ULLMANN'S

Industrial Toxicology

Volume 1

Toxicology in Occupational and Environmental Setting

ULLMANN'S

Industrial Toxicology

Volume 1
Toxicology in Occupational and Environmental Setting

Volume 2
Toxic Agents, Pharmacological and Medical Fundamentals

Further Ullmann's Publications

WILEY-VCH (Ed.)
Ullmann's Encyclopedia of Industrial Chemistry
Sixth, Completely Revised Edition, 40 Volumes
2003, ISBN 3-527-30385-5

Wiley-VCH (Ed.)
Ullmann's Processes and Process Engineering, 3 Volumes
2004, ISBN 3-527-31096-7

Wiley-VCH (Ed.)
Ullmann's Chemical Engineering and Plant Design
2004, ISBN 3-527-31111-4

WILEY-VCH (Ed.)
Ullmann's Electronic Release 2005
Online plus CD-ROM
2005, ISBN 3-527-31097-5
http://interscience.wiley.com/ullmanns

WILEY-VCH (Ed.)
Industrial Inorganic Chemicals and Products
An Ullmann's Encyclopedia, 6 Volumes
1999, ISBN 3-527-29567-4

WILEY-VCH (Ed.)
Industrial Organic Chemicals – Starting Materials and Intermediates
An Ullmann's Encyclopedia, 8 Volumes
1999, ISBN 3-527-29645-X

ULLMANN'S

Industrial Toxicology

Volume 1

Toxicology in Occupational and Environmental Setting

WILEY-VCH Verlag GmbH & Co. KGaA

Library of Congress Card No.:
applied for

British Library Cataloguing-in-Publication Data:
A catalogue record for this book is available from the British Library.

Bibliographic information published by Die Deutsche Bibliothek
Die Deutsche Bibliothek lists this publication in the Deutsche Nationalbibliografie; detailed bibliographic data is available in the Internet at http://dnb.ddb.de

Typesetting: Steingraeber Satztechnik GmbH, Dossenheim

Printing: and Binding: betzdruck GmbH, Darmstadt

Binding: Schäffer GmbH, Grünstadt

Cover Design: Gunther Schulz, Fußgönheim

Printed in the Federal Republic of Germany
Printed on acid-free paper

ISBN-10: 3-527-31247-1
ISBN-13: 978-3-527-31247-4

Preface

Since the entire 40-volume Ullmann's Encyclopedia is inaccessible to many readers – particularly individuals, smaller companies or institutes – all the information on industrial toxicology, ecotoxicology, process safety as well as occupational health and safety has been condensed into this convenient 2-volume set.

Based on the latest online edition of Ullmann's containing articles never been before in print, this ready reference provides practical information on applying the science of toxicology in both the occupational and environmental setting, and explains the fundamentals necessary for an understanding of the effects of chemical hazards on humans and ecosystems. The detailed and meticulously edited articles have been written by renowned experts from industry and academia, and much of the information has been thoroughly revised. Alongside explanations of safety regulations and legal aspects, this set covers food additives, toxic and mutagenic agents as well as medical and therapeutical issues. Top-quality illustrations, clear diagrams and charts combined with an extensive use of tables enhance the presentation and provide a unique level of detail. Deeper insights into any given area of interest is offered by referenced contributions, while rapid access to a particular subject is enhanced by both a keyword and author index.

We are convinced that these handy two volumes are a convenient one-stop resource for health protection professionals, environmental scientists, safety engineers and all those involved in the field of industrial toxicology.

Summer, 2005
Weinheim

The Publisher

Contents

Volume 1

Toxicology in Occupational and Environmental Setting

Volume 2

Toxic Agents, Pharmacological and Medical Fundamentals

Ullmann's Industrial Toxicology

ISBN: 3-527-31247-1

Symbols and Units

Symbols and units agree with SI standards (for conversion factors see page XI). The following list gives the most important symbols used in the encyclopedia. Articles with many specific units and symbols have a similar list as front matter.

Symbol	Unit	Physical Quantity
a_B		activity of substance B
A_r		relative atomic mass (atomic weight)
A	m^2	area
c_B	mol/m^3, mol/L (M)	concentration of substance B
C	C/V	electric capacity
c_p, c_v	$J\,kg^{-1}K^{-1}$	specific heat capacity
d	cm, m	diameter
d		relative density (ϱ/ϱ_{water})
D	m^2/s	diffusion coefficient
D	Gy (= J/kg)	absorbed dose
e	C	elementary charge
E	J	energy
E	V/m	electric field strength
E	V	electromotive force
E_A	J	activation energy
f		activity coefficient
F	C/mol	Faraday constant
F	N	force
g	m/s^2	acceleration due to gravity
G	J	Gibbs free energy
h	m	height
$\hbar$	$W \cdot s^2$	Planck constant
H	J	enthalpy
I	A	electric current
I	cd	luminous intensity
k	(variable)	rate constant of a chemical reaction
k	J/K	Boltzmann constant
K	(variable)	equilibrium constant
l	m	length
m	g, kg, t	mass
M_r		relative molecular mass (molecular weight)
n_D^{20}		refractive index (sodium D-line, 20 °C)
n	mol	amount of substance
N_A	mol^{-1}	Avogadro constant (6.023×10^{23} mol^{-1})
p	Pa, bar *	pressure
Q	J	quantity of heat
r	m	radius
R	$J\,K^{-1}mol^{-1}$	gas constant
R	Ω	electric resistance
S	J/K	entropy
t	s, min, h, d, month, a	time

Symbols and Units (Continued from p. IX)

Symbol	Unit	Physical Quantity
t	°C	temperature
T	K	absolute temperature
u	m/s	velocity
U	V	electric potential
U	J	internal energy
V	m^3, L, mL, µL	volume
w		mass fraction
W	J	work
x_B		mole fraction of substance B
Z		proton number, atomic number
α		cubic expansion coefficient
α	$W\,m^{-2}K^{-1}$	heat-transfer coefficient (heat-transfer number)
α		degree of dissociation of electrolyte
$[\alpha]$	$10^{-2}deg\,cm^2g^{-1}$	specific rotation
η	Pa · s	dynamic viscosity
θ	°C	temperature
$\varkappa$		c_p/c_v
λ	$W\,m^{-1}K^{-1}$	thermal conductivity
λ	nm, m	wavelength
μ		chemical potential
ν	Hz, s^{-1}	frequency
ν	m^2/s	kinematic viscosity (η/ϱ)
π	Pa	osmotic pressure
ϱ	g/cm^3	density
σ	N/m	surface tension
τ	Pa (N/m^2)	shear stress
φ		volume fraction
χ	Pa^{-1} (m^2/N)	compressibility

* The official unit of pressure is the pascal (Pa).

Conversion Factors

SI unit	Non-SI unit	From SI to non-SI multiply by
Mass		
kg	pound (avoirdupois)	2.205
kg	ton (long)	9.842×10^{-4}
kg	ton (short)	1.102×10^{-3}
Volume		
m^3	cubic inch	6.102×10^{4}
m^3	cubic foot	35.315
m^3	gallon (U.S., liquid)	2.642×10^{2}
m^3	gallon (Imperial)	2.200×10^{2}
Temperature		
°C	°F	$°C \times 1.8 + 32$
Force		
N	dyne	1.0×10^{5}
Energy, Work		
J	Btu (int.)	9.480×10^{-4}
J	cal (int.)	2.389×10^{-1}
J	eV	6.242×10^{18}
J	erg	1.0×10^{7}
J	$kW \cdot h$	2.778×10^{-7}
J	$kp \cdot m$	1.020×10^{-1}
Pressure		
MPa	at	10.20
MPa	atm	9.869
MPa	bar	10
kPa	mbar	10
kPa	mm Hg	7.502
kPa	psi	0.145
kPa	torr	7.502

Powers of Ten

E (exa)	10^{18}	d (deci)	10^{-1}
P (peta)	10^{15}	c (centi)	10^{-2}
T (tera)	10^{12}	m (milli)	10^{-3}
G (giga)	10^{9}	μ (micro)	10^{-6}
M (mega)	10^{6}	n (nano)	10^{-9}
k (kilo)	10^{3}	p (pico)	10^{-12}
h (hecto)	10^{2}	f (femto)	10^{-15}
da (deca)	10	a (atto)	10^{-18}

Abbreviations

The following is a list of the abbreviations used in the text. Common terms, the names of publications and institutions, and legal agreements are included along with their full identities. Other abbreviations will be defined wherever they first occur in an article. For further abbreviations, see page IX, Symbols and Units; page XVI, Frequently Cited Companies (Abbreviations), and page XVII, Country Codes in patent references. The names of periodical publications are abbreviated exactly as done by Chemical Abstracts Service.

abs.	absolute
a.c.	alternating current
ACGIH	American Conference of Governmental Industrial Hygienists
ACS	American Chemical Society
ADI	acceptable daily intake
ADN	accord européen relatif au transport international des marchandises dangereuses par voie de navigation interieure (European agreement concerning the international transportation of dangerous goods by inland waterways)
ADNR	ADN par le Rhin (regulation concerning the transportation of dangerous goods on the Rhine and all national waterways of the countries concerned)
ADP	adenosine 5′-diphosphate
ADR	accord européen relatif au transport international des marchandises dangereuses par route (European agreement concerning the international transportation of dangerous goods by road)
AEC	Atomic Energy Commission (United States)
a.i.	Active ingredient
AIChE	American Institute of Chemical Engineers
AIME	American Institute of Mining, Metallurgical, and Petroleum Engineers
ANSI	American National Standards Institute
AMP	adenosine 5′-monophosphate
APhA	American Pharmaceutical Association
API	American Petroleum Institute
ASTM	American Society for Testing and Materials
ATP	adenosine 5′-triphosphate
BAM	Bundesanstalt für Materialprüfung (Federal Republic of Germany)
BAT	Biologischer Arbeitsstoff-Toleranz-Wert (biological tolerance value for a working material, established by MAK Commission, see MAK)
Beilstein	Beilstein's Handbook of Organic Chemistry, Springer, Berlin – Heidelberg – New York
BET	Brunauer – Emmett – Teller
BGA	Bundesgesundheitsamt (Federal Republic of Germany)
BGBl.	Bundesgesetzblatt (Federal Republic of Germany)
BIOS	British Intelligence Objectives Subcommitee Report (see also FIAT)
BOD	biological oxygen demand
bp	boiling point
B.P.	British Pharmacopeia
BS	British Standard
ca.	circa
calcd.	calculated
CAS	Chemical Abstracts Service
cat.	catalyst, catalyzed
CEN	Comité Européen de Normalisation
cf.	compare
CFR	Code of Federal Regulations (United States)
cfu	colony forming units
Chap.	chapter
ChemG	Chemikaliengesetz (Federal Republic of Germany)
C.I.	Colour Index
CIOS	Combined Intelligence Objectives Subcommitee Report (see also FIAT)
CNS	central nervous system
Co.	Company
COD	chemical oxygen demand
conc.	concentrated
const.	constant
Corp.	Corporation
crit.	critical

CTFA The Cosmetic, Toiletry and Fragrance Association (United States)
DAB 9 Deutsches Arzneibuch, 9th ed., Deutscher Apotheker-Verlag, Stuttgart 1986
d.c. direct current
decomp. decompose, decomposition
DFG Deutsche Forschungsgemeinschaft (German Science Foundation)
dil. dilute, diluted
DIN Deutsche Industrie Norm (Federal Republic of Germany)
DMF dimethylformamide
DNA deoxyribonucleic acid
DOE Department of Energy (United States)
DOT Department of Transportation – Materials Transportation Bureau (United States)
DTA differential thermal analysis
EC effective concentration
EC European Community
ed. editor, edition, edited
e.g. for example
emf electromotive force
EmS Emergency Schedule
EN European Standard (European Community)
EPA Environmental Protection Agency (United States)
EPR electron paramagnetic resonance
Eq. equation
ESCA electron spectroscopy for chemical analysis
esp. especially
ESR electron spin resonance
Et ethyl substituent ($-C_2H_5$)
et al. and others
etc. et cetera
EVO Eisenbahnverkehrsordnung (Federal Republic of Germany)
exp (...) $e^{(\ldots)}$, mathematical exponent
FAO Food and Agriculture Organization (United Nations)
FDA Food and Drug Administration (United States)
FD & C Food, Drug and Cosmetic Act (United States)
FHSA Federal Hazardous Substances Act (United States)
FIAT Field Information Agency, Technical (United States reports on the chemical industry in Germany, 1945)
Fig. figure
fp freezing point
Friedländer P. Friedländer, Fortschritte der Teerfarbenfabrikation und verwandter Industriezweige, Vol. 1 – 25, Springer, Berlin 1888 – 1942
FT Fourier transform
(g) gas, gaseous
GC gas chromatography
GefStoffV Gefahrstoffverordnung (regulations in the Federal Republic of Germany concerning hazardous substances)
GGVE Verordnung in der Bundesrepublik Deutschland über die Beförderung gefährlicher Güter mit der Eisenbahn (regulation in the Federal Republic of Germany concerning the transportation of dangerous goods by rail)
GGVS Verordnung in der Bundesrepublik Deutschland über die Beförderung gefährlicher Güter auf der Straße (regulation in the Federal Republic of Germany concerning the transportation of dangerous goods by road)
GGVSee Verordnung in der Bundesrepublik Deutschland über die Beförderung gefährlicher Güter mit Seeschiffen (regulation in the Federal Republic of Germany concerning the transportation of dangerous goods by sea-going vessels)
GLC gas-liquid chromatography
Gmelin Gmelin's Handbook of Inorganic Chemistry, 8th ed., Springer, Berlin – Heidelberg – New York
GRAS generally recognized as safe
Hal halogen substituent ($-F$, $-Cl$, $-Br$, $-I$)
Houben-Weyl Methoden der organischen Chemie, 4th ed., Georg Thieme Verlag, Stuttgart
HPLC high performance liquid chromatography
IAEA International Atomic Energy Agency
IARC International Agency for Research on Cancer, Lyon, France

IATA-DGR International Air Transport Association, Dangerous Goods Regulations
ICAO International Civil Aviation Organization
i.e. that is
i.m. intramuscular
IMDG International Maritime Dangerous Goods Code
IMO Inter-Governmental Maritime Consultive Organization (in the past: IMCO)
Inst. Institute
i.p. intraperitoneal
IR infrared
ISO International Organization for Standardization
IUPAC International Union of Pure and Applied Chemistry
i.v. intravenous
Kirk-Othmer Encyclopedia of Chemical Technology, 3rd ed., J. Wiley & Sons, New York – Chichester – Brisbane – Toronto 1978 – 1984; 4th ed., J. Wiley & Sons, New York – Chichester – Brisbane – Toronto 1991 – 1998
(l) liquid
Landolt-Börnstein Zahlenwerte u. Funktionen aus Physik, Chemie, Astronomie, Geophysik u. Technik, Springer, Heidelberg 1950 – 1980; Zahlenwerte und Funktionen aus Naturwissenschaften und Technik, Neue Serie, Springer, Heidelberg, since 1961
LC_{50} lethal concentration for 50 % of the test animals
LCLo lowest published lethal concentration
LD_{50} lethal dose for 50 % of the test animals
LDLo lowest published lethal dose
ln logarithm (base e)
LNG liquefied natural gas
log logarithm (base 10)
LPG liquefied petroleum gas
M mol/L
M metal (in chemical formulas)
MAK Maximale Arbeitsplatz-Konzentration (maximum concentration at the workplace in the Federal Republic of Germany); cf. Deutsche Forschungsgemeinschaft (ed.): Maximale Arbeitsplatzkonzentrationen (MAK) und Biologische Arbeitsstoff-Toleranz-Werte (BAT), WILEY-VCH Verlag, Weinheim (published annually)
max. maximum
MCA Manufacturing Chemists Association (United States)
Me methyl substituent ($-CH_3$)
Methodicum Chimicum Methodicum Chimicum, Georg Thieme Verlag, Stuttgart
MFAG Medical First Aid Guide for Use in Accidents Involving Dangerous Goods
MIK maximale Immissionskonzentration (maximum immission concentration)
min. minimum
mp melting point
MS mass spectrum, mass spectrometry
NAS National Academy of Sciences (United States)
NASA National Aeronautics and Space Administration (United States)
NBS National Bureau of Standards (United States)
NCTC National Collection of Type Cultures (United States)
NIH National Institutes of Health (United States)
NIOSH National Institute for Occupational Safety and Health (United States)
NMR nuclear magnetic resonance
no. number
NOEL no observed effect level
NRC Nuclear Regulatory Commission (United States)
NRDC National Research Development Corporation (United States)
NSC National Service Center (United States)
NSF National Science Foundation (United States)
NTSB National Transportation Safety Board (United States)
OECD Organization for Economic Cooperation and Development
OSHA Occupational Safety and Health Administration (United States)

p., pp. page, pages
Patty G. D. Clayton, F. E. Clayton (eds.): Patty's Industrial Hygiene and Toxicology, 3rd ed., Wiley Interscience, New York
PB report Publication Board Report (U.S. Department of Commerce, Scientific and Industrial Reports)
PEL permitted exposure limit
Ph phenyl substituent ($-C_6H_5$)
Ph. Eur. European Pharmacopoeia, 2nd. ed., Council of Europe, Strasbourg 1981
phr part per hundred rubber (resin)
PNS peripheral nervous system
ppm parts per million
q. v. which see (quod vide)
ref. refer, reference
resp. respectively
R_f retention factor (TLC)
R. H. relative humidity
RID règlement international concernant le transport des marchandises dangereuses par chemin de fer (international convention concerning the transportation of dangerous goods by rail)
RNA ribonucleic acid
R phrase (R-Satz) risk phrase according to ChemG and GefStoffV (Federal Republic of Germany)
rpm revolutions per minute
RTECS Registry of Toxic Effects of Chemical Substances, edited by the National Institute of Occupational Safety and Health (United States)
(s) solid
SAE Society of Automotive Engineers (United States)
s.c. subcutaneous
SI International System of Units
SIMS secondary ion mass spectrometry
S phrase (S-Satz) safety phrase according to ChemG and GefStoffV (Federal Republic of Germany)
STEL Short Term Exposure Limit (see TLV)
STP standard temperature and pressure (0° C, 101.325 kPa)
T_g glass transition temperature
TA Luft Technische Anleitung zur Reinhaltung der Luft (clean air regulation in Federal Republic of Germany)
TA Lärm Technische Anleitung zum Schutz gegen Lärm (low noise regulation in Federal Republic of Germany)
TDLo lowest published toxic dose
THF tetrahydrofuran
TLC thin layer chromatography
TLV Threshold Limit Value (TWA and STEL); published annually by the American Conference of Governmental Industrial Hygienists (ACGIH), Cincinnati, Ohio
TOD total oxygen demand
TRK Technische Richtkonzentration (lowest technically feasible level)
TSCA Toxic Substances Control Act (United States)
TÜV Technischer Überwachungsverein (Technical Control Board of the Federal Republic of Germany)
TWA Time Weighted Average
UBA Umweltbundesamt (Federal Environmental Agency)
Ullmann Ullmann's Encyclopedia of Industrial Chemistry, 6th ed., Wiley-VCH, Weinheim, 2002, 5th ed., VCH Verlagsgesellschaft, Weinheim, 1985 – 1996; Ullmanns Encyklopädie der Technischen Chemie, 4th ed., Verlag Chemie, Weinheim 1972 – 1984
USAEC United States Atomic Energy Commission
USAN United States Adopted Names
USD United States Dispensatory
USDA United States Department of Agriculture
U.S.P. United States Pharmacopeia
UV ultraviolet
UVV Unfallverhütungsvorschriften der Berufsgenossenschaft (workplace safety regulations in the Federal Republic of Germany)
VbF Verordnung in der Bundesrepublik Deutschland über die Errichtung und den Betrieb von Anlagen zur Lagerung, Abfüllung und Beförderung brennbarer Flüssigkeiten (regulation in the Federal Republic of Germany

	concerning the construction and operation of plants for storage, filling, and transportation of flammable liquids; classification according to the flash point of liquids, in accordance with the classification in the United States)
VDE	Verband Deutscher Elektroingenieure (Federal Republic of Germany)
VDI	Verein Deutscher Ingenieure (Federal Republic of Germany)
vol	volume
vol.	volume (of a series of books)
vs.	versus
WGK	Wassergefährdungsklasse (water hazard class)
WHO	World Health Organization (United Nations)
Winnacker-Küchler	Chemische Technologie, 4th ed., Carl Hanser Verlag, München, 1982-1986; Winnacker-Küchler, Chemische Technik: Prozesse und Produkte, Wiley-VCH, Weinheim, from 2003
wt	weight
$	U.S. dollar, unless otherwise stated

Frequently Cited Companies (Abbreviations)

Air Products	Air Products and Chemicals
Akzo	Algemene Koninklijke Zout Organon
Alcoa	Aluminum Company of America
Allied	Allied Corporation
Amer. Cyanamid	American Cyanamid Company
BASF	BASF Aktiengesellschaft
Bayer	Bayer AG
BP	British Petroleum Company
Celanese	Celanese Corporation
Daicel	Daicel Chemical Industries
Dainippon	Dainippon Ink and Chemicals Inc.
Dow Chemical	The Dow Chemical Company
DSM	Dutch Staats Mijnen
Du Pont	E.I. du Pont de Nemours & Company
Exxon	Exxon Corporation
FMC	Food Machinery & Chemical Corporation
GAF	General Aniline & Film Corporation
W.R. Grace	W.R. Grace & Company
Hoechst	Hoechst Aktiengesellschaft
IBM	International Business Machines Corporation
ICI	Imperial Chemical Industries
IFP	Institut Français du Pétrole
INCO	International Nickel Company
3M	Minnesota Mining and Manufacturing Company
Mitsubishi Chemical	Mitsubishi Chemical Industries
Monsanto	Monsanto Company
Nippon Shokubai	Nippon Shokubai Kagaku Kogyo
PCUK	Pechiney Ugine Kuhlmann
PPG	Pittsburg Plate Glass Industries
Searle	G.D. Searle & Company
SKF	Smith Kline & French Laboratories
SNAM	Societá Nazionale Metandotti
Sohio	Standard Oil of Ohio
Stauffer	Stauffer Chemical Company
Sumitomo	Sumitomo Chemical Company
Toray	Toray Industries Inc.
UCB	Union Chimique Belge
Union Carbide	Union Carbide Corporation
UOP	Universal Oil Products Company
VEBA	Vereinigte Elektrizitäts- und Bergwerks-AG
Wacker	Wacker Chemie GmbH

Country Codes

The following list contains a selection of standard country codes used in the patent references.

AT	Austria	ID	Indonesia
AU	Australia	IL	Israel
BE	Belgium	IT	Italy
BG	Bulgaria	JP	Japan *
BR	Brazil	LU	Luxembourg
CA	Canada	MA	Morocco
CH	Switzerland	NL	Netherlands *
DE	Federal Republic of Germany (and Germany before 1949) *	NO	Norway
		NZ	New Zealand
DK	Denmark	PL	Poland
ES	Spain	PT	Portugal
FI	Finland	SE	Sweden
FR	France	US	United States of America
GB	United Kingdom	ZA	South Africa
GR	Greece	EP	European Patent Office *
HU	Hungary	WO	World Intellectual Property Organization

* For Europe, Federal Republic of Germany, Japan, and the Netherlands, the type of patent is specified: EP (patent), EP-A (application), DE (patent), DE-OS (Offenlegungsschrift), DE-AS (Auslegeschrift), JP (patent), JP-Kokai (Kokai tokkyo koho), NL (patent), and NL-A (application).

Periodic Table of Elements

element symbol, atomic number, and relative atomic mass (atomic weight)

1A "European" group designation and old IUPAC recommendation
1 group designation to 1986 IUPAC proposal
IA "American" group designation, also used by the Chemical Abstracts Service until the end of 1986

1A **1** IA	2A **2** IIA	3A **3** IIIB	4A **4** IVB	5A **5** VB	6A **6** VIB	7A **7** VIIB	8 **8** VIII	8 **9** VIII	8 **10** VIII	1B **11** IB	2B **12** IIB	3B **13** IIIA	4B **14** IVA	5B **15** VA	6B **16** VIA	7B **17** VIA	0 **18** VIIIA
1 **H** 1.0079																	2 **He** 4.0026
3 **Li** 6.941	4 **Be** 9.0122											5 **B** 10.811	6 **C** 12.011	7 **N** 14.007	8 **O** 15.999	9 **F** 18.998	10 **Ne** 20.180
11 **Na** 22.990	12 **Mg** 24.305											13 **Al** 26.982	14 **Si** 28.086	15 **P** 30.974	16 **S** 32.066	17 **Cl** 35.453	18 **Ar** 39.948
19 **K** 39.098	20 **Ca** 40.078	21 **Sc** 44.956	22 **Ti** 47.867	23 **V** 50.942	24 **Cr** 51.996	25 **Mn** 54.938	26 **Fe** 55.845	27 **Co** 58.933	28 **Ni** 58.693	29 **Cu** 63.546	30 **Zn** 65.409	31 **Ga** 69.723	32 **Ge** 72.61	33 **As** 74.922	34 **Se** 78.96	35 **Br** 79.904	36 **Kr** 83.80
37 **Rb** 85.468	38 **Sr** 87.62	39 **Y** 88.906	40 **Zr** 91.224	41 **Nb** 92.906	42 **Mo** 95.94	43 **Tc*** 98.906	44 **Ru** 101.07	45 **Rh** 102.91	46 **Pd** 106.42	47 **Ag** 107.87	48 **Cd** 112.41	49 **In** 114.82	50 **Sn** 118.71	51 **Sb** 121.76	52 **Te** 127.60	53 **I** 126.90	54 **Xe** 131.29
55 **Cs** 132.91	56 **Ba** 137.33		72 **Hf** 178.49	73 **Ta** 180.95	74 **W** 183.84	75 **Re** 186.21	76 **Os** 190.23	77 **Ir** 192.22	78 **Pt** 195.08	79 **Au** 196.97	80 **Hg** 200.59	81 **Tl** 204.38	82 **Pb** 207.2	83 **Bi** 208.98	84 **Po*** 208.98	85 **At*** 209.99	86 **Rn*** 222.02
87 **Fr*** 223.02	88 **Ra*** 226.03		104 **Rf*** 261.11	105 **Db*** 262.11	106 **Sg**	107 **Bh**	108 **Hs**	109 **Mt**	110 **Uun**[a]	111 **Uuu**[a]	112 **Uub**[a]		114 **Uuq**[a]		116 **Uuh**[a]		

[a] provisional IUPAC symbol

57 **La** 138.91	58 **Ce** 140.12	59 **Pr** 140.91	60 **Nd** 144.24	61 **Pm*** 146.92	62 **Sm** 150.36	63 **Eu** 151.97	64 **Gd** 157.25	65 **Tb** 158.93	66 **Dy** 162.50	67 **Ho** 164.93	68 **Er** 167.26	69 **Tm** 168.93	70 **Yb** 173.04	71 **Lu** 174.97
89 **Ac*** 227.03	90 **Th*** 232.04	91 **Pa*** 231.04	92 **U*** 238.03	93 **Np*** 237.05	94 **Pu*** 244.06	95 **Am*** 243.06	96 **Cm*** 247.07	97 **Bk*** 247.07	98 **Cf*** 251.08	99 **Es*** 252.08	100 **Fm*** 257.10	101 **Md*** 258.10	102 **No*** 259.10	103 **Lr*** 260.11

* radioactive element; mass of most important isotope given.

Toxicology in Occupational and Environmental Setting

Toxicology

WOLFGANG DEKANT, Institute of Toxicology, University of Wuerzburg, Germany
SPIRIDON VAMVAKAS, Institute of Toxicology, University of Wuerzburg, Germany

Ullmann's Industrial Toxicology

ISBN: 3-527-31247-1

Abbreviations:

Ah-R	arylhydrocarbon receptor
AP	apurinic/apyrimidinic site
APS	adenosine 5′-phosphosulfate
BHK	baby hamster kidney
BIBRA	British Industrial Biological Research Association
CoA	Coenzym A
DDT	1,1′-(2,2,2-trichloroethylidene)bis-(4-chlorobenzene)
DHHS	U.S. Department of Health and Human Services
DHP	delayed hypersensitive response
ECETOC	European Chemical Industry Ecology and Toxicology Centre
ED	effective dose
ELISA	enzyme-linked immunosorbent assay
FCA	Freund's complete adjuvant
FAD	flavine adenine dinucleotide
GABA	γ-aminobutyrate
GC/MS	gas chromatography/mass spectroscopy
GOT	glutamic acid oxalacetic transaminase
GSH	glutathione
GSSG	glutathione disulfide
GST	glutathione *S*-transferase
GTP	guanosine 5′-triophosphate
HGPRT	hypoxanthine – guanine phosphoribosyltransferase
IPCS	International Programme on Chemical Safety
LDH	lactate dehydrogenase
LOAEL	lowest-observed-adverse-effect level
LOEL	lowest-observed-effect level
MIF	migration inhibition factor

mRNA	messenger RNA
MTD	maximum tolerated dose
NADPH	nicotinamide dinucleotide phosphate (H)
NOEL	no-observed-effect-level
NTP	National Toxicology Program
PAPS	3′-phosphoadenosine-5′-phosphosulfate
PG	prostaglandin
rRNA	ribosomal RNA
SHE	Syrian hamster embryo
SMART	somatic mutation and recombination test
T, or TCDD	2,3,7,8-tetrachlorodibenzodioxin
TD	tumor dose
TK	thymidine kinase
tRNA	transfer RNA
UDP	uridine diphosphate
UDPG	uridine diphosphate glucose
UDPGA	uridine diphosphate glucuronic acid
UDS	unscheduled DNA synthesis

1. Introduction

1.1. Definition and Scope

Chemicals that are used or of potential use in commerce, the home, the environment, and medical practice may present various types of harmful effects. The nature of these effects is determined by the physicochemical characteristics of the agent, its ability to interact with biological systems (hazard), and its potential to come into contact with biological systems (exposure).

Toxicology studies the interaction between chemicals and biological systems to determine the potential of chemicals to produce adverse effects in living organisms. Toxicology also investigates the nature, incidence, mechanisms of production, factors influencing their development, and reversibility of such adverse effects. Adverse effects are defined as detrimental to the survival or the normal functioning of the individual. Inherent in this definition are the following key issues in toxicology:

1) Chemicals must come into close structural and/or functional contact with tissues or organs to cause injury.
2) All adverse effects depend on the amount of chemical in contact with the biological system (the dose) and the inherent toxicity of the chemical (hazard). When possible, the observed toxic effect should be related to the degree of exposure. The influence of different exposure doses on the magnitude and incidence of the toxic effect should be quantitated. Such dose-response relationships are of prime importance in confirming a causal relationship between chemical exposure and toxic effect (for details, see Section 1.7).

Research in toxicology is mainly concerned with determining the potential for adverse effects caused by chemicals, both natural and synthetic, to assess their hazard and risk of human exposure and thus provide a basis for appropriate precautionary, protective and restrictive measures. Toxicological investigations should permit evaluation of the following characteristics of toxicity:

1) The basic structural, functional, or biochemical injury produced
2) Dose-response relationships
3) The mechanisms of toxicity (fundamental biochemical alterations responsible for the induction and maintenance of the toxic response) and reversibility of the toxic effect
4) Factors that modify response, e.g., route of exposure, species, and gender

For chemicals to which humans may potentially be exposed, a critical analysis, based on the pattern of potential exposure or toxicity, may be necessary in order to determine the risk-benefit ratio for their use in specific circumstances and to devise protective and precautionary measures. Indeed, with drugs, pesticides, food additives, and cosmetic preparations, toxicology testing must be performed in accordance with government regulations before use.

1.2. Fields

Toxicology is a recognized scientific discipline encompassing both basic and applied issues. Although only generally accepted as a specific scientific field during this century, its principles have been appreciated for centuries. The harmful or lethal effects of certain chemicals, mainly present in minerals and plants or transmitted venomous animals, have been known since prehistoric times. In many countries, toxicology as a discipline has developed from pharmacology. Pharmacology and toxicology both study the effect of chemicals on living organisms and have often used identical methods. However, fundamental differences have developed. Years ago, only the dependence on dose of the studied effects separated pharmacology and toxicology. Pharmacology focused on chemicals with beneficial effects (drugs) at lower doses whereas toxicology studied the adverse health effects occurring with the same chemicals at high doses. Today, the main interest of research in toxicology has shifted to studies on the long-term effects of chemicals after low-dose exposure, such as cancer or other irreversible diseases; moreover, most chemicals of interest to toxicologists are not used as drugs.

The variety of potential adverse effects and the diversity of chemicals present in our environment combine to make toxicology a very broad science. Toxicology uses basic knowledge from clinical and theoretical medicine and natural sciences such as biology and chemistry (Fig. 1). Because of this diversity, toxicologists usually specialize in certain areas.

Any attempt to define the scope of toxicology must take into account that the various subdisciplines are not mutually exclusive and frequently are heavily interdependent. Due to the overlapping mechanisms of toxicity, chemical classes, and observed toxic effects, clear divisions into subjects of equal importance are often not possible.

The professional activities of toxicologists can be divided into three main categories: descriptive, mechanistic, and regulatory. The *descriptive toxicologist* is concerned directly with toxicity testing. Descriptive toxicology still often relies on the tools of pathology and clinical chemistry, but since the 1970s more mechanism-based test systems have been included in toxicity testing [1]. The appropriate toxicity tests in experimental animals yield information that is extrapolated to evaluate the risk posed by exposure to specific chemicals. The concern may be limited to effects on humans (drugs, industrial chemicals in the workplace, or food additives) or may encompass animals, plants, and other factors that might disturb the balance of the ecosystem (industrial chemicals, pesticides, environmental pollutants).

The *mechanistic toxicologist* is concerned with elucidating the mechanisms by which chemicals exert their toxic effects on living organisms. Such studies may result in the development of sensitive predictive toxicity tests useful in obtaining information for risk assessment (see Chap. 4). Mechanistic studies may help in the development of chemicals that are safer to use or of more rational therapies for intoxications. In addition, an understanding of the mechanisms of toxic action also contributes to the knowledge of basic mechanisms in physiology, pharmacology, cell biology, and biochemistry. Indeed, toxic chemicals have been used with great success as mechanistic tools to elucidate mechanisms of physiological regulation. Mechanistic toxicologists are often active in universities; however, industry and government institutions are now undertaking more and more research in mechanistic toxicology.

Regulatory toxicologists have the responsibility of deciding on the basis of data provided by the descriptive toxicologist and the mechanistic toxicologist if a drug or chemical poses a sufficiently low risk to be used for a stated purpose. Regulatory toxicologists are often active in government institutions and are involved in the establishment of standards for the amount of chemicals permitted in ambient air in the environment, in the workplace, or in drinking water. Other divisions of toxicology may be based on the classes of chemicals dealt with or application of knowledge from toxicology for a specific field (Table 1).

Forensic toxicology comprises both analytical chemistry and fundamental toxicologic principles. It is concerned with the legal aspects of the harmful effects of chemicals on humans. The expertise of the forensic toxicologist is invoked primarily to aid in establishing the cause of death and elucidating its circumstances in a postmortem investigation. The field of *clinical toxicology* recognizes and treats poisoning, both chronic and acute. Efforts are directed at treating patients poisoned by chemicals and at the development of new techniques to treat these intoxications. *Environmental toxicology* is a relatively new area that studies the effects of chemicals released by man on wildlife and the ecosystem and thus indirectly on human health.

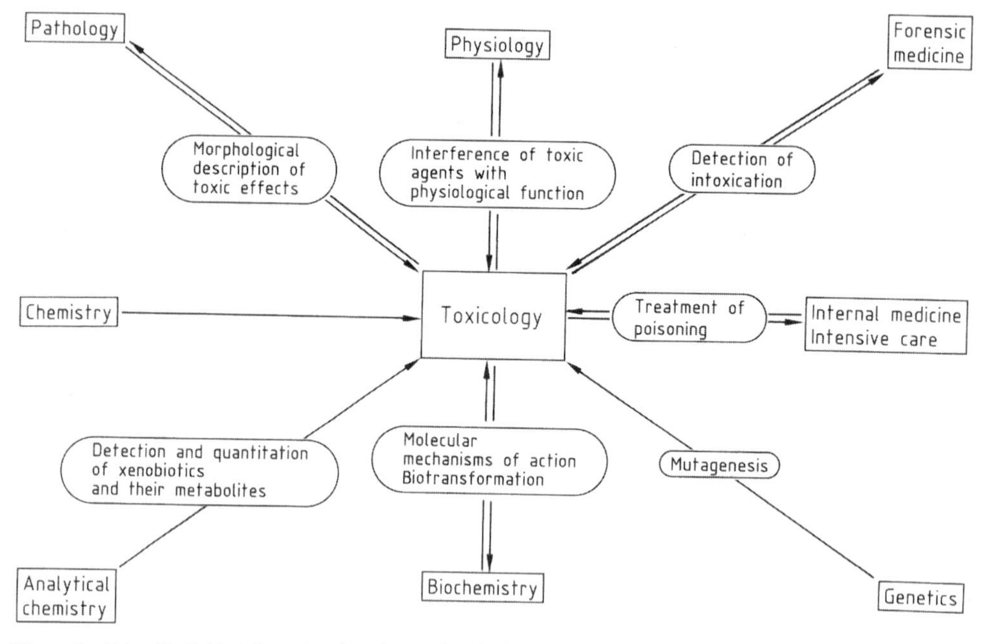

Figure 1. Scientific fields influencing the science of toxicology

Table 1. Areas of toxicology

Field	Tasks and objectives
Forensic toxicology	diagnoses poisoning by analytical procedures
Pesticide toxicology	studies the safety of pesticides, develops new pesticides
Occupational toxicology	assesses potential adverse effects of chemicals used in the workplace, recommends protective procedures
Drug toxicology	studies potential effects of drugs after high doses, elucidates mechanisms of sideeffects
Regulatory toxicology	develops and interprets toxicity testing programs and is involved in controlling the use of chemicals
Environmental toxicology	studies the effects of chemicals on ecosystems and on humans after low-dose exposure from the environment

Drug toxicology plays a major role in the preclinical safety assessment of chemicals intended for use as drugs. Drug toxicology also elucidates the mechanisms of side effects observed during clinical application. *Occupational toxicology* studies the acute and chronic toxicity of chemicals encountered in the occupational environment. Both acute and chronic occupational poisonings have exerted a major influence on the development of toxicology in general. Occupational toxicology also helps in the development of safety procedures to prevent intoxications in the workplace and assists in the definition of exposure limits. *Pesticide toxicology* is involved in the development of new pesticides and the safety of pesticide formulations. Pesticide toxicology also characterizes potential health risks to the general population caused by pesticide residues in food and drinking water.

1.3. History

Toxicology must rank as one of the oldest practical sciences because humans, from the very beginning, needed to avoid the numerous toxic plants and animals in their environment. The presence of toxic agents in animals and plants was known to the Egyptian and Greek civilisations. The papyrus Ebers, an Egyptian papyrus dating from about 1 500 B.C., and the surviving medical works of HIPPOCRATES, ARISTOTLE, and THEOPHRASTUS, published during the period 400–250 B.C., all included some mention of poisons.

The Greek and Roman civilizations knowingly used certain toxic chemicals and extracts for hunting, warfare, suicide, and murder. Up to the Middle Ages, toxicology was restricted to the use of toxic agents for murder. Poisoning was developed to an art in medieval Italy and has remained a problem ever since, and much of the earlier impetus for the development of toxicology was primarily forensic. There appear to have been few advances in either medicine or toxicology between the time of GALEN (131–200 A.D.) and PARACELSUS (1493–1541). The latter laid the groundwork for the later development of modern toxicology. He clearly was aware of the dose–response relationship. His statement that "All substances are poisons; there is none that is not a poison. The right dose differentiates a poison and a remedy," is properly regarded as a landmark in the development of the science of toxicology. His belief in the value of experimentation also represents a break with much earlier tradition. Important developments in the 1700s include the publication of RAMAZZINI's *Diseases of Workers*, which led to his recognition as the father of occupational medicine. The correlation between the occupation of chimney sweepers and scrotal cancer by POTT in 1775 is also noteworthy.

ORFILA, a Spaniard working at the University of Paris, clearly identified toxicology as a separate science and wrote the first book devoted exclusively to it (1815). Workers of the later 1800s who produced treatises on toxicology include CHRISTISON, KOBERT, and LEWIN. They increased our knowledge of the chemistry of poisons, the treatment of poisoning, the analysis of both xenobiotics and toxicity, as well as modes of action and detoxication. A major impetus for toxicology in the 1900s was the use of chemicals for warfare. In World War I, a variety of poisonous chemicals were used in the battlefields of France. This provided stimulus for work on mechanisms of toxicity as well as medical countermeasures to poisoning. Since the 1960s, toxicology has entered a phase of rapid development and has changed from a science that was almost entirely descriptive to one in which the study of mechanisms has become the prime task. The many reasons for this include the development of new analytical methods since 1945, the emphasis on drug testing following the thalidomide tragedy, the emphasis on pesticide testing following the publication of Rachel Carson's *Silent Spring* and public concern over environmental pollution and disposal of hazardous waste.

1.4. Information Resources

Because of the complexity of toxicology as a science and the impact of toxicological investigations on legislation and commerce, a wide range of information on the toxic effects of chemicals is available. No single, exhaustive source of toxicological data exists; several sources are required to obtain comprehensive information on a particular chemical. Printed sources are often quicker and easier to use than computer data bases, but interactive online searching can rapidly gather important information from the huge number of sources present.

The information explosion in toxicology has resulted in a comprehensive volume dedicated to toxicological information sources:

P. Wexler, P. J. Hakkinen, G. Kennedy, Jr. F. W.Stoss, *Information Resources in Toxicology*, 3rd ed., Academic Press, 1999.

Textbooks. The easiest way to obtain information on general topics in toxicology and secondary references are a range of textbooks available on the market. Only a few selected books are listed below:

C. D. Klaasen, *Casarett and Doull's Toxicology; The Basic Science of Poisons*, 6th ed., McGraw-Hill, New York, 2001.

G. D. Clayton, F. E. Clayton (eds): *Patty's Industrial Hygiene and Toxicology*, Wiley, New York, 1993.

J. G. Hardman, L. E. Limbird, *Goodman and Gilman's, The Pharmacological Basis of Therapeutics*, 10th ed., McGraw-Hill, New York, 2001.

W. A. Hayes, *Principles and Methods of Toxicology*, 3rd ed., Raven Press, New York, 2001.

E. Hodgson (Ed.): *Textbook of Modern Toxicology*, 3rd ed., Wiley Interscience, 2004.

T. A. Loomis, A. W. Hayes, *Loomis's Essentials of Toxicology*, 4th ed., Academic Press, San Diego, 1996.

The huge volume by N. I. Sax and R. J. Lewis, *Dangerous Properties of Industrial Materials*, 7th ed., Wiley, New York, 1999, contains basic toxicological data on a large selection of chemicals (almost 20 000) and may serve as a useful guide to the literature for compounds not covered in other publications.

Monographs. The best summary information on toxicology is published in the form of series by governments and international organizations. Most of these series are summarizing the results of toxicity studies on specific chemicals. The selection of these chemicals is mainly based on the extent of their use in industry (e.g. trichloroethene), their occurrence as environmental contaminants (mercury) or their extraordinary toxicity (e.g. 2,3,7,8-tetrachlorodibenzodioxin):

American Conference of Governmental Industrial Hygienists, Threshold Limit Values and Biological Exposure Indices (Cincinnati, OH). Published annually.

MAK-Begründungen, VCH Publishers, Weinheim, Federal Republic of Germany. This German series includes detailed information on the toxicity of chemicals on the German MAK list (ca. 150 reports are available; the series is continuously expanded).

The Commission of the European Communities publishes the Reports of the Scientific Committee on Cosmetology and the Reports of the Scientific Committee for Food.

The Environmental Protection Agency (EPA) publishes a huge number of reports and toxicological profiles. They are indexed in "EPA Publications. A Quarterly Guide."

The European Chemical Industry Ecology and Toxicology Centre (ECETOC) issues "Monographs" (more than 20 have been published) and "Joint Assessments of Commodity Chemicals."

The monographs of the International Agency for Research on Cancer are definitive evaluations of carcinogenic hazards. The "Environmental Health Criteria" documents of the International Programme on Chemical Safety (IPCS) assess environmental and human health effects of exposure to chemicals, and biological or physical agents. A related "Health and Safety Guide" series give guidance on setting exposure limits for national chemical safety programs.

The National Institute for Occupational Safety and Health (NIOSH), has published 50 "Current Intelligence Bulletins" on health hazards of materials and processes at work.

The technical report series of the National Toxicology Program (NTP) reports results of their carcinogenicity bioassays, which include summaries of the toxicology of the chemicals studied. A status report indexes both studies that are under way and those that have been published. The program also issues an "Annual Review of Current DHHS [U.S. Department of Health and Human Services], DOE [U.S. Department of Energy] and EPA Research" related to toxicology.

A large number of internet-based resources are also available to collect information on toxic effects of chemicals and methods for risk assessment. Some information sites containing large amounts of downloadable information are listed below:

US Environmental Protection Agency (EPA), Integrated Risk Information System (IRIS), http://www.epa.gov/iris/index.html

US Environmental Protection Agency (EPA), ECOTOX Database, http://www.epa.gov/ecotox/

Organisation for Economic Co-operation and Development (OECD), test guidelines, http://www.oecd.org

Agency for Toxic Substances and Disease Registry (ATSDR), toxicological profile information sheet http://www.atsdr.cdc.gov/toxprofiles/

European Chemicals Bureau, http://ecb.jrc.it/

National Toxicology Programm, http://ntp-server.niehs.nih.gov/htdocs/liason/-Factsheets/FactsheetList.html

United Nations Environment Programm, Chemicals http://www.chem.unep.ch/

Journals Results of toxicological research are published in more than 100 journals. Those listed below mainly publish research closely related to toxicology, but articles of relevance may also be found in other biomedical journals:

Archives of Environmental Contamination and Toxicology
Archives of Toxicology
Biochemical Pharmacology
Chemical Research in Toxicology
CRC Critical Reviews in Toxicology
Clinical Toxicology
Drug and Chemical Toxicology
Environmental Toxicology and Chemistry

Table 2. Toxic effects of different chemicals categorized by time scale and general locus of action

Exposure	Site	Effect	Chemical
Acute	local	lung edema	chlorine gas
	systemic	liver damage	carbon tetrachloride
		narcosis	halothane
Subchronic	local	sensitization	toluene diisocyanate
	systemic	neurotoxicity	hexane
Chronic	local	bronchitis	sulfur dioxide
		nasal carcinoma	formaldehyde
	systemic	bladder carcinoma	4-amino-biphenyl
		kidney damage	cadmium

Food and Chemical Toxicology
Fundamental and Applied Toxicology
Journal of the American College of Toxicology
Journal of Analytical Toxicology
Journal of Applied Toxicology
Journal of Biochemical Toxicology
Journal of Toxicology and Environmental Health
Neurotoxicology and Teratology
Pharmacology and Toxicology
Practical In Vitro Toxicology
Regulatory Toxicology and Pharmacology
Reproductive Toxicology
Toxicology
Toxicology and Applied Pharmacology
Toxicology and Industrial Health
Toxicology In Vitro
Toxicology Letters

Databases and Databanks. Electronic sources, such as computer data bases or CD-ROM are a fast and convenient way to obtain references on the toxicity of chemicals. Since on-line searching of commercial data bases such as STN-International may be expensive, CD-ROM-based systems are increasingly being used. The major advantages are speed, the ability to refine searches and format the results, and non-text search options, such as chemical structure searching on Beilstein and Chemical Abstracts.

Useful information about actual research on the toxicology of chemicals may be obtained by searching Chemical Abstracts or Medline with the appropriate keywords. Specific data banks covering toxicology are the Registry of Toxic Effects of Chemical Substances, which gives summary data, statistics, and structures; Toxline (available in DIMDI) gives access to the literature.

1.5. Terminology of Toxic Effects

Toxic effects may be divided according to timescale (acute and delayed), general locus of action (local, systemic, organ specific), or basic mechanisms of toxicity (reversible versus irreversible). *Acute toxic effects* are those that occur after brief exposure to a chemical. Acute toxic effects usually develop rapidly after single or multiple administrations of a chemical; however, acute exposure may also produce *delayed toxicity*. For example, inhalation of a lethal dose of HCN causes death in less than a minute, whereas lethal doses of 2,3,7,8-tetrachlorodibenzodioxin will result in the death of experimental animals after more than two weeks. *Chronic effects* are those that appear after repetitive exposure to a substance; many compounds require several months of continuous exposure to produce adverse effects. Often, the chronic effects of chemicals are different from those seen after acute exposure (Table 2). For example, inhalation of chloroform for a short period of time may cause anesthesia; long-term inhalation of much lower chloroform concentrations causes liver damage. Carcinogenic effects of chemicals usually have a long latency period; tumors may be observed years (in rodents) or even decades (in humans) after exposure.

Toxic effects of chemicals may also be classified based on the type of interaction between the chemical and the organism. Toxic effects may be caused by reversible and irreversible interactions (Table 3). When reversible interactions are responsible for toxic effects, the concentration of the chemical present at the site of action is the only determinant of toxic outcome. When the concentration of the xenobiotic is decreased by excretion or biotransformation, a parallel decrease of toxic effects is observed.

Table 3. Reversible and irreversible interactions of chemicals with cellular macromolecules as a basis for toxic response

Mechanism	Toxic response	Example
Irreversible inhibition		
of Esterase	neurotoxicity	tri-*o*-cresylphosphate
Covalent binding to DNA	cancer	dimethylnitrosamine
Reversible binding to		
Hemoglobin	oxygen deprivation in tissues	carbon monoxide
Cholinesterase	neurotoxicity	carbamate pesticides

After complete excretion of the toxic agent, toxic effects are reduced to zero (see below). A classical example for reversible toxic effects is carbon monoxide. Carbon monoxide binds to hemoglobin and, due to the formation of the stable hemoglobin–carbon monoxide complex, binding of oxygen is blocked. As a result of the impaired oxygen transport in blood from the lung, tissue oxygen concentrations are reduced and cells sensitive to oxygen deprivation will die. The toxic effects of carbon monoxide are directly correlated with the extent of carboxyhemoglobin in blood, the concentration of which is dependent on the inhaled concentration of carbon monoxide. After exhalation of carbon monoxide and survival of the acute intoxication, no toxic effect remains (Fig. 2).

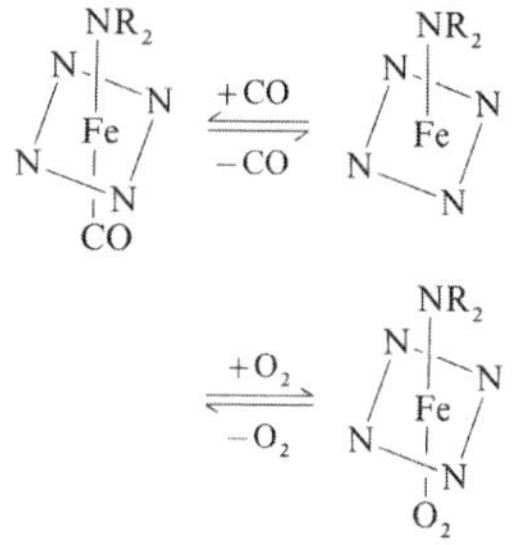

Figure 2. Reversible binding of carbon monoxide to hemoglobin and inhibition of oxygen transport

Irreversible toxic effects are often caused by the covalent binding of toxic chemicals to biological macromolecules. Under extreme conditions, the modified macromolecule is not repaired; after excretion of the toxic agent, the effect persists. Further exposure to the toxic agent will produce additive effects; many chemicals carcinogens are believed to act through irreversible changes (see Section 2.5.6).

Another distinction between types of effects may be made according to the general locus of action. *Local toxicity* occurs at the site of first contact between the biological system and the toxic agent. Local effects to the skin, the respiratory tract, or the alimentary tract may be produced by skin contact with a corrosive agent, by inhalation of irritant gases, or by ingestion of tissue-damaging materials. This type of toxic responses is usually restricted to the tissues with direct contact to the agent. However, life-threatening intoxications may occur if vital organs like the lung are damaged. For example, inhaled phosgene damages the alveoli of the lung and causes lung edema. The massive damage to the lung results in the substantial mortality observed after phosgene intoxication.

The opposite to local effects are *systemic effects*. They are characterized by the absorption of the chemical and distribution from the port of entry to a distant site where toxic effects are produced. Except for highly reactive xenobiotics, which mainly act locally, most chemicals act systemically. Many chemicals that produce systemic toxicity only cause damage to certain organs, tissues, or cell types within organs. Selective damage to certain organs or tissues by systemically distributed chemicals is termed organ- or tissue-specific toxicity [2]; the organs damaged are referred to as target organs (Table 4).

Table 4. Organ-specific toxic effects induced by chemicals that are distributed systemically in the organism

Chemical	Species	Target organ
Benzene	humans	bone marrow
Hexachlorobutadiene	rodents	damage to proximal tubules of the kidney
Paraquat	rodents, humans	lung
Tri-*o*-cresylphosphate	humans	nervous system
Cadmium	humans	kidney
1,2-Dibromo-3-chloropropane	humans, rodents	testes
Hexane	rodents, humans	nervous system
Anthracyclines	humans	heart

Major target organs for toxic effects are the central nervous system and the circulatory system followed by the blood and hematopoietic system and visceral organs such as the liver or the kidney. For some chemicals, both local and systemic effects can be demonstrated; moreover, chemicals producing marked local toxicity may also cause systemic effects as secondary responses to major disturbances in homeostasis of the organism.

1.6. Types of Toxic Effects

The spectrum of toxic effects of chemicals is broad, and their magnitude and nature depend on many factors such as the physiocochemical properties of the chemical and its toxicokinetics, the conditions of exposure, and the presence of adaptive and protective mechanisms. The latter factors include physiological mechanisms such as adaptive enzyme induction, DNA repair, and others. Toxic effects may be transient, reversible, or irrversible; some are deleterious and others are not. Toxic effects may take the form of tissue pathology, aberrant growth processes, or altered biochemical pathways. Some of the more frequently encountered types of injury constituting a toxic response are described in the following.

Immune-mediated hypersensitivity reactions by antigenic materials are toxic effects often involved in skin and lung injury by repeated contact to chemicals resulting in contact dermatitis and asthma. Inflammation is a frequently observed local response to the application of irritant chemicals or may be a component of systemic injury. This response may be acute with irritant or tissue damaging materials or chronic with repetitive exposure to irritants. Necrosis, that is, death of cells or tissues, may be the result of various pathological processes resulting from biochemical interactions of xenobiotics, as described in Chapter 3. The extent and patterns of necrosis may be different for different chemicals, even in the same organ. Chemical tumorigenesis or carcinogenesis (induction of malignant tumors) is an effect often observed after chronic application of chemicals. Due the long latency period and the poor prognosis for individuals diagnosed with cancer, studies to predict the potential tumorigenicity of chemicals have developed into a major area of toxicological research. Developmental and reproductive toxicology are concerned with adverse effects on the ability to conceive, and with adverse effects on the structural and functional integrity of the fetus. Chemicals may interfere with reproduction through direct effects on reproductive organs or indirectly by affecting their neural and endocrine control mechansims. Developmental toxicity deals with adverse effects on the conceptus through all stages of pregnancy. Damage to the fetus may result in embryo reabsorption, fetal death, or abortion. Nonlethal fetotoxicity may be expressed as delayed maturation, decreased birth weight, or structural malformation. The most sensitive period for the induction of malformation is during organogenesis; neurobehavioral malformations may be induced during later stages of pregnancy.

1.7. Dose–Response: a Fundamental Issue in Toxicology

In principle, a poison is a chemical that has an adverse effect on a living organism. However, this is not a useful definition since toxic effects are related to dose. The definition of a poison thus also involves quantitative biological aspects. At sufficiently high doses, any chemical may be toxic. The importance of dose is clearly seen with molecular oxygen or dietary metals. Oxygen at a concentration of 21% in the atmosphere is essential for life, but 100% oxygen at atmospheric pressure causes massive lung injury in rodents and often results in death. Some metals such as iron, copper, and zinc are essential nutrients. When they are present in insufficient amounts in the human diet, specific disease patterns develop, but in high doses they can cause fatal intoxications. Toxic compounds are not restricted to man-made chemicals, but also include many naturally occurring chemicals. Indeed, the agent with the highest toxicity is a natural poison found in the bacterium *Clostridium botulinum* (LD_{50} 0.01 μ/kg).

Therefore, all toxic effects are products of the amount of chemical to which the organism is exposed and the inherent toxicity of the chemical; they also depend on the sensitivity of the biological system.

The term “dose” is most frequently used to characterize the total amount of material to which an organism is exposed; dose defines the amount of chemical given in relation to body weight. Dose is a more meaningful and comparative indicator of exposure than the term exposure itself. Dose usually implies the exposure dose, the total amount of chemical administered to an organism or incorporated into a test system. However, dose may not be directly proportional to the toxic effects since toxicity depends on the amount of chemical absorbed. Usually, dose correctly describes only the actual amount of chemical absorbed when the chemical is administered orally or by injection. Under these circumstances, the administered dose is identical to the absorbed dose; other routes of application such as dermal application or inhalation do not define the amount of agent absorbed.

Different chemicals have a wide spectrum of doses needed to induce toxic effects or death. To characterize the acute toxicity of different chemicals, LD_{50} values are frequently used as a basis for comparisons. Some LD_{50} values (rat) for a range of chemicals follow:

Ethanol	12 500
Sodium bicarbonate	4 220
Phenobarbital sodium	350
Paraquat	120
Aldrin	46
Sodium cyanide	6.4
Strychnine	5
1,2-Dibromoethane	0.4
Sodium fluoroacetate	0.2
2,3,7,8-Tetrachlorodibenzodioxin	0.01

Certain chemicals are very toxic and produce death after administration of microgram doses, while others are tolerated without serious toxicity in gram doses. The above data clearly demonstrate that the toxicity of a specific chemical is related to dose. The dependence of the toxic effects of a specific chemical on dose is termed dose–response relationship. Before dose–response relationships can be appropriately used, several basic assumptions must be considered. The first is that the response is due to the chemical administered. It is usually assumed that the responses observed were a result of the various doses of chemical administered. Under experimental conditions, the toxic response usually is correlated to the chemical administered, since both exposure and effect are well defined and can be quantified. However, it is not always apparent that the response is the result of specific chemical exposure. For example, an epidemiologic study might result in discovery of an "association" between a response (e.g., disease) and one or more variables including the estimated dose of a chemical. The true doses to which individuals have been exposed are often estimates, and the specificity of the response for that chemical is doubtful.

Further major necessary assumptions in establishing dose–response relationships are:

- A molecular site (often termed receptor) with which the chemical interacts to produce the response. Receptors are macromolecular components of tissues with which a chemical interacts and produces its characteristic effect.
- The production of a response and the degree of the response are related to the concentration of the agent at the receptor.
- The concentration of the chemical at the receptor is related to the dose administered. Since in most cases the concentration of an administered chemical at the receptor cannot be determined, the administered dose or the blood level of the chemical is used as an indicator for its concentration at the molecular site.

A further prerequisite for using the dose–response relationship is that the toxic response can be exactly measured. A great variety of criteria or end points of toxicity may be used. The ideal end point should be closely associated with the molecular events resulting from exposure to the toxin and should be readily determined. However, although many end points are quantitative and precise, they are often only indirect measures of toxicity. For example, changes in enzyme levels in the blood can be indicative of tissue damage. Patterns of alterations may provide insight into which organ or system is the site of toxic effects. These measures usually are not directly related to the mechanism of toxic action. The dose–response relationship combines the characteristic of exposure and the inherent toxicity of the chemical. Since toxic responses to a chemical are usually functions of both time and dose, in typical dose–response relationships, the maximum effect observed during the time of observation is plotted against the dose to give time-independent curves. The time-independent dose–response relationship may be used to study dose–response for both reversible and irreversible toxic effects. However, in risk assessments that consider the induction of irreversible effects such as cancer, the time factor plays a major role and has important influences on the magnitude or likelihood of toxic responses. Thus, for this type of mechanism of toxic action, dose–time–response relationships are better descriptors of toxic effects.

The dose–response relationship is the most fundamental concept in toxicology. Indeed, an understanding of this relationship is essential for the study of toxic chemicals.

From a practical point of view, there are two different types of dose–response relationships. Dose–response relationships may be quantal (all or nothing responses such as death) or graded. The graded or variable response involves a continual change in effect with increasing dose, for example, enzyme inhibition or changes in physiological function such as heart rate. Graded responses may be determined in an individual or in simple biochemical systems. For example, addition of increasing concentrations of 2,3,7,8-tetrachlorodibenzodioxin to cultured mammalian cells results in an increase in the concentration of a specific cytochrome P450 enzyme in the cells (for details of mechanisms, see Section 2.5.4.1). The increase is clearly dose related and spans a wide range (Fig. 3). An example for a graded toxic effect in an individual may be inflammation caused by skin contact with an irritant material. Low doses cause slight irritation; as the amount increases, irritation turns to inflammation and the severity of inflammation increases.

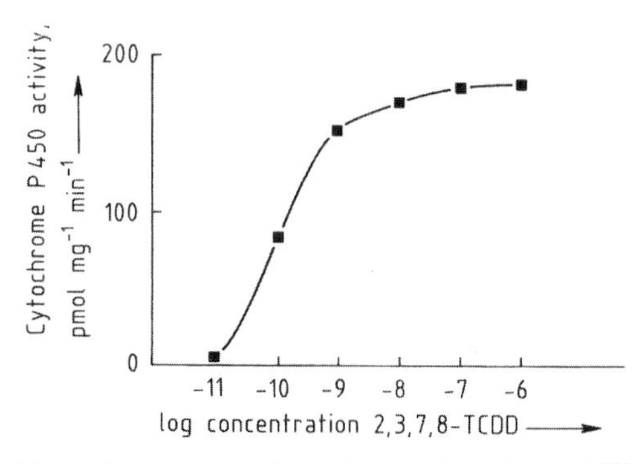

Figure 3. Dose-dependent induction of cytochrome P450 1A 1 protein in cultured liver cells treated with 2,3,7,8-tetrachlorodibenzodioxin [3]

In dose–response studies in a population, a specific endpoint is also identified and the dose required to produce this end point is determined for each individual in the population. Both dose-dependent graded effects and quantal responses (death, induction of a tumor) may be investigated. With increasing amount of a chemical given to a group of animals, the magnitude of the effect and/or the number of animals affected increase. For example, if an irritant chemical is applied to the skin, as the amount of the material increases, the numbers of animals affected and the severity of inflammation increases. Quantal responses such as death induced by a potentially lethal chemical will also be dose-dependent. The dose dependency of a quantal effect in a population is based on individual differences in the response to the toxic chemical. A specific amount of the potentially lethal xenobiotic given to a group of animals may not kill all of them, but as the amount given increases, the proportion of animals killed increases.

Althought the distinctions between graded and quantal dose–response relationships are useful, the two types of responses are conceptually identical. The ordinate in both cases is simply labeled response, which may be the degree of response in an individual, or the fraction of a population responding, and the abscissa is the range of administered doses.

1.7.1. Graphics and Calculations

Even with a genetically homogenous population of animals of the same species and strain, the proportion of animals showing the effect will increase with dose (Fig. 4A). When the number of animals responding is plotted versus the logarithm of the dose, a typical sigmoid curve with a log-normal distribution that is symmetrical about the midpoint, is obtained (Fig. 4B).

When plotted on a log-linear scale, the obtained normally distributed sigmoid curve approaches a response of 0% as the dose is decreased, and 100% as the dose is increased, but theoretically never passes through 0 or 100%. Small proportions of the population at the right- and left-hand sides of the curve represent hyposusceptible and hypersusceptible members. The slope of the dose–reponse curve around the 50% value, the midpoint, gives an indication of the

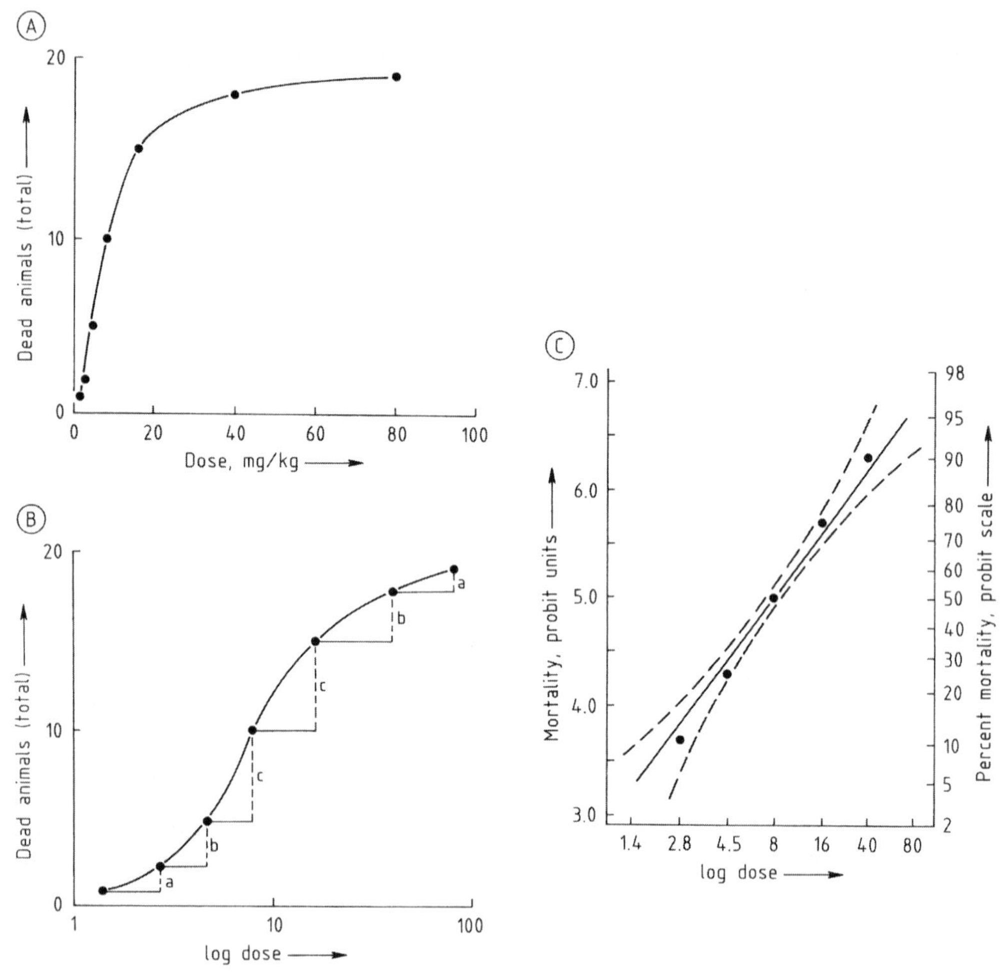

Figure 4. Typical dose – response curves for a toxic effect
Plots are linear – linear (A); log – linear (B); and log – probit (C) for an identical set of data

ranges of doses producing an effect. A steep dose–response curve indicates that the majority of the population will respond over a narrow dose range; a shallow dose–response curve indicates that a wide range of doses is required to affect the majority of the population. The curve depicted in Fig. 4B shows that the majority of the individuals respond about the midpoint of the curve. This point is a convenient description of the average response, and is referred to as the median effective dose (ED_{50}). If mortality is the endpoint, then this dose is referred as median lethal dose (LD_{50}).

Death, a quantal response, is simple to quantify and is thus an end point incorporated in many acute toxicity studies. Lethal toxicity is usually calculated initially from specific mortality levels obtained after giving different doses of a chemical; the 50% mortality level is used most frequently since it represents the midpoint of the dose range at which the majority of deaths occur. This is the dose level that causes death of half of the population dosed. The LD_{50} values are usually given in milligrams of chemical per kilogram of body weight (from the viewpoint of chemistry and for comparison of relative potencies of different chemicals, giving the LD_{50} in moles of chemical per kilogram body weight would be desirable). After inhalation, the reference is to LC_{50} (LC = lethal concentration), which, in contrast to LD_{50} values, depends on the time of exposure; thus, it is usually expressed as X-hour LC_{50} value. The LD_{50} or LC_{50} values usually represent the initial information on the toxicity of a chemical and must be regarded as a first, but not a quantitative, hazard indicator that may be useful for comparison of the acute toxicity of different chemicals [3].

Similar dose–effect curves can, however, be constructed for cancer, liver injury, and other types of toxic responses. For the determination of LD_{50} values and for obtaining comparative information on dose–response curves, plotting log dose versus percent response is not practical since large numbers of animals are needed for obtaining interpretable data. Moreover, other important information on the toxicity of a chemical (e.g., LD_{05} and LD_{95}) cannot be accurately determined due to the slope of sigmoid curve. Therefore, the dose–response curve is transformed to a log-probit (probit = probability units) plot. The data in the Fig. 4B form a straight line when transformed into probit units (Fig. 4C). The EC_{50} or, if death is the end point, the LD_{50} is obtained by drawing a horizontal line from the probit unit 5, which is the 50% response point, to the dose–effect line. At the point of intersection a vertical line is drawn, and this line intersects the abscissa at the LD_{50} point. Information on the lethal dose for 90% or for 10% of the population can also be derived by a similar procedure. The confidence limits are narrowest at the midpoint of the line (LD_{50}) and are widest at the two extremes (LD_{05} and LD_{95}) of the dose–response curve. In addition to permitting determination of a numerical value for the LD_{50} of a chemical with few groups of dosed animals, the slope of the dose–response curve for comparison between toxic effects of different chemicals is obtained by the probit transformation [4].

The LD_{50} by itself, however, is an insufficient index of lethal toxicity, particulary if comparisons between different chemicals are to be made. For this purpose, all available dose–response information including the slope of the dose–response line should be used. Figure 5 demonstrates the dose–response curves for mortality for two chemicals.

The LD_{50} of both chemicals is the same (10 mg/kg). However, the slopes of the dose–response curves are quite different. Chemical A exhibits a "flat" dose–response curve: a large change in dose is required before a significant change in response will be observed. In contrast, chemical B exhibits a "steep" dose–response curve, that is, a relatively small change in dose will cause a large change in response. The chemical with the steep slope may affect a much larger proportion of the population by incremental increases in dose than chemicals having a shallow slope; thus, acute overdosing may be a problem affecting the majority of a population for chemicals with steeper slopes. Chemicals with shallower slopes may represent a problem for the hyperreactive groups at the left-hand side of the dose–response curve. Effects may occur at significantly lower dose levels then for hyperreactive groups exposed to chemicals with a steep dose–response.

While the LD_{50} values characterize the potential hazard of a chemical, the risk of an exposure is determined by the hazard multiplied by the exposure dose. Thus, even very toxic chemicals like the poison of *Clostridium botulinum* pose only a low risk; intoxications with this compound are rare since exposure is low. Moreover, acute intoxications with other highly toxic agents such as mercury salts are rarely seen, despite detectable blood levels of mercury salts in the general population, since the dose is also low. On the other hand, compounds with low toxicity may pose a definite health risk when doses are high, for example, constituents of diet or chemicals formed during food preparation by heat treatment.

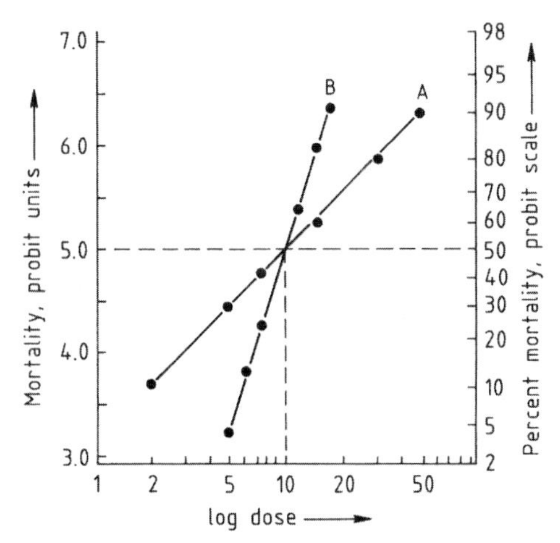

Figure 5. Comparison of dose – response relationships for two chemicals (log – probit plot)
Both chemicals have identical LD_{50} values, but different slopes of the dose – response curve

Therefore, for characterizing the toxic risk of a chemical, besides information on the toxicity, information on the conditions of exposure are necessary. When using LD_{50} values for toxicity characterisation, the limitations of LD_{50} values should be explicitly noted. These limitations include methodological pitfalls influenced by

1) Strain of animal used
2) Species of animal used
3) Route of administration
4) Animal housing

and intrinsic factors limiting the use of LD_{50} values

1) Statistical method
2) No dose–response curve
3) Time to toxic effect not determined
4) No information on chronic toxicity

The most serious limitation on the use of LD_{50} values for hazard characterization are the lack of information on chronic effects of a chemical and the lack of dose–response information. Chemicals with low acute toxicity may have carcinogenic or teratogenic effects at doses that do not induce acute toxic responses. Other limitations include insufficient information on toxic effects other than lethality, the cause of death, and the time to toxic effect. Moreover, LD_{50} values are not constant, but are influenced by many factors and may differ by almost one order of magnitude when determined in different laboratories.

1.8. Dose-Response Relationships for Cumulative Effects

After chronic exposure to a chemical, toxic response may be caused by doses not showing effects after single dosing. Chronic toxic responses are often based on accumulation of either the toxic effect or of the administered chemical. Accumulation of the administered chemical is observed when the rate of elimination of the chemical is lower than the rate of administration. Since the rate of elimination is dependent on plasma concentrations, after long-term application an equilibrium concentration of the chemical in the blood is reached. Chemicals may also be stored in fat (polychlorinated pesticides such as DDT) or bone (e.g., lead). Stored chemicals usually do not cause toxic effects because of their low concentrations at the site of toxic action (receptor). After continuous application, the capacities of the storage tissues may become saturated, and xenobiotics may then be present in higher concentration in plasma and thus at the site of action; toxic responses result. Besides cumulation of the toxic agent, the toxic effect may also cumulate (Fig. 6).

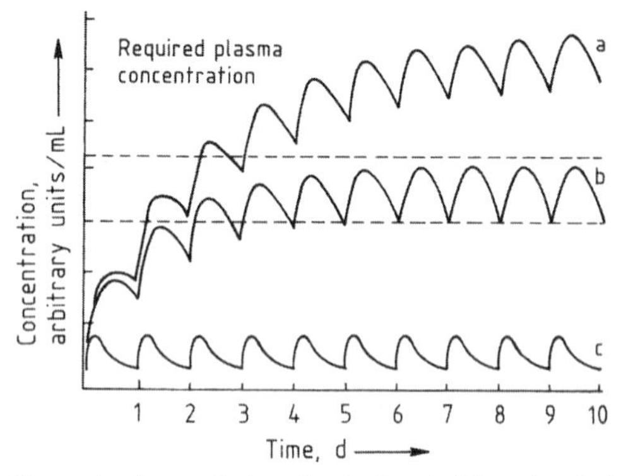

Figure 6. Accumulation of toxic chemicals based on their rate of excretion
a) The rate of excretion is equal to the rate of absorption, no accumulation occurs; b) Chemical accumulates due to a higher rate of uptake and inefficient excretion; the plasma concentrations are, however, not sufficient to exert toxic effects; c) The plasma concentrations reached after accumulation are sufficient to exert toxicity

For chemicals which irreversibly bind to macromolecules, the magnitude of toxic responses may be correlated with the total dose administered. In contrast to chemicals which act reversibly, the effect is not dependent on the frequency of dosing. Effect accumulation is often observed with carcinogens and ionizing radiation. In Figure 7 accumulation of effects is exemplified by the time- and dose-dependent induction of tumors by 4-(dimethylamino)azobenzene, a potent chemical carcinogen [5]. The TD_{50} values (50% of the treated animals carry tumors) are used to characterize the potency. Identical tumor incidences were observed after high doses and a short exposure time or after low doses and long exposure; the tumor incidence was only dependent on the total dose administered.

Reversibility of toxic responses also depends on the capacity of an organ or tissue to repair injury. For example, kidney damage by xenobiotics is often, after survival of the acute phase of the intoxication, without further consequence due to the high capacity of the kidney for cell proliferation and thus the capacity to repair organ damage [6]. In contrast, injury to the central nervous system is largely irreversible since the differentiated cells of the nervous system cannot divide and dead cells cannot be replaced.

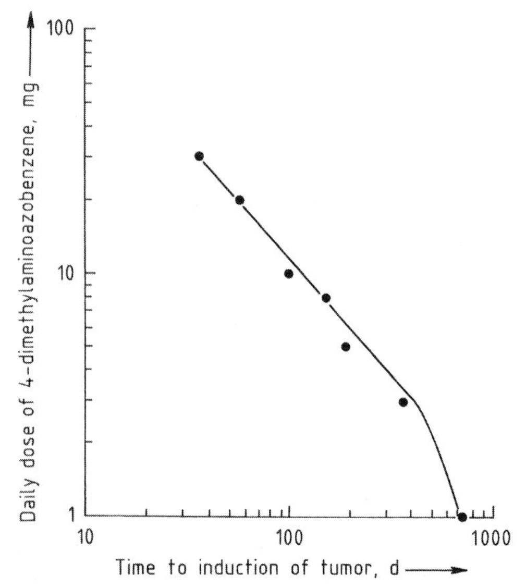

Figure 7. Time-dependent induction of tumors after different daily doses of 4-dimethylaminoazobenzene in rats [5]

1.9. Factors Influencing Dose–Response

In animals and humans, the nature, severity, and incidence of toxic responses depend on a large number of exogenous and endogenous factors [7]. Important factors are the characteristics of exposure, the species and strain of animals used for the study, and interindividual variability in humans [8]. Toxic responses are caused by a series of complex interactions of a potentially toxic chemical with an organism. The type and magnitude of the toxic response is influenced by the concentration of the chemical at the receptor and by the type of interaction with the receptor. The concentration of a chemical at the site of action is influenced by the kinetics of uptake and elimination; since these are time-dependent phenomena, toxic responses are also time-dependent. Thus, the toxic response can be separated into two phases: toxicokinetics and toxicodynamics (Fig. 8).

Toxicokinetics describe the time dependency of uptake, distribution, biotransformation, and excretion of a toxic agent (a detailed description of toxicokinetics is given in Section 2.5). Toxicodynamics describes the interaction of the toxic agent with the receptor and thus specific interactions of the agent (see below). Toxicokinetics may be heavily influenced by species, strain, and sex and the exposure characteristics [9–13]. Differences in toxic response between species, route of exposure, and others factors are often dependent on influences on toxicokinetics. Since toxicodynamics (mechanism of action) are assumed to be identical between species, this provides the basis for a rational interspecies extrapolation of toxic effects when differences in toxicokinetics are defined.

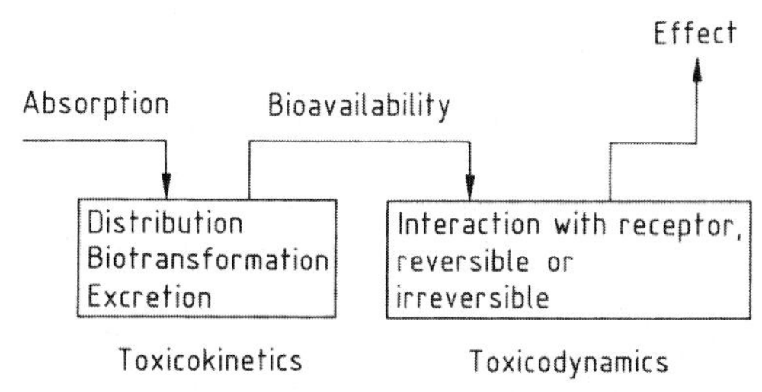

Figure 8. Toxicokinetics and toxicodynamics as factors influencing the toxic response

1.9.1. Routes of Exposure

The primary tissue or system by which a xenobiotic comes into contact with the body, and from where it may be absorbed in order to exert systemic toxicity, is the route of exposure. The frequent circumstances of environmental exposure are ingestion (peroral), inhalation, and skin contact. Also, for investigational and therapeutic purposes, intramuscular, intravenous, and subcutaneous injections may also be routes of exposure.

The major routes by which a potentially toxic chemical can enter the body are – in descending order of effectiveness for systemic delivery – injection, inhalation, absorption from the intestinal tract, and cutaneous absorption. The relationship between route and exposure, biotransformation, and potential for toxicity, may be complex and is also influenced by the magnitude and duration of dosing (Table 5).

The route of exposure has a major influence on toxicity because of the effect of route of exposure on the bioavailability of the toxic agent. The maximum tissue levels achieved, the time to maximum tissue levels, and thus the duration of the effect are determined by the rate of absorption and the extent of distribution within the system.

Table 5. Toxicity of chemicals applied by different routes of exposure (data taken from [13])

Chemical	Species	Route of application	LD_{50}, mg/kg
DDT	rat	intravenous	68
	rat	oral	113
	rat	skin contact	1931
Atropine sulfate	rat	intravenous	41
	rat	oral	620
1-Chloro-2,4-dinitrobenzene	rat	oral	1070
	rat	intraperitoneal	280
	rabbit	skin contact	130
Dieldrin	rat	oral	46
	rat	intravenous	9
	rat	skin contact	10

Direct injection into veins is usually restricted to therapeutic applications, but it is important for the toxicology of intravenously injected drugs in addicts. Chemicals applied by intravenous injection are rapidly distributed to well-perfused organs in the blood and thus may result in the rapid induction of toxic effects. The rapid dilution of a chemical after intravenous injection by venous blood permits even the injection of locally acting or corrosive chemicals which are well tolerated. The likelihood of toxicity from inhaled chemicals depends on a number of factors, of which the physical state and properties of the agent, concentration, and time and frequency of exposure are important. Major influences on the absorption and disposition of xenobiotics are exerted by species peculiarities since the anatomy of the respiratory tract and the physiology of respiration show major differences between rodents and humans. The water solubility of a gaseous xenobiotic has a major influence on penetration into the respiratory tract. As water solubility decreases and lipid solubility increases, penetration into deeper regions of the lung, the bronchioli, and the alveoli becomes more effective. Water-soluble molecules such as formaldehyde,are effectively scavenged by the upper respiratory tract and may have toxic effects on the eye and throat. In contrast, gases with low water solubility such as phosgene may penetrate through the bronchii and bronchioli to the alveoli. Damage to the alveolar surface may initiate a series of events that finally results in lung edema. The degree to which inhaled gases, vapors, and particulates are absorbed, and hence their potential to produce systemic toxicity, depend on their diffusion rate through the alveolar mebrane, their solubility in blood and tissue fluids, the rate of respiration, and blood flow through the capillaries.

Uptake through the alimentary tract represents an important route of exposure for xenobiotics accumulated in the food chain, for natural constituents of human diet, and, drugs. Absorption from the gastrointestinal tract is dependent on the lipophilicity of a chemical, the molecular mass of the xenobiotic, and the presence of certain dietary constituents may influence the extent and rate of absorption. Chemicals absorbed from the gastrointestinal tract are transported to the liver via the portal vein; hepatic metabolism ("hepatic first-pass effect") may efficiently reduce the concentration of the xenobiotic available in the systemic circulation after oral uptake. Compounds undergoing bioactivation in the liver usually exhibit greater toxicity when given orally than when absorbed across the respiratory tract, due to the high proportion of material passing through the liver. In contrast, chemicals causing toxicity to extrahepatic, well-perfused organs such as the kidney often show a lower degree of toxicity to extrahepatic target organs when given orally.

Skin contact is an important route of exposure in the occupational and domestic environments. Local effects may include acute inflammation and corrosion, chronic inflammatory responses, immune-mediated reactions, and neoplasia. The percutaneous absorption of materials may also be a significant route for the absorption of systemically toxic materials. Factors influencing the percutaneous absorption of substances include skin site, integrity of skin, temperature, formulation, and physicochemical characteristics, including charge, molecular mass, and hydro- and lipophilicity.

1.9.2. Frequency of Exposure

The exposure of experimental animals may be categorized as acute, subacute, subchronic, and chronic. Acute exposures usually last less than 24 h, and all above-mentioned routes of exposure may be applied. With chemicals of low toxicity, repeated exposures may be used. Acute inhalation exposure is usually less than 24 h; frequently 4–8 h is chosen as timescale. Repeated exposure refers to application of the chemical for less than one month (subacute), one to three months (subchronic), and more than three months (chronic). Chronic exposures to detect specific toxic effects (carcinogenicity of a chemical) may span most of the lifetime of a rodent (up to two years). Repeated exposure may be by any route; the least labor intensive route is oral, by mixing the chemical with the diet; only for specific chemicals or to simulate likely routes of exposure for humans are application in drinking water, by gastric intubation, and by inhalation applied. These are more labor-intensive and require skilled personnel and/or sophisticated techniques and thus are more expensive.

The toxic effects observed after single exposure often are different form those seen after repeated exposure. For example, inhalation of high concentration of halothane causes anesthesia in animals and humans. In contrast, long-term application of halothane in lower doses causes liver damage in sensitive species The frequency of exposure in chronic studies is important for the temporal characterisation of exposure. Chemicals with slow rates of excretion may accumulate if applied at short dosing intervals, and toxic effects may result (see Section 1.6). Also, a chemical producing severe effect when given in a single high dose may have no detectable effects when given in several smaller doses. Interspecies and strain differences in susceptibility to chemical-induced toxicity may be due to heterogeneity of populations, species specific physiology (for example of the respiratory system), basal metabolic rate, size- and species-specific toxicokinetics and routes of metabolism or excretion (Table 6). In some cases, animal tests may give an underestimate, in others an overestimate, of potential toxicity to humans [14].

1.9.3. Species-Specific Differences in Toxicokinetics

Species-specific differences in toxic response are largely due to difference in toxicokinetics and biotransformation. Distribution and elimination characteristics are quite variable between species. Both qualitative and quantitative differences in biotransformation may effect the sensitivity of a given species to a toxic response (Table 7).

Table 6. Comparative LD_{50} values for four different chemicals in different animal species and estimated LD_{50} for humans

Chemical	Species	LD_{50}, mg/kg
Paraquat	rat	134
	mouse	77
	guinea pig	41
	human	32 – 48
Ethanol	rat	12 500
	mouse	8000
	guinea pig	5500
	human	3500 – 5000
Acetaminophen	rat	3763
	mouse	777
	guinea pig	2968
	human	42 800
Aspirin	rat	1683
	mouse	1769
	guinea pig	1102
	human	3492

Table 7. Species and sex differences in the acute toxicity of 1,1-dichloroethylene after oral administration and inhalation in rats and mice (data from World Health Organization, Geneva, 1990)

Species	Dosing criteria	Estimated LD_{50}/LC_{50}
Rat, male	inhalation/4 h	7000 – 32 000 mg/L
Rat, female	inhalation/4 h	10 300 mg/L
Mouse, male	inhalation/4 h	115 mg/L
Mouse, female	inhalation/4 h	205 mg/L
Rat, male	gavage	1550 mg/kg
Rat, female	gavage	1500 mg/kg
Mouse, male	gavage	201 – 235 mg/kg
Mouse, female	gavage	171 – 221 mg/kg

For example, the elimination half-live of 2,3,7,8-tetrachlorodibenzodioxin in rats is 20 d, and in humans it is estimated to be up to seven years [15]. An example for quantitative difference in the extent of biotransformation as a factor influencing toxic response is the species differences in the biotransformation of the inhalation anesthetic halothane. Both rats and guinea pigs metabolize halothane to trifluoroacetic acid, a reaction catalyzed by a specific cytochrome P450 enzyme [16–18]. As a metabolic intermediate, trifluoroacetyl chloride is formed, which may react with lysine residues in proteins and with phosphatidyl ethanolamine in phospholipids (Fig. 9).

This interaction initiates a cascade of events finally resulting in toxicity. The metabolism of halothane in guinea pigs occurs at much higher rates than in rats, so guinea pigs are sensitive to halothane-induced hepatotoxic effects and rats are resistant. Qualitative differences in biotransformation are responsible for apparent differences in the sensitivity of rats and guinea pigs to the bladder carcinogenicity of 2-acetylamidofluorene. In rats, 2-acetylamidofluorene is metabolized by *N*-oxidation by certain cytochrome P450 enzymes. The *N*-oxide is further converted to an electrophilic nitrenium ion which interacts with DNA in the bladder; this biotransformation pathway explains the formation of bladder tumors in rats after long-term exposure to 2-acetylamidofluoren. In guinea pigs, 2-acetylamidofluorene is metabolized by oxidation at the aromatic ring; since nitrenium ions cannot be formed by this pathway, guinea pigs are resistant to the bladder carcinogenicity of 2-acetylamidofluorene (Fig. 10).

With some chemicals, age may significantly affect toxicity, likely due to age related differences in toxicokinetics. The nutritional status may modify toxic response, likely by altering the concentration of cofactors needed for biotransformation and detoxication of toxic chemicals. Diet also markedly influences carcinogen-induced tumor incidence in animals [19] and may be a significant factor contributing to human cancer incidence.

The toxic response is influenced by the magnitude, number, and frequency of dosing. Thus, local or systemic toxicity produced by acute exposure may also occur by a cumulative process with repeated exposures to lower doses; also, additional toxicity may be seen in repeated-exposure situations. The relationships for cumulative toxicity by repetitive exposure compared with acute exposure toxicity may be complex, and the potential for cumulative toxicity from acute doses may not be quantitatively predictable. For repeated-exposure toxicity, the precise profiling of doses may significantly influence toxicity.

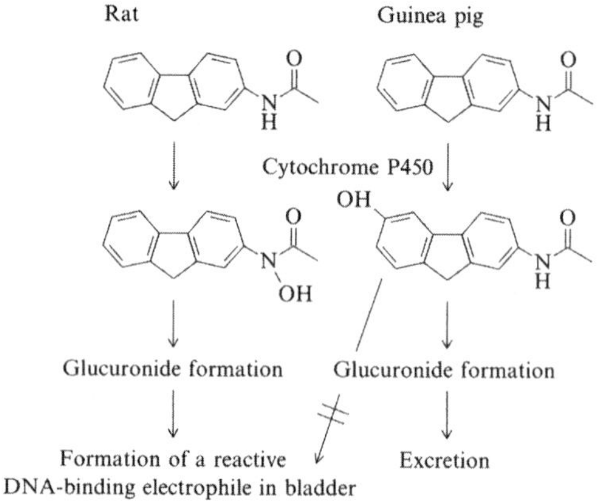

Rat
Guinea pig
Cytochrome P450
OH
Glucuronide formation
Glucuronide formation
Formation of a reactive DNA-binding electrophile in bladder
Excretion

Figure 10. Biotransformation pathways of 2-acetylamidofluorene in rats and guinea pigs

1.9.4. Miscellaneous Factors Influencing the Magnitude of Toxic Responses

A variety of other factors may affect the nature and exhibition of toxicity, depending on the conditions of the study, for example, housing conditions, handling, volume of dosing, vehicle, etc. Variability in test conditions and procedures may result in significant interlaboratory variability in results of otherwise standard procedures. For chemicals given orally or applied to the skin,

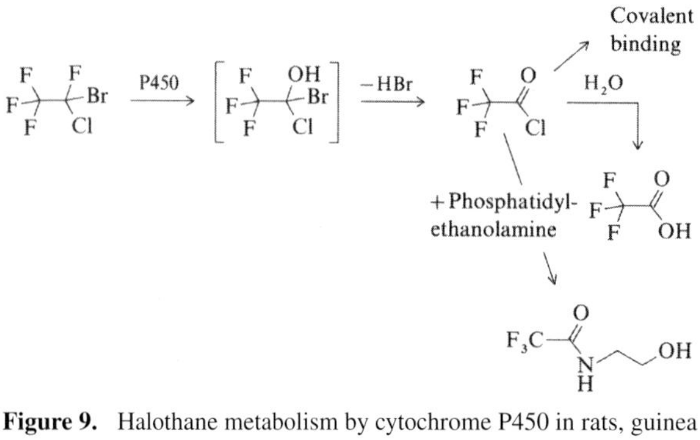

P450
−HBr
Covalent binding
H_2O
+Phosphatidyl-ethanolamine
F_3C

Figure 9. Halothane metabolism by cytochrome P450 in rats, guinea pigs, and humans

toxicity may be modified by the presence of materials in formulations which facilitate or retard the absorption of the chemicals. With respiratory exposure to aerosols, particle size significantly determines the depth of penetration and deposition in the respiratory tract and thus the site and extent of the toxic effects.

1.10. Exposure to Mixtures

In experimental animals most data on the toxic effects of chemicals are collected after exposure to a single chemical; in contrast, human exposure normally occurs to mixtures of chemicals at low doses. Moreover, prior, coincidential, and sucessive exposure of humans to chemicals is likely. Interactions between the toxic effects of different chemicals are difficult to predict, effects of exposure to different chemicals may be independent, additive, potentiating (ethanol and carbon tetrachloride), antagonistic (interference with action of other chemical, e.g., as seen with antidotes administered in case of intoxications), and synergistic. Ethanol exerts a potentiating effect on the hepatotoxicity of carbon tetrachloride. In rats pretreated with ethanol, the hepatotoxic effects of carbon tetrachloride are much more pronounced than in control animals. This potentiation is due to an increased capacity for bioactivation (see Section 2.4) of carbon tetrachloride in pretreated rats due to increased concentrations of a cytochrome P450 enzyme in the liver [20]. Thus, an important considerations for the assessement of potential toxic effects of mixtures of chemicals are toxicokinetics and toxicodynamic interactions. Toxicokinetic interactions of chemicals may influence absorption, distribution, and biotransformation, both to active and inactive metabolites. Mixtures of solvents often show a competitive inhibition of biotransformation. Usually, one of the components has high affinity for a specific enzyme involved in its biotransformation, whereas another component has only a low affinity for that particular enzyme. Thus, preferential biotransformation of the component with the high affinity occurs. Different outcomes of enzyme inhibition are possible: if the toxic effects of the component whose metabolism is inhibited is dependent on bioactivation, lower rates of bioactivation will result in decreased toxicity; if the toxic effects are independent of biotransformation, the extent of toxicity will increase due to slower rate of excretion. Toxic effects of mixtures may also not be due to a major component, but to trace impurities with high toxicity. For example, many long-term effects seen in animal studies on the toxicity of chlorophenols are believed to be due to 2,3,7,8-Tetrachlorodibenzodioxin, which was present as a minor impurity in the samples of chlorophenols used for these studies.

2. Absorption, Distribution, Biotransformation and Elimination of Xenobiotics

2.1. Disposition of Xenobiotics

The induction of systemic toxicity usually results from a complex interaction between absorbed parent chemical and biotransformation products formed in tissues; the distribution of both parent chemical and biotransformation products in body fluids and tissues; their binding and storage characteristics; and their excretion.

The biological effects initiated by a xenobiotic are not related simply to its inherent toxic properties; the initiation, intensity, and duration of response are a function of numerous factors intrinsic to the biological system and the administered dose. Each factor influences the ultimate interaction of the xenobiotic and the active site (Section 1.9). Only when the toxic chemical has reached the specific site and interacted with it can the inherent toxicity be realized. The route a xenobiotic follows from the point of administration or absorption to the site of action usually involves many steps and is termed toxicokinetics. Toxicokinetics influence the concentration of the xenobiotic or its active metabolite at the receptor. In the dose–response concept outlined in Section 1.9 and 1.7, it is generally assumed that the toxic response is proportional to the concentration of the xenobiotic at the receptor. However, the same dose of a chemical administered by different routes may cause different toxic effects. Moreover, the same dose of two different chemicals may result in vastly different concentrations of the chemical or its biotransformation products in a particular target organ. This differential pattern is due to differences in the disposition of a xenobiotic (Fig. 11).

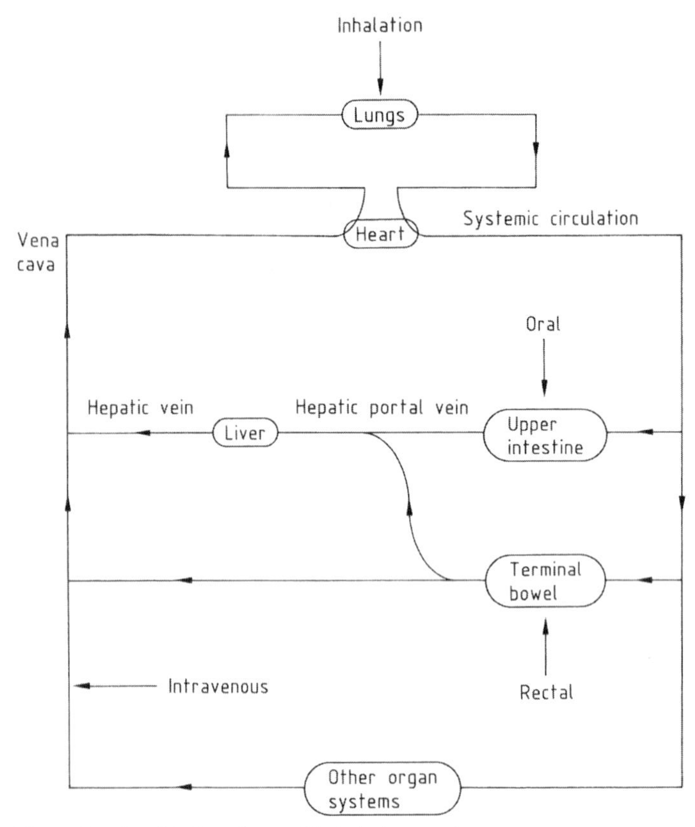

Figure 11. Possible fate of a xenobiotic in the organisms

The disposition of a xenobiotic consists of absorption, distribution, biotransformation, and excretion, which are all interrelated. The complicated interactions between the different processes of distribution are very important determinants of the concentration of a chemical at the receptor and thus of the magnitude of toxic response. They may also be major determinants for organ-specific toxicity.

For example, in the case of absorption of a xenobiotic through the gastrointestinal tract, the chemical proceeds from the intestinal lumen into the epithelial cells. Following intracellular transport, it passes through the basal membrane and lamina propria and enters the blood or lymph capillaries for transport to the site of action or storage. At that site, the xenobiotic is released from the capillaries, into an interstitial area, and finally through various membranes to its site of action, which may be a specific receptor, an enzyme, a membrane, or many other possible sites.

2.2. Absorption

The skin, the lungs, and the cells lining the alimentary tract are major barriers for chemicals present in the environment. Except for caustic chemicals, which act at the site of first contact with the organism, xenobiotics must cross these barriers to exert toxic effects on one or several target organs. The process whereby a xenobiotic moves through these barriers and enters the circulation is termed absorption.

2.2.1. Membranes

Because xenobiotics must often pass through membranes on their way to the receptor, it is important to understand membrane characteristics and the factors that permit transfer of foreign compounds. Membranes are initially encountered whether a xenobiotic is absorbed by the dermal, oral, or vapor route. These membranes may be associated with several layers of cells or a single cell. The absorption of a substance from the site of exposure may result from passive diffusion, facilitated diffusion, active transport, or the formation of transport vesicles (pinocytosis and phagocytosis). The process of absorption may be facilitated or retarded by a variety of factors; for example, elevated temperature increases percutaneous absorption by cutaneous vasodilation, and surface-active materials facilitate penetration. Each area of entry for xenobiotics into the organism may have specific peculiarities, but a unifying concept of biology is the basic similarity of all membranes in tissues, cells, and organelles.

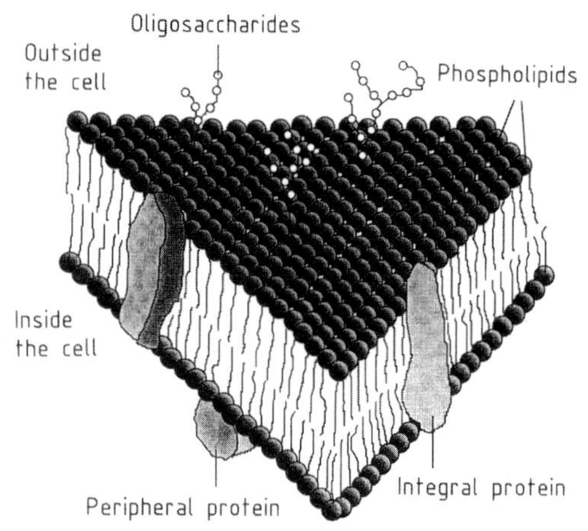

Figure 12. Simplified model of the structure of a biological membrane

All membranes are lipid bilayers with polar head groups (phosphatidylethanolamine, phosphatidylcholine). The polar groups predominate at the outer and inner surfaces of the membrane; the inner space of the membrane consists of perpendicularly arranged fatty acids [21]. The fatty acids do not have a rigid structure and are fluid under physiological conditions; the fluid character of the membrane is largely dominated by the fatty acid composition. The width of a biological membrane is approximately 7–9 nm. Figure 12 illustrates the concept of a biological membrane (fluid-mosaic model).

Proteins are intimately associated with the membrane and may be located on the surface or inside the membrane structure, or extend completely through the membrane. These proteins may also form aqueous pores. Hydrophobic forces are responsible for maintaining the structural integrity of both proteins and lipids within the membrane structure. The ratio of lipid to protein in different membranes may vary from 5:1 (e.g,. myelin) to 1:5 (e.g., the inner membrane of mitochondria). Usually, pore diameters in membranes are small and permit only the passage of low molecular mass chemicals. However, some specialized membranes such as those found in the glomeruli of the kidney, which can have pore sizes of up to 4 nm, also permit the passage of compounds with molecular mass greater than 10 000.

The amphipathic nature of the membrane creates a barrier for ionized, highly polar compounds; however, changes in lipid composition, alterations in the shape and size of proteins, and physical features of bonding may cause changes in the permeability of membranes [22].

2.2.2. Penetration of Membranes by Chemicals

A chemical can pass through a membrane by two general processes: passive diffusion and active transport. Passive diffusion is described by Fick's law and requires no energy. Active transport processes involve the consumption of cellular energy to translocate the chemical across the membrane. Active transport may also act against a concentration gradient and result in the accumulation of a xenobiotic in a specific organ, cell type or organelle.

Diffusion of Chemicals through Membranes. Many toxic chemicals pass membranes by simple diffusion. Their rates of diffusion depend on their lipid solubility and are often correlated with the partition coefficient (solubility in organic solvents/solubility in water). Lipophilic chemicals may diffuse directly through the lipid domain of the membrane. However, a certain degree of water solubility seems to be required for passage since many poorly lipid soluble chemicals have been shown to penetrate easily. Once initial penetration has occurred, the molecule must necessarily traverse a more polar region to dissociate from the membrane. Compounds with extremely high partition coefficients thus tend to remain in membranes and to accumulate there rather than pass through them. Polar compounds that are insoluble in the nonpolar, fatty-acid-containing inner space of the membrane often cannot penetrate membranes, although some low molecular mass polar chemicals may slowly penetrate through the aqueous pores of the membranes.

The rates of movement of nonpolar xenobiotics through membranes can be predicted based on the assumptions from Fick's law of diffusion. Polar compounds and electrolytes of low molecular mass are believed to behave similarily. A first-order equation appears to be applicable to the majority of xenobiotics. The rate of diffusion of a xenobiotic is related to its concentration gradient across the membrane ($C_1 - C_2$), the surface area available for transfer A, the diameter of the membrane d, and the diffusion constant k. The latter is related to the size and structure of the molecule, the spatial configuration of the molecule, and the degree of ionization and lipid solubility of the xenobiotic.

$$\text{Rate of diffusion} = k\,\frac{A\,(C_1 - C_2)}{d}$$

As the xenobiotic is rapidly removed after absorption, C_2 can usually be ignored. and a log/linear plot of the amount of unpenetrated chemicals present over time should be linear. When relatively comparable methods have been used, calculation of the half-time of penetration $t_{1/2}$, is useful. The rate constant of penetration k is derived from

$$k = \frac{0.693}{t_{1/2}}$$

When the half-time of penetration after oral and dermal administration of several environmental contaminants were compared, rates were found to vary considerably. Clearly, rates of penetration by different routes in mammals show little or no correlation.

Ionization becomes particularly important when xenobiotics are introduced into the gastrointestinal tract, where a variety of pH conditions are manifest (see Section 2.2.4.2). Although many drugs are acids and bases and thus potentially ionizable form, most xenobiotics are neither acids nor bases and thus are unaffected by pH. The amount of a xenobiotic in the ionized or unionized form depends upon the pK_a of the xenobiotic and the pH of the medium. When the pH of a solution is equal to the pK_a of the dissolved compound, 50% of the acid or base exists in the ionized and 50% in the unionized form. The degree of ionization at a specific pH is given by the Henderson–Hasselbalch equation:

$$pK_a - \text{pH} = \log \frac{[\text{nonionized}]}{[\text{ionized}]}$$

$$pK_a - \text{pH} = \log \frac{[\text{ionized}]}{[\text{nonionized}]}$$

Since the unionized, lipid-soluble form of a weak acid or base may penetrate membranes, weak organic acids diffuse most readily in an acidic environment, and organic bases in a basic environment. There is some degree of penetration even when xenobiotics are not in the most lipid-soluble form, and a small amount of absorption can produce serious effects if a compound is very toxic.

2.2.3. Mechanisms of Transport of Xenobiotics through Membranes

Filtration. Passage of a solution across a porous membrane results in the retention of solutes larger than the pores. This process is termed filtration. For example, filtration of solutes occurs in the kidney glomeruli, which have large pores and retain molecules with molecular masses greater than 10 000. Elsewhere in the body, filtration by pores may only result in the passage of relatively small molecules (molecular mass ca. 100), and most larger molecules are excluded. Thus, uptake of xenobiotics through these pores is only a minor mechanism of penetration.

Special Transport Mechanisms. Special transport processes include active transport, facilitated transport, and endocytosis (Table 8). Often, the movement of chemicals across membranes is not due to simple diffusion or filtration. Even some very large or very polar molecules may readily pass through membranes.

Active transport systems have frequently been implicated in these phenomena. *Active transport* may be effected by systems that help

Table 8. Special transport processes involved in the passage of xenobiotics through biological membranes

Type of transport	Carrier molecule required	Examples of substrates	Energy required	Against concentration gradient
Active transport	yes	organic acids in the kidney	yes	yes
Facilitated transport	yes	glucose	yes	no
Endocytosis	no	proteins	yes	?

transport endogenous compounds across membranes. Such processes require energy and transport xenobiotics against electrochemical or concentration gradients. Active transport systems are saturable processes and exhibit a maximum rate of transport; they are usually specific for certain structural features of chemicals. A carrier molecule (likely a protein) associates with the chemical outside the cell, translocates it across the membrane for ultimate release inside the cell. This is particularly important for compounds that lack sufficient lipid solubility to move rapidly through the membrane by simple diffusion. Active transport plays a major role in the excretion of xenobiotics from the body, and major excretory organs such as the liver or the kidney have several transport systems which may accept organic acids, organic bases, or even metal ions as substrates.

In contrast to other special transport processes, some carrier-mediated processes do not require energy and are unable to move chemicals against a concentration gradient. These processes are termed *facilitated transport*. Facilitated transport is particulary beneficial for compounds which lack sufficient lipid solubility for rapid diffusion through the membrane. Facilitated transport is more rapid than simple diffusion up to the point at which concentrations on both sides of the membranes are equal. For example, the transport of glucose through a variety of membranes occurs by facilitated transport. The mechanisms by which facilitated transport occurs are not well understood.

Pinocytosis (liquids) and *phagocytosis* (solids) are specialized processes in which the cell membrane invaginates or flows around a xenobiotic, usually present in particulate form, and thus enables transfer across a membrane. Although of importance once the xenobiotic has gained entry into the organism, this mechanism does not appear to be of importance in the initial absorption of a xenobiotic.

2.2.4. Absorption

Absorption is the process whereby xenobiotics cross body membranes and are translocated to the blood stream. The primary sites of absorption of environmental contaminants are the gastrointestinal tract (gastrointestinal absorption), the skin (dermal absorption), and the lung (respiratory absorption). Absorption of chemicals may also occur from other sites such as muscle, the subcutis, or the peritoneum after administration by special routes. In clinical medicine, many drugs are injected directly into the bloodstream to circumvent the problems of absorption posed by the peculiarities of the different routes.

2.2.4.1. Dermal Absorption

Human skin can come into contact with many potentially toxic chemicals. Skin is relatively impermeable to aqueous solutions and most xenobiotics present as ions. Therefore, it is a relatively good barrier separating the human body from the environment. However, skin is permeable in varying degrees to a large number of xenobiotics, and some chemicals may be absorbed through the skin in sufficient amounts to cause a toxic response [23]. A striking example of the significance of absorption through the skin is the large number of agricultural workers who have experienced acute poisoning from exposure to parathion (dermal $LD_{50} \approx 20$ mg/kg) during application or from exposure to vegetation previously treated with this pesticide.

The human skin is a complex, multilayered tissue with approximately 18 000 cm^2 of surface in an average human male. Chemicals to be absorbed must pass through several cell layers before entering the small blood and lymph capillaries in the dermis. Transport in blood and lymph then distributes absorbed chemicals in the body. The human skin consists of three distinct layers (Fig. 13) and a number of associated appendages (sweat and sebaceous glands, hair follicles).

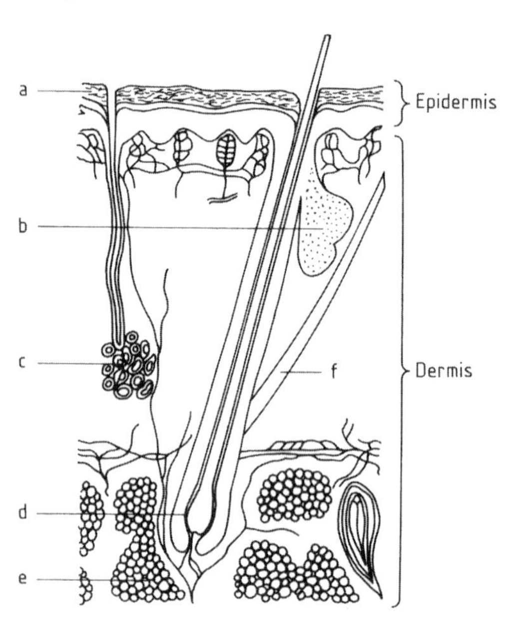

Figure 13. Cross section of human skin
a) Stratum corneum; b) Sebaceous gland; c) Sweat gland; d) Hair follicle; e) Fat; f) Muscle

The *epidermis* is a multilayered tissue varying in thickness from 0.15 (eyelids) to 0.8 mm (palms). This tissue appears to be the greatest deterrent to the absorption of xenobiotics. The epithelial tissues of the skin develop and grow divergently from other tissues. Proliferative layers of the basal cells (stratum germinativum) differentiate and gradually replace cells above them as surface cells deteriorate and are sloughed from the epidermis. Cells in this layer produce fibrous, insoluble keratin that fills the cells, and a sulfur-rich amorphous protein that comprises the cell matrix and thickened cell membrane. This cell layer, the stratum corneum, provides the primary barrier to the penetration of foreign compounds. It consists of several layers of flattened, stratified, highly keratinized cells. These cells are approximately 25–40 μm wide and have lost their nuclei. Although highly water retarding, the dead, keratinized cells of the stratum corneum are highly water absorbent (hydrophilic), a property that keeps the skin supple and soft. A natural oil covering the skin, the sebum, appears to maintain the water-holding capacity of the epidermis but has no appreciable role in retarding the penetration of xenobiotics. The rate-determining barrier in the chemical absorption of xenobiotics is the stratum corneum.

The *dermis* and *subcutaneous tissue* offer little resistance to penetration, and once a substance has penetrated the epidermis these tissues are rapidly traversed. The dermis is a highly vascular area that provides ready access to blood and lymph for distribution once the epithelial barrier has been passed. The blood supply in the dermis is subjected to complex, interacting neural and humoral influences whose temperature-regulating function can have an effect on distribution by altering blood supply to this area. Therefore, the extent of absorption of a chemical through the skin may be influenced by temperature, and relative humidity [24].

The *skin appendages* are found in the dermis and extend through the epidermis. The primary appendages are the sweat glands (epicrine and apocrine), hair, and sebaceous glands. These structures extend to the outer surface and therefore may play a role in the penetration of xenobiotics; however, since they represent only 0.1 to 1% of the total surface of the skin, their contribution to overall dermal absorption is usually minor.

Percutaneous absorption can occur by several routes, but the majority of unionized, lipid-soluble xenobiotics appear to move by passive diffusion directly through the cells of the stratum corneum. Important arguments for the importance of transepidermal absorption are that epidermal damage or removal of the stratum corneum increases permeability, the epidermal penetration rate equals whole-skin penetration, epidermal penetration is markedly slower than dermal, and the epidermal surface area is 100–1000 times the surface area of the skin appendages. Very small and/or polar molecules appear to have more favorable penetration through appendages or other diffusion shunts, but only a small fraction of toxic xenobiotics are chemicals of this type. Polar substances, in addition to movement through shunts, may diffuse through the outer surface of the protein filaments of the hydrated stratum corneum, while nonpolar molecules dissolve in and diffuse through the nonaqueous lipid matrix between the protein filaments.

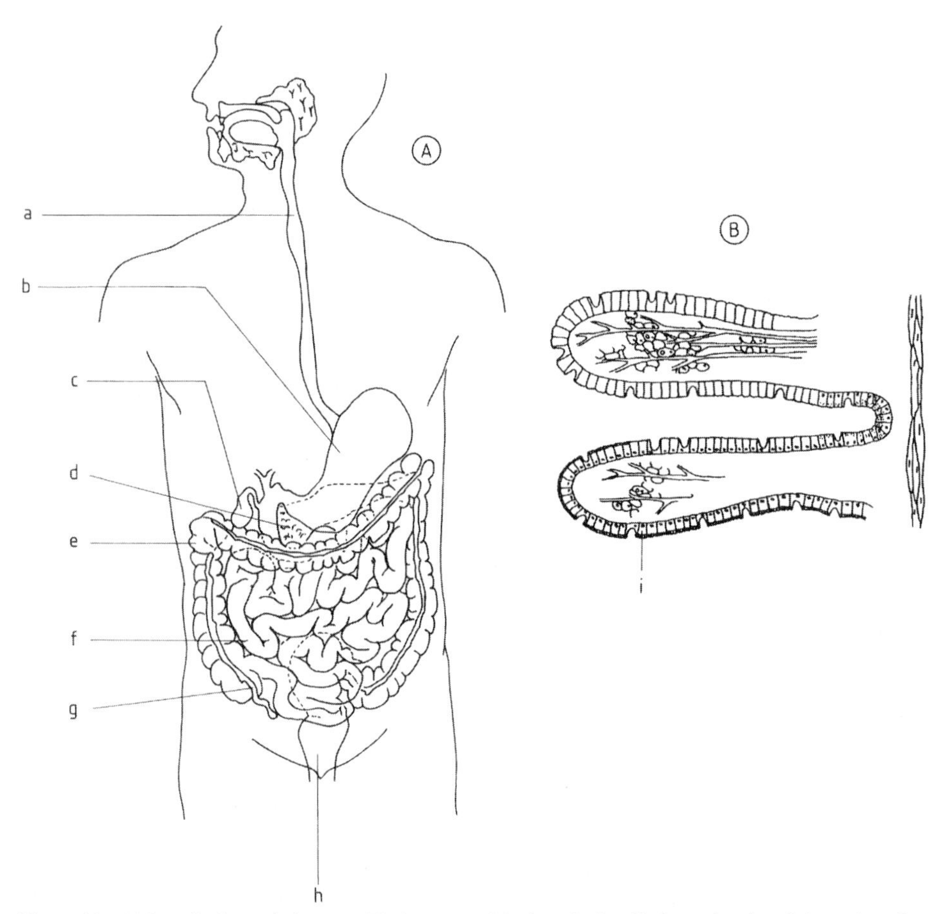

Figure 14. A) Intestinal tract in humans; B) Anatomy of the intestinal wall, the major site of absorption of xenobiotics
The lining of the small intestine is highly folded and has a special surface structure (brush-border membrane) to give a large surface available for the efficient uptake of nutrients.
a) Esophagus (4 – 7.2); b) Stomach (1.0 – 3.0); c) Duodenum (4.8 – 8.2); d) Pancreas; e) Colon (7.9 – 8.0); f) Jejunum (7.6); g) Ileum (7.6); h) Rectum (7.8); i) Brush-border membrane
Numbers in brackets represent pH in different parts of the intestinal tract.

Human stratum corneum displays significant differences in structure from one region of the body to the other, which affect the rate of absorption. Penetration at certain body regions thus varies according to the polarity and size of the molecule, but it is generally accepted that for most unionized xenobiotics the rate of penetration is in the following order: scrotal > forehead > axilla = scalp > back = abdomen > palm and plantar. The palm and plantar regions are highly diffuse, but their much greater thickness (100–400 times that of other regions) introduces an overall lag time in diffusion.

The condition of the skin greatly influences the absorption of xenobiotics. Damage to or removal of the stratum corneum cause a dramatic increase in the permeability of the epidermis for xenobiotics. Caustic and corrosive chemicals such as acids or alkali or burns will greatly enhance dermal absorption and thus influence the toxicity of a xenobiotic applied to the skin. Soaps and detergents are among the damaging substances routinely applied to skin. Whereas organic solvents must be applied in high concentrations to damage skin, 1% aqueous solutions of detergents increase the rate of penetration of solutes through human epidermis dramatically. Alteration of the stratum corneum by organic solvents may also be the cause of increased penetration.

Organic solvents can be divided into damaging and nondamaging categories. Damaging solvents include methanol, acetone, diethyl ether, hexane, and some solvent mixtures. These solvents and mixtures can extract lipids and proteolipids from tissues and are thus expected to alter permeability. Although the mechanical strength of the stratum corneum is unaltered, delipidization produces a more porous, nonselective surface. Solvents such as higher alcohols, esters, and olive oil do not appear to damage skin appreciably. On the contrary, the penetration rate of solutes dissolved in them is often reduced. Surprisingly, lipid-soluble xenobiotics may be markedly resistant to washing, even a short time after application. For example, 15 min after application, a substantial portion of parathion cannot be removed from contaminated skin by soap and water.

When comparisons across species are made, human skin appears to be more impermeable, or at least as impermeable, as the skin of the cat, dog, rat, mouse, or guinea pig. The skin of pigs and guinea pigs in particular serves as a useful approximation to human skin, but only after a comparison has been made for each specific chemical.

Temperature, surface area of applied dose, simultaneous application of another xenobiotic, relative humidity, occlusion, age, and hyperthermia are among a number of chemical, physical, and physiological factors that may alter skin penetration.

2.2.4.2. Gastrointestinal Absorption

The oral route of entry into the body is specially important for accidental or purposeful (suicide) ingestion of poisonous materials. Food additives, food toxins, environmental xenobiotics accumulated in the food chain, and airborne particles excluded from passage to to alveoli are also introduced into the digestive system. The penetration of orally administered xenobiotics is primarily confined to the stomach and intestine [25].

The gastrointestinal tract may be viewed as a tube traversing the body. It consists of the mouth, esophagus, stomach, small and large intestine, colon, and rectum (Fig. 14). The digestive tract is lined by a single layer of columnar cells, usually protected by mucus, which do not present a barrier to penetration. The circulatory system is closely associated with the intestinal tract (30–50 μm from membrane to vasculature), and once xenobiotics have crossed the epithelium of the intestinal tract, entry into capillaries is rapid. Venous blood flow from the stomach and intestine rapidly removes absorbed xenobiotics and introduces them into the hepatic portal vein, which transports them to the liver.

Absorption of chemicals may take place along the entire gastrointestinal tract, but most xenobiotics are absorbed in the stomach and the small intestine. A major factor favoring absorption in the intestine is the presence of microvilli that increase the surface area to an estimated 100 m^2 in the small intestine (see Fig. 14) Because the intestinal area thus offers maximal opportunity for absorption, it is generally accepted that absorption of xenobiotics is greatest in this area of the gastrointestinal tract. Although the gastrointestinal tract has some special transport processes for the absorption of nutrients and electrolytes, most xenobiotics seem to enter the body from the gastrointestinal tract by simple diffusion. Exeptions are some heavy metals such as thallium and lead, which mimic the essential metals iron and calcium, respectively. They are thus absorbed by active transport systems developed for the uptake of these nutrients.

The gastrointestinal tract has areas of highly variable pH, which can markedly change the permeability characteristics of ionic compounds. For example, passive diffusion is greatly limited except for unionized, lipid-soluble chemicals. Although variable according to secretory activity, the pH of the stomach is ca. 1–3 and that of the intestine ca. 7. The measured pH of the intestinal contents may not be the same as the pH of the epithelium at the site of absorption, and this explains the entrance of compounds whose p*K*a would suggest less favored absorption. The variations in pH in the different sections of the intestinal tract may influence the absorption of acids and bases. Since most xenobiotics are absorbed by diffusion, only the unionized, membrane-permeable form may be absorbed. Weak organic acids are mainly present in the unionized, lipid-soluble form in the stomach, and predominantly in the ionized form in the intestine. Therefore, organic acids are expected to be more readily absorbed from the stomach than from the intestine. In contrast, weak organic bases are ionized in the stomach but present in the lipid-soluble form (unionized) in the intestine. Absorption of such compounds should therefore predominantly occur in the intestine rather than in the stomach.

However, other factors determining the rate of membrane penetration such as surface area available for diffusion, blood flow (influencing concentration gradients), and the law of mass action also influence the site of absorption of acids or bases from the gastrointestinal tract. For example, although only 1% of benzoic acid is present in the lipid-soluble, unionized form in the small intestine, the large surface area and the rapid removal of absorbed benzoic acid with the blood result in its efficient absorption from the small intestine.

Other factors contribute to gastrointestinal absorption Clearly a xenobiotic must be dissolved before absorption can take place. Particle size, organic solvents, emulsifiers, and rate of dissolution thus also effect absorption. In addition, the presence of microorganisms and hydrolysis-promoting pH offer opportunities for the biotransformation of many xenobiotics. Other factors affecting gastrointestinal absorption include binding to gut contents, intestinal motility, rate of emptying, temperature of food, effects of dietary constituents, health status of the individual, and gastrointestinal secretion.

2.2.4.3. Absorption of Xenobiotics by the Respiratory System

The respiratory system is an organ in direct contact with environmental air as an unavoidable part of living. A number of xenobiotics exist in gaseous (carbon monoxide, nitric oxides), vapor (benzene, carbon tetrachloride), and aerosol (lead from automobile exhaust, silica, asbestos) forms and are potential candidates for entry via the respiratory system. Indeed, the most important cause of death from acute intoxication (carbon monoxide) and the most frequent occupational disease (silicosis) are caused by the absorption or deposition of airborne xenobiotics in the lung.

The respiratory tract consists of three major regions: the nasopharyngeal, the tracheobronchial, and the pulmonary (Fig. 15). The nasopharynx begins in the mouth and extents down to the level of the larynx. The trachea, bronchii, and bronchioli serve as conducting airways between the nasopharynx and the alveoli, the site of gas exchange between the inhaled air and the blood. The human respiratory system is a complex organ containing over 40 different cell types. These cell types contribute to the pulmonary architecture and function over various zones of the lung, although to some extent, individual cell types can be found in several zones. The tracheobronchial system comprises airways lined with bronchial epithelium with associated submucosal glands and several different tissues with specific function and the lung vasculature.

The absorption of xenobiotics by the respiratory route is favored by the short path of diffusion, large surface area (50–100 m^2), and large concentration gradients. At the alveoli (site of gas exchange), the membranes are very thin (1–2 µm) and are intimately associated with the vascular system. This enables rapid exchange of gases (ca. 5 ms for CO_2 and ca. 200 ms for O_2). A thin film of fluid lining the alveolar walls aids in the initial absorption of xenobiotics from the alveolar air. Simple diffusion accounts for the somewhat complex series of events in the lung regarding gas absorption. The sequences of respiration, which involve several interrelated air volumes, define both the capacity of the lung and factors important to particle deposition and retention. Among the elements important in total lung capacity is the residual volume, that is, the amount of air retained by the lung despite maximal expiratory effort. Largely due to slow release from this volume, gaseous xenobiotics in the expired air are not cleared immediately, and many expirations may be necessary to rid the air in the lung of residual xenobiotic. The rate of entry of vapor-phase xenobiotics is controlled by the alveolar ventilation rate, and a xenobiotic present in alveolar air may come into contact with the alveoli in an interrupted fashion about 20 times per minute. The diffusion coefficient of the gas in the fluids of pulmonary membranes is another important consideration, but doses are more appropriately discussed in terms of the partial pressure of the xenobiotic in the inspired air. On inhalation of a constant tension of a gaseous xenobiotic, arterial plasma tension of the gas approaches the tension of gas in the expired air. The rate of entry is then determined by blood solubil-

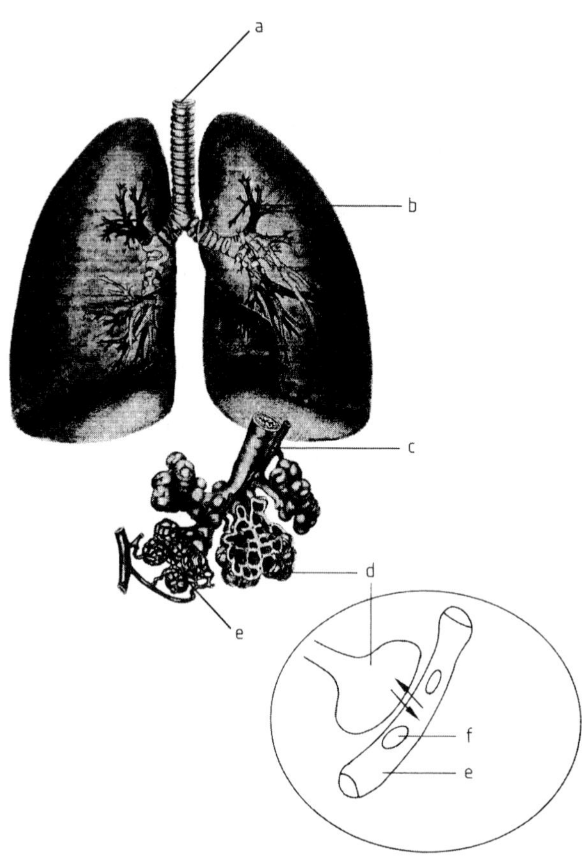

Figure 15. Anatomy of the human respiratory system
a) Trachea; b) Bronchii; c) Bronchioli; d) Alveoli; e) Capillary; f) Erythrocyte

ity of the xenobiotic and blood flow. For a high blood/gas partition coefficient, a larger amount must be dissolved in the blood to raise the partial pressure. Chemicals with a high blood/gas partition coefficient require a longer period to approach the same tension in the blood as in inspired air than less soluble gases.

Aerosols and Particulates. The entry of aerosols and particulates is affected by a number of factors. A coal miner inhales ca. 6000 g of coal dust particles during his occupational lifetime, and only ca. 100 g are found postmortem; therefore, effective protective mechanisms are operative. The parameters of air velocity and directional changes in air flow favor impaction of particles in the upper respiratory systems. Particle characteristics such as size, chemistry of the inhaled material, sedimentation and electrical charge are important to retention, absorption, or expulsion of airborne particles. In addition to the other aforementioned lung characteristics, a mucous blanket propelled by ciliary action clears the respiratory tract of particles by directing them to the gastrointestinal system (via the glottis) or to the mouth for expectoration. This system is responsible for 80% of lung particulate clearance. The deposition of various particle sizes in different respiratory regions is summarized in Figure 16, which shows that particle size is important for disposition and particles larger than 2 μm do not reach the alveoli [26].

The direct penetration of airborne xenobiotics at alveolar surfaces or in the upper respiratory tract is not the only action of toxicological importance. Both vapors and particulates can accumulate in upper respiratory passages to produce irritant effects. Irritant gases may be deposited in the respiratory tract depending on their water solubility and may cause localized damage characterized by edema, swelling, mucus production, and increased d vascular permeability. If major airways are obstructed by these processes or important anatomical structures of the respiratory tract like the alveoli are damaged, life-threatening or deadly intoxications may be caused by the inhalation of irritant gases.

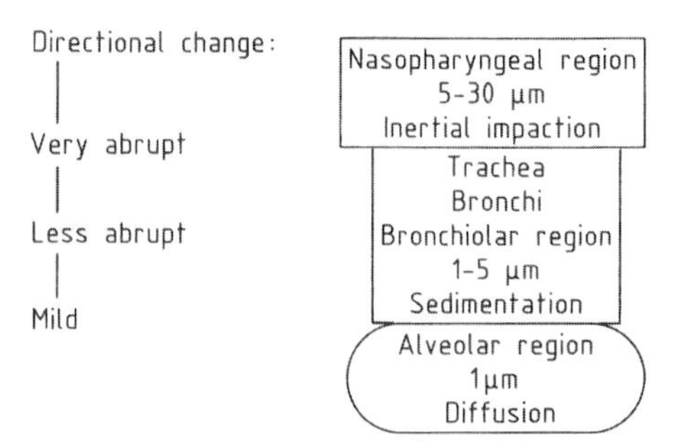

Figure 16. Effect of size on the disposition and sedimentation of particulates in the respiratory tract
The site of particle sedimentation is determined largely by particle size; only very fine particles are deposited in the alveoli; larger particles do not reach the lung but are deposited in the nasopharynx.

Despite the effectiveness of ciliary movement and phagocytosis, the cumulative effects of silica, asbestos, or coal dust ultimately cause chronic fibrosis even though direct absorption is of minor importance. Thus, phagocytosis prevents acute damage but may contribute to chronic toxicity. There is little evidence for active transport in the respiratory system, although pinocytosis may be of importance for penetration. The lung is an area of extensive metabolic activity; enzymes present in the lung may catalyze both activation and detoxication of xenobiotics (see Section 2.4).

2.3. Distribution of Xenobiotics by Body Fluids

After entering the blood by absorption or by intravenous administration, xenobiotics are available for distribution throughout the body. The initial rate of distribution to organs and tissues is determined by the blood flow to that organ and the rate of diffusion of the chemical into the specific organ or tissue. Uptake of xenobiotics into organs or tissues may occur by either passive diffusion or by special transport processes. Within tissues binding, storage, and/or biotransformation may occur. After reaching equilibrium, the distribution of a chemical among organs and tissues is largely determined by affinity; blood flow determines distribution only during the initial phase shortly after uptake.

Body fluids are distributed between three distinct compartments: vascular water, interstitial water and intracellular water. Plasma water and interstitial water are extracellular water. Plasma water plays an important role in the distribution of xenobiotics. Human plasma accounts for about 4% of the total body weight and 53% of the total volume of blood. By comparison, the interstitial tissue fluids account for 13% of body weight, and intracellular fluids account for 41%. The concentration of a xenobiotic in blood following exposure will depend largely on its apparent volume of distribution. If the xenobiotic is distributed only in the plasma, a high concentration will be achieved within the vascular tissue. In contrast, the concentration will be markedly lower if the same quantity of xenobiotic were distributed in a larger pool including the interstitial water and/or intracellular water.

Among the factors that affect distribution, apart from binding to blood macromolecules, are the route of administration, rate of biotransformation, polarity of the parent xenobiotic or biotransformation products, and rate of excretion by the liver or kidneys. Gastrointestinal absorption and intraperitoneal administration provide for immediate passage of a compound to the liver, whereas dermal or respiratory routes involve at least one passage through the systemic circulation prior to reaching the liver. The metabolism of most xenobiotics results in products that are more polar and thus more readily excreted than the parent molecules. Therefore, the rate of metabolism is a critical determinant in the distribution of a compound, since compounds that are readily metabolized are usually readily excreted, and thus are proportionally less prone to accumulate in certain tissues. The same principle applies to polarity, since very polar xenobiotics will be readily excreted. Chemicals may circulate either free or bound to plasma protein or blood cells; the degree of binding and factors influencing the equilibrium with the free form may influence availability for biotransformation, storage, and/or excretion [27].

Patterns of xenobiotic distribution reflect certain physiological properties of the organism

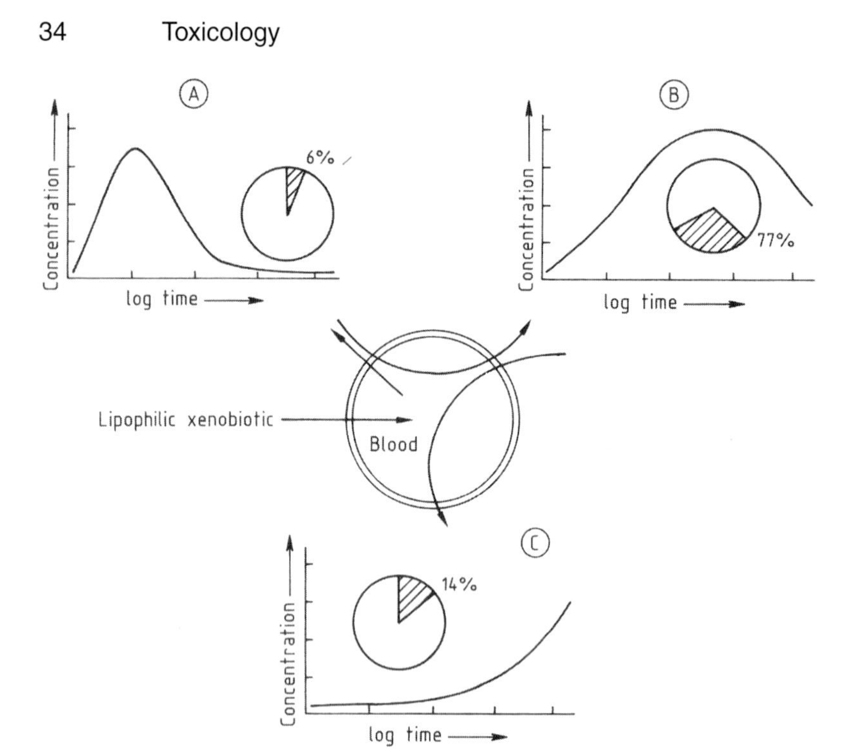

Figure 17. Uptake and redistribution with blood of lipophilic xenobiotics
Lipophilic xenobiotics in the blood are first distributed to well-perfused organs (A); after some time, they are redistributed to organs with lower blood flow representing a larger fraction of the body weight (B, C)

and the physicochemical properties of the xenobiotics. An initial phase of distribution may be distinguished that reflects cardiac output and blood flow to organs. Heart, liver, kidney, brain, and other well-perfused organ- receive most of a lipophilic xenobiotic within the first few minutes after absorption. Delivery to the smooth muscles, most viscera, and skin is slower, and the time to reach a steady-state concentration of a xenobiotic in these organs may be several hours. A second phase of xenobiotic distribution may therefore be distinguished; it is limited by blood flow to an organ or tissue and involves a far larger fraction of body mass than the first phase of distribution (Fig. 17).

Only a limited number of xenobiotics have sufficient solubility in blood to account for simple dissolution as a route of distribution; the distribution of many xenobiotics occurs in association with plasma proteins. The binding of drugs to plasma proteins is of key importance in transport. Many organic and inorganic compounds of low molecular mass appear to bind to lipoproteins, albumins, and other proteins in plasma and are transported as protein conjugates. This binding is reversible. Cellular components may also be responsible for transport of xenobiotics, but such transport is rarely a major route. The transport of xenobiotics by lymph is usually quantitatively of little importance since the intestinal blood flow is 500–700 times greater than the intestinal lymph flow.

A large number of studies on binding of drugs by plasma protein have demonstrated that binding to serum albumin is particularly important for these chemicals. Only few studies on the reversible binding of toxic xenobiotics have been performed, but available evidence suggests a significant role of lipoproteins in plasma. These plasma proteins may bind xenobiotics as well as some physiological constituents of the body. Examples for plasma proteins which may bind xenobiotics are albumin, α- and β-lipoproteins, and metal-binding proteins such as transferrin. Lipoproteins are important for the transport of lipid-soluble endogenous chemicals such as vitamins, steroid hormones, and cholesterol, but they may also bind lipophilic xenobiotics. If a xenobiotic is bound to a protein, it is immobilized remote from the site of action. The extent of binding to plasma proteins varies considerably among xenobiotics. While some are not at all bound, for others more than 90% of administered dose may be bound to plasma proteins. These ligand–protein interactions are reversible and provide a remarkably efficient means for transport of xenobiotics to various tissues.The xenobiotic–protein interaction may be simply described according to the law of mass action as:

$$[T]_F + [\text{free sites}] \underset{k_2}{\overset{k_1}{\rightleftharpoons}} [T]_B$$

where $[T]_F$ and $[T]_B$ are the concentrations of free and bound xenobiotic molecules, respectively, and k_1 and k_2 are the rate constants for association and dissociation; k_2, which governs the rate of binding to the protein, dictates the rate of xenobiotic release at a site of action or storage. The ratio k_1/k_2 is identical with the dissociation constant K_{diss}. Among a group of binding sites on proteins, those with the smallest K_{diss} for a given xenobiotic will bind it most tightly.

In contrast to the covalent binding to proteins seen with many xenobiotics or their electrophilic metabolites (see Section 2.5.6.6), the interaction of xenobiotics with plasma proteins is most often noncovalent and reversible. Noncovalent binding is of primary importance with respect to distribution because of the opportunities to dissociate after transport. Binding of xenobiotics to plasma proteins may be due to several types of interactions which are summarized in the following.

Ionic Binding. Electrostatic attraction occurs between two oppositely charged ions. The degree of binding varies with the chemical nature of each compound and the net charge. Dissociation of ionic bonds usually occurs readily, but some transition metals exhibit high association constants (low K_{diss} values), and exchange is slow. Ionic interactions may also contribute to binding of alkaloids with ionizable nitrogen groups and other ionizable xenobiotics.

Hydrogen Bonding. Generally, only the most electronegative atoms form stable hydrogen bonds. Protein side chains containing hydroxyl, amino, carboxyl, imidazole, and carbamyl groups may form hydrogen bonds, as can the nitrogen and oxygen atoms of peptide bonds. Hydrogen bonding plays an important role in the structural configuration of proteins and nucleic acids.

Van der Waals forces are very weak interactions between the nucleus of one atom and the electrons of another atom, i.e., between dipoles and induced dipoles. The attractive forces are based on slight distortions induced in the electron clouds surrounding each nucleus as two atoms come close together. The binding force is critically dependent upon the proximity of interacting atoms and diminishes rapidly with distance. However, when these forces are summed over a large number of interacting atoms that "fit" together spatially, they can play a significant role in determining specificity of xenobiotic–protein interactions.

Hydrophobic Interactions. When two nonpolar groups come together they exclude the water between them, and this mutual repulsion of water results in a hydrophobic interaction. In the aggregate they represent the least possible disruption of interactions among polar water molecules and thus can lead to stable complexes. Some consider this a special case of van der Waals forces. The minimization of thermodynamically unfavorable contact of a polar group with water molecules provides the major stabilizing effect in hydrophobic interactions.

Consequences of the binding to plasma proteins are reduced availability of the free xenobiotic in the cells and a delayed excretion. The xenobiotic bound to plasma protein cannot cross capillary walls due to its high molecular mass. The fraction of dose bound is thus not available for delivery to the extravascular space or for filtration by the kidney. It is generally accepted that the fraction of xenobiotic that is bound may not exert toxic effects; however, many xenobiotics and endogenous compounds appear to compete for the same binding site, and thus one compound may alter the unbound fraction of another by displacement, thereby potentially increasing toxic effects. Plasma proteins that can bind endogenous chemicals and xenobiotics are listed below, together with examples of bonded xenobiotics:

α-Lipoproteins	vitamins A, K, D steroid hormones dieldrin
Albumin	salicylate tetracyclines phenols vitamin C

Binding of xenobiotics to plasma proteins is mostly reversible; the bound xenobiotic is in equilibrium with free xenobiotic, and thus binding usually slows excretion or delivery to cellular sites of action. Toxicological consequences of the reversible binding of a xenobiotic to plasma proteins may arise after saturation of the binding capacities of plasma proteins and by displacement of the bound xenobiotic by another chemical with higher affinity, which increases the free fraction of the formerly bound xenobiotic. This will result in an increased equilibrium concentration of the xenobiotic in plasma and in the target organ, with potentially harmful consequences.

2.4. Storage of Xenobiotics in Organs and Tissues

Absorbed xenobiotics may be concentrated in specific organs or tissues. The concentration of a xenobiotic in a specific tissue may cause toxic effects to that particular tissue; some xenobiotics actually attain their highest concentration at the site of toxic action. However, other xenobiotics may be concentrated in tissues without harmful consequences. Some tissues have a high capacity to accommodate certain xenobiotics and may release them only slowly. The compartment or tissue in which a chemical is concentrated can also be considered as a storage depot for this xenobiotic. If a chemical is stored in a depot and thus removed from the site of action (e.g., polychlorinated biphenyls in fat or lead in bone), no immediate manifestation of toxicity may be observed, even though a potential for adverse effects exists. For example, lead stored in bone does not cause a toxic response but has the potential for mobilization and thus for migration into soft tissues; toxic effects may appear after mobilization. As the xenobiotic in storage depots is in equilibrium with the free xenobiotic in plasma, mobilization is constant, and exposure of the target organ to low concentrations of the xenobiotic is constant. Some storage depots for specific chemicals follow:

Lead	bone
Fluoride	bone, teeth
Cadmium	kidney
Iron	transferrin (a blood protein)
Polychlorinated pesticides such as DDT	fat
Arsenic	skin

Liver, kidney, fat, bone, and plasma proteins may serve as storage depots for absorbed xenobiotics. Both liver and kidney have a high capacity to store xenobiotics and are major storage sites for a multitude of chemicals. Accumulation of circulating xenobiotics form the blood by active transport systems and binding to certain tissue constituents are major mechanisms involved in the renal and hepatic storage.

Several thiol-rich proteins present in liver and kidney have a high affinity for xenobiotics [28]. The binding protein ligandin, which as a glutathione *S*-transferase also has enzymatic activity (Section 2.5.5.6) and thus participates in xenobiotic biotransformation, binds organic acids, some azo dyes, and corticosteroids. Metallothionein, a cysteine-rich protein present in liver and kidney, serves as a binding and storage protein for metals such as cadmium and zinc. Its biosynthesis increases after exposure to metals,and this may result in storage of a considerable percentage of cumulative metal dose as metalothionein complex in liver and kidney.

Highly lipophilic chemicals rapidly penetrate membranes and are thus efficiently taken up by tissues. Lipophilic substances that are inefficiently biotransformed accumulate in the most lipophilic environment in the organism, fat. Most xenobiotics seem to accumulate by physical dissolution in neutral fats, which may constitute between 20 and 50% of the body weight in human males. Large amounts of lipophilic xenobiotics may therefore be present in fat; for xenobiotics that do not undergo biotransformation (e.g., 2,3,7,8-tetrachlorodibenzodioxin), determination of the concentration in body fat is a good measure for exposure.

Usually, xenobiotics stored in fat also do not induce toxic responses, since the xenobiotic is not readily available at the target site. However, during rapid mobilization of fat during disease or starvation, a sudden increase in the plasma concentration, and thus toxic effects in target organs, may occur. For example, signs of organochlorine pesticide intoxication have been observed after starvation in animals pretreated with persistent organochlorine pesticides.

Some xenobiotics have a high affinity for bone and may accumulate in the bone matrix. For example, 90% of the lead and a major part of the strontium present in the body after chronic exposure are stored in the skeleton [29]. Lead and strontium accumulate in bone due to their similarity with calcium; inorganic fluoride, which is also a "bone-seeker", accumulates in bone due to similarities in size and charge to the hydroxyl ion. The storage of xenobiotics in bone may or may not be responsible for toxic effects. Lead stored in the skeleton is not toxic to bone, but both stored fluoride and stored strontium cause toxic effects to bone (fluorosis, osteosarcoma). Xenobiotics stored in bone are also in equilibrium with the unbound xenobiotic circulating in plasma and may thus be released.

Certain plasma proteins have a high affinity for xenobiotics, binding of a chemical to plasma proteins may constitute a both transport form and a storage form. Globulins such as transferrin (involved in iron transport) and ceruloplasmin (copper) and α- and β-lipoproteins (lipophilic xenobiotics and endogenous chemicals) may be involved in binding.

Storage in tissue may greatly alter the rate of excretion of a xenobiotic. Only xenobiotics present in plasma are available for distribution, biotransformation, and excretion. However, excretion or biotransformation alters the plasma concentration of the xenobiotic, and some of the stored chemical is released into plasma from the site of storage. Owing to this mechanism, the rate of excretion of a xenobiotic stored in tissues can be very low.

2.5. Biotransformation

Most xenobiotics entering the body are lipophilic. This property enables them to penetrate lipid membranes, to be transported by lipoproteins in the blood, and to be rapidly absorbed by the target organ. However, the efficient excretory mechanisms of the organism require solubility of the xenobiotic in aqueous media and thus a certain degree of hydrophilicity is required for efficient excretion. Lipophilic substances can only be excreted efficiently by exhalation, but this is restricted to volatile xenobiotics. In the absence of efficient means for excretion of nonvolatile chemicals, constant exposure or even intermittent single exposures to a lipophilic chemical could result in accumulation of the xenobiotic in the organism.

Therefore, animal organisms have developed a number of biochemical processes that convert lipophilic chemicals to hydrophilic chemicals and thus assist in their excretion. These enzymatic processes are termed biotransformation, and the enzymes catalyzing biotransformation reactions are referred to as biotransformationenenzymes. The biotransformation enzymes differ from most other enzymes by having a broad substrate specifity and by catalyzing reactions at comparatively low rates. The low rates of biotransformation reactions are often compensated by high concentrations of biotransformation enzymes. For example, ca. 5 wt% of the protein in rat liver consist of cytochromes P450, which are major biotransformation enzymes.

The broad specifity of the biotransformation enzymes likely has evolutionary reasons. Biotransformation enzymes have evolved to facilitate the excretion of lipophilic chemicals present in the diet of animals. The broad substrate specificity helped to adjust to new dietary constituents and thus led to evolutionary advantages.

Biotransformation is generally the sum of several processes by which the structure of a chemical is changed during passage through the organism. The metabolites formed from the parent chemical are usually more water soluble; the increased water solubility reduces the ability of the metabolites to partition into membranes, restricts renal and intestinal reabsorption, and thus facilitates excretion with urine or bile.

2.5.1. Phase-I and Phase-II Reactions

Xenobiotic metabolism is catalyzed by a number of different enzymes. For solely operational purposes, the biotransformation enzymes are separated into two phases. In phase-I reactions, which involve oxidation, reduction, and hydrolysis, a polar group is added to the xenobiotic or is exposed by the biotransformation enzymes. Phase-II reactions are biosynthetic and link the metabolite formed by phase-I reactions to a polar endogenous molecule to produce a conjugate. Various endogenous molecules with high polarity and are utilized for conjugation; the resulting conjugates are often ionized at physiological pH and thus highly water soluble. Moreover, the moieties used for conjugation are often recognized by specific active transport processes, which assist in their translocation across plasma membranes and thus further enhance the rate of excretion.

The fate of a particular chemical and the participation of the various phase-I and phase-II biotransformation enzymes is determined by its chemical structure; biotransformation is usually complex and often integrated. Many chemicals bearing functional groups undergo conjugation without prior phase-I biotransformation, whereas others are oxidized or reduced prior to conjugation. However, chemicals lacking functional groups may also undergo phase-II biotransformations without being subjected to a prior phase-I reaction (examples are 1,2-dibromoethane and perchloroethene; see Section 2.5.4).

2.5.2. Localization of the Biotransformation Enzymes

The biotransformation enzymes are localized mainly in the liver. A significant fraction of the blood from the splanchnic area, which also contains xenobiotics absorbed from the intestine, enters the liver. Therefore, the liver has developed the capacity to enzymatically modify most of these chemicals before storage, release, or excretion. However, most other tissues also have the capacity to catalyse biotransformation reactions; indeed, most tissues tested have shown the presence of enzymes which can catalyze biotransformation reactions. The contribution of extrahepatic organs to the biotransformation of a chemical depends on many factors, including chemical structure, dose, and route of administration. However, biotransformation of a chemical within an extrahepatic tissue may have toxic effects on this specific tissue and may thus have important toxicological consequences.

Inside cells, phase-I enzymes are mainly present in the endoplasmatic reticulum, a myriad of lipoprotein membranes extending from the mitochondria and the nucleus to the plasma membranes of the cell. When an organ is homogenized, the endoplasmatic reticulum is broken and membrane fragments are sealed of to form microvesicles. These microvesicles can be sedimented by differential centrifugation, and the materials thus obtained is known as a microsomes. They are highly enriched in vesicles from the endoplasmatic reticulum and retain active biotransformation enzymes. Microsomes are often used to study the enzymatic biotransformation of xenobiotics in vitro.

The presence of phase-I enzymes within membranes has important implications, since lipophilic chemicals will preferentially distribute into lipid membranes; thus, high concentrations of lipophilic xenobiotics are present at this site of biotransformation. In contrast to phase-I enzymes, phase-II enzymes are often soluble, non-membrane-associated, and present in the cytoplasm of the cell. They are found in the supernatant (cytosol) obtained by ultracentrifugation of homogenized tissues. The subcellular localizations of enzymes responsible for biotransformation afollow:

Phase-I enzymes	
–Cytochrome P450	microsomal
–Flavin-dependent mono-oxygenase	microsomal
–Prostaglandin synthase	microsomal
–Epoxide hydrolase	microsomal/cytosolic
Phase-II enzymes	
–UDP-glucuronyl-transferases	microsomal
–Sulfotransferases	cytosolic
–*N*-acetyltransferases	cytosolic
–Glutathione *S*-transferase	cytosolic/microsomal

2.5.3. Role of Biotransformation in Detoxication and Bioactivation

The general purpose of biotransformation reactions is detoxication, since xenobiotics should be transformed to metabolites which are more readily excreted. However, depending on the structure of the chemical and the enzyme catalyzing the biotransformation reaction, metabolites with a higher potential for toxicity than the parent compound are often formed. This process is termed bioactivation and is the basis for the toxicity and carcinogenicity of many xenobiotics with a low chemical reactivity (see Section 2.5.6). The interaction of the toxic metabolite initiates events that ultimately may result in cell death, cancer, teratogenicity, organ failure, and other manifestations of toxicity. Formation of reactive and more toxic metabolites is more frequently associated with phase-I reactions; however, phase-II reactions and combinations of phase-I and phase-II reactions may also be involved in toxication.

2.5.4. Phase-I Enzymes and their Reactions

Phase-I reactions are catalyzed by microsomal monooxygenases and peroxidases, cytosolic and mitochondrial oxidases, reductases, and hydrolytic enzymes. All these reactions add or expose functional groups which can be conjugated later.

2.5.4.1. Microsomal Monooxygenases: Cytochrome P450

Microsomal monooxygenases are the cytochrome P450 enzymes and the mixed-function amine oxidase or flavin-dependent monooxygenase. Both enzyme systems add a hydroxyl moiety to the xenobiotic. Cytochrome P450, a carbon monoxide binding hemoprotein in microsomes, is the most important enzyme system involved in phase-I reactions. The name cytochrome P450 is a generic term applied to a group of hemoproteins defined by the unique spectral property observed when reduced cytochrome P450 (Fe^{2+}) is treated with carbon monoxide. The complex formed has a maximum absorption at 450 nm imparted by the presence of an axial thiolate ligand on the heme iron atom. This spectral characteristic is only present when the protein is intact and catalytically functional. Denatured cytochrome P450 shows, like other heme proteins, an absorbance maximum at 420 nm.

Cytochrome P450 enzymes are a coupled enzyme system composed of the heme-containing cytochrome P450 and the NADPH-containing cytochrome P450 reductase [30]. This flavoprotein has a preference for NADPH as its cofactor and transfers one or two electrons from NADPH to cytochrome P450. Cytochrome P450 and the reductase are embedded into the phospholipid matrix of the endoplasmatic reticulum. The phospholipid matrix is crucial for enzymatic activity since it facilitates the interaction between the two enzymes. The importance of the phospholipid matrix is indicated by the following: Both cytochrome P450 and cytochrome P450 reductase can be purified to apparent homogeneity; the enzymatic activity of the purified and recombined enzymes is dependent on the addition of phospholipids.

In vertebrates, the highest concentrations of cytochrome P450 are found in the liver, but cytochrome P450 enzymes are also present in lung [31–35], kidney, testes, skin, and gastrointestinal tract [36]. The presence of several forms of cytochrome P450 with different substrate specificity and different rates of biotransformation for certain xenobiotics was indicated by studies in the 1970s. In the early 1980s, several different cytochrome P450 enzymes from rodents were purified to apparent homogenicity. Moreover, a large number of cytochrome P450 enzymes have been purified from human organs. All these cytochrome P450 enzymes share the heme, but they differ in both the composition and thus the structure of the polypeptide chain and in the reactions they catalyze [37–40].

The individual enzymes are regulated in their expression by a variety of factors such as treatment with xenobiotics, species, organ, sex, and diet. Because of the multitude of enzymes present, the term "superfamily" of cytochromes P450 is frequently used. In mammals, two general classes of cytochrome P450 exist: six families involved in steroid metabolism and bile acid biosynthesis, and four families containing numerous individual cytochromes P450, mainly responsible for xenobiotic biotransformation. A complex nomenclature, based on amino acid sequence similarity, has been developed to designate individual cytochromes P450. The genes for the individual enzymes are named by the root CYP followed by a number designating the family, a letter for the subfamily, and another number denoting the individual enzyme (see Table 9).

Table 9. Mammalian cytochromes P450 involved in xenobiotic biotransformation

Family	Number of subfamilies	Number of forms
CYP1	1	2
CYP2	8	59
CYP3	2	11
CYP4	2	10

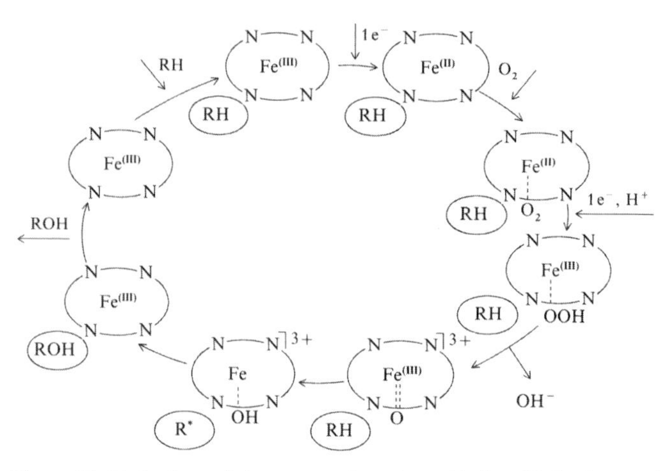

Figure 18. Mechanisms of electron transfer and xenobiotic oxidation by cytochrome P450

Table 10. Oxidations catalyzed by cytochrome P450

Type of reaction	Examples
Aliphatic hydroxylation	$R{-}CH_2{-}CH_2{-}CH_3 \xrightarrow[P450]{O_2} R{-}CH_2{-}CHOH{-}CH_3$
N-Dealkylation	$(R)(R^1)N{-}CH_3 \xrightarrow{P450} [(R)(R^1)N{-}CH_2{-}OH] \longrightarrow (R)(R^1)NH + (H)(H)C{=}O$
O-Dealkylation	$O_2N{-}C_6H_4{-}OCH_3 \longrightarrow [{-}O{-}CH_2{-}OH] \xrightarrow{P450} O_2N{-}C_6H_4{-}OH + (H)(H)C{=}O$
Epoxidation	$H_2C{=}CH_2 \xrightarrow{P450}$ ethylene oxide (epoxide ring with O)

Reactions catalyzed by cytochrome P450. Cytochromes P450 are monooxygenases. They utilize one of the oxygen atoms of molecular oxygen and incorporate it into the xenobiotic RH:

$$RH + O_2 + NADPH + H^+ \rightarrow ROH + H_2O + NADP^+$$

The other oxygen atom is reduced to water with consumption of NADPH as reducing cofactor. The likely mechanisms of electron transfer and xenobiotic oxidation is shown in Figure 18.

In the first step of the catalytic cycle, the xenobiotic combines with the oxidized form of cytochrome P450 (Fe^{3+}) followed by a one-electron reduction by NADPH-cytochrome P450 reductase to form a reduced (Fe^{2+}) cytochrome P450-substrate complex. This complex then combines with molecular oxygen, and another electron from NADPH is accepted. In a series of further steps, which are not completely understood, an oxygen atom from the intermediate is transferred to the substrate, while the other oxygen atom is reduced to water. In the last step of the catalytic cycle, the oxidized substrate dissociates and regenerates the oxidized form of cytochrome P450. Examples for oxidation reactions catalyzed by cytochromes P450 are shown in Table 10.

Cytochromes P450 may catalyze the hydroxylation of carbon–hydrogen bonds to transform hydrocarbons to the corresponding alcohols. In larger aliphatic chains, the $(\omega - 1)$ position is often a favored point of attack. Oxidative *N*-, *O*-, or *S*-dealkylation and oxidative dehalogenation are similar in mechanism to aliphatic hydroxylation, but, due to further reactions of the intermediate products, give different end products. Olefins are also oxidized by cytochrome P450, and with some substrates, epoxides are formed as products. The reaction, however, does not proceed in a concerted manner, but involves discrete ionic intermediates, which may also rearrange to products other than epoxides, as shown for chloroolefins in Figure 19.

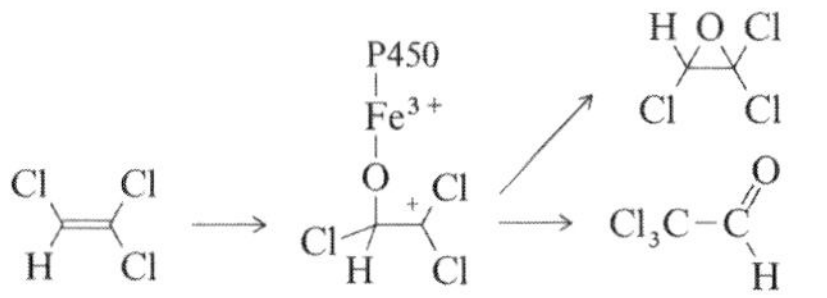

Figure 19. Mechanism of oxidation and rearrangement of trichloroethylene to chloral and trichlorooxirane, respectively

Oxidation at sulfur or nitrogen occurs by the addition of oxygen at the unshared electron pair on the sulfur or nitrogen atom. The products formed (sulfoxides or hydroxylamines) may be stable (many sulfoxides), may be further oxidized by other enzymes in the organism (e.g., hydroxylamines), or may decompose to sulfur and the corresponding oxo compound.

The above-mentioned reactions may be catalyzed by most cytochromes P450 involved in xenobiotic biotransformation; the type of reaction catalyzed seems to be more influenced by steric factors regarding the substrate-binding site of individual cytochromes P450 than by electronic factors. As shown in Table 11, which gives an overview of human cytochrome P450s involved in the biotransformation of xenobiotics and drugs, steric factors are likely major determinants of the substrate specificity of cytochrome P450 enzymes.

In addition to promoting oxidative metabolism, cytochrome P450 may also catalyze reductive biotransformation reactions [41, 42]. These reaction are favored under reduced oxygen pressure or occur with xenobiotics lacking oxidizable C–H bonds or olefinic moieties. In these cases,instead of oxygen, the xenobiotic accepts one or two electrons from NADPH-cytochrome P450 reductase or from cytochrome P450. Reductive biotransformation catalyzed by cytochrome P450 has been demonstrated with some azo dyes and several aromatic nitro compounds. The double bond in azo compounds may be progressively reduced to give amine metabolites; aromatic nitro groups may also be reduced via the nitrone and the hydroxylamine to the corresponding amine.

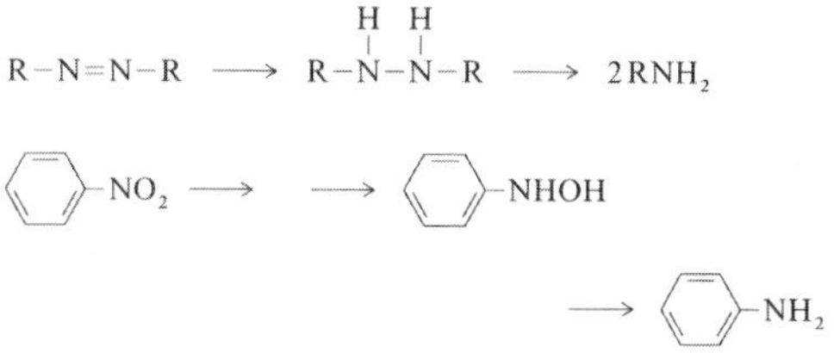

The reductive biotransformation of polyhalogenated alkanes is exemplified by the one-electron reduction of carbon tetrachloride to the trichloromethyl radical and chloride; reductive biotransformation of carbon tetrachloride by a two-electron reduction results in formation of chloroform:

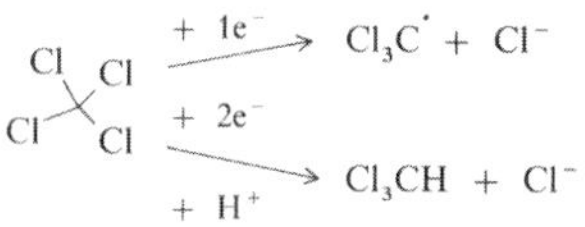

$$R{-}N{=}N{-}R \longrightarrow R{-}NH{-}NH{-}R \longrightarrow 2\,RNH_2$$

$$C_6H_5{-}NO_2 \longrightarrow \longrightarrow C_6H_5{-}NHOH \longrightarrow C_6H_5{-}NH_2$$

$$CCl_4 \xrightarrow{+\,1e^-} Cl_3C^{\cdot} + Cl^-$$

$$CCl_4 \xrightarrow[+\,H^+]{+\,2e^-} Cl_3CH + Cl^-$$

2.5.4.2. Microsomal Monooxygenases: Flavin-Dependent Monooxygenases

Tertiary amines and sulfur-containing drugs have been known to be metabolized to *N*-oxides or sulfoxides by a microsomal monooxygenase which is not dependent on cytochrome P450 [43, 44]. This enzyme, which is historically referred to as mixed-function amine oxidase, is a flavoprotein that is present in the endoplasmatic reticulum. It is capable of oxidizing nucleophilic nitrogen and sulfur atoms in xenobiotics. However, this enzyme shows a catalytic mechanism different from those of other heme- or flavin-containing enzymes. Like other monooxygenases, flavin-containing monooxygenases require molecular oxygen and NADPH as cofactors for oxygenation. In contrast to the other monooxygenases, flavin-dependent monooxygenases do not contain heme or iron, and the binding of the substrate is not required for the generation of the enzyme bound oxygenating intermediate (Fig. 20).

The active, oxygenating form of the enzyme is present in the cell, and any soft, oxidizable nucleophile that can gain access to the enzyme-bound oxygenating intermediate will be oxidized. Precise fit of substrate to the enzyme is not necessary. This property seems to be largely responsible for the broad substrate specificity of flavin-dependent monooxygenases. Flavin-

Table 11. Human cytochromes P450 identified as major catalysts in the biotransformation of specific xenobiotics that seem to play major roles in the oxidation of substrates listed

Cytochrome P450 1A1	1A2	2E1	3A4
Benzo[*a*]pyrene	phenacetin	vinyl chloride	aflatoxin B_1
Other polycyclic hydrocarbons	1-aminofluorene	trichloroethylene	17-β-estradiol
	2-amino-3-methylimidazo-[4,5-*f*]quinoline	halothane benzene	6-aminochrysene sterigmatocystine
	2-naphthylamine	dimethylnitrosamine	nifedipine
		acetaminophen	ethinylestradiol

dependent monooxygenase catalyzes the oxidation of a wide variety of xenobiotics with few, if any, common structural features at maximum rate (Table 12).

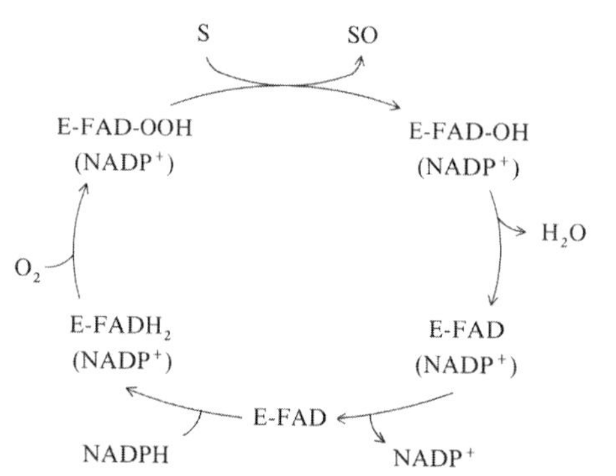

Figure 20. Mechanism of xenobiotic oxidation by the flavin-dependent monooxygenase
FAD = flavineadenine dinucleotide

Many essential xenobiotics also bear functionalities that are oxidized by flavin-dependent monooxygenases. However, these enzymes apparently discriminate between physiologically essential and xenobiotic soft nucleophiles and seem to exclude the former.

As with cytochrome P450, species- and tissue-specific forms of flavin-dependent monooxygenase have been described. Species differences in hepatic flavin-dependent monooxygenase seem to be quantitative rather than qualitative, whereas tissue specific forms in the same species are clearly distinct enzymes. For example, hepatic and pulmonary flavin-dependent monooxygenases in rabbits exhibit distinct, but overlapping substrate specificities and are different gene products. Recent studies have shown that several isoforms of flavin-dependent monooxygenase exist; all isoforms show differences in their distribution in species and in organs within species [45].

Table 12. Oxidations catalyzed by flavin-dependent monooxygenases

Type	Examples
Amine oxidation	$C_6H_5\text{-}NH_2 \rightarrow C_6H_5\text{-}N^+H_2\text{-}O^-$
Hydroxylamine oxidation	$RN(OH)\text{-}CH_3 \longrightarrow RN^+(O^-)\text{=}CH_2$
Thioamide oxidation	$R\text{-}C(\text{=}S)\text{-}NH_2 \rightarrow R\text{-}C(\text{=}S\text{-}O^-)\text{-}NH_2$
Thiol oxidation	$2R\text{-}SH \longrightarrow R\text{-}S\text{-}S\text{-}R$
Disulfide oxidation	$R\text{-}S\text{-}S\text{-}R \longrightarrow 2R\text{-}SO_2$

The regulation of expression of flavin-dependent monooxygenase seems to be complex. Xenobiotics which were shown to increase the concentration of cytochrome P450 in mammals did not influence flavin-dependent monooxygenase concentrations. Recent evidence suggest that soft nitrogen and sulfur nucleophiles in the diet may act as inducers of flavin-dependent monooxygenases. Since the dietary inducers are continously taken up, flavin-dependent monooxygenases are present in maximal concentrations in rats on commercial rat chow.

2.5.4.3. Peroxidative Biotransformation: Prostaglandin-synthase

During the biosynthesis of prostaglandins, a polyunsaturated fatty acid, arachidonic acid, is oxidized to yield prostaglandin G_2, a hydroperoxy endoperoxide. This is further transformed to prostaglandin H_2. Both the formation of prostaglandin G_2 and its further transformation into prostaglandin H_2 are catalyzed by the same

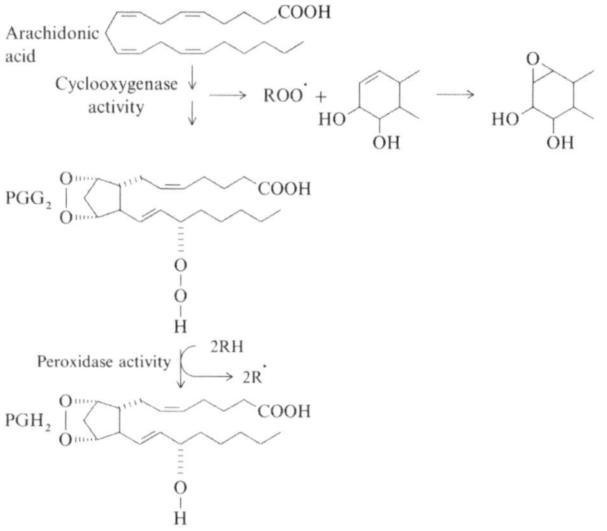

Figure 21. Cooxidation of xenobiotics during the biosynthesis of prostaglandins

enzyme, prostaglandin synthase [46–48]. This enzyme is a glycoprotein with a molecular mass of approximately 70 000 Dalton and contains one heme per subunit. The enzyme is found in high concentrations in germinal vesicles and renal medulla, but also in several other tissues such as skin and adrenals [49]. The cyclooxygenase and peroxidase activity of prostaglandin synthase generate enzyme- and substrate-derived free-radical intermediates (Fig. 21).

Biotransformation of xenobiotics may be associated both with the cyclooxygenase and the peroxidase activity of prostaglandin synthase (Fig. 21). During the cyclooxygenase-catalyzed conversion of arachidonic acid to prostaglandin G_2 peroxy radicals are formed as intermediates. These lipid peroxy radicals represent a source of reactive oxygen metabolites (see below) and can in turn biotransform xenobiotics. Oxidation by the cyclooxygenase activity of prostaglandin synthase is important in the oxidation of diols derived from carcinogenic polycyclic aromatic hydrocarbons and transforms these diols to diol epoxides (Fig. 21, top).

During the reduction of the peroxide prostaglandin G_2 to prostaglandin H_2, the peroxidase undergoes a two-electron oxidation. To return to the ground state the enzyme requires two consecutive one-electron reductions, which are achieved by abstracting electrons from available donors. In addition to endogenous substrates, xenobiotics may act as electron donors and may thus be oxidized to radicals (Fig. 21, bottom). This process is termed cooxidation of xenobiotics. Classes of xenobiotics that undergo cooxidation during prostaglandin syntheses are aromatic amines, phenols, hydroquinones, and aminophenols. The role of prostaglandin synthase in the biotransformation of xenobiotics is somewhat unclear, since many of the end products of prostaglandin-synthase-mediated cooxidation of xenobiotic are identical to those formed by cytochrome P450. Therefore, it is assumed that prostaglandin synthase may contribute to the oxidative biotransformation of xenobiotics in tissues low in monooxygenase activity [50].

In addition to prostaglandin synthase, other peroxidases may also participate in the oxidation of xenobiotics [51–53]. For example, mammary gland epithelium contains lactoperoxidase and leucocytes contains myeloperoxidase. The general reaction catalyzed by this enzymes involves the reduction of hydroperoxide coupled to the oxidation of the substrate:

$$ROOH + XH \rightarrow ROH + XOH$$

The availability of peroxides in tissues likely controls the extent of peroxidative biotransformation; however, the availability of hydrogen peroxide is usually low due to efficient scavenging by catalase and glutathione peroxidase. Therefore, peroxidative metabolism occurs mainly in tissues which can maintain an oxidizing environment.

2.5.4.4. Nonmicrosomal Oxidations

Several enzymes located in mitochondria or the cytoplasm of the cell may also catalyze the oxidation of xenobiotics. In contrast to cytochromes P450 with their broad substrate specificity, most of the nonmicrosomal oxidases have a more narrow substrate specificity and accept only xenobiotics bearing specific functional groups as substrates. Alcohol dehydrogenases catalyze the oxidation of alcohols to aldehydes or ketones:

$$RCH_2OH + NAD^+ \rightarrow RCHO + NADH + H^+$$

The enzyme is mainly found in the soluble fraction of liver, but also in other organs such as the kidney and lung and is responsible for the oxidation of ethanol. The expression of alcohol dehydrogenase is under genetic control, which gives rise to a number of variants with differing activities. Usually, the oxidation of alcohols to aldehydes is reversible, since the reduction of aldehydes is also efficiently catalyzed by aldehyde reductases. However, in vivo, the reaction proceeds in the direction of alcohol consumption since aldehydes are further oxidized by aldehyde dehydrogenases. These enzymes catalyze the formation of acids from aliphatic and aromatic aldehydes:

$$RCHO + NAD^+ \rightarrow RCOOH + NADH + H^+$$

The reaction may be catalyzed by aldehyde dehydrogenase, which has broad substrate specificity, and several isozymes of aldehyde dehydrogenase are found in liver cytosol, mitochondria, and microsomes with characteristic substrate specificities. Other enzymes in the soluble fraction of liver that can oxidize aldehydes are the flavoproteins aldehyde oxidase and xanthine oxidase.

Monoamine oxidases are a family of flavoproteins present in many tissues including liver, kidney, brain, and intestine. These mitochondrial enzymes have a broad and overlapping substrate specificity and oxidize a varietyof amines:

$$RCH_2NH_2 + O_2 + H_2O \rightarrow RCHO + NH_3 + H_2O$$

The monoamine oxidase found in the central nervous system is concerned primarily with neurotransmitter turnover.

2.5.4.5. Hydrolytic Enzymes in Phase-I Biotransformation Reactions

Many tissues contain enzymes with carboxylesterase and amidase activity. These enzymes are located both in microsomal and soluble fraction and hydrolyse ester and amide linkages in xenobiotics.

$$R{-}C(=O)OCH_3 \xrightarrow[+H_2O,\ -CH_3OH]{\text{Esterase}} R{-}C(=O)OH$$

$$R^1{-}C(=O)NR^2R^3 \xrightarrow[+H_2O,\ -HNR^2R^3]{\text{Amidase}} R^1{-}C(=O)OH$$

Although esterases and amidases were thought to be different enzymes, all purified esterases have been demonstrated to have amidase activity; similarly, all amidases have esterase activity. In general, esters are cleaved more rapidly than amides. The expression of many esterases is under genetic control; thus, extremes of high/low esterase activity and resistance/sensitivity to toxic effects mediated by esterases are known.

Epoxide hydrolase is an important enzyme cleaving aliphatic and aromatic epoxides. The enzyme hydrates arene oxides and aliphatic epoxides to the corresponding *trans*-1,2-diols [54]. Water is required as cofactor, and the catalytic mechanism of epoxide hydrolases involves ester formation of the oxirane with a carboxylic acid function at the active site of the enzyme and hydrolysis of this ester by water; no metals or other cofactors are required.

$$E{-}C(=O)OH \xrightarrow{\text{epoxide}} E{-}C(=O)O{-}R(OH) \xrightarrow{+H_2O} E{-}C(=O)OH + \text{HO, OH (trans-diol)}$$

Microsomal epoxide hydrolases are thought to be present in close proximity to the microsomal cytochromes P450; in most cases, the conversion of the epoxide to the less reactive diol is considered to represent an important detoxication reaction for metabolically formed oxiranes. Epoxide hydrolases are found in many tissues such as liver, kidney, testes, and intestine. Their distribution is heterogenous between different cell types in a specific organ; in addition, several forms of microsomal epoxide hydrolases with broad substrate specificity have been found in different animal species. Moreover, in animals, in addition to membrane-bound epoxide hydrolase, a soluble epoxide hydrolase is present in cytoplasm of several tissues [55].

2.5.5. Phase-II Biotransformation Enzymes and their Reactions

Products of the phase-I biotransformation reactions carrying functional groups such as hydroxyl, amino, or carboxyl often undergo a conjugation reaction with an endogenous substrate. The endogenous substrates may include sugar derivatives, sulfate, amino acids, and small peptides (glutathione). The conjugation products are usually more polar and thus more readily excreted than their parent compounds.

In contrast to phase-I biotransformation reactions, phase-II reactions are biosynthetic and require energy to drive the reaction. Energy is usually consumed to generate a high-energy cofactor or an activated intermediate and then utilized as cosubstrate. Thus, depletion of the cofactor or general interference with cellular energy status may interfere with the ability of cells to conduct phase-II biotransformation reactions.

2.5.5.1. UDP-Glucuronyl Transferases

Glucuronidation represents one or the main phase-II biotransformation reactions in the conversion of both endogenous and exogenous compounds to water soluble products [56 – 59]. The formed glucuronides are excreted with bile or urine. The formation of an activated glucuronide (uridine diphosphate glucuronic acid, UDPGA), is required for glucuronide formation. UDPGA is formed in a sequential reaction from uridine and glucose-1-phosphate.

$$\text{Uridine triphosphate} + \text{Glucose 1-phosphate} \xrightarrow[\text{Pyrophosphorylase}]{\text{UDPG}} \text{Uridine diphosphate glucose} + \text{Pyrophosphate}$$

$$\text{Uridine diphosphate glucose} + 2\,NAD^{+} + H_2O \xrightarrow{\text{Dehydrogenase}} \text{UDPGA} + 2\,NADH_2$$

The enzymes that carry out the coupling of the xenobiotic with UDPGA are termed UDP-glucuronyl transferases. They couple D-glucuronic acid with a wide variety of xenobiotics carrying functional groups to give β-D-glucuronides [57]. These glucuronides are highly polar and ionized at physiological pH, and hence are rapidly excreted. The membrane-bound UDP-glucuronyl transferases are found in highest concentration in the liver, but also in most other tissues studied. The reaction catalyzed involves a nucleophilic displacement (S_N2) of the functional group of the substrate with Walden inversion (Fig. 22).

UDP-glucuronyl transferases, like cytochrome P450, represent a family of enzymes, and at least ten individual forms are known. The various forms respond differently to inducers and have preferences for certain classes of chemicals. UDP-glucuronyl transferases catalyze the conjugation of numerous functional groups of xenobiotics with glucuronic acids. Some typical examples are shown in Table 13.

Glucuronides formed in the liver are excreted with urine or bile. Aglycones with molecular masses higher than 300 are transformed to glucuronides that surpass the molecular mass

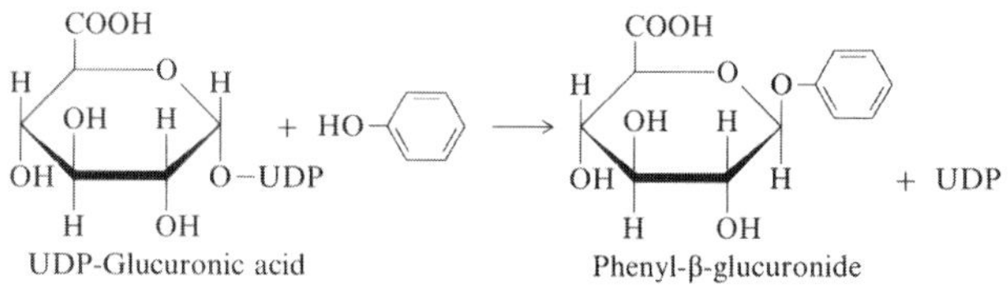

UDP-Glucuronic acid + HO–C₆H₅ → Phenyl-β-glucuronide + UDP

Figure 22. Conjugation of phenol to phenyl glucuronide catalyzed by UDP-glucuronyltransferase (UDP = uridine diphosphate)

Table 13. Typical conjugation reactions catalyzed by UDP-glucuronyltransferases

Type of reaction	Examples
O-Glucuronide formation	1-Naphthol (OH) + UDPGA —UDP-glucuronyl-transferase→ O-glucuronide (COOH, HO, OH, OH) + UDP
N-Glucuronide formation	2-Naphthylamine (NH_2) + UDPGA —UDP-glucuronyl-transferase→ N-glucuronide (N–H, COOH, HO, OH, OH) + UDP
S-Glucuronide formation	Thiophenol (SH) + UDPGA —UDP-glucuronyl-transferase→ S-glucuronide (S, COOH, HO, OH, OH) + UDP

threshold for biliary excretion (see Section 2.5.8.2) and will thus be excreted with bile into the intestine. Glucuronides may by cleaved there by β-glucuronidase present in intestinal microflora to the respective aglycone, which may be reabsorbed from the intestine and translocated back to the liver with the blood. The resulting cycle is called enterohepatic circulation; compounds that undergo enterohepatic circulation are only slowly excreted and usually have a longer half-life in the body.

2.5.5.2. Sulfate Conjugation

The formation of water-soluble sulfate esters is observed with many xenobiotics carrying functional groups such as alcohols, phenols, and arylamines. These reactions are catalyzed by sulfotransferases, a large group of soluble enzymes found in many tissues [60–62]. Sulfotransferases catalyze the transfer of a sulfate group from the "active sulfate" 3′-phosphoadenosine-5′-phosphosulfate to hydroxyl groups and amines. The resulting products are referred as sulfate esters of sulfamates.

The products of this reaction are ionized at physiological pH and may therefore be rapidly excreted in urine; thus, sulfate conjugation is an effective mechanism to enhance the rate of excretion of many xenobiotics. Sulfotransferases are also a family of enzymes, and at least four different classes of sulfotransferases are involved in biotransformation reactions: sulfotransferases, hydroxysteroid sulfotransferases, estrone sulfotransferases, and bile salt sulfotransferases; each class again has been divided into several distinct forms differing in substrate specificity, optimum pH and immunological properties.

Sulfate conjugation also requires a sulfate donor, 3′-phospho-adenosine-5′-phosphosulfate. 3′-Phospho-adenosine-5′-phosphosulfate is likely synthesized in the cytosol of most mammalian cells by a two-step reaction consuming ATP and utilizing inorganic sulfate originating from the catabolism of cysteine or from diet. In the first step of this sequence the sulfation of ATP to adenosine 5′-phosphosulfate is catalyzed by an ATP sulfurylase. Adenosine 5′-phosphosulfate is further transformed to 3′-phosphoadenosine 5′-phosphosulfate by an adenosine 5′-phosphosulfate kinase. However, the equilibrium concentration of 3′-phosphoadenosine 5′-phosphosulfate in mammalian cells may be low, and due to the tight coupling of two enzymes in 3′-phosphoadenosine 5′-phosphosulfate, biosynthesis may proceed rapidly.

$$SO_4^{2-} + ATP \xrightarrow{\text{Sulfurylase}} APS + \text{Pyrophosphate}$$

$$APS + ATP \xrightarrow{\text{APS-phosphokinase}} PAPS + ADP$$

Despite the rapid synthesis, the sulfation of xenobiotics may by limited by reduced availability of 3′-phosphoadenosine 5′-phosphosulfate. The availability of the cofactor of the synthesis reaction, sulfate, may be limited by consumption due to sulfation or the limited availability of free cysteine for transformation to sulfate. Therefore, the sulfation capacity for certain xenobiotics is dependent on dose. Following administration of low doses the compound may be excreted as sulfate; after high doses, the capacity of sulfate conjugation may be saturated, and other biotransformation reactions such as glucuronide formation may become more important for biotransformation.

2.5.5.3. Methyl Transferases

A large number of alcohols, phenols, amines, and thiols present both as endogenous and exogenous compounds may be methylated by several *N*-, *O*- and *S*-methyl transferases. The most common donor for the methyl group is *S*-adenosyl methionine, which is formed from methionine and ATP. Often, these reactions do not increase the water solubility of a xenobiotic, but they are regarded as phase-II reactions since they mask potentially toxic functional groups and thus may serve as detoxication reactions.

A large variety of enzymes catalyze methylations of xenobiotics. The more important enzymes involved in methylation reactions are usually found in many tissues and are present in the soluble fraction of tissues. Some of the enzymes have a high specificity for certain endogenous compounds such as histamine or noradrenaline; others, such as catechol *O*-methyl transferase, metabolize both endogenous catechols and certain xenobiotics carrying aromatic rings with catechol functionalities [63].

2.5.5.4. *N*-Acetyl Transferases

Aromatic amines, hydrazines, sulfonamides, and certain aliphatic amines are biotransformed into amides in a reaction catalyzed by *N*-acetyl transferases. Enzymes that catalyze the acetylation of amines are designated as acetyl CoA: amine *N*-acetyl transferases. These enzymes utilize acetyl coenzyme A as cofactor [64]. The acetylation reaction of arylamines occurs in discrete steps. In the first step, the acetyl group from acetyl coenzyme A is transferred to the *N*-acetyl transferase, which then acetylates the arylamine, thus regenerating the enzyme and forming the amide (Fig. 23).

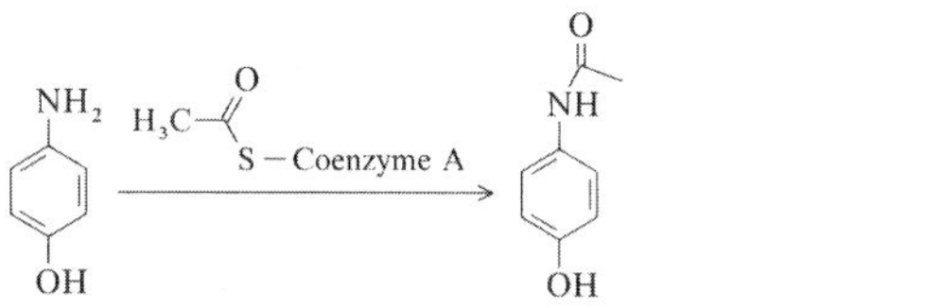

Figure 23. Mechanism of the *N*-acetylation of amines by *N*-acetyl transferase

N-Acetyl transferases are found in a number of different forms in cytosol of many tissues. In many species, the expression of *N*-acetyl transferases is under genetic control, and polymorphism for the expression of *N*-acetyl transferase has been found in several animal species and in humans. The transfer of an acetyl group to amines is reversible, and deacetylation of amides occurs in many species; as noted above, there are large differences between strains, species, and individuals in the extent of expression of amidases [65].

2.5.5.5. Amino Acid Conjugation

Exogenous carboxylic acids are conjugated with a variety of amino acids to form amides. Substrates for conjugation are mainly carboxylic acids containing aromatic rings. Glutamate and glycine appear to be the most common amino acids involved in these conjugation reactions in mammals; in other species such as reptiles and birds, ornithine is involved. The reaction proceeds in two steps and is catalyzed to two different enzymes. In the first step, the carboxylic acid is activated to form a coenzyme A derivative in a reaction involving coenzyme A and ATP. The enzymes that catalyze this reaction are called ATP-dependent acid coenzyme A ligases and are present in mitochondria. They appear to be identical to the intermediate-chain-length fatty acid:coenzyme A synthetase. The thus-formed coenzyme A thioester then transfers its acyl group to the amino group of the acceptor amino acid. This reaction is catalyzed by an *N*-acetyl transferase (Fig. 24). The enzymes catalyzing both steps in amino acid conjugation exist in several forms with different substrate specifities.

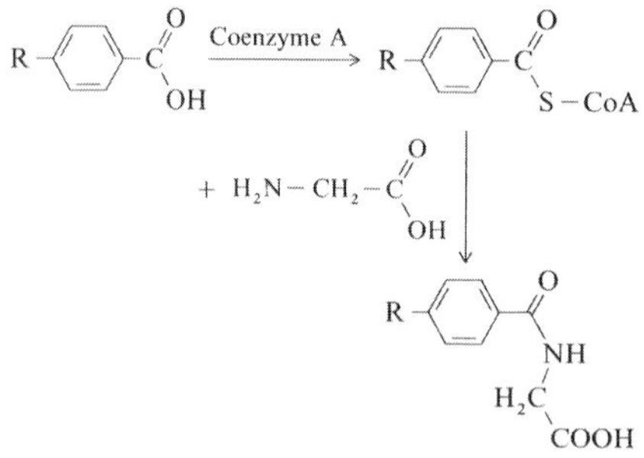

Figure 24. Amino acid conjugation of a xenobiotic carboxylic acid by ATP-dependent acid- coenzyme A ligases followed by *N*-acetyltransferase

2.5.5.6. Glutathione Conjugation of Xenobiotics and Mercapturic Acid Excretion

The conjugation of xenobiotics or their metabolites with the tripeptide glutathione is an important conjugation reaction. Glutathione is composed of the amino acids cysteine, glutamic acid, and glycine (γ-glutamylcysteinylglycine) and is present in many cells in high concentrations (up to 10 mM in liver cells) [66]. Since glutathione conjugation captures reactive electrophiles and transforms them into stable, often nontoxic thioethers, the formation of glutathione conjugates protects cells from the harmful effects of these electrophiles and thus serves as a major detoxication reaction (Section 2.5.6.5). Glutathione conjugation is catalyzed by a family of enzymes termed glutathione *S*-transferases, which are present in the highest concentration in the liver, but are also found in high activity in the kidney, testes, and lung [67]. Glutathione *S*-transferases exist in both membrane-bound and soluble forms; with most substrates, the activity of soluble glutathione *S*-transferase is higher than that of microsomal glutathione *S*-transferase [68].

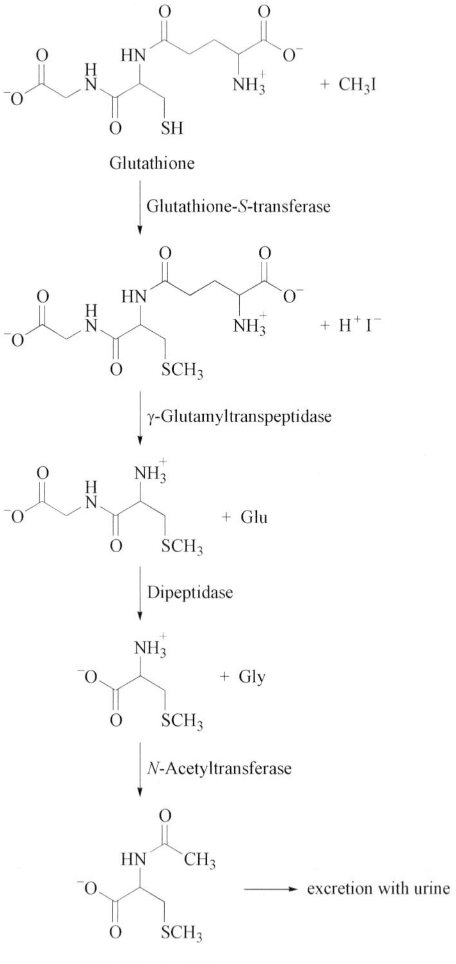

Figure 25. Formation of mercapturic acids by processing of glutathione *S*-conjugates as exemplified by the metabolism of methyl iodide

Cytosolic glutathione *S*-transferases exist in numerous different isoforms, each species being a dimer differing in subunit composition [67, 69]. The glutathione *S*-transferase gene family exists of at least six different families. In contrast to the multiple forms of soluble glutathione *S*-transferases thus possible, only one form of the membrane-bound enzyme is known. The glutathione *S*-transferases catalyze the reaction of the sulfhydryl group of glutathione with chemicals containing electrophilic carbon atoms (Table 14).

Thioethers are formed by reaction of the thiolate anion of glutathione with the electrophile; a spontaneous reaction, albeit at low rates, of

Table 14. Substrates for mammalian glutathione S-transferases

Type of reaction	Examples
Aryltransferase	1-Chloro-2,4-dinitrobenzene (Cl, NO_2, NO_2 substituted benzene) + GSH* ⟶ (SG^{**}, NO_2, NO_2 substituted benzene) + HCl
Arylalkyl-transferase	$C_6H_5CH_2Cl$ + GSH* ⟶ $C_6H_5CH_2SG^{**}$ + HCl
Alkylene-transferase	$CHCOOC_2H_5{=}CHCOOC_2H_5$ + GSH* ⟶ $CH_2COOC_2H_5$–$CH(SG^{**})COOC_2H_5$ (**GS–$CHCOOC_2H_5$)
Epoxide-transferase	C_6H_5–O–CH_2–CH–CH_2 (epoxide O) + GSH* ⟶ C_6H_5–O–CH_2–CH(OH)–CH_2–SG^{**}

GSH = Glutathione. ** GS = Glutathionyl residue.

the electrophile with glutathione without assistance by glutathione *S*-transferases is required for enzymatic catalysis. Glutathione thioethers formed in the organism are not excreted, but further processed to excretable mercapturic acids. Mercapturic acids are thioethers derived from *N*-acetyl-L-cysteine. Mercapturic acid formation is initiated by conjugation of the xenobiotic or an electrophilic metabolite with glutathione (Fig. 25).

This is followed by transfer of the glutamate by γ-glutamyltranspeptidase, an enzyme specifically recognizing γ-glutamyl peptides and found in high concentrations in the kidney and other excretory organs. Dipeptidases catalyze the loss of glycine from the intermediary cysteinylglycine *S*-conjugate to give the cysteine *S*-conjugate which, in the final step of mercapturic acid formation, is *N*-acetylated by a cysteine-conjugate-specific *N*-acetyltransferase using acetyl coenzyme A as cofactor. The mercapturic acids formed are readily excreted into urine by active transport mechanisms in the kidney [70].

Glutathione conjugation is one of the most important detoxication reactions for reactive intermediates formed in organisms. Usually, metabolically formed intermediates are efficiently detoxified, but under specific circumstances, glutathione conjugation may be overwhelmed by high concentrations of electrophiles, which result in covalent binding of intermediates to cellular macromolecules, disruption of important cellular functions, and cell death and necrosis (see Section 2.5.4.5).

2.5.6. Bioactivation of Xenobiotics

Many xenobiotics with low chemical reactivity (e.g., the solvent carbon tetrachloride, the environmental contaminant hexachlorobutadiene, and the heat-exchanger fluid tri-*o*-cresyl phosphate) cause toxic effects. These toxic effects are initiated by covalent binding to macromolecules of metabolites formed in the organism by biotransformation enzymes. This process is termed bioactivation. With many chemicals, reactive metabolites formed during bioactivation may be efficiently detoxified; thus, toxic effects only occur when the balance between the production of reactive metabolites and their detoxication is disrupted. For example, toxic effects may be observed with a certain chemical only when the formation of reactive intermediates is enhanced or when the capacity for detoxication is diminished.

Table 15. Basic mechanisms involved in the bioactivation of xenobiotics based on chemical reactivity of intermediates formed

Mechanism	Structure and reactivity of the intermediate	Examples
Biotransformation to stable but toxic metabolites	different structures, selective interaction of formed metabolite with specific acceptors, or disruption of specific biochemical pathways	dichloromethane, acetonitrile, parathion
Biotransformation to electrophiles	reactive electrophiles	dimethylnitrosamine, acetaminophen, bromobenzene
Biotransformation to free radicals	radicals	carbon tetrachloride
Formation of reactive oxygen metabolites	radicals	paraquat, aromatic nitro compounds

The mechanisms of bioactivation of xenobiotics may be classified into four categories describing the basic types of reactive intermediates formed and their potential reactivity (Table 15) [71–73].

2.5.6.1. Formation of Stable but Toxic Metabolites

This mechanisms is limited to a few selected chemicals because few xenobiotic metabolites are both stable and toxic. The bioactivation of the solvents *n*-hexane and dichloromethane are examples of this mechanism. *n*-Hexane produces a characteristic neuropathy and peripheral nerve injury after chronic exposure. The same typical manifestations of toxicity are also observed when the *n*-hexane metabolites 2-hexanone and 2,5-hexanedione are administered to animals. The mechanism of *n*-hexane neuropathy thus involves oxidation of *n*-hexane by cytochromes P450 at both ends of the carbon chain (ω–1 hydroxylation) and further oxidation of the thus-introduced alcoholic function. The 2,5-hexanedione formed reacts with critical lysine residues in axonal proteins by Schiff base formation followed by cyclization to give pyrroles [74]. Oxidation of the pyrrole residues then causes crosslinking between two *n*-hexane-modified proteins; the resulting changes in the three-dimensional structures of proteins perturb axonal transport and function and cause damage to nerve cells (Fig. 26).

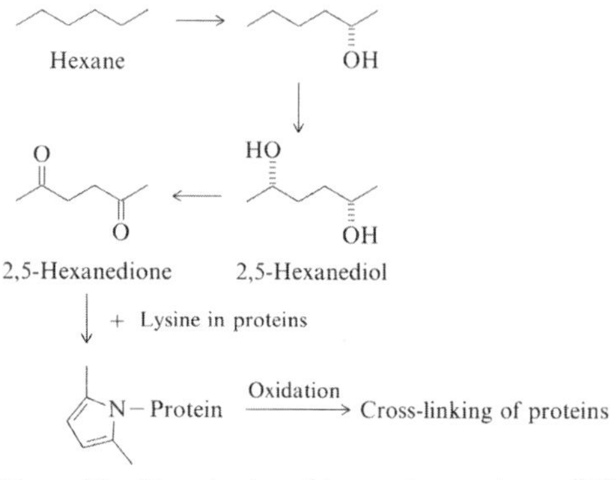

Hexane → OH
2,5-Hexanedione ← 2,5-Hexanediol
+ Lysine in proteins
N – Protein —Oxidation→ Cross-linking of proteins

Figure 26. Bioactivation of hexane by cytochrome P450 to 2,5-hexanedione
Hexanedione reacts with lysine groups in proteins to form pyrroles; oxidation of two neighboring pyrrole residues causes the cross-linking of proteins.

Carboxyhemoglobin formation is observed after human exposure to dichloromethane. Dihalomethanes are oxidized by cytochrome P450, likely by P450 2E1, to carbon monoxide, which, due to its high affinity for iron(II)-containing porphyrins, binds to hemoglobin and interferes with oxygen transport in the blood. Other examples for the formation of stable, but toxic metabolites include the oxidation of acetonitrile to cyanide

$$H_3CCN \xrightarrow{P450} H_2C(OH)CN \longrightarrow CN^- + HCHO$$

and the oxidative desulfuration of parathion.

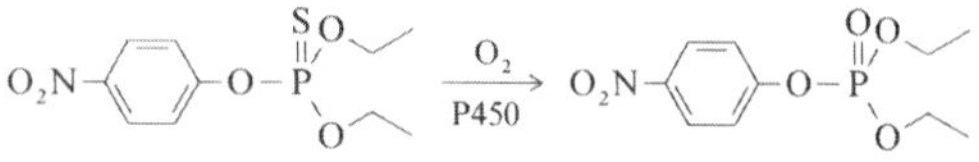

Parathion —O_2, P450→ Paraoxon

Parathion is a potent insecticide acting as an inhibitor of cholinesterase, but itself is only a weak cholinesterase inhibitor. Biotransformation of parathion by oxidative desulfuration to give the potent cholinesterase inhibitor paraoxone is responsible for the high insecticidal potency.

2.5.6.2. Biotransformation to Reactive Electrophiles

Biotransformation to reactive electrophiles is the most common pathway of bioactivation. The cytotoxicity and carcinogenicity of many chemicals is associated with the formation of electrophiles and the ensuing alkylation or acylation of tissue constituents such as protein, lipid, or DNA. Reactive intermediates include such chemically diverse functionalities as epoxides, quinones, acyl halides, carbocations, and nitrenium ions. The metabolic formation of electrophiles may be catalyzed by many different enzymes, although the majority of cases elucidated to date involve cytochrome P450-mediated oxidations.

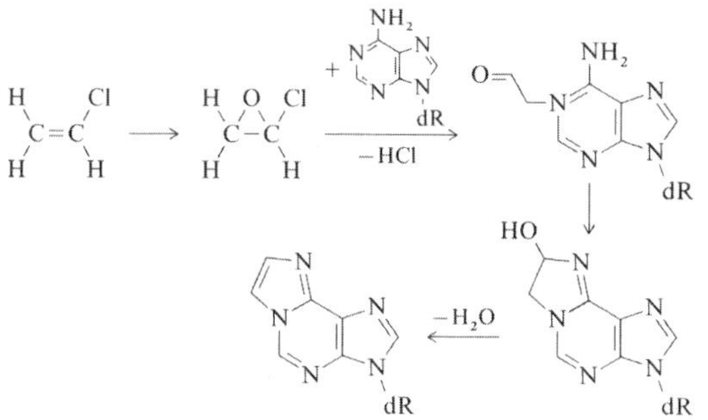

1,N^6-Ethenodeoxyadenosine

Figure 27. Bioactivation of vinyl chloride to chlorooxirane and reaction of the epoxide with critical macromolecules in the cell
dR = Deoxyribose

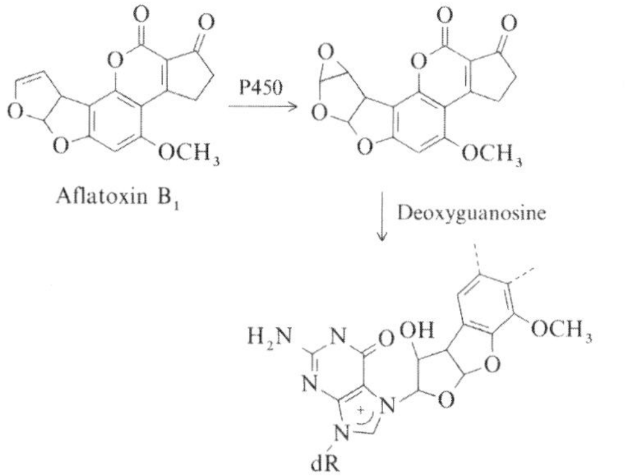

Aflatoxin B_1

2,3-Dihydro-2-(deoxyguan-7-yl)-3-hydroxyaflatoxin B_1

Figure 28. Bioactivation of aflatoxin B_1 to an electrophilic oxirane which results in the formation of DNA adducts and is believed to initiate tumor induction by aflatoxin B_1 in the liver

Cytochrome P450 catalyzes the transformation of olefins to reactive and electrophilic oxiranes. For example, the carcinogenicity of the industrial intermediate vinyl chloride (Fig. 27) and the fungal toxin vaflatoxin B1 (Fig. 28) are dependent on their transformation to electrophilic oxiranes [75, 76].

Carbocations are formed during the cytochrome P450-mediated oxidation of dialkyl nitrosamines. For example, the mutagen and potent carcinogen dimethylnitrosamine is hydroxylated by cytochrome P450 followed by loss of formaldehyde. Monomethylnitrosamine thus formed is unstable and rearranges to release an electrophilic carbocation (Fig. 29).

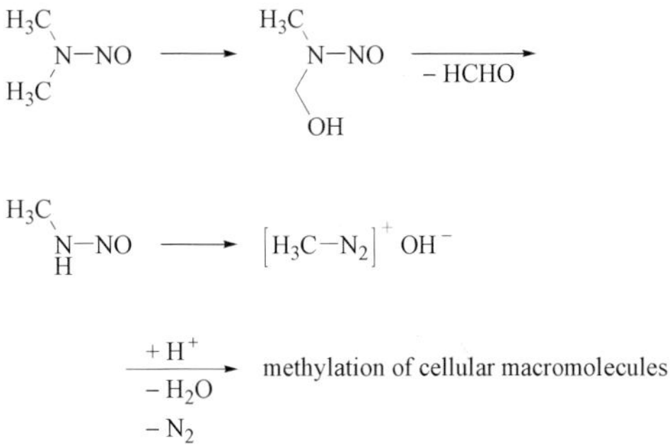

Figure 29. Bioactivation of dimethylnitrosamine to a methylating agent by cytochrome P450

Acyl halides are formed by the oxidation of carbon atoms bearing at least two halogen atoms. The initially formed products are unstable α-halohydrins, which lose hydrogen chloride and thus give reactive acyl halides. An example for the formation of acyl halides as reactive intermediates is the cytochrome P450-mediated oxidation of chloroform to phosgene [77] (Fig. 30).

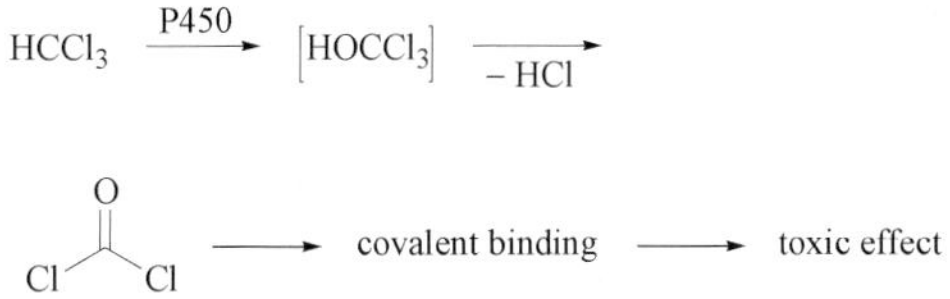

Figure 30. Bioactivation of chloroform by cytochrome P 450 mediated hydroxylation of a C–H bond

However, besides cytochromes P450, other monooxygenases such as flavin-dependent monooxygenase and of phase-II biotransformation enzymes such as UDP-glucuronyl transferases, sulfotransferases, or even the glutathione *S*-transferases may catalyze the bioactivation of xenobiotics [78]. For example, *N*-acetylamidofluorene is oxidized to *N*-hydroxyacetylamidofluorene by cytochrome P450. However, this metabolite is not electrophilic and requires further biotransformation via sulfate conjugation to the highly reactive *O*-sulfate ester. This sulfate ester fragments to a reactive intermediate (a nitrenium ion) which covalently binds to tissue constituents such as DNA (Fig. 31).

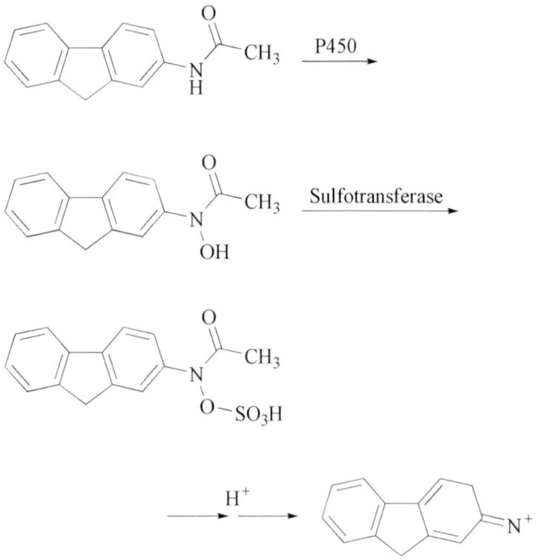

Figure 31. Bioactivation of acetylamidofluorene by cytochrome P450 and by UDP glucuronyl transferases. The glucuronide formed is acid-labile and decomposes to a nitrenium ion.

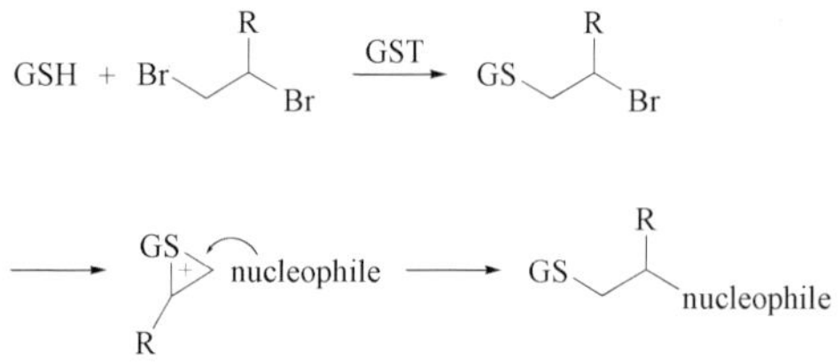

Figure 32. Bioactivation of 1,2-dibromoethane by glutathione conjugation to a reactive and electrophilic episulfonium ion

Some glutathione *S*-conjugates which are biosynthesized to detoxify electrophiles are toxic and mutagenic [79, 80]. 1,2-Dibromoethane is metabolized by glutathione conjugation to *S*-(2-bromoethyl)glutathione. Intramolecular displacement of the bromine on the adjacent carbon atom gives a highly strained, electrophilic episulfonium ion (Fig. 32).

Other toxic glutathione *S*-conjugates require processing by the enzymes of mercapturic acid formation to give electrophiles. A minor pathway in perchloroethene biotransformation results in *S*-(1,2,2-trichlorovinyl)glutathione [81]. This glutathione *S*-conjugate is cleaved by γ-glutamyl transpeptidase and dipeptidases to *S*-(1,2,2-trichlorovinyl)-L-cysteine, which is a substrate for renal cysteine conjugate β-lyase and transformed to pyruvate, ammonia, and a reactive thioketene, binding of which to cellular macromolecules is likely responsible for the renal toxicity of perchloroethene (Fig. 33).

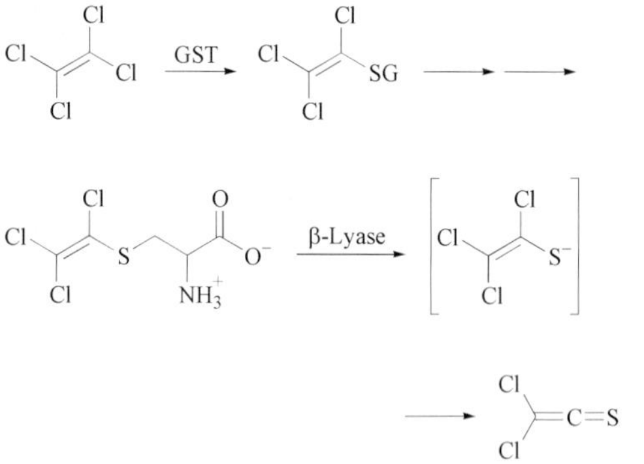

Figure 33. Bioactivation of perchloroethene by glutathione conjugation. The conjugate *S*-(1,2,2-trichlorovinyl)glutathione is biosynthesized in the liver, translocated to the kidney to be processed by γ-glutamyl transpeptidases and dipeptidases, and finally cleaved by cysteine-conjugate β-lyase to give dichlorothioketene.

Due to high concentrations of *S*-(1,2,2-trichlorovinyl)-L-cysteine obtained by active transport to the kidney, covalent binding of the dichlorothioketene formed via this pathway occurs only in the kidney; despite the presence of cysteine conjugate β-lyase in many other organs.

2.5.6.3. Biotransformation of Xenobiotics to Radicals

Free radicals are chemical species that may be formed by a one-electron oxidation to give a radical cation, by a one-electron reduction to give a radical anion, or by homolytic fission of a σ-bond to give a neutral radical.

Free radicals are highly reactive and, when formed in biological systems, are expected to react with a variety of tissue molecules. Radicals may abstract hydrogen atoms, undergo oxidation-reduction reactions, dimerizations and disproportionation reactions. Radicals may also participate in a chain mechanism, which is initiated by a reaction causing a free radical and propagated by a subsequence of reactions causing further radicals as products.

The toxic and tumorigenic solvent carbon tetrachloride is the outstanding example of a bioactivation reaction to a free radical. Carbon tetrachloride is biotransformed by a one-electron reduction to yield the trichloromethyl radical and chloride:

$$Cl_4C + e^- \rightarrow Cl_3C^\bullet + Cl^-$$

The trichloromethyl radical may abstract hydrogen atoms from tissue macromolecules to give chloroform, a proven metabolite of carbon tetrachloride, or may dimerize to give hexachloroethane, which is also a metabolite of carbon tetrachloride. Toxic effects of radicals formed during biotransformation reactions are lipid peroxidation and oxidative modification of proteins (see Section 2.5.6.4). Formation of radicals has been implicated in the bioactivation of many xenobiotics. Radicals may be formed by NADPH-dependent cytochrome P450 reductase, nitroreductases, or one-electron oxidations catalyzed by peroxidases such as prostaglanding synthetases. Formation of free radicals from tissue constituents also plays an important role in the toxic effects of ionizing radiation [82–85].

2.5.6.4. Formation of Reactive Oxygen Metabolites by Xenobiotics

Xenobiotic-induced formation of reduced oxygen metabolites such as the superoxide radical anion, hydrogen peroxide, and the hydroxyl radical has been implicated as a mechanism of producing cell damage, so-called oxidative stress [86–89]. The biotransformation of certain xenobiotics that are involved in redox cycles or undergo enzyme-catalyzed oxidation/reduction reactions may be associated with the production of reduced oxygen metabolites. 2-Methylnaphthoquinone (Menadione) has been intensively used to study the formation and cellular reactions of reduced oxygen metabolites. Menadione and other quinones undergo enzymatic redox cycling; these one-electron oxidation reactions are associated with the formation of the superoxide radical anion ($O^{\cdot 2-}$) by one electron reduction of triplet oxygen. In aqueous solution, superoxide is not particularly reactive, but dismutation or further reduction of superoxide may give rise to hydrogen peroxide (Fig. 34).

Hydrogen peroxide is also a poor oxidant in biological systems, but sufficiently stable to cross biological membranes. The toxicity of hydrogen peroxide is attributed to the formation of the hydroxyl radical by the Fenton reaction, catalyzed by metal ions such as Fe^{2+} (M = transition metal):

$$M^n + H_2O_2 \rightarrow M^{(n+1)} + HO^\bullet + HO^-$$

The highly reactive hydroxyl radical may then initiate cellular damage by radical-based mechanisms. Besides menadione, oxidative stress may also be initiated by other xenobiotics such as the bis-pyridinium herbicide paraquat and nitroheterocycles. Moreover, the formation of reduced oxygen metabolites plays an important role in host defense against infectious agents and in the initiation and propagation of certain diseases such as arteriosclerosis and polyarthritis.

Since oxygen radicals are also formed in low concentrations during cellular respiration, efficient mechanisms for their detoxication exist (see Section 2.5.6.5). Oxidative stress is thus only observed when the eqilibrium between oxidants and reductants is disturbed and detoxication mechanisms are overwhelmed.

2.5.6.5. Detoxication and Interactions of Reactive Metabolites with Cellular Macromolecules

Reactive intermediates formed inside cells may react with low and high molecular mass cellular constituents. These interactions may result in formation of less reactive chemicals and thus in detoxication, or may perturb important cellular functions and thus result in acute and/or chronic toxic effects such as necrosis or cancer. Usually, the interaction with low molecular mass constituents in the cell results in detoxication, whereas the irreversible interaction with cellular macromolecules results in adverse effects [70, 90 – 92].

Detoxication of reactive intermediates may be due to hydrolysis, glutathione conjugation, or interactions with cellular antioxidants. The reaction of electrophilic xenobiotics with the nucleophile water, present in high concentrations

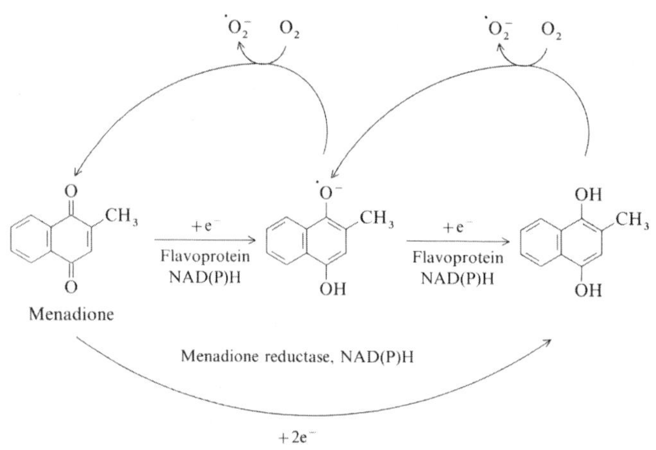

Figure 34. Biotransformation of menadione and induction of oxidative stress by reduction of triplet oxygen to the superoxide radical anion

in all cells, is the simplest form of detoxication. Many of the products thus formed are of low reactivity and may be rapidly excreted. For example, acyl halides formed by the oxidation of olefins such as perchloroethylene are hydrolyzed rapidly to halogenated carboxylic acids; only minor amounts of the intermediate acyl halide reacts with protein and lipids (Fig. 35).

Glutathione-dependent detoxication is an important mechanism for metabolically formed electrophiles, free radicals, and reduced oxygen metabolites [93–96]. Electrophiles react with the nucleophilic sulfur atom of glutathione in a spontaneous or enzyme-catalyzed reaction. Spontaneous reactions are only observed at appreciable rates with soft electrophiles (glutathione is a soft nucleophile); the conjugation of hard electrophiles with glutathione requires enzymatic catalysis; usually, the rates of conjugation catalyzed by glutathione *S*-transferase differ between hard and soft electrophiles; soft electrophiles are conjugated more efficiently. For example, the hard electrophile aflatoxin B_1 8,9-oxide does not spontaneously react with glutathione; only in the presence of a certain glutathione *S*-transferase enzyme is a glutathione *S*-conjugate of aflatoxin B_1 8,9-oxide formed. Species differences in the tumorigenesis of aflatoxin B_1 may serve to illustrate the important role of glutathione *S*-transferases in the expression of toxicity and carcinogenicity. Aflatoxin B_1 is a potent liver carcinogen in rats; in mice, aflatoxin B_1 is only weakly carcinogenic. The liver of mice contains a glutathione *S*-transferase which efficiently detoxifies aflatoxin B_1 8,9-oxide. This glutathione *S*-transferase enzyme is not present in rat liver; thus, the binding of aflatoxin B_1 8,9-oxide to rat liver DNA and liver carcinogenicity of aflatoxin B_1 are much higher in rats than in mice.

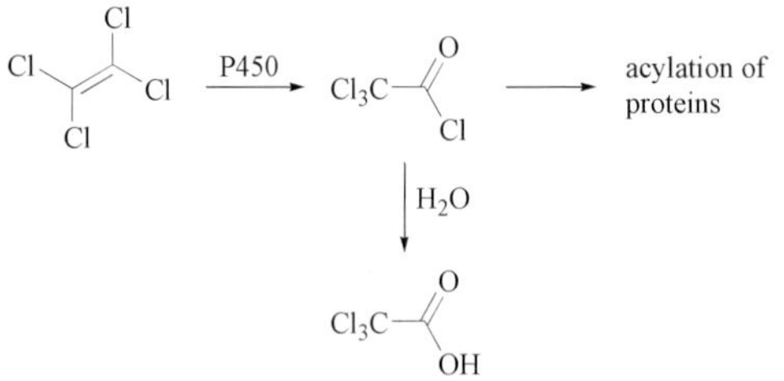

$$Cl_2C{=}CCl_2 \xrightarrow{P450} Cl_3C{-}COCl \longrightarrow \text{acylation of proteins}$$
$$Cl_3C{-}COCl \xrightarrow{H_2O} Cl_3C{-}COOH$$

Figure 35. Biotransformation of tetrachloroethylene to trichloroacetyl chloride followed by hydrolysis to trichloroacetic acid, the major urinary metabolite formed from tetrachloroethylene. Only a small amount of the acyl halide formed reacts with proteins.

Glutathione also plays a major role in the detoxication of reactive oxygen metabolites and radicals. Selenium-dependent glutathione peroxidases are important enzymes catalyzing the detoxication of hydrogen peroxide. In the glutathione peroxidase catalyzed reaction, two moles of glutathione are oxidized to glutathione disulfide:

$$H_2O_2 + 2\,GSH \underset{\text{Glutathione reductase}}{\overset{\text{Glutathione peroxidase}}{\rightleftarrows}} GSSG + H_2O$$

Glutathione can be recycled by the reduction of glutathione disulfide by glutathione reductase.

The copper- and zinc-dependent cytosolic and - manganese-dependent mitochondrial superoxide dismutases detoxify superoxide radical anions. Hydrogen peroxide formed by dismutation of superoxide is converted to water and oxygen and thus detoxified by catalase:

$$2\,H_2O_2 \xrightarrow{\text{Catalase}} 2\,H_2O + \tfrac{1}{2}\,O_2$$

Several cellular antioxidants also play a role in the detoxication of radicals. α-Tocopherol is an important lipophilic antioxidant, whose presence in lipid membranes prevents damage to lipid constituents (e.g. unsaturated fatt acids) by radicals. The hydroxyl radical, the superoxide radical anion, and peroxy radicals react with α-tocopherol to yield water, hydrogen peroxide, and hydroperoxides, which may be detoxified further by catalase and glutathione peroxidase. α-Tocopherol is transformed during these reactions to give a stable radical of comparatively low reactivity. Ascorbic acid is an important antioxidant present in the cytoplasm of the cell and may also participate in the detoxication of radicals.

2.5.6.6. Interaction of Reactive Intermediates with Cellular Macromolecules

Although a substantial body of information is available on the biotransformation of xenobiotics to reactive metabolites and the chemical nature of those metabolites, considerably less is known about how reactive intermediates interact with cellular constituents and how those interactions cause cell injury and cell death. The reaction of toxic metabolites may result in the formation of covalent bonds between the molecule and a cellular target molecule, or they may alter the target molecule without formation of a covalent bond, usually by oxidation or reduction [97].

Electrophilic metabolites may react with different nucleophilic sites in cells. Nucleophilic sites in cellular macromolecules are thiol and amino groups in proteins, amino groups in lipids, and oxygen and nitrogen atoms in the purine and pyrimidine bases of DNA. The formation of a covalent bond may permanently alter the structure and/or activity of the modified macromolecule and thus result in a toxic response. The complexity of the reaction of electrophilic metabolites with the various nucleophilic sites in cells may be interpreted on the basis of the concept of hard and soft electrophiles and nucleophiles (hard and soft acids and bases). The donor atom of a soft nucleophile is of high polarizability and low electronegativity, and is easily oxidized; the donor atom of a hard nucleophile is of low polarizability and high electronegativity. Hard electrophiles carry a high positive charge and have a small size; soft electrophiles are of low positive charge and large size. Soft electrophiles react predominantly with soft nucleophiles, and hard electrophiles with hard nucleophiles [98]. Thus, hard electrophiles formed during a biotransformation reaction (e.g., carbocations formed from dialkylnitrosamines) predominantly react with hard nucleophiles such as the oyxgen and nitrogen atoms of DNA, In contrast, soft electrophiles such as α,β-unsaturated carbonyl compounds (e.g., acrolein, benzoquinone) react predominatly with soft tissue nucleophiles such as the sulfhydryl groups of cysteine in proteins (Table 16).

Covalent interactions of xenobiotics with proteins occur with several nucleophilic nitrogen atoms; both alkylation and acylation reactions of amino acids have been reported as consequences of formation of reactive intermediates in cells. Besides the sulfur atom of cysteine, nitrogen atoms in the amino acids lysine, histidine, and valine are frequent targets for electrophilic metabolites. Consequences of the modifications may be inactivation of enzymes important for cellular function, changes in the tertiary structure of proteins, or changes in gene expression. Alkylation of the sulfhydryl-dependent enzymes of mitochondrial respiration is thought to play an important role in the initiation of mitochondrial dysfunction and thus cell damage.

Some modified proteins may also serve as immunogens, and hypersensitivity reactions, formation of immune complexes, and delayed hypersensitivities may be the consequences of protein adduct formation. Indeed, many drug- and chemical-related hypersensitivity reactions observed in clinical medicine are based on the formation of covalent protein adducts and their recognition as "foreign" by the immune system [99].

Oxidative stress produces mixed disulfides of proteins with low molecular mass thiols such as glutathione and thus alters protein structure

Table 16. Metabolically formed electrophiles and their prime targets for covalent binding in cells

Soft		→		Hard
		Nucleophile		
SH of cysteine or glutathione	sulfur of methionine	primary or secondary nitrogen atoms in peptides (lysine, arginine, or histidine)	amino groups of purines and pyrimidines in RNA and DNA	oxygen in purines and pyrimidines in DNA and RNA
		Electrophile		
α,β-Unsaturated carbonyls, quinones	epoxides, alkyl sulfates, alkyl halides	nitrenium ions	benzylic carbocations	aliphatic and aromatic carbocations

and function. In addition, oxidants and radicals promote the oxidation of amino acids in proteins, which may increase the susceptibility of these proteins to proteolysis [100, 101]. Increased protein oxidation has been implicated in cellular aging and in the mechanisms of toxicity of several redox-active transition metals.

Radicals formed during the biotransformation of xenobiotics may abstract hydrogen atoms from cellular components [82]. The abstraction of hydrogen atoms from polyunsaturated fatty acids of lipids results in a process termed lipid peroxidation. The fatty acid radicals thus formed may react with molecular oxygen to give peroxy radicals and further to hydroperoxides. The initiated radical chain reactions cause the cleavage of carbon–carbon bonds in the fatty acids to short fragments such as α,β-unsaturated carbonyl compounds [102, 103] (see Section 3.3.1).

The disruption of membranes and the formation of toxic hydroperoxides and α,β-unsaturated carbonyl compounds may cause disruptions in cellular calcium homeostasis and thus cause biochemical changes that ultimately lead to cell death [104, 105]. The reaction of electrophilic metabolites with DNA constituents results in the formation of altered purine and pyrimidine bases or other DNA damage such as DNA strand breaks or loss of single bases from the double helix. Many of these modifications are "premutagenic lesions". After gene expression, these lesions may be translated into mutations [106]. Mutations in certain genes are considered to be the basis for the evolution of neoplastic cells and cancer and thus play a major role in chemical carcinogenesis. Other types of DNA damage may result in the activation of genes important for cellular differentiation or other regulatory functions. Electrophilic intermediates alkylate the nitrogen and oxygen atoms of the purine and pyrimidine bases in DNA; deoxyguanosine is often preferentially alkylated. The site of alkylation of a certain base in DNA is again dependent on the electrophilicity of

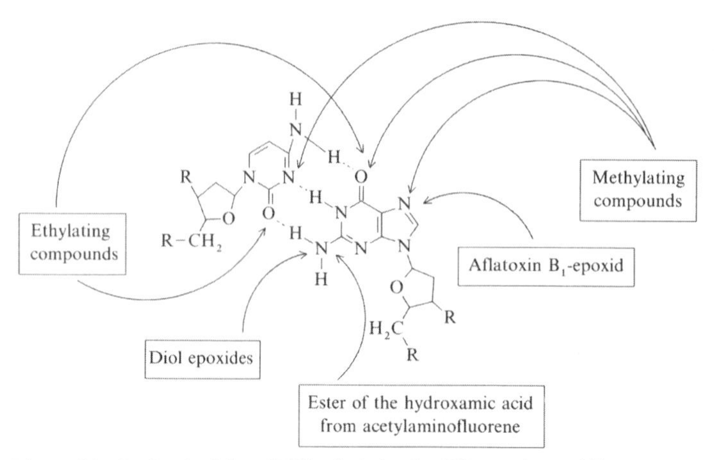

Figure 36. Regioselectivity of DNA alkylation by different electrophiles

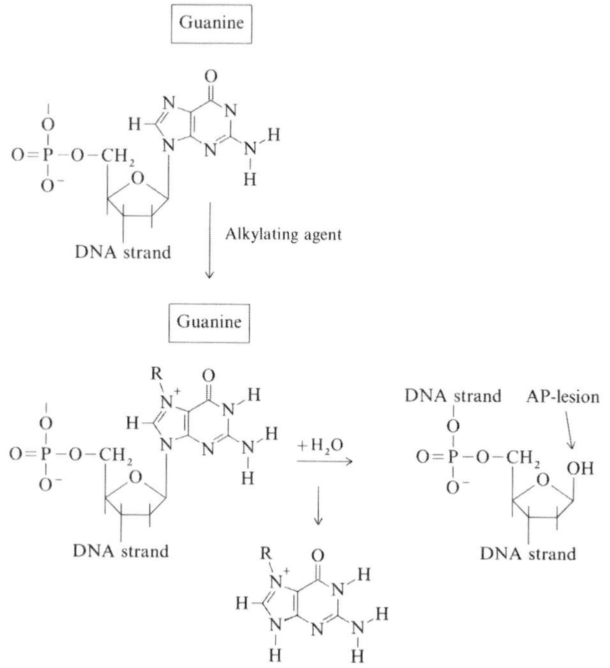

Guanine

Alkylating agent

DNA strand

Guanine

$+H_2O$

DNA strand AP-lesion

Figure 37. Labilization of the bond of guanosine to the DNA backbone by alkylation of the N-7 position resulting in the loss of the modified guanosine from DNA and the formation of an apurinic site (AP)

the alkylation agent; hard electrophiles preferentially react with the oxygen atoms of guanosine, while soft electrophiles alkylate the exocyclic amino groups (Fig. 36).

The pattern of base alkylation is different when DNA is modified in biological systems or when isolated DNA, nucleosides, and nucleotides are treated with the xenobiotic or its metabolite(s). Regioselectivity is further modified by solvents, buffer salts, and concentration of reactants. Certain modifications of deoxyguanosine result in the labilization of the glycosidic bond; loss of the deoxyguanosine derivate results in an "apurinic site" in DNA other chemical modifications may result in labilization of the five-membered ring and ring opening after reaction with water (Fig. 37).

Radicals formed as reactive intermediates may also cause DNA damage. Besides DNA strand breaks, which have been frequently observed, the reaction of oxygen-derived radicals may also result in the oxidation of purine and pyrimidine nucleotides. Due to the development of sensitive techniques for the characterisation of oxidative modifications in DNA, a number of modified bases have been identified:

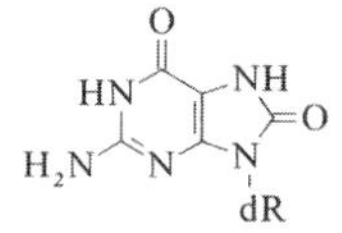

8-Hydroxydeoxyguanosine
(7,8-Dihydro-8-oxodeoxyguanosine)

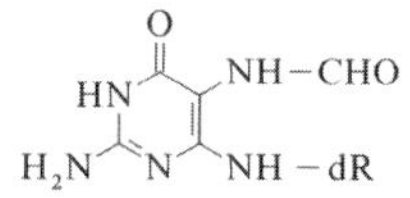

Formamidopyrimidine-deoxyguanosine

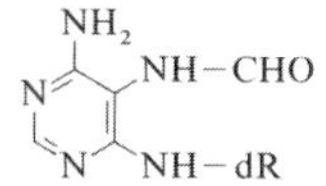

Formamidopyrimidine-deoxyadenosine

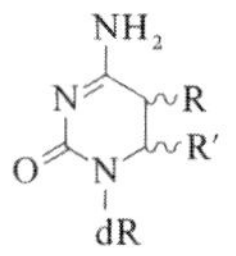

R,R′ = H, OH, OOH
dR = Deoxyribose

5,6-Dihydrodeoxycytidine

8-Hydroxydeoxyguanosine, a premutagenic modification, is considered as one of the more important lesions induced by oxidative DNA damage; because sensitive methods are available for its quantification, it can serve as a marker for the extent of oxidative DNA modification caused by a xenobiotic or by other processes. DNA oxidation has also been implicated in aging; an increase in oxidative DNA modifications may occur with age due the decreased availability of antioxidants in cells of aging mammals. Several theories suggest a correlation between increased oxidative DNA damage and the increased incidence of tumors in the aged population [107, 108].

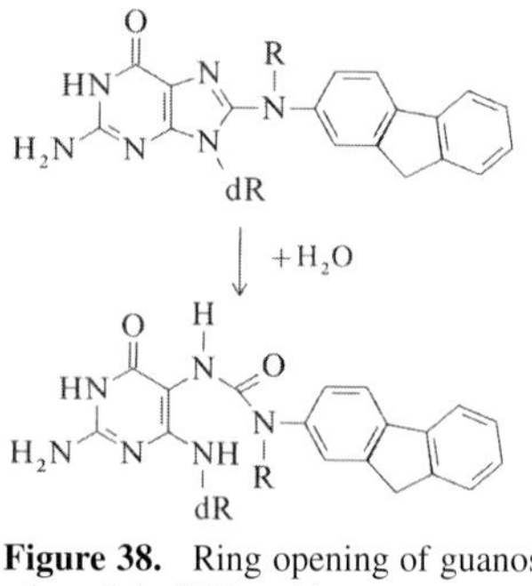

Figure 38. Ring opening of guanosine in DNA by alkylation of the N-7 position

2.5.7. Factors Modifying Biotransformation and Bioactivation

The biotransformation of xenobiotics may be modified by a variety of factors both intrinsic and extrinsic to the normal functioning of the organism. The changes in the extent of biotransformation may have profound effects on the toxicity of a specific chemical. When biotransformation results in detoxication and rapid excretion, increased rates of biotransformation will decrease toxicity. On the other hand, the toxicity of a chemical bioactivated to reactive intermediates will increase on enhancing biotransformation. A great variety of factors have been shown to influence the extent of biotransformation; many of the effects listed below have been primarily decribed in experimental animals. However, observations in humans (e.g,. after drug treatment) indicate that similar effects, albeit not of the same magnitude or duration, must occur in humans.

2.5.7.1. Host Factors Affecting Biotransformation

Enzyme Induction. The activity of biotransformation enzymes can be enhanced by pretreatment with a range of structurally different chemicals. These chemicals can be drugs, pesticides, natural products, environmental contaminants, and even ethanol. The enhanced enzyme activities and the increased enzyme concentrations may results from increased de novo synthesis of the protein, reduced degradation, or from other, often unknown effects. An increase in the concentration of an biotransformation enzyme in the organism, a certain organ or cell type is termed "enzyme induction" [109, 110]. Several hundred different chemicals have been demonstrated to increase the biotransformation of other xenobiotics and to act as enzyme inducers. The majority of these studies focused on the induction of microsomal monooxygenases, mainly cytochrome P450 enzymes; however, other membrane bound enzymes such as UDP-glucuronyl transferases may also be induced. Glutathione *S*-transferases are the only cytosolic biotransformation enzymes whose activities may be increased by the administration of inducers to experimental animals (Table 17).

Table 17. Inducers of the enzymes of biotransformation and enzymes whose cellular concentrations are increased by pretreatment

Inducing agent	Induced enzymes
2,3,7,8-Tetra-chlorodibenzodioxin	cytochrome P450, UDPglucuronyltransferase
Ethanol	cytochrome P450
Phenobarbital	cytochrome P450, epoxide hydrolase, UDP-glucuronyltransferases
trans-Stilbenoxide	epoxide hydrolase
3-Methylcholanthrene	cytochrome P450, UDPglucuronyltransferases

The onset, magnitude, and duration of increases in the concentration of biotransformation enzymes after the administration of an inducer and the associated biochemical and morphological effects depend on the chemical nature of the inducing agent, dose, and time of administration. For example, the time required for maximum induction of specific cytochrome P450 enzymes by the classical inducers phenobarbital and 3-methylcholanthrene are different. Moreover, besides increases in the activity of hepatic monooxygenases, phenobarbital administration results in marked hepatic hypertrophy and proliferation of the smooth endoplasmatic reticulum; these effects are absent in animals treated with 3-methylcholanthren. Induction by parenteral application of 3-methylcholanthren results in maximum enzyme concentrations in the liver within 48 h, whereas maximal induction by parenteral application of hypnotic doses of phenobarbital requires up to 5 d.

Enzyme induction is reversible after withdrawal of the inducing agent, and the enzyme activities return to basal levels over a characteristic time span. Again, this time span is dependent on the chemical nature of the inducing agent. For example, cessation of phenobarbital treatment will result in a decline of enzyme activities to basal levels within one to two weeks. The mechanisms of enzyme induction are complicated and only partially understood. Apparently, different chemicals influence the activities of the biotransformation enzymes by different mechanims; even the effect of a specific chemical on different enzymes may be due to separate mechanisms (Table 18) [111].

Table 18. Mechanisms of cytochrome P450 enzyme induction by different xenobiotics

Cytochrome P450	Inducing agent	Mechanism of induction
1A1	2,3,7,8-tetra-chlorodibenzodioxin	increased gene transcription
1A2	3-methylcholanthrene	stabilization of messenger RNA
2B1, 2B2	phenobarbital	increased gene transcription
2E1	ethanol, acetone	protein stabilization
3A1	dexamethasone	increased gene transcription, independent of glucocorticoid receptor
3A1	triacetyloleandomycin	protein stabilization
4A1	clofibrate	increased gene transcription, receptor mediated

Only a few mechanisms are well understood. Modulation of gene expression seems to be the basis for many inductive effects. For example, induction of cytochrome P450 1A1 is prevented by inhibitors of protein synthesis. Moreover, studies using the potent inducing agent 2,3,7,8-tetrachlorodibenzodioxin identified a high-affinity binding protein with the properties of a receptor for 2,3,7,8-tetrachlorodibenzodioxin in rat liver cytosol [112, 113]. Binding of 2,3,7,8-tetrachlorodibenzodioxin to this protein results, after further interaction with other proteins, in the translocation of the formed complex from the cytosol to the nucleus. This translocation is followed by interaction with specific recognition sites on the genome, transcription, and translation of the specific gene for cytochrome P450 1A1 and for other biotransformation enzymes such as UDP-glucuronyl transferases (Fig. 39).

In contrast, the mechanisms of enzyme induction by phenobarbital and, for example, ethanol are not defined. A specific receptor for phenobarbital could not be demonstrated, but some experiments suggest involvement of the glucocorticoid receptor in phenobarbital-mediated enzyme induction. Ethanol and other inducers seem to stabilize the cytochrome P450 2E1 protein against degradation by an unknown mechanism.

Enzyme Inhibition. The decrease in the activity of specific biotransformation enzymes is termed inhibition of biotransformation. As noted above, inhibition of biotransformation may increase or decrease the toxicity of a xenobiotic. For example, the inhibition of cytochrome P450 by 2-(diethylamino)ethyl-2,2-diphenylpentanoate (SKF-525A) causes an increase in hexobarbital sleeping time, but a decrease in the hepatoxicity of carbon tetrachloride.

Several mechanisms responsible for inhibition of xenobiotic metabolizing enzymes are operative [114, 115]. Besides inhibition of protein synthesis and thus synthesis of the enzyme, xenobiotics may irreversibly bind to the active site of the enzyme. This process is termed suicide inhibition [116]. Following activation of the xenobiotic by the enzyme, the reactive intermediate formed reacts with constituents of the enzyme at or near the active site, thus blocking further catalytic activity. This effect has been demonstrated with several xenobiotics which are inhibitors of cytochrome P450, such as alkenes and compounds containing allylic and acetylenic derivatives. For example, ethylene oxide, the reactive metabolite formed by cytochrome P450 catalyzed oxidation of ethylene, may alkylate the pyrrole nitrogen atoms in the heme moiety and thus result in heme destruction.

During exposures to mixtures, chemicals with high affinity to certain biotransformation enzymes will be preferentially metabolized, and thus the biotransformation of other constituents

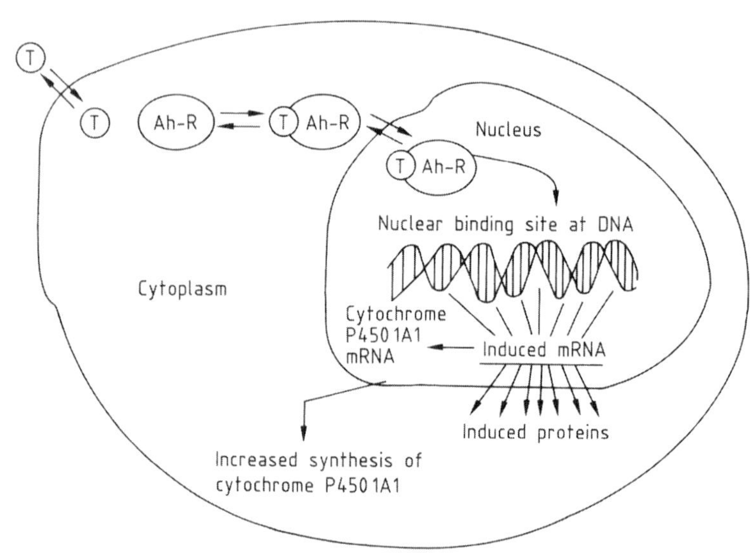

Figure 39. The Ah receptor (Ah-R) and mechanism of enzyme induction by 2,3,7,8-tetrachlorodibenzodioxin (T)

of the mixture will be reduced or even totally inhibited. This reduced biotransformation may also alter the toxicity of a chemical present in a mixture compared with that of the pure chemical. For example, coadministration of ethanol markedly reduces the toxic effects of methanol (metabolic acidosis, reversible or even permanent blindness). These toxic effects are caused by the oxidation of methanol to formic acid as toxic metabolite; formic acid accumulates in the organism and damages the optical nerve. Oxidation of methanol to formaldehyde is competitively blocked by administration of ethanol; under these circumstances, the toxic metabolite formic acid can not be formed; methanol is excreted from the organism unchanged in the urine and by exhalation.

Inhibition of some biotransformation enzymes may also be caused by effects on the tissue levels of necessary cofactors. The availability of glutathione for conjugation is reduced by blocking glutathione biosynthesis; diethyl maleate and some other chemicals deplete intracellular glutathione concentrations by reacting with glutathione to give a glutathione *S*-conjugate. Pretreatment of experimental animals with these chemicals followed by the application of a xenobiotic which requires glutathione for detoxication will result in an increased toxic response. Moreover, large doses of nontoxic chemicals metabolized by sulfotransferases may deplete the cofactor for sulfate conjugation and may thus alter the disposition and, probably the toxicity, of other xenobiotics that undergo sulfate conjugation.

Genetic Differences in the Expression of Xenobiotic Metabolizing Enzymes. The ability of different animal species to metabolize xenobiotics is related to evolutionary development and therefore to different genetic constitution; thus, major species differences in the extent and pathways of biotransformation exist. These variations may be divided into qualitative and quantitative differences. Qualitative differences involve metabolic pathways and are related to species defects or peculiar reactions of a species. For example, guinea pigs do not have the enzymatic capacity to catalyze the last step in mercapturic acid formation, the *N*-acetylation of cysteine *S*-conjugates, and therefore excrete cysteine *S*-conjugates as end products of this pathway. Certain species such as cats do not have the capability to form glucuronides from xenobiotics.

Quantitative variations are often due to species differences in gene and enzyme regulation. For example, interindividual differences have been decribed in humans in the biotransformation of many drugs. The *N*-acetylation of the tuberculostaticum Isoniazid has a genetic basis. Some individuals are homozygous for a recessive gene, and this may result in the absence of isoniazid *N*-acetyltransferase, that is, they are "slow acetylators". In normal homozygotes or heterozygotes, "rapid acetylators", Isoniazid is rapidly transformed to the *N*-acetyl derivative. The *N*-acetyltransferase polymorphisms are correlated with different responses to Isoniazid-induced toxicities. This genetic polymorphism is also seen in human, 80% of the Japanese and Eskimos are "rapid acetylators"; whereas in some European populations only 40 to 60% are "rapid acetylators" [117]. Polymorphisms in the expression of cytochrome P450 1A1 and a specific glutathione *S*-transferase in human lung have been implicated in an increased rate of lung cancer in individuals expressing high cytochrome P450 1A1 and deficient in this glutathione *S*-tansferase. Cytochrome P450 1A1 bioactivates aromatic hydrocarbons present in cigarette smoke to yield electrophiles; glutathione *S*-transferase detoxifies these metabolites.

Influence of sex on biotransformation reactions. Sex differences in the extent and pathways of biotransformation may be based on sex-dependent expression of certain biotransformation enzymes. For example, adult male rats metabolize many xenobiotics at higher rates than females; both phase-I and phase-II biotransformation reactions seem to be influenced by sex-dependent factors. With cytochrome P450, at least three different hepatic enzymes have been demonstrated to be under the control of sex-hormones (Table 19).

Table 19. Sex hormone-dependent hepatic cytochrome P450 in the rat

Enzyme	Sex specificity	Remarks
P450 2D	female	expressed at a hormone-independent basal rate, stimulated by estrogen, suppressed by androgen
P450 2C	male	
P450 3A	male	neonatally imprinted by androgen

In addition to the liver, sex differences in biotransformation are also found in extrahepatic tissues such as the kidney and may be responsible for sex-specific toxic effects of xenobiotics in these organs. For example, the kidneys of male mice contain a cytochrome P450 enzyme which bioactivates chloroform, acetaminophen, and 1,1-dichloroethene to reactive intermediates. The enzyme is present only in much lower activity in female mice, which are thus not susceptible to the renal toxicity of these chemicals.

In experimental animals, sex differences in the expression of biotransformation enzymes usually become apparent at puberty and are maintained throughout adult life. Despite the relatively large sex-dependent variations seen in animal studies, sex seems not to have a profound influence on the biotransformation of chemicals in humans.

Dietary constituents and the biotransformation enzymes. Nutritional factors influencing biotransformation may be mineral deficiencies, vitamin deficiencies, protein content, starvation, and natural substances in the diet. Mineral deficiencies (calcium, copper, zinc) have been shown to reduce the activities of cytochromes P450. On the other hand, an excess of dietary iron has been observed to increase monooxygenase activity. Dietary cobalt, calcium, and manganese may increase the hepatic levels of glutathione and may thus influence glutathione *S*-conjugate formation. Several vitamins are directly or indirectly involved in the regulation of cytochrome P450. For example, diets deficient in vitamins C and E reduce the activity of monooxygenases, whereas deficiencies in other vitamins increase monooxygenase activity. Moreover, several vitamins serve as important cellular antioxidants and influence the energy and redox state of the cell and thus also affect biotransformation reactions.

Low-protein diets generally reduce the activity of cytochrome P450 and certain phase-II biotransformation reactions. Thus, the nutrient status may also modify the toxicity of xenobiotics. For example, dimethylnitrosamine is a potent hepatocarcinogen in rats kept on a high-protein diet; but almost without effect in rats kept on a low-protein diet. Food deprivation reduces the hepatic concentration of glutathione by as much as 50% due to reduction of glutathione biosynthesis. Thus, xenobiotics detoxified by glutathione conjugation are more toxic in starved than in fed animals. Fasting has also been shown to increase the levels of cytochrome P450 2E1, but decrease the levels of cytochrome P450

2C11. Several natural ingredients in the diet of laboratory animals such as indoles, diallyl disulfid,e and psoralens may increase the activities of cytochrome P450. However, some of these compounds may selectively increase some cytochrome P450 enzymes, but inhibit others. For example, diallyl disulfide, a constituent of garlic, has been shown to induce cytochrome P450 2B in rats, but inhibits cytochrome P450 2E1. Vegetable ingredients present in broccoli are potent inducers of phase-II biotransformation enzymes and thus increase the capacity of the organism to detoxify reactive intermediates. These natural ingredients are thought to play a major role in the anticarcinogenicity of diets rich in those vegetables.

2.5.7.2. Chemical-Related Factors that Influence Biotransformation

Xenobiotic-related factors influencing biotransformation are the physiocochemical properties (i.e., chemical structure including the presence of functional groups) and dose. The major determinant of the rate of biotransformation is the concentration of the substrate at the active site of the enzyme. This concentration is determined by structure and lipophilicity and by dose. Lipophilic xenobiotics readily cross cell membranes and are rapidly absorbed and distributed in the organism. Moreover, lipophilic xenobiotics show a higher partitioning into lipid membranes. These factors contribute to higher concentrations of lipophilic xenobiotics at the active center, especially of membrane-bound enzymes. The presence of functional groups also influences rates and routes of biotransformation. Certain functional groups may compete for the same substrate for conjugation; also, specific functional groups may undergo different reactions, as indicated in Figure 40 for *p*-aminobenzoic acid.

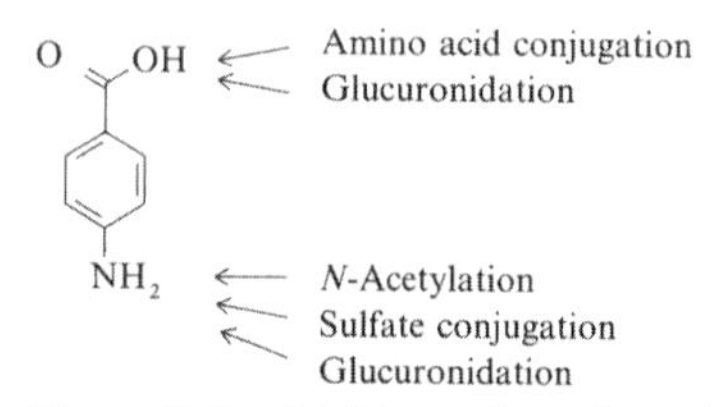

Figure 40. Possible biotransformation pathways for *p*-aminobenzoic acid

The presence of other specific functional groups may have a major effect on biotransformation and its regioselectivity. For example, the presence of trifluoromethyl groups in alkanes renders the adjacent methylene carbon atom almost inert to hydroxylation and strongly influences the regioselectivity of enzymatic hydroxylations on aromatic rings.

Dose is one of the most important factors determining rate and route of biotransformation for more complex molecules. Certain biotransformation enzymes have a high affinity but low capacity for a specific chemical, while others have a high capacity but low affinity. As dose increases, high-affinity, low-capacity enzymes will become staturated, and low-affinity, high capacity pathway(s) will biotransform a larger percentage of dose [118].

2.5.8. Elimination of Xenobiotics and their Metabolites

The evolution of complex forms of life necessitated the development of specialized mechanisms to eliminate waste products formed from endogenous compounds and to prevent the accumulation of toxic xenobiotics present taken up in the diet. Excretion of wastes by the earlier forms of aquatic life was largely passive and involved the loss of large volumes of water and nutrients. For land-living animals, conservation of water, minerals, and nutrients was necessary for survival. Therefore, complex mechanisms for the elimination of both endogenous chemicals and xenobiotics evolved. A wide variety of xenobiotics can be handled by evolved excretory mechanisms and can thus be efficiently eliminated from the body.

Xenobiotics may be excreted as the parent compound, as metabolites, and/or as conjugates formed in phase-II biotransformation reactions. A major route of excretion of xenobiotics is via the kidney, and in some cases the urinary elimination of parent compound or metabolite can be used to determine absorbed dose. The kidney is the only organ which functions almost exclusively as an organ of elimination. The cells of the liver have more varied functions than those found in the kidney; however, the liver also plays an important role in the excretion of chemicals not effectively eliminated by the kidneys. Chemicals may be eliminated from the liver into bile and thus be finally excreted with feces. Active transport mechanisms, that is, transport against a concentration gradient, play a major role in renal and hepatic excretion of xenobiotics. In contrast, in most other organs which may serve as excretory systems for xenobiotics, passive excretion mechanisms are operative. For example, volatile chemicals and metabolites may be eliminated in expired air; this route is quantitatively significant for some solvents and inhalation anesthetics. Specific xenobiotics may also be excreted in sweat, saliva, and milk.

2.5.8.1. Renal Excretion

The kidneys are the only organs that are primarily designed for excretion. The function of these organs accounts for the elimination of most of the byproducts of normal metabolism and most of the polar xenobiotics and metabolites of lipophilic xenobiotics to which humans and experimental animals are exposed [119]. The kidney is a complex structure which consist of a number of different cell types [119 – 121]. Essentially, the kidney filters the bood and all components present in blood with a molecular mass of less than 50 000 (depending on structure and charge) enter the tubular system; there, important nutrients and most of the filtered water are recovered. Only a small fraction of the primary filtrate is excreted as urine (one to two liters per day). The human kidney consists of approximately two million nephrons, which are the functional units that filter the blood and the recover essential nutrients. The structure and components of a single nephron are shown in Figure 41.

Glomerular Filtration. Renal excretion is the product of three complex and interactive processes: glomerular filtration, tubular reabsorption, and tubular secretion. Glomerular filtration is the passive filtering of the plasma as a result of its passage through glomerular pores (7–10 nm in diameter) under hydrostatic pressure generated by the heart. The average rate of glomerular filtration in adults is 125 ml/min or almost 200 liters/d. Glomerular filtration shows little specificity other than molecular size, and free solutes in the plasma that pass through the glomerular pores will all appear in the filtrate. Only protein-bound low molecular mass xenobiotics will not appear in the filtrate and remain in blood. Glomerular filtration is influenced by factors that affect the hydrostatic pressure or integrity of the glomerulus; thus, these factors may result in elevated plasma concentrations of excretory products formed from endogenous chemicals and from xenobiotics.

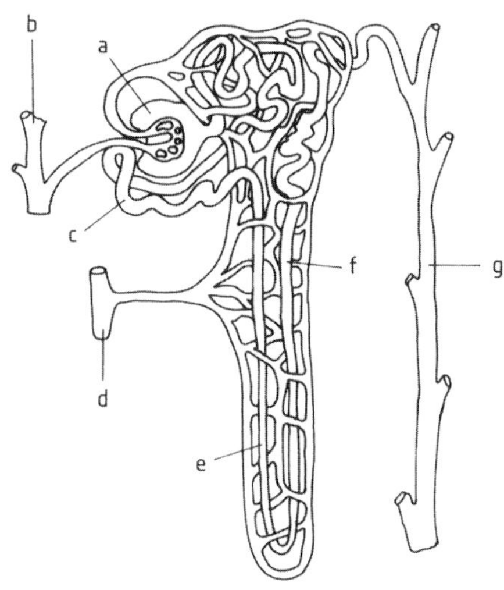

Figure 41. Structure of a nephron
a) Glomerulus; b) Renal artery; c) Proximal tubule; d) Renal vein; e) Loop of Henle; f) Distal tubule; g) Collecting duct

Tubular Reabsorption. The daily volume of glomerular filtrate exceeds that of the total body water by a factor of four and contains many necessary nutrients such as glucose, amino acids, and salt; therefore, most of the glomerular filtrate must be recovered. Thus, the second major process occuring in the kidney is tubular reabsorption. A number of discrete mechanisms, both active and passive and of varying degrees of specificity, are involved in tubular reabsorption. Many of these reabsorptive mechanisms are located in the cells of the proximal segments of the tubules. These cells account for the reabsorption of 65–90% of the glomerular filtrate. Glucose, certain cations, low molecular mass proteins, amino acids, and organic acids are actively reabsorbed. Water and chloride are passively reabsorbed as a result of the osmotic and electrochemical gradients generated by the active transport of sodium and potassium. The osmolarity of the fluid in the collecting duct is regulated in the loops of Henle; most of the remaining water and ions are reabsorbed in the distal tubules and collecting ducts. The rate of reabsorption in these segments of the proximal tubule is regulated to maintain the osmolar concentration of the blood. Most xenobiotics are also reabsorbed after glomerular filtration by passive diffusion during passage through the nephron. Passive tubular reabsorption of lipophilic xenobiotics is therefore greater than the reabsorption of polar xenobiotics or endogenous wastes.

Tubular Secretion. Xenobiotics present in blood may also be excreted by the kidney by tubular secretion. This secretion transports xenobiotics from the peritubular fluid (blood) to the lumen (urine) in the tubule. Tubular secretion is often selective; active transport mechanisms account for the secretion of many organic acids, including glucuronides and sulfates, and strong organic bases.

The secretion of weak bases and some weak acids may also occur by a passive mechanism that utilizes pH differences between peritubular fluid and urine. At the pH of the tubular lumen, these compounds become ionized and do not diffuse back across the cell wall.

Factors Affecting Renal Excretion of Xenobiotics. Xenobiotics are excreted by the same mechanisms which eliminate endogenous wastes; highly polar xenobiotics in plasma water are removed primarily by glomerular filtration and excreted in the urine with minimal involvement of tubular reabsorption or secretion. The rate of renal elimination of most of these xenobiotics is largely dependent on the rate of glomerular filtration. Since lipophilic compounds cross cell membranes more readily, they distribute into a much larger tissue volume than polar compounds, which are more likely to be restricted to the vascular volume. However, lipophilic xenobiotics metabolized to more polar compounds are usually rapidly returned to the circulation and are readily excreted. Therefore, the rate of metabolism of a xenobiotic may also play an important role in its rate of excretion.

2.5.8.2. Hepatic Excretion

Nutrients and xenobiotics are delivered from the gastrointestinal tract to the liver by the portal vein to be biotransformed there. Thus the liver is located between the intestinal tract and the general circulation and ideally located for the biotransformation of nutrients and xenobiotics taken up by this route. Besides participating in the biotransformation of xenobiotics, the liver is a major excretory organ and contributes to the excretion of many xenobiotics by eliminating them with bile into the gut and thus into feces [122–125].

The bulk of the liver consists of cells arranged in plates two cells thick. These plates are arranged radially around the terminal branches of the hepatic veins and are exposed to blood from the portal vein and hepatic artery flowing through interconnecting spaces referred to as hepatic sinusoids. The sinusoidal walls are freely permeable even to relatively large particles; special transport may only play a role in the uptake of certain anions from the blood. The epithelial hepatic cell is the smallest unit of the liver and accounts for most of the varied functions of this organ, including storage, secretion, biotransformation, and excretion (Fig. 42).

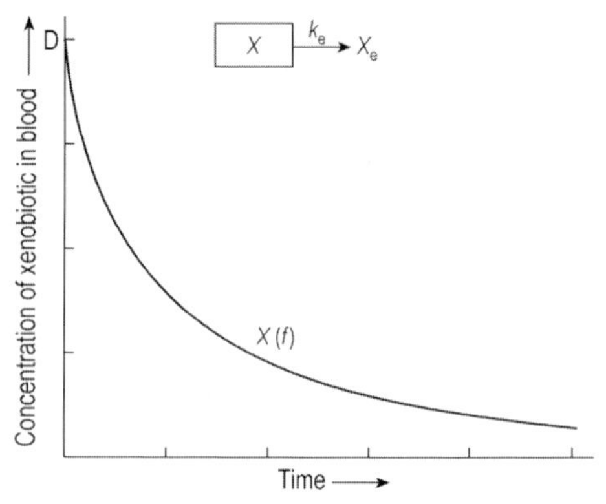

Figure 42. One-compartment model with first-order elimination and instantaneous absorption

Bile formation is thought to be the result of active transport of certain ionized compounds and passive transport of other solutes and water, which follow a concentration or electrochemical gradient. Active secretion of anions and cations appears to be controlled by different mechanisms, but the compounds actively excreted are usually amphipathic molecules and have both polar and nonpolar portions in their structures. Bile acids are the classical example of endogenous amphipathic molecules. Conjugates of lipophilic xenobiotics are also examples of amphipathic molecules, and many of these conjugates are ionized, a fact that facilitates excretion from the hepatocyte by active transport mechanisms.

As a result of both active and passive secretion into the bile, xenobiotics excreted into bile may be classified into different groups. Solutes found in bile may be divided according to their concentration in bile versus blood. For example, the excretion of Na^+, K^+, Cl^-, and glucose, for which the bile/blood ratio is close to unity, is thought to be passive. Conjugates of xenobiotics and bile acids have a bile/blood ratio of greater than 10 and are thought to be actively excreted from liver into bile. Other compounds such as proteins, inulin, sucrose, and phosphates do not cross the canalicular membrane of hepatocytes and have a bile/blood ratio of much less than one. Bile secreted by the liver cells into the bile canaliculi flows into the narrowest branches of the bile duct, the cholangioles, then into the hepatic duct, which carries the bile to the gallbladder and finally into the intestine. Lipophilic xenobiotics may appear in bile at low concentrations prior to metabolism. However, a compound absorbed from the stomach or intestine after ingestion is likely reabsorbed from the intestine if it is secreted in bile prior to structural modifications by biotransformation. Moreover, conjugates excreted into bile, such as glucuronides, may be hydrolyzed by enzymes present in the bacteria of the intestinal microflora, and the aglycone may be reabsorbed. Most xenobiotics reabsorbed from the intestine are returned to the liver. The process of excretion into bile, reabsorption from the intestine, and return to the liver is termed enterohepatic circulation. Enterohepatic circulation serves as an efficient physiological recovery mechanism for bile acids and certain hormones. When xenobiotics are trapped in this cycle, their rate of excretion from the body may be significantly reduced and their toxicity may be significantly increased.

Effect of Molecular Mass on Hepatic Excretion of Xenobiotics. From the liver cell, xenobiotics may be excreted into bile or returned to blood. The molecular mass of nonvolatile organic xenobiotics or their metabolites determines the primary route by which they are excreted from the hepatocyte. In the rat, xenobiotics, their metabolites and conjugates formed in phase-II reactions with molecular masses greater than 500 are most often excreted into bile, while xenobiotics and their metabolites with molecular masses of less than 350 return from the hepatocyte to blood and are thus delivered to the general circulation and to the kidney for excretion. In humans, the critical molecular mass threshold for biliary excretion is approximately 500.

2.5.8.3. Xenobiotic Elimination by the Lungs

Any xenobiotic present in blood with sufficient volatility will pass from the blood across the alveolar membrane into the air space of the lung and may be exhaled. The rate of elimination from blood of a volatile xenobiotic is dependent on the solubility of the xenobiotic in blood, the rate of respiration, and the blood flow to the lung. Xenobiotics like diethyl ether, which are highly soluble in blood, are rapidly eliminated by exhalation; their elimination may also be efficiently increased by a forced increase in the rate of respiration (hyperventilation). In contrast, volatile xenobiotics with low solubility in blood are only slowly cleared from the lung and thus from the body by exhalation, and their rate of exhalation may not be markedly influenced by hyperventilation. The proportionality among xenobiotic volatility, blood solubility, and the concentration of a volatile xenobiotic in the blood is utilized to quantitate blood alcohol content and thus estimate sobriety by breath analysis.

2.6. Toxicokinetics

The toxic response to chemical exposure depends particularly on the magnitude, duration, frequency, and route of exposure. These determine the amount of material to which an organism is exposed (the exposure dose) and hence the amount of material which can be absorbed (the absorbed dose). The latter determines the amount of material available for distribution and toxic metabolite formation, and hence the likelihood of inducing a toxic effect. Absorption and metabolite accumulation are opposed by elimination. All these factors define the disposition of the xenobiotic. The modeling and mathematical description of the course of disposition of a potentially toxic xenobiotic in the organism with time is termed toxicokinetics. Most of the methodologies and principles applied in toxicokinetics were first used to model the kinetics of chemicals applied as drugs (pharmacokinetics).

The goal of both toxicokinetics and pharmacokinetics is to quantitate the dynamic course of xenobiotic absorption, distribution, biotransformation, and elimination processes in living organisms with time. Both the whole process of disposition or individual steps such as elimination may be characterized.

2.6.1. Pharmacokinetic Models

A pharmacokinetic or toxicokinetic model is a functional representation that has the ability to describe the movement of a xenobiotic over time in a real biological system [126–128]. A common way to describe the kinetics of drugs is to represent the body as a number of interconnected compartments which may or may not have an anatomical or physiological reality. These compartments represent all tissues, organs, and fluids in the body that are kinetically indistinguishable from each other. A compartment might be represented by a cluster of cells within an organ, an organ, the blood, or the whole body taken together. More recently, so called physiologically based pharmacokinetic and toxicokinetic models have been developed. These models permit the modeling of the time course of the concentration of a chemical in a tissue on the basis of physiological considerations and may be particulary useful in interspecies extrapolations such as are necessary in risk assessment.

2.6.1.1. One-Compartment Model

This simplest toxicokinetic model depicts the body as a single homogeneous unit within which a xenobiotic is uniformly distributed at all times. The toxicokinetics of a xenobiotic may be analyzed by a one-compartment model if the determined plasma concentration after a single dose decreases exponentially (Fig. 43; the plot of the logarithm of the plasma concentration versus time yields a straight line). In this model, elimination of a chemical from the body occurs by first-order processes. A mathematical description of the first-order process is

$$\frac{\mathrm{d}X}{\mathrm{d}t} = -k_{\mathrm{e}}X$$

where X is the amount in the body at time t and k_{e} is the rate constant for first-order elimination. In compartmental analysis, k_{e} is often referred to as an apparent first-order rate constant, to emphasize that the underlying processes may in reality only approximate first-order kinetics. For example, the xenobiotic might actually be eliminated by active biliary secretion, for which zero-order saturation kinetics would be observed under suitable conditions.

If the xenobiotic is not present in the organism and a known amount X_0 is rapidly administered, the total amount of the xenobiotic initially in the body will be approximately X_0 = total dose. The amount of the xenobiotic present in the organism at a certain time after administration is then

$$X(t) = X_0 \times \mathrm{e}^{-kt}$$

The first-order rate constant of elimination can be determined from the slope of the plot of the logarithmic plasma concentration versus time and can be used to estimate the half-life of elimination for the xenobiotic:

$$t_{1/2} = \frac{0.693}{k_{\mathrm{e}}}$$

Within seven half-lives, a xenobiotic is almost completely eliminated (99.2%), although theoretically complete elimination will never be achieved. Important characteristics of the first-order elimination of a xenobiotic according to the one-compartment model are:

1) The half-life of the xenobiotic is independent of dose; the semilogarithmic plot of the plasma concentration of the xenobiotic versus time yields a straight line.
2) The concentration of the xenobiotic in plasma decreases by a constant fraction per time unit.

Xenobiotics responsible for toxic effects are usually not injected into blood and intake is thus not instantaneous compared to distribution and elimination and a lag period before maximal concentration in plasma is observed. If intake also approximates a first-order absorption process, then the rates of absorption and elimination

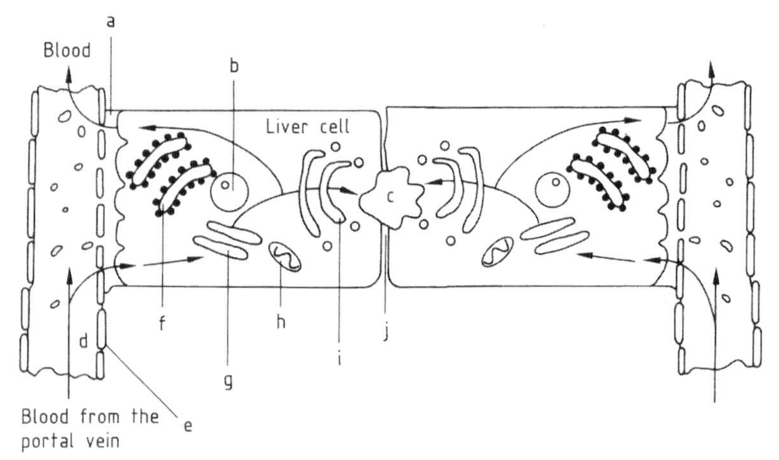

Figure 43. Structure of a liver cell (hepatocyte) and possible pathways for the uptake and elimination of xenobiotics and their metabolites
a) Space of Disse; b) Nucleus; c) Bile; d) Liver capillary; e) Sinusoidal wall; f) Rough endoplasmic reticulum; g) Smooth endoplasmic reticulum; h) Mitochondria; i) Golgy complex; j) Bile duct

will determine the time course of the plasma concentration (Fig. 44).

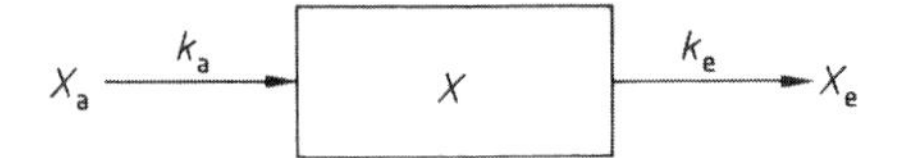

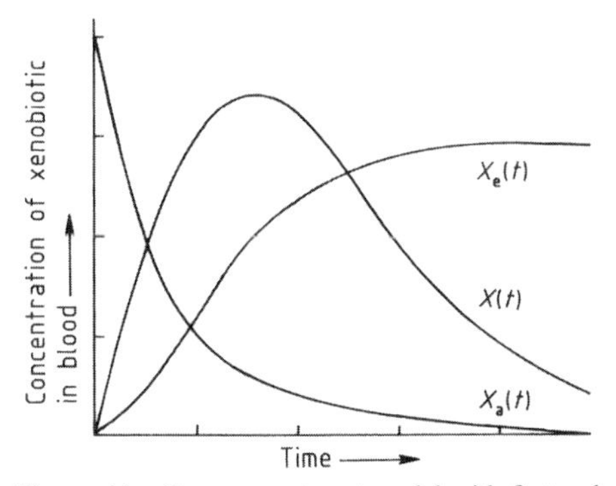

Figure 44. One-compartment model with first-order absorption and first-order elimination

The rate of change in the concentration of a xenobiotic in a one-compartment model with first-order absorption and elimination may be described by:

$$\frac{\mathrm{d}X}{\mathrm{d}t} = -k_{\mathrm{e}}X_{\mathrm{e}}+k_{\mathrm{a}}X_{\mathrm{a}}$$

where k_a is the rate constant of absorption, X_a is the amount of the xenobiotic present at the site of absorption, and k_e is the rate constant of elimination.

The rate constant of elimination can be experimentally determined by treating the elimination phase as if it occurred after the intravenous injection described above by extrapolation of the terminal straight line of the semilogarithmic plot to the ordinate at a putative X_0. By determing the plasma concentration of the xenobiotic at time points shortly after administration, the rate constant for absorption can be determined by plotting the difference between the early plasma concentrations and the extrapolated portion of the elimination curve.

2.6.1.2. Two-Compartment Model

If a xenobiotic does not distribute and equilibrate throughout the body rapidly, a two-compartment model provides a better description of the kinetics of disposition. In this case, the semilogarithmic plot of plasma concentration versus time does not yield a straight line. In two-compartment models, the main or central compartment is assumed to represent the blood and highly perfused organs and tissues such as the liver, the heart, and the kidneys, which are in rapid distribution equilibrium with the blood, while the second or peripheral compartment corresponds to poorly perfused tissues such as muscles and fat. This model is depicted in Figure 45 and is called a two-compartment open pharmacokinetic model.

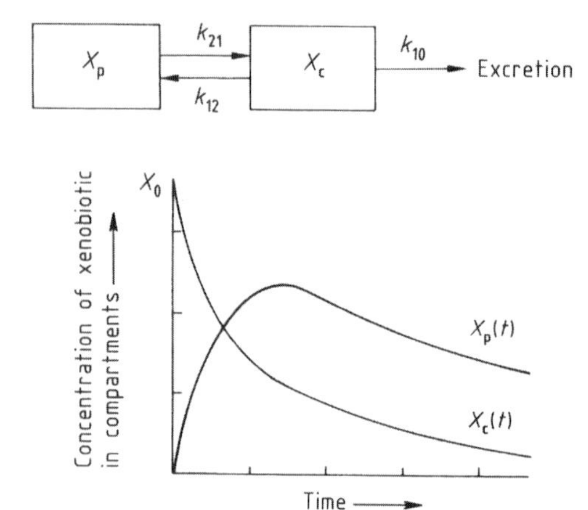

Figure 45. Two-compartment model for the distribution of a xenobiotic
X_c = concentration in the central compartment; X_p = concentration in the peripheral compartment; X_0 = initial dose; k_{21}, k_{12}, k_{10} = rate constants

The time course of the plasma concentration of the xenobiotic can be decribed by two overlying monoexponential terms of the type:

$$X = Ae^{-kt}+Be^{-k't}$$

where A and B are proportionality constants and k and k' are rate constants. Thus, the pharmacokinetics of a two-compartment model describe a biexponential decay in the amount of xenobiotic in the body. Analogous to a one-compartment model, a distribution equilibrium is assumed to exist within each compartment that can be expressed in terms of blood concentration. According to this equation, a plot of the logarithm of blood concentration versus time will yield a biphasic decay from which the constants A, B, k, and k' can be estimated either graphically or by nonlinear regression analysis. Once these experimental constants have been determined, the pharmacokinetic parameters can be calculated. The numerical values of these parameters can aid in assessing the relative importance of tissue distribution and elimination to the disposition of the xenobiotic.

2.6.2. Physiologically Based Pharmacokinetic Models

Although compartmental analysis models are convenient to use and provide useful descriptions of the overall time course of xenobiotic disposition, the practical limitations of curve fitting to experimental data generally restrict such models to one or two compartments. Alternatively, it is often possible to develop a compartmental model from physiological principles and thereby circumvent the inherent limitations of curve-fitting analysis. A physiological pharmacokinetic or toxicokinetic model is a mathematical description of the disposition of a xenobiotic in the organisms or in a part thereof (e.g., a specific organ) Such a model is constructed by using physiological and biochemical parameters such as blood flow rates, tissue and organ sizes, binding, and biotransformation rates [129–131]. These are generally more complex and require the specification of many parameters; thus, physiological compartmental models are still simplified representations of biological systems. In addition, the accurate determination of physical and biochemical parameters is often both difficult and inaccurate. However, the physiological framework provides several advantages: the physical definition of compartments and transfer rates facilitates the incorporation of existing knowledge about the quantitative behavior of biological systems into the model; physiological changes with time during chronic exposure to a xenobiotic, such as those due to physical growth, or induction of biotransformation and changes in excretion rates, can be introduced (Fig. 46).

Most important, however, a reasonable basis exists for extrapolating the kinetics of a xenobiotic to predict the disposition of a xenobiotic following various types and patterns of exposure, and to extrapolate from experimental data obtained in one species of experimental animals to other animal species and humans.

Physiology-based models use the concepts of mass balance and flow-limited transport under the following definitions: Blood serves to distribute a xenobiotic from the site of absorption to the other parts of the body. In the normal sequence of events, a chemical entering the bloodstream is often rapidly distributed within the blood, and its blood concentration can be considered to be essentially uniform.

The chemical enters and leaves the compartment with the blood flow and diffuses or is transported from blood to tissue and back. The chemical may also undergo a variety of physical interactions, such as binding to macromolecules, and the result is a partitioning between tissue and blood that depends on the affinity of the chemical for each medium. Diffusion or transport directly between adjacent compartments, enzymatic biotransformations, and excretion may also occur. The net result of all of these changes is expressed by a mass-balance differential equation, which is simply a mathematical statement of the conservation of mass. Frequently, these mass balance equations can be greatly simplified, since many of the terms required may not apply to a particular compartment.

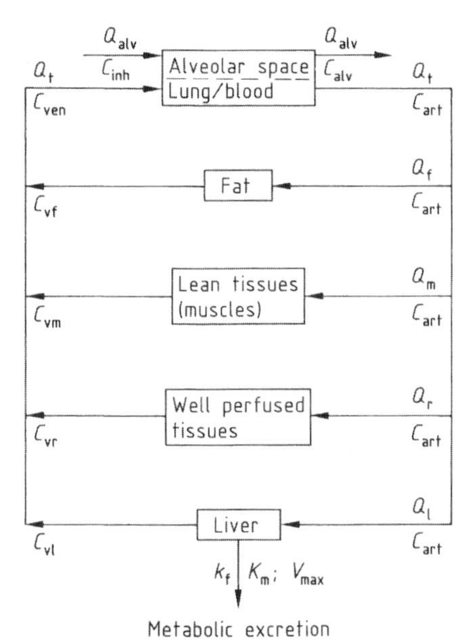

Figure 46. Schematic representation of a physiologically based toxicokinetic model

3. Mechanisms of Acute and Chronic Toxicity and Mechanisms of Chemical Carcinogenesis

3.1. Biochemical Basis of Toxicology

Since the 1960s, toxicology has moved from observing and classifying the harmful effects of chemicals in animals with the tools of pathology (descriptive toxicology), to a discipline able to explain the mechanisms of the basic changes in cell function responsible for toxic effects (mechanistic toxicology). This progress resulted from the widespread application of techniques and concepts from a range of basic sciences, most notably biochemistry and cell biology. Mechanistic explanations of toxic phenomena aid in the prevention of chemical or biological toxicity and provide a rational basis for the use of animal data to assess the anticipated consequences of human exposure to a particular chemical.

Many toxic compounds are chemically stable and produce their characteristic effects by interference with biochemical or physiological homeostatic mechanisms. Many adverse events are the consequence of disturbance of normal physiology and do not result in cell death. A xenobiotic may, for example, activate plasma membrane receptors and induce physiological signal transduction pathways. Induction of a normal cascade at the wrong time or to the wrong extent (too much) may cause undesirable or harmful effects. Toxic compounds interfering with homeostatic mechanisms do not necessarily cause cell death; however, the induced changes may have a harmful impact both on the altered cells and on the involved tissue or on the entire organ. Therefore, it is critical to have a proper knowledge on biochemical and molecular sites of action of the xenobiotics.

Cytotoxicity resulting in cell death is often the consequence of exposure to a harmful xenobiotic, but the number of cells which must be killed before the function of a tissue or organism is noticeably impaired is highly variable. Some cell types like the epithelia of the kidney and the liver have the ability to regenerate in response to damage, while others like neurons can not. Furthermore, some organs, such as the liver, lung, and kidney, have a substantial functional reserve capacity in excess of normal requirements, and normal function can be maintained even in the presence of extended necrosis.

In addition to cell death, disturbances in the regulation of cell division induced by toxic xenobiotics may have harmful long-term consequences for the organism affected. Nonlethal alterations in the genome of somatic cells can result in mutations and can lead to malignant transformation and tumor formation [132]. More recently, it has become clear that compounds not directly interacting with the genomic DNA can also produce cancer by so-called epigenetic mechanisms [133]. These may involve a proliferative response of epithelial cells to cytotoxicity, as is suggested to occur with high-dose carcinogens such as allyl isothiocyanate and chloroform, or a more direct action that enhances the rate of cell division in the absence of cytotoxicity, as is seen with 2,3,7,8-tetrachlorodibenzodioxin. Increased cell replication, whatever the cause, is accompanied by increased chance of unrepaired DNA lesions that may be fixed as mutations. Hyperplasia has long been suspected of preceding neoplasia, but the inevitability of such a progression has never been established on a pathological basis alone. Even in absence of foreign compounds, DNA is damaged to a considerable extent by reactive oxygen species formed during different biochemical processes in the cell, it may thus be speculated that nongenotoxic carcinogens act by enhancing the likelihood of this normal DNA damage being fixed as a mutation and leading to cancer [134].

Having established the fundamental differences between toxic chemicals which act through physiological inbalance, through cytotoxicity, or by causing alterations in cell proliferation, we now consider some sites or chemical reaction mechanisms of toxic action. The following major mechanisms are neither comprehensive nor mutually exclusive.

3.2. Receptor-Ligand Interactions

3.2.1. Basic Interactions

Some xenobiotics can interact with physiological receptors due to structural similarities to endogenous compounds. The toxic effects of these chemicals are related to their ability to interfere with normal receptor–ligand interactions either as agonists or as antagonists of the physiological ligand. Receptors are macromolecular components, most often proteins, of tissues which interact with specific endogenous ligands or structurally related xenobiotics to induce a cascade of biochemical events and produce characteristic biological effects. The binding between a receptor (R) and a ligand (L) is usually reversible and can be described by the equilibrium reaction:

$$\mathrm{R} + \mathrm{L} \underset{k_2}{\overset{k_1}{\rightleftharpoons}} \mathrm{RL}$$

The dissociation constant K_d that describes this relationship is thus given by:

$$K_\mathrm{d} = \frac{k_1}{k_2} = \frac{[\mathrm{L}]\,[\mathrm{R}]}{[\mathrm{LR}]}$$

where [L], [R], and [LR] are the concentrations of ligand, unbound receptor, and ligand-bound receptor, respectively. The affinity of the ligand to a certain receptor is proportional to $1/K_\mathrm{d}$. The ligand may be an endogenous substance that interacts with the receptor to produce a normal physiological response, or it may be a xenobiotic that may either elicit (agonist) or block (antagonist) the response.

Receptor–ligand interactions are generally highly stereospecific. Usually, small changes in the chemical structure of the ligand can Drastically influence its capability to bind to the receptor and thus reduce or completely abolish the effect elicited by the ligand–receptor interaction. In other cases, changes in the chemical structure may have an important impact on the response without altering the binding of the xenobiotic to the relevant receptor. Differences in the extent of activity are not only observed with structurally distinct chemicals but also with chemicals that are chiral and thus may be present as racemic mixtures of stereoisomers. Synthetic chemicals with chiral centers generally contain both enantiomers, often in a 1:1 ratio, yet in many instances only one of the enantiomers is biologically active. In pharmacological or toxicological studies, the inactive or weakly active enantiomer should be viewed as an "impurity" which may confound the interpretation of results obtained with the racemic mixture. Because the enzymes responsible for biotransformation of xenobiotics also contain active sites with specific steric requirements, stereoisomers may be differentially biotransformed. The selective biotransformation of only one enantiomer may markedly alter the potency and efficacy of one enantiomer compared with another. Stereoselective differences in action among enantiomers should not be surprising, if one considers that these chemicals are mirror images, much as the right hand is of the left, and that receptors have a physical orientation that can be compared with a glove. Although the left-hand and right-hand gloves look remarkably similar, they do not fit both hands equally well. However, these theoretical considerations do not imply that stereoisomers always elicit different responses. Since the chiral center of a stereoisomer is not necessarily involved with the active site of the receptor, chirality does not always results in differences in toxic response among enantiomers.

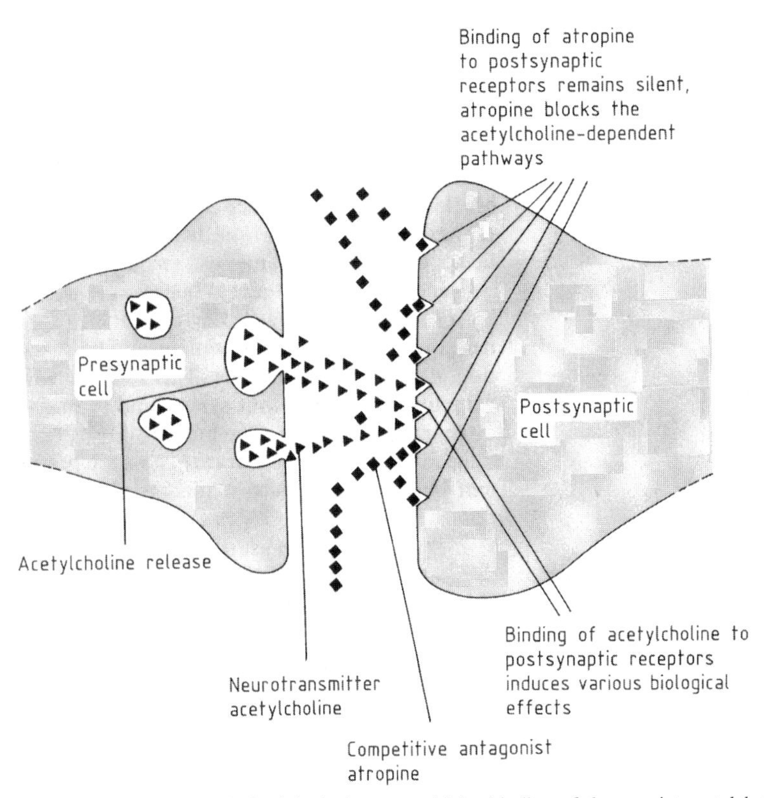

Figure 47. Induction of physiological response(s) by binding of the agonist acetylcholine to postsynaptic receptors; in contrast, binding of the antagonist atropine to the same receptors does not induce biological effects

The acute toxic effects of many xenobiotics are related directly to their ability to interfere with normal receptor–ligand interactions. This is most clearly the case with neurotoxins, acting within and outside the central nervous system (CNS). The proper function of the nervous system is highly dependent on a diverse array of receptor–ligand interactions. For example, the belladonna alkaloids atropine and scopolamine bind to and block the cholinergic receptors, the binding site of the physiological agonist acetylcholine (Fig. 47). The interactions of acetylcholine and atropine are an excellent example demonstrating that interference of a xenobiotic with physiological regulatory mechanisms may result both in desired (pharmacological) and an undesired (toxic) effects.

Since atropine itself can not elicit the physiological responses mediated by acetylcholine binding to this receptor, toxic effects may result. In clinical therapy, atropine is often used as an antispasmodic to reduce the hypermotile state of the gastrointestinal tract or the urinary bladder. Its anticholinergic effects are also utilized in cardiovascular pharmacology to decrease pathologically elevated heart rates. At the same time, however, blockage of acetylcholine-mediated physiological cascades may cause dry mouth, blurred vision, and constipation. In addition, in the CNS toxic effects of atropine include restlessness, confusion, hallucinations, and delirium. Blockage of the receptors for the neutral amino acid glycine by strychnine represents a further example of the important and, in this particular case, life-threatening agonist–antagonist interactions. Binding of glycine to its receptor induces an increased permeability of the plasma membrane to chloride ions resulting in hyperpolarization and reduced activity of nerve cells, that is, glycine exerts an inhibitory effect on the nervous system. Blockage of the glycine receptors by the antagonist strychnine, that does not elicit a response, results in hyperactivity and severe convulsions.

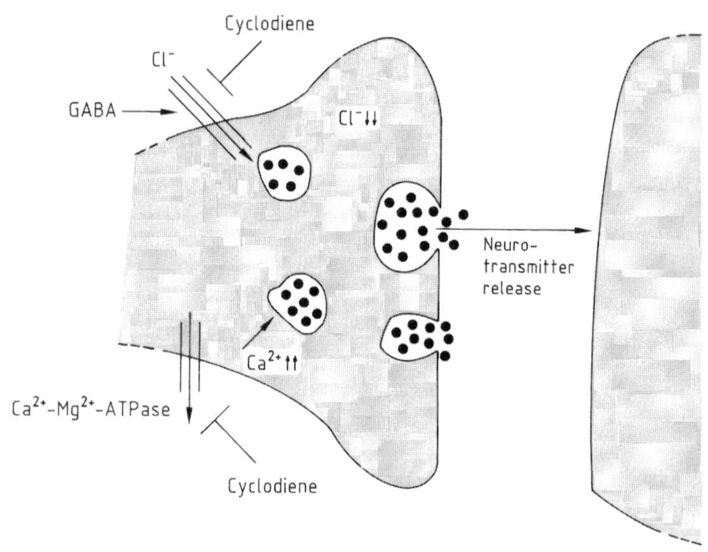

Figure 48. Sites of action of chlorinated cyclodiene insecticides on the plasma membrane of nerve cells

Table 20. Chemicals causing toxic response by interaction with receptors (Ah = aryl hydrocarbon)

Type of receptor	Chemical
Muscarinic receptor	atropine
Glycine receptor	strychnine
Ah receptor	TCDD
Peroxisome proliferator receptor	peroxisome proliferators (clofibrate, diethylhexylphthalate)

In addition to acute toxicity, chronic influences exerted by xenobiotics on steroid hormone homeostasis may result in adaptive and proliferative responses in certain tissues. The long-term consequences may involve impairment or loss of the physiological function. Moreover, binding to a specific receptor may cause changes in gene expression and contribute to tumor formation. For example, binding to the Ah receptor (the aryl-hydrocarbon receptor, named on the basis of its interaction with planar, polycyclic aromatic compounds) mediates many of the effects observed with the highly toxic and carcinogenic 2,3,7,8-tetrachlorodibenzodioxin (TCDD). A cytosolic binding protein (cytosolic receptor) has also been identified for peroxisome proliferators, such as clofibrate and diethylhexyl phthalate [135, 136]. This protein has DNA-binding domains and appears to activate gene transcription in the nucleus in a manner analogous to the steroid hormones. The increased cell proliferation caused in target organs such as the liver may be responsible for the carcinogenicity of many peroxisome proliferators. Characteristic examples of receptor–xenobiotic interactions with harmful acute or chronic consequences are summarized in Table 20.

3.2.2. Interference with Excitable Membrane Functions

The maintenance and stability of excitable membranes is essential to normal physiology. Excitable membranes are critical to the function of nerves and muscles due to their ability to generate and propagate action potentials. Action potentials are elicited by the exchange of ions between the intra- and extracellular compartments, and hence they depend on the normal activity of ion channels and membrane ion pumps.

Xenobiotics may perturb excitable membrane functions in many different ways. Xenobiotics can block sodium channels resulting in toxic effects such as paralysis. For example, tetrodotoxin from the the puffer fish irreversibly blocks the sodium channel along the nerve axon and thus prevents the inward sodium current of the action potential while leaving the outward potassium current unaffected. The marine toxin saxitoxin, which is structurally quite different from tetrodotoxin, also produces its paralyzing effects by blocking sodium channels in excitable membranes in essentially the same manner. The insecticide dichlorodiphenyltrichloroethane (DDT) produces its neurotoxic action by interfering with the closing of sodium channels, which impairs the physiological repolarization of excitable membranes. This results in hyperactivity of the nervous system and repetitive discharges of neurons. The most striking symptoms in poisoned insects or mammals are persistent tremor and convulsive seizures. Other neurotoxins such as cyclodiene insecticides and picrotoxin interact with the γ-aminobutyrate (GABA) receptor in neurons. Blockage of the GABA receptor by these antagonists impairs the transport of chloride ions into the cell, which results in partial depolarisation of the membranes in the absence of adequate signals. The clinical symptoms of the pathophysiological changes can be summarized as a state of uncontrolled excitation. In addition, with cyclodiene insecticides this hyperexcitability is augmented by the inhibition of Ca^{2+}–Mg^{2+}–ATPase, which results in accumulation of intracellular free calcium and increased release of neurotransmitters from storage vesicles (Fig. 48).

Organic solvents have a depressant effect on the CNS that results in narcosis by nonspecific alterations in membrane fluidity due to their lipid solubility rather than by interference with specific ion channels. This depressant effect is not based on certain structural features or specific interactions, that is, it is also observed after inhalation of the inert, but highly lipophilic gas xenon.

3.2.3. Interference of Xenobiotics with Oxygen Transport, Cellular Oxygen Utilization, and Energy Production

Continuous, sufficient production of energy in form of adenosine triphosphate (ATP) in specialized cellular organelles (mitochondria) is essential for cell survival and function. The process of cellular energy production requires oxygen, cofactors, and a wide array of specialized enzymes acting in concert. Many chemicals produce their toxic effects by interfering with the transport and cellular utilization of oxygen or with the oxidation of carbohydrates, which is coupled to the synthesis of ATP by oxidative phosphorylation.

The interference with cellular energy production may occur at different sites in the uptake, transport, and cellular utilization of oxygen. Oxygen transport from the lung to tissues may be reduced or blocked by xenobiotics which compete with oxygen for the binding site in the transport protein hemoglobin or which chemically modify this binding site. Carbon monoxide competitively blocks the binding of oxygen to the transport protein hemoglobin. Chemical oxidation of the iron in hemoglobin also impairs oxygen transport. Some xenobiotics oxidize the Fe^{2+} at the oxygen binding site in hemoglobin to Fe^{3+}. Hemoglobin with Fe^{3+}, known as, methemoglobin, can not reversibly bind oxygen. Thus, xenobiotics causing methemoglobinemia (e.g., nitrites, aromatic amines) effectively block oxygen transport.

The cellular utilization of oxygen in the tissues is blocked by cyanide, hydrogen sulfide, and azide because of their affinity for cytochrome oxidase. Cyanide exerts its toxic effects by interrupting electron transport in the mitochondrial cytochrome electron-transport chain. In addition, interferences with the enzyme systems producing ATP is a well-defined mechanism of toxicity. The ultimate formation of ATP in the cell by the oxidation of carbohydrates may also be blocked at other sites. For example, rotenone and antimycin A interfere with specific enzymes in the electron-transport chain, nitrophenols uncouple oxidative phosphorylation, and sodium fluoroacetate inhibits the citric acid (Krebs) cycle. Nitrophenols interfere with the production of high-energy phosphates in mitochondria. They prevent the phosphorylation of adenosine diphosphate (ADP) to ATP, but electron flow and oxygen consumption continue. In addition to inducing loss of cell function due to depletion of ATP, this type of uncoupler also causes a marked elevation of body temperature due to excess heat production. Toxicity resulting from blockage of the tricarboxylic acid cycle occurs in organs heavily relying on the availability of a continuous energy supply and results in cardiac and nervous system toxicity. A classic example for a mechanism inhibiting the Krebs cycle is termed "lethal synthesis" and is exemplified by fluoroacetate, which is incorporated into the Krebs cycle as fluoroacetyl coen-

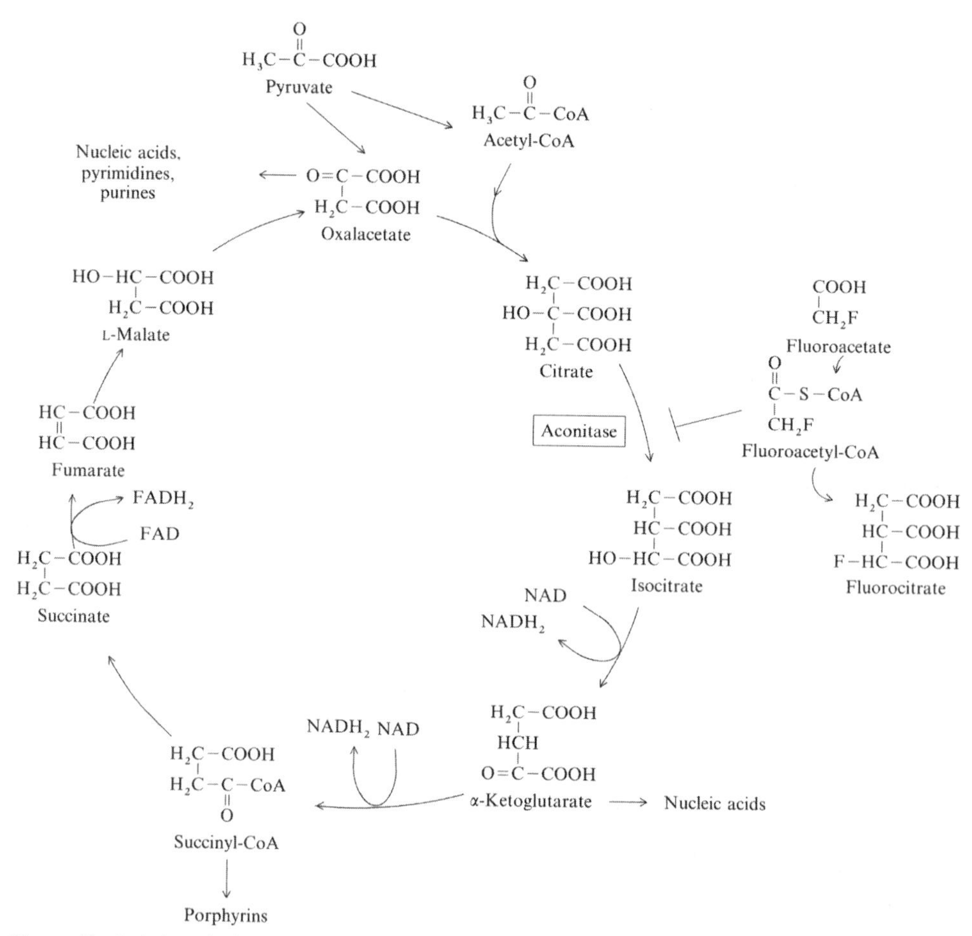

Figure 49. Lethal synthesis
Incorporation of fluoroacetate in the Krebs cycle; formation of fluorocitrate and inhibition of aconitase results in blocking of the tricarboxylic acid cycle and interruption of cellular energy production

zyme A, inhibits the aconitase-catalyzed conversion of citrate to isocitrate, and interrupts the Krebs-cycle (Fig. 49). The Krebs cycle is the major degradative pathway for the generation of ATP and also provides intermediates for biosynthesis. For example, the majority of the carbon atoms in porphyrins comes from succinyl CoA and many of the amino acids are derived from α-ketoglutarate and oxalacetate. Hence, interruption of the Krebs cycle by the incorporation of fluoroacetate may be lethal for the cell.

The consequences of ATP depletion include impairment of membrane integrity, ion pumps, and protein synthesis. Depending on its extent, energy depletion will inevitably lead to loss of cell function and cell death.

3.3. Binding of Xenobiotics to Biomolecules

Many toxic chemicals exert their effects by covalent linkage of reactive metabolites to essential macromolecules of the cell. Xenobiotics may become covalently bound to the active site of enzymes or to other macromolecules whose function is critical to the cell. These include proteins (Section 3.3.1) and lipids (e.g., as membrane structural elements; see Section 3.3.1 and 3.3.2) and nucleic acids (see Section 3.3.1 and 3.3.2 and 3.3.3). The linkage generally involves binding of electrophilic metabolites to nucleophilic sites such as thiol, amino, and hydroxyl groups in the side chains of proteins and is essentially irreversible; it depends only on the turnover of the

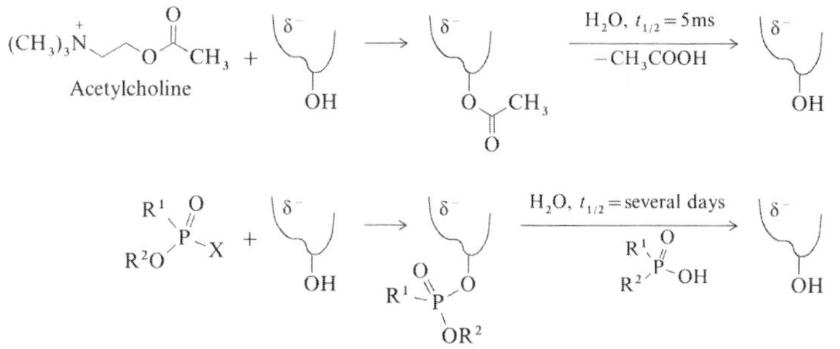

Figure 50. Mechanisms of cholinesterase inhibition by organophosphate pesticides
Compare the $t_{1/2}$ of the enzyme complex with the physiological substrate acetylcholine (a few milliseconds) to the $t_{1/2}$ of the enzyme complex with organophosphate pesticides (several days).

macromolecule in question for the repair of the lesion. In addition, other toxins may bind to proteins and impair their normal function without first being converted to reactive intermediates.

3.3.1. Binding of Xenobiotics or their Metabolites to Cellular Proteins

Many toxic substances exert their effects via binding to the active sites of enzymes or other proteins that are critical to cellular function. For example, hydrogen cyanide binds with high affinity to the Fe^{3+} ion in cytochrome oxidase and thus blocks the last step in the mitochondrial electron-transport chain, which is important for cellular energy production. This single and highly specific site of action is responsible for the rapid induction of the often fatal toxic effects of cyanide. Carbon monoxide binds to the reduced form of iron in hemoglobin, whose physiological function is to bind, transfer, and deliver oxygen to tissues. Since carbon monoxide has a 210-fold higher affinity to this binding site than oxygen, very low concentrations of carbon monoxide in the atmosphere are sufficient to displace the physiological ligand and produce severe toxic effects. Thus, 0.1% carbon monoxide in the atmosphere can occupy roughly 50% of the available hemoglobin binding sites, because on the basis of the 210-fold higher affinity to hemoglobin compared with oxygen, 0.1% carbon monoxide is equivalent to 20% oxygen in the atmosphere.

Another example for the importance of protein binding in toxicity is the binding of metal ions to protein thiol groups. Many toxic metals such as lead, mercury, cadmium, and arsenic bind to proteins with free sulfhydryl groups, which contributes by as-yet largely unknown mechanisms to the toxicity of these metals. Induction of porphyria by lead, mercury, and other metals is based in part on the inhibition of specific enzymes of heme biosynthesis and results in the accumulation of specific intermediates in heme synthesis. The binding of cadmium to the sulfhydryl-rich protein metalothionein results in active concentration of the metal-metalothionein complex in the proximal tubules of the kidney. Catabolism of the metal complex in the lysosomes of proximal tubule cells with concomitant liberation of the toxic metal results in nephrotoxicity, a common effect of cadmium exposure in humans and experimental animals [137].

Binding to active sites in enzymes may result in the inhibition of biochemical pathways vital to the cell. The induction of toxic effects may be due to accumulation of a specific enzyme substrate or insufficient amount of substrate available for normal physiological function. For example, many organophosphate pesticides inhibit cholinesterases by covalently binding to the active site, the amino acid serine of cholinesterase. Since the phosphorylated cholinesterase is stable, the covalent binding results in inhibition of enzymatic activity (Fig. 50). Due to the impaired cleavage, acetylcholine accumulates at cholinergic synapses and neuromuscular junctions. Organophosphates thus produce the typical signs of acetylcholine poisoning, such as increased salivation and lacrimation, abdominal cramps and diarrhea, cough, bronchoconstriction and breathing impairment, mental confusion, headaches, tremor, and coma.

Protein modifications may also be the basis for immunosuppressive effects and chemical-induced allergy. For many chemicals causing necrosis, binding to cellular macromolecules is essential for the expression of toxic effects. However, in most cases the temporal sequence of events that is triggered by covalent binding and the cause–effect relationships between these events are not fully understood.

3.3.2. Interaction of Xenobiotics or their Metabolites with Lipid Constituents

Lipid peroxidation in biological membranes by free radicals initiates a series of events finally causing cellular dysfunction and cell death. The formation of radicals during peroxidation is a self-propagating process; the reaction may be started by organic radicals formed during biotransformation or by oxygen radicals formed by disruption of cellular energy metabolism. The initiation of lipid peroxidation by interaction of free radicals with polyunsaturated fatty acids to form lipid peroxy radicals, which then produce lipid hydroperoxides and other lipid peroxy radicals, has been proposed as a critical step leading to cell injury and death. Peroxidative damage to plasma membrane lipids may cause impairment of membrane integrity and, finally, rupture of the plasma membrane. In addition, breakdown of the membranes of subcellular organelles such as those of the mitochondria, the endoplasmic reticulum, and lysosomes may also contribute to induction of cell death. The end products of the breakdown of membranes, mainly unsaturated aldehydes, may also produce toxicity in distal tissues. Induction of lipid peroxidation is involved in the toxicity of many chemicals which are converted to free radicals. A well-studied example is carbon tetrachloride, which is converted by cytochrome P450 to the trichloromethyl radical (CCl_3) and the trichloromethylperoxyradical (CCl_3O_2). Radicals and reactive oxygen species may interact with the major soluble cellular thiols, glutathione (GSH), and thiol-containing proteins. Depletion of cellular glutathione and modification of thiol-containing proteins by oxidation or mixed disulfide formation with glutathione and other low molecular mass thiols may results in oxidative stress in the cell.

An important function of reduced glutathione is to protect the sulfhydryl groups of proteins by keeping them in the reduced state. Depletion of intracellular glutathione stores appears to be a prerequisite for the onset of significant oxidative stress. Many critical enzymes in the cell depend on reduced thiol groups (SH groups) to maintain their activity; hence, concentrations of reactive oxygen species in excess of that necessary to deplete intracellular glutathione can induce oxidation of protein thiols to form disulfide linkages, thereby impairing enzymatic activity. Although the direct covalent interaction of electrophilic chemicals with protein thiols may contribute to enzyme inhibition; it appears that reversible oxidation of the thiol group as a result of oxidative stress plays a more important role. One group of thiol-containing enzymes whose impairment as a result of oxidative inactivation may play a critical role in cell injury and death are the Ca^{2+}-transporting membrane systems.

3.3.3. Interactions of Xenobiotics or their Metabolites with nucleic Acids

Electrophilic compounds, usually formed by oxidative biotransformation of xenobiotics, may also interact with various nucleophilic sites in DNA, principally O-6, N-7, N-2, and C-2 of guanine. Moreover, other types of DNA damage may occur as a consequence of covalent binding of electrophiles or by interaction of DNA with reactive oxygen metabolites (Table 21).

These interactions may alter gene expression (see Sections 3.5 and 3.7). The changes in gene expression may be quantitative, that is, formation of the wrong amount of a protein in an inappropriate time period of cellular life, or qualitative, that is, formation of a protein with altered or impaired properties. In both cases, the changes may cause the death of the cell. However, interaction of xenobiotics with nucleic acids is more important in generating somatic mutations which can be the initiating event for a process ultimately leading to malignant transformation of cells and tumor growth. Alkylation of the O-6 position of guanine appears important in the mutagenicity and carcinogenicity of nitrosamines and other chemicals that readily form methyl carbonium ions. In addition, other sites such as the N-7, N-2, and C-2 positions of guanine may

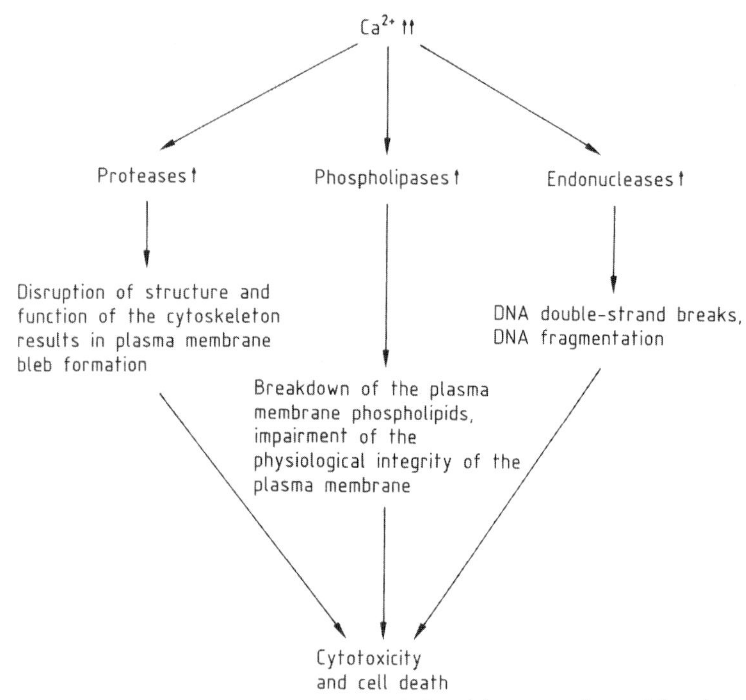

Figure 51. Increased concentrations of intracellular calcium may activate calcium-dependent degradative enzymes, resulting in the destruction of proteins and lipids and in DNA fragmentation, thus causing cell death

also play an important role in DNA adduct formation with other electrophilic chemicals. Ribonucleic acid (RNA) also contains nucleophilic sites, and thus critical intracellular functions of RNA such as protein synthesis may be perturbed by covalent interaction of electrophilic chemicals with RNA.

Table 21. Types of DNA damage induced by xenobiotics

Type of DNA-damage	Examples
Modifications of purines and pyrimidines	alkylation by electrophiles, oxidation to 8-hydroxyguanosine and other oxidized purines, UV-induced formation of thymidine dimers
DNA strand breaks	oxygen radicals, activation of calcium-dependent endonucleases
AP lesions (apurinic, apyrimidinic sites)	loss of DNA bases through labilization after covalent binding of xenobiotics to specific sites in purine and pyrimidines
DNA cross-links	bifunctional electrophiles react with nucleophilic sites in both DNA strands (interstrand cross-link) or with two bases on one strand (intrastrand cross-link)

3.4. Perturbation of Calcium Homeostasis by Xenobiotics or their Metabolites

Intracellular calcium concentrations are rigorously maintained at around 10^{-7} M against an extracellular concentration of more than 10^{-3} M. Three principal buffer systems are important in maintaining this steep gradient in virtually all cells: the plasma membrane, the mitochondria, and the endoplasmic reticulum. All systems require ATP to transport calcium, directly or indirectly. Exposure of freshly isolated or cultured cells to numerous toxins such as *tert*-butyl hydroperoxide, quinones, paracetamol, and carbon tetrachloride induces a rapid and sustained rise in cytosolic calcium concentrations, which correlates well with the subsequent loss of cell viability [138, 139]. Prevention of this rise by omitting calcium in the extracellular buffer or by addition of calcium chelators may prevent cell death. A central event in calcium-induced toxicity is the activation of calcium-dependent degradative enzymes such as proteases, phospholipases, and endonucleases (Fig. 51).

Cytoskeletal integrity is important for a number of cellular functions including motility, shape, secretion, and division. Activation of calcium-dependent proteases results in disruption of cytoskeletal components and formation of plasma membrane blebs (protrusions). In addition, activation of calcium-dependent phospholipases may induce membrane phospholipid breakdown, and calcium-dependent endonucleases may lead to DNA fragmentation by the formation of DNA double-strand breaks. These toxic biochemical cascades may act in concert to cause cell death [104].

3.5. Nonlethal Genetic Alterations in Somatic Cells and Carcinogenesis

The induction of cancer by chemicals is a complex multistep process involving interactions between environmental and endogeneous factors. Tumors are formed as a result of aberrant tissue growth due to loss of control mechanisms of cell division. A model traditionally used for an operational description of carcinogenesis is the initiation–promotion model. Initiation, the first stage, requires a genotoxic event such as binding of an electrophilic xenobiotic to DNA causing a premutagenic lesion in a single cell [140, 141]. After DNA replication, the premutagenic lesion may be transformed into a heritable mutation (i.e. by base-pair substitution; see Section 3.7). In the promotion stage, several, primarily nongenotoxic, mechanisms facilitate (promote) the preferential proliferation of the initiated cell and finally resulting in the formation of a tumor. Initiators are usually genotoxic agents, whereas promoters generally act by interfering with extranuclear sites and processes; most promoters increase cell growth and cell proliferation.

In contrast to initiators, which are believed to result in irreversible changes in the cellular genome without thresholds, promoters show some reversibility, and thresholds for promoters can be discussed. This stepwise nature of carcinogenesis was first shown in mouse skin, and this model has now been well characterized (Fig. 52). In a typical experiment, an initiating chemical such as dimethylbenzanthracene is applied to mouse skin at a low dose so that very few tumors, if any, are produced in the animals lifetime. After an interval of one week to one year, the treated (initiated) skin is exposed to multiple applications of a promoter such as the phorbol esters found in croton oil. Tumors begin to appear within 5 to 6 weeks after application of the promotor, and all mice carry tumors by 10 to 12 weeks after the start of application (Fig. 52).

The initiation–promotion experiments on mouse skin and similar experiments in the rodent liver have led to the following general rules of carcinogenesis:

1) The initiator must be given first; no tumors or very few tumors result if the promoter is given first.
2) The initiator, if administered once at a subcarcinogenic dose, does not produce tumors during the time of observation; however, repeated doses of the initiator may cause tumors even in the absence of the promoter (the initiator is a complete carcinogen in this case).
3) The action of the initiator is irreversible; tumors result in nearly the same yield if the interval between initiation and promotion is extended from one week to one year.

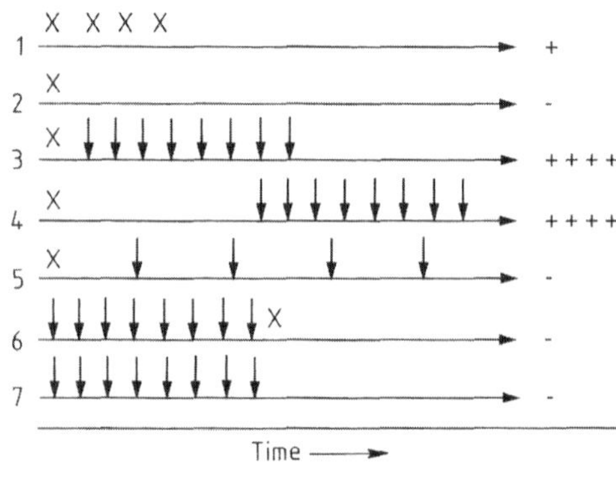

Figure 52. Schematic presentation of the initiation–promotion model of chemical carcinogenesis in the mouse skin

Promoters reduce the latency period and increase the number of tumors only when applied after the initiator (3, 4) The application frequency required for promoting effects to occur depends on the initiating carcinogen and the target tissue. If the application frequency of the promoter is too low the promoting effects are not observed (i.e., no tumors are formed or the tumor yield is not increased, 5). This is also the case, when the promoter is applied prior to the initiator (6) or alone without initiating agent (7). Tumors can also be produced by the initiating agent alone if it is applied at sufficient concentration or above a certain application frequency (1). In this case, the initiator is termed a complete carcinogen.

4) The initiator is an electrophile, or is bioactivated to an electrophile, which binds covalently to DNA causing a mutation after DNA replication.
5) In contrast, there is no evidence of covalent binding of promoters or their metabolites to DNA.
6) The action or the promoter is reversible at an early stage and usually requires repeated exposure; thus, there is probably a threshold level of exposure to promoters. However, threshold levels can not be reliably defined, as long as the biochemical mechanism involved in tumor promotion is not precisely known. In contrast to the established target of the initiation process (DNA), the molecular mechanisms of tumor promotion are still largely unknown.

In addition to promotion of skin carcinogenesis by the phorbol esters, there are known or suspected promoters for tumors in other organs [142, 143]. Bile acids are known to be promoters of colon carcinogenesis in experimental animals. In humans, there is a strong association between high intake of dietary fat and cancer of the colon; since ingestion of fat increases the amount of bile acids in the colon, the increased incidence of colon cancer may be due to the promoting effect of the bile acids on intestinal epithelia. In rat bladder, saccharin and cyclamate are promoters for tumors initiated by a single dose of dimethylnitrosourea; tryptophane is a promoter for urinary bladder tumors in dogs treated with an initiating dose of 4-aminobiphenyl or 2-naphthylamine. Hormones are also known modifiers of chemical carcinogenesis. Oral or intravenous administration of dimethylbenzanthracene produces mammary tumors in susceptible female mice. Prolactin will increase and accelerate tumor development, whereas ovariectomy results in reduced tumor yield.

In addition to these two stages a third stage, termed progression, has since been established as an integral part of the carcinogenic process. In this last stage additional mutational events increase the malignancy of the tumor, that is, the tumor grows in an invasive manner, destroying the surrounding tissues and forming metastases in other organs (Fig. 53).

The understanding of the molecular mechanisms on route from DNA interactions to a clinically observable tumor requires some knowledge on the basic roles of DNA biochemistry and biology in cellular function and replication. Therefore, a short chapter on DNA biochemistry and biology follows. Further and in-depth information is presented in textbooks on biochemistry and molecular biology.

3.6. DNA Structure and Function

3.6.1. DNA Structure (→ Nucleic Acids, Chap. 2.1)

With the exception of certain viruses, the genetic information of all cells is contained in deoxyribonucleic acid (DNA), whose structure and constituents permit the accurate storage of a vast amount of information. In the cell nucleus, the

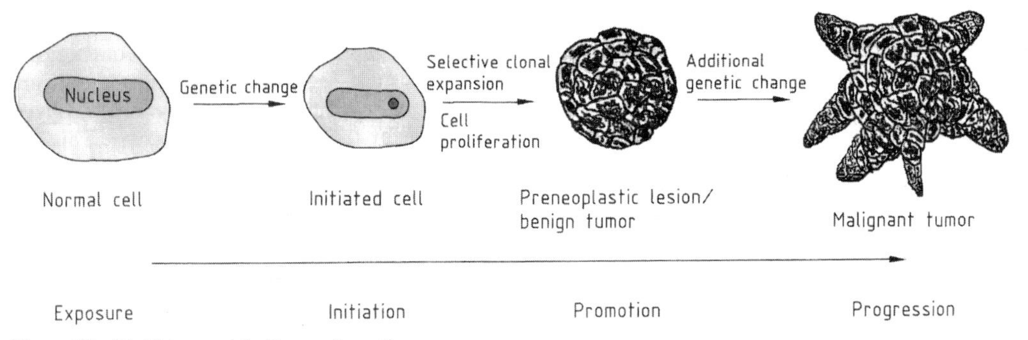

Figure 53. Multistep model of tumor formation
Genotoxic events resulting in heritable mutations cause the formation of an initiated cell (initiation). In the following stage of promotion, nongenetic (epigenetic) events contribute to preferential proliferation of the initiated cell. In the third stage of progression, additional genetic events increase the malignancy of the tumor tissue (i.e., its growth becomes increasingly destructive and metastases are formed in other organs).

DNA is packaged with proteins to form chromatin.

The genetic code of the DNA is denoted by four letters: two pyrimidine nitrogenous bases, thymine (T) and cytosine (C), and two purine bases, guanine (G) and adenine (A). These are are functionally arranged in codons (or triplets). Each codon consists of a combination of three letters and codes for a specific amino acid.

The bases on one strand are connected together by a sugar (deoxyribose) phosphate backbone. DNA can exist in a single-stranded or double-stranded form. In the latter state, the two strands are held together by hydrogen bonds between the bases. The adenine on one strand binds to thymine on the sister strand, and guanine pairs with cytosine. Two hydrogen bonds are involved in the binding between adenine and thymine, while three hydrogen bonds are involved in the binding of guanine to cytosine.

Double-stranded DNA has the unique property that it can make identical copies of itself when supplied with precursors and the required enzymes. In simplified terms, two strands begin to unwind and separate as the hydrogen bonds are broken. This produces single-stranded regions. Complementary deoxyribonucleotide triphosphates then pair with the exposed bases under the control of the enzyme DNA polymerase.

The information in DNA is assembled in structural genes. A structural gene is a linear sequence of codons, which contain the information for a functional protein, consisting of a sequence of amino acids. Individual proteins may function as structural components of the cell, as enzymes, or may regulate important cellular functions. The DNA of eukaryotic cells contains repeated sequences of some genes. Also, eukaryotic genes, unlike prokaryotic (i.e., bacterial) genes, have noncoding DNA regions called introns between coding regions known as exons. This property means that eukaryotic cells have an additional processing mechanism at transcription.

3.6.2. Transcription (→ Nucleic Acids, Chap. 4.2.1)

The linkage between the DNA in the nucleus and proteins in the cytoplasm is not direct (Fig. 54). The information contained in the DNA molecule is transferred to the protein-synthesizing machinery of the cell via another informational nucleic acid, called messenger RNA (mRNA), which is synthesized complementary to the relevant DNA sequence by RNA polymerase. Although similar to DNA, mRNAs are single-stranded, and contain the base uracil instead of thymine, and ribose instead of deoxyribose. The mRNA molecules act as transport vehicles for the information contained in the genes being expressed.

In eukaryotic cells, the initial mRNA copy contains homologues of both the intron and exon regions. The intron regions are then removed and the exon regions are spliced together to form the active mRNA molecules, which are then transported through the pores of the nuclear membrane to the cytoplasm.

3.6.3. Translation (→ Nucleic Acids, Chap. 4.2.2)

The next process involves the translation of mRNA molecules into polypeptides. This procedure requires many enzymes and two further types of RNA: transfer RNA (tRNA) and ribosomal RNA (rRNA). There is a specific tRNA for each amino acids. The tRNA molecules are involved in the transport and coupling of amino acids into the resulting polypeptide (Fig. 54). Each tRNA molecule has two binding sites, one for the specific amino acid, the other containing a triplet of bases (the anticodon) which is complementary to the appropriate codon on the mRNA.

The rRNA is complexed with protein to form subcellular globular organelles called ribosomes. Ribosomes can be regarded as the reading heads, which allows the linear array of mRNA codons each to base-pair with an anticodon of an appropriate incoming tRNA amino acid complex.

3.6.4. Regulation of Gene Expression

All cells possess the same genetic information, but different types of cells exhibit distinct gene transcription patterns. These differences in gene expression are critical to the morphological and biochemical properties of the many thousands of

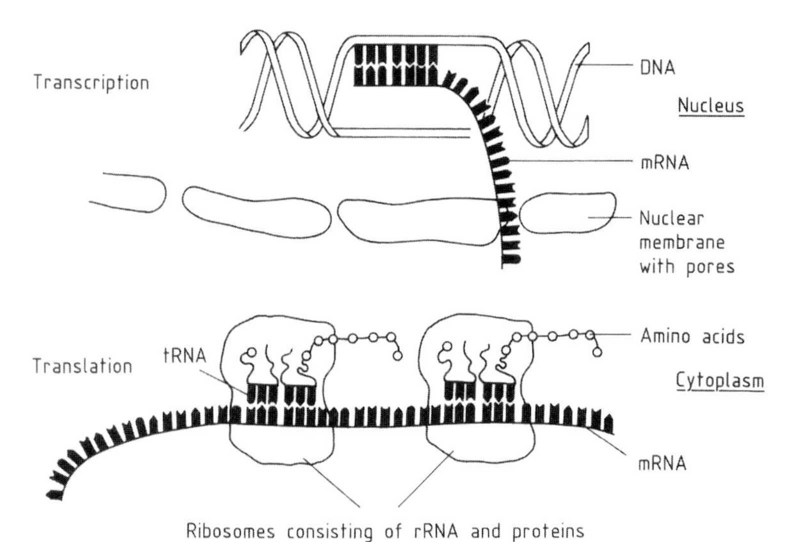

Figure 54. Scheme of the important steps in gene transcription and protein synthesis

cell types of the human and animal body. Hence, mechanisms are required that regulate gene expression, that is, determine which genes are expressed and to what extent and which genes are not expressed in a certain cell type at a particular time. The mechanisms involved in regulation of gene transcription are not entirely understood. The transcription of structural genes is regulated by a special set of codons, in particular, promotor sequences, the initial binding sites for RNA polymerase before transcription begins. Different promoter sequences have different affinities for RNA polymerases. Additional regulatory genes called operators regulate the activity of several genes or gene groups (operons). The activity of the operator itself is further controlled by a repressor protein, which stops the transcription of the whole operon by binding to the operator sequence. Due these regulatory mechanisms cells are able to express only the genes required at a given moment for their specialized function. This not only helps to conserve cellular energy, but is also critical for correct cellular differentiation, tissue pattern formation and function, and maintenance of the physiological integrity of the entire organism.

3.6.5. DNA Repair

All living cells possess several efficient DNA repair processes. DNA repair is crucial in protecting cells from spontaneous and exogenous lethal and mutating effects such as heat-induced DNA hydrolysis, UV radiation, ionizing radiation, DNA-reactive endogenous chemicals, free radicals, and reactive oxygen species. Among the various DNA repair mechanisms, the most comprehensively studied mechanism in eukaryotes is the excision repair pathway. This mechanism involves a group of enzymes acting cooperatively to recognize DNA lesions, remove them, and correctly replace the damaged sections of DNA.

The excision repair pathway is regarded as error-free and does not lead to the generation of mutations. However, this pathway may become saturated after excessive DNA damage. In this case, the cell may be forced to activate other repair mechanisms which do not operate error-free. Several of these mechanisms, such as error-prone repair, have been well characterized in bacteria, but their counterparts, if any, in mammalian cells have not been identified yet.

3.7. Molecular Mechanisms of Malignant Transformation and Tumor Formation

3.7.1. Mutations

Mutations are hereditary changes in genetic information, resulting from spontaneous or xenobiotic-induced DNA damage. The term mutation can be applied to point mutations, which are qualitative changes involving one or a few bases within one gene, and to larger changes involving parts of the chromosome detectable by light microscopy or even whole chromosomes and thus many thousands of genes (Table 22).

Table 22. Types of mutations

Gene mutations	base-pair substitutions, deletions, insertions, gene rearrangements, gene amplifications
Chromosomal mutations	
Structural	breaks, translocations
Numerical	loss or gain of an entire chromosome

Thymine : adenine

Cytosine : guanine

Thymine : O^6-methylguanine

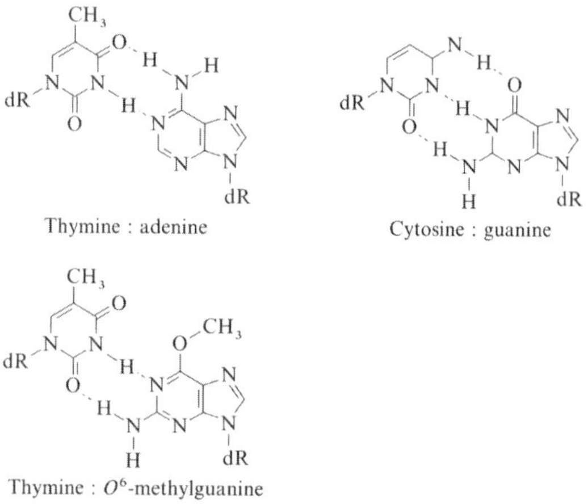

Figure 55. Formation of a base substitution
Guanine normally pairs with cytosine and adenine with thymine (upper part); in contrast, O^6-methylguanine (a frequent DNA modification induced by methylating agents such as dimethylnitrosamine) pairs with thymine (lower part), resulting in a hereditary change of the genetic information

Point mutations can occur when one base is substituted for another (base substitution) or when base pairs are deleted or inserted (deletions/insertions). Substitution of another purine for a purine base or of another pyrimidine for pyrimidine is called a transition, while substitutions of purine for pyrimidine or pyrimidine for purine are called transversions. Very small alterations in the chemical structure or the DNA bases may be sufficient for a base-pair substitution to occur. Guanine, for example, normally pairs with cytosine, while O^6-methylguanine (a frequent DNA modification seen with methylating agents such as dimethylnitrosamine) pairs with thymine (Fig. 55).

These changes in certain codons may cause insertion of the wrong amino acid into a relevant polypeptide. In this case, the changes are named missense mutations. Such proteins may have dramatically altered properties if the new amino acid is close to the active center of an enzyme or affects the three-dimensional structure of an enzyme or a structural protein. Hence, the alterations may result in marked changes in the differentiation and proliferative characteristics of the affected cells. A base substitution can also result in the formation of a new inappropriate stop (or nonsense) codon. The result of nonsense mutations is the formation of a shorter and, most likely, inactive protein. Owing to the redundancy of the genetic code, about a quarter of all possible base substitutions will not result in amino acid replacement and will be silent mutations.

Bases can be also deleted or added to a gene. As each gene is of a precisely defined length, these changes, if they involve a number of bases that is not a multiple of three, result in a change in the reading frame of the DNA sequence and are known as frameshift mutations. Such mutations often have a dramatic effect on the polypeptide of the affected gene, as most amino acids will differ from the point of insertion or deletion of bases in the DNA strand onwards.

Some forms of unrepaired alkylated bases are lethal due to interference with DNA replication. Others, such as O^6-methylguanine lead to mutations if unrepaired. These differences indicate that not all DNA adducts are of equivalent importance. In fact, some adducts appear not to interfere with normal DNA functions or are rapidly repaired, others are mutagenic, and yet others are lethal. The most vulnerable base is guanine, which can form adducts at several of its atoms (e.g., N-7, C-8, O-6 and exocyclic N-2, see Section 2.5.6.6).

Intrastrand and Interstrand Cross-Links. Xenobiotics with bifunctional alkylating properties can also form links between adjacent bases on the same strand (intrastrand cross-links) or between bases on different strands (intrastrand cross-links). The induction of frameshift muta-

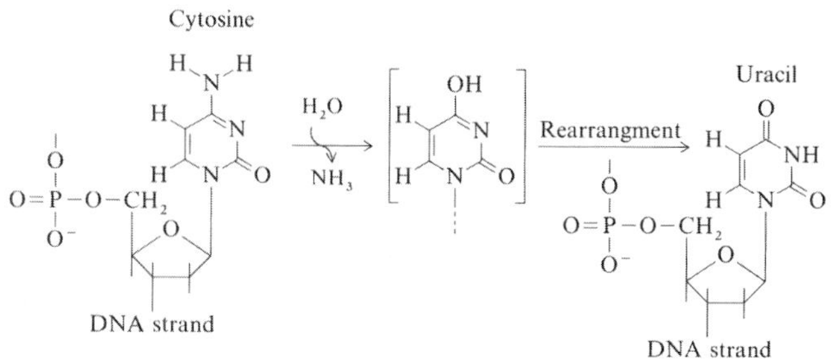

Figure 56. Deamination of cytosine to uracil
dR = Deoxyribose

tion does not necessarily require formation of covalent adducts. Some compounds that have a planar structure, particularly polycyclic aromatic hydrocarbons, can intercalate between the strands of the DNA double strand. The intercalated molecules may interfere with DNA repair or replication and cause insertions and/or deletions of base pairs. The precise molecular event is still unclear, although several mechanisms have been proposed. Hot spots for frameshift mutations often involve sections of the DNA strand taht contain a run of the same base (e.g., the addition of a guanine to a run of six guanine residues).

DNA strand breaks result from the hydrolysis of the sugar–phosphate bond or a nucleotide. In a double-strand, both a single- and double-strand breaks may occur. DNA strand breaks are often induced by hydroxyl radicals, which are formed at high rates both spontaneously during normal cell life and in the presence of exogenous chemicals. AP lesions (apurinic/pyrimidinic sites) in the DNA strand result from spontaneous hydrolysis of the glucosidic bond and loss of the DNA base (see Fig. 37). Similar to the situation with DNA strand breaks, AP lesions are a common spontaneous event; however, the hydrolysis can be dramatically increased by various types of DNA adducts, such as N^7-substituted purines, a common target of alkylating chemicals (see Section 2.5.6.6). Another common spontaneous event is the desamination of cytosine to uracil (Fig. 56); approximately 100 desaminations take place in each cell every day (→ Nucleic Acids, Chap. 4.2.1).

3.7.2. Causal Link between Mutation and Cancer (see also → Carcinogenic Agents; → Mutagenic Agents)

The change from cells undergoing normal, controlled cell division and differentiation to cells that are transformed, divide without control, and are undifferentiated or abnormally differentiated does not occur in a single step. Malignant transformation is a multistage process. Evidence for the involvement of multiple stages comes from in vitro studies, animal models, and epidemiological observations. In humans, the latent period between exposure to a chemical carcinogen and the appearance of a tumor in the target tissue is approximately 10–25 years. Modern molecular-biology techniques enable thorough investigations of the genome of malignant cells in comparison with the genome of their normal counterparts. These studies clearly show that a single mutation is not sufficient to induce malignant transformation. The number of genetic changes varies between two and seven in different tumor types. Also, several types of mutations are usually formed in a malignant transformed cell (i.e., base-pair substitutions, gene rearrangements, chromosomal breaks, and deletions.

3.7.3. Proto-Oncogenes and Tumor-Suppressor Genes as Genetic Targets

Why should mutations be causally linked to cancer? The answer to this question has increasingly become clear since the 1980s with the study of proto-oncogenes and tumor-suppressor genes [144, 145]. It is now appreciated that normal control of cell division and differentiation is based on the interplay of two sets of genes, the proto-oncogenes and the tumor-suppressor genes. Abnormal activation of proto-oncogenes and/or inactivation of tumor-suppressor genes eventually leads to malignant transformation.

Oncogenes were originally discovered in the genome of transforming retroviruses and were therefore named v-oncogenes [144, 146]. Subsequent studies showed that these viral genes were originally derived from the mammalian genome. In the normal cell, these proto-oncogenes have important functions in signal transduction pathways.

3.7.4. Genotoxic versus Nongenotoxic Mechanisms of Carcinogenesis

Oncogene activation and tumor-supressor gene inactivation induced by mutations provide strong evidence for the involvement of genotoxic mechanisms in tumor formation. However, it has been recognized for many years that cancers can arise without direct or indirect interaction between a chemical and cellular DNA, that is, in the absence of direct mutations. The distinction between nongenotoxic and genotoxic carcinogens was more sharply defined following the identification of a comparatively large number of nongenotoxic carcinogens by the U.S. National Toxicology Program [147]. These include a wide range of chemicals acting by a variety of mechanisms, such as disruption of normal hormonal homeostasis in hormone-responsive tissues, and peroxisome proliferation and proliferation of urothelial cells of the urether and urinary bladder following damage by kidney stones.

Genotoxic carcinogens tend to induce tumors in several tissues of both males and females in both rats and mice. In contrast, nongenotoxic carcinogens usually induce tumors only at high doses, in one tissue, in one sex, or only in one species. The experimental evidence available so far does not support the existence of real thresholds for DNA-reactive carcinogens, although very low concentrations may exist for which practically no clinically manifest tumors may be observed in the animal (or human) lifespan. However, these concentrations can not be considered as thresholds. In the case of carcinogens that operate via other biological effects, the carcinogenic activity would parallel dose–response relationships of the relevant biologic effects, a very important aspect for human risk assessment. Treatment regimens or exposure scenarios that do not elicit biological effects would not promote tumor formation. There are, however, two major problems: very few dose–response studies have been performed with nongenotoxic carcinogens, and in most cases the biochemical mechanisms responsible for the tumor promoting action are not understood.

Usually, non-enotoxic carcinogens are divided into two major categories. The first includes compounds that induce cytotoxicity and regenerative cell proliferation, e.g., 2,2,4-trimethylpentane and other branched-chain hydrocarbons in the proximal tubules. The second group of nongenotoxic carcinogens induce cell proliferation in the absence of cytotoxicity (they are directly mitogenic), relevant examples in this group are carcinogenic hormones or peroxisome proliferators such as di(2-ethylhexyl) phthalate and clofibrate. Induction of cell proliferation is involved in both categories of nongenotoxic carcinogens and may contribute to malignant transformation by increasing the number of spontaneous genetic errors, since DNA replication does not occur with 100% fidelity. Furthermore, in rapidly proliferating cells, DNA damage has a higher chance of being converted to heritable mutations. However, in all these cases cell proliferation is the final result of an as-yet unidentified molecular mechanism.

Genotoxic and nongenotoxic mechanisms are not mutually exclusive events. Rather, they cooperate in tumor formation, as can be seen with many genotoxic carcinogens. In most cases genotoxic carcinogens induce tumors only after applications of high doses, concomitantly causing cytotoxicity, cell death, and regenerative proliferation. Hence, a tumor is the final outcome of a complex, multistep interplay between genotoxic and extranuclear events.

3.8. Mechanisms of Chemically Induced Reproductive and Developmental Toxicity

The term reproductive toxicity covers any detrimental effect on the male and female reproductive system due to exposure to toxic chemicals.

Developmental toxicity refers to detrimental effects produced by exposure to developing organisms during embryonic, fetal, and neonatal stages of development. The main phases of reproduction are listed below:

Germ cell production
Spermatogenesis (man)
Oogenesis (woman)

Preimplantation phase
Fertilization
Formation of the blastocyst
Implantation

Embryonic phase
Organogenesis (in humans the first 12 weeks, in rat the first 2 weeks of gestation)

Fetal phase
Functional maturation and growth of the organs, in humans week 12 to 30

Peri- and postnatal phase
Last week of gestation, birth and first period after birth

Such effects can be irreversible or reversible. Embryolethal effects are incompatible with survival and result in resorption or spontaneous abortion. Irreversible effects that are compatible with survival may cause structural or functional abnormalities in the offspring, and these are called teratogenic. Embryotoxic chemicals may also cause overall growth retardation or delayed growth of certain organs.

3.8.1. Embryotoxicity, Teratogenesis, and Transplacental Carcinogenesis

For an agent to be classified as a developmental toxicant, it must be harmful to the developing organism at exposure levels that do not induce severe toxicity in the mother, such as substantial reduction in weight gain, persistent emesis, or convulsions. Adverse effects on the developing organism under severe maternal toxicity may be secondary to perturbations in the maternal system. For practical purposes, however, the test compounds can be initially administered at maternally toxic doses to determine the threshold level for adverse effects on the offspring. At these exposure levels conclusions can be qualified to indicate that adverse effects of the conceptus were obtained at maternally toxic exposure levels, and may not be indicative for selective developmental toxicity.

The susceptibility of the developing organism to xenobiotic insults varies dramatically within the narrow time span of the major developmental stages (the preimplantation, embryonic, fetal, and perinatal periods), because developing organisms undergo rapid and complex changes within this short period. The major morphogenic events occuring during preimplantation development are formation of a compact mass of cells (the morula) and of the blastocyst. The latter already exhibits a certain degree of cellular differentiation. Considerable similarity exists in the timing of preimplantation development across several mammalian species, regardless of the total length of gestation [148]. At the time of blastocyst formation, cell division and metabolic capacity increase dramatically. During the preimplantation period, biochemical changes under progesterone and estrogen control render the endometrium sensitive to the blastocyst implantation. One important sign of blastocyst implantation is a prostaglandin-dependent increase in endometrial vascular permeability. Alterations in the hormonal milieu or direct excretion of specific xenobiotics into the uterine epithelia during this period can impair implantation and cause embryolethality. Limited data suggest that the preimplantation embryo appears to be susceptible to lethality but rarely to induction of structural aberrations (teratogenicity) with chemical insults. Sublethal exposures of preimplantation embryos have not yet been adequately explored.

Following implantation, organogenesis takes place, which is characterized by the division, migration, and association of cells into primitive organs. The most characteristic susceptibility of the embryo to xenobiotics during the organogenesis period is the induction of structural birth defects (terata). Within the organogenesis period (embryonic period), individual organ systems possess highly specific periods of vulnerability to teratogenic insult (Fig. 57). Administration of a teratogen on day 9 of rat gestation would result in a high level of brain and eye defects, while on day 15 structural abnormalities of the kidney and urinary bladder would predominate. As shown in Figure 57, critical periods for the susceptibility of different organs overlap, so that exposure to teratogens usually results in a spectrum of more or less severe malformations in a number of organ systems [149].

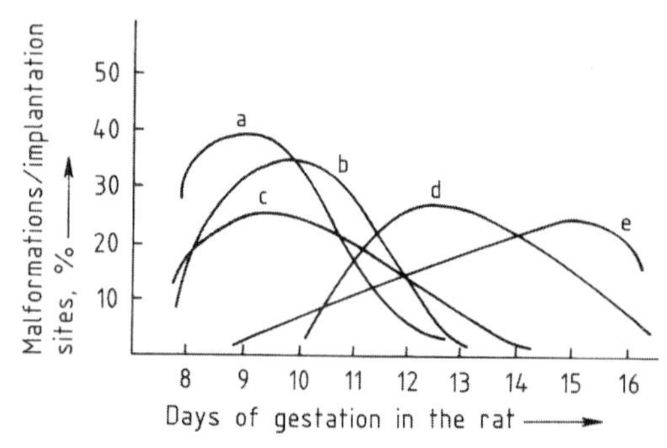

Figure 57. Degree of susceptibility of various organ rudiments of the developing organism to teratogenic xenobiotics (data from [144])
a) Eye; b) Brain; c) Heart and axial skeleton; d) Palate; e) Urogenital system

The critical phase for inducing anomalies in individual organ systems may be as short as one day or may extend throughout organogenesis. Urogenital defects, for example, can result from drug treatment from day 9 to 16 of gestation in the rat. However, the structural defects of the urogenital system depend upon the day(s) of exposure to the teratogenic antibiotic. The development of the urogenital system is multiphasic, and individual stages may have different sensitivities to chemical insult. The mechanisms governing embryonic differentiation are not well understood, but they are certainly involved in intrinsic susceptibility of individual organs to teratogenic insult.

Functional maturation and growth are the major processes occurring after organogenesis, during the fetal and perinatal periods. Insult at these late developmental stages leads to growth retardation or to more specific functional (but not structural) disorders and transplacental carcinogenesis. The fetal and perinatal period of life is highly susceptible to carcinogenesis, due to the high cellular replication rates, presence of xenobiotic biotransforming enzymes in the fetus, and immaturity of the immune system in the developing organism. Several childhood tumors occur so early after birth that prenatal origin is considered likely. These include acute lymphocytic leukemia, Wilms tumor (nephroblastoma), and neuroblastoma. Studies with direct-acting transplacental carcinogens such as ethylnitrosourea indicate that susceptibility to carcinogens begins after completion of the organogenesis period in rodents. Tumors in offspring occurred primarily when ethylnitrosourea was given during the fetal period, whereas birth defects and embryolethality predominated with exposures in the embryonic phase [150, 151]. However, this does not imply that teratogenesis and carcinogenesis are mutually exclusive processes. Teratogenesis and carcinogenesis can be regarded as graded responses of the embryo to injury, with teratogenesis representing the more gross response involving major tissue necrosis in early, relatively undifferentiated embryos, combined carcinogenicity–teratogenicity damage in older embryos, and finally, carcinogenicity alone in the fetus.

3.8.2. Patterns of Dose–Response in Teratogenesis, Embryotoxicity, and Embryolethality

The major toxic effects of prenatal exposure observed at the time of birth are embryolethality, malformations, and growth retardation. The relationship between embryolethality, malformations, and growth retardation is quite complex and depends on the type of agent, the time of exposure, and the dose. Some developmental toxins may cause malformations of the entire litter at exposure levels that do not cause embryolethality (Figure 58). If the dose is increased, embryolethality can occur, often in combination with severe maternal toxicity. Malformed fetuses are often more or less retarded in growth, and the curve for growth retardation is often parallel to and slightly displaced to the right from the curve for teratogenicity. Such a pattern of response is indicative of agents with high teratogenic potency.

A more common dose–response pattern involves embryolethality, malformations, and growth retardation of surviving fetuses. Exposure to these chemicals results in a combination of resorbed, malformed, growth-retarded, and "normal" fetuses. Depending on the teratogenic potency of the agent, lower doses may cause predominantly malformations. As the dosage increases, however, embryolethality predominates until the entire litter is resorbed. Growth retardation can precede both these outcomes or parallel the teratogenicity curve.

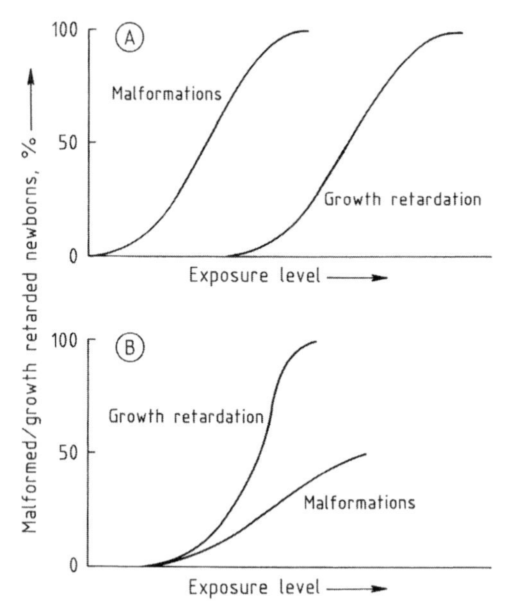

Figure 58. Possible dose – response relationship of teratogens.
A) Teratogens interfering with specific events in differentation; B) Teratogens acting via general cytotoxicity and induction of cell necrosis

A third dose–response pattern consists of growth retardation and embryolethality without structural abnormalities. Growth retardation of surviving fetuses usually precedes significant embryolethality. Agents producing this pattern of response would be considered embryotoxic but not teratogenic, and are also toxic to the maternal organism. In contrast, potent and specific teratogens often show only weak toxicity to the maternal organism.

The best known example is the hypnotic sedative thalidomide, which was rarely associated with severe undesired effects in adult humans but induced malformation in thousands of children whose mothers had taken the drug as a sleeping aid at the recommended therapeutic doses during gestation. The existence of these three general patterns of response indicate that for some agents embryolethality and teratogenicity are different degrees of manifestations of the same primary insult. For other agents, there is a qualitative difference in response, and the primary insult leads to embryotoxicity and embryolethality alone.

For practical purposes (for exact experimental procedures, see Section 4.7.2) a relatively small number of pregnant rodents (approximately eight per group) are exposed on days 6 through 15 of gestation to the test agent at doses up to those causing limiting maternal toxicity and/or severe embryotoxicity (death, severe growth retardation). The purpose of this dose-range finding study is to obtain a qualitative yes/no signal about the potential developmental toxicity of the agent, and information on doses causing severe impairment of the maternal organism. For the main evaluation of developmental toxicity the highest dose should cause measurable but slight maternal toxicity (i.e., significant depression of weight gain) or embryotoxicity (i.e., significant depression of fetal body weight, increased embryolethality, and/or structural malformations), and the low dose should cause no observable effects.

4. Methods in Toxicology

4.1. Toxicological Studies: General Aspects

The aim of toxicology is the assessment and management of potential hazards from exposure of humans and the general environment including animals and plants to chemicals. To achieve this objective, detailed knowledge on the inherent hazard of a xenobiotic (for definition, see Section 5.2.1), that is, its acute and chronic toxicity, its no observed effect level (NOEL), and its teratogenic, mutagenic and carcinogenic effects, is required. This information can not be obtained from a single experiment. A battery of in vivo and in vitro toxicity tests must be utilized. As required by law in most industrialized countries, all toxicity testing must be performed under the rules of good laboratory practice with exact documentation of all relevant conditions and results. Adequate planning of toxicity tests for obtaining optimal information from the experiments may greatly improve the basis for the risk assessment of a chemical and may reduce the number of animals needed and the financial expense associated with toxicity studies. Therefore, all test batteries should be part of an integrated approach to toxicity studies and include not only methods to determine the toxic effects of a chemical

Table 23. OECD guidelines on short- and long term toxicity testing in vivo

No.	Title	Original adoption	Updated
401	Acute Oral Toxicity	12 May 1981	20 Dec. 2002*
402	Acute Dermal Toxicity	12 May 1981	24 Feb. 1987
403	Acute Inhalation Toxicity	12 May 1981	
404	Acute Dermal Irritation/Corrosion	12 May 1981	24 April 2002
405	Acute Eye Irritation/Corrosion	12 May 1981	24 April 2002
406	Skin Sensitization	12 May 1981	17 July 1992
407	Repeated Dose 28-Day Oral Toxicity Study in Rodents	12 May 1981	27 July 1995
408	Repeated Dose 90-Day Oral Toxicity Study in Rodents	12 May 1981	21 Sept. 1998
409	Repeated Dose 90-Day Oral Toxicity Study in Non-Rodents	12 May 1981	21 Sept. 1998
410	Repeated Dose Dermal Toxicity:28-Day	12 May 1981	
411	Subchronic Dermal Toxicity: 90-Day	12 May 1981	
412	Repeated Dose Inhalation Toxicity: 28/14-Day	12 May 1981	
413	Subchronic Inhalation Toxicity: 90-Day	12 May 1981	
414	Prenatal Developmental Toxicity Study	12 May 1981	22 Jan. 2001
415	One-Generation Reproduction	Toxicity	26 May 1983
416	Two-generation Reproduction Toxicity Study	26 May 1983	22 Jan. 2001
417	Toxicokinetics	4 April 1984	
418	Delayed Neurotoxicity of Organophosphorus Substances Following Acute Exposure	4 April 1984	27 July 1995
419	Delayed Neurotoxicity of Organophosphorus Substances: 28-Day Repeated Dose Study	4 April 1984	27 July 1995
420	Acute Oral Toxicity – Fixed Dose Procedure	17 July 1992	17 Dec. 2001
421	Reproduction/Developmental Toxicity Screening Test	27 July 1995	
422	Combined Repeated Dose Toxicity Study with the Reproduction/Developmental Toxicity Screening Test	22 March 1996	
423	Acute Oral Toxicity–Acute Toxic Class Method	22 March 1996	17 Dec. 2001
424	Neurotoxicity Study in Rodents	21 July 1997	
425	Acute Oral Toxicity: Up-and-Down Procedure	21 Sept. 1998	17 Dec. 2001
426	Developmental Neurotoxicity Study	Draft New Guideline, October 1999	
427	Skin Absorption: In vivo method	Expected, Approved by WNT (May 2002)	
428	Skin absorption: In vitro method	Expected, Approved by WNT (May 2002)	
429	Skin Sensitization: Local Lymph Node Assay	24 April 2002	
430	In Vitro Skin Corrosion: Transcutaneous Electrical Resistance Test (TER)	Expected, Approved by WNT (May 2002)	
431	In Vitro Skin Corrosion: Human Skin Model Test	Expected, Approved by WNT (May 2002)	
432	In Vitro 3T3 NRU phototoxicity test	Expected, Approved by WNT (May 2002)	
433	Acute Inhalation Toxicity: Fixed Dose Procedure	Draft New Guideline, October 1999	
451	Carcinogenicity Studies	12 May 1981	
452	Chronic Toxicity Studies	12 May 1981	
453	Combined Chronic Toxicity/Carcinogenicity Studies	12 May 1981	

*Date of deletion.

and their dose dependence, but also toxicokinetics, biotransformation, and mechanisms of action. This chapter provides an overview of the currently used methods for the assessment of a chemical's toxic profile. For details, specific guidelines on practical aspects of toxicity studies and types of data required can be obtained from web sites of national and international organisations (see Section 5.2.1 and 1.4). OECD guidelines for the toxicity testing of chemicals in vivo are listed in Table 23.

For the evaluation of a new chemicals toxic effects in laboratory animals two types of studies are carried out: acute-toxicity and repeated-dosing studies.

Acute Toxicity. Following administration of a single dose of the test substance or of multiple doses given over a period of up to 24 h, potentially adverse effects are usually monitored during the following 14 d. Acute toxicity studies in animals aim to assess the human risk from single exposure to high doses, for example, in industrial accidents, after drug overdoses, or after suicide attempts.

Repeated-Dosing Studies: Subacute, Subchronic, and Chronic Toxicity. The purpose of repeated daily doses of a chemical for part of the animal's life span is to study subchronic and chronic effects. Studies on *subacute toxicity* are carried out for two to four weeks, while studies on *subchronic toxicity* usually last for a period of three months. These studies are helpful in assessing the human risk resulting from frequent exposure to household or workplace chemicals and from intake of chemicals used for therapeutic purposes. Studies to determine *chronic toxic effects* are carried out for at least six months; studies aiming to investigate the carcinogenic effects of a test compound are carried out over the animal's entire lifetime. Lifetime exposure of humans may occur to widespread environmental pollutants, food additives, or residues of agricultural chemicals in food.

In addition to the acute and repeated-dose toxicity studies, the reproductive and developmental toxicity as well as the genotoxicity of a new chemical must be investigated in separate experiments.

Many thousands of new and potentially toxic compounds are synthesized every year. It would be a waste of money, resources and manpower if the entire battery of toxicity tests were automatically performed for every new chemical. Therefore, toxicity testing is rather undertaken on the basis of a decision-point approach in several stages, as shown in Figure 59. At the end of every stage, the decision must be met, if the development will be continued or if, on the basis of the toxicity data available so far, the potential human risk of the exposure to this chemical is unacceptable. If the latter is true, the development and consequently the toxicity testing is stopped.

Animal Husbandry. The use of standardized conditions for the housing of animals plays a major role in the planning, evaluation, and interpretation of toxicity tests. Animals must be kept in a controlled environment, i.e., constant temperature of 22 ± 3°C, sufficient ventilation, relative humidity between 30 and 70% and a 12 h light/dark cycle. Diet composition and quality of drinking water must also be standardized and controlled throughout the experiment. Only healthy young adult animals should be enrolled in the studies, and the animals should be allowed to acclimatize to the experimental conditions for at least one week prior to first dosing. After the acclimatization period, animals with poor health or body weights varying by more than 20% of the the group's mean body weight are either excluded from the studies or randomized to ensure a homogenous population in the different control and treatment groups. The basic guidelines – choice of species, number of animals, dosing regimens, duration and frequency of observation, assessment of specific body functions – for acute, subchronic and chronic toxicity tests are summarized in Table 24.

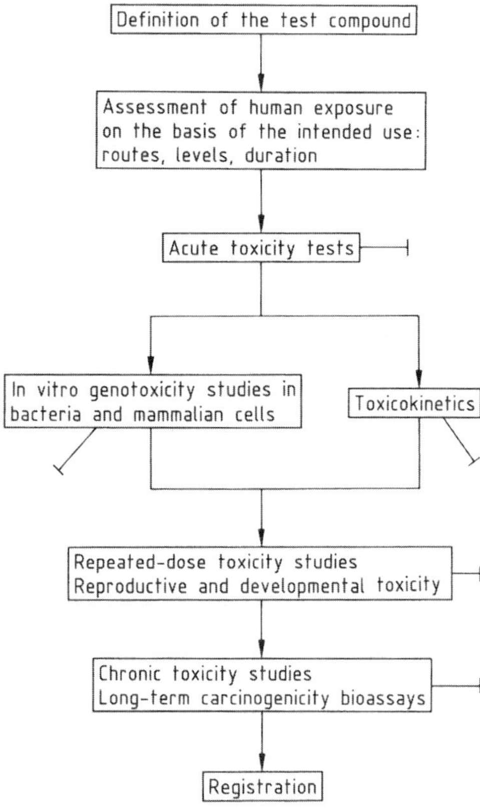

Figure 59. Evaluation of the toxicity profile of a new chemical compound on the basis of a decision point approach At every step of toxicity testing, further development may be interrupted if, on the basis of data collected up to that point, the human risk is considered unacceptable.

A variety of in vitro methods are in development to reduce the numbers of animals used in toxicity testing. Some of the developed methods have gained regulatory acceptance, and some may be used for a priority-determining process for further testing. All well-evaluated methods focus on local effects such as skin and eye irritation, where most progress in the development of nonanimal methods has been made. Regarding replacement of toxicity studies on systemic effects after repeated exposure by nonanimal

Table 24. Basic guidelines for acute, subchronic, and chronic oral toxicity tests

	Acute oral	Subchronic oral	Chronic oral
Animals	rats preferred	rodent and nonrodent species	rodent and nonrodent species
Sex	males and females equally distributed per dose level		
Age	young adult, weight variation within 20 % of mean	rodents, 6 weeks; dogs 4 – 6 months old	
Number of animals	at least 10 (5 per sex)	at least 20 for rodents (10 per sex)	50 per sex group for rodents
Number of treatment groups	3; mortality rates between 10 and 90 % should be produced	3; mortality should not exceed 10 % in high-dose group	3; low dose should reflect expected human exposure and high dose must produce not more than 10 % mortality
Untreated control	not necessary	yes	yes
Vehicle control	yes, if vehicle of unknown toxicity is used	yes	yes
Dosing	gavage; single dose, same dose of vehicle; if necessary use divided doses over 24 h	diet, gavage, drinking water	diet, gavage, drinking water
Duration of study	at least 14-d observation period	90 d	6 – 24 months in rats
Body weight determination	before dosing, weekly thereafter, and at death	weekly and at termination	weekly for first 13 weeks; every 2 weeks thereafter and at termination
Necropsy	all animals	all test animals; organ weights of liver, kidney, heart, lungs, brain, gonads, adrenals, and spleen	
Histopathology	examination of organs showing evidence of gross pathological change	all tissues high-dose and control groups; liver, kidney, heart, lungs, target organs, and any gross lesion in mid- and low-dose groups	all tissues of animals
Frequency of cage-side observations	frequently during day of dosing; once each morning and late afternoon thereafter	daily	daily
Observations, assessments	nature, onset, severity, and duration of any effect observed		
		ophthalmoscopy: pretest and at termination in control and high-dose groups	
		hematology/clinical chemistry:	
		pretest, dosing midpoint, termination	pretest and at 3, 6,12, 18, 24 months
		urine analysis: dosing midpoint, termination	pretest and at 3, 6,12, 18, 24 months

methods, due to the complexities of interactions resulting in toxic responses, acceptable nonanimal tests equaling the predictive power of animal testing are unlikely to be available in the near future.

4.2. Acute Toxicity

4.2.1. Testing for Acute Toxicity by the Oral Route: LD_{50} Test and Fixed-Dose Method

The objectives of acute toxicity tests are

1) To assess the intrinsic toxicity of the test compound
2) To identify target organs of toxicity affected by the xenobiotic
3) To provide information concerning the dose selection and treatment regimens for repeated-dose studies
4) To provide information for human risk assessment after a single high-dose exposure to the chemical
5) To provide essential data for the classification, labeling, and transportation of the chemical (regulatory view report)

LD_{50} Test. The determination of the mean lethal dose (LD_{50}) is still often considered as the first step in the evaluation of the acute oral or inhalation toxicity of a new chemical; with the present knowledge and recent experiences, the formal determination of the LD_{50} is no longer considered as necessary, and alternative methods that are also to be used for classification and labeling have been developed. However, since the test is still widely used, it will be briefly described before treating the newer methods used in testing of acute toxicity and the reasons that led to these changes.

In the LD_{50} test, groups of animals (usually female rats) are treated with graduated doses of the test compound, and by using mathematical models the dose which causes death in 50% or more of the population is determined. Internationally accepted guidelines recommend the use of at least three dose groups with five males and five females for each dose or the use of three dose groups with five animals of one sex and one dose with five animals of the other sex [152]. The LD_{50} is then determined from data obtained as described in Section 1.7. Chemicals with LD_{50} values ≤ 25 mg/kg are considered very toxic, between 25 and 200 mg/kg as toxic, and between 200 and 2000 mg/kg as harmful. Some potentials and limitations of the LD_{50} test follow:

Potential

1) Useful as a first approximation of hazards in the workplace
2) Basis for the design of subchronic studies
3) Properly conducted test may give useful information on other relevant toxicity parameters
4) Rapid completion

Limitations and Problems

1) Lethality only criterion applied, other toxic effects not considered
2) Animal welfare is major point of concern because large number of animals are required to obtain statistically acceptable values
3) Large variations in LD_{50} in different laboratories with identical chemicals, many influencing factors
4) Species and strain differences cause difficulties in extrapolation
5) No information on chronic toxicity obtained (chronic effects are more important for regulating exposure)

The scientific significance of the LD_{50} test has been repeatedly questioned, not only because the lethal dose is not relevant for human risk assessment but also on the basis of the variability of the test results and last not but least for reasons of animal welfare. Comparative assessment of LD_{50} values in 60 laboratories under controlled conditions resulted for example in considerably different LD_{50} values (by a factor of up to 14). Therefore, numerous alternatives to the LD_{50} test have been proposed for the evaluation of acute toxicity that rely on signs of toxicity rather than on mortality. One of this procedures, the fixed-dose method – has recently gained acceptance by the OECD and the EU.

Fixed-Dose Method. The fixed-dose method relies on the observations of clear signs of toxicity developed at one of a series of fixed-dose levels (i.e. 5, 50, 300, and 2000 mg/kg of the chemical oper kilogram body weight). The dose levels at which signs of toxicity but no deaths are detected are used to classify the test compounds according to their toxic potential (Table 25).

Table 25. Classification of toxicity of a xenobiotic with the fixed-dose method

Dose (oral), mg/kg	Results	Classification
5	less than 90 % survival	very toxic
	90 % or more survival but evident toxicity	toxic
	90 % or more survival, no evident toxicity	retest at 50 mg/kg
50	less than 90 % survival	toxic, retest at 5 mg/kg
	90 % or more survival, but evident toxicity	harmful
	90 % or more survival, no evident toxicity	retest at 500 mg/kg
500	less than 90 % survival or evident toxicity and no death	harmful, retest at 50 mg/kg
	no evident toxicity	retest at 2000 mg/kg
2000	less than 90 % survival	harmful
	90 % or more survival, with or without evident toxicity	unclassified, does not represent a significant acute toxic risk if swallowed, no further testing necessary

As can be seen in Table 25, the fixed-dose method allows a classification identical to that previously obtained in the LD_{50} test. Moreover, comparative investigations utilizing both the LD_{50} test and the fixed-dose procedure revealed that in the majority of the test compunds (80–90%), the toxicity class assigned by determining the LD_{50} was identical to that determined by the fixed-dose method [152, 153].

In the fixed-dose method, at least ten animals (five per sex) are used for each dose investigated. The initial dose chosen (5, 50, 300, or 2000 mg/kg body weight) is one that is judged likely to produce evident toxic effects, but no mortality. When such a judgement can not be made due to lack of information on the potential toxic effects of the xenobiotic, an initial "sighting" study should be carried out. If clear signs of toxicity do not occur at the starting dose of 300 mg/kg during the two weeks observation period, the dose is increased to the next level. A careful clinical examination of the animals is performed at least twice on the day of administration and once daily thereafter for the next two weeks. Animals obviously in pain or showing severe signs of distress or toxicity are humanely killed. Cage-side examinations include skin and fur, eyes and mucous membranes, respiratory system, blood pressure, somatomotor activity, and behavior (for procedures, see Section 4.11). Particular attention is directed to observation of tremors, convulsions, hypersalivation, diarrhoea, and coma as indices of neurotoxicity. Food consumption and weight development are also monitored constantly. At the end of the observation period, all animals in the study are killed and subjected to gross autopsy. Organs showing macroscopic evidence of gross pathology are further subjected to histopathological examination.

The fixed-dose method offers several important advantages as compared with the traditional LD_{50} test:

1) The available evidence suggest that the fixed-dose method produces more consistent results without substantial interlaboratory variation.
2) It provides information on the type, time of onset, duration, and consequences of toxic effects. This information is more relevant for assessing the risks of human exposure to the chemical than the mean lethal doses of the LD_{50} test.
3) It requires fewer animals than the LD_{50} test (roughly 50%) and subjects the animals to less pain and distress.
4) It enables the classification of chemicals according to regulatory requirements.

For the "standard" acute oral and dermal tests the LD_{50} should be determined, except when the substance causes no mortality at the limit dose (usually 2000 mg/kg). Similarly, for an acute inhalation toxicity study the LC_{50} should be determined, unless no mortality is seen at the limit concentration (5 mg per L per 4 h for aerosols and particulates, 20 mg per L per 4 h for gases and vapors). In the fixed-dose procedure, the discriminating dose (the highest of the preset dose levels which can be administered without causing mortality) should be determined. For the acute toxic class and the up-and-down methods the final dose used in the study should be determined following the testing protocol, except when the substance causes no mortality at the limit dose.

Whichever approach is used in determining acute toxicity critical information must be derived from the data to be used in risk assessment. It is important to identify the dose levels at which signs of toxicity are observed, the relationship of the severity thereof with dose, and the level at which toxicity is not observed (i.e. the acute NOAEL). However, note that a NOAEL is not usually determined in acute studies, partly because of the limitations in study design.

4.2.2. Testing for Acute Skin Toxicity

Irrespective of whether a substance can become systemically available, it may cause changes at the site of first contact (skin, eye, mucous membrane/gastrointestinal tract, or mucous membrane/respiratory tract). These changes are considered local effects. A distinction can be made between local effects observed after single and after repeated exposure. For local effects after repeated exposure, see Section 3.9. Only local effects after single ocular, dermal, or inhalation exposure are dealt with in this section. Substances causing local effects after single exposure can be further classified as irritant or corrosive substances, depending on the (ir)reversibility of the effects observed.

Irritants are noncorrosive substances which through immediate contact with the tissue can cause inflammation. Corrosive substances are those which can destroy living tissues with which they come into contact.

Knowledge on the dermal toxicity of a new chemical is one of the prerequisites for assessment of the risks associated with human exposure to the chemical, because skin contact may represent a very important route of exposure in the occupational setting and in the home. Testing for dermal toxicity is usually performed in rabbits. Three types of application of the test chemical are employed: nonocclusive, semiocclusive, and occlusive. The test compound is applied uniformly to the back or a band around the trunk (clipped free of hair); approximately

10% of the body surface of the animal should be covered. Solid substances are pulverized and moistened to a paste with physiological saline or another appropiate solvent whose effects have been fully evaluated prior to the skin test. For occlusive or semiocclusive testing, the application site is covered with a plastic sheet (or other impervious material) or with a porous gauze dressing, respectively. For unocclusive exposure, the application site should be as close to the head as possible to prevent ingestion of the chemical by the animal licking the site of application. The duration of exposure varies between 4 and 24 h. If no test-chemical-related toxic effects on the skin or systemic toxicity are observed up after doses of up to 2 g/kg body weight, testing at higher doses is unnecessary. At the end of the exposure period, the compound is removed with cotton wool soaked in an appropiate solvent and the skin irritation is scored according to the Draize scoring system as shown in Table 26. In addition, any adverse systemic effects caused by percutaneous absorption of the test compound are monitored.

However, substantial differences exist in skin anatomy between humans and experimental animals. In general, the penetration of chemicals through the human skin is similar to that of pig, miniature swine, and squirrel monkey and clearly slower than that of the rat and rabbit. For example, administration of the insecticides lindane and parathion to rabbit skin results in an absorption of 51.2 and 99.5% of the dose, respectively; the corresponding absorption rates for human skin are 9.3 and 9.7%.

Since the 1980s, in vitro studies using human skin samples have been increasingly conducted to estimate percutaneous absorption of chemicals. The following experimental design is commonly used: A piece of excised human skin is attached to a diffusion apparatus that has a top chamber for the test compound, an O-ring to hold the skin in place, and a bottom chamber to collect samples for analysis. The flow of a chemical across the skin can be calculated with models based on chemical thermodynamics, taking into consideration the octanol/water partition coefficients, the saturated concentration in aqueous solution, and the molecular mass of the test compound [154]. However, for routine applications, the method has not been sufficiently evaluated.

Table 26. Evaluation of skin reactions according to the Draize scoring system

Erythema	Score
No erythema	0
Very slight erythema (barely perceptible)	1
Well-defined erythema	2
Moderate to severe erythema	3
Severe erythema (beet redness)	4
No edema	0
Very slight edema (barely perceptible)	1
Slight edema (edges of area well defined by definite raising)	2
Moderate edema (area raised ca. 1 mm)	3
Severe edema (raised more than 1 mm and extended beyond area of exposure)	4

Furthermore, an increasing number of toxicokinetic models for estimating the extent of percutaneous absorption of chemicals has appeared in the literature. Among them, a physiologically based toxicokinetic model was recently developed to describe the percutaneous absorption of volatile and lipid-oluble organic contaminants in dilute aqueous solution [155]. This toxicokinetic model considers both physiological parameters such as volumes of body compartments and blood flow rate, as well as the properties of the test compound. Presently, these models do not play a role in regular toxicity testing and are therefore not discussed in depth here.

More recently extensive progress has been also made in developing in vitro systems for evaluating the dermal irritation potential of chemicals. An overview of the systems that have been evaluated so far for a range of compounds by comparison of their predictive accuracy with animal test results is presented in Table 27 (for a review see [156]).

Table 27. In vitro test systems for detection of dermal irritation potential

System	End point
Mouse skin organ culture	leakage of LDH* and GOT**
Human epidermal keratinocytes	release of labeled arachidonic acid, cytotoxicity
Cultured BHK21/C13 cells	growth inhibition, cell detachment
SKINTEX – protein mixture	protein coagulation

* LDH = Lactate dehydrogenase.
** GOT = Glutamic acid oxalacetic transaminase.

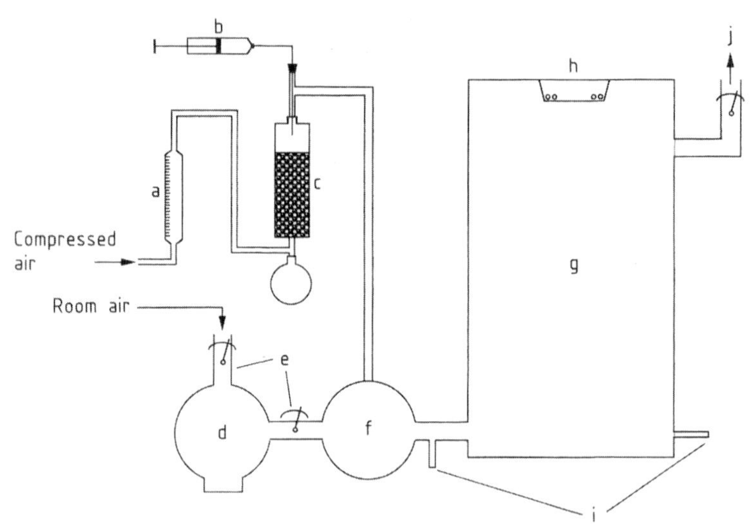

Figure 60. Example of an open inhalation chamber for exposure to volatile liquids
a) Meter; b) Syringe coupled to infuser; c) Heated beads of silica or glass; d) Compressor; e) Metering valves; f) Mixing chamber; g) Exposure chamber; h) Ventilation; i) Sampling valves for determination of atmospheric concentration; j) Exhaust

When evaluating these studies, attention should be given to the occurrence of persisting irritating effects, even those which do not lead to classification. Effects such as erythema, oedema, fissuring, scaling, desquamation, hyperplasia, and opacity which are not reversible within the test period may indicate that a substance will cause persistent damage to the human skin and eye.

4.2.3. Testing for Acute Toxicity by Inhalation

Studying the toxicity of a chemical by inhalation exposure requires a considerable technological input. Therefore, inhalation exposure is usually not tested if this absorption pathway is not expected to occur because the test chemical is not volatile or the physicochemical properties of solids do not allow the generation of respirable particles. Particles with diameters greater than 100 µm are unlikely to be inhaled, because they settle too rapidly. Particles with diameters of 10–50 µm are likely retained in the nose and the upper parts of the respiratory tract, while particles with diameters of less than 7 µm can reach the alveoli of the human lung. When performing toxicity studies with inhalation exposure, the differences in respiratory physiology between humans and the small laboratory rodents must be considered. In contrast to humans, the rat is an obligate nose breather with a complex nasal turbinate structure which filters many small particles. Therefore, the upper size limit for particles reaching the alveolar region in rats is in the range of 3–4 µm in diameter.

The duration of exposure in acute inhalation studies is usually 4–6 h, and they may be performed either as whole-body or head(nose)-only procedures in specific exposure chambers. A number of important considerations should be taken into account in planning and evaluating inhalation studies. The frequently made assumption that on the basis of their physical form, gases and vapors will be absorbed uniformly throughout the respiratory tract is incorrect. Many gases or vapors, such as ammonia, formaldehyde, and sulfur dioxide have high solubility in water and are rapidly absorbed by the humid epithelial surface of the upper respiratory tract. Therefore, toxic effects observed after inhalation of this type of chemical will generally be confined to these regions, especially to the nasal passages. In contrast, chemicals with low solubility in water such as nitrogen dioxide, phosgene, and ozone will penetrate readily to the low pulmonary regions, even at relatively low concentrations in the respiratory air. In mixed atmospheres containing both a gas or vapor and particulates, the vapor or gas may be absorbed on the particulate fraction, so that the deposition pattern of the vapor or gas is governed by the size of the particulate fraction and not by the water solubility of the vapor or gas.

The inhalation exposure of experimental animals may be performed in a dynamic or static mode. In dynamic systems, the test atmosphere is continuously renewed, ensuring atmospheric stability and constant concentrations of the chemical in the gas phase. This mode of inhalation exposure is more complicated and requires larger amounts of the test chemical and suitable systems to ensure complete mixing of the continuously applied test compounds with the stream of air flowing through the exposure system (Fig. 60).

Static systems are sealed, and the atmosphere is circulated. The concentration of the chemical in the gas phase decreases during exposure due to its uptake and biotransformation by the experimental animals. Static systems are used predominantly in acute toxicity studies and in research laboratories because they are relatively inexpensive and consume only a small amount of the chemical in comparison to the amounts need to generate a dynamic exposure. In analogy to the LD_{50} values obtained in the acute oral studies the mean lethal air concentrations LC_{50} are assessed in acute inhalation studies, usually for an exposure time of 4 h (Fig. 61).

4.3. Repeated-Dose Toxicity Studies: Subacute, Subchronic and Chronic Studies

Repeated-dose toxicity studies assess the effects resulting from the accumulation of a compound or its toxic effects in the organism and, unlike acute studies, they can also reveal toxic effects that appear after a latency period. A classic example is the delayed neuropathy caused by some organophosphorous insecticides and by cresyl phosphates, which is manifested several weeks after the first administration of the test compound. In contrast to the marked clinical symptoms observed in the course of delayed neuropathy, these compounds hardly cause any acute symptoms immediately after the first administration. Hence, false negative results may be obtained if an assessment were based only acute toxicity testing. The major aims of repeated-dose toxicity studies are the identification of starting points for the extrapolations required in human risk assessment such as NOAEL and benchmark doses, and the identification of critical end points to be carried over into the risk assessment process.

Testing for subacute toxicity is usually performed over a time period of 2–4 weeks at three dose levels as an aid in selecting the dose levels for the subchronic studies. Studies to determine subchronic effects are usually performed in rats and dogs over 10% of the animals life-span (3

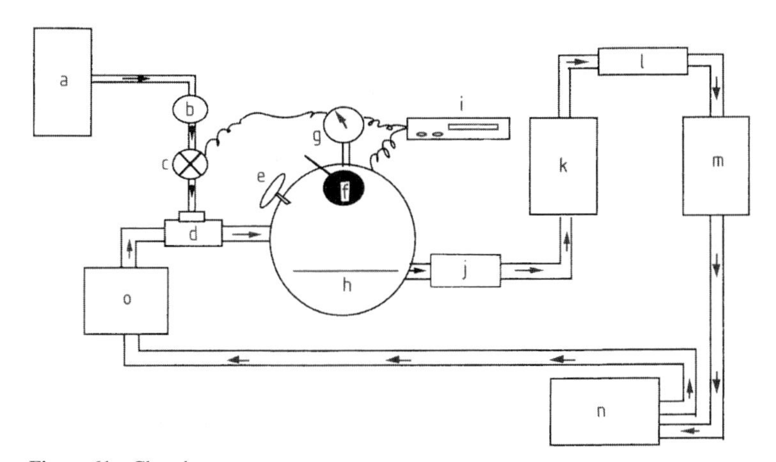

Figure 61. Closed exposure system
a) Oxygen cylinder; b) Metering valve; c) Solenoid valve; d) Mixing chamber; e) Thermometer; f) Oxygen sensor; g) Pressure gauge; h) Exposure chamber; i) Oxygen monitor; j) Injection port; k) Condenser; l) Flow meter; m) Carbon dioxide absorber; n) Gas chromatograph; o) Pump

months in rats, 12 months in dogs). At the start of the study, rodents should be 6–8 weeks old and dogs 4–6 months old. The animal numbers enrolled are between 10 and20 rats and 6 and 8 dogs per sex per dose group. Ideally, the lowest dose of the chemical should not induce toxic effects, the intermediate dose should induce slight toxicity, and the high dose should induce clear signs of toxicity without causing death in more than 10% of the animals.

In both subchronic and chronic toxicity studies (see below), the test compound is often incorporated into the diet or added to the drinking water. Food consumption varies from weanling to maturity, with younger animals consuming more food on a bodyweight basis. Therefore, it is necessary to predict the changes in body weight and food consumption on a weekly basis and adjust the concentration of the test compound in the diet in order to ensure constant dosing throughout the study. Compounds not stable in diet or water or not accepted by the animals may also be applied by gavage. Application by gavage directly into the stomach ensures constant dosing, but gavage studies require skilled personnel, and gavage-related trauma may reduce survival in all groups. Oral application with the feed is usually performed on a 7 days per week basis, while a 5 days per week scheme is frequently used for gavage administration, skin application, and inhalation studies.

Studies on the chronic toxicity of chemicals are usually performed for at least six months in rodents and 12 months in dogs, the chronic toxicity studies can be combined with a carcinogenicity study. Dose levels are usually selected on the basis of the results of studies on acute and subacute toxicity. The highest dose applied should be toxic, i.e., suppress body weight up to 10% (maximum tolerated dose, MTD). The two other dose-levels are usually 1/4 and 1/8 of the MTD. Xenobiotics showing no adverse effects in the short-term studies are usually tested at doses which are 100–200 times higher than the expected human exposure.

In both subchronic and chronic studies, cage-side examination and clinical chemistry are performed routinely during the test period. After termination of the study, much emphasis is placed on the histopathological evaluation of treatment-induced adverse effects.

Cage-side observations during the study

1) Body weight, food and water consumption
2) Skin and fur, eyes, mucous membranes
3) Respiration and blood circulation
4) Motor activity and behavioral pattern

Clinical chemistry during the study

1) *Blood:* erythrocyte, leukocyte, and differential leukocyte counts; hemoglobin concentration; hematocrit, platelet, and reticulocyte counts; electrolytes; inorganic phosphorus and alkaline phosphatase; glucose, protein, albumin, creatinine, urea, lipids, enzymes
2) *Urine:* volume and coloration/turbidity, osmolality and pH, glucose and protein, urine enzymes and cytology

Toxicologic pathology after termination of the study and in animals dying during the study

1) Organ weights and macroscopic evaluation
2) Histopathological examination of brain, liver, kidney, spleen, testes, and every organ with macroscopic changes

4.4. Ophtalmic Toxicity

The majority of injuries to the eyes by direct contact with a chemical occur with substances which are handled in an uncontrolled manner, e.g,. by children in the home; and this type of injury is easily prevented at the occupational setting by using simple protective procedures. However, many irritant gases and vapors may also produce ophtalmic toxicity, and these toxic effects are of practical importance in occupational medicine.

The conventional in vivo eye irritation test in the rabbit was formalized by Draize some 50 years ago and still remains the only fully validated method to assess ophtalmic toxicity. Increasing criticism primarily based on the discomfort and the deliberate injury caused to the animals led to the development of several in vivo and in vitro alternative methods, which will be described at the end of this chapter.

In the conventional in vivo test the chemical (0.1 ml of liquid or 100 mg of solid chemical) is instilled into one eye of each of the test rabbits, the contralateral eye serving as control. Eyes are then examined periodically (usually after 1, 24, 48, 72 h and 7 dd) and the ocular lesions are scored essentially according to Draize et al.:

Cornea

A)	Opacity	
	No opacity	0
	Scattered or diffuse area, iris details clearly visible	1
	Easily discernible translucent areas, iris details slightly obscured	2
	Opalescent areas, iris details not visible, pupil size barely discernible	3
	Opaque, iris invisible	4
B)	Area of cornea involved	
	One-quarter or less, but not zero	1
	Greater than one-quarter, but less than half	2
	Greater than half, but less than three-quarters	3
	Greater than three-quarters, up to whole area	4

Corneal score = (A) × (B) × 5 (maximum total score = 80)

Iris

A)	Normal	0
	Folds above normal, congestion, swelling, circumcorneal injection (any or all), iris still reacting to light	1
	No reaction to light, hemorrhage, gross destruction	2

Iris score = (A) × 5 (maximum total score = 10)

Conjunctivae

A)	Vessels normal	0
	Vessels definitely injected, above normal	1
	Diffuse, deep crimson red, individual vessels not readily discernible	2
	Diffuse beefy red	3
B)	No chemosis (swelling)	0
	Any swelling above normal (includes nictitating membrane)	2
	Obvious swelling with partial eversion of lids	2
	Swelling with lids about half closed	3
	Swelling with lids about half to completely closed	4
C)	No discharge	0
	Any amount of discharge different from normal	1
	Discharge with moistening of lids and hairs adjacent to lids	2
	Discharge with considerable moistening around eyes	3

Conjunctival score = [(A) + (B) + (C)] × 2 (maximum total score = 20)

Total maximum score (cornea + iris + conjunctiva) = 110

There is increasing evidence that a volume of 0.01 ml of liquid xenobiotics is as sensitive as the conventionally used 0.1 ml and is probably more appropiate for comparison with human exposure situations. Ocular toxicity testing for exposure to gases, vapors, and aerosols is carried out in appropiate exposure chambers.

A number of in vivo and in vitro alternatives to the conventional eye irritation test have been suggested. The in vivo alternatives aim at reducing discomfort of the animals by employing lower doses of the test material and increasing the sensitivity of the test by using noninvasive objective measurements. Among them, the assessment of corneal thickness and of the intravascular pressure seem to be sensitive parameters to identify mild to moderately irritant chemicals. In spite of the large number of suggested in vitro tests, currently no single in vitro test has proved effective in predicting the eye irritation. Therefore, in vitro tests can not replace the rabbit eye test yet. However, in vitro assays are useful as screens for product development to reduce the number of tests performed in animals later. A number of these tests are presented in [157].

4.5. Sensitization Testing

Chemicals that have the potential to elicit allergic reactions are continously introduced into the human environment. Therefore, allergic reactions of the skin are becoming an increasingly important problem, especially in the workplace. Allergic contact dermatitis is one of the most common occupational diseases and may become debilitating unless the causative agent is identified and exposure stopped. While irritant dermatitis is generally produced by direct interaction of the chemical with skin constituents, allergic dermatitis is the result of a systemic immune reaction which in turn induces effects in the skin. An important characteristic of allergic reactions that must be taken into account when testing for allergenic potential, is that allergic responses usually have a biphasic course. The induction period between initial contact with the causative agent and the development of skin sensitivity may be as short as two days for strong sensitizers such as poison ivy extract, or may require several years for a weak sensitizer such as chromate; for most of the chemical compounds with allergenic potential the induction period usually takes from 10 to 21 d. After this initial development of sensitivity to a certain allergenic chemical, the time between reexposure to this agent and the occurrence of clinical allergic symptoms is generally between 12 and 48 h; in animal testing, this period is called the challenge phase.

Table 28. Guinea pig sensitization tests

Test	Induction: route/number of applications	Challenge: route/number of applications
Draize	intradermal/10	intradermal/1
Open epicutaneous	epidermal open/20	epidermal open/1
Buehler	epidermal occlusive/3	epidermal occlusive/1
FCA*	intradermal in FCA*/3	epidermal open/1
Split adjuvants	epidermal occlusive/4 + FCA* intradermal/1	epidermal occlusive/1
Optimization	intradermal + FCA*/10	intradermal/1 epidermal occlusive/1
Maximization	intradermal + FCA*/1 epidermal occlusive/1	epidermal occlusive/1

* FCA = Freund's complete adjuvant.

The general objectives are to determine whether there are indications from human experience of skin allergy or respiratory hypersensitivity following exposure to the agent and whether the agent has skin sensitization potential based on tests in animals. There are two methods currently described in EU Annex V and OECD guidelines for skin sensitization in animals: the guinea pig maximisation test (GPMT) and the Buehler test. The GPMT is an adjuvant-type test in which the allergic state (sensitization) is potentiated by the use of Freund's Complete Adjuvant (FCA). The Buehler test is a non-adjuvant method involving topical application for the induction phase rather than the intradermal injections used in the GPMT (Table 28; for reviews see references [158, 159]. Although they differ by route and frequency of treatment, they all utilize the guinea pig as test species. In general, for the induction phase the chemical is administered to the shaved skin intradermally, epicutaneously, or by both routes several times over a period of two to four weeks. Freund's complete adjuvant (FCA, a mixture of heat-killed *Mycobacterium tuberculosis*, paraffin oil, and mannide monooleate) is often included to increase the immunological response. During the challenge phase, a nonirritating concentration of the chemical is applied. The concentration of the test chemical and the application route (epi- or intradermal) are often different between the two phases. Sensitization is assessed by examining the skin reactions (edema, erythema) following the challenge phase and comparing them with any skin reactions observed immediately after the induction phase; the latter reactions are considered to result from direct irritating (toxic) properties of the test chemical. Hence, the difference between the symptoms observed after the induction and after the challenge phase is attributed to the allergenic effects of the chemical.

The guinea pig maximization test is the most widely used and is considered to be very sensitive. The first part of the induction phase includes simultaneous injection of FCA alone, the test compound in saline, and the test compound in FCA into three different areas in close proximity to each other. The second part of the induction phase 7 d later employs epicutaneous application of the chemical on a filter paper, which is occluded and left in place for 48 h. The challenge phase is conducted epicutaneously for 24 h, two weeks after the induction phase. The maximization test is very sensitive and may produce false positive results. The original procedure (injection of the test compound) does not allow testing of final product formulations. Therefore, a modified procedure has been developed. In the first week, the FCA is injected four times and the test product formulation is administered epidermally, as in the second induction week.

Both the GPMT and the Buehler test have demonstrated the ability to detect chemicals with moderate to strong sensitization potential, as well as those with relatively weak sensitization potential. These guinea pig methods provide information on skin responses, which are evaluated for each animal after several applications of the substance, and on the percentage of animals sensitized.

The murine local lymph node assay (LLNA) is another accepted method for measuring skin sensitization potential. It has been validated internationally and has been shown to have clear animal welfare and scientific advantages compared with guinea pig tests. In June 2001, the OECD recommended that the LLNA should be adopted as a stand-alone test as an addition to the existing guinea pig test methods.

Respiratory hypersensitivity is a term that is used to describe asthma and other related respiratory conditions, irrespective of the mechanism by which they are caused. When directly considering human data in this document, the clinical diagnostic terms asthma, rhinitis, and alveolitis have been retained.

There are currently no internationally recognised test methods to predict the ability of chemicals to cause respiratory hypersensitivity. Potentially useful test methods based on allergic mechanisms are the subject of research and development. However, there are currently no test methods under development which are designed specifically to identify chemicals that cause respiratory hypersensitivity by nonimmunological mechanisms.

4.6. Phototoxicity and Photosensitization Testing

The biologically active spectrum of light can be divided into UV (220–400 nm) and visible light (400–760 nm). The UV spectrum is further divided into UVA (315–400 nm), UVB (280–315 nm), and UVC (220–280 nm); the last-named is absorbed in the stratosphere and does not reach the surface of the earth. The primary source of toxic effects on the skin is UVB, although UVA may also play a critical role in some reactions.

Xenobiotics localized within the skin may be activated by UVB and induce phototoxicity and/or photosensitization (photoallergy). Photoallergy is similar, both mechanistically and clinically, to allergic contact dermatitis, the only difference being that the chemical must react with light to become allergenic. Photoallergic reactions are not necessarily dose-dependent and show great variability between individuals. In analogy, phototoxicity may be compared with irritant dermatitis. Many phototoxic reactions may be caused by the formation of free radicals followed by lipid peroxidation and localized inflammation. In addition to phototoxicity and photoallergy, light-induced activation of chemicals may cause depigmentation, induction of an endogenous photosensitizer or of a disease characterized by photosensitization such as lupus erythematodes or pellagra. However, these reactions are comparatively rare. The tests to identify photoallergenic and phototoxic chemicals are performed in analogy to the tests for skin sensitization and irritation, respectively.

Photoallergenic potential is evaluated by repeated application of the test compound on the skin of guinea pigs and exposure of the treated area with UV light after every application; the UV treatment should cause a very slight erythema. Several days after this induction phase, the challenge phase is conducted by treatment with a low dose of the test compound together with UV light. Phototoxic reactions can usually by observed after the first exposure to the test compound together with UV light; in addition to guinea pigs, mice and rabbits are also utilized in tests for phototoxicity.

4.7. Reproductive and Developmental Toxicity Tests

The general objectives of reproductive and developmental toxicity testing are to establish whether exposure to the chemical may be associated with adverse effects on reproductive function or capacity, and whether administration of the substance to males and/or females prior to conception and during pregnancy and lactation causes adverse effects on reproductive function or capacity. Another focus of these studies are induction of nonheritable adverse effects in the progeny and whether the pregnant female is potentially more susceptible to general toxicity.

Reproductive and developmental toxicity is a very broad term including any adverse effect on any of the following aspects:

1) Male or female sexual structure and function (fertility)
2) Development of the new organism through the period of major organ formation, organogenesis (embryotoxicity and teratogenicity)
3) Development of the new organism during the peri- and postnatal periods

The field of reproductive toxicology has become increasingly important with the recognition that viral and bacterial infections and xenobiotic chemicals can produce severe and irreversible defects in the offspring. The first reports on malformations due to rubella virus infections, ionizing radiation, hormones, dietary deficiencies, and chemicals appeared in the 1930s and 1940s, but the potential impact of reproductive toxicity on public health was only recognized more than two decades later. In 1960, a large increase in newborns with specific limb malformations, which are rarely seen otherwise, was recorded in Germany and in other parts of the world. One year later, the sedative/hypnotic drug thalidomide was recognized as the causative agent. The thalidomide epidemic resulted in

10 000 malformed children and subsided after the drug was withdrawn from the market at the end of 1961. One important consequence of the thalidomide disaster was the introduction of requirements for the testing of potential new drugs for reproductive toxicity. All test batteries required include detailed tests for reproductive toxicity to prevent a repetition of the thalidomide disaster.

The first specific reproductive toxicity test to be conducted is usually the two-generation study. which should be initiated after the rat 90-d subchronic repeated-exposure study, since the results obtained may provide information necessary for selecting dose levels for the two-generation study. Additionally, repeated-exposure studies that can provide information relevant to reproductive toxicity should be used in the design of the two-generation study. For example, the observation of neurological effects may indicate the need to evaluate developmental neurotoxicity.

The first developmental toxicity study is normally performed after completion of the two-generation study. The design of the developmental toxicity study should use all information derived from the repeated-exposure and two-generation studies, in particular dose–response relationships and information on maternal toxicity. The preferred species for the two-generation study is the rat; the necessity of a developmental toxicity study in the rabbit is dependent on the outcome of the first study.

4.7.1. Fertility and General Reproductive Performance

Segment I experiments are usually conducted in rats (20 animals of each sex per dose) with three doses of the test chemical, most often administered with diet. The treatment must not cause general systemic toxicity in the parental organism; therefore, in dose selection the dose levels are chosen according to observations in studies of subacute and subchronic toxicity, which are usually performed before testing for reproductive toxicity. Young adult male rats are treated for 60–80 d prior to mating to cover a whole period of spermatogenesis. Female rats are pretreated for 14 d to cover three estrous cycles. Treatment of both sexes is continued during the mating period and that of females throughout pregnancy. Half of the females are sacrificed just before term, and the numbers of resorbed and dead fetuses as well as structural abnormalities in the developed fetuses are assessed. In the United States, pregnancy is interrupted midterm in half of the females. Treatment of the remaining females is continued through parturition and lactation until weaning of the newborns, usually 21 d after birth. The young animals (F_1 generation) are reared without receiving the test compound until sexual maturity, when their fertility is assessed. During the rearing period, the development of the young animals is monitored with cage-side carefully clinical observations. If an adverse effect on fertility, pregnancy, or development of the offspring is observed, it is necessary to evaluate whether the effect is due to toxicity to the male or female reproductive system or both. This information can be obtained by separate mating of treated males with untreated females and vice versa.

Furthermore, evaluation of toxic effects on the male and female reproductive systems with specific test systems may be required. The effect of the test compound on male reproductive performance may be evaluated by monitoring mating behavior (e.g., frequency of copulation). Structural and functional impairment of the male reproductive organs is assessed by conducting gross pathology and histology of the testes and sperm analysis (viability, motility, and morphology). Histological examination of the ovaries plays an important role in assessment of toxic effects on the female reproductive system.

4.7.2. Embryotoxicity and Teratogenicity

Segment II studies assess adverse effects during the period of organogenesis. Xenobiotics interfering with the developing organism during this extremely sensitive period may cause severe and irreversible structural malformations. These studies are carried out in two species, usually rats (20 per dose) and rabbits (10 per dose); in most cases, two dose levels and an untreated control group are included. Pregnant animals are tested during the period of organogenesis: days 6 to 15 for rats and 6 to 18 for rabbits. The fetuses are delivered by cesarean section one day prior to the estimated time of delivery: day 21 for rats and day 31 for rabbits. The main reason for avoiding natural delivery is to prevent loss of deformed or dead fetuses by cannibalism, which happens in rodents and rabbits. The uterus of the maternal animal is excised, weighed, and examined for implantation sites and resorbed fetuses. The pups are weighed, and one-half of each litter is usually examined for skeletal defects and the remaining one-half for soft-tissue defects.

4.7.3. Peri- and Postnatal Toxicity

For segment III studies, treatment of pregnant rats with three dose levels (10–12 animals per dose) begins on day 16 of gestation and is carried on through delivery and lactation, until weaning of the offsprings, normally on day 21 postpartum. Treatment during parturition and lactation is also performed in segment I studies; however, segment III studies may offer additional information since higher doses can be used than in segment I. The peri- and postnatal segment evaluates effects on birth weight and survival as well as development of the offspring in the postnatal period. However, extrapolation of results of segment III studies from the rat to the human situation should be performed with care and consideration of the specific circumstances in each species. In contrast to the human situation, where in addition to the mother the social environment takes care of the newborns, young rats depend completely on the functional integrity of the maternal organism. Furthermore, the organism of newborn rats is significantly less mature by the time of delivery than that of newborn humans. Hence, treatments that affect the maternal organism simply by causing sedation or fatigue may significantly impair the development of newborn rats in the first 21 days of life. Such effects should not be automatically interpreted as relevant for postnatal toxicity in humans.

4.7.4. Multigeneration Studies

Multigeneration studies assess the cumulative effects of continuous application of the test chemical on reproduction and development during two or three generations. This application mode is relevant for long-term exposure to chemicals in the environment, such as pesticide residues in food or contamination of drinking water with agricultural chemicals or nonbiodegradable solvents.

The two-generation study is a general test which allows evaluation of the effects of the test substance on the complete reproductive cycle including libido, fertility, development of the conceptus, parturition, postnatal effects in both dams (lactation) and offspring, and the reproductive capacity of the offspring. The two-generation study is preferable to the one-generation study because the latter has some limitation regarding assessment of post-weaning development, maturation, and reproductive capacity of the offspring. Thus, some adverse effects such as oestrogenic- or antiandrogenic-mediated alterations in testicular development may not be detected. The two-generation study provides a more extensive evaluation of the effects on reproduction because the exposure regime covers the entire reproductive cycle, permitting an evaluation of the reproductive capabilities of offspring that have been exposed from conception to sexual maturity. The prenatal developmental toxicity study only provides a focused evaluation of the potential effects on prenatal development.

Three dose levels are usually given to groups of 25 female and 25 male rats shortly after weaning at days 30 to 40 of age. In the multigeneration study, these rats are referred to as the F_0 generation. The F_0 generation is treated throughout breeding, which occurs at about 140 d of age, and the female animals also during pregnancy and lactation. Hence, the offspring (F_1 generation) has been exposed to the test compound in utero, via the maternal milk, and thereafter in the diet. In many protocols, the F_1 generation is standardized to include certain numbers of animals, e.g., eight animals per litter. In analogy to the F_0 generation, the F_1 generation is bred at about 140 d of age to produce the F_2 generation. In some of the parents (F_0 and F_1 generations), gross necropsy and histopathology is conducted with greatest emphasis on the reproductive organs. In addition, necropsy and histopathology are carried out in all animals dying during the study.

The percentage of F_0 and F_1 females that become pregnant, the number of pregnancies carried to full term, the litter size, and number of resorptions, stillborns, and live births are recorded. Viability counts and pup weights are recorded at birth, and at days 4, 7, 14, 21 and 28 of age. With these data, the following parameters are calculated for assessment of the long-term reproductive toxicity of the test compound:

$$\text{Fertility index (\%)} = \frac{\text{Number of pregnancies}}{\text{Number of matings}} \times 100$$

$$\text{Gestation index (\%)} = \frac{\text{Number of litters}}{\text{Number of bred females}} \times 100$$

$$\text{Birth index (\%)} = \frac{\text{Number of pregnancies resulting in live off spring}}{\text{Number of pregnancies}} \times 100$$

$$\text{Viability index (\%)} = \frac{\text{Number of animals alive at day 4 after birth}}{\text{Number of new borns}} \times 100$$

$$\text{Lactation index (\%)} = \frac{\text{Number of animals alive at day 28 after birth}}{\text{Number of animals alive at day 4 after birth}} \times 100$$

4.7.5. The Role of Maternal Toxicity in Teratogenesis

If an agent with selective developmental toxicity is administered throughout the organogenesis period (days 6 to 15 in the rat), identifying the most sensitive target organs becomes difficult. In addition, teratogenic effects may be masked by embryolethality with repeated dosing during the organogenesis period. Developmental toxicity in the form of increased resorption and decreased fetal body weight is generally accepted to occur at maternally toxic dose levels. The role of maternal toxicity in causing congenital malformations, however, is not clear. Doses causing maternal toxicity, as indicated by reduced maternal body weight, clinical signs of toxicity, or death, commonly cause reduction in fetal body weight, increased resorption, and rarely, fetal deaths. Three patterns of association between maternal toxicity and malformations can be observed: (1) for some compounds, maternal toxicity is not associated with malformations; (2) for others, maternal toxicity is associated with a diverse pattern of malformations, often including cleft palate; and (3) the maternal toxicity of still others is associated with a characteristic pattern of malformations.

Compounds in the second category are the most difficult to classify in terms of teratogenic potential. Cleft palate is the principal malformation resulting from food and water deprivation during pregnancy in mice; however, cleft palate is also a malformation specifically induced in mice by a number of teratogens, most notably the glucocorticoids, without apparent maternal toxicity. Complete determinations of food and water consumption, maternal body weights, and impairment of the maternal organism are necessary to distinguish between cleft palate caused by the teratogenic effect of a chemical on the embryo and that resulting from systemic maternal toxicity, which secondarily affects embryonic development. The association of maternal toxicity with major malformations, such as exencephaly and open eyes, is not generally accepted, although most investigations agree that maternal toxicity can cause minor structural abnormalities such as variants in the ribs.

4.7.6. In Vitro Tests for Developmental Toxicity

Models to elucidate the mechanism of embryogenesis have been under development for several decades, and therefore developmental toxicology is a fields in which alternative methods to animal experimentation are available. However, because of the complicated, multistep nature of the development of a new life, none of the in vitro systems presently available can replace the animal tests. In vitro tests rather serve for screening purposes, i.e., to preclude the extensive traditional whole-animal test protocol for compounds with marked toxicity on reproduction and development. The existing alternative test systems fall into six groups: lower organisms, cell-culture systems, organ-culture systems, whole-embryo cultures, embryos, and others (Table 29). Since none of the in vitro methods is sufficiently validated for a set of compounds for which the effects on humans or animals are known and the field is much too extensive to be comprehensively reviewed here, the reader is referred to two comprehensive reviews of this field [160].

Table 29. In vitro test systems for developmental toxicity

Group	Test system/organisms	End points monitored
Lower organisms and small animals	sea urchins	growth
	drosophila	
	trout, medaka (fish species)	
	plania	
	brine shrimp	
	animal virus	
Cell culture	pregnant mouse and chick lens epithelial cell	protein synthesis
	avian neural crest	cell differentiation
	neuroblastoma	cell differentiation
Organ culture	frog limb	regeneration
	mouse embryo limb bud	morphological and biochemical differentiation, toxicity
	metanephric kidney organ culture from day 11 mouse embryos	morphological and biochemical differentiation
Whole embryo cultures	chick embryo	embryotoxicity, malformations
	frog embryo teratogenesis assay	lethality, no observed-effect-level, development stage, attained growth, motility, pigmentation, gross anatomical malformations
	rat embryo culture (postimplantation embryo)	viability, growth and macromolecular content, gross structural and histological abnormalities

Table 30. Basic procedures of rodent carcinogenicity bioassays

Species	rat (Fischer 344, Sprague – Dawley, Wistar) mouse (B6C3F1, CD)
Age of animals at the beginning	4 to 6 weeks (shortly after weaning)
Number of animals	50 per sex per dose for carcinogenicity 10 – 20 for additional studies during the course of experiment
Dose	at least three doses and vehicle control maximum tolerated dose intermediate dose nontoxic dose
Duration	24 months
Application	gavage, in feed, drinking water, inhalation (only if absolutely necessary)
Toxicologic pathology	all animals: gross necropsy weight of all important organs histopathology of all tissues (ca. 40) and all tumors and preneoplastic lesions by two independent pathologists

4.8. Bioassays to Determine the Carcinogenicity of Chemicals in Rodents

Despite the many available short-term in vivo and in vitro tests to determine the genotoxic and carcinogenic potential of chemicals and the vast amount of literature on the subject, the lifelong carcinogenicity bioassay remains the main instrument for reliable evaluation of the carcinogenic properties of a xenobiotic. The principal guidelines for performing bioassays were established some 25 a ago by the U.S. National Cancer Institute and have essentially been adopted with slight alterations by all regulatory authorities (Table 30).

A number of factors may interfere with the analysis and interpretation of data from animal carcinogenicity studies. A variety of statistical techniques has been developed to adjust for confounding factors and to estimate confidence intervals and significance of results. Significance tests are used to assess neoplastic response in treated groups as compared to control groups or historical controls (cancer incidence in the identical strain and species observed in control groups for other cancer bioassays under identical housing conditions in the same facility). To estimate the absolute cancer risk posed by a specific chemical, background or spontaneous cancer incidences (induction of neoplasms not related to the administration of the test chemical) must be well defined. In general, high background incidences of cancer such as liver cancer observed in specific strains of mice requires larger number of animals in the treatment groups to detect increases in cancer incidence induced by the administration of the test chemical and to obtain statistically significant results. The demonstration of a dose–response curve for the cancer incidence in groups of animals treated with different doses of the carcinogen will increase the confidence in positive results of an animal cancer bioassay. The same holds for identical results observed with cancer as an endpoint in an independent study.

However, despite the importance of animal cancer bioassays for characterizing chemical carcinogens, this approach has been criticized recently (see also Section 5.2.4.4). Due to the influence of rodent carcinogenicity assay on the development of new chemicals and pharmaceutical drugs, the pros and cons of this type of assay are discussed in depth in the following.

Practically all chemicals identified as human carcinogens produce tumors in the rodent bioassay. Hence, the test has a very good predictive value, and every chemical exerting carcinogenicity in rodents should be handled as a potential human carcinogen. Considering the many new chemicals developed each year, two major disadvantages of the life-long assay are its high cost and long duration. A two-year gavage study in only one species amounts to approximately €10^6 and takes 3–4 years or longer for complete evaluation. For optimized evaluation of carcinogenic properties, it is important to use animal species that are closest to humans with regard to biotransformation and toxicokinetics of the test compounds. However, for practical and financial reasons, only rodents can be used. Long-term carcinogenic bioassays in dogs or primates, for example, require seven to ten years for completion and are much more costly.

The application of the maximum tolerated dose (MTD) in rodent bioassays has been the subject of much controversy [161–167], but the experimental design limitations of in vivo studies make the application of high doses necessary. For example, if a specific dose of a chemical causes a 0.5% increase in human cancer incidence, this would result in several hundred thousands of additional cancer cases in a country such as Germany each year and would thus definitely pose an unacceptable risk. However, the identification of this 0.5% increase in cancer incidence with statistical confidence in the rodent bioassay would require a minimum of 1000 animals, provided the incidence of spontaneous tumors is zero. Therefore, there seems to be general agreement that the use of the MTD, although not an optimum solution, is necessary for risk assessment. According to the U.S. National Cancer Institute, the MTD is defined as "the highest dose that can be predicted not to alter the animals' normal longevity from effects other than carcinogenicity". In practical terms "MTD is the dose which, in the subchronic three-months toxicity study causes not more than a 10% weight decrement as compared to the control groups and does not produce mortality, clinical signs of toxicity, or pathological lesions other than those which may be related to a neoplastic response that would be predicted to shorten an animal's life span". As stated above, the MTD is determined in the preliminary three-month studies on subchronic toxicity, where it fulfills the above requirements. Ideally, this is exactly what should also happen in the 24-month carcinogenicity study. However, due to cumulation of toxic effects and/or alterations in toxicokinetics of the xenobiotics during the study, for example, induction of toxification or detoxification pathways by application of the xenobiotic in high doses, the MTD dose group often shows reduced survival rates in the life-long bioassay. This may invalidate the study, that is, make it inadequate for evaluation of the carcinogenic potential. Indeed, this is not a rare event in carcinogenicity studies. The opposite effect may also occur: due to toxicokinetic differences between the three- and 24-month studies, the MTD chosen may turn out too low in the long-term bioassay. In spite of all these problems and because of the absence of a satisfactory alternative solution, the use of the MTD is currently the only method to compensate for the fact that in relation to the human population exposed to potential carcinogens, the numbers of rodents used in the carcinogenicity bioassay are extremely low. The legitimate argument against the MTD is that any chemical given at a sufficiently high dose level will induce adverse effects. This understanding, which is beyond dispute in toxicology, has been tentatively generalized by several scientists in recent years by the notion that carcinogenic effects obtained at the MTD may exclusively result f rom target-organ toxicity, and the increased cell proliferation may contribute to tumor formation by increasing the rate of spontaneous mutations, since DNA replication does not take place with 100% fidelity. Furthermore, during increased cell turnover, the time available to repair DNA damage is reduced, so that an increased number of damaged DNA sites may be converted to heritable mutations. Although this may be the mechanism underlying the carcinogenic effects of some nongenotoxic chemicals, it can not be generalized to every tumor observed at the MTD. Toxicity and cell proliferation do not necessarily result in tumor formation. Table 31 summarizes the important differences between rodent bioassays and human exposure to carcinogens.

Table 31. Some important differences between carcinogenicity tests in rodents and human exposure to potential carcinogens

Rodent carcinogenicity test	Human exposure
High doses	(usually) low doses
Continuous exposure	(often) infrequent or not regular exposure
Single compound, no interactions	simultaneous exposure to several carcinogenic chemicals, interactions probable
Homogeneous population	heterogeneous population

In addition to these problems, which are inherent in the bioassay procedure, the evaluation of the toxicological pathology has repeatedly become an issue of debate, since differences in evaluations between pathologists are frequent. This does not necessarily indicate incompetence of one of the pathologists. The different evaluations may be the result of difference in terminology. Also, sometimes evaluations are conducted years apart, and in the interim period understanding of the pathogenesis of lesions may have changed. Thus, even the same pathologist may not come to the same conclusion when reevaluating tissue slices several years after the first examination.

Due to the uncertainties of rodent bioassays and the extremely high costs and personnel requirements, a multitude of short-term tests has been developed in recent years. These tests aim to predict the carcinogenic potential of chemicals. Most of these in vitro tests are based on damage to the genetic material (genotoxicity) by the chemical or its metabolites. Genotoxicity is without doubt the field in toxicology with the best established and validated in vivo and in vitro short-term tests.

4.9. *In Vitro* and *In Vivo* Short-term Tests for Genotoxicity

Genetic toxicology a comparatively new field of research that has rapidly grown since the 1960s, deals with mutagenicity and genotoxicity.

Mutagenicity is the induction of permanent transmissible changes in the genetic material of cells or organisms. Changes may involve a single gene or gene segment, a block of genes, or whole chromosomes. Effects on whole chromosomes may be structural and/or numerical.

Genotoxicity is a broader term and refers to potentially harmful effects on genetic material which may not be associated with mutagenicity. Thus, tests for genotoxicity include systems which give an indication of damage to DNA (no direct evidence of mutation). End points determined here are unscheduled DNA synthesis (UDS), sister-chromatid exchange (SCE), DNA strand breaks, formation of DNA adducts and mitotic recombination.

Evidence has increasingly accumulated that many carcinogens are also mutagenic, and a large number of short-term in vitro and in vivo tests were developed as predictive tools. Most of these tests are well validated and aim to assess

Table 32. OECD guidelines on genetic toxicology testing and guidance on the selection and application of assays

No.	Title	Original adoption	Updated
471	Bacterial Reverse Mutation Test	26 May 1983	21 July 1997
472	Genetic Toxicology: *Escherichia coli*, Reverse Assay	26 May 1983	21 July 1997*
473	In Vitro Mammalian Chromosome Aberration Test	26 May 1983	21 July 1997
474	Mammalian Erythrocyte Micronucleus Test	26 May 1983	21 July 1997
475	Mammalian Bone Marrow Chromosome Aberration Test	4 April 1984	21 July 1997
476	In Vitro Mammalian Cell Gene Mutation Test	4 April 1984	21 July 1997
477	Genetic Toxicology: Sex-Linked Recessive Lethal Test in *Drosophilia melanogaster*	4 April 1984	
478	Genetic Toxicology: Rodent dominant Lethal Test	4 April 1984	
479	Genetic Toxicology: In Vitro Sister Chromatid Exchange assay in Mammalian Cells	23 Oct. 1986	
480	Genetic Toxicology: *Saccharomyces cerevisiae*, Gene Mutation Assay	23 Oct. 1986	
481	Genetic Toxicology: *Saccharomyces cerevisiae*, Mitotic Recombination Assay	23 Oct. 1986	
482	Genetic Toxicology: DNA Damage and Repair, Unscheduled DNA Synthesis in Mammalian Cells In Vitro	23 Oct. 1986	
483	Mammalian Spermatogonial Chromosome Aberration Test	23 Oct. 1986	21 July 1997
484	Genetic Toxicology: Mouse Spot Test	23 Oct. 1986	

*Date of deletion (method merged with TG 471).

genotoxic properties of chemicals. Today, the majority of potential carcinogens are first identified as mutagens or chromosome-damaging agents in short-term tests and subsequently as carcinogens in the rodent carcinogenicity bioassay. The numerous in vivo and in vitro assays can be categorized into two major groups:

1) Short-term tests detecting gene mutations
2) Short-term tests detecting structural and/or numerical chromosomal aberrations

A number of these in vitro test procedures have gained regulatory acceptance for toxicology testing (Table 32).

In vitro tests to detect gene mutations can be categorized into two groups: microbial and mammalian cell assays. An important step in the history of modern genetic toxicology was the development of genetically precisely defined strains of bacteria carrying mutations in particular genes coding for enzymes involved in the biosynthesis of amino acids. Among the numerous tests, the most widely used and best validated is the assay in *Salmonella typhimurium* developed by AMES et al.

4.9.1. Microbial Tests for Mutagenicity

4.9.1.1. The Ames Test for Bacterial Mutagenicity

The most common method to detect mutations in microorganisms is selecting for reversions in strains that have a specific nutritional (i.e., amino acid) requirement differing from wild-type members of the species; the tester strains are "auxotroph" for this particular nutrient. The *Salmonella typhimurium* mutant strains developed by AMES can not synthesize histidine, because each strain carries one of a number of mutations in in the operon (group of genes) coding for histidine biosynthesis [168, 169]. The result of this mutation is that the tester strains can not grow and form colonies in histidine-free medium. The mutation may revert to the wild-type sequence or a functionally equal sequence either spontaneously (a rare event) or by exposure of the tester strains to genotoxic compounds. The revertant colonies are, like the wild-type bacteria, capable of synthesizing histidine and form colonies in histidine-free medium. For the common tester strains, the DNA sequence at the site of the original mutation in the relevant histidine gene has been determined. According to the type of mutation leading to the inability to synthesize histidine, the strains can be categorized into two groups: base-substitution and the frameshift strains. The difference between these two categories can be illustrated with the following sentence, in which each letter represents a DNA base, each word a triplet coding for an amino acid, and the whole sentence a gene coding for a protein (i.e. enzyme of histidine biosynthesis). The correct sentence represents the gene in the wild-type strain.

THE NUN SAW OUR CAT EAT THE RAT	original sentence (wild-type)
Base-pair substitution:	
THE **S**UN SAW OUR CAT EAT THE RAT	missense mutation coding for a wrong amino acid
THE N**S**N SAW OUR CAT EAT THE RAT	nonsense mutation resulting in interruption of gene transcription
Frameshift mutation:	
THE NUN SA**S** WOU RCA TEA TTH ERA T	+ 1-frameshift mutation

This example illustrates that a base substitution can be reverted by another base substitution and in analogy a frameshift mutation can be reverted by another frameshift mutation. Hence, the Ames test not only provides information on the genotoxic potential of the test compound but also on the nature of the DNA damage.

Genetic Makeup of the *Salmonella typhimurium* Strains Used in the Ames Test. *Histidine mutations of the tester strains.* The base-pair substitution strains can be categorized into two families.

1) The *Salmonella typhimurium* strains TA100 and TA1535 carry the sequence CCC (for leucine) instead of the wild-type sequence CTC (for proline). These missense mutations may be efficiently reverted by mutagens with alkylating properties.
2) The second group of base-pair substitution strains carry a nonsense mutation (TAA instead of CAA). These strains (TA2638, TA100) may detect mutations induced by radicals or oxidizing agents such as hydrogen peroxide and reactive oxygen metabolites.

The commonly used frameshift strains TA98 and TA1538 carry a +1 frameshift mutation near

Table 33. Genetic makeup of commonly used Salmonella typhimurium tester strains

Strain	Histidine mutation		Additional genetic alteration
	Location	Route	
TA100	his G46	CCC instead of CTC base substitution results in proline instead of leucine	*uvrB* *rfa* *pkM*101
TA1535	his G46	CCC instead of CTC base substitution results in proline instead of leucine	*uvrB* *rfa*
TA2638	his G428 his G8476	ATT nonsense mutation results in interruption of transcription	*rfa* *pkM*101
TA102	his G428	multiple copies of plasmid pAQ1 carry the revertible nonsense mutation ATT	*rfa* *pkM*101
TA98	his D3052	-1-frameshift mutation with a GCGCGCGC sequence	*uvrB* *rfa* *pkM*101
TA1538	his D3052	-1-frameshift mutation with a GCGCGCGC sequence	*uvrB* *rfa*

a GCGCGCGC sequence. This strain may be used to detect frameshift mutagens such as polycyclic aromatic hydroxycarbons, certain aromatic amines, and certain aromatic nitro compounds.

In addition to the mutation in one of the genes of histidine biosynthesis, the Ames strains carry additional genetic alterations that increase their sensitivity to detect mutagens (an overview of the genetic make-up of the *Salmonella typhimurium* tester strains is provided in Table 33).

rfa Mutations. The *rfa* mutation results in a defective lipopolysaccharide membrane and thus increases the permeability of the cell wall to bulky hydrophobic chemicals. In addition, strains with defective cell walls are not pathogenic to experimental animals and humans.

uvrB Deletion. Wild-type bacteria possess several effective DNA repair systems, that operate practically error free and can repair DNA damage without allowing mutations to occur. To overcome the problem of DNA repair and thus the decreased sensitivity of the test system, Ames constructed a series of strains with a deletion of the *uvrB* gene, which codes for a subunit of an important enzyme (the cor-endonuclease-I) of the error-free excision repair system. This change increases the sensitivity of the tester strains to mutagens by several orders of magnitude.

The plasmid pkm101. Wild-type *Salmonella typhimurium* strains do not process an "error-prone" DNA repair which is found for example in *Escherichia coli* and some other members of the Enterobacteriaceae. In contrast to the above-mentioned error-free excision repair pathway, the error-prone system operates with low fidelity and introduces new mutations into the genome while repairing a damaged DNA site. To overcome this deficiency and to increase sensitivity to mutagens, the gene for the error-prone repair system has been introduced into some of the *Salmonella typhimurium* tester strains with the plasmid pkm101. This plasmid also carries the genetic information for ampicillin resistance, an important property for monitoring the physiological integrity of the tester strains in the course of the experiments.

The plasmid pAQ1. The *Salmonella typhimurium* strain TA 102 carries the revertible histidine mutation on multiple copies (approximately 30) of the plasmid pAQ 1 and not on the chromosome. Reversion of one of these copies returns the bacteria's capability to synthesize histidine. This increase in sensitivity is partly offset because strain TA 102 has an intact excision repair system (it does not have the *uvrB* deletion). This strain has been constructed because DNA interstrand cross-linking agents such as the quinone mitomycin C or the combination of psoralens und UV light require an intact excision repair system to generate mutations. DNA interstrand cross-links must first be removed from one strand along with a small number of adjacent bases by excision repair. The gap left behind is repaired, and the remaining broken cross-link attached to the other strand is a premutagenic lesion that may give rise to a mutation by error-prone repair.

The Problem of Bioactivation in the Ames Test. The majority of the mutagenic xenobiotics require bioactivation (toxification) to reactive electrophiles to induce DNA damage. The main disadvantage of bacterial mutagenicity assays is that the tester strains do not express many of the enzymes that bioactivate xenobiotics in mammals. For example, cytochrome P450 activity is not detectable with most substrates in *Salmonella typhimurium*. Therefore, there is a need to simulate the biotransformation occurring in the intact animal by supplementing the test system with the enzymes of bioactivation and the necessary cofactors. For this purpose, numerous exogenous metabolic systems have been used in the last decades. They can be grouped into cell-free and cell-based systems, and among the cell-free systems the 9000 *g* supernatant (called S-9 fraction, the S-9-fraction with the cofactors necessary for enzyme activity is called S-9 mix) from rat liver is the most widely used and best validated.

Preparation of S-9 Fraction and S-9 Mix. S-9 fraction is usually prepared from the liver of male adult rats. Preparations from uninduced animals may contain only low activities of important enzymes of bioactivation such as cytochrome P450 1A1. Since these deficiencies could limit the use of the S-9 fraction as exogenous activating systems in bacterial assays, the rats are usually pretreated with enzyme inducers. Among them Aroclor 1254, a mixture of polychlorinated biphenyls, is the most widely used. Pretreatment of the rats with Aroclor 1254 results in induction of a broad range of cytochrome P450 enzymes. For specific purposes, more selective inducers of cytochrome P450 enzymes such as phenobarbital (cytochrome P450 2A1, P450 2B1) and 3-methylcholanthrene (cytochrome P450 1A1) may be also used.

The livers are removed from the animals after sacrifice at specific time intervals after application of the inducer, minced, homogenized, and centrifuged at 9000 *g* for 15 min. The monooxygenases contained in this S-9 supernatant require NADPH as cofactor, which is normally generated by glucose-6-phosphate dehydrogenase from glucose-6-phosphate by reducing $NADP^+$. Therefore, the S-9-fraction is also supplemented with these two cofactors and with magnesium and potassium salts to yield the final activating system, the S-9 mix. This standard S-9 mix is capable of performing phase-I biotransformation reactions (e.g. oxidations, reductions) but is deficient in most phase-II systems (conjugation reactions). The latter are often involved in detoxification reactions, while phase-I enzymes usually result in toxification of xenobiotics to more electrophilic metabolites. Hence, the S-9-mix may efficiently simulate bioactivation of xenobiotics in the liver, but not detoxification. This may be regarded as an advantage, since it usually increases the sensitivity of the system. On the other hand, the discrepancy between phase-I and phase-II enzymatic activities may be a source of false positive results because the genotoxicity observed in vitro may not reflect the in vivo situation, and detoxification may predominate in the intact animal. The deficiency in conjugation reactions may be partly overcome by adding appropiate cofactors for phase-II conjugation reactions, such asglutathione. The S-9 fraction from other organs may also be prepared and used for the bioactivation of organ specific carcinogens, for example, S-9-fraction of the renal cortex for compounds inducing renal cell tumors. Alternatively, cell-based systems may be used to bioactivate xenobiotics in bacterial mutagenicity assays. Freshly isolated hepatocytes or hepatocytes in primary culture retain the activity of the phase-I and phase-II enzymes of the intact liver and do not require the addition of cofactors for enzyme activity. However, besides greater technical difficulty in obtaining hepatocytes of good quality compared with S-9 fraction, additional problems arise when using intact cells because the electrophiles formed may have a very short half-life and may be trapped and react with macromolecules before diffusing out of the hepatocytes, resulting in false negative results.

An recent elegant approach to overcoming the deficiencies of the bacteria in biotransformation reactions is to clone genes of mammalian biotransformation enzymes into plasmids and to introduce these into tester strains. Although this approach has not been widely used so far, it seems to be a promising method for the future. For example, introducing the gene for *N,O*-acetyl transferase, an enzyme important in the bioactivation of aromatic amines, into *Salmonella typhimurium* TA 98 and TA 100 resulted in an approximately 100-fold increase in the corresponding enzymatic activity in the newly engineered strain. In contrast to the original strains, the *N,O*-acetyl transferase-proficient strains permitted efficient detection of the mutagenic activity of several nitroarenes and aromatic amines. A similar increase in the mutagenicity of bromo- and chloroalkanes was abtained with tester strains expressing a specific human glutathione *S*-transferase isoenzyme [170, 171]. Specific potentials and limitations of these tests follow:

Potentials

- High reliability when testing known carcinogens and noncarcinogens
- Extensive data base available
- High sensitivity
- Information about the mechanism of mutations may be obtained by comparing results in different strains
- Rapid and inexpensive
- May be used as a bioassay to detect mutagenic components in complex mixtures

Limitations

- Some important carcinogens are not active due to specific mechanisms of action (metals, particles, asbestos)
- Data are only qualitative
- Sample must be sterile
- Problems in testing bactericidal chemicals

Experimental Procedure for the Ames Test. Presently three different protocols are widely used for the Ames test: the plate-incorporation assay, the preincubation assay and the fluctuation test.

Plate Incorporation Assay. The basic procedure of the plate incorporation assay is illustrated in Figure 62. Briefly, 2.0 mL aliquots of soft agar overlay medium (Top Agar, 0.6% agar and 0.5% sodium chloride in distilled water) containing a trace of histidine and excess biotin and maintained in the liquid state at 45°C, 100 μL of the tester strain, 20–100 μL of the test compound and, when necessary, 500 mL S9-mix (or another bioactivation system) are added. After mixing, the solution is poured onto dried Vogel–Bonner minimal medium plates. The plates carrying untreated, solvent, and positive (UV light or established mutagen) controls are incubated for 2 d at 37°C. At least ten concentrations of the test compound are usually tested with two plates per concentration.

Preincubation Assay. Some mutagens, particularly those metabolized to short-lived reactive electrophiles, may not be detected in the standard plate incorporation assay due to reaction of the electrophile with constituents of the medium. This type of xenobiotic may be detected more efficiently by using the preincubation assay protocol, in which the bacteria are preincubated with the test compound and S-9 mix in suspension for 30–120 min in the absence of top agar. After the end of this preincubation period, 2.0 mL of soft agar is added to each tube and the reaction mixture is poured into Vogel–Bonner plates, which are incubated as described above.

Fluctuation Test. In the fluctuation test, the number of mutants in a series of small independent replicate cultures is detected. Overnight bacterial cultures are incubated with the test compound and the appropriate metabolizing system (when necessary) in the presence of a trace of histidine. As in the other protocols, this trace of histidine allows a few replication cycles, which are necessary for expression of the mutations following the initial DNA damage. The above mixture is then divided into a large number (usually 50–100) of test tubes or microtiter plates. When the histidine trace is consumed, only revertant cells can grow. Test tubes or wells of the microtiter plates containing revertants become turbid, and the media turn acidic as a result of acid release during growth. This pH drop to 5.2–6.8 can be demonstrated by the color change of a pH indicator such as bromoethyl blue.

Scoring for Colonies in the Plate Incorporation and Preincubation Assay The limited growth of nonrevertant colonies due to the trace of histidine added to enable mutational expression of DNA damage results in a slight background lawn of growth. Therefore, before scoring the plates for revertant colonies, the background lawn should be examined either macroscopically or under the low magnification of a light microscope. At toxic concentrations of the test chemical, the plates appear clear. Also, at toxic concentrations of the test chemical, tiny colonies (minicolonies) may be formed. This happens when, at bactericidal concentrations of the test chemical, very few bacteria survive. The

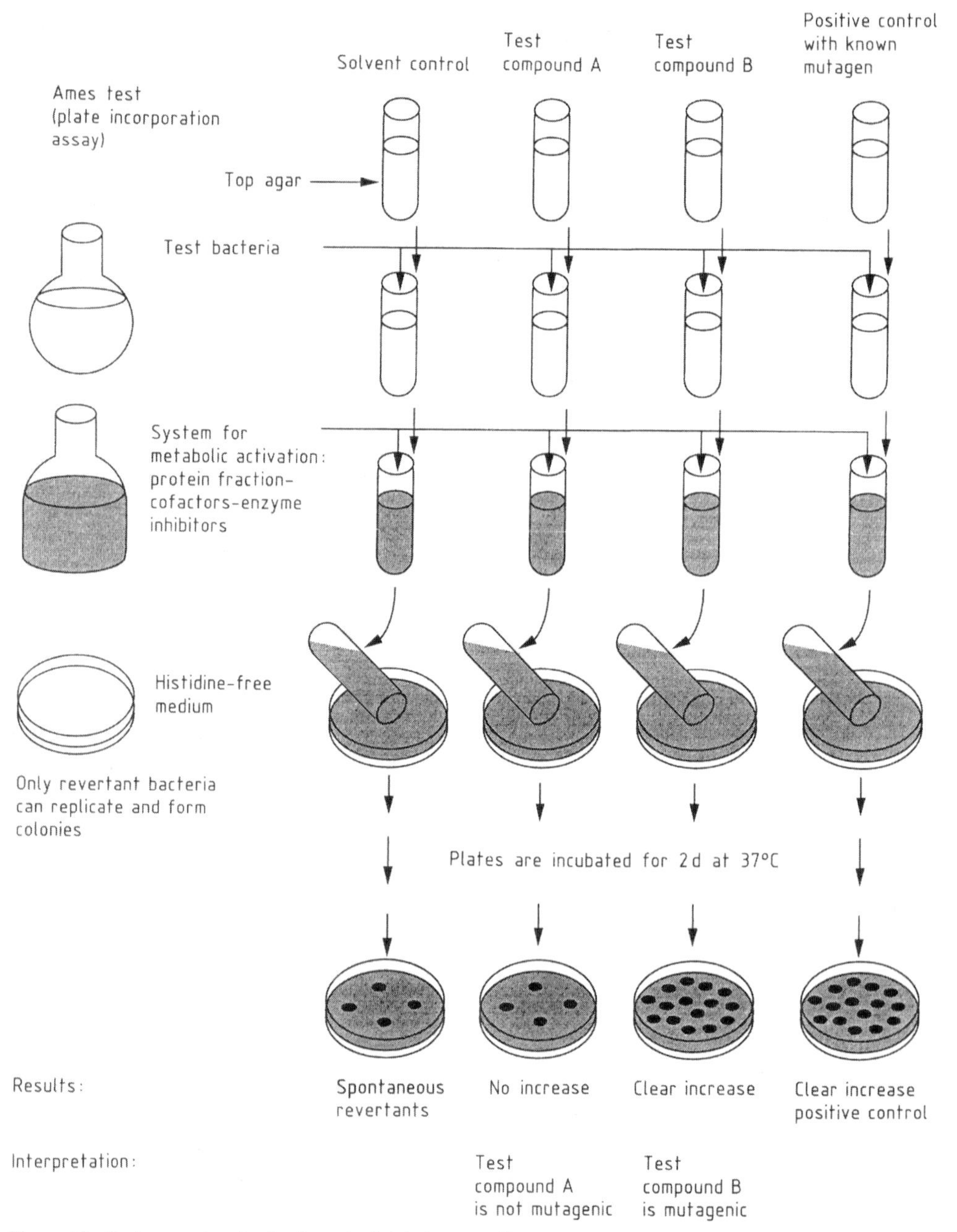

Figure 62. Basic procedure for the Ames test/plate incorporation assay
50 to 200 plates are usually prepared and evaluated per test. At least ten concentrations of the test compound are examined with two plates per concentration level

histidine trace usually added to the system may then be sufficient for formation of these mini-colonies (pseudorevertants). Revertant colonies resulting from gene mutations are clearly larger and can be counted either by hand or with an automatic colony counter.

Evaluation of Results. A typical mutagenic dose–response curve is illustrated in Figure 63. From the linear part of the dose–response curve, the mutagenic potency of the test compound can be calculated and is usually expressed in number of revertants per nanomole of test compound. At higher, toxic concentrations of the test chemical, the curve may flatten. At concentrations which are very toxic to the bacteria, the number of revertant colonies may start to decline after reaching a maximum. The significance of the results is analyzed with usual statistical methods.

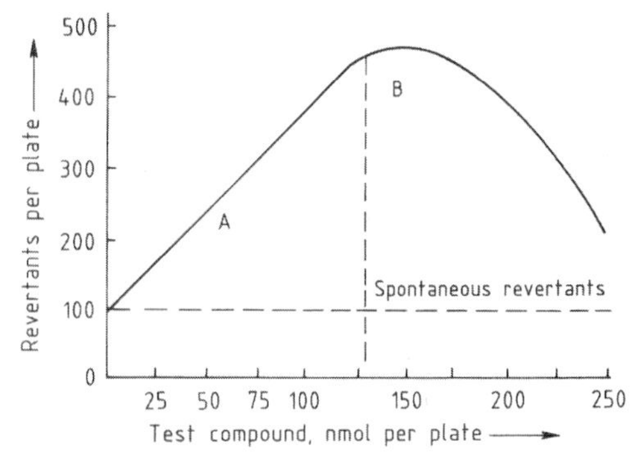

Figure 63. Typical dose – response curve in the Ames test
A) At nontoxic mutagen concentrations the number of revertants increases in a dose-dependent manner; B) Higher, toxic concentrations inhibit growth, which is reflected in flattening of the curve followed by decreasing numbers of obtained revertants

Role of the Ames Test in Evaluating Carcinogenic Properties.. The Ames test is a system to detect mutagenicity which has gained great practical importance as a short-term predictive test for potential carcinogenicity of a xenobiotic. Hence, the Ames test may predict the carcinogenic properties of genotoxic carcinogens, but not of nongenotoxic carcinogens. The very high correlation between mutagenicity and carcinogenicity found in previous reports (values of 90–95% were repeatedly estimated) are certainly not correct. One reason for this overestimation may be that in previous decades mainly structurally alerting chemicals (xenobiotics carrying functional groups present in other carcinogens and known or suspected to be metabolized to electrophiles) were tested for long-term rodent carcinogenicity, and most of these structurally alerting carcinogens are mutagenic. In contrast, more recently, more important environmental pollutants and occupational hazardous chemicals have increasingly been tested for carcinogenicity, independent on their chemical structure. As a result, an increasing number of carcinogenic compounds that do not react directly with DNA to cause mutations have been found. These xenobiotics are thought to act via additional and (unfortunately) only partly defined so,called epigenetic mechanisms (i.e., not involving genetic changes). Nongenotoxic carcinogens cannot be detected in the Ames test. In contrast to the genotoxic carcinogens, which are usually active in two species and/or at multiple sites, nongenotoxic carcinogens are often active in one species and at a single site, for example, branched chain hydrocarbons in the male rat kidney or peroxisome proliferators in the mouse liver. This is confirmed by comparing the carcinogenicity results with the bacterial mutagenicity. While more than 70% of the two-species/multiple-site carcinogens are positive in the Ames test, only 40% of the one-species/single-site carcinogens exert bacterial mutagenicity. Therefore, a reasonable evaluation of the results obtained in bacterial mutagenicity systems can only be conducted by taking into consideration additional information on the chemistry, toxicokinetics, biotransformation, and biological effects of the test compounds.

Presently, the Ames test is an important part of a screening battery for possible carcinogenic properties in the course of the development of new pharmaceutical drugs and chemicals. Depending upon the intended use of the chemical, the development may be interrupted if a compound is clearly positive in the Ames test and in an additional short-term test, often the in vivo micronucleus test (see Section 4.9.5.3). With this "decision-point approach", the number of unnecessary life-long bioassays in rodents is significantly reduced and large amounts of money and time may be saved.

4.9.1.2. Mutagenicity Tests in *Escherichia coli*

The *E. coli* tester strain WP2 and additional tester strains subsequently developed originate from the wild-type *E. coli* B strain; all tester strains are tryptophan(trp)-auxotrophic. For example, the *E. coli* WP2 trp E group has a terminating mutation at an AT base pair of the trp E gene which codes for anthranilate synthetase, an enzyme of the tryptophan biosynthetic pathway. Hence, the *E. coli* mutagenicity assays, like the Ames test, detect reversion of the deficiency to synthetize tryptophan. The *E. coli* tester strains have umuDC$^+$ genes coding for the error-prone repair system, and their membrane structure permit the passage of many large molecules. The experimental procedures and the inherent advantages and limitations are practically identical to those of the *Salmonella typhimurium* test. However, the *E. coli* test systems are less well validated than the Ames test, and this renders the selection of appropiate strains for the investigation of the mutagenicity of a specific xenobiotic and the interpretation of the obtained results more difficult.

4.9.1.3. Fungal Mutagenicity Tests

The most widely used species is *Saccharomyces cerevisiae*; the test detects reversion of the isoleucine auxotrophy induced by exposure to mutagens, in analogy to the Ames test.

4.9.2. Eukaryotic Tests for Mutagenicity

Although prokaryotic systems are fast, inexpensive and versatile, there is a requirement for mammalian test systems due to the important differences between prokaryotic and eukaryotic organisms in structure and function of the genome and in biotransformation and transport of xenobiotics. The eukaryotic systems for detecting mutations include tests in the fruit fly *Drosophila melanogaster*, in many different types of cultured mammalian cells, and in the intact animal in vivo.

4.9.2.1. Mutation Tests in *Drosophila melanogaster*

The fruit fly *Drosophila melanogaster* has a short generation time of 10 d and a similar cellular and chromosomal structure and function to mammalian cells. In addition, the genome of the fly is well characterized, and *Drosophila* species can perform many of the phase-I and phase-II reactions occuring in mammals. The *Drosophila* sex-linked recessive lethal test was a popular screening system in the 1970s and was used to screen many hundreds of compounds. However, since then the popularity of the test has waned drastically because of its poor performance in international collaborative trials that investigated the utility of mutagenicity assays as predictive tools. Another *Drosophila* test, devised in the 1980s, the somatic mutation and recombination test (SMART) seems promising but has not been sufficiently validated. Since both tests are not widely used for screening, the experimental procedures are not be described here (the interested reader may find them in reference [156]). Some potentials and limitations of the mutation tests in *Drosophila melanogaster* follow:

Potentials
- Large number of test organisms can be raised
- Metabolic activation endogenous
- Genotoxic effects evaluated in germ cells
- Large data base on mutagens

Limitations
- Some known carcinogens such as polycyclic hydrocarbons respond poorly
- Problems with toxic compounds (insecticides)

4.9.2.2. In Vitro Mutagenicity Tests in Mammalian Cells

Many short-time mutagenicity tests in mammalian cell lines have been developed. These systems have the advantage of determining genotoxic effects in relevant cell types and may permit the study of mechanisms of carcinogenicity. The potentials and limitations of these in vitro tests follow.

Potentials
- Full range of mutational responses may be observed
- Low costs when compared to in vivo assays
- Respond to particulates and other types of compound not detected in the Ames test
- More relevant end point for mammals

Limitations
- Most cell lines used are transformed cells and do not represent "normal" cells
- Only specific loci from the entire genome are monitored
- Most cell lines used are deficient in biotransformation enzymes

In contrast to popular bacterial assays that detect reversion mutations, the commonly used mammalian mutagenicity assays are based on the detection of forward mutations. A large number of cells is treated with the test compound. After a certain period of time, the cells are exposed to a toxic agent that is lethal to all cells not carrying mutations (i.e., only mutated cells can survive). Cultured mammalian cells are normally diploid and have two copies of each gene. Hence, recessive mutations may be missed if a normal copy is present on the homologous chromosome because the probability that a mutagenic compound alters genes is very low. Therefore, mutations are assessed in genes on the X chromosome in male cells, where only one copy of the gene is present. Alternatively, heterozygous cells are used, in which one copy of the gene is already inactivated.

Mammalian cells have genes that allow the cell to salvage nucleotides from the surrounding medium. These genes, although not essential for cell survival, save cellular energy, since the cell need not synthesize these molecules from simple precursors by energy-consuming pathways. If the medium is supplied with altered nucleotides, the normal (not mutated) cell will incorporate them into the DNA, and this will result in cell death. However, if this salvage property is lost due to mutation, the mutated cells are not able to incorporate the toxic nucleotides from the surrounding medium and will survive and form colonies which can be detected as a parameter of genotoxicity. Although a gene may be inactivated by a mutation, the mRNA and the corresponding enzyme produced prior to the mutational event may be present for some time after exposure to the mutagenic test compound. Therefore, the cells have to be left for some time before challenging with the toxic nucleotide, this period is called expression time.

Basically, two genes of the salvage pathway are utilized in mutagenicity assays: the hypoxanthine–guanine phosphoribosyltransferase (HGPRT) gene and the thymidine kinase (TK) gene. The HGPRT gene is located on the X chromosome in humans and in the Chinese hamster, from which many useful cell lines have been developed. The genetic changes detected in the HGPRT gene as the target site are mainly point mutations because chromosomal deletions usually extend to flanking regions of the target gene, which may contain essential genes. Since only one copy of these essential genes exists in male cells, this results in cell death so that the cells in which the HGPRT gene is inactivated by deletion are usually lost for the test. This is not the case with the TK gene, because it is autosomal and heterozygous cells (TK^+/TK^-) are used in the assay. Here, both point mutations and changes at the chromosomal level can be detected. The HGPRT gene is the target gene in the Chinese hamster CHO and V 79 cell lines, while mouse lymphoma L 5178 Y $TK^{+/-}$ cells detect mutations at the TK locus. The mechanistic background of these tests is shown in Figure 64.

The use of Chinese hamster cell lines for mutagenicity screening is limited due to low sensitivity. Chinese hamster cell lines grow on monolayers and, owing to metabolic cooperation, only a relatively small cell population can be used. In contrast, L5178Y $TK^{-/-}$ cells grow in suspension, and the system is not impaired by the problems of metabolic cooperation, because intercellular bridges do not occur. As described above, a wide variety of genetic events including gene mutations, recombinations and mitotic nondisjunction can result in the formation of the $TK^{+/-}$ genotype from the heterozygote $TK^{+/-}$. Two protocols have been devised for mutation assays with mouse lymphoma L5178Y cells:the plate test in soft agar and the fluctuation test in suspension. For dose-finding, a preliminary cytotoxicity assay is usually conducted when testing new compounds. In this assay, the cloning efficiency (for the plate test) or the relative suspension growth (for the fluctuation test) is determined. The highest test concentration for the mutation assay is usually the concentration that reduces cloning efficiency or suspension growth to approximately 10–20% of the control values. In addition, a moderate concentration causing survival reduction to 20–70% and a low concentration ($> 70\%$ survival) are also used. After treatment with the test compound (3–7 h; in the presence or absence of an exogenous metabolic system, e.g., S-9 mix) the cells are incubated for 2 d for mutation expression before exposure to trifluorothymidine for 10–12 d. In the presence of trifluorothymidine, only mutant cells can survive. Thus, the relative growth of the cells at the end of the experiment can be used as a parameter for mutagenicity.

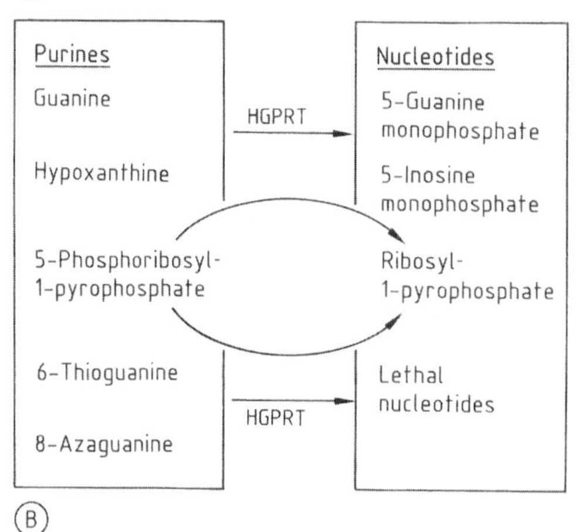

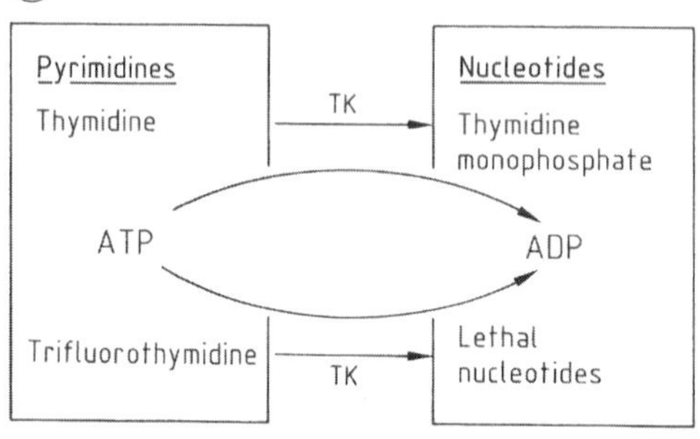

Figure 64. Assessment of mutagenicity in mammalian cells
A) The V79/HGPRT (hypoxanthine-guanine phosphoribosyltransferase) assay; B) The L5178Y TK$^{+/-}$ (thymidine kinase) mouse lymphoma assay
Mutant cells do not form the lethal nucleotides and survive in the presence of 6-thioguanine or trifluorothymidine.

4.9.3. *In Vivo* Mammalian Mutation Tests

In contrast to the well-established and widely used micronucleus tests in rodents (see Section 4.9.5.3), the existing in vivo tests to detect gene mutations are not used for routine genetic toxicology testing; therefore, only brief descriptions are given here (for exact experimental procedure, see [156]).

4.9.3.1. Mouse Somatic Spot Test

The mouse somatic spot test detects the inactivation of genes at a set of heterozygous loci controlling hair pigmentation. Inactivation can be induced by a wide spectrum of genetic alterations ranging from gene mutations to major chromosomal changes.

4.9.3.2. Mouse Specific Locus Test

The mouse specific locus test consists of treating parental mice homozygote for a marker gene controlling coat pigmentation, intensity, and pattern or size of the external ear. Treated mice are mated with a tester stock that is homozygous recessive at the marker loci. The resulting F_1 generation is normally heterozygous at the marker genes and therefore expresses the wild-type phenotype. In the case of mutations at any of these genes, the F_1 offspring express the recessive phenotype. The mouse specific locus test has a comparatively low sensitivity, thus requiring that many thousands of newborns be scored. However, it remains an important method for investigating heritable mutations.

4.9.3.3. Dominant Lethal Test

The dominant lethal test assesses embryonic death resulting from genetic changes in parental germ cells. The procedure is based on observing the viability of uterine implantations. Implantations that die at an early stage form a deciduoma or mole. The genetic event resulting in early death is predominantly chromosomal damage in the parental germ cells resulting in a dominant lethal mutation. Although the assay was popular in the 1960s and 1970s as a screen for germ cell damage, it is not widely used at present because it is considered relatively insensitive.

4.9.4. Test Systems Providing Indirect Evidence for DNA Damage

4.9.4.1. Unscheduled DNA Synthesis (UDS) Assays

Two major DNA repair mechanisms have been identified in mammalian cells. The first mechanism involves direct reversal of DNA damage, such as cleavage of pyrimidine dimers that have been induced by UV light or removal of potentially mutagenic methyl groups from the O^6 position of guanine by O^6-methylguanine-DNA methyltransferases. The second mechanism recognizes damaged DNA bases, removes them after nicking the DNA backbone, fills the gap formed with the correct bases using the opposite strand as a template, and finally seals the second strand break.

The most widely used method to study DNA repair is the "unscheduled" DNA synthesis assay. The principle of the UDS assay consists in detection of the incorporation of radioactive DNA bases or chemical analogues of nucleotides such as bromodeoxyuridine into DNA in the course of excision repair after damaged bases have been removed. This unscheduled DNA synthesis induced by DNA damage must be distinguished from the semiconservative DNA synthesis during DNA replication. DNA replication can occur in the same cells that undergo repair or, alternatively, in other cells of the test population and usually contributes to an incorporation of radiolabeled base or nucleotide analogue in concentrations several orders of magnitude higher than repair. Three approaches are currently available to overcome this problem and to detect selectively DNA repair-dependent incorporation.

1) Use of quiescent cell populations, that do not replicate
2) Suppression of DNA replication by incubation in serum-reduced and arginine-depleted medium and pretreatment with the ribonucleotide reductase inhibitor hydroxyurea
3) Evaluation of radioactivity by autoradiography of single cells, which allows the distinction between heavily labeled replicating cells and lightly labeled DNA-repairing cells

The UDS assay can be conducted in vitro and in vivo. The following description briefly outlines the in vitro assay in cultured mammalian cells. After inhibition of DNA replication by serum reduction, arginine depletion, and treatment with hydroxyurea (if replicating cell lines are used), the cells are exposed to the test compound in presence of radiolabeled thymidine. Among the various cell types used, primary cultures of rat hepatocytes are often preferred to permanent cell lines, because their biotransformation capabilities are closer to those of liver cells in vivo. In addition, primary cultures of hepatocytes do not divide, so that no suppression of cell proliferation is necessary.

Radioactivity incorporation can be measured by scintillation counting of the isolated DNA or by autoradiography of whole cells. For determination of UDS in vivo, the animals are treated with the test compound and radiolabeled thymidine (most often applied by implanted minipumps). Radioactivity in DNA is determined after isolation of the DNA. Alternatively, a combined in vivo/in vitro approach may be used. In this case, in vivo treatment of the animals is followed by isolation of cells from the appropiate organs and labeling of the cells with the radioactive nucleotide in vitro. Assessment of radioactivity incorporation after in vivo treatment is usually conducted by autoradiography.

Several factors influence the interpretation of UDS data. Detection of excision repair demonstrates that the test compound has caused repairable damage to DNA, but does not answer the question whether this has any significance for the mutagenic potential of the chemical. This is very important, because measurements of UDS are averages over all cells of the study population, whereas mutation is a rare event in an individual cell. In addition, extranuclear events such as mitochondrial toxicity resulting in oxidative stress and increased production of reactive oxygen species or increased calcium concentrations in the nucleus and activation of endonucleases may also influence indirect DNA damage and induction of UDS.

4.9.4.2. Sister-Chromatid Exchange Test

Sister- chromatid exchanges are reciprocal exchanges between sister chromatids thought to occur at homologous loci. Although detection of sister-chromatid exchanges can also be conducted after in vivo treatment, the test is widely applied in established cell lines in vitro (Chinese hamster cell lines, human fibroblast cell lines) and in freshly isolated human lymphocytes. Any cell type that is replicating or can be stimulated to divide can be used. The test is carried out essentially according to the following basic procedure. Cells in exponential growth are usually exposed simultaneously to 5-bromo-2′-deoxyuridine and the test compound for a period equivalent to approximately two cell cycles (incubation time depends on the cell type used). In the final 1–2 h of incubation, a spindle inhibitor (colchicine or demecolcine) is added to arrest the cells in metaphase. For detection of sister-chromatid exchanges, the two chromatids are differentially stained with a fluorescent dye such as Hoechst 33258 plus Giemsa. With this method, sister-chromatid exchanges appear as color change on one chromatid (Figure 65).

The sister-chromatid exchange test is one of the quickset, easiest, and most sensitive methods for detecting genetic damage. A large number of compounds have been evaluated with this method, which is usually part of the test battery for genotoxicity screening.

The problem is that the precise genetic event detected by this test is basically not known. sister-chromatid exchanges are obviously not related to chromosomal aberrations since they often fail to detect potent clastogens. They relate more to gene mutations, although there are examples of compounds that clearly induce sister-chromatid exchanges in the absence of mutations as well as the converse. Despite the uncertainty resulting from the unknown mechanism, the sister-chromatid exchange test is very useful because it detects important groups of potentially carcinogenic compounds such as alkylating agents and nucleotide analogues, as well as compounds inducing DNA single-strand breaks or acting through DNA binding.

4.9.5. Tests for Chromosome Aberrations (Cytogenetic Assays)

4.9.5.1. Cytogenetic Damage and its Consequences

In addition to gene mutations, most tumor cells investigated thus far exhibit structural and/or numerical chromosomal aberrations. Chromosomal aberrations such as deletions and translocations may result in activation of proto-oncogenes or inactivation of tumor-suppressor genes. In certain cancers of the lymphatic system (human Burkitt's lymphoma and mouse plasmocytoma), the normal proto-oncogene *c-myc*, which is involved in the regulation of cell proliferation, is translocated to the immunoglobulin locus. Immunoglobulin genes show very high transcription rates during development of B lymphocytes. The translocated *c-myc* is subjected to the same control mechanisms, resulting in inappropriately elevated expression of the gene and increased cell proliferation. In addition to these quantitative changes, chromosome aberrations may result in alterations of structure and function of cellular proteins. In many cases of human chronic myeloid leukemia, a piece of chromosome 9 carrying the proto-oncogene *c-abl* undergoes reciprocal translocation with a piece of chromosome 22. This results in joining of the proto-oncogene *c-abl* with a gene on chromosome 22 and production of a fusion protein encoded by both DNA sequences.

During cell division, chromosome segregation depends upon the functional integrity of the proteins of the spindle apparatus, that correctly divides the replicated genome into the two new nuclei. Impairment of this process, i.e. by chemical damage to the spindle apparatus, may result in nondisjunction, which means that both

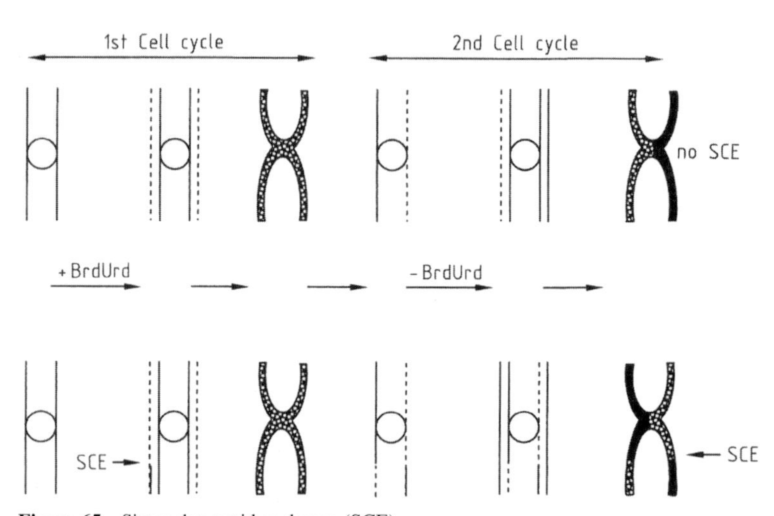

Figure 65. Sister-chromatid exchange (SCE)
The exchange is visualized by treating cells with 5-bromo-2′-deoxyuridine (BrdUrd) during two DNA replication cycles. The distribution of BrdUrd between chromatids is then determined by staining with a fluorescent dye and UV irradation.

copies of a particular chromosome move into one daughter cell while the other receives none. Approximately 10% of the tumor types investigated so far are monosomic or trisomic for a specific chromosome. Aneuploidy is also an important cause of severe birth defects. Trisomy or monosomy of large chromosomes usually leads to early death of the embryo; trisomy of smaller chromosomes may allow the embryo to survive, but the newborn shows severe anatomical abnormalities and major physiological impairment. The most common syndrome of this group is Down's syndrome, which results from trisomy of chromosome 21.

Hence, structural and numerical chromosome aberrations may have detrimental effects on the health of the affected persons, and the following sections describe assays to detect chromosome aberrations in vitro and in vivo.

4.9.5.2. In Vitro Cytogenetic Assays

In the in vitro cytogenetic assays, proliferating cultured cells are treated with the test compound, and chromosomal aberrations are investigated after one or more cell cycles. Although cytogenetic assays can, in principle, be carried out in any cell type, Chinese hamster cell lines and human peripheral blood lymphocytes from healthy donors are generally used. In contrast to the spontaneously dividing Chinese hamster cell lines, peripheral blood lymphocyte cultures must be stimulated to divide by a mitogenic agent, such as phytohemagglutinin. Treatment is performed about 44 h after phytohemagglutinin stimulation, when the cells are proliferating. At least three doses of the test compound should be investigated, with the highest being in the low cytotoxic range. The usual recommendation is to treat the cells for ca. 1.5 normal cycles times, which is 15 h for Chinese hamster cells and 12–14 h for human peripheral lymphocytes. Many authors recommend longer treatment periods because some chemicals induce a mitotic delay at clastogenic doses (i.e., they increase the duration of the cell cycle). At the end of treatment, cells are induced to accumulate in the metaphase stage of mitosis with a spindle inhibitor such as colchicine or demecolcine, and at least 200 cells per treatment group are scored under the microscope for chromosomal aberrations. Chromosome breaks, fragments, and exchanges are considered specific structural aberrations, whereas gaps are usually excluded from the quantitation because the mechanism underlying their formation is not understood. Chromosome pulverization (complete fragmentation in small pieces) may also be regarded not as a parameter of clastogenicity, but rather as an indicator of severe cytotoxicity and cell death induced by the test chemical. Chemicals that are clastogenic in vitro at noncytotoxic concentrations are likely to be clastogenic in vivo. When bias resulting from lack of or insufficient biotransformation in the in vitro situation can be excluded, negative results in the in vitro assay provide a strong indication for absence of in vivo clastogenesis.

4.9.5.3. *In Vivo* Cytogenetic Assays

Induction of chromosome aberrations can be detected in intact animals either by examination of metaphases as described above for cultured cells or by the formation of micronuclei. Rats and Chinese hamsters are usually employed in metaphase analysis, while mice are commonly preferred in the widely used micronucleus test in bone marrow cells.

Rodent Micronucleus Test. The rodent micronucleus assay is an important part of the test battery for genotoxicity and is usually included in the toxicity evaluation of both pharmaceuticals and chemicals. The assessment of micronuclei is commonly conducted in polychromatic erythrocytes of bone marrow after treatment of young mice for six to eight weeks. Induction of numerical chromosome aberrations or chromosome breaks in the immature erythroblast of the bone marrow results in the formation of a micronucleus after cell division because the chromosome fragments are not participating in the chromosome segregation accomplished by the spindle apparatus. In the course of the physiological maturation of the red blood cells, the main nucleus is extruded from the erythroblast, while the micronucleus remains in the cell and can be detected in the young polychromatic erythrocyte after staining (Figure 66). The lack of nucleus in these bone marrow cells facilitates scoring for micronuclei under the microscope;

however, peripheral blood cells and liver cells have also been used.

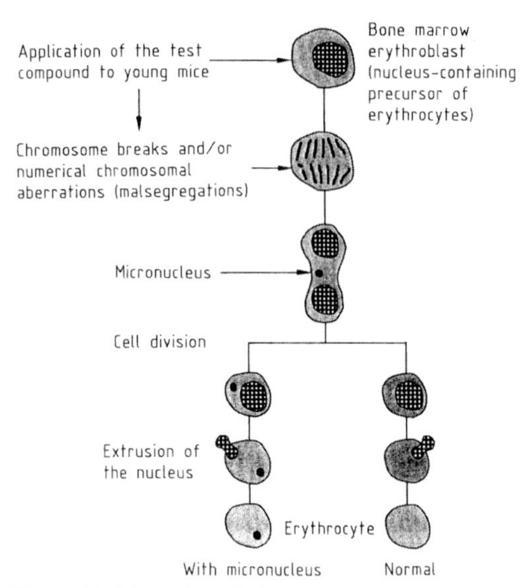

Figure 66. Mechanism of micronucleus formation in bone marrow erythroblasts

Usual treatment protocols include application of a single dose of the test compound and harvesting of bone marrow cells after 24, 48, and 72 h; or, alternatively, multiple applications (two or three doses 24 h apart) followed by a single harvest another 24 h after the last dose. The bone marrow is aspirated with a needle from the proximal end of the femur shafts and flushed with serum to obtain a homogeneous cell suspension, and after staining, the number of micronucleated immature polychromatic erythrocytes (not the number of micronuclei) is scored under a microscope. The immature erythrocytes stain bluish and can be distinguished clearly from the orange-red mature erythrocytes. Only tests with positive and negative control values within the historical control value of the laboratory are considered adequate. The minimum number of cells scored depends on the spontaneous frequency of micronuclei; 2000 cells should be scored if the control incidence is $< 0.2\%$ [172]. The ratio of polychromatic immature erythrocytes : monochromatic mature erythrocytes indicates the integrity of proliferation and differentiation processes in the bone marrow and is used as a measure of general toxicity. If the test compound has been administered at the maximum tolerated dose (as determined in the subacute toxicity test) and no increase in micronucleated erythrocytes can be detected, it is classified as negative for clastogenic activity in vivo. Clear positive results in the in vivo assay indicate clastogenic potential also in humans.

Rodent Bone Marrow Metaphase Analysis. Metaphase analysis is usually carried out in rats at 6, 24, and 48 h after administration of a single dose or alternatively at 6 and 24 h after administration of multiple doses at 2 to 4 h prior to sampling. The animals are treated with spindle inhibitor such as colchicine and demecolcine to induce accumulation of cells in the metaphase. Metaphase analysis is much more time- consuming than scoring for micronuclei; therefore, only 50 cells per animal (500 per experiment with five male and five female animals) are usually analyzed [172]. In contrast to the rodent micronucleus test, the bone marrow metaphase analysis is not automatically included in the general screening scheme for genotoxicity; it is performed only when specific questions arise or for research purposes.

4.9.6. Malignant Transformation of Mammalian Cells in Culture

The term transformation in the present context means that a cultured cell line has taken on one or more morphological and/or malignant changes to give tumor cells. Among the various changes observed by exposure of cell lines to carcinogenic compounds, the following are two clearly established parameters indicating morphological transformation:

1) Inducation of anchorage independence (i.e., cells become capable of forming colonies in soft agar)
2) Loss of contact inhibition resulting in a tendency to grow in a piled-up criss-crossed pattern and to form foci

The term malignant transformation indicates that a cell line is capable of producing an invasive tumor in a suitable host. Hence, not every morphologically transformed cell line has necessarily undergone malignant transformation. On the other hand, human cancer cells are by definition malignantly transformed because they may metastasize under appropriate conditions.

The vast majority of cell transformation studies have been carried out with Syrian hamster cells [i.e., Syrian hamster embryo (SHE) cells and baby hamster kidney (BHK) cells] and with human fibroblasts.

SHE cells, isolated from 13-day-old embryos, can be induced to form piled-up criss-crossed colonies (foci) by treatment with carcinogens. Cells derived from these foci can produce tumors in athymic mice after extended subculturing (35–70 population doublings). Hence, the induction of malignantly transformed tumorigenic cells is a laborious, time- consuming procedure with several inherent problems (e.g., contamination of the plates in the course of the multiple passaging steps over several months). Therefore, the tendency developed to establish morphological transformation that can be induced in a comparatively short time as a short-term test for carcinogens. However, anchorage-independent cells showing criss- crossed growth pattern are usually not carcinogenic in animals. In addition, interlaboratory comparative elaboration studies indicated that scoring may be arduous and subjective, and in many cases the lack of dose-response and difficulties in obtaining consistently reproducible results in repeated assays may occur, even in the same laboratory [173].

Malignant transformation of cultured cells is currently an important tool to investigate the molecular mechanisms, i.e. mutations in oncogenes and tumor suppressor genes likely involved in the carcinogenic process. However, the use of cell transformation assays for screening purposes and for human risk assessment is very limited.

Two in vitro tests are required to provide a base level of information on the mutagenic potential of a chemical. These are a gene mutation test in bacteria and an in vitro mammalian cell test capable of detecting chromosome aberrations. For chemicals with significant toxicity to bacteria, an *in vitro* mammalian cell gene mutation test can be used as an alternative first test.

There are various options for selection of further test procedures in mammalian cells. An in vitro chromosome aberration test, i.e., a cytogenetic assay for structural chromosome aberrations using metaphase analysis will povide information on potential aneugenicity by recording the incidence of hyperdiploidy, polyploidy, and/or modification of mitotic index (e.g., mitotic arrest). A mouse lymphoma assay (L5178Y cells, TK locus) may detect gene mutations and structural chromosome aberrations but is not sufficiently sensitive for the detection of aneugens. These systems may be combined with an in vitro micronucleus tests, which is capable of detecting structural chromosome aberrations as well as aneuploidy.

4.9.7. In Vivo Carcinogenicity Studies of Limited Duration

In vivo tests of limited duration provide evidence for carcinogenicity in the whole animal in a short period (i.e., 52 weeks or less) compared to the lifelong bioassay. The experimental procedure usually involves administration of several doses of a known initiating carcinogen for a specific target organ to investigate promoting effects of the subsequently administered test compound. Alternatively, a known potent promoter for a specific tumor can be used to facilitate detection of initiating properties of the test compound. Also, many limited bioassays currently established or being developed aim to predict carcinogenicity by detecting cell lesions that consistently precede the appearance of the relevant tumor (i.e., induction of altered foci in the liver). In vivo carcinogenicity studies of limited duration are not automatically included in the usual test battery for carcinogenicity and are not accepted by regulatory authorities as a replacement for the conventional lifelong carcinogenicity study. The decision to carry out limited bioassays is made individually according to the available and required information on the test compound. Therefore, the present chapter includes only a brief description of the limited bioassays currently available.

4.9.7.1. Induction of Altered Foci in the Rodent Liver

In the course of rat liver carcinogenesis, cell foci exhibiting specific biochemical alterations precede the formation of liver tumors. These foci have, for example, abnormal concentrations of γ-glutamyl transpeptidase, glucose-6 phosphatase, adenosine triphosphatase, and the placental form of glutathione S-transferase and, unlike normal liver parenchyma, do not accumulate iron following iron loading. These parameters allow reliable and objective histochemical identification of preneoplastic foci as soon as three weeks after the initiation of treatment with known carcinogens. The yield is usually highest by the 12th to 16th week of exposure; therefore, the recommended approach is exposure for 12 weeks to the test compound with subcutaneous injection of iron during last two weeks to produce the iron load [174].

4.9.7.2. Induction of Lung Tumors in Specific Sensitive Strains of Mice

Certain strains of mice (e.g., the A/Heston) exhibit a high spontaneous incidence of lung tumors and are extremely sensitive to pulmonary carcinogens such as polycyclic aromatic hydrocarbons, certain nitrosamines, aflatoxin B_1 ethyl carbamate, hydrazines, and certain alkylating agents. The sensitivity of the test is highest 30 to 35 weeks after exposure initiation. Extension of treatment for a longer period is not recommended, because after 35 weeks the spontaneous lung tumor incidence increases rapidly in control animals as well, resulting in decreased sensitivity of the test.

4.9.7.3. Induction of Skin Tumors in Specific Sensitive Strains of Mice

The carcinogenic activities of several chemicals and crude mixtures can be readily revealed by their continuous application to the skin of a highly sensitive mouse strain, the Sencar mouse. Tars from coal, petroleum, or tobacco, as well as the pure polycyclic aromatic hydrocarbons and their congeners contained in such mixtures, give clearly positive results in this limited bioassay; in contrast, some arylamines and other established carcinogens do not elicit a positive response in this mouse skin test. These discrepancies are due mainly to differences in toxification or detoxification reactions between the systemic and the local applications. The mouse skin contains specific cytochrome P450 enzymes and peroxidases that can bioactivate polycyclic aromatic hydrocarbons, hence the differences in response. Therefore, this test cannot be considered a reliable predictor of the potential carcinogenicity of a xenobiotic or of human risk.

4.9.8. Methods to Assess Primary DNA Damage

4.9.8.1. Alkaline Elution Techniques

Many biochemical and analytical procedures are available to detect and quantitate the damage to DNA after contact with xenobiotics. Some of these procedures are very sensitive and thus permit detection and quantitation of DNA damage or DNA modifications after application of relevant doses of carcinogens to animals and after occupational or environmental exposure of humans.

The alkaline DNA filter elution was developed based on the observation that the rate at which large DNA single-strands pass through a membrane filter under alkaline denaturing conditions depends on the length of the strands. A broad spectrum of DNA damage types – both direct and indirect effects caused by effects of the toxicant on cellular function – can be detected with this techniques. The indirect effects include DNA single-strand breaks, DNA–protein crosslinks, and interstrand crosslinks (DNA–DNA).

The basic procedure (Fig. 67) operates essentially as follows: After treatment of cells in vitro or isolation of cells from an organ or a tissue of an animal treated in vivo, cell samples are placed onto membrane filters and lysed with a detergent- containing solution. This lysis solution is allowed to flow through the filter, thus removing most of the cellular protein and RNA, the intact DNA of cellular chromatin being retained on the filter. An elution solvent with a pH generally >12.0 is then pumped slowly through the filter to disrupt the hydrogen bonds between DNA strands. Treatment-induced DNA single-strand breaks produce DNA fragments with reduced molecular masses, thus increasing the rate at which DNA passes through the filters, whereas DNA–DNA or DNA–protein crosslinks decrease the rate compared to untreated controls. The amount of DNA eluted can be quantified either by using radiolabeled cell populations (by pretreating the cells with [^{3}H]thymidine) or by fluorogenic DNA-reactive compounds. The major limitations of the test are essentially the same as described in Section 4.9.4. DNA strand breaks do not necessarily indicate mutagenicity or interaction of a xenobiotic with DNA; they may also result from extranuclear damage such as increased production of reactive oxygen species due to impairment of mitochondrial functions. Hence, although a vast amount of data exists on the ability of mutagens and carcinogens to induce single-strand breaks, the test has only limited value as a screening method for mutagenicity.

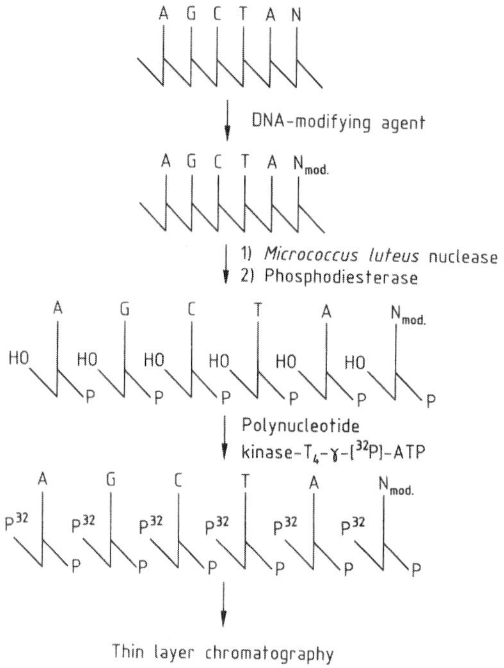

Figure 67. The ^{32}P-postlabeling procedure for detection of DNA damage ($N_{mod.}$ = modified deoxynucleotide)

4.9.8.2. Methods to Detect and Quantify DNA Modifications

The increased sensitivity of analytical methods has resulted in the development of several sensitive and selective methods to detect and quantify modifications of DNA bases induced by xenobiotics. These methods often rely on sophisticated chromatographic techniques to separate unchanged DNA constituents from modified DNA bases.

High-performance liquid chromatography (HPLC) with fluorescence detection may be used to detect and quantify DNA adducts formed from highly fluorescent xenobiotics, such as aflatoxin$_{B1}$ and polycyclic aromatic hydrocarbons. In addition, some DNA adducts of xenobiotics are highly fluorescent, such as alkylations of the N^7 atom of guanosine [175] and cyclic derivatives formed from vinyl chloride [176]. HPLC with electrochemical detection may detect oxidative DNA modifications, such as 8-hydroxydeoxyguanosine with high sensitivity and has been used to study DNA damage by cellular aging [177].

Gas chromatography/mass spectrometry (GC-MS) techniques are widely used in the detection and quantitation of DNA modifications. After derivatization to form volatile derivatives, DNA adducts formed by methylating and ethylating agents can be detected by GC-MS with selected-ion monitoring [178]. Due to the low detection limit (attomole range), GC-MS coupled with chemical ionization and negative-ion detection (after electrophore labeling with. e.g. pentafluorobenzyl bromide) has been applied in DNA adduct monitoring [179, 180]. However, the procedures are very time consuming and prone to artefact formation. At present, due to much simpler sample workup and high sensitivity LC-MS/MS methods promise to be useful for detecting DNA modifications.

A widely used method to detect DNA adducts of bulky organic substituents, which can not be transformed to volatile derivatives, is the ^{32}P-postlabeling technique [181]. The method can be widely used because it does not require the use of radiolabeled xenobiotic and sophisticated and expensive mass spectrometry. Due to the availability of ^{32}P-labeled adenosine triphosphate with high specific activity and the possibility of concentrating DNA adducts, this procedure may detect adduct frequences as low as one adduct in 10^{10} nucleotides. The method involves isolation of DNA from an animal or cells treated with the xenobiotic, enzymatic hydrolysis of the DNA to the 3′-nucleoside monophosphates and enzymatic phosphorylation to the 3′,5′-nucleoside diphosphates with ^{32}P-labeled adenosine triphosphate as phosphate donor. The obtained mixture of ^{32}P-labeled nucleosides and modified nucleosides is then separated by multidimensional TLC. Modified nucleosides are detected by placing the TLC plates on radiation-sensitive film; quantification can be performed by liquid scintillation spectrometry after cutting the adduct spots from TLC plates. Formation of artefacts and the poor resolution of the employed TLC method are problematic.

Several procedures to detect DNA adducts based on immunological methods have also been developed. These procedures have the advantage of being rapid and simple, but they are usually selective only for a specific type of adduct and require a time-consuming procedure to generate the antibody. In addition, sensitivity is often insufficient, quantification of adducts is complicated, and cross-reactivity generates artefacts.

Modern LC-MS/MS methods have sufficiently low detection limits to detect DNA modifications in a concentration range relevant for animal toxicity studies and do not require time-consuming sample preparation. Therefore, these methods have major potential for widespread use in the detection and quantitation of DNA modifications.

4.9.9. Interpretation of Results Obtained in Short-Term Tests

Although no single rule applies in every case and each test compound should be evaluated individually, some general recommendations may be useful for a reasonable interpretation of results from short-term tests. Positive results in one in vitro test for mutagenicity (i.e., the Ames test) and one in vivo test (i.e., the rodent bone marrow micronucleus assay) indicate potential carcinogenicity. The test compound is highly suspect for potential carcinogenicity if clear-cut, dose-dependent evidence for genotoxicity in more than one in vivo and one in vitro test has been obtained, especially if the compound or its metabolites are structurally alerting for DNA reactivity.

This is supported by a recent evaluation on the activity of known human carcinogens in the *Salmonella* mutagenicity test and the rodent bone marrow micronucleus test [182]. As shown in Table 34, most of the human carcinogens identified so far are positive in these popular tests. Apart from hormones, no nongenotoxic organic chemical has been shown to cause cancer in humans thus far. In contrast, since the late 1980s an increasing number of organic chemicals has emerged that are clearly carcinogenic in rodents but are not mutagenic in *Salmonella* or clastogenic in rodent bone marrow. In addition, most of these nongenotoxic rodent carcinogens (and their metabolites) do not exhibit structural alerts for DNA reactivity. The reason for the differences between the epidemiological data in humans and the chronic rodent bioassays are not

Table 34. Mutagenicity in Salmonella typhimurium and clastogenicity in rodent bone marrow of known human carcinogens [191]

Human carcinogens*	*Salmonella* mutagenicity	Rodent bone marrow chromosomal aberrations or micronuclei
Organic compounds		
Aflatoxins	+	+
4-Aminobiphenyl	+	+
Analgesics containing phenacetin	+	+
Azathioprine	+	+
Benzene		+
Benzidine	+	+
Betel quid and tobacco	+	+
Bis(chloromethyl) ether	+	(+)
Chlorambucil	+	+
Chlornaphazine	+	+
Cyclophosphamide	+	+
Melphalan	+	+
Mustard gas	+	ND(+)
Myleran	+	+
2-Naphthylamine	+	+
Tobacco, smokeless	+	+
Tobacco, smoke	+	ND(+)
Treosulphan	+	+
Vinyl chloride	+	+
Soot, tars, and oils		
Coal tar pitch	+	ND
Coal tar	+	ND
Mineral oil (untreated and mildly treated)	+	ND
Soot	+	ND(+)
Metals		
Arsenic compounds		+
Chromium compounds (hexavalent)	+	+
Nickel and nickel compounds		ND

+ = positive response; (+) = predicted positive response; ND = not tested so far.

* The carcinogenic hormones (essentially natural and synthetic estrogens) and fibers (asbestos, erionite, and talc containing asbestiform fiber) are not included; the members of these groups tested so far gave negative or inconclusive results in the *Salmonella* mutagenicity and rodent bone marrow clastogenicity assays.

known. However, these considerations also suggest that although positive results in the short-term assays for genotoxicity indicate carcinogenic potential and may give reason to interrupt further development of a compound, negative results cannot be taken as a guarantee of the absence of carcinogenic activity in vivo, but rather as a green light to proceed to the lifelong rodent bioassay.

4.10. Evaluation of Toxic Effects on the Immune System

In the last twenty years experimental and epidemiological evidence has accumulated that the mammalian immune system may be altered (i.e. suppressed or induced) by a wide range of environmental chemicals and drugs. Well-known examples are the polychlorinated and polybrominated biphenyls; dibenzo-*p*-dioxins; benzene; isocyanates; metals such as chromium, lead, and nickel; and organometallics such as di-*n*-octyltin chloride and tri-*n*-butyltin oxide.

Immunotoxicity is the ability of a substance to adversely affect the immune system: the immune response of affected individuals is altered. Immunotoxic responses may occur when the immune system is the target of the chemical insult; this in turn can result in either immunosuppression and a subsequent decreased resistance to infection and certain forms of neoplasia, or immune disregulation, which exacerbates allergy or autoimmunity. Alternatively, toxicity may arise when the immune system responds to an antigenic specificity of the chemical as part of a specific immune response (i.e., allergy or autoimmunity). Changes in immunological parameters may also be a secondary response to stress resulting from effects on other organ systems. Therefore, it must be recognized that in principle all chemical substances may be able to influence parameters of the immune system if administered at sufficiently high dosages. However, an immunotoxic effect should only be discounted when a thorough investigation has been performed. Although the immune system is considered as a target organ with regard to systemic toxicology, it consists of several different organ systems. A very large number of different cell types, present in practically all tissues and compartments of the human or animal body, participate in the immune response. Due to the complexity, it is not possible to describe the immune system in the context of the present chapter. Only a very brief description will be given to help the understanding of the current practical approaches of the evaluation of adverse effects on immune function induced by chemicals (see also → Immunotherapy and Vaccines, Chap. 1).

The overall immune system can be categorized in two major subsystems: the humoral and the cell-mediated subsystems. The effects of the humoral system are mediated by B lymphocytes producing antibodies that react with antigenic (usually foreign) material (i.e., antibodies attack bacteria and viruses before they can enter the host cell). The cell-mediated system involves primarily the mobilization of phagocytic leucocytes (macrophages) to ingest foreign organisms such as bacteria and the activation of T lymphocytes.

The two systems do not function independently in different situations; rather, they interact by complex feedback mechanisms that are only partly understood. One of the main properties of the immune system is the rapid production of a large number of cells capable to react with a specific antigen when this antigen is presented (again) to the organism. This important property is based on the presence of a wide variety of memory cells that were specifically adapted to the antigen at the time of initial contact.

Immunocompetent cells are required for host resistance, and thus exposure to immunotoxicants can result in increased susceptibility to any type of toxicity and disease including cancer.

The evaluation of adverse effects on the immune system can be carried out in two tiers [183, 184]. The first tier evaluates immune-related parameters (haematology, blood chemistry, organ weights and histopathology) that are included in revised standard testing protocols of repeated-dose (28 and 90 d) toxicity studies (Table 35).

Compounds showing some immunotoxic properties in this first tier and also chemicals suspected of having immunotoxic effects based on information from prior studies or structure–activity relationships are further evaluated by functional assays that assess competence of immune cells (Table 36) For a detailed description of the experimental procedures see [196].

Table 35. Immunological parameters that can be included in repeated-dose toxicity studies in rats for compounds for which no prior immunotoxic potential has been identified (tier I)

Parameters assessed	Tests
Hematology and blood chemistry	differential white blood cell counts
	bone marrow cellularity
	albumin : globulin ratio
	serum immunoglobulin classes
Organ weights	thymus, spleen, lymph nodes
Histopathology	thymus, spleen, lymph nodes, bone marrow, Peyer's patches (if oral administration)
	bronchus-associated lymphoid tissue (in case of pulmonary administration)

Table 36. Selected functional assays to assess immunotoxicity of compounds for which some immunotoxic properties have been implied in tier I evaluation (see Table 35) or in prior studies (tier II)

Parameter assessed	Tests
Cell-mediated immunity	mixed leukocyte response to antigenic determinants; induction of cell proliferation by the T-lymphocyte mitogen concanavalin A; T-lymphocyte cytotoxicity; delayed hypersensitivity response (DHR) to keyhole limpet hemocyanin (in contrast to the first three assays, that are carried out in cultured cells in vitro, the DHR is induced in vivo)
Antibody-mediated immunity	antibody plaque-forming cell response; serum antibody titer after exposure to specific antigens determined by enzyme-linked immunosorbent assay (ELISA)
Natural and induced host resistance	natural killer cell cytotoxicity; macrophage-mediated phagocytosis and intracellular killing; assessment of macrophage and T-lymphocyte competence by inoculation of *Listeria monocytogenes*; host resistance to melanoma cells

4.11. Toxicological Evaluation of the Nervous System

Neurotoxicity is the induction by a chemical of adverse effects in the central or peripheral nervous system, or in sensory organs. It is useful for the purpose of hazard and risk assessment to differentiate effects specific to sensory organs from other effects which lie within the nervous system. A substance is considered neurotoxic if it induces a reproducible lesion in the nervous system or a reproducible pattern of neural dysfunction.

The identification and characterization of neurotoxic properties of chemicals is one of the essential goals of every toxicity screening programm. Neurotoxicity has traditionally been associated with structural pathological modification of nervous system constituents (i.e. neurons, glial cells, and endothelial cells). Hence, the most important tool to assess neurotoxic effects in the past has been the histopathological examination of the nervous system of animals acutely or chronically exposed to xenobiotics. Over the last ten or twenty years, however, several reasons have called for the development and use of functional tests in neurotoxicity screening. First, detailed histopathological analysis of the nervous system is very time-consuming and requires experienced neuropathologists. Second, many chemical compounds disturb the nervous system without causing identifiable structural lesions. This clear need for functional tests to assess neurotoxicity has led to the development of a functional observational battery and an automated test to detect locomotor activity in rats based on methods that have been used by neuropharmacologists for many decades for the evaluation of psychoactive neurologic and autonomic pharmaceutical compounds.

4.11.1. Functional Observational Battery

The functional observational battery can be carried out both in acute and repeated-dose toxicity studies. The sequence of tests is usually arranged to progress from the least to the most interactive with the animal. The assessment begins with home cage observations followed by measurement made while handling the animals and assessment of activities in the open field. Assessment of reflexes, as well as physiologic and neuromuscular parameters is carried out at the end of the examination. Specific parameters assessed at these different stages are summarized in Table 37, and more detailed descriptions of some important tests are given below (reviews: [197–199]).

Catalepsy. Catelpsy can be measured by placing the rat on four corks (35 mm high, 40 mm in diameter, 100 mm between fore- and hindfeet, 60 mm between right and left feet).

Duration of immobility at this position is measured for a period up to 60 s. For the assessment of catalepsy, a number of experimental modifications exist (i.e. placing the rats on a horizontal bar 12 cm above the ground). However, the four-corks procedure allows the best distinction between cataleptic and heavily sedated animals.

Table 37. General parts of the functional observational battery to assess neurotoxicity and specific examples of parameters recorded

A.	home cage observations	posture palpebral closure (eyelids wide open to completely shut) convulsions (clonic, tonic) biting
B.	observations while handling the animal	ease of removal from cage and of handling in hand, lacrimation, salivation, piloerection, fur appearance
C.	open-field activity (observations over 3-min period)	time to first step number of rears (supported and unsupported) number of urine pools and defacations mobility and gait tremors, ataxic gait, convulsions
D.	reflexes	approach response (e.g., to a pencil) touch response click response tail response pupil and eye blink response forelimb and hindlimb extension righting reflex: hold rat in supine position, drop from approximately 30 cm and note ease of landing
E	physiologic observations	catalepsy body temperature body weight
F	neuromuscular observations	rotarod performance grip strength hindlimb extension strength hindlimb foot splay

Grip Strength. The rat is allowed to grip a triangular ring with its forepaws and is pulled back along a platform until its grip is broken. As the pulling back continues, the animal's hindpaws reach a T-shaped rear limb grip bar, which it is allowed to grasp and then forced to release by continued pulling. Special devices are used to measure the maximum strain required to break forelimb and hindlimb grip.

Rotarod Performance (Rotating-Rod Test). This test requires preliminary training of the animals to walk on a rotating rod (7 cm in diameter, 5–12 rpm rotation rate). During the training period, the containers underneath the rod are filled with water to prevent the rats from jumping off. For testing, the time each rat remains on the rotarod is measured up to 2 or 3 min. Due to its objectivity and reproducibility this test is one of the most popular for the evaluation of adverse effects to the nervous system.

Hindlimb Foot Splay. The hind feet are painted, animals are dropped from a horizontal position 30 cm above a table onto paper and the distance between the middle of the ink spots is measured.

4.11.2. Locomotor Activity

Locomotor activity is not automatically part of the screening battery for neurotoxicity, the decision to carry out the test being rather made individually based on preliminary observations. Locomotor activity is assessed in special test chambers using a photocell detection procedure. Animal movement inside the chamber interrupts the photobeams and is translated into activity counts. Data on animal activity are usually recorded over three 5 min intervals.

In addition, a number of further methods to investigate neurotoxicity are available and standardized approach for an evaluation may not be given, but need to be decided on a case-by-case basis (Table 38).

Table 38. Methods for investigation of neurotoxicity

Effect	Methods available
Morphological changes	Neuropathology Gross anatomical techniques Immunochemistry Special strains
Physiological changes	Electrophysiology (e.g. nerve conduction velocity, NCV) Electroencephalogram (EEG, evoked potentials)
Behavioral changes	Functional observations Sensory function tests Motor function tests (e.g., locomotor activity) Cognitive function tests
Biochemical changes	Neurotransmitter analyses Enzyme/protein activity Measures of cell integrity

A major problem in neurotoxicity screening is that the general behaviour of rats is not stable and uniform, even when they are kept under strictly controlled conditions and not exposed to neurotoxic compounds. Therefore, many authors suggest to carry out the tests without knowledge of the treatment protocols. Furthermore, since many observations are subjective, it is important to use experienced and well trained scientists and technicians in the evaluation.

4.12. Effects on the Endocrine System

Although endocrine disruption is often regarded as a specific end point in toxicity testing, it is just a mechanism by which a chemical may induce adverse effects. Many endocrine-dependent toxicities will be detected in the course of the already available methods for toxicity testing. However, a number of specific tests for endocrine effects of chemicals are under discussion for introduction into toxicity testing.

The endocrine system consists of a number of glands such as the thyroid, gonads, and the adrenals, and the hormones they produce such as thyroxine, oestrogen, testosterone and adrenaline. These hormones may influence development, growth, reproduction and behaviour of animals and humans.

Endocrine disrupters are defined as:

- Chemicals that have properties that might be expected to lead to endocrine disruption in an intact organism, its progeny, or (sub)populations
- Chemicals that alter functions(s) of the endocrine system and consequently may cause adverse health effects in an intact organism or its progeny

Endocrine disrupters may to interfere with the endocrine system by several mechanisms:

- By mimicking the action of a naturally produced hormone such as oestrogen or testosterone and thereby inducing similar chemical reactions in the body
- By blocking the hormone receptors in cells and thus preventing the action of normal hormones;
- By affecting the synthesis, transport, metabolism, and excretion of hormones and thus altering their concentrations.

A variety of test systems to characterize the potential of chemicals for endocrine disruption, ranging from receptor-binding assays to complex in vivo measurements, have been proposed, but a specific testing approach has not yet been agreed.

Some testing guidelines (e.g., the 28-d study guideline) detect effects on endocrine function, but there are no test strategies/methods available which can detect all possible effects that may be linked to the endocrine disruption mechanism.

5. Evaluation of Toxic Effects

One of the major environmental and occupational issues of concern to both scientists and administrators is the control of potential health hazards to humans due to the production, use, and disposal of chemicals. The concern arises from the increasing numbers of chemicals in production and use and the increasing numbers of chemicals demonstrated to exert toxic effects in one or several of the sensitive toxicity testing systems available. This situation has afforded growing legislative control of the production and application of chemicals to ensure adequate protection of human health. Control measures based on the recognition of potential adverse health effects may limit the presence of hazardous chemicals in the environment or regulate the use of hazardous chemicals, thus reducing the potential health risks to humans (Table 39).

Table 39. Possible measures to reduce human exposure to hazardous chemicals

Application or exposure to chemical in question	Measures to reduce exposure
Industrial chemicals	reduction or cessation of application; protective measures in the workplace; alternative chemicals with lower hazard
Pharmaceuticals	cost – benefit analysis
Alcohol, smoking, drugs of abuse	education
Environmental chemicals	quantitation of exposure, strategies for avoidance or reduction of environmental pollution

The assessment of potential human health risks resulting from the exposure to chemicals provides the basis for appropriate regulatory and control measures. The health risk assessment determines whether a xenobiotic may cause adverse health effects, at what level and frequency of exposure, and the probability that adverse health effects will occur. The term "risk assessment" is increasingly used in the context of potentially toxic chemicals. Scientific risk assessment considers the available data on the toxicology of a specific chemical when judging which agents potentially pose a significant risk to the human population. Toxicology focuses on the identification and quantitation of potential hazards by using animal studies as surrogates for humans. Permissible exposure levels for humans are derived from the results of the animal studies by using margins of safety or defining "acceptable" incidences of adverse health effects in exposed humans [185].

Health risk assessment and its use in regulatory decisions have recently generated intense controversy. The debate over risk assessment is politically and emotionally charged, and creates an adverse atmosphere heightened by the extraordinary sums of money at stake. Industry complains that the costs of complying with possible overregulation based on inappropriate risk assessments may be excessive; moreover, lawsuits on potential environmentally caused diseases, especially in the United States, involve huge sums of money. On the other hand, environmentalists claim that risk assessment practices and policies do not adequately protect human health; moreover, health care costs for the treatment of environmentally caused diseases may also be very high. These considerations have led to an intensive rethinking of the health risk assessment process and have increased the awareness that in many cases, the scientific foundation for risk assessment is weak. This rethinking led to the conclusion that resolution of the controversies by the development of effective prevention strategies and rational priority setting may be achieved only by strengthening the scientific background and available data by research and by developing better methods to estimate risks due to chemical exposures [185–187].

Before considering the practice of health risk assessment, several terms frequently used and misused in risk assessment and its perception should be clarified. In discussions on health effects of potentially toxic chemicals, the terms "hazard" and "risk" are often used with an identical meaning, although they are clearly different. Hazard defines the intrinsic toxicity of a chemical and is not identical to risk. Risk is the estimated or measured probability of injury or death resulting from exposure to a specific chemical. Risk may be described either in semiquantitative terms such as high or low risk or in quantitative terms such as one person experiencing an adverse effect per 10 000 persons exposed. Risk may also be described in absolute terms (probability of adverse effects due to a specific chemical exposure) or in comparative terms by comparing the probability of adverse effects between a population exposed to an agent and an unexposed population.

The health risks due to the contact with potentially toxic chemicals are dependent on the conditions of exposure, since not only the intrinsic toxicity of a chemical determines the magnitude ot the adverse effect but also the dose. As noted in Chapter 1, the magnitude of the toxic effects is the product of the intrinsic toxicity of a chemical multiplied by the dose taken up by exposed animals or humans; thus, all toxic effects are dose-dependent and even very toxic chemicals may not cause toxic effects when the dose is low. If the dose is zero, despite a very high intrinsic toxicity of a specific chemical, the toxic effect and the risk of adverse health effects will be zero. On the other hand, chemicals with low intrinsic toxicity may induce toxic effects when the dose is high and may thus pose a significant risk. In toxicological terms, risk is therefore the product of the intrinsic toxicity of a chemical and the exposure characteristics.

5.1. Acceptable risk, Comparison of Risks, and Establishing Acceptable Levels of Risk

In earlier phases of risk assessment, the basic belief was that few chemicals are toxic and all of these toxic chemicals are derived from synthetic processes. To achieve a zero risk, chemical exposure must be reduced below a threshold level, under which it causes absolutely no risk. However, where such a threshold cannot be demonstrated, one must assume that a finite risk may occur at any exposure level, consequently, absolute control of risk is possible only if the source of exposure is eliminated altogether. These considerations resulted in the zero-risk concept. The Delaney Clause of the Federal Food and Drug Act in the United States is an example of a zero-risk approach in the regulation of food additives. This law states that no xenobiotic whose carcinogenic potency in animals has been demonstrated may be used as a food additive.

However, the more widespread testing of chemicals for toxicity, the increased sensitivity of analytical instruments to determine chemicals in the environment and at the workplace, and the developments in the science of toxicology put the basic assumption of the zero risk concept – that is, only synthetic chemicals are toxic – into question.

However, the more widespread testing of chemicals for toxicity, the increased sensitivity of analytical instruments to detect chemicals in the environment and the workplace, and developments in the science of toxicology put the basic assumption of the zero-risk concept – that only synthetic chemicals are toxic – in question. These developments led to the recognition that zero risk was unachievable and, perhaps, unnecessary for the regulation of chemicals. The observation was based mainly on a few facts: (1) all chemicals, both of synthetic and natural origin, are toxic under specific exposure conditions; (2) most of the hazardous chemicals routinely encountered by humans are of natural rather than synthetic origin; (3) most of the exposure to hazardous synthetic chemicals cannot be avoided entirely or be eliminated from the environment without changing profoundly the way of life in many countries; and (4) in the case of cancer risk assessment, DNA damage and mutations, assumed to be of major significance in the process of carcinogenesis, occur spontaneously, albeit at a low rate. Examples of endogenously occurring DNA damage are hydrolytic deamination, depurination, oxidative modification, and endogenously formed DNA adducts. Well-known examples of hazardous synthetic chemicals are benzene, which is present in the environment as a result of its emission from motor vehicles, cigarette smoking, and other sources, or 2,3,7,8-tetrachlorodibenzodioxin formed in forest fires. The effect of naturally occurring chemicals and chemical exposure due to life-style factors is best exemplified by the estimated contribution of different factors to the incidence of avoidable cancers in humans (Fig. 68).

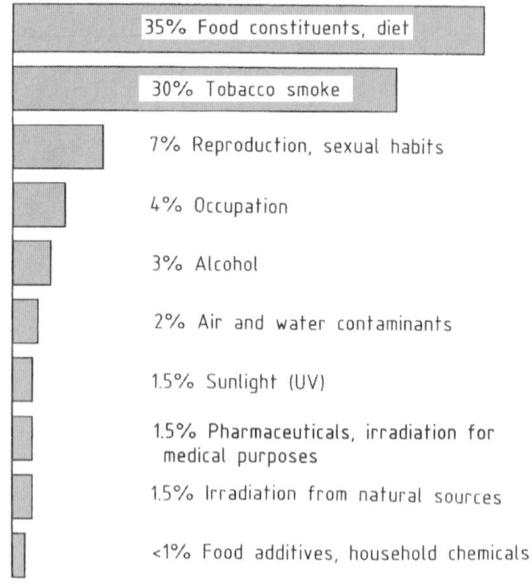

Figure 68. Contribution of chemical exposure and life-style factors to the incidence of avoidable cancer in human [200]

According to the large epidemiological study of Doll and Peto (1981), natural chemicals in diet and chemicals inhaled by cigarette smoking are the major causative agents in human cancer. Occupational and environmental exposure to synthetic chemicals constitutes only a minor causative factor.

Given these facts, the acceptable risk concept was developed as an alternative. The acceptable risk concept realizes that it is not possible to eliminate all potential health risks associated with chemical exposure due to the life style. According to the concept of acceptable risk, safety – the reciprocal of risk – is no longer an absolute term but is redefined as a condition of certain, but very low and thus acceptable, risk. This conceptual change improves the ability to deal with potentially very low risks identified by the increased sensitivity of analytical instrumentation and with increasingly sensitive scientific methods to detect potential adverse effects of xenobiotics. The concept of acceptable risk also permits the definition of limits for the exposure to toxic chemicals that can be considered to have a negligible impact on the incidence of adverse effects in an exposed population. Risk assessment is therefore unavoidable and must implicitly or explicitly involve a balance of risk and benefit [185–189]. Some of the main factors considered in establishing acceptable risk levels for exposure to a chemical follow:

Beneficial aspects
- Economic growth
- Employment
- Increased standard of living
- Increased quality of life
- Taxes generated

Detrimental aspects
- Decreased quality of life
- Health effects
- Lawsuits
- Loss of environmental resources
- Loss of work
- Medical expenses

5.2. The Risk Assessment Process

Several individual elements make up the risk assessment process. In the first step, the potential adverse health effects of a chemical, a mixture of chemicals or a specific technical process are evaluated by the application of toxicity tests (for details, see Chap. 4) and, if the chemical is already in widespread use and humans are exposed, by considering the data from epidemiological studies. The second step in risk assessment determines the dose–response for the observed adverse effects. In parallel, the exposure of humans to this xenobiotic is quantitated by analytical procedures or, if the chemical is not yet in widespread use, by the estimation of likely exposure scenarios. In the third step, the results obtained in the toxicity studies are extrapolated. This involves an extrapolation both from adverse effects seen in experimental animals to humans, and often an extrapolation from the effects seen after high doses in animals to the much lower doses humans usually encounter. The last step of risk assessment, risk characterization, involves the combination of steps one to three to judge the existence and magnitude of the public health problem and characterizes the uncertainties inherent in the risk assessment process

5.2.1. Hazard Identification Techniques

Hazard identification is the step in which the adverse effects of the xenobiotic are determined. Evaluation of both acute and chronic toxicity is performed by using animals as experimental models for humans. The use of animals as surrogate for humans is based on the following assumptions: (1) xenobiotics with a likely adverse effect in humans will manifest some degree of toxicity in other living systems when the dose is sufficiently high; (2) if a sufficient number of animal species are dosed with the xenobiotic, at least one should exhibit a similar pattern of biotransformation and toxicokinetics to that seen in humans; and (3) if a sufficient number of different animal species are dosed with the xenobiotic, at least one is likely to exhibit the toxic responses and clinical symptoms occurring in humans.

Typical end points in studies aimed at hazard identification in intact animals include mortality, reproductive and developmental effects, target organ toxicity, and cancer. Hazard identification studies at present also include the determination of a range of biochemical end points related to specific toxic effects such as toxicokinetics, routes and extent of biotransformation, structure of reactive intermediates, and binding of reactive intermediates to cellular macromolecules. In addition, many short-term tests for specific toxic effects such as DNA damage, mutagenicity, or clastogenicity are increasingly included in hazard identification procedures.

The acute toxicity of xenobiotics is evaluated by a number of procedures from which the LD_{50} may be calculated. With the more recently recommended fixed-dose method information on target organs affected and types of toxic effects may also be obtained. Repeated-dose toxicity studies last between two weeks (subacute toxicity studies) and 6–24 months (chronic toxicity studies), the lifespan of the animals, including post-mortem examination, histopathology, clinical chemistry, and hematology at termination and at specified time points during the study. From the chronic toxicity studies, the lowest observed effect level (LOEL), also referred to as the lowest observed adverse-effect level (LOAEL), and the no observed effect level (NOEL) are obtained for noncancer endpoints. The NOEL is the highest does administered that does not induce observable toxic effects. The NOEL may not be identical to the no-effect level if insensitive methods are applied or the wrong end point is chosen. Moreover, the value obtained for the NOEL will depend on the number of animals used in the study and the spacing of the applied doses.

Short-Term Tests for Specific Toxic Effects *In Vivo* and *In Vitro*. A variety of short-term tests has been developed for the detection and quantification of toxic effects. Most of the more established and well-evaluated test systems are designed to evaluate the genotoxic activities of xenobiotics and employ well-defined genetic changes (DNA damage, gene mutations, chromosome defects, cell transformation) in prokaryotes, lower eukaryotes, and mammalian cells. Knowledge that a genotoxic chemical is active in vivo in the target organ of carcinogenesis enhances confidence that the genotoxicity of the chemical is important in the process of cancer induction by that chemical. Before any short-term test can be used with confidence to assess potential toxic effects, its validity should be thoroughly evaluated and its major drawbacks should be explicitly noted. Moreover, it should be kept in mind that most available short-term tests are well designed to give qualitative information and their major use for risk assessment purposes is therefore the confirmation or exclusion of a specific toxic response. The magnitude of toxic response in intact animals or humans depends on both toxicokinetics and toxicodynamics. The toxicokinetic phase of the toxic response is not considered in most of the in vitro short-term tests. Therefore, short-term tests in vitro may be used for hazard identification only in combination with studies on adverse effects in animals and studies on the toxicokinetics of a xenobiotic.

Biotransformation and Toxicokinetics. Studies on the extent of biotransformation including structural identification of the metabolites formed from the xenobiotic both in intact animals and in appropriate in vitro systems such as organ homogenates or fractions with enzymatic activity also contribute to hazard identification [190]. The structures of metabolites formed and the presence and extent of covalent binding of metabolites to macromolecules such as protein and DNA indicate the formation of electrophilic metabolites and thus a potential hazard.

For example, the structure of excreted mercapturic acids may give information on the structure and reactivity of the electrophilic metabolite and sites of cellular interactions [191]. Information on the rate of absorption and elimination may indicate a possible accumulation of the xenobiotic in humans, with the consequence of potential adverse effects. These studies should be performed on at least two animal species in vivo. For in vitro studies, human tissue samples should be included to confirm that biotransformation reactions are identical to those, observed in animals occur in humans. Different mechanisms of toxicity may operate at different dose levels; in these instances, toxicokinetic data may help in understanding dose-dependent mechanistic differences. The toxicokinetics of a xenobiotic in humans may be extrapolated by physiologically based pharmacokinetic models from the results obtained in experimental animals. Information on potential pathways of biotransformation may also be made by computerized structure analysis of the xenobiotics with specific computer programs [147, 192]. These are designed for predicting routes and rates of biotransformation based on the presence of functional groups in the molecule; however, at present, the available programs are far from perfect, and they should only be used in conjunction with expertise and as a basis for experimental planning.

Structure – Activity Relationships and Chemical Structure Analysis. Predictive data on the potential of xenobiotics, mainly organic chemicals, to induce adverse effects may be derived from relationships between chemical structure (physicochemical properties, presence of functional groups, atomic configuration) and biological activity, termed structure–activity relationships.

Two approaches to hazard prediction utilize structure–activity relationships. The first is essentially a qualitative approach and involves the comparison of the structure of the xenobiotic with that of other compounds already known to cause specific toxic effects. The comparison with known structures and the knowledge of biotransformation reactions and mechanisms of toxicity permits the identification of toxophores in the structure of xenobiotics. Toxophores are functional groups present in the molecule which are likely converted to toxic metabolites or posess chemical reactivity related to mechanisms of toxicity. Examples of toxophores are olefinic moieties, which may be oxidized to epoxides, and terminal carbon atoms in aliphatics bearing two halogen atoms, which may be oxidized to acyl halides. The scope and usefulness of structure–activity relationships depend on the availability of a sufficiently large database on the toxic effects of chemicals with common subgroups and structures [147], [192]. The main limitation of this approach is the qualitative nature of the resulting estimate of potential toxicity. The second approach involves quantitative structure–activity relationships and relies on the computer-aided analysis of databases on toxic effects of chemicals. A basic feature of the applied techniques is the use of pattern-recognition schemes or substituent weighting factors coupled with regression analysis. Some useful predictions have been made with these techniques, but due to the complex nature of toxic effects and the multitude of factors governing the toxic response, there are still severe limitations to the applicability of quantitative structure–activity relationships for predicting toxicity profiles.

Clinical and Epidemiological Studies. In principle, the best evidence for the toxic effects of a chemical in humans is derived from clinical and epidemiological studies. These studies assess effects in the species of interest for risk assessment and at relevant concentrations. Potentially confounding extrapolations from high to low dose and from animals to humans are not required. However, several weaknesses limit the applicability of these studies to hazard identification. In most cases, reliable exposure data are lacking, consequently, dose–response relationships cannot be established. In addition, the sensitivity of epidemiological studies to detect health problems is comparatively low. Unless the toxic effect of a particular xenobiotic is very unusual in control (unexposed) groups, it may pass unnoticed in a normal survey. Examples are the identification of asbestos exposure as a cause of mesothelioma or vinyl chloride as a cause of hemangiosarcoma, very rare forms of cancer in humans not exposed to asbestos or vinyl chloride, respectively. Moreover, the results of epidemiological studies, especially on cancer risk, may reflect the risk associated with exposure to chemicals decades ago because of the long latency period. Other limitations are confounding variables such as smoking and concomitant exposure to other xenobiotics, which often impair the interpretation of carcinogenicity data in humans. Thus, evidence based on epidemiological observations has identified only a limited number of chemicals as human carcinogens; many of the identified compounds are used in cancer chemotherapy and have the intrinsic property of genotoxicity (Table 40).

Table 40. Examples of established human carcinogens based on epidemiological observations

Chemical or agent	Site of tumor formation
Aflatoxin	liver
Alcoholic drinks	mouth, esophagus
4-Aminobiphenyl	bladder
Benzidine	bladder
2-Naphthylamine	bladder
Arsenic	skin, lung
Asbestos	lung, pleura, peritoneum
Azathioprine	reticuloendothelial system
Benzene	bone marrow
Bis(chloromethyl) ether	lung
Cadmium	prostate
Chlorambucil	bone marrow
Chlornaphazine	bladder
Chromium	lung
Cyclophosphamide	bladder
Bis(2-chloroethyl) sulfide	larynx, lung
Nickel compounds	nasal cavity, lung
Estrogens	endometrium, vagina
Phenacetin	kidney and lower urinary tract
Polycyclic aromatic hydrocarbons	skin, scrotum, lung
Steroid hormones	liver
Tobacco	mouth, pharynx, larynx, esophagus, lung, bladder
Vinyl chloride	liver

5.2.2. Determination of Exposure

The quantification of exposure, both in individuals and in populations, is a prerequisite for the quantification of risk. Reliable data on exposure are needed to assess the adverse effects of the xenobiotic and to recognize specific risk factors such as occupation, life style, and social status. The dimensions of exposure include intensity, frequency, route, and duration; in addition, the nature, size, and makeup of the exposed population should be characterized. The assessment of exposure is a difficult and complex task, and is often neglected. Typically, estimations and field measurements are required. The estimation of human exposure to a particular xenobiotic involves an initial estimation of the possible sources of the chemical and the possibilities for exposure. A good inventory of sources may provide important information on critical pathways of exposure, populations at particular risk, and the levels of exposure.

In many cases, the duration and level of exposure, especially after chronic contact, may only be estimated from ambient levels of the xenobiotic in the environment, and estimations may thus be crude; owing to the large numbers of potentially exposed persons, only in special situations (e.g., occupational exposure, after disasters), will exposure data, including determination of the internal dose, be available.

Specific procedures to detect exposure to a certain xenobiotic and procedures to estimate exposure by determining biological effects may be used. These include:

- Direct measurement of the chemical in environmental samples such as water, air, and soil.
- Measurement of the chemical, its metabolites or products of the interaction of the chemical or its biotransformation products with cellular macromolecules (protein and/or DNA) in body fluids and tissues (biomonitoring).

Biological end points in exposure assessment may be:

- Assessment of biochemical indicators for specific adverse effects known to be caused by the xenobiotic, e.g., inhibition of specific enzymes such as cholinesterase activity in persons exposed to organophosphate pesticides.
- Observation of pathological evidence of exposure such as cytogenetic changes in lymphocytes from workers exposed to chromosome-damaging chemicals at the workplace. However, the use of biological end points for exposure assessment lacks the resolving power to discriminate between endogenous changes and the effects of xenobiotics, and therefore chemical specific indices of exposure should be favored for quantitation of exposure [178, 193].

Due to the time-consuming and cost- and labor-intensive procedures required, data on exposure to xenobiotics are usually limited. Difficulties in identifying concomitant exposures, interactions with other xenobiotics or activities, special risk groups such as the very old or very young and pregnant women, and patterns of exposure result in a high degree of uncertainty in exposure assessment in human populations.

A stepwise approach to exposure assessment using the following hierarchy can be applied. The most reliable exposure assessments are based on measured data, including the quantification of key exposure determinants. When these are not available, appropriate surrogate data may be used. Modeling may also be used in the absence of useful data, but the limitations of the modeling approach should be clearly stated.

5.2.3. Dose-Response Relationships

The establishment of the dose–response relationship for adverse effects in animals is the decisive step in risk assessment. This step quantifies the relationship between received or administered dose and biological response and may be performed on an individual or population basis. Dose–response assessment includes exposure intensity and duration and factors modifying toxic response such as sex, age, health status, and route of administration. Since the dose–response assessment can only rarely be performed in humans, extrapolation from data obtained in animals to humans is usually required. Moreover, many animal experiments, particularly carcinogenicity bioassays, are performed with high doses to increase the sensitivity of the assay. Therefore, besides species extrapolation, an extrapolation from effects seen after high doses in animals to the low doses usually encountered by humans is necessary. This extrapolation step from high, sometimes toxic doses in animals to low doses in humans is controversial.

The extrapolations form high to low dose are performed differently depending on the type of toxic response elicited by the xenobiotic. As explained in Chapter 1, toxic effects of a chemical may be caused by both reversible and irreversible interactions of the xenobiotic or its metabolite(s) with macromolecules in the organism. Many acute toxic responses such as carbon monoxide poisoning are based on reversible interactions and are associated with thresholds. Threshold doses are doses below which the probability of a response is zero. The biological basis for thresholds is well founded and may be demonstrated on a mechanistic basis. On the other hand, many chronic toxic responses, particulary chemical carcinogenesis, are often considered nonthreshold effects. Since a negative can never be proven, the absence of thresholds cannot be demonstrated by experiments and is based on consideration of the mechanisms of chemical carcinogenesis. Different approaches have thus been developed in establishing acceptable levels of exposure to threshold and nonthreshold responses (see below).

5.2.4. Risk Characterization

The hazard identification, the exposure assessment, and the dose–response assessment merge into risk characterisation. Risk characterisation estimates the incidence of expected adverse health effects in exposed populations. As noted in Section 5.2.1, risk characterization and the establishment of acceptable exposure levels are handled differently for carcinogenic and noncarcinogenic xenobiotics. For chemicals that cause adverse effects by mechanisms with thresholds, the safety factor approach was developed. For nonthreshold responses such as cancer, both quantitative and qualitative risk assessment procedures are used.

5.2.4.1. The Safety-Factor Methodology

When the safety assessment of a chemical is based on animal toxicity testing, a different evaluation of the data is required compared to safety evaluations relying primarily on human observations. Since animal experiments to determine health hazards of chemicals were widely used in the 1940s and human health risks had to be extrapolated from these data, the idea of a safety factor, which was to be applied to the results from animal studies, was developed. The safety factor approach was first proposed in the 1950s by Arnold Lehmann of the FDA as a protective measure for human health and is intended to compensate for uncertainties in the extrapolations of animal data to humans. This concept was applied to the determination of acceptable daily intakes (ADIs) by the World Health Organization from the 1960s onward. The ADI was defined as "the daily intake of a chemical which, during an entire lifetime, appears to be without appreciable risk on the basis of all known facts at that time."

The safety factor, or as it is sometimes called, the uncertainty factor, has been introduced to consider both interspecies and interindividual differences in response to potential toxic effects of the chemical under consideration. The major purpose of the safety factor is the protection of human health by establishing safe exposure levels; the exposure levels defined do not mean that exposure above these levels will result in adverse effects. Despite several limitations and criticisms, the safety factor approach has been used for many years in Western Europe and the United States, and has proven useful and reliable. It was thus adopted internationally as the standard procedure for assessing the ADI.

This risk assessement approach is based on the establishment of a point of departure for the required extrapolation, which may be an NOEL or NOAEL or a benchmark dose (e.g. ED_{05}) in animal studies on chronic toxicity, and then defining permissible human exposures by the application of a safety factor. In this context, the NOEL is defined as the lowest dose of the xenobiotic in an animal experiment which produced no detectable effect in the most sensitive animal species treated. Once a NOEL has been determined, a safety factor for human exposure is introduced and, often, a permissible level of exposure of one hundredth of the NOEL in animals is defined for humans. The 100-fold safety factor is justified on the basis of a 10-fold difference to reflect an interspecies difference in susceptibility and a 10-fold difference to reflect possible interindividual variations in susceptibility in humans [188]. The acceptable daily intake is then obtained by dividing the NOEL from the study by the safety factor.

The safety-factor approach assumes that toxic effects exerted by the chemical exhibit a dose–response curve with a threshold, that the results of the toxicity studies in animals are relevant to humans, and that extrapolation of the dose is reliable.

A similar approach is the margin-of-safety (MOS) approach which determines by how much the derived NOAEL exceeds the determined or estimated exposure. The size of the MOS (< 100 or > 100) determines the extent of concern of a specific exposure.

As noted above, safety factors of 100 are frequently applied. However, when reliable data for adverse effects of the chemicals in humans are available, a safety factor of only 10 may be applied. On the other hand, a safety factor of more than 100 is appropriate when no or only limited data on the toxicity of the chemical in animals are available. Note that the safety factor is not based on scientific evidence. Additional data on the mechanisms of toxic effects in animals and on the toxicokinetics and biotransformation of the xenobiotic in animals and humans are required and add scientific credibility to the proposed safety factors. Therefore, the choice of a safety factor should consider, in addition to the NOEL, a series of further qualitative parameters:

Evaluation of animal toxicity studies
- Number of studies and effects observed
- Type of toxic effects
- Time course for toxic effects
- Tumorigenicity

Evaluation of biochemical end points
- Biotransformation and toxicokinetics
- Mechanism of action, covalent binding to macromolecules
- Short-term test for genotoxicity and other nonthreshold effects

Evaluation of species differences
- Interspecies variations in biotransformation and toxicokinetics
- Influence on anatomical and physiological differences of toxic effects between species

Application of the most appropiate safety factor requires careful analysis of the data available on the toxic effects for every chemical, on a case-by-case basis. Consideration should be given to the quality and completeness of the data and the number and spread of the dose levels used. Besides the derived NOEL, all further information obtained in the long-term animal toxicity studies is valuable for characterizing toxic effects and determining safety factors. The type of toxic effect and the shape of the dose–response curve should also be taken into account in setting the size of the safety factor.

Also important are considerations of the ability to extrapolate from toxic effects seen in animals to human exposure scenarios and the results of human epidemiology studies, when available. Because of the complexity of information required, no universally accepted guidelines can be developed for determination of the precise magnitude of safety factors; therefore, expert judgment on an individual basis is a major factor contributing to the size of safety factors.

The safety factor approach has several problems. In experiments to determine the NOEL in animals, the experimental group size and the spacing of dose levels are major determinants of the numerical value of the NOEL obtained; the smaller the size of the individual dose groups and the larger the spacing of doses, the less likely is it that an effect will be observed. This phenomenon has the effect of rewarding poor experimental design because small group sizes in experiments will tend to produce higher NOELs. Moreover, nonthreshold effects may not be detected due to the low number of animals enrolled or the dose levels applied. In addition, the slope of the dose-response curve is often not considered or may not be determined with sufficient accuracy, and chemicals with steep and shallow dose–response curves are treated alike. Advantages and disadvantages of safety factors in the risk assessment process follow:

Advantages
- Simple application
- Ease of understanding
- Flexibility of use
- Use of expert judgment

Disadvantages
- Uncertainties of threshold values and size of safety factor
- No risk comparison possible
- Slope of dose–response curve not adequately considered
- Experimental NOELs are dependent on group size in the animal toxicity testing and end point selected

One of the most promising alternatives to the use of NOAELs or NOELs is the benchmark concept [194]. In the benchmark approach, a dose–response curve is fitted to the complete experimental data for each effect parameter. On the basis of the fitted curve, the lower confidence limit of the dose at which a predefined critical effect size is observed (i.e., the dose at which adverse effects start to arise or where 5% of the animals are predicted to be affected, effective dose, or ED_{05}) is defined as the benchmark dose. Advantages of this approach over the NOAEL are:

- The benchmark dose is derived by using all experimental data and gives a better reflection of the dose–response curve.
- The benchmark dose is independent of predefined dose levels and spacing of dose levels.
- The benchmark approach makes more reasonable use of sample size, and better study designs result in higher benchmark doses.

A disadvantage of this method is the uncertainty with respect to the reliability of the approach when results are obtained from toxicity studies performed according to the requirements defined in current guidelines. For the derivation of reliable dose–response relationships, the classical study design of three dose groups and a vehicle control group is limited, since adverse effects may only be observed at the highest dose level. An improved benchmark model fit could be achieved by increasing the number of dose groups without changing the total number of animals in the test.

At present, the determination of a NOAEL is mandatory for risk assessment in the EU. Nevertheless, the benchmark dose method can be used in parallel when a NOAEL cannot be established for the selected toxicological end point because only a LOAEL is available. In this case, benchmark modeling preferred over LOAEL–NOAEL extrapolation, which uses more or less arbitrary assessment factors. Benchmark dose software (BMDS) is available from the US EPA internet site (www.epa.gov).

5.2.4.2. Risk Estimation Techniques for Nonthreshold Effects

The procedures outlined below are most often applied to the low-dose risk estimation of human or animal carcinogens. Since dose–response data are not available for effects in animals at doses relevant to human exposure, extrapolations are required for determining the potential human cancer risk. The methodology employs mathematical modeling to characterize the relationship between exposure and response or to place an upper bound on the dose–response relationship. Dose–response data, available from specific study situations (mainly animal bioassays using high doses of the chemical and, sometimes, heavily exposed population groups), are extrapolated to the often much lower exposures of the general population in order to calculate the possible risk. Therefore, cancer risk assessment generally involves extrapolating risk from the relatively high exposure levels employed in animal studies, or occupational studies where cancer responses can be measured, to risks at the relatively low exposure levels that are of environmental concern. However, since the majority of carcinogenicity experiments use only two or three doses, it is impossible to assess the shape of the dose-response with a reasonable degree of precision. Risk assessment must therefore rely on some arbitrary assumptions about the shape of the dose-response relationship at low doses. Risk estimates thus obtained are not true or actual risk but values obtained by extrapolation well below the range of experimental observations. A summary of the advantages and disadvantages of quantitative risk assessment follows:

Advantages

- Gives numerical values on risk that may be used for setting exposure limits.
- Permits the comparison of risks due to different chemicals.
- Provides a reasonable basis for setting exposure limits by identifying compounds with high risk.

Disadvantages

- Extrapolation of data obtained at high doses to low doses relevant for human exposure by means of mathematical models which are not based on cancer biology and pathophysiology.
- Mechanistic and kinetic data are not used for the risk estimation process.
- Expensive and time- consuming lifelong bioassays are required.

The risk extrapolation techniques used are based on several conservative default assumptions, some unsupported by any direct empirical evidence. Conservatism is introduced to ensure maximum protection for those exposed to presumed chemical carcinogens. These assumptions were adopted to achieve some consistency in the application of risk-estimating techniques. The major assumptions are (1) carcinogenic risks are estimated from the data obtained in the most sensitive animal model only using positive responses (data from bioassays that do not show a treatment-related increase in cancer incidence are ignored); (2) linear, nonthreshold dose-risk models are applied; (3) statistical upper confidence limits are used rather than best estimates; and (4) a linear dose–response curve is assumed at low doses.

5.2.4.3. Mathematical Models Used in High- to Low-Dose Risk Extrapolation

Mathematical models for quantifying human cancer risk from exposure to carcinogens were first developed in the 1950s. These models were based on the one-hit or the multistage model of chemical carcinogenesis. A probit model was proposed in 1961 for assessing low-dose risk by extrapolation. An improved probit model was introduced in the 1970s by the FDA for computing the level of carcinogenic chemicals permissible in food. This procedure included the suggestion that the dose causing a very low risk (one additional cancer in 1 000 000 exposed people) to be considered as a "virtually safe dose." Introduction of this mathematical model into the risk assessment process stimulated the development of a variety of other mathematical models for carcinogen risk assessment. The basis of all the above methods is to apply a mathematical model to the tumor incidence observed in a long-term animal bioassay.

A major problem in dose– response and risk extrapolation is the determination of an appropriate mathematical model to predict effects at hypothetical low levels of exposure. Several models have been developed, and most of them contain analytical functions that appear to fit the experimental dose range quite well and also contain a dose–response functionality. Most models differ in the functions used to estimate response in the very low dose range (i.e., as dose approaches zero). The mathematical models in common use in carcinogen risk assessment are

- The linear model.
- Statistical or distribution models: log–probit, Mantel–Bryan, logit, Weibull.
- Mechanistic models: one-hit (linear), multihit, multistage (Armitage–Doll), linearized multistage, Moolgavkar model.
- Other models: statisticopharmacokinetic, time-to-tumor.

Linear Model. Linear extrapolation involves the intersection of a straight line between the origin (zero dose) and the upper confidence limit of the response at the single, lowest experimental dose. This model is based on the assumption that the increase in tumor incidence by the applied xenobiotic augments an already proceeding process.

Distribution models are based on mathematical functions of presumed population characteristics i.e. on the assumption that every member of a population has a critical dosage (threshold) below which the individual will not respond to the exposure in question. The probit model assumes that log dose–responses have a normal distribution. This model serves as the basis for the Mantel-Bryan risk extrapolation procedure. Other distribution models on which carcinogenicity dose–response models have been based include the logit and Weibull models.

Mechanistic models are based on the currently presumed mechanisms of chemical carcinogenesis. Each model reflects the assumption that a tumor originates from a single cell. The concept underlying the one-hit model is that a tumor can be induced by exposure of DNA to a single molecule of a carcinogen. This model is essentially equivalent to assuming that the dose–response is linear in the low-dose region and as a consequence tends to produce very low calculated "virtually safe doses" compared with the other currently applied models. The multihit model is a generalized version of the one-hit model which assumes that more than one hit is required at the cellular level to initiate carcinogenesis.

The biological justification for the multistage (Armitage–Doll) model is that cancer is assumed to be a multistage process that can be approximated by a series of multiplicative linear functions. It assumes that the effect of a chemical carcinogen occurs in multiple steps and that the effect of each step is additive. The dose–response predicted by this model is approximately linear at low doses and it results in estimates of potential risk that are similar to those of the one-hit model.

The Moolgavkar–Venson–Knudson model attempts a more comprehensive consideration of the biologic processes of cancer formation than the other mechanistic models and may provide a more accurate estimate of human risk by reducing some of the interspecies uncertainty. It is based on a two-stage growth model and considers the birth and death of cells, the effect of cell proliferation on the number of available cells for malignant transformation, and assumes two specific, irreversible, and rate-limiting mutational events to be necessary for cancer development. This model may quantitatively consider major phenomena influencing cancer formation such as cell proliferation, initiation/promotion, genetic predisposition, and synergism/antagonism. Moreover, model parameters may be obtained experimentally. At present, the major limitation for the application of this model is the lack of availability of many of the important biological parameters.

Other Models. The carcinogenicity of many chemicals is based on their biotransformation to reactive metabolites. The statisticopharmacokinetic model arises from consideration of competing metabolic activation and deactivation processes (e.g., detoxification and DNA repair) and estimates the "effective dose", i.e., the level of reactive metabolites formed and interacting with critical macromeolecules, rather than the administered dose.

A modification of the Probit model relates it to the time at which a tumor is detected. For risk assessement, this time-to-tumor model uses the time to observance (latency) in addition to the proportion of animals bearing tumors at each dose.

5.2.4.4. Interpretation of Data from Chronic Animal Bioassays

The results of experimental animal studies may provide a variety of data in addition to the simple indication of the presence or absence of cancer. This information may be useful for the evaluation of potential human cancer risk and includes the number of neoplasms per animal, the number of different types of neoplasms observed, and the number of species affected. The organ or target tissue in which the carcinogenic response occurs is also important, as some rodents have extremely high and variable spontaneous incidence of certain tumor types. Where a chemical increases the number or accelerates the formation of neoplasms which spontaneously occur in high and variable incidence, the response carries less weight than the appearance of tumors with very low spontaneous rates. The time to development of tumor will also give an indication of potency. Further considerations on the interpretation of the obtained data are given in Section 4.8.

5.2.4.5. Problems and Uncertainties in Risk Assessment

The practice of using tumor incidences obtained in long-term animal experiments with high doses (MTD, maximum tolerated dose, see Section 4.8) for human cancer risk assessment and several of the conservative default assumptions used in the extrapolation processes have become the subject of major criticism. Some of the assumptions made when using experimental animal data for human risk assessment are not directly testable experimentally. Nevertheless, these assumptions are widely used because risk assessment would be difficult or impossible without them [188].

However, recent results on the mechanisms of carcinogenesis have revealed a complex biological process with many variables; a comprehensive consideration of all these parameters by a single, generally applicable mathematical model is not possible. Moreover, observations such as nonlinear toxicokinetics and delineated mechanisms of action for nongenotoxic carcinogens have indicated that a more scientifically based approach to carcinogen risk assessment, likely in a case-by- case examination by expert panels, may be more appropriate for defining actual risk than mathematical modeling. The major points of controversy in risk extrapolation from animal experiments and possible solutions to reduce uncertainties are described in Table 41.

Table 41. Uncertainties in quantitative risk assessment and the application of scientifically based methods for reduction of these uncertainties

Uncertainty	Reduction by
Extrapolation between species	physiologically based toxicokinetic modeling, comparative studies on toxicokinetics
Extrapolation from high to low dose	toxicokinetic modeling, quantitative determination of dose – response for biochemical effects of xenobiotics (e.g., DNA adducts, genetic changes), mechanisms of carcinogenesis
Extrapolation from controlled experimental conditions to variable human situations	none
Conservatism	elucidation of mechanism of actions
Toxicity and increased cell proliferation	dose – response for cell proliferation and cytotoxic effects
Dosimetry	toxicokinetic modeling and physiologically based toxicokinetic modeling
Mixtures	research on mechanisms of interactions
Bioassay-inherent factors in risk assessment	expert judgment

Extrapolation Between Species. In the absence of information on the possible mechanism of carcinogenesis by a particular xenobiotic, the tumor incidences observed in an animal experiment are often assumed to be useful in assessing human risk. However, the differences in the physiology and anatomy of laboratory animals and humans are well recognized. For example, the life span of the laboratory rodent is approximately two years whereas that of humans is approximately seventy years. Cancer appears to develop in rodents over a time scale that is proportional to the life span, and it is generally assumed that this will apply to all chemicals being assessed. The assumption that humans will respond in a similar fashion to laboratory animals is frequently shown to be inappropriate. Mechanistic studies demonstrate that qualitative and quantitative aspects of toxicokinetics and biotransformation, DNA repair and tissue susceptibility, and immune and other defense mechanisms may explain observed differences in the response of laboratory species and humans to exposure to carcinogens [195]. Expert judgement may be required to assess the nature of the end point or the mechanisms of carcinogenic action of the chemical in the experimental animal and to decide whether they are relevant to man.

Extrapolation from High to Low Dose. The use of very high doses in animal cancer bioassays and the required extrapolation form high to low dose are a major point of critisim. Most cancer bioassays are performed in relatively small groups of rodents (between 50 and 100) and with administration of high doses. The highest dose selected is usually the "maximum tolerated dose" (MTD) in order to have maximum sensitivity and to ensure that the results obtained do not overlook a carcinogenic response simply because the dose was too low.

The following information that is obtained in bioassays with high doses in animals:

- Definite identification of compounds with carcinogenic activity in the species.
- Information about relative potency of different chemicals.
- Information about the carcinogenic activity of the test chemical, when administered alone.
- Characterization of tumor types, target organs, and presence or absence of dose–response relationship, which permit comparison of different chemicals and help in establishing structure–activity relationships to improve predictive capabilities.
- Information on the lack of carcinogenicity of many chemicals to assist priority setting in public health.

However, they also have the following limitations:

- No direct information about effects at doses lower than dose studied.
- No information on the mechanism of carcinogenicity.
- No information on the effects of the test compound, when administered together with other chemicals (synergistic/antagonistic effects).
- Use of high doses that may cause unspecific toxic effects contributing to carcinogenicity.
- Acute toxic effects that may prohibit long-term administration of specific chemicals in sufficiently high doses to cause tumors.

Human exposure to carcinogens usually occurs at doses several orders of magnitude lower than those used in the experiment. Clearly, the possible shape of the dose–response relationship is vitally important in establishing the likely effects at doses substantially below those in the observable range. Mathematical models used for extrapolation only give the upper limits of risk, the real risk may be somewhere between zero and the number calculated. Since the dose–response relationship can not be determined experimentally, application of different mathematical models results in markedly different slopes of the curves in the low-dose range (Fig. 69), and this has major implications for risk assessment and the establishment of ADI values.

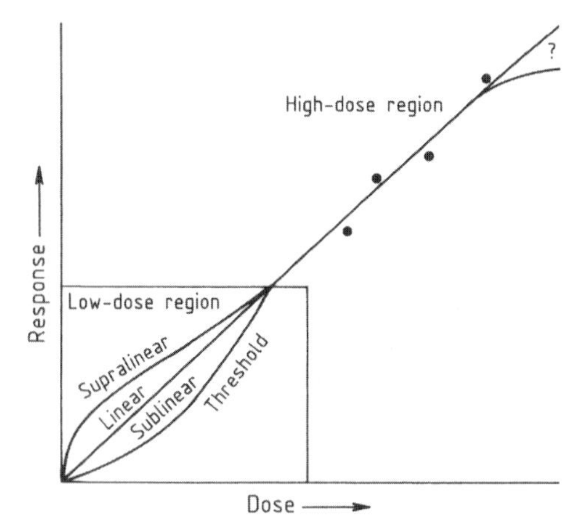

Figure 69. Possible slopes of dose – response curves in the very low dose range below the ability of experimental determination in cancer bioassays
• = Experimental data point

Only one study to determine the actual shape of the dose–response curve at low doses with group sizes large enough to constitute statistical significance was performed, the so-called ED_{01} study using the potent carcinogen 2-acetylaminofluoren as a model compound administered in the diet to mice. Urinary bladder and liver neoplasms were found to be related to 2-acetylaminofluorene administration. The resulting tumor incidences showed that even within one species and with one specific carcinogen, there are major differences in the shape of the dose–response curve at low doses taht depend on the target organ affected. The incidence of bladder tumors at all time points suggested the presence of a threshold for this end point, whereas the dose–response data for the liver tumors was nearly linear at all time points and did not indicate a threshold (Fig. 70).

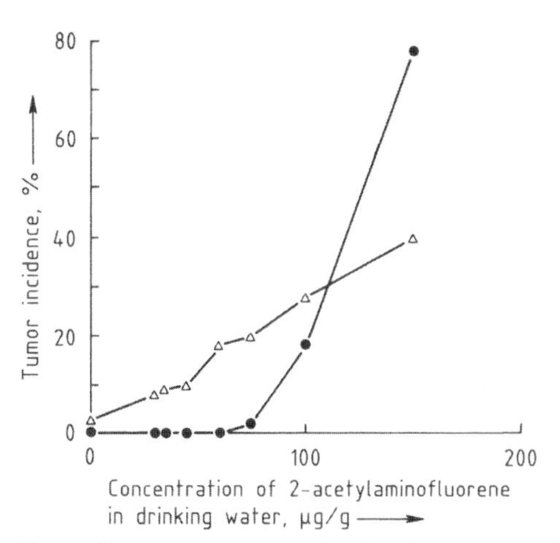

Figure 70. Dose – response data for the induction of liver (Δ) and urinary bladder (•) tumors induced by 2-acetylaminofluorene in mice after 24-month administration in diet

Extrapolation from Controlled Experimental Conditions to Variable Human Situations. In most animals studies, all critical factors are carefully controlled to guarantee the reproducibility of experimental observations. This is in contrast to the sometimes very large differences and time-dependent changes in the environmental exposure and other circumstances of the human population, which cannot be simulated in animal experiments. The complexity of these circumstances (i.e., changing life-style factors or workplaces) would necessitate an assessment of each human individually. Thus, risk assessment normally extrapolates directly from the animal data unless good evidence exists to suggest that an important confounding factor has been introduced. Besides the variable exposure situations, the human populations exposed to chemicals often differ in age, sex, and ethnic background and will certainly have a more heterogeneous genetic makeup than experimental animals. For the majority of carcinogenic risk assessments, individual variability can be taken into account only by the application of safety factors to compensate for these uncertainties.

Dosimetry. The problems of estimating environmental and/or occupational exposure to potential carcinogens have been described above. There are equally difficult problems in extrapolating the doses applied in animal experiments to those which might be encountered by humans due to environmental or occupational exposure and to distinguish between administered dose and effective dose.

Doses can be extrapolated from animal experiments to humans by five methods:

1) Expression of dose as a function of body weight (mg/kg or mmol/kg).
2) Expression of dose as a concentration in food or water (usually parts per million, i.e., μg/g or μL/L).
3) Expression of dose as a concentration in inhaled air (usually parts per million, i.e., μL/L).
4) Correction of dose for surface area. This is achieved in a complicated process by first raising the body weight of the laboratory animal to the power of 2/3 or 3/4. The correction converts doses that are expressed as concentrations in air or food to milligrams per kilogram of body weight in the laboratory species. The dose is then corrected for body surface area and converted back to concentrations in air or food by using appropriate conversion factors for humans. Corrections using surface area are based on the observation that metabolic rate is proportional to body surface area. For acute toxicity, this method may have some merit, but the rate of biotransformation may differently affect the potency of a carcinogen, depending on the role of biotransformation in detoxification or activation. Using correction factors for surface area provides data suggesting that humans may be more susceptible to potential carcinogens than laboratory animals compared to the data obtained by extrapolation on the basis of body weight.
5) Extrapolation assuming that the tissue dose is the primary determinant of carcinogenic response. This approach requires a study of the pathways of biotransformation and the kinetics of the pathway that generates the ultimate electrophile. In vitro studies may be necessary to obtain appropriate human data.

Further problems occur if the experiment is carried out by a protocol that is completely dissimilar from the human experience, for example, if animal exposure is carried out for a lifetime and human exposure is for a shorter period or if the animal experiment uses an exposure route that is not relevant for humans. As a general rule for chemical carcinogens, where the target tissue dose is presumably the most important determinant of the carcinogenic potency, the total body burden must be computed from the various routes of exposure to assess overall dose.

Mixtures. Cancer bioassays in animals are most often performed with a single chemical of definite high purity; exposure of experimental animals to other chemicals or other confounding factors such as vector-based disease is carefully avoided. In contrast, humans are continuously exposed to mixtures of chemicals and other agents. Thus, even when animals or humans are exposed to two or more chemicals, further complications of data interpretation are introduced. In general, the response of the animal and the human will depend on whether the chemicals' activities are interactive or not and whether the effects of the chemicals are similar or not. If the chemicals are not interactive and the response is similar, an additive effect may be assumed to occur. If they are not interactive, the overall response may be less than additive. In addition, a variety of metabolic or infectious diseases in the course of a human's life may have major impact on the development of a certain tumor due to exposure to the chemical under question. It is not feasible to simulate these real-life factors in the experimental situation [196].

Bioassay-Inherent Factors in Risk Assessment. A variety of biological factors exist that influence the assessment of carcinogenic hazard. Among these are the quality of the experiment, including the quality of the pathology, environmental control of animal facilities, estimation and standardization of chemical administration, purity of the chemical administered, and many other factors that together make up compliance with good laboratory practice. A further problem may be encountered if differences occur in outcome from different experiments. In the majority of cases, some aspect of an exceptional experiment may explain why the response was different (e.g., use of a different strain of animal). Since only positive data are used and negative data are ignored for carcinogenic risk assessment purpose to provide a conservative estimate of risk, this factor may also contribute to an overestimation of risk.

Conservatism and the Mechanisms of Chemical Carcinogenesis. To provide maximum protection, a conservative approach to risk assessment is preferred. This approach assumes that a single molecule of a carcinogen can interact with DNA and produce cancer; therefore, there can be no threshold for chemical carcinogenesis. Moreover, selection of the most sensitive response, irrespective of mechanism of action, and the assumption that the dose–response is linear at low doses may overestimate risk. This overestimation may be due to the unknown degree of conservatism at each step and to the amplification of previous bias in the assumption by the next step. The magnitude of overstatement of risk by these conservative assumptions is unknown, but is claimed to amount to several orders of magnitude.

Role of Toxicity and Increased Cell Proliferation. Critics point out that the high doses used in rodent carcinogenicity studies may have nonspecific effects such as an increase in the rate of cell proliferation due to cytotoxic effects [197]. These effects can be unique to high doses, and mitogenesis may itself be mutagenic in numerous ways, either by errors in replication or by conversion of endogenous DNA damage (e.g., by oxidative processes) and exogenous DNA damage to mutations before repair can occur. Moreover, sustained increases in the rate of cell proliferation may also yield secondary mutational events and could be important in the promotional phase of carcinogenicity by increasing the clonal expansion of initiated cells and thus increasing the chance that multiple critical mutational events will occur [198]. Thus, cell division will increase the chance of tumor formation. In this case, the tumor incidence for the same chemical applied at much lower doses is likely to be much lower than a linear model would predict and may even be zero [165, 199, 200]. Therefore, it would be important to add methods for determining cell division to animal cancer bioassays and apply the obtained results to estimate low-dose risks more adequately. Cell proliferation has been implicated as a major contributor to the cancerogenicity of several chemicals such as phenobarbital, 1,4-dichlorobenzene, *d*-limonene, and peroxisome proliferators.

Selection of Mathematical Model. The various models used in risk extrapolation fit the experimentally observed data well; however, they predict widely differing potential risks at low doses. Depending on the data used in the calculation, the predicted risks may differ by several orders of magnitude. These differences are inherent in the application of mathematical models, since only two or three doses are used in the experimental dose groups.

5.3. Future Contributions of Scientifically Based Procedures to Risk Assessment and Qualitative Risk Assessment for Carcinogens

Since quantitative cancer risk assessment can neither claim to be a scientific basis for development of historical models nor attempt to incorporate the large amount of scientific data on the mechanism of carcinogenesis into the assessment of risk, qualitative approaches considering all relevant data may be the best available solution for risk assessment. This approach, referred to a "weight-of-evidence determination of risk" is increasingly emphasized by regulatory authorities. The weight-of-evidence approach includes critical evaluation of the animal bioassay and all other available information on adverse effects of the chemical together with biotransformation, toxicokinetics, and expert judgment [201].

The criticims outlined above have demonstrated the need for further refinement of the extrapolation procedures by toxicokinetics and studies on the mechanisms of tumor formation. The role of animal experiments in predicting the potential risk of human carcinogen exposure will

Table 42. Descriptive dimensions proposed as a framework to facilitate the use of mechanistic data in evaluation of carcinogenic risk to humans

Data set	Example of information required
Evidence of genotoxicity	DNA adduct formation, mutagenicity, bioactivation
Evidence of effects on the expression of genes relevant to the process of carcinogenesis	alterations of the structure or quantity of product of a proto-oncogene or suppressor gene
Evidence for effects on cell behavior	mitogenesis, cell proliferation, hyperplasia
Evidence of time and dose – response relationships and interactions	initiation, promotion, progression

likely remain an important step of the risk assessement process despite the obvious limitations of this approach [197, 202]. However, the data obtained with animals must be interpreted with caution and in light of other data, both quantitative and qualitative, on the adverse effects of the compound (Table 42).

Application of Toxicokinetic Models in Risk Assessment. Information on the way in which a chemical is absorbed, biotransformed, and excreted may be critical in extrapolating the relevance of the results obtained in experimental animals to humans [130, 203]. For the purposes of risk extrapolation, it is generally assumed that the administered dose is proportional to the effective dose. However, many chemicals are known to be carcinogenic only after they have been activated to reactive electrophiles by enzymatic reactions. The amount of reactive metabolite formed might not be directly related to dose because saturable enzymatic processes are involved in both bioactivation and detoxication. The extent of bioactivation and detoxication may therefore be highly dose dependent, and the relationship between administered dose and target dose may not be linear at all administered doses. For example, when high-affinity, low-capacity enzymes catalyze the detoxication of a xenobiotic and low-affinity, high-capacity enzyme catalyze its bioactivation, the amount of reactive metabolite formed is likely highly dependent on the administered dose (i.e., a tenfold higher administered dose may result in a 100-fold higher effective target dose). Moreover, there may be depletion of cosubstrates required by the enzymes catalyzing bioactivation or detoxication. After the reactive metabolite is formed, it is often deactivated by a second enzyme, such as epoxide hydrolase or glutathione *S*-transferase. These enzymes can also be saturated. The reactive metabolites that are not destroyed by these detoxication pathways may bind to DNA. The metabolites bound to DNA can be removed by saturable DNA repair systems. These effects are referred to as nonlinear toxicokinetics. Nonlinear kinetics are also seen with chemicals inhibiting or inducing drug-metabolizing enzymes. Application of a carcinogen at high doses may result in the induction of enzymes catalyzing its bioactivation and may leave detoxifying enzymes unaffected. Thus, the expected steady-state concentrations of reactive intermediates present in the cell and capable of binding to DNA are expected to be disproportionately higher after application of high doses which induce biotransformation enzymes than after low doses which leave the levels of biotransformation enzymes unaffected. Moreover, the metabolism of a xenobiotic may be changed as a consequence of the effects of long-term administration of the chemical [204].

To estimate the dose-dependent relationship between administered dose and the effective dose, toxicokinetic models incorporating saturable processes have been applied. The examples in Figure 71 show the theoretically derived relationships between administered dose and effective dose for the same chemical in animals by varying the kinetic parameters for bioactivation, detoxication, and DNA repair [205]: saturation of enzymatic bioactivation (A), saturation of detoxication and activation (B), and saturation of activation, detoxication, and DNA repair (C).

The hockey-stick shape of the dose-response curve may remain unnoticed in animal experiments with high dose. The nonlinear correlation between effective dose and administered dose in the low, relevant dose region for human exposure may result in a decrease in the potential risk of exposure when compared to risk estimation based on experiments with high doses. Effective doses may be determined by measuring the amount of DNA adducts formed after administration of a carcinogen based on the assumption that for genotoxic agents the carcinogenic response is related to the extent of DNA adduct formation in target tissues. Unfortunately, nonlinear dose–response curves for effective doses have not yet been observed experimentally; all attempts to determine the dose-dependent formation of DNA adducts of potent carcinogens have shown linear relationships between administered dose and effect [206]. However, the increased sensitivity of analytical instruments currently available offers the opportunity to study the dose–effect curve for a wide range of carcinogens transformed to intermediates with different electrophilic reactivity in the low-dose range by quantifying the dose-dependent concentrations of DNA lesions. Quantitation of xenobiotic–hemoglobin adducts and chromosomal abnormalities in lymphocytes (e.g., sistechromatid exchange frequency) offers other examples of biological markers that may prove useful in the definition of dose–response curves. The application of these procedures offers the advantage of obtaining quantitative information on a nonstochastic effect thought to be relevant in the mechanisms of carcinogenesis and thus in risk assessment.

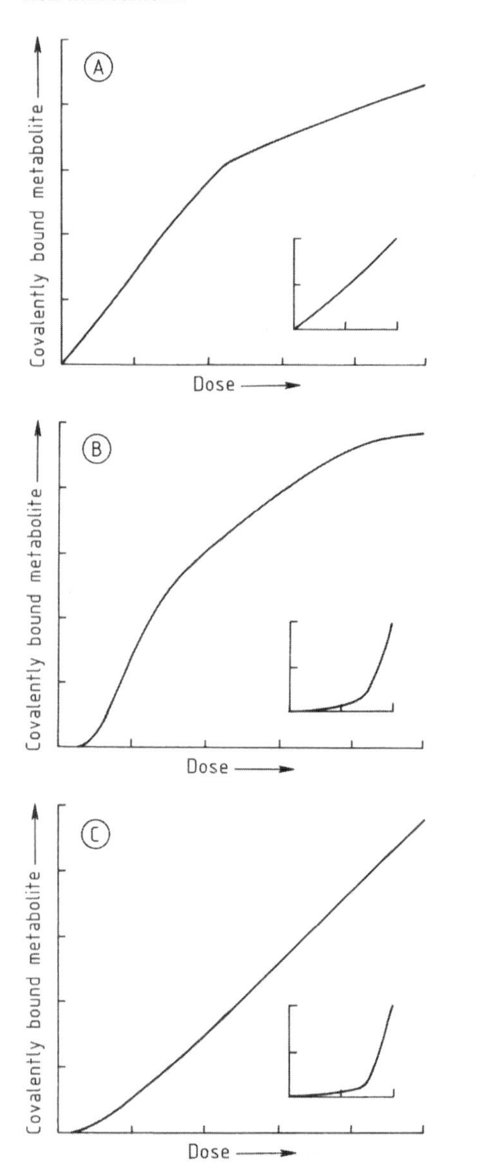

Figure 71. Possible relationships between administered dose and effective dose for the same data set including correction for nonlinear toxicokinetics
A) Saturation of enzymatic bioactivation; B) Saturation of detoxification and activation; C) Saturation of activation, detoxification, and DNA repair
The shapes of the curves in the low-dose range are shown in detail in the small sections.

As noted in Section 5.2.4.5, dosimetry may be a major cause of uncertainties in risk extrapolation form animals to humans. Besides saturable enzymatic reactions, the effective dose of a carcinogenic chemical may also be influenced by species differences in absorption, distribution and elimination, which may also be influenced by the dose administered.

To account for such factors, physiologically based pharmacokinetic models are increasingly used in the process of risk assessment. The principal purpose of the application of physiologically based pharmacokinetic models is to predict the concentration of carcinogen at the target site and describe the relationship between administered dose and target dose over a range of concentrations. By application of these models, a more accurate dose extrapolation is possible over broad ranges and may also incorporate nonlinearities in bioactivation and detoxification. Since the principles employed in the development of these models apply across species, the definition of the relationship between administered dose and effective dose and the important rate processes that cause a deviation from linearity permits a more accurate species extrapolation. The application of known physiological parameters may also enable the prediction of target organ concentrations in humans when direct measurements are not possible and may permit the comparison of different routes of application for effective dose. Physiologically based pharmacokinetic models are increasingly used to support quantitative risk assessments. Due to the risk estimates based on effective dose rather than external dose levels and the consideration of potential nonlinear relationships, an overestimation of risk obtained by linear extrapolations may be overcome and result in more scientifically founded risk assessments.

Mechanisms of Carcinogenicity and the Risk Assessment Process. Research on the molecular effects of particular agents has increased our understanding of the mechanisms of carcinogenicity. A range of biological processes has been implicated in carcinogenesis. Some of these mechanisms may be common to most carcinogens, others may be restricted to particular classes of chemicals or specific circumstances. In spite of the incomplete knowledge on carcinogenesis and a limited understanding of many processes, mechanism.based decisions are increasingly being introduced into risk assessment [207].

The emerging field of molecular toxicology may hold promise for the future of human risk assessment, and the application of methods developed for molecular biology may also offer more accurate ways of assessing human risk. The most promising application of mechanistic results for risk assessment are non-genotoxic carcinogens [208]. A number of xenobiotics are carcinogenic in animals but do not cause any detectable mutagenicity in in vitro studies. In contrast to other chemical carcinogens, which affect multiple organs or both sexes of both rat and mouse, many nongenotoxic carcinogens affect only a single organ in a single sex of a single species [133]. There are numerous examples of xenobiotics or treatment regimens in which production of tumors in animal experiments unlikely involves interaction of the xenobiotic or its metabolite with DNA. Examples include the induction of subcutaneous sarcomas by the repeated injection of glucose or saline solutions, the induction of skin cancer after chronic skin damage, and the induction of bladder cancer by the implantation of solid materials [209]. Recent experimental evidence also suggests that a number of xenobiotics that do not damage DNA may induce cancer in specific organs after systemic administration. Examples include chemicals binding to the circulating protein α_{2u}-globulin [210], peroxisome proliferators, and several structurally unrelated compounds interacting with receptors [211] (Table 43).

Table 43. Examples of nongenotoxic carcinogens and their presumed mechanisms of action

Class of nongenotoxic carcinogen	Typical examples	Presumed mechanism of action
Peroxisome proliferators	diethylhexylphthalate	receptor-mediated increase in gene transcription, oxidative stress
α_{2u}-Globulin-binding agents	unleaded gasoline, limonene, decalin	regenerative cell proliferation due to cytotoxicity
Phenobarbital		mitogenic activity
Chlorinated dioxins	2,3,7,8-tetrachlorodibenzodioxin	receptor-mediated increase in gene transcription
Hormones	17-α-ethynylestradiol	receptor-mediated increase in gene transcription

The terms "genotoxic" and "nongenotoxic" were defined by Butterworth [133]: "A genotoxic agent is one for which a primary biological activity of the chemical or a metabolite is alteration of the information encoded in the DNA. These can be point mutations, insertions, deletions or changes in chromosome structure or number. Chemicals exhibiting such activity can usually be identified by assays that measure reactivity with the DNA, induction of mutations, induction of DNA repair or cytogenetic effects. Nongenotoxic chemicals are those that lack genotoxicity as a primary biological activity. While these agents may yield genotoxic events as a secondary result of other induced toxicity, such as forced cellular growth, their primary action does not involve reactivity with the DNA."

Xenobiotics such as unleaded gasoline, limonene, and 1,4-dichlorobenzene induce renal cancer in male rats, but not in female rats or in either sex of mice [212]. Metabolites of these xenobiotics bind to the circulating protein α_{2u}-globulin, whose synthesis in the rat is under the control of androgens. The modified α_{2u}-globulin is concentrated in the kidney proximal tubules and is not degradable by the processes responsible for degradation of unmodified α_{2u}-globulin. Thus, accumulation of the modified protein causes cytotoxicity and regenerative cell proliferation, which has been implicated as the cause of tumor formation in the kidney.

Tumors in organs whose function is regulated by the endocrine system are often observed after hormonal therapies in humans and hormone treatment in animals and, due to the absence of genotoxicity of most hormones, nongenotoxic processes involving receptor-mediated transcription and activation or repression of specific genes have been implicated in hormonal carcinogenesis [213].

For both of the above cited examples, threshold mechanisms may be postulated. Kidney cancer induced by α_{2u}-globulin-binding agents requires cytotoxicity and cell death. Nontoxic low concentrations of these xenobiotics applied may not be tumorigenic. For receptor-mediated processes, a disproportional relationship between receptor occupancy and hormonally mediated cancer is also considered likely [214–216]. Hence, any assessment of carcinogenic risk of these chemicals must take into account the mechanisms by which they produce their effects.

The use of molecular toxicology in risk assessment is also advocated by the International Agency for Research on Cancer [207]. A monograph on the identification of cancer risks by mechanistic investigations suggested that cancer risks by mechanistic investigations suggests that "when available data on mechanisms are thought to be relevant to evaluation of the carcinogenic risk of an agent to humans, they should be used in making the overall evaluation, together with the combined evidence for animal and/or human carcinogenicity." The consensus report further states that no definite guidelines for the inclusion of mechanistic data in the evaluation of carcinogens can be elaborated. However, a range of options are available: "First, information concerning mechanisms of action may confirm a particular level of carcinogen classification as indicated on the basis of epidemiological and/or animal carcinogenicity data. Second, for a particular agent, strong evidence for a mechanism of action that is relevant to carcinogenicity in humans could justify 'upgrading' its overall evaluation. Third, an overall evaluation of human cancer hazard on the basis of animal carcinogenicity data could be downgraded by strong evidence that the mechanism responsible for tumor growth in experimental animals is not relevant to humans. In keeping with the goal of public health, priority must be given to the demonstration that the mechanism is irrelevant to humans."

5.4. Risk Assessment for Teratogens

The timing of exposure and patterns of dose–response from animal studies have important implications for extrapolating animal data to humans. A wide spectrum of end points can be produced, even under the controlled conditions of timing and exposure that can be achieved in animal studies. In some cases, the spectrum includes a continuum of responses, with depressed birth weight or functional impairment occurring at low doses, birth defects at intermediate doses, and lethality at high doses. Less commonly, birth defects alone or lethality alone are produced. Therefore, in estimating human risk, all exposure-specific adverse outcomes must be taken into consideration, and not just birth defects.

A similar response pattern is observed in humans exposed to developmental toxicants. The spectrum of responses is determined by the time and duration of exposure, magnitude of exposure, and interindividual differences in sensitivity. Hence, manifestations of developmental toxicity can not be expected to be identical across species; that is, an animal model can not be expected to forecast exactly the human response to a given exposure. For instance, an agent that induces cleft palate in the mouse may elevate the frequency of spontaneous abortion or retard growth in humans. However, any manifestation of exposure-related developmental toxicity in animal studies can be regarded as indicative of a spectrum of response in humans.

Epidemiological data suggest that the majority of human embryos with chromosomal and/or morphologic abnormalities are lost through early miscarriage and that relatively few survive to term. Consequently, determination of malformations or growth retardation at the time of birth alone (malformations, stillbirths, low birth weight) is likely to result in a substantial underestimate of the true risk, since the onset of embryolethality would be missed.

The sensitivity (ability to detect a true positive response in humans) and specificity (ability to detect a true negative response in humans) of laboratory animal studies have been evaluated [217]. Of 38 compounds with demonstrated or suspected teratogenic activity in humans, all but one (tobramycin, which causes otological deficits in humans) tested positive in at least one animal species. Approximately 80% of the compounds were positive in multiple species. A positive response was elicited to 85% in the mouse, 80% in the rat, 60% in the rabbit, 45% in the hamster, and 30% in the monkey. These findings indicate that the usual laboratory animal species are highly sensitive for detecting human teratogens.

In contrast to the high sensitivity, laboratory animals show low specificity for predicting human teratogenesis; of 165 test compounds with no evidence of teratogenic activity in humans 65 (41%) were positive in more than one animal species. The high percentage of false positive results (i.e., compounds with no evidence for human teratogenicity inducing malformations in animals may be in part accounted for by the high doses usually applied in the animal tests. Also, human studies determine effects from the time of birth onward, which may result in underestimation of the true risk.

6. References

1. D. Henschler, *Trends Pharmacol. Sci. Fest. Suppl* (1985) 26 – 28.
2. W. Dekant, H.-G. Neumann: *Tissue Specific Toxicity: Biochemical Mechanisms*, Academic Press, London 1992.
3. D. O. Chanter, R. Heywood, *Toxicol. Lett.* **10** (1982) 303 – 307.
4. Y. Alarie, *EHP Environ. Health Perspect.* **42** (1981) 9 – 13.
5. H. Druckrey, D. Schmähl, W. Dischler, A. Schildbach, *Naturwissenschaften* **49** (1962) 217 – 228.
6. P. H. Bach, C. P. Ketley, I. Ahmed, M. Dixit, *Food Chem. Toxicol.* **24** (1986) 775 – 779.
7. P. D. Anderson, L. J. Weber, *Toxicol. Appl. Pharmacol.* **33** (1975) 471 – 483.
8. W. W. Weber, D. W. Hein, *Pharmacol. Rev.* **37** (1985) 25 – 79.
9. T. Green, M. S. Prout, *Toxicol. Appl. Pharmacol.* **79** (1985) 401 – 411.
10. D. A. Smith, *Drug Metab. Rev.* **23** (1991) 355 – 373.
11. J. H. Lin, *Drug Metab. Dispos.* **23** (1995) 1008 – 1021.
12. D. F. Lewis, C. Ioannides, D. V. Parke, *Environ. Health Perspect.* **106** (1998) 633 – 41.
13. M. R. Juchau, H. Boutelet-Bochan, Y. Huang, *Drug Metab. Rev.* **30** (1998) 541 – 68.
14. E. J. Calabrese, *J. Pharm. Sci.* **75** (1986) 1041 – 1067.
15. M. VandenBerg, J. DeJongh, H. Poiger, J. R. Olson, *CRC Crit. Rev. Toxicol.* **24** (1994) 1 – 74.
16. R. C. Lind, A. J. Gandolfi, P. d. I. M. Hall, *Anesthesiology* **70** (1989) 649 – 653.
17. R. A. van Dyke, J. A. Gandolfi, *Drug Metab. Dispos.* **2** (1974) 469 – 476.
18. I. G. Sipes, A. J. Gandolfi, L. R. Pohl, G. Krishna, B. R. Brown, *J. Pharmacol. Exp. Therap.* **214** (1980) 716 – 720.
19. A. Turturro, R. Hart, *Exp. Gerontol.* **27** (1992) 583 – 592.
20. K. O. Lindros, Y. A. Cai, K. E. Penttila, *Hepatology (N.Y.)* **12** (1990) 1092 – 1097.
21. S. J. Singer, G. L. Nicolson, *Science* **175** (1972) 720 – 731.
22. M. Eisenberg, S. M. McLaughlin, *Residue Rev.* **31** (1976) 1 – 8.
23. R. C. Scott, J. D. Ramsey, R. I. Ward, M. A. Thompson, C. Rhodes, *Br. J. Dermatol.* **115** (1986) 47 – 48.
24. P. H. Dugard, *Food Chem. Toxicol.* **24** (1986) 749 – 753.
25. T. R. Bates, M. Gibaldi in J. Swarbick (ed.): *Current Concepts in the Pharmaceutical Sciences: Biopharmaceutics*, **vol. 1**, Lea and Febiger, Philadelphia 1970.
26. A. R. Brody, *EHP Environ. Health Perspect.* **100** (1993) 21 – 30.
27. J.-P. Tillement, E. Lindenlaub: *Protein Binding and Drug Transport*, F. K. Schattauer Verlag, Stuttgart 1986.
28. M. P. Waalkes, P. L. Goering, *Chem. Res. Toxicol.* **3** (1990) 281 – 288.
29. R. A. Goyer, *EHP Environ. Health Perspect.* **100** (1993) 177 – 187.
30. F. P. Guengerich, T. L. Macdonald, *Acc. Chem. Res.* **17** (1984) 9 – 16.
31. G. S. Yost, A. R. Buckpitt, R. A. Roth, T. L. McLemore, *Toxicol. Appl. Pharmacol.* **101** (1989) 179 – 195.
32. F. M. Guengerich, *Annu. Rev. Pharmacol. Toxicol.* **29** (1989) 241 – 264.
33. F. P. Guengerich, *J. Biol. Chem.* **266** (1991) 10019 – 10022.
34. F. P. Guengerich, *Life Sci.* **50** (1992) 1471 – 1478.
35. F. P. Guengerich, *FASEB J.* **6** (1992) 745 – 748.
36. L. S. Kaminsky, M. J. Fasco, *CRC Crit. Rev. Toxicol.* **21** (1992) 407 – 422.
37. A. J. Paine, *Hum. Exper. Toxicol.* **14** (1995) 1 – 7.
38. F. P. Guengerich, *Chem. Res. Toxicol.* **14** (2001) 611 – 50.
39. F. P. Guengerich, *Drug Metab. Rev.* **34** (2002) 7 – 15.
40. X. Ding, L. S. Kaminsky, *Annu. Rev. Pharmacol. Toxicol. 43* (2003) 149 – 73.
41. D. R. Koop, *FASEB J.* **6** (1992) 724 – 730.
42. P. B. Danielson, *Curr. Drug Metab.* **3** (2002) 561 – 597.
43. D. M. Ziegler, *Drug Metab. Rev. 34* (2002) 503 – 511.
44. J. R. Cashman, *Curr Opin. Drug Discov. Dev.* **6** (2003) 486 – 93.
45. R. E. Tynes, R. M. Philpot, *Mol. Pharmacol.* **31** (1991) 569 – 574.
46. T. E. Eling, D. C. Thompson, G. L. Foureman, M. F. Hughes, *Annu. Rev. Pharmacol. Toxicol.* **30** (1990) 1 – 45.

47. B. J. Smith, J. F. Curtis, T. E. Eling, *Chem. Biol. Interact.* **79** (1991) 245 – 264.
48. C. Vogel, *Curr. Drug Metab.* **1** (2000) 391 – 404.
49. W. L. Smith, L. J. Marnett, *Biochim. Biophys. Acta.* **1083** (1991) 1 – 17.
50. B. H. Bach, J. W. Bridges, *CRC Crit. Rev. Toxicol.* **15** (1985) 217 – 329.
51. R. J. Duescher, A. A. Elfarra, *J. Biol. Chem.* **267** (1992) 19859 – 19865.
52. P. D. Josephy, T. E. Eling, R. P. Mason, *Mol. Pharmacol.* **23** (1983) 461 – 466.
53. A. J. Kettle, C. C. Winterbourn, *J. Biol. Chem.* **267** (1992) 8319 – 8324.
54. H. Thomas, F. Oesch, *ISI Atlas Sci.: Biochemistry* **1** (1988) 287 – 291.
55. S. A. Sheweita, A. K. Tilmisany, *Curr. Drug Metab.* **4** (2003) 45 – 58.
56. B. Burchell, W. H. Coughtrie, *Pharmacol. Ther.* **43** (1989) 261 – 289.
57. T. R. Tephly, B. Burchell, *Trends Pharmacol. Sci.* **11** (1990) 276 – 279.
58. R. H. Tukey, C. P. Strassburg, *Annu. Rev. Pharmacol. Toxicol.* **40** (2000) 581 – 616.
59. J. K. Ritter, *Chem. Biol. Interact.* **129** (2000) 171 – 193.
60. K. W. Bock, B. S. Bock-Hennig, G. Fischer, L. W. D. Ullrich in H. Greim et al. (eds.): *Biochemical Basis of Chemical Carcinogenesis*, **vol. 1**, Raven Press, New York 1984, pp. 13 – 22.
61. G. J. Mulder: *Sulfation of Drugs and Related Compounds*, CRC Press, Boca Raton, Fla., 1981.
62. M. W. Duffel, A. D. Marshal, P. McPhie, V. Sharma, W. B. Jakoby, *Drug Metab. Rev.* **33** (2001) 369 – 395.
63. E. Y. Krynetski, W. E. Evans, *Pharm Res.* **16** (1999) 342 – 349.
64. D. A. Evans, *Pharmacol. Ther.* **42** (1989) 157 – 234.
65. P. Meisel, *Pharmacogenomics* **3** (2002) 349 – 366.
66. I. M. Arias, W. B. Jakoby: *Glutathione: Metabolism and Function*, Raven Press, New York 1976.
67. B. Ketterer, D. J. Meyer, A. G. Clark in H. Sies, B. Ketterer (eds.): *Glutathione Conjugation: Mechanisms and Biological Significance*, Academic Press, London 1988, pp. 73 – 135.
68. R. Morgenstern, G. Lundqvist, G. Andersson, L. Balk, J. W. DePierre, *Biochem. Pharmacol.* **33** (1984) 3609 – 3614.
69. J. L. Cmarik, P. B. Inskeep, M. J. Meredith, D. J. Meyer, B. Ketterer, F. P. Guengerich, *Cancer Res.* **50** (1990) 2747 – 2752.
70. J. Caldwell, W. B. Jakoby: *Biological Basis of Detoxification*, Academic Press, New York 1983
71. M. W. Anders: *Bioactivation of Foreign Compounds*, Academic Press, Orlando 1985.
72. M. W. Anders in I. M. Arias, W. B. Jakoby, H. Popper, D. Schachter, D. A. Shafritz (eds.): *The Liver: Biology and Pathology*, 2nd ed., Raven Press, New York, 1988, pp. 389 – 400.
73. J. G. Bessems, N. P. Vermeulen, *CRC Crit. Rev. Toxicol.* **31** (2001) 55 – 138.
74. G. M. B. StClair, V. Amarnath, M. A. Moody, D. C. Anthony, C. W. Anderson, D. G. Graham, *Chem. Res. Toxicol.* **1** (1988) 179 – 185.
75. F. P. Guengerich, D. C. Liebler, *CRC Crit. Rev. Toxicol.* **14** (1985) 259 – 307.
76. D. C. Liebler, F. P. Guengerich, *Biochemistry* **22** (1983) 5482 – 5489.
77. L. R. Pohl, R. V. Branchflower, R. J. Highet, J. L. Martin, D. S. Nunn, T. J. Monks, J. W. George, J. A. Hinson, *Drug Metab. Dispos.* **9** (1981) 334 – 339.
78. M. W. Anders, W. Dekant: *Conjugation-dependent Carcinogenicity and Toxicity of Foreign Compounds*, Academic Press, San Diego, 1994.
79. T. J. Monks, M. W. Anders, W. Dekant, J. L. Stevens, S. S. Lau, P. J. van Bladeren, *Toxicol. Appl. Pharmacol.* **106** (1990) 1 – 19.
80. M. W. Anders, W. Dekant, S. Vamvakas, *Xenobiotica* **22** (1992) 1135 – 1145.
81. W. Dekant, S. Vamvakas, M. W. Anders, in W. Dekant, H.-G. Neumann (eds.): *Tissue Specific Toxicity: Biochemical Mechanisms*, Academic Press, London, 1992, pp. 163 – 194.
82. S. D. Aust, C. F. Chignell, T. M. Bray, B. Kalyanaraman, R. P. Mason, *Toxicol. Appl. Pharmacol.* **120** (1993) 168 – 178.
83. B. D. Goldstein, B. Czerniecki, G. Witz, *EHP Environ. Health Perspect.* **81** (1989) 55 – 57.
84. B. D. Goldstein, G. Witz, *Free Radicals Res. Commun.* **11** (1990) 3 – 10.
85. J. P. Kehrer, B. T. Mossman, A. Sevanian, M. A. Trush, M. T. Smith, *Toxicol. Appl. Pharmacol.* **95** (1988) 349 – 362.
86. P. A. Cerutti, B. F. Trump, *Cancer Cells* **3** (1991) 1 – 7.
87. E. Chacon, D. Acosta, *Toxicol. Appl. Pharmacol.* **107** (1991) 117 – 128.
88. R. H. Burdon, V. Gill, C. Rice-Evans, *Free Radicals Res. Commun.* **11** (1990) 65 – 76.

89. R. W. Pero, G. C. Roush, M. M. Markowitz, D. G. Miller, *Cancer Detect. Prev.* **14** (1990) 551 – 561.
90. W. B. Jakoby, D. M. Ziegler, *J. Biol. Chem.* **265** (1990) 20715 – 20719.
91. H. Sies, *Naturwissenschaften* **76** (1989) 57 – 64.
92. D. M. Ziegler in W. B. Jacoby (ed.): *Enzymic Basis of Detoxification*, **vol. I**, Academic Press, New York 1980, pp. 201 – 227.
93. E. Boyland, L. F. Chasseaud, *Adv. Enzymol. Relat. Areas Mol. Biol.* **32** (1969) 173.
94. T. A. Fjellstedt, R. H. Allen, B. K. Duncan, W. B. Jakoby, *J. Biol. Chem.* **248** (1973) 3702 – 3707.
95. R. C. James, R. D. Harbison, *Biochem. Pharmacol.* **31** (1982) 1829 – 1835.
96. L. I. McLellan, C. R. Wolf, J. D. Hayes, *Biochem. J.* **258** (1989) 87 – 93.
97. J. A. Hinson, D. W. Roberts, *Annu. Rev. Pharmacol. Toxicol.* **32** (1992) 471 – 510.
98. S. D. Nelson, P. G. Pearson, *Annu. Rev. Pharmacol. Toxicol.* **30** (1990) 169 – 195.
99. L. R. Pohl, *Chem. Res. Toxicol.* **6** (1993) 786 – 793.
100. E. R. Stadtman, C. N. Oliver, *J. Biol. Chem.* **266** (1991) 2005 – 2008.
101. E. R. Stadtman, *Science* **257** (1992) 1220 – 1224.
102. H. Esterbauer, H. Zollner, R. J. Schaur, *ISI Atlas Sci.: Biochemistry* **1** (1988) 311 – 317.
103. H. Kappus, *Biochem. Pharmacol.* **35** (1986) 1 – 6.
104. J. L. Farber, *Chem. Res. Toxicol.* **3** (1990) 503 – 508.
105. J. P. Kehrer, D. P. Jones, J. J. Lemasters, J. L. Farber, H. Jaeschke, *Toxicol. Appl. Pharmacol.* **106** (1990) 165 – 178.
106. K. Hemminki, *Carcinogenesis* **14** (1993) 2007 – 2012.
107. B. N. Ames, *Mutat. Res.* **214** (1989) 41 – 46.
108. R. Adelman, R. L. Saul, B. N. Ames, *Proc. Natl. Acad. Sci. USA* **85** (1988) 2706 – 2708.
109. K. W. Bock, H.-P. Lipp, B. S. Bock-Hennig, *Xenobiotica* **20** (1990) 1101 – 1111.
110. D. R. Koop, D. J. Tierney, *Bioassays.* **12** (1990) 429 – 435.
111. A. H. Conney, *Annu. Rev. Pharmacol. Toxicol.* **43** (2003) 1 – 30.
112. L. Wu, J. P. J. Whitlock, *Proc. Natl. Acad. Sci. USA* **89** (1992) 4811 – 4815.
113. J. P. J. Whitlock, *Chem. Res. Toxicol.* **6** (1993) 754 – 763.
114. B. Testa, P. Jenner, *Drug Metab. Rev.* **12** (1981) 1 – 117.
115. B. Testa, *Xenobiotica* **20** (1990) 1129 – 1137.
116. J. R. Halpert, F. P. Guengerich, J. R. Bend, M. A. Correia, *Toxicol. Appl. Pharmacol.* **125** (1994) 163 – 175.
117. M. Blum, D. M. Grant, W. McBride, M. Heim, U. A. Meyer, *DNA Cell Biol.* **9** (1990) 193 – 203.
118. M. E. Andersen, *CRC Crit. Rev. Toxicol.* **11** (1981) 105 – 150.
119. M. W. Anders, W. Dekant in M. W. Anders, W. Dekant, D. Henschler, H. Oberleithner, S. Silbernagl (eds.): *Renal Disposition and Nephrotoxicity of Xenobiotics*, Academic Press, San Diego 1993, p. 155 – 183.
120. R. M. Brenner, F. C. J. Rector: *The Kidney*, W. B. Saunders Co., Philadelphia 1981.
121. E. C. Foulkes, *Proc. Soc. Exp. Biol. Med.* **195** (1990) 1.
122. B. A. Saville, M. R. Gray, Y. K. Tam, *Drug Metab. Rev.* **24** (1992) 49 – 88.
123. C. D. Klaassen, J. B. Watkins, *Pharmacol. Rev.* **36** (1984) 1 – 67.
124. O. Fardel, L. Payen, A. Courtois, L. Vernhet, V. Lecureur, *Toxicology* **167** (2001) 37 – 46.
125. Y. Kato, H. Suzuki, Y. Sugiyama, *Toxicology* **181–182** (2002) 287 – 290.
126. J. G. Filser, *Arch. Toxicol.* **66** (1992) 1 – 10.
127. M. E. Andersen, *Toxicol. Lett.* **138** (2003) 9 – 27.
128. R. Dixit, J. Riviere, K. Krishnan, M. E. Andersen, *J. Toxicol. Environ. Health. B Crit. Rev.* **6** (2003) 1 – 40.
129. M. E. Andersen, M. L. Gargas, H. J. Clewell III, K. M. Seceryn, *Toxicol. Appl. Pharmacol.* **89** (1987) 149 – 157.
130. M. E. Andersen, D. Krewski, J. R. Withey, *Cancer Lett.* **69** (1993) 1 – 14.
131. R. B. Conolly, M. E. Andersen, *Annu. Rev. Pharmacol. Toxicol.* **31** (1991) 503 – 523.
132. S. M. Cohen, L. B. Ellwein, *Science* **249** (1990) 1007 – 1011.
133. B. Butterworth, *Mutation Res.* **239** (1990) 117 – 132.
134. Q. Chen, J. Marsh, B. Ames, B. Mossman, *Carcinogenesis* **17** (1996) 2525 – 2527.
135. R. A. Roberts, S. Chevalier, S. C. Hasmall, N. H. James, S. C. Cosulich, N. Macdonald, *Toxicology* **181–182** (2002) 167 – 170.
136. M. S. Denison, S. R. Nagy, *Annu. Rev. Pharmacol. Toxicol.* **43** (2003) 309 – 34.
137. J. Abel, D. Hohr, H. J. Schurek, *Arch. Toxicol.* **60** (1987) 370 – 375.
138. S. Orrenius, P. Nicotera, *Arch. Toxicol. Suppl.* **11** (1987) 11 – 19.

139. S. Orrenius, D. J. McConkey, G. Bellomo, P. Nicotera, *Trends Pharmacol. Sci.* **10** (1989) 281 – 285.
140. E. C. Miller, J. A. Miller, *Pharmacol. Rev.* **18** (1966) 805 – 838.
141. E. C. Miller, J. A. Miller, *Cancer* **47** (1981) 2327 – 2345.
142. P. A. Cerutti, *Science* **227** (1985) 375 – 381.
143. *International Agency for Research on Cancer: Models, Mechanisms and Etiology of Tumour Promotion*, IARC Sci. Publ., Lyon, 1984.
144. J. M. Bishop, *Science* **235** (1987) 305 – 311.
145. J. M. Bishop, *Cell* **64** (1991) 235 – 48.
146. J. M. Bishop, *Cell* **42** (1985) 23 – 38.
147. J. Ashby, R. W. Tennant, *Mutation Res.* **257** (1991) 229 – 306.
148. R. L. Brinster in T. H. Shepard, J. R. Miller, M. Marois (eds.): *Methods for Detection of Environmental Agents that Produce Congenital Defects*, Elsevier New York, 1975, pp. 113 – 124.
149. J. M. Manson, L. D. Wise, in M. O. Amdur, J. Doull, C. D. Klaassen (eds.): *Casarett and Doull's Toxicology. The Basic Science of Poisons*, Pergamon Press New York 1991, pp. 226 – 254.
150. R. P. Bolande in J. G. Wilson, F. C. Fraser (eds.): *Handbook of Teratology*, **vol. 2**, Plenum Press New York 1977, pp. 293 – 328.
151. S. Ivankovic: "Perinatal Carcinogenesis, DHEW Publication no. (NIH) 7a–1063", *Nat. Cancer Inst. Monogr.* **51** (1979) 103 – 116.
152. E. Schlede, U. Mischke, R. Roll, D. Kayser, *Arch. Toxicol.* **66** (1992) 455 – 470.
153. M. J. Van den Heuvel, *Human Exptl. Toxicol.* **9** (1990) 369.
154. B. Berner, E. R. Cooper in A. F. Kydonieu, B. Berner (eds.): *Transdermal Delivery of Drugs*, CRC Press, Boca Raton 1987, pp. 107 – 130.
155. J. A. Shatkin, H. S. Brown, *Environ. Res.* **56** (1991) 90 – 108.
156. B. Ballantyne, T. Marrs, P. Turner: *General and Applied Toxicology*, Macmillan Press, New York 1993.
157. A. P. Worth, M. Balls: *Alternative (non-animal) Methods for Chemical Testing: Current Status and Future Prospects*, FRAME, Nottingham, UK, 2002.
158. P. A. Botham, D. A. Basketter, T. Maurer, D. Mueller, M. Potokar, W. J. Bontinck, *Food Chem Toxicol.* **29** (1991) 275 – 86.
159. G. M. Henningen in D. W. Hobson (ed.): *Dermal and Ocular Toxicology. Fundamentals and Methods*, CRC Press, Boca Raton 1991, pp. 153 – 192.
160. W. E. Ecvam in European Center for the Validation of alternative methods *Alternative (non-animal) Methods for Chemical Testing: Current Status and Future Prospects*, **vol. 30**, Suppl. 1, FRAME, Nottingham, UK 2002, pp. 95 – 102.
161. D. B. Clayson, F. Iverson, R. Mueller, *Teratog. Carcinog. Mutagen* **11** (1991) 279 – 296.
162. J. K. Haseman, A. Lockhart, *Fund. Appl. Toxicol.* **22** (1994) 382 – 391.
163. D. G. Hoel, J. K. Haseman, M. D. Hogan, J. Huff, E. E. McConnell, *Carcinogenesis* **9** (1988) 2045 – 2052.
164. B. N. Ames, R. Magaw, L. S. Gold, *Science* **236** (1987) 271 – 80.
165. B. N. Ames, L. S. Gold in T. Slaga, B. Butterworth (eds.): *Chemically Induced Cell Proliferation: Implications for Risk Assessment*, Wiley-Liss, New York 1991, pp. 1 – 20.
166. L. S. Gold, T. H. Slone, B. R. Stern, N. B. Manley, B. N. Ames, *Science* **258** (1992) 261 – 265.
167. D. J. Waxmann, L. Azaroff, *Biochem. J.* **281** (1992) 577 – 592.
168. B. N. Ames, J. McCann, E. Yamasaki, *Mutation Res.* **31** (1975) 347 – 64.
169. D. M. Maron, B. N. Ames, *Mutat. Res.* **113** (1983) 173 – 215.
170. M. Watanabe, M. J. Ishidate, T. Nohmi, *Mutat. Res.* **234** (1990) 337 – 348.
171. T. P. Simula, M. J. Glancey, C. R. Wolf, *Carcinogenesis* **14** (1993) 1371 – 1376.
172. M. Richold, A. Chaudley, J. Ashby, D. G. Gatehouse, J. Bootman, L. Henderson in D. J. Kirkland (eds.): *UKEMS Sub-Committee on Guidelines for Mutagenicity Testing, report part Revised Basic Mutagenicity Tests, UKEMS Recommended Procedures*, Cambridge University Press, Cambridge 1990, pp. 115 – 141.
173. J. J. McCormick, V. M. Mahler, *Environ. Molec. Mutag.* **14**(S 15) (1989) 105 – 113.
174. G. M. Williams, J. H. Weisburger, *Mutat. Res.* **205** (1988) 79 – 90.
175. N. Ozawa, F. P. Guengerich, *Proc. Natl. Acad. Sci. USA* **80** (1983) 5266 – 5270.
176. H. Bartsch, A. Barbin, M.-J. Marion, J. Nair, Y. Guichard, *Drug Metab. Rev.* **26** (1994) 349 – 372.
177. C. Richter, J.-W. Park, B. N. Ames, *Proc. Natl. Acad. Sci. USA* **85** (1988) 6465 – 6467.
178. P. B. Farmer, H.-G. Neumann, D. Henschler, *Arch. Toxicol.* **60** (1987) 251 – 260.

179. J. A. Swenberg, N. Fedtke, F. Ciroussel, A. Barbin, H. Bartsch, *Carcinogenesis* **13** (1992) 727 – 729.
180. I. A. Blair, *Chem. Res. Toxicol.* **6** (1993) 741 – 747.
181. E. Randerath, T. F. Danna, K. Randerath, *Mutat. Res.* **268** (1992) 139 – 153.
182. M. D. Shelby, E. Zeiger, *Mutat. Res.* **234** (1990) 257 – 261.
183. J. G. Vos, *Arch. Toxicol Suppl.* **4** (1980) 95 – 108.
184. A. D. Dayau, R. F. Hertel, E. Hesseltine, G. Kazautzis, E. Smith, M. T. Van der Venne: *Immunotoxicity of Metals and Immunotoxicology*, Plenum Press, New York 1990.
185. D. M. Maloney, *Environ. Health Perspect.* **101** (1993) 396 – 401.
186. C. C. Harris, *Science* **262** (1993) 1980 – 1981.
187. J. V. Rodricks, *Environ. Health Perspect.* **102** (1994) 258 – 264.
188. F. R. Johannsen, *CRC Crit. Rev. Toxicol.* **20** (1990) 341 – 367.
189. A. M. Jarabek, W. H. Farland, *Toxicol. Ind. Health* **6** (1990) 199-.
190. D. W. Nebert, *Mutat. Res.* **247** (1991) 267 – 281.
191. R. T. H. van Welie, R. G. J. M. van Dijck, N. P. E. Vermeulen, N. J. van Sittert, *CRC Crit. Rev. Toxicol.* **22** (1992) 271 – 306.
192. R. W. Tennant, J. Ashby, *Mutat. Res.* **257** (1991) 209 – 227.
193. A. C. Beach, R. C. Gupta, *Carcinogenesis* **13** (1992) 1053 – 1074.
194. K. Crump, *CRC Crit. Rev. Toxicol.* **32** (2002) 133 – 53.
195. C. C. Travis, *Toxicology* **47** (1987) 3 – 13.
196. H. Vainio, M. Sorsa, A. J. McMichael: *Complex Mixtures and Cancer Risk*, International Agency for Research on Cancer (WHO), Lyon 1990.
197. J. A. Swenberg, R. R. Maronpot in (ed.): *Chemically Induced Cell Proliferation: Implications for Risk Assessment*, Wiley-Liss, New York 1991, pp. 245 – 251.
198. S. M. Cohen, L. B. Ellwein, *Cancer Res.* **51** (1991) 6493 – 6505.
199. B. N. Ames, L. S. Gold, *Science* **249** (1990) 970 – 1.
200. B. N. Ames, L. S. Gold, *Mutat. Res.* **250** (1991) 3 – 16.
201. G. M. Gray, J. T. Cohen, J. D. Graham, *Environ. Health Perspect.* **101** (1993) 203 – 208.
202. J. C. Barrett, *Environ. Health Perspect.* **100** (1993) 9 – 20.
203. M. E. Andersen, H. J. Clewell, 3rd, M. L. Gargas, F. A. Smith, R. H. Reitz, *Toxicol. Appl. Pharmacol.* **87** (1987) 185 – 205.
204. T. Green, *Xenobiotica* **20** (1990) 1233 – 1240.
205. D. G. Hoel, N. L. Kaplan, M. W. Anderson, *Science* **219** (1983) 1032 – 1037.
206. W. K. Lutz, *Carcinogenesis* **11** (1990) 1243 – 1247.
207. H. Vainio, P. N. Magee, D. B. McGregor, A. J. McMichael: Mechanisms of Carcinogenesis in Risk Identification, IARC Scientific Publications No. 116, Lyon 1992.
208. C. Shaw, H. B. Jones, *Trends Pharmacol. Sci.* **15** (1994) 89 – 93.
209. P. Grasso, M. Sharratt, A. J. Cohen, *Annu. Rev. Pharmacol. Toxicol.* **31** (1991) 253 – 87.
210. J. A. Swenberg, *Environ. Health Perspect.* **101** (1993) 39 – 44.
211. L. Poellinger, M. Göttlicher, J.-A. Gustafsson, *Trends Pharmacol. Sci.* **13** (1992) 241 – 245.
212. S. J. Borghoff, B. G. Short, J. A. Swenberg, *Annu. Rev. Pharmacol. Toxicol.* **30** (1990) 349 – 67.
213. R. M. McClain, *Toxicol. Lett.* **64/65** (1992) 397 – 408.
214. A. M. Tritscher, J. A. Goldstein, C. J. Portier, Z. McCoy, G. C. Clark, G. W. Lucier, *Cancer Res.* **52** (1992) 3436 – 3442.
215. C. Portier, A. Tritscher, M. Kohn, C. Sewall, G. Clark, L. Edler, D. Hoel, G. Lucier, *Fund. Appl. Toxicol.* **20** (1993) 48 – 56.
216. J. P. V. Heuvel, G. C. Clark, M. C. Kohn, A. M. Tritscher, W. F. Greenle, G. W. Lucier, D. A. Bell, *Cancer Res.* **54** (1994) 62 – 68.
217. V. H. Frankos, *Fund. Appl. Toxicol.* **5** (1985) 615 – 622.

Ecology and Ecotoxicology

OTTO FRÄNZLE, Christian-Albrechts-Universität, Kiel, Federal Republic of Germany [(Sections1.1, 1.2, 2.1 (in part), 2.2, 2.3 (in part), 2.4, 3.1, 3.2 (in part)]

MILAN STRAŠKRABA, Akademie věd České Republiky, Biomatematická laboratoř, Entomologický institut, České Budějovice, Czech Republic (Sections 2.1.4, 2.3.5, 3.2.3)

SVEN ERIK JØRGENSEN, Danmarks Farmaceutiske Højskole, Copenhagen, Denmark (Section 3.3)

Ullmann's Industrial Toxicology

ISBN: 3-527-31247-1

Abbreviations:

a species-specific maximum interception
A constant; evaporating surface
ALT (*t*) leaf senescence function
b constant; density of vegetation cover; coefficient adjusting evaporation coefficient to other substances than oxygen
B natality
BCF bioconcentration factor
C connectance of the web
C_{dis} concentration at which long-lasting disturbances of essential soil functions occur
C_{eff} concentration at which low-level adverse effects first appear
CHAID chi-square automatic interaction detection
D disturbance; Simpson's diversity index; standardized damage level
2,4-D 2,4-(dichlorophenoxy)acetic acid
DOC dissolved organic carbon
e efficiency; vapor pressure
E evenness (equitability); rate
E (*N*) photosynthetic efficiency of a particular needle age class
ECETOC European Chemical Industry Ecology and Toxicology Centre
eff specific effect constant
EFF average photosynthetic efficiency
EM ecological magnification (concentration trophic level *n*/concentration trophic level $n-1$)
$E_{\mathrm{o}}(\lambda)$ scalar irradiance
ES solar radiation
ET evapotranspiration
Ex excretion coefficient
f fugacity
F amount of food uptake per day
F_{m} site factors
$F*$ rate of infiltration
F_{k} correction factor
f_{oc} fraction of organic carbon in the solid being considered
F_{oc} distribution coefficient for organic matter
F_{s} distribution coefficient
GIS geographic information system
GW groundwater
GWP global warming potential
GWR groundwater run-off
h soil water suction; vector (Eq. 52)

H Shannon's diversity index; Henry's law constant
H_a sensible heat transfer through air
HC_p protective hazard concentration
H_s heat flow in soil
i cumulative infiltration at time t; flux
INT total intake of toxic substance per day
IRPTC International Register of Potentially Toxic Chemicals
J rate of irreversible phenomena (e.g., heat flux)
k von Karman's constant; rate constant
k_b ($=\mu_m/Y$) biodegradation constant
K hydraulic conductivity; rate constant, buffer constant; sediment – water partition coefficient (Eq. 80)
K vertical eddy diffusion coefficient
$K_{a,\lambda}$ specific absorption rate
K_G gas film mass-transfer coefficient
K_L liquid film mass-transfer coefficient; evaporation coefficient (Eq. 89)
K_{OW} n-octanol – water partition coefficient
K_s saturated hydraulic conductivity, half-saturation constant (Fig. 10); concentration of substrate to support half-maximum specific growth rate (Eq. 50)
L specific coefficient (e.g., of thermal conductivity or diffusion)
$L(N)$ leaf mass of particular needle age class
LAI leaf area index
$LEAF$ total leaf mass
m expectation (Eq. 52); adsorbent mass
M mortality
n number of individuals per species; number of molecules adsorbed
n_m number of molecules required for a complete unimolecular surface film
N precipitation; rate of transfer; total number of individuals in the community
$N_i(R)$ net interception
NJ number of needle age classes
N_o above-canopy precipitation
NOEC no-observed-effect concentration
O_n organisms
ODP ozone-depletion potential
p concentration of chemical in solution
P precipitation; rate of absorption (photosynthesis); equilibrium vapor pressure
PCB polychlorinated biphenyl
PCP pentachlorophenol
PEC predicted environmental concentration
p_i proportion of individuals or biomass contributed to the total of the sample
PNEC predicted no-effect concentration
POCP photochemical ozone formation potential
p_o saturation vapor pressure
p_r amount of rain required to pond the soil
q instantaneous value of water vapor concentration; water vapor content at level
QSAR quantitative structure – activity relationship
r resistance factor; increased number of organisms per unit time and unit population (Eq. 81)
r_{ET} transfer rate of latent heat due to evapotranspiration
r_p rainfall intensity
R recovery; universal gas constant
R_c long-wave counterradiation of the lower atmosphere
R_n net radiant flux
R_s surface runoff
s sorptivity; specific leaf surface
S total number of species in the community; sorptivity; nutrient concentration; saturation concentration or solubility at fixed temperature; net source – sink term (Eq. 51); entropy; standardized damage level
$S*$ pondage capacity
SM soil moisture, volumetric soil water content
S_s IR absorption strength in the interval 800 – 1200 cm^{-1}
SW perched water
T safety factor (Eq. 65)
TCO_2 percentage of total CO_2
2,4,5-T 2,4,5-(trichlorophenoxy)acetic acid
TF throughflow
TJ season
To equilibrium concentration in water
TOD theoretical oxygen demand
T_r average residence time due to reactions
u velocity
U wind speed at level
v rate of infiltration (Eq. 11)
v_z vertical flow of water at depth z (Eq. 8)
Vd vapor density
w momentary vertical wind speed
W accumulation; body weight (dry or wet matter)
$W(t)$ accumulated pollutant at time t

x amount of substance adsorbed
X thermodynamic "force" (e.g., gradient of temperature or concentration); biomass per unit volume
Y biomass produced per unit amount of substrate consumed; random function (Eq. 52)
z depth, depth of lowest point of measurement (Eq. 8); elevation above surface (Eq. 21)
Z fugacity capacity
α measure of the effectiveness of Br in ozone depletion with respect to Cl; reflectivity (albedo)
β average interaction strength
β_{ij} effect of species *j*'s density on species *i*'s rate of increase
γ coefficient of absorption of photosynthetically active radiation
ε_λ molar absorptivity at wavelength λ
$\varepsilon\sigma T^4$ long-wave radiation from the earth's surface
μ specific growth rate
μ_m maximum specific growth rate
ϱ constant in Hammett – Taft equation; air density
σ Hammett constant; reflectivity
τ_s atmospheric lifetime
Φ_d quantum yield
Ψ soil water potential

1. Ecology and Environment

1.1. Humans and Environment

In the course of their evolution, humans have exerted an ever-increasing influence on the terrestrial and aquatic environments that ensured their livelihood. In early Paleolithic times, *Homo erectus* was dispersed over large geographical ranges of the Old World continents and had already attempted to circumvent environmental limitations, via culture, in expressing dietary preferences different from the youngest australopithecines and in colonizing both drier and colder environments. Because of their food preferences, these hunters and gatherers probably required larger territories and thus comparatively few Acheuleans could be supported in any one area; consequently, mobility and the periodicity of seasonal activities were considerable [1], [81]. The sites occupied indicate a preference for open, grassy environments with large herds of gregarious herbivores, which means that human influence on ecosystems was by no means different from that of hunting or gathering animals.

Things changed drastically during the tenth millennium B.C., when the first agricultural societies arose. The new subsistence system, based on crop planting, livestock raising, and permanent settlements, had been developed to the point that long-term population increases became possible. Significant ecosystem changes, often associated with disturbances of the hydrological cycle and soil erosion [82], resulted when native floras and faunas began to be replaced by domesticated stock and cultigens. Thus, not only were the largely prevailing agrarian and woodland ecosystems profoundly changed in species composition, but natural recycling processes also constantly lost importance.

In the present millennium, characterized by an unprecedented population increase in the last two centuries, the formerly "isolating" woodlands or uncultivated areas around the agrarian ecosystems and settlements have been constantly reduced, which means that the ecological buffering capacity of these nonagrarian areas, which had such an important role in the intricate ecosystem interplay of the past, no longer exists in densely populated regions of the world. Consequently, both aquatic and terrestrial types of ecosystems are now open to human impact on the regional and continental levels, and present-day analytical knowledge provides ample evidence that even the most remote areas on earth bear more or less marked imprints of human activity.

Among them, chemical imprints have attracted particular attention, because their importance has grown since the middle of the last century with ever-increasing speed, and the development is not likely to change in principle, albeit sectoral differentiations are evident. A comparison of high-volume chemicals illustrates for the United States that petroleum derivatives represent 10 % of the total number of entries in the inventories, but account for 55 % of the total production, whereas inorganics represent both 12 % of the materials and 12 % of the production. Another 6.9 % of the production is due to materials

Table 1. Estimated global release of organic reference chemicals in 10^3 t/a (after [84])

Chemical	Release	Chemical	Release
1,2,3-Trichlorobenzene	0.09	1,2-Dichloropropane	15
1,3,5-Trichlorobenzene	0.09	Diethyl phthalate	23.1
1,2,4,5-Tetrachlorobenzene	0.15	Pentachlorophenol	27
4-Chloroaniline	0.16	*p-p′*-DDT	28.5
2-Chloroaniline	0.17	Aniline	32.7
1,2-Dichloroethylene	0.3	Dichloroisopropyl ether	45
1,1,2,2-Tetrachloroethane	0.45	Nitrobenzene	46.5
4-Nitroaniline	0.55	Malathion	47.4
Hexachlorobutadiene	0.6	*o*-Xylene	63
N,N′-Dimethylaniline	0.72	1,4-Dichlorobenzene	66
3,4-Dichloroaniline	1.2	Hexachlorobenzene	80
4-Methylphenol	1.2	Tetrachloromethane	84
Dimethoate	1.8	*p*-Xylene	100
1,1,2-Trichloroethane	2.4	Atrazine	111
1,3-Dichlorobenzene	3	Alachlor	123
Dieldrin	3	Phenol	180
Tri-*N*-butyl phosphate	3	Styrene	210
1,2,4-Trichlorobenzene	3.3	Di-*N*-butyl phthalate	230
m-Xylene	3.6	Trichlorofluoromethane	291
Endrin	6	Trichloroethylene	326
Chlorobenzene	6.6	Di-2-ethylhexyl phthalate	402
Dimethyl phthalate	6.6	Tetrachloroethane	470
Aldrin	7.5	1,1,1-Trichloroethane	537
Heptachlor	8.1	1,2-Dichloroethane	547
1,2-Dibromoethane	9	Ethylbenzene	600
Parathion (methyl and ethyl)	9.6	Cyclohexanol	1 200
Chlorpyrifos	10	Toluene	1 800
1,1-Dichloroethane	12	Benzene	11 000
γ-HCH (lindane)	14.1		

that are residues from the processing of ferrous metals. Saturated hydrocarbons are responsible for 6.7 %. Well-defined organic substances are the most numerous, and cause major concern in testing, because they account for about 34 % of the inventory sample, but only 6 % of the total production. Polymers and plastics represent 24 % of the number of materials and 3 % of the total production [83]. Table 1 provides insight into the estimated global release of organic reference chemicals. The figures listed indicate that humans, owing to the accelerated growth of the chemical industry since the 1950s, produce and use many synthetic compounds that, in localized exposure at least, no longer appear to be negligible in comparison to natural fluxes and concentrations. Thus, the ecological role of humans has changed profoundly in little more than a century of their long cultural history, because they are now capable of and responsible for introducing new substances or increasing the concentration of natural ones in the environment to an extent unthought of in the past [2].

Because the products of the chemical industry are also a most essential part of both our economic and our social life, governments and industry alike have become increasingly concerned about the potential, unintended consequences that the use of man-made chemicals could have on both human health and the environment. Therefore, a number of key chemical-producing countries have passed, or are enacting, general substance control legislation (→ Legal Aspects, Chap. 2., 6., 7.). Common to all these legal instruments is the preventive aspect (i.e., the notification of chemicals prior to marketing), which entails the presentation of data derived from laboratory investigations together with additional information to permit the evaluation of potential hazards.

1.2. Scope of Ecology and Ecotoxicology

Environments are highly complex systems whose evolution is determined by complicated networks of positive and negative feedback loops. Human interference, particularly when inadvertent, is therefore reflected in an ever-increasing number of environmental problems that cause major concern in today's societies. To better understand and resolve these problems, knowledge of the basic earth and life sciences unified under a comprehensive ecological perspective is essential, and a consideration of the bewildering number of interrelationships from a generalized system point of view appears indicated.

Ecological thinking began to emerge in the 17th and 18th centuries [3], while the very term "ecology" as derived from the Greek *oikos* (home) was coined and introduced into biological terminology by HAECKEL [4]. Paraphrasing HAECKEL, ecology can be defined as the scientific study of the interactions between organisms and their environment or, more informatively: "Ecology is the scientific study of the interactions that determine the distribution and abundance of organisms" [5]. The environment of an organism consists of all those physical and chemical (abiotic) or biotic (other organisms) factors and phenomena outside the organism that influence it.

At the *level of the organism*, ecology deals with how individuals interact with their biotic and abiotic environments (autecology). At the *population level*, ecology determines the presence or absence of particular species, their abundance or rarity, and trends and fluctuations in their numbers (demecology). Two approaches can be distinguished at this level: one deals first with the attributes of individual organisms and then considers the way in which they combine to define the characteristics of the population. The other deals directly with the characteristics of populations and relates these to relevant aspects of the environment. *Community ecology* deals with the structure of communities and with the pathways and fluxes of energy, water, nutrients, and other chemicals through them. This implies adequate consideration of human influence on nature (e.g., pollution and global warming).

Since ecology must deal explicitly with these three levels of the biological hierarchy and the overwhelming multitude of interactions between organisms and their physical, chemical, and biological environments, it is particularly subtle and complex. Yet ecology also has the distinction of being confronted with uniqueness in a very peculiar way: it must consider millions of different species, each composed of frequently countless numbers of genetically distinct individuals, all living and interacting in a highly varied and ever-changing world. Thus, "the beauty of ecology is that it challenges us to develop an understanding of very basic and apparent problems, in a way that recognizes the uniqueness and complexity of all aspects of nature but seeks patterns and predictions within this complexity rather than being swamped by it" [6].

Ecology of such a comprehensive scope cannot be an easy science but must be a complex interdisciplinary structure grouping several dozens of sciences on the basis of a hierarchically structured system of hypotheses. For instance, one of the major present-day ecological research schemes, the long-term ecosystem research project in the Bornhöved Lake District (Schleswig-Holstein) comprises no less than 25 disciplines, ranging from mathematics to social sciences, around a nucleus of physical, chemical, biological, and earth sciences [85].

Ecotoxicology may be defined as "a science which is concerned with the toxic effects of chemicals and physical agents on living organisms, especially on populations and communities within defined ecosystems; it includes the transfer pathways of those agents and their interactions with the environment" [7]. In formal analogy with the objectives of classical toxicology, which has evolved as an aut- or demecologically oriented discipline, ecotoxicology aims at defining toxic effects in terms of reactions of ecosystems or representative compartments of such systems. In view of the complicated structure of ecosystems and their tremendous variability in time and space, however, particular problems arise. While many experimental approaches exist to predict the hazards of drugs, food additives, and contaminants for humans (→ Toxicology, Chap. 3., 5.), which are based on extrapolation from animal or other data, the analogous ecotoxicological procedures are much more complicated and hitherto far less satisfactory [8].

The very crux of the ecotoxicological extrapolation problem is the difficulty that an eventual decision-making process will lead to either "false negatives or false positives" [86]. In this connection, a false negative implies that a chemical that appears acceptable on the basis of laboratory data may cause considerable damage in practical circumstances. If, for instance, the prediction of possible environmental exposure levels on the basis of standardized laboratory tests turns out to underestimate the actual levels occurring in "real-world" environments, populations of certain species will be much more at risk than expected. Thus, species with mono- or oligomaniacal feeding behaviors (i.e., food specialists), run a much higher risk of bioaccumulation effects if specific chemicals are applied against their favorite or only prey than less specialized predators.

In contrast, the false positive involves rejection of chemicals because laboratory tests indicate intolerable hazards for the environment, while in fact such undesirable effects are not likely to occur. Pesticides may be labeled toxic to fish on the basis of standardized laboratory trials, although this does not necessarily imply that they will always cause mortality in fish when applied in, or in the immediate neighborhood of, their natural aquatic habitat. The framework of tsetse fly control in West Africa amply demonstrated that the potentially fish-toxic insecticides endosulfan and deltamethrin could actually be used without having any noticeable effects on fish [87]. The effective dose rate for killing tsetse flies was minimized by careful selection of the insecticide formulation on the one hand and application methods on the other.

Thus, the hazards of pesticides usually depend on such factors as the amount applied, formulation, methods and times of application, and intensity of use, which are – in their extremely variable specific combinations – virtually impossible to simulate. Attempts must be made, therefore, to consider the ecological characteristics of target areas as well as possible nontarget effects in connection with the use pattern envisaged in order to avoid unrealistic predictions. A compound that appears unacceptable in a certain application in one place need not be so in another. Hence, basic physical – chemical data such as those described in the *Collection of Minimum Pre-Marketing Sets of Data* of the OECD Working Group on Exposure Analysis [88] have to be matched to data on the relevant properties of typified environments into which the compounds are ultimately released. From this particular perspective, the essential structural and dynamic characteristics of both chemicals and ecosystems that are relevant to ecotoxicology are described in the following chapters.

2. Principles of Ecology and Environmental Chemistry

2.1. Structure and Dynamics of Terrestrial Ecological Systems

2.1.1. Structure of Communities and Ecosystems

In a classic paper, Tansley [89] defined the term *ecosystem* as "... the whole *system* (in the sense of physics), including not only the organism-complex, but also the whole complex of physical factors forming what we call the environment of the biome – the habitat factors in the widest sense. Though the organisms may claim our primary interest, when we are trying to think fundamentally we cannot separate them from their special environment, with which they form one physical system. It is the systems so formed which, from the point of view of the ecologist, are the basic units of nature on the face of the earth .... These ecosystems, as we may call them, are of the most various kinds and sizes. They form one category of the multitudinous physical systems of the universe which range from the universe as a whole down to the atom."

From the definition of an ecosystem as an arbitrary unit with respect to both its spatial extent and the phenomena considered, ecosystems are often portrayed as food webs or graphs in which "all" the trophic relations between component species are given [9], [10], [90]. Species can also be grouped into a hierarchy of trophic levels or size spectra [11], [91], [92], which facilitates an understanding of the flows of energy and matter through a community and underscores biological differences in interactions within and between levels. Because of the impacts of pollution and environmental change on the living world [2], the concept of stability has also been germane to the argument of how best to depict ecosystems (Section 2.3.1).

Figure 1 illustrates that an ecosystem can be viewed as a composite of the lower levels of organization (i.e., communities, populations, and individuals). Coexisting individuals of a single species possess characteristics such as density, sex ratio, age – class structure, rates of natality and mortality, and immigration or emigration that are unique to *populations*. The *community* is an assemblage of species populations that occur together in space and time.

In proceeding from the level of physiological and behavioral ecology – dealing primarily with individuals – to the ecology of populations and communities, collective properties such as species diversity, community biomass, and productivity can be identified. Organisms of the same or different species interact with each other in highly diversified processes of mutualism, parasitism, predation, and physical or chemical competition. Thus, the nature of a higher organizational level is obviously more than just the sum of its constituent parts; it is their sum plus the interactions among them. Thus, *emergent properties* appear only when increased complexity (i.e., different levels of organizational structure) is the focus of attention. A primary aim of community and ecosystem ecology is to determine whether repeating patterns in collective and emergent properties exist, even when the differences in the particular species that happen to be assembled together are great. Discussions need not be segregated as ecosystem rather than community topics, since these phenomena depend explicitly on fluxes between living and nonliving components of ecosystems. Thus, for instance, detailed knowledge of the role that communities play in biogeochemical cycling is essential for understanding and adequately combating the effects of acid rain or the increasing levels of atmospheric CO_2 on forests, fields, greenlands, streams, lakes, and other communities.

2.1.1.1. Description of Community Structure

The apparently simplest way to characterize a community is to list its species, but in practice this may become surprisingly difficult, because of taxonomic problems and because usually only a subsample of the organisms in an area can be counted. Thus, the *species richness* of different communities can properly be compared only if investigations are based on the same sample sizes in terms of the area of habitat studied, the time devoted to sampling, or the number of individuals or modules included in the samples. Statistics in general, and geostatistics in particular, are essential to ensure valid sampling (see, e.g., [12], [13], [93]).

The simplest measures that take into account both species richness and abundance or biomass patterns are diversity indices. As examples, the Simpson, McIntosh, and Shannon diversity indices are given:

The Simpson index [94] is

$$D = \left(\sum_{i=1}^{S} p_i \right)^{-1} \qquad (1)$$

where D is the diversity index, S the total number of species in the community, and p_i is the proportion of individuals or biomass that it contributes to the total in the sample. An important further specification is expressed by the term evenness (or equitability, E), i.e., the proportion of the maximum possible value D would assume if individuals were completely evenly distributed among species:

$$E = D \cdot D_{\max}^{-1} \qquad (2)$$

Evenness assumes a value between 0 and 1.

The McIntosh [95] index is given as

$$D = \left[N - \left(\sum_{i=1}^{n} n_i^2 \right)^{\frac{1}{2}} \right] \cdot \left(N - N^{\frac{1}{2}} \right)^{-1} \qquad (3)$$

where N is the total number of individuals in the community and n is the number of individuals per species. In Section 2.3.3, the McIntosh index is used in the framework of stability-oriented analyses of biodiversity.

Shannon's diversity index H [96] again depends on an array of p values; thus,

$$H = -\sum p_i \ln p_i \qquad (4)$$

and

$$E = H \cdot H_{\max}^{-1} = H \, (\ln S)^{-1} \qquad (5)$$

Since an infinite number of logarithmic systems exist, different authors use other logarithms, in particular those with bases 2 and 10, which should be specified when calculating H.

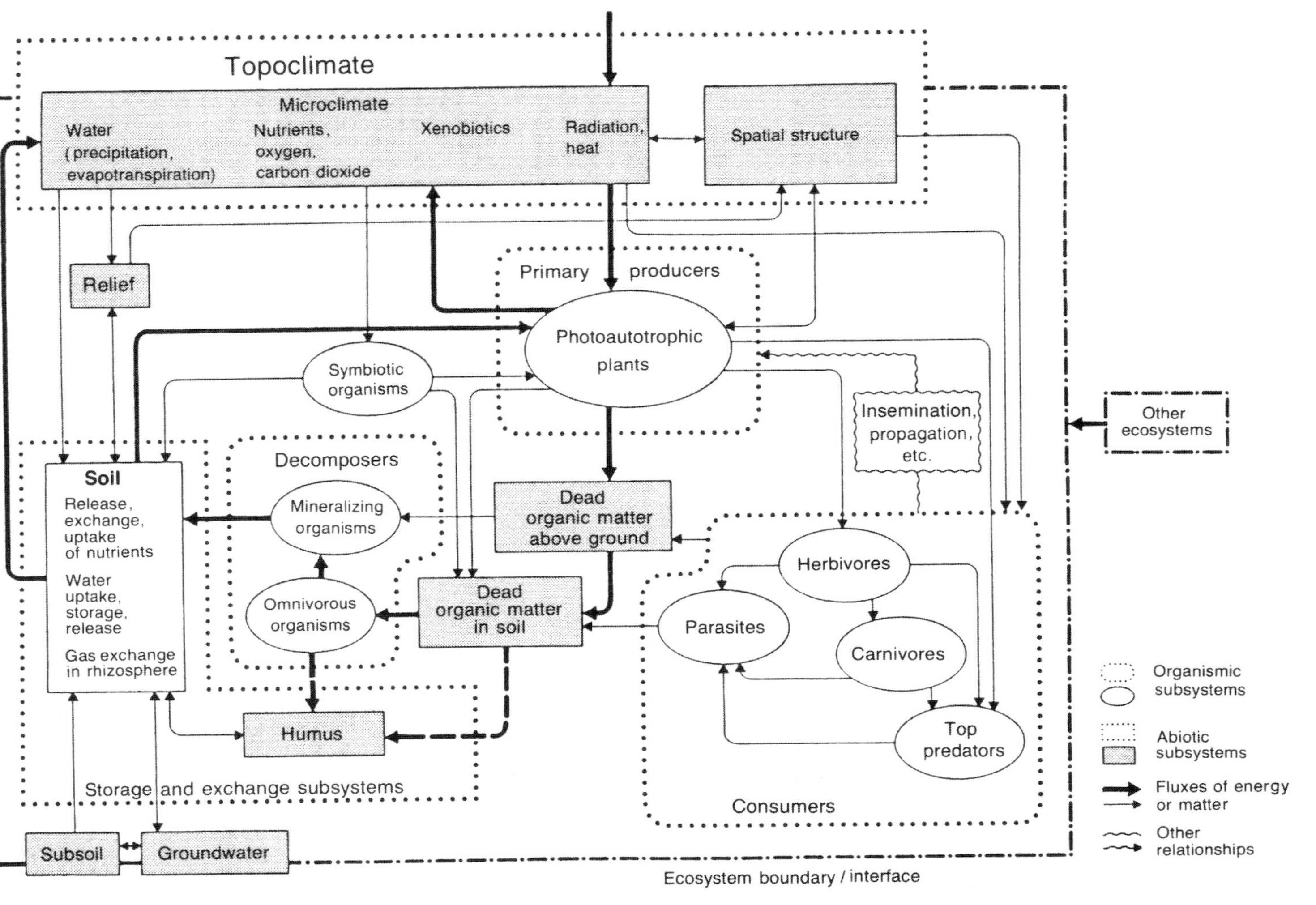

Figure 1. Graph model of an ecosystem (after [90], modified)

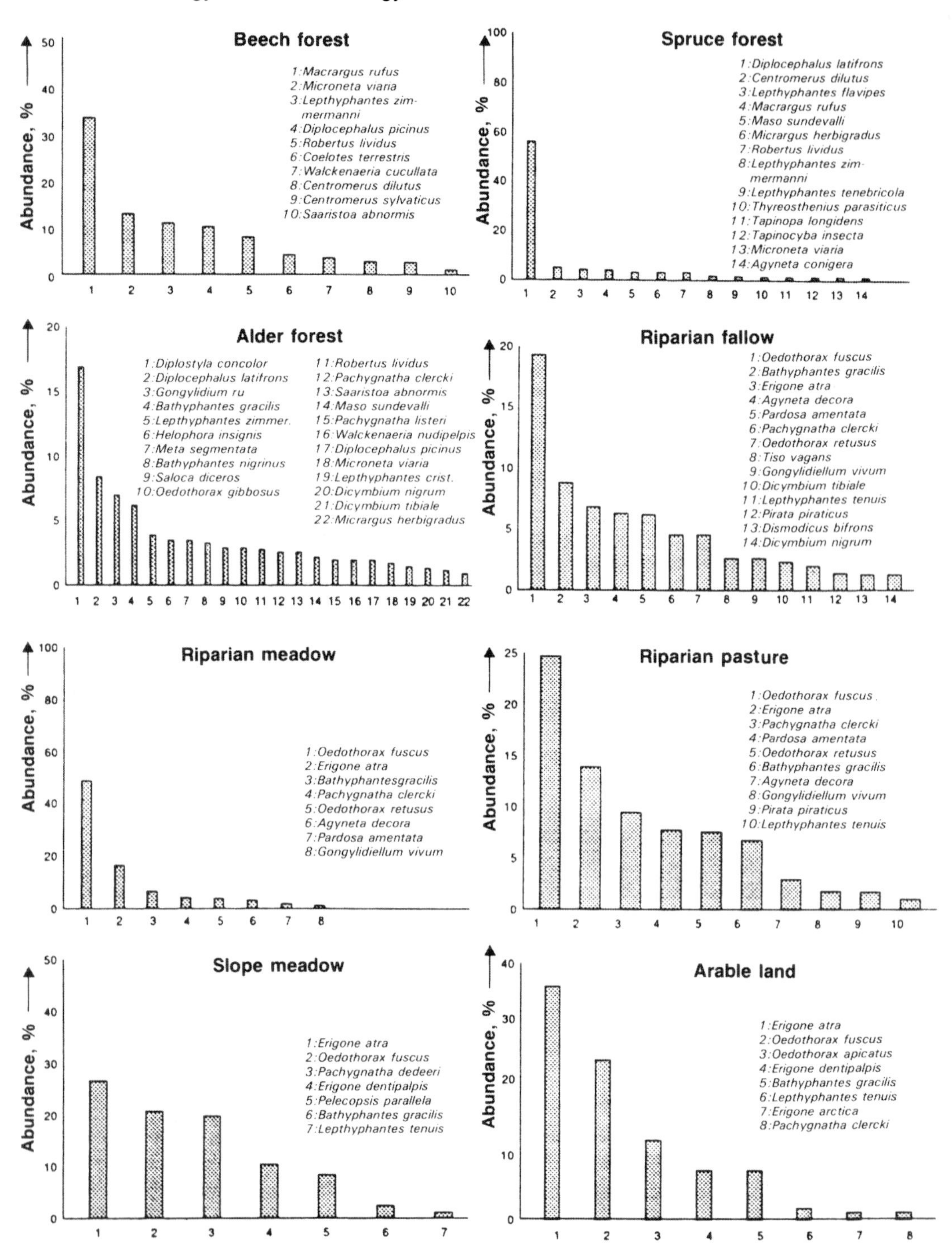

Figure 2. Rank – abundance histograms of epigeic spider communities from various ecosystems of the Bornhöved Lake District (Schleswig-Holstein) (after [85])

Rank – abundance diagrams provide a more complete picture of the distribution of species abundances in a community since they make use of the full array of p_i values by plotting p_i against rank. A rank – abundance diagram can be drawn for the number of individuals, for the area of ground covered by sessile species, or for the biomass contributed to a community by various species (see Fig. 2).

Other classical statistical techniques to reduce the subjectivity of community description are ordination and classification.

Ordination organizes communities on a graph so that those that are most similar in both species composition and relative abundance appear closest together. To this end, the axes of the graph are derived mathematically solely from the species composition of the various stands. Interpretation of the resultant patterns in terms of environmental variables is the second step, in which the scatter of points in the ordination is analyzed to see if the axes correspond to ecologically relevant gradients. Since the latter can be brought about by a host of interactions and a priori determination of the important ones is not possible, the same critical remarks relative to gradient analysis (see below) apply to the interpretation of ordination diagrams.

Classification, in a sense opposed to ordination, produces groups of related communities by a process conceptually related to taxonomic classification. Communities with similar species compositions are grouped together in subsets, and similar subsets may be further combined in a sequential manner by means of clustering algorithms [14], [97]. Figure 3 provides an illustration of such a community analysis (Fig. 3 A) and classification (Fig. 3 B), in which the primary characterization of the individual phytosociological units and the related Collembola faunas was effectuated by means of Renkonen indices.

Gradient analysis and canonical correspondence analysis are uni- and multivariate approaches to detect patterns in communities. In the first case, the investigator searches (sometimes with a considerable amount of subjectivity) for some feature of the environment that appears to matter to the organism. Then, data about the species concerned are organized along a gradient of that factor, which is not necessarily a priori the most appropriate factor. The fact that individual species from a community can be arranged in a sequence along a gradient may imply only that the factor chosen is loosely correlated with whatever really matters in the lives of the species considered. Thus, gradient analysis is a minor step on the way to the objective description of communities.

Canonical Correspondence Analysis. Much better suited to unravel the intricate interplay of environmental factors and to define their relative importance more precisely are multivariate methods (for a comparative analysis, see [98]). The canonical correspondence analysis [99] compares optimum and pessimum curves of species with Gaussian distributions on the basis of multivariate data obtained in the field. Figure 4 illustrates the results of such an analysis for the Collembola fauna of several stands of the Bornhöved Lake District (Schleswig-Holstein). In analogy to biplot analysis described below, in Figure 4 the orthogonal projection of the points marking species on the various vectors (environmental factors) defines the relative importance of these factors for the distribution of the Collembola species explored.

Biplot analysis [100], [101] permits analysis of the inherent structure of comprehensive geographic data matrices, representing, for instance, organisms (O_n) in relation to metrically defined site factors (F_m).

$$[\boldsymbol{M}\,(O_n, F_m)]_{1<n,m} \qquad (6)$$

Such an $n \times m$ matrix is equivalent to a cluster in m-dimensional space, where m is the number of factors as related to n organisms considered. This equivalence means that any matrix with $n \times m$ elements may be represented by one vector for a row and another vector for each column such that the elements of the matrix are the inner products of the vectors. When the matrix is of rank 2, or if a higher-rank matrix can be closely approximated by a matrix of this rank by using singular-value decomposition procedures, the vectors may be plotted and the resulting matrix representation (i.e., the biplot), inspected visually. Thus, a biplot not only permits viewing the individual data and their differences, but

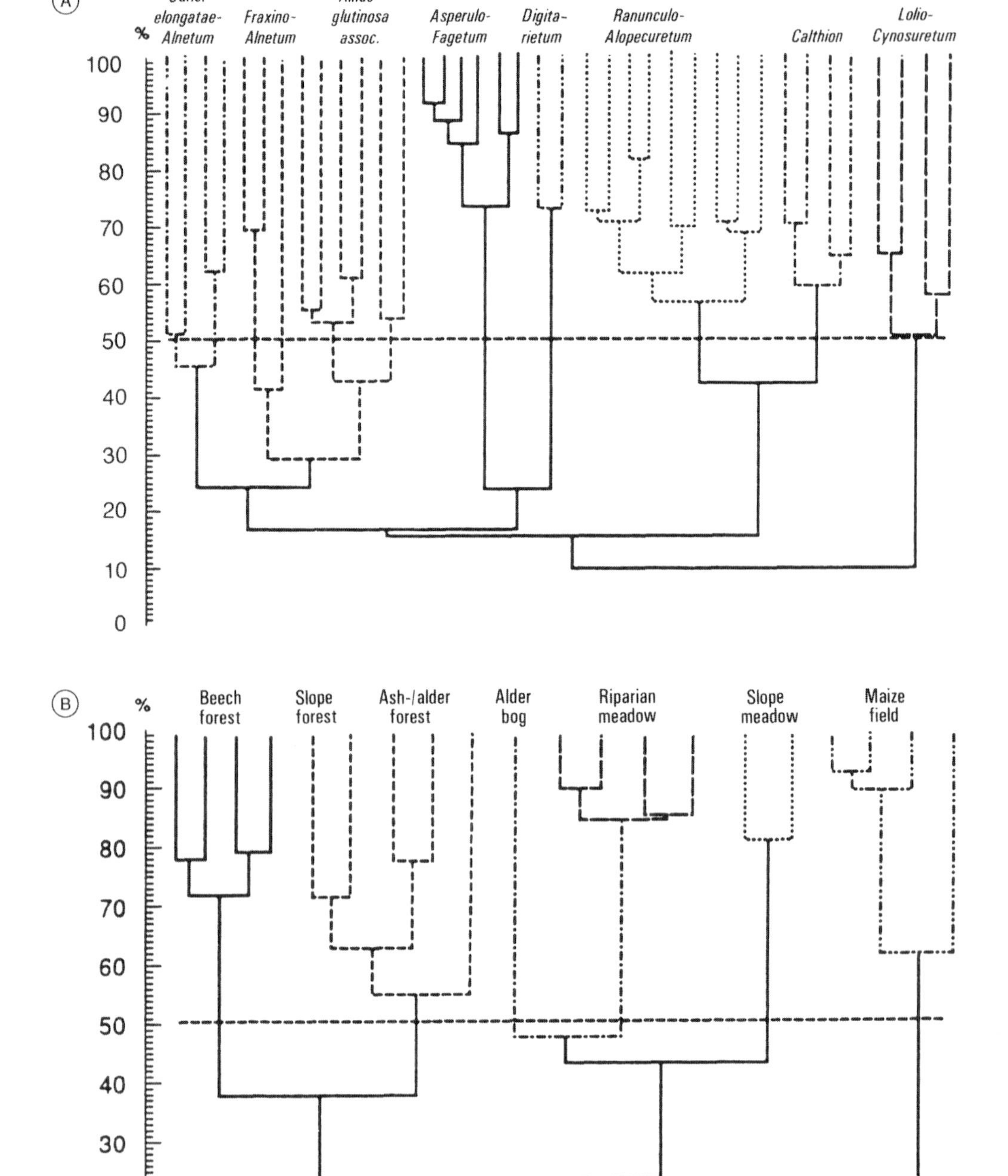

Figure 3. Cluster analysis of plant communities and related Collembola faunas in the Bornhöved Lake District (after [85]) A) Group analysis; B) Formation of subsets and clusters

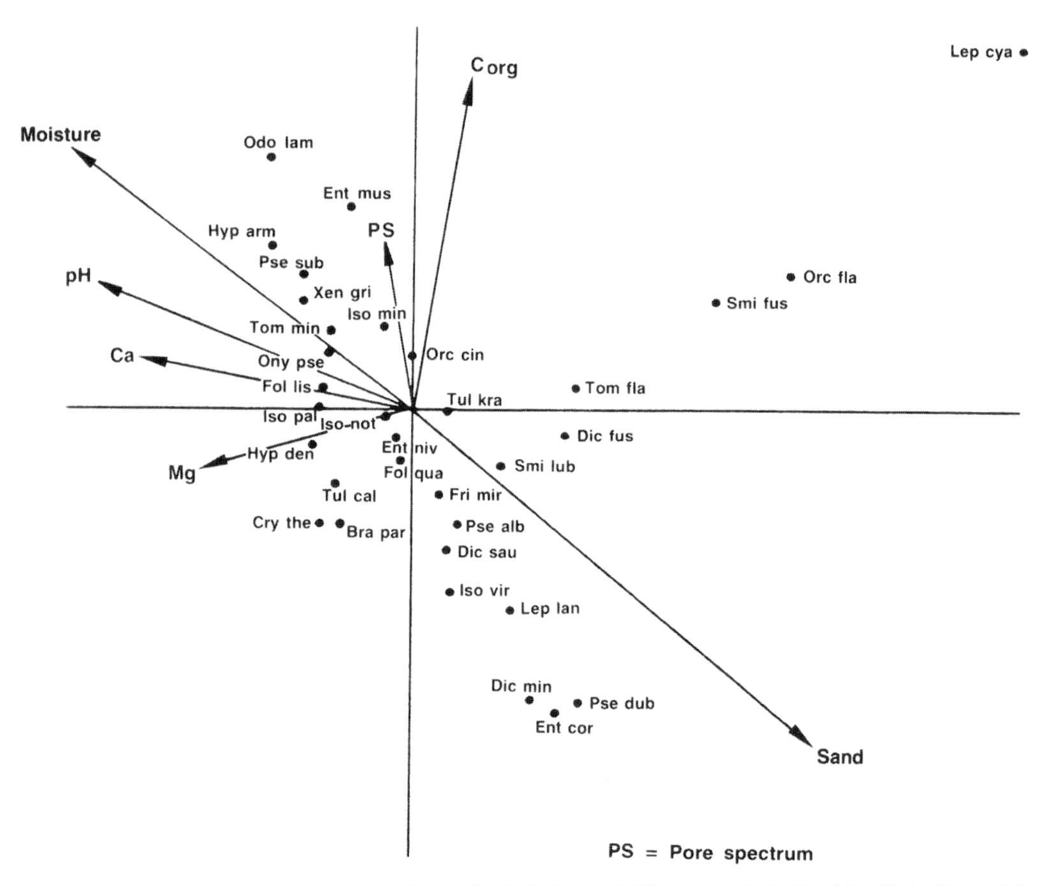

Figure 4. Canonical correspondence analysis of the Collembola faunas of different sites in the Bornhöved Lake District (after [85])
Species: *Brachystomella parvulu, Cryptopygus thermophilus, Dicyrtoma fusca, Dicyrtoma minuta, Dicyrtoma saundersi, Entomobrya corticalis, Entomobrya muscorum, Entomobrya nivalis, Folsomia listeri, Folsomia quadrioculata, Frisea mirabilis, Hypogastrura armata, Hypogastrura denticulata, Isotoma notabilis, Isotoma viridis, Isotomiella minor, Isotomurus palustris, Lepidocyrtes cyaneus, Lepidocyrtes lanuginosus, Odontella lamellata, Onychiurus pseudovanderdrifti, Orchesella cincta, Orchesella flavescens, Pseudachorutes dubius, Pseudachorutes subcrassus, Pseudosinella alba, Sminthurus fuscus, Sminthurus lubbocki, Tomocerus flavescens, Tomocerus minor, Tullbergia callipygos, Tullbergia krausbaueri, Xenylla grisea*

further allows scanning the standardized differences between units and inspecting variances, covariances, and correlations of the variables.

Chi-Square Automatic Interaction Detection. In addition to the above analytical procedures for metric data, chi-square automatic interaction detection (CHAID) is a hierarchically disaggregative classification procedure for nominal and ordinal data [98]. As a multivariable procedure, CHAID subdivides the base set of data automatically in relation to χ^2 values and significance levels into exhaustive subsets. Because of mathematical advantages and the easy interpretability of the resultant dendrogram presentation of data, it has found wide application in forest dieback studies in Germany [102], [103]. Figure 5 illustrates the results of a comprehensive CHAID analysis of environmental factors at 895 sites in North Rhine-Westphalia that proved to be of decisive importance for forest damage.

2.1.1.2. Successions

The relative importance in the community of many organisms (particularly those with relatively short lifetimes) changes with time of year because of seasonal change. In addition, interannual variations in the abundance of species oc-

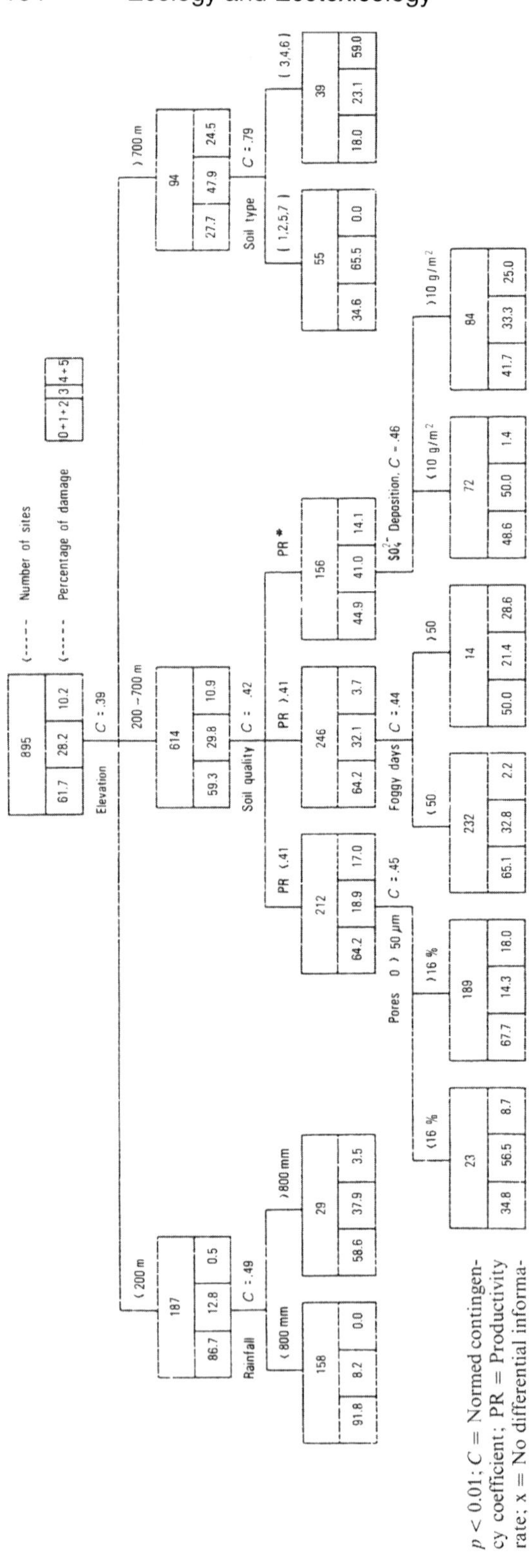

Figure 5. CHAID dendrogram of site factors controlling forest damage in North Rhine-Westphalia (after [103])
$p < 0.01$; C = Normed contingency coefficient; PR = Productivity rate; x = No differential information

cur because individual populations respond to a multitude of environmental factors that influence their reproduction and survival rates. On still higher time scales, different processes of succession occur, which may be defined "as the nonseasonal, directional, and continuous pattern of colonization and extinction on a site by species populations" [6]. On any time scale, successions are of particular ecotoxicological interest since the varying communities normally have different stability, resilience, and consequently sensitivity properties.

The underlying mechanisms of such serial replacements may be quite different. *Degradative successions* usually occur over a relatively short time scale of months or years [104]. Any packet of dead organic matter – for example, litter, the body of an animal, or a fecal deposit – is exploited by microorganisms and detritivores. Normally different species invade and disappear in turn, as degradation of the organic matter uses up part of the resources and makes others available. Ultimately, degradative successions come to an end because the resource is completely metabolized and (partly) mineralized [15], [16].

In addition, autogenic and allogenic successions can be distinguished. *Allogenic successions* occur as a result of changing external environmental conditions (e.g., changes in habitat quality due to erosion or deposition, modifications of soil pH). *Autogenic successions* usually occur on newly exposed landforms, and one of the principal forces driving this type of serial replacement is the change in soil conditions caused by the early colonists. These may alter the availability of resources (e.g., nutrients) in a habitat, in such a way that the entry of new species is facilitated. The process of *facilitation* [105] may be particularly important in primary successions where conditions are initially severe [e.g., on till laid bare after the retreat of a glacier or inland ice (see Section 2.3.7)]. Another mechanism underlying autogenic succession is inhibition – specific modification of the environment by early species that makes it less suitable for recruitment of "late-successional" species.

Connell and Slatyer [105] proposed three models whose fundamental characteristics are summarized in Figure 6. Hemprich [106] modeled the changes in community composition on the basis of comparative phytosociological analyses in the Bornhöved Lakes area. Figure 7 is an example of successions in grassland communities that typically take from several decades to hundreds of years to run their course.

2.1.1.3. Patch Dynamics

Disturbances that open up gaps in woods may be caused by the death of a tree, high winds, lightning, animals, or lumberjacks. Perturbations in grassland communities may be due to frost, to grazing or burrowing animals, and to the feet and dung of grazers. The formation of such gaps of highly variable size is of importance to sensitive or sedentary species with a requirement for open space, whereas it matters comparatively little in the lives of mobile animals for whom space is not a limiting factor.

Viewing the evolution of communities or succession from such a "patch-dynamics" point of view leads to an important class of models consisting of a number of cells that are colonized at random by individuals of different species. Although species interactions within cells proceed according to Lotka – Volterra principles [107], the stochastically cellular structure of the community and the manifold opportunities in each generation for migration between cells allow species-rich communities to evolve and persist. Thus, patch-dynamics models are much more realistic than equilibrium models with resource partitioning and niche differentiation (e.g., [108–110]) or nonequilibrium models incorporating temporal variation in conditions (e.g., [111]), which both deal with closed systems. Patch-dynamics models including competitive interactions and dispersal are of particular interest in an ecotoxicological context. Dhôte and Hollier [112] considered the dynamics of forest stands from two complementary points of view, namely, their intrinsic biological behavior and their management. This involves explicit consideration of different spatial and temporal scales, levels of organization, and model aggregation. Yodzis [113] distinguished two different types of community organization depending on the type of competition between component species. The case in which some species are competitively superior is called "dominance controlled", while the other with

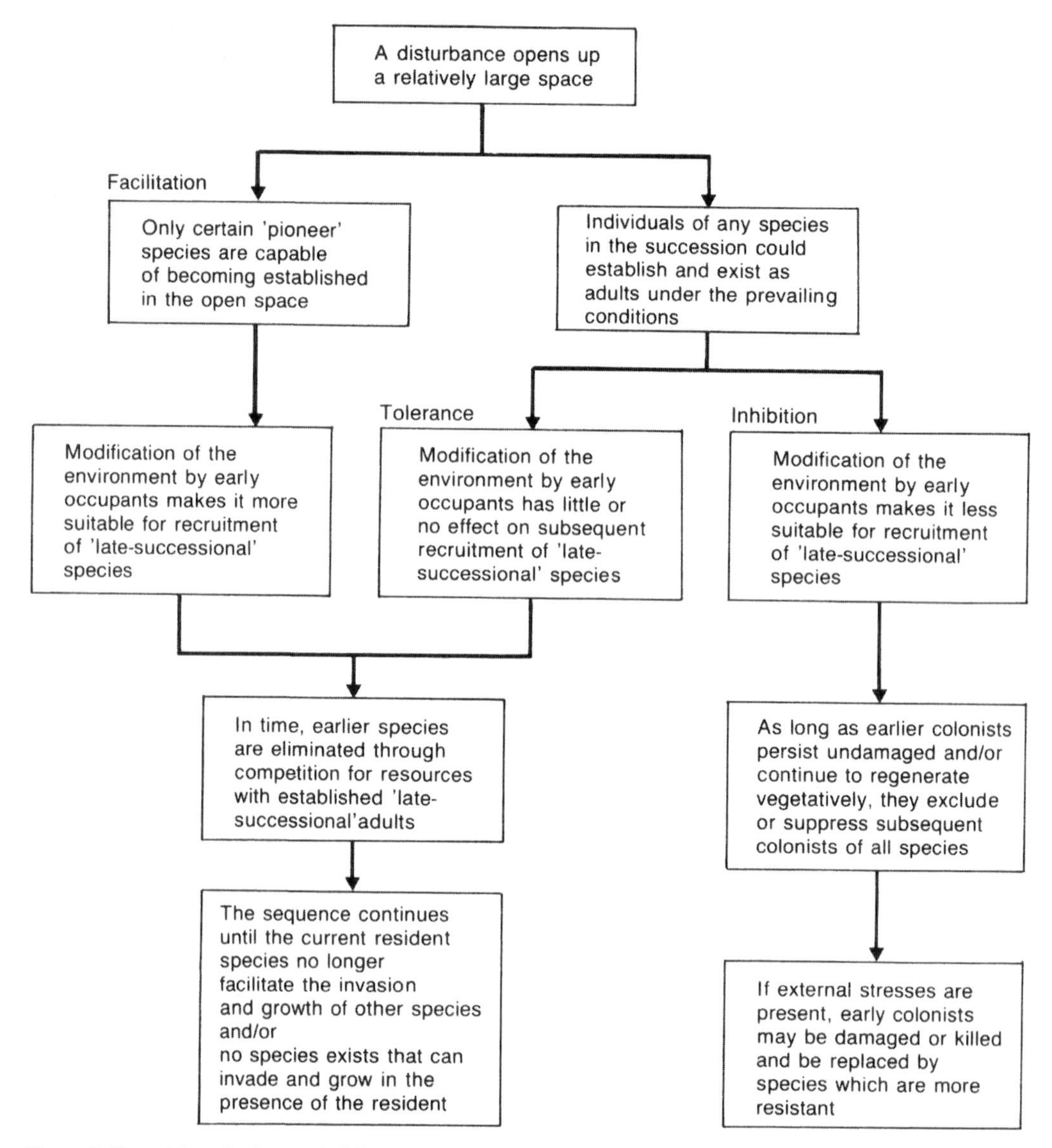

Figure 6. Essential mechanisms underlying terrestrial plant successions (after [105])

all species having similar competitive abilities is "founder controlled."

Dominance-Controlled Communities. In dominance-controlled communities, disturbances that open up gaps lead to relatively predictable species sequences, where early species are good colonizers and fast growers, whereas later species can tolerate lower resource levels and grow in the presence of early species, eventually outcompeting them. Thus, the community regains the climax stage when the most efficient competitors oust their neighbors. Disturbances may be synchronized, or phased, over extensive areas (e.g., bush or forest fires). Other disturbances are much smaller and thus produce a mosaic of patches, particularly if these disturbances are unphased. The resultant climactic macrosystem has a much higher species diversity than an extensive area undisturbed for a very long period and occupied by just a few dominant species.

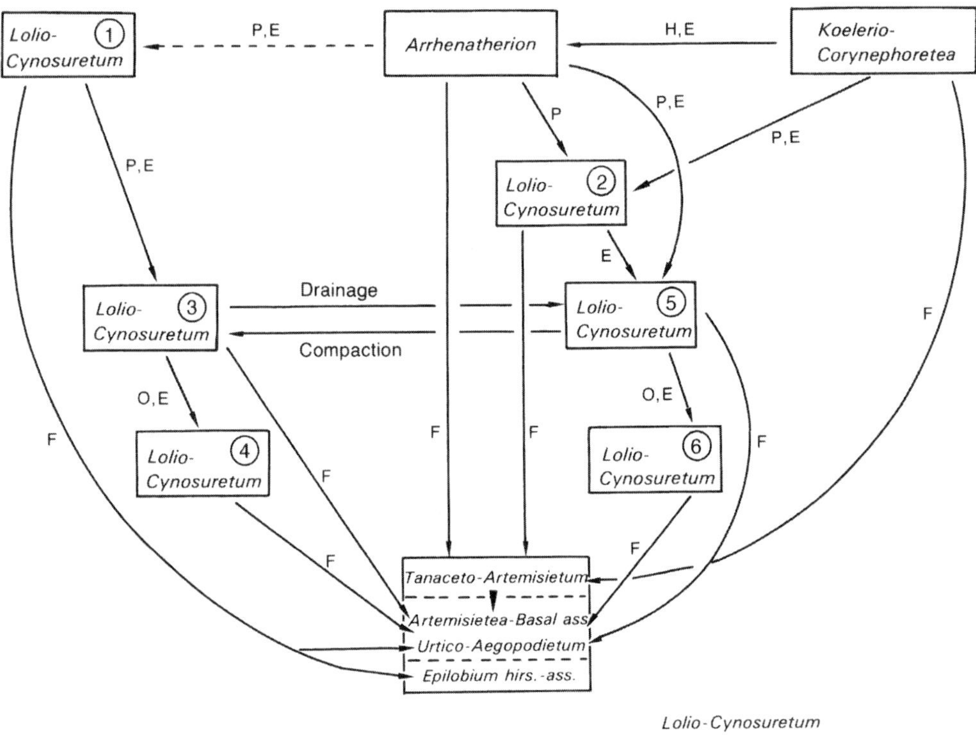

Lolio-Cynosuretum

1: *Lotus uliginosus* sub-association
2: Typ. sub-ass., variant of *Hypochoeris radicat*
3: Typ. sub-ass., variant of *Agrostis stolonifera*
4: Typ. sub-ass., variant of *A. stol.*, with *Rume*
5: Typ. sub-ass., typ. variant
6: Typ. sub-ass., typ. variant, with *Rumex obt.*

Figure 7. Successions in grassland communities of the Bornhöved Lake District (after [106], modified)
E = Eutrophication; P = Pasture; O = Overgrazing; H = Hay harvest; F = Fallow

Gap size and structure (defined in terms of relief, soil, climate, etc.) are important factors of community structure because they induce contrasting mechanisms of recolonization. The central parts of very large gaps are likely to be colonized by species propagules that can travel greater distances, whereas the smallest gaps may be filled by simple lateral movements of individuals around the periphery. Regrowth results from seeds, plants established prior to disturbance, and lateral growth of branches from surrounding trees. Nearer ground level, colonization is frequently due to saplings that were already present but suppressed [114]. Also, animals can be overrepresented in gaps, as LEVEY [115] has shown in a study of tropical rain forest in Costa Rica. Here, nectarivorous and frugivorous birds were much more abundant in treefall gaps where understory plants tend to produce distinctly more fruit over longer periods than conspecies fruiting under an undisturbed canopy.

Founder-Controlled Communities. Whereas in dominance-controlled communities colonizing ability and competitive status are inversely related, founder-controlled communities are characterized by species that are both good colonists and essentially equal competitors. If a large number of species nearly equivalent in their capacity to invade gaps, and equally tolerant of the abiotic conditions, can "defend" the gaps against all comers, then the probability of competitive exclusion is much reduced in communities where gaps open continually and randomly. On each occasion an organism dies, a gap is reopened for invasion. Thus, all conceivable replacements are possible and biodiversity will be maintained at a high level. An additional factor is the unpredictability of neighboring individuals in space and time, which contributes to reducing the likelihood of competitive exclusion of one species by another in the cause of succession. In any case, however, numerous feedback relationships exist between local and regional successional dynamics and the energy and water balances of the ecosystems affected.

2.1.2. Water Balance of Ecosystems

A number of models for plant – environment interactions, and particularly for the utilization of water and energy by plants, have been developed since the 1950s. Several of these attempt to understand plant growth and water use as related to specific physiological and environmental parameters [116], [117]. These models, however, cannot be applied to situations for which few data are available unless a number of simplifying assumptions are used. On the regional and geographical levels, other models of a predominantly qualitative character have been suggested [118–120]. As a consequence of this dichotomous development, attempts have been made to unify these two approaches with a view to simplifying the comprehensive models, so as to make them applicable to regional use in areas with limited data, without introducing misleading oversimplifications.

Examples of soil – water balance studies that appear to partly fill in this middle ground are those of Fuchs and Stanhill [121], Slatyer [122], Benecke and Van Der Ploeg [123].

The soil – water balance equation is generally written in the form:

$$P-R_{\mathrm{S}}-(GWR+TF)-ET+\Delta\,(SM+SW+GW)=0 \qquad (7)$$

where P denotes precipitation, R_{S} the surface runoff, GWR the groundwater runoff, TF the throughflow, ET evapotranspiration, SM soil moisture, SW the perched water (above a compact horizon), and GW groundwater. For the sake of greater convenience, $(GWR+TF)$ is termed deep drainage, and $\Delta\,(SM+SW+GW)$ is the change in soil water storage (initial minus final) during the period and for the depth of measurement. For a number of cases, deep drainage may be adequately defined as the amount of water passing beyond the root zone or, for experimental purposes, as the amount passing below the lowest point of measurement. All symbols have dimensions of length, for example, millimeters. Equation (7) can be written more specifically as

$$\int_{t_1}^{t_2} [(P-R_{\mathrm{S}})-(ET)-v_z]\,\mathrm{d}t = \int_{t_1}^{t_2}\int_{0}^{z} \frac{\partial\,(SM)}{\partial t}\,\mathrm{d}z\,\mathrm{d}t \qquad (8)$$

where (t_2-t_1) is the time interval over which the measurements are made (s), z is the depth to the lowest point of measurement (cm), v_z is the net downward flux of water at depth z (cm/s), and SM is the soil moisture, i.e., the volumetric soil water content (cm^3 of water per cm^3 of soil). P, R_{S}, and ET are in units of mm/s or $g\,cm^{-2}\,s^{-1}$. In the absence of a water table near the surface, v_z is generally positive. Compared to above-ground measurements of estimates of the water vapor flux, the soil – water balance approach has the advantages of ease of data processing and integration, since the soil – water reservoir (SM) automatically integrates extraction rates between observations. The disadvantages are associated largely with a somewhat lower level of measurement accuracy and the difficulty of adequately assessing evapotranspiration during periods of rainy weather. Therefore, its applicability is, to a considerable degree, restricted to regions of relatively high potential evaporation rates and sufficiently well-defined alternations of rainy and dry weather.

2.1.2.1. Precipitation and Interception

The measurement of precipitation N at a site or in a region is generally considered simpler and more straightforward than that of the other terms in the water balance equation. The technical requirements are comprehensively described by Von Hoyningen-Huene and Nasdalack [124] and by Sevruk [125]. The identification of unequivocal spatial structures should be controlled by means of variogram analysis, which is particularly important in vegetation stands (see Section 3.1.4.3).

Here, marked differences in the pattern of precipitation actually reaching the ground normally develop in many plant communities because of the gross interception of precipitation by the vegetation. Subsequently, precipitation is partly transferred to the soil by channeling down the main streams ("stemflow") and partly by dripping from branches, twigs, and foliage ("canopy leaching"), or it may be lost by evaporation from the wet surface. This latter proportion constitutes the term "net interception." Further differences in the amount of precipitation reaching the ground between plants ("throughfall") are due particularly to the disturbed wind structure and are most noticeable in the case of snow.

In view of the complicated physical nature of net interception, attempts to assess it by means of measurement and by indirect estimations are quite numerous. In the framework of a comprehensive agroclimatological model, BRADEN [126] developed the following interception estimate:

$$N_i\,(R) = a \cdot LAI \cdot \left(1 - \frac{1}{1 + N_0 b / aLAI}\right) \tag{9}$$

where (in the author's notation) $N_i(R)$ is the net interception; $a \cdot LAI$ the saturation parameter, dependent on leaf area index (LAI) and a species-specific maximum interception a; N_0 the above-canopy precipitation; and b the density of vegetation cover. The BRADEN approach has the double advantage of mathematical simplicity and physical foundation, and its validity has been widely tested [127].

With rain, drip or stemflow is usually observed after an area rain total of about 2 mm has been received. However, with freezing rain or snow under conditions favoring retention on the leaves, twigs, branches, and stems (i.e., low wind, temperatures a few degrees below freezing), several times this amount may be accumulated. Stemflow is enhanced by a smooth bark and by branches and leaves that are inclined upwards. Thus, in deciduous (beech) and evergreen (spruce) forests the amounts vary considerably, beech providing much more stemflow and much less interception loss, while oak would have an intermediate position (for greater details, see [2]).

2.1.2.2. Infiltration and Surface Runoff

Horton Overland Flow. Surficial runoff in the sense of Horton overland flow occurs whenever the rate of effective precipitation (i.e., precipitation net interception) exceeds the rate of infiltration ($F*$) and the resultant accumulation of surface water exceeds the pondage capacity ($S*$) at the point of measurement. The most important regulator is $F*$. It is useful in applied hydrology and in relation to pollutant transport to characterize the dynamics of infiltration by a small number of parameters. PHILIP developed a simple physical model of infiltration, which is, however, closely related to more precise diffusion descriptions of infiltration [128].

$$i = St^{0.5} + At \tag{10}$$

where i is the cumulative infiltration at time t, and the constants S ("sorptivity") and A have a physical meaning related to the diffusion analysis of infiltration. The first term on the right-hand side of Equation (10) describes the contribution to infiltration due to capillarity; the second term represents mainly the contribution due to gravity. The differential form of Equation (10) is Equation (11), where v is the rate of infiltration (cm/s):

$$v = \frac{1}{2} St^{-1/2} + A \tag{11}$$

Some values of the infiltration rate (mm/h) for particular soils in specified conditions are given below:

Sand, loess, silt	11 – 7
Sandy loam	7 – 4
Clayey loam, soils poor in organic matter	4 – 1
Clays, alkaline soils (solonets)	<1

These values are much lower than those determined by means of a field rain simulator in the Schönbuch Nature Reserve (Baden-Württemberg), where the use of buffered cylinder infiltrometers proved impossible because of the extremely high spatial variability of the infiltration-relevant soil characteristics. The experimental conditions were such that a constant surface runoff rate was produced by a constant amount of rainfall when the soil had reached its saturation point. Since evaporation is negligible during the short duration of the experiment in comparison to the rainfall applied (100 – 250 mm/h) and the increment in soil and water storage is assumed to be zero, the difference between precipitation and surface runoff may be equated to infiltration [129]. For two clay soils tested, the minimum infiltration capacity amounted to 58 and 60 mm/h; for sandy soils, 79 mm/h was measured, while the infiltration rates of loamy soils varied between 63 and 76 mm/h. These results are, on the one hand, indicative of a marked enhancement of infiltration due to organic matter and, in particular, desiccation cracks and root voids; on the other hand, they point to high throughflow rates close to the surface (e.g., piping). As a consequence, no differences in runoff characteristics could generally be found on soils in coniferous, broadleaf, or mixed forests. Variation in the runoff rates was high when the soils were initially dry, but low when the soils were wet. A comparison of these minimum infiltration rates with the maximum net precipitation rates recorded within the last hundred years leads to the conclusion that surface runoff is an exceedingly rare phenomenon in temperate forests due to preferential seepage.

Horton overland flow appears to be a common process in semiarid and arid regions, where precipitation intensities are high and the infiltration capacity of the sparsely vegetated soil is low. It is further caused or intensified by the development of a crust on the soils because the surface layer becomes compacted and the pores blocked as a result of the redistribution of soil particles following raindrop impact. Crust formation due to lateral iron translocation is a particularly widespread phenomenon in ferric luvisols of the semihumid tropics, where it largely contributes to enhance pediplanation processes.

Saturated Overland Flow. In temperate environments with normally modest precipitation rates and well-structured soils, Horton overland flow is the exception rather than the rule, except under certain conditions of cultivation and when the ground is frozen. In temperate environments, all the pore spaces may become filled with water after a period of prolonged rainfall, thus saturating the soil. At this point the water table has risen to the surface and the effective infiltration capacity is consequently reduced to zero. Subsequent rainfall runs off directly across the surface of the slope as saturated (or saturation) overland flow. This situation is likely to come about toward the base of a slope or in microtopographical depressions on a slope where both local infiltrations and throughflow received from higher up the slope contribute to soil moisture. SMITH and PARLANGE [130] describe simple relationships that enable saturating or ponding times to be estimated from values of saturated hydraulic conductivity K_s and sorptivity s. In Morocco IMESON [131] found that the amount of rain required to pond the soil p_r could be estimated reasonably well with one of these equations, namely,

$$\int_0^{tp_r} p_r \, dt = \frac{A}{K_s} \ln \frac{r_p}{r_p - K_s} \tag{12}$$

where A is 0.5 s^2 and r_p is the rainfall intensity.

Because of the number of boundary conditions operative in runoff, the latter varies considerably with the amount, intensity, and duration of precipitation, as well as with slope configuration and soil fabric, which determine the degree and extent to which pondage can occur [129], [132], [133]. In natural situations, the slope is rarely constant and, while runoff tends to reduce soil water recharge at the top of the slope and increase it at the bottom, minor changes of slope generally modify the slope – runoff interrelation. Table 2 provides some runoff figures from two sites in the Federal Republic of Germany. Bochow is representative of loamy luvisols developed from Weichselian till, while Müncheberg is characterized by a podzolized cambisol on coversand overlying Saalian till.

Table 2. Hydrometeorological characteristics of two ploughland sites in northeastern Germany for 1961 – 1965 (after [2])

Characteristic	Müncheberg	Bochow
Number of erosion-inducing showers	30	39
Precipitation, mm		
Median value	14.8	11.8
Mean extreme value	23.6	18.4
Absolute extreme value	34.8	25.2
Runoff, mm		
Median value	1.2	2.1
Mean extreme value	3.3	5.1
Absolute extreme value	10.2	8.1

2.1.2.3. Deep Drainage

The deep-drainage term in balance Equation (7) comprises throughflow and groundwater flow and can be equated to a vertical flow that, in turn, may be calculated from hydraulic conductivity and soil water potential data. The normal equation for vertical flow of water v_z is

$$v_z = K + K\frac{\delta h}{\delta z} \quad (13)$$

where K is the hydraulic conductivity (cm/s) and $(\delta h/\delta z)$ is the rate of change of soil water suction (h in cm) with depth (z in cm). Soil water suction is derived from the soil water potential ψ (dyn/cm^2) by the relationship $h = -\psi/\varrho_W\, g$, where ϱ_W is the density of water and g the vertical acceleration due to gravity. Unless h is very small, $\delta h/\delta z$ is usually much greater than unity so the K term in Equation (13) is often negligible [122]. Under these circumstances, the deep-drainage term of the water balance equation is given by

$$GWR + TF = \int_{t_1}^{t_2} v_z \cdot \mathrm{d}t \quad (14)$$

where $(t_2 - t_1)$ is the time between observations [134], [135].

In other situations, a net upward flux of soil water into the root zones can occur from wetter underlying soil horizons or, in particular, from a water table closer to the surface. For comprehensive reviews of methods available for the determination of deep drainage with a particular emphasis on groundwater recharge, see [136–140].

Under certain soil conditions, diffuse water movement through the intergranular pore spaces and voids may be supplemented by concentrated turbulent throughflow in networks of pipes. These result from large voids that exist in many soils and are enlarged by soil fauna (e.g., mice, rats, hamsters, moles, weasels, ground squirrels) and the growth and decay of roots. Frequently, soil pipes develop at the interface between organic soil and the underlying mineral soil. Discharge in completely filled pipes varies depending on pressure and gravity potentials, and in partly filled pipes in response to the gradient of the water surface. Usually, therefore, pipe flow velocity is much more rapid than that of matrix flow. Table 3 quotes estimates of flow velocities. From these figures, pipe flow may be seen to attain considerable importance in chemical transport, although the distances covered are normally small in comparison to channel flow. Piping is perhaps more strongly associated with semiarid areas than with humid regions. Drainage and slope development in badlands all over the world are frequently dominated by piping.

Table 3. Flow velocities along different routes in a catchment (after [141])

Type of flow	Flow route	Velocity, m/h
Surface	channel flow	300 – 10 000
	overland flow	50 – 500
Soil flow	pipe flow	50 – 500
	matrix throughflow	0.005 – 0.3
Groundwater flow	limestone (jointed)	10 – 500
	sandstone	0.001 – 10
	shale	10^{-8} – 1

2.1.2.4. Soil Water

Measurement of changes in soil water storage is conducted most accurately by the use of weighing or hydraulic lysimeters, provided they are properly designed and sited [142]. Lysimeters cannot be used, however, when the nature of the species composition, the spatial structure of the vegetation cover, the depth and ramification of the root system, or other factors make it impossible to simulate the natural environment inside the lysimeter itself. In such cases, determinations of changes in soil water storage at different points in the plant community provide the only technique for evaluating $\Delta(SM + SW + GW)$. Soil water sensing equipment may still fall short of operator requirements, although marked advances have been made since the 1980s. Probably the most commonly used techniques, at the present time, are those of neutron moderation and tensiometers or tensiographs.

2.1.3. Energy Balance of Ecosystems

The disposition of radiant energy at the surface of the earth is of prime importance for understanding soil – water balances and the related chemical transport and transformation processes. A formalized transcription of the energetic relationships yields the following form of the energy balance equation:

$$R_n + H_s + H_a + r_{ET} = 0 \quad (15)$$

In this equation, which summarizes the solar energy cascade with its numerous regulation and transformation components, any form of energy that is flowing toward the surface is considered positive and any form that is moving away from it is considered negative. R_n is the *net radiant flux*, H_S is the *heat flow* in soil at the surface, H_a is the *sensible heat* transfer through air at the surface, and r_{ET} is the transfer rate of *latent heat* due to evapotranspiration. Conventionally, all terms are expressed in joules per minute and square centimeters; 4.19 J cm^{-2} min^{-1} equals ca. 1 mm of water depth evaporated per hour.

2.1.3.1. Net Radiation

The earth's surface receives short-wave and long-wave radiation from the sun – some of which is lost through reflection – and long-wave counterradiation from the atmosphere, while itself emitting long-wave radiation. Whereas photochemical reactions are controlled mostly by short-wave radiation, the diurnal variability of surface temperatures and the related energy fluxes are very much under the influence of long-wave radiation. Thus, the net radiant flux R_n may be derived from the short- and long-wave radiation balances:

$$R_n = ES\,(1-\alpha) + R_c - \varepsilon\sigma T^4 \quad (16)$$

where the solar radiation *ES* is the sum of direct beam radiation ES_S and diffuse radiation ES_D, each term comprising short-wave and solar long-wave components; R_c is the long-wave counterradiation of the lower atmosphere; $\varepsilon\sigma T^4$ the long-wave radiation from the earth's surface (Stefan – Boltzmann radiation); and α the reflectivity. Regionally, the net radiant flux of the earth's surface varies considerably with a general trend toward a positive balance or surplus in low latitudes and a negative balance or deficit in high latitudes (for details, see [17]). Factors such as cloudiness and atmospheric humidity, which in conjunction with aerosols affect the transmission of both solar and terrestrial radiation, and the radiative properties of the soil and vegetation canopy create the spatial variation in radiation balance over the earth's surface.

2.1.3.2. Soil Heat Flow

The rate at which heat is transferred downwards into the soil and subsurface substrate (H_s) is directly related to the nature and efficiency of the distribution mechanisms. In solids, heat is redistributed by conduction, and the flow rate depends on *thermal conductivity*. Units are watts per meter per degree Celsius (W m^{-1} $°C^{-1}$) or joules per centimeter per second per degree Celsius (J cm^{-1} s^{-1} $°C^{-1}$). Under steady-state conditions, the flow of heat ($\partial Q/\partial t$) is related to the temperature gradient ($\partial\Theta/\partial x$) and the thermal conductivity λ by

$$H_s = \frac{\partial Q}{\partial t} = \lambda\frac{\partial\Theta}{\partial x} \quad (17)$$

A major problem in soils is that steady-state conditions are rarely achieved and thermal conductivity is a complicated function of granulometry, mineralogical composition, compaction, and water content. The conventional determination of thermal conductivity is consequently fairly difficult. An alternative parameter, *thermal diffusivity*, is used, which is given by $\lambda/(c\varrho)$, where c denotes the specific heat and ϱ the density. Units are square meters per second. For a homogeneous medium, thermal diffusivity defines the rate at which temperature changes ($\partial\Theta/\partial t$) take place:

$$\frac{\partial\Theta}{\partial t} = \frac{\lambda}{c\varrho}\cdot\frac{\partial^2\Theta}{\partial x^2} \quad (18)$$

2.1.3.3. Sensible Heat Flow in the Lower Atmosphere

The magnitude of the heat flow into the air at the surface H_a can be obtained by a difference when the other components of the energy balance in Equation (16) are measured, or it can be determined directly. Among the latter approaches, two methods, the aerodynamic method and the

Sverdrup – Albrecht method merit particular attention.

The *aerodynamic method* is based on very precise determinations of the vertical temperature and wind profiles, and involves a horizontal homogeneity of the surface over considerable distances in the luff of the measuring station.

The *Sverdrup – Albrecht method* determines either H_a or ET as components of the energy balance in Equation (19)

$$H_a = \frac{H_a}{ET}(R_n - H_s) \cdot \left(1 + \frac{H_a}{ET}\right)^{-1} \quad (19)$$

The ratio H_a/ET (Bowen ratio) can be estimated from measurements of the vertical gradients of temperature ($\Delta\Theta/\Delta z$) and vapor pressure ($\Delta e/\Delta z$) above the surface.

2.1.3.4. Latent Heat Flux

The latent heat flux can be estimated as a result of the vaporization or condensation of water. Consequently, five general methods are used to evaluate the water vapor flux caused by evapotranspiration.

Analogous to Equation (19), the *Sverdrup – Albrecht method* may be used to determine ET:

$$ET = (R_n - H_s)\left(1 + \frac{H_a}{ET}\right) \quad (20)$$

THORNTHWAITE and HOLZMAN [143] suggested

$$ET = \frac{\varrho k (U_2 - U_1)(q_1 - q_2)}{[\ln (z_1/z_2)]^2} \quad (21)$$

where q is the average water vapor content, ϱ the air density (g/cm^3), U the average wind speed (cm/s), z the average elevation above surface (cm), and k von Karman's constant (0.4).

A *continuous record of water vapor convection by means of eddy-correlation techniques* as described by DYER [144] yields estimates of ET by

$$ET = \frac{1}{t}\int_0^t \varrho w q \mathrm{d}t \quad (22)$$

where ϱ denotes air density, w the momentary vertical wind speed, and q the instantaneous value of water vapor concentration. The problem of measuring ET as the time average of the vertical flux of water vapor thus becomes one of designing instrumentation capable of measuring the turbulent air motion and structure of q. This involves equipment whose response time is sufficiently short to take account of all frequencies in the turbulent spectrum contributing significantly to the flux. This requirement depends on both the height of measurement and the stability of the air, which means, in general terms, that the sensing elements should respond adequately to signals with a period of 1 s.

If all terms of the energy budget are known except the flux due to evaporation (or evapotranspiration), the latter can be obtained by a *difference*. Finally the latent heat flux can be found by *direct measurement*, i.e., by means of weighing lysimeters on the local scale or as the difference term of the water budget of a drainage system on the regional scale.

2.1.3.5. Efficiency of Photosynthesis

In comparison to the above components of the energy balance summarized in Equation (15), the absorption of photosynthetically active radiation in the plant cover plays a quantitatively minor role. Yet, it furnishes useful information on the amounts of solar energy available for photochemical reactions at the surface of soil and vegetation under natural conditions. A quantitative description of the absorption process is provided by

$$dP = \gamma (ES)_v \, s dz, \quad (23)$$

where P is the rate of absorption, γ is the coefficient of absorption of photosynthetically active radiation (i.e., wavelength ranges from 400 – 480 nm to 640 – 690 nm); $(ES)_v$ is the incident direct beam and diffuse short-wave radiation; s is the specific leaf surface (i.e., the surface of leaves in a unit volume); and z is the vertical coordinate counted upward [18].

According to MONTEITH and SZEICZ [145], the daily energy rates required for gross assimilation are about 28.5, for respiration 6.7, and for net assimilation 21.3 J cm^{-2} d^{-1}. This means that normally the photosynthetic efficiency of the vegetation cover is in the order of magnitude of 1 – 2 % of the total incident radiation. These findings were corroborated by numerous authors. Examples of studies from different parts of Central Europe are, for instance, [19] and [146]. From the purely quantitative budgetary point of view this may appear little, but with regard to bioproductivity and photochemical transformation reactions the situation is quite different. A number of compounds (e.g., various chlorinated aromatics) show distinctly higher conversion rates if adsorbed on particulate matter than if deposited as solids or on thin films [147]. This can be attributed to a bathochromic shift, changes in the relative extinction, the appearance of new absorption bands as a consequence of adsorption, and finally fixation of the dispersed molecules in the adsorbed phase resulting in an intensified compound – oxygen contact.

2.1.4. Nutrient Cycles and Productivity of Terrestrial and Aquatic Ecosystems

Speaking about nutrient cycles means considering how the same atoms of elements move through different components of nature or different biota, and participate in different kinds of reactions (Fig. 8). In principle, these are always chemical reactions; however, the conditions driving them vary highly if embedded in biochemical, biological, or ecological structures. A typical feature of the organic world is the hierarchy constructed of levels of increasing complexity (Fig. 9). In such a hierarchy, the control of processes is both bottom-up, from the low levels to high ones, and top-down, from the higher to lower levels. Biological reactions are controlled by several mechanical, chemical, and neurological mechanisms, which depend on the internal and external environment of biota. An example is the formation of urine in kidneys. In ecological reactions, the conditions for biological, biochemical, and chemical reactions are driven by population cycles of organisms, by mutual reactions of the environment and biota, and by interactions among biota themselves.

Biological production is based on the uptake of nutrients, but external conditions such as light, temperature, and oxygen have major effects. Production is the output of organic matter formation by plants, the primary producers, and of transformation by other groups of biota. Biomass is the organic mass of organisms, which can be expressed as fresh weight, as dry weight, as protein content, or in energy equivalents. Biomass is the standing stock, which is not necessarily identical with production. Deciduous forests in temperate regions, for example, produce yearly the biomass of leaves that then decay in fall. The next year, a new leaf biomass is formed.

In the following, nutrients and biological production are described separately for the sake of simplicity. However, they are shown to be inseparably interrelated. A distinction is made between terrestrial ecosystems (those inhabiting land) and aquatic ecosystems (those inhabiting water).

The effect of human activities, particularly of industry, on the environment is enormous. Humans are at present the dominant driving force on earth. They must take care that this force is not misused and is not driving humanity to extinction.

2.1.4.1. Nutrients

Nutrients are defined as those chemical elements that are essential for the construction of organisms. Depending on the proportion of the organism biomass that they represent, nutrients are usually divided into two classes: (1) macronutrients and (2) micronutrients or trace elements. *Macronutrients* represent major components of the biological material and are needed in large amounts. Usually, C, N, P, K, S, Ca, and Mg are considered macronutrients. *Micronutrients* are needed only in small amounts; micronutrients include Si, Mo, Cu, Zn, Mn, Fe, B, and Cl. The needs of organisms for elements belonging to the same class differ by a factor of ten. An indication of the ratio of the first three macronutrients in organisms is given by the stoichiometric ratio of the organisms' bodies. The average elemental composition (by weight) of aquatic autotrophs, for example, is about C : N : P = 106 : 16 : 1. In land plants the average ratio of C : N : P corresponds to 250 : 10 : 1 according to data in [21], and the entire set of macro-elements will give a C : N : P : K : Mg : S ratio of 250 : 10 : 1 : 5 : 5 : 1 : 0.5. Micronutrients are present in land plants in lower amounts (Fe 100, Mn 50, B 20, Zn 20, Co 6, and Mo 0.2 μg per gram dry weight). However, their ratios can vary

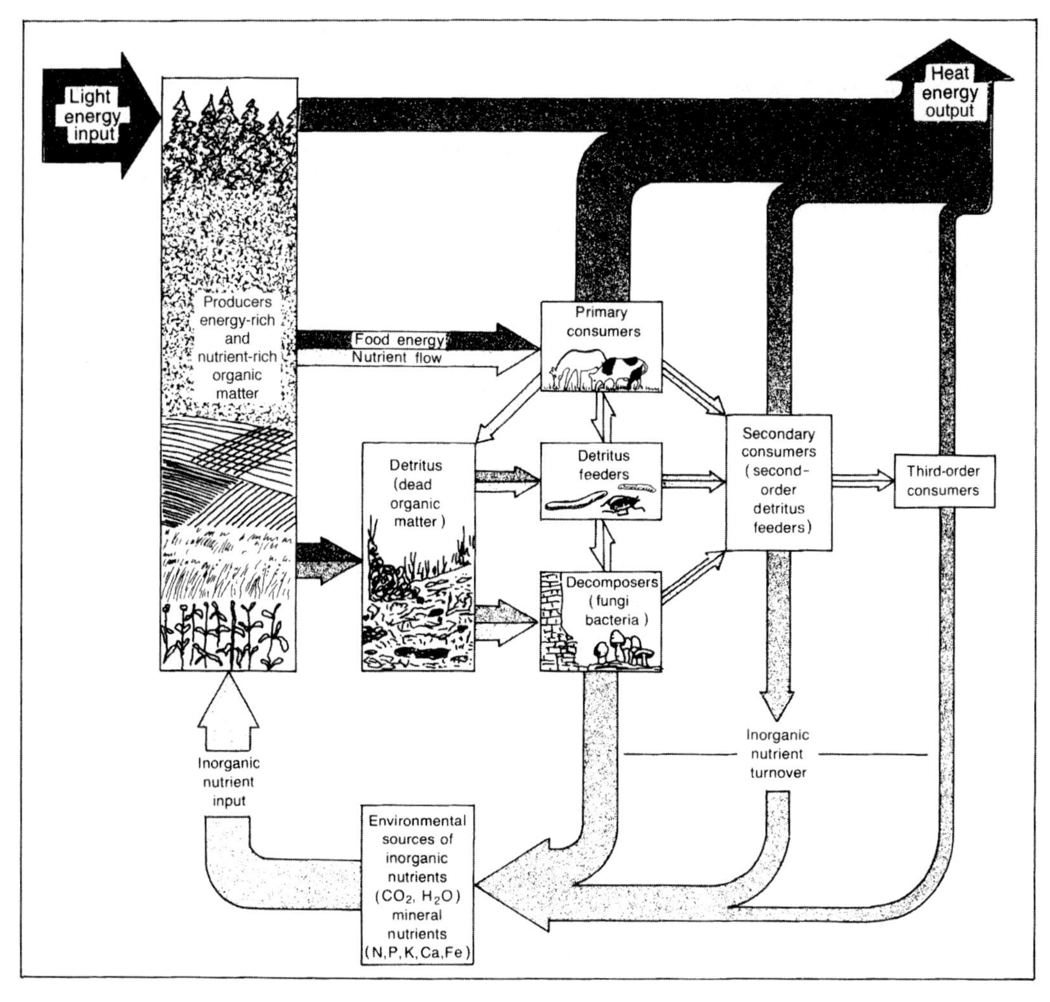

Figure 8. Nutrient cycling and productivity of ecosystems production decreases in relation to successive steps of the trophic chain (producers > herbivores > predators) (after [20], modified)

widely in individual plants (see, e.g., [148] for aquatic plants).

Chemical elements usually are not taken up as such by plants; rather, chemical compounds (e.g., C in the form of CO_2) are. Plants are now recognized to optimize their uptake. For example, in selection between nitrates and ammonia, the energetically more favorable ammonia is taken up first. Only some compounds of phosphorus are bioavailable [22].

A typical curve for the kinetics of nutrient uptake is given in Figure 10, which also shows that an element essential for the life of organisms may have a negative effect if it is present in excess quantities. In agriculture this is known as overfertilization. Plants need at least the first three macronutrients (C, N, P) simultaneously. The theory of limiting nutrient states that the nutrient in shortest supply represents the bottleneck. The uptake of all macronutrients is driven by the limiting one, the rest being taken up proportionally to the total needs.

Inorganic nutrients are taken up only by plants with photosynthetic pigments that use solar energy for the construction of highly structured organic matter from more simple inorganics. Such plants are called *autotrophs*; *heterotrophs* are organisms bound to organic matter formed primarily by autotrophs, but possibly transformed by other heterotrophs (e.g., carni-

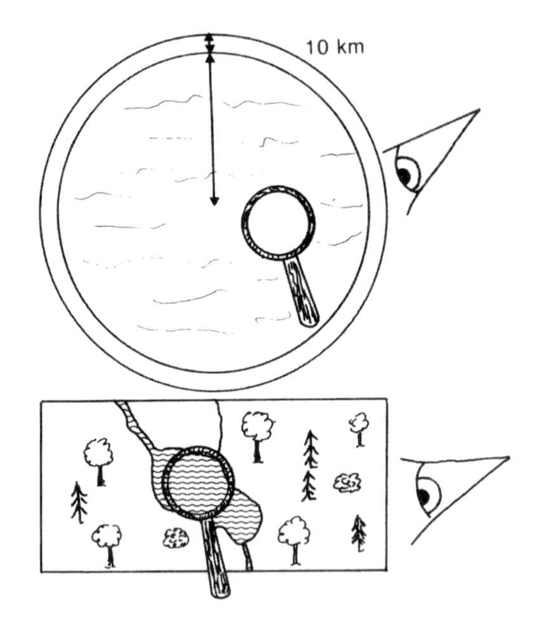

World

On the globe we see only the position of land masses and of major biomes. The atmosphere extent is highly magnified - in real proportions it would represent only much less than a millimeter.

Biome

The basic vegetation cover types occupying vast areas of the global surface are called biomes. Tundra biomes represented schematically.

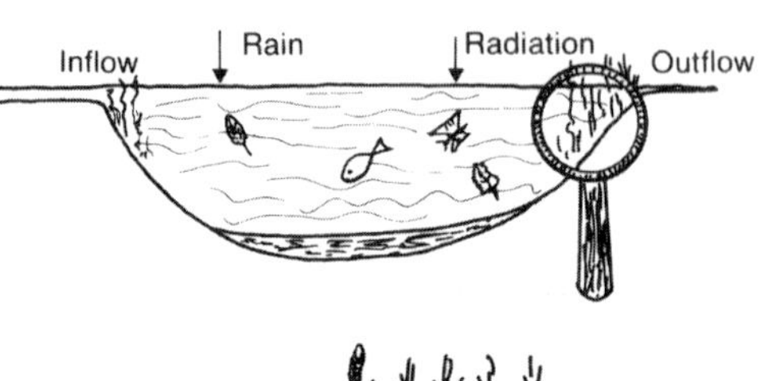

Ecosystem

An ecosystem is defined as a relatively isolated part of nature like a forest or a lake, with its abiotic components and biota mutually interacting.

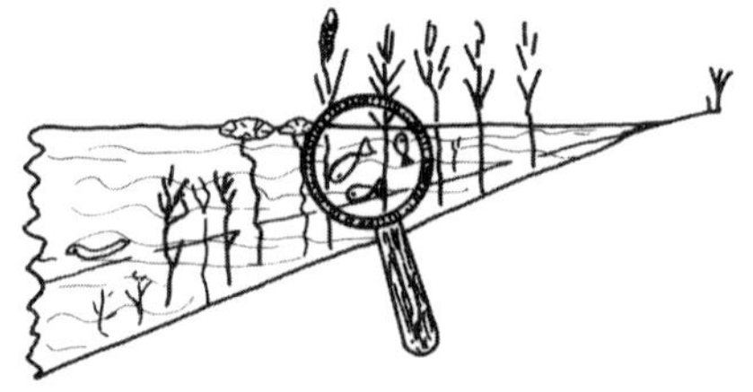

Biocenosis

An assemblage of plants and animals living in certain specific types of environment. The example depicts the nearshore (littoral) biocenosis of a lake.

Population

One-species population consisting of different age groups which may also appear successively.

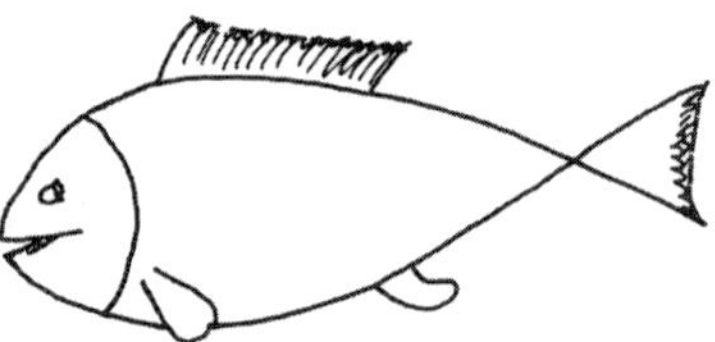

Individuum

Within the population different individua possess different capabilities.

Figure 9. Hierarchy of biological organization; looking at selected objects of nature with successively larger or smaller magnification involves a simultaneous change of view

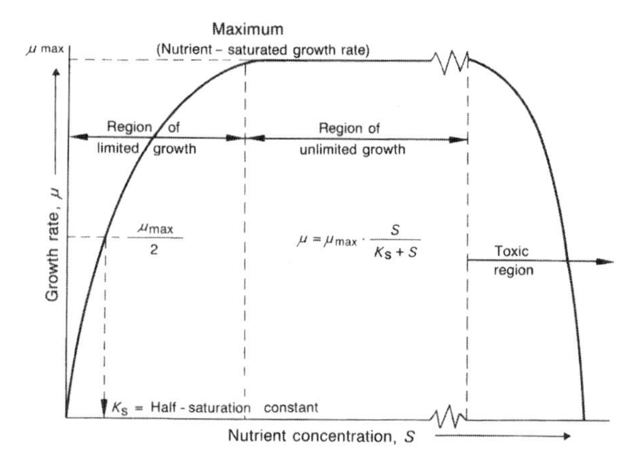

Figure 10. Kinetics of nutrient uptake by organisms
Organismic growth depends predictably on nutrient concentration, following a relationship known from enzyme kinetics.

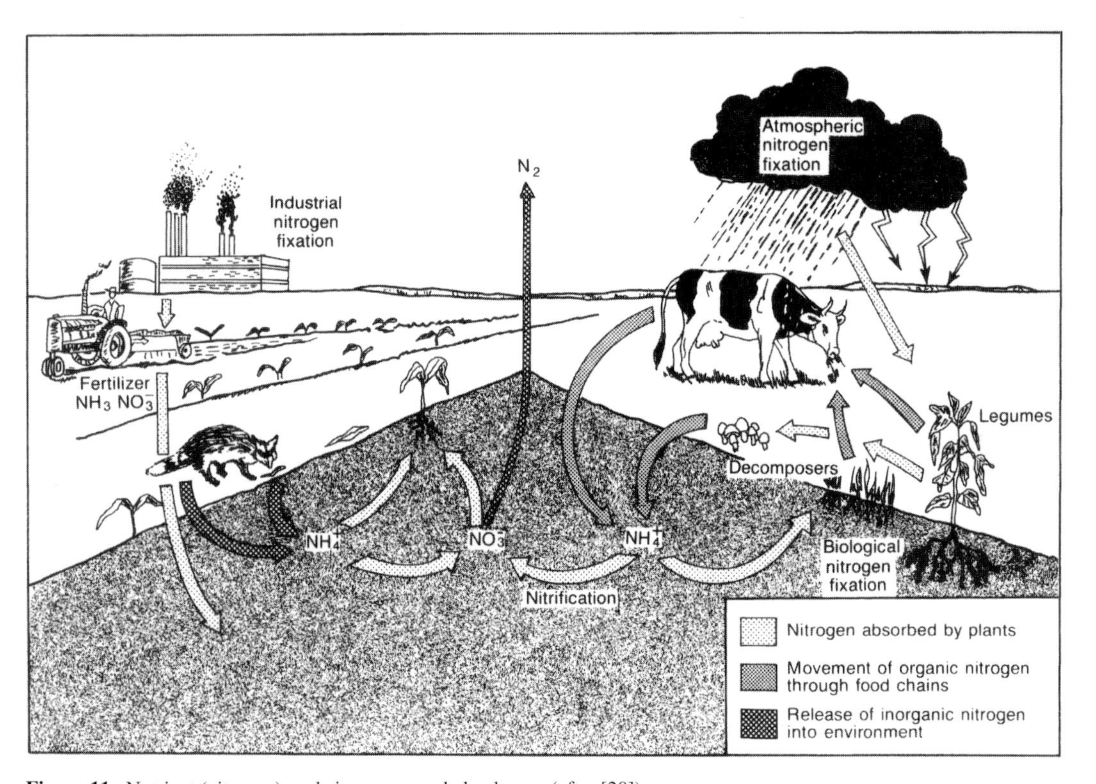

Figure 11. Nutrient (nitrogen) cycle in a man-made landscape (after [20])

vores). Bacteria, in this connection, are known as *decomposers*. They decompose matter to mineral components; bacteria belong to a larger group of destruents, degrading dead organic material, the so-called detritus.

2.1.4.2. Nutrient Cycling in Ecosystems

Two basic types of cycles can be distinguished: the short-circuit cycle in natural ecosys-tems (see Fig. 8) and the open cycle in agricultural ecosystems (Fig. 11). In natural ecosystems, nutrients circulate rapidly among the organisms, soil, and groundwater. Soil constantly coveredby vegetation retains water with the dissolved nutrients. The vegetation cover also retains humidity. Evaporation creates backflow from the atmosphere as rain. Soil is rich in organic matter, which keeps the water capacity high. Therefore, nutrient loss from vegetated soils is low. Simultaneously, soil erosion and washout of nutrients are minimized.

In today's agricultural and inhabited landscapes, the situation is drastically different. The soil is barren or loosely covered, at least for a part of the year. Because of lack of rhizosphere, it is subject to high erosion and flushing out of nutrients to streams and rivers and, consequently, to lakes or sea. The soil is continuously deprived of organic matter, which decreases its water retention capacity. Instead of a short-circuit, an open cycle is observed, with considerable nonreversible losses. In the United States, soil erosion was estimated in 1968 and 1980 to represent 36×10^9 kg/a, about one-half of which originated from agriculture.

The opening of the cycle of organics and nutrients occurred in the Middle Ages when the increasing accumulation of people in towns caused diseases to spread. The construction of sewers disrupted the natural cycle, the original backflow of organic remnants and nutrients to the soil being diverted.

An experiment showing which major changes are produced by deforestation was obtained on the Hubbard Brook Ecosystem [4], [23]. Trees were logged within the entire watershed, and the outflow and its chemical composition were carefully measured. An increase in soil erosion and leaching of nitrogen, phosphorus, and sulfur to groundwater and streams was observed. Because water flux through the soil has increased, the soil loses its fertility. The organic forest floor rapidly decomposes and eventually disappears. The variability of flow in streams increases, with much more transport of the streambed produced during larger flows. However, the average flow decreases. Nitrogen released to the atmosphere during burning of remains has negative effects on the atmosphere. Reradiation from barren soil creates ground-level frosts on clear nights. Soil temperature is significantly increased on hot days.

Carbon Cycle. The global cycling of carbon is gaining in importance, not only because of plant nutrition but mainly because of its increasing concentrations in the atmosphere and the oceans. Because of the increase of several "greenhouse" gases – among which CO_2 plays the leading role – a global change in world weather conditions is expected. Carbon is essentially exchanged among the four reservoirs shown in Figure 12, where the past (in the atmosphere) and present reserves and flows are also given. The importance of forests and oceans as balancing elements in the CO_2 budget is enormous.

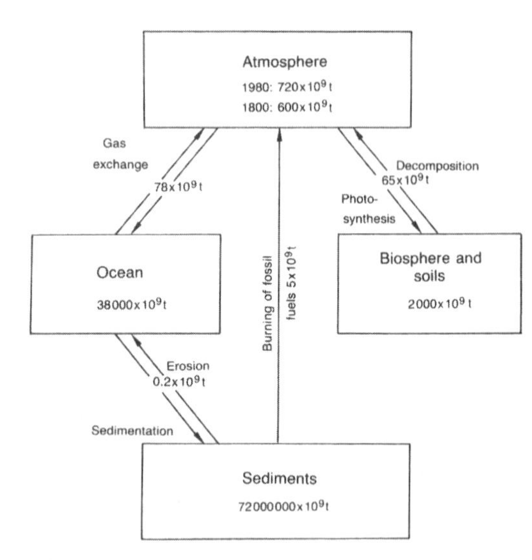

Figure 12. Major carbon reservoirs (after [24])

Nitrogen Cycle. Basic transformation processes in the terrestrial cycle of nitrogen are shown in Figure 13. Two pathways lead out of the short-circuiting between soil, crops, and animals: to the atmosphere and to water. Losses to the atmosphere have two dominant sources: industry, which releases NO_x; and agriculture, specifically animal husbandry, which produces NH_3. Both gases contribute to the greenhouse effects. Losses of nitrogen to waters also have negative consequences. Figure 14 shows the enormous rise in concentration of nitrates in water bodies in intensively cultivated landscapes of Europe and North America. Since the pristine state, the concentration has increased by a fac-

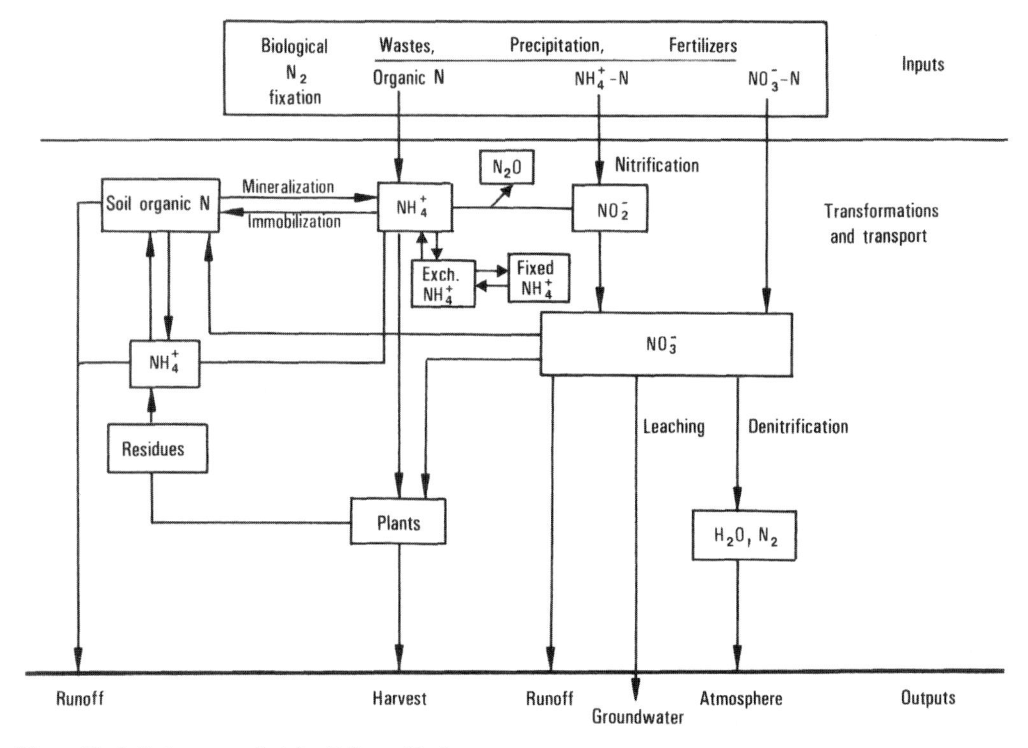

Figure 13. Soil nitrogen cycle (after [25], modified)

tor of ten and is occurring in both surface and groundwater. Most of this nitrogen increase is attributable to agriculture, for which it represents enourmous economic losses because of unutilized fertilizers. World fertilizer use increased from about 5 kg per capita in 1950 to 28 kg in 1985 [151]. The consequence of this fertilization of waters is an increased growth of aquatic vegetation, particularly of nuisance algae. In drinking water supply, excess nitrates result in hygienic problems, among them the deadly infant disease methemoglobinemia. For these reasons, every country has standards for the maximum permissible concentrations of nitrates in drinking water. The problem of cheap extraction of nitrogen from water is far from being solved on a global scale, although technological nitrogen-elimination processes are known. The nonsensicality of losing amounts of costly nutrients from agriculture on the one hand and eliminating them from waters at high costs on the other seems evident.

Phosphorus Cycling. A popular representation of the phosphorus cycle is given in Figure 15. Similarly to C and N, short-circuit cycling is characteristic of natural undisturbed vegetated landscapes. An open cycle with losses, mainly to waters, typifies agricultural and town environments. The difference from the previous cycles is that no phosphorus gas phase occurs. In organisms, phosphates are oxidized to produce energy in cellular respiration, and the phosphate is released in urine or similar waste back to the environment. When phosphorus compounds (particulate P, adsorbed P, dissolved orthophosphates, inorganic polyphosphates, organic phosphorus dissolved in the water phase) enter waterways, they enrich – indeed, overenrich – aquatic ecosystems. In waters, the processes of phosphorus uptake by aquatic flowering plants and microflora as well as processes of P immobilization and microbial synthesis are inadequate to utilize these amounts. Besides nuisance growth, considerable amounts of P-rich sediments accumulate in estuaries, lakes, and reservoirs. From water, little natural return to the land occurs. The sources of phosphorus for water are listed below:

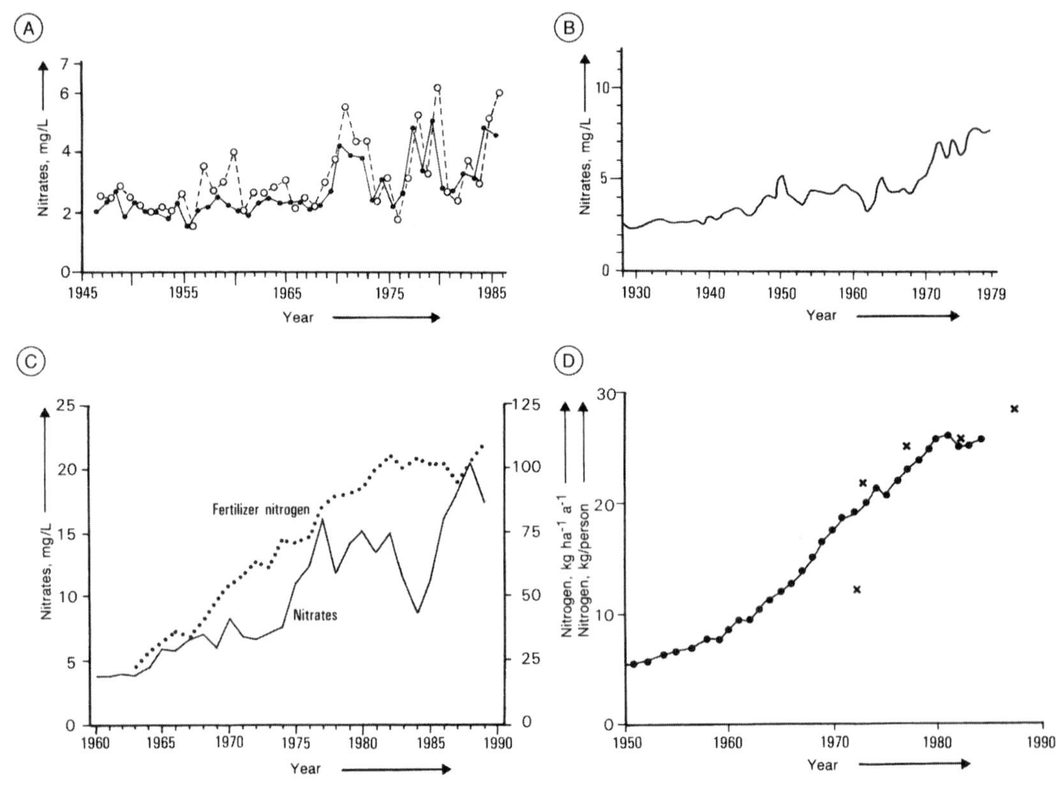

Figure 14. Trends of nitrate in freshwater and fertilizer use
A) Mean winter – spring concentrations of nitrate in Esthwaite Water and Blelham Tarn (Lake District, United Kingdom); B) Mean annual nitrate concentrations at Walton on Thames (United Kingdom), after [149]; C) Mean annual nitrate concentrations in the Slapy Reservoir (River Vltana, Czech. Republic), and related fertilizer use in the drainage basin [150]; D) Mean global fertilizer use per capita in 1950 – 1984 [151]; crosses represent newer data [26]

Natural surface load	0.07 – 0.12 kg/ha
Atmospheric load	0.4 – 1.2 kg/mm
From agricultural areas	up to 1.8 kg/ha
Livestock sewage	up to 13 kg per capita cattle
Sewers	0.85 kg per capita dweller
Septic tanks	0.14 kg per septic tank
Municipal sewage	1.1 kg per hectare of town area

Although great differentiation exists locally, two sources of phosphorus seem to dominate in the developed world: detergents and soil erosion. Phosphorus applied as fertilizer does not go into waterways in significant amounts unless some scattering from stores or spreading during aircraft application occurs.

2.1.4.3. Productivity of Ecosystems

Major biomes of the world are represented in Figure 16. The total biomass of plants on the globe is estimated to be 1.84×10^{15} kg [28]. The value is rapidly diminishing because of forest clearing, soil salinization, and desertification. In tropical countries of America, Africa, and Asia, the rate of loss represented 37, 48, and 40 %, respectively, between 1980 and 1985 [29]. Typical feedback between the nutrients and organisms is shown in Figure 17, which illustrates the dynamics of natural systems.

However, nutrients are not the only decisive agents for biological production. As evident from the summary photosynthetic equation

$$6\,CO_2 + 6\,H_2O \xrightarrow{\text{Energy}} C_6H_{12}O_6 + 6\,O_2 \qquad (24)$$

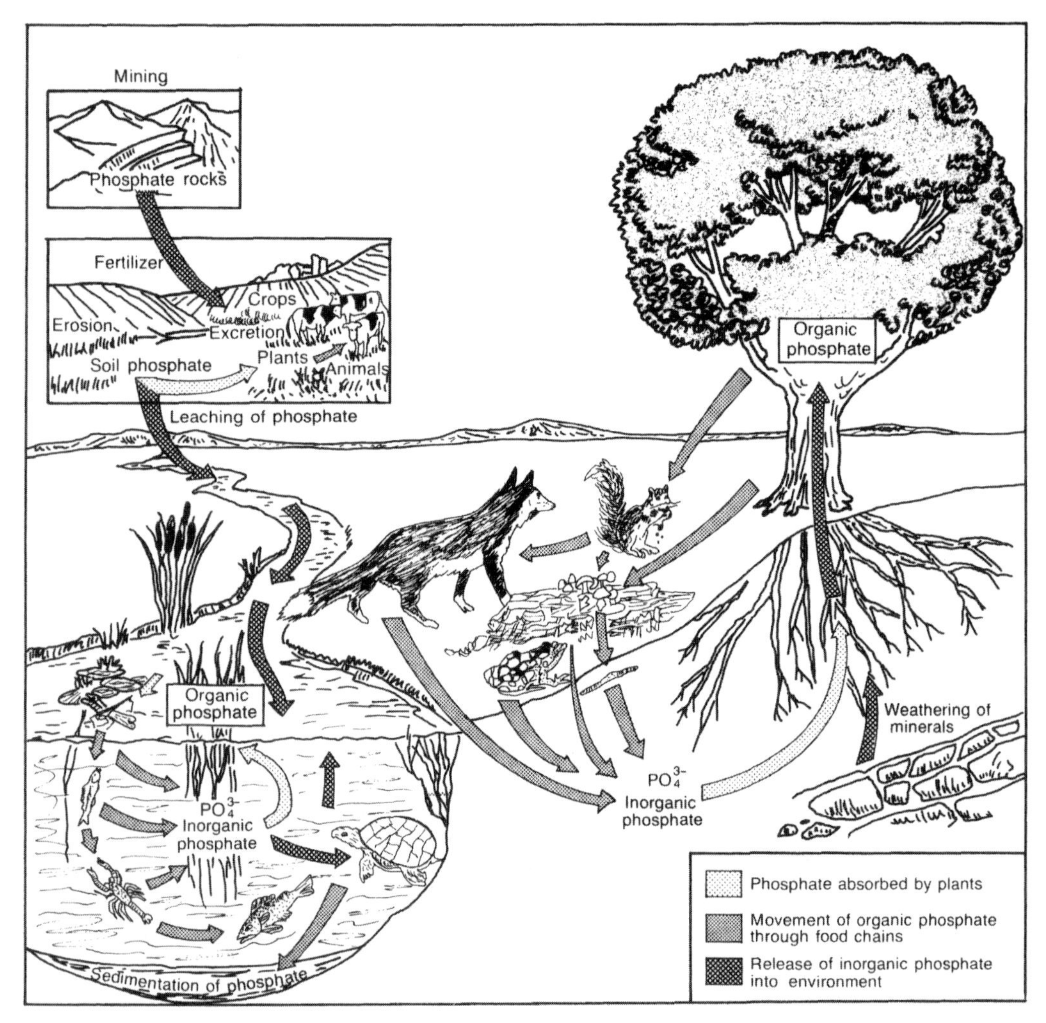

Figure 15. Phosphorus cycle (after [20])

energy (usually solar energy) and water are necessary. Like chemical reactions, biological reaction rates are also affected by temperature. The difference is that life is confined only to some −50 to +50 °C. Characteristic shapes of light and temperature dependence of organismic growth are given in Figure 18. For the above reasons the fate of nutrients is highly dependent on geographical position as well as seasonal and daily changes, as shown in Figure 19.

Productivity of Terrestrial Ecosystems. A distinction is made between natural and the agricultural production, although temperate forests are now also cultivated. Quantitative data on primary production around the world were elaborated in [28], [152]; data on sites where primary and secondary production was measured simultaneously are summarized in [30], [153]. The results of these studies are given in Table 4. Such estimates are methodologically very difficult, and some of them might be highly biased. Only the net aboveground production was measured; belowground production is known to represent up to five times as much as aboveground [31]. Net production is distinguished from gross production in that it does not include organic matter produced but used immediately for maintenance (respiration, excretion). The range of values for 104 sites was es-

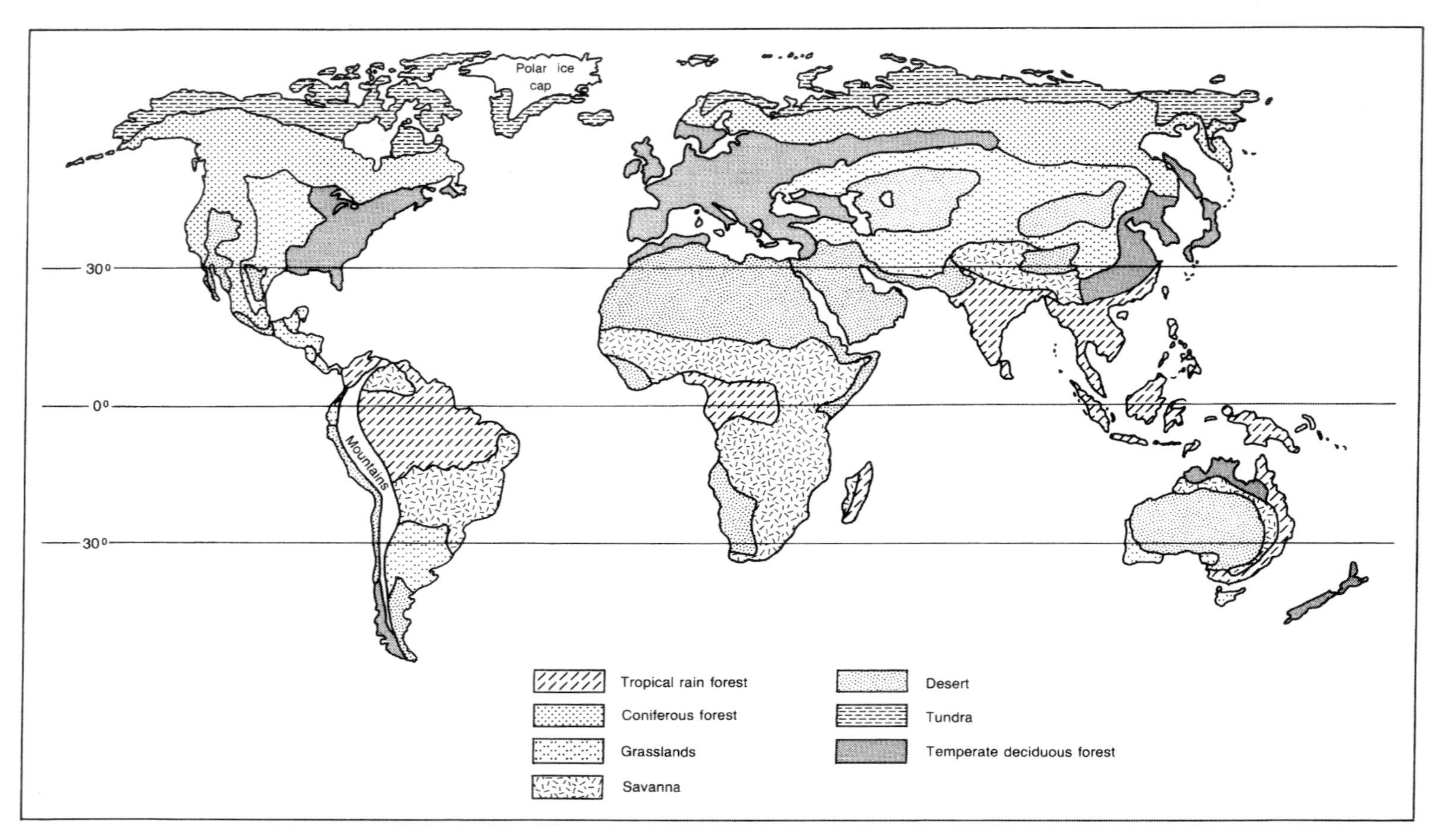

Figure 16. Major biomes of the world (after [27])

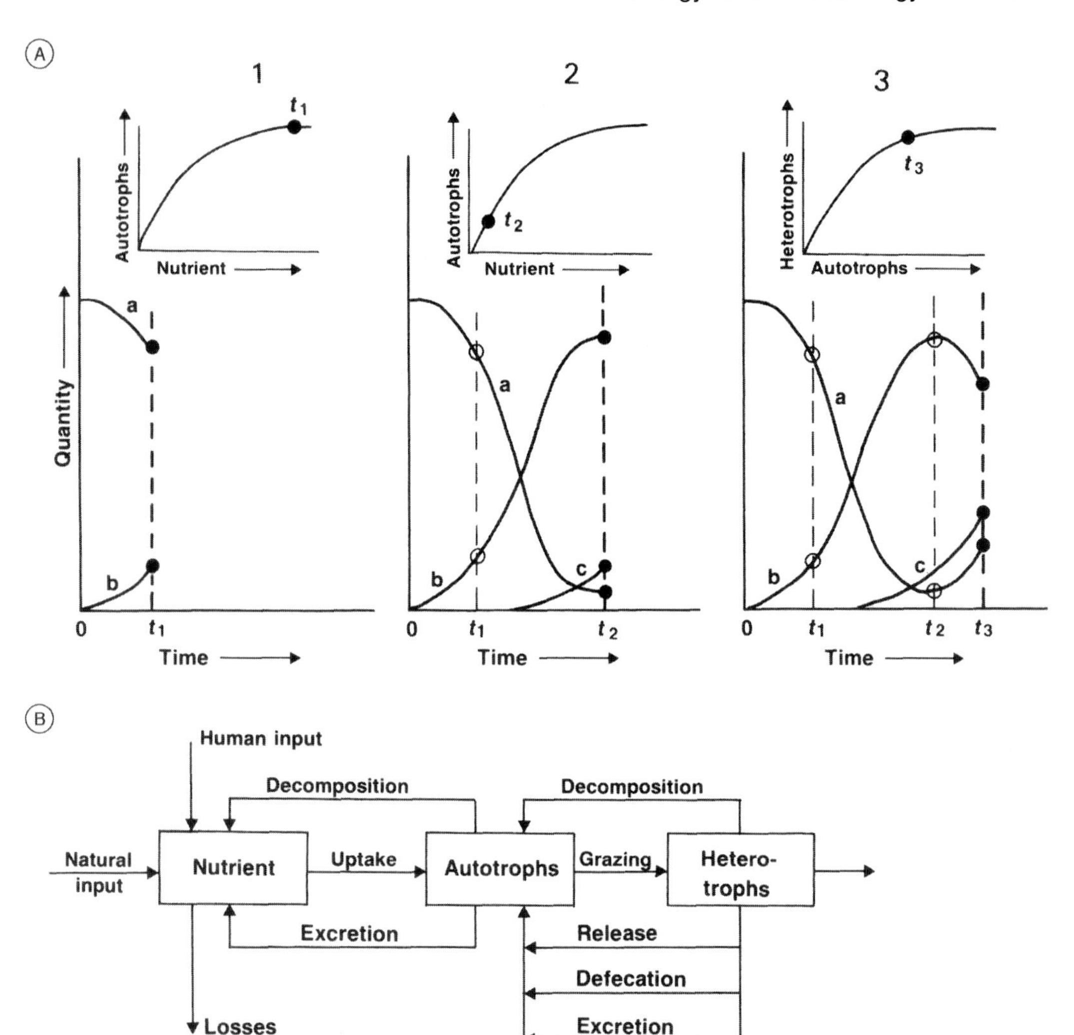

Figure 17. Essential processes of ecosystem dynamics
A) Time development of a population of autotrophs and heterotrophs in a nutrient-limited environment 1) At time t_1 (e.g., in winter) nutrients (curve a) are high and autotrophs (curve b) low, due to low temperatures. When autotrophs start to appear, they grow quickly and take up nutrients rapidly. 2) At time t_2, growth ceases, being severely limited by the decreasing quantity of nutrients (see inset). 3) Heterotrophs (curve c) appear with retardation, because their life cycles are long and at the beginning their abundance is limited by low food supply (i.e., low amounts of autotrophs on which they feed). At time t_3 autotrophs decrease, being heavily grazed on by herbivores (heterotrophs). Nutrients are regenerated both from senescent and decaying autotrophs and from excretion, defecation, and nutrient release (as in Fig. 17 B)
B) Main processes in the nutrient-productivity cycle. Losses are represented by flushing nutrients into groundwater and streams.

timated as 30 g fresh weight per square meter and year for a desert to 7050 g m^{-2} a^{-1} for a tropical forest. The average estimates for various ecosystems range from 0 – 10 g dry weight per square meter and year for extreme desert through 10 – 250 for deserts and 400 – 2000 for boreal forests up to 1000 – 3500 for tropical rain forest and 800 – 6000 g dry weight per square meter and year for swamps and marshes. Table 4 shows that forests in general are the most productive ecosystems. Column 3 also shows that a high proportion of the primary production in nature is consumed by herbivores, be they insects, small rodents, or large ungulates. On average, 1.3 to 64 % of primary production is consumed in different biomes, with by far the highest pro-

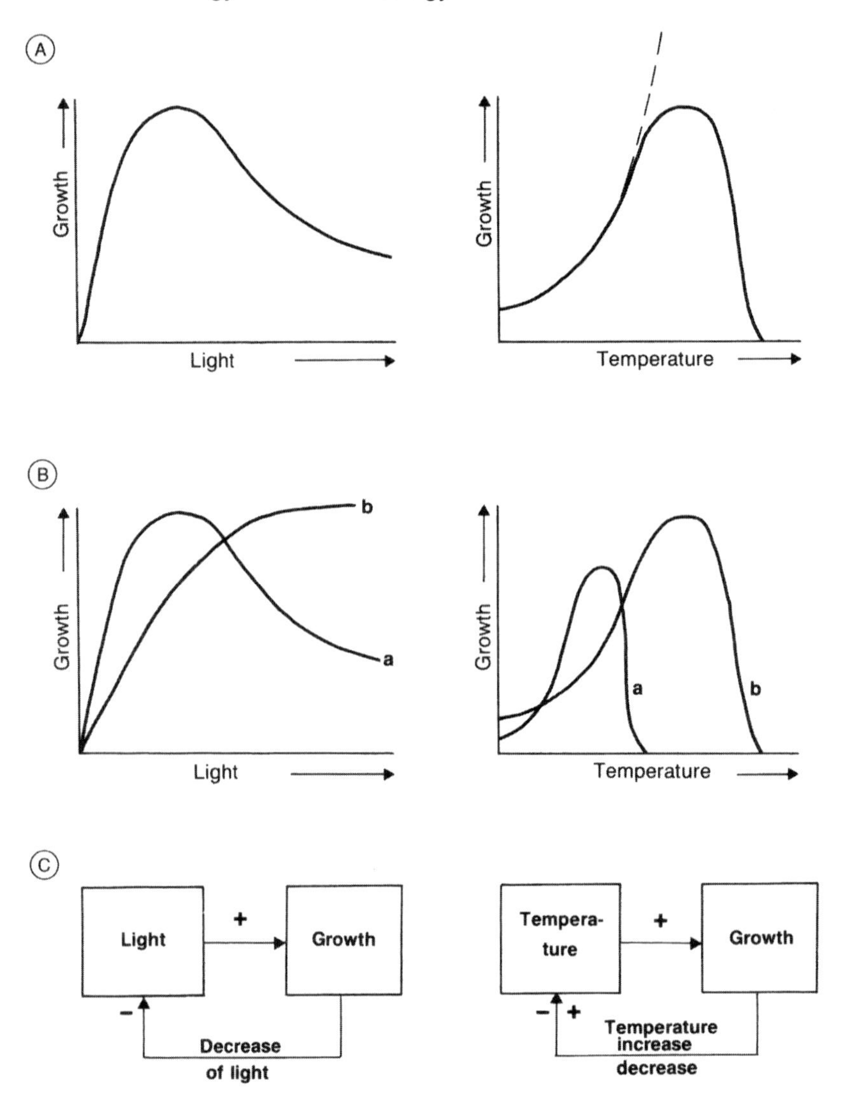

Figure 18. Light and temperature dependence of organismic growth and related feedback mechanisms
A) Typical shapes (equations not shown) of the dependence of plant growth on light and temperature. The dotted line shows the dependence typical for a chemical reaction (Arrhenius curve). B) In contrast to chemical reactions, which follow a fixed curve, the organisms are capable of adaptation to changing conditions. The two curves show changes for low (a) and high (b) light and low (a) and high (b) temperature conditions, respectively. Within the pertinent range of conditions the adapted organisms have the same high performance. However, full compensation is not always possible, as the temperature case shows. C) Feedback effect of growth on light (left) and temperature (right). In the case of light the feedback is negative; for temperature, either positive (increase of temperature, e.g., in water where more radiation is absorbed by microscopic algae) or negative feedback relationships (decrease of temperature by shading in forests) are possible

portion in tropical grassland. The primary production taken up is converted to herbivore production with an ecological efficiency (expressed as the amount of primary production converted to herbivore production) of about 0.35 %. The efficiency of herbivore utilization of primary production calculated from columns 3 and 4 results in an average value of 0.93 %, with a range between 0.09 for tropical grasslands and 13.3 % for salt marshes. Two types of ecological cycles in terrestrial ecosystems are distinguished in [31]: a short one, where most of the primary produc-

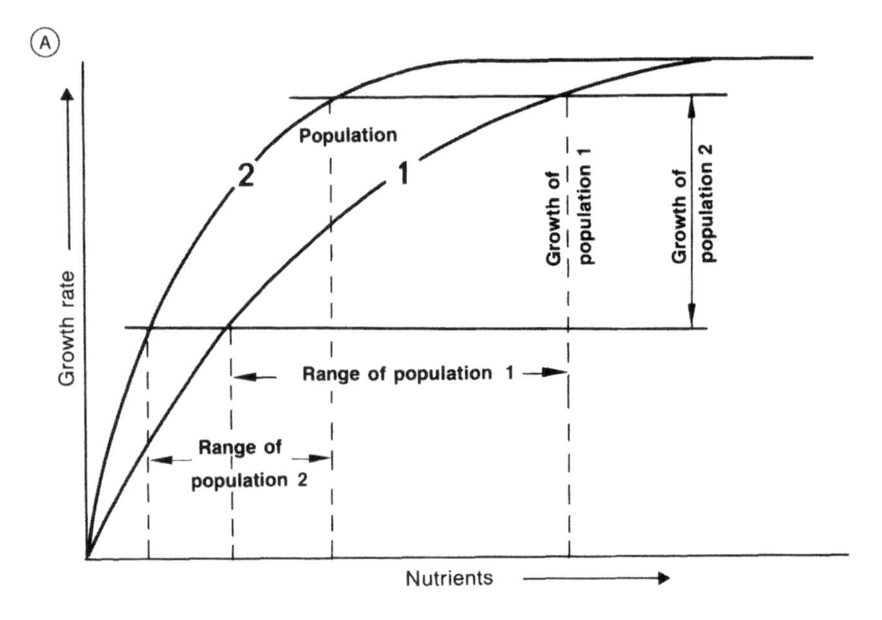

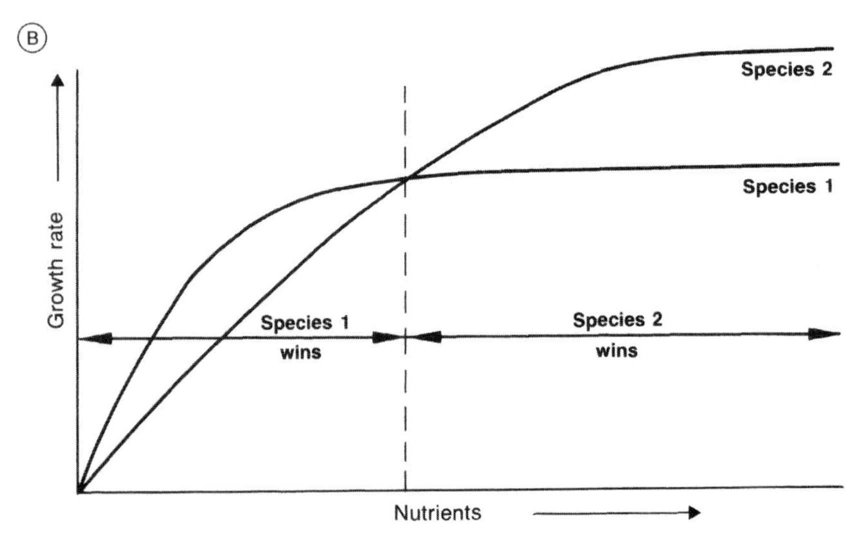

Figure 19. Adaptation of populations to nutrient supply and competition for nutrients
A) Populations adapt to low-nutrient conditions, as indicated by increasing slope of the curve. Population 1 living in high-nutrient conditions shows a lower initial slope of the curve, while population 2 living in low concentrations shows a high initial slope, enabling it to utilize low concentrations more efficiently
B) Species compete for nutrients. In a certain condition range, one or the other species dominates, depending on the range in which the curve reaches the highest values.

tion is immediately decomposed, while in the long one, herbivores not only consume a large part of primary production but also are preyed upon by carnivores and a complex food web originates. In Table 4, column 3, only temperate forests show the short cycle, with only about 1 % of primary production being consumed by herbivores. In all other instances the primary production consumed is of the order of magnitude of 10 %. For tropical grassland, seemingly the most complex system, more than half of the primary production is consumed by herbivores.

Table 4. Data on primary and secondary (herbivore) production of natural ecosystems in different biomes (after [153], modified) [a]

Ecosystem	Prim. prod. [b]	Biomass [c]	Consumption [d]	Herbivores [e]	Efficiency [f]
Desert	403	1	32 (7.9)	0.1	0.02
Tundra	257	0.5	16 (6.2)	0.2	0.07
Temperate grassland	1181	1	161 (13.6)	14.6	1.23
Temperate old field	1352	2.2	95 (7.0)	10.3	0.76
Tropical grassland	2694	8	1725 (64.0)	1.5	0.06
Temperate forest	5032	1.2	64 (1.3)	0.8	0.02
Tropical forest	5400	17	591 (10.9)		
Salt marsh	4328		345 (8.0)	46.0	1.06
Average	2276	8.1	861 (37.9)	8.0	0.35

[a] Values are given in grams of fresh weight per square meter and year.
[b] Prim. prod. = net aboveground annual plant production.
[c] Herbivore biomass in grams of fresh weight per square meter and year.
[d] Amount of primary production consumed by herbivores. The percentage of primary production being consumed by herbivores is given in brackets.
[e] Annual production by herbivores.
[f] Ecological efficiency, i.e., the percentage of primary production converted to herbivore production.

Table 5 gives examples of net aboveground primary production of agrarian ecosystems and secondary production of domestic livestock compared to game.

Table 5. Examples of net aboveground primary production of agricultural eosystems and production of domestic livestock per area as compared to wild animals (data from [10], [23])

Primary production, kg dry weight per square meter and year		
Crop	World average	Maximum per country
Wheat	0.4	1.3
Corn	0.7	1.3
Rice	0.6	1.5
Potatoes	2.5	4.6
Soybean	0.35	0.6
Sugarcane	0.9	3.3
Sugar beet	2.0	
	Secondary production	
Type of animals	Weight gain, kg fresh weight per day	Biomass, kg/km^2
Sheep	0.05	
Cattle	0.14	
Thomson's gazelle	0.06	
Topi	0.20	
Wildebeest	0.24	
Domestic livestock		10 000 – 27 000
Domestic livestock in East Africa		10 000 – 14 000
Wild ungulates in East Africa		61 000 – 87 000
Bison		12 000 – 17 000
Black tailed deer in California		9 200

Surprisingly, comparison of Table 5 with Table 4 shows that natural production and agricultural production are not much different, if one ignores the extreme energy needs to achieve high agricultural yields [154]. Often the fossil energy input is greater than the energy content of the food harvested [32]. According to [33], for an eightfold increase in yield of major crops, a twentyfold increase in energy applied is needed. The idea that much more output is obtained when artificially cultivating domestic livestock is not supported by these data. The use of natural vegetation by natural livestock is far from inefficient. That was recognized in tropical countries and became a political issue in opposition to what was exported from the developed countries as the "green revolution." The green revolution appears to lead to the continuing poverty of poor countries. They are made dependent on Western technology and on energy sources not available at home, without adequate rewards. The simultaneous trend for sustainable development leads to major revisions of prevailing ideas concerning agricultural production.

Productivity of Aquatic Ecosystems. The primary production of fresh waters may be divided into two groups: (1) production of flowering plants living in shallow water of marshes, swamps, and along the shores of lakes and slow-flowing streams; and (2) microflora inhabiting primarily the open lake and slow river central regions. Microflora consists of several groups of microscopic algae and species of cyanobacteria (also called blue-green algae). These free-floating organisms are called phytoplankton.

The transition between land and water belongs to the most productive plant environments: no water deficit exists; nutrients are obtained both from rich aquatic soil and from water. Under comparable geographic conditions, the highest production among all vegetation is obtained in aquatic macroflora. The dependence of aquatic primary production on geographical position is due to a combination of effects such as light and temperature. For phytoplankton, conditions of mixing in lakes dependent on geographical latitude and altitude are decisive. During the annual cycle, lakes undergo periods of intensive deep mixing to the bottom, as well as periods of stratification. During summer stratification, upper layers become warmer than deeper ones, and intensive mixing is confined to several top meters. The density difference between warm top and cold bottom layers is usually sharp. At a certain depth, a jump in temperature occurs between the two basic layers called the mixing layer (thermocline) and the hypolimnion. The depth of the mixing layer affects primary production. Algae moved within this layer periodically encounter high light intensity when taken to the water surface and reduced or no light when mixed to depths. Greater lakes are usually clearer and mixed to greater depths due to higher wind stress; their primary productivity is lower than that of shallow lakes with comparable nutrients.

The most important determinant of lake productivity is nutrient input. In the 1960s, controversy existed – mainly between representatives of the soap and detergent industry and limnologists studying lakes – whether carbon, nitrogen, or phosphorus is the nutrient most often limiting aquatic primary production. The situation is now clear: depending on the mutual relation among C, N, and P any of them can be limiting, as can several micronutrients. However, in most lakes at present, phosphorus is the limiting agent, since nitrogen is usually supplied from human activities in excess. Carbon can limit in instances of an excess of both nitrogen and phosphorus; carbon deficiency is due mainly to limited exchange of CO_2 between water and atmosphere. This is the case in highly fertilized fish ponds. A systematic difference exists between fresh water on the one hand and brackish and marine waters on the other: in the latter, nitrogen limitation prevails. Instances of limitation by micronutrients are reported rarely, particularly from regions without human activities, where the lack of some elements is geologically conditioned [34].

Table 6. Primary and secondary production in fresh waters, based on different sources and data obtained mainly during the International Biological Program [5] (for more detailed data see [31])

	Macroflora	
Type of water body	Aboveground primary production, kg dry weight per square meter and year	Average biomass, kg dry weight per square meter and year
Lakes and streams	0.1 – 3.0	0.02
Swamps and marshes	0.8 – 8.5	1.5
	Microflora (phytoplankton)	
Phosphorus and nitrate concentrations, μg/L	Primary production, kg dry weight per square meter and year	Average biomass, g dry weight per square meter
Low P (<5) and N (<250)	0.015 – 0.09	0.4 – 2.0
Medium P (<30), N (<500)	0.075 – 0.3	2 – 5
High P (>30), N (>500)	>0.43	>5
	Type of water body fish production,* kg fresh weight per square meter and year	
Small lakes	0.002 – 0.18	
Great lakes	1 – 7	
Tropical lakes	2 – 250	
European fish ponds	500 – 850	
Chinese fish ponds	500 – 7500	

* Fish production is estimated from harvested production. According to recent estimates harvested production represents in lakes about 10 % of total production and in culture about 50 %.

Table 6 gives some ideas of the primary production of fresh water depending on phosphorus and nitrogen concentrations. In addition, there is a geographical differentiation of phytoplankton primary production at similar conditions of nutrients. In Arctic and Antarctic lakes values between 0.002 and 0.07 kg dry weight per square meter and year were observed. In alpine lakes the lowest values observed are similar to those in the Arctis but maximum values up to 0.2 are reached. For temperate and tropical lakes the ranges are 0.004 – 1.8 and 0.06 – 5.0, respectively. Some data on herbivorous production, represented by fish, are also indicated. The biomass of phytoplanktonic algae is about three factors of ten lower than in macrophytes, whereas production is similar. This is due to the short life cycle and high turnover rate of microscopic algae. Under intensive artificial cultivation, some algal species have produced up to 8 kg dry weight per square meter and year.

2.2. Terrestrial, Lotic, Lentic, and Wetland Ecotones

A boundary in nature may be created by discontinuities in the external influences that control the ecosystems or by discontinuities in the functional network of the systems themselves. Consequently, a wide spectrum of equilibrium system responses to values of environmental gradients exists.

In Figure 20 A the response of the ecosystem as characterized by one indicator variable is as smooth and linear as the changes in gradient; Figure 20 B depicts an abrupt break in response, probably due to a threshold response in one of the components of the ecosystem. In Figure 20 C the response has a twist, indicating that the system can have two stable equilibrium states. In these cases the ecosystem state and behavior with respect to driving variables depend on the history of the systems (hysteresis). Hysteresis plays a major role in catastrophe theory. In Figure 20 D the system has two stable states and can display multiple internal reactions to take on one or the other.

In more abstract or unifying terms the individual response curves of Figure 20 can be considered as a continuum of ecosystem responses to an environmental variable. Thus, the various responses grade into one another depending on internal reactions and autonomous features of the ecosystem. The environmental gradients can be continuous or discontinuous to form a variety of pattern that, together with the response of the ecosystems, control the nature and pattern of the transition zones between adjacent systems.

These transition zones or *ecotones* [156] have a set of characteristics uniquely defined by space and time scales and by the strength of interaction between adjacent ecosystems (see Section 2.3). Because ecotones usually represent relatively steep gradients in both abiotic and biotic variables, they facilitate the ecological analysis of mechanisms contributing to biodiversity and flow of energy or materials across the landscape [157].

2.2.1. Terrestrial and Lotic Ecotones

In view of the ecological importance of ecotones, they can be divided into three broad classes.

Terrestrial ecotones may be pragmatically defined as discontinuities between systems. Consequently, the most obvious ecotones in terrestrial ecosystem complexes involve transitions between plants of communities dominated by different life forms. Large-scale examples are transitions from tundra to taiga along thermal gradients or transitions from prairie to forest along continental moisture gradients; a frequent small-scale type is the transition from field to woodlot in agricultural landscapes. This pragmatic view of terrestrial ecotones depends highly on the contrast between the different elements contained in the definition of the landscape, and the degree of contrast is necessarily a function of the variables used by the observer to categorize the landscape elements [138], [158]. Thus, a more precise and generally applicable method of defining ecotones is based on areally valid measurements of ecologically relevant parameters and subsequent multivariate cluster analysis. The results are operationally homogeneous spatial units (patches) separated by distinct boundaries whose location and geometry depend on the variables used to characterize the primary measuring points or elementary spatial units, respectively; the fusion level of clustering adopted; and finally the scale selected [159], [160].

Lotic ecotones can be defined, in an analogous way, as fluvial boundaries. This definition indicates that resource patches are separated by both longitudinal (i.e., upstream – downstream) and lateral (landward) ecotones that operate over highly variable temporal and spatial scales. This type of ecotone appears to be highly sensitive to landscape changes caused by physical and biotic disturbances. Examples of direct influences of disturbances are removal of riparian vegetation by floods or landslides, and edaphic or vegetative modifications by animals and, in particular, humans. Indirect influences, such as changes in concentration gradients of dissolved chemical compounds and pathways of chemical reactions,

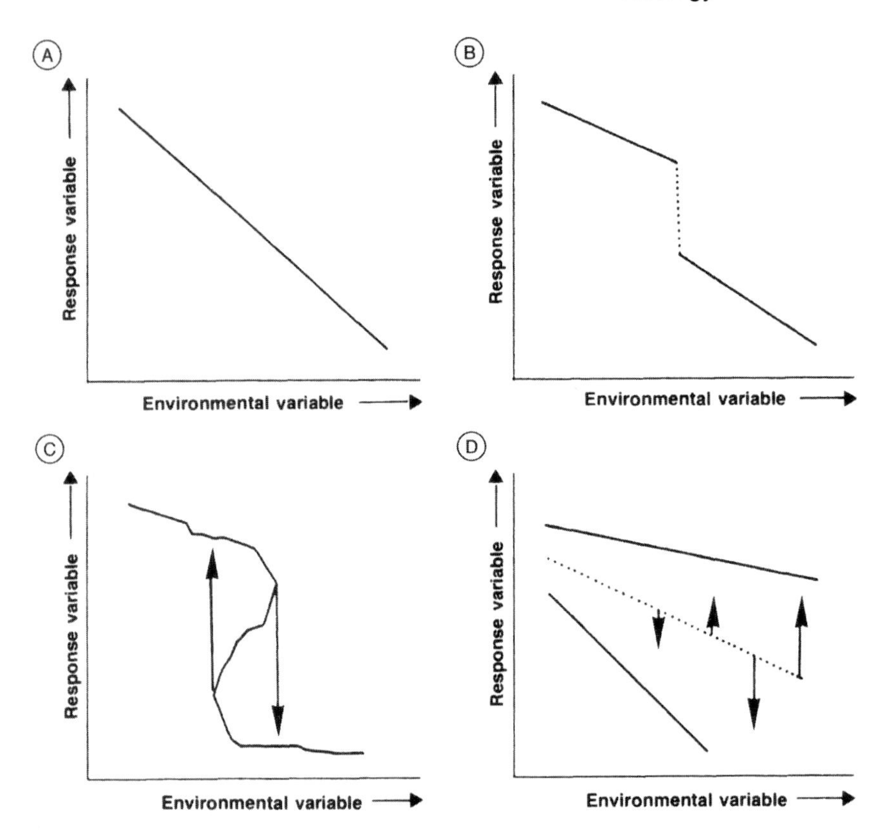

Figure 20. Ecosystem response to different types of environmental gradients (after [155], modified)
A) Case in which the response of the ecosystem is smooth with respect to the environmental variable; B) Case in which the response is discontinuous; C) Case in which the response is folded and results in hysteresis; D) Case in which the system can have multiple responses

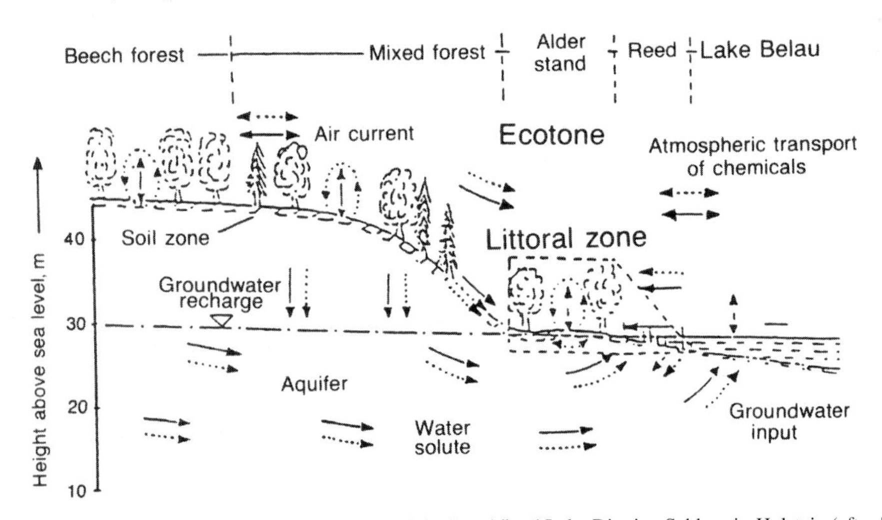

Figure 21. Exchange processes in ecotones of the Bornhöved Lake District, Schleswig-Holstein (after [157], modified)
a.s.l. = above sea level

may result from changes in edaphic and vegetative properties following direct impacts [2], [35], [161], [162].

2.2.2. Lentic and Wetland Ecotones

Lentic ecotones. (i.e., boundaries between water bodies such as lakes, ponds, or reservoirs and adjacent terrestrial patches) also play an important role in coupling ecosystems. With their steep gradients and typically great heterogeneity, lentic ecotones can contribute considerably to mitigating unfavorable changes in aquatic ecosystems. Furthermore, they regulate the landscape mosaic by controlling energy and nutrient flow between adjacent patches in the ecosystems [157], [163].

Wetlands are defined as lands transitional between terrestrial and aquatic ecosystems where the water table is at or near the surface or the land is seasonally covered by shallow water. Therefore they support, at least periodically, hydrophytes, and their substrate is classified predominantly as undrained hydric soil [164]. A clear-cut ecological distinction separates *tidal* and *inland wetlands*, which may be subdivided further [36]. Wetlands have internal and external boundaries separating distinct vegetation patches [165], [166]. They can be categorized into various types according to the flows that dominate in the tidal and inland environments. In both tidal and inland wetlands, vertical and horizontal transfers occur across a series of surficial and lateral boundaries. Flows across *surficial boundaries* (e.g., graded sediment beds or soil horizons) include transfer from aerobic to anaerobic soils, from soils to surficial vegetation and then to litter and vice versa, and from open water to the atmosphere.

Transfers across *lateral boundaries* include material flows from the upland to the wetland (upland – wetland ecotones) or from the wetland into adjacent open water (wetland – openwater ecotones), from groundwater aquifers into soils, or across vegetation patches with each community dominated by different species (wetland – wetland ecotones).

Figures 21 and 22 illustrate a combined lentic – wetland – upland ecotone situation in the Bornhöved Lake District (Schleswig-Holstein). The figures show that lentic ecotones form unusually differentiated habitats [136]. Various microenvironments are formed within the shore zone of the water body, and similar microenvironments occur in certain sections of water bodies contrasting in size, depth, productivity, and catchment area [167]. Lentic ecotones have rich plant and animal communities characterized by a shifting proportion of terrestrial and aquatic organisms.

Allochthonous and autochthonous material is the source of organic matter in ecotonal shore zones: matter originating from primary production by macrophytes and algae and accumulated material of terrestrial origin supplied by the drainage basin. A fast decomposition rate in several ecotonal habitats makes this material available to higher trophic levels much more quickly than in most land and water habitats [136], [163].

The majority of lentic ecotones in densely populated areas of the world is under visible human impact. Lake eutrophication, air pollution, sewage input, agricultural practices, and tourist activities often damage them. Disappearance of terrestrial and aquatic vegetation is the most common response to such unfavorable conditions.

2.3. Sensitivity of Ecosystems and Ecotones

In ecology, the sensitivity of biota or ecosystems to disturbances may be defined in terms of community stability. Questions of persistence and the probability of extinction are of particular importance in this connection; hence, these measures are also frequently defined as aspects of stability. Stability denotes the ability of a system to return to an equilibrium state after a limited disturbance, while persistence is an indication of resilience in the sense defined by HOLLING [168]: i.e., the capacity to absorb changes of state and driving variables and parameters. The more rapidly a system returns to equilibrium with the least fluctuation, the more stable it is, which means in terms of thermodynamics that stability is coupled with a (relative) minimum of entropy production [2], [169]. Further differentiations lead to various notions of community stability and nondemographic stability measures.

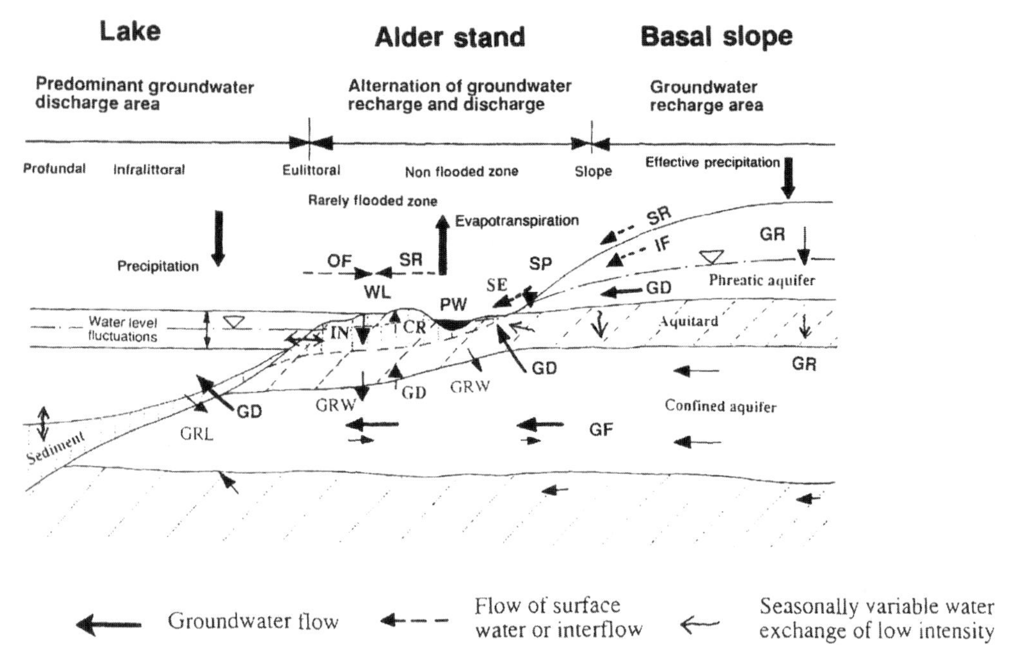

Figure 22. Water balance of ecotones adjacent to Lake Belau, Schleswig-Holstein (after [157], modified)
CR = Capillary rise; GD = Groundwater discharge; GF = Groundwater flow; GR = Groundwater recharge; GRL = Groundwater recharge by lake water; GRW = Groundwater recharge by wetland water; IF = Interflow; IN = Infiltration; OF = Overflow; PW = Ponding water; SE = Seepage on slope; SP = Spring at foot of slope; SR = Surface runoff; WL = Waterlogging

In the light of these definitions, a sensitive ecosystem is one that responds readily to a particular stress; an insensitive ecosystem may be oblivious to the stress. Sensitivity has also a temporal component: a system that responds more rapidly than another is considered more sensitive. Because of the complex character of system sensitivity to stress, various measurable aspects of sensitivity are recognized, which are summarized below:

- *Stability* how an (eco)system responds to perturbation and how readily it recovers
- *Resistance* rate at which (eco)system indicators change in response to a given stress – more resistant systems change more slowly
- *Recovery* how the system responds following removal of the stress. Recovery has two components: how rapidly and how effectively. (The temporal aspect is characterized as the system's resilience.)
- *Resilience* the inverse of the length of time required to return to near normal. According to WESTMAN [170], four components of resilience can be distinguished:
  - *Elasticity:* time required to restore the system to its initial steady state
  - *Amplitude:* zone of deformation from which the system will still return to its initial state
  - *Hysteresis:* extent to which the paths of degradation under chronic disturbance, and of recovery when disturbance ceases, are mirror images of each other
  - *Malleability:* degree to which the new steady state established following recovery differs from the original steady state

Most of the above definitions explicitly or implicitly assume that the system is in a steady state before a stress is applied. This is not necessarily true because ecosystems are generally continuously changing and developing. Therefore, the system need not return to a steady state after recovery, merely to a "near-normal" state.

In some instances the ecosystem can switch with relatively minor disturbances into another steady state. This is called bifurcation: rapid transition at a critical point into a new steady state. Examples of bifurcation of aquatic ecosystems are, for example, the rapid change of lakes from a clean-water state with low amounts of harmless algae to a highly polluted state with highly noxious blue-green algae, or the switching of prairie lakes with healthy fish populations to a state with fish-kills.

2.3.1. Different Notions of Community Stability

A first distinction is between local and global stability. *Local stability* or stability in the vicinity of an equilibrium point describes the tendency of a community to return to its original state when subjected to a minor perturbation. *Global stability* describes this tendency when the community is subjected to a major disturbance. The graphic visualization of solutions to the equations of population dynamics for an m-species community may be represented on some m-dimensional surface, where each point marks a set of populations. Then, equilibrium situations are in principle characterized by those points at which the surface is flat; by equating the configuration with that of a landscape, in customary geographical notation this means on hilltops and in valley bottoms. The hilltop equilibria, however, are obviously labile (i.e., unable to survive the smallest displacement); only the valley bottoms are stable configurations (see Fig. 23 A). In looking beyond the realm of linearized stability in the immediate neighborhood of equilibrium points for which straightforward mathematical tools exist the situation becomes more complicated. Appropriate representation of the *global stability* of a system implies recourse to nonlinear equations of population biology (see Fig. 23 B). Again in geographical terms, such a global analysis aims at comprehending the stability of an entire landscape, not just the immediate vicinity of equilibrium points. This involves appropriate recognition of the fact that real-world environments are uncertain and stochastic, which means that the corresponding environmental parameters in the model equations exhibit random fluctuations [6], [171], [172].

This leads to a yet more general meaning of stability, termed *structural stability*, which refers to the qualitative effects on solutions of the model equations that are due to gradual variations in the model parameters. Thus, a system may be considered structurally stable if these solutions change in a continuous manner. Conversely, a system is structurally unstable if gradual changes in the system parameters (e.g., alterations in site factors of a community) produce qualitatively discontinuous effects [2], [37], [172].

2.3.2. Biodiversity and Stability of Model Communities

An important issue of theoretical ecology is how biodiversity relates to community structure and which community-level properties emerge from the disparate interactions of organisms, populations, and site qualities. Possible examples of such “emergent properties” include trophic and guild structures, stability, resilience, and successional stages. A generalized offshoot of the concept of emergent properties is the notion that biological systems are hierarchically organized, with new properties at each level of the hierarchy [11], [38], [173]. According to this view, diversity is better understood if ecosystems or biota are decomposed hierarchically so that each process can be viewed as a stabilizing or disruptive factor at each level in a hierarchy of time and space scales.

During the 1950s and 1960s, the “conventional wisdom” in ecology [6] was that increased complexity within or diversity of a community lead to increased stability [39], [174]. May [175], however, came to contrasting conclusions by means of model food webs comprising a numberof interacting species. The term β_{ij} was used to measure the effect of species j's density on species i's rate of increase. Thus β_{ij} would be zero when there was no effect, whereas both β_{ij} and β_{ji} would be negative for two competing species, and β_{ij} would be positive and β_{ji} negative for a predator (i) and its prey (j). Setting all self-regulatory terms (β_{ii}, β_{jj}) in his randomly connected networks at -1, May distributed all other β-values at random, including a certain number of zeros. Thus, the cybernetic webs serving as ecosystem models could be described by

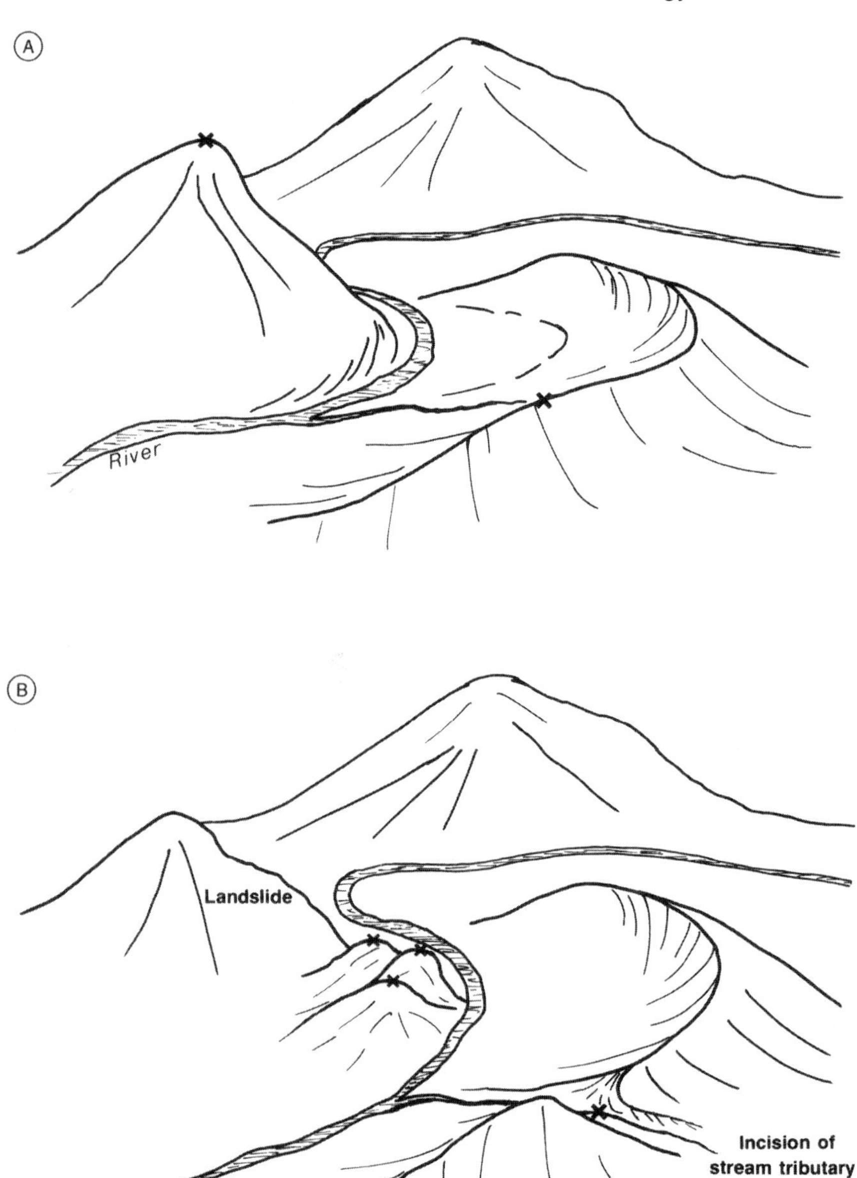

Figure 23. A) Visualization of system dynamics and stability conditions, equating phase – space characteristics with a landscape – labile equilibria (i.e., both low local and global stability); B) Illustration of global stability conditions by means of a landscape configuration, indicating different states of low local, but generally high stability

three parameters: S, the number of species; C, the connectance of the web, i.e., the fraction of all possible pairs of species interacting directly ($\beta_{ij} \neq 0$); and β, the average interaction strength, i.e., the average of nonzero β-values, regardless of sign.

Comparative analysis of these networks showed that they were likely to be stable (i.e., the populations represented would return to equilibrium after a small disturbance) only if

$$\beta\,(SC)^{1/2} < 1 \quad (25)$$

In other words, MAY's model suggests, like others, that "too rich a web connectance or too large an average interaction strength leads to instability. The larger the number of species, the more pronounced the effect." This inference is clearly in contradiction to the above conventional wisdom, but it indicates that no general, unavoidable connection exists between complexity or diversity, respectively, and community stability. Another question is, how much is MAY's result an artifact arising from particular characteristics of the model and the interpretative techniques applied [6], [37].

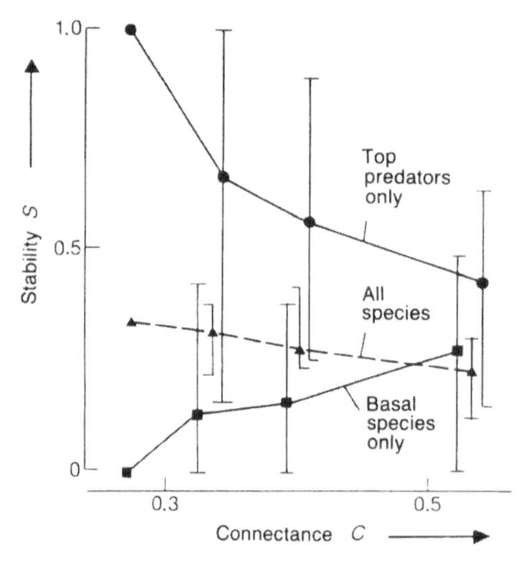

Figure 24. Species-deletion stability plotted against interspecific connectance (after [6])

The picture also alters if, instead of focusing on local stability and correspondingly minor disturbances, larger perturbations are considered or "species-deletion stability," which is particularly interesting from an ecotoxicological point of view. Under the assumption of the deletion of one species of a community owing to a persistent impact, the system is said to be species-deletion stable if all of the remaining species are retained at locally stable equilibria. In the simple six-species community of Figure 24 containing two top and two intermediate predators and two basal species (either plants or categories of dead organic matter), interaction strengths were defined at random, while connectance was varied systematically; thus, complexity could be equated with connectance. Again, stability generally decreased with increasing diversity, but this trend was reversed when basal species or elements were eliminated. However, within the simulated small subset of stable communities, resilience actually increased with complexity.

Other theoretical approaches to the stability problem are due to MCMURTRIE [176], PIMM [10], and ULANOWICZ [40], who established criteria for the stability of ecosystems under certain restrictive conditions, such as zero-sum games. In particular, they have shown that ecosystems can become unstable against weak perturbations when the complexity increases beyond a certain value, although the transition is not always sharp [177].

BACHAS and HUBERMAN [178] related the complexity of hierarchical structures, as measured by their diversity [179], to dynamical behavior. The result suggests that there might be a quantifiable relation between the diversity of a hierarchical structure and its stability. Furthermore, the results of DOREIAN [180] and of BRIAND and COHEN [181] suggest that the effective dimensionality of the space in which the interactions occur, and the overlaps in food webs, contribute to the nature of the structural representation of the community. Hence, it may be concluded that their stability will likewise be affected, which is in agreement with the results of a comparative model analysis of unstructured cooperative interactions between arbitrary species on the one hand and ecosystems with pyramidal organization on the other [182].

In returning to the original problem set out by MAY [175], this author derived a condition of stability for more general systems as a function of species diversity and the strength of interactions. The stability of such systems is determined by the behavior of the largest eigenvalue of matrices governing the response of the system to small perturbations. As a result, MAY and coworkers showed in the case of nonhierarchical organizations how the removal of a zero-sum game condition can lead to further reduction of stability. Examining hierarchical ecologies, they demonstrated that these hierarchical structures are intrinsically more stable than unstructured ones [182].

The conflicting results among models developed so far must be considered in the light of model structure, particularly the number and composition of elements (i.e., species) simulated. In comparison to real communities with hundreds or thousands of species [136] and highly variable degrees of interaction, most models are extremely "impoverished" in species. Furthermore, the models quite often refer to randomly constructed communities, whereas real communities and ecosystems are far from being randomly constructed and normally display a complex hierarchical structure [2], [173]. Unstable communities are liable to collapse when they experience environmental conditions that reveal their instability, but the range and predictability of conditions may vary markedly from place to place. Under stable and predictable site conditions, a community will experience only limited fluctuations; thus, even a dynamically fragile one may still persist. By contrast, in a variable and largely unpredictable environment, only dynamically robust communities are likely to persist.

2.3.3. Stability-Oriented Biodiversity Analyses of Real Communities

Following MAY's approach, studies have been performed on the general importance of dynamic stability constraints by examining the relationship among S, C, and β in real communites. It follows from the inequality of Equation (25) that increases in S will lead to decreased stability unless compensatory decreases occur in C or β. Since data on interaction strengths for entire communities are unavailable, β is usually assumed to be constant. In such a case, communities with more species would retain stability only with a corresponding decline in average connectance, which means that the product SC should remain approximately constant.

REJMANEK and STARY [183] studied geographically distinct plant – aphid – parasitoid communities in Central Europe and found a decrease of C with an increase in S, with SC lying between 2 and 6. Evaluating the literature, BRIAND [184] analyzed 40 terrestrial, freshwater, and marine food webs and calculated a single value for connectance on the basis of both predator – prey and presumed competitive interactions (i.e., C_{max}). Once again, connectance decreased in an approximately hyperbolic way with species number, and SC was about 7.

MCNAUGHTON's field test [185] of MAY's inference, carried out in 17 grassland stands in Tanzania's Serengeti National Park, showed that "both average interaction strength and connectance declined as species richness of the grassland increased. The correlation was somewhat stronger for interaction strength than for connectance." Hence, if the grassland communities analyzed are representative biocenoses, species-poor stands are likely to be characterized by strong interactions among species, while species interacting with many others do so relatively weakly. Diffuse competition obviously increases with the number of species; in addition, MCNAUGHTON's findings suggest that communities have a blocklike or patchy organization consisting of species interacting among themselves but only little with species in other blocks.

Table 7. Influence of nutrient addition on species richness, equitability (H/ln S), and diversity in two fields; and influence of grazing by African buffalo on species diversity of two vegetation plots (after [185])

	Control plots	Experimental plots	Statistical significance
Nutrient addition			
Species richness per $0.5\ m^2$ plot			
Species-poor plot	20.8	22.5	n.s.*
Species-rich plot	31.0	30.8	n.s.*
Equitability			
Species-poor plot	0.660	0.615	n.s.*
Species-rich plot	0.793	0.740	$p < .05$
Diversity			
Species-poor plot	2.001	1.915	n.s.*
Species-rich plot	2.722	2.532	$p < .05$
Grazing			
Species diversity			
Species-poor plot	1.069	1.357	n.s.*
Species-rich plot	1.783	1.302	$p < .005$

* n.s.=not significant.

McNaughton [185] also tested the prediction that complex communities are less likely to return to their state prior to perturbation. In the first experiment, the perturbation resulted from addition of plant nutrients; in the second, it involved the action of grazing buffalo. The results summarized in Table 7 indicate that each perturbation significantly reduced the diversity of the species-rich but not the species-poor community.

According to these analyses and studies on the species diversity of tropical rain forest ecosystems in particular, a climax community is more likely to have both low fluctuations in composition (i.e., high stability) and low resilience as the homogeneity of its environment in space and time increases. Figure 25 summarizes the diversity and abundance spectra of numerous neo- and paleotropical rain forest communities as related to soil properties [2]. The diagram clearly shows that in most cases, rain forest stands on highly nutrient-depleted ferralsols, acrisols, and podzols of tropical lowlands attain diversity indices greater than 90 % of the theoretical maximum. However, a further differentiation according to local nutrient status and water budget is not possible with the data available. Nevertheless, stand 38 (Borneo) is of particular interest since it is representative of a forest that developed on a tropical bog; its exceptionally low index value allows the conjecture that its nutrient supply has fallen below a critical value, which is discussed in a broader thermodynamic context in Section 2.3.7.

In terms of May's stability criterion (Eq. 25), these findings would indicate a very low degree of connectance and average interaction strength between plants of the same species. With regard to nutrient uptake from the soil, this means a well-balanced low-grade intra- and interspecific competition involving a marked selection of K strategies (i.e., care for the offspring in the widest sense) and the formation of numerous ecological niches. Therefore, highly complex rain forest communities with their relatively constant environments appear to be more susceptible to outside, unnatural disturbances than the simpler, more robust communities of ectropical regions.

The example of Vietnam's inland forests is illuminating in this respect, since they still bear the scars of a ten-year herbicide spraying program by the U.S. Air Force during the Vietnam war. According to official U.S. figures, some 10.3 % of these forests, in addition to 36.1 % of mangrove forests, 3 % of cultivated land, and 5 % of other land, were sprayed. Thus, some 90 000 t of herbicides and antiplant chemicals was used between 1962 and 1971 [186]. Most of the spraying was with Agent Orange [i.e., a 1 : 1 mixture of (2,4-dichlorophenoxy)acetic acid (2,4-D) and (2,4,5-trichlorophenoxy)acetic acid (2,4,5-T)]; Agent White (2,4-D and picloram); and Agent Blue (cacodylic acid). Some 20 years after the spraying ended, little regrowth had occurred in areas sprayed three or four times. The fertility of the soils is rapidly reduced further if forests are replaced with grassland or bamboo, as has happened in a large part of the sprayed region. In particular, a loss of minerals and nitrogen has occurred, and in some cases a decrease in soil pH, which all inhibit diversified recolonization.

Defoliation has clearly affected the faunal assemblages, not only in the areas totally defoliated and subsequently converted into grassland but also in those regions less frequently sprayed where the plants in the upper reaches of the forest were damaged. In both areas, the number of animals and birds has dropped dramatically, and animals are also at considerable risk in some of the forest areas that have been isolated from the main body by herbicide application. Furthermore, the concern is that Agent Orange contained a high concentration of the highly toxic contaminant 2,3,7,8-tetrachlorodibenzo-*p*-dioxin, which is both teratogenic and carcinogenic in animals, and toxic to humans as well.

In comparison to tropical lowland rain forests where species diversity and nutrient content of the soil are negatively correlated, the situation in temperate regions is different. The composition of 132 plant communities from all parts of the continental United States (except Alaska) evaluated in relation to climate and soils shows that the amount of water available for evapotranspiration is of decisive importance for species diversity. This is further corroborated by the highly significant correlation between diversity and the temporal distribution of rainfall. Conversely, the analysis of 194 plant communities in northern Germany led to the conclusion that, most frequently, high diversity coupled with medium to low proportional abundances occurs on sites

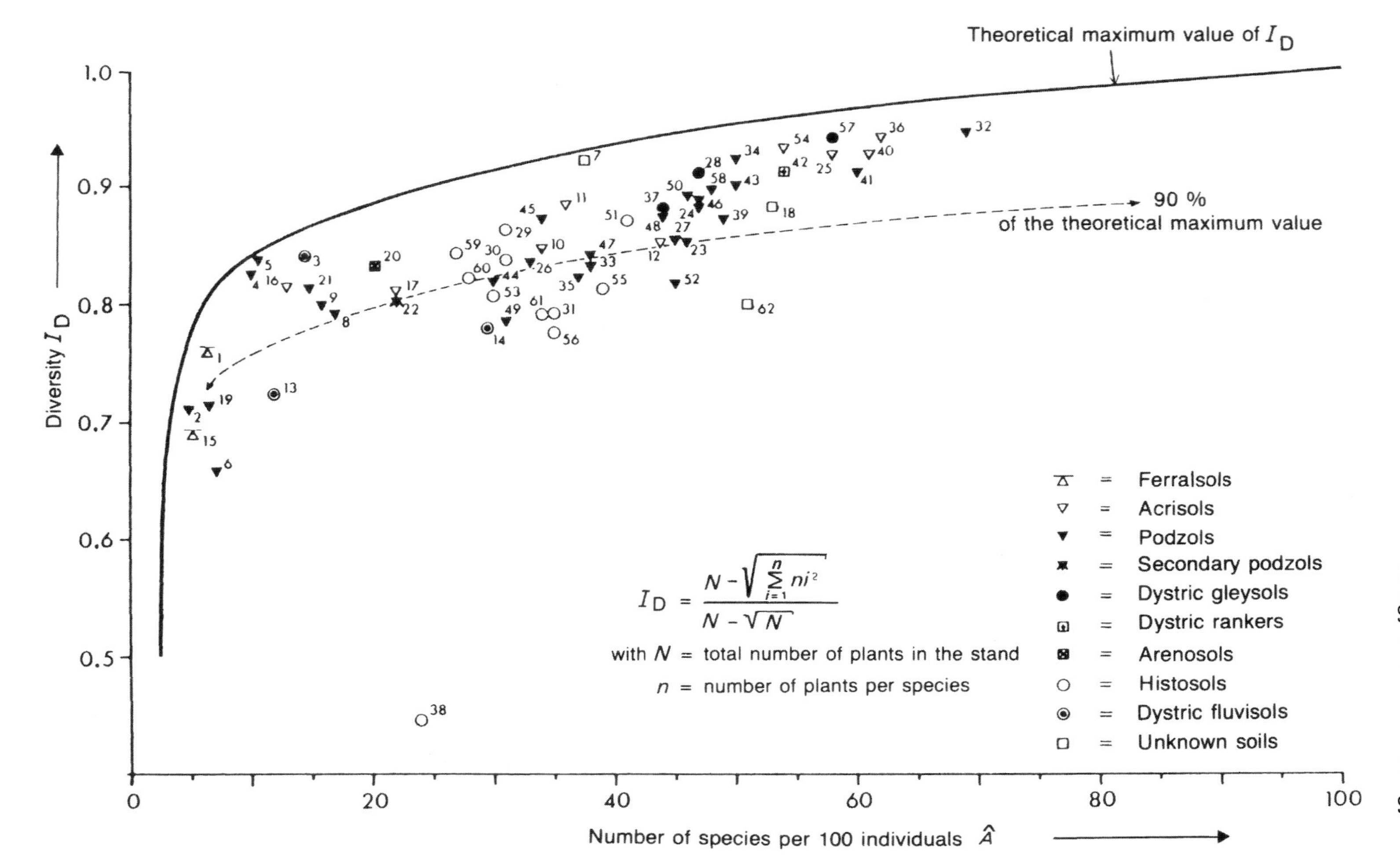

Figure 25. Diversity and abundance spectra of neo- and paleotropical rain forest communities

with high nutrient supply and medium to low soil moisture, whereas the combination of both high diversity and high abundance with high nutrient and medium to low moisture status is comparatively rare. These inverse diversity –nutrient relationships in tropical and ectropical environments, on the one hand, and the fact that soil moisture controls diversity in the United States but not in northern Germany, on the other, lead to the final question of whether a unifying interpretation of these apparently incoherent results is possible. This is given in Section 2.3.7 by considering ecological communities as dissipative structures.

2.3.4. Nondemographic Measures of Stability

The introductory remarks on stability and resilience indicate that community and ecosystem stability can be viewed from perspectives other than demographic ones. The highly different aspects of functioning (e.g., primary productivity, nutrient cycles, energy and water balance) and the composite aspects of structure such as standing-crop biomass appear to be particularly important [6].

McNaughtons's [185] studies on Serengeti grassland communities showed that species-rich associations responded to perturbations with a significant drop in species diversity, while species-poor communities did not. In terms of *primary productivity*, however, grazing as a natural perturbation reduced the standing-crop biomass in the species-poor grassland much more than that of species-rich communities. Similarly, several studies have revealed a marked influence of *food web* structure on community resilience in response to perturbations of energy and nutrient supplies. O'Neill [187] considered a three-compartment system consisting of active (photoautotrophic) plant tissue, heterotrophs, and dead organic matter, and found that the rate of change in standing crop depends on the energy transfer among these compartments. With regard to the autotrophic compartments, it depends on the input variable (net primary productivity) and two output variables (fraction consumed by heterotrophs and fractions lost to soil by litterfall) in addition to a great variety of translocation processes [85]. The rate of change of the heterotrophic compartment depends on two inputs (consumption of living plant biomass and dead organic matter) and two outputs (respiratory heat loss and defecation). Finally, as Figure 26 shows, the energetic exchange rate of soil organic matter depends on two inputs (litterfall and defecation) and two outputs (physical transport into neighboring ecosystems and consumption by heterotrophs).

On the basis of this simple configuration, and inserting real data from communities representing a pond, a spring and tundra, tropical forest, a temperate deciduous forest, and a salt marsh, O'Neill [187] subjected the models of these communities to a standard perturbation that consisted of a 10 % decrease in the initial standing-crop vegetation. The rates of recovery toward equilibrium thus monitored and plotted as a function of the *energy input* per unit standing crop of living tissue are presented in Figure 27. It shows that the flux of energy through the different subsystems has an important influence on community resilience; the higher the flux is per unit standing crop, the more quickly will the effects of perturbation be compensated. This, in turn, seems to depend in part on the relative importance of heterotrophs in the system considered. The pond as the most resilient system has a heterotrophic biomass 5.4 times that of autotrophs, reflecting the high turnover rate of phytoplankton; the least resilient tundra has a heterotroph : autotroph ratio of only 0.004. De Angelis [188] has come to analogous conclusions in terms of residence times of various nutrients in different compartments of woodland communities. Thus, a unit of nitrogen exists in soil for an average of 109 years, is taken into forest biomass for 88 years, remains in the litter for up to 5 years, and passes through the detrivore compartment in a few days. In contrast to nitrogen, calcium is a less tightly cycled element whose residence times in soil and biomass amount to only 32 and 8 years, respectively.

More detailed long-term investigations in the forest and agroecosystems of the Bornhöved study area (Schleswig-Holstein, Germany) corroborate these findings. Here, the nitrogen deposition amounted to 19.9 and 18.2 kg ha^{-1} a^{-1} in 1989 and 1990, respectively, with an ammonia fraction of 47 % (60 %) and 23 % (7 %) of organic nitrogen-compounds, while nitrate deposition yielded 6.1 and 5.9 kg nitrogen per hectare

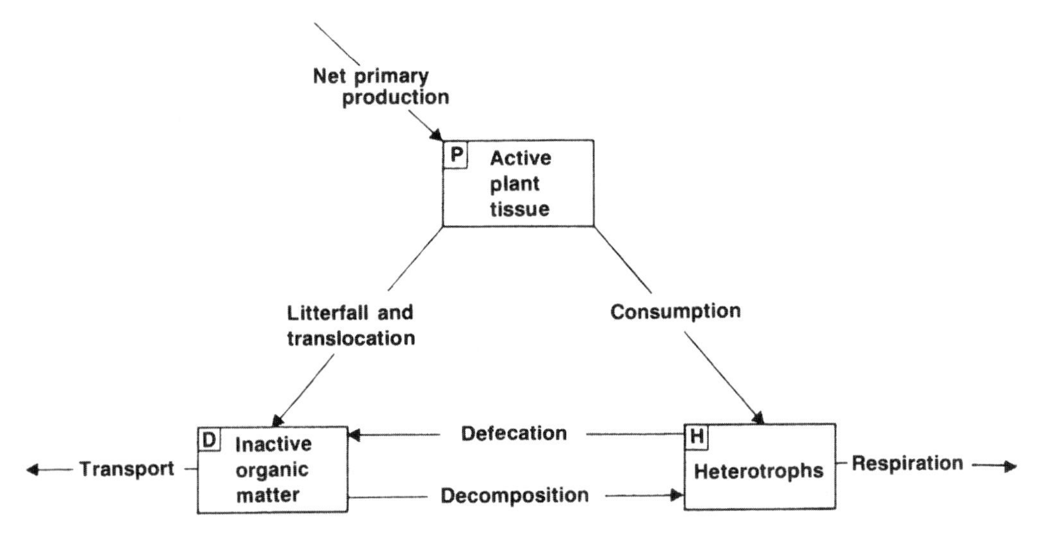

Figure 26. Three-compartment model of an ecosystem with typical energy transfers (after [187], modified)

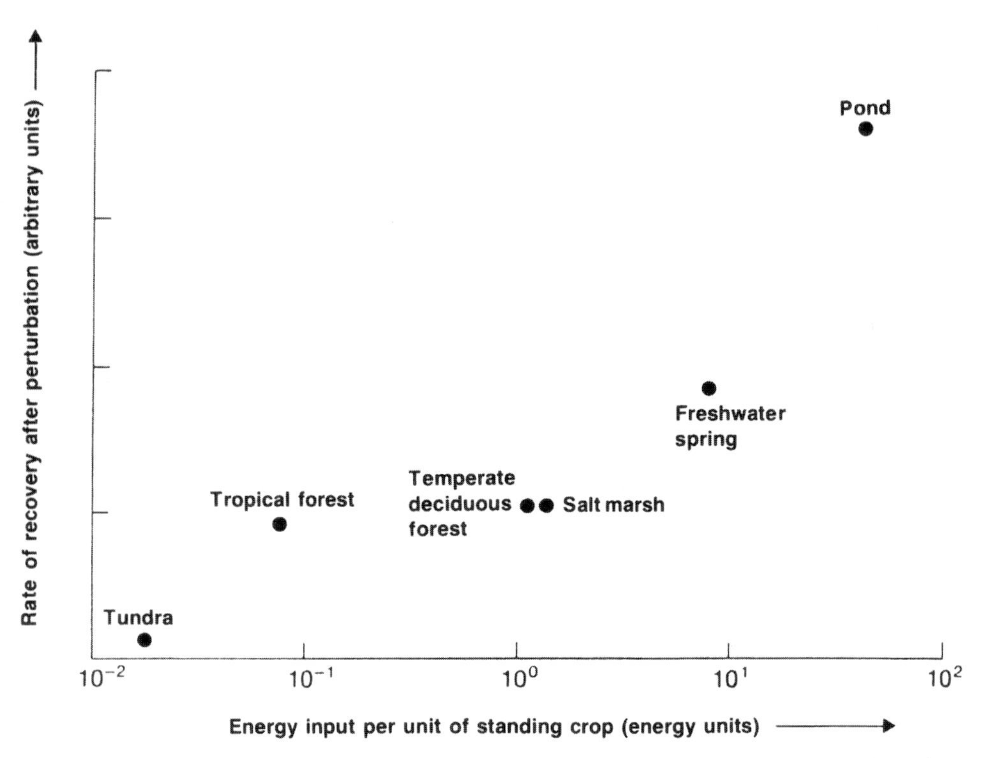

Figure 27. Recovery rate as an index of resilience after perturbation as a function of energy input per unit standing crop (after [187])

and year, respectively. The nitrogen balance of a field is summarized in Table 8.

Table 8. Nitrogen balance of a field in the Bornhöved Lake District

Process	Amount of nitrogen (measured, kg ha^{-1} a^{-1})	
	1989	1990
Initial N content	51 [a]	38
Total N input	384	222
NO_3-N	30	14
NH_4-N	118	64
N_{org}	236	144.5
Mineralization	221 [b]	215 [b]
Nitrification		
Plant uptake	206	121
N × leakage	110	129
Other N losses [c]	7	7
Residual N_{min} content	39	4.5
Balance	+59	+70

[a] Deduced value. [b] Net mineralization. [c] Denitrification and volatilization of ammonia.

The wet deposition of nitrates via canopy flow in a neighboring beech stand (*Asperulo–Fagetum*) of relatively uniform age structure (95 years) amounted to 30.3 kg nitrogen per hectare and year in 1989 and 37.5 kg in 1990, again with a major ammonia fraction. Comparably high were the deposition rates to the soils of a neighboring alder stand (*Carici elongatae –Alnetum*), where additional nitrogen fixation is due to the symbiotic activity of *Frankia*.

With regard to internal nitrogen fluxes the beech and alder stands differ considerably. In the first case, litterfall transported 78 kg nitrogen per hectare and year to the soil surface, where the chemical and microbial transformation processes induce seasonal changes in N_{min} content as illustrated by Figure 28. In the light of the site characteristics of this stand (Dystric Cambisol), the figures are indicative of rather limited nitrogen export from the community, which is further corroborated by measurements of the nitrogen fluxes in the related seepage water. In the distinctly more resilient alder stand, the nitrogen input due to litterfall amounted to 109.4 kg ha^{-1} a^{-1} in 1989 and 104.5 kg ha^{-1} a^{-1} in 1990, respectively. In comparison to the above beech stand, microbial degradation of the litter is very high in the *Alnetum*, and within 18 months almost all of the leaf tissue is degraded., Chap. 1.

2.3.5. Sensitivity of Aquatic Ecosystems

Aquatic ecosystems are natural collectors of materials from their surroundings, and water washes matter out from the land surface. Human activities increase this natural washout enormously by destroying the natural vegetation cover of the landscape due to agricultural, mining, construction, and forest-cutting activities. For organic matter of natural origin, with its contents of nutrients necessary for further growth of vegetation, a relatively closed cycle originally existed: the remains of organic matter consumed by humans and their domestic animals were returned to fields. In medieval times, increasing population concentrations in towns created local accumulations of decaying organic matter leading to the spread of infectious diseases. This problem consequently was solved by disruption of the feedback cycle. Organics with their content of elements essential for life were directed to streams, rivers, lakes, and ultimately the sea. The result was continuous deprivation of nutrients in soil and a need to subsidize them to maintain and enhance agricultural productivity. The open flow of nutrients from soil and deeper earth layers started to create unpleasant conditions by increasing aquatic production not only in streams, lakes, and ponds but also in the greatest lakes of the world and marginal seas such as the Adriatic. In addition, humans pollute water by applying various chemicals to the environment – both at home and in industry – and by extracting materials from the earth and transporting them over large distances, in part by means of waterways. Tanker catastrophes are becoming more and more common, and road transport is no less hazardous to waters.

Pollution sources are usually classified in two categories: point and nonpoint sources. Although *point sources* coming from a factory or town are technically much easier to handle, *nonpoint* sources are not only difficult to detect but still more difficult to eradicate. Typical nonpoint pollution sources originate from fertilizers and other chemicals spread on fields. Another source of water pollution is deposition of airborne pollutants, some of which reach water bodies in remote regions. The findings of elevated concentrations of DDT in Antarctica show the global scale of pollution problems.

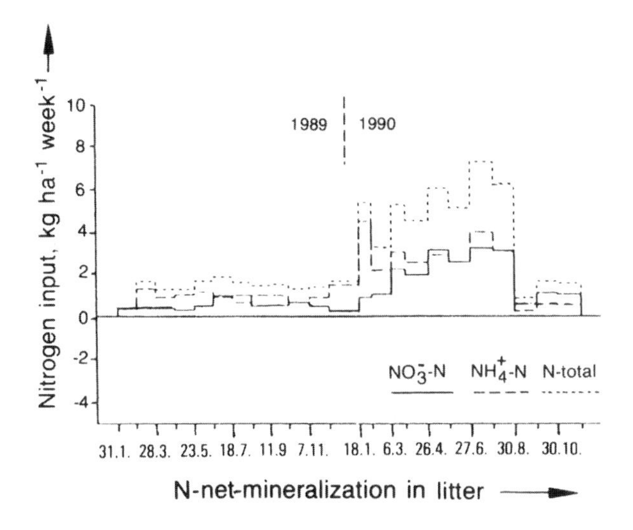

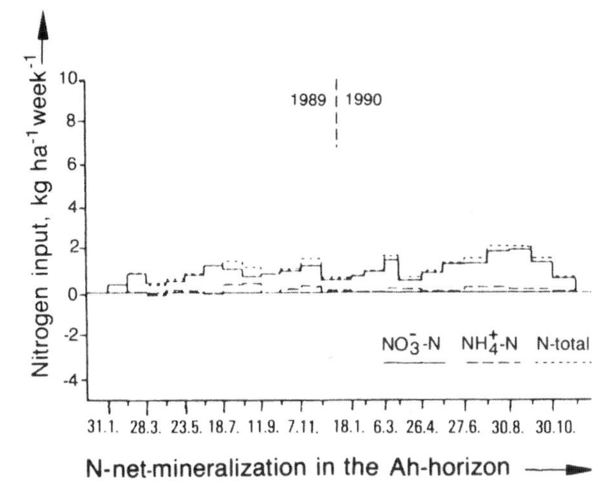

Figure 28. Seasonal variation of the N_{min} content of beech litter of a Schleswig-Holstein *Asperulo – Fagetum* stand in 1989 and 1990
Ah-horizon = part of the surface soil

The sensitivity of aquatic ecosystems to disturbances differs depending on the type and size of the respective ecosystem and the nature of the disturbance. In scientific terminology, all disturbing factors are stressors and the term "sensitivity to stress" is used. Stress is usually defined as a measurable alteration of the state of organisms as induced by an environmental change that renders the individual, population, or community more vulnerable to further environmental change.

2.3.5.1. Reactions of Aquatic Ecosystems to Stress

Aquatic ecosystems under stress undergo both changes of structure and changes of function. *Changes of structure* are manifest by changes in the composition of populations of different species in the ecosystem. *Changes of function* are represented by differences in the organic matter productivity of the respective ecosystem and in the rates of release and utilization of different gases and minerals. Thus, both a sensitivity of ecosystem structure and a sensitivity of ecosystem function occur. No uniformity of opinion exists as to whether the structure or function of ecosystems in general is more sensitive to various stressors. Several researchers feel that functional variables, especially those that are substrate limited, will always be less sensitive than structural measures because of functional redundance in the community. They suggest that any loss of functional capacity by one organism is immediately compensated by increased activity of another organism. Other scientists feel that functional capacity can be affected before compensatory mechanisms operate. This is especially likely when the compensatory mechanisms themselves are adversely affected by the stress or when they operate on a more lengthy time scale relative to the functional measure [189]. These authors have also summarized various functional reactions of ecosystems:

1) Community respiration increases
2) Productivity – respiration ratio becomes unbalanced
3) Productivity – biomass ratio increases as energy is diverted from growth and reproduction into acclimation and compensation
4) Importance of auxiliary energy increases (import becomes necessary)
5) Export of primary productivity increases
6) Nutrient turnover and nutrient loss increase
7) One-way transport increases and internal cycling decreases
8) Life span decreases; turnover of organisms increases
9) Trophic dynamics shift, food chains shorten, and functional diversity declines
10) Efficiency of resource use decreases
11) Condition declines

A well-documented example of greater sensitivity of functional variables of a terrestrial ecosystem is presented by VAN VORIS et al. [190], who measured structural and functional changes in small patches of a meadow subjected experimentally to cadmium stress. Functional changes such as spectral complexity and calcium export were affected at concentrations that did not affect structural measures such as the presence or absence of certain species.

In the direction of slowly and rapidly reacting ecosystems at opposite ends of the scale are pelagic lake and reservoir communities composed of very short-living (few hours to few days) species that react to any stress by their rapid appearance or disappearance, and forests with very long-living tree species that are incapable of rapid presence – absence types of reaction. Benthic communities of rivers and fish populations are intermediate: their life span is measured in months to years.

In aquatic ecosystems, many reports have shown greater relative sensitivity of structural than functional variables. For example, during his extensive comparative studies of Canadian Shield lakes, SCHINDLER [191] found no significant changes in decomposition or nutrient cycling in whole lakes treated with acid, but species composition of phytoplankton was among the earliest indicators of change. CROSSEY et al. [192] found in impaired rivers that measures of production and respiration were more variable than the macroinvertebrate composition. CRUMBY et al. [193] studied the biological reaction of Roaring River in north-central Tennessee to stress caused by construction around the river and by inadequate agricultural practices in the watershed. Species composition changes were reflected in a general decline in numbers of intolerant species and simultaneous increase of tolerant species.

However, many reports of the greater sensitivity of functional variables of aquatic ecosystems also exist. RODGERS et al. [194] found that process rate changes were more sensitive than biomass or chlorophyll concentration in detecting the effects of diverse chemicals on periphyton in artificial streams. When dealing with enrichment, functional measures are often a good warning indicator. UHLMANN et al. [195] and GNAUCK [196] summarized structural and functional changes in aquatic ecosystems and gave examples of experiments equivalent to those of VAN VORIS et al. [190]. Figure 29 compares the degree of variability (expressed by a relative index of instability) during a 50-d experiment with an experimental purification pond for several structural and a few functional variables. The greatest instability equivalent to the highest sensitivity is observed (with the exception of the alga *Scenedesmus obliquus*) for the biomass of individual species of organisms (index values between 0.85 and 1.45), while functional variables such as oxygen concentration, primary production, and turbidity and the summarizing biomass of functional groups of organisms such as zooplankton or phytoplankton or chlorophyll-a levels have an index close to or below 0.5. The least sensitive appeared to be the organic carbon elimination capacity of the pond, which characterizes, from the human point of view, its most important function.

Once the stress has ceased, two possibilities exist:

1) Reversible changes are induced
2) The changes prove irreversible

The deforestation in the coastal hills of Venezuela is a drastic example of irreversible changes of an extensive ecosystem. The soil structure, seed sources, and local physical environment have changed to such a degree that forests have not returned even after the areas were abandoned by humans. Perhaps the example merely reflects an exceedingly long recovery period, and the system will eventually recover. Yet for practical purposes, these ecosystem changes are permanent. A dried lake and a watershed with homogeneously distributed amounts of toxic substances are examples of irreversible changes of aquatic ecosystems.

If the disturbance is sufficiently regular and of near-natural origin, components of the ecosystem may adapt and eventually require disturbance to maintain a normal, resilient system (tidal pools require daily exchange with the sea, pastures require mowing or grazing, many forests are adapted to periodic fires). In aquatic ecosystems, this system of reaction is seen in periodic pools, which are inhabited by animals adapted to desiccation. Their characteristic fauna will disappear when water is permanently present because they are inferior in competition with longer-lived species not needing desiccation.

For chemical stress, the recovery of aquatic ecosystems depends mainly on the degree of accumulation of the respective chemical in the environment and the rate of flow. This conclusion can be corroborated by comparing the findings in lakes with those in rivers. The recovery of eutrophic or highly polluted lakes after the sources of organic pollution and phosphorus causing lake eutrophication have been reduced is very slow. Recovery takes as much as ten to twelve years. The delayed reaction is due to the enormous quantities of decomposable organic matter and phosphorus stored in the bottom mud. Oxygen at the bottom is consumed during the decomposition of organic matter and, under such conditions, phosphorus is continuously released. This phenomenon is called internal P load, which indicates that when the external load ceases, an internal one plays a major role in continuing eutrophication. At the other end of the spectrum are rivers with high flushing. YOUNT et al. [197] made an overview of the recovery of rivers after a stress and found that almost full recovery may occur within about two years.

2.3.5.2. Sensitivity to Different Stressors

The sensitivity of marine and freshwater fish worldwide to recent changes in the environment was determined by means of changed species diversity [41]. The numbers of extinct, endangered, or threatened species and species of special concern represent 21 – 69 % of the total fauna in different regions of the world, with the exception of only 9 % in Latin America. The highest numbers were found in California and South Africa; low figures in Iran, Australia, Sri Lanka, and also Arkansas. Major causes of this situation are the extraction of water for various human uses, physical alteration of fish habitats (mainly by channelization, construction of weirs and reservoirs, siltation from mass erosion and degradation of wetlands), and pollution (mainly municipal and industrial point-source pollution, agricultural pollution, and acid rain). The sensitivity of aquatic ecosystems to different specific stressors depends on the type of ecosystem and the agent. For chemical agents the degree of interference with the structure and metabolism of organisms is important. However, even such components as structural macroelements of organisms may be harmful, once their concentrations in the environment exceed certain levels. This is the case, for example, with plant nutrients. When present in large excess, they decrease life functions (e.g., plant growth; see Section 2.1.4.3). Sensitivity to toxic chemicals is discussed in Section 2.4. Sensitivity to acidification as an environmental process is described in Section 2.3.5.3 together with other signs of global change.

The sensitivity of aquatic ecosystems to *ionizing radiation* is not well known. However, some reactions parallel to those observed for terrestrial ecosystems are expected to be valid.

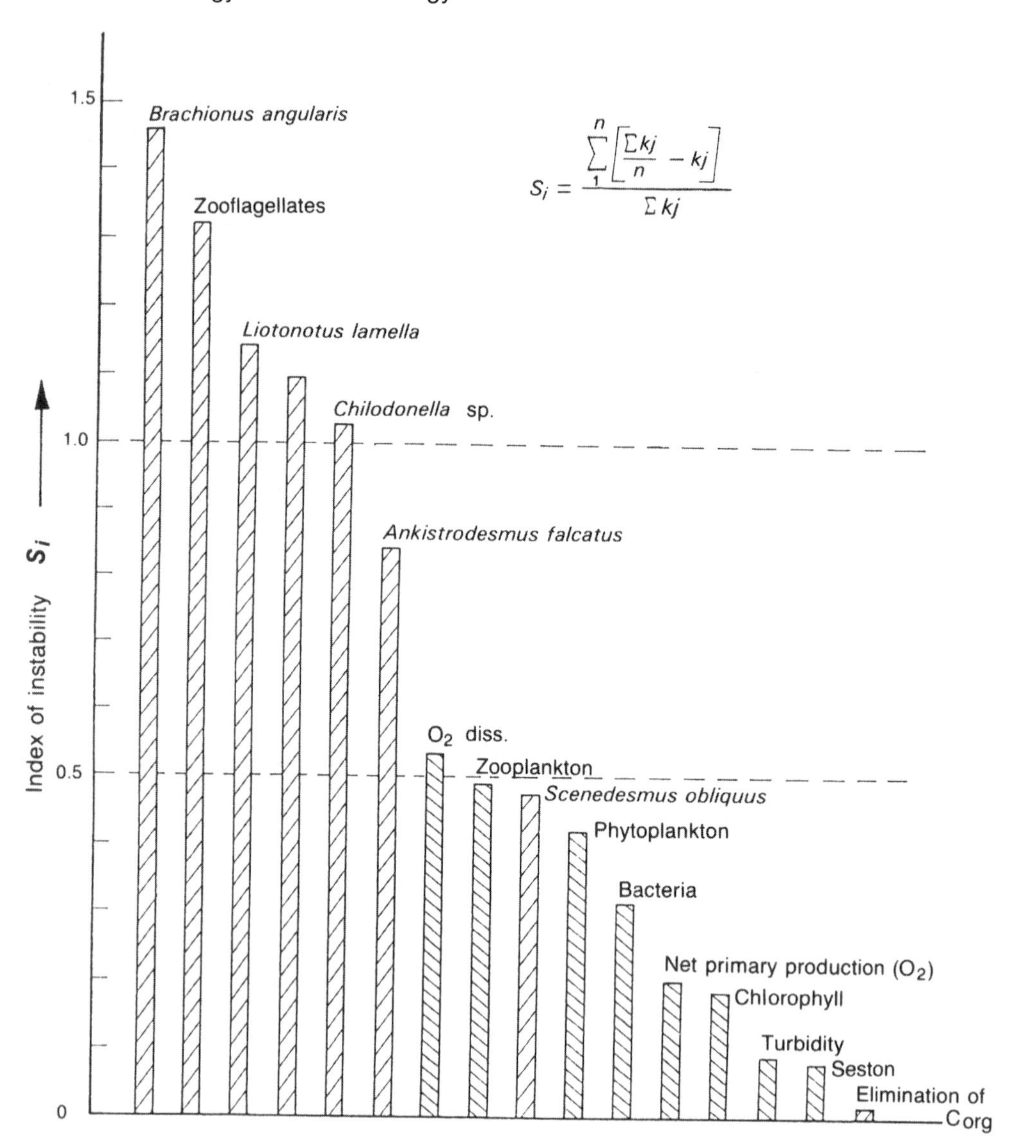

Figure 29. Comparison of the degree of variability, expressed in terms of an instability index, during a 50-d experiment with an artificial purification pond (after [195], modified)

The latter were studied in detail by WOODWELL [198]. He demonstrated that in a mixed forest – shrub – grassland exposed to different levels of ionizing radiation, the diversity of species is more sensitive than organic production. The sensitivity increases with the size of the organisms. The reason is that increased size puts greater demand on the photosynthetic mechanism for maintenance, leaving less for growth and repair. Sorting of the sensitivity was also observed on the basis of chromosome size: plants with large chromosomes were more sensitive.

Sensitivity to *communal organic pollution* is well known for both stream and lake ecosystems; this is the classical type of pollution studied by quantitative methods since the beginning of this century, the methods being continuously improved and new methods being continuously developed. From the pollution point of view, organic matter is not characterized as a chemical species but in a summarizing way, namely, the total of easily decomposable or refractory matter. The difference between the two categories is arbitrary: easily degradable matter is decomposed by the natural population of microorganisms within five days' incubation. In streams, systematic changes of the composition of benthos are observed with certain amounts of organic pollution. This is the basis for the bioindicator approach to detect the degree of organic pollution. However, geographic differences in the benthic fauna of different regions; dependence on stream characteristics such as size, rate of flow, and bottom type; and combination of pollution types all complicate the evaluation.

2.3.5.3. Sensitivity to "Global Change"

One of the effects anticipated by part of the scientific community to cause major changes in world climate and consequently local weather and food production within the next decades is "global change." Since the 1940s, compounds such as CO_2, chlorofluorocarbons (CCl_2F_2 and CCl_3F), methane (CH_4), and dinitrogen monoxide (N_2O) have displayed notable and deleterious increases or decreases in the case of tropospheric ozone (O_3). These gases (except O_3) are classified as "greenhouse gases" because they have a greenhouse effect on earth's atmosphere.

About a hundred years ago, CO_2 values of 280 ppm (mL/m^3) were measured in the atmosphere; (in the 1990s), concentrations are about 25 % higher. Carbon is both fixed and released by oceans, biota, and human activities. Approximately 30 % of the global annual carbon fixation by plants occurs in the seas, which also form an integral part of the nitrogen flux of the earth. An estimate for 1980 gave the following figures for the fluxes of carbon among atmosphere, ocean, biosphere, and human activities [199]: 78×10^{12} kg C per year (flux ocean $\leftrightarrow$ atmosphere), 65×10^{12} kg C (flux biosphere $\leftrightarrow$ atmosphere), and 5×10^{12} kg C (flux due to human activities, mainly burning of fossil fuels and land clearance $\leftrightarrow$ atmosphere). Although the contribution by human activities is far less than the natural fluxes, it shifts the total balance significantly.

In the United States, the sources of CO_2 from human activities were estimated for 1987 as follows [42]: 11 % residential, 24 % industry, 30 % transportation, and 35 % electric and power generation. Industry is therefore an important CO_2 source, even when the significant portions of electric power generation and transport associated with industry are neglected.

The climate has changed throughout the last few thousand years due to natural causes; the influence of human activities superimposed on these variations has also contributed significantly. Scientific documentation demonstrates that the overall rise in CO_2 has already had marked effects on the biosphere. Comparison of different recent plants with herbarium specimens collected many years ago disclosed that the leaf area and the nitrogen content of leaves in 15 species of trees, shrubs, and herbs decreased over the last 240 years. Also, a systematic increase in the growth rate of bristlecone pines and a decrease in stomatal density of the leaves of plants over the last 200 years was observed [200]. Although no aquatic plants were included in the studies, they are expected to be subject to similar effects. The reported changes may seem unimportant; however, this may be only the easily observable tip of the iceberg. More relevant parameters cannot be compared with those of earlier times when no scientific measurements existed. The detection of changes increases the credibility of predictions by demonstrating that biological changes have already taken place and are not merely scientists' anticipation.

Despite many uncertainties in predictions, a certain consensus exists among part of the scientific community that the increase in CO_2 and the other greenhouse gases due to human activities will warm the climate by the middle of the next century between 1 and 5 °C ("global warming"). Greater temperature increases are expected at high latitudes, especially in winter, with consequences for ice melting and sea-level rise. Consensus also exists that changes of this magnitude in temperature will be accompanied by substantial regional anomalies in temperature, rainfall, evaporation, cloudiness, windiness, drought, and so forth that could have substantial impact on agriculture, water supply, human and animal health, biological diversity, energy demand, and sea level [199], [201].

The changes expected in aquatic environments due to these increasing quantities of greenhouse gases are of both a physical and a chemical nature. Physical changes mainly involve global warming and an increased level of ultraviolet radiation at the earth's surface. Ultraviolet radiation in its present spectral composition is readily adsorbed in water, and only organisms living in the uppermost (few centimeters) water layers can be damaged. However, alterations in spectral composition of ultraviolet radiation expected as a consequence of global changes can markedly increase the layers affected.

One indirect effect of global warming on streams is expected in connection with increased variability in stream flow, associated with increased rates of sediment transport and sedimentation. Sediments are expected to fill in some pools and generally cover surfaces. However, streams may be flushed of material periodically. These conditions would favor invertebrates with shorter life cycles adapted to softer sediments, while larger invertebrates could be adversely affected by these conditions because they graze rock surfaces [43].

Indirect effects on aquatic ecosystems are also expected due to the high sensitivity of forest ecosystems to environmental variables. The sensitivity of forests to climate changes was estimated from changes observed during the geological past and from some geographical dependencies observed more recently [202]. The results indicate that during global change, shifts will occur in the geographical ranges of tree species, along with changes in forest productivity. Because of the interrelation of soil nutrients, soil moisture, groundwater, and tree growth, changes in forest ecosystems will have major consequences for aquatic ecosystems, primarily rivers. River flow will be changed and, with it, leaching of minerals and organics from soil will occur to varying extents, with consequences for river inhabitants [44]. Additional indirect effects will be caused by changes in agricultural production.

The probable effect of the expected climate warming on fish of the Great Lakes Basin was estimated by MEISNER et al. [203]. They concluded that no full extirpation of species is to be expected, but the populations of the most valued fish, salmonids and coregonids, will decrease, whereas the species less valued by fishermen (cyprinids, esocids, centarchids, and ictalurids) will probably increase. Thus, a decrease of fishery yields of preferred species is expected.

Other global changes in aquatic ecosystems are already observed locally in acidified regions. They result from direct and indirect chemical effects caused by increasing CO_2 levels. Doubling of atmospheric CO_2 levels, which is expected to occur at the beginning of the next century, is supposed to decrease the pH of rainwater by about 0.4 unit, which means a pH drop in some localities to 4.0. The lower pH will increase the solubility of carbonate minerals and increase the calcium, magnesium, bicarbonate, and sulfate content of rivers draining carbonate terrain. The greatest secondary effects on organisms are caused by the increased solubility of aluminum at low pH. Aluminum would be released, predominantly in bedrock terrain. The speciation of aluminum in river water is important in determining the toxicity of elevated aluminum levels as well as the acidity of soil solutions. Inorganically complexed or free aluminum is more toxic than organically complexed aluminum. Lakes with present pH levels of 6.5 are expected to become acidified to pH <5, with considerable effects on inhabiting organisms. Major impoverishment of lakes due to acidification has already occurred in northern Europe and Canada, and in mountainous lakes of more southern localities [45].

2.3.6. Sensitivity of Ecotones

Like communities and ecosystems, ecotones also differ in physical stability or resistance to changes (resilience) caused by disturbances. In contrast to lotic ecotones, adjacent terrestrial ecosystems and ecotones may be viewed as occupying distinctly more stable positions in a landscape where natural disturbances due to geomorphic processes are less frequent than the time for patch adjustment to them (see Sections 2.3.5 and 2.3.7).

In distinguishing reactions, recovery and persistence phases, and recurrence intervals as characteristic parameters of system response to physical disturbance [46], the recovery and disturbance recurrence intervals are of particular ecological importance. The ratio of the recovery to disturbance recurrence intervals *R*/*D* suggests differences in recovery or sensitivity characteristics of both ecotones and ecosystems. For unstable systems, the *R*/*D* ratios exceed 1. Typically, this is the case for predominantly transient systems such as many *lotic ecotones* [162]. Normally, deposition and erosion interact with riparian vegetation at varying intensities and frequencies so that communities are kept in the early stages of succession. A decrease in seasonal flooding of the riparian vegetation results in changes via several possible successional pathways, which may lead to the formation of purely terrestrial communities when the system is no longer influenced by flooding.

Examples of this type support the assertion of NAIMAN et al. [155] that a paradox of modeling interfaces is predicting when and where ecotone parameters become unpredictable relative to adjacent homogeneous sites. Chaos theory may be useful in dealing with this uncertainty because ecotones are not simple averages of adjacent systems; the spatiotemporal pattern of environmental factors and the interactions between patches are as important in determining ecotone dynamics on different scales as are the characteristics of adjacent patches.

2.3.7. Stability and Resilience of Ecosystems in Evolution

On a very general level, the apparently unusual – because inverse – relationship between the nutrient status of soil and the diversity of related tropical rain forest communities described in Section 2.3.3 may be interpreted in terms of biological thermodynamics [169]. Biotic communities are open systems in exchange of energy and matter with their environment; their entropy production dS thus comprises the terms $\mathrm{d_i}S$ and $\mathrm{d_e}S$, where $\mathrm{d_i}S$ denotes the entropy production within the system while $\mathrm{d_e}S$ is a flux term describing entropy "export" into the environment:

$$dS = \mathrm{d_i}S + \mathrm{d_e}S \tag{26}$$

The term $\mathrm{d_i}S$ can only be >0, but $\mathrm{d_e}S$ can also be negative. By identifying entropy with disorder, Equation (26) shows that an isolated system can evolve only toward greater disorder. For an open system, however, the "competition" between $\mathrm{d_e}S$ and $\mathrm{d_i}S$ permits the system, subject to certain boundary conditions, to adopt new states or structures. These are stationary if

$$dS = 0 \tag{27}$$

or

$$d_e S = -\mathrm{d_i}S < 0 \tag{28}$$

$\mathrm{d_i}S$ can be expressed in terms of thermodynamic "forces" X_i, and rates of irreversible phenomena J_i [165]; X_i may be gradients of temperature or concentration; the corresponding "rates" are then heat flux and chemical reaction rate. Hence,

$$\frac{d_i S}{\mathrm{d}t} = \sum_{i=1}^{n} J_i X_i \tag{29}$$

Around equilibrium, the relationship between fluxes and forces is linear

$$J_i = \sum_{j=1}^{m} L_{ij} X_j \tag{30}$$

where L_i represents specific coefficients (e.g., coefficients of thermal conductivity or diffusion).

Provided the reservoirs of energy and matter in the environment of the open system are sufficiently large to remain essentially unchanged, the system can tend to a nonequilibrium stationary state far beyond the domain of linear thermodynamics. This state may be associated with dissipative structures [204] (i.e., structures resulting from a dissipation of energy rather than from conservative molecular forces). In considering phytocenoses from the viewpoint of stationary dissipative structures, the relationship of $\mathrm{d_i}S$ and $\mathrm{d_e}S$ as expressed in Equation (28) and the specific boundary conditions controlling entropy production and flux rates appear to be particularly important. A consequence of Equation (28) is that, thermodynamically speaking, stability or the capacity to maintain a nonequilibrium steady state is coupled with a (relative) minimum of total entropy production dS. Clearly this can be accomplished by either minimizing $\mathrm{d_i}S$ or maximizing $\mathrm{d_e}S$ or by a combination of both strategies.

Concentration processes involved in the normal metabolic activities of living systems play an important role in this connection, as can be seen from

$$\Delta G^{\circ} = R \cdot T \cdot \ln \frac{C_2}{C_1} \tag{31}$$

where ΔG° is the difference in standard free energy, $R = 8.31\ J\ mol^{-1}\ K^{-1}$, T = temperature in kelvin, and C_2, C_1 are higher and lower thermodynamic concentrations, respectively. Changes in concentration are a physical prerequisite for the production of a great many compounds, and an absolutely cogent one if substances are produced whose free energy is higher than that of the corresponding "raw materials."

On the nutrient-depleted soils of Amazonia (Fig. 25), for example, forest stands with their mosaic-like structure owing to the patchiness of rejuvenation processes (see Section 2.1.1) developed specific adaptations to facilitate these concentration processes. A dense root mat is formed over the soil in intimate contact with litter, with root tips growing upward into the fallen litter. Mycorrhizal associations, as the most common plant root – soil microorganism symbioses, benefit their host plants by effectively increasing the absorptive surface area of the root system, thus providing for increased uptake of nutrients. The dense fabric of predominantly fine roots also plays an important part in exchange and adsorption of nutrients from throughfall water. In addition, the structure of the foliage favors the use of nutrients during a long active life, the retransport of certain nutrients before leaf shedding, and a high polyphenol content and coriaceous nature, which both reduce herbivory. Another factor of great importance is the multilayered structure of the forest and the activities of epiphytes and microorganisms on much of the exposed surfaces, which together form highly efficient filtering systems scavenging nutrients from rainwater. Finally, biological nitrogen fixation in the root – humus – soil interface appears to play an important role in the nutrient budget [205]. Taking into account that even in the simplest cells the normal metabolic pathways imply several thousand complex chemical reactions that must be coordinated by means of an extremely sophisticated functional network, means that hierarchical order in both functional and spatiotemporal respects constitutes a further and most powerful negentropic factor. It characterizes every living system from the submicroscopic level to gigantic rain forest biomes such as the Amazonian Hylaea.

The effectiveness of these negentropic processes is further enhanced by most efficient entropy fluxes related to the transpiration and nocturnal respiration of plants. The molal entropy of H_2O increases from $63\ J\ mol^{-1}\ K^{-1}$ (liquid) to $189\ J\ mol^{-1}\ K^{-1}$ (gas) in the course of evaporation, and CO_2 has a molal entropy of $214\ J\ mol^{-1}\ K^{-1}$. Consequently, the reverse process, the photosynthetic fixation of CO_2, is of comparable importance for the negentropy balance in the light of Equations (26) and (28).

As a consequence of these mechanisms, the results of the above comparative diversity analyses may be given a unified interpretation:

1) Species diversity is not a monotonous function of the nutrient status of soils. If soils form on basic or intermediate parent materials, pedogenic nutrient supply and species diversity may both increase during periods of $10^3 - 10^4$ (10^5) years, provided the water and energy factors are not limiting. In Figure 30, the period of ascending evolution is termed phase I. Shorter fluctuations of variable but generally decreasing amplitude are likely to be superimposed on this long-term trend. In phase II, which can last for several hundred thousand years, soils degrade in regard to nutrient supply, but species diversity keeps increasing for negentropic reasons (see the tropical rain forest stands in comparison to the ectroprical associations described in Section 2.3.3). Clearly such an evolution toward and including the following phase requires a sufficient degree of climatic and geomorphic stability (as has been the case in central Amazonia and considerable parts of Malesia), or else the continuity of pedogenesis will be interrupted by truncation processes that rejuvenate the soils (as happened in the major part of the Congo Basin). Phase III marks the eventual and comparatively acute decrease in diversity of phytocenoses once the nutrient status of soils has fallen below a critical level.
2) Spatial heterogeneity as a characteristic feature of the sites during phase I results in variability in the number of species and plant individuals. With this variability the associations can simultaneously retain genetic and behavioral types that can maintain their existence in low populations together with others capitalizing on opportunities for dramatic increase. The more homogeneous the environment in space and time (i.e., during phase II,

and probably also phase III), the more likely is the system to have low fluctuations (i.e., high stability). Tropical rain forests represent climatically buffered and (largely) self-contained systems with relatively low (natural) variability on highly impoverished soils of the ferralsol, acrisol, and podzol classes. Temporary external disturbances are consequently likely to affect a proportionally greater number of species than in phase I populations, where the interspecific differences in ecological potency are usually distinctly higher. Hence, resilience of mature biocenoses is low during phase II. In contrast, biocenoses may be highly resilient in phase III of evolution, although unstable.

The greatly reduced resilience of phase II stands is particularly well illustrated by the influence of clearing activities on tropical forests. The soil, with its low retention capacity, is irrevocably impoverished after a brief period of fertilization due to sudden release of nutrients in the biomass. The brief period of fertility is used by shifting cultivators to produce crops, but even this system, developed by small native populations "in close contact with the environment" [206], can induce severe losses if forced to fit a market-oriented economy or to feed larger populations [207], [208].

2.3.8. Appraisal

The analysis of real communities and the comparative assessment of ecological models [2], [37] show that real communities are not those homogeneous and temporally invariant systems described by simple Lotka – Volterra equations and exemplified by most laboratory microcosms. Population dynamics of real-world ecosystems are spatially distributed in general, associated with a marked temporal variation, providing a multitude of ways in which the probability of coexistence is enhanced and biodiversity increased. Coexistence under stochastic, nonequilibrium conditions as described by patch-dynamics models at different scales can be just as strong and stable as that occurring under a deterministic, niche-differentiation model [209].

In a single patch, species extinction can occur as a result of competitive exclusion, overexploitation, and other destabilizing interspecific interactions or be due to environmental instability (e.g., unpredictable disturbances or changes in conditions).

De Angelis and Waterhouse [210] and Remmert [211] have emphasized that the macroscale integration of unstable patches that are out of phase with each other can lead to persistent species-rich communities.

An important parallelism seems to exist between the properties of a community and the properties of its component populations, which will be subject to a (relatively) high degree of K selection in stable environments, while r selection (selection in the direction of high numbers of offspring) predominates in variable environments. The K-selected populations with their high competitive capacity, high inherent survivorship, and low reproductive rates are normally resistant to disturbances but, once perturbed, have little possibility of recovery (i.e., low resilience). The r-selected populations, by contrast, have less resistance but distinctly higher resilience. This inverse relationship between resistance and resilience is of particular importance in the ecotoxicological context (Section 3.2).

The above stability considerations mean that food chains should be shorter in "fluctuating" environments, since only the most resilient food webs would be expected to persist and short food chains are more resilient. An analysis of Briand's findings [184], however, shows that further critical tests of these logical ideas are required because the average maximal food chain lengths for 40 communities from both unstable and "constant" environments did not differ significantly. In conclusion, most communities are probably organized by a temporally and spatially varying mixture of " forces" – namely, competition, predation, and disturbances, with competition and predation being presumably less important in more disturbed environments [2], [212], [213]. Consequently, no single stability or sensitivity measure exists for a community. Stability varies with the aspect of the community or ecosystem under study and the nature of the disturbance. Without pertinent semantic specifications relating to defined temporal and spatial scales (duration and dimensionality of observations), indicator variables and the nature

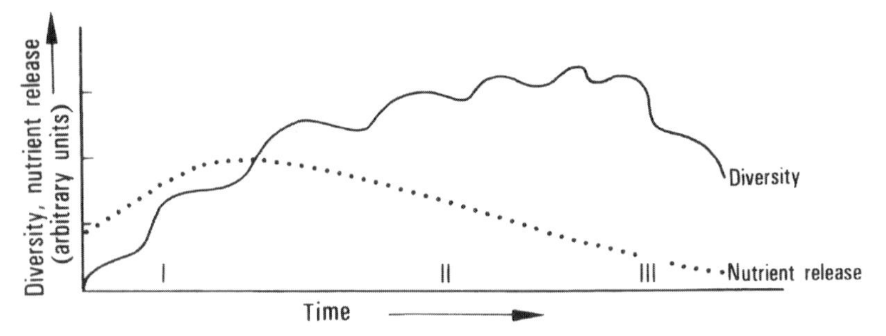

Figure 30. Generalized model of long-term ecosystem evolution in reaction to nutrient supply of soil

of perturbation statements about system stability would be meaningless in a general ecological and a more specific ecotoxicological context.

2.4. Exposure and Effect Criteria of Chemicals

Methodologies for estimating fluxes, transformations, and resultant environmental concentrations of chemicals depend on the knowledge of how rapidly the chemicals are discharged to a specific compartment in the environment and how rapidly they are transformed or removed by physical, chemical, and biological processes. Apart from the specific structure of the receiving compartment, particularly its physical and chemical characteristics, these processes are largely controlled by the inherent molecular properties of the chemicals released. These may be appropriately classified as sets of equilibrium and kinetic constants that define the exposure potential, and a complementary set of data that characterizes the effect potential, i.e., the chemical, biotic, and particularly, ecotoxicological effects of a substance.

In the following sections the essential exposure and effect criteria are described. Other chemical and physical properties such as flash points, flammability limits, and autoignition temperature are omitted because they are not of direct concern to the environmentalist. A comprehensive review of these and other dangerous properties of chemicals is given in [214]. With regard to the following quantitative specifications, however, note that chemicals are never 100 % pure and that the very nature and quantity of the impurities can have a significant impact on many environmental qualities. In particular, the following parameters are very sensitive to the presence of impurities: water solubility, odor characteristics and threshold values, biological oxygen demand, and toxicity [47].

2.4.1. Equilibrium Constants for Chemical Distribution

Density, Molar Mass, and Structure. The density of a substance, particularly in comparison to that of water or air, is of special environmental importance since it largely controls dispersion processes. Hydrocarbons are usually lighter than water. Figure 31 illustrates the relationships of molar mass and density.

Melting and Boiling Points. Normally, melting involves an increase in volume by about 10 %; exceptions are water, bismuth, and gallium. According to the sign of this change in volume, melting temperature rises or falls by 0.01 – 0.1 K/bar, which means that, under environmental conditions, pressure has little effect on melting point.

The melting and boiling points of the members of homologous series of hydrocarbons increase with increasing molar mass. When comparing organic compounds containing different halogens, however, the mass effect can be (partly) counterbalanced by polarization effects (e.g., in the case of CH_3F and CH_3Br). Substitution of another atom or group for hydrogen in an alkane results in an elevation of the boiling point. Thus, alkyl halides, alcohols, aldehydes, ketones, and acids boil at higher temperatures than ("elementary") hydrocarbons with the same carbon skeleton. The introduction of groups promoting association brings about a marked rise in boil-

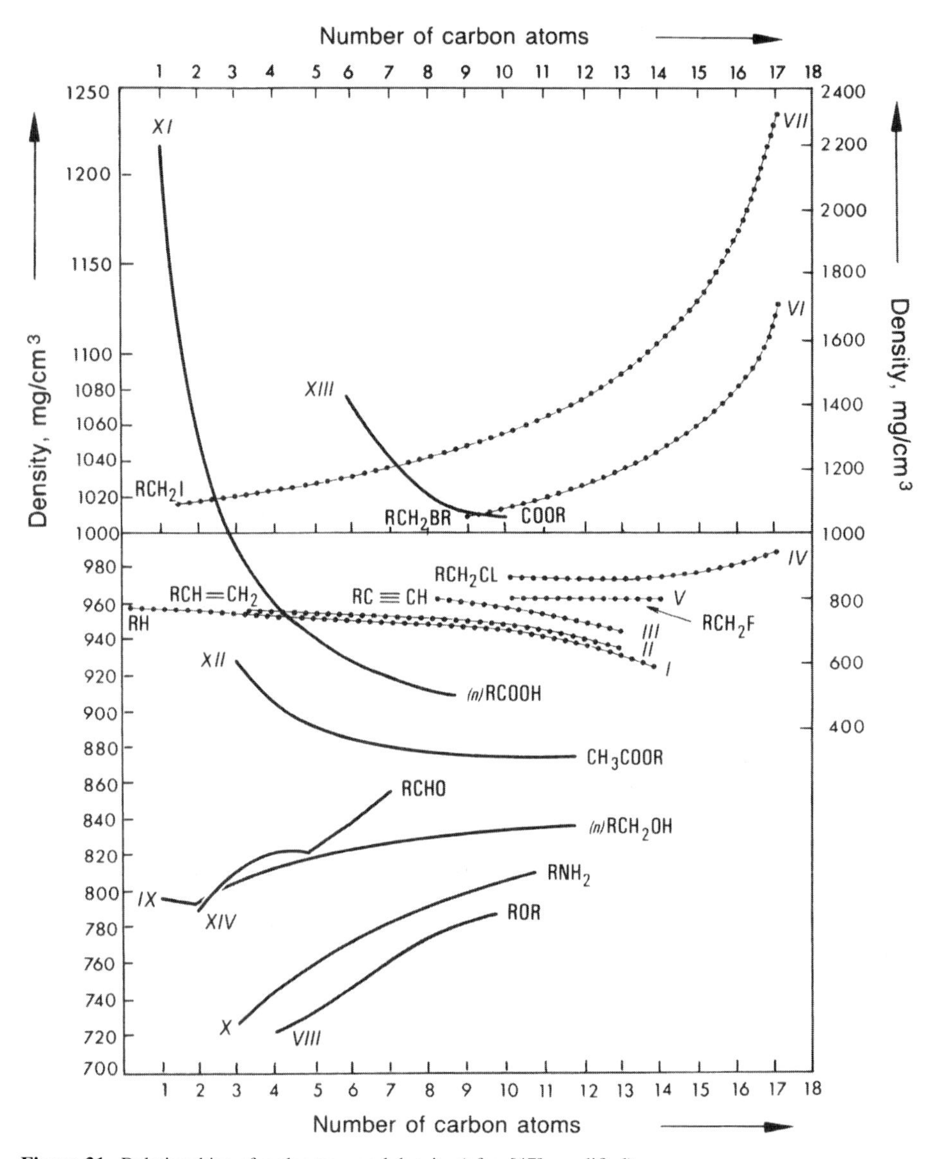

Figure 31. Relationships of molar mass and density (after [47], modified)
Dotted lines *I – VII* refer to right scale, lines *VIII – XIV* to left-hand scale

ing point, which is especially pronounced in alcohols and acids since hydrogen bonds can be formed [47].

Vapor Pressure, Vapor Density, and Volatility. The *vapor pressure* of a liquid or solid is the pressure exerted by the gas in equilibrium with the liquid or solid at a given temperature. The evaporative loss of a substance, volatilization, is a very important source of material for airborne transport and subsequent transformation and deposition processes. Vapor pressure data permit, in combination with solubility data, calculations of rates of evaporation of dissolved compounds from water by using Henry's law constant H, which is normally defined as

$$H = PS^{-1} \quad (32)$$

where P is the equilibrium vapor pressure of the pure chemical and S its saturation concentration or solubility at a fixed temperature.

Vapor density, compared to the density of air (=1), indicates whether a gas will be transported along the ground, possibly subjecting ecosystems to high exposure, or whether it will disperse rapidly. Vapor density is related to equilibrium vapor pressure through the fundamental state equation for a gas:

$$PV = nRT \tag{33}$$

If the mass of the substance and its molar mass are substituted for the number of moles n, this yields for the vapor density Vd [47]:

$$Vd = PM\,(RT)^{-1} \tag{34}$$

where P is the equilibrium vapor pressure (bar), $R = 8.31\ \mathrm{J\,mol^{-1}\,K^{-1}}$, M = molar mass (g/mol), and T = absolute temperature (K).

Volatilization of chemicals from water is an important transport process for a number of substances that have low water solubility and low polarity. Despite their very low vapor pressure, many compounds can volatilize rapidly because of their very high activity coefficients in solution, and by using the mathematical approach of MACKAY and PATERSON [215], [216] the rate constant for volatilization from water is given by

$$K_{\mathrm{vw}} = \frac{A}{V}\left(\frac{1}{K_{\mathrm{L}}} + \frac{RT}{HK_{\mathrm{G}}}\right)^{-1} \tag{35}$$

where A is the evaporating surface ($\mathrm{cm^2}$), V the liquid volume ($\mathrm{cm^3}$), H Henry's law constant (bar L/mol), K_{L} the liquid film mass-transfer coefficient (cm/h), K_{G} the gas film mass-transfer coefficient (cm/h), R the gas constant ($8.31\ \mathrm{J\,mol^{-1}\,K^{-1}}$), and T the temperature (K).

Volatilization processes from soil surfaces are complicated by variable contributions due to volatilization of the chemical from the water at the surface, evaporation of water itself, and the related capillary rise, which transports more water and dissolved chemical to the surface [217].

Water Solubility. Water solubility is an important parameter for the environmental assessment of all solid or liquid chemicals. In general, molecules with highly polar bonds will be relatively strongly attracted to one another; consequently, water and alcohol are potent solvents for ionic compounds such as salts, as well as polar nonionic molecules such as sugar. In contrast, nonpolar molecules such as tetrachloromethane, benzene, or gasoline are more weakly attracted to one another. They mix easily and are good solvents for nonpolar molecules, such as fats, grease, and paraffins, which are not soluble in water. Some solids are so firmly bonded that they are not significantly soluble in any common solvents except those with which they react chemically, whereby the nature of the molecule being dissolved is altered. Quartz is of this type: its solubility in pure water is of the order of 5 mg/L, but if the water contains a 0.01 M concentration of orthodiphenols, quartz is transformed into soluble esters and the solubility is increased to 40 mg/L [218].

The composition of natural waters varies considerably in space and time (see, e.g., [48]). Variables such as pH; water hardness; and concentration of cations, anions, and organic substances such as gelatin or fulvic and humic acids, all affect the solubility of chemicals in water.

Solubility of Mixtures. Mixtures of compounds behave differently from the individual compounds when brought in contact with water because each component partitions between the aqueous phase and the mixture. Compounds with a high water solubility tend to move into the aqueous phase, while the less soluble components have a greater tendency to remain in the mixture phase. Thus, the fractional composition of the water-soluble fraction differs from the original composition of the mixture, and the concentrations of the individual components of the water-soluble fraction are generally lower than their maximum solubilities [219].

***n*-Octanol – Water Partition Coefficient.** The *n*-octanol – water partition coefficient (K_{OW}) is defined as the ratio of the equilibrium concentration C of a dissolved substance in the two-phase system *n*-octanol and water (two solvents that are largely immiscible):

$$K_{\mathrm{OW}} = \frac{C_{\mathrm{octanol}}}{C_{\mathrm{water}}} \tag{36}$$

Usually the decadic logarithm of K_{OW} is given ($\log K_{\mathrm{OW}}$). Because K_{OW} is a dependent variable with regard to water solubility simultaneous use of both in structure – activity relationships or multivariate classification procedures will give spurious correlations. Experimental determination of both the aqueous solubility and the n-octanol – water partition coefficient led to the following regression [220]:

$$\log K_{\mathrm{OW}} = 5.00 - 0.670 \log S \quad (37)$$

where S is the solubility in water (µmol/L). When solubility is expressed in mg/L, Equation (37) becomes

$$\log K_{\mathrm{OW}} = 4.5 - 0.75 \log S \quad (38)$$

The regression covers many classes of both liquid and solid compounds with highly different polarities, from hydrocarbons and organic halides to aromatic acids, pesticides, and polychlorinated biphenyls (PCBs). The corresponding correlation coefficient is 0.97, which normally allows an estimation within one order of magnitude of the partition coefficient of a given chemical from its solubility in water. However, the scatter increases considerably for solubilities greater than 100 mg/L, as the example of L-tyrosine shows, where the experimental value of $\log K_{\mathrm{OW}}$ is -2.26 while calculations using Equation (37) yield values of +2.7 or +2.5.

Furthermore, whether the above equations would also apply to chemicals with high dissociation constants, such as salts and strong acids or bases, is questionable since their activities in solution cannot be approximated by their concentration. Basically, the same restriction holds for aliphatic acids or bases whose partition coefficients can vary drastically with changes in pH [47], [221]. Nevertheless, the n-octanol – water partition coefficient has proved useful as a simple means of predicting biological uptake, lipophilic storage, and biomagnification factors (see Sections 2.4.2, 3.3.5 and Fig. 57), whereas extrapolations of the multifactorial soil adsorption processes are much more difficult.

2.4.2. Kinetic Constants for Environmental Processes

Every chemical reaction takes place at a definite rate depending on the specific conditions under which it occurs. These boundary conditions are the concentrations or pressure of the reacting substances, temperature, presence or absence of a catalytic agent, and radiation. Chemical kinetics deals with the rates and mechanisms of these reactions whose understanding is rendered difficult by the fact that both time and intermediate products are involved, compared to the subject of chemical equilibrium, which is concerned only with the initial and final states of reactants. Because of the complexity of these subjects, the following presentation is limited to determination of the essential rate constants for environmental processes from structure alone by using structure – activity relationships (see also → Toxicology, Chap. 2.).

2.4.2.1. Hydrolysis

Hydrolysis of compounds usually results in the introduction of a hydroxyl function and is most commonly associated with the elimination of a leaving group X:

$$\mathrm{RX} + \mathrm{H_2O} \rightarrow \mathrm{ROH} + \mathrm{HX} \quad (39.1)$$

In water, the reaction is catalyzed mainly by H^+ and OH^- ions; whereas in moist soil, loosely complexed metal ions such as copper or calcium may also constitute important catalysts for the hydrolysis of several types of chemical structures. In addition, sorption (→ Soil) of the chemical may increase its reactivity toward hydronium or hydroxyl ions [222], [223].

Mabey and Mill [224] reviewed kinetic data for hydrolysis of a variety of organic compounds in aquatic systems in relation to the chemical characteristics of most freshwater systems. Quantitative structure – activity relationships, used as predictive test methods to develop the essential kinetic data, generally take the form of linear free-energy relationships such as the Hammett and Taft

$$\log K_x / K_o = \varrho \sigma \quad (39.2)$$

or Brønsted

$$\log K_x / K_o = m \mathrm{p}K_a + C \quad (39.3)$$

approaches, where ϱ is a constant for a given reaction under a given set of conditions, and σ is the Hammett constant, characteristic of a certain substituent; K_o and K_x are the rate constants for the unsubstituted and substituted structures, respectively, while the dissociation constant pK_a refers specifically to the acidity of the leaving group.

Highly variable hydrolysis reactions have been observed in soil and sediment. This appears to be the result of unusual pH relationships at the surface of soil constituents, possible occurrence of metal-ion-catalyzed processes, and general acid- and base-catalyzed reactions, possibly involving phenolic, amine, or sulfide groups in soils [223]. In some cases (e.g., in the presence of clay minerals), hydrolysis rates of organic chemicals are accelerated considerably compared to bulk solution, while adsorption to humic components frequently impedes hydrolysis. A detailed understanding of the mechanisms involved is limited, however, and structure – activity relationships appear to be available for only a few compounds.

2.4.2.2. Photolysis

Direct Photolysis. Natural direct photolysis results from absorption of photons from the solar spectrum above 290 nm by the chemical. The specific absorption rate ($K_{a,\lambda}$) is a measure of the overlap between the solar spectrum and the electronic absorption spectrum of the compound (i.e., an inherent property of a photoreactive chemical) [225]:

$$K_{a,\lambda} = 2.303 E_0 (\lambda) \varepsilon_\lambda \quad (40)$$

where $E_0 (\lambda)$ is the scalar irradiance and ε_λ the molar absorptivity at wavelength λ.

Since pollutant concentrations are low under most environmental conditions, photolysis can be described by a first-order expression; i.e., the rate constant of direct photolysis k_d is directly proportional to pollutant concentration. From Equation (40), then

$$k_d = 2.303 \Phi_d \int E_0 (\lambda) \varepsilon d\lambda \quad (41)$$

where integration is over the range of $Eo(\lambda)$, which is absorbed by the compound, and Φd denotes the quantum yield, which may attain a maximum value of 1. However, rapid direct photolysis can occur even with compounds that react with very low quantum efficiencies [225].

Since more than one primary reaction may be involved in a photochemical reaction, the primary quantum yield may be defined with respect to a specific reaction. Moreover, the sum of the quantum yields of all primary reactions, including deactivation, must be unity. The overall quantum yield, which is the quantity most often determined, on the other hand, does not distinguish between the effects of primary and secondary reactions and it may exceed unity.

Modeling photochemical reactions in natural waters is still in its infancy. No proven general model applicable to every kind of natural water is presently available for estimating the rates of consumption or production of chemical species through light-induced reaction pathways. A lot of fundamental and essential information for constructing a precise and realistic model is still not available. For example, the chemical composition of natural waters and their temporal and spatial variabilities are still inadequately known. Thus, the reactants that may participate directly and indirectly in photochemical reactions in a competitive, synergistic, or catalytic way cannot yet be fully identified. The rate at which a particular photochemical reaction may proceed in this poorly defined medium, and to which extend it will proceed cannot be quantified extensively by empirical data. Besides, as stated previously, even in a well-defined medium, the detailed reaction pathways of photochemical reactions can still rarely be identified definitively.

Indirect photolysis may occur if a natural material can absorb solar photons and transfer part of the energy to a chemical. Thus, compounds that do not absorb light directly can still undergo mediated photolysis, and chemicals that photolyze directly may photolyze much more rapidly in natural waters containing organic substances in dissolved or particulate form. Despite these difficulties, general rate expressions for sensitized photolysis can be defined. The rate of light absorption is

$$I^s_{a,\lambda} = 2.3 \varepsilon_{\lambda,s} E_o (\lambda) [S] \quad (42)$$

where $\varepsilon_{\lambda,s}$ is the molar absorptivity of the sensitizer and [S] is its concentration [225]. Thus, from Equation (42), the rate constants for indirect photolysis should increase with increasing [S] if $E_o (\lambda)$ is held constant. When the concentration of pollutant P is very dilute, the quantum yield is directly proportional to this concentration. By denoting C_λ as the proportionality constant, the total rate expression becomes

$$(Rate)_\lambda = 2.3E_o(\lambda)\,\varepsilon_{\lambda,s}C_\lambda\,[S]\,[P] \quad (43)$$

Provided the sensitizer concentration is constant, which is usually the case in nature, the rate expression follows first-order kinetics (→ Air, Chap. 1.).

2.4.2.3. Oxidation and Reduction

Direct and indirect phototransformation as described in the preceding section are possible removal pathways only for those chemicals that, either by their very nature or due to fixation onto light-absorbing carrier substances, can absorb radiation in the range of the solar spectrum. However, the main transformations that lead to the removal of chemicals from the atmosphere and water involve reactions with photochemically generated *oxidants* such as the hydroxy radical ($OH\cdot$), ozone (O_3), hydroperoxy radical ($HO\cdot_2$), organic peroxy radicals ($RO\cdot_2$), singlet oxygen ($O_2\,\Delta_g$), and nitrate radical ($NO\cdot_3$). Table 9 summarizes the reactivity of different classes of organic compounds toward some of the above oxidants.

Table 9. Rates of oxidation in air and water (after [222])

Type of oxidant	Class or structure	Half-life of reactant
$HO\cdot$ radical (in air)	*n*-, iso-, cycloalkanes	1 – 9 d
	alkenes	0.05 – 1 d
	halomethanes	0.2 – 100 a
	alcohols, ethers	1 – 3 d
	ketones	0.2 – 6 d
	aromatics	0.1 – 3 d
$RO\cdot_2$ radical (in water)	alkanes, alkenes	220 – 2000 a
	alkyl derivatives	220 – 2000 a
	phenols, arylamines	1 d
	hydroperoxides	150 min
	polyaromatics	10 d
$O_2\,\Delta_g$ (in water)	aliphatic compounds	100 a
	substituted or cyclic alkenes	8 – 40 d
	alkyl sulfides	1 d
	dienes	19 h
	enamines	15 min
	furans	1 h

The *reduction* of organic compounds requires an environment of low redox potential, i.e., eutrophic or deeper levels of aqueous systems and a soil depth at which oxygen is limited [226]. Reduction of organics can follow various mechanisms such as hydrogenation, halogenation, or coupled oxidation – reduction, many of which require catalysts. Like with many other chemical reductions, rates of reductive alteration of organics are temperature and pH sensitive [2]. Since the most common mechanism for reduction of organic molecules in the environment is via microbial attack, separating biologically induced from straight chemical reduction in situ is difficult. From the viewpoint of prospective hazard assessment, however, identifying those compounds that are potentially susceptible to reduction and the transformation products that might be deposited in the environment as a result is more important.

2.4.2.4. Sorption and Ion Exchange

Further processes of fundamental importance for the distribution of environmental chemicals in water and soil are sorption and ion exchange. Adsorption is the adhesion of ions or molecules to surfaces or to solid – liquid, liquid – gas, and liquid – liquid interfaces, causing an increase in the concentration of chemicals on the surface or interface over the concentration in solution. Adsorption occurs as a result of a variety of processes with a multitude of mechanisms when the concentration of chemicals lowers the free interfacial energy. The higher the free surface energy at an interface, the greater is the number of substances capable of lowering it.

Physical Adsorption. Molecules of a given material may be adsorbed from the gas or liquid phase to a solid by van der Waals forces. This process resembles a condensation that reaches equilibrium in a time limited chiefly by molecular diffusion rates. The weakness of van der Waals forces is reflected in the low adsorption enthalpy of physiosorptive (or unspecific) bonding; it amounts to hardly more than 40 kJ/mol [227]. Reversibility is a characteristic of physical adsorption that results from rapid equilibration; i.e., at a given temperature and pressure the same amount of material is adsorbed regardless of the direction from which equilibrium pressure and temperature have been approached. From the environmental point of view, however, the fact that physical adsorption of gases by highly porous media such as silica gel, charcoal, and soil often exhibits "hysteresis," a form of irreversibility due to condensation of liquid in pores is very important. Analogous phenomena occur with chemicals bonded at solid – liquid interfaces.

Chemical Adsorption. Molecules of a given substance may also be adsorbed from the liquid or gas phase by a solid through the activation of valence forces. Consequently, they may be considered as reacting with atoms or molecules on the surface of the solid in a way analogous to any other chemical reaction, and the products of this reaction are usually conveniently considered to be species quite different from the original reacting molecules of the atoms (chemisorption or specific bonding in terms of soil chemistry). Temperatures at which chemisorption occurs are related not to the boiling point of the adsorbate (material being adsorbed), as is the case with physical adsorption, but to the stability of the surface compound and its rate of formation. Therefore, rates of chemisorption vary enormously with material and temperature, as do those of ordinary chemical reactions, and may be very much slower than rates corresponding to diffusion of gases or solutes to the surface. Furthermore, the rates are likely to vary strongly with the amount of material adsorbed. Because of these rate characteristics, the amount of adsorbate at a given pressure or concentration and temperature – when observed over practical time periods – may depend strongly on the way these boundary conditions have been reached. Therefore, irreversibility is extremely common in chemisorption, which is characterized by enthalpy values >100 kJ/mol [227].

Ion Exchange. The presence and extent of ionization also have a great influence on the chemical behavior of a substance, particularly on solubility, sorption, toxicity, and other biological characteristics. Ion exchange as an important sorption mechanism for substances prone to ionization is viewed as an exchange with some other ion that initially occupies an adsorption site on the solid [226]. For example,

$$H^+ \overset{Ca^{2+}\,Mg^{2+}}{\underset{K^+\,Na^+}{\boxed{}}} Al^{3+} + 10\,NH_4^+ \rightleftharpoons NH_4^+ \overset{4\,NH_4^+}{\underset{2\,NH_4^+}{\boxed{}}} 3\,NH_4^+ + H^+ + K^+ + Na^+ + Ca^{2+} + Mg^{2+} + Al^{3+} \quad (44)$$

The ion-exchange property of a soil results mostly from the clay and silt fractions, organic matter, and hydrous oxides of iron. Soil particles have an amphoteric character but normally carry a net negative charge. Positive charges, as indicated by the amphoteric nature of the clay fraction, may originate from hydrous oxides of iron (ferrihydrites), aluminum, and manganese, and from exposed octahedral groups that react as bases by attracting protons from the surrounding soil solution. The negative charge increases and the positive charge decreases with rising pH as a result of increasing ionization of the acid groups and decreasing protonation of the basic groups. For decreasing pH, the corresponding change occurs in the opposite direction. The carboxyl groups of humus ionize under acid conditions, the phenolic hydroxyl groups mainly above pH 6. Because of the highly irregular shape of many clay minerals (and clay – humus complexes, in particular) and the nonuniform distribution of charges in the particles, the surface charge density is quite variable.

To describe ion-exchange processes, several formulas have been proposed in the course of exchange studies. The first group is essentially empirical and intended mainly to provide mathematical formulations best fitting the experimental data. Among this group, the Freundlich [49], Langmuir, and BET (i.e., Brunauer – Emmett – Teller) are particularly important [228]. The second group of formulas is related to the law of mass action.

Empirical Adsorption Formulas. In the *Freundlich equation*, the concentration of a substance in solution is related to that adsorbed as follows [49]:

$$\frac{x}{m} = kc^{\frac{1}{n}} \quad (45)$$

where x is the amount of substance adsorbed; m the mass of the adsorbent; c the concentration of substance remaining in solution; and k, n are specific constants.

The Freundlich formula is of a parabolic type and therefore cannot give a maximum adsorption value as does the *Langmuir theory*. According to the latter, adsorption is considered proportional to the pressure of gas (or concentration of a solute) and to the amount of surface not yet covered with adsorbed molecules. The rate of desorption (or evaporation) of adsorbed molecules is considered proportional to the number of such molecules. At equilibrium, these two rates must be equal; thus,

$$\frac{n}{n_{\mathrm{m}}}=\frac{bp}{1+bp} \tag{46}$$

This equation, in which n is the number of molecules adsorbed, n_{m} the number of molecules required for a completely unimolecular surface film, p the concentration of a chemical in solution, and b a constant at a given temperature, is the *Langmuir adsorption equation* (*Langmuir isotherm*).

A third theory of adsorption widely used for representing the adsorption of gases or vapors on solids was advanced by Brunauer, Emmet, and Teller [228]. In contrast to Langmuir's theory, the *BET theory* makes allowance for adsorption of multimolecular layers as is commonly observed when nonporous solids adsorb vapors at pressures approaching saturated vapor pressure. The BET equation is therefore

$$\frac{n}{n_{\mathrm{m}}}=\frac{b\frac{p}{p_0}}{\left(1-\frac{p}{p_0}\right)\left[1+(b-1)\frac{p}{p_0}\right]} \tag{47}$$

where p_0 denotes the saturation vapor pressure, while the other quantities have the same meaning as in Equation (46). The equation is designed for physical adsorption and predicts a very sharp increase in the amount of material adsorbed as the saturated vapor pressure is approached, and this is found to be the case [50].

2.4.2.5. Biotransformation

Biotransformation is a more accurate and general term than biodegradation, because many chemicals are transformed to products of comparable molecular complexity instead of being degraded. Chemically speaking, biotransformation in water and soil systems includes hydrolysis, oxidative cleavage, and reduction, even in aerobic systems. Kinetics of biotransformation are normally based on the Monod model [51], which couples the rate of loss of the chemical to the growth of organisms involved. In the natural environment, however, growth is usually controlled by nutrients present in comparatively constant amounts, and the kinetic expression for biotransformation processes simplifies to a second- or pseudo-first-order reaction:

$$d\,[C]\,/dt=k_{\mathrm{bt}}\,[\mathrm{C}]\,[\mathrm{B}]=k'_{\mathrm{bt}}\,[\mathrm{C}] \tag{48}$$

where d [C]/dt is the rate of transformation of C; [B] is the number of organisms per unit volume; k_{bt} the second-order rate constant for biotransformation; and k'_{bt} the pseudo-first-order rate constant. The half-life of the chemical under transformation ($t_{1/2}$ at a given cell concentration) can then be calculated as

$$t_{1/2}=\frac{\ln 2}{k_{\mathrm{bt}}\,[\mathrm{B}]} \tag{49}$$

In deriving this equation, the microbial community in water, sludge, or soil is assumed to already be acclimated to the chemical, so no lag time is involved in production of the necessary level of biotransforming organisms or appropriate mutants or enzymes.

Because of the remarkable versatility of microorganisms in their catabolic activity, many substances are subject to microbial mineralization or cometabolism in water and soil. If, however, microorganisms have no enzymes active in a necessary catabolic sequence or if the substrate is protected from microbial attack because of special properties of the environment, the compound tends to persist. In some instances, persistence may be attributed to absence of the requisite enzymes or to the lack of permeability of microorganisms to the substrate [229]. Other mechanisms of recalcitrance may be inherent in the molecule itself; in some cases, biotransformable compounds are degraded quickly in one environment but are partially or totally refractory in another.

2.4.2.6. Chemical Structure and Biodegradation

Biodegradation is the most important transformation mechanism for organic compounds in nature. It can occur under conditions in which oxygen is present (aerobic) or absent (anaerobic). In aerobic processes, organic carbon is oxidized to CO_2, while in anaerobic processes it may ultimately be reduced to CH_4. In either case, the more important environmental variables affecting the rate and extent of biodegradation are temperature, pH, salinity, oxygen concentration, concentration of the chemical involved, ionic strength, composition of microbial associations and concentration of viable microorganisms, quantity and quality of nutrients (other than xenobiotic substances), trace metals, and vitamins.

In general, microbial degradation is impeded by ether bonds and by triple-substituted or condensed phenyl groups–particularly, if the latter have methoxyl, sulfone, nitro, or chlorine substituents in the *meta* position. Among the nitrophenols and chloroanilines, the *ortho*-isomers are relatively recalcitrant. Multiple substitution increases the recalcitrance of aromatic molecules considerably because their aerobic degradation is effected by relatively specific mono- and dioxygenases [229], [230]. Halogen substituents delay the electrophilic attack of these oxygenases considerably by reducing enzyme affinity. Therefore, multiple chlorine-substitued aromatics form relatively persistent compounds, the more so because the delayed cometabolic attack yields "dead-end" metabolites, which cannot be used for cell-building purposes or as an energy source.

Anaerobic Processes. As a consequence, reliable conclusions about biodegradation are not generally possible on the basis of chemical structure alone [47]; therefore, many laboratory methods have been developed for studying the aerobic biodegradative potentials of chemicals [231–234]. In comparison, tests for assessing anaerobic biodegradation have received less attention [84]. Nevertheless, anaerobic degradation is an important process for substances that are not aerobically biodegradable or whose physicochemical properties are such that their occurrence in an aerobic environment is restricted (e.g., chemicals that are strongly adsorbed or insoluble). For these substances, anaerobic biodegradation can be the major process responsible for breakdown in the environment.

Anaerobic biodegradation is complex and usually considered to occur in at least three consecutive main stages, the first of which involves hydrolysis of complex organic molecules such as carbohydrates, proteins, and lipids by the action of extracellular enzymes. In the subsequent stage (acidogenic step), the hydrolysis products are fermented to yield mainly short-chain fatty acids, alcohols, hydrogen, and carbon dioxide. Alcohols and acids are then converted in the acetogenic step. The last degradation step is effected by methanogenic bacteria (methanogens), which utilize acetate and hydrogen to form methane.

Anaerobic processes tend to be self-inhibitory because the hydrolysis of complex substrates such as fats, proteins, and carbohydrates yields volatile organic acids. By lowering the pH of the system, these may inhibit the growth of methanogens unless the system has sufficient buffer capacity [84]. Inhibition of gas production as a result of the toxicity of chemicals is also possible. Finally, some indications have been found that certain compounds (e.g., nitrilotriacetic acid) are degraded anaerobically only when aerobic organisms acclimatized to the compound are incorporated in the system [235].

When the organic compound liable to degradation or enzymatically induced metabolic transformations is utilized as a carbon source, the growth rate of the responsible microorganism depends on the concentration of the former. The rate of substrate utilization then becomes [52]

$$-dC/\mathrm{d}t = \mu X/Y = (\mu_\mathrm{m}/Y)\cdot CX/(K_\mathrm{s}+C) = k_\mathrm{b}\cdot CX/(K_\mathrm{s}+C) \quad (50)$$

where μ is the specific growth rate; X the biomass per unit volume; μ_m the maximum specific growth rate; K_s the concentration of the substrate to support half-maximum specific growth rate ($0.5\,\mu_\mathrm{m}$); $k_\mathrm{b} = \mu_\mathrm{m}/Y$, the biodegradation constant; and Y is the biomass produced from a unit amount of substrate consumed. The constants μ_m and K_s depend on temperature, pH, salinity, oxygen concentration, concentration of chemical, composition of microbial associations, and concentration of viable microorganisms, quantity and quality of nutrients (other than xenobiotic substances), trace metals, and vitamins.

2.4.2.7. Bioconcentration, Bioaccumulation, and Ecological Magnification

Bioconcentration is the concentration of a chemical in an organism due to direct uptake from the environment (e.g., ambient air or water), which does not include contaminated food. *Bioaccumulation* includes the latter pathway of concentration, while *biomagnification* considers only concentration processes via food uptake. *Ecological magnification* defines the increase in concentration of a substance in an ecosystem or a food web when passing from a lower trophic level to a higher one [53].

Bioconcentration or bioaccumulation is the result of both kinetic (diffusional transport and biodegradation) and equilibrium (partitioning) processes. It thus involves the following set of fundamental events [47]:

1) Partitioning of the xenobiotic between the environment and some surface of the organism
2) Diffusional transport of the molecules across cell membranes
3) Transport processes mediated by body fluids, such as exchange between blood vessels and serum lipoproteins
4) Concentration of the xenobiotic in various tissues depending on its affinity for specific biomolecules such as nerve lipids
5) Metabolism, cometabolism, or biodegradation of the xenobiotic

The relative importance of the above processes is likely to differ considerably as a result of both the physicochemical properties of the xenobiotic and the morphological and physiological structure of the target organisms. Therefore, an essential ecotoxicological requirement is that the determination of bioconcentration factors (*BCF*s) be complemented by an assessment of biomagnification values. Under real-world conditions, these would provide deeper insight into the complex bioaccumulation processes of xenobiotics as a result of transport and partitioning into higher levels of the food web, which may ultimately result in toxic concentrations [235].

On the basis of bioconcentration studies in different test organisms, correlations with both water solubility and *n*-octanol – water partition coefficients have been established. In some cases, the results proved excellent; in others, the correlations yielded only satisfactory structure – activity relationships (i.e., estimates of *BCF*s to within an order of magnitude for chemicals of log K_{ow} ranging from 1 to nearly 7).

3. Ecotoxicology

The production, use, and disposal of chemicals lead to their dispersion in the environment. The way in which substances are released depends on their physical and chemical properties (see Section 2.4.1), the production process (→ Plant and Process Safety), including extraction of natural resources and manufacturing; and the use pattern and means of disposal.

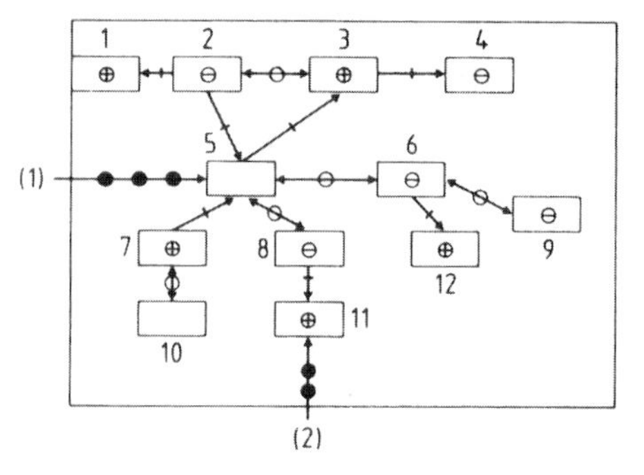

Figure 32. Component-specific influence of two xenobiotics on an ecosystem (after [159], modified)
Substance (1) affects a functionally central component, thus causing breakdown of the whole ecosystem. Substance (2) affects a peripheral component; even its deletion leaves the ecosystem practically unchanged.

Compartments of ecosystems have a limited capacity to assimilate chemicals by breakdown or dilution; if this capacity is exceeded, damage may ensue. Thus, the hazard posed by a single substance or a mixture of chemicals is a function of its inherent toxicity to organisms and the concentration attained. At the community and ecosystem levels, the sensitivity has to be considered specifically (Section 2.3), i.e., the functional position of the organisms affected with regard to the entire system. Figure 32 illustrates schematically the different influence of xenobiotics, affecting two elements of an ecosystem. In case (2), only a peripheral element (in the functional sense of the term) is damaged or deleted, which leaves the system relatively intact. In case (1), however, a component of central importance is eliminated, which leads to breakdown of the entire system.

Thus, the assessment of whether a substance presents a hazard to organisms in the environment is based on a comparison of its predicted environmental concentration (PEC) with the predicted no-effect concentration (PNEC) for organisms in the ecosystem affected. Whenever the PEC : PNEC is <1.0, a substance may be concluded not to present an environmental hazard under the boundary conditions considered. In the opposite case, hazard assessment should be complemented by risk assessment approaches, which define the probability that a hazard will occur. Figure 33 illustrates the temporal and spatial scales that must be considered when analyzing the impact of environmental chemicals on biota or ecosystems.

PEC values can be measured in the field. The identification of unequivocal spatial structures related to emission, transport, and sedimentation normally involves the construction of isopleth maps by means of metric interpolation. A pertinent statistical method that permits testing the validity of interpolation procedures is variogram analysis, normally applied in conjunction with kriging methods (i.e. methods in geostatistics for valid derivation of isopleths) [236], [237] (Section 3.1.4). Alternatively, the predicted environmental concentration can be estimated, if the manner and quantity in which a chemical enters the environment, and its subsequent distribution and transformation, are known (see Section 2.4.2; [2], [238]). This normally involves the application of models that are available at different levels of complexity, ranging from empirical models for practical purposes to rather abstract ones aiming at generalized insights via simulation.

Effects assessment establishes a maximum concentration of a substance that does not cause adverse effects. The PNEC values may be derived from whatever ecotoxicological data are available, by using simple mono- and oligospecific test systems, or micro- and mesocosms, and ideally from toxicity measurements at the ecosystem level. The lower the level of assessment approach, the more appropriate are assessment factors to compensate for restrictions in the data.

3.1. Exposure Assessment

Exposure assessment involves the identification of point or diffuse emission rates and the subsequent distribution tendencies, based on predictive models. In the first assessment step, the initial concentration ($PEC_{initial}$) in each compartment of concern must be estimated. If necessary, a revision in light of relevant new information may lead to a revised $PEC_{initial}$ and a PEC_{local} or a $PEC_{regional}$. The PEC_{local} is the estimated concentration of a substance when released to a particular environmental compartment, with further fate processes, (e.g., volatilization, sorption, biodegradation) taken into account. The $PEC_{regional}$ denotes the predicted concentrations of a substance in all relevant compartments, when the further distribution and fate of the chemical are taken into account.

3.1.1. Release Estimation

For realistic exposure assessments, basically the entire life cycle of a chemical should be analyzed and PEC values calculated for each relevant process. To standardize the release estimation for screening purposes [239], the EEC and OECD have defined three different categories of use pattern: main, industrial, and use. The main category classifies substances in four groups (see Table 10).

Table 10. Release fractions in the screening phase of hazard assessment of chemicals related to the main category (after [239])

Main category	Release fraction of production volume	Examples
Closed system	0.01	chemical intermediates
Enclosed in a matrix	0.1	plastic additives
Nondispersive	0.2	photochemicals
Wide dispersive use	1.0	solvents, plant protection products, detergents

In addition to the main categories, there are 15 different "industrial" categories and 55 different "use" categories [240]. This classification scheme allows a stepwise approach in release estimation. In the screening phase a rough estimate based on the main category may be sufficient, whereas in the confirmatory phase more

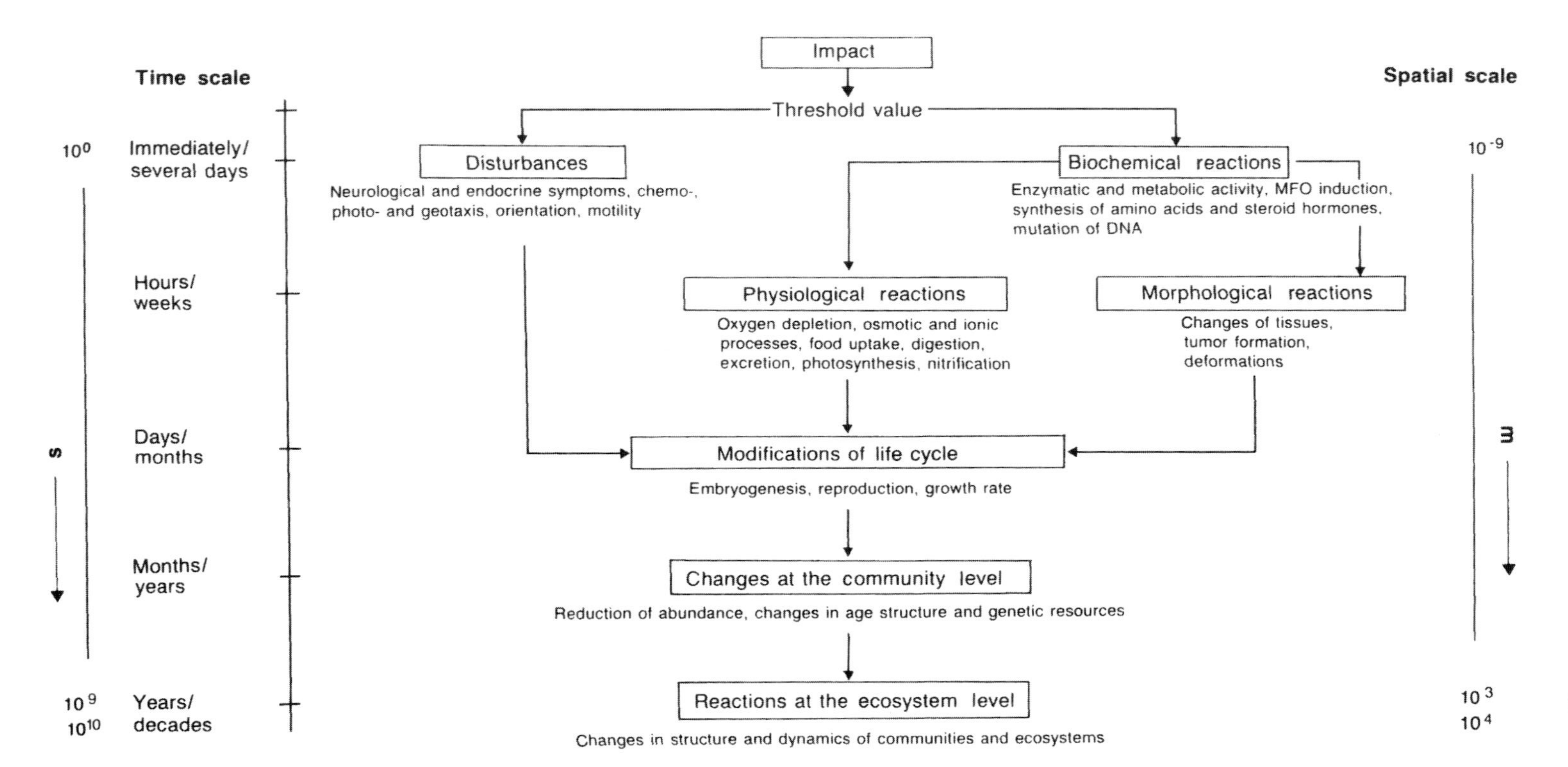

Figure 33. Temporal and spatial scales of the impact of chemicals on biota (after [53], modified)
MFO = monooxygenases

refined release data (e.g., based on other categories) are needed. In the investigation phase, finally, detailed information on specific releases of the substance is required (→ Chemical Products: Safety Regulations, Chap. 3.).

3.1.2. Elimination

A reduction in the concentration of chemicals in gaseous or aqueous discharges prior to their release to the environment is termed elimination. Elimination may be due to physical, chemical, or biochemical processes. The most significant elimination mechanisms are biochemical processes, especially aerobic biodegradation by bacteria or an alternation of aerobic and anaerobic phases (see Section 2.4.2) [2], [35]. Physical processes transfer a substance from one phase to another. In the case of volatile substances, the aeration process normally enhances their removal by stripping them from the solid or liquid phases to the atmosphere [47]. Substances from exhaust gas streams may be eliminated by scrubbing, e.g., by adsorption on a suitable material or passage through a suitable aqueous solution.

3.1.3. Dispersion

Pollution transport and dispersion are directly related to wind speed or flow velocity of rivers, etc. The higher the velocity, the greater is the volume of air passing a point source per unit time, and the lower is the resultant concentration per unit volume. Higher wind speed also means greater turbulence, which involves fluctuations of the local wind vector on various scales, and these eddies act to diffuse a plume. In addition to this "stretching" phenomenon [241], wind speed is important for the transport distance. In strong winds, pollutants may be transported long distances, but the resultant dilution over 10 – 50 km is usually such that most individual plumes have lost their identity and contribute to a more general (background) contamination of the lower atmosphere. Consequently, weaker regional wind systems may exhibit much greater potential for pollution since both horizontal transport and eddy diffusion are reduced.

This simple fact becomes all the more important because, under these conditions, local wind systems tend to develop that are difficult to predict with any accuracy [54]. In so far as they are closed smaller-scale circulations (e.g., land and sea breezes, mountain and valley winds, city winds), they often produce recirculation of pollutants coupled with a progressive increase in pollutant loading with time. The city thermal wind system operating mainly in the urban canopy layer is particularly dangerous; it is produced by microscale processes in the streets [242–244]. Here, flow converges on the city (or its variable thermal subcenters) from all directions, rises, diverges, and then moves outward to subside on the urban – rural fringe, where it rejoins the inflow.

The worst conditions for dispersion, however, occur in the case of temperature inversion and consequent formation of a stable boundary layer. With the warmer air layer acting as a lid on the mass of cooler air below, pollutants cannot escape from the cooler zone and thus increase in concentration in the confined space. This situation can be brought about by cooling (usually radiative) from below, by warming (usually adiabatic) from above, or by advection of warmer or cooler air.

Since the 1970s, modeling of dispersion processes in air has become more and more sophisticated owing to the progressive consideration of pertinent boundary layer effects and concentration or removal processes ([245–248], e.g.). This involves not only a considerable increase in computer time but also, and often more prohibitive, much more detailed input data. In the light of these circumstances, a question of practical importance is whether simpler models are not likely to yield comparable results on the condition that their data base can be made proportionally more comprehensive.

Box models, including dry and wet deposition and chemical transformations, are likely to provide calculated concentrations that are in reasonable or good agreement with measured values on a local scale as well as with transport data over a distance of 1000 km. An exemplary prototype whose fundamentals have often been used was developed by [249], describing SO_2 and NO_x concentrations in the industrial areas of North Rhine – Westphalia along the Rhine and Ruhr rivers.

Statistical models define immission concentrations by means of multiple regression relating immissions to emissions, meteorological parameters, topography, or other dispersion-controlling factors. Models of this type vary considerably with respect to both regression analytical approach and selection of data.

Observations of looping, coning, or (to a lesser degree) fanning plumes on an instantaneous basis reveal that the concentration is very peaked, with a marked maximum in the plume center. Averaged over growing time intervals, an increasingly smooth and wide plume envelope appears, which contains all of the short-period plume fluctuations. The wider spread results in a flatter concentration curve, with a correspondingly lower peak centered on the time mean axis of the plume (i.e., the mean wind direction). Analogous bell-shaped or Gaussian concentration profiles also characterize the vertical plane through the plume, although there may be a difference between the width of the horizontal and vertical profiles indicating that turbulence is greater in one or another direction.

Eulerian or Grid Model. The Eulerian or grid model is a mathematical representation of the pertinent transport processes and chemical reactions occurring in a grid of cells yielding the spatial and temporal variations of the ground cell concentrations of primary and secondary pollutants. It is characterized by fixed coordinate systems with respect to the ground and consists of a set of nonlinear coupled partial differential equations expressing the conservation of mass of each pollutant considered. Inputs to the model are numerical representations of the meteorological and emission conditions as a function of time of day, initial concentrations, and boundary conditions of inflow.

Lagrangian Model. In the trajectory or Lagrangian model, the time behavior of the concentration c of a pollutant is described by a differential equation of the type

$$\frac{\partial c}{\partial t} = \frac{\partial}{\partial z}\left(\underline{K} \cdot \frac{\partial c}{\partial z}\right) + S \qquad (51)$$

where c depends only on height above ground z and the time of travel t of an air parcel, while S refers to the net source – sink term representing the combined effects of emissions and chemical processes during transport, deposition, and flow across the upper boundary. $\underline{K}$ represents a vertical eddy diffusion coefficient whose values characteristically depend on time of day, height, and location. In comparison with the above Eulerian models, trajectory models have commendable mathematical advantages resulting from their moving coordinate systems with respect to a fictitious vertical air column moving horizontally with the advecting wind [250]. Considered as the present state of the art in operational long-range transport modeling, the Lagrangian approach has been used in numerous studies in the United States and Canada relating to transboundary air pollution.

3.1.4. Local and Regional Concentration Estimates

After emission, the environmental fate of a substance is governed by both transport and transformation processes that may be assessed by measurements or modeling approaches of which there are two basic types. The *local* or *monocompartment approach* is focused on the distribution of a substance within a particular environmental compartment, which leads to a PEC_{local}. The *regional* or *multicompartment approach* is directed toward the overall distribution, where transport occurs in all relevant environmental compartments (i.e., air, water, or some other medium) leading to $PEC_{regional}$ values.

The first approach is appropriate if the degradation rate is (distinctly) greater than the intercompartmental transfer rate; otherwise a multicompartment or multimedia approach is indicated (Section 3.3) [251]. Irrespective of the type of model to be used, the environmental parameters such as compartment volumes or compartment ratios must be defined; they are sensitive points for the final result. For substances that enter the environment from diffuse emission sources, a realistic PEC estimate cannot be calculated by means of the habitual point-source models. In such cases, the exposure assessment must be based on scale considerations and spatial modeling [252].

3.1.4.1. PEC Estimates in the Air Compartment

For the primary exposure assessment, models of advection and related dispersion processes may be used that compute concentrations of a substance at different receptor compartments. These require different minimum data sets consisting of discharge emission rates, effective height of source, and meteorological conditions, (stability classes of atmosphere, wind speed etc.). For regulatory purposes, the TA Luft model [253] is widely used (see also Section 3.1.3).

For subsequent revisions of the exposure assessment, information on the long-term fate of the chemical in the atmosphere is required. This involves the appropriate consideration of direct and indirect photodegradation and other chemical reactions [245], [254]. The reactivity of the hydroxy radical formed predominantly via photolysis of ozone by reaction of electronically excited oxygen atoms with water vapor is of fundamental importance in this connection.

A balance must be struck between simple models that do not represent the real-world situation and models that are too difficult to use. Thus, level 1 (equilibrium, conservative, steady state) and level 3 models (nonequilibrium, nonconservative, steady state) in the sense of the MACKAY et al. classification [251] are considered useful simple approaches. Level 2 models (equilibrium, nonconservative, and steady state) usually are of minor interest, whereas level 4 models (nonequilibrium, nonconservative, nonsteady state) should be used only for estimating disappearance of chemicals from the environment [238].

An important drawback of these "unit world" models is that they compute only one concentration value for each compartment, whereas measurements normally indicate that actual concentrations in the environment may range over several orders of magnitude. Thus, they are suited only to provide an indication of concentrations in places far from the source (i.e., background levels). Improvements are possible by performing a sensitivity analysis, by using statistical approaches, or by adapting the unit world to the region of interest [255], [256]. An alternative method is the application of dispersion models, provided that appropriate input data, representative dispersion parameters, and sufficient meteorological information are available (see Sections 3.1.3, 3.3).

3.1.4.2. PEC Estimates in Soil and Subsoil

Xenobiotics in soil and subsoil generally result from agricultural or industrial activities. The main problems in recent years were related to the application of manure and fertilizers, pesticides, sewage sludge, and to contaminated sites [2], [48]. Table 11 illustrates, on a comparative basis, heavy-metal concentrations in Schleswig-Holstein soils. In the light of the very high spatial variability of the soil cover, representative sampling is particularly important for concentration measurements. Regionalized variables should be applied to adequately characterize the chemical soil properties under consideration; consequently, variogram analysis in combination with kriging procedures is the appropriate method [159], [236].

3.1.4.3. Variogram Analysis

Mathematical Concept. A regionalized variable can be considered the realization of a random function Y. Usually, this function is assumed to be stationary; this implies the following:

1) The expectation (m) of Y at any point of x is constant and independent of x.

$$E\left[Y\left(x\right)\right]=\left(m\right) \quad (52)$$

2) The covariance function of any pair of points $\boldsymbol{x}$ and $\boldsymbol{x}+\boldsymbol{h}$ depends exclusively on the vector $\boldsymbol{h}$ and is independent of $\boldsymbol{x}$

$$E\left[Y\left(\boldsymbol{x}\right)Y\left(\boldsymbol{x}+\boldsymbol{h}\right)\right]-m^2=K\left(\boldsymbol{h}\right) \quad (53)$$

In many cases, only the increments of the functions are supposed to be stationary. The intrinsic hypothesis for a vector $\boldsymbol{h}$ concerning expected value and variance is

$$E\left[Y\left(\boldsymbol{x}+\boldsymbol{h}\right)-Y\left(\boldsymbol{x}\right)\right]=0 \quad (54)$$

$$Var\left[Y\left(\boldsymbol{x}+\boldsymbol{h}\right)-Y\left(\boldsymbol{x}\right)\right]=2\gamma\left(\boldsymbol{h}\right) \quad (55)$$

Table 11. Average concentration of selected heavy metals in Schleswig-Holstein soils in mg/kg (after [257])

Substrate	As		Cd		Co		Cr		Cu		Hg		Ni		Pb		Zn	
	A	C	A	C	A	C	A	C	A	C	A	C	A	C	A	C	A	C
S	2.61	**2.67**	**0.36**	**0.38**	1.96	**3.33**	2.55	**5.00**	**4.63**	**3.75**	0.082	**0.010**	8.97	**11.53**	**21.25**	**7.49**	**15.89**	**14.97**
Su2			0.46				5.67		7.70	5.61					23.91		25.95	16.41
Su3			0.31				18.83		13.26						34.72		43.26	
Su4																		
Su	7.13		0.37	0.42	3.30		11.45	16.03	**9.88**	4.75			11.02		29.00	6.29	**32.66**	15.61
Slu																		
Sl2			0.35				11.42		11.12				9.06		29.61		35.04	
Sl3	8.11		0.40	0.26	5.40		**22.73**	28.43	**9.21**	11.82			10.61	22.19	**30.67**	13.20	**43.91**	34.20
Sl4	7.21	**6.52**	0.51	**0.32**	6.73	**6.63**	**34.52**	**34.75**	**12.56**	**12.63**			**15.98**	19.65	**30.31**	**16.66**	**50.69**	**41.26**
Sl	**7.57**	**6.32**	0.43	**0.28**	**5.75**	**6.37**	**24.42**	**31.75**	**10.69**	**11.88**			**13.09**	20.46	**31.51**	**15.04**	**43.54**	**38.57**
Ls2																		
Ls3	7.05	**6.99**	0.70	**0.63**	7.84	**8.21**		**41.26**	13.83	**13.61**		0.022	19.61	**24.41**	31.99	**17.10**	63.59	**50.38**
Ls4				0.52				30.26		13.74				31.44		13.57		39.75
Ls	6.68	**6.68**	0.68	**0.60**	7.15	8.00	40.45	**39.07**	13.19	**13.63**	0.100	**0.022**	19.08	**25.44**	30.00	**16.49**	58.53	**48.38**
Lts		**6.45**		**1.15**		**11.62**				**15.35**		**0.023**		**31.09**		**15.58**		**52.92**
Lt2																		
Lt3																		
Ltu																		
Lt		**6.53**		**1.14**		**11.48**		59.71		**17.10**		**0.024**		**34.98**		**16.00**		**57.72**
Total	**6.78**	**5.46**	**0.45**	**0.55**	**5.34**	**7.24**	**21.04**	**30.20**	**10.05**	**10.70**	**0.085**	**0.019**	**13.63**	**22.90**	**29.04**	**13.34**	**39.64**	**36.75**

A = Average of A horizons (0 – 20 cm)
C = Average of C horizons
Total = Average heavy-metal content except marshland and peat
PNEC: As 20 mg, Cd 3 mg, Co 50 mg, Cr 100 mg, Cu 100 mg, Hg 2 mg, Ni 50 mg, Pb 100 mg, Zn 300 mg
0.00 for $n \geq 10$ and < 30
0.00 for $n \geq 30$

Su = Silty sand
Su2 = Low silt content
Su3 = Medium silt content
Su4 = High silt content

Slu = Loamy-silty sand
Sl = Loamy sand
Sl2 = Low loam content
Sl3 = Medium loam content
Sl4 = High loam content

S = Sand
Sl = Loamy sand
Ls2 = Low sand content
Ls3 = Medium sand content
Ls4 = High sand content

Lts = Clayey – sandy loam
Lt = Clayey loam
Lt2 = Low clay content
Lt3 = Medium clay content
Ltu = Clayey – silty loam content

The so-called semivariogram (designated "variogram" further on), i.e., the function of the vector $\boldsymbol{h}$ is defined as

$$\gamma(\boldsymbol{h}) = 1/2Var\,[Y(\boldsymbol{x}+\boldsymbol{h})\,Y(\boldsymbol{x})] \quad (56)$$

From Equations (55) and (56),

$$\gamma(\boldsymbol{h}) = 1/2E\,[Y(\boldsymbol{x}_i+\boldsymbol{h}) - Y(\boldsymbol{x}_{\mathrm{i}})]^2 \quad (57)$$

In case of discontinuous data the expected value ($\boldsymbol{h}$) can be estimated by the formula

$$(h) = \frac{1}{2n}\sum_{i=1}^{n}[Y(\mathrm{x}_i+\mathrm{h}) - Y(\mathrm{x}_i)]^2 \quad (58)$$

where n is the number of pairs of points.

The points are situated in either one-, two- or three-dimensional space. In two-dimensional space as dealt with here, the coordinates h_1 and h_2 determine the vector $\boldsymbol{h}$. Hence, the variance of measured values depends on the distance and the direction of the difference vector $\boldsymbol{h}$.

In practice, the variogram is usually computed for four main axes (Fig. 34) to account for possible directional effects. Distinctions of the range in different directions (anisotropy) then enable a more detailed interpretation. A mean variogram that is independent of such directional effects is usually computed on the basis of the four directional variograms. A variogram is geometrically characterized by two boundary criteria called "sill" and "range" (Fig. 35). The latter denotes the maximum extent of influence, while the former is the analogous limiting value of influence as measured on the ordinate.

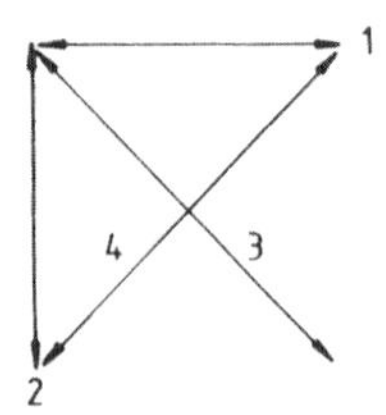

Figure 34. Principal axes of a variogram

Interpretation of the Variogram. *Zone of Influence*. The foregoing indicates that a variogram does not necessarily have a maximum or a level of stabilization. Because of this, two basic types of variograms are distinguished (Fig. 36). In Figure 36 A the maximum range is reached when the correlation between $Y(\boldsymbol{x})$ and $Y(\boldsymbol{x}+\boldsymbol{h})$ becomes nil

$$\gamma(\mathrm{h}) = 1/2Var\,[Y(\mathrm{x+h}) - Y(\mathrm{x})] \quad (59.1)$$

$$= 1/2\,\{Var\,[Y(\mathrm{x+h})] + Var\,[Y(\mathrm{x})]\} \quad (59.2)$$

$$= \frac{2\sigma^2}{2} = \sigma^2 \quad (59.3)$$

This means that a mathematically meaningful interpolation is possible if and only if the maximum distance between neighboring points of measurement is shorter than the distance indicated by the range value of the variogram. In Figure 36 B the zone of influence extends beyond the area examined.

Mathematical Models of the Curve. To best fit the curve of the variogram under construction to the sequence of points obtained primarily from computation of the variance function, various mathematical models are used, the most important of which are described briefly (Fig. 37). Power functions (Fig. 37 A):

$$\gamma(\boldsymbol{h}) = C\,|\boldsymbol{h}|^{\gamma} \ \text{with} \ 0<\gamma<2 \quad (60)$$

$\gamma(\boldsymbol{h}) = C—\boldsymbol{h}—$ is the special case of a linear model. Spherical model (Fig. 37 B):

$$\gamma(\mathrm{h}) = C\left[3/2\frac{\mathrm{h}}{a} - 1/2\frac{|\mathrm{h}|^3}{a^3}\right] \ \text{for} \ |\mathrm{h}| \leq a \quad (61)$$

$$\gamma(\mathrm{h}) = c \ \text{for} \ |\mathrm{h}| > a$$

Exponential model (Fig. 37 C):

$$\gamma(\mathrm{h}) = C\left[1 - e^{\frac{-|\mathrm{h}|}{a}}\right]; \ a \simeq 1/3 \ \text{of range} \quad (62)$$

Gaussian model (Fig. 37 D):

$$\gamma(\mathrm{h}) = C\left[1 - e^{\frac{-|\mathrm{h}|^2}{a^2}}\right]; \ a \simeq 0.58 \times \text{range} \quad (63)$$

Often, for a more exact fit the models must be combined. The variance of the sample is approximately the same as the value of the sill as far as bounded variograms are concerned. The model of the curve is important primarily for determination of the "nugget effect," a phenomenon that must be dealt with when the curve's behavior near the origin is considered.

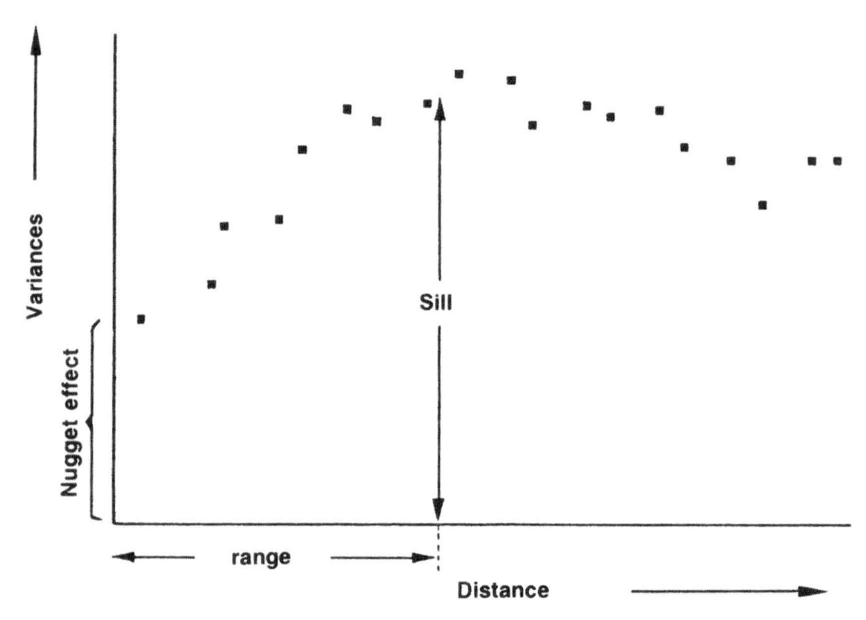

Figure 35. Sill and range of an experimental variogram

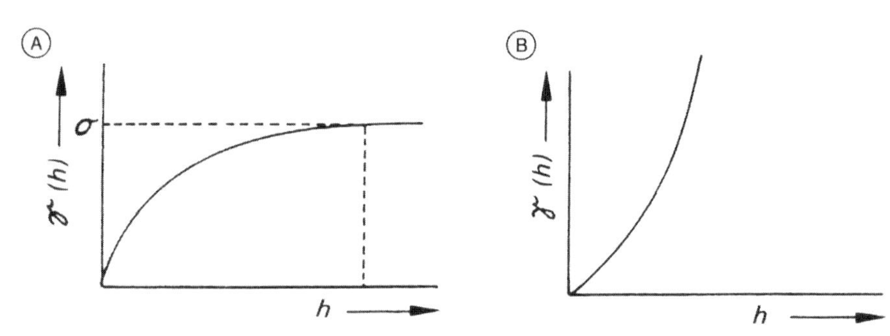

Figure 36. Bounded (A) and unbounded (B) variogram

Behavior of the Curve near the Origin.

1) A parabolic shape shows a high degree of continuity of the regionalized variable. It is differentiable.
2) A linear shape shows continuity "in average" according to MATHERON [236].
3) The curve does not intersect the abscissa at the origin. The nugget effect reveals great irregularity and can be caused fundamentally either by an extremely discontinuous distribution in the immediate neighborhood of the sample taken or by observational errors. Supplementary information is consequently needed.
4) A straight line parallel to the abscissa indicates that no correlation exists between any points $Y(\boldsymbol{x}+\boldsymbol{h})$ and $Y(\boldsymbol{x})$ whatever their distance might be. The sample does not show any spatial structure.

Anisotropies. Distributions characterized by different variabilities in different directions are reflected in the resulting variogram and called anisotropy.

1) Geometrical or affine anisotropy exists whenever elliptical zones of influence can be deduced in a two-dimensional space
2) In a three-dimensional space, variations might also appear in the vertical direction; this is the case of stratified anisotropy

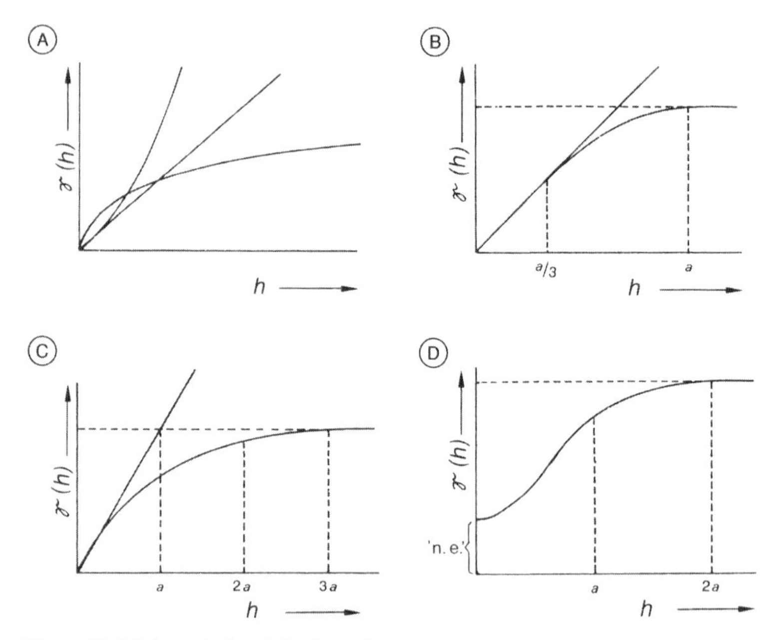

Figure 37. Mathematical models of a variogram
A) Power function model; B) Spherical model; C) Exponential model; D) Gaussian model with "nugget effect"

3.1.4.4. Modeling Chemical Distribution in Soil and Subsoil

Assessing chemical concentrations in soil by means of models normally involves two basically complementary approaches. One type of model simulates chemical movement in the unsaturated zone, (i.e., from the soil surface down to the water table); the other describes chemical transport in the saturated zone or aquifer. BONAZOUNTAS [258], FRÄNZLE [2], and MATTHESS [48] provide state-of-the-art reviews of these models.

Figures 38 and 39, representing one of the most comprehensive hierarchical models developed so far [252], illustrate that the two most important factors influencing chemical behavior in the unsaturated soil zone are sorption and biotransformation. The essential boundary conditions and processes involved (Section 2.4) are indicated in the model that is free from limiting meso- and macro-scale assumptions such as homogeneity (i.e., homogeneous compartments and uniform distribution of chemicals) and stationarity (i.e., large-scale equilibria in extensive compartments). Extended versions of this model serve specific purposes, e.g., modeling of carbon and nitrogen fluxes through ecosystems (Fig. 40).

Small-Scale Assessment of Chemical Distribution. Fate and behavior of chemicals can be described approximately by means of table functions, see Table 12 [259]. In combination with geographic information systems, they permit the production of smaller-scale single-species sensitivity or distribution maps. Six parameters, defined in score form on a 0 (practically none) to 5 (very much, very high) scale, are employed to describe chemical reactions in soil:

1) Solubility in mg/L (20°C) 0: <0.1, 1: 0.1 – 1, 2: 1 – 10, 3: 10 – 100, 4: 100 – 1000, 5: >1000
2) Adsorption, as defined by the Freundlich k values for organic matter (k_{oc}) and clay (k_c) at a soil – solution ratio of 1 : 2 0: <0.5, 1: 0.5 – 50, 2: 50 – 150, 3: 150 – 500, 4: 500 – 5000, 5: >5000 pH influence + or − (i.e., adsorption increases with increasing or decreasing pH)
3) Degradability, defined in terms of time required to produce an approximately 90 % transformation at a soil temperature ranging

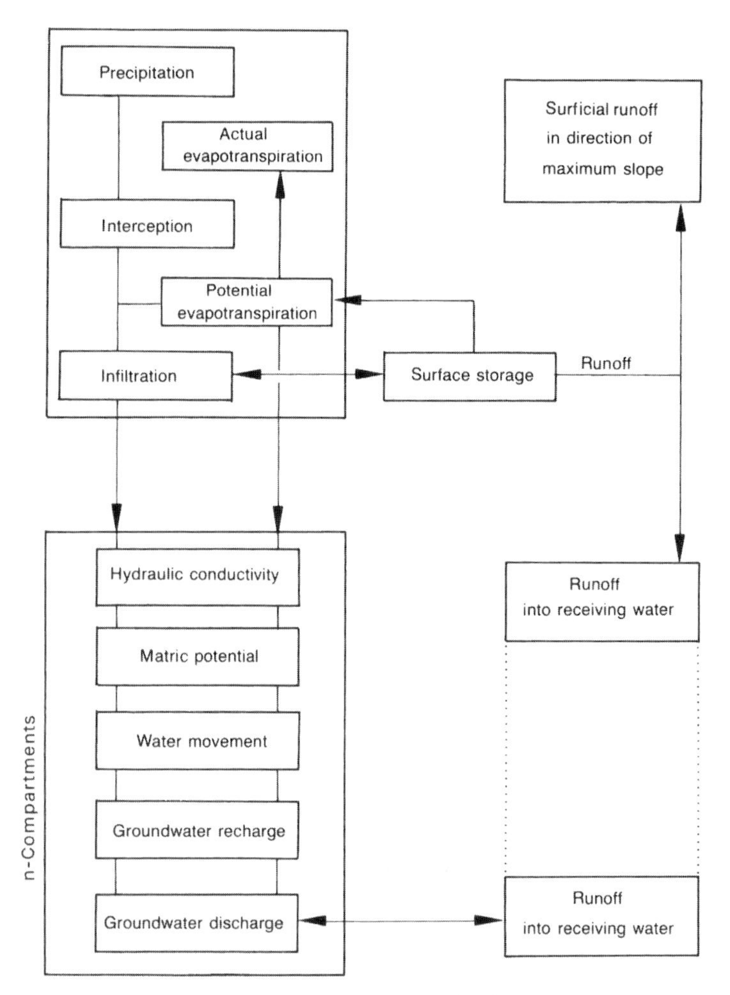

Figure 38. Soil moisture module of the integral WASMOD/STOMOD model (Wasserhaushaltsmodell/Stoffflußmodell, after [252], modified)

from 11 to 16 °C: 1: >3 years, 2: 1 – 3 years, 3: 18 weeks to 1 year, 4: <18 – 6 weeks, 5: <6 weeks

4) Volatility, defined by the vapor pressure of the pure substance in (hPa) at 20 – 25 °C: 1: $<10^{-3}$, 2: 10^{-3} – 100, 3: 100 – 500, 4: >500 or Henry constants 1: $<4\times10^{-6}$, 2: 4×10^{-4}, 3: 4×10^{-4} – 0.04, 4: >0.04

In considering the influence of organic matter content on the adsorption process, the average scores of Table 12 must be defined more precisely according to Table 13. Further refinements relate to texture, pH value, temperature, etc. (Tables 14, 15, 16).

3.1.4.5. PEC Estimates in Aquatic Compartments

Establishing valid monitoring networks or sampling grids poses distinctly lesser problems in surface waters than in soils, but the inherent difficulties should not be underestimated. For instance, determination of the surficial nitrate concentration pattern in a dimictic – holomictic lake of about 1-km^2 size requires 40 – 50 stations [167]. For other chemical species, network densities must be higher still in the light of variogram-analytical evaluations.

In the light of these technical difficulties, modeling approaches play a considerable role

Table 12. Behavior of organic chemicals in soil (after [259], for explanation of figures see)

Chemicals	Aqueous solubility	Adsorption to		pH influence	Degradation		Volatility	Concentrations*, mg/kg		
		Organic matter	Clay		Aerobic	Anaerobic		A	B	C
Chlorinated alkanes and alkenes										
Dichloromethane	4	1	1	0	3–4	2	4	0.1	5	50
Trichloromethane	4	2	2	0	2–3	1–2	4	0.1	5	50
Tetrachloromethane	4	1–2	2	0	2–3	1–2	4	0.1	5	50
Dichloroethane	4	2–3	2	0	3	2	4	0.1	5	50
Monochloroethylene	4	1	1	0	4	4	3			
Trichloroethylene	4	2	1	0	3	2	4	0.1	5	50
Tetrachloroethylene	3	1–2	1	0	3	2	3	0.1	5	50
Carbocyclic compounds										
Benzene	4	1	1	0	3	2	4	0.01	0.5	5
1,2-Dichlorobenzene	3	3	2	0	2	2	3	0.05	1	10
1,2,4-Trichlorobenzene	2	3–4	1–2	0	2–3	2–4	3	0.05	1	10
Phenol	4	1–2	1–2	–	4	3	2	0.05	1	10
2,5-Dichlorophenol	4	3	2	–	3	4		0.01	0.5	5
2,4,5-Trichlorophenol	4	3	1	–	4	4		0.01	0.5	5
Pentachlorophenol	2	4	2	–	3–4	4	2	0.01	0.5	5
Aniline	4	1–2	1	+	4–5	4	2			
4-Chloroaniline	4	3	2–3	+	2	2	2			
Toluene	3	1	1	0	4	2–3	3	0.05	3	30
Toluidine	4	2–3	1	+	2	1				
Xylene	3	2–3	1–2	0	3		2	0.05	1	10

* A = reference level, can be regarded as unpolluted; B = indicative value for further investigation; C = clean-up necessary.

Table 13. Sorption capacity of soil related to organic matter content (after [259])

Organic matter content of soil, %	Relative sorption capacity (Table 12)				
	1	2	3	4	5
0.5–1	0	0	1	1.5	2
1–2	0	0.5	1.5	2	3
2–8	0.5	1	2	3	4
8–15	0.5	1.5	2.5	3.5	4.5
>15	1	2	3	4	5

* Average value of the uppermost 30 cm of soil (i.e., the Ah or Ap horizons).

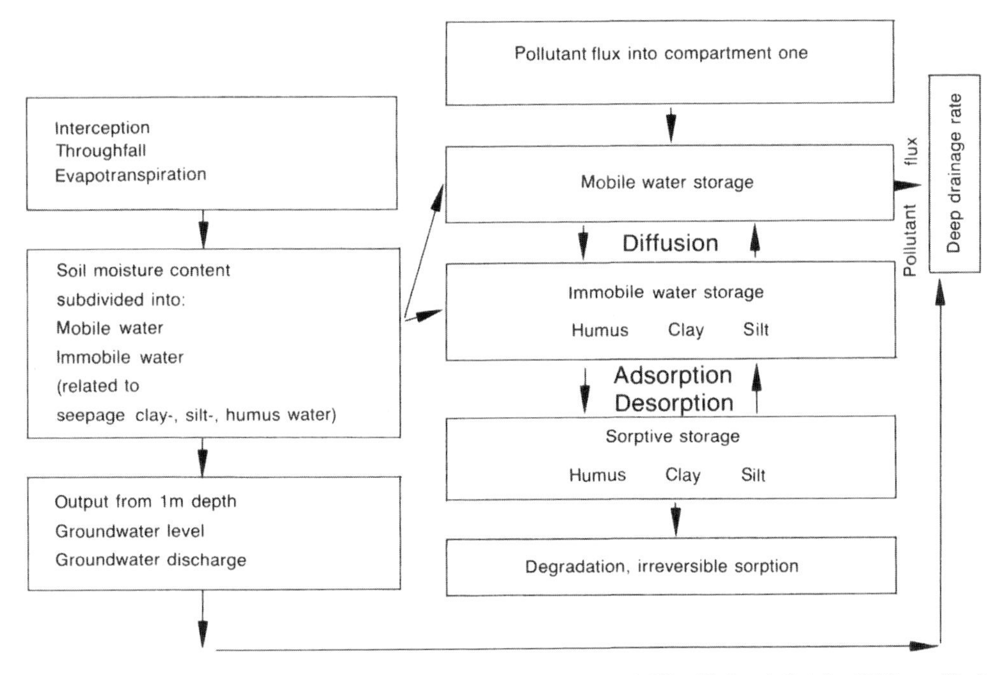

Figure 39. Solute module of WASMOD/STOMOD (Wasserhaushaltsmodell/Stoffflußmodell) (after [252], modified)

Table 14. Sorption capacity of soil related to texture (after [259])

Texture classes*	Relative sorption capacity (Table 13)			
	1	2	3	4
1. S Ss′2*	0	0.5	1	1.5
2. Sc2 Sl2, 3 Ss′ 3, 4 S′s S′	0.5	1	1.5	2
3. Sls′ Sl4 S′l2, 3 S′ls Ls Ls′ Sc3 Cs4 S′C2, 3	0.5	1	2	3
4. S′l4 Cl Cs2, 3 C′s Lcs Lc2, 3 Lcs′ S′c4	1	1.5	2.5	3.5
5. C	1	2	3	4

* Medium texture of the uppermost 30 cm of soil (i.e., the Ah or Ap horizons). S = Sand, S′ = Silt, L = Loam, C = Clay, s = Sandy, s′ = Silty, l = Loamy, c = Clayey; 2 = low, 3 = medium, 4 = high content.

Table 15. pH control of relative sorption capacity of soil (after [259])

pH influence	pH ($CaCl_2$)			
	>6.5	6.5 – 5.5	5.5 – 4	<4
+	+0.5	0	−0.5	−1
−	−0.5	0	+0.5	+1

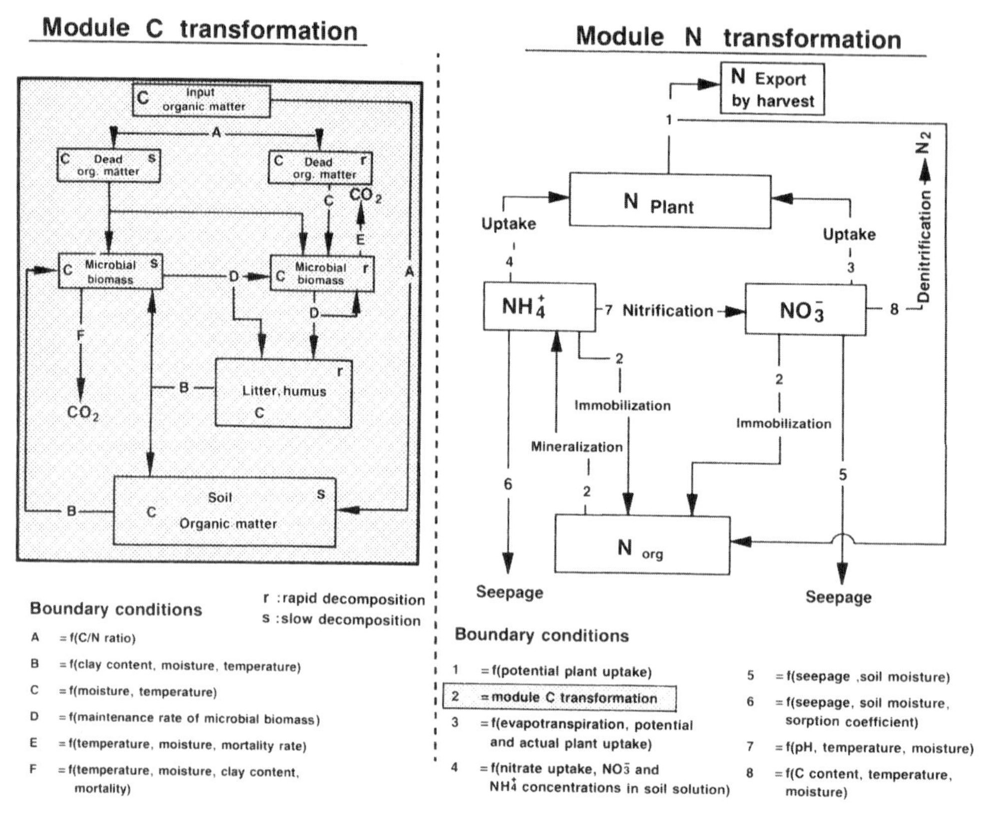

Figure 40. Carbon and nitrogen modules of the extended WASMOD/STOMOD (Wasserhaushaltsmodell/Stoffflußmodell) model (after REICHE, unpublished, modified)

Table 16. Degradability of organic chemicals related to mean air temperature (after [259])

Degradability	Mean temperature of vegetation period, °C			Mean annual temperature, °C		
(Table 12)	21–16	16–11	11–6	12–9	9–6	6–3
1	1.5	1	0.5	0.5	0.5	0
2	2.5	2	1.5	1.5	1	0.5
3	3.5	3	2.5	2.5	2	1.5
4	4.5	4	3.5	3.5	3	2.5
4–5	5	4.5	4	4	3.5	3

[238]. Differentiation between large-scale dilution models operating at the regional level, (e.g., a whole river), and site-specific models requiring a more specific characterization of the projected site is indicated. Simple dilution models assume a homogeneous distribution of chemicals in a water body; i.e., for a given set of stream or river and sewage treatment (→ Wastewater, Chap. 1.) or discharge conditions, an average concentration is calculated. Site-specific models may yield concentration profiles of a chemical in the water body. Also concentrations of chemicals released into the river at many points can be estimated by dilution modeling. The simulation then comprises a relatively large region, typically a major catchment or an entire country. When a sufficient amount of data on dilution factors is available, chemical exposure can be expressed in the form of PEC frequencies in the rivers considered.

The data required to run a surface water model relate to the character of the chemical, the hydrological structure of the catchment, and the discharge pattern [238]. The chemical data include molar mass, melting and boiling points, water solubility, ionization constant, and kinetic parameters relating to sorption, volatilization, photolysis, hydrolysis, oxidation, and biotransformation processes (see Section 2.4). Environmental data vary according to the complexity of the model, beginning with a stream dilution figure for a whole region and ending with more sophisticated approaches based on data on the number and discharge flows of sewage treatment plants and the flow regimes of rivers of a major catchment. Additional site-specific input data may include hydraulic geometry of channels, sediment load, terms of the catchment water balance, physical and chemical characteristics of the water body and its suspended and bottom sediments, etc. In addition, basic dilution models require data on the fraction of the chemicals released to the drain and its removal in different types of sewage treatment. Site-specific models require additional data on chemical loading for each simulated compartment. For spill models, volume and density of the spill are needed [260], [261].

For specific purposes, aquatic equilibrium modeling may become necessary to predict environmental concentrations more reliably. To this end, fast reactions of inorganic species dissolved in aqueous media can be classified under the following categories:

1) Reactions with solvent molecules, in particular dissociation
2) Substitution reactions with solvent or dissolved species
3) Redox reactions with dissolved gases or other ionic species

These reactions can lead to the formation of new complex ions with different ligands in the coordination sphere, ions with different oxidation state of the metal centers, or still other ions. The reactions are essential, since the medium in which the particular inorganic substance is dissolved largely determines the speciation of the particular metal ion. In addition, upon mixing of solutions of these metals with the environment, chemical modification of the species will occur initially by thermodynamically favored rapid reactions and subsequently by slower reactions [262–264]. The following slower reactions are important in aqueous media [258]:

1) Ligand substitution reactions of kinetically relatively inert ions
2) Electron-transfer reactions involving inner-sphere mechanisms for relatively inert ions and some outer-sphere reactions
3) Reactions with dissolved gaseous species or bacterially induced reduction
4) Precipitation of solids by formation of insoluble species through substitution reactions
5) Formation of metal hydroxides and sulfides by oxidation reactions

These slower reactions impact the speciation and concentration in the aqueous phase and determine both type and extent of further interactions in a way comparable to fast reactions. Therefore, the time frame for some of the above reactions may indicate that characterizing the speciation in the original aqueous phase will also define the chemical behavior of the compound in the environment, whereas in other cases, understanding speciation implies exact knowledge of environmental interactions. To this end, a larger number of speciation models have been developed to address a series of specific situations. They are based on chemical thermodynamic reactions and principles, and also account for additional processes such as redox reactions [265], adsorption [266], complexation [267–269], ligand exchange [270], and others [271].

3.2. Effects Assessment

The assessment of possible effects of chemicals on the environment involves the determination of no-effect concentrations (PNECs) for aquatic species, edaphic species, and for species that may be affected by ecological magnification, i.e., by accumulation of substances through the food web. In addition, assessment procedures are indicated for substances that may have a negative impact on natural physicochemical processes in the atmosphere (for definitions and terminology of risk, see Appendix, Chap. 4).

Hitherto the predictive assessment of dose – effect relationships between chemicals and biotic elements in the environment has usually started with laboratory tests. Their results need extrapolation to the complex real-world situation, which proves rather difficult, as a rule, since the experiments are performed with defined amounts of chemicals in contact with a single species or restricted group of organisms. In contrast, the integral approach to assessing the effects of potential pollutants in the natural environment has to deal with the manifold interactions and complicating factors or boundary conditions, which are carefully avoided or controlled in the laboratory situation (see also → Toxicology, Chap. 3., 4., 5.).

The following are most prominent of these complicating factors in the natural environment [272]:

1) Uneven distribution of the pollutant in an ecosystem compartment because of gradients or currents, dilution, volatilization, sorption processes, etc. Bioaccumulation in successive elements of the food web may cause considerable variations in the individual load of different species.
2) Differences in specific sensitivity may lead to shifts in the abundance and diversity proportions of the biocenoses affected. Interspecific relationships, synergistic relations, and food web effects can be analyzed only in multispecies systems.
3) In a natural ecosystem, the effect of a pollutant on the biocenosis will be greatly influenced by weather, diurnal and seasonal variations, simultaneously occurring variable loads such as thermal and chemical loads, along with intermittent discharges.

To a certain extent, the influence of these complications can be accounted for by simulation techniques in the laboratory, but practical limitations necessarily exist. On the other hand, retrospective field experiments alone cannot be relied on since the doses are not known exactly, intentional pollution for experimental purposes is frequently prohibited, and biological observations are time-consuming and expensive. As a consequence, comparative tests should be carried out with selected model chemicals under both laboratory and natural conditions to determine the relative representativity of the test procedure. This validation of the tests should also pay appropriate attention to the important question of how far the test system or the experimental field plot is representative from a regional point of view [160].

In comparing this comprehensive, normative list of ecotoxicological models with the (still) predominant present-day practice, a fairly marked tendency exists to favor the bioindicator approach and aquarium or microcosm tests or technically comparable procedures [273], [274]. This has the appreciable advantage of relative simplicity and ensures a fairly high amount of reproducibility, but also poses the above problems as to rationally and reliably extrapolating to the complex real-world situation.

3.2.1. Test Systems for Aquatic Ecosystems

3.2.1.1. Level 0

Test procedures on the infraorganismic level (i.e., organells, cells, tissues) are omitted here and the hierarchical system suggested in the OECD Chemicals Testing Programme is followed [275]; consequently three successive levels of sophistication labeled 0, 1, and 2 are distinguished. The results of level 0 tests together with an increase in production of the chemical considered and its type of use pattern should indicate that subsequent tests on higher levels have to be carried out.

The following set of studies is considered relevant for the level 0 dossier:

1) Acute toxicity study with *Daphnia magna*
2) Acute toxicity study with fish
3) A static biodegradation test

According to the 7th Amendment to the Directive 67/548/EEC, Annex VII A, the minimal data basis (base set) for effects assessment purposes comprises acute tests for fish, *Daphnia*, and algae. In the OECD testing hierarchy, the latter forms part of level 1 enquiries.

Toxicity Test with *Daphnia magna*. The water flea *Daphnia magna* (Daphniidae, Cladocera) is a suitable test organism for static and semistatic tests and for flow-through systems. The results of these tests are relevant for initial hazard assessment because of the trophic importance of daphnids as primary consumers, their sensitivity to chemicals, and their relationship to various other ecologically important aquatic invertebrates. Also, automatic biomonitoring devices are feasible. The tests usually consist of two phases:

1) An acute phase that gives the 24-h EC_{50} value, the highest concentration that still causes no immobilization (EC_0), and the lowest concentration that causes 100 % immobilization (EC_{100})
2) A reproduction phase yielding the EC_{50} and LC_{50} values at 24, 48, and 96 h; 7 and 14 d; and the end of the test (further information is provided by time span until the emergence of the first brood, the number of live and dead young, their condition, etc.)

The slope of the time – response curve depends largely on the speed and degree of absorption, inherent detoxification, and excretion mechanisms. While a steep slope usually indicates rapid absorption and toxic effects appearing within a short period of exposure, a shallow slope may be due either to low absorption or to toxic effects delayed beyond the period of observation.

The 14-d effect level in relation to the 24-h LC_{50} is, to a certain extent, indicative of the potential for chronic toxicity of a compound. Since the test duration is limited to two weeks, however, it may provide misleading results if the chemicals considered are taken up slowly or have delayed toxic activity. Yet the *Daphnia* test is in any case useful for determining whether further ecotoxicity studies are necessary to indicate which parts of an ecosystem or which species are most sensitive. A low effect concentration may trigger further tests while, conversely, a relatively high *Daphnia* effect concentration may guide any further testing away from invertebrates [276], [277].

Acute Toxicity Studies with Fish. Fish are of particular importance in aquatic environments, and their survival in waters bears witness to the absence of toxic substances or adverse conditions for the species in question. Consequently, toxicity tests with fish have been carried out in laboratories for about a century, but not infrequently the results obtained have differed widely because of lack of standardization of test organisms and boundary conditions.

Therefore, considerable numbers of studies have been undertaken since the 1970s to define the parameters involved in performing toxicity tests: taxonomy, age, and size of fish; composition of environments, in particular pH value, buffering capacity, mineral and organic content, temperature, and oxygenation; duration of contact with toxic substances; etc. The result has been the perfection of numerous techniques that can be applied in fresh water or seawater; they yield satisfactory results, depending on their objectives, although they cannot always be applied readily to other cases in view of major differences in the experimental conditions and the sensitivity of species selected.

Species frequently used in such studies are *Brachydanio rerio*, *Carassius auratus*, *Leuciscus idus*, *Phoxinus phoxinus*, *Rasbora heteromorpha* (Cyprinidae); *Jordanella floridae*, *Oryzias latipes* (Cyprinodontidae); *Poecilia reticulata* (Poeciliidae); and *Oncorhynchus mykiss* (Salmonidae). They comprise species that occur in aquatic environments into which the chemicals under study are released and so-called aquarium fish, which (frequently) are not representative of natural environments but are more suitable for screening tests.

The acute toxicity test on fish [275] is specifically designed to estimate the LC_{50} with observations after 24, 48, 72, and 96 h. However, the longer durations are reported to yield possibly misleading results because test organisms are not fed during the experiment [278]. The resulting data are used to prepare concentration – mortality curves with LC_{50} values, which also permit estimation of the maximum concentration causing no mortality and the minimum concentration causing 100 % mortality. Where possible, LC_{50} values are also estimated for each of the recommended observation times, thus giving a more precise indication of the time – toxicity curve.

As for Daphniidae, comparison of the LC_{50} values with the potential environmental concentrations of a chemical will provide a first crude estimate of the potential hazard to fish. When the slope of the curve is steep and soon tends to be asymptotic to the time axis, longer exposure is not likely to result in a significant increase in mortality unless noxious effects that result from different mechanisms are delayed. Conversely, where the slope is shallow, cumulative toxicity and subsequent chronic effects may be indicated and an extension of the test period could result in a considerably lower LC_{50} estimate. For a more detailed review of the increasing range of types of acute toxicity tests on fish, see [47],

[275], [278], [279]. In view of the numerous tests suggested, however, one should note that environmentally acute lethal toxicity is not the major problem. Acute lethal poisoning is easily detectable and should be readily controllable as well. Thus, subchronic and chronic exposures are likely to be the sources of problems. Consequently, obtaining information on the approximate magnitude of acutely lethal concentrations over an exposure period of up to 96 h might appear sufficient: "The greater the range of water qualities in which the acute LC_{50} is measured and the greater the variety of fish species used, the more likely is it that sufficiently cautious estimates can be arrived at. In this field of toxicity testing it is more important to get some measure of the variability existing, and ever perhaps try to account for it, than it is to mask any real differences which exist by conveniently standardizing tests. (Standard tests with fish appear, in fact, not to be concerned with the assessment of toxicity as such but rather with reproducibly bioassaying the concentration of a given chemical in some particular quality of water)" [280].

Tests for Biodegradation. Assessment of the environmental hazard of a particular chemical involves a prediction of interactions that may lead to degradation of the chemical or its eventual accumulation in particular environmental compartments. In the framework of the OECD Chemicals Testing Programme [275], the following test procedures *defining ready biodegradibility* have been suggested:

1) Modified AFNOR test
2) Modified Sturm test
3) Closed bottle test
4) Modified OECD screening test

These tests provide limited opportunities for biodegradation and acclimatization. Hence, a chemical giving a positive result may be assumed to biodegrade in the environment rapidly and, therefore, may be classified as "readily biodegradable."

Modified AFNOR Test. The modified AFNOR test requires a test material that must at least be soluble at the concentration tested [40 mg dissolved organic carbon per liter (DOC/L)] have negligible vapor pressure, not be inhibitory to bacteria, and not significantly adsorb on glass surfaces. The test substance is degraded by chemicoorganotrophic microorganisms added in low concentrations with the products used as the sole source of carbon and energy. The organic carbon remaining in solution after 3, 7, 14, 28 (and 42) d is measured, and the corresponding level of biodegradation calculated. A detailed description of the test, which is sensitive to an order of magnitude of $\pm$ 10 %, is given in AFNOR T-90-302 Test and the OECD Guideline 301 A [231].

Modified Sturm Test. The modified Sturm test has the same requirements as the preceding one, but uses the amount of CO_2 produced as indicator variable [281]. To this end, a chemically defined liquid medium, essentially free of other organic carbon sources, is spiked with the test substance and inoculated with a low concentration of sewage microorganisms. The CO_2 released is trapped as $BaCO_3$. By reference to suitable blank controls, the total amount of CO_2 produced by the test compound is determined and calculated as the percentage of total CO_2 (TCO_2) that the chemical tested could have theoretically produced in relation to its carbon content. In the absence of toxic effects on the inoculum, the reproducibility of the Sturm test is $\pm$ 5 %. Unless the test is adapted to handle ^{14}C-labeled compounds, test chemical concentrations cannot be lower than 5 mg/L [231].

Closed Bottle Test. The purpose of the closed bottle test is measurement of the biodegradability of organic compounds in an aerobic, aqueous medium at a test concentration of 2 (i.e., standard concentration) to 10 mg/L of active material. The degradation or biotransformation is given as the biochemical oxygen demand (BOD) within 28 d as a percentage of either the theoretical oxygen demand (TOD) or the chemical oxygen demand (COD) [231]. Most data elaborated with this test pertain to water-soluble compounds, but volatile compounds and those of low water solubility may also be tested, at least in principle. In the closed bottle test, a predetermined amount of test substance is dissolved in a mineral nutrient solution to provide the above concentration of active substance per liter. The solution is inoculated with a small number of microorganisms from a mixed population (derived from unfertilized garden soil and activated sludge) and kept in closed bottles in the dark in a constant temperature bath at 20 $\pm$ 1 °C. Degradation is followed by oxygen analyses over a

28-d period. A control with inoculum, but without test material, is run parallel for the determination of oxygen blanks. The reproducibility of the test is appropriate for screening purposes but not sufficient for a final decision regarding biodegradability. Although most experience has been gathered with water-soluble chemicals, the structure of the tests also permits, in principle at least, application to volatile and insoluble compounds.

Modified OECD Screening Test. The modified OECD screening test aims at measuring the ultimate biodegradability of water soluble, nonvolatile organic compounds in an aerobic, aqueous medium with a starting test concentration of 5 – 40 mg DOC per liter. Degradation is calculated as the percentage DOC removal within 28 d with respect to the test material. The compound to be tested is dissolved in an inorganic medium (i.e., a mineral nutrient solution, fortified with a trace element and essential vitamin solution), providing the above concentration. The solution is inoculated with a small number of microorganisms from a mixed population and aerated at 20 – 25 °C in the dark. Degradation is determined by DOC analysis over the 28-d period, which is checked by means of a standard. A control with inoculum, but without test material or standard, serves to determine DOC blanks [231]. Data obtained from these and similar tests are applicable only to aerobic biodegradation. Since a more comprehensive evaluation of the possible fate of a chemical must also take anaerobic situations into account, other test procedures may become necessary.

Other Tests for Ready Biodegradability. At least some 20 additional methods have been suggested to evaluate biodegradation in surface waters or sewage treatment plants [47]. On the basis of comparative degradation tests, GERIKE [282] came to the conclusion that determination of DOC removal usually yields the most reliable and relevant results and could therefore constitute the major source of primary information on ready biodegradability.

Tests for Inherent Biodegradability. Substances that are not readily biodegradable in terms of the above or comparable tests may be examined with the aim of assessing their intrinsic potential for biodegradation (i.e., their liability to microbial breakdown under favorable conditions). At present, the modified Zahn – Wellens test and several methods for assessing the biodegradation potential of soil are available for determining inherent biodegradability [283].

The *Zahn – Wellens test* [284] has a number of advantages. As a batch die-away test, it employs activated sludge, mineral nutrients, and the test substance in solution as the only source of carbon. The mixture is agitated and aerated in a 1 – 4-L glass under diffuse illumination for up to 28 d. Degradation is monitored by determining the DOC (or COD) values in the (filtered) solution at regular intervals. The ratio of eliminated DOC (or COD) after each interval to the initial concentration is expressed as percentage biodegradation; plotted versus time it gives the biodegradation curve. This test provides a good indication of inherent biodegradability and permits acclimatization phenomena to be detected from the degradation – time relationships recorded. Its drawback lies in the high initial concentration of the test substance, which may exert toxic effects on the inoculum [284]. Like some others, the Zahn – Wellens test is suitable only for water-soluble, nonvolatile compounds.

The principle of the inherent biodegradability test in soil is described in OECD Test Guideline 304 A [234]. A small (50-g) sample of soil is treated with the ^{14}C-labeled test chemical in a biometer flask apparatus. The amount of $^{14}CO_2$ released from the test substance is absorbed in a 0.1 N KOH solution and determined by liquid scintillation counting. The test is one example of a class of indirect estimations of microbial degradation processes that generally rely on some unique characteristic of the compound in question, such as the release of $^{14}CO_2$ or of chloride from organics containing organically bound chlorine. Destruction of aromatic compounds in model systems can also be detected in the loss of ultraviolet light absorbance as the aromatic ring is cleaved [285].

Direct Measurement of Ecotoxic Substances. Among the direct measurement methods, gas chromatography provides a rapid, simple, and highly sensitive tool, especially in combination with mass spectrometry and in the form of multidimensional headspace capillary gas chromatography. The latter method increased the reliability of the qualitative and quantitative analysis of volatile compounds by the simultaneous chromatographic separation of samples in two capillary columns of different polarity. It is not affected by disturbances due to salts and compounds with higher molar mass and avoids effects otherwise exerted by the extractant used. By virtue of its high degree of mechanization, headspace injection can be automated easily.

Bioassays. In bioassays for the degradation of toxic compounds, a plant, insect, microorganism, or other suitable indicator organism that is sensitive to the treated sample is exposed at regular intervals after initial modification of the soil. The substance or its inhibitory derivatives are assumed to remain as long as the susceptible species or strain fails to grow or develops atypically, and the length of incubation of the soil required before the indicator organism starts developing in the test sample as well as in untreated soil is considered a reflection of the time at which the toxic component has dissipated [286].

3.2.1.2. Level 1

Subject to positive results from the acute toxicity tests of level 0, an increase in production of the compound, and a risk estimate associated with the use pattern intended, the following test procedures are foreseen for the level 1 dossier:

1) Chronic toxicity study with *Daphnia magna*
2) Subacute toxicity study with fish
3) Algal test
4) Test for species accumulation
5) Additional biodegradation test

Chronic Toxicity Study with Daphnia Magna. The chronic toxicity study with *Daphnia magna* is thought expedient if the LC_{50} appeared lower than 100 mg/L, and it should also include determination of the "no-toxic-effect levels" for both reproduction and lethality. For higher LC_{50} values, such a test appears justified only on the condition that biodegradation has not been proved by the relevant test carried out on level 0.

Subacute Toxicity Study with Fish. The subacute toxicity study with fish is a medium-term test with an exposure time of at least 14 d. As such, it is generally dynamic and based on the substantiation of lethal or sublethal effects on growth, behavior, or reproduction, including the determination of "threshold levels." If the LC_{50} for fish has been assessed at values lower than 1000 mg/L, the same species is to be used for the subacute toxicity study. For higher LC_{50} values, such a test appears necessary only if the previous biodegradation study yielded inconclusive results [287].

Algal Test. An algal test [275] enables a determination of the effects of a substance on the growth of unicellular green algae under laboratory conditions over a period of several cell generations. Since algae, as primary producers, are at the very base of aquatic food webs, their increase or decrease may influence higher trophic levels in natural aquatic ecosystems. In this respect, the possible toxic influences of a chemical on algal growth are highly relevant for the assessment of potential effects in the natural environment. The value of the test is, however, necessarily limited by the inherent difficulty of extrapolating laboratory conditions to the natural situation with its involvement of many other biotic and abiotic factors.

To obtain reproducible results, the algae (frequently *Chlorella pyrenoidosa*, *Scenedesmus quadricauda*, or *Scenedesmus subspicatus*) are exposed to the test substance at defined concentrations for 96 h. The algal growth, measured by the increase in biomass or cell number, is determined at least 24, 48, 72, and 96 h after the start of the test, and the corresponding concentrations per milliliter are calculated. The EC_{50} value is calculated graphically or by computation.

For assessment purposes, both the intrinsic growth rate (i.e., slope of the curve during logarithmic growth) of the control and experimental algal population and the total yield have proved useful. Even if the 96-h EC_{50} value does not seem to reflect high toxicity, this may be due to the fact that the test chemical has degraded considerably in the course of the test.

Test for Species Bioaccumulation. The potential of a chemical to bioaccumulate in aquatic organisms is determined by means of a sequential static test for species bioaccumulation, for which fish are preferably the animals of choice. Bioaccumulation is expressed by the bioconcentration of the chemical in the test organism relative to that in water under steady-state conditions [234]. In the first step of the test procedure, the time interval necessary to reach the steady-state concentration is determined by exposing one group of fish to one concentration of the test chemical without renewal of the test solution. In the second step, the dependence of bioaccumulation on various (i.e., at least three) concentrations of the test chemical is determined, and the result obtained is expressed in form of a sorption isotherm. In the third step of the procedure, the duration rate is measured. To this end, fish exposed to the maximum test compound concentrations in the preceding step are transferred to noncontaminated water from which the depurated compound is continually removed. The residual concentration in the fish is determined periodically until about 80 % depletion is reached or for a period of up to ten days, whichever comes first.

The test for species bioaccumulation was established with ^{14}C-labeled pesticides such as 2,4-D, atrazine, DDT, and others; consequently, such compounds can be appropriately used for standardization. The use of nonlabeled chemicals as standards requires highly sensitive (ppb-range) and compound-specific analytical methods. With respect to specificity, the test is suitable for nonionized organics that are not readily biodegradable, are relatively stable in the aquatic environment (i.e., half-life >4 d), are soluble in water at <2000 ppm (μg/g), and have an n-octanol – water partition coefficient (K_{ow}) >1000 and a volatility (i.e., water – air partition coefficient) >100.

With regard to direct bioaccumulation in individual species, it clearly correlates with the degree of lipophilicity, expressed in terms of the n-octanol – water partition coefficient (see Section 2.4.1). Exceptions normally relate to cases where K_{ow} overestimates the bioconcentration factor. The higher the K_{ow} of a chemical (i.e., the more lipophilic it is) the slower is it depurated (desorbed) from the organism. Since this process does not always follow first-order kinetics, however, apparent half-lifes should not be extrapolated beyond the ranges measured.

In addition, reactions of chemicals in living organisms may result in derivatives that are more lipophilic than the parent compound or in binding (parts of) the chemical to special cell constituents. The latter applies, in particular, to inorganic or organometallic chemicals. Clearly, in either case, the K_{ow} value of the parent compound as such is not an appropriate indicator of the bioaccumulation potential. For organic compounds ionizing under physiological conditions (i.e., in the pH range 3 – 9), K_{ow} is predictive only if the chemical has no specific reactivity to cell constituents.

If the studies carried out on level 0 did not yield a satisfactory amount of information about biodegradation, another *static biodegradation test* can be chosen. If its result is negative again, a dynamic test in a flow-through system on the basis of lower concentrations, together with another inoculum, should then be used.

3.2.1.3. Level 2

The results of studies carried out on test level 1 are preferably to indicate which tests are to be made on level 2; this level is characterized by

1) An additional test for bioaccumulation
2) A chronic toxicity study with fish
3) An acute and a subacute toxicity test with another organism

Additional Bioaccumulation Studies. In the framework of supplemental bioaccumulation studies, the water solubility and n-octanol – water partition coefficient of the chemical in question merit particular attention (Section 2.4.1). Depending on the results of the previous accumulation test on level 1 and the physical and chemical properties of the compound, a flow-through test may be appropriate. It provides an opportunity for more adequate regulation of test conditions, including actual exposure concentrations, and is not limited in duration, thus allowing more comprehensive assessment of the significance of exposure time in evaluating bioaccumulation. The test substances are dissolved, if necessary with surfactant, in water. The solution is introduced continuously into an aquarium with constantly flowing water where the test fish are kept. The concentrations of chemicals in both water and test fish are analyzed, and the BCF is calculated. Since the variability of bioconcentration factors in several fish species is fairly high, it may be useful to achieve a higher degree of normalization by relating BCFs to the lipid content of the fish.

A technical advantage of the easily applicable test [288] is the possibility of automatic control.

The dissolved oxygen concentration, pH, water temperature, etc. can be automatically and continuously measured and recorded.

Chronic Toxicity Study. A chronic toxicity study with fish is indicated if the "no-effect level" for reproduction is more than a factor 10 lower than the no-effect level for lethality of *Daphnia magna*. The same applies if the threshold level for fish as determined under level 1 is 10 times lower than the LC_{50} for fish. According to the relevant OECD Test Guideline [287], fish such as the aforementioned *Poecilia*, *Oryzias*, and *Jordanella* species should preferably be used as test organisms. On the basis of present limited experience, the tendency is to believe that the results of a toxicology study with a second species are not likely to change the estimated "no-toxic-effect" level to a great extent. Therefore, acute and subacute toxicity studies with another aquatic or terrestrial species, for instance the Japanese quail, should be performed only if the need is really indicated.

Representatives of the trophic levels described are listed below. Standard organisms are underlined. The assignment of an organism to a trophic level is based on the energy balance of the ecosystem considered and does not depend primarily on the species. Therefore, a population may represent more than one trophic level in dependence on the nutrition available for the species. In addition, earlier life stages often live on completely different nutrition compared to adults of the same species.

Primary Producers. Primary producers photo- or chemoautotrophically synthesize organic compounds using inorganic precursors. They include

1) Chlorophyll-containing species of vascular plants
2) Algae (e.g., green algae: *Selenastrum*, *Scenedesmus*, *Chlorella*; blue-green algae: *Microcystis*)
3) Purple sulfur bacteria, chlorobacteria
4) Chemoautotrophic bacteria (nitrifying bacteria, sulfur bacteria)

Primary Consumers. They live mainly on living or dead autotrophic organisms or on microorganisms. Representatives of this trophic level are especially plant-eating animals (i.e., species that are not carnivorous) of the following taxonomic groups:

1) Protozoa (e.g., *Uronema, Entosiphon, Tetrahymena*)
2) Annelids (e.g., *Tubifex, Enchytraeus*)
3) Crustacea (e.g., *Artemia*, *Daphnia sp.*, *Copepoda*, *Gammarus*, *Asellus*)
4) Mollusks (e.g., *Dreissensia, Mytilus, Ostrea*; several gastropods: *Patella, Viviparus*)
5) Insects (some insect larvae that are not carnivorous)
6) Nematodes (species that live in water)

Secondary Consumers. They live mainly on primary consumers. Among them are

1) Predatory insects and larvae of insects (e.g., *Chaoborus*)
2) Carnivore protozoa
3) Rotatoria
4) Coelenterates (e.g., *Hydra*)
5) Predatory copepods
6) Fish (Teleostei: e.g., *Cyprinus carpio*, *Brachydanio rerio*, *Poecilia reticulata*, *Oryzias latipes*, *Pimephales promelas*, *Lepomis macrochirus*, *Oncorhynchus mykiss* (former: *Salmo gairdneri*) *Leuciscus idus melanotus*, *Cyprinodon*, *Carassius*
7) Amphibians (e.g., *Rana*, *Xenopus*)

Decomposers. Organisms of this trophic level break down dead organic material to inorganic constituents. Representatives are

1) Bacteria (most species)
2) Fungi

Test organisms for benthic ecosystems

Biological group/test organism	Parameter tested
Benthic invertebrates	
Panagrellus redivivus	survival
Caenorhabditis elegans	survival
Tubifex tubifex	survival
Stylodrilus heringianus	survival, repellency, rate of transformation of the sediment, growth
Hyalella azteca	survival, growth, reproduction
Pontoporeia hyi (*Diporeia sp.*)	survival, repellency
Corbicula fluminea	survival, growth
Anodonata imbecilis	survival
Chironomus tentans	survival, growth, hatching
Chironomus riparius	survival, growth
Hexagenia limbata	survival, frequency of exuviation
Macrobenthic biocenoses	indices for biocenoses and populations
Macrophytes *Hydrilla verticilata*	length of shoots and roots, dehydrogenase activity, chlorophyll-a, peroxidases

3.2.1.4. PNEC Values and Assessment Factors

The function of hazard or risk characterization and assessment is the overall protection of the environment. With regard to aquatic ecosystems, two habitual assumptions are therefore made to extrapolate from the above single-species short-term toxicity data to ecosystem effects: (1) ecosystem sensitivity depends on the most sensitive species, and (2) protecting community structure protects community function. While the latter assumption is hardly questionable, the former needs further examination in light of the introductory statements to Chapter 3.

For many existing and most new substances, the pool of data on ecosystem effects is very limited; i.e., only short-term toxicity data are available as a rule. This implies the application of assessment factors that permit one "to predict a level at or above which the balance of probabilities suggests an environmental effect will occur. It is not intended to be a level below which the chemical is considered safe. However, again, the probability is that there will be no effects" [289]. When defining assessment factors, a number of uncertainties must be taken into account in attempting to extrapolate from mono- or oligospecific laboratory test systems to multispecies ecosystems: (1) intra- and interspecies variations, (2) acute to chronic toxicity extrapolation, and (3) extrapolation of laboratory data to field impact. Additional uncertainty results from the different testing methods themselves or interlaboratory variation in toxicity data.

When reviewing data from laboratory test systems the assessment factors of Table 17 are proposed. Instead of using assessment factors on the lowest-effect data, statistical extrapolation methods may be applied to a (relatively large) set of long-term data as an alternative approach. The underlying assumptions are [292]:

1) The distribution of species sensitivities follows a theoretical distribution function
2) The groups of species tested in the laboratory are a random sample of this distribution

This implies the use of no-observed-effect concentrations (NOECs) for identical parameters (e.g., reproduction) of the species considered [293], especially if only limited data from different species are available [294]. Normally long-term toxicity data are log-transformed and fitted according to the distribution function, and a prescribed percentile (e.g., 95 % [292]) of this distribution is used as a criterion. This means that the NOEC may be exceeded for 5 % of the species in the community. The validity of some of the above assumptions may be questioned (see also [294], [295]). First of all, it must be taken into account that for most compounds few longterm NOECs are available so that most tests on distribution functions have low power. Second, the NOECs available are not a random sample. Emans et al. [296] compared NOECs derived from multiple-species experiments for aquatic systems with extrapolated values [297], [298] and came to following conclusions:

1) Validation is rendered difficult by the lack of reliable data from multiple-species tests and long-term monospecific tests
2) Results from both statistical extrapolation approaches do not differ significantly
3) Results of the statistical extrapolation with 5 % confidence are in good agreement with results from (semi-) field tests

Table 17. Assessment factors for aquatic toxicity testing (after [289–291])

Toxicity data	Factor		
	EC	OECD	ECETOC*
One short-term toxicity value: algae, daphnia, or fish		1000	
Short-term toxicity: algae, daphnia, and fish	1000 [lowest $L(E)C_{50}$]	100	200
Long-term toxicity: fish or daphnia	100 (NOEC)		
Long-term toxicity: algae, daphnia, and fish	10 (lowest NOEC)	10	5
Long-term toxicity: two species in two taxonomic groups	50 (lowest NOEC)		
Field or mesocosm toxicity	case-by-case evaluation		1

* ECETOC = European Chemical Industry Ecoloy and Toxicology Centre.

3.2.2. Testing Strategies for Terrestrial Ecosystems

Soil constitutes a central compartment of terrestrial ecosystems. Chemicals introduced into soil can be adsorbed on soil constituents or transformed by soil organisms; furthermore, they may be taken up by plants, washed out by rain or irrigation water, or evaporated in gaseous form. The extent of adsorption is controlled by various soil properties, including composition and content of organic matter, type and content of clay minerals, ion-exchange capacity, and acidity; it is also closely connected to the physical and chemical properties of the substance in solution (see Section 2.4.2). In view of this complex situation, the empirical knowledge of the behavior of environmental chemicals is still fairly limited, and the possible danger of excessive fertilizer or manure application, pesticides, and heavy-metal contamination of soils and crops has been recognized only in the last decades, mostly in relation to the agriculture use of sewage sludge or in connection with atmospheric deposition of organics or heavy metals.

Given the fundamental importance of soil for biological conservation, fertility, element fluxes, biodegradation of organic materials, and retention and filtration of groundwater, broad agreement exists as to the necessity of specific, use-oriented risk assessment procedures. As for risk characterization in aquatic ecosystems, the predicted no-effect concentration in soil ($PNEC_{soil}$) must be compared with the predicted environmental concentration in soil and subsoil (PEC_{soil}) as described in Section 3.1.4.

3.2.2.1. PNEC Estimates for Soil and Subsoil

Terrestrial Tests and Assessment Factors. A substance discharged into soil can have noxious effects on both the soil's physicochemical properties and its biocenoses. If the pore structure of a soil is affected by the chemical, the soil might silt up and be compacted; thus, the percolation rate is reduced considerably, the water balance is impaired (see Section 2.1.2), and the air supply is impeded. Since soil acts as a pH buffer system and a reservoir for bound substances, impairment of the soil adsorption capacity by xenobiotics can infer exhaustion of the buffer capacity. Impairment of microbial activity implies a reduction of the catabolic and metabolic capacity of the soil, which may lead to retarded liberation of nutrients and accumulation of mainly organic substances; this implies further effects on soil-dwelling organisms [35]. A chemical represents a risk to the soil compartment of an ecosystem if its concentration exceeds the degradation potential of the soil or if it changes the soil properties – in particular, the filtration, buffering, and transformation functions – in such a way that adverse effects result for the environment [299]. Thus, data on geoaccumulation as obtained in the framework of exposure assessment may give a first indication of possible changes in soil properties. The $PNEC_{soil}$ can be derived from the results of toxicity tests on soil and subsoil organisms by the application of assessment factors. If no ecotoxicological data are available for such organisms, the equilibrium partitioning method may be applied as a first approximation. It assumes that the bioavailability and therefore the toxicity of chemicals are determined by their concentration in the pore water of soil. The applicability of the method has been tested less for soil- than for sediment-dwelling organisms; VAN GESTEL [300] has demonstrated its validity forearthworm toxicity for several chlorophenols, chlorobenzenes, and chloroanilines. Whether this also holds true for other compounds or species is questionable, however (e.g., lipophilic compounds or organisms that are exposed primarily via food).

For few existing chemicals, short-term terrestrial tests are available with microorganisms, earthworms, and higher plants; in addition, a long-term test is used with springtails. For new substances, two standardized short-term tests with higher plants and earthworms are required at level 1. If no acute tests are available, tests on primary producers, consumers, and decomposers should be applied for the preliminary stage of the concentration-effect assessment [301–304]. The Umweltbundesamt [290] suggests use of the same assessment factors for terrestrial as for aquatic ecosystems, depending on the type of investigations, (i.e., acute or long-term toxicity test) and the number of trophic levels studied (see Table 18).

Table 18. Assessment factors for terrestrial ecosystems (after [290])*

Information available	Assessment factor
L(E)C_{50} for preliminary stage toxicity tests (e.g., acute toxicity to plants and earthworms or microorganisms)	1000
NOEC for one additional long-term toxicity test	100
NOEC for additional long-term toxicity tests for two species of two trophic levels	50
NOEC for additional long-term toxicity tests for three species of three trophic levels	10
Field data/data of model ecosystems	case by case

* The $PNEC_{soil}$ is calculated on the basis of the lowest effect value measured.

Standardized long-term tests for earthworms [305] and for springtails [304] are available, both of which analyze effects on reproduction. The standardization of two long-term tests is close to completion: the test on staphylinids (Coleoptera) registers degree of parasitism, hatching rate, and reproduction [306]; the test on *Enchytraeus* (Annelida) can replace the reproduction test on earthworms [307]. VAN STRAALEN and VAN GESTEL [308], STAVOLA [309], and PEDERSEN and SAMSØE-PETERSEN [310] discuss test methods with various terrestrial species and different degrees of standardization.

Evaluation of Stress Tolerance of Soils as Ecosystem Compartments. To comply better with the exigencies of a provision-oriented environmental policy, the critical loads of soil with regard to nutrients, heavy metals, and xenobiotics require a generalized ecological evaluation, in addition to estimates of human exposure to contaminants in soil (see → Toxicology, Chap. 5.). This implies the rational definition of soil standards that aim at sustaining appropriate physicochemical and biotic soil functions.

A recent approach of this type is based on the regionalized determination of geogenic and anthropogenic concentrations of potentially hazardous chemicals, and the assessment of different degrees of adverse chemical effects on soil and related biota [299] (see Table 19). Thus, the preventive threshold value C_p characterizes chemical concentration levels not affecting any essential ecological soil function. The C_{eff} value is indicative of first and low-level adverse effects and is defined by means of ecotoxicological test procedures and determination of the physicochemical behavior of xenobiotics in soil. By contrast, the C_{dis} value indicates long-lasting or permanent disturbance of essential biotic and abiotic soil functions. Among the extrapolation methods for determining ecologically relevant effect concentrations [293], [311–316], the last mentioned appears most appropriate. It defines a hazard concentration (HC_p) that ensures the protection of a certain number of soil-living animals. Thus, the approach is based explicitly on resilience considerations (see Section 2.1), using NOECs as reference values. The hazard assessment involves four steps:

1) Determination of at least five NOECs
2) Standardization of NOECs values with regard to a standard soil from at least three taxonomic groups according to the following equation

$$\mathrm{NOEC}' = \mathrm{NOEC}\,(L,H)\,\frac{R\,(25,10)}{R\,(L,H)} \qquad (64)$$

where NOEC′ is the NOEC of standard soil; NOEC (L, H) ist the experimental NOEC with given clay (L) and organic matter content (H); $R\,(25, 10)$ is the reference value of standard soil with 25 % clay and 10 % organic matter; and $R\,(L, H)$ is the reference value on the basis of the clay and organic matter content of the test soil.

3) Determination of a safety factor T in relation to the different sensitivity of indicator species

$$T = \exp\left[\frac{3S_{\mathrm{m}}d_{\mathrm{m}}}{\pi^2}\ln\left(\frac{1-\delta_1}{\delta_2}\right)\right] \tag{65}$$

where S_{m} is the standard deviation of ln NOECs; d_{m} is a factor dependent on the number of test species; δ_1 is the proportion of nonprotected species, related to defined hazard concentration (HC_{p}); δ_2 is the probability of overestimating HC_{p}; and T is a safety factor for NOEC average.

4) Based on this safety factor the hazard concentration HC_{p} is defined as

$$HC_{\mathrm{p}} = \frac{\exp\,(x_{\mathrm{m}})}{T} = \frac{\overline{\mathrm{NOEC}}}{T} \tag{66}$$

where m ist the number of test species; x_{m} the mean value of ln NOECs; and $\overline{\mathrm{NOEC}}$ the geometric mean value of NOECs.

The exemplary application of Equation (66) to cadmium and a test system of *Dendrobaena rubida, Lumbricus rubellus, Eisenia foetida, Helix aspera, Porcellio scaber, Platynothrus peltife*, and *Orchsella cincta* would yield an HC_5 (i.e., $\delta_1 = 0.05$) of 0.16 mg Cd per kilogram of soil. In comparison, the present A level of the Dutch "Leidraad Bodemsanering" is equivalent to 0.80 mg Cd per kilogram of standard soil.

As long as the C_{eff} value, defined by means of the above prodecure, is not exceeded, no essential reduction of ecological soil functions occurs and the long-term multifunctionality of soil is sustained. In contrast to this threshold, the C_{dis} value indicates a distinctly higher concentration of pollutants in soil that exerts long-lasting or irreversible effects on biotic and regulative soil functions. For assessment purposes, basically the same approach is foreseen as for the definition of C_{eff} (i.e., multispecies determination of HC_{p} with a p around 50). Unter these circumstances, neighboring ecosystems may also be affected due to export of primary pollutants or their metabolites. Thus, the C_{dis} value characterizes a threshold that necessitates protective or rehabilitation measures. Depending on both quality and intensity of land use, these may be quite different, including, e.g., limitations of crop production or reduction of fertilizer and pesticide application [299].

Complex Biotic Indicator Systems for PNEC Estimates. Toxic effects of chemicals in a biocenosis change according to the species composition of the community exposed to them. This is particularly important when a community is continually exposed to a concentration of a substance since its species are either least sensitive to this substance or adapted to it. Consequently, the effects of the substance on a selected laboratory test organism or a monospecific assemblage not previously exposed to that particular substance may be quite different. In consequence of these and other limitations of the aforementioned tests, artificial laboratory microecosystems (microcosms that simulate some of the characteristics of natural ecosystems) are more frequently used in toxicological studies [288], [317].

Model Ecosystem Approach. The model ecosystem approach uses physical models of natural ecosystems to simulate a selected subset of the relevant physical, chemical, and biotic characteristics of a potentially affected environment. These models may vary in complexity from a bacterial culture in a test tube to a net cage isolating a part of an ecosystem. Normally, the model ecosystem involves an aquarium or other small container in which soil, sediment, water, and several trophic levels of organisms are kept. Thus, METCALF and coworkers [318] studied the effects of chemicals in a terrestrial – aquatic interface to determine their potential to accumulate in food chain organisms and to degrade. Tests with the *Metcalf microcosm* have been standardized to permit comparison of different chemicals. Although this system does not allow for prediction of absolute concentrations of a substance expected or observed in a wide variety of actual environmental conditions, it has shown positive relationships among physical, chemical, and biotic properties, (e.g., water solubility and bioconcentration potential). However, it has not been used typically to determine rates of dominant processes, nor has it sufficiently incorpo-

Table 19. Complex ecological threshold values for soil quality characterization [299]

Natural background "clean soil"	Range of tolerable disturbance no further chemical stress	Range of permanent heavy disturbances changes in land use, hazard assessment, and restoration indicated
C_p	C_{eff}	C_{dis}
No reduction of ecological soil functions		
Multifunctionality: optimum range of ecological soil functions	first low-level adverse effects on ecological soil functions	long-lasting or permanent disturbances of essential ecological functions
No detrimental fluxes of chemicals	no detrimental fluxes of chemicals	detrimental fluxes of hazardous chemicals into neighboring ecosystems
Optimum conservation of species	almost complete conservation of species (HC_5)	species deletion (HC_{50})
Site-specific quasi-natural climax communities	$\longrightarrow$	possibilities of land use reduced
Any form of ecologically indicated land use possible	gradually increasing reduction of soil functions	biotic and abiotic soil functions reduced

rated many of the dynamic properties found in other aquatic environments.

FREHE's *greenhouse approach* [319] is an example of microcosm research related to agriculture. Each microcosm, measuring $207 \times 193 \times 200\,cm^3$, was fixed on a lysimeter containing a soil mixture (loess loam, sand, and peat) with an earthworm (*Lumbricus terrestris*) population; an association of pea (*Pisum sativum*), vetch (*Vica sativa*), oat (*Avena sativa*), and mustard (*Sinapis alba*) was grown in it. The greenhouses were provided with ventilation and sprinkling systems permitting the application of di-(2-ethylhexyl) phthalate for testing purposes in aerosol form.

Studies in such small-scale laboratory systems are valuable tools to demonstrate some principles of concern in ecotoxicology and have to be included in the general knowledge of the fate of chemicals in natural large-scale ecosystems. They are indispensable additions to single-species experiments in order to more reliably evaluate the influence of interspecific interrelationships on the toxic potential of compounds. An example is zinc toxicity with respect to marine diatoms and dinoflagellates, which, in the majority of cases, proved to be definitely more pronounced in five-species tests than in monospecific cultures [320]. Hence, an abbreviated model ecosystem approach may prove useful in environmental concentration – fate studies by providing data on the rate constants of some of the processes that govern the transport and transformation of chemicals. Thus limited-scope model ecosystems contribute to determine whether a mathematical model accurately or approximately predicts concentrations in the various environmental compartments included in the model ecosystem. Expanded types of microcosms may be envisaged to include a substantially higher amount of that variety of physical, chemical, and biotic properties that a chemical is likely to encounter in the real environment. In terms of structural diversity, these comprehensive, and consequently more expensive, *mesocosms* lead over to biocenotic experiments under open-air conditions, which constitute the highest assessement level possible.

Biocenotic Experiments. Individual *populations* display marked changes in abundance as a direct consequence of the exposure pattern and an indirect consequence of the induced variations of intra- and interspecific competition and resource availability. Thus, almost all collembolan species exposed to cypermethrin displayed a longer increase in population density after an initial reduction, followed by a secondary decrease, probably due to excessive exploitation of food resources or an increase of predators [321]. In numerous experiments, complete annihilation of species was the exception; recedent taxa were mostly affected. Yet in a few cases, populations with dominant species were almost completely suppressed (e.g., the predatory mite *Rhodacarellus silesiacus* and the collembolan species *Isotomina bipunctata* on experimental plots treated with aldicarb [322]). Also, nontarget organisms experienced changes in their abundance spectra that lasted for several years [323].

At the *community level*, chemicals nearly always induce changes in the abundance spectra, whereas changes in species composition occur less frequently, particularly as a result of persistent multifactorial stress [324], [325].

Ecosystem PNEC Assessment. Integral PNEC indicators at the *ecosystem level* are

1) Fluxes of macro- and micronutrients (K, Ca, Mg, N, P, S and Fe, Mn, Cu, Zn, Mo, B, Cl)
2) Duration of biochemical cycles
3) Energy flux rates and entropy production of phytocenoses
4) Changes in biodiversity and population dynamics
5) Structure of trophic networks or experimental food chains

Examples of the first groups of indicators are the influence of pentachlorophenol (PCP), 2,4-D, 2,4,5-T, and aldicarb on the nitrogen balance and litter degration of beech, pasture, and fallow ecosystems [326], [327]. In either case, degradation was impeded and the nitrogen content of the litter bags reduced. Shortly after application of the chemicals, the release of inorganic nitrogen compounds increased considerably at the expense of microflora and fauna. Furthermore, the retention capacity and the availability of many nutrients were diminished, and these effects lasted for more than four years depending on the chemical character and application rate of the compounds tested.

3.2.2.2. Complex Assessment and Modeling of Vegetation Damage

Pollutant gases such as sulfur dioxide and nitrogen oxides are frequently emitted in combination and normally with other pollutants such as soot and heavy-metal particles; in many places, photooxidants are emitted simultaneously. In these cases, the effects of combinations of pollutants are not simply the sum of the effects of individual components, but the relative concentrations of components or the specific sensitivity of the plants to one of them may determine the response. Thus, pollutants in combination may display antagonistic (reduced), additive (sum of individual pollutant effects), or synergistic (superadditive) effects.

In the past, most experiments on exposure to combinations of pollutants were carried out under laboratory conditions using higher concentrations of gases than are typical in the field. Newer fumigation techniques, for instance open-top chambers [328] or field fumigation systems [329], have now enabled the exposure of plants under field conditions with increasingly realistic concentration ranges. These experiments indicate that pollutant combination effects depend on pollutant concentrations; SO_2 and O_3 are often antagonistic at levels where the individual gases cause severe leaf injury, while super-additive effects predominate at lower concentrations (i.e., $SO_2 < 140\,\mu g/m^3$, $O_3 < 100\,\mu g/m^3$), particularly near the threshold dosage [330]. In addition, the injury symptoms for SO_2 and O_3 in combination are not a simple mixture; frequently, the symptoms of one pollutant occur more clearly or earlier than those of the other [331].

With regard to combinations of NO_2, SO_2, and O_3, effects were produced in many plants by lower concentrations of NO_2 and the other gases than if they acted alone, especially at the lower exposure levels [332], [333]. Synergistic effects of NO_2 in combination with SO_2 or O_3 apparently result from a decrease of the nitrate and nitrite reductase activities. While the latter tend to rise in response to NO_2 acting on its own, activation of these reducing enzymes is limited or even suppressed by the simultaneous presence of SO_2 or O_3 [334]. This adverse effect is further enhanced by the formation of free radicals. An inhibitory influence on the regulation of transpiration occurred at less than $20\,\mu L/m^3$ $SO_2 + 20\,\mu L/m^3$ NO_2 [335]. The combination of NO_2 with O_3 is likely to endanger vegetation more than $NO_2 + SO_2$ in combination and should receive greater attention since these pollutants form essential components of photochemical smog. The threat to vegetation may increase still further if all three pollutants interact simultaneously [332], [333], [336], [337].

Combination effects, however, also depend on the temporal order in which the pollutants affect the plants. If they are first exposed to NO_2 and thereafter to SO_2, the usual SO_2 effects are reduced, while the opposite sequence causes superadditive effects because SO_2 affects nitrite reductase [334], [338]. Attention must also be drawn to the combination of gaseous SO_2 with acid precipitation; the effects of this combination are discussed in connection with the new types of forest damage.

Critical Levels. In conclusion, and with some reservations resulting from the above remarks, critical levels for short-term and long-term exposures may be defined. In view of the fact that most horticultural and agricultural plants do not show adverse effects at *sulfur dioxide concentrations* <30 μg/m^3, while particularly sensitive species of trees, moss, lichens, and bushy and grassland vegetation are adversely influenced by concentrations of ca. 20 μg/m^3, the critical level for SO_2 acting on its own is to be set at 20 μg/m^3 (≈0.007 mL/m^3) as the annual mean value. Experimental fumigation trials that indicate first adverse effects from 70 μg/m^3 SO_2 (≈0.025 ppm) for cultivated plants form the basis for setting the short-term value at this concentration [331].

Leaf necroses were the predominant effects of short-term, mainly single exposures of plants to >1 mL/m^3 *nitrogen dioxide*; lower concentrations caused alterations in photosynthesis and respiration as well as in organelle ultrastructures. The particular sensitivity of plants at night must also be taken into account, and the fact that high NO_2 concentrations are often associated with high NO levels. For long-term exposures, a seasonal differentiation of sensitivity is indicated. Thus, for summer, when the physiologically more active plants can rapidly metabolize NO_2, a half-year mean of 0.03 mL/m^3 appears tolerable. In winter, however, when plants are particularly sensitive, a half-year mean of 0.02 mL/m^3 must not be exceeded to protect both short-lived plants and long-lived arboreal flora.

Experimental investigations of *ozone damage* [337] permit definition of the following critical levels for the protection of vegetation against O_3 as a single pollutant. For longer-lasting exposures, 50 μg/m^3 (≈0.025 mL/m^3) O_3 appears indicated as a vegetation period average (composed of 7-h daylight mean values).

For *NO_2 in combination with O_3 and SO_2*, the World Health Organization has adopted an Air Quality Guideline of 30 μg/m^3 NO_2 as the arithmetic annual mean for areas in which SO_2 and O_3 may occur in annual and vegetation period means of up to 30 μg/m^3 SO_2 and 60 μg/m^3 O_3 [333]. However, nitrogen oxides do not only have direct effects, they may also affect plants indirectly via the soil, with additional nitrogen input greatly endangering nutrient-poor plant formations such as heath and moor communities. An amount of 10 μg NO_2 per cubic meter may suffice to damage the particularly sensitive ombrotrophic mires [339].

On the basis of exposure – effect relationships as determined for a variety of plants of widely differing sensitivity by various authors, POSTHUMUS [340] deduced the following critical levels for *ammonia*:

100 μg/m^3 (≈0.14 mL/m^3) for monthly mean concentration
600 μg/m^3 (≈0.86 mL/m^3) for 24-h mean concentration
10 mg/m^3 (≈14.3 mL/m^3) for 1-h mean concentration

Impacts on Forest Productivity. Since the 1970s the indirect effects of acid precipitation on forest trees and crops has been a major area of both scientific and economic concern. In Germany, forest decline attributed to flue gases, particularly SO_2, was described in the neighborhood of industrial plants more than a century ago. The situation did not change essentially until the late 1960s, when a new type of forest injury became evident that was no longer related to local emitters but occurred to an increasing extent in presumably clean air areas far away from conurbation and industrial centers [341–345]. In the mid-1970s, severe dieback of silver fir (*Abies alba*) was observed in southern Germany. In the early 1980s, other conifers, especially Scotch pine (*Pinus sylvestris*) and Norway spruce (*Picea abies*) showed similar symptoms over rapidly increasing areas, and deciduous species such as beech (*Fagus sylvatica*) and oak (*Quercus robur*) have recently been affected. According to the 1985 damage inventory, 52 % (i.e., 3.8×10^6 ha) of the forest areas in the 58 growth districts of the Federal Republic of Germany were injured. Level 1 of damage (reduction of vitality) characterizes about two-thirds of this portion, while medium to high damage (levels 2 – 4) occurs on 19 % of the forested area [346].

Comprehensive studies [347–350] show that the lack of nutrients, climatic factors [351], mistakes in forest management, infections, and pests may play an important role in the present complex forest diseases; but a rapidly growing number of findings indicate that the above anthropogenic air pollutants, and possibly others such as triethyl lead or nitrophenols [352], [353], and related soil acidification are the essential direct and indirect causes of damage [328], [354–357].

At the cellular level, pollutants may lead to a disturbance of photosynthesis and respiration and the following primary and secondary metabolisms. In some cases, impairment of photosynthesis occurs before visible damage can be observed [358]. At the same time, photooxidants and acid rain cause weathering of cuticular structures, destruction of membrane systems, and leaching of nutrients from leaves. Quantitative and qualitative changes in photosynthesis [359] lead to reduced primary production, increased respiration, and a changed assimilate allocation to the root – mycorrhiza system, stem, branches and leaves, and also to the reproduction and phytogenous defense system. Such stressed trees emit unsaturated hydrocarbons such as ethylene, isoprene, and terpenes, which react with secondary pollutants (e.g., ozone) and highly phytotoxic tertiary pollutants in the immediate neighborhood of the plants. Weakening by reduced assimilate allocation to the roots leads to reduced uptake of nutrients from soil that has already been disturbed by acidification (i.e., impoverished in potassium and magnesium).

MURACH [360] determined a calcium deficiency in the roots of declining spruce and fir, but the same levels of aluminum as in healthy trees; REHFUESS et al. [361] noted magnesium and possible calcium deficiencies by foliar analyses even on base-rich soils. Contrary to ULRICH's [362], [363] contention that a change in fine root biomass is paralleled by toxically high Al : Ca ratios, REHFUESS points out that the two phenomena are not necessarily synchronized, so that marked decreases in fine root biomass can precede the increase in soil solution aluminum.

Mineral deficiency due to leaching and soil acidification is frequently coupled with a relative surplus of nitrogen with respect to assimilated carbon as a consequence of enhanced deposition of nitrate and ammonia and the intensity of subsequent nitrification processes. KRIEBITZSCH [364], who studied nitrification in many types of acidic forest soils, divided them into groups A, B, C, and D. In group A, no nitrification occurred and ammonium was the only nitrogen source. In soil of groups B and C, partial nitrification occurred, while group D soils had complete nitrification. Former heathland soils in Germany now cultivated with various *Pinus* species and *Pseudotsuga* belong mainly to group A, where the nitrate level is low, whereas the ammonium concentration is high.

Quantitative and qualitative changes in the living biomass of forest trees will bring about changes in the food web of the forest ecosystem; the population of some consumer (e.g., insects) or decomposer organisms (e.g., fungi) may grow explosively while others are completely suppressed. The disturbance of soil biocenoses may then induce a reduced litter decomposition that leads to a further shortage of minimal nutrients. Trees subject to these stress factors are in general extremely sensitive to secondary damage by frost, drought, or insect infestation.

Visible *injury to crops* as a result of ambient acidic precipitation has not been reported from modern high-productivity agrarian ecosystems. Therefore, attempts to evaluate impacts of acidic deposition must be based on simulated precipitation in field experiments or controlled environments. Yet most of the crop cultivars studied in the field and the crop varieties studied in controlled environments exhibited no effect on growth or yield as a result of exposure to simulated rain usually up to ten times more acidic than ambient [365]. Likewise, out of five cultivars of soybeans studied, only one exhibited reduction in yield [366]. Genetic variations, possibly in combination with characteristics of the soil or simulated rain applied in these experiments, may account for the observed variability.

3.2.2.3. Statistical Analysis of Forest Dieback

Descriptive or static models of forest decline are analyses of the spatial correlation pattern of site qualities and damage levels [348], [367] that corroborate the importance of pedogenic nutrient supply and acidification processes for forest decline. Figure 41 illustrates, on the basis of comprehensive cross-tabulations of several hundred German stands, the extent of forest damage in terms of relevant site factors. The standardized damage level *S* (i.e., proportion of injured : unaffected forest area) correlates best with the pedogenic nutrient supply, frequency of fog situations, and elevation, while the correlation with other site factors is also significant on the 99.9 % level although of little statistical relevance; $S = f(Sq)$ is a monotonous function, while $D = f\ (Nf)$ has a marked maximum. Thus, stands on members of the ranker, cambisol, and podzol soil groups with pH <4, marked nutrient deficiency, and concomitant reduction of buffering capacity down to the aluminum and iron ranges are most liable to dieback. Among airborne pollutants, SO_2 is particularly important. In combination with other pollutants, its effect on both vegetation and soil is enhanced by fog (contingency coefficient after PEARSON $C = 0.53$). The amount of sulfur deposition is controlled mainly by precipitation ($C = 0.80$). Comparison of the pH and sulfur deposition curves ($C = 0.89$) summarizes the major influence of gaseous and particulate pollutants on forests as a result of the above mechanisms. Their inherent complexity implies the recognition that critical levels for tolerable pollutant concentrations can be defined adequately only with regard to the entire set of concomitant stress factors affecting a plant community.

A more sophisticated analysis of the data of forest decline monitoring programs can be brought about by the CHAID approach (see Section 2.1.1.1) [103], [368]. CHAID divides the total data set successively into subsets that are, at each level of partitioning, characterized by the very variable that best explains the distribution and intensity of damage. Thus, the explicative value of different combinations of decline-inducing factors can be tested on high levels of statistical significance. A further advantage is the easy interpretability of the CHAID dendrograms.

Thus, a dendrogram of forest decline of about 500 stands in North Rhine-Westphalia clearly indicates the predominant role of the factors age, crown density, and composition of stand as boundary conditions of forest decline, while climatic factors such as annual precipitation and anthropogenic pollutant gases play a distinctly lesser role. These results are corroborated by comparable findings in other parts of Germany [103].

Dynamic Models of Forest Dieback. Simulation models of forest decline have to deal with two types of change – (1) change in complexity through increasing or decreasing differentiation, and (2) adjustment to extrasystemic changes. Negative and positive feedback processes can be correlated with systemic continuity or discontinuity, respectively. Continuity results when deviation-correcting mechanisms operate to ensure that structure and function are kept viable within the given parameters of the system. Quantization occurs when deviation is amplified to the point at which no deviation-correcting mechanism can prevent rupturing of the basic systemic framework (i.e., when resilience mechanisms are no longer able to contain and canalize the energies and thrust that have been generated). The process alters the relationships of the system to its environment, creating new spatiotemporal, structural, and functional boundaries – in short, the system is transformed to a new level of internal organization and environmental integration. Therefore, and specifically with regard to forest decline, models must employ system concepts to explain the circumstances in which quantization can occur from one developmental stage to another. Hence, they have to account for both systematic levels of biocenotic organization and cybernetic processes that demonstrate (1) systemic self-stabilization within a given organizational level and (2) systemic transformation resulting in a biocenotic quantum jump across an environmental frontier.

Models with Nonlinear Dose – Effect Relationships. Following models of reduced photosynthesis and leaf formation, and increased leaf aging and shedding under air pollution stress [369], [370], KOHLMAIER and PLÖCHL [371] developed a more comprehensive model concept for damage of Norwegian spruce. It includes the root system with reduced nutrient uptake and higher feeder root turnover, thus coupling the canopy and the soil root system. Reduced photosynthesis and changed assimilate allocation are simulated during the entire exposition to airborne pollutants, which may be a part of or the whole life span of a tree. Assimilate demand and supply have a hierachic structure with regard to, e.g., respiration, growth of foliage, roots, and living wood. The effect of an air pollutant on leaves and roots depends on both the

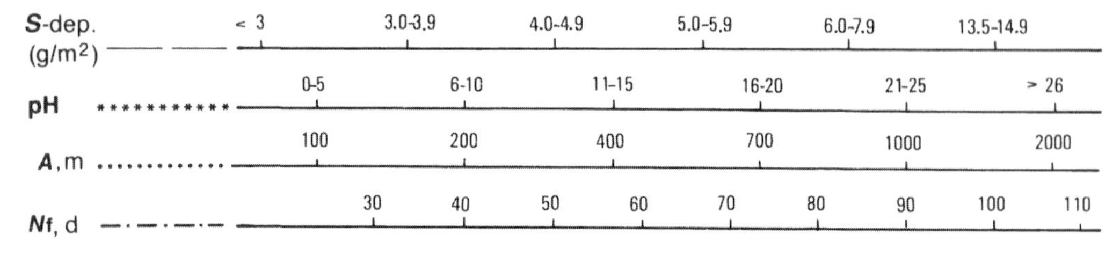

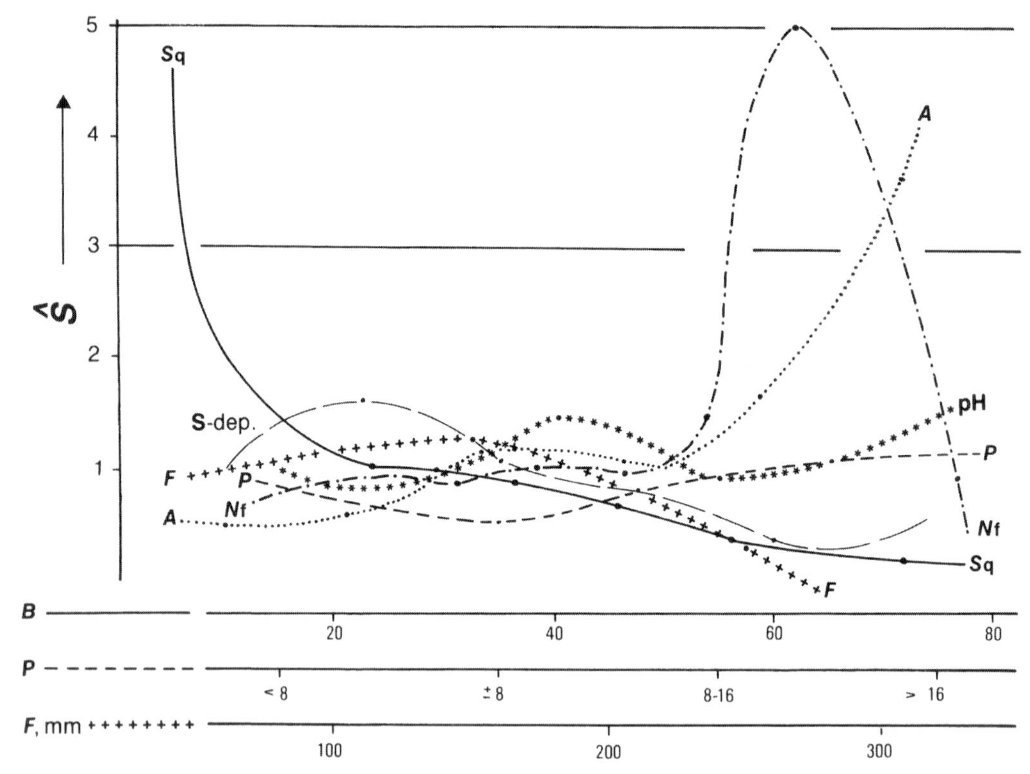

Figure 41. Forest decline as related to selected site factors
$\hat{S}$ = Ratio of damaged : undamaged forest area; A = Altitude (m); Nf = Number of foggy days; Sq = Soil qualitiy (nutrient supply); P = Porosity (percentage of pores >50 μm); F = Plant available field capacity (mm); pH = Percentage frequency of acid rain (i.e., pH <4.5); S-dep. = Sulfur deposition

specificity of instantaneous effects and the cumulative effects during the entire exposure time involving elimination and repair mechanisms of the plant. The complicated interaction pattern of these response mechanisms normally results in nonlinear dose – effect relationships with respect to concentration and accumulated pollutant and damage level of the plant. During one time step, the model first calculates the photosynthesis rate as dependent on external factors such as light intensity and air temperature, and internal parameters such as photosynthetic leaf efficiency; water uptake by roots is the limiting factor. Pollutants can enter the system by two different paths (i.e., leaves or needles and roots); in the first case, the photosynthetic efficiency of a particular needle age class is reduced; in the second, fine root turnover and consequently assimilate demand are increased. The momentary pollutant flux into the leaf is described in terms of gas exchange between the intercellular air and the ambient air by Fick's first law:

$$I_{\text{in}}^{\text{leaf}} = (c_{\text{a}} - c_{\text{i}}) / r_{\text{ai}} \quad (67)$$

where the flux $I_{\text{in}}^{\text{leaf}}$ is proportional to the concentration difference $(c_{\text{a}} - c_{\text{i}})$ between ambient and intercellular air, and r_{ai} is a specific resistance factor for a particular pollutant. The corresponding flux per unit canopy or unit ground area is then obtained by multiplication with the leaf area index *LAI*. For the boundary condition $c_a \gg c_i$ Equation (67) reduces to

$$I_{in} = (r_{ai}/LAI)\, c_a = u_{dep} c_a \quad (68)$$

which is approximately equivalent to the dry deposition rate, with u_{dep} being the deposition velocity.

The pollutant load in the leaf or at its surface may be assumed to be eliminated either by physical processes (e.g., litterfall), by transport to other parts of the plant, or by metabolic processes. It appears more appropriate, however, to model the elimination process according to Michaelis – Menten kinetics of the type

$$d\bar{c}_i/dt = u_{dep} c_a - k_{out}\, \bar{c}_i / [1 + (\bar{c}_i/K_c)] \quad (69)$$

where $\bar{c}_i$ is the concentration per unit ground area and k_{out} the characteristic elimination constant, while the other symbols have the above meaning. The mode of interaction and the effect of a pollutant inside the plant are likely to be specific for the particular pollutant *j*. In the simplest case, the effect, above the specific no-effect level, will be proportional to the interior concentration $\bar{c}_i^j$:

$$eff^j\,(t) = a_j \bar{c}_i^j\,(t) \quad (70)$$

where a_j is a specific effect constant relating to a particular process (e.g., photosynthesis). The effect variable *eff* (*t*) may be composed of linear and higher-order terms of *n* interacting species:

$$eff\,(t) = \sum_{j=1}^{n} a_j \bar{c}_i^j\,(t) + \sum_{j=1}^{n} \sum_{k=1}^{n} a_{jk} \bar{c}_i^j \bar{c}_i^k \quad (71)$$

where synergetic effects are included in the second term with $j \neq k$.

According to Bossel et al. [370], the effect variable affects the leaf senescence function

$$ALT\,(t) = 0.1\,[1 + eff\,(t)] \quad (72)$$

with $0 \leq eff \leq 1.5$ and $0.1 \leq ALT \leq 0.25$, implying that the value 0.1 is the natural senescence rate without pollutant damage. The senescence rate influences the number of needle age classes:

$$NJ = 0.8/ALT\,(t) \quad (73)$$

and both determine the leaf's photosynthetic efficiency. Hence, the photosynthetic efficiency of a particular needle age class is

$$E\,(N) = 1 - ALT\,(t)\,[(N-1) + TJ] \quad (74)$$

where $0 < TJ < 1$ characterizes the season and $1 < N < J$ the age of a needle class.

The average photosynthetic efficiency is then expressed by

$$EFF = \sum E\,(N)\, L\,(N)\, / LEAF \quad (75)$$

where $L\,(N)$ is the leaf mass of a particular needle age class and *LEAF* the total leaf mass. In unaffected systems, *EFF* ranges between 0.55 and 0.65 [371].

Acidification of soil and related leaching of mono- and bivalent cationic nutrients leads to higher feeder root turnover rates while a gradual loss of feeder root mass occurs. Following a titration curve of a buffered chemical system, this effect may be described by a logistic curve of the type

$$W_{to}\,(t) = f\,[W\,(t)] = W_{max} / \left[1 + F_k \left(3^{-Klt}\right)\right] \quad (76)$$

where $W\,(t)$ is the accumulated pollutant at time t; W_{max} the maximum accumulation; F_k the correction factor $(W_{max} - 1)$; K the buffer constant; l the constant input; and $W_{to}\,(t)$ the fine root turnover factor at time t [370].

This supply – demand model of photosynthetic efficiency and assimilate allocation appropriately describes the decline of a forest ecosystem under the influence of chronic air pollution stress. It clearly indicates that nonlinear effects may drastically accelerate the observed dieback phenomena, but more experimental data on both the biochemical and the physiological levels are required to better account for the combined effect of two or more pollutants [55], [372–374].

Time Series Analysis and Mapping. A common feature of the various methods of time series analysis of dynamic systems such as ecosystems is that they involve the observation of a complete set of state variables. This may constitute a major drawback in studies of empirical systems whose inherent dynamics are not known a priori. Under these circumstances, however, use can be made of the fact that in dynamic systems with a finite number of state variables, information on the momentary value of all state variables can be replaced by information on the recent history of a part of the variables. GROSSMANN et al. [375] therefore used time series analysis in combination with geographic information systems (GISs) to develop a scenario that describes forest damage in sequential form by means of maps. A geographic information system is a computer hardware and software system designed to collect, manage, analyze, and display spatially referenced data. Geographic information systems are emerging as the major spatial data-handling tools for solving complex natural resource planning problems, and the use of GIS technology has revolutionary implications for conducting research and presenting research results.

The GROSSMANN et al. approach POLLAPSE is a combination of dynamic models writing information into the data bank of a GIS, which thus produces a time series of maps of forest growth or damage [375]. These maps can be compared with the actual development so that deviations become readily discernible. The hypotheses underlying this type of regionalized modeling are

1) Photochemical oxidants cause foliar damage, and subsequent attack of airborne pollutants leads to leaching, eventually resulting in nutrient deficiency [376]
2) Ulrich's soil acidification hypothesis [362]
3) A combination of the two preceding assumptions

The predictive quality of these hypotheses was tested in the Pfaffenhofen area (Bavaria) on the basis of 462 forest stands, and the POLLAPSE methodology showed that the combined hypothesis (3) yielded the relatively best interpretation of forest damage phenomena observed in terms of both spatial pattern and degree of damage. The above methodology of time series analysis and mapping also offers excellent possibilities for the management of forests or other biotic systems as a base for sustained viability.

3.2.3. Impact of Xenobiotics on Aquatic Ecosystems

Xenobiotics released to aquatic environments have a great impact on biocenoses inhabiting streams, wetlands, and lakes. This adversely affects not only the respective aquatic ecosystems, but also humans who are in contact with this water. Aquatic organisms show how far-reaching consequences can be caused by effects that are difficult to detect on a short-term scale and therefore easily overlooked. Similar to effects on aquatic biocenoses, in most instances the effects of xenobiotics on humans cannot be recognized, not only because the necessary analyses cannot be made, but also because small, difficultly detectable effects can grow into major and irreparable population damage. The following describes how difficult it is to foresee the effect of a chemical on an aquatic ecosystem merely from (easy to make) short-term tests. This is a warning that care for human health and life means that society cannot wait until some clearly seen negative effect is proved beyond any doubt but must react on early warnings obtained by sensitive methods. For more detailed information, see [56–64].

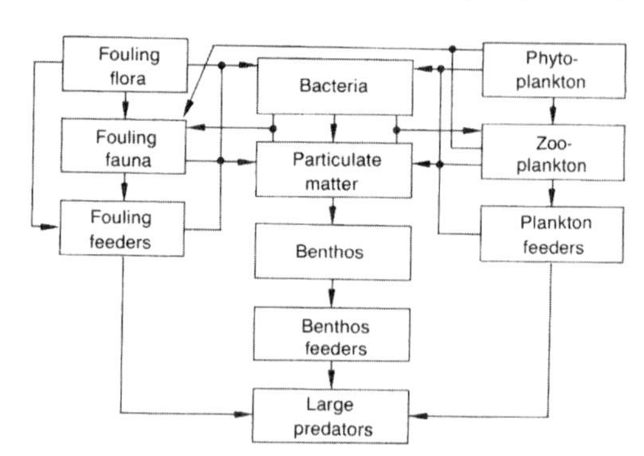

Figure 42. Food chains or food webs as the organization of organisms within an ecosystem according to food selection; arrows indicate directions of nutrient flow

3.2.3.1. Approaches and Methods of Studying the Impact of Xenobiotics on Aquatic Life

Aquatic life is organized into several hierarchic levels (Fig. 9 in Section 2.1.4):

1) Individual organisms
2) Monospecific populations
3) Communities of different species inhabiting the same environment, e.g., the pelagic communities freely swimming in water and benthic communities inhabiting the bottom of rivers and standing waters (ponds, lakes, reservoirs)

4) Ecosystems with their nonliving components interacting with organisms; within an ecosystem, organisms are organized into food chains (or more exactly, food webs) according to their selection of food items (Fig. 42)

Detection of the effects of xenobiotics is performed at all hierarchic levels of biological organization. In addition to this organizational hierarchy, a hierarchy of time scales of the effects can be distinguished, from very short ones measured in minutes up to very long ones measured in human generations [377], [378].

Methodology. The most common approaches to detect the impact of xenobiotics on aquatic life are listed below:

1) Classical toxicity tests on test individuals or cultivated populations of aquatic organisms to detect short-term lethal effects
2) Methods for detecting longer-term sublethal effects on test individuals and populations
3) Methods for detecting genotoxicity
4) Use of biomarkers, a relatively new approach for detecting stress effects of long-term exposure of individual organisms living in contaminated environments
5) Microcosm and mesocosm studies
6) Monitoring of long-term consequences of contaminants in nature by following the species diversity and loss of sensitive species in biocenoses and ecosystems

Other approaches are, e.g., predictions of toxicity based on molecular structure of the respective substances using quantitative structure – activity relationships (QSARs, see, e.g., [10], [220]), and tests for changes of morphologic or morphometric characteristics of organisms and detection of pathological individuals.

Classical Short-Term Toxicity Bioassays. Water organisms employed for these tests are usually some species of easily cultivated algae, water flies, and fish (see Section 3.2.1). Benthic, bottom-dwelling organisms commonly used are worms of the family Tubificidae living in the bottom mud of streams and lakes that are fairly polluted with organic matter of natural origin. In the simplest version, the tests consist of adding different concentrations of xenobiotics to jars (aquaria) containing the respective organisms and determining the time until half of the population dies (LC_{50} or LD_{50}). Recently, much care has been taken to standardize aquatic toxicity tests to maximize their major advantage – the rapid comparative evaluation of many xenobiotics [65].

Long-Term Sublethal Effects of Xenobiotics. In the aquatic environment, long-term sublethal effects are studied by measuring different reactions of test organisms. An example is the behavioral test with the freshwater bivalve mollusk *Dreissensia polymorpha* [379]. Its filtering rate, which is determined anemometrically in this test, represents a vital function, providing food and therefore vitality, growth, and reproduction of the organism. Detection limits of long-term sublethal tests are several orders of magnitude higher than those of short-term toxicity tests.

Genotoxicity. Compounds such as diethylstilbestrol that disrupt the endocrine systems of fish, wildlife, and humans were found to induce large-scale dysfunction at the human population level [66] by causing genetic and reproductive changes. Genotoxicity is studied by observations of the offspring of the affected population.

Determination of Biomarkers. Measurements of body fluids, cells, or tissues are carried out that indicate in biochemical or cellular terms the presence of contaminants or the magnitude of the host response [63]. Examples are, e.g., determination of marker enzyme activity in cells of target tissues, serum enzyme activity, appearance or loss of reaction products of an enzyme; major changes in the concentration of a cellular component such as glycogen, fat, mucus, or metallothionein; and measurements of damage to DNA.

Microcosm and Mesocosm Studies (see Section 3.2.2.1). Microcosms are laboratory model ecosystems consisting of mixed natural populations of organisms transferred from nature to a laboratory jar or aquarium, sometimes an arrangement that imitates a stream. Mesocosms are of the same origin, but of much larger volume and usually kept outdoors in their natural environment (e.g., [59], [380–382]). When xenobiotics are added, the changes in some common properties are measured (e.g., species composition, diversity, photosynthesis, or respiration).

Studies of Changes in Structure and Function of Communities and Ecosystems. Toxic chemicals can affect independently both the structure and the function of aquatic ecosystems. Low levels of perturbation may cause changes in community composition or structure without a change in function. Alternatively, a process, the rate of which is changed by stress (such as respiration or photosynthesis), may be reduced under sublethal stress, but the community structure is left intact. According to [383], the most predictable changes seem to be the disappearance of sensitive species from communities under environmental stress.

The integrated bioindicator approach currently gives the best results. It involves monitoring a suite of selected exposure and effects indicators at several levels of biological organization from the biomolecular to the community level [63], [384].

Critical Evaluation of the Above Approaches to Aquatic Ecosystems. For aquatic environments, *short-term lethal toxicity tests* have many advantages, as well as many disadvantages. Their major advantage is the ability to rapidly compare the relative toxicity of specific chemicals or specific mixtures. The limitations become particularly obvious when trying to use their results for higher levels of biological organization in the aquatic ecosystem. Thus, they do not account for the effects of

1) Chemical speciation in the environment
2) Kinetics and hysteresis in sorption of chemicals to sediments
3) Bioaccumulation through the food chain from the lowest level of algae and plants to the highest level of fish
4) Modes of toxic action on survival, growth, and reproduction of individual species populations

For aquatic sediments, an additional inadequacy is that sediment toxicity tests often rely on equilibration of chemicals in a sediment – water mixture or employ benthic organisms whose survival, growth, or reproductive success may depend not only on the presence of the toxicant but also on physical characteristics of the sediment.

The application of results of (laboratory) toxicity tests to natural populations is especially risky for the following reasons:

1) The way in which laboratory organisms are exposed to pollutants differs from exposure in the environment
2) Laboratory tests deal with single chemicals, and organisms are exposed to complex mixtures in the environment
3) Criteria for effects in the laboratory are not important functional endpoints in population and systems dynamics

The inadequacy of short-term tests for the aquatic environment may be shown by conclusions drawn from careful monthly bioassay results with the survival and reproduction of a water fly (*Ceriodaphnia*) and larvae of fathead minnow. These indicated that the effluent of an energy plant at Oak Ridge (Tennessee) causes no observed effect. However, community surveys demonstrated clear effects on diversity and species richness downstream of the power plant [63].

The use of *biomarkers* is more relevant to conditions in nature than acute toxicity methods. Biomarkers provide evidence of exposure to rapidly metabolized contaminants and in this way integrate pharmacodynamic and toxicological interactions. Correlations between biomarkers of exposure and ecological effects are expected to be better than those between populationmonitoring parameters and indicators of exposure.

Observations in nature can be performed only in rare instances when the concentration of the xenobiotic in nature is increased for a prolonged period. These observations cannot be used for testing new materials. Therefore an integrated system of xenobiotic hazard evaluation has been developed (see Sections 3.2.1, 3.2.3.4).

3.2.3.2. Behavior of Xenobiotics in the Aquatic Environment and their Influence on Aquatic Organisms and Indirectly on Humans

A chemical of potentially xenobiotic character follows a complicated flow from production site to the environment, as shown in Figure 43. The situation in fresh water depends not only on the direct emission of the respective chemical to the water body but also on its emissions and fates in soil and groundwater (both associated with vegetation growth and decay), on the chemical and biological processes in the water body, and on the bioaccumulation in different organisms of a food chain. Environmental processes connected with local hydrometeorology, with precipitation and flow, and with temperature and light conditions modify the fate of xenobiotics. Figure 44 illustrates the behavior of hydrophobic xenobiotics in standing waters, particularly the exchange between the pelagic and benthic environment (i.e., between water and sediments). The participating major processes are shown: sedimentation, resuspension, and mud – water exchange. For hydrophilic xenobiotics, another important process is bioaccumulation (i.e., accumulation inside the organism's tissues and organs, mainly in fat and liver). In addition to these processes of an individual chemical compound, complex interactions occur between different chemical species in nature.

Many organic contaminants are readily taken up and *bioaccumulated* by aquatic organisms [61], [68] if the chemicals are dissolved in water. The bioavailability is much lower if the same compounds are sorbed to soil or sediment particles. Many toxic chemicals do not bioaccumulate but rather are metabolized. In many cases the metabolites are more toxic than the parent compounds present in the environment. Bioaccumulation, the process by which a substance is taken up by an aquatic organism from water and through food, is found to depend strongly on the size (weight) of the respective organisms (see, e.g., for pesticides [56], [387]). This is due to the dependence of most physiological processes on size (e.g., [59]). Biomagnification within aquatic food chains (i.e., the increasing concentration of toxic substances in higher organisms) was found important only for some chemicals (e.g., DDT [388]).

Humans are affected by the processes in water in different ways, as a result of their direct and indirect contact with water, but also through air, soil, and food. Large differences exist in the biological availability of contaminants associated with different environmental media, as well as individual and species-specific differences in pharmacodynamic disposition.

In most instances, one thinks of direct effects of the xenobiotic on organisms. However, recent ecological findings demonstrate that in nature the indirect effects are more important [389]. They are caused, e.g., by changes in food composition when some organisms become rare or disappear because of toxic effects and by changes in competition and other possible relations between organisms. An easily understandable example is the eradication of an aquatic weed species. Indirectly, this will cause a complete change of abiotic conditions not only for other species that depend on the weed directly, such as invertebrates living on it, but also for fish species finding their shelter.

Adaptation. Organisms are capable of adapting to the effects of xenobiotics to a certain degree, and this can have both positive and negative effects on individual organisms (including humans) and on the total environment, as well as on environmental management. Adaptive acquisition of degradative abilities for some organic compounds or of resistance to heavy metals has been demonstrated for bacteria in laboratory ecosystems [390]. Such an adaptive response may be caused by different molecular and biochemical processes: (1) induction of specific enzymes in members of the community, resulting in an increase in the observed degradative capacity of the total community; (2) growth of a specific subpopulation of a microbial community able to take up and metabolize the substrate; and (3) selection of mutants that acquire altered enzymatic specificity or novel metabolic activities at the onset of exposure of the community to the introduced compounds.

The *negative effects of adaptation* may be exemplified by pests (e.g., insects or weeds) in the framework of eradication programs [67]. In West Africa, the widespread parasitic disease filariosis is caused by the *Filaria.* This parasite is transmitted to humans by blackflies. To eradicate the flies, the organophosphate Temephos was applied on a large scale in about 18 000 km of rivers in the region during 1974 – 1980. However, resistance to this xenobiotic developed soon in one region and spread rapidly. Therefore, other means of eradication were sought that would act more persistently and not show the negative effects of the insecticide on other organisms in the rivers, since Temephos causes up to 100 % mortality of many insect species living on the river bottom. No toxicity to fish was observed. However, the insecticide accumulated in fish tissues,

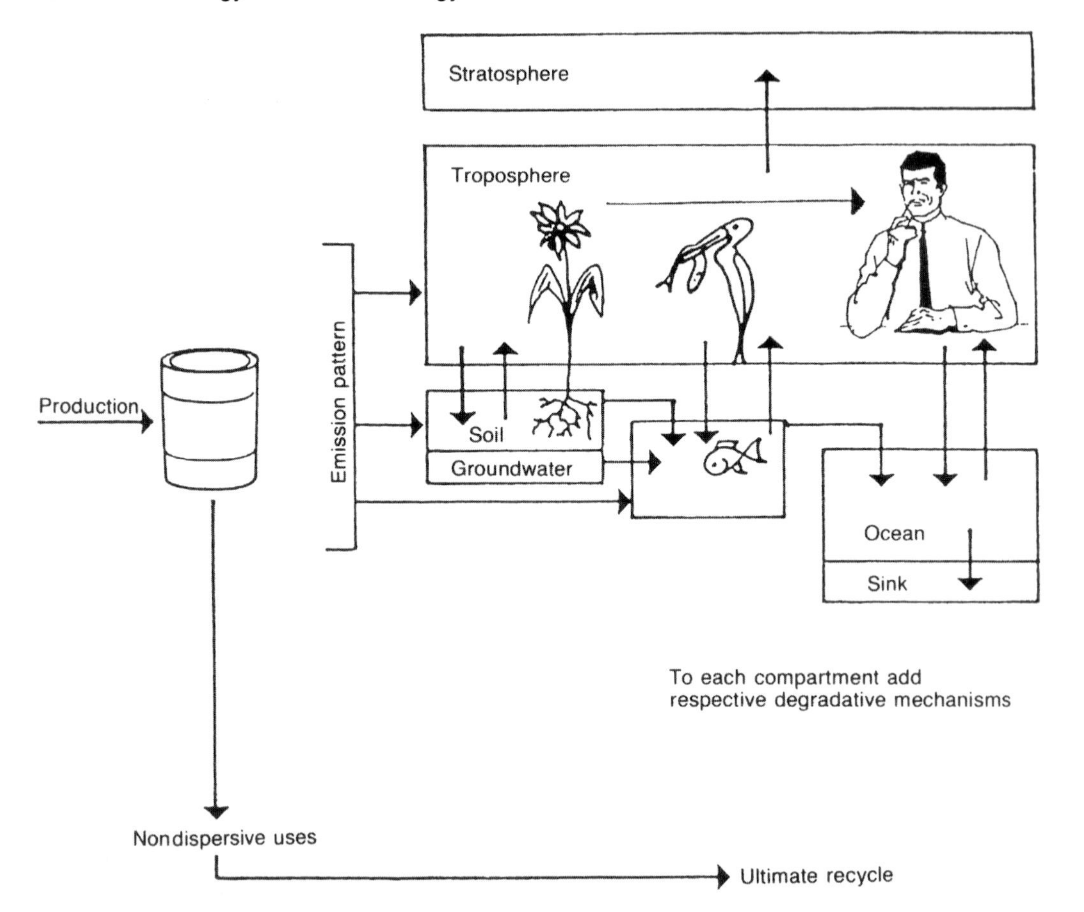

Figure 43. Schematic representation of the flow of a chemical from production to the environment (after [385])

which may have negative effects on humans that consume the fish.

In addition to the use of another more toxic organophosphate (chlorophoxime), a biological method of eradication was developed. This method demonstrated the general advantage of using biological agents. The bacterium *Bacillus thuringiensis* appeared to be very selective in killing only *Simulium* and few other species, and no adaptation of the fly populations was observed. Therefore, the application of a biological agent has the following advantages: (1) it is selective for the target goal; (2) no negative side effects are observed; (3) no rapid adaptation of pests takes place; and (4) the cost of application is lower.

Because of adaptation and other differences, the applicability of basic ecotoxicological principles derived in the temperate region to warm, arid, and cold climates is not without problems [67]. No simple straightforward relation exists for assessing regional differences of toxic effects due to differences in temperature and associated climatic factors. The general belief that metabolic rates are much higher in tropics has not been supported; nevertheless, an increased risk of toxic effects in tropics is to be expected.

3.2.3.3. Impact of Xenobiotics on Pelagic and Benthic Biocenoses

The data bank of the International Register of Potentially Toxic Chemicals (IRPTC) of UNESCO contains a record of some 8000 chemicals, and the number is growing rapidly. Therefore, only a few illustrative examples are described in the following, with an emphasis on related problems. The quantitative aspects of the

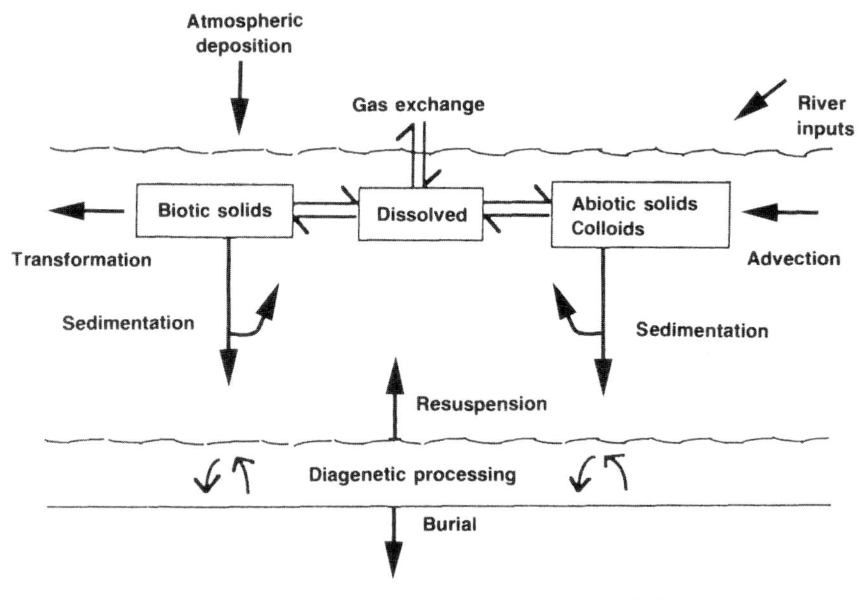

Figure 44. Behavior of hydrophobic xenobiotics in standing waters (after [386])

processes and predictions of the impacts of xenobiotics are described in Section 3.3, dealing with ecotoxicological modeling.

An example of the dependence of toxicity not just on one element, but on the ratio of several, can be demonstrated by the case of arsenic. REUTHER [391] found for Swedish lakes that, when added as arsenate, arsenic was readily accumulated in plankton at concentrations close to those found in nature. If the ratio P : As was low, arsenic proved toxic at much lower concentrations.

For the trace metals cadmium, lead, copper, and zinc, the effect of external variables on the amounts present in the environment was specific for each element [392]; especially important for Cd and Pb were sediment- and water-related abiotic variables. For the essential trace metals, the variance expressed by the effect of the common physical variables of the surrounding environment was very low: from 25 % for Zn up to 37 % for Cd and Cu. The quantities of the common xenobiotic cadmium in the aquatic environment can be demonstrated by the example of the river Ruhr in Germany (Fig. 45). A long-term effect of Cd on aquatic biocenoses was measured by using the microcosm method [380]. Figure 46 shows a steady decrease of the number of species in the microcosm with increasing cadmium concentration.

Some detergents greatly affect nature not only because of the amount of phosphates they contain (leading to eutrophication of water bodies), but also because of their toxicity. The detergent FIT was found to be highly toxic to frogs and their larvae [394].

Evidence exists of the sometimes close relation between the occurrence of some xenobiotics and the amounts produced in the respective region. Figure 47 shows how the quantities of PCBs, DDT, mirex, and chlorobenzene accumulated per year in layers of some U.S. lake sediments follow the sales or production rates of the respective years.

A systematic study of 50 chemical compounds demonstrated that their joint toxic effects can be considered simply as additive [395]. The authors proposed a quantitative structure – activity relationship (QSAR) technique for predicting the concentrations of the components of mixtures that would cause 50 % inhibition of the toxic effect by joint action. How far the effect of other compounds can act synergistically is not clear.

Range of cadmium concentrations of different waters in the Ruhr area

	Concentration, μg/L
Groundwater	0.05 – 0.15
Drinking water	0.05 – 0.2
Falling rain at Essen	0.2 – 1.1
Storm water flow from paved areas	1–10
Seepage water from sewage sludge deposits	<1
Seepage water from industrial sludge deposits	5–50

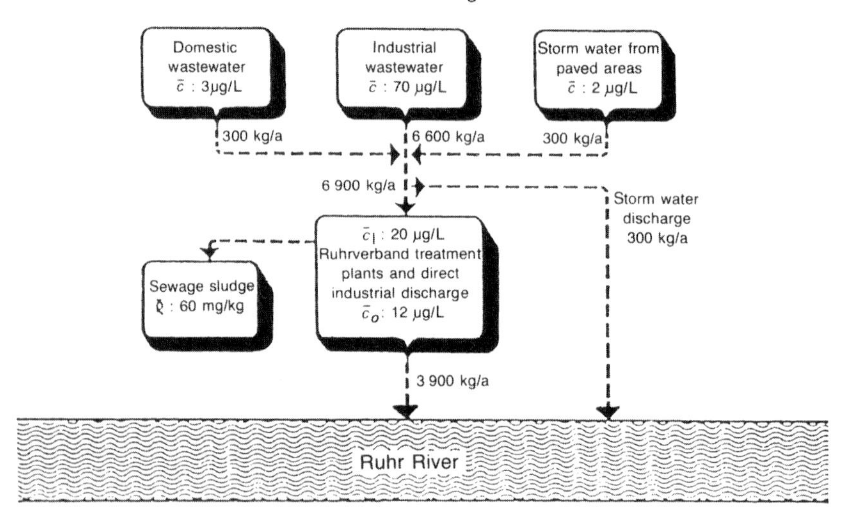

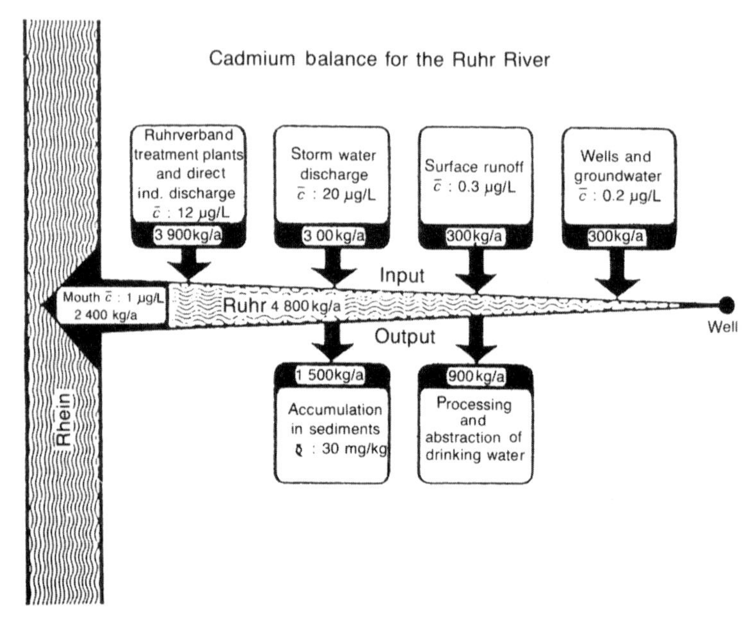

Figure 45. Origin, fate, concentrations, and annual input–output figures for cadmium in the river Ruhr (Germany) (after [393])

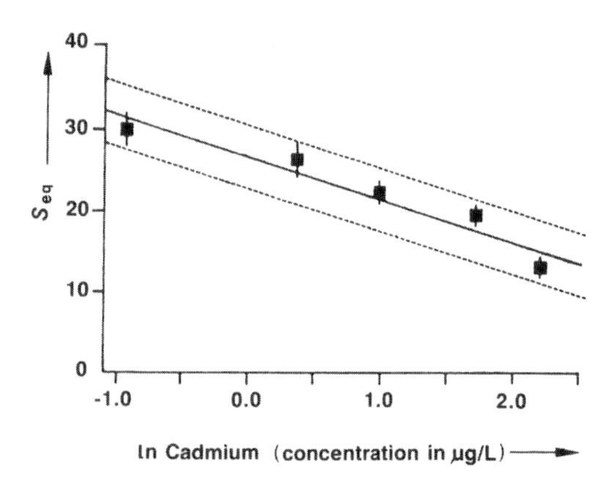

Figure 46. Microcosm analysis of the long-term effect of cadmium on the equilibrium number of species (S_{eq}) in aquatic biocenoses (after [380])

3.2.3.4. Ecological Risk Assessment of Xenobiotics

Ecological risk is a condition in which the normal functions of a population, ecosystem, or entire landscape are threatened by external forces or stress factors that, at present or in the future, may diminish the health, productivity, genetic structure, economic value, or aesthetic quality of the system [395]. In addition to xenobiotics, ecological risk may be caused by excessive inputs of nutrients; disturbance of normal energy flow by thermal pollution or global warming due to greenhouse gases; drastic changes in ecosystem structure from drying peatlands; and physical destruction of ecosystems by excessive grazing of rangelands, burning of forests, or covering with heavy loads of industrial dust.

Ecosystem risk assessment is defined as a set of procedures for measuring risk to the environment associated with the use of xenobiotics through an objective and probabilistic exercise based on empirical data and scientific judgment. The results of such a risk assessment can then be used to provide a consistent means of estimating the limiting concentrations of substances that will produce no unacceptable negative effects on ecosystems that may be exposed to the substance (see Section 3.2.5). The phases of making and reviewing an environmental risk assessment are presented in Figure 48. More about the method can be found in [68], [397], [398].

Concluding Remarks. The consequences of massive application of xenobiotics in Vietnam are drastic. Pesticide damage to nature involved the death of millions of trees and often their ultimate replacement by grasses, maintained in turn by subsequent periodic fires; deep and lasting inroads into the mangrove habitat; widespread site debilitation via soil erosion and loss of nutrients in solution; decimation of terrestrial wildlife, primarily through destruction of their habitat; losses in freshwater fish, largely because of reduced availability of food species; and a possible contribution to the decline of offshore fishery. Humans are suffering both direct and indirect effects. In addition to a variety of health problems of exposed humans, the abovementioned widespread damage to the inland and coastal forest ecosystems has had major indirect effects due to large-scale devastation of crops [67].

In the future, additional complications must be expected in the framework of xenobiotic application. These will be connected not only with the continuing adaptation of target organisms but also with changes in the total environment that are expected to be global. These include global warming accompanied by shifts in species composition, in food crops and production, and in conditions for human life. A mistake has been made in conceding to chemicals the right to be judged innocent until proven guilty. Therefore, a new strategy toward the application of xenobiotics is necessary, based on the following rules:

1) Minimize the number of newly invented chemicals in production and commercial use
2) Minimize the number of chemicals used in social and economic activities
3) Give priority to naturally occurring rather than synthetic chemicals
4) Give priority to the production of solid substances over liquids and liquids over gases
5) Minimize the spatial scope and time of their presence in the environment

3.2.4. Testing Approaches to Atmospheric Effects

Biotic Effects. For atmospheric hazard assessment, biotic and abiotic approaches are

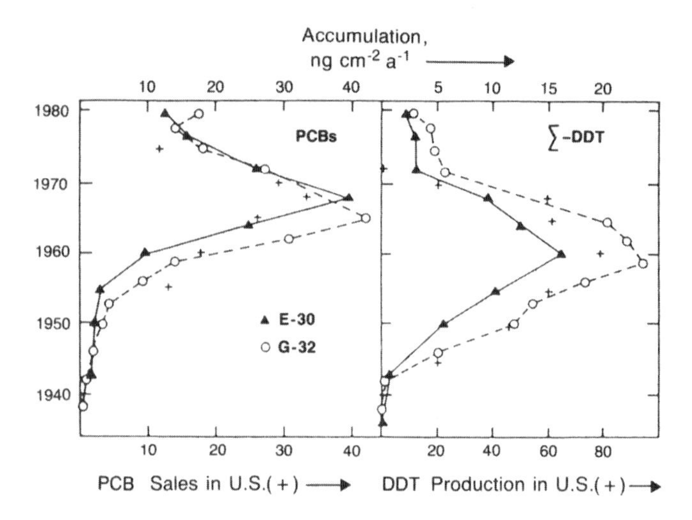

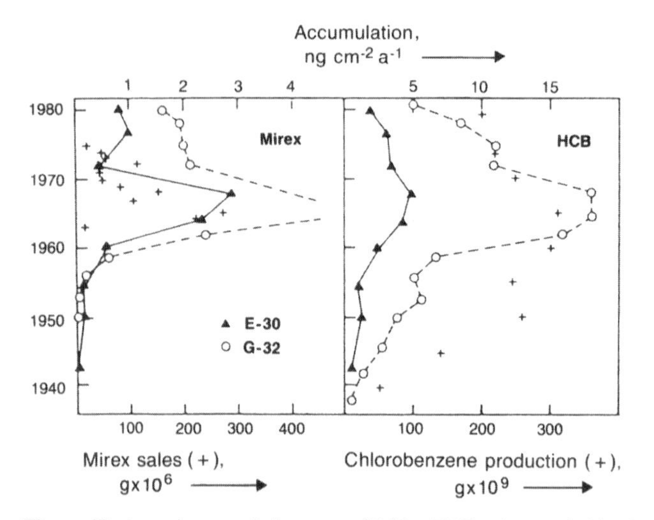

Figure 47. Annual accumulation rates of PCBs, DDT, mirex, and chlorobenzene in sediments of selected U.S. lakes (after [386])
E-30 and G-32 represent two different sampling stations (cores)

possible, although internationally accepted test guidelines to more precisely define the effects of airborne contaminants on organisms have not yet been developed. In most cases, only data on acute mammalian toxicity are available, which can be used for rough estimation of the risk a chemical poses for animals. However, atmospheric concentrations are normally not high enough to cause acute toxic effects. Therefore, data on long-term or chronic toxicity or on mutagenic and teratogenic effects would be more appropriate.

Concerning the toxicity to terrestrial plants [2], no standardized test procedures are available for chemicals applied directly in gaseous or deposited form since a guideline for tests with herbaceous species has not yet been accepted. For some chemicals, enquiries into invertebrate toxicity – mostly insects (e.g., *Apis mellifera, Syrphus corollae*) – are available, which were conducted according to guidelines for the testing of plant protection agents.

Abiotic Effects. As regards abiotic effects, consideration of the following atmospheric

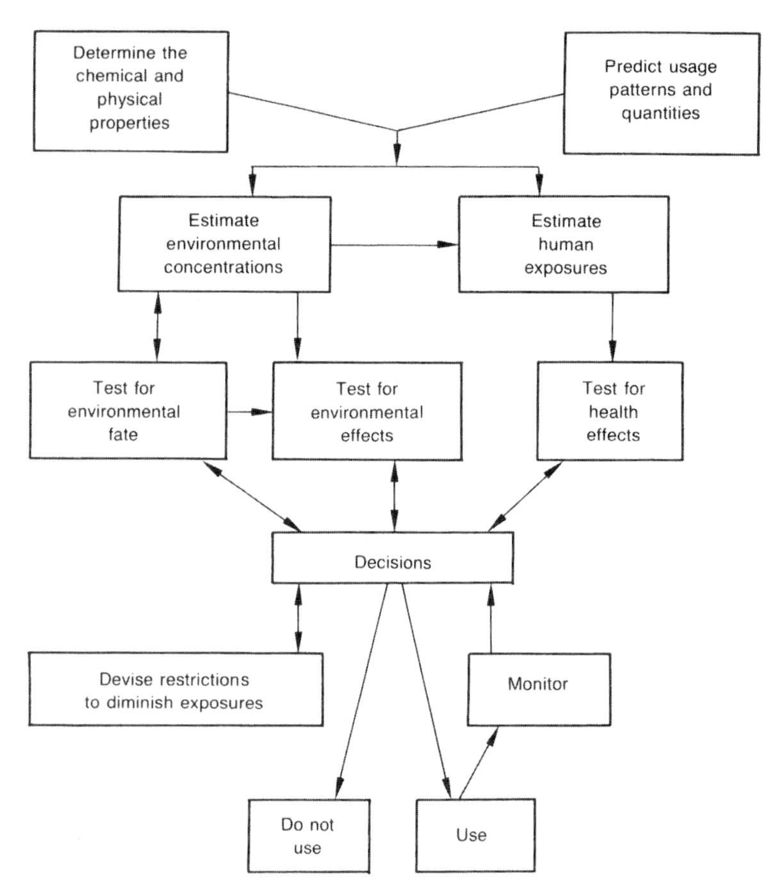

Figure 48. Environmental risk assessment (after [396])

processes is indicated [399]: global warming, stratospheric ozone depletion, tropospheric ozone formation, and acidification. A potential greenhouse gas has absorption bands in the 800–1200-cm^{-1} atmospheric window. Its *global warming potential* (GWP) is defined as the ratioof calculated warming for each mass unit of a gas emitted into the atmosphere relative to the calculated warming for a mass unit of the reference gas CFC 11 (CCl_3F). An approximate GWP value of a chemical S is given by

$$GWP = \frac{\tau_s M_{CFC11} S_s}{\tau_{CFC11} M_s S_{CFC11}} \quad (77)$$

where τ_s is the atmospheric lifetime ($\tau_{CFC11} =$ 60 a); M_s the molar mass ($M_{CFC11} = 137$ g/mol); and S_s the IR absorption strength in the interval 800 – 1200 cm^{-1} ($S_{CFC11} = 2389$ cm^{-2} bar).

In the above formula, the atmospheric lifetime is defined as

$$\tau = \frac{1}{k_d + k_w + k_c} \quad (78)$$

where k_d is the removal rate for dry deposition; k_w the removal rate for wet deposition; and k_c the pseudo-first-order chemical transformation rate.

A substance may have an impact on *stratospheric ozone* when its atmospheric lifetime is sufficiently long to allow for transport to the stratosphere and it contains chlorine or bromine substituents [290]. The ozone-depletion potential (ODP) is then defined as the ratio of calculated ozone change for a unit mass of gas emitted into the atmosphere relative to the depletion determined for an equal mass of the reference gas CFC 11 (ODP = 1). Thus, the ODP value can be approximated by

$$\mathrm{ODP} = \frac{\tau_{\mathrm{s}} M_{\mathrm{CFC11}} n_{\mathrm{Cl}} + \alpha n_{\mathrm{Br}}}{\tau_{\mathrm{CFC11}} M_{\mathrm{s}} 3} \quad (79)$$

where τ_{s} is the atmospheric lifetime ($\tau_{\mathrm{CFC11}} = 60$ a); M_{s} the molar mass of gas considered ($M_{\mathrm{CFC11}} = 137$ g/mol); $n_{\mathrm{Cl}}, n_{\mathrm{Br}}$ the numbers of Cl and Br atoms, respectively, per molecule; and α a measure of the effectiveness of Br in ozone depletion with respect to Cl ($\alpha = 30$).

The generation of *tropospheric ozone* depends on

1) The reactivity of the substance and the degradation pathway
2) The meteorological situation, with highest O_3 concentrations expected at high temperatures, intense solar radiation, and low wind speeds
3) The concentration of air pollutants, where the concentration of nitrogen oxides has to exceed several μL/m^3

Less reactive species (eg., CH_4 or CO) are important for long-term O_3 concentrations, while highly reactive species such as alkylenes or aldehydes contribute significantly to ozone peaks.

In analogy to the above depletion potential a photochemical ozone formation potential (POCP index) has been defined, based on ethylene, which has a POCP value of 100. Comparative studies showed, however, considerable variability in the POCP values assigned to each organic compound, which is difficult to explain by recourse to the basic chemical characteristics alone [2]. Therefore, ozone formation is best derived from an approximate reactivity scale based on the rate constant for the (hydrocarbon + OH·) reaction and molar mass [290]:

$$\mathrm{OH}^{\cdot}\ scale = \frac{k_{\mathrm{s}} M_{\mathrm{ethylene}}}{M_{\mathrm{s}} k_{\mathrm{ethylene}}} \cdot 100 \quad (80)$$

where k_{s} is the rate constant at $T = 298$ K for the reaction with OH radicals ($k_{\mathrm{ethylene}} = 8.5 \times 10^{-12}$ cm^3 mol^{-1} s^{-1}) and M_{s} is the molar mass ($M_{\mathrm{ethylene}} = 28$ g/mol). For a presentation of pollutant impact on metals, organic and inorganic nonmetallic materials, see [2].

3.2.5. Appraisal

Hazard assessment of chemicals is based on stepwise determinations of predicted environmental concentrations and ecotoxicological no-effect concentrations followed by comparison of the two values. Figures 49 and 50 [239] indicate that whenever the PEC/PNEC ratio is <1.0, the chemical may be concluded not to present an environmental hazard under the conditions analyzed.

In addition to and as a substitute for specifically designed networks yielding spatially representative data, fate and exposure models are useful tools for estimating concentrations and are increasingly applied in many situations. In view of the considerable number of models available, careful selection is indicated to comply with specific problems. The fundamental problem is to define the set of questions a model is expected to answer and accuracy required at a given scale. Even if a suitable model has been chosen and valid input data have been obtained, the results can still be misinterpreted if the underlying principles are not understood properly. Thus, both the assumptions and the limitations of different modeling approaches must be recognized (see Section 3.3).

The first field of model application is the assessment of the fate of chemicals at a specific location (i.e., predicting concentrations related to well-defined sources of emission). The second is the screening of chemicals whose purpose is the evaluation of potential exposure, typically at larger or even global scales. Therefore, consideration of an average environment appears justified in the latter case.

As in the case of monitoring networks [2], the networks providing input data for the calibration and validation of models also require a geostatistical control to allow for interpolation and spatial extrapolation. Only in the case of site-specific modeling is strict validation possible, while for screening models proper validation cannot be ensured, because no specific situation is available as a reference [238]. Validation of probabilistic models is possible, but this requires a substantial amount of valid environmental concentration measurements in different locations to allow geostatistical analysis of the primary data.

As summarized in Figure 49, the principle of effects assessment is to use toxicity data from

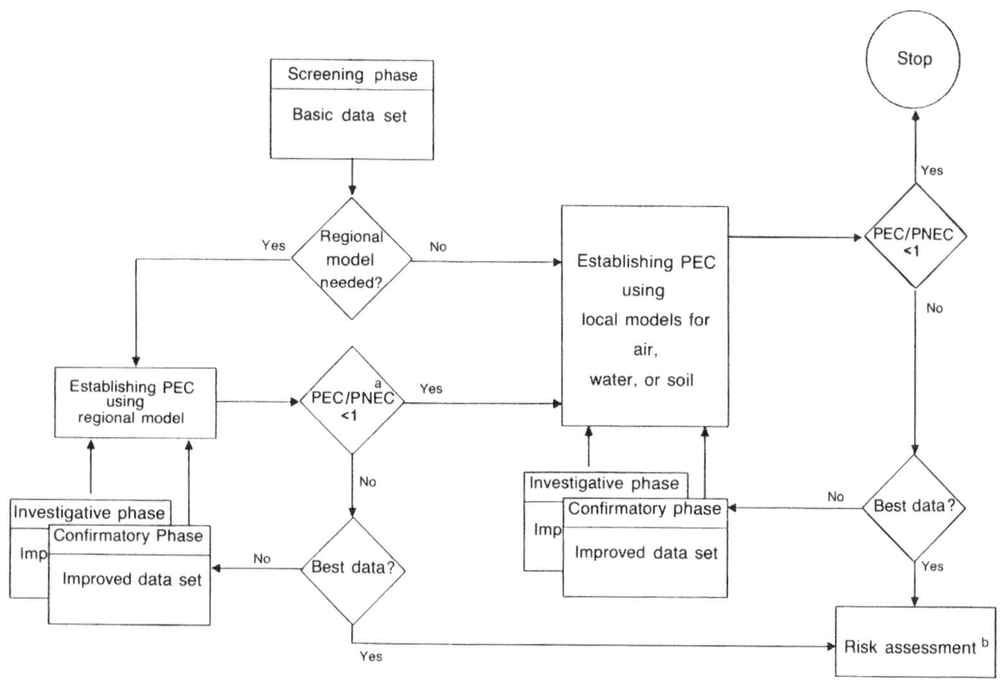

Figure 49. Exposure assessment scheme (after [239])

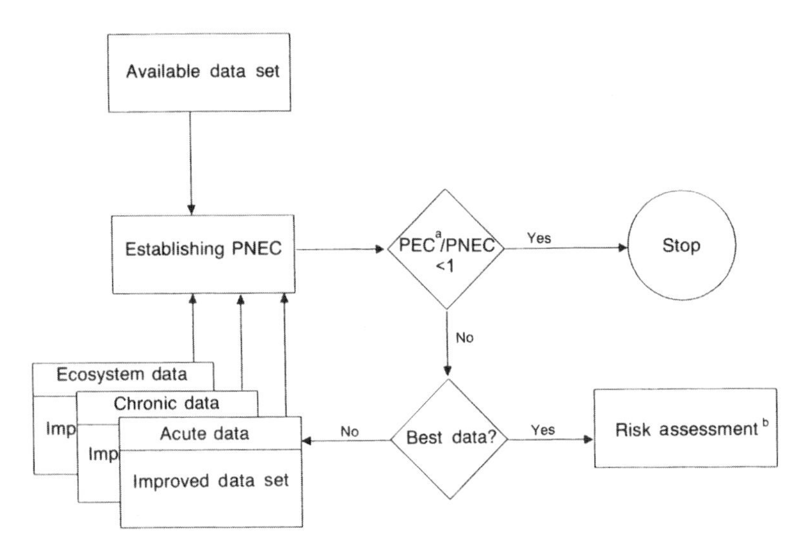

Figure 50. Effects assessment scheme (after [239])

short- and long-term experiments and appropriate assessment factors to calculate a PNEC value. In principle, this is accomplished separately for each environmental compartment, but practical difficulties may arise from the fact that most of the ecotoxicity data on existing chemicals were obtained for the aquatic compartment only.

Aquatic Hazard Assessment. According to the base set definition [289], [290], acute effect data for fish, *Daphnia*, and algae should be available. In addition, chronic toxicity investigations with standard or nonstandard organisms (see Section 3.2.1.3) may also be available. Multispecies tests, micro- and mesocosm studies, and field tests are comparatively rare for most chemicals [296]. The standard organisms used represent three trophic levels – primary producers and primary and secondary consumers – but in view of the great variability of aquatic phyto- and zoocenoses, their functional representativeness should be evaluated more precisely. In addition, a given population may represent more than one trophic level depending on the amount and spectrum of nutrition for the species; earlier life stages not infrequently live on completely different food than adults of the same species.

Therefore, ecotoxicological data collected in laboratory experiments can be difficult to translate into accurate predictions of effects in ecosystems. This applies in particular if the duration of a test is different from that of standard tests or if the test parameters (e.g., end points) are not comparable to those used in standardized test procedures (e.g., investigations of photosynthesis or behavior, studies conducted on the cellular or subcellular level).

In general, the use of laboratory test data will lead to an overestimate of chemicals' effects on complex aquatic ecosystems, which implies very conservative assessment factors, for the following reasons:

1) Bioavailability of chemicals is maximized in filtered-water laboratory systems, while in natural waters the higher concentrations of dissolved organic matter and suspended solids may reduce it substantially as a result of adsorption or complexation [291].
2) In laboratory tests, exposure concentrations are kept constant as far as possible, whereas in natural waters, exposure is likely to vary over time and degradation is greater. An overestimate is obvious when relatively long-term laboratory toxicity studies are the basis for extrapolation.
3) Laboratory experiments are normally conducted with a limited number of individuals or monospecific communites (e.g., *Daphnia magna*) and therefore exclude the various homoeotic mechanisms operating in complex natural multispecies communities.

The following features of laboratory experiments are likely to lead to an underestimate of toxic effects in ecosystems:

1) Few test species are used in laboratory test systems (see Section 3.2.1.3); therefore, they cannot be expected to encompass the most susceptible species in the field. In addition, the definition of sensitivity is dependent largely on the choice of the respective end point.
2) Information on the toxicity of substances normally relates to individual substances, while in nature complex mixtures are often present. Their effects may be additive, but the extent to which toxicant concentrations lower than the NOEC can contribute to toxicity is not yet clear [400–402].
3) In the laboratory, efforts are normally made to remove stressors other than the substance tested. In the field, however, a variety of stressors (e.g., temperature, drought, or disease) may contribute to increase the susceptibility to a chemical well above that found in the laboratory.
4) The use of fixed assessment factors between acute EC_{50} values and chronic NOECs excludes the possibility that differences in toxic mechanisms exist between acute and chronic exposure.
5) The same applies to fixed assessment factors between chronic NOECs and ecosystem NOECs since the history of the ecosystem in a monitoring study is not always known precisely enough. Therefore, an observed "anomaly" in ecosystem behavior cannot be conclusively attributed to the presence of a substance at the time of monitoring.

Hazard Assessment for Sediments. In comparison to aquatic hazard assessment, analogous approaches for sediments have not yet reached the same level of standardization. Various techniques (e.g., equilibrium partitioning, determination of interstitial water quality, or spiked sediment toxicity) are under discussion [292]. As long as ecotoxicological effect data on sediment-dwelling organisms are lacking, the $PNEC_{sed}$ may be approximated by the PNEC for aquatic taxa and the sediment – water partitioning coefficient K according to the following formula:

$$PNEC_{sed} = K \cdot \mathrm{PNEC}_{\mathrm{water}} \tag{81}$$

Terrestrial Hazard Assessment. Hitherto, hazard assessment schemes have been based predominantly on data obtained on aquatic organisms, while the amount of valid data on sediment and soil organisms, particularly of data obtained under realistic exposure conditions, is distinctly lower. Therefore, data obtained for aquatic test organisms should be used as an inital approach only to indicate whether terrestrial tests must be performed.

Aquatic organisms might display a similar susceptibility to chemicals compared to soil organisms exposed mainly to the pore water of the soil, but the transferability of aquatic effect values to pore water organisms requires further examination on the basis of comparative tests with chemicals of various groups. These have to make proper allowance for the fact that most chemicals relevant for soil either are resistant to degradation, show a strong adsorption to soil, or have a certain bioaccumulation potential. However, even readily biodegradable chemicals such as detergents may pose a problem for the terrestrial environment if applied in considerable amounts. Therefore, in many cases an adsorption potential may be more a supposition of biodegradibility than an exclusion criterion.

If no acute tests are available, one test with higher plants and one test with earthworms or microorganisms merit particular attention since they are involved in the transformation of organic matter and contribute considerably to the transformation potential of a terrestrial ecosystem.

Extrapolation of test results obtained for vertebrates (i.e., mice, rats, or birds) on the effects of a chemical in the field is difficult because of behavior- or stress-related individual or specific responses of the test organisms [255]. Furthermore, the assessment of critical environmental concentrations must be based on concentration estimates in the soil, which lead to effects via ingestion [403].

Fuzzy Set and Fuzzy Logic Applications in Ecological Research. A fuzzy set and a fuzzy logic approach [404] can be used to solve some problems of uncertainty in ecotoxicological research and practice, namely, the uncertainty of data and uncertainty of expert knowledge. Compared with the conventional methods of information processing, the application of the fuzzy set theory offers a better applicability of imprecise data and vague knowledge. This involves

1) The representation and processing of imprecise data in the form of fuzzy sets
2) The representation and processing of vague knowledge in the form of linguistic rules with imprecise terms defined as fuzzy sets

Two main applications of the fuzzy set theory in ecological and ecotoxicological research are data analysis and knowledge-based modeling. Conventional clustering methods definitely placed an object within only one cluster. With *fuzzy clustering*, an object need not be definitely placed within one cluster since the membership value of this object can be split up between different clusters. In comparison to conventional clustering methods, distribution of the membership value thus provides additional information from which the membership value can be interpreted as the degree of similarity between properties of a particular object and properties represented by particular clusters.

Kriging belongs to the most popular methods of spatial interpolation of data, but its application is sometimes restricted owing to an insufficient amount of data. The amount of available measurement data is often too low for conventional kriging methods, or exact data may be unavailable at some locations in the study area. In such a case, the data set can be completed by using additional imprecise data estimated subjectively by an expert. *Fuzzy kriging* is a modification of the conventional kriging procedure that utilizes exact (crisp) measurement data as well as imprecise estimates obtained from an expert. These imprecise data are defined as so-called fuzzy numbers (a fuzzy number is a fuzzy set whose membership function is convex and takes its maximal value 1 for only one parameter value). Fuzzy kriging can be used to estimate interpolated values in the form of fuzzy numbers reflecting the imprecision of the input data.

Fuzzy knowledge-based modeling can be particularly useful when no analytical model of the relations to be examined exists or data are insufficient for statistical analysis [405]. To facilitate the construction of fuzzy knowledge-based models of complex ecological or toxicological systems, a modeling support system based on fuzzy logic (FLECO) has been developed [406–408]. The FLECO procedure employs fuzzy logic to handle inexact reasoning and fuzzy sets to handle the uncertainty of data. The main tasks of this system are

1) To facilitate the creation of the fuzzy knowledge base
2) To simplify changes within the fuzzy knowledge base
3) To create and facilitate the connections between submodels
4) To offer a set of inference methods
5) To facilitate the simulation process

3.3. State of the Art of Modeling in Ecotoxicology

3.3.1. Characteristics of Ecotoxicological Models

An increasing interest in management of pollution by toxic substances has emerged during the 1980s, and this has caused an equally large interest in toxic substance modeling. Models for toxic substances attempt to model the fate and effect of these substances in ecosystems. These models are management models, deterministic models, more frequently compartment models than matrix models, holistic models, and lumped models (because the difficulty of determining parameters is greater than for ecological modeling in general). Toxic substance models are most often biogeochemical models because they attempt to describe the mass flows of the toxic substances considered, although models of population dynamics also exist, which include the influence of toxic substances on birthrate or mortality and therefore should be considered toxic substance models.

Toxic substance models differ from other ecological models by the following:

1) The need for parameters to cover all possible toxic substance models is great, and general estimation methods are therefore used widely. Section 3.3.5 is devoted to this question.
2) The safety margin should be high when, for instance, expressed as the ratio between the actual concentration and the concentration that produces undesired effects.
3) The possible inclusion of an effect component, which relates the output concentration of a toxic substance to its effect, is easy. However, finding a well-examined relationship to base it on is often a problem.
4) The need is for simple models because of items (1) and (2), and because of limited knowledge of process details, parameters, sublethal effects, antagonistic and synergistic effects.

Several questions must be clarified before a toxic substance model can be developed in accordance with the general procedure used for development of ecological models. The following information must be available:

1) The best possible knowledge about the processes in which the toxic substance considered is involved in the ecosystem (as far as possible, knowledge about the quantitative importance of the processes should be obtained)
2) Obtaining parameters from the literature and/or from experiments (in situ or in the laboratory)
3) Estimation of all parameters by the methods presented in Section 3.3.5
4) Comparison of the results from items (2) and (3) and an attempt to explain the discrepancies
5) Estimation of which processes and state variables should be included in the model (if the slightest doubt exists at this early stage, too many processes and state variables should be included rather than too few)

6) Use of sensitivity analysis to evaluate the significance of individual processes and state variables, which in many cases will lead to further simplification

Section 3.3.2 reviews the characteristic features of ecotoxicological models and tries to classify them. Section 3.3.3 gives an overview of some of the ecotoxicological models published during the last 10 – 15 years. The description of the chemical, physical, and biological processes in the models will, in general, be in accordance with equations generally applied for physicochemical and biological processes. Some submodels of particular ecotoxicological interest are presented in Section 3.3.4. Section 3.3.5 deals with parameter estimation methods, which are of particular importance in ecotoxicological models. Section 3.3.6 is devoted to fugacity models in general, including their basic equations, and a case study in which the fugacity modeling approach is used is presented.

3.3.2. Classification of Ecotoxicological Models

Ecotoxicological models differ from ecological models in general by

1) Most often being more simple
2) More parameters are unknown, cannot be found in the literature
3) Wider use of parameter estimation methods
4) Possible inclusion of an effect component

Ecotoxicological models can be divided into six classes. The classification presented here is based on differences in the modeling structure. The decision on which model class to apply will be based on the ecotoxicological problem to be solved. The definitions of model classes are given below, along with the most appropriate use of each model type.

Class 1. Food Chain or Food Web Dynamic Models. Class 1 models consider the flow of toxic substances through the food chain or food web. Such models are relatively complex and contain many state variables. The models furthermore contain many parameters, which often have to be estimated by one of the methods presented in Section 3.3.5. These types of model are used when a great number of organisms are affected by the toxic substance, or the entire structure of the ecosystem is threatened by the presence of a toxic substance. Because of their complexity, these models have not been used widely. They are similar in many aspects to the more complex eutrophication models that consider the flow of nutrients through the food chain or even through the food web. In some instances, they are even constructed as submodels of a eutrophication model (see, for instance, [409]). Figure 51 shows a conceptual diagram of an ecotoxicological food chain model for lead. Lead from atmospheric fallout and wastewater enters an aquatic ecosystem, where it is concentrated through the food chain. Simplification is rarely possible for this model because the aim of the model is to describe and quantify the bioaccumulation and ecological magnification through the food chain.

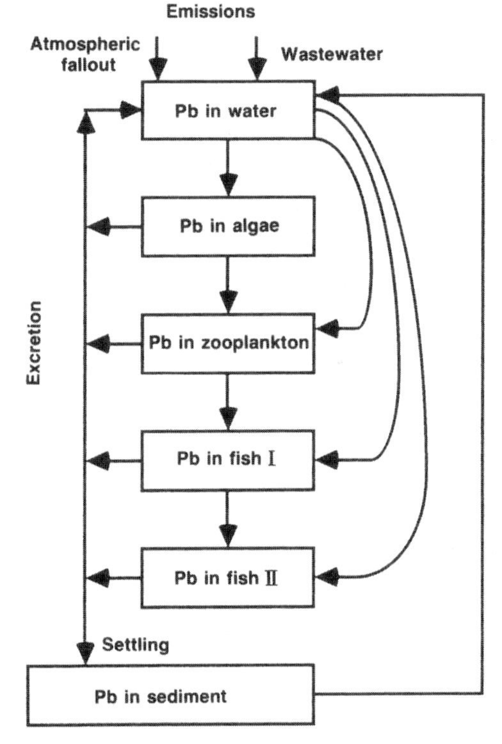

Figure 51. Conceptual diagram of lead accumulation in the food chain of an aquatic ecosystem

Class 2. Static Models of the Mass Flow of Toxic Substances. If seasonal changes are minor, or of minor importance, a static model of the mass flows is often sufficient to describe the situation and even indicate the expected changes if the input of toxic substances is reduced or enlarged. This type of model is based on a mass balance as shown in the example in Figure 52. It often, but not necessarily, contains more trophic levels, and the modeler often is concerned with the flow of the toxic substance through the food chain. If some seasonal changes occur, this type, which in most cases is simpler than class 1, can still be used advantageously, for instance, if the modeler is concerned with the worst case of seasonal changes.

Class 3. Dynamic Model of a Toxic Substance in a Trophic Level. Often only the toxic substance concentration in one trophic level is of concern. This includes the zero trophic level, which is understood as the medium soil, water, or air. Figure 53 shows a model of copper contamination in an aquatic ecosystem as an example. The main concern is the copper concentration in the water, which may reach a toxic level for phytoplankton. Zooplankton and fish are much less sensitive to copper contamination, so the "alarm clock" rings first at the concentration level that is harmful to phytoplankton. However, only dissolved copper ions are toxic, and therefore the partition of copper in ionic, complex-bound, and adsorbed forms must be modeled. The exchange between copper in the water phase and in sediment is also included because sediment can accumulate relatively large amounts of heavy metals. The amount released from sediment may be significant under certain circumstances, for instance, at low pH.

Figure 54 gives another example. Here, the main concern is the DDT concentration in fish, which may be so high that, in accordance with WHO standards, the fish are unfit for human consumption. The model can therefore be simplified by including not the entire food chain but only fish. Some physicochemical reactions in the water phase are, however, important and are included in the diagram.

The so-called HESP model [260] estimates exposure, or the total concentration of a toxic substance for individuals living at a contaminated site. A number of possible exposure routes are identified and subsequently quantified in the model. The model can be used to initiate a hazard assessment process, which can provide a sound basis for decisions regarding possible measures.

As seen from these examples, simplifications are often feasible when the problem is well defined. Simplifications take into account which component is most sensitive to toxic matter and which processes are most important for concentration changes.

Class 4. Ecotoxicological Models in Population Dynamics. Population models are biodemographic models and therefore have numbers of individuals or species as state variables. The simple population models consider only one population. Population growth is the difference between natality and mortality:

$$dN/\mathrm{d}t = B \cdot N - M \cdot N = r \cdot N \quad (82)$$

where N is the number of individuals; B is the natality (i.e., the number of new individuals per unit of time and unit of population); M is the mortality (i.e., the number of organisms that died per unit of time and unit of population); and r is the increase in number of organisms per unit of time and unit of population (i.e., $B-M$). The concentration of a toxic substance in the environment or in the organisms influences natality and mortality, and if the relations between the concentration of a toxic substance and these population dynamic parameters are included in the model, it becomes an ecotoxicological model of population dynamics.

Population dynamic models may include two or more trophic levels, and ecotoxicological models will include the influence of the concentration of the toxic substance on natality, mortality, and interactions between these populations. In other words, an ecotoxicological model of population dynamics is a general model of population dynamics with the inclusion of the relation between toxic substance concentrations and some of the model parameters.

Class 5. Ecotoxicological Models with Effect Components. Although class 4 models include relations between concentrations of toxic substances and their effects, these are limited to population dynamic parameters. In comparison, class 5 models include more comprehensive relations between concentrations of toxic substances and effects. These models may include not only lethal or sublethal effects but also effects on biochemical reactions or on the enzyme

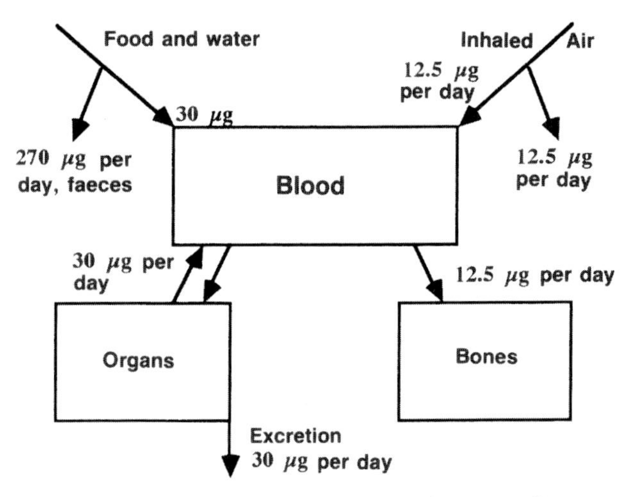

Figure 52. Static model of the daily lead uptake of an average Dane

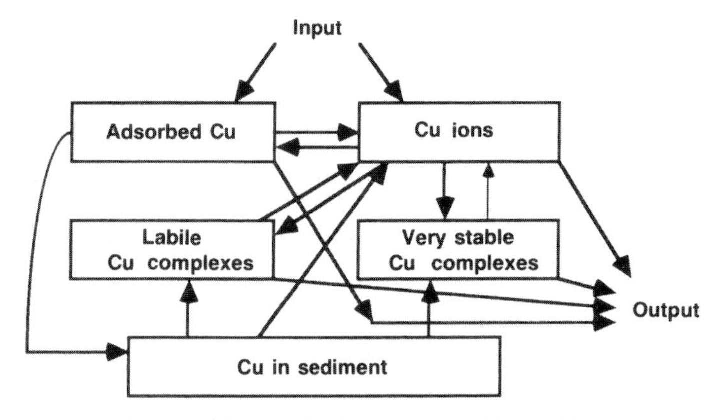

Figure 53. Conceptual diagram of a simple copper partition model

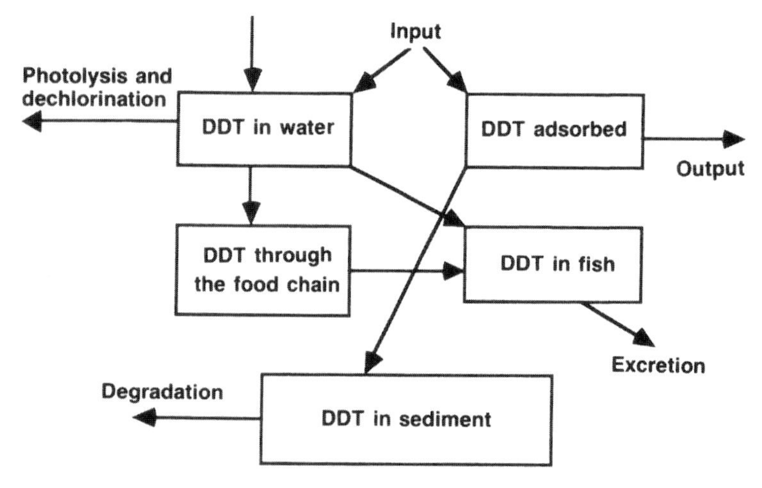

Figure 54. Conceptual diagram of a simple DDT partition model

system. As Figure 55 shows, the effects on various levels of the biological hierarchy may be considered from cells to ecosystems.

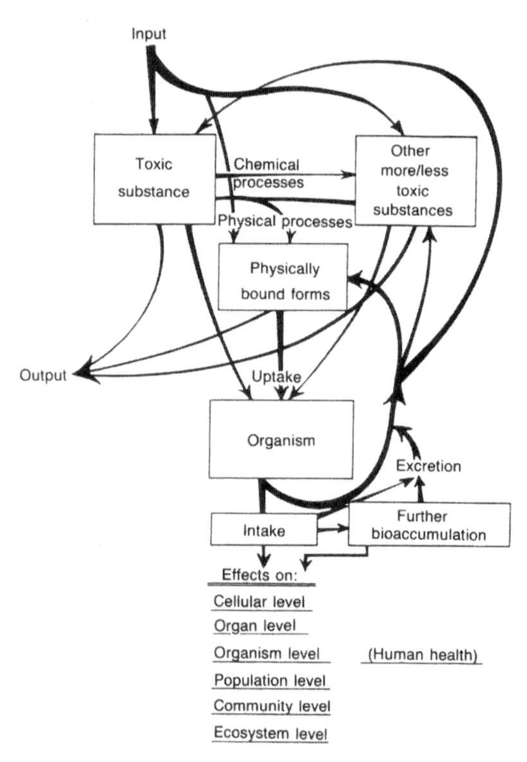

Figure 55. Conceptual diagram of ecotoxicological models with effect components on various levels of biological complexity

In many problems, more details about the effects may be required to answer the following relevant questions:

1) Does the toxic substance accumulate in the organism?
2) What will the long-term concentration in the organism be when uptake rate, excretion rate, and biochemical decomposition rate are considered?
3) What is the chronic effect of this concentration?
4) Does the toxic substance accumulate in one or more organs?
5) What is the transfer between various parts of the organism?
6) Will decomposition products eventually cause additional effects?

A detailed answer to all of these questions may require a model of the processes that take place in the organism and a translation of concentrations in various parts of the organism into effects. This implies, of course, that the intake = (uptake by the organism) × (efficiency of uptake) is known. Intake may be from either water or air, which can also be expressed by the use of concentration factors (the ratios between the concentration in the organism and that in air or water).

However, if all of the aforementioned processes are considered for just a few organisms, the model will become too complex, contain too many parameters to calibrate, and require more detailed knowledge than can be provided. Therefore, most models in this class do not consider too many details of the partition of toxic substances in organisms and their corresponding effects, but rather are limited to simple accumulation in the organisms and their effects. In most cases, accumulation is rather easy to model, and the following simple equation is often sufficiently accurate:

$$dC/\mathrm{d}t = (e_\mathrm{f} \cdot C_\mathrm{f} \cdot F + e_\mathrm{m} \cdot C_\mathrm{m} \cdot V)/W - Ex \cdot C = (INT)/W - Ex \cdot C \qquad (83)$$

where C is the concentration of the toxic substance in the organism; e_f and e_m are the efficiencies for the uptake from the food and medium, respectively (water or air); C_f and C_m are the concentration of the toxic substance in the food and medium, respectively; F is the amount of food uptake per day; V is the volume of water or air taken up per day; W is the body weight as either dry or wet matter; and Ex is the excretion coefficient (d^{-1}). As seen from the equation, INT covers the total intake of toxic substance per day.

This equation has a numerical solution:

$$C/C\,(\mathrm{max}) = \{INT \cdot [1 - \exp\,(Ex \cdot t)]\}/(W \cdot Ex) \qquad (84)$$

where C (max) is the steady-state value of C:

$$C\,(\mathrm{max}) = INT/(W \cdot Ex) \qquad (85)$$

i.e., the concentration of the toxic substance approximates a steady-state value after a certain time interval.

Class 6. Fate Models with or without a Risk Assessment Component. The last type of ecotoxicological models focuses on the fate of toxic substances: Where in the ecosystem will the toxic substance be found? In what concentration? These models are sometimes called multimedia models because they focus on the distribution of toxic substances in the four spheres. The complete solution of an ecotoxicological problem requires in principle four (sub)models, of which the fate model may be considered the first model in the chain; see Figure 56. The four components are [410] (1) a fate or exposure model; (2) an effect model, translating the concentration into an effect (see class 5); (3) a model for human perception processes; and (4) a model for human evaluation processes. The first two submodels are in principle "objective," predictive models, whereas the latter two are value oriented. The development of submodels (1) and (2) is based on physical, chemical, and biological processes. They are very similar to other environmental models and are based on mass transfer; mass balances; and physical, chemical, and biological processes.

The second submodel requires good knowledge of the effects of the toxic components. Submodels (3) and (4) differ from the generally applied environmental management models and are presented in some details below in the discussion of risk assessment.

Fugacity models (see Section 3.3.6) are a special type of fate model. They attempt to answer the following questions: In which of these six compartments (air, water, soil, sediment, suspended sediment, and biota) can the greatest problem be expected for chemicals emitted to the environment? What concentration will be expected in each compartment? What are the implications of these concentrations? The fugacity models are to a large extent based on physicochemical parameters and the use of these parameters to estimate other required parameters. The applicability of the fugacity model is therefore very dependent on the use of the estimation methods presented in Section 3.3.5.

The risk assessment component, associated with the fate model comprises human perception and evaluation processes (see Fig. 56). These submodels are explicitly value laden but must of course build on objective information about concentrations and effects. Factors that may be important to consider in this context are

1) Magnitude and time constant of exposure
2) Spatial and temporal distributions of concentrations
3) Environmental conditions determining the process rates and effects
4) Translation of concentrations into magnitude and duration of effects
5) Spatial and temporal distribution of effects
6) Reversibility of effects

The uncertainties relating to the information on which the model is based and the uncertainties related to the development of the model, are crucial in risk assessment. The uncertainty in risk assessment may be classified in one or more of the following five categories:

1) Good direct knowledge and statistical evidence on the important components (state variables, processes, and interrelations of the variables) of the model are available.
2) Good knowledge and statistical evidence on the important submodels are available, but the aggregations of the submodels are less certain.
3) No good knowledge of the model components for the considered system is available, but good data are available for the same processes from a similar system, and these data are considered applicable for use directly or with minor modifications, to the model development.
4) Some, but insufficient, knowledge ist available from other systems. Attempts are made to use these data without the necessary transferability. Attempts are made to eliminate gaps in knowledge by the use of additional experimental data as far as possible within the limited resources available for the project.
5) The model is to a large extent based on the subjective judgment of experts.

Acknowledgment of the uncertainty is of great importance and may be taken into consideration, either qualitatively or quantitatively. Another problem is of course: Where and how should the uncertainty be taken into account? Should the economy or the environment benefit from the uncertainty? If the results of the management strategy are to be ensured, the un-

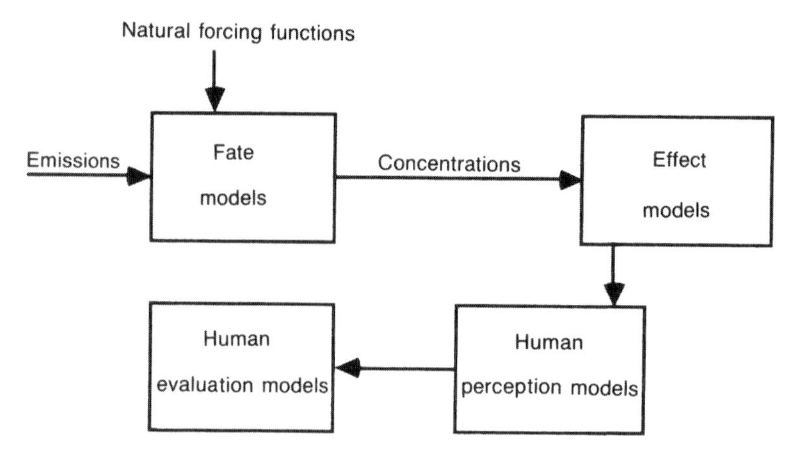

Figure 56. Components and general structure of comprehensive ecotoxicological modeling systems

certainty must be used to reduce the emission standards.

Until 10 – 15 years ago, researchers had very little understanding of the processes by which people actually perceive the exposures and effects of toxic chemicals, but these processes are just as important for risk assessment as the exposure and effect processes. The following list of risk characteristics summarizes the concerns:

1) Is the risk voluntary or involuntary?
2) Are the levels known to the exposed people or only to science?
3) Is the risk novel, or old and familiar?
4) Is it common or dreaded (e.g., does it involve cancer) ?
5) Does it involve death?
6) Are mishaps controllable?
7) Are future generations threatened?
8) Is the risk global, regional, or local?
9) Is the risk a function of time? How does it depend on time (e.g., increase or decrease) ?
10) Can it be reduced easily?
11) Is the expected effect immediate or delayed?
12) Are many or a few people involved?
13) Are the effects observable immediately?

A factor analysis has been performed by Slovic et al. [411] that shows, among other results, a not surprising emphasis on people's perception of dreadful and unknown risks. Broadly speaking, two methods of selecting the risks exist. The first may be described as the *rational actor model*, involving people that look systematically at all the risks they face and make choices about which they will live with and at what levels. For decision making, this approach would use some single, consistent, objective functions and a set of decision rules. The second method may be called the *political – cultural model*. It involves interactions among culture, social institutions, and political processes for the identification of risks and the determination of those that people will live with and at what level.

Both methods are unrealistic, since they are both completely impractical in their pure form. Therefore, a strategy for risk abatement must be selected that is founded on a workable alternative based on the philosophy behind both methods.

Several risk management systems are available. Here, some recommendations should be given for the development of risk management systems:

1) Consider as many of the characteristics listed above as possible and include human perception of these characteristics in the model.
2) Do not focus too narrowly on certain types of risks; this may lead to suboptimal solutions. Attempt to approach the problem as broadmindedly as possible.
3) Choose strategies that are pluralistic and adaptive.
4) Benefit – cost analysis is an important element of the risk management model, but it is far from being the only important element, and the uncertainty in evaluation of benefit

and cost should not be forgotten. The variant of this analysis applicable to environmental risk management may be formulated as follows :

Net social benefit = social benefits of the project minus "environmental" costs of the project

5) Use multiattribute utility functions, but remember that people in general have trouble thinking about more than two to three (at most, four) attributes in each outcome.

Presentation of the six classes of models above clearly shows the advantages and limitations of ecotoxicological models. The simplifications used in classes 2, 3, and 6 (at least without risk assessment components) often offer great advantages. They are sufficiently accurate to give a very applicable picture (overview) of the concentrations of toxic substances in the environment, because of the use of large safety factors. Application of the estimation methods presented in Section 3.3.5 makes feasible the construction of such models, even though knowledge of the parameters is limited. The estimation methods obviously have a high uncertainty, but the great safety factor helps in accepting this uncertainty. On the other hand, knowledge about the effects of toxic substances is very limited, particularly at the organism and organ level. Therefore, models with effect components must not be expected to give more than a first rough picture of what is known today in this area.

Because of the character of ecotoxicological models, a few questions should be clarified before beginning the modeling procedure:

1) Obtain the best possible knowledge about the processes of the toxic substance under consideration
2) Attempt to get parameters of the processes in which the toxic substance is involved in the environment from the literature
3) Estimate all parameters using methods presented in Sections 3.3.4 and 3.3.5
4) Compare the results from items (2) and (3) and attempt to explain discrepancies, if present
5) Use sensitivity analysis widely to estimate which processes and state variables would be feasible and relevant to include in the model

3.3.3. An Overview: Application of Models in Ecotoxicology

Some toxic substance models are reviewed in Table 20 to give an idea of the types of models available today. Most models reflect the proposition that good knowledge of the problem and the ecosystem can be used to make reasonable simplifications. Model characteristics indicated in Table 20 are state variables or processes considered in the model. Note the number of modeled toxic substances and the processes taken into account. The model class, according to the classification 1 to 6 discussed in Section 3.3.2, is given in brackets after the toxic substance. Only a few class 4 models are included in Table 20. Ecological modeling has been approached from two sides: population dynamics and biogeochemical flow analysis. Since the second approach has been most involved in environmental issues, the toxic substance problem has also been approached from this angle. The few class 4 models, including the example given in Table 20, are population dynamic models, with some additional equations to account for the influence of toxic substances on natality and mortality. Provided these relations are available, constructing this type of model should be relatively easy.

The most difficult part of modeling the effect and distribution of toxic substances is obtaining relevant knowledge about the behavior of toxic substances in the environment and using this knowledge to make feasible simplifications. The modeler of ecotoxicological problems is particularly challenged to select the proper, balanced complexity, and many examples can be found of rather simple ecotoxicological models that are able to solve the basic problem.

It can be seen from the overview in Table 20 that most ecotoxicological models have been developed during the last decade. Before ca. 1975, toxic substances were rarely associated with environmental modeling, since the problems seemed straightforward. The many pollution problems associated with toxic substances were believed to be easily solved simply by eliminating the source of the toxic substance. During the 1970s, the environmental problems originating from toxic substances were acknowledged to be very complex because of the interaction of

Table 20. Examples of toxic substance models

Toxic substance, model class	Model characteristics	Reference
Cadmium (1)	food chain similar to a eutrophication model	[412]
Mercury (6)	6 state variables: water, sediment, suspended matter, invertebrates, plant, fish	[413]
Vinyl chloride (3)	chemical processes in water	[414]
Methyl parathion (1)	chemical processes in water and microbial degradation and adsorption of benzothiophene, 2–4 trophic levels	[415]
Methyl mercury (4)	a single trophic level: food intake, excretion, metabolism, growth	[416]
Heavy metals (3)	concentration factor, excretion, bioaccumulation	[417]
Pesticides in fish, DDT and methoxychlor (5)	ingestion, concentration factor, adsorption on body, defecation, excretion, chemical decomposition, natural mortality	[418]
Zinc in algae (3)	concentration factor, secretion, hydrodynamic distribution	[419]
Copper in the sea (5)	complex formation, adsorption, sublethal effect of ionic copper	[420]
Lead (5)	hydrodynamics, precipitation, toxic effects of free ionic lead on algae, invertebrates, and fish	[421]
Radionuclides (3)	hydrodynamics, decay, uptake, and release by various aquatic surfaces	[69]
Radionuclides (2)	in grass, grains, vegetables, milk, eggs, beef, and poultry–state variables	[422]
SO_2, NO_x, and heavy metals on spruce forests (5)	threshold model for accumulation effect of pollutants (air and soil)	[369]
Toxic environmental chemicals in general tests (6)	hazard ranking and assessment from physicochemical and a limited number of laboratory data	[423]
Heavy metals (3)	adsorption, chemical reactions, ion exchange	[424]
Polycyclic aromatic hydrocarbons (3)	transport, degradation, bioaccumulation	[424]
Persistent toxic organic substances (3)	groundwater movement, transport and accumulation of pollutants in groundwater	[425]
Cadmium, PCB (2)	hydraulic overflow rate (settling), sediment interactions, steady-state food chain submodel	[409]
Hydrophobic organics (3)	gas exchange, sorption/desorption, hydrolysis compounds, photolysis, hydrodynamics	[426]
Mirex (3)	water – sediment exchange processes, adsorption, volatilization, bioaccumulation	[427]
Toxins (aromatic hydrocarbons, Cd) (3)	hydrodynamics, deposition, resuspension, volatilization, photooxidation, decomposition, adsorption, complex formation, (humic acid)	[428]
Heavy metals (2)	hydraulic submodel, adsorption	[429]
Oil slicks (3)	transport and spreading, influence of surface tension, gravity, and weathering processes	[430]
Acid rain (soil) (3)	aerodynamics, deposition	[70]
Acid rain	C, N, and S cycles and their influence on acidity	[431]
Persistent organic chemicals (6)	fate, exposure, and human uptake	[432]
Chemicals, general (6)	fate, exposure, ecotoxicity for surface water and soil	[433]
Toxicants, general (4)	effect on populations of toxicants	[434]
Chemical hazard (6)	basin-wide ecological fate	[435]
Pesticides (4)	effects on insect populations	[436]
Insecticides (2)	resistance	[437]
Mirex and lindane (6)	fate in Lake Ontario	[427]
Pesticides (3)	degradation in soil	[438]
Pesticides (3)	degradation in soil	[439]
Pesticides (3)	leaching to groundwater	[440]
Acid rain (5)	effects on forest soils	[70]
Acid rain (5)	cation depletion of soil	[441]
pH, calcium, and aluminum (4)	survival of fish populations	[442]
Photochemical smog and NO_x (6)	fate and risk	[443]
Nitrate (3)	leaching to groundwater	[444]
Oil spill (6)	fate	[445]
Toxicants (4)	effects on populations	[446]
Chromium (2)	distribution and accumulation in mussels	[71]
Toluene, benzene, Zn, Ni, Pb (3)	assess exposure levels of soil contaminants for humans	[447]

many sources and of many processes and components simultaneously. Several accidental releases of toxic substances into the environment have reinforced the need for models. Thus, various ecotoxicological models have been developed since the late 1970s. The list in Table 20 represents a comprehensive survey of available ecotoxicological models, but it should not be considered complete since it is not the result of an exhaustive literature review.

Table 21. Regression equations for estimation of concentration, bioconcentration, and ecological magnification factors

Indicator	Relationship	Correlation coefficient	Range (indicator)
K_{ow}	$\log CF = -0.973 + 0.767 \log K_{ow}$	.76	$2.0\times10^{-2} - 2.0\times10^{6}$
K_{ow}	$\log CF = 0.7504 + 1.1587 \log K_{ow}$	.98	$7.0 - 1.6\times10^{4}$
K_{ow}	$\log CF = 0.7285 + 0.6335 \log K_{ow}$	.79	$1.6 - 1.4\times10^{4}$
K_{ow}	$\log CF = 0.124 + 0.542 \log K_{ow}$	.95	$4.4\times100 - 4.2\times10^{7}$
K_{ow}	$\log CF = -1.495 + 0.935 \log K_{ow}$	.87	$1.6\times100 - 3.7\times10^{6}$
K_{ow}	$\log CF = -0.70 + 0.85 \log K_{ow}$	.95	$1.0 - 1.0\times10^{7}$
K_{ow}	$\log CF = 0.124 + 0.542 \log K_{ow}$	.90	$1.0\times100 - 5.0\times10^{7}$
S, µg/L	$\log BCF = 3.9950 - 0.3891 \log S$	.92	$1.2 - 3.7\times10^{7}$
S, µg/L	$\log BCF = 4.4806 - 0.4732 \log S$	.97	$1.3 - 4.0\times10^{7}$
S, µmol/L	$\log BCF = 3.41 - 0.508 \log S$	.96	$2.0\times10^{-2} - 5.0\times10^{3}$

3.3.4. Ecotoxicological Processes

Ecotoxicological processes differ from other ecological processes, and a review of the relevant processes and their quantitative description is therefore necessary as basis for model development. A survey of the most crucial ecotoxicological processes and the most applied mathematical descriptions of these processes is given below.

Concentration Factor. The concentration factor relates the concentration of the xenobiotic in the medium to the concentration in organisms, sediment, or suspended matter. The concentration in organisms is a result of uptake by respiration, and the concentration in sediment or suspended matter is caused by adsorption and sedimentation processes. The concentration factor CF can often be described with sufficient accuracy as a constant independent of the concentration in the medium (water):

$$CF = C_S / C_W \tag{86}$$

where C_S is the concentration in the organisms, suspended matter, or sediment and C_W is the concentration in water (or air). If the uptake includes the indirect route through food, the term bioconcentration factor BCF is often used ($BCF \geq CF$), which can be described by the same equation.

Table 21 gives equations relating water solubility and the n-octanol – water partition coefficient to CF and BCF. The n-octanol – water partition coefficient K_{ow} is defined in Section 2.4.1. An overview on relevant literature references dealing with the determination of the n-octanol –water partition coefficient of various chemicals in fish is given in Table 22. If the n-octanol –water partition coefficient is not known, it can be estimated from the water solubility of the corresponding substance (see Section 3.3.5 and Fig. 58).

Table 22. Determination of n-octanol – water partition coefficients in fish

Animal	Number of chemicals	References
Fish species	36	[448]
Mosquito fish	9	[449]
Mosquito fish	11	[450]
Trout	8	[451]
Fish species	26	[448]
Fathead minnow	59	[452]
Fathead minnow, bluegill	59	[453]
Fish species	13	[448]
Fish species	50	[448]
Mosquito fish	9	[449]
Fish species	36	[448]
Mosquito fish, whole	15	[450]

Biomagnification describes the process whereby pollutants are passed from one trophic level to the next and increasingly concentrated in organisms with higher trophic status (see Section 2.4.2.7). The magnification (i.e., the increasing concentration through the food chain) is caused by greater retention of toxic substances than of general food components by the organisms. The process may be described by the biological or ecological magnification factor, which indicates the magnification from one level in the food chain to the next.

Further important processes in ecological modeling are adsorption (see Section 2.4.2.4) and desorption.

Excretion of toxic substances by an organism can be shown to follow first-order kinetics to a good approximation, and the rate coefficient depends on the chemical compound and the size of the organism considered (see [37]). This process is in most cases insignificant for the entire mass balance of the components considered but inclusion of the excretion process may be important for a model of a toxic substance on the organism level.

Biodegradation. It is rather difficult to estimate the reaction rate of biodegradation in ecotoxicology because a large number of factors influence the reaction and thereby also the concentration of microorganisms. The most important of these factors are: temperature, sorption, redox potential, ionic strength, adaptation of microbial populations, macronutrients and micronutrients available [454], [455].

Distribution of a Chemical Compound. Distribution of a chemical compound in the four spheres is of particular interest in fate models.

Distribution air – water

$$\frac{C^*}{To^*} \equiv F_{\mathrm{L}} = \frac{H18}{1000RT} \text{ (approximation)} \tag{87}$$

or at (103.33 kPa) and 20 °C

$$F_{\mathrm{L}} = 7.49 \cdot 10^{-4} H \tag{88}$$

where H is Henry's constant (bar); $C*$ and $To*$ are equilibrium concentrations (mass/volume) in air and water, respectively.

Rate of evaporation (distribution liquid – vapor phase)

$$Ev = -D \cdot A \frac{\mathrm{d}To}{\mathrm{d}x} = bK_{\mathrm{L}} \cdot A \cdot (To - To^*) \tag{89}$$

where D is the molecular diffusion coefficient (area/time), A is the area, and b is a coefficient that adjusts the evaporation coefficient K_{L} (length/time) to substances other than oxygen ($b = 1.0$ for oxygen); b can be found in [72] for other compounds. Several empirical equations for calculation of K_{L} can be found in [73], [74].

$$b = \frac{K_{\mathrm{L}}\text{ (compound)}}{K_{\mathrm{L}}\text{ (oxygen)}} = \frac{D\text{ (compound)}}{D\text{ (oxygen)}} \tag{90}$$

For lakes and streams,

$$To(t) = To(0) \cdot \mathrm{e}^{-b \cdot K_{\mathrm{L}} \cdot \mathrm{a}} \cdot t \tag{91}$$

where

$$a = \frac{A}{V} = \frac{\text{area}}{\text{volume}} \tag{92}$$

Distribution water – solid (sediment)

A distribution coefficient may be used at low concentrations (see also Langmuir's and Freundlich's adsorption isotherms):

$$F_{\mathrm{S}} = \frac{C^*}{To^*} \tag{93}$$

where $C*$ and $To*$ are the concentrations in the solid and water phases at equilibrium. Since the dimension of $C*$ is mass/mass (e.g., mg per gram dry matter) and $To*$ has the dimension of mass/volume (mg/L), F_{S} has the dimension volume/mass (L/g or L/kg).

F_{S} may be found from

$$F_{\mathrm{S}} = F_{\mathrm{oc}} \cdot f_{\mathrm{oc}} \tag{94}$$

where F_{oc} is the distribution coefficient for organic matter and f_{oc} is the fraction of organic carbon in the solid being considered.

The following relationship between F_{oc} and K_{ow} exists [456]:

$$\log F_{\mathrm{oc}} = -0.006 + 0.937 \log K_{\mathrm{ow}} \tag{95}$$

MATTER-MÜLLER et al. [457] found the following relationship for activated sludge:

$$\log F_{\mathrm{S}} = 0.39 + 0.67 \log K_{\mathrm{ow}} \tag{96}$$

where F_{S} has the units L/kg.

F_{oc} and F_{S} values for selected chemical compounds can be found in the literature.

Complex Formation. As for other chemical processes, the law of mass action is valid for complex formation reactions. For the process

$$Me^{n+} + \mathrm{L}^{m-} = \mathrm{MeL}^{(n-m)+} \tag{97}$$

(where Me denotes a metal and L a ligand), the law of mass action has the following form:

$$\frac{[MeL^{(n-m)+}]}{[\mathrm{Me}^{n+}]\,[\mathrm{L}^{m-}]} = K \tag{98}$$

K is termed the stability constant if the process – as in this case – is the formation of a complex.

The formation of coordination complexes plays a major role in modeling the distribution as well as the effect of metal ions in aquatic ecosystems because of the following influences:

1) Increase in metal solubility

$$MeY\ (S) + L = MeL + Y \quad (99)$$

where Me represents a metal that is bound in the sediment in the compound MeY; L denotes a ligand

2) Alteration of the distribution between oxidized and reduced forms of metals. The equilibrium constant for the process

$$Me^{n+} + me^{-} = \mathrm{Me}^{(n-m)+} \quad (100)$$

differs from the one for

$$MeL^{n+} + me^{-} = \mathrm{MeL}^{(n-m)+} \quad (101)$$

if the complexes, MeL^{n+} and $\mathrm{MeL}^{(n-m)+}$, have different stability constants, which often is the case

3) Alleviation of toxicity due to alteration of the availability to aquatic life
4) Change in adsorption and ion exchange on sediment and suspended matter
5) Change in stability of the metal-containing colloids

The presence of ligands in water generally increases the transfer of metals from sediment, soil, and suspended matter and thereby increases the solubility. However, since complexes show lower toxicity than metal ions, the harm to aquatic life in most cases does not increase in proportion to the increased solubility.

The application of coordination chemistry to environmental problems is very complex because many ligands are present in aquatic ecosystems and all ligands compete simultaneously for formation of complexes with metal ions.

Photolysis. The *Stark – Einstein law* states that if a species absorbs radiation, one particle is excited for each quantum of radiation absorbed. The energy of the particle is limited by the energy of the photon absorbed. This energy is given by

$$E = h\nu \quad (102)$$

where E is the energy of the photon, h is Planck's constant, and ν is the frequency of the photon. Thus, depending on the nature of the incident light and the property of the chromophore, different excited states of the chromophore may be formed. Although not exclusively true, in general only light absorption that brings about electronic excitation of the chromophore will result in photochemical reactions. Excitation of the vibrations and rotations of the absorbing species in its ground electronic state is less likely to induce chemical changes.

The effectiveness of photochemical reactions in the consumption or production of a given chemical species is given by the *overall quantum yield* of that process (see Section 2.4.2.2). The overall quantum yield is the number of molecules of a given reactant consumed for each photon absorbed. It may be further subdivided into the primary quantum yield (molecules consumed by primary reactions) and the secondary quantum yield (molecules consumed by secondary reactions).

If the quantum yield is not strongly wavelength dependent, for an incident beam of light with a certain spectral characteristic, the complete rate law for direct photolysis is

$$-\left(d\,[CA]\,/\mathrm{d}t\right) = k_{\mathrm{a}}\,[CA] \quad (103)$$

where *CA* is the concentration of the reactant and k_{a} is the sum of the k_{a} values of all the wavelengths represented by the photons in the incident light. Models for estimating the rates of direct photolysis in natural waters under restrictive conditions have been attempted by several investigators [458–466].

Acid – base reactions are of a great environmental interest because almost all processes in the environment depend on pH. A few illustrative examples are included in the following list:

1) Ammonia is toxic to fish and the ammonium : ammonia ratio depends on pH.
2) Carbon dioxide is toxic to fish and the ratio of hydrogencarbonate to carbon dioxide depends on pH.
3) The fertility of fish and zooplankton eggs is highly pH dependent.
4) All biological processes have a pH optimum, which is usually in the range 6 – 8. This implies that algal growth, microbiological decomposition, nitrification, and denitrification are all influenced by pH.

5) The release of heavy-metal ions from soil and sediment increases very rapidly with decreasing pH. Heavy-metal hydroxides have a very low solubility product, which implies that most heavy-metal ions are precipitated at pH 7.5 or higher.

Acid rain illustrates the problems related to a low pH value, but too high a pH is also of environmental concern because of items (1) and (4) above.

Other Chemical Processes. In addition, hydrolysis and redox processes may also be important for the speciation and general presence of toxic substances. Many metal hydroxides may be dissolved by hydrolysis, and the release of metal ions from lake sediments is highly dependent on the redox potential at the interface between water and sediment.

Organic compounds can also undergo hydrolysis reactions. Many compounds react with water, with introduction of a hydroxyl group into the molecule. Most toxic organic compounds are oxidized by oxygen, and carboxyl groups may be formed. For all of these reactions, a chemical pathway should be found and the application of simple first-order reaction kinetics is often sufficient to describe the process in an ecotoxicological model.

3.3.5. Parameter Estimation

About 50 000 chemicals are produced on a large scale (1000 kg or more per year), but only about 4000 chemicals are produced at a quantity of $\geq 1 \times 10^6$ kg/a. Approximately one-third of these chemicals are well-defined organics. Their environment-relevant properties are listed below:

Physical properties
vapor pressure, water solubility, Henry's constant, adsorption coefficient, partition coefficient octanol – water, mass-transfer coefficient, density, viscosity, diffusion coefficient in air and water, boiling and melting point

Chemical properties
rate of hydrolysis, rate of oxidation or reduction, rate of photolysis, acid dissociation coefficient(s)

Biological properties
bioconcentration factor, rate of biodegradation in soil and water under aerobic or anaerobic conditions

Most of the biological parameters are needed for all species or at least a representative fraction of the species on earth. This implies that for construction of all possible ecotoxicological models, about 20 000 parameters are needed for 50 000 chemicals (i.e., a total of ca. 1×10^9 parameters). A review of the literature reveals that hardly 1 % of these parameters, or 10×10^6 values are known. Providing the 9.9×10^8 parameters lacking will be an enormous task – a task that will take several decades to accomplish. How can ecotoxicological models be constructed under these circumstances? The answer is to estimate the parameters, which is acceptable, because ecotoxicological models do not require a high accuracy, or rather, one is forced to interpret the model results by the use of large safety factors, which makes it less necessary to demand great accuracy of the modeling results.

Many estimation methods are available for ecotoxicological parameters, and too many pages would be required to present even an overview of these methods. JØRGENSEN [37] presents a rather simple method that can be applied easily and still give sufficient accuracy for development of most ecotoxicological models.

The *physical properties* of interest in ecotoxicology are vapor pressure P, water solubility S, partition coefficient K_{ow}, and Henry's constant H; for determination of the transfer between air and water, adsorption isotherms K_{ac} and diffusion coefficients D are also necessary. For organic compounds, all these properties can be estimated just on the basis of their chemical structure. Some estimation methods are based on estimation of the boiling point and critical properties, which are then used later to estimate the aforementioned physical properties. Experience shows that generally better estimates can be given if the boiling point is known and not found by estimation methods. Simultaneous use of two or more estimation methods will always give a better overall estimation (an average value is used) than if only one method is applied. All relevant equations for the described estimation system are given in [37]. The system itself

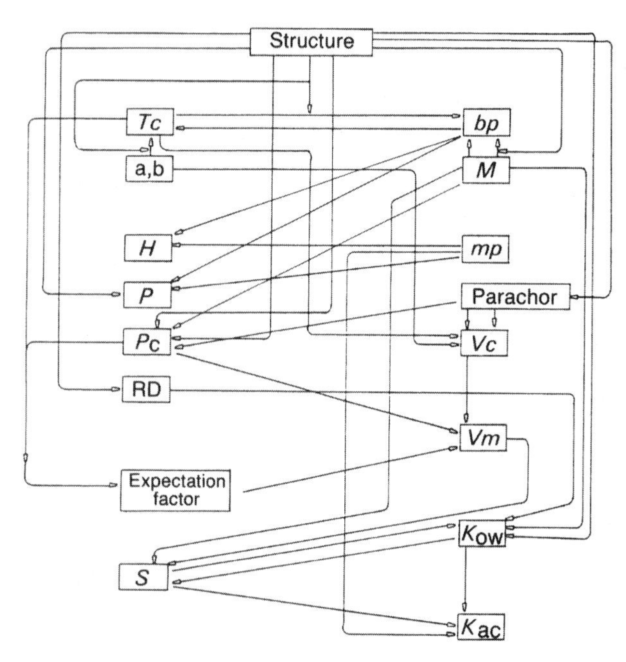

Figure 57. Conceptual diagram of an estimation system for ecotoxicological parameters (after [37])

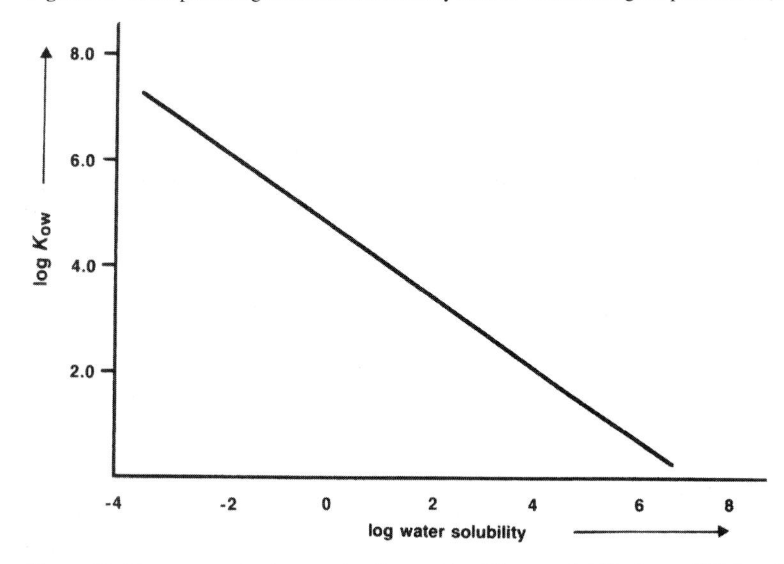

Figure 58. Linear regression between log K_{ow} and log S (water solubility in µmol/L), based on a range of organic compounds including aliphatic and aromatic hydrocarbons, aromatic acids, organochlorines, organophosphorus insecticides, and PCBs (after [467])

is shown conceptually in Figure 57 while the correlation between the partition coefficient *n*-octanol – water and the water solubility is given in Figure 58.

Chemical properties (i.e., rate of hydrolysis, rate of oxidation or reduction, and acid dissociation coefficients) can also be estimated.

Obviously, *biological properties* are very important in ecotoxicological modeling, and since these properties are generally even less well known than physical and chemical properties, the estimation methods are even more important. Figure 59 shows how to estimate CF, BCF, and the ecological magnification EM by use of solubility data S and partition coefficients K_{ow}, see also the relation in Figure 60. From the structure, furthermore, biodegradation rate coefficients and toxicity can be estimated. Some of the methods work via molar mass M and estimations of mp and bp (see Fig. 59).

The estimation methods used in practical ecotoxicological modeling have been tested widely on compounds with known properties and have been found to have a standard deviation of 10 – 60 % for most parameters, although the standard deviations are higher for toxicological parameters. This is an acceptable accuracy since large safety factors are recommended and applied in ecotoxicological modeling. Although the number of relevant parameters that can be found in the literature is limited, review of the literature is advantageous to find the parameters that actually have been determined. Furthermore, a comparison of the parameters that have been estimated by various methods is helpful to get an idea of the accuracy of these methods for the compounds considered. The following references may be used as primary sources: [37], [47], [71], [72], [75].

3.3.6. Fugacity Models

The application of fugacity in environmental modeling has been discussed by MACKAY and PATERSON [215], [216]. Fugacity has the dimension of pressure and may be considered a measure of "escaping tendency" from the phase. When phases are at equilibrium, their fugacities are equal. Fugacity f is related to concentration by the following equation:

$$f = C/Z \tag{104}$$

where C is the concentration (mol/m^3) and Z is the fugacity capacity (mol m^{-3} Pa^{-1}). Equilibrium involves

$$C_a/C_w = Z_a/Z_w = K_p \tag{105}$$

where the indices a and w indicate air and water, respectively, as examples of the phases; K_p is a partition coefficient. Transfer rates between two phases by diffusion are expressed by the following equation (models per unit of area and time):

$$N = D \cdot \Delta f \tag{106}$$

where N is the rate of transfer, D is the diffusion coefficient, and Δf is the difference in fugacity. D is the total resistance for the transfer consisting of the resistances of the two phases in series. Note that D may be determined as $K \cdot Z$, where K is the transfer coefficient and Z is the fugacity capacity defined in Equation (104).

The so-called unit world model consists of six compartments: air, water, soil, sediment, suspended sediment, and biota. This simplified model aims at identification of the partition of toxic substances emitted to the environment among these six compartments. The conceptual diagram of the unit world model is shown in Figure 61.

The fugacity model may be used at four levels. The *first level* calculates the equilibrium distribution of a chemical between phases. It assumes that each compartment is well mixed, and no reaction or advection into or out of the system occurs.

If M is the total amount of a chemical in moles, m_i and V_i are the amounts of the chemical and the volumes in each compartment, then:

$$M = \sum m_i = \sum C_i \cdot V_i = \sum f_i \cdot Z_i \cdot V_i = f \cdot \sum Z_i \cdot V_i \tag{107}$$

Since M is known, f may be calculated and the amount of substance in each compartment is given by

$$m_i = f \cdot V_i \cdot Z_i \tag{108}$$

The percentage distribution is found as m_i/M and the concentrations are found by

$$C_i = Z_i \cdot f \tag{109}$$

Level two considers equilibrium but includes reaction and advection as well. Reactions comprise photolysis, hydrolysis, biodegradation, oxidation, and so on. All the processes are assumed to be first-order reactions. If the rate constants are not known or cannot be estimated, they are set to zero. Since process rates are expressed in first-order form, they are additive and the total

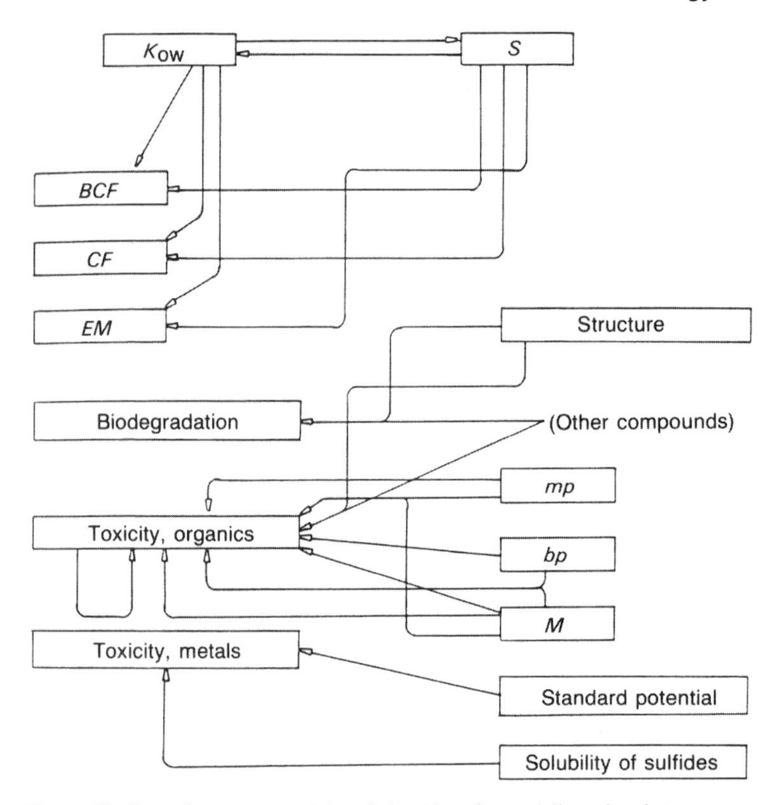

Figure 59. General structure – activity relationships of potentially toxic substances

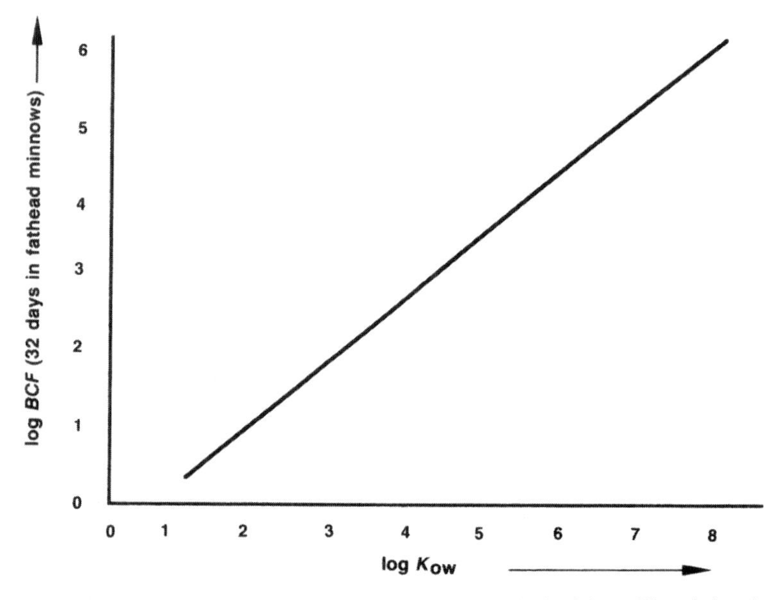

Figure 60. Relationship between log K_{ow} and log BCF, valid for fish ca. 25 cm in length

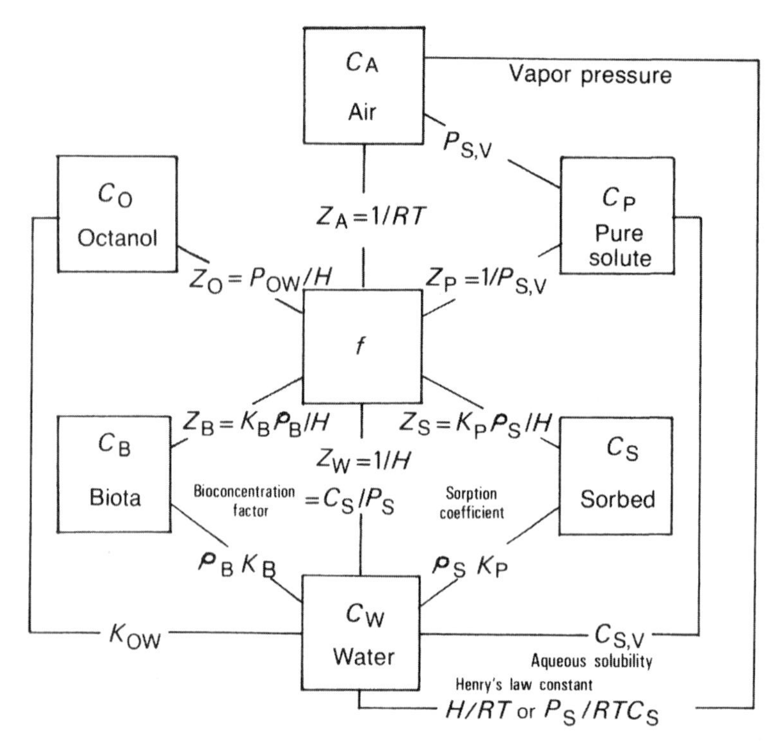

Figure 61. Conceptual diagram of fugacity models

removal rate of a chemical from a compartment is

$$\sum V_\mathrm{i} \cdot C_\mathrm{i} \cdot \sum k_j = \sum V_\mathrm{i} \cdot C_\mathrm{i} \cdot k_\mathrm{i} \quad (110)$$

where k_j and k_i denote the rate constants of conversion of substances *j* and *i*.

As in level one, a common fugacity *f* prevails, and if $f \cdot Z$ is substituted for *C* the total rate *E* is

$$E = f \cdot \sum V_i \cdot Z_i \cdot k_i \quad (111)$$

The average residence time due to reactions t_r may be found by use of the following equation:

$$t_\mathrm{r} = M/E \quad (112)$$

and the overall rate constant *K* is *E*/*M* or $1/t_\mathrm{r}$.

The *third level* is devoted to a steady-state, nonequilibrium situation, which implies that the fugacities are different in each phase. Equation (105) is used to account for the transfer. The *D* values may be calculated from quantities such as interface areas, mass-transfer coefficients (as indicated above, *D* is the product of the transfer coefficient and the fugacity capacity: $D = K \cdot Z$), release rate of chemicals into phases such as biota or sediment, and *Z* values, or by use of the estimation methods presented in Section 3.3.5.

Level four involves a dynamic version of level three where emissions, and thus concentrations, vary with time. This implies that differential equations must be applied for each compartment to calculate the change in concentrations with time, for instance:

$$V_i \cdot \mathrm{d}C_i/\mathrm{d}t = E_i - V_i \cdot C_i \cdot k_\mathrm{i} - \sum D_{ij} \cdot \Delta f_{ij} \quad (113)$$

This model level is similar in concept to the EXAMS model [251].

Levels one and two are sufficient in most cases, but if the environmental management problem requires the prediction of (1) the time required for a substance to accumulate to a certain concentration in a phase after emission has started, or (2) the length of time for the system to recover after the emission has ceased, the fourth level must be applied.

This approach has been used widely, and a typical example is given by MACKAY et al. [251]. It concerns the distribution of PCB between air and water in the Great Lakes. In this case, H was 49.1 bar and the air – water distribution coefficient (=H/RT) 0.02. The fugacity capacity for water (=$1/H$) was found to be 0.0204 bar and the fugacity capacity for air (=$1/RT$) 0.000404 bar. The distribution coefficient of PCB between water and suspended matter in the water was estimated to be 100 000. Since the concentration of suspended matter in the Great Lakes has been found to be 2×10^{-6} on a volume basis (approximately 4 mg/L) the fraction dissolved was $1/(1+0.2)=0.833$.

The PCB concentration in water of the Great Lakes was 2 ng/L, and in the air 2 ng/m^3. The fugacity can be calculated in water and air as C/Z, and the fugacity in water was found to be 17 times higher than in air, which implies that volatilization will occur. If the transfer coefficient in water is assumed to be 10^{-5} m/s and in air 10^{-3} m/s, the volatilization rate can be calculated from the traditional two-resistance model, using the relation $D=K\cdot Z$ to find the overall diffusion coefficient, D:

$$1/D = 1/10^{-5}\cdot 0.0204\,\mathrm{bar} + 1/10^{-3}\cdot 0.000404\,\mathrm{bar} \qquad (114)$$

D is 1.36×10^{-7}; N is calculated by use of Equation (106):

$$N = D\,(f_{\mathrm{w}} - f_{\mathrm{a}}) = D\,(2.8\times10^{-7}\,\mathrm{Pa} - 1.53\times10^{-8}\,\mathrm{Pa}) \qquad (115)$$

N is found to be 35.9×10^{-15} mol m^{-2} s^{-1}.

The transfer with precipitation is negligible compared to the volatilization rate, while the washout of particles and dry deposition are important processes. If these processes are considered, the net flux to the atmosphere becomes about 75 % of the flux found above.

4. Appendix [291]

Risk Assessment Directive 93/67/EEC	Document XI/730/89 rev. 3 (Ispra)	Notes
Risk Assessment		
A generic term describing an administrative and technical process that entails some or all of the elements below	no equivalent term	much wider meaning of the term than at Ispra
Effects Assessment		
1) Hazard Identification	Hazard Identification	these two terms are synonymous
Identification of the adverse effects a substance has an inherent capacity to cause; and, where possible or appropriate, assessment of a particular effect		they describe inherent properties
2) Dose – response assessment		
Estimation of the relationship between dose (or level of exposure) and the incidence and severity of an effect		
Exposure assessment	Hazard assessment (part only)	
Determination of the emissions, pathways, and rates of movement of a substance and its transformation or deradation in order to estimate the concentrations/doses to which human populations or environmental compartments are or may be exposed	= "environmental exposure"	this part of hazard assessment (Ispra) is synonymous with exposure assessment (draft directive)
		leads to PEC
	Hazard assessment (part only)	
	= "effect data with reference to the environmental compartment of concern"	this is aided by a knowledge of hazard identification (Ispra)
		eads to PEC
Effects assessment plus exposure assessment	Hazard assessment	synonymous terms
		comparison of PEC and PNEC
Risk characterization	Risk assessment (see below) plus hazard assessment	synonymous terms, i.e., PEC : PNEC ratios against a probability scale
Estimation of the incidence and severity of the adverse effects likely to occur in a human population or environmental compartment due to actual or predicted exposure to a substance and may include		
Risk estimation	Risk assessment	synonymous terms
Quantification of the likelihood of the incidence and severity of the adverse effects	estimation of the probability that a substance causes adverse effects as a result of its presence in the environment at a given concentration	
Risk reduction	Risk management	synonymous terms , e.g., special use and disposal instructions, emission control measures, restrictions on types of use, total ban of use
Measures that would enable the risks for man or the environment in connection with the marketing of the substances to be lessened	the taking of measures appropriate at least to diminish significantly the presence of a substance in the environmental compartments of concern	

5. References

General References

1. A. Leroi-Gourhan: *Le Fil du Temps,* Fayard, Paris 1983.
2. O. Fränzle: *Contaminants in Terrestrial Environments,* Springer Verlag, Berlin 1993.
3. L. Trepl: *Geschichte der Ökologie,* Athenäum, Frankfurt/Main 1987.
4. E. Haeckel: *Generelle Morphologie der Organismen,* Reimer, Berlin 1866.
5. C. J. Krebs: *Ecology,* Harper & Row, New York 1972.
6. M. Begon, J. L. Harper, C. R. Townsend: *Ecology: Individuals, Populations and Communities,* Blackwell, Boston – Melbourne 1990.
7. G. C. Butler (ed.): *Principles of Ecotoxicology,* J. Wiley, Chichester 1978.
8. H. Parlar, D. Angerhöfer: *Chemische Ökotoxikologie,* Springer Verlag, Berlin 1991.

9. H. Ellenberg: *Ökosystemforschung,* Springer Verlag, Heidelberg 1973.
10. S. L. Pimm: *Food Webs,* Chapman & Hall, London 1982.
11. R. V. O'Neill, A. R. Johnson, A. W. King "A Hierarchical Framework for the Analysis of Scale," *Landscape Ecology* **3** (1989) 193 – 205.
12. K. Dierßen: *Einführung in die Pflanzensoziologie,* Wissenschaftliche Buchgesellschaft, Darmstadt 1990.
13. W. Schröder, L. Vetter, O. Fränzle (eds.): *Neue statistische Verfahren und Modellbildung in der Geoökologie,* Vieweg, Wiesbaden 1994.
14. H. G. Gauch: *Multivariate Analysis in Community Ecology,* Cambridge University Press, Cambridge 1982.
15. W. Ziechmann: *Huminstoffe,* Verlag Chemie, Weinheim 1980.
16. P. Schachtschabel et al.: *Lehrbuch der Bodenkunde,* Enke Verlag, Stuttgart 1989.
17. K. Y. Kondratyev: *Radiation Processes in the Atmosphere,* WHO, Genever 1972.
18. M. I. Budyko: "Solar Radiation and the Use of it by Plants". Agroclimatological Methods, *Proc. Reading Symp.,* UNESCO, Paris 1968, 39 – 53.
19. H. Ellenberg, R. Meyer, J. Schauermann (eds.): *Ökosystemforschung. Ergebnisse des Solling-Projekts,* Enke Verlag, Stuttgart 1986.
20. B. J. Nebel: *Environmental Science. The Way the World Works,* Prentice Hall, Englewood Cliffs, N.J., 1990.
21. M. G. Barbour, J. H. Burk, W. D. Pitts: *Terrestrial Plant Ecology,* Benjamin/Cummings, Menlo Park, CA, 1980.
22. BUA (Beratergremium für umweltrelevante Altstoffe der Gesellschaft Deutscher Chemiker): *Umweltrelevante Alte Stoffe III – Prioritätensetzung und eingestufte Stoffe der dritten Stoffliste,* VCH Verlagsgesellschaft, Weinheim 1992.
23. O. Hutzinger: *The Handbook of Environmental Chemistry,* 5 vols., Springer Verlag, Berlin 1989 – 1991.
24. T. Rosswall, R. R. Woodmanse, P. G. Risser (eds.): *Scales and Global Change,* J. Wiley, Chichester 1988.
25. F. W. Schaller, G. W. Bailey (eds.): *Agricultural Management of Water Quality,* Iowa State University Press, Ames, IA, 1983.
26. *FAO Yearbook: Fertilizer,* **vol. 39**. Statistical Analysis Service, Statistics Division, Food and Agriculture Organization, Rome 1989.
27. D. K. Northington, Y. R. Goodin: *The Botanical World,* Times Mirror/Mosby College Publ., St. Louis, MO, 1984.
28. M. Lieth, R. H. Whittaker: *Primary Productivity of the Biosphere,* Springer Verlag, Berlin 1975.
29. L. R. Pomeroy, J. J. Alberts: *Concepts of Ecosystem Ecology. A Comparative View,* Springer Verlag, New York 1988.
30. J. Cole, G. Lovett, S. Findlay (eds.): *Comparative Analyses of Ecosystems,* Springer Verlag, New York 1991.
31. H. Walter, S. W. Breckle: *Ökologie der Erde,* **vol. 1** , "Ökologische Grundlagen in globaler Sicht," G. Fischer Verlag, Stuttgart 1991.
32. D. Pimentel (ed.): *Handbook of Energy Utilization in Agriculture,* CRC Press, Boca Raton, FL, 1980.
33. K. E. F. Watt: *Understanding the Environment,* Allyn and Bacon, Boston 1982.
34. R. G. Wetzel: *Limnology,* Saunders College Publ., New York 1983.
35. H.-P. Blume (ed.): *Handbuch des Bodenschutzes,* Ecomed, Landsberg/Lech 1992.
36. W. J. Mitsch, J. G. Gosselink: *Wetlands,* Van Nostrand Reinhold, New York 1986.
37. S. E. Jørgensen: *Modelling in Ecotoxicology,* Elsevier, Amsterdam 1990.
38. T. H. F. Allen, T. B. Starr: *Hierarchy,* Univ. Chicago Press, Chicago 1982.
39. C. Elton: *The Ecology of Invasion by Animals and Plants,* Methuen, London 1958.
40. R. E. Ulanowicz. *Growth and Development: Ecosystems Phenomenology,* Springer Verlag, New York 1986.
41. P. B. Moyle, R. A. Leidy: "Loss of Biodiversity in Aquatic Ecosystems: Evidence from Fish Faunas," in P. S. Fielder, K. J. Subodh (eds.): *From Conservation Biology: the Theory and Practice of Nature, Conservation, Preservation and Management,* Chapman and Hall, New York 1992, pp. 127 – 169.
42. J. C. White (ed.): *Global Climate Change Linkages,* Elsevier, Amsterdam 1967.
43. A. K. Ward: "Geological Mediation of Stream Flow and Sediment and Solute Loading to Stream Ecosystems due to Climate Change," in P. Firth, S. G. Fischer (eds.): *Global Change and Freshwater Ecosystems,* Springer Verlag, Berlin 1991.
44. F. H. Borman, G. E. Likens: *Pattern and Process in a Forested Ecosystem,* Springer Verlag, New York 1979.

45. C. E. W. Steinberg, R. F. Wright (eds.): *Acidification of Freshwater Ecosystems. Implications for the Future,* J. Wiley, Chichester 1994.
46. R. J. Chorley, S. A. Schumm, D. E. Sudgen: *Geomorphology,* Methuen, London 1984.
47. K. Verschueren: *Handbook of Environmental Data on Organic Chemicals,* Van Nostrand Reinhold, New York 1983.
48. G. Mattheß: *Die Beschaffenheit des Grundwassers,* Borntraeger, Berlin – Stuttgart 1994.
49. H. Freundlich: *Kapillarchemie,* Akad. Verlagsgesellschaft, Leipzig 1930.
50. J. N. Israelachvili: *Intermolecular and Surface Forces. With Applications to Colloidal and Biological Systems,* Academic Press, London 1985.
51. S. L. Faust, J. V. Hunter (eds.): *Kinetics of Biologically Mediated Oxidation of Organic Compounds in Aquatic Environments,* Marcel Dekker, New York 1978.
52. J. W. Moore, S. Ramamoorthy: *Organic Chemicals in Natural Waters,* Springer Verlag, Berlin 1984.
53. F. Korte (ed.): *Lehrbuch der ökologischen Chemie,* Thieme Verlag, Stuttgart 1987.
54. R. Geiger: *Das Klima der bodennahen Luftschicht,* Vieweg & Sohn, Braunschweig 1961.
55. B. Ulrich (ed.): *Internationaler Kongreß Waldschadensforschung: Wissensstand und Perspektiven,* Kernforschungszentrum Karlsruhe, Karlsruhe 1990.
56. R. Dallinger, P. S. Rainbow: *Ecotoxicology of Metals in Invertebrates,* Lewis Publ., Boca Raton, FL, 1992.
57. J. A. Foran, L. E. Fink: *Regulating Toxic Substances in Surface Waters,* Lewis Publ., Boca Raton, FL 1993.
58. P. H. Howard: *Handbook of Environmental Fate and Exposure Data for Organic Chemicals,* **vols. II – III**, CRC Press, Boca Raton, FL, 1991.
59. K. C. Jones: *Organic Contaminants in the Environment,* Elsevier Applied Science, New York 1991.
60. R. A. Lewis: *Environmental Toxicology,* CRC Press, Boca Raton, FL, 1991.
61. M. Mansour (ed.): *Fate and Prediction of Environmental Chemicals in Soils, Plants and Aquatic Systems,* Lewis Publ., Boca Raton, FL, 1993.
62. S. Matsui, B. F. D. Barrett, J. Banerjee, (eds.): *Guidelines of Lake Management,* vol. 4, "Toxic Substances Management in Lakes and Reservoirs," International Lake Environment Committee Foundation, Otsu, Japan 1991.
63. J. F. McCarthy, L. R. Shugart: *Biomarkers of Environmental Contamination,* CRC Press, Boca Raton, FL, 1990.
64. R. G. Tardiff (ed.): "Methods to Assess the Impact of Pesticides on Non-target Organisms," SCOPE **49** (1992) IPCS Joint Symposia 16, SGOMSEC 7.
65. A. M. V. M. Soares, P. Calow (eds.): *Standardization of Aquatic Toxicity Tests,* Lewis Publ., Boca Raton, FL, 1993.
66. T. Colborn, C. Clement: *Chemically-induced Alterations in Sexual and Functional Development: The Wildlife/Human Connection. Advances in Modern Environmental Toxicology,* **vol. XXI**, Princeton Scientific Publ. Co., Princeton, N.J., 1993.
67. P. Bourdeau, J. A. Haines, W. Klein, C. R. Krishna Murti (eds.): "Ecotoxicology and Climate with Special Reference to Hot and Cold Climates," SCOPE **38** (1989).
68. S. M. Bartell: *Toxicological Risk Assessment in Aquatic Ecosystems,* CRC Press, Boca Raton, FL, 1991.
69. M. J. Gromiec, E. F. Gloyna: "Radioactivity Transport in Water," Final Report no. 22 to U.S. Atomic Energy Commission, Contract AT (11-1)-490, 1973.
70. P. Kauppi et al.: "A Model for Predicting the Acidification of Forest Soils: Application to Acid Deposition in Europe," IIASA Research Report 1984.
71. S. E. Jørgensen, S. Nors-Nielsen, L. A. Jørgensen: *Handbook of Ecological Parameters and Ecotoxicology,* Elsevier, Amsterdam 1991.
72. S. E. Jørgensen (ed.): *Modelling in Environmental Chemistry,* Elsevier, Amsterdam 1991.
73. S. E. Jørgensen, M. Gromiec: *Mathematical Submodels of Water Quality Systems,* Elsevier, Amsterdam 1989.
74. S. E. Jørgensen, I. Johnsen: "Principles of Environmental Science and Technology," *Stud. Environ. Sci.* **33** (1989).
75. Y. Samiullah: *Prediction of the Environmental Fate of Chemicals,* Elsevier Applied Science, London 1990. U. Schlottmann (ed.): *Prüfmethoden für Chemikalien,* S. Hirzel Verlag, Stuttgart 1994.

76. BUA (Beratergremium für umweltrelevante Altstoffe der Gesellschaft Deutscher Chemiker): *Umweltrelevante Alte Stoffe – Auswahlkriterien und Stoffliste,* VCH Verlagsgesellschaft, Weinheim 1986.
77. BUA (Beratergremium für umweltrelevante Altstoffe der Gesellschaft Deutscher Chemiker): *Umweltrelevante Alte Stoffe II – Auswahlkriterien und Zweite Stoffliste,* VCH Verlagsgesellschaft, Weinheim 1988.
78. H. Bick: *Ökologie,* G. Fischer Verlag, Stuttgart 1989.
79. M. Straškraba, A. Gnauck: *Freshwater Ecosystems. Modelling and Simulation,* Elsevier, Amsterdam 1985.
80. J. Gilbert, M.-J. Dole-Olivier, P. Marmonier, P. Vervier in R. J. Naiman, H. Décamps (eds.): *The Ecology and Management of Aquatic-Terrestrial Ecotones,* UNESCO, Paris; Parthenon, Park Ridge, 1990, pp. 199 – 225.

Specific References

81. L. G. Freeman, *Am. Anthr.* **38** (1973) 3 – 44.
82. J. Seymour, H. Girardet: *Fern vom Garten Eden,* Fischer Verlag, Frankfurt/Main 1985.
83. E. H. Blair, *Proc. of the Workshop on the Control of Existing Chemicals under the Patronage of the Organisation for Economic Co-operation and Development,* Umweltbundesamt, Berlin 1981, pp. 252 – 260.
84. ECETOC: "Concentrations of Industrial Organic Chemicals measured in the Environment," Technical Report no. **29,** Brussels 1988.
85. Leitungsgremium des Projektzentrums "Ökosystemforschung im Bereich der Bornhöveder Seenkette" (ed.): "Arbeitsbericht 1988 – 91," *EcoSys,* **vol. 1,** Kiel 1992.
86. I. H. Koeman, *Ecotoxicol. Environ. Saf.* **6** (1982) 358 – 362.
87. W. Takken, F. Balk, R. C. Jansen, J. H. Koeman, PANS **24** (1978) 455 – 466.
88. Umweltbundesamt: Collection of Minimum Premarketing Sets of Data Including Environmental Residue Data on Existing Chemicals, Umweltbundesamt, Berlin 1982.
89. A. G. Tansley, *Ecology* **16** (1935) 284 – 307.
90. L. Finke: *Landschaftsökologie,* Westermann, Höller und Zwick, Braunschweig 1986.
91. C. Elton: *Animal Ecology,* Macmillan, New York 1927.
92. R. L. Lindeman, *Ecology* **23** (1942) 399 – 413.
93. O. Fränzle in B. Markert (ed.): *Environmental Sampling for Trace Analysis,* VCH Verlagsgesellschaft, Weinheim 1994, pp. 305 – 320.
94. G. G. Simpson: *The Major Features of Evolution,* Columbia University Press, New York 1953.
95. R. P. McIntosh, *Ecology* **48** (1967) 392 – 404.
96. C. E. Shannon, *Bell Syst. Tech. J.* **27** (1948) 379 – 423, 623 – 653.
97. O. Fränzle, W. F. Killisch, *Kieler Geographische Schriften* **50** (1979) 211 – 245.
98. L. Vetter: "Evaluierung und Entwicklung statistischer Verfahren zur Auswahl von repräsentativen Untersuchungsobjekten für ökotoxikologische Problemstellungen," Dissertation, Universität Kiel 1989.
99. R. M. G. Jongman, C. F. R. Ter Braak, O. F. R. Van Tongeren: "Data Analysis in Community and Landscape Ecology," PUDOC, Wageningen 1987.
100. K. R. Gabriel, *Biometrika* **58** (1971) 453 – 467.
101. O. Fränzle, W. F. Killisch in W. Schröder, L. Vetter, O. Fränzle (eds.): *Neuere statistische Verfahren und Modellbildung in der Geoökologie,* Vieweg & Sohn, Braunschweig 1994, pp. 129 – 143, 255 – 281.
102. O. Fränzle et al.: "Synoptische Darstellung möglicher Ursachen der Waldschäden: Untersuchungen zu Langzeitwirkungen von Kompensationskalkungen auf Buchen-, Fichten- und Kiefernforst-ökosysteme," Forschungsbericht 108 03 046/13 im Umweltforschungsplan des BMU, Kiel 1989.
103. W. Schröder: "Ökosystemare und statistische Untersuchungen zu Waldschäden in Nordrhein-Westfalen: Methodenkritische Ansätze zur Operationalisierung einer wissenschaftstheoretisch begründeten Konzeption," Dissertation, Universität Kiel 1989.
104. J. H. R. Gee, P. S. Giller (eds.): *Organization of Communities: Past and Present,* Blackwell, Oxford 1987.
105. J. H. Connell, R. O. Slatyer, *Am. Nat.* **111** (1977) 1119 – 1144.
106. G. Hemprich: "Landschaftsökologische Untersuchungen im Bereich des Belauer Sees und des Schmalensees," Diplomarbeit Geographie, Universität Kiel 1991.
107. V. Volterra: Variations and Fluctuations of the Numbers of Individuals in Animal Species Living Together (1926) (reprinted in R. N. Chapman: *Animal Ecology,* McGraw Hill, New York 1931).

108. J. H. Connell, *Am. Nat.* **122** (1983) 661 – 696.
109. P. Grubb, *Biol. Rev. Cambridge Philos. Soc.* **52** (1977) 107 – 145.
110. L. R. Lawlor, *Am. Nat.* **116** (1980) 394 – 408.
111. G. Müller, E. Foerster, *Acker-Pflanzenbau* **140** (1974) 161 – 174.
112. J.-F. Dhôte, F. Houllier in J.-D. Lebreton, B. Asselain (eds.): *Biométrie et Environnement,* Masson, Paris 1993, pp. 175 – 204.
113. P. Yodzis in J. Diamond, T. J. Case (eds.): *Community Ecology,* Harper & Row, New York 1986, pp. 480–4 91.
114. C. D. Canham, *Ecology* **69** (1988) 786 – 795.
115. D. J. Levey, *Ecology* **69** (1988) 1076 – 1089.
116. J. L. Monteith, *Symp. Soc. Exp. Biol.* **29** (1965) 205 – 234.
117. C. T. de Wit: *Simulation of Assimilation, Respiration and Transpiration of Crops,* PUDOC, Wageningen 1978.
118. C. W. Thornthwaite, *Geogr. Rev.* **38** (1948) 85 – 94.
119. L. Turc, *Ann. Agrar.* **5** (1954) 491 – 596.
120. B. L. Dzerdzeevskii in: *Climat et Microclimatologie. Actes du Colloque de Canberra,* (Arid Zone Research/Recherches sur la Zone Aride XI), UNESCO, Paris 1958, pp. 315 – 325.
121. M. Fuchs, G. Stanhill, *Isr. J. Agric. Res.* **13** (1963) 63 – 78.
122. R. O. Slatyer in: *The Use of Soil Water Balance Relationships in Agroclimatology* (Natural Resources Res. VII) UNESCO, Paris 1968, pp. 73 – 87.
123. P. Benecke, R. R. van der Ploeg, *Forstarchiv* **49** (1978) 1 – 7, 26 – 32.
124. J. v. Hoyningen-Huene, S. Nasdalack, *Landschafts-ökol. Messen Auswerten* **1** (1985) 107 – 116.
125. B. Sevruk: "Methodische Untersuchungen des systematischen Meßfehlers des Hellmann-Regenmessers im Sommerhalbjahr in der Schweiz," Dissertation, ETH Zürich 6789, 1981.
126. H. Braden, *Mitt. Dtsch. Bodenkd. Ges.* **42** (1985) 294 – 298.
127. F.-J. Löpmeier: *Verhandlungen 45. Dtsch. Geographentages,* Steiner Verlag, Stuttgart 1987, pp. 434 – 438.
128. J. R. Philip in E. S. Hills (ed.): *Water Resources, Use and Management,* Melbourne University Press, Melbourne 1964, pp. 257 – 275.
129. O. Schwarz in G. Einsele et al. (eds.): *Das landschaftsökologische Forschungsprojekt Naturpark Schönbuch,* VCH Verlagsgesellschaft, Weinheim 1986, pp. 161 – 179.
130. R. E. Smith, J. Y. Parlange, *Water Resour. Res.* **14** (1978) 533 – 538.
131. A. C. Imeson, *CATENA Suppl.* **4** (1983) 79 – 89.
132. G. Einsele, G. Agster, M. Elgner in G. Einsele (ed.): *Das landschaftsökologische Forschungsprojekt Naturpark Schönbuch,* VCH Verlagsgesellschaft, Weinheim 1986, pp. 209 – 234.
133. J. Drescher, R. Horn, M. de Boodt (eds.): "Impact of Water and External Forces on Soil Structure", *CATENA Suppl.* **11** (1988).
134. E. Petzold, *Verh. Ges. Ökol.* **12** (1984) 487 – 498.
135. M. Renger, O. Strebel, *Wasser Boden* **32** (1980) 362 – 366.
136. Arbeitskreis Grundwasserneubildung, *Geol. Jahrb. Reihe C* **C 19** (1977) 3 – 98.
137. O. Fränzle et al.: "Auswahl der Hauptforschungsräume für das Ökosystemforschungsprogramm der Bundesrepublik Deutschland," Forschungsbericht 101 04 043/02 im Umweltforschungsplan des BMU, Universität Kiel 1987.
138. H. J. Heckmann, K. F. Schreiber, R. Thöle, *Mitt. Dtsch. Bodenkd. Ges.* **41** (1985) 353 – 356.
139. H. Sager, Informationsbericht, part 1, Bayer. Landesamt für Wasserwirtschaft, München 1983, pp. 59 – 72.
140. P. Greminger: "Physikalisch-ökologische Standortuntersuchung über den Wasserhaushalt im offenen Sickersystem Boden unter Vegetation am Hang," *Mitt. Eidgen. Anst. forstl. Versuchswesen* **60** (1985) no. 2.
141. I. D. White, N. D. Mottershead, S. J. Harrison: *Environmental Systems,* Allen & Unwin, London 1984.
142. M. Schroeder, *Dtsch. Gewässerkd. Mitt.* **20** (1976) 8 – 13.
143. C. W. Thornthwaite, B. Holzman, *Tech. Bull. U.S. Dep. Agric.* **817** (1942).
144. A. I. Dyer, *Q. J. R. Meteorol. Soc.* **87** (1961) 401 – 412.
145. J. L. Monteith, G. Szeicz, *Q. J. R. Meteorol. Soc.* **86** (1960) 205 – 214.
146. S. v. Stamm: "Untersuchungen zur Primärproduktion von Corylus avellana an einem Knickstandort in Schleswig-Holstein und Erstellung eines Produktionsmodells," *EcoSys,* **suppl. vol. 4** , Kiel 1992.

147. S. Gäb, H. Parlar, F. Korte, *Chemosphere* **3** (1974) no. 5, 187 – 192.
148. M. Straškraba, *Mitt. Int. Ver. Limnol.* **14** (1968) 212 – 230.
149. J. Blake in K.-H. Zwirnmann (ed.): "Nonpoint Nitrate Pollution of Municipal Water Supply Resources: Issues of Analysis and Control," *IIASA Collab. Proc. Ser.* **CP-82-S4** (1982) 231 – 245.
150. L. Procházková, P. Blažka, personal communication, 1993.
151. L. R. Brown: *State of the World,* W. W. Norton & Co., New York 1985.
152. H. Lieth, E. Box in R. Y. Mather (ed.): *Publications in Climatology,* **vol. 25,** C. W. Thornthwaite Associates, New Jersey 1972, pp. 37 – 46.
153. S. McNaughton, M. Oesterheld, D. A. Frank, K. J. Williams in [30, pp. 120 – 139].
154. I. Aselmann, H. Lieth, *Prog. Biometeorol.* **3** (1983) 101 – 119.
155. R. J. Naiman, H. Décamps, J. Pastor, C. A. Johnston, *J. North Am. Benthological Soc.* **7** (1988) 289 – 306.
156. F. E. Clements: *Research Methods in Ecology,* University Publ. Comp., Lincoln 1905.
157. W. Kluge, O. Fränzle, *Ges. Ökol.* **21** (1992) 401 – 407.
158. D. L. Urban, R. V. O'Neill, H. H. Shugart, *BioScience* **37** (1987) 119 – 127.
159. O. Fränzle in E. Bayer, H. Behret (eds.): "Bewertung des ökologischen Gefährdungspotentials von Chemikalien," *GDCh-Monogr.* **1** (1994) 45 – 90.
160. O. Fränzle in B. Markert (ed.): *Sampling of Environmental Materials for Trace Analysis,* VCH Verlagsgesellschaft, Weinheim 1994, pp. 217 – 227.
161. W. M. Stigliani: "Changes in Valued 'Capacities' of Soil and Sediments as Indicators of Nonlinear and Time-delayed Environmental Effects," IIASA Ecoscript **35** (1988), International Institute for Applied Systems Analysis, Laxenburg.
162. R. C. Wissmar, F. J. Swanson in R. J. Naiman, H. Décamps (eds.): *The Ecology and Management of Aquatic-terrestrial Ecotones,* UNESCO, Paris; Parthenon, Park Ridge 1990, pp. 171 – 198.
163. E. Pieczinska in R. J. Naiman, H. Décamps (eds.): *The Ecology and Management of Aquatic-terrestrial Ecotones,* UNESCO, Paris; Parthenon, Park Ridge 1990, pp. 103 – 140.
164. U. Schleuß: "Böden und Bodenschaften einer norddeutschen Moränenlandschaft," *EcoSys,* **suppl. vol. 2,** Kiel 1992.
165. M. M. Holland, D. F. Whigham, B. Gopal in R. J. Naiman, H. Décamps (eds.): *The Ecology and Management of Aquatic-terrestrial Ecotones,* UNESCO, Paris; Parthenon, Park Ridge 1990, pp. 171 – 198.
166. D. Scholle: "Vegetationskundliche Untersuchungen im Raum Bornhöved und deren Auswertung mit Hilfe eines Geographischen Informationssystems (GIS)," Diplomarbeit, Universität Saarbrücken, 1991.
167. G. Schernewski: "Raumzeitliche Prozesse und Strukturen im Wasserkörper des Belauer Sees," Diplomarbeit Geographie, in *EcoSys,* **suppl. vol. 1**, Kiel 1992.
168. C. S. Holling in E. Jantsch, C. H. Waddington (eds.): *Evolution and Consciousness,* Addison-Wesley, Reading, Mass., 1976, pp. 73 – 92.
169. I. Prigogine in E. Jantsch, C. H. Waddington (eds.): *Evolution and Consciousness,* Addison-Wesley, Reading, Mass., 1976, pp. 93 – 133.
170. W. E. Westman, *BioScience* **28** (1989) 705 – 710.
171. A. Gigon, Ber. Geobot. Inst. Eidg. Techn. Hochsch., Stift. Rübel Zürich **50** (1983) 149 – 177.
172. E. van der Maarel, *Ber. Dtsch. Bot. Ges.* **89** (1976) 415 – 443.
173. O. T. Solbrig, G. Nicolis: *Perspectives in Biological Complexity,* IUBS, Paris 1991.
174. R. H. MacArthur, *Ecology* **36** (1955) 533 – 536.
175. R. M. May, *Nature (London)* **238** (1972) 413 – 414.
176. R. E. McMurtrie, *J. Theor. Biol.* **50** (1975) 1 – 11.
177. J. E. Cohen, C. M. Newman, *Ann. Probab.* **12** (1984) 283 – 310.
178. C. P. Bachas, B. A. Huberman, *J. Phys. A Math. Gen.* **20** (1987) 4995 – 5014.
179. H. A. Ceccato, B. A. Huberman, *Physica Scr.* **37** (1988) 145 – 150.
180. P. Doreian, *J. Soc. Biol. Struct.* **9** (1986) 115 – 139.
181. F. Briand, J. E. Cohen, *Science (Washington D.C. 1883)* **238** (1987) 956 – 960.
182. T. Hogg, B. A. Huberman, J. M. McGlade, *Proc. R. Soc. London B* **237** (1989) 43 – 51.
183. M. Rejmanek, P. Stary, *Nature (London)* **280** (1979) 311 – 313.
184. F. Briand, *Ecology* **64** (1983) 253 – 263.
185. S. J. McNaughton, *Am. Nat.* **111** (1977) 515 – 525.

186. A. Hay, *Nature (London)* **302** (1983) 208 – 209.
187. R. V. O'Neill, *Ecology* **57** (1976) 1244 – 1253.
188. D. L. de Angelis, *Ecology* **61** (1980) 764 – 771.
189. J. Cairns Jr., B. R. Niederlehner, *Environ. Professional* **15** (1993) 116 – 124.
190. P. van Voris, R. V. O'Neill, W. R. Emanuel, H. H. Shugart, *Ecology* **6** (1980) 1352 – 1360.
191. D. W. Schindler, *Can. J. Fish. Aquat. Sci.* **44** (1987) Suppl 1, 6 – 25.
192. M. J. Crossey, T. W. LaPoint, *Hydrobiologia* **162** (1988) 109 – 121.
193. W. D. Crumby, M.-A. Webb, F. J. Bulow, H. J. Cathey, *Trans. Am. Fish. Soc.* **119** (1990) 885 –893.
194. J. H. Rodgers Jr., K. L. Dickson, J. Cairns, Jr., in R. G. Wetzel (ed.): "Methods and Measurements of Periphyton Communities," *ASTM,* Philadelphia, PA, 1980, pp. 142 – 167.
195. D. Uhlmann, H. Mihan, A. Gnauck, *Acta Hydrochim. Hydrobiol.* **6** (1978) 421 – 444.
196. A. Gnauck: "Strukturelle und funktionelle Änderungen in aquatischen Ökosystemen," Kongreß- und Tagungsberichte, Martin-Luther-Universität Halle-Wittenberg, 1982, pp. 335 – 344.
197. J. D. Yount, G. J. Niemi: "Recovery of Lotic Communities and Ecosystems from Disturbance: Theory and Applications," *Environ. Manage. (N.Y.)* **14** (1990) no. 5, 132 – 137.
198. G. M. Woodwell, *Science (Washington, D.C. 1883)* **156** (1967) no. 3774, 461 – 470.
199. T. Rosswall, R. G. Woodmanse, P. G. Risser: *Scale and Global Change: Spatial and Temporal Variability in Biospheric and Geospheric Processes,* J. Wiley, Chichester 1988.
200. J. Peñuelas in J. D. Ross, N. Prat (eds.): "Hommage to Ramón Margalef: or Why There is Such Pleasure in Studying Nature," Universitat de Barcelona Publications, Barcelona 1991, 367 – 385.
201. S. H. Schneider, N. J. Rosenberg in N. J. Rosenberg, W. E. Easterling, P. J. Crosson, J. Darmstadter (eds.): "Greenhouse Warming: Abatement and Adaptation," Proceedings of a Workshop held in Washington D.C., June 14 – 15, 1988, Resources for the Future, Washington 1989, pp. 7 – 34.
202. J. S. Clark in R. L. Wyman (ed.): *Global Climate Change and Life on Earth,* Routledge, Chapman and Hall, New York 1991, pp. 65 – 98.
203. J. D. Meisner et al., *J. Great Lakes Res.* **13** (1987) no. 3, 340 – 352.
204. P. Glansdorff, I. Prigogine: *Thermodynamic Theory of Structure, Stability, and Fluctuations,* Wiley-Interscience, New York 1971.
205. E. J. Fittkau: "Tropische Regenwälder – Ökologische Zusammenhänge," *Bensberger Protokolle* **66** (1991) 27 – 63 (Thomas-Morus-Akademie, Bensberg).
206. R. Herrera, C. F. Jordan, E. Medina, H. Kluge, *Ambio* **10** (1981) 109 – 114.
207. H. Sioli: *Amazonien. Grundlagen der Ökologie des größten tropischen Waldlandes,* Enke Verlag, Stuttgart 1980.
208. C. F. Jordan (ed.): *An Amazonian Rain Forest: the Structure and Function of a Nutrient Stressed Ecosystem and the Impact of Slash and Burn Agriculture,* UNESCO, Paris, Parthenon, Carnforth 1989.
209. D. L. de Angelis, *Ecology* **56** (1975) 238 – 243.
210. D. L. de Angelis, J. C. Waterhouse, *Ecol. Monogr.* **57** (1987) 1 – 21.
211. H. Remmert, *Verh. Ges. Ökol.* **16** (1987) 27 – 34.
212. H. M. Wilbur, *Ecology* **68** (1987) 1437 – 1452.
213. C. R. Townsend, *J. North Am. Benthological Soc.* **8** (1989) 36 – 50.
214. N. I. Sax: *Dangerous Properties of Industrial Materials,* Van Nostrand Reinhold, New York 1975.
215. D. Mackay, S. Paterson, *Environ. Sci. Technol.* **15** (1981) 1006.
216. D. Mackay, S. Paterson, *Environ. Sci. Technol.* **16** (1982) 654 A.
217. W. F. Spencer, W. J. Farmer in R. Haque (ed.): *Dynamics, Exposure and Hazard Assessment of Toxic Chemicals,* Ann Arbor Sci., Ann Arbor 1980, pp. 143 – 161.
218. O. Fränzle, *Z. Geomorphol. N.F.* **15** (1971) 212 – 235.
219. M. C. Lee et al., *Water Res.* **13** (1979) 1249 – 1258.
220. ECETOC: "Structure-activity Relationships in Toxicology and Ecotoxicology: an Assessment," *ECETOC Monogr.* **8** (1986).
221. A. Leo, C. Hansch, D. Elkins, *Chem. Rev.* **71** (1971) 525.
222. T. Mill in Umweltbundesamt (eds.): "Proc. Workshop Control of Existing Chemicals under the Patronage of the Organisation for Economic Cooperation and Development," June 10 – 12, 1981, Umweltbundesamt, Berlin 1981, pp. 207 – 227.

223. G. Mattheß et al.: "Der Stofftransport im Grundwasser und die Wasserschutzgebietsrichtlinie W 101 – Statusbericht und Problemanalyse," *Ber. Umweltbundesamt Ger.* **7** (1985).
224. W. Mabey, T. Mill, *J. Phys. Chem. Ref. Data* **7** (1978) 383 – 415.
225. ECETOC: "Experimental Assessment of the Phototransformation of Chemicals in the Atmosphere," Technical Report no. 7, Brussels 1983.
226. P. Schachtschabel, H.-P. Blume, G. Brümmer et al.: *Lehrbuch der Bodenkunde,* Enke Verlag, Stuttgart 1989.
227. R. Brdicka: *Grundlagen der physikalischen Chemie,* Deutscher Verlag der Wissenschaften, Berlin 1971.
228. G. Sandstede, *Z. Phys. Chem. (Munich)* **29** (1961) 120 – 133.
229. M. Alexander in R. Haque (ed.): *Dynamics, Exposure and Hazard Assessment of Toxic Chemicals,* Ann Arbor Sci., Ann Arbor 1980, pp. 179 – 190.
230. J. C. G. Ottow, *Landwirtsch. Forsch.* **35** (1982) 238 – 256.
231. OECD Guidelines for Testing of Chemicals, no. 301 A – E, OECD, Paris 1981.
232. OECD Guidelines for Testing of Chemicals, no. 302 A, OECD, Paris 1981.
233. OECD Guidelines for Testing of Chemicals, no. 303 A, OECD, Paris 1981.
234. OECD Guidelines for Testing of Chemicals, no. 304 A, OECD, Paris 1981.
235. H. Bernhardt (ed.): *NTA: Studie über die aquatische Umweltverträglichkeit von Nitrilotriacetat (NTA),* Richarz, Sankt Augustin 1984.
236. A. G. Journel, C. J. Huijbregts: *Mining Geostatistics,* Academic Press, London 1978.
237. G. Verly, M. David, A. G. Journel, A. Maréchal (eds.): "Geostatistics for Natural Resources Characterization," *NATO Adv. Sci. Inst. Ser.,* Reidel, Dordrecht 1984.
238. ECETOC: "Estimating the Environmental Concentrations of Chemicals Using Fate and Exposure Models," Technical Report no. 50, Brussels 1992.
239. ECETOC: "Environmental Hazard Assessment of Substances," Technical Report no. 51, Brussels 1993.
240. HEDSET (1992): Harmonized Electronic Data Input Set for the Council Regulation on the Evaluation and Control of Existing Substances and the OECD, Existing Chemicals Programme (05/07/92).
241. T. R. Oke: *Boundary Layer Climates,* Methuen, London 1978.
242. T. R. Oke, *J. R. Meteorol. Soc.* 1982, 1 – 24.
243. D. O. Lee, *Prog. Phys. Geogr.* **8** (1984) 1 – 31.
244. E. Jauregui, *Erdkunde* **41** (1987) 48 – 51.
245. G. M. Hidy, C. S. Burton in G. M. Hidy et al. (ed.): *The Character and Origin of Smog Aerosols,* Wiley, New York 1980, pp. 385 – 433.
246. T. Schneider, L. Grant (eds.): *Air Pollution by Nitrogen Oxides,* Elsevier Sci., Amsterdam 1982.
247. B. Henderson-Sellers: *Modeling of Plume Rise and Dispersion – The University of Salford Model: U.S.P.R.,* Springer Verlag, Heidelberg 1987.
248. J. Pankrath, *Staub Reinhalt. Luft* **47** (1987) 239 – 244.
249. W. Klug in [255, pp. 243 – 248].
250. A. Eliassen, Ø. Hov, I. S. A. Isaksen, J. Saltbones in [255, pp. 347 – 356].
251. D. Mackay, S. Paterson, W. Y. Shiu, *Chemosphere* **24** (1992) 695 – 717.
252. E. W. Reiche: "Entwicklung, Validierung und Anwendung eines Modellsystems zur Beschreibung und flächenhaften Bilanzierung der Wasser- und Stickstoffdynamik in Böden", *Kieler Geogr. Schr.* **79** (1991) 1 – 150.
253. TA Luft, Aug. 28, 1974, GMBl I, pp. 426, 525, GMBl 1983, p. 98.
254. BUA (Beratergremium für umweltrelevante Altstoffe): "OH-Radikale in der Troposphäre: Konzentration und Auswirkung," BUA-Stoffbericht 100, Hirzel Verlag, Stuttgart 1993.
255. OECD: "Compendium of Environmental Exposure Assessment Methods for Chemicals," *Environ. Monogr.* **27** (1989).
256. OECD, Workshop on the Application of Simple Models for Environmental Exposure Assessment, Berlin, Sep. 11 – 13, 1991.
257. S. Wiegmann: "Die geogenen Schwermetallgehalte der Böden Schleswig-Holsteins," Diplomarbeit Geographie, Universität Kiel 1994.
258. M. Bonazountas in H. Barth, P. L'Hermite (eds.): *Scientific Basis for Soil Protection in The European Community,* Elsevier, London 1987, pp. 487 – 566.
259. DVWK (Deutscher Verband für Wasserwirtschaft und Kulturbau e.V.): *Abschätzen des Verhaltens organischer Chemikalien in Böden,* DVWK, Bonn 1989.

260. ECETOC: "Hazard Assessment of Floating Chemicals after Accidental Spill at Sea," Technical Report no. 35, Brussels 1990.
261. D. G. Farmer, M. J. Rycroft (eds.): *Computer Modelling in the Environmental Sciences,* Clarendon Press, Oxford 1991.
262. A. Dahmke: "Lösungskinetik von feldspatreichen Gesteinen und deren Bezug zu Verwitterung und Porenwasser-Chemie natürlicher Sander-Sedimente." *Berichte – Reports Geol. Paläont.-Inst. Universität Kiel* **20** (1988).
263. A. Petersen: "Laboruntersuchungen zum Einfluß organischer Komplexbildner auf die Kinetik der Feldspatverwitterung," *Berichte – Reports Geol. Paläont.-Inst. Universität Kiel* **21** (1988).
264. D. Boening: "Untersuchungen zur Modellierung des Stofftransportes in einem durch Deponiesickerwasser verunreinigten Grundwassergerinne," *Berichte – Reports Geol.-Paläont. Inst. Universität Kiel* **30** (1989).
265. R. D. Lindberg, D. D. Runnels, *Science* (Washington, D.C. 1883) 425 (1984) 925.
266. D. Langmuir in E. A. Jenne (ed.): "Chemical Modeling in Aqueous Systems," *ACS Symp. Ser.* **93** (1979).
267. D. J. Kirkner, T. L. Theis, A. A. Jennings, *Adv. Water Resour.* **7** (1984) 120.
268. P. G. Daniele, A. de Robertis, C. de Stefano, S. Sammarfano, *J. Chem. Soc. Dalton Trans.* 1985, 2353 – 2361.
269. R. Katlein: "Bestimmung von Komplexkapazitäten mit der Mangandioxid-Methode," Dissertation, Universität Marburg 1986.
270. W. Stumm, R. Kummert, L. Sigg, *Croat. Chem. Acta* **48** (1980) 291 – 312.
271. R. V. James, J. Rubin in E. A. Jenne (ed.): "Chemical Modeling in Aqueous Systems," *ACS Symp. Ser.* **93** (1979).
272. H. J. Hueck, D. M. M. Adema, W. C. de Kock, J. Kuiper, *Ber. Umweltbundesamt Ger.* **10** (1978) 159 – 167.
273. H. Biek, D. Neumann (eds.): "Bioindikatoren," *Decheniana Beih.* **26** (1982).
274. L. Steubing, H. J. Jäger (eds.): *Monitoring of Air Pollutants by Plants. Methods and Problems,* Junk, The Hague 1982.
275. OECD Guideline for Testing of Chemicals, no. 305 E, OECD, Paris 1981.
276. W. Ernst, *Decheniana Beih.* **26** (1982) 55 – 66.
277. E. A. Nusch, *Decheniana Beih.* **26** (1982) 87 – 98.
278. B. Hamburger, *Decheniana Beih.* **26** (1982) 78 – 81.
279. W. K. Besch, *Decheniana Beih.* **26** (1982) 67 – 77.
280. C. C. Brown in G. C. Butler (ed.): *Principles of Ecotoxicology,* J. Wiley, Chichester 1978, pp. 115 – 148.
281. P. Gerike, in [283, pp. 139 – 143].
282. R. N. Sturm, *J. Am Oil. Chem. Soc.* **50** (1973) 159 – 167.
283. R. Zahn, M. Wellens, *Chem. Ztg.* **98** (1974) 228 –232.
284. Umweltbundesamt: "Tests for the Ecological Effects of Chemicals," *Ber. Umweltbundesamt Ger.* **10** (1978).
285. J. S. Whiteside, M. Alexander, *Weeds* **8** (1960) 204 – 213.
286. A. Krotzky et al., *Z. Pflanzenernähr. Bodenkd.* **146** (1983) 634 – 642.
287. OECD: Data Interpretation Guides for Initial Hazard Assessment of Chemicals, Provisional OECD, Paris 1984.
288. P. Friesel, P. D. Hansen, R. Kühn, J. Trénel: "Überprüfung der Durchführbarkeit von Prüfungsvorschriften und der Aussagekraft der Stufe 1 und 2 des Chemikaliengesetzes," part VI, Umweltforschungsplan des Bundesministers des Innern, F+E-Vorhaben 106 04 001/08 (1984).
289. European Communities (EC): Technical Guidance Documents in Support of the Risk Assessment Directive (933/67/EEC) for New Substances Notified in Accordance with the Requirements of Council Directive 67/548/EEC, Brussels 1993.
290. Umweltbundesamt: "Technical Guidance on Environmental Risk Assessment of Existing Chemicals in Accordance with the Requirements of Council Regulation (EEC)" no. 793/93, draft, Berlin 1994.
291. ECETOC: "Aquatic Toxicity Evaluation," Technical Report no. 56, Brussels 1993.
292. OECD: "Report of the OECD Workshop on the Extrapolation of Laboratory Aquatic Toxicity Data on the Real Environment," *Environ. Monogr.* **59** (1992).
293. N. M. van Straalen, C. A. J. Denneman, *Ecotoxicol. Environ. Saf.* **18** (1989) 241 – 251.
294. H. Lokke in M. H. Donkers, H. Eijsackers, F. Heimbach (eds.): "Ecotoxicology of Soil Organisms," *SETAC Special Publication Ser.,* Lewis, Michigan, 1994, pp. 425 – 441.
295. T. L. Forbes, V. E. Forbes, *Funct. Ecology* **7** (1993) 249 – 254.

296. H. J. B. Emans et al., *Environ. Toxicol. Chem.* **12** (1993) 2139 – 2154.
297. T. Aldenberg, W. Slob, *Ecotoxicol. Environ. Saf.* **25** (1993) 48 – 63.
298. C. Wagner, H. Lokke, *Water Res.* **25** (1991) 1237 – 1242.
299. O. Fränzle et al.: "Grundlagen zur Bewertung der Belastung und Belastbarkeit von Böden als Teilen von Ökosystemen," Texte 59/93 des Umweltbundesamtes, Berlin 1993.
300. C. A. M. van Gestel in H. Becker et al. (eds.): *Ecotoxicology of Earthworms,* Intercept, Andover 1992, pp. 45 – 54.
301. OECD Guideline for Testing of Chemicals, no. 207, OECD, Paris 1984.
302. OECD Guideline for Testing of Chemicals, no. 208, OECD, Paris 1984.
303. NEN (Nederlandse Norm): Soil-determination of the Influence of Chemicals on Soil Nitrification, no. 5795, Nederlands Normalisatie Instituut, Delft 1988.
304. BBA (Biologische Bundesanstalt für Land- und Forstwirtschaft): "Bestimmung der Reproduktionsleistung von *Folsomia candida* (WILLEM) in künstlichem Boden," (Draft), Braunschweig 1990.
305. ISO (International Organisation for Standardisation): Soil Quality Effects of Pollutants on Earthworms (Eisenia foetida) – part 2: Method for the Determination of Effects on Reproduction. Draft Int. Standard 1993.
306. L. Moreth, E. Naton: Richtlinie zur Prüfung der Nebenwirkung von Pflanzenschutzmitteln auf *Aleochara bilineata* Gyll. (Col., Staphylinidae). Unveröff. Richtlinienvorschlag. 1992.
307. J. Römbke: "Entwicklung eines Reproduktionstests an Bodenorganismen – Enchytraen," UBA F+E-Vorhaben no. 106 03 051/01. Batelle-Institut, Frankfurt/Main 1991.
308. N. M. van Straalen, C. A. M. van Gestel: "Ecotoxicological Terrestrial Test Methods Using Terrestrial Arthropods," Detailed Review Paper for the OECD Test Guideline Programme, Amsterdam 1992.
309. A. Stavola: Detailed Review Paper on Terrestrial Ecotoxicology Test Guidelines, OECD Updating Programme, Periodical Review. OECD, Paris 1990.
310. F. Pedersen, L. Samsøe-Petersen: Discussion Paper Regarding Guidance for Terrestrial Effects Assessment (Draft), Water Quality Institute, Horsholm, Denmark, 1993.
311. H. Blanck, *Ecol. Bull.* **36** (1984) 107 – 119.
312. C. A. J. Denneman, C. A. M. van Gestel: "Bodemverontreiniging en bodemecosystemen: voorstel voor C-(toetsings)waarden op basis van ecotoxicologische risico's," Rijksinstituut voor Volksgezondheid en Milieuhygiene, Bilthoven 1990.
313. EPA (Environmental Protection Agency): "Pesticide Assessment Guidelines," Office of Pesticides and Toxic Substances, Washington D.C., 1982.
314. S. A. L. M. Kooijman, *Water Res.* **21** (1987) no. 3, 269 – 276.
315. W. Sloof, J. A. M. v. Oers, D. Zwart, *Environ. Toxicol. Chem.* **5** (1986) 841 – 852.
316. C. E. Stephan et al.: "Guidelines for Deriving Numerical National Water Quality Criteria for the Protection of Aquatic Organisms and their Uses," U.S. Environmental Protection Agency, Washington 1985.
317. K. Terytze: Qualitative and quantitative Aspekte der Bewertung von Böden. Manuskript 1994.
318. R. L. Metcalf, *Environ. Qual. Saf.* **5** (1976) 141 – 151.
319. C. Frehe, *Verh. Ges. Ökol.* **12** (1984) 507 – 510.
320. H. Kayser, *Helgol. Wiss. Meeresunters.* **30** (1977) 682 – 696.
321. A. M. Albert in B. Scheele, M. Verfondern (eds.): "Auffindung von Indikatoren zur prospektiven Bewertung der Belastbarkeit von Ökosystemen," Endberichte der geförderten Vorhaben, Teil 2, Jül.-Spez. 503, A 1 – A 116, Jülich 1989.
322. G. Weidemann, K. Mathes, H. Koehler in B. Scheele, M. Verfondern (eds.): "Auffindung von Indikatoren zur prospektiven Bewertung der Belastbarkeit von Ökosystemen," Jül-Spez. 439, 7 – 223, 1988.
323. L. Beck, K. Dumpert, U. Franke, W. Schönbor in B. Scheele, M. Verfondern (eds.): "Auffindung von Indikatoren zur prospektiven Bewertung der Belastbarkeit von Ökosystemen," Jül-Spez. 439, 548 – 702, Jülich 1988.
324. T. Basedow, H. Rzehak, W. Liedtke in B. Scheele, M. Verfondern (eds.): "Auffindung von Indikatoren zur prospektiven Bewertung der Belastbarkeit von Ökosystemen," Jül-Spez. 439, 223 – 369, Jülich 1988.
325. G. Kneitz, W. J. Kloft in B. Scheele, M. Verfondern (eds.): "Auffindung von Indikatoren zur prospektiven Bewertung der

Belastbarkeit von Ökosystemen," Jül.-Spez. 439, 702 – 907, Jülich 1988.
326. G. Weigmann et al.: "Wirkung und Verbleib von PCP-Na in einem Brachland-Ökosystem," Abschluß-bericht 037 288, 037 289, 037 294 eines im Programm: "Auffindung von Indikatoren zur prospektiven Bewertung der Belastbarkeit von Ökosystemen" geförderten Verbundvorhabens, 1986.
327. M. Runge et al. in B. Scheele, M. Verfondern (eds.): "Auffindung von Indikatoren zur prospektiven Bewertung der Belastbarkeit von Ökosystemen," Endberichte der geförderten Vorhaben, Teil 2, Jül.-Spez. 503, B 1 – B 218, Jülich 1989.
328. R. Guderian (ed.): *Air Pollution by Photochemical Oxidants,* Springer Verlag, Berlin – Heideberg – New York – Tokyo 1985.
329. L. Steubing, A. Fangmeier, *Environ. Pollut. Ser. A* **44** (1987) 297 – 306.
330. L. W. Kress, J. M. Skelly, *Plant Dis.* **66** (1982) 1149 – 1152.
331. H.-J. Jäger, E. Schulze: "Critical Levels for Effects of SO2," ECE Critical Levels Workshop, Bad Harzburg, March 14 – 18, 1988, Draft Report part II, pp. 15 – 50, UN Econ. Comm. Europe, Geneva 1988.
332. R. Guderian: "Critical Levels for Effects of NOx," ECE Critical Levels Workshop, Bad Harzburg, March 14 – 18, 1988. Draft Report part II, pp. 79 – 104, UN Econ. Comm. Europe, Geneva 1988.
333. R. Guderian: "Critical Levels for Effects of Ozone (O3)," ECE Critical Levels Workshop, Bad Harzburg, March 14 – 18, Draft Report part II, pp. 51 – 78, UN Econ. Comm. Europe, Geneva 1988.
334. D. C. Robinson, A. R. Wellburn, *Environ. Pollut. Ser. A* **32** (1983) 109 – 120.
335. E. A. Wright, P. W. Lucas, D. A. Cottam, T. A. Mansfield, *CEC Air Pollut. Res. Rep.* **4** (1986) 187 –200.
336. J. Mooi, *Forst. Holzwirt* **39** (1984) 438 – 444.
337. R. Guderian, K. Küppers, R. Six, *VDI-Ber.* **560** (1985) 657 – 701.
338. A. R. Wellburn, C. Higginson, D. Robinson, C. Walmsley, *New Phytol.* **88** (1981) 223 – 237.
339. J. A. Lee, M. C. Press, S. J. Woodin in CEC: "Study on the Need for a NO2 Long-term Limit Value for the Protection of Terrestrial and Aquatic Ecosystems," Final Report, EUR 10 546 EN, 1985.
340. A. C. Posthumus: "Critical Levels for Effects of NH_3 and NH_{+4}," ECE Critical Levels Workshop, Bad Harzburg, March 1988, pp. 117 – 127, UN Econ. Comm. Europe, Geneva 1988.
341. J. Bauch, P. Klein, A. Frühwald, H. Brill, *Eur. J. For. Pathol.* **6** (1979) 321 – 331.
342. S. Athari, *Mitt. Forstl. Bundesversuchsanst. Wien* **39** (1981) 7 – 27.
343. P. Nogler, *Allg. Forst Z.* **36** (1981) 709 – 711.
344. S. Athari, H. Kramer, *Forst. Holzwirt.* **38** (1983) 204 – 206.
345. G. K. Kenk, *Forst. Holzwirt* **39** (1984) 435 – 438.
346. Bundesministerium für Ernährung, Landwirtschaft und Forsten: Waldschadenserhebung 1985, Bonn 1985.
347. Bundesminister für Forschung und Technologie: "Umweltforschung zu Waldschäden," 2nd report, Bonn 1985.
348. O. Fränzle, W. Schröder, L. Vetter: "Saure Niederschläge als Belastungsfaktoren: Synoptische Darstellung möglicher Ursachen des Waldsterbens," Umweltforschungsplan des Bundesministers des Innern, Forschungsbericht 106 07 046/13, Berlin 1985.
349. Umweltbundesamt (eds.): "Wissenschaftliches Symposium zum Thema Waldschäden 'Neue Ursachenhypothesen'," *Schriftenr. "Texte" Umweltbundesamtes* **19** (1986).
350. Umweltbundesamt (eds.): "IMA-Querschnittsseminar zur Waldschadensforschung 'Belastung und Schäden auf Ökosystemebene und ihre Folgen'," *Schriftenr. "Texte" Umweltbundesamtes* **18** (1986).
351. H. H. Cramer in [348, pp. 308 – 312, 423 – 426].
352. H. Faulstich in [348, pp. 22 – 27, 381 – 382].
353. G. Rippen et al., *Environ. Technol. Lett.* **8** (1987) 475 – 482.
354. B. Ulrich, *Umschau* **11** (1984) 348 – 355.
355. B. Prinz, *Umschau* **18** (1984) 544 – 549.
356. E. F. Elstner, W. Osswald, R. J. Youngman, *Experientia* **41** (1985) 591 – 597.
357. A. Hüttermann, *Allg. Forst Jagdztg.* **4** (1978) 67 – 70.
358. U. Arndt, M. Kaufmann, *Allg. Forst Jagdztg.* **156** (1985) 16 – 20.
359. P. Benner, A. Wild, *J. Plant Physiol.* **129** (1987) 59 –72.
360. D. Murach, *Göttinger Bodenkd. Ber.* **77** (1984).
361. K. E. Rehfuess, C. Bosch, E. Pfankuch: "Nutrient Imbalances in Coniferous Stands in Southern Germany," Int. Workshop on Growth Disturbances of Forest Trees, Oct. 1982, IUFRO/FFRJ-Jyvaskyla, Finland.

362. B. Ulrich, *Forstwiss. Centralbl.* **100** (1981) 228 – 236.
363. B. Ulrich, *Allg. Forst Z.* **38** (1983) nos. 26/27, 670 –677.
364. W. U. Kriebitzsch: "Stickstoffnachlieferung in sauren Waldböden Nordwest-Deutschlands," *Scr. Geobot.* (1978).
365. R. Linthurst, *VDI Ber.* **500** (1983) 175 – 185.
366. L. S. Evans et al., *Water Air Soil Pollut.* **16** (1981) 469 – 509.
367. W. Schröder, O. Fränzle, L. Vetter, *Allg. Forst Z.* **41** (1986) no. 22, 543 – 544.
368. W. Saager: "Multivariate Bestimmung waldschadensdisponierender Faktoren in Schleswig-Holstein auf der Grundlage der terrestrischen Waldschadensinventur," Diplomarbeit Geographie, Universität Kiel 1990.
369. G. H. Kohlmaier et al., *Ecol. Modell.* **22** (1984) 45 – 65.
370. H. Bossel, W. Metzler, H. Schäfer (eds.): *Dynamik des Waldsterbens,* Springer Verlag, Berlin – Heidelberg – New York 1985.
371. G. H. Kohlmaier, M. Plöchl in [349, pp. 41 – 54].
372. H. Bossel, *Ecol. Modell.* **34** (1986) 259 – 288.
373. D. L. Godbold, A. Hüttermann, *Water Air Soil Pollut.* **31** (1986) 509 – 515.
374. R. Schultz: "Vergleichende Betrachtung des Schwermetallhaushalts verschiedener Waldökosysteme Norddeutschlands," Berichte des Forschungszentrums Waldökosysteme/Waldsterben, Reihe A **32** (1987).
375. W.-D. Grossmann, J. Schaller, M. Sittard, *Allg. Forst Z.* **39** (1984) no. 38, 837 – 843.
376. B. Prinz, G. H. M. Krause, H. Stratmann, *LIS Ber.* **28** (1982).
377. S. M. Adams et al., *Mar. Environ. Res.* **28** (1989) 459 – 464.
378. S. Gerlach: *Marine Pollution: Diagnosis and Therapy,* Springer Verlag, Berlin 1981.
379. A. Moubadan, J.-C. Pihan, *Hydroécol. Appl.* **5** (1993) 97 – 109.
380. J. Cairns, Jr., B. R. Niederlehner in J. Cairns, Jr., B. R. Niederlehner, D. R. Orvos (eds.): *Advances in Modern Environmental Toxicology,* **vol. XX**: "Predicting Ecosystem Risk", Princeton Scientific Publ. Co., Princeton, New Jersey, 1992, pp. 327 – 343.
381. P. H. Pritchard in R. A. Conway (ed.): *Environmental Risk Analysis for Chemicals,* Van Nostrand Reinhold, New York 1991, pp. 257 – 353.
382. P. H. Pritchard, H. W. Bourquin in H. H. White (ed.): *Concepts in Marine Pollution Measurements,* University of Maryland, 1984, pp. 117 – 138.
383. R. W. Howarth in J. Cole, G. Lowett, S. Findlay (eds.): *Comparative Analyses of Ecosystems,* Springer Verlag, New York 1991, pp. 169 – 195.
384. J. Cairns, Jr., J. R. Pratt: "Multispecies Toxicity Testing Using Indigenous Organisms – A New, Cost effective Approach to Ecosystem Protection," *Environmental Conference, TAPPI Proceedings,* TAPPI Press, Atlanta, Georgia, 1985, pp. 149 – 153.
385. W. B. Neely: *Chemicals in the Environment,* Marcel Dekker, New York 1980.
386. D. I. Swackhamer, S. J. Eisenreich in K. C. Jones (ed.): *Organic Contaminants in the Environment,* Elsevier Applied Sci., London 1991, pp. 33 – 86.
387. S. Griesbach, R. H. Peters, S. Youakim, *Can. J. Fish. Aquat. Sci.* **39** (1982) 727 – 735.
388. K. J. Macek, S. R. Petrocelli, B. H. Sleight in L. L. Marking, R. A. Kimerle (eds.): "Aquatic Toxicology," *Am. Soc. Test. Mater.* (1979) 251 – 268.
389. B. C. Patten in M. Higashi, T. P. Burns (eds.): *Theoretical Studies in Ecosystems: The Network Perspective,* Cambridge University Press, Cambridge 1991, pp. 288 – 351.
390. J. Roelof van der Meer, W. M. DeVos, S. Harayama, A. J. B. Zehnde, *Microbiol. Rev.* **56** (1992) 677 – 694.
391. R. Reuther, *Sci. Total Environ.* **115** (1992) no. 3, 219 – 237.
392. B. van Hattum, K. R. Timmermans, H. A. Govers, *Environ. Toxicol. Chem.* **10** (1991) 275 – 292.
393. R. R. Imhoff, P. Koppe, E. A. Nusch in S. Matsui, B. F. D. Barrett, J. Banerjee (eds.): "Toxic Substances Management in Lakes and Reservoirs," *Guidelines of Lake Management,* **vol. 4,** International Lake Environment Committee, UNEP, Shiga, Japan, 1991, pp. 127 – 159.
394. J. Plotner, R. Gunther, *Int. Rev. Gesamten Hydrobiol.* **72** (1987) no. 6, 759 – 771.
395. N. Nirmalakhandan et al., *Water Res.* **28** (1994) 543 – 552.
396. A. W. Maki, M. W. Slimak in W. Grodzinski, E. B. Cowling, A. I. Breymeyer (eds.): *Ecological Risks – Perspectives from Poland and the United States,* National Academy Press, Washington, D.C., 1990, pp. 77 – 87.

397. J. Cairns, J. R. Pratt, *Water Sci. Technol.* **19** (1987) no. 11, 1 – 12.
398. C. N. Haas, *Water Qual. Int.* **4** (1993) 30 – 32.
399. F.-A. A. M. de Leeuw, *Chemosphere* **27** (1993) no. 8, 1313 – 1328.
400. J. S. Alabaster, *Chem. Ind.* **1** (1981) 529 – 534.
401. J. Hermens, E. Broekhuyzen, H. Canton, R. Wegman, *Aquat. Toxicol.* **6** (1985) 209 – 217.
402. J. W. Deneer: "The Toxicity of Aquatic Pollutants: QSARs and Mixture Toxicity Studies, Chap. VI: The Joint Acute Toxicity to *Daphnia magna* of Industrial Organic Chemicals at Low Concentrations," Proefschrift 81 – 89, Rijsuniversiteit Utrecht, 1988.
403. ECETOC: "Assessment of Non-occupational Exposure to Chemicals," Technical Report no. 58, Brussels 1994.
404. L. A. Zadeh: "Fuzzy Sets," *Inf. Control* **8** (1965) 338 – 353.
405. A. Salski, *Ecol. Modell.* **63** (1992) 103 – 112.
406. A. Salski, P. Kandzia, *Springer Informatik Fachber.* **296** (1991) 303 – 310.
407. M. Schepers: "Ein Unterstützungssystem zur wissensbasierten Modellierung und Simulation von ökologischen Prozessen," Diplomarbeit, Inst. f. Informatik und Prakt. Mathematik, Universität Kiel 1991.
408. M. T. Bui: "Weiterentwicklung des Unterstützungssystems zur wissensbasierten Modellierung unter dem Einsatz der Fuzzy-Set-Theorie," Diplomarbeit, Inst. f. Informatik und Prakt. Mathematik, Universität Kiel 1991.
409. R. V. Thomann, *Ecol. Modell.* **22** (1984) 145 – 170.
410. M. G. Morgan in J. V. Rodricks, R. G. Tardiff (eds.): "Assessment and Management of Chemical Risks," *ACS Symp. Ser.* **239** (1984) chap. 8.
411. R. C. Schwing, W. A. Albers (eds.): *Societal Risk Assessment: How Safe is Safe Enough?* Plenum Press, New York 1982.
412. R. V. Thomann et al., *Water Res.* **8** (1974) 841 – 851.
413. D. R. Miller in G. C. Butler (ed.): "Principles of Ecotoxicology," *SCOPE* **12** (1979) 71 – 90.
414. J. W. Gillet et al.: *A Conceptual Model for the Movement of Pesticides Through the Environment,* National Environmental Research Center, U.S. Environmental Protection Agency, Corvallis, OR 1974, Report EPA 660/3-74-024, pp. 79.
415. R. R. Lassiter: *Principles and Constraints for Predicting Exposure to Environmental Pollutants,* U.S. Environmental Protection Agency, Corvallis, OR, 1978, Report EPA 118-127 519.
416. T. Fagerstrøm, B. Aasell, *Ambio* **2** (1973) 164 – 171.
417. I. Aoyama, Yos. Inoue, Yor. Inoue, *Water Res.* **12** (1978) 837 – 842.
418. D. K. Leung: *Modelling the Bioaccumulation of Pesticides in Fish,* Center for Ecological Modelling, Polytechnic Institute, Troy, NY, 1978, Report 5.
419. K. L. Seip, *Ecol. Modell.* **6** (1978) 183 – 198.
420. G. T. Orlob, D. Hrovat, F. Harrison, *Adv. Chem. Ser.* **189** (1980) 195 – 212.
421. D. C. L. Lam, T. J. Simons in J. O. Nriago (ed.): *Metals Transfer and Ecological Mass Balances. Environmental Biochemistry,* **vol. 2,** Ann Arbor Sci., Ann Arbor 1976, pp. 537 – 549.
422. T. B. Kirchner, F. W. Whicker, *Ecol. Modell.* **22** (1984) 21 – 44.
423. F. Bro-Rasmussen, K. Christiansen, *Ecol. Modell.* **22** (1984) 67 – 85.
424. S. M. Bartell, R. H. Gardner, R. V. O'Neill, *Ecol. Modell.* **22** (1984) 109 – 123.
425. C. G. Uchrin, *Ecol. Modell.* **22** (1984) 135 – 144.
426. R. P. Schwarzenbach, D. M. Imboden, *Ecol. Modell.* **22** (1984) 171 – 213.
427. E. Halfon, *Ecol. Modell.* **22** (1984) 213 – 253.
428. J. R. W. Harris et al., *Ecol. Modell.* **22** (1984) 253 – 285.
429. N. Nyholm, T. K. Nielsen, K. Pedersen, *Ecol. Modell.* **22** (1984) 285 – 324.
430. J. C. J. Nihoul, *Ecol. Modell.* **22** (1984) 325 – 341.
431. P. A. Arp, *Ecol. Modell.* **19** (1983) 105 – 117.
432. S. Paterson, D. Mackay, *Ecol. Modell.* **47** (1989) 85.
433. M. Matthies et al., *Ecol. Modell.* **47** (1989) 115.
434. J. T. de Luna, T. G. Hallam, *Ecol. Modell.* **35** (1987) 249.
435. T. Morioka, S. Chikami, *Ecol. Modell.* **31** (1986) 267.
436. G. B. Schaalje, R. L. Stinner, D. L. Johnson, *Ecol. Modell.* **47** (1989) 223.
437. B. C. Longstaff, *Ecol. Modell.* **43** (1988) 303.
438. D.-S. Liu, S.-M. Zhang, Z.-G. Li, *Ecol. Modell.* **41** (1988) 75.
439. D.-S. Liu, S.-M. Zhang, *Ecol. Modell.* **37** (1987) 131.
440. R. F. Carsel, L. A. Mulkey, M. N. Lorber, L. B. Baskin, *Ecol. Modell.* **30** (1985) 101.

441. G. R. Gorban, E. Bosatta, *Ecol. Modell.* **40** (1988) 25.
442. J. E. Breck, D. L. de Angelis, W. van Winkle, S. W. Christensen, *Ecol. Modell.* **41** (1988) 1.
443. D. S. Wratt et al., *Ecol. Modell.* **64** (1992) 185 – 204.
444. G. Wuttke, B. Thober, H. Lieth, *Ecol. Modell.* **57** (1991) 263 – 276.
445. J. K. Cronk, W. J. Mitsch, R. M. Sykes, *Ecol. Modell.* **51** (1990) 161 – 192.
446. T. C. Gard, *Ecol. Modell.* **51** (1990) 273 – 280.
447. ECETOC: Hazard Assessment of Chemical Contaminants in Soil, Technical Report no. 40, Brussels 1990.
448. E. E. Kenaga, C. A. I. Goring: "Relationship Between Water Solubility, Soil Sorption, Octanol-water Partitioning and Bioconcentration of Chemicals in Biota," Special Technical Publication (STP) 707, ASTM, Philadelphia, PA, 1980.
449. R. L. Metcalf, G. K. Sangha, I. P. Kopoor, *Environ. Sci. Technol.* **5** (1975) 709 – 713.
450. P.-Y. Lu, R. L. Metcalf, *Environ. Health Perspectives* **10** (1975) 269 – 284.
451. W. B. Neely, D. R. Branson, G. E. Blau, *Environ. Sci. Technol.* **8** (1974) 1113 – 1115.
452. G. D. Veith, D. L. Defoe, B. V. Bergstedt, *J. Fish. Res. Board Can.* **36** (1979) 1040 – 1048.
453. R. R. Lassiter: "Modelling Dynamics of Biological and Chemical Compartments of Aquatic Ecosystems," U.S. Environmental Protection Agency, Washington, D.C., 1975, EPA-660/3-75-012.
454. R. J. Larson, R. L. Perry, *Water Res.* **15** (1981) 697.
455. P. H. Pritchard, A. W. Bourquin, H. L. Fredrickson, T. Maziarz in A. W. Bourquin, R. H. Pritchard (eds.): "Proceedings of the Workshop: Microbial Degradation of Pollutants in Marine Environments," U.S. Environmental Protection Agency, Gulf Breeze, FL, 1979, p. 251.
456. D. S. Brown, E. W. Flagg, *J. Environ. Qual.* **10** (1981) 382 – 386.
457. C. Mater-Müller, W. Gujer, W. Giger, W. Stumm, *Water Technol. Sci.* **12** (1980) 299 – 314.
458. O. C. Zafiriou, M. B. True, *Mar. Chem.* **8** (1979) 9 – 32.
459. O. C. Zafiriou, M. McFarland, *J. Geophys. Res.* **86** (1981) 3173 – 3182.
460. R. G. Zika, L. T. Gidel, D. D. Davis, *J. Geophys. Res.* **11** (1984) 353 – 356.
461. C. van Baalen, J. E. Marler, *Nature (London)* **211** (1966) 951.
462. R. M. Baxter, J. H. Carey, *Freshwater Biol.* **12** (1982) 285 – 292.
463. W. R. Haag, J. Hoigne, E. Grassmann, A. M. Braun, *Chemosphere* **13** (1984) 631 – 640.
464. P. G. Zepp, *Environ. Sci. Technol.* **12** (1978) 327 –329.
465. P. G. Zepp in J. Calkins (ed.): *The Role of Solar Ultraviolet Radiation in Marine Ecosystems,* Plenum Press, New York 1982, pp. 293 – 307.
466. P. G. Zepp in O. Hutzinger (ed.): *The Handbook of Environmental Chemistry,* **vol. 2,** part B, Springer Verlag, Berlin 1982, pp. 19 – 40.
467. C. T. Chiou, V. H. Freed, D. W. Schmedding, R. L. Kohnert, *Environ. Sci. Technol.* **11** (1977) 475 – 478.

Occupational Health and Safety

LEOPOLD W. MIKSCHE, Bayer AG, Leverkusen, Federal Republic of Germany (Chap. 1, Section 4.6)

GUNTER MEYER, IG-Chemie-Papier-Keramik, Hannover, Federal Republic of Germany (Section 2.1)

JAN J. KOLK, Arnhem, The Netherlands (Sections 2.2, 2.3 in part, 4.1, 4.2, 4.5)

PHILIP NOORDAM, The Hague, The Netherlands (Sections 2.2, 2.3 in part)

DONALD W. LAMB, Miles Inc., Pittsburgh, United States (Sections 2.4, 3.10, 4.8)

DAN BOSATRA, Dow Europe S. A., Horgen, Switzerland (Section 3.1–3.9)

NIKO KIESSELBACH, Bayer AG, Leverkusen, Federal Republic of Germany (Section 4.3)

JOACHIM GRUBER, Bayer AG, Leverkusen, Federal Republic of Germany (Section 4.4)

RAINER DOSCH, Bayer AG, Leverkusen, Federal Republic of Germany (Section 4.7)

Ullmann's Industrial Toxicology

ISBN: 3-527-31247-1

1. Introduction [1–3]

Work has always involved certain health hazards and risks of accidents. With progressing industrialization the pattern of potential threats to human health at work has been changing. At present, in a highly industrialized society, health protection and accident prevention at work are issues of high ethical and socioeconomic priority.

1.1. History

Work-related health impairments are not a discovery of modern times.

Signs of dust disease of the lungs have been discovered in prehistoric corpses. Scripts from ancient Egypt noted that farmworkers suffered from intense heat and dust on dry soil, whereas working in the muddy grounds along the Nile induced severe rheumatic disorders and chronic arthritis, which still can be seen in mummies. Reports and documentation of occupation-related diseases can be traced throughout ancient history. In ancient Greece ARISTOTELES and PLATO described certain diseases found in specific professions such as "tanners, porters of heavy loads, and couriers." Around 400 B.C. HIPPOCRATES noted: "there are many crafts and arts which cause a number of illnesses and adverse health conditions for those who work in these specific professions."

He also requested that physicians investigate thoroughly the professional history of their patients. Signs of pneumoconiosis and arthrosis of the knee in miners were described by him, as well as the toxicity of lead in lead miners.

Silicosis and diseases caused by lead and mercury or by certain melting procedures in gold mining are reported most frequently in the early European literature.

GEORGIOS AGRICOLA described miners diseases in the early 1500s, and PARACELSUS wrote a book on occupational diseases in Germany at the same time.

The Italian physician BERNARDINO RAMAZZINI (1633 – 1714) has been called the father of occupational medicine. In *De Morbis Artificum Diatriba* he described different diseases of workmen and their treatment, and stressed the importance of prevention of occupational diseases and accidents to make work "a love rather than a curse." His book includes information on afflictions caused by inhaling noxious gases and dusts, as well as such "modern" issues as illnesses resulting from disorderly motion and improper posture of the body.

With increasing industrialization plant physicians began to take care of workers and their health in the early 1800s. In the second half of the 1800s legislation on health protection and technical inspection of factories was established, for example, in England and Germany.

Some important events in the history of occupational health are summarized in the following:

B.C.

1500	Description of back pain and deformities in quarrymen (Papyros Ebers).
400	Hippocrates (460 – 377 B.C.) observed an occupational disease (colic) in slaves working in lead mines.
40	Pliny the Elder wrote about lead, mercury, sulfur, and zinc as occupational hazards. He also noted that in dusty conditions workers tied bladders over their mouths to breathe through, and that workers who used asbestos to make textiles became ill.

A.D.

250	Galen identified the danger of acid mists in copper mines.
1473	Ellenbog identified dangers of metal fumes in a report on gold miners. He also discussed carbon monoxide, lead, mercury, and nitric acid, and provided instructions in preventative measures.
1500	Paracelsus (1493 – 1541) wrote a book on occupational diseases, identifying diseases of miners, which he attributed to vapors from the metals. He advised that the vapors be avoided.
1500s	Agricola (1494 – 1555) published *De Re Metallica*, a series of books on the hazards of mining. He recommended the use of ventilation in mines and of protective masks to limit exposure to dusts.
1700s	*De Morbis Artificum*, a book dealing, among others, with mercury, lead, and the pathology of silicosis, was published by Bernardino Ramazzini, the "father of occupational medicine". He advised physicians to ask patients what their trade was, and to study work environments to learn about occupational diseases.
1775	Percival Pott (1714 – 1788), a London surgeon, identified cancer of the scrotum in chimney sweeps as the first occupational cancer.
1833	The English Child Labour Laws were passed, to limit childrens' working hours, and to provide a mechanism for inspections in certain industries.
1884	In Germany the law on accident insurance for workers was passed.
1925	The first official list of occupational diseases was passed in Germany.

1.2. Differentiation of Tasks

Historically health protection and provision of safety at the workplace has been taken care off mostly by physicians.

They tried to detect work-related health impairments and to discover the causes for these diseases in order to prevent them by changing work hygiene or other conditions at the workplace. In the 1800s close cooperation between physicians and technicians was necessary to improve production processes and working technology to prevent diseases and work-related accidents.

Government inspectors began supervising factories and mines in the second half of the 1800s. At the turn of the century, and specially in the first half of the 1900s, additional groups of specialists developed for specific tasks in the field of health and safety at work.

Nowadays there is a separation of certain tasks covered by different specialists (see Section 1.5). The occupational health physician is responsible for health protection, prevention of occupational illnesses and diseases, accident prevention, and medical surveillance of the workforce. Provision and control of technically safe workplaces, machinery, and working conditions and the technical side of accident prevention is dealt with by safety engineers in most countries. In a number of countries a third group is involved in health and safety at the workplace: the industrial hygienists. Their task is the maintenance of hygienic conditions at the worksite and surveillance of the working environment to identify chemical and physical hazards and give advice on their control and minimization.

In Germany the profession of industrial hygienist has so far not become established. These tasks are taken care off by occupational physicians and safety engineers in close cooperation.

In countries where the occupational health system is not completely separate from general health care, provision of general health care and preventive measures for non-work-related diseases are partially covered by occupational physicians.

1.3. Hazards in the Chemical Industry

Throughout the development of the chemical industry the number of available chemical compounds has grown tremendously. This has led in turn to a broader range of potential health risks caused by chemical compounds. In earlier centuries mostly dusts, fumes, and heavy metals were described as disease-causing agents. Since the mid-1800s organic compounds gathered increasing importance as causative agents for work-related illnesses and diseases.

The specific hazards to human health by chemicals are derived from their general toxicity and from specific properties such as mutagenicity, teratogenicity, carcinogenicity.

The major organ systems affected by chemicals are the eyes (irritation), the skin (chemical burns, allergy), the respiratory system (dust diseases, allergy, irritation), the central and peripheral nervous system (neurotoxicity), the liver, and the kidneys.

The second group of hazardous factors are physical agents such as noise, vibration, heat, dust, radioactive radiation, high pressure, low pressure, and magnetic fields.

The third group of hazardous agents are biological agents, which are gaining in importance with the development of biotechnology and gene technology. Whereas in former times pathogenic bacteria, viruses, and fungi have been known as risk factors in the health-care industry and the pharmaceutical industry during production of vaccines and antibiotics, nowadays microorganisms used in genetic engineering are causing concern as potential health risks.

Developments in the last decades have also attached greater importance to possible health impairments caused by physical conditions such as lifting of heavy loads, monotonous repetitive motions, and improper posture of the body.

Diseases caused by these conditions were already described by RAMAZZINI. But while in former times humans had to adapt to fit the working conditions – as is demonstrated by the employment of children in certain mining industries – it is nowadays the general philosophy to adjust equipment, machinery, and work-place conditions to the physiological capacity of the human body.

Besides chemical, physical, and biological agents, other factors that have come to be regarded as probable hazards to health in the working population in the last few decades include mental stress and socioeconomic conditions.

The question of occupational disease caused by stress at work is under serious consideration.

1.4. General Outline of Goals

The joint efforts of all parties involved in occupational health and safety at present can be summarized in the following:

1) Prevention of occupational diseases, health impairments, and illnesses caused by influences at work
2) Prevention of work-related accidents
3) Early detection of potential new hazards for human health in the working environment
4) Elimination of identified health hazards and accident sources
5) Adaptation of working conditions, tools, and working environment to the physiological capability of the workers
6) Assistance in providing a socioeconomic work environment to prevent effects on mental health

The means to reach the above-listed goals are:

1) Thorough inspection of the working environment and provision of necessary improvements
2) Provision of a safe technical environment
3) Regular health surveillance and preplacement health examinations
4) Provision of a high standard of hygiene at the worksite
5) Provision of health-supporting socioeconomic conditions

To reach these goals of health and safety provision at the workplace close cooperation of the professional specialists with the employer, employees, and employee representatives is essential.

1.5. Experts and their Training

The tasks of health protection and safety at the workplace are carried out by experts whose professional training and field of expertise varies from country to country.

Occupational Health Physicians. In some countries occupational medicine is a state-licenced speciality of the medical profession. This means that physicians receive additional training in occupational health after several years of medical practice.

In several European countries, training in occupational medicine takes two years, including several months of theoretical study at a university or a special academy. In some countries this additional training leads to certification in occupational medicine, which in Germany includes a state exam.

In the United States a specific training program and residency of two or three years – depending on the university – has been established, leading to the degree of master in public health (MPH). This degree is normally achieved by physicians but can also be acquired by other medical professionals.

In the United States and Great Britain, special training for nurses leads to the degree of occupational health nurse; this is unknown in other countries. The occupational health nurse performs part of the occupational health service in the plant.

Safety Experts. Safety experts normally hold a technical degree, mostly in engineering or chemistry. The special additional education as safety expert or safety engineer differs widely in different countries. Whereas in some countries it is a special academic training with a masters degree or the degree of safety engineer, other countries require only a minor amount of theoretical training parallel to practical experience on the job.

Industrial Hygienists. The profession of industrial hygienist is not yet established in all European countries. It was first developed in the United States and Great Britain, where industrial hygienists hold a specific academic degree in that field. In other countries, such as Germany, industrial hygiene activities are taken care of partially by occupational health physicians or by safety engineers who then are mostly either chemical engineers or chemists by profession.

Worldwide there is wide variability in provision of health and safety services in industry as well as in training and experience of the health and safety specialists.

Many countries are still understaffed and are developing their occupational health and safety systems, with the advise and assistance of more advanced countries. The future goal according to WHO and ILO (International Labour Office) is to provide occupational health and safety services to the entire working population.

2. Legal Background [4–8]

2.1. Laws and Regulations in Germany

The Federal Republic of Germany is a confederation of states. The responsibilities of the government are divided between the Federal Government and the states (Länder) as described below.

Laws, decrees, and guidelines are passed at a national level, the states contributing through the Bundesrat. The states are responsible for the enforcement of laws and for supervising observance of government regulations via their trade supervisory office. In addition to the central government and the states, the legal accident insurance institutions are also involved in the industry-wide occupational safety system. Statutory accident insurance is a branch of social security and is a compulsory insurance. The legal basis of the accident insurance is anchored in the book on social welfare legislation and in the "Reichsversicherungsordnung" (RVO).

The statutory accident insurance institutions are the trade associations. Trade associations are organized according to the specific trade. They have the right to pass accident prevention regulations (UVV).

These regulations serve the trade-specific concretization of governmental laws and decrees.

The relationship between the central government and states on the one hand and the trade associations on the other hand correspond to the principle of subsidiarity. To control the observance of accident prevention regulations, the trade associations have a technical control service (TAD) which assumes control and consultation functions in the factories parallel to the governmental trade supervisory office.

2.1.1. State Legislation and Controls

The enforcement and control of federal laws on occupational safety are the responsibility of the Länder. The State Ministry of Labor (Landesarbeitsministerium) is the highest state authority and is superordinate to the state institute for occupational safety and industrial medicine and the trade supervisory offices. In the application and further development of occupational safety regulations, close cooperation is required between the participants of the industry-wide occupational safety system and also between the industry-wide and the company occupational safety system.

To guarantee smooth cooperation between the governmental trade supervisory office and the trade associations, the Federal Minister of Labor has passed an appropriate administrative regulation. Apart from the central government, states, and trade associations, private organizations also have important functions in occupational safety.

In Germany, there is a long tradition that private organizations fulfill by means of definitions (e.g., "generally recognized rules of technology") the requirements of federal laws and decrees. These include the standards of the German Institute of Standardization (DIN), guidelines of the Association of German Engineers (VDI), regulations of the Association of German Electrical Engineers (VDE), etc. This is intended to guarantee that occupational safety laws are linked to the further development of technology.

The German Institute of Standardization plays a special role because as far as standardization within the European Community is concerned, it is the direct partner of the European standardization institute CEN/CENELEC.

Private organizations are also active in supervising industrial safety, e.g., the Technical Control Board (TÜV) in the inspection of plant.

The Most Important Laws. The government regulations pertaining to occupational safety can be divided into six areas:

Workplaces, Including Industrial Hygiene. As early as 1891, the first general basic demands on the workplace were laid down in the industrial code. With the workplace decree of 1975, obligatory regulations for the construction and equiping of workplaces were drawn up for all companies.

The individual control areas of the workplace regulations are summarized in the workplace guidelines:

Ventilation
Room temperature
View outside
Artificial lighting
Safety lighting
Floors
Transparent walls
Doors and gates
Glass doors, doors with glass inserts
Protection against falling doors and gates
Automatic doors
Protection against falling objects
Fire extinguishing equipment
Roads
Escalators
Seating
Pause rooms
Rest rooms
Changing rooms
Washrooms
Opportunities for washing outside of designated washrooms
Toilets
Medical room
First-aid equipment
Tempory accommodation for building sites
Washrooms for building sites

Machines, Equipment, and Technical Facilities. Apart from the regulations governing factory premises, a book of rules also exists for the equipping of the premises with machines, instruments, and technical systems.

In particular, the law passed in 1968 governing technical working devices (Equipment Safety Law), the decrees on systems that require monitoring, the Accident Regulations, and the Accident Prevention Regulations should be mentioned here.

The range of the Equipment Safety Law is very wide. It covers not only technical working devices which are used in industrial, manual, and agricultural concerns as well as in the administrative sector, but also machines and equipment used in the private sector, e.g., for the household, leisure activities, recreation, and sports.

The manufacturer and importer are required by this law to put only such equipment into circulation which is safely built. A safety label (GS: tested safety) has been introduced as an orientation aid for the buyer. All machines and instruments carrying this test label have been tested for safety by an officially recognized testing institute.

The most important regulations for facilities that require monitoring are shown in Figure 1.

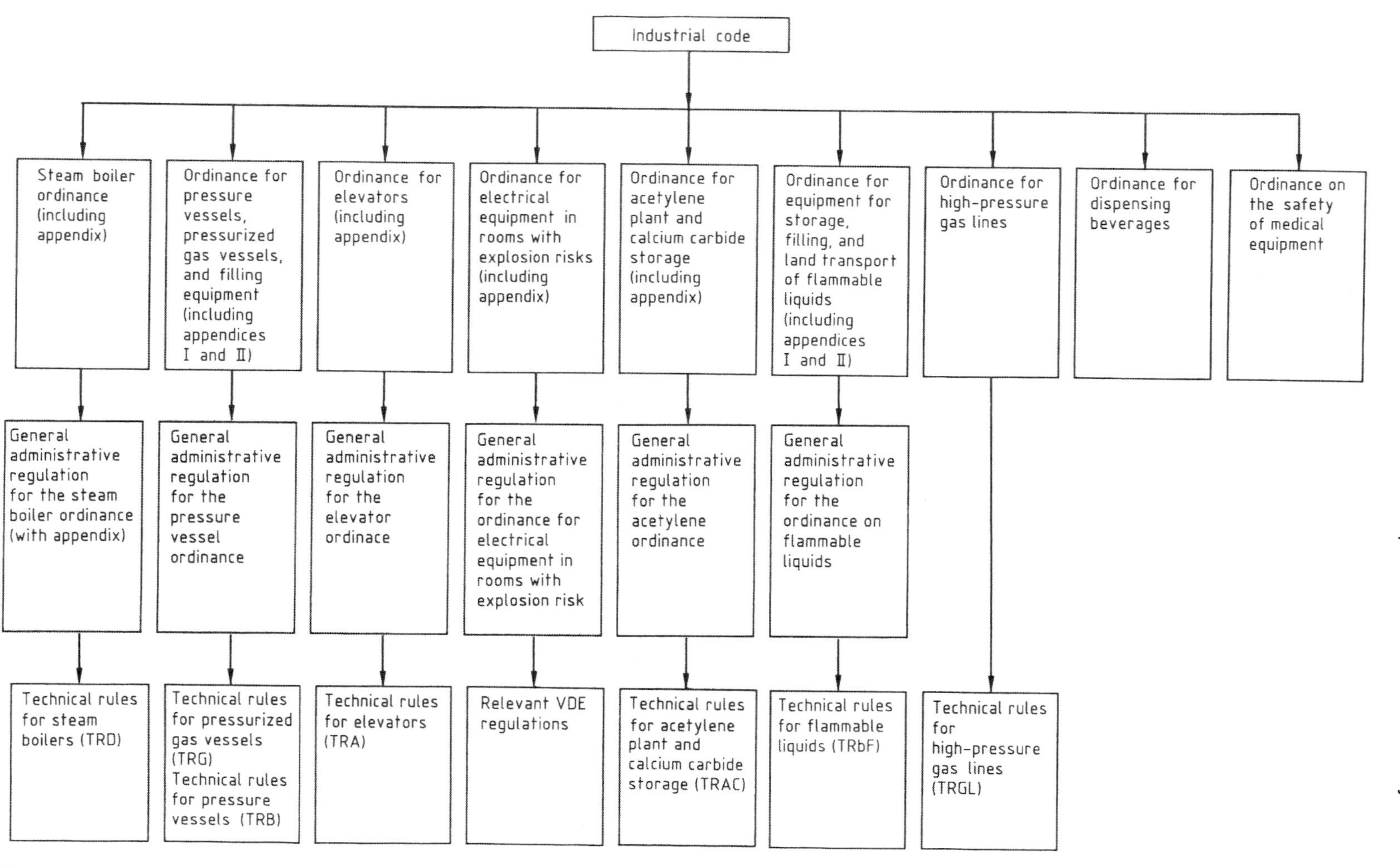

Figure 1. Important regulations for facilities that require monitoring

The emergence of the regulations for facilities that require monitoring can be understood from the historical development. On examination of complex processing plants today, it is clear that these regulations often cover parts of the plant only.

Therefore, occupational safety in processing plants requires supplementary regulations which take the system behavior into account. Based on the federal emissions protection law, the decree pertaining to industrial accidents (12th BImSchV) covers not only environmental protection requirements, but also occupational safety measures in the case of disruption of normal operation.

This is significant because the operating personnel can be directly affected by the onset, control, and possibly also by the elimination of the results of a disturbance.

Accident prevention regulations occupy the fourth part of the area "machines, equipment, and technical facilities". They contain safety requirements for special machines, equipment, and technical facilities which are used in certain branches of industry.

Hazardous Substances. A modern industrial society is inconceivable without the use of chemicals. Today, it is estimated that more than 100 000 chemicals are used worldwide to manufacture products.

Some of these chemicals are classified as dangerous. Therefore, they may be used and put into circulation only after certain safety precautions are taken.

The Chemicals Law (Chemikaliengesetz) and the decree on dangerous substances (Gefahrstoffverordnung) are the most important regulations.

The Chemicals Law covers not only occupational safety, but also environmental protection and general health protection. Nevertheless, occupational safety is the main focus of the law because in most cases the potential hazard of a substance is first observed at the workplace.

The main emphasis of the Chemicals Law is a testing obligation and a compulsory notification for each new substance introduced to the market. Apart from containing special regulations for introduction into the market, labelling, packing, storage, and passing on of information on occupational safety and environmental protection to users of chemicals, the decree on dangerous substances primarily regulates the handling of chemicals within the company.

This decree demands, in particular, a compulsory inquiry (how and which chemicals are handled within the company), compulsory control (checking the maintenance of stipulated air and biological threshold values, etc.), protective measures to be taken, and prohibition or limitation of use.

To further concretize the decree on dangerous substances, the Federal Ministry of Labor has set up a sociopolitical committee which has the task of elaborating the technical regulations for dangerous substances (TRGS). This committee for dangerous substances (AGS) has passed technical regulations for the following areas:

1) Technical and organizational safety measures (e.g., asbestos: demolition, renovation, and maintenance work)
2) Technical guide concentrations (threshold values for carcinogenic substances)
3) Substitutes and substitute processes (e.g., substitutes for cleansing agents containing chlorinated hydrocarbons).

Regulation of Working Hours. Regulations pertaining to working hours are contained in the working hours code, trading and industrial code, the law regulating the closing time of shops, and the special regulations for the working hours of professional drivers.

The regular daily working hours must not exceed 10 h. Working on Sundays and holidays is basically forbidden. However, there are a series of exceptions:

1) Work that must be done without delay in emergencies or in the public interest
2) Work that is required to prevent the spoiling of raw materials or the failure of production
3) Work that is required to prevent excessive damage when an unforeseeable necessity arises

Protection of Certain Groups of People. The law protecting mothers protects the working mother and the fetus from dangers and health impairments. It protects the mother from excessive demands at the workplace, financial losses, and loss of her job due to pregnancy and delivery.

The law for the protection of children and young persons applies to the employment of persons under 18 years of age. The minimum working age has been fixed at 15 years.

Apart from the laws protecting mothers and young persons, there are also regulations for the disabled and for working at home.

Occupational Safety Organization in the Company. The employer is responsible for the enforcement of occupational safety in the company. He is obliged by law to take all measures required for the protection of his employees in all the occupational safety areas mentioned above, including occupational safety organization in the company. Here, the Workers' Protection Law, the rules for the prevention of accidents, and the labor management act are of special importance.

The Workers' Protection Law requires that the companies engage doctors and qualified personnel for occupational safety. The qualified personnel and the company doctors have to support the employer with regard to occupational safety and accident prevention in all questions of health protection and working safety, including the designing of humane working conditions.

The qualification requirements, profile of duties, and working hours are further concretized in the rules for the prevention of accidents, "Safety Engineers and other Specialists for Occupational Safety" (VBG 122) and "Company Doctors" (VBG 123).

Responsibilities of the Federal Government and the Länder. At the federal level, the Federal Ministry of Labor is in overall charge of occupational safety. In this ministry, drafts of new laws and decrees are elaborated and EC guidelines are translated into German law.

Furthermore, via lower authorities, this ministry supports research and model projects and provides counseling and information.

Supplementary focal points are taken on by other ministries, e.g., the program "Work and Technology" at the Federal Ministry of Research and Technology.

Institutions and Control Functions. The Federal Ministry of Labor is supported by the Federal Institution of Industrial Safety and the Federal Institution of Industrial Medicine.

The duties of the Federal Institution of Industrial Safety include:

1) Cooperation with state officials and with statutory accident insurance institutions.
2) Observation and analysis of occupational safety, health situation, and working conditions in companies and offices.
3) Development of solutions to problems by the application of knowledge of technical safety, industrial medicine, ergonomics, and other industrial sciences. For this purpose, the Federal Institution of Industrial Safety conducts research or awards research grants.
4) Promotion of the application of acquired knowledge through publications, assistance with the defining of rules, development of materials for training and further education, model-like implementation and counseling, exhibitions and professional meetings.

The Federal Institution of Industrial Medicine also belongs to the Federal Ministry of Labor. It emerged from the Central Institute of Industrial Medicine of the former German Democratic Republic and has the following duties:

1) To observe and evaluate the effects of working conditions on health and make contributions to prevention.
2) To conduct and support research on subjects in industrial medicine. The main emphasis is placed on biotechnology and genetic engineering, allergies, stress, hazardous substances, noise, vibration, radiation, dust, and carcinogenic, mutagenic, and teratogenic substances.
3) To collect and disseminate knowledge of industrial medicine as well as to draw up concepts for the care given by company doctors and for health promotion in the company.

The trade supervisory authority that belongs to each state ministry is responsible for technical occupational safety.

This trade supervisory authority is equipped with all the required official powers, especially the right to inspect and check plants at any time. The trade supervisory authority checks that laws, decrees, and other regulations are complied with. Legal occupational safety covers the following:

1) Machines, instruments, plants
2) Hazardous substances, radiation
3) Working hours
4) Social occupational safety (protection of special groups of people)

5) Accident prevention and safety technology

Each trade supervisory office is responsible for all the companies in its area – independent of the branch of industry – and supervises not only the observation of government safety regulations, but also environmental protection. It is also responsible for the granting of exemptions.

The trade medical service also comes under the state ministries and is responsible for medical occupational safety. It is concerned with preventive health protection, such as evaluation of the workplace and medical checkups.

With the help of medical inquiries, this service should elaborate information on possible health hazards caused by the type of work, substances, etc. In addition, it is involved in proceedings for the acknowledgement of occupational diseases.

2.1.2. Trade Associations

Origin, Structure, and Organization. The fundamental activities of trade associations are specified by the Accident Insurance Law of 1884. These are:

1) The private liability of the employer for his employees is replaced by the public law system of statutory accident insurance.
2) Employers in the same or related branches of industry join together to form trade associations. The demands of accident victims are henceforth directed towards the appropriate trade association.
3) The benefits provided by law are granted independently of who caused the accident. The injured person is entitled to this benefit even if he caused the accident himself by thoughtlessness, negligence, or unlawful action. He is denied the benefit only if he intentionally caused the accident.
4) The trade associations are obliged to pursue accident prevention by all suitable means. The employer, who remains fully responsible for occupational safety in his company, receives professional advice and help from his trade association.

The common basis for the social, health, and accident insurance was created by the "Reichsversicherungsordnung" (RVO) of 1911.

At present, the legal accident insurances are divided into one marine, 21 agricultural, and 34 industrial trade associations.

The statutory accident insurance institutions are the federal government, the states, the commune and insurance associations, certain municipalities (cities with their own accident insurance), and the fire department accident fund. The accident insurance institutions are public corporations. They have the right of self-administration, i.e., they carry out the duties assigned to them by law on the responsibility of their own honorary autonomous organs, but under state supervision.

The autonomous organs are the assembly of representatives and the board of directors, elected every six years. They are made up of equal numbers of representatives of the insured persons and of the employers. In the case of agricultural trade associations, representatives of the insured persons and of the employers and of the self-employed make up one-third each.

The assembly of representatives establishes the rules and other autonomous rights of the insurance institutions (e.g., accident prevention regulations). The expenditure of the trade associations is covered by the companies. The employees make no contribution.

This regulation of the fees is based on the fact that the statutory accident insurance is primarily meant to replace the civil liability of the employer for his employees with regard to occupational diseases and to accidents at the workplace and on the road to work.

Further duties of the trade associations include prevention and rehabilitation.

Accident Prevention Regulations. The regulations for the prevention of accidents are the rules of the particular statutory insurer passed according to the RVO and confirmed by the Federal Ministry of Labor.

The employers united in the trade associations and their employees are obliged to observe the accident prevention regulations.

At present, more than 150 regulations for the prevention of accidents exist with different contents. These regulations must be adapted by the individual trade associations to each sphere of responsibility and then passed. Thus, more than 1800 individual accident prevention regulations exist. These regulations control facilities, orders and measures taken for the prevention of accidents, rules of action for employees, requirements for medical examinations, and demands made on the company safety organization. They are supplemented by numerous guidelines, safety rules, leaflets, and brochures issued by the trade associations to facilitate enforcement within the company.

The trade associations have their own umbrella organizations for the areas of industry and agriculture. In the civil service, there is also a head organization: the statutory accident insurance institution.

The main organization of the industrial trade associations looks after common interests and primary duties of the trade associations. These include:

1) Occupational safety, industrial medicine, and first aid
2) Medical and professional rehabilitation
3) Research, testing, and counseling
4) Training and further education of the members of the trade associations
5) Provision of information material
6) Implementation of interstate social insurance agreements

The main organization looks after the common duties of the trade associations in the region through the individual state organizations.

Duties and Control Functions. Each trade association has its own technical control service which supervises the observation of accident prevention regulations. In addition, it counsels companies with regard to industrial safety and health protection. For instance, it can give valuable information on occupational safety prior to the procurement of new machines or the use of new processes. Moreover, the members of the technical control service also work on the further development of the book of technical safety rules in Germany and in the EC. In addition, they advise manufacturers on the safety and ergonomic development of working facilities through the expert committees of the trade associations. The trade associations are also active in the area of occupational medical counseling. Companies that do not have their own doctor can fall back on the doctors of the trade associations' medical service.

In addition, qualified safety personnel, safety representatives, employers and executives, works and staff council members, and company workers are trained in occupational safety and health protection by the trade associations (see Section 2.1.3).

2.1.3. Company Occupational Safety System

The employer is responsible for ensuring that occupational safety regulations are observed. His facilities, measures, and instructions must provide safe and healthy working conditions in the company.

Management must ensure that, as part of their authority to issue directives, all executives and senior employees must regard the practice of occupational safety and health protection as one of their most important executive duties.

In spite of this transfer of duties, the employer alone takes the entire responsibility.

The Specialists. Legal regulations (e.g., workers' protection law) and the rules of the trade associations require companies to provide doctors who are specialists in industrial medicine. The requirements are stipulated in the accident prevention regulation: "Company Doctors" VBG 123 (regulations of the main organization of the trade associations). Company doctors counsel the employer on all matters relating to occupational and health protection:

1) Planning, implementation, and maintenance of production facilities and of social and sanitary facilities
2) Procurement of technical working devices and the introduction of working methods and substances
3) Selection and testing of protective equipment
4) Occupational physiological, occupational psychological and other ergonomic and occupational hygienic questions, which include working rhythm, working hours and pauses, design of workplace, course of work, and working surroundings

Their duties also comprise organization of first aid in the company and questions of changing the workplace and integration of the handicapped into the working process.

Company doctors also perform medical checkups and provide medical care for employees.

The working hours of company doctors are regulated by the trade associations and the public accident insurance institutions in their accident prevention regulations. They depend on the potential hazard of the particular branch and the number of employees.

Company doctors can be employees of the company in question, employees of an industry-wide occupational medical service, or self-employed. The type of medical care is stipulated in agreement with the employer and the works council.

The employer is also required by the workers' protection law to hire qualified personnel for occupational safety. These are engineers, master craftsmen, or technicians with additional training in occupational safety. Apart from counseling duties, similar to those of the company doctors (see above), they are responsible for the technical safety supervision of production facilities and working devices and processes, and for observing occupational safety and accident prevention in practice. Like the doctors, they must inspect the workplaces at regular intervals, report shortcomings to the employer, and recommend measures for their elimination.

Similar to the doctors, the qualified safety personnel has no authority to issue directives. In their work, however, they are free of directives, i.e., not subject to the opinions of company executives when they make demands or have complaints in the interest of occupational safety.

The working hours of the qualified safety personnel are stipulated in the Accident Prevention Regulation: "Safety Engineers and other Qualified Personnel for Occupational Safety" (VBG 122).

Like the company doctors, the qualified safety personnel can be employees of the company in question, employees of an industry-wide technical safety service, or self-employed.

The work of the qualified safety personnel is supplemented by the safety representatives in the company.

In companies with more than 20 employees, safety representatives are appointed by the employer in accordance with the RVO. In their field of activity, they have to pay attention to safety at the workplace. The number of safety representatives depends on the branch, number of employees, and hazard potential of the particular company.

Apart from company doctors, qualified personnel for occupational safety, and safety representatives, other occupational safety experts can also work in the company, e.g., representatives for radiation protection and representatives for biological safety. The experts within the company also include recognized specialists who are authorized to check plants that require monitoring (§ 24 c of the industrial code).

The company representatives for environmental protection (waste, wastewater, and transport of hazardous materials) are also to be taken into account because their duties can partially overlap with occupational safety.

Employees have to support all measures that serve occupational safety. They are obliged to comply with directions of the employer for the purpose of occupational safety. Furthermore, they have to use the personal protective equipment placed at their disposal. Employees must use company equipment only for purposes that are in accordance with the regulations and they need not follow directions that are adverse to safety.

Before taking up employment and with each assignment to a new workplace, the employee is to be informed about the accident and health hazards involved in his work and about the measures and equipment available to avert these risks. These safety instructions can be delivered verbally or in writing. However, the mere handing over of leaflets – without additional explanations – is not sufficient.

If an employee discovers that some equipment is not perfectly safe, he must correct the defect immediately. If this does not belong to his duties or if he does not have the necessary knowledge, he should report the defect to his superior immediately.

If an employee makes use of his right to lodge a complaint about inadequate safety measures with his employer, no disadvantages should accrue to him therefrom. After checking the complaint, the employer must inform the employee about the results and, if necessary, immediately rectify unsafe conditions.

The employee can also lodge a complaint with the works council, either immediately or after he has unsuccessfully complained to the employer. The works council must check the complaint and, in case it is justified, urge the employer to take remedial action.

Duties of the Works Council. To combat accident and health hazards in the company, all those who can contribute are required to work together. In this connection, the employees have a special role. The German Industrial Relations Act regulates the right of information, participation, and codetermination of employees. The workforce elects and is represented by the works council. The main concern of the Industrial Relations Act is to involve the works council (and the workforce) in all occupational and health protection decisions taken by the employer. With respect to occupational safety, the works council has the mandate to control and organize, the right of codetermination, the obligation to support the authorities, and the right of information and participation.

The works council should ensure that not only the laws, decrees, and accident prevention regulations passed in favor of the employee, but also the negotiated industrial agreements and employment agreements (see Section 2.1.4) are observed and enforced.

The mandate to monitor applies to the occupational safety measures to be taken by the employer and company executives as well as to those to be taken by employees. The works council has the right and the duty to point out shortcomings in occupational safety and accident prevention to the employer or responsible company executives and to press for their elimination with all means at its disposal. The supervisory mandate also includes the obligation to urge the employees to comply with the regulations passed for their protection during the time they spend in the company.

The supervisory mandate is supplemented by the organizational mandate. The works council is required to propose measures that serve the company and the workforce. It can deal with all matters that serve the protection from accidents and health hazards. It can demand from the employer that appropriate measures are taken to eliminate hazards.

According to the Labor Management Act, the works council has the right of codetermination in the settlement of matters of occupational safety and accident prevention. This means that the consent of the works council to the safety measure proposed by the employer is required. It is not sufficient for the works council to simply take note of the intended occupational safety measure. In accordance with the Workers' Protection Law, the appointment of qualified safety personnel and company doctors is also subject to this codetermination.

Based on their obligation to support the industrial safety authorities (trade supervisory authority and trade association), the works council has the duty to support the authorities in their work in the company by providing suggestions, advice, and information.

On the other hand, the authorities are obliged to involve the works council in all company inspections, accident investigations, and in meetings on subjects of occupational safety and to hold discussions, if necessary. Apart from the control services, the works council can also take advice from the qualified safety personnel and from the company doctors. In agreement with the employer, the works council can also consult other experts outside the company. The works council can do justice to its mandate to supervise, organize, codetermine, and support only if it is adequately informed. For this reason, according to the Labor Management Act, the works council is granted an extensive right of information. This right of information is based on the principle that the employer and the works council should have available the same level of information on occupational safety and accident prevention in order to fulfill their duties.

Although the works council has the right to control and participate, the employer carries the sole responsibility for occupational safety. The works council cannot independently enforce safety measures.

The Occupational Safety Committee. All experts for occupational safety in the company are obliged to cooperate. Therefore, the employer is responsible for forming a suitable occupational safety organization within the company.

The central body of the company occupational safety organization is the occupational safety committee. Occupational safety committees are to be formed in all companies in which company doctors and qualified safety personnel are appointed. The following are members of the committee: the employer or his representative, two members selected by the works council or staff council, company doctors, qualified safety personnel, and safety representatives.

The occupational safety committee is required to discuss matters of concern for occupational safety. It meets at least once every three months. Apart from routine duties, such as dealing with accident statistics or the general supervisory function and coordination, the occupational safety committee should take into account the conditions in the particular company and concentrate its efforts on, e.g.,

1) Counseling and recommendations for safety programs in the company
2) Discussing suggestions for putting into effect safety investments of the company
3) Regular evaluation of the frequency and severity of occupational diseases
4) Analysis of the results of technical safety controls of work processes and the introduction of new working methods or new substances
5) Discussions of proposals for company participation in industry-wide accident prevention measures

In companies with more than three safety representatives and an existing occupational safety committee, it is also required that the safety representatives meet regularly.

In these meetings, the technical safety problems of the individual areas are to be discussed. The safety representatives for the occupational safety committee are to be named.

In smaller companies which have no qualified safety personnel and company doctors, a safety committee is to be formed if more than three safety representatives are appointed. With the participation of the works council, the employer should meet at least once a month with the safety representatives or the safety committee.

2.1.4. Role of the Social Partner and Associations

The function of the unions and the employers' associations are not regulated by law. However, some of their powers are legally described, especially the right to negotiate industrial agreements and to send representatives to certain occupational safety bodies.

Occupational safety and health protection of workers have always been a goal of the unions. Progress has been made in some areas, e.g., a reduction of the number of accidents at the workplace or on the road to work and an increase in the recognition of industrial diseases. However, new problems arise at many workplaces due to job reorganization (e.g., due to stress, performance demands, and new substances).

To attain their occupational safety and health protection goals, the unions work on the further development of the books of rules for occupational safety in both parliamentary and extraparliamentary bodies (occupational safety bodies). Their efforts are concentrated in the sociopolitical bodies of the Federal Ministry of Labor (e.g., committee for hazardous substances), in the advisory councils of the Federal Institution of Industrial Safety and the Federal Institution of Industrial Medicine, and in the aid program "Work and Technology" of the Federal Ministry for Education, Science, Research, and Technology. Moreover, occupational safety councils have been founded in some states with the participation of the unions. However, the most extensive field of action of the unions is the self-government of the trade associations, which comprise equal numbers of representatives of the employer and of the employees. In this connection, unions can be represented by working members or by full-time officials.

Via the autonomous organs (assembly of representatives and the board) and their committees (e.g., accident prevention committee, pension committee), the unions can directly influence the book of rules of the trade associations, help to plan the essential features of the activity of the technical control service (information, counseling, control), and assist in the assessment of occupational diseases and accidents at the workplace or on the road to work.

The representatives of the employer who work in the autonomous bodies of the trade associations can also be the representatives of employers' associations. The resulting common planning platform in occupational safety of trade associations is supplemented by agreements or industrial contracts between the social partners.

These contracts can contain regulations which go beyond the book of rules of the government or the trade associations or concretize these rules in a branch- or company-specific manner. For the employees, industrial contracts of this type can be completed only be the trade associations. For the employers, however, both the employers' associations (joint agreements) as well as each individual employer (company or house agreements) may negotiate. The workman's guild and the guild associations are also entitled to complete contracts by virtue of special legal regulations. Where occupational safety was a part of joint industrial agreements in the past decades, it usually applied to the classical areas of working hours, protection of special groups of people, and special allowances.

On the other hand, company agreements contain detailed regulations regarding, e.g., the participation of employees and members of the works council in the enforcement of the Workers' Protection Law and the decree on dangerous substances, or the role of the safety representative.

2.1.5. Future Developments

As a result of the formation of the European Common Market in 1 January 1993, Germany is required to translate European guidelines into German law. Of special significance is the guideline 89/391/EEC: "EEC guideline for the enforcement of measures to improve the safety and health protection of the employee at work". At present, it is being translated into a basic German occupational safety law. The political consultations are not yet completed.

The important points of this EC guideline or basic occupational safety law will be:

1) Extension of the understanding of occupational safety including health protection in the company and, thus, improvement of prevention
2) All workplaces and work processes are to be subjected to risk assessment
3) Employees are to be involved in the shaping of working conditions
4) The control, counseling, and compensation mandate of the trade associations will be extended to include prevention of all health hazards due to work

Due to the "centralization" of the occupational safety legislation in the EC, adaptation of the accident prevention regulations of the trade associations will become necessary. National standards (DIN, VDI, VDE) for machines, equipment, and facilities will be replaced gradually by European standards by CEN/CENELEC.

In Germany, the statutory health insurance companies will also play a new role in future occupational safety and health protection in the company. As a branch of social security, the health insurance companies have been assigned duties in health protection at the workplace in accordance with Section 4, § 20 of the social code. These duties are:

1) To counsel and inform insured persons about health hazards in general and about the prevention of diseases in particular
2) To pursue the causes of health hazards and to work on their elimination
3) To assist in the prevention of health hazards caused by work and to cooperate with the statutory accident insurance institutions (the trade associations)

2.2. Legal Requirements in Other European Countries

Most European countries have specific regulations to protect workers from harm due to work. For some of these countries the regulations are described in more detail below (situation as of December 1994).

United Kingdom. The principal legislative framework is the Health and Safety at Work Act 1974. This sets out general duties of employers and others and is supplemented by specific regulations with more detailed requirements. In the case of chemicals the principal regulations are:

1) Notification of New Substances Regulations 1993 (NONS)
2) Chemicals (Hazard Information and Packaging) Regulations 1993 (CHIP)
3) Control of Substances Hazardous to Health Regulations 1994 (COSHH)

Strictly speaking, NONS and CHIP, both of which are derived from EC directives, deal with the supply of hazardous substances. COSHH, by contrast, deals with the use of hazardous substances, and with assessing and managing the health risk to employees. If the assessment indicates that there may be risks to health, COSHH requires the employer to:

1) Prevent (by substitution or elimination) or adequately control exposure
2) Make sure that precautions are used and controls maintained
3) Monitor, if necessary, exposure and/or carry out health surveillance
4) Inform, instruct, and train employees as necessary

COSHH contains specific provisions to deal with carcinogens; these implement the EC Carcinogens Directive.

Other regulations deal with more specific risks such as from lead, asbestos, and flammable liquids.

A list of occupational exposure limits (OELs) for use with COSHH is published annually. The list contains two types of OEL. Maximum exposure limits (MEL) are maximum concentrations of airborne substances, averaged over a reference period, to which employees may be exposed by inhalation under any circumstances. The occupational exposure standard (OES) is the concentration of an airborne substance, averaged over a reference period, at which, according to current knowledge, there is no evidence that it is likely to be injurious to employees if they are exposed daily by inhalation to that concentration.

Denmark. The general regulation of chemicals at the workplace is laid down in Order No. 540 of Substances and Materials, which came into force in 1983. (In practice "materials" are considered to be the same as "preparations").

The order establishes some very general provisions for working with chemicals. More specifically it also defines the substances and preparations that are to be considered as hazardous. This Danish definition for workplace purposes is broader than the EC definition, laid down in the 1967 Substances Directive and in the 1988 Preparation Directive.

Included in the definition of "hazardous" are:

1) All substances and preparations classified as dangerous according to the EC directives
2) All substances on the Danish list of OELs
3) All substances on a recommended list (an annex to the list of the OELs) of volatile organic solvents and preparations containing more than 0.5 % of such a substance
4) All substances on a list of agents considered to be carcinogenic and preparations containing more than 0.1 % of such a substance; the list consists of EC carcinogens and IARC carcinogens (Groups 1, 2 A, and 2 B)

The extension compared to the EC definition is primarily for organic solvents. For carcinogens the difference to the EC definition is only small but was substantial until 5 years ago.

For the substances and preparations classified as hazardous specific provisions are given:

1) Substitution/replacement by less dangerous substances if technically feasible
2) Safety data sheets prepared by the supplier, including obligatory information on specific Danish restrictions for use
3) Safety data sheets prepared by the employer (more detailed than those from the supplier)

In more detailed governmental orders, specific rules are laid down for minimization of exposure and for specific types of preparations such as paints.

A specific order prohibits employers ordering persons under 18 years of age to work with preparations containing more than 0.5 % of a volatile solvent.

In general, legislation is similar to the latest EC directives.

Before use of a new substance or new preparation at the workplace the supplier must notify the Product Register at the Danish Working Environment Service. Among other information the notification must contain a full quantitative and qualitative description of the substance/preparation and the amount expected to be sold per annum in Denmark.

Finland. Finnish occupational chemical safety regulations are based on two framework acts: the Labor Protection Act (299/58, amendments) and the Chemicals Act (744/89, amendments). The Act on the Supervision of Labor Protection and Appeal Procedure in matters concerning Labor Protection (131/73, amendments) lays down the rights and powers of the labor protection authorities. In addition to these, the Occupational Health Care Act (743/78) regulates the occupational health care services which are to be provided to workers.

The Chemicals Act prescribes the requirements concerning chemical substances and preparations as products. The purpose of the Act is to prevent damage to health and environment. Being a framework act, it contains quite general provisions on the duties of manufacturers, importers, and suppliers of chemicals.

With some minor exceptions the regulations concerning chemical substances and preparations as well as safety data sheets are identical with the corresponding EU directives. Under Article 12 of the Preparations Directive, Finnish manufacturers and importers of chemicals hazardous to health are obliged to send fairly detailed information to the Ministry of Labor, which maintains a registry of chemical products. The EU directives regarding restrictions on the marketing and use of certain dangerous substances and preparations have also been implemented in Finnish legislation.

Protection of workers against the risks related to chemical agents is covered by the Labor Protection Act. It also contains general provisions on protection of workers against other hazardous factors at the workplace. More detailed regulations are given in decisions of the Council of State and Ministry of Labor. Detailed regulations have been issued, e.g., on benzene (355/82), lead (1154/93), asbestos (886/87, 1413/91), carcinogenic substances (1182/92, 838/93), and teratogenic substances (1043/91, 1044/91).

The most important regulations concerning exposure to chemicals, however, are contained in the Council of State Decision on the Protection of Workers from the risks related to exposure to chemical agents (920/92, 727/93). These regulations cover widely the obligations of the employer, from assessment of exposure and risk to preventive and protective measures, as well as instructions and information to be given to the employees. The limit values for workplace air contaminants are also specified. Basically, there are two types of limit values. The Council of State issues "binding limit values for workplace air contaminants" which are directly legally binding (e.g., the values for asbestos, benzene, lead, and vinyl chloride). The other type of limit values are the "indicative limit values of workplace air contaminants" (HTP values). The philosophy of the indicative limit values is very close to the concept of the EU indicative limit values. Employers have to take the indicative limit values into account when assessing the quality of workplace environment, exposure of workers, and the meaning of the measurement results.

Spain. Spanish law concerning health and safety at work derives from a number of different sources with varying levels of importance. The Constitution of 1978 requires the public authorities to safeguard workers' safety and health. International obligations such as ILO conventions are included in national law. The Spanish Parliament has adopted a number of specific acts dealing with health and safety, in particular the Workers' Statute of 1980, which states that workers have a general right to be protected by safety rules and the responsibility to perceive the rules adopted by the employer. There are also elaborate administrative regulations establishing the Labor Inspectorate and laying down detailed and technical health and safety rules.

The general principles as laid down in the law are worked out in regulations. The most important of the regulations concerning health and safety at work is the general Ordinance Concerning Safety and Health at Work of 1971, being a tool for the Ministry of Labor to make rules, to initiate research, to give technical guidance, to lay down the functions of the Labor Inspectorate, to establish provincial occupational safety and health councils, to lay general duties on both employers and workers, to set out the functions of occupational safety and health committees and of safety officers, and to make provision for administrative fines.

At the same time there are a number of other regulations which set out specific technical standards, e.g., on protection against lead, asbestos and ionising radiations.

New legislation is designed to bring Spain into line with EC legislation, especially the Framework Directive. Until now provisions have been adopted to comply with the directives on vinyl chloride, lead, asbestos, and banning of certain specified agents and activities. This new law will probably be adopted in 1995.

France. The French laws concerning the risks of chemicals is mainly based on EC directives. For the workplace this means:

1) Risk classification of substances and preparations
2) Notification of new chemicals and preparations
3) Material safety data sheets
4) General prevention rules for workplaces
5) Quality control procedures for workplace air
6) Prevention of cancer risks

There are specific rules and protection measures for women and younger persons. Basis for taking prevention measures is the risk assessment of dangerous substances at the workplace.

Occupational exposure limits (not binding) give indications for protection of workers. At the moment about 520 indicative limit values are available.

The Netherlands. The Working Environment Act (*Arbeidsomstandighedenwet*), adopted in 1980, deals with safety, health, and well-being during work and is meant as a framework law. It comprises basic requirements and responsibilities of employers and employees to be elaborated in decrees and regulations for special topics.

With respect to chemical substances the following regulations are in force:

Relevant decrees are the *Decree on Industry and Workshops* (1938) and the *Decree on Commercial and Noncommercial Services* (1990). For substances a central point is that effective measures must be taken to prevent harm, damage, or nuisance to workers exposed to chemicals during their work. The hierarchy of preventive measures to be taken reads:

1) Prevention of all emissions at source
2) Ventilation
3) Separation of workers and source
4) Use of personal protective equipment

The measures to be taken have to be effective with respect to health protection and prevention of nuisance, but feasibility aspects may also be taken into account. Only in the ultimate case may personal protective equipment be used.

The Minister of Social Affairs has the right to set legal limit values for exposure during work. In March 1994 an initial list of 96 substances with legally binding limit values was published. All these chemicals have passed the Dutch three-step procedure for setting limit values. Additionally, legally binding limit values for 15 carcinogenic chemicals have been covered by a specific decree on carcinogenic substances. In addition to the binding values, administrative occupational exposure limits (MAC values) are in force, adopted from the United States (ACGIH), Germany, Sweden, and the United Kingdom.

The *Decree on Specific Harmful Chemical Compounds* (1991) prohibits the production, storage, and use of certain aromatic amines (2-naphthylamine, 4-aminodiphenylamine, benzidine, and their salts, 4-nitrodiphenylamine). Also, preparations containing more than 0.1 wt % fall under this regulation. It is an implementation of EC Directive 88/642.

The *Decree on Carcinogenic Substances and Processes* (1994) is an implementation of EC Directive 90/394 on the protection of workers from harmful effects due to exposure to carcinogenic chemicals. It covers all chemicals that fall within the criteria for carcinogenicity in EC Directives 67/548 and 88/379.

The decree covers:

1) Registration and evaluation of carcinogenic substances
2) Restrictive working conditions to minimize exposure
3) Setting legal limit values for carcinogenic substances (so far 15 have been published)
4) Medical surveillance, including environmental and biological monitoring and medical investigations
5) Keeping personal log-books of workers involved, which are to be kept for at least 40 years

Moreover, a number of acts have since been published, all of which are appended to the Working Conditions Law. The *Propane Sultone Act* (1976) prohibits the use, production, and storage of this suspect carcinogen.

The *Asbestos Act* (1977) is an implementation of EC Directive 83/477. This Act has been amended several times; the 1993 version prohibits the use and production of all types of asbestos. In situations where significant exposure occurs (e.g., demolition) strict rules and standards apply.

The *Benzene Act* (1977) is an implementation of ILO Treaty 136 of 1971 and restricts the use of this solvent. The *Vinyl Chloride Act* (1987) implements EC Directive 78/610. The *Lead Act* (1988) includes biological limit values for exposure.

2.3. European Legislation

Due to the increasing mobility of the labor force throughout Europe, it is becoming more and more desirable to harmonize regulations and rules concerning working conditions.

As in other matters, the EU Member States try to come to a joint standpoint by mutual discussions. Directives related to working conditions are drafted by the Directorate General V (DG V, Health and Safety) in Luxemburg. These drafts are elaborated in the relevant EC Council working group, leading to a common position. After adoption by the Council of Ministers the new EC legislation must be implemented in national legislation within a certain time period. Since EC workplace legislation concerns minimum requirements (Art. 118 A of the EC Treaty), each Member State has the right to set stricter regulation than specified in EC directives.

Since the 1970s a number of directives on working conditions have been adopted. With respect to chemical substances the following directives, dealing with the protection of workers against health risks of exposure to chemical agents during work, are noteworthy:

Directive 78/610 – vinyl chloride
Directive 80/1107 – a framework directive for chemical, physical and biological agents with general principles of protective measures
Directive 82/605 – exposure to metallic lead and its compounds
Directive 83/477, amended by 91/382 – handling of asbestos
Directive 88/642 – binding and indicative limit values
Directive 88/364 – relating to the banning of certain agents
Directive 90/394 – the carcinogens directive
Directive 91/322 – establishing indicative limit values for 27 substances

Since 91/322 a multistep procedure has been followed to develop additional indicative limit values. As a first step the Scientific Expert Group (SEG) of DG V establishes a health-based recommended value; in the next step the socioeconomic and technical feasibility of the SEG value is investigated by the Commission, the Member States, and the tripartite Advisory Committee on Safety, Hygiene, and Health Protection at Work. In the third and last step the committee of Member State representatives votes on the Commission proposal for a definite indicative limit value.

As of December 1994 a number of SEG recommendations has been published. The next steps in the procedure, however, have stagnated seriously.

A Commission proposal [COM (93) 155 final – SYN 459] for a council directive on the protection of the health and safety of workers from the risks related to chemical agents at work is under discussion in the Council working group.

100 A-Directives. In addition to the specific 118 A-workplace directives, a number of 100 A Common Market Directives also have important implications for workers.

Most prominent is Council Directive 67/548/EEC of 27 June 1967 relating to the classification, packaging and labelling of dangerous substances, as amended by Council Directive 92/32/EEC of 30 April 1992. The sixth amendment (Directive 79/831/EEC) concerns notification of new substances. In the seventh amendment (1993) the risk assessment paragraph has been worked out extensively.

Further Council Directive of 26 June 1978 relating to the classification, packaging, and labelling of dangerous preparations (pesticides), and Council Directive 88/379/EEC of 7 June 1988 relating to the classification, packaging, and labelling of dangerous preparations, as amended by Council Directive 90/492/EEC, defining and laying down a system of specific information on dangerous substances and preparations, in the form of safety data sheets, are principally intended to enable industrial users to take the measures necessary to ensure the protection of the safety and health of workers.

2.4. Regulations in the United States

Industrial safety and health in the early stages in America changed from work-related accident and illness being the worker's responsibility to a team approach of worker and management. Some industrialists realized early that unhealthy or unsafe work conditions cost money, with lowered work quality, interrupted work schedules, and compensation payments to injured or sick workers. Others refused to recognize the inherent hazards of their business, and this required a series of regulatory requirements for adequate controls to be introduced.

The Occupational Safety and Health Act (OSHA) of 1970 authorized the Department of Labor to set and enforce mandatory occupational and health standards for businesses. The act provided for research and assistance with the discovery of latent data to establish any connection between diseases and the work environment. While other laws gave the authority to regulate food additives, drugs, and pesticides for public exposure, no employer was previously required to assure worker health in production operations. OSHA is unique because it does not have authority to pre-approve products entering the market, unlike FDA and EPA. Furthermore, OSHA cannot require mandatory testing of products.

The Act created the National Institute of Occupational Safety and Health (NIOSH) under the Department of Health and Human Services, with responsibility for research and recommendations for reports on employee exposure and the authorization of medical examinations and tests where desirable to protect employees.

OSHA health standards typically set maximum limits on employee exposure and prescribe procedures or equipment to achieve this level. The Environmental Protection Agency (EPA) under the Toxic Substances Control Act (TSCA) requires disclosure of safety information and testing to develop data for health and environmental effects for chemicals intended for manufacture, import, or distribution. Testing is initiated by the finding that a chemical presents a potential risk or that humans will be exposed to substantial quantities. With the different regulatory agencies involved (federal and state), there is potential for inconsistency in government standards.

The Interagency Testing Committee, now the Regulatory Council, had the goal of securing agreement on requiring chemical-specific testing and common policies. Agencies have also achieved collaboration in the National Toxicology Program (NTP) to coordinate numerous federal efforts to improve methods and performance of animal studies on selected chemicals, funded by public taxation.

Another joint effort is the Agency for Toxic Substances and Disease Registry (ATSDR) under the Public Health Service, where the EPA publishes toxicological profiles of chemicals determined by the collaborators to be the most significant potential threat to human health.

In summary, the Federal Government's Occupational and Health Administration, in cooperation with state agencies, sets and enforces standards for production and process workers. Other agencies (EPA, FDA, and Consumer Product Safety Commission) have various laws on packaging, transportation labels, and pre-marketing requirements to protect consumer health and the environment. To combine efforts, OSHA has published a standard on Hazard Communications (1986), which includes requirements for material safety data sheets and labels for hazardous products.

3. Industrial Hygiene [9–28]

3.1. Introduction and History

Introduction. Industrial Hygiene (IH) is a term given in the United States primarily to the discipline broadly concerning the recognition, evaluation, and control of the chemical, physical, and biological stressors that may cause discomfort to or impair the health or well-being of employees in the workplace. Such stressors may result from exposure to:

1) Chemical gases, vapors, fumes, particulate matters, etc.
2) Physical agents such as noise, vibration, heat, and cold
3) Ionizing and non-ionizing radiation
4) Biological materials or microorganisms
5) Ergonomic stress

In a number of European countries this discipline is defined more appropriately as Occupational Hygiene (OH), and this term, originating from the United Kingdom, is generally used in these Sections. In other countries (e.g., Germany) the term industrial hygienist or occupational hygienist is less well-known, and the general functions, defined above, are generally performed by specialized safety or medical personnel.

The main objective of occupational hygiene is the prevention of undesirable effects to health by assessing and adequately controlling harmful exposures to chemical, physical, and biological agents in the workplace. While safety is more broadly dedicated to the prevention of personal injuries and loss of assets and equipment resulting from sudden events and failures, OH focuses on potential hazardous exposures arising from short- and long-term working conditions.

Occupational medicine, also concerned with health protection, monitors for any early signs of illness by means of a health surveillance program on individual workers. Proper interaction and communication between occupational hygiene and occupational medicine guarantees a comprehensive and truly preventive occupational health program.

History. Occupational diseases were identified more than two millennia ago by the Greek physician, HIPPOCRATES, who described lead toxicity in mining (400 B.C.). It is only since the early 1900s, however, that the dimensions of the problem of health and safety of workers have raised general interest and stimulated legislative actions in some industrialized countries.

During the long intermediate period, occupational health remained dormant, with the exception of a few, far-sighted persons who studied and recorded the problem in now famous treatises. Among them are AGRICOLA (1500s), who in his *De Re Metallica* described a lung disease caused by the inhalation of silica (silicosis); PARACELSUS (1567), the father of modern toxicology; and RAMAZZINI (1700s), who first outlined preventive measures for various occupational hazards in *De Morbis Artificum*.

Despite the enactment of some well-meaning legislation and the establishment of public health services in the United States and many European countries, working conditions improved little, and statistics continued to show high incidences of both injury and illness at work until the end of World War II. It was during this period that the first occupational hygiene services and associations were formed in the United States. With the exception of the United Kingdom, occupational hygiene in Europe came later, and a leading role in this field was played by occupational physicians from public services, universities, and industry.

The explosive technical progress in all scientific fields since the 1960s and the enormous development of different industrial activities and services (extractive, chemical, pharmaceutical, agrochemical, automotive, transport, nuclear, electronic, etc.) have required a new and more specialized effort in the increasingly complex field of health and safety at work.

A strong safety culture has been established by industry during this period. This resulted in programs which broadly incorporated health and environmental protection, and aimed at protecting key assets (people, property, and technology). Workplace conditions have progressively improved, as demonstrated by positive trends in safety records for occupational injuries and illnesses. New factors have more recently emerged which have modified public awareness of health and environmental matters, changed priorities and expectations on quality of life, and, in general, reduced trust in industry and its ethics.

The scientific knowledge of the effects of chemical, physical, and biological agents on human health gained by toxicologists and health professionals has increased dramatically, together with the analytical ability to measure smaller and smaller amounts of substances in any media (air, water, soil, food, biological fluids, etc.). Scientific and social standards of acceptability for both workplace and environmental pollutants have continuously been lowered. In addition public perception of the risk arising from chemicals changed after recent industrial disasters which damaged and threatened entire populations (Seveso, Bhopal, etc.). Regulations have become increasingly stringent and sometimes bureaucratic and complex. Industry has recognized these concerns and is working to rebuild confidence and trust with more transparent and coherent environmental, health, and safety (EH & S) programs. In particular, the chemical industry has relaunched its commitments by voluntary worldwide implementation of the Responsible Care program, with excellent and measurable performance in integrated EH & S as the main objective.

In this new climate the role and function of cost-effective occupational hygiene services can significantly contribute to the goal of protecting and promoting the health and well-being of workers and the community.

3.2. Requirements for an Effective Program

There are no consolidated and internationally accepted requirements for establishing an effective occupational hygiene (OH) program, and different approaches have been adopted in various countries, appropriate to national legislation, cultural background, and needs. Attempts to harmonize OH practices are being made by international associations and standardization bodies [e.g., ILO (International Labour Office), WHO (World Health Organization), EC, etc.] but the quality of service varies considerably, particularly in Europe.

Public OH services exist in many countries. They are usually part of the governmental health and safety institutions, which may have an advisory role and/or assure compliance with legislation. Other services are part of academia and may act as consultants for the government and for companies.

Most large companies have in the last decade established an internal OH service, usually as part of the environmental, health, and safety organization. Private consultancies have developed in some countries, and also represent a resource for medium-sized and small companies that cannot afford an in-house service.

Experience gained from well-run and successful services, particularly in multinational chemical and petrochemical companies, indicates that a number of key basic elements and systems should be considered and implemented. The philosophy and detailed practical guidelines for each element are described in the following sections of this chapter.

3.3. Objectives and Standards

Independent of the nature and size of the OH service, the first and most important factors to be defined with full agreement and committed support of top management, is the definition of mission of the service, together with long- and short-term objectives.

The mission statement together with well-defined objectives will clearly indicate the core values the service is aiming at and provide directions, means, and ethical provisions which will be adopted to achieve the mission.

In addition, the OH service should define the level of standards strived for in the various program elements. In some cases the reference standard can simply be national legislation and practices; in other cases, for large multinational companies, stringent internal requirements are set to guarantee high-quality and consistent working conditions, independent of the local country's economic and social structure, legislation, authorities, and verification of compliance.

Corporate codes of conduct and guidelines should also be issued to harmonize practices and to help implement the programs. Periodic review and adjustment of the objective should also be instituted, together with an auditing system to measure the performance of the service against the targets.

3.4. Organization and Staff Management

The second essential element for an OH service is an adequate organization, sufficiently staffed and well managed, in which allocation of responsibilities is clearly defined. It is clearly specified in all legislations that the health and safety of employees is the responsibility of the employer, and not of the safety, medical, or OH services. The role of the service is to provide management and employees with technical and scientific opinions, expert advice, and authoritative input for implementation and interpretation in the OH field. This is a clear advisory function in which the occupational hygienist should maintain his professional independence and ethics.

Typically, an occupational hygienist has a university degree in science (e.g., chemistry, biology) or engineering, or a specific degree in occupational hygiene or environmental health, plus postgraduate specialization in industrial hygiene/occupational hygiene and accumulated practical experience in this field. This broad, basic knowledge in a variety of disciplines, which should also include elements of toxicology, occupational medicine, safety, environmental protection, ergonomics, analytical chemistry, and statistics, allows the occupational hygienist to perform well in a complex workplace environment.

Codes of ethics for this profession have been established by some national and international associations [e.g., AIHA (American Occupational Hygiene Association), IOH (Institute of Occupational Hygiene), and IOHA (International Occupational Health Association)]. In addition, there are well-known, highly reputed, professional certification schemes in the United States and the United Kingdom [ABIH (American Board of Industrial Hygiene) and BERBOH (British Examining and Registration Board in Occupational Hygiene)], with the aim of elevating the level of professionalism and guaranteeing high standards. These independent institutions provide certification by examination at various levels of competence and specialization. The ABIH also requires periodic re-training and qualification in order to keep the certification.

It is important that the professional OH service not become disconnected from workplace practices or be regarded as a consultative body. Continuous links with management can be established by implementing a so- called OH unit contact concept. Well-trained persons can be nominated as OH contact in the production unit or plant where they work. In this function they carry out specific routine tasks on a daily basis under the technical assistance of the professional OH staff. These tasks can include sampling and analytical activities, workplace observation, and self inspection. Good cooperation and balanced approach between OH contacts and the professional OH organization is instrumental in the effectiveness of the program. A similar approach with a part-time person, eventually trained and supported by external services, can be adopted by small to medium-sized companies, for which the hiring of a professional OH is economically infeasible.

Continuous communication and good teamwork is needed between the two main health professionals – the industrial hygienist and the occupational physician. This is particularly important in areas where knowledge of each other's findings and data is essential for a truly preventive occupational health program (work history; job class; exposure data; biological monitoring; first aid, acute and chronic cases; health surveillance trends, etc.). Good communication and cooperation should also be sought with other related expert functions, such as safety and environmental protection due to the overlapping and often conflicting fields of interest (e.g., communication and training of employees, personal protective equipment, job safety analysis and working procedures, auditing, ergonomics, accident investigation, air pollution, and emergency responses).

3.5. Anticipation of Health Hazards

The early anticipation of health problems is probably the most cost-effective practice of occupational hygiene. It consists of identifying and correcting sources of exposure or potential health-affecting situations early in the design stage of new facilities and when modifying existing ones. It is cheaper to make changes at the design stage than to retrofit processes when in operation. Although the concept is very straightforward, it is not easy for the engineering/construction group to foresee and minimize all possible factors adversely affecting the project (e.g., process yields, quality, performance, reliability, environment, health, and safety). If the occupational hygienist is involved, for example, at the research and development stage, he can review the properties and intended use of new products and processes. He can help determine additional toxicological testing if necessary, in order to make a preliminary hazard assessment.

Similar assistance should be provided when a material new to a facility is introduced or handled. In this case one should also search for existing occupational exposure limits (OELs) and the availability of adequate sampling and analytical methods in case quantitative evaluation is required. Advice on ergonomically designed equipment, workplaces and procedures, requirements for appropriate hygiene and facility decontamination, and selection and testing of machinery based on specifications driven by health hazard considerations, are other examples of anticipation. Major maintenance work or shut-down operations, which currently represent the conditions with the highest risk level, are also cases where a program should be specially designed to prevent major exposure problems during execution. Consultation of OH before acquiring an existing facility should also be considered by large companies.

3.6. Occupational Exposure Assessment

3.6.1. Introduction

Research and development (R & D) activities, manufacturing operations and related services (e.g., maintenance, utilities), transport and distribution, end-use and disposal of products are all stages at which, in spite of technological progress, the personnel involved may be exposed to workplace pollutants. This section describes the most traditional and fundamental function of an occupational hygienist: the assessment of exposure to chemical, physical, and biological agents. The more accurate the assessment, the more valid the determination of related health risks that the occupational hygienist and other health professionals can then make.

Appropriate risk assessment allows the risk manager to take the most effective measures to control hazardous exposure and ultimately assure a safe and healthy work environment throughout the working life of all employees. Exposure assessment can be carried out by various methods, which often reflect the different legislative emphasis and OH traditions. The strategy for an exposure assessment described here is a general and simplified one which reflects current accepted good practice and is founded on a preventive philosophy. This strategy is outlined in Figure 2 as a cyclic sequence of steps which are broken down into subsets and described in detail in the following sections. The entire process starts again when a re-assessment is needed due to significant process and procedure modifications, or simply because of ageing of the unit. It is good practice to consider reassessment every four to five years if exposure conditions have not changed prior to that time.

3.6.2. Characterization and Identification

The first step in occupational exposure assessment is collecting and evaluating the relevant information, documents, and reports that allow full recognition and characterization of the factors and conditions leading to hazardous exposures. This is the preparatory stage upon which the entire assessment is built. The various factors that influence the degree and frequency of exposures depend on the type of work and working environment, how the personnel operate, and the nature of the agents. All these intimately interconnected aspects, once properly described and examined, provide the basic information for the assessment process.

To properly conduct this step, the occupational hygienist should become very familiar with the unit/plant under investigation, and spend sufficient time collecting information, studying, holding discussions with supervisors and workers, and observing what happens in the process. Naturally, previous OH data and experience, if available, are a fundamental source of information. The information, answers, and factors that generally have to be gathered during this preparatory phase are listed in the following:

Work environment and process:	Type of process and technology, chemistry and operating conditions
	Type of production and variation (continuous, batch, etc.)
	Layout and building characteristics
	Equipment and machinery
	Degree and type of environmental controls
	General and local ventilation systems
	Main emission sources and propagation
	Seasonal variation
	Maintenance requirements
	Hygiene facilities
Workforce and human factors:	Number of people, age, sex
	Shift system
	Awareness and training
	Personal protective equipment
	Workclothes
	Food and drink consumption
	Routine operations and tasks
	Infrequent and critical tasks
	Typical length and frequency
	"Dirty jobs"
	Special groups at risk
	Contractors
Agents and their characteristics :	List of chemical substances (raw materials, intermediates, byproducts, products, etc.)
	Physical and warning properties
	Noise, vibration, heat stress, and other physical agents
	Toxicological profile and main health adverse effects
	Acute, chronic, carcinogen, mutagen, reproductive
	Immediate or delayed
	Reversible and irreversible
	Asphyxiant, irritant, allergenic, etc.
	Biological agents
	Route of exposure (inhalation, ingestion, skin contact)
	Applicable occupational exposure limits (OELs)

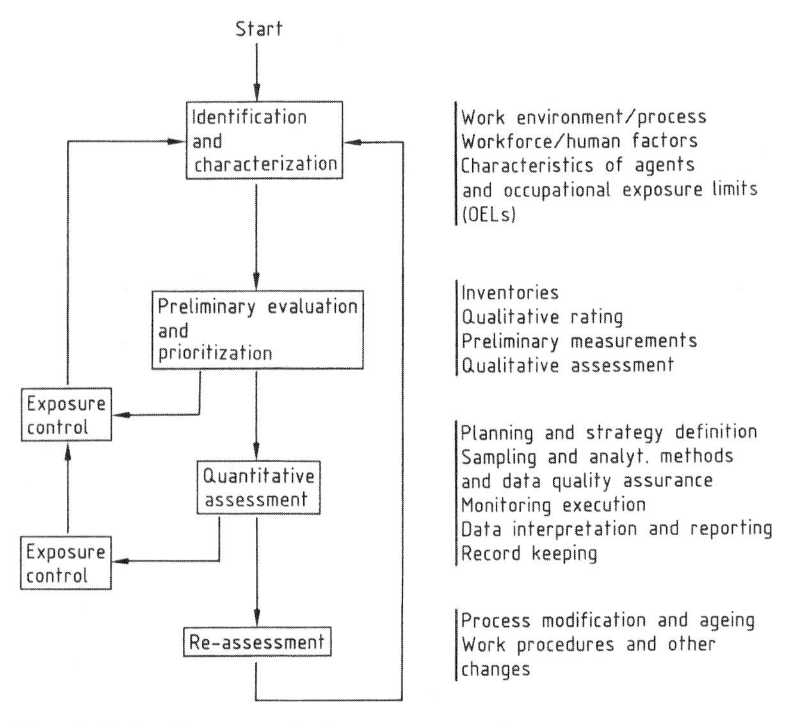

Figure 2. Outlines for an occupational exposure assessment

A list of typical areas, operations, and conditions where experience has shown that higher risk of exposure can occur is as follows:

Warehouse	Sampling raw materials
	Fork lift trucks
Unloading/loading stations	Ship, barge, rail car, tank truck unloading/loading operations
	Handling bags/sacks/drums
Laboratories	Handling incoming samples
	Storage and handling of reagents/solvents
	Lab waste disposal/glassware washing
Process area	Tank vents, pumps, centrifuges, filters, dryers, extruders, grinders, mill furnaces, process sampling, adding raw materials, filling line, packaging and bagging operation, waste handling
Maintenance and mechanical shop	Decontamination and line breakup
	Welding, soldering, cutting, grinding, machinery degreasing, polishing, sand blasting, vessel entry, and confined space operation

The list is not exhaustive and refers to chemical production facilities, but can be extrapolated to other operations.

3.6.3. Preliminary Evaluation and Prioritization (Qualitative Assessment)

By using a systematic approach and documenting the preliminary evaluation as described in recent publications or recommended by some codes of practice based or legislation (Germany, United Kingdom, EC, etc.), the preliminary evaluation becomes a valid qualitative assessment of health risks. In addition, it can constitute the basis for a system for prioritizing potential OH concerns so that resources can effectively be directed to those conditions that pose the greatest risk. This prioritization scheme is determined by the qualitative assessment and monitoring results, but can also be used to establish priorities in other important areas, such as exposure control, communication, and training of employees.

Preliminary assessment consists of estimating the nature and degree of exposure, combining considerations of type and extent of health effects, and finally deriving a sound judgment of the overall related risk. The degree of exposure is estimated by reviewing the information collected during basic characterization with one or more of the following tools: previous monitoring data, experience from similar conditions, exposure modeling techniques, semi-quantitative field monitoring, and spot measurements. Qualitative ratings or indices can be established for typical exposure categories for tasks performed by homogeneous exposure groups: 0 = none or negligible, 1 = low and infrequent, 2 = moderate and infrequent, 3 = high and frequent.

Similarly, toxicological properties, related health effects, OELs, or other parameters can be used to qualitatively classify the hazards of agents: 0 = minimal/low effects, 1 = moderate/reversible, 2 = moderate/high, 3 = very high/irreversible. Lack of OELs does not prevent assessment but makes it more complicated. A useful approach is to assign to the agent a more stringent priority category where doubts exist.

By combining the degree of exposure ranking and the hazard classification of agents, it is possible to assign an overall qualitative risk index and prioritization to the various homogeneous exposure groups or working areas. In most modern workplaces, which are characterized by an enclosed and fully automated process, the assessment normally indicates negligible risk and low OH priority; few are normally judged unhealthy and require control (Fig. 2); some others cannot be properly evaluated with a qualitative method and require further, quantitative assessment. The above rating refers to routine, normal operating, or planned conditions and cannot be used for emergency and unplanned operations.

3.6.4. Occupational Exposure Limits

Occupational exposure limits (OELs) are the most traditional and widely used criteria to assess the acceptability of workplace conditions. The various types of OELs for chemical and physical agents are established by means of different approaches, purposes, definitions, and updating systems by various national and international regulatory bodies. Most are guidelines for evaluating health hazards and assisting in other preventive activities (adequacy of control, design standard, selection of PPE, etc.). Others have legal status and serve for demonstration of compliance. OELs are based on current knowledge of toxicological and health effects of the substance on the working population, and on the best judgment of health scientists and occupational health professionals.

In general, an OEL for a chemical substance refers to the airborne concentration to which most workers can be repeatedly exposed by inhalation for a full working day and working lifetime without adverse short- or long-term health effects. However, OELs do not guarantee complete protection of all workers and cannot be used as an absolute dividing line between safe and harmful concentrations, but only as a guideline.

OELs have a long tradition worldwide, and the first recommendations were published in Germany by LEHMANN in 1889. A systematic development of OELs started in the United States in the mid-1940s with the ACGIH-TLV system which has remained the most widely used and highly reputed system to today. Germany started an independent MAK system in 1965, and after this other nations started their own national schemes (Sweden, United Kingdom, France, The Netherlands, Denmark). In 1977 the International Labour Office (ILO) in Geneva published its "air quality limits" for the workplace, followed in 1980 by the WHO with "safe limits" for occupational exposure. The ILO and WHO OELs are health-based limits established solely by considerations of the health of the working population. Most other OELs, especially those with regulatory compliance status, incorporate other technical considerations (technical feasibility, analytical methods) as well as socioeconomic factors.

In light of the differences that exist between OELs from country to country in Europe, which is also influencing the principle of free movement of labor, the EC Commission began in the 1980s to tackle the problem by offering a scheme to harmonize OELs in the Member States and to coordinate the activities of the national bodies. More than 1500 of the most important chemical substances now have an official OEL established by combining the activities of the various agencies around the world, and many hundreds more are set internally by international enterprises. Although these limits are relevant for many key occupational health concerns, the very large number of chemicals on the market (100 000) represents a complex challenge for both authorities and companies.

When no OEL has been established for an existing chemical (particularly common in the case of process intermediates) or for a novel substance, assessment becomes more complicated, and assistance from other health professionals (occupational physicians and toxicologists) is necessary. A program to obtain a minimum set of toxicological data with the aim of establishing a preliminary OEL or internal guideline is required.

Different categories of OELs are established to take into account the toxicological properties and related health effects of the substances. The most common category is the TWA concentration for a defined period of work (normally 8 h). A defined number of excursions above the TWA are allowed per shift, taking into account the short-term effects of the substance (Short-Term Exposure Limits, STELs). For certain substances, where exposure can result in acute health effects, there is a ceiling limit that describes the concentration beyond which workers should not be exposed under any circumstances.

For a limited number of chemical substances, a biological exposure limit has been established. This provides criteria for carring out and evaluating a biological monitoring program which assesses the overall internal exposure to the chemical by analysis of biological samples. It should complement the traditional OH airmonitoring program.

3.6.5. Quantitative Assessment – Occupational Hygiene Monitoring

The preliminary assessment step indicates, in order of priority, which exposure conditions, related agents, groups of workers, and homogeneous job groups need to be assessed by an appropriate monitoring program.

The objective of this section (see also Fig. 2) is to analyze the elements and provide guidance on how to plan, conduct, document, and maintain records of an investigation evaluating exposure of workers to chemicals.

The technique for evaluating exposure in the workplace has changed in many ways over the years. Work situations in which workers are clearly and grossly exposed have declined; in most cases, the levels of exposure are quite low. A quantitative assessment of very low levels of complex chemical mixtures is a challenge requiring professionalism, experience, and a carefully thought out working model. Before the start of any monitoring, the preparation of written plans for appropriate monitoring (protocol) is the first step to assure the required level of quality, reliability, and effectiveness of the entire program.

3.6.6. Monitoring Planning and Strategy Definition

There are many reasons for conducting an OH monitoring program besides evaluating personnel exposure: to demonstrate regulatory compliance, verification of applied control, selection of PPE, continuous air monitoring for leak detection and air emissions, etc. Each requires a different strategy. The first elements of a monitoring protocol should state clearly the purpose of the program and define the most appropriate strategy to accomplish these objectives. A detailed and in-depth review of information and data collected during the preliminary assessment, specifically related to the agents and workplace prioritized for the monitoring, is then required. Defining a monitoring strategy simply means answering the basic questions what, when, how, how long, and how frequently sampling should be carried out.

Typical workplaces are complex, and concentrations of environmental agents vary temporally and spatially within and between workdays. Workers move through these variations in concentration in variable patterns, and they also influence the concentration and the degree of workplace contamination by the tasks that they perform. The exposure routes by which workers can absorb a substance must be carefully studied in relation to its physical, chemical and toxicological properties. Inhalation is usually the primary mode of exposure, and this depends on the vapor pressure of the substance, temperature and pressure of the process, etc. Absorption via skin contact can also be significant, especially if the substance readily penetrates the skin in toxic amounts. Exposure via ingestion is generally negligible, but it is necessary to watch for potential intake via food, beverage consumption, and smoking.

The toxicological properties of the substance and its potential health effects should be reviewed in detail: how it exerts its acute properties; the type of chronic toxicity; the target organs; the carcinogenic, reproductive, and neurotoxic potential; the reversibility or irreversibility of the effects; and the biological half-life, etc. A careful review of the documentation of the OEL of the substance is necessary in order to understand the rationale and the safety factors adopted.

There are two major categories of sampling strategy: traditional and statistical. In practice there is a third category that results from specific regulatory requirements, where approach, methods, procedure, etc. are indicated by the legislation. The traditional diagnostic approach is based on past experience and professional judgment of the industrial hygienist. It is normally an effective combination of the most common types of sampling: personal, task, and stationary sampling. TWA personal sampling is performed by taking samples with miniaturized devices in the breathing zone of a worker for the entire shift (ca. 8 h). This is the preferred and most effective technique to evaluate exposure.

Task-related sampling is carried out, for example, at the workers' breathing zone during potentially critical operations and tasks. Stationary sampling, carried out in strategically fixed locations of the process, is generally useful in identifying the source of exposure and can be used to estimate worker exposure if other sampling is impractical. Continuous air monitors are generally sophisticated stationary sampling systems equipped with alarms, which are installed to measure one or more chemical agents in various selected parts of the process area. They can be used as leak detectors and for monitoring background concentrations.

There are various statistical approaches reported in the literature and specifically recommended for regulatory compliance. The selection and application of the most appropriate and powerful statistical model, backed up by sound professional judgment, can result in enhancement of the overall quality and credibility of the program.

Frequency of sampling is dictated by statistical considerations, and largely depends on the variability of the conditions being sampled. The more variable the conditions are, the more samples are required. However, 5 – 8 sampling repetitions is a good rule of thumb, valid for most conditions. Different shifts should be sampled for the same homogeneous exposure group. Work load, seasonal, and evironmental conditions should be considered. The conditions during sampling should be representative of the conditions that one intends to evaluate.

Estimation of potential for skin contact can be made by measuring surface or item contamination by wipe-tests. Choosing an adequate analytical method to match the selected sampling strategy is the next step, and often requires help from analytical chemists.

The final important elements of a monitoring protocol include a careful timetable and cost estimate for the entire program. Extensive monitoring can become extremely expensive and may not be necessary and/or add very little valuable information to the assessment. A proper balance between needs and scope of monitoring and cost should be the goal.

3.6.7. Sampling and Analysis; Data Quality Assurance

Sampling equipment and analytical techniques for air monitoring have undergone marked changes over the past decades, with improved detection sensitivity, miniaturization, and automation.

For the most important chemical agents, specific and reference methods are published by the National Institute of Occupational Safety and Health (NIOSH), and DFG (Deutsche Forschungsgemeinschaft). These methods generally describe sampling, analytical procedure, and instrumentation and specify the important parameters that have been determined in laboratory and field tests to satisfy good practice for trace monitoring. Generally the range of applicability, stability, recovery, effect of humidity, interferences, limits of detection, accuracy, and precision are indicated. When a reference-validated method is available and its field of application satisfies the needs of the planned monitoring strategy, the normal quality assurance tests, such as instrument calibration, blanks, and known standard samples should be run to verify that the sampling and analytical conditions chosen correspond to those of the validated method. If no method is available, a plan for its development should be designed in cooperation with analytical chemists. Air monitoring techniques, which are highly developed because OELs are generally based on inhalation exposure, can be conducted by:

1) Collecting contaminated air in glass bulbs, plastic bags, etc. for later analysis
2) Instruments and techniques that directly measure the pollutants in air
3) Removal of the chemical agent from the air for later analysis

Direct-measurement techniques range from relatively simple colorimetric detectors to highly sophisticated portable infrared detectors or gas chromatographs (GC). They are extremely useful as preventive tools when equipped with alarm systems, or in diagnostic evaluation.

The most widely used techniques fall into the third category, where a volume of air is drawn by a pump (active system) or diffusion (passive) into an appropriate collecting medium, either solid (e.g., charcoal tubes) or liquid (e.g., solvent). This is generally used for vapors and gases, which are then extracted and analyzed. Similarly, filters or membranes are used to remove airborne particulate matter (dust and aerosols) from pumped contaminated air. When a specific fraction of the particles (e.g., inhalable fraction) is to be measured, an appropriate size-selecting device is placed before the membrane.

Common analytical techniques include GC, Atomic Absorption (AA), HPLC, and GC – MS. Classification and counting under the phase-contrast microscope is carried out for airborne fibers (e.g., asbestos).

3.6.8. Sampling and Analysis Execution

Good communication and arrangements should exist with the unit and analytical lab supervision prior to the sampling and analytical execution in order to take into account production schedule changes, etc. This is particularly important when monitoring specific and infrequent tasks. Sampling preparation often requires calibration of sampling devices and instrumentation, and labelling of tubes, badges, vials, and bags.

During sampling execution, it is advisable to use predefined sampling record forms, in which the key information and observations noted during sampling are written down (humidity, setup conditions, use or not of personal protective equipment, etc.). Handling, temporary storage, and transportation of collected sample to the laboratory should be done according to planned procedure to minimize errors and prevent invalidation of results.

3.6.9. Data Interpretation and Reporting; Record Keeping

Once the analytical results have been converted into appropriate units (e.g., ppm), they must be grouped into meaningful categories (TWA personal data of each homogeneous exposure group, the STEL results of similar operations, etc.) and then studied by adding the information noted during sampling. Appropriate statistical tools are then used to describe the central tendency of samples (e.g., arithmetic mean) and their distribution (range, geometric standard deviation, etc.). Ultimately, the best estimate of exposure is compared with the OEL to determine the acceptability of the assessed exposure conditions.

Another possible interpretation is that it is not possible to make a decision with the data available, and further monitoring is necessary. If the sampling strategy includes a specific statistical analysis, additional and more precise conclusions can be drawn (e.g., chance of exposure above the OEL).

The main elements of a quantitative assessment should be documented in an appropriate survey report. This should include purpose, information on the process/unit and homogeneous exposure groups, OELs and criteria of evaluation, sampling and analysis, data quality assurance, tables of results, conclusions, and recommendations.

Because much of the information is already available in the monitoring protocol, proper use should be made of existing documentation and/or references to avoid duplication of efforts. The recommendations, if related to exposure control methods, should be made in consultation with management and workers' representatives if necessary.

The OH survey report is a document that must be completed in accordance with the retention requirements of the company and/or the general legislative retention time of 30 – 40 years. Other OH documents that should be kept are: chemical and physical inventories with qualitative assessment; work histories; homogeneous exposure group and other OH documents that, together with health surveillance and biological monitoring data, can serve future epidemiologial or legal needs.

3.7. Exposure Control Methods

3.7.1. General Concepts

The most important function of a sound occupational health program is controlling and reducing exposure to hazardous agents to acceptable levels as part of risk management. The entire function consists of the selection and application of the most appropriate control measure and subsequent verification of its effectiveness.

Even if he is not directly responsible, the specific knowledge and experience of an occupational hygienist, particularly in certain control methods [e.g., industrial ventilation, personal protective equipment (PPE)], can contribute significantly to the success of the entire process. When conclusions drawn from qualitative and quantitative assessments (see Fig. 2) indicate that reduction of exposure is necessary, a number of factors should be considered by management in selecting the most appropriate method. Such factors should include urgency, cost-effectiveness, reliability, maintenance, and comfort of employees. Methods of exposure control can be divided into different categories. The order in which these categories are listed below corresponds to the hierarchical order of the risk manager's options; he should make his choice in line with good OH practice and in the spirit of current legislation.

1) Substitution (e.g., replacement of a toxic substance)
2) Containment (e.g., total enclosure)
3) Removal at source (e.g., local exhaust ventilation)
4) Administrative/procedural (e.g., reduced length of exposure)
5) Personal protective equipment (e.g., dust mask)

The above approach is based on the fact that elimination of exposure should be considered first, while PPE should be adopted as the last line of defense when it is neither possible nor practical to use any of the other methods. Using a material with better toxicological properties or different emission characteristics is clearly the best method of controlling exposure, although this is generally quite difficult because it requires extensive research and investment.

Control by containment means enclosing the process so that the characteristic emission of the substance cannot reach the worker. An automated feed system in processes and glove boxes in labs are typical examples. Removal at source normally implies the use of a local exhaust ventilation (LEV). The principles behind control of exposure by LEV are as follows:

1) Design the maximum degree of enclosure around the operation that still allows the work to be done efficiently. Decide on the necessary capture velocity to entrain emissions at the remaining enclosure opening (e.g., 0.5 m/s for vapors released into still air).
2) Calculate the necessary volume flow rate to achieve the required capture velocity.
3) Design, install, test, and maintain the system.

3.7.2. Personal Protective Equipment

Where it is not feasible or practical to adopt one of the other methods, including procedural measures, personal protective equipment (PPE) must be considered. Management has the responsibility to ensure that correct PPE is available and used, while workers must ensure its proper use. PPE includes a variety of devices for head, eyes, face, hands, ears, body, feet and lungs, providing protection against a wide variety of workplace health and safety hazards (chemical, physical, mechanical, biological, electrical, falls, fire, etc.). A sound program for the proper use of PPE taking into account its limitations, should be established and should address in general the following points:

1) Identification of health and safety conditions and hazards

2) Selection of the most appropriate PPE types
3) Procedures for use, storage, cleaning, maintenance, and disposal
4) Emergency procedure and use of PPE
5) Medical fitness and physical stresses
6) Education, training, and retraining of wearers
7) Inspections
8) Documentation

The Commission of the European Community (CEC) has promulgated a directive (89/656/EEC) which lays down minimum requirements for PPE used by workers at work. This directive, together with related CEC brochures and, above all, the standardization of technical specifications for PPE in CEN standards will harmonize manufacturing, selection and proper use of PPE in the EU countries, which up to now was regulated by national standardization schemes.

The safety officer normally administers the entire program. However, OH skills and knowledge are particularly significant when adequate respiratory protection and skin protection from chemicals are to be assured during routine tasks, as well as during special emergency operations.

The definition of the most suitable types of respirators derive from the analysis of the hazards and the assessment of risks to personnel, and then comparing merits and limitations of the various types and classes available (air-purifying respiration, air-supplying respirators, tight or loose fitting face piece, negative or positive pressure mode, escape only, etc.)

Consideration of the nominal protection factor (PF) assigned to a certain class of respirator, the required level of protection, the specific face-fit of the wearer, etc. are very important for an appropriate selection and are vital in critical conditions when personnel with facial hair may also be involved.

The protection factor is the ratio of ambient concentration outside the respirator to that inside the respirator facepiece. Multiplying the PF by the OEL of the substance under consideration gives the maximum use concentration.

The selection of adequate protective clothing as a barrier against chemical penetration also requires specialized knowledge. Interpretation and judgment of extensive and often conflicting data from suppliers and literature on chemical permeation testing is needed, together with considerations of comfort and hand sensitivity.

3.8. Physical Agents

Various physical agents may be encountered in the workplace. Some, such as noise and vibrations, are very diffuse and affect many workers; others, such as compression and decompression, are typical of work environment where the atmosphere is not at normal pressure and temperature (e.g., divers).

Other physical agents include:

1) Stress due to heat and cold.
2) Sources of intense optical radiation, such as ultraviolet and infrared lights and lasers, which may produce eye and skin damage.
3) Electromagnetic fields, for which certain chronic health effects are still under investigation.
4) Ionizing radiation, typical of the nuclear industry and in various other applications (e.g., medicine, gauging equipment). Exposure to ionizing radiation causes damage to living cells.

Each physical agent has its own characteristics, mode of entry, health effects, etc., and therefore requires a specific approach and technology if it is to be assessed and controlled in the workplace. Specialized literature, international guidelines, and legislative requirements for physical agents have been established in some countries.

Occupational exposure to *noise* is by far the most diffuse, most widely studied, and has the greatest impact of all physical agents. It has been estimated that $> 13\times10^6$ workers in the EU states are exposed to excessive noise for most of their working time. Although for many years noise at work has been known to cause hearing loss, it was only during the last three decades that major scientific studies have allowed full understanding and quantification of relationship between noise exposure and risk to the human auditory system.

Noise, or an unwanted sound, is described as oscillations in pressure above and below atmospheric pressure which evoke an auditory sensation. Prolonged exposure to sufficiently high levels of noise can produce permanent damage to the hearing mechanism. Other nonauditory effects (e.g., physiological) can be attributed to noise exposure at lower levels. Such effects are less well-known and also occur in many human leisure activities. The level of exposure to noise for which there is international consensus that the risk of hearing damage is limited and at which most legislations set the limit for 8 h of exposure is 85 dB (A) as a time weighted average. Impulse noise should also not exceed 120 dB at any time.

Assessment of noise exposure is the first important element of a complete hearing conservation program. The other integral steps are: technical measures and abatement techniques; administrative measures (reducing time of exposure, area delimitation, noise specifications, etc.); selection and use of hearing protection (various types of ear muffs and ear plugs); education or training of personnel; and preliminary and routine audiometric testing to check the degree of hearing impairment.

There are two basic methods of assessing noise exposure. The first consists of fixed-position measuring of noisy areas or equipment by a sound level meter which can also measure impulse noise, and calculating the total noise exposure, taking into account the time spent by the worker in each measured area. In the second method, the worker has a personal audiometer, capable of continuously monitoring noise levels throughout the working shift. The use of both techniques at appropriate intervals can be considered the most effective approach for assessing noise exposure.

3.9. Communication and Training

The final essential element of an OH program is providing information, training, and education in the field of IH and hazard communication for personnel at all levels. This is a basic requirement of all health and safety legislation. Regular communication of accomplishments, performance, new relevant information, issues, concerns, and regulatory development to management is needed to maintain high awareness and support.

Potentially more exposed personnel who operate equipment and handle materials can contribute with personal control and attention in minimizing exposure if they understand the nature of the hazard and how work procedure affects emission and contamination.

OH training is normally done in cooperation with the supervisor and other departments (e.g., safety and medical) as part of the initial routine health and safety training. Topics that should be included are:

1) Physical, toxicological, and warning properties of substances and their possible consequences and health effects upon exposures
2) Ergonomic factors and recommendations
3) Results and interpretation of occupational exposure assessment
4) The proper use and limitations of PPE

Many specific sources of such information are available at international and national level. The most appropriate approach and method to provide such training should be designed according to the background and level of education of the people it is intended for.

Verification of the effectiveness of this program, for example, by means of a questionnaire, should be considered. Training records should be made according to legislative requirements and practices.

3.10. Industrial Hygiene in the United States

Although some education programs existed previously, the first industrial hygiene section was organized by the Public Health Association in 1914.

In 1915 the U.S. Public Health Service organized a division of Industrial Hygiene and Sanitation. The *Journal of Industrial Hygiene* was first established in 1919 and in 1925 became the *American Industrial Hygiene Association Journal*. The present Harvard School of Public Health had a Department of Industrial Hygiene in 1922. Chemists, engineers, physicians, toxicologists, and other specialized graduates have applied their knowledge to the growing scientific profession of industrial hygiene. The American Hygiene Association was organized in 1939. Since then improvements in plant design, machine design, and plant layout have seen the worker come to expect a safe workplace, in part because of the activities of the industrial hygienist.

With progressing industrial and workplace safety, today's industrial hygienist is still involved with control of dust hazards, skin irritants, noise levels, personal protective equipment, classification of hazards, and training. However, with additional data, better monitoring methods, and more precise measurements, chronic occupational problems are now of greater importance. Information has been developed on chemical exposure limits in workplace air by several organizations such as the National Institute for Occupational Safety and Health (NIOSH) with recommended exposure limits (RELs). The Occupational Safety and Health Administration has proposed permissible exposure limits (PELs) for several hundred chemicals. If accepted by the legal process, these could become regulatory standards. As with all OSHA standards, the PELs are being strongly challenged in the courts.

The American Conference of Government Industrial Hygienists (ACGIH) have developed guidelines or recommendations called threshold limit values (TLVs). These recommendations and guidelines all assist the industrial hygienist with the control of potential health hazards. In addition, for different working conditions, these values are expressed as time weighted averages (TWAs) and as threshold limit values- ceiling (TLV-C) for exposures not to be exceeded even instantaneously.

All of these guides and values represent levels to which nearly all workers can be exposed without adverse health effects. With the help of data automation, the industrial hygienist sets up sampling monitoring and analytical programs to measure and correct exposure conditions. Physical agents such as temperature, stress, radiation, vibration, and pressure variations are more recent areas of study where practicing industrial hygienists can continue their strategy of occupational safety.

Today's industrial hygienist must draw on the knowledge of a broad range of disciplines in the medical, biological, and technological fields to prevent work strain from reducing performance or becoming pathological. The primary objective is to provide a workplace free of recognized hazards.

4. Occupational Medicine

4.1. Health Surveillance [29–32]

One of the main goals of occupational health care is to protect workers from harmful effects due to work or the working environment.

The various means that can be used to achieve this goal can be summarized under the heading of health surveillance.

In general terms health surveillance comprises four sequential steps:

1) Monitoring/measuring
2) Evalution
3) Interpretation
4) Corrective actions

The first step in health surveillance consists of measuring parameters of exposure and uptake or effects. Secondly, the results must be evaluated and interpreted with respect to the question of whether any health impairment exists or is to be expected.

When it is concluded from the monitoring results that inadmissible exposure or uptake is occurring, corrective measures should be taken to prevent further health risks. Which measures have to be taken depends on the particular situation and is decided by a case-to- case approach.

Health surveillance in the framework of occupational health care may be aimed at:

1) Specific occupational branches, such as health care services, construction and building industry, chemical industry
2) Special occupational cohorts, such as busdrivers, railway personnel, operators in the process industry
3) Particular occupational diseases, such as hepatitis in nurses and skin conditions in bricklayers
4) Occupational health problems due to exposure to particular biological, physical, or chemical agents

The discussion in this chapter is restricted to examinations dealing with potential hazards from occupational exposure to chemicals.

4.1.1. Monitoring

Monitoring is "the repetitive and continued observation, measurement and evaluation of health and/or environmental or technical data for defined purposes, according to pre-arranged schedules in space and time and using comparable methodologies for sensing and data collection" [30]. This description is still valid.

The most essential criteria for reliable monitoring procedures are:

1) The activity should serve a defined objective
2) Continuous or repeated observation
3) Comparison of the data observed with a reference level [31]

Monitoring implies a multidisciplinary activity in testing, watching, and observing situations and/or persons. In the occupational setting this means examinations of the working environment and/or the workers.

Which mode of monitoring will be preferred depends primarily on the goal of the examination and a number of factors, which are discussed below.

Monitoring Strategy. Before any health surveillance action is started, a decision has to be made about the nature and design of a prearranged schedule, adopted to meet a defined objective. This decision depends mainly on the answers to a number of questions to be posed, certain criteria to be met, and the availability of appropriate monitoring methods.

The most important and crucial questions deal with the aims and means of a health surveillance program:

Why carry out an examination? What are the objectives and what do we want to know? Is exposure of workers actually taking place?

The answer to these key questions generally implies the answers to the next questions.

Who must be included in the surveillance program? What is the actual population at risk? Which criteria are valid for assessing this risk group?

What will be measured and monitored?

The answer to this question presupposes some knowledge about the nature of the exposure.

What is the most suitable way and the most relevant place to monitor and which parameters need to be measured?

The answer to this question presupposes fairly good knowledge of the biological fate of a chemical in the human body.

Which sampling procedures and analytical methods are available and suitable in a particular situation?

When and how often must monitoring be performed?

In some countries detailed guidelines or strict rules are given for the methods and frequency with which health surveillance is to be performed. In Germany this is regulated by the "Berufsgenossenschaftliche Grundsätze für Arbeitsmedizinische Vorsorgeuntersuchungen" [31]. Guidelines are given in the United Kingdom by the Health and Safety Executive (HSE) and in the United States by the Occupational Safety and Health Administration (OSHA).

Monitoring Criteria. Apart from the above mentioned questions, some criteria must be met for proper design of a health surveillance program.

In 1975 WHO issued criteria for periodical medical examination that are still valid. The most important of these criteria are:

1) It should not involve undue expenditure of time, equipment, or staff.
   In other words, the cost and discomfort for the individual and for society should be in balance with the health benefits to be expected.
2) It should not be inconvenient to workers.
   A liver biopsy might be considered to be the best diagnostic means in a particular situation. But, as this operation could be fatal, one is very reluctant to perform it for the medical surveillance of healthy people.
   This criterion can also play a role in the choice of biological medium. Thus taking a blood sample or collecting a 24-h urine sample could be experienced as difficult and inconvenient.
3) It should be an early reversible predictor before overt health effects are impending. Surveillance activities which can be classified under biological monitoring and biological effect monitoring meet this criterion.
4) There should be a quantitative relationship with exposure and with health risk.

For most effects caused by chemicals, a dose – response relationship exists. Measuring a particular parameter is only significant when this dose – response relation and/or dose – effect relationship can be expressed numerically. Preferably, the effect measured should have a steep dose – response curve which enables the intensity of effects to be more easily distinguished.

For proper evaluation and interpretation, valid reference data are essential.

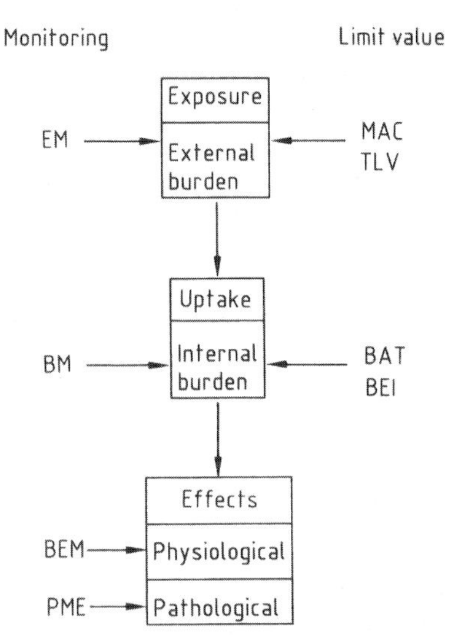

Figure 3. Instruments for health evaluations of workers
EM = Environmental monitoring; BM = Biological monitoring; BEM = Biological effect monitoring; PME = Periodical medical examination
MAC = Maximum allowed concentration; TLV = Threshold limit value; BAT = Biologische Arbeitsstofftoleranz; BEI = Biological exposure indices

A very important criterion is that the tests should have high validity and effectivity.

Thus the aim is the highest possible specificity and sensitivity. Low sensitivity leads to an unsatisfactory number of false negatives (positives that wrongly score negative). The higher the specificity of a test is, the lower is the number of false-positives.

An essential condition for performing meaningful health surveillance programs for workers is that the management responsible is prepared to improve working condition if this is indicated. Also employees must be willing to improve their working habits and hygiene when necessary. If these requirements are not met, any health surveillance program is completely redundant and a waste of time for everyone involved.

Modes of Monitoring. The following modes of health surveillance can be distinguished:

Estimation of the health risk:
  Environmental monitoring (EM),
  Biological monitoring (BM),
Estimation of the state of health:
  Biological effect monitoring (BEM)
  Periodical medical examinations (PME).

In Figure 3 a scheme is given of these "instruments" for health evaluations.

These different monitoring methods can be exemplified on the basis of lead exposure.

When workers are exposed to lead fumes or dust, the following can be investigated:

1) The concentration of the fumes or dust in the working area (environmental monitoring, EM).
2) The content of lead in blood or urine of the exposed workers (biological monitoring, BM).
3) Indications of biological changes in blood production by measuring the content of δ-aminolevulinic acid dehydrogenase (δ-ALAD) or zinc protoporphyrine (ZPP) in the blood, or δ-aminolevulinic acid in the urine. This method is known as biological effect monitoring (BEM).
4) Indications of impaired health by physical medical examination of symptoms and signs of specific adverse health effects: anemia, lead line on the gums, abdominal cramps, obstipation, wrist drop, impaired renal function, and blood changes.

Environmental monitoring (EM) in this context means measuring exposures in the air at the workplace. These measurements are generally specific for a defined chemical. EM is used to determine the intensity of exposure, expressed as the concentration of a single chemical or group of chemically related compounds in the working environment.

However, exposure to a chemical or group of chemicals, does not necessarily mean that uptake into the human body will happen or has already taken place. Whether, in what way, and to what extent uptake results from exposure depends on a number of factors.

Air sampling close to the mouth and nose of an individual gives the best reflection of the actual nature and intensity of the chemicals inhaled during work. This type of EM is called personal air sampling (see Section 3.6.6).

In biological monitoring (BM) a chemical compound or one of its metabolites is determined in a biological medium, mostly urine or blood. BM can be regarded as a person-related, specific measurement of so-called biomarkers, which reflect the internal burden of a compound in the human body, irrespective of the route of entry.

Therefore, BM is a relevant method of health surveillance when individual screening is the goal. However, when it is intended to check or to control the working environment, environmental monitoring is the first method of choice. BM is discussed in more detail in Section 4.2.

4.1.2. Periodical Medical Examination

Periodical medical examination (PME) is the periodically repeated medical examination of individuals for health effects that may be related with their work or working circumstances. The main aim of these examinations is to detect early, reversible (pre-)clinical signs of health impairment.

PME is individual-related and aimed at the detection of nonspecific anatomical and/or functional disturbances of organs or tissues, irrespective of the cause.

Besides occupation, a number of intrinsic and nonoccupational extrinsic factors, including eating, drinking and smoking habits, may influence the nature and intensity of health effects. This means that in most cases it only can be estimated or inferred indirectly from occupational histories that a specific chemical is the causative agent.

A periodical medical examination implies the drafting of a medical history, a physical examination and, when indicated, additional tests.

Elements of a periodical medical examination are:

Medical History
- current state of health
- previous ailments
- medication
- life style

Physical examination
Additional examinations
- blood or urine tests
- lung function
- neurological/electrophysiological
- chromosomes
- sperm

Example 1. The presence in blood of a particular quantity of carbon monoxide bound to hemoglobin (CO-Hb), may be caused by the exposure and uptake of carbon monoxide itself, but can also be due to exposure and uptake of the solvent dichloromethane. Heavy smoking has to be considered as a confounding factor.

Example 2. A large number of pesticides, chiefly organophosphates, act in the body by blocking the enzyme acetylcholinesterase, which regulates nerve – muscle action by hydrolyzing acetylcholine; thus, excessive amounts of this muscle stimulator accumulate. However, the clincal symptoms of inhibition of this enzyme do not indicate which particular acetylcholinesterase inhibitor is the causal agent.

Example 3. Headache may be caused by a large number of etiological factors, unrelated to any occupational activity. But even when a relationship with occupation cannot be ruled out, it is often difficult to determine whether this symptom is caused by the exposure and uptake of a neurotoxic organic solvent or is due to mental stress at work.

Example 4. A disorder of liver function may be of bacterial, viral, or chemical origin. But even when it is obvious that it has been induced by a chemical agent, it may have been the result of a lifelong excessive drinking of alcoholic beverages during leisure hours, the use of particular pharmaceuticals, or exposure to and uptake of chlorinated hydrocarbons during work.

In the early days periodical medical examination of individual workers consisted almost entirely of detection of clinically apparent symptoms. Nowadays, increasingly subtle diagnostic methods are used, enabling the physician to detect early reversible effects which may be considered to be not yet adverse or pathological.

A number of early effects can be regarded as forming part of the physiological defense mechanism of the human organism. ZIELHUIS and HENDERSON [32] use the term biological effect monitoring for this kind of examination and defined it as: "the measurement and assessment of early biological effects, irrespective of their significance to health, in exposed workers and health risk compared to an appropriate reference."

However, the line between physiological and pathological effects is often hard to draw because the distinction may be indistinct.

For some effects it is still unclear whether they should be considered adverse or harmful, or neither. Examples of effects that are still under discussion between experts are enzyme induction (see Section 4.5) and some electrophysiological phenomena in nerves or muscles.

Apart from whether a particular effect can be regarded as a biological or pathological response of the human organism, there is another crucial criterion for deciding whether a specified effect is tolerable or not. This criterion is the reversibility of an effect after exposure to the causative chemical agent ceases.

Sometimes it is a matter of academic or political dispute whether a particular person-related measurement is called BM or BEM. Although this does not change measurement method, it can have a major impact on the establishment of biological limit values and the setting of standards.

For example, the determination of carbon monoxide – hemoglobin complexes in the peripheral blood may be regarded as an indication of the total body burden of CO (which is BM), but also as a parameter of BEM because it reflects the proportion of hemoglobin that is blocked and not available for oxygen transport from the lungs to the tissues.

4.1.3. Evaluation and Interpretation of Data

The most important step in the entire health surveillance (HS) process is the evaluation and interpretation of the test results.

For interpretation of the monitoring results, the data must be compared with reference data. Therefore, it should be clear before starting any health surveillance program which references can and should be used.

The availability of valid reference data is an inherent part of the considerations during the structuring of a surveillance scheme.

Evaluation. Before the test results of a health surveillance examination can be interpreted properly, it should be determined whether the collected data are appropriate and reliable with respect to health.

It is essential that relevant and proper methods be used with appropriate sensitivity and specificity, and that they be performed with correct sampling, storage, and analysis. For this reason some method of quality control for accuracy and precision should be incorporated [e.g., good laboratory practice (GLP) or round robin tests].

For intra- and interindividual comparison with previous test results, adjustment of data may be indicated. Thus, data from urine samples have to be adjusted for specific gravity or creatinine excretion.

Another consideration is whether – and if so which – intrinsic and extrinsic factors could have confounded the results. These factors can include smoking, drinking, diet, medication, and others.

For correct evaluation and interpretation proper choice of reference data is essential. The reference data used depend on the objectives of the HS program and the conclusions to be achieved: Do we have to make comparisons with figures for the general national or local population? Will we draw conclusions for groups of workers or individuals? Do we want a cross-sectional impression of the current overall health state of a group of workers or a longitudinal view of the progression of specific parameters throughout the years for individuals or groups of workers? In longitudinal studies, the current data can be compared with results from previous examinations. For individual workers, reference data from the pre-employment examination may be of value. When no reliable reference data are available, the best estimate should be made from epidemiological and animal studies (see also Section 4.5).

Reference data for physical (pre-)clinical symptoms and signs of impaired health depend on the knowledge and skills of the physician who performed the periodical medical examinations. Data from laboratory tests for estimating the state of health depend on the methods used and the specificity and their sensitivity.

As HS is meant to determine preclinical reversible signs in healthy people before these must be considered adverse, in general the laboratory tests have to be more sensitive and the reference data more subtle than those used in hospital for patients with impaired health.

For a number of chemicals, occupational threshold limit values for exposure are set. For a limited number of chemicals biological limit values for uptake have also been established. In the United States biological exposure indices (BEI) are used, and in Germany Biologische Arbeitsstofftoleranzwerte (BAT values). For details, see Sections 3.6 and 4.2.

To ensure that all essentials are considered, the use of a checklist for the evaluation of health surveillance data is helpful:

Methods and sampling
  appropriate (sensitivity, specificity) ?
  correctly performed?
Adjustment for correct interpretation
Confounding factors?
  intrinsic (age, enzyme deficiencies, etc.)
  extrinsic (life-style, pharmaceuticals, etc.)
Reference data available?
  individual
  group
  general population
Occupational limit values available?
  guideline/advisory
  legal/obligatory

Interpretation. Once the data have been evaluated it has to be established whether the HS results indicate impairment of health.

Universally fixed "normal" values do not exist, because geographical, ethnic, age, and gender differences may be apparent. Even between individuals within a quite homogeneous group of people and, even within the same individual, fluctuations can be found, due to the range of physiological variability. Differences in basic body burden of individuals may be due to extrinsic factors such as life-style (eating, smoking, etc.) and leisure activities.

With respect to pre-clinical reversible biological effects, it has to be considered whether these are medically acceptable or not. This is completely a matter of professional knowledge and skills. The final decision however, on what is tolerable, is not only a medical but also a political issue and will include a balance between technical and economical feasiblity and acceptability by society. This can differ from country to country and between social groups.

4.1.4. Reporting Health Surveillance Results

After completion of the health surveillance program, it has to be decided to whom, when, and which information about the results will be given.

The workers involved as well as the responsible management must be provided with the relevant information.

With respect to BM and BEM data, it should be emphasized that these monitoring methods are person-related. This implies that individual workers have the right to be informed of their own personal test results. To enable the worker to put these data into perspective and to understand the uses and limitations of BM, some basic information on the dose – effect relation, reference data, etc., should be provided as well. The individual data must be considered as confidential and only accessible to the worker and physician involved. They are only to be provided to workers' representives or management with the explicit consent of the individual worker.

The employer is given an anonymized compilation of the group data, to allow him to judge the working conditions and the effectiveness of preventive measures already taken.

4.1.5. Corrective Actions

The results of health surveillance indicate whether corrective preventive actions are necessary or advisable.

These actions may be restricted to individuals in a group who showed aberrant and unwanted signs of chemical overload or a health threat, compared with the group mean and/or the average in a group of nonexposed people.

It must be investigated why these individuals differ from the others: are they at risk due to higher exposure or uptake, or are they subject to individual hypersensitivity?

Individuals with abnormal HS results should be temporarily withdrawn from their work until the figures are back to normal. Sometimes, permanent withdrawal from specific work conditions has to be considered; for instance, after (pre-)clinical signs or symptoms of asthma due to isocyanate exposure or when skin conditions due to a susceptibility for sensitizing agents have been diagnosed.

Moreover, certain medical conditions unrelated to work may be incompatible with certain job activities. Restrictive and obstructive lung diseases prohibit exposure to irritating gases, vapors, and dusts. Exposure to a number of chemicals is contraindicated for workers with functional liver and kidney disorders.

When the group mean of the examined workers appears too high compared with nonexposed people, or higher than the group mean in a previous examination, measures at the job or working environment level should be considered.

A critical survey of current work systems and practises can often lead to improvements which are simple and cheap to achieve and lead to reduction of emissions, exposures, and uptake.

As has been mentioned before, TLV values may be not always protect every worker from adverse health effects under all circumstances. Therefore, for each group of workers and for each specific working condition, a critical appraisal should be made about the time-weighted average concentration of chemicals that can be considered to be safe in a defined situation. This holds too for chemicals that are not included in official lists of TLV values.

A reduction of emission can be achieved by substitution of potentially toxic chemicals with less hazardous ones, or modification of manufacturing processes, technology, and installations. Also changes in procedures and organization can improve the working conditions.

The intensity of uptake can be reduced by personal protective equipment (see Section 3.7.2).

4.2. Biological Monitoring [33–38]

Biological monitoring (BM) is one of the methods for the health surveillance of individual workers.

BM can be described as: "Measurement and assessment of workplace agents or their metabolites either in tissues, secreta, excreta or any combination of these, to evaluate exposure and health risk, compared to an appropriate risk" [33].

4.2.1. Basic Considerations

BM is a person-related specific way of monitoring to estimate the health risk due to exposure to a particular chemical or group of chemicals. The aim is to obtain a measure of the internal burden of a defined chemical in the body of an individual after absorption, irrespective of the route, time, or duration of intake. BM is especially important for chemicals that are readily absorbed through intact skin, such as benzene, acrylamide, dimethylformamide, dinitro-*o*-cresol, and methyl parathion.

With respect to health surveillance, BM is far more relevant than environmental monitoring, which only reflects the possible exposure by inhalation during work.

When appropriate methods are available, BM is always the method of choice for the health surveillance of individual workers, because it reflects the individual uptake by all routes and is almost always substance-specific.

Thanks to steady progress in knowledge about the biological fate of chemicals in the human body and improved specific and sensitive analytical methods, there is a growing number of chemicals for which BM methods have become available.

Although a specific response is measured by means of BM, it is not a reflection of the way the chemical has entered the body. Mercury in the urine can be caused by ingestion of metallic mercury, inhalation of mercury vapor, or transdermal and transmucosal absorption of an organomercury compound. Carbon monoxide, determined in the blood, might be due to smoking, gas fumes in the home due to poor ventilation, occupational exposure to carbon monoxide in a blast-furnace plant, or stripping paints with methylene chloride. Hippuric acid in the urine might be due to the metabolism of toluene taken up during work or to digestion of various food constituents.

As BM always reflects a compilation of past exposures, it is generally not suitable for determining peak exposures during work. It also will give no indication of contact of skin or mucous membranes with chemicals that are not absorbed through them, although they may exhibit local effects.

Moreover BM results must be evaluated with care. Figures that exceed certain reference values may not necessarily be considered as harmful to health and therefore unacceptable. And levels below reference values may not always be acceptable for every individual or in all circumstances.

4.2.2. Prerequisites

BM can only be used properly when the following conditions have been met:

1) Before any measuring strategy can be designed the aims of health surveillance must be defined. For example: Are we interested in current or past exposures? Do we want individual figures or group means? Only when these conditions have been met is the design of a proper and sensible BM scheme possible. In general this means that in a BM program adapted to the conditions chosen, the most suitable parameter should be measured in the most suitable biological medium in the right way at the right moment in the right population.
   In general, determination of concentrations in blood better mimics current exposure, whereas urine samples give a better indication of past exposures.
2) As the parameters to be measured are intended to the way the human body deals with the chemical involved, relevant knowledge of pharmacokinetics, uptake, distribution and the route and rate of elimination is essential.
3) To consider possible health effects additional information about a quantifiable relationship between the parameters to be measured and apparent clinical or biological effects is needed.
4) Appropriate and reliable analytical methods must be available, and must meet the requirements for sensitivity, specificity and reproducibility.
5) The considerations for a proper BM surveillance scheme are the same as for other methods of health surveillance (see Section 4.1).

4.2.3. Biological Media

In principle, various media can be used:

1) Body fluids such as whole blood, plasma, and serum
2) Body tissues, e.g., hairs, nails, teeth, fat tissues, placenta
3) Excreta: urine, sweat, saliva, feces, expiratory air, mother milk.

However, the choice of biological medium depends upon a number of considerations, including aims, feasibility, availability of proper analytical methods, specificity, sensitivity, and cost.

Urine and blood are still the prevalent media in common practice. The advantages and disadvantages of the most frequently used media are discussed below:

Urine is the most suitable and easy obtainable test medium for water-soluble chemicals or metabolites that are excreted in the urine. It can give an impression of the average uptake in the past.

A drawback of the determination of chemicals in the urine is that the quantity and composition of the urine varies over a 24-h period, as does the content of the chemical to be determined. Sampling over a 24-h period overcomes this disadvantage to some extent, but it is technically difficult to achieve. For correct interpretation of the analysis of one single sample a correction for specific gravity or creatinine excretion is indicated.

Blood is the most important means of transport and distribution for chemicals in the body. Biomarkers determined in blood reflect the circulating quantity of the chemical itself or of a metabolite. Based on the half-life of a chemical in the body, the data give an impression of recent uptakes, and sometimes peak exposures can even be distinguished.

Some chemicals can be found in the red blood cells (e.g., chromium) whilst others are bound to plasma proteins. Chemicals such as carbon monoxide, hydrogen sulfide, cyanides, and nitrites form complexes with hemoglobin.

Sampling of blood by venopuncture may often act as an emotional barrier for people to cooperate.

Expiratory air is especially useful for volatile compounds which are metabolized slowly, are poorly soluble in body fluids, and are largely exhaled. Sampling and analysis of expiratory air is often simple. The interpretation is complicated by inter-individual differences in respiratory patterns.

Analysis of chemicals in *faeces* is suitable for chemicals that are taken up orally or are excreted into the intestines via the biliary system. The interpretation with respect to exposure or contact and to health is difficult or even impossible.

For most chemicals, determination in hair, nails, saliva, mother milk, etc. does not provide additional useful information with respect to health risks for exposed people. Determination of chemicals in mother milk may give an indication of the amount that could be taken up by breast-fed babies.

4.2.4. Biological Occupational Limits

For a number of chemicals, reference values are available for some relevant biomarkers. Until now, only in Germany and the United States are lists of this type of limit value published. The figures may differ between these two countries due to differences in definition.

In the United States biological exposure indices (BEI) are published by the American Conference of Governmental Industrial Hygienists [34]. These indices are intended as guidelines for evaluation of potential health hazards. They "represent the levels of determinants which are most likely to be observed in specimens collected from a healthy worker who has been exposed to the same extent as a worker with inhalation exposure to the TLV."

The German BAT value (tolerated biological value for occupational exposure to chemicals) refers to the maximal concentration of a chemical or metabolite which, based on current knowledge, does not impair the health of individual workers [35].

For the current values refer to the lists mentioned in the literature reference list. In Table 1 parameters for BM and BEM are given for some chemicals.

Except for inorganic lead, no biological exposure limits are set as legal standards. Until now they only serve as guidelines and recommended reference data in the health surveillance of workers.

For inorganic lead, a CEC Directive was adopted in 1982, which all Member States are committed to incorporate in the national legislation.

The Directive sets action levels and limits both for lead in air and for biological indicators.

Based on the CEC Directive, a resolution on lead ("Loodbesluit", 1988) came into force in The Netherlands with the same action levels and limits. However, the Dutch Expert Commission for Occupational Standards recommended a biological acceptable limit of 400 μg lead/L blood for men and 300 μg lead/L blood for women.

For health surveillance of workers exposed to lead, it is recommended to start with a simple determination of ZPP (zinc protoporphyrin) in blood with a hematofluorometer. When the ZPP content of blood exceeds 20 μg/g hemoglobin, additionally PbB should be determined.

Action levels:
PbB > 400 μg/L ⟶ information of workers
PbB > 500 μg/L ⟶ full application of the directive, including obligation for periodical medical examination
Limit values:
PbB max. 700 μg/L
PbB 700 – 800 μg/L is acceptable when:
ALAU > 20 mg/g creatinine
ZPP > 20 μg/g hemoglobine or,
ALAD < 6 Eur. Units

where PbB = lead in blood, ALAU = δ-amino-levulinic acid in urine, ZPP = zinc protoporphyrin in blood, ALAD = δ-aminolevulinic acid dehydrogenase in blood.

4.2.5. Interpretation of Biological Monitoring Data

From the point of view of health, BM is much more relevant than EM. Even when the relation between BM and EM results is fairly well known on a group basis, this relationship may differ significantly on an individual basis.

Therefore, BM programs should be performed under the supervision of an (occupational) physician. Interpretation of BM data for groups or individuals should be based on medical-toxicological expertise. The physicians involved must have the expertise to evaluate the data with respect to intrinsic and extrinsic individual factors which might explain interindividual variability, in order to come to the correct conclusion about the state of health of the individuals under surveillance.

Table 1. Biological monitoring and biological effect monitoring for some chemicals

Compound	Biological monitoring		BEI**	BAT**	Biological effect monitoring	
	Parameter	Medium*			Parameter	Medium*
Acetone	acetone	U				
	formic acid	U				
	acetone	B				
	acetone	E				
Acrylonitrile	thiocyanate	B				
	acrylonitrile	B				
Aniline	*p*-aminophenol	U	X		methemoglobin	B
	aniline	U		X		
Antimony	antimony	U				
Arsenic	arsenic	U				
Benzene	benzene	B			leucocytes	B
	benzene	E	X		hemoglobin	B
	phenol	B	X			
Bromine (salts)	bromide	U				
Butylglycidyl ether	butoxyacetic acid	U				
Cadmium	cadmium	U	X		β-microglobulin	U
	cadmium	B	X			
Chloroform	chloroform	E				
	chloroform	B				
Chromium	chromium	U	X			
Cyanides	thiocyanate	U			cyanohemoglobin	B
	thiocyanate	B				
	cyanide	B				
2-Ethoxyethanol	2-ethoxyacetic acid	U		X		
Ethylbenzene	mandelic acid	U	X			
	ethylbenzene	E	X			
Ethylene glycol	oxalic acid	U				
Fluorides	fluoride	U	X	X		
Formaldehyde	formic acid	U				
	formaldehyde	B				
n-Hexane	2,5-hexanedion	U	X	X	electromyography	
Hydrogen cyanide	thiocyanate	U			cyanohemoglobin	B
	thiocyanate	B				
	cyanide	B				
Isopropylbenzene	2-phenyl-2-propanol	U				
Lead	lead	U	X	X	zinc protoporphyrine	B
	lead	B	X		δ-aminolevulinic acid	U
Manganese	manganese	U				
	manganese	B				
Mercury	mercury	U	X	X	renal functions	
	mercury	B	X			
Methanol	methanol	U	X	X		
	formic acid	U	X			
Methylchloroform	methylchloroform	E	X			
	trichloroacetic acid	U	X			
	trichloroethanol	U	X			
	trichloroethanol	B	X			
Methylene chloride	methylene chloride	B		X	CO – hemoglobin	B
Methyl ethyl ketone	methyl ethyl ketone	U	X			
Monochlorobenzene	4-chlorocatechol	U	X	X		
	p-chlorophenol	U	X	X		
Nitrates					methemoglobin	B
Nitrites					methemoglobin	B

Table 1. (Continued)

Compound	Biological monitoring		BEI**	BAT**	Biological effect monitoring	
	Parameter	Medium*			Parameter	Medium*
Nitrobenzene	*p*-nitrophenol	U	X		methemoglobin	B
	aniline	B		X	methemoglobin	B
Organophosphates					cholinesterase	B
Parathion	*p*-nitrophenol	U	X	X	cholinesterase	B
Pentachlorophenol	pentachlorophenol	U	X			
Phenol	phenol	U	X	X		
Pyrene	hydroxypyrene	U				
	1-pyrenol	U				
Styrene	mandelic acid	U	X	X		
	styrene	B	X			
	styrene	E				
	hippuric acid	U				
	phenylglyoxylic acid	U	X			
Tetrachloroethene	tetrachloroethene	E	X			
	tetrachloroethene	B	X	X		
	trichloroacetic acid	U	X			
Toluene	hippuric acid	U	X			
	toluene	E	X			
	toluene	B	X	X		
1,1,1-Trichloroethane	trichloroacetic acid + trichloroethanol	U				
Trichloroethene	trichloroacetic acid	U		X		
	trichloroethene	E				
	trichloroethanol	B	X	X		
Xylene	methylhippuric acid	U	X	X		
	xylene	B		X		
Carbon monoxide	carbon monoxide	E	X		CO−hemoglobin	B
	carbon monoxide	B	X	X		
Carbon disulfide	thiothiazolidine-4-carboxylic acid	U	X			

* B = Blood; U = Urine; E = Expired air. ** BEI = Biological exposure index (ACGIH); BAT = Biologische Arbeitsstofftoleranz.

Among other things, inter-individual differences in enzyme configurations can influence the possible health effects of chemicals. For instance, due to a difference in biotransformation, people with inherited low acetylating enzyme profiles ("slow acetylators") are at greater risk of intoxication with methemoglobin forming compounds such as aniline or nitrobenzene than those who have large amounts of the acetylating enzyme.

Note, that BM cannot always replace EM, and sometimes a combination of both methods is indicated. Especially when the test results from BM exceed the reference data or recommended biological occupational limits, EM is indicated to estimate whether air levels at the work place are responsible, or other work-related factors such as poor hygiene at work, or endogenic or nonoccupational exogenic factors must be considered.

This question is particularly important for the decision whether corrective changes in the working conditions must be made or not. BM may also serve as a check for the effectivity of measures taken to reduce the exposure of workers.

4.3. Occupational Epidemiology (General references: [39–43])

4.3.1. Introduction

Going back to ancient Greek roots, epidemiology can be defined as the science (*logos*) of what comes upon (*epi*) the people (*demos*).

In modern terms it should be understood as the study of the occurrence and patterns of disease in groups of people. Whereas daily occupational medical routine looks at individual persons, occupational expidemiology is concerned with groups or cohorts. The basic requirements for any analytical epidemiological study are careful collection of relevant data, appropriate statistical analysis, and qualified interpretation of the results. The type of study chosen to answer a particular question should not only suit the size of the group under investigation but also the quality and quantity of data that can be obtained. Although occupational epidemiology can be regarded as one of several scientific instruments for the assessment of a particular occupational health hazard, it must be remembered that no epidemiological study can prove or disprove cause and effect relationships. How close to the "truth" a particular study will come largely depends on the quality of the study (i.e., the quality of the data), on careful performance of the study, and on expert interpretation of the results. The results should be interpreted in the context of the outcome of similar studies. Successful occupational epidemiological studies demonstrate the magnitude of risk associated with a particular level of exposure to a particular agent at the workplace.

4.3.2. Basic Requirements – Data Collection

4.3.2.1. Data on Exposed Individuals in the Study Group

For occupational cohort studies, data are needed to identify individuals in the cohort (e.g., full name, sex, ethnic origin, date of birth, personnel or social security number).

Also needed are data that connect these individuals to the cohort, such as date of entry into the company or plant where the individual is employed during the period of study, and where applicable, also the date of leaving employment.

For mortality studies the date of death or the vital status on the closing date of the study should be known. Also the cause of death must be known and it should be stated from which source (e.g., death certificate, autopsy report, etc.) this information was obtained.

For morbidity studies (concerning persons suffering from a disease) the date on which the key diagnosis was made should be known, as well as its diagnostic certainty (e.g., in the case of cancer whether the diagnosis was made by clinical means only or by histology of surgical material).

If other, confounding factors influence the occurrence of a disease (e.g., smoking in the case of lung cancer) then they must also be considered.

During data collection, it is essential to link personal data to occupational or work-environment data if a study on a specified occupational hazard is to make any sense. For subsequent statistical evaluation and publication, the data can be anonymized without any loss of quality to the study.

In many countries restrictive laws on medical confidentiality and/or data protection severely hamper the conduction of sound epidemiological studies.

Data may be very difficult to obtain when studies deal with exposures in the past because personnel files or occupational medical records may have been destroyed. In case of foreign labor many cases may be lost to the study because of workers having returned home.

In many countries death certificates are not available for occupational epidemiological studies; therefore, data on cause of death must be obtained by the best available means. If the study population is compared with the general population of the country by using the official death statistics, which are based on death certificates, then this differing source of information must be taken into consideration, as it can influence the result of the study.

For morbidity studies (e.g., the occurrence of certain types of cancers) the obstacles can even be greater. Fully functioning cancer registries, as well as national population registries, are often lacking. Some cancer registries contain little or no information on the occupational history of the cases or only include the last occupation. Even if all cancer cases were known in a particular population under study the comparison with another (e.g., the general population) would be difficult if not impossible. Thus it is not only important to very carefully collect data but also to validate the source from which they are obtained.

4.3.2.2. Technical Data on the Occupational Environment

The occupational environmental exposure of an individual worker stems from the local environment (the workplace), the activity performed (the job), and the factors present such as chemicals or physical agents. It is thus essential that information on these items be available. For studies concerned with past exposure (retrospective studies) this may be very difficult. For example, the time between the beginning of exposure to a bladder carcinogen such as 2-naphthylamine and the development of a bladder tumor (the latency period) may be as long as 20 – 40 years. Quantitative exposure data may not be available after such a long period, or may never have existed. The only source of information may be the detailed occupational history taken from those persons who are still available for questioning by the occupational physician in charge of the study. In some cases an intelligent guess can be made with the help of the industrial hygienist about the exposure level.

For studies concerned with future exposure (prospective studies) occupational exposure recording systems should be established in production plants where a potential health hazard is present or suspected and should be linkable to the individuals working in that particular plant or production site. Two types of exposure data can be used: exposure information relating to the workplace or to the job. For some chemicals biological monitoring yields a third type of exposure data specific for individual workers.

Information on the Workplace. Descriptions of industrial processes and working conditions should be available and should be updated when changes occur. Lists of chemicals or raw materials should be available in an accessible and comparable form, indicating substances present by means of appropriately defined workplace codes (e.g., by the CAS code).

Records should be kept on the quantities used and produced. Exposure measurements at the workplace (ambient monitoring) should be based on standardized sampling and analytical methods, backed by quality control programs to confirm the validity of the data produced. Results of personal monitoring should be stored with cross-references to the individual concerned to permit dose – effect and/or case studies to be performed.

Information on the Job. Job exposure information can be obtained from data representative of the activity performed (job) as opposed to that of a particular workplace. For this an appropriate description of the job is necessary (job exposure profile), which should be exposure orientated. Wherever possible the job description should be complemented with job-specific exposure measurements. Problems arise when a worker cannot be assigned to one specific workplace (e.g., working in different parts of the production or doing maintenance work), when mixtures of several chemical substances are handled, or production and products change very frequently.

Biological monitoring gives information on individual exposure (inhalatory and skin absorption), and helps to identify exposed workers.

4.3.3. Descriptive and Analytical Study Types

Descriptive studies are concerned with the observation of the distribution and the progress of a disease in groups of exposed persons. They are limited to the determination of incidence (number of new cases occurring per unit time relative to the number of persons at risk), prevalence (the number of existing cases per unit population relative to the number of persons in the population at a certain point of time), or mortality rates for certain diseases.

Analytical studies are concerned with a hypothesis about causal (etiological) factors of the disease, developed from a descriptive study or from individual occupational medical observations. The purpose of an analytical study is to determine an association between a suspected causal or risk factor (e.g., a carcinogen) and an effect (e.g., the occurrence of a specific cancer). Under favorable circumstances the strength of this association can be quantified (the "power" of the study).

The several types of analytical studies are described below.

Cross-Sectional Studies. Cross-sectional (also called prevalence) studies concentrate on a defined population at a particular point in time. They are set up to determine prevalences of disease in the exposed and nonexposed groups. Cross-sectional studies are relatively simple and quick to carry out and may be used for preliminary testing of a hypothesis. But since they merely examine new cases or drop outs they give rise to a selection effect.

Case Control Studies. In case control studies two groups of individuals are selected: the cases (suffering from the disease) and the controls (not suffering from the disease). The two groups are then compared with respect to present or past characteristics thought to be of possible relevance to the etiology of the disease. This type of study is always retrospective, i.e., the effects are already present at the beginning of the study. The conditions of exposure must therefore be reconstructed from the past.

A case control study has the advantage of being easily carried out in a relatively short time and does not need large numbers. Indeed, for the study of rare diseases this may be the only useful method. The great disadvantage, however, is the difficulty of collecting past data for both cases and controls. Interviewing cases in the knowledge that they are suffering from the disease may bias data collection, and it may be difficult to find appropriate controls.

Cohort Studies. The cohort study compares the incidence of a disease in two or more groups (cohorts) which are observed in parallel for a given period of time. An exposed group is composed of individuals with known exposure to a suspected risk factor. The control group is not exposed, but otherwise comparable for as many criteria as possible.

This type of study is also called longitudinal. It can be carried out in a prospective way, i.e., starting with apparently healthy individuals and observing them for a defined period. It can also be carried out in retrospective (historical) way, i.e., a study date is set from which the observation is retrospectively carried out to a starting date. For both types of cohort studies standardized rates of mortality or morbidity are used and compared with standardized rates of a suitable reference population.

Prospective studies have the great advantage of being more reliable and of better validity than retrospective studies, but they may take a very long time and be very costly. Retrospective studies can yield a much quicker result and cost less but often are less reliable and of lower validity. The "power" of cohort studies tends to be dependent on fairly large numbers, particularly if the disease searched for has a low incidence.

Intervention studies are concerned with the observation of an effect on a population, e.g., by reducing a certain exposure, or by altering any other relevant factor.

4.3.4. Usefulness of Occupational Epidemiological Studies

An occupational epidemiological study may be initiated to describe the health status of a particular work population. It may also be used to analyze possible associations of a disease with the place of work or the work environment. It could be used to answer specific questions resulting from findings in animal toxicological studies (mutagenicity, carcinogenicity, etc.), or to assess the validity of existing or planned limit or threshold values for certain substances at the place of work or in the work environment. Under favorable circumstances occupational epidemiological studies can even act as a monitor for the efficiency of protective measures. Occupational medical questions that might be answered by epidemiology are:

What is the morbidity or mortality for workers exposed to a particular substance or agent?

Is there an elevated risk of morbidity or mortality due to a particular disease?

What working history is found for workers who suffer from a particular disease?

What is the significance of certain abnormal laboratory or biological monitoring findings in certain workers, with, for example, elevated urine or blood levels of a particular substance or its metabolites?

The interpretation of an occupational epidemiological study eventually has to be compared with experimental animal and laboratory data, which often may give more precise and reproducible data on dose and effect. Since epidemiology is a descriptive science it concerns itself with humans and helps to evaluate possible effects at exposure levels that are actually experienced in the work environment.

But any occupational epidemiological study can only be as good as the basic data used for it. The interpretation of an epidemiological study is rarely a simple task and should be made within the epidemiological framework, taking into consideration the findings of similar studies. Therefore, so-called negative studies should also be published whenever possible.

4.4. Ergonomics

4.4.1. Introduction; Definitions

The physiology of work is based on the principles, experiences, and findings of applied physiology. It is the systematic study of the interactions between humans and their work. In particular, the effects of various forms of human work on individual body functions or on the entire human organism are analyzed. The factors studied include time, duration, hardness, difficulties, amount, and monotony of work as well as influence of the surroundings. These effects are then compared with the prerequisites for the performance of the working person.

The performance prerequisites can be differentiated according to the capabilities as structural characteristics of the person:

1) Prerequisites for physical performance, general condition, constitution, acquired physical and mental prerequisites, such as knowledge, experience, skills, and degree of practice and training
2) Prerequisites relating to work, such as powers of vision and hearing, dexterity, etc.
3) Willingness to work as driving and guiding characteristics: advancement orientation, intro- and extroversion, perseverance, satisfaction of needs, striving for a sense of achievement, varying physiological readiness to work as a result of the circadian rhythm, etc. [44], [45]

The goal of the physiology of work is to acquire scientific principles, findings, and rules in order to adapt the various forms of human work to the capabilities of the working person in the sense of preventing strain-induced health damage.

Ergonomics is regarded as an interdisciplinary core area of industrial science. A broad definition is given in [46]: "Industrial science deals with the analysis and organization of work systems and working facilities. The starting point and the goal of this study are the individual and social relationships of the working person to the other elements of the work system."

Therefore, industrial science is the science of:

1) Human work, especially from the standpoint of cooperation with others and interaction with working facilities or objects
2) The prerequisites and conditions under which the work is carried out
3) The consequences and effects of work on people, their behavior, and hence their productivity
4) The factors by which work, its conditions, and effects can be influenced in a humane fashion

Thus, the organization of work according to knowledge of industrial science includes all measures by which the system human and work can be humanely influenced.

These diverse and multifaceted questions can be solved only by the cooperation of the various scientific fields, by using findings based on human work. These fields include:

1) Medicine, especially physiological, hygienic, and pathological aspects
2) Social sciences, especially psychology, sociology, and education
3) Technical sciences
4) Economics and jurisprudence

The terms ergonomics and industrial science are sometimes used synonymously because clear and generally recognized boundaries do not exist, and opinions on the definition of "ergonomics" differ.

However, it can be stated that ergonomics is an integral part of industrial science without completely encompassing the entire field. Numerous problems of social science and sociology and a series of medical areas are not addressed by ergonomics. On the whole, industrial science is regarded as a pure science and ergonomics is more of an applied science.

Ergonomics is increasingly becoming a human-centered systematic science. The results are based on humans and the constructive consequences for technology are also derived from people [47]. The term ergonomics is derived from the Greek *ergon* (work, act) and *nomos* (law, rule, teaching).

The term ergonomics has spread internationally, largely due to the journals *Ergonomics* (since 1957) and *Applied Ergonomics* (since 1969) [48], [49].

The International Ergonomics Association (IEA) was founded in 1959 and represents about 20 national associations [50]. Today, ergonomics includes various branches of industrial science, e.g., anatomic, physiologic, psychologic, medical, and scientific aspects.

Ergonomics has increasingly opened up different spheres of activity and areas of application, including: household articles, traffic including safety aspects, sport, and leisure activities. Only areas of ergonomics that are relevant to industrial medicine are discussed here.

4.4.2. Scope of Ergonomics

The following focal points of ergonomics are worth mentioning:

1) Analysis, measurement, and assessment of human – machine/human – work systems and the determination of the limits of overworking and underworking. In this connection, the stress – strain concept plays an important role.
2) Organization of human – machine/human – work systems with the purpose of adapting the work to the person. This area includes the ergonomic planning and construction of the workplace, working facilities, working surroundings, and hours and structure of work.
3) In an inter-individually and intra-individually varying range (variable time-dependent prerequisites for performance of an individual person), the adaptation of the person to the work is also possible by means of education, practice, habit, training, and rehabilitation measures.

4.4.3. Stress Factors

The stress – strain concept of ergonomics mentioned above includes a large number of influencing factors which act on the working person during the work under investigation. The following factors can contribute to the burden of the working person.

Workplace:	Spatial design, visual conditions, acting forces (e.g., when lifting and carrying loads)
Working facilities:	Indicators and signals, screens, working chairs, handles and actuators, working aids and tools
Work surroundings:	Sounds, vibrations, light (e.g., lighting), and air (e.g., ventilation)
Working hours:	Shift system, flexibility, regulation of breaks
Work structure:	Work content (e.g., enrichment of work) distribution of work (e.g., group work)

The above sequence corresponds to experience gained in production in the processing industries, predominantly in mechanical engineering, the motor industry, electrical engineering, and precision engineering. However, it can also be applied to other areas of work, such as the chemical industry. It has proved to be a suitable basis for examining and developing the main aspects for the ergonomic organization of work [50].

4.4.4. Assessment and Testing Criteria

According to [51], prior to a detailed examination, the following criteria and questions can be used for the assessment of working conditions as a prerequisite for the organization of work.

1st	Criterion:	Feasibility
	Question:	Can the planned work be carried out with the designated persons?
2nd	Criterion:	Endurability
	Question:	Is the work on a permanent basis endurable for the designated persons?
3rd	Criterion:	Reasonableness
	Question:	Are the work and the expected working conditions reasonable?
4th	Criterion:	Satisfaction
	Question:	Will the persons designated for the work be satisfied with the work and the working conditions to be expected [51] ?

An examination of the above criteria is helpful in assessing whether all persons designated for the work are capable of carrying out the specific task planned.

For instance, a certain task cannot be performed if it involves work that must be performed beyond arms' reach. If the strength required exeeds that of the designated persons, this can be a reason for a task being not performable [50].

The criterion bearableness includes the term "strain". Here, the effect of a specific task on humans is examined. For instance, a given stress does not produce the same strain in all persons.

The demand made on the particular worker depends on individual capabilities, which can vary widely due to, e.g., age, constitution, sex, personal abilities, and skills. Consequently, strain is the individual, psychophysical reaction of the organism to a stress [52]. For the assessment of individual strain, evaluation criteria from industrial medicine are of considerable importance. This applies in particular to handicapped employees. ROHMERT describes strain as a varying degree of exhaustion of individually different capabilities by a given stress [53].

Strain can be technically measured by physiological means, e.g., cardiac rate, changes in blood pressure, body temperature, electromyogram, electrooculogram, or by psychophysiological means, e.g., flicker fusion frequency or selection reaction test. Psychological and sociological strain can also be detected by widening the stress – strain concept, e.g., by means of a questionnaire, interview, strain scaling, and group observation.

This can be explained with the help of a few examples:

Stress by lifting and carrying of weights will lead to strain of the musculo-skeletal and the cardio-vascular systems. The intensity of strain depends on, e.g., posture, individual muscular power, quality of cartilage and intervertebral tissue, the axial conditions of the strained section of the skeleton, e.g., post traumatic faulty alignment, individual patterns of movement, movements required by the specific task (e.g., frequent overhead movements), and on the individual cardio-vascular performance.

The thermoregulation system is strained by the stress resulting from the action of heat when emission of body heat is obstructed, e.g., by working in fully protective clothes, or heat is supplied to the body from the outside. Other factors to be taken into account include the heaviness of the work, working time under the action of heat, the interval between and duration of cooling breaks, and the provision of special clothing. Moreover, the intensity of strain in this example depends on the degree of acclimatization to the working conditions.

A control and supervisory task can require high attention and concentration, especially if the product quality essentially depends on this work. The intensity of strain can depend on a series of individual performance prerequisites, e.g., reactions, experience, routine with regard to the control and supervisory task, and motivation (e.g., avoidance of standstill periods and achievement of the best possible quality).

In the sense of motivation or inhibiton, certain factors can be linked to individual abilities and skills. These factors include working conditions resulting from the subordination or dependency of workers, conditions resulting from the organization of the company, working atmosphere, wages and social benefits, and social and company systems of standardization, control, and sanctioning.

Negative social relationships can represent an additional stress which, because of the resulting strain, can possibly lead to the overtaxing of the individual [52], [54].

In the case of screen work, strain can depend on the amount of work, arrangement and design of the keyboard, the office chair, the quality of the screen, and glare in the field of vision. Here, the individual factors can be further differentiated from the standpoint of ergonomics. For instance, increased demand on the eyes due to impeded focusing (accommodation) can be caused by low resolution of the characters, insufficient contrast between the characters and the background, instability of the characters, reflections on the glass surface of the display, and unsuitable design of characters and typeface [55].

The spatial design of a computer-terminal workplace must take into account the variations of body size. To optimize the design of the workplace, a graphic scale model of the workplace is used into which the schematic depiction of human figures with the different expected body sizes can be inserted. This method is known as somatography [56]. In the meantime, somatographic models and computer aided methods are available, making perspective representations of the spatial situation possible [50].

In many cases, ergonomic findings have already been embodied in law, e.g., the decree and guidelines on workplaces, acknowledged rules of technology in numerous clauses of DIN and

VDI, guidelines of the EC Council on the minimum regulations with regard to safety and health protection, e.g., for working with visual display units and for manual handling of loads (see Sections 2.1, 2.2, 2.3, and 2.4).

The application of ergonomic rules in industrial medicine requires ergonomic assessment of the working conditions and medical assessment of the working person.

To assess working conditions, numerous methods for the analysis of work and stress can be used. These include ergonomic test lists for the evaluation of the state of organization [54], questionnaires for the ergonomic description of working conditions [57], and representations in text books and monographs [47,54,55,58–65]. Moreover, the legally embodied regulations mentioned above must be considered. To assess the working person, industrial medical criteria are to be used. These must be supplemented by the application of biomechanical, neurophysiological, and functional anatomic principles as well as physiological and psychological findings.

4.4.5. Objectives

In health protection at the workplace, ergonomics supports the concept of prevention and helps avoid work-related illness and disease [66].

Attempts should be made to replace corrective ergonomics (subsequent consideration of ergonomic findings and industrial scientific evaluation criteria when, e.g., ergonomic findings have been violated in the planning process) by the concept of prospective ergonomics.

In this concept, LAURIG [66] summarizes two meanings of the term "prospective":

1) The possible prediction of future effects of working conditions and their assessment (e.g., the prediction of the limits for the endurability of the effects of vibration)
2) The application of ergonomic findings to the improvement of work productivity (e.g., the design of lighting conditions to improve quality)

In principle, for the application of the concept of prospective ergonomics, it is true that both ergonomic and industrial-medical organizational objectives must always be established in connection with the definition of production engineering planning objectives. In this process, the regulations of public occupational safety and of the trade association must be taken into account (see Sections 2.1, 2.2, 2.3, and 2.4).

Consequently, this priority can also result in restrictions on decisions preferred from a production engineering standpoint if these cannot be made compatible with the criterion of harmlessness.

Therefore, ergonomic findings should always be taken into account in the planning of human – machine/human – work systems. Only in this way can it be guaranteed that a workplace need not to be changed after it has been set up [50].

4.5. Evaluation of Health Risks [67–72]

The American National Academy of Sciences has defined health evaluation of persons exposed to chemical compounds as:

"The scientific activity of evaluating the toxic properties of a chemical and the condition of human exposure to it in order both to ascertain the likelihood that exposed humans will be adversely affected, and to characterize the nature of the effect they may experience" (1982).

Before the health risk of a particular task or exposure can be evaluated, terms such as "health" and "risk" must be defined, to arrive at a reliable balance and conclusion.

4.5.1. Basic Considerations

The state of health of an individual is in no way a static entity. It is a dynamic balance between the inborn potential of the human body and external, environmental factors (i.e., the balance between capacity and burden of an individual). Every biological alteration may disturb or change the balance. For instance, age is a biological factor that continuously influences the capacity of an individual.

A second problem in the evaluation of health risks is the meaning of the phrase "risk". Risk is a combination of the chance and nature of health impairment due to exogenic factors, caused by a particular action or event. It is difficult to outline a universally applicable framework for the phrases "risk" and "safety", because they are largely determined for individual cases.

Since there is no life without risk, which kind and degree of risk is accepted or tolerated as inherent to normal life, and therefore not be considered as risk, differs from person to person and situation to situation. Each person will, intentially or intuitively, create his own framework of safety that embodies the acceptance of a particular chance and extent of disability or harm.

Thus inter-individual evaluation of a particular hazard may vary in nature and intensity, according to whether it is an activity performed by oneself or another person. Smokers accept the risk of dying early from lung cancer, whilst passive smokers do not accept harmful effects from inhaling cigarette smoke produced by other persons.

A sensible definition of health with respect to medical checks on individuals might be: "the absence of objectively and/or subjectively perceivable signs that are considered to be harmful".

However, when using this operational definition, another problem is introduced: the question to what extent a particular finding must be considered harmful or not.

Also, from a medical point of view, not every response of the body due to a particular external physical and chemical agent should by definition be considered as harmful or impairing to health. The human body responds in one way or another to all types of endured burden. The nature and intensity of the burden is largely decisive for the nature and intensity of the response.

Increased physical burden on performing strenuous physical work, for instance, induces increased oxygen demand of the muscular tissue. This is compensated for by faster and more intense transport of oxygen from the lungs to the tissues. This compensation mechanism includes the intensivation of respiration and increased heartbeat. As soon as the burden becomes too high, these mechanisms become insufficient, and functional decompensation follows. The incidence of decompensation must be regarded as medically unacceptable and harmful, unless it is readily reversible when exposure ceases, and reconvalescence can be expected within a short time.

The working capacity of the human body mainly depends on the intrinsic, inborn properties but also on other, external stresses.

This principle of individual capacity and load also holds for chemical exposures. Contact with and exposure to chemicals may cause local and/or systemic effects.

Local effects can be caused by physical contact of the skin, eyes, or mucous membranes with certain chemicals. Contact with acids and alkalis, in particular, may cause such local effects, but also other compounds, such as aldehydes (e.g., formaldehyde) and ketones may cause local harm.

Systemic effects can result from uptake of a compound via the lungs, the digestive tract, or through the intact skin and mucous membranes.

The nature and extent of uptake depend mainly on compound-related properties such as volatility and solubility and the exposure circumstances. Some examples that illustrate this are:

1) During spraying of the volatile soil and space fumigant bromomethane large amounts can be inhaled if proper respiratory protection devices are not used. But even when inhalation is not possible, large amounts of liquid and vapor can be absorbed through unprotected skin.
2) The oily liquid aniline with only minor volatility, is easily absorbed through intact skin.
3) Dimethylformamide is a high-boiling bipolar solvent with hepatotoxic (livertoxic) potential. The percutaneous absorption rate is fast and high. Thus, exposure to DMF vapors far below the threshold limit values do not guarantee the absence of harmful effects when handling this solvent.

4.5.2. Difficulties in Interpretation

The systemic response, which may or may not be clinically apparent, can also be regarded as a physiological response as long as the extent of uptake of the chemical does not exceed the compensatory capacity of the human body. The most important compensation mechanisms for chemical agents are detoxification and elimination.

A sharp distinction between biological response and harmful, pathological effects cannot always be made. Therefore, interpretation differences between experts concerning evaluation of data are possible.

For some phenomena, such as chromosome aberration, the biological composition of semen, and some electrophysiological responses of muscle tissue and peripheral nerves, insufficient knowledge is available about normal inter- and intra-individual biological variations. An increase in the number of chromosome aberrations and a decrease of the number of healthy spermatozoa in the semen in response to the uptake of a particular chemical must not à priori be considered harmful to health.

Increased production of an enzyme that interacts with a particular chemical may follow prolonged exposure to that chemical. This enzyme induction can be regarded as a physiological response to a particular chemical burden. In some cases enzyme induction leads to better and faster detoxification and is an efficient defense mechanism against a chemical burden. However, there is still difference of opinion between experts whether enzyme induction should be regarded as an unacceptable effect or a normal response of the human organism to exogenic chemical stresses.

To arrive at a consensus, it is imperative to attain uniformity in the judgement of health risks. When this cannot be realized, a compromise should be accepted by all parties than can be expressed in a operational limitation between physiology and pathology.

Another difference of opinion between experts in the assessment of health risks might arise about the significance of impairment of health. Also, consensus should be sought about the nature and intensity of impairment that will be considered harmful.

4.5.3. Dose/Effect/Polymorphism

The intrinsic harmful properties of a chemical are determined by its inherent properties. Most effects on the human organism due to exposure to chemical compounds satisfy the principle of dose-dependency of effect. This holds that for almost every response to the uptake of a chemical, a no effect level and/or a no adverse effect level exists. It implies that in principle every chemical, under defined conditions, in a defined quantity and route of administration, and in some or all exposed individuals may cause harmful effects. This principle was discovered and published by the alchemist PARACELSUS as early as the 1500s.

An exception to this rule concerns genotoxic chemicals. For a proper preventive policy, when no contradicting data are available, it must be assumed that all concentrations of these particular chemicals, even in molecular quantities, may initiate a carcinogenic or mutagenic process after entering the cell.

However, in reality this appears to be too stringent. Firstly, it may only be valid for chemicals that have not been detoxified before reaching the organ cells. And secondly, living species have a great capacity for repairing DNA damage in cells.

The toxicological risk for an individual of contact with a chemical depends on (1) the intrinsic properties of the chemical, (2) the impairing potential (toxicity), (3) the exposure conditions, and (4) a number of person-related factors. In particular factors such as age, gender, and life-style (smoking, alcohol consumption, eating habits) may influence both the nature and intensity of an effect. This also holds for genetically determined differences in the availability of particular enzymes or in tissue susceptibility.

It is well known that a fairly large part of the global population has insufficient amounts of the enzyme *N*-acetyltransferase to guarantee optimal detoxification of aromatic amines via noncarcinogenic metabolic pathways.

Another example is the congenital deficiency of glucose-6-phosphate dehydrogenase (G 6 PD) in some ethnic groups. This enzyme is necessary for the metabolic reduction of organic peroxides. G 6 PD deficiency can lead to serious damage to the membranes of red blood cells, followed by extensive internal hemorrhage, after uptake of peroxide-forming chemicals, because these peroxides are reduced too slowly.

Although a number of inter-individual differences in biochemical capacity can be well explained, the cause of a number of these differences is still unclear. Therefore, in every estimate of health risks, the possibility should be reckoned with that a small number of a population will respond to exposure to a particular chemical, quantitatively or qualitatively, in a manner other than the majority.

Especially in setting exposure limits for chemicals, the existence of particular risk factors or risk groups, should always be considered.

The estimate of health risks of a particular activity or exposure to a particular harmful agent during work passes through the following phases [68]:

1) Hazard identification
2) Risk estimation
3) Risk evaluation
4) Risk limitation

4.5.4. Hazard Identification

The initial phase of a health risk assessment includes inventarization and description of all components of an activity to establish qualitatively whether there is a health risk due to conditions and influences at work.

A detailed study should be made of the nature of work and of the workplace. The inventory should include a full list of chemicals to be used, their physical and chemical characteristics (e.g., volatility, solubility, flammability, explosivity) and toxicity.

Toxicity. Information on toxicity can be derived from

1) Human experience with the compound
2) Results from animal experiments

Data on the state of health of humans who have been exposed to a particular compound in the past must be carefully collected and evaluated to obtain relevant information. Human experimentation is restricted, stringently regulated, and rarely acceptable. So, almost all human effects due to chemical exposure are only observational and descriptive.

The aim of this kind of investigation is to obtain knowledge about dose – effect relationships. For this reason, knowledge as accurate as possible concerning the extent of past exposures is essential. Unfortunately, this condition causes major problems because in the past measurements have not been performed at all, not on a regular basis, or by means of an unreliable or insensitive method (see Section 4.3).

As reliable data on previous exposures are often not available for estimating the toxicity, the assessment must be based on the results of animal experiments.

Animal experiments have the advantage that the exposure to a single compound can be controlled with respect to nature, dose, and duration. Thus, confounding influences, such as simultaneous exposures to other chemicals and life-style factors, can be avoided.

A major drawback, however, may be differences in susceptibility between human and animal species. Therefore, the use of an animal species that best mimics human metabolism is recommended. However, in many cases the most elementary knowledge for making a reliable choice is lacking.

Due to metabolic differences between species and sometimes in different strains of the same species, fundamental differences in detoxification and elimination mechanisms may be possible.

This can be illustrated by the following examples:

1) The human carcinogen β-naphthylamine does not give a carcinogenic response in assays with rats and rabbits
2) Mice respond with malignant liver and lung tumors in long-term inhalation assays with dichloromethane, whereas in rats only males show benign fibro-adenoma in mammary glands
3) B 6 C 3 F 1 and C 3 H mice strains are highly susceptible to spontaneous liver carcinoma. In males hepatic tumor incidence is ca. 30 % and in females ca. 6 %. With the same treatment in mice of the C 57 BL/6 J strain, induction of liver tumors does not occur [72].

Until now no generally accepted method for bridging inter- and intra-species differences is available. Only when a good qualitative and quantitative understanding of the toxicokinetics of a compound exists with respect to the metabolic pathway can a valid translation from animal to human be made. In case of doubt, uncertainty in the extrapolation can be compensated by using safety factors or uncertainty factors. The magnitude of these factors can vary from zero to several thousand, depending on the intended extent of certainty that no harmful effects will occur. The choice of magnitude of a safety factor is arbitrary, but is always related to the nature, quantity, and quality of the available data. The greater the knowledge gap, the greater is the safety factor.

4.5.5. Risk Estimation

The qualitative inventory of risk factors is followed by a quantification of these factors. For this phase a clear understanding of the dose – effect relationship is indispensible. As sufficient relevant and valid data are often lacking, the risk estimation is often an educated guess or "guestimate".

A second activity during this phase is the assessment of the nature of the exposure. Is it a single, short-term or a long-lasting, chronic exposure? What is the magnitude of the exposure?

Also the nature of interaction between simultaneously acting chemicals should be determined, as well as the influence of other environmental conditions and individual- or grouprelated factors. For this reason, the nature of the environmental air at the working place and the magnitude of the concentrations should be estimated as accurately as possible. For details on analysis and monitoring strategies, see Section 3.3.

4.5.6. Risk Evaluation

The third phase of a risk assessment procedure deals with the compilation of the data collected during the two previous phases. This compilation may lead to a conclusion about the potential health damage of a defined population due to a defined activity and about the chance that the damage will actually happen. It is a matter of political considerations whether this estimated risk is acceptable or not. In a number of situations it can be estimated whether the risk is in compliance with actual or optional limit values.

4.5.7. Risk Limitation

Based on knowledge of the likelihood, the nature, and seriousness of potential health risks a program can be designed to prevent harmful health effects completely or to reduce the chance of health risks as far as possible.

The nature and the extent of preventive measures depend on the technological possibilities on the one hand, and on political, social, and economical considerations on the other. Tightening the limits of risk acceptance always results in more expensive preventative programs. Therefore, a balance between technical and economical feasability is indicated. Which precautionary measures can be taken is discussed in Section 4.1.

4.6. Occupational Diseases and Work-Related Illnesses [73–76]

4.6.1. Introduction; Definitions

A variety of factors at the workplace can effect human health and cause health impairment, illness, and disease.

More than 4000 years ago the ancient Egyptians described certain illnesses linked to specific jobs or professions (see Chap. 1). Since then knowledge about the negative effects of factors from the working environment on human health has increased tremendously. The continuous advance in medical science and the ongoing improvement in diagnostic techniques has fostered the identification of occupational diseases and work-related illnesses.

An occupational disease is "a disease caused by any physical factor, chemical substance, or biological agent encountered in the course of work under a contract of employment" [74]. It must be established by scientific means that the disease can be caused by the factor under consideration and the worker must have been exposed seriously to this factor during his work. The exposure generally must persist over a long period of time, and the adverse health condition develop as a chronic disease.

The term work-related illnesses covers all situations of ill health that are triggered or aggravated by influences at work. They may be acute and curable, or chronic. In practice, the term is used to characterize illnesses that are not covered by the term occupational disease.

4.6.2. Occupational Diseases

Occupational diseases are classified in various subgroups.

Each disease caused by a specific agent exhibits characteristic symptoms which are the basis for its identification.

The largest group of causative agents are chemical substances. Chemicals that are known to be capable of causing occupational disease include the following elements and their compounds

Arsenic
Beryllium
Cadmium
Chromium VI
Cobalt
Fluorine
Lead
Manganese
Mercury
Nickel
Phosphorus
Vanadium
Zinc

and the following compounds

Aromatic amines
Benzene and homologes
Carbon disulfide
Carbon monoxide
Halogenated aliphatic hydrocarbons
Halogenated aromatic hydrocarbons
Hydrogen sulfide
Nitro derivatives of aromatic hydrocarbons
Phenols, their homologues, and halogenated derivatives

Occupational lung diseases caused by dust or fiber exposure are regarded as a special group. Occupational diseases caused by dusts and fibers are:

Asbestos	asbestosis
Quartz	silicosis
Certain metal dusts	pneumoconiosis

Certain physical specifications are required (specific fiber size for asbestos, crystalline dust for quartz).

Here the disease is caused either by chemical effects, mechanical impact, and in some cases allergic reactions, or by a combination of these effects.

Another specific group of chemically induced illnesses are asthmatic disorders and allergic skin effects. The most prominent causative agents for occupational allergies and their target organs are summarized in the following:

Chromium	skin
Animal hair	respiratory tract
Flour	skin, respiratory tract
Isocyanates	skin, respiratory tract
Nickel	skin
Penicillin	skin, respiratory tract

Most compounds can cause both skin and respiratory sensitization.

With allergic occupational diseases, early detection of the cause – effect relationship and consequent avoidance of any further exposure to the allergen is of major importance for the further course of the disease. The severity and irreversibility of the health impairment is directly correlated with the duration of exposure after the onset of allergy.

Since the first description of scrotal skin cancer in chimney sweeps by Pott in 1775 a number of chemicals have been identified that can cause occupational cancer.

The connection between exposure to chemical agents at the workplace and cancer development in humans has been definitely established for the following substances:

4-Aminodiphenyl	bladder cancer
Arsenic	skin cancer
Asbestos	lung cancer, mesothelioma
Benzene	leukemia
Benzidine	bladder cancer
Bischloromethyl ether	lung cancer
Calcium chromate	lung cancer
Coal tars and pitches	skin cancer
2-Naphthylamine	bladder cancer
Vinyl chloride	liver cancer (hemangiosarcoma)

Occupational diseases can also be caused by physical agents such as noise, ionizing radiation, high pressure, heat, and vibration. The related diseases are:

Ionizing radiation	bone marrow damage, skin changes
Noise	hearing deterioration
Overpressure	divers' disease
Vibration	white finger syndrome

Another group of occupational diseases can be caused by biological agents. This group includes infectious diseases and parasitic disorders acquired during work with infectious agents or in areas where infection is enhanced by climate or other conditions. Bacteria, viruses, and parasites can cause diseases in this group.

For most occupational diseases no specific therapy is available except for those caused by biological agents, where therapy with antibiotics can often cure the disease.

Some of the diseases continue to progress even after the exposure to the causative agent has been stopped. Classical examples for this are some of the pulmonary diseases caused by dust, such as beryllosis and silicosis.

Occupational cancer normally develops after a latency period of 15 – 40 years after first exposure to the causative agent. Workers with known exposure to human carcinogens should therefore be observed even after retirement in order to detect the cancer as early as possible.

In most European countries and North America, as well as in some other countries throughout the world, occupational diseases are compiled in a national list, which is generally the basis for possible compensation [73], [74].

4.6.3. Prevention

The prevention of occupational diseases is one of the most important tasks in occupational health and safety.

All groups of professional experts must cooperate with employers and employees to avoid adverse conditions at work that could cause occupational diseases. Of special importance is the medical surveillance (see Section 4.1) and the control and improvement of hygiene at the workplace.

Technical means and substitution of known hazardous chemicals (e.g., carcinogens) as well as providance and proper use of personal protective equipment are important tools in prevention of occupational diseases.

4.6.4. Compensation for Occupational Diseases

In the states throughout Europe and North America, as well as in an increasing number of countries in other parts of the world, recognition of occupational diseases and compensation for resulting disability is regulated by national laws.

An important basis for compensation is the respective national list of occupational diseases. Within the European Union a proposal for a European list of occupational diseases has been issued, and attempts to harmonize the legal basis for occupational diseases throughout the European Union will continue [75].

The World Health Organisation (WHO) has also issued a list of occupational diseases, which can be used by states that have not yet established a national list. The existing national lists of occupational diseases show large variations in classification, scope, and number of diseases listed.

The basic condition for the recognition of a disease as occupational is

1) Proof of sufficient exposure to the causative agent
2) Scientific proof that the disease in question can be caused by that agent

The mode of compensation differs from one country to another. In some countries standardized rules govern the amount of compensation.

In Germany, for example, compensation for an occupational disease is only granted if it causes a disability of 20 % or more. Below that margin the occupational disease is recognized as such but without compensation [73].

In the United States compensation is granted under diversified rules in the different states, based upon state legislation. The U.S. federal list of occupational diseases is modified at state level. Thus a wide variety of conditions governs the question of compensation for occupational diseases [76].

In several states, including the United States, Germany and other European countries, diseases can be recognized and compensated as occupational even if they are not contained in the national list, provided that:

1) The disease is causally related to a hazard at the workplace
2) The claimant has been exposed to the causative agent by reason of his employment
3) The incidence of the disease is substantially greater in that specific industry/occupation than in the general population

By this provision new occupational diseases can be discovered and compensated, and later included in the national list.

The most important tool for discovery of new occupational diseases is epidemiology (see Section 4.3).

Compensation for occupational diseases and related disability is provided in different countries under the national legislation either by a specific insurance fund, by government funds, or directly from the employer. Depending upon the respective legislation it can be payed as a lump sum or provided as a monthly pension.

For several occupational diseases compensation is granted only if the employee is not able to continue work in his former profession. Under German legislation this is the case of irritative and allergic disorders of the pulmonary tract and of the skin as well as for the newly established occupational diseases of the cervical and lumbar spine.

Under these circumstances the statutory accident insurance institutions in Germany (Berufsgenossenschaft) provide the financial means for training in a new profession.

This possibility is also available early on even if the occupational disease is not yet compensable but is expected to progress if work under the causative conditions continues.

4.6.5. Trends

The continuing development in medical diagnostics and epidemiology is providing new information on the possible cause – effect relationship for illnesses or diseases so far not recognized as occupational disease.

Areas under research and discussion are certain orthopedic diseases like chronic diseases of the spine; repetitive motion trauma (RMT), e.g., tendovaginitis; neurological disorders in connection with a number of organic chemicals (mainly solvents); and stress-related health impairments, including myocardial infarction triggered by stress at work. Further research and epidemiological studies will help to clarify these questions.

4.6.6. Work-Related Illnesses

Whereas occupational diseases mostly develop under chronic influence of the causative agent and are of persistant nature, work-related illnesses can be caused by acute or chronic exposures and are often fully reversible with adequate therapy and cessation of exposure. The scope of illnesses covered under this term is under discussion and covers acute infections as well as acute effects by chemical agents and acute disorders of the musculo-skeletal system, or health impairments related to shift work. Aggravation of preexisting ailments by influences at work is also covered by this term.

So far there is no systematic categorization or compensation. In some instances repeated acute illnesses of the same kind can lead to chronic occupational disease and in turn to compensation. Again the prevention of adverse influences at work is the best way to avoid adverse health effects. Frequent workplace inspections and evaluation of working conditions by the occupational physician and the industrial hygienist are of major importance for prevention, as is occupational medical surveillance for early detection of work-related illnesses.

4.7. First Aid

Basic Principles. In the case of occupational accidents, the medical care given must follow the same principles as at the communal level. The individual therapy steps taken to prevent life-threatening emergencies must be adapted to the particular requirements and must be well coordinated. The prerequisite for this is a well-functioning rescue chain consisting of the following four links:

1) Initial care at the site of the accident
2) Observation during transport
3) Initial care in the clinic
4) Final care in the clinic

This section discusses the measures necessary at the accident site.

Basic Requirements in the Companies. In Germany, the accident insurance institutions (trade associations) demand certain precautionary measures from the employers. Producing companies must train 10 % of their workers in first aid. In the case of shift work, one person trained in first aid should be present on each shift. In clerical work, 5 % of employees should be trained in first aid. This training is conducted by organizations authorized by the trade associations. The basic course includes heart–lung resuscitation. A refresher course must be taken after two years at the latest.

Companies must have alarm systems to call the rescue units immediately and lead them to the site.

In the case of contact with chemical products, the victims must be showered immediately. Companies must be equipped with the necessary emergency showers and eye-washing devices. First aid for accident victims includes the application of an emergency dressing. Depending on the number of employees, first aid boxes of various sizes must be kept ready in suitable containers that are quickly reached and readily accessible.

Rescue of Accident Victims. When freeing workers who are trapped or have fallen, the actual rescue measures are taken by the fire fighters. The fire brigades must have available all the technical facilities required for rescuing accident victims. Workers contaminated by chemical substances must be removed from the danger zone as quickly as possible. In case of a gas cloud, for example, they must be removed at right angles to the wind direction. The rescuers may be coworkers or firemen who have arrived at the site; they must protect themselves from contact with the substances.

The rescue equipment includes not only breathing masks and compressed air masks, but also protective clothing. Exhalation of toxic substances, such as hydrogen cyanide, carbon monoxide, or solvents, can be dangerous for the rescuer during mouth-to-nose respiration. Accident victims must be immediately freed of clothing contaminated with the product. The rescuers must wear gloves to protect themselves from skin contact. Victims should be undressed under running water, if possible. Contaminated clothes must be safely disposed of.

Maintenance of Vital Functions. The vital functions of each accident victim must be checked before he is transported. In case of respiratory arrest and/or circulatory arrest, reanimation measures must be taken immediately by the colleagues trained in first aid, with the support of medically trained firemen.

The doctor on emergency call continues the life-saving therapy, determines when the patient can be moved, and accompanies the victim to the hospital.

Product Removal. In case of skin contact, chemical substances must be removed immediately. Every second counts, especially if the victim has been burned with, e.g., hot phenol or hydrofluoric acid.

Workers must be informed about how and where they can reach the next washing facility. The primary measure is washing with copious water for several minutes, possibly with the additional use of soap. In all chemical companies which work with hazardous substances, shower facilities must be installed at certain key positions, where the accident victims can shower immediately. This helps to reduce the damage caused by all products, especially if water-soluble substances are involved. However, even sparingly water soluble products can be washed off by this purely mechanical measure.

For certain products, special neutralizing agents should be made available:

Hydrofluoric acid	calcium gluconate gel, calcium gluconate solution
Aromatic nitro and amino compounds	poly(ethylene glycol) 400
Phenols, cresols	mixture of ethanol and poly(ethylene glycol) 300 (1 : 2)
Chromium	ascorbic acid

In case of exposure to such products, a thorough aftertreatment/neutralization must be carried out in the company-owned accident department.

For neutralization, acetic acid (0.3 %) can be used for alkaline substances, and sodium acetate solution (0.3 %) for acids. Repeated testing with pH indicator paper is required to ensure that neutralization is complete.

The hair, finger nails, and ears deserve special attention. Hardened plastics must be mechanically removed.

The eyes are severely endangered in chemical accidents. The eyes must be rinsed directly at the site of the accident, and rinsing continued during transport.

Water should generally be used as the rinsing agent because of the wide range of products in most companies. In the company accident department, this preliminary treatment of eye burns is continued after the application of anesthetizing eye drops. A special neutralizing solution can be used.

Treatment. In treating victims of chemical accidents, symptomatic measures must generally be taken. Special antidotes exist only for very few chemicals. The following antidotes are permitted in Germany:

Chemical substance	*Antidote*
Acrylonitrile	*N*-acetylcysteine
Aromatic nitro and amino compounds	vitamin C and toluidine blue
Hydrogencyanide and cyanides	sodium thiosulfate and 4-dimethyl-aminophenol
Carbamates	atropine
Chromic acid, dichromates	vitamin C
Hydrofluoric acid	calcium gluconate, magnesium sulfate
Methanol	ethanol and folic acid
Phosphoric acid esters	atropine and toxogonin
Irritant gases	glucocorticoids
Hydrogen sulfide	4-dimethylaminophenol

Toxic pulmonary edema caused by inhalation of irritant gases such as phosgene must be treated directly in the inital stage. In addition to the administration of glucocorticoids, positive pressure ventilation improves the success of the treatment. The clinical latent phase, i.e., the interval between the accident and the appearance of clinical symptoms, is of varying length. In case of individual pulmonary irritants, it can last up to 24 h.

Toxic pulmonary edema can be radiologically detected in the early phase, between 6 and 8 h after the inhalative trauma, before clinical symptoms will appear. The duration of the latent phase mainly depends on the dose inhaled. The higher the dose, the shorter is the latent phase.

Prevention. First aid measures must be well organized in each company. Furthermore, the rescue chain must function smoothly. It must be constantly checked whether the necessary precautions have been taken in the company. Regular practice with the simulation of possible accidents improves the effectivity. The antidotes for certain types of poisoning should be kept ready in special therapy containers to minimize the time between the accident and the start of treatment. Regular schooling of employees, efficient training of medical staff, and technically perfect equipment help to minimize damage.

4.8. Occupational Medicine in the United States

Occupational medical programs in the United States function similarly to those in Europe with the exception that U.S. programs extend beyond the workplace conditions. Overall general health and wellness programs are included in addition to health examinations and biological monitoring.

Biological exposure indices provide a tool for assessing workers exposure to chemicals, in addition to workplace air monitoring. Biological monitoring can be required by regulatory standards or recommended guidelines in the evaluation of potential health hazard. Biological exposure indices (BEIs) provided by the American Conference of Government Industrial Hygienists are reference values usually measured in exhaled air, urine, and blood.

Biological monitoring is subject to normal biological variability and requires multiple samples over a period of time to reduce the effects of variable factors. Interpretation of biological monitoring should be with professional judgement and compared to workplace air monitoring and nonwork activities. Legal considerations are necessary when designing such monitoring programs in the United States. Health risk assessment is increasingly important and the basis for recent developments of major regulations and legislative proposals in the United States.

The American Industrial Health Council (AIHC) has made fundamental recommendations for preserving the scientific credibility of risk assessments to improve the role of science in decision-making for protection of workers' health and public health.

AIHC's proposals include having an active process to update current existing guidelines with full weight-of-the-evidence to incorporate new data and methodology. Also, the legislation should promote input from all interested parties including the general public, and should lead to meaningful communications to managers, workers, and the public.

Government-wide risk assessment practices should be considered under the Office of Science and Technology Policy to avoid different regulatory agencies having different interpretations from a single data base. The International Program on Chemical Safety (IPCS) is also discussing issues related to risk assessment harmonization. IPCS is coordinating the attainment of a common frame for risk assessment and is working to improve risk assessment methodology in the areas of reproductive/development toxicity, genotoxicity, and carcinogenicity (see Section 4.5).

Epidemiological studies have a special relationship to risk assessment. These studies are normally developed to investigate associations between health effects and a contributing factor. Comparisons between exposed and nonexposed groups can establish associations between exposure and health effects (see Section 4.3).

However, many epidemiological studies do not have the adequately quantified exposures which are necessary for risk assessment methods.

Retroactive studies have limitations because of confounding exposures and numerous health effects. Animal studies can be evaluated for similarity in response, but special attention must be given to species susceptibility, response of target organ mechanisms, and dose response.

Well-designed epidemiological studies are always preferable to animal studies for the risk assessor, due to the advantage of observing dose – response results in humans rather than extrapolating animal results to humans.

In general, for prevention of occupational diseases (risk management) the recommended concentrations for threshold chemical effects use safety factors, while zero threshold chemicals require establishing acceptable concentrations with risk assessment methods.

5. References

General References

1. A. H. Murke, H. Rodegra: "Geschichte der Arbeitsmedizin," in J. Konietzko, H. Dupuis (eds.): *Handbuch der Arbeitsmedizin,* (I-1), ecomed Verlag, Landsberg 1988.
2. W. N. Rom: *Environmental and Occupational Medicine,* Little, Brown and Company, Boston 1983.
3. D. J. Hansen: *The Work Environment,* vol. 1, Lewis Publisher, Chelsea, Mich. 1992.

General References

4. Bundesminister für Arbeit und Sozialordnung (ed.): *Übersicht über das Recht der Arbeit,* Bonn 1994.
5. Arbeitskammer des Saarlandes (eds.): *Das Arbeitssicherheitsgesetz,* Saarbrücken 1992.
6. Arbeitskammer des Saarlandes (eds.): *Arbeitsschutzvorschriften, Unfallverhütungsvorschriften, Vorsorgeuntersuchungen,* Saarbrücken 1993.
7. Bundesanstalt für Arbeitsschutz (eds.): *Arbeit menschlicher machen,* Dortmund 1992.
8. BASF: *Mehr Erfolg durch Sicherheit, sichere Betriebsführung – Leitfaden für Führungskräfte,* Ludwigshafen 1994.

General References

9. *The Industrial Environment. Its Evaluation and Control,* NIOSH Publication, U.S. Govt Printing Office, Washington 1973.
10. R. S. Brief: *Basic Industrial Hygiene – A Training Manual,* AIHA Publication, Fairfax, VA, 1975.
11. R. W. Allen et al.: *Industrial Hygiene,* Prentice Hall, Englewood Cliffs, N.J., 1976.
12. L. Parmeggiani: *Encycl. Occup. Health Saf.,* 3rd ed., ILO, Geneva 1983.
13. G. D. Clayton, F. E. Clayton (eds.): "General Principles," *Patty's Industrial Hygiene and Toxicology,* **vol. I,** Wiley-Interscience, New York 1978.
14. J. T. Garrett, L. V. Cralley, L. J. Cralley: *Industrial Hygiene Management,* Wiley & Sons, Chichester 1988.
15. Leidel, Busch, Lynch: *Occupational Exposure Sampling Strategy Manual,* NIOSH Publication, U.S. Govt Printing Office, Washington 1977.
16. R. R. Langner et al.: "Two Methods Establishing Industrial Hygiene Priorities," *Am. Ind. Hyg. Assoc. J.* **40** (1979) 1039 – 1045.
17. N. Hawkins, S. Norwood, J. Rock: *A Strategy for Occupational Exposure Assessment,* AIHA, Fairfax, VA, 1991.
18. BOSH Technical Guide, N. 11: Sampling Strategies for Airborne Contaminants in the Workplace, 1993.
19. CEFIC: *Report on Occupational Exposure Limits and Monitoring Strategy,* Brussels.
20. CEFIC: *Symposium on Occupational Exposure Limits and Harmonisation in the Setting and Control of OELs for the Protection of Workers,* Brussels 1984.

21. L. V. Cralley, L. J. Cralley: *Industrial Hygiene Aspects of Plant Operations,* MacMillan, New York 1987.
22. ACGIH: *Documentation of the Treshold Limit Values,* 5th ed., Cincinnati, 1986.
23. ACGIH: *Threshold Limit Values 1993 – 1994,* Cincinnati.
24. *Occupational Exposure Limits Worldwide,* AIHA Publication, Akron, OH, 1987.
25. *NIOSH Manual of Analytical Methods,* 3rd ed., Cincinnati 1984.
26. E. H. Berger et al.: *Noise and Hearing Conservation Manual,* 4th ed., AIHA, Fairfax, VA, 1986.
27. ACGIH: *An Industrial Ventilation – A Manual of Recommended Practices,* 20th ed., Cincinnati 1988.
28. K. Hallenbeck, K. M. Cunnigham: *Quantitative Risk Assessment for Environmental and Occupational Health,* Lewis Publishers, London 1987.
29. P. H. L. Williams, J. L. Burson: *Industrial Toxicology,* Van Nostrand Reinhold, New York 1985.
30. WHO: Early Detection of Health Impairment in Occupational Exposure to Health Hazard, *Tech. Rep. Ser.* **571** (1975).
31. BG Chemie: *Berufsgenossenschaftliche Grundsätze für Arbeitsmedizinische Vorsorge-Untersuchungen,* Gentner Verlag, Stuttgart 1971.
32. R. L. Zielhuis, P. Th. Henderson: "Definitions of Monitoring Activities and their Relevance for the Practice of Occupational Health," *Int. Arch. Occup. Environ. Health* **57** (1986) 249 – 257.
33. A. Berlin, R. E. Yodaiken, B. A. Henman (eds.): *Assessment of Toxic Agents at the Workplace. Roles of Ambient and Biological Monitoring,* Nÿhoff, Boston 1984.
34. ACGIH: *Threshold Limit Values and Biological Exposure Indices,* Cincinnati 1991/1992.
35. *List of MAK and BAT Values 2005,* Wiley-VCH, Weinheim 2005.
36. R. Roi et al.: *Occupational Health Guidelines for Chemical Risk*, CEC, Luxemburg 1983.
37. L. Allessio et al.: *Human Biological Monitoring of Industrial Chemicals Series,* CEC, Luxemburg 1983.
38. The MAk Collection for Occupational Health and Safety, PART IV: *Biomonitoring Methods*, Wiley-VCH, Weinheim.
39. J. M. Last (ed.): *A Dictionary of Epidemiology,* Oxford University Press, Oxford 1988.
40. M. Alderson: *An Introduction to Epidemiology,* Mac Millan Press, London 1976.
41. R. R. Monson: *Occupational Epidemiology,* CRC Press, Boston 1990.
42. M. Alderson: *Occupational Cancer,* Butterworths, London 1986.
43. R. Doll, R. Peto: *The Causes of Cancer,* Oxford University Press, Oxford 1981.
44. W. Hacker: *Arbeitspsychologie. Psychische Regulation von Arbeitstätigkeiten,* Huber Verlag, Bern 1986.
45. R. Tielsch: "Ansatz, Kriterien und Ergebnisse zur Ermittlung subjektiver Belastung und Beanspruchung an industriellen Arbeitsplätzen," *Fortschr. Ber. VDI Reihe 17* (1987).
46. Gesellschaft für Arbeitswissenschaft: Denkschrift "Arbeitswissenschaft in der Gesetzgebung," Rationalisierungskuratorium der deutschen Wirtschaft, 2nd ed., Europäische Verlagsanstalt, Frankfurt/Main 1974.
47. H. Schmidtke: *Handbuch der Ergonomie,* 2nd ed., Hanser Verlag, München 1989.
48. *Ergonomics,* an international journal of research and practice in human factors and ergonomics, Taylor & Francis, London – Washington.
49. *Appl. Ergonomics,* journal on the technology of mans' relations with machines, environment and work systems, Butterworth Heinemann, Oxford.
50. W. Laurig: "Einführung in die Ergonomie," in J. Konietzko, H. Dupuis (eds.): *Handbuch der Arbeitsmedizin,* 6th ed., ecomed Verlagsgesellschaft, Landsberg 1991.
51. W. Rohmert: "Aufgaben und Inhalt der Arbeitswissenschaft," *Die berufsbildende Schule* **24** (1972) 3 – 14.
52. K. H. Norpoth: *Einführung in die Arbeitsmedizin,* ecomed Verlagsgesellschaft, Landsberg 1991.
53. W. Rohmert: "Das Belastungs-Beanspruchungs-Konzept," *Zentralbl. Arbeitswissensch.* **38** (1984) 193 – 199.
54. REFA: *Methodenlehre der Betriebsorganisation. Grundlagen der Arbeitsgestaltung,* Hanser Verlag, München 1991.
55. E. Grandjean: *Physiologische Arbeitsgestaltung,* 4th ed., ecomed Verlagsgesellschaft, Landsberg 1991.

56. P. Jenik: "Maschinen menschlich konstruiert: Somatographie als arbeitswissenschaftliche Methode für Konstrukteure," *Maschinenmarkt/MM-Industriejournal* **78** (1972) 87–90.
57. W. Laurig, K. Wieland, H.-D. Mecheln: "Arbeitsplätze für Behinderte – I. Dokumentation technischer Arbeitshilfen," Forschungsbericht no. 233 der Bundesanstalt für Arbeitsschutz und Unfallforschung Dortmund, Wirtschaftsverlag NW, Bremerhaven 1980.
58. W. Laurig: *Grundzüge der Ergonomie, Erkenntnisse und Prinzipien,* 4th ed., Beuth Verlag, Berlin 1992.
59. W. Rohmert, J. Rutenfranz: *Praktische Arbeitsphysiologie,* 3rd ed., Thieme Verlag, Stuttgart 1983.
60. J. Rutenfranz, P. Knauth: *Schichtarbeit und Nachtarbeit: Probleme – Formen – Empfehlungen,* 2nd ed., Bayerisches Staatsministerium für Arbeit und Sozialordnung, München 1987.
61. K. Ruppe, K. Nienerowski: *Methoden der Arbeitshygiene. Erkennen – Messen – Bewerten – Gestalten,* Verlag Volk und Gesundheit, Berlin 1988.
62. H. Schmidtke, J. Jastrzebska-Fraczek, H. Ruhmann: *Ergonomische Prüfung von technischen Komponenten, Umweltfaktoren und Arbeitsaufgaben: Daten und Methoden,* Hanser Verlag, München 1989.
63. K. Wieland, W. Laurig: "Arbeitsplätze für Behinderte: Handbuch technischer Arbeitshilfen zur Arbeitsgestaltung," Forschungsbericht no. 375 der Bundesanstalt für Arbeitsschutz und Unfallforschung Dortmund, 4th ed., Wirtschaftsverlag NW, Bremerhaven 1987.
64. K. Landau, W. Rohmert: *Recent Developments in Job Analysis,* Taylor & Francis, London 1989.
65. W. Rohmert, K. Landau: *Das arbeitswissenschaftliche Erhebungsverfahren zur Tätigkeitsanalyse (AET),* Huber Verlag, Bern 1979.
66. W. Laurig: *Prospektive Ergonomie – Utopie oder Wirklichkeit?* Arbeitswissenschaft des AGV Metall, Köln, no. 8, Arbeitgeberverband der Metallindustrie, Köln 1984.
67. BG Chemie: *Toxikologische Bewertungen,* Springer Verlag, Berlin 1970.
68. ECETOC: *Risk Assessment of Occupational Carcinogenes,* Monograph no. 3, Brussels 1982.
69. ECETOC: Joint Assessment Commodity Chemicals Reports,
70. ECETOC: Monographs Enquiries ECETOC, Av. E. van Nieuwenhuyse 4, Box 6, B 1160 Brussels, Belgium
71. Arbeidsinspectie "Health based Recommended Occupational Exposure Limit" reports. Edited by the Dutch Expert Committee for Occupational Standards. Ed. Min. Sociale Zaken en Werkgelegenheid/Dir. Generaal van de Arbeid, P.O.Box 90 804, 2509 LV The Hague, Netherlands.
72. T. R. Fox, T. L. Goldsworthy: "Molecular Analysis of the H-ras Gene. An Understanding of the Mouse Liver Tumor Development," *CIIT-Activities* **13** (1993) 1–2.
73. E. Perlebach, G. Mehrtens: *Die Berufskrankheitenverordnung (BeKV),* Ergänzbare Sammlung der Vorschriften, Merkblätter und Materialien, Erich Schmidt Verlag, Berlin.
74. Institute of Occupational Health: *Act on Occupational Diseases and Ordinance on Occupational Diseases,* Helsinki 1989.
75. Empfehlungen der Kommission betreffend die Annahme einer Europäischen Liste der Berufskrankheiten. Empfehlung der Kommission vom 22.05.90 – (90/326/EWG), Amtsblatt der Europäischen Gemeinschaften Nr. L 160/39
76. U.S. Department of Labor: *Recordkeeping Guidelines for Occupational Injuries and Illnesses,* Washington 1986.

Plant and Process Safety

VOLKER PILZ, Bayer AG, Leverkusen, Federal Republic of Germany (Chap. 1, Sections 3.1 – 3.3)

HERBERT BENDER, BASF Aktiengesellschaft, Ludwigshafen, Federal Republic of Germany (Section 2.1)

MICHAEL MÜLLER, Bayer AG, Leverkusen, Federal Republic of Germany (Section 2.2.1, Section 3.5.3 in part)

DIETRICH CONRAD, Berlin, Federal Republic of Germany (Section 2.2.2)

CLAUS-DIETHER WALTHER, Bayer AG, Leverkusen, Federal Republic of Germany (Section 2.2.2)

WERNER BERTHOLD, BASF Aktiengesellschaft, Ludwigshafen, Federal Republic of Germany(Section 2.2.3)

MARTIN GLOR, Ciba-Geigy AG, Basel, Switzerland (Section 2.2.3)

PETER-ANDREAS WANDREY, Bundesanstalt für Materialforschung und –prüfung, Berlin, Federal Republic of Germany (Section 2.2.4)

KARL-HEINZ MIX, Bayer AG, Leverkusen, Federal Republic of Germany (Section 2.2.4)

JÖRG STEINBACH, Schering AG, Berlin, Federal Republic of Germany(Section 2.3)

ALBERT EBERZ, Bayer AG, Leverkusen, Federal Republic of Germany(Section 2.3)

FRANCIS STOESSEL, Ciba-Geigy AG, Basel, Switzerland(Section 2.3)

HANS HAGEN, Bayer AG, Leverkusen, Federal Republic of Germany (Sections 3.4.1, Section 5.1 in part)

HELMUT SCHACKE, Bayer AG, Leverkusen, Federal Republic of Germany (Section 3.4.2)

RICHARD VIARD, Bayer AG, Leverkusen, Federal Republic of Germany (Section 3.4.2)

BERND SCHRÖRS, Bayer AG, Leverkusen, Federal Republic of Germany (Section 3.4.3)

STEPHAN WEIDLICH, Hoechst Aktiengesellschaft, Frankfurt, Federal Republic of Germany (Section 3.4.3)

STEFAN DREES, Bayer AG, Leverkusen, Federal Republic of Germany (Section 3.5.1)

GÜNTER HESSE, Bayer AG, Brunsbüttel, Federal Republic of Germany (Section 3.5.2)

HANS FÖRSTER, Physikalisch-Technische Bundesanstalt, Braunschweig, Federal Republic of Germany (Section 3.5.3 in part)

KLAUS BARTELS, Berufsgenossenschaft der chemischen Industrie, Heidelberg, Federal Republic ofGermany (Section 4.1)

ULRICH WIDMER, Sandoz, Basel, Switzerland (Section 4.2.1)

ADRIAN GEIGER, Sandoz, Basel, Switzerland (Section 4.2.1)

KLAUS NOHA, Hoechst Aktiengesellschaft, Frankfurt, Federal Republic of Germany (Section 4.2.2)

EDMUND MÜLLER, BASF Aktiengesellschaft, Ludwigshafen, Federal Republic of Germany (Section 4.3)

DIETER GRENNER, Bayer AG, Dormagen, Federal Republic of Germany (Sections 4.4, 4.5)

NIKOLAUS SCHULZ, Bayer AG, Dormagen, Federal Republic of Germany (Sections 4.4, 4.5)

JÜRGEN ZIMMERMANN, Bayer AG, Dormagen, Federal Republic of Germany (Sections 4.4, 4.5)

JÜRGEN HARBORDT, Bayer AG, Leverkusen, Federal Republic of Germany (Section 5.1 in part)

MATTHIAS WALPER, Bayer AG, Brunsbüttel, Federal Republic of Germany (Sections 5.2, 5.3)

Ullmann's Industrial Toxicology

ISBN: 3-527-31247-1

1. Safety Problems in Chemical Plants

1.1. Types and Sources of Hazards

1.1.1. Fundamentals

Large-scale chemical production is one of the most important industrial activities. Its products play key roles in human nutrition, health, quality of life, and welfare. Nevertheless, many people regard chemical production as dangerous, even though chemical processes are associated with accident frequencies which are much lower than in other industries – and very few such accidents have anything to do with chemistry [2].

Despite the good accident statistics of chemical plants, grave accidents have occurred, endangering human life and the environment, and causing considerable damage. The names

Flixborough (1974) [3], Seveso (1976) [4], Bhopal (1984) [5], and Basel (1986) [6] call spectacular events to mind.

Three types of event are traditionally associated with the chemical industry:

Releases and spills (Seveso, Bhopal)
Fires (Basel)
Explosions (Flixborough)

"Explosion" as used here refers to an event in which large amounts of energy are released very rapidly (ultimately, in the form of expansion energy), producing a pressure wave that is (at least) audible and propagates at high velocity, decaying over distance.

Damage caused by the above mentioned three types of event comes about through two distinct processes:

Direct action of chemicals on humans and the environment
Indirect action of liberated energy

In the Bhopal accident, many people died after inhaling a dangerous gas that had escaped from a tank storage facility [5]. Most of the damage from the Basel warehouse fire was due to fire fighting water carrying starting material, decomposition products, and combustion products out of the stored material, harming the ecosystem of the Rhine [6]. The consequences of the Flixborough explosion occurred largely through the destructive action of the pressure wave, initiated by ignition of a cloud of escaped flammable gases [3].

Generally, chemical reactions underlie such harmful effects. When substances act directly on living matter, a chemical reaction may alter or destroy cellular tissue (toxic effect). In an explosion, energy is liberated by an exothermic chemical reaction.

It is obvious that the dangers arising from chemical plants have much to do with the nature of the substances being processed, the way they are treated in the plant, and their tendency to take part in chemical reactions under these conditions.

A conflict arises between process needs and chemical safety. On the one hand, chemistry requires reactive substances; on the other hand, this reactivity of the substances is a key aspect of the danger they pose.

A chemical process is impossible without substances that show hazardous properties and effects. The substances must therefore be reliably contained in the process equipment, and their reactivity must be governed so that uncontrolled chemical reactions cannot take place.

1.1.2. Causes and Effects

Releases and Spills. An uncontrolled release from a closed facility presupposes an undesired opening of the containment (e.g., of a valve), or damage to the walls of vessels, process equipment, fittings, or piping (e.g., as a consequence of corrosion, seal failure, or rupture).

If liquids and solids are released, people outside the plant are, as a rule, not directly endangered (or at worst after some time delay) and they are thus more easily protected. The material, on the other hand, can penetrate into the soil, contaminating the groundwater or passing through sewers into surface waters, posing a danger to the environment. Precautions inside the facility include erection of barriers (liquidproof floors, seals on sewers, and retention systems; see Section 5.1).

If the substances are gaseous or finely dispersed (aerosols), they are transported by the motion of the ambient air after their release, so that they can be spread over large areas, diluted by convection and diffusion, separated from the air by sorption and possibly sedimentation, and gradually degraded through chemical reactions with the environment.

Above certain concentration levels in air, pollutants can harm people, fauna, and flora in downstream regions. The amount of damage depends on the exposure time (dose). Usually, such a drifting gas or aerosol cloud causes a short-term burden on individuals exposed to varying concentrations. Toxicology must provide information about these harmful effects (see Section 2.1).

Explosions. An explosion involves the rapid, almost instantaneous release of heat and expansion energy resulting in the formation of large gas volumes and a (destructive) pressure wave, which propagates at high velocity. At a fixed point, the wave causes a sudden rise in pressure followed by a pressure drop, all within a fraction of a second.

The pressure wave arises because the volumes of gas formed cannot escape quickly enough by mass motion. The destructive action of the wave results from the pressure force and the rapid transfer of momentum to objects in the region of wave propagation.

Energy-supplying processes in an explosion can be chemical or purely physical. There are three different types of chemical explosion. The best-known is the instantaneous decomposition of sensitive chemical compounds to gaseous species occupying a large volume, because it figures in the use of explosives. The process is initiated by external excitation. The volume required for the gaseous products is markedly increased by the heat of decomposition evolved in the process. Decomposition can be excited mechanically by shear forces (friction) or impact; thermally by heating; or by the chemically generated shock wave from an igniter (see Section 2.2.4).

The second chemical explosion process is also fairly well known: rapid oxidation of gases or finely dispersed particles in oxidizing gases (see Sections 2.2.2, 2.2.3). A highly exothermic reaction (usually combustion) heats the gases to high temperature so quickly that the pressure in confined systems generally increases by a factor of 6 – 10 (depending on the situation); in unconfined systems, the increase in volume can bring about a pressure wave with a peak overpressure of up to 100 kPa, with a positive pressure phase lasting for a few to a few hundred milliseconds (depending on the quantity of material that takes part in the spontaneous reaction, how well the fuel gas is mixed with air, how much turbulence there is to accelerate the combustion process, and how well the reacting mixture is confined). This type of explosion can be prevented "classical" anti-explosion measures (see Section 3.4.2).

The third type of chemical explosion, less well known, at least among the general public, is the "thermal runaway" explosion, initiated by a homogeneous exothermic reaction that goes out of control.

If the heat of an exothermic reaction cannot be removed, it further raises the temperature of the system. The reaction velocity increases, more heat is produced per unit time, and the system continues to heat up more and more rapidly until the maximum possible reaction velocity has been reached. The accompanying temperature rise can lead to a dangerous pressure buildup in two ways: by raising the vapor pressure of reactants and solvents, or by initiating another reaction (e.g., a decomposition) that liberates gases (see Section 2.3). If the pressure increase leads to overstressing of the reactor, it may burst and the expansion of its contents may generate a pressure wave.

In principle, an explosion can also be brought about by a purely physical process. A sudden liberation of expansion energy is possible after gases have been compressed or when vaporization is abruptly initiated, e.g., if a hot, condensed substance is quickly mixed with another substance that vaporizes readily. So far, this phenomenon does not seem to have played a major role in chemical plant explosions, and it is not discussed here; but its possibility must be considered in relation to plant safety.

Fires. Fires in chemical plants can occur if flammable substances are released and come in contact with an ignition source. Section 2.2.1 deals with the substance properties and ambient conditions that must be present for fires to start; Section 3.4 discusses measures to control fires.

The effects of fires relate to the action of heat (energy evolved mainly as thermal radiation) and the release of pollutants (combustion gases and decomposition products).

1.2. Requirements for Safe Chemical Plants

The phenomenological discussion in Section 1.1 showed that most of the dangers in the operation of chemical plants have to do with the reactivities of the substances present. The principal hazards are caused by releases and spills of substances outside the plant, and by uncontrolled chemical reactions between substances. The danger is greater, the larger the quantities of substances and energy released.

Plant and process safety efforts must have as their essential goals:

1) To minimize the quantities of substances (Section 3.1)
2) To control the potential risks that remain

Objective (2) requires:

Reliably confining hazardous substances in process equipment that can withstand the anticipated stresses (due to pressure, temperature, corrosive attack, etc.).

Insuring that process parameters (e.g., pressure, temperature, concentration) do not take on values such that the substances can undergo uncontrolled reactions.

Some help is gained from the fact that substances, as a rule, react spontaneously with one another only if they have been suitably prepared (e.g., by treatment to increase their surface area); mixed and concentrated; and excited (e.g., by heat).

Experimental laboratory tests are performed, e.g., to determine the temperature and concentration conditions under which reactions can get so far out of control that they can lead to an explosion, or to establish the initiation conditions that can cause dangerous reactions.

A point of special interest is where the regions begin in which uncontrolled reactions cannot occur. The characteristic (critical) parameters found in such experiments are called safety ratings. They include such figures as the ignition temperature and the explosion limits. Safety ratings form the basis for the safety measures in chemical plants. All of Section 2 concerns how safety ratings are determined; the account is broken down by the event classes of Section 1.1.

At the planning and design stage, plant and process safety requires taking steps so that critical concentrations, temperatures, and pressures are not reached. This is achieved by appropriate process design and process control engineering (Chap. 3). Safety also entails preventing the occurrence of potential ignition sources that could cause an undesired reaction (e.g., hot surfaces or sparks generated by mechanical or electrical equipment); great care needs to be taken in the layout of machinery and process equipment (Section 3.4).

Critical process parameters and hazardous potential ignition sources must not occur in the plant as a result of process upsets or human error. These become an additional concern of plant safety, requiring a painstaking cause-and-effect analysis of all possible errors and malfunctions, and the institution of measures to prevent or neutralize situations that could lead to an unsafe condition. Such measures may be technical or organizational; the second group includes operating instructions, inspections, etc. Chapters 3 and 4 present a fuller treatment of these points.

The complete absence of all possible hazards – absolute safety – is not possible, for several reasons.

1) First, it cannot be ruled out that several safety measures will fail simultaneously, so that a potential hazard may become an actual one
2) People make errors from time to time, and they can misjudge things, assess them wrongly, even fail to notice them at all

The fact that absolute safety is impossible is important; two inferences follow from it. Knowledge has to be continually advanced, and precautions have to be taken to avert risks in case of failure (Chap. 5).

1.3. Definitions

The fact that absolute safety is not possible leads to two key questions:

1) What does plant safety mean in the first place?
2) How safe is safe enough?

These intimately related questions can be answered in two different ways. In the first approach, safety is defined in terms of risk; a plant is said to be safe if the risk created by it is acceptable.

Here "risk" means the possibility of harm, defined in terms of the probability of the harm during the lifetime of the plant (or the frequency of the harm) and its anticipated severity.

In the second approach, a plant is said to be safe if it complies with the appropriate regulations and codes. Then it is also said to be safe enough.

While the first method is based on a probabilistic concept, risk, the second employs a deterministic principle. In both cases, what is safe is clearly a matter of convention.

The German standard DIN 31 000, Part 2, defines safety as "a state of affairs in which the risk is not greater than the greatest acceptable risk due to the technical process or condition under consideration" [7].

The standard states that this risk is generally not quantifiable, since only in rare cases can it be expressed as R, the product of a frequency F and a measure of severity S:

$$R = FS \qquad (1)$$

The standard treats danger as the diametric opposite of safety, where the risk of a process is greater than the acceptable limiting risk (Fig. 1).

A useful notion in plant and process safety is the hazard potential, a measure of the greatest harm that can occur in the worst possible event in a plant or plant subdivision. It is reasonable to use this concept in assessing safety measures in a plant: the greater the hazard potential, the more and better safety measures are needed to lower the probability of occurrence of the undesired event to the point that the level of risk is at or below the acceptable risk level.

Safety measures may include intrinsic measures and conditions [8], which insure a priori that a hazard potential can become real only in the event of a relatively improbable combination of multiple independent failures.

Where product quality considerations make it necessary to design the process and process control system so as to prevent an exothermic reaction getting out of control, safety or protective measures can be built up on the basis of this intrinsic safety in order to lower the risk to the acceptable level.

1.4. Magnitude of Hazard Potentials and Risks

The risks created by chemical plants are often overestimated if their principles are not known.

1.4.1. Risk According to Previous Experience

An anticipated value for the risk posed by chemical plants to employees or uninvolved third parties can be derived in a relatively simple way by statistical analysis of historical data.

Consider the risk of death incurred by a chemical worker due to a typical chemical accident (poisoning, chemical burn, explosion). In the Federal Republic of Germany, this risk can be determined by analysis of the annual reports of the mutual accident insurance association of chemical industry, the so-called *Berufsgenossenschaft der chemischen Industrie* [2]. When the number of persons per year suffering death from poisoning, chemical burn, fire, or explosion is divided by the total number of persons employed in the chemical industry, the annual individual lethal risk averaged over the period 1983 – 1992 is ca. 7×10^{-6} a^{-1}, i.e., statistically, 7 persons in 10^6 die every year owing to an on-the-job chemical accident. This risk is a factor of 20 less than the risk of dying in a traffic accident in Germany (ca. 1.5×10^{-4} a^{-1} [9]) and is comparable to the risk of drowning (ca. 8×10^{-6} a^{-1} [9]).

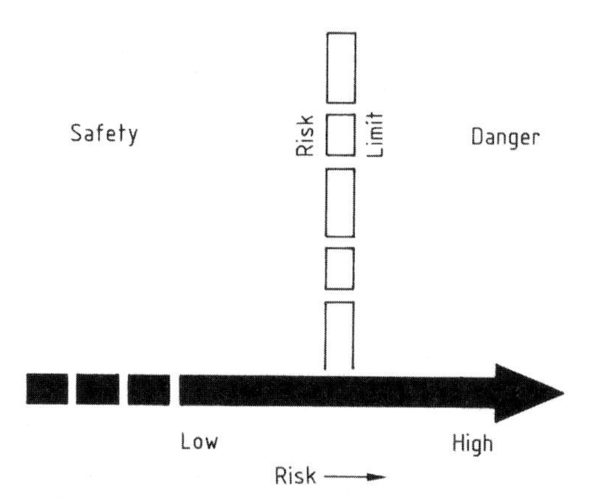

Figure 1. Risk chart [7]

Figure 2 illustrates the situation. The vertical axis shows the probability of any employee dying in the next year from one of the causes cited (individual lethal risk). For comparison, values for a range of risks, [9–12], are also plotted. In terms of a statistical average, a chemical worker is much more likely to be killed in a traffic or domestic accident than in an on-the-job accident. The worker is also more likely to be a victim of violent crime than to die in an industrial chemical accident. His on-the-job risk is low compared with other risks threatening his life.

Those living nearby and others outside the chemical plant are even safer from chemical effects, because the effects of the infrequent incidents in chemical plants fall off quite rapidly with distance. It can be assumed that this risk is, at most, of the same order as the risks due to natural catastrophes [10]. In Germany, the past 50 years have seen no identifiable serious personal injuries or deaths outside a chemical plant site resulting from accidents inside. This shows that the German chemical industry, like those in many other industrialized countries, operates very safely.

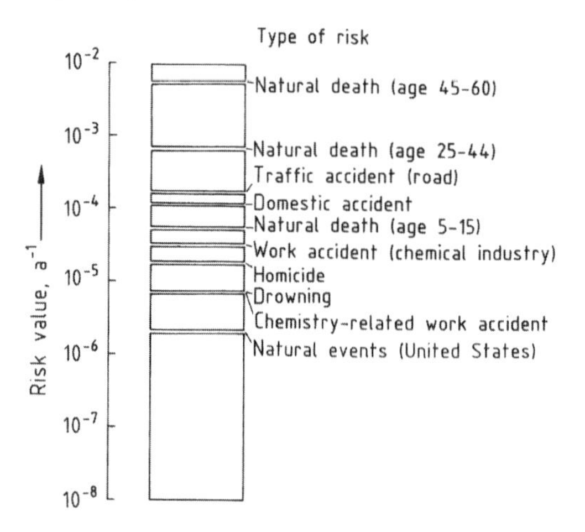

Figure 2. Comparison of lethal risks to humans (per person per year) in the Federal Republic of Germany [2], [9–11]

However, it is true that a low risk may well conceal high hazard potentials, when the probability of occurrence is low. It is therefore advantageous to consider the size of hazard potential in chemical plants.

Analysis of hazard potentials is broken down by type of accident:

Spills and releases
Explosions

1.4.2. Hazard Potential from Release of a Volatile Substance

When a volatile substance posing a health hazard is released, the hazard arises because the liberated gas is taken up by air currents, and propagates in the atmosphere as it is steadily diluted. The size of the impact area, the region in which concentration levels of the substance rise above a critical value φ_H, is one measure of the hazard potential; ultimately, however, the hazard potential is governed by the number of persons who stay in this impact area without protection.

To assess the hazard potential, it is necessary to determine the greatest mass of gas that can be released (in accordance with the model).

Consider a vessel that contains a mass M_L of a substance in the liquid state under its vapor pressure p_v at a temperature T_α, above the boiling point T_b. The mass of vapor in the vessel is M_v. If the containment fails, the contents of the vessel expand to ambient pressure; the liquid continues to vaporize and cool until the remaining liquid reaches its boiling point T_b at atmospheric pressure.

Because this process is very fast, it approximates to an adiabatic expansion. The mass of substance vaporized plus the original mass of vapor present are released into the atmosphere; this is the quantity that determines the hazard potential. A material and energy balance, with some simplifications, gives, for the mass vaporized:

$$\Delta M_v = M_{F(\alpha)} \left[1 - \exp\left(-c_L \Delta T / \Delta h_v\right)\right] \quad (2)$$

where $\Delta T = T_\alpha - T_b$, c_L is the specific heat of the liquid, and Δh_v is the heat of vaporization.

If the temperature change is small, i.e., $(c_L/\Delta h_v)\Delta T \ll 1$, a series expansion leads to the approximation:

$$\Delta M_v \approx M_{F(\alpha)} \left(c_L \Delta T / \Delta h_v\right) \quad (3)$$

A typical value for $c_L/\Delta h_v$ is $5\times10^{-3}\,K^{-1}$. For ΔT, a value of ca. 50 °C, corresponding to an overpressure of 0.5 – 1.0 MPa, is not unusual for the storage of liquefied gases. Calculation shows that with these assumptions 20 – 30 % of the liquid contents of the vessel vaporizes spontaneously. A small additional amount vaporizes because of heat supplied from the surroundings, especially if liquid escapes from the vessel in finely divided form entrained with the vapor. In any case, this simple analysis shows that only a fraction of liquefied mass under its vapor pressure in a vessel can generally be spontaneously released into the atmosphere on failure of confinement, even when substances are kept under high pressure well above their boiling point. Only if the contents are superheated by some hundreds of degrees will the released fraction approach 100 %.

To evaluate the acute hazard created by such a process, atmospheric propagation and dilution of the gas cloud have to be calculated, but these processes show an extremely strong dependence on the motion and turbulence of the atmosphere. Results can differ by orders of magnitude; what is more, initial dilution (e.g., dilution due to momentum of released vapor), height of release, and gas density all play roles. Exact prediction is impossible, but a rough estimate is possible [13]: under unfavorable conditions, ca. 1000 m^3 (reduced to standard temperature and pressure) or several tonnes of gas have to be released to establish maximum concentrations of ca. 10 ppm out to a range of several kilometers from the point of release. Under propagation conditions of higher probability, the concentrations drop below such levels at a range of less than 1 km [13], [19].

As a rule of thumb, a hazard to humans due to a volatile toxic gas release at a range of more than a few hundred meters from the source presupposes catastrophic failure of a vessel, and the vessel must contain several tonnes of the gas in releasable form (compressed or liquefied). Such apparatus can be found in chemical plants, but the kind of catastrophic damage under discussion is extremely improbable because such vessels are designed with safety factors against stress, and, they are subject to regular inspection.

From the safety standpoint, it is still important to assess hazard potentials relative to one another and to know how they can be influenced by process parameters.

The following argument can be of help. The simple equations describing gas propagation show [13] that the maximum concentration in a gas cloud (plume) at distance x downwind of the source is directly proportional to the quantity released M_{rel} and inversely proportional to the nth power of x:

$$\varphi = \text{const}\ (M_{\text{rel}}/x^n) \quad (4)$$

where $2 < n < 3$ in general.

In the case of a release near the surface, the downwind impact area, the region swept over by a concentration higher than φ_H, is commonly cigar-shaped. It becomes broader and shorter, the more turbulent the atmosphere.

For a given state of atmospheric turbulence (i.e., fixed n and constant of proportionality), the (maximum) width of the impact region is always proportional to its length. Therefore, the impact region, as a measure of the hazard potential (HP), is proportional to the square of the distance x from the source to the point where the concentration falls below φ_H:

$$HP \sim x^2 = \text{const}\ (M_{\text{rel}}/\varphi_{\text{H}})^{2/n} \quad (5)$$

For the case under discussion (spontaneous vaporization and escape of a substance posing a health hazard and having a critical concentration level φ_H) the hazard potential is:

$$\text{HP} = K \left\{ \frac{c_{\text{L}}\,\Delta T}{\Delta h_{\text{v}}} \frac{M_{\text{F}(\alpha)}}{\varphi_{\text{H}}} \right\}^{2/n} \quad (6)$$

Nevertheless, this is not a well-defined quantity, because K and n depend on the ambient conditions (state of motion of the atmosphere) and can only be derived from the known gas dispersion formula if a special case is assumed [13], [19].

The hazard potential is also a function of the quantity of substance, its properties, and its thermodynamic state inside the vessel. The hazard potential increases with the mass of substance present and its superheating (above boiling point), i.e., with pressure.

From this, it is possible to devise ways to reduce the hazard potential, even to the point of inherent safety.

1.4.3. Hazard Potential from an Explosion

The hazard potential of an explosion is due to the pressure wave, and is determined by the size of the (ideally circular) area around the explosion center in which there is significant damage, e.g., collapsing walls or broken windows. Window glass typically breaks at a peak overpressure in the 1 kPa range and masonry walls begin to fail at overpressures ≥ 30 kPa [14], [15]. BRASIE and SIMPSON speak of serious damage at peak overpressures > 15 kPa [15].

The task of evaluating the hazard potential can thus be reduced to determining at what distance from the explosion center the peak pressure just falls below a critical value.

Suppose the mass M of substance being handled is such that its sudden decomposition liberates an enthalpy of reaction Δh_r. The total energy evolved is:

$$E = M\,\Delta h_{\text{r}} \quad (7)$$

HOPKINSON and CRANZ independently found that for two distinct "charges" (releasing energies E_1 and E_2) to produce equal effects, the radii (distances R_1 and R_2 from the source point) must be related as the cube roots of the energies [16]:

$$\frac{R_1}{R_2}=\frac{E_1^{1/3}}{E_2^{1/3}} \tag{8}$$

This result led SACHS to postulate that the dimensionless peak overpressure and the dimensionless momentum of a pressure wave (the quantities that govern the destructive action of the wave) can each be represented as a unique function of a dimensionless distance from the origin [16], [17]. The dimensionless peak overpressure $p*$ is obtained by dividing by the ambient pressure; the dimensionless range $R*$, from an expression involving the energy and ambient pressure:

$$p^*=\frac{p_\mathrm{P}-p_0}{p_0} \tag{9}$$

$$R^*=\frac{R_\mathrm{P}}{(E/p_0)^{1/3}} \tag{10}$$

The dimensional formula corresponding to Equation (10) is:

$$\frac{\mathrm{m}}{\left[\left(\frac{\mathrm{Nm\cdot m^2}}{\mathrm{N}}\right)^{1/3}\right]}=\frac{\mathrm{m}}{\mathrm{m}}$$

A diagram presented by BAKER and co-workers (Fig. 3) shows that these parameters make possible a relatively accurate dimensionless representation of the characteristic pressure-wave parameters, at least for many explosives [16].

For a provisional estimate, consider 1 t of a typical explosive, TNT ($\Delta h_\mathrm{r}=4230$ kJ/kg). The energy liberated in an explosion is:

$$E=(4230\mathrm{kJ/kg})\times 1000\mathrm{kg}=4.23\times 10^6\mathrm{kJ}$$
$$=4.23\times 10^9\mathrm{Nm}$$

If $p_0=100$ kPa, Equation (10) yields:

$$R_\mathrm{P}=\left(\frac{E}{p_0}\right)^{1/3}R^*=\left(\frac{4.23\times 10^9\mathrm{Nm}}{10^5\mathrm{N/m^2}}\right)^{1/3}R^*$$

$$R_\mathrm{P}=\left(42300\mathrm{m^3}\right)^{1/3}R^*$$

$$R_\mathrm{P}=(35\mathrm{m})\,R^* \tag{11}$$

Figure 3 yields the following dimensionless ranges:

Δp, kPa	$R*$
30	ca. 1.5
15	ca. 2
1	ca. 20

Thus, Equation (11) implies that an explosion corresponding to the detonation of 1 t of TNT would break windowpanes (Δp=1 kPa) out to a range of several hundred meters (R_P = (35 m)×$R*$ = 700 m) and cause serious damage (Δp = 15 kPa) in a circle radius 50 – 100 m (R_P = 70 m).

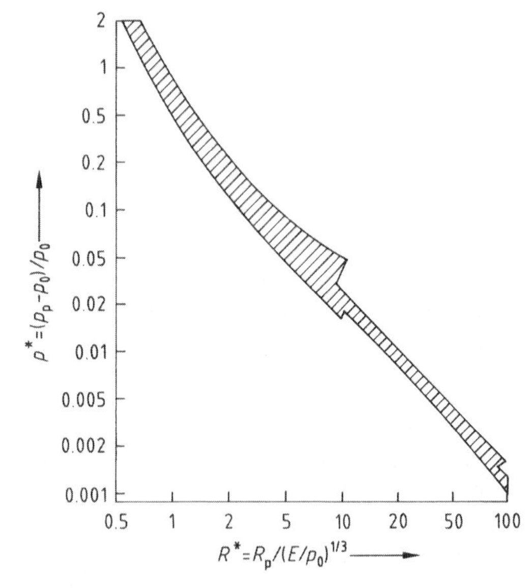

Figure 3. Dimensionless peak overpressure in an explosion pressure wave as a function of dimensionless distance from the source [16]
The expression for the dimensionless distance involves the explosion energy and the initial pressure

Explosives and similar materials are not commonly used in chemical plants. Reactions that can lead to explosion-like phenomena (Sections 1.1.2) are governed by other laws and differ strongly in kinetics, enthalpies of reaction, and efficiencies (in terms of energy conversion). For this reason, no direct comparison with the explosion just described is possible.

Even so, a number of workers have sought correlations [4], [18]. MARSHALL [1.3, p. 256], working with event analyses for explosions of unconfined gas clouds, arrives at the conclusion that for less than ca. 1 t of fuel gas forming an unconfined explosive cloud, peak overpressures on ignition are negligibly low. He estimates that for an unconfined cloud with $> 10\,t$ fuel gas, about 5 % of the liberated energy (of combustion) can contribute to a TNT-equivalent pressure wave, but with significantly lower pressures ($< 100\,kPa$) in the near region.

Hence, 2 t fuel gas reacting to completion would produce the same effect as 1 t TNT (with allowance for the fact that the enthalpy of combustion of a fuel gas is a factor of 10 greater than the energy of TNT decomposition).

In a real unconfined gas cloud formed through damage to a vessel or in a plant, only a fraction of the fuel gas is ever in a region where the concentration is high enough for sudden combustion – in the worst case. It must therefore be supposed that the effect produced by the explosion of 1 t TNT can be matched only by the spontaneous release and ignition of ca. 10 t fuel gas. This could not happen without serious damage to a large vessel.

The hazard potential of an explosion is directly proportional to the area within which a certain level of damage occurs. According to Equations (8) or (10) and (7):

$$HP \sim R^2 \sim E^{2/3} = (M \Delta h_r)^{2/3} \quad (13)$$

i.e., the hazard potential increase is less than directly proportional to the energy release and the mass involved.

1.5. Conclusions

The analyses given here show, as does experience, that chemical plants may occasionally represent high hazard potentials, especially when volatile substances of a flammable or health-endangering nature are processed in large quantities, and when exothermic (decomposition and combustion) reactions are possible.

It is also clear that, while hazards can be identified and analyzed, the magnitude of hazard potentials cannot be quantified precisely. For the design of safe processes and plants, a qualitative order-of-magnitude estimate, a relative comparison with other hazard potentials, and an identification of key independent variables is sufficient, because such considerations give a point of reference such that hazard potentials can be reduced in the most efficient way, and safety measures and equipment can be designed to handle the remaining potentials (Chap. 3).

Experience has led to the low risk values cited for chemical plants and shown that major chemical plant accidents have been infrequent, especially in view of the scope of industrial activity in this field. The advanced chemical industry has obviously had much success in controlling its hazard potentials.

2. Determination and Evaluation of Hazardous Properties of Substances (Safety Ratings)

2.1. Harmful Effects of Substances

According to the EC classification criteria [20], substances are described in terms of 15 hazard-related characteristics, which can be broken down into four groups:

1) *Acute Toxicity:*
   Very toxic substances
   Toxic substances
   Harmful substances
   Corrosive substances
   Irritant substances
2) *Specific Toxic Properties:*
   Sensitizing substances
   Carcinogens
   Substances with effects on reproduction
   Substances with heritable effects
3) *Physicochemical Properties:*
   Extremely flammable substances
   Highly flammable substances
   Flammable substances
   Oxidizing substances
   Explosive substances
4) *Environmental Impacts:*
   Substances harmful to the environment

While acute and specific toxic properties can cause direct harm to health, physicochemical properties lead to indirect harm. Substances can take any of three routes into the body:

Oral: by mouth directly into the stomach
Dermal: through the skin
Inhalation: via the respiratory organs

The actions of a substance can differ greatly depending on the route of exposure. Because of the acidic conditions in some regions of the gastrointestinal tract (pH 1 – 5), hydrolyzable substances can be broken down after oral intake. Chemical reactions can lead to both more toxic substances (metabolic activation) and less toxic ones (detoxification). While acidic compounds are preferentially absorbed in the stomach, basic and lipophilic substances are preferentially taken up in the intestinal tract. Chemicals not resorbed in either the stomach or the intestines can be excreted directly without realizing their toxic potential (e.g., mercury, barium sulfate).

The skin protects the body against external effects. It performs very effectively against substances readily soluble in water. Fat-soluble (lipophilic) substances, on the other hand, commonly diffuse well through the skin into the body. Dermal resorption is extremely effective for polar molecules with lipophilic and hydrophilic groups (e.g., many organic solvents such as dimethyl formamide, dimethyl sulfoxide, ethyl acetate, toluene, chloroform). Corrosive chemicals resorbed through the skin (e.g., phenol) are taken up very quickly and effectively by this route.

Highly corrosive substances are those that destroy the skin (necrosis) if allowed to act for a maximum of 3 min; corrosive substances have the same effect within 4 h. Examples of highly corrosive substances are inorganic acids and alkalies. In general, the corrosive action of alkaline substances is stronger than that of acids.

Irritant substances are characterized by inflammation of the skin after a maximum of 4 h exposure (e.g., dilute acids and alkalies, weak organic acids, acrylates, alcohols, amines). While the irritant action of substances is reversible, corrosive action implies irreversible injury. The mucosa of the eyes and respiratory passages are particularly sensitive to irritation.

Highly water-soluble substances, when taken up by inhalation, are generally absorbed by the mucosa in the upper part of the trachea, and do not reach the deeper layers of the lung. Because many receptors are present, irritation reactions such as coughing and sneezing may be initiated by agents such as ammonia, hydrogen chloride, sulfur dioxide, and vapors of acids and alkalies.

Compounds that are less water-soluble can readily penetrate into the bronchi. Partial diffusion through the thin bronchial tissue is possible and can lead to lung injury (e.g., chlorine, bromine, iodine, ozone, phosphorus chlorides).

If the tissue of the lung suffers local corrosion, liquid can seep into the alveoli (pulmonary edema) and the oxygen transfer required for life can be hindered for hours to days after exposure (latent action; e.g., phosgene, nitrogen dioxide, ozone, some isocyanates). Lipophilic compounds, in particular, can cross over via the alveoli when respired air is exchanged with the blood.

Substances with systemic action, after inhalation or oral or dermal exposure, are distributed throughout the body by the circulatory system. In this way they can reach all organs.

If injury takes place only at the point of exposure, the substance is said to have a local action.

With regard to acute toxic effects, substances are classed as very toxic, toxic, or harmful. This description is based on the median lethal dose LD_{50}: the dose at which half of the experimental animals die. Toxicity is reported as oral, dermal, or by inhalation. Oral and dermal LD_{50} values are given in milligrams of the substance per kilogram of the experimental animal for one-time administration; LD_{50} values by inhalation are given in milligrams of the substance per liter of respired air (exposure time 4 h). Because oral toxicity values are more common, they alone are discussed here.

The lethal dose represents only the end point of toxic effects. The toxic action, however, begins to appear even when much smaller quantities are present in the body. The dose at which no biologically relevant effect can be detected is called the no adverse effect level (NOAEL).

Table 1 shows how substances are classified in accordance with EU guidelines [20]. Substances are arranged by acute toxicity properties; the table also gives danger symbols and indications along with the "R phrases" (which specify the principal risk more closely). In some non-European countries, the boundaries between classes differ slightly from the values shown.

Table 1. LD_{50} values (oral)

Attribute	LD_{50}, mg/kg	Symbol	R phrases
Very toxic	<25	T+	26, 27, 28
Toxic	25 – 200	T	23, 24, 25
Harmful	200 – 2000	Xn	20, 21, 22

Substances with LD_{50} (oral) values over 2 g/kg do not, according to past experience, require special precautions; these are no longer classed as hazardous substances.

Sensitizing substances are those that evoke an allergic reaction in a large number of individuals, either on the skin (skin or contact allergy; e.g., formaldehyde, glutaraldehyde, alkyl acrylate) or in the respiratory passages (respiratory allergy; e.g., phthalic anhydride, isocyanates). Sensitizing reactions of chemicals cannot be distinguished from allergic reactions to agents such as flour, pollen, and animal hair.

Substances that threaten the ability to reproduce are termed reproductive toxins or substances with effects on reproduction. Such effects are classified as:

Developmental Injuries (symbol R_E). These effects include birth defects (teratogenic effects) and embryo- or fetotoxic effects such as low weight and growth or developmental disorders of the unborn child.

Impaired Fertility (symbol R_F). The EU classification criteria [20] divide substances with effects on reproduction into three categories:

Category 1. Substances known to have effects on reproduction in humans.

Category 2. Substances that should be regarded as having effects on reproduction on the basis of animal studies.

Category 3. Substances suspected of having effects on reproduction.

The labeling requirements in countries of the European Union are summarized in Table 2 [20].

Table 2. Classification of reproductively toxic substances according to the EU guidelines [20]

	Category 1	Category 2	Category 3
Hazard symbol	T	T	Xn
Classification	toxic	toxic	harmful
R phrase	60 (R_F), 61 (R_E)	60 (R_F), 61 (R_E)	62 (R_F), 63 (R_E)

Some countries have regulations different from the EU scheme. For example, the MAK Commission in Germany classifies substances with respect to the MAK value (*maximale Arbeitsplatzkonzentration* = maximum occupational exposure concentration) [21]:

Pregnancy group A. A risk of fetal injury has been demonstrated. Injury cannot be ruled out, even if the MAK value is complied with.

Pregnancy group B. A risk of fetal injury cannot be ruled out, even if the MAK value is complied with.

Pregnancy group C. A risk of fetal injury need not be feared if the MAK value is complied with.

If a health hazard is posed, not by one-time exposure to the substance, but by quantities administered over a longer period, the terms subacute, subchronic, and chronic are employed, depending on the exposure time. Carcinogenicity is merely one specific chronic effect.

Carcinogenic substances can initiate tumors. A tumor is an uncontrolled cell growth that penetrates surrounding cells in infiltrative fashion and destroys them. The latency period (time between exposure and externally detectable cancer growth) is typically 20 – 40 a for chemical carcinogens. According to the EU criteria, carcinogenic substances are classified as follows [20]:

Category 1. Known to be carcinogenic in humans.

Category 2. Unambiguously carcinogenic in animal trials.

Category 3. Suspected of being carcinogenic.

Labeling requirements under the EU scheme [20] are listed in Table 3.

Table 3. Classification of carcinogenic substances according to the EU guidelines [20]

	Category 1	Category 2	Category 3
Hazard symbol	T	T	Xn
Classification	toxic	toxic	harmful
R phrase	45 (gases, vapors) 49 (dusts)	45 (gases, vapors) 49 (dusts)	40

The best-known substances in Category 1 are asbestos, benzene, benzidine, β-naphtylamine, and vinyl chloride; in Category 2, acrylonitrile, butadiene, and *N*-nitrosamines.

The classification of the German MAK Commission [21], [261] is very similar to the EU scheme (Table 4).

Table 4. Comparison of classifications of carcinogenic substances

	MAK Commission	EU Classification
Letter codes	K	C
Human carcinogens	1	Category 1
Animal carcinogens	2	Category 2
Suspected carcinogens (classification is temporary)	B	Category 3
Carcinogens without genetoxic effect	4	–
Carcinogens with genetoxic effect	5	–

MAK = maximum workplace concentration (Germany), *GefStoffV* = hazardous materials regulation

From the standpoint of toxicological action, substances causing heritable genetic damage do not differ fundamentally from carcinogens; the injury they cause, however, is to the genetic information of the germ cells rather than to that of the body cells. The classification again includes Categories 1 – 3. No Category 1 chemicals causing heritable genetic damage are known.

The classification of hazardous substances on the basis of thermal properties employs the flash point (Table 5). Substances are characterized as oxidizing if they can sustain combustion with the exclusion of oxygen. The symbol O is used in their labeling (e.g., permanganates, bromates, hypochlorites, chlorites, nitrates).

Table 5. Thermal properties according to EC guidelines [20]

Property	Flashpoint, °C
Extremely flammable*	<0
Highly flammable	<21
Flammable	21

* *bp* must be <35 °C.

Substances with a greater explosive tendency than dinitrobenzene (reference substance) are described as explosive, and must be labeled with the symbol E (e.g., picrates, pentaerythritol tetranitrate, many organic peroxides).

In order to assess the workplace, the concentration of hazardous substances must be determined and compared with the applicable workplace concentration limits (see § 18 of the *German Hazardous Materials Regulation* [22]). The most important limits for workplace air are the German MAKs [21] and the American TLVs (threshold limit values) [23]. If these limits are maintained, no damage to health is to be feared, given a daily exposure time of 8 h and a weekly exposure time of 40 h for healthy adult workers over a working lifetime (50 a). The OEL values (occupational exposure levels) set by the EU Commission [24] are also determined primarily on scientific grounds. Because no health hazard is present if these limits are satisfied, they are less than the NOAELs mentioned earlier. The limits cited are defined as the maximum permissible concentrations averaged over one working day (shift averages). For this reason, exposure peaks must be limited. Substances are classified in exposure peak groups according to their toxicological action profiles. According to, say, the provisions of the German MAK Commission, maximum exposure levels are 2 – 10 times the MAK. Substance-specific STELs (short-term exposure levels) are also established in the United States. For a few substances, "ceiling values" are also stated; it is prohibited to exceed these at any time [23].

Because NOAELs cannot be established for carcinogens and substances with heritable effects, some countries have not established medically and toxicologically justified limits for them. In Germany, for example, TRK values (*Technische Richtkonzentration* = technical guideline concentration) can be set for carcinogens, instead of MAKs [25]. A TRK represents the exposure level that can be reached at the state of the art. If the TRK is observed, acute health hazards are eliminated; the cancer risk is minimized, but not completely eliminated. These TRKs are also 8h or shift averages.

For important industrial chemicals not governed by limits, the chemical industry has created internal guidelines on its own initiative. The results are called PELs (permissible exposure limits) in the United States and ARWs (*Arbeitsplatzrichtwerte* = workplace guideline values) in Germany [26]. All workplace exposure levels are now summarized in Germany in a single guideline [27].

With regard to protective practices, technical measures always take priority over personal protective measures (see, e.g., § 19 of the *GefStoffV* [22]). Only when the use of enclosed apparatus is not reasonable can other measures consistent with the protective objective be used, e.g., the application of suction at the source in sampling, filling, and draining operations. The use of respiratory protection is permitted only temporarily, for short-term activities, and employment as a long-term measure is not permitted.

In order to prevent dermal intoxication, suitable body protection must always be worn when working with chemicals. Polar compounds, in particular, commonly diffuse very rapidly through ordinary glove materials; breakdown times reported by manufacturers must always be observed. For further details, see [28].

In the preparation of the safety analysis pursuant to the *Störfallverordnung* (The Seveso Directive of the EU), the hazards due to product release in an emergency must be determined. With the aid of generally accepted propagation calculations (e.g., VDI 3783), the concentration in the region near a plant is calculated on the basis of a plausible leakage rate. "Accident rating values" are employed in assessing the health risk. These limits, worked out by the German chemical industry, represent concentrations such that there is no immediate danger to life, given a maximum exposure time of 0.5 h. Temporary impairment of health cannot, however, be completely ruled out.

2.2. Ratings of Flammable and Explosive Substances

2.2.1. Flammability Ratings

2.2.1.1. Introduction

A variety of data and experimental techniques can be used to characterize flammability and combustibility. Requirements as to fire behavior and fire resistance apply not just to chemicals themselves, but also to the materials and structures used in buildings where chemicals are handled.

2.2.1.2. Combustibility/Flammability of Chemicals

Combustion is a reaction that takes place in the gas phase and involves oxygen or another oxidizing agent present in the ambient gas phase. The flame phenomenon is characteristic of the reaction for liquids, gases, and most solids; constituents previously vaporized or carbonized and mixed with oxygen by diffusion oxidize in the flame. Any residue in, e.g., coke-like form, is heated to incandescence in this process and then reacts with oxygen present at the surface of the solid and diffusing into its pores. Substances such as pure carbon or iron, which do not vaporize sufficiently at the temperatures under consideration, react by glowing [29].

The speed of a self-sustaining combustion process depends on the rate of reaction, the heat liberated in the reaction, and the balance of reaction heat, heat loss to the surroundings, and heat required for melting/vaporization.

Combustibility of Liquids. The safety parameters flash point and ignition point are used to characterize the flammability of liquids.

The flash point is the temperature of a liquid, tested in a certain apparatus and under certain conditions, at which the gas phase above the liquid can burst into flame when an ignition source is brought near. The flash point is measured in various test methods [30], [31], e.g., DIN 51 755, DIN 51 758, and DIN 53 213 for Germany; GOST 12.1.044-84 for the Russian Federation; BS 2000 Part 34 and Part 170 for the United Kingdom; and ASTM D 56-87, ASTM D 3278-89, and ASTM D 93-90 for the United States. All test setups and specifications allow for the fact that the flash point is a rough measure because of the tolerated uncertainty of measurement and the variation of material properties (e.g., viscosity).

The two basic types of apparatus are the open and closed cups. Methods differ in the access of oxygen and in the quality of mixing or accumulation of fuel over the liquid. Methods can also be classified as thermodynamic equilibrium or nonequilibrium, and a given test substance may show different flash points in these two kinds of test.

In spite of national and international differences, all the methods yield data sufficiently reliable for safety assessments.

Figure 4 shows one setup for flash point determination.

If the temperature of a liquid exceeds the flash point, flammable mixtures with air may form over the liquid. As a rule, combustion of this mixture does not result in self-sustaining combustion of the liquid; this is possible only when its temperature is raised (by a few degrees celsius) to the combustion point. Even then, it is necessary that the temperature rise does not cause any further inert constituents (e.g., water in spirits of very low alcohol content) to go into the gas phase.

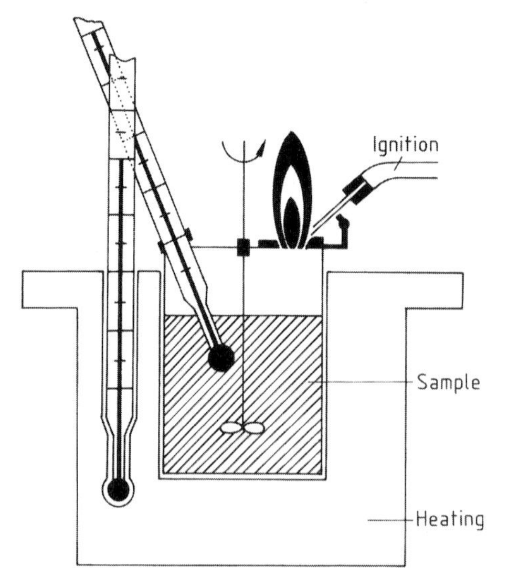

Figure 4. Flash point test equipment
Examples of flash point values (°C): Diethyl ether, gasoline − 20; ethanol 12; rum (40 vol %) − 27; diesel fuel 55; glycerine 160

Internationally, the flash point is the basis for placing substances in hazard classes with regard to ignitability. Under international transportation law [30], a liquid is classed as flammable if it has a flash point < 60.5 °C (closed cup) or 65.6 °C (open cup) or if it is handled at a temperature higher than the flash point. Self-sustaining combustion is tested in a separate procedure [32]. If a substance with flash point > 35 °C does not continue to burn after ignition in this test, it is not placed in Class 3 (flammable liquids); this rule takes into account the reduced risk of fire.

According to the German regulations (e.g., the VbF (Verordnung brennbarer Flüssigkeiten) or Flammable Liquids Regulation [31]), a substance having flash point ≤ 100 °C is classified as combustible. Liquids miscible (extinguishable) with water make up Hazard Class B; those not miscible with water, Hazard Class A. Details of the classification are as follows:

Hazard Class A
Liquids with flash point ≤ 100 °C which do not have the water solubility characteristics of Hazard Class B

A I: flash point < 21 °C
A II: flash point 21 – 55 °C
A III: flash point 55 – 100 °C

Hazard Class B
Liquids with flash point < 21 °C which are soluble in water at 15 °C or whose combustible components are soluble in water at 15 °C

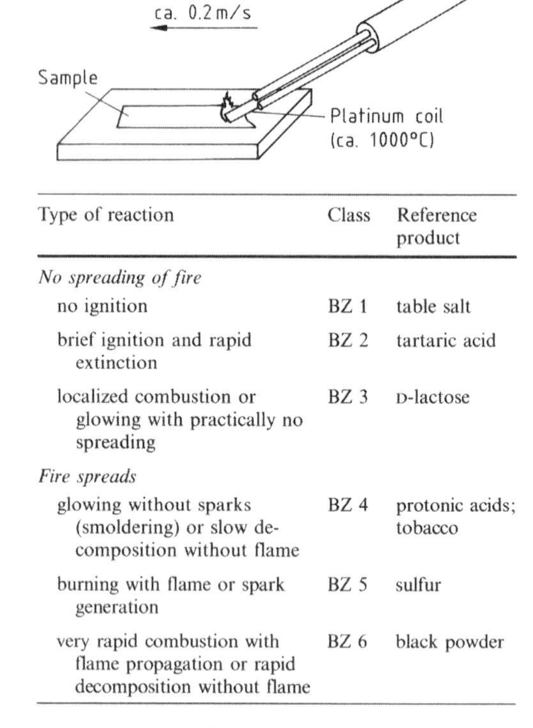

Type of reaction	Class	Reference product
No spreading of fire		
no ignition	BZ 1	table salt
brief ignition and rapid extinction	BZ 2	tartaric acid
localized combustion or glowing with practically no spreading	BZ 3	D-lactose
Fire spreads		
glowing without sparks (smoldering) or slow decomposition without flame	BZ 4	protonic acids; tobacco
burning with flame or spark generation	BZ 5	sulfur
very rapid combustion with flame propagation or rapid decomposition without flame	BZ 6	black powder

Figure 5. Flammability index test equipment

Combustibility of Solids (Dusts). The combustibility/flammability of solids is characterized by the flammability index [33] and the burning rate [33], [34] of an uncompacted layer of dust. The flammability index gives a qualitative description of how such a bed of dust behaves when a strong ignition source (igniting flame, incandescent coil) is applied to it (Fig. 5). A substance with flammability index > 3 tends to keep burning after local ignition; a burning-rate test must be performed on such a substance. For flammability index ≤ 3, the substance generally shows no tendency to rapid flame propagation. On the basis of a burning-rate test (flame path 100 mm, time limit 45 s), solids are classified as highly flammable or not highly flammable, but these terms are misleading because they both relate to the rate at which a fire spreads once started, not the tendency of a substance to catch fire.

Beside ignition from an outside source, another concern with dusts is the spontaneous flammability in air or other gaseous oxidizing medium due to self-heating. VDI 2263 describes the Grewer accelerated test (Fig. 6) and the heat-aging tests used to characterize the spontaneous flammability of dusts. Volumes of the dust ranging from 5 to 1000 mL are subjected to a programmed variation in temperature in the presence of (usually flowing) oxidant, and the behavior of the dust sample is recorded. Maximum permissible storage and surface temperatures can be derived from the results with the aid of mathematical models and appropriate safety margins. Similar test methods, with other volumes of dust tested in other containers (square, cylindrical), are used in Transportation Class 4.2 testing under international transportation law [30], and in spontaneous flammability testing pursuant to the *Chemikaliengesetz* (Chemicals Act) [34].

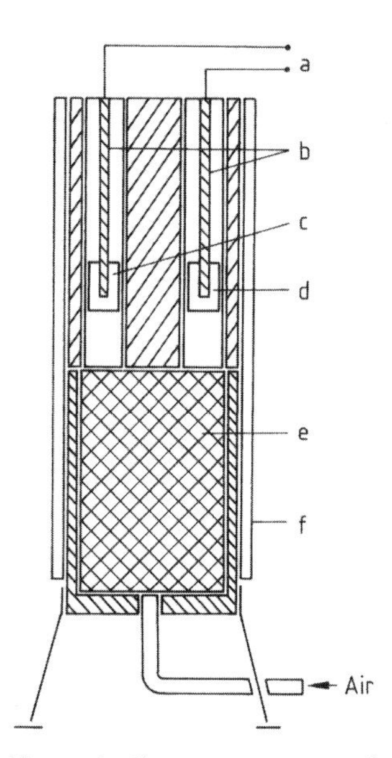

Figure 6. Grewer test apparatus for testing spontaneous flammability of dusts (VDE 2263)
a) Temperature recording; b) Thermocouples; c) Test substance (8-cm^3 wire basket); d) Reference substance; e) Packing; f) Temperature-controlled housing

Applicability of Flammability Data. The combustibility measures described up to this point are based on standard test methods. The data cannot be extended to practical conditions unless possible variations in the ambient conditions are taken into account. The flash point, for example, depends heavily on pressure as well as oxidizing medium. Changes in these lead to changes in fire behavior. The temperature of spontaneous ignition can also fall markedly below the ignition temperature of DIN 51 794 if organic liquids (including heat-transfer media) penetrate into open-pored insulants such as rock wool.

Assessment of dusts is likewise based on the experimental conditions (air as oxidizing medium, adequate heat removal from combustion zone). If the partial pressure of oxygen is increased, a stronger oxidizing agent is employed, the heat-removal conditions are made worse (storage in large drums, silos, etc.), or the ambient temperature is higher, an increased risk of fire must be expected.

In the case of solids or ointments containing solvents, it has to be decided case by case whether the flash point or the burning rate/flammability index better describes the situation.

2.2.1.3. Oxidizing Properties of Chemicals

Oxidizing substances in contact with combustible substances, give off oxygen or other oxidizing agents so as to increase the danger of fire or the vigor with which the combustibles burn. As a rule, oxidizing substances are not themselves combustible. In the pan-European market area, testing is done pursuant to Directive 84/449/EEC Annex A 17 [34]. Various mixtures of the test substance with cellulose powder are prepared and the burning rates are determined; the results are compared with the burning rate of a reference mixture of barium nitrate and cellulose. A substance is classified as oxidizing if the burning rate of a test mixture is greater than that of the reference mixture. As a rule, incorrect predictions are made if this test procedure is applied to liquids or to melting and highly flammable (combustible) substances, so its use with these substances is ruled out. A comparable test method is employed internationally for the transportation of hazardous substances [30]. In some instances, the two procedures give different classifications for the same substance.

Oxidizing properties make a chemical a hazardous substance, and safety regulations apply to its storage and transportation.

Table 6. Classification and test method for building materials [35]

Building material class	Building inspection designation	Test method
Class A	noncombustible	
A 1 *		furnace test 750 °C
A 2		Brandschacht smoke density to ASTM D 2843-70 and DIN E 53 436/37 toxicity to DIN 53436 calorific potential to DIN 51 900 Part 2 and heat release to DIN 4102 Part 8 or furnace test 750 °C
Class B	combustible	
B 1 **	low flammability	Brandschacht and small burner test
		special case of floor coverings, radiant panel test (NBSIR 75-950) including evaluation of smoke density
B 2	moderately flammable	small burner test
		special case of textile floor coverings (DIN 54 332)
B 3	highly flammable	no tests, no compliance with *B 2*

* *Class A 2* requirements must also be satisfied.
** *Class A 2* requirements must also be satisfied.

2.2.1.4. Flammability of Construction Materials

For fire precautions to be effective, they must take into account not only the combustibility of chemicals, but also the fire behavior of construction materials and structures, floor coverings, and other items within the production facility.

In Germany, structures and construction materials are tested as set forth in DIN 4102 [35]. Evaluated along with the fire endurance are the flammability, the burning rate, the toxicity of the combustion gases, and (for structures) the mechanical strength. The fire endurance is stated in minutes, for example F 30, F 60, F 90. DIN 4102 requires a fire endurance of F 90 for firewalls. Construction materials are classified as combustible or noncombustible according to the test results (Table 6).

Outside the Federal Republic of Germany, standards for assessing the combustibility of construction materials and other substances vary from country to country. A comprehensive survey of national regulations can be found in [35].

2.2.2. Explosion Data for Gas Mixtures

(General References: [36–40])

2.2.2.1. Introduction

In any operation on combustible and/or unstable gases, explosive systems may be formed and explosions initiated. This danger can be effectively countered only if the safety characteristics of the gases involved are well known. These characteristics are not constants in the physical sense, for they are influenced by the method used to determine them. There is no generally accepted method for estimating and calculating them, although for a few characteristics empirical formulas exist for specific groups of substances [41]. In other words, recourse must be had to the experimental methods described in the following sections. Laboratory values must be of such a quality that they can serve as a basis for explosion-protection measures in full-scale plants. This is possible only if the method of determination is standardized or meets other requirements derived from experience. The link between parameters and explosion-protection measures must be taken into account, even in bench-scale determination; this link essentially has to do with the type of plant and the process parameters, regardless of whether explosion protection is achieved through precautions or constructional measures.

2.2.2.2. Explosion Limits; Limiting Oxygen Concentration

A gas explosion is a rapid chemical reaction (oxidation) in the gas phase, which propagates through the explosive mixture in a self-sustaining process. Because the reaction proceeds rapidly, the heat released leads rapidly to high temperatures, and can result in the buildup of high pressures, especially in enclosed vessels.

A gas explosion requires the presence of an explosive mixture consisting of a finely dispersed fuel in an oxidizer (oxidizing medium) and the simultaneous presence of an effective ignition source. Fuels include all combustible gases (e.g., hydrogen, methane, carbon monoxide, ethylene) as well as combustible vapors such as those of the volatile hydrocarbons. The oxidizer is commonly atmospheric oxygen, but there are other oxidizing gases, the most important being chlorine, fluorine, ozone, nitrous oxide, nitric oxide, and nitrogen dioxide. Inert gases such as nitrogen and carbon dioxide play an indirect role in explosion reactions.

Any gaseous system containing mixed fuel and oxidizer is explosive only within a certain concentration range, the "explosion range", governed by the properties of the substances involved and by pressure and temperature. In practice, there is often only one fuel and air is the oxidizer; here the explosion range can be characterized simply by stating the concentration limits of the fuel in air. These bounds, the lower explosion limit (LEL) and the upper explosion limit (UEL) have been tabulated for a number of combustible gases and vapors [42], [43].

Nonexplosive mixtures in the range below the LEL are said to be "too lean" and those above the UEL are said to be "too rich."

The explosion limit itself is defined quite generally as the fuel concentration such that an explosive reaction initiated by an outside source just fails to propagate through the mixture in a self-sustained fashion. In practice, the explosion limit is recognized by the fact that the visible flame (present in virtually all gas explosions and directly tied to the reaction zone) just fails to detach from the ignition source; in other words, the velocity of flame propagation becomes zero. This flame propagation velocity or combustion rate can take values from a few centimeters per second up to several kilometers per second in the case of detonations. It is not an important parameter, except in certain technical systems with precisely defined boundary conditions (e.g., the internal combustion engine).

One basic explosion-protection measure is to prevent the formation of explosive mixtures, i.e., to keep the mixture composition outside the explosion range (see also Section 3.4.2). This requires an accurate knowledge of the explosion limits for the system.

Any apparatus for determining explosion limits must satisfy definite minimum requirements, since the limits relate not just to thermal conditions, but also to reaction kinetics.

A set-up for measuring the explosion limits of gases should therefore comprise a heated cylindrical glass vessel (ca. 1 L) with inner diameter ≥ 6 cm, set with its axis vertical, and an ignition source located on-axis in the bottom third (Fig. 7). The boundary of the ignition vessel is thus sufficiently far from the ignition point and has only a slight deactivating (explosion-hindering) action; the propagation or detachment of the flame can be observed visually and is not hindered by the burned gases, which are free to escape upward.

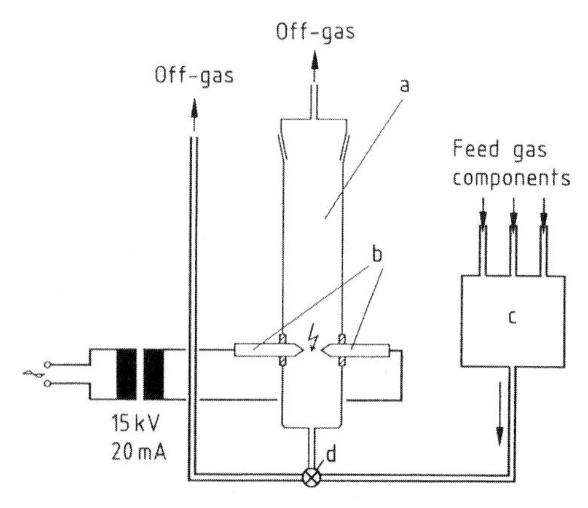

Figure 7. Apparatus for determining the explosion limits of gases
a) Ignition vessel; b) Ignition electrodes; c) Mixture preparation; d) Three-way valve

The criterion for a self-sustaining progressive combustion reaction (explosion) is the visually observed detachment of a flame from the ignition source (Fig. 8, [44]). Each ignition trial must be done in a stationary mixture, since movement (e.g., flow) of the unburned mixture, especially at the explosion limits, leads to additional transport of heat and free radicals out of the ignition volume, and this can interfere with the formation of the explosion reaction.

A proven way of delivering energy to initiate the reaction is with an induction spark (e.g., that generated with a 15 kV transformer) lasting ca. 0.2 – 0.5 s; the spark jumps a gap of 5 mm between two pointed rod electrodes ca. 3 mm diameter.

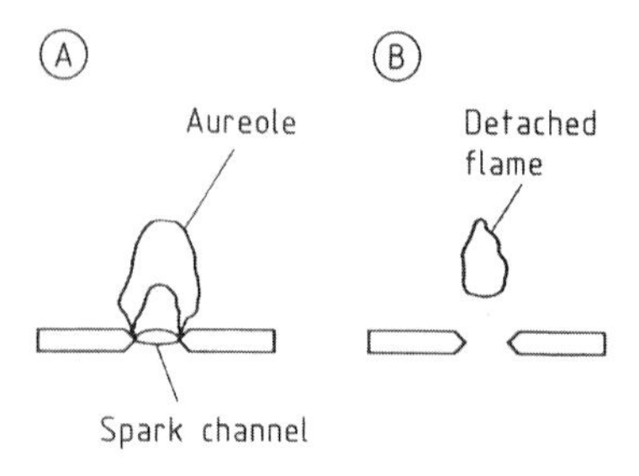

Figure 8. Flame detachment criterion
A) No flame detachment; B) Flame detachment

To determine the explosion limits accurately, the preparation of the mixture and its transfer into the test vessel must introduce as little error as possible. Calibrated devices must be used to meter the volume flow rates (of gases) and mass flow rates (of totally vaporized liquid mixtures). Temperature control, especially of the ignition vessel, should be as gradient-free as possible.

Gas-metering pumps with volumetric delivery have been found satisfactory for preparing pure gas mixtures. The test vessel must be adequately purged with the test gas mixture before every ignition trial, to remove burned or unburned residues and so that the vessel wall is sufficiently conditioned. To insure constant mixture composition, it is useful to avoid interrupting the flow of mixture during the trial; a three-way valve can be used to bypass the mixture around the vessel.

In pure gas mixtures, the explosion limits are generally stated in percentage by volume of fuel gas relative to the total mixture (vol %). This is appropriate when the mixture is prepared with a metering device with volumetric delivery. Otherwise, the limits should be based on the molar ratio (mol %).

If care is taken in operating the mixture preparation and explosion devices, explosion limits can be determined to an accuracy of ca. ± 0.1 vol % [44].

When the UEL is being determined, it should be kept in mind that some substances, such as acetylene and ethylene oxide, can decompose explosively without the presence of oxidizer; this can result in an observable flame detachment or pressure rise not attributable to an oxidative gas explosion, and it may not be possible to state an UEL. Because the effects of explosive decomposition (e.g., the maximum pressure) may actually surpass those of an oxidative gas explosion, a UEL of 100 vol % is reported and the decomposability of the substance is noted [42].

In the case of an explosive gas mixture containing several fuel species, if the concentrations and the explosion limits of the individual fuels are known, the explosion limit of the fuel mixture can be calculated, provided certain mixing rules (e.g., that of Le Chatelier [45]) are known to hold (see also [46]). As a rule, this gives satisfactory results only for the LEL.

If explosion limits must be determined for an elevated initial pressure, the tests must be carried out in a closed, pressure-tight apparatus, such as that used for measuring the maximum explosion pressure and the maximum rate of pressure rise (see Section 2.2.2.3).

Along with tables of LELs and UELs for combustible gases and vapors in air [42], [43], a triangular plot (explosion diagram) can also be used for systems of three or more components. The triangular plot displays the LEL, UEL, limiting oxygen concentration [O_{2max}], and characteristic concentration ratios needed to avoid explosive mixtures. Figure 9 presents such a diagram for a hydrocarbon/nitrogen/oxygen system; the explosive region is shaded. Each concentration (fuel on the base of the triangle, inert gas on the right leg, oxygen on the left leg) is shown increasing in the counterclockwise direction. In systems with more components, those whose concentration ratio is held constant can be plotted on one axis; for example, air ($[N_2]/[O_2]=3.76$) can be plotted as one species if CO_2 is introduced as an additional (inert) component. The so-called air line, joining 0 vol % N_2 (point B) and 21 vol % O_2 connects all mixture compositions in which the concentration ratio $[N_2]/[O_2]$ is constant at 3.76. Points L and U where the boundary of the explosive region intersects this line give the lower and upper explosion limits in air. The tangent to the explosive region boundary parallel to the N_2-axis cuts the O_2-axis at point D, which defines an O_2 concentration below which no explosion can take place because of the deficiency of oxygen, regardless of how much fuel or N_2 is present (partial inertization). This O_2 concentration (point D) is called the limiting oxygen concentration [O_{2max}]; it is a key parameter in the design of inert-gas blanketing (including that for plant-scale equipment), and the oxygen level must never be permitted to exceed it. A detailed explanation of the triangular diagram has appeared in [47]. Limiting oxygen concentrations for some substances are listed in [48].

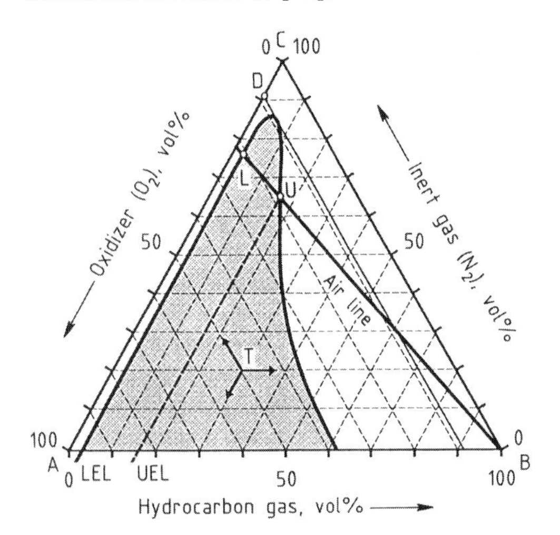

Figure 9. Triangular plot for a hydrocarbon – oxygen – nitrogen system

In practice, inerting is often best achieved by monitoring just one component, commonly the oxygen level at partial inertization, because ratio monitoring is generally more involved and expensive.

Since the explosion reaction depends on pressure and temperature, so do the explosion limits. Generally as pressure increases and temperature rises, the explosion range becomes larger. This change is much more strongly marked near the UEL than at the LEL, as Figure 10 illustrates for ethylene. The limiting oxygen concentration decreases as the pressure increases. The explosion limits and the limiting oxygen concentration vary similarly with temperature [49]. An exception is the pressure dependence of the explosion limits of hydrogen; with increasing pressure the explosion range for hydrogen does not become wider but narrower [50]. A similar behavior was found for carbon monoxide.

In the design of explosion-protection measures, it must be kept in mind that the safety-relevant parameters employed are those that apply to the conditions (pressure, temperature) of the particular application.

At present, explosion limits can be calculated only by fitting experimental data, because simulation models do not offer an adequate (substance-specific) description of the physicochemical conditions at the explosion limit.

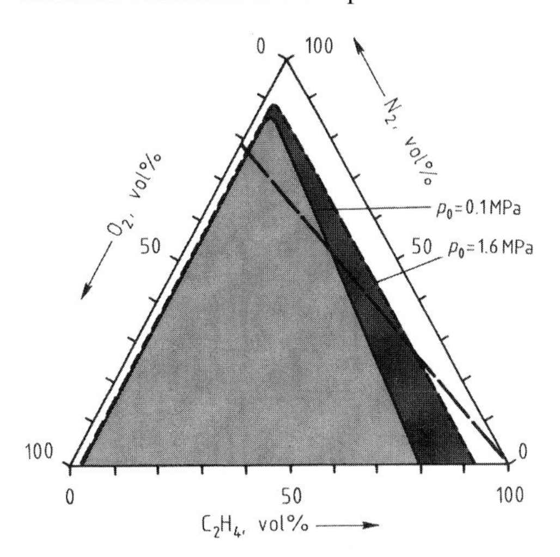

Figure 10. Pressure dependence of the explosion range

2.2.2.3. Maximum Explosion Pressure and Maximum Rate of Pressure Rise

In chemical process engineering, it is frequently not possible to avoid working with reactive gas systems in closed equipment. Avoiding potential ignition sources is often not sufficient to insure safe working conditions. The hazardous effects of an explosion on the equipment and its surroundings, including people, can be averted only through design measures [48]. The elimination of ignition sources is necessary, if only for economic reasons, as a supplementary measure. Before protective practices are planned, the following question must be elucidated: can the gas system become capable of detonation (occurrence of shock waves) as a result of the operating conditions (pressure, temperature, composition) or the geometry of the apparatus (long vessels, piping) ? If so, the discussion that follows does not apply. If the occurrence of detonations can be ruled out with confidence, it is possible to design vessels and piping such that they can withstand a deflagrative explosion. The principle of pressure-resistant or pressure-shock-resistant design is employed in such cases.

In pressure-resistant design, the anticipated explosion pressure must not exceed the design pressure; in shock-resistant design, the pressure is allowed to go above the design pressure, and deformations are allowed, provided the system remains tight (deformed parts must be replaced after such an event).

For either of these design approaches, the maximum anticipated explosion pressure generated by the gas system under the design operating conditions must be known. In simple systems, a theoretical calculation is possible, but in other cases the maximum explosion pressure has to be determined experimentally.

If economics dictates that pressure-resistant or shock-resistant design not be used, pressure relief is still a design option for drastically reducing the pressure loads. The design of rupture disks and explosion doors requires knowledge of (among other parameters) the maximum rate of pressure rise in the anticipated explosion. This cannot be calculated theoretically and must be determined experimentally for every application [51–53].

When pressure relief is employed for protection – an approach calling for a high degree of experience – it must not be forgotten that the release of products into the atmosphere may infringe air pollution control regulations. Both the maximum explosion pressure p_{max} and the maximum rate of pressure rise $(dp/dt)_{max}$, are measured in a single laboratory trial. There is no single prescribed method for determining these parameters. The requirement that bench-scale measurements must be suitable for the design of much larger equipment is best met by using a spherical autoclave with the ignition source at the center of the explosion vessel. This choice stems from the knowledge that the flame front of a gas explosion propagates through the mixture as a sphere, centered at the point of initiation [54]. To allow the reactions to proceed in virtually adiabatic fashion (i.e., to minimize heat loss through the autoclave wall in comparison with the total heat of reaction), the distance between the igniter and the wall must be $\geq$ 10 cm. Otherwise, excessive heat loss causes the measured values to be too low and yields poor estimates of the safety ratings [55]. An autoclave with a pressure rating of ca. 100 MPa at 200 °C covers the pressure and temperature ranges of practical interest in such studies. Proven types of initiators are wire igniters and surface-discharge plugs.

Energies, which depend not only on the electrical parameters, but also on the gas and the pressure, are 50 – 80 J [56].

The following instrumentation chain is recommended for recording the time dependence of pressure during reactions:

1) Pressure transducer with appropriate range; pressure-sensing membrane must make a tight seal in the autoclave wall
2) Amplifier adapted to the pressure pickup
3) Digital storage oscilloscope or computer with measuring card

For undistorted transmission of the pressure signal, the measuring chain must have a rise-time short in comparison with the pressure rise-time in the explosions under study.

Other internal features of the autoclave are a thermocouple to monitor the initial temperature and a mixer to homogenize the gas mixture. The mixer is dispensable if gas mixtures are prepared in a separate mixing vessel. Generally, the individual components are charged into the evacuated vessel in accordance with their partial pressures; the quantities are governed by the mixture composition and the initial pressure.

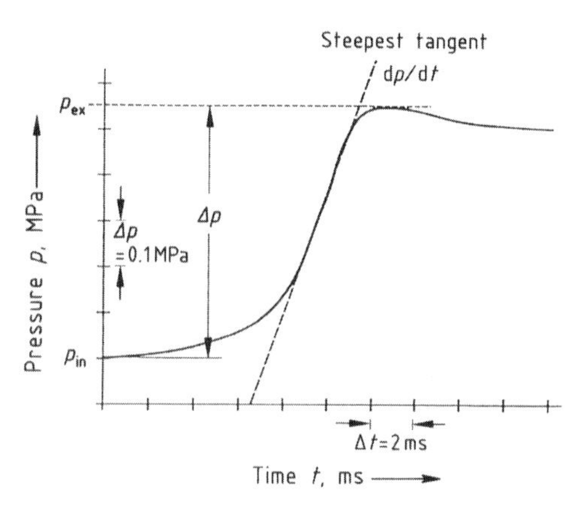

Figure 11. Time dependence of pressure in a gas explosion p_{in} = initial pressure; Δp = explosion overpressure; p_{ex} = explosion pressure; dp/dt = rate of pressure rise

Initiation trials are performed in still gas mixtures (mixer turned off) with various fuel concentrations. Analysis of each separate run with a given fuel-gas content yields an explosion pressure value (maximum on the $p-t$ curve) and a rate of pressure rise (slope of the steepest tangent to this curve, Fig. 11). Analysis of the digitally stored curves can conveniently be done by computer. Values from individual trials, plotted against the fuel-gas content, yield two curves (Fig. 12) whose maxima are the parameters sought, p_{max} and $(dp/dt)_{max}$ for the system under study for the specified initial pressure and temperature conditions. The maximum explosion pressure thus obtained remains virtually unchanged when the volume is increased.

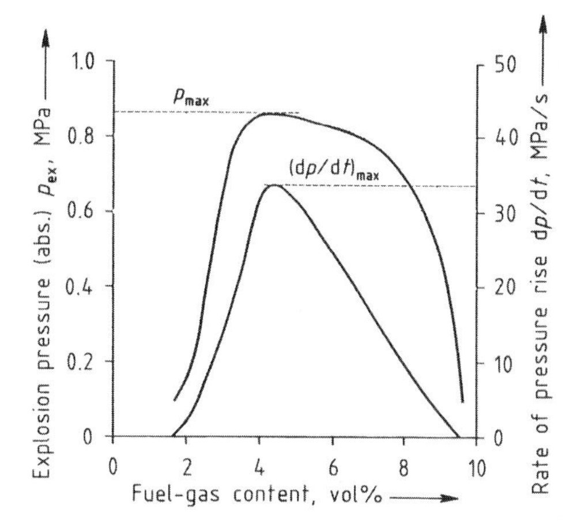

Figure 12. Illustrative curves of the explosion pressure and the rate of pressure rise for a fuel – air mixture, plotted versus the fuel-gas content
Tests took place in a still gas mixture in a 14-L spherical autoclave.

If the maximum pressure is divided by the initial pressure p_{in} the result is called the maximum pressure increase factor:

$$F_{max}=p_{max}/p_{in} \tag{14}$$

Typical values for explosive mixtures with air are $F_{max} = 5 - 10$. F_{max} is more or less independent of the initial pressure, but decreases with rising temperature.

The value found for the maximum rate of pressure rise cannot be extended directly to larger volumes. The only practical way to get an acceptable value for volumes larger than laboratory autoclaves is to apply the so- called cube law, but it must not be forgotten that the validity of the law is restricted; it can be used only for volumes that do not differ too greatly in their geometry. Such a transformation is not possible if the turbulence characteristics differ, e.g., when the laboratory value is determined in the still gas mixture, as it commonly is. This must be taken into consideration in practical situations, especially when the pressure is to be relieved with blast pipes; full-scale tests may be necessary.

According to the cubic law, the maximum rate of pressure rise multiplied by the cube root of the test volume is a constant K_G:

$$K_G = (dp/dt)_{max,V_L}\, V_L^{1/3} \tag{15}$$

where V_L is the volume of the laboratory autoclave. The maximum rate of pressure rise can be calculated for a different volume V:

$$(dp/dt)_{max,V} = K_G V^{-1/3} \tag{16}$$

The pressure increase factor and K_G value can now serve as a basis for analyzing design practices, regardless of the fuel content in the explosive mixture in the actual plant. (Note that the spherical autoclave can be replaced by a different ignition vessel if the new vessel has the same volume/area ratio as the sphere. For example, a cylindrical vessel can be employed if its diameter and height are equal.)

2.2.2.4. Pressure Limit of Stability for Unstable Gases

There are gaseous compounds whose thermodynamic and/or chemical properties enable them to decompose explosively, without any air or oxygen, if an ignition source is present. A thermodynamically unstable compound can be prepared by direct synthesis from the elements only if energy (specifically, the free enthalpy of formation) is supplied. If such a compound (e.g., acetylene, nitrous oxide, ethylene) decomposes to its elements, energy equal to the free enthalpy of formation is released [57]. Thermodynamically stable compounds with a negative free enthalpy of formation can also decompose to compounds of lower molecular mass if the sum of the free enthalpies of formation of the decomposition products is more negative than the free enthalpy of formation of the starting substance (e.g., ethylene oxide, tetrafluoroethylene). The features responsible for the ability of thermodynamically stable compounds to decompose are "weak bonds" between atoms in a molecule, where the molecule can dissociate if its vibrations are strongly excited (e.g., by absorption of external energy). Organic molecules with multiple bonds between carbon atoms are susceptible because of the diminished binding forces at these sites. Thus there are two factors, one thermodynamic and one structural that suggest an ability to decompose. A gas with both kinds of instability (e.g., acetylene) is most likely to decompose.

Fortunately, an unstable gas does not decompose spontaneously. The free enthalpy of activation ΔG^{A}, must be supplied to initiate the reaction. Only when the energy threshold ΔG^{A} has been overcome (Fig. 13) the reaction can proceed, with the release of the free enthalpy of reaction ΔG_{r} [58]. Such reactions become dangerous if the enthalpy of reaction is evolved so rapidly that it can no longer be absorbed by the surroundings, but remains in the system in some form. The reaction products can be heated up so that the pressure rises (in the same way as in a gas explosion involving oxygen). The energy threshold ΔG^{A} is overcome more easily the higher the pressure and/or temperature of the gas. For safe handling of gases tend to decompose, it is urgently necessary to know the pressure and temperature limits such that the energy threshold cannot be exceeded, even by a strong ignition source. These limits can be determined only by experiment.

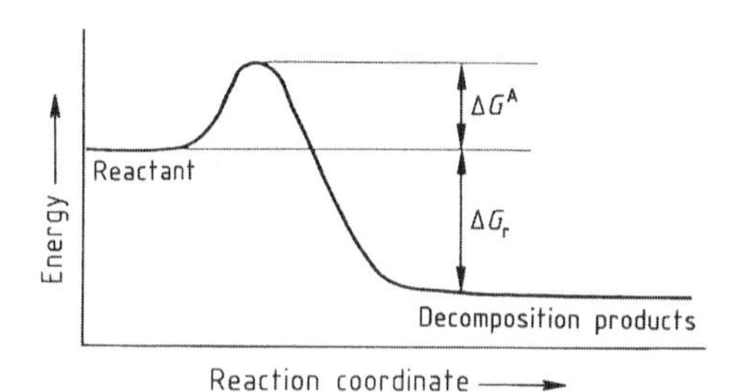

Figure 13. Energy diagram of a decomposition reaction

The apparatus described in Section 2.2.2.3 is well suited to studies of this kind. By performing a series of ignition trials, the researcher seeks to determine the pressure, as a function of temperature, at which it just becomes impossible to initiate decomposition. The result is the pressure limit of stability as a function of temperature. To provide at least approximate statistical support for the limit, the test is replicated five times. Figure 14 shows the pressure limit of stability for ethylene at 100 – 200 °C [59].

To deal with problems relating to the safe transportation of unstable gases, studies of the kind just described are run, up to 70 °C. Pressure limits of stability at even higher temperatures are of interest in view of operations in chemical plants. It is also often desirable to find out how the pressure limit of stability can be modified by the addition of a stable gas (not necessarily inert) or what amount of foreign gas must be added to render the system stable under the operating conditions. All these questions can be answered by experiments with the laboratory apparatus described.

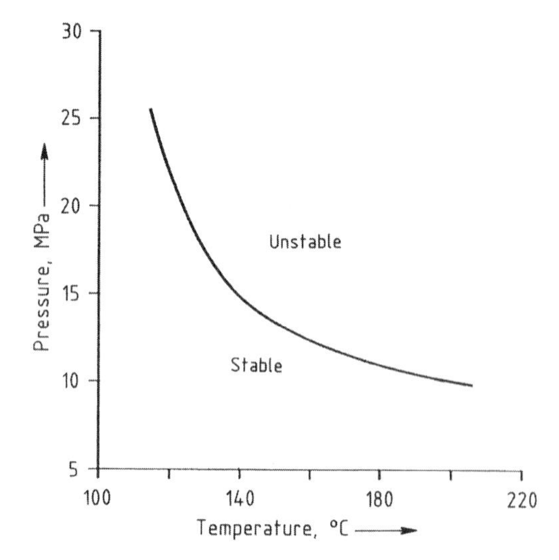

Figure 14. Pressure limit of stability for ethylene as a function of temperature, ignition power 1 kW

In practice, an adequate margin of safety relative to the pressure limit of stability must always be maintained when working with unstable gases.

2.2.2.5. Ignition Temperature, Ignition Energies

Explosive systems are characterized by certain parameters (LEL, UEL, $[O_{2max}]$, p_{max}, $(dp/dt)_{max}$, K_{G}) that can be employed in the design of explosion-protection measures. Two approaches to explosion protection are to prevent explosive mixtures from forming, and to insure safe handling of such mixtures by explosion-proof design with or without pressure relief. In a similar way, the sensitivity of these systems to ignition by certain ignition sources is described in terms of parameters that can be used in safety design (elimination of ignition sources).

It is also important in practice to find a parameter that characterizes fuels with regard to the propagation of explosions, for example, through narrow gaps. The propagation of an explosion from a part of the plant featuring explosion-resistant design into upstream or downstream sections that are not so designed, or that are not vented, must be effectively prevented.

One such parameter is the ignition temperature, which makes it possible to assess the effectiveness of a hot surface (e.g., a motor winding) as an ignition source. Others are the minimum ignition energy, used for evaluating electrostatic (capacitive) discharges, and the minimum ignition current, for evaluating electrical (inductive) sparks. Finally, the ability of exploding gas mixtures to propagate through a narrow gap is characterized by the maximum experimental safe gap (MESG), which can be used in the design of flame arresters or "pressure-resistant" enclosed electrical equipment (see EN 50 018).

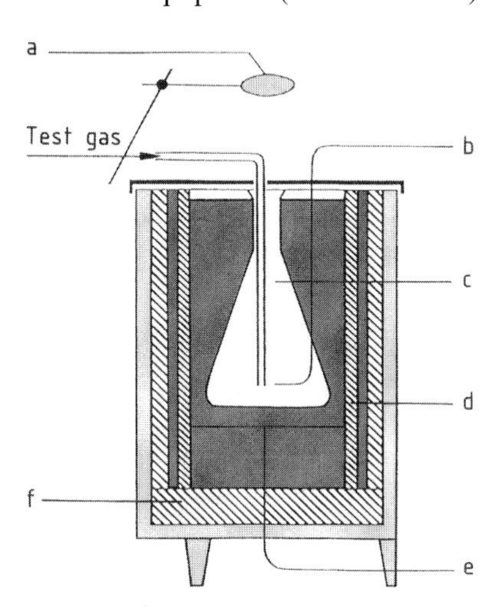

Figure 15. Apparatus for determining the ignition temperature (DIN 51 794)
a) Observation mirror; b) Thermocouple, FeCo; c) 200-mL Erlenmeyer flask; d) Heating; e) Resistance thermometer (Pt-100); f) Thermal insulation

The ignition temperature T_I is the lowest temperature of a hot surface at which a fuel – air mixture of the most easily ignitable composition can just be stimulated to burn with a visible flame. Two crucial points for determination of T_I are the geometry of the test vessel, which must provide a temperature field as closed as possible, and the formation of steep fuel concentration gradients when the fuel is admitted to the oxidizing gas phase already present in the vessel. The standard method of IEC 79-4 (embodied in DIN 51 794 and ASTM D 2155) takes these points into account. The glass ignition vessel (a 250-mL Erlenmeyer flask) is charged with air or another oxidizer. The fuel is admitted to the vessel through an inlet tube (for gaseous substances) or dropwise (for liquids; Fig. 15). The criterion for ignition is a flame forming in the vessel, which can be observed with mirrors.

On the basis of the ignition temperatures in air, fuels are placed in temperature classes (EN 50 014/5.78, DIN VDE 0165; Table 7). Ignition temperatures and temperature classes have been tabulated [42]. The maximum surface temperature T_{max} associated with a given temperature class must not be exceeded by explosion-resistant electrical (and other) equipment belonging to the class (EN 50 014, pr EN 1127).

Table 7. Temperature classes according to EN 50 014

Ignition temperature*, °C	Temperature class	Maximum surface temperature, °C
$450 < T_I \leq 600$	T 1	450
$300 < T_I \leq 450$	T 2	300
$200 < T_I \leq 300$	T 3	200
$135 < T_I \leq 200$	T 4	135
$100 < T_I \leq 135$	T 5	100
$85 < T_I \leq 100$	T 6	85

* Measured according to IEC 79-4 (e.g., DIN 51 794).

When determining the ignition temperatures T_I of fuel mixtures, it must be kept in mind that there is no linear relationship between the concentration ratio (or mole fraction) of the fuels and the ignition temperature for the mixture [46], [60]. T_I values for fuel mixtures must therefore be determined for every individual case. Ignition temperatures for a given fuel in oxidizing media other than air may sometimes be quite low. This is especially so for pure oxygen and chlorine, in which T_I may be lowered by as much as hundreds of kelvins: T_I (toluene, air) = 535 °C, T_I (toluene, Cl_2) = 175 °C. This effect with pure O_2 can be accounted for by the lack of a diluting inert gas; in other cases, e.g., Cl_2, the ignition temperature is lower because the Cl_2 molecule has a much smaller dissociation energy than O_2. Also, T_I generally decreases with increasing when initial pressure [60].

The minimum ignition current, used as a measure of the igniting ability of electrical (inductive) sparks, is a fuel-specific parameter. It is the smallest current flowing in a circuit of a certain inductance, such that the spark produced when the current flow is interrupted (in an explosive mixture of optimal ignitability) just ignites the mixture. If the minimum ignition current is known, electrical circuits can be designed so that no igniting sparks are created by switching, or on failures such as a short-circuit. This design approach is employed in explosion-resistant electrical equipment rated “intrinsically safe” (i; see EN 50 020).

The minimum ignition energy is the smallest amount of energy, stored in a capacitor, that is just sufficient, when discharged across a spark gap, to ignite the most ignitable explosive mixture. It is a fuel-specific parameter. As a measure of the effectiveness of capacitive sparks to ignite mixtures, it is particularly useful in the evaluation of electrostatic ignition sources.

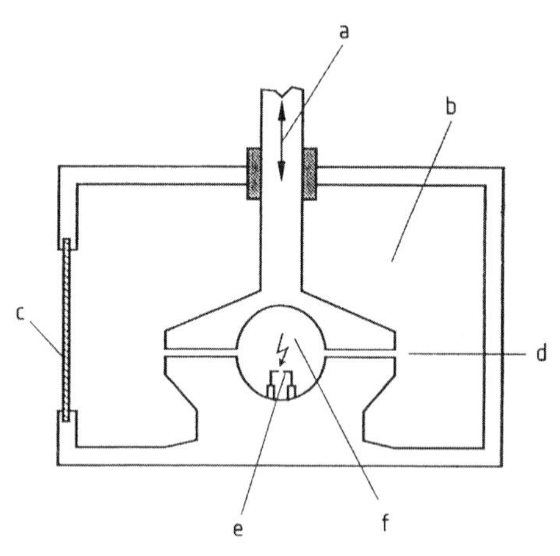

Figure 16. Apparatus for determining the maximum experimental safe gap
a) Gap adjustment; b) Outer chamber; c) Observation window; d) Gap; e) Ignition source; f) Inner chamber

The minimum ignition current, the minimum ignition energy, and the maximum experimental safe gap (MESG) make it possible to rank fuels with allowance for the ability of explosive mixtures to propagate initiating action through a narrow gap.

The apparatus for determining the MESG is shown in Figure 16. It consists of two chambers connected by an annular gap of constant length (25 mm) and variable width. The two chambers and the gap are filled with an explosive mixture of the most sensitive composition. After ignition in the inner chamber (ca. 20 cm^3), the presence or absence of a flame breaking through into the outer chamber is determined visually (cf. IEC 79-1 A).

On the basis of the MESG, fuels are classed into explosion groups (EN 50 014, DIN VDE 0165): I (CH_4; only relevant for hazards due to firedamp); II A (e.g., C_2H_6); II B (e.g., C_2H_4); II C (H_2, C_2H_2, CS_2).

This knowledge makes it possible to design objects such as flame arresters, which prevent the transmission of explosions between sections of the plant, or to build explosion-resistant electrical equipment with the “flame-proof enclosure” (d) rating (EN 50018). A lamp switch housing so rated is constructed so that ignition of the mixture inside the housing does not propagate outside it; this requires the housing also to be explosion-pressure resistant.

The great advantage of being able to rank fuels is that a knowledge of the explosion group is enough for the selection of suitable explosion-resistant equipment for ignition-source elimination or explosion decoupling of certain plant sections; there is no need for costly experimental studies.

When a gas mixture contains species belonging to different explosion groups, it is necessary to determine whether the properties of the sensitive components (e.g., H_2 or other Group II C substances) govern the explosion propagation behavior of the mixture as a whole.

2.2.3. Explosion Indices of Dust-Air Mixtures

2.2.3.1. Introduction

All substances that burn in the solid state may “explode” when finely dispersed in air in the form of a dust cloud. The violence of such a dust explosion, which is best characterized by its pressure – time diagram, is similar to that of the explosion of a homogeneous gas – air mixture. In the latter, fuel dispersion takes place on a molecular level whereas in dust – air mixtures the size of the dispersed particles is typically 10 – 200 μm. Historically, dust explosions have been associated with coal mining and grain milling. Further industrialization has led to dust explosions in the chemical, pharmaceutical, metallurgical, food, and wood processing industries. All these industries have supported research to investigate measures to prevent dust explosions, or to limit their effects [61–66].

The probability of a dust explosion is highest for very fine dusts (particle diameter < 63 μm). The upper size limit for particles forming an explosible dust cloud is ca. 400 μm. Handling or processing of coarser product may lead to the accumulation of fines (e.g., by abrasion) and thus to the formation of an explosible dust cloud.

Dust explosions may occur within apparatus in which combustible powders are handled, transported, or processed, and in which dust clouds at explosible concentrations are difficult to avoid. They may also occur in rooms with deposits of combustible powders on the floor or other surfaces when this powder is taken up in the blast wave of a primary explosion, or by some other shock wave, in the presence of an ignition source. Layers of deposited powder as thin as 0.2 mm may be sufficient to fill a room homogeneously with an explosible dust cloud.

In contrast to gas – air mixtures, which are sufficiently identified by the nature and concentration of the flammable gas, dust – air mixtures need more parameters for their exact description. Owing to gravity the particles do not stay long in suspension; the large particles settle rather quickly. Therefore, to maintain a dust cloud for a certain amount of time, and to improve homogeneity, the particles should continue to move around. A dust cloud is not normally a stationary system, but maintains a degree of turbulence which is one of the additional parameters. Others are particle size distribution (or median of the mass distribution in the case of rather monodisperse powder), water content of solid particles, and dust concentration. Ignition and explosion characteristics of dust clouds may change drastically when a flammable gas or vapor is also present (hybrid mixture).

Because of the complexity of dust – air mixtures, it is necessary to standardize methods to determine the explosion indices which describe ignition and explosion behavior.

2.2.3.2. Ignition Sensitivity of Dust Clouds

Minimum Ignition Temperature (MIT). The MIT is the lowest temperature of a hot surface at which a dust – air mixture is ignited on contact. For determination of the MIT the Godberg –Greenwald furnace or the BAM oven are used [65–67]. The Godberg – Greenwald furnace consists of an electrically heated vertical tube (Fig. 17). A dust sample is dispersed by a blast of air from a chamber containing pressurized air, and flows through the heated tube. The temperature of the tube is varied until flames are seen to emerge from the heated tube. The Godberg – Greenwald furnace is in the process of becoming an IEC standard.

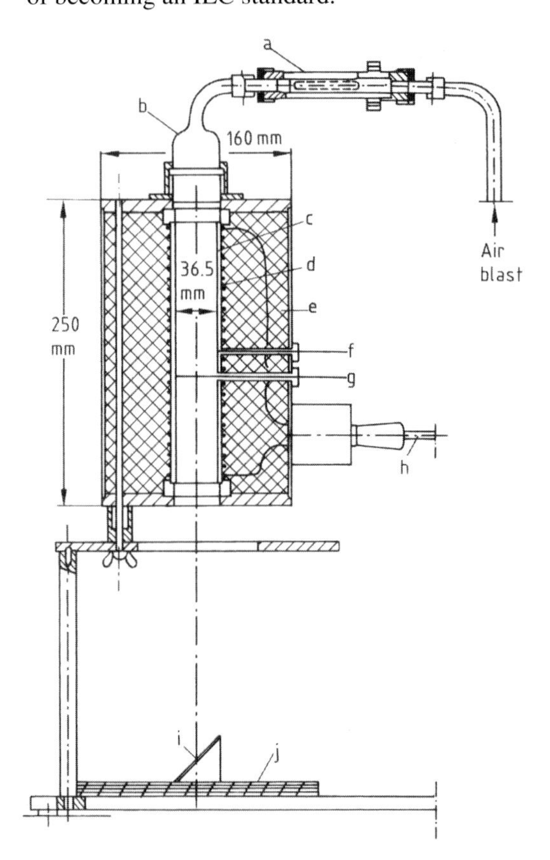

Figure 17. Godbert – Greenwald furnace
a) Dust sample chamber; b) Glass adaptor; c) Ceramic tube; d) Heating coil (1 kW); e) Thermal insulation; f) Control thermocouple; g) Measurement thermocouple; h) Power; i) Mirror; j) Heat-resistant plate

The BAM oven consists of an electrically heated horizontal pipe (Fig. 18). Close to the center of the pipe there is a cap shaped impact plate to which the dust sample is blown from the outside with air. Starting at 600 °C, the temperature is gradually decreased until ignition no longer occurs.

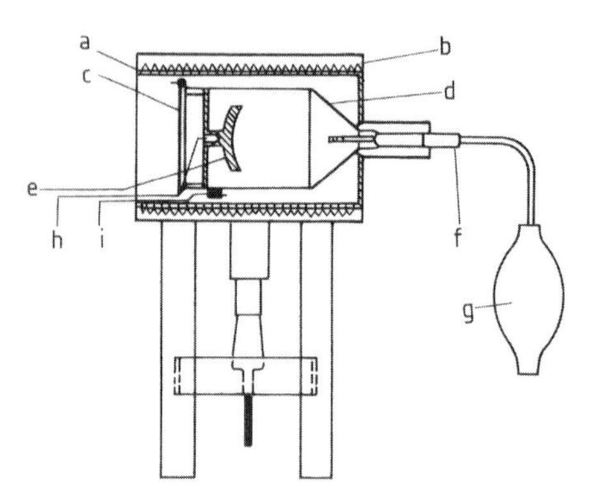

Figure 18. BAM oven
a) Oven; b) Heating coil (1.5 kW); c) Flap; d) Test chamber; e) Deflecting surface; f) Air inlet pipe; g) Rubber bulb; h) Measurement thermocouple; i) Control thermocouple

Minimum Ignition Energy (MIE). The MIE is the lowest value of the energy stored in a capacitor which, when released in a spark discharge, is just sufficient to ignite the most readily ignitable dust – air mixture at atmospheric pressure and room temperature. No internationally accepted test method exists for determination of the MIE of dispersed powder, but an IEC standard is being prepared [68]. The dust is dispersed in a modified Hartmann apparatus (Fig. 19), a 1-m^3 vessel (Fig. 20) [66], [67], or a 20-L apparatus (Fig. 21) [66], [67]. The spark gap must be adjusted to 6 mm and the discharge circuit should meet special requirements. The inductance of the circuit should be 1 – 2 mH, except when data are to be used for assessment of electrostatic ignition hazards, when it should be $\leq 25\ \mu H$. On varying the ignition delay time

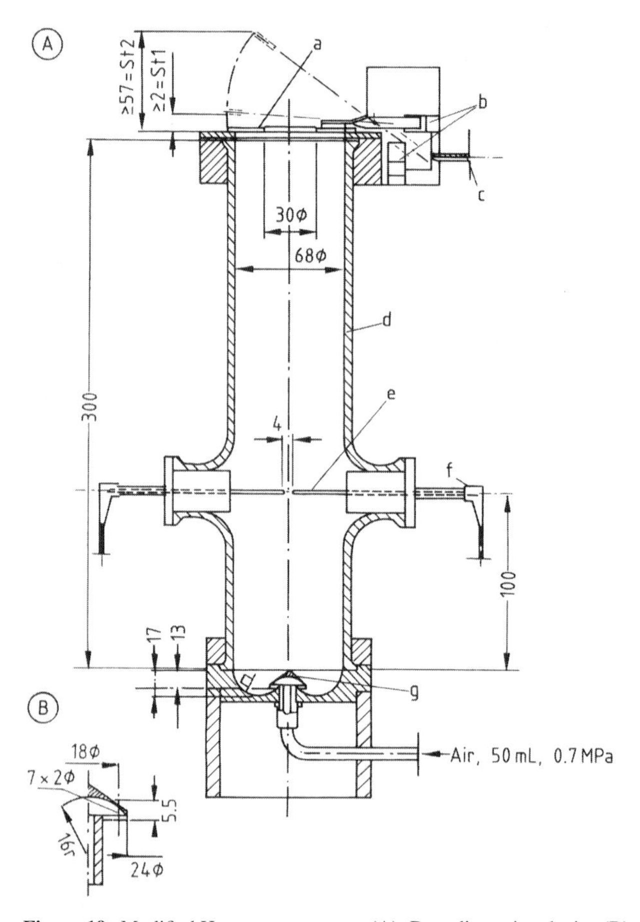

Figure 19. Modified Hartmann apparatus (A); Dust dispersion device (B)
a) Cover (Al, 17.2 g); b) Induction coil; c) Connection for digital indicator; d) Pyrex tube; e) Electrodes; f) Connection to transformer; g) Mushroom-type dust dispersion nozzleDimensions given in millimeters.

(between dust dispersion and activation of the ignition source) in the modified Hartmann apparatus, as well as the dust concentration (in all types of enclosure) the energy is always decreased by the same factor until no ignition occurs in a defined number of successive trials. The actual MIE lies between this lowest energy and the next higher energy at which ignition persists. Both values are usually reported.

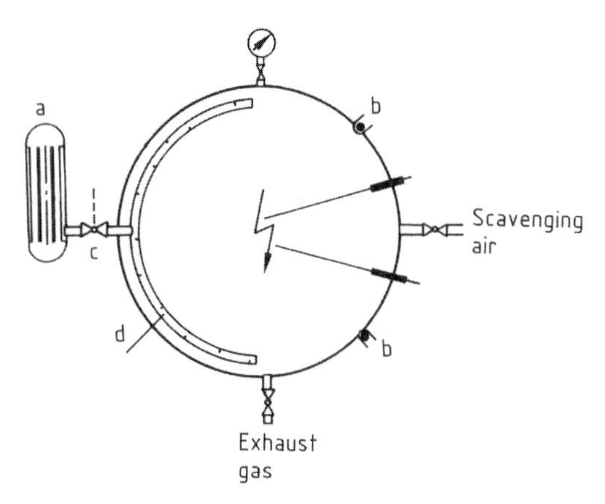

Figure 20. 1-m^3 Vessel
a) Dust container; b) Pressure sensor; c) Detonator-activated valve; d) Semiannular dispersion nozzle

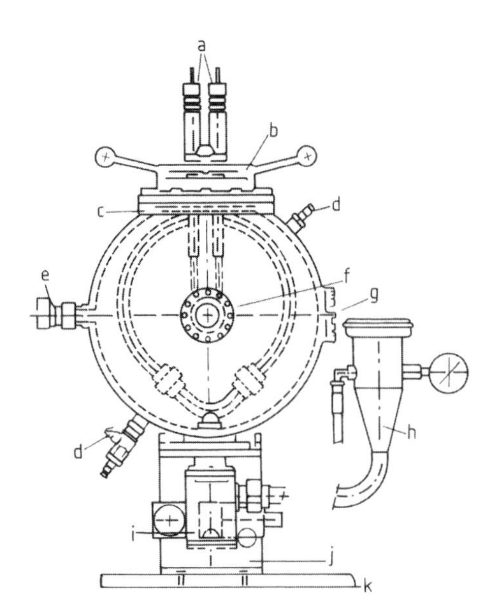

Figure 21. 20-L Apparatus
a) Ignition leads; b) Bayonet closure; c) Flange; d) Thermostatically controlled heating/cooling; e) Vent; f) Measuring flange; g) Sight glass; h) Dust storage chamber; i) Outlet valve; j) Support; k) Base plate

The MIE of fine organic powders is typically ca. 1 – 1000 mJ. It decreases with increasing temperature.

2.2.3.3. Dust Explosibility

The fact that a dust cloud is not normally stationary has implications for the course of the flame front after ignition. From studies of flammable gases, it is known that any disturbance of the flame front results in an increase of the reaction interface and thus leads to an increase of the rate of pressure rise. The same is true for combustible dusts. The rate of pressure rise measured in any test apparatus therefore depends on the dust dispersion procedure and the exact moment of ignition. Knowledge of the rate of pressure rise is most important when designing explosion venting or explosion suppression. It is necessary to use a procedure for the generation of the dust cloud which can easily be reproduced, and which represents the worst turbulence characteristics that may occur in practice.

The ISO dust dispersion procedure, developed in Germany in the 1960s, meets these requirements. This procedure is used worldwide, in combination with 1-m^3 or 20-L explosion chambers. In combination with the 1-m^3 vessel, it has become an ISO standard [69]. In Germany, this dust dispersion procedure in combination with the 1-m^3 vessel or the 20-L apparatus is recommended in the VDI Guideline 2263 [67]. At CEN level, work is in progress on standardization of dust explosion indices on this basis. The dust sample is kept at high air pressure (2 MPa) in a storage container connected to the explosion chamber. For generation of the dust cloud, the powder is released into the explosion chamber via a rapid-action valve and a specially designed dust dispersion system. After a well-defined interval (ignition delay time) a chemical ignitor, located at the center of the explosion chamber, is activated, and the pressure within the explosion chamber is recorded as a function of time.

For many years the Hartmann apparatus has been used for dust explosibility tests (e.g., the closed Hartmann apparatus and the modified Hartmann apparatus). The latter (Fig. 22) has a hinged top cover. When an explosion occurs, the cover is lifted to a certain degree, depending on the violence of the explosion (i.e., the pressure rise per unit time). The closed Hartmann apparatus is almost obsolete, but the modified Hartmann apparatus is still useful as a screening instrument. However, for reliable dust explosion indices (lower explosion limit, LEL, limiting oxygen concentration LOC, maximum explosion pressure p_{max}, maximum rate of pressure rise $(dp/dt)_{max}$, or the constant K_{max}, required for the design of safety measures such as inerting, explosion venting [70] or explosion suppression [71], it is necessary to use results from tests performed in accordance with the ISO standard test procedure in the 20-L apparatus or the 1-m^3 vessel.

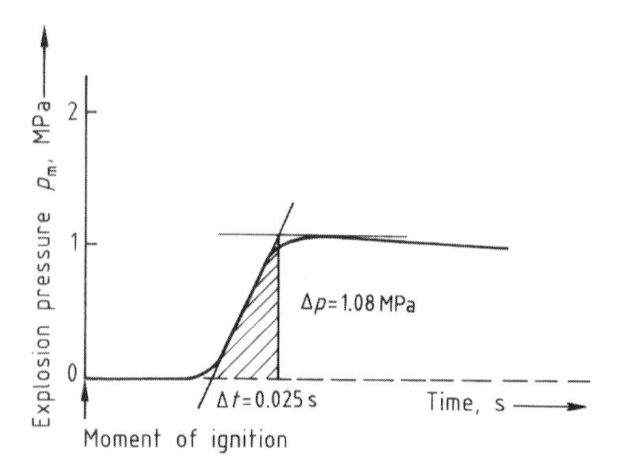

Figure 22. Explosion pressure in an enclosure as a function of time

$$\left(\frac{dp}{dt}\right)_m = \frac{1.08\,\mathrm{MPa}}{0.025\,\mathrm{s}} = 43.2\,\mathrm{MPa/s}$$

Explosion Limits (EL). The explosion limits describe the range of dust concentrations in air within which an explosion is possible. Usually, only the lower explosion limit (LEL) is determined. Owing to the rather fast settling of dust particles, the upper explosion limit (UEL) is usually not significant. It is important to note that the LEL loses its importance if deposits of fine powder are present which may become dispersed in an uncontrollable way. For the determination of the LEL of powders the 1-m^3 vessel or the 20-L apparatus are used, the former in combination with chemical ignitors of total energy 10 kJ, the latter with chemical ignitors of total energy 1, 2, or 10 kJ (the exact value is still a matter of debate). Starting with an explosible concentration, the concentration is gradually reduced until there is no longer an explosion. The dust concentration (g/m^3) at which ignition just fails over a defined number of successive trials is stated as the LEL.

The LEL can also be calculated from the heat content of the dust [72–74]. For organic powders it is typically ca. 15 – 60 g/m^3. The LEL decreases with increasing temperature and increases proportionally with increasing absolute pressure.

Limiting Oxygen Concentration (LOC). The LOC is the experimentally determined maximum residual oxygen concentration in a mixture of air and inert gas at which a dust explosion does not occur. It depends on the dust and the type of inert gas. For determination of the LOC of powders, the 1-m^3 vessel or the 20-L apparatus are used (the former in combination with chemical ignitors of total energy 10 kJ, the latter with chemical ignitors of total energy 1, 2, or 10 kJ). Starting from an oxygen concentration, at which explosion occurs, the oxygen concentration is gradually reduced, varying the dust concentration until no explosion occurs. The oxygen concentration (vol %) at which no explosion occurs over a defined number of successive trials is stated as the LOC.

For organic powders and nitrogen as inert gas, the LOC is usually 9 – 14 vol %; for fine metal powders values down to 5 vol % are reported [66]. It decreases with increasing temperature; the pressure has only a minor influence on the LOC measured in vol %.

Maximum Explosion Pressure (p_{max}) and constant K_{max}. p_{max} is the maximum explosion pressure of a combustible powder, in a closed system at optimum concentration. In spherical or cubic enclosures with the ignition source located at the center, p_{max} is independent of volume. K_{max} is the rate of pressure rise per unit time, measured in a 1-m^3 enclosure at optimum concentration. The dependence of the rate of pressure rise on the volume V is known as the cubic law (see also Section 2.2.2.3):

$$K_{max} = \left(\frac{dp}{dt}\right)_{max} V^{1/3} \qquad (17)$$

This relationship is used to calculate K_{max} from results obtained in test apparatus of any volume.

Both, p_{max} and K_{max} can be determined in the 1-m^3 vessel or the 20-L apparatus with chemical ignitors of total energy 10 kJ located at the center. The dust cloud is generated according to the ISO standard procedure and ignited after a well-defined ignition delay time. The explosion pressure p_m is measured as a function of time (Fig. 22). Such explosion tests are performed over a wide range of dust concentration until no further increase in either p_m or $(dp/dt)_m$ is observed. From a plot of p_m and $(dp/dt)_m$ against dust concentration, the final quantities p_{max} and $(dp/dt)_{max}$ are determined (Fig. 23).

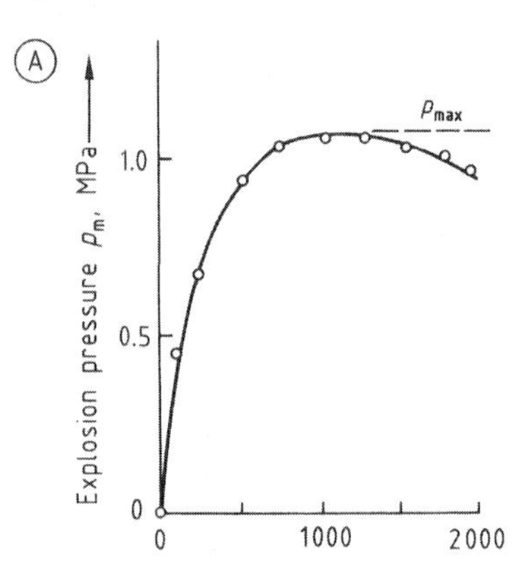

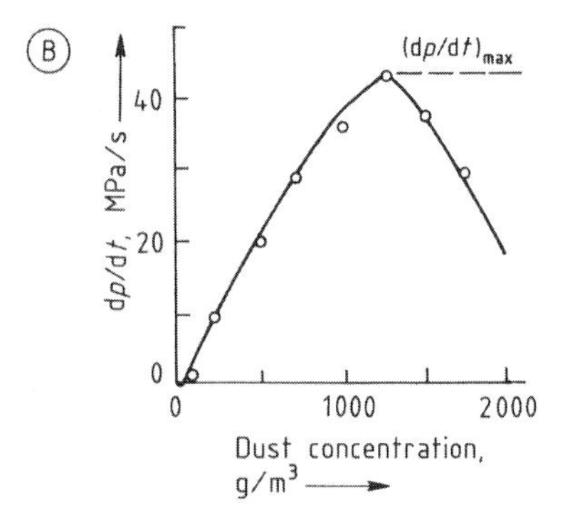

Figure 23. Explosion pressure p_m (A) and $(dp/dt)_m$ (B) as a function of dust concentration
Maximum explosion pressure 1.08 MPa; maximum rate of pressure rise 43.2 MPa/s.

Fine organic powders usually have p_{max} ca. 0.6 – 0.9 MPa, and K_{max} ca. 5 – 30 MPa m s^{-1}. Very fine aluminum powder may have p_{max} up to 1.3 MPa and $K_{max} > 100$ MPa m s^{-1}. Substances are classified in accordance with their explosion violence, as expressed by K_{max}, tabulated as follows:

K_{max}, MPa m s^{-1}	Dust explosion class
<0 – 20	St 1
20.1 – 30	St 2
>30	St 3

Determination of p_{max} and K_{max} by the standard procedure described is the decisive test of whether a powder can be ignited at all in the form of a dust cloud.

p_{max} increases proportionally with increasing absolute pressure at least up to 1.1 MPa initial absolute pressure. For K_{max} this is true only up to initial pressure 0.2 – 0.4 MPa; p_{max} decreases with increasing temperature, and the effect of temperature on K_{max} cannot generally be predicted.

2.2.4. Characterization of Explosive Condensed Substances

2.2.4.1. Introduction

Explosions can occur in the gaseous phase as well as in the liquid or solid phase. Three types of explosions have been defined in Chapter 1.

- Explosions due to spontaneous decomposition
- Explosions due to spontaneous combustion
- Thermal explosions

Gas-phase explosions (due to spontaneous combustion) and the thermal explosions are described in Sections 2.2.2, 2.2.3, and 2.3. This section discusses condensed explosive substances which are likely to undergo explosions due to spontaneous decomposition.

2.2.4.2. Chemical Nature and Definition of an Explosion

A condensed-phase explosion is defined as an exothermic chemical reaction, during which gases and vapors are formed at such a high rate as to have destructive effects on their surroundings. Fragments, projectiles, ground and air shock waves are produced by the high temperature, pressures, and rates of pressure rise. The extent of damage depends on the type and the mass of the explosive substance as well as on the nature of the surroundings.

An explosive reaction can be initiated only in substances which have a high positive heat of formation or a high negative heat of decomposition. Such substances usually have a defined chemical structure. Being organic in nature and possessing, in most cases, reactive groups with available oxygen, they can undergo intramolecular oxidation of the combustible part. Examples of oxidizing reactive groups are nitro, nitroso, and peroxy groups, but organic compounds containing azide, acetylenic, or diazonium groups may also be powerfully explosive substances. Explosive compositions may also be produced by mixing inorganic oxidizing substances with combustible materials; well-known examples are: mixtures of potassium nitrate, sulfur, and carbon (black powder); ammonium nitrate and fuel oil (ANFO); nitration mixtures containing nitric acid; and oxidation mixtures with hydrogen peroxide. BRETHERICK [75] and KING [76] give guidance for identifying substances or mixtures as potentially explosive because of the presence of reactive groups. When a substance or mixture contains reactive groups, it depends on the molecular mass and the type and number of reactive groups, or on the oxygen balance, whether and with what intensity explosive properties are exhibited. An important difference between gas-phase and condensed-phase explosions is that in the latter the participation of oxygen from the air is not necessary.

The power of an explosion is closely related to the dynamics of gas production and energy release, as well as to the amount of gas produced and the total heat of decomposition. The energy release when a substance explodes is difficult to determine because it depends on the resulting explosion products. There may be considerable uncertainty in reported heats of decomposition of explosive substances. A slow reaction at low temperature, due to a low degree of adiabatic efficiency under the test conditions, may result in a less energetic decomposition measurement. It is known [77] that the decomposition energy measured can be an underestimate of the energy actually available. On the other hand, computer programs such as CHETAH (chemical thermodynamic and hazard evaluation [78]) can be used to predict the maximum heat of decomposition for any substance or mixture of known chemical formula and structure. As a rough guide, substances with decomposition energy > 500 J/g may have explosive properties, and those > 800 J/g may be detonable. Although this is often a good qualitative prediction of explosivity, reliable assessment of an explosive substance necessarily requires experimentally determined data.

2.2.4.3. Grouping of Explosive Substances

Explosive substances are grouped according to their intended use and/or the type of hazard connected with handling them.

When substances, because of their energy content, are used to produce explosive, initiating, propellant, or pyrotechnic effects, they are called explosives. They may be used, e.g., for blasting in rocks, coal, or salt, as initiating explosives in blasting caps (detonators), as propellant powders in ammunition, or as components of pyrotechnic articles.

All explosives are regulated internationally with respect to handling, storage, and transport. Except for international transport, this is not so for explosive substances used for chemical, scientific, or technical purposes.

According to the type of associated hazard, explosive substances (especially explosives and articles with explosive properties) are divided into groups representing:

1) The hazard of a mass explosion (mass detonation)
2) The hazard of explosions (detonations) together with the production of fragments, without the hazard of mass explosion
3) The hazard of a mass fire
4) Ordinary fire hazard

This grouping is the basis for all modern regulations and safety measures during transport and storage (e.g., safe distances).

2.2.4.4. Explosion Mechanisms

Commonly, "explosion" is used collectively for the three distinguishable mechanisms by which a substance can explode. The thermal explosion proceeds, usually in the liquid phase and more or less homogeneously, as a temperature-controlled (Arrhenius law) self-accelerating reaction (see Section 2.3). When, during the homogeneous decomposition, a temperature gradient develops, a deflagration may start at the hottest spot. It passes through the substance as a reaction front in which the heat of reaction and the reaction products are liberated. The deflagration is propagated by conductive, radiative, or convective heat transfer into the unreacted materials. If a substance is liable to deflagrate, this reaction can be initiated locally, e.g., by flame, heat, impact, or friction. The deflagration velocity increases with the energy content, temperature, and porosity of the substance, and exponentially with pressure. Typical values are 0.003 – 100 m/s.

The pressure dependence of the reaction velocity explains why deflagration of a small quantity of material may proceed slowly, whereas under confinement by pressure buildup, or in even larger masses by self-confinement, it occurs with explosive violence. When the speed of the gaseous deflagration products reaches sonic velocity, a shock wave develops which propagates supersonically into the unreacted substance; a deflagration to detonation transition (DDT) has taken place. Compression combined with strong heating initiates chemical reaction of the substance. Typical detonation velocities are 1000 –8000 m/s; detonation pressures reach values up to several thousand mega pascales. Detonation velocity increases with the energy content of the substance and its density. The detonation reaction may be initiated locally by heat, impact, friction, DDT, or a shock wave from other detonating substances, leading to mass explosion, and producing disastrous damage. Primary (initiating) explosives give rise to direct immediate detonation by flame, impact, or friction in quantities of a few milligrams. Secondary explosives may be detonated by primary explosives or by DDT in larger quantities. The nature of detonation is further described in the thermohydrodynamic theory of detonation [79–81].

2.2.4.5. Testing of Explosive Substances

To determine the risks associated with handling an explosive substance, experimental investigation into its individual explosive properties is imperative. Explosive properties refer to the mechanisms by which an explosive reaction can proceed, to the types of stresses and the ease with which explosions can be initiated (sensitivity), and to the power of the explosion once it takes place. Understanding the initiation, propagation, and the possibility of terminating explosive reactions, and recognition of any destructive potential form the basis for safe handling of an explosive substance [82]. The sensitivity of an explosive substance is of special importance since, if it is too high, it may preclude some or all modes of technical handling.

For several decades, explosive properties have been described by test data determined by standardized test methods [83], [84]. The latest and most comprehensive collection of tests, covering all relevant properties together with evaluation criteria for classifying explosive substances for the purpose of transport, is given in the UN Test Manual [85]. In Part I, the Manual contains test methods and criteria, together with a classification flowchart, for deciding whether a substance or mixture is an explosive substance or article of Class 1 of the UN Recommendations on the Transport of Dangerous Goods [86], and which division of Class 1 best reflects its risks during transport. Part II of the Manual presents tests methods, criteria, and a flowchart for evaluating organic peroxides and self-reactive substances (Divisions 5.2 and 4.1 of the UN Recommendations) for explosive properties.

The Manual includes ca. 70 test methods on laboratory and field scale for the substances (packed as for transport). The test methods are related to the properties to be identified:

- Detonability by shock wave, including sensitivity to detonation shock of variable strength, detonation velocity, and sensitization by cavitation (gas bubbles)
- Deflagration after ignition in an open vessel or under confinement, determining the linear deflagration velocity or the rate of pressure rise
- DDT after ignition under confinement
- Thermal sensitivity to heating under variable defined degrees of confinement, the sensitivity being characterized by the limiting diameter of a pressure-relief vent, or by the maximum pressure and rate of pressure rise
- Sensitivity to mechanical stresses (impact and friction), determination of the sensitivity limits
- Explosive power after initiation with a blasting cap or thermal decomposition, measuring the work performed or the specific energy
- Thermal stability of explosives on storage at 75 °C, or of organic peroxides and self-reactive substances on storage under adiabatic, isothermal, or heat-accumulation conditions, determining the self-accelerating

decomposition temperature (SADT) related to the packed substances

The European Community has provisionally agreed on three methods for evaluating a substance as explosive or not. These screening methods include testing the sensitivity of substances to impact, friction, and heating under partial confinement [87]. The EC regulates the classification, packaging, and labeling of such substances for all modes of handling except transport.

2.2.4.6. Conclusions

Use of standardized test methods resulting in criteria for assessing the safety of chemical procedures and installations involving explosive substances should take special note of certain factors limiting their applicability.

It requires considerable experience to perform the tests, to modify them if necessary, and to evaluate observations correctly. Each test result should be assessed for plausibility in connection with all other results. Sensitivity does not depend on the energy content of the substance, as does the explosive power. To increase the reproducibility of test results, standardized test procedures often reflect idealized conditions. The heating rate, degree of filling of the apparatus, and the strength of confinement may greatly influence the results of thermal tests. Problems of scaling must also be addressed.

Any test result corresponds to the conditions under which it was obtained, and its validity for substances in a particular plant should be thoroughly checked. Special care should be taken before neglecting a given detonability of a substance. With respect to production in a chemical plant (excluding the explosives industry), it is imperative that initiation of a detonable substance under the conditions of handling should be impossible.

2.3. Exothermic and Pressure-Inducing Chemical Reactions

2.3.1. Introduction

The overwhelming majority of chemical reactions are accompanied by heat release, i.e., the overall change in enthalpy between the starting materials and products is negative (exothermic).

The heat produced heats up the material itself, the container (e.g., the reaction vessel), and/or the surroundings. According to the law of conservation of energy, these heat flows equilibrate in a heat balance (Fig. 24). The degree of heat dissipation depends on the heat capacity of the system itself and the heat removal capacity provided by the reaction vessel design and/or the latent heat of phase transitions. If the heat production rate of a chemical process exceeds the heat removal capacity of the system, and if there are no self-stabilizing boundary conditions, e.g., reaching of a boiling point, self-acceleration will occur, resulting in a "runaway" reaction. As runaway reactions are rather hazardous, it is necessary to understand the underlying mechanism, as well as the systematic experimental test procedures to assess the hazard potential for a runaway, and to know the design criteria for safe processes in order to prevent runaway scenarios from occurring.

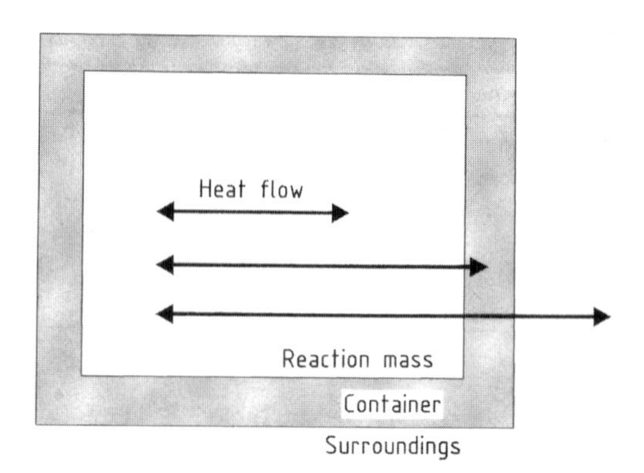

Figure 24. Schematic representation of heat balance [88]

2.3.1.1. Exothermic Reactions: Runaway Potential

To assess the runaway potential of a chemical system it is necessary to understand the basic physicochemistry and its interaction with the transport phenomena of heat, mass, and momentum, which determine the behavior of the system, including the reactor. The hazardous character of a chemical reaction is determined by the

Table 8. Typical heats of reaction for common chemical processes [89]

Reaction type	ΔH, kJ/mol	Reaction type	ΔH, kJ/mol
Neutralization (HCl)	− 55	Hydrogenation (nitroaromatics)	−560
Neutralization (H_2SO_4)	−105	Amination	−120
Diazotization	− 65	Combustion (hydrocarbons)	−900
Sulfonation	−105	Diazo-decomposition	−140
Nitration	−130	Nitro-decomposition	−400
Epoxidation	− 96		

overall heat of reaction (thermodynamics), and the rate of heat production (strongly influenced by the kinetics). Typical values for heats of reaction are summarized in Table 8. The total heat output of a real system can only be influenced by the concentration of the reacting materials. In the sense of a primary method to prevent runaway reactions it can be stated that the more dilute a system the smaller is its runaway potential. The disadvantage of this rule, besides its effect on the economics of the process, is the environmental consequence of a higher demand for auxiliary materials and energy consumption for their recovery. Consequently, an inherent runaway potential has to be accepted, but must be adequately controlled, either kinetically (e.g., by choice of process temperature), or technically (by reactor design).

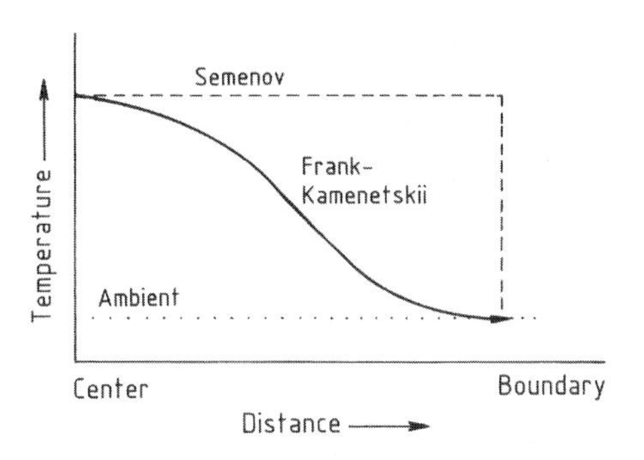

Figure 25. Semenov (uniform) and Frank-Kamenetskii (nonuniform) temperature profiles

As already mentioned, control of chemical heat release depends on the ability of the system to dissipate this energy. Mathematically this is formulated as a heat balance, including an accumulation term as well as the heat generation and dissipation terms. Appropriate description of the dissipation terms depends strongly on the heat transport mechanism (conductive or convective, free or forced). For a uniform temperature distribution within the reaction mass, which is usually the case in well-mixed gases or Newtonian liquid, heat transfer is governed by the boundary, e.g., across a reactor jacket. This situation was originally discussed by SEMENOV [90].

The other extreme, nonuniform temperature distribution and heat loss governed by conduction through the bulk, is a good model for large, unstirred masses, solids, and powders and was first discussed by FRANK-KAMENETSKII [91]. Both cases are shown in Figure 25. For more detailed discussion of self-heat models see [92].

For phenomenological discussion of runaway reaction systems the Semenov model [90] is the more adequate. It assumes a chemical reaction $A \rightarrow B$ with an indefinite amount of initial material, so that its consumption may be neglected. It is further assumed that the temperature dependence follows the Arrhenius relationship. Based on this pseudo-zero-order kinetic approach, the heat production rate $\dot{Q}_R$ is governed by an exponential dependence on temperature:

$$\dot{Q}_R = V\,(-\Delta_R H)\,k_\infty c_{A,\infty} \exp\left(-\frac{E}{RT}\right) \qquad (18)$$

The rate of heat removal $\dot{Q}_C$, assuming the coolant has Newtonian properties, depends linearly on the driving temperature difference between the uniform reaction mass temperature and ambient (jacket) temperature:

$$\dot{Q}_C = U A\,(T_R - T_A) \qquad (19)$$

The controllability of the heat production rate can best be explained by plotting the two heat flows (Eq. 18 and 19) as a function of temperature (Fig. 26). Three different cases can be discussed: with two, one, or no intersections. These three cases can be obtained either by varying the ambient temperature (Fig. 26) or by varying the slope of the heat removal lines (equal to the product of heat transfer area and overall heat transfer coefficient). The intersections represent pseudo-steady-state conditions. In case

1 (low T_A), small deviations from the steady state, represented by the lower of the two intersections, automatically result in a return to the origin, as can easily be deduced from a comparison of the relative magnitudes of the two heat flow terms. This operating point is rated "stable". With respect to the upper steady state, once a temperature deviation occurs, the original operating conditions are never reached again. In the case of a temperature decrease, the process quickly approaches the lower steady state; for a temperature increase, the heat production rate always exceeds the heat removal capacity of the system. This leads to an unhindered self-acceleration of the reaction rate and thereby of the heat production rate (thermal explosion or runaway reaction). The same is true for all operating conditions of case 3 (high T_A). Case 2 (medium T_A) represents the limiting case of the first occurrence of an unstable operating point, characterized by the equality of rates and their temperature derivatives. From these characteristics, it can be shown mathematically that, as long as the following relation holds:

$$\frac{RT_R^2}{E} > T_R - T_A \qquad (20)$$

runaway conditions need not be expected. This may be regarded as a first rule of thumb for the safety assessment of chemical processes. More sophisticated design criteria, presented in 2.3.1.4 and Section 2.3.3.1, extend the stability discussion of chemical processes from this pseudo-steady-state approach to a discussion of dynamic stability.

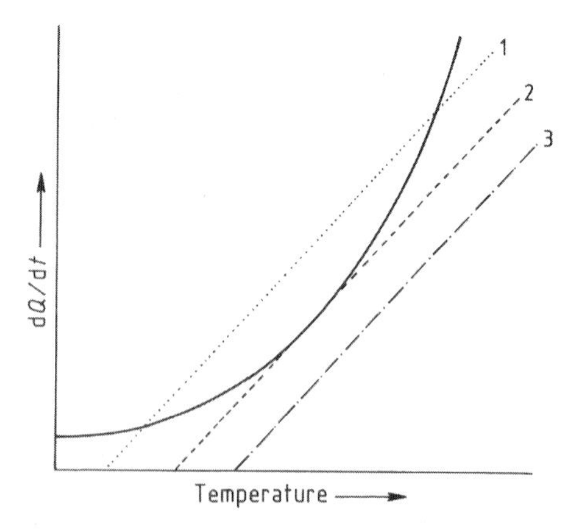

Figure 26. Semenov plot
1) Low T_A; 2) Medium T_A; 3) High T_A

2.3.1.2. Causes and Consequences of Overpressure-Inducing Exothermic Reactions

If the defined operating conditions do not include extensive overpressure, as is common for a number of petrochemical processes and hydrogenations, the sudden and unexpected occurrence of overpressure must be regarded as hazardous. The discussion of causes and consequences of a pressure increase must begin with a distinction between effects related to gas-generating chemistry and those related to process design.

For chemistry-related effects, the prime cause of overpressure is the formation of noncondensable gases. These may be reaction products of the desired process or of secondary reactions initiated at the elevated temperatures due to a runaway. For the first category, the basis of a suitable safety assessment is a fundamental understanding of the chemical process. Many common chemical reactions produce large amounts of gases, e.g., reductions with $LiAlH_4$ or substitutions of NH_2 groups by halogens via diazonium intermediates. Less obvious processes are those where the gas is formed in parallel reactions, e.g., the formation of gaseous ethylene oxide from 2-chloro-1-ethanol in the presence of a strong base. Provided that the reaction has been properly identified, these kind of gas-generating processes are controllable by conventional means, as their generation rate is kinetically controlled.

The most important class of gas-generating processes are decomposition reactions. These are initiated, e.g., when a runaway reaction raises the operating temperature so that the decomposition reaction overwhelms the desired reaction, or when cooling medium comes into contact with the reaction mixture, owing to a jacket or pipe rupture. Many of such undesired reactions stoichiometrically produce more than 1 mol gas per 1 mol initial material, which accelerates the pressure increase dramatically. In these cases, the process cannot be controlled and only mitigating measures can help reduce the consequences. It is therefore important to assess the process in all conceivable fault conditions, in order to initiate appropriate process or plant modifications to make the process inherently safe, or to define suitable preventive measures as the second best choice.

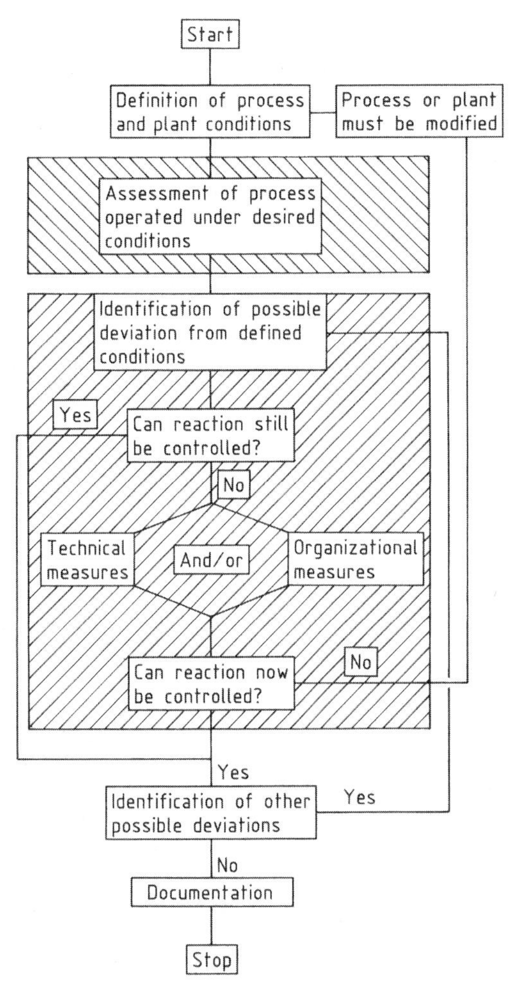

Figure 27. Process hazard and safety evaluation (PHASE)

2.3.1.3. Processes Hazard Assessment and Safety Evaluation for Exothermic Reactions

Process hazard assessment and safety evaluation (PHASE) [93] is a mandatory procedure to assure safe manufacturing of chemicals. The PHASE procedure has two characteristics:

1) It has to follow predefined systematic pathways to optimize its efficiency to identify as many hazardous situations as possible, and to assure the implementation of recommended safety measures
2) It is not performed once and for all, but should be used repeatedly during the lifetime of a plant or process

PHASE has to be applied to differing degrees, depending on the status of development of a process. Preliminary assessments should be performed on laboratory scale. The first fully comprehensive application of PHASE is recommended before the first run, on pilot plant scale. Later it should be repeated whenever a process scale-up or process/plant modifications are intended.

PHASE is a stepwise iterative procedure (Fig. 27). It starts with an assessment of all chemicals, together with operating and plant conditions, for the desired process. If hazardous situations are identified at this stage, process or plant modifications have to be initiated. Experimental methods and evaluation criteria for this purpose are the subject of Sections 2.3.2 and 2.3.3. If the desired process is rated as safe, possible fault conditions and process deviations have to be identified and evaluated. If the implementation of technical or organizational preventive measures cannot assure safe operation, further process or plant modifications become mandatory, and assessment of the normal operating conditions has to be repeated, accounting for the changed parameters. This is the iterative part of PHASE. For the second step in PHASE, tools such as HAZOP, fault trees, or FMEA are recommended (see Section 3.3).

2.3.1.4. Hazard Characteristics of Exothermic Processes due to Process Design

Chemical reaction engineering principles form the basis of an understanding of reaction hazards originating from process design. Basically three different designs can be distinguished: continuous, semicontinuous, and batch. A detailed discussion of reactor design principles is given in the standard textbooks [94], [95]. This section reviews those characteristics which have a direct impact on the hazard assessment of exothermic processes.

Design and safety criteria depend strongly on whether the system is homogeneous or heterogeneous. Heterogeneous processes depend on a much larger number of parameters (e.g., particle size distribution, surface area, and porosity for solid – liquid systems, or mass transfer parameters for liquid – liquid systems). For a detailed discussion of heterogeneous processes see [94], [95]. The following discussion is for homogeneous systems only.

Continuous processes are performed either in continuously stirred tank reactors (CSTR) or plug-flow tube reactors (PFTR). Phenomenologically, the longitudinal spatial coordinate of the PFTR is equivalent to the time coordinate of a batch process. Therefore, with respect to continuous processes, the following discussion focuses on CSTR. Continuous processes have two major advantages: the reaction rate can be controlled by the reactant feed stream; the reaction products and the solvent immediately dilute the initial materials entering the reactor. Both characteristics influence the effective concentration of reacting materials directly and reduce the thermal potential. The problem areas with CSTR are the correct design of the start-up procedure and the avoidance of dynamic instabilities. Additionally (but not of primary importance to hazard assessment), continuous processes require a larger reaction volume than discontinuous processes to achieve the same space – time yield. It is recommended to start up a CSTR as a semicontinuous process, up to the point where the expected stationary extent of reaction is achieved, and then to start the second feed stream. To avoid ignition/extinction and oscillatory phenomena, design criteria are available in [94], [95], which can easily be applied, provided the thermal kinetics of the overall process have been determined appropriately, e.g., by reaction calorimetry.

The true batch process represents the other extreme of reactor design. Batch reactors are often preferred in the fine chemicals and pharmaceutical industries because of the smaller amounts of substance handled. From a hazard evaluation viewpoint, batch processes are the most difficult, as all reacting materials are charged initially, so that the total hazard potential is present in the reactor right from the beginning. Depending on plant or process design, batch reactors (BR) can be operated either isothermally, under computer control, or isoperibolically, i.e., with a coolant kept at constant temperature (Fig. 28). In the first case, the assessment must focus on the maximum heat rate developed, ideally at $t=0$. If the coolant temperature can be rapidly lowered by the control system, quasi-isothermal conditions can be achieved. On the other hand, sensitivity analyses by SEMENOV and others suggest, that the cooling temperature difference should not be too great [96]. There are four engineering numbers, which can be used to give guidance on safe operability: the adiabatic temperature increase ΔT_{ad}; the thermal reaction number B; the dimensionless reaction rate Da (Damkoehler number); and the modified Stanton number St (dimensionless cooling capacity of the system):

$$\Delta T_{\mathrm{ad}} = \frac{(-\Delta_{\mathrm{R}} H)\, c_{\mathrm{A0}}}{(-\nu_{\mathrm{A}})\, r c_{\mathrm{P}}} \tag{21}$$

$$B = \frac{E \Delta T_{\mathrm{ad}}}{R T^2} \tag{22}$$

where ν_{A} is the stoichiometric coefficient of the limiting component.

$$Da = \frac{(-\nu_{\mathrm{A}})\, r_0 t}{c_{\mathrm{A0}}} \tag{23}$$

$$St = \frac{U A t}{V r c_{\mathrm{P}}} \tag{24}$$

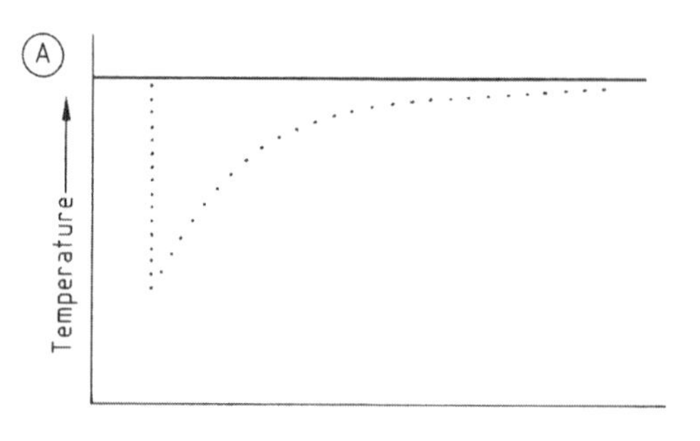

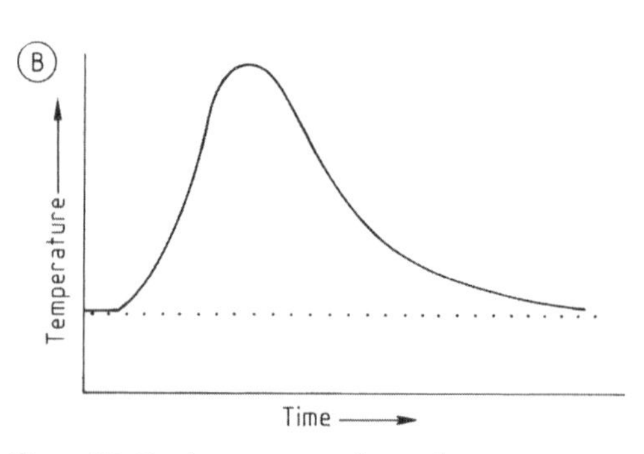

Figure 28. Batch reactor operating modes
(A) Isothermal; B) Isoperibolic
Full lines, internal temperature; dotted lines, jacket temperature.

If $\Delta T_{\mathrm{ad}} < 50$ K and $B < 5$, isothermal batch process can be operated safely. Such processes may be regarded as inherently safe, acknowledging that possible plant hazards are excluded from this assessment.

Under isoperibolic operating conditions, the hazard assessment focuses on the maximum temperature difference, which, under these conditions, occurs some time into the reaction. Additionally it has to be assessed whether or not the maximum reaction temperature exceeds a limiting value, which represents the temperature level at which undesired reactions become dominant. It is also necessary to assess the sensitivity of this operating point to the normal variability of other operating parameters, e.g., coolant temperature or overall heat transfer coefficient. If $Da/St << 1$ and thereby $(T_{max} - T_A) < 0.15\,\Delta T_{ad}$, such batch processes can be regarded as safe.

Semicontinuous processes are performed by charging one reacting component initially and feeding the second at a constant rate. Processes performed in these semi-batch reactors (SBR) combine the advantages of CSTR, i.e., control of the heat production rate by the feed rate, and of BR, i.e., the relatively small reactor volumes. These positive factors become effective only if the problem inherent to SBR design – the risk of reactant accumulation – can be prevented. Accumulation occurs if the rate of addition exceeds the rate of reaction. A SBR can be operated safely if $Da/St > 1$. Provided this is assured by the process design, accumulation of the added component does not pose a hazard [97].

2.3.2. Methods of Investigation and their Systematic Application

It is the aim of safety studies either to determine whether or not a chemical process may be carried out under predefined conditions (e.g., optimal conditions with regard to yield and time), or to establish safe operating conditions with respect to temperature, pressure, concentrations, and time. For this purpose, it is necessary to make experimental investigations by appropriate test methods. Theoretical considerations may be helpful (e.g., estimation of the hazard potential of a process from the chemical equation and molecular structures), but they cannot replace experiments. Several screening methods and more detailed test methods are used in the chemical industry to obtain information necessary for the assessment of exothermic or pressurizing chemical processes [98].

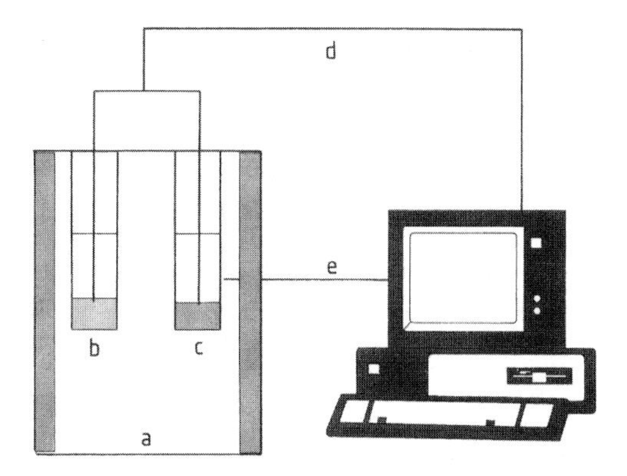

Figure 29. Differential thermal analysis (DTA)
a) Oven; b) Sample; c) Reference; d) Temperature difference signal; e) Oven temperature signal

2.3.2.1. Screening Methods

Thermal Analysis. Differential thermal analysis (DTA) is a common method for the investigation of chemical reaction hazards. A sample of the material to be tested, which may be a pure compound, a reaction mixture, a residue, or a waste sample, is exposed to thermal stress. For this purpose, the sample and a thermally inert reference material (e.g., Al_2O_3) is placed in an oven. Thermal stress can be applied either dynamically by ramping up the oven temperature at a fixed rate, or isothermally in thermal aging at elevated temperatures. The difference between sample and reference temperature is recorded as a function of oven temperature or time. The recorded signal is converted to a heat flow by a calibration function. For screening purposes, linear heating rates of 1 – 10 K/min, from room temperature up to 300 – 500 °C, are widely used. To suppress the endothermic effect of vaporization and catalytic effects of the vessel material, sealed glass ampoules are often used. A more sophisticated method is differential scanning calorimetry (DSC), where the heat flux between the sample and its surroundings is recorded directly. The resulting thermogram shows exothermic and/or endothermic transitions, originating from physical or chemical effects. Integration of the signal peaks provide the heat content of the observed transitions. The measurable beginning of the transition is called the onset temperature (Figs. 29, 30).

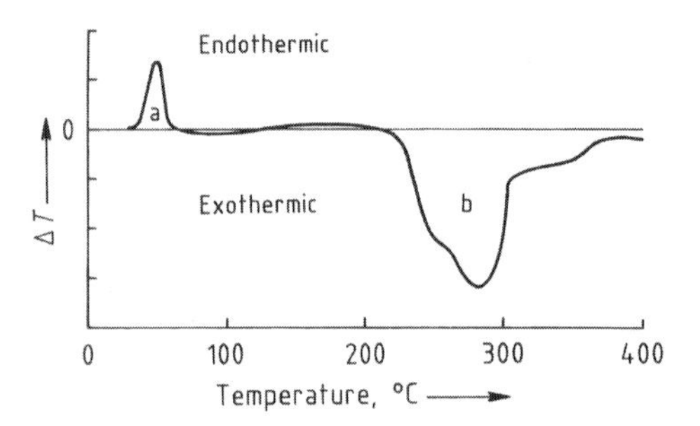

Figure 30. Typical thermogram of a solid
a) Melting (endothermic); b) Decomposition (exothermic)
Heating rate 3 K/min

The advantages of DTA/DSC are the use of small sample masses and the short time (1 – 3 h) required for a run. Onset temperature, peak area, and peak shape give hints on the hazard potential of a sample. It is not permitted to perform a direct scale-up of the measured values to the true chemical process or unit operation conditions. The recorded onset temperature depends critically on the measuring conditions, e.g., heating rate and sensitivity of the apparatus. Furthermore, there is no analytical control on the extent of reaction. So the area of a peak due to a chemical reaction cannot be interpreted as equivalent to the molar heat of a defined reaction. As the sample mass cannot be stirred during the experiment, the measuring conditions are not representative for heterogeneous systems. In particular, thermograms of such samples must be interpreted carefully.

Other Screening Methods. Isoperibolic calorimetry on the scale of several grams is another screening method for testing the thermostability of chemical systems, and may be used as an alternative or in addition to DTA/DSC. The calorimeter consists of a jacketed sample vessel (the jacket temperature may be programmed). The temperature difference between the sample mass and the jacket is measured and recorded as a function of jacket temperature or time. The recorded curves are evaluated similarly to DTA/DSC curves. The use of a larger sample mass (compared with DTA/DSC) and the option of stirring allows the investigation of heterogeneous systems with more success. The low mechanical stability of the sample vessel may be a disadvantage.

For a comprehensive assessment of the hazard potential of chemical systems, not only the thermal behavior, but also pressure effects in a closed vessel, or the rate of material loss in an open system, are of interest. Both effects can be measured with the help of these screening tools as a function of the sample temperature (programmed) for sample masses of ca. 50 – 1000 mg. For the interpretation of results, restrictions similar to those for thermograms apply. The results must not be scaled up directly to the true process and plant conditions.

2.3.2.2. Thermal Aging and Heat Accumulation Storage Tests

Data from thermal aging tests can be used directly for plant-scale assessment, provided the test conditions are representative of actual process conditions (e.g., quality of solid or liquid sample identical with the actual plant material, temperature and testing time identical with the process temperature and residence time). By varying the sample amount it is possible to extrapolate the experimental results for larger quantities. A sample vessel of defined size (e.g., a glass cylinder 100 – 500 mL) is stored in a drying oven at fixed temperature. Temperature sensors inside the sample detect the onset of an exothermic reaction. It is possible to eliminate the dependence of the exothermic onset on the sample mass by storing the test sample under adiabatic conditions, using so-called heat accumulation storage tests (open or under pressure). For adiabatic tests, the sample is stored in a Dewar flask with an insulating lid. A higher degree of adiabaticity is achieved by reducing heat losses to the surroundings. For this purpose, the Dewar flask is placed in a controllable oven, thereby assuring an ambient temperature equal to the sample temperature (Fig. 31).

The heat accumulation storage test provides the sample temperature – time profile under adiabatic conditions, starting at a fixed storage temperature. In evaluating the curves, it is necessary to account for the "thermal inertia" of the experimental system. Thermal inertia expresses the fact that the heat produced by the chemical reaction system is partially absorbed by the sample container and internal components (temperature sensor, stirrer, etc.). From a heat balance viewpoint, the extent of this effect depends on the ratio of the heat capacity of the complete measuring system to that of the reaction mixture, the so- called Φ-factor:

$$\Phi = \frac{m_{R} c_{PR} + m_{A} c_{PA}}{m_{R} c_{PR}} \quad (25)$$

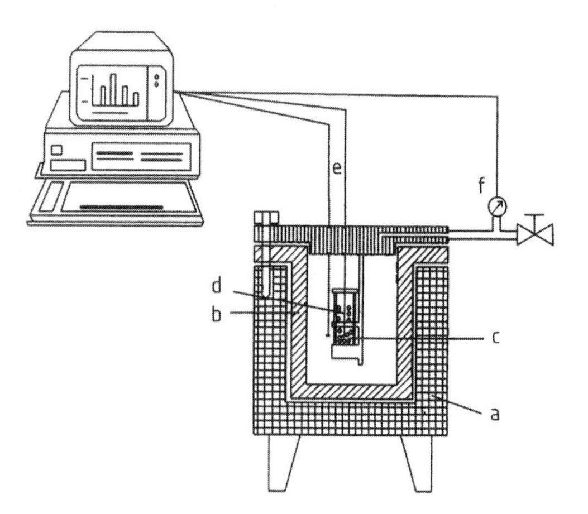

Figure 31. Adiabatic Dewar test setup
a) Oven; b) Autoclave; c) Dewar; d) Sample; e) Temperature signal; f) Pressure signal

Generally, Φ approaches unity with increasing mass of the reaction mixture. This is important as this is both demand and justification for model based mathematical correction to data obtained in small experimental tools. Figure 32 shows typical temperature – time curves for the decomposition of a substance, starting the test at different initial temperatures.

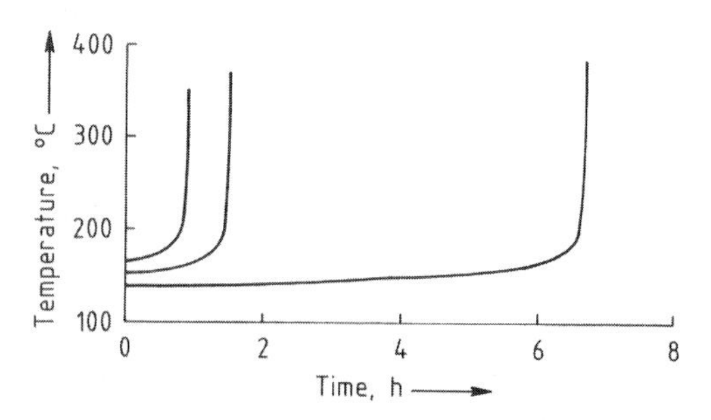

Figure 32. Typical recording of decomposition of a substance from adiabatic Dewar experiments

The handling of toxic reaction products and the repair problems are the disadvantages of the thermal aging and heat accumulation storage tests when a decomposition reaction occurs. The problem can be reduced by using smaller sample masses (< 10 g) in metallic bombs with high pressure stability. As such apparatus has a high heat-loss factor, it is necessary to use an efficient oven with accurate temperature control to minimize this heat loss.

With modern equipment, it is not necessary to predefine a fixed starting temperature for the adiabatic test. The measuring system detects the temperature at which a threshold value of the heat production rate (e.g., 1 – 2 W/kg) is exceeded, by a "search-and-wait" procedure. Evaluation of the test run yields the temperature and pressure rates as a function of temperature, as well as a thermal kinetic interpretation of the exothermic reaction under adiabatic conditions, and the adiabatic temperature increase. The temperature/pressure rate – temperature data of such an experiment (Fig. 33) can be used for designing a rupture disk or a safety relief valve.

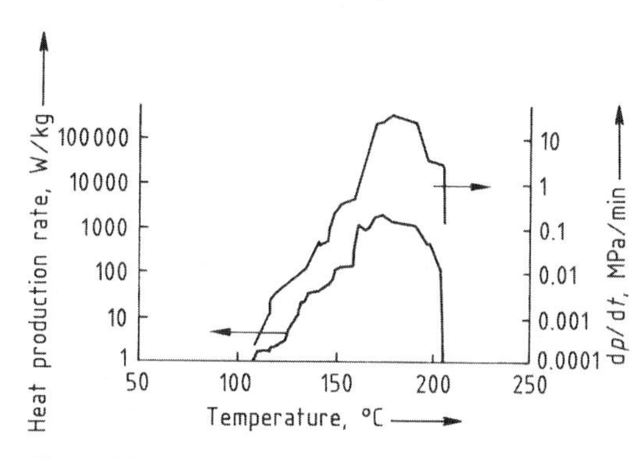

Figure 33. Experimental self-heat and pressure rate data from an adiabatic study of an exothermic reaction

2.3.2.3. Calorimetry

The methods described usually provide information on the thermal and/or adiabatic runaway potential of batch systems. They are not suitable for studying reactions under their predefined, normal or upset operating conditions. For this purpose, reaction calorimetry is the recommended method. In a reaction calorimeter, batch, semibatch, and continuous processes can be studied, and complex reaction procedures can be simulated (e.g., pH control, temperature or pressure control, reflux, distillation). Moreover, the effect of deviating operating conditions can be tested (e.g., by varying the amounts of reaction components, catalyst, or solvent, temperature, loss of agitation, simulation of adiabatic conditions).

The equipment consists of a stirred glass reactor with a cooling jacket. For the investigation of reactions under pressure, special test vessels are also available. The reactor lid accommodates several measuring instruments and input/output devices. The experiment is controlled by a data processing/acquisition unit. The calorimeter can be operated isothermally (reaction temperature kept constant throughout the run), in an isoperibolic mode (jacket temperature is kept constant), in a free-programmable mode (linear temperature ramps for the internal or the jacket temperature can be defined), or in an adiabatic mode. Calorimetric evaluation is based on temperature measurements and calibration runs, to characterize the heat transfer. The evaluation yields the overall heat production rate over time of the reaction system as investigated, as well as the heat of reaction. Special evaluation techniques allow determination of the thermal accumulation potential in the reactor due to unreacted material.

Evaluation can be based on heat-flow or heat-balance algorithms. In the first case, the heat flux between the inside of the reactor and the surrounding jacket is determined by measuring the temperature difference between the reaction mass and the coolant. This kind of evaluation may be significantly in error if the overall heat transfer coefficient changes markedly during the reaction in a nonlinear way. This is especially the case for polymerization reactions, as they are usually accompanied by major changes in viscosity. This systematic error can be eliminated by running the calorimeter in the isoperibolic mode, and evaluating the heat balance for the jacket independently. However, this evaluation method is rather complex.

If the calorimetric experiment is evaluated according to the heat-balance algorithm, the liquid flow rate of the coolant and its heat capacity must be well defined and measured; otherwise, significant errors may occur.

If reactions are investigated under reflux or distillation, it is necessary that all heat flows related to the vapor phase are accounted for appropriately. Figure 34 shows a typical apparatus for reaction calorimetry. The information obtainable from reaction calorimetry is of great use, not only for process safety, but also for process development and optimization.

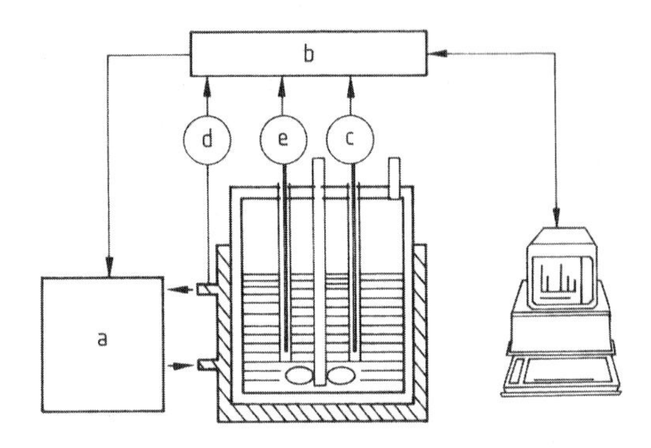

Figure 34. Typical reaction calorimetry setup
a) Thermostat; b) Controlling unit; c) Reaction temperature; d) Jacket temperature; e) Quality measurement

2.3.2.4. Systematic Testing

Some 25 different experimental techniques are commercially available for the investigation and characterization of exothermic reactions. It has to be emphasized that there is no comprehensive method; it is sensible to apply more than one method to obtain the necessary PHASE data. On the other hand, PHASE demands a systematic procedure, so a certain ranking of the different techniques is required. This can be achieved by considering the life cycle of a chemical process. Researchers usually carry out their experiments on the 1-g scale; thorough experimental hazard characterization is not required, as possible consequences of events are limited to the fume cupboard, provided occupational health and safety recommendations have been strictly followed.

Process development on the 1-kg scale represents the first milestone of the process life cycle, where the preliminary PHASE should be performed, based on screening test data. Before the first run of a process in a pilot plant, the first comprehensive PHASE is required. This can only be performed reasonably if it is based on data which are not subject to large model-based, scale-up corrections; reaction calorimetry and Dewar tests are the recommended techniques [99].

For intended plant or process modifications, selection of the appropriate experimental test methods depends on each individual case, and no general recommendations can be given.

2.3.3. Assessment Criteria

The incident risk is defined as the product of incident severity and its probability of occurrence. Hence, risk assessment includes both severity and probability. In the following, a generalized procedure for risk assessment of runaway reactions with special emphasis on evaluation criteria is presented.

2.3.3.1. Runaway Scenario

It is assumed that a batch reactor is being operated under normal operating conditions, and a cooling failure occurs. Provided that unconverted material is still present in the reactor, the temperature increases, owing to the heat output related to completion of the reaction. This temperature increase is proportional to the amount of unreacted material. The temperature reached at the end of this period is called the maximum temperature of the synthesis reaction (MTSR). At this level, a secondary decomposition reaction may be initiated (Fig. 35). The heat produced by this undesired reaction leads to a further increase in temperature [100].

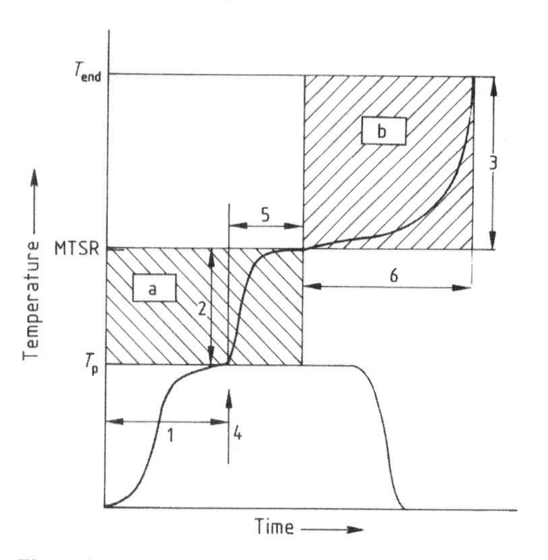

Figure 35. Runaway scenario
a) Runaway of desired process; b) Runaway of undesired process
1 = time to cooling failure; 2 = ΔT_{ad} due to unreacted material (for desired reaction only); 3 = ΔT_{ad} due to decomposition; 4 = $T_p - T_{initial}$; 5 = time to reach MTSR; 6 = adiabatic induction period

The following questions help to characterize the runaway scenario and to provide data for risk assessment:

1) Can the process temperature be controlled by the cooling system?
2) What temperature can be reached after runaway of the desired reaction?
3) What temperature can be reached due to runaway of the decomposition reaction?
4) Which moment of a cooling failure will have the worst consequences?
5) How fast is runaway of the desired reaction?
6) How fast is runaway of the decomposition at MTSR ?

This type of scenario should be worked out for the different possible process deviations, e.g., charging errors or higher accumulation due to loss of agitation. It allows assessment of the sensitivity of the process to these deviations.

2.3.3.2. Severity: Adiabatic Temperature Increase

Most reactions in the fine chemicals industry are exothermic. In the case of a cooling failure, or if heat removal is not sufficient to compensate for the heat production, the temperature increases proportionally to the heat of reaction. Thus, the reaction energy is a direct measure of the severity of a runaway, i.e., the destructive potential. It is usually expressed as the adiabatic temperature increase ΔT_{ad}.

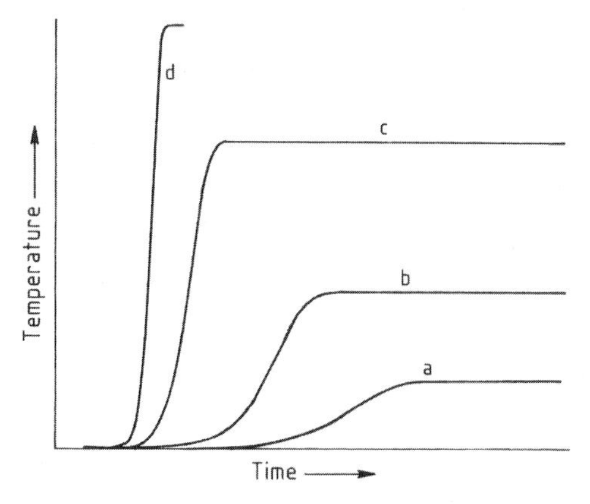

Figure 36. Adiabatic runaway temperature profiles
a) – d) increasing rates of temperature increase with increasing energy

Independent of its meaning for characterization of the hazard potential and determination of attainable temperature levels, the adiabatic temperature increase assists estimation of the dynamics of a runaway. As a general rule, high energies result in fast runaway or thermal explosion, while lower energies ($\Delta T_{ad} < 100$ K) result in slower temperature increase rates (Fig. 36), given the same activation energy and initial heat release rate [89].

The heats of common industrial synthesis reactions are often of the order of − 100 kJ/mol, whereas decomposition reactions may reach far higher levels, e.g., −400 kJ/mol for mononitrated aromatic compounds. In many cases, the desired reactions are not themselves inherently dangerous, but decomposition reactions may lead to dramatic effects. The consequences of a runaway event can be manifold. One possibility is solvent evaporation with the subsequent possibility of a vapor cloud explosion, if the boiling point of the system is reached. Another possibility is formation of gaseous products from a decomposition reaction, which leads to a pressure increase and the risk of vessel rupture. In practice, three levels – low, medium, and high – are sufficient for assessment the severity:

Low:	$\Delta T_{ad} < 50$ K, no pressure buildup
Medium:	$50\ \mathrm{K} < \Delta T_{ad} < 200$ K
High:	$\Delta T_{ad} > 200$ K

2.3.3.3. Probability and Kinetics of a Runaway

There is no direct quantitative measure for the probability of occurrence of an incident, or in the case of thermal process safety, for the occurrence of a runaway reaction. But, if a runaway, anticipated to occur within several minutes, is compared with one expected to occur over several hours, it is obvious that in the first case there is very little time to take preventive measures, whereas in the second case there is time to regain control of the reaction. The probability of runaway is higher with a fast than with a slow temperature increase. While quantification of probabilities is not easy, at least a comparison is feasible.

To assess the probability of occurrence of a decomposition reaction, it is necessary to characterize its kinetics. The concept of time to explosion or $\mathrm{TMR_{ad}}$ (time to maximum rate under adiabatic conditions) is of great help [98]. It can be estimated by the following equation:

$$\mathrm{TMR_{ad}} = \frac{c_p R T_0^2}{q_0 E_a}\ (\mathrm{s}) \qquad (26)$$

where c_p is the heat capacity of reaction mixture (J kg^{-1} K^{-1}); R is the gas constant, 8.31431 J mol^{-1} K^{-1}; T_0 is the initial temperature of the runaway (K); q_0 is the heat release rate at T_0 (W/kg); and E_a is the activation energy (J/mol).

This equation was derived for zero-order reaction kinetics, but it can be used for reactions of higher order, provided the influence of concentration on the reaction rate can be neglected. This approximation is particularly valid for fast, highly exothermic reactions.

The isothermal mode of DSC provides an easy way to measure the kinetic parameters in the $\mathrm{TMR_{ad}}$ equation [89]. A set of isothermal experiments is run at different temperatures. The natural logarithm of maximum heat release rate, determined on each thermogram, is plotted against the reciprocal temperature in an Arrhenius diagram (Fig. 37). Runaway curves and the corresponding time to maximum rate can also be measured by adiabatic calorimetry.

In practice, three levels are sufficient for the assessment of the probability. For discontinuous chemical reactions on an industrial scale, a probability can be considered low if the time to maximum rate of a runaway reaction under adiabatic conditions is > 24 h. The probability becomes high if the time to maximum rate is < 8 h (a working shift).

These timescales are only orders of magnitude, and depend on a lot of organizational and plant design factors, e.g., degree of automation, operator training, frequency of electrical power failures, reactor size. This scaling of probabilities is valid only if safety measures in proportion to the known severity are implemented.

2.3.3.4. Assessment of Criticality

For reactions presenting a thermal potential, criticality ranking can be based on the relative values of four temperatures.

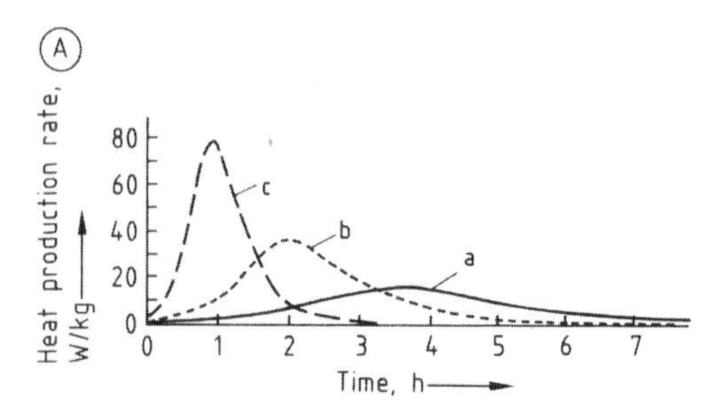

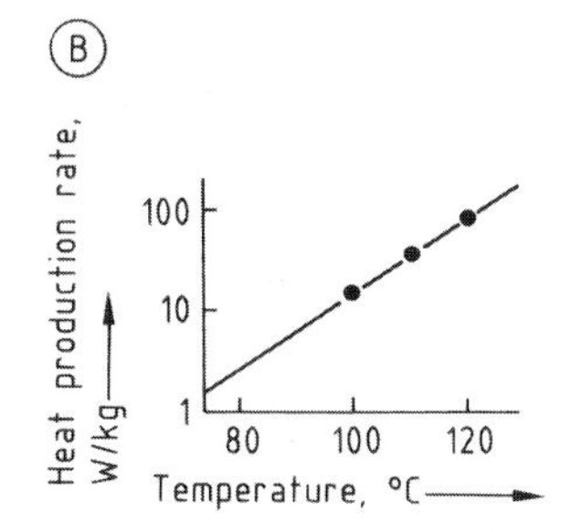

Figure 37. Determination of TMR_{ad} by differential scanning calorimetry (DSC)
A) Isothermal DTA traces: a) $T = 100\,°C$; b) $T = 110\,°C$; c) $T = 120\,°C$
B) Pseudo-Arrhenius diagram for evaluation of the isothermal runs

- $T_{process}$ (process temperature) is the initial temperature in the cooling failure scenario. In nonisothermal processes, it is defined as the temperature level at which a cooling failure produces the most severe consequences (worst case).
- MTSR (maximum temperature of synthesis reaction) depends essentially on the degree of accumulation of unconverted reactants, and therefore depends strongly on the process design [101].
- ADT 24 (temperature at which TMR_{ad} = 24 h) is defined by the thermal stability of the reaction mixture.
- MTT (maximum temperature of technical reasons), in an open system, is the boiling point. For a closed system, it is the temperature at which the pressure reaches the maximum permissible value, i.e., the set pressure of a safety valve or rupture disk.

These four temperatures permit classification of the scenarios into five different classes, from the least critical (1) to the most critical (5) [102] (Fig. 38). They are defined as follows:

1) In the case of loss of control of the desired reaction, the boiling point cannot be reached and the decomposition reaction cannot be triggered.
2) The situation is very similar to scenario (1), but if the reaction mass is kept under heat accumulation conditions for a longer time, the decomposition reaction can be triggered and the boiling point of the system can be reached.
3) In the case of loss of control of the desired reaction, the boiling point of the system is reached, but the decomposition reaction cannot be triggered. The safety of the process depends on the heat production rate of the desired reaction at the boiling point [103].
4) In the case of loss of control of the desired reaction, the boiling point is reached, and the decomposition reaction can, theoretically, be triggered. The safety of the process depends on the heat production rate of both the desired and decomposition reactions at the boiling point. Evaporative cooling may serve as a safety barrier in open systems.
5) In the case of a loss of control of the desired reaction, the decomposition reaction is triggered and the boiling point is reached during runaway of the decomposition reaction. It is very unlikely that evaporative cooling can serve as a safety barrier. The heat production rate of decomposition at the boiling point determines the thermal safety of the process. It is the most critical of all scenarios.

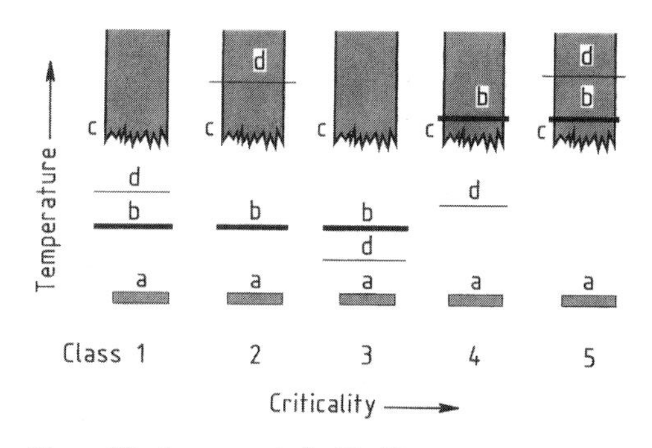

Figure 38. Assessment of criticality
a = T_p; b = MTSR; c = range where decomposition becomes relevant; d = *bp*

2.3.4. Measures

Three categories of measures, with different applications, are distinguished:

1) Hazard-preventing measures, which suppress the potential, e.g., dilution, change of synthesis route. These measures render the process inherently safe [101].
2) Hazard-controlling measures, e.g., change from batch to semibatch or continuous operation, interlocks in the process control system (Section 3.4.3) etc. These measures render the process fail-safe.
3) Consequence-mitigating measures, e.g., dumping, quenching, venting (Sections 3.5, 5.1).

Depending on the criticality, different measures can be applied to prevent or control the runaway or to mitigate its consequences:

1) No special measures are required, but the reaction mass should not be kept under heat accumulation conditions for long periods. Evaporative cooling serves as an additional safety barrier.
2) No special measures are required, but the reaction mass should not be kept under heat accumulation conditions for long periods.
3) The measure of choice is to use the evaporative cooling to keep the reaction mass under control. A distillation system must be designed for that purpose, and has to work effectively, even in cases of utility failures. A backup cooling system, dumping of the reaction mass, or quenching can also be used. These measures must be designed adequately and must be available immediately after the failure occurs.
4) Similar to (3). The same measures apply, but the additional heat production rate due to the secondary reaction has also to be taken into account.
5) In this case, the boiling point is very unlikely to serve as a safety barrier. Therefore only quenching or dumping can be used. Since, in most cases, the decomposition reactions release very high energies, particular attention has to be paid to the design of safety measures. It is worth considering an alternative process design, in order to reduce the severity or at least the probability. The following possibilities should be considered: reduction of concentration, change from batch to semi-batch, optimizing semibatch operating conditions to minimize accumulation [89], [104], changing to continuous operation, etc.

3. Development, Design, and Construction of Safe Plants

3.1. Objectives, Regulations, and Concerns

The goal of all plant safety effort is to eliminate or reduce the possible hazards described in Chapters 1 and 2. According to Section 1.3, this means limiting the risk created by the facility to an acceptable level. At the same time, the relevant regulations and engineering codes must be complied with. In many areas, these implicitly establish the required level of safety (or acceptable risk).

All the industrialized countries have adopted regulations to protect workers, uninvolved third parties, and the environment [105]. These regulations employ various combinations of the following approaches:

1) Setting standards for the quality of engineering facilities and their safe operation
2) Establishing approval or licensing procedures for the erection and operation of plants and for substantial modifications to them
3) Requiring regular inspections of the plant and its technical equipment by the operators, government agencies, and (sometimes) independent third parties
4) Assigning civil liability and legal responsibility for damage caused

Because there was initially little experience relating to the hazards of technology, the first regulations were written for individual types of apparatus regarded as dangerous (steam boilers, pressure vessels, machinery with moving parts) and aimed chiefly to insure occupational safety and to protect third parties. Later, regulations for other types of equipment were added, attempts were made to extend protection to new areas such as surface waters, and new regulatory principles such as the principle of precautionary action were introduced. It is not unusual that regulations are created by many independent bodies, with very different jurisdictions. As a result, in Germany (for example) there is a rather complicated system of regulations for chemical plants, with overlaps and multiple rules, and the density and the degree of detailing of regulation differ widely from one regulation to another.

The safe design and safe operation of chemical plants in Germany are governed by regulations stemming from the following fields of law [106]:

1) Building and construction law
2) Occupational safety law (including the Equipment Safety Act, regulations defining facilities subject to supervision, and the Accident Prevention Regulations)
3) Hazardous substances law (including the Chemicals Act and the Hazardous Materials Regulation)
4) Water pollution control law (including the Regulation on Facilities with Water-Contaminating Substances)
5) Air pollution control law (including the Major Hazards Directive [107])

This complicated situation is not likely to be fundamentally changed by harmonization efforts in the European Single Market, for two reasons: First, under the EEC Treaty (Article 100 a) [108], only requirements on the quality of technical devices are subject to harmonization. Second, regulations in the European Union (EU) continue to be created, adopted, and enacted into law by the member countries according to the old pattern (e.g., for specific types of facilities).

As a rule, the regulatory system is structured vertically [109]. Legislation at high level sets forth the scope and objectives as well as the solution approaches in general form, while engineering codes and standards at lower levels are largely concerned with technical details.

Regulations, especially at sublegislative levels, are designed deterministically in that the required safety is deemed to be attained if the plant satisfies certain requirements on design, sizing and equipment. An exception can be found in a Dutch regulation under which a chemical plant, to be approved, must operate at a risk below a value described in terms of fatal accident probabilities [110].

Plant builders and operators view plant safety in terms of the types of hazard arising from the process. For them and their dealings with the authorities, it would be simpler if there were a single consistent set of regulations, broken down by hazard type (releases and spills, fires, explosions) and – instead of setting fixed technical solutions for protective tasks – such a code would use fixed solutions merely as examples of ways to achieve the stated objective.

This approach would have the advantage that the required safety could be achieved by the most advantageous technical resources in each instance. In addition to that, the "plant" orientated view would no longer require multiple specific regulations covering occupational safety and health, air and water pollution control, since one criterion for the plant (e.g., tightness) would simultaneously take care of related problems in the other areas.

The main safety tasks for the process designer and the designer and builder of process plants are:

1) To identify and correctly assess all types of hazard
2) To take appropriate steps to reduce and control these hazards

The key hazard types are those listed in Section 1.1:

Releases and spills
Fires
Explosions

The safety tasks can be broken down into the process itself and into the safe design and operation of the technical facility required for the process. The specific tasks can then be listed as follows:

1) Achieve safe process design by: a) identifying all types of hazards; b) assessing their hazard potentials; c) minimizing the hazard potentials; d) deactivating the hazard potentials
2) Achieve safe plant design and operation by: a) systematically analyzing danger sources; b) evaluating their probabilities of occurrence (usually qualitatively); c) minimizing sources of trouble and error; d) employing fault-tolerant design

In detail (see also Section 1.2), this means:

1 a) For all substances to be processed, safety ratings and toxicologically and ecologically relevant data must be acquired (see Chapter 2). A comparison of process parameters and design data with safety ratings, step by step through the process, reveals where danger sources exist, or may arise.
1 b) The magnitude of each hazard potential is evaluated through analyses, e.g., those described in Sections 1.4.2 and 1.4.3.
1 c) It must be determined whether the hazard potentials can be reduced through suitable process design.

The practices recommended here lead to inherent safety [111], [112]. To achieve this goal the planner should replace hazardous substances by less hazardous ones wherever possible. Large inventories should be avoided as far as possible, e.g., by using process steps that can be carried out quickly in a small volume, by introducing continuous operations in place of batch operations, and by eliminating large buffer volumes.

1 d) Any hazard potentials that remain must be deactivated in such a way that they cannot manifest themselves in the process. Lowering the temperature far below critical values or diluting substances and handling them in solution (to lower the vapor pressure) are the approved methods. They also help make the process inherently safe [111], [112].

If the plant is to be conceived for a process that has been safety optimized in this way, two analytical tasks, followed by two design tasks, must still be performed:

2 a) The system plant must be systematically searched for danger sources, i.e., possible defects and failures that can activate the deactivated hazard potential.
2 b) When possible faults are identified, their frequencies or probabilities of occurrence must be evaluated so that appropriate safety measures can be taken.

The last two tasks can now be undertaken with an eye to the magnitude of each hazard potential (task 1 b) and a (qualitative) measure of the probability of a defect that would activate the hazard potential:

2 c) All possibilities for minimizing sources of trouble and error must be exhausted [112].
2 d) As far as possible, the facility must be designed and equipped so that faults are "forgiven" without resulting in harm. One way to accomplish this is to use redundant (multiple) safety devices [112] (see Section 3.4.3).

In carrying out these tasks, some of the steps (2 a – 2 d) may have to be done more than once, recursively, because any change in the system due to the new measures can introduce new danger sources.

For this reason, and because the development of a process and the associated plant is done step by step, with concomitant advances in understanding a "holistic" procedure, segmented by time, technical specialty, and logical relationships, must be adopted in the development, design, construction, and operation of a chemical plant.

Examples of such methods are presented in the Section 3.2.

3.2. Procedure for Designing and Constructing Safe Plants

Chemical processes and their associated technical facilities are developed in steps. Process development in the laboratory is followed by testing of the process on a pilot scale before the project goes through the various planning stages (preliminary, draft, and detailed design). The planning process culminates in the purchase of equipment and erection of the plant. After a test and commissioning phase, the facility is put into production. The left-hand side of Figure 39 shows these stages [113].

Each phase involves questions as to the safety of the process and the plant, each of which must be answered immediately or, at the latest, before the process goes on to the next phase. The most expedient way of creating a safe plant is thus to plan for safety studies. At each step in process and plant development, safety analyses must be done (ideally integrated into the development) in order to pose the right questions and immediately seek solutions to the problems identified.

These ideas are the basis for the procedure chosen by Bayer, a large German chemical company (Fig. 39). Safety analyses are broken down into four sections, each concluding with a certificate prepared by safety experts. This certificate

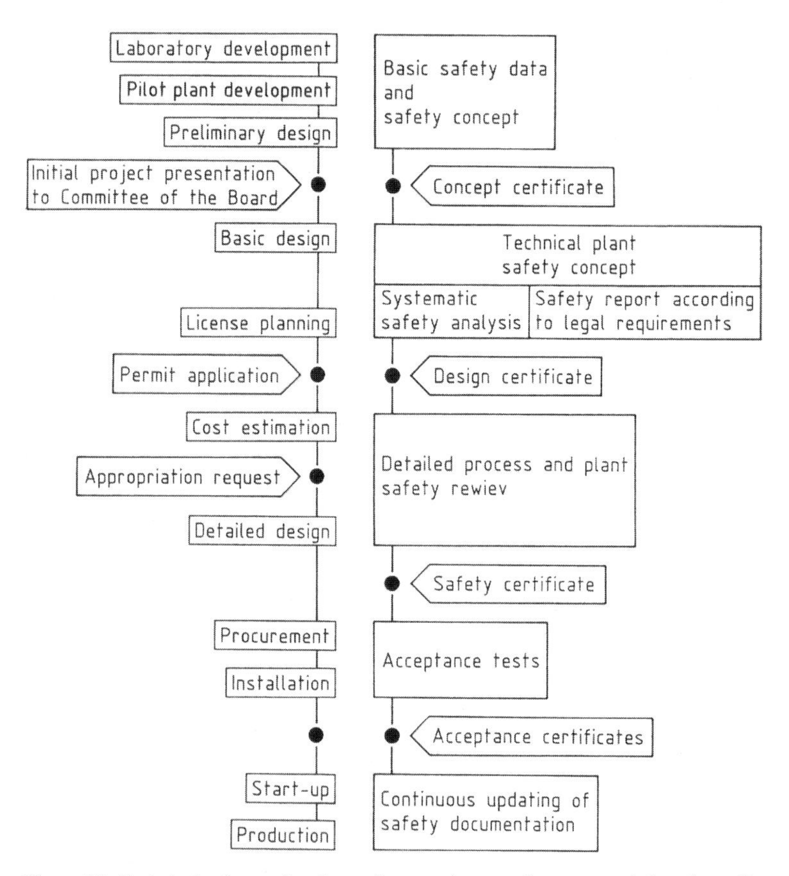

Figure 39. Technical safety audit scheme for new plants or changes to existing plants (Bayer) [113]

states that the required studies have been performed, and the correct conclusions drawn from them.

The four phases of safety analysis can be listed as follows:

A 1) To create safety principles by: compiling and determining safety, toxicological, and ecological data; identifying sources of danger in the process; examining possible safety solutions; establishing the safety concept for the process

A 2) Defining the safety concept for the plant by: performing systematic analysis; identifying technical protective measures

A 3) Performing a detailed safety analysis by: analyzing all plausible forms of trouble as to cause, effect, and corrective measures; adopting the final detailed safety concept

A 4) Conducting the safety acceptance of the plant by: doing a nominal/actual comparison (concept/implementation); carrying out functional tests

This procedure is organized in an obvious way, and has the advantage that a completion certificate must be prepared at the end of each phase, before the next phase begins. The project is not released unless the planning certificate is ready, and the plant does not go on stream unless there is a safety acceptance certificate. These rules are set forth in an internal directive [114].

Similar procedures are common in other companies. They are not always so formal, but they embody the same principles and use the same resources.

BASF, for example, uses a three-phase approach for these studies [115]. Phase 1 generates a basic safety concept with provisional solutions for the main possible hazards in the process. Phase 2 involves elaboration of the safety concept, so that it can be submitted to the authorities along with the approval application (safety analysis, safety report). Detailed safety verification takes place in phase 3.

Hoechst utilizes the principle that certain safety analysis and design results, once created, become important documents that must accompany the process and the plant throughout their lives [116]. A substance data report and transmission report have been developed for this purpose; they function to gather key safety information and communicate it within the company; a checklist is used in the safety assessment of production facilities.

A number of publications on the performance of safety analyses have been issued by ESCIS, the Expert Commission on Safety in the Swiss Chemical Industry [117], [118]. These stress principles and organization for this kind of work, establish key points to be analyzed at various times, and include methodological aids.

Essential in all these procedures is that nothing is overlooked and the proper specialists are brought in at the proper time. All such procedures can therefore be said to regulate three things – who does what and when.

The methods employed by Degussa [119] and by Lurgi, a plant construction company [120], assign great importance to these aspects.

Over the plant's economic life, all the results of safety work are used again and again for guidance, e.g., during maintenance and plant modifications, or when new personnel are being trained. Close attention must be paid to clear and complete documentation of the safety analyses; the recommended form for this is a concise, uniform, computerized database keyed to process, process step, plant, and plant section, and covering all danger sources with causes, effects, and corrective measures (including brief justifications). It can be used directly in the writing of safety reports pursuant to the Major Hazards Regulation (Seveso Directive) [107] and can easily be consulted from the plant operating personnel whenever necessary.

By way of summary: the design and construction of safe plants calls for a highly structured and organized procedure clearly setting forth what has to be done by whom and in what way, and focusing on the creation and routing of documents. This is called a safety management system.

In accordance with this principle, it has recently been suggested in both the United States [121] and Germany [122] that "safety management" should be implemented on the model of quality management following the well known international standards ISO 9000 ff. This idea makes sense, provided the procedures recommended for quality assurance are reinterpreted for this new purpose. This tends toward the procedures discussed above, which are better suited to the legal situation, at least in Germany.

In order that nothing should be overlooked in the analyses, it is desirable to provide methodological aids to the specialists as they deal with various problems. Section 3.3 concerns such aids.

3.3. Methodological Aids

The safety concept for a chemical plant must be complete, with an answer ready for any safety question. All necessary questions which can help reveal potential sources of danger must be posed ahead of time.

The problem of a comprehensive and complete analysis comes up twice in the course of process and plant development.

First, for the process (task 1 a of Section 3.1) the safety figures for the various substances (toxicity, flammability, explosibility) must be determined. Second, for the plant (task 2 a of Section 3.1) defects and malfunctions in the system that can activate the hazard potentials must be identified.

Appropriate safety engineering involves assessing the hazards as to both possible scope (task 1 b of Section 3.1) and probability of occurrence (task 2 b of Section 3.1).

Methodological aids available for use in these tasks [123], [124] are all characterized by clear and easily understood structures and systematic procedures.

The specific methods differ in:

Objective of the analysis
Functional principle employed
Basic knowledge required
Appropriate time of use
Aids needed
Results achievable
Documentation
Cost/benefit ratio

If we go back to the four chief tasks of safety analysis

1) Completely identifying hazards due to substances (task 1 a)
2) Assessing hazards as to potential scope (determining the hazard potential, task 1 b)
3) Identifying danger sources in the plant (identifying defects and possiblities for malfunctions, task 2 a)
4) Evaluating the possible hazards with respect to probability of occurrence (task 2 b)

the methods can be placed in four groups, two relating to the identification of problems and two to their assessment.

Table 9 summarizes the most important methods. The literature refers to many more methods than those in the table. In most cases, the additional names do not refer to formal methods, e.g., "safety audit", "what-if-method", and "preliminary hazard analysis", as defined in [124]; or they are synonyms for known methods: "human error analysis", "action error analysis", and "human reliability analysis", all concerned with human errors and using checklists, keywords, tables, or event trees to study their effects; or they are combinations of two methods, e.g., "cause – consequence analysis", a combination of incident sequence analysis and fault tree analysis [123–125].

The methods of Table 9 increase in difficulty and cost from top to bottom; with the exception of consequence analysis, they are also arranged roughly in chronological order to use in process and plant development. In accordance with this ordering, the amount of input information required, the level of complexity, and the amount of special knowledge required also increase from top to bottom. Therefore assessment methods must generally be carried out by specialists, while the identification methods should be among the tools used by every chemist and engineer in the process industries, particularly the chemical industry.

The identification methods used in safety analysis are summarized below.

Checklists and Relationship Charts. A checklist enumerates points that, according to experience, are associated with hazards in the handling of substances and mixtures of substances, or in the performance of a technical process. It can be as detailed as desired, and must be suitable for the kind of analysis being carried out. For example, checklists can be developed and used for thorough analysis of all safety-relevant material properties. If the concern is whether explosions can be initiated by undesired reactions, the checklist comprises such questions as:

- Are the substances stable?
- What happens if foreign substances are admitted?
- What happens if the process parameters (temperature, concentration) change?

The formal method is to work through such lists of questions and to determine whether:

1) A hazard is possible (yes or no)
2) Further study is needed
3) Safety measures are called for

The results of the analysis can be documented directly on the checklist or on a summary form reflecting its organization.

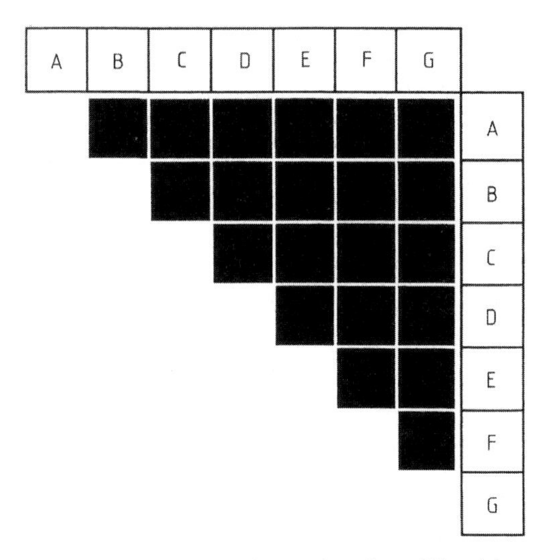

Figure 40. Reaction matrix showing all possible pairings of substances (tests for unwanted reactions)

Checklists can be used to insure the completeness of the safety concept in later phases of a plant project, e.g., in the design phase, to insure that possible events such as power outage, cooling outage, or stirrer outage are included in the safety concept. The creation and use of checklists is also customary in plant operation. They have the obvious advantage that they can

Table 9. Safety analysis methods

Task	Aim	Working principle	Method
1) Identification of hazards	1) To complete the safety concept	1) Memory-jogging	1) Checklists 2) Relationship charts
		2) Use of search aids and tables	3) Failure modes and effects analysis 4) Action error analysis 5) Hazard and operability studies
2) Evaluation of hazards according to probability of occurrence	2) To optimize safety systems with regard to reliability and availability	3) Representation of interconnections of failures in graphical form and evaluation of probability	6) Incident sequence analysis (inductive) 7) Fault tree analysis (deductive)
3) Evaluation of hazards according to consequences	3) To minimize the hazard potential and devise optimum protective measures	4) Mathematical analysis of physicochemical processes	8) Hazard consequence analysis

be adapted to any problem; their drawback is that things not included may not always be recognized and dealt with.

The same limitation applies to the second "memory-jogging" method, the use of compatibility charts [124] or relationship charts [125]. One use for this technique is to check the compatibility of all the substances that can come into contact in a process. If, for example, there are substances A to G, then half of a two-dimensional matrix (Fig. 40) gives all possible pairings of substances.

By systematically scanning over the interaction fields, it is possible to check whether mixing two substances creates a hazard, and whether further experimental study is called for. The disadvantage of this technique is obvious: mixtures of three or more components do not appear. Caution is therefore indicated, and matrices are more often used simply to illustrate safety situations, e.g., causes and effects or effects and corrective actions [117].

Hazard and Operability Studies (HAZOP) and Similar Methods. The second group of problem-identifying methods includes:

- Failure modes and effects analysis [123], [124]
- Action error analysis [123], [124]
- HAZOP studies (called PAAG methods in Germany) [127], [128]

These methods are so similar that they can be described together. All can be characterized as "deviation analysis" in that they look for possible hazards that can arise in the process, in the plant, or in plant operation if an error occurs, or the state or sequence of actions deviates from the prescribed state of sequence. (The assumption is that there are no hazards in the prescribed state or sequence.)

In this analysis of deviations, similar forms are employed for all three cases. As a rule, the form covers the following aspects:

- Deviation (error, failure)
- Cause
- Effect
- Corrective action

In many analysis, the following are also included:

- Error detection (how?)
- Frequency (often/seldom)
- Severity of effects

The third trait common to all three methods is the use of search aids. It is here and in the objects of study that the methods differ.

Failure modes and effects analysis (FMEA) is oriented to items of equipment and machinery, each having a certain function in the plant. It employs the working hypothesis that this function is not performed, i.e., each piece of equipment is examined for the effect its failure has, and what corrective action may be required.

The search aids in action error analysis and HAZOP studies are keywords describing deviations from nominal conditions or a nominal sequence. Keywords used to characterize human errors include:

- Too early

- Too late
- Not
- Wrong object
- Wrong sequence

Typical keywords for the HAZOP method [128] include:

- More
- Less
- No
- Different from

Much has been written about this and the other methods [123–128]; Table 10 presents a few examples of analyses by all three methods for a stirred-tank reactor.

All that remains is to indicate the proper time for applying these methods. In the design of a new plant, they should come in at the detailed safety design phase, when most of the piping and instrumentation diagrams are ready. These methods are also useful in the inspection of plants already built and on stream.

In the author's view, the possibility of creating suitable keywords for all parameters, functions, and action sequences in a plant makes the HAZOP method so flexible that it can cover the full range of applications of the other techniques.

Just two identification methods are therefore of any practical importance in the development of a complete safety concept: checklist methods and the HAZOP method.

Assessment of Hazards by Consequences. The concept of the hazard potential was introduced in Section 1.3 as a way of evaluating the magnitude or severity of a safety problem. Section 1.4 described how the hazard potential can be determined and what difficulties and imponderable factors must be taken into account.

If the scope and quality of safety practices are to be suited to the hazard potential so that a generally comparable and acceptable risk level is achieved, it is necessary to get at least a rough idea of the magnitude of the hazard potential (see also Section 3.1).

On the other hand, if effect-limiting safety devices are to be custom designed, e.g., an emergency pressure-relief system or a scrubber to handle off-gases in case of an emergency, it is necessary to model the physical and chemical processes taking place during the accident and to investigate their effects. Protective systems can be tailormade, and hazard potentials can be minimized (tasks 1 b and 1 c of Section 3.1).

Evaluation of Hazards by Probability of Occurrence: Incident Sequence Analysis and Fault Tree Analysis. Both methods in this class examine links between faults, represent them graphically, and (in principle) can assign probabilities to them. Incident sequence analysis starts with a single fault, and observes how it may develop. The working hypothesis is that every safety measure can succeed or fail with a certain probability.

Figure 41 shows a simple event tree for a temperature rise in an exothermic semibatch chemical reaction where three safety measures (heating shut-down, feed cut-off, and emergency cooling) have been instituted to prevent the reaction getting out of control.

The figure illustrates the three-stage safety concept. The probability of the undesired event can be calculated if the probability of the initial event and the failure probabilities of the several safety measures are known.

More careful analysis of the sequence of events reveals that the model is far too simple for the process under discussion. Whether the safety actions succeed or fail depends heavily on the intensity and rate of temperature rise, as well as the dynamic response of each action, and these are quantities that cannot be simply described in probability terms. Therefore, the only utility of event tree analysis for such processes is that it can make relationships easy to understand in simple cases.

Similar limitations apply to fault tree analysis, at least when used for typical processes in chemical systems (Fig. 42). The analysis goes in the opposite direction to that in incident sequence analysis: deductively from an undesired event. Faults are connected from top to bottom in chains that can lead to the top event. There are two main kinds of link:

1) The blocking AND junction: more than one event must occur in order to open the path upward
2) The passing OR junction: one event from a set of several events is sufficient to lead to the next higher event

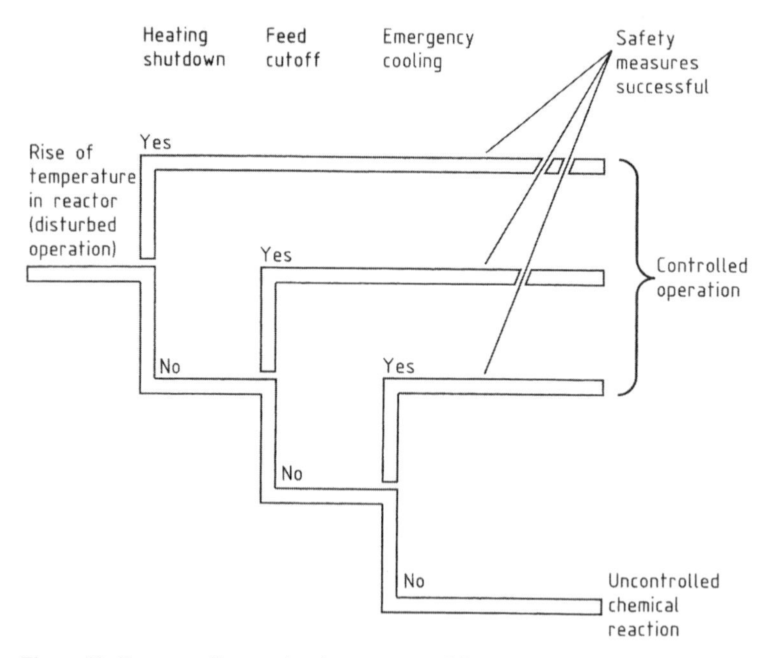

Figure 41. Event tree diagram showing success or failure of safety measures

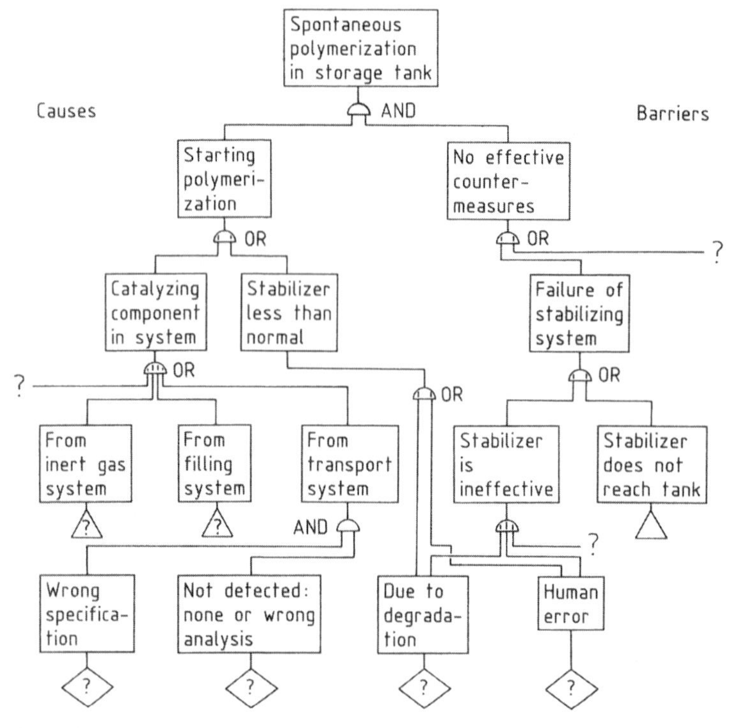

Figure 42. Fault tree diagram

Table 10. Examples of safety analysis methods applied to a stirred-tank reactor

Method	Deviation	Cause	Consequences	Safety measure
Failure modes and effects analysis (failure of equipment)	stirrer: does not turn, turns wrong way:	no electricity, motor defective, wrongly installed	heat transfer impeded, solids precipitate	revolution counter, feed shut-off, control unit
	steam valve: does not close	jammed	temperature increase in reactor	feed shut-off
Action error analysis (human error)	"too late": steam off, cooling water in	operator not paying attention, overworked	temperature increase, danger of uncontrolled reaction	alarm, feed shut-off, emergency cooling
	"not": steam off	operator not paying attention		automatic shut-off
	"incorrect": product	mix-up	unknown reaction	labeling analysis
Hazard and operability studies (deviations in functions and process variables)	"more": steam, heat	pressure reduction fails, no cooling	temperature increase, danger of uncontrolled reaction	steam shut-off by hand, feed shut-off, emergency cooling
	"none": cooling water, stirring	valve does not open, driving mechanism defective	heat dissipation impeded, temperature increase	feed shut-off, emergency cooling
	"reverse": direction of stirrer	wrongly installed	solids precipitate	direction indicator, control unit

Both fault tree analysis and event tree analysis pose stringent requirements on correct application and interpretation. Both call for experts. The methods have been fully described in the literature [123–126] and have been standardized in Germany [129], [130]; with respect to chemical safety, however, they have so far not played a significant role, for the reasons discussed earlier. Their main area of application is in the assessment and optimization of complicated technical systems in which the components display a yes/no failure behavior with assignable probabilities, and where the failure rates are known (e.g., in aviation and space flight). For further discussion see [123], [124], [131].

It can be maintained that these methods do not provide a satisfactory basis for assessing hazards in chemical processes as to probability or frequency of occurrence. The process variables in the analyses are also functions of time and intensity, and there is not a sufficiently large number of comparable states for statistical analysis. Estimates of probabilities therefore have to be established qualitatively or semiquantitatively on the basis of experience and the number and quality of safety measures.

For the same reasons, attempts to evaluate risks from chemical plants on the basis of occurrence probabilities of single events and process and component failure rates have not yielded useful results [131]; but this is not a fatal drawback because the risk due to chemical production (Section 1.4.1) can easily be derived from historical data of long-time experience and has proved low in the past.

3.4. Safe Processing: Strategies

3.4.1. Fire Protection

3.4.1.1. Fire Prevention

The key factors governing the occurrence and impact of fires are the quantity of combustible substances and possible ignition sources present in the plant. Both factors must be determined and taken into consideration for effective design of fire-protection measures.

First, it is necessary to ascertain the form and quantity of combustible, flammable, and explosive substances located inside battery limits. As well as products, auxiliary substances, and packaging materials, thus evaluation must also include energies, construction materials, and combustible parts of the plant.

Next, these combustible substances are rated according to their key fire parameters (see Section 2.2.1). The analysis must include the state of aggregation in which these substances are held in the plant under normal or process conditions, as well as the extent to which a possible fire, by raising the temperature, can initiate further critical reactions. A crucial point is whether fire loads are unprotected or are protected from the direct action of ignition sources by noncombustible packaging or enclosures, e.g., steel tanks or reactors [132–134].

The first important measure to prevent fires is to minimize the fire loads. In industrial plants, this can be done only to a limited extent, so possible ignition sources must also be identified, eliminated, or isolated from the existing combustible substances by bulkheads. Some examples of ignition sources are [135]:

- Open flames
- Sparks
- Compensating currents
- Spontaneous ignition
- Hot surfaces
- Electrostatic charges
- Chemical reactions
- Vibrations, shock waves
- Thermal manifestations of mechanical energy (e.g., friction)
- Thermal manifestations of electrical energy (e.g., overheating)

Ignition sources present during maintenance and repair operations also merit special attention. Sparks from welding, burning, and abrasive cutting have repeatedly caused fires. It is therefore desirable to prescribe the necessary safety precautions in safety certificates (see Section 4.4).

Smoking must be expressly forbidden in plants and warehouses. The creation of special smoking areas is preferable to concealed smoking in no-smoking areas. Appropriate containers, e.g., self-extinguishing wastepaper baskets, must be furnished for the disposal (daily if possible) of combustible wastes.

To avoid ignition by lightning, all buildings must have lightning protection. Potential compensation between large-area metal structures should be kept in mind, as should the use of overvoltage protection.

Critical effects can also follow from reactions between combustible substance and a fire-extinguishing agent. If incompatibilities lead to restrictions on extinguishing agents, these must be identified and integrated into the planning of organizational and technical fire-fighting procedures. For example, the reaction of fire-fighting water with a combustible substance may release not only flammable gases (from potassium, alkali metal compounds, etc.) but also corrosive vapors (HCl from chlorosilanes, acetyl chloride, etc.). Ignition by static discharge has caused accidents where CO_2 or medium-expansion foam has been improperly used [133], [136].

3.4.1.2. Limitation of Fire Effects

The principal effects of fire have to do with the toxic action of the smoke, and the high temperatures. To minimize harm to humans and animals, the environment, and property, a fire-protection concept must be developed for every industrial facility. The following sections identify the measures that must be included in such a concept.

Design und Construction. Access to the plant must be guaranteed by linking it to roads, streets, and fire service entries. Staircases – their number and placement dictated by the permissible lengths of escape and rescue routes – must give access to interior areas of the building. Escape and rescue routes simultaneously serve as lines of attack for firefighters, and must therefore be built to remain safely usable for an extended time. In modern industrial buildings, this is achieved by erecting massive stair towers (turrets) or administrative service wings, which are more or less independent of the supporting structure in the production building proper [137–139].

The availability of fire-fighting water is insured by a main supply designed to serve the needs of the entire developed area, and a plant supply designed to meet the special needs of the facility.

Fireproofing of the supporting beams of the building should be suited to the possible thermal loads in the anticipated fire. Simple approximations can be used for rough estimates (e.g., DIN 18 230), or more exact thermal-balance calculations can be performed. Regardless of the result, the use of noncombustible structural materials, or at least materials that burn only with difficulty or are difficult to ignite, is preferable.

An effective way to reduce temperature stress is to build in smoke and heat exhausts. Smoke extraction devices, which have to be opened mechanically, serve chiefly to remove cold smoke, and thus protect escape and rescue routes; heat exhaust systems are intended to remove hot fumes, and thus reduce the thermal stress on the supporting members of the building. Materials that fuse in a short time can be employed effectively and economically in these devices [140], [141].

The subdivision of the plant building into fire compartments or fire-fighting compartments should prevent unacceptable spreading of the fire. This is accomplished by walls and ceilings rated for a specified fire endurance (commonly 90 min). Openings in these members must be closed off appropriately; this includes doors, windows, ventilation ducts, cable and pipe penetrations, openings for tracked conveying systems, and other weak points [139], [142].

When a building is compartmentalized in this way, care should be taken to separate areas allocated to uses, e.g., administration, production, technical zones, and storage. Storage of starting materials and end products in production areas should be avoided.

Retention of contaminated fire-fighting water must form part of the design concept for fire protection. Retention is implemented, appropriate to the purpose of the building, the water contamination class of the substances, and the protective measures employed. If facilities are available for intercepting and removing contaminated fire-fighting water, normal water-retention facilities can be erected for more than one plant or storage area [143].

Fire-Protection Equipment. When fire compartments must be very large, for reasons dictated by production aspects, or when very large quantities of combustible substances or high fire risks are present, fire-protection equipment must be installed. The selection of such equipment is governed by the fire risk. The fundamental distinction is between room-protective equipment, which covers the entire enclosed region or fire compartment, and local protective equipment, whose action is restricted to a particular unit [144], [145]:

Early fire-detection devices are connected to a continuously staffed central office (that of the fire-fighting service if possible) and give a prompt alarm to the personnel assigned to damage control, so that the time allowed for the fire to develop undisturbed is cut short. Depending on the combustile material present, smoke, heat, and flame detectors can be used.

Automatic fire-extinguishing systems involve more than mere early detection. They are connected to a continuously staffed central office, and not only send an alarm to fire-fighting personnel, but also take immediate fire-fighting action. A properly designed and constructed system can suppress a nascent fire or block its progress, so that the fire-fighting service has a better chance of success.

The design of the fire-extinguishing system must include the proper extinguishing agents carefully matched to the type and quantity of combustible substances. Restrictions on extinguishing agents, e.g., because of critical reactions of stored substances with water, must be observed.

A sprinkler system consists of fixed piping with spray nozzles, closed off by heat-sensitive elements (liquid-filled glass vessels or fusible closures, with triggering temperatures ca. 50 – 260 °C). If a fire starts, the elements above the fire site experience a thermal load, which opens the outlets of the spray nozzles so that extinguishing agent is supplied to the fire site. Because only the nozzles above the actual site of the fire open, damage by the extinguishing agent is limited. Applications include piece-goods storage areas and production areas with large quantities of combustible substances. Water sprinkler systems are suitable mainly for Class A solid fuels. Sprinkler systems with a mixture of water and foaming medium can be employed with Class A solid fuels and with combustible liquids (Class B).

A water-spray or deluge system consists of a fixed pipe grid with open nozzles. In contrast to the sprinkler system, extinguishing agent is delivered through all the nozzles at once, over the entire area of the system. The extinguishing action is stronger than that of sprinklers, but the possibility of secondary damage is greater. This type of system is generally used only where the fire spreading rate is extremely high, e.g., transformer installations or warehouses containing substances that burn very rapidly or explosively. Water and mixtures of water with foaming media can be used.

Carbon dioxide fire-extinguishing systems deliver CO_2 through piping systems with spray or "snow" nozzles. The high concentration of extinguishing agent required means that the compartment must be isolated by bulkheads in case of fire, so systems of this type are preferred for smaller spaces.

Their use is essentially restricted to combustible liquids and electrical equipment. Carbon dioxide does not penetrate to a great depth, and so there are problems in putting out solid (smoldering) fires by this method. CO_2 is useless against metal and carbon fires, because the high temperatures that may occur in such fires can break down CO_2 to carbon monoxide and oxygen, and the liberated oxygen can promote the development of the fire. Certain metals such as magnesium can reduce CO_2 to carbon, e.g.:

$$CO_2 + 2\,Mg \longrightarrow 2\,MgO + C$$

so these metals can continue to burn, even after CO_2 has displaced oxygen.

The toxicity of the extinguishing agent means that special measures must be taken for personnel safety (redundant warning systems). CO_2 fire-fighting installations are increasingly used for local protection, e.g., in laboratory fume hoods or computer centers.

Powder extinguishing systems can be used primarily as local protective devices for special applications, e.g., to extinguish compressed gases or organometallic liquids. This type of system has the advantage that its use is largely independent of temperature (-50 to $+60\,°C$). Powder extinguishing agents are based on sodium hydrogen carbonate, potassium hydrogen carbonate, potassium sulfate, and other chemicals. Sodium chloride or potassium chloride powder can be used against metal fires.

Halon extinguishing systems can be used only with special approval because of current bans on these halogenated hydrocarbons; their application is limited to military uses, and on aircraft, spacecraft, and ships.

Semistationary, nonautomatic fire-extinguishing systems have recently come into increasing use alongside automatic systems. In such a device, the extinguishing agent is delivered through fixed piping from mobile supply units operated by the fire-fighting service. This kind of system is effective only when it is carefully matched to the fire service's equipment; it is used chiefly where the plant has its own fire-fighting unit. The same extinguishing media can be used as in automatic systems.

Dry or wet standpipes are almost always required because they markedly improve the performance of the fire-fighting service.

Fire-Fighting Methods. Industrial fire-fighting includes both the training of employees to carry out immediate response, and the installation of fire extinguishers or wall hydrants to deliver suitable extinguishing agents.

Establishing a properly trained in-plant fire-fighting service improves efficiency; training and equipment should be similar to those of the municipal fire services. Professional industrial fire-fighting services differ from in-plant services in that they are recognized by government agencies, and are trained, certified, and equipped to the same level as municipal fire departments. They represent the highest level of fire protection in industrial plants. Their value stems from local knowledge of the personnel, special equipment tailored to the plant, and the fact that they can work continually to provide high-quality fire-prevention and fire-protection services.

An essential aspect of industrial fire precautions is continuous firefighter training, together with combined exercises with plant personnel [145].

3.4.2. Explosion Prevention and Protection

3.4.2.1. Introduction

Flammable substances (gases, liquids, and solids), if in the form of gas, mist, or dust and mixed with a gaseous oxidizer (most commonly atmospheric oxygen, but other substances such as pure oxygen and chlorine can serve), can form an explosive mixture. The explosion of such systems is generally a fast, exothermic oxidation – a gas-phase chain reaction with chain branching [146]. Special aspects are presented by substances that are not only oxidizable in the gas phase, but can experience exothermic, explosive decomposition reactions without other reactants; examples are ethylene, ethylene oxide, and acetylene.

A gas-phase explosion is initiated by an ignition source, delivering sufficient energy and having a suitable energy distribution. The reaction initiated propagates spontaneously through the mixture. Such combustion reactions are accompanied by the release of a large quantity of energy with increases in temperature and pressure (and often also by the formation of dangerous reaction products). The hazards associated with an explosion are thus governed by three factors :

Occurrence of an explosive mixture
Presence of an effective ignition source
Effects of an explosion

These factors provide a logical structure for explosion prevention and protection in the following order:

Hazard identification
Risk assessment
Identification of countermeasures, risk reduction

3.4.2.2. Hazard Identification

(see also Sections 2.2.2, 2.2.3)

In gas-phase explosions, it is not the flammable material on its own that represents the potential hazard, but its contact or mixing with an oxidizer. The first step in identifying the hazard is therefore to define the combustion properties of the substances. The result indicates whether, and under what conditions, the substances can give rise to an explosive gas mixture. These conditions are characterized by data such as flashpoint, lower and upper explosion limits, and limiting oxygen concentration.

The second step is to determine the (minimum) requirements for the explosion hazard to be activated, i.e., the ignition characteristics of the system. These include the minimum ignition temperature for a dust layer, the ignition temperature of a flammable gas or liquid, the minimum ignition temperature of a dust cloud, and the minimum ignition energy.

The third step, focusing on the behavior of the explosive system after ignition, provides information on the expected physical explosion effects. Of interest are the heat of combustion, the explosion pressure, the rate of pressure rise, and the maximum experimental safe gap.

After the characteristics of the substances are known, this information has to be linked to the dimensions and parameters of the real chemical process and plant. Since substance data are obtained under standard laboratory conditions, it is important to know how they depend on factors such as temperature, pressure, volume, and volume of the process and plant under consideration. The basic transformation rules are qualitatively summarized in Table 11 [147].

It is necessary to compare substance properties with parameters of the chemical process and plant, not only for normal operating conditions, but also for failure states of the process and of the plant. Such analysis is particularly important for identification of process and plant conditions that may lead to the release of energy (i.e., creation of an ignition source) into the chemical system being handled (see the discussion of ignition hazards in Section 3.4.2.3).

3.4.2.3. Explosion Risk

The necessary condition for an explosion in the gas phase is the simultaneous occurrence of an explosive mixture and an effective ignition source. The likelihood of an explosion may therefore be considered as the product of the probability of the occurrence of an explosive mixture and the probability of the presence of an effective ignition source. The risk is defined as the combination of the (relative) frequency of an event occurring and the expected extent of damage; this means that explosion effects must also be considered.

Likelihood of Explosive Mixtures. Even when the chemical and physical relationships are understood in detail and years of practical experience have been accumulated, the likelihood that explosive mixtures may arise cannot be quantitatively evaluated, and it is doubtful whether such an evaluation would be of any use. It is more to the point to define a few qualitative categories that take in the limiting cases. In these categories, the probability of the formation or presence of a hazardous explosive mixture is described verbally [148]. The following zone definitions have been worked out in the course of European harmonization of explosion prevention and protection regulations [149]:

1) The explosive atmosphere (i.e., mixture under normal conditions) is present continuously, for an extended time, or frequently

Table 11. Transformation rules

Substance characteristics *	Real plant parameters			
	Temperature	Pressure	Dispersion	Volume
T_f (flashpoint)		+		
LEL (lower explosion limit)	−	−	−	0
UEL (upper explosion limit)	+	+	(+)	0
LOC (limiting oxidator concentration)	−	−	−	0
MIT (minimum ignition temperature)		−	−	−
MIE (minimum ignition energy)	−	−	−	0
p_{max} (maximum explosion pressure)	−	+	0	0
$(dp/dt)_{max}$ (maximum rate of explosion pressure rise)	+	+	+	−
MESG (maximum experimental safe gap)	−	−	(−)	(0)

* Laboratory data
+, −, 0 indicate parallel, inverse, or neutral effects; i.e., increase in value of parameter results in increase, decrease, or no change in substance property compared with laboratory data.

(i.e., for most of the time in the process under consideration). This zone is, as a rule, limited to the interior of apparatus: Zone 0 for flammable gases, vapors, and mists; Zone 20 for flammable dusts.

2) The explosive atmosphere may be present occasionally during normal operation: Zone 1 for flammable gases, vapors, and mists; Zone 21 for flammable dusts.
3) The explosive atmosphere is present only rarely and for a short time only (e.g., in the case of an infrequent malfunction or upset): Zone 2 for flammable gases, vapors and mists; Zone 22 for flammable dusts. (Zone 22 also covers deposits of dust that can lead to an explosive atmosphere in the infrequent event of the dust becoming agitated in air.)
4) The "zone-free area" where the occurrence of explosive mixtures is not anticipated, even in infrequent upsets (nonhazardous area).

In countries whose codes are based on United States practice, the classification differs slightly, but the rating arrived at is similar [150].

A mixture of flammable substances with air is explosive only if the concentration of flammable substances lies between the lower and upper explosion limits (see Sections 2.2.2 and 2.2.3 for definitions). With flammable liquids, if the processing temperature is above the flashpoint, an explosive mixture can be produced above the free liquid surface. If a flammable liquid is atomized (to form a mist or aerosol), the degree of dispersion can be sufficient to generate an explosion hazard, even at temperatures below the lower explosion point. The probability of occurrence of explosive mixtures for a given process in a given plant is not constant for all points in the system:

1) Inside the apparatus, this likelihood depends on the substances present, the prevailing process conditions, and the possibility of upsets
2) Flammable gases, vapors, liquids, or dusts can enter the outside atmosphere at openings and leaks (release points) and form explosive mixtures with air (explosive atmosphere)

The frequency of a release in the vicinity of apparatus depends on the procedures used; it is important to retain flammable substances in closed vessels, not just on grounds of explosion prevention, but also for reasons of exposure and environmental protection. Accordingly, portions of the plant containing flammable substances, as well as equipment and piping associated with these plant sections, are designed to be technically tight in relation to the mechanical, chemical, and thermal stresses anticipated under the intended operating conditions. The following examples illustrate the differences between various forms of release [151]:

- Continuous sources (continuous release) include the permanent free surface of a flammable liquid, open to the atmosphere (open vessel or separator, immersion bath, conveyor belts carrying products moist with solvent) at or above the flashpoint.
- Primary sources (occasional release) include expected leaks from flanges, shaft seals, apparatus and connections opened in normal

operation, venting to the atmosphere during filling or as a result of overpressure, occasional upsets such as the failure of an unmonitored cooling system, sampling, and drainage.
- Secondary sources (infrequent release) include cracks in apparatus walls or piping, defective gaskets in flange joints, defective shaft seals, and emergency pressure-relief devices (e.g., safety valves).

Dilution in the air causes the concentration of flammable substances to decline with distance from the release point. The size of an explosion hazard region thus depends on the rate of release and the propagation conditions. The rate of release depends in turn on the geometry of the source and the substance and process parameters, all of which control the mechanism of release (gas release, flash evaporation, spray jet, pond evaporation).

A central problem is to determine the maximum anticipated leak rate [151]. Catastrophic leaks, e.g., a large leak in a pipe or complete pipe rupture, do not fall within the scope of current explosion regulations.

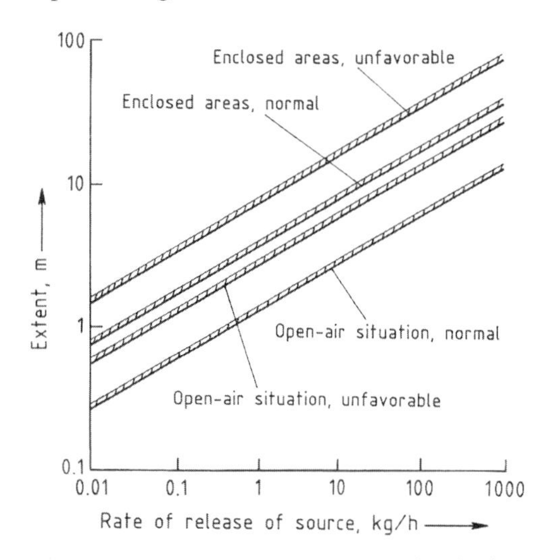

Figure 43. Hypothetical extent of an explosive cloud as a function of release rate of flammable substances for different ventilation conditions
Enclosed areas: normal, $C/f = 5$; unfavorable, $C/f = 0.2$
Open-air situation: normal, $C/f = 100$; unfavorable, $C/f = 4$
Extent $= 2r_z$ (see Fig. 44 for definitions); assumed typical LEL ca. 50 g/m^3

Other expected malfunctions generally lead to leaks up to ca. 0.01 – 1 kg/h. The failure of a seal (leak size in the mm^2 range) results in a release rate up to ca. 10 kg/h for gases or 100 kg/h for liquids, but in liquids it may be that only a fraction of this amount contributes to the explosion hazard.

The main factors governing propagation of released substances are the density of the fluid, the momentum of the release, and the climatic conditions and topography of the surroundings (see Section 1.4.2).

Ventilation conditions can contribute greatly to reducing explosion hazards:
- *Local ventilation.* If the atmosphere is exhausted from the release point (e.g., around the margins of open vessels), flammable substances can be kept from spreading into the compartment.
- *Dilution.* The concentration of flammable substances is lowered by dilution in air so that the concentration may fall below the lower explosion limit.

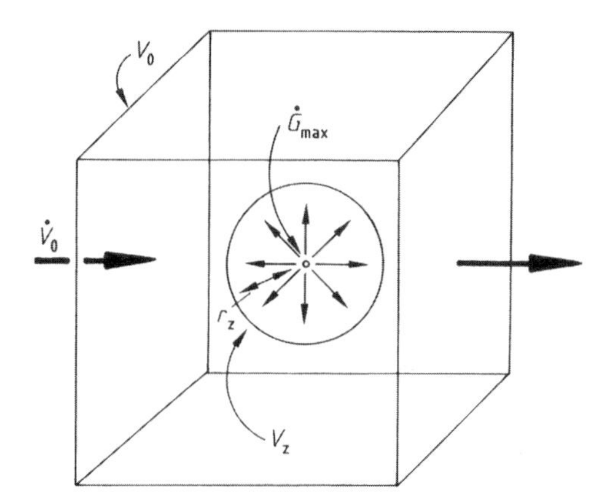

Figure 44. Ventilation model
V_0 volume considered; $\dot{G}_{max}$ release rate of flammables (mass/time); $\dot{V}_0$ fresh airflow (volume/time); $C = \dot{V}_0/V_0$ air change (1/time); f ventilation quality factor (1 – 5); $V_z = f\dot{V}_{min}/C$ hypothetical residual sphere of explosive mixture with radius r_z on dilution by fresh airflow; $\dot{V}_{min} = \dot{G}_{max}/k \cdot$ LEL required minimum fresh airflow to dilute below LEL with safety factor k

In buildings above ground with no special air intake or exhaust openings, an airflow of at least 1 h^{-1} is normally maintained. In typical chemical plants with natural ventilation, the temperature differences commonly present between different process apparatus units produce an airflow > 3 h^{-1}. Below ground, ventilation conditions are generally less favorable; most flammable substances are denser than air, so they tend to collect in places such as drains or underfloor spaces.

Artificial ventilation moves larger amounts of air, increasing dilution. Air movement can be designed in an artificial ventilation system, with allowance for explosive mixtures with density different from that of air. In an outdoor installation, air movement produces a much more rapid turnover.

Figure 43, based on a simple ventilation model (Fig. 44) [148], [151], shows the characteristic size of an explosive cloud as a function of release rate for various airflows. For a given release rate, an explosive cloud in a closed building is a factor of 3 – 6 larger than in an outdoor plant.

Initially, a hazardous area is created around any potential release source. Figure 45 shows the size of the hazardous areas around a stirred tank opened for charging in a ventilated room. The areas where an explosion hazard exists are identified in the ground plan. Where these areas overlap, the zone of higher probability prevails. For clarity, the hazardous areas have been smoothed to simple geometric forms containing the actual regions. The result is an area classification document. Figure 46 shows such a document for a simplified facility in a compartment with artificial ventilation. In practice, especially in smaller compartments, it is normal to extend, say, Zone 2 out to the boundaries of the room [151], [152].

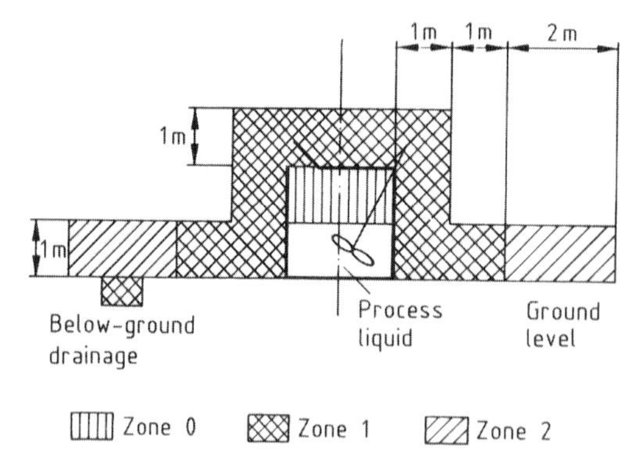

Figure 45. Type, shape, and extent of hazardous areas for a stirred vessel opened during normal operation [151]

Ignition Hazards. In view of the various ways energy can be introduced into an explosive system, 13 types of ignition sources have been defined [148]:

Hot surfaces
Flames, hot gases
Mechanical sparks
Electrical installations
Transient currents
Lightning
Static electricity
Electromagnetic waves (high frequency)
Electromagnetic waves ($3\times10^{11} - 3\times10^{15}$ Hz)
Ionizing radiation
Ultrasonics
Adiabatic compression, shock waves
Chemical reactions

The conditions for such ignition sources to become effective can be generalized in only a few cases. The effectiveness of ignition sources involved in process operation or the working of apparatus depends strongly on the properties of the substances present (see Section 3.4.2.2).

Hot surfaces may be present in normal operation (heated piping) or may arise through malfunctions (increased friction between moving parts). A large hot surface can initiate an explosion if its temperature exceeds the ignition temperature of the fuel that is present. With regard to small (cm^2 scale) hot surfaces, studies have shown that these must be at temperatures much higher than the ignition temperature before they can initiate an explosion [153].

Flames and hot gases are among the most effective ignition sources.

Visible sparks produced mechanically (friction, impact, and grinding) can be attributed to the combustion of metal particles heated to 1000 °C on separation from solid materials. The effectiveness of such sparks as ignition sources depends on the material as well as the ignition temperature of the fuel. For most optimal mixtures of common flammable gases and vapors with air, sparks from friction against steel are effective ignition sources; however, these sparks are effective on dust – air mixtures only if the minimum ignition energy and ignition temperature are low. At relative speeds of ca. 1 m/s, visible frictional sparks no longer occur (Fig. 47). Impact sparks may be effective especially with the combination of light metal and non-stainless steel.

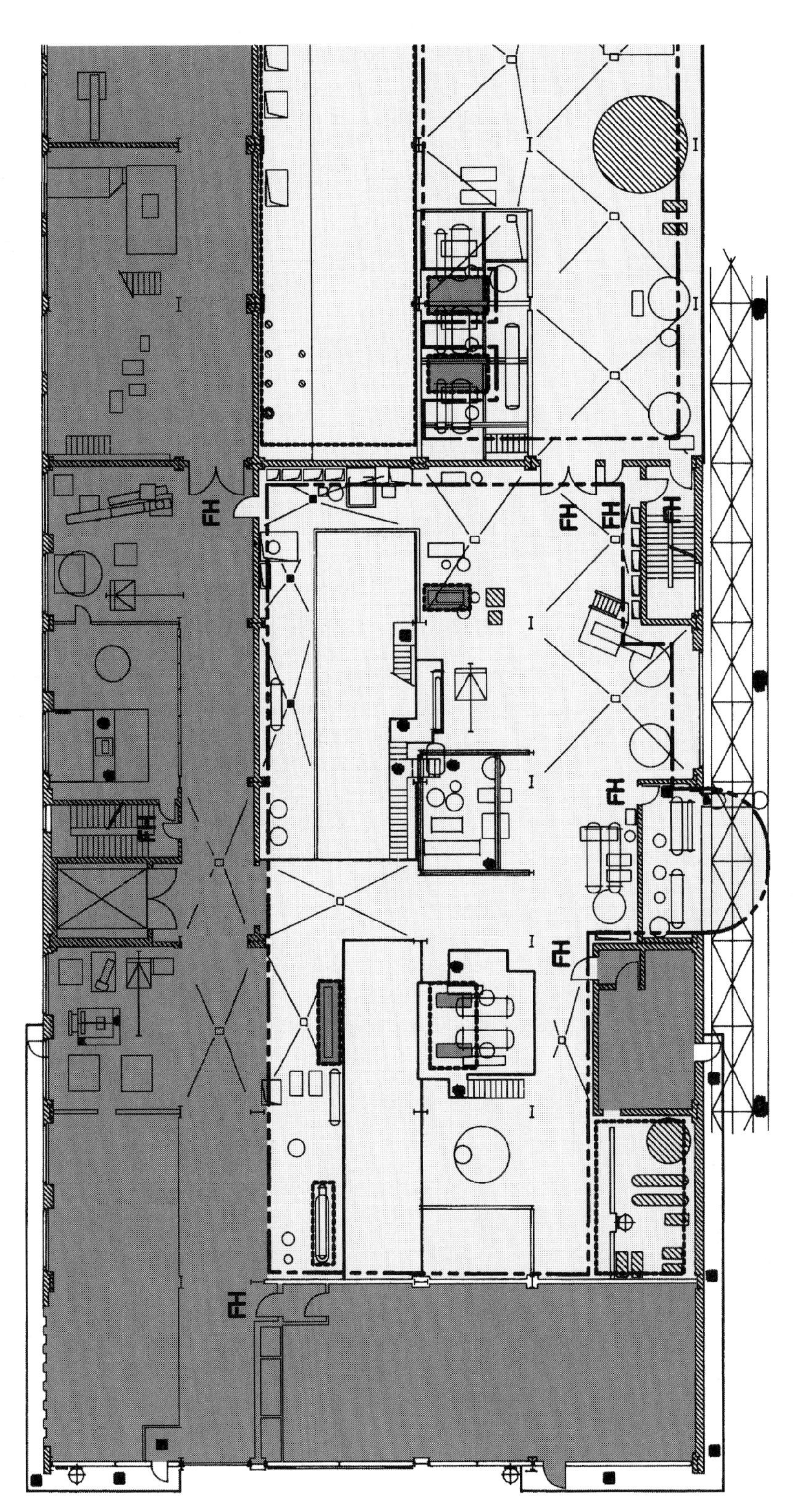

Figure 46. Plan view of the area classification of an in-house chemical plant Zone 1 enclosed by boundary; Zone 2 by boundary.

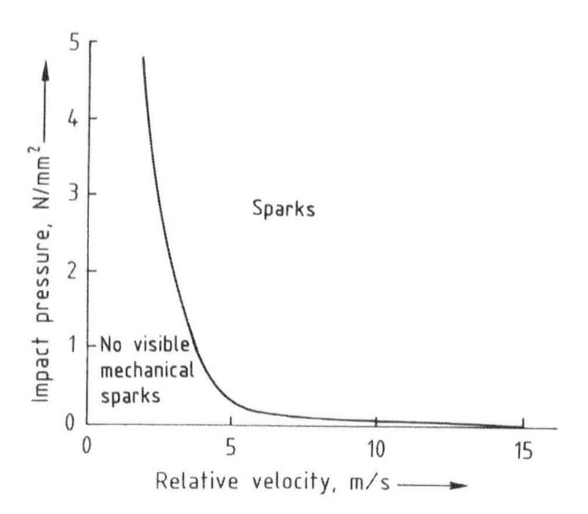

Figure 47. Generation of mechanical sparks
Typical curve for steel construction material [153]

Electrical sparks and hot surfaces may occur as ignition sources in electrical installations. Explosive mixtures can be ignited, even at low voltages. Electrical equipment is considered a priori safe if none of the following values is exceeded: voltage 1.2 V, current 0.1 A, energy 20 μJ, power 25 mW. The use of electrical devices in hazardous areas is strictly (and internationally) regulated (Fig. 48), and as a rule such equipment is rated on the basis of type approval tests or comparable technical documentation, e.g., a manufacturer's certificate [155].

Static electricity as an ignition source is a typical example of how a realistic assessment demands an analysis of the interaction of substance properties and process/plant parameters [156], [157]. Electrostatic discharges occur primarily with substances of very low electrical conductivity; discharges result from fast separation processes at interfaces. Solids with surface resistances $< 10^9\ \Omega$ and liquids with conductivities $> 10^8$ S/m cannot be charged to dangerously high potentials, provided they are grounded. In the case of a liquid capable of holding a charge, charging (e.g., in stirring) can be significantly increased if the liquid is a multiphase system. Dusts often acquire an electrostatic charge during handling. Gases do not become charged, but solid or liquid contaminants or solid or liquid species formed by condensation in flowing gases can lead to an electrostatic charge.

If, say, a pipe wall is sufficiently highly charged, the dielectric strength of the ambient air is exceeded and a gas discharge ensues. The effectiveness of the discharge as an ignition source depends on the form of the discharge as well as the sensitivity of the explosive mixture.

Figure 48. Explosion-proof electrical equipment, tested and certified, with the required explosion-resistant marking and data of testing house (with kind permission of the manufacturers STAHL and ABB CEAG)
A) Limit switch, flame-resistant enclosure "d", suitable for use in zone 1, explosion hazard due to products of Group II C and Temperature Class T6 (e.g., CS_2);
B) Electronic device for PCI, intrinsic safety "i(a)", suitable for use even with zone 0, explosion hazard due to products of Group II C and Temperature Class T 6

Spark discharges take place when charged, conductive objects are brought close together and a discharge channel forms between these "electrodes." Such discharges are effective ignition sources if the energy stored capacitively is greater than the minimum ignition energy of the mixture, as is commonly the case with large objects. Conductive substances (resistivity

$< 10^4\ \Omega \cdot m$), if isolated, represent an increased ignition hazard.

Brush discharges can occur from charged nonconductive substances. The counterelectrode can be a conductive, grounded piece of equipment or a (grounded) person. The effectiveness of bruhs discharges as ignition sources depends on the charged area; they can ignite gaseous mixtures with a minimum ignition energy of up to ca. 3 mJ (dust – air mixtures cannot generally be ignited by brush discharges).

At pointed electrodes, brush discharges give way to corona discharges; this weakest form of discharge, in its pure form, is not usually an effective ignition source.

If a grounded, conductive body has a thin, chargeable coating, charge can be drained off, and the danger of ignition by brush discharge may be greatly reduced. If rapid charging takes place, as in the pneumatic conveying of dusts, the charge density in the coating can become so high that propagating brush discharges with high energy content can occur (Fig. 49).

Figure 49. Propagating brush discharge from a chargeable surface [154]

Chemical reactions, by producing heat, can raise the temperature of any substances present, and thus become ignition sources. In highly reactive systems, this can happen through spontaneous exothermic reactions; such systems are said to be auto-igniting or unstable. Self-heating does not occur in less reactive systems under normal operating conditions. If heat removal is blocked, e.g., by (oxidation-sensitive) solid deposits, the temperature can rise to the point that the conditions for ignition are attained.

Other conceivable types of ignition source are less important in practice and cannot be discussed here.

Ignition sources can be categorized in a similar way to explosion hazards (the rating of explosion prevention measures is discussed later in this section).

Explosion Effects. Structures that may rupture in an explosion present a danger to the surroundings, owing to overpressure, underpressure, flames, and expelled fragments. In the field of explosion protection, design data for apparatus are compared with the anticipated maximum explosion pressure. If the possible explosion effects exceed the design limits of the apparatus, and product releases or flying debris from ruptured equipment cannot be ruled out, substantial consequences have to be reckoned with. In general, severe effects will not extend beyond a few tens of meters (see Section 1.4.3).

Rating of Explosion Prevention Measures. In the chemical industry, a distinction is made between cases involving appreciable impacts (i.e., severe personal injury or environmental harm) and those where damage to plant and equipment is the most that can be expected. If no appreciable impacts are anticipated, then economic considerations alone (plant availability, product loss, damage) dictate that an analysis be done to determine how tolerable is the occurrence probability of an explosion. If appreciable impacts cannot be eliminated, the occurrence probability of events must be made low enough that such events are regarded as positively prevented. The following commonly accepted rating system should be applied to such cases:

A) Where an explosive atmosphere is anticipated most of the time (Zone 0 or Zone 20), the occurrence of an ignition source is never tolerable, not even under extreme fault conditions.

B) Where an explosive atmosphere occurs occasionally (Zone 1 or Zone 21) the only ignition sources that are tolerable are those that result from a very infrequent malfunction, e.g., during run-up of a mill rotor (mechanical sparks) or a short- circuit (electrical sparks).
C) Where an explosive atmosphere can occur only infrequently (Zone 2 or Zone 22) continuous ignition sources or those present for long periods or for most of the time, cannot be tolerated. Sources occurring only as a result of a (simple) upset, e.g., a hot surface due to abnormal overheating of a heat exchanger, can be tolerated.
D) In an area where an explosive atmosphere cannot occur at any time, even under extreme upset conditions (i.e., nonhazardous area), the question of possible ignition sources is irrelevant. Even continuous ignition sources remain ineffective, and can be tolerated (e.g., the flame in an off-gas flare system).

For these four categories, Table 12 uses keywords to summarize complementary measures to restrict the formation of explosive mixtures and to avoid ignition sources.

The prerequisite for such a rating (categories B and C) is that the occurrence of an explosive atmosphere and the occurrence of an ignition source are mutually independent; i.e., both do not result from a common malfunction in the process or plant (common mode failure).

The explosion rating of a system states what zone exists for the situation under consideration, and in what zone the ignition sources found would still be tolerable. This safety balance can take different forms, depending on the specific question posed:

1) The ignition source appears to be tolerable in the ascertained category of occurrence of explosive mixtures. The system properties are so favorable that the probability of occurrence of an explosion is sufficiently low. Such a system is said to be intrinsically safe. No further precautions are necessary [158].
2) There is a gap between the category in which the ignition source would still be tolerable and the ascertained zone. Precautions are necessary, but they need only complement the existing intrinsic safety. Three cases can be observed:

a) For a reduction by one level (e.g., explosion hazard lowered from Zone 0 to Zone 1), a "simple" measure is sufficient
b) For reduction by two levels (e.g., from Zone 0 to Zone 2), the measure must satisfy extensive requirements
c) For a three-level reduction (e.g., from Zone 0 to "zone-free"), the reliability of the measure(s) taken must be enhanced even more, e.g., by redundancy

3.4.2.4. Explosion Risk Reduction

There are two options for lowering the explosion risk: measures to prevent an explosion, or measures to limit its effects.

If explosion prevention is chosen, measures are instituted to reduce the explosion hazard (preferable) or the ignition hazard; combinations of both approaches are possible. Limitation of the explosion hazard establishes a zone. If the limitation of the ignition danger meets the requirements for this zone, prevention of an explosion is insured.

Reduction of Explosion Hazard. The avoidance or restriction of explosive mixtures should always be the first priority. Especially outside closed systems, this rule also applies on grounds of occupational safety and health, as well as environmental protection. A simple way is to replace flammable substances with nonflammable ones (e.g., a flammable solvent by water).

Inside apparatus and equipment, the formation of explosive mixtures can be prevented by:

1) Keeping the concentrations of flammable substances in the gas phase outside the explosion range. Gas systems can be operated at overpressure, to prevent entry of air. Suitable purging is needed when such a system is placed in service [159]. In waste-gas streams, the total concentration of flammable components can be held above the upper explosive limit, e.g., by adding flammable gases (enrichment). Conversely, a stream can be diluted with air to reduce the concentration of flammable gases below the lower explosion limit. With liquids, the formation of

Table 12. Keywords for explosion risk assessment

Category	Occurrence of explosive mixtures	Keyword	Tolerance of ignition sources	Keyword
A	continuously/for long periods/frequently	permanently	not even in very rare situations	never
B	occasionally in normal operation	occasionally	not even in rare situations	rarely
C	in rare situations and with short duration	rarely	not in normal operation	occasionally
D	not even in very rare situations	never	in normal operation	permanently

explosive mixtures can be avoided by keeping the temperature sufficiently far below the flashpoint. Substances with low flashpoints can often be replaced by substances whose flashpoints are a safe margin above the ambient and process temperatures.

2) Displacing or replacing air (or other gaseous oxidizing agent) with an inert gas until the oxidizer concentration drops below a case-specific critical level (see Section 2.2.2). A dryer for flammable dusts can be operated with combustion off-gases no longer having oxidizing properties, or a solvent tank can be purged with nitrogen.

In the vicinity of process equipment, the occurrence of explosive mixtures can be largely controlled by:

- Using closed systems to avoid release of flammable substances during normal operation. Examples include gas equilization lines for storage tanks, and the use of lock systems for tank filling and emptying.
- Installing apparatus that has more integrity in terms of possible leaks. Examples include: welded piping to avoid flange joints; O-ring seals instead of stuffing boxes; and materials less susceptible to fracture, such as metal instead of glass.
- "Good housekeeping," i.e., the immediate removal of all flammable liquids and dusts released in the vicinity of plant equipment.
- Creating suitable ventilation to insure the instant dilution of any gas mixtures. This practice can be established throughout the plant (e.g., open-air plant). Selective efforts can also be made to avoid formation of explosive mixtures in areas where sparks can occur, e.g., the installation of an open cage on stirrers and pumps insures that vapors escaping at the shaft seal are immediately diluted with air.

Reduction of Ignition Hazard. The ignition hazard from each type of ignition source listed in Section 3.4.2.3 can be reduced by specific practices. Hot surfaces can be avoided by matching the cooling capacity to the energy dissipated in the apparatus, and providing overtemperature protection. If necessary, insulation can be installed as a form of shielding.

Possible ignition sources involving frictional heat or sparks can be dealt with by suitable design:

- Use of adequately sized components with sufficiently long service lives
- Avoidance of improper material combinations, use of materials with favorable dry friction properties
- Monitoring of temperatures, and of lubrication and cooling systems
- Maintenance of adequate clearance between moving parts

Electrical ignition sources can be avoided by limiting electrical characteristics such as energy and power so as not to produce sparks capable of initiating an explosion ("intrinsic" safety, as in instrumentation circuits) or by preventing extreme conditions and overload in equipment that is free of ignition sources under normal conditions ("enhanced" safety, as in electric power distribution and lighting).

The chief way of reducing ignition hazards due to electrostatic discharges is to use (grounded) conductive construction materials, and to ground conductive objects, liquids, and personal (resistance to ground $< 10^6\ \Omega$). The use of objects made of chargeable, nonconductive substances is generally permitted in Zone 2, but not in Zones 0 and 1. Ignition hazards can be avoided by limiting the dimensions of such objects, depending on the zone, and the ignition sensitivity of substances present; e.g., chargeable surfaces can be kept below $100\ cm^2$ for substances classified in explosion groups II A and II B in Zone 1. Similarly, the thickness of chargeable coatings on conductive substrates can be restricted (e.g., < 2 mm for substances belonging to explosion groups II A and II B) and ignition thus avoided.

When working with chargeable liquids, dangerous charging can be avoided by limitations on filling rates (e.g., < 7 m/s for hydrocarbons in a pipe with nominal diameter 50 mm) and by systematic under-level filling.

Ignition hazards due to chemical reactions are avoided by excluding pyrophoric systems from hazardous areas, and maintaining safe process conditions (e.g., layer thickness, temperature, residence time, and thermal load in drying processes).

Limitation of Explosion Effects. Design practices can insure that, if an explosion does take place in an apparatus, it does not have appreciable impacts. Maximum possible pressures generated by an explosion can be determined to acceptable accuracy by thermodynamic calculations, or measured experimentally.

Explosion-pressure resistant design consists in designing apparatus to withstand the anticipated explosion pressure, generally less than 10 times the initial pressure for fuel – air mixtures. If the safety factor relative to yield strength is smaller, the term used is explosion pressure shock resistant design. In piping where detonations are possible, the standard design for PN 10 includes significant hidden reserves that are adequate for nominal diameters ≤ 200 mm; for nominal diameters > 200 mm, the piping should be designed to PN 16 at least. In interconnected vessels, an explosion in one portion can raise the pressure in the remainder; in this way, a (consequent) explosion can be induced, with a higher initial pressure. Explosion protection in such a case may entail decoupling of volumes.

Pressure-relief devices (rupture disks or explosion valves) open as the explosion pressure rises and reaches their response pressure. They insure that the apparatus experiences only a reduced explosion pressure. Such systems are designed especially for protection against dust explosions, where the rate of pressure rise is generally fairly low [160], [153]. Naturally, pressure relief must not create new hazards (see Section 3.5.1).

In an explosion suppression system, a complex of detectors respond at a very early stage in the explosion, and an extinguishing agent (e.g., ammonium-phosphate-based powder) is injected on a scale of milliseconds in order to quench the explosion. The protected apparatus can then be designed for a lower pressure, e.g., 200 kPa. The amount of extinguishing agent needed depends on the violence with which the substance explodes, and the volume of apparatus to be protected (typically 50 kg per 25 m^3 for St 1 dusts; see also Section 2.2.3).

Isolation ("decoupling") in the context of explosion protection means separation of an explosion-resistant section (where an explosion can take place without danger) and a non-explosion-resistant section (where explosions must be positively prevented). A number of devices have been created to prevent flame-front breakthrough; these employ several principles:

- Fast mechanical closure and lock-type features
- Extinguishing of flames in narrow gaps
- Use of flame-extinguishing agents (as in explosion suppression)
- Flame arresting by high counterflow velocity
- Flame arresting by liquid barriers

Equipment selection must take into account substance properties; key aspects include sensitivity to ignition and propagation velocity (see Section 3.5.3). Narrow gaps are also employed in the design of explosion-proof equipment, e.g., fans, gas pumps for recycle systems, submerged pumps, and internal combustion engines [161]. This principle, when used in electrical equipment design, results in the rating "flame-proof enclosure."

3.4.2.5. Legal Aspects

In the domain of European legislation, EU directives set the framework for explosion protection at continental and national levels. The principal directives are ATEX 100 a [162], ATEX 118 a [163] and portions of the Machinery Directive [164].

The ATEX 100 a directive, aimed at manufacturers of "devices and protective systems intended for use in explosion hazard regions," governs the quality and certification of such devices and systems. (For electrical equipment intended for use in Zones 0 and 1, type testing and a certificate of conformity issued by an authorized agency have long been required [165], according to earlier harmonized European standards.) For Zone 2 equipment, the manufacturer's declaration of conformity is sufficient. Similar rules apply for equipment to be used in areas where there is a dust explosion hazard (this equipment is still classified according to earlier zone definitions with just two types, i.e., Zone 10 and Zone 11, substantially comparable to the "new" zones 20 and 21 or zone 22, respectively).

The ATEX 100 a directive, in contrast to its predecessor, now includes equipment and devices as well as protective systems "which are intended to halt incipient explosions immediately and/or to limit the effective range of explosion flames and explosion pressures," and makes such equipment and systems subject to certification. The pertinent standards are being drawn up or are under revision by the European standards institutions CEN (e.g., committees TC 305 and TC 114) and CENELEC (e.g., TC 31), partly in collaboration with international standards bodies such as ISO and IEC.

The ATEX 118 a directive (in preparation) applies to "the minimum requirements for improving the safety and health protection of workers potentially at risk from explosive atmospheres." It is aimed at plant and process operation and represents a synthesis of earlier national regulations, reflecting the basic principles of explosion protection.

3.4.3. Safety Techniques Based on Process Control

Safety objectives in chemical process plants can be achieved with systems based on process engineering and process control engineering (PCE). Technical or organizational measures, or a combination, can be employed. Alternative solutions that are equivalent from the safety standpoint can be found, so that the most appropriate solution from the process engineering standpoint can be selected.

Safety concepts for specific safety objectives set down in technical regulations must take first priority.

This section deals with PCE equipment for safety in process plants. The discussion is based on NAMUR Recommendation NE 31 [166], VDI/VDE 2180 [167], and provisional standard DIN V 19250 [168], with some points from the book *Praxis der Sicherheitstechnik* (Practical Safety Engineering), vol. 1, *Anlagensicherung mit Mitteln der PLT-Technik* (Safety Techniques Based on Process Control Engineering) [169].

The PCE safety measures suggested in this section are tailored to the risk component that would arise from a process plant in the absence of such measures. This way of assessing risks is only qualitative, and must be carried out by case. Other procedures leading to a comparable safety level are possible.

As used in this section, "safety objective" refers to preventing injuries to persons, major environmental damage, and major equipment damage to property. Safety precautions for electrical installations and equipment, occupational safety practices, or measures taken to safeguard machinery are excluded.

3.4.3.1. Integration of PCE into the Safety Concept

As a rule, safety requirements under DIN V 19250 and the measures required are more stringent the greater the risk that must be covered. The risk must be reduced at least to the acceptable limit (DIN VDE 31000 [170]; see Section 1.3) by non-PCE or PCE safety measures. Non-PCE safety measures and equipment may include elimination of ignition sources, isolation to control explosions, pressure-proof design, safety valves, interception spaces, and retention systems.

PCE safety measures commonly reduce only a portion of the risk resulting from a unit. The risk on which the following discussion is based is just the portion to be covered by PCE safety measures. The remainder can be dealt with by non-PCE safety measures.

Safety tasks allotted to PCE vary widely in character and significance. Similarly diverse are the safety requirements on PCE equipment and practice (technical; nontechnical, e.g., organizational).

The use of PCE equipment for plant safety, the tasks assigned, and their performance are among the matters to be decided in a general safety review. The following are determined in the safety conference:

- The safety objectives (personnel, environment, property)
- The PCE equipment to be used for plant safety
- The classification of PCE equipment by task:
  Operating systems
  Monitoring systems
  Safety systems
  Damage minimizing systems
- The required technical and organizational measures (e.g., cycle of functional testing)

The safety review is the basis for the task-driven planning, construction, and operation of PCE safety systems, at a reasonable expense and with clearly defined functional scope.

The selection of measures for maximum simplicity and direct impact generally leads to a safe and economic solution. PCE safety measures find use when other approaches are not applicable, inadequate, or uneconomic, given a comparable level of risk reduction. The economic comparison takes in not only one-time investment costs, but also recurring maintenance expenses.

The requirements for PCE safety and monitoring systems are derived from this definition. Inadvertent triggering of the PCE safety and monitoring systems during any phase of the process must not lead to an unacceptable fault condition of the plant.

3.4.3.2. Classification of PCE Systems

In chemical process plants, PCE systems are classified as operating systems, monitoring systems, and safety systems (Fig. 50), in relation to the ranges of a process variable:

- Specified operation, subdivided into
  Normal operating range
  Admissible error range
- Nonspecified operation – nonadmissible error range

Region 1 of Figure 50 corresponds to the situation in which factors inherent in the process keep the process variable from reaching the nonadmissible error range. A monitoring system is sufficient. The process variable is brought into the normal operating range by automatic or (after a status signal) manual action. Region 2 represents the case where the process variable exceeds the limit for the nonpermissible range. Because another protective device is present (safety valve, rupture disk, rapid-opening or rapid-closing valve), a PCE device connected ahead of it to signal or limit the increase in the process variable is classified under monitoring system. In region 3, the PCE device keeps the process variable from entering the nonpermissible range, and is therefore classed as a safety system.

PCE Operating System. PCE operating systems are used in specified operation of the plant in its normal operating range. These devices implement the automation functions required for production: measurement and control of all variables relevant to operation, including related functions such as logging and report generation. High-level control algorithms, complex control sequences, automated recipe processing, and optimization strategies are increasingly employed. Many binary, digital, and analog signals have to be processed if all these tasks are to be performed. Because the functions of PCE operating systems are called on continuously or frequently in operation, these devices are subject to plausibility checking by plant personnel, so that failures and malfunctions can be detected immediately.

PCE Monitoring System. When the plant is in specified operation, PCE monitoring systems responds to conditions in which one or more process variables are outside the normal operating range, but there is no safety reason to discontinue operation; i.e., these monitoring devices respond at the boundary between normal operating and admissible error ranges of process variables.

Acceptable fault conditions of the plant are reported, in order to evoke heightened awareness or direct action by operating personnel; the monitoring system may even initiate action itself to bring process variables back into the normal operating range.

Also included are PCE devices connected ahead of PCE or non-PCE safety systems in order, if possible, to prevent them from responding.

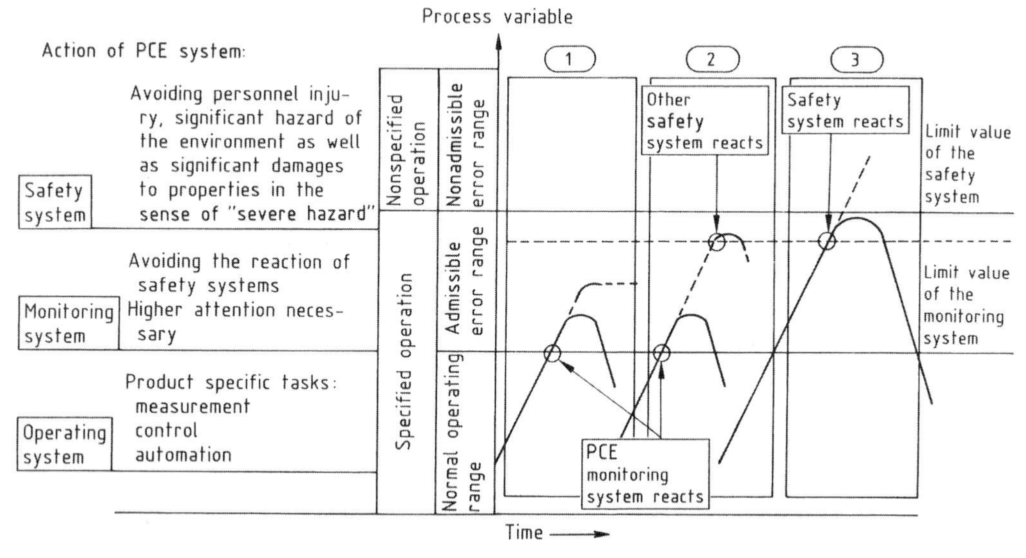

Figure 50. Functions of PCE systems

PCE Safety Systems. In contrast to the functions of PCE operating and monitoring systems, PCE safety devices have the function of preventing nonadmissible error state of the plant.

A PCE safety system is necessary if its absence would allow the plant to reach states that could lead directly to personal injury, major environmental damage, or major equipment damage.

The task of a PCE safety system is usually to monitor a process safety variable and take one of the following actions if the variable goes outside the admissible error range:

1) Initiate a control process
2) Notify the operating personnel (who are always present) so that the proper action can be initiated ahead of time

The functions of PCE safety systems always take priority over those of PCE operating and monitoring systems, and must be executed at a level near the process level with minimum complexity.

The functions of PCE safety systems, in contrast to those of PCE operating equipment, are called on extremely seldom, because the probability of occurrence of the undesired event is low, and because PCE operating, monitoring, and safety systems are often in a staggered configuration (Fig. 50).

Because they are called on so infrequently, it may be desirable to allow sharing of PCE safety components, such as actuators, by PCE operating systems, to enhance availability and facilitate plausibility checking. Such common components must meet the requirements for PCE safety equipment.

PCE safety systems employed for prevention of major equipment damage are designed solely on economic grounds and are not discussed here.

PCE Damage Minimizing Systems. PCE damage minimizing systems come into action: when the plant is in nonspecified operation; when an undesired event occurs. They limit the impacts on persons and the environment, limiting the extent of harm under these extremely rare circumstances.

If PCE equipment is used to detect the undesired event, what is monitored is not a process variable (e.g., pressure, temperature), but some other quantity such as the concentration of gases released in the atmosphere. Effectors put into action do not influence the process, but have to do with the threatened region outside the vessels, apparatus, and piping (e.g., initiation of a water curtain to prevent the spread of ammonia). PCE damage control devices are often combined with non-PCE damage minimizing devices and organizational measures.

3.4.3.3. Requirements for PCE Equipment for Process Plant Safety and Design Principles

PCE Monitoring Systems. No special requirements apply to PCE monitoring devices; they are treated like operating equipment.

PCE Safety Systems. If the interdisciplinary safety review establishes the need to employ a PCE safety system, the following procedure is employed [168]:

1) Qualitative estimation of risk
2) Setting requirements for PCE safety system
3) Definition of technical and organizational measures

In particular, the following steps are identified and documented:

- Task statement, safety problem
- Function of PCE safety system
- Technical design (principle)
- Nature and frequency of scheduled function testing
- Other organizational measures (e.g., scheduled maintenance; see Section 4.4)

Two points must be considered in the selection of PCE safety devices: the minimum risk reduction aimed for; and the safety-related availability of PCE safety systems.

Fundamental Requirement. PCE safety systems must be designed and operated in such a way that if one passive fault occurs, the protective function is still performed.

In the design of PCE safety systems, the safety-relevant availability must therefore be chosen such that the risk is reduced to a residual level lower than the limiting risk, even if one passive fault occurs.

The safety-relevant availability of PCE safety systems depends on:

- The failure rate due to passive faults
- The mean time to detect and remedy passive faults
- The degree of redundancy of the PCE safety system

The safety-relevant availability can thus be enhanced by the following practices, or a combination of them:

- Redundancy: the presence of more operable technical means than are required to fulfill the intended basic functions (e.g., special safety or reliability requirements)
- Homogeneous redundancy: redundant design of a device or parts thereof such that the redundant channels are identically structured, and operate by identical physical processes
- Heterogeneous redundancy or diversity: redundant design of a device or parts thereof such that the redundant channels operate by different physical processes or are differently structured
- Fail-safe quality: ability of a safety device to hold a process safety variable in a safe state, or to modify the variable directly into another safe state, on the occurrence of certain faults in the safety device
- Self-monitoring: a protective device is self-monitoring if, aside from certain faults, it is so constructed that all other faults are detected by the self-monitoring, and thus the safe state is achieved

To meet the requirements on PCE safety systems after risk estimation, distinction is made between high and low risk application [166].

Low Risk Application

Requirement: One passive fault must be detected and remedied within a time interval in which no violation of specified operation is anticipated.

Response: Single-channel PCE safety device with:

1) Short fault-detection time (e.g., high frequency of function testing, continuous plausibility checking); or
2) Low probability of passive faults of the PCE safety device.

A reduced risk may be present if non-PCE safety measures of a technical or organizational nature are employed.

High Risk Application

Requirement: One passive fault must not impair the ability of the PCE safety device to carry out the safety function. Independently of the process behavior, such a fault must be detected and remedied within a time interval in which the simultaneous occurrence of a second, independent fault is not anticipated.

Response: Redundancy of PCE safety device. In general, one-out of-two or (for simultaneously high production availability) two-out of-three design combined with regular functional checking is sufficient. Diversity (e.g., in acquisition of measurements) does not automatically enhance safety over homogeneous redundancy. It is a supplementary measure for preventing possible systematic faults.

Fail-safe or self-monitoring systems are equivalent to redundant devices.

Within a PCE safety system, combinations of these measures are possible depending on the particular availability requirements; e.g., acquisition of measured values may be redundant, control fail-safe, and actuator single-channel.

There are several ways to estimate risk. Quantitative evaluation is not possible for the chemical production plants under consideration here (see also Sections 1.3 and 3.3), because the wide variety of processes and the comparatively short lives of plants mean that adequate statistical material cannot be obtained. This holds equally for the safety relevant availability of PCE safety devices.

Accordingly, the qualitative scheme of Table 13 is used in the design of PCE safety systems. It involves measurements of the risk to be dealt with as well as the safety-relevant availability of the PCE safety systems (the availability of the single-channel device when there is redundancy).

Table 13. Grading of PCE safety systems by risk level and safety-related availability of PCE safety system (availability of single-channel device when there is redundancy)

Safety availability	Risk to be covered	
	Lower	Higher
Higher	I	II
Lower	II	II

Damage Minimizing Systems. Because the undesired event is expected to occur extremely seldom, damage minimizing systems are commonly single-channel units, and must be tested for function in regular intervals.

Principles for Design and Construction of PCE Safety Systems. A number of important principles must be considered in the design and construction of PCE safety systems:

- Proven, reliable hardware and installation methods must be employed.
- The PCE safety device must be simple in construction. Fault effects (e.g., secondary or sequential faults in the PCE safety device) should, if possible, be limited by suitable barriers to fault propagation: high-impedance decoupling, short-circuit strength, galvanic isolation, etc.
- Harmful effects due to environment and products, e.g., vibration, impact, static strain forces, thermal action, corrosion, contamination, wear, and electromagnetic effects (including those resulting from lightning, ripple content in power supply, grid malfunctions, grid noise, etc.), must be taken into account.
- Fail-safe properties of equipment must be utilized (actuator with spring return to safe position, closed circuit to reset, etc.).
- When operating and monitoring systems are shared with PCE safety systems: the safety function must take priority over other functions; and the shared elements must be rated as for the safety device.
- The measurement of process safety variables, processing operations, and the implementation of the safety function must be done with accuracy and speed suited to the safety problem.
- The measurement ranges of process safety variables must be chosen so as to insure adequate resolution. Limiting values must be far enough from range end points so that resolution is still guaranteed if measurement errors are within tolerance.
- The correct setting of limiting values must be protected against inadvertent change.
- As a rule, automatic reactivation after initiation of the safety function should be disabled.
- Whenever possible, process safety variables should be selected such that they can be measured directly, simply, and by a proven method. Indirect derivation of process safety variables by combining measurement signals should be employed only when direct measurement is impossible or unreliable.
- It is desirable to record process safety variables.
- Analog process safety variables should be displayed, together with the limiting values, in the monitoring (control) room, or at local

control panels. In this way, operating personnel can do plausibility checking, so that fault detection times are kept short and the limiting value setting can be checked easily.
- The design of PCE safety systems must also take account of maintenance and start-up needs. Ease of inspection and accessibility of all components of the PCE safety system are important, even in the planning phase.
- Manual overrides can be provided to allow inspection or repair of PCE safety systems during plant operation.
- In isolated cases, redundant PCE safety systems should be examined to determine whether fire hazard or the possibility of mechanical damage necessitates: a split construction; or a protected and/or separate power supply and spatial separation of cable runs for both channels.

Use of Programmable Electronic Systems (PES). Most PCE safety systems are hardwired. The main arguments for this are:

1) The setup must be simple and straightforward.
2) The PCE safety device is seldom, or never, modified while in service.
3) Only a few of the PCE devices in a process plant are safety devices.

In special cases (more complex protective devices), if the use of PES is more economic, the following principles apply:

1) Certified systems can be used for both single-channel and multichannel safety devices in the applications, and under the conditions, set forth in the test certificate.
2) Noncertified systems may be used for safety measures only as part of a multichannel safety device, and then for at most one channel. The other channels must be either hardwired (as required for PCE safety devices) or constructed with certified systems. The following minimum requirements apply to the noncertified system:
   - The system must be proven in former applications
   - Safety and nonsafety functions must be implemented separately whenever possible (in software and, if applicable, in hardware) to avoid influencing the safety section.
   - Along with the standard system monitoring diagnostics (e.g., "watchdog" circuits), which return outputs to the safe state in malfunctions or failures, further system-specific requirements (fan monitoring, climate control, etc., as instructed by the manufacturer) are to be implemented.
   - Shutdowns may be implemented only via binary outputs. Resetting analog outputs to 0/4 mA is not acceptable as the sole shutdown action.
   - Reports of a safety function response must be distinct from reports of operating and monitoring equipment (e.g., different colors).
   - Application software must be created only by trained personnel, written in an easy-to-follow form (structured, modular software), and easy to test.
   - The safety portion of the application program must be subjected to initial examination by the persons involved, and by a specialist, along with the planning documentation.
   - When major changes are made in software, even in the non-safety-relevant portion, a functional test of the safety-relevant functions should be performed.
   - Scheduled functional testing should also include the current software status (application software and firmware), including software documentation.

Labeling. All important components of the PCE safety system must be labeled as such in the documentation, locally in the instrument and the control room.

Testing Before Commissioning. Before the PCE safety system is first placed in service, a test must be performed to determine whether its design and function comply with the provisions of the safety review. At this time, the documentation of the PCE safety system and the test instructions for periodic testing must be available.

Testing must be done in such a way that proper functioning is demonstrated in the interplay of all components. The first functional test must be documented in writing.

3.4.3.4. Operation of PCE Safety System

Organizational Measures. Organizational measures are required for the operation of the PCE safety systems. They fall into four groups:

Continuous monitoring (surveillance)
Inspection (functional testing)
Maintenance
Repair

Continuous Monitoring. Malfunctions of PCE safety systems must be detected by regularly observing process variables, and checking them for plausibility. Such observations are done by qualified personnel.

Outwardly noticeable defects or damage to the PCE safety equipment or their installations are identified by regular visual inspections, and must be immediately repaired.

Functional Testing. Functional tests are needed in order to reveal passive failure. Test instructions must be prepared, summarizing the nature and extent of periodic testing. These instructions must contain information on the nominal state and nominal performance of the safety device, along with a description of the properties and functions to be tested. This includes, in particular, statements of limiting values and measurement ranges, and other specified features to be inspected (such as actuating times for valves, time lags for trigger signals, and other properties important for the performance of the safety task). The testing procedure must be described in test instructions (e.g., checklists) that the checking personnel can understand. The limiting values must be documented in writing by the operations manager.

The test cycle is set in the general safety review. Differences in availability may dictate that some parts of a protective device be tested more often than others.

If no comparable experience exists, an appropriately short testing interval should be set at first. If the tests reveal sufficient safety-related availability, the test interval can be lengthened as the time in service increases.

By analogy with pertinent technical guidelines, inspection of the entire PCE safety system (from sensor to actuator) must take place at least once a year. The operations manager is responsible for seeing that functional tests are performed.

It is desirable that testing be done under conditions corresponding to the demand case and with the least possible modification of the PCE safety system. Whenever modifications are necessary to perform the check, special care should be taken to restore the PCE safety system to its proper state. If manual overrides are installed in multichannel PCE safety systems, only one channel at a time may be bridged.

Care must be taken that neither safety nor availability is substantially impaired during the test. The method used should start with the assumption that faults may be present in the safety device; thus suitable (e.g., organizational) measures should be taken to insure that the plant remains in specified operation.

Further details on testing, especially testing methods for PCE safety devices, are explained in VDI/VDE 2180, Sheet 4.

In addition, tests should be performed after long shutdowns and repairs to the device.

Maintenance. When service conditions are severe, or in the case of certain measurement techniques (e.g., process analytical instruments), scheduled maintenance may be necessary. Work schedules must be prepared summarizing the nature and extent of periodic maintenance, and the required permits from plant management. Maintenance operations are performed by qualified personnel in accordance with these work schedules, and the work must be documented.

Repair. Repair of PCE safety devices must be performed without delay by qualified personnel whenever defects are found in the devices, and there is no alternative that will maintain the level of safety.

Documentation. Testing, maintenance, and repair of PCE safety and damange control devices must be documented. In particular, documentation of functional tests must include at least the following information:

- Identification of test objective
- Results of test, with detailed information on faults corrected
- Date of test
- Signature of tester
- Signature of operations manager

The signatures of the tester (inspector) and operations manager confirm the release and acceptance of the functional PCE safety or damage control device. The test report must be retained for at least five years so that performance of specified tests can be proved.

Fault Analysis. To improve the reliability of protective devices, any faults discovered must be carefully analyzed; longitudinal documentation of test results reveals weak points. If the same causes are seen to produce faults again, this indicates a weak point and calls for improvement or testing at shorter intervals.

Decommissioning, Restart, and Change of Limiting Values. *Decommissioning and Short-Time Overrides.* If a PCE safety device must be temporarily taken out of service or bypassed, e.g., during plant start-up, this must be documented in writing. Other technical or organizational measures must be put into effect to insure safety while the device is out of service or bypassed. The decommissioning or override status must be clearly identifiable.

Appropriately labeled technical devices, e.g., key switches, can be provided when a safety device has to be bypassed repeatedly. A clear signal must be given to indicate bypassing; an automatic interlocking involving the bypassed PCE safety device must be released if necessary.

There are three options for restart, and these must likewise be documented:

1) Restart without functional testing (e.g., after overriding for calibration)
2) Restart with partial or special testing (e.g., after replacement of a unit, cable, or data line, or after correction of a malfunction). A complete test (as under 3) should be performed as soon as possible.
3) Restart with full functional testing, as provided in the testing specification (e.g., after prolonged interruption of service)

Restart after Initiation of Safety Function. Automatic restart after initiation of the safety function is to be "disabled," i.e., restarting must be prevented when the plant is placed in operation again. This applies particularly when the process safety variable has already returned to the normal operating range. Until the plant state has not been checked portions of the plant affected by the initiation of PCE safety devices cannot be returned to service by manual action. PCE devices used for restoring the PCE safety device to service are considered PCE operating equipment.

Setting Limiting Values. Limiting values for PCE safety systems may be changed only on the instructions of the operations manager, who must also determine the extent to which this action affects the results of the safety review.

A report must be prepared to document the change in limiting value (time of change, authority, person performing the change), and the correct setting of the new limiting value must be verified.

3.4.3.5. Summary

In the technical safety concept, non-PCE and PCE safety measures are employed to reduce the risk arising from a chemical process plant to a value below the limiting risk. The risk to be countered is evaluated qualitatively, and the result is used in "grading" protective measures by scope. Technical measures are implemented as safety systems.

The task of a PCE safety system is to reduce the risk at least to the limiting risk, possibly in cooperation with non-PCE safety systems or organizational measures. From the risk assessment and the associated requirements, safety requirements on PCE safety systems are first determined, and from these are derived the technical and nontechnical measures needed to fulfill the safety function. In a second step, graded measures are devised to meet the requirements. The requirements are more stringent the greater the risk to be covered by the PCE safety system.

3.5. Special Safety Equipment

3.5.1. Pressure-Relief Devices

3.5.1.1. Introduction

Liquids and gases in chemical plants are frequently held in vessels that can be closed. A variety of mechanisms (Section 3.5.1.4) can cause overpressure to develop in such vessels. Initially, the vessel walls withstand this overpressure, but if it exceeds some limit, the "maximum allowable working pressure," which depends on vessel shape, wall thickness, and material properties, the vessel bursts and the fluid contained is released, unless some sort of relief device limits the rising pressure to the maximum allowable working pressure.

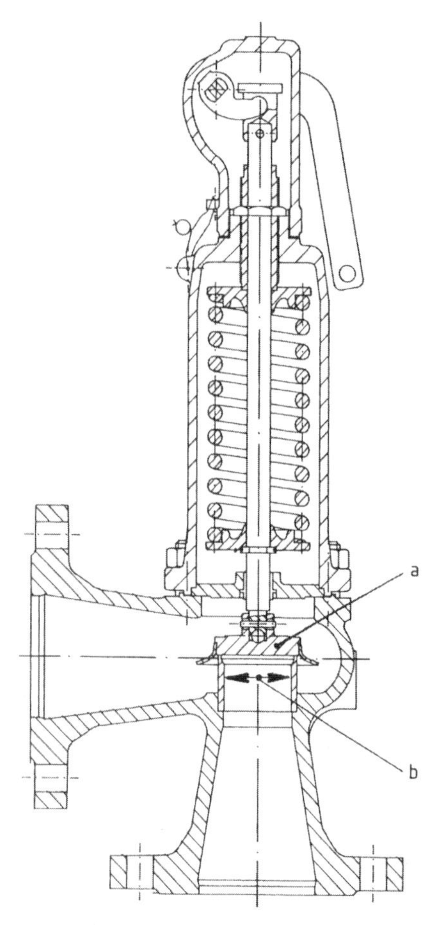

Figure 51. Safety valve
a) Valve disk; b) Flow area

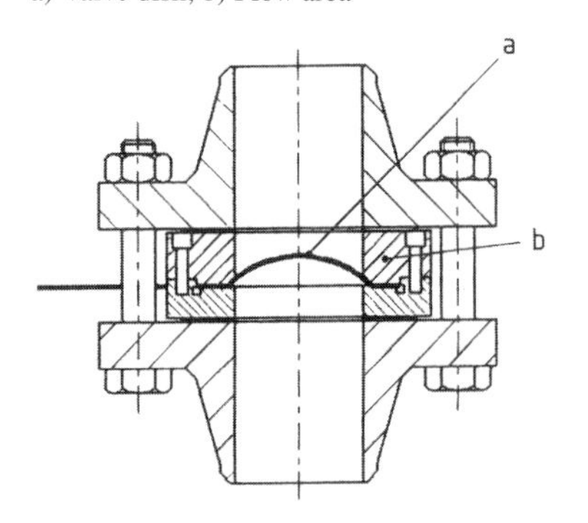

Figure 52. Bursting disk
a) Disk; b) Holder

Classical devices that act automatically to prevent excessive pressure in pressurized compartments are the safety valve and the bursting disk (Figs. 51 and 52) [171–176].

Vents on tanks are operating equipment, not actual safety devices. Not until modern plants were built did their function change as artificial ventilating and gas-displacement systems were installed on tanks.

3.5.1.2. Safety Valves

A safety valve opens automatically when the set pressure p_0 of the valve is exceeded. The valve disk is lifted by an amount h and makes the flow area available for pressure relief. In general, the full lift is attained within a pressure rise of no more than 10 % above the set pressure. The disk reseats when the pressure has dropped below the set pressure. In this way, the amount of vessel contents released is only that needed for instantaneous pressure limitation.

The "opening characteristic" of a safety valve is described by the lift of the valve disk as a function of the pressure p in the vessel. This characteristic depends on the interaction of opening and closing forces on the disk. Opening forces are the pressure force on the disk and, after the disk begins to lift, the fluid dynamic forces exerted by the escaping fluid. These fluid dynamic forces are a function of the lift and strongly influenced by the geometry of the disk, which is designed to achieve a suitable opening characteristic. Closing forces are a constant weight or a lift-depending spring force and the back pressure on the valve disk, which is the sum of an external "superimposed" pressure and the pressure generated by the escaping fluid in the discharge line after the valve opens.

Figure 53 illustrates the interaction of opening and closing forces. Where the opening forces, as a function of disk lift at constant pressure, increase more slowly than the spring force, the valve can open only when the pressure rises further. In this range (A – B in Fig. 53), it opens continuously. If the opening forces increase more rapidly than the closing force (to the right of B), the valve opens abruptly along the constant-pressure curve, up to the next equilibrium point D or the full lift limit C. Similarly, as the pressure declines, the valve reacts by closing continuously from C (or D) to E and on to F. It then closes abruptly (F – G) because the remaining opening forces at constant pressure and decreasing lift (to the left of F) are smaller than the closing force exerted by the spring. The pressure increase to achieve the full lift between A and C (or D) is the "opening pressure difference." The pressure drop between A and G is called the "reseating pressure difference," or "blowdown."

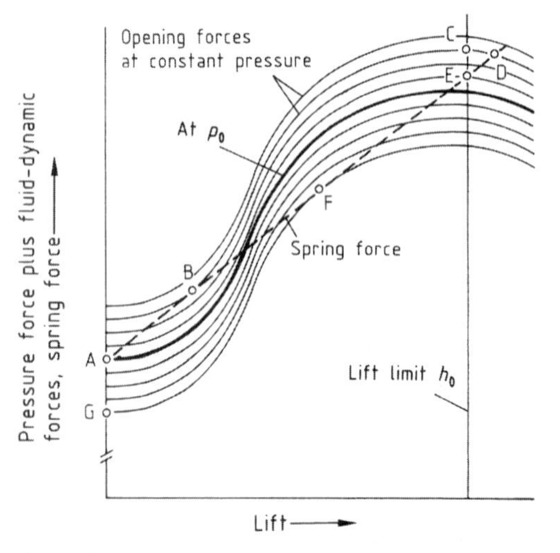

Figure 53. Lift – force – pressure diagram
See text for details

Specially designed safety valves have supplementary loads, which act as a function of pressure to reinforce the closing force until the set pressure is reached, or reinforce the opening force when the set pressure is exceeded. These loads can completely replace weights or springs. In the bellows-type safety valve, a bellows between the disk and housing largely eliminates the effect of back pressure on the discharge side.

Safety valves are classified by opening characteristic and loading principle. The *full-lift safety valve* opens almost abruptly to full lift within a pressure rise of $\leq 5\,\%$ above the set pressure, thereby attaining maximum discharge capacity very quickly. It also closes abruptly. The *proportional safety valve*, in contrast, opens continuously (though not necessarily linearly) as the pressure increases above the set pressure; it also closes gradually. There are no special requirements on the opening characteristic of the *standard safety valve*, and its performance falls between those of the full-lift and proportional types. With *weight-loaded* and *spring-loaded safety valves*, the set pressure must be significantly higher than the normal working pressure in the protected compartment so the valve remains tight during normal operation. The working and set pressures can be much closer together in *supplementary loaded* and in *controlled safety valves*, because the (controlled) closing and opening forces at the setpoint can change almost discontinuously. *Pilot-operated safety valves* have to be classed as controlled safety valves because of their construction and functional principles.

The full-lift safety valve is preferred when large volumes of gas have to be discharged suddenly. The proportional type adapts better to a fluctuating mass flow, and is well suited to low volume flow rates such as those resulting from thermal expansion, and also to variable flows of mostly incompressible fluids (liquids). Supplementary loaded and controlled safety valves are used chiefly in power plants, where large mass flow rates have to be handled and the generally constant service conditions make it possible to run the plant continuously at the limit of its design pressure. Because it permits good sealing of the interior space against the surroundings, the bellows valve is preferred with fluids that must never be released to the environment. It is suitable where back pressure in the discharge line is high or fluctuating, because it restricts the back-pressure effect.

The mechanical reliability and discharge capacity of safety valves are commonly determined by type testing. The discharge capacity is characterized by the certified coefficient of discharge α_w, which holds only for a range of set pressures established by type testing. Normally $\alpha_w = 0.6–0.8$ for full-lift valves and gaseous discharge fluids; for proportional valves and liquids $\alpha_w = 0.1 – 0.35$. Against the actual discharge performance, α_w includes a safety margin of ca. 10 %. Unauthorized changes made on a safety valve, affect the basis of the type testing and the discharge capacity is no longer certified. In particular, the installation of an incorrect spring can keep the valve from fully opening, even though the set pressure is correct, so that the required discharge capacity is not available.

For proper functioning, a safety valve must have the correct set pressure, and must be able to discharge the maximum possible mass flow on opening. Spring-loaded full-lift valves tend to go into self-excited oscillation (chattering) when there are abrupt pressure changes close to the set pressure; the valve opens and closes at short intervals, and the discharge capacity is markedly reduced. Chattering of the valve disk damages the seat and possibly the entire valve, and tightness is impaired. To prevent chattering, the pressure drop in the feed line during discharging must not exceed 3 % of the set pressure. Chattering can also occur if the safety valve is too large for the mass flow rate it is to discharge. If chattering cannot be eliminated, the safety valve can be equipped with a vibration damper. A design modification to limit the full lift is also possible in oversized valves.

The discharge capacity is also reduced if the back pressure in the discharge line is too high. It should therefore not exceed 15 % of the set pressure, or 50 % for bellows safety valves.

3.5.1.3. Bursting Disks

A bursting disk is a closure element that fails when its "bursting pressure" is reached, leaving open the flow area of a discharge line. The design is governed by the construction materials.

Disks can be classified as follows:

Membrane-type disks are loaded by membrane stresses. They fail when the yield point is reached. *Brittle plates* are loaded by bending stresses. They fracture when the breaking stress is reached. The yield point and breaking strength are functions of temperature, so the bursting pressure varies with service temperature. *Shells of revolution* fail by buckling when a stability limit, governed by the modulus of elasticity, is reached (reverse-acting disks). The stability limit is less temperature dependent than the yield point and breaking strength. The bursting of the shell after buckling can be aided by special devices, e.g., a fixed knife in the form of a cross, located behind the bursting disk. The serviceability of bursting disks is determined by bursting-pressure tests on samples.

A bursting disk causes only a slight increase in the flow resistance of the discharge line; it opens spontaneously, and frees a larger flow area than does a safety valve of the same nominal diameter. It is thus particularly well suited to rapidly rising pressures and large volume flow rates. Bursting disks are less sensitive than safety valves to gumming and polymerizing media. They have the disadvantage that they do not reclose, and if a disk fails a large quantity of product can be released. Depending on the design, bursting disks are often sensitive to underpressure in the protected compartment and back pressure on the discharge side.

Membranes are employed chiefly for large flow areas and low overpressures. Panels are used when the corrosiveness or other properties of the medium dictate the use of chemically resistant materials, e.g., graphite, which are not appropriate for the fabrication of thin shells.

3.5.1.4. Sizing of Safety Valves

The protection of a plant section with a pressure-relief device such as a safety valve involves five steps:

1) Definition of scenario
2) Calculation of relief cross section
3) Valve selection
4) Determination of mass flow capacity
5) Impact analysis

The first step in defining the scenario is a systematic safety analysis of the plant section under consideration, e.g., in a HAZOP study. These scenario can be divided into two groups:

1) An energy or material flow is supplied from outside (physical action)
2) The energy or material flow is generated in the system (chemical reaction)

A simple systems analysis makes it possible to break the first group down further:

1) The system is heated. Pressure builds up as a result of thermal expansion, rising vapor pressure, or evolution of dissolved substances. The mass flow rate to be relieved depends on the maximum quantity of heat per unit time that can be supplied to the system.
2) A substance flows from a higher-pressure system into the system under consideration. The mass flow rate to be relieved depends on the rate at which mass can be delivered to the system.

3) A flow of matter is supplied to the system by a pressurizing device, e.g., a pump. The rate at which mass must be discharged depends on the delivery capacity of the pressurizer.

The second group includes systems in which an unacceptable rise in pressure would result if a reaction went to completion. The pressure rise can result from the increase in vapor pressure as the temperature increases, and/or from the liberation of gas, e.g., by decomposition of the condensed phase. Possible causes and mechanisms of such reactions are described further in Section 2.3, while Section 2.3.2 lists methods by which the progress of an exothermic reaction can be determined experimentally. As a rule, studies of this kind are essential in the design of a pressure-relief device. Useful findings include the rate of pressure rise and the rate of heat production. This forms the basis for determining the rate of mass discharge, and hence for designing the safety valve, with the help of fundamental hydrodynamic equations [171], [172], [177].

If the required minimum cross section of the valve is known, a valve is selected such that its smallest cross section is larger than the required value. The cross section gives the maximum rate of discharge through the valve, which makes it possible to calculate the pressure drops on the high-pressure and discharge sides of the valve. These are compared with the limits discussed in Section 3.5.1.2, which must not be exceeded.

It must also be determined whether two-phase discharge can occur. Correlations useful for this purpose have been published by DIERS (Design Institute of Emergency Relief Systems) [178].

Figure 54 illustrates what happens in a pressure-relief event with two-phase flow [179]. When the pressure in the gas space of a vessel is relieved, single-phase discharge occurs first; the pressure drops rapidly. The liquid cannot immediately react to this change of state with a corresponding drop in temperature, and becomes superheated. To restore thermodynamic equilibrium, intensified formation of vapor bubbles begins (after a boiling delay time). The growing and ascending bubbles both displace the liquid and entrain it, setting the contents of the vessel in upward motion. If the free surface of the liquid reaches the discharge opening, a two-phase flow results.

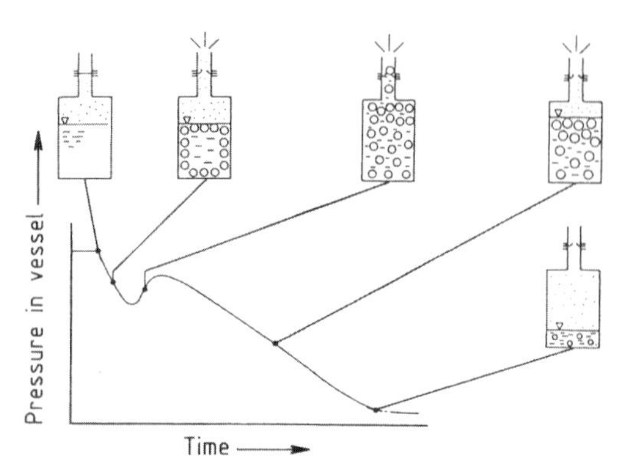

Figure 54. Pressure variation and behavior of vessel contents when pressure on the vapor space above a single-component mixture is relieved

This process depends on both the properties of the reactor contents (viscosity, surface, tension, foaming tendency) and the reactor geometry, fill level, and discharge capacity of the valve [178]. Thus low surface tension and high viscosity promote two-phase flow, while low fill level and a pressure-relief opening small in comparison with the vessel cross section help suppress two-phase flow.

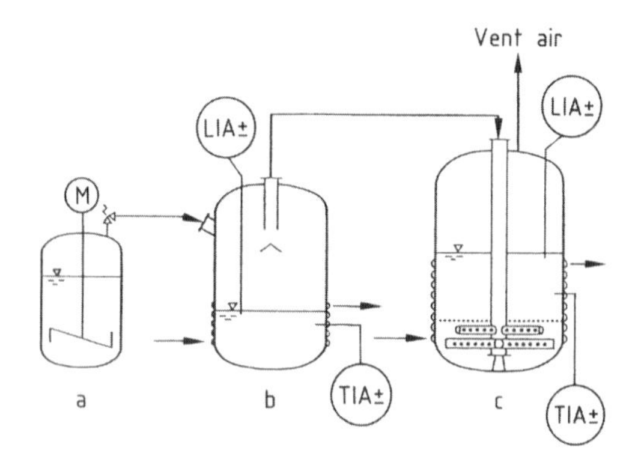

Figure 55. Retention system with liquid separator and direct condenser
a) Reactor; b) Liquid separator with retention volume and cooling jacket; c) Direct-contact condenser with dip tube and cooling jacket

The development of two-phase flow has the consequences that more liquid is removed, more total mass is removed, and the mass flow rate is increased substantially. On the other hand, the gas/vapor mass flow rate that has to be discharged for safe pressure relief is smaller than in single-phase flow. This means that a valve designed for single-phase discharge can be undersized for safe pressure relief under two-phase discharge conditions, and a new test must be performed for this case (the pressure drops on the pressure and discharge sides of the valve must also be remeasured).

The occurrence of two-phase flow can result in markedly higher inertial forces and momentum transfer.

An iterative procedure is usually followed until all constraints are met. A thorough discussion of models to describe two-phase flow has appeared in [180].

3.5.2. Blowdown Systems

3.5.2.1. Procedure

The substances discharged for pressure relief must be handled in such a way that they do not cause personal injury or environmental damage [181]. In general, it is necessary first to determine [178] whether the discharge flow is single-phase, gas/vapor, or two-phase. In the two-phase case, it is expedient to separate the liquid phase [182] and retain it in order to suppress any further reactions. When separation is accomplished in separate vessels, e.g., impingement separators or cyclones, the reaction is stopped by a large quantity of cold water, containing a reaction inhibitor if necessary. On grounds of space and cost, it is desirable to use one apparatus for both functions, if possible (Fig. 55) [183].

The safe removal of substances in gas/vapor form must also be insured. Three approaches to gas/vapor handling are possible, depending on the total quantity of the substances discharged and their properties [184]:

Discharge into the atmosphere
Retention by treatment systems
Retention in closed recovery systems

Substances may be released only if it has been demonstrated for the particular case, e.g., by a propagation calculation [185], that the impact of the released substances on humans and the environment stays within acceptable limits, otherwise, retention in treatment systems is necessary. Industrial treatment systems include thermal cleanup systems, flare systems, scrubbers, and dip-tube and other condensation units.

When systems of these kinds are employed, the first point to check is that the risk potential is reduced to an adequate degree, since both untreated residues and newly formed degradation products are released into the environment. The second point is whether the risk potential is merely shifted; a scrubber may clean up the exhaust air, but then generate highly contaminated wastewater. Finally, it must be demonstrated that the disposal systems are not harmed by the reaction forces and momentum transfer occurring during discharge.

If no satisfactory answers are obtained to these three questions, the substances must be delivered to a closed retention system [183], [186]. In the simplest case, they can be captured in a static vessel, or in systems that can be blown up, but devices of this type may quickly run into practical limits because the volumes of gas to be intercepted are often very large.

One way to minimize the volume needed is to dissolve or condense the discharged substances, or to fix them chemically. Often, this can be done by introducing the gases or vapors into a water-filled vessel [187]. The degree of condensation or absorption can be enhanced, and the necessary volume minimized, by spraying the substances into the vessel through nozzles or distributor pipes [188], [189]. Other variables affecting the quality of condensation or absorption are the temperature of the liquid in the vessel, the fill level, and the pressure in the retention system, which should be as high as possible, consistent with the requirements of the pressure-relief device (Fig. 55).

3.5.2.2. Alternatives

The safe design of a pressure-relief system combined with suitable facilities for handling the discharge can be a time-consuming iterative process. When costly retention or treatment units appear necessary, economic and environmental factors dictate that alternative pressure-relief arrangements must also be evaluated.

For example, pressure relief can be dispensed with if the vessel is constructed so that neither physical nor chemical processes can generate a pressure higher than its maximum allowable working pressure.

In addition, no pressure-relief device is needed if appropriate instrumentation and control [190] (Section 3.4.3) can insure that an unacceptable pressure rise cannot occur. These features must reliably control the reaction, or provision must be made to inject an emergency suppressor or emergency coolant to stop the reaction immediately [191].

Which solution is preferable can be decided only when the specific constraints of the application are known.

3.5.3. Flame Arresters and Explosion Barriers [192–201]

3.5.3.1. Introduction

The initiation and propagation of explosions in (chemical) plants where explosive mixtures are present can lead to significant harm to persons, the environment, and property. One feasible, and often economic, way to control and prevent such explosions in closed systems is to separate parts of the plant with flame arresters. These block the propagation of an explosion, and help limit and control its impacts. This type of isolation can be applied to mixtures of flammable gases/vapors and oxidants as well as explosive dust – oxidant mixtures.

3.5.3.2. Flame Arresters for Mixtures of Vapors (Gases) and Air

In normal plant operation, flame arresters permit the free passage of explosive mixtures. If ignition occurs, however, they block the passage of the flame, preventing the ignition from being transmitted to the protected side. The extinguishing action in present-day flame arresters, regardless of design, is based on one or more of three mechanisms:

Flame Quenching in Narrow Channels. Through intimate contact with the cold walls of a filter element comprising many narrow channels, heat and free radicals are withdrawn from the combustion process, and the flame is extinguished. This mechanism is employed in dry flame screens (e.g., the crimped-ribbon arrester shown in Fig. 56).

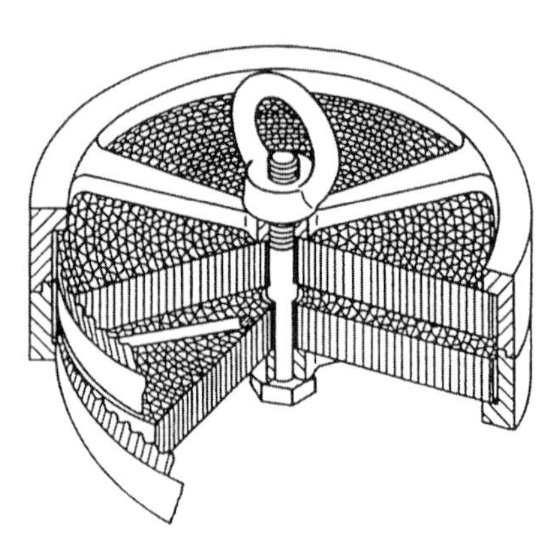

Figure 56. Cutaway drawing of a crimped-ribbon-type flame arrester [201]

Blocking by Liquids. There are two main types of liquid-type flame arresters, the liquid seal in the form of a siphon inserted in a liquid-conveying pipe, and the hydraulic arrester in which a stream of gas passes through a dip tube and is divided into noncoalescing bubbles. The siphon, by virtue of the configuration, is continuously filled with the conveyed medium (Fig. 57). Water is the most common barrier liquid in the hydraulic flame arrester.

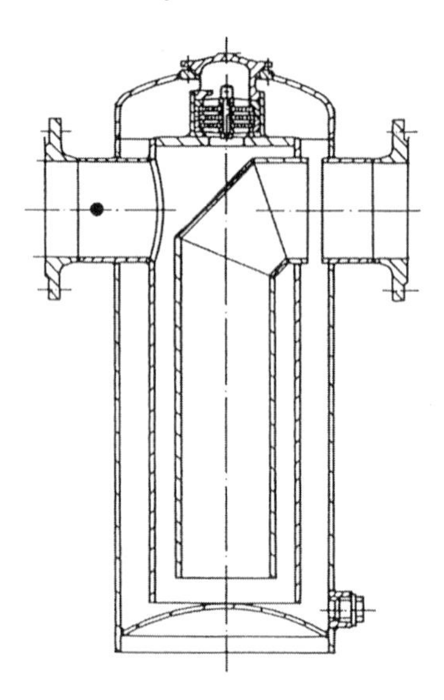

Figure 57. Wet-type flame arrester (siphon arrester) [201]

Dynamic Blocking. The geometry of the flame arrester includes a narrow clearance, in which the flow velocity is always higher than the turbulent flame velocity (combustion velocity) in the explosive mixture. Upstream propagation of the reaction is thus impossible.

Use of Flame Arresters in Practice. The selection of an arrester is governed by a variety of constraints. Only exact knowledge of the system to be protected can provide the basis for choosing suitable and economical units.

First, it must be established what parts of the plant are to be safeguarded against what effects, i.e.,

What are the potential ignition sources?
Is ignition due to external or internal sources?

This is crucial in fixing the location, installation direction, and layout for the arrester. Two applications must be considered for the layout: deflagration and detonation (i.e., coupled combustion and shock waves with rapid pressure rise); the type of protection chosen will depend on which installation is used. Under some circumstances, a run of a few meters of pipe is enough to allow detonation, so that detonation-type arresters are commonly selected for installation in piping. Deflagration-type arresters can be employed where run-up distances are very short, e.g., where tank vents to the atmosphere are to be protected.

In a second step, the construction of the flame arrester is considered. The properties of the medium being transported are important: contamination with solids, solidification temperatures (heatability, frost protection), and condensation or solidification of impurities from the stream.

Quenching-type flame arresters in some services may become contaminated and suffer corrosion; these devices need regular inspection and cleaning. It makes sense to install several such flame arresters in parallel so that one at a time can be bypassed. If even a few of the passages in an arrester become enlarged, or if the flame filter is incorrectly inserted in the holder, the unit may fail.

In the case of pipework not continuously filled with liquid, the wet types of arrester are desirable on grounds of maintenance and availability; these can be obtained in a variety of designs.

Hydraulic arresters are often used to protect contaminated exhaust air streams. Safe operation depends on correct sizing and design, as well as the use of suitable instrumentation for continuous monitoring of important parameters. An automatic alarm should be provided, together with automatic initiation of emergency response when limiting values are violated. Safety-relevant parameters of this type of arrester include a sufficiently high liquid level over the end of the dip tube, limitation of maximum air flow rate, monitoring of liquid temperature (frost protection), and monitoring of temperature in the gas space of the arrester. As a rule, buildup of contaminants and condensing species from the air stream necessitates continuous cycling of liquid in and out of the arrester. To prevent the formation of a continuous explosion channel by the bursting of explosive gas bubbles in the ascending cloud of bubbles – which would allow an ignition to get through the arrester – it is important to have a hydraulic arrester designed for maximum air flow rate and minimum dip depth.

For the correct design of the flame arrester, it is essential to know the maximum experimentally safe gap for the substance being handled. This parameter serves as the basis for classifying substances into Explosion Groups II A, II B, and II C. In Germany, for cases covered by the Flammable Liquids Regulation (VbF), flame arresters must be type-tested (type-approved) for the appropriate explosion group, and for one of the two layouts (deflagration or detonation). Type approval tests are done with appropriate mixtures of fuel and air. Any change in oxidant (e.g., oxygen or chlorine instead of air) and any increase in oxygen partial pressure can lead to a change in performance, which can be assessed only after further experimental tests. Where the VbF does not apply, Germany does not yet have a type-testing requirement.

The quenching of flame in an arrester presupposes adequate removal of heat from the reaction or flame zone. In some arrester designs, handling flowing explosive mixtures, heat removal may be hindered so that a flame can stabilize on the arrester. The filter element and piping may be heated to the ignition point of the flowing mixture, and the flame can be propagated through the arrester. As a rule, the only way to insure adequate heat removal is by using end-of-line arresters. Experimental demonstration of the "endurance burning safety" (ability to prevent flame transmission for an unlimited time) is required. Any flame arrester that does not have this property must be expected to fail after stabilizing the flame for a certain time unless countermeasures are taken. Therefore continuous temperature monitoring (Fig. 58) is required, along with initiation of emergency functions (e.g., cutting off the gas stream).

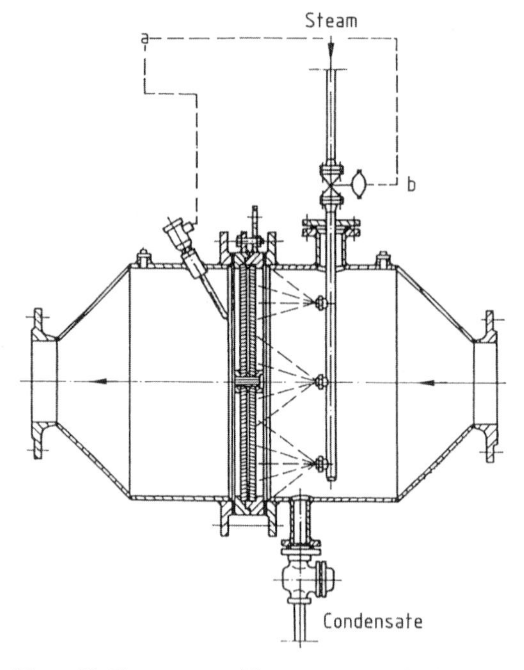

Figure 58. Flame arrester with temperature monitoring and steam injection [201]
a) Electronics; b) Solenoid valve

The use of a flame arrester leads to quenching of the flame at the arrester. However, the pressure generated in the explosion must be safely accomodated through design features (pressure-proof design) in conjunction with adequate provisions for pressure relief.

Safety Concepts. The use of flame arresters to separate plant sections that are not proof against explosion pressure shocks or explosion pressures is just one of a range of concepts applicable to explosion protection.

As is usual in explosion protection, all protective actions are based on risk analyses, which call for countermeasures with appropriate availability, depending on the probability of occurrence of explosive atmospheres (classified in zones) and potential ignition sources. Even when flame arresters are used, the zone and the ignition probability must still be analyzed. As a result of this interplay of occurrence probabilities redundancy of flame arresters is necessary. The important factor is the number of independent protective measures (Table 14).

Table 14. Number of measures to prevent flame transmission

Ignition sources anticipated in plant	Number of precautionary measures		
	Zone 0	Zone 1	Zone 2
Normal operation (e.g., burner flame)	3	2	1
Failures that are likely to occur	2	1	0
Failures that are not likely to occur	1	0	0

German regulations mandate the use of flame arresters to protect non-pressure-proof plant sections in storage facilities and filling points for flammable liquids. Further details can be found in the Flammable Liquids Regulation [196] and the Engineering Code on Flammable Liquids (TRbF).

3.5.3.3. Flame Arresters for Dust – Air Mixtures

The flame arresters that have been discussed are not designed for use in plant areas and piping where there is a danger of dust explosion. Other approaches must therefore be employed in such cases. Among the means used for isolation are star-wheel feeders, active barriers with extinguishing media, rapid-closing valves (explosion valves) and doors, and explosion vents. These methods are described more fully in VDI Guideline 2263 [200].

The use of star-wheel feeders requires prior testing to determine the ability to prevent flame propagation, as well as to withstand the explosion pressure. In order that no smoldering product is transported into downstream parts of the plant, the star wheel must be automatically cut off (e.g., with a pressure monitor) in the event of an explosion.

When active barriers and rapid-closing doors are used, injection of extinguishing agent into the flame front or closing of the doors is initiated in a few milliseconds by a flame detector. Specialists must perform tests to determine the effectiveness (speed from detection to response) in the real plant geometry, and the ability to withstand pressure. Explosion valves, acting without auxiliary sources of energy, close at the higher flow velocities occurring during explosive pressure relief, automatically blocking off pipe cross sections, and preventing the explosion from propagating.

When explosion vents are used, the formation of long flames and the release of large amounts of product, possibly of environmental relevance, must be anticipated.

4. Safe Plant Operation

4.1. Safe Handling of Chemicals

4.1.1. Introduction

Substances and formulations in the form of raw materials, intermediates, and finished products are employed in many branches of industry. Some substances and formulations have dangerous properties (see also Sections 2.1 and 2.2); these are referred to as hazardous substances.

When hazardous substances are present, safety precautions must be taken so that concentrations stay below the limiting values in the workplace air, so that employees do not come into direct contact with hazardous substances, and so that fires and explosions do not occur.

Measures must be taken to prevent upsets in operation. If upsets come about anyway, the threat to employees must be limited.

Systematic safety analyses should be performed to determine the measures required [202]. Hazard sources and the conditions for their activation are systematically and comprehensively examined so that the hazard potentials can be ascertained. The knowledge gained makes it possible to identify the necessary measures case by case.

Points to be considered in establishing safety practices include provisions of legislation, engineering codes, industry and factory standards, and safety information relating to specific processes.

Technical measures, including the selection of processes with the lowest possible hazard potential, should be favored over administrative measures whenever possible. Preference should go to practices that safeguard the employee or mitigate risks, regardless of employee behavior.

Despite many technical measures, ranging up to "enclosed" plants, contact of employees with hazardous substances cannot be completely prevented. Normal operation entails filling, transferring, and emptying media, taking samples, and cleaning and inspecting vessels. In abnormal operation, for example, substances can get into work areas by leakage and when pressure-relief devices operate.

Selected practices for the safe handling of chemicals are discussed in what follows. These should prevent the release of hazardous substances and threats to employees and the environment.

4.1.2. Normal Operation

Transferring, Filling, and Emptying. In the chemical industry, the handling of liquids is part of daily routine. Liquids must be drained from tanks, drums, and reactors, or masterbatch tanks and reactors must be filled with liquids. It is often necessary to meter exact quantities in specified times [203].

Preventing Hazardous Concentrations of Substances in the Gas Phase. Some ways to prevent hazardous concentrations are:

1) Transferring liquids in systems that are enclosed or can be equalized (e.g., fixed piping from storage tank to plant); to be employed especially when transferring carcinogenic, highly toxic, and toxic liquids
2) Using gas-displacement devices for pumping
3) Extracting vapors at the point of escape
4) Providing adequate ventilation at the workplace

Avoiding Escape of Liquids. While being transferred, liquids can escape by splashing, dripping, spilling, or overflowing. Some ways to prevent the escape of liquids are:

1) Restricting transfer operations
2) Selecting and properly installing suitable filling devices
3) Using appropriate designs for piping disconnections

4) Employing process-specific practices as well as instrumentation and control, examples being:
   Emptying into calibrated vessels
   Emptying through volumetric meters
   Running drain lines to installed scales

Avoiding Electrostatic Charging. Liquids can become electrostatically charged while flowing next to walls or when being sprayed, filtered, or stirred. Charges transported along with the liquid can charge vessels to the extent that static electricity is discharged at the liquid surface; this can result in the ignition of flammable gas – air or vapor – air mixtures. The amount of charging depends strongly on the conductivity, the flow velocity, and the quantity of liquid transported.

A high level of charging must always be expected when interfaces are present in nonconductive liquids, e.g., when immiscible liquids are transported, or when liquid mists or vapors are generated.

Liquids with conductivities $> 10^4$ pS/m do not acquire a charge while flowing.

Avoiding Mix-up of Liquids and of Containers. Dangerous reactions may occur if substances are mistaken for others. Mix-up can be avoided by:

1) Design practices; connections for vessels containing different liquids can differ in diameter from one liquid to another or be made with incompatible flange constructions (grooves, lugs)
2) Unambiguous labeling; vessels, piping, and connections for individual liquids can be labeled, e.g., with substance names, colors, or numbers
3) Identity checking before liquids are drained

Avoiding, Restricting, or Assisting Manual Transport Operations. Ways of avoiding transport operations include:

1) Fixed piping between storage tanks and consumer (reactor, tap); the preconditions for this is that pipe runs be as short as possible, i.e., all parts of the plant are arranged compactly
2) Mechanization of operations

Ways of restricting transport operations include:

1) Purchase of liquids in larger vessels, e.g., 1000-L containers instead of 200-L drums
2) Suitable placement of vessels and other measures to minimize the need for transfers

If manual transfers are unavoidable, offthe-shelf or plant-designed equipment can make them easier and thus less dangerous.

4.1.3. Sampling

Samples are required for the following reasons:

1) Determining identity and quality of feedstocks
2) Monitoring and controlling chemical reactions and other processes in the plant
3) Assessing quality of intermediate and end products

Employees conducting sampling operation may be endangered by hazardous substances and the sampling conditions [204].

Safe Design of Sampling Devices. The following points must be considered in the design of sampling stations:

1) The sample should be withdrawn at a point in the plant where the pressure and temperature are as low as possible.
2) The cross-sectional area of the sampling device must be kept as small as possible.
3) Sampling must be designed such that large quantities of hazardous substances cannot escape from a plant or part of a plant, e.g., because of a malfunctioning or damaged sampling device.
4) The inevitable pre-sample liquid must be returned to the closed system, as far as possible. Two points merit special attention: the sampling point should be designed such that there is no pre-sample flow, or the unavoidable pre-sample flow is safely returned to the processing system.

When the sample is transferred into the sample container, splashing, vaporization, dripping, overflowing, and escape of hot liquids must be prevented. It may be necessary to take steps to prevent static charging, such as:

If possible, avoid sampling from stirred vessels through the manhole (especially if the vessel contains flammable substances). Otherwise, turn off stirrer, sprayers, and other charge-producing components before sampling and wait for a "calming" time.

Do not allow to spray flammable, chargeable liquids into open sample containers, especially if partial vaporization of the liquid is anticipated.

If possible, conduct sampling of flammable, chargable liquids in a closed system only, e.g., in a nitrogen-purged sample vessel.

Limit the outflow velocity of the liquid; rule-of-thumb ca. 1 m/s.

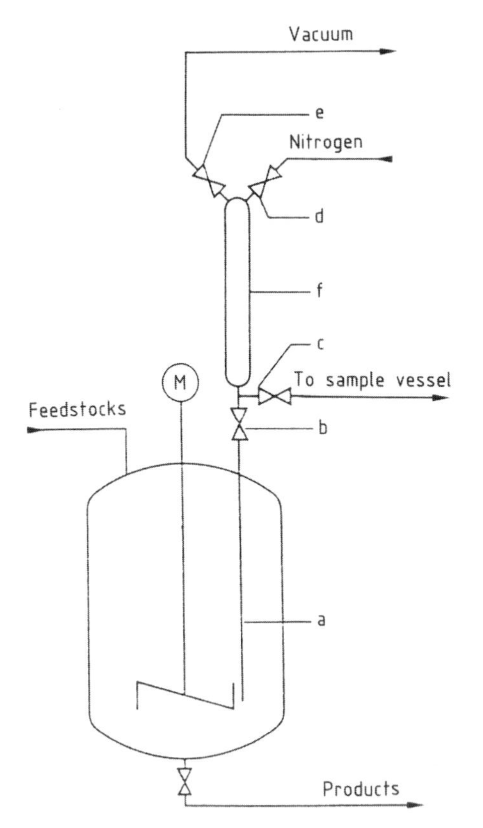

Figure 59. Sampling with vacuum
a) Dip tube; b) Stopcock; c), d), e) Fittings; f) Container

Sampling with Vacuum. Figure 59 shows a setup for taking samples with no pre-sample flow. The procedure is shown by the numbers in the diagram:

1) Open ball valve (b)
2) Blow dip tube (a) empty through fitting (d) (e.g., with nitrogen)
3) Close fitting (d)
4) Apply vacuum to fitting (e) in order to draw product into container (f)
5) Close ball valve (b)
6) Close fitting (e)
7) Apply nitrogen pressure through fitting (c) to transfer sample into suitable vessel
8) Close fitting (c)

Nonreturn valves in the vacuum line prevent the working fluids in the vacuum generator from getting into the product vessel.

4.1.4. Cleaning of Vessels

Cleaning methods should be designed so that worker health is not endangered by residues, contaminants, or cleaning agents [205]. Before cleaning is begun, the status of the vessels to be cleaned must be determined. It must be known what substances, with what hazardous properties, are or have been present (solid deposits, liquids, vapors, gases).

Hazards can arise from:

1) Reaction of the cleaning liquid with residues and contaminants
2) Flammable cleaning agents
3) Static charging when liquids are sprayed under high pressure
4) Materials unstable with respect to the substances, pressures, temperatures, and mechanical stresses present

The proper precautions must be ascertained. Cleaning must be performed by trained personnel under expert supervision, in accord with written information. The cleaning method must be described in an instruction manual.

Whenever possible, closed cleaning systems should be used.

4.1.5. Safety Practices when Working with Hazardous Substances Under Abnormal Operation

Safety practices must prevent employees being exposed to excessive concentrations of hazardous substances, even in abnormal or upset operation. Pressure vessels and pressure piping must be designed, constructed, and operated in such a way that personnel and third parties are not endangered. The action of pressure-relief devices must not threaten employees or third parties. The quantity and nature of the substances present may make it necessary to conduct releases to the atmosphere or to provide supplementary features, e.g., scrubbers, flare systems, or blowdown tanks.

Leaks represent a further source of danger to personnel and the environment.

Seal systems installed on shafts of pumps, drives, mixers, stirrers, etc. must be selected in accord with the media present and the regulatory requirements [206].

Along with proper design, the effectiveness of the seal systems must be tested regularly, and any leaks must be detected. When media are under pressure, splash deflectors must be used.

Technical protective practices must also be coordinated with administrative and personnel practices, especially for the case of abnormal operation.

4.2. Safety in Batch and Continuous Processes

4.2.1. Safety in Batch Processes [207–210]

4.2.1.1. Introduction

Safe production imposes stringent requirements on the owners and operators of chemical plants, who must observe and implement a body of legislation, rules, and regulations. What is more, the manufacturer of chemical products must make compromises between productivity, quality, safety, and environmental protection, but safety must take priority.

The principles of safety and environmental protection are established by management in accordance with the pertinent laws. Every chemical process involves a special combination of chemicals, equipment, and process conditions.

Safe operation makes it essential to study each process separately. The written set of instructions, the operating manual, occupies a central place in the safety effort.

4.2.1.2. The Operating Manual

Principles. The operating manual is a comprehensive document that fully describes a particular chemical or physical process used in making a product. It must guarantee that the process can be carried on in a safe, environmentally benign, economical, quality-oriented way.

A fundamental requirement is that the manual must describe in full scope and depth the steps that must be performed in the process. It must be unambiguous in that the employee at the workplace knows at all times precisely what actions are to be taken, in what order (e.g., numbering of steps). In extreme cases, every detail of a manipulation may have to be described.

While the manual must be exact and complete, it must also be easily understood. It may be expedient to write separate SOPs (standard operating procedures) when complicated and extended actions must be repeated for each batch (e.g., preparation of a hydrogenator).

Aids. The operating manual is based on a wide range of documents and other information sources (Fig. 60). For reasons of bulk, it cannot include all this information down to the last detail; a vital task of the author is to set priorities and to pick out specific points about a process and set them apart from what is obvious. Operators of multipurpose plants face a special challenge, for they must diagnose continuously varying conditions in the same apparatus and decide what process-specific information and actions are appropriate for each case.

Electronic data processing systems are nearly universal today, and the use of standard forms makes it possible to prepare the operating manual in a consistent way and present it in an easily understood fashion. Modifications and additions can be done quickly (e.g., if production is to be shifted to other plants).

The author of the manual must harmonize process data with plant-specific features while observing all guidelines on safety, environmental protection, occupational health, etc. Many documents and other information sources can be used as aids in describing the process, but it is not enough merely to have these to hand. They must be correctly interpreted and applied. A plant manager often finds it difficult to read documents written by specialists in other disci-

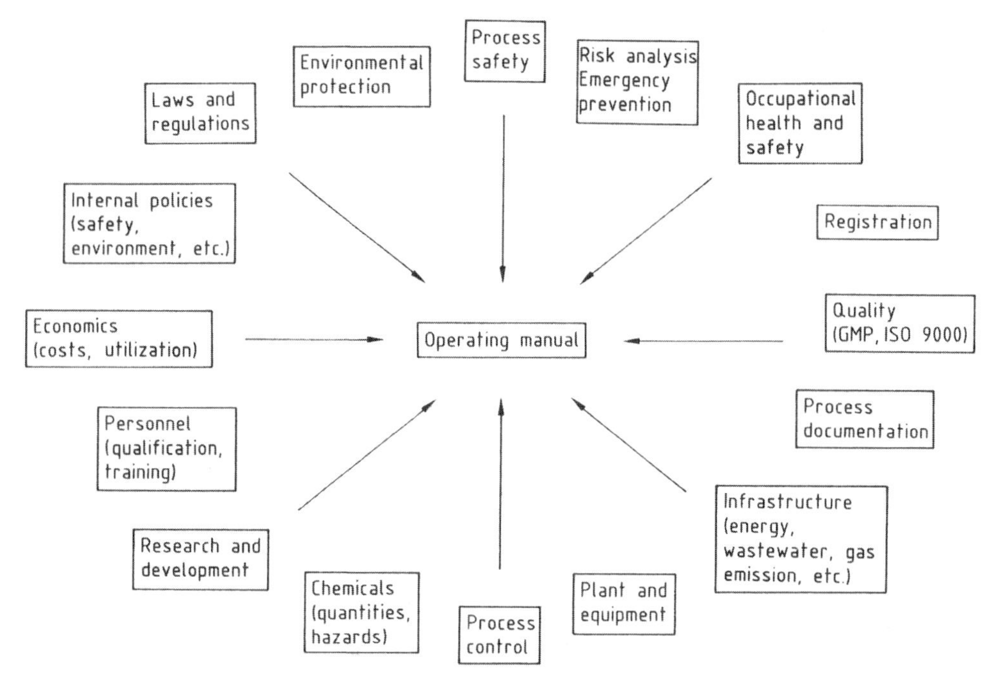

Figure 60. Operating manual: documents and information sources

plines (e.g., some chemists find technical plant flowsheets incomprehensible). Close cooperation with these specialists is needed to insure safe operation. The following information must be checked for completeness:

- Correctly acquired and interpreted material and process safety data
- Knowledge about the plant (cooling capacities, construction materials, plant status, etc.)
- Problems of high-technology facilities (automation, measurement and control, process control engineering systems)
- Limitations of technical safeguards (safety equipment)
- Utilization of infrastructure (energy supply, alarm systems, emergency systems, gas emission, wastewater, etc.)
- Correct and complete risk analysis (process and plant) and transfer of results into practice

For some plants or chemical processes, further points may have to be added to this list.

A good operating manual is never the product of one individual; it can be created only based on effective collaboration among specialists. As soon as the definitive operating manual is in being, the involved departments involved should apply their special knowledge to prepare opinions on the process. It is desirable to set up a fixed routing (toxicology/occupational health, safety, environment, registration, etc.) and have each activity signed off on the manual.

Contents and Organization. The organization of the operating manual is dictated by company policy, and there is no universal rule. The following three-part breakdown has been tested in practice:

1) Information section: intended chiefly for management; a quick overview of the process, presenting all safety-relevant aspects
2) Operations section: focused on the operating procedures; containing all information needed for the practical conduct of the process in the plant
3) Appendixes: data, details, and special information on the process; can be used to gain a deeper understanding of the process

A possible table of contents reflecting this overall breakdown is shown below.

Overall Organization of Operating Manual
Information Section

A. Changes from Precedent Operating Procedures
B. Chemical Scheme of Process
C. Materials (Chemicals, Yield)
D. Environmental and Safety Aspects
  D.1. Safety Data
  D.2. Summary of Risk Analysis
  D.3. Waste Disposal
  D.4. Environmental Impacts (Ecogram)
E. Plant and Equipment
E.1. Flowsheet (Streams in Apparatus)
E.2. Apparatus List
F. Material Data
G. Safety and Occupational Health
  G.1. General Provisions
  G.2. Process-Specific Provisions
H. Response to Accidents, Fire Alarms, Power Outages
Operations Section
I. Process Performance
  I.1. Step-by-Step Description
  I.2. Process Instructions
  I.3. Process Deviations, Remarks
J. Intermediate and End Product Inspection
K. Packaging and Storage
L. Equipment Cleaning on Product Changeover
Appendixes
M. Safety Tests, Wastewater Analyses, Operating Instructions for Individual Apparatus, Distribution List

Safety-relevant items are discussed as follows.

D.2, *Summary of Risk Analysis* gives a quick overview of the risks involved in the chemical process and the apparatus used. Known major risks are described along with key safety practices for the process. These include administrative practices, special technical safeguards, and important limit values (temperature, pressure, etc.).

G.1, *Safety and Occupational Health: General Provisions* lists general directives and rules to be followed in the chemical process. These may relate to reporting of deviations from the operating specifications, grounding, inert gas blanketing, etc. The instructions and rules must be formulated so that they can be observed in practice.

G.2, *Safety and Occupational Health: Process-Specific Provisions* gives detailed information and practices derived from the risk analysis. It may include special characteristics of equipment, special protective measures, handling of hazardous chemicals, increased risks, temperature limits, etc. This section is crucial for the operating personnel, telling how this process differs from others. It is especially important in multipurpose plants where the hazard potential varies continuously.

H, *Response to Accidents, Fire Alarms, Power Outages* is one of the hardest sections to write because problems often occur where they are least expected. With the help of the risk analysis, the steps in the process are analyzed, device by device; the analysis suggests what problems may occur and need to be considered in the planning of emergency response practices. Critical quantities are temperatures, times, and mixing; these conditions have to be determined separately for a variety of potential events. Practices are defined for each step in the process so that any kind of deviation can be responded to and the process returned to a safe condition.

I, *Process Performance* is the heart of the operating manual. The sequence of operations is detailed and described in chronological order. Sequential numbers are assigned to the process steps, which must be formulated in a clear and unambiguous way. The process conditions to be maintained and the quantities of chemicals to be used are written down as nominal values; the person controlling the process must acknowledge these in the form of actual values. When key reactants are present, or when process conditions are especially safety-relevant, it is desirable to confirm maintenance of the nominal values by having a second person (possibly a supervisor) take readings and sign.

In multipurpose plants, the base settings of the apparatus take on central significance for safety. The frequent changes of conditions in a given apparatus mean that adjustments must be made as a function of process conditions (temperature limits, stack gas downstream of scrubber, or stack gas cleanup unit, etc.), and some apparatus may have to be modified (cartridge filters, charging hoppers, etc.). Before a piece of equipment is placed in service, these base values must be set and checked.

The operations to be performed may be identical for a number of apparatus items (e.g., charging of a centrifuge). In such cases, reference may be made to separate operating instructions or unit operation instructions (SOPs). When this is done, all that has to be set down step-by-step are the parameters specific to the process (e.g., speed of a centrifuge).

I.3, *Process Deviations, Remarks* is a form, to be completed on site for each batch. It helps in documenting irregularities in the process. The plant manager thus has a powerful tool for early detection of safety-relevant excursions; a shift log serves the same purpose.

L, *Equipment Cleaning on Product Changeover* is very important. From the safety standpoint, it is essential to describe the individual process steps chronologically and in detail, precisely as in I, *Process Performance*. It is a mistake to think that cleaning with water is completely harmless; if, for example, warm water is run into a reactor thought to be empty, solvent residues in the vapor phase may reach an explosive concentration. The mechanical removal of tenacious crusts with various tools can be critical if solvent residues are still present. It is therefore important that all safety principles are observed, even when cleaning the equipment.

Appendixes give standard operating procedures (SOPs), checklists, piping and electrical diagrams, etc. The exact distribution list is also important, because such a list is the only way to make certain that changes are made in all existing copies of the operating manual.

Introduction of a New Operating Manual. Careful, comprehensive documentation is just one important condition for safety in chemical production. Equally important is that employees properly understand the instructions. Often, the use of technical language and foreign words in the operating manual places it at too high a linguistic level for the workers. It is also remarkable to see the wide range of interpretations different people can put on imprecisely formulated directions! Another problem arises when some employees are not fluent in the language of the instructions, and thus fail wholly or in part to understand an exact description. Any new operating manual must be presented and explained, one step at a time, by oral instruction. Actions that are difficult to describe in speech or writing must be demonstrated in the plant. The objective of such training is not simply that the staff understand the instructions; as comprehension grows, so do acceptance and safety awareness.

The instruction program must end in a test to ascertain whether the operating manual has been understood. This can take the form of a game or competition, or can be administered as a moderated discussion. Another proven method is to have the employees make a display, illustrating and explaining the steps in the process.

4.2.1.3. The Human Aspect of Safety

Responsibility. The chemical process employee works in an environment with a substantial hazard potential. Every day brings the need to work with harmful substances, dangerous reactions, and large quantities of chemicals. Each employee thus bears a high degree of responsibility for co-workers, the environment, and the plant. Even small mistakes can lead to grave accidents. In multipurpose plants, the situation is even more critical: Processes differing in their demands on work methods and safety are performed in succession. Close concentration and an ability to adapt quickly to new conditions are required.

One of the most urgent tasks for management is to provide each employee with the conditions for safe work and the resources for performing the job. Occupational safety becomes a task for management, to insure that workers are safety-motivated, and to aid them in carrying out their demanding tasks with discipline and self-confidence – and without accidents.

Training. The training of plant operating personnel is a continuing task for plant managers. Thorough training greatly enhances on-the-job safety by helping workers to deal better with the pressure of responsibility, and to react calmly and correctly in stress situations. Training must be related to practice in the plant; an ideal example is discussion of the operating manual before production is started.

Special attention must be focused on the training of new employees, who possess good technical knowledge, but not the specific skills for their jobs. During their first months, these employees must be given technical orientation and support so they can perform in accordance with safety rules.

A particularly important point in training is how to respond in abnormal situations. In the context of the operating manual, responses must be explained in the clearest possible way. Practical and theoretical emergency training (e.g., with simulator programs) help establish a routine for dealing with abnormal conditions. It is important to instill in the worker the self- confidence that will lead to calm, considered actions in a real emergency.

This training is documented and evaluated. Feedback, including worker self-assessment, helps reveal gaps so that the appropriate steps can be taken.

Motivation. People are not inherently motivated to work safely; external motivation must be provided. Repeated use of convenient, but unsafe, work methods, not immediately leading to an accident, gives rise to bad habits. Administrative measures (e.g., checkups) and technical precautions can evoke safe working methods, and make unsafe behavior difficult or impossible. Correct behavior should be associated with all possible advantages: recognition, easier job performance, higher qualifications, etc.

Cooperation. In the course of time, process workers accumulate much practical experience. It is essential for plant management to make use of this knowledge by integrating the workers in the safety process. It is an important distinction whether the wearing of protective clothing and respirators is ordered or self-prescribed by a group within the plant working on a safety problem. Safety awareness, and above all the acceptance of key safety practices, can be markedly enhanced in this way.

4.2.1.4. Internal Organization and Policies

The processes that follow one another in a multipurpose facility call for intensified administrative policies. Without clear direction and clear structures, an operation where products are changed over, say, weekly slides into chaos, and an accident becomes inevitable. In what follows, some points are listed that should help avoid such problems.

Weekly Program. The weekly program provides a quick survey of planning for the next week. It yields information on personnel allocations, production program, maintenance activities, and special events. The weekly program is presented orally to the operating personnel and is posted in the plant. Every worker can then make the optimal preparations for coming assignments. Surprises and improvisation are largely eliminated.

Weekly Conference. All employees take part in a brief conference every week. This is an occasion for comments on the preceding week and the coming one, and is also very important for exchanging safety information. Absentees must be identified and informed later.

Shift Change. A formalized shift change must be instituted, so that the chemical process can be carried on without pause. The procedure should include explicit allocation of tasks, provided in writing if necessary (e.g., in a shift log).

Responsibility. Clear structures in work allocation is essential. The chain of command (hierarchical levels) must be preserved. Unless a situation arises calling for prompt intervention, the plant manager should issue instructions to process personnel only through supervisors and team leaders.

This avoids, the confusion generated by instructions and counterinstructions, and by indistinct areas of responsibility.

Monitoring of Administrative Practices. Procedures on paper are not enough; they must be carried out. Spot checks should be done to find out whether selected points in the operating manual are being complied with. Above all, misunderstandings and gradual safety-relevant changes in work procedures must be detected promptly.

Information Wall. An information or notice wall can be used to promote safety awareness, but this can fulfill its purpose only if it is actually read, so it must be prominently located. Attractive presentation (generous use of color, graphics, etc.) makes it more likely that employees take an interest in the information, but no information wall can replace the oral communication of important instructions.

4.2.1.5. Safety in Production Practice

Before the Start of Production. The saying "as you make your bed, so you must lie on it" takes on great significance when preparations are being made for a new process. The decisions made at this stage must lead to safe, accident-free operation later. Serious preparation is particularly important in multipurpose plants where chemical processes change frequently, so that the plant must continually be modified to fit changed process conditions and operations. Plant personnel must also adapt. There must be a distinct "breathing space" between finished and following production. Overlap allows the possibility of confusion, and increases risk.

Central importance attaches to the production conference with all employees concerned. The operating manual is reviewed step by step and supplementary instructions are given by management. This production conference is an intensified form of the conference discussed under the heading *Introduction of a New Operating Manual* at the end of Section 4.2.1.2. Its timing has much to do with its success; it should be held immediately before the start of production, so that employees can concentrate on the new process without distraction, and get themselves mentally set for it.

The preparations for a production changeover are extensive, but identical in their main features. It is worth standardizing these procedures; ideally with a checklist.

Checklist for Production Preparation

1) Information for leadership
   - Scope and timing of production
   - Study of process documentation
   - Leadership conference
2) Operating manual
   - Check operating manual (validity, changes)
   - Make available report forms
   - Post necessary documents (e.g., chemical data sheets) in plant
3) Feedstocks
   - Verify availability of reactants required
   - Verify quality of reactants
4) Training
   - Conduct process-specific training
   - Discuss process with all employees concerned
5) Plant/facilities
   - Check maintenance completed
   - Carry out functional tests
   - Install additional safety features
   - Make piping hookups according to diagrams
   - Perform leak tests
   - Check cleanness of equipment
6) Trial run
   - Carry out and document trial run with solvent and prepare report
7) Analysis
   - Prepare analytical sampling
8) Administrative
   - Order necessary placards and labels
   - Prepare folders and forms for process documentation
9) Wastes
   - Check disposal system
10) Storage
    - Prepare packages/containers
    - Set aside storage capacity
11) Safety
    - Verify completeness of safety documentation
    - Consider safety and occupational health checks

During Production. In a correctly prepared start-up, the management concentrates on inspections, monitoring work procedures, and verifying process safety. This task includes:

Surveillance of Action on the Shop Floor. It is obvious that a safety problem exists when the author of an operating manual is convinced that the instructions are being strictly followed while, for whatever reason, something entirely different is going on. The comment, "That's what it says in the book, but we've always done it differently," must not be tolerated. Discrepancies should be identified and immediately corrected.

Detection of Irregularities. A permanent task for both leadership and operating personnel is to identify process irregularities promptly; an important aid is a form for reporting process excursions. Prompt response to irregularities is a key element in accident prevention.

Assessment of Process Modifications. If deviations from the process instructions are necessary (e.g., because of defective apparatus), the plant manager must consider carefully the safety consequences of such a change. If a distillation is carried out in a different piece of equipment under the same conditions, but the reactor jacket offers worse heat transfer, or the stirrer is less effective, the operation may last longer than specified. This type of change can lead to a dangerous condition.

Feedback. To obtain feedback on the practicability of an operating manual, it is essential to make contact with the workers at the workplace. An amazing amount of information and suggestions can be conveyed in a short, relaxed conversation.

Working Conditions. Noise, heat, uncomfortable posture, and other conditions can strongly affect human performance; working conditions should continually be examined and improved.

At Termination of Production. When a process is terminated, extensive process documentation is generated: production reports, analysis reports and certificates, printouts, etc., as well as a report summarizing all vital aspects of the process. This documentation yields information on potentially safety-relevant deviations in process chemistry and technology.

Process Documentation

1) Title/production stage
   Product, reactant, duration
2) Operating manual
   Applicable operating manual, deviations from valid procedure
3) Production
   Technical and chemical feasibility
4) Yields
   Comparison with earlier productions
5) Analysis/quality
   Remarks on product and reactants
6) Modifications/remodelling
   Summary comment
7) Economic aspects
   Calculation
8) Documentation
   Identification of changes, defects, and missing or outdated documents (e.g., safety tests); all changes made in flowsheets and other schemes in the course of repairs
9) Conference notes on termination of production
   Proposals for improving safety and process conditions

Good process documentation guarantees continuity of production with regard to work procedures and safety. Reading these documents when the same process is resumed is an absolute must.

Finally, the operating manual must be revised on the basis of the process documentation and, if necessary, rewritten. This insures that the manual reflects the reality of the plant and is ready to use for a new production at any time.

4.2.2. Safety in Continuous Processes

4.2.2.1. Introduction

A continuously operating production plant usually makes a single product. The steps involved can include both chemical reactions and physical separation operations. The quantity of material involved in both cases can be extraordinarily large, and special care is required, particularly when the substances have dangerous properties (flammable, explosive, toxic, or environmentally harmful). Continuous production plants are highly automated, and are seldom run without process control systems. Apart from start-up and shut-down modes, they operate according to fixed, specified process parameters. Intervention by operating personnel is necessary only during indicated or infrequent deviations from normal operating conditions. Because the process does not change, the danger of neglecting to readjust safety-relevant process variables on a change of product is not crucial. Typical continuous processes are:

- Refinery processes
- Cracking processes for olefin production
- Polymerizations
- Processes for making organic and inorganic bulk products

4.2.2.2. Analogies with Batch Processes

While the objectives, procedures, and plant sizes in the batch case are much different from those in the continuous case, quite a few of the basic statements in Section 4.2.1 also apply to continuous processes.

For example, exact knowledge of the process steps is important for process safety, in particular the acceptable deviations from the specified process parameters. The safety parameters and their acceptable ranges, established before the process plant is designed, must be set forth in an operating manual, and embodied in safe procedures for use by the operating personnel. In the context of risk or safety analysis (Section 3.3), the conditions for the occurrence of conceivable upsets must be identified, the consequences must be analyzed in terms of occurrence probability and severity, and appropriate safety practices must be instituted and covered in detail by bulletins and training of operating personnel. These studies must include not only the failure of technical devices (compressors, pumps, mills, and centrifuges) along with malfunctions of instrumentation and control devices, but also errors made by operating personnel, and partial and total power outages.

There must be a plant emergency plan, prescribing how workers should respond to a variety of abnormal conditions and accidents. A higher-order emergency and hazard-control plan dealing with major accidents, i.e., those that can pose a serious danger to employees, the immediate neighborhood of the plant, or the environment, must establish internal report routing and hazard-control measures. This plan must also designate those accidents in which external assistance, e.g., the public fire brigade or disaster response teams, must be requested, and what public institutions, e.g., the police and regulatory authorities, are to be informed, immediately or later.

Before production is started, operating personnel must receive training and orientation based on:

Process documentation:
- Process flowsheet
- Chemical reactions
- Physical steps in the process
- Parameter values to be maintained
- Safety rules
- Functioning of the process monitoring system

Safety information:
- Working with dangerous substances
- Use of personal protective equipment
- Periodic activities in the plant

Operating and safety instructions:
- Proper operation of apparatus and machinery
- Repair procedures
- Response to abnormal events

This information must be presented to all employees before they take up work in a plant, and at regular intervals thereafter. It is essential to test whether the knowledge presented has been understood and whether it can be promptly and correctly put into practice in stress situations. When the plant is being shut down on purpose, it is recommended that the shut-down process be combined with a simulated emergency.

Another vital activity in continuous plants is the keeping of a shift log. This helps inform later shifts about all important events in the plant. It is documentary in character, and must contain the following items:

- Bulletins from management
- Process upsets
- Deviations from normal operation
- Defects found in the plant
- Repairs needed and those completed
- Availability of standby equipment and machinery
- Warnings from supplier and customer plants

4.2.2.3. Special Features of Continuous Processes

Usually, large quantities of materials are involved in continuous processes. Large quantities, especially of flammable gases or toxic liquids or solids, represent high hazard potentials that must be taken into consideration when starting up the process for the first time or restarting after a temporary halt, during operation, and in the shut-down phases.

A continuous process has the advantage of inherent safety. While reactors have large dimensions and high throughputs, the quantities of products directly involved in the reaction and the holdup in the reactor are small; the hazard potential due to the reaction is minimal, especially for strongly exothermic reactions.

Continuous plants must be carefully inspected before they are first brought on stream:

1) All installation work, including subordinate work such as steam-trace lines and thermal insulation, has been completed
2) Scaffolding has been removed from the plant
3) Craft workers and assistants not actually needed in starting up the process must have left the plant
4) The entire system (logically subdivided into functional units) has been pressurized with gases (air or nitrogen) or liquids (water or harmless products) and checked for leaks
5) Auxiliary loops, coolers and chillers, special product or scrubber circuits, cleaning loops, and the wastewater and waste air cleanup units have been put through successful test runs
6) All instrumentation and control equipment, particularly their safety-relevant and fail-safe functions have been subjected to functional testing by qualified specialists. These functional checks must include not only the operation of alarms and safety interlocks on partial or total power outage, but also the emergency shut-down systems for plant sections or the entire plant

Almost all large continuous plants are now run with computer-assisted process control systems. Operating personnel must be thoroughly trained in the use of these systems so that they can take the proper risk-reducing actions in case of trouble, even in stress situations. Safety-related circuits are commonly hard-wired in parallel with the process control system. Switch points (temperatures, pressures, quantities, flow rates) are then fixed; they may be changed only after plant management and/or a safety expert has performed a safety analysis and issued a written directive. Where safety circuits are integrated into the process control system, technical or administrative policies must insure that safetyrelevant switch points cannot be altered by operating personnel on their own authority.

When starting up a continuous chemical process, special attention should be focused on "priming" the reaction. The pressure, temperature, quantities, and concentrations of reactants or catalysts must meet the criteria specified for the start-up phase. To avoid dangerous conditions due to delayed reactions and the associated buildup of unreacted feeds, the reaction start should be precisely monitored for rates of change (e.g., temperature or pressure rise) or with real-time analytical methods. If irregularities are seen, the reaction start should be interrupted, the fault identified and remedied, and the plant readied for another attempt. Many process control systems have sequence controls programmed to halt start-up automatically if an upset occurs.

During on-stream operation, the plant personnel continuously observe the process parameters, comparing them with the acceptable nominal ranges. Analyses of intermediate and end products not only help keep the commerical product within specification, but also provide information for use in maintaining the process. In particular, buildup of undesired byproducts due to recycles within a process or in closed-cycle scrubber units must elicit a high level of operator attention, and lead to countermeasures if appropriate.

Important process parameters that are recorded or stored by the process control system must be retained for up to ten years (depending on legislative provisions). These can be used to demonstrate that the process was run properly, or to prove compliance with emission limits and other air and water pollution control standards during regulatory audits; they are also indispensable in post-accident reconstruction.

As a rule, a continuous plant is shut down once a year for cleaning, inspection, and major repairs. After the process is stopped and before these operations are started, the plant must be carefully drained, purged, and cleaned according to the operating manual. For especially hazardous work (entry into vessels and tight spaces, or welding in plants that process flammable or explosive substances) plant management must establish safety practices that must be specified in detail in written work permisssions given to the employees performing the jobs. Exact instructions for the performance of repairs must be given (see Section 4.4).

When several craft teams are working at the same time, a coordinator must be designated to prevent them from endangering each other. The coordinator plans the work in time and space and is authorized to issue instructions to the craft teams.

Large continuous plants where explosive, flammable, acutely toxic, or environmentally harmful substances are handled in such quantities as to create a high latent hazard potential present a special concern. Such a plant should have a rapid-response service, organized at each key level of the hierarchy, which can provide expert advice to emergency workers, to minimize the consequences of an accident.

4.3. Technical Inspection [211–217]

Introduction. In what follows, the term "technical inspection" denotes the inspection of technical devices by independent third parties; its purpose is to find out whether the equipment meets requirements set forth in legislation, regulations, or standards. In this way, unacceptable risks to employees working with the equipment – and to third parties and the environment – are prevented.

Plant hardware subject to technical inspection includes pressurized systems (vessels, piping); storage facilities for flammable substances or water pollutants; elevating, hoisting, and conveying equipment for the transportation of persons or freight; electrical equipment; nuclear facilities; means of transport such as automobiles and trucks, railroads, ships, and aircraft; power plants; petrochemical plants; and chemical plants.

Experts. As a rule, inspection is performed by an expert recognized by the responsible authority. Conditions for recognition include not only personal integrity, but also appropriate specialist qualifications (through education and continuing education), and it must be determined that the expert is not subject to instruction or restraint having to do with the inspection activity, i.e., independent.

The specialist qualification is generally a natural science or engineering degree. In many cases, a specified period of employment in the field is also required.

The expert may also be required to take part in an exchange of experience, which has two purposes: to assure consistency of technical inspection, and to identify crucial points of concern for a certain type of technical equipment. The results of such an exchange can also provide a basis for technical progress, and serve as input for standards development.

The role of the expert in technical inspection has long been established in Germany. The high point of the independent expert tradition was the field of law dealing with facilities "subject to surveillance." The experts recognized in this area are usually comprised in Technical Inspection Boards (TÜVs), which have in turn combined (along with some industrial firms that support their own inspection services, staffed by recognized experts) in the Association of Technical Inspection Boards.

Official recognition as an expert used to be limited to an individual on the German system. At European level, organizations will be entrusted with the performance of safety tests in the context of equipment manufacturing. Such organizations, after gaining accreditation at national level, can apply to the European Commission for "notification". These inspection organizations or "notified bodies" are to be engaged by manufacturers, especially for certain conformity tests pursuant to EU directives. The requirements on notified bodies are set forth in the Directives and relate chiefly to independence, personnel and hardware resources, and employee training. While expert inspection in Germany has been handled on a regional basis, notification as a testing service is valid throughout Europe, i.e., a laboratory accredited by a European Union member country and notified by the Commission can also do the pertinent conformity tests in the other member countries.

German Federal Water Quality and Pollution Control Act (Wasserhaushaltsgesetz) is following a similar path. Where tests by experts used to be prescribed, jurisdiction is shifting to inspection organizations, whose recognition in one Federal State is valid in all states.

Legal Background. The purpose of technical inspection is to make certain that the equipment inspected satisfies the quality and operational requirements set forth in regulations and standards.

Regulations are based on legislation that designates goods for protection and identifies technical devices to be regulated. The legislature thus authorizes regulatory bodies to set requirements, e.g., quality, operation, and inspection. Administrative actions such as permits and approvals are also authorized in this way.

A regulation commonly defines the scope of application, defines exceptions, states what technical devices are covered, establishes the scope and intervals of inspection, designates the persons or institutions that perform the inspections, and states what agencies have surveillance authority. The requirements set forth in such regulations are then implemented in instructions to the agencies or in engineering codes developed by private or public standardization bodies.

The harmonized European standards promise to be of great importance in this respect. These standards implement the basic safety requirements stated in European directives. Although their application is voluntary, they allow the presumption that a technical device complying with the harmonized European standards also fulfills the essential safety requirements of the pertinent directives.

Expert Inspection. Inspection and testing by the expert usually takes place in two phases. In the first, the existence and completeness of all necessary documents for the technical device are verified (official approvals, type approvals, type test certificates, certificates covering the inspection and use of materials, technical drawings and plans, operational descriptions, and reports from individual tests already performed). The second phase comprises the actual testing of physical facts.

The process of expert inspection can also be broken down in time. The first point examined is whether the design being inspected is suitable for the intended service. In the second phase, it is determined whether the technical device as fabricated complies with the submitted and examined design.

The next step is to verify that the finished technical device is operated in such a way that no stresses are imposed on it other than those considered in the first phase. An integrated system analysis may be called for here, especially if interaction with other technical devices or within the system as a whole must be taken into account. Finally, after a set inspection interval has passed, so-called periodic tests are conducted. These verify whether changes in the technical device due to its service, e.g., those resulting from wear, suggest that the stresses experienced in the next inspection cycle may no longer be tolerated. Another question at this stage is whether there have been safety-relevant changes within the system in which the component is integrated.

In addition to inspections before and during service, technical devices may also have to be inspected after they are taken out of service, if significant danger can result from improper operation.

When technical devices or their service conditions are modified in such a way that there may be safety-relevant effects, a new expert inspection generally has to be performed.

Finally, the responsible agency can instruct the expert to assess departures from the regulations or to investigate accidents.

The methods used in technical inspection range from simple visual examination to complex, specialized testing methods such as nondestructive ultrasonic tests or electron microscopy in connection with accident investigations. Such tests are generally performed, and the results interpreted, by specialists acting in support of the expert, who evaluates the results and cites them in the final report.

Integrated Plant Inspection. The interdisciplinary efforts of specialists in assessing safety problems is particularly important in the inspection of facilities subject to approval under the German Federal Pollution Control Act (*Bundesimmissionsschutzgesetz*). "Whole-plant inspection" means a comprehensive system inspection and monitoring of a plant with regard to safety functions.

This procedure takes in material properties, process conditions, interaction between apparatus and other plant components, interaction between human beings and hardware, administrative policies, and the effect of external safety-relevant factors. Inspections in such a program are concerned with whether the operation complies with legislative requirements (pollution control, occupational safety and health, water quality, soil conservation), pertinent regulations, and recommendations.

These inspection and surveillance tasks concern all those concerned. The three pillars of whole-plant inspection are thus:

1) Supervision of the plant as part of the operator's responsibility
2) Inspection and surveillance by public agencies

3) Safety inspections by experts

Such a procedure can also serve as a model for the inspection of plants, as proposed in a draft European Council Directive for avoiding the dangerous consequences of serious accidents with hazardous substances.

4.4. Maintenance

A plant operating in a trouble-free manner is in its safest condition. However, because technical components are subject to mechanical and process-related wear, upsets can occur. As a result, maintenance is necessary to keep the plant in a safe condition or restore it to such a condition. On the other hand, maintenance can create new dangers in the plant.

This maintenance-related risk can be minimized only by good safety practices in the planning and execution of maintenance work.

In the past, mistakes in the preparation or execution of maintenance were a significant cause of accidents in the chemical industry. The gas explosion at Flixborough in 1974 was traced back to a makeshift repair procedure [218], and many smaller and less spectacular events have resulted from maintenance errors [219], [220]. In a representative time interval, 30 % of all events occurred during maintenance [220].

Therefore, the value of safe maintenance in the chemical industry cannot be overstated. This is not surprising if the cost of maintenance is analyzed. It amounts to 3 – 5 % of the plant replacement value per year; in large companies, this can add up to more than DM 10^9 a year [221], [222]. These costs arise from a large number of small operations, and are made up mainly of personnel costs (including those incurred by vendors). A high level of administrative and logistical expenditure is called for. Maintenance actions, especially repair work, are only partially recurrent; many repair jobs must be planned and carried out individually.

These points imply the following three fundamental goals:

1) Plants should be designed and constructed in such a way that they are easy to maintain, i.e., easy to repair (good cleaning and draining capabilities, proper access, etc.); require little maintenance; and are made up of the most reliable and long-lived components possible
2) The proportion of unforeseen repairs caused by damage should be reduced in favor of scheduled (including preventive) operations
3) Safe procedures must be adopted and enforced for the performance of maintenance work, to reduce the risk arising from maintenance

4.4.1. Actions During Plant Design and Construction

The task ensuring safety during maintenance work is to create the technical and administrative conditions for safe performance of the work. This requires a systematic procedure (Fig. 61).

A good maintenance strategy begins with the design of a low-maintenance, easy-to-maintain plant. Every intervention in the plant can lead to danger during maintenance work and on restarting. The achievement of long life and high reliability at this stage is also economically important.

Systematic safety analysis (e.g., HAZOP) can identify the consequences of malfunctions and their possible causes at the design stage. Two types of event must be distinguished:

- Events that follow immediately from the technical failure and the resulting deviation of the process parameters from the design conditions
- Events that can occur during a maintenance operation

The objective in event analysis is to minimize hazards:

- To achieve the highest possible reliability in terms of failures, and to limit the consequences of failures
- To minimize the repair risk, i.e., to reduce the number of maintenance actions and to perform them as safe as possible

Selection Criteria for Apparatus and Machinery. High reliability is obtained by designing and selecting components on the basis of their requirements. Key criteria are:

Construction materials

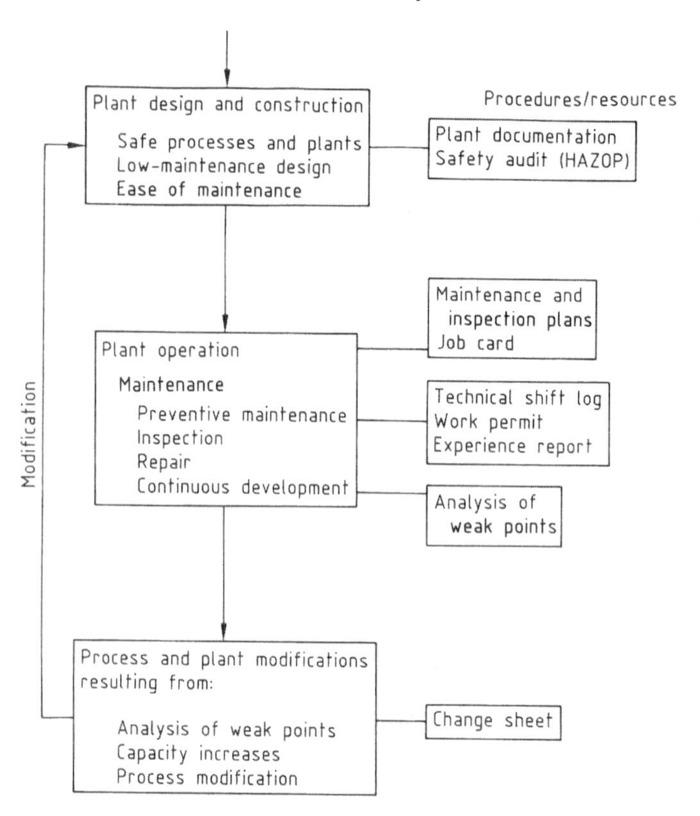

Figure 61. Maintenance strategy: central significance in a chemical plant

Functional principle
Robustness

Secondary points are hardware safeguards against improper operating conditions, e.g., flow monitors on pumps, which help to prevent damage.

Because a component may fail anyway, the consequence of such a failure must be analyzed. Even if the probability of occurrence is low, an event (failure mode) with serious consequences can influence the above criteria and make it necessary to seek a different solution.

Once these criteria are set, quality assurance systems must be instituted to monitor and guarantee compliance with the standards during the subsequent fabrication and installation of apparatus and machinery.

Design Criteria. Because maintenance and repair are inevitable, danger arising from them can be most effectively minimized by good installation and design practices. The criteria at this stage include clear layout (accessibility), easy and safe disassembly, and draining and pressure-relief capabilities. In a chemical plant, special importance attaches to draining. Some maintenance jobs can be made safer by the use of assembly aids, local spot-vent systems, etc.

Aids for Monitoring Plants and Performing Safe Maintenance. The documents needed for reliable operation and safe maintenance should be created at the design stage. These documents can usually be assembled from documents required for plant design and construction, e.g., manufacturers' data sheets, internal experience reports, and piping and instrumentation diagrams.

Periodic inspections are supported by maintenance and inspection schedules, which must be present when the plant is commissioned. Test methods involving no opening of equipment or piping in the plant (e.g., ultrasonic inspection) are preferred. Where safety-relevant process control equipment is not redundant, self-monitoring is recommended. Inspections, including functional tests and plausibility checks, help todiscover concealed defects and to identify needed repairs. In some situations, automatic monitoring devices such as vibration monitors can also be used.

Anticipated, i.e., recurrent repairs must be thought out ahead of time. Detailed job instructions, supported by extracts from technical drawings (job cards, Figs. 62, 63), cut the number of mistakes during a repair or preparation for one.

Other documents needed later for the proper execution of maintenance tasks include: piping and instrumentation diagrams with piping and seal specifications; piping and fitting lists; apparatus specifications; measurement station data sheets (to ensure compliance with standards); and functional diagrams of the process control system with lists of setpoints, limits, and alarm settings (unambiguous transfer of process parameters to process control system).

4.4.2. Actions During Plant Operation

Figure 64 illustrates the maintenance strategy for this phase. The maintenance task is to keep the plant in its nominal condition so that requirements on product quality, safety, and availability are always met. According to DIN 31 051, maintenance is defined in the following way:

1) Preventive maintenance: monitoring and keeping plant in nominal state
2) Inspection: determination and assessment of actual state
3) Repair: restoration of nominal state

There is a fundamental distinction between scheduled maintenance and breakdown maintenance. An attempt is made to minimize the proportion of breakdown maintenance caused by unforeseen failures. Scheduled maintenance is particularly concerned with components having safety and quality relevance; it includes upkeep and inspection as well as measures to obviate repairs or cut their cost [223], [224].

Preventive maintenance and inspection are usually done according to predetermined schedules and are thus systematic (timing and scope set in advance). Maintenance and inspection schedules are used for this purpose. Inspections are scheduled on the basis of the anticipated wear of the components involved, known either from manufacturers' data or from operating experience.

Preventive maintenance and inspection are done both while the plant is on stream, nominal/actual comparisons being made continuously and documented during regular rounds, and during scheduled shutdowns. Maintenance and inspection schedules are a reliable and efficient way to organize the latter. Unforeseen failures of plant equipment occur despite scheduled maintenance, and these must be covered by administrative policies (usually operating procedures) providing for later repairs. This "breakdown" maintenance is also supported by repair instructions and aids rendering the work more reliable and faster.

4.4.2.1. Maintenance Activity

The procedures for scheduled and breakdown maintenance are similar (Fig. 64). The key step in minimizing risk of breakdown is early detection of a deviation from nominal conditions, which can be apparent in:

- Obvious defects: drips, noise, etc.
- Concealed defects, e.g., gradual slippage of process parameters

Regular rounds by shift personnel under a fixed plan can aid detection of the problem and thus contribute to minimizing risks. The maintenance and inspection schedule for every technical component contains information on:

- Test criteria applied in maintenance and inspection
- Methods practiced
- Timing

The action ends with a finding, documented in a report signed by the inspector.

If problems are identified, but cannot be remedied immediately, they are recorded in the plant shift log. The description should be oriented to the concrete observation, and should not contain speculation about the cause. The proper remedy cannot be established until the malfunction or the damage has been jointly assessed by plant management, the shift supervisor, the shop supervisor, and in some cases a specialist. A meeting must be held to determine how and with

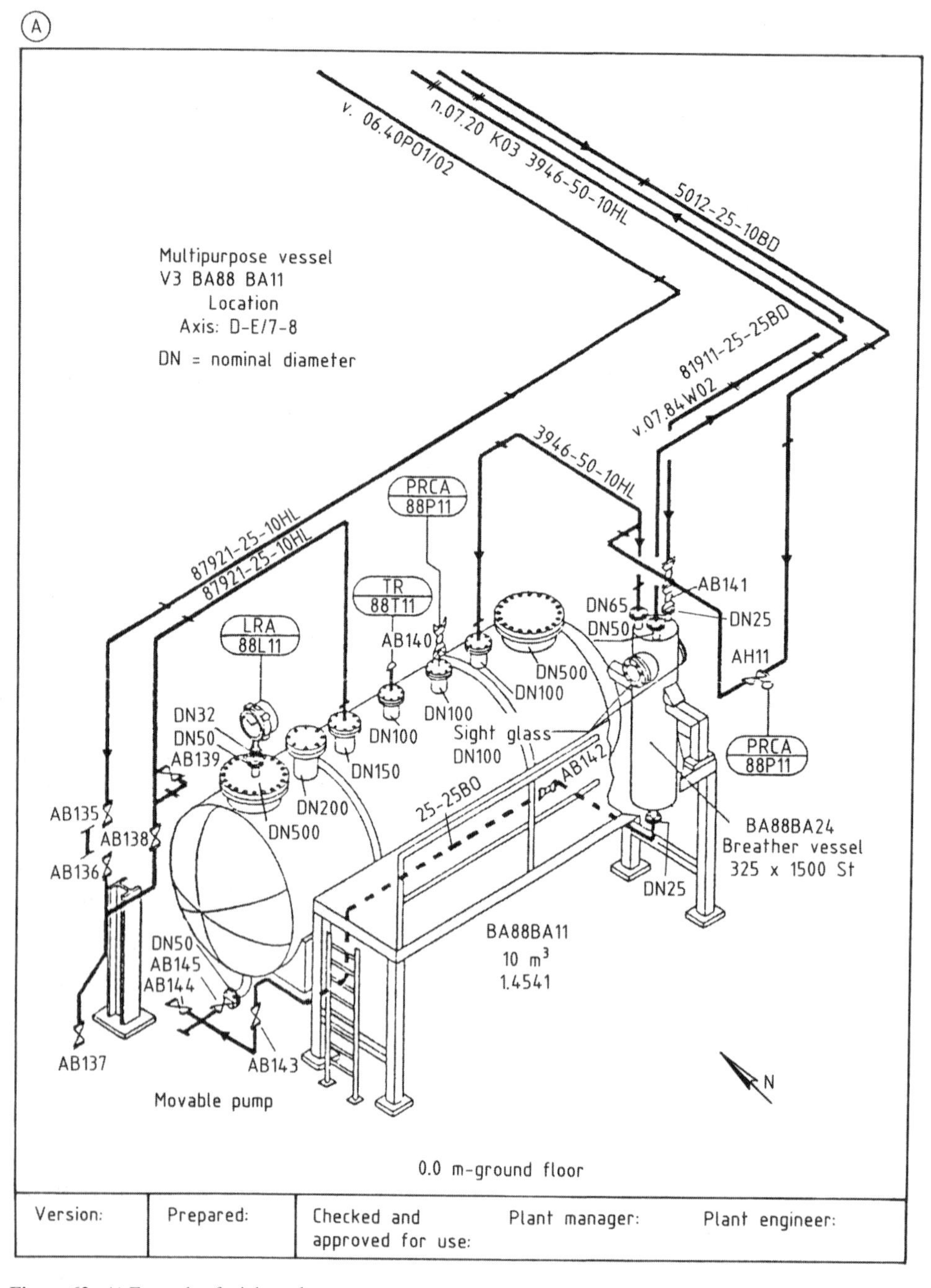

Figure 62. A) Example of a job card

(B)

	Multipurpose vessel					
MID Plant	B577 Building	22 Works	SIS Plant compl.	V 3 Unit	BA88 Unit sect.	BA11 Techn. device
Shop Job No.		Short text		Recurrent Job No.		Job No.

Work permit Yes ☐ No ☐	Documentation Yes ☐ No ☐	No. in shift log	Exhibitor	Tel.No.	Date

Job instructions:

Inspections:		Medium	Pressure, bar IR	RDDR	AR	RUDR		Acceptance test requested:
Pressure test	☐							Safety department ☐
Leak test	☐							Voluntary ☐
Intl. exam.	☐							Acceptance test performed:
Endoscopy	☐							Safety department ☐
Ultrasonic wall thickness meas.	☐							Material testing ☐
— — — —	☐							SK ☐
Cl.No. — — — —	☐							

No.	DN	PN	Material	Bolts	Remarks

Dept.	X	Crafts	Week MO,TU,WE, TH,FR,SA,SU	Week	Week	Week	Week
Maintenance		Plant crafts					
		Measurement					
		Pool					
		PCS B					
Production		Operating					
Contractor		Central assembly					
		Piping fabrication					
		Steel construction					
		Erection/cranes					
		Apparatus fabrication					
Civil		Scaffolding					
		Insulation					
		Painting					
Engineering		PAT 3.2					

Prepared:
Name: Date:

Figure 63. B) Example of a job card

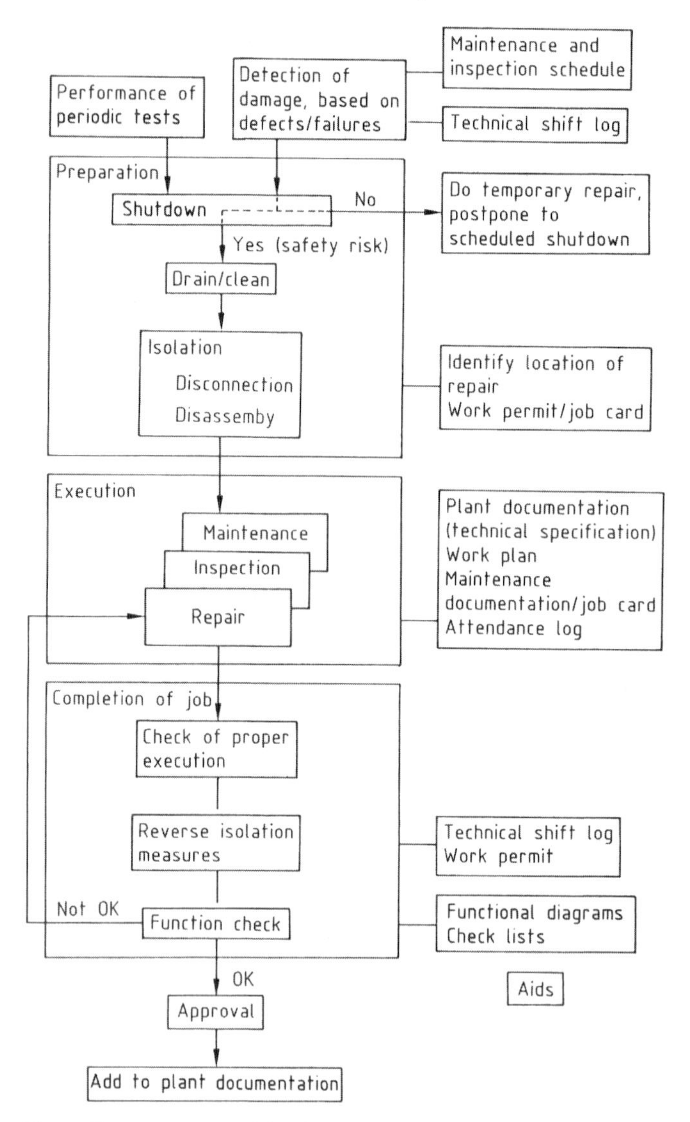

Figure 64. Maintenance activity chart

what precautions the repair work is performed (Section 4.2.2.2).

A further aid to proper reporting of technical problems by operating personnel to maintenance personnel is the technical shift log, which simultaneously serves as a tool for evaluating damage; the shift personnel enter the following information:

- Trouble during the shift
- Exact location (i.e., identification of component)
- Nature of trouble or malfunction observed
- Time of discovery

The log should also show what maintenance action was taken; the person responsible must sign it.

The purpose of the technical shift log is to state clearly what was done, and who was responsible, to prevent ill-considered and uncoordinated repair efforts. With the help of the technical shift log – including the signature of the responsible person – it is always possible to verify that necessary action has been done properly.

The technical shift log performs two other functions:

- Identification of temporary and makeshift repairs
- Evaluation in terms of recurrent problems (analysis of weak points)
- Computer support is recommended so that:
- Temporary repairs are not left in place permanently
- Statistical analysis can be done and appropriate countermeasures instituted to extend service life and reduce risk by cutting the number of repair projects [225]

Safety procedures must be followed whether the maintenance job is:

- Anticipated when preventive maintenance, inspection, and repairs are to be carried out during a scheduled shut-down (scheduled maintenance); or
- Unplanned, i.e., caused by a failure (breakdown maintenance)

In either case, the procedures must be designed to cope with the risks occurring at every stage.

If a problem arises, e.g., failure of a safety device, it may directly affect plant safety. This aspect is not pursued here; the proper response is always set out in the context of the safety analysis.

As a rule, the plant or sections of it must be put in a safe condition by a shut-down procedure when a problem occurs; this must not be confused with the "emergency off" or "panic" button for a section of the plant. While planned or "checklist" shutdown of the plant is a frequent procedure, the failure of a technical component required for shutdown may lead to the isolation of specific apparatus or sections of the plant, or to overriding the safety interlocks. Examples include bypassing a leaking agitated vessel and releasing an interlocked fitting that normally cannot be opened except at a certain point in the reaction.

Shutdown is not always necessary. In many instances, countermeasures can be taken that do not lead directly to a solution of the problem. There are three possibilities:

1) Postponing the proper repair (e.g., temporary plugging of leaks)
2) Using planned fallbacks (e.g., putting standby systems on stream)
3) Taking alternative measures (e.g., blowing a vessel down instead of pumping it dry)

Alternative actions are temporary modifications; such makeshifts may be more desirable than immediate shutdown of the plant, e.g., in a large continuous process where operation can be maintained until the next planned shutdown. Any modification must be checked first by a systematic safety analysis.

4.4.2.2. Performance of Maintenance

Scheduled and breakdown maintenance differ, especially in the documentation existing at the start of production, which describes the specific work to be done and takes the form of instructions:

- Scheduled maintenance is covered by preventive maintenance and inspection schedules describing the nature and scope of the work. Repair instructions (e.g., job cards) are generally used. Plans for shutdown maintenance are of this type, especially in continuous process plants.
- For breakdown maintenance, the problem and the corrective action taken are recorded in the technical shift log. Special repair instructions are not always available, and the documentation is then limited to the written plant documentation, including updates.

There are essentially two further possibilities:

1) Maintenance work while the plant is on stream
2) Maintenance work during a shutdown

In what follows, only maintenance during shutdown is discussed. There is no fundamental difference between total and partial shutdown. The only routine actions that should be taken in an operating plant are those that raise no safety concerns.

Maintenance can be broken down into three phases:

Preparation
Execution
Post- completion measures

When danger can arise during the performance of this work, all workers involved must follow special safety practices, which are written up in a "work permit". The purpose of this is twofold: to make certain that the proper safety procedures are followed; and to ensure that responsibility for the components of the work are clearly defined by signature.

Work Permit **No.**

Equipment: **Polymerization reactor 1 (R 3.02)**

Description of work (incl. to whom assigned)
Reattachment of submerged product feed pipe by fitters

1. **Isolation of plant section/component:**
   Depressurize all product lines connected; open pipe unions and close off with blank flanges; disconnect N_2 lines and close off with blank flanges. Disconnect external steam tracing. Disconnect power to agitator motor. Release interlock on bottom drain valve. Completely empty vessel through R 2.03, and leave valve open.
2. **Actions on isolated plant section/component:**
   Using high-pressure cleaning unit, clean vessel with water through manhole. Close bottom drain valve and fill with water. Drain both cleaning waters through R 2.04 for analysis. Have specialist team perform radiometric measurement of level from outside pipes. Suspend ventilation device in vessel. Close bottom drain valve.
   Date: Time: Signature:
3. **Action before beginning work:**
   Instruct fitters that no welding or driving is permitted.
   Instructions given to: Signature:
4. **Occupational safety measures:**
   Introduce conductors safely into tank; post lookout at manhole; ventilation unit; protective clothing for fitters in vessel; only ordinary work clothing (note: hardhat, protective goggles, safety shoes, . . .).
5. **Actions after completion of work:**
6. **Management approval (by plant manager) for performance of work:**
   From 1 March 1993, 10 : 30 am, to 1 March 1993, 4 : 30 pm
   Date: Time: Signature:
7. **Signature of shop supervisor responsible for compliance with items 4 and 5**
   Date: Time: Signature:
8. **Approval to proceed on site by engineering supervisor of plant**
   Date: Time: Signature:
9. **After completion of work: approval by plant supervisor**
10. **All- clear and technical acceptance**
    All clear:
    Date: Time: Signature of shop supervisor:

    Technical acceptance:
    Date: Time Signature:
11. **Waiver of stated procedures (see 1 and 2)**
12. **Acceptance before restart**

Preparatory steps include isolating single components or whole sections of the plant. This requires proper shutdown of the section concerned to a safe condition. Where redundant components exist for the malfunctioning components, plant management and/or the set operating instructions may allow these to be placed on stream. Other temporary actions such as:

- Blowing down instead of pumping dry
- Changing the preset process control switching or alarm values
- Interfering with interlocks

are permitted only after a special safety check, unless they are provided for in special operating procedures.

After shutdown, the critical danger points are release of substances by opening up the system to drain it (danger of poisoning or fire/explosion), and the possible introduction of undesired substances during cleaning.

Ideally, a plant should have emptying devices and holding tanks for the contents of the equipment, so the plant does not have to be opened for drainage. This concept is desirable for substances with high hazard potentials, but the cost may be substantial. Regardless of whether fixed holding tanks are installed, emptying by gravity should be possible.

Normally it is necessary to drain only certain portions of the plant. Such areas can be safely separated from other, product- containing sections by closing some fittings and inserting blanks. Residual quantities of product remain in the drained parts of the plant, however, especially in fittings, filters, and pumps. The individual draining features of these must be used to free them of product. Drainage is frequently aided by nitrogen pressure. Subsequent cleaning is not required with highly volatile substances. The vented air from such an operation must be considered; a vent system may be necessary, along with downstream treatment of the gases.

When cleaning is required, the agents used must be compatible with the product and, physiologically, as innocuous as possible. If inert media cannot be used, the following become particularly important:

- Medium adequately diluted in the shortest possible time

- Agent reactive only with the small quantity of product found in the drained segment of the plant
- Cleaning liquid safely separated from sections of the plant containing larger amounts of product

Failure to observe these conditions has frequently led to events, often because, in a nonroutine cleaning operation, the flow direction, pressure, and valve configuration were different from the normal conditions, and so were interpreted wrongly. All connections to other sections of the plant must therefore be examined, including those not ordinarily filled with product (e.g., off-gas piping).

When maintenance work calls for disconnection or disassembly of a plant section, personal injury hazards can arise if the repair site is not rendered safe, e.g., if the electric power to drives is not disconnected or product-filled pipes are not safely closed with a blank flange after disassembly of a subsystem. The latter point is crucial when equipment such as pumps and fittings are disassembled, because the safety measures are carried out right at the disassembly site; in larger maintenance operations, entire piping systems are generally made product-free.

Engineering codes, accident prevention regulations, and special safety instructions aid in the proper execution of these operations. To avoid misunderstandings at the site, particularly in a complex installation, the defective item must be identified, and its isolation must be indicated visually. The area can be further marked with a tag stating which craft has jurisdiction and what kind of work is being done.

The key requirement for safe repairs is up-to-date plant documentation. On the basis of the technical specifications (complete and unambiguous description of performance data, media, construction materials, and design data), it must be ensured that parts and materials used (e.g., greases) are such that no danger arises when the affected part of the plant is on stream.

Hazards in this sense are leaks due to improper materials (e.g., gaskets) or incorrect temperature/pressure design; unwanted reactions due to the catalytic action of construction materials or other media present; and parameters outside safe ranges owing due to the installation of unsuitable equipment (e.g., a pump with too large an impeller generating an excessive pressure). Similar hazards must be considered when replacing or repairing process control system components. A typical mistake involving an instrument is the use of the wrong range.

Just as a repair or inspection site must be labeled to prevent confusion, replacement parts or newly delivered repaired components must be labeled so they can be recognized as equipment meeting the specifications. Often there is no outward difference between two parts, e.g., when only the materials inside the device differ.

These practices are intended to ensure that the on-site activity can be finished in the shortest possible time, without interruption or risk. Safety and economy are not in conflict.

Virtually every maintenance activity comprises many single activities performed by different crafts. Because economics does not permit craft specialists to be employed especially for all such tasks, workers unfamiliar with the special hazards of the production plant have to be brought in; the requirement of up-to-date plant documentation must be supplemented by explicit instructions to the craftspeople at the work site, and a clear task description (see the job card in Fig. 62).

It is necessary to establish both the plant documentation needed for maintenance, and the scheduling of various maintenance activities in a plant area. Only by matching and skillfully coordinating maintenance activities to the rhythm of the process they can be prevented from threatening each other. An aid in this respect is a working plan, which describes the sequence of tasks and their duration from their starting times (e.g., a bar chart). In an area where welding, driving, and caulking work is done, for example, explosive mixtures are not permitted. Production not connected to the maintenance action may have to be interrupted. The same holds when flammable or explosive substances are present in the course of maintenance, e.g., when a protective coating is applied to a surface.

Some kinds of work also require a second person to be present, even if the job requires only one craft worker. This is the case when a hazard such as the escape of toxic or explosive mixtures cannot be positively ruled out, or when work has to be done inside vessels.

Maintenance jobs can extend over more than a day, so interruptions cannot always be avoided.

The area of the plant where repairs are being done should be marked off with signs throughout the job, especially to prevent an inadvertent return to service before the work is complete.

Approval to restart can be given only by persons whose functions include this task, e.g., the responsible plant engineer. Only a single person should be made responsible for maintenance. This person is the contact point throughout maintenance, otherwise there is a danger that someone who does not have an overview of the work may make a wrong decision.

To avoid the danger of premature startup, interruptions of the work must be recorded, e.g., in an attendance log, and completion must be documented in the technical shift log. For work that requires a permit, the permit includes a space for approval after completion of the job.

A maintenance action cannot be considered finished until all the unit activities have been checked and accepted by the coordinator of the overall action. This includes both inspection for correct execution of a task, and functional testing. Proper execution means both the use of the components provided (correct temperature and pressure ratings, material, etc.) and their functionally correct installation. Functional testing is then done to make certain that all technical components work together in the way necessary for the process. The simplest example is leak testing of a repaired section of the plant; a very complicated one is the comprehensive testing of a process control system.

Where preparatory action has been taken, e.g., isolation of sections of the plant, these must be reversed in the same way as temporary modifications. Approval can then be given for production. In some cases, the plant is restarted a step at a time, extra functional tests with reduced throughput or enlarged crew being done first. Such precautions are needed especially when extensive maintenance work has been carried out.

4.4.3. Continuing Plant Development (Analysis of Weak Points)

The elimination of problems by repair is a necessity, but it also offers the plant operators a chance to analyze recurring problems, and to use the results in improving the system. Analysis of maintenance actions for continuing development presupposes that the technology and the operating conditions are adequately documented (this includes unambiguous nomenclature in the plant). Entries in the technical shift log aid in evaluating the frequency of problems. If plant documentation is computerized, the cost of such a "weak point analysis" can be substantially reduced. The goal is to use technically feasible and economically acceptable means to optimize sections of the plant, i.e., to minimize the frequency and/or severity of trouble. Before such an analysis, the possibility of human error causing the problem must be eliminated.

This analysis of weak points must not be limited to the technology. Often, the technical defect can be eliminated more simply by changing the process conditions. The weak point found may not be the cause, but rather the effect, of defects in the process, e.g., process conditions or product specifications. In such cases, the analysis of weak points becomes a process analysis (Section 4.5).

The causes of problems may well be hard to identify. A final statement of causes often has to be based on the vigilance of the operating personnel on the spot, as well as the years they have spent learning the details. The plant manager must therefore make a continuing effort to motivate the personnel to be observant.

4.5. Modification of Plants

4.5.1. Reasons for Modifications

While the objective of maintenance is to keep the plant in its nominal condition, modifications to the process and plant involve changing the documented nominal state of either, or both. A variety of factors can lead to process and plant modifications:

- Increases in capacity
- Change in process to lower manufacturing costs and/or improve product quality, or to modify the product
- Measures to eradicate weak points
- Upgrading to the state of the art (technology or safety)

Any change in the nominal state means that the corresponding parts of the process and plant documentation must be revised or rewritten.

In this respect, modification differs in scope, but not in basic execution, from a new design project. The reason it is so important to treat a modification like a new design is that there is no way to know ahead of time how widely the modification will ramify itself. Even minor changes, e.g., replacing one apparatus by another, or raising the temperature in a unit operation, can cause a violation of the design or process control parameters. The systematic methods already described for evaluating new plants must be applied here too, because they cover the entire field of possible hazards. First, all areas of concern, e.g., explosion hazards, are studied to determine whether the proposed change affects them at all. Because a change often relates to only a few danger spots, the number of areas of concern quickly becomes small, as does the number of aspects to be considered; detailed analysis is needed for only a few aspects in most cases. Knowledge of the existing safety concept is essential for assessment of all new practices. Any change represents a "compatible" expansion of the whole concept.

4.5.2. Procedure for Modification

A suitable procedure must be established if modifications are to be carried out correctly. As soon as a major change becomes necessary in the plant, e.g., because a new unit operation is being introduced, the project is implemented in the same way as any other design project, right up to the construction of a new plant (Section 3.2).

In the case of minor changes, the amount of investigative effort required is generally small, so a simplified procedure is useful in such cases. The principle remains the same, but the effort invested in safety analysis and documentation updating is matched to the scope of the planned change. A useful aid is the change sheet, which has two functions: to describe the objective and the implementation of the planned change; and to insure that necessary safety questions have been asked and properly answered. The change sheet also documents the work and its approval, and is used in identifying the parts of the plant documentation affected by the change (Section 4.4.1). A signature is required for each approval and each inspection of each individual step in the process. The work permit already mentioned complements this task by listing actions to be taken during its execution.

The items contained in the change sheet are:

1) Identification of what is to be changed (including nature of change and other ordering criteria; suitable tools for this purpose are job cards, piping and instrumentation diagrams, and operating manuals)
2) Purpose of the change
3) Description of the change
4) Identification of essential danger spots with the help of a checklist
5) Discussion with specialist departments on the danger spots relevant to each
6) Inquiry into necessary updating of safety analysis
7) List of plant and process documents affected, with possibility of monitoring progress of the change; testing steps after completion (e.g., acceptance test, test run)
8) Questions as to possible effects on product quality

Important consultations with the specialist departments and other safety information must be recorded, and the safety analysis must be attached.

The following fictitious examples show how small changes can have wide-ranging consequences.

1) An additive used in a thermally sensitive mixture is purchased from another vendor; the composition appears to be only slightly different, but it lowers the safe temperature limit. The mixture is to be subjected to distillation at high temperature. The result is thermal decomposition of the bottoms, faster heat evolution, and a rise in pressure in the reboiler.

Corrective action: Perform safety study in the lab with the new feedstocks; lower the temperature set point.

1) A pump delivering acid from a storage tank must fill a vessel more rapidly. On the basis of the characteristic curve, the old pump was selected so that the vent line of the vessel cannot be completely flooded. As a result, releases of substances are prevented without a level-limiting feature having to be installed. The new pump, with a larger impeller, attains a higher pressure. It is now possible to flood

the vent completely so that acid is released. The danger is not noticed as long as overfilling does not occur. The inherent safety formerly present no longer exists.

Corrective action: Select pump according to proper safety documentation; see change sheet.

4.5.3. Summary of Plant Maintenance and Process Modification

The systematic procedure described, with the associated aids, contributes to plant safety from design to operation and process upgrades. By itself, however, it is not sufficient to guarantee plant safety. While it cannot replace the knowledge of experienced workers, it offers a guide and support for daily work. The reasons for the increasing volume of technical documentation and administrative support are:

- Plants themselves have become far more complex (process control engineering and other advances)
- Maintenance work is being done by a larger number of specialized teams
- There is increased public interest in readily understandable safety practices

5. Hazard Control

5.1. Means of Limiting Accident Impacts

The responsible care of every plant-operating company to take active measures to protect employees, third parties, and the environment against potential harm means that the operator must take responsibility for technical and administrative precautions to limit the impacts of minor and major accidents in the facility. Legal and regulatory provisions independently require limitation of the consequences of accidents in plants; e.g., the "Seveso Directive" [226], [227] and the implementing regulations in EC member countries.

Action to limit accident impacts is also mandated by other regulations, e.g., the German Pressure Regulation [228]. The release of substances from pressure-relief features of a plant (safety valves, rupture disks, emergency blowdown devices) must present "no hazard" to employees, third parties, and the environment [228], [229].

Whether such a release is permissible must be assessed in view of the properties of the substances, the location of the plant, the propagation behavior of the substances, the duration of the release, and the occurrence of hazardous concentrations. If a release is not possible without hazard, retention or disposal equipment must be provided to limit the consequences.

German water law also requires that plants be designed, built, and operated in such a way that water pollutants cannot escape. Substances that escape anyway must be retained quickly and reliably so that contamination of receiving waters does not become a concern [230], [231].

The basic idea of a two-barrier system is contained in these provisions. The enclosing shell of a storage tank, e.g., for fuel oil, or the pressure-resistant construction of a reactor in the chemical industry, is the first barrier; it must safely withstand the mechanical, thermal, and chemical stresses imposed on it in normal operation, and in the event of accidents.

If a release occurs in spite of the preventive measures, a second barrier must limit the consequences of the release in such a way that employees and third parties, as well as the neighborhood of the plant and the environment, cannot be threatened, or that such threat is held to an acceptable level.

The chemical industry uses a variety of means to limit the impacts of accidents. As a rule, these also have functions relating to occupational safety, pollution control, and water and soil conservation. In addition to retaining the product in liquid, solid, or gaseous form in case of an upset in the plant, they also provide for retention of contaminated wastewater or cooling water, and the retention of water contaminated by fire fighting.

The necessity of retention systems and the way in which they are implemented depends chiefly on the properties of the substances being handled, or those produced in the plant during an accident, and on the actions of such substances on persons and the environment – not just in terms of the immediate effect of exposure, but also the effects produced by a reaction (fire, explosion) of such substances. The hazard potential of a substance is assessed on the basis of its physical, chemical, and biological properties. The water-polluting effect of a substance is generally evaluated from its assignment to a pollutant class [232], [233], based on the acute toxicity of the substance to mammals, bacteria, and fish (sometimes to algae and Daphnia), its degradability, long-term effects, and distribution.

The principal ways to limit the impacts of potential accidents are:

- Retention systems, catch wells
- Quick-closing valves, emergency compartmentalization systems
- Emergency drain and collection systems
- Blowoff and disposal systems
- Spray-curtain systems
- Partial and complete containment
- Fire protection
- Retention of water contaminated by fire fighting

The choice of the means to limit accident consequences must be based on the operating conditions (e.g., pressure, temperature, substance properties), the type of plant, its location, and the situation in its vicinity.

5.1.1. Retention Systems, Catch Wells

The best-known retention system for liquids is the catch pan or well. In Germany, the Engineering Code for Flammable Liquids (TRbF) prescribes this way of retaining any liquid that leaks from tanks located in the well [234], [235].

Catch wells not only prevent harm to soils and receiving waters by flammable liquids, but also limit the spread of fires and emissions. If the liquid surface is completely covered (e.g., with a low-expansion fire-fighting foam), fires and emissions can be almost entirely suppressed.

In addition to product retention, a catch pan or well can also hold back contaminated wastewater and water contaminated by fire fighting.

Special requirements apply to the design and selection of material of chemically resistant, leakproof retention systems such as catch wells, pans, and baseplates. These standards are chiefly dictated by the stresses occurring, for example:

- Mechanical stress due to loads, impacts, friction, building subsidence, internal stresses, and contraction
- Thermal stresses due to ambient temperature cycles (day – night, summer – winter) or hot media
- Chemical stresses dependent on water pollutant class, intermittent attack, penetration, or aggressive properties of retained substances relative to construction materials

The resistance to chemical attack needed in a retention system, and hence the material chosen, generally depend on the water pollutant classification and chemical nature of the substances present. Chemical stresses are classed by severity:

1) Low load: no water pollutants present in normal operation; retention of water contaminated by fire fighting
2) Moderate load: leaks with action limited in time (nonrecurring or recurring but brief)
3) High load: intermittent contact with water-polluting liquids; regular contact with heavily laden cleaning water

The duration of exposure must be determined for each component. If the loads are low or moderate, the design can be based on a fairly low-probability event, occurring once and of limited duration. Infrastructural administrative or technical measures, e.g., provision for prompt detection and remedy of leaks, may also be taken [236–239].

If required by the load assessment, retention systems can be rendered tight and chemically resistant by special coatings (resins, paints, etc.) or plastic or metal foil liners.

A large industrial plant can set up a further barrier for retention of chemicals and contaminated fire-fighting water by connecting production equipment or storage tanks to biosewers leading to the plant's wastewater treatment system.

A special type of retention system involves double-wall construction of vessels, pipes, etc. The retention space is attached to the containing barrier (in the case of internally jacketed vessels, placed inside it) in such a way that the space between the two walls catches any leaks and can be fitted with an alarm [239–242].

Double-wall design is not always safer than single-wall construction. The reasons have to do with fabrication cost and additional stresses (e.g., those due to constrained expansion at penetrations and supports) as well as stability against denting and buckling. Inspection and maintenance of a double-wall component is difficult or impossible.

5.1.2. Rapid-Closing Valves, Emergency Compartmentalization Systems

In plants containing large amounts of products where the release of a substance would represent a high hazard potential, it may be useful to subdivide the plant with rapid-closing valves and fittings. These can be actuated after the detection of a release, e.g., by a gas warning system [243], [244], and after the scope of the release has been evaluated. They can be activated manually or automatically. The effect is to isolate the area of the leak from the rest of the facility. The quantity released at a leak can thus be limited to a value estimated in advance. The use of a rapid-closing valve system presupposes detailed advance planning and design of the plant sections, knowledge of the amounts of products contained, the way in which a leak is detected, etc. The analysis may also include the emergency shut-down of the entire plant if dangerous conditions in the plant can be created by blocking individual sections [245], [246].

5.1.3. Emergency Drain and Collection Systems

The product in a plant or in an isolated portion of the plant may be drained or released into a retention system (e.g., a catch tank or blowdown tank permanently installed in the plant) if a leak or other hazardous condition arises (e.g., fire). The retention system may be fitted with a separator (e.g., a cyclone) if the properties of the substance and the system pressure demand separation of the phases of the substance. It is also useful to configure the system in such a way that the plant drains by gravity. A pressure-relief system may also be needed to reduce the pressure of the gas phase [247], [248].

5.1.4. Blowoff and Disposal Systems

In plants handling gases, flares, scrubbers, absorbers, and condensers are useful for pressure relief and gas disposal. Systems of this type function not only as emergency pressure-relief systems, but are also used for the disposal of gases and vapors arising during normal operation, or when the plant is up or shut down, e.g., where large amounts of low-quality gas are produced for a short time. If these gases or vapors are flammable, the plant design must also meet the safety standards for explosion protection; other safeguards against spontaneous ignition and fire may also be necessary, e.g., with absorption on activated- carbon filters [249], [250].

5.1.5. Spray-Curtain (Drench) Systems

Spray curtains are often employed in plants with critical gases or vapors, where a release at an unpredictable location cannot be ruled out and may lead to a hazard to employees, third parties, or the environment. The plant section in question, or the whole plant, is surrounded by a pipe system fitted with closely spaced spray nozzles. If a pollutant escapes, a high-pressure mixture of water and steam issues from the nozzles, surrounding the plant in a dense water mist that can condense or capture the released substances. Neutralizing agents can also be added to the water; e.g., ammonia for phosgene plants. Due to the high-pressure and thermodynamic effects the effectiveness of these systems can be maintained, even in windy weather.

Mobile equipment such as articulated booms and water cannons operated by the plant fire fighting service can be used to create spray curtains in smaller plants, or as a supplementary measure in larger ones [251].

5.1.6. Partial and Complete Containment

A technique well known from the nuclear industry is to place the entire plant inside a pressure-tight containment, enclosed on all sides, but with personnel entry points. This approach has also been discussed for chemical plants and portions of plants where very dangerous substances are handled.

This approach may be correct from the environmental protection standpoint, but in practice there are limits, since some chemical plants cannot be remotely controlled and cannot be run without operating personnel.

For reasons of occupational safety and health, a chemical plant containment cannot be hermetically sealed from the outside world while the plant is in normal operation. Entry for operation and maintenance must be possible, and continuous ventilation is required, both for waste heat removal and on occupational health grounds. A release would have to be detected very promptly in order for all openings to be shut and the stack-gas cleanup system turned on so that retention could be achieved.

A further problem arises with flammable substances. Even a small cloud of an explosive mixture released because of a malfunction is diluted much more slowly in the containment than in an open plant, so it remains flammable over a long time. Containment also exacerbates the effect of an explosion; because there are no pressure-relief openings, the pressure wave and heat of an explosion affect humans and structures many times more than in the open case.

A large fire in a containment plant inevitably causes a huge amount of heat to accumulate in the building. Thermal stresses on the structure and plant equipment result, possibly causing their complete failure. Mobile fire fighting is far more difficult in such a facility.

It is possible that an initially minor event in a containment plant may lead to substantial personal injury and a dangerous release of substances.

A variety of plant subdivisions, where special process or design conditions bring about a severe hazard potential for the environment, are already being built with double walls or enclosed in boxes or compartments. Aside from this form of enclosure and double-walled construction, whole-plant containment cannot be recommended for general use. Complete containment is an option only in rare cases, e.g., where there are no flammable or explosive substances; even in such cases it is necessary to strike a careful balance between advantages and disadvantages (pollution control, industrial safety, fire and explosion protection, plant reliability, and safe maintenance practices) [251–253].

5.1.7. Fire Protection; Retention of Water Contaminated by Fire Fighting

Effects on employees, third parties, and the environment must also be expected when fire occurs in the plant. These can result from the fire itself (heat, combustion gases, etc.), or from the extinguishing agents (e.g., the Sandoz accident at Basel). Fire protection legislation requires that the fire fighting service seek to extinguish a fire virtually always and under all circumstances. From the environmental protection standpoint, it may be preferable not to extinguish a fire that has already passed a certain size, but to let it burn at a high temperature while keeping it from spreading to the surroundings; the amount of environmentally harmful combustion gases generated is smaller, and no contaminated extinguishing agent is produced.

There must, however, always be fire protection measures to prevent the start and growth of a fire, or to safeguard the surroundings from the effects of fire (Section 3.4.1). Fixed facilities for the retention of water contaminated by fire fighting must be provided, depending on the substances involved, their water pollution classifications, the construction of the first barrier, and other factors [254].

Mobile retention systems can also be used as a barrier. A number of options, e.g., mobile folding vessels and rapid-set-up tanks, are now on the market. Even simple devices such as magnetic plates and compressed-air bags to close off sewer intakes, have their place in contaminated fire-fighting water retention.

5.2. Hazard Control Plans at Plant Level and Beyond

5.2.1. Hazard Control Plans

In what follows, plans established by a plant or works to control hazards inside the battery limits are referred to as plant alarm and hazard control plans. In Europe, plants subject to the "Seveso Directive" (implemented in Germany as the Industrial Emergencies Regulation) are required to create an alarm and hazard control plan [255], [256].

An alarm and hazard control plan describes administrative and technical actions to minimize the impacts of accidents; relates to various levels of organization (e.g., constituent plants and the larger facility to which they belong) and is prepared with an eye to the immediate surroundings of each unit.

The actions prescribed include:

Management
Personnel
Technical resources
Communications
Documentation
Procedures and jurisdictions

In facilities containing a large number of varied units, alarm and hazard control plans are prepared at "plant" (smaller unit) and "works" (larger unit) levels [257].

5.2.1.1. Plant Hazard Control Plan

The plant alarm and hazard control plan sets forth the plant safety organization. It contains information necessary to plan and specify the emergency response.

The coordinated administrative measures taken in response to an accident are such as to afford the highest possible level of protection to humans and facilities.

The hazard control plan identifies all potential threats within the plant (e.g., those due to fire, explosion, or other accident) and external to the plant (e.g., trouble in nearby plants, flooding), lists safety equipment in the plant, and prescribes the desired actions of plant personnel in a dangerous situation.

The plant hazard control plan is aimed primarily at the people in charge (e.g., plant managers and engineers, laboratory and workshop managers), who must ensure that their personnel are properly instructed as to how to act in a particular hazard situation. The plan is a document for use in plant personnel instruction and drills, as provided in the Industrial Emergencies Regulation. The plant hazard control plan serves as a guide for people in charge, giving instructions for actions in a particular hazard situation. It lays down guidelines both for measure within the plant and for measures by technical task forces and service groups (e.g., fire safety, occupational safety, plant security, maintenance). The plan is also an informational aid and guide to emergency action for management, and provides a pattern for the allocation of technical resources.

5.2.1.2. Works Hazard Control Plan

The alarm and hazard control plan of the works relates to a higher organizational level than the plant, and documents the emergency response organization adopted by the works when an event occurs.

It defines, for all personnel involved:

The essential tasks to be performed
The way in which efforts are coordinated

It also establishes duties relating to information and prescribes the forms of cooperation with:

External organizations
Government agencies

5.2.1.3. Required Contents of the Plan

Many suggestions have appeared concerning what alarm and hazard control plans should contain [258], [259]. In the German state of North Rhine – Westphalia, a working group, including independent experts as well as agency and industry representatives, has developed a model plant alarm and hazard control plan [257], which can serve as a pattern for plant or works plans. Major headings in the model plan are:

Introduction. Mailing address, scope, revision (update) status

General Information on the Facility and its Surroundings. Orientation for emergency workers on:

1) Access routes and staging areas
2) Information on the work force
3) Danger sources, major hazards
4) Safety equipment and plans in place

Alarm Plan. The routing of alarms in the event of an accident is summarized, as is mandatory internal and external reporting.

Hazard Control by Internal Activities. This lists departments in the works (works manager, plant, works fire service, central safety office, industrial medicine, safety engineer, etc.) and defines their jurisdictions and assignments.

Warnings. Options for informing persons in the works or plant about possible hazards are discussed, along with ways of delivering warnings and other information to those living nearby.

Instructions for Particular Events. In view of local circumstances, precautions and responses are described for a variety of events (flood, smog, power outage, product release, need for outside assistance, etc.).

Notification of Government Agencies and the Public. This describes the internal location, where all information required to be reported externally can be retrieved.

Resources and Technical Specialists. A table lists all current technical documents, devices, equipment, and technical specialists, and records their availability and contact information.

Telephone List. Telephone numbers that may be vital for government agencies (for inquiries within the works) and for responsible works management personnel (e.g., for reports to agencies) must be listed here.

Definitions, Regulations, Conventions. This lists concepts and procedures defined in regulatory documents; key passages may be quoted verbatim.

Appendixes. These include maps, keyword index, and other items.

5.2.2. Off-Battery Hazard Control Plans

An example of an off-battery hazard control plan is the special accident response plan, created jointly by the disaster control agency and the plant operator. Such a plan relates to a plant or works in which, because of the production process, storage, and other features, hazards to nearby inhabitants may occur and disasters cannot be ruled out. The special accident response plan repeats portions of the works alarm and hazard control plan that are vital for response by the agency; these include:

1) Description of the facility
2) Routing of reports and alarms
3) Immediate response to accident/disaster
   Identification of threat
   Identification of threatened area
   Measurement of pollutant levels
   Warning/notification of populace
4) Subsequent actions

The disaster response plan of a municipality covers all events that may figure in a disaster. It is not restricted to any one plant or works.

5.3. Public Awareness and Responsible Care

Responsible Care. The chemical industry's worldwide responsible care program obliges companies to strive for steady improvements in health, safety, and environmental protection. It mandates that corporations seek a stronger dialog with the public, to enhance public knowledge of product and manufacturing safety.

The centerpiece of an active environmental philosophy, as the Association of the German Chemical Industry made binding on its member companies in 1992, consists of ten Guiding Principles and six Management Practice Codes intended to implement them. CEFIC (the European Chemical Industry Council) has published similar guiding principles for member societies [260].

Member companies are bound by the following guiding principles:

1) To acknowledge the public interest in chemical products and the activities of manufacturers, and to respect this interest
2) To develop and manufacture only those products that can be safely produced, transported, used, and disposed of
3) In the planning of new products and production processes and the improvement of existing products and processes, to assign a high priority to health, safety, and environmental aspects
4) In any chemical-related health and environmental activity, to provide full information to government agencies, employees, customers, and the public, and to recommend suitable protective practices
5) To advise customers on the safe use and transportation of chemical products, as well as their safe disposal
6) To operate production facilities so as to safeguard the environment as well as the health and safety of employees and the public

7) To conduct research to increase knowledge about the possible health, safety, and environmental impacts of products, production processes, and waste products
8) To cooperate with other companies in solving problems occasioned in the past by the handling and disposal of hazardous substances
9) To cooperate with the government and government agencies in devising responsible legislation, procedures, and standards that enhance the safety and protection of the public, workers, and the environment
10) To promote the principles and practices of the responsible care program by sharing experience with all the other companies manufacturing, using, transporting, and disposing of chemical products, offering all kinds of support to them

The six Management Practice Codes deal with:

Information and emergency response
Process and plant safety
Environmental protection
Transportation safety
Occupational health and safety
Product responsibility

CEFIC is promoting and developing the responsible care concept throughout Europe.

Public Awareness. Article 13 of the Seveso Directive for Europe in general, and § 11a of the Industrial Emergencies Regulation in Germany, mandates that the operators of accident-relevant plants provide information on safety practices and proper accident response to persons who may be affected by an accident as well as to the public at large. This information must be made available in suitable form, and without having to be solicited [255].

The objective is to create a bulletin, understandable to the general public, containing the following information:

- Name of plant operator and location of plant
- Name and title of the person giving information
- Brief description of the nature and purpose of the plant
- Identification of substances or preparations that can cause an accident, along with their essential hazardous attributes
- Nature of dangers in an accident, including possible impacts on humans and the environment
- Nature of warning given to affected persons and follow-up
- Correct behavior and actions of affected persons in the event of an accident
- Pertinent safety practices
- Internal and external hazard control plans
- Coordination of plans between plant operator, municipality, and agencies responsible for hazard control

In areas of high industrial concentration, a joint bulletin may be issued by the operators of plants where emergencies may occur.

6. References

General References

1. "Das Sicherheitskonzept für die chemische Technik," *Dechema Monogr.* **88** (1980) no. 1818 – 1835. BASF (eds.): *Sicherheit in der Chemie,* Verlag Wissenschaft und Politik, Berend von Nottbeck, Köln 1979. Bayer AG (eds.): *Sichere Chemietechnik,* Leverkusen 1982. F. P. Lees: *Loss Prevention in the Process Industries,* **vols. 1 and 2,** Butterworths, London 1980. G. L. Wells: *Safety in Process Plant Design,* J. Wiley & Sons, New York – Toronto 1980. S. Lange: *Ermittlung und Bewertung industrieller Risiken,* Springer Verlag, Berlin – Heidelberg 1984. "Fortschritte der Sicherheitstechnik I," *Dechema Monogr.* **107** (1987). D. A. Crowl, J. F. Louvar: *Chemical Process Safety: Fundamentals with Applications,* Prentice Hall Inc., New Jersey 1990. J. Whiston: *Safety in Chemical Production,* Blackwell Sci. Publ., Oxford 1991. H. Pohle: *Chemische Industrie – Umweltschutz, Arbeitsschutz, Anlagensicherheit,* VCH, Weinheim 1991. G. Strohrmann: *Anlagensicherung mit Mitteln der MSR-Technik,* 2nd ed., R. Oldenburg Verlag, München – Wien 1983. DECHEMA (ed.): "Anlagensicherung mit Mitteln der MSR-Technik," *Praxis der Sicherheitstechnik,* **vol. 1,** Frankfurt/Main 1988. NAMUR (eds.): *NAMUR-Empfehlung: Anlagensicherung mit Mitteln der Prozeßleittechnik,* Erstausgabe 30. 10. 92, Vertrieb: NAMUR-Geschäftsstelle. Berufsgenossenschaft der chemischen

Industrie (BG Chemie) (eds.): *Ratgeber Anlagensicherheit; Grundlagen und Anwendungshilfen zur Anlagensicherheit,* Verlag Kluge, Berlin 1992.

Specific References

2. Berufsgenossenschaft der chemischen Industrie (BG Chemie), Heidelberg, annual reports 1983 – 1992.
3. Her Majesty's Stationery Office (HMSO), *The Flixborough Disaster – Report of the Court of Inquiry,* London 1975.
4. V. C. Marschall: *Major Chemical Hazards (Ellis Horwood series in chemical engineering),* Ellis Horwood, Chichester 1987.
5. H. Steen et al.: *Das Bhopal-Unglück 1984,* Umweltbundesamt (eds.), Berlin 1987.
6. Bundesminister für Umwelt, Naturschutz und Reaktorsicherheit (eds.): "Rhein-Bericht, Bericht der Bundesregierung über die Verunreinigung des Rheins durch die Brandkatastrophe bei der Sandoz AG/Basel," Bonn, Feb. 12, 1987.
7. DIN/VDE 31 100,part 2 (Deutsche Norm), Begriffe der Sicherheitstechnik, Beuth Verlag, Berlin 1987.
8. H. Schacke et al.: "Redundanz im Staubexplosionsschutz? – Konzept komplementärer Schutzmaßnahmen," *Staub Reinhalt. Luft* **53** (1993) 453 – 459.
9. Statistisches Bundesamt (eds.): *Statistisches Jahrbuch 1993 für die Bundesrepublik Deutschland,* Wiesbaden 1993.
10. T. A. Jäger: "Zur Sicherheitsproblematik technologischer Entwicklungen," *Qualität Zuverlässigkeit* **19** (1974) no. 1, 2 – 9.
11. Polizeiliche Kriminalstatistik des Bundeskriminalamtes, Wiesbaden (für die Jahre 1985 bis 1991).
12. Amtliche Mitteilungen der Bundesanstalt für Arbeitsschutz, no. 2, Dortmund, April 1993.
13. V. Pilz: "Grundlagen für die Vorhersage der Auswirkungen von Störfällen," *VFDB 2.* (1981) no. 3, 116 – 125.
14. H. Giesbrecht et al.: "Analyse der potentiellen Explosionswirkung von kurzzeitig in die Atmosphäre freigesetzten Brenngasmengen," *Chem. Ing. Tech.* **52** (1980) no. 2, 114 – 122 (part 1); **53** (1981) no. 1, 1 – 10 (part 2).
15. W. C. Brasie, D. W. Simpson: "Guidelines for Estimating Damage Explosion," Reprint 21 A, Symp. Loss Prevention Process Ind., *63rd Nat. Meeting AIChE,* St. Louis, Mo, 1968.
16. W. E. Baker et al.: *A Short Course on Explosion Hazards Evaluation,* Southwest Research Institute, San Antonio, Tx, 1976.
17. W. E. Baker: *Explosions in Air,* Wilfried Baker Engineering, San Antonio 1973, (2nd Printing 1983).
18. R. A. Strehlow, W. E. Baker: *The Characterization and Evaluation of Accidental Explosions,* NASA CR 134 779, Urbana 1975.
19. VDI-Richtlinie, VDI 3783, Blatt 1, Dispersion of Emissions by Accidental Releases, Beuth Verlag, Berlin 1987.
20. Directive 92/32/EEC, ABIEG Nr. L 154 vom 5. 6. 1992, 7. Amendment to Framework Directive 67/548/EEC.
21. Deutsche Forschungsgemeinschaft, Senatskommission zur Prüfung gesundheitsschädlicher Arbeitsstoffe, 31. Mitteilung, Sep. 1, 1995.
22. Verordnung zum Schutz vor gefährlichen Stoffen (Gefahrstoffverordnung), Sep. 19, 1994, BGBl. I, p. 2557.
23. American Conference of Governmental Industrial Hygienists, Threshold Limit Values for Chemical Substances and Physical Agents and Biological Exposure Indices, ACGIH Technical Information Office, Cincinatti 1995.
24. Framework Directive 80/1107/EEC, Amendment Directive 88/642/EEC, Directive for Occupation Exposure Levels: 91/322/EWG.
25. Technische Regel für Gefahrstoffe (TRGs) 102, "Technische Richtkonzentration für gefährliche Stoffe," BArbBl. (1994) no. 2, p. 71.
26. Arbeitsplatzrichtwerte, BArbBl. no. 3/1991, p. 69;
27. Technische Regel für Gefahrstoffe (TRGS) 900, "Grenzwerte," BArbBl (1995) no. 4, p. 47.
28. H. Bender: *Sicherer Umgang mit Gefahrstoffen,* VCH Verlagsgesellschaft, Weinheim 1995.
29. H. H. Freytag: *Handbuch der Raumexplosionen,* Verlag Chemie, Weinheim 1965.
30. United Nations: *Recommendations on the Transport of Dangerous Goods,* 8th rev. ed., New York 1993.
31. VbF, Verordnung über Anlagen zur Lagerung, Abfüllung und Beförderung brennbarer Flüssigkeiten zu Lande mit Technischen Regeln für brennbare Flüssigkeiten (TRbF), C. Heymanns Verlag, Köln.
32. K. Nabert, G. Schön: *Sicherheitstechnische Kennzahlen brennbarer Gase und Dämpfe,*

2nd rev. ed., Deutscher Eichverlag, Braunschweig 1980.
33. VDI-Kommission Reinhaltung der Luft, VDI 2263, Blatt 1, Untersuchungsmethoden zur Ermittlung von sicherheitstechnischen Kenngrößen von Stäuben, Düsseldorf 1990.
34. Richtlinie 84/449/EWG, Prüfmethoden für physikalisch chemische Eigenschaften, Anhang A, veröffentlicht im Amtsblatt der Europäischen Gemeinschaft L 251/89.
35. J. Troitzsch: *International Plastics Flammability Handbook,* 2nd ed., Hanser Verlag, München 1990.
36. W. Jost: *Explosions- und Verbrennungsvorgänge in Gasen,* Springer Verlag, Berlin 1939.
37. B. Lewis, G. v. Elbe: *Combustion, Flames and Explosions of Gases,* 2nd ed., Academic Press, New York 1961.
38. H. H. Freytag: *Handbuch der Raumexplosionen,* Verlag Chemie, Weinheim 1965.
39. D. A. Franck-Kamenetzkii: *Stoff- und Wärmeübergang in der chemischen Kinetik,* Springer Verlag, Berlin 1959.
40. W. Bartknecht: *Explosionsschutz – Grundlagen und Anwendung,* Springer Verlag, Berlin 1993.
41. F. Funk: "Berechnung sicherheitstechnischer Kennzahlen," *Chem. Tech.* (Leipzig) 29 (1977) no. 9, 494–497.
42. K. Nabert, G. Schön: *Sicherheitstechnische Kennzahlen brennbarer Gase und Dämpfe,* Deutscher Eichverlag, Braunschweig 1980/89 mit Nachträgen.
43. H. F. Coward, G. W. Jones: "Limits of Inflammability of Gases and Vapors," *Bull. U.S. Bur. Mines* **279** (1952) no. 503.
44. DIN 51 649,part 1, Bestimmung der Explosionsgrenzen von Gasen und Gasgemischen in Luft, Beuth Verlag, Berlin 1986. pr EN 1839, Determination of Explosion Limits of Gases, Vapours and their Mixtures.
45. H. Le Chatelier, *Ann. Mines.* **(Ser. 8) 19** (1891) 388–395.
46. W. Berthold, U. Löffler: *Lexikon Sicherheitstechnischer Begriffe in der Chemie,* Verlag Chemie, Weinheim 1981.
47. D. Conrad: "Inertisierung explosionsfähiger Gassysteme als Maßnahme des primären Explosionsschutzes," *2. Int. Kolloquium der IVSS,* Frankfurt 1973.
48. Hauptverband der gewerblichen Berufsgenossenschaften, *Explosionsschutz-Richtlinien (EX-RL),* C. Heymanns Verlag, Köln 1990.
49. M. Müller, H. Schacke, C.-D. Walther: Explosion Characteristics of Carbon Monoxide at High Temperatures, Loss Prevention and Safety Promotion in the Process Industries, *Proc. Int. Symp. 6th,* Oslo, June 1989, vol IV, p. 104/104-12, European Federation of Chemical Engineering, 1989.
50. D. Conrad, R. Kaulbars: "Druckabhängigkeit der Explosionsgrenzen von Wasserstoff," *Chem. Ing. Tech.* **67** (1995) no. 2, 185 – 188.
51. H.-J. Heinrich: "Bemessung von Druckentlastungsöffnungen zum Schutz explosionsgefährdeter Anlagen in der chemischen Industrie," *Chem. Ing. Tech.* **38** (1966) 1125 – 1133.
52. H.-J. Heinrich: "Zur Bemessung von Druckentlastungsöffnungen bei Gas- und Staubexplosionen," Wissenschaftliche Berichte aus der Bundesanstalt für Materialprüfung (BAM), Berlin 1973, pp. 277–280.
53. S. Crescitelli, G. Russo, V. Tufano, *J. Occup. Accid.* **2** (1979) 125 – 133.
54. W. Jost: *Probleme der Flammenfortpflanzung,* report no. 226, Deutsche Versuchsanstalt für Luft- und Raumfahrt e.V., Oberpfaffenhofen 1962, pp. 3 – 24.
55. J. Nagy et al.: "Explosion Development in Closed Vessels," *Rep. Invest. U.S. Bur. Mines* **7507** (1971) 1 – 50.
56. D. Conrad, S. Dietlen: "Untersuchungen zur Zerfallsfähigkeit von Distickstoffoxid," *BAM-Forschungsbericht,* no. 89, Verlag für neue Wissenschaft, Bremerhafen 1983.
57. H. R. Christen: *Thermodynamik und Kinetik chemischer Reaktionen,* Diesterweg u. Salle, Frankfurt 1974.
58. D. Conrad: "Vermeidung von Gefahren beim Umgang mit zerfallfähigen Gasen," *2. Sicherheitstechnische Vortragsveranstaltung über Fragen des Explosionsschutzes,* PTB Braunschweig, March 1983.
59. D. Conrad, R. Kaulbars: "Untersuchungen zur chemischen Instabilität von Äthylen," *Chem. Ing. Tech.* **47** (1975) 265.
60. D. Oberhagemann: "Zündtemperaturen von Ein- und Mehrkomponentensystemen," *VDI-Fortschrittber.* , Reihe 3, no. 185, VDI Verlag, Düsseldorf 1989.
61. P. Field: *Dust Explosions,* Elsevier Sci. Publ., Amsterdam 1982.
62. J. Cross, D. Farner: *Dust Explosions,* Plenum Press, New York 1982.

63. W. E. Baker et al.: *Explosion Hazards and Evaluation,* Elsevier Sci. Publ., Amsterdam 1983.
64. M. Glor: *Electrostatic Hazards in Powder Handling,* Research Studies Press, Letchworth, 1983.
65. R. K. Eckhoff: *Dust Explosions in the Process Industries,* Butterworth-Heinemann, Oxford 1991.
66. W. Bartknecht: *Explosionsschutz, Grundlagen und Anwendung,* Springer Verlag, Berlin 1993.
67. VDI-Guideline 2263, part 1, Test Methods for the Determination of the Safety Characteristics of Dusts, Beuth Verlag, Berlin 1990.
68. W. Berthold: "Bestimmung der Mindestzündenergie von Staub/Luft-Gemischen," *VDI Fortschrittber.* **134, Reihe 3,** VDI-Verlag, Düsseldorf 1987.
69. ISO 6184/1- 1985 (E) Explosion Protection Systems, part 1, Determination of Explosion Indices of Combustible Dusts in Air.
70. VDI-Guideline 3673, part 1, Pressure Release of Dust Explosions, Beuth Verlag, Berlin 1979 (Draft 1992).
71. ISO 6184/4- 1985 (E) Explosion Protection Systems, part 4, Determination of Explosion Suppression Systems.
72. J. Zehr: "Anleitungen zu den Berechnungen über die Zündgrenzwerte und die maximalen Explosionsdrücke," *VDI-Ber.* **19** (1975) 62.
73. J. Schönewald: "Vereinfachte Methode zur Berechnung der unteren Zündgrenze von Staub-Luft-Gemischen," *Staub Reinhalt. Luft* **31** (1971) no. 9, 376.
74. DIN 55 990,part 6, Pulverlacke, Berechnung der unteren Zündgrenze.
75. L. Bretherick: *Handbook of Reactive Chemical Hazards,* 4th ed., Butterworths, London 1990.
76. R. King: *Safety in the Process Industries,* Butterworths-Heinemann, London 1990.
77. T. Grewer, O. Klais: *Exotherme Zersetzung – Untersuchung der charakteristischen Stoffeigenschaften,* VDI-Verlag, Düsseldorf 1988.
78. W. H. Seaton, E. Freedman, D. N. Treweek: "CHETAH: The ASTM Chemical Thermodynamic and Energy Release Potential Evaluation Program," *ASTM Data Ser.* **DS 51** (1974).
79. M. A. Cook: *The Science of High Explosives,* Robert E. Krieger Publ. Corp., Huntington, New York, 1971.
80. J. Taylor: *Detonation in Condensed Explosives,* Clarendon Press, Oxford 1952.
81. K. K. Andreyev, A. F. Belyayev: *Theorie der Explosivstoffe* (Translation from Russian into German by Svenska Nationalkommittén för Mekanik, Sektionen för Detonik och Förbränning, Stockholm 1964).
82. T. M. Groothuizen, J. W. Hartgerink, H. J. Pasman in C. H. Buschmann (ed.): "Phenomenology, Test Methods and Case Histories of Explosions in Liquids and Solids," *First International Loss Prevention Symposium Hague/Delft, May 28 – 30, 1974,* Elsevier Sci. Publ. Corp., Amsterdam – London – New York 1974.
83. H. Koenen, K. H. Ide, K.-H. Swart: "Sicherheitstechnische Kenndaten explosionsfähiger Stoffe," *Explosivstoffe* **9** (1961) 4, 30, 195.
84. K.-H. Swart, P.-A. Wandrey, K. H. Ide, E. Haeuseler: "Sicherheitstechnische Kenndaten explosionsfähiger Stoffe," *Explosivstoffe* **13** (1965) 339.
85. Recommendations on the Transport of Dangerous Goods – Test and Criteria, 3rd ed., United Nations, New York 1995.
86. Recommendations on the Transport of Dangerous Goods, 9th ed., United Nations, New York 1995.
87. EG A 14 Explosionsgefahr, Richtlinie 92/69/EWG, Amtsblatt der Europäischen Gemeinschaft L 383 A, 12/95.
88. J. Barton, R. Rogers: Guide on Chemical Reaction Hazards, Institute of Chemical Engineers (IChmE), Rugby 1993.
89. ESCIS-Booklet 8, 1993: Thermal Process Safety Data, Assessment Criteria, Measures (available from SUVA, CH-6001 Lucerne).
90. N. N. Smenov, *Z. Phys. Chem.* **48** (1928).
91. D. A. Frank-Kamenetskii, *Dokl. Akad. Nauk SSSR* **18** (1938) 1.
92. P. C. Bowes: *Self-heating: Evaluating and Controlling the Hazard,* Her Majesty's Stationery Office (HMSO), London 1984.
93. N. Gibson, Conference on Chemical Reaction Hazards, London 1993.
94. O. Levenspiel, *Chemical Reaction Engineering,* J. Wiley & Sons, New York 1972.
95. K. R. Westerterp, W. P. M. van Swaaij, A. A. C. M. Beenackers: *Chemical Reactor Design and Operation,* J. Wiley & Sons, New York 1984.
96. J. Steinbach, Habilitationsschrift, TU Berlin 1994.
97. P. Hugo, *Chem. Ing. Tech.* **52** (1980) 712.

98. T. Grewer et al., *Dechema Monogr.* **111** (1987) 41 – 65.
99. J. Steinbach: *Chemische Sicherheitstechnik,* 1st ed., VCH Verlagsgesellschaft, Weinheim 1995.
100. J. Steinbach, *Dechema Monogr.* **107** (1986) 133 – 152.
101. D. I. Townsend, J. C. Tou, *Thermochim. Acta* **37** (1980) 1.
102. J. Wiss, G. Killé, F. Stoessel, *Chimia* **47** (1993) 417–423.
103. F. Stoessel, *Chem. Eng. Prog.* **89** (1993) no. 10, 68 – 75.
104. P. Hugo, J. Steinbach, F. Stoessel, *Chem. Eng. Sci.* **43** (1988) no. 8, 2147 – 2152.
105. DECHEMA: *Safety Standards and Regulations for Chemical Plant in Europe: A Comparison,* Frankfurt/Main 1988 (Praxis der Sicherheitstechnik, vol. 2).
106. V. Pilz: "Der Technische Ausschuß für Anlagensicherheit – Wegbereiter für ein ganzheitliches Regelwerk zur Sicherheitstechnik aus immissionsschutzrechtlicher Sicht," *UMWELT* (1992) no. 11, 415 – 416.
107. European Council Directive on the Major Accident Hazards of Certain Industrial Activities (the "Seveso Directive"), 82/501/EEC, June 24 1982, European Community, Brussels.
108. "EWG-VERTRAG, Grundlage der Europäischen Gemeinschaft," Europa-Union Verlag, Bonn 1988.
109. P. Marburger: "Die Regeln der Technik im Recht," C. Heymanns Verlag, Köln 1979.
110. J. M. Ale: "Risk Analysis and Risk Policy in the Netherlands and the EEC," *J. Loss Prev. Process Ind.* **4** (1991) 58 – 64.
111. T. A. Kletz: "Inherently Safer Plants: An Update," *Plant/Oper. Prog.* **10** (1991) no. 2, 81 – 84.
112. V. Pilz: "Integrierte Sicherheit bei verfahrenstechnischen Anlagen", *UMWELT* **19** (1989) no. 5, D 27 – D 30.
113. V. Pilz: "Bayer's Procedure for the Design and Operation of Safe Chemical Plants," paper presented at the *XIII World Congress on Occupational Safety and Health,* New Delhi, India, April 1993.
114. Bayer: "Directive Process and Plant Safety," 2nd ed., Leverkusen 1989.
115. BASF: *Sicherheitsbetrachtungen bei der Planung von Chemieanlagen,* 2nd ed., Ludwigshafen 1989.
116. Hoechst: *Richtlinie zur Sicherheitsüberprüfung von Verfahren und Produktionsanlagen (Überprüfungsrichtlinie),* Nov. 1989, Frankfurt.
117. "Einführung in die Risikoanalyse," *Schriftenreihe der Expertenkommission für Sicherheit in der chemischen Industrie der Schweiz (ESCIS),* 2nd ed., **no. 4,** Basel 1986.
118. "Behelf für die Durchführung von Sicherheits-überprüfungen (Safety Audits)," *Schriftenreihe der Expertenkommission für Sicherheit in der chemischen Industrie der Schweiz (ESCIS),* no. 9, Basel 1991.
119. K. A. Ruppert: "Sicherheitsanalytische Vorgehensweise für Alt- und Neuanlagen," *Chem. Ing. Tech.* **62** (1990) no. 11, 916 – 927.
120. Lurgi: *Sicherheitsleitfaden,* Frankfurt, Nov. 1990.
121. A. W. Bickum, A. E. Barnette: "Common Documentation Systems for ISO 9000 and Process Safety Management," *Preprints Int. Process Safety Management Conference and Workshop (AIChE-CCPS),* San Francisco, Sept. 1993.
122. H. W. Adams: "Erhöhung der Sicherheit durch Qualitätssicherung bei Planung, Bau und Betrieb von chemischen Produktionsanlagen," *Forschungsbericht 10 409 221,* Umweltbundesamt, Texte 18/93.
123. V. Pilz: "Safety Analyses for the Systematic Checking of Chemical Plant and Processes – Methods, Benefit and Limitations," *Ger. Chem. Eng. (Engl. Transl.)* **9** (1986) 65 – 74; *Chem. Ing. Tech.* **57** (1985) no. 4, 289 – 307.
124. "Guidelines for Hazard Evaluation Procedures," 2nd ed., Center for Chemical Process Safety (CCPS) of AIChE, New York 1992.
125. G. L. Wells: *Safety in Process Plant Design,* J. Wiley & Sons, New York 1980.
126. F. P. Lees: *Loss Prevention in the Process Industries,* **vol. 1,** Butterworths, London 1980, chaps. 8 and 9.
127. K. Bartels et al.: *Risikobegrenzung in der Chemie, PAAG-Verfahren (HAZOP),* BG Chemie/IVSS, Heidelberg 1990.
128. H. G. Lawley: "Operability Studies and Hazard Analysis," *Chem. Eng. Progr.* **70** (1974) no. 4, 45 –56.
129. DIN 25 419,Ereignisablaufanalyse, Beuth Verlag, Berlin, Nov. 1985.
130. DIN 25 424,Fehlerbaumanalyse, Beuth Verlag, Berlin, Sep. 1981.

131. V. Pilz: "Risk Analysis for Chemical Production Processes? – Some Remarks on Meaningful Applications of Available Methods and Limitations Thereof," *Angew. Systemanalyse* **2** (1981) no. 4, 175 – 178.
132. H. Dembeck: *Chemie-ABC für Feuerwehr- und Sicherheitsfachkräfte,* Kohlhammer Deutscher Gemeindeverlag, Stuttgart 1981.
133. A. Rempe, G. Rodewald: *Brandlehre,* Kohlhammer Deutscher Gemeindeverlag, Stuttgart 1988.
134. J. Troitzsch, *Brandverhalten von Kunststoffen,* Hanser Verlag, München 1982.
135. S. Bussenius, *Brand- und Explosionsschutz in der Industrie,* Staatsverlag der DDR, Berlin 1982.
136. F. Kaufhold, A. Rempe: *Feuerlöschmittel,* Kohlhammer Deutscher Gemeindeverlag, Stuttgart 1990.
137. *Vereinigung zur Förderung des deutschen Brandschutzes, VFDB* **2** (1986).
138. W. M. Predtetschenski, A. I. Milinski: *Personenströme in Gebäuden,* Verlagsgesellschaft R. Müller, Köln 1971.
139. K. Klingsohr, *Vorbeugender baulicher Brandschutz,* Kohlhammer Deutscher Gemeindeverlag, Stuttgart 1986.
140. H. P. Morgan, *Rauchschutzmethoden in ein- oder mehrgeschossigen, geschlossenen Einkaufszentren: Eine Entwurfsübersicht,* Colt, Kleve 1988.
141. Richtlinien des Verbandes der Sachversicherer, Köln, zu: Brandmeldeanlagen; Rauch- und Wärmeabzugsanlagen; Sprinkleranlagen; CO2-Feuerlöschanlagen.
142. Fire Resistance Tests, European standards EN 1363 – EN 1366, draft 1994.
143. Verband der Sachversicherer: *VdS-Fachtagung 1990: Lagerung von gefährlichen Stoffen,* Köln 1990.
144. Seminarberichte des Instituts für Baustoffe, Massivbau und Brandschutz, Brandschutz im Industriebau bzw. Brandschutz – Forschung und Praxis, Braunschweig 1987 – 1992.
145. J. Steinmetz, E. Merz: *Umweltschutz und Gefahrenabwehr als betriebliches Gesamtkonzept,* Boorberg Verlag, Stuttgart 1992.
146. H. H. Freytag: *Handbuch der Raumexplosion,* Verlag Chemie, Weinheim 1965.
147. H. Schacke: "Fact-finding and Basic Data/Fire and Explosion, Safety in Chemical Production," *Proceedings of the First IUPAC Workshop on Safety in Chemical Production,* Blackwell Scientific Publications, Oxford 1991, pp. 21 – 26.
148. Berufgenossenschaft der chemischen Industrie: *Explosionsschutz-Richtlinien (EX-RL),* issue 9/94, Druckerei Winter, Heidelberg 1994.
149. prEN 1127-1 Safety of Machinery, Fire and Explosions, Part 1, Explosion Prevention and Protection/Draft Standard CEN TC 114 WG 16, Apr. 1995.
150. NFPA 497 A-1986 Standard-National Electrical Code (USA).
151. prEN 50 145 Electrical Apparatus for Potentially Explosive Gas Atmospheres/Classification of Hazardous Areas, draft, March 1993.IEC-SC 31 J Revision of publication 79-10 transforming the publication in IEC Standard: Electrical apparatus for explosive atmospheres/Area classification and installation requirements/Classification of hazardous areas, to be issued as a DIS, 1995.
152. A. W. Cox, F. P. Lees, M. I. Ang: *Classification of Hazardous Locations,* Institution of Chemical Engineers, Rugby, UK 1990.
153. W. Bartknecht: *Explosionsschutz – Grundlagen und Anwendung,* Springer Verlag, Heidelberg 1993.
154. K. Ritter: "Mechanisch erzeugte Funken als Zündquellen," *VDI-Ber.* **494** (1984) 129 – 144.
155. DIN VDE 0165/02.91, Errichten elektrischer Anlagen in explosionsgefährdeten Bereichen, Beuth Verlag, Berlin 1991.
156. G. Lüttgens, M. Glor: *Elektrostatische Aufladungen begreifen und sicher beherrschen,* expert Verlag, Ehningen 1993. G. Lüttgens, M. Glor: *Understanding and Controlling Static Electricity,* expert Verlag, Ehningen 1989.
157. Berufgenossenschaft der Chemischen Industrie: Richtlinien für die Vermeidung von Zündgefahren infolge elektrostatischer Aufladungen, issue 10/89, ZH 1/200, C. Heymanns Verlag, Köln 1989.
158. R. Viard, H. Schacke, C.-D. Walther: "Structurized Explosion Prevention and Protection – An Adequate and Economical Way to Satisfy Safety Requirements," *Loss. Prev. Saf. Promot. Process Ind. Int. Symp.* **8th** (1995) 527 – 538.
159. Berufsgenossenschaft der chemischen Industrie: Unfallverhütungsvorschrift Gase (VBG 61), Jedermann-Verlag Dr. O. Pfeffer, Heidelberg 1995.
160. VDI 3673, Pressure Venting of Dust Explosions, part 1, Beuth Verlag, Berlin 1995.

161. Safety of Machinery – Industrial Trucks – Operation in Potentially Explosive Atmospheres, CEN TC 150 WG 7, draft standard, May 1994.
162. Directive 94/9/EC of the European Parliament and the Council.
163. Draft Proposal for a Directive Concerning the Minimum Requirements for Improving the Safety and Health Protection of Workers Potentially at Risk from Explosive Atmospheres, EU-Commission, Doc. No. 427/4/93.
164. Directives 89/392/EEC and 91/368/EEC of the European Parliament and the Council.
165. Directives 76/117/EEC, 79/196/EEC, 82/130/EEC, and 90/487/EEC of the European Parliament and the Council.
166. NAMUR (eds.): NAMUR-Empfehlung "Anlagensicherung mit Mitteln der Prozeßleittechnik," NE 31, NAMUR-Geschäftsstelle Bayer AG, Leverkusen, Jan. 1993.
167. VDI/VDE-Richtlinie 2180, Sicherung von Anlagen der Verfahrenstechnik mit Mitteln der Meß-, Steuerungs- und Regelungstechnik, "Einführung, Begriffe, Erklärungen," Blatt 1, April 86; "Berechnungsmethoden für Zuverlässigkeitskenngrößen von Sicherungseinrichtungen," Blatt 2, April 86; "Klassifizierung von Meß-, Steuerungs- und Regelungseinrichtungen," Blatt 3, Dec. 84; "Ausführung und Prüfung von Schutzeinrichtungen," Blatt 4, July 88; "Bauliche und installationstechnische Maßnahmen zur Funktionssicherung von Meß-, Steuerungs- und Regelungseinrichtungen in Ausnahmezuständen," Blatt 5, Dec. 84.
168. DIN V 19 250,Grundlegende Sicherheitsbetrachtungen für MSR-Schutzeinrichtungen, Mai 1994.
169. DECHEMA: "Praxis der Sicherheitstechnik, ein Leitfaden für Planung, Bau und Betrieb chemischer Produktionsanlagen," *Anlagensicherung mit Mitteln der MSR-Technik,* vol. 1, Frankfurt/Main 1988.
170. DIN VDE 31 000,Allgemeine Leitsätze für das sicherheitsgerechte Gestalten technischer Erzeugnisse, Begriffe der Sicherheitstechnik, part 2, Dec. 1987: Grundbegriffe.
171. AD-Merkblatt A 1, Berstsicherungen, C. Heimanns Verlag, Köln 1992.
172. AD-Merkblatt A 2, Sicherheitsventile, C. Heimanns Verlag, Köln 1993.
173. DIN 3320,part 1, Sicherheitsventile – Begriffe Größenbemessung Kennzeichnung, 1984.
174. ISO 4126,part 1, Safety Valves, draft 1994.
175. ISO 4126,part 5, Controlled Safety Pressure Relief Systems, draft 1994.
176. ISO 6718,Bursting Discs and Bursting Disc Devices, 1991.
177. R. Bozòki: *Überdrucksicherungen für Behälter und Rohrleitungen,* Verlag TÜV Rheinland, Köln 1986.
178. C. M. Sheppard: "Diers Bubbly Disengagement Correlation Extended to Horizontal Cylinders and Spheres," *J. Loss Prev. Ind.* **7** (1994) no. 1, 3 – 5.
179. E. Molter: "Druckentlastung von Gas-/Dampf-Flüssigkeitsgemischen," PhD Thesis, Universität Dortmund 1991.
180. F. Mayinger: *Strömung und Wärmeübergang in Gas/Flüssigkeitsgemischen,* Springer Verlag, Berlin 1982.
181. Hauptverband der gewerblichen Berufsgenossenschaften, ZH 1/621 Technische Regeln Druckbehälter, TRB 404, 6.1 – 6.3, C. Heimanns Verlag, Köln, Apr. 1989, p. 27.
182. S. Muschelknautz: *Mechanische Phasentrennung bei Entspannungsverdampfung, VDI-Forschrittsber.* **72** (1990) Reihe 15, Umwelttechnik.
183. R. Spatz, E. Molter, H. Schoft: "Auffangsysteme zur Entsorgung der aus Druckentlastungseinrichtungen abgeblasenen Stoffströme," *Chem. Ing. Tech.* **63** (1991) no. 3, 233 – 236.
184. Technischer Ausschuß für Anlagensicherheit beim Bundesminister für Umwelt, Naturschutz und Reaktorsicherheit, Gesellschaft für Reaktorsicherheit GSR, Köln. Sicheres Rückhalten von gesundheitsgefährlichen Stoffen aus Druckentlastungseinrichtungen, TAA-GS-06.
185. VDI-Richtlinie, VDI 3783, Ausbreitung von störfallbedingten Freisetzungen – Sicherheitsanalyse, Beuth Verlag, Köln.
186. V. Pilz: "Auslegung, Einsatz und Wirkungsweise von Blow-Down-Systemen als Sicherheitseinrichtungen von Chemieanlagen," *Chem. Ing. Tech.* **49** (1977) no. 11, 873 – 883.
187. K. Beher, A. Steiff, P.-M. Weinspach: "Direktkondensation inertgashaltiger Dämpfe mit Hilfe von Tauchvorlagen," *Chem. Ing. Tech.* **62** (1990) no. 2, 136 – 137.
188. K. Herrmann, H.-G. Schecker, H. Schoft: "Direktkondensation notentspannter Dampf/Gas-Gemische in einem

Strahlkondensator," *Chem. Ing. Tech.* **64** (1992) no. 2, 183 – 184.
189. H. Schoft, R. Spatz: "Direktkondensation in Strahlapparaten zur Niederschlagung entlasteter Dampf/Gas-Gemische," *Chem. Ing. Tech.* **65** (1993) no. 6, 735 – 739.
190. AD-Merkblatt A 6, C. Heymanns Verlag, Köln.
191. P. Walzel, H. Schoft: "Einmischen von Notabstoppern in chemische Reaktoren," *Chem. Ing. Tech.* **65** (1993) no. 4, 447 – 449.
192. K. Schampel: *Flammendurchschlagsicherungen,* Expert Verlag, Ehningen 1988.
193. D. Lietze, *J. Occup. Accid.* **5** (1983) 17 – 37.
194. G. G. Börger, W. Hüning, H. U. Nentwig, M. Schweitzer, *Chem. Ing. Tech.* **55** (1983) 396.
195. H. Förster, *MariChem93 Conference Papers,* Sect. 8.1, Gastech RAI Ltd., London 1993.
196. VbF, Verordnung über Anlagen zur Lagerung, Abfüllung und Beförderung brennbarer Flüssigkeiten zu Lande mit Technischen Regeln für brennbare Flüssigkeiten (TRbF), C. Heymanns Verlag, Köln 1994.
197. H. H. Freytag: *Handbuch der Raumexplosionen,* Verlag Chemie, Weinheim 1965.
198. H. Förster, H. Leinemann: *Neuester Stand der Explosionsschutzmaßnahmen mit flammendurchschlagsicheren Einrichtungen,* Unterlagen zum Lehrgang der Technischen Akademie Esslingen, 1990.
199. K. Nabert, G. Schön: *Sicherheitstechnische Kennzahlen brennbarer Gase und Dämpfe,* 2nd ed., Deutscher Eichverlag, Braunschweig 1980.
200. VDI-Richtlinie 2263, VDI Kommission Reinhaltung der Luft, Düsseldorf, Nov. 1986.
201. Braunschweiger Flammenfilter GmbH, Braunschweig, catalog.
202. V. Pilz, *Chem. Ing. Tech.* **57** (1985) no. 4, 289 – 307; *Ger. Chem. Eng. (Engl. Transl.)* **9** (1986) 65 – 74.
203. Berufsgenossenschaft der chemischen Industrie: Technisches Merkblatt T 025, Sicherer Umgang mit Flüssigkeiten, part 1: Umfüllen.
204. Berufsgenossenschaft der chemischen Industrie: Technisches Merkblatt T 026, Sicherer Umgang mit Flüssigkeiten, part 2: Probenahme.
205. Berufsgenossenschaft der chemischen Industrie: Technisches Merkblatt T 006, Reinigen von Behältern.
206. Berufsgenossenschaft der chemischen Industrie, Verein Deutscher Sicherheitsingenieure: Ratgeber Anlagensicherheit, Verlag Kluge, Berlin.
207. Berufsgenossenschaft der chemischen Industrie: *Betriebsanweisung für den Umgang mit Gefahrstoffen,* Merkblatt A 010, Jedermann-Verlag Dr. Otto Pfeffer, Heidelberg 1992.
208. T. A. Kletz: "Managing Risk in Chemical Manufracture," *Proc. Int. Conf.,* Cambridge, Jul. 13 – 16, 1992, R. Soc. Chem. 1992, 92 – 105.
209. R. Leicht, K. A. Ruppert: "Sicherheitstechnische Dokumentation in Chemieanlagen," *Tech. Überwach.* **34** (1993) no. 1, 22 – 25.
210. Y. Sakami: "Safety Management at Chemical Plants in Japan," *Proceedings of Second IUPAC-Workshop on Safety in Chemical Production,* Yokohama, May 31 – June 4, 1993 (in preparation).
211. W. E. Hoffmann: *Unabhängig und neutral. Die TÜV und ihr Verband VdTÜV,* Wirtschaftsverlag, Wiesbaden 1986.
212. G. Wiesenack: *Wesen und Geschichte der Technischen Überwachungsvereine,* C. Heymanns Verlag, Köln 1971.
213. W. E. Hoffmann: *Erfahrungsaustausch in der Technischen Überwachung, Technische Eigenüberwachung in der Chemie,* Verlag Wissenschaft und Politik, Köln 1982.
214. H. Pohle: *Chemische Industrie Umweltschutz, Arbeitsschutz, Anlagensicherheit,* VCH Verlagsgesellschaft, Weinheim 1991.
215. W. Schoch: *Selbstverwaltung der Wirtschaft auf dem Gebiet der Technischen Überwachung, Sicherheit in der Chemie,* Verlag Wissenschaft und Politik, Köln 1979.
216. Vergleich des Prüf- und Sachverständigenwesens in Großbritannien, Frankreich und Deutschland am Beispiel der Druckgeräte, Arthur D. Little International, Wiesbaden 1992.
217. DECHEMA: *Safety Standards and Regulations for Chemical Plant in Europe: A Comparison,* Frankfurt/Main 1988.
218. V. C. Marshall, *Major Chemical Hazards,* Ellis Harwood, Chichester 1987.
219. N. M. Azarkin: "Unflagging Attention to the Preparation of Chemical Equipment for Maintenance Operations, in which there is Hazard of Fire," *Sov. Chem. Ind.* (Engl. Transl.) 20 (1988) no. 7, 59 –61.
220. *Dangerous Maintenance, Health and Safety Executive,* Library and Information Services, Sheffield 1987.

221. K. Schwab, “Praxisbericht, Vorbeugende Instandhaltung (VI) und Qualitätssicherung (QS)”, *Konferenz Sicherheits- und Instandhaltungsmanagement in der chemischen Industrie,* VCI, Darmstadt 1993.
222. U. Olsen, “Inspektion und Wartung,” in *Die zustandsorientierte Maschineninstandhaltung in der Praxis, 5. Instandhaltungs-Forum,* Reihe Praxiswissen für Ingenieure – Instandhaltung, TÜV Rheinland, Köln 1989, pp. 95 – 123.
223. H.-H. Rauschhofer: “Vorbeugende Instandhaltung ist ein Garant auch für Sicherheit,” *Maschinenmarkt/MM Industriejournal* **87** (1981) no. 37, 737 –740.
224. S. Pradham: “Apply Reliability Centered Maintenance to Sealless Pumps,” *Hydrocarbon Process.* **72** (1993) no. 1, 43 – 47.
225. N. Hering, Verbesserung der Voraussetzungen zum Optimieren von Sicherheit und Wirtschaftlichkeit im Industrieanlagenbetrieb, *VDI-Z.* **134** (1992) 22 –28.
226. Amt für amtliche Veröffentlichungen der Europäischen Gemeinschaft, Luxemburg: Richtlinie des Rates über die Gefahren schwerer Unfälle bei bestimmten Industrietätigkeiten (82/501/EWG) (mit 1. und 2. Änderungsrichtlinie), ABl. Nr. L 230 vom 05. 08. 1982 (Abl. Nr. L 85 vom 28. 03. 1987, Abl. Nr. L 336 vom 07. 12. 1988).
227. Amt für amtliche Veröffentlichungen der Europäischen Gemeinschaft, Luxemburg: Vorschlag für eine Richtlinie des Rates zur Abwehr der Gefahren schwerer Unfälle mit gefährlichen Stoffen vom 26. 01. 1994, Katalog-Nr. CB-CO-94-016-DE-C.
228. Verordnung über Druckbehälter, Druckgasbehäl-ter und Füllanlagen (Druck-behälterverordnung –DruckbehV)/Technische Regeln zur Druckbehälterverordnung (z. B. TRB 404 Nr. 6, TRB 600 Nr. 3.4), C. Heymanns Verlag, Köln.
229. J. Steinmetz, E. Merz: *Umweltschutz und Gefahrenabwehr als betriebliches Gesamtkonzept,* Boorberg Verlag, Stuttgart 1992.
230. § 19 g ff Wasserhaushaltsgesetz (WHG), *Bundesgesetzblatt,* 1986, part 1, pp. 1536 ff.
231. Verordnung über Anlagen zum Umgang mit wassergefährdenden Stoffen und über Fachbetriebe (VAwS) – (des jeweiligen Bundeslandes, z. B. NW, Gesetz und Verordnungsblatt für das Land Nordrhein-Westfalen – no. 56, pp. 676 ff).
232. Katalog wassergefährdender Stoffe Bekanntmachung des BMI, March 1, 1985 – U III 6523074BGemeinsames Ministerialblatt, 36. Jahrgang, no. 11, pp. 175 ff und deren Fortschreibungen.
233. Allgemeine Verwaltungsvorschrift v. 09. 03. 1990 über die nähere Bestimmung wassergefährdender Stoffe und ihre Einstufung entsprechend ihrer Gefährlichkeit – VwV wassergefährdender Stoffe (VwVwS), Gemeinsames Ministerialblatt, 41. Jahrgang, no. 8, pp. 114 ff.
234. Verordnung über Anlagen zur Lagerung, Abfüllung und Beförderung brennbarer Flüssigkeiten zu Lande (Verordnung über brennbare Flüssigkeiten – VbF) und Allgemeine Verwaltungsvorschrift mit den zugehörigen Technischen Regeln für brennbare Flüssigkeiten (TRbF), C. Heymanns Verlag, Berlin – Köln.
235. Verband der Technischen Überwachungs-Vereine e.V., VdTÜV-Merkblatt “Auffangräume für die Lagerung brennbarer und nichtbrennbarer wassergefährdender Flüssigkeiten,” Essen 1992.
236. Verband der Chemischen Industrie: Sicherheitskonzept für Anlagen zum Umgang mit wassergefährdenden Stoffen, Frankfurt/Main 1987.
237. B. Wittke: “Lager- und Abfüllanlagen für Gefahrstoffe,” *VDI-Ber.* no. 726 (1989).
238. V. Papenhausen: “Standardisierung von Ableitflächen,” *VDI-Ber.* no. 869 (1991).
239. DIN Deutsches Institut für Normung e.V.:Richtlinie des DAfStB – Unbeschichtete Betonbauteile beim Umgang mit wassergefährdenden Stoffen, Berlin 1994.
240. DIN 6608 Teil 2,Liegende Behälter (Tanks) aus Stahl, doppelwandig, für die unterirdische Lagerung wassergefährdender, brennbarer und nichtbrennbarer Flüssigkeiten, Beuth Verlag, Berlin.
241. TRbF 501, Richtlinie/Bau- und Prüfgrundsätze für Leckanzeigegeräte für Behälter.
242. TRbF 502, Richtlinie/Bau- und Prüfgrundsätze für Leckanzeigegeräte für doppelwandige Rohrleitungen.
243. Sicherheitsregeln für Anforderungen an Eigenschaften ortsfester Gaswarneinrichtungen für den Explosionsschutz, ZH 1/8. Hauptverband der gewerblichen Berufsgenossenschaften, Fachausschuß Chemie, C. Heymanns Verlag, Köln.

244. Umgebungsüberwachung von verfahrenstechnischen Freiluftanlagen durch Gasdetektoren, Abschlußbericht des F+E-Vorhabens Nr. 104 09 215 des Umweltbundesamtes vom 12. 08. 1991.
245. S. Muschelknautz, K. Nießer: "Den Schaden begrenzen," *Chem. Ind.* no. 9 (1993) 31 – 33.
246. Sicherheitstechnische Hinweise und Anforderungen an Abschott- und Entlastungssysteme aus Sicht der Störfall-Verordnung (Entwurf), Ausarbeitung eines Arbeitskreises bei der Landesanstalt für Immissionsschutz NRW, Essen.
247. V. Pilz: "Auslegung, Einsatz und Wirkungsweise von Blow-Down-Systemen als Sicherheitseinrichtung bei Chemieanlagen," *Chem. Ing. Tech.* **49** (1977) no. 11, 873 – 883.
248. Leitfaden Rückhaltung von gefährlichen Stoffen aus Druckentlastungseinrichtungen, Abschlußbericht eines Arbeitskreises des Technischen Ausschusses für Anlagensicherheit beim BMU, TAA-GS-06.
249. J. Seeger, E. Marx: "Hochfackeln zur Verbrennung von Industriegasen," *Gas Wärme Int.* **29** (1980) nos. 2/3, 91 – 98.
250. Hauptverband der gewerblichen Berufsgenossenschaften, Fachausschuß Chemie, Sicherheitsregeln für Anlagen zum Entfernen von Gasen und Dämpfen organischer Lösemittel aus der Abluft nach dem Adsorptionsverfahren, ZH 1/595, C. Heymanns Verlag, Köln.
251. Landesanstalt für Immissionsschutz NRW, LIS-Info Anlagensicherheit Nr. 2, Anforderungen an phosgenführende Anlagenteile, Essen.
252. V. Pilz: internal report, Bayer AG, Leverkusen, Mai 1992.
253. DOW Öffentlichkeitsarbeit: Die Herstellung von Polymeric-MDI . . . ein sicherer chemischer Prozeß.
254. Richtlinie zur Bemessung von Löschwasser-Rückhalteanlagen beim Lagern wassergefährdender Stoffe (LöRüRL), Ministerialblatt für das Land Nordrhein-Westfalen, no. 71 vom 20. 11. 1992, pp. 1719 ff
255. "Seveso-Richtlinie" 82/501/EWG, 1982, überarbeiteter Entwurf von 1992 in H. J. Uth: *Störfall-Verordnung,* Kommentar, 2nd ed., Bundesanzeiger Verlagsgesellschaft, Köln 1994.
256. Bekanntmachung der Neufassung der 12. Verordnung zur Durchführung des Bundes-Immissionsschutzgesetzes (Störfall-Verordnung), Sep. 20, 1991, BGBl. I, p. 1891.
257. NRW Arbeitskreis "Alarm- und Gefahrenabwehrplan": Betrieblicher Alarm- und Gefahrenabwehrplan, Stand 23. 08. 1993, Düsseldorf.
258. Leitlinie zur Erstellung betrieblicher Alarm- und Gefahrenabwehrpläne, VCI, Frankfurt/Main, Mar. 1988.
259. W. Steuer: Fachveranstaltung "Der chemische Unfall," Haus der Technik, Essen 1991.
260. Responsible Care, A Chemical Industry Commitment to Improve Performance in Health, Safety and the Environment, CEFIC Schrift 5/1993.
261. DFG MAK- und BAT-Werte-Liste, 2001, Wiley-VCH Weinheim

Transport, Handling, and Storage

CLEMENS STROETMANN, Bad Honnef, Federal Republic of Germany (Chap. 1)

EMILE BERSON, Ministère des Transports, Paris, France (Chap. 2)

NADIA BERSON, Paris, France (Chap. 2)

DETLEF RENNOCH, Bundesanstalt für Materialforschung und -prüfung, Berlin, Federal Republic of Germany (Chap. 3)

BERNHARD DROSTE, Bundesanstalt für Materialforschung und -prüfung, Berlin, Federal Republic of Germany (Chap. 4)

WIEGER JOHANNES VISSER, Nederlandse Spoorwegen, Utrecht, The Netherlands (Sections 5.1 – 5.2.1)

FRITS WYBENGA, International Standards Coordinator, Office of Hazardous Materials Safety, Washington D.C. 20590, United States (Section 5.2.2)

YOSHIO YASOGAWA, Nippon Kaiji Kentei Kyokai, Tokyo, Japan (Section 5.2.3)

HEINZ W. HÜBNER, Bundesanstalt für Materialforschung und -prüfung, Berlin, Federal Republic of Germany (Section 6.1)

KURT-JOSEPH DOKTOR, Bayer AG, Leverkusen, Federal Republic of Germany (Sections 6.2.1.1, 6.2.1.2, 6.3, 6.4)

KLAUS JOCHEM DÖHRN, Germanischer Lloyd, Hamburg, Federal Republic of Germany (Section 6.2.1.3)

JOSEF HAWIGHORST, Bayer AG, Leverkusen, Federal Republic of Germany (Sections 6.2.1.1, 6.2.1.2)

ANNETTE ROLLE, Bundesanstalt für Materialforschung und -prüfung, Berlin, Federal Republic of Germany (Section 6.2.2)

HANS KLUSACEK, Bayer AG, Dormagen, Federal Republic of Germany (Section 6.3)

HANS ASCHENBRENNER, Bayer AG, Leverkusen, Federal Republic of Germany (Section 6.4.2)

HANS-PETER LÜHR, Technische Universität, Institut für wassergefährdende Stoffe, Berlin, Federal Republic of Germany (Chap. 7)

HEINRICH FREIHERR VON LERSNER, Umweltbundesamt, Berlin, Federal Republic of Germany (Chap. 8)

Ullmann's Industrial Toxicology

ISBN: 3-527-31247-1

1. Introduction

At present, there is a lack of comprehensive data on the transport of dangerous goods, in terms of transport volume, transport capacity, carriers, dangerous goods (including dangerous wastes), accident frequency and causes, nature and extent of releases of material because of accidents, leakages, and recovery, etc. The same is true for the area of storage and transport-dependent intermediate storage of dangerous materials and for comparative surveys, e.g., strategies for transport avoidance and traffic deflection as well as site contingency plans. The EC Green Paper on the Effect of Traffic on the Environment makes the following points:

1) An enormous increase is expected in total carriage, average distances, frequencies of consignments, and speed [1]
2) From 1990 to 2010, a 42 % increase of the carriage of goods by road and a 33 % increase is expected by rail [2,3]
3) Transport of dangerous goods involves extensive risks, especially potential effects on the environment [4]
4) A package of measures for the reduction or avoidance of risk during the transport of dangerous goods has been prepared [5], which sets out the environmental, traffic policy, and economic dimensions of these trends

Comparable forecasts for individual Member States, e.g., Germany, predict a large increase both of goods traffic volume and of transport capacity over the next two decades [6], the proportion of dangerous goods in rail traffic being estimated as ca. 12.4 – 13.8 % and in road traffic as 9.1 – 9.6 % [7]. A breakdown of shipments of dangerous goods by transport sector shows that ca. 27.4 % go by inland waterways, 25 % by road transport (long-distance and short hauls across frontiers), ca. 24 % by sea, and 23.5 % by rail [8].

A forecast of the traffic volume and its distribution, which shows the percentage change compared with 1988, assumes a growth in long-distance road transport of goods of 72.9 % by the year 2010; in inland waterway transport 35.3 %, but in rail transport (less polluting than by road) a reduction of 7.3 %.

Moves toward economic integration of Europe should influence the scale of macroeconomic activity in the short, medium, and long term; this should result in a rise in the demand for transport services. With the increase of freight traffic volume, the number of shipments of dangerous goods inevitably rises, and with it the risk of accidents and release of hazardous materials. There is a consequent rise in other external costs (e.g., destruction of the landscape, pollution, and noise) which differ from each other according to the mode of carriage.

The following global goals can be derived from the forecasts outlined above:

1) Increase of transport safety, including transport avoidance strategies
2) A consistent emphasis on precautionary traffic measures.

A multitude of detailed regulations, constantly updated and amended, are aimed at minimizing the special risks of the transport of dangerous goods. However, most regulations – aside from the UN Recommendations – are specific to a particular mode of transport. In some cases, they have different legal status, different member or signatory states, different geographic areas of application, and to some extent different content. Above all, implementation of regulations in different countries may depend on national interest, the external environment, and geographic region.

However, these regulations do not address the long-term transport policy and environmental challenges of goods outlined above, and are often incomprehensible to the nonexpert and difficult to enforce. Even the UN Recommendations fail to deal with the necessity of transport avoidance and the transfer of consignments of dangerous goods to safer or environmentally beneficial means of transport. The basic issue of classification of dangerous materials and goods is still far from being resolved. The worldwide harmonization of dangerous goods regulations is to be encouraged, but it must not lead to a downgrading of safety and environmental standards, especially under the influence of commercial competition. Undifferentiated harmonization can threaten national instruments, in some cases tried and true, for improving transport safety and guaranteeing precautionary environmental protection, e.g., traffic deflection regulations and the allocation of special routes for certain dangerous goods.

It is also necessary to eliminate distortions of competition between carriers, frequently at the expense of safety and environmental considerations, and to resolve conflicts of aims, e.g., between the freedom of service-rendering and legitimate safety interests, or between interests of transport and environmental policy (assistance or avoidance strategies for transport).

The development of a sufficiently differentiated, yet nationally and internationally coordinated safety culture oriented toward prevention is therefore essential.

2. Safety

Although modern technology has overcome many natural dangers (predators, famines, diseases), it has introducted dangers that were unknown in the past. For example, it is estimated that there are about 100 000 dangerous substances, increasing by 1000 every year.

Furthermore, while relying more and more on technology, people are increasingly less prepared to tolerate the risk that this involves. The problem of safety is therefore difficult and ongoing.

2.1. Risk

Risk is an uncertain, possible event, and, as regards safety, often involves damage.

Damage occurs when a harmful agent, finding a medium favorable to its production and/or propagation, reaches a receptive site, i.e., a target.

Safety objectives are designed to prevent the formation of harmful agents, to confine them so as to prevent their dispersal, or to ensure that receptive sites are beyond their reach. Safety measures often combine these objectives. Elaboration of these measures involves analysis of the various parameters and their evolution over time. This risk identification requires a constant questioning approach, e.g., what can a harmful agent do? What can be done to prevent it?

All possibilities should be explored: normal operation, incident, and accident, up to a possible serious catastrophe; in each case measures have to be devised to prevent the situation becoming worse and to reduce damage. Such a method is effective only if it covers all the stages (e.g., in an industrial installation the design, execution, and operation) and involves all factors in these stages.

Risk identification should be supplemented by an evaluation of the probability that the risk will occur. A risk is fully defined only by the conjunction of its detrimental effects and its probability; this enables risks to be compared. Statistical methods enable such an evaluation to be made, and although these methods are still imperfect and limited, especially in new fields, they are improving all the time.

One method that applies particularly to fixed installations is that based on experimental feedback. Starting from an analysis of incidents by the method of event trees, it evaluates the probability that such events may degenerate into a serious accident, and by adding these probabilities over a period of time and referring the result to the number of installations of the same type, it can sometimes be shown that a serious accident becomes probable after a certain period. The importance of this method is that it is able to show that the safety of a system has become compromised, even though no accident has occurred and those responsible may honestly believe that the system is operating satisfactorily.

How can these principles be applied to the transport and storage of dangerous materials? First, this field concerns public safety, on account of the extent and potential consequences of accidents; in all developed countries, it is regulated. Regulations are generally national or local as regards production and storage or national or international as regards transport.

All regulations include an identification of dangerous materials and their classification according to the nature of the danger they present; classification consists of a list of products, drawn up on the basis of experience, and often supplemented by criteria and tests for new materials. These lists and criteria are not necessarily the same for storage and transport (e.g., carcinogenicity or toxicity by bioaccumulation are of little importance in connection with transport). Listing of a product means that the safety regulations applicable to the material and to the activity in question must be observed.

The regulations almost always include the following provisions:

1) Declaration and marking of materials: transport document (manifest) or storage declaration, and labeling of containers
2) Confinement: regulations covering the construction and testing of containers and tanks, safety systems, and warning systems
3) Operations: safety instructions for normal operation and in an accident
4) Environmental factors: distances of storage sites from places of habitation, regulations governing the construction and movement of transport vehicles

2.2. Transport Safety

Transport is an important stage in the life cycle of a dangerous material, which extends from its production to its use. Dangerous materials are scarcely handled at all during transport, except during loading and unloading; the materials are packed and not normally in direct contact with potential targets (humans, animals, the environment).

Accidents due solely to the material itself are therefore fairly rare; the risk comes from the medium in which the material is found, namely traffic. Those responsible for transport safety have to deal specifically with traffic accidents and the consequences arising from the presence of dangerous materials, attempting to reduce the probability of:

Traffic accidents
Escape of material in the event of an accident
Contact between the unconfined material and a target

The best way of preventing a transport accident is to avoid transport. This is why transport of some extremely dangerous materials (e.g. nondesensitized nitroglycerine) is prohibited. Manufacturing industries have to adapt their production methods to these demands.

Measures intended to reduce the probability of escape of material involve the use of the strongest possible containers (crates, boxes, tanks), not forgetting that, except in certain cases (e.g., containers for certain radioactive materials), they are unable to withstand major accidents, and that, if there is a fire, containers that are too strong can cause high pressures, resulting in explosions. Particularly for large containers, such as tanks, it is better to concentrate on their ability to deform without bursting when subjected to violent shock, rather than their overall strength. The ductility of the material and the amount of energy it absorbs on deformation are important safety factors, as is the provision of structures which distribute the deformation (rings, collars, baffles, etc.).

A compromise has to be reached in the solution of these complex problems, depending on the hazard level of the material, type of transport, and industrial technology, which is reflected in national and international transport regulations that lay down provisions and tests for containers.

In addition tanks, large bulk containers, and gas cylinders are subjected to periodic checks. Also, limits are placed on the amount of material per container in order to reduce the amount of material that can escape in the event of an accident. Some materials are not allowed to be transported in bulk. In limiting the amount of material carried per vehicle, it has to be remembered that this increases the number of vehicles transporting this material, and thus the probability of an accident. This type of provision is restricted almost exclusively to explosives.

Materials that can react with one another, causing a new additional danger, e.g., fire, explosion, release of toxic gas, must not be packed together. Similarly, it is prohibited to transport in the same vehicle materials (explosives and organic peroxides) that can cause an explosion with other dangerous materials, since dispersal of these materials by an explosion exacerbates their effects and hinders their recovery.

In order to reduce the probability of a traffic accident, the safety of vehicles is continually being improved, e.g. by improving braking (ABS system, heavy-duty brakes) or by lowering the center of gravity of tanks. Attempts are also being made to improve driver behavior by awareness and reflex training.

To reduce the probability of traffic accidents, and to prevent contact between unconfined material and a target, is more problematic, sometimes contradictory. For example, in the case of land transport there may be a choice between rail and road. It might seem preferable to chose rail since the overall probability of an accident is very low, but railway lines in general pass through built-up areas, whereas bypasses on trunk roads often avoid these.

Furthermore, railway stations, with their points crossings and intersections (i.e., in populated areas) involve the greatest probability of an accident. The amounts of dangerous materials transported in a train are much greater than those in a lorry, and much more varied, which gives rise to the possibility of much more serious accumulations and dangerous synergistic effects. The decisions taken by some countries in favor of rail transport are often politically or commercially motivated. Modes of transport can be compared only from case to case, between a specified starting point and destination.

The same applies when choosing one particular route over another. Local authorities are sometimes inclined to prohibit movement of certain dangerous materials within their built-up areas. Such a decision, which shifts the danger elsewhere, should be taken only on the basis of a comparative analysis of the risk of transport subject or not subject to such a prohibition. The decision should specify a recommended alternative route for transporting dangerous goods. The choice of such a route is complex, and should take account of the environment, avoid possible proximity of industrial activities capable of exacerbating a potential road accident, and avoid populated areas, taking account of daily journeys and travel (the commercial and business center of a large town contains many people during the day, whereas the suburbs are almost empty; the reverse is true at night).

Other types of more general prohibition or restriction exist, often decided at government level and concerning the days, or even hours, when road traffic is very heavy and/or the probability of accidents is very high (beginning and end of holidays, weekends, festivals, etc.). All these measures tend to channel the movement of dangerous goods into periods and zones where the probability and consequences of an accident are a minimum.

Another way to tackle the consequences of an accident is to improve the speed and efficiency of the emergency services. Studies are being carried out on the possibilities of satellite surveillance and warning.

2.3. Storage Safety

With regard to the production or storage of dangerous goods the drafting of the safety measures should take account of the agent, medium, and target. Identification of the agent appears simple. However, with regard to production as well as storage, when the length or conditions of storage can affect stability, all the possible forms and transformations of a material have to be examined. The major accident in a chemical factory in Seveso led to a European Community Regulation, the so- called Seveso Directive. The product, which was distributed over a large area by the accident, was an extremely toxic dioxin. No measures had been envisaged to control this compound, since it was not the end product of the process, but was formed only as an intermediate and converted immediately into a less dangerous chemical. No one had considered that an accident resulting in escape could occur during the brief existence of the dioxin. This graphically illustrates the need for extreme care in identifying the danger.

Unlike transport, the medium is geographically fixed, which simplifies matters, though its study should take into account all the possible variations: time of day; seasonal climatic conditions; wind speeds and directions, as well as all the normal, incidental, and unplanned operating conditions in the installation itself, so as to evaluate the forseeable consequences and formulate appropriate safety measures.

These measures vary, depending on the nature of the goods that are stored and the industrial activity, but always aim to reduce the risk of escape of dangerous products. They include restricting or splitting up amounts of materials, strengthening containers, verifying that they are in good condition by periodic checks, using devices to contain leaks, etc. It should also be ensured that the installation can withstand foreseeable natural catastrophes, including earthquakes, tornadoes, etc. Special attention should be paid to the operation of the installation. Any handling of the product is a dangerous operation on account of the risk of error. The human factor, despite technical progress, remains a decisive element in the uncertainty of risk calculations; there is an increasing tendency to reduce the human element by automation, and to create intrinsically safe systems which can "excuse" an operator error, and even restore normal operation in the case of an incident. Despite this, humans continue to be indispensable, which is why information and workforce training are necessary to reduce risk. Active participation of all those involved in designing safety measures should be encouraged since this is the best way of achieving a safety culture in the enterprise where people are no longer the weak link in the chain, but become the central positive element of the safety system.

In most cases of manufacture or storage of dangerous substances, the consequences of a serious accident may extent far beyond the confines of the enterprise. Its siting is therefore of primary importance, the ideal being to locate it as far as possible from populated areas. Care should also be taken to ensure that, in the case of toxic gases, the prevailing winds are not in the direction of such areas, and that, in the event of accidental spillage, the impermeability of the ground prevents pollution of underground water.

The location of dangerous sites in relation to built-up areas often presents a problem. In most European countries, demographic trends in the last two decades have led to an often uncontrolled expansion of towns and cities, which is why previously well-isolated industrial installations are now within urban boundaries, with a frightening increase in risk. Even where new enterprises are sited in rural areas, it is sometimes found that public bodies responsible for land management do little to resist pressure from landowners who wish to divide into plots and sell land that has been developed for industrial purposes, sometimes to allow company employees to live near their place of work. It is very important that any enterprise involved in the manufacture or storage of dangerous materials be enclosed by an adequate safety perimeter, including the zone where the presence of houses and dwellings presents a serious risk in the event of an accident. The safety measures adopted by the company should include alerting and informing the population involved as well as the public authorities.

This is also prescribed in European Directive 82-501 (the Seveso Directive). This covers industrial establishments (other than nuclear, military, and waste disposal) that produce, use, or store amounts of dangerous materials above the levels specified in a list appended to the directive. The directive makes the manufacturer responsible for determining the risk, taking appropriate safety measures, and informing and training the workforce. The manufacturer should also notify the relevant authorities and keep an up-to-date list of the type and amount of dangerous materials being produced and stored; describe the installation and production processes; internal emergency procedures; and provide all the necessary information for the formulation of an external emergency plan by the official safety bodies. The manufacturer should notify the authorities of all major accidents, including the circumstances, materials involved, measures adopted and to be adopted; notification has to be followed by the setting up of a special enquiry.

The safety principles governing dangerous installations in Europe are fairly well defined and regulated. An important feature of these measures is that they require regular safety checks to be carried out by an independent outside body.

Any safety system, however well designed, will tend to deteriorate over time. Not only must the safety equipment and systems be regularly checked, serviced, and repaired, but instructions to the workforce and their level of awareness and training also need to be periodically updated. There are two periods when the risk associated with an installation is greatest: during start-up, since the safety systems have not yet been tested; and after a long incident-free period, when the risk has been minimized or even forgotten by managers and workers. This is why it is essential for a disinterested observer, if possible outside the company or at least completely independent of the production organization, to make periodic visits to evaluate the safety system and update the objectives, without however interfering in the choice of measures to be adopted to achieve these objectives, which should remain the responsibility of the management.

2.4. Security

Safety may be defined as the complete absence of risk. Zero risk is an objective which one must strive for, but which is never reached. In many languages, safety and security are virtually synonymous.

Economic imperatives must be taken into account in this constant striving to improve safety. Safety has its price. No reliable safety exists if the means, and thus the necessary funding, have not been allocated for this purpose. This implies that it should not be left unsupervised in the hands of those responsible for production.

Contrary to a widespread misconception, safety cannot be achieved regardless of cost. It is therefore unacceptable that safety regulations be drawn up by a remote bureaucratic system. Even in high-level international bodies, such a system tends to add new constraints without taking into account the probability of the risk that these constraints are designed to combat, their economic effects, or sometimes even their feasibility. It is possible to arrive at provisions that are so complex and difficult to apply that they are not implemented.

The safety costs involved in the production, transport, and storage of a dangerous product cannot exceed a certain fraction of the internationally accepted value of the product. On the other hand, the industrial safety field is so complex that there are always several ways of achieving the same result. Accordingly, the aim of experts in this field, whether they be regulatory bodies or industrial managers, should be to avoid implementing measures indiscriminately, even if they are ostensibly aimed at security, but instead choose the measures that have the best cost/benefit ratio.

Industrial safety, whether of persons or the environment, has become a political football for decision-makers and an economic debate among manufacturers. There may thus be a temptation for the former to emphasize the "public relations" impact to the detriment of efficiency, and for the latter to promote eye- catching features in order to maximize profit, all this often being supported by the media, which tend to seek sensationalism rather than objective information. Sometimes, under the pretext of safety, measures may be accepted that would otherwise have been rejected, or decisions are made purely on economic grounds.

Although these comments may appear rather pragmatic, it should not be forgotten that safety research is eminently worthwhile, even if it demands a great deal of perseverance, attention to detail, and humility. One should never believe the experts when they say, "We have done everything we can," one must continue to question. Safety techniques improve at least as quickly as production techniques, which, in the end, is very reassuring.

3. Dangerous Goods and Substances

Test methods, classification, labeling, and assigning to packaging groups of dangerous goods should be harmonized worldwide for all modes of transport (sea, rail, road, and air).

The UN Recommendations cover all modes of transport. Where less stringent requirements can be applied to a particular mode, this is indicated only in special circumstances. For air transport, more stringent requirements may occasionally apply. The Recommendations are internationally implemented through regulations and agreements covering rail, road, sea, and air transport. The parts of the Recommendations which are relevant to this chapter are the definitions of classes, principles and criteria for classification, test methods, labeling, and the *List of Dangerous Goods Most Commonly Carried* (approximately 3000 entries). There is a great degree of flexibility in the data on which the classification is based, because the ultimate classification, i.e., designation of class and inclusion in the UN Recommendations list of substances (Chap. 2) is done by UNCETDG [9] which passes judgement on the data and the proposals for classification submitted to it. The UN system uses the notion of primary hazard, which determines the hazard class. This is reflected in the recommendation that "in general not more than one danger class label should be affixed to a package". If a substance meets the definition of more than one class, a rule of precedence applies, and a distinction is made between primary and subsidiary risks (see Sections 3.1 and 3.1.10).

Principles Underlying the Regulation of the Transport of Dangerous Goods. Transport of dangerous goods is regulated in order to prevent, as far as possible, accidents to persons or property and damage to the means of transport employed or to other goods. At the same time, regulations should be framed so as not to impede the movement of such goods, other than those too dangerous to be accepted for transport. With this exception, the aim of regulations is to make transport feasible by eliminating risks or reducing them to a minimum.

Classification and Definitions of Classes of Dangerous Goods. The classification of goods by type of risk involved has been drawn up to meet technical conditions while at the same time minimizing interference with existing regulations. Classifications are made on consideration of data submitted to the Committee by governments, intergovernmental organizations, and other international organizations.

Classification of Dangerous Wastes. Wastes should be transported under the requirements of the appropriate class, considering their hazards and the criteria of the Recommendations. Wastes not otherwise subject to the Recommendations, but covered under the Basel Convention [10] may be transported under Class 9.

Many of the substances listed in Classes 1 – 9 are deemed, without additional labeling, to be environmentally hazardous.

In Germany, the classification of "dangerous wastes", "special wastes", "dangerous residuals", "oils from wastes", and "chemical wastes" can be made on the basis of the *Abfall-Transport-Berater*, which adopts a general view on the character of the wastes/residual substances and their classification in the ADR/RID (GGVS/GGVE), taking into account the *Abfallartenkatalog*. It enables the sender, to classify such substances without too much effort [11].

Determination of the Physical State. Unless there is an explicit or implicit indication to the contrary in the UN Recommendations, dangerous goods with a melting point or initial melting point of 20 °C or lower at a pressure of 101.3 kPa are considered to be liquids. A viscous substance for which a specific melting point cannot be determined should be subjected to the ASTM D 4359-90 test; or the test for determining fluidity (penetrometer test) prescribed in Appendix A.3 of Annex A of the *European Agreement Concerning the International Carriage of Dangerous Goods by Road* (ADR) with the modifications that the penetrometer should conform to ISO 2137: 1985, and that the test should be used for viscous substances of any class.

Goods too Dangerous to be Carried Without Special Restrictions. Inherent instability of goods may involve different hazards, e.g., explosion, polymerization with intense evolution of heat, or emission of toxic gases. In most substances, such tendencies can be controlled by correct packing, dilution, stabilization, addition of an inhibitor, refrigeration, or other precautions.

Where (bearing in mind the manner in which such items are generally packed) precautionary measures are laid down for a given substance or article (e.g., that it should be "stabilized" or "inhibited") such substance or article should not normally be carried when these measures have not been taken.

Collective Entries. National or international regulations may list, either as single or as suitable collective entries, substances or articles which do not appear in the Recommendations. A "generic" or "not otherwise specified" entry (N.O.S. entry) may be used to permit the transport of substances or articles which do not appear specifically by name in the list of dangerous goods. Such a substance or article may be transported only after its dangerous properties have been determined. Any substance or article having or suspected of having explosive characteristics should first be considered for inclusion in Class 1. Some collective entries may be of the generic or N.O.S. type, provided the regulations contain provisions ensuring safety, both by excluding extremely dangerous goods from normal transport, and by covering all subsidiary risks inherent in some goods.

Classification of Solutions and Mixtures. A mixture or solution containing a dangerous substance identified by name in the Recommendations and one or more substances not subject to the Recommendations should be treated according to the requirements given for the dangerous substance, provided the packaging is appropriate to the physical state of the mixture or solution, unless:

1) The mixture or solution is specifically identified by name in the Recommendations; or
2) The entry in the Recommendations indicates that it applies only to the pure substance; or
3) The hazard class, physical state, or packing group of the solution or mixture is different from that of the dangerous substance; or
4) There is significant change in the measures to be taken in emergencies.

For a solution or mixture of which the hazard class, physical state, or packing group differs from that of the listed substance, the appropriate N.O.S. entry should be used, including its packaging and labeling provisions.

3.1. Transport Classification of Dangerous Goods

UN Classes and Divisions are summarized as follows:

Class 1	Explosives
1.1	Substances and articles which have a mass explosion hazard
1.2	Substances and articles which have a projection hazard, but not a mass explosion hazard
1.3	Substances and articles which have a fire hazard and either a minor blast hazard or a minor projection hazard, or both, but not a mass explosion hazard
1.4	Substances and articles which present no significant hazard
1.5	Very insensitive substances which have a mass explosion hazard
1.6	Extremely insensitive articles which do not have a mass explosion hazard
Class 2	Gases: compressed, liquefied, dissolved under pressure, or refrigerated
Class 3	Flammable liquids
Class 4.1	Flammable solids: readily combustible solids, self-reactive substances, wetted explosives
Class 4.2	Substances liable to spontaneous combustion: pyrophoric substances, self-heating substances
Class 4.3	Substances, which in contact with water, emit flammable gases
Class 5.1	Oxidizing substances
Class 5.2	Organic peroxides
Class 6.1	Toxic substances
Class 6.2	Infectious substances
Class 7	Radioactive material
Class 8	Corrosives
Class 9	Miscellaneous dangerous substances

General Provisions of ADR/RID. Annex A of ADR or Annex I of RID specifies the dangerous goods to be excluded from international carriage by road and the dangerous goods to be accepted for such carriage under certain conditions. It groups the dangerous goods in restrictive and nonrestrictive classes. Of the dangerous goods covered by the restrictive classes (Classes 1, 2, and 7) those which are listed in the clauses concerning these classes are to be accepted for carriage only under the conditions specified; others are excluded from carriage. Some of the dangerous goods covered by the nonrestrictive classes (Classes 3, 4.1, 4.2, 4.3, 5.1, 5.2, 6.1, 6.2, 8, and 9) are excluded from carriage; of the other goods covered by the nonrestrictive classes, those which are mentioned in the clauses concerning these classes are to be accepted for carriage only under the conditions specified; those not mentioned or covered by one of the collective headings are not deemed to be dangerous goods and are to be accepted for carriage without any special conditions.

ADR/RID Classes and Divisions. Hazard categories and terms used in the Regulations concerning the International Carriage of Dangerous Goods by Road and Rail (RID/ADR) [12] are summarized below.

Class 1	Explosive substances and articles	Restrictive
Class 2	Gases: compressed, liquefied, or dissolved under pressure	Restrictive
Class 3	Flammable liquids	Nonrestrictive
Class 4.1	Flammable solids	Nonrestrictive
Class 4.2	Substances liable to spontaneous combustion	Nonrestrictive
Class 4.3	Substances which, in contact with water, emit flammable gases	Nonrestrictive
Class 5.1	Oxidizing substances	Nonrestrictive
Class 5.2	Organic peroxides	Nonrestrictive
Class 6.1	Toxic substances	Nonrestrictive
Class 6.2	Infectious substances	Nonrestrictive
Class 7	Radioactive material	Restrictive
Class 8	Corrosive substances	Nonrestrictive
Class 9	Miscellaneous dangerous substances and articles	Nonrestrictive

From 1 January 1995, the classification criteria concerning flammable liquids are the same worldwide; the upper flash point limit is ≤ 61 °C c.c. (closed cup test).

Hazard Categories and Terms used in the International Maritime Dangerous Goods Code (IMDG-Code) [13]. Regulation 2 of part A of Chapter VII of the *International Convention for the Safety of Life at Sea, 1974*, as amended, sets out the various classes of dangerous goods. For the purposes of the Code, it has been found necessary to subdivide a number of these classes and to define in greater detail the characteristics and properties of the substances, materials, and articles which fall within each class or division. In accordance with the criteria for selection or marine pollutants for the purposes of Annex III of the *International Convention for the Prevention of Pollution from Ships, 1973*, as modified by the Protocol of 1978 (MARPOL 73/78), a number of dangerous substances in the various classes have also been identified as substances harmful to the marine environment (Marine Pollutants, see 3.1.11).

IMO Classes and Divisions. For the following classes, the IMDG-Code differs from the UN Recommendations and the RID/ ADR:

Class 3. Flammable liquids are defined as liquids, mixtures of liquids, or liquids containing solids in solution or suspension (e.g., paints, varnishes, lacquers, but not including substances which, on account of other dangerous characteristics, have been included in other classes), which give off a flammable vapor at or below 61 °C (141 °F) closed cup (c.c.) test (corresponding to 65.6 °C (150 °F) open cup test).

In the IMDG-Code, Class 3 is further subdivided:

Class 3.1. Low-flash-point liquids; flash point < −18 °C (0 °F) c.c.

Class 3.2. Intermediate-flash point liquids; flash point from − 18 °C (0 °F) up to, but not including, 23 °C (73 °F) c.c.

Class 3.3. High-flash-point liquids; flash-point from 23 °C (73 °F) up to, and including, 61 °C (141 °F) c.c.

Substances with flash point > 61 °C (141 °F) c.c., are not considered dangerous by virtue of their fire hazard, where the flash point is indicated for a volatile liquid it may be followed by the symbol "c.c.", representing determination by a closed cup test, or by "o.c.", representing an open cup test. A reference to these tests is given in Section 6 of the General Introduction of the IMDG-Code.

3.1.1. Explosive Substances and Articles (Class 1)

UN Recommendations. Class 1 comprises:

1) Explosive substances (a substance which is not itself an explosive, but which can form an explosive atmosphere of gas, vapor, or dust is not included in Class 1), except those that are too dangerous to transport, or those where the predominant hazard is appropriate to another class;
2) Explosive articles, except devices containing explosive substances in such quantity of or such a character that their inadvertent or accidental ignition or initiation during transport causes no effect external to the device either by projection, fire, smoke, heat, or loud noise; and
3) Substances and articles not mentioned under (1) and (2) which are manufactured with a view to producing a practical, explosive, or pyrotechnic effect.

Transport of explosive substances which are unduly sensitive or so reactive as to be subject to spontaneous reaction is prohibited.

Under the UN Recommendations, the following definitions apply:

1) An explosive substance is a solid or a liquid substance (or a mixture of substances) which is in itself capable by chemical reaction of producing gas at such a temperature and pressure and at such a speed as to cause damage to the surroundings. Pyrotechnic substances are included even when they do not evolve gases.
2) A pyrotechnic substance is a substance or mixture of substances designed to produce an effect by heat, light, sound, gas, or smoke, or a combination of these as the result of nondetonative self-sustaining exothermic chemical reactions.
3) An explosive article is an article containing one or more explosive substances.

Class 1 is divided into six divisions and 13 compatibility groups.

Division 1.1. Substances and articles which have a mass explosion hazard (a mass explosion is one which affects almost the entire load virtually instantaneously).

Division 1.2. Substances and articles which have a projection hazard, but not a mass explosion hazard.

Division 1.3. Substances and articles which have a fire hazard and either a minor blast hazard or a minor projection hazard or both, but not a mass explosion hazard. This division comprises substances and articles: which give rise to considerable radiant heat; or which burn one after another, producing minor blast or projection effects, or both.

Division 1.4. Substances and articles which present no significant hazard. This division comprises substances and articles which present only a small hazard in the event of ignition or initiation during transport. The effects are largely confined to the package and no projection of fragments of appreciable size or range is to be expected. An external fire should not cause virtually instantaneous explosion of almost the entire contents of the package.

Division 1.5. Very insensitive substances which have a mass explosion hazard. This division comprises substances which have a mass hazard, but are so insensitive that there is very little probability if initiation or of transition from burning to detonation under normal conditions of transport. The probability of transition from burning to detonation is greater when large quantities are carried in a ship.

Division 1.6. Extremely insensitive articles which do not have a mass explosion hazard. This division comprises articles which contain only extremely insensitive detonating substances, and which demonstrate a negligible probability of accidental initiation or propagation; the risk is limited to the explosion of a single article.

Class 1 is unique in that the type of packaging frequently has a decisive effect on the hazard, and therefore on the assignment to a particular division. The correct division and compatibility group and the packaging are determined by the method outlined in Chapter 4 of the UN Recommendations.

The definition and the classification in divisions and compatibility groups of the UN Recommendations corresponded the those of IMO and RID/ADR.

3.1.2. Gases (Class 2)

UN Recommendations. A gas is a substance which:

1) At 50 °C has a vapor pressure > 300 kPa;
2) Is completely gaseous at 20 °C at a standard pressure of 101.3 kPa.

The transport condition of a gas is described according to its physical state as:

1) Compressed gas: a gas (other than in solution) which, when packaged under pressure for transport, is entirely gaseous at 20 °C;
2) Liquefied gas: a gas which when packaged for transport is partially liquid at 20 °C;
3) Refrigerated liquefied gas: a gas which when packaged for transport is partially liquid because of its low temperature; or
4) Gas in solution: compressed gas which when packaged for transport is dissolved in a solvent.

The class includes compressed gases, liquefied gases, gases in solution, refrigerated liquefied gases; mixtures of gases, mixtures of one or more gases with one or more vapors of substances from other classes, articles charged with a gas, tellurium hexafluoride, and aerosols.

Class 2 substances are assigned to one of three divisions based on the primary hazard during transport.

Division 2.1. Flammable gases. Gases which at 20 °C and a standard pressure of 10.3 kPa:

1) Are ignitable in a mixture of 13 vol % or less with air; or
2) Have a flammable range with air of at least 12 %, regardless of the lower flammability limit. Flammability should be determined by tests or by calculation in accordance with ISO 10 156: 1990 (where insufficient data are available, tests by a comparable method recognized by a national competent authority may be used).

Division 2.2. Non flammable, nontoxic gases. Gases which are transported at a pressure not less than 280 kPa at 20 °C, or as refrigerated liquids, and which:

1) Are asphyxiant gases, which dilute or replace the oxygen normally in the atmosphere; or
2) Are oxidizing gases, which may, generally by providing oxygen, cause or contribute to the combustion of other material more than air does; or
3) Do not come under the other divisions.

Division 2.3. Toxic gases. Gases which:

1) Are known to be so toxic or corrosive to humans as to pose a hazard to health; or
2) Are presumed to be toxic or corrosive to humans, because they have $LC_{50} \leq 5000\ mL/m^3$ (ppm) when tested in accordance with Chapter 6 subparagraph 6.5 (c) of the Recommendations. Gases meeting the above criteria owing to their corrosive properties are classified as toxic with a subsidiary corrosive risk.

Mixtures of Gases. For the classification of gas mixtures into one of the three divisions (including vapors of substances from other classes) the principles of Chapter 1 of the UN Recommendations can be used.

Hazard Precedence. Gases and gas mixtures with hazards associated with more than one division take the following precedence:

1) Division 2.3 takes precedence over all other divisions
2) Division 2.1 takes precedence over Division 2.2

ADR/RID Definitions and Divisions of Class 2. Substances with critical temperature $< 50\,°C$, or at 50 °C, a vapor pressure greater than 300 kPa (3 bar) are deemed to be Class 2 substances. Substances and articles of Class 2 are classified as follows:

A) Compressed gases with critical temperature $< -10\,°C$
B) Liquefied gases with critical temperature $\geq -10\,°C$:
   a) Liquefied gases with critical temperature $\geq 70\,°C$
   b) Liquefied gases with critical temperature $\geq -10\,°C$, but $< 70\,°C$;C.
C) Deeply refrigerated liquefied gases
D) Gases dissolved under pressure
E) Aerosol dispensers and nonrefillable containers of gas under pressure
F) Gases subject to special requirements
G) Empty receptacles and empty tanks

The substances and articles of Class 2 are subdivided according to their chemical properties, as follows:

(a) Nonflammable
(at) Nonflammable, toxic
(b) Flammable
(bt) Flammable, toxic
(c) Chemically unstable
(ct) Chemically unstable, toxic

Unless otherwise specified, chemically unstable substances are considered to be flammable.

IMO Definitions and Divisions of Class 2 are the same as those in the UN Recommendations.

3.1.3. Flammable Liquids (Class 3)

UN Recommendations. Flammable liquids are liquids, or mixtures of liquids, or liquids containing solids in solution or suspension (paints, varnishes, lacquers, etc., but not including substances otherwise classified on account of their dangerous characteristics) which give off a flammable vapor at temperatures of not more than 60.5 °C c.c., or not more than 65.6 °C o.c., normally referred to as the flash point. However such liquids with a flash point of more than 35 °C, which do not sustain combustion need not be considered as flammable liquids for the purposes of the UN Recommendations. Liquids offered for transport at temperatures at or above their flash point are in any case considered as flammable liquids. Flammable liquids also include substances that are transported or offered for transport at elevated temperatures in a liquid state, and which give off a flammable vapor at or below the maximum transport temperature.

Liquids are considered to be noncombustible if they have passed a suitable combustibility test, if their fire point according to ISO 2592: 1973 is greater than 100 °C, or if they are water-miscible solutions with a water content of more than 90 wt %.

Viscous substances with flash point $< 23\,°C$ may be placed in Packing Group III. Substances classified as flammable liquids owing to their being transported or offered for transport at elevated temperatures are included in Packing Group III.

ADR/RID Definitions and Divisions of Class 3.

1) Flammable liquids within the meaning of ADR, are those that are liquid at the maximum temperature of 20 °C, or for viscous substances for which a specific melting point cannot be determined, and which are highly viscous according to the criteria of the penetrometer test (see Appendix A.3 of ADR and Appendix III of RID), or are liquid according to the ASTM D 4359-90 test method
2) Have at 50 °C a vapor pressure ≤ 300 kPa (3 bar)
3) Have a flash point of $\leq 61\,°C$ c.c.

Class 3 substances, are assigned to one of the following groups designated (a), (b), or (c) according to their degree of danger:

(a) Very dangerous substances: flammable liquids with boiling point or initial boiling point not exceeding 35 °C, and flammable liquids with flash point < 23 °C which are either highly toxic according to the criteria of Class 6.1 or highly corrosive according to the criteria of Class 8
(b) Dangerous substances: flammable liquids with flash point < 23 °C which are not classified under (a), with the exception of viscous substances
(c) Substances presenting a minor danger: flammable liquids with flash point 23–61 °C, and some viscous substances

Methods used for grouping, and combustibility testing of flammable liquids as well as methods for grouping of flammable viscous substances with flash point < 23 °C can be found in Appendix A.3 of ADR and Appendix III of RID, respectively. Viscosity and flash point data are summarized in Table 1.

Table 1. Viscosity and flash point data of Class 3 substances

Kinematic viscosity ν (extrapolated), at near-zero shear rate at 23 °C, mm^2/s	Flow time t (ISO 2431-1984), s	Jet diameter, mm	Flash point, °C
20–80	20–60	4	>17
80–135	60–100	4	>10
135–220	20–32	6	> 5
220–300	32–44	6	>−1
300–700	44–100	6	>−5
>700	>100	6	≤ −5

IMO definitions and divisions of Class 3 are described in Section 3.1.

3.1.4. Flammable Solids, Substances Liable to Spontaneous Combustion, and Substances which in Contact with Water Emit Flammable Gases (Class 4)

UN Recommendations. Class 4 comprises:
Division 4.1. Flammable solids. Solids which, under conditions encountered in transport, are readily combustible or may cause or contribute to fire through friction; self-reactive and related substances which are liable to undergo a strongly exothermic reaction; desensitized explosives which may explode if not diluted sufficiently.

Division 4.1 comprises the following types of substances:

(a) Flammable solids
(b) Self-reactive and related substances
(c) Desensitized explosives

Division 4.2. Substances liable to spontaneous combustion under conditions encountered in transport, or to heating up in contact with air, and which are then liable to catch fire;
Division 4.2 comprises:

(a) Pyrophoric substances
(b) Self-heating substances

Division 4.3. Substances which in contact with water emit flammable gases. Substances which, by interaction with water, are liable to become spontaneously flammable or to give off flammable gases in dangerous quantities.

3.1.4.1. Flammable Solids [Division 4.1 (a)]

Properties. Flammable solids are readily combustible solids and solids which may cause fire through friction. Readily combustible solids are powders or granules, ignited by brief contact with an ignition source, such as a burning match, provided the flame spreads rapidly. The danger may come not only from the fire, but also from toxic combustion products. Metal powders are especially dangerous because of the difficulty of extinguishing a fire since normal extinguishing agents such as carbon dioxide or water can increase the hazard.

Classification of flammable solids should be carried out according to Chapter 14 of the UN Recommendations.

3.1.4.2. Self-Reactive and Related Substances [Division 4.1 (b)]

Definitions. Self-reactive substances are thermally unstable substances liable to undergo a strongly exothermic decomposition, even without participation of oxygen (air). Substances should be excluded from Division 4.1, if:

1) They are explosives, according to the criteria of Class 1

2) They are oxidizing substances, according to the assignment procedure of Division 5.1
3) They are organic peroxides, according to the criteria of Division 5.2
4) Their heat of decomposition is < 300 J/g
5) Their self-accelerating decomposition temperature (SADT) is $> 75\,°C$ for a 50-kg package

Heat of decomposition can be determined by any internationally recognized method, e.g., differential scanning calorimetry, adiabatic calorimetry. Any substance which shows the properties of a self-reactive substance should be classified as such, even if this substance gives a positive test result for inclusion in Division 4.2. Substances related to self-reactive substances are distinguished from the latter by having a self-accelerating decomposition temperature $> 75\,°C$.

Properties. The decomposition of self-reactive substances can be initiated by heat, contact with catalytic impurities (e.g., acids, heavy-metal compounds, bases), friction, or impact. The rate of decomposition increases with temperature and varies with the substance. Decomposition, particularly if no ignition occurs, may result in the evolution of toxic gases or vapors. For certain self-reactive substances, the temperature should be controlled. Some self-reactive substances may decompose explosively, particularly if confined. This characteristic may be modified by the addition of diluents or by the use of appropriate packaging. Some self-reactive substances burn vigorously.

Self-reactive substances include: aliphatic azo compounds; organic azides; diazonium salts; *N*-nitroso compounds; aromatic sulfohydrazides.

Classification of self-reactive and related substances should be carried out according to Chapter 14 of the UN Recommendations.

Temperature Control Requirements. Self-reactive substances should be subject to temperature control if their self-accelerating decomposition temperature (SADT) is $\leq 55\,°C$. Test methods for determining the SADT are given in the current edition of the Recommendations on the Transport of Dangerous Goods, Manual of Tests and Criteria. The test selected should be conducted in a manner which is representative, both in size and material, of the package to be transported.

3.1.4.3. Desensitized Explosives [Division 4.1 (c)]

Properties. Desensitized explosives are substances which are wetted with water or alcohol or diluted with other substances to suppress their explosive properties.

3.1.4.4. Substances Liable to Spontaneous Combustion [Division 4.2]

Properties. Self-heating of substances, leading to spontaneous combustion, is caused by reaction of the substance with oxygen (in the air), conduction of the heat developed not being rapid enough to avoid combustion. Spontaneous combustion occurs when the rate of heat production exceeds the rate of the heat loss and the autoignition temperature is reached. Two types of substance can be distinguished with spontaneous combustion properties:

(a) Substances, including mixtures and solutions (liquid or solid), which even in small quantities ignite within 5 min of coming into contact with air; these substances are the most liable to spontaneous combustion and are called pyrophoric substances.
(b) Other substances which in contact with air, without addition of energy, are liable to self-heating; these substances ignite only when present in large amounts (kilograms) and after long periods of time (hours or days), and are called self-heating substances.

Test methods for pyrophoric solids and liquids as well as for self-heating substances can be found in Chapter 14 of the UN Recommendations.

3.1.4.5. Substances which in Contact with Water Emit Flammable Gases (Division 4.3)

Properties. Certain substances in contact with water may emit flammable gases that can form explosive mixtures with air. Such mixtures are easily ignited by all ordinary sources of ignition, e.g., naked lights, electrical sparts, or unprotected light bulbs. The resulting blast wave and flames may endanger people and the environment. A suitable test should be used to determine whether the reaction of a substance with water produces a dangerous amount of gases which may be flammable; it should not be applied to pyrophoric substances.

Methods of classification and assignment to packaging groups can be found in Chapter 14 of the UN Recommendations.

3.1.5. Oxidizing Substances and Organic Peroxides (Class 5)

UN Recommendations. This class comprises:

Division 5.1. Oxidizing substances which, while in themselves are not necessarily combustible, may, generally by yielding oxygen, cause, or contribute to, the combustion of other material.

Division 5.2. Organic peroxides which contain the bivalent – O – O – group and may be considered as hydrogen peroxide derivatives, with one or both of the hydrogen atoms replaced by organic radicals. Organic peroxides are thermally unstable substances, which may undergo exothermic self-accelerating decomposition.

Methods of classification and assignment to packaging groups can be found in Chapter 11 of the UN Recommendations.

Assignment of Organic Peroxides to Division 5.2. Any organic peroxide should be considered for classification in Division 5.2, unless the formulation contains:

1) Not more than 1.0 % available oxygen from the organic peroxides, when containing not more than 1.0 % hydrogen peroxide; or
2) Not more than 0.5 % available oxygen from the organic peroxides, when containing more than 1.0 %, but not more than 7.0 % hydrogen peroxide.

Allocation of new organic peroxides, or new formulations, or mixtures of currently assigned organic peroxides to a generic entry should be made by the competent authority of the country of origin, on the basis of a test report.

Methods for classification and assignment to packaging and packaging groups of organic peroxides can be found in Chapter 11 of the UN Recommendations. Suitable test methods with pertinent evaluation criteria are given in the current edition of Recommendations on the Transport of Dangerous Goods, Manual of Tests and Criteria, Part III.

Temperature Control. To avoid unnecessary confinement, metal packagings meeting the test criteria of Packing Group I should not be used. Organic peroxides are assigned to Packing Group II (medium danger).

3.1.6. Toxic and Infectious Substances (Class 6)

UN Recommendations. This class comprises:

Division 6.1. Toxic substances are liable either to cause death or to harm human health if swallowed, or inhaled, or by skin contact.

Division 6.2. Infectious substances contain viable microorganisms, including a bacterium, virus, Rickettsia, parasite, fungus, or a recombinant, hybrid or mutant, that are known or are reasonably believed to cause disease in animals or humans.

Genetically modified microorganisms and organisms which do not meet the definition of an infectious substance should be considered for classification in Class 9 and assignment to UN 3245.

Toxins from plant, animal, or bacterial sources which do not contain infectious substances or organisms, or which are not contained in them should be considered for classification in Division 6.1 and assignment to UN 3172.

3.1.6.1. Toxic Substances (Division 6.1)

Criteria for Defining Toxicity. Substances of Division 6.1, including pesticides, should be allocated among the three packing groups according to their degree of toxic hazard in transport as follows:

Packing Group I:	substances and preparations presenting a very severe toxicity risk
Packing Group II:	substances and preparations presenting a serious toxicity risk
Packing Group III:	substances and preparations presenting a relatively low toxicity risk

In making this grouping, account should be taken of experience in instances of accidental poisoning in humans, and of special properties possessed by any individual substance, such as liquid state, high volatility, any special likelihood of penetration, and special biological effects.

In the absence of human experience, grouping should be based on data obtained from animal experiments. Three possible routes of administration should be examined: oral ingestion; dermal contact; and inhalation of dust, mist, or vapor. Appropriate animal tests for the various routes of exposure are described in Chapter 6 of the UN Recommendations. When a substance exhibits a different order of toxicity by two or more of these routes of administration, the highest degree of danger indicated by the tests should be assigned.

The criteria for grouping a substance according to the toxicity it exhibits by all three routes of administration are presented in Chapter 6 of the UN Recommendations.

3.1.6.2. Infectious Substances (Division 6.2)

Infectious substances contain viable microorganisms: bacterium, virus, *Rickettsia*, parasite, fungus, or a recombinant, hybrid or mutant, that are known or reasonably believed to cause disease in animals or humans. However, they are not subject to the recommendations for this division if the spread of disease to humans or animals exposed to such substances is considered unlikely.

Infectious substances should be considered for classification in Division 6.2 and for assignment to UN 2814 (UN-No. 3245: genetically modified microorganisms) or UN 2900 after allocation to four risk groups, based on criteria developed by the World Health Organization.

Genetically modified microorganisms and organisms are microorganisms and organisms in which genetic material has been altered by genetic engineering.

Biological products are either finished biological products for human or veterinary use, manufactured in accordance with the requirements of national public health authorities, transported, if required, under special approval or license from such authorities; or biological products transported prior to licensing for development or investigation; or finished products for experimental treatment of humans or animals, manufactured in compliance with the requirements of national public health authorities. They also cover unfinished biological products, prepared in accordance with procedures of specialized government agencies.

Some licensed vaccines may present a biohazard in certain parts of the world only. In that case, competent authorities may require these vaccines to comply with the requirements for infectious substances, or may impose other restrictions.

Diagnostic specimens are any human material including, but not limited to, excreta, secreta, blood and its components, tissue and tissue fluids, transported for diagnostic or investigation purposes, but excluding live infected animals.

Wastes transported under UN 3291 (UN-No. 3291: clinical waste, unspecified, n.o.s. or (bio) medical waste, n.o.s. or regulated medical waste, n.o.s.) are wastes derived from the medical treatment of animals or humans or from bioresearch where there is a relatively low probability that infectious substances are present. Waste infectious substances which can be specified should be assigned to UN 2814 (UN-No. 2814: infectious substances, affecting humans) or to UN 2900 (UN-No. 2900: infectious substances, affecting animals only). Decontaminated wastes which previously contained infectious substances should be considered nondangerous, unless the criteria of another class are met.

These requirements are also applicable to RID/ADR and IMO from 1 January 1995.

3.1.7. Radioactive Material (Class 7)

UN Recommendations. A radioactive material is defined as any material with specific activity > 70 kBq/kg (0.002 μCi/g). In this context, specific activity means the activity per unit mass of radionuclide or, for a material in which the radionuclide is essentially uniformly distributed, the activity per unit mass of material.

Regulations regarding the transport of radioactive material have been prepared by the International Atomic Energy Agency (IAEA) in consultation with the United Nations, specialized agencies, and IAEA Member States.

3.1.8. Corrosive Substances (Class 8)

UN Recommendations. These are substances which, by chemical action, cause severe damage on contact with living tissue, or, in the case of leakage, materially damage, or even destroy, other goods or the means of transport; they may also cause other hazards.

Special Recommendations Relating to Class 8. Substances and preparations of Class 8 are divided among three packing groups, according to their degree of hazard in transport as follows:

Packing Group I:	very dangerous substances and preparations
Packing Group II:	substances and preparations presenting medium danger
Packing Group III:	substances and preparations presenting minor danger

Allocation of substances to the packing groups in Class 8 has been on the basis of experience, taking into account such additional factors as inhalation risk and reactivity with water (including the formation of dangerous decomposition products). [A substance or preparation meeting the criteria of Class 8 having an inhalation toxicity of dusts and mists (LC_{50}) in the range of Packing Group I, but toxicity through oral ingestion or dermal contact only in the range of Packing Group III or less, should be allocated to Class 8.] New substances, including mixtures, can be judged by the contact time necessary to produce full-thickness destruction of human skin. Substances judged not to cause full-thickness destruction of human skin should still be considered for their potential to cause corrosion of metal surfaces.

In making this grouping, account should be taken on experience in instances of accidental exposures to humans. In the absence of such experience grouping should be based on data obtained from animal experiments in accordance with OECD Guideline 404 [15].

The test criteria for the three groups in this class are:

Packing Group I (very dangerous substances): Substances that cause full-thickness destruction of intact skin tissue within an observation period of 60 min, starting after an exposure time of 3 min or less.

Packing Group II (substances presenting medium danger): Substances that cause full-thickness destruction of intact skin tissue within an observation period of 14 d, starting after an exposure time of more than 3 min, but not more than 60 min.

Packing Group III (substances presenting minor danger):

(a) Substances that cause full-thickness destruction of intact skin tissue within an observation period of 14 d, starting after an exposure time of more than 60 min, but not more than 4 h.

(b) Substances which are judged not to cause full-thickness destruction of intact skin tissue, but which exhibit a corrosion rate on steel or aluminum surfaces exceeding 6.25 mm/a at 55 °C. For the purposes of testing steel, type P 3 [ISO 2604 (IV): 1975] or a similar type, and for testing aluminum, non-clad types 7075-T 6 or AZ 5 GU-T 6 should be used.

3.1.9. Miscellaneous Dangerous Substances and Articles (Class 9)

UN Recommendations. These are substances and articles which during transport present a danger not covered by other classes, including substances that are transported or offered for transport at tempertures ≥ 100 °C in a liquid state or at temperatures ≥ 240 °C in a solid state.

IMO Definitions. Class 9 comprises:

1) Substances and articles not covered by other classes which experience has shown, or may show, to be of such a dangerous character that the provisions of part A Chapter VII of the International Convention for the Safety of Life at Sea, 1974, as amended, should apply
2) Substances not subject to the provisions of part A of Chapter VII of the aforementioned Convention, but to which the regulations of Annex III of the International Convention for the Prevention of Pollution from Ships, 1973, as modified by the Protocol of 1978 (MARPOL 73/78), apply

ADR/RID Definitions. Class 9 covers substances and articles which, during carriage, present a danger not covered by the headings of other classes; it includes:

Substances which, on inhalation as fine dust, may endanger health
Substances and apparatus which in the event of fire may form dioxins
Substances evolving flammable vapor
Lithium batteries

From 1 January 1995, a new subsection "Environmentally hazardous substances" should be implemented, including UN 3077 (Environmentally hazardous substances, solid, n.o.s.), and UN 3082 (Environmentally hazardous substances, liquid, n.o.s).

These UN numbers take into account inland water pollution effects as well as genetically modified microorganisms and organisms.

3.1.10. Estimation of Major Hazards when there are Several Hazardous Properties

Precedence of Hazard Characteristics of the UN Recommendations. Table 1.44 of the UN Recommendations may be used as a guide in determining the class of a substance, mixture, or solution having more than one risk, when it is not named in the list of dangerous goods in the UN Recommendations. For goods having multiple risks which are not specifically listed, the most stringent packing group assigned to the respective hazards of the goods takes precedence over other packing groups, irrespective of the precedence of the hazard table. The precedence of hazard characteristics of the following have not been dealt with in the table, as these primary characteristics always take precedence:

Class 1 substances and articles
Class 2 gases
Division 4.1 self-reactive and related substances and desensitized explosives
Division 4.2 pyrophoric substances
Division 5.2 substances
Division 6.1 substances with a Packing Group I inhalation toxicity, except for substances or preparations meeting the criteria of Class 8 having an inhalation toxicity of dusts and mists (LC_{50}) in the range of Packing Group I, but toxicity through oral ingestion or dermal contact only in the range of Packing Group III or less, which should be allocated to Class 8
Division 6.2 substances
Class 7 material

3.1.11. Seawater Pollutants and Inland Water Pollutants

In Germany, the individual federal states are responsible for regulating the transport of inland water pollutants, as reflected in the *Wasserhaushaltsgesetz* (WHG Paragraph 19 G). From 1 January 1995 the transport of inland water pollutants is regulated by the new section F of Class 9 marginals 11 and 12 ADR/RID.

Seawater Pollutants. At the International Conference of Marine Pollutants, 1973, the need was recognized to preserve the marine environment. Provisions are contained in Annex III of the *International Convention for the Prevention of Pollution from Ships, 1973*, as modified by the Protocol of 1978 (MARPOL 73/78). The Marine Environment Protection Committee, at its 21st session in 1985, decided that Annex III should be implemented through the International Maritime Dangerous Goods Code (IMDG-Code). This decision was endorsed by the Maritime Safety Committee at its 51st session in 1985.

Substances, materials, or articles are identified in the individual schedules of the IMDG-Code as Marine Pollutants. Substances, articles or materials that are so identified, but do not possess any other hazard, are listed in the appendices to the schedules for UN 3077, or UN 3082, in Class 9.

Solutions, Mixtures, and Isomers. A solution or mixture containing $\geq 10\ \%$ of one or more Marine Pollutants is itself a Marine Pollutant.

Certain Marine Pollutants have an extreme pollution potential and are identified as Severe Marine Pollutants in the individual schedules. A solution or mixture containing 1 % or more of these Severe Marine Pollutants is a Marine Pollutant.

A solution or mixture which does not fall within the criteria of Classes 1 – 8, but which meets the criteria for Marine Pollutants should be offered for transport as an Environmentally Hazardous Substance, Solid, N.O.S, or as an Environmentally Hazardous Substance, Liquid, N.O.S, under Class 9, even though it is not listed by name.

An isomer of a substance identified as a Marine Pollutant and covered by a generic entry in Classes 1 – 8, but which does not fall within the criteria of these classes, should be offered for transport as an Environmentally Hazardous Substance, Solid, N.O.S, or as an Environmentally Hazardous Substance, Liquid, N.O.S, under Class 9, even though it is not listed by name.

Packages containing Marine Pollutants should be durably marked with the Marine Pollutant Mark; marking should not apply to:

1) Packages containing Marine Pollutants in inner packagings with contents of: 5 L or less for liquids; or 5 kg or less for solids
2) Packages containing Severe Marine Pollutants in inner packagings with contents of: 0.5 L or less liquids; or 500 g or less for solids

In the case of loss or likely loss overboard of marked Marine Pollutants into the sea, the master or other person in charge of the ship concerned must report the particulars of such loss or likely loss overboard, wherever it occurs, by the fastest telecommunication channel available with the highest possible priority to the nearest coastal state.

Inland Water Pollutants. In 1995 new requirements for the inclusion of Environmentally Hazardous Substances (Aquatic Pollutants) came into force. The criteria are consistent with the final text for substances pollutant to the aquatic environment adopted by the EC in Commission Directive No. 91/325/EEC, Section 5 [16].

In Appendix III/A.3 of RID/ADR a new Section G has been added test methods to determine the ecotoxicity, persistance and bioaccumulation of substances in the aquatic environment for classification in Class 9. The test methods used shall be those adopted by the OECD and EC. If other methods are used, they shall be internationally recognized, be equivalent to the OECD/EC tests and be referenced in the test report.

Tests to determine aquatic pollutant effects are as follows:

Acute toxicity for fish
Acute toxicity for *Daphnia*
Algal growth inhibition
Tests for bioaccumulation potential

3.2. Storage of Dangerous Goods

In Germany, the Technical Regulations for Dangerous Substances (*Technische Regeln für Gefahrstoffe*, TRGS) established by the Committee of Dangerous Substances (*Ausschuß für Gefahrstoffe* (AGS) reflect, albeit never definitively and completely, the state of regulations and knowledge mentioned in § 17 of the Directive of Dangerous Substances (*Gefahrstoff V*), especially for handling purposes. This also involves legal considerations of the storage of dangerous goods and substances.

The regulations to be applied are as follows:

- TRGS 511: storage of ammonium nitrate
- TRGS 514: storage of severely toxic and toxic substances in packaging and in portable tanks
- TRG 515: storage of oxidizing substances in packaging and in portable tanks

3.2.1. Explosive Substances (Class 1)

The storage of explosive substances is regulated by the Second Directive to the Explosive Law (*2. SprengV*) in its latest edition of 5 September 1989, the Technical Regulation for the Storage of Explosives, which are subject to 2. SprengV and the section on Accident Prevention Regulation (*Unfallverhütungsvorschrift*, UVV) 46 a *Explosive Substances and Articles with Explosive Substances*, General rule (VBG 55 a) established by the Employers' Liability Insurance Association of Chemical Industries (*Berufsgenossenschaft der chemischen Industrie*). In 2. SprengV in §1, item (2), No. 1 and 2, the area of application is clearly described. The Directive is not applicable to explosive substances during transportation by road, rail, water, or air, and on vehicles mentioned in § 1, item (2), No. 1 and during loading or unloading.

Storage Groups. Explosives and explosive articles are classified in four storage groups, by analogy to the transport Divisions 1.1 – 1.4. There are no storage groups corresponding to Divisions 1.5 and 1.6.

The assignment of substances and articles to storage groups is normally based on results of tests which are carried out by the Federal Institute for Materials Research and Testing (BAM). These tests and the criteria for the assignment to storage groups are described in the Rules for the Storage of Explosives (*Sprengstofflagerrichtlinie*, Spreng LR 010). Of most importance are the properties of the substances and articles in their packagings, in connection with their degree of risk to the public. Explosive substances from the chemical industry are also covered by the German Explosives Law, but for transportation purposes are normally placed in classes other than Class 1 (e.g., Classes 4.1 or 5.2). Such explosive chemicals are also assigned to certain storage groups on the basis of experimental procedures described the Spreng LR 011.

Compatibility Groups. According to certain properties of the substances or articles, a compatibility group is determined (A, B, C, D, E, F, or G). The definitions of the compatibility groups correspond to those used for transport.

3.2.2. Flammable Liquids (Class 3)

For storage of flammable liquids in Germany specific Technical Regulations (*Technische Regeln für brennbare Flüssigkeiten*, TRbF) are valid. Their objective, as part of the according Directive (*Verordnung über brennbare Flüssigkeiten*, VbF), is to regulate filling, loading, and unloading of flammable liquids.

The regulations are divided as follows:

- TRbF 001: general provisions, structure, and applicability
- TRbF 002: standards, etc.
- TRbF 003: testing of flammable liquids
- TRbF 100/200: general safety requirements
- TRbF 111/211: places of filling, emptying
- TRbF 120/220: general provisions for metallic or nonmetallic fixed tanks
- TRbF 121/221: specific provisions for metallic or nonmetallic fixed tanks

According to the VbF, flammable liquids are those which are liquid or viscous at a temperature not exceeding 35 °C and which have vapor pressure not exceeding 300 kPa (3 bar) at a temperature of 50 °C.

3.2.3. Oxidizing Substances (Class 5.1)

The storage of oxidizing substances in packaging and in portable tanks should be in accordance with the provisions of the technical rule TRGS 515. This rule is not valid for substances in process, handling, storage under the provision for the transport of dangerous goods, and amounts of substances stored up to 200 kg. From 1 October 1993, the storage and handling of organic peroxides is regulated by a new Accident Prevention Regulation (VBG 58) published by the Employer's Liability Insurance Association of Chemical Industries Peroxides are assigned, according to their danger, to storage groups, similar to those for explosive substances.

3.2.4. Toxic Substances (Class 6.1)

The storage of toxic substances in packaging and in portable tanks should be in accordance with the provisions of the technical rule TRGS 514. This rule is not valid for substances in process, handling, storage under the provision for the transport of dangerous goods, and amounts of substances stored up to 200 kg (50 kg for very toxic substances).

3.2.5. Radioactive Material (Class 7)

Within the nuclear fuel cycle, the safe storage of different types of radioactive material is necessary. An important product for nuclear fuel fabrication is uraniumhexafluoride UF_6 which is shipped and stored in steel cylinders designed to withstand a test pressure of 2.8 MPa (28 bar).

Spent nuclear fuel must be stored after removal from the reactor, first in water ponds within the nuclear power plant. This first stage of storage (when most of the decay heat is dissipated) is followed by reprocessing or interim storage for a longer period, up to 50 years. There are several interim storage options for spent fuel or high-level waste:

- Dry storage in an inert gas atmosphere in casks, silos, or vaults
- Wet storage in water ponds

In every case, the technical measures have to satisfy the following protection aims:

Shielding against radiation
Containment of the radioactive inventory
Nuclear safety
Heat removal

The back-end of the nuclear fuel cycle is the disposal of the radioactive waste products, or the direct disposal of spent nuclear fuel.

In some countries, low-level or non-heat-generating waste is disposed of in above-ground repositories (United States, France); in other countries in deep underground repositories (Sweden, Germany). High-level waste or spent fuel should be disposed of only in deep geological formations. The general safety principles for the disposal of radioactive materials are:

- Immobilization of radioactive isotopes (e.g., by vitrification, cementation)
- Containment by technical and geotechnical barriers during the service period of the repository
- Long-term containment by geological barriers

4. Safety Measures

To minimize the risk in transportation, handling, and storage of hazardous substances appropriate safety measures must:

1) Assure safe service conditions
2) Reduce the probability of incidents or accidents
3) Reduce the consequences of accidental release

Safety measures can be classified as:

1) Measures to assure safe confinement of the hazardous substance during normal service conditions, abnormal service conditions (incidents), and accidents (i.e., accident prevention measures)
2) Measures to reduce the consequences of accidental losses (i.e., accident limitation measures)

Besides this "horizontal" differentiation, a "vertical" division of safety measures into:

Technical safety measures
Operational safety measures

is possible. Table 2 shows the combination of the main components of safety measures. To ensure that all measures are met effectively and within specified limits, qualitiy assurance is necessary, at every stage.

The most interesting problem in safety science and technology is the correct choice (i.e., the selection of the appropriate combination and technical level) of measures to provide a level of safety, accepted on commercial considerations by the supplier, on safety grounds by the workers, and on environmental aspects by the public. Figure 1 shows the correlation between incident probability and consequences. If the specific hazard potential of substance 2 (case 2) is bigger than that of substance 1 (case 1), only a lower "consequence grade" (e.g., a lower release rate) in case of an accident can be accepted; the consequences for substance 2 have to be restricted to the site area, but for substance 1 an accident consequence may be accepted outside. With respect to the safety measures necessary to reduce the incident probability to an acceptable level, the minimum safety in case 2 has to be higher than in case 1.

Table 2. Combination of the main components of safety measures

Safety measures	Normal service safety measures	Accident prevention measures	Accident limitation measures
Technical measures	design of containment and service equipment according to normal conditions	accident-safe design with special equipment (e.g., double containment, fire protection, alarms, shut-down) special provisions (e.g., fail-safe, fire-safe, redundancy, diversity)	separation, size limitation, safety distance, hold-back measures
		marking, labeling	
Operational measures		experience, training	
	handling and testing instructions, maintenance		emergency response, recovery action
		Quality assurance measures	

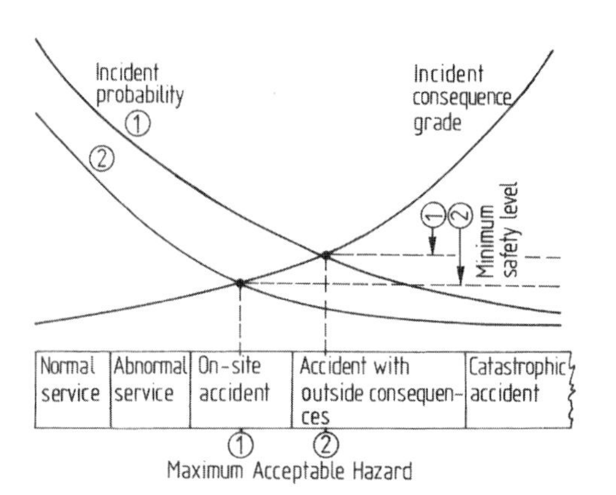

Figure 1. Correlation between incident probability, consequences, and safety level for two cases of acceptable hazard

4.1. Technical Safety Measures

Technical safety measures have to assure safe service conditions or (if they are so designed) safety in accidents. The design of a technical system has to consider the right configuration of "function units" with components made from suitable materials, correctly dimensioned, and equipped with all necessary devices. The design or construction process has to consider all relevant design loads:

Design load
Mechanical impacts
Pressure
 internal and/or external over-pressure
 design pressure
 working pressure
 test pressure
 hydrostatic pressure
 pump pressure
Vibration
Operational loads
Wind, snow floods
Accident impact
 drop
 crush
 puncture
 ripping
 missile
Thermal impact
Substance temperature
Ambient temperature
Fire
 torch fire
 full engulfment
 partial engulfment
 radiation from external fire
Corrosion
Substance
Environment
Time-induced effects
Aging
Creep
Stress relaxation

Construction/design features
Purpose
Lifetime
Materials
Dimensions
Equipment
Configuration of units and components
Safety principles
 safety level
 single or multiple barriers
 redundancy

diversity
fail-safe design
fire-safe design

The following examples illustrate the interdependance of design features, design loads, and purpose.

4.1.1. Design and Equipment of Tanks and Packages

Transport and storage containers exceeding a capacity of 3 m^3 are usually designed as pressure vessels. Commonly, calculation of the resistance to internal over-pressure is based on the maximum vapor pressure of the dangerous substance under service conditions; this pressure depends on the kind of substance and the reference temperature of the vapor pressure. The vessel's resistance to internal over-pressure is determined by the wall thickness and the maximum allowable shell stress value, i.e., a specified proportion of the yield or tensile strength of the tank material. The ratio of strength to allowable stress is the safety factor: the material's ability to withstand mechanical impacts should only be consumed partially under normal service conditions, leaving a reserve in the case of accidents. The calculation of the minimum shell thickness for a cylindrical pressure vessel is given by:

$$s = \frac{D_a p}{20\frac{K}{S}v + p} + c_1 + c_2$$

with s = minimum tank shell thickness (mm); p = design pressure (bar = 100 kPa); D_a = external diameter of the cylindrical tank shell (mm); K = material strength at design temperature (N/mm^2); S = safety coefficient at design pressure; v = "weld coefficient", representing the absorption of the allowable design stress in a welded joint (v = 1 if a 100 % nondestructive test of the welded joint is required); c_1 = allowable tolerances of the tank shell material's dimensional standards; and c_2 = possible effects of corrosion.

Other components of a pressure vessel (heads, manholes, flanges, pipes, valves, etc.) are designed according to similar calculations.

The whole vessel and piping design, including calculation, material selection, equipment, manufacturing, and testing must be carried out according to an accepted pressure vessel design code, e.g., ASME [17], or *AD-Merkblätter* [18].

The safety level expressed by these codes is adequate only up to the pressure hazard potential, usually expressed as the product of pressure and volume. For a given material, the wall thickness may be taken as indication for the safety level of a pressure vessel; every parameter determining this wall thickness (for a given design pressure) depends on the type of vessel material. These parameters have been established on the basis of experience for the most common structural materials, and are defined in the codes. For example, Table 3 shows safety coefficients for some ductile metals; interpretation of these data yields some valuable safety principles:

1) The better the ductility of the material, the lower the safety coefficient to be applied (plastification before fracture can give significant energy absorption before rupture of the vessel)
2) Vessels of forged or rolled steels and aluminum alloys, i.e., metals with reasonably high ductility, are stressed at the final inspection, and in reinspection tests, at a test pressure 1.3 times the maximum allowable working pressure (MAWP); the tank shell is stressed up to 90 % of the material's strength, usually the yield strength
3) Vessels of metals with limited ductility (e.g., cast iron) have to withstand a test pressure which is twice the MAWP

An additional possibility to influence the safety level via the tank shell thickness is the determination of the design pressure of the tank. Usually, the design pressure is the maximum over-pressure that may occur during operation. For liquids of low vapor pressure, the MAWP may be mainly determined by the filling pump pressure; for high vapor pressures (e.g., in transportation and storage of gases) the design pressure is the vapor pressure at the maximum operational temperature. These "reference temperatures", together with the relations between MAWP and test pressure, and the definitions of the maximum allowable stresses for different modes of transport and for storage are given in Figure 2.

Table 3. Safety coefficients for ductile metals

Material	Safety coefficient S at design temperature and maximum working pressure	Safety coefficient S' at test pressure
Forged and rolled steels	1.5	1.1
Cast steels	2.0	1.5
Ductile cast iron (DIN 1693)		
GGG 60/70	5.0	2.5
GGG 50	4.0	2.0
GGG 40	3.5	1.7
GGG 40.3 (with guaranteed toughness)	2.4	1.2
Aluminum and Al alloys	1.5	1.1

For the same tank size and material, calculation for a pressure-liquified gas such as propane shows that a road transportation tank with sun protection has to be designed with a shell 1.33 times thicker than that of an above-ground storage vessel. Transportation regulations generally have to consider a higher probability of accidents impacting the tank from the outside, e.g., in crashes. For hazardous liquids, this is achieved by setting minimum shell thickness standards or "fictive" high design pressures, to design a tank that is more resistant to external forces.

According to the specific hazard potential of substances, land transportation regulations (RID/ADR) require tank shell thicknesses based on "fictive" design over-pressure steps of 0.4, 1, 1.5, and 2.1 MPa; sea transportation regulations (IMGD-Code) require minimum shell thickness values of 6, 8, 10, or 12 mm (mild steel, tank diameter > 1.8 m). These regulatory shell thickness requirements qualitatively enhance tank safety, but only in a very general manner, and do not consider quantitatively safety against external impact. Attempts to increase the safety of tanks by technical measures and material selection are made in risk assessment studies [19,20], or by investigations of the global or local resistance of tanks or tank components to accidental mechanical forces [21,22].

The safety of packagings and intermediate bulk containers (IBC) has to be proved mainly by performance tests. According to the UN Test Requirements for Packagings (with the exceptions of Class 2 and Class 7 substances) the following tests have to be carried out on every design type:

Drop Test. The number of test samples and the drop orientations depend on the kind of packaging; the packaging must be dropped onto a rigid, nonresilient, flat, horizontal target. The drop height depends on the packing group (PG) of the substance, i.e; the specific hazard potential: 1.8 m for PG I, 1.2 m for PG II, and 0.8 m for PG III.

Leakproofness test at an air pressure not less than 30 kPa (PG I) or not less than 20 kPa (PG II, III).

Internal pressure (hydraulic) test at a test pressure of 250 kPa (PG I) or not less than 1.5 times the vapor pressure at 55 °C (minus 10 kPa), but at least 10 kPa (PG II, III).

Stacking test with minimum stacking height of 3 m.

A very special case is the set of performance test criteria for so- called type-B packages for radioactive materials. These criteria include tests of the ability to withstand severe accident conditions in transport. A package sample is subjected to the cumulative effects of mechanical tests (9 m drop onto an unyielding target, 1 m drop onto a steel bar, and a crush test by a 500 kg mass dropped from a height of 9 m) as well as a thermal test (30 min full fire engulfment, e.g., in a kerosene pool fire with flame temperatures exceeding 800 °C). An additional requirement is a water immersion test (150 kPa external gauge pressure of 2 MPa for packages with a radioactive inventory exceeding a specified limit). These severe performance tests, after which a specified activity release rate has to be guaranteed, lead to a package design that is safe, even under the worst accident conditions.

4.1.2. Fire Protection Measures

The biggest losses in industrial sites are often caused by fires. Storage sites or tanks with large amounts of hazardous materials have to be carefully fire protected. The usual classification is:

"Passive" fire protection measures, such as buildings, building components, insulation layers, and coatings resistant over a specified duration to a specified fire scenario, e.g., a furnace test fire [23]

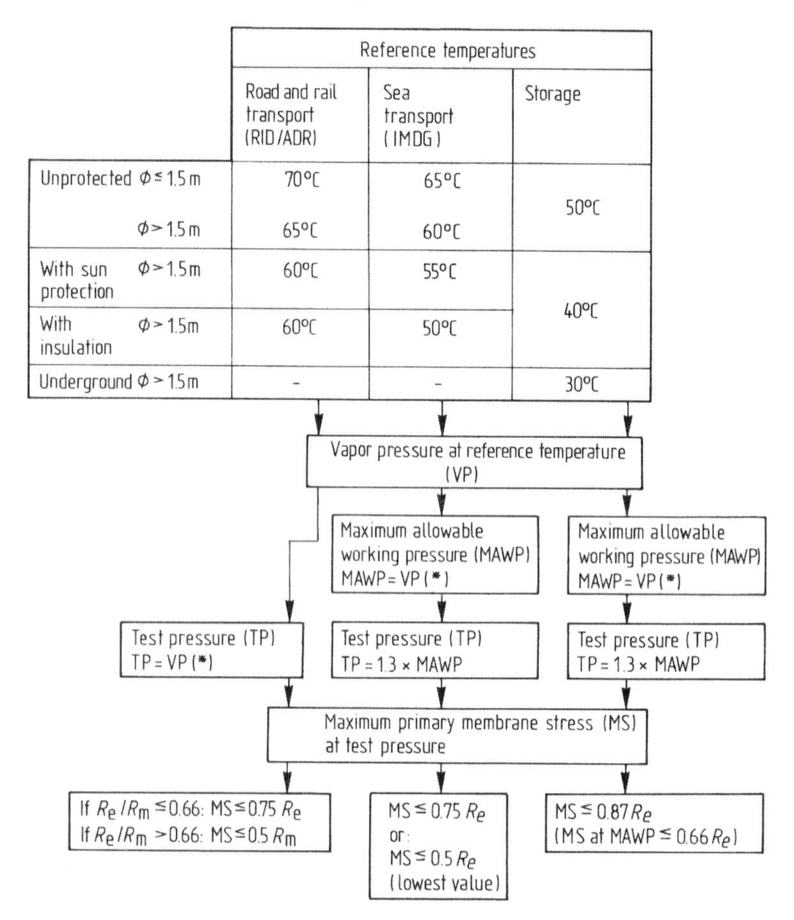

Figure 2. Pressure design criteria for gas tanks
(∗) Set pressure of the pressure-relief device.

"Active" fire protection measures, such as fire extinguishing systems, sprinkler and water spray systems

For effective fire protection of dangerous goods tanks, and for fire fighting in sites where the tanks are on fire or close to it understanding the behavior of tanks engulfed by a fire is essential. For LPG tanks Figure 3 shows the relation between fire duration and vessel-bursting pressure (upper curves), and the internal over pressure (lower curves). If the tank is equipped with a pressure-relief valve, and protected by an effective coating, which may be a layer of insulation, intumescent or subliming material [24], the internal over-pressure is always well below the bursting pressure. An unprotected tank, although equipped with a high-capacity pressure-relief device, will burst after a reasonably short fire duration if the tank is fully engulfed. If the tank is only partially engulfed this will happen just as fast if the fire engulfs the tank shell in the vapor space, where the shell is heated very rapidly; the shell wetted by the liquid is not heated so rapidly because of efficient heat transfer to the liquid. Figure 3 also shows cases where the tank has little or no pressure-relief apparatus; vessel bursting may occur very rapidly, in the case of LPGs leading to a boiling liquid expanding vapor explosion (BLEVE) with desastrous consequences for the environment (tank fragmentation, fire-ball heat radiation, and explosion overpressure).

Fire extinguishing systems, commonly used in closed-storage facilities, put the storage space under inert gas (e.g., CO_2) or foam to extinguish a fire. Sprinkler and water spray systems oper-

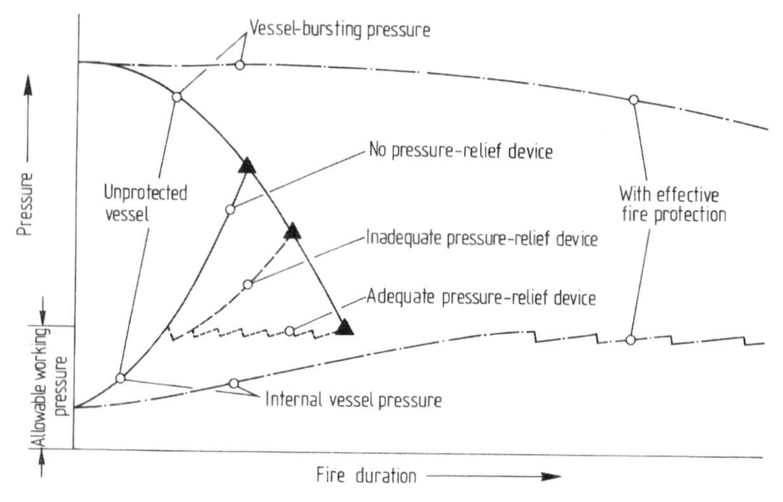

Figure 3. Liquefied petroleum gas (LPG) tank behavior in a fire
△ = BLEVE (boiling liquid expanding vapor explosion)

ate by cooling. All these systems have to be initiated, usually automatically by fire detectors; their reliability depends on the proper functioning of many components, and their full efficiency is reached only after a lag time. Therefore, inherently safe fire protection measures, e.g., a fire-resistant construction, is often preferred for sites with major risks. Details of fire protection methods can be found in industrial handbooks e.g., [25,26].

4.2. Quality Assurance Measures

Quality assurance (QA) programs should be established for the design, manufacture, testing, documentation, use, maintanance, inspection, transport, handling, and storage operations ofall containments for dangerous goods to ensure compliance with the relevant safety regulations and the provisions of approval certificates. In this context, "quality" is the compliance of a tank or package at any stage of its lifetime with the regulations, so that an approved safe condition is achieved.

For a package, a brief survey of QA is shown in Figure 4.

Quality assurance is based on a quality policy, quality management (within all institutions that design, manufacture, operate, test, examine, inspect, or approve), and a system of quality controls and audits.

Basic ISO QA standards (with comparable European standards) are given in:

- ISO 9000 (EN 29 000) Quality Management and Quality Assurance Standards – Guidelines for Selection and Use
- ISO 9001 (EN 29 001) Qualitiy Systems – Model for Quality Assurance in Design/Development, Production, Installation and Servicing
- ISO 9002 (EN 29 002) Qualitiy Systems – Model for Quality Assurance in Production and Installation
- ISO 9003 (EN 29 003) Quality System – Model for Quality Assurance in Final Inspection and Test
- ISO 9004 (EN 29 004) Quality Management and Quality System Elements – Guidelines

Quality can be produced only by skilled engineering, manufacture, and operation, by well-educated and trained personnel. A very sophisticated QA program is able to filter out failures only to a certain extent.

4.3. Safety Distances

Accidental release should be prevented by technical and operational safety measures as described above, but within a safety analysis for

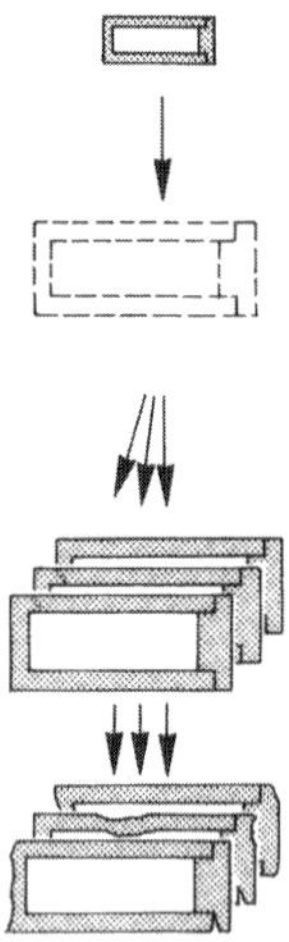

Phase	Item	Quality assurance tasks	Quality assurance instruments
Design control and approval	Test object	Quality control of: test object properties test parameters	Test program, test records, examination and documentation of test object properties, test report
	Package design	Validation of design safety evaluation, considering: transferability of the test results, assessment of quality standards	Safety evaluation report, including design specifications (drawings, specifications of materials and components)
Manufacture	Package produced	Assessment of fabrication and test procedures Quality control to assure design specifications Evaluation of deviations from specification	Quality assurance program QA manual fabrication and inspection plans including: fabrication instructions test instructions responsibility of inspectors/experts according to a classification system final inspection plan QA records
Operation		Maintenance, safe operation, and quality control to assure design specifications	Operation and inspection plan Maintenance and reinspection test instructions } Transport Acceptance test criteria Storage manual } Storage

Figure 4. Quality assurance

any site where a major risk exists, the consequences of a potential hazardous material release have to be determined.

Estimation of consequences may be necessary for the planning of emergency response action, as well as for plant construction or storage size limitation to provide appropriate distances between the hazard source and other plant areas (storage spacing) or public facilities. First, a realistic accident scenario has to be considered to derive a source term (instantaneous release of a certain amount of dangerous goods, or a specified continuous release rate). Depending on the behavior of the substance in mind, the subsequent reaction may be ignition, explosion, release of hazardous reaction products, or dispersion only. Depending on the substance-specific hazard potential (specific toxicity, flammability, explosibility, causticity, radioactivity, etc.), and other influences (e.g., metereologic and plant conditions) the hazard spreads, up to a certain distance. The distance up to which the hazard is at an unacceptable level, e.g., 50 % lethal dose toxicity (LD_{50}), or heat radiation or explosion over-pressure causing severe injury or building damage is termed the “safety distance.”

The effects of the release of flammable liquids and gases (heat flux, over-pressure and tank missile flight after a BLEVE, vapor cloud explosion, or fire) can be estimated from formulae in the appropriate guides and handbooks, e.g., [28, 30]. Somewhat more difficult is the calculation of gas dispersion to estimate, e.g., the lower flammability distance in LPG releases or toxic concentration distances (e.g., for ammonia or chlorine releases) because of the effects of metereologic circumstances; calculation standards for these are also available [31].

If the existing safety distance is less than an estimated unacceptable hazard distance, additional safety measures have to be taken. These may be either source-term-reducing measures, such as smaller vessels, thinner or shorter pipes between shut-off valves, or hazard-distance-reducing measures, such as gas-tight walls or confinements, or water curtains to reduce the gas concentration.

4.4. Emergency Response

Emergency response action is necessary as a back-up, to limit the consequences of an accidental release of hazardous substances in case of a failure of all technical barriers. The correct and durable marking, labeling, and placarding of dangerous goods containments is the first important measure for corrective action. Transport media for dangerous goods must carry additional labels or instructions identifying the primary and subsidiary risks of the substance, and indications of the initial emergency response. More details are necessary for the fire fighters at the accident site. Some countries have set up special task-forces with appropriate equipment for dangerous goods transport accidents; these task-forces can get additional support from quick-response hazardous substances data banks from specialist advisers, or from technical equipment supplied by the chemical industry. Support is organized either on a national basis, e.g., in Germany by TUIS, a transportation accident information and emergency response system run by the chemical industry [32], or internationally, e.g., in Europe by ICE run by the European Chemical Industry Council, CEFIC [33]. For major-risk storage sites, a detailed alarm and emergeny response plan that regulates all necessary action inside the plant, and the activation of external help by local authorities and fire fighters must be approved and regularly up-dated during operation.

5. Regulations and Standards

5.1. World Transport Regulations

The purpose of all regulations for the transport of dangerous goods is to minimize the associated risks, while limiting the costs for the industry concerned to an acceptable level.

The basis of all regulations is a classification system for dangerous goods by type of hazard and degree of danger.

According to the predominant hazard of the dangerous substance, classification is the same for all regulations for the transport of dangerous goods:

Class 1	Explosives
Class 2	Gases
Class 3	Flammable liquids
Class 4.1	Flammable solids
Class 4.2	Substances liable to spontaneous combustion
Class 4.3	Substances which in contact with water emit flammable gases
Class 5.1	Oxidizing substances
Class 5.2	Organic peroxides
Class 6.1	Toxic substances
Class 7	Radioactive material
Class 8	Corrosive substances
Class 9	Miscellaneous dangerous substances

About 2500 individual substances and articles have been classified; substances, not specifically listed in the regulations, but which are dangerous according to the criteria, have to be classified, to be acceptable for carriage under the same conditions as substances presenting a comparable hazard. Some substances are too dangerous for transport, and may be carried only in a stabilized form or in solutions or mixtures. All regulations specify the conditions of carriage for the dangerous substances, such as:

Conditions of packing
Marking and danger labels on packages
Information in the transport document
Conditions for tank transport
Mixed loading and segregation
Special conditions for different modes of transport

5.1.1. UN Recommendations on the Transport of Dangerous Goods

Since 1956 a Committee of Experts on the Transport of Dangerous Goods established by the United Nations Economic and Social Council has been developing Recommendations on the Transport of Dangerous Goods [34]: Although the UN Recommendations have no legal force, they aim at presenting a basic scheme to allow international regulations for the various modes of transport to develop in a uniform fashion.

All bodies responsible for the international regulations have committed themselves to harmonize their regulations as far as possible with the principles of the UN Recommendations.

The most important part of the Recommendations is the *List of Dangerous Goods Most Commonly Carried*. A four-digit number is assigned in the list to each substance or article. Unfortunately, this UN number gives no information about the hazards of the substance or article.

Classification of dangerous substances not individually listed in the Recommendations is possible under a great number of "generic" or "not otherwise specified" (N.O.S.) entries. Such classifications should be based on the results of tests, and on experience [35].

The UN Recommendations also cover provisions for packing, multimodal tank transport, consignment procedures, marking, and labeling.

The Committee of Experts gives no recommendations for the transport of radioactive materials. In 1959, the UN Economic and Social Council asked the International Atomic Energy Agency (IAEA) to prepare Regulations for the Safe Transport of Radioactive Material [36]. These Regulations, which are referred to in the UN Recommendations, have been introduced in all international regulations for the transport of dangerous goods.

5.1.2. The IMDG-Code

The International Maritime Organization in London has prepared requirements for the safe transport of dangerous goods by sea. These requirements were originally based on the *International Convention for the Safety of Life at Sea* (SOLAS) of 1960. To prevent marine pollution by the transport of dangerous substances, the requirements also include provisions based on the *International Convention for the Prevention of Pollution from Ships* (MARPOL) of 1973.

The requirements are laid down in the *International Maritime Dangerous Goods Code* (IMDG-Code, [37]). The IMDG-Code is recommended to governments for adoption as the basis for national regulations in pursuance of their obligations under the SOLAS Convention.

The IMDG-Code follows very closely the UN Recommendations with regard to the classification, packing, and tank transport of dangerous goods.

A special feature of the IMDG-Code is the presentation of provisions for packing: an individual page for each substance or entry makes the IMDG-Code the most voluminous of all international regulations for the transport of dangerous goods.

In addition to the provisions in accordance with the UN Recommendations, the IMDG-Code sets out special requirements for the transport of dangerous goods in ships, such as:

- Provisions for freight container traffic
- Stowage and segregation of dangerous goods on ships
- Fire precautions
- Dangerous goods on roll-on, roll-off ships
- Transport of solid bulk materials.

Furthermore, a supplement to the IMDG-Code includes publications of the International Maritime Organization relating to the IMDG-Code, such as:

- The Emergency Procedures for Ships carrying Dangerous Goods (EmS)
- The Medical First Aid Guide (MFAG)
- The Code of Safe Practice for Solid Bulk Cargoes
- Reporting procedures
- The Guidelines for Packing Cargo Transport Units
- Recommendations on the use of pesticides in ships.

5.1.3. The ICAO Technical Instructions

The International Civil Aviation Organization in Montreal is responsible for regulating the air transport of dangerous goods. In 1984, the first edition of the ICAO *Technical Instructions for the Safe Transport of Dangerous Goods by Air* [38], based on Annex 18 to the Convention on International Civil Aviation was introduced. Before then, the transport of dangerous goods by air had been regulated by the International Air Transport Association (IATA), also in Montreal in the IATA Restricted Articles Regulations.

The ICAO Technical Instructions are also fully harmonized with the UN Recommendations as regards the classification and the provisions for packing, labeling, and documentation.

There are, however, many restrictions on air transport: many substances and articles from the UN Recommendations, e.g., explosives, are forbidden for air transport, and there are many restrictions on the net quantity per package. Restrictions are more severe for passenger than for cargo aircraft.

Most of the conditions of carriage in the ICAO Technical Instructions are included in the alphabetical Table of Dangerous Goods, which also refers to packing instructions and special provisions.

In addition to the provisions in accordance with the UN Recommendations, the ICAO Technical Instructions cover special requirements for air transport, such as:

Provisions for magnetized material
Loading of dangerous goods on aircraft
Information to flight crews
Training of shippers and operators and all other persons handling dangerous air cargo
Variations of the Technical Instructions, notified by states and airline operators.

5.2. Regional Transport Regulations

5.2.1. Europe

The first regulations for international transport of dangerous goods in Europe were developed over 100 years ago for rail transport. Regulations for road transport and transport by inland waterway were developed after World War II, and are all based on the regulations for rail transport (RID). Considerable effort has been made to harmonize all European regulations on aspects such as classification, packing, labeling, and marking as well as the requirements for tank wagons, tank containers, and tank vehicles.

All European regulations are being harmonized with the UN Recommendations. This is difficult, because there are still fundamental differences in safety principles underlying the UN Recommendations and the European regulations.

The Regulations Concerning the International Carriage of Dangerous Goods by Rail (RID). The RID (*Règlement Concernant le Transport International Ferroviaire des Marchandises Dangereuses* [39]) is Annex I to the International Convention concerning the Carriage of Goods by Rail (CIM). The RID is applicable to transport between most European countries and some countries of northern Africa and the Middle East, i.e., the Member States of the Intergovernmental Organization for the International Carriage by Rail (OTIF), based in Bern.

The first regulations for the transport of dangerous goods by rail came into force in 1893. Development of the regulations is the responsibility of the RID Safety Committee, which works closely with road transport experts in the joint RID/ADR Meeting.

Although the criteria for classification of dangerous goods in RID are harmonized with the UN Recommendations, there is a special feature in RID and other European regulations. The conditions for carriage are set out for the substances of each class separately, starting with a systematical list of substances and articles; all substances and articles are grouped under collective headings according to their physical and chemical properties and their degrees of hazard. The appropriate N.O.S. entry from the UN Recommendations concludes every group of substances under the collective heading, to accommodate classification of new substances and articles. This classification system ensures that the same conditions of carriage apply to substances under one collective heading or item; the format of the regulations can be kept simple, without the need for amendment of the regulations for every new substance.

The conditions of carriage in each class cover the following subjects:

- Conditions of packing, including marking and labeling
- Particulars in the consignment note
- Conditions for wagons and loading, including marking and labeling

The RID also includes appendices with provisions for testing substances, packaging (including intermediate bulk containers), for radioactive material, and provisions for tank containers and tank wagons.

The European Agreement Concerning the International Carriage of Dangerous Goods by Road (ADR). The ADR (*Accord Européen Relatif au Transport International des Marchandises Dangereuses par Route* [40]) was concluded in 1957. Most European states are contracting parties to the ADR. Development of the ADR is the responsibility of a Working Party, established by the Inland Transport Committee of the United Nations Economic and Social Council.

Annexes A and B cover the same subjects as RID. Annex B also covers some provisions important for road transport, such as:

Technical equipment of road vehicles

General service requirements, driver training

Provisions concerning loading, unloading, and handling

Provisions concerning the operation of vehicles

The European Provisions Concerning the International Carriage of Dangerous Goods by Inland Waterway (ADN). It is the intention of the UN Economic Commission for Europe to additionally put into force an agreement for the transport of dangerous goods by inland waterways. At present the ADN exists only in the form of a draft.

The Regulations for the Carriage of Dangerous Goods on the Rhine (ADNR). The ADNR (ADN-Rhine) is based on the existing drafts for the ADN and has been developed by the Central Commission for the Navigation on the Rhine in Strasbourg. The plan is for the ADNR to be subsumed by the ADN.

For the transport of packaged goods, the ADNR refers to other regulations for the transport of dangerous goods; it mainly regulates the transport of dangerous goods in bulk, and therefore consists of provisions for the construction, equipment, and operation of vessels.

Directives of the European Communities with Regard to the Transport of Dangerous Goods. It is the intention of the Commission of the European Communities to extend the implementation of the existing international regulations for the transport of dangerous goods to national transport in the Member States of the Community.

5.2.2. United States

The transport of dangerous goods (called hazardous materials in the United States) is subject to Title 49 of the United States Code of Federal Regulations Parts 107 – 180 (49 CFR Parts 107 – 180), except that bulk transport in tank ships and tank barges on the navigable waters of the United States is covered by other regulations. The Associate Administrator for Hazardous Materials Safety of the Research and Special Programs Administration (RSPA) is responsible for maintaining the regulations and is also the United States competent authority for purposes of the ICAO Technical Instructions on the Safe Transport of Dangerous Goods by Air and the IMDG-Code.

In many respects the contents of the United States Hazardous Materials Regulations (HMR) are similar to the international requirements for dangerous goods transport contained in the ICAO TI and the IMDG-Code. This is largely the result of amendments made to 49 CFR in December 1990 and 1994. These amendments substantially aligned United States requirements for classification, transport documents, nonbulk packaging, labeling, package marking, and placarding with the UN Recommendations on the Transport of Dangerous Goods. In addition, the HMR also include vehicle construction and operating requirements for transporting hazardous materials by rail and highway, operating requirements for transporting hazardous materials by air, and the procedures under which an exemption to the regulations may be obtained.

In the case of consignments by air or water, two provisions in the Regulations are particularly important. Subject to specified conditions, 49 CFR 171.11 authorizes transport of hazardous materials on aircraft and any associated highway transport in accordance with the ICAO Technical Instructions. Similarly, 49 CFR 171.12 authorizes transport of hazardous materials by ship and any associated land transport in accordance with the IMDG-Code if part of the transport of the material involves transport by water. A number of exceptions apply to transport in accordance with the ICAO and IMO requirements and the reader is advised to consider these provisions which are identified in 49 CFR 171.11 and 171.12. Some of the more important conditions include:

Environmentally hazardous substances are identified in the Regulations and must be transported under UN 3077 or UN 3082 if they do not meet the criteria of hazard Classes 1 – 8, and special hazard communication requirements apply to environmentally hazardous substances identified in the regulations.

Transport of hazardous materials in portable tanks is subject to 49 CFR which contains its own tank construction requirements.

Materials that meet the criteria of Division 6.1 on the basis of their toxicity by inhalation are referred to as "poison by inhalation" (PIH) materials which are subject to special packaging, marking, labeling, and transport document requirements. It should be noted that a number of substances have been found to meet the criteria of PIH materials, even though this is not reflected by their UN classification.

Gas cylinders used in transport must meet construction requirements in 49 CFR.

Explosives, including materials that exhibit explosive properties, require special authorization by RSPA.

Emergency response information must accompany hazardous materials shipments, and the transport document must include a telephone number where additional information on the material may be obtained at any time while the material is in transport.

Training requirements apply to shippers and carriers that bring hazardous materials into the United States

Additional restrictions pertaining to the transport of radioactive materials are included.

The requirements also include special provisions relating to air transport.

Under 49 CFR 171.12 (c), fewer special provisions apply when hazardous materials transported in accordance with the IMDG-Code are transported within a single United States port area.

5.2.3. Japan

Japanese regulations for transport of dangerous goods by rail and road exhibit some differences from those of European countries and the United States. International carriage of dangerous goods by sea and/or air is subject to the domestic regulations based on the IMDG-Code and ICAO Technical Instructions, but for inland transport by rail and/or road, the current regulations are not based on United Nations or other international recommendations. Dangerous goods can be transported from foreign countries to Japan by ship or aircraft under common classification, packaging, labeling, marking, etc., in compliance with IMDG-Code or ICAO Technical Instructions, but for inland transport by rail and/or road, carriage is regulated by Japanese requirements. In Japan, dangerous goods are regulated en bloc for manufacturing, storage, transportation, consumption, etc., except for regulation of carriage by ship and aircraft.

Japanese regulations for transport of dangerous goods are summarized as follows.

Sea transport is regulated by the following laws:

1) Ships Safety Law. Transport requirements for dangerous goods by ship are provided in the Regulations for Carriage and Storage of Dangerous Goods by Ship as an order of the Ships Safety Law, in which almost all of the IMDG-Code has been adopted.
2) Port Regulation Law. This provides for berthing and cargo handling of a ship carrying dangerous goods in port, but not packaging, labeling, marking, etc., which must be in accordance with the Ships Safety Law.

Air transport is regulated by the Civil Aeronautics Law. This prescribes transport requirements regarding classification, packaging, labeling, marking, etc., and complies the requirements of ICAO Technical Instructions.

Inland transport of dangerous goods is implemented in Japan under comprehensive regulations regarding manufacturing, storage, transport, consumption, etc. Classification, packaging, etc. for transport are not based on international recommendations.

1) Fire Service Law. This provides requirements for storage, road tank vehicles, means of road transport, emergency procedures, etc. It covers Classes 3, 4 and 5 of the UN Recommendations, with some differences in classification criteria.
2) Railway Operation Law. Requirements for carriage of dangerous goods by rail are regulated by this law, but classification, packaging, labeling, marking, etc. are subject to other inland transport regulations, such as the High-Pressure Gas Control Law and the Explosives Control Law. It covers all classes of the UN Recommendations, with some differences in classification criteria.

3) High-Pressure Gas Control Law. Storage, packaging, transport, etc. of gases on land are regulated by this law. Classification criteria are different from the UN Recommendations.
4) Explosives Control Law. Manufacturing, storage, transport, sale, consumption, etc. of explosives are regulated by this law. The Classes and Divisions of the UN Recommendations are not adopted.
5) Poisonous and Deleterious Substances Law. Import, manufacturing, sale, handling, transport, etc. of poisonous and irritant substances are regulated by this law. It covers Classes 6.1 and 8 of the UN Recommendations, with some differences in classification criteria.
6) Law for the Regulation of Nuclear Source Materials, Nuclear Fuel Materials, and Reactors. This provides land transport requirements, e.g., packages, test procedures, radiation protection, for nuclear materials. It adopts IAEA Regulations (SS 6).
7) Law for the Prevention of Damage from Radioactive Isotopes. This law provides land transport requirements, e.g., packages, test procedures, radiation protection, for radioactive materials other than nuclear materials. It adopts IAEA Regulations (SS 6).

6. Containment Technology

6.1. Transport of Piece Goods

Shipment of dangerous goods in packed form, i.e., in packagings or intermediate bulk containers (IBC), is still the most common means of carriage from one site to another by all modes of transport.

6.1.1. UN Packagings, "Packagings by Description", Intermediate Bulk Containers

In its *Recommendations on the Transport of Dangerous Goods* [41], the UN Committee of Experts on the Transport of Dangerous Goods has issued a set of provisions to provide as much safety for the transport of dangerous goods as possible with regard to the needs of all parties concerned, e.g., administrators, commercial companies, or the public.

The recommendations on the packing of dangerous goods form the basis for existing international [42–45] and national regulations. Account is also taken of a prevailing trend to replace the detailed specification/description by tests designed to ensure that the packaging or IBC can withstand normal conditions of transport, and to ensure a desirable level of safety.

6.1.1.1. UN Packagings

Types of Packagings. For the so- called UN Packagings, the types of packaging are defined as follows:

Bags: flexible packagings made of paper, plastic film, textiles, woven material, or other suitable materials.

Boxes: packagings with complete rectangular or polygonal faces, made of metal, wood, plywood, reconstituted wood, fiberboards, plastics, or other suitable material.

Drums: flat-ended or convex-ended cylindrical packagings made of metal, fiberboard, plastics, plywood, or other suitable material. Wooden barrels of jerricans are not covered by this definition.

Jerricans: metal or plastics packagings of rectangular or polygonal cross-section.

Wooden barrels: packagings made of natural wood, of round cross-section, having convex walls, consisting of staves and heads, and fitted with hoops.

There are two more complex forms of dangerous goods packagings:

Combination packagings: consist of one or more inner packagings secured in an outer packaging in such a way that they cannot break, be punctured, or leak their contents into outer packagings (usually boxes).

Composite packagings: consist of an outer packaging and an inner receptacle, constructed so as to form an integral packaging. Once assembled, it remains thereafter an inseparable unit; it is filled, stored, transported, and emptied as such.

General packing requirements applicable to the packing of substances of all classes other than Classes 2 and 7 define the design and behavior of packagings. It is required that these should be constructed and closed so as to prevent any leakage which might occur under normal conditions of transport, by vibration, or by changes in temperature, humidity, or pressure. Parts of packagings which are in direct contact with dangerous substances should not be affected by chemical or other action of those substances, or should be provided with a suitable inner coating or treatment, e.g., a liner.

The packagings should be manufactured and tested under a quality assurance program which satisfies the competent authority that each manufactured packaging meets the requirements of the design type.

Testing and Approval. Requirements in the regulations for package strength are expressed as performance standards or tests. The regulations prescribe what must be achieved, rather than how itis done. Therefore, the most important item is that each packaging, except inner packagings of combination packagings, should conform to design types successfully tested, as set out in the regulations and in accordance with procedures established by the competent authority. The testings consist of drop tests (for all design types) and stackings tests (for all design types except bags and nonstackable composite packagings). Packagings intended to carry liquids, i.e., drums, jerricans, additionally have to pass hydraulic pressure tests and leakproofness tests; if they are made of plastics they also have to prove the compatibility of the plastics materials with the contents, i.e., prestorage for six months with the liquid(s) intended or with so-called standard liquids: wetting solution, acetic acid, *n*-butyl acetate/*n*-butyl acetate-saturated wetting solution, mixtures of hydrocarbons (white spirit, nitric acid, or even water). The latter standard liquids only apply to defined high molecular mass polyethlene packagings. These are the so-called compatibility tests.

Tests must be carried out on packagings and packages prepared as for despatch, including inner packagings of combination packagings. The packagings shall be filled to not less than 95 % of their capacity for solids or 98 % for liquids. A given number of test samples have to pass the tests in such way that the weakest part of the packaging is tested. The severity of the test conditions depends on the packing group(s) (PG) for which the packaging is intended.

There are drop tests from three different heights: 1.8 m for PG I, 1.2 m for PG II, or 0.8 m for PG III. Where the relative density of a liquid to be carried is greater than 1.2 g/cm^3, the drop height (in meters) is calculated on the basis of the relative density by multiplying it by 1.5 for PG I, by 1.0 for PG II, and by 0.67 for PG III. The drop tests are carried out on a target with a rigid, nonresilient, flat, and horizontal surface, having mass 50 times that of the test specimen (see ISO 2248).

There are two different air pressures for the leakproofness test, i.e., not less than 30 kPa for PG I or not less than 20 kPa for PG II and III.

The hydraulic (or internal) pressure test to be applied depends on the vapor pressure of the filling substance, normally at 55 °C, but must not be less than 100 kPa.

The stacking test must prove that the test samples can withstand an additional mass equivalent to a stacking height of at least 3 m.

There are additional tests such as the permeability test for plastic drums and jerricans when used for flammable liquids with flash point < 55 °C, or a supplementary test for wooden barrels.

The competent authority may also establish procedures for the approval of packagings, now common practice in countries which have adopted the UN Recommendations or international regulations derived therefrom into national law.

Coding System and Marking. In order to identify the packagings with their design and tested properties, a system of coding and marking has been developed. The code consists of an arabic numeral indicating the kind of packaging, e.g., 1 for a drum, 3 for a jerrican, 4 for a box, or 5 for a bag, followed by one or more capital letters indicating the nature of the material, e.g., A for steel, G for fiberboard, H for plastics, etc., followed where necessary by an arabic numeral indicating the category of packaging, e.g., 1 when the packaging is designed with a nonremovable head or a closure with diameter ≤ 70 mm, 2 for a removable head or a closure with diameter > 70 mm. There are additional coding provisions for composite packagings and packagings manufactured to different specifications.

The complete marking of a dangerous goods packaging, having met the UN Recommendations and been approved by a competent authority, normally consists of the UN symbol followed by the packing code; a code in two parts, i.e., the letter X, Y, or Z designating the packing groups I, II, or III for which the design type has been approved and the relative density (liquids) or the maximum gross mass in kilograms; either a letter S (solids or inner packagings) or, where a hydraulic pressure test has been passed, the test pressure in kPa; the year of manufacture (last two digits); the mark of the state in which the approval was issued; and an identification mark specified by the competent authority.

Typical markings are for example:

UN 1A1/X/250/94/D/BAM 1234-00, where 1A1 means a steel drum with openings of diameter < 70 mm, submitted to drop tests from a height of 1.8 m, and tested at an internal hydraulic pressure of 250 kPa. The certificate of approval has been issued for PG I by the *Bundesanstalt für Materialforschung und -prüfung* (BAM) in Berlin, being the competent authority in Germany, the registration number is 1234, and the drum was manufactured in 1994 by a company with registration letters 00. This steel drum can be used for liquids of packing groups PG I, II, and III.

Or another example:
UN 4G/Y 20/S/94/D/BAM 9876-QQ, where 4G means a fiberboard box, submitted to drop tests from 1.2 m. The box can be used for solids of PG II and III or inner packaging(s) with substances of PG II and III. Here too the certificate of approval has been issued by the BAM in Germany; the registration number is 9876, and the box was manufactured in 1994 by the manufacturer with the registration letters QQ.

6.1.1.2. Packagings by Description

Though the system of tested and approved design types of packagings has been established for dangerous goods of Classes 1, 3, 4.1, 4.2, 4.3, 5.1, 5.2, 6.1, 6.2, 8, and 9, there are still so-called described packagings, especially for Class 2, and in some cases 7. There are provisions for materials properties, e.g., aluminum alloy receptacles for certain gases of Class 2, or steel receptacles. The same applies for packagings intended for the shipment of low-specific-activity radioactive material (LSA-I) of Class 7, where provisions for general conditions of packagings also apply with no particular requirements for testing.

Class 2 is under revision, and it is planned to establish provisions for the testing of the appropriate packagings.

6.1.1.3. Intermediate Bulk Containers (IBC)

An intermediate bulk container (IBC) is a rigid or flexible portable packaging, other than specified in Section 6.1.1.1, that has capacity ≤ 3000 L (≤ 250 L for sea transport) is designed for mechanical handling, and is resistant to the stresses produced in handling and transport as determined by specified tests. Typically, IBCs are intended for multimodal transport of PG II and III substances.

Types of IBC. *Metal IBCs* consist of a metal body together with appropriate service and structural equipment.

Flexible IBCs consist of a body constituted of film, woven fabric, or any other flexible material, or combination thereof, together with any appropriate service equipment and handling devices.

Rigid plastics IBCs consist of a rigid plastics body, which may have structural equipment together with appropriate service devices.

Composite IBCs with plastics inner receptacles consist of structural equipment in the form of a rigid outer casing enclosing a plastics inner receptacle, together with any service or other structural equipment. They are so constructed that the inner receptacle and outer casing once assembled form, and are used as such, an integrated single unit to be filled, stored, transported, or emptied as such.

Fiberboard IBCs consist of a fiberboard body with or without separate top and bottom caps, if necessary an inner liner (but no inner packagings), and appropriate service and structural equipment.

Wooden IBCs consist of a rigid or collapsible wooden body, together with an inner liner (but no inner packagings) and appropriate service and structural equipment.

The IBCs must be so constructed and closed that none of the contents can escape under normal conditions of transport. All service equipment must be so positioned or protected as to minimize the risk of leakage owing to possible damage during handling and transport. The IBCs must be designed, manufactured, and tested under a quality assurance program which satisfies the competent authority, in order to ensure that each manufactured IBC meets the given requirements.

Testing and Approval. The design type of each IBC must be tested and approved by the competent authority. For each design type, a single IBC must pass the required tests before such an IBC is used. An IBC design type is defined by the design, size, material, and wall thickness, manner of construction, and means of filling and discharging; it may include various surface treatments.

Test shall be carried out on IBCs prepared for despatch, i.e., filled either with the contents intended to be carried or with other substances having at least equivalent physical properties, e.g., mass, grain size, relative density. For the latter, the same calculation procedure used for packagings can be applied.

The main test provisions are:

- A bottom lift test for each design type to be handled in this way, except for flexible IBCs
- A top lift test for metal, flexible, rigid plastics or composite IBCs
- A tear test for flexible IBCs
- A stacking test for all design types
- A leakproofness test for IBCs intended to carry liquids
- An internal hydraulic pressure test for IBCs intended to carry liquids
- A drop test for all design types; another IBC of the same design may be used
- A topple test for flexible IBCs only
- A righting test when a flexible IBC is designed to be lifted from the top or from the side

Coding System and Marking. The code system for IBCs consists of two arabic numerals indicating the type of IBC, a capital letter indicating the nature of the material and, where necessary, an arabic numeral indicating the category of IBC within the type. For composite IBCs, two capital letters are used; the first indicates the material of the inner receptacle, and the second that of the outer packaging.

Rigid IBCs for solids are coded 11 if loaded and/or unloaded by gravity, 21 if loaded and/or unloaded under pressure > 10 kPa, and 31 if intended to carry liquids. Flexible IBCs are coded 13.

The nature of the material is indicated by A for steel (all types and surface treatments), B for aluminum, C for natural wood, D for plywood, F for reconstituted wood, G for fiberboard, H for plastics, L for textiles, M for paper (multiwall), and N for metals other than steel or aluminum; i.e., the letters are the same as for the materials of packagings.

The IBC code is followed in the marking by a letter indicating the packing group(s), i.e., Y for PGs II and III, Z for PG III. The use of IBCs for PG I substances is still under discussion, and up to now scheduled only for certain solids.

Deviating from the marking system of packagings, there is a so-called primary marking consisting of: the UN symbol; the design type code; the letter for the packing group(s); the month and year (last two digits) of manufacture; the mark of the state in which the approval was issued; the name or symbol of the manufacturer, or any other identification as specified by the competent authority; the stacking test load in kilograms; and the maximum permissible gross mass or, for flexible IBCs, maximum permissible load in kilograms.

Additional marking follows the primary one, namely the tare mass (kg) for all IBCs other than flexible IBCs; for metal IBCs, rigid plastics IBCs, and composite IBCs the capacity (L at 20 °C); the date of the last leakproofness test (month and year), if applicable; date of last inspection (month and year); and the maximum filling/discharge pressure (kPa), if applicable. For metal IBCs only, the marking is completed by indicating the body material and its minimum wall thickness (mm) and the serial number of the manufacturer. For rigid plastics IBCs and composite IBCs, the test pressure (gauge) in (kPa), if applicable, concludes the marking.

A typical marking of a metal IBC could be: UN 11A/Y/0394/NL/Van Hoog 001/5500/1500, where 11A means a steel IBC for solids discharged, e.g., by gravity, for PG II and III;

manufactured in March 1994; approved by The Netherlands; manufactured by van Hoog in conformity with a design type to which the competent authority has allocated the serial number 001; load used for the stacking test 5500 kg; maximum permissable gross mass 1500 kg.

Operational Requirements. Before being filled and handed over for transport, every IBC must be inspected to ensure that it is free from corrosion, contamination, or other damage, and with regard to proper functioning of any service equipment.

When filled with liquids, sufficient ullage shall be left to ensure that, at the mean bulk temperature of 50 °C, the IBC is not filled to more than 98 % of its water capacity.

During carriage, no dangerous residue should adhere to the outside of an IBC. Also, IBCs must be securely fastened or contained within the transport unit so as to prevent lateral or longitudinal movement or impact, and to provide adequate external support.

After discharging, an empty IBC must be treated as for a filled IBC, unless it has been purged of the residue of the dangerous substance.

Where IBCs are used for the transport of liquids with a flash point ≤ 60.5 °C or powders liable to dust explosion, measures must be taken to prevent a dangerous electrostatic discharge.

6.1.2. Packagings for Radioactive Material

Required packaging for radioactive material (RAM) is based on the assurance that the level of safety built into such a packaging meets the potential hazards of the contents being shipped. Because of the specific dangers arising from RAM, there are special regulations to ensure their safe transport [46,47].

There are four primary types of package, relating to the activity and physical form of their radioactive contents: excepted packagings, industrial packagings, type A packagings, and type B packagings.

An *excepted package* contains quantities of RAM sufficiently small (e.g., smoke detectors) to allow their exception from most design and use requirements. Such packages must, however, meet certain requirements which ensure that their contents can be identified when opened, and that they can be safely handled and transported. These include design requirements to assure safe handling and stowage of the package, and to exclude adverse effects of shock, vibration, collection and/or retention of water, and chemical or radiolytic degradation of the packaging materials.

Because of the extremely low hazard posed by the contents of excepted packages, they may be transported with limited administrative controls, and without externally visible warning labels. Local postal regulations may permit certain packages of this type to be shipped by mail.

Industrial packages are used to transport materials known as either low specific activity (LSA) or surface- contaminated objects (SCO). Materials of low specific activity, having little activity per unit mass, and certain nonradioactive objects, having low levels of surface contamination, are safe for two reasons: either the contained activity is very low, or the material is in a form which is not easily dispersed and which presents only a low internal radiation hazard.

Materials of low specific activity, such as radioactive ores, can sometimes be transported unpackaged. Otherwise, these materials are transported in industrial packages.

In addition to meeting the requirements for excepted packages, industrial packages must meet requirements relating to normal conditions of transport which include minor mishaps. Industrial package integrity is graded in relation to the hazard posed by its contents. There are three types:

Industrial packages type 1 (IP-1) must meet certain temperature and pressure requirements beyond those imposed on excepted packages

Industrial packages type 2 (IP-2) also have to pass free drop and stacking tests

Industrial packages type 3 (IP-3) must in addition meet water spray and penetration test requirements

Type A packages are intended to provide a safe, economical means to transport relatively small quantities of radioactive material: they are expected to retain their integrity under "normal" abuse, i.e., quite likely to occur during transport:

Falling from vehicles or being dropped from similar heights
Being struck by a sharp object which may penetrate the surface
Being exposed to rain
Having other cargo stacked on them

These packages must be designed to satisfy all the requirements imposed on an IP-3 package, and must also meet more stringent test requirements if the radioactive content is liquid or gaseous. They must also satisfy stringent dimensional, ambient environment, internal pressure, and containment specifications which are not imposed on industrial packages.

It is assumed that Type A packages may be damaged in a severe accident, and that a fraction of their contents may be released. The regulations therefore prescribe limits on the maximum amounts of radionuclides that can be transported in such packages. These limits ensure that in the event of a release the risks from external radiation or contamination are low.

In addition to normal environmental conditions, such as high and low pressure, and to reduced atmospheric pressure as sustained by the excepted and industrial packages, Type A packages must be able to retain their contents and shielding integrity if subjected to:

1) The free drop test, where the test specimen is dropped onto a nonresillient hard target so as to cause maximum damage to the safety features to be tested. The normal drop height is 1.2 m. Heavier packages, of more than 1.5 t are dropped from a lower height ($\geq$0.3 m). Where a package is intended to carry liquids or gaseous contents the drop height is 9 m.
2) The compression (stacking) test, where the package is subjected for a period of 24 h to a load equal to or greater than five times the mass of the actual package.
3) The penetration test, where the packaging is placed on a rigid, flat, horizontal surface, and a bar of mass 6 kg and 3.2 cm diameter with a hemispherical end is dropped from a height of 1.0 m (1.7 m if the package is intended to carry a liquid or gas), with its longitudinal axis vertical, so that it falls onto the center of the weakest point of the package.

Type B packages are used to carry larger amounts of radioactive material. They must be able to withstand the effects of servere accidents; tests for resistance to immersion, representing hypothetical accident conditions, are required. Additionally, each design must be approved by the competent authority of the country in which the package was designed; and under some conditions, by the competent authority of each country through or into which it is shipped.

Type B packages are used for carrying radioisotopes for industrial radiography, irradiated nuclear fuel, nuclear wastes, and similar highly radioactive material. Type B packages must be capable of withstanding both Type A and Type B tests.

Packages Designed for Fissile Materials. This type of package is required if the contents are capable of sustaining a nuclear chain reaction. Special assessments and controls are required for such packages. And each design must be approved by the competent authority of each country from, through, or into which it is shipped.

Type A and B packages and packages for fissile materials also have to meet specific requirements. These are:

Mechanical tests, where any given packaging must pass two tests. In most cases, an impact (or drop) and a penetration (or puncture) test are required. These consist of dropping a test specimen onto two different targets so to cause maximum damage. In one drop test, the packaging is dropped from a height of 9 m onto a unyielding surface, in the other it falls onto a mild steel bar, 15 cm in diameter, from a height of 1 m. The bar is rigidly mounted, perpendicular to the target surface, and must project at least 20 cm above the base. An alternative to the 9 m drop test for some lightweight Type B packages is the so- called *crush-test*. This consists of placing the packaging on the unyielding target in such a way as to experience the maximum possible damage when a 500 kg steel mass is dropped onto it from a height of 9 m.

Thermal test, where the packaging is fully engulfed in a thermal environment of at least 800 °C for 30 min. Any combustion of the packaging's components must be allowed to continue self-extinction.

Water immersion test, where most packagings are immersed in water at a pressure equivalent to a depth of at least 15 m for not less than 8 h. A water immersion test applicable to some packagings for irradiated nuclear fuel requires immersion in water at a pressure equivalent to a depth of at least 200 m for not less than 1 h resulting in no leakage.

Means for identifying shipments of radioactive material are: labels (with the trefoil symbol); markings assigned by the competent authority – both labels and markings must be visible on the packaging; and/or placards (with the trefoil symbol) on vehicles, rail cars, freight containers, and portable tanks.

As outlined earlier, safety in transport depends primarily on the integrity of the packaging. In particular, packages designed for dangerous goods and especially those for the transport of materials with high potential hazards are designed and manufactured to the highest engineering standards to meet the requirements of the transport regulations. The regulations for the transport of radioactive materials were the first to contain specific provisions on quality assurance (QA) for all transport operations; QA is defined as any systematic evaluation and documentation of performance judged against regulatory requirements. Where approval of the competent authority is required for a design or shipment, such must be contingent on the adequacy of a QA program.

6.1.3. Consignments, Means of Loading, and Loading Security

The shipment of dangerous goods from one site to another on public routes involves a number of measures. These are covered by the so- called consignment procedures. The whole cycle from the manufacturing of the substance or product to its final use is a complex interaction of all parties and measures involved.

The manufacturer has to inform the consignor of any dangerous properties. The packer has to choose the admissible packaging or IBC, to take care of any prohibition of mixed loading, and to be responsible for the correct marking and labeling of the packages or IBCs. The consignor has to ensure proper documentation for the shipment, including the transport document containing at least: a description of the goods, including the substance identification number (where available); the class; necessary initials (if any); the number and a description of the packages or IBCs; the gross mass; the names and addresses of the consignor and consignee(s); and/or declaration required by the terms of any special arrangement. The carrier has to take care that all relevant and necessary information about the load and the transport documents be handed to the driver, that no damaged packages have been loaded, that the necessary testing certificates for tanks or tank containers are available, that filling limits have been observed, and all necessary labeling of the vehicle or container has been applied. There are additional requirements for the driver and assistant (if any) as well as for the vehicle itself [44].

The various components of a dangerous goods load must be properly stowed on the vehicle and adequately secured to prevent them from being displaced in any way in relation to each other and to the walls of the vehicle. Minor displacements may be permitted, provided they are of a stabilizing nature, and do not lead to constant frequent motions of the load in the cargo space. If the load comprises goods from different categories, the packages of dangerous goods have to be separated from the others. The load has to be uniformly distributed over the loading space, and/or sufficiently secured against slipping backwards, sideways, or falling out. Packages must be stowed so they are protected from any penetration or impact, and from being squeezed, punctured, or exposed to a pressure which might break the package. This can be achieved by clamping beams, brackets, pads, or dunnage baulkings. Cargo spaces and packages or IBCs must be clear of oil, ice, or other covering which may reduce friction. Most accidents in transportation are due to improper loading security measures, and carriers or shippers of dangerous goods must ensure appropriate loading security.

6.2. Bulk Transport

A bulk shipment is the transport of a loose, unpacked material (gas, liquid, or solid) in a means of transportation (Fig. 5).

	Bulk transport	
	Liquids, gases	Solids
Road and rail transport	Tank containers Tank trucks Rail tank wagons	Containers, tank containers, bulk containers Tank trucks, silo trucks Sheeted or open, covered vehicles with box-type construction
Marine transport	Ships	
Air transport	Not permitted	

Figure 5. Overview of packagings for bulk transport

Different definitions of the "packed state" of a material apply for the individual carriers.

For shipment by rail and road, dangerous goods are considered packed when they are present in tested and permitted packagings according to Appendix A.5 of ADR [48] or Appendix V RID [49] (max. volume 450 L) or in tested and permitted intermediate bulk containers according to Appendix A.6 ADR or A.VI RID (max. volume 3000 L). For ocean and inland-waterway transport, road and rail vehicles as well as containers already represent a so-called bulk packaging [50].

However, because of the large quantities which are shipped in one packaging, shipment in bulk represents a higher potential of release. Therefore, only less dangerous substances are permitted for this mode of carriage or the packaging for this mode of carriage or the packaging must be designed to be adequately safe (e.g., double wall). The logistic chain is often greatly simplified by bulk shipment, the number of shipments required is greatly reduced, and savings by orders of magnitude are made in packing material.

6.2.1. Bulk Transport of Liquids and Gases

6.2.1.1. Tank Containers [42,51–56]

In the transport of gaseous and liquid goods (as well as granular materials and powders, see also Section 6.2.2), the tank container is one of the most widely used means of transport equipment.

Tank containers are equally suitable for all transport routes (water, road, rail) and therefore, as a multimodal means of transport, offer a variety of possible uses.

These possibilities range from transport via intermediate storage up to long-term storage or stockpiling of goods. With regard to the design and manufacture of the tank, and its testing and operation, regulations (IMDG-Code, based on the UN recommendations) and standards (ISO) valid and accepted worldwide must be observed.

It must be demonstrated by calculation that the tank container withstands the planned loads, i.e., internal and external overpressure and the load which results from the inertia of the tank's contents during movement in transport. The tank, including its attachment fittings, must be able to absorb the following forces at the highest permissible weight of its contents:

1) Twice the total weight in the direction of travel
2) The total weight horizontally perpendicular to the direction of travel (if the direction of travel is not clearly defined, twice the total weight applies in each direction)
3) The total weight vertically upwards
4) Twice the total weight vertically downwards

With these forces acting, there must also be a safety factor of at least 1.5 relative to the elastic limit.

The tank material must be resistant to the cargo or be protected by a resistant lining or coating. Items of equipment on the tank container must be so mounted that during transport or handling they are secure against tearing off or damage. They must guarantee the same safety as the tank, and remain leak-free if the tank container turns over.

Tanks with discharge underneath must be provided with two independent closures in series, the first of the two closures being internal.

Each tank container licensed under the IMDG-Code must be fitted with a pressure-compensating device. This can consist of a melting fuse, a bursting disk, or a pressure-relief valve and must be approved by the responsible authority.

The total amount blown off by the presssure-compensating device from a tank completely surrounded by fire must be sufficient to limit the pressure in the tank to a value of 20 % above the response pressure of the pressure-compensating device.

Each tank container is subject to a type license by the national licensing authority concerned.

Tank containers built according to this license, including their items of equipment, have to be tested before being brought into service and subsequently at regular intervals by experts recognized by the authorities.

In the operation of tank containers the maximum degree of filling must be observed. In tanks of capacity >7500 L, to avoid forces arising from the surging of liquids, a minimum degree of filling of 80 % must be observed, or surgeobstructing arrangements provided.

As a result of the regulations valid and applied worldwide it is guaranteed that tank containers satisfy a uniformly high standard of safety engineering.

6.2.1.2. Tank Trucks and Railroad Tank Cars [51,52,57–64]

In addition to the transport of dangerous materials in tank containers, transport in tank trucks and railroad tank cars is of great importance.

A tank truck is a vehicle with one or more tanks fastened to it for the transport of liquids and gases, as well as powders and granular solids (see also Section 6.2.2).

For *tank trucks,* aside from the constructional and equipment requirements of the actual tank, which are comparable with those of tank containers, the following requirements with regard to the electrical and nonelectrical equipment of the vehicle must be observed.

The rear of the vehicle must be protected against collision by a bumper over the whole width of the tank.

In vehicles that transport liquids with a flash point of <61 °C or flammable gases, the vehicle engine, the exhaust pipes, and the loading pump (if present) must be constructed such that danger to the load resulting from heating or ignition is avoided.

The vehicles' fuel tank must be so arranged that it is protected as well as possible in a collision. If fuel nevertheless escapes, it should be able to run off directly to the ground. Positioning of the fuel tank directly above the exhaust arrangement is not permitted.

In many countries the quality of the vehicles' braking system is subject to testing at regular intervals.

Tank trucks carrying dangerous materials must be equipped with at least one portable fire extinguisher, mainly intended for extinguishing burning tires.

Special requirements must be met by the electrical equipment of a vehicle if liquids with a flash point of <61 °C, combustible gases, or certain explosives are transported:

1) Sufficient dimensions of the cables to avoid overheating
2) Adequate insulation
3) Protection of the leads from impacts, stones, and the heat from the exhaust arrangement
4) Fitting of a battery isolation switch in the driver's cabin or outside on the vehicle, which breaks all circuits in the event of danger

The use of safety valves to avoid unacceptably high pressure in the tank is not uniformly prescribed. In land traffic, it is mainly tightly sealed tanks that are transported.

Arguments against fitting with safety valves are, in particular:

1) Possible operation of the valves (e.g., in tunnels or when driving through built-up areas), whereby the operation is possibly not noticed
2) Possible gumming-up of valves with product and, therefore, failure to function
3) Experience with valves "flaring off" has shown that after a short time the tank and valve material is annealed and loses strength, which can then lead to a tank explosion

Arguments for using safety valves are:

1) In the case of overfilling, overstressing of the tank by hydraulic pressure is avoided
2) The noise of the valve blowing off serves as an indication of danger

As is prescribed in many countries, the drivers of tank trucks should possess a certificate demonstrating that they have taken part in training and have passed an examination on the special requirements to be met in the transport of dangerous materials.

Railroad tank cars consist of an assembly of one or more tanks and the underframe. The basic requirements for the tanks with regard to construction, equipment, and testing are comparable with those for tank containers and tank trucks. Requirements specific to the railroad are set for the underframe, consisting of running carriage, suspension, drawgear, and buffer gear, and for the brakes.

Within the scope of application of ADR and RID, as described in Section 6.2.1.1, both tank trucks and railroad tank cars are subject to type approval by the responsible authorities and regular testing by officially recognized experts.

6.2.1.3. Marine Tankers [65–73]

Whereas the transport of liquids in tank vehicles and tank containers by land is considered to be bulk transport, in shipping, such transport in units regarded as "bulk packaging" is carried out in accordance with the rules for packed dangerous goods. The term bulk cargo is used if unpacked material is taken into the ship's tanks.

The requirements for such ships for the field of ocean shipping are contained in the SOLAS (*International Convention for the Safety of Lifes at Sea*) Conventions and MARPOL (*International Convention for the Prevention of Pollution from Ships*) as well as in the IBC (*International Bulk Chemical Code*) and IGC (*International Gas Carrier Code*) Codes, and the requirements for inland navigation in the ADN and/or ADNR Regulations. Supplementary dimensioning and construction regulations are contained in the appropriate classification and construction rules of internationally recognized classification corporations (e.g., Germanischer Lloyd).

In view of the increasing oil pollution of the oceans and environmental protection in general, measures for raising the standard of safety both in inland-waterway and ocean shipping have been introduced stepwise in the above-mentioned regulations, or have come into force in 1995 for inland shipping and for existing ships in ocean shipping.

Inland Shipping. Inland navigation dispenses with its own list of substances and adopts, with a few additions, that of the land traffic regulations by direct reference in Marginal no. 6002 1a to Appendix A of ADR in its current version. Problems in the transport of packed dangerous goods on transfer from land transport to inland shipping are therefore excluded.

Bulk transport in tankers is permitted only for products listed in Appendix B 2. This list at present contains ca. 270 entries for directly named substances and N.O.S. products of Classes 3 and 8 to which a UN number has been allocated. The list is to be regarded as open, i.e., it can be extended after examination of a request to transport. The time until acceptance is bridged by exemptions regulations limited to one year, according to a precisely defined procedure.

As a result of the requirements included in the substance list for the products and further substance-related regulations in Part II of Appendix B 2 of ADNR, the tanker regulations of the new ADNR are now related to substance rather than type of ship.

In spite of substance-related construction regulations, there are still three types of ship, which however can now be regarded only as a rough classification. The allocation of any product to one of the three types of ship

Type G tanker (gas tanker)
Type C tanker (chemicals tanker)
Type N tanker (normal tanker) is given in the substance list.

The *Type G Ship* is a tanker intended for the transport of gases. Transport is normally carried out in the liquefied state, either under pressure at ambient temperature or at cryogenic temperatures and atmospheric pressure. Partial cooling and transport in pressure tanks is also possible.

For the transport of liquefied gases under pressure, the hull must be constructed in the cargo area as follows:

1) As a double-hull ship with wing walls and double bottom minimum wing-wall width 0.8 m minimum double-bottom height 0.5 m
2) As a single-hull ship with reinforced side-wall structures minimum distance cargo tank – side wall 0.8 m minimum distance cargo tank – bottom 0.6 m
3) Other designs, provided a collision energy of at least 22×10^6 Nm is attained.

The maximum permissible tank capacity depends on the size of the ship (LBH value), but must in no case exceed 380 m^3 per tank. A minimum number of tanks is no longer prescribed, but proof must be provided of sufficient damage stability when two adjacent compartments spring a leak under precisely defined damage conditions. When the main engine room springs a leak, only the single-compartment status has to be proved.

The *Type C Ship* is a tanker designed for the transport of especially dangerous liquids, such as poisons of Class 6.1 or combustible liquids of Class 3 with additional dangerous properties such as carcinogenicity and toxicity. This is a flush-deck, double-hull ship with wing walls and double bottom, the cargo tanks in the standard version being formed by the hull.

The use of independent cargo tanks also is permissible if they are installed in the inner holds formed by wing walls and double bottoms.

The minimum width of the wing walls is 1 m, but if certain collision reinforcements are carried out it can be reduced to 0.8 m. The average double-bottom height must be 0.7 m, and this dimension must not be less than 0.6 m at any point.

Apart from the requirement of a maximum permissible tank capacity, a maximum cargo tank length must also be observed. The tank capacity depends on the size of the ship, but must not in any case exceed 380 m^3.

Sufficient damage stability when two adjacent compartments spring a leak has to be demonstrated, whereby certain minimum assumptions of the extent of damage in the event of a leak are made.

Here also, the proof of single-compartment status is sufficient for the main engine room.

There is no need to assume damage to the inner longitudinal bulkhead of the wing walls; likewise the crown of the tank is to be regarded as undamaged, provided the double-bottom height is at least 0.6 m.

Type C double-hull ships with closed tanks must be designed for pressure ratings that rule out a response of the safety valves during the transport of toxic and carcinogenic substances.

A reduction of the safety valve setting can be obtained by additional sprinkling, provided pressure sensors ensure that the sprinkler system comes into operation early enough.

The *Type N Ship* is a conventional tanker for the transport of liquids. These are usually single-hull ships of various pressure ratings for the transport of combustible liquids of Class 3 (comparable with the former tanker Types II to V, but with altered classification criteria).

Acid and alkali tankers for the transport of goods of Class 8 are of this type, even though they are mostly double-hull ships or ships with inserted tanks. With these ships special attention must be paid to material compatibility and it should be noted that acids and alkalis generally have a relatively high density.

Particular distances of tanks from the ship's hull or from the bottom for safety reasons are not required for these ships, nor is proof of damage stability. Sufficient intact stability is, however, prescribed (see Marginal no. 331 214).

In addition to the maximum permissible tank capacity (environmental protection), a maximum cargo tank length must be complied with. The former requirement for a minimum number of tanks has been dispensed with.

There are few restrictions regarding the construction of Type N Ships, so the following alternatives exist, unless they are restricted by the invidual requirements of the substance list:

1) Flush-deck, single hull ship
2) Single-hull ship with trunk
3) Flush-deck, double-hull ship
4) Double-hull ship with trunk
5) Ship with inserted tanks
6) Ship with inserted pressure tanks

The basic safety requirements are the same for all Type N Ships. However, Marginal no. 210 287 (Appendix B 2, Chap. 1) exempts tankers of Type N, open and with and without flame arresters from the observance of a number of requirements.

Thus, the requirements for tankers of pressure rating 0, i.e., those having tanks with an open venting system, are effectively not increased. Only the number of products that can be transported in a tanker with flame arrester has been restricted.

A simplified comparison of the Type N Ships with the former tanker Types I to V is shown in Figure 6.

Ocean shipping has its own list of substances in the IMDG-Code, which in spite of having the same classification, deviates from that of land transport in the ADR, although both

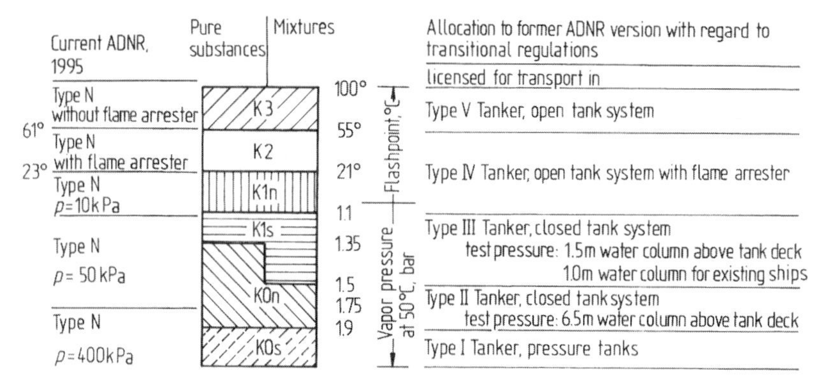

Figure 6. Simplified comparison for Type N Ships with former Type I to V Tankers

originate from the list of substances in the UN recommendations.

The basic regulations for the bulk transport of liquids, as far as dangerous goods are concerned, can be found in Chapter VII of SOLAS and in the MARPOL Convention. These are supplemented by the IBC and the IGC Code.

MARPOL 73/78 is the International Convention of 1973 for the *Prevention of Pollution from Ships* in the version of the protocol of 1978. Appendix 1 of MARPOL 73/78 contains the rules for the prevention of oil pollution by ships and came into force, in part retroactively, on 2 October 1983.

"Oil" here means all petroleum products, including crude oil, fuel, oil sludge, oil residues, and refinery products such as fuel oil and gasoline. "Oil tankers" are defined as ships that transport oil as bulk cargo.

At the 31st MEPC committee meeting in July 1991, the text of the new Rule 13 F for oil tankers was formulated and adopted. It applies to new oil tankers of >600 t load whose construction contract was concluded after 6 July 1993.

Rule 13 F requires for oil tankers of ≥5000 t capacity that the whole cargo tank length is protected as follows by ballast tanks or holds that are neither cargo nor fuel tanks:

1) By wing tanks or spaces over the whole depth to the deck of the ship or from the double bottom to the topmost deck with a width in meters of

$$w = 0.5 + \mathrm{DW}/2000,$$

where DW is the load of the ship in tonnes deadweight (tdw). The minimum value for $w = 1$ m, and the maximum required value is $w = 2.0$ m.

2) By a double bottom with a height in meters of

$$h = B/15,$$

where the minimum value for $h = 1.0$ m and the maximum required value is $h = 2.0$ m (B = breadth of ship).

3) In addition, the total capacity of the ballast tanks must be sufficient for a safe voyage in ballast with the screw fully immersed.

For oil tankers of <5000 t load, equivalent solutions are provided.

As a precursor of MARPOL Rule 13 F, Rule 13 E for crude oil tankers above 20 000 tdw and for product tankers above 30 000 tdw, in effect from 1 June 1979, required a partial double hull by having segregated ballast tanks in a protective arrangement. "Segregated" means that these ballast tanks must be completely separated from the oil cargo and fuel systems; i.e., they can only be used for ballast water. This partial double hull can be formed by a double bottom and/or wing tanks.

The degree of protection J was 0.45 for tankers of 20 000 tdw and 0.3 for tankers of 200 000 tdw. Intermediate values are found by linear interpolation. For tankers of >200 000 tdw J could be reduced to as low as 0.2 according to Rule 23/24, depending on the ratio of the hypothetical to the permissible oil outflow.

In Rule 13 G measures are provided for existing tankers for the prevention of oil pollution in cases of collision and stranding.

Various measures are provided for crude oil tankers of $\geq$20 000 tdw and product tankers of $\geq$30 000 tdw for raising the safety standard of existing tanker fleets. These measures can be summarized as follows:

1) Stricter inspections. In addition keeping a complete record of the inspection reports, which is available to the responsible authorities of each of the signatory states.
2) Conversion to a double-hull ship or laying-up after a ship's lifetime of 25 a. For ships that fully satisfy Appendix I of MARPOL 73/78, i.e., those for which the construction contract was concluded after 1 June 1979 (so-called MARPOL tankers), a period of 30 a is laid down.

The requirements of Rule 13 G came into force on 6 July 1995.

It can be gathered from the above that according to Rule 13 F new oil tankers in ocean shipping with a load >5000 t must be built over the whole load range with wing tanks and double bottom, and for those with a load <5000 t equivalent regulations must be observed. However, according to Rule 13 G, for the existing tanker fleet a stricter scope of inspection is required only for crude oil tankers of $\geq$20 000 tdw and product tankers of $\geq$30 000 tdw, together with conversion to double-hull ships or laying-up after 25 or 30 years, respectively.

For existing smaller tankers, corresponding precautionary measures have not been agreed upon, so that a definite final date for the existence of small "single-hull tankers" depends on their lifetime, since new ones may not be built.

Apart from the crude oil tankers and product tankers that correspond to the Type N Tankers in inland shipping, ocean shipping has according to Chapter VII of the SOLAS Convention declared the International Bulk Chemical (IBC) Code and the International Gas Carrier (IGC) Code to be binding regulations.

The IBC Code contains material-related regulations for the transport of dangerous liquid chemicals in bulk, allocated to the three types of ship:

1) A *Type 1 Ship* is a chemical tanker intended for the transport of materials that constitute a very great danger with regard to environment and safety and require the highest possible level of safety precautions against the escape of cargo
2) A *Type 2 Ship* is a chemical tanker intended for the transport of materials that constitute a great danger with regard to environment and safety and require considerable safety precautions against the escape of cargo
3) A *Type 3 Ship* is a chemical tanker intended for the transport of materials that constitute a fairly great danger with regard to environment and safety and require a certain measure of safety precautions to improve the buoyancy under damage conditions

Arrangement of the cargo tanks:

Type 1 Ships:	*B*/5 or 11.5 m from the side (the smaller value is required)
	B/15 or 6 m from the bottom (the smaller value is required)
Type 2 Ships:	*B*/15 or 6 m from the bottom (the smaller value is required)
	760 mm from the side
Type 3 Ships:	no requirements (i.e., Type 3 Ships are single-hull ships)

Apart from intact stability, damage stability and buoyancy in the leaking state under specified damage conditions must be demonstrated.

The regulations relevant to substances can be obtained from the substance list, comprising about 800 products, as well as from the special chapters: "Special Requirements," "Additonal Measures for Protection of the Marine Environment," and "Operating Instructions."

The IGC Code governs the sea transport of gases liquefied under high pressure and by cooling. Depending on the potential danger originating from the gas, one of the following specific types of ship is prescribed: Type 1 G Ship, Type 2 G Ship, Type 2 PG Ship, or Type 3 G Ship.

A Type 1 G Ship is a liquefied gas tanker intended for the transport of materials with the highest potential danger, while Type 2 G/2 PG and Type 3 G Ships are provided for the transport of materials with lower potential danger. Accordingly, a Type 1 G ship should be able to survive the severest assumed damage and its cargo tanks have to be arranged at the greatest prescribed distance from the side plating.

If a ship is intended to transport more than one substance of those listed in Section 19 of the IGC Code, the assumptions about damage must correspond to the substance for which the highest demands are made on the type of ship.

The requirements for the arrangement of the individual cargo tanks, on the other hand, refer to the substance actually to be transported in each case.

The Code contains precise specifications with reference to the arrangement of the cargo tanks.

The damage stability and the buoyancy under prescribed damage conditions must be demonstrated, in addition to the intact stability.

6.2.2. Solid Bulk Cargo

Bulk cargo is a loose material (other than liquid or gas) in pourable form [74].

It is characteristic of bulk material that the components or particles are disordered and mutually displaceable, so that the material as a whole has a variable shape and is fluid or pourable. Gravitational force can therefore be exploited for moving bulk material.

Characteristic physical quantities of bulk material are cohesion and adhesion, bulk density, angle of repose, and grain shape and grain size d:

Dust	$d < 0.06$ mm
Fine	$0.5 < d < 10$ mm
Coarse grained	$10 < d < 60$ mm
Medium	$60 < d < 160$ mm
Lump	$d > 16$ mm

These material quantities – together with the hazard properties of the bulk material such as toxicity, gas evolution, corrosivity, infectiousness, flammability, and explosiveness – form the basis for choosing a suitable packaging.

6.2.2.1. Solid Bulk in Road and Rail Transport

Definition. In road and rail transport, the shipment of dangerous material in bulk is defined as "the shipment of a solid material without packaging" [75,76]. This definition is not entirely correct, since by using the term "material" it excludes an important area of bulk cargo shipment: the disposal of uncleaned packages, contaminated objects, and equipment. In general therefore, one should refer to the shipment of solid goods without packaging.

Permitted Goods. The road and rail dangerous goods regulations name three modes of shipment of goods: shipment in bulk, shipment in containers, and shipment in tanks.

This system is illogical and contradicts the definition of shipment in bulk. According to ADR/RID the "bulk" mode of shipment is permitted only for certain materials [76]:

Class	Number	Materials
4.1	4 c)	wastes
	6 c)	all materials and wastes
	11 c)	all materials and wastes
	12 c)	all materials and wastes
	13 c)	all materials and wastes
	14 c)	all materials and wastes
4.2	1 c)	all materials
	2 c)	all materials
	3 c)	all materials
	12 c)	2793 iron metal swarf
	16 c)	1376 iron oxide, used; iron oxide sponge, used
4.3	11 c)	all materials
	12 b)	1405 calcium silicide pieces
	12 c)	all materials
	13 b)	3170 aluminum sweepings
	13 c)	all materials
	14 c)	all materials
	15 c)	all materials
	17 b)	all materials
	20 c)	all materials
5.1	11	all materials and wastes
	12	all materials and wastes
	13	all materials and wastes
	16	all materials and wastes
	18	all materials and wastes
	19	all materials and wastes
	21	all materials and wastes
	22 c)	all materials and wastes
6.1	44 b)	all materials and wastes
	60 c)	all materials and wastes
	63 c)	all materials and wastes
	all in c)	all wastes
6.2	1	all
	2	all
	3	all
	5	all
	9	all
8	1 b	sulfuric-acid-containing lead sludge
	23	all materials and wastes
	all in c)	all wastes
9	4 c)	2211 foamable polymer beads
All		empty (uncleaned) packagings

For all unlisted materials it is forbidden. The selection of materials is not subject to any system but rather is governed by need. Owing to the assumed low safety level of the packings in question, however, the less dangerous materials (c materials) predominate. For some special areas (e.g., spent accumulators, contaminated soil), there are exceptional national regulations or regulations specific to federal states or special agreements with individual ADR/RID treaty states.

A considerably greater number of bulk materials is permitted according to ADR/RID for shipment in tanks, since relevant constructional, equipment, and testing requirements guarantee a high safety level of the packaging "tank" (see Sections 6.2.1.1 and 6.2.1.2).

Means of Transport and Requirements. For the shipment of bulk material by land, the following means of transport are mainly used (Fig. 7):

1) Open, sheeted, or covered vehicles with box-type construction (dumper trucks)
2) Tank trucks with pressure discharge, or silo vehicles
3) Containers/bulk containers according to ISO 1496/ Part 4 [77]; so-called logistical containers, e.g., put-down containers, according to DIN 30 723 [78] and roller containers according to DIN 30 722 [79]; tank containers

The bodywork of the vehicles is not subject to test requirements. According to the definitions in ADR/RID, a covered vehicle is simply a vehicle whose body can be closed. A sheeted vehicle is one that is provided with a canvas cover to protect the load. The design of the bodywork, like that of containers, remains largely unrestricted in respect of mechanical stability and watertightness.

If materials permitted according to ADR/RID for shipment in bulk are shipped in tank trucks or tank containers, the regulations for construction, equipment and operation of tanks are not applicable [80], i.e., for tank shipment also there are no minimum requirements for stability, watertightness, and equipment.

For the shipment of a number of flammable or spontaneously flammable materials a metallic vehicle construction, and in some cases the use of impermeable or flame-resistant sheeting is also required.

The vehicle bodywork or container for spontaneously flammable materials or materials that evolve flammable gases on contact with water must be capable of airtight closure. For ignitable materials (Class 4.1) and nauseating materials (Class 6.2), sufficient ventilation is prescribed. In the shipment of materials with igniting (oxidizing) effect (Class 5.1), contact with wood or other combustible materials must be prevented.

Figure 7. Bulk container
Top: 12-m silo tank container for the shipment of toxic dust; Bottom: 4-m^3 silo tank for shipment of toxic lead compound

In various fields, particularly in the disposal of industrial wastes, there is a requirement for packages for the collection and shipment of waste, which can guarantee safe disposal of even dangerous wastes, sometimes as heterogeneous mixtures, in large amounts ($>3\ m^3$), and whose construction is adapted to the logistical requirements of an optimum waste-disposal chain (e.g., large openings for filling and emptying).

There have already been a number of special developments whose design is based partly on requirements from the tank field and partly on the regulations for IBCs, as well as additional material requirements (Fig. 8). For these shipment concepts, a special license from the responsible authority is necessary.

Figure 8. Special bulk container for shipment of hazardous waste (Courtesy of HIM)

Starting from the assessment that the current regulations for bulk shipment in land transport do not correspond to the state of the art and do not take sufficient account of the current requirements, a rearrangement of this shipment field is being contemplated that should lead to systematic and clear regulations and objective test requirements.

6.2.2.2. Solid Bulk in Marine Transport

Definition. The regulation for marine and inland-waterway transport define bulk solids as "goods, other than liquids and gases, that can be loaded directly without any intermediate packing into the hold of a ship." Goods are included that are loaded in a barge carrier's lighter on a barge carrier. In addition, the term "bulk package," which refers to packaging (containers, road or rail vehicles) that contain bulk cargo, has been introduced in the field of marine transport [50].

Permitted Materials, Requirements. Solid materials in bulk packages may be shipped in marine transport if, according to the relevant substance page of the IMDG-Code, bulk shipment is permissible. The permissible range of materials corresponds approximately to that in land transport. The materials must be shipped in closed containers or closed road or rail vehicles with superstructures or freight car bodies of metal (dust-tight wooden floors are permissible) or in mobile tanks. The relevant land transport regulations are also to be observed.

For some materials there are also special requirements for the bulk packages. Solid materials that are shipped in bulk in the cargo hold of a ship and from which, because of their chemical or physical properties, a danger originates during transportation are listed in Appendix B of the Code on the Safe Handling of Bulk Cargoes (BC Code).

Some of the materials listed therein represent no substantial danger if they are transported in packagings. However, shipped in bulk, they represent a hazard. These substances include:

1) Aluminum wastes
2) Fly ash
3) Activated carbon, coal, coke
4) Direct-reduced iron, iron phosphide
5) Fluorspar
6) Quick lime
7) Magnesium
8) Pitch
9) Sawdust, wood shavings, wood pellets
10) Silicomanganese
11) Vanadium ore

For materials classified as dangerous goods, the corresponding substance pages of the IMDG Code contain a reference to the BC Code. Before each shipment in bulk of an unlisted material, comprehensive information on the physical and chemical properties of the material can if necessary be obtained by request to the responsible authority.

The prescribed safety measures gathered in the "instructions for the safe handling of bulk cargoes in transportation by ocean ships" are adequate to match the potential hazard of bulk cargoes of dangerous material on ships.

6.3. Cargo Storage

For the storage of goods in packages, logistic considerations, the type of store, and safety requirements all have to be harmonized.

The logistic plan is determined by the type and number of goods, the packing, the storage units (pallet units), the turnover rate, and the type of carrier.

The kinds of storage that must be distinguished are block storage, for large lots of stock units of the same type, and shelf storage, for the storage of single packages and pallet units.

In the design of the store, logistic and safety aspects have to be taken into account:

1) Unroofed open-air storage consists of a paved area, impermeable to liquids, and sufficiently firm. Products can be stored here that are insensitive to temperature and have weather-resistant packing. Individual storage zones are separated by safe distances.
2) In roofed open-air storage the stored goods are shielded from the direct effects of the weather by a self-contained roof structure, but the store has no continuous external walls. The stored goods must be insensitive to temperature. Individual storage zones are separated by safe distances or by walls.
3) In storage buildings, a distinction is made between single-level, multilevel, and racked stores. Products with different safety requirements are accommodated in different store rooms. Storage zones are separated by fire-resistant walls or ceilings. The store is suitably divided into loading zone, handling area, and stocks on hand.

In siting the store facility, sufficient distances must be maintained from residential areas, busy traffic routes, and special protected areas such as drinking water extraction areas. Areas with risk of flood, earthquake, landslide, and avalanche must be avoided. The infrastructure must include good traffic connections, good accessibility for the fire brigade, and a supply of water for fire fighting.

The safety plan for a store must take into account the potential dangers to employees, the neighborhood, and the environment. Control measures serve to avoid a danger arising, to prevent the danger spreading, and to fight the danger. Rules and prohibitions governing storage together of incompatible substances must be observed.

The release of dangerous substances, especially toxic and combustible liquids and gases, is prevented by the choice of corrosion-resistant, stable packing drums, by prohibition of inadmissible overstacking, and by safety devices to prevent falling from shelves.

Escaped product is retained in durable collecting troughs or in the storeroom. The collecting devices must be liquid-tight and product-resistant. With readily volatile liquids and gases, it may be necessary to close openings to the building and switch off the artificial ventilation.

Measures for explosion protection and ventilation are used for the avoidance of fires and explosions, especially with combustible liquids and gases. Ignition sources must be avoided on principle.

The spreading of fire is hindered by constructional arrangements (fireproof walls, ceilings, and load-bearing parts of the building) and by specific store organization (separation of the storage zones by walls and spacings, buffer zones with nonflammable materials). In fairly small storerooms, the spreading of fire can also be hindered by a gas-tight mode of construction and automatic closure of openings (ventilation-controlled fire).

The fire is fought initially by the store personnel, using hand fire extinguishers and wall hydrants. The fire alarm is given by press-button alarm, by telephone, or by automatic fire alarm installations connected to the responsible fire department. The nature, equipment, and speed of the fire brigade must match the danger potential. If the danger potential is considerable, it is necessary to equip the store with automatic extinguishers.

The fire protection plan, including selection of the extinguishing agent, is drawn up with the local fire department. The nature of the fire protection methods also influences the arrangements for retaining contaminated fire-fighting water.

In addition to constructional and technical measures, organizational rules are necessary for the safe and economic operation of a store. In particular:

1) A storage plan and a stock list; both used for rapid information and orientation in the event of danger
2) An internal alarm and accident prevention plan
3) Written instructions for the store personnel on handling dangerous substances, and on rules of conduct in the event of danger

6.4. Bulk Storage

6.4.1. Storage Concept [82–95]

When storage concepts are being developed, the dangerous properties of substances must be taken into account. The optimum amount stored for a store of a given scale follows from production and sales considerations. This can raise an area of conflict between reliability of supply and profitability, on the one hand, and the protection requirements of employees and third parties on the other. Protection of third parties includes neighborhood and environmental protection.

The following forms of stationary tank are customary for the storage of bulk liquids: horizontal and vertical cylindrical tanks, spherical tanks, and flat-bottomed cylindrical tanks in a collecting space. The following types of installation can be considered: above-ground: in the open or in a containment; underground: corrosion-protected; in a collecting space or double-walled or soil- covered with an open end.

The following substance-related risk factors exist: amount stored, pressure, energy potential (product of pressure and capacity), temperature, explosion danger, combustibility, hazard to health, chemical reactivity (e.g., ethylene oxide), corrosion risk (e.g., hydrogen-induced stress corrosion cracking by ammonia in ferritic steels), and danger to water sources. Further intrinsic risk factors arise from the design of the tank system and from processes in which the tank takes place.

The following environment-related risk factors are important: site conditions, fire loads, ignition sources, storage together (layout), foundation (e.g., in mining or earthquake districts), traffic, natural risk factors (e.g., flood danger) and artificial ones. Operational risk factors include human errors in filling, emptying, sampling, cleaning, and maintenance, as well as leaks during operation.

The inclusion of a store in an existing infrastructure reduces risk, e.g., with regard to the reliable supply of energy and water as well as to the immediate engagement of a works fire department.

Changes to existing safety rules must follow the same protective goals. The goals can also be met by other appropriate methods, with consideration of the balance between cost and safety gain.

In this connection, control of disturbances by the third barrier (organizational measures) and the fourth (limitation of consequential damage) must be adequately evaluated when designing the first barrier (tight enclosure) and the second (sufficient safe drainage of any leakage). In works with a well-developed infrastructure, especially in the chemical, petrochemical, and petroleum industries, this third barrier is always readily available.

Principles of a safety plan comprise:

1) Keeping stored amounts small: reducing total amounts stored to the quantity necessary for reliable supply.
2) Lowering the hazard potentials: storage in chemically stabilized or dilute form, mixed with inert carrier material, in solvents, or at low temperature.
3) Distribution: amounts of substance distributed between discrete sections of the store, separated by fire walls.
4) Storing substances separately: this requires that dangerous properties (explosiveness, combustibility, hazard to health) be itemized. For combustible substances it is important to know whether they are readily flammable or even spontaneously flammable; such substances should be stored separately. It is advisable to classify substances into groups according to the transport regulations, extended to include chronic toxicity, carcinogenicity, crop-damaging and genetically modifying properties, and potential danger to water sources. Combustible substances should not be stored together with igniting or spontaneously flammable substances, or with toxic substances, or gases that are compressed, liquefied, or dissolved under pressure.

Avoiding Environmentally Dependent Hazards. It is important to check the site; eliminate fire loads; avoid ignition sources where combustible substances are concerned; separate substances of different danger classes; secure foundations (taking note of soil surveys, earthquake zones); protect against mechanical damage; provide protective distances (e.g., explosion protection) and safe distances (e.g., from hospitals and schools).

Optimizing Organizational Protective Arrangements. The following requirements must be set: operating instructions (including avoidance of human errors, in the initial filling and in the event of trouble); qualified personnel; regular training; cleaning and maintenance permits; safe infrastructure (energy supplies, water reservoir, fire protection, waste disposal); and practices of alarm and accident prevention plans.

Avoiding Hazard Activation: Maintaining Primary Safety (First Safety Barrier). The leak-tightness of the plant (tank, piping, equipment, feeding devices, seals) must be continuously monitored. The following protective goals are important:

- Equipment to be designed for pressure, temperature, and other stresses
- Compatibility to be preserved between equipment materials and the stored substance
- Confusion of substances to be avoided (control of substances delivered)
- Materials to be sufficiently resistant to or protected from weathering or other external influences
- Protective equipment to be reliable (redundant, e.g., over-filling protective devices LS^+A^+ and level indicators LIA^+), fail-safe, or self-monitoring
- Alarms and monitors to be provided
- Automatic leak-testing to be provided at separable connections
- Protection to be provided against heat, fire, and cold
- Adequate grounding to be provided
- Protection to be provided against mechanical action from outside, e.g., guide planks
- Arrangement of the installations to be clear and understandable
- Filling systems to be situated outside the storage area
- Human errors to be reduced by organizational measures (operating instructions), e.g., rules for acknowledgment action taken, and by technical measures (e.g., over-filling safety devices with automatic disconnection LS^+A^+)
- Parts subject to wear (e.g., hoses) to be replaced in good time
- Safety equipment (e.g., over-filling safety devices) to be given regular operational tests with appropriate documentation
- Regular leakage checks with appropriate documentation to be carried out
- Periodic tests to be carried out on vessels, lines, hoses, devices for avoiding electrostatic charging, and for protection from lightning and fire

Limiting the Effects of Malfunctions (Second Safety Barrier). Malfunctions can be detected early by regular monitoring (e.g., patrols). The following protective measures can be suitable:

In store rooms (housing), regular monitoring to be supplemented by automatic early warning equipment (e.g., gas warning devices), optionally with suction into a destruction system.

Components to be designed according to the "equal safety" principle (structures of the same strength) and "leak before break": fracture of a pipe connection and catastrophic failure is then impossible, and need not be included in the considerations.

Emergency shutdown systems with quick-acting gate valves to be applied when transport tanks roll away, and optionally automatically effective in case of fire.

Emergency nitrogen and power supplies to be available if the fail-safe principle is not used.

Pipe-break safety devices or quick-acting gate valves to be incorporated in lines that convey liquid.

Quick-separation systems to be used in ship reloading.

Full-hose systems to be used at reloading or filling stations.

In connection with filling with hoses or articulated arms: quick-acting fittings to be used on both sides, in above-ground installation of tanks for combustible substances in the liquid state: good ventilation to be applied; soil areas below the tanks to be liquid-tight and drained by a safe route.

Tanks to be installed in accordance with the properties of substances, e.g., very toxic gases in the liquid state in collecting spaces, double-walled tanks with leak monitoring, or as single-walled accommodation with suction; very-low-temperature storage with collecting space for toxic and combustible substances; as a rule, above-ground installation in the open air is best with regard to safety and cost, since the tanks can be monitored and maintained in the best possible way throughout their life.

Electrical and mechanical explosion protection to be provided.

In the case of danger to water sources, arrangements to be made for the collection of leaks (e.g., collecting spaces or double-walled tanks with leak monitoring).

No gully holes and ignition sources to be present in the protected zone, especially for separable connections.

Water sprinkling to be provided against the effect of solar radiation.

Safety valves to have a safe discharge line.

Possibility of covering the collecting space with foam or, in the case of combustible substances, filling the collecting space, e.g., with gravel.

For confined liquid volumes, e.g., in pipelines, relief valves to be present, in the case of toxic substances with optional downstream flash vessels and upstream bursting disks and buffer vessels.

Fire protection plan to be coordinated, e.g., sprinkling or flooding with water; water sprinkling by the works fire department, but only when the heat is exclusively radiant; protective gap; protective wall; fire insulation; insulation or soil cover. For installation above ground with water sprinkling, a protected extinguishing agent supply and fire protection equipment to be provided; for remote sites without continuous monitoring, optional water sprinkling or flooding with fusible fire safety device and automatic response; amounts of water sprinkling must be laid down as a function of tank size and type; for large spherical tanks ($>1000\ m^3$), sprinkling densities of $\geq 100\ L\,m^{-2}\,h^{-1}$ are sufficient.

Measures to be provided against fire under a tank and spontaneous combustion.

Personal safety equipment to be provided.

Alarm and accident prevention plans to be drawn up.

The fire protection plan in large works complexes with corresponding infrastructure includes: active explosion protection, qualified operators always present, specially trained works fire department always available, access to the basic pieces of equipment by agreement with the fire department for cooling and, if the need arises, for extinguishing, stock of water always available, maintenance, preventive maintenance, protected energy supplies.

Before commissioning, the safety plan of the plant is to be checked (plant monitoring). The important aspects are the following: substance data, piping and instrumentation flow charts, space-assignment plan, notice of approval, and, if occasion arises, safety analysis and operating instructions.

The acceptance inspection is concerned with operation according to the regulations and possible disturbance of this operation.

The conformity test is based on the observance of the applicable safety regulations. It assumes that there are appropriate vessels, piping, fittings, pressure generators, warning, alarm, and safety equipment, as well as seals, in accordance with the documentation used for checking of manufacturers' specifications as well as the piping and instrumentation flow charts. In the space-assignment test, the following considerations have to be assessed: fire protection, explosion protection including electrostatic charging, health protection, water protection, and infrastructure.

Requirements and design of the safety equipment are based on the process-engineering relationships, the characteristic quantities for safety engineering (properties of the substances), and the local conditions, e.g., safe discharge.

An operational test is to be made of the warning, alarm, and safety equipment. Appropriate documentation on the plant, based on the tests carried out, is to be drawn up.

The special conditions to be taken into account for individual cases, e.g., installation in a water protection area or in the neighborhood of sensitive equipment outside the works, must also be subjected to appraisal in the storage plan. In this connection, the fourth safety barrier is of decisive importance.

In some countries, detailed regulations exist on the handling and storage of dangerous substances. In Germany, the *Technische Regeln für Gefahrstoffe* (TRGS) incorporate the guidelines of the European Community, especially on the classification, packing, and labeling of dangerous industrial substances according to the level of technology employed.

6.4.2. Tanks [91,92,96–102]

In almost every branch of industry tanks are used worldwide for the storage of a variety of liquids, while bulk materials and silage fodder are stored in silos. Tanks are mostly made of ferritic steels, but also of high-grade steels, nonferrous metals, and plastics, e.g., glass-reinforced plastics (GRP). For economic reasons, ferritic sheets with cladding of high-grade steel or nonferrous metal are frequently used in tank construction. Good alternatives are ferritic steel tanks with plastic internal coatings, rubber linings, enameling, and internal coverings applied by means of vacuum (between covering and steel) to the supporting walls. Depending on the construction, it is normal to refer to flat-bottomed tanks (up to more than 100 m in diameter), cubic tanks, cylindrical tanks, double-walled tanks, double-bottomed tanks, and spherical tanks (spherical containers). Another characteristic is the permissible over-pressure, e.g., quasi-pressureless tanks ($p < 10$ kPa) detonation-resistant tanks (these may deform plastically in an explosion, but not develop leaks), and tanks with internal over-pressure ($p > 10$ kPa). For the storage of large quantities of gases liquefied under high pressure, spherical tanks (spherical containers) are preferable, which rest on legs, columns, or eggcup-shaped bases. Flat-bottomed tanks are mostly made with a fixed roof; above ca. 60 m diameter they are built as floating-roof tanks for economic reasons. For the storage of gases liquefied at low temperature (e.g., ammonia, oxygen, nitrogen, argon), the tanks are built with ca. 1.5 m thick special insulation and for permissible operating temperatures down to $-200\,^{\circ}$C or below. For substances that are not pumpable at ambient temperature, the tanks are provided with direct or indirect heating. The usual tank equipment includes: liquid-level indicators, over-filling safety devices, air-inlet and air-removal connections, gas displacement connections, filling and withdrawal connections with shut-offs. For the storage of combustible liquids with flash point $< 55\,^{\circ}$C, the tanks must be earthed and have flame arrestors in the respirator lines. In the case of flat-bottomed tanks for combustible liquids, the weld seam between cylinder and roof must be fabricated as a rated-rupture seam, so that with an inadmissible over-pressure only the roof tears off, and the stored liquid remains in the tank. The tanks are erected, preferably above ground (simple checking of corrosion damage and gas-tightness from outside as well as better accessibility during cleaning, repair, and alteration work), standing vertically or horizontally on stable concrete foundations. In the case of tanks erected above ground, catwalks, stairs, piping, and similar equipment must be so arranged that if there are changes of length as a result of temperature changes or tank settlements, they exert no inadmissible additional forces on the tanks.

Tanks for combustible and water-polluting liquids must be erected within bund walls. Flat-bottomed tanks for water-polluting liquids must be so erected or equipped that leakages in the bottom of the tank can be detected reliably and quickly. This can be achieved, e.g., by:

1) Erection on continuous concrete foundations with plastic or metal foil separating layers or the detector of a leak-detection system between the tank bottoms and the foundations
2) Erection on strip footings
3) Double-bottomed tanks with an inspection hole or leak-detection device

In Germany, steel tanks for the underground storage of water-polluting liquids must be double-walled and the leak-tightness of the interspace must be monitored by a suitable leak-detection device. In the case of underground steel tanks, the use of a cathodic corrosion protection system is advisable. For reasons of explosion-proofness and environmental protection, newly erected tanks are increasingly operated closed, i.e., with no air admission to the atmosphere and with inert gas blanketing (e.g., nitrogen).

Depending on the properties and the amount of stored liquid, the tanks are subject in many countries to national regulations for construction and equipment as well as for the environment, in particular for the protection of water sources. The detailed requirements regarding tank construction, equipment, erection, and testing contained in these regulations and in technical rules and standards associated with them are on the whole comparable as far as above-ground storage is concerned.

Tanks in operation must be tested at regular intervals in accordance with the national regulations by experts recognized by the authorities.

These experts are personally responsible for the operational safety of the tanks tested. Testing intervals are determined by the requirements of the authorities, or by the experience of the operator or builder. Testing must cover the whole tank structure, including all associated equipment. After being taken out of service, the tanks must be cleaned in a professional and environmentally acceptable way, and disposed of likewise.

7. Pipelines

7.1. Types of Pipeline

Pipelines transport and distribute substances in bulk, and are used for a variety of purposes in industry and technology. Their design specifications, layout, and operation are therefore covered by a range of technical and legal codes and standards.

Pipelines can be used as transport systems between two points (line systems), or as network systems for transport and distribution. This applies for long-distance pipelines and intra-plant systems. Typical applications for pipeline systems are for oil, gas, or water transport over long distances. Typical networks are intra-company distribution systems, communal water supply or sewerage systems, and airport fueling lines.

Pipeline systems should always be viewed as a whole. This holistic view of the transport line or network system should include the plant and facilities involved at both ends of the pipeline (Fig. 9), such as tanker ships, rail tank wagons, or road tank lorries. At the input end this may also include the source (e.g., the oil well) where this feeds into the pipeline, either directly or via a collecting tank. At the output end there may be tankage and transport plants, as well as user facilities (filling and transfer plant). The pipeline also includes all fittings, pumps valves, etc. along its length.

Pipeline systems can be laid overground or underground, or as a mixture of both. They are used to supply substances, or to remove unwanted substances (notably sewage). It is important to distinguish between pressure pipelines and gravity pipelines. Gravity pipelines are usually used only for the transport of water and sewage.

Pipelines are used to transport and distribute fluids such as oil and its derivatives, water, sewage, brine, and fluids for production, such as natural gas, hydrogen, acetylene, as well as suspensions of fine-grained solids such as cement, coal, ore, or sulfur. It is important to distinguish between flammable and nonflammable substances, since construction and operating requirements differ considerably between these.

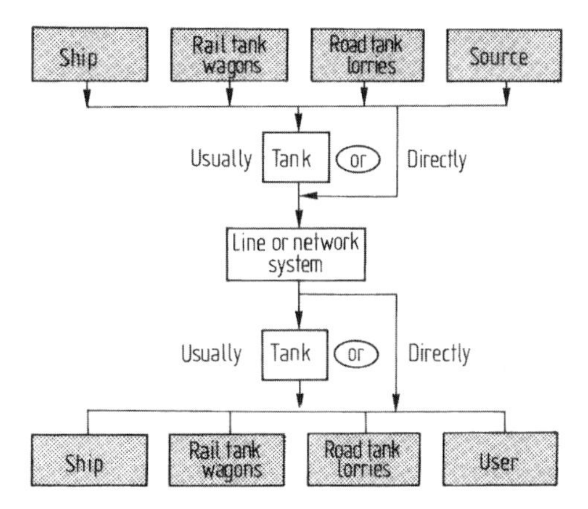

Figure 9. Components of a pipeline system

The numerous possibilities are covered by extensive technical codes, standards, and laws, and the relevant national requirements must be taken into consideration when applying for operating licenses.

7.2. Requirements for Pipeline Systems

Pipeline systems must meet two basic requirements:

1) They must be designed for continuous operation, to ensure supply or removal of the substances involved (which is in the interests of the operators).
2) To obtain permission to operate a pipeline, whether for long-distance transport or for intra-plant operation, it must be shown that the construction and mode of operation exclude the possibility of uncontrolled release of substances into the environment during normal operation, and in the event of damage or disturbances. This requires a redundant safety system, both to protect public safety

(avoiding fires and explosions) and to pollution of soil, ground and surface waters.

The redundant safety system consists of two barriers. The primary barrier is that directly enclosing the substances, with all safety and auxiliary fittings. The secondary barrier serves to contain substances released in the event of damage or disturbance, and also any unavoidable leak and drip losses. This barrier is generally designed as a leakproof system with appropriate safety features and warning systems. Special attention needs to be paid to the transfer points at both ends of the pipeline system, since these represent weak points. This is particularly important for systems involving the discharging or loading of ships.

Extensive technical and operating requirements have been introduced, aimed at preventing unacceptable environmental damage. It is important to establish the substances which are to be transported by a system and the volumes involved, since this influences the choice of materials for the primary and secondary barriers, as well as the range of possible environmental pollution.

The toxic properties of transported substances are of vital importance. The greater the potential threats posed by these substances to the environment, the more care needs to be taken in the design and operation of the pipeline.

In addition to direct toxic effects of substances, whether to humans or more generally in the environment, other environmental effects may also need to be taken into account. Natural gas, for example, is toxic only to the extent that it carries traces of substances such as the carcinogen benzene. However, if released underground it can lead to the displacement of oxygen, with adverse effects on roots and soil microorganisms. Accompanying changes in soil conditions such as pH and redox potentials can also lead to the mobilization of heavy metals.

7.3. Design and Operation

7.3.1. Routing

A well-considered route for a pipeline can go a long way toward meeting the safety requirements for the system. This applies to both external influences on the pipeline and the effects of the pipeline on its surroundings. It is important, for example, to avoid heavily populated areas, areas of importance for water resource management, and mining areas. Where this is not possible special precautions must be taken.

Where roads or railways cross a pipeline, in addition to soil pressure it is subject to further static and dynamic loads from passing vehicles. If the soil cover is less than 1.5 m the traffic loads dominate. For cross-country pipes under railway embankments or paved roads, it is possible to calculate soil pressures and traffic loads, and hence the loads on the pipeline. The additional loads must be taken into account when determining the dimensions and properties of the pipes; special measures may be necessary.

Local conditions and climate can be particularly important for safety. In cold climates it is necessary to protect the pipeline against frost, and to elevate it on stands to prevent it from sinking and breaking during thaws. In areas subject to earthquakes the bearings must allow compensation for horizontal and vertical displacements to avoid deformation.

7.3.2. Pipeline Design

Line or network pipeline systems consist of various elements:

Pipes
Pumps
Shut-off devices (valves, stopcocks, slide gates)
Fittings
Purging cocks (for gas lines)
Condensation traps
Monitoring systems
Expansion joints

A pipeline system viewed as a whole should always be designed as a double-barrier system, capable of being monitored and repaired. The primary barrier with all its elements encapsulates the transported substance. Its freedom from leaks depends mainly on the materials used in its construction in relation to the mechanical loads to which it is subjected, and the chemical properties of the transported substances. Unavoidable losses (drip leaks at pumps, valves, and gates), should also be contained by a second, physical barrier.

Wherever possible, there should be a continuous secondary barrier for the entire pipeline system, in the form of jacket pipes, collecting basins, and channels. In some cases this is not possible (e.g., because it would impede cathodic corrosion protection, or because differential expansion of the pipe and outer jacket threaten the system). The secondary barrier can then only consist of intensive organizational and operative back-up measures: shorter inspection periods; shorter distances between shut-off points to limit possible losses; more elaborate provisions for leak detection; continual remote monitoring of the pipeline system (both internally and externally); equipment for tackling emergency events and catastrophes; contingency plans, etc.

Such precautions are particularly important in endangered areas. Where there are threats of subsidence, it is important to carry out regular geodesic measurements along the route, and to measure the pipes for possible stretching. In densely populated areas and in important water catchment areas, the following safety measures are necessary:

- Higher specifications on the pipelines (design, materials, testing); additional supervision during pipe-laying
- Installation of additional measuring points to monitor cathodic corrosion protection
- Arrangement of shut-off fittings to limit losses
- Groundwater monitoring wells or local use of leak detectors
- Construction of safety embankments
- Laying warning bands above the overhead transmission line
- Special marking for the routes of cross-country lines
- Monitoring the route at short intervals
- Protection measures or adjacent installations

A frequent cause of incidents is damage due to external causes, e.g., excavation work. It is therefore important to provide underground pipelines with a suitable cover. For the Transalpine (TAL) Pipeline, which passes Lake Constance on its way from Genoa to Ingolstadt, the city of Stuttgart included an ozonization system in the purification plant for the water it draws from the lake, in order to be able to tackle oil leakages effectively.

7.3.3. Materials

The safety of the pipeline is affected to a considerable degree by the properties of the materials used in its construction and by quality assurance during pipe production. The measures adopted are continually being improved, e.g., better steel quality through continuous casting, secondary metallurgy, and thermomechanical rolling; improved quality control documentation (quality planning and monitoring); use of computer-aided production methods and optimum nondestructive testing procedures.

Further development of raw materials has resulted in the use of thermomechanical rolled steels and the reduction of sulfur and phosphorus levels.

When selecting and dimensioning the materials, the following criteria are important:

- Amount per unit time
- Internal pressure [constant operating pressure, shock waves (hammering), pulsating stress]
- External pressure (soil and traffic loads, bending loads)
- Temperature
- Internal and external corrosion

Selection of raw materials depends on the substances to be transported (Table 4).

7.3.4. Pipe Laying

Transport and Storage. Damage to pipes and their insulation can be avoided by suitable transport, careful loading and unloading, and proper interim storage. By training and informing personnel, careful supervision of work, and spot checks, it is possible to reduce the frequency of errors, and to detect those which do occur in time to take counter-measures.

Welding. Before welding begins in situ, it is important to check that pipes comply with delivery specifications. Only permitted filler materials may be used. Random nondestructive tests must be carried out on welded joints along the finished pipeline section (ultrasound, irradiation, or both in combination).

Table 4. Selection of raw materials

Substance group	Special properties	Materials usually used
Water/wastewater	no special requirements (exception: corrosive water)	steel, earthenware, concrete
Gases	high mechanical strength at all pressures; resistance to corrosion by sulfur compounds and humidity in the gas	steel
Oil	high requirements for mechanical strength and expansion; high temperature stability	steel
Chemicals	high resistance to internal corrosion; high requirements for tensile strength and temperature stability	steel, copper, special alloys
Solids in suspension	high resistance to mechanical scouring; high mechanical strength	steel

Installation. In underground pipeline systems, the pipe trench and its floor must be examined before the pipe is laid. The trench must be deep enough to allow the necessary cover over the pipe, and it must be possible to install the pipe in the trench without damaging the jacket. The trench floor requires special attention to ensure that the pipe is not damaged or subjected to impermissible bending loads. The pipeline must be bedded on sand, and a compactible material must be used for filling in the trench.

7.3.5. Operation

To ensure that the pipeline transports substances at the desired rate and with constant quality, a number of measures are necessary; these depend on the necessary safety provisions, on the mode of operation, and on local conditions.

General Safety Measures. Appropriate safety equipment must be provided for all plant. The safe operation of a pipeline requires trained personnel. Organization and communication are particularly important, especially when an alarm is raised or in the event of catastrophes.

Control. The most important process information parameters, which must be continually passed to the operations center (including during interruptions), are pressure, pressure flow, volume, density, and temperature. Updates on the current state of all fittings, such as pumps, gates, relief systems, and other plant, must be passed to the operations center, together with monitoring data, in particular for the detection and location of leaks and for levels of corrosion.

In pressure pipelines, monitoring the line pressure is important. Changes in pressure can give rise to pressure waves or shock waves (hammering) which can destroy the pipeline; it is necessary to ensure continuous monitoring of shut-off valves, pressure-release devices, and safety valves.

Provisions must also be made to precisely detect, quantify, and localize losses along a pipeline due either to damage or to leaks developing under normal operating conditions. Methods are available which make use either of changes in pressure (pressure-drop or shock-wave methods) or the measurement of mass. Gradual leaks can be monitored during interruptions (pressure – temperature, pressure difference, pressure – volume methods) or during operation (leak-detection rabbits, drains, or cables).

Special rabbits or pigs can be used to identify precisely the nature, location, and size of safety threats. Distortion, corrosion, and cracks can also be identified by ultrasound, magnetic fringing flux, and high-frequency and impulse eddy current methods.

It is also important to monitor safety along the route of the pipeline, by patrols or aerial observation.

Maintenance and Repair. Regular inspection and repair of defective components serves both general safety and the operability of the pipeline systems. In addition to routine inspections on the basis of a servicing plan, a general inspection of all functions should be carried out at specific intervals (usually annually).

7.4. Incidents

No statistics on incidents are available which cover all pipeline systems and all fields of application. In Western Europe, detailed data has been published regularly since the early 1970s by

CONCAWE (oil companies' European organization for environmental and health protection) for 65 operators of cross- country oil pipelines.

There were 14 oil spillage incidents in 1991 (4 in 1990 and an annual average of 13.3 since 1971). Of these, 12 were from pipelines and 2 from pumpstations (Table 5). All incidents required cleaning up operations. The level of soil contamination caused by the incidents varied: 2 caused no contamination; 5 were classified as causing slight contamination (< 1000 m^2 soil affected); 7 were classified as serious. Two petrol/diesel incidents polluted watercourses.

Table 5. Causes and numbers of incidents since 1971

	1991	Average 1971 – 91
Mechanical failure	7	3.3
Operation		0.9
Corrosion	4	4.4
Natural disaster		0.6
Damage by others	3	4.1
Total	14	13.3

The net oil loss into the environment in 1991 was 902 m^3 (0.00015 % of the total volume transported). The gross loss was 1346 m^3 (0.00023 %). A total of 444 m^3 (33 % of the gross loss) was recoverable.

7.5. Economic Aspects

Pipeline systems provide continuous transport and distribution of substances in bulk. Features of pipelines, in comparison with other means of transport, are low transport costs (low energy requirements), virtually loss-free transport, high operational reliability, and ease of maintenance. Pipeline systems allow greater flexibility in the choice of sites for industrial development, since the raw materials and fuels can be transported to the location, rather than the plant having to be located near the source of the raw material. Pipelines can thus be of considerable economic advantage to a country or region, since they allow the continued use of existing industrial infrastructures.

Advantages over other means of transport include shorter routes, which are usually more direct, fully automated operation (with correspondingly low costs for operation and personnel), as well as reduced environmental impact. The distances involved must be suitably large, and demand should be evenly distributed over the year, to warrant the relatively large investment costs involved.

No global summary is available for all bulk pipeline systems for oil and its derivatives, gas, water, wastewater, solids, and chemicals. In Western Europe, companies operate some 210 different service pipelines which, at the end of 1991, had a combined length of 21 000 km. In 1991, 593×10^6 m^3 of crude oil and refined products was transported, ca. 8 % more than in 1990, most of the increase being in crude oil. This gives some idea of the importance of pipelines for the economy as a whole.

8. Future Developments

With the liberalization of shipments, world freight traffic continues to increase. In Europe this tendency is reinforced by the implementation of the Common Internal Market and by the opening up of Eastern Europe.

In the past, the transport of dangerous goods developed in parallel with freight traffic as a whole. In Germany at present about 15 % of freight volume involves the transport of dangerous goods; in the European Community the proportion is similar [103, 104]. The proportion of road transport has grown steadily, whereas rail and inland waterway transport have been pushed back (Fig. 10).

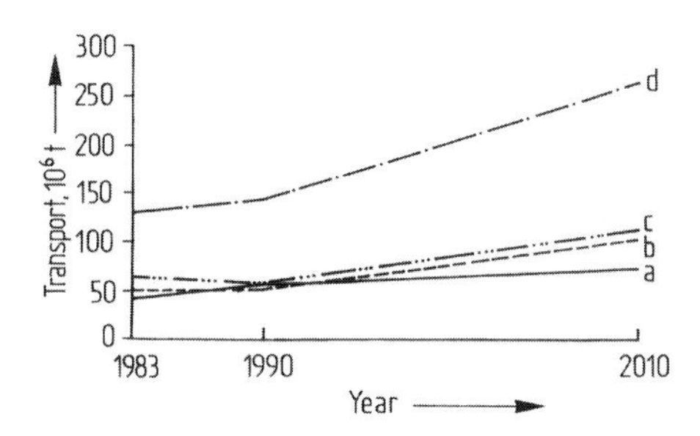

Figure 10. Transport of dangerous goods in the Federal Republic of Germany
a) Road; b) Rail; c) Ship (inland waterway); d) Total

Also to be expected in future are further increases in the transport of dangerous goods, in haulage distances, and in traffic across frontiers, and a growth in the proportion of road transport of dangerous goods (Fig. 10).

These developments entail an inevitable increase in environmental pollution as a result of total pollutant emissions from the operation of transport systems, even if the emissions in relation to distance covered can be reduced. In addition there are waste disposal problems, such as the cleaning of freight containers. In Germany, the release of dangerous goods in traffic accidents and from storage has not increased to the same extent as the total freight traffic volume. Nevertheless, the aim must be to prevent any release of dangerous materials into the environment.

Long-distance road transport has the greatest danger potential. Figure 11, for transport of water-endangering materials in Germany, shows that long-distance road transport accounts for more than seven times as many accidents, relative to transport volume, as rail transport. The volume of water-endangering materials leaked and not recovered, relative to the transport volume, is almost three times as large. Inland waterway transport has lower accident figures than road transport, but the amount of water-endangering materials entering the environment as a result is distinctly higher than for the other carriers.

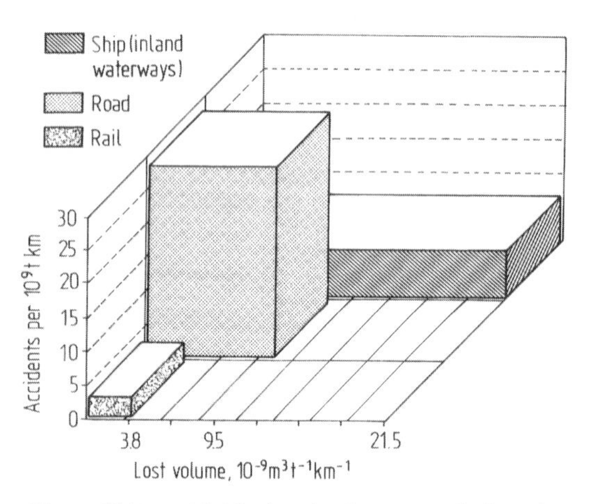

Figure 11. Potential risk of carriers for water-polluting substances

Not all shipments of dangerous goods can be transferred to rail, for capacity reasons. Besides, it may be preferable not to send a particular shipment of dangerous goods by rail if the route leads through districts that are densely populated or particularly worthy of protection, if road transport permits a harmless detour. Preconditions for a greater transfer of shipments of dangerous materials to rail are freight traffic centers on the margins of the more densely populated zones, as provided in the German Railroads plan [105], and an area-wide presence of the railroad in short-haul transport.

The existing waterways in Germany could absorb an additional freight transport volume of about 20 % [106], but the new construction or development of waterways, e.g., as investigated for the Elbe, is problematic for ecological reasons [107].

8.1. International Harmonization of Dangerous Goods Regulations

Although concern about the safety of shipments of dangerous goods has led to international agreements, questions of environmental damage have so far played rather a minor role.

On 1 January 1995 criteria for the classification of environmentally relevant substances were introduced for the first time into the international agreements on cross-border transport of hazardous materials existing since 1957 [108, 109], taking into consideration the expert recommendations of the UN.

The EU has adopted these regulations in framework directives, which remain to be introduced by the Member States. From 1 January 1997 onwards the directives will also apply to intrastate transport in the EU. Up to now the criteria deal only with aspects of water pollution; hazards to soil and air are not included.

Furthermore, there is a need for action regarding harmonization of the classification systems of the individual transport routes. Thus it is possible that goods transported by ship are classified, packaged, and labeled differently to those carried by rail or road.

In addition a uniform classification system in the field of transport should be harmonized with the classification schemes in other legal areas. Such a harmonization has already been commissioned:

In 1992 the Conference for Environment and Development (UNCED) adopted Agenda 21 in which the signatories commit themselves in Chapter 19 to work out an internationally coordinated classification system.

8.2. Transport Avoidance and Regulatory Measures

Previous national measures, like international ones, were aimed at making the transport of dangerous goods safer: the transport volume itself was not the subject of regulation. The safest strategy for dangerous goods, however, is undoubtedly one in which they are not transported at all. Shipments of dangerous goods can become superfluous, e.g., by abandoning products involving the use of dangerous substances, or by rearranging production processes.

The best-known example of an international environmental protection agreement that resulted, as a side effect, in a rearrangement of production processes and a withdrawal of products, is the 1987 Montreal Agreement for the protection of the ozone layer. The prohibition of the production and use of specific chlorofluorocarbons that, aside from their action on the stratospheric ozone layer, are also water-endangering, has required rearrangements of production [110].

Conversion of the Basel agreement for controlling the export of wastes across frontiers to an EC statutory regulation [111] is likely to lead to an increase in the transport of wastes. This anomaly arises from the provisions in the statutory regulation, for the acceptance of new lists and reclassification and deletion of existing lists, graded according to the degree of surveillance, the vagueness of the definition of waste, and the tendency to liberalization in the worldwide movement of goods under the influence of rising costs of local waste disposal.

Aside from regulatory means, incentives for transport avoidance, and the traffic deflection, can also be achieved by economic instruments. It has been estimated that consistent application of the "polluter pays principle", which requires internationalization of external costs, would add $46\times10^9 - 250\times10^9$ DM per year to transportation costs in the federal states of pre-unification Germany. These are low estimates as they take no account of important external costs such as the effects of climatic change [112, 113].

In many cases, external costs can not be allocated unequivocally to the polluter. In addition, subsequent internalization of external costs is impossible if irreversible environmental damage has occurred [112].

The level of external costs shows that it would be possible, e.g., by a general tax on the transport of dangerous goods, to create inducements to rearrange production and contribute to transport avoidance.

A first step in this direction is represented by a proposed EC directive in which the external costs would be charged to heavy commercial vehicles (environmental loads are expressly mentioned as external costs) [114].

9. References

1. EC Green Paper on the Effect of Traffic on the Environment, Publication of the European Commission, KOM (92) 46th, final version, IV b no. 97.
2. In [1], IV a no. 87.
3. In [1], IV a no. 89.
4. In [1], III d nos. 80 – 82, III e no. 84.
5. In [1], III d nos. 83 – 85.
6. Der Rat von Sachverständigen für Umweltfragen, Umweltgutachten 1994 – Für eine dauerhaft umweltgerechte Entwicklung, Wiesbaden 1994, p. 423.
7. S. Nicodemus, H. Eberlein in Statistisches Bundesamt (ed.): *Schätzung von Umfang und Struktur des Transportaufkommens gefährlicher Güter,* im Auftrage des Bundesministers für Verkehr, Wiesbaden 1993, p. 46.
8. In [7], pp. 30 – 33.
9. The Recommendations are based on those originally prepared by the United Nations Committee of Experts on the Transport of Dangerous Goods (ST/ECA/43-E/CN.2/170) considered by the Economic and Social Council at its twenty-third session (Resolution 645 G (XXIII) of 26 April 1957). The Recommendations have been amended and updated by succeeding sessions of the Committee of Experts and published in accordance with subsequent resolutions of the Economic and Social Council.
10. Basel Convention of the Control of Transboundary Movements of Hazardous Wastes and their Disposal, 1989.
11. D. Rennoch: *Abfall-Transport-Berater,* K. O. Storck Verlag, Hamburg 1993.
12. European Agreement concerning the International Carriage of Dangerous Goods by Road (ADR) and protocol of signature done at Geneva on 30 September 1957, United Nations

Publication, ECE/TRANS/110, vols. I and II New York – Geneva.Ordnung für die internationale Eisenbahnbeförderung gefährlicher Güter (RID), Zentralamt für den internationalen Eisenbahnverkehr, Bern.
13. International Maritime Dangerous Goods Code (IMDG-Code), International Maritime Organization, London 1995.
14. *Laboratory Biosafety Manual,* WHO, Geneva 1983.
15. OECD Guidelines for Testing of Chemicals, no. 404, Acute Dermal Irritation/Corrosion, Paris 1992.
16. *Official Journal of the European Communities* **34** (1991) 180.
17. American Society of Mechanical Engineers: *Boiler and Pressure Vessel Code,* Section VIII, Division 1, "Rules for the Construction of Pressure Vessels," ASME, New York 1992.
18. *AD-Merkblätter,* Beuth Verlag GmbH, Berlin 1994.
19. B. Jäger et al.: *Die Auswirkungen des Sicherheitsrisikos von Lagerung und Transport gefährlicher Stoffe auf die Entwicklung verbesserter Transporttechnologie,* BMFT-Report RGB 8010, Verlag TÜV Rheinland 1983.
20. Health and Safety Commission: *Major Hazard Aspects of the Transport of Dangerous Substances,* Library and Information Services, London 1991.
21. M. Arras, *TU* **34** (1993) 84.
22. E. Lehmann et al., *TU* **34** (1993) 79.
23. Underwriters Laboratories Inc.: Outline of Proposed Investigation of Structural Steel Protected for Resistance to Rapid-Temperature-Rise-Fires, UL 1709, London, Sep. 1983.
24. B. Droste, *Fire Technol.* **28** (1992) 257 – 269.
25. *Allianz-Handbuch der Schadenverhütung,* VDI-Verlag, Düsseldorf 1984.
26. *Handbook of Industrial Loss Prevention,* McGraw-Hill Book Co., New York 1967.
27. W. E. Baker et al.: Workbook for Estimating Effects of Accidental Explosions in Propellant Ground Handling and Transport Systems, NASA Contractor Report 3023, National Aeronautics and Space Administration, 1978.
28. Control of Urban Development Around High-Risk Industrial Sites, Secretary of State to the Prime Minister for the Environment and the Prevention of Major Technological and Natural Risks – DEPPR, France, Oct. 1990.
29. G. Uelpenich, *Brandschutz* **47** (1993) 696 – 702.
30. Verein Deutscher Ingenieure (ed.): VDI 3783, part 1, Dispersion of Emissions by Accidental Releases – Safety Study, 1987; part 2, Dispersion of Heavy Gas Emissions by Accidental Releases – Safety Study, 1990, Beuth Verlag, Berlin 1987, 1990.
31. Verband der Chemischen Industrie e.V. (VCI), TUIS, Transport-Unfall-Informations- und Hilfeleistungs-System, Frankfurt/Main, Oct. 1992.
32. CEFIC, ICE News, Brussels, Belgium, Sep. 1992 and May 1993.
33. H. Wefers et al.: Hinweise zur Erstellung und Prü-fung von betrieblichen Alarm- und Gefahrenabwehrplänen nach der Störfall-Verordnung, LIS-Berichte no. 83, Essen 1988.
34. Recommendations on the Transport of Dangerous Goods, United Nations Publication, New York.
35. Recommendations on the Transport of Dangerous Goods, Tests and Criteria, United Nations Publications, New York.
36. Regulations for the Safe Transport of Radioactive Material, International Atomic Energy Agency, Vienna.
37. International Maritime Dangerous Goods Code, Volumes I to IV and Supplement, International Maritime Organization, London.
38. Technical Instructions for the Safe Transport of Dangerous Goods by Air, International Civil Aviation Organization, Montreal.
39. Règlement concernant le transport international ferroviaire des marchandises dangereuses (RID), Office Central des Transports Internationaux par Chemins de Fer, Bern.
40. European Agreement concerning the international carriage of dangerous goods by road (ADR), United Nations, Publication, ECE/TRANS/100, New York.
41. Recommendations on the Transport of Dangerous Goods, ST/SG/AC. 101/Rev 8, latest edition United Nations Publication, New York.
42. International Maritime Dangerous Goods Code (IMDG-Code), 27th Amendment, International Maritime Organization, London 1995.
43. European Agreement Concerning the International Carriage of Dangerous Goods by Road (ADR), United Nations Publication, ECE/TRANS/100, vols. I and II, New York – Geneva, 1994.

44. European Agreement Concerning the International Carriage of Dangerous Goods by Rail (RID), unofficial edition by Department of Transport, HMSO Publications Centre, London 1995.
45. Technical Instructions for the Safe Transport of Dangerous Goods, International Civil Aviation Organization, DOC 9284-AN/905, latest edition, Montreal.
46. Regulations for the Safe Transport of Radioactive Materials, 1985 Edition (as amended 1990), Safety Series no. 6, STI/PUB/517, International Atomic Energy Agency, Wien.
47. Advisory Material for the IAEA Transport Regulations for the Safe Transport of Radioactive Material, 1985 Edition, Safety Series no. 37, STI/PUB/589, International Atomic Energy Agency, Wien 1990.
48. Europäisches Übereinkommen vom 20. Sept. 1957 über die internationale Beförderung gefährlicher Güter auf der Straße (ADR) in der Fassung der Bekanntmachung vom 15. Januar 1992 (BGBl: II, S. 95) in der Fassung der 11. Änderungsverordnung vom 4. März 1993, BGBl. II S. 234 – Anlagenband zum BGBl. Teil II Nr. 9 vom 23. März 1993.
49. Ordnung für die internationale Eisenbahnbeförderung gefährlicher Güter (RID), Anlage I zu Anhang B des Übereinkommens über den internationalen Eisenbahnverkehr vom 9. Mai 1980 (COTIF-Übereinkommen) zuletzt geändert durch die 4. RID-Änderungsverordnung vom 25. 06. 1993 (BGBl. II Nr. 19).
50. International Maritime Dangerous Goods Code (IMDG-Code), 25th Amendment, International Maritime Organization, London 1989; B. Anz. Nr. 69, 8 April 1992.
51. Verordnung über die innerstaatliche und grenz-überschreitende Beförderung gefährlicher Güter auf Straßen (GGVS); Europäisches Übereinkommen über die internationale Beförderung gefährlicher Güter auf der Straße (ADR).
52. Verordnung über die innerstaatliche und grenzüberschreitende Beförderung gefährlicher Güter mit Eisenbahnen (GGVE); Ordnung über die internationale Eisenbahnbeförderung gefährlicher Güter (RID).
53. H. Engels: "Tankcontainer minimieren Risiko," *Chem. Ind.* (1988) no. 10, 28 – 32.
54. E. Mühling: "Safety First," *Chem. Ind.* (1990) no. 1, 12 – 14.
55. B. Schulz-Forberg, B. Droste: "Tankbau – was, wie und warum," *Gefährliche Ladung* **36** (1991) no. 11, 506 – 512.
56. K. Wissenbach: "Gefährliche Güter sicher transportieren," *Maschinenschaden* **62** (1989) no. 5, 181 – 185.
57. Recommendations on the Transport of Dangerous Goods, 5th ed., United Nations Publication, New York 1988.
58. K. Wissenbach: "Gefährliche Güter sicher transportieren," *Maschinenschaden* **62** (1989) no. 5, 181 – 185.
59. M. Hegenauer: "Vorschriften für die Beförderung gefährlicher Güter," *Chem.-Tech.* **14** (1985) no. 6, 153 – 164.
60. K. Ridder: "Transportvorschriften für gefährliche Güter," *Entsorgungs-Technik,* Jan./Feb. 1990, 36 –38, March/April 1990, 38 – 40
61. V. Pilz: "Technische Maßnahmen zur Verhinderung von Schadstoff-Freisetzung aus Chemie-Anlagen und beim Chemikalien-Transport," *VFDB Z.* **35** (1986) no. 4, 160 – 163.
62. H. J. Mayer: "Für und Wider von Überdruckventilen an Tanks zur Beförderung gefährlicher Güter," *TU* **22** (1981) no. 6, 245 – 248.
63. R. J. Janssen: "Machen Überdruckventile Druckgaskesselwagen sicherer?" *Gefährliche Ladung* **25** (1980) no. 9, 376 – 378.
64. G. F. List: "Modeling and Analysis for Hazardous Materials Transportation," *Transp. Sci.* **25** (1991) no. 2, 100 – 114.
65. International Maritime Organization (IMO), SOLAS-Convention 74/78 as amended.
66. IMO, MARPOL-Convention 73/78 as amended.
67. Zentralkommission für die Rheinschiffahrt (ZKR), ADNR Regulations – Version prior to 1995.
68. ZKR, ADNR Regulations – Version 1995.
69. Germanischer Lloyd (ed.): "Inland Waterway Vessels," part 2 in "I Ship Technology," *Rules for Classification and Construction.*
70. Germanischer Lloyd (ed.): "Seagoing Ships," part 1 in "I Ship Technology," *Rules for Classification and Construction,* Chap. 1.
71. Germanischer Lloyd (ed.): "Seagoing Ships," part 1 in "I Ship Technology," *Rules for Classification and Construction,* Chaps. 6, 7.
72. K. J. Döhrn: "Das neue ADNR/ADN, Bauvorschriften für Binnentankschiffe," Germanischer Lloyd, Hamburg.

73. M. Böckenhauer: "Die aktuellen MARPOL-Vorschriften zur Verbesserung der Tanksicherheit," Germanischer Lloyd, Hamburg..
74. VDI-Richtlinie 2700, "Ladungssicherung auf Straße," Mai 1990.
75. G. Großmann et al.: *Technologie für Transport, Umschlag und Lagerung im Betrieb,* VEB Verlag Technik, Berlin 1983.
76. N. Müller: "Neuland für Stückgut-Fahrer," *Gefährliche Ladung* **2** (1993) 66.
77. ISO 1496 – 4:1991, Series 1 freight containers – specification and testing – Part 4: Non-pressurized containers for dry bulk.
78. DIN 30 723,Absetzkipperfahrzeuge, Juli 1986.
79. DIN 30 722,Abrollkipperfahrzeuge, Abrollbehälter, November 1990.
80. Richtlinien zur Durchführung der Gefahrgutverordnung Straße vom 24. März 1994, VkBl., no. 7, 1994, p. 281.
81. Richtlinien für die sichere Behandlung von Schüttladungen bei der Beförderung mit Seeschiffen vom 30. August 1990, B Anz., Nr. 226 a vom 6. Dez. 1990, zuletzt geändert durch die im B Anz., Nr. 23 vom 04. Feb. 1993 veröffentl. Bekanntmachung vom 18. Dez. 1992.
82. P. Schmitt: *Lagerung gefährlicher Güter in der Europäischen Gemeinschaft,* Verlag TÜV Rheinland, Köln 1991.
83. P. Roberts: "Application of Hazard and Risk Analysis to Gas Making and Gas Storage Facilities," International Conference on Safety and Loss Prevention in the Chemical and Oil Processing Industries, 1990.
84. J. E. S. Venart: "Major Hazards in the Transport and Storage of Pressure Liquefied Gases," *J. Hazard. Mater.* **12** (1988) 3 – 392.
85. V. Pilz: "Sichere Lagerung von Stoffen in der chemischen Industrie – safe storage of chemicals," *Chem. Ing. Tech.* **60** (1988) no. 6, 452 – 463.
86. V. Marshall: *Major Chemical Hazards,* Ellis Horwood Ltd., Chichester 1987.
87. V. Pilz: "Technische Maßnahmen zur Verhinderung von Schadstoff-Freisetzungen aus Chemie-Anlagen und bei Chemikalien-Transporten," *VFDB Z.* **35** (1986) no. 4, 160 – 163.
88. K. J. Doktor, A. Borchard: "Abnahmeuntersuchung an Druckbehälteranlagen," *TU* **30** (1989) nos. 7/8, 294 – 296.
89. K. J. Doktor: "Gase sicher lagern," *TU* **30** (1989) no. 6, 245 – 249.
90. D. L. Blomquist: "Neue Normen für Schadenverhütung bei Flüssiggasen (API 2510 and 2510 A)," *Hydrocarbon Process.* **11** (1988) 58 – 61.
91. API Standard 2510 – Design and Construction of Liquefied Petroleum Gas (LPG) Installations, 6th ed., American Petroleum Institute, Washington, D.C. 1989.
92. API Standard 2510 A – Fire Protection Considerations for the Design and Operation of Liquefied Petroleum Gas (LPG) Storage Facilities, 1st ed., American Petroleum Institute, Washington, D.C. 1989.
93. K. Strohmeier: "Abschätzung des Gefährdungspotentials druckverflüssigter Gase," *Chem. Ing. Tech.* **65** (1993) no. 4, 410 – 413.
94. A. Schwierczinski, R. Herrmann: "Praxisbezogene Ermittlung von Sicherheitsabständen bei der Ausbreitung freigesetzter schwerer Gase," Siemens AG (KWU), Offenbach 1990.
95. "Ermittlung sicherheitstechnischer Kriterien zur Flüssiggastechnologie und Herleitung geeigneter Sicherheitsstandards," Bundesanstalt für Materialforschung und -prüfung (BAM), Berlin 1988, Report 01501, VDI-Technologiezentrum Physikalische Technologien.
96. DIN 4119,Oberirdische zylindrische Flachboden-Tankbauwerke aus metallischen Werkstoffen, Teil 1 Grundlagen, Ausführung, Prüfungen, Teil 2 Berechnung, Beuth Verlag GmbH, Berlin 1979/1980.
97. Verordnung über brennbare Flüssigkeiten – VbF mit ihren Anhängen und den dazugehörigen Technischen Regeln für brennbare Flüssigkeiten (TRbF), Carl Heymanns Verlag KG, Berlin 1992.
98. BS 7777 – Flat-bottomed, Vertical, Cylindrical Storage Tanks for Low Temperature Service; part 1, Guide to the general provision applying for design construction, installation and operations; part 2, Specification for the design and construction of single, double and full containment metal tanks for the storage of liquified gas at temperatures down to $-165\,^{\circ}$C; part 3, Recommendations for the design and construction of prestressed and reinforced concrete tanks and tanks foundations; and the design and installation of tank insulation, tank liners and tank coatings; part 4, Specification for the design and construction of single containment tanks for the storage of liquid

oxygen, liquid nitrogen and liquid argon, British Standard Institution, London 1992.
99. Technische Regeln zur Druckbehälterverordnung (TRB), Carl Heymanns Verlag KG, Köln, loose leafe collection.
100. O. Scharz: *Glasfaserverstärkte Kunststoffe, kurz und bündig,* Vogel-Verlag, Würzburg 1975.
101. BS 4994, Design and Construction of Vessels and Tanks in Reinforced Plastics, British Standard Institution, London 1987.
102. K. Krekeler, G. Wick: "Polyvinylchlorid," *Kunststoffhandbuch,* vol. I. Carl Hanser Verlag, München 1993.
103. S. Nicodemus: *Transport von gefährlichen Gütern 1982 – 1991,* Statistisches Bundesamt, Wiesbaden 1992.
104. A. Coco: "Grand Prix d'Harmonisation," *Gefährliche Ladung* **7** (1991) 278.
105. Die Deutschen Bahnen, Masterplan GVZ Deutschland, Zentrale der DR/DB, Berlin/Frankfurt a.M. 1993.
106. G. Hulsmann: "Die Binnenschiffahrt in großer Not," *Frankfurter Allgemeine Zeitung,* June 18, 1993.
107. Öko-Projekt Elbe-Raum, Elbe – Fluß oder Kanal, Tagungsband, Dresden 1992.
108. ADR, Accord européen relativ au transport international des marchandises Dangereuses par voie de Route; BGBl, 1969 II, p. 1489.
109. (RID), Office Central des Transports Internationaux par Chemins de Fer, Bern; BGBl, 1985 II, p. 666. Réglement International concernant le transport feroviaire des marchandises dangereuses.
110. Umweltbundesamt: *Verzicht aus Verantwortung: Maßnahmen zur Rettung der Ozonschicht,* Berichte 7, Erich-Schmidt-Verlag, Berlin 1989.
111. Beschluß des Rates vom 1. Febr. 1993 zum Abschluß des Übereinkommens über die Kontrolle der grenzüberschreitenden Verbringung von gefährlichen Abfällen und ihrer Entsorgung (Basler Übereinkommen), 93/98/EWG.
112. E. Dogs, H. Platz: *Externe Kosten des Verkehrs – Schiene, Straße, Binnenschiffahrt,* Planco Consulting GmbH, Essen 1990.
113. D. Teufel et al.: *Ökologische und soziale Kosten der Umweltbelastung in der Bundesrepublik Deutschland im Jahr 1989,* Umwelt- und Prognose-Institut, Heidelberg 1991.
114. Änderung des Vorschlags für eine Richtlinie des Rates zur Anlastung der Wegekosten an schwere Nutzfahrzeuge, Drucksache der Europäischen Kommission, KOM (92) 405 endg., vom 30.09.1992.

Chemical Products: Safety Regulations

GOFFREDO DEL BINO, European Commission, Brussels, Belgium (Chap. 1)

BERNARD BROECKER, formerly Hoechst Aktiengesellschaft, Frankfurt/Main, Federal Republic of Germany (Chaps. 2 and 4)

ROBERTO BINETTI, Istituto Superiore di Sanità, Rome, Italy (Chap. 3)

NORMAN KING, Department of the Environment, London, United Kingdom (Chaps. 5 and 6)

1. Introduction

About 30 years ago, governments started to adopt measures to control chemicals with the aim of protecting people and the environment. Since then, rules and regulations have been introduced in many industrialized countries to assess and manage the risks of chemicals. It was recognized from the outset that harmonization of controls is important if the need for protection was to be reconciled with the aim of avoiding technical barriers to trade. The European Union (EU) took the first step in 1967 with the adoption of Directive 67/548 on classification, packaging, and labeling of dangerous substances, since then amended seven times to introduce appropriate risk assessment measures.

In the United States, the Toxic Substances Control Act (TSCA) became law in 1976. It introduced pre-manufacturing/pre-marketing assessment of the effects of chemicals.

In the 1970s, the OECD, grouping for 24 most industrialized countries in the world, launched a Chemicals Programme which is still the most

Ullmann's Industrial Toxicology

ISBN: 3-527-31247-1

important international initiative in the field of chemicals control.

All chemicals are toxic. The difference between them is the dose required to cause toxic effects: sugar and salt, if consumed by the kilogram, may be as lethal as milligrams of cyanide or strychnine. In addition to their toxicity, chemicals may be explosive, flammable, or corrosive. Others may have deleterious effects on our environment, e.g., chlorofluorocarbons.

Given that all chemicals are potentially dangerous, it is reasonable that precautions be taken to assess each chemical and to ensure that appropriate steps are taken to reduce the potential risks associated with its use to acceptable levels.

Control of chemicals includes the different steps illustrated in Figure 1.

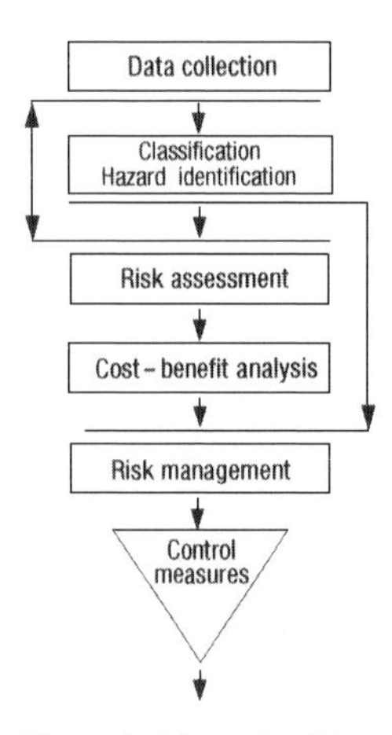

Figure 1. Schematic of the process of chemicals control

In recent years, governments and industry have come to realize the part that consumer choice can play in protecting the environment from chemicals and other hazards. The need to develop more environmentally friendly products is now well recognized, and this article also deals with the objectives and function of ecolabeling systems.

2. Definitions

Risk assessment is a stepwise, iterative process, comprising the following steps:

1) Hazard identification: the identification of the adverse effects which a substance has an inherent capacity to cause
2) Dose (concentration) – response (effect) assessment: estimation of the relationship between the dose, or level of exposure to a substance, and the incidence and severity of its effect
3) Exposure assessment: determination of the emissions, pathways, and rates of movement of a substance, and its transformation or degradation in order to estimate the concentrations/doses to which human populations or environmental compartments are or may be exposed
4) Risk characterization: estimation of the incidence and severity of the adverse effects likely to occur in a human population or environmental compartment owing to the actual or predicted exposure to a substance
5) Risk estimation: quantification of the estimated likelihood in a risk characterization

3. Hazard Identification

The risk assessment of a chemical substance is a complex procedure based on comparison of its potential adverse effects with the reasonably foreseeable exposure of people and the environment to that substance. The identification of the potential adverse effects of a substance is the first step of this procedure [1].

Hazard identification is mainly based on a critical evaluation of the intrinsic properties of a substance, i.e., the physicochemical, toxicological, and ecotoxicological properties. For each of these groups of properties it is possible to define a number of specific end points, and for each of them internationally recognized guidelines [2], [3], to determine specific parameters. The interpretation of such parameters leads to the hazard identification of the chemicals.

3.1. Significance of Parameters

Intrinsic properties characterize the behavior of a chemical substance. They are directly linked to the structure of the molecule, although sweeping conclusions about behavior and properties cannot be drawn simply because a substance belongs to a chemical class or family. It is enough to substitute one radical by another to induce specific properties which are not shared by the whole family.

Consequently, although the relationship between structure and activity is now an area of great scientific interest and ever-growing endeavors are being made to replace compulsory, costly, and long-term experimental studies by mathematical models, it is still necessary to use experimental methods to determine the basic properties from which the potential effects of a given substance can be deduced.

The following paragraphs deal with physicochemical, toxicological, and ecotoxicological properties, attempting briefly to clarify the significance of each of these and the principles underlying the methods of determining them, with direct reference to existing internationally recognized methods.

3.1.1. Physicochemical Properties

Physicochemical properties play an important role in the potential of a substance to produce adverse effects. There is a close correlation between physicochemical properties and the occurrence of toxic and environmental effects.

Each physicochemical property plays a specific role. For instance, the higher the molecular mass, the lower the possibility of absorption by the organism; the higher the melting point, the greater the physical stability of the substance; the lower the boiling point, the more readily the substance can be inhaled; liposoluble substances can penetrate more readily through the skin, facilitating their bioaccumulation, while hydrosoluble substances have a preference for gastric and pulmonary channels; substances able to lower surface tension are more readily absorbed and promote the absorption of substances which are difficult to absorb on their own. Similar considerations apply to the various sectors of the environment since substances can enter, move among, persist, and bioaccumulate in natural species.

Molecular Mass. The size of the molecule gives an idea of whether the substance is bioavailable, i.e., can be absorbed, assimilated, and metabolized by the organism. A comparatively small molecule can enter the organism and produce its effects much more readily than a large molecule. A high polymer cannot generally be absorbed and assimilated and is therefore much more inert as regards interaction with the human organism or other species.

Melting Point. The melting point characterizes the physical state of the substance at ambient temperature, and normal pressure. If the substance is solid at room temperature, it is less mobile than a substance in the liquid or gaseous state.

Boiling Point. Like the melting point, the boiling point characterizes the physical state of the substance under normal pressure conditions, providing important information on its ability to affect, for instance, channels of inhalatory exposure. In case of diffusion into the environment, it qualitatively indicates how easily the substance can be transferred into the air, and to what extent it diffuses into the environment. The higher the boiling point, the smaller the risk of inhalatory effects.

Relative density represents the mass ratio between a solid or liquid substance and water or, in the case of gaseous substances, between the substance and air.

When referred to water, it is normally expressed as d_4^{20} (mass of the substance measured at 20 °C and mass of water measured at 4 °C). If the substance does not dissolve in water, its density indicates whether the substance can be expected to stratify on the surface (density <1) or to sink to the bottom (density >1), which is important in relation to mobility in the environment and the diffusion of pollution into different compartments, including indirect human exposure.

When referred to air, substances which are denser than air (density >1) tend to stratify around the source of emission, constituting a major risk of exposure at relevant concentrations, while at density <1, diffusion into the air is easier, and the risk is generally reduced since it is less persistent in the vicinity of the emission site.

Vapor pressure is also an important way of finding out whether the substance, at ambient temperature, may readily become volatile from a solution, another liquid, or solid matrix. It is defined as the pressure of saturation above a solid or a liquid. A substance with a high vapor pressure chiefly affects inhalatory and dermal channels of exposure. As regards diffusion into the environment, higher values are linked to an increased transfer potential. As regards physical danger, higher vapor pressure may be associated with risks of explosion and the need for greater precautions as regards storage, packaging, and transportation. In brief, greater risks are attached to high vapor pressure values.

Surface Tension. The ability of a substance to lower the surface tension of water is closely linked to its increased potential for absorption, permeation of barriers (membranes), and diffusion into water, even when the substance is scarcely hydrosoluble. A substance which can lower the surface tension of water may also make it easier for liposoluble substances which are insoluble or only slightly soluble in water to dissolve in water. A typical example is provided by surfactants which facilitate absorption of substances such as polycyclic aromatic hydrocarbons which are not hydrosoluble.

The surface tension value which is of interest is not that of the substance per se, but of its solution in water. The greatest risks are associated with low surface tension values measured in this solution.

Water solubility is defined as the concentration of saturation in water at a defined temperature. It is generally expressed as mass per unit volume of solution.

Hydrosolubility has direct implications as regards both potential toxicological and ecotoxicological effects. Hydrosolubility shapes adsorption by gastric or pulmonary channels, the possibility of dissolution in surface waters, and therefore the mobility in water and soil, as well as the precipitation of substances onto the soil in rain.

Fat solubility (liposolubility) expresses a substance's potential to dissolve and eventually accumulate in human and natural species' fats. The more liposoluble a substance, the more easily it can penetrate through the skin and produce local or systemic effects. It is normally measured using standard triglycerides with defined fatty acids content. Alternatively, the solubility in organic (possibly nonpolar) solvents may be useful information.

Partition Coefficient *n*-Octanol/Water. The problem of bioaccumulation may reach very high proportions in natural species. Aquatic species, which filter enormous quantities of water, accumulate liposoluble substances on a scale which increases with the species' physical size and life expectancy, playing a part in "magnifying" pollution of the trophic chain.

An idea of the bioaccumulation capacity of a substance may be gained from the *n*-octanol/water partition coefficient. This is obtained by establishing the ratio at which a given substance is distributed when brought into contact with a system of two immiscible phases formed by normal octyl alcohol (*n*-octanol) and water. The *n*-octanol simulates the fatty phase: if the substance dissolves to a greater extent in the aqueous phase the partition coefficient assumes negative values (it is normally expressed as a decimal logarithm, $\log P_{ow}$), showing that the substance is more hydrosoluble than liposoluble; positive values in the case of substances which dissolve to a greater extent in the octanol phase, indicating that the substance can bioaccumulate (normally this requires $\log P_{ow} > 3$).

The partition coefficient is particularly important for substances which are scarcely soluble in water and fats. It might be thought that low liposolubility entails a low level of risk of bioaccumulation, whereas if the substance is more liposoluble than hydrosoluble (i.e., high partition coefficient), bioaccumulation may also occur, but to a lower extent because of the low absolute liposolubility. The partition coefficient is independent of the absolute values for liposolubility and hydrosolubility, expressing the ratio between them.

Flammability. This property, like the following physicochemical properties, provides direct information on the possibility of a so-called physical effect (it is better to talk of physicochemical effects, since oxidation is a typical chemical effect). While the previously described properties "characterize" the substance, giving indirect information on the potential bioavailability and mobility into the environment, the properties regarding physical effects provide information on immediate and direct risks. Most damaging accidents are linked to physical effects, often with consequences which are disastrous in terms of their size, persistent effects, and number of casualties.

The risk of flammability may derive from a low vapor ignition temperature in the presence of an ignition source (primer) or from the possibility of the substance igniting spontaneously in air with no ignition point (autoflammability).

For liquid products, the flash point is the parameter which allows quantification of the flammability risk. It is defined as the lowest temperature, at a corrected pressure of 101.325 kPa, at which a liquid emits vapors in such quantity as to produce a flammable vapor/air mixture. Experimentally, the substance is slowly and progressively heated and, at regular intervals, its vapors are exposed to a spark: the lowest temperature at which the vapors/air mixture flashes is the flash point. This type of test is appropriate for liquids and low-melting-point substances. A different type of test is carried out for solid substances, yielding a simple positive/negative response, by evaluating the substance against a reference flammable product.

If the substance decomposes during heating, the flash point is still of interest, as the ignition of the vapors may be due to the mixture of decomposition product vapors rather than the substance per se. Moreover, account has to be taken of the case in which the substance per se is not flammable, but releases flammable gases in contact with other substances, e.g., water.

Some gaseous substances ignite in the presence of a primer at low or very low temperature; in this case it is useful to know the range of concentration in air at which flame propagation occurs.

Information on autoflammability can prevent serious accidents, especially during long-term storage of chemical products. It is vital to know whether a substance may autoignite at temperatures which are foreseeable under bad storage conditions.

Explosive Properties. A substance is considered to have explosive properties if, following heating, shock, or friction, it may decompose abruptly, releasing large quantities of energy and/or gas. Explosive properties are tested by methods which take account of the three different possibilities (heat, shock, friction). If the substance is shown to have explosive properties similar to or greater than those of a reference substance (e.g., dinitrobenzene) it is considered explosive.

Oxidizing Properties. A substance with oxidizing properties may, when brought into contact with chemically oxidizable organic or inorganic substances, bring about dangerous reactions causing fire, explosion, or the formation of other hazardous substances. Oxidizing properties are not just related to the presence of oxygen in the molecular structure, but to its potential for acting more generally as an oxidant in an oxidation – reduction reaction. In many cases, specific tests are not necessary to determine whether or not it is oxidizing; examination of its structural formula may provide predictive information. Organic peroxides, for instance, are generally considered as oxidizing, even in the absence of test results. The only doubt is whether they should also be considered as explosive; for this, experimental evidence is needed.

3.1.2. Toxicological Properties

(See also → Toxicology)

Toxicological properties are of direct interest in assessing the risk of chemical substances. They include a range of effects understood either as general poisoning of the organism or aimed at one or more particular target organs, and whose intensity is generally proportional to the dosage absorbed and to specific effects, which may be independent of dose, e.g., mutagenic and carcinogenic effects. There may also be immediate effects, generally acute, or effects deferred in time, linked to the intake of small, continued doses, or with a period of latency before the effects appear.

The dosage administered and the duration of toxicological tests are inversely proportional with reference to acute, subacute, subchronic, and chronic studies. When studying acute effects, the substance is usually administered to test animals as a single dose, at a level sufficiently high to cause death. When studying subacute effects, lower dosages are used, but for a longer and continuous period (28 d). For subchronic toxicity, even smaller dosages are used, but over a period corresponding to ca. 10 % of the animal's life; for chronic toxicity the test is extended throughout the animal's life, usually at even lower dosages.

Some basic properties need to be differentiated for assessment of the various types of risk connected with toxicological properties.

Acute toxicity includes those effects which are or may be produced following a single administration. In terms of acute effects, it is common to distinguish between systemic effects (lethal effects following oral, cutaneous, or inhalatory administration) and local effects, such as irritation of the eyes and skin, or corrosive and sensitizing properties.

Each of these effects needs to be examined in slightly more detail.

LD_{50} and LC_{50}. The acute lethal effects are generally evaluated through the LD_{50} (dose killing 50 % of animals tested), valid for oral and cutaneous administration, and the LC_{50} (concentration killing 50 % of animals tested), valid for inhalatory administration. Other routes of administration, e.g., peritoneal, intramuscular, or intravenous, although of some interest, have no direct correlation with the study of chemical substances in relation to personal, domestic, professional, or environmental exposure of humans.

A basic consideration underlies the determination of the oral LD_{50}, cutaneous LD_{50}, and inhalatory LC_{50}: the greater the toxicity of the substance, the smaller the value obtained.

The oral LD_{50} is generally determined by administering a known quantity of the substance to the animal (the rat is taken as a reference) by gastric probe (gavage), possibly dispersed in an appropriate vehicle, depending on the solubility of the substance in water or oil, or a different inert vehicle for insoluble substances. After administration, the animal must be observed for up to 14 d. The LD_{50} is assumed to be the dosage, expressed in milligrams substance per kilogram animal body weight (mg/kg), able to cause, during the observation period, death of 50 % of the test animals. In order to determine the LD_{50}, it is necessary, after preliminary tests, to use groups of animals to which a range of increasing dosages are administered, so that the LD_{50} value may be interpolated. The test should not be limited to establishing the number of animals killed in relation to those tested, but should also be used to show, through appropriate histopathological studies, the organs chiefly affected, and thereby to gather information on the possible mechanism of action of the substance.

The cutaneous LD_{50} is determined by applying the substance, possibly dissolved or dispersed in an inert vehicle, to the shaved skin of the test animal (rat or rabbit) and ensuring that it remains in contact for a period of time (24 h or periods depending on the substance's potential corrosive or irritant effects). As mentioned above, liposoluble substances can readily penetrate through the skin, but this possibility is not significant for nonliposoluble substances. Observation must be continued for 14 d after the removal of the bandage, and subsequent investigations should endeavor to gather any other useful information. The cutaneous LD_{50} is also expressed in mg/kg.

The inhalatory LC_{50} is determined by exposing the animal (rat) for 4 h to an atmosphere in which the substance is uniformly dispersed at a constant concentration. The dosage capable of killing 50 % of the animals is expressed as a concentration, in milligrams per liter of air.

For some substances in the form of dusts, it may be difficult or impossible to disperse a concentration of the substance sufficient to cause fatal effects. In such cases, it is possible only to indicate that the LC_{50} is higher than the maximum concentration tested.

Irritant or Corrosive Properties. The irritant or corrosive properties of a test substance are determined with respect to the skin and the eyes. Two separate tests are conducted on rabbits and provide qualitative or semiquantitative information.

Irritant or corrosive effects on the skin are tested by leaving the substance in contact with the intact, shaved skin of the animal, and observing the effects produced after 4 h; these effects may range from the absence of effects to erythema, the formation of eschar or edema, or tissue corrosion. Corrosion occurs when the whole thickness of cutaneous tissue is destroyed.

Irritant effects on the eyes are tested by instilling a small quantity of the substance into one of the animal's eyes and using the other eye as a reference. The severity of the effects is compared with a reference scale, looking in particular at the effects on the cornea, iris, and conjunctivae.

Sensitizing Properties. Sensitization is an allergic cutaneous or inhalatory reaction occurring when the organism, hypersensitive to a chemical substance, is exposed to the substance, even at very low concentrations. This property may be experimentally evaluated only for skin sensitization; no methods exist for the evaluation on test animals of inhalatory sensitization. The test substance is administered to the animal (guinea pig) by cutaneous injection (induction phase) and after an appropriate period the administration is repeated (challenge phase), observing whether the animal is subject to sensitization phenomena.

Sensitizing effects on the respiratory tract are considered a more severe hazard. However, given the lack of a practicable test, a substance is considered to be sensitizing by inhalation on the basis of epidemiological evidence.

Subacute, Subchronic, and Chronic Toxicity. Medium- or long-term toxicity tests provide essential additional information on the more hidden effects which the substance may cause by exposure to low dosages over a period of time. This type of test can also be conducted via the various routes of exposure (oral, inhalatory, cutaneous) in accord with the main route of human exposure.

A basic aim of these tests is to identify the maximum dose level at which the substance does not present toxic effects. This makes it necessary to test several increasing dose levels, so as to determine the dosage at which toxic effects start to appear. This makes it possible to show whether the toxic phenomenon is dose dependent, i.e., whether the intensity of the effect is proportional to the dose administered.

Tests results in the literature contain a range of notation:

1) NEL (no effect level), expressing the maximum level where there are not even minor effects
2) NOEL (no observed effect level), expressing the maximum level at which no effects have been observed
3) NOAEL (no observed adverse effect level), expressing the maximum level at which no adverse effects have been observed

Between NEL and NOEL, generally the latter is preferred, expressing the actual experimental dose level at which no effects have been observed; NEL is a more theoretical value, situated somewhere above the NOEL, more precisely between the dose level corresponding to the NOEL and the subsequent dose level. For identification of the NOAEL, the problem is to characterize exactly what "adverse effect" means. There is the need to distinguish between severe and minor effects, and the debate is still open, although the internationally recognized guidelines have attempted to define, at least in general terms, an adverse effect.

In any case, the NOEL (or NEL, or NOAEL) should be expressed in milligrams substance per kilogram animal body weight and day ($mg\,kg^{-1}\,d^{-1}$).

If the test is conducted by administering the substance to the animal in its food, the results may also be expressed as ppm in the diet (mg/kg of food). It is possible to convert, within an acceptable approximation, one value to another. In the rat, for instance, this conversion consists of dividing the concentration of the substance in the food by 20. Animals other than the rat, with metabolism closer in some ways to that of humans, are mice, dogs, and monkeys.

Subacute toxicity tests are designed to reproduce repeated exposure over short periods. Subchronic toxicity tests may, in some cases, replace chronic toxicity tests; if subchronic toxicity tests, which last 3 months in the case of rats (i.e., 10 % of their life expectancy), do not reveal major effects at sufficiently high dosages and it is assumed that the chronic toxicity tests are not likely to add relevant information, longer term tests may not be necessary, thereby reducing suffering of the animals and study commitments. Chronic toxicity tests are conducted for the animal's entire life and are designed to reproduce continuous, prolonged exposure.

Various macroscopic (body weight, general condition, food intake) and chemical and clinical measurements (blood, urine, and other specific tests) are also conducted during the main test. Small groups of animals may be killed at predetermined times for histopathological studies which are in any case carried out at the end of the trial with a detailed examination of the individual organs and glands.

The NOEL value is a basic factor in assessing the risks connected with the toxicity of a given substance as a result of medium- or long-term exposure. The smaller the NOEL value, the more hazardous the substance, with the result that the risks are inversely proportional to the numerical value of this parameter.

Mutagenesis. A substance is considered mutagenic if its adsorption may produce heritable genetic damage or increase their frequency (→ Mutagenic Agents). Assessing a substance's ability to cause mutation in humans, at germinal or somatic level, is very difficult on the basis of current knowledge. Several in vivo and in vitro tests are designed to predict the presence of a potential ability to cause mutations. The comparative simplicity and brevity of these tests, together with the knowledge that many carcinogenic substances are also mutagenic, has promoted increasing interest in a wide-ranging set of mutagenesis tests, each of which can provide specific information on mutagenic potential, bearing in mind that general conclusions cannot be drawn from a single test. It is therefore necessary to carry out a battery of tests to obtain a minimum set of information.

In vivo tests obviously have a greater significance than in vitro tests, although the latter are more sensitive. In vivo studies are usually performed only when a battery of in vitro tests gives positive results.

A number of categories for mutagenic substances have been established for operating purposes and for risk assessment. The EU and other international and national organizations have formulated criteria defining different categories of danger with different levels of mutagenic potential. The allocation of a substance to one or other of these categories provides an overall qualitative assessment of its possible mutagenic potential.

The EU, for instance, has adopted the following classification:

Category 1: substances known to be mutagenic to humans. There is sufficient evidence to establish a causal association between human exposure to a substance and heritable genetic damage.

Category 2: substances which should be regarded as mutagenic to humans. There is sufficient evidence to provide a strong presumption that human exposure to the substance may result in the development of heritable genetic damage, generally on the basis of appropriate animal studies or other relevant information.

Category 3: substances which cause concern, owing to possible mutagenic effects, but in respect of which the available information does not satisfactorily demonstrate heritable genetic damage. There is evidence from mutagenicity studies, but this is insufficient to place the substance in Category 2.

As in the case of other properties of chemical substances, it is possible in mutagenesis to use the structure – activity relationship for an initial assessment of possible mutagenic effects. Extremely reactive substances which are free radical donors are likely to give positive results in mutagenesis tests.

Carcinogenesis. Substances are considered carcinogenic if, after exposure (generally long-term) by inhalation, ingestion, or skin contact, they may produce cancer or increase its frequency (→ Carcinogenic Agents). Test protocols have been drawn up at international level for studies of carcinogenic potential of chemical substances.

These long-term tests, conducted using the exposure routes (oral, inhalatory, cutaneous) assumed to be most representative of normal exposure conditions for humans, include detailed histopathological investigations intended to show the number, type, and location of tumors; data need to be statistically processed in comparison with control groups to take account of the natural occurrence of tumors in the particular strain of animals studied, and to establish significance.

Several, substantially similar, international classifications have also been formulated for carcinogens; substances are allocated to different categories in relation to the carcinogenic potential demonstrated.

The following three categories, in decreasing order of risk, have been formulated by the EU:

Category 1: substances known to be carcinogenic to humans. There is sufficient evidence to establish a causal association between human exposure to a substance and the development of cancer.

Category 2: substances which should be regarded as carcinogenic to humans. There is sufficient evidence to provide a strong presumption that human exposure to a substance

may result in the development of cancer, generally on the basis of appropriate long-term animal studies or other relevant information.
Category 3: substances which cause concern owing to possible carcinogenic effects, but in respect of which the available information is not adequate for satisfactory assessment. There is some evidence from animal studies, but this is insufficient to place the substance in Category 2.

Also in the case of carcinogenesis, the fact that a substance belongs to a chemical family whose carcinogenic properties are known should pave the way for detailed specific studies.

Embryotoxicity. Teratogenic effects, i.e., nonheritable birth defects, produced by a chemical substance, are now covered by a broader class of effects, covering all possible effects on reproduction; embryotoxicity seems to be the most appropriate term. Two main subcategories of effects exist: possible effects on human fertility, and developmental toxicity.

Substances are considered dangerous to human fertility if they may produce or increase the frequency of damage either to reproductive functionality or to the reproductive capacity of males and females. Substances are considered toxic to development if they may produce adverse nonheritable effects in the offspring, or increase their frequency. There are also specific test protocols to predict potential embryotoxicity on the basis of animal studies.

The present EU scheme distinguishes three categories for each kind of effect:

1) Substances dangerous to human fertility
   Category 1: substances known to impair fertility in humans. There is sufficient evidence to establish a causal relationship between human exposure and impaired fertility.
   Category 2: substances which should be regarded as impairing fertility in humans. There is sufficient evidence to provide a strong presumption that human exposure may result in impaired fertility on the basis of appropriate animal studies or other relevant information.
   Category 3: substances which may cause concern for human fertility, essentially on the basis of animal studies providing sufficient evidence to cause a strong suspicion of impaired fertility in the absence of toxic effects.
2) Substances dangerous to human development
   Category 1: substances known to cause developmental toxicity in humans. There is sufficient evidence to establish a causal relationship between human exposure and subsequent developmental toxic effects in the progeny.
   Category 2: substances which should be regarded as causing developmental toxicity to humans. There is sufficient evidence to provide a strong presumption that human exposure to the substance may result in developmental toxicity, generally on the basis of animal studies or other relevant information.
   Category 3: substances which cause concern owing to possible developmental toxic effects, generally on the basis of animal studies which provide sufficient evidence to cause a strong suspicion of developmental toxicity in the absence of signs of marked maternal toxicity.

3.1.3. Ecotoxicological Properties

(→ Ecology and Ecotoxicology)

As toxicological properties, studied by tests on animals, are indicators of potential effects in humans, so ecotoxicological properties, studied through tests on natural species, are indicators of the potential effects which a substance may have on the environment. Some species, considered to be "biological indicators" are taken to be representative of the various compartments (water, soil, air).

It is difficult to make a clear distinction between these compartments since air, soil, and water interact. Soil contamination can be assessed directly through toxicity tests on species which live in the soil, e.g., plants and earthworms, but soil may be permeated by water and washed by atmospheric and surface waters. This leads to contamination of the water compartment and aquatic species, both in situ and in watercourses and catchment areas. Diffusion of a substance in the air may lead to precipitation and dissolution in water. Diffusion in the soil or in water may lead to volatilization and thus to contamination of the air. Direct contamination of rivers or lakes may pave the way for diffusion through the soil. Contamination of soil may contaminate land species (flora and fauna) which may become the prey of birds.

The aim of these tests is to identify sufficiently representative biological indicators of the toxicity of a substance to ecosystems.

Toxicity in Fish. Toxic effects on fish may occur following acute or prolonged exposure to concentrations which may be very low, bearing in mind the ability of aquatic organisms to recycle very high quantities of water and, in many cases, to bioaccumulate toxic substances.

Acute toxicity is evaluated by keeping the fish for 48 – 96 h in water containing a known concentration of the substance. A number of tests have to be conducted in parallel to determine the concentration at which no lethal effect occurs (LC_0), the concentration which is fatal for all tested animals (LC_{100}) and the median concentration which is fatal for 50 % (LC_{50}). The latter is normally calculated by interpolation among the other data or at least between the two extreme values. LC_{50} is expressed in milligrams per liter of water (mg/L).

The smaller the LC_{50} value, the greater the toxicity of the substance for the species tested.

Long-term toxicity in fish is of interest, even if their sensitivity to extraneous substances in water is very high, and the LC_{50} can be taken as a very representative parameter. The LC_{50} and the LC_{100} are often very close numerically. It is important to look for long-term toxicity in substances which are not highly hydrosoluble since ingestion of very small quantities of substances in the aqueous medium may present delayed toxic effects.

The species recommended for the test are:

Brachydanio rerio (zebra fish)
Pimephales promelas (fathead minnow)
Cyprinus carpio (common carp)
Oryzias latipes (red killifish)
Poecilia reticulata (guppy)
Lepomis macrochirus (bluegill)
Salmo gairdneri (rainbow trout)
Leuciscus idus (golden orfe)

On the basis of tests specifically designed to measure long-term toxicity, it is possible to identify an NOEC (no observed effect concentration).

Toxicity in Daphnia. *Daphnia* is a very small crustacean that makes rapid and frequent swimming movements. The *Daphnia* test is designed to detect the immobilization effect, which is taken as an index of the adverse effects caused by the substance.

The substance is dissolved in water, over a range of increasing concentration and the *Daphniae* are exposed for 48 h. The mean effective concentration (EC_{50}) is generally interpolated between the EC_0 and EC_{100}. All data are expressed in mg/L.

As for fish, the lower the EC_{50} value, the greater the acute toxicity of the substance for *Daphnia.* There may be problems with substances which are scarcely hydrosoluble in acute toxicity tests with *Daphnia.*

Effects on reproduction may be evaluated through long-term tests (21 d) on *Daphnia magna*. An NOEC can be identified.

Acute Toxicity in Birds. No acute toxicity tests for birds formulated at international level are yet available, but the literature provides toxicity data for a number of species (quails, ducks, pigeons, chickens).

Administration generally takes place orally through the diet, linking the quantity ingested with body weight. Use is therefore made of the dose level which is lethal for 50 % of the birds in the test (LD_{50}), expressed in mg/kg.

Toxicity in Higher Plants. Plants are biological indicators of major importance. The damage to higher plants by environmental pollution, for instance, is commonly measured. Up to now there have not been any standard methods, given the wide range of species and reactions which may be obtained. The tests available are based on the effect which a given substance may have on the germination and growth of the plants tested, expressed as the EC_{50}.

Herbicides, which by their nature and function, have toxic effects on plants (phytotoxic effects) may be considered as reference substances. In the absence of accepted international methods, the substance can only be generally considered as phytotoxic, moderately phytotoxic, or nonphytotoxic.

Toxicity in Earthworms. The earthworm is an important indicator of soil life following contamination by chemical substances. This information, together with data from toxicity tests on plants, relates therefore to the soil compartment.

The test already defined at international level uses *Eisenia foetida* kept in standard artificial loam soil treated with different concentrations of the substance. After 14 d the soil is examined and the surviving earthworms counted. The result is expressed as LC_{50} with respect to the soil.

Effects on Algae. Algae are important biological indicators. Unfortunately, the available data focus largely on the detection of effects inhibiting algal gowth, while experience has shown that major adverse effects on the environment may result from the abnormal increase of algae as a result of contaminants in the water.

The tests are carried out in water and the EC_{50} is understood as the concentration which inhibits growth by 50 % (also referred to as inhibitory concentration, IC_{50}). It is more convenient to refer to the substance's ability to produce both positive and negative effects, i.e., changes in normal development.

Persistence. The lifetime of a chemical substance may vary as a function of external conditions. A substance which is kept dry in a dark, sterile, oxygen-free atmosphere can remain unchanged indefinitely.

Once introduced in the environment, a substance is subject to different kinds of aggression which, depending on the molecular structure, may produce total or partial degradation of the substance. Such aggression may be divided in two types: biotic and abiotic.

The more rapidly decomposition takes place, biotically or abiotically, the lower the chance of the substance having any residual adverse effect. Nevertheless, unless the substance is totally decomposed to simple compounds (H_2O, CO_2, NO_x, SO_x, PO_x), identification of the degradation products is essential for complete evaluation of residual effects; the degradation products may be more dangerous than the parent molecule.

Biotic degradation is caused by bacteria and microorganisms which use the substance for their own development, causing it to decompose. Organic chemicals are largely formed from carbon and other elements such as hydrogen, oxygen, nitrogen, sulfur, and phosphorus, and they may become involved in the various cycles of use and conversion by bacteria and microorganisms.

The methods formulated for the determination of biodegradation are therefore based largely on determining the quantity of oxygen needed for the oxidation, in due time, of the organic carbon contained in the test substance. Bearing in mind the various environmental conditions in which the substance may be dispersed, the analytical methods set out different conditions for the addition of bacterial inocula, e.g., in activated sludge and sewage, and different operating methods which attempt to ascertain whether the substance can be decomposed under particular conditions by the action of bacteria and microorganisms.

An initial measurement (ready biodegradability) is based on determining the percentage of the substance degrading over 28 d; a substance which degrades by more than 70 % is generally considered biodegradable.

The biochemical oxygen demand (BOD) is related to the chemical oxygen demand (COD), the latter representing the amount of oxygen needed for complete oxidation of the substance by chemical methods. The BOD/COD ratio is a useful factor for assessing to what extent the substance degrades by biotic media.

Abiotic Degradation. The degradation of a chemical substance may also take place abiotically as a result of hydrolysis if it comes in contact with water or moisture and/or as a result of exposure to sunlight (photodegradation).

Decomposition by hydrolysis is usually assessed by determining the concentration of the substance in an aqueous solution as a function of time and pH, to simulate a range of possible acid or basic conditions in the receptor medium. The "half-time" $t_{1/2}$ is calculated as the time for the concentration of the substance to reduce to half the initial concentration.

Photodegradation is ascertained by exposing the substance to light of a given wavelength, under specific environmental conditions, and determining the residual fraction as a function of time. Here too, the aim of the test is the evaluation of $t_{1/2}$.

3.2. Interaction of Parameters

Each of the parameters examined in Section 3.1 gives information on the risk potential for humans and the environment of a chemical substance. Such parameters are, however, not completely independent. Some physicochemical properties, besides being clear indicators of mobility in the various environmental sectors, directly influence the biological behavior.

3.2.1. Hydrosolubility

Hydrosolubility is an important predictor of the behavior of a chemical substance, both in terms of mobility and biological activity. A substance with high hydrosolubility behaves in a largely predictable manner with respect to possible effects on human health and the environment.

A first consequence of high hydrosolubility is, in general, reduced liposolubility; there is almost always a predominance of either lipophilic or hydrophilic groups in a molecule. Only in some types of chemicals (e.g., surfactants) are both types of group represented to such an extent that the substance has hydrosolubility very similar to its liposolubility. Another direct influence is its *n*-octanol/water distribution coefficient log P_{ow}; since high hydrosolubility often corresponds to a low liposolubility, log P_{ow} is negative, resulting in a slight bioaccumulation potential in living organisms.

From the point of view of possible direct effects on humans, a hydrosoluble substance is easily absorbed through the gastrointestinal tract or the pulmonary alveolae; but it is less likely to be absorbed through the skin. In toxicological studies on animal species, it is necessary to concentrate on the oral route and, depending on the volatility (liquids) or particle size (solids), on inhalation. In the initial analysis, the cutaneous route can be ignored for possible systemic effects, though it remains relevant for possible local (corrosive/irritant) effects.

From the environmental aspect, hydrosolubility is fundamental in evaluating possible effects on aquatic species. Experiments on biological indicators such as fish, *Daphnia*, and algae are useless unless the substance has significant hydrosolubility. The same also applies to biotic and abiotic degradation rates.

3.2.2. Liposolubility

As hydrosolubility and liposolubility are generally inversely proportional, it may be expected that a substance with a high liposolubility will behave in a complementary manner to a substance with a high hydrosolubility.

A substance with high liposolubility generally has a low hydrosolubility and a high octanol/water distribution coefficient, with a significant bioaccumulation potential. From the point of view of the effects on humans, the cutaneous route becomes important as a liposoluble substance can readily be absorbed through the skin to produce immediate or subsequent systemic effects, the latter of which are particularly important because of the potential for bioaccumulation.

It is often difficult to evaluate the potential environmental danger of liposoluble substances since, at least as regards aquatic behavior, the tests are performed in aqueous solution without the addition of auxiliary solvents which can affect the hydrosolubility of the substance. Important is the real concentration, not the nominal concentration (this refers to the amount of substance added to the aqueous medium, and it may not be completely dissolved). If the hydrosolubility is negligible, acute toxicity studies in fish, *Daphnia*, and algae are irrelevant. It is wrong to believe that a highly liposoluble substance does not constitute a risk to the aquatic environment; incidents connected with the accidental discharge of petroleum products (crude oil in particular) reinforce this point. On the other hand, the presence of extremely low concentrations in water may constitute a danger for aquatic species, given their bioaccumulation capacity.

3.2.3. Corrosion/Irritation and Oxidizing Power

Chemically aggressive substances such as oxidizing agents can produce local effects on the skin. High reactivity is often synonymous with low persistence; it is particularly important to investigate the acute effects. It seems certain that chemically aggressive substances can influence genotoxicity.

Corrosive substances can present problems in interpretation of results, if they are tested to evaluate possible systemic effects. If a substance produces local effects, this does not mean that it cannot also produce systemic effects. It is precisely through the local effect that the cutaneous barrier is weakened, and the substance can enter the circulation and produce effects on organs or systems remote from the entry site. In spite of this, in order to understand the exact mechanism of action, it is important to distinguish mortality due to the local effect (e.g., perforation of the esophageal tract due to the corrosive effect) from mortality due to the systemic effect.

3.3. Classification Schemes: The European Union System

A chemical substance is characterized by a critical examination of its physicochemical, toxicological, and ecotoxicological properties. However, it is not feasible to require that every time a chemical substance (per se, or mixed with other substances) is handled, transported, or stored, it should be subjected to a comprehensive evaluation. Those who may come into contact with a chemical substance, whether laboratory staff or domestic users, hardly ever have sufficient knowledge to make such an evaluation; this work is therefore entrusted to committees of experts who systematically examine the relevant data in order to draw up concise guidelines. It is vitally important that such guidelines be expressed in clear and understandable terms, so as to provide essential information not only on the risk potential of the substance, but above all on the measures to minimize such risks.

Such ideas are implemented in practice via the danger classification in conjunction with the danger guidelines and information on reducing risk, generally using one or more danger symbols accompanied by standard wording describing, very clearly and in terms comprehensible to all, the risk associated with a possible use of the chemical substance, and the measures to be adopted to prevent exposure and reduce the consequences of accidents.

These classification systems generally tend to cover all possible effects of a substance, and are intended to inform users directly through the labeling. More specialized types of classification exist, aimed at specific end points, e.g., carcinogenesis; these classifications are drawn up by national or international agencies, (e.g., EPA, IARC) and are regularly updated, with the aim of protecting human health, particularly that of exposed workers.

Besides these "specific" classifications, there are the far more widely used "generic" classifications. The two systems that are most widely used are:

1) The UN classification system for the transport of dangerous goods
2) The EU classification system for the introduction of dangerous substances onto the market

3.3.1. Transportation

National or international transport of goods can be by air, rail, road, sea, or inland waterways. For each of these modes of transport, standards authorities have issued international norms regulating the technical details of the transport of dangerous goods. The term "goods" is extremely general, but covers both substances and dangerous preparations. Such norms (e.g., RID for rail transport, ADR for road transport) differ from one another as regards the packaging characteristics and the safety measures to be adopted, because the particular features of each type of transport have to be taken into account. However, as regards the criteria for classifying goods, they all follow the system defined by the UN and published in the "Orange Book" [4].

The Orange Book not only sets out classification criteria for transport, but for each class lists and regularly updates the products officially recognized as belonging to the class; such lists are not exhaustive, but they are supplemented by guideline criteria which increase their scope. Such guideline criteria cover a total of nine danger classes, these are treated in detail in →Transport, Handling, and Storage, Chap. 3.

Each of the nine classes is characterized by a specific danger symbol, applied to the external packaging of the dangerous product being transported (Table 1). The general criteria are then defined in more detail according to the technical demands of each type of transport.

Table 1. Danger symbols for transport of dangerous goods

Class 1

Explosives
Symbol (exploding bomb): black
background: orange

Class 1—*Division 1.4*
(*Except safety explosives*)

Class 1—*Division 1.5*

Explosives
Background: orange; figures: black; numerals must be ca. 30 mm in height and ca. 5 mm wide (for a label measuring 10 × 10 cm)

Class 2

Compressed Nonflammable Gases
Included in Class 2
Symbol (gas cylinder): black; background: green

Class 3

Flammable Liquids
Symbol (flame): black;
background: red

Class 4—*Division 4.1*

Flammable Solids
Symbol (flame): black;
background: white with
vertical red stripes

Class 4—*Division 4.2*

Substances Liable to Spontaneous
Combustion
Symbol (flame): black;
background: upper half white,
lower half red

Class 4—*Division 4.3*

Substances which, in Contact
with Water, emit Flammable Gases
Symbol (flame): black;
background: blue

Class 5

Oxidizing Substances; Organic
Peroxides
Symbol (flame over circle): black;
background: yellow

Class 6—*Division 6.1*

Poisonous (toxic) Substances
Danger groups: I and II
Symbol (skull and cross-bones):
black; background: white

Class 6—*Division 6.1*

Poisonous (toxic) Substances
Danger group: III
The bottom part of the label
should bear the inscription:
Harmful stow away from foodstuffs
Symbol (St. Andrew's cross
over an ear of wheat): black;
background: white

Class 6—*Division 6.2*

Danger Infectious Substances
The bottom part of the label should
bear: INFECTIOUS SUBSTANCE
(Optional) and the inscription: **In case of damage or leakage immediately notify public health authority** (optional)
Symbol (three crescents superimposed
on a circle) and inscription: black;
background: white

Table 1. (Continued)

Class 7

Radioactive Substances
(a) Category "white"; background: white with one vertical red stripe in the bottom half. Text (mandatory) black in bottom half of label: **Principal radioactive content... Activity of contents... curies...**
Symbol (trefoil): black

Radioactive Substances
(b) Category "yellow"; background: top half yellow, with two or three red stripes in the bottom half white. Text (mandatory) black in bottom half of label: **Principal radioactive content... Activity of contents... curies... Transport index...**
Symbol (trefoil): black

Class 8

Corrosives
Symbol (acids spilling from two glass vessels and attacking a hand and a metal): black; background: upper half white, lower half black with white border

The importance of the transport classification is also due to the fact that, in some countries, it is the only classification system and is therefore used, apart from the specific field of transport, as a generic characterization method.

3.3.2. The EU Classification for the Marketing of Dangerous Substances

In the European Union, a system of classification has been developed since the 1970s, covering the labeling to be applied when dangerous substances and preparations are marketed; this includes both professional and domestic use. In the EU, classification norms and labeling for transport are carefully distinguished from those relating to commercial use; e.g., a container or crate intended for transporting tins of paint is labeled according to the norms for transport, whereas the tins of paint are labeled according to the norms for commercial use.

Only in cases involving a common single form of packaging, which is used for both transport and marketing, is it permitted to use a mixed label, consisting of the danger symbol specified by the transport norm and the danger wording and safety instruction wording specified by the marketing norm. EEC Directive 67/548 [5] presents the Community's legal basis for the whole range of norms issued for classification and labeling of dangerous substances and preparations. A group of experts from the Member States is engaged in classifying substances considered to be a priority at the community level; ca. 1500 substances have been officially classified, as reported in Commission Directive 93/73 [6]. Since the basic directive, in an updated version [7], also provides that substances not officially classified should be evaluated by the manufacturer or importer for the purposes of possible classification and labeling (regarded as provisional), the Community system has to clarify the criteria for danger classification in order to provide industry with a clear, unambiguous framework that avoids as far as possible subjective evaluation and differing interpretation of the data. The most up-to-date version (1998) is Commission Directive 98/98/EG [8]. Application of these criteria enables substances to be allocated to one or more of the following danger categories:

Explosive
Extremely flammable
Readily flammable
Flammable
Oxidizing
Very toxic
Toxic
Harmful
Corrosive
Irritant
Sensitizing
Carcinogenic
Mutagenic
Toxic to reproduction
Hazardous to the environment

A substance can be characterized by a maximum of three danger symbols; one for physicochemical effects, one for biological effects, and one for environmental effects. Figure 2 shows the 10 danger symbols used to identify the 15 danger categories enumerated above.

There is no exact correspondence between the danger categories and the danger symbols, in the sense that in several cases the same symbol is used to characterize more than one danger category (e.g., the skull-and-crossbones used for very toxic, toxic, carcinogenic, mutagenic, and substances toxic to reproduction; or the St. Andrew's cross for harmful substances, irritants, substances producing sensitization on inhalation, Category 3 carcinogens, Category 3 mutagens, and Category 3 substances toxic to reproduction). The symbol thus provides only preliminary generic information on the type of risk; the real danger of the substance is indicated only by the danger indications and risk phrases (R). The safety instructions (S phrases) provide information on the specific precautions to be taken during handling, and first-aid measures in case of accidental contamination or ingestion.

The classification criteria provide a way of utilizing the physicochemical, toxicological, and ecological data on a substance most effectively for assigning danger symbols, risk and safety phrases.

Classification Based on Physicochemical Effects. A substance is classified as explosive if it falls within at least one of the three criteria: shock, friction, or heating. A substance is extremely flammable if it is a liquid with flash point $<0\,°C$ and boiling point $\leq 35\,°C$, or if it is gaseous and becomes flammable on contact with air at ambient temperature and pressure. A substance is highly flammable if it is a liquid with flash point 0 – 21 °C, or if it is a solid capable of igniting and continuing to burn after the source of ignition is removed. A substance is flammable if it is a liquid with flash point 0 – 55 °C.

A substance is an oxidizing agent if it responds positively to a specific test designed to check its potential to increase flame propagation velocity in a sample of combustible substance.

Classification Based on Biological Effects. With regard to the lethal acute effects the LD_{50} or LC_{50} values for the appropriate species are used; normally, oral and inhalatory studies on rats and cutaneous studies on rabbits or rats are preferred.

Table 2 gives reference values for danger classification and, within the scope of such a classification, the distinction between the categories of very toxic, toxic, and harmful. For the inhalatory route, two separate sets of reference values are given, depending on the physical state of the substance; in the majority of cases it is practically impossible with powdery solids and high boiling point liquids to prepare suspensions or aerosols of concentrations >5 mg/L.

In order to reduce the number of animals subjected to experimentation, the "fixed-dose" method has recently been developed for the determination of acute toxicity, based on determining not the lethal dose, but the "discriminating dose", i.e., the dose that produces toxic, but not lethal effects. Reference values (mg/kg body weight) are:

Very toxic	<5
Toxic	5
Harmful	50 – 500

As regards long-term effects, a substance is classified as toxic or harmful if repeated or prolonged exposure can produce serious damage (evident functional disturbances or morphological changes of toxicological significance) below specified dose levels. With regard to subchronic toxicity studies, i.e., at 90 d, reference values are:

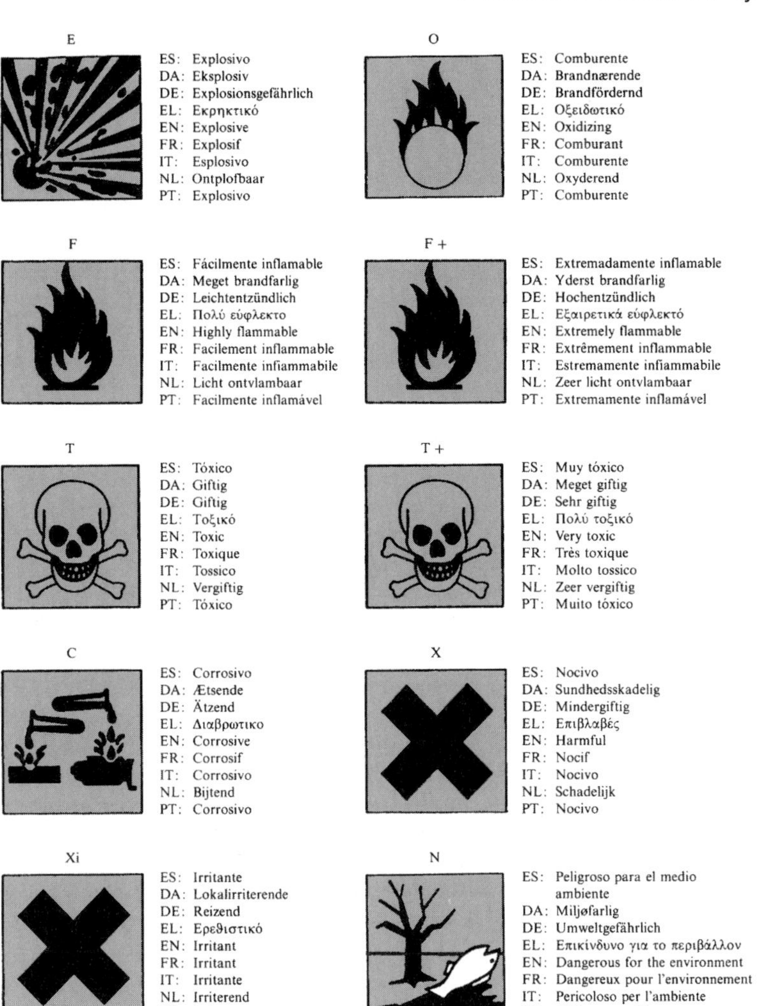

Figure 2. Symbols used to identify danger categories within the EU classification system

Table 2. Reference values for danger classification based on animal tests

Category	LD_{50}, oral in rat, mg/kg	LD_{50}, dermal in rat or rabbit, mg/kg	LC_{50}, (inhalatory) in rat, mg/L for 4 h	
			Aerosols or particulates	Gases and vapors
Very toxic	≤ 25	≤ 50	≤ 0.25	≤ 0.5
Toxic	25 – 200	50 – 400	0.25 – 1	0.5 – 2
Harmful	200 – 2000	400 – 2000	1 – 5	2 – 20

Oral route, rat: $\leq 50\,mg\,kg^{-1}\,d^{-1}$
Cutaneous route, rat or rabbit: $\leq 100\,mg\,kg^{-1}\,d^{-1}$
Inhalatory route, rat: ≤ 0.25 mg/L, 6 h per day.

For subacute toxicity studies, i.e., at 28 d, these values are multiplied by 3.

In a chronic toxicity study, which normally lasts the whole life of the animal (2 years for rats), it is not possible to give individual reference values, and the evaluation is made case by case.

When an individual reference value is given, it is possible to distinguish between toxic and harmful effects from the dose level at which the toxic effect is observed; if it is slightly below the reference value, the substance is classified as harmful. However, if it is significantly below this value the substance can be classified as toxic. The EU classification reserves the category of very toxic for acute lethal or nonlethal effects.

With regard to local effects (corrosive and irritant) criteria are defined for assigning specific risk wording depending on the type of effect and its severity. For corrosive effects, it is possible to use two different risk wordings, depending on the time it takes a substance to destroy the whole thickness of the cutaneous tissue (between 3 min and 4 h).

A substance is classified as a skin irritant if it produces significant cutaneous inflammation that lasts for at least 24 h following a maximum exposure time of 4 h; such inflammation is regarded as substantial if the scores obtained separately for the occurrence of erythema, eschar, or edema, are greater than the arbitrarily specified limiting values.

With regard to effects on the eyes, a distinction may be made between serious ocular lesions and ocular irritation, depending on the scores obtained for four specific end points:

Opacity of the cornea
Lesion of the iris
Redness of the conjunctiva
Edema of the conjunctiva

It is also possible to classify a substance as a respiratory tract irritant, but since no experimental models are available to evaluate such an effect on laboratory animals, it has to be based on practical observations mainly connected with the professional use of the substance.

For cutaneous sensitization, a reference method (maximization test) exists based on the guinea pig; a positive response in at least 30 % of the animals tested classified the substance as a skin sensitizing agent. There is no experimental model for sensitization by inhalation: in this case tests have to demonstrate reliably that the substance can produce a sensitization reaction in subjects by inhalation, at a frequency greater than that encountered in a control population.

There are three danger categories with regard to carcinogens:

Category 1: substances known to be carcinogenic in humans on the basis of epidemiological studies
Category 2: substances for which there is sufficient evidence of carcinogenesis in experimental animals (as a precaution, such substances are labeled as being in Category 1)
Category 3: substances to be regarded as suspect with regard to possible carcinogenic effects in humans on account of a positive, but inconclusive response in experimental animals

Three danger categories are defined for mutagenesis:

Category 1: substances known to have mutagenic effects in humans. There is sufficient evidence to establish a causal relationship between human exposure and the occurrence of hereditary changes (up to now no substances fulfilling such criterion are known). Moreover, it is extremely difficult to obtain reliable information from studies on the incidence of mutations in the human population, or on a possible increase in frequency.
Category 2: substances for which there is sufficient evidence to say that it is probable that human exposure will lead to the development of hereditary genetic changes. It is necessary to obtain positive results from studies demonstrating the mutagenic effect on mammalian germ cells in vivo or the mutagenic effect on somatic cells, which together clearly demonstrate that the substance, or one of its metabolites, is capable of reaching the germ cells.
Category 3: substances to be considered as suspect with regard to possible mutagenic effects, based on positive test results that demonstrate the mutagenic effect on mammalian somatic cells in vivo, normally supported by positive results of mutagenesis tests in vitro.

For reproductive toxicity, the effects on both fertility and development must be borne in mind. Three danger categories are defined for each of these effects.

1) Effects on fertility:

Category 1: substances capable of reducing human fertility, as demonstrated by epidemiological studies
Category 2: substances capable of affecting human fertility, based on clear evidence from animal studies of reduced fertility in the absence of toxic effects
Category 3: substances that give rise to concern about possible effects on human fertility, based on animal studies demonstrating a decrease in fertility, but not providing sufficient evidence to justify classification in Category 2

2) Effects on development:
Category 1: substances capable of producing harmful developmental effects in human progeny as shown by epidemiological studies
Category 2: substances that can produce harmful developmental effects in human progeny, based on unequivocal results in suitable animal studies, in the absence of maternal toxicity
Category 3: substances that give rise to concern on account of the possible effects on human progeny based on animal studies showing harmful developmental effects in the absence of maternal toxicity, but not providing sufficient evidence to justify classification in Category 2

Classification Based on Environmental Effects. Criteria have been defined at EU level for classifying chemical substances according to their potential environmental hazard. Such criteria essentially refer to the aquatic compartment, at present considered representative of the whole ecosystem. Three parameters are taken into consideration:

Effects on aquatic species
Bioaccumulation capacity
Degradation

Effects on Aquatic Species. Normally these refer to acute effects, even if under some conditions the long-term toxicity data can modify the classification obtained from the acute effects. The species used are:

Fish (LC_{50}, 96 h);
Daphnia (EC_{50}, 48 h);
Algae (IC_{50}, 72 h);

The cut-off toxicity values that define the various subcategories are 1, 10, and 100 mg/L, intended as concentrations in the aquatic medium.

Bioaccumulation Capacity. Normally, a substance is regarded as capable of bioaccumulation if $\log P_{ow} \geq 3$, unless the experimental bioconcentration factor $BCF \leq 100$.

Degradation. A substance is regarded as rapidly degradable if one of the following criteria is satisfied:

1) In a 28-d degradation study, a 70 % degradation is obtained if the test is based on the measurement of dissolved organic carbon, or 60 % of the theoretical maximum value if the test is based on measuring the loss of oxygen or formation of carbon dioxide. These biodegradation levels should be reached within 10 d of the start of degradation, i.e., from the moment at which 10 % of the substance is degraded.
2) In cases in which only the COD and BOD_5 are available, the ratio between the two $BOD_5/COD \geq 0.5$.
3) Data exist, demonstrating that the substance can be degraded (in a biotic or abiotic manner) in an aquatic environment to levels >70 % within 28 d.

3.4. Hazard Identification of Preparations: Calculation Systems for Different Use Categories (EU Approach)

For intentional mixtures of chemical substances (preparations), the problem of hazard identification may be particularly acute given the widespread commercial availability of chemical products, definable as preparations, intended for a wide variety of uses, including domestic environments, where those particularly at risk (children, the elderly) can come into contact with such products and suffer harmful effects. The possibility of human exposure is extremely variable, depending on the intended use of the product. The potential danger of a preparation depends on the danger characteristics of its components and their concentration in the product. Such a danger can essentially be evaluated in two ways:

1) By subjecting the preparation to a series of experimental tests to determine the toxicological and ecotoxicological properties and physical effects, in the same way as for a substance
2) By employing calculation methods that take account of the danger of the various components and their concentrations, normally expressed in wt % (assuming the absence of synergistic or antagonistic phenomena)

Some categories of preparations, because of their intended use, are subject to extremely stringent regulations that require a series of experimental studies to determine the potential danger to humans and, in some cases, to the environment (e.g., for pharmaceuticals and agricultural pesticides). In the majority of cases, however, evaluation of the potential danger is the responsibility of the manufacturer or importer, who is basically free to choose between the experimental and calculation options, with a clear preference for calculation, given the extremely high cost of performing a set of experimental studies. A series of norms has recently been approved at international level that aim to discourage, or at least minimize, experimental studies on animals, for humanitarian reasons.

In the past, a series of specific technical norms has been defined at EU level for particular areas of use (solvents, paints, inks, pesticides), which besides sanctioning the principle of alternative between biological testing and application of a calculation system, define specific mechanisms for establishing the danger classification of the preparation, depending on the nature and proportions of its components. Directive 88/379/EEC [9] is a first step in the unification and rationalization of such systems to arrive at a single evaluation mechanism regardless of the intended use of the product.

3.4.1. Classification Based on Physical and Chemical Effects

No calculation system exists that enables the flammability, explosiveness, and oxidizing power of a preparation to be evaluated, depending on the presence of components exhibiting such properties. The characteristics of the preparation are largely influenced by the chemical nature of the other components. For example, if a liquid preparation contains 20 – 30 % of a flammable substance, it is likely that the preparation will exhibit such characteristics if the other components are nonflammable liquid organic substances, whereas if the preparation contains a large amount of water it is highly probable that this will inhibit flammability. Therefore, in the case of chemical and physical effects, the preparation should be tested as if it were a substance, and eventually be classified according to the experimental results. Not all preparations are to be tested, only those that contain components exhibiting such properties above specific limiting concentrations.

3.4.2. Classification Based on Biological Effects

Given that it is possible to evaluate biological end points by means of experimental tests on the preparation, it is permitted to use a conventional calculation system. Such a system is based on the assumption that the components of the preparation have previously been evaluated and, if considered dangerous, have been correctly classified and labeled. The starting point of the classification of preparations is therefore classification of its components; if a substance is classified in different ways by different manufacturers and importers, such variation will be reflected in preparations containing that substance.

Another basic feature of the calculation system of Directive 88/379 is that each specific effect is evaluated separately, with the possibility of assigning to a substance not just an overall percentage limit, but various limits depending on the specific effects of the substance, and reflected in the risk wording assigned to the latter. For each substance, at the time of its official classification by the group of EU experts it is possible to define specific percentage limits, depending on the various effects of the substance, which are applied when the substance is present in a preparation. Figure 3 shows a file card relating to acrylonitrile, characterized by various types of effect [6]. The card includes identification data, structural formula, chemical name in the various Community languages, classification, labeling, and the various percentage limits

Cas No 107-13-1

No 608-003-00-4

NOTA D

NOTA E

$CH_2=CH-CN$

ES : acrilonitrilo

DA : acrylonitril

DE : Acrylnitril

EL : ακρυλονιτρίλιο

EN : acrylonitrile

FR : acrylonitrile

IT : acrilonitrile

NL : acrylnitril

PT : acrilonitrilo

Clasificación, Klassificering, Einstufung, Ταξινόμηση, Classification, Classification, Classificazione, Indeling, Classificação

F ; R 11	Carc. Cat.2 ; R 45	T ; R 23/24/25	Xi ; R 38

Etiquetado, Etikettering, Kennzeichnung, Επισήμανση, Labelling, Étiquetage, Etichettatura, Kenmerken, Rotulagem

Límites de concentración, Koncentrationsgrænser, Konzentrationsgrenzwerte, Ορια συγκέντρωσης, Concentration limits, Limites de concentration, Limite di concentrazione, Concentratiegrenzen, Limites de concentração

C ≥ 20 %	T; R 45-23/24/25-38
1 % ≤ C < 20 %	T; R 45-23/24/25
0,2 % ≤ C < 1 %	T; R 45-20/21/22
0,1 % ≤ C < 0,2 %	T; R 45

Figure 3. Example of a file card published by the EU, defining classification, labeling and concentration limits for a substance (in this case, acrylonitrile) in preparations

Table 3. Concentration limits for substances as components of preparations

Gas-phase preparation, vol %			Substance classification	Solid or liquid preparation, wt %		
T+	T	Xn		T+	T	Xn
		5	R 20, 21, 22, Xn			25
	5	0.5	R 23, 24, 25, T		25	3
1	0.2	0.02	R 26, 27, 28, T+	7	1	0.1
		5	R 40/20, 21, 22, Xn			10
	5	0.5	R 39/23, 24, 25 T		10	1
1	0.2	0.02	R 39/26, 27, 28, T+	10	1	0.1
		5	R 48/20, 21, 22, Xn			10
	5	0.5	R 48/23, 24, 25, T		10	1
		0.2	R 42, 42/43, Xn			1
	0.1		R 45 carc. 1, 2		0.1	
	0.1		R 46 muta. 1		0.1	
		0.1	R 46 muta. 2			0.1
		1	R 40 muta. 3			1
	0.2		R 60 reprotox. 1 R 61		0.5	
	0.2		R 60 reprotox. 2 R 61		0.5	
		1	R 62 reprotox. 3 R 63			5
Gas-phase preparation, vol %			Substance classification	Solid or liquid preparation, wt %		
C R 35	C R 34	Xi*		C R 35	C R 34	Xi*
1	0.2	0.02 (R 37)	R 35, C	10	5	1 (R 36, 38)
	5	0.5 (R 37)	R 34, C		10	5 (R 36, 38)
		5 (R 36, 37 38)	R 36, 37, 38, Xi			20 (R 36, 37, 38)
		0.5 (R 36)	R 41, Xi			5 (R 36)
		5 (R 41)				10 (R 41)
						1 (R 43)
			R 43, Xi			

* Applying R ... given in parenthesis.

to be adopted when the substance is present in a specific concentration in a preparation.

Only a minority of substances have been assigned specific percentage limits, but Directive 88/379 allows the calculation system to be employed, having defined a series of generic limits (established according to the toxicological significance of each effect and independent of the chemical nature of the substance). Table 3 summarizes the percentage limits to be employed, depending on the classification of the components of the preparation. For gaseous preparations, the percentage figures are preferably given by volume rather than by weight.

3.4.3. Classification Based on Environmental Effects

The Directive does not provide for the possibility of classifying a preparation according to its potential environmental danger. A proposal to introduce such criteria is currently under examination. The aim is to define specific percentage limits depending on the risk phrase assigned to the substances. The possibility of determining experimentally the characteristics of the preparation would be limited to the effects on aquatic species, since it is technically impossible to determine log P_{ow} and biodegradability of a preparation. In order to demonstrate experimentally that the preparation is not toxic to aquatic species, it is necessary to carry out tests on all the three species, fish, *Daphnia*, and algae.

4. Risk Assessment

4.1. General Methodology

Risk assessment comprises one or more of the five steps defined in Chapter 2. The fundamental principle is that risk depends on the inherent hazardous properties of a substance and the extent of exposure to that substance. Consequently, a risk assessment can be concluded at a relatively early stage if it has been demonstrated that a substance exhibits no hazardous properties or that neither humans nor the environment will be exposed to it. The following scheme illustrates the steps and the relationships between them.

Information gathering

Effects	*Exposure*
Hazard identification; dose (concentration) – response (effect) assessment, if appropriate; toxicokinetics	Human exposure assessment (workers, consumers, via the environment, as appropriate); environmental exposure assessment (water, soil, air, as appropriate)

↓

Risk characterization

Human health	*Environment*
	Evaluation of effects data and comparison with exposure data

↓

Outcome of risk assessment

A risk assessment should be amenable to revision in the light of any new information.

4.2. Qualitative and Quantitative Aspects: Toxicological and Ecotoxicological Properties (→ Toxicology, → Ecology and Ecotoxicology)

The definitions in Chapter 2 of risk characterization and risk estimation illustrate the distinction between the qualitative and quantitative aspects of risk assessment. It is sometimes useful to be able to quantify a risk as a numerical probability; equally, it can sometimes be misleading and often impossible. A qualitative characterization of risk, however, should be included in any assessment in which data on intrinsic properties have been compared with data on exposure.

4.2.1. Hazard Identification

The principal toxicological and ecotoxicological properties which should be considered in a hazard identification are described in Chapter 3. If it can be demonstrated that a substance is not inherently hazardous, it is appropriate to conclude that it poses no risk without proceeding to an exposure assessment.

4.2.2. Dose (Concentration) – Response (Effect) Assessment

This step cannot always be separated from hazard identification. Ideally, it should be a quantitative exercise in which an assessor determines the dose or concentration at or below which an adverse effect is unlikely to occur.

In relation to toxicological properties, such a threshold level can be determined for subacute, subchronic, or chronic toxicity, for reproductive toxicity, and for nongenotoxic carcinogenicity. For most other effects, it is usually impossible or inappropriate to determine such a level, and only a qualitative assessment of a substance's inherent capacity to cause such effects can be made.

In relation to ecotoxicity, particularly aquatic toxicity, it is possible to calculate a predicted no-effect concentration (PNEC) by applying an assessment factor to an experimentally derived toxicity value, such as the LC_{50} or NOEC (see Section 3.1.3). An assessment factor is an expression of the degree of uncertainty in extrapolation from test data on a limited number of species to the environment; it should be adjusted according to the extent and quality of the data.

4.2.3. Exposure Assessment

Ideally, exposure assessment should result in the estimation of an exposure level or predicted environmental concentration (PEC). For chemical substances and products already on the market and in use, it should be possible to measure actual exposure levels and concentrations. When conducting a risk assessment before marketing or use, it is usually necessary to predict exposure on the basis of information such as the properties of the substance, its use pattern, and the expected frequency and duration of exposure. Sometimes it is not possible to achieve a quantitative estimation and a qualitative estimation must be made instead.

Exposure assessment should take account of the three principal human populations (workers, consumers, and those exposed indirectly via the environment) and the three principal environmental compartments (water, soil, and air). A risk assessment can be concluded and a substance deemed to pose no risk if there is evidence that neither humans nor the environment will be exposed to it. If it can be demonstrated that a particular population or compartment is not likely to be exposed to the substance, exposure assessment and risk characterization for that compartment or population are unnecessary.

4.2.4. Risk Characterization

If dose (concentration) – response (effect) assessment and exposure assessment have been conducted quantitatively, it should be possible to compare a human exposure level with a threshold level for toxicological effects or a PEC with a PNEC and to determine an exposure/effect ratio. If only a qualitative assessment of effects or exposure has been made, the comparison will have to be made in more qualitative terms.

Where a threshold level for toxicological effects has been determined, it may be appropriate to apply an assessment factor to that level before comparing it with the exposure level. The assessment factor should take account of the following elements:

1) Nature and severity of the effect
2) Uncertainties in experimental data, including interspecies variations
3) Different susceptibilities in exposed humans
4) Extent to which exposure can be controlled and monitored in different human populations (workers, consumers, those exposed indirectly)

On the basis of the comparison of effects and exposure, whether quantitative or qualitative, the assessor should establish the degree of concern associated with the substance and whether further action is necessary, possibly including recommendations that further information on effects or exposure be acquired so that the risk assessment can be repeated and the risk characterized more precisely. As indicated above, risk estimation (quantification of the likelihood of adverse effects) is not always possible or appropriate.

4.2.5. Physicochemical Properties

As indicated in Section 3.1.1, flammability, explosiveness, and oxidizing potential are the three physicochemical properties which correspond to direct adverse effects. The procedures for assessing risk in relation to those properties are not entirely analogous to those for toxicity and ecotoxicity. The dose (concentration) – response (effect) step is not applicable. The concept of an exposure level is meaningless in relation to physicochemical effects, so exposure assessment amounts to determining the reasonably foreseeable conditions of use; risk characterization then entails evaluation of the likelihood that an adverse effect will occur under those conditions.

4.3. Risk – Benefit Considerations

The outcome of a risk assessment can be that measures should be taken to reduce the risk posed by a substance. The process of deciding precisely which measures are appropriate falls outside the scope of risk assessment. Socioeconomic and political considerations may influence decisions on risk reduction; it is therefore important that risk assessment, based solely on technical data, should be clearly separated from that process.

This applies especially where the withdrawal of a substance or product from the market is being considered. It may be appropriate to weigh the risk which would thus be removed against the benefits which would be forgone. A substance may be necessary to the safe operation of an industrial process for which no satisfactory alternative is available; withdrawal of that substance to remove a toxic risk identified in the risk assessment may create a new risk to worker safety. In other cases, the benefits of the risk reduction itself may be compared with its costs in order to assess whether they are proportionate; if the costs are judged to outweigh the benefits, it may be decided not to reduce the risk or to employ risk reduction measures (e.g., exposure limits) which stop short of withdrawal of the substance.

The allocation of a financial cost to a risk is very difficult and likely to entail subjective judgment. Decisions not to reduce a risk or to take only limited risk reduction measures imply the concept of "acceptable risk."

4.4. Definition of Acceptable Risk

It is extremely difficult, if not impossible, to achieve a general definition of acceptable risk;

acceptability is inherently subjective and variable. In individual cases, it may be useful to quantify the risk (see Section 4.2.4).

Perception of a risk by those subject to it is a highly significant factor. There is evidence that high risk may be considered acceptable if it is perceived as controllable and dependent on personal choice (e.g., car driving, smoking). The risks due to exposure to toxic substances in chemical products, even if quantified as very low, may be considered unacceptable by the public because they are perceived as being beyond its control.

The level of probable damage to humans and the environment by a substance must be viewed in relation to the advantages which its use offers to society. This is difficult, since the groups who have to bear the risk are not the same as those who draw benefit from the product. An abstract definition of the benefit is not as a rule possible, since such a definition varies greatly between individuals, and also depends on sociopolitical conditions. The process of weighing up risks and benefits, which ultimately leads to a definition of the acceptance of a certain risk, is thus often extremely blurred and determined by political motives. It is frequently the case that bodies are entrusted with this task in which relevant, social groups are represented, but it is not feasible for such an assessment to be carried out for every chemical substance. Conventions are, therefore, often established for this evaluation, but the reasons for the final decision are nevertheless frequently of a purely political nature.

Most experience gained in the field of risk–benefit examination has been in pharmaceuticals, since the benefit can usually be very clearly defined, and it is known that harmful side effects cannot be completely avoided. Things are far more problematical in the industrial chemicals field, substances with irreversible effects and without a NOEC being the most difficult to deal with.

A classic example is formaldehyde, which is known to be capable of causing a sensitizing and carcinogenic effect. Because it is indispensable for technical purposes, it is accepted that below certain statutory limits it may continue to be used, even in consumer articles, although adverse effects cannot be precluded, at least in particularly sensitive persons. As a rule, an attempt is made to focus on risks to which humans or the environment are unavoidably exposed. It was calculated, for example, that the occurrence of ethylene oxide as a micropollutant in cosmetic raw materials does not result in a significant increase in this substance in the human metabolism, and that these traces can be regarded as harmless.

Examples in the ecological field are chelating agents such as ethylenediaminetetraacetic acid (EDTA) and nitrilotriacetic acid (NTA), for which it is feared that heavy metals present in sediments in surface waters may be remobilized. An attempt is being made by monitoring programs to ensure that these substances do not accumulate in surface waters in concentrations at which such an effect might arise.

Since many of these effects are highly speculative and elude quantitative assessment, any risk–benefit examination inevitably remains blurred and open to considerations of political expediency. Bans or restrictions on chemical products, which would result in considerable economic consequences, have to a large extent not been imposed.

4.5. Summary of Risk Assessment

The present state of the scientific art – even in toxicology with its rather long tradition, and even more in ecotoxicology – does not allow a real quantified risk characterization which might be used as basis for the definition of an acceptable risk, as practiced in particular in the United States. Attempts to compare risks, identified and characterized as outlined above, with other risks in normal life are seldom successful because society has a different perception of various risks.

A pragmatic approach is appropriate, and should embrace the following elements:

1) Identifying chemicals for which the risk, due to minimal hazardous effects or to minimal exposure, is obviously negligible
2) Refinement of data bases for which this decision cannot be taken on the basis of available data
3) Quantification of risk characterization, as far as possible case by case by expert judgment, for chemicals which show a significant risk potential
4) Decisions on risk reduction measures by political bodies on the basis of acceptability of

risk, taking account of economic and social considerations

5. Risk Management

Risk management should flow logically from risk assessment, but is separate from it. Risk assessment should have characterized the risks, have identified who or what is at risk, and the levels, sources and pathways of exposure. The function of risk management is to reduce the risks and to mitigate the consequences if an adverse event occurs.

It may be possible to eliminate a particular risk, but more usually a residual risk remains, and the challenge is to reduce this to an acceptable level. Science has a role in defining this level, but socioeconomic and other factors are more important, and outside the scope of this article. Those wishing to explore the subject further will find useful starting points in references [10], [11].

There can be no single system of risk management. The key to successful risk management is the selection of the most appropriate set of tools for particular situations and their acceptance by those causing the risks and by those most likely to be affected by them.

Although this article is principally concerned with products control, there are instances where products are only one source of exposure and the risk assessment has to consider the overall impact arising from all sources. Lead is an example of this; there is concern about the effects, particularly on children, of uptake of lead into the body from a range of sources through various pathways. Products control can reduce exposure but cannot eliminate the problem. A much larger range of measures is needed.

In the context of chemicals control, hazard is an inherent property of a chemical but risk depends on the extent of exposure. Exposure is a function of variable, controllable elements (how the chemical is made, how it is incorporated into a product, and how that product is used).

The more important tools for the management of risks from chemical products are dealt with below.

5.1. Supplying Information (Risk Communication)

This is a fundamental requirement in all strategies for managing risks. Unless all those making, taking, or regulating the risks, i.e., the "stakeholders", are aware of the risks, their consequences, and what they can do to reduce these, the strategy will be flawed. Thus:

1) Those creating risks are often best able to take steps to reduce them. They need to inform themselves about the risks associated with their activities and products, and about the possible consequences of the risks. They also need to know what can be done to reduce the risks and to mitigate the consequences.
2) Those who might be affected are more likely to find the risks acceptable if they know what the risks are and what measures have been taken to minimize them. They also need to know what they themselves can do to reduce the risks and the possible consequences.
3) Those responsible for regulating the activities or products which create the risks need to have as much information as possible relating to the risks and about the tools available to reduce them.

Ensuring safety during the manufacture of chemicals, and products containing them, is primarily a matter for industry and the appropriate regulators: safe use in industry and commerce involves management and workers. Outside the workplace, individual behavior becomes the most important factor, and is greatly influenced by personal knowledge, experience, and perception of the risks.

Much environmental, consumer safety, and worker protection legislation requires those producing, importing, and supplying products to give information on the hazardous properties and risks associated with their products, and there is provision for such information to be passed down the manufacturing and supply chain. This is a feature in much EC, Scandinavian, and North American legislation. The most obvious manifestation is the information that appears on labels or is supplied as safety advice with the product. Labels usually carry information of different kinds and give only very basic information on risks, but they may be the only means of influencing user behavior, which is often important in determining whether a risk becomes manifest. Effective labeling is very important.

Labeling requirements are generally linked to a classification system. These systems cover a range of hazards, e.g., toxicity, irritancy, flammability, effects on the environment, and set criteria to define degrees of hazard, e.g., very toxic, toxic, harmful. The information required on the label, often in the form of standard "risk phrases," follows from this classification. Packaging requirements, which can be important in reducing risk (e.g., through accidental exposure) may also be linked to this classification, as in the EU. Labels are primarily intended to warn and protect the user but, to an increasing extent, include information on environmental safety, in particular on safe disposal.

Labeling is frequently supplemented by data sheets which provide more detailed information. Chemical manufacturers often supply data sheets to their workers, to their customers who process chemicals into products, and to the emergency services. Legislation in the EU on the safe handling of chemicals requires the provision of information in this form. The preparation of international data sheets, principally for worker protection, is a feature of the International Programme for Chemical Safety, a collaborative program involving the World Health Organization, International Labour Office, and United Nations Environment Programme.

Access to information on the hazardous properties of chemicals and on exposure has improved in recent years in most countries. Some countries (e.g., the United States) have legislation requiring such information to be made available. The EC Directive on access to environmental information [12] should ensure greater availability and easier access within the EU to information held by public authorities. Public interest groups have campaigned for greater freedom of access to information, and often act as alternatives to official or industrial sources of information on risks (often seen as less biased or more independent).

5.2. Codes of Practice

Since the way in which products are used can by very important in terms of protecting human health and the environment, codes of good practice are often developed for the more dangerous substances to help ensure safe use. These give guidance on good practice in the use of products and the safe disposal of any wastes. In some countries, demonstration that such codes have been followed is a valid defense in the event of legal proceedings. Failure to comply with a code may lead to prosecution.

5.3. Product Standards

These are usually intended to ensure that a product delivers the required performance, but they can also reduce the impact on human health and the environment, e.g., by specifying maximum levels of dangerous impurities. Conversely, by specifying a performance which can be met only through the use of a dangerous substance, or by failing to revise obsolete standards which encourage the continuing use of dangerous substances, needless risks can be created or perpetuated.

5.4. Control of Supply

This includes control of production and importation. Prevention of entry into the manufacturing and supply chain is among the most powerful and effective forms of product control. Control of production has been used to control the supply and hence the eventual release of chlorofluorocarbons, but control at the point of sale is more common. The latter is widely used as part of approval schemes for pesticides and pharmaceuticals, etc. (e.g., particularly dangerous substances and formulations may be supplied only to professional users). It is also common in EU Member States to implement EU controls on marketing and use, e.g., wood preservatives containing pentachlorophenol cannot be sold to the public in the EU, and the supply for professional use is strictly limited. Lead-based paints may be supplied only for use on specific historic buildings in the United Kingdom.

Attaching conditions to the supply of a chemical rather than banning it outright can be an effective means of control, while allowing appropriate uses. Many of the pesticides available to the public must be supplied in small packs or in diluted form.

Controls on supply are fairly easy to enforce (suppliers can be readily identified, and most are already subject to other forms of control, e.g., inspection by environmental, consumer protection, or labor protection authorities). However, controlling supply has no impact on products already in the control of users. This is why it is often seen as a weaker form of control than control of use (see Section 5.5).

5.5. Control of Use

Although potentially a more powerful form of control, control of use is more difficult than controlling supply. Control of use by an individual may involve changing behavior or interfering with personal rights, preventing the individual from obtaining the product in the first place is a more effective option. But control of use, if it can be enforced effectively, addresses the control of products already in the possession of the user. Many countries have laws which seek to control use by individuals. To help prevent illegal use or disposal, public information and recovery schemes may be introduced to encourage those holding stocks of the product to return them to the supplier or to a body designated for their safe disposal.

Controlling use by industry and business is more feasible. They can be more readily identified, targeted, and the controls enforced by labor or other inspectors.

5.6. Voluntary Agreements

Regulations can be expensive to develop and implement, and often follow rather than force change. By the time regulation takes effect it is likely that industry has adjusted its processes, withdrawn problematic products, and is marketing alternative products. Voluntary agreements sometimes offer a more cost-effective alternative. Paint manufacturers in the United Kingdom agreed to stop using lead driers in paints intended for home use: it is claimed that the voluntary agreements associated with the Toxic Release Inventory in the United States have significantly reduced releases of chemicals to the environment. All the principal manufacturers or suppliers must be parties to such agreements, which must be clear, set targets and timetables, and include systems for monitoring progress. They should be public documents. Even if voluntary agreements are replaced by regulations, they can provide valuable experience which helps avoid problems in implementing the regulations.

5.7. Economic Instruments: Harnessing Market Forces

There is a growing interest in measures to achieve environmental ends through market mechanisms; these can either replace or complement regulation. There are few examples in the field of chemicals control, but tax differentials on motor fuels containing lead additives or to encourage the use of diesel fuels have been adopted in a number of countries.

The attractions of economic instruments are that they highlight that the "polluter is paying," encourage the market to allocate the substance concerned to more essential uses, and provide an incentive to develop substitutes.

Probably the most significant examples of the power of market forces for environmental benefit do not involve government intervention at all. "Green consumer" campaigns can claim many successes, including reducing the use of chlorofluorocarbons in domestic products. Many retailers have recognized the selling power of "environmentally friendly" products and require their suppliers to produce products which exploit this. The ecolabeling schemes now being introduced in the EU and elsewhere will further enhance this trend.

5.8. Integration of Pollution Control and Chemicals Control

Chemicals control deals with problems arising from chemicals as products–pollution control is concerned with chemicals (and other agents) in waste streams. Insofar as the overall problem is from chemicals released to the environment, chemicals control and pollution control complement one another.

In 1991, the OECD adopted a Council Recommendation on "Integrated Pollution Prevention and Control" which commits Member

States to "take into account the effects of activities and substances on the environment as a whole, and the whole commercial and environmental life cycles of substances when assessing the risks they pose and when developing and implementing controls to limit their release." Based on this Recommendation, the Commission of the European Communities is preparing a Union proposal which should lead to greater coordination of the control of waste streams from the chemical and related industries. It should also facilitate the integration of pollution control and chemicals control. Some European countries, e.g., United Kingdom, The Netherlands, Sweden, have already enacted such legislation, but none have gone as far in that direction as the OECD Recommendation implies.

Where total exposure to a substance is important (e.g., lead) and there is a significant contribution from industrial emissions, more effective control of industrial waste streams must clearly be part of the overall risk management strategy.

5.9. Monitoring Effectiveness

The principal requirement is for systems which monitor the effectiveness of the measures taken. It is important that the initial risk assessment be used to define a baseline against which progress can be assessed, and that risk management strategies provide a framework within which progress can be measured. Such a framework should set clear objectives, targets, and timetables which address questions such as:

- Are the measures reducing or containing the risks?
- By how much and to what timescale?
- What are the costs–to all the stakeholders?
- Does more need to be done or can efforts be relaxed?

It may be possible to measure directly whether or not risks have been reduced, e.g., from accident statistics, measures of mortality and morbidity, recovery of species and ecosystems. In many cases, reduction in exposure of the "at-risk group" will be measurable, implying a reduction in the risk, in other cases reduction of inputs to pathways and environmental compartments (air, water, etc.) will be the best that can be achieved. Monitoring programs such as these will answer the first two questions. Normal financial practices, if set up appropriately, can begin to answer the third question, but it is important that all the costs are exposed, including the costs of not taking action, for all those affected. The various stakeholders may differ in their ability to monitor and calculate their costs, and placing a monetary value on environmental damage or improvement remains a difficult problem for economists (some costs are readily monitored while others remain difficult). Nevertheless, cost considerations cannot be avoided.

The final question can only be answered following an assessment of the kinds of information referred to above. The answers represent the outcome of the "feedback loop" which should be an integral part of all risk management strategies.

6. Ecolabeling

6.1. Philosophy

There has been a growing recognition over the past decade that it is not just for public authorities to set rules for environmental behavior. Companies should include the environment in their strategies, over and above the minimum regulatory requirements. Consumers can also play their part by stimulating (by individual choice) the production and marketing of more environmentally friendly products.

The objectives underlying ecolabeling schemes are generally:

1) To provide purchasers with guidance on choosing products with least impact on the environment
2) To encourage the development of products with less impact on the environment, but which still meet the customers' requirements on costs and effectiveness
3) To harness purchasing power to complement regulatory and other controls aimed at protecting and improving the environment

The German "Blue Angel" (Blauer Engel) scheme, set up in 1978, is an early example of a successful ecolabeling scheme, but in the last few years there has been an upsurge in interest in such schemes. Public awareness of environmental issues has been raised, and more and more of the public want to play a part in preventing or dealing with environmental problems; industry has come to realize that being seen to be "environmentally friendly" helps to sell products, and governments have recognized that consumer purchasing power can be useful in securing environmental improvement. However, the proliferation of labeling schemes and unsubstantiated claims by manufacturers about the "greenness" of their products may confuse, rather than inform, the public. There has also been concern that national schemes might become barriers to international trade. An EC Regulation was agreed in March 1992 [13] and the members of the Nordic Council, Finland, Norway, and Sweden, agreed to set up a joint scheme in November 1989. In North America, the Canadian Government set up its "Environmental Choice Program" in 1988 which, for the first time, adopted a "cradle-to-grave" approach–a concept subsequently adopted by the EU and by the Nordic countries. This approach requires a life-cycle analysis of the product (the EU scheme uses the indicative assessment matrix, Fig. 4).

All existing schemes have restrictions on the types of product covered, e.g., food, drink, and pharmaceuticals are excluded from the EU scheme. Products falling within a scheme are divided into product categories or groups (e.g., washing machines, paints, detergents) and criteria to evaluate the environmental impact of each category are published. Manufacturers who believe that their products meet the criteria may submit them for approval and, if successful, can display the appropriate label or logo on the product and packaging and use it in advertisements and promotional material.

The criteria for award of an ecolabel can be separated into those of a general nature (e.g., legal health and safety requirements) and those that are specific to a particular kind of product. Chemicals per se are unlikely to qualify for any of the product groups in any of the schemes, but they are components of a great many of the products which are eligible for consideration (e.g., adhesives, paints, detergents), and there are many examples where the chemical content can have a bearing on whether or not a product is eligible for the award of an ecolabel. Under the Nordic scheme, no product can be awarded an ecolabel if it contains substances which do not decompose or which bioaccumulate. The Canadian criteria for adhesives prohibit the award of a label to products containing aromatic hydrocarbons, formaldehyde, or chlorinated solvents. The Nordic criteria for detergents limit the content of phosphates, EDTA, NTA, and sodium perborate to less than 0.2 %, and exclude products containing carcinogenic, allergenic, and teratogenic substances.

6.2. The EU Scheme

The scheme was officially launched on June 30, 1993 with the adoption of the first sets of ecological criteria relating to washing machines and dishwashers.

The objectives of the scheme are:

1) To promote the design, production, marketing, and use of products which have a reduced environmental impact throughout their entire life cycle
2) To provide consumers with better information on the environmental impact or products

All consumer products, in principle, are eligible under the scheme, with the exception of food, drink, and pharmaceuticals. Consumers in this context are defined broadly to include business purchases.

The award of the ecolabel means that successful applicants are permitted to use the official logo (Fig. 5) on their approved products. A consumer faced with a choice of, say, washing machines of different makes can pick out, by means of this logo, those which have been shown to achieve a high standard of environmental performance. The same logo is used regardless of the type of product.

Figure 5. The EU official logo for ecolabeling

Environmental fields	Product life-cycle				
	Pre-production	Production	Distribution (including packaging)	Utilisation	Disposal
Waste relevance					
Soil pollution and degradation					
Water contamination					
Air contamination					
Noise					
Consumption of energy					
Consumption of natural resources					
Effects on eco-systems					

Figure 4. Assessment matrix for the EU ecolabeling scheme

The strength of the label is its European dimension. Once approved by one Member State, it can be used throughout the Community. This avoids having to make an application in every country where a national label exists, and thus avoids time-consuming and costly procedures. More importantly perhaps, the success of the EU scheme avoids a proliferation of national schemes, which might confuse the public, serve as barriers to trade and hamper the development of the Single Market.

The scheme should gradually improve the environmental performance and impact of products by giving recognition to those which are the most favorable to the environment. By providing the consumer with information essential to making a sensible purchasing decision, ecolabeling is beginning to make an effective contribution at EU level in the joint interest of consumers and the environment.

7. References

1. Directive 93/67/EEC of the Commission defining the principles for evaluating the risks for man and the environment of substances notified within the meaning of Directive 67/548/EEC of the Council, July 20, 1993.
2. OECD Guidelines for Testing of Chemicals.
3. Directive 92/69/EEC of the Commission July 31, 1992, regarding the 17th adaption to the technical progress of Directive 67/548/EEC of the Council regarding the harmonization of legislative, regulatory, and administrative provisions relating to the classification, packaging and labeling of dangerous substances, July 31, 1992.
4. Transport of Dangerous Goods – Recommendations prepared by the Committee of Experts on the Transport of Dangerous Goods, United Nations, New York.
5. Directive 67/548/EEC of the Council concerning the harmonization of legislative, regulatory and administrative provisions relating to the classification, packaging, and labeling of dangerous substances, June 6, 1967.
6. Directive 93/72/EEC of the Commission regarding the 19th adaption to the technical progress of Directive 67/548/EEC of the Council concerning the harmonization of legislative, regulatory, and administrative provisions relating to the classification, packaging, and labeling of dangerous substances, Sep. 1, 1993.
7. Directive 92/32/EEC of the Council regarding the 7th adaption of Directive 67/548/EEC concerning the harmonization of legislative, regulatory, and administrative provisions relating to the classification, packaging, and labeling of dangerous substances, April 30, 1992.
8. Directive 98/98/EG of the Commission regarding the 25th adaption to the technical progress of Directive 67/548/EEC of the Council concerning the harmonization of legislative, regulatory, and administrative provisions relating to the classification,

packaging, and labeling of dangerous substances, 1998.
9. Directive 88/379/EEC of the Council concerning the harmonization of legislative, regulatory, and administrative provisions of the Member States relating to the classification, packaging, and labeling of dangerous preparations, June 7, 1988.
10. Health and Safety Executive (ed.): *The Tolerability of Risk from Nuclear Power Stations,* HMSO Publications Centre, London 1992.
11. Royal Society: "Risk; Analysis, Perception and Management," report of a Royal Society Study Group, The Royal Society, London 1992.
12. Directive 90/313/EEC of the Council on the freedom of access to information on the environment, June 23, 1990.
13. Regulation 880/92/EEC of the Council on a Community ecolabel award scheme, April 11, 1992.

Legal Aspects

JEAN-MARIE DEVOS, CEFIC, Brussels, Belgium (Chaps. 1, 3)

LUDWIG KRÄMER, Head of Legal Services, DG XI, European Commission, Brussels , Belgium (Chap. 2, Chap. 5 in part)

GABRIELLE H. WILLIAMSON, LeBoeuf, Lamb, Greene & MacRae, Brussels, Belgium (Chap. 4)

PASCALE KROMAREK, Elf Aquitaine, Direction Environment, Paris-La-Defense, France (Chap. 5)

TURNER T. SMITH, JR., Hunton & Williams, Washington, D.C., United States (Chap. 6)

MUNETO ITODA, JCIA, Tokyo, Japan (Chap. 7)

Ullmann's Industrial Toxicology

ISBN: 3-527-31247-1

1. Industrial Chemistry and Law

1.1. Law, Society, Chemistry, and Ethics

Law is the normative instrument that society uses to determine in a peaceful way the rules governing the relations between people (private law); people and authorities (public law); and states (international public law). It reflects to a large extent the values and ethics of a given society at a given time Normative activities have – at an early stage – addressed the question of humans and their interface with industrial and trade activities either to promote such activities, to control or restrict them, or in some cases, even to prohibit them.

Although the requirements of an organized society, as reflected in law, are supposed to be integrated, complementary, and consistent, they might actually conflict or at least appear contradictory. For example, the legal system might aim at economic objectives – for example, the free movement of products in an economic area – as reflected in one of the four basic freedoms of the European Communities Treaties. On the other hand, society and lawmakers might impose severe restrictions on the movement of products because of their impact on organized collectivities and the environment (see Section 4.3).

Law might actually impose restrictions and administrative burdens on research, production, transport, and distribution activities, either because of the "dangerous" nature of such activities or because they involve products that are, to various degrees, considered dangerous or simply perceived as negative or hostile to people or the environment. The end of the twentieth century has seen the emergence of the "environmental society". With the growing satisfaction of basic needs in the developed world and the collapse of traditional ideologies, growing awareness of environmental problems has become a major factor influencing sociology, politics, law, and economics. In parallel to the new "environmental" awareness, and in apparent contrast, industry and trade activities are more and more characterized by increased competition and "globalization." After the success of the General Agreement on Tariffs and Trade (GATT) round in Uruguay and the creation of a World Trade Organization, the dominant economic concepts and trade realities are clearly hostile to trade barriers and restrictions. The resolution of potential and actual conflicts between these two approaches should lie in a coherent and humanistic vision, based on human values, on responsibility, and on confidence in sustainable progress and science.

The main objective of this chapter is to focus on a few key trends which affect, or will influence, the socio-legal framework in which industry has to operate. As shown in the various chapters of the keyword Legal Aspects, innovation, production, distribution, and marketing strategies must integrate environmental and other ethical values which determine their legal framework.

At the same time, industry's response should be based on sound and credible science, on economic sense and on appropriate communication.

In other words, modern *legal systems* tend to address production, distribution activities and products from "cradle to grave" and respond to a growing social concern about perceived risks generated by "dangerous" activities and products. This is particularly relevant in the area of chemicals which are subject to comprehensive sets of rules in the various regions of the world (see Chaps. 3, 6, 7). Perceived risks may be real and have to be seriously addressed. They may also result from fears generated by some media or pressure groups. In contrast, *science* has never been more advanced but, in a way, has become a victim of its own success as technology allows society to better measure even minor risk factors (or simply traces) which are then deemed to be unacceptable. [See for example the progress of detection limits of modern analytics using ultrasensitive equipment and the debate on the revision of the Drinking Water Directive [1]. The key issues are the allowance of greater flexibility for nonhealth parameters and reliance on toxicological data in setting maximum admissible concentrations (MAC) for health-related parameters. Progress in toxicological, analytical, and epidemiological knowledge and methods may imply that MAC should evolve, in line with WHO guidelines.]

The importance of innovation as a driving force of modern economy and society was defined by SCHUMPETER at the beginning of the 20th century as the entire process from the birth of an idea through to its widespread application in society. Innovation implies changed technologies, economic and social evolutions, and adaptability. The driving forces of innovation and changes are rooted in human curiosity and needs and are influenced by their evolving hierarchy.

Social psychology based on MASLOW's approach, distinguishes five hierarchical levels of needs

1) Physical survival and physiological needs
2) Safety
3) Social belonging
4) Respect and recognition
5) Self-fulfilment

According to classical marketing sciences, innovation was responding to "technology pushes" on the one hand and almost unlimited "market demands" on the other hand. Until the lower needs are satisfaction of human needs as described in MASLOW's ladder in industrialized countries and with the achievement of high standards of living, perceived priorities have evolved. Public awareness has become more and more attentive, and critical of "side effects" of innovation.

In other words, the better and the more social needs are fulfilled, the more acute is the perception and concern of the society about the instruments developed to respond to such needs and about their effects. This phenomenon has had a direct and massive impact on the lawmaking process and has led to a vast number of regulations, particularly in the health, safety, and environment area.

The Rio United Nations Conference on Environment and Development (UNCED) Declaration is probably the widest international effort to address environmental concerns on a global basis, although reflecting different Regional sensitivities and priorities. The growing environmental awareness has generated many "law – regulatory"-driven technologies and market developments (e.g., the compulsory introduction of catalytic converters for automobiles) and has led to the phased withdrawal of products considered as harmful to the environment (as for example, the phasing out of CFCs, initiated by the 1987 Montreal protocol on substances that deplete the ozone layer). Market and competition effects are far-reaching.

In the face of these developments and in parallel to the international, European and national lawmaking process, industry has gradually developed comprehensive "pro-active" programs. Such programs are meant to meet the public's and the authorities' concerns and to offer a sound managerial and operational response from industry. Such a response may take the form of general policy declarations or more operational programs. [See for example the "Business Charter for Sustainable Development" (November 1990) of the International Chamber of Commerce. The Charter lists sixteen principles for Environmental Management.] Typically, one leading feature of such policies is the improvement of information flows between industry, the public and the authorities.

On a more sectorial basis, the "Responsible Care" program developed by the chemical industry on a worldwide and a Regional basis reflects this sector's commitment towards achieving continuous and measurable progress in the health, safety, and environment areas. Association members of the International Council of Chemical Associations, such as the European Chemical Industry Council (CEFIC), the Japanese Chemical Industry Association (JCIA), and the U.S. Chemical Manufacturers Association (MA) have identified Responsible Care as a priority and a contribution to sustainable development.

1.2. Policy and Legal Principles Affecting Industrial Chemistry

Years have seen the emergence of two major policy principles that influence and will continue to impact with special significance the lawmaking process affecting the exercise of activities related to chemicals.

1.2.1. Precautionary Principle

The "precautionary principle" has been used more and more frequently and quoted in public statements adopted in several international forums and according to some opinions has become a principle of (customary) international law [2–5]. Examples of international forums include:

1) Bergen Conference on Action for a Common Future, Norway, May 8 – 16, 1990. The Bergen Conference was organized by the government of Norway in cooperation with the UN Economic Commission for Europe as part of the preparation of the UN Rio Conference on Environment and Development. The Bergen Conference established that "in order to achieve sustainable developments, policies must be based on the Precautionary Principle. Environmental measures attack the causes of environmental degradation."
2) Principle 15 of Rio Declaration on Environment and Development, Rio de Janeiro, Brazil, June 13, 1992. "In order to protect the environment, the precautionary approach shall be widely applied by States according to their capabilities. When there are threats of serious or irreversible damage, lack of full scientific certainty shall not be used as a reason for postponing cost-effective measures to prevent environment degradation."

The precautionary principle is part of the wider concept of sustainable development and has been integrated in the European Communities' law since its introduction in the Maastricht Treaty [6]. Article 130 r, § 2, of the Treaty explicitly refers to the precautionary principle: "Community policy on the environment shall aim at a high level of protection .... It shall be based on the 'Precautionary Principle' and on the principles that preventive action should be taken, that environmental damage should as a priority be rectified at source and that the polluter should pay ...."

The European Constitution is a major example of the fast recognition in a legally binding instrument of the two major principles for future environmental policies.

The precautionary principle institutionalizes caution when (sufficient) evidence exists that an activity or substance is likely to cause unacceptable harm to the environment. Its actual meaning, interpretation, and definition, however, are much debated because the implications of the principle could be far reaching, if not devastating. Definitions vary widely from encouraging a cautious and responsible attitude, based on scientific evidence, not to generate or discharge in an uncontrolled way elements hostile to humans and the environment, to a much more radical "anti-industrialization" attitude, which is leading to the prohibition of substances or activities even in the absence of conclusive scientific evidence. According to the precautionary principle, as defined by part of the doctrine and in several marine pollution conventions, a substance of activity posing a threat to humans or to the environment should therefore be prohibited (i.e., banned or restricted) even if there is no conclusive scientific proof of a causal link between that particular substance, or activity, and humans or environmental damage.

This rather radical approach was first used in the Ministerial Declaration of the Second North Sea Conference [7]. "The Parties agree to ... accept the principle of safeguarding the marine ecosystem of the North Sea by reducing polluting emissions of substances that are persistent, toxic and liable to bioaccumulate at source by the use of the best available technology and other appropriate measures."

This applies especially when there is reason to assume that certain damage or harmful effects on the living resources of the sea are likely to be caused by such substances, even where there is no scientific evidence to prove a causal link between emissions and effects ("the principle of precautionary action").

The overall goal was to achieve a substantial reduction in the total quantity of persistent, toxic and bioaccumulative substances reaching the North Sea. Actual policies of Member States are aiming at a target of 50 %. The national interpretations of the London Declaration have been far from homogeneous, most of them requiring some certainty of (scientific) evidence regarding environmentally damaging effects (i.e., France) or calling for actual evidence before taking action.

Those variations clearly reflect the embarrassment of public authorities who are confronted by complex situations in which multiple causation factors make scientific assessment and conclusions difficult or uncertain. [This embarrassment was reflected in the Third International Conference on the Protection of the North Sea, The Hague, March 7 – 8, 1990 (Third North Sea Conference). The final declaration confirmed the continuous application of the precautionary principle but limited it to those substances that are persistent, toxic, and liable to bioaccumulate.] A radical interpretation of the precautionary principle as reflected in the London Declaration is seriously objectionable – from both legal and scientific standpoints – because it conceptually opposes and rejects scientific evidence and causality as instruments for sound policy actions.

Such an interpretation necessarily leads, at the extreme, to an impasse and might be translated into a "paralysis principle." Indeed, a requirement based on the existence of "reasons to believe" that damage or harmful effects are "likely" to be caused becomes meaningless if opposed to or separated from scientific evidence. For example, a mere trace of a substance in the environment would equal pollution of that environment leading to possible prohibitons. The progress of analytical methods today allows the detection of minute quantities of a given substance (e.g., a piece of sugar in a lake). The progress of scientific and analytical methods without parallel progress of impact and risk assessment techniques could actually multiply purely "emotionally"-based policies. The exclusion of scientific evidence as a determining factor for shaping environmental policy might actually lead to the opposite of a precautionary and reasonable attitude. At the extreme, it might lead to decision-making processes entirely dominated by "beliefs" and "opinions" dominated by ideology without due regard for an objective and scientifically sound analysis of facts.

The radical interpretation of the precautionary principle seems to be disregarded by provisions of Article 130 r of the Maastricht Treaty. While endorsing the concept of the precautionary principle, the Treaty does not define it and obliges the Community to take account of available scientific and technical data [Article 130 r (§ 3)].

Realistically, complex situations and interactive systems might lead to a degree of uncertainty where scientific evidence is incomplete or even appears contradictory. The lack of full scientific evidence might encourage policy makers to take measures "where there are threats of serious or irreversible damage." Such measures must, however, be based on sufficient scientific evidence.

It remains to be seen if and how a moderate and reasoned interpretation will finally emerge in the international and national spheres, where scientific evidence and causation will actually serve and precondition the applicability of the precautionary principle.

Clean Water Initiative Submitted to Congress, 1 February 1994. While aiming at a strategy for prohibiting, reducing, and substituting the use of chlorine and chlorinated compounds, the administration proposed a time schedule allowing federal agencies and "other experts" to comprehensively assess the use the environmental and health impacts of chlorine and chlorinated compounds, and availability and relative efficiency and safety of substitutes. Within 30 months, a plan "for our appropriate actions" should be developed. The National Academy of Sciences is expected to complete a study for the Congress on the current knowledge of chemicals "that exhibit endocrine immune and nervous system health effects in human and wild life." The highly political and emotion-driven nature of this issue is demonstrated by the more moderate approach of the administration, which resulted from fierce reactions from many circles, including industry.]

The sole answer for policy makers is to develop more, not less, science and technology assessment and to assist policy and decision makers to take the most appropriate decisions.

1.2.2. The Polluter-Pays Principle

Historically, an economic and preventive principle, the "polluter-pays principle" is gradually influencing public and administrative law, or civil law.

The polluter-pays principle was initially adopted by the OECD in its Recommendations [8]. The 1991 OECD Recommendation [9] extended the polluter-pays principle to the "cost of damage" generated by pollution.

The Parliamentary Assembly of the Council of Europe summarized the extension of the polluter-pays principle to compensation of damage as follows:

"States, industry and all persons shall be liable to pay for the environmentally harmful consequences of their actions and development programmes ... The principle of the polluter's liability, or the "Polluter Pays Principle" as it is commonly known, shall be strictly applicable." [Recommendation 1130 (1990)]

The polluter-pays principle has been introduced into the European Communities Constitution via Article 130 r of the Single European Act [10] and later in the Treaty on the European Union. The extension of the polluter-pays principle as an environmental law principle has received priority consideration by the OECD. In its preventive form, the principle implies that those whose emissions could create a risk to the environment should bear the full cost of reducing or eliminating such emissions so as to prevent this risk from materializing.

The principle is thus also a means of internalizing environmental costs [11]. However, it is not yet clear what impact it will have in specific areas such as civil liability for damage, although it may be argued that it is based on a similar concept, i.e., the author of a damage, to persons or to property, has to compensate for such damage.

One of the important, and complex, questions raised by the polluter-pays principle within civil liability mechanisms is linked to the very question of "what is a damage?" since there is a tendency to extend the notion of damage to "damage to the environment", or "ecological damage". These are extremely difficult to define and to quantify. (For example, how is the degradation of the ozone layer to be evaluated in terms of damage and to which cause can it be rationally attributed? Who is the polluter? The "global climatic change" issue raises a number of complex questions of this nature.)

Other difficulties are linked to the establishment of causation, when several sources of pollution exist, and to the establishment of evidence. Should the plaintiff bring the full evidence that damage was caused by a given attributable incident?

Or is "prima facie" evidence sufficient? Should the burden of the proof be reversed with the risk of jeopardizing the very concept of civil liability? These, and other questions, have been debated in the context of the "Council of Europe" Convention and are still much argued within the European Union.

Efforts have also been made to make civil liability rules not only a legal instrument for compensation of injury to people and damage to property but also to use them as a (limited) tool for environment remediation. It is sometimes argued that strict civil liability could play a role in the prevention of environmental degradation and therefore be used as an instrument of environmental policy.

1.3. Economic Integration and Environmental Protection

The interface between economic integration and environmental protection is a major emerging issue at the end of the 20th century. Both on the world trade scene and within the European Union, law and policymakers have to reconcile potential conflicts between the logic of economic integration implying and enforcing similar rules and policies to facilitate trade and competition and, on the other hand, the logic of national/regional prohibitions and restrictions on products and activities. Global issues call for global solutions. Achieving cohesion is far from obvious, as illustrated by the present debates concerning the tensions between the Internal Market of the Community and environmental policies, but also in the debate within the World Trade Organization.

International trade of products which might have an impact on the environment is increasingly subject to severe restriction and/or to authorization and information procedures. For example, the case of waste movement restrictions and free trade is illustrative of such tensions. The EC Regulation on Control of Waste of February 1, 1993 [12] is intended to implement the Basel Convention on the "Control of Transboundary Movements of Hazardous Waste and their Disposal" (adopted and opened for signature on March 22, 1989). This Regulation completes and tightens the monitoring system already applicable in the Community. It covers five categories of transport: transfer between Member States, transfer within Member States, export to non-EC countries, imports from outside the EC and waste transfers. The role of the OECD, in seeking a workable solution especially for waste for recovery should be stressed.

The Regulation provides for a prior authorization system, recognizes a broad right for Member States to object to shipment and implements the self-sufficiency and the proximity principles which both aim at encouraging local waste disposal. Paradoxically, a piece of Community legislation, supposed to have direct effects, is opposing one of the fundamental freedoms of the EC Treaty, i.e., free circulation of goods. This has raised much criticism within the Community and the legal debate continues on the (impossible?) reconciliation of free movement of goods principles and restrictions of transboundary shipments of waste and recyclable materials (secondary raw materials) (see Chap. 4). Rather than concentrating its policy on the proximity principle, which may actually create inefficiencies and additional unnecessary costs, the Community should develop sound and uniform criteria for operating waste facilities and movements of waste.

Another example of restrictions to trade motivated by environmental concerns is the prior informed consent (PIC) procedure. The PIC procedure is based on the principle that international shipment of a chemical that is banned or severely restricted in order to protect human health and the environment should not proceed without the agreement, or contrary to the decision of, the designated national authority in the importing country [13–17]. These developments raise the question of their interface and compatibility with world trade liberalization agreements under the GATT. There is no doubt that the whole "trade and environment" debate will constitute a major global policy issue of the end of this century. For a comprehensive review, see [18].

1.4. Other General Trends of Health, Safety and Environmental Laws Affecting Industrial Chemistry

While legislations dealing with environmental aspects of industrial activities have to be situated within general public and civil law, the evolution of health, safety and, especially, environmental (HSE) laws has been characterized by the following elements:

1) A more autonomous, systematic system of law is gradually developing both in the national, European, and international spheres. The fast development of a European Community comprehensive environmental law is based on policy programs which aim at fully integrating environmental policies in other Community policies and in setting long-term objectives and targets [19], [20]. The primacy of Community law over national legislation, where sufficient, clear, and precise provisions exist, and the direct effect into national legal systems have far-reaching implications.
2) The gradual emergence of a "right to know" both in the European Union and in the United States is a good example of the impact of environmental law concepts on public and administrative law. This right to know gives citizens broad access to information obtained by public authorities in the environmental area, subject to some limitations [21].
3) The impact of environmental issues on competition policy. In a growing number of cases, "environmental benefits" are claimed in industrial and marketing strategies both vis-à-vis consumers and vis-à-vis public authorities. This development raises many questions about ethics and (fair) competition. The eco-label concept is an illustration of this evolution [22].

Other aspects are linked to state aids given for environmental purposes which might distort competition in a given industrial sector.

The European Commission has issued guidelines describing its policy on state aid for environmental protection. Such guidelines would authorize up to 30 % state subsidy for (voluntary) investment related to the environment, while investment aimed at compliance with environmental legislation could be supported up to 15 % of the cost [23].

1.5. Scope of this Keyword

The keyword Legal Aspects does not aim at providing a comprehensive or exhaustive review of European, national, or international rules affecting chemicals. For a publication relating to international aspects of chemical control see [24]. The mere coverage of, for example, European Community legislation would require several volumes of Ullmann's. Many specific aspects have already been addressed in other keywords dealt with in volumes B 7/B 8 or in relevant sections of them. The main objective of the keyword Legal Aspects is to make better understood the interface between the legal system in general and chemical-related activities and to present in this respect the major trends characterizing the evolution of law on the eve of the year 2000.

2. European Community Legislation Affecting Chemicals

Definition of Terms Used in this Chapter. The European Union (EU) was set up by the Treaty of Maastricht, which came into force in November 1993. Part of the Treaty on the EU are the Treaties on the European Communities (EC), the EURATOM Treaty, and the Coal and Steel Treaty.

The EU does not have legal personality, but the EC has. Therefore, any legal instrument is adopted not by the EU but by the EC. Whenever references to legal texts are made, the term EC is correct, EU may only be used in a political context.

Prior to the coming into force of the Maastricht Treaty the correct name for the EC was European Economic Community (EEC). Thus, all legal texts adopted before November 1993 have in their title the reference EEC.

2.1. The European Community Framework

The European Community has, under different Treaties, the objective of creating an integrated economic community in which industrial frontiers are no longer economic frontiers, and which includes the free circulation of goods and services, capital, and labor as well as a high level of health, safety and environmental protection. The mandate to establish and monitor the Internal Market for goods and services, capital, and labor is in the Treaties given to Community institutions: Council and Commission, the European Parliament, the Economic and Social Committee, and the European Court of Justice (ECJ). Member States have the objective of facilitating achievement of the Community's tasks, and abstaining from any measures that could jeopardize the attainment of Community objectives [25]. In contrast to that there is the joint responsibility between the European Community and its Member States acting in isolation to ensure an appropriate protection of the environment [26]. This principle of subsidiarity, enshrined since autumn 1993 in the general provisions of the EC Treaty, has existed since the 1987 amendment of the environmental chapter of the EEC Treaty. Proposals for legislation made by the EC Commission have been adopted or rejected without express recurrence to this principle.

Chemical legislation in the European Community must find its way through this framework of responsibilities. In very broad terms, past legislation on product standards was generated mainly at Community level and thus had a considerable degree of uniformity. Legislation banning products or restricting their use was more often adopted at a national level; with the completion of the internal Community market, a marked trend exists towards integrated rules for bans and restrictions, but progress is slow. With regard to production-orientated legislation, Community legislation in the past has been unsystematic and limited to framework rules. The tendency has developed to arrive at a more systematic approach, although the responsibility for rule making remains shared between Member States and the Community, the Community being limited to umbrella legislation.

2.2. Product-Oriented Legislation

Community legislation on chemical substances and preparations as well as on specific chemical products – food additives, pharmaceuticals, cosmetics, fertilizers – started early in the 1960s, shortly after the creation of the EEC. This legislation was based almost entirely on Article 100 of the EEC Treaty, which reads:
"The Council shall, acting unanimously on a proposal from the Commission, issue Directives for the approximation of such provisions laid down by law, Regulations or administrative action in Member States which directly affect the establishment or functioning of the common market...".

Total or Optional Harmonization? The approximation of national legislation aimed at diminishing the differences between national rules that created, or were capable of creating, barriers to the free circulation of goods. Since in all the above-mentioned sectors of chemical products, questions of health and safety considerably influenced rule setting, the EEC did not attempt the legislative technique called "optional harmonization" but rather used a "total harmonization" approach.

Under *optional harmonization*, Member States fix standards for national production at their discretion and at a level they consider appropriate. However, they are obliged to admit on their market those products from other Member States that comply with Community standards. This kind of approach was, for instance, largely used by the EEC in the area of technical legislation.

Under the *total harmonization* approach, which was used in the area of chemical products, national legislation is replaced by Community legislation so that, finally, one rule exists within the Community. This approach is obviously more ambitious, more delicate, and more time consuming. Since questions of health and safety are central to all Member States' national policy, progress in replacing the different national rules by one single Community rule has been slow and even slower in the areas of food and pharmaceuticals than in other areas.

In the area of chemical substances and preparations, national legislation in the 1960s was not very sophisticated. Therefore, attempts to create uniform rules for chemicals were relatively successful. Community legislation on chemical substances [27], solvents [28], pesticides [29], and paints and varnishes [30] was adopted soon after the transition period for setting up the EEC. This legislation set uniform rules for the classification, labeling, and packaging of substances and preparations, and ensured that chemicals conforming to these requirements were allowed to be circulated freely within the Community. Since this legislation in the 1960s and 1970s had to be adopted unanimously, Member States were able to block the adoption of Community legislation that did not conform to their national rules on health and safety, until their concern was met.

A Dynamic Cooperative System under the Control of the Commission. The possibility of blocking legislation served to protect the interests of Member States in terms of health and safety, before such legislation was adopted. However, this need could also arise after the adoption of Community legislation, for instancce, when an accident in a Member State demonstrated that the Community rule was not strict enough or where new scientific evidence was found. Since an amendment of Community legislation was slow and required an initiative from the EEC Commission, this mechanism was not considered able to adequately cover all cases. Therefore, the Community legislation on chemicals contained a safety clause: where a Member State found that a chemical, although complying fully with Community rules, presented a risk to health and safety, the Member State was allowed to take provisional measures to counteract that risk. These measures could even go so far as to remove the chemical from the market. The Member State that took such a measure had to inform the Commission and other Member States of its action and give a reason for its decision. Then, a Community procedure was started to decide whether the national measure should be extended to the entire Community. Until such a decision, the national safeguard measure was maintained. By way of illustration, the text of such a safeguard clause is given here [31]:

"Where a Member State has detailed evidence that a substance, although satisfying the requirements of this Directive, constitutes a hazard for man or the environment by reason of its classification, packaging and labelling, it may provisionally prohibit the sale of that substance or subject it to special conditions in its territory. It shall immediately inform the Commission and the other Member States of such action and give reasons for its decision. The Commission shall consult the Member States concerned within six weeks, then give its view without delay and take the appropriate measures. If the Commission considers that technical adaptations to this Directive are necessary, such adaptations shall be adopted, either by the Commission or by the Council in accordance with the procedure laid down in Article 21; in such a case, the Member State which has adopted safeguard measures may maintain them until the adaptations enter into force."

In the mid-1970s, two elements were added to this situation – attempts to introduce Community-wide bans and restrictions on chemicals, and rising environmental concerns. Clearly, the feeling was that legislation on chemicals that tried to open up national frontiers and set uniform rules was insufficient since national prohibitions and restrictions for use continued to exist, and with growing concerns for health, safety, and the environment, new restrictions were being introduced at the national level. To ensure uniform standards, the EC felt the need to adopt Community-wide restrictions or bans. This legislation [32] was adopted on an ad hoc basis: whenever nationally – or indeed in international bodies – the need was felt to ban or restrict the use of a chemical product, the European Economic Community tried where appropriate, to reach Community-wide standards. The difficulty of reaching a consensus on such restrictions under the unanimity rule had, as a consequence, the adoption of a relatively limited number of such Community rules. Several attempts to accelerate the decision-making procedure at the Community level were unsuccessful, probably mainly because Member States were reluctant to see responsibility for health and safety questions shift more and more to the Community. Thus, between 1976 and 1992 some 30 substances were banned or severely restricted in use, covering only a part of the national bans or restrictions.

The second element was upcoming environmental concern in legislation. This concern gradually led some Member States to legislate in the area of chemicals, in order to protect, preserve, or improve the quality of the environment. The EEC reacted by including environmental considerations into its existing legislation on chemicals. The evolution of Directive 67/548/EEC, adopted in 1967 at Community level, is characteristic in this regard: in the mid-1970s, France informed the Commission of its intention to introduce new chemical legislation that took into a account environmental considerations. The Commission was of the opinion that such legislation would be capable of creating new obstacles to trade within the Community. Therefore, it suggested an amendment to Directive 67/548/EEC that finally led to the adoption of Directive 79/831/EEC, which, for the first time in such precise terms, talked of the need "to protect man and the environment" [33]. Since environmental policy is based, among other factors, on the principle that preventive action should be taken [34], that Directive introduced the principle that before new chemical substances were placed on the market, information had to be provided to competent authorities, together with test results on the substances, in order to enable the authorities to evaluate the foreseeable risk for humans and the environment. This evaluation is made Community-wide in concert between national and Community administrations. Conditions for marketing of new chemical substances based on health, safety, or environmental considerations are fixed for the entire Community.

A similar approach regarding notification prior to placing a substance on the market was used when the Community adopted legislation for genetically modified organisms on the market (deliberate release) [35]. Again, evaluation of the risk of such deliberate release is made Community-wide prior to the release.

However, this approach is by no means generalized; it is not applied to all chemicals, i.e., chemical preparations can be placed freely on the market without prior provision of test results and other information to competent authorities.

In 1987, the EC Treaty was amended. Specific rules were introduced to ensure achievement of the Internal Market by the end of 1992. For that purpose, a new Article 100 a provided that measures with the objective of achieving the Internal Market could be adopted by majority decision of Member States. At the same

time, a new Article 100 a (4) was included in the Treaty, which reads:
"If, after the adoption of a harmonization measure by the Council acting by a qualified majority, a Member State deems it necessary to apply national provisions on grounds of major needs referred to in Article 36, or relating to protection of the environment or the working environment, it shall notify the Commission of these provisions. The Commission shall confirm the provisions involved after having verified that they are not a means of arbitrary discrimination or a disguised restriction on trade between Member States ...".

Furthermore, Community environmental policy was considerably strengthened, by the introduction of an environmental chapter, Articles 130 r to 130 t, into the Treaty. Environmental measures were, in principle, to be adopted unanimously (Art. 130 s). Article 130 t stipulated, "The protective measures adopted in common pursuant to Article 130 s shall not prevent any Member State from maintaining or introducing more stringent measures compatible with this Treaty."

The borderline between the provisions of Article 100 a and Article 130 s seems, with regard to chemical products, to be rather clear: since chemical preparations are products and should –under the concept of an Internal Market, where national frontiers are no longer economical frontiers – circulate freely within the Community, the legal foundation for rules on their classification, labeling, or packaging – including their ban or restrictions on use – should be based on Article 100 a. Article 130 s would be reserved to production-related legislation that did not affect the establishment or the functioning of the Internal Market.

The majority of Community chemical legislation did indeed follow this approach. However, not all legislation respected this pattern. When the Council adopted legislation on chlorofluorocarbons (CFCs) and other ozone-depleting substances, this legislation was based on Article 130 s [36]; the same approach was followed on waste legislation, although, from the point of view of the environment, there exists little difference between poisoning by chemicals or by waste [37]. Thus, because of national initiatives remaining possible under Article 130 t, uniformity of rules within the Community was not achieved.

Even where new legislation on chemicals was based on Article 100 a, the establishment of uniform rules was questioned. Indeed, certain interpretations of the above-mentioned clause of Article 100 a (4) are equivalent to the wording of Article 130 t. In other words, this interpretation would allow Member States, after the adoption of Community standards, to introduce new, diverging rules at the national level, not merely to continue to apply national legislation that existed prior to the adoption of Community rules. Whether Article 100 a really has the same meaning as Article 130 t or should be interpreted more narrowly, remains to be decided by the European Court of Justice, which, until now, has not had occasion to pronounce itself on this question.

Pending this question and more generally, Member States remain responsible, together with Community institutions, for the protection of health, safety, and the environment. They have thus not lost the responsibility for banning or restricting the use of chemical products, as long as no comprehensive legislation setting up Community-wide bans and restrictions exists. Article 100 a, seen in more general context, enables the Community to draft uniform rules for chemicals that are based on a high level of protection of health, safety, and the environment. To the extent that the Community manages to elaborate, adopt, and permanently adapt such rules to technical and scientific progress, recurrence to national bans and restrictions will not be necessary; thus Article 100 a (4) would lose practical importance.

The Maastricht Treaty on European Union has not changed the principles on rule making at the Community level. Thus, the trends described above are likely to remain applicable in the coming years.

2.3. Legislation on Chemical Installations

As stated above, Community legislation on chemical installations has not been very systematic in the past. The establishment and functioning of a common market were not in the past perceived to require the elaboration of Community-wide standards for industrial installations. Thus, rules affecting chemical installations were set up only with the introduction of an environmental policy at the Community level and under environmental protection auspices. Industrial installations that discharge dangerous substances into waters or into groundwater or emit pollutants into the air need specific authorization for such emission [38]. Such permitting rules were, however, far from covering all installations or all pollutants.

An industrial accident in Seveso, Italy, provoked Community legislation on the prevention of major accident hazards of certain industrial activities [39]. This so- called Seveso Directive, adopted in 1982 and amended several times, requested installations that process or stock chemicals to take measures to prevent industrial accidents, to inform the adjacent population of potential risks, and to elaborate on-site and off-site emergency plans. This legislation is described in detail in Chapter 5.

In 1985, the EEC adopted a Directive that required an environmental impact assessment for a number of public or private projects, before planning consent was given [40]. In 1993, the EEC adopted a Regulation on the voluntary participation in a system of eco-auditing [41]. This Regulation aims at the improvement of eco-management by industrial installations. The Regulation will take effect in spring 1995. Also, in 1993, the Commission proposed a legislative measure at the Community level on integrated pollution control, which aims at a generalized, integrated permitting system for industrial installations, in order to improve, rationalize, and harmonize authorizations for industry and adequately protect the environment [42].

These Community rules on chemical installations were either umbrella rules or left Member States relatively large discretion in complementary rule making at the national level. This reflects the fact that while chemical products should circulate freely within the EC, the chemical installations that produced these products are not movable. Therefore, where legislation on installations was adopted prior to the 1987 amendment of the EC Treaty, it was based on Articles 100 and 235 of the EC Treaty at the same time. After the single Act 1987 took effect, Community legislation on installations was based on Article 130 s; the one exception until now – legislation harmonizing national programs on disposal of waste from the titanium dioxide industry, which was finally based on Article 100 a of the EEC Treaty [43] – seems isolated. In general, legislation on industrial installations will for the foreseeable future probably be based on Article 130 s of the EEC Treaty. This is likely also to apply to accessory environmental legislation, such as emission of pollutants into the air or soil.

The consequence of the use of Article 130 s is that Member States, under Article 130 t, will remain entitled to maintain or introduce more protective legislation with regard to chemical installations. Different national legislation, however, might lead to different production costs and gradually influence the competitive situation of industrial installations. This, in turn, might have an influence on the free circulation of goods produced in such installations. Where Community rules on installations have a significant impact on the cost of production, good reasons exist to consider whether the establishment of an Internal Market does not require equivalent standards over the entire Community territory. This would suggest that for such rules, Article 100 a be chosen as a legal basis, since standards that are laid down on the basis of Article 130 s will – because of the existence of Article 130 t – never be capable of achieving uniformity. For example, Directive 82/501/EEC on the prevention of major accidents, which affects mainly industrial installations that produce or process chemical products, might well request such considerable investments for accident prevention that rules within different Member States would distort competition for the chemical products that are placed on the market. According to the content of such a revised Directive, good reason might thus exist to base such a Directive on Article 100 a and not on Article 130 s of the Treaty. Indeed, the precise content of the Directive will determine whether one can argue that the objective of such an accident prevention Directive was principally the protection of humans and the environment and whether the establishment of local competitive conditions among installations was only an accessory objective.

3. Towards the Adoption of New Civil Liability Rules: A New Challenge for the Chemical Industry

"Toutes les pertes et tous les dommages qui peuvent arriver par le fait de quelque personne, soit imprudence, légèreté, ignorance de ce que l'on doit savoir, ou autres fautes semblables, si légères qu'elles puissent être, doivent être réparées par celui dont l'imprudence ou autre faute y a donné lieu. Car c'est un tort qu'il a fait, quand même il n'aurait pas eu l'intention de nuire." Extract from "Lois Civiles, II" by Jean Domat (1625 – 1696). The original French text can be translated as follows: "All losses and all damage caused by the deeds of any person, through negligence, thoughtlessness, ignorance of what one should know or similar shortcomings, however trivial they may be, must be compensated by the person whose fault or negligence has created the damage. Because he generated a tort, even though this was unintentional."

3.1. Introduction

Significance of Civil Liability and its Evolution. A significant number of research, production, and processing activities are carried out by a host of individuals and companies. Once manufactured, the most diverse products – from the simplest to the most technologically complex – are put on the market and are subject, among other things, to transport, packaging, labeling, and storage operations. All of these activities may cause damage either to persons, to goods, or to the environment. Thus, one can easily imagine the importance of "civil liability law" or "third-party liability" and its economic impact on industry. "Manufacturer's liability" must be considered in the framework of general legal systems of third-party liability, or "responsabilité civile ou quasi-délictuelle" in French civil code systems.

Basics of Civil Liability. Traditionally speaking, civil liability for damage means an obligation, under certain conditions, to compensate for damage caused to persons, property, or proprietary rights. The concept of environmental damage is relatively new and refers only to the nature of the damage. It basically addresses damage to the unowned environment, for example, disappearance of a fish species in a river. The plaintiff has to prove the existence of the following factors: (1) that he/she has suffered damage to his/her person or property; (2) that the damage was caused by an offense of negligence by the defending party (i.e., was based on fault); and (3) that a causal link exists between the damage and the incident.

These fundamental principles of civil law are similar in most European countries but are subject to evolution as described below. Liability instruments of direct interest to industry have been drafted in relation to

1) Defective products
2) Transport of (hazardous) goods
3) Industrial activities

Chapter 3 addresses liability, and compensation questions, in the context of industrial and marketing or trade activities. In the following material, the major trends affecting the evolution of civil liability are outlined, and various major international legal instruments recently adopted or still under discussion in this area are reviewed.

3.2. General Trends and Debates Affecting Modern Civil Liability Systems

3.2.1. New Trends

When can a Person or Company Be Held Liable? In most European countries, existing civil liability laws are based on fault ("continental law systems") or on the "negligence concept" ("common law systems"). However, in some countries new laws with a "strict" liability regime as a basis for liability have been enacted or are under consideration. The strict liability regime usually applies to damage caused by activities that are deemed especially dangerous, such as handling hazardous waste, dangerous substances, or nuclear energy. In other cases, the case law (theoretically based on the fault concept), has led to a similar result. Under common law, the "duty of care" has been considerably extended and case law developed in the direction of strict liability. Similarly, fault-based liability is interpreted by courts and jurisdictions in such a stringent manner that the difference between these regimes has shrunk. Even without such specific law, the case law of most countries has developed a strict liability regime for damage caused by dangerous activities, such as handling explosives or operating dangerous installations.

At European level, the Directive on defective products (see Section 3.3) is based on strict liability of producers for defective products put on the market. A similar trend toward strict liability has characterized international conventions in the context of land and maritime transport of dangerous goods [the so- called International Institute for the Unification of Private Law (UNIDROIT) convention, and the present "Hazardous Noxious Substances" (HNS) draft convention of the International Maritime Organization].

Caution is needed and "labels" can be misleading because a fault-based liability regime can actually be more severe for potentially liable parties than a well-defined strict liability system. Enormous conceptual and practical differences also exist between the so-called strict liability systems.

For example, little common ground exists between the "operator's liability" approach, which is the basis of the Council of Europe Convention (see Section 3.7), and the absolute "generator's liability," which was once proposed in the context of a draft protocol on liability for waste to the Basel Convention [44]. Both texts are supposed to be based on strict liability. The operator's liability approach focuses on actual responsibility of the person in control of certain activities and substances; the generator's liability approach merely indicates who should bear the damage consequences of products (here waste) whether this person is, or is not, in control. The generator can be held liable even for acts committed by third parties, professional or not.

Arguments in Defense. Certain arguments may be used to defend oneself. One cannot be held liable if one proves the existence of "force majeure" (i.e., the damage was caused by an act of war or hostilities or by a natural phenomenon of an exceptional, inevitable, and irresistible character). A similar defense is, for instance, a claim that the damage was caused by an intentional act by a third party or that the person who suffered the damage had by his own fault contributed to the damage.

Additionally, different laws contain defenses of a special nature typical of the area concerned. The state-of-the-art defense is one example of regulatory exclusion of liability. According to the Directive on defective products, producers are not held liable if they prove that the state of scientific and technical knowledge at the time they put the product into circulation was not such as to enable the existence of a defect to be discovered.

In the context of liability related to industrial activities, compliance with permits is taken into consideration although it does not lead to an absolute defense. Emissions within "tolerable levels under local circumstances" (Council of Europe Convention) also offer possible defenses depending on legislation or case laws.

The Emergence of the Polluter-Pays Principle in Civil Liability Law (see Section 1.2.2). The polluter-pays principle was introduced in the European Communities Constitution via Article 130 r of the Single European Act and later in the Treaty on the Economic Union. Scientific literature on the polluter-pays principle is considerable but in what concrete and practical manner it will be applied is still unclear because interpretations of questions such as "Who is the polluter?" still differ.

3.2.2. Recognition of Ecological Damage in International and Community Law of Civil Liability

One of the most significant aspects of the present debates and current work in the international sphere is the extension of the concept of damage (death, personal injury, damage to property) to ecological damage. This same ecological damage is defined in various ways, depending on the instruments analyzed. Some common features can, however, be identified: (1) environmental damage concerns the community, not the individual; (2) environmental damage is difficult to quantify. For example, what is the value of a variety of flowers that has disappeared because of pollution? U.S. jurisdictions and legal professions have made considerable efforts to quantify the ecological damage. Another interesting source is the case law developed in connection with major maritime pollution cases (e.g., the Amoco Cadiz decision).

The European Commission Green Paper on Remedying Environmental Damage of 1993 (see Section 3.2) envisages (1) the harmonization of civil liability rules for environmental damage and (2) the possibility of remedying environmental damage not met by the application of civil liability principles. The Green Paper, being a discussion document only, does not propose any definition of environmental damage but refers to definitions contained in the Council of Europe Convention and in the amended proposal for a Council Directive for damage caused by waste. The fundamental importance of and difficulties related to the definition are emphasized.

The 1991 amended proposal for a Council Directive on civil liability for damage caused by waste (see Section 3.4) defines "impairment of the environment" as any significant physical, chemical or biological deterioration of the environment in so far as this is not considered to be damage within the meaning of "damage to property." "Damage" as such, in turn, is traditionally defined to mean death or physical injury and damage to property. The compensation obligation under the Directive would apply for both damage and impairment of the environment. The Council of Europe Convention refers to "loss or damage by impairment of the environment ... [which] shall be limited to the costs of measures of reinstatement actually undertaken or to be undertaken." This definition includes the definition of the "CRTD Geneva Convention" of October 10, 1989 (see Section 3.5).

The word environment is itself defined very broadly in the Council of Europe Convention; it covers natural abiotic and biotic resources, such as air, water, soil, fauna, and flora, and the interaction among the same factors, assets making up the cultural heritage, and the characteristic aspects of the landscape. These last two elements of the definition illustrate the erosion of the traditional concept of damage, because it is now a matter of covering damage to collective property by civil liability mechanisms. Such damage is particularly difficult to quantify. The Commission, unlike the Council of Europe, has avoided giving a definition of the word environment as such in the mentioned proposal for the Directive on liability for waste.

3.2.3. International Harmonization in the Context of Liability for Dangerous Activities

Opinion is growing that civil liability rules should be addressed at the international level not only in relation to products and goods, but also in respect of industrial activities. International rules are said to be necessary because pollution is no respecter of national boundaries and Member States have been establishing different Regulations in this field that could distort competition and hamper the smooth operation of a single market. Protection of the environment is also one of the Community aims, equivalent to creation of the single market, as enshrined in the Treaty of Rome Articles 130 r – t.

This approach was much debated in the early 1990s. Although harmonization of civil laws is conceptually attractive from the point of view of the Internal Market, it is effectively desirable for industry only if it would simplify existing systems and if new international instruments do not simply add another layer of possible legal actions. Harmonization efforts are also seriously challenged by the existence of different procedural rules linked to national judicial systems.

The two most important efforts to introduce harmonization of liability for dangerous activities have been developed within the Council of Europe and the EEC. The Commission Green Paper on Remedying Environmental Damage envisages harmonization of (strict) civil liability rules for environmental damage, as well as the possibility of remedying environmental damage not met by the application of civil liability principles. The latter means creating compensation funds to be used when no liable party can be identified or in case of the insolvency of the liable party. Civil liability is also considered the domain of Community legislation because it is a legal principle closely linked to general European enviromental policy principles, such as the prevention of damage and the polluter pays.

Another important international document is the already-mentioned Council of Europe Convention on civil liability for damage resulting from activities dangerous to the environment. The convention was influenced by major pollution of the Rhine river by Sandoz that affected several countries. It is based on the belief that issues relating to civil liability and compensation for damage resulting from dangerous activities need to be addressed at the international level.

3.2.4. Actions and Requests by Organizations

The right to sue is normally given only to the party with a legal (i.e., direct and certain) interest in recovering compensation for injuries or for damage suffered to property. Where damage occurs to property that is not owned (i.e., environmental damage), no injured party with the right to bring legal action can be identified. At the European level the tendency is to grant such a right – although limited and under the control and at the discretion of the courts – to environmental associations. This right to protect the "public interest" normally rests with the public authorities that operate under public law and democratic institutions.

For example, according to Article 4(3) of the amended proposal for a Council Directive on civil liability for damage caused by waste, common interest groups or associations, which have as their objective the protection of nature and the environment, are given the right either to seek remedy or to join in legal proceedings that have already been brought. Also the Council of Europe Convention gives certain restricted rights to associations seeking to protect the environment; namely, the group may request of courts or jurisdictions (1) that an unlawful, dangerous activity, which poses a grave threat of damage to the environment be prohibited; (2) that the operator be ordered to take measures to prevent an incident or, after an incident, to prevent damage; or (3) that the operator be ordered to take measures of reinstatement. However, any party to the convention may reserve the right not to apply this provision.

3.2.5. Financial Security Schemes

Several proposals suggest compulsory insurance or other financial security schemes to cover the liability under various pieces of legislation. Article 11 of the amended proposal for a Council Directive on civil liability for damage caused by waste provides that the liability of the producer, who in the course of a commercial or industrial activity produces waste, and of the eliminator of waste, be covered by insurance or other financial security. The amended proposal for a Council Directive on the landfill of waste, in turn, contains an obligation on the operator of the landfill to have a financial guarantee or any other equivalent, whose purpose is to cover the estimated costs of the closure procedures and after-care operations of the landfill (Art. 17).

3.2.6. Essential Requirements

The trends described above represent a real evolution in fundamental civil liability rules. However, some of them are still quite flexible, and others imply more theoretical than practical changes compared to the existing situation. An example of this is the discussion of whether liability should be fault based or strict.

Nevertheless, some essential requirements exist for any legislation on civil liability, one of the most significant of which is maintaining the establishment of causation. It is of utmost importance to the entire system that the causal link between the incident and the damage suffered be shown by the plaintiff. Likewise, in the context of damage caused by industrial activities, causation must be proved to exist between the operator in charge of the activity and the incident that led to the damage.

3.3. Liability for Defective Products and European Community Law

3.3.1. Background and Legal Basis

The Directive on defective products was approved by the Council on July 25, 1985 [45]. European Community Member States had to implement the Directive by July 1988. In early 1994, most Member States had effectively introduced the Directive into their internal law. Only France and Spain had not yet enacted the Directive, which led the Commission to open legal proceedings in the European Court of Justice for the noncompliance of these two countries with their obligations under the EC Treaty.

The declared objective of approximation of the laws has been partially attained since the text allows Member States to introduce some variations to three specific provisions. Article 100 of the Treaty of Rome provides the legal basis for the Directive and lays down the following conditions:

1) Unanimous council decision, based on a proposal from the Commission
2) Compulsory consultation of the assembly (European Parliament) and the economic and social committee if there is a need to amend national laws. Both institutions were consulted and expressed their opinion.
3) Grounds for the Directive: According to indent one, the approximation of the laws of the Member States concerning the liability of producers for damage caused by the defectiveness of products is necessary because existing differences may distort competition, affect the free movement of goods within the Common Market, and lead to differing degrees of consumer protection.
4) Social – political justification: The preamble assumes that "liability without fault on the part of the producer is the sole means of adequately solving the problem, peculiar to our age of increasing technicality, of a fair apportionment of the risks inherent in modern technological productions."

The industry has challenged strict liability and expressed preference for liability regimes based on fault or negligence in accordance with traditional principles of the law of torts. This battle was lost for the chemical industry, however, when even Italy, once a strong opponent of strict liability, accepted this regime in the political compromise that led to the adoption of the Directive in 1985. Was this a real source of concern to industry and should strict liability a priori be perceived as hostile by industry? The answer very much depends on three interrelated factors:

1) The content of actual provisions of the Directive
2) The content of national laws implementing the Directive
3) Case law, including that of the EC Court of Justice, that could play an important role in clarifying ambiguities

As stressed in Chapter 5, more important than the "labels" or names used to qualify liability regimes are their specific features. Other elements are also important: the practice of the courts, the judicial organization and its cost, and the general attitude of the public toward the liability and compensation system. In this connection, the Directive, although far from perfect from the point of view of legal certainty, was probably a reasonable compromise acceptable to governments, industry, and consumers (although some would not share this view, both in consumer circles and in industry).

As shown below, the question of recognition of the state-of-the-art defense was an important aspect of the "package" finally adopted and illustrates quite well the conceptual tensions and arguments that prevailed during the European Commission and the Council of Ministers discussions.

3.3.2. Nature of the Liability Regime

Principle: No-fault Liability. The Directive provides for the following system: the injured party has to prove the damage, the defect of a product, and the causal relationship between the defect and the damage (Art. 4). Thus, negligence and fault are no longer relevant as sources of producer's liability. Suppression of the "fault" standard should ease the task of the victim. However, it should be stressed that the case law of most countries had already introduced a softening of the victim's position, for example, in reversing the burden of proof or in extending considerably the concept of fault or "negligence" of the producer. According to common law concepts, the duty of care is now very strict indeed, with the important exception of the state-of-the-art defense.

Lines of Defense and Exemptions from Liability (Art. 7 of the Directive). The producer must prove that his/her is one of the cases of exemption provided for, namely,

1) That he/she did not put the product into circulation
2) That the defect did not exist when the product was put into circulation or that it came into being afterwards
3) That the product was neither manufactured for sale or any other form of distribution, nor manufactured or distributed by him/her in the course of his/her business

4) That the defect is due to compliance of the product with mandatory Regulations issued by the public authorities
5) State-of-the-art defense ("development risk", see Section 3.3.3)
6) In the case of a manufacturer of a component, that the defect is attributable to the design of the product in which the component has been fitted or to instructions given by the manufacturer of the product

This last point is very important and was stressed by the chemical industry. All sectors of the chemical industry should be fully aware of this defense. It will be more and more important to have clear agreements between purveyors and buyers about the purity and characteristics of component products.

3.3.3. The Three Options of the Directive

A fundamental question that had to be addressed by the authors of the Directive was the so- called harmonization question. Should the EEC Directive provide for strict harmonization ("maximum Directive") or should it create a "minimum system" allowing the Member States to establish different and additional standards, depending on their concepts of consumer protection? Real harmonization being one of the raisons d'être of this text was preferable. The EEC Member States delegations recognized that a maximum Directive would be the best solution from the Treaty of Rome point of view, but at the level of details, compromises became necessary and the maximum approach turned into a limited approximation of laws.

State-of-the-Art Defense. A highly important question is, should the manufacturer of a defective product be held liable if the state of scientific and technological development did not enable him to detect the existence of a defect at the time the product was put in circulation. The inclusion of development risks in the scope of liability was predicted to interfere with technological innovation. Until now, in the vast majority of Western legal systems, no party has had to accept the risk of the unknown and unforeseeable. As CEFIC wrote in a position paper addressed to the EEC Council:

"While it is perfectly understandable that a system of liability should oblige producers to take all possible precautions against any possible injury to the consumer, it seems totally unreasonable to require that these same producers be liable for the unknown, and for risks over which they have no control whatsoever. Injury caused by a defect in a product cannot be a source of liability if the defect was undetectable and unforeseen at the time of its being put into circulation because of the state of the art."

After long and difficult debates, the EC Council finally introduced in the Directive what is known as the state-of-the-art defense. Accordingly, the producer has the possibility of proving that the state of scientific and technical knowledge when the product was put into circulation was not such as to enable the existence of the defect to be discovered. This defense may therefore be introduced in the legislation of Member States and was effectively introduced in most national laws of implementation.

However, the Community regime allows individual Member States to maintain the liability of the producer for "development risk" if it already existed in their internal laws when the Directive was approved (i.e., July 1985). Only Germany was affected by this provision, because its Pharmaceutical Act did not allow producers to use the state-of-the-art defense.

The Community regime also allows Member States to introduce derogations to that principle in new legislation, although this possibility is tempered by a clause of Community "standstill" and the obligation to follow a specific procedure under the supervision of the Commission [Art. 15 (1) of the Directive]. The importance of recognizing a state-of-the art defense for sectors such as the chemical and related industries is clear if its impact on research and development is considered, which in certain sensitive areas might otherwise be discouraged. Evidence already exists, especially in the United States, showing that the liability regime based on absolute liability and disregarding important factors such as the state of technology and the plaintiff's behavior has had adverse effects on new products.

In the United States, the impact of certain court cases based on "absolute producer liability" on insurance costs and the (non)availability of coverage has been tremendous. In the Federal Republic of Germany, the Pharmaceutical Law has not – thus far – generated similar experiences. However, nobody can predict what the consequences would be for the German pharmaceutical industry of a serious "developmental risk" case similar to the thalidomide affair. The compromise achieved in the Directive on the state-of-the-art defense seems to preserve a fair balance and efficiency.

Derogation in Respect of Financial Limitation of Liability. The Directive makes no provision for a Community system of financial ceilings to liability. It does, however, allow Member States to introduce on an individual basis a liability "ceiling" for personal injury.

The Commission has been instructed to submit a report to the Council ten years after the date of notification of the Directive (1995). Acting in the light of this report, the Council is to decide whether or not to cancel the option of introducing ceilings. As with the state-of-the-art issue, such a provision will mirror the Council's political embarrassment at the time of its adoption. The Council will, however, be free to decide to maintain the state-of-the-art and the ceilings option if a political majority so wishes.

Derogation in Respect of Agricultural Products. The Directive normally excludes agricultural products from its scope. However, Member States are authorized to include them within their laws and to apply the regime of the Directive.

3.3.4. Other important Aspects of the Directive

The Directive covers not only personal injury but damage to property.

Definition of Producer. "Producer" is extensively defined in the Directive as

1) The manufacturer of a finished product
2) The producer of any raw material
3) The manufacturer of a component part
4) Any person who, by putting his name, trademark, or other distinguishing feature on the product, presents himself as its producer

Clearly, the concern of the Directive's drafters has been to avoid the possibility of an injured person finding himself in a situation for which no one is responsible. Hence, care is taken in defining the producer. So true is this that the Directive provides for treating the importer into the Community as a producer within the meaning of the Directive and, should it prove impossible to identify the actual producer, treating the supplier as the producer.

Product Defectiveness. For the purposes of the Directive, a product is defective not according to objective or material criteria but with reference to the "safety which a person is entitled to expect." Thus, § 6 of the recital states that defectiveness does not mean the product's lack of fitness for its intended use but the lack of safety that the public at large is entitled to expect. Such a reference creates difficulties of interpretation. In reality, it will be for the judge to assess. However, the Directive provides that all circumstances be taken into account in assessing the safety the product should provide.

Some of the criteria to be considered in assessing the safety of a product are: (1) the presentation of the product; (2) the use to which it could reasonably be expected to be put; and (3) the time when the product was put in circulation. Safety is to be assessed by taking into account the expected use of the product, its marketing, the information given and user instructions, and similar factors related to the product.

Conclusion. The new Community regime on liability for defective products contains both pros and cons and seems generally balanced. Despite imperfections and weaknesses, it genuinely contributes to the progressive establishment of similar legislations in Europe: one major weakness is that it leaves open the question of diverging procedural rules in the judicial systems of the Member States.

Another important question is the possible impact of other more recent Community legislation, i.e., the Product Safety Directive and the proposed Directive on Liability of Suppliers of Services on the Product Liability System.

The Product Safety Directive was adopted on June 29, 1992 and had to be implemented by the Member States in June 1994. The Directive was challenged on grounds of the Maastricht Treaty "subsidiarity" principle by the Federal Republic of Germany. A very wide and strict "general safety requirement" is introduced, although the Directive does not apply to sectors that are already subject to specific legislation. The Directive introduces a general obligation to market "safe products" only, which is not the same concept as a (non)defective product of the Product Liability Directive. Other provisions address the following issues:

1) Monitoring of products
2) Provision of information to consumers concerning possible risks involved in use of the products
3) Adoption of measures to ensure information of consumers
4) Ensuring uniform management of tests, checks, warnings, prohibitions, and market withdrawals

In practice, this legislation should be seen in relation to the already existing comprehensive laws on safety and should contribute to the highest safety standard for products marketed in the European Union.

The proposed Directive on the Liability of Suppliers of Services would make the supplier liable for damage to the health and physical integrity of persons or to property caused by a fault committed by the supplier. However, the proposal "revises the burden of proof" since the supplier would have to establish the absence of fault. The prospects of the Directive being adopted were not very good in 1994 because most Member States were reluctant to accept Community legislation in this area.

The practical impact of the new Community regime, particularly on insurance costs, has so far not been dramatic, but it is still uncertain and will depend on the content of national laws and their application by national courts under the control of the European Court of Justice.

3.4. Community-Proposed Directive on Liability for Waste

Implementation of the polluter-pays principle provides for consideration of the scope of civil liability to better define responsibility in the environmental field. So as to answer this requirement, also expressed in the previous Directive on the control of transfrontier movements of waste (84/631/EEC) and several other Community documents, the Commission in 1989 submitted a proposal for a Directive on civil liability for damage caused by waste. This proposal was amended two years later, in 1991 [46].

The proposal raised a number of arguments, one of them being that the Community Directive on civil liability should be compatible with the UNIDROIT Convention and, on a broader level, with the approach developed in the Council of Europe Convention. With the content of the amended proposal, this compatibility did not seem ensured. For example, contrary to the international conventions mentioned, the proposal channelled liability onto the waste producer regardless of whether or not the producer was in control of waste. To create a system that is workable and practicable, all other equivalent instruments must be studied and considered. Although the proposal was put aside pending the results of comprehensive discussions on the basis of the Green Paper on Remedying Environmental Damage, the main aspects of the draft Directive are described.

3.4.1. The "Channelling" of Liability

Article 3 of the proposal from 1991 poses the principle of the liability of the waste producer for damage and impairment of the environment caused by waste, irrespective of fault on the part of the producer. Simultaneously, the proposal is intended to implement the polluter-pays principle. However, this principle – the legal ambit of which remains ill-defined – does not necessarily imply a strict liability regime. A liability regime based on the polluter's fault would be compatible with this principle if the regime ensures that the polluter actually suffers the consequences of the pollution. The Commission's choice of a strict liability regime channelled onto the producer is explained by the concern to use civil liability as an instrument to eliminate or reduce waste. Such an objective, however, has no direct relation to the function of civil liability, which is to ensure compensation of damage arising from wrong use or handling of products whether they are waste or not.

A positive feature of the proposal is Article 2.2.C., providing a transfer of liability whenever the producer delivers waste to a licensed or approved treatment or disposal facility. Applying the same logic, it is difficult to see how a similar transfer of liability would not occur upon delivery to a licensed carrier.

3.4.2. Definition of the Producer

The Commission proposal refers to any person who, in the course of a commercial or industrial activity, produces waste and/or anyone who carries out preprocessing, mixing, or other operations resulting in a change in the nature or composition of this waste. Article 2(2), however, states that the following are to be deemed the producer of the waste:

"The person who imports the waste into the Community;

The person who had actual control of waste when the incident giving rise to the damage to or impairment of the environment occurred if the producer is not identified;

The person responsible for the installation, establishment or undertaking where the waste was lawfully transferred to such installation, establishment or undertaking licensed pursuant to Directive 75/442/EEC on waste (amended thereafter), 75/439/EEC on waste oils or 78/319/EEC on hazardous wastes (amended thereafter)."

The method chosen in the draft leads to a complex definition based on a system of references to previous Directives, all of them providing for treatment, storage, and disposal operations subject to licensing. This leads to a contradiction in the underlying philosophy when it "channels" liability onto the producer while acknowledging the responsibility of the operator or the person having "actual control" of the waste, but only in cases where the producer has not been identified.

3.4.3. The Concept of Waste

The Commission proposal refers to the definition of the original waste Directive 75/442/EEC, which was amended in 1991. Waste is defined under Article 1 (a) of the amended Directive (91/156/EEC) as any substance or object in the categories set out in Annex 1 that the holder discards or intends or is required to discard. The concept of waste is rather broad by definition but excludes, for instance, gaseous effluents emitted to the atmosphere, radioactive waste, waste arising from mining operations, and decommissioned explosives. Given the fact that the definition is also fundamentally based on a subjective approach – the intent of the holder is decisive – judging whether or not a certain material is waste is far from simple. The categories set out in the mentioned annex do not make this easier, giving only rough examples of what should be considered waste.

Problems related to the definition of waste would complicate implementation of the Directive proposal. Different interpretations of waste in various Member States might also distort competition. Problems related to the definition have been under discussion for years, but a satisfying solution has not yet been found. Finally, one more question is still to be raised: Why is a harmonization of civil liability laws required especially for damage caused by waste? Would that not create an uneven situation when other possible sources of damage are left untouched?

3.4.4. Insurance and Compensation Fund

Article 11 of the draft Directive states the obligation of the producer, who produces waste in the course of a commercial or industrial activity, and of the eliminator of waste to be covered by insurance or other financial security. The article also provides for the Commission to study the feasibility of establishing a European fund for compensating damage caused by waste when the person liable cannot be identified or is incapable of providing full compensation.

3.5. Civil Liability for Damage Caused during the Carriage of Dangerous Goods by Land: The Convention of October 10, 1989

The International Institute for the Unification of Private Law and the United Nations Economic Commission for Europe have actively cooperated in framing an international instrument on liability and compensation for damage caused during the carriage of dangerous goods by road, rail, and inland waterways. Prepared within the

Institute of Rome, the draft was subsequently forwarded to the UN Economic Commission for Europe. CEFIC, along with other international organizations representing transport and insurance interests, participated as an observer in the work of UNIDROIT and the United Nations. The draft convention was finalized at the extraordinary session of the Inland Transport Committee of the UN Economic Commission for Europe, held in Geneva from October 2 – 10, 1989.

The main thrust of the Convention rightly establishes the linkage between civil liability and operational responsibility. This last, dated October 10, 1989, has been open for signature by the states since February 1990. Five ratifications are required for it to come into force. Among the countries that expressed at that time their active support are West European countries (especially The Netherlands, Denmark, the Federal Republic of Germany, and France with some reservations, Finland, and Austria) and also Central and Eastern European countries, and the United States.

The salient features of the Convention are as follows:

1) A "strict" liability regime is in effect.
2) For the purposes of the Convention, carriage comprises the period from the beginning of loading of goods to the completion of unloading operations.
3) Liability is channelled onto the carrier (save in the case of loading or unloading when such operations are carried out under the control of another person).
4) Reference is made to the ADR list for the definition of substances to which the Convention will apply (including hazardous wastes).
5) The principle of compulsory insurance (or any system of equivalent financial security) exists.
6) Total financial liability of the carrier (road, rail) is limited to a ceiling of 30×10^6 € per vehicle (18×10^6 € for personal injury; 12×10^6 € for damage to property and the environment). In the event of the "personal injury" ceiling being used up, the amounts of unmet claims for personal injury could compete with claims for damage to property within the "damage to property" ceiling. The liability ceiling specified for inland waterway transport is 15×10^6 € (8×10^6 € for personal injury; 7×10^6 € for other damage).
7) The carrier (road, waterway) is defined as being the person having use of the vehicle. The carrier is presumed to be the person on behalf of whom the vehicle was registered. This presumption may be reversed.
8) The rail carrier is defined as being the person(s) operating the railway on which the incident giving rise to damage occurred.
9) Grounds for exemption from liability are based on:
   - Lack of information by the shipper or any other person as to the nature of the goods
   - The fact that a third party acted or failed to act with intent to cause damage
   - Deliberate fault of the injured party
   - Occurrence of certain exceptional events: act of war, civil war, or natural phenomena of an exceptional and irresistible nature
10) The definition of "damage" is extended to cover ecological damage in addition to the more traditional categories of "death or personal injury" and "damage to property." Ecological damage is defined as "any loss or damage by contamination of the environment caused by dangerous substances, provided that the compensation for impairment of the environment ... is limited to costs of reasonable measures of reinstatement actually undertaken and to be undertaken."

The states will, however, be able to introduce some reservations if they so wish (particularly concerning defenses and the limiting of liability). From the economic and practical standpoints, it does not look as though the Convention will result in upheaval in land or rail transport. The Conseil des Bureaux (international insurers' body designed to facilitate the settlement of damage caused by automobile traffic – "green card system") has stated that the "green card" system will adapt fairly easily to the introduction of the Convention into the laws of Member States. As far as rail is concerned, the vast majority of railway companies have their own systems of insurance coverage. The Convention is, by contrast, a novelty for waterway transport, for which no compulsory insurance regime has existed up to now.

The authors of the draft have forwarded a resolution to the International Maritime Organization for a similar convention to be adopted for maritime transport, thereby complementing the work carried out in regard to land transport.

3.6. Liability in Connection with the Carriage of Dangerous Substances by Sea (HNS)

The adoption of an international draft convention on civil liability has for some years been included in the agenda of the IMO Legal Committee. An initial draft based on a "two-tier" system of liability – both complicated and costly – failed to achieve adoption in 1984. Despite this setback, the International Maritime Organization still sees the issue as a priority. The IMO Legal Committee is discussing several alternatives:

Alternative 1:
Sole liability of the shipowner coupled with a general raising of amounts of the total limitation stipulated in the 1976 Convention on the Limitation of Liability for Maritime Claims (CLLMC).

Alternative 2:
Sole liability of the shipowner coupled with an additional specific amount provided for in the CLLMC Convention for damage caused by HNS substances.

Alternative 3:
Shipowner liability limited in accordance with the provisions of the CLLMC and complemented by compulsory insurance, the cost of which would be borne by the shipper.

Alternative 4:
Shipowner liability limited in accordance with the provisions of the CLLMC and complemented by a fund financed by the shipper through either a pre- or a post-event financing mechanism.

Other options are possible, including the adoption, envisaged at the beginning of 1994, of a maritime carrier liability regime complemented by setting up an international compensation scheme, to come into operation beyond the limits provided by the shipowners' liability coverage or when the person responsible is not solvent. The system would be based on financial contributions calculated as a function of the type of cargoes and their hazard. Various options have been discussed, including the possibility of requesting importers, not shippers, to contribute. In any event, users of transport services would be asked to contribute.

The issue of adoption by the IMO of an international convention on liability and compensation is expected to be addressed in a 1996 diplomatic conference and will remain an important priority in years to come.

3.7. The Council of Europe Convention on Compensation for Damage Caused by Dangerous Activities

3.7.1. Background

In the 1986 Oslo Conference of European Ministers of Justice, the decision was taken to consider options to extend the reach of civil law to promoting environmental protection. Accordingly, the European Council of Ministers Committee set up a committee of experts in 1987 to propose measures for compensation for damage caused to the environment.

After its sixth meeting, the committee of experts instructed the Secretariat General and the Legal Affairs Directorate of the Council of Europe to prepare an interim activity report. This report was to be submitted to the European Committee on Legal Cooperation (CDCJ), a senior Council of Europe body in the legal field, and to Member States of the Council. The report enclosed preliminary draft rules on compensation for damage resulting from dangerous activities. The content of the Convention was adopted by the European Council of Ministers on March 24, 1993. The Convention was opened for signature on June 21, 1993 in Lugano and was signed by seven Council of Europe Member States (Cyprus, Greece, Finland, Italy, Liechtenstein, Luxembourg, and The Netherlands). Other countries were expected to follow suit, and debate within the European Union concerning the

Commission's Green Paper has directly influenced this process.

The main aim of the Convention is to ensure prompt and effective compensation of damage to people, property, and the environment during the exercise of dangerous activities carried out in a professional capacity in installations or on sites. As stated in Article 1 of the Convention, it aims at providing adequate compensation for damage resulting from activities that are dangerous to the environment and also provides means of prevention and reinstatement.

3.7.2. Key Features of the Council of Europe Convention

3.7.2.1. Liability Regime

The Convention provides for a regime involving strict (no-fault) liability of "the operator of a dangerous activity." The operator's responsibility is not based on the concept of fault, but on the causal link existing between the damage caused and the incident occurring at the time or during the period when the operator was in charge of a dangerous activity. This general approach is in line with the one persistently advocated by the author and CEFIC, which have always wanted to see a strict link maintained between civil liability for damage and operational responsibility. The explanatory report to the Convention rightly points out major reasons for making the person in charge of the activity from which the damage resulted bear the cost of damage; the activity of this person is the source of the damage, and this person is best placed to prevent the damage and limit its extent.

3.7.2.2. Concept of Operator

The operator, as already discussed above, is defined as being "the person who exercises the control of a dangerous activity." Essential in determining who is in charge of a certain activity is to consider who has the power to decide on the way in which the activity is carried out.

3.7.2.3. Incident

Incident is defined as any sudden occurrence or any series of occurrences having the same origin that cause damage or create a grave and imminent threat of damage.

3.7.2.4. Concept of Dangerous Activity

Dangerous activity is defined as being "one or more of the following activities provided that it is performed professionally, including activities conducted by public authorities:

1) The production, handling, storage, use, or discharge of one or more dangerous substances or any operation of a similar nature dealing with such substances;
2) The production, culturing, handling, storage, use, destruction, disposal, release, or any other operation dealing with one or more:
   - Genetically modified organisms which as a result of the properties of the organism, the genetic modification and the conditions under which the operation is exercised, pose a significant risk for man, the environment or property;
   - Microorganisms which as a result of their properties and the conditions under which the operation is exercised pose a significant risk for man, the environment or property, such as those microorganisms which are pathogenic or which produce toxins;
3) The operation of an installation or site for the incineration, treatment, handling, or recycling of waste, such as those installations or sites specified in Annex II, provided that the quantities involved pose a significant risk for man, the environment or property;
4) The operation of a site for the permanent deposit of waste.

Dangerous substances , in turn, are defined as follows:

1) Substances or preparations which have properties which constitute a significant risk for man, the environment, or property. A substance or preparation which is explosive, oxidizing, extremely flammable, highly flammable, very toxic, toxic, harmful, corrosive, irritant, sensitizing, carcinogenic, mutagenic, toxic for reproduction, or dangerous for the environment within the meaning of Annex I, Part A to the Convention shall in any event be deemed to constitute such a risk;

2) Substances specified in Annex I, Part B to the Convention. Without prejudice to the application of sub-paragraph a above, Annex I, Part B may restrict the specification of dangerous substances to certain quantities or concentrations, certain risks or certain situations."

The operators of waste installations and sites for permanent waste deposit are explicitly subject to the new system. However, specific transitional provisions and liability rules are to be applied in the case of permanent waste deposit.

3.7.2.5. Concept of Damage

The concept of damage is defined as follows:

1) Loss of life or personal injury
2) Loss of or damage to property other than to the installation itself or property held under the control of the operator, at the site of the dangerous activity
3) Loss or damage by impairment of the environment in so far as this is not considered damage within the meaning of the preceding subparagraphs, and provided that compensation for impairment of the environment, other than for loss of profit from such impairment, is limited to the costs of measures of reinstatement actually undertaken or to be undertaken
4) The costs of preventive measures and any loss or damage caused by preventive measures, to the extent that the loss or damage referred to in subparagraphs (1) to (3) arises out of or results from the hazardous properties of the dangerous substances, genetically modified organisms, or waste.

The Convention covers legal categories traditionally included under the concept of "damage" in civil law (i.e., loss of life or personal injury). Additionally, the text introduces the concept of loss or damage by impairment of the environment, provided compensation is limited to the costs of measures of reinstatement actually undertaken or to be undertaken. Since "the environment" is defined in a remarkably broad way, the Convention potentially goes beyond the precedents established in the UNIDROIT Convention or in the EC draft Directive on liability for waste.

3.7.2.6. Environment

As already mentioned, the word "environment," in the confines of the Convention, includes the following factors:

1) Natural resources both abiotic and biotic, such as air, water, soil, fauna, and flora and the interaction among these factors
2) Property that forms part of the cultural heritage
3) Characteristic aspects of the landscape

The definition contained in the Convention is not exhaustive. On the contrary, the explanatory report to the Convention explicitly states that damage to any other aspect of the environment may also give rise to liability under the Convention. Even without this note, the definition is considered broad. Point (1) covers the factors traditionally understood to constitute "the environment." Points (2) and (3) refer to more immaterial values.

A legal definition of the environment (or damage to the environment) will influence the process of determining the type and scope of the necessary remedial action. The Commission Green Paper on Remedying Environmental Damage refers to this definition of the environment and also notes the problems related to the matter.

3.7.2.7. Right of Action of Private Groups

Article 18 of the Convention to some extent authorizes actions by private associations that according to their statutes, aim at the protection of the environment, while it restricts the scope of such actions before the courts:

1) Requests for prohibition of any unlawful dangerous activity posing a grave threat of damage to the environment
2) Requests for an order to the operator to prevent an incident or damage
3) Requests for an order after an incident to prevent damage or reinstate the environment

3.7.2.8. Causal Link and Administration of Proof

According to Article 10, while maintaining the requirement of a causal relationship between the incident giving rise to damage and the activity of the operator, the court "shall take due account of the increased danger of causing such damage inherent in the dangerous activitiy."

3.7.2.9. Financial Security Scheme

According to Article 12 of the Convention, each Member State shall ensure that where appropriate, taking due account of the risks of the activity, operators conducting a dangerous activity on its territory be required to participate in a financial security scheme or to have and maintain a financial guarantee up to a certain limit, of such type and terms as specified by internal law to cover the liability under this Convention.

The article leaves a lot of freedom for national legislators, because it provides for a financial security scheme for operators covered by the Convention where appropriate, of the type and to such a limit as wanted, respectively. This solution is to be welcomed as leaving Member States free in a matter that does not require strictly prescribed international rules.

3.7.2.10. Access to Information

A system of access to information held by authorities or operators is also contained in the Convention. The same system explicitly recognizes the right to protect the confidentiality of data. The system is based broadly on the provisions of Community law governing the subject.

3.8. The 1993 European Commission Green Paper on Remedying Environmental Damage

In addition to international organizations, new civil liability regimes have been enacted by various EC Member States. In its Green Paper on Remedying Environmental Damage of 1993, the European Commission assumed that different approaches adopted might lead to distortion of competition and hamper the smooth operation of the single market. For this reason, the Commission suggested that a Community system of civil liability in relation to activities dangerous to the environment was necessary. Civil liability is also considered to belong to the domain of the Community legislation, since it presents many interfaces with EC environmental policy.

The purpose of the Green Paper, adopted in March 1993, is to inspire discussion on whether and how requirements to remedy environmental damage might be introduced appropriately and effectively within the Community to recover the costs of such restoration [47]. As its name envisages, the communication is a discussion document that does not formally propose a certain regime on the matter. Instead, it presents alternatives for the legislator and summarizes different arguments for and against different solutions. Nonetheless, the document outlines the main principles as follows:

1) The Commission aims at harmonization of civil liability systems for environmental damage in cases when a liable party can be identified, and it expresses a clear preference for strict (objective) liability, e.g., as defined in the Council of Europe Convention
2) Civil liability is not seen as a means for compensating damage when no liable party can be found; thus, the Green Paper looks for another kind of compensation mechanism, such as a collective compensation fund

Therefore, the Commission aims at a strict liability regime combined with a nonliability-related system (i.e., the cost of repair could be shared among several economic sectors).

The Green Paper suggests that such a joint compensation mechanism is needed in the event of the following:

1) Damage is unbounded or latent
2) Cumulative acts or incidents have occurred
3) Liable parties cannot be identified
4) No basis exists for liability
5) No causal link is determinable
6) No party has a legal interest to bring action

Questions to be answered are on whom the burden to contribute to such funds should rest and whether such funds are necessary in addition to existing mechanisms. The Green Paper does not provide the answer but asks whether the polluter-pays principle should be respected to the greatest possible degree so that contributions are requested from "the economic sectors most closely linked to the type of damage needing restoration." However, a number of difficulties exist in interpretation of the principle and in identification of those sectors. For example, should the "sector" that is today most closely linked to the damage pay for damage caused by operators in the past? In addition, the identification of sectors that are most closely linked to the damage raises serious principle and technical objections. It can be argued that postpollution, unrelated to faulty or illegal pollution, is the result of a given stage of development of society. Therefore, collective and public law answers and compensation are more suitable for such "historic" pollution.

Apart from the U.S. Superfund [Comprehensive Environmental Response Compensation Liability Act 1980 (CERCLA)] and the funds established by the oil industry [Brussels Convention on the Establishment of an International FUND for Compensation for Oil Pollution Damage (1971) and parallel financial structures, such as TOVALOP (Tankers Owners Voluntary Agreement Concerning Liability for Oil Pollution), CHRISTAL (Contract Regarding Interim Supplement to Tanker Liability for Oil Pollution), and OPOL (The Offshore Pollution Liability Agreement)], such compensation schemes are still unusual and of minor significance. In the Netherlands, a fund on damage from air pollution was created in 1972. In France, a fund for noise, which compensates persons living around Paris airports, was created in 1973. The U.S. Superfund, as is widely agreed, has proved inefficient and unfair and therefore should not serve as a model. Solutions for the oil industry, in turn, serve only for purposes characteristic of that industry (i.e., those resulting from major transport accidents).

The Green Paper considers a number of issues related to civil liability and aims at stimulating discussions among all interested parties. Therefore, it has a major role in opening discussions on the fundamental principles of Community environmental policy and also on the Community's role and need for harmonization in this area. The Council of Europe Convention recently adopted will, without doubt, have a great influence on the discussions since although it is binding on only a few states so far, it is the first existing international legal instrument covering civil liability in the context of dangerous activities for environmental damage as well as damage to persons and property.

The communication does not suggest any definition of environmental damage but merely refers to definitions contained in the Council of Europe Convention and in the amended proposal for a Council Directive for damage caused by waste. However, the fundamental importance of and difficulties related to the definition are emphasized.

3.9. Economic Aspects and Questions Bearing on the Insurance of the Industrial Risk

In the last few years, industry has encountered growing difficulties in renewing its "civil liability operation" insurance policies, in terms of both the cost of premiums and the extent of risks covered. Insurance companies, as well as the reinsurance market, have been very hesitant, chiefly for the following reasons:

1) The impact of the U.S. liability crisis, characterized by the number and magnitude of claims for damages
2) Fear of a similar situation developing in Europe, particularly in view of the introduction of new liability laws and increasingly stringent technical Regulations
3) The potential liability of businesses involved in actions understood to be based, among other things, on damage to the environment whose cause is uncertain
4) Anticipated development at the national and international levels of new laws that would deal specifically with liability for pollution

This situation has prompted industry to study new additional methods of risk financing. In the United States, chemical and insurance companies alike have created insurance pools to increase available capacity.

In Europe, industry has found it increasingly difficult to find appropriate coverage, especially for "gradual pollution." Studies are therefore being undertaken to assess the "feasibility" of and arrangements for setting up a European Industrial Insurance Pool designed to provide added capacity and better conditions. Coverage of major industrial accident risks and of the risk of environmental pollution is reportedly being envisaged in particular. However, the actual adoption and operation of a pool depend largely on the development of the traditional insurance market. In many cases, conditions surrounding the renewal of policies can be said to have improved in recent years. One reason for this favorable trend could be that the envisaged creation of an insurance pool under direct industry control has served to make the insurance market more buoyant. The concept of a pool has rightly been perceived as mirroring the grave concern of industry and as offering an alternative to insurance mechanisms that no longer appear to be meeting industry's demand adequately.

The insurance crisis has not disappeared, however. Its acuteness has merely been mitigated a little. Major incidents or legislative initiatives unreasonably extending the scope of industrial civil liability could adversely affect the world insurance and reinsurance market, especially in the area of pollution.

3.10. Conclusions

Seeking harmonization of civil liability laws for all European Community member countries was the basis of the "product liability" Directive. Harmonization, or at least approximation of laws regarding liability for dangerous activities, was offered as a working hypothesis in the Green Paper published by the European Commission in 1993.

The issue could, however, be addressed in an even wider context, given that other economic or geographic areas might be affected by the transfrontier effects of pollution, as illustrated by major industrial accidents outside EC territory. With regard to the European Community, a harmonization of national laws is to be supported, if it eliminates trade distortions and simplifies existing provisions. A harmonization of this kind could, moreover, be achieved in a broader framework than that of the Community. However, real harmonization – a simplification of laws and a reduction of potential conflicts – is a most difficult and challenging task for lawmakers. No genuine harmonization can be achieved in this area without parallel progress in harmonization of the procedural rules governing civil liability. Therefore, this is the beginning, not the end, of a complex process.

The U.S. liability and insurance crisis has demonstrated that the dilution of basic legal requirements related, for example, to causation and evidence can lead to complete economic and legal failure. The question of need, justification, and possible creation of nonliability-related compensation mechanisms will have to be addressed carefully in the coming years, especially in relation to compensation for "historic" gradual nonattributable pollution damage and old dumping sites. Public authorities, the chemical industry, the insurance world, and the public at large have a joint interest in finding consensus-based solutions in a most important societal issue as the 20th century draws to a close.

4. Waste Shipments in the European Union: The Interface between Environmental Regulation and the Internal Market

4.1. Introduction

The last few years have seen an increase in the interaction between international trade policies and environmental protection. The first international convention on this subject was the 1933 Convention Relative to the Preservation of Fauna and Flora in Their Natural State. However, until the late 1980s, environmental regulators were developing their policies in large part separately from economic or trade considerations. This situation is changing rapidly, and today almost every international, intergovernmental organization has a program on trade and the environment. Moreover, environmental considerations are addressed as integral components of trade policy decisions, and trade or economic implications of environmental regulations are, conversely, being seen as key components of environmental regulatory decisions as well.

What is responsible for this dramatic change? First has been an increased recognition of the need to focus on transnational pollution issues such as the question of carbon dioxide emissions, chlorofluorocarbons (CFCs), and industrial accidents, for example, the Sandoz Rhine river pollution case. These transnational issues are increasingly addressed by multilateral environmental agreements, which also contain trade provisions. One example is the Montreal Protocol on CFCs of the United Nations Environment Programme (UNEP, 1987). Another example is the Basel Convention on Control of Transboundary Movements of Hazardous Wastes (UNEP, 1989).

In addition, focus has increased on the impact on competitiveness caused by some countries' having strict environmental standards or by attempts by such countries to erect barriers against imports produced in countries having less stringent environmental standards. This has led international, intergovernmental organizations, such as the General Agreement on Tariffs and Trade (GATT), to address the danger of countries taking unilateral actions under the guise of protection of environment, health, or safety, with the effect (and perhaps even the intention) of discriminating against imports. Included here would be such unilateral actions as discriminatory product standards, testing/certification requirements, and environmental subsidies.

A series of GATT disputes have been related directly to environmental regulation brought under the GATT nondiscrimination and national treatment provisions. In addition, the GATT, now the World Trade Organization (WTO), has instituted a Trade and Environment Program to facilitate the enforcement of "free" trade principles while simultaneously encouraging sustainable development.

Related subjects involve the effect of environmental policies of the developed or "Northern" countries on developing or "Southern" countries. Complicated issues of market access, technology transfer, commodity pricing of natural resources, debt levels, paternalism, and protectionism must be addressed here.

Regrettably, writings on the subject of free trade and environmental regulation tend to represent one polarized view or the other, either environmental protection or free trade. On the *free trade side*, the position is that environmental objectives do not justify violating the multilateral trading system rules of national and nondiscriminatory treatment. Moreover, the free traders believe that the sovereignty of countries in setting their own environmental priorities and requirements should be respected; that environmental policy should not impede the free flow of goods, services, persons, and capital; and that international environmental issues, while perhaps a component of trade policy, should be subservient to trade policy in the event of a dispute.

In contrast, *environmentalists* generally believe that economic expansion leads to environmental degradation; that harmonized standards reduce environmental protection; that unilateral trade-restrictive actions are valid or even desirable to promote domestic or global environmental objectives; and that environmental subsidies and other incentives or disincentives should be encouraged. The environmentalist perspective is not concerned with what it perceives as "outdated" issues of nondiscriminatory trade. Protection of the environment justifies almost any measures necessary.

4.2. European Community/Union Measures

In the European Community/Union, the interrelationship between trade and environmental regulation is best seen in the development of the Internal Market.

4.2.1. The Internal Market, "1992"

The Treaty of Rome was enacted in 1957 to establish the European Economic Community and to guarantee, among members of the Community, four basic freedoms: the free movement of goods, persons, services, and capital between Member States.

The concept of 1992, "the Internal Market," was one of the best public relations coups in recent political history. Under the Treaty of Rome and other early documents of the European Economic Community, a single market and the elimination of physical, technical, and fiscal barriers (explanation see below) between Member States of the Community had already been anticipated. However, discouraged by the slow progress toward such a single market, the Community developed the 1992 program to accelerate its progress.

The first major document in the 1992 process was the White Paper on Completing the Internal Market [49]. It was published in 1985 and included a variety of proposals for legislation and timetables in which the 300-plus measures necessary to eliminate barriers to the free movement of persons, goods, services, and capital in the Community had to be enacted by the end of 1992. The barriers the Community was trying to eliminate among its Member States were physical, technical, and fiscal ones.

Physical barriers are such things as customs services, passports, and transport formalities. *Technical barriers* include health and safety requirements, standards, and product packaging. *Fiscal barriers* are among the most difficult to eliminate. Included here are harmonization of general tax rates and value-added tax (VAT) rates. Of the 300-plus measures necessary to implement the Internal Market, between one-third and one-half were environmental, health, and safety initiatives.

4.2.2. The Single European Act (see Section 2.1)

The Single European Act took effect on July 1, 1987 (Single European Act, signed at Luxembourg on February 17, 1986, and at The Hague on February 28, 1986). It contains numerous provisions amending the Treaty of Rome, many of which were necessary for the 1992 process to be successful. The most significant provisions affecting trade and waste shipments are the following:

1) Inclusion of a new Article 100 A in the Treaty of Rome providing that harmonization measures can generally be adopted by the Council by a qualified majority vote, rather than by unanimity as before
2) Inclusion for the first time in the Treaty of Rome of provisions specifically relating to environmental protection
3) Revision of the consultation procedure with the Parliament, giving Parliament a stronger role in the policy-making process

New Article 100 A of the Treaty of Rome (Article added by Article 18 of the Single European Act) provides that measures can be adopted by the Council by qualified majority, if their objective is the approximation of the provisions of national laws in order to further the establishment and functioning of the Internal Market, i.e., harmonization measures (see Section 2.1).

In the past, Council decisions had to be reached on a unanimous basis. As result, many Council actions, especially controversial ones, were deadlocked because one or another Member State could not agree with the rest. To facilitate action on measures necessary to implement the Internal Market, a new Article 100 A was included to permit the Council to act on Internal Market measures by qualified majority vote. (The votes of the Council are weighted. The total number of votes among the 12 Member States is 76; a qualified majority consists of 54 votes.) Exemptions from qualified majority voting exist here for measures related to fiscal provisions; worker rights issues; as well as health, safety, environmental, or consumer protection issues. Here, unanimity still had to prevail under the Single European Act.

Enactment of Community legislation by unanimous Council votes is usually more complicated and time consuming than reaching qualified majority decisions. One of the early steps in the EC legislative process under the Single European Act, therefore, always involves "jockeying for position" between the European Commission and Member States to determine the legislative basis of the measure in question. A dispute on the legal basis of a piece of Community legislation concerning titanium dioxide went as far as the European Court of Justice [50]. In this case, the Court of Justice ruled on the legal basis of Community legislation that was aimed both at the achievement of the Internal Market and at environmental protection. The Court upheld the legal basis of the legislation as an Internal Market, rather than environmental, measure and annulled the Directive. The Court ruled that the choice of legal basis should be governed by the aim and content of the legislation in question.

As stated above, the Single European Act also specifically provides that protection of the environment is a valid area of Community legislation. In practice, the Community began legislating in the environmental area in the late 1960s. However, the original Treaty of Rome is primarily a trade agreement and does not specifically provide authority for such legislation. Under the

Single European Act, the Community must now consider environmental implications in adopting its other legislation, as well [51]:

"Action by the Community relating to the environment shall be based on the principles that preventive action should be taken, that environmental damage should as a priority be rectified at source, and that the polluter should pay. Environmental protection requirements shall be a component of the Community's other policies."

The Member States were rather suspicious of this amended Treaty language, since environmental, health, and safety regulation had traditionally been within their own purviews. Therefore, the Treaty language now also includes two other provisions: (1) the issue of subsidiarity (see Section 2.1), and (2) unanimous voting on environmental matters. "Subsidiarity," a term that has received much attention in the context of the Maastricht Treaty, in this environmental context, means merely that environmental legislation or enforcement should be undertaken at the lowest feasible regulatory level.

New Article 130 T of the Treaty of Rome [51] specifies that Member States will not be prevented from adopting more stringent protective environmental measures than those enacted at the Community level, provided these measures are compatible with the Treaty of Rome (see Section 2.2). This is very useful for Member States, such as Denmark, the Netherlands, and Germany, that periodically adopt more stringent legislation. At the Community level, there has also been a trend to "harmonize up" (i.e., to harmonize toward more stringent environmental regulation). Invariably, the Community has usually taken existing legislation from one of the more stringently regulated Member States as a model for its own Community-wide legislation.

A number of procedural changes also occurred as a result of the Single European Act, the majority of which increased the power of the European Parliament. The general procedure of parliamentary advisory review for Community actions is one of the Commission proposing legislation and the Council acting on it. Prior to such Council action, Parliament and ECOSOC review the proposal and issue advisory opinions, which the Commission and/or Council can accept or reject.

The Single European Act introduced another procedural step in certain situations, the so-called Cooperation Procedure. Under this procedure, two readings in the Parliament and two Council reviews (under new Article 149, as replaced by Article 7 of the Single European Act) take place.

4.2.3. The Maastricht Treaty – Environmental Provisions [52]

Mention should also be made of the environmental provisions of the Maastricht Treaty. The Maastricht Treaty was agreed to by the Heads of State and Government of the European Community Member States in December 1991, to revise the Treaty of Rome to provide for such changes as gradual political integration, a common currency, and a Single European Bank.

The Maastricht Treaty is notable for its further integration of environmental considerations into all other policies of the European Community; the text goes even further than that of the Single European Act. (Since the entry into force of the Maastricht Treaty changed the name "European Community" to "European Union," henceforth, reference is to either the European Union or the EU, even though the measures described may have been initiated in EC times.) It also contains language on the interrelationship between transportation, energy, and agricultural policies and environmental protection. Moreover, the Treaty deals with the need to consider trade or competition and economic development issues within the framework of environmental considerations. In fact, protection of the environment is explicitly contained in the introductory section of the Maastricht Treaty (i.e., impact on the environment should be considered in the development of all other EU policies and the EU has competence to regulate in this area). The concept of "sustainable development" is also found in this Treaty.

The Treaty also eliminates the need for unanimous votes for environmental measures and adds a new "Cooperation Procedure" which expands the role of the European Parliament. In addition, the question of the meaning of "subsidiarity" is supposedly clarified. However, confusion about the exact definition of this term, and when it is to be applied, is still present under this new Treaty.

Finally, the Maastricht Treaty provides for the creation of a "Cohesion Fund" to assist the least financially sound of the European Member States, Ireland, Spain, Portugal, and Greece (including to help them improve environmental protection and enforcement measures).

4.2.4. Other Applicable Provisions of the Treaty of Rome

Notification and Standstill Procedure on Technical Standards and Regulations. Significant European Court of Justice case law exists on "mutual recognition"; i.e., goods lawfully produced and marketed in one EU Member State must be accepted in all other Member States (except where refusal is justified by Articles 30 and 36). Directive 83/189/EEC [53] specifies notification procedures for technical standards and regulations in the EU. The procedure was developed to discourage national, unilateral actions in this area. Member States must inform the Commission on each draft technical standard or regulation. The Commission then informs the other Member States. The Commission and the other Member States have the option of consenting to the draft technical standard or regulation text. The acting Member States must take their positions into account. Waiting periods for adoption of the measure(s) are also specified.

Article 30. Article 30 of the Treaty of Rome provides that "quantitative restrictions on imports and all measures having equivalent effect shall be prohibited." Quantitative restrictions have never constituted a serious obstacle to intra-EU trade. Therefore, the emphasis in practice has been on measures having an "equivalent effect." All legislative rules and administrative provisions, as well as administrative practices that form a barrier to importation or exportation, are governed. Included here are provisions and practices that render importation or exportation more expensive or difficult compared to sales of home products on the domestic market. The *Cassis de Dijon* decision [54] established the basic principles to be applied in judging the legitimacy of measures adopted by EU Member States that hinder the free movement of goods among them. In the absence of harmonized EU law, Member States are free to establish rules affecting the free movement of goods.

However, goods lawfully produced or marketed in one Member State must be accepted in other Member States, except where refusal is based on the "rule of reason" or on exceptions provided in Article 36. The rule-of-reason exceptions are applicable only to noneconomic matters, i.e., (1) environmental protection, (2) consumer protection, (3) prevention of unfair commercial practices or (4) effectiveness of fiscal supervision – provided, however, that the national measure in question is not unreasonable, disproportionate, or protectionist. Exceptions provided in Article 36 include national restraints on trade, if necessary to protect (1) public morality, public policy, or public security; (2) health and life of persons, animals, or plants; (3) national treasures of artistic, historic, or archaeological value; and (4) industrial or commercial property.

However, not all trade restrictions based on the "rule of reason" or on Article 36 exemptions are acceptable. A restriction may not, for example, constitute a means of arbitrary discrimination or disguised restriction on trade between Member States. These concepts are referred to as the principles of "nondiscrimination" and "proportionality;" the restriction must apply equally to nationally produced goods and goods imported from other Member States and must be necessary to achieve legitimate, proportional objectives.

For environmental cases, there should also be a balance between environmental and free trade purposes. A regulatory measure must be necessary to achieve a stated goal, and no less restrictive measures can exist to achieve this goal. A case in point is the Danish Bottles case [55], which demonstrates the complexities of justifying the compatibility of environmental measures with rules on the free movement of goods. In question was a Danish law providing that mineral water, soft drinks, and beer could be sold only in refillable containers. A deposit and return system to ensure refilling of a large portion of containers was also included. The result of this law was that beverages lawfully sold in other Member States could not be sold in the same containers in Denmark, a clear violation of Article 30 of the Treaty of Rome.

The European Commission took Denmark to the European Court of Justice to contest this legislation. The Court found that the Danish system of mandatory deposit on containers for mineral water, soft drinks, and beer, although placing a greater burden on non-Danish producers, did not infringe Article 30. The Court held that environmental protection is an objective that may justify restraints on trade. The Danish deposit system appeared necessary to attain the desired objective, and no other less restrictive means existed.

A more recent example of one State's unilateral trade action, ostensibly to protect the environment, is the German Packaging Waste Regulation [56]. Under it, a complicated take-back and recycling system for packaging and packaging waste (set up and funded by industry because of the constraints of the Regulation) imposes, in practice, significantly disproportionate problems on importers into Germany than on the domestic industry.

A number of complaints have been filed with the European Commission alleging Germany's violation of Article 30 of the Treaty of Rome. One benefit of the German Regulation is that it has resulted in the EC's accelerating the development of its own Directive on Packaging and Packaging Waste [57] to harmonize all EU Member State rules on this subject. Germany, however, has steadfastly opposed the text of this EC draft Directive and refused to amend its own Regulation. As a result, the European Commission decided, on July 6, 1994, to initiate legal proceedings against Germany because its unilateral packaging Regulation creates an unjustified barrier to trade between EU Member States. In all likelihood, this is merely a political ploy to get Germany to drop its opposition to the EC's draft Packaging and Packaging Waste Directive and to bring the German regulatory requirements for waste packaging into line with the EC Directive's requirements (i.e., reduce the German recycling and reuse targets and build up domestic German recycling capacity more rapidly).

4.3. Waste Shipments: Conflict between European Union Free Movement of Goods Principles and Environmental Regulation

4.3.1. The New Regulation 259/93: Introduction

Recent regulatory initiatives on waste shipments provide perhaps the most dramatic example of the conflict between the European Union's principles of free movement of goods and environmental regulation. On May 6, 1994, the European Union Regulation "On the Supervision and Control of Shipments of Waste within, into and out of the European Community" [58] became effective (hereinafter Regulation 259/93). Regulation 259/93 imposes a complex system of shipment notification and consent on all shipments of waste, not just hazardous waste, that cross borders in the European Union, as well as all waste imports into and waste exports from the EU. Although the notification and consent procedures vary depending on the type of shipment, its destination, and its purpose; waste shipments for recovery or recycling operations are covered, as is waste for disposal.

Regulation 259/93 must be viewed in the context of other key EC waste legislation, most notably, Directive 75/442 EEC on Waste [59], as amended by Directive 91/156/EEC [60]; and Directive 91/689/EEC on Hazardous Waste [61], replacing Directive 78/319 on Dangerous Waste [62]. For these pieces of legislation, see → Waste, Chap. 2. International commitments of the EU on this subject (e.g., Basel Convention) are also interrelated.

On the specific topic of transfrontier movements of waste, the EEC had Regulations as early as 1984, with Council Directive 84/631/EEC, as amended [63], "On the Supervision and Control within the European Community of the Transfrontier Shipments of Hazardous Waste." Directive 84/631 required Member States to take measures to ensure the supervision and control of transfrontier shipments of hazardous waste within, into, and out of the Community to protect human health and the environment.

Under Directive 84/631, all cross-border movements of hazardous waste were subject to a notification requirement to be effected by a uniform Consignment Note and typically made to the Competent Authority of the Member State of destination. Procedures for acknowledgment of and objections to the shipment, if any, were provided. However, given the narrow definition of "hazardous waste," Directive 84/631 applied to only a limited number of transfrontier shipments of waste. According to 1992 figures of the Organization for Economic Cooperation and Development, ca. 200×10^6 t of waste cross OECD borders each year (140×10^6 t cross EU borders). Of these 200×10^6 t, < 1 % was estimated by the OECD to involve so- called hazardous waste.

In addition, since this 1984 legislation was a Directive, EC Member States tended to implement it inconsistently. To fill this regulatory gap, address implementation inconsistencies, and adapt the rules regulating transfrontier movements of hazardous wastes to the requirements of the March 22, 1989 Basel Convention on Transboundary Movements of Hazardous Waste and Their Disposal and the Lomé Convention of December 15, 1989 (the Fourth ACP – EEC Convention), the European Council adopted the Regulation "On the Supervision and Control of Shipments of Waste Within, Into and Out of the Community" (Regulation 259/93). Since this legislation is in the form of a Regulation, as opposed to a Directive, it is directly applicable and binding on EU Member States.

Regulation 259/93 was also drafted to ensure consistency with the OECD, March 1992 Council Decision Concerning the Control of Transfrontier Movements of Waste Destined for Recovery Operations, specifically its "Green," "Amber," and "Red" Lists of Waste. These lists are categorized according to hazard ("Green" being least hazardous, with "Red" most hazardous) and appear at Annexes II through IV of Regulation 259/93. Regulation 259/93 was also deemed necessary to ensure continued and improved environmental protection in EU waste management, in view of completion of the Internal Market and its resultant removal of border controls.

Finally, the issues of waste management in general in the EU, and waste shipments in particular, are highly emotional and political ones. Regulation 259/93 is at least partially the result of successful political pressure by environmental groups, some Member State regulatory authorities, and "the public" for stricter and harmonized European Union legislation on this subject.

The overriding purpose of the Regulation is, of course, to ensure prior notification and consent for waste shipments to Competent Authorities, Member States, enabling them to be informed of the particular type, movement, and disposal or recovery of waste, so that they can take all necessary measures for the protection of health and the environment, including the possibility of raising reasoned objections to the shipments (e.g., de facto "PIC" or prior informed consent).

Regulation 259/93 provides for a Consignment Note notification system for the covered waste shipments to the Competent Authorities of destination, dispatch, and transit and to the consignee (see Sections 4.3.2.1 and 4.3.2.2). The content of the notification and the subsequent regulatory avenues available to the Competent Authorities to "control" these shipments depend on the nature and composition of the waste (i.e., its hazard); the shipment's purpose (disposal or recovery); and its destination (to or through other countries in the EU, to non-EU OECD Decision signatories, to other parties to the Basel Convention, to countries having equivalent bilateral or multilateral agreements with EU Member States – the possibilities are numerous).

The bases on which Member States may object to a waste shipment for disposal include the principles of proximity (waste should be disposed of near the point of generation – a policy encouraging local waste disposal); self-sufficiency (to ensure that each Member State can dispose of its waste in its own territory); and priority for recovery. Differing positions have been taken by commentators as to whether the first two of these principles can or should be extended to waste shipments for recovery under the Regulation. Political pressures by local authorities, NIMBY ("not-in-my-backyard")-ism by the neighboring public if such an interpretation is upheld, and resulting negative competitive impacts on industry in the EU can be anticipated if this interpretation is sustained.

4.3.2. Specific Provisions of Regulation 259/93

Exclusions from the scope of Regulation 259/93 are:

1) Offloading to shore of waste generated by normal operation of ships and offshore platforms, including wastewater and residues, provided that such waste is subject to a specific binding international instrument
2) Shipments of civil aviation waste
3) Shipments of radioactive waste
4) Shipments of waste already covered by other relevant legislation
5) Shipments of waste into the EU subject to the environmental protocol of the Antarctic Treaty
6) Shipments of waste contained on Annex II (the OECD Green List of Waste) destined only for recovery, subject to certain conditions

The main text of the Regulation is divided into sections, depending on whether the shipments are between EU Member States or involve imports or exports and, in each section, whether disposal or recovery is involved.

4.3.2.1. Shipments of Waste between EU Member States

Shipments of Waste for Disposal. If a notifier intends to ship waste for disposal from one EU Member State to another or pass it in transit through one or several other Member States, the Competent Authority of destination must be notified, and a copy of the notification must be sent to the Competent Authority of dispatch and of transit and to the consignee. Notification must also cover any intermediate stage of the shipment from the place of dispatch to its final destination. The notification is to be effected by means of a Consignment Note, which is to be issued by the Competent Authority of dispatch. In making the notification, the notifier must complete the Consignment Note and must supply any additional information and/or documentation requested by the Competent Authorities.

Moreover, the notifier must enter into a contractual agreement with the consignee for disposal of the waste. Specific provisions are included as acknowledgments of the Consignment Note and authorization of the shipments. The shipment may be effected only after the notifier has received authorization from the Competent Authority of destination.

To implement the principles of (1) proximity, (2) priority for recovery, and (3) self-sufficiency at the EU and national levels, Member States may take appropriate measures generally or partially to prohibit shipments of waste for disposal.

Shipments of Waste for Recovery. The notification requirements of Regulation 259/93 with respect to shipments of waste between Member States for recovery are generally consistent with the requirements described above for shipments between Member States for purposes of disposal, with certain key deviations. First, the content of the Consignment Note differs. Moreover, tacit consent for a shipment is deemed to exist if 30 days have passed following notification, and no objection has been lodged.

In addition, Competent Authorities with jurisdiction over specific recovery facilities may decide, notwithstanding the notification and consent requirements of the Regulation, that they will not raise objections concerning shipments of certain types of waste to specific recovery facilities. Such decisions may be limited for a specific time period and may be revoked at any time. The Commission must be notified of any such special arrangements; it will inform the Competent Authorities of the other EU Member States and the OECD. However, Member States of transit or dispatch may object.

Shipments of Waste within one Member State. The notification and consent provisions described above do not apply to shipments of waste within one Member State. Here, the individual Member State in question has the role of establishing an appropriate system for the supervision and control of shipments of waste, "coherent with" the system established by Regulation 259/93. The Member States must inform the Commission of these individual systems. However, shipments between Regions are covered if the Region has jurisdiction over waste management (see also Section 4.3.3.1).

4.3.2.2. Exports of Waste from the EU

Exports for Disposal. Exports of waste from the EU for disposal are prohibited under Regulation 259/93, except to European Free Trade Area (EFTA) countries, which are also parties to the Basel Convention. Exports of waste for disposal to an EFTA country may also be banned if the EFTA country of destination prohibits imports of such waste, if it has not given its written consent to the specific import in question, or if the Competent Authority of dispatch in the EU has reason to believe that the waste will not be managed in accordance with environmentally sound methods in the EFTA country of destination.

The notification and consent requirements with respect to exports of waste from the EU for disposal generally track those described above for intra-Union shipments for disposal, with certain deviations. For example, the notification and consent time periods are significantly expanded here. The Regulation also contains language as to obligations of EU Customs offices of departure.

Exports of Waste for Recovery. Exports from the EU of waste destined for recovery are prohibited under Regulation 259/93, except for those exports to countries to which the OECD Decision applies and other countries that are parties to the Basel Convention (or with which the EU or the EU and its Member States have concluded bilateral, multilateral, or regional agreements/arrangements compatible with the Basel Convention and EU legislation). The Regulation also provides specific requirements that such agreements or arrangements must meet.

In addition to the procedures above, exports of waste from the EU for recovery to countries to which the OECD Decision, Basel Convention, and/or bilateral or multilateral agreements or arrangements apply may also be banned if the country in question prohibits all imports of such waste, if it has not given its consent to the specific import in question, or if the European Union Competent Authority of dispatch has reason to believe that the waste will not be managed in accordance with environmentally sound methods in the country of destination.

Generally, waste specified on Annex II of Regulation 259/93 (OECD Green List of Waste) for recovery is not subject to the notification and consent requirements of the Regulation, again with certain exceptions. First, the Commission is required to inform every country to which the OECD Decision does not apply of the waste categories contained in Annex II and request written confirmation that such waste is not subject to control in the country of destination and that the latter will accept such waste to be shipped without recourse to the control procedures. Annex II waste must be exported for recovery operations within a facility that is locally authorized and operating in the importing country. Moreover, such exports will be subject to surveillance.

As to exports of Annex III and IV waste for recovery (OECD Amber and Red Lists, respectively), the notification and consent procedures discussed earlier with respect to intra-Union shipments generally apply.

Finally, the Regulation provides that all exports of waste to ACP (African Caribbean, Pacific) States will be prohibited. This prohibition does not prevent an EU Member State to which an ACP State has decided to export waste for processing from returning that processed waste to the ACP State of origin.

4.3.2.3. Imports of Waste into the EU

Imports for Disposal. Imports of waste into the EU for disposal will be prohibited, except those from EFTA countries that are Parties to the Basel Convention, other countries Party to the Basel Convention, or countries that have concluded appropriate bilateral or multilateral agreements or arrangements with the EU as a whole or with individual Member States. Such agreements or arrangements must be notified to the European Commission, which will review them.

As in the case in the intra-EU and export shipment situations discussed above, notification of imports of waste for disposal will be subject to the Consignment Note procedure. Notification must be made to the Competent Authority of destination, and the Consignment Note will be issued by that Competent Authority. Again, specific deadlines for notification, objections, and decisions are provided.

Imports for Recovery. Imports of waste into the EU for recovery are prohibited under Regulation 259/93, with the exception of imports from countries to which the OECD Decision applies and, again, countries Party to the Basel Convention or with which the EU/Member States have concluded appropriate bilateral, multilateral, or regional agreements or arrangements.

The notification and control procedures of Regulation 259/93 also apply to waste imports into the EU for recovery from countries and through countries to which the OECD Decision applies for waste listed on Annexes III and IV of the Regulation or for waste that has not yet been assigned to an Annex.

4.3.2.4. Transit of Waste from Outside and Through the EU for Disposal or Recovery Outside the European Union

Here again, the Consignment Note and control procedures apply. The notification is to be effected by means of the Consignment Note sent to the last Competent Authority of transit within the EU, with copies to the consignee, the other Competent Authorities concerned, and the Customs offices of entry into and departure from the EU.

The shipment may be admitted into the European Union only if the notifier has received the written consent of the last Competent Authority of transit, which authority will also issue the Consignment Note. Specific time deadlines for notifications and decisions are also provided. Transits of waste listed on Annexes III and IV for recovery through one or more EU Member States from a country, and transit of waste for recovery to a country to which the OECD Decision applies require notification of all Competent Authorities of transit in the Member States concerned. Such shipments may be effected only in the absence of objections.

4.3.2.5. Common Provisions

Regulation 259/93 contains certain common provisions applicable to all shipments, whether they are for disposal or recovery, or involve waste imports, intra-EU shipments, or exports from the European Union. These cover general notification possibilities, illegal traffic, financial guarantees and insurance, inability to facilitate recovering disposal and return, regulatory authority designation, documents retention, etc.

4.3.3. The Trade – Free Movement and Environmental Protection Conflicts in Complying with EC Regulation 259/93

4.3.3.1. Definitions

One of the most difficult issues to anticipate in compliance with this new EC Regulation is a definitional one. Even though much progress has been made in the definitions of "waste" and "hazardous waste" in the OECD and in recent EC Directives on the subject, clear EC definitions have still not been developed, and differing EU Member State definitions continue to exist. Therefore, regulatory inconsistencies on this subject can be expected to remain.

One case in point is the German distinction between something having commercial value (Wirtschaftsgut) and waste, and the German treatment of secondary raw materials. However, the new German Federal Comprehensive Waste Reform Law (Kreislaufwirtschaftsgesetz) intends to change German waste definitions to comply with EU definitions. This law was finally passed by the Bundesrat before its summer 1994 recess and is expected to come into effect, at the latest, by 1996.

The EC Waste and Hazardous Waste Directives attempt to define waste in their texts. The definition of waste in the 1991 Waste Directive (91/156/EEC) is as follows:

"Waste" shall mean any substance or object in the categories set out in Annex I which the holder discards or intends or is required to discard" (Article 1).

The "hazardous" waste definition under the 1991 Hazardous Waste Directive (91/689/EEC) is much more complicated and includes various references to Annexes, origin, composition, properties, and a final umbrella category of "any other waste" considered by a Member State to be hazardous, in effect negating the usefulness of the harmonized European-level definition.

In addition, both the 1991 Waste and the Hazardous Waste Directives mandate the compilation of lists or catalogues of waste under their domain. The EU has published its Waste Catalogue [64] under the 1975 and 1991 Waste Directives [65], and the so-called Waste Catalogue II or List of Hazardous Wastes under the 1991 Hazardous Waste Directive was published on December 31, 1994 [66].

The process of developing both European Commission lists has been highly controversial. Officials at the Commission have taken varying positions as to whether these should be listings of products and substances or whether they should, instead, be criteria-based. Moreover, a number of key Member States, most notably Germany and France, are still lobbying to have the new European List of Hazardous Waste be similar to those in place in their countries (i.e., a listing of substances or products that have been found to be hazardous; therefore, waste containing any of these products or substances in whatever quantities would, likewise, be categorized as hazardous).

As was the case with Waste Catalogue I, the scope and content of the EU Hazardous Waste List (Waste Catalogue II) are of major importance to a number of industries and companies operating in Europe. The content of this new List will affect such commercial questions as whether and where a material can be recycled or remediated, whether it can be shipped across borders, and what other regulatory constraints and liabilities will apply vis-à-vis manufacturers, generators, and transporters. Sales price and contractual and marketing questions of customers are intricately involved here as well.

The new List of Hazardous Waste Pursuant to Article 1 (4) of Directive 91/689/EEC on Hazardous Waste (January, 1994), while still product related, deals only with broad product categories and is keyed to the thresholds and concentrations developed for hazardous properties under EU chemicals legislation [e.g., the 0.1 % concentration for category 1 and 2 carcinogens in the 1988 Preparations Directive (88/379/EEC)]. The inclusion of these thresholds and concentrations has been the subject of some controversy.

Wallonian Waste Decree. As illustrated in the German "Wirtschaftsgut" example above, another definitional issue of concern in complying with Regulation 259/93 is whether a material is a "product" or "good," hence coming under the Treaty of Rome's free movement of goods principles, or whether it is something else entirely. Under the new EC Regulation 259/93 and such recent legal interpretations as the July 1992 ECJ Judgment on Wallonian Waste Imports [67], waste should be granted special treatment for environmental protection purposes, tantamount to at least a partial exemption from the European Union's free movement of goods and Internal Market principles. This case was particularly interesting in that the Wallonian Decree at issue not only impeded the flow of waste across EU Member State borders, but also impeded intra-Belgian trade on a regional basis. While the ECJ held that the challenged Wallonian Decree did not conform to the 1984 Hazardous Waste Movements Directive (the precursor to Regulation 259/93), it found that the Decree's banning of the disposal of wastes in Wallonia that were generated outside Wallonia was not "discriminatory" within the meaning of the Treaty of Rome. This decision, was based on the "particularité" of wastes, in that Article 130 S of the Treaty of Rome requires that, as a priority, "environmental damage" should be rectified at the source. The Court also referred to the Basel Convention's principle of waste self-sufficiency.

The ECJ appears to have erroneously interpreted the Treaty of Rome provisions here. The Wallonian Decree clearly violates not only the 1984 Hazardous Waste Shipments Directive, but also the free movement of goods provisions of the Treaty of Rome. Relying on Article 130 provisions is also questionable since a Directive that deals with shipments of materials and border transfers within the EU should more appropriately be considered an Internal Market measure and, therefore, judged under Article 100 A of the Treaty of Rome, not Article 130 S.

In addition, because the Wallonian Judgment did not address all of the issues on this subject raised by Regulation 259/93, a definitive legal interpretation is needed on the free trade versus environmental protection argument for waste shipments in the EU.

Article 130 S vs. Article 100 A. The debate continues – the ECJ has just rejected the European Parliament's request to annul Regulation 259/93 because it was improperly based on Article 130 S (environment) of the Treaty of Rome. The Parliament said that this Regulation was essentially an Internal Market (trade) measure, hence based on Article 100 A of the Treaty of Rome and subject not only to the Treaty's free movement of goods provisions, but also to Article 113 of the Treaty dealing with external trade (It is not just happenstance that an Article 100 A determination for the new Regulation would give the Parliament greater powers as to the form and content of the Regulation.) However, the ECJ rejected the Parliament's arguments and upheld the Article 130 S basis of the Regulation. Waste, it seems, is indeed special and, at least for now, will not be called a "good" in the EU.

The European Court of First Instance has also just rejected as inadmissible a request by two French waste management companies and one Luxembourg one that it annul Article 4 of Regulation 259/93, which permits individual Member States to ban imports of waste, even nonhazardous waste, from other EU countries. (At issue was a French Decree banning imports into France of household waste for disposal in landfill.) The Court of First Instance did not address the substance of the plaintiffs' complaint. Instead, the case was dismissed on the procedural ground of lack of "standing" of the plaintiffs, since they were not "directly concerned." In this ruling, the President of the Court said that Article 4 of the 1993 Council Regulation was not aimed at companies, "except in their objective quality as economic operators in the waste management and transport sector, the same as for any other operators in an identical situation." The Court's President also rejected the request for damages and interest. Case T-475/93, SA Buralux, SA Satrod, SA Ourry vs. Council of the European Union, judgment published in O.J.E.C. C188, July 9, 1994.

Disposal and Recovery. Another definitional concern with Regulation 259/93 is seen in the question of how to define "disposal" and "recovery." Member States of the European Union have, to date, taken inconsistent positions as to whether operations to be carried out are either of the two. This is obviously a critical point for purposes of defining the type of notification and the content of information to include in the Consignment Note under Regulation 259/93. Defining an operation as "recovery" or "recycling/further use," as opposed to "disposal," will also have a critical effect on how the proximity principle would apply. This issue is particularly difficult for operations involving energy generation and incineration.

Perhaps the most difficult practical problem resulting because of these vague definitions is in the area of waste for recovery. The impact on byproducts, coproducts, secondary raw materials, and product or container take-back programs must be clarified.

The practical, commercial difficulties posed by all of the above definitional issues for EU companies in attempts to adhere to Regulation 259/93 are discussed in Sections [**?**]. These problems are magnified even more in the international arena.

In addition to the dilemmas caused by inconsistencies between the EU-level definitions and those in some EU Member States, OECD and Basel Convention strictures, obviously, also have to be adhered to. It is already evident that definitional divergencies between these two organizations and the EU may cause problems in the future.

In addition, many affected companies are multinational enterprises, with manufacturing, waste-generating plants in and outside the EU. Company attempts to impose one comprehensive waste management scheme throughout all of such companies' international activities will be stymied. A case in point is seen in the differences in the U.S. and EU definitions of "waste," "hazardous waste," "disposal," and "recovery," and the shipment restrictions imposed. For example, unlike the ECJ, the U.S. Supreme Court has determined that waste is an article of commerce (i.e., a "good") meriting constitutional protection against discriminatory state and local Regulations that impede its free movement and that nothing special or distinctive exists in waste that should permit discriminatory restriction of its flow across state lines to apply in the United States.

A second example is seen in the fact that the United States attempts to define waste in a way that encourages beneficial recycling (exemption of certain kinds of materials destined for certain recycling). The EU definition follows a less desirable, inconsistent approach: It covers all secondary materials as "waste" as well, regardless of whether disposed or recycled, but then differs in the regulatory treatment of conventional waste and recyclable waste.

Some parties, frustrated by these definitional differences and the length and complexity of the debate believe that "waste prevention" should be the focus in the future, instead of waste shipment restrictions, especially when the words "waste" and "hazardous waste" cannot even be defined properly.

4.3.3.2. Liabilities

An attempt must also be made to address liability issues under Regulation 259/93 at the EU level to ensure some degree of consistency in determining liable parties and in enforcement. From past experience, this issue cannot be left solely to the Member States. Adoption of EU legislation on Civil Liability for Damage Caused by Waste is a first step (see Section 3.4), but continued Member State inconsistencies on this subject will further confuse companies in compliance.

4.3.3.3. Administrative Issues

European industry, while favoring harmonized regulation of waste shipments throughout the EU, has raised some of the issues noted in the previous sections. However, it appears to be focusing currently on the failure of the EU and its Member States to provide the administrative structures and procedures to allow industry to comply with Regulation 259/93 by its effective date of May 6, 1994, or shortly thereafter.

Among the most active EU-level industry associations on this subject have been UNICE, Eurométaux (which has particular concerns about disruption of shipments of scrap metals as secondary raw materials), the European Chemical Industry Council (CEFIC), and the European Federation of Waste Management (FEAD). According to a UNICE discussion paper on the implementation of Regulation 259/93,

"... the adminstrative backup necessary for the proper implementation of the Regulation is widely lacking in most of the Member States. It is both paradoxical and disastrous for a Community Regulation, directly applicable according to Community law to be paralyzed in this way ...."

The concern about the lack of administrative systems in the Member States to implement the Regulation and the enormous burden to be placed on Competent Authorities in the Member States in homogeneous or integrated implementation is well justified. As noted in Sections [**?**], the notification and approval system under Regulation 259/93 is very complex and inconsistent, depending on the types of shipments, their purposes and destinations, and quite bureaucratic and paper-intensive. It is questionablewhether the Competent Authorities will be able to deal with the large number of notifications to be expected in the time anticipated by the Regulation. Given the large degree of discretion still left to the Member States in implementation, it is also questionable whether a truly harmonized system will evolve. Therefore, it cannot be expected that a significant degree of increased environmental protection will ensue as a result of Regulation 259/93. (Of course, on the other hand, trade, will be severely hampered.)

Moreover, given the fact that Regions having responsibility for waste management will also be in a position of appointing Competent Authorities, an additional administrative question is raised. It is rare that the proper Regional infrastructure exists in each of the Regions throughout the European Union which may appoint such Competent Authorities.

If this administrative situation is not solved quickly, it could lead at worst to failure of the entire system or, at a minimum, to the need to store a large number of wastes prior to disposal or recovery, pending clearance. Such storage may result in more environmental, health, and safety problems than the new requirements in the Regulation will solve.

Other administrative concerns include the fact that on the eve of implementation of the Regulation, the standard Consignment Note to be drafted by the Commission was not yet available; the list of competent authorities to be designated by Member States was also not available (indeed, most of the Member States had not even submitted their Competent Authority designations to the Commission); the list of customs offices called for had not yet been published; and the list of wastes on the OECD Green List that will be subject to specific controls in certain Member States was not available, nor was information on which third countries have bilateral agreements with which Member States or the EU authorizing such trade.

Thus, according to one industry observer, "chaos" will result as companies attempt to comply with the terms of the Regulation, with clearly inadequate and inconsistent administrative structures and procedures in place.

The Commission Staff, while admitting certain transitional administrative problems, does not appear to believe that all of these concerns are fully justified, and international environmental organizations believe that the administrative and compliance concerns of industry are excessive. Reports that illegal traffic in toxic waste, particularly in a West-to-East direction in Europe and to Asian countries, is increasing have led to growing environmental group pressure to strengthen EC legislation even further on this subject. Greenpeace, in particular, is reportedly concerned with increased shipments to Asian countries, as markets in Africa and in Central and Latin America become increasingly difficult to enter, and with reports of mislabeling waste shipments as bound for recycling or recovery facilities, when, in fact, they are being sent to developing countries for disposal.

4.3.3.4. Trade and Competition Issues

Since broad disparities exist among EU Member States on the administrative structures and procedures to implement Regulation 259/93 and definitional and liability disputes continue, competitive differences will arise state-to-state in the EU. In addition, possible uses and abuses of the proximity and self-sufficiency principles by Competent Authorities could give rise not only to political problems, but also to major anticompetitive ones as well.

Although disposal of waste at sites as close as possible to the point of generation is generally preferable, in some circumstances moving waste over greater distances to ensure its proper disposal may be desirable. Moreover, many companies in Europe have one or a limited number of waste disposal or recycling sites to ensure the proper disposal or recovery of their company's waste throughout the EU. Sending waste to these sites to be disposed of will involve movement, possibly over considerable distances, from one site to another and perhaps across national borders in the EU. This growing tendency for large, usually multinational, companies to be as self-sufficient as possible in the integrated management of their own waste should be encouraged rather than penalized by possibly inappropriate application of the proximity principle.

How the growing number of contractual arrangements involving manufacturers of products taking back contaminated products or packaging from customers (under a life-cycle philosophy) will be merged with the requirements of Regulation 259/93 is also unclear. Therefore, under this new Regulation, industry could well face disruption of its waste disposal and recovery programs, delays in operations, and the need to store waste at significant cost and potential environmental risk, pending administrative resolution of these questions.

In addition, some would argue that it is unwise for the EU, given the current economic situation in particular, to adopt more stringent legislation on this subject than its major trading partners have and than is required by its international commitments.

The Parties to the Basel Convention agreed on March 25, 1994, to adopt the EU Council Common Position to amend the Basel Convention to impose additional restrictions on exports of hazardous waste to developing countries (potentially non-OECD states), for disposal immediately and for recycling or recovery as of December 31, 1997. As a result, at the Second Conference of the Parties to the Basel Convention (Basel COP 2) recently held in Geneva, the Basel signatory countries agreed, in Decision II/12, to ban trade of hazardous waste between OECD and non-OECD countries. The U.S. Government and U.S. industry oppose this action and estimate that such a ban on trading waste for recycling between OECD countries and non-OECD (developing) countries could cost the United States about $\$ 2.2\times10^9$ a year in commodity trade.

A major indusry sector affected by this ban would be the scrap metal industry. In addition, many non-OECD countries are desperate for raw materials to fuel their economies and may, in response, redefine what is considered a Basel waste in order to circumvent the Treaty.

In the European Union, Greenpeace has pressured the European Commission immediately to bring it into conformance with these Basel Convention commitments. The Commission, on the other hand, does not believe it is necessary to modify the Regulation and certainly not at this time.

Such a ban on hazardous waste shipments between OECD and non-OECD countries, especially for recycling or recovery, would arguably not be compatible with the GATT. In addition, if the hazardous waste definition is overly broad, valuable goods not usually considered waste – for example, chemical byproducts, secondary raw materials, and used electronic goods that constitute legitimate trade between the EU and non-OECD countries – would be blocked. Again, all of these issues and the differing ways in which the EU and its trading partners handle them will likely result in competitive disadvantages for European industry vis-à-vis its non-EU competitors.

4.4. Conclusion

With its Regulation 259/93, the European Community has undertaken an ambitious regulatory program in a much needed area of harmonized EC legislation, with laudable environmental protection goals. However, the scope and complexity of the Regulation, without adequate administrative implementation procedures in place at the EC level and in the Member States, will give rise to significant legal, trade and competition, and administrative compliance problems for European industry. Moreover, the undesirable inconsistencies on this subject among EU Member States are likely to continue to exist in the application and enforcement of this Regulation. In addition, despite ECJ determinations to the contrary, Regulation 259/93 is primarily a free movement of goods/Internal Market measure, not an environmental one. Wastes, especially waste for recovery, should be treated as "goods" under the free movement provisions of the Treaty of Rome. Failure to do so for intra-Community movements of waste violates the letter and spirit of the Treaty of Rome, as amended – without providing additional harmonized environmental protection.

Finally, inconsistent ratifications and implementation of Basel Convention and OECD commitments by the EU and its trading partners will further complicate the competitive problems facing European industry. Recent Basel Convention amendments on this subject, especially with respect to shipments of waste for recovery, arguably violate GATT rules. It will be interesting to see whether, and how, this issue will be resolved in the GATT's – WTO's new Trade and Environment Programme or perhaps in a new GATT "Green Round."

5. The Seveso Directive: An Example of the Interface between Community Law and National Implementation Rules

5.1. Historical Background and Content

Council Directive 82/501/EEC of June 24, 1982 on the major accident hazards of certain industrial activities owes its existence to a number of industrial accidents. Subsequent to accidents in Flixborough in 1974 (United Kingdom), Beek (The Netherlands) in 1975, and in particular at an ICMESA plant in Seveso (Italy) in July 1976, the Commission undertook to draft legislation for the prevention of major industrial accidents, having been encouraged by the intentions of the United Kingdom, The Netherlands, France, and Italy to adopt national legislation in order to improve accident prevention.

The Commission's proposal was submitted to the Council in July 1979, about three years after the Seveso accident. The proposal was largely inspired by measures that existed for nuclear energy installations to avoid nuclear accidents. While the general lines of the proposed Directive were agreed to relatively quickly by the Council, adoption of the Directive was held up because France, in particular, was afraid that information on risks and safety measures to be given to other Member States might create difficulties for the siting and activity of its nuclear installations. Under pressure from the European Parliament, which had from its beginnings been in favor of the adoption of the Directive, and the Benelux countries, a compromise was finally found that enabled the Directive to be adopted in June 1982.

At the end of 1984, a leak of methyl isocyanate from a Union Carbide plant in Bhopal, India caused the death of more than 2000 persons, and some 20 000 persons were injured. Following that accident, the Commission proposed an amendment to Directive 82/501/EEC, which was adopted in early 1987 (Directive 87/216/EEC). The Parliament, in its opinion on this proposal, had suggested a wide-ranging review of the Directive. However, the accident of Sandoz in Basel in October 1986 and the public concern raised by it necessitated immediate action. The Commission submitted in spring 1988 a proposal for a further amendment that considerably enlarged the existing provisions for accident prevention related to the storage of dangerous substances. That proposal was adopted within six months, a record time, as Directive 88/610/EEC.

The announced general revision of Directive 82/501/EEC was inserted into the Commission's legislative program 1993 but has, as of April 1995, not yet been adopted. Directive 82/501/EEC aims at preventing major accidents and limiting the consequences thereof. Approximating the legislation of the Member States is a means of attaining this objective.

Industrial activity is defined in Article 1 of this Directive. Industrial activity means any operation in an industrial installation listed in Annex I to the Directive that involves one or more dangerous substances and is capable of presenting major accident hazards, as well as storage of dangerous substances under conditions laid down by Annex II of the Directive.

Major accidents are also defined in Article 1: "an occurrence such as a major emission, fire or explosion resulting from uncontrolled developments in the course of an industrial activity, leading to a serious danger to man, immediate or delayed, inside or outside the establishment, and/or to the environment, and involving one or more dangerous substances."

The Commission thought it necessary to publish guidelines for the interpretation of this concept. Directive 82/501/EEC provides for a number of obligations:

1) Manufacturers who process or store certain chemicals have a general duty to prevent major accidents and limit their consequences for humans and the environment (Art. 3 and 4). Should a major accident occur, manufacturers have a duty to report it (Art. 10). The Competent Authorities of Member States have a general duty to organize inspections or other methods of control [Art. 7(2)].
2) Manufacturers, furthermore, have specific duties that apply either when any of 180 dangerous substances are used in quantities above specified threshold levels or when any of 28 dangerous substances are stored. In such cases, the manufacturer must produce and update periodically a safety report (Art. 5) and an on-site emergency plan.
3) The Competent Authorities must ensure that an off-site emergency plan is produced (Art. 5 and 7). They must further ensure that "persons liable to be affected by a major accident" are informed of safety measures and of the correct behavior to adopt in case of accident.

5.2. Implementation Requirements

Member States had to adapt their national legislation to the requirements of the Directive by January 8, 1984. This national legislation and the above mentioned amendments of the Directive phased in gradually. The following shows how three Member States – France, Germany, and Greece – have complied with the requirements under Directive 82/501/EEC.

Situation in France. On August 16, 1982 (i.e., only seven weeks after the adoption of the Directive), France issued an internal circular, by which it drew the attention of the Commissaires de la Republique (the Préfets) to the existence of the Directive and requested that specific care be given to the information requirements contained in its Article 8. It listed the installations that came under the Directive and were, in France, the subject of the legislation on "classified installations" (loi sur les installations classées) of July 19, 1976 and of several decrees issued under that legislation, in particular a decree of September 21, 1977.

The legislation of 1976 provides that all types of installations, listed in a specific register, that might present a risk must obtain an administrative license for operation. The corresponding list includes more than 300 activities and thus goes far beyond the number of installations that are covered by Directive 82/501/EEC. To fully adapt its existing legislation, in 1987, France adopted legislation on the organization of civil protection, forest protection against fire, and the prevention of major accidents. This legislation requested the establishment of national, regional, or departmental offices for organizing help in cases of accidents and natural or man-made catastrophes. Emergency plans among others were to be prepared by the departmental prefects around certain industrial installations.

These are the essential pieces of French legislation to which workers' safety provisions must be added in order to cover the area of the Directive. What is striking is the fact that the legislation on industrial installations is, to a large extent, umbrella legislation. The major part of the obligations that the Directive places on the individual manufacturer is to be determined by individual administrative decision. Some 500 000 French plants are registered and subject either to a system of declaration or, for some 55 000, to a system of authorization procedure under the legislation of 1976. Most of the elements of Directive 82/501/EEC are found in the 1976 legislation, although this legislation does not seem to cover all elements of Articles 3 and 4.

The requirement for safety reports (risk analysis) was enlarged after adoption of the Directive to cover existing installations. These safety reports under Article 5 of Directive 82/501/EEC refer to some 320 types of installations. On-site emergency plans were, prior to adoption of the Directive, mandatory only for oil industry installations; since the first circulars of 1982 and 1983, they have also been required for installations that come under Article 5 of the Directive. Off-site plans have been requested gradually since 1985, subsequent to legislation adopted that year.

The French system is a clear case of existing means being amended in order to comply better with a Directive that postdates the basic national legislation. Furthermore, the French system gives considerable authority to the departmental, regional, and national administrations, which operate via individual administrative decisions.

Situation in Germany. Germany effected Directive 82/501/EEC and its subsequent amendments by a great number of provisions, under essentially four different categories:

1) Legislation on workers' safety
2) Legislation on chemical substances and preparations
3) Legislation on catastrophes
4) Environmental legislation

Overall, more than 40 pieces of legislation, adopted at the federal or regional level, apply to the subject matter covered by the Directive.

The most relevant provisions to Directive 82/501/EEC are contained in the so-called "Störfallverordnung," the 12th Regulation on the implementation of the Federal Act on Quality Standards (Bundesimmissionsschutzgesetz). For political reasons, Germany thought it necessary to have its national legislation in place when Community legislation on this subject was adopted. This led the German government to submit a proposal for the 12th Regulation on February 20, 1980, while the Commission proposal for Directive 82/501/EEC dates from July 16, 1979. In 1980, the 12th Regulation was adopted in Germany, while the Community Directive was politically agreed to, though not yet formally adopted. The same coincidence occurred after the Sandoz accident in 1986, when both the German government and the Commission submitted proposals. This time the Community acted more quickly; Directive 88/610/EEC was adopted in 1988, while the German amendment was finally adopted in 1991.

The multiplicity of legal texts makes it difficult to identify discrepancies between the 12th Regulation and Directive 82/501/EEC, without being confronted with the objection that other legal provisions explicitly or implicitly cover the aspect in question. The text of the German 12th Regulation is more precise than that of the French legislation; it gives greater legal certainty to industrial installations and less discretion to the administration.

The manufacturers' general obligations are listed specifically, as well as the obligation to establish an on-site emergency plan and a safety analysis. The obligation to inform persons outside the plant has been placed on manufacturers. Off-site measures are to be agreed to with the competent regional or federal authorities. Generally, the threshold levels in German legislation are more stringent than those in the Directive. Also, the list of substances that make the provisions applicable is longer than that of Directive 82/501/EEC.

Situation in Greece. Greece is a Member State that, at the time of its accession to the Community in 1981, had relatively little legislation on the environment. Accident prevention measures for installations were a rather new item, and the Greek administration was not really accustomed to monitor manufacturers' behavior with regard to safety. These rather sketchy lines might help explain why national measures to transpose the amendment of Directive 82/501/EEC took several years. A similar attitude is to be observed regarding the amendment made by Directive 88/610/EEC: only in 1993 after the Commission had filed an application to the European Court of Justice for noncommunication of national implementation measures, Greece transformed the requirements of Directive 88/610/EEC into national law.

The national legislation reproduces the requirements of Directive 82/501/EEC and its amendments correctly. However, the responsibility of monitoring the legislation is given to no less than four bodies [environmental and industrial ministries, prefects at local – regional level, and a Coordinating Interministerial Committee (CIC)]. These rather dispersed responsibilities might have contributed to certain doubts being raised about effective application of the legislation. Indeed, several accidents suggest that effective application, safety analysis, and on-site or off-site emergency plans have not been made for all installations that come under the Directive, nor is regular inspection always made. One reason for this situation is that the different central and local – regional administrations did not receive any supplementary financial or human resources, when the new monitoring tasks were assigned to them in 1988. Thus, the administration does not act as counterpart interlocutor to manufacturers. Since the requirement of informing the public of measures to be taken in case of accident has been placed on prefects, a manufacturer does not incur too serious risks regarding compliance with the Directive's requirements.

5.3. Conclusions

If one looks horizontally at Directive 82/501/EEC and its implementation, a number of observations can be made:

1) The Directive owes its existence and its tightening-up to industrial accidents and public concern; these accidents made possible the adoption of preventive provisions that had not been acceptable at the national or Community level prior to the accidents.
2) Of the three Member States considered, France undoubtedly gives the greatest discretion to the administration and Germany the greatest legal certainty to manufacturers.
3) Effective application of the requirements of the Directive requires flexible cooperation between manufacturers and administration regarding inspection, monitoring, safety analysis, and risk assessment. The administration itself needs to be trained and experienced to be able to contribute to accident prevention without unduly hampering manufacturers' activity. The absence of a trained, effective, and active administration that effectively monitors the practical application seems to be the greatest handicap in Greece.
4) The Directive is not entirely clear on the question of who has to inform the public on safety measures and on behavior in case of accident (Art. 8) – the administration or the manufacturer. This sometimes leads to practical difficulties. Furthermore, if this kind of information is too detailed, it might provoke negative reactions among the population concerned.
5) The cost of prevention might be significant, while the benefits – the absence of major industrial accidents – are very difficult to estimate. Both France and Germany had, prior to adoption of the Directive, taken a number of steps to prevent major accidents, although these measures focused more on human health than on environmental protection.
6) It is not clear under the Directive whose task it is to set up the off-site emergency plan, to ensure its consistency with the on-site emergency plan, and to manage off-site emergency operations. Obviously, the manufacturer is not capable of providing for all details of the off-site plan such as blocking roads or evacuating the population. Therefore, active cooperation by the administration is necessary, not only in case of accident, but also to learn about cooperative mechanisms and gain experience for any accidental event.

7) French administration has a tendency to ask for measures to prevent accidents in the worst possible scenario, while German administration normally is satisfied with the degree of accident prevention that is reached by using the best available technology.
8) German administration seems to attach limited importance to the question of siting of industries, whereas this aspect plays an increasing role in the French administration's approach to accident prevention.
9) Safety reports play an essential role among all measures required by the Directive. They are considered a means to obtain more precise information on installations, to identify their accident potential, and to prepare preventive measures more adequately. However, which role the administration has to play in this context is not made clear in the Directive. The differences among Member States seem to be important. The administration in Greece seems to be easily satisfied with the safety reports submitted. In France, the authorities examine the safety reports and sometimes require other preventive measures, but never "accept" the report, thus leading to a kind of uncertainty by manufacturers. It might be useful to systematically check, as German administration does, the reports' completeness, correctness, and their effective monitoring and updating.

The hope is that further amendments of the Directive will not be the result of new industrial accidents.

6. U.S. and European Environmental Legislation

6.1. Introduction

This Chapter discusses the nature and role of environmental Regulation, particularly as it applies to the chemical industry. It makes no attempt to summarize the vast and rapidly evolving body of current U.S. and EC environmental legislation. Other sources can be found for this (see, e.g., [68], [69]). The environmental legal provisions applicable to all phases of the chemical industry are now so extensive, detailed, and complex that they cannot be treated exhaustively under a single keyword of Ullmann's.

Rather, the attempt here is to step back from the system and compare briefly the style and context found in the world's two major centers of regulatory activity, the United States, and Europe, to note some of the key features of the legislation in both systems; and to comment briefly on what may lie ahead.

6.2. The Environmental Regulatory System [70]

The United States and Western Europe have had significant environmental laws on the books since the early 1970s. The United States has erected an elaborate edifice of Regulations. At the EC level and in many European countries (Germany being a noteworthy exception), however, a great deal has not happened until recently. Over the past several years, the European Community and its Member States have been generating a flow of environmental initiatives of ever-increasing importance. Now developing in Europe is an extensive framework of environmental regulatory and liability law (see Chaps. 3, 4) that presents business operating in Europe with many of the same kinds of risks found in the United States. The ways these laws are developed in the European Community and in the United States differ fundamentally.

6.3. Methods of Lawmaking in the United States and the European Community

United States. In the United States, the law and lawyers have been central to the development of environmental policy, and highly structured efforts at the federal level have led the way. U.S. Regulation relies heavily on generic, federal administrative rule making to achieve specificity and uniformity, and to gather the technical, scientific, and economic data that form the premise for public decision making. Administrative rule making proceedings provide public, formal, and structured opportunities for citizens and industry to participate in the development of Regulation through comments, public hearings, and sometimes judicial challenges to rules.

The "political" aspects of the U.S. environmental regulatory process take the form of lobbying directed at the U.S. Congress, as well as less formal efforts to lobby the administrative agencies that establish, implement, and enforce Regulations.

In short, in the United States, contending interests use legal processes in the judicial arena and before administrative agencies to influence the formulation and execution of policy. This is particularly true of environmental groups who, with some frequency, use citizen suit provisions in federal environmental legislation as a means of influencing environmental policy.

The result of this dynamic, contentious law and rule making process is a highly articulated, but often inflexible, regulatory system, i.e., a regulatory system that is at times too rigid, complex, and confusing.

EC Regulation. Regulation in Europe, on the other hand, has traditionally been regarded as "technical" or "political," rather than "legal" in nature, although this attitude appears to be changing. European industry, as a whole, has not in the past looked to legal counsel to assist it in handling environmental matters, but has relied primarily on its technical staff, who often operate through national and European-wide trade associations. The EC regulatory process, furthermore, is largely nonadversarial and is the result of a slow, nonpublic, lawmaking structure designed to achieve political consensus among the Member States. A close industry – government relationship exists. Industry operates through contacts with national authorities; through "expert" advisory committees at the Commission of the European Community; and through lobbying of the Commission, Parliament, and COREPER (the permanent representation of Member States, which assists the Council on day-to-day matters). Proposed Directives are generally made publicly available and are commented on by Parliamentary committees, but until recently, this was not required at the early stages. This situation has been changed, at least in theory, in 1992 – 1993 with the adoption of Council and Commission Resolutions calling for early consultation of interested bodies, including industry associations. No structured open administrative process for promulgating Directives or Regulations exists, however, such as the U.S. notice and comment rule making process.

Implementation of EC Directives is left to the authorities of individual Member States. The degree to which Directives are effectively implemented varies widely from country to country. While some Member States do not give full effect to Directives until long after adoption, other states adopt even stricter measures than those existing at the Community level. For example, Denmark, The Netherlands, and Germany have well-developed national environmental policies. Ireland, Greece, Portugal, Spain, and to some extent, Italy, are in earlier stages of development. Belgium, France, and the United Kingdom are somewhere in the middle, although they have had environmental policies for many years. British law is now developing especially rapidly. As a result of the wide disparity in Member State national environmental laws, chemical companies operating in Europe must be familiar with both Community and national environmental legislation in order to minimize the risk of unexpected environmental Regulation or liability.

6.4. U.S. and European Environmental Laws

Environmental law, to date, has progressed chiefly in two broad streams – one regulating the manufacturing process and the other regulating the products themselves (→ Chemical Products: Safety Regulations). A third stream of natural resource law regulates extraction of the oil, gas, coal, and various minerals that provide the raw material for the chemical manufacturing process. (A number of minor subjects in environmental law deal, for example, with the special protection of certain types of resources, such as wetlands, endangered species, wild and scenic rivers, archaeological sites, and historic preservation. They are not dealt with here since they are less important to and less specific to the chemical industry.) The focus in this section is on chemical process and chemical product Regulation.

Chemical process Regulation encompasses chiefly pollution control Regulation – the classic areas of air emissions, water effluents, and solid and hazardous wastes. It centers on the chemical manufacturing plant, although it typically also applies to all related commercial or industrial facilities (e.g., storage depots, means of transport, and disposal facilities). Generally, the basic tools are construction or operating permits that impose relevant operating constraints, monitoring and reporting requirements, etc. Beyond these standard provisions, however, environmental impact assessment laws, accident planning laws, and public information laws have also been developed, sometimes implemented in permits or the permit process and sometimes separately.

6.4.1. Environmental Impact Assessment Laws

Both the United States and the EC have environmental impact assessment (EIA) laws [71]. These apply chiefly during the licensing of new chemical or other manufacturing facilities or the major expansion of existing plants. They require a systematic, detailed, and written analysis of the likely environmental impacts of the proposed project. This analysis is designed to force both chemical industry management and government regulators to consider environmental factors in their project decision making. In both the United States and the EC, however, the EIA laws themselves mandate no particular result. Rather, they are procedural laws premised on the assumption that a better decision-making process leads to better decisions. U.S. law goes beyond EC law in many ways. It requires consideration of all reasonable alternatives to the project, and has more detailed guidance and "teeth" than its EC counterpart.

6.4.2. Accident Planning and Public Information Laws

The United States and the EC likewise have accident planning and public information laws of great significance for the chemical industry. The EC developed broad-gauged legislation on this subject earlier than the United States, and most European countries have a more systematic approach to risk assessment and analysis built into the new (and modified) plant licensing process than does the United States to date. The EC law (the Seveso Directive, see Chap. 5) is designed to prevent and limit the consequences of major accidents resulting from industrial operations involving dangerous substances [72].

The corresponding U.S. law, the Emergency Planning and Community Right-to-Know Act, requires (1) emergency planning, (2) emergency release notification, (3) community right-to-know reporting, and (4) toxic chemical release inventory reporting [73]. While the EC does not yet have an analogue to the U.S. toxic chemical release inventory requirements, one is on the drawing board. For many years, the United States has also had separate oil spill legislation requiring the well-known Spill Prevention, Control and Countermeasure (SPCC) plans [74] and has much better developed legislation on aboveground and belowground storage tank Regulation than either the EC or most European countries (Germany being an exception in this regard).

6.4.3. Air and Water Pollution Legislation

The air and water pollution legislation in both the United States and the EC is extensive [75]. In both cases, the legislation uses "best-technology"-type substantive emission and effluent standards (emission limits in German terms) as well as case-specific emission and effluent limits to be set at levels necessary to attain air and water quality requirements (the latter called immission standards in German terms).

U.S. water pollution legislation has for years built its control program around a comprehensive permit system, and U.S. air pollution legislation is now adopting the same technique. The EC legislation has had no comprehensive permit system requirements to date, although the Commission is now developing a multimedia permit system modeled after the British Integrated Pollution Control system. The U.S. air and water legislation has regulated new plants more stringently than existing sources through technology-based new source performance standards for industry categories. The legislation has evolved industry-specific technology-based limits for existing sources on the water side and is now moving to do likewise on the air side, and has developed complex antidegradation requirements (in effect, special limits on degradation of an air or water resource that meets existing ambient requirements). The EC legislation has tended, in practice, chiefly to apply technology-based limits for new sources in a few large industrial categories, and the pervasive air and water quality-based Regulation of existing sources called for by EC legislation has never been put in place. Indeed, not even the air and water quality monitoring systems necessary to implement such ambient-based Regulation have been developed. The Commission is still struggling to come up with a politically and administratively feasible vehicle for setting air and water technology-based limits for the industry, such as are pervasive in the United States and in Germany. Antidegradation requirements, to the extent theoretically present under EC law, have been largely ignored.

In both the United States and the EC, efforts are now being made to move beyond Regulation of the few parameters dealt with in both the air and the water context in the early years and to tackle the difficult and complex job of effectively regulating a multitude of toxic pollutants.

6.4.4. Solid and Hazardous Waste Regulation

U.S. Regulations. In the field of solid and hazardous waste Regulation, the United States has for years had a much more effective regulatory system in place for hazardous waste than the EC (although the German waste regulatory system is comprehensive, except for a loophole in the case of "recyclable" wastes, now to be closed by the new German Eco-Cycle Waste Law of 1994) [76]. The U.S. cradle-to-grave regulatory system uses registration requirements, permits, and transport manifests to regulate virtually all aspects of the generation, storage, transport, treatment, and disposal of a broad range of hazardous wastes (including many destined for recycling), under an extremely complex set of rules and both written and unwritten regulatory practices. All types of treatment, storage, and disposal facilities must meet specific design and operating performance standards, including statutory minimum technology and groundwater monitoring requirements and a prohibition on the land disposal of untreated hazardous wastes. The U.S. system has complex rules for mixtures of wastes; for wastes derived from other wastes; and for distinguishing wastes from products, coproducts, byproducts, chemical intermediates, and secondary raw materials. It regulates some wastes destined for recycle or reuse and does not regulate others. Despite all this hazardous waste Regulation, however, the U.S. system has virtually no comprehensive or systematic Regulation of solid waste at the national level, this area being left almost entirely to state and local governments.

EC Regulations (see Section 4.3). The EC waste management framework is more coherent, but less effective, than that in the United States. The European system is built on a basic framework Directive that deals with all forms of waste and is supplemented by a more specific Directive dealing with hazardous waste [77]. The entire system is now in flux (see Section 4.3).

Further, the European waste regulatory system has depended, from the outset, on Regulation of the *shipment of waste* (see Section 4.3.1). Both the United States and the EC countries have Regulations governing how hazardous materials must be transported, but EC waste shipment Regulation governs whether hazardous waste can be shipped across national borders at all, and under what conditions. As early as 1984, the Directive on Transfrontier Shipments of Hazardous Wastes regulated hazardous waste shipments across Member State borders and into and out of the EC by creating prior notification, manifesting, and packaging requirements [78]. More recently, the community has adopted a new Regulation on Shipment of Waste that implements the Basel Convention in Europe (the Basel Convention governs shipment of hazardous waste on an international scale) and that imposes more stringent requirements (including prior authorization in some cases) on the shipment of all forms of waste, both hazardous and normal, across Member State borders and into or out of the EC (see Section 4.3.1). Shipments of waste for recovery operations would, depending on how they were listed in three categories (Red, Amber, and Green; see Section 4.3.1) on lists prepared by the "Committee Procedure," be (1) largely excluded from Regulation (green), (2) subject to a prior notification regime (amber), or (3) subject to a prior authorization regime (red; waste not listed would be subject to the authorization system). This Regulation could balkanize the EC on waste matters, with local authorities lodging objections motivated by protectionism of local disposal or recovery facilities and NIMBY-ism. Chemical facilities may be faced with disruption of their waste disposal and recovery arrangements, and those dependent on off-site waste disposal or recovery (and having only limited on-site waste storage capacity) may find their operations at risk of interruption.

To implement the EC's Waste and Hazardous Waste Directives as recently amended, the Commission has drawn up a comprehensive list of solid and hazardous wastes (the European Waste Catalogue). A separate list has been adopted by the Council, which will be critical to EC waste legislation because it will determine the threshold question, for particular wastes, of whether they must be regulated as "solid" or as "hazardous" (see Section 4.3.3). The Commission is also having great difficulty establishing and implementing the definition of "waste" found in relevant Directives.

U.S. Superfund and EC Packaging Legislation. Two idiosyncratic areas of waste Regulation remain to be discussed – one in the United States and one in the EC. In the United States *Superfund legislation* uses government-funded cleanup backed up by a civil liability scheme to deal with soil and groundwater contamination [79]. Originally designed for old dump sites, this broad-reaching legislation extends in fact to all instances of such contamination, both old and new and both on-site at chemical manufacturing facilities and off-site. The government-funded cleanup process has been slow and has been required by Congress to use extremely stringent cleanup levels. The liability regime imposes retroactive, joint and several, strict liability regarding site cleanup liability and responsibility for natural resources. This liability is imposed on virtually everyone connected with the contaminated site since its contamination, including lenders in some cases. The primary Superfund program has been widely criticized and drives broad secondary efforts of private remediation at sites where the potentially responsible parties hope to avoid becoming entangled in the main cleanup and liability program. The European Commission has been considering legislation to deal with soil and groundwater contamination in Europe, and the Council of Europe has adopted a convention on this subject (see Chap. 3).

The second piece of idiosyncratic legislation is the effort by the EC and a number of Member States to regulate one form of solid waste, *packaging*, through product take-back (or advance collection, sorting, recycle, or disposal fees) and thus to develop a new paradigm for regulating all forms of solid waste . The Commission, reacting to the so-called Töpfer Ordinance on packaging waste in Germany and to similar efforts in France and differently structured efforts in other Member States, has proposed a Directive on Packaging and Packaging Waste, which is aimed at eliminating and preventing trade barriers arising from varying national packaging waste initiatives. The essence of this legislation is the ability to impose product charges to force those in the packaging chain, from manufacturer to consumer, to begin to internalize some elements of the costs of collection, sorting, recycle, or disposal of solid waste. The German system achieves the same end by imposing on the commercial parts of the packaging chain a legal "product take-back" obligation. By forcing internalization of some of the costs of the end-of-life disposition of products, the scheme aims at correcting market conditions so as to favor recycling, while letting the market rather than direct Regulation sort out how much recycling is warranted and in which cases. Particularly at the EC level, however, extensive direct regulatory measures, which may well not be environmentally wise or economically efficient, are also being considered.

6.4.5. Regulations on Chemical Products

EC Regulations. Environmental legislation regarding chemical products was one of the first areas pursued by the European Community, although initially its motivations had more to do with the Common Market than with environmental protection. The first major piece of chemical legislation was adopted in 1967, and established a common Community system for the classification, packaging and labeling of dangerous chemical substances [80]. This basic Directive was modified in 1979 by the so-called Sixth Amendment, which established a premarketing clearance requirement [81]. Under this amendment, companies marketing new "dangerous substances" are required to submit the new chemical substance to a prior notification procedure that involves testing the substance for toxicity and eco-toxicity and preparing technical dossiers to provide information on the substance's characteristics (see Section 2.2). In 1988, the Community adopted similar requirements for dangerous preparations (i.e., mixtures of dangerous substances) [82].

The Commission has also recently adopted material safety data sheet (MSDS) rules and has extended its Regulation to include the marketing and use of certain dangerous substances and preparations, as well as enacting Directives in the field of biotechnology designed to control the handling and release of genetically modified organisms. The Commission is also moving to impose on manufacturers and importers of existing chemical substances some significant data reporting obligations for assessment of their risk characteristics.

U.S. Regulations. In the United States, chemical product Regulation on a comprehensive basis began with the Toxic Substances Control Act in 1976 [83]. That Act places on manufacturers the responsibility of providing data on the health and environmental effect of chemical substances and mixtures and gives the Environmental Protection Agency comprehensive authority to regulate the manufacture, use, distribution in commerce, and disposal of chemical substances. The central regulatory mechanism is the requirement for premanufacture review of new chemical substances prior to their commercial production and introduction to the marketplace. The Environmental Protection Agency also has authority to require testing of chemicals, although no requirement for a "base level" of tests exists as a precondition to premanufacture review as in the Sixth Amendment to the EC Regulation.

Regulation of environmental harms from chemicals in the workplace takes place in the United States under the Occupational Safety Health Act. Under that act, the Hazard Communication Standard requires that employees be informed about the potential chemical exposures involved in their work and be trained to deal with them [84]. Chemical manufacturers and importers are required to evaluate the hazards of the chemicals they produce or import, and to use MSDS to transmit this information to their own employees and to employers to whom they ship the chemicals. Further, chemicals shipped must contain labels on the containers with similar information.

Beyond these core areas, both the U.S. and the European systems regulate use of polychlorinated biphenyls, asbestos, and agricultural chemicals. Further, hazardous chemicals are regulated to varying degrees and in varying ways in consumer products, foods, drugs, and other end uses such as cosmetics and medical devices as well.

6.5. Some Contrasts in U.S. and EC Environmental Legislation [70]

For obvious constitutional and political reasons, the scope, specificity, and rigor of EC legislation do not equal those of U.S. federal legislation.

6.5.1. Regulatory Gaps

Important regulatory gaps exist in the EC (e.g., on wetlands, underground storage tanks, and until recently, municipal sewage), along with weaknesses (e.g., Member State ability to determine, on a relatively subjective basis, the applicability of certain EC Directives on fish protection and bathing water). Nonetheless, in some areas, EC provisions or proposals go beyond present U.S. practices or requirements (e.g., eco-labeling, eco-auditing, packaging waste, and economic and fiscal measures such as a carbon or fuel tax).

6.5.2. Implementation of Legislation

Implementation of the EC Directives is spotty and frequently ineffective. Directives, particularly the early ones, tend to be general policy statements, with little of the detailed standards and permit requirements of their U.S. counterparts. For example, the European Community as yet has no specific standards for surface impoundments (e.g., pits, ponds, lagoons), underground tanks or pipes, waste piles, and various forms of waste treatment. Where Member State implementation takes place, standards and permit conditions are often negotiated on a case-by-case basis, and systematic monitoring and reporting obligations are not always imposed.

United States. In the United States, the implementation of legislation at the state level is reasonably well assured by U.S. federal legislation through a series of devices lacking in EC legislation. First, a well-staffed, trained, and funded agency (the EPA) can be found at the federal level in the United States. Also available is a great deal of public, comparable monitoring data and other information, as well as effective freedom-of-information legislation at the federal and state levels. Second, the EPA may as a formal matter withdraw a state's right to implement many of the federal regulatory programs if it judges the proposed implementation to be inadequate. The EPA also normally has direct prior approval authority over state-implemented Regulations, standards, and individual permits before those measures can go into effect under federal law (although they can and do sometimes go into effect under state law without regard to EPA approval at the federal level). In the absence of an EPA-approved state implementation program or Regulations, standards, or permits, the EPA is frequently required to promulgate directly applicable federal provisions. Should it fail to supervise state implementation where it has a mandatory duty to do so, it may be sued in federal court by citizens and environmental groups and forced to act. Lastly, the EPA has frequently threatened to withhold federal funding on which the state programs depend if state implementation is deemed inadequate.

6.5.3. Enforcement of Legislation

The enforcement of EC legislation is perhaps the weakest element of the EC regulatory framework. All the European Community can really do is insist that Member States enact legislation (i.e., transpose EC legislation into national law). The European Court of Justice has new power to impose penalties on Member States for failing to implement EC environmental law, but it remains to be seen how effective this tool will prove. An enforceable obligation on the part of Member States to undertake effective, concrete implementation through more specific regulatory standards and requirements, comprehensive permit or approval systems, and an adequately staffed and trained bureaucracy with sufficient funding and legal authority is generally lacking in many countries. Furthermore, the European Environment Agency, which is only just getting started has only a data collection and dissemination function. It does not, for the time being, have enforcement powers.

Enforcement in the United States is vigorous and takes a number of forms. Nearly all of the major U.S. environmental regulatory statutes require some form of publicly available self-monitoring and reporting. The system thereby automatically highlights lapses in compliance and the need for enforcement. When the EPA deems state enforcement inadequate, it normally has the authority at any time to override the state's failure to enforce and to issue an administrative order or, in some cases, to assess civil penalties on an administrative basis. In either case, the EPA may enforce its action in federal court or take the violator directly into federal court for injunctive relief and civil or criminal penalties. Should the EPA fail to do so, citizens and environmental groups can themselves normally take the violators into federal court.

Further Development in the EC. The EC institutions currently have none of these tools at their disposal. At the level of enforcement of specific standards, requirements, or permit conditions against an individual regulated facility, the Community plays virtually no role. Such enforcement is left up to the Member States. Enforcement is difficult, even when attempted, since self-monitoring and reporting obligations have not always been imposed either by Regulation or by permit conditions. Government failure to prosecute generally cannot effectively be challenged except through publicity and the political process.

The Commission is now engaged in a series of moves to surmount the difficulties it has encountered in trying to develop an effective "command-and-control" EC regulatory system. It is considering the use of fiscal and economic measures instead of administratively driven control requirements. It is trying to make greater use of directly applicable Regulations in the place of Directives that must be implemented by Member States; and in any event, EC legislation, overall, is becoming more specific, thereby providing the detail necessary for ensuring the consistent implementation that has been absent in earlier legislation. The Commission is drafting an Integrated Pollution Control Directive, modeled on British legislation, that would impose comprehensive permitting requirements covering all environmental media, i.e., air, water, soil, groundwater (but, initially, applying only to a restricted list of larger chemical, metallurgical, and paper plants). This Directive would impose substantive and management requirements as permit conditions and would implement on a case-by-case basis the existing media-specific air and water quality legislative requirements. The Commission has also adopted an eco-audit Regulation that provides a (presently "voluntary") comprehensive, universal system that is a partial, periodic substitute for the self-monitoring and reporting provisions that are lacking in existing Directives. These eco-audits would have to be verified by independent outside auditors, with public disclosure of certain information. The likely result of these eco-audits will be public and political pressure for more effective standard setting, implementation, and enforcement.

At a fundamental level, for the first time, the Treaty on Political Union (the Maastricht Treaty), would provide the Court of Justice with limited powers to sanction with penalties those Member States failing to implement EC law. Also, the Court of Justice has been developing the following legal principles to increase the significance of EC law: (1) obligations on national bureaucrats at all levels to apply EC Regulations, even where the national legislature has failed to implement that EC Regulation adequately (i.e., "direct applicability"); (2) "direct effect" of provisions in EC Directives that are sufficiently precise and unconditional, even where the national legislatures have failed to implement adequately; and (3) state responsibility for harm to individuals arising from the government's failure to implement EC legislation adequately. The cumulative effect of these developments will be to increase the importance of EC environmental law and to render more likely its effective, if not consistent, implementation and application throughout the European Community.

6.6. Outlook [70]

While the basic framework of U.S. environmental law has been fairly settled (the U.S. Clean Air Act and new initiatives by the first Republican-controlled Congress in 40 years being exceptions), environmental law in Europe is changing rapidly. The past and present roles of European environmental law, of course, indicate that the EC environmental proposals may not necessarily have more than limited importance for business. As is well known, a gap has existed between the rhetoric and the reality of European environmental law. The European Community and many Member States have a good deal of environmental "law" on the books, but much of the EC legislation requires little real action. Moreover, an inability at the EC level and, in some cases, a reluctance at the Member State level to implement and enforce existing environmental law, have been evident. Further, a remarkable lack of comparable data to document many European environmental problems, as well as a lack of data on government actions and enforcement in regard to them, are noticeable.

The politics of the environment appear to be changing, however, As a result of such crises as Seveso, Chernobyl, and Sandoz, and of the palpable, long-neglected environmental problems facing Western and Eastern Europe, public attitudes toward environmental risks in the European Community are evolving quickly. Environmental groups, although perhaps not yet as sophisticated and well funded as their U.S. counterparts, are becoming increasingly well organized and active. Adoption of the EC Directive on Freedom of Information on the Environment and creation of the European Environment Agency, by improving the quantity and quality of publicly available environmental information, will in all likelihood result in heightened public awareness, increased effectiveness of environmental groups, and consequently, increased political pressure for consistent and effective environmental legislation and enforcement.

7. Environment, Safety, and Chemical Safety in Japan

7.1. Current Situation Concerning Environmental Protection

Since the early 1970s, Japan has made a strong effort to tackle the problem of environmental preservation. Various measures, such as reinforcement of pollution control Regulations and amplification of pollution control facilities have been implemented.

As a result, some serious environmental conditions have been alleviated and the situation has generally improved since the 1980s. Moreover, because the social demand for environmental preservation, particularly in metropolitan areas, tends to be more and more diversified and sophisticated, the Environment Agency is promoting various measures in response to this situation.

In addition, finding solutions to the recent environmental problems that are developing into international and global issues and global climate problems are considered important political subjects by the Japanese government as in other countries. In November 1993 the basic law for environmental pollution control was put in force in response to the result of the UNCED held in June 1992.

7.1.1. Air Pollution Control

Regulatory standards to protect human health (environmental standards) have been set for SO_2, NO_2, CO, photochemical oxidants, and airborne particles based on the Basic Law for Environmental Pollution Control. For Regulation of exhaust gas emission accompanying industrial activities, emission control Regulations are in force for SO_x, NO_x, cadmium and its compounds, Cl_2 and HCl, F_2, HF and SiF_4, lead and its compounds, and exhaust dust. The Regulations are based on the Air Pollution Prevention Law. With regard to SO_x and NO_x, Regulations on total amounts of emissions are also valid within the designated area in the country. In addition to these nationwide uniform standards, special emission standards must be met by newly constructed or expanded facilities in the specified area. In certain areas, additional emission standards have been set by local governments (in municipalities designated by government ordinance).

To meet these standards, industries have taken measures such as desulfurization of exhaust gas [176×10^6 m^3/h (STP) at 1810 facility units], denitration of exhaust gas [142×10^6 m^3/h (STP) at 379 units], development of combustion processes that emit lower amounts of NO_x, absorption of HCl and HF, electrostatic dust separation, etc.

The regulatory system for total NO_x emission control for each industrial installation was introduced into three areas with a high population density, and the system was implemented in Yokohama and in Kawasaki in April 1982, and in Osaka and Tokyo in November 1982.

However, NO_2 concentration in the environment shows an increasing trend, and reinforcing the Regulations on NO_2 mobile emission sources such as automobiles is considered necessary.

In June 1989, the Air Pollution Prevention Law was amended. Fibrous materials such as asbestos that are hazardous to human health have been treated as specific dangerous forms of dusts in this amendment.

7.1.2. Control of Water Pollution

Wastewater Regulation standards were implemented in 1970 based on the Water Pollution Prevention Law. Together with additional standards implemented by the respective local governments (in municipalities designated by government ordinance), water pollution has been controlled with regard to the following items:

1) **Substances Posing a Health Hazard (Toxic Substances).** Concentration limits are set for cadmium and its compounds, cyanides, organic phosphorus compounds, lead and its compounds, chromium(VI) compounds, arsenic and its compounds, mercury, alkylmercury and other mercury compounds, polychlorinated biphenyls, trichloroethylene, and tetrachloroethylene.
2) **Items Affecting Living Environment.** Limits have been set on pH range, BOD, COD, suspended solids, concentration of oily substances, phenols, copper, zinc, soluble iron,

soluble manganese, chromium, fluorine, and a number of *Escherichia coli* colonies.

Industries have promoted various measures such as wastewater treatment and process rationalization. As a result, the environmental standards on toxic substances have been satisfied by 99.9 %, and a general trend of improvement is observed for household effluents. However, in closed water areas such as rivers, inner bays, lakes, and swamps, pollution was still proceeding and eutrophication was advancing. Since the cause of such situations was the insufficent installation of sewage systems for treatment of household wastewater, the Water Pollution Prevention Law was revised.

The Law on Special Measures for Preservation of Water Quality of Lakes and Swamps was enacted in July 1984, and an environmental standard and wastewater discharge control standard for nitrogen and phosphorus concentration were introduced.

The Regulation on the total quantities of contaminants in water was implemented in 1978 for three closed sea areas, namely, Tokyo Bay, Ilse Bay, and Seto Inland Sea, and all-around measures for reduction of the total COD were implemented. In January 1991 the third Regulation for the reduction of total quantities of contaminants was prepared. The countermeasures for water quality control of the closed sea will become more important in the future because tremendous growth of algae resulting from eutrophication will seriously damage fishery by causing red tide.

To counteract groundwater pollution, trichloroethylene and tetrachloroethylene were classified as hazardous substances under the Ordinance for Enforcement of the Water Pollution Prevention Law in April 1989; in June, upon amendment of the Water Pollution Prevention Law, percolation of specific water containing hazardous materials into the groundwater was prohibited. In this connection, the ordinances for reinforcement of the Law related to the Disposal and Cleaning of Wastes were also revised.

7.1.3. Revision of Compensation Systems for Patients with Pollution-Related Diseases

The compensation system for patients suffering from pollution-related diseases was implemented in September 1974, and compensation has been paid, at the cost of polluting parties, to the approved patients in Class no. 1 (directed to the four designated diseases caused by air pollution) and Class no. 2 (directed to diseases caused by water pollution or soil contamination). For Class no. 1, the Central Pollution Study Committee submitted a report in October, 1986 proposing a system that reflects a substantial improvement in environmental pollution.

Based on the above report, a revision bill of the Law on Compensation of Pollution-Related Patients was approved by Parliament. The contents of the revision are (1) all-out cancellation of class I area designation due to the improvement of the environmental situation; (2) continued compensation to approved patients; and (3) promotion of comprehensive environmental hygiene measures (establishment of a fund). As a result, a long-awaited rational revision of this system was implemented in March 1988. By the end of 1993 the number of approved patients was 84 471 and expenses totaled 93.4×10^9 Yen.

7.1.4. Measures for Waste Disposal

Since 1970, the Law on Waste Treatment and Cleaning has been in force, with the two groupings of "Domestic Wastes" and "Industrial Wastes." Section 3 of the law stipulates that an entrepreneur shall be responsible for, e.g., self-accountability, appropriate treatment, discharge reduction, not making products difficult to treat, etc. With respect to industrial waste, the law provides for standards on collection, transportation, and disposal and Regulations concerning the industrial waste treatment industry, industrial wastes treatment facilities, and hazardous industrial wastes.

As far as domestic wastes are concerned, awareness is growing of the need for waste reduction, including recycling of waste. This is due to the fact that the quantities of wastes discharged have been increasing in recent years, existing incineration and final treatment facilities have reached their maximum capacity, and need exists to utilize limited resources more effectively. The Law Concerning the Utilization of Recycled Resources was published in April 1991. The bill for revision of the Law on Waste

Treatment and Cleaning was deliberated by Parliament in September 1991.

7.1.5. Response to Global Environmental Problems

The Law Concerning Protection of Ozone Layers by Regulation of Specified Substances was enforced in Japan in May 1988. The law specifies measures that are to be taken to regulate the manufacture of specific substances, to restrict their discharge, and to rationalize the use of these substances.

In June 1990, the Montreal Protocol was revised to strengthen Regulations by placing a complete ban on the use of certain fluorocarbons by the year 2000 and adding to the list substances such as carbon tetrachloride and 1,1,1-trichloroethane. The related industries in Japan are making efforts to meet these requirements by developing alternative substances for use in place of existing ones.

In August 1990, IPCC of United Nations Environment Program (UNEP)/WHO (the intergovernmental panel regarding climate fluctuation) issued a report about the forecast influences and countermeasures of global warming, which had been studied for three years since 1988. In Japan, corresponding studies have been conducted under the leadership of relevant government authorities. The issue of global warming is also of utmost importance to industries.

It is estimated that ca. 244×10^6 t/a of CO_2 (in terms of C) is discharged in Japan, accounting for 4.3 % of the amount discharged worldwide. Of this amount, about 4.9 % originates from the chemical industry. Since the chemical industry has curtailed energy consumption by about 40 % since the 1980s, little room is available for further energy saving. Thus, reduction of CO_2 discharge appears to be a difficult problem.

7.1.6. Basic Policy of the Japanese Chemical Industry Association about Environment and Safety

The Executive Committee of the International Council of Chemical Associations (ICCA) has set up the ICCA Principles of Responsible Care with the aim of promoting a program for the improvement of voluntary environmental and safety measures.

The Japan Chemical Industry Association concluded that the chemical industry clearly realizes its social responsibility and continues its endeavors to undertake voluntary measures in order to achieve sound growth of the chemical industry at large, at a time when humans are faced with new international problems such as global-scale environmental issues, waste disposal, and the safety of chemical substances. Japan's chemical industry has already made great progress in environmental pollution countermeasures and is at a very advanced level compared with to its U.S. and European counterparts.

7.2. Current Situation Concerning Safety and Fire Prevention Measures

As a result of a series of explosions and fires that occurred in 1973 – 1974 in chemical and oil refinery complexes in Japan, the chemical and related industries have implemented safety and fire prevention measures. Due to such efforts, large-scale explosion accidents in the plants and works have decreased drastically since 1978.

However, the number of accidents is gradually leveling off, and further safety and fire prevention measures should be taken and promoted.

7.2.1. Revision of Fire Service Law

Ignitable and flammable materials are classified as Hazardous Materials based on the Fire Service Law.

On May 24, 1988, the law for partial revision of the Fire Service Law (Law no. 55 of 1988) was promulgated, and in this relation, the revision of the government and ministerial ordinances. The essential part of this revision is the "review of the scope of hazardous materials" to define the hazardous material more clearly and introduce methods to test whether a substance is hazardous or not. The tests should be performed and judged by the enterprises themselves, following and considering these laws and Regulations.

Another important point is a review of the scope of the term "dangerous articles."

1) The term dangerous article is clearly defined, and judgment of a system by testing is introduced. The test methods and criteria are established by taking into account the definition and international test methods.
2) A designated quantity is determined by government ordinance by taking into account the danger of the articles concerned.
3) Quasi-hazardous materials and special combustible materials are regulated under the category of combustible materials.

The provisions relating to the revision of the scope of dangerous materials in this law were enforced as of May 23, 1990.

7.2.2. Safety Measures for the Transport of Dangerous Goods

For the air and marine shipments involved frequently in international transport, international Regulations have been introduced so that Japanese domestic legislation is consistent with international transport standards.

7.2.2.1. Marine Transport Standards for Hazardous Substances

Breakbulk Transport. The Ministry of Transport provided Rules for the Marine Transport and Storage of Dangerous Goods with respect to the breakbulk transport. The rules were determined with reference to the International Marine Dangerous Goods Code (IMDG Code) set by the International Maritime Organization. The notice of revision incorporating the 25th revision of the IMDG code was enforced in January 1991. With regard to container testing, the test system and UN label indication system have been internationally enforced since January 1991, and the use of qualified containers having UN labels is obligatory for the transport of dangerous goods.

Transportation in Bulk. With regard to the marine transport in bulk of hazardous liquids by chemical tankers and liquid gas tankers, the Rules for the Massive Transport and Storage of Dangerous Substances were amended as of July 1986, and the International Bulk Chemical Code (IBC) and International Gas Carrier Code (IGC) of the IMO were officially adopted.

The amendment of IBC Codes was adopted in March 1989. In response to this amendment, the Ocean Pollution Prevention Law, the Regulations on the Marine Transport and Storage of Dangerous Materials, etc. were revised and enforced in October 1990. In particular, as for the Ocean Pollution Prevention Law, it is not necessary to make prior assessment for pollution classification of goods mixed with pollutants already assessed or that contain pollutants already classified by the International Marine Organization.

7.2.2.2. Standards for Air Transport of Hazardous Cargoes

With respect to air transport, standards for the transport of hazardous cargoes (notification) in Japan are determined on the basis of the Aviaiton Act. These standards are in accordance with the Regulations on hazardous cargoes (International Civil Aviation Convention Annex no. 18) and the technical guidelines of the International Civil Air Organization (ICAO). These standards (notice) were revised on December 20, 1990, and enforced on January 1, 1991. Testing of containers and the use of containers with UN labels have been obligatory since 1991 as in the case of marine transport.

7.2.2.3. Standards for Transportation Containers for Poisonous and Toxic Substances

The Control Act on Poisonous and Toxic Substances stipulates the handling of poisonous and toxic substances.

The Pharmaceutical and Supply Bureau of the Ministry of Health and Welfare issued the standard on transportation containers for poisonous and toxic substances (no. 3; a standard for small-size transportation containers having an inner volume of 450 L or less). The standard went into effect on October 1, 1991.

This standard is in line with the United Nations Hazardous Substances Experts Committee Recommendation-no.6 Edition from the standpoint of international harmonization. Also on

October 1, notice was given of the establishment and revision of standards concerning each of the poisonous and toxic substances with respect to emergency measures in transportation accidents as well as disposal.

7.2.2.4. Safety Measures in the Use of Gases Under High Pressure

Safety Regulations with respect to high-pressure gas are in accordance with the High-Pressure Gas Control Act and governmental ordinances associated therewith.

The Industrial Location and Environmental Protection Bureau of the Ministry of International Trade and Industry has been holding meetings of the High-Pressure Gas Safety Policy Committee since September 1989 to study the best possible way of organizing high-pressure gas safety administration in the future. In June 1991, a report was published, proposing the following:

1) To promote self-security (i.e., self-protection measures and attitude for security)
2) To strengthen safety measures in response to diversification of applications
3) To review import Regulations from the standpoint of internationalization
4) To promote international cooperation in the field of safety
5) To strengthen execution systems by making full use of private-sector inspection organizations
6) To enhance systems supporting voluntary security activities

Since July 1991, the Ministry of International Trade and Industry has been holding meetings of the High-Pressure Gas and Explosives Safety Review Committee to study the future administration of high-pressure gas safety, including revision of the High-Pressure Gas Control Act.

7.3. Present Situation Regarding Safety Measures for Chemical Substances

In Japan, two laws exist for maintaining the safety of chemical substances under the notification system for new chemical substances prior to the manufacture thereof: (1) The Law Concerning Examination and Regulation of Manufacture, etc. of Chemical Substances, which is designed to regulate substances that are hazardous to human health because of contamination of the environment and (2) the system of Investigation of Toxicity of Chemical Substances under the Industrial Safety and Health Law, which is to regulate the concentration of carcinogenic substances in the workplace.

7.3.1. The Law Concerning Examination and Regulation of Manufacture, etc. of Chemical Substances

The Law Concerning Examination and Regulation of Manufacture, etc. of Chemical Substances had been, until its revision in April 1987, implemented effectively for the 14 years since its enforcement in 1973 to regulate PCBs and similar substances.

However, as Regulations became stricter in Europe and the United States, the need arose to coordinate Regulations on chemical substances in various countries and to expand Regulations on harmful chemical substances that are slowly degraded in the environment. The Japanese government then enacted a Law for the Partial Revision of the Law Concerning Examination and Regulation of Manufacture, etc. of Chemical Substances in May 1986 and implemented it in April 1987. The gist of the revised law is as follows:

1) The system of prior examination of new chemical substances was reinforced by demanding that toxicity tests, namely, mutagenicity tests (Ames test, chromosonal aberrations test) and repeated dose tests are to be conducted for 28 days, in addition to the biodegradability and accumulation tests now in force. The tests must be performed, at the time the government is notified of the chemical substances, prior to their manufacture and import.
2) Substances that are difficultly degradable and are suspected to have chronic toxicity are to be termed "Designated Chemical Substances" and must be put under surveillance with respect to their production volume and application after manufacture and import.

3) When such "Designated Chemical Substances" are suspected to be contaminating the environment and be hazardous to human health, the manufacturer or importer will be advised to perform a study on hazards of such substances.
4) If such studies reveal that the substances are hazardous, they are termed "Class II Specified Chemical Substances," and their production volume and handling methods must be regulated.
5) "Specific Chemical Substances" with respect to the old Law Concerning Examination and Regulation of Manufacture, etc. of Chemical Substances are now termed "Class I Specified Chemical Substances," and the same control measures, such as ban on their manufacture in principle and on their use in open systems, are used as in the past.
6) As a result of tests conducted by the government on existing chemical substances (which correspond to a prior examination for new substances), some of them may be termed "Designated Chemical Substances."

7.3.2. Situation Regarding the Enforcement of Law Concerning Examinations and Regulations of Manufacture, etc. of Chemical Substances ("Chemical Substances Law")

Under the old Chemical Substances Law (until March 1987) the reported number of new chemical substances admitted for manufacture and import was 3864; under the new Chemical Substances Law (April 1987 – December 1993), which applies more stringent Regulations, it is 1492.

The designated chemical substances as of the end of September 1993 amounted to 97, including chlorohydrocarbons and organotin compounds. Among them are 135 substances (excluding the 23 substances that were regrouped to the "Class II Specified Chemical Substances") whose manufacture and import quantities must be reported. The 23 specified chemical substances inlcude chlorohydrocarbons (trichloroethylene, tetrachloroethylene, carbon tetrachloride) and organotin compounds (e.g., triphenyl- and tributyltin compounds). Furthermore, safety measures for each substance such as the preparation of proper use manuals and labeling of containers are being taken by individual industry organizations.

As far as existing chemical substances are concerned, the government continues to conduct systematic safety monitoring. The Ministry of International Trade and Industry checked the safety of 970 substances from 1974 through 1993. The result was that 358 substances decomposed in solvents to nonhazardous compounds, 594 substances showed low and 18 substances high accumulation of hazardous compounds upon decomposition. Of these, the nine substances or classes of substances listed below are designated Class I Specified Chemical Substances. In addition, the Ministry of Health and Welfare conducts toxicity testing of each substance.

Polychlorinated biphenyls
Polychlorinated napththalenes
Hexachlorobenzene
Aldrin
Dieldrin
Endrin
DDT
Chlordane
Bis(tributyltin)oxide

More careful handling and discharge, not only of those substances that are regulated specifically but of all chemical substances in open systems, is desired.

7.3.3. International Cooperation on Chemical Substance Regulations

In Japan, the Good Laboratory Practice (GLP) system was introduced in 1986, and the coordination of testing methods with those of OECD and the mutual acceptance of test data were promoted.

Moreover, the OECD started to promote the work of international cooperation for investigating the safety of existing chemical substances. Specifically, the preparation of data bases (EXICHEM) on the testing status in member countries and joint investigation of chemical products of high production volume are being put in force. Concerning the joint investigation of chemical products with a high production volume, 53 substances were allocated in April 1990 to each country as Phase 1 in accordance with the investigation priority substances (P 1) list. Japan took charge of nine substances and collected data on them. Based on the result of this investigation, tests in 1991 were conducted with respect to characteristics on which no data were available or the data were indefinite. In addition, investigation of about 100 substances (Phases 2 and 3) was initiated toward the end of 1991. In this program Japan is to evaluate data on 24 substances. Furthermore, until the year 2000 about 5000 chemical substances of high production volume including the above figures will be surveyed. Japan will evaluate one-fourth of those substances.

8. References

1. Council Directive 80/778/EEC Relating to the Quality of Water Intended for Human Consumption, OJL 229, August 30, 1980, p. 11.J.
2. J. Cameron, J. Aubouchay: "Precautionary Principle: a Fundamental Principle of Laws and Policy for the Protection of the Global Environment," Boston College International and Comparative Law Review, 1991.
3. E. Rehbinder, "The Precautionary Principle in an International Perspective," in Miljørettens grundspøgsmål – bidrag til en nordisk forskeruddannelse – G-E-C Gad, København 1994.
4. M. Rémond-Gouilloud, "Le risque de l'incertain: la responsabilité face aux avancées de la science" – La Vie des Sciences, *Comptes rendus, série générale* vol. 10 (1993) no. 4, 341 – 357 – Archives de l'Académie des Sciences, Paris.
5. Oslo and Paris Conventions, Paris, 21 – 22 Sep. 1992, vol. 14 (1991) no. 1, p. 16.
6. Treaty on European Union, Maastricht, 7 Feb. 1992.
7. "London Declaration," *2nd Int. Conf. on the Protection of the North Sea,* 24 – 25 Nov. 1987.
8. [C(72)128], OECD 1972; [C(74)223, OECD 1974.
9. [C(90)177 (final)], OECD 1991.
10. O.J.E.C., L 169 1 – 29, June 26, 1987.
11. H. Smets: "Le principe pollueur payeur, un principe de l'environment," *Rev. Gen. Droit International Public* **2** (1993) April/June.
12. Regulation 93/259/EEC on the Supervision and Control of Shipments within, into and out of the Community.
13. International Information Exchange and the PIC Procedure (UNEP/FAO).
14. UNEP (1989): London Guidelines for the Exchange of Information on Chemicals in International Trade (1987, amended 1989), UNEP, Nairobi.
15. FAO (1989): International Code of Conduct on the Distribution and Use of Pesticides (1985, amended 1989, FAO, Rome).
16. FAO/UNEP (1991): Guidance for Governments: Operation of the Prior Informed Consent Procedure for Banned or Severely Restricted Chemicals in International Trade, FAO/UNEP, Rome/Geneva.
17. Council Regulation (EEC) No 2455/92 or July 23, 1992 (amended on January 11, 1994) concerning the Export and Import of Certain Dangerous Chemicals.
18. Rege, "GATT Law and Environment-Related Issues Affecting the Trade of Developing Countries," in Journal of World Trade Law (June 1994).
19. "Towards Sustainability," European Commission Fifth Environmental Programme [*COM* (92) 93 final]
20. Resolution of the Council and the representatives of the Governments of the Member States on the fifth Community Policy and Action Programme for the Environment and Sustainable Development, adopted on March 18, 1992, OJC 138, May 17, 1993.
21. Council Directive 90/313/EEC On the Freedom of Access to Information on the Environment of June 7, 1990 and Communication to the Council, Parliament and the Economic and Social Committee 93/C 156/05 on Public Access to the Institutions' Documents.
22. Council Regulation 880/92/EEC on a Community Award Scheme for an Eco-Label (OJL 99, April 11, 1992)
23. OJC 72/3, March 10, 1994.
24. R. Lönngren: *International Approach to Chemical Control: A Historic Overview,* Kemi, The National Chemical Inspectorate, Stockholm 1992.
25. EEC Treaty, Article 5.
26. EEC Treaty (1987 – 1993), Article 130 r (4). Maastricht Treaty on European Union, Article 3b.
27. Directive 67/548/EEC Relating to the Classification, Packaging and Labelling of

Dangerous Substances, O.J. 1967, no. 196, p. 1 with later amendments.
28. Directive 73/173/EEC relating to the Classification, Packaging and Labelling of Dangerous Preparations (Solvents), *O.J.* 1973, no. L 189, p. 7, with later amendments.
29. Directive 78/631/EEC Relating to the Classification, Packaging and Labelling of Dangerous Preparations (Pesticides), *O.J.* 1978, no. L 206, p. 13, with later amendments.
30. Directive 77/728/EEC Relating to the Classification, Packaging and Labelling of Paints, Varnishes, Printing Inks, Adhesives and Similar Products, *O.J.* 1977, no. L 303, p. 23 with later amendments.
31. Example taken from Directive 67/548/EEC (note 3) in its version of the 6th amendment, Directive 79/831/EEC, *O.J.* 1979, no. L 259, p. 10.
32. Directive 76/769/EEC on the Approximation of the Laws, Regulations, and Administrative Provisions of the Member States Relating to Restrictions on the Marketing and Use of Certain Dangerous Substances and Preparations, *O.J.* 1976, p. 201 with later amendments.
33. Directive 79/831 (note 7), recital 1.
34. EEC Treaty, Article 130 r (2).
35. Directive 90/220/EEC on the Deliberate Release in to the Environment of Genetically Modified Organisms, *O.J.* 1990, no. L 117, p. 15.
36. Regulation 594/91/EEC on Substances that Deplete the Ozone Layer, *O.J.* 1991, no. L 67, p. 1; now replaced by Regulation 3093/94, O.J. 1994 no. L 333, p. 1.
37. Directive 91/156/EEC on Waste, amending Directive 75/442/EEC, O.J. 1991, no. L 78, p. 32; Directive 91/689/EEC on Hazardous Waste, *O.J.* 1991, no. L 377, p. 20.
38. Directive 76/464/EEC on Pollution caused by Certain Dangerous Substances Discharged into the Aquatic Environment of the Community, O.J. 1976, no. L 129, p. 23; 80/68/EEC on the Protection of Groundwater Against Pollution Caused by Certain Dangerous Substances, *O.J.* 1980, no. L 20, p. 43; 84/360/EEC on the Combating of Air Pollution from Industrial Plants; *O.J.* 1984, no. L 188, p. 20.
39. Directive 82/501/EEC on the Major Accident Hazards of Certain Industrial Activities, *O.J.* 1982, no. L 230, p. 1 with later amendments.
40. Directive 85/337/EEC on the Assessment of the Effects of Certain Public and Private Projects on the Environment, *O.J.* 1985, no. L 175, p. 40.
41. Regulation 1836/93 allowing Voluntary Participation by Companies in the Industrial Sector in a Community Eco-management and Audit Scheme, *O.J.* 1993, no. L 168, p. 1.
42. Commission Proposal for a Council Directive on Integrated Pollution Prevention and Control, O.J. 1993, no. C 311, p. 6.
43. Directive 92/112/EEC, *O.J.* 1992, no. L 409, p. 11.
44. UNEP, "Liability and Compensation for Damage resulting from the transboundary Movement of Hazardous Wastes and their Disposal." Basel Convention, Montevideo, Dec. 1992, draft articles.
45. Directive on Civil Liability caused by Defective Products, *O.J.E.C.* no. L 210, August 7, 1985.
46. Amended proposal for a Directive on Civil Liability for Damage caused by Waste, *O.J.E.C.,* no. L 192/6 (July 23, 1991).
47. Communication to Council and Parliament: Green Paper on Remedying Environmental Damage, *COM* (93), 47, March 1993.
48. Sections of this chapter have been taken, in amended form, from two previous publications by G. H. Williamson: "Environmental Protection versus Free Trade: A European Community Perspective," in: *Private Investments Abroad,* Southwestern Legal Foundation, Dallas, Texas, 1993, Chap. 3. G. H. Williamson: "The European Union's New Regulation on Waste Shipments: Its Content and Practical Implementation Concerns," *Environment Watch, Western Europe,* May 6, 1994.
49. "Completing the Internal Market: White Paper from the Commission to the European Council," Commission of the European Communities, June 1985.
50. Commission vs. Council, Judgment of June 11, 1991, on Dir. 89/428 of the Council.
51. Paragraph 2, new Article 130 R, added to Part Three of the Treaty by Article 25 of the Single European Act.
52. Treaty on European Union, signed at Maastricht on Feb. 7, 1992 (*O.J.E.C.* C191, July 25, 1992).
53. *O.J.E.C.* 1983, L 109/8, amended by Directive 88/182/EEC, O.J.E.C. 1988, L 81/75.
54. Rewe-Zentral AG vs. Bundesmonopolverwaltung für Branntwein (Case 120/78), Judgment of Feb. 20, 1979. ECR (1979) p. 649.

55. Commission vs. Denmark (Case 302/86), Judgment of Sep. 20, 1988, ECR (1988) p. 4607;1 CMLR 408.
56. Verordnung über die Vermeidung von Verpackungsabfällen, Bundesgesetzblatt 1991, part I, p. 1234.
57. Common Position No. 13/94 with a View to Adopting a European Parliament and Council Directive on Packaging and Packaging Waste, 94/C/137/08, *O.J.E.C.* C 137/65, May 19, 1994.
58. Regulation 259/92, *O.J.E.C.* L 30/1, Feb. 2, 1993.
59. Directive 75/442/EEC, *O.J.E.C.* L 194/39, July 25, 1975.
60. Directive 91/156/EEC, *O.J.E.C.* L 78/32, March 26, 1991.
61. Directive 91/689/EEC, *O.J.E.C.* L 377/21, Dec. 31, 1991, as amended.
62. Directive 78/319/EEC, *O.J.E.C.* L 84/43, March 31, 1978.
63. Directive 84/631/EEC, *O.J.E.C.* L 326/31, Dec. 13, 1984.
64. Commission Decision 94/3/EC, *O.J.E.C.* L 5/15, Jan. 7, 1994.
65. 75/442/EEC and 91/156/EEC.
66. 91/689/EEC; List published at O.J.E.C. L 356/14 seq., Dec. 31, 1994.
67. Commission vs. Belgium, E.C.J. Case no. C-2/90, decided July 9, 1992.
68. S. M. Novick et al. (eds.): *Law of Environmental Protection,* Environmental Law Institute, Washington, D.C. 1987.
69. T. T. Smith, Jr., R. D. Hunter (eds.): *European Community Deskbook,* Environmental Law Institute, Washington, D.C. 1992.
70. T. T. Smith, Jr., et al.: "Understanding European Environmental Regulation," *Conference Board Monograph,* report no. 1026, New York 1993.
71. 42 U.S.C. §§ 4321-70C; Directive 85/337, O.J.E.C. L 175/40, July 5, 1985.
72. Directive 82/501, *O.J.E.C.* L 230/1, Aug. 5, 1982.
73. 42 U.S.C. $ 11001 (a) – (c).
74. 33 U.S.C. § 1321.
75. The Clean Air Act, 42 U.S.C. §§ 7401 ff, as amended; the Clean Water Act, 33 U.S.C. § 1251 – 1376. The EC Air and Water Legislation is Summarized and Cited in Ref. [69].
76. The Resource Conservation and Recovery Act, 42 U.S.C. §§ 6901 ff. The EC Waste Legislation is Summarized in Ref. [69] pp. 24 – 28.
77. Directive 75/442, *O.J.E.C.* L 194/26 (July 25, 1975), as amended; Directive 78/319, *O.J.E.C.* L 78/32 (Mar. 31, 1978), as amended.
78. Directive 84/631, *O.J.E.C.* L 326/31 (Dec. 13, 1984), as amended.
79. 42 U.S.C. §§ 9601 et seq.
80. Directive 67/548, *O.J.E.C.* L 196/1 (Aug. 16, 1967), as amended.
81. Directive 79/831, *O.J.E.C.* L 259 (Oct. 15, 1979).
82. T. T. Smith, Jr., R. D. Hunter (eds.):*European Community* in [69] pp. 28 – 31.
83. 15 U.S.C. §§ 2601 – 29.
84. 29 C.R.R. § 1910.1200, issued under 5 U.S.C. 553.
85. T. T. Smith, Jr., et al.: in [70] p. 7 – 10.
86. T. T. Smith, Jr., et al.: in [70] p. 10.

ULLMANN'S

Industrial Toxicology

Volume 2

Toxic Agents, Pharmacological and Medical Fundamentals

ULLMANN'S

Industrial Toxicology

Volume 1
Toxicology in Occupational and Environmental Setting

Volume 2
Toxic Agents, Pharmacological and Medical Fundamentals

Further Ullmann's Publications

WILEY-VCH (Ed.)
Ullmann's Encyclopedia of Industrial Chemistry
Sixth, Completely Revised Edition, 40 Volumes
2003, ISBN 3-527-30385-5

Wiley-VCH (Ed.)
Ullmann's Processes and Process Engineering, 3 Volumes
2004, ISBN 3-527-31096-7

Wiley-VCH (Ed.)
Ullmann's Chemical Engineering and Plant Design
2004, ISBN 3-527-31111-4

WILEY-VCH (Ed.)
Ullmann's Electronic Release 2005 Online plus CD-ROM
2005, ISBN 3-527-31097-5
http://interscience.wiley.com/ullmanns

WILEY-VCH (Ed.)
Industrial Inorganic Chemicals and Products
An Ullmann's Encyclopedia, 6 Volumes
1999, ISBN 3-527-29567-4

WILEY-VCH (Ed.)
Industrial Organic Chemicals – Starting Materials and Intermediates
An Ullmann's Encyclopedia, 8 Volumes
1999, ISBN 3-527-29645-X

ULLMANN'S

Industrial Toxicology

Volume 2

Toxic Agents, Pharmacological and Medical Fundamentals

WILEY-VCH Verlag GmbH & Co. KGaA

Library of Congress Card No.:
applied for

British Library Cataloguing-in-Publication Data:
A catalogue record for this book is available from the British Library.

Bibliographic information published by Die Deutsche Bibliothek
Die Deutsche Bibliothek lists this publication in the Deutsche Nationalbibliografie; detailed bibliographic data is available in the Internet at http://dnb.ddb.de

Typesetting: Steingraeber Satztechnik GmbH, Dossenheim

Printing: and Binding: betzdruck GmbH, Darmstadt

Binding: Schäffer GmbH, Grünstadt

Cover Design: Gunther Schulz, Fußgönheim

Printed in the Federal Republic of Germany
Printed on acid-free paper

ISBN-10: 3-527-31247-1
ISBN-13: 978-3-527-31247-4

Preface

Since the entire 40-volume Ullmann's Encyclopedia is inaccessible to many readers – particularly individuals, smaller companies or institutes – all the information on industrial toxicology, ecotoxicology, process safety as well as occupational health and safety has been condensed into this convenient 2-volume set.

Based on the latest online edition of Ullmann's containing articles never been before in print, this ready reference provides practical information on applying the science of toxicology in both the occupational and environmental setting, and explains the fundamentals necessary for an understanding of the effects of chemical hazards on humans and ecosystems. The detailed and meticulously edited articles have been written by renowned experts from industry and academia, and much of the information has been thoroughly revised. Alongside explanations of safety regulations and legal aspects, this set covers food additives, toxic and mutagenic agents as well as medical and therapeutical issues. Top-quality illustrations, clear diagrams and charts combined with an extensive use of tables enhance the presentation and provide a unique level of detail. Deeper insights into any given area of interest is offered by referenced contributions, while rapid access to a particular subject is enhanced by both a keyword and author index.

We are convinced that these handy two volumes are a convenient one-stop resource for health protection professionals, environmental scientists, safety engineers and all those involved in the field of industrial toxicology.

Summer, 2005
Weinheim

The Publisher

Contents

Volume 1

Toxicology in Occupational and Environmental Setting

Volume 2

Toxic Agents, Pharmacological and Medical Fundamentals

Ullmann's Industrial Toxicology

ISBN: 3-527-31247-1

Symbols and Units

Symbols and units agree with SI standards (for conversion factors see page XI). The following list gives the most important symbols used in the encyclopedia. Articles with many specific units and symbols have a similar list as front matter.

Symbol	Unit	Physical Quantity
a_{B}		activity of substance B
A_{r}		relative atomic mass (atomic weight)
A	m^2	area
c_{B}	$\mathrm{mol/m}^3$, mol/L (M)	concentration of substance B
C	C/V	electric capacity
c_p, c_v	$\mathrm{J\,kg^{-1}K^{-1}}$	specific heat capacity
d	cm, m	diameter
d		relative density ($\varrho/\varrho_{\mathrm{water}}$)
D	$\mathrm{m^2/s}$	diffusion coefficient
D	Gy (= J/kg)	absorbed dose
e	C	elementary charge
E	J	energy
E	V/m	electric field strength
E	V	electromotive force
E_{A}	J	activation energy
f		activity coefficient
F	C/mol	Faraday constant
F	N	force
g	$\mathrm{m/s^2}$	acceleration due to gravity
G	J	Gibbs free energy
h	m	height
$\hbar$	$\mathrm{W \cdot s^2}$	Planck constant
H	J	enthalpy
I	A	electric current
I	cd	luminous intensity
k	(variable)	rate constant of a chemical reaction
k	J/K	Boltzmann constant
K	(variable)	equilibrium constant
l	m	length
m	g, kg, t	mass
M_{r}		relative molecular mass (molecular weight)
n_{D}^{20}		refractive index (sodium D-line, 20 °C)
n	mol	amount of substance
N_{A}	$\mathrm{mol^{-1}}$	Avogadro constant ($6.023\times10^{23}\ \mathrm{mol^{-1}}$)
p	Pa, bar *	pressure
Q	J	quantity of heat
r	m	radius
R	$\mathrm{J\,K^{-1}mol^{-1}}$	gas constant
R	Ω	electric resistance
S	J/K	entropy
t	s, min, h, d, month, a	time

Symbols and Units (Continued from p. IX)

Symbol	Unit	Physical Quantity
t	°C	temperature
T	K	absolute temperature
u	m/s	velocity
U	V	electric potential
U	J	internal energy
V	m^3, L, mL, μL	volume
w		mass fraction
W	J	work
x_B		mole fraction of substance B
Z		proton number, atomic number
α		cubic expansion coefficient
α	$W\,m^{-2}K^{-1}$	heat-transfer coefficient (heat-transfer number)
α		degree of dissociation of electrolyte
$[\alpha]$	$10^{-2} deg\,cm^2 g^{-1}$	specific rotation
η	$Pa \cdot s$	dynamic viscosity
θ	°C	temperature
$\varkappa$		c_p/c_v
λ	$W\,m^{-1}K^{-1}$	thermal conductivity
λ	nm, m	wavelength
μ		chemical potential
ν	Hz, s^{-1}	frequency
ν	m^2/s	kinematic viscosity (η/ϱ)
π	Pa	osmotic pressure
ϱ	g/cm^3	density
σ	N/m	surface tension
τ	Pa (N/m^2)	shear stress
φ		volume fraction
χ	Pa^{-1} (m^2/N)	compressibility

* The official unit of pressure is the pascal (Pa).

Conversion Factors

SI unit	Non-SI unit	From SI to non-SI multiply by
Mass		
kg	pound (avoirdupois)	2.205
kg	ton (long)	9.842×10^{-4}
kg	ton (short)	1.102×10^{-3}
Volume		
m^3	cubic inch	6.102×10^{4}
m^3	cubic foot	35.315
m^3	gallon (U.S., liquid)	2.642×10^{2}
m^3	gallon (Imperial)	2.200×10^{2}
Temperature		
°C	°F	$°C\times1.8+32$
Force		
N	dyne	1.0×10^{5}
Energy, Work		
J	Btu (int.)	9.480×10^{-4}
J	cal (int.)	2.389×10^{-1}
J	eV	6.242×10^{18}
J	erg	1.0×10^{7}
J	$kW\cdot h$	2.778×10^{-7}
J	$kp\cdot m$	1.020×10^{-1}
Pressure		
MPa	at	10.20
MPa	atm	9.869
MPa	bar	10
kPa	mbar	10
kPa	mm Hg	7.502
kPa	psi	0.145
kPa	torr	7.502

Powers of Ten

E (exa)	10^{18}	d (deci)	10^{-1}
P (peta)	10^{15}	c (centi)	10^{-2}
T (tera)	10^{12}	m (milli)	10^{-3}
G (giga)	10^{9}	µ (micro)	10^{-6}
M (mega)	10^{6}	n (nano)	10^{-9}
k (kilo)	10^{3}	p (pico)	10^{-12}
h (hecto)	10^{2}	f (femto)	10^{-15}
da (deca)	10	a (atto)	10^{-18}

Abbreviations

The following is a list of the abbreviations used in the text. Common terms, the names of publications and institutions, and legal agreements are included along with their full identities. Other abbreviations will be defined wherever they first occur in an article. For further abbreviations, see page IX, Symbols and Units; page XVI, Frequently Cited Companies (Abbreviations), and page XVII, Country Codes in patent references. The names of periodical publications are abbreviated exactly as done by Chemical Abstracts Service.

abs. absolute
a.c. alternating current
ACGIH American Conference of Governmental Industrial Hygienists
ACS American Chemical Society
ADI acceptable daily intake
ADN accord européen relatif au transport international des marchandises dangereuses par voie de navigation interieure (European agreement concerning the international transportation of dangerous goods by inland waterways)
ADNR ADN par le Rhin (regulation concerning the transportation of dangerous goods on the Rhine and all national waterways of the countries concerned)
ADP adenosine 5′-diphosphate
ADR accord européen relatif au transport international des marchandises dangereuses par route (European agreement concerning the international transportation of dangerous goods by road)
AEC Atomic Energy Commission (United States)
a.i. Active ingredient
AIChE American Institute of Chemical Engineers
AIME American Institute of Mining, Metallurgical, and Petroleum Engineers
ANSI American National Standards Institute
AMP adenosine 5′-monophosphate
APhA American Pharmaceutical Association
API American Petroleum Institute
ASTM American Society for Testing and Materials
ATP adenosine 5′-triphosphate
BAM Bundesanstalt für Materialprüfung (Federal Republic of Germany)
BAT Biologischer Arbeitsstoff-Toleranz-Wert (biological tolerance value for a working material, established by MAK Commission, see MAK)
Beilstein Beilstein's Handbook of Organic Chemistry, Springer, Berlin – Heidelberg – New York
BET Brunauer – Emmett – Teller
BGA Bundesgesundheitsamt (Federal Republic of Germany)
BGBl. Bundesgesetzblatt (Federal Republic of Germany)
BIOS British Intelligence Objectives Subcommitee Report (see also FIAT)
BOD biological oxygen demand
bp boiling point
B.P. British Pharmacopeia
BS British Standard
ca. circa
calcd. calculated
CAS Chemical Abstracts Service
cat. catalyst, catalyzed
CEN Comité Européen de Normalisation
cf. compare
CFR Code of Federal Regulations (United States)
cfu colony forming units
Chap. chapter
ChemG Chemikaliengesetz (Federal Republic of Germany)
C.I. Colour Index
CIOS Combined Intelligence Objectives Subcommitee Report (see also FIAT)
CNS central nervous system
Co. Company
COD chemical oxygen demand
conc. concentrated
const. constant
Corp. Corporation
crit. critical

CTFA The Cosmetic, Toiletry and Fragrance Association (United States)
DAB 9 Deutsches Arzneibuch, 9th ed., Deutscher Apotheker-Verlag, Stuttgart 1986
d.c. direct current
decomp. decompose, decomposition
DFG Deutsche Forschungsgemeinschaft (German Science Foundation)
dil. dilute, diluted
DIN Deutsche Industrie Norm (Federal Republic of Germany)
DMF dimethylformamide
DNA deoxyribonucleic acid
DOE Department of Energy (United States)
DOT Department of Transportation – Materials Transportation Bureau (United States)
DTA differential thermal analysis
EC effective concentration
EC European Community
ed. editor, edition, edited
e.g. for example
emf electromotive force
EmS Emergency Schedule
EN European Standard (European Community)
EPA Environmental Protection Agency (United States)
EPR electron paramagnetic resonance
Eq. equation
ESCA electron spectroscopy for chemical analysis
esp. especially
ESR electron spin resonance
Et ethyl substituent ($-C_2H_5$)
et al. and others
etc. et cetera
EVO Eisenbahnverkehrsordnung (Federal Republic of Germany)
exp (...) $e^{(\cdots)}$, mathematical exponent
FAO Food and Agriculture Organization (United Nations)
FDA Food and Drug Administration (United States)
FD & C Food, Drug and Cosmetic Act (United States)
FHSA Federal Hazardous Substances Act (United States)
FIAT Field Information Agency, Technical (United States reports on the chemical industry in Germany, 1945)
Fig. figure
fp freezing point
Friedländer P. Friedländer, Fortschritte der Teerfarbenfabrikation und verwandter Industriezweige, Vol. 1 – 25, Springer, Berlin 1888 – 1942
FT Fourier transform
(g) gas, gaseous
GC gas chromatography
GefStoffV Gefahrstoffverordnung (regulations in the Federal Republic of Germany concerning hazardous substances)
GGVE Verordnung in der Bundesrepublik Deutschland über die Beförderung gefährlicher Güter mit der Eisenbahn (regulation in the Federal Republic of Germany concerning the transportation of dangerous goods by rail)
GGVS Verordnung in der Bundesrepublik Deutschland über die Beförderung gefährlicher Güter auf der Straße (regulation in the Federal Republic of Germany concerning the transportation of dangerous goods by road)
GGVSee Verordnung in der Bundesrepublik Deutschland über die Beförderung gefährlicher Güter mit Seeschiffen (regulation in the Federal Republic of Germany concerning the transportation of dangerous goods by sea-going vessels)
GLC gas-liquid chromatography
Gmelin Gmelin's Handbook of Inorganic Chemistry, 8th ed., Springer, Berlin – Heidelberg – New York
GRAS generally recognized as safe
Hal halogen substituent ($-F$, $-Cl$, $-Br$, $-I$)
Houben-Weyl Methoden der organischen Chemie, 4th ed., Georg Thieme Verlag, Stuttgart
HPLC high performance liquid chromatography
IAEA International Atomic Energy Agency
IARC International Agency for Research on Cancer, Lyon, France

IATA-DGR International Air Transport Association, Dangerous Goods Regulations
ICAO International Civil Aviation Organization
i.e. that is
i.m. intramuscular
IMDG International Maritime Dangerous Goods Code
IMO Inter-Governmental Maritime Consultive Organization (in the past: IMCO)
Inst. Institute
i.p. intraperitoneal
IR infrared
ISO International Organization for Standardization
IUPAC International Union of Pure and Applied Chemistry
i.v. intravenous
Kirk-Othmer Encyclopedia of Chemical Technology, 3rd ed., J. Wiley & Sons, New York – Chichester – Brisbane – Toronto 1978 – 1984; 4th ed., J. Wiley & Sons, New York – Chichester – Brisbane – Toronto 1991 – 1998
(l) liquid
Landolt-Börnstein Zahlenwerte u. Funktionen aus Physik, Chemie, Astronomie, Geophysik u. Technik, Springer, Heidelberg 1950 – 1980; Zahlenwerte und Funktionen aus Naturwissenschaften und Technik, Neue Serie, Springer, Heidelberg, since 1961
LC_{50} lethal concentration for 50 % of the test animals
LCLo lowest published lethal concentration
LD_{50} lethal dose for 50 % of the test animals
LDLo lowest published lethal dose
ln logarithm (base e)
LNG liquefied natural gas
log logarithm (base 10)
LPG liquefied petroleum gas
M mol/L
M metal (in chemical formulas)
MAK Maximale Arbeitsplatz-Konzentration (maximum concentration at the workplace in the Federal Republic of Germany); cf. Deutsche Forschungsgemeinschaft (ed.): Maximale Arbeitsplatz-konzentrationen (MAK) und Biologische Arbeitsstoff-Toleranz-Werte (BAT), WILEY-VCH Verlag, Weinheim (published annually)
max. maximum
MCA Manufacturing Chemists Association (United States)
Me methyl substituent ($-CH_3$)
Methodicum Chimicum Methodicum Chimicum, Georg Thieme Verlag, Stuttgart
MFAG Medical First Aid Guide for Use in Accidents Involving Dangerous Goods
MIK maximale Immissionskonzentration (maximum immission concentration)
min. minimum
mp melting point
MS mass spectrum, mass spectrometry
NAS National Academy of Sciences (United States)
NASA National Aeronautics and Space Administration (United States)
NBS National Bureau of Standards (United States)
NCTC National Collection of Type Cultures (United States)
NIH National Institutes of Health (United States)
NIOSH National Institute for Occupational Safety and Health (United States)
NMR nuclear magnetic resonance
no. number
NOEL no observed effect level
NRC Nuclear Regulatory Commission (United States)
NRDC National Research Development Corporation (United States)
NSC National Service Center (United States)
NSF National Science Foundation (United States)
NTSB National Transportation Safety Board (United States)
OECD Organization for Economic Cooperation and Development
OSHA Occupational Safety and Health Administration (United States)

p., pp. page, pages
Patty G. D. Clayton, F. E. Clayton (eds.): Patty's Industrial Hygiene and Toxicology, 3rd ed., Wiley Interscience, New York
PB report Publication Board Report (U.S. Department of Commerce, Scientific and Industrial Reports)
PEL permitted exposure limit
Ph phenyl substituent ($-C_6H_5$)
Ph. Eur. European Pharmacopoeia, 2nd. ed., Council of Europe, Strasbourg 1981
phr part per hundred rubber (resin)
PNS peripheral nervous system
ppm parts per million
q. v. which see (quod vide)
ref. refer, reference
resp. respectively
R_f retention factor (TLC)
R. H. relative humidity
RID règlement international concernant le transport des marchandises dangereuses par chemin de fer (international convention concerning the transportation of dangerous goods by rail)
RNA ribonucleic acid
R phrase (R-Satz) risk phrase according to ChemG and GefStoffV (Federal Republic of Germany)
rpm revolutions per minute
RTECS Registry of Toxic Effects of Chemical Substances, edited by the National Institute of Occupational Safety and Health (United States)
(s) solid
SAE Society of Automotive Engineers (United States)
s.c. subcutaneous
SI International System of Units
SIMS secondary ion mass spectrometry
S phrase (S-Satz) safety phrase according to ChemG and GefStoffV (Federal Republic of Germany)
STEL Short Term Exposure Limit (see TLV)
STP standard temperature and pressure (0° C, 101.325 kPa)
T_g glass transition temperature
TA Luft Technische Anleitung zur Reinhaltung der Luft (clean air regulation in Federal Republic of Germany)
TA Lärm Technische Anleitung zum Schutz gegen Lärm (low noise regulation in Federal Republic of Germany)
TDLo lowest published toxic dose
THF tetrahydrofuran
TLC thin layer chromatography
TLV Threshold Limit Value (TWA and STEL); published annually by the American Conference of Governmental Industrial Hygienists (ACGIH), Cincinnati, Ohio
TOD total oxygen demand
TRK Technische Richtkonzentration (lowest technically feasible level)
TSCA Toxic Substances Control Act (United States)
TÜV Technischer Überwachungsverein (Technical Control Board of the Federal Republic of Germany)
TWA Time Weighted Average
UBA Umweltbundesamt (Federal Environmental Agency)
Ullmann Ullmann's Encyclopedia of Industrial Chemistry, 6th ed., Wiley-VCH, Weinheim, 2002, 5th ed., VCH Verlagsgesellschaft, Weinheim, 1985 – 1996; Ullmanns Encyklopädie der Technischen Chemie, 4th ed., Verlag Chemie, Weinheim 1972 – 1984
USAEC United States Atomic Energy Commission
USAN United States Adopted Names
USD United States Dispensatory
USDA United States Department of Agriculture
U.S.P. United States Pharmacopeia
UV ultraviolet
UVV Unfallverhütungsvorschriften der Berufsgenossenschaft (workplace safety regulations in the Federal Republic of Germany)
VbF Verordnung in der Bundesrepublik Deutschland über die Errichtung und den Betrieb von Anlagen zur Lagerung, Abfüllung und Beförderung brennbarer Flüssigkeiten (regulation in the Federal Republic of Germany

	concerning the construction and operation of plants for storage, filling, and transportation of flammable liquids; classification according to the flash point of liquids, in accordance with the classification in the United States)
VDE	Verband Deutscher Elektroingenieure (Federal Republic of Germany)
VDI	Verein Deutscher Ingenieure (Federal Republic of Germany)
vol	volume
vol.	volume (of a series of books)
vs.	versus
WGK	Wassergefährdungsklasse (water hazard class)
WHO	World Health Organization (United Nations)
Winnacker-Küchler	Chemische Technologie, 4th ed., Carl Hanser Verlag, München, 1982-1986; Winnacker-Küchler, Chemische Technik: Prozesse und Produkte, Wiley-VCH, Weinheim, from 2003
wt	weight
$	U.S. dollar, unless otherwise stated

Frequently Cited Companies (Abbreviations)

Air Products	Air Products and Chemicals
Akzo	Algemene Koninklijke Zout Organon
Alcoa	Aluminum Company of America
Allied	Allied Corporation
Amer. Cyanamid	American Cyanamid Company
BASF	BASF Aktiengesellschaft
Bayer	Bayer AG
BP	British Petroleum Company
Celanese	Celanese Corporation
Daicel	Daicel Chemical Industries
Dainippon	Dainippon Ink and Chemicals Inc.
Dow Chemical	The Dow Chemical Company
DSM	Dutch Staats Mijnen
Du Pont	E.I. du Pont de Nemours & Company
Exxon	Exxon Corporation
FMC	Food Machinery & Chemical Corporation
GAF	General Aniline & Film Corporation
W.R. Grace	W.R. Grace & Company
Hoechst	Hoechst Aktiengesellschaft
IBM	International Business Machines Corporation
ICI	Imperial Chemical Industries
IFP	Institut Français du Pétrole
INCO	International Nickel Company
3M	Minnesota Mining and Manufacturing Company
Mitsubishi Chemical	Mitsubishi Chemical Industries
Monsanto	Monsanto Company
Nippon Shokubai	Nippon Shokubai Kagaku Kogyo
PCUK	Pechiney Ugine Kuhlmann
PPG	Pittsburg Plate Glass Industries
Searle	G.D. Searle & Company
SKF	Smith Kline & French Laboratories
SNAM	Societá Nazionale Metandotti
Sohio	Standard Oil of Ohio
Stauffer	Stauffer Chemical Company
Sumitomo	Sumitomo Chemical Company
Toray	Toray Industries Inc.
UCB	Union Chimique Belge
Union Carbide	Union Carbide Corporation
UOP	Universal Oil Products Company
VEBA	Vereinigte Elektrizitäts- und Bergwerks-AG
Wacker	Wacker Chemie GmbH

Country Codes

The following list contains a selection of standard country codes used in the patent references.

AT	Austria	ID	Indonesia
AU	Australia	IL	Israel
BE	Belgium	IT	Italy
BG	Bulgaria	JP	Japan *
BR	Brazil	LU	Luxembourg
CA	Canada	MA	Morocco
CH	Switzerland	NL	Netherlands *
DE	Federal Republic of Germany (and Germany before 1949) *	NO	Norway
		NZ	New Zealand
DK	Denmark	PL	Poland
ES	Spain	PT	Portugal
FI	Finland	SE	Sweden
FR	France	US	United States of America
GB	United Kingdom	ZA	South Africa
GR	Greece	EP	European Patent Office *
HU	Hungary	WO	World Intellectual Property Organization

* For Europe, Federal Republic of Germany, Japan, and the Netherlands, the type of patent is specified: EP (patent), EP-A (application), DE (patent), DE-OS (Offenlegungsschrift), DE-AS (Auslegeschrift), JP (patent), JP-Kokai (Kokai tokkyo koho), NL (patent), and NL-A (application).

Periodic Table of Elements

element symbol, atomic number, and relative atomic mass (atomic weight)

1A "European" group designation and old IUPAC recommendation
1 group designation to 1986 IUPAC proposal
IA "American" group designation, also used by the Chemical Abstracts Service until the end of 1986

1A **1** IA	2A **2** IIA	3A **3** IIIB	4A **4** IVB	5A **5** VB	6A **6** VIB	7A **7** VIIB	8 **8** VIII	8 **9** VIII	8 **10** VIII	1B **11** IB	2B **12** IIB	3B **13** IIIA	4B **14** IVA	5B **15** VA	6B **16** VIA	7B **17** VIA	0 **18** VIIIA
1 **H** 1.0079																	2 **He** 4.0026
3 **Li** 6.941	4 **Be** 9.0122											5 **B** 10.811	6 **C** 12.011	7 **N** 14.007	8 **O** 15.999	9 **F** 18.998	10 **Ne** 20.180
11 **Na** 22.990	12 **Mg** 24.305											13 **Al** 26.982	14 **Si** 28.086	15 **P** 30.974	16 **S** 32.066	17 **Cl** 35.453	18 **Ar** 39.948
19 **K** 39.098	20 **Ca** 40.078	21 **Sc** 44.956	22 **Ti** 47.867	23 **V** 50.942	24 **Cr** 51.996	25 **Mn** 54.938	26 **Fe** 55.845	27 **Co** 58.933	28 **Ni** 58.693	29 **Cu** 63.546	30 **Zn** 65.409	31 **Ga** 69.723	32 **Ge** 72.61	33 **As** 74.922	34 **Se** 78.96	35 **Br** 79.904	36 **Kr** 83.80
37 **Rb** 85.468	38 **Sr** 87.62	39 **Y** 88.906	40 **Zr** 91.224	41 **Nb** 92.906	42 **Mo** 95.94	43 **Tc*** 98.906	44 **Ru** 101.07	45 **Rh** 102.91	46 **Pd** 106.42	47 **Ag** 107.87	48 **Cd** 112.41	49 **In** 114.82	50 **Sn** 118.71	51 **Sb** 121.76	52 **Te** 127.60	53 **I** 126.90	54 **Xe** 131.29
55 **Cs** 132.91	56 **Ba** 137.33		72 **Hf** 178.49	73 **Ta** 180.95	74 **W** 183.84	75 **Re** 186.21	76 **Os** 190.23	77 **Ir** 192.22	78 **Pt** 195.08	79 **Au** 196.97	80 **Hg** 200.59	81 **Tl** 204.38	82 **Pb** 207.2	83 **Bi** 208.98	84 **Po*** 208.98	85 **At*** 209.99	86 **Rn*** 222.02
87 **Fr*** 223.02	88 **Ra*** 226.03		104 **Rf*** 261.11	105 **Db*** 262.11	106 **Sg**	107 **Bh**	108 **Hs**	109 **Mt**	110 **Uun**[a]	111 **Uuu**[a]	112 **Uub**[a]		114 **Uuq**[a]		116 **Uuh**[a]		

[a] provisional IUPAC symbol

57 **La** 138.91	58 **Ce** 140.12	59 **Pr** 140.91	60 **Nd** 144.24	61 **Pm*** 146.92	62 **Sm** 150.36	63 **Eu** 151.97	64 **Gd** 157.25	65 **Tb** 158.93	66 **Dy** 162.50	67 **Ho** 164.93	68 **Er** 167.26	69 **Tm** 168.93	70 **Yb** 173.04	71 **Lu** 174.97
89 **Ac*** 227.03	90 **Th*** 232.04	91 **Pa*** 231.04	92 **U*** 238.03	93 **Np*** 237.05	94 **Pu*** 244.06	95 **Am*** 243.06	96 **Cm*** 247.07	97 **Bk*** 247.07	98 **Cf*** 251.08	99 **Es*** 252.08	100 **Fm*** 257.10	101 **Md*** 258.10	102 **No*** 259.10	103 **Lr*** 260.11

* radioactive element; mass of most important isotope given.

Toxic Agents, Pharmacological and Medical Fundamentals

Carcinogenic Agents

WOLFGANG PFAU, Department of Toxicology and Environmental Medicine of the Fraunhofer Society and Department of Toxicology, Hamburg University Medical School, Hamburg, Germany

1. Introduction

Cancer incidences are increasing worldwide and in the industrialized countries malignant tumors are a leading cause of death, second only to cardiovascular diseases. Improvements in cancer treatment, surgery, pharmacological and radiation therapy, have not been able to reduce the threat that this illness is posing to the general public.

The term cancer covers a broad spectrum of malignant diseases that affect the blood-forming system in the bone marrow (leukemia, 4 % of human cancers), soft tissues and bones (sarcoma, 2 %), the lymphatic system (lymphoma, 5 – 9 %) and, most often, epithelial tissues including the skin, lung, gastrointestinal tract, and glands (carcinoma, 85 %).

Intensive investigations into the genesis of neoplasms have revealed a number of clues at the molecular level that may help in prevention and treatment of cancer.

2. Carcinogenesis

In most cases the induction of cancer is the consequence of low dose chronic exposure to carcinogenic agents and the process of carcinogenesis may take several years or even decades. A broad spectrum of mechanistic variations has been discovered. However, generally molecular biological investigations have confirmed the concept of carcinogenesis as being a multi-step process which is generally divided in the phases of initiation, promotion, and progression. On each of

Ullmann's Industrial Toxicology

ISBN: 3-527-31247-1

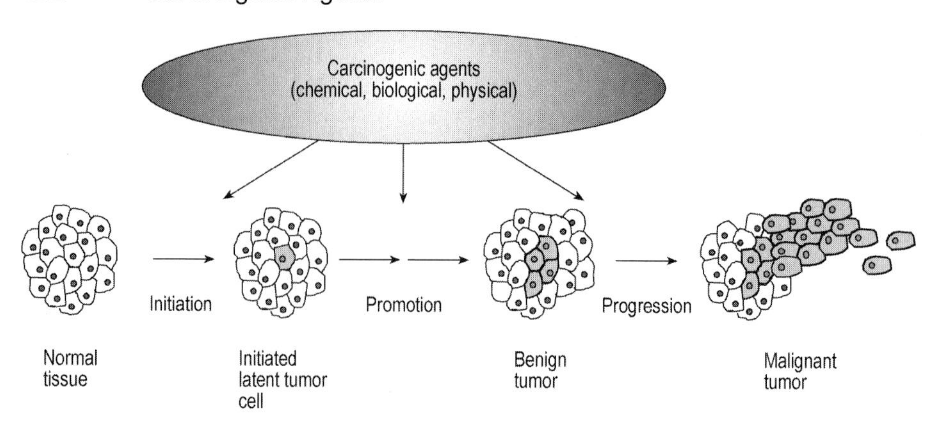

Figure 1. Carcinogenesis is a multi-step process. At any stage of this development carcinogenic agent may have an impact

these phases carcinogenic agents may exert their specific activity and it has been possible in a few types of tumors to assign specific genetic changes in the process of transforming a normal, healthy cell into a malignant clone of tumorous cells (Fig. 1; see above).

Knowledge of the mechanisms of carcinogenesis may help to intervene in this process and reduce the incidence rates of cancer.

It is a biological fact that an increase in the number of tumors induced and/or a decrease of the latent period is observed with increasing dose of a carcinogen. For carcinogenic agents with a genotoxic mechanism, however, a no-effect dose cannot be defined. On the other hand, for carcinogens that act via nongenotoxic mechanisms thresholds are assumed. Thus, for the purpose of risk assessment it is pivotal to understand mechanisms of action in order to define possible threshold levels of exposure [1].

3. Genotoxic Mechanisms

The majority of carcinogenic agents acts as genotoxic agents, i.e., by damaging cellular DNA. These agents are termed mutagenic (see → Mutagenic Agents) or clastogenic.

3.1. Metabolic Activation

With few exceptions (see Section 7.9) carcinogenic organic chemicals (termed procarcinogens) are metabolically activated in the body to reactive metabolites (Fig. 2; see next page). These metabolic transformations are considered accidental in the general metabolism which improves hydrophilicity of foreign agents to facilitate their excretion from the body.

Reactive metabolites formed may include radicals and most commonly electrophilic species (proximate or ultimate carcinogens). These may intermediately release positively charged ions that may react with nucleophilic components in the cell, such as H_2O or glutathione (a tripeptide containing cystein) resulting in a detoxification. These reactive species may also bind to cellular proteins leading to fatal cellular injury or the formation of immunogenic antigens. However, these electrophilic metabolic intermediates may also covalently modify the DNA in the cell nucleus.

3.2. DNA Binding

Several nucleophilic sites in the DNA have been identified as targets for covalent modification by electrophilic agents. DNA bases may be modified with preferential binding sites of alkylating agents (A), aromatic amines (AA) or polycyclic aromatic hydrocarbons (PAH).

The most prominent target of DNA binding is the base guanine where binding has been detected at positions N^7, O^6, N^2, and C^8, depending on the carcinogenic agent involved. These covalent modifications have been termed DNA adducts and may yield mutations in the genome upon DNA replication. The ability of an agent to form covalent DNA adducts is, thus, an

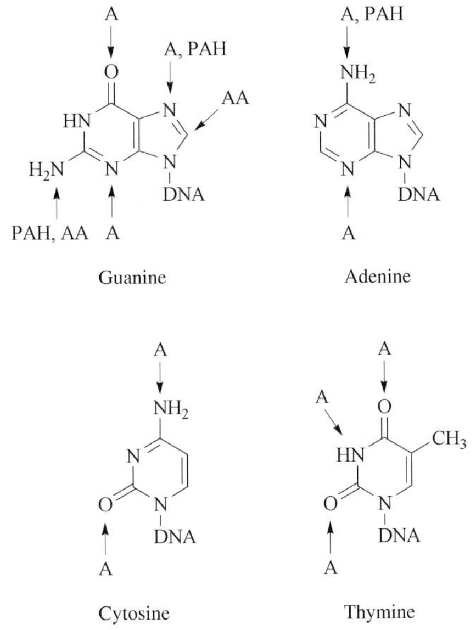

Guanine

Adenine

Cytosine

Thymine

indicator of its genotoxic activity and this damage may lead to the development of neoplasms. Although the formation of these DNA adducts is not sufficient for tumor development, all genotoxic agents identified as human carcinogens have been shown to bind covalently to DNA.

A form of indirect DNA damage is produced by reactive oxygen species (e.g., hydroxyl radicals) which may be formed by oxidants (Section 7.8) or by ionizing radiation (see Section 7.17).

3.3. DNA Repair

The integrity of the genome is pivotal for the survival of a cell. Therefore, a highly sophisticated system of removal and repair of DNA lesions is at work in a cell to identify and remove modifications. Indeed, mechanisms have been identified (involving the tumor suppressor gene *p53*) that arrest DNA replication and cell division in order to allow sufficient time for this repair to occur. Deficient DNA repair is often associated with an increased risk for developing cancer.

3.4. Molecular Targets in the Genome

Genetic targets for carcinogenic agents have been proposed that are termed proto-oncogenes or tumor suppressor genes. A number of these have been found to be frequently altered in tumor cells.

DMNA

2-AN

2-NN

DBP

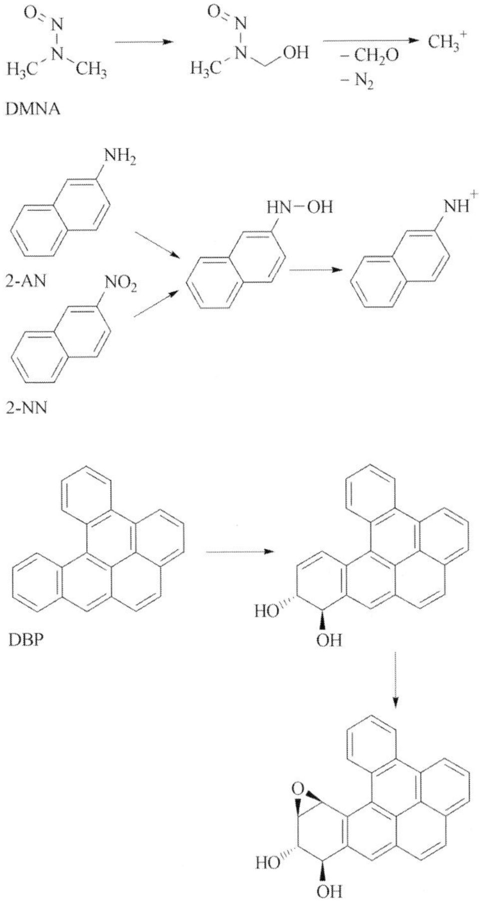

Figure 2. Metabolic activation of chemical carcinogens: Formation of electrophilic ultimate carcinogens. *N*-Nitrosodimethylamine (dimethylnitrosamine, DMNA) is activated via α-hydroxylation to a carbenium ion; polycyclic aromatic amines (2-aminonaphthalene, 2-AN) are oxidized and polycyclic aromatic nitro compounds (2-nitronaphthalene, 2-NN) are reduced to form a nitrenium ion; polycyclic aromatic hydrocarbons (dibenz[a,l]pyrene, DBP) are oxidized to form a dihydrodiol derivative and subsequently vicinal dihydrodiol epoxides in the fjord region of the molecule

3.4.1. Oncogenes

Proto-oncogenes (→ Toxicology, Chap. 3.7.3.) are genes that encode for proteins implicated in vital functions of a cell. These gene products, e.g., growth factors, protein kinases, and receptors, are involved in signal transduction mechanisms. Certain mutations induced in these genes result in an activation of these proto-oncogenes: point mutations (change of one letter in the DNA code), amplifications (enhanced expression, i.e., production, of the gene product) or translocations (also resulting an increase in expression).

3.4.2. Tumor-Suppressor Genes

When mutations in a tumor-suppressor gene of a cell lead to a deletion of the encoded protein or the production of a nonfunctional protein this may result in the cell's expansion and unrestrained growth [2] (→ Toxicology, Chap. 3.7.).

4. Nongenotoxic Mechanisms

There are a number of carcinogenic compounds that induce cancer without possessing genotoxic activity. These agents act by various mechanisms and are summarized as *epigenetic carcinogens*. In contrast to the DNA damaging (genotoxic) agents, for epigenetic carcinogens doses can be established below which no adverse effects are expected to occur. These agents are often carcinogenic via mechanisms that involve acute toxic effects, thus, threshold levels of exposure can be defined.

4.1. Tumor Promoters

Tumor promoters are not carcinogenic on their own, but when applied subsequently to a (genotoxic) tumor initiator increase the number of tumors per animal or shorten the latency period of tumor development in an experimental setting. Thus, tumor promotion is an operational definition rather than a mechanistic concept. The biological phenomenon of tumor promotion is very specific for the animal species tested and the organ site of tumorigenesis, where mouse skin and rat liver are the most commonly tested organ sites. Mechanisms under consideration are mitogenesis (increased cell proliferation) and inflammatory processes.

Different from genotoxic mechanisms, promotion is considered to be reversible and threshold values for promoters are assumed to exist.

Prominent promoting agents are phorbol ($4\beta, 9\alpha, 12\beta, 1\alpha$,20-pentahydroxy-1,6-tigliadien-3-one) esters [such as tetradecanoyl phorbol acetate (TPA) isolated from *Euphorbiaceae*], phenobarbital, saccharin, halogenated hydrocarbons (including DDT and other organochloro pesticides, polyhalogenated biphenyls (PCBs) and polyhalogenated dibenzodioxins/dibenzofurans (TCDD and congeners).

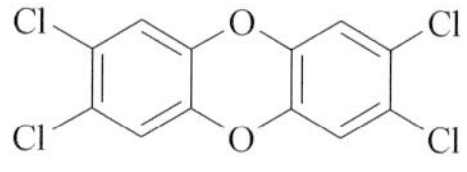

TCDD

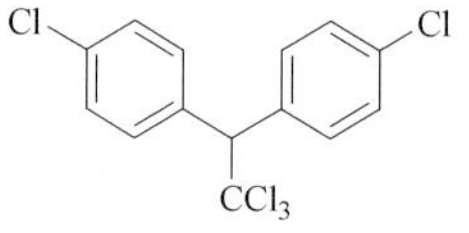

DDT

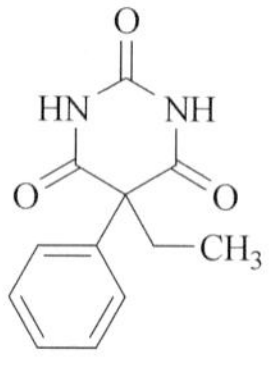

Phenobarbital

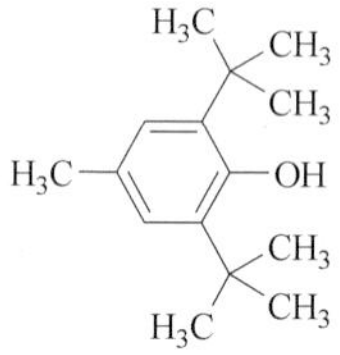

BHT
2,6-di-*tert*-butyl-4-methylphenol
(butylared hydroxytoluene)

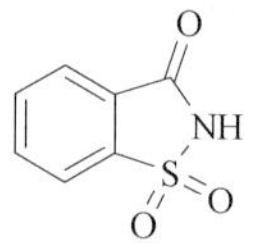

Saccharin

4.2. Hormones

Sexual hormones, mainly estrogens and progesterons, have been shown in animal experiments to increase the number of tumors at certain organ sites such as mammary gland, prostate, and endometrium. Tumor incidences in these organs are dependent on the sex; therefore, it is believed that a promoting effect of sex hormones is involved. The lifetime estrogen dose is considered a major factor for human breast cancer risk.

Steroid hormones are essential for the growth, differentiation, and function of many organs. Thus, the induction of cell proliferation is believed to be the mechanism by which steroid hormones act as carcinogenic agents.

In addition, genotoxic mechanism have been reported for estrogen, its metabolites, and related drugs (diethylstilbestrol, tamoxifen, cyproterone acetate).

A number of anthropogenic environmental contaminants have been discovered to have affinity to the estrogen or androgen receptor and to either mimic (agonists) or inhibit hormonal activity (antagonist). These xenoestrogens (hormonal disrupters) include polyhalogenated biphenyls, halogenated dibenzodioxins, and nonylphenol. Additionally, estrogen, its metabolites and anticontraceptive drugs (e.g., ethinyl estradiol) have been detected in sewage effluents.

Moreover, it has been shown that naturally occurring compounds may bind to the estrogen receptors. These compounds include flavonoids such as genistein and daidzein.

However, these exogenous hormones have generally a much lower affinity to the receptor than the endogenous hormones. Currently, it is unclear if these low levels of xenoestrogens and phytoestrogens have an impact on human endocrine homeostasis. A possible contribution to human cancer risk is a matter of speculation.

4.3. Peroxisome Proliferators

A considerable number of nongenotoxic agents that induce hepatic tumors in the rat have been shown to increase the number of peroxisomes in the rat liver. These cell organelles are responsible for the catabolism of long-chain fatty acids and cholesterols. During catabolism reactive oxygen species are produced in the hepatic cell and these may be the ultimate carcinogenic agents. Whether this mechanism may also play a role in human liver is a matter of dispute. Peroxisome proliferators include plasticizers (e.g., phthalates), halogenated organic solvents (1,1,2-trichloroethylene) and blood lipid lowering drugs (clofibrate, fenofibrate).

5. Identification of Carcinogenic Agents

In order to reduce cancer risk it is necessary (1) to identify the relevant agents and (2) to minimize exposure to these agents. The different approaches to identify the carcinogenic activity of an agent are outlined below. Ideally, both, experimental and human epidemiological data are available.

5.1. Animals Tests (in vivo)

In the National Toxicology Program in the United States and as a result of legislative requirements for new drugs and chemicals of a production volume of more than 1000 t/a in Europe, long term animal carcinogenicity studies are carried out. These two-year animal experiments are generally performed with rats, using several doses of the compound, including the maximum tolerated dose (MTD) which induces marginal toxic effects in the test animals. In the 1990s this concept has been challenged since this dosing regimen is generally several orders of magnitude higher than the anticipated human exposure and may produce many false positives because of toxic effects induced at the high doses [3].

Since this type of long term study is both labor- and cost-intensive, several approaches have been developed to introduce short-term in vitro or in vivo assays that may point towards a possible carcinogenic activity of a test substance.

5.2. In vitro Assays

Although the process of carcinogenesis may generally not be mimicked in a test tube or petri dish, several in vitro assays that are very sensitive in detecting genotoxic activity have been developed. These tests include microbial mutagenicity assays (e.g., the Ames test using various strains of *Salmonella typhimurium*, → Toxicology, Chap. 4.9.1.1.) and genotoxicity tests in mammalian cells. These in vitro assays are generally used as prescreening tests and may give hints on a possible carcinogenic activity of a test compound. They may also help in elucidating the mechanism of action of a rodent carcinogen (→ Mutagenic Agents).

More closely resembling the in vivo tumor induction are cell transformation tests. Here, the transformation of apparently normal cells into tumor cells is recorded as endpoint. These transformation tests are capable of detecting both, genotoxic and nongenotoxic carcinogens.

5.3. Modeling

Computer-assisted analysis of the results from several hundred rodent bioassays on the carcinogenic potency of organic chemicals has helped to identify certain structural elements as carcinogenic alerts [4]. These functional elements ("alerting structures") are listed in Table 1 (see next page).

However, even highly sophisticated calculation methods are presently not capable to replace biological testing.

5.4. Epidemiological Studies

Since experimental studies with possibly carcinogenic compound can, for obvious ethical reasons, not be performed with humans, data on human carcinogenicity have to be deduced by extrapolation from animal experiments, in vitro experiments (using human tissue or cells) and epidemiological studies.

Exposures may be occupational, environmental or accidental (see Chap. 9). Together with experimental studies such results may also provide insights into the mode of carcinogenic activity and biological plausibility. This combined approach is the most secure way of identifying a potential carcinogenic hazard [5].

5.5. Molecular Epidemiology

Conventional epidemiological investigations have been refined by mathematical improvements. However, as was pointed out by TAUBES [6] in a controversial paper these studies are only suitable to detect low risks for the general population and cannot be employed for individual risk assessment. A novel approach that takes into account the individual exposure levels and predisposing (e.g., genetic) factors has been termed molecular epidemiology [7]. By analyzing the internal dose (such as serum levels of a given compound) and/or metabolic capacities of individuals, this approach may help to identify both, subpopulations at risk and agents that pose a significant risk for human cancer.

6. Predisposing Factors

Several types of cancer are inheritable diseases. Since most cancer incidences occur at a late age past the reproductive age, the selective mechanisms of evolution do not help to eradicate these predispositions.

These inheritable forms of cancer account for less than 10 % of all cancer cases. More importantly, the predisposition to develop a malignant tumor may be inherited but is clinically manifest only when other factors such as exposure to specific carcinogenic agents occur. The mechanisms leading to an elevated individual risk include certain enzyme activities in the pathway of metabolic activation such as an enhanced enzyme activity with regard to toxification, a reduced activity in detoxification steps, or defects in DNA repair capacity. Individual risk is also a function of the levels of exposure to carcinogenic agents, the duration of this exposure, and the time in life when this exposure occurs (higher risk at young age) [7].

7. Classes of Carcinogenic Agents

7.1. *N*-Nitroso Compounds

Included in this group are the directly acting *N*-nitrosourea derivatives and the *N*-nitrosodialkylamines. The latter group requires metabolic activation (α-hydroxylation) to release an alkylating species that modifies the cellular DNA (see Fig. 2). *N*-Nitrosamines are among the most potent carcinogenic agents ever tested and *N*-nitrosodimethylamine has been detected in a variety of foods, industrial products, and the environment.

N-Nitrosamines may also be formed endogenously in the acidic environment of the stomach by nitrosation of alkylamines with nitrite. *N*-Nitroso derivatives of nicotine and related alkaloids contained in tobacco are formed upon combustion of tobacco (termed tobacco specific nitrosamines) and are considered major contributors to the carcinogenic potential of tobacco smoke.

Table 1. Potential structural alerts for mutagenicity

Structural alert	Structure	Structural alert	Structure
Alkyl esters of phosphonic or sulfonic acids	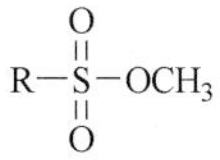 $R-S(=O)_2-OCH_3$	Halogenated methanes (X = H, F, Cl, Br, I)	CX_4
Aromatic nitro groups	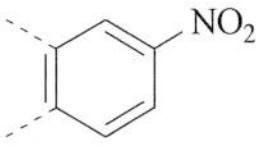 NO_2	*N*- or *S*-mustard (*β*-haloethyl)	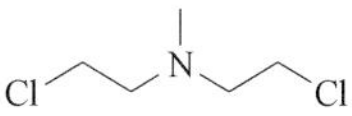 Cl, N, Cl
Aromatic amino groups	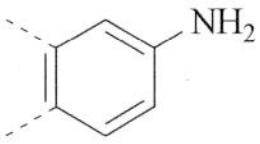 NH_2	*N*-Chloramines	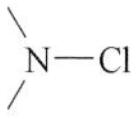 N—Cl
Aromatic hydroxylamines (+ esters)	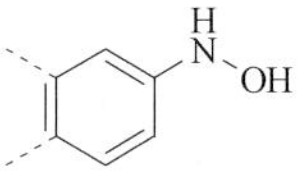 H, N, OH	Propiolactone	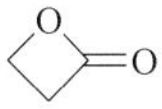 O, =O
Aromatic ring-N-oxides	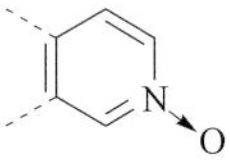 N, O	Aromatic aziridinyl derivatives	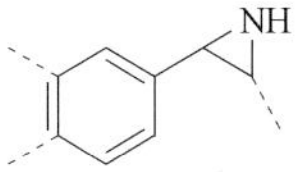 NH
Aromatic azo groups	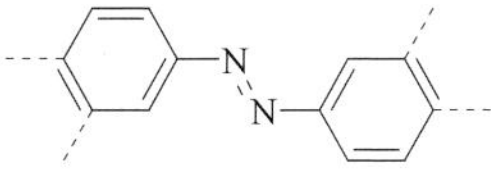 N, N	Primary alkyl halides	CH_2Cl
Aromatic azides	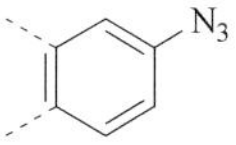 N_3	Carbamates	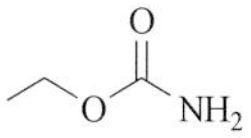 O, O, NH_2
Aromatic mono- or dialkyamino groups	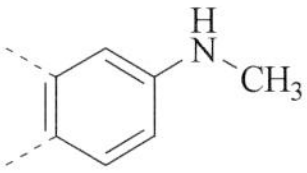 H, N, CH_3	*N*-Nitroso alkylamines	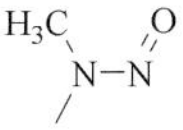 H_3C, O, N—N
Alkyl hydrazines	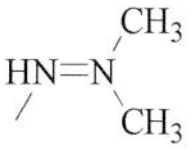 HN=N, CH_3, CH_3	Michael reactive a,*β*-unsaturated aldehydes	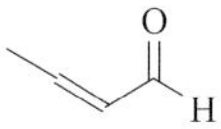 O, H
Alkyl aldehydes	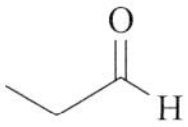 O, H	Polycyclic aromatic hydrocarbons (bay regions)	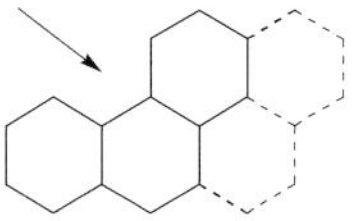
N-Hydroxymethyl groups	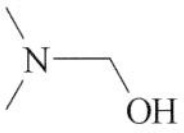 N, OH	Polycyclic aromatic hydrocarbons (fjord region)	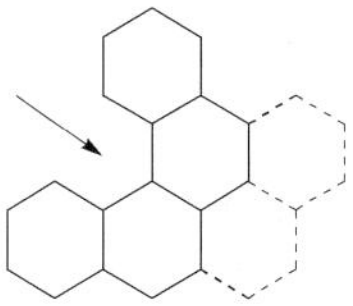
Monohaloalkenes	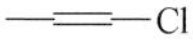 Cl	Aliphatic and aromatic epoxides	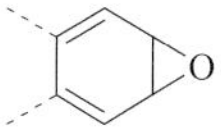 O

7.2. Benzene

Benzene is a constituent of crude petroleum and motor gasoline and is produced in large quantities in the chemical industry as a synthetic intermediate. Both experimental and epidemiological studies have identified benzene as a human carcinogen inducing different forms of leukemia.

7.3. Polycyclic Aromatic Hydrocarbons

When organic matter such as fossil fuels, wood, tobacco, or waste is combusted, polycyclic aromatic hydrocarbons (PAH) are formed, especially under oxygen deficient conditions.

Members of this substance group were among the first individual compounds to be identified as carcinogenic agents. The carcinogenic potency varies among different PAHs depending on the number of aromatic rings and structural features. The most potent PAH are among the five- to six-ring compounds with benzo[*a*]pyrene being the prototype of a carcinogenic PAH. It was shown that PAHs require a multi-step metabolic activation process which results in the formation of dihydrodiol epoxide derivatives (Fig. 2). Genotoxic potency of these dihydrodiol epoxides is enhanced when activation occurs in a *bay region* or in a *fjord region* of a PAH molecule (Table 1). Thus, PAHs with these structural features such as benzo[*a*]pyrene, dibenz[*a,h*]anthracene, or dibenz[*a,l*]pyrene have been shown to be most potent carcinogenic PAHs. Furthermore, biological activity of these diol epoxides varies with on the conformation of the different hydroxyl substituents.

7.4. Nitroaromatic Compounds

Associated with the occurrence of PAHs are their nitro derivatives. Relatively high levels of these compounds are observed in emissions of diesel-fueled motor vehicles. These compounds also require metabolic activation, the enzymatic process involves reduction of the nitro group to a hydroxyl function that results in a nitrenium ion as ultimate carcinogenic species (Fig. 2).

7.5. Aromatic Amines

Similar to aromatic nitro compounds aromatic amines are also transformed metabolically to hydroxyl derivatives, albeit via an oxidative pathway (Fig. 2). This class of carcinogenic compounds includes the monocyclic aromatic amines (aniline and its derivatives), polycyclic aromatic amines (e.g., 4-aminobiphenyl, ABP) and heterocyclic aromatic amines (e.g., aminoimidazo quinoxalines). A number of these aromatic amines are found in tobacco smoke and polycyclic amines such as 2-aminonaphthaline (β-naphthylamine), benzidine, and 4-aminobiphenyl have been associated with the increased risk of urinary bladder cancer observed among cigarette smokers. The high incidence of bladder cancer among workers in the dye industry has also been attributed to polycyclic impurities in aniline, namely 2-aminonaphthalene and 4-aminobiphenyl, but recently ortho-toluidine has also been considered as etiologic agent in these occupational cancers.

7.6. Halogenalkyls

Halogenated aliphatic hydrocarbons are widely used as solvents and industrial intermediates. These include halomethanes like CCl_4 and $CHCl_3$ that have been shown to induce liver tumors in rodents via nongenotoxic mechanisms. Dichloromethane induces tumors in lung and liver of mice; the significance of these findings for human cancer risk has been questioned.

Halogenated unsaturated hydrocarbons induce tumors in the kidneys of exposed rodents. A genotoxic mechanism involving a β-lyase mediated metabolic activation has been discovered.

Vinyl chloride (chloroethylene) the monomer of poly(vinyl chloride) (PVC) is an established liver carcinogen inducing hemangiosarcoma of the liver of exposed individuals.

Polyhalogenated cyclic hydrocarbons including pesticides (e.g., DDT and hexachlorocyclohexane), polyhalogenated biphenyls, polyhalogenated dibenzodioxins and dibenzofurans are exceptionally persistent in the environment. They are accumulated in the fatty tissues of animals and are enriched in the food chain. These compounds may act by various mechanisms as (nongenotoxic) tumor promoting agents.

Haloethers, such as bis(chloromethyl) ether or bis(chloropropyl) ethers, have been detected in substantial amounts in the environment as persistent contaminants. These ethers have a considerable carcinogenic potential.

7.7. Aldehydes

Formaldehyde is an ubiquitous chemical intermediate used in several fields of health sciences, in the production of synthetic resins, plastics and chip- and fiberboard. Formaldehyde is mutagenic and induces tumors in the rat (nasal cavity) following high level exposure. It has been suggested that the carcinogenic effect of formaldehyde in rodents is due to high-dose cytotoxic effects and there is inadequate epidemiological evidence of a carcinogenic potential in humans. Acetaldehyde and glutaraldehyde must be viewed similarly.

Unsaturated aldehydes such as crotonaldehyde may react with DNA bases to form cyclic DNA adducts and have been shown to be strongly carcinogenic in rodents.

7.8. Oxygen Damage

Reactive oxygen species are formed as a result of physiological cellular function. Glutathione and several enzymatic systems detoxify these reactive species (e.g., $OH^{\bullet}$) and guard the cellular function up to a certain level. Additional oxidative stress is imposed by chemicals that undergo one-electron oxidation or reduction in the cell forming radical species. This may lead to oxidative damage of DNA (e.g., 8-oxoguanine) that is usually effectively repaired but can also result in mutations. Indirectly, the oxidative damage of cellular lipids may lead to premutagenic lesions via modification of DNA by lipid peroxidation products. Oxidative damage of DNA has been proposed as a mechanism of heavy metal carcinogenesis. The carcinogenic effect of ionizing radiation is believed to be effected by the formation of reactive oxygen species.

Antioxidative compounds (such as α-tocopherol, vitamin E) may have an anticarcinogenic effect because they detoxify the reactive oxygen derivatives.

7.9. Alkylating Agents

Mustard gas (sulfur mustard) was used in large quantities in chemical warfare. Both experimental data and studies from occupational exposure indicate this DNA alkylating agent to be carcinogenic to humans. Similarly, there is sufficient evidence that alkylating antineoplastic drugs (cyclophosphamide → Cancer Chemotherapy, Chap. 3.3., melphalan → Cancer Chemotherapy, Chap. 3.2.) increase human cancer risk (urinary bladder, leukemia) because the modification (alkylation) of cellular DNA is not restricted to the tumor but causes genotoxic damage in various organs of the patient.

7.10. Metals

There is indisputable evidence that beryllium, chromium, nickel, and arsenic represent a carcinogenic hazard to humans (pulmonary carcinoma and cancer of other organ sites), when present in certain well defined physicochemical forms. The evidence for the carcinogenicity of cadmium is less clear (probable targets are lung and prostate) although in the laboratory chronic inhalative exposure resulted in lung tumors in rats. Either way, cadmium compounds are very toxic in other respects (kidney and bones) so that they should be used with particular care.

The mechanism of metal carcinogenesis has tentatively been described as a genotoxic pathway since metal salts have been shown to be mutagenic in several in vitro short term tests.

Earlier misconceptions were based on a lack of mutagenic activity of these metals in the Ames test, the most widely used mutagenicity test. Other metals that are carcinogenic in laboratory animals include cobalt, iron, lead, titanium, and zinc.

With the exception of arsenic, metal carcinogenesis seems not to be a problem of environmental contamination at trace levels but is relevant to occupational exposure.

7.11. Natural Compounds

The pollution of the environment with synthetic chemicals has supported the notion that these anthropogenic substances are the major cause of human cancer. Natural compounds are generally considered as safe by the general public. However, as has been brought forward rather pointedly by AMES and coworkers there is a considerable number of substances of natural origin that have a strong carcinogenic potency. These authors hypothesize that natural agents are indeed far more relevant to human carcinogenesis than environmental contamination with synthetic substances.

Aflatoxins are produced by molds (*Aspergillus flavis*). Among these, aflatoxin B_1 is one of the most potent liver carcinogens in the rat and acts via a genotoxic pathway: after metabolic activation the epoxide reacts with guanine residues in the DNA. Molecular epidemiological studies have shown that in certain regions of east Africa (The Gambia and Senegal) and provinces of China the endemic high incidences of liver cancer are associated with aflatoxin-contaminated food in a synergistic combination with hepatitis B viral infection.

Carcinogens have also been detected in plants, e.g., phorbol esters [such as tetradecanoyl phorbol acetate (TPA) isolated from *Euphorbiaceae* or Aquilid A from bracken fern (*Pteridium aquilinum*)] and herbal drugs such aristolochic acid derived from birthwort (*Aristolochia* species), pyrrolizidine alkaloids (*Senecio ssp*), or anthraquinones (e.g., senna leaves). Nicotine in tobacco is a constituent of carcinogenic tobacco smoke (see Section 7.5).

When proteinaceous food is grilled, fried, or broiled, heterocyclic aromatic amines are formed via the Maillard reaction. Among these, the aminoimidazo quinoxalines exhibit extremely strong mutagenic responses in the Ames test and have also been shown to induce tumors at several organ sites in rodents. Thus these heterocyclic amines are considered to be involved in human cancer at organ sites (colon, breast, prostate, pancreas) that are associated with a diet high in fried or grilled meat.

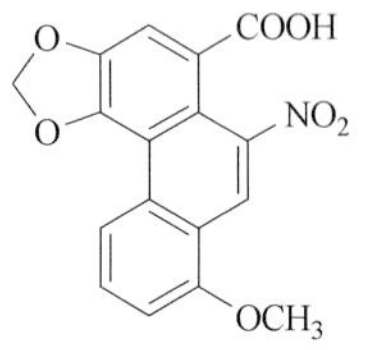

Aristolochic acid I

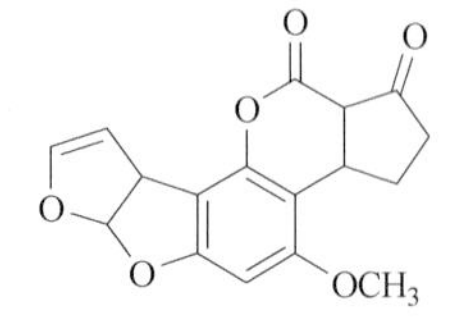

Aflatoxin B_1

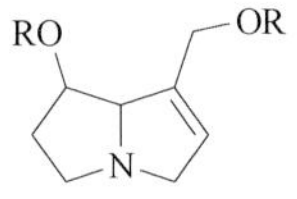

Pyrrolizidine

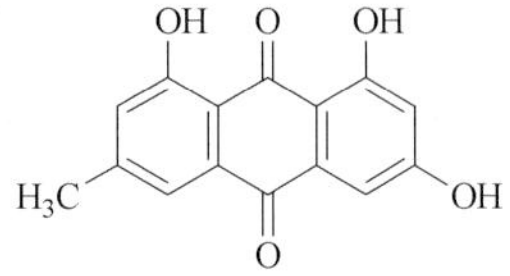

Emodin (anthraquinone derivative)

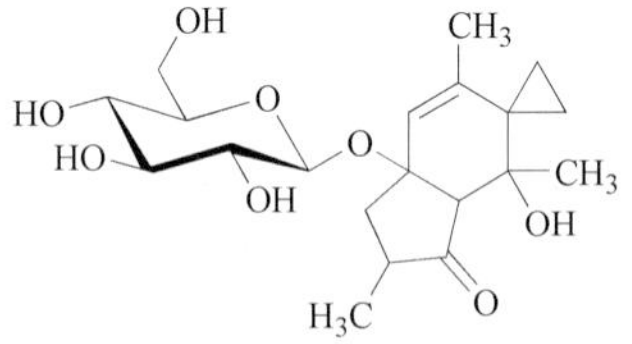

Aquilid A

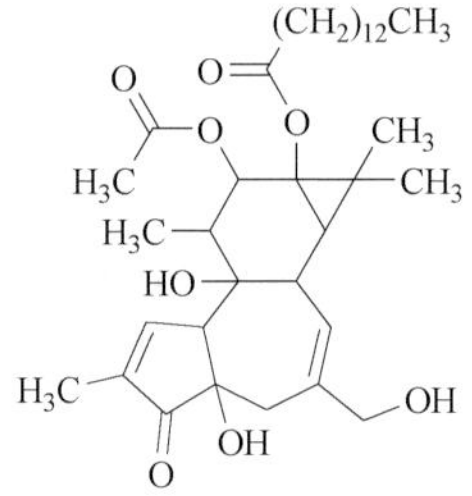

TPA

7.12. Hormones

See also Section 4.2.

Hormones include both endogenous sexual hormones, hormonal drugs, and environmental compounds with affinity to hormonal receptors. The latter can be of natural origin (phytohormones) or originate from anthropogenic environmental contaminants. The affinity to the hormonal receptors varies between different compounds by several orders of magnitude.

7.13. Mixtures

Humans are most often exposed to mixtures of carcinogenic agents, such as mixtures of polycyclic aromatic hydrocarbons in smoke or coal tar, mixtures of heterocyclic amines in fried meat or even complex mixtures of different types of chemical agents. These may act in an additive, synergistic (overadditive), or inhibitory manner. Since the composition of these complex mixtures is variable (for example, tobacco smoke or environmental pollution) it is difficult to predict quantitatively the biological potency of such complex mixtures. One of the most complex and most important chemical mixtures is the human diet.

7.14. Mineral Fibers

Dust of the naturally occurring rock asbestos, particularly blue asbestos (crocidolite) can induce mesothelioma (malignant tumors of the pleura) following inhalative exposure. The physical properties of the needle-like fibers (length > 5 μm, diameter < 3 μm) and the very long biological half-life are crucial for the activity. Thus, other fibers (erionite, a zeolite mineral) with similar properties have also carcinogenic potential.

Other synthetic inorganic fibers (rock wool, glass fibers, ceramic fibers) are less persistent in the body; probably for this reason the carcinogenic potential of these is less certain.

7.15. Viruses and Other Biological Agents

Viral infection is a major risk factor for human cancers, chronic infection with specific viruses being implicated in the genesis of up to 20 % of tumors.

Retroviruses.

HTLV-I. has been implicated as a cause of adult T-cell leukemia/lymphoma; less than 4 % of individuals infected with this virus develop the tumor, and then only several decades after the initial infection.

HIV-1, HIV-2. The increased incidence of lymphoma and Karposi's sarcoma in patients infected with human acquired immune deficiency syndrome (AIDS) viruses is almost certainly the result of the chronic immunosuppression that occurs in the disease, rather than a direct oncogenic effect.

DNA and RNA Viruses. Chronic infection with human papilloma virus (HPV), acquired as a result of sexual transmission , has been implicated as a cause of cervical, anal, and vulval cancer. HPV is present in more than 60 % of cervical carcinomas, and additional risk factors such as herpes simplex virus and smoking, seem to be necessary.

Hepatitis B Virus (HBV) and Hepatitis C Virus (HCV). Epidemiological evidence links chronic infection with HBV or HCV to the development of hepatocellular carcinoma. It has been suggested that these viruses may induce liver cancer indirectly, acting as a promoter by causing chronic inflammatory liver disease.

Herpes Viruses. Epstein Barr virus (EBV) may be involved in the pathogenesis of a number of human tumors, including Burkitt's lymphoma, Hodgkin's disease and salivary gland, urogenital or nasopharyngeal carcinoma. In AIDS patients, lymphomas (immunoblastic and nervous system) are also associated with this virus.

Parasites. Schistosomiasis, a parasitic infection with different types of the flatworm *Schistosoma* is endemic in different tropical regions of the world and is associated with an increased risk of cancer of the urinary bladder. Chronic inflammation of the bladder and metabolic changes leading to increased endogenous formation of *N*-nitrosamines is being considered as mechanism of action.

Bacterial Infection. Infection with *Heliobacter pylori* is believed as causally involved in human stomach cancer.

7.16. Nonionizing Radiation

Skin cancer is the most common neoplasm in Caucasians in the United States with a lifetime risk nearly equal to that of all other cancers combined. Sun exposure is the major environmental agent implicated in induction of nonmelanoma skin cancer (basal cell carcinoma or squamous cell carcinoma).

The initiating event is the DNA damage induced by UV radiation: cross-linking of two thymidine bases to form a cyclobutane ring (2 + 2-cycloaddition). This hypothesis is corroborated by the observation that patients with the rare inherited disorder *Xeroderma pigmentosum* have a skin cancer risk 1000 times higher than that of the general population, probably due to a defect in DNA nucleotide excision repair.

With the reduction of the stratospheric ozone layer, UV exposure is increasing and skin cancer rates are rising. Preventing severe sunburn especially at young age and regular medical checkups are important to minimize the risk.

7.17. Ionizing Radiation

Long term follow-up of survivors of the atomic bomb releases in Hiroshima and Nagasaki, as well as of patients that received extensive radiation therapy provided epidemiological evidence that ionizing radiation can induce cancer in any organ in which cancer occurs naturally. In particular the rate of leukemia, cancer of the breast, lung, and thyroid are increased with latency periods up to 20 years. Most pronounced effects were observed when the exposure to a single high dose of radiation occurred at a young age.

Radiation may cause DNA damage indirectly: formation of reactive radicals by radiochemical homolysis of water molecules and induction of DNA strand breaks by these reactive intermediates.

The relationship between cancer and low-level radiation is a subject of much controversy. The considerable level of natural background radiation has been said to exceed the low levels of exposures such as in the vicinity of nuclear power or reprocessing plants. Recently, the observation of inheritable genomic instability induced by low amounts of radiation has been discussed.

7.18. Electromagnetic Fields

The increase of electrical appliances in the home, the widespread use of mobile phones, high tension lines and the constant exposure to computer monitors and television sets results in an exposure of the human body to electromagnetic fields. These fields are not without biological effects; however, the proposed carcinogenic (promoting) effects of extremely low frequency (0 – 300 Hz) electromagnetic fields have not been definitely confirmed experimentally.

8. Human Carcinogens

The evidence provided by epidemiological data is currently taken as the only acceptable proof for a given compound to be a human carcinogen. It is a matter of controversy whether or to what extent experimental data, e.g., from rodent assays may be regarded as adequate data to classify a compound a human carcinogen.

It has been pointed out that standard testing at the MTD results in a number of false positives, e.g., many agents are labeled carcinogenic only because of the cytotoxic action at unrealistic high doses [1], [3]. The International Agency for Research on Cancer (IARC) has evaluated a number of agents [5]. These are listed below and are classified according to the IARC categorization scheme:

Carcinogenic to humans (Group 1): a causal relationship has been established between the exposure and human cancer

Chemicals and groups of chemicals

Aflatoxins
4-Aminobiphenyl
Arsenic and Arsenic compounds
Asbestos
Benzene
Benzidine
Bis(chlormethyl)ether
Chromium compounds (hexavalent)
Erionite
Mustard gas (sulfur mustard)
2-Naphthylamine
Nickel compounds
Radon and its decay products
Talc containing asbestiform fibers
Vinyl chloride

Mixtures

Alcoholic beverages
Betel quid with tobacco
Coal-tar pitches
Coal tar
Mineral oils
Shale oils
Soots
Tobacco products, smokeless

Tobacco smoke
Drugs
Analgesic mixtures containing phenacetin
Azathioprine
N,N-Bis(2-chloroethyl)-2-naphthylamine (Chlornaphazine)
1,4-Butanediol dimethanesulfonate (myleran)
Chloroambucil
1-(2-Chloroethyl)-3-(4-methylcyclohexyl)-1-nitrososurea (Methyl-CCNU)
Cyclophosphamide
Cyclosporin
Diethylstilbestrol
Melphalan
8-Methoxypsoralen plus UV radiation
Estrogen replacement therapy
Estrogens, nonsteroidal
Estrogens, steroidal
Thiotepa
Treosulfan
Biological and physical factors
Hepatitis B virus
HTLV 1
Ionizing radiation
Ultra-violet radiation

Probably carcinogenic to humans (Group 2A): a positive association has been observed, a causal interpretations credible, but chance bias or confounding factors cannot be ruled out with confidence; sufficient evidence of carcinogenicity in animals);

Chemicals and groups of chemicals
Acrylonitrile
Benzo[*a*]anthracene
Benzidine-based dyes
Benzo[*a*]pyrene
Beryllium and beryllium compounds
Cadmium and cadmium compounds
p-Chlorotoluidine
Dibenz[*a,h*]anthracene
Diethyl sulfate
Dimethylcarbamoyl chloride
Dimethyl sulfate
Epichlorohydrin
Ethylenedibromide
Ethylene oxide
N-Ethyl-*N*-nitrosourea
Formaldehyde
4,4'-Methylene-bis(2-chloroaniline) (MOCA)
N-Methyl-*N*'-nitro-*N*-nitrosoguanidine (MNNG)
N-Methyl-*N*'-nitrosourea
N-Nitrosodiethylamine
N-Nitrosodimethylamine
Propylene oxide
Silica, crystalline
Styrene oxide
Tris(2,3-dibromopropyl)phosphate
Vinyl bromide
Mixtures
Creosotes
Diesel engine exhausts
Polychlorinated biphenyls
Drugs
Adriamycin
Anabolic steroids
Azacytidine
N,N'-bis(2-chloroethyl)-*N*-nitrosourea (BCNU)
Chloramphenicol
N-(2-chloroethyl)-*N*'-cyclohexyl-*N*-nitrosourea (CCNU)
Chlorototocin
Cisplatin 5-Methoxypsoralen
Nitrogen mustard
Phenacetin
Procarbazine hydrochloride
Tris(1-aziridinyl)phosphine sulfide (Thiotepa)
Biological factors
Clonorchis sinensis
Epstein Barr virus
Herpex simplex virus 2
Human immunodeficieny viruses
Human papilloma virus
Opisthorchis viverini
Schistosoma haematobium

Possibly carcinogenic to humans (Group 2B): sufficient evidence of carcinogenicity in animals, but no adequate data on cancer in exposed humans)

Not classifiable as to its carcinogenicity to humans (Group 3)

Probably not carcinogenic to humans (Group 4)

Other classification schemes by other organizations exist including those of the EPA, the EC, or the Chemical Manufacturers Association. These evaluations agree with respect to the classification of compounds that are known human carcinogens but place different emphasis on the results of animal or in vitro studies. These lists of agents with an established carcinogenic effect may help to avoid or at least reduce exposure to these substances. However, it cannot be assumed that substances not included in these lists are nonhazardous; it may simply be a lack of data on humans.

9. Forms of Exposure

9.1. Occupational Exposure

Historically, human occupational exposure to relatively high doses of chemical carcinogens has provided insight into the etiology of human cancer. The earliest observations were reported by POTT (1775) on an increased risk of scrotal cancer in chimney sweeps or by REHN (1897) who observed an elevated incidence of tumors in the urinary bladder among workers of the dye industry [8].

Whereas for high exposure levels at the workplace (asbestos, metals) often a clear epidemiological evidence exists for the carcinogenicity of certain substances, such causal relationships are difficult to verify for environmental (low level) exposures. In the developed countries, but less so in the developing or newly industrialized countries [9], safety precautions in the industry have been implemented by law and have resulted in a reduced levels of exposure and subsequently in lower incidences of occupational cancer.

9.2. Environmental Exposure

Many carcinogenic agents identified in an occupational setting can also be found in the general environment at lower levels. One example of these is asbestos. Whether the low levels of carcinogenic compounds in the general environment pose a significant cancer risk is a matter of intensive debate. It has been brought forward that the risk calculated from animal carcinogenicity data and environmental exposure levels is very low and may statistically effect only few people. On the other hand, individual differences in susceptibility and combination effects resulting from exposure to complex mixtures must be considered.

For genotoxic compounds the general assumption is that there is no threshold. For nongenotoxic carcinogenic agents biological considerations, that take into account the mechanisms of action, point towards the existence of a threshold dose. However, determination of threshold levels for chronic biological effects is a formidable task [1], [5].

9.3. Lifestyle

Perhaps the most common cause of human cancer is tobacco smoking and other forms of tobacco abuse (chewing, sniff). It has been estimated that over 90 % of lung cancer and significant proportions of cancers at other organ sites (urinary bladder, pancreas, gastrointestinal tract, upper respiratory passages) are attributable to excessive tobacco smoking. Overall, about one-third of all cancer deaths are believed to result from this habit [1], [5], [10]. Accordingly, it has been convincingly demonstrated that smoking cessation reduces the cancer risk significantly.

The carcinogenic effect of excessive intake of alcoholic beverages is not as readily apparent. However, the risk for cancer of the oral cavity or the larynx increases markedly when an individual smokes tobacco and abuses alcoholic beverages. Similarly, a synergistic effect has been observed for liver cancer when individuals infected with hepatitis B virus consume alcoholic beverages excessively [5].

Viral infections (e.g., HIV, HPV) are associated with promiscuity. Here, the use of protective measures upon sexual intercourse has been shown to reduce the risk of an infection (see Section 7.15)

9.4. Diet

Dietary factors have been considered to be most important among the different contributors to human cancer [10], [11]. In the diet, carcinogenic agents can be found such as PAH and heterocyclic amines in grilled meat, carcinogenic plant products (safrol) or mycotoxins (see Section 7.11). A high calorie diet and/or a diet high in fat (especially saturated animal fat) has been shown to increase tumor formation in experimental animals but epidemiological evidence is insufficient. There are also speculations on factors with anticarcinogenic activity such as antioxidants, vitamins, and fibers.

9.5. Drugs

Among the cytostatic drugs used in the chemotherapy of neoplasms several classes of compounds exist that have been shown to induce tumors in rodents or even lead to the induction of secondary tumors in treated patients. These include alkylating agents [mitomycin C, cyclophosphamide and nitrosourea derivatives (BCNU)], and anthracyclines (doxorubicin).

Genotoxic activity has been reported for hormone drugs (cyproterone acetate, tamoxifen) while a postulated increase in tumor incidences as a result of continued use of contraceptive drugs was not confirmed by epidemiological studies. Diethylstilbestrol, a synthetic estrogen used in the 1960s to prevent spontaneous abort, induced carcinoma in the children of treated patients.

Various drugs such as phenacetin that have been identified as carcinogenic agents have been withdrawn from the market. A number of herbal drugs contain carcinogenic components including aristolochic acid, anthraquinones, and pyrrolizidine alkaloids (see Section 7.11).

9.6. Exposure Levels in Developing Countries

To some extent, industry is being forced to develop cleaner and safer processes and products in North America and Europe. However, in the rapidly industrializing countries of the Third World, corporate responsibility is not compelled by public awareness, regulation, and compensation laws. It has been pointed out not only by industry critics but also the International Labor Office and the United Nations Center on Transnational Corporations that global corporations are operating more polluting and dangerous plants in Third World countries [12]. Carcinogenic or otherwise toxic chemicals, e.g., pesticides such as DDT, banned for certain uses or voluntarily withdrawn from markets in the United States or Europe, are being produced and marketed in Africa, Asia, and Latin America. In small-scale industries that account for the bulk of industrial enterprises in developing and newly industrialized countries the standards of industrial hygiene, safety and, waste management are low [9].

With the industrialization the western lifestyle is adopted in these countries; thus, developing and newly industrialized countries are the fastest growing markets for the tobacco industry. It has been predicted that here the rates of lung cancer are going to increase in an exponential way in the years to come [13].

9.7. Safe Handling of Carcinogenic Agents

For laboratory handling of chemical carcinogens, the U.S. National Research Council has provided useful guidelines [14]. Because the toxicological properties of most laboratory chemicals are undetermined, virtually all operations should be performed in well-ventilated fume-cupboards. Care should be taken to avoid contact with the skin and to avoid ingestion even of traces. Additional precautions are to be taken for the handling of chemicals with high carcinogenic potency. National or federal legislation regulates the appropriate labeling of carcinogens, suspected carcinogens (see Table 1) and other hazardous chemicals; these should be handled only by trained persons.

10. References

1. L. Tomatis et al.: "Avoided and avoidable risks of cancer," *Carcinogenesis* **18** (1997) 97 – 105.
2. T. Jacks, R. A. Weinberg: "Cell-cycle control and its watchman," *Nature* **381** (1996) 643 – 644.
3. B. N. Ames, L. S. Gold: "Chemical Carcinogenesis: Too many rodent carcinogens," *Proc. Natl. Acad. Sci. USA* **87** (1994) 7772 – 7776.
4. C. D. Klassen (ed.): *Casaret and Doull's Toxicology,* 5th ed., Pergamon Press, New York 1996.
5. L. Tomatis (ed.): *Cancer, Causes, Occurrence and Control,* IARC Scientific Publications 100, Lyon 1990.
6. G. Taubes: "Epidemiology faces its limits," *Science* **269** (1995) 164 – 169. E. L. Wynder, *Am. J. Epidem.* **143** (1996) 747 – 749.
7. F. P. Perera: "Molecular epidemiology: Insights into cancer susceptibility, risk assessment, and prevention," *J. Natl. Cancer Inst.* **88** (1996) 496 – 509.

8. H. Marquardt, S. G. Schäfer (eds.): *Lehrbuch der Toxikologie,* BI Wissenschaftsverlag, Mannheim 1994.
9. D. C. Christiani et al.: "Occupational health in developing countries: Review of the research needs," *Am. J. Ind. Med.* **17** (1990) 393 – 401.
10. R. Doll: "Nature and nurture: possibilities for cancer control," *Carcinogenesis* **17** (1996) 177 – 184.
11. World Cancer Research Fund: *Food, Nutrition and the Prevention of Cancer: a global perspective* American Institute for Cancer Research, Washington 1997.
12. B. I. Castleman: "The migration of industrial hazards," *Int. J. Occup. Environ. Health* **1** (1995) 85 – 96.
13. S. J. Jay: "The emerging tobacco epidemic in China," *J. Amer. Med. Ass.* **279** (1998) 1346 – 1347.
14. National Research Council: *Prudent Practices for Handling Hazardous Chemicals in Laboratories.* National Academy Press, Washington, DC, 1981, pp. 30 – 56.

Mutagenic Agents

WOLFGANG PFAU, Umweltmedizin Hamburg e.V. and Institute of Toxicology, Hamburg University Medical School, Hamburg, Germany

1. Introduction

Mutations are changes in the composition of the genetic information that are passed on to the daughter cells. Mutagenic agents are, thus, chemical or physical agents that damage the genome and are also referred to as *genotoxic agents*.

However, changes in the genome are also the basis of evolution and may, in rare occasions, confer a selective advantage. Since evolution is a slow process, this may be observed in our time scale only in rapidly proliferating organisms, such as bacteria or viruses, which may acquire resistance towards antimicrobial or antiviral drugs by mutations. The overall majority of mutations are, however, either silent or detrimental.

There are a number of hereditary diseases that have been assigned to specific mutations in relevant genes (such as sickle cell anemia, hereditary breast cancer). Also, hereditary enzyme deficiencies have been associated with a mutated genetic code for the respective enzymes.

As an instrument in molecular biology site directed mutagenesis is a means to elucidate proteins by systematically introducing changes in the amino acid sequence.

Because mutagenicity is linked to carcinogenicity it is import to determine if a substance has a mutagenic potential. Genotoxic carcinogens are generally considered to have no safe dose, no threshold, and therefore stringent regulations are imposed on mutagenic agents minimizing human exposure to these agents to the lowest (reasonably) achievable level.

2. Mutagenesis

Mutations in a *germ line cell* result in changes in the genetic code, that are manifest in all cells of the off-spring and will be hereditary. However, these are very rare events. More common are mutations in *somatic cells* which may, in the worst case, lead to cancer or, if introduced in the developing embryo, to teratogenic effects.

Mutagenic agents may induce premutagenic lesions. These are transformed into mutations during DNA replication or by incorrect DNA repair processes (see below).

There are different types of mutations resulting from different types of damage in the DNA: gene mutations (point and frameshift mutations), chromosomal aberrations, and genome mutations.

Ullmann's Industrial Toxicology

ISBN: 3-527-31247-1

2.1. Gene Mutations

Gene mutations are small changes in the DNA at the level of the bases. Upon replication a wrong nucleotide may be incorporated into the daughter strand and as a consequence the mutation is then passed on to the following generations of cells.

2.1.1. Point Mutations

By substitution of nucleotides the sequence of the bases is changed. This may be of no consequence when this mutation occurs in a nontranscribed region of the genome or when a silent mutation is introduced (because the genetic code is degenerate, different codons coding for the same amino acid). However, changing a codon may result in changes of the amino acid sequence of the resulting protein or the introduction of a stop codon.

Endogenous or exogenous agents may induce chemical changes in the bases, for example oxidative damage may result in deamination of adenine by nitrous acid. This deamination results in the incorporation of cytosine during replication and within two cell divisions a A–T base pair is converted into a G–C pair.

The most common damage induced by mutagenic agents is the covalent binding of a reactive, electrophilic metabolite to the nucleophilic DNA bases. Small *alkylating agents* bind to the O^6- or N7-position of guanine, the O^4-position of thymidine, or the N6, N3 or N7 of adenine. These small DNA adducts may lead to a mispairing upon replication and, thus, to the introduction of a point mutation (see Fig. 1.).

Polycyclic aromatic amines and nitroarenes are metabolically transformed and the ultimate electrophilic agents, the nitrenium ions bind to the C8-position of guanine and sometimes to the N^2- or N^6-exocyclic amino groups of the purine bases (Figure 2) (→ Carcinogenic Agents, Chap. 3.1).

Polycyclic aromatic hydrocarbons bind generally, upon metabolic activation, to the exocyclic amino groups of purine bases (Figure 2) but may also, after a one-electron oxidation, bind to the N7-position. Binding to the N7-position results in a weakening of the *N*-glycosidic bond and may lead to a loss of the base, resulting in an apurinic site.

Bivalent electrophilic agents, such as unsaturated carbonyl compounds, dicarbonyl compounds, or dihalogenated aliphatic agents may form cyclic adducts, binding to both exocyclic and endocyclic nitrogen atoms in the purine bases or cytidine.

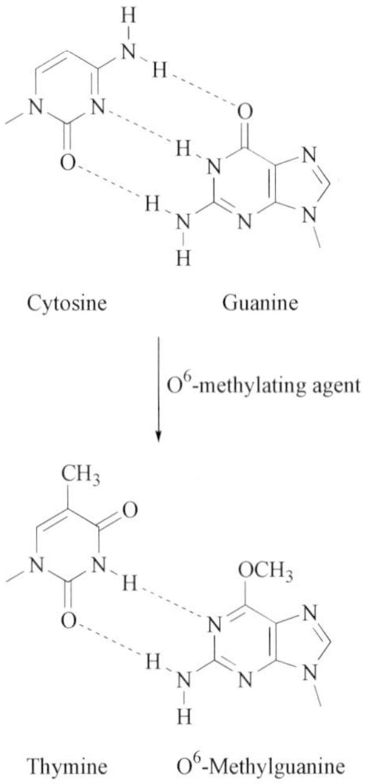

Cytosine Guanine

O^6-methylating agent

Thymine O^6-Methylguanine

Figure 1. Normally guanine pairs with cytosine. Methylation at the O^6-position freezes the enolic form of guanine and may result in introduction of thymine in the opposing strand. Upon replication one of the daughter cells will, thus, contain a TA base pair instead of CG.

2.1.2. Frameshift Mutations

Addition or elimination of bases to the sequence may result in a change of the reading frame. Generally, this leads to more serious changes in the resulting protein. A classical example is the congenital phenylketonuria, where the enzyme responsible for the conversion of phenylalanine to tyrosin is inactive due to a (hereditary) frameshift mutation.

Chemicals with a planar aromatic structure may intercalate into the DNA helix. This, but also covalent DNA adducts may result in a slip or an addition of one base during replication or DNA repair.

2.2. Chromosomal Aberrations

When parts of a chromosome end up in a different location in the same or a different chromosome, are lost or amplified, this is called chromosomal aberration. These types of damage are caused by breakages, deletions, or rearrangements during the cell cycle. The places where breaks or exchanges occur are usually not randomly distributed over the chromosome.

Chromatid-type aberrations include micronuclei, deletions, and interchromatid exchanges. *Chromosome-type aberrations* include bicentric chromosomes, terminal deletions, and acentric rings.

Microscopically, chromosomal aberrations may be observed after fixation and staining of the chromosomes during the metaphase stage of the cell cycle.

Agents inducing chromosomal aberrations are called *clastogens*, these include ionizing radiation and numerous chemical agents.

An example of a genetic disease resulting from structural chromosome aberrations is the Philadelphia chromosome. Here, a portion of the long arm of chromosome 22 is translocated to chromosome 9, an aberration, that is observed in the bone marrow cells of 90% of patients with chronic myeloid leukemia.

2.3. Genome Mutations

Genome mutations are also referred to as numerical chromosomal aberrations and involve a change in the number of chromosomes. *Euploidy* refers to a multiple of the complete set of chromosomes (haploid, diploid, triploid). *Aneuploidy* describes the situation when the number of some chromosomes is different from the normal diploid set or exact multiples of it. This may result from non-disjunction of chromosomes at anaphase or chromosome loss during cell division with an uneven distribution of chromosomes between daughter cells.

Aneuploidy may be induced by agents affecting the spindle apparatus, such as the drugs colchicine, taxol and vinca alkaloids. Also aneugenic environmental chemicals have been identified, for example high levels of the pesticide trichlorfon in the diet of mothers may be responsible for a cluster of congenital abnormalities including Down's syndrome, observed in a Hungarian village in 1990.

The frequency of aneuploidy in humans appears to be high: At conception a rate of about 16–50% is estimated. Most of the affected embryos are not viable and are spontaneously aborted. About one in 200 newborns displays aneuploidy, and a strong increase in this frequency is observed with the mother's age.

Trisomy of chromosome 21 results in Down's syndrome characterized by physical abnormalities, physical and mental growth retardation. This most common autosomal abnormality occurs with a frequency of about 1 in 600 children.

3. DNA Repair

A large variety of DNA repair enzymes continuously scan the DNA and replace damaged nucleotides. Therefore, despite the frequent occurrence of spontaneous or induced lesions (estimated to amount to several thousand in a typical mammalian cell each day) the number of mutations occurring is low and the genetic information is largely maintained intact.

DNA repair depends on the presence of a separate copy of the genetic information in each strand of the DNA double helix. A lesion on one strand can therefore be cut out by a repair enzyme and a good strand resynthesized using the undamaged strand as a template. There are a number of repair systems that identify and amend specific lesions in the DNA strand prior to the replication. For example, alkyl residues at the guanine O^6-position are being removed by a dedicated methyl transferase. In *base excision repair* an altered base is removed by a DNA glycosylase enzyme, followed by excision of the resulting sugar phosphate. In *nucleotide excision repair* a small region of the strand surrounding the damage is removed from the DNA helix as an oligonucleotide. In both cases the small gap left in the DNA helix is filled in by the sequential action of DNA polymerase and DNA ligase.

While some of these repair enzymes are constitutively expressed, others are inducible, in an adaptive response these enzymes are expressed when a DNA damage occurs.

DNA double strand breaks may be repaired by homologous recombination using the intact homologous chromosome as a template (this may lead to sister-chromatid exchanges) or by rejoining the ends of two DNA strands thereby losing several nucleotides at the joining point and introducing mutations.

In order to transform a DNA lesion into a mutation it is necessary that the DNA polymerase system is by-passing this lesion and is introducing a (wrong) base opposite the lesion in the daughter strain. The molecular mechanisms of these translesion DNA synthesis with a family of error-prone polymerases have been identified most recently.

4. Mutations and Cancer

Hereditary predispositions towards certain types of cancer have been linked to mutations in one of the two copies (alleles) of specific genes that humans carry in every cell. For example, a tumor suppressor gene may be damaged, such as BRCA1 in hereditary breast cancer. Induction of a mutation that damages the intact copy of this tumor suppressor will lead to the development of a tumor at a specific organ site.

Similarly, it is believed that mutations in oncogenes may lead to the development of malignant tumors. Generally, several genetic changes are necessary for a malignant tumor to develop. This high incidence of specific mutagenic events is possible only because in tumor cells the systems ensuring the integrity of the genome are also lost or damaged.

Also, chromosomal mutations (aneuploidy) play a major role in the induction and development of malignant neoplasms.

It has been proposed that the number and pattern of mutations, the so-called mutation spectrum, may be an indicator of the damaging agent. For example the fingerprint of mutations detected in several thousand of cancer cases in a specific gene (p53) is indicative of polycyclic aromatic hydrocarbons in cases of lung cancer, whereas in cases of skin cancer the majority of mutations resembled the damage induced by UV-irradiation.

5. Mutagenicity Tests

A large number of test systems can be employed to determine the mutagenic potential of an agent. Basically, these systems include in vitro tests in bacteria, yeast, or mammalian cells and in vivo short-term tests in experimental animals. The extent of testing required for new substances depend on the intended use (pesticide, drug, food additive, etc.) and the production volume.

Several tests have been standardized and the testing procedures are laid down in OECD Guidelines, EU test guidelines, or by the Environmental Protection Agency of the USA.

The principles of these tests are described in detail in → Toxicology.

5.1. Microbial Tests

Most commonly the so-called *Ames test* is performed. Here gene mutations (point mutations or frameshift mutations) are recorded in a battery of mutated *Salmonella typhimurium* strains that require histidine. With or without addition of external metabolic activation (added homogenate of rat liver) reverse mutations are detected by regaining histidine-prototrophy (OECD Guidelines → Toxicology, Chap. 4.).

5.2. Eukaryotic In Vitro Tests

Whereas microbial tests are easier to conduct, assays involving mammalian cells are generally considered more relevant for the prediction of animal or human in vivo effects and also allow detection of clastogenic or aneugenic potency.

5.2.1. Mammalian Gene Mutagenicity Tests

Most commonly the V79 cell line (Chinese Hamster) or the mouse lymphoma cell line L5178Y $TK^{+/-}$ are employed. These tests may be conducted with or without external metabolic activation. In principle, mutations in the hypoxanthine-guanine phosphoribosyl transferase gene (HPRT) are detected in V79 cells. This gene, which is a available only in one copy because it is located on the X chromosome, confers susceptibility towards purine derivatives such as 6-thioguanine. Similarly, the mouse lymphoma cell line L5178Y $TK^{+/-}$ possesses only one copy of the gene coding for thymidine kinase, an enzyme involved in nucleotide metabolism and leading to susceptibility towards the pyrimidine derivative trifluorothymidine. Inactivation of these enzymes (by mutation) leads to insensitivity towards these agents. Cells with a mutation survive and may be scored as growing clones.

In the mouse lymphoma test small mutant colonies are considered the result of chromosome damage while large mutant colonies are considered the result of gene mutations (OECD Guidelines → Toxicology, Chap. 4.).

5.2.2. In Vitro Cytogenetic Assays

Chromosomal aberrations are analyzed microscopically in cell populations in metaphase. Mostly cells of the Chinese hamster such as V79 and CHO are utilized but also human lymphocytes and mouse lymphoma cells. Metaphase chromosomes are stained and inspected for rearrangements. In the *sister-chromatid exchange assay*, chromosomes are differentially stained and inspected for the presence of exchanges between sister chromatids.

In recent years the *micronucleus assay* has become more popular: Micronuclei (MN) may be formed during the cell division and may arise from a whole lagging chromosome (aneugenic event leading to chromosome loss) or an acentric chromosome fragment detaching from a chromosome after breakage (clastogenic event). The MN test may be performed with different cultured mammalian cells and involves the treatment of cells with the actin polymerization inhibitor cytochalasin B to identify cells that divide as binucleate cells. Micronuclei are detected microscopically. By means of kinetochore-antibody staining or fluorescence in situ hybridization they may be differentiated into micronuclei containing a whole chromosome or an acentric chromosome fragment, indicating aneugenic or clastogenic activity, respectively, of the test compound (OECD Guidelines → Toxicology, Chap. 4.).

5.3. In-Vivo Tests

5.3.1. Micronucleus Assay

As the in vitro micronucleus test, this assay tests mutagenic effects on the chromosome and genome level, detecting clastogenic and aneugenic agents. The in vivo micronucleus test is generally performed in mice, and the tissues most often assessed for frequency of micronuclei are bone marrow and peripheral blood. Erythrocytes are the cells that are scored in the bone marrow or the blood for presence of micronuclei. Because they are constantly produced from precursor cells and the nucleus is expulsed from the cell immediately after full differentiation is accomplished, erythrocytes are an ideal cell type for a micronucleus test. If a stem cell is damaged by a chemical and a micronucleus is formed as a consequence of this damage, the micronucleus remains in the cell after the main nucleus has been expulsed and can be observed microscopically.

Mouse bone-marrow micronucleus tests typically employ one to three treatments with the chemical under study; with doses extending up to the maximum tolerated dose. The route of administration in these short-term tests is usually either intraperitoneal injection or oral gavage. Based on the cell cycle and maturation times of the erythrocytes, harvesting of the bone marrow usually occurs 24 h after the final dosing. At that time, about 50% of the erythrocytes in the bone marrow are immature, newly formed erythrocytes, and these are the cell types that are checked for presence of micronuclei. The bone marrow is flushed from the femurs and spread onto slides. The slides are dried, fixed, and stained with a fluorescent DNA-specific stain that easily illuminates any micronuclei that may be present. 2000 polychromatic erythrocytes (PCEs, or immature erythrocytes) are scored per animal for frequency of micronucleated cells. In addition, the percentage of PCEs among the total erythrocyte population in the bone marrow is scored for each dose group as a measure of toxicity. Short-term studies that give positive results are repeated to confirm the response.

Critical parameters in analyzing micronucleus test data are the number of animals per dose group (a minimum of three is required); dose levels and number of doses administered; route of administration; tissue and cell type analyzed; sample time (interval between last dosing and harvesting of cells for analysis);, frequencies of micronucleated cells in the negative and positive controls; and the results of the statistical analyses. In the case of a negative result it is important to establish that the test chemical reached the target organ (bone marrow), which may be verified by signs of bone-marrow toxicity.

5.3.2. Indicator Tests

Assays that measure damage to DNA may be performed either in vitro or in experimental animals. These tests include detection of *unscheduled DNA synthesis* (UDS), indicating the repair of damaged DNA. The test is most commonly performed in primary hepatocytes in vitro or in hepatocytes following in vivo treatment (OECD test guideline 486, EU test guideline B39).

The formation of DNA adducts may be analyzed by ^{32}P-postlabeling analysis. Here covalent modifications are analyzed chemically with a sophisticated procedure involving enzymatic DNA hydrolysis, enzymatic labeling with ^{32}P-phosphate and chromatographic separation of labeled modified nucleotides (→ Toxicology, Chap. 4.9.4.)

The analysis of DNA strand breaks may be accomplished by single cell electrophoresis (comet assay). Here nuclei are exposed to an electric field under DNA denaturing conditions in an agarose gel. Damage to DNA results in DNA fragments that move faster in the gel. This can be visualized upon fluorescence staining and microscopically scored.

The advantage of short-term in vivo assays such as ^{32}P-postlabeling analysis and comet assay is that a number of organs may be assayed in parallel and genotoxic activity testing is not confined to one organ such as the liver in UDS tests or bone marrow in the case of the micronucleus assay.

The more common application of these, liver and bone marrow specific, tests is based on the assumption that mutagens exhibit a general genotoxicity which would be detectable in all organs. However, this is contradicted by the organ specificity observed by some chemical carcinogens.

5.3.3. Transgenic Animals

Assays based on transgenic animals such as Muta Mouse or Big Blue Mouse (and rats) allow determination of genotoxic potency in any organ. These commercially available animals have been genetically modified so that each cell contains a shuttle vector containing several copies of the lactose operon of *E. coli*, lacZ or lacI, respectively. These serve as indicators of gene mutations. Briefly, these animals are treated with a test compound, animals are sacrificed, DNA is isolated after a certain delay period, and the shuttle vector is transferred to *E. coli*. Detection of mutations in the shuttle vector sequences is accomplished by white/blue selection of growing bacteria in plates containing appropriate indicators.

These assays show a high degree of correlation with conventional in vitro gene mutation assays and also mutagenic activity was detected in target organs of numerous carcinogens. While these assays are detecting mutations in nongenomic DNA and may not detect chromosomal or genome mutations, there are new developments under way such as the gpt delta mouse, for example, that may be capable of detecting large deletions induced by clastogens.

6. Testing Strategies

Depending on the type of substance there are guidelines and legislation, for example in the European Union, that regulate the requirements for genotoxicity testing.

Basic testing for genotoxicity includes one gene mutation assay (mostly a bacterial assay) and one in vitro mammalian assay (most commonly the in vitro chromosomal aberration test). If these tests are negative, there is no further testing required for existing chemicals and cosmetic ingredients.

A third in vitro test (mammalian cell gene mutation test) is required for plant protection products, biocides, food additives, food contact materials, and human pharmaceuticals.

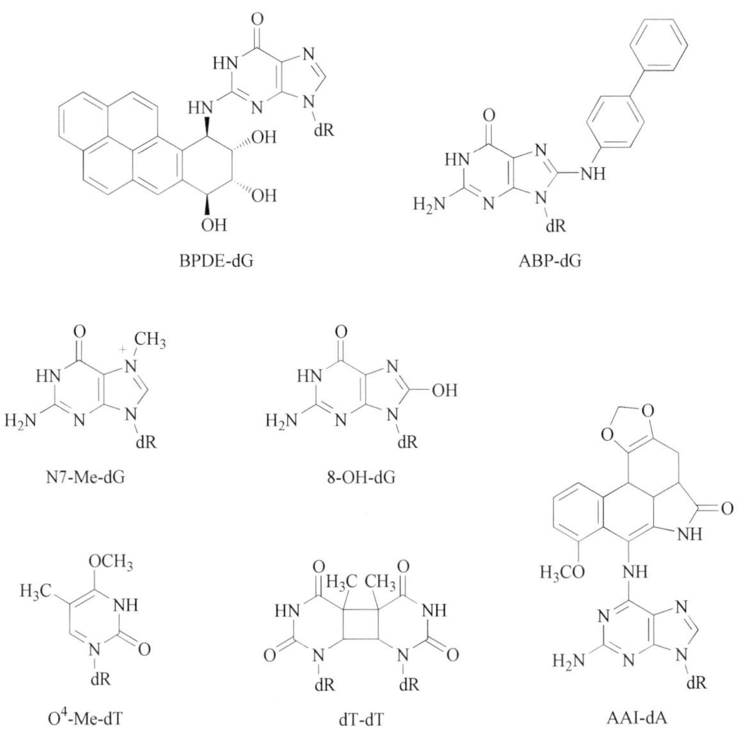

Figure 2. Prototypes of DNA adducts: Major deoxyguanosine-adducts formed by benzo[a]pyrene (via its diol-epoxide, BPDE-dG), 4-aminobiphenyl (C8-ABP-dG), hydroxyl radicals (8-OH-dG) and *N,N*-dimethylnitrosamine (N7-Me-dG). Alkylating agents may also modify thymidine such as O^4-Me-dT. The thymidine dimer formed is the major lesion formed by UV-light and the N^2-deoxyadenosine adduct is induced by the herbal carcinogen aristolochic acid I (AAI-dA).

For human pharmaceuticals and veterinary drugs for food producing animals, basic testing includes a bacterial gene mutation assay, mammalian cell assay (mouse lymphoma assay or in vitro chromosome aberration), and an in vivo assay (testing chromosomal aberration or micronucleus formation).

New chemical substances require testing depending on tonnage of production.

If these basic tests are consistently negative the test substance is considered nonmutagenic. However, further testing is required in the case of positive findings in one or all basic assays conducted.

Generally, in vitro tests are preferred to tests in animals in line with animal welfare considerations. However, in vivo test are more important in the final assessment of a test substance, for examples a positive result in a microbial assay is generally overruled by negative outcome of one or more in vivo assays.

In order to assess a test substance the judgement of an experienced toxicologist is required.

7. Metabolic Activation

Strictly speaking most so-called mutagenic chemicals are not mutagenic themselves but, in order to exhibit their genotoxic activity require metabolic activation. By enzymatic transformation reactive metabolites are formed, such as α-hydroxy derivatives of *N*-nitrosamines, hydroxylamines from aromatic amines and nitroarenes, epoxides from hydrocarbons, or radicals that are capable of binding to the DNA. Covalent binding to DNA results in DNA adduct formation (Fig. 2.). These adducts are mostly removed by repair mechanisms but may result in the introduction of mutations (→ Carcinogenic Agents)

UV- or visible light may also serve as an activating agent. In assays investigating photomutagenicity treated cells are exposed in vitro to a source of UV(and/or visible) light. A number of examples of agents inducing gene mutations upon photoactivation have been reported. These agents include polycyclic aromatic hydrocarbons, and fluoroquinolones. Also clastogenic activity may be induced by photoactivation for example in the cases of 8-methoxypsoralen or chlorpromazine. These effects are especially relevant for ingredients of cosmetic products.

8. Types of Mutagenic Agents

The most prominent mutagenic agents are also carcinogenic and are described in Carcinogenic Agents. Computer programs are available that, based on biological data, assess chemical structures for functional elements ("alerting structures"), which may act as mutagens. (→ Carcinogenic Agents, Chap. 5.3). These programs are useful to scan large libraries of chemicals being tested early in high throughput drug development, but even highly sophisticated calculation methods are presently not able to replace biological testing.

Members of the following classes of chemical compounds have been identified as mutagenic: see → Carcinogenic Agents

Chap. 7.1 *N*-Nitroso compounds
Chap. 7.2 Benzene
Chap. 7.3 Polycyclic aromatic hydrocarbons
Chap. 7.4 Nitroaromatic compounds
Chap. 7.5 Aromatic amines
Chap. 7.6 Halogenalkyls
Chap. 7.7 Aldehydes
Chap. 7.9 Alkylating agents
Chap. 7.10 Metals

In addition, a number of natural compounds and mixtures show mutagenic potential. Mutations can also be induced by oxygen damage. *Physical mutagenic agents* include nonionizing and ionizing radiation.

Aneugenic agents with no direct genotoxic activity are listed below. These may act by inhibition of repair enzymes, cell cycle control or apoptosis related proteins, tubulins of the mitotic or meiotic spindle or interference with detoxification or activating enzymes:

Benomyl
Cadmium chloride
Carbendazim
Chloralhydrate
Colchicine
Diethylstilbestrol
Griseofulvin
Hydroquinone
Vinblastine

9. Conclusions

In addition to the application as short-term test to detect the mutagenic potency of test chemicals, assays such the micronucleus test, comet assay or ^{32}P-postlabeling are applied as biomarkers in molecular epidemiological studies. Increased levels of DNA adducts have been observed in cohorts occupationally exposed, e.g., to polycyclic aromatic hydrocarbons or styrene, to aflatoxins in the diet, to tobacco smoke, or to genotoxic drugs (including chemotherapeutic agents or herbal drugs).

In vivo animal experiments reflect the human situation only to a certain extent but generally the predictive value of animal assays is believed to be quite high. In contrast, in vitro assays are highly artificial and several aspects such as metabolic activation and toxicokinetics are not included. Therefore, data from in vivo assays are considered more relevant for the evaluation of a test chemical. In vitro assays are, however, indicative of the mechanism of action of a mutagen.

Antimutagenic agents inhibit or reduce the activity of mutagens. These may act by inducing detoxifying mechanisms, by antioxidative or receptor antagonism.

Co-mutagens are not mutagenic but may in combination with other agents increase mutagenic activity. Here, induction of activating enzymes, inhibition of detoxifying mechanisms, or formation of mutagenic metabolites is involved.

There are also compounds that are positive in several short-term tests for mutagenicity but failed to induce tumors in animal carcinogenicity studies. Examples are flavonoids such as quercetine or *p*-phenylendiamine. This may relate to high doses, that have been tested in vitro but are too toxic to be administered in vivo. However, these cases are highly disputed in the scientific community.

On the other hand there are carcinogenic agents that act via a nongenotoxic mechanism of action, such as mitogenic agents, tumor promoters, or hormones (→ Carcinogenic Agents).

Most genotoxic agents inducing gene mutations are also clastogenic. There are aneugenic agents that act solely by inhibiting the synthesis and assembly of the mitotic spindle, with no direct activity on the DNA, such as the methyl benzimidazole carbamate pesticides benomyl and carbendazim. Other indirect mechanism may involve inhibition of DNA repair (e.g., action of cadmium).

Based on theoretical considerations, *indirect genotoxins* are believed to be thresholded, i.e., there exists a safe exposure level. DNA damaging agents (*direct genotoxins*) are considered to act without a threshold, i.e., already a single molecule could induce mutation and cancer albeit with a exceedingly small but finite probability. Therefore, it is important to identify the mechanism by which a mutagenic agent is exerting its activity.

10. References

1. H. Marquardt, S. G. Schäfer, R. O. McClellan, F. Welsch (eds.): *Toxicology*, Academic Press, San Diego 1999.
2. Niesink, DeVries, Hollinger (eds.): *Toxicology*, CRC Press, Boca Raton 1996.
3. BgVV: Strategies for genotoxicity testing of substances, 2002.
4. Kirsch-Volders et al., *Toxicol. Lett.* **140–141** (2003) 63 – 74.
5. Parry et al. *Mutagenesis* **17** (2002) 509 – 521.

Food Additives

ERICH LÜCK, formerly Hoechst Aktiengesellschaft, Frankfurt, Federal Republic of Germany

GERT-WOLFHARD VON RYMON LIPINSKI, formerly Hoechst Aktiengesellschaft, Frankfurt, Federal Republic of Germany

1. Introduction

Food additives are substances that are intentionally added to food, without themselves being considered food in the ordinary sense of the term. The term "food additive" is defined differently by the food laws of individual countries. As examples, the definitions used in the Codex Alimentarius, in the United States, in Japan, and in the Federal Republic of Germany are given in Chapter 6. Food additives are indispensable for the production and processing of many foods. Some are essential for the economic production

Ullmann's Industrial Toxicology

ISBN: 3-527-31247-1

and distribution of foods. Together with continued improvements in food technology, additives ensure the general availability of high-quality food with a satisfactory shelf life. The four basic reasons for using additives are

1) to influence the nutritive value of food,
2) to improve the stability of food,
3) to affect the sensory properties of food, and
4) to make certain technological processes possible.

Food additives are sometimes used to produce food more cheaply. However, they may not be used to deceive the consumer, for example, about the character or freshness of food. Furthermore, additives must not adversely affect food quality.

The properties and production methods of some additives are described in other articles; the reader is advised to consult the cumulative index for specific products. The following discussion is, therefore, confined to more general and universal aspects.

1.1. History

Two preservatives, smoke and salt, are probably the oldest food additives. They were also used to give food a good taste. Like spices, they have been used since prehistoric times to improve the taste of food from plant or animal sources. These foods were either eaten raw or prepared over an open fire. Until the 18th century, the use of additives remained low. However, with the beginning of industrialization in the 19th century, an increasing number of people moved to towns. This meant that they had less time available to provide and process their food. More and more food was produced either by specific craftsmen or industrially. Additives became increasingly necessary; especially those that protected food against spoilage. Some food additives introduced in the 19th century are still being used today, e.g., baking powder, benzoic acid, and saccharin. However, others in use at that time do not comply with current toxicological requirements and have thus disappeared from food processing.

In the 20th century, the industrial production of food developed further and the importance of additives increased. Thus, the invention of processed cheese in 1913 would have been impossible without the development of emulsifying salts, such as citrates, phosphates, or similar substances. During this time emulsifiers were developed which improved and simplified the production of margarine. Additives used in certain food for physiological reasons have also become important, e.g., trace elements, vitamins, and essential amino acids. More recently, the importance of aroma and flavor enhancers, as well as flavoring compounds, has increased substantially.

Hygienic and legal controls have also been tightened considerably in this century. One example of this is in coloring agents. Many substances used in the first half of the 20th century have not survived modern toxicological requirements.

Despite all the advances, unresolved problems still remain in the area of food additives, e.g., finding substitutes for two specific types of preservative, nitrite and sulfite, which are considered undesirable for various reasons. The search also continues for substances or processes that can be used to reduce the undesirably high levels of sugar and common salt in some foods, without impairing their quality. The financial risks involved in developing new food additives are high because development of a new food additive is now almost as expensive as developing a new drug. Even when additives have been evaluated successfully and necessary authorization under food law has been obtained, the cost of development can be recovered only if the additive can be used widely.

1.2. Safety Aspects

Some people still harbor an aversion to food additives. This can be explained by earlier, uncontrolled use of chemicals in food processing. Today, food additives are authorized only if no harmful effects of any kind can be shown after extensive toxicological testing. Many additives have undergone more rigorous testing than some foods or food components.

To date, no internationally binding standards exist for the toxicological testing of additives. The most widely accepted guidelines are those laid down by expert committees of the WHO

[1]. The U.S. regulations [2] are also observed widely.

Additives are tested for their acute, subchronic, and chronic toxicity, as well as for carcinogenicity, mutagenicity, teratogenicity, and biochemical behavior. Before testing begins, the substance must be chemically and physically homogeneous and pure. As a matter of principle, a food additive should not be pharmacologically active at the concentration at which it is used. For obvious reasons, the testing of additives is carried out largely on animals. The tendency away from animal testing is increasing, but at present, this is only possible in a few areas.

Acute toxicity, expressed as LD_{50}, and subchronic toxicity, established by the so-called 90-day test, are only crude measures of the toxicological properties of a substance. The acute toxicity of food additives should always be as low as possible to prevent poisoning even in the case of accidental overdosing or misuse.

Chronic toxicity is more important (→ Toxicology). The specifications for food additives are far more strict in this respect than those for drugs, because additives may be consumed continuously for a very long period of time. Also, in contrast to drugs, side effects are not considered an acceptable risk with food additives.

Food additives must not be mutagenic or teratogenic (→ Mutagenic Agents). Because they are absorbed exclusively via the gastrointestinal tract, carcinogenicity is tested mainly by feeding.

Biochemical tests are performed to determine the extent to which a food additive is absorbed by the organism, the factors affecting its absorption, its distribution within the body, and the manner in which it is excreted. Four basic possibilities exist:

1) A substance is excreted quickly and unchanged. Examples of this are the sweeteners saccharin [*128-44-9*] and acesulfame-K [*55589-62-3*].
2) A substance is quickly metabolized within the body and excreted in this metabolized form. An example is the preservative benzoic acid [*65-85-0*], which is converted in the body to hippuric acid [*495-69-2*], by reaction with glycine [*56-40-6*], and excreted in this form.
3) A substance is utilized quickly as a nutrient. Some examples are the preservative sorbic acid [*110-44-1*] and the sweetener aspartame [*22839-47-0*].
4) A substance is excreted more slowly than it is absorbed and thus accumulates in the body. An example of this is boric acid, used as a preservative in the past. Such substances are considered undesirable and are no longer used as food additives.

Table 1. Acceptable daily intake (ADI) values of some food additives

Additive	CAS registry no.	ADI, mg/kg per day
Acesulfame-K	[*55589-62-3*]	0–9
Alginates		0–50
Amaranth	[*915-67-3*]	0–0.75 (temporary)
Ascorbic acid	[*50-81-7*]	0–15
Aspartame	[*22839-47-0*]	0–40
Azo Rubine	[*3567-69-9*]	0–4
Benzoic acid	[*65-85-0*]	0–5
Butylated hydroxyanisole	[*25013-16-5*]	0–0.5
Butylated hydroxytoluene	[*128-37-0*]	0–0.5
Caramel (ammonia process)		0–100 (temporary)
Cellulose ethers		0–25
Cyclamate	[*100-88-9*]	0–11
Erythrosine	[*16423-68-0*]	0–2.5
Ethyl maltol	[*4940-11-8*]	0–2
Fumaric acid	[*110-17-8*]	0–6
Gallates		0–0.2
Nitrates		0–5
Nitrites		0–0.2 (temporary)
Phosphates		0–70
Quinoline Yellow	[*8004-92-0*]	0–0.5 (temporary)
Saccharin	[*128-44-9*]	0–2.5 (temporary)
Sorbic acid	[*110-44-1*]	0–25

Animal trials are conducted to determine the concentration to which an additive remains nontoxic. Expert WHO committees use this value to define the *acceptable daily intake* (ADI) by reducing it by a factor of 100 (except in special cases). This takes into consideration the following facts: the food intake of some animals, relative to their body weight, is greater than that of humans; that some people (e.g., children, the sick, and the elderly) show special metabolism; and the ADI applies over the entire life-span.

The ADI is related to kilograms of body weight and is the amount of additive in milligrams that can be consumed safely per day (this includes the reduction by a factor of 100). The ADI values of some food additives are given in Table 1. The actual consumption of food additives is generally far lower than the ADI. For preservatives in the Federal Republic of Germany, consumption is about 1 – 10 % of the ADI. Only sulfur dioxide is consumed in larger amounts [3].

2. Substances with Nutritive and Other Dietary Effects

The human organism requires regular supplies of about 40 – 50 nutrients if all bodily functions are to proceed smoothly. If even one of them is missing or present in insufficient amounts, deficiency symptoms result. A particularly well-known example of this is scurvy, i.e., vitamin C deficiency.

All nutrients and nutritive supplements needed by the body are normally present in sufficient amounts in food. However, in exceptional cases or with unbalanced diets, the amount of these substances may be insufficient. Some can be destroyed or lost during processing or storage. In such cases, these nutrients can be added in a controlled manner.

Substances with nutritive and dietetic functions are added to food to increase its nutritive quality or, in special cases, to reduce it in a controlled way. Some of these substances may also ameliorate metabolic disorders with a dietary or physiological cause and correct various deficiencies.

Incorporating these substances homogeneously is sometimes difficult because they must be added at very low concentrations to relatively large quantities of food. In such cases, special premixes are used to allow easier incorporation into the appropriate food. Some substances with nutritive and dietary effects are particularly susceptible to degradation, especially by oxidation. Therefore, they must be stabilized. In practice, this is achieved by the synthesis of especially stable molecules which will remain physiologically and dietetically effective, by the use of stabilizing additives, or by encapsulation with suitable carrier materials (e.g., microencapsulation).

2.1. Vitamins and Provitamins

Vitamins are used as food additives for dietary reasons, when sufficient vitamins are not supplied by the food itself or when losses have occurred during processing. Natural variation in the vitamin content of food can also be adjusted in this way. Vitamins may also be added when the body is unable to absorb sufficient amounts of the vitamins contained in food [4–6]. The controlled addition of vitamins to food is called *vitaminization,* and the addition to compensate for processing losses is *revitaminization. Enrichment* or *fortification* means the addition of vitamins beyond the natural content.

Standardization means adjustment of natural variations. Oversupply is harmful only with the fat-soluble vitamins A and D, because it can lead to so-called hypervitaminoses.

Examples of *revitaminization* are the addition of vitamins A and D to skimmed milk and the addition of B vitamins to cereal. Examples of *standardization* or *enrichment* are the addition of vitamin C to beverages or fruit products and the addition of vitamin D to milk. Vitamins are added to some foods that naturally contain few or no vitamins because they constitute a particularly suitable carrier for a particular vitamin. An example of this is the addition of fat-soluble vitamins A and D to margarine.

Vitamins and provitamins can also be added to foods for technological reasons. For example, ascorbic acid [*50-81-7*] and tocopherols have an antioxidative effect (see Section 3.2), and some carotenoids with vitamin functions are important food colorants (see Section 4.1).

2.2. Amino Acids (→ Amino Acids)

Some amino acids are added to foods for technological reasons, e.g., flavoring or flavor enhancers (Chap. 4) or baking aids (Section 5.11). Their use as food additives for *dietetic reasons* is restricted to the essential amino acids, i.e., those which the body cannot synthesize. These are added to food for the same reason as vitamins. During food processing, lysine [*56-87-1*], in particular, can react with carbohydrates by the Maillard reaction to form compounds from which it is no longer physiologically available. Deficiencies in low-protein foods which also have a low lysine ratio (e.g., some cereals) can be adjusted by enrichment with lysine. Mixtures of amino acids are sometimes administered with food in special diets, especially low-fiber diets. The importance of amino acids as human food additives is far less than in animal feed.

2.3. Minerals and Trace Elements

Sufficient amounts of minerals and trace elements are generally present in the average diet. In addition, they are not damaged by industrial processes so they are added to foods only in special cases. Iron salts are of some importance because iron is sometimes not absorbed in sufficient quantity from food. The same is true of calcium, magnesium, copper, and zinc salts [7]. The importance of magnesium for the cardiovascular system encouraged the addition of magnesium to food. In areas where the iodine content of drinking water is low, a sufficient iodine intake can be achieved by addition of iodides and iodates to common salt. A difference of opinion exists about the need for fluorine in human nutrition. In principle, fluorine can be added to common salt or to drinking water in the form of fluorides.

2.4. Bulking Agents

Bulking agents are inert substances with little or no nutritive value. They are used in reduced-calorie foods to generate a feeling of satiety.

The cheapest bulking agents are water and air. If these are to be incorporated into foods, the addition of emulsifiers (Section 3.6) or thickening agents (Section 3.7) is usually necessary. Other bulking agents include crystalline cellulose, polydextrose (a polycondensate of sucrose with sorbitol) and the disaccharide alcohol isomalt [*64519-82-0*], which have low calorie values.

3. Substances with Stabilizing Effects

Products with increased shelf life are in demand for many reasons: the increasing extent to which food is produced industrially, higher food-quality specifications, contemporary shopping and consumption patterns, as well as certain health reasons. Therefore, prevention of food spoilage–whether due to microbiological, chemical, biochemical, or physical factors–is a necessity. Improved shelf life can often be achieved by physical methods such as heating. However, some of these methods change the structure or taste of food, which is the reason they are sometimes supplemented or reinforced by the use of additives.

3.1. Preservatives

Preservatives are compounds that delay, or ideally prevent, microbiological spoilage of food [8], [9]. Not only do these act against visible spoilage of food by yeast, molds, and bacteria, they also prevent the formation of toxins, especially those produced by bacteria and molds. Today, this effect is more important than the protection of food against economic losses. For example, the antimicrobial agent nitrite destroys *Clostridia* which excrete botulinum toxin, and sorbic acid [*110-44-1*] acts against molds that can form carcinogenic aflatoxins.

The use of preservatives, also referred to as chemical preservation of food, does not compete with physical methods of preservation, e.g., sterilization and chilling; rather, it complements them. Sometimes the combined use of preservatives and physical measures makes possible a reduction of either the intensity of physical measures or the concentration of preservatives. A typical example is the production of intermediate-moisture, partially dried, plums protected by sorbic acid from mold development. Their quality is superior to that of fully dried products.

The combination of physical and chemical preservation processes of any kind is referred to as the "hurdle concept" [10]. The addition of preservatives also protects food against subsequent spoilage once packages are opened or if they are not hermetically sealed. This is independent of, and in addition to, any physical preservation measures that may be used.

3.1.1. Mode of Action

In practice, a distinction is often drawn between bacteriostatic and bactericidal (antibacterial effect) or between fungistatic and fungicidal (antiyeast and antimold effect), i.e., between germ-inhibiting and germ-killing effects. Strictly speaking, this distinction is not justified [11]. Preservatives present in low-germ foods at sufficiently high dosages kill the microorganisms present more or less rapidly. If no reinfection occurs, this final state is achieved in a period of several days to several weeks. This is the crucial difference between preservatives and disinfectants (→ Disinfectants). The latter must destroy microorganisms in a very short period of time. The course per unit time of microbial death due to preservatives corresponds to a first-order reaction [12]:

$$k = \frac{1}{t} \cdot \ln \frac{Z_0}{Z_t} \quad \text{or} \quad Z_t = Z_0 \cdot e^{-kt}$$

where k is the death constant, t is time, Z_0 is the number of viable cells at the start of the preservative action, and Z_t is the number of viable cells after time t.

The *antimicrobial effect* of preservatives can essentially be explained by the following phenomena:

1) interference with the genetic microstructure of the protoplast,
2) inhibition of protein synthesis,
3) inhibition of enzyme activity,
4) damage to the cell membrane, and
5) damage to the cell wall.

Substances that lower the *water activity* of a substrate and thereby impede microbial growth work differently. Packages, coatings, oil, and some packaging gases block access of oxygen to foods and thus inhibit the growth of aerobic microorganisms. As a rule, none of these substances kill microorganisms but, at best, damage them and stop them from multiplying. When the influence of these inhibitors is removed, the contaminating germs start multiplying again, and the product eventually spoils. Finally, preservatives such as acetic and lactic acid lower the pH of food and thereby inhibit the growth of some microorganisms, especially bacteria.

No preservative is equally effective against all microorganisms. Most substances used today are more effective against yeasts and molds than against bacteria. Unlike antibiotics, preservatives do not lead to the development of organisms resistant to them [13], [14].

Occasionally, two or more preservatives are used in tandem. In the past, such combinations were widely used commercially. However, their effectiveness has been overrated. Synergism of sufficient practical importance is unknown for preservatives. However, a combination of preservatives can sometimes broaden the spectrum of activity, e.g., the combined use of substances that are individually effective against molds or bacteria.

3.1.2. Methods of Evaluation

The results of tests based on nutrient media, commonly used in microbiology, are difficult to transfer to a food. This is because substrate characteristics influence the effect of preservatives in a variety of ways. For example, the pH and the redox potential of a food, the presence or absence of fats, and its water activity are of great practical importance for the effectiveness of a preservative. Some preservatives (e.g., nitrite) react with food components.

In practice, the preservative is added in different concentrations to batches of the food. These are then packaged commercially, and their shelf life is monitored under realistic storage conditions. Here, the microbiological status of the food is observed as are any changes in its sensory qualities. A common problem is inoculation with appropriate microorganisms. Not only is this a question of inoculating with suitable types of microorganisms, but the number of organisms must also represent the actual conditions found in practice. The role of preservatives is not to combat very high numbers of germs but rather to attack the relatively small population that can be expected to be present in food produced under hygienic conditions.

3.1.3. Uses and Individual Substances

Preservatives are used in cheese, meat products, fruit-based products, beverages, and baked goods. Depending on the type of food and the spoilage expected, different preservatives are used. For example, meat products are preserved mainly with common salt and nitrite; fruit-based products, with sulfur dioxide; beverages, with sorbic acid; and baked goods, with sorbic acid or propionates.

The most important preservatives commonly used in foods are summarized in Table 2. In addition to these, the following have some importance as preservatives although they are used partially for other reasons, such as flavoring: sodium chloride, nitrogen, silver, ozone, hydrogen peroxide, chlorine, ethanol, sucrose [*57-50-1*], and smoke.

3.2. Antioxidants (→ Antioxidants)

Antioxidants delay oxidative spoilage of food. They interfere with the early stages of oxidative and autoxidative processes to prevent formation of unwanted reaction products. Broadly speaking, antioxidants are substances that react quickly with oxygen and thereby quench it.

During food processing and storage, autoxidation is particularly troublesome with fats that contain unsaturated fatty acids, which can form highly odoriferous, unsaturated aldehydes. These cause the fats to smell and taste unpleasant; i.e., they become rancid. The formation of polymeric reaction products is also undesirable.

3.2.1. Mode of Action

The autoxidation of fats proceeds via a radical mechanism. Hydroperoxides are formed as intermediates, which then decompose to form radicals and can thereby initiate further reactions. The chain reactions can be terminated by scavengers that react with radicals to form inert compounds [15–17]. Radical scavengers constitute, therefore, the most important group of antioxidants.

If a mixture of two or more antioxidants is used, a stronger effect can be achieved than with a single antioxidant. To optimize this synergism, the necessary ratio of the two antioxidants must be established experimentally. In addition, some antioxidants inhibit lipoxygenase and, by this side effect, can prevent the enzymatic oxidation of fats.

Antioxidants have the desired effect only within a certain concentration range. At excessive concentration, they may even act as prooxidants. If the aim is to protect fats in the finished food product after it has been heated, antioxidants must be used which maintain their effectiveness after such processing. For example, certain phenolic antioxidants show this carry through effect [18].

Antioxidants, in the widest sense, scavenge oxygen and thereby prevent the occurrence of enzymatic reactions that require oxygen. Sometimes they also inhibit the enzymes involved such as polyphenol oxidases.

3.2.2. Methods of Evaluation

The most important means of evaluating antioxidants used for fats are the Swift test, the Barcroft–Warburg test, and the thiobarbituric acid test. In the *Swift test,* samples to which antioxidants have been added are aerated at 97 °C under defined conditions. The peroxide content is measured and compared to that of a sample without antioxidants. In the *Barcroft–Warburg test,* a sample is held under oxygen in a closed system and the oxygen uptake is monitored manometrically. In the *thiobarbituric acid test,* the presence of aldehydes is measured. Under suitable test conditions, this test is said to correlate highly with sensory properties of the fat [17].

3.2.3. Uses and Individual Substances

Antioxidants are added mainly to fats or fat emulsions and to foods containing fats in a more or less fine distribution. Antioxidants can also prevent unwanted oxidative changes in essential oils, aroma compounds, and chewing gum. In practice, antioxidants are combined with synergists to increase their effectiveness. Fruit- and potato-based products, in particular, can be protected against discoloration through addition of ascorbic acid [*50-81-7*], sulfur dioxide, and sulfites. Sulfur dioxide or sulfites alone, and in combination with ascorbic acid, considerably delay undesirable oxidation processes in beverages, e.g., the madeirization of wine [19]. However, here sulfur dioxide and sulfites are used because

Table 2. Concentrations of preservatives found in foods

	Preservative	Concentration, ppm
E* 200 – E 203	sorbic acid [*110-44-1*] and sorbates	500 – 2000
E 201 – E 213	benzoic acid [*65-85-0*] and benzoates	500 – 1000
E 214 – E 219	4-hydroxybenzoic acid esters (parabens) and their sodium derivatives	500 – 1000
E 220 – E 227	sulfur dioxide and sulfites	200 – 2000
E 230	biphenyl (diphenyl) [*92-52-4*]	50 – 70
E 231 – E 232	-phenylphenol [*90-43-7*]	10 – 12
E 233	thiabendazole [*148-79-8*]	3 – 6
E 236 – E 238	formic acid [*64-18-6*] and formates	3000 – 4000
E 239	hexamethylenetetramine [*100-97-0*]	20 – 200
E 249 – E 250	nitrites	50 – 100
E 251 – E 252	nitrates	200 – 600
E 260 – E 263	acetic acid [*64-19-7*] and acetates	5000 – 30 000
E 270	lactic acid [*50-21-5*]	5000 – 10 000
E 280 – E 283	propionic acid [*79-09-4*] and propionates	2000 – 3000
E 290	carbon dioxide	

* Serial number for use on labels

of their preservative effect and for other reasons as well, e.g., to scavenge fermentation byproducts.

The most important antioxidants commonly used in foods are listed below. In addition, the following substances are important antioxidants in some countries: 3,3′-thiodipropionic acid [*111-17-1*], dilauryl thiodipropionate [*123-28-4*], trihydroxybutyrophenone [*52262-23-4*] (THBP), ethoxyquin [*91-53-2*], and *tert*-butylhydroquinone [*1948-33-0*] (TBHQ).

E 220 – E 227	sulfur dioxide and sulfites
E 300 – E 304	ascorbic acid [*50-81-7*] and ascorbates
	isoascorbic acid [*89-65-6*] (erythorbic acid)
E 306 – E 309	tocopherols
E 310 – E 312	gallates
E 320	butylated hydroxyanisole [*25013-16-5*] (BHA)
E 321	butylated hydroxytoluene [*128-37-0*] (BHT)

3.3. Synergists and Sequestrants

Sequestrants are compounds that form complexes with metal ions and thereby convert them into an inactive form. Of special importance is the inactivation of traces of heavy metals, such as copper and iron, which accelerate oxidative changes catalytically.

Sequestrants also inactivate certain enzymes by binding the metal ions needed to activate them and thereby support the microbiological stabilization of foods. Sequestrants can sometimes suppress discoloration caused by interaction between metals and amines or mercapto compounds in delicatessen products and fish or shellfish preserves. Undesirable cloudiness in beverages, such as that caused in wine by iron and copper, is avoided by using sequestrants to convert the appropriate metal ions into ineffective complexes.

Sequestrants used in combination with antioxidants are called *synergists* because their reaction with metallic ions indirectly supports the antioxidative effect. Compounds that regenerate spent antioxidants are also called synergists.

The most important synergists and sequestrants commonly used in foods are summarized below. In addition, the following substances are of some importance as synergists and sequestrants: phytic acid [*83-86-3*] and its salts, calcium gluconate [*299-28-5*], calcium polyphosphate, and potassium ferrocyanide [*13943-58-3*].

E 270	lactic acid [*50-21-5*]
E 322	lecithin [*8002-43-5*]
E 324 – E 327	lactates
E 330 – E 333	citric acid [*77-92-9*] and citrates
E 334 – E 337	tartaric acid [*147-71-7*] and tartrates
E 338 – E 341	phosphoric acid [*7664-38-2*] and phosphates
E 342	calcium and sodium salts of ethylenediaminetetraacetic acid [*60-00-4*] (EDTA)

3.4. Packaging Gases

Packaging gases are used essentially to exclude oxygen from stored food [20] and thus primarily protect against oxidative changes. Furthermore, they indirectly act as preservatives by displacing the oxygen which is vital for strictly aerobic microorganisms.

The main packaging gases used in commercial food processing are nitrogen, carbon dioxide, and mixtures of these. Besides displacing oxygen, carbon dioxide at higher concentration specifically inhibits some microorganisms, especially molds.

3.5. Stabilizers

Stabilizers are substances that protect from, or counteract, changes in the structure of food which, in a wider sense, can be considered spoilage. Stabilizers include emulsifiers, thickeners and gelling agents, foam stabilizers, humectants, anticaking agents, and coating agents.

3.6. Emulsifiers

Emulsifiers are substances that facilitate, or may be essential to, the production of emulsions. Furthermore, they improve the structure of foods that naturally contain small amounts of fat, e.g., dough for baked goods. Emulsifiers improve and stabilize the consistency of foods and sometimes their viscosity, texture, and mouthfeel as well. They improve the shelf life of some food such as baked goods.

Thickeners (see Section 3.7) often stabilize emulsions by greatly increasing the viscosity of the aqueous phase. The formation and stabilization of emulsions in foods that contain protein and fat can also be achieved by addition of phosphates. The properties and effects of phosphates depend on the degree of condensation. Phosphates cause hydration and swelling of proteins, thereby increasing the viscosity of the system.

3.6.1. Mode of Action

A stable emulsion can be achieved only with emulsifiers or emulsifying systems that are specifically tailored for the components involved. Not every emulsifier is suitable for every food. The HLB system can be used to help select a suitable non-ionic emulsifier [21–23]. Fat-soluble, hydrophobic emulsifiers have HLB values between 0 and 9; water-soluble compounds have HLB values between 11 and 20. Ionic emulsifiers do not fit into this system. For water-in-oil emulsions (W/O emulsions), emulsifiers or emulsifier mixtures with HLB values of 3 – 8 are suitable; for oil-in-water emulsions (O/W emulsions), HLB values between 8 and 12 are good. Compounds with HLB values of 12 to 18 can be used as solution promoters (solubilizers).

Emulsifier mixtures are preferred when the object is to achieve a particularly good emulsification. The HLB values of such mixtures are obtained by addition from the proportions of individual emulsifiers contained in the mixture. A suitable emulsifier is not the only requirement for stable food emulsions; the emulsification conditions and the components of the phases to be emulsified also strongly influence the quality of an emulsion.

3.6.2. Uses and Individual Substances

Fat spreads with high water content (e.g., margarine, a water-in-oil emulsion), cannot be produced unless a suitable emulsifier is used. In fat emulsions for frying, emulsifiers prevent spattering, which is why they are referred to as antispattering agents. For mayonnaise and salad dressings (oil-in-water emulsions), the emulsifying effect of the lecithin contained in egg yolk is often exploited. When these products are made with a reduced egg content or without any egg yolk, other emulsifiers must be added.

Particularly effective emulsifiers are needed for ice cream manufacture. Ice cream is the foam of an oil-in-water emulsion, in which the aqueous phase is a colloidal solution of milk proteins and, at the same time, a true solution of various sugars and salts as well as a suspension of ice crystals. Here, emulsifiers have a major effect on stability and melting behavior.

In yeast-leavened baked goods, certain emulsifiers such as monoglycerides delay staling through clathrate formation with starch. This inhibits the retrogradation of the starch and delays softening of the crust. Interaction between emulsifiers and proteins in the dough increases the elasticity of the dough and thereby improves its gas-retaining capacity. The finished baked goods have a more uniform pore size and distribution and a larger volume [24], [25].

Shortenings are special baking fats with added emulsifiers. In pastry, emulsifiers cause the formation of foamy oil-in-water emulsions and thereby affect its pore structure.

In chocolate manufacture, emulsifiers lower the viscosity of the chocolate mass and delay bloom development in the finished chocolate. Instant foods containing fats require the addition of emulsifiers. This leads to a particularly fine distribution of the fat and good flowability of the finished product.

In meat products, especially sausages, emulsifiers improve the distribution of meat, fat, and water. In this context, the stability of emulsions on heating is also important. The sausage binders used to make sausage-type products are emulsifiers in a wider sense; they are used in the manufacture of frankfurter type sausages. They influence the swelling properties of the protein and thus improve the water- and juice-retention capacity of the meat so that it can be processed like freshly slaughtered meat. Under their influence, the denatured muscle protein coagulates uniformly during heat treatment, so that the final product can be sliced. Sausage binders are mixtures of common salt with salts of acetic, lactic, tartaric, or citric acid, and especially, with certain phosphates [26], [27].

The most important emulsifiers commonly used in foods are summarized below:

E 322	lecithins [*8002-43-5*]
E 339 – E 341	phosphates
E 450 a – c	diphosphates, triphosphates, and polyphosphates
E 470	sodium, potassium, and calcium salts of fatty acids
E 471	mono- and diglycerides of fatty acids
E 472 a – f	esters of mono- and diglycerides of fatty acids
E 473	sucrose esters of fatty acids
E 474	sucroglycerides
E 475	polyglycerol esters of fatty acids
E 477	propane-1,2-diol esters of fatty acids
E 481 – E 482	sodium and calcium stearoyl-2-lactylate
E 483	stearyl tartrate

3.7. Thickeners

Thickeners are hydrocolloids that are soluble in water or can be readily hydrated or dispersed. They form viscous solutions. The thickeners commonly used in food are polysaccharides of plant, microbial, or semisynthetic origin, e.g., milled seeds, extracts of sea algae, and plant extracts or exudates. Thickeners are used to maintain or improve the viscosity and flow characteristics of food [28–32].

3.7.1. Mode of Action

In contrast to emulsifiers, thickeners are not fat-soluble. They interact with the water in food through their fibrous or cross-linked structure and their many polar groups, especially hydroxyl groups. Water molecules, which are polar, orient themselves around the polar groups of the thickeners. Formation of a hydration layer of this kind, often accompanied by unfolding of the molecules, limits the mobility of the water so that the viscosity of the system increases.

Thickeners compete for the available water with other food components that can also be readily hydrated. They therefore influence the swelling of other food components, such as proteins. Different thickeners can associate among themselves or with other high molecular mass food components to give a particularly strong increase in viscosity.

3.7.2. Methods of Evaluation

Viscosity measurements are used to characterize and evaluate swollen or dissolved thickeners. At commonly used concentrations, solutions and suspensions of hydrocolloids exhibit non-Newtonian, specifically pseudoplastic, flow behavior. However, the measured viscosities are material constants only for liquids with Newtonian flow behavior. As a practical rule, relative measurements within a particular class of substances are sufficient. For this reason, measuring apparatus and experimental conditions must be standardized.

3.7.3. Uses and Individual Substances

Thickeners are generally added under defined conditions to water or to a food component with a high water content. For successful application, the method by which they are incorporated, the water temperature, and the stirrer speed are very important.

In fermented milk products such as yogurt, the addition of thickeners leads to a very finely flocculated precipitation of the casein and thus a particularly uniform structure. In the manufacture of ice cream, thickeners prevent the formation of large ice and sugar crystals and thereby lead to uniform melting behavior. In addition, the ice cream mixture can be whipped more readily into a foam because of stabilization of the finely dispersed air. In fat emulsions, thickeners can exercise an indirect emulsifying effect or improve the effect of emulsifiers, which leads to improved temperature stability. Thickeners also serve as clouding (Section 3.10) and bulking agents for reduced-calorie food (see Section 2.4).

The most important thickeners commonly used in foods are listed below:

E 400 – E 405	alginic acid [*9005-32-7*], alginates, propane-1,2-diol alginate [*12698-40-7*]
E 406	agar [*9002-18-0*]
E 407	carrageenan (Irish moss extract) [*9000-07-1*]
E 410	carob gum (locust bean gum) [*9000-40-2*]
E 412	guar gum [*9000-30-0*]
E 413	tragacanth gum [*9000-65-1*]
E 414	acacia gum (gum arabic) [*9000-01-5*]
E 415	xanthan gum [*11138-66-2*]
E 440 a – b	pectin [*9000-69-5*] and amidated pectin
E 461 – E 466	cellulose ethers, starch, starch esters, starch ethers, modified starches

3.8. Gelling Agents

Gelling agents combine with water to form pseudogels or gels. They maintain or improve the structure, consistency, or elasticity of a food. Gelling agents facilitate development of the gel structure desired in some food. They are used in confectionery, desserts, jams, preserves, and coating for fish and meat products. They also increase the freeze – thaw stability of frozen food. The action of some gelling agents depends on the calcium ion content and pH of the food. The most important gelling agents used in food are gelatin [*9000-70-8*], alginates, agar [*9002-18-0*], carrageenan [*9000-07-1*], and pectin [*9000-69-5*].

3.9. Foam Stabilizers

Foam stabilizers are used particularly in the baking and confectionery industries. They impart greater stability to foamy preparations. In sugar-containing preparations, hydrocolloids like methyl cellulose [*9004-67-5*] and tragacanth [*9000-65-1*] can be used; in foamy water-in-fat emulsions, emulsifiers may be used instead. Phosphates stabilize whipped protein foam. The foam of soft drinks and beer can also be stabilized by certain hydrocolloids.

3.10. Clouding Agents

Clouding agents are thickeners that increase the viscosity of a beverage and, thereby, retard or prevent the deposition of finely suspended particles. Certain emulsifiers are also used to prevent the separation of essential oils in beverages.

3.11. Humectants

Humectants are added to food to maintain a predetermined moisture level. They thereby prevent excessive drying out, or any changes in texture associated with this, and hardening. In intermediate-moisture food intended to have a long shelf life without expensive packaging, the role of humectants is to standardize water activity to 0.85 or less. This prevents the multiplication of bacteria which, as a rule, require water activity of 0.90 or more for growth. Growth of molds and yeasts is still possible below this limit, but it can be suppressed easily by addition of sorbic acid [8].

In confectionery, humectants can prevent undesirable crystallization of the sugar. They thereby contribute to the softness or chewiness of confectionery.

The primary humectants are sorbitol [*50-70-4*] and glycerol [*56-81-5*]. Common salt and sugar are added to intermediate-moisture foods to decrease water activity.

3.12. Anticaking Agents

Anticaking agents maintain the flowability of formed, powdered, or fine-grained products by preventing caking and adhesion. Mold-release agents (see Section 5.6), which keep food from sticking to molds or other containers, are different from anticaking agents.

Anticaking agents are inert organic or inorganic compounds which are generally insoluble in water. They coat the particles to be protected with a thin layer, increase the distance between them, and sometimes reduce their cohesion. In addition, they may prevent electrostatic attraction between particles with opposite charges.

The primary uses of anticaking agents are in common salt; spices; vegetable, beverage, and fruit powders; powdered soups; powdered sauces; leavening agents; and confectionery such as hard candy. The most important anticaking agents used in food are calcium stearate [*1592-23-0*] and magnesium stearate [*557-04-0*], silicon dioxide, silicates, talc [*14807-96-6*], flour, starch, and for common salt, alkali-metal ferrocyanides.

3.13. Coating Agents

Coating agents are compounds that are used to cover or coat foods, or that are added to food surfaces for their protection. They are distinct from packaging materials, which are not food additives. Coating agents are used to protect foods or their surfaces against undesirable changes, such as drying out and loss of aroma. In some food, coating agents may prevent oxidation or the growth of undesirable microorganisms on the surface. Other coating agents give an attractive look to the surface of a food that would not otherwise look very appealing.

Coatings are used mainly for citrus fruit, confectionery, meat products, and cheese. The most important coating agents used in food are waxes, resins, oils, cellulose esters or acetic acid esters of the monoglycerides of edible fatty acids, and talc.

4. Substances with Sensory Effects (Organoleptic Substances)

Not only should food have a nutritive value and be stable, it should stimulate the appetite by its appearance, taste, and aroma. Some substances have properties that influence positively the senses of smell, taste, and vision before, during, and after the consumption of food.

Foods naturally contain sensorially active compounds. However, many of these are volatile or unstable. They may, therefore, be lost during food processing and storage. The main purpose of food additives that produce sensory effects is to compensate for such losses as authentically as possible. In addition, some substances stimulate the secretion of saliva and digestive fluids in the gastrointestinal tract. Others may impart sweet, salty, or sour taste nuances or may tone down or mask undesirable odor and taste. Under no circumstances should such food additives be used to mask unsatisfactory quality or spoilage. Excessive flavoring and unnaturally high coloring are objectionable and are not in the interest of the consumer.

4.1. Coloring Agents

An attractively colored food stimulates the appetite more than a discolored one. Apparently, there is a relationship between the eye and the gustatory nerves. Another purpose of food coloring is to provide a more variable range of products, which is especially important in the confectionery industry. Despite some controversy in this area, the coloring of food is unavoidable [33–35].

Coloring fresh food is not permitted as a matter of principle. As a rule, coloring is used only for processed food with no color of its own or in which only residual amounts of color remain. Food must not be colored to simulate a higher level of nutritionally important components or, worse, to mask poor quality or spoilage.

4.1.1. Specifications

Coloring is generally added to food at a very early stage in processing. Coloring agents must therefore be stable to heating, cooling, acid, or oxygen. In particular, they must remain stable during the storage of the food when they are often exposed to light. Many natural colorants lack this stability so that, despite other advantages, their use is limited. Sulfur dioxide, which is used in a number of foods, can destroy many colors.

4.1.2. Uses and Individual Substances

Coloring agents are distinguished on the basis of their solubility, i.e., insoluble coloring, pigments, and water-soluble or fat-soluble coloring. *Pigments* are used mainly for surface coloration of confectionery; *fat-soluble colorants,* for fatty foods such as margarine and cheese, and *water-soluble colorants,* for food with a high water content such as fruit products and beverages.

Depending on their origin, coloring agents may be classified as self-coloring food, natural coloring, or synthetic coloring. Examples of *natural coloring* are extracts of beetroot and turmeric, as well as carotene and other carotenoids, although as a rule the latter are produced synthetically [36]. Among purely *synthetic coloring* agents, triphenylmethane and azo dyes are used most often because they are the most stable.

The main uses for food coloring are in confectionery, cakes, fruit-based products, beverages, margarine, cheese, and certain fish products.

The most important colorants commonly used in foods are summarized below. In contrast to other food additives, major differences exist in the food-coloring laws of the EEC countries, Japan, and the United States (see Chap. 6).

E 100	curcumin [*458-37-7*]
E 101	riboflavin (lactoflavin) [*83-88-5*]
E 102	tartrazine [*1934-21-0*]
E 104	Quinoline Yellow [*8004-92-0*]
E 110	Sunset Yellow FCF [*2783-94-0*]
E 120	cochineal (carminic acid) [*1260-17-9*]
E 122	carmoisin (Azo Rubine) [*3567-69-9*]
E 123	amaranth [*915-67-3*]
E 124	Ponceau 4R (Cochineal Red A) [*2611-82-7*]
E 127	Erythrosin [*16423-68-0*]
E 131	Patent Blue V [*129-17-9*]
E 132	Indigo Carmine (indigotine) [*860-22-0*]
E 140 – E 141	chlorophyll [*1406-65-1*] and its copper complexes
E 142	Green S (Acid Brilliant Green BS or Lissamine Green) [*3087-16-9*]
E 150	caramel
E 151	Black PN (Brilliant Black BN) [*2519-30-4*]
E 153	carbon black (vegetable carbon)
E 160	carotenoids (carotene, annatto [*1393-63-1*], bixin [*6983-79-5*], norbixin [*542-40-5*], capsanthin [*465-42-9*], capsorubin [*470-38-2*], lycopene [*502-65-8*], β-apo-8′-carotenal and its ethyl ester)
E 161 a – g	flavoxanthin [*512-29-8*], lutein [*127-40-2*], cryptoxanthin [*472-70-8*], rubixanthin [*3763-55-1*], violaxanthin, [*126-29-4*], rhodoxanthin [*116-30-3*], canthaxanthin [*514-78-3*]
E 162	Beetroot Red (betanin) [*7659-95-2*]
E 163	anthocyanins
E 171	titanium dioxide [*13463-67-7*]
E 172	iron oxides and hydroxides
E 173 – E 175	metals (aluminum, silver, gold)
E 180	Pigment Rubine (Lithol Rubine BK) [*5281-04-9*]

4.2. Color Stabilizers

Color stabilizers help foods retain their natural color during processing and storage and prevent undesirable discoloration. Color stabilizers are especially important with meat products. The red pigment myoglobin can be oxidized by oxygen in the air to form metmyoglobin which is brown to grayish brown. Nitrates or nitrites stabilize the desired red meat color through formation of nitrosomyoglobin. This complex has good storage, cooking, and baking stability [37].

4.3. Bleaching Agents

Discoloration in light-colored food can be removed in some cases with bleaching agents. Sulfur dioxide and sulfites are effective color stabilizers for enzymatic oxidation processes because they are enzyme inhibitors.

Bleaching agents not only remove undesirable discoloration but also destroy vitamins. Because of this, they are added only to products in which this side effect is unimportant.

4.4. Intense Sweeteners

Intense sweeteners are compounds with a far more intensely sweet taste than sugar [38–40]. They can be synthetic or derived from plants. Because of their intense sweetness, they need be added only in small amounts and do not contribute (or do not contribute significantly) to the caloric value of food. The sweeteners can be tolerated by diabetics.

Intense sweeteners are judged by two criteria, their sweetening power and the quality of their sweetness. The sweetening power is the factor by which the substance tastes sweeter than sucrose; it is concentration dependent. Data quoted in the literature usually refer to a comparison with 0.1 mol/L sucrose solution.

Intense sweeteners are used in foods for three main reasons: because of their lack of calories, because they are not harmful to diabetics, and because they are cheaper than sugar.

In the early 1970s, the two traditional sweeteners, saccharin and cyclamate, were suspected of being carcinogenic. Because of this, their use as food additives was prohibited in some countries. In other countries their use was restricted or made difficult because labels warning of their potential danger were required. This led to the development of new sweeteners; among these, aspartame, other peptides, and acesulfame-K have achieved some importance. Possible sweeteners currently being discussed include chlorinated sugars.

4.5. Nutritive Sweeteners

Nutritive sweeteners are noncarbohydrate materials with a degree of sweetness similar to that of sugar; they can be used as substitutes for sucrose, glucose, and other sugars [38–40]. Like sugars they are metabolized. However, in contrast to sucrose and glucose, they can be metabolized independently of insulin and are, therefore, suitable for diabetics. Nutritive sweeteners increase osmotic pressure, sometimes to an even greater degree than sugars. When consumed in large amounts, they can therefore act as laxatives.

Nutritive sweeteners are used largely in place of sucrose or glucose in foods designed for diabetics. Because unlike sugars, they are not readily fermented, oral bacteria are unable to convert them into acids. Consequently, the risk of caries is lower with confectionery made from nutritive sweeteners than with that made from sugars.

The most frequently used nutritive sweeteners are the polyols sorbitol [*50-70-4*], xylitol [*87-99-0*], and mannitol [*69-65-8*]. In some countries, fructose is included among the nutritive sweeteners because it is tolerated by diabetics.

4.6. Acidulants

Acidulants are used to impart a sour taste to foods in a controlled way. Besides being sour, most of them have a characteristic taste of their own. This can be desirable, for example, in the case of citric acid. Acids that taste only sour or nearly so are malic acid [*6915-15-7*] and orthophosphoric acid. Acidulants can highlight or suppress other aroma components of a food. The subjective perception of a sour taste is strongly influenced by salts with a buffering effect, by sweet compounds, and by other flavorings. In special cases, an acidulant may be chosen according to the speed with which the sour taste is registered or with which it fades.

Acidulants are added to beverages, especially nonalcoholic soft drinks, various fruit products, desserts, confectionery, pickles, mayonnaise-type products, salads, and fish products [41].

Vinegar and acetic acid are the most commonly used acidulants for pickles and mayonnaise-type or fish products. Orthophosphoric acid is used specifically to acidify cola beverages. Citric acid [*77-92-9*] is very widely used. Besides these, the most important acidulants are tartaric acid [*147-71-7*], lactic acid [*50-21-5*], malic acid [*6915-15-7*], succinic acid [*110-15-6*], and fumaric acid [*110-17-8*].

4.7. Substances with a Salty Taste

Common salt is the most important substance with a salty taste used in food. Because it has been used for thousands of years, in many cases it is no longer considered a food additive in legal terms. Besides its sensory function, salt is very important as a preservative (see Section 3.1).

Salt is an essential taste component of many foods such as meat products, certain fish products, vegetables, ready-to-eat meals, and most of all, bread. Without salt, many foods would taste bland or, in the view of some, be unpalatable.

If the intake of sodium must be restricted because of illness, substitutes for table salt, such as salts of potassium, calcium, magnesium, or choline with organic or inorganic acids, can be used.

4.8. Substances with a Bitter Taste

A bitter or slightly bitter taste in foods is desirable only in certain cases, generally in beverages. Quinine is used as an additive for tonic; hops resins, and extracts from hops and hops flowers are used as additives in beer. Caffeine, which is added to some cola beverages, also has a slightly bitter taste.

4.9. Substances with an Alkaline Taste

Among flavorings with an alkaline taste, the only one that has even a limited practical role is sodium hydroxide solution. It is used in some countries in the manufacture of pretzels.

4.10. Flavor Enhancers

Flavor enhancers are compounds that particularly enhance certain tastes or reduce undesirable flavors without having an especially strong taste of their own. They harmonize taste components and make food preparations more palatable. The Japanese expression *Umami* substances has come into general use [42].

Important Umami substances include glutamic acid [*56-86-0*] and glutamates, as well as purine-5′-ribonucleotides, especially inosine [*4691-65-0*], guanosine [*5550-12-9*], and adenosine [*61-19-8*] 5′-monophosphates. The latter are effective at far lower concentrations than glutamates. Lately, thaumatin [*53850-34-3*] has been of interest; this substance was originally developed as a sweetener. Common salt is also a flavor enhancer in the wider sense.

The main uses for flavor enhancers are in meat, soups, and vegetable products; gravies; ready-to-eat meals; seasoning; confectionery; beverages; and fruit products. Maltol [*118-71-8*] and ethyl maltol [*4940-11-8*] are flavor enhancers in confectionery, soft drinks, and fruit products.

4.11. Spices and Flavorings

In addition to the substances with a certain taste already described, certain materials, mostly of plant origin, have been used from ancient times to improve the flavor of food. These include spices, herbs, herb extracts, and essences. Aroma compounds are substances with a more or less pronounced odor that impart a specific aroma to foods [43–45]. Chemically pure compounds, as well as plant extracts and microbial metabolites or extracts, are used. Essential oils and preparations made from them are especially important. Raw materials of animal origin play only a subordinate role in flavoring food.

Aroma compounds are perceived mainly through the nose; compounds with a taste and substances that affect food consistency are perceived more via the tongue. However, these groups of substances cannot always be considered separately. Aromagrams are now being discussed as a systematic method for overall evaluation of odor and taste. An aromagram is a form that can be used to sketch the aroma of any food and that allows the interaction between taste and odor to be illustrated, while the effect of consistency is also taken into account. The areas representing compounds that have overall sensory impact are shaded. In addition, the key substance (i.e., the primary aroma component) is given. Figure 1 shows the aromagram for Emmentaler cheese [43].

Flavoring is classified as natural, nature-identical, and artificial. *Natural flavorings* are obtained through purely physical processes, usually from plant material, or through microbiological processes. In many countries, synthetically produced substances which are chemically identical to the natural ones are called *nature-identical. Artificial flavorings* are substances synthetically produced that have not been found in any natural products suitable for human consumption. In some countries, only artificial flavorings are considered additives within the meaning of the food law.

Some aroma compounds are formed by the controlled use of technological methods. These may be strengthened or simulated by additives. An example involves flavor compounds produced by the Maillard reaction, which develop during the roasting of meat or baking of bread.

Both naturally occurring flavors and flavors produced during roasting and baking, as a rule, consist of numerous individual components. So far it has not been possible to synthetically reproduce natural flavors in their entirety in such a way that they are fully identical with the natural product. However, flavor impressions of a very high quality can be generated with synthetically produced flavoring.

4.12. Chewing Gum Bases

Chewing gum bases, also called masticatory substances, are the underlying components of chewing gum. At body temperature, they must be plastic, but they must also offer a certain bite resistance. Important chewing gum bases include gums, natural and synthetic resins, synthetic polymers, paraffins, and waxes.

5. Processing Aids

The Codex Alimentarius defines *processing aids* as "any substance or material, not including apparatus or utensils, and not consumed as a food

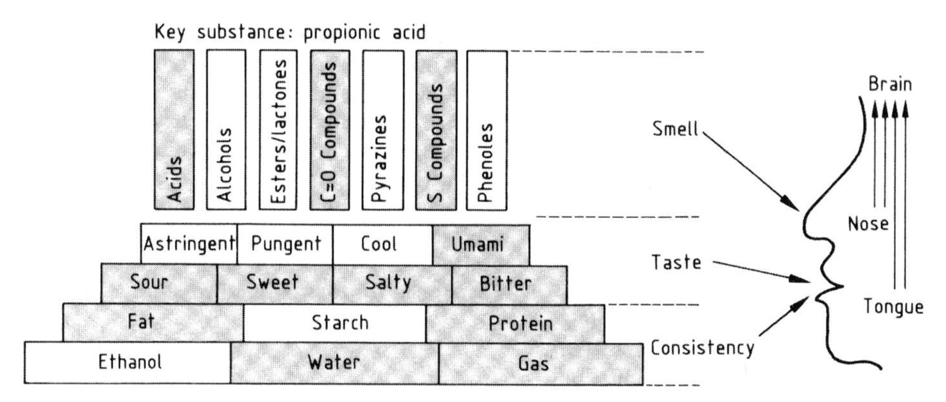

Figure 1. Aromagram of Emmentaler cheese

ingredient by itself, intentionally used in the processing of raw materials, foods or its ingredients, to fulfil a certain technological purpose during treatment or processing and which may result in the non-intentional but unavoidable presence of residues or derivatives in the final product" [46]. Processing aids have a special status among additives in the food laws of practically every country. Some countries do not consider them additives at all.

5.1. Extractants

Extractants are solvents used to extract certain components of food. This is done so that these components may be used separately or so that the food can be freed of certain unwanted components. As a rule, these solvents are removed from the food when extraction is complete, unless the solvent is necessary as a diluent for the extract.

The most commonly used solvents are water, ethanol, and other lower aliphatic alcohols, as well as hexane, other hydrocarbons, and in special cases, dichloromethane. Supercritical carbon dioxide has been used increasingly as an extraction solvent because of its good solvent properties and because it leaves no residue. Extractants are used in fat extraction from fatty raw materials, in the extraction of aroma compounds from plant materials, and in the extraction of caffeine from unroasted coffee beans.

Carrier solvents are distinct from extractants and are used to work small amounts of other additives into a food, e.g., to dissolve vitamins, antioxidants, or aroma compounds. Unlike extractants, carrier solvents are food additives in the true sense.

5.2. Clarifying Agents

Clarifying agents speed up the separation of suspended matter from those beverages in which clarity is desired, such as wine and beer. They also attract colloids and precipitate them. In addition to adsorbents, colloids of opposite charge can be effective against other colloids. If added in the right amounts, these precipitate together with the substances to be eliminated. Clarifying agents must be removed as far as possible with the suspended particles, mostly by sedimentation and ultrafiltration. The most important clarifying agents for food are bentonite [*1302-78-9*], activated charcoal, gelatin [*9000-70-8*], tannin, polyvinylpyrrolidone [*9003-39-8*], casein [*9000-71-9*], and caseinates.

Other clarifying agents include substances that form compounds of poor solubility with traces of soluble heavy metals. These can then be filtered readily. Thus, potassium ferrocyanide [*13943-58-3*] forms insoluble precipitates with traces of iron and copper. This process is especially useful in removing traces of iron from wine and is called blue or Moeslinger fining.

5.3. Filter Aids

To facilitate filtration, on suitable supports filter aids form a filter layer of appropriate pore size and maximum filtering capacity. These materials must be inert and should not contaminate the liquid to be filtered with any soluble compounds. In settling filters, mixtures of filter aids and adsorbents are sometimes used, especially with activated charcoal.

The most important filter aids used in the food industry are kieselguhr and cellulose. Asbestos, much used in the past, has been largely abandoned for toxicological reasons.

5.4. Propellants

Propellants are used in the production of aerosols. Whipping cream, frying and baking and mold release oils, as well as spice extracts are used in aerosol form. Carbon dioxide, mixtures of propane and butane, dinitrogen monoxide, and certain fluorohydrocarbons and mixtures of these are used as propellants.

5.5. Cooling Agents and Cryogens

Cooling agents and cryogens are substances with a low boiling point which, when in direct contact with foods, extract heat from them and thereby cool them. The oldest *cooling agent* is crushed ice, and it continues to be important in chilling fish and other marine products at sea. The use of cryogens is more recent. The most important of these is liquid nitrogen which is sprayed onto food in special freezing tunnels. The advantage of liquid nitrogen is that the freezing process is particularly fast when oxygen is absent.

5.6. Mold-Release Agents

Release agents facilitate the removal of food such as confectionery or baked goods from molds. The most common mold-release agents are lecithins, stearic acid [*57-11-4*], calcium and other stearates, and waxes.

5.7. Antifoaming Agents

Antifoaming agents prevent unwanted foam development that may occur during food processing. Dimethylpolysiloxane, fatty acid esters of sorbitan and ethoxylated sorbitan, and various emulsifiers are suitable for this purpose.

5.8. Acidity Regulators

Acidity regulators generate a pH that is favorable or even essential for certain food processes. They are used, for example, to optimize the activity of enzymes or preservatives. The action of some gelling agents also depends on a particular range of pH. The most important regulators are the acids and acid salts described in Section 3.1. In addition, glucono-δ-lactone [*90-80-2*] may be used; it is gradually converted to gluconic acid, something that is occasionally desirable.

5.9. Emulsifying Salts

Emulsifying salts are necessary for the production of processed cheese. They inactivate calcium which is important for the stability of the cheese gel. The action of emulsifying salts causes the insoluble paracasein gel to be converted to homogeneously flowing paracasein sol. After the melt has cooled down, the sol reconverts into a gel which, however, remains homogeneous and is stable. Some emulsifying salts are antimicrobial; thus, processed cheese is less susceptible to mold attack than natural cheese [47].

The most important emulsifying salts used in food are citrates and ortho-, di-, or polyphosphates. They are used mainly as mixtures that are tailored to the type and maturity of the original cheese to be processed.

5.10. Dough Conditioners and Flour Improvers

Dough conditioners and flour improvers enhance the processing and baking properties of flour and dough. Basically, they affect the gluten in flour and balance variations in the flour, especially in enzyme activity. Both oxidizing and reducing compounds can be used as dough conditioners. The effectiveness of *reducing* dough conditioners is attributed to their splitting of hydrogen or disulfide bridges. This breakdown improves the rheological properties of doughs so that less energy is used in working them.

Important dough conditioners include cysteine hydrochloride [*52-89-1*] and ascorbic acid [*50-81-7*]. Some countries, also allow azodicarbonamide [*123-77-3*] and bromates to be used. Certain emulsifiers and amylolytic enzymes can also improve the properties of dough.

5.11. Leavening Agents

Leavening agents lighten baked products, mostly by liberating carbon dioxide. Carbon dioxide can be generated biologically (e.g., by yeast or sourdough) or chemically (by baking powder). The latter is a mixture of sodium hydrogen carbonate and an acid, e.g., a solid organic acid such as tartaric or citric acid, or sour salts such as certain phosphates.

5.12. Enzymes [55]

Enzymes are widely distributed throughout food of plant or animal origin. They are involved in food spoilage and, therefore, must often be inactivated during food processing, usually by heat. On the other hand, some enzymes are deliberately added to foods to convert or remove certain components in a controlled and careful way, or to generate particularly desirable end products [48–51]. Often, enzymes make possible processes that either could not be carried out at all without them or would lead to profound and usually detrimental changes in food. The action of an enzyme is greatly influenced by its concentration, its activity, the length of time of application, pH optimum, the possible presence of enzyme inhibitors, and the process temperature. An advantage of using enzymes is that simple heating to 50 – 70 °C inactivates the enzyme and stops the reaction instantaneously. Enzymes generally have marked specificity and, when optical isomers are involved, usually react with only one enantiomer.

Uses and Individual Substances. Hydrolases, lyases, and isomerases are the most important classes of enzymes used as food additives. Enzymes are added to food either as standardized powders or in solution.

The addition of *pectinases* to fruit increases juice yields by destroying the cellular structure of the fruit. Pectin-containing suspended matter, which may impede the filtration of fruit juice, can be flocculated by use of suitable pectinases and then removed. For this reason, such enzymes are called carifying enzymes.

Amylases hydrolyze starch both during dough production and in the manufacture of glucose syrup and glucose. They are used in alcohol and beer manufacture to break down starch so as to obtain readily fermentable oligosaccharides. Of the enzymes that hydrolyze disaccharides, lactase and invertase have practical importance. *Lactase* (E.C. 3.2.1.108) [*9031-11-2*] cleaves lactose which has poor solubility and low sweetness, and cannot be tolerated by many people. It thereby improves milk and dairy products. *Invertase* (E.C. 3.2.1.26) [*9001-57-4*] cleaves sucrose into glucose and fructose (inversion). The addition of invertase makes possible the production of sweets with a solid coating and a soft center. Under the action of invertase, an originally solid sucrose-containing mass with a suitable coating takes on a soft, creamy, or liquid consistency. *Glucose isomerase* (E.C. 5.3.1.18) [*9055-00-9*] isomerizes glucose to fructose. Because of the greater sweetness of fructose compared to sucrose and glucose, its use leads to sweeter products. *Naringinase* [*9068-31-9*] makes orange and grapefruit juices more palatable by cleaving their bitter component, naringin, into substances that do not taste bitter.

Among the *proteolytic enzymes,* exogenous proteases are used to tenderize meat. Like the proteases that occur naturally in meat, they split proteins into shorter chain proteins and peptides. Papain (E.C. 3.4.22.2) [*9001-73-4*], bromelain (E.C. 3.4.22.4) [*37189-34-7*], and ficin (E.C. 3.4.22.3) [*9001-33-6*] are proteases of plant origin that also break down proteins of connective tissues. In beverages, proteases can prevent flocculation of proteins and clouding due to protein – tannin complex formation.

The milk-clotting enzyme rennin (E.C. 3.4.23.4) [*9001-98-3*], obtained from the mucous membrane of the stomach of young calves has great commercial importance. It is indispensable for the manufacture of many cheeses. Microbial proteases are also used in cheese manufacture.

5.13. Microbial Cultures

Cultures of microorganisms resemble enzymes in their application and mode of action [51]. Frequently, their activity is due to the enzymes released by these cultures into food. Cultures of microorganisms may be used if they do not produce toxins or other undesirable substances.

Commercial preparations are tested routinely for this.

From the earliest times, yeast has been very important as a dough-leavening agent, and pure cultures of certain yeasts are used in wine making. Certain mold cultures are used to make blue cheese, and pure cultures of certain molds are employed in the manufacture of salami. So-called starter cultures are used in certain areas of dairy and meat technology.

6. Legal Aspects

6.1. Food Law Authorizations

In many countries, into the 20th century food law was limited to prohibiting the use of additives that were damaging to health. This was achieved partly by listing certain substances on so-called negative lists. In the long run, this procedure did not protect consumers and now has generally been replaced by so-called positive lists. According to these, all substances are prohibited unless specifically permitted by decree. Generally, the law also states which substances may be used for which foods. Sometimes, the maximum amount permitted in a food and other conditions are also fixed. Both authorization and maximum amount are based on technological necessities, with calculation of a sensible safety margin. An obvious prerequisite is that the substance be innocuous.

6.2. Definitions

In the following material, the term "food additive" is defined according to the Codex Alimentarius and the food legislation of the United States, Japan, and the Federal Republic of Germany.

6.2.1. Codex Alimentarius

In the Codex Alimentarius, *food additive* means any substance not normally consumed as a food by itself, and not normally used as a typical ingredient of the food, whether or not it has nutritive value, the intentional addition of which to food for a technological (including organoleptic) purpose in the manufacture, processing, preparation, treatment, packing, packaging, transport or holding of such food results, or may be reasonably expected to result (directly or indirectly) in it or its by-products becoming a component of or otherwise affecting the characteristics of such foods. The term does not include "contaminants" or substances added to food for maintaining or improving nutritional qualities [46].

6.2.2. United States of America

In the United States, a food additive is any substance the intended use of which results or may reasonably be expected to result, directly or indirectly, in its becoming a component or otherwise affecting the characteristics of any food (including any substance intended for use in producing, manufacturing, packing, processing, preparing, treating, packaging, transporting, or holding food; including any source of radiation intended for any such use), if such substance is not generally recognized, among experts qualified by scientific training and experience to evaluate its safety, as having been adequately shown through scientific procedures (or, in the case of any substance used in food prior to January 1, 1958, through either scientific procedures or experience based on common use in food) to be safe under the conditions of its intended use; except that such term does not include a pesticide chemical in or on a raw agricultural commodity, or a pesticide chemical to the extent that it is intended for use or is used in the production, storage, or transportation of any raw agricultural commodity, or a color additive, or any substances used in accordance with a sanction or approval granted prior to the enactment of this paragraph pursuant to this Act, the Poultry Products Inspection Act or the Meat Inspection Act of March 4, 1907, as amended and extended [52].

Thus in the United States, the term additive is more widely defined than anywhere else in the world, yet food coloring and GRAS (generally recognized as safe) substances are not classified as food additives. These GRAS substances include salt, sugar, vinegar, and baking powder, as well as many substances classified as food additives in other countries, which are considered to be especially safe, such as citric and sorbic acids.

However, regulations covering the use of coloring and GRAS substances in the United States are very similar to those for food additives. In the United States, additives include practically all materials that come in contact with food in any way.

Distinction is made between three types of food additive. (1) Food additives permitted for direct addition to food for human consumption are those substances which, in other countries, are considered food additives as such. (2) Secondary direct food additives permitted in food for human consumption include polymers and polymer adjuvants for food treatment, enzyme preparations and microorganisms, solvents, lubricants, release agents and related substances, and other products that come in contact with food only temporarily. (3) Indirect food additives are adhesives, components of coatings, food-packaging materials, components of paper and paperboard, polymers, processing aids, and sanitizers.

6.2.3. Japan

In Japan, the term "additive" means anything added to, mixed with, permeating, etc., food in the process of manufacturing, processing, or preserving it [53].

In Japanese food law, synthetic and naturally occurring additives are treated differently. The latter, in particular naturally occurring flavors and vitamins, do not require any special permission for use. This explains, for example, why sweeteners isolated from plants must be specifically allowed as additives everywhere else in the world, while they can be used freely in Japan.

6.2.4. Federal Republic of Germany

For the purpose of German food legislation, additives are substances destined to be added to foods to influence their character or to achieve particular properties or effects. This definition does not include substances of natural origin or those which are chemically identical to natural ones and are, according to general commercial practice, used predominantly because of their nutritive value, aroma, or flavor characteristics, or which are used as stimulants. Drinking or table water is also included in this category. The following are equivalent to additives:

1) minerals and trace elements and their compounds except common salt,
2) amino acids and their derivatives,
3) vitamins A and D and their derivatives,
4) nutritive sweeteners except fructose,
5) intense sweeteners,
6) adipic acid,
7) nicotinic acid and nicotinamide,
8) nitrite salt for curing,
9) substances for use in the manufacture of edible coatings,
10) substances to be added to nonedible food surfaces,
11) substances to be used in the treatment of food in such a way as to find their way into or onto the food, and
12) propellants or similar substances to be applied by pressure to food whereby they come in contact with it [54].

A peculiarity of the food law in the Federal Republic of Germany is that substances of natural origin or their chemically identical synthetic analogues are not considered additives if, by common commercial practice, they are used predominantly because of their nutritive, aroma, or taste values. This means, for example, that in the Federal Republic of Germany, flavorings, including nature-identical ones as well as citric and acetic acids, do not constitute additives within the framework of food law.

6.2.5. Regulations and Specification

In most countries, additives must be listed with the ingredients.

In many countries, all additives permitted are summarized in a single list. Examples are relevant regulations in the Scandinavian countries, the Federal Republic of Germany, and Switzerland. In countries such as the United Kingdom, authorization of food additives is controlled by individual regulations, with reference to specified groups of additives. Some countries have authorizations within product regulations for certain foods. In addition, mixed systems exist, e.g., in the United States.

The international consensus is largely that, in principle, authorization for food additives should be given only if justified technologically. This means that without such additives, the required effect could only be achieved either uneconomically or not at all. In every country, food additives are subject to purity specifications which are fixed by law.

7. References

1. World Health Organization Geneva: "Principles for the Safety Assessment of Food Additives and Contaminants in Food" *Environ. Health Criter.* **70** (1987).
2. US Food and Drug Administration Bureau of Foods: *Toxicological Principles for the Safety Assessment of Direct Food Additives and Color Additives Used in Food,* Washington 1982.
3. E. Lück, K.-H. Remmert: *Z. gesamtes Lebensmittelrecht* **3** (1976) 115 – 143.
4. Hoffmann-La Roche: *Vitamin-Compendium,* 2nd ed., Grenzach–Wyhlen 1980.
5. R. A. Morton: *Fat-soluble Vitamins,* Pergamon Press, Oxford 1970.
6. J. N. Counsell, D. H. Hornig: *Vitamin C (Ascorbic Acid),* Applied Science Publ., London 1981.
7. H. Zumkley: *Spurenelemente–Grundlagen–ätiologieDiagnose–Therapie,* Thieme Verlag, Stuttgart–New York 1983.
8. E. Lück: *Antimicrobial Food Additives,* Springer Verlag, Berlin–Heidelberg–New York–Tokyo 1980; *Chemische Lebensmittelkonservierung,* 2nd ed., Springer Verlag, Berlin–Heidelberg–New York–Tokyo 1986.
9. A. L. Branen, P. M. Davidson: *Antimicrobials in Foods,* Marcel Dekker, New York–Basel 1983.
10. L. Leistner, W. Röder, K. Krispien in L. B. Rockland, G. F. Stewart (eds.): *Water Activity Influences on Food Quality,* Academic Press, New York–London–Toronto–Sidney–San Francisco 1981, p. 855.
11. M. von Schelhorn, *Arch. Mikrobiol.* **19** (1953) 30 – 44.
12. M. von Schelhorn: *Fette Seifen Anstrichm.* **56** (1954) 221 – 224.
13. D. A. A. Mossel: *Proc. 21st Symp. Soc. Gen. Microbiol.,* 1971, pp. 177 – 195.
14. E. M. Lukas: *Zentralbl. Bakteriol. Parasitenkd. Infektionskr. Hyg., Abt. 2 Naturwiss. Allg. Landwirtsch. Tech. Mikrobiol.* **117** (1964) 485 – 509.
15. J. C. Allan, R. J. Hamilton: *Rancidity in Foods,* AVI, Westport 1977.
16. R. R. Hiatt: *CRC Crit. Rev. Food Sci. Nutr.* **7** (1975) 1 – 12.
17. WHO: *A Review of the Technicological Efficacy of Some Antioxidants and Synergists.* WHO: Food Additives Series No. 3, Geneva 1972.
18. E. R. Sherwin: *J. Am. Oil Chem. Soc.* **49** (1972) 468 – 472.
19. L. C. Schroeter: *Sulfur Dioxide,* Pergamon Press, Oxford 1966.
20. G. Dinglinger: *Z. Lebensmitteltechnol. Verfahrenstechn.* **35** (1984) 49, 50, 52, 54, 55.
21. G. Schuster: *Emulgatoren für Lebensmittel,* Springer Verlag, Berlin–Heidelberg–New York–Tokyo 1984.
22. *Emulgatoren–Ihre Wirkung in Lebensmitteln,* B. Behr's, Hamburg 1983.
23. N. Krog: *J. Am. Oil Chem. Soc.* **54** (1977) 124 – 131.
24. V. Bade: *Getreide Mehl Brot* **28** (1974) 296 – 299.
25. F. C. Greene: *Baker's Dig.* **49** (1975) no. 3, 16 – 18, 20 – 24, 26.
26. R. H. Ellinger: *Phosphates as Food Ingredients,* CRC Press, Cleveland 1972.
27. *Phosphate–Anwendungen und Wirkung in Lebensmitteln,* B. Behr's, Hamburg 1983.
28. H. D. Graham: *Food Colloids,* AVI, Westport 1977.
29. W. Burchard: *Polysaccharide,* Springer Verlag, Berlin–Heidelberg–New York–Tokyo 1985.
30. M. Glicksman (ed.): *Food Hydrocolloids,* **vol. I,** vol. II, CRC Press, Boca Raton 1982, 1983.
31. H. Neukom, W. Pilnik (eds.): *Gelier- und Verdikkungsmittel in Lebensmitteln,* Forster, Zürich 1980.
32. *Hydrokolloide–Stabilisatoren, Dickungs- und Geliermittel in Lebensmitteln,* B. Behr's, Hamburg 1984.
33. J. N. Counsell (ed.): *Natural Colours for Food and Other Uses,* Applied Science, London 1981.
34. *Food Technol. (Chicago)* **34** (1980) 77 – 84.
35. *Farbstoffe für Lebensmittel,* Boldt, Boppard 1978.
36. J. C. Bauernfeind: *Carotenoids as Colorants and Vitamin A Precursors,* Academic Press, New York 1981.
37. G. G. Giddings: *CRC Crit. Rev. Food Sci. Nutr.* **9** (1977) 81 – 114.

38. C. A. M. Hough, K. J. Parker, A. J. Vlitos (eds.): *Developments in Sweeteners–1,* Applied Science, London 1979.
39. T. H. Grenby, K. J. Parker, M. G. Lindley: *Developments in Sweeteners–2,* Applied Science, London 1983.
40. L. O'Brien Nabors, R. C. Gelardi (eds.): *Alternative Sweeteners,* Marcel Dekker, New York–Basel 1986.
41. M. H. M. Arnold: *Acidulants for Foods and Beverages,* Food Trade Press, London 1975.
42. J. A. Maga: *CRC Crit. Rev. Food Sci. Nutr.* **18** (1983) 231–312.
43. K. H. Ney: *Lebensmittelaromen,* B. Behr's, Hamburg 1987.
44. W. Auerswald, B. M. Brandstetter: *Natürliche, natur-identische und synthetische Geschmacksstoffe und deren lebensmittelrechtliche Aspekte,* Maudrich, Wien 1973.
45. R. C. Lindsay: *Food Technol. (Chicago)* **38** (1984) 76–81.
46. Joint FAO/WHO Food Standards Programme. Codex Alimentarius Commission: Codex Alimentarius, vol. 14 *Food Additives,* Food and Agriculture Organization of the United Nations, Rome 1983.
47. A. Meyer: *Joha-Schmelzkäsebuch,* Benckiser-Knapsack, Ludwigshafen 1970.
48. S. Schwimmer: *Source Book of Food Enzymology,* AVI, Westport 1981.
49. G. G. Birch, N. Blakebrough, K. J. Parker (eds.): *Enzymes and Food Processing,* Applied Science, London 1981.
50. *Enzympräparate–Standards für die Anwendung in Lebensmitteln,* B. Behr's, Hamburg 1983.
51. *Starterkulturen und Enzyme für die Lebensmitteltechnik,* VCH Verlagsgesellschaft, Weinheim 1987.
52. Y. H. Hui: *United States Food Laws, Regulations and Standards,* J. Wiley & Sons, New York–Chichester–Brisbane–Toronto 1979.
53. The Federation of Food Additives Associations: *Food Sanitation Law,* "Food Additives in Japan," Japan 1981.
54. Gesetz über den Verkehr mit Lebensmitteln, Tabakerzeugnissen, kosmetischen Mitteln und sonstigen Bedarfsgegenständen (Lebensmittel- und Bedarfsgegenständegesetz) vom 15. 8. 1974.
55. W. Aehle (ed.): *Enzymes in Industry*, 2nd ed., Wiley-VCH, Weinheim 2004.

Antioxidants

PETER P. KLEMCHUK, CIBA-GEIGY Corporation, Ardsley, New York 10502, United States

1. Introduction

Organic matter has a strong tendency to react with oxygen and oxidize. This is true for most commercial organic materials, e.g., plastics, elastomers, fibers, fuels, lubricants, and foods. It is also true for humans, if aging is considered to be, at least in part, an oxidation process.

In 1861, A. W. VON HOFMANN first made the connection between oxygen and the deterioration of polymer properties in his studies on the aging of gutta-percha [1]. The concept of antioxidants and the discovery that small amounts of reducing agents could protect materials susceptible to oxidation were attributed to S. L. BIGELOW (1898) for sodium sulfite solution [2]; to P. SISLEY (1903 and 1904) for the stabilization of dyed silk to oxidation by hydroquinone (among other compounds); to A. LUMIÈRE, L. LUMIÈRE, and A. SEYEWETZ (1905) for the protection of photographic developers (these authors first used the term antioxidant); and to CH. MOUREU and CH. DUFRAISSE (1920) for the protection of acrolein.

Virtually all polymers that have become commercial products since these early years are dependent on stabilizing additives, antioxidants, which permit the polymers to withstand the oxidative stresses of processing and fabrication and to be used in so many different ways.

Oxidation of organic materials takes place by a number of processes:

autooxidation
biooxidation
combustion
photooxidation

The ultimate result of these processes is the conversion of carbon and hydrogen in organic compounds to carbon dioxide and water, with the accompanying release of large amounts of energy for each mole of carbon dioxide and water formed. In the autooxidation of commercial organic materials, oxidation to carbon dioxide and water is seldom complete.

This article primarily describes the *autooxidation* of organic materials and its prevention by antioxidants, with emphasis on the autooxidation of commercially produced polymers. *Biooxidation* and *combustion* are important processes but are not covered here. *Photooxidation* is discussed only briefly. Although conventional antioxidants, such as hindered phenols, thio-

Ullmann's Industrial Toxicology

ISBN: 3-527-31247-1

synergists, and phosphites, also are moderate stabilizers of organic materials against photooxidation, other, more effective additives, especially ultraviolet absorbers, hindered amines, and nickel chelates, are required to give adequate light stability.

Antioxidants, also called inhibitors (of oxidation), are organic compounds that are added to oxidizable organic materials to retard autooxidation and, in general, to prolong the useful life of the substrates. Relatively few chemical classes are effective as antioxidants. Those in common use are hindered phenols, secondary aromatic amines, certain sulfide esters, trivalent phosphorus compounds, hindered amines, metal dithiocarbamates, and metal dithiophosphates.

Another class of stabilizers, the metal deactivators, reduces markedly the rate of oxidation of polyolefin insulation on copper conductors.

Some antioxidants are capable of functioning alone as *primary stabilizers.* These are the hindered phenols, aromatic amines, metal dithiocarbamates and phosphates, and hindered amines. The phosphites and sulfide esters are *secondary stabilizers* and are used most commonly in conjunction with primary stabilizers. Stabilizer practices are discussed Chapter 5.

Many of the secondary aromatic amines derived from *p*-phenylenediamine are active as both antioxidants and antiozonants. *Antiozonants* are added to unsaturated elastomers to reduce the rate of ozone cracking that arises from the reaction of ozone with carbon – carbon double bonds in the polymer. This reaction ultimately leads to chain cleavage and to cracking of the elastomer when sufficient numbers of bonds are broken.

2. Oxidation

Oxygen is a key factor in oxidation. In the ground state, it exists as a biradical (**1**) and reacts readily with carbon free radicals, even at ambient temperatures, to yield *peroxy radicals*

$$RCH_2\cdot + \underset{\mathbf{1}}{\cdot O-O\cdot} \longrightarrow RCH_2OO\cdot$$

The reaction of carbon free radicals with oxygen has such a low-activation energy that it is the preferred pathway when not limited by oxygen diffusion. The resulting peroxy radicals abstract hydrogen from the substrate, generating another carbon free radical and *a hydroperoxide*:

$$RCH_2OO\cdot + RH \longrightarrow CH_2OOH + R\cdot$$

This reaction has severe consequences for the organic material because: (1) it is a chain reaction that has a significant length and can cause extensive localized oxidation before termination; (2) the hydroperoxide can yield *radical initiators* of oxidation by unimolecular and bimolecular processes:

$$RCH_2OOH \longrightarrow RCH_2O\cdot + HO\cdot$$

$$2\,RCH_2OOH \longrightarrow RCH_2OO\cdot + RCH_2O\cdot + H_2O$$

Unless these radical reactions are inhibited with antioxidants, the organic substance can quickly deteriorate. The physical properties and appearance of polymers and elastomers are reduced, fuels and lubricants sour, and foods lose their flavor and become rancid.

Ions of certain *metals,* e.g., copper, iron, and cobalt, which can exist in two oxidation states, act as catalysts to decompose hydroperoxides:

$$M^{n+} + ROOH \longrightarrow M^{(n+1)+} + RO\cdot + OH^-$$

$$M^{(n+1)+} + ROOH \longrightarrow M^{n+} + ROO\cdot + H^+$$

$$2\,ROOH \longrightarrow RO\cdot + ROO\cdot + H_2O$$

Hydroperoxides are also susceptible to *photolytic cleavage* by near ultraviolet light (< 400 nm but particularly ca. 300 nm):

$$RCH_2OOH \xrightarrow{h\nu} RCH_2O\cdot + HO\cdot$$

Temperature plays a key role in the rate of oxidation of organic materials. Most materials are subjected, however briefly during some stage, to relatively high temperatures; in these intervals significant oxidation can occur and significant amounts of oxidation initiators can be formed. Commercial polymers (thermoplastics, elastomers, synthetic fibers) are dried, processed, and fabricated at elevated temperatures; lubricants are employed at elevated temperatures; foods are cooked at high temperatures. Therefore, processed and fabricated materials will contain significant, but varying, amounts of hydroperoxide and peroxide functionalities, depending on the degree to which the material is stabilized with antioxidants and on the degree of thermo-oxidative exposure.

With plastics and synthetic fibers, compounding to incorporate stabilizer, filler, lubricant, pigment, etc., requires several minutes exposure at 175 – 300 °C. Subsequent fabrication into final articles usually requires < 10 min at temperatures that vary according to the fabrication process:

Fabrication process	Temp., °C
Injection molding	175 – 340
Film blowing	175 – 260
Film extrusion	175 – 260
Blow molding	175 – 230
Fiber spinning	230 – 350
Rotomolding	280 – 340

Processing at high temperature can have a significant impact on a polymer. For example, polypropylene was extruded a total of five times at 260 °C in a 2.54 cm laboratory extruder. Table 1 shows the melt flow rates for pellets retained from the first, third, and fifth extrusions. The large increase in melt flow rate of the unstabilized sample reflects considerable polymer chain cleavage during processing. Melt flow rate determinations and viscosity measurements are convenient and sensitive methods for following changes in average molecular masses of polymers.

Table 1. Multiple extrusion of polypropylene at 260° C

Antioxidant (AO) stabilizer	Melt flow rate (g/10 min) after extrusion number:			
	0	1	3	5
None	4*	8	14	23
0.1 % phenolic AO	–	6	7	9

* With 0.5 % phenolic AO

The influence of processing on the thermal stability of a polymer can be seen in Table 2. polypropylene samples (0.6 mm plaques), after one, three, and five extrusions at 260 °C, were oven aged at 150 °C; the deleterious effect of additional processing on thermal stability is clearly evident.

Table 2. Influence of processing on thermal stability of polypropylene

Antioxidant (AO) stabilizer	Hours to failure after extrusion number:		
	1	3	5
None	<20	<20	<20
0.1 % phenolic AO	510	400	330

The *ease of oxidation* of organic materials is related to the tendency of the substrate to participate in free-radical reactions with oxygen. Substrates that contain carbon – carbon unsaturation with allylic hydrogen atoms are particularly prone to oxidation. For this reason, unsaturated polymers, notably unvulcanized elastomers, and foods with unsaturated fats and oils oxidize easily. The impact of unsaturation is shown by the following comparative polymer data (0.1 % phenolic AO, 100 °C, 0.6 mm plaques)

Polymer	Hours to failure	
polypropylene	10000 [a]	↓
Impact polystyrene	500 [b]	Increasing
Styrene – butadiene rubber	69 [c]	degree of unsaturation
Polybutadiene rubber	15 [c]	↓

[a] Disintegration.
[b] Color formation.
[c] Gel formation.

3. Reaction Mechanisms

3.1. Autooxidation Mechanisms

The work on autooxidation carried out at the Natural Rubber Producers' Research Association in England in the 1940s and early 1950s was a major contribution to understanding autooxidation and stabilization reactions. The work of Bateman [3], Bolland [4], Gee, and Ten Have, among others, showed the oxidation of hydrocarbon materials to:

1) be free radical in nature,
2) have a period of inhibition,
3) have the properties of an autocatalytic chain reaction,
4) yield hydroperoxides as primary oxidation products.

Autooxidations have three distinct steps:

Initiation:

$$ROOH \xrightarrow{k_1} RO\cdot + HO\cdot$$

Propagation:

$$ROO\cdot (RO\cdot, HO\cdot) + RH \xrightarrow{k_2} ROOH + R\cdot \quad (1)$$

$$R\cdot + O_2 \xrightarrow{k_3} ROO\cdot$$

Termination:

$$2\,ROO\cdot \xrightarrow{k_4} [ROOOOR] \longrightarrow ROOR + O_2 \quad (2)$$

$$2\,R\cdot \xrightarrow{k_5} RR$$

$$R\cdot + ROO\cdot \xrightarrow{k_6} ROOR$$

In most instances, when the oxygen concentration is not limiting, termination involves mainly peroxy radicals.

In his work on the oxidation of ethyl linoleate, a model for unsaturation in elastomers and, incidentally, also for fatty foods, BOLLAND [5] identified hydroperoxides as the main products of oxidation. He also showed ethyl linoleate hydroperoxide to be an initiator of ethyl linoleate oxidation. The rate expression in this instance is:

$$\frac{dc_{\mathrm{ROOH}}}{dt} = k_2 c_{\mathrm{RH}} \sqrt{\frac{k_1 c_{\mathrm{ROOH}}}{k_4}}$$

The termination of secondary peroxy radicals proceeds through a tetroxide intermediate to yield *sec*-alcohols and ketones [6]:

$$2\,R_2CHOO\cdot \longrightarrow R_2CHOH + R_2C{=}O + O_2$$

Tertiary peroxy radicals, on the other hand, do not terminate directly. In this case the metastable tetroxide intermediate decomposes to yield alkoxy radicals and oxygen [7–9]:

$$2\,ROO\cdot \longrightarrow [ROOOOR] \longrightarrow 2\,RO\cdot + O_2 \quad (3)$$

where R contains a *tert*-carbon atom attached to oxygen. With cumylperoxy radicals, about 10 % of termination takes place by Reaction (2) to yield dicumyl peroxide, and 90 % proceeds by Reaction (3) [9]. The cumylalkoxy radical reacts further and eventually terminates as a methylperoxy radical:

$$RO\cdot \longrightarrow CH_3\cdot + C_6H_5COCH_3$$

$$CH_3\cdot + O_2 \longrightarrow CH_3O_2\cdot$$

$$CH_3O_2\cdot + RO_2\cdot \longrightarrow CH_2O + ROH + O_2$$

where R is $C_6H_5C(CH_3)_2$.

Oxidation can start in a hydrocarbon at ambient temperature because most commercial organic materials are subjected to high temperature, e.g., during processing, fuel and lubricant refining, and food preparation. Small but detectable quantities of peroxides form and serve as initiators of autooxidation. The main product of autooxidation, hydroperoxide, is an initiator of autooxidation. This is the reason autooxidation is autocatalytic.

The plot of oxygen uptake (Fig. 1) for high-density polyethylene (HDPE) at 100 °C exhibits the autocatalytic nature of autooxidation. During an induction period of about 35 h, little oxygen is absorbed. Once initiator concentration is sufficiently high and oxidation inhibitors, if any, are consumed, however, oxygen consumption increases rapidly to a steady rate. This behavior is characteristic of organic materials undergoing autooxidation.

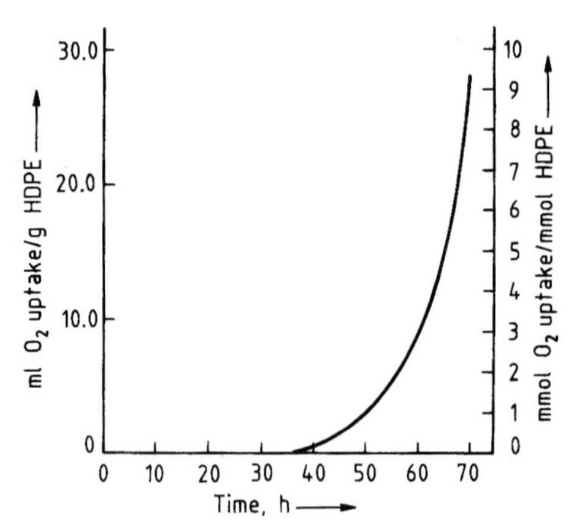

Figure 1. Oxygen uptake curve of Du Pont Alathon 5496 HDPE at 100 °C and 101.3 kPa

3.2. Antioxidant Mechanisms

Antioxidants are classified as either radical trapping (chain breaking) or peroxide decomposing, terms that describe the mechanisms by which they function.

3.2.1. Radical Trapping

Radical trapping antioxidants (AH) function by reaction with (trapping) the propagating *oxygen free radicals* in autooxidation:

$$ROO\cdot + AH \longrightarrow ROOH + A\cdot \quad (4)$$

The antioxidant thus competes with the organic substrate (RH) as a source of hydrogen for peroxy radicals in the usual autooxidation sequence (Eq. 1). For effective radical trapping antioxidants, Reaction (4) is much faster than Reaction (1). In addition, the radical A · must be stable and not abstract hydrogen atoms from the organic substrate; i.e., it should not function as a chain-transfer agent:

$$A\cdot + RH \not\longrightarrow AH + R\cdot \quad (5)$$

Otherwise, the antioxidant does not become a chain terminator but only transfers the propagation of oxidation from one molecule to another.

Radical trapping antioxidants, e.g., hindered phenols and secondary aromatic amines in particular, react with oxygen radicals (peroxy and alkoxy) almost exclusively. That they usually do not trap carbon free radicals efficiently is a distinct disadvantage. Trapping a *carbon free radical* (Eq. 6) is the preferred radical trapping reaction because it reduces to zero the oxidative chain length (the average number of oxygen molecules consumed in a chain reaction), and markedly reduces hydroperoxide formation. The usual radical-trapping reaction is represented by Equation (4).

$$R\cdot + AH \longrightarrow RH + A\cdot \quad (6)$$

A new class of radical-trapping stabilizers, the hindered amines trap alkyl radicals (see Section 4.5).

Even though they do not reduce the oxidative chain length to zero, as would be preferred, radical-trapping antioxidants are very effective in reducing the oxidative chain length. For example, 1,4-benzohydroquinone at 0.03 % (mole fraction of hydroquinone to ester 9×10^{-4}) reduced the oxidative chain length for the benzoylperoxide-initiated, uninhibited oxidation of ethyl linoleate from 241 to one [10]. This means that in the uninhibited reaction, 241 molecules of oxygen were consumed and 241 molecules of hydroperoxide were formed in the chain reaction following the generation of one initiating radical. With the antioxidant present, only one molecule of oxygen was consumed and one molecule of hydroperoxide was formed per initiating radical. Both the uninhibited and the inhibited oxidative chain lengths vary with the substrate; however, stabilization by antioxidants always results in significant lowering of chain lengths.

3.2.2. Peroxide Decomposing

Peroxide decomposing antioxidants reduce the hydroperoxidic and peroxidic products of oxidation to more innocuous compounds, usually alcohols or ethers. Because peroxides are initiators of oxidation, their removal from organic materials is important in the inhibition of oxidation. The two major classes of peroxide decomposing stabilizers are trivalent phosphorus compounds and thiosynergists (see Sections 4.3 and 4.4).

Trivalent phosphorus compounds– phosphites and phosphonites – function as peroxide decomposers by abstracting peroxidic oxygen from hydroperoxides and peroxides, and reducing them. The phosphorus compounds are oxidized to phosphates or phosphonates:

$$(RO)_3P + R'OOH \longrightarrow (RO)_3PO + R'OH$$

Thiosynergists, mainly esters of thiodipropionic acid, function mainly at elevated temperatures (> 100 °C) both by direct reaction with hydroperoxides and by the nonradical decomposition of hydroperoxides catalyzed by the acidic oxidation products of the thiosynergists. This latter reaction is similar to the acid-catalyzed decompositions of cumene hydroperoxide used to manufacture phenol and acetone.

3.3. Metal Deactivators

Metal deactivators are effective in reducing the particularly deleterious effects that copper in wires and cables has on polyolefin insulation. Copper is a particularly severe agent for degrading polymers by thermal oxidation. Metal deactivators are postulated to function by deactivating copper to oxidation by complexation at the polymer – metal interface (see Section 4.6).

4. Antioxidant Classes

4.1. Hindered Phenols

Hindered phenols are a major class of antioxidants. They are presently used as radical-trapping agents in plastics, elastomers, synthetic fibers, fuels, lubricants, and foods. Steric hindrance is provided by bulky substituents in positions on the ring ortho to the hydroxyl group. These bulky substituents influence the specificity of the phenols by blocking phenoxyl radicals from abstracting hydrogen atoms from organic substrates.

2,6-Di-tert-butylphenol [*128-39-2*] is the most widely used functionality in commercial hindered phenolic antioxidants. This group is effective and easily made at moderate cost by the ortho alkylation of phenol by olefins using an aluminum alkoxide alkylation catalyst [11]:

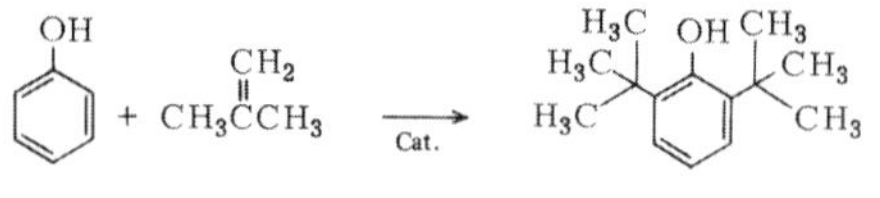

Structures of common phenolic antioxidants are included in Table 3. A more comprehensive list of phenolic and other antioxidants can be found in [12].

Substituents in the 4-position of 2,6-di-*tert*-butylphenol influence the solubility of the antioxidant. Moreover, high-molecular-mass substituents influence the migration and extractability of antioxidants, reducing their extraction from food-packaging plastics, and making such antioxidants suitable for use as indirect food additives. The 4-substituent can also influence the reactivity of the hindered phenol, making some compounds more suitable for certain substrates, such as unsaturated elastomers.

The mechanism of the reaction of hindered phenols with alkylperoxy radicals, as a model for chain propagating radicals, was investigated. In oxidizing organic substrates [13], [14]. In solution each hindered phenolic moiety consumed two peroxy radicals:

$$\mathrm{ROO\cdot + AH \longrightarrow ROOH + A\cdot} \qquad (7)$$

$$\mathrm{ROO\cdot + A\cdot \longrightarrow ROOA}$$

4-Alkylperoxycyclohexadienones were isolated:

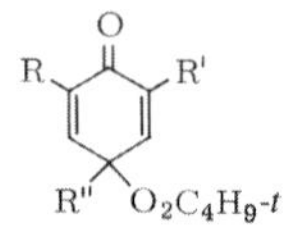

where R, R′, R″ are CH_3 or t-C_4H_9.

Compounds with 4-methyl substituents also yielded *stilbenequinone* derivatives (**15**), arising from the following reaction sequence:

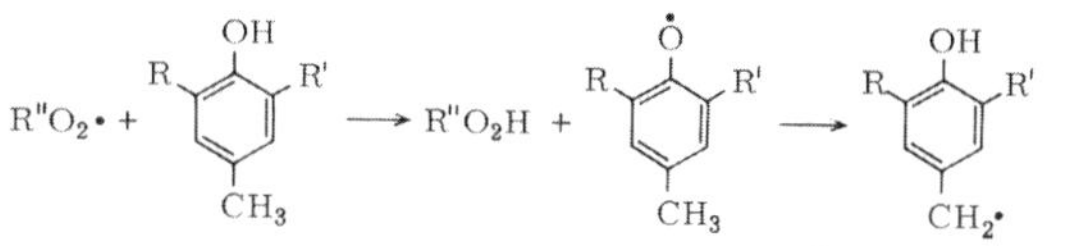

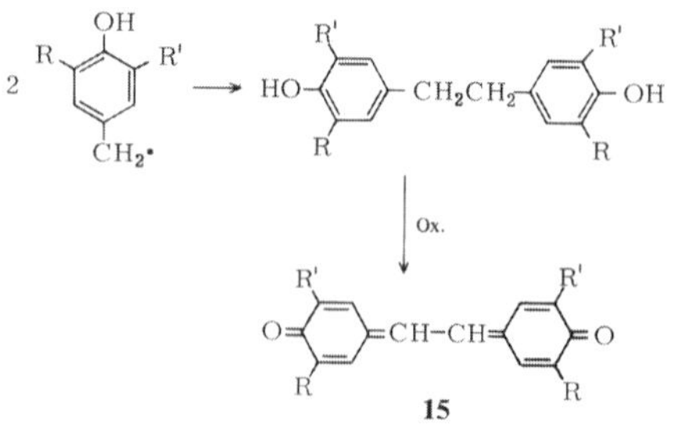

Stilbenequinones have very high-extinction coefficients in the visible portion of the spectrum, and can contribute significant color to the materials in which they are incorporated. Stilbenequinones are serious problems, particularly in plastics stabilization because amounts as small as 5×10^{-6} parts are sufficient to impart a yellow color to the plastic, which is objectionable in colorless or white plastic articles.

The antioxidant efficiency of phenolic antioxidants as well as the degree of chain transfer have been measured [14]. The *antioxidant efficiency* is defined as $2\,k_a/k_r$, where k_a is the rate constant for Equation (4) and k_r is the rate constant for Equation (1). Efficiencies were found to be 32 for 2,4,6-tri-*tert*-butylphenol and 33.5 for 2,6-di*tert*-butyl-4-methylphenol. No chain transfer (Eq. 5) was found with either antioxidant.

The 4-alkylperoxy-4-alkyl-2,6-di-*tert*-butylcyclohexadienones have been formed in the presence of relatively high concentrations of alkylperoxy radicals. Because that situation does not usually exist in inhibited oxidation of organic materials under conditions of use, the normal reaction in plastics is the disproportionation of two phenoxy radicals. This reaction regenerates starting phenol and quinone methide (**16**):

$$2\ \text{(phenoxyl radical, } CH_2R) \longrightarrow \text{(phenol, } CH_2R) + \textbf{16}\ \text{(quinone methide, } =CHR)$$

Quinone methides have limited stability and react to yield a complex mixture of products [15], [16].

4.2. Aromatic Amines

Secondary aromatic amines for use in elastomer stabilization were among the first antioxidants. The main classes are diarylamines, amine – ketone condensation products (e.g., aniline – acetone condensates), and substituted *p*-phenylenediamines. Most are colored and form highly colored oxidation products. Thus, their uses are generally limited to those for which discoloration is not a drawback, e.g., carbon-black-filled, vulcanized elastomers are a major application.

In spite of their long history of use, the *mechanisms of action* of aromatic amine antioxidants are not as well understood as are those of hindered phenolic antioxidants. To a great extent this is because of the mixture of oxidation products obtained from aromatic amines, and the complex composition of some of the aromatic amine antioxidants. Nevertheless, the general features of the mechanisms are known.

The secondary amines serve as a source of hydrogen atoms, which are abstracted by peroxy radicals (Eq. 4) to yield relatively stable aminyl radicals. The same competitive reactions as found with hindered phenols are the key to the activity of amines as antioxidants.

The *antioxidant efficiencies* of secondary aromatic amines were found to be comparable to those of hindered phenols in the sensitized oxidation of 9,10-dihydroanthracene [17]. The intermediate aminyl radicals had a low tendency to abstract hydrogen atoms from polymers. They were similar in this behavior to hindered phenoxyl radicals. Tertiary amines showed no oxidation retarding properties. *N,N′*-Di-*sec*butyl-*p*-phenylenediamine was particularly efficient in trapping peroxy radicals; it trapped two peroxy radicals per diamine molecule, leading to the formation of quinone imide.

In the diphenylamine-inhibited oxidation of cumene at 68.5 °C, diphenylaminyl radicals were converted in significant amount to nitroxyl radicals; about 20 % of the initial diphenylamine was isolated as diphenyl nitroxide [18].

$$R\dot{N}R + R'O_2^{\bullet} \longrightarrow RN(O^{\bullet})R + R'O^{\bullet}$$

The *N*-oxyl may eventually trap an alkyl radical:

$$RN(O^{\bullet})R + R'^{\bullet} \xrightarrow{k_7} RN(OR')R \qquad (8)$$

The ratio of the rate constants, k_8/k_7, is about ten in the oxidation of styrene:

$$R'^{\bullet} + O_2 \xrightarrow{k_8} R'O_2^{\bullet} \qquad (9)$$

Reaction (9) therefore is favored in the presence of significant oxygen. However, in the virtual absence of oxygen, Reaction (8) is important, as evidenced by the ability of 4,4-dimethoxydiphenyl nitroxide to function as an inhibitor for styrene polymerization initiated by azobisisobutyronitrile (AIBN) at 60 °C [19].

4.3. Organosulfur Compounds

The usefulness of organosulfur compounds as stabilizers for organic materials was recognized early. Metal salts of dialkyldithiocarbamic acids, zinc dialkyldithiophosphates, and esters (dilauryl, dimyristyl, distearyl) of thiodipropionic acid have become important commercial antioxidants for the stabilization of diverse materials: lubricants, elastomers, and plastics. Several comprehensive reviews have been published on these compounds as antioxidants [20–23].

Mechanisms of Action. Organosulfur antioxidants decompose hydroperoxides by *nonradical reactions.* The effectiveness of various organosulfur compounds, i.e., alkyl- and arylsulfides, alkyl- and arylthiols, and a zinc dialkyl-

Table 3. Major commercial hindered phenolic antioxidants

Structure	Chemical name	CAS registry no.	Trade names
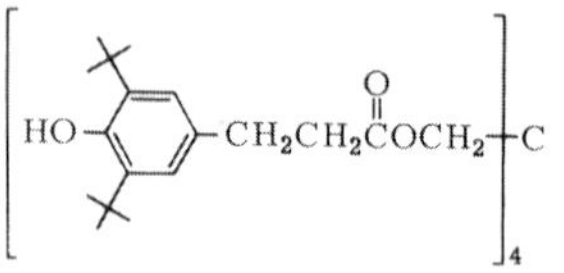 **2**	tetrakis [methylene (3,5-di-*tert*-butyl-4-hydroxyhydrocinnamate)] methane	[*6683-19-8*]	Irganox 1010
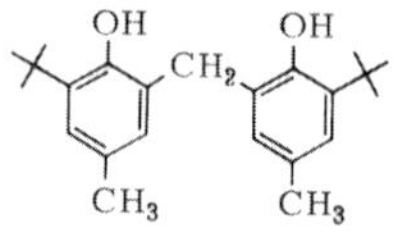 **3**	2,2′-methylenebis-(4-methyl-6-*tert*-butylphenol)	[*119-47-1*]	Cyanox 2246
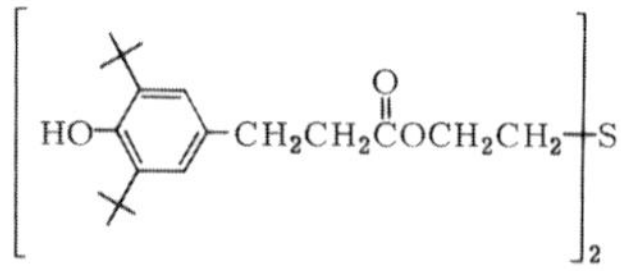 **4**		[*41484-35-9*]	Irganox 1035
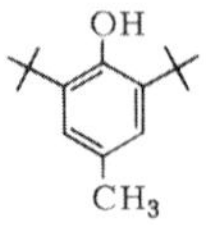 **5**	2,6-di-*tert*-butyl-4-methylphenol	[*128-37-0*]	Butylated hydroxytoluene (BHT)
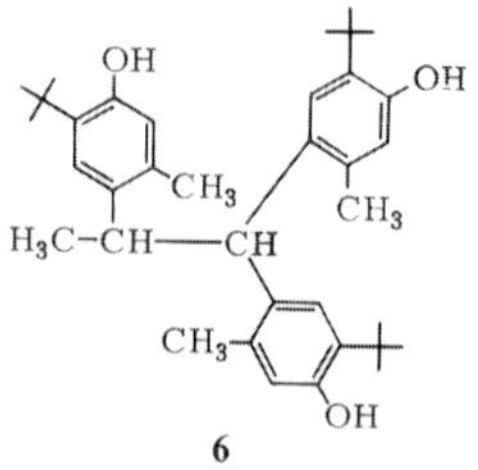 **6**		[*1843-03-4*]	Topanol CA
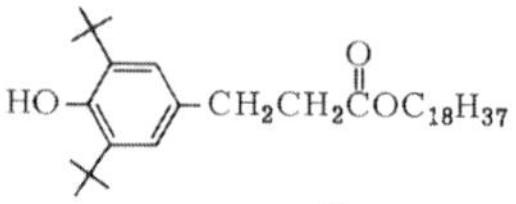 **7**		[*2082-79-3*]	Irganox 1076
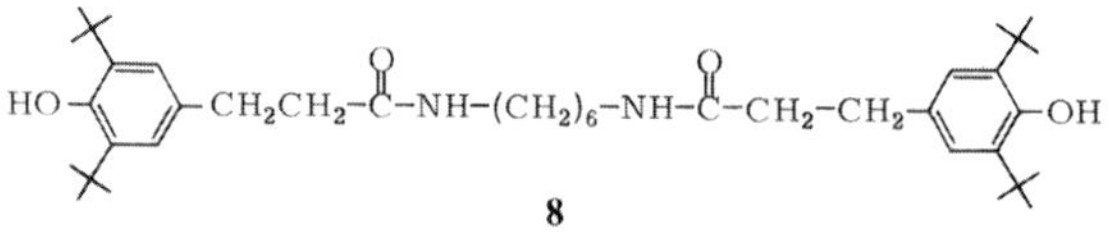 **8**	*N,N′*-1,6-hexamethylene-bis-3-(3,5-di-*tert*-butyl-4-hydroxyphenyl) propionamide	[*23128-74-7*]	Irganox 1098

Table 3. Continued.

Structure	Name	CAS no.	Trade name
9	4,4′-Butylidenebis-(6-*tert*-butyl-3-methylphenol)	[*85-60-9*]	Santowhite powder
10		[*40601-76-1*]	Cyanox 1790
11		[*27676-62-6*]	Good-rite 3114
12		[*34137-09-2*]	Good-rite 3125
13		[*1709-70-2*]	Ethanox 330 Irganox 1330
14	4,4′-Thiobis(2-*tert*-butyl-5-methylphenol)	[*96-69-5*]	Santonox R

dithiophosphate in decomposing cumene hydroperoxide was determined in mineral oil at 150 °C [24]. The most effective compounds were zinc di(4-methylpentyl)-2-dithiophosphate, *n*-dodecylthiol, and a sulfonic acid. All reactions gave high yields of phenol, implying that they proceeded by acid catalysis [25].

Oxidation and decomposition products of the organosulfur compounds are responsible for inhibiting the oxidation of organic materials by decomposing the hydroperoxides. Sulfur dioxide, one of these terminal oxidation products, is a highly effective hydroperoxide decomposer [26]. Sulfides and disulfides react with hydroperoxides to yield sulfoxides and thiosulfinates:

$$RSR + R'OOH \longrightarrow RS(O)R + R'OH$$
$$RSSR + R'OOH \longrightarrow RS(O)SR + R'OH$$

Sulfenic acid, RS(O)H, and thiosulfoxylic acid, RSS(O)H, are decomposition products of sulfoxides and thiosulfinates. Each molecule of these acids is capable of decomposing many hydroperoxide molecules [27]. In addition higher oxidation products of sulfenic and sulfoxylic acids, i.e., sulfonic, thiosulfurous, thiosulfuric, and sulfuric acids, and sulfur dioxide are also catalytic hydroperoxide decomposers [27]. A large number of active hydroperoxide decomposers are generated by the oxidation of sulfides and disulfides with hydroperoxides.

Dilauryl thiodipropionate (**17**) [*123-28-4*], (DLTDP) is a common thiosynergist for polyolefin stabilization. Its mechanism of action is similar to relatively simple aromatic and aliphatic sulfide and disulfide model compounds [28]. The DLTDP sulfoxide (**17a**) was thermally unstable at 150 °C and decomposed to a gas (SO_2, H_2S, H_2O), a liquid (containing lauryl acrylate), and a solid (a complex mixture of several compounds, including the sulfenic acid, $H_{25}C_{12}O_2CCH_2CH_2SOH$) [28]. In addition to hydroperoxide decomposition, a second mechanism for DLTDP activity involves the reaction of DLTDP (**17**) with the peroxycyclohexadienone reaction product of a hindered phenol. The hindered phenol was regenerated in 41 % yield [28]:

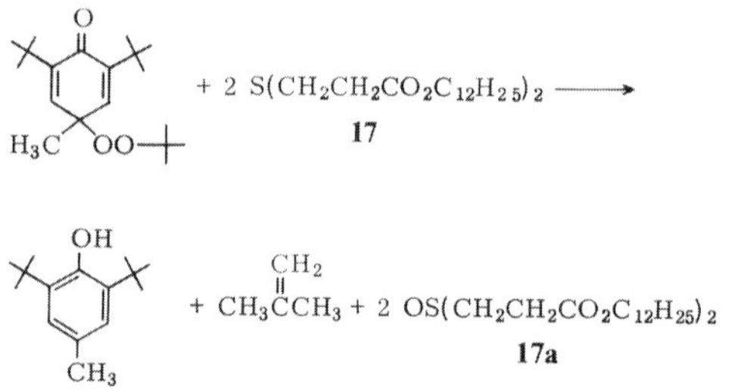

$$+ 2\ S(CH_2CH_2CO_2C_{12}H_{25})_2 \longrightarrow$$
17

$$+ CH_3\overset{\displaystyle CH_2}{\overset{\|}{C}}CH_3 + 2\ OS(CH_2CH_2CO_2C_{12}H_{25})_2$$
17a

This mechanism can be important in oxidations where significant amounts of peroxycyclohexadienone products are formed from hindered phenols, and in the performance of 2,6-diphenyl-4-octadecyloxyphenol [29].

A third mechanism for thiosynergists comes from the observation that zinc dialkyldithiophosphates, zinc dialkylxanthates, and zinc dialkyldithiocarbamates, inhibit the azobisisobutyronitrile oxidation of squalene at 60 °C [30], [31]. The inhibition is the result of peroxy radical trapping involving electron transfer from sulfur in the zinc salts to the peroxy radicals. Disulfides and zinc peroxides are the products:

$$[(RO)_2P(S)S]_2Zn + 2\,R'OO\cdot \longrightarrow$$
$$(RO)_2P(S)S{-}S(S)P(OR)_2 + Zn(OOR')_2$$

Hydroperoxide decomposition is generally the main mechanism of stabilization by organosulfur compounds. The relative importance of other mechanisms, peroxy radical trapping and hindered phenol regeneration, depends on the stabilizers and circumstances of use.

Thiosynergism. A *synergism of action* exists between radical trapping antioxidants (hindered phenols or secondary aromatic amines) and peroxide-decomposing antioxidants (zinc dialkyldithiophosphate or dialkyl sulfide) in the stabilization of mineral oil at 155 °C and 180 °C [24]. Sulfoxides inhibit oxidation only after the absorption of a small amount of oxygen. Sulfoxides, RS(O)R, and thiosulfinates, RS(O)SR, were especially effective inhibitors for the oxidation of squalene at 75 °C [32].

Esters of thiodipropionic acid play a key role in the high-temperature stabilization of polyolefins by hindered phenols. Figure 2 shows the considerably longer lifetimes obtained with the combination of a hindered phenol and DLTDP [33]. All samples contained 0.4 wt% of stabilizer. Curves a and b depict the lifetimes obtained with DLTDP and hindered phenol, respectively; c is calculated from a and b; and d is a plot of the experimental values from actual compositions of hindered phenol and DLTDP tested at 140 °C. The markedly superior performance shown by the actual mixtures is a result of synergistic interaction between the two stabilizers, the one basically radical trapping, the other hydroperoxide decomposing.

Thioesters are effective primarily when polymers are in the solid state and are at temperatures $< 160\,°C$. This is in contrast to phosphites (see Section 4.4), another class of hydroperoxide decomposers, which are effective in protecting polymers in the molten state during processing and fabrication, usually at temperatures $\gg 160\,°C$.

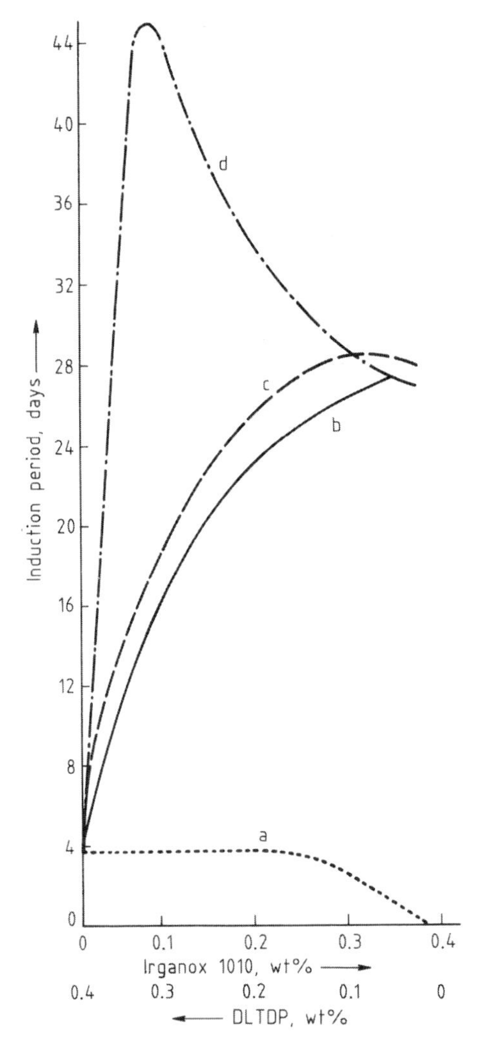

Figure 2. Synergists in heat stabilization
a) DLTDP (dilauryl thiodipropionate); b) Irganox 1010; c) Calculated; d) Experimental

4.4. Phosphorus Compounds

Trivalent phosphorus compounds, i.e., phosphites and phosphonites, are practically the only types of phosphorus compounds being used commercially as antioxidants. The only other major class is zinc dialkyldithiophosphate, which finds use primarily in lubricants. Its mode of action is discussed in Section 4.3.

Mechanisms of Action. Trivalent phosphorus antioxidants generally function by reducing hydroperoxides and peroxides to alcohols and ethers:

$$(RO)_3P + R'OOH \longrightarrow (RO)_3PO + R'OH$$

Neither the phosphate nor the alcohol is an oxidation initiator. The reaction between phosphites or phosphonites and peroxy compounds takes place at moderate temperature but, unfortunately, is stoichiometric. Therefore the effectiveness ceases once all the trivalent compound is oxidized.

Phosphites also function as alkoxy and peroxy radical trappers, as in reactions (10) and (11). Stabilization results if (**18**) is a hindered phenoxy radical with a low tendency to abstract hydrogen from polymer, i.e., if it is comparable in activity and stability to radicals generated from effective hindered phenolic antioxidants [34]:

$$(RO)_3P + R'OO\cdot \longrightarrow (RO)_3\dot{P}OOR' \qquad (10)$$
$$(RO)_3\dot{P}OOR' \longrightarrow (RO)_3P{=}O + R'O\cdot$$

$$R'O\cdot + (RO)_3 \longrightarrow (RO)_3\dot{P}OR' \qquad (11)$$
$$(RO)_3\dot{P}OR' \longrightarrow RO)_2POR' + RO\cdot$$

18

Certain cyclic phosphites, notably catechol phosphites, are catalytic hydroperoxide decomposers, their corresponding phosphates functioning as strong acids [35]:

NO_2 (o, p-OR) $\xrightarrow{\text{red.}}$ NH_2 (o, p-OR)

where R is alkyl

Thiophosphites, thiophosphates, and easily hydrolyzable phosphates also are catalytic hydroperoxide decomposers. The thiophosphorus compounds are oxidized to acidic products that function as the active agents [36].

Uses. Phosphites, such as tris(nonylphenyl) phosphite (TNPP) (**19**), have been used commercially for about 40 years as stabilizers for unsaturated elastomers. In spite of limited stability to hydrolysis and attendant corrosion problems, TNPP continues to be used for low-cost antioxidant applications. More sophisticated trivalent phosphorus compounds have been marketed primarily to minimize molecular mass changes in polyolefins during compounding and fabrica-

Table 4. Influence of phenolic antioxidant, phosphite, and thiosynergist on polypropylene processing stability at 260°C

Multiple extrusions, Profax 6501	Melt flow (extrusions)			YI color 3 mm plaques (extrusions)		
	1st	3rd	5th	1st	3rd	5th
None	7 [a]	17	28	6	10	13
0.1 % AO 1 [b]	6	8	9	12	16	19
0.1 % AO 1 + 0.05 % P 1 [c]	4	6	9	12	13	16
0.1 % AO 1 + 0.05 % P 2 [d]	4	6	7	8	11	14
0.1 % AO 1 + 0.05 % S 1 [e]	6	8	9	10	13	16

[a] Melt flow was 3 for unprocessed polymer with 0.5 % BHT.
[b] Irganox 1010.
[c] Phosphite 168.
[d] Sandostab P-EPQ.
[e] Distearyl thiodipropionate.

tion. Several of these newer products are regulated by the FDA as indirect food additives for plastic food packaging. They are used with hindered phenolic antioxidants not only to minimize molecular mass changes, but also to reduce the color development on high-temperature processing and subsequent use that is usually associated with hindered phenols (Table 4; see above). The data in Table 5 show the reduction in concentration of phosphites during high-temperature processing of polypropylene. These results emphasize the sacrificial role of phosphite stabilizers and suggest they are not suitable candidates for long-term stabilization. However, by inhibiting polyolefin thermooxidative degradation during high-temperature processing, phosphites help prolong polymer life.

Table 5. Influence of processing on additive concentration; multiple extrusions at 260°C, Profax 6501

Stabilizers *	Additive conc., %	
	1st extrusion	5th extrusion
None	–	–
0.1 % AO 1	0.08	0.06
0.1 % AO 1 +	0.09	0.07
0.05 P1	0.04	0.02
0.1 % AO 1 +	0.09	0.05
0.05 % P2	0.04	<0.001
0.1 % AO 1 +	0.10	0.08
0.05 % S1	0.05	0.05

* Stabilizers identified in Table 4.

Table 6 (see next page) gives the major commercial trivalent phosphorus compounds used as antioxidants for plastics and rubbers.

4.5. Hindered Amines

Hindered amines are used primarily as light stabilizers but were also recognized early as stabilizers against thermooxidative degradation. The following data [37] summarize the antioxidant activity of hindered amine stabilizer in oven aged (120 °C) polypropylene multifilaments (8000/396 den):

Stabilizer	Days to embrittlement
Hindered amine stabilizer, O.1 %	47
Hindered amine stabilizer, 0.25 %	80
Hindered amine stabilizer, 0.5 %	110
Phenolic antioxidant-1, 0.1 %	14
Phenolic antioxidant-2, 0.1 %	20

In certain instances, particularly at moderate temperatures (< 120 °C), some hindered amines are more effective than hindered phenols for the stabilization of polypropylene stretched tapes [38]. These results are in line with the generally accepted *mechanisms of action* of hindered amines:

1) trapping of carbon free radicals by hindered amine nitroxyls [39]

$$>NO\cdot + R\cdot \longrightarrow >NOR$$

2) trapping of peroxy radicals by hydroxyl amine ethers (reaction products from trapping carbon free radicals by hindered amine nitroxyls) with the regeneration of nitroxyls and formation of peroxides [40]

Table 6. Trivalent phosphorus antioxidants

Structure	CAS registry number	Trade names
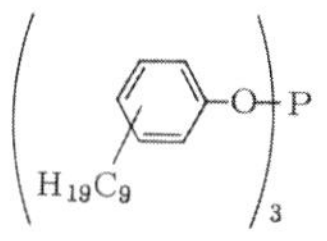 $(H_{19}C_9\text{-}C_6H_4\text{-}O)_3P$ **19**	[*26523-78-4*]	TNPP
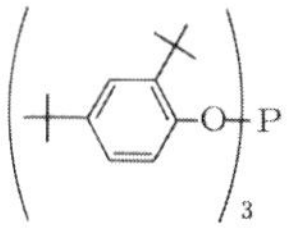 (2,4-di-*tert*-butylphenyl-O)$_3$P **20**	[*31570-04-4*]	Phosphite 168
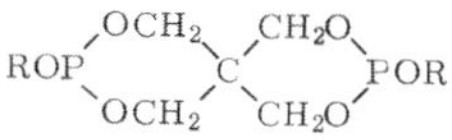 $ROP(OCH_2)_2C(CH_2O)_2POR$ **21** R = $C_{18}H_{37}$,	[*3806-34-6*]	Weston 618
22 R = 2,4-di-*tert*-butylphenyl	[*26741-53-7*]	Ultranox 626
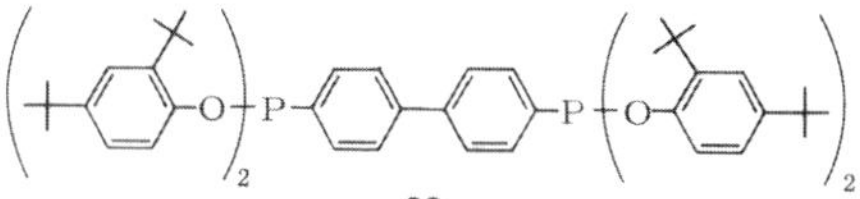 (2,4-di-*tert*-butylphenyl-O)$_2$P-C$_6$H$_4$-C$_6$H$_4$-P(O-2,4-di-*tert*-butylphenyl)$_2$ **23**	[*38613-77-3*]	Sandostab P-EPQ Irgafos P-EPQ

$$>NOR + R'O_2\cdot \longrightarrow >NO\cdot + R'OOR$$

3) complexing of hydroperoxides by hindered amines and hindered amine nitroxyls [41].

Trapping carbon free radicals is preferable to trapping peroxy radicals for purposes of stabilization because the former assures the oxidative chain length will be zero (no hydroperoxide formed), whereas trapping peroxy radicals yields one or more hydroperoxide products.

The complexation of hydroperoxides by stabilizer molecules is also a real advantage because it effectively increases the stabilizer concentration in the vicinity of the main species responsible for initiating oxidation.

Table 7 contains the main hindered amines available commercially.

4.6. Metal Deactivators

Metal deactivators were developed primarily to increase the lifespan of polyolefin insulation in contact with copper conductors in communications wire. Phenolic antioxidants alone are not effective in providing long-term stability. Copper is a prooxidant in polyolefin oxidation. It functions as a catalytic decomposer of hydroperoxides, generating alkoxy and peroxy radical initiators of oxidation:

$$ROOH + Cu^{+} \longrightarrow RO\cdot + Cu^{2+} + OH^{-}$$
$$ROOH + Cu^{2+} \longrightarrow ROO\cdot + Cu^{+} + H^{+}$$
$$2\,ROOH \longrightarrow ROO\cdot + RO\cdot + H_2O$$

In the absence of metal deactivators, copper penetrates polyethylene via oxidation reactions, yielding copper carboxylates among many products [42]. Metal deactivators markedly reduce the rate of copper penetration and are postulated to function by complexing copper at the polymer–metal interface and within the polymer [42]. Table 8 (see page after next) shows the severe effect of copper on the heat stability of polyethylene and the effectiveness of two different metal deactivators in reducing thermal degradation.

Table 7. Major commercial hindered amine stabilizers

Structure	CAS registry number	Trade names
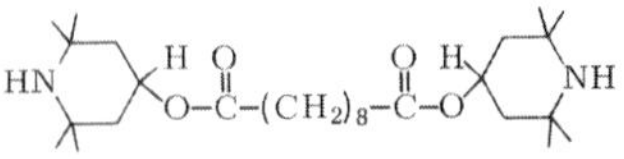 HN, H O, O H, NH, O–C–(CH$_2$)$_8$–C–O **24**	[*52829-07-9*]	Tinuvin 770
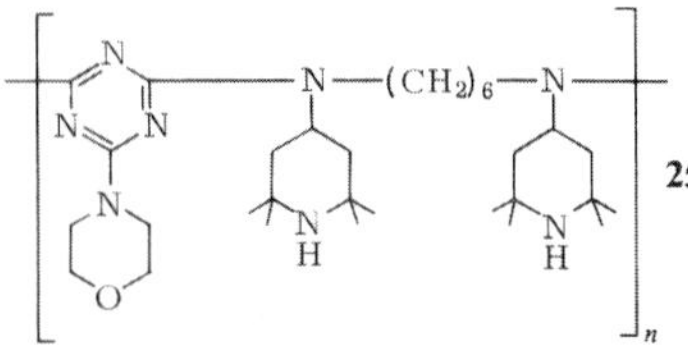 N, N, N, N–(CH$_2$)$_6$–N, N, O, N H, N H **25** *n*	[*82451-48-7*]	Cyasorb UV-3346
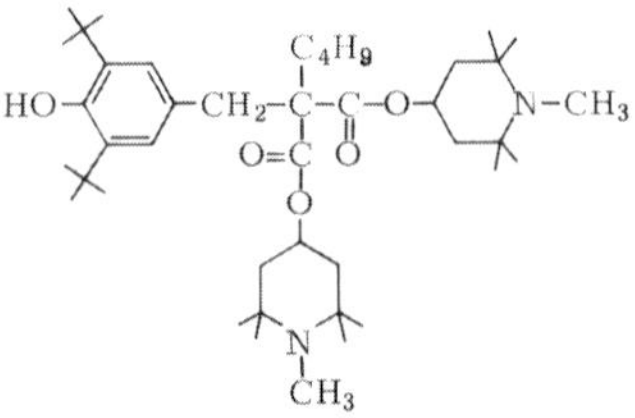 HO, CH$_2$–C(C$_4$H$_9$)–C(=O)–O, N–CH$_3$, O=C, O, N, CH$_3$ **26**	[*63843-89-0*]	Tinuvin 144
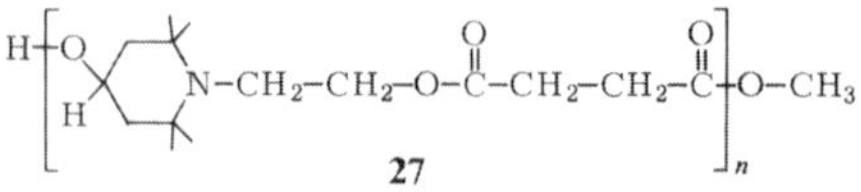 H–[O, H, N–CH$_2$–CH$_2$–O–C(=O)–CH$_2$–CH$_2$–C(=O)]$_n$–O–CH$_3$ **27**	[*65447-77-0*]	Tinuvin 622LD
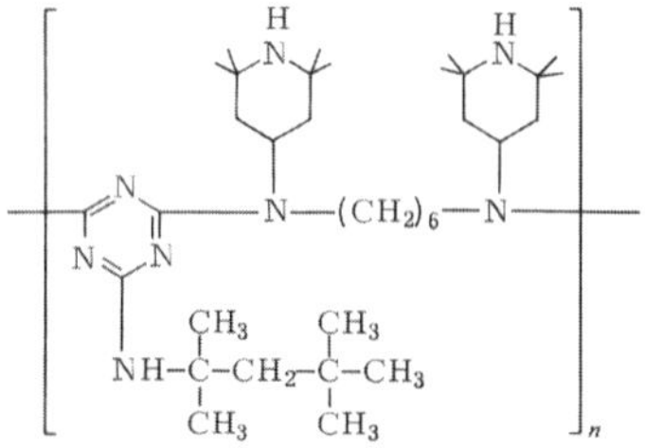 H N, H N, N, N, N, N–(CH$_2$)$_6$–N, NH–C(CH$_3$)$_2$–CH$_2$–C(CH$_3$)$_2$–CH$_3$ **28** *n*	[*70624-18-9*]	Chimassorb 944LD
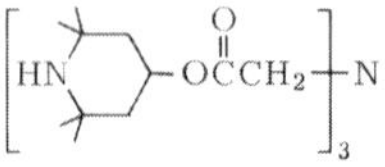 [HN, –OC(=O)CH$_2$]$_3$–N **29**	[*64022-57-7*]	Mark LA 55
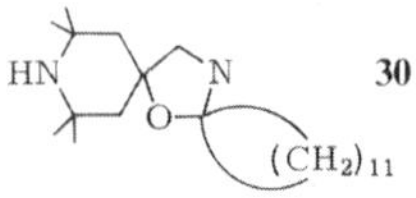 HN, N, O, (CH$_2$)$_{11}$ **30**	[*81406-61-3*]	Hostavin TMN 20
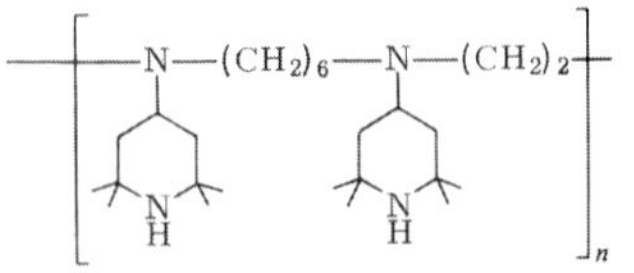 [N–(CH$_2$)$_6$–N–(CH$_2$)$_2$]$_n$, N H, N H **31**	[*61269-61-2*]	Spinuvex A-36

Table 8. Thermal stability of poly(propylene – ethylene) compression molded plaques containing embedded copper screen

0.2 % Irganox1010 +	Hours to failure				
	120 °C	40 °C		150 °C	
	Cu	No Cu	Cu	No Cu	Cu
No metal deactivator	460	1710	360	850	100
0.2 % Irganox MD 1024	1320	1870	570	760	250
0.2 % Eastman OABH	1360	1780	500	760	250

The major commercially available metal deactivators are Irganox MD-1024 [*32687-78-8*], (**32**); Eastman Inhibitor OABH [*6629-10-3*], (**33**); Naugard XL-1 [*70331-94-1*], (**34**).

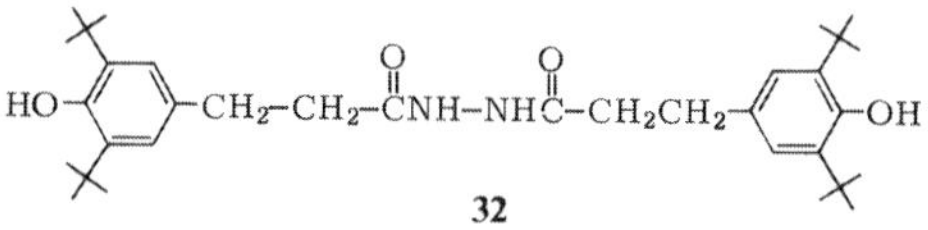

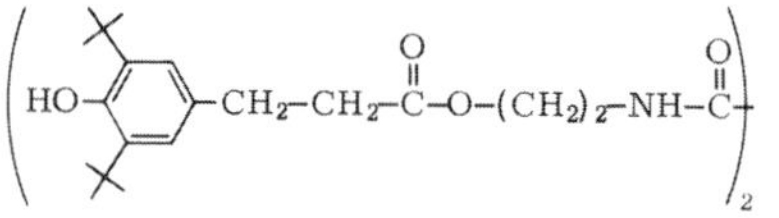

The amide or hydrazide functionalities complex copper, and are responsible for the inhibition activity. Normally, a combination of a hindered phenol and a metal deactivator is used to stabilize polyolefins in the presence of copper. However, because Irganox MD 1024 and Naugard XL 1 contain both phenolic and nitrogen groups, they can be used as the sole stabilizers for that application.

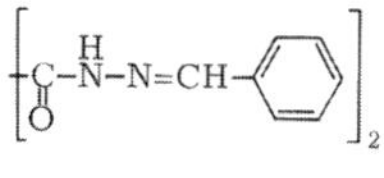

Metal deactivators are also used with hindered phenols for the stabilization of filled polyolefins, where metallic impurities in fillers adversely affect the thermooxidative stability of polymers [43].

5. Practices in Stabilization with Antioxidants

Specialized practices have evolved for each of the four major substrate classes that are stabilized with antioxidants:

1) foods,
2) fuels,
3) lubricants, and
4) polymers.

Some antioxidants are used in more than one substrate class. For the most part, however, each class has its own stabilizer types and amounts. Some antioxidant practices are common to all substrate classes: antioxidants for each class are selected on the basis of effectiveness, solubility, and cost. In the United States, antioxidants used in food and food-packaging plastics are limited to ones regulated by the Food and Drug Administration (FDA) for these uses.

5.1. Stabilization of Food

The antioxidants most commonly used to stabilize food against autooxidation are butylated hydroxytoluene (**5**), see Table 3; butylated hydroxyanisole (BHA) (**35**) [*25013-16-5*], mixture of 2- and 3-*tert*-butyl-4-methoxyphenol, Tenox BHA (Kodak); and *n*-propyl gallate (**36**) [*121-79-9*], 3,4,5-trihydroxybenzoic acid propyl ester, Tenox PG (Kodak).

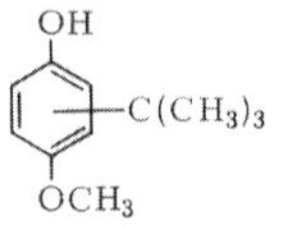

These antioxidants may be used directly in foods governed by FDA regulations. They are used to protect fats and oils during frying and retain activity during storage. In general they are used to prevent oxidation of foods, which results in deterioration of flavor and odor. Butylated hydroxytoluene is used in packaging materials for cookies, cold cereals, etc., where the relatively high-vapor pressure of BHT permits it to transfer from the packaging material to the contained food.

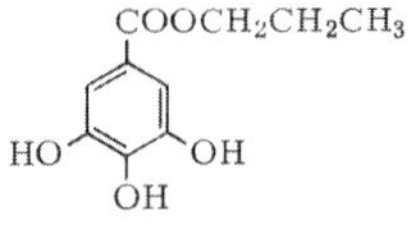

Use of BHT and BHA as direct food additives has been questioned as a result of published toxicity studies showing a higher than normal incidence of benign, precancerous, and cancerous tumors in test animals having been fed the antioxidants. However, detailed evaluations of the studies have always been resolved in favor of the continued use of both antioxidants for food. A number of studies have supported the safety of BHT for the stabilization of food [44].

5.2. Stabilization of Fuels

Hydrocarbon fuels, gasoline, jet fuel, and fuel oil, are subject to autooxidation. Autooxidation of fuels causes deposits that can foul carburetors, valves, pumps, and other precision equipment. Acidic oxidation products corrode critical metal parts in fuel systems. Hydroperoxide oxidation products are undesirable in gasoline because their rapid homolysis under combustion conditions leads to preignition, thereby lowering octane performance.

The addition of antioxidants to fuels effectively stabilizes them against autooxidation during storage. Both phenolic and aminic antioxidants are used, sometimes in combination. The main phenolic antioxidants for fuels are BHT (**5**); 2,4-dimethyl-6-*tert*-butylphenol, and 2,6-di*tert*-butylphenol. Among the aminic antioxidants, *N,N′*-di-*sec*-butyl-*p*-phenylenediamine has been of the greatest importance historically and continues to be used in fuels. However, other aminic antioxidants, some proprietary, including *p*-phenylenediamine with *N*-isopropyl or *N*-heptyl substituents, are also used.

Use levels of antioxidants for fuel stabilization are presently on the order of 0.5 – 1 kg per 1000 barrels of fuel (1 U.S. barrel = 159 L). This level is markedly lower than the 5 – 7 kg and more per 1000 barrels used in the past. The lower stabilizer requirement has come about because the composition of, and the production processes for fuels, especially gasoline, have increased the oxidative stability of base fuels.

5.3. Stabilization of Lubricants

Lubricating oils usually operate at relatively high temperature in gas turbines, steam turbines, jet engines, and crankcases of internal combustion engines. The performance requirements of stabilizers for these conditions are very demanding. Antioxidants minimize the deterioration of these lubricants by retarding viscosity increase, metal corrosion (usually controlled in conjunction with a corrosion inhibitor), and formation of acid, sludge, resins, and lacquers.

Mineral oil lubricants (gas turbines, steam turbines) are stabilized with 0.5 – 1.0 % of zinc dialkyldithiocarbamates in combination with 0.5 – 1.0 % of a phenolic antioxidant (BHT or 2,6-di-*tert*-butylphenol) or with an aminic antioxidant (octylated or nonylated diphenylamine, or phenyl α-naphthylamine or its derivatives).

Mineral crankcase oils may contain 0.5 – 1.0 % zinc dialkyldithiophosphates, which function as radical-trapping and peroxide-decomposing antioxidants in addition to their antiwear activity. These additives often are supplemented with sulfurized olefins (obtained by reacting sulfur with olefins), and aminic and phenolic antioxidants. The trend in the United States to unleaded gasolines and catalytic convertors has put pressure on the lubricant manufacturer to remove phosphorus-containing additives from crankcase oils, but so far with little success.

Mineral lubricants for diesel engines are stabilized with 0.5 – 1.0 % zinc diaryldithiophosphates, which are more thermally stable than the dialkyl derivatives and occasionally are supplemented with octylated or nonylated phenyl-α-naphthylamine.

Synthetic lubricants, both polyester and polyolefin types, are stabilized with 0.5 – 1.0 % alkylated phenyl-α-naphthylamine and/or alkylated diphenylamine.

5.4. Stabilization of Polymers

Virtually all commercial organic polymers, including elastomers, thermoplastics, synthetic fibers, and adhesives, are susceptible to oxidative degradation during processing and end use, with resulting reduction in physical properties and discoloration. Most polymers would not be useful without the small amounts of stabilizing additives, antioxidants, and processing stabilizers that are added to permit them to survive high-temperature processing and other thermooxidative conditions.

Antioxidant properties that influence performance include compatibility or solubility in the polymer or other substrate, volatility, discoloration resistance, and thermal stability. Antioxidants are selected on the basis of how important these properties are for the specific application and how well antioxidants meet them.

Additional factors in selecting antioxidants include the desired lifetime, the product thickness (samples of small cross section are more difficult to stabilize than thick samples); other components of the formulation, i.e., fillers, pigments, and flame retardants that may influence the effectiveness of the antioxidant, use conditions, FDA status, and cost performance. The last is a key consideration because it is not the selling price of the antioxidant that should be determinant but the cost of stabilization to get the job done.

5.4.1. Elastomers

Many different stabilization practices have evolved for polymers over the years. Elastomers, which commonly exhibit carbon – carbon unsaturation in the backbone, are particularly sensitive to oxidative degradation. They require stabilization during drying and storage in the uncompounded state as well as subsequently in the vulcanized forms. Both nonstaining and staining antioxidants are used. *Nonstaining, hindered phenolic antioxidants,* such as BHT (**5**), *phosphites*, such as tris(nonylphenyl phosphite) (**19**), and *nonstaining amines,* such as alkyl-substituted diphenylamines and alkyl-substituted phenyl-α-naphthylamines, are used when the final formulations are relatively light colored. However, because much of the production of elastomers is used in carbon-black-filled, vulcanized formulations, *staining antioxidants* may be used. These antioxidants contain dark-colored components capable of migrating into light-colored objects in contact with the vulcanized elastomer, causing objectionable discoloration or staining. Aryl- and alkyl-substituted *p*-phenylenediamine derivatives are commonly used as staining antioxidants for elastomers because they provide antioxidant as well as antiozonant activity. Other staining aminic antioxidants, such as aniline-acetone condensation products, also are used as antioxidants for elastomers.

Stabilization of elastomers against oxidation was historically among the earliest uses of polymer antioxidants [2]. Most were staining antioxidants could not be directly applied to other polymers. However, nonstaining antioxidants used in elastomers, hindered phenols, and phosphites particularly, and nonstaining amines, to a limited degree, are used also in other polymers.

5.4.2. Thermoplastics

Stabilization in the early history of thermoplastics revolved around hindered phenols, such as BHT (**5**); 2,2′-methylenebis(4-methyl- 6-*tert*-butylphenol) (**3**), 4,4′-methylenebis(2,6-di-*tert*-butylphenol), 4,4′-thiobis(2-*tert*-butyl-5-methylphenol) (**14**), and other relatively simple phenolic antioxidants. These antioxidants were generally adequate for the limited stabilization demands of early thermoplastics. However, stabilization requirements became more demanding with the advent of polystyrene containing unsaturated elastomeric constituents for impact modification and especially with the introduction of polyolefins from Ziegler-Natta catalysis in the mid-1950s. More effective antioxidants were needed for these demanding applications. Today, sophisticated, relatively high-molecular-mass antioxidants, stabilize the bulk of commercial polymers (see Table 3). Although BHT continues to be used for the stabilization of thermoplastics, its relatively high-vapor pressure limits its use to low-demand applications.

Hydrocarbon thermoplastic polymers, which are the main consumers of antioxidants, *polystyrenes, polyolefins,* and *thermoplastic elastomers* are stabilized with phenolic antioxidants, phosphites or phosphonites, and thiosynergists; the composition and concentrations depend on life stage and end use, as shown in Table 9. These stabilizers are used to retard the loss of physical properties and to minimize discoloration of polymers that accompany thermooxidative degradation. The stability problems associated with these polymers are increased by recycling scrap from processing operations. Thus, some of the polymers may be processed one or more times, depending on the amount of scrap recycled.

Table 9. Practices in polymer stabilization

Life stage of polymer	Stabilizer	Usual concentration, ppm
Drying	antioxidant	<250
Storage	antioxidant	<250
Compounding	antioxidant	500 – 1000
	phosphite	500 – 1000
Fabrication	antioxidant	500 – 1000
	phosphite	500 – 1000
End-use		
low stress	same as compounding and fabrication	
thermally stressing	antioxidant	1000 – 5000
	thiosynergist	1000 – 5000
	phosphite	500 – 1000
weathering	antioxidant	0 – 1000
	phosphite	500 – 1000
	hindered amine	0 – 10000
	ultraviolet absorber	0 – 10000

Polyoxymethylene polymers are sensitive to hydrolysis so that in addition to 0.1 – 0.5 % hindered phenolic anti oxidants, 0.1 – 0.5 % of an acid – formaldehyde scavenger, such as cyanoguanidine, calcium ricinoleate, or terpolyamide, is also used.

Polyamides discolor and their physical properties deteriorate as a result of thermal oxidation. Early in their history, polyamides were stabilized with copper halides and simple phosphites, sometimes even phosphorous acid. More recently, hindered phenols, notably *N,N′* – 1,6-hexamethylene-bis-3-(3,5-di-*tert*-butyl- 4-hydroxyphenyl)propionamide (**8**), have been used at 0.1 – 0.5 % to stabilize nylon 6 and nylon 66. Polyamides with longer aliphatic chains, such as nylon 12, are stabilized with phenolic antioxidants, such as tetrakis[methylene(3,5-di-*tert*butyl-4-hydroxyhydrocinnamate)]methane (**2**).

Several polymers, notably *polyesters, acrylics,* and *polycarbonates,* are relatively thermally stable and require only small amounts of phosphites or hindered phenolic antioxidants to reduce discoloration.

5.4.3. Synthetic Fibers

Synthetic fibers (polyamides, polyesters, and polyolefins) are stabilized in accordance with the practice for the corresponding thermoplastics, with the exception that the fibers are usually stabilized against photooxidation as well as thermooxidative degradation. Stabilization against photooxidation usually entails the use of hindered amines and ultraviolet absorbers as well as hindered phenolic antioxidants and phosphites.

5.4.4. Economic Aspects

The estimated consumption of antioxidants for plastics in the United States is given in Table 10 [45]. Table 11 shows the estimated consumption by plastic in the United States [46].

Table 10. Estimated consumption of antioxidants for plastics in the United States [45]

	$t \times 10^3$		
Antioxidant	1981	1982	1983
BHT	3.5	3.1	3.9
Other hindered phenols	5.6	5.2	7.3
Phosphites	3.2	2.9	4.5
Thioesters	2.7	2.4	3.1
Other	0.9	0.8	1.1
Total	15.9	14.4	19.9
	$\$ \times 10^6$		
Antioxidant	1981	1982	1983
BHT	12.8	10.2	15.9
Other hindered phenols	52.9	47.9	93.2
Phosphites	8.1	7.4	15.7
Thioesters	10.2	8.9	15.3
Other	4.0	3.4	6.7
Total	88.0	77.8	146.8

Table 11. Estimated consumption of antioxidants by plastic in the United States [46]

Plastic	Consumption, $t \times 10^3$		
	1981	1982	1983
ABS	4.6	3.7	4.4 (29 %)
Polyethylene	3.2	2.9	3.0 (20 %)
polypropylene	4.4	4.1	4.5 (29 %)
Polystyrene	2.2	1.9	2.0 (13 %)
Other	1.5	1.3	1.4 (9 %)
Total	15.9	13.9	15.3 (100 %)

6. Food Safety Considerations

The safety of antioxidants used as both direct and indirect food additives is monitored rigorously by food agencies in various countries, such as the FDA in the United States. In order to be permitted for use as stabilizers for food packaging, the safety of an indirect food additive needs to be demonstrated in subacute feeding studies with test animals, usually rats and dogs, and by migration studies with food-simulating solvents. The latter studies permit the estimation of levels of additives that may migrate into various types of foods from plastic packaging materials during filling, sterilization, and storage.

These data and the use profiles of the plastic packaging materials permit the calculation of *estimated daily intakes* (EDI) of the additive. Safety assessment is made by examining the EDI in light of the results of the feeding studies, especially the no-effect levels established by the feeding studies. For more details of the food safety and toxicologic aspects of antioxidants, see [47], → Food Additives.

7. Test Methods to Evaluate Polymer Thermal Oxidative Stability

Antioxidants are used selectively for specific polymers and for specific applications. Therefore, selecting the best antioxidant is dependent on the performance characteristics of the antioxidants, their limitations, the formulation methods, and the oxidation behavior of the organic material to be stabilized. Evaluations must be conducted on a small scale in a finite time to predict field results, sometimes of thirty or more years.

7.1. Melt Processing

Melt processing is an important operation in the preparation of polymer formulations and of molded objects. However, the temperatures and high-shear rates involved in this processing degrade many polymers. The efficacy of stabilizers in preventing the degradation is usually evaluated by repeated melt processing.

The polymeric material is extruded and injection or blow molded. Then the processed material is chopped and reprocessed anywhere from 2 to 20 times, depending on the polymer and the application. Changes in melt viscosity are usually monitored after specific numbers of processing cycles.

Generally, little or no change in melt viscosity on repeated processing is desired, and additives are evaluated relative to one another. In addition, the color of samples on repeated processing is of considerable importance. Additives that prevent discoloration and are not themselves discolored are especially desirable.

7.2. Oven Aging

Oven aging at elevated temperature is another important test method used in the plastics industry. By this method, oxidative degradation is accelerated and the results obtained in relatively short times. However, to minimize the risks of extrapolation, *temperature* for oven aging is selected as close to the end-use requirement as is feasible from a testing time point of view. Results from at least three test temperatures permit extrapolation of failure times in apparent Arrhenius plots to a use temperature at which a life span estimate is desired.

The two most important variables in oven aging are the temperature and *air velocity* in the oven. Tests carried out in ovens without air movement can give misleading results because additives of low-molecular mass do not volatilize in static air ovens to the degree that they do in actual use and to the degree that they do in ovens with significant air velocities. The measurement of oven air velocities is difficult because considerable variations exist depending on the location in an oven. One method of overcoming this difficulty involves continuous rotation of specimens within the oven during the test period.

7.3. Oxygen Uptake

Another useful method for studying accelerated thermal oxidative degradation of polymers is oxygen uptake. Polymer samples are enclosed in a system with an oxygen atmosphere, and the rate of oxygen consumption is measured. Oxygen uptake methods have the advantage of being relatively rapid but require more specialized equipment. In addition, the results from oxygen uptake investigations are more difficult to relate to end-use conditions because of the problem in extrapolating results in an oxygen atmosphere to normal atmospheric conditions.

7.4. Thermal Analytical Methods

Differential scanning colorimetry (DSC) and thermal gravimetric analysis (TGA) are useful methods that can provide information relatively quickly on the thermal stability of polymers. However, these measurements are usually carried out at temperatures above 180 °C, where many polymers are molten. Extrapolation of the results to performance in solid polymers at ambient temperature is questionable. On the other hand, DSC is very useful for quality control purposes once a formulation has been selected. Also TGA is particularly applicable in assessing the thermal properties of polymers, polymer formulations, and additives. Measurements by TGA are carried out using both programmed temperature increases and isothermal conditions. Temperatures at 1 %, 5 %, 10 %, and 50 % loss in sample mass, for example, are noted in the programmed temperature mode. In the isothermal mode, time to these mass losses are recorded.

7.5. Tests to Monitor Polymer Properties

Color and physical property measurements are the two most important ways of assessing performance of stabilizer compositions.

Color. Discoloration of plastic articles is usually undesirable. Color is generally monitored initially and after various exposure periods. Color may be assessed visually or measured instrumentally. Instruments generally read three parameters: L, the brightness of the specimen; a, the redness or greenness of the specimen; and b, the blueness or yellowness of the specimen; and tristimulus values X, Y, and Z. Various relationships of L, a, and b have been used to express specimen color. Yellowness index, a commonly used expression for relative color levels, is calculated for transparent samples from the tristimulus values:

$$\frac{1.28X-1.06Z}{Y}\times 100$$

Table 12. Specific test details for selected polymers

Polymer	Property	Test conditions
ABS	color color impact strength	repeated processing oven aging oven aging
Polyacetal	weight loss weight loss	TGA 200 – 250 °C oven aging
Polyamides	solution viscosity color tensile retention of fiber	oven aging oven aging oven aging
Polycarbonate	color impact strength	oven aging oven aging
Polyester	color impact strength	oven aging oven aging
Polyethylene, low density	M_r – melt index gloss and clarity of film	repeated processing film blowing
Polyethylene, low density, cross-linked	tensile strength and elongation	oven aging
Polyethylene high density	M_r – melt index physical integrity embrittlement	repeated processing oven aging to
polypropylene	M_r – melt index physical integrity	repeated processing oven aging to embrittlement
Polystyrene, impact	color color impact strength	repeated processing oven aging oven aging
Polyvinyl chloride	color	oven aging
Polybutadiene rubber	M_r – Mooney viscosity gel content	oven aging oven aging
Thermoplastic elastomers	color tensile and elongation	oven aging oven aging

Physical Properties. Among physical test methods commonly used for evaluating the oxidative stability of polymers are tensile strength and elongation, and impact strength. *Elongation* is a more useful measure of oxidative degradation than tensile strength. *Tensile strength* may not be markedly affected by sample embrittlement, whereas elongation is much more sensitive and can provide evidence of degradation more readily. However, because both measurements are made simultaneously, it is not necessary to choose one over the other prior to testing.

Impact strength is measured in different ways depending on the end-use performance of articles made from the polymers in question. Among the methods used are unnotched Izod, notched Izod, tensile-impact, chip impact, falling dart, and high-speed-impact testing. Each has disadvantages and no single ideal impact test method exists. However, high-speed-impact testing with instrumented tups has several key advantages and would very likely be more universally adopted were it not for the relatively high cost of equipment, in the neighborhood of $ 50000 – $ 60000.

Monitoring of molecular mass changes on oxidative degradation is also a frequently employed technique. The melt index determination (Section 7.1) is a standard technique in the plastics industry. Molecular mass is also monitored by solution viscosity measurements, less frequently by gel permeation chromatography, and, in the case of elastomers, by Mooney viscosity measurements.

In undergoing oxidation, hydrocarbon polymers develop carbonyl groups (i.e., ketones, aldehydes, acids, esters, etc.). The monitoring of carbonyl absorbance by infrared spectroscopy is a useful technique for following the course of polymer oxidation.

For the specific test used to assess various polymer properties, see Table 12.

8. References

1. A. W. von Hofmann, *J. Chem. Soc.* **13** (1861) 87.
2. A. Seyewetz, P. Sisley, *Bull. Soc. Chim. De France* **31** (1922) 672.
3. L. Bateman, *Q. Rev. Chem. Soc.* **8** (1954) 147.
4. J. L. Bolland, *Q. Rev. Chem. Soc.* **3** (1949) 1.
5. J. L. Bolland, *Proc. Royal Soc. London Ser. A* **186** (1946) 218.
6. G. L. Russell, *J. Am. Chem. Soc.* **79** (1957) 3871.
7. P. D. Bartlett, T. G. Traylor, *J. Am. Chem. Soc.* **85** (1963) 2407.
8. T. G. Traylor, *J. Am. Chem. Soc.* **85** (1963) 2411.
9. T. G. Traylor, C. A. Russell, *J. Am. Chem. Soc.* **87** (1965) 3698.
10. J. L. Bolland, P. Ten Have, *Trans. Faraday Soc.* **43** (1947) 201.
11. A. J. Kolka, J. P. Napolitano, A. H. Filbey, G. C. Ecke, *J. Org. Chem.* **22** (1957) 642.
12. T. J. Henman: *World Index of Polyolefin Stabilizers,* Kogan Page, London 1982.
13. A. F. Bickel, E. C. Kooyman, *J. Chem. Soc.* 1953, 3211.
14. A. F. Bickel, E. C. Kooyman, *J. Chem. Soc.* 1956, 2215.
15. J. Pospisil, *Adv. Polym. Sci.* **36** (1980) 69.
16. J. Pospisil in N. S. Allen (ed.): *Advances in Polymer Photochemistry-2,* Applied Science Publishers Ltd., Essex, England 1981, p. 53.
17. A. F. Bickel, E. C. Kooyman, *J. Chem. Soc.* 1957, 2217.
18. J. R. Thomas, C. A. Tolman, *J. Am. Chem. Soc.* **84** (1962) 2930.
19. I. T. Brownlie, K. U. Ingold, *Can. J. Chem.* **45** (1967) 2427.
20. J. R. Shelton in W. L. Hawkins (ed.): *Polymer Stabilization,* Wiley-Interscience, New York 1972, p. 84.
21. J. R. Shelton in G. Scott (ed.): *Developments in Polymer Stabilization-4,* Applied Science Publishers, Essex, England 1981, p. 23.
22. G. Scott in G. Scott (ed.): *Developments in Polymer Stabilization-6,* Applied Science Publishers, Essex, England 1983, p. 29.
23. S. Al-Malaika, K. Chakraborty, G. Scott in G. Scott (ed.): *Developments in Polymer Stabilization-6,* Applied Science Publishers, Essex, England 1983, p. 73.
24. G. W. Kennedy, W. L. Patterson, Jr., *Ind. Eng. Chem.* **48** (1956) 1917 – 1924.
25. H. Hock, S. Lang, *Ber. Dtsch. Chem. Ges.* **77 B** (1944) 257.
26. W. L. Hawkins, H. Sautter, *Chem. Ind.* 1962, 1825 – 1826.
27. J. R. Shelton, E. R. Harrington, *Rubber Chem. Technol.* **49** (1976) 147.
28. N. P. Neureiter, D. E. Bown, *Ind. Eng. Chem. Prod. Res. Dev.* **1** (1962) 236 – 241.
29. C. R. H. I. DeJonge, P. Hope in G. Scott (ed.): *Developments in Polymer Stabilization-3,* Applied Science Publishers Ltd., Essex, England, 1980, p. 21.
30. T. Colcough, J. I. Cunneen, *J. Chem. Soc.* 1964, 4790.
31. A. J. Burn, *Tetrahedron* **22** (1966) 2153 – 2161.
32. D. Barnard, L. Bateman, M. E. Cain, T. Colclough, J. I. Cuneen, *J. Chem. Soc.* 1961, 5339.
33. D. Rysavy, *Kunststoffe* **60** (1970) 120.
34. K. Schwetlick, *Pure Appl. Chem.* **55** (1983) 1629 – 1636.

35. K. J. Humphris, G. Scott, *J. Chem. Soc.* 1974, 617.
36. J. Holcik: *4th International Conference on Advances in the Stabilization and Controlled Degradation of Polymers,* Lucerne, Switzerland, June, 1982.
37. A. Tozzi, G. Cantatore, F. Masine, *Text. Res. J.* **48** (1978) 433.
38. CIBA-GEIGY, unpublished data.
39. K. Murayama, S. Morimura, Y. Toshioka, *Bull. Chem. Soc. Jpn.* **42** (1969) 1640.
40. Y. B. Shilov, F. T. Denisov, *Vysokomol. Soed. Ser. A* **16** (1974) 2313.
41. D. W. Grattan, A. H. Reddock, D. J. Carlsson, D. M. Wiles, *J. Polym. Sci., Polym. Lett. Ed.* **16** (1978) 143.
42. D. L. Allara, C. W. White, R. L. Meek, *J. Polym. Sci.* **14** (1976) 93.
43. H. Mueller in R. Gaechter, H. Mueller (ed.): *Plastic Additives,* Hanser-Verlag, München 1985, p. 75.
44. William H. Lederer: “Status of Butylated Hydroxylated Toluene,” International Life Sciences Institute, Washington, D.C., Sept., 1982.
45. R. Lacallade, *Chem. Eng. News* 1983 (October 24) 18.
46. *Mod. Plast.* **1983** (September) 1967.
47. R. Leimgruber in R. Gaechter, H. Mueller (ed.): *Plastics Additives,* Hanser-Verlag, München 1985, p. 685.

Antiseptics → Disinfectants

Disinfectants

HANS-P. HARKE, Schülke & Mayr, Hamburg/Norderstedt, Federal Republic of Germany

1. Introduction

Systematic pioneering studies in the 19th century, especially those of LOUIS PASTEUR (1822 – 1895) and ROBERT KOCH (1843 – 1910), showed that diseases are caused by microorganisms. Simultaneously, microorganisms were found to cause spoilage of food, cosmetics, drugs, and other materials. Preventing microorganisms from causing disease or destroying products led to the concept of disinfection.

Even prior to Koch's studies, IGNAZ PHILIP SEMMELWEIS, working in Vienna, Austria in 1847, dramatically reduced maternal mortality in his hospital when he ordered physicians to wash their hands with chlorinated lime prior to assisting in birth. Although misunderstood by many of his colleagues, SEMMELWEIS was successful in introducing hand disinfection without realizing the importance of bacteria [1], [2].

JOSEPH L. LISTER, who was influenced by Pasteur's studies, was equally successful in England in 1867, when he introduced the use of phenol solutions for treating skin wounds, thereby preventing severe infection [3].

Thereafter, successes followed rapidly. In 1888, PAUL F. FÜRBRINGER introduced the use of alcohol for disinfecting hands. The first proprietary disinfectant was Lysol (Schülke & Mayr), which appeared on the market in 1892 and proved its efficacy during the 1893 cholera epidemic in Hamburg, Germany. The disinfective properties of formaldehyde were recognized ca. 1900. This was followed in 1912 by the introduction of less toxic phenolic compounds, which superseded the use of cresols and tar oils for more delicate disinfection procedures.

Consequently, disinfectants and antiseptics rank high and they are becoming more and more important as protection against infections in the medical field and for production hygiene, particularly in food processing.

The discovery and development of antibiotics in the second half of this century seemed to have banned the threat posed by infectious diseases so that the significance of disinfectants rapidly declined for physicans and in public opinion. This development went the wrong way and has changed meanwhile.

1) All over the world nosocomical infections (i.e., in hospital acquired infectious diseases) play an increasingly important role not only regarding mortality and morbidity of patients but also in view of economy. The German Hospital Society reported 638 000 – 706 000 cases of nonsocomial infections in 1987 for Germany on average. These patients had to stay in hospitals 10 days longer than normal [4]. In the United States additional costs of US $ 2100 per nosocomical infection have been calculated [5]. In Germany it is estimated that nowadays of all patients being hospitalized for different reasons 1×10^6 suffer from nosocomical infections in addition to their original disease. The most important prophylactic measure is the compliance with hygienic and disinfection plans.

Ullmann's Industrial Toxicology

ISBN: 3-527-31247-1

2) Many well known pathogens which were supposed to have been overcome have developed a resistance against antibiotics. Infectious diseases caused by multiresistant microorganisms are difficult to treat. Their transmission can be restricted or inhibited by specific disinfection measures [6].
3) New infective agents are appearing or identified for the first time, e.g., *Helicobacter pyloris* causing gastritis or gastric carcinoma, Hepatitis-C-virus against which vaccination is still not possible, or *Escherichia coli* type EHEC (*E. coli* 0157:H7), causing the hemolytic uremic syndrom through food poisoning [7]. A careful hygiene and disinfection help to significantly reduce the danger emanating from such pathogens.

2. Definitions

The goal of *disinfection procedures* is to prevent transmission or spread of organisms that cause disease or spoilage. The following definition is frequently quoted in international publications: "Disinfection is the specific elimination of undesirable microorganisms for the purpose of preventing their transmission by interfering with their structure or metabolism irrespective of their functional state" [8]. This rather broad statement has been replaced by more manageable definitions, all of which are basically similar but which address only the medical aspect of this term. Disinfection has been defined as:

1) killing or inactivating all disease-causing organisms by chemical agents or physical procedures (Roche Lexikon, 1984) [9]
2) removing or destroying pathogenic microorganisms (Oxford Companions to Medicine, 1986) [10]
3) killing or irreversibly inactivating all causative agents of transmissible diseases [11].

In some circumstances, notably in the United States, the term "disinfectant" is customarily used only for active agents applied to inanimate objects. According to the EPA definition: "Disinfectants destroy or irreversibly inactivate infectious or other undesirable bacteria, pathogenic fungi, or viruses on surfaces or inanimate objects." Agents for application to living tissue are referred to as "antiseptics." In this article, "disinfectant" refers to active agents for both living tissue and inanimate objects.

Common to all definitions is the fact that disinfection is directed only against organisms that are considered harmful. The irreversible inactivation of the causative agent is considered to be essential by most definitions, except the British one, which also permits removal. Reber [8] is the only one to extend the definition of disinfection to include product protection.

Distinction should be made among disinfection, sterilization, and preservation. *Sterilization* refers to the complete destruction or irreversible inactivation of all microorganisms. *Preservation* is aimed at inhibiting multiplication of microorganisms in materials that are likely to spoil, with activity extending over a relatively long period.

Applying disinfectants means a careful selection of the range of microorgansims that will be inactivated by the corresponding preparation. The contained active agents as well as use concentrations and contact times are decisive.

In general it can be stated that disinfectants are formulated to inactivate bacteria and fungi. Efficacy against viruses, in particular against uncovered viruses, can be limited. Not all preparations inactivate bacterial spores, high concentrations of active agents are absolutely necessary.

3. Application

The aim of a disinfection measure will only be achieved if an appropriate disinfectant for the intended purpose is chosen observing the recommended use concentration and time and provided all parts of the body or the object to be treated come into contact with the preparation or its solution.

Typical application fields for disinfectants are in hospitals and other medical institutions in veterinary medicine; in many industrial areas for maintenance of a good production hygiene (e.g., in the pharmaceutical, food processing and cosmetics industry, in catering establishments, etc.); and in public facilities (swimming pools, prisons, barracks etc.).

Disinfection is important in agriculture for disinfection of barns used for intensive animal husbandry and disinfection of working surfaces and equipment used in greenhouses (to maintain a monoculture free of outside germs).

In public services, sanitation of drinking water and swimming pools must be considered.

In private households disinfection measures may be advisable for the care of seriously ill patients or for food processing.

Within the mentioned application fields specific disinfection procedures with different types of preparations are necessary for different purposes. What is meant by "different purposes" may be demonstrated in the field of medical applications.

1) Skin and mucous membrane disinfection of patients, e.g., before surgical treatment, injections, punctures, catheterization of the bladder, or for antiseptic wound treatment. With these measures a transmission of microorganisms in primary sterile areas will be avoided.
2) Hygienic hand disinfection to inactive microorganisms on the hands of medical staff to avoid the transmission of microorganisms from one patient to the other via the hands or from contaminated surfaces and equipment to patients via the hands.
3) Surgical hand disinfection is executed before surgical treatment by the operating team before gloves are put on. The endogenic mixed flora, too, must be inactivated in order to prevent microorganisms from entering the opened body during operation in case the gloves are defect.
4) Instrument disinfection by immersion of used instruments into disinfection solutions with the aim to protect the staff against infections before further handling (e.g., cleaning, sterilization, etc.). Patients are also protected if thermolabile material is handled (e.g., endoscopes).
5) Thermochemical instrument disinfection in (washing) machines at 50 – 60 °C instead of immersion into disinfection solutions.
6) Surface disinfection by rubbing and wiping the surfaces of treatment tables in consulting rooms, the surfaces of equipment for treatment and diagnosis, etc., or floors.
7) Laundry disinfection, disinfection of excrements and other special applications.

4. Efficacy Testing Methods, Requirements for Efficacy

Up to now many countries have their own test methods to evaluate a disinfectant. These methods may be found compiled in national guidelines for normalization or in guidelines published by scientific associations. An example are the test methods of the DGHM (German Society for Hygiene and Microbiology) [12]

Suspension tests and germ carrier tests are most important for testing disinfectants.

In *suspension tests* a suspension of the microorganisms is mixed with the disinfectant in a given test concentration. After the desired exposure time the effect of the disinfection is interrupted by suitable methods (dilution or chemical neutralization), surviving germs are counted and deducted from the amount of germs that remained without any influence of the disinfectant. From both figures the logarithmic germ count reduction is calculated. The suspension test also allows to determine the influence of disturbing substances such as blood, or albumin on the efficacy.

Germ carrier tests allow to simulate practice-relevance conditions. Microorganisms are brought onto germ carriers made of glass, metal, plastic, textiles, etc. After drying, the carriers are placed into the disinfection solution or they are wetted with the solution. The further procedure is similar to the suspension test.

For hand, skin or mucous membrane disinfectants tests on the corresponding epithelia are carried out. These tests have to be organized like chemical trials.

Since 1990 a European group of experts associated to CEN (European Committee for Standardization) in Brussels has been establishing European standards for testing disinfectants and antiseptics and an obligatory profile for tests and requirements. After tests in phase 1 where it is evaluated whether a formulation has disinfecting properties at all, the suitability for a particular application field is investigated in phase 2.

Corresponding standards have already been released for the application field hand disinfection in the medical area. Standing for all other test methods, they are introduced as follows: In a suspension test EN 12054 the bactericidal effect is investigated against a range of four bacterial strains in short exposure times. These bacteria represent all other bacteria. If the test requirements are met, the efficacy on hands is investigated in a test under practical conditions. In case the formulation is intended for hygienic hand disinfection, the hands of volunteers are heavily contaminated with *E. coli* and afterwards treated with the disinfectant or with 2-propanol (60 vol %) in a reference procedure (EN 1500). The obtained germ count reductions are compared. The test preparation must be at least as effective as 2-propanol in the reference procedure.

The efficacy of a formulation for surgical hand disinfection is tested on the normal hand, i.e., without special contamination and is compared with a reference procedure using 1-propanol (60 vol %) EN 12791.

For the application field hygienic hand disinfection two different methods have been accepted throughout Europe. In Germany, Austria, and in parts of Switzerland alcohol preparations are rubbed into the hands, in other countries aqueous washing preparations with a bactericidal effect are used. The aforementioned EN 1500 refers to such rubb-in preparations. EN 1499 is effective for washing preparations. In the reference procedure (EN 1499) potassium soap is used. The test preparation must have a better effect than the soap in a test under practice-relevance conditions on the hand. If necessary, efficacy tests against mycobacteria, fungi, or viruses must also be added. Corresponding standards are being worked out.

For all other application fields for disinfectants and antiseptics test methods and requirement profiles have also been issued as ENs or are being developed.

All valid standards can be obtained at DIN (in German language), BSI (in English language) or at AFNOR (in French language).

5. Formulations

Disinfectant formulations mostly contain a combination of active ingredients to satisfy user demands, and to fulfill the appropriate test procedures. In addition, additives must be incorporated to achieve good skin tolerance, materials compatibility, cleaning power, and other goals (see Table 1).

The specific spectrum of effectiveness of the active ingredients is discussed in Chapter 6. However, because closely related chemical substances commonly exhibit marked differences in potency, referring globally to aldehydes, phenols, etc. as disinfectants is inappropriate.

Currently, the preferred active ingredients are from the following classes of chemicals: aliphatic and aromatic alcohols, aldehydes, peroxy acids, phenols, cation-active compounds, amphoteric surfactants, alkylamines and inorganic oxidizing agents.

The benefits and risks associated with commercial products can only be assessed on the basis of test results for the complete preparation. Definitive conclusions are impossible if they are based solely on the properties of individual ingredients.

Examples of some preparations used for different applications are given in Table 2. Some preparations are used without further dilution, whereas others must be diluted prior to use. Many products contain active ingredients from several different chemical classes.

6. Active Ingredients

6.1. Alcohols

Ethanol [*64-17-5*] and the two propanols (1-propanol [*71-23-8*] and 2-propanol [*67-63-0*]) are essentially the only aliphatic alcohols used, even though they are effective only at high concentration. Their advantage lies in their water miscibility, rapid evaporation, and favorable pharmacologic and toxicologic properties [13]. Higher alcohols are more effective as antimicrobial agents, but their low volatility, unpleasant odor, and danger of resorption exclude their use.

The concentration of primary alcohol effective within 2 min against gram-negative and gram-positive bacteria, when tested in a suspension test, is as follows [14]:

Chain length	C_1	C_2	C_3	C_4	C_6	C_{12}
Concentration, %	60	50	20	10	5	15

Maximum effectiveness is reached with alcohols having six to eight carbon atoms. 2-Ethylhexanol [*104-76-7*] appears to be the most

Table 1. Common disinfectant additives

Additive	Disinfectant use [a]			
	Hand	Skin	Instrument	Surface
Germicides	+	+	+	+
Germistatics	+	+	–	(+)
Surfactants [b]	–	–	+	+
Wetting agents	+	+	–	–
Anticorrosives	–	–	+	+
Indicator dyes	–	+/–	–	(+)
Fragrances [c]	+	–	(+)	(+)
Solvents	+	+	+	+

[a] Key: + Useful; – Unnecessary; (+) Occasionally useful.
[b] Cleansing agents.
[c] Fragrances.

Table 2. Typical disinfectants for various applications

Application and trade names	Chemical class	Producer	Application form
Hand disinfection			
Desderman N	alcohol – phenol	Schülke & Mayr	undiluted
Sterillium	alcohol – cationic	Bode Chemie	undiluted
Skin disinfection			
Betaisodona solution	PVP * – iodine	Mundipharma	undiluted
Braunol 2000	PVP * – iodine	Braun Melsungen	undiluted
Cutasept	alcohol – cationic	Bode Chemie	undiluted
Kodan Tinktur forte	alcohol – phenol	Schülke & Mayr	undiluted
Spitaderm	alcohol – cationic	Henkel	undiluted
Mucous membrane antiseptics			
Betaisodona (range)	PVP * – iodine	Mundipharma	undiluted
Octenisept	Octenidine – aromatic alcohol	Schülke & Mayr	undiluted
Medical instrument disinfection			
Alhydex compact	aldehyde	Johnson & Johnson	undiluted
Desoform	aldehyde	Lysoform	diluted
Gigasept FF	aldehyde	Schülke & Mayr	diluted
Kohrsolin ID	aldehyde	Bode Chemie	diluted
Korsolex FF	aldehyde – amine	Bode Chemie	diluted
Lysetol AF	aromatic alcohol – cationic – alkylamine	Schülke & Mayr	diluted
Sekusept plus	alkylamine – aminoacid	Henkel	diluted
Surface disinfection			
Buraton 10 F	aldehyde	Schülke & Mayr	diluted
Dettol	phenol	Reckitt & Colman	diluted
Incidin perfekt	aldehyde – cationic	Henkel	diluted
Kohrsolin	aldehyde	Bode Chemie	diluted
Lysoformin	aldehyde	Lysoform	diluted
Perform	peroxy acid	Schülke & Mayr	diluted
Terralin	aromatic alcohol – cationic	Schülke & Mayr	diluted

* PVP = Polyvinylpyrollidone

effective. When combined with solvents and emulsifiers, a 0.2 % solution of 2-ethylhexanol kills gram-negative organisms within 15 min [15]. Despite its odor, these favorable properties have led to limited use of 2-ethylhexanol in combination with other agents in concentrated surface disinfectants that are diluted with water before use.

The most important application of short-chain alcohols is for disinfecting skin and hands. The alcohols are mostly combined with substances having germistatic activity. These alcohols are not suitable for disinfection of mucous membranes. Another application is for surfaces that must be disinfected rapidly, e.g., examination tables and working surfaces.

The maximum potency of ethanol is as a 70 % aqueous solution. For example, in a germ carrier test using gauze as the substrate, *Staphylococcus aureus* is killed by a 50 or 80 % solution within 10 min, but is killed within 5 min by a 60 – 75 % solution [16]. The alcohols are ineffective against spores of bacteria. Ethanol in concentrations below 75 – 80 %, 2-propanol, and 1-propanol are ineffective against unenveloped viruses like the polyvirus as a representative for enteroviruses. However, lipophilic viruses like the herpesvirus are rapidly and effectively inactivated by all three alcohols [17], [19]. The effectiveness of an ethanol-based hand disinfectant against polio virus within 10 – 60 s has been proven in a clinical trial. 2-propanol (60 vol %) in comparison was absolutely ineffective even after 60 s [18].

Among aromatic alcohols, benzyl alcohol [*100-51-6*], $C_6H_5CH_2OH$, phenethyl alcohol [*60-12-8*], $C_6H_5CH_2CH_2OH$, and phenoxyethanol [*122-99-6*], $C_6H_5OCH_2CH_2OH$, and the mixture of isometric phenoxypropanols (1-phenoxypropanol-2 [*770-35-4*] and 2-phenoxypropanol-1 [*4169-04-4*] find limited application, generally in combination with other active agents. These compounds have germicidal as well as germistatic activity (Table 3). They are more toxic than short-chain aliphatic alcohols [20]. An overview of the properties of alcohols as disinfectants has been published [21].

6.2. Aldehydes

Certain aldehydes have proved to be excellent agents for disinfecting instruments and surfaces. However, aldehydes are not used as skin disinfectants mainly because their local tolerance is unacceptable. Within the large group of aldehydes only very few have microbiocidal properties. For example, succinaldehyde and glutaraldehyde are extremely effective whereas adipaldehyde and malonaldehyde are ineffective. On the other hand the few active aldehydes are effective against a broad spectrum of infective agents provided the contact times and concentrations are chosen correctly.

The most commonly used aldehydes are formaldehyde [*50-00-0*], HCHO; succinaldehyde [*638-37-9*], $HCOCH_2CH_2CHO$; glutaraldehyde [*111-30-8*], $HCO(CH_2)_3CHO$; 2-ethylhexanal [*123-05-7*], $CH_3(CH_2)_3CH(CH_2CH_3)CHO$; and glyoxal [*107-22-2*], HCOCHO. As with alcohols, the most effective monoaldehydes have seven or eight carbon atoms, but their pungent odor limits their use. In this connection formaldehyde must be regarded differently. This might be due to its pronounced tendency to form oligomeres on the one hand and hydrates on the other hand. Thus, marked differences in efficacy are noticeable when the results of surface and suspension tests are compared [15]. In the suspension test, a 1 % solution of formaldehyde is effective in less than 30 min against a wide range of fungi and bacteria and, thus, is considerably less effective than, e.g., 2-ethylhexanal, a 0.1 % solution of which kills all microorganisms within 2.5 min. On the other hand, a 0.3 % solution of formaldehyde kills *Staphylococcus aureus* on surfaces within 4 h, whereas an equally strong solution of 2-ethylhexanal is inadequate even after contact for 6 h.

These results clearly demonstrate the advisability of combining active ingredients to formulate effective and acceptable preparations. This is especially important because the effects of combination preparations are frequently not only additive, but also synergistic. For example, a product (Buraton 10 F), which at its working strength contains 0.045 % formaldehyde, 0.01 % glutaraldehyde, 0.0025 % 2-ethylhexanal, and 0.07 % glyoxal, meets the DGHM specification for a surface disinfectant because it is effective within 1 h both in the suspension test and the surface test. Often aldehydes are combined with cationic compounds in commercial preparations. So the concentration of aldehydes can be lowered and the acceptability is increased (reduction of smell).

Among the dialdehydes, glyoxal was ineffective against the usual spectrum of microorganisms in the suspension test (30 min exposure, 1 % solution). On the other hand, a 0.2 % solution of succinaldehyde or glutaraldehyde was effective within 15 and 2.5 min, respectively (if the effect on *Candida albicans*, which required a longer contact time, is excluded). The same concentration was also effective within 1 h against *Staphylococcus aureus* on surfaces; for glyoxal this effect required only a 0.1 % solution. Therefore, glyoxal is useful as a surface disinfectant only when combined with other active ingredients.

Aldehydes are effective virucidal agents against lipophilic viruses and enteroviruses, such as the Coxsackie viruses or polioviruses. They are to some extend less effective against papovaviruses such as the SV 40 virus [22], [23].

Succinaldehyde and glutaraldehyde have a low tendency to be adsorbed onto plastic materials. This property, combined with their generally broad activity and excellent materials

Table 3. Effective concentration and contact time to kill germs with aromatic alcohols [15]

Aromatic alcohol	Concentration, %	Contact time, min			
		Escherichia coli	*Pseudomonas aeruginosa*	*Proteus mirabilis*	*Staphylococcus aureus*
Benzyl alcohol	1	>30	>30	>30	>30
Phenethyl alcohol	1.25	2.5	2.5	2.5	>30
	2.5	2.5	2.5	2.5	5
Phenoxyethanol	1.25	15	2.5	2.5	>30
	2.5	2.5	2.5	2.5	>30

compatibility, has led to their use as agents capable of disinfecting (contact period 15 min) or almost sterilizing (contact period 10 h) delicate thermolabile instruments, e.g., endoscopes. The best known products are Gigasept/Gigasept FF (succinaldehyde – dimethoxytetrahydrofuran) and Alhydex – Cidex (glutaraldehyde). In U. S. journals such preparations are defined as "sterilizer". The treatment of an instrument with the preparation (immersion, rinsing, drying) does not guarantee sterility of the instrument. Therefore, this term is not used in Europe, and the term "disinfection" is preferred. A survey on the properties of glutaraldehyde is given in [24].

6.3. Peroxy and Peroxo Acids

Peroxycarboxylic acids and inorganic peroxo acids contain active oxygen and are extremely effective disinfectants. Because these compounds also act as strong oxidants, formulating stable, liquid preparations that contain these active ingredients combined with additives that have cleaning, skin protective, anticorrosive, or other properties is difficult.

Peroxycarboxylic acids may be prepared in solution from reactive carboxylic acid derivatives (e.g., acid chlorides, acid anhydrides, and possibly esters or amides) and excess H_2O_2. The direct reaction of carboxylic acids with H_2O_2 requires the presence of a strong mineral acid such as sulfuric acid. Reviews on peroxycarboxylic acids as disinfectants have been published [25], [26].

Solid reactive acid derivatives can be mixed with solid hydrogen peroxide sources (e.g., peroxycarbonates, peroxyborates, or similar substances). In this case, the peroxycarboxylic acid is not formed until water is added. Other additives may be included in such a formulation, because the final solution must remain stable for only a few hours. Such solutions also exhibit a broad spectrum of antimicrobial activity [27], [28].

Aliphatic Peroxycarboxylic Acids.

Peracetic acid [*79-21-0*], CH_3CO_3H, was described as a germicidal substance as early as 1902 [29].

In spite of its effectiveness at low concentration against all types of microorganisms (Table 4), peracetic acid has not found wider acceptance because of its corrosive properties, penetrating odor, the danger of explosive reactions, and its limited stability. There are also some discussions about possible toxicity of peracetic acid [30], [31]. Peracetic acid is used for disinfection of artificial kidneys, linen, for treating ion exchange resins or for use in food industry.

Monoperoxyglutaric acid [*3851-97-6*], $HO_2C(CH_2)_3CO_3H$, exhibits a broad spectrum of antimicrobial activity. The compound is odorless, essentially nonvolatile, and quite stable in solution [35].

Monoperoxysuccinic acid [*3504-13-0*], $HO_2CCH_2CH_2CO_3H$, is generated only during the preparation of the aqueous application solution as described previously. The compound is relatively unstable. It inactivates viruses not only by capsid alteration, but also by RNA fragmentation [36].

Peroxyformic acid and peroxypropionic acid resemble peracetic acid in their effectiveness, but both are difficult to apply and are used only occasionally.

Aromatic Peroxycarboxylic Acids. These compounds are frequently much more stable than the aliphatic acids. However, their low water solubility and limited stability in alcohols have prevented their practical application. The antimicrobial properties of some members of this class, e.g., peroxybenzoic acid, peroxyanisic

Table 4. Contact time to kill germs with four dilutions of peracetic acid [33], [34]

Microorganism	Contact time, min			
	0.5 %	0.05 %	0.005 %	0.0005 %
Bacillus subtilis (spores)	1	>30	>30	>30
Escherichia coli	1	1	1	10
Staphylococcus aureus	1	1	5	10
Trichophyton mentagrophytes	5	>30	>30	>30

acid, and chloroxyperbenzoic acid, have been described [37].

Monoperoxyphthalic acid is commercially available as the hydrated magnesium salt [38]. The compound is used in a surface disinfectant which is a white granulate.

Inorganic Peroxo Acids. Peroxomonosulfuric acid [*7722-86-3*], HO_3SOOH, and its alkali-metal salts have been found to be as effective as organic peroxy acids . These compounds also have good overall effectiveness against viruses [39]. Viruses including papovaviruses which are resistant to other disinfectants are deactivated. The absence of toxic vapors that might endanger personnel or patients is noteworthy. However, peroxomonosulfuric acid can oxidize chloride to chlorine. So, if the solutions are prepared with tap water which contains relatively high amounts of chloride, the chlorine smell may influence the user in favor of the compound.

6.4. Phenols

Phenol [*108-95-2*], C_6H_5OH, the simplest member of this class, was introduced as an antiseptic in 1867 by JOSEPH L. LISTER. Systematic studies were undertaken around the turn of the century to determine the effect of substitution on disinfectant properties [40], [41]. Alkyl substituents were found to increase efficacy, with those having 5 – 7 carbon atoms being the most effective. Substitution by chlorine or bromine exerted an even greater effect. These relationships have been discussed in detail [42].

In general, phenols are effective germicidal and germistatic agents against bacteria, fungi, and lipophilic viruses, but are not effective against spores and hydrophilic viruses. They are markedly more effective in a weakly acidic rather than in an alkaline environment. However, their water solubility under acidic conditions is low. Surfactants may impair the efficacy of phenols.

Phenols are frequently incorporated into commercial soap concentrates; the diluted solution is used to disinfect surfaces and instruments, and occasionally skin and hands. Phenol derivatives are also incorporated as the active germistatic agents in alcohol-based products.

Increased discussions on pharmacological, toxicological, and ecological properties of phenols have reduced use to relatively few, preferably nonhalogenated compounds. This is especially true for Middle European and Scandinavian countries, but not for English speaking countries. Polychlorinated phenols, such as pentachlorophenol and even hexachlorophene, are no longer used.

The efficacy and pharmacologic and chemical properties of the phenols have been described in detail [43].

Nonhalogenated Phenols. *2-Phenylphenol* [*90-43-7*], $C_6H_5C_6H_4OH$, M_r 170.21, (2-hydroxybiphenyl, 2-hydroxydiphenyl, *ortho*-phenylphenol) *mp* 55.5 – 57.5 °C, *bp* 280–284 °C, vapor pressure 70 Pa at 100 °C, solubility in water 0.07 g in 100 g, forms colorless to pale yellow crystalline flakes.

This substance is more effective against gram-negative bacteria and fungi than against gram-positive cocci (Table 5) [44]. No effect was observed on enteroviruses when exposed for 10 min at concentrations up to 12 % [45]. However, adenoviruses and vacciniaviruses were rapidly inactivated by 0.12 % solutions. A combination of 2 % 2-phenylphenol, 18 % 1-

Table 5. Germicidal concentration after 30 min contact with 2-phenylphenol

Microorganism *	Germicidal concentration, %		
	pH 4.5	pH 7	pH 8.5
Gram-negative bacteria	0.005 – 0.01		0.0125 – 0.025
Staphylococcus aureus	0.1		0.25
Candida albicans	0.01 – 0.0125	0.025	

* Initial concentration 10^9/mL.

propanol – 2-propanol, and soap was effective against hepatitis B viruses [46].

Based on its well-researched and generally favorable toxicologic properties, 2-phenylphenol is currently the most widely used phenol. It is approved in many countries for treating citrus fruit. The WHO has set the acceptable daily intake to 0.2 mg/kg and limited maximum daily intake to 1 mg/kg.

Thymol [*89-83-8*], 2-[$(CH_3)_2CH]C_6H_3$-5-$(CH_3)OH$, M_r 150.22, 2-isopropyl-5-methylphenol, *mp* 51.5 °C, *bp* 232 °C, vapor pressure 12.7 Pa at 40 °C, solubility in water 0.098 g in 100 g at 25 °C, forms colorless crystals.

Thymol is a component of many natural essential oils and, probably for this reason, has been incorporated as a component of antimicrobial mouth and throat antiseptics. Solutions containing 0.03 – 0.05 % are effective against fungi [47].

Eugenol [*97-53-0*], 4-$(H_2C{=}CHCH_2)C_6H_3$-2-$(OCH_3)OH$, M_r 164.20, 2-methoxy-4-(2-propenyl)phenol, *mp* – 9 °C, *bp* 225 °C, is a clear liquid.

Eugenol, also a natural substance, is quite effective against bacteria and fungi. Its tendency to form complexes with metal ions, its sensitivity to oxidation, and its intensive smell of cloves limit general use of this compound.

4-tert-Amylphenol [*80-46-6*], $C_2H_5C(CH_3)_2C_6H_4OH$, M_r 164.25, 4-*tert*-phenylphenol, 4-(2-methylbutyl)phenol *mp* 88–89 °C, *bp* 225 °C, is extremely effective against gram-positive bacteria and for this reason is still used in the United States in surface disinfectants.

Although they are quite effective, cresols and xylenols are no longer extensively used because of their toxic properties and strong odor. They were once used extensively in cresol – soap solutions as described in various pharmacopoeias.

Halogenated Phenols.

4-Chloro-3-methylphenol [*59-50-7*], $ClC_6H_3(CH_3)OH$, M_r 142.59, 4-chloro-*m*-cresol, *mp* 55.5 °C, *bp* 235 °C, vapor pressure 700 Pa at 100 °C, solubility in water 0.4 g in 100 g, is a colorless powder.

4-Chloro-3-methylphenol is currently an important chlorophenol; it is used as a commercial preservative as well as an active agent in disinfectants. In suspension tests, chlorocresol was found to be effective within 30 min against dermatophytes at concentrations of 0.005 – 0.01 % (pH 4.5) and at 0.0125 – 0.025 % against *Candida albicans*.

2-Benzyl-4-chlorophenol [*120-32-1*], $C_6H_5CH_2C_6H_3(Cl)OH$, M_r 218.69, Clorophene, *mp* 48.5 °C, *bp* 160 – 162 °C at 466 Pa, vapor pressure 13 Pa at 100 °C, solubility in water 0.007 g in 100 g, occurs as colorless crystals.

Chlorophenol is extremely effective. In the suspension test, a concentration of 0.025 % is adequate to kill all types of microorganisms within 30 min [44].

4-Chloro-3,5-dimethylphenol [*88-04-0*], $(CH_3)_2ClC_6H_2OH$, M_r 156.61, chloroxylenol, *mp* 115.5 °C, *bp* 246 °C, solubility in water 0.05 g in 100 g, occurs as colorless prisms or needles.

Chloroxylenol is used in combination with other phenols. It is more effective against gram-positive bacteria than against gram-negative bacteria [48].

Other halogenated phenols used as disinfectants include chlorothymol, 2-chloro-6-methyl-4-benzylphenol, 4-bromo-2,6-dimethylphenol,

2,3,4,5-tetrabromo-6-methylphenol, and 2,4,4′-trichloro-2′-hydroxydiphenyl ether (Triclosan). Triclosan is used as active agent in germicidal soap.

6.5. Cation-Active Compounds

Cation-active compounds were introduced as antimicrobial agents in 1935 by GERHARD D. DOMAGK. This class of disinfectants includes compounds of widely differing structures, such as quaternary ammonium compounds and guanidinium and pyridinium derivatives. The first generation of cation-active compounds showed effectiveness against gram-positive bacteria at low concentration, low activity against gram-negative bacteria, and ineffectiveness against mycobacteria. Continuous development and testing of new structures for activity have produced the relatively well-balanced compounds currently available. However, even such substances are inactive against mycobacteria.

Benzalkonium chloride [*68391-01-5*], $[C_6H_5CH_2N(CH_3)_2C_nH_{2n+1}]^+Cl^-$, is a mixture of alkyldimethylbenzylammonium chlorides. The alkyl chains have a length of 8 – 18 carbon atoms. According to the European Pharmacopeia 1997, the mean molecular mass should be ca. 354 [49].

The effect of alkyl chain length of benzalkonium chloride on gram-negative and gram-positive bacteria has been tested in a suspension test [50]. The effective concentration against *Staphylococcus aureus* and *Pseudomonas aeruginosa* is shown in Table 6. This test indicates that gram-negative bacteria are less susceptible. The most effective benzalkonium chlorides are compounds in which the alkyl residues have 12 or 14 carbon atoms.

Benzalkonium chloride is used as an active ingredient in surface disinfectants. Its lack of activity against mycobacteria would disallow its use as an instrument disinfectant.

Benzalkonium chloride has low oral toxicity [51]. Because it is effective in low concentration, this agent is frequently chosen as an ingredient for disinfectants to be used in the vicinity of food. However, benzalkonium chloride is inactivated, as are other cation-active compounds, by many anions, including anionic detergents.

Table 6. Concentration of benzalkonium chloride effective within 10 min against *Staphylococcus aureus* and *Pseudomonas aeruginosa*

Chain length	Concentration, ppm	
	Pseudomonas aeruginosa	*Staphylococcus aureus*
C_8	6000	3000
C_{10}	1200	450
C_{12}	120	45
C_{14}	40	15
C_{16}	200	30
C_{18}	1000	450

Didecyldimethylammonium chloride [*2390-68-3*], is an interesting alternative to benzalkonium chloride. Toxicological and ecotoxicological properties have been thoroughly investigated (Lonza AG) with favorable results.

Poly(hexamethylenebiguanide) hydrochloride [*32289-58-0*] is also used as a surface disinfectant and is alleged to be suitable for skin disinfection. This compound has a slow effect and does not meet the practical requirements for prophylactic antiseptics in this respect. Although it is somewhat less effective than benzalkonium chloride (Table 7), it is sometimes used instead of benzalkonium because it is less foam producing under use conditions.

$$\left[-HN-(C(=NH)NH)_2-(CH_2)_6- \right]_n \cdot HCl$$

Poly(hexamethylenebiguanide) hydrochloride

Cetylpyridinium chloride [*123-03-5*] is used for treatment of mucous membranes and as throat medicine. Furthermore the substance is applied as preservative due to its excellent bacteria inhibiting property. In this respect it is of some interest that gram-negative bacteria are less sensitive by a factor of 4 than gram-positive bacteria [53]. The substance does not inactivate polioviruses or adenoviruses [54].

$$n\text{-}C_{16}H_{33}-N^+C_5H_5 \quad Cl^-$$

Cetylpyridinium chloride

Chlorhexidine [*55-56-1*], a bisbiguanide, is a strong base that is only slightly soluble in water (solubility 0.008 % at 20 °C). It is used worldwide, especially in skin and hand disinfectants, generally in the form of the digluconate, as a bacteriostatic. A group of experts at the German BGA summarized the available results and cer-

Table 7. Required exposure time as determined in suspension test [52]

Microorganism	Exposure time, min			
	A *	A * plus 10 % serum	B **	B ** plus 10 % serum
Staphylococcus aureus	5	5	15	15
Pseudomonas aeruginosa	30	60	15	15
Proteus vulgaris	15	15	60	60
Candida albicans	5	15	15	30

* 0.05 % Benzalkonium chloride.
** 0.05 % Poly(hexamethylenebiguanide) hydrochloride.

tified that chlorhexidine has above all an excellent bacteriostatic efficacy against gram-positive bacteria. The minimum inhibiting concentration is ≥ 1 µg/L. Significantly higher concentrations are given for gram-positive bacteria and fungi (10 to > 73 µg/mL). In the presence of blood or protein the efficacy is reduced by a factor of 100 to 1000 [55].

$$Cl-C_6H_4-(HNC(=NH))_2-NH-(CH_2)_6-(HNC(=NH))_2-C_6H_4-Cl \cdot 2\,C_6H_{12}O_7$$

Chlorhexidine digluconate

4 % solution of chlorhexidine digluconate is frequently used in hand wash lotions. Application of such a lotion leads to ca. 90 % germ reduction, which can be increased to 98 % after 6 applications [56]. However, it does not meet specifications established nowadays in Europe for a hand disinfectant. Under the test conditions of the European Standard EN 1499 for the efficacy testing of hygienic wash preparations it turned out that a logarithmic reduction of 3.1 is reached with a 4 % chlorhexidine solution which is customary in trade. There is no significant difference in the efficacy between this solution and soap [57]. In a multicenter study performed in six European countries which included 28 hospitals it was observed that the rate of wound infection had not been influenced by total body washings with a chlorhexidine-containing detergent the night before operation. In this study more than 2800 patients were included [58]. This result may be regarded in connection with investigations showing that a 0.5 % chlorhexidine solution reduces the number of organisms by only ca. 60 % [59], [60].

Chlorhexidine has bacteriostatic effects at low concentration and, like other cation-active compounds, remains on the skin. These properties suggest a combination of chlorhexidine with alcohol. Preparations containing both agents act rapidly, and the presence of chlorhexidine assures a residual effect.

Chlorhexidine is ineffective against polioviruses and adenoviruses. The effectiveness of a 0.1 % solution against herpes viruses has not yet been established unequivocally [54].

Octenidine [*71251-02-0*], a new cation-active agent, is a bispyridine that was used for the first time in an approved disinfectant in 1987 [61], [62]. It may be even more suitable as a remanent active ingredient than chlorhexidine combined with alcohols because it is more effective and better balanced (Table 8). Residual octenidine, after skin disinfection with an alcohol-based, octenidine-containing preparation, was capable of inactivating micrococci introduced later [63].

$$n\text{-}C_8H_{17}-N=C_5H_4N-(CH_2)_{10}-NC_5H_4=N-C_8H_{17} \cdot 2\,HCl$$

Octenidine dihydrochloride

Octenidine has been applied as active agent in a mucous membrane antiseptic since 1990 [64], [65]. Today it is one of the preparations prevalently used in many European countries for the treatment of the urogenital tract, buccal cavity [66], and wounds. Octenidine is absorbed neither through the skin nor through mucous membranes nor via wounds and does not even pass the placental barrier [67].

Other cation-active compounds that are used include polymeric quaternary ammonium salts (Buckman Laboratories or Onyx Chemical Co.), and guanidine [68].

6.6. Amphoteric Compounds

Amphoteric compounds, which are derived from glycine by replacing one amino hydrogen atom with an alkylamine group, exhibit antimicrobial activity. Solutions containing such agents are commercially available under the trade name Tego (Goldschmidt, Essen, Federal Republic of Germany) and have the following basic formula:

$RNH(CH_2)_2NH(CH_2)_2NHCH_2COOH$

Solutions of 1 % are effective within a short period against gram-positive bacteria and to a certain extent against gram-negative bacteria as well [69]. Lipid viruses are also inactivated but hydrophilic viruses, such as the poliovirus, are not [70].

The efficacy of these amphoteric compounds is less affected by a high protein burden than other agents generally are. Therefore, such compounds are well suitable for use in the food industry to disinfect working surfaces and equipment. These compounds have surface-active properties similar to those of quaternary ammonium compounds. They have low oral toxicity [71], and toxicologically significant amounts are apparently not absorbed through the skin.

Table 8. Comparison between octenidine and chlorhexidine determined by the suspension test *

	Effective concentration, %	
	Octenidine dihydrochloride	Chlorhexidine digluconate
Staphylococcus aureus	0.025	>0.2
Escherichia coli	0.025	0.1
Proteus mirabilis	0.025	>0.2
Pseudomonas aeruginosa	0.025	>0.2
Candida albicans	0.01	0.025

* Exposure time 5 min.

6.7. Alkylamines

It has already been known for quite some time that selected alkylamines have antimicrobial properties. Their activity against sulfate-reducing bacteria made them of particular interest for the oil-producing industry. Cocospropylenediamine [*61791-63-7*], e.g., is applied in this area. The importance of these compounds in other fields of disinfection increased after a successful strengthening of their antimicrobial properties by further derivatization and/or improvement of their compatibility [72].

N,N-bis-(3-aminopropyl)laurylamine [*2372-82-9*] has a surprisingly high effect against mycobacteria. This compound can be used instead of aldehydes for formulating tuberculocidal preparations [73]. 1.5 % solutions are capable to kill a great deal of mycobacteria in the presence of 0.5 % bovine albumine.

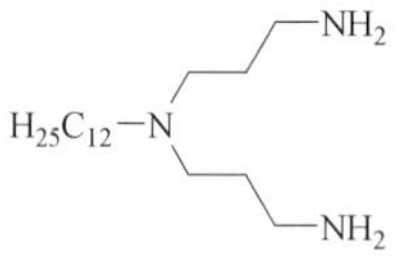

N,N-bis-(3-aminopropyl)laurylamine

Glukoprotamine [*2372-82-9*], a reaction product of cocospropylene-1,3-diamine and L-glutamine acid is present as a mixture of the amide and the corresponding ammonium salts [74]. Glukoprotamine has a well balanced microbiological effect and also kills mycobacteria. An 0.5 % solution is effective in a germ carrier test at pieces of cotton within 60 min against *Mycobacterium terrae*. Enteroviruses are not inactivated.

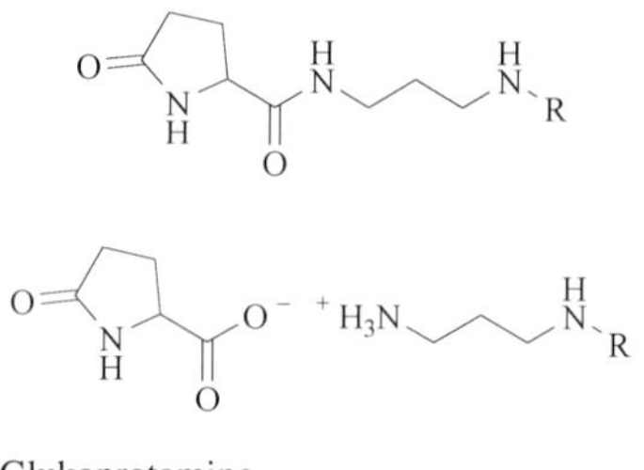

Glukoprotamine

The effectiveness of both compounds remains for cocospropylenediamine guanidinium diacetate [*67-63-0*], whereas the mycobactericidal effect is restricted.

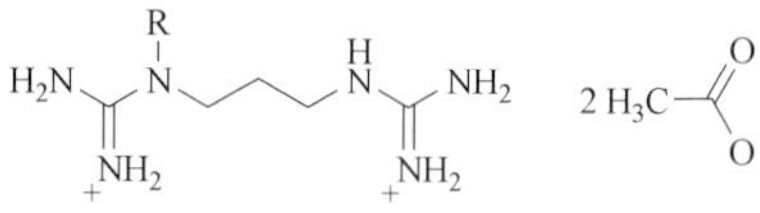

Cocospropylenediamine guanidinium diacetate

6.8. Inorganic Compounds

Chlorine. Chlorine [*7782-50-5*] is used in large quantities in some countries and has long been known as a disinfectant. The term chlorine is frequently used rather imprecisely. In the present context, the term refers generally to an aqueous solution of chlorine, hypochlorite, or hypochlorous acid. Occasionally, even chlorine-releasing compounds, such as *N*-chlorosulfonamides (e.g., Chloramin T) or *N*-chloroisocyanuric acid, and their salts are included in this group.

Aqueous solutions of chlorine and hypochlorite are extremely effective against a broad spectrum of microorganisms. Frequently, a concentration of < 1 ppm of available chlorine is sufficient to kill bacteria. Viruses are also inactivated under similar conditions [75]. Higher concentration is required to kill spores and mycobacteria.

Because chlorine is so chemically reactive, contaminants, especially organic substances, rapidly impair its effectiveness, which is one of the reasons that chlorine cannot be used universally for disinfection [76].

Other conditions may also significantly alter chlorine effectiveness. One example is pH. Spores of *Bacillus macerans* were reduced by 99 % in 8.5 min when 15 ppm chlorine was used at pH 6, but 42 min were required at pH 8 [77].

Chlorine is used mainly to disinfect drinking water, swimming pools, and food industry installations. Chlorine for household use or for wastewater treatment is questionable for ecological reasons, as the reaction between chlorine and organic material leads to the formation of undesirable organic chlorine compounds.

Iodine. Like chlorine, iodine [*7553-56-2*] has a broad spectrum of effectiveness. It was used mainly to disinfect wounds, skin, and mucous membranes. However, a number of adverse reactions made its use questionable and led to changes in its mode of application. Adverse reactions include local effects, primarily skin irritation, pain when applied to wounds, and the possibility of sensitization. In addition, systemic reactions, notably those affecting the thyroid gland, are known.

Since the mid-1950 PVP – iodine [*25655-41-8*], (providone – iodine), an iodine – polyvinylpyrrolidone complex with 10 % available iodine is being used instead of the old iodine tincture.

This change in the galenic of the iodine preparation had the effect that some of the above-mentioned disadvantages of the iodine tincture were significantly reduced, including sensitization, primary skin irritation, or pain.

Around 1980 it was observed that the PVP – iodine use solutions did not have the same efficacy as diluted test solutions [79], [78]. Very soon it was discovered that this phenomenon was due to the fact that aqueous solutions of 10 % PVP – iodine contain only a few ppm of free iodine [80]. The contents of free iodine increases when diluted with water and in consequence efficacy increases, too. However, at the same time the effectiveness decreases when the PVP – iodine solution is loaded with blood, protein, etc. to an inadmissible point because the contents of free available iodine decreases.

By optimizing the galenics it was possible to increase the level of free iodine even in 10 % PVP–iodine solutions to such a point that even undiluted the desired effect was obtained [81].

The use of PVP – iodine is counterindicated in pregnant women and in newborns because the thyriod gland of a fetus or a newborn is not fully functional.

Like chlorine, iodine kills vegetative bacteria rapidly and effectively in the ppm range [82]. Organic materials severely impair its activity. Spores are inactivated only at high concentration of free iodine and longer contact times are used. The use of PVP – iodine on skin for this purpose is questionable. In addition, PVP – iodine is ineffective against enteroviruses, but is active against lipophilic viruses [83], [84].

Acids and bases. Strong alkali-metal and alkaline-earth hydroxides are occasionally employed for disinfecting excretions or as disinfectants in animal husbandry. The concentration must be chosen carefully so that the hydrolytic reactions of microorganisms as well as of organic contaminants proceed at ambient temperature. The same applies to nonoxidizing inorganic acids like sulfurous acid or SO_2.

Metals. Metals, especially heavy metals, have been used for years because of the so-called oligodynamic effect; that is, they are toxic to bacteria at low concentration. The metals react with thiol (SH) or amine (NH) groups of enzymes or proteins, a mode of action somewhat different from that of the other active agents and one to which microorganisms may develop resistance. Such resistance may be transmitted by plasmids [85–87].

Only mercury and silver compounds are used currently, and then just for specific applications. For example, silver sulfadiazine [*22199-08-2*] is used as a prophylactic antiseptic ointment for extensive burns. An equilibrium dispersion of colloidal silver with dissolved silver ions can be used to purify drinking water at sea [88].

Phenylmercuric borate [*8017-88-7*] or acetate [*62-38-4*] were used for disinfecting mucous menbranes at an effective concentration of 0.07 % in aqueous solutions [89]. Due to toxicological and ecotoxicological reasons phenylmercury salts are no longer applied nowadays. Nevertheless, surgeons use mercurochrom [*129-16-8*] even today and despite toxicological objections. The benefit, however, may be more an adstringent effect than an antimicrobial one.

7. Toxicology

When the toxicology of disinfectants is evaluated, their widely varying methods of application must be considered. Important factors include the final use concentration, whether direct body contact is intended, and the volatility of the substance with the accompanying danger of inhalation. Because coverage of the many possible conditions would be difficult, the present discussion is restricted to a few general comments, and references to toxicologic studies of the active agents.

Short-chain alcohols are considered safe, provided they are properly applied to skin or small areas. They are not absorbed by the skin. *Phenols* used in disinfectants have commonly been well investigated with regard to their pharmacological toxicological properties. For 2-phenylphenol [*90-43-7*] cancerogenicity and teratogenicity tests are available [90]. Even in the applied high dosages neither cancerogenic nor teratogenic effects were observed. These dosages are significantly and by more than a factor 1000 higher than those amount of substances being applied for hand disinfection (Desderman N). Noteworthy is the fact that phenols are readily absorbed through the skin.

Aldehydes may cause local irritation and may elicit sensitization. A formaldehyde concentration as low as 1 ppm in air irritates the upper respiratory tract and the mucous membrane of the eye. This may be an advantageous warning that the facility needs to be ventilated. When used correctly as a surface disinfectant for washing and scrubbing floors, the maximum concentration attained for a short time is just above 0.1 ppm [91]. Concentrations of formaldehyde that cause mucous membrane carcinomas in animals are not likely be reached when appropriate disinfectants are used properly [92], [93].

Modern preparations containing *active oxygen* are characterized by low volatility. They are a genuine improvement over aldehyde-based preparations, especially considering their broad range of activity. Their acute oral toxicity is low and they cause only mild local irritation.

Cation-active and amphoteric compounds are not readily absorbed through the skin or the mucous membrane. These compounds cause local irritation and are extremely poisonous when administered parenterally [51]. Specific toxic effects have been observed occasionally as, for example, effects of chlorhexidine on hearing and balance [94].

With *iodine*, the systemic effect and the effect on thyroid function are important. With *metals*, central nervous system involvement may be a consideration.

In summary, the toxic properties of the products must be made known by the manufacturers, and proper precautions must be taken by the consumers. If directions for use are followed, most preparations can be handled without presenting any danger.

8. References

1. S. W. B Newsom, *J. Hospital Infect.* **23** (1993),175 – 187.
2. M. L. Rotter, *Hyg. Med.* **22** (1997) 332 – 339.
3. N. V. Nussbaum, *Ärztliches Intelligenzblatt* **5** (1875)43 – 47.
4. H. Rüden, F. Daschner, M. Schumacher in Bundesministerium für Gesundheit (ed.): *Nosokominale Infectionen in Deutschland,*

vol. 56, Nomos Verlag, Baden-Baden 1995,pp. 13 – 14.

5. W. J. Martone, W. R. Jarvis, D. H. Culver, R. W. Haley: "Incidence and nature of endemic and epidemic nosocomical infections," in J. V. Benett, P. S. Brachman (eds.) *Hospital Infection,* Little Brown and Co., Boston 1992, pp. 577 – 596.
6. M. Kresken, A. Jansen, B. Wiedemann, *J. Antimicrob. Chemother.* **25** (1990) 1022 – 1024.
7. D. Bitter-Suermann, M. Exner, K.-O. Gundermann, H.-P. Harke: *The Threat Posed by Infectious Diseases,"* Rudolph Schülke Stiftung, mhp-Verlag, Wiesbaden 1996.
8. H. Reber, *Zentralbl. Bakteriol. Mikrobiol. Hyg., Abt. 1 Orig. B* **157** (1973) 421 – 438.
9. Hoffmann-La Roche (eds.): *Roche Lexikon Medizin,* Urban & Schwarzenberg, München-Wien-Baltimore 1984, p. 340.
10. J. Walton, P. B. Beeson, R. B. Scott (eds.): *The Oxford Companion to Medicine,* Oxford Univ. Press, Oxford-New York 1986, p. 314.
11. J. Borneff: *Hygiene,* 4th ed., Georg Thieme Verlag, Stuttgart-New York 1982, p. 416.
12. Desinfektionsmittelkommission der DGHM (ed): *Prüfung und Bewertung chemischer Desinfektionsverfahren,* Status 12., July 1991, mhp-Verlag, Wiesbaden 1991.
13. H.-P. Harke, *Hyg. Med.* **3** (1978) 76 – 80.
14. H. Eggensperger, *Hosp.-Hyg., Gesundheitswes. Desinfekt.* **67** (1975) 321 – 329.
15. H. Eggensperger, *Hosp.-Hyg., Gesundheitswes. Desinfekt.* **68** (1976) 39 – 55.
16. C. Harrington, H. Walker, *Boston Med. Surg. J.* **148** (1903) 548 – 552.
17. H. J. Eggers, *Zbl. Bakt.* **273** (1990) 36 – 51.
18. A. Böse-Eggert, J. Steinmann, Evaluation of Desderman N against Polivirus Type 1 (Sabin) using Fingerpads of Adult Volunteers, Report Bremen 1997.
19. W. Schürmann, H. J. Eggers, *Antiviral Res.* **3** (1983) 25 – 41.
20. H.-P. Harke, *Hyg. Med.* **4** (1979) 401 – 403.
21. E. L. Larson, H. E. Morton in S. S. Block (ed.): *Disinfection, Sterilization and Preservation,* 4th ed., Lea & Febiger, Philadelphia 1991, pp. 191 – 203.
22. O. Thraenhart, E. Kuwert, *Zentralbl. Bakteriol. Mikrobiol. Hyg., Abt. 1 Orig. B* **161** (1975) 209 – 232.
23. O. Thraenhart, E. Kuwert, *Zentralbl. Bakteriol. Mikrobiol. Hyg., Abt. 1 Orig. B* **164** (1977) 22 – 44.
24. E. M. Scott, S. P. Gorman in S. S. Block (ed.): *Disinfection, Sterilization and Preservation,* 4th ed., Lea & Febiger, Philadelphia 1991, pp. 596 – 614.
25. B. Tichacek, V. Merka, A. Kramer in W. Weuffen, G. Berencsi, D. Gröschel, B. Kemter et al. (eds.): *Handbuch der Antiseptik,* **vol. 2,** part 2, G. Fischer Verlag, Stuttgart-New York 1983, pp. 160 – 176.
26. A. Kramer, W. Weuffen, V. Merka, B. Tichacek in [25], pp. 177 – 187.
27. H. Eggensperger, *Dtsch. Apoth.-Ztg.* **118** (1978) 1073 – 1076.
28. H. Eggensperger, *Dtsch. Apoth.-Ztg.* **122** (1982) 2599 – 2602.
29. P. C. Freer, F. G. Novy, *Am. Chem. J.* **27** (1902) 161.
30. A. Kramer et al., *Hyg. Med.* **16** (1991) 279 – 286.
31. *Bundesgesundheitsblatt* 1997, 72 – 74.
32. W. Weuffen, A. Kramer, V. Adrien, *Hyg. Med.* (1987) in press.
33. V. Merka, F. Sita, V. Zikes, *J. Hyg. Epidemiol., Microbiol., Immunol. (Prague)* **9** (1965) 220 – 226.
34. V. Merka, J. Dvorak, *J. Hyg. Epidemiol., Microbiol., Immunol. (Prague)* **12** (1968) 115 – 121.
35. Schülke & Mayr, DE 2 654 164, 1976 (H. Eggensperger, W. Beilfuss).
36. J. Sporkenbach-Höffler, K. J. Wiegers, R. Dernick, *Zentralbl. Bakteriol. Mikrobiol. Hyg., Abt. 1 Orig. B* **177** (1983) 469 – 481.
37. Schülke & Mayr, DE-OS 2 653 738, 1976 (H. Eggensperger, W. Beilfuss, W. Zerling).
38. Interox Chemicals Ltd., Technisches Merkblatt,Cheshire, UK.
39. Schülke & Mayr, DE 3 046 769, 1980 (J. Sporkenbach, H. Eggensperger, L. Bücklers, H. H. Ehlers et al.).
40. H. Bechthold, P. Ehrlich, *Z. Physiol. Chem.* **47** (1906) 173 – 199.
41. K. Laubenheimer: *Phenol und seine Derivate als Desinfektionsmittel,* Urban & Schwarzenberg, Berlin 1909.
42. D. O. O'Connor, J. R. Rubino in S. S. Block (ed.): *Disinfection, Sterilization and Preservation,* 4th ed., Lea & Febiger, Philadelphia 1991, pp. 204 – 224.
43. W. Beilfuss, L. Bücklers, U. Eigener, H.-P. Harke et al., [25], part 2.
44. W. Beilfuss, H. Nolte, Personal communication, 1987.
45. M. Klein, A. Deforest, *Soap Chem. Spec.* **39** (1963) 70 – 97.

46. H.-P. Harke, *GIT Labor-Med.* **5** (1982) 35 – 36.
47. K. H. Wallhäusser: *Sterilisation Desinfektion Konservierung,* G. Thieme Verlag, Stuttgart 1978, p. 348.
48. W. B. Hugo, A. D. Russel: *Pharmaceutical Microbiology,* Blackwell Scientific Publications, Oxford 1977, pp. 163 – 169.
49. *European Pharmacopeia,* 3rd. ed., Council of Europe, Strasbourg; official german edition: Deutscher Apotheker Verlag, Stuttgart and Govi Verlag, Eschborn 1997.
50. R. A. Cutler, E. B. Cimijotti, T. J. Okolowich, W. F. Wetterau: *Chemical Specialties Manufacturers Association (CSMA) Proceedings of the 53rd Annual Meeting,* 1966, pp. 102 – 113.
51. R. A. Cutler, H. P. Drobeck in E. Jungermann (ed.): *Cationic Surfactants,* Marcel Dekker, New York 1970.
52. W. Beilfuss, Personal communication, 1986.
53. K.-H. Wallhäusser: *Praxis der Sterilisation Desinfektion Konservierung,* G. Thieme Verlag, Stuttgart–New York 1995,p. 597.
54. B. Eckhoff, I. Grell-Büchtmann, J. Steinmann, *Hyg. Med.* **11** (1986) 328 – 330.
55. *Bundesanzeiger no. 159,* 28.08.1994, p. 9122.
56. K. Askgaard, *Ugeskr. Laeg.* **137** (1975) 2515 – 2518.
57. M. L. Rotter: "Hand Washing and Hand Disinfection", in: C. G. Mayhall (ed.): *Hospital Epidemiology and Infection Control,* Williams & Wilkins, Baltimore 1996, pp. 1052 – 1068.
58. M. L. Rotter et al., *J. Hosp. Infect* **11** (1988)310 – 320.
59. E. J. L. Lowbury, H. A. Lilly, *Br. J. Surg.* **61** (1974) 19.
60. G. A. J. Ayliffe, J. R. Babb, K. Bridges, *J. Hyg.* **75** (1975) 259 – 274.
61. D. M. Bailey, C. G. DeGrazia, S. J. Hoff, P. L. Schulenberg et. al., *J. Med. Chem.* **27** (1984) 1457 – 1464.
62. H.-P. Harke, *Zbl. Hyg.* **188** (1989) 188 – 193.
63. U. Eigener, U. Behrens, *Hyg. Med.* **10** (1985) 475 – 478.
64. G. Wewalka et al., *Hyg. Med.* **16** (1991) 335 – 345.
65. V. Hanf, M. Dominguez, P. Heeg, H. R. Tinneberg, *Fertilität* **7** (1991)140 – 143.
66. A. Kramer et al., *Zent.bl. Hyg. Umweltmed.* **200** (1998)443 – 456.
67. E. R. Weissenbacher et al., *Int. J. Experimt. Clinical Res.* **10** (1997) Supl. 1.
68. K. H. Wallhäusser: *Praxis der Sterilisation Desinfektion Konservierung,* 5th ed., Georg Thieme Verlag, Stuttgart–New York 1995, pp. 598–604.
69. S. S. Block in S. S. Block (ed.): *Disinfection Sterilization and Preservation,* 4th ed., Lea & Febiger, Philadelphia–London 1991 pp. 263 – 273.
70. P. G. Micheletti, R. Ponti, C. Cantoni, *Arch. Lebensmittelhyg.* **29** (1978) 94 – 96.
71. H. Edelmeyer, A. Laqua, M. Wiemann, *Arch. Lebensmittelhyg.* **29** (1978) 62 – 65.
72. H. W. Rossmoore in S. S. Block (ed.): *Disinfection Sterilization and Preservation,* 4th ed., Lea & Febiger, Philadelphia–London 1991, pp. 290 – 321.
73. Lonza AG, EP 0 343 605 B 1, 1992 (S. Güller, J. Fritschi, F. Lichtenberg).
74. K. Disch, "Glucoprotamin – ein neuer antimikrobieller Wirkstoff", *Hyg. Med.* **17** (1992) 529 – 534.
75. N. A. Clarke, S. L. Chang, *J. Am. Water Works Assoc.* **51** (1959) 1299 – 1317.
76. J. Peters, *Bundesgesundheitsblatt* (1987) in press.
77. W. A. Mercer, I. I. Sommer, *Adv. Food Res.* **7** (1957) 129 – 160.
78. H.-P. Werner, *Hyg. Med.* **7** (1982) 205 – 212.
79. K. O. Gundermann, H.-P. Harke, D. Horn, G. Koppensteiner et al., *Hyg. Med.* **8** (1983) 175 – 178.
80. D. Horn, W. Ditter in G. Hierholzer, G. Görz (eds.): *PVP Jod in der operativen Medizin,* Springer Verlag, Berlin-Heidelberg-New York-Tokyo 1984, pp. 7 – 19.
81. E. Pinter, H. Rackur, R. Schubert, *Pharm. Ind.* **46** (1984) 640 – 645.
82. W. Gotthardi in S. S. Block (ed.): *Disinfection Sterilization and Preservation,* 4th ed., Lea & Febiger, Philadelphia–London 1991 152 – 166.
83. J. Sporkenbach, *Hyg. Med.* **5** (1980) 357 – 362.
84. C. H. Carter, *Proc. Soc. Exp. Biol. Med.* **165** (1980) 380.
85. E. J. L. Lowbury, J. R. Babb, K. Bridges, D. M. Jackson, *Br. Med. J.* **1** (1976) 493 – 496.
86. R. J. Pinney, *J. Pharm. Pharmacol.* **30** (1978) 228 –232.
87. A. O. Summer, S. Silver, *J. Bacteriol.* **112** (1972) 1228 – 1236.
88. G. E. Pierce, J. J. Brimm, E. E. Reiber, J. H. Litchfield, *Dev. Ind. Microbiol.* **20** (1979) 455 – 461.
89. A. Cremieux, *Ann. Pharm. Fr.* **37** (1979) 165 – 174.

90. National Institute of Health: *Toxicology and Carcinogenesis Studies of ORTHO-Phenylphenol,* NIH Publication 86-2557, 1986.
91. U. Knecht, H.-J. Woitowitz, *Oeff. Gesundheitswes.* **41** (1979) 715 – 723.
92. W. D. Kerns, K. L. Pavkov, D. J. Donofrio, E. J. Gralla et al., *Cancer Res.* **43** (1983) 4382.
93. J. K. McLaughlin, *Int. Arch. Occup. Envion Health* **66** (1994) 295 – 301.
94. T. Morizono, B. M. Johnstone, E. Hadjar, *J. Oto-Laryngol. Soc. Aust.* **3** (1983) 550.

Nucleic Acids

HELMUT BURTSCHER, Boehringer Mannheim GmbH, Penzberg, Federal Republic of Germany (Chaps. 1–6)

SIBYLLE BERNER, Boehringer Mannheim GmbH, Tutzing, Federal Republic of Germany (Chap. 7)

RUDOLF SEIBL, Boehringer Mannheim GmbH, Penzberg, Federal Republic of Germany (Chap. 8)

KLAUS MÜHLEGGER, Boehringer Mannheim GmbH, Tutzing, Federal Republic of Germany (Chap. 9)

1. Introduction

Nucleic acids are high molecular mass compounds found in all living cells and viruses. Their name originates from their discovery in the nuclei of eucaryotic cells. They can be chemically degraded to yield phosphoric acid, pentoses, and nitrogen-containing heterocycles (bases). Nucleic acids can be divided into two main classes depending on the sugar they contain: *deoxyribonucleic acids* (DNA) contain 2-deoxy-D-ribose and *ribonucleic acids* (RNA) contain D-ribose.

Nucleic acids are long, unbranched chains of sugar and phosphate (Fig. 1, see next page): the C-3′ atom of each sugar is linked by a phosphodiester bond to the C-5′ atom of the neighboring sugar. Either a purine (adenine, guanine) or a pyrimidine (cytosine and thymine in DNA; cytosine and uracil in RNA) is attached to C-1′ of the sugar by a β-glycosidic bond. For a detailed description of purines and pyrimidines, see → Purine Derivatives, → Pyrimidine and Pyrimidine Derivatives.

Although nucleic acids have been known since the second half of the nineteenth century it was only in the 1940s that their importance as the carrier of genetic information became clear. Genetic engineering and improved physical and biochemical methods of analysis have led to enormous progress in the understanding of the structure of DNA, DNA–protein interactions, and gene organization, expression, regulation, and transfer. The importance of nucleic acids became even more obvious after the discovery that they can have other functions in addition to their ability to store and transfer genetic information. It is widely assumed that in the course

Ullmann's Industrial Toxicology

ISBN: 3-527-31247-1

of evolution first RNA and then DNA came into being [36,37].

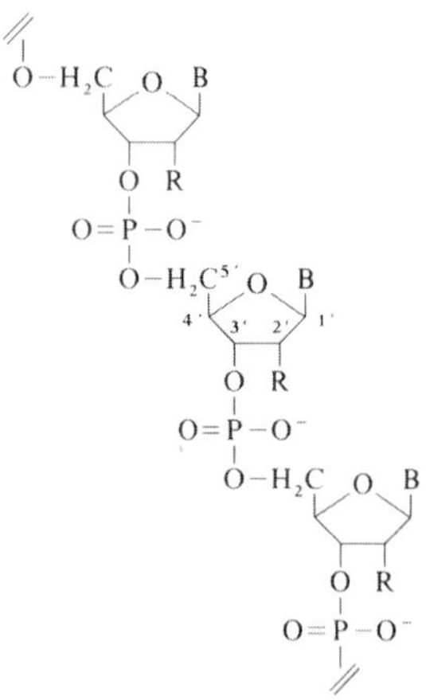

Figure 1. Structure of DNA (R = H) and RNA (R = OH)
B = base (adenine, guanine, thymine or uracil, cytosine)

2. Structure

2.1. Structure of DNA

The joining of the DNA building blocks by 5′- and 3′-phosphodiester bonds gives the molecule polarity (Fig. 1); base sequences are always written starting with the 5′-terminus, i.e., in the $5' \rightarrow 3'$ direction. The specific base sequence of DNA and its ability to form double-stranded structures according to precisely defined rules are of utmost importance for the storage of genetic information and for interactions with other nucleic acids and proteins.

From X-ray analysis data, CRICK and WATSON proposed a double-stranded structure for DNA in 1953 in which two antiparallel (i.e., $5' \rightarrow 3'$ and $3' \rightarrow 5'$) polynucleotide chains form a right-handed helix (i.e., looking along the axis of the helix, the strands are coiled clockwise). Naturally occurring DNA usually consists of right-handed helices with a major and a minor groove (Fig. 2). The hydrophobic bases are located inside the helix and the sugar – phosphate "backbone" on the outside [38]. Bases that are opposite each other are paired according to defined rules as a result of hydrogen bond formation: adenine always pairs with thymine or uracil and guanine with cytosine. Complementary bases can be bound by the more common Watson – Crick pairing (Fig. 3 A) or by Hoogsteen base pairing (Fig. 3 B). The double-stranded structure is further stabilized by hydration of the phosphate groups and hydrophobic interactions between the aromatic ring systems that result in stacking of the bases.

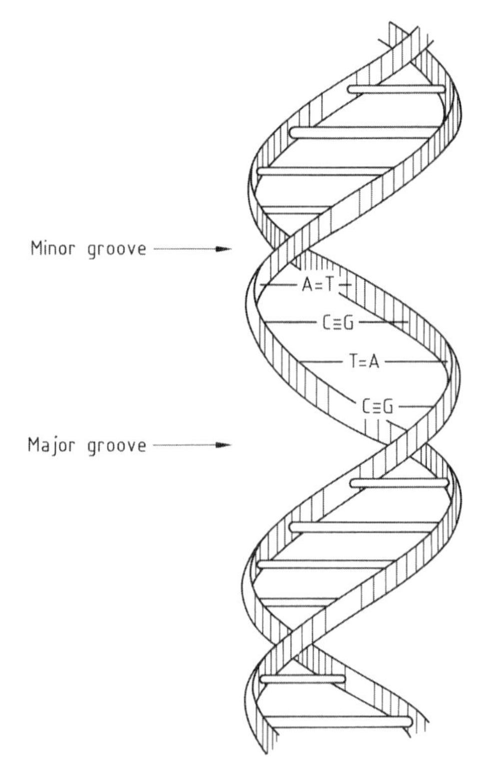

Figure 2. Right-handed double-helix of DNA
A = adenine, C = cytosine, G = guanine, T = thymine

In double-stranded DNA, the bases are densely stacked and there is a cooperative effect between hydrogen bonding and stacking. Internal bases can be continuously paired and unpaired; double-stranded regions open and form single-stranded "bubbles" ("breathing" of DNA). Breathing is more frequent in regions rich in A – T pairs and could be important for interactions with proteins. DNA helices can exist in various forms (A, B, C, D, and Z) [39] some of which are interconvertible depending on the concentration and type of salts present. The helices always exhibit a degree of microheterogeneity that plays an important part in genetic regulation mechanisms.

The DNA helices can exist in linear form (e.g., in the chromosomes of higher organisms)

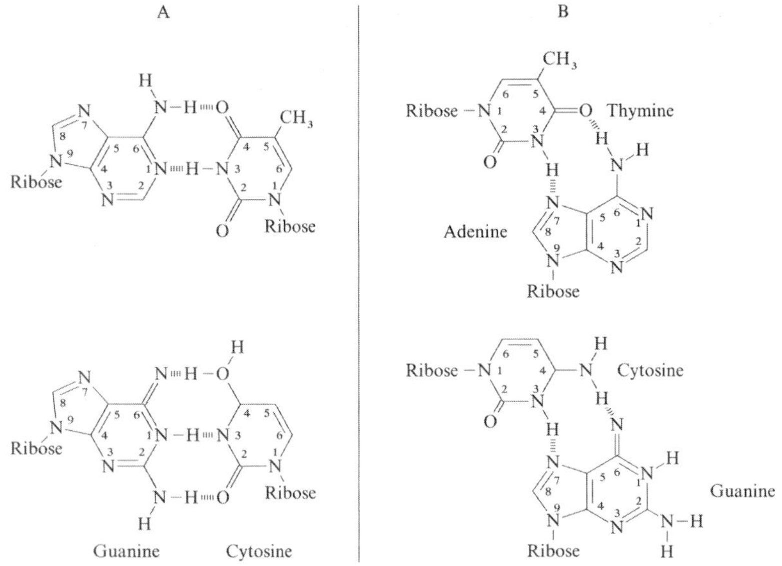

Figure 3. Watson – Crick base-pairs (A), and Hoogsteen base-pairs (B)

or as closed rings (e.g., in *Escherichia coli*); the molecules can also be twisted (superhelicity or supercoiling). In order to accommodate the large amount of DNA present in living cells, it must be packaged as compactly as possible with the help of proteins and RNA. Proteins can recognize specific binding sites on the DNA. The grooves of the DNA helix are large enough to allow proteins to come into contact with the bases [40]. Defined regions in DNA can also be recognized with the help of the methylation pattern of the bases (see Section 4.3).

Forms of DNA.

A-DNA can be observed in X-ray analyses at 66 % relative humidity. It has 11 base pairs per turn of the helix, the planes of the base pairs are tilted away from the vertical helical axis (19°), the helix is right-handed and has a diameter of ca. 2.3 nm.

B-DNA is the classical Watson – Crick form. It represents the structure of DNA at a relative humidity of > 92 % and largely corresponds to that found under physiological conditions. The helix is also right-handed with about 10.2 – 10.4 base pairs per turn and a diameter of ca. 2 nm. Single unpaired bases can be "looped out" of the helix and barely disturb the rest of the structure [41,42]. Protein – DNA interactions usually require recognition of nucleotide sequences in the major groove of the B-DNA double helix.

C-DNA helices can be observed at a relative humidity of 44 – 66 % in the presence of lithium salts. The helix is also right-handed and similar to the B form, but with 9.3 base pairs per turn.

D-DNA occurs in nature only in sequences with alternating adenine and thymine residues and in the DNA of the bacteriophage T 2 (T-DNA). The helix is also right-handed and has 8 base pairs per turn.

The left-handed conformation of *Z-DNA* has an alternating sequence of pyrimidines and purines and is formed in vitro at high salt concentrations (> 2 mol/L NaCl) or in the presence of divalent cations (Mg^{2+} > 0.7 mol/L). Unlike the right-handed helices (which have two grooves), this structure forms a single, very deep groove that penetrates the helix axis. The sugar – phosphate backbone assumes a zig-zag arrangement (therefore Z-DNA) with 12 base pairs per turn of the helix. Z-helices can form in vivo at physiological salt concentrations. They are less stable than B-DNA, but are stabilized by supercoiling, proteins, special ions, and methylation [43]. Torsional stress of DNA in vivo can favor the formation of Z-DNA [44]. Z-DNA and B-DNA are interconvertible; part of a DNA molecule may exist in the B form and another part in the Z form.

Supercoiling. Circular DNA and DNA between fixed sites can be twisted to supercoils. The term *supercoiling* refers to the curvature of the double helix axis. Supercoiled (superhelical) DNA was discovered in the 1960s in polyoma virus [45]. Rotation in the direction of winding is called *positive supercoiling* and rotation in the opposite direction is called *negative supercoiling*.

Torsional stress due to negative supercoiling can be overcome by the formation of DNA structures other than the B form. Negative supercoiling is a strong driving force for the stabilization of Z-DNA. Supercoiling makes DNA more compact, which is very important in DNA packaging. Almost all naturally occurring superhelical DNAs are underwound (i.e., have negative superhelices) but overwound DNAs also exist [46]. The strain produced by over- or underwinding can be accommodated by the formation of local single-stranded regions which tends to increase with increasing temperature. "Breathing" of the DNA (see third paragraph in Section 2.1) plays an important part here too. A sequence with $> 90\,\%$ A–T can exist permanently unpaired in a superhelical molecule. This is important for many reactions of DNA. Supercoiling influences transcription (see Section) and vice versa. Positive supercoils are formed in front of the transcription apparatus and negative supercoils behind it; these supercoils are controlled by enzymes [47].

Bending. The base sequence of DNA is of tremendous importance for its structure [44]. In a right-handed helix, the twist angle between two bases changes depending on the sequence. This may result in the bending of a linear double helix. Bending can also be caused by proteins. Bending is of significance for the packaging of DNA and for many of its biological reactions [48,49].

Intrinsically bent DNA is formed when special base sequences or structural motives are repeated in phase with the DNA helical repeat; homopolymeric A tracts being the best example [50,51]. Protein-induced DNA bending plays an important role in recombination, initiation of transcription, and replication [52,53]. Bends are also important structural features; indeed, regulatory protein binding sites can be replaced by an intrinsic bend [49,54].

Special Structural Elements. Short sequences are frequently repeated in regulatory regions. *Repeats* can be recognized by DNA-binding proteins. Owing to DNA breathing, double-stranded regions (hairpins or stem-loops and cruciform structures) can be formed at repeats within a single strand (Fig. 4). This rarely happens in double-stranded DNA because stem loops are energetically less favorable than linear double strands. However, it is encountered frequently in single-stranded DNA and RNA. Supercoiling can promote the formation of cruciform structures, whereas transcription inhibits it [55]. Hairpins can play a part in replication, transcription, and RNA processing [56].

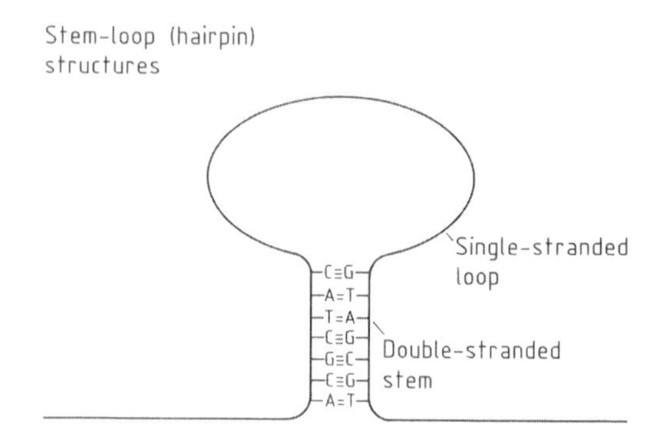

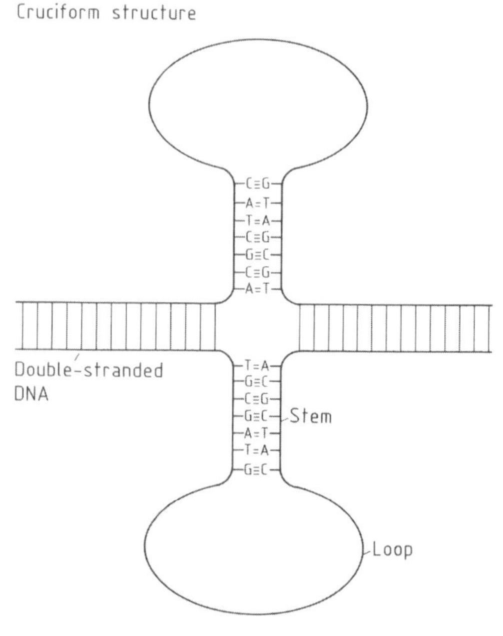

Figure 4. Special structural elements in nucleic acids

Homopyrimidine – homopurine runs are frequently found in regulatory regions of eucaryotic genes and are especially sensitive to nucleases [57]. There is a high tendency to form right-handed structures other than B-DNA in such regions [58].

Although the chains of double-stranded DNA are normally antiparallel, *parallel double-stranded oligomers* have also been found in vitro [59]. They form a right-handed helix and are even recognized by several enzymes. They are less temperature-stable than the corresponding oligomers.

At homopurine – homopyrimidine sections of the DNA, a homopyrimidine oligonucleotide can attach itself parallel to the homopurine strand in the major groove and form a *triple helix* [58]. Structures of this type can be used for specific strand cleavage with the help of coupled ellipticine derivatives or metal chelates [60,61].

Centromeres are important compact DNA structures of the eucaryotic chromosome that are rich in adenine and thymine. Their exact structure is not known but they are important for the attachment of the spindle fibers during mitosis.

The ends (*telomeres*) of linear chromosomes (as in eucaryotic DNA) pose a special problem. DNA polymerases synthesize DNA from a DNA template and always require an RNA primer to start replication. Cleavage of this primer then results in a small 5′-gap which cannot be closed by the polymerase. Under normal replication conditions, the ends should therefore beocme shorter with every cycle of DNA replication (see Section 4.1). Special enzymes (telomerases) are responsible for adding telomere repeats to the chromosome ends to maintain constant length (Fig. 5): repeats can fold back and provide a 3′-OH group which serves as a primer for copying the last segment of a linear DNA molecule. Disturbances in telomeres can lead to aging phenomena [62,63] and a role in carcinogenesis is also being discussed. Broken ends of chromosomes that are no longer protected by telomeres are very susceptible to fusion with other DNA ends and to degradation by nucleases [64]. The antiparallel structure and function of telomeres are highly conserved in all eucaryotes and are species specific. They consist of simple, tandemly repeated sequences with clusters of G residues [65,66]. The G-rich strand is aligned in the $5' \rightarrow 3'$ direction towards the end of the chromosome and has a single-stranded 3′-end containing 12 – 16 nucleotides. Telomeres can associate to form stable, parallel, four-stranded structures (G 4-DNA) [67].

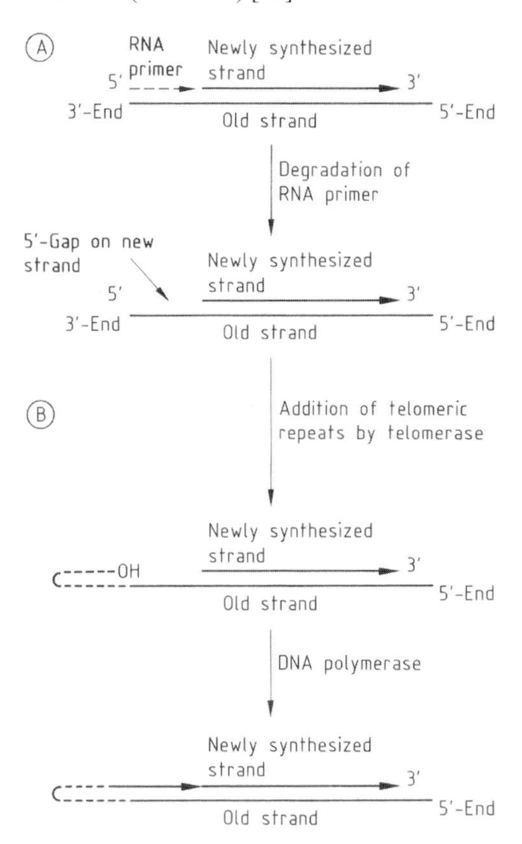

Figure 5. The importance of telomeres
Without telomere addition (A) newly synthesized DNA strands become shorter; with telomere addition by telomerase (B) constant length can be maintained.

2.2. Structure of RNA

RNA is an unbranched single-stranded polymer with many intramolecular double-stranded sections that may account for 50 – 67 % of the molecule. As in DNA, the backbone of RNA consists of 3′,5′-phosphodiester bonds (Fig. 1); however the sugar is ribose (and not deoxyribose) and uracil replaces thymine. Double-stranded RNA cannot form a B-helix because of steric hindrance caused by the 2′-OH groups of ribose; helices of the A type are, however, possible.

The functional groups of the nucleotides in the major groove of the A type of double helix found in RNA are not easily accessible to proteins [68]. Protein binding to RNA probably occurs via interaction with single-stranded regions.

Four functional RNA families exist: messenger RNA (mRNA), ribosomal RNA (rRNA), transfer RNA (tRNA), and small nuclear RNA (only in eucaryotes). The structure of tRNAs has been studied most extensively; about half of the ca. 75 – 90 nucleotides within the tRNA molecule are paired, resulting in a secondary structure with a stem and three loops similar to that of a cloverleaf [69].

RNA has many different biological functions and exhibits a spectrum of flexible structures that more closely resemble those of proteins rather than those of the chemically related DNA [70]. RNAs have secondary structures–double-stranded sections, hairpins, internal loops, and bulged bases. With unpaired nucleotides pronounced tertiary structures are formed in addition to the secondary structure. Examples of tertiary structure motives are pseudo knots [71], produced by folding back in a hairpin and formation of a second stem – loop structure [72].

Formation of DNA – RNA hybrids is of importance in the replication and transcription of DNA and in the reverse transcription of viral RNA. Such hybrids can form secondary structures but they are considerably more polymorphous than DNA alone [73].

3. Properties

3.1. Physical and Chemical Properties

The size of naturally occurring DNA varies from a few thousand to 10^9 base pairs. The length of such molecules (micro- to centimeter range) can easily be measured under the electron microscope.

DNA absorbs UV light at 260 – 280 nm due to its bases. Aqueous DNA solutions are very viscous; viscosity depends on DNA length, DNA concentration, and temperature. Heating to a critical temperature is accompanied by a decrease in viscosity because the hydrogen bonds responsible for base pairing are disrupted and the helix structure collapses. This process is called *thermal denaturation or melting of DNA*. The temperature at which one-half of the base pairs is disrupted is denoted the *melting temperature*. It depends on the base composition (G – C pairs are more stable than A – T pairs). Double-stranded DNA ranging in size from 100 to > 100 000 base pairs melts at ca. 90 °C. In shorter double strands a gradual decrease in the melting temperature is observed. The melting temperature increases with increasing salt concentrations because the solubility of the bases decreases and hydrophobic interactions are increased. Chemicals that compete with hydrogen bond formation, such as urea or formamide, lower the melting temperature of DNA. Methanol has a similar effect; it increases the solubility of the bases and increases the interaction with water. The "melting" of double-stranded DNA is also facilitated by solvents such as ethylene glycol, dimethylformamide, dimethyl sulfoxide; low ionic strength; or extreme pH values. DNA can be denatured at an alkaline pH because the keto – enol equilibria of the bases are shifted preventing these groups from participating in hydrogen bonding.

Since the stacked bases in the double-stranded helix are not as easily excited by UV light as in single strands, absorption at 260 nm is lower for double-stranded DNA than for single strands. Increase in UV absorption can thus be used to measure DNA denaturation. At 260 nm solutions containing 50 µg/mL of double-stranded DNA, 50 µg/mL of single-stranded DNA, and 50 µg/mL of free bases have absorptions of ca. 1.00, 1.37, and 1.60, respectively.

Denaturation can also occur in the presence of proteins that destabilize the helix (melting proteins). Such proteins are required to unwind the helix during replication and to facilitate interaction between single strands during genetic recombination.

The reassociation (renaturation) of thermally denatured DNA is a spontaneous process but only occurs if the solution is cooled slowly below the melting temperature. Renaturation can take several hours, depending on the size of the molecule, because it initially relies on random base pairing (hybridization); it is, however, a cooperative process. Rapid cooling of denatured DNA at salt concentrations > 50 mmol/L produces a very compact molecule in which about two-thirds of the bases are hydrogen bonded or stacked. At salt concentrations below

10 mmol/L the DNA remains denatured even after cooling.

The length of RNA varies greatly: tRNA has a length of 75 – 90 nucleotides and mRNA can be up to several thousand nucleotides long. Denaturation effects are rarely observed because RNA has few truely double-stranded regions; it is most likely to be observed in tRNA.

Because they are extremely long, DNA molecules are extremely sensitive to mechanical influences (shearing forces, e.g., vigorous stirring) and easily break into small fragments (ca. 1000 base pairs). Ultrasonic treatment of DNA in solution produces fragments of ca. 100 – 500 base pairs owing to disruption of hydrogen bonds and single-strand and double-strand breaks in the sugar – phosphate backbone [74]. Nucleic acids are sparingly soluble in water (depending on the molecular mass). They are negatively charged and acidic at physiological pH and form water-soluble alkali and ammonium salts that can be precipitated with ethanol.

RNA and DNA are insoluble in cold acid. DNA is more sensitive to acid hydrolysis than RNA. At pH < 1, however, both DNA and RNA break down into the free bases, phosphoric acid, and (deoxy)ribose. Acid hydrolysis can be used to determine the base composition of nucleic acids (e.g., total hydrolysis can be achieved by heating DNA in 90 % formic acid at 180 °C for 30 min). The β-glycosidic linkage between the N-9 of purines and the C-1 of deoxyribose is selectively cleaved at ca. pH 4, resulting in apurinic sites. Anhydrous hydrazine cleaves the pyrimidine residues.

DNA is stable at pH 13, only 0.2 of 10^6 phosphodiester bonds are broken per minute at 37 °C. In contrast, RNA is rapidly hydrolyzed at alkaline pH.

DNA can be both specifically and nonspecifically cleaved by a variety of enzymes [deoxyribonucleases (DNases)]. RNA is cleaved by ribonucleases (RNases). Some of these cleavage reactions are exploited for sequencing RNA [75, 76].

3.2. Interaction with Proteins

In bacteria, DNA occurs as a complex with RNA and proteins that is bound to but not surrounded by a membrane. The DNA often has a closed circular form and is organized in a series of superhelical loops.

The DNA of higher cells is enclosed within the nuclear membrane as morphologically distinct units of varying size (chromosomes); it is associated with basic proteins called histones. The number and size of the chromosomes are species specific (karyotype). Two full turns of the DNA double helix (146 base pairs) are wound around a histone octamer (diameter ca. 8.6 nm) to form a nucleosome. The width of the grooves varies due to the periodic arrangement of A – T trinucleotides on the inside and G – C trinucleotides on the outside of the nucleosome at intervals of about ten base pairs [53]. Nucleosomes can become condensed into fibers of 10 or 30 nm (super superhelices, solenoids).

Eucaryotic cellular organelles (e.g., mitochondria, chloroplasts) possess closed circular DNA that is not associated with histones.

4. Biosynthesis and Biological Function

4.1. DNA Replication

The genetic information of all cellular organisms is stored in double-stranded DNA (viruses may, however, also have single-stranded DNA or RNA, as well as double-stranded RNA). It is extremely important that the transfer of biological information in DNA (i.e., its base sequence) occurs with a very high degree of accuracy. Because of perfected proofreading and repair mechanisms (see Section 4.5) DNA replication has an error level of $10^{-8} – 10^{-11}$ [77], i.e., for every $10^8 – 10^{11}$ bases in newly synthesized DNA only one is incorrectly incorporated. The replication of DNA is carried out by DNA polymerases which require a single strand of DNA as a template and a short double-stranded piece of nucleic acid (formed with the help of a primer) for initiation (Fig. 6).

The DNA is synthesized from deoxyribonucleotide triphosphates which are polymerized on the single-stranded DNA template with the release of pyrophosphate; the cleavage of pyrophosphate by a pyrophosphatase provides the energy required for DNA biosynthesis. The addition of new nucleotides always takes place at the 3′-OH group of the sugar, therefore all biologically synthesized nucleic acids grow in the $5' \rightarrow 3'$ direction. Divalent cations such as Mg^{2+} or Mn^{2+} are important cofactors. Some polymerases contain $3' \rightarrow 5'$ exonuclease activity and can remove bases that have been incorrectly incorporated. The primer in vivo is usually RNA, which is synthesized by a special RNA polymerase (primase) and removed later by an exonuclease (e.g., DNA polymerase with $5' \rightarrow 3'$ exonuclease activity). The various activities are combined in a multienzyme complex.

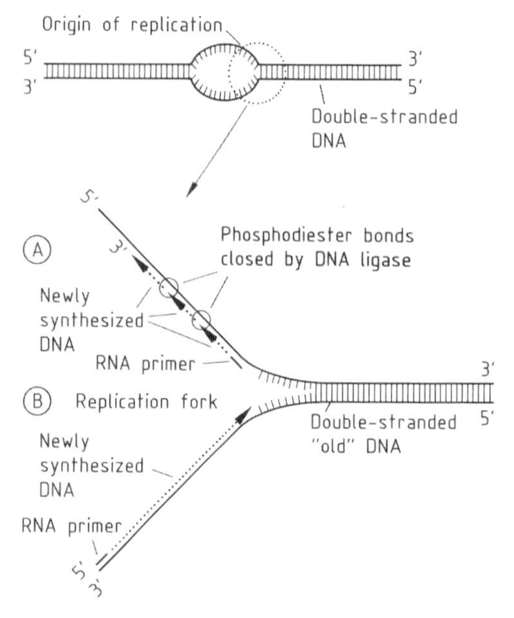

Figure 6. Simplified scheme of DNA replication
A) Lagging strand synthesis: after primer addition by primase a stretch of DNA is synthesized by DNA polymerase, the primer is removed and after synthesis of the adjacent stretch of DNA the phosphodiester bond between the two stretches is closed by a DNA ligase.
B) Leading strand synthesis: after primer addition by primase the new strand is synthesized by DNA polymerase. Arrows denote direction of synthesis.

The replication of DNA is a semiconservative process, i.e., one strand of each of the two new daughter molecules of DNA is an old strand and the other a newly synthesized one [78]. Replication starts with the creation of replication forks at the "origin of replication", and proceeds in opposite directions along the DNA (see Fig. 6).

The DNA of procaryotes is replicated from a single replication origin. The main replication enzyme in *Escherichia coli*, DNA polymerase III, consists of at least ten different subunits. Replication is probably a membrane-bound process [79].

Because DNA strands are antiparallel, one strand (lagging strand) must be synthesized "backwards" to keep the replication complex together. The direction of synthesis of this strand is contrary to the direction of movement of the replication fork and occurs discontinuously. Segments (Okazaki fragments) are formed which are then joined by a DNA ligase. The strand, that is synthesized "forward" is called the leading strand. Replication in eucaryotes starts at several points along the DNA and the replication complexes are fixed to the nuclear matrix [80]. Viroids are the smallest self-replicating structures known. These single-stranded, circular, protein-free RNAs are a few hundred nucleotides long and occur as pathogens in higher plants [81].

4.2. Gene Expression

The base sequence of DNA constitutes the genetic information in the form of discrete units (genes). A sequence of three successive bases (codon) acts as a code for a certain amino acid (e.g., GCT, GCC, GCA, or GCG code for alanine) or as a stop signal (TAG, TGA, or TAA) in protein synthesis. Of the 64 ($=4^3$) possible triplets, 61 code for amino acids and 3 specify stop signals. Since this genetic code is almost universal, eucaryotic proteins can be synthesized in procaryotes from eucaryotic DNA. Some organisms and organelles differ in a few codons from the universal code [82]. The genetic information contained in DNA is normally used as a template for the synthesis of mRNA which itself then acts as a template for protein synthesis.

The regulation of gene expression in living organisms is based primarily on the recognition of nucleic acid sequences by proteins. However, nucleic acids can also be involved. "Antisense RNA" for instance, can hybridize with mRNA, leading to a block of translation and to the degradation of the mRNA by RNase. The termination of transcription can occur in a similar way. This principle may possibly be exploited for therapeutic applications [83].

4.2.1. Transcription

The process by which genetic information is transferred from DNA to mRNA is called transcription. This process is catalyzed by DNA-dependent RNA polymerases and involves the synthesis of mRNA from ribonucleotide triphosphates in the presence of Mg^{2+} or Mn^{2+} and a single- or double-stranded DNA template. The direction of synthesis is $5' \rightarrow 3'$ and no primer is required. As in DNA synthesis, energy is obtained from the cleavage of pyrophosphate. The accuracy of transcription is lower than that of DNA replication (ca. 10^{-4}) because there are no repair processes [77]. Most RNA polymerases are complex enzymes consisting of several subunits. Bacteria generally have one, but eucaryotes have three.

Due to the specific base pairing in the double strand, the base sequence information of DNA determines the base sequence of the mRNA synthesized by RNA polymerases. Messenger RNA acts as an intermediate for conveying the information required for protein synthesis from the DNA to the protein-synthesizing structures, i.e., the ribosomes (see below). For interaction with the ribosome, mRNA possesses a ribosomal binding site shortly before the translation initiation sequence. This site is complementary to rRNA sequences [84]. The average length of *Escherichia coli* mRNA is about 1200 base pairs, eucaryotic mRNA can be much longer. Some types of mRNA are very rapidly degraded and exist in relatively low concentrations. Other types are more stable and accumulate in considerable amounts in the cell. The ends of mRNA are important for its stability.

Modifications in the form of loop structures [85] in procaryotes, and poly(A) at the 3′-terminus and a cap (formed by addition of 7-methyl-GTP followed by methylation) at the 5′-terminus in eucaryotes protect the ends from nonspecific degradation by nucleases (Fig. 7). Capping and the addition of poly(A) usually occur before splicing. Random degradation of mRNA is also prevented by cellular ribonuclease inhibitors. The mRNA transcript is finally exported to the cytoplasm where it acts as a template for protein synthesis.

Transcription is increased or decreased by sequence elements on the DNA that act as binding sites for protein factors [86–88]. Transcription also involves initiation and termination factors, some of which are ribonucleoproteins. Structural features also affect transcription, e.g., bent DNA can activate the process [52].

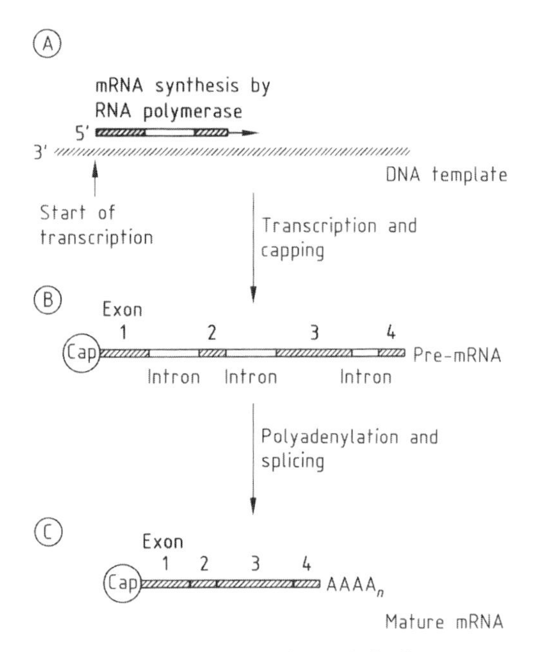

Figure 7. Simplified scheme of transcription in eucaryotes A DNA template is transcribed into pre-mRNA by RNA polymerase (A) and a cap is added to the 5′-end of the mRNA (B). A stretch of A-residues is then added to the 3′-end. During splicing intervening sequences (introns) are removed and the sequences carrying the information for protein synthesis (exons) are joined to form mature mRNA (C).

RNA polymerase binds to a promoter region in the DNA template and then transcribes the DNA into mRNA. In most eucaryotic genes protein-coding sequences (exons) are interrupted by intervening noncoding sequences (introns). Many exons code for protein domains; exon rearrangements in DNA can facilitate the exchange of domains (exon shuffling). Lower eucaryotes have a higher content of intron-free genes than higher eucaryotes. Introns are rarely found in bacterial genes [89]. Introns are removed from eucaryotic precursor mRNA and the exons are then joined to give mature mRNA in a process known as splicing (see Fig. 7). The length of introns varies from 30 to 100 000 base pairs, the average exon length is 50 – 300 base pairs. Splicing demands a high degree of accuracy and is carried out by a multicomponent complex known as a spliceosome that consists of protein and small RNAs (snRNAs = small nuclear RNAs). The main components of the spliceosome are ribonucleoproteins (snRNPs = small nuclear ribonucleoproteins) [90,91]. Introns usually start with GU (5′-splice site) and end with a pyrimidine-rich section followed by AG (3′-splice site) [92]. The half-life of introns in mRNA varies from a few seconds to 10–20 min. Splicing always takes place in the nucleus. Unspliced RNA remains in the nucleus and is degraded.

The splicing mechanism is highly conserved: a mammalian cell extract can process yeast RNA. The number of possible transcription products can be increased by alternative splicing (i.e., not all available exons of a gene are used to create one mature mRNA).

In protozoa RNA is also "edited", the primary transcript is converted into its functional form not only by post-transcriptional removal of certain bases, but also by insertion of others [93,94].

In some RNA viruses (retroviruses), replication of the genetic material proceeds via a DNA intermediate in a process known as reverse transcription. The information in RNA is transcribed back to DNA with suitable enzymes (reverse transcriptases). The DNA is then used as a template for RNA synthesis. Reverse transcription can also be used in vitro to produce intron-free complementary DNA (cDNA) from eucaryotic mature mRNA. Telomerases are special reverse transcriptases with an internal RNA template [95,96].

4.2.2. Translation

Protein synthesis takes place on the ribosomes, which are complexes of rRNAs and more than 50 different proteins. Protein factors are involved in all phases of translation. Various types of RNA are involved in protein synthesis: mRNA, the carrier of genetic information from the DNA; rRNA, component of ribosomes and directly involved in almost all stages of protein synthesis [97]; and tRNA, which makes activated amino acids available for protein synthesis. Translation is an irreversible process with error rates of $10^{-3}-10^{-4}$ [77].

The initiation of protein synthesis is a very intricate multistep process [98,99]. A translation initiation complex is formed from ribosomes, mRNA, a special tRNA, initiation factors, and GTP. The mRNA start codon for protein biosynthesis is usually AUG but GUG, UUG, or AUU may also be used [89]. After initiation of translation, the protein is synthesized step by step (elongation) from amino acids that are delivered and activated by tRNA. Each amino acid has a specific tRNA.

The tRNAs act as adapters between the codons of mRNA and the corresponding amino acids. Apart from the "usual" bases, tRNA contains several rare bases (e.g., 5-hydroxymethylcytosine). Transfer RNAs have an amino acid attachment site and a template recognition site (anticodon) for the mRNA. Amino acids are attached by an ester bond either to the 3′-or 2′-OH group of the terminal adenosine of tRNAs. Loading the tRNAs with the correct amino acids is crucial for exact translation of genetic information. This step is catalyzed by specific enzymes (aminoacyl-tRNA synthetases). Each of these enzymes must be able to recognize a specific amino acid and its associated tRNA.

The amino acid sequence of the protein product is specified by the mRNA codons and thus ultimately by the base sequence of DNA. A stop codon (TAA, TAG, TGA) on the mRNA leads to termination of protein synthesis and the release of the protein. Differences between translation complexes in procaryotes and eucaryotes permit the selective interruption of procaryotic protein synthesis (e.g., by some antibiotics).

4.3. Modification and Degradation

DNA is exposed to many damaging influences and can be modified and cross-linked by radiation and chemicals. Spontaneous hydrolysis of DNA bases may also occur.

Mutations change the normal base sequence of DNA. Erroneous bases can appear as a result of replication errors or chemical modification (→ Mutagenic Agents). Intercalating agents such as benzopyrenes, aflatoxins, or ethidium bromide can cause deletions. Insertion of individual bases is also possible. Insertion of larger DNA segments can occur through mobile DNA elements (transposable elements, jumping genes). The insertion site is usually arbitrary. This process (transposition) can lead to gene activation or inactivation.

Methylation of DNA is used by cells in different ways [100, 101], e.g., to distinguish between newly synthesized strands (unmethylated) and old strands (methylated). Methylation occurs at defined sites: in bacteria at the N-6 position of adenosine and the C-5 position of cytosine, and in eucaryotes, physiologically only at cytosine. The methylation pattern also differentiates between endogenous and foreign DNA and is recognized by restriction – modification systems (in procaryotes): endogenous DNA is methylated at specific sites and foreign DNA is digested. Methylation can be used to switch genes on or off in eucaryotes. C-Methylation influences the equilibrium between B- and Z-DNA [102] and can lead to altered DNA – protein interactions by the modulation of DNA topology. Methylation patterns are stable and hereditary. Methylation can have a mutagenic effect: the deamination of 5-methylcytidine to thymine occurs spontaneously.

Degradation of nucleic acids can be effected in vivo and in vitro by a series of enzymes. Some DNases are specific for single strands and others for double strands, some cleave bonds located within a double strand (endonucleases) and others require a free DNA end (exonucleases). Exonucleases act either in the $3' \rightarrow 5'$ or the $5' \rightarrow 3'$ direction; exonuclease activity is often associated with DNA polymerases for proofreading in DNA replication. Restriction endonucleases are a special class of nucleases. RNases can also be divided into exo- and endonucleases. RNase II is a bacterial exonuclease; RNase H is an endonuclease that digests RNA in DNA – RNA hybrids.

4.4. Recombination

In recombination two DNA molecules interact and exchange segments to produce two new molecules that contain genetic information from both parental DNAs. DNA recombination is by no means limited to genetic engineering, but occurs widely in nature. This process is catalyzed by enzymes and usually requires the presence of homologous regions (homologous recombination). In rarer cases, nonhomologous, illegitimate recombination can take place. Transposition (see Section 4.3) is also a recombination phenomenon [103].

4.5. DNA Repair

Although changes in base sequences create the genetic variety required for evolution, these changes are only rarely beneficial for individual organisms. Despite many damaging influences (see Section 4.3), nucleic acids are replicated with amazing accuracy. This exact conservation of genetic material is one of the fundamental requirements for genetic success. Damaged DNA that is not repaired can lead to mutations and cell death, and give rise to tumors. Living organisms have therefore developed a variety of enzymatic mechanisms for eliminating DNA damage. Most damage is recognized by repair systems in the cell and repaired either by directly reversing the damage or by using the undamaged second strand as a template. *Escherichia coli* has been used as a model for studying repair mechanisms [104]. The best studied mechanism is nucleotide excision repair. This multistep process involves recognition of the error, nicking of the damaged strand, removal of nucleotides at the damaged site, repair DNA synthesis, and ligation. It requires the participation of multienzyme complexes [105, 106].

4.6. Nucleic Acids as Enzymes

Ribozymes are RNA molecules that act as catalysts: sequence-specific RNA endonucleases, nucleotidyl transferases, phosphatases, kinases, and glucanotransferases [37, 107]. The sequence-specific cleavage of oligodeoxyribonucleotides and single-stranded DNA can also be catalyzed by ribozymes [108]. Previously, only proteins (enzymes) were thought to have activities of this kind. Ribozymal properties were discovered in the early 1980s in the self-splicing introns of *Tetrahymena* [109] and ribonuclease P [110], which cleaves tRNA precursors and contains RNA as the catalytic subunit. In eucaryotes, small nuclear RNAs (snRNAs) are required for the splicing of pre-mRNA; they are part of the spliceosome, an RNA – protein complex [90].

The smallest RNA structure that exhibits catalytic (cleavage) activity is 19 nucleotides long and forms a hammer head structure with its target [111]. No pure DNA catalysts have been discovered yet [112]. However, since a few of the ribonucleotides in ribozymes can be replaced by deoxyribonucleotides, the three-dimensional structure that permits nucleic acid catalysis may not necessarily be limited to RNA [113, 114].

In all proven examples of RNA catalysis, the substrate attacked by the ribozyme is itself a nucleic acid. However, there are indications that rRNA also attacks other molecules (i.e., is an RNA enzyme). RNase P is the only known naturally occurring ribozyme that does not cleave itself, but attacks other molecules [72, 115].

5. Isolation, Purification, and Transfer

Chromosomal DNA from bacteria or eucaryotes has a very high molecular mass and is extremely fragile. Thus to isolate DNA, cells must be gently lysed with enzymes (pronase, protease K, lysozyme) and mild detergents. Accompanying RNA can be removed by treatment with enzymes which digest RNA or by gel filtration [116]. Proteins associated with DNA are denatured with phenol and extracted [117].

Plasmid DNA is relatively small and has a closed circular and supercoiled form. It is more stable than chromosomal DNA when subjected to shearing forces and can be renatured much more easily after denaturation; isolation and purification of plasmid DNA is achieved relatively easily (e.g., by equilibrium centrifugation in cesium chloride – ethidium bromide gradients at 150 000 – 400 000 g for 5 – 48 h). Kits are also available which permit the isolation of pure plasmid DNA in a short time without ultracentrifugation using binding to glass powder (glass milk) and/or gel filtration. Similar methods are used for the *isolation of RNA*. However, degradation by ribonucleases must be prevented (RNA is much more sensitive to nucleases than DNA). Either proteases or chloroform are employed to remove protein; with chloroform protein precipitates at the phase boundary. Phenol is not normally used because a small part of the RNA also goes into the phenol phase. Oligo(dT) cellulose (cellulose-bound oligonucleotides containing only thymine) can be used for the purification of eucaryotic mRNA [118] taking advantage of its poly(A) tails.

Nucleic acids can also be purified by ion-exchange chromatography (e.g., by HPLC) [119]. Suitable separating media are hydroxyapatite and anion exchangers. Reversed phase methods and gel filtration are also used.

Microgram amounts of DNA fragments can easily be isolated and purified by techniques such as electroelution from agarose gels, binding to anion-exchange paper, dialysis, or binding to glass milk (powdered glass suspension). After purification, DNA can easily be concentrated by precipitation with ethanol or 2-propanol.

DNA can be transferred relatively easily in vivo between related organisms (especially in bacteria) via endogenous mechanisms (e.g., conjugation, some viruses). Several other methods are available for the transfer of DNA in the laboratory: application as a $CaCl_2$ precipitate, exposure to an electric field (electroporation), and the use of viral vectors.

6. Analysis of Nucleic Acids

Sequencing. Computer data banks such as EMBL [120] and GenBank [121] contain sequence information in the order of millions of nucleotides.

Single or double-stranded DNA can be sequenced [122]. The most widely used methods are the Sanger method [123] and the Maxam and Gilbert method [124]. In the Maxam and Gilbert method the DNA is cleaved with chemicals statistically and specifically at one of the four bases. The Sanger method employs enzymatic synthesis of a complementary copy of a single-stranded DNA template and random chain termination with dideoxynucleotides. Both methods produce DNA fragments that are separated by gel electrophoresis.

Both methods initially made use of radioactively labeled nucleotides and subsequent exposure on X-ray films for identification of the fragments. More recent variations use fluorophores [fluorescein, tetramethylrhodamine, Texas red,

4-chloro-7-nitrobenzo-2-oxa-1-diazole(NBD)] which are more stable, easier to use, less dangerous, and permit the automatic detection of all four nucleotides on a single track. Sequences of >450 bases can be determined per gel run and track [125–128]. An automatic sequencing machine allows determination of sequences of up to 10 000 bases per day. Further acceleration by combination with automatic DNA preparation in robotic workstations may be possible [129].

More recent techniques aim at faster sequence determinations (several hundred bases per second [130]) with smaller amounts of DNA. Variations of the Sanger method permit determination of sequences up to 10 000 bases without subcloning [131].

The Sanger method can also be used for sequencing RNA [132] and sequences up to 150 bases can be determined in a single run. Reverse transcriptase is used to synthesize DNA with RNA as a template. The RNA sequence is then deduced from the DNA sequence. Another method for RNA sequencing involves the use of specific RNases [75, 76].

X-Ray Analysis. X-ray analysis is of considerable importance for elucidating the structure of nucleic acids and their building blocks [44]. Useable crystals can be obtained from aqueous solutions for molecules up to the size of tRNA (75 – 90 base pairs). Nucleic acids of higher molecular mass are too heterogeneous to produce crystals, quasi-crystalline fibers drawn from concentrated aqueous solutions are used instead.

Nuclear Magnetic Resonance (NMR). The best method for determining the structure of relatively short (10 – 20 bases or base pairs) DNA or RNA fragments in solution is ^{1}H-, ^{13}C-, ^{15}N-, or ^{31}P-NMR (one- or two-dimensional) [133, 134]. NMR is also used to study short triple helices [135], DNA – RNA hybrids in solution [136], DNA – drug complexes, mismatches, and intercalating agents. Torsion angles and the mobility of base pairs and the sugar phosphate backbone can be determined.

Electron Microscopy. Electron microscopy permits the determination of the length of DNA. DNA heteroduplexes, binding of large protein factors [137], and supercoiled DNA [138] can be visualized. A special application is DNA-hybridization electron microscopy [139, 140]: biotinylated DNA probes are hybridized to selected regions of DNA or RNA and then reacted with avidin and visualized by electron microscopy.

Scanning tunneling microscopy (STM) allows the observation of single-stranded DNA adsorbed on graphite with atomic resolution. It opens up the possibility of structural studies on DNA and of direct sequencing [141–143]. The profiles show excellent correlation with the results of X-ray analysis.

Centrifugation. Ultracentrifugation in cesium chloride gradients can be used to measure the molecular mass of DNA and also for the experimental identification of covalently closed circles. Sedimentation at alkaline pH (>11.3) causes disruption of the hydrogen bonds and unwinding of the DNA molecules. In the case of linear DNA, two single strands are obtained which have a higher sedimentation coefficient than native DNA at salt concentrations >0.3 mol/L. Covalently closed circles cannot be denatured into single strands, they collapse and have a sedimentation coefficient that is three times higher than that of native DNA.

In equilibrium centrifugation in cesium chloride, ethidium bromide is firmly bound to DNA in salt solutions by intercalating between base pairs. This lowers the density of the DNA by about 0.15 g/cm^3 and the DNA unwinds with increased binding of ethidium bromide. Unlike linear DNA, covalently closed molecules have a limited absorption capacity for ethidium bromide. Linear DNA is consequently lighter than covalently closed circular DNA. Use of a cesium chloride density gradient with ethidium bromide therefore permits the separation of supercoiled and linear (relaxed) DNA of the same size.

Staining and Quantification. Nucleic acids can be stained with ethidium bromide [144], Stainsall, 1-ethyl-2-[3-(1-ethylnaphtho[1,2 *d*] thiazoline-2-ylidene)-2-methyl-1-propenyl]-naphtho[1,2 *d*] thiazolium bromide [145], or silver [146–148]. DNA stained with ethidium bromide can be quantified in electrophoresis gels after being photographed [144, 149]. Silver staining followed by a densitometer scan is equally sensitive (1 – 2 ng of DNA can be detected) [146]. A quick and sensitive estimation of the amount of nucleic acid can be obtained by measuring the UV absorption at 260 nm: an absorption of 1.00 corresponds to a concentration of ca. 50 µg/mL of double-stranded DNA.

DNA quantification in the picogram range has been reported with a silicon sensor, that responds to the interaction between DNA-binding proteins and DNA [150]. Quantification in the picogram range is also possible after specific hybridization (see Chap. 8).

Fluorescence imaging of DNA in picogram to nanogram amounts is made possible by laser excitation of a DNA – ethidium homodimer [5,5′-diazadecamethylene) bis(3,8-diamino-6-phenylphenanthridinium) dichloride dihydrochloride] complex at 488 nm [151].

Electrophoresis. Since DNA is negatively charged because of its phosphate groups, it migrates in an electrical field (usually at a pH of ca. 7). The electrophoretic mobility of DNA fragments in solution is independent of the molecular mass (constant linear charge density). However, electrophoresis of DNA in gel matrices (agarose or polyacrylamide) at a constant field strength is an effective method of separating DNA molecules according to molecular mass. The mechanism of separation is not well understood [152].

Important experimental parameters are gel concentration, field strength, temperature, and running time. Gels of 0.5 – 2 % are usually employed for the separation of 50 – 20 000 base pairs at field strengths of 2 – 3 V/cm [153]. The mobilities of linear and supercoiled DNA of equivalent molecular mass are different. Estimation of the molecular mass of supercoiled DNA molecules based on linear standards is therefore not possible.

Polyacrylamide gels (4 – 12 %) give a sharp separation of polynucleotides up to a length of about 600 base pairs. Special gels can separate polynucleotides with up to 2500 base pairs [148]. Mobility varies logarithmically with the size of the DNA at field strengths below 1 V/cm [154]. Separation efficiency is limited in the range of 10 000 – 50 000 base pairs, there is no resolution above 50 000 base pairs.

Denaturing gels (e.g., with urea) are used in the sequencing and analysis of single-stranded nucleic acids. Under special conditions, the behavior of individual DNA molecules in an electric field can be observed with a fluorescence microscope [155].

Pulse-field electrophoresis is used to separate larger linear DNA molecules. Here the direction of the electric field is changed at intervals. The DNA molecules therefore have to change their orientation, the time required for this change depends on the size of the DNA [44, 156]. Unlike conventional electrophoresis (duration 30 – 60 min), a pulse-field run takes 20 – 140 h. Usually, 1.5 % agarose gels are used and 10 – 20 µg of DNA can easily be separated at field strengths of 2.5 – 10 V/cm. Separation can be achieved up to ca. 12×10^6 base pairs [157]. Separation is affected by DNA topology and sequence, pulse time (seconds to minutes), field geometry, field strength, gel composition, sample concentration, temperature, and running time [154, 158].

7. Chemical Synthesis

The chemical synthesis of oligonucleotides began in 1955 when A. M. MICHELSON and A. R. TODD prepared the first dinucleoside monophosphate [159]. They used the so-called *triester method* to show that oligonucleotides of defined sequence can be prepared by purely chemical methods. In 1956 KHORANA et al. successfully synthesized the same nucleotide by the *diester method* [160]. This process dominated oligonucleotide synthesis until the mid-1970s, and was also used by KHORANA and coworkers for the first total synthesis of a gene (alanine-specific tRNA from yeast) [161]. The importance of synthetic oligonucleotides was generally underestimated because the synthesis was time-consuming and expensive. It was the progress in genetic engineering that resulted in a dramatic increase in the demand for synthetic oligonucleotides, which, in turn, led to the development of new synthesis concepts and to the automation of DNA and RNA synthesis.

7.1. Synthesis Strategy

The chemical synthesis of DNA consists of three steps:

1) Separate synthesis of the two complementary strands
2) Hybridization of the two strands (formation of hydrogen bonds between A and T, and between C and G)
3) Enzymatic joining of the DNA molecules to give larger DNA units

Three standard methods are used for the synthesis of the 3′,5′-internucleotide phosphodiester bond. They differ in the type of monomer building blocks used and are known as the phosphate triester, phosphoramidite (phosphite), and the *H*-phosphonate methods (Fig. 8, Section 7.4).

An oligomer can be synthesized either by stepwise addition of individual monomer building blocks or by joining oligomer blocks (e.g., dimers or trimers). The former concept is employed in solid-phase synthesis, which was developed almost simultaneously by R. B. MERRIFIELD [162] for peptides and by R. L. LETSINGER [163] for oligodeoxyribonucleotides. The first nucleoside (sugar and base) is covalently bound to an insoluble support at the 3′-OH group of the sugar. Stepwise synthesis of the chain then proceeds by condensation of the appropriate nucleoside monomers. Since the reaction product remains immobilized on the support, it can easily be freed of other reactants by washing; time-consuming chromatographic purification is not required. Synthesis is usually carried out on the 0.2 – 1 µmol scale and all

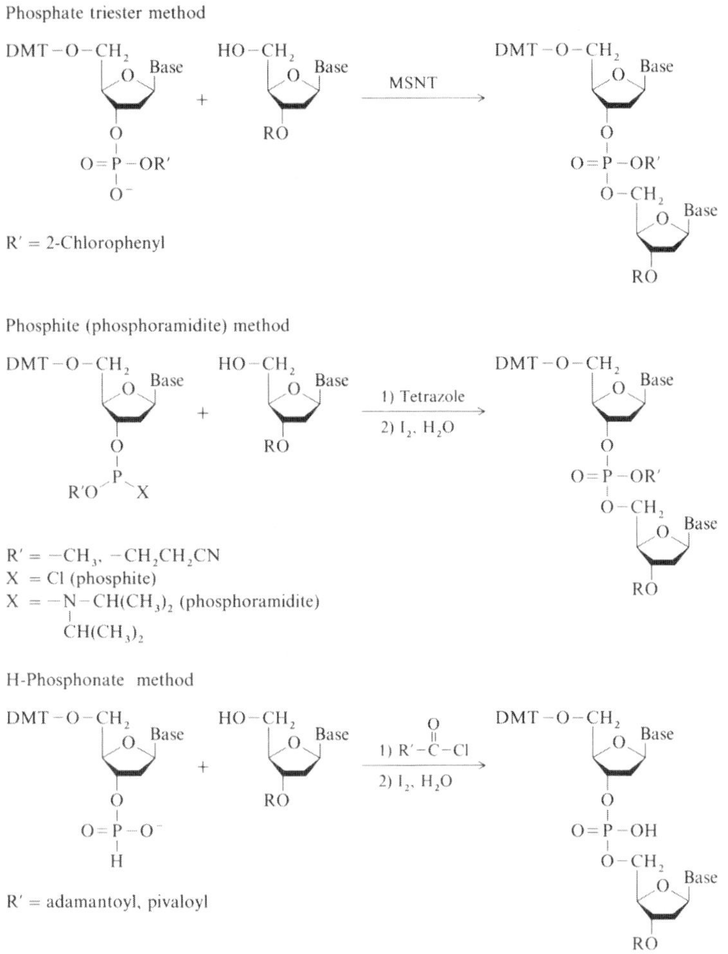

Phosphate triester method

DMT–O–CH_2 Base, O, O, O=P–OR′, O^- + HO–CH_2 Base, O, RO —MSNT→ DMT–O–CH_2 Base, O, O, O=P–OR′, O–CH_2 Base, O, RO

R′ = 2-Chlorophenyl

Phosphite (phosphoramidite) method

DMT–O–CH_2 Base, O, O, P, R′O X + HO–CH_2 Base, O, RO —1) Tetrazole 2) I_2, H_2O→ DMT–O–CH_2 Base, O, O, O=P–OR′, O–CH_2 Base, O, RO

R′ = $-CH_3$, $-CH_2CH_2CN$
X = Cl (phosphite)
X = $-N(CH(CH_3)_2)-CH(CH_3)_2$ (phosphoramidite)

H-Phosphonate method

DMT–O–CH_2 Base, O, O, O=P–O^-, H + HO–CH_2 Base, O, RO —1) R′–C(=O)–Cl 2) I_2, H_2O→ DMT–O–CH_2 Base, O, O, O=P–OH, O–CH_2 Base, O, RO

R′ = adamantoyl, pivaloyl

Figure 8. Chemical synthesis of oligodeoxynucleotides
DMT = 4,4′-dimethoxytrityl residue; MSNT = mesitylene sulfonylnitrotriazole; R = polymer support

reagents are applied in large excess. Solid-phase synthesis of oligonucleotides can be divided into the following steps:

1) Preparation of monomer units that are either protected or can be activated
2) Functionalization of the support
3) Stepwise synthesis of the desired nucleotide sequence by one of the three methods listed above
4) Cleavage of the protecting groups and the support
5) Isolation and purification of the desired sequence

7.2. Protecting Groups

In order to avoid unwanted side reactions during oligonucleotide synthesis, functional groups that do not participate in the linking reaction must be suitably protected. Only a few protective groups have gained acceptance and are based on the work of KHORANA et al. [164].

Two types of groups can be distinguished: permanent and intermediary protecting groups.

Permanent protecting groups are maintained throughout the synthesis and protect the exocyclic amino functions of adenine, cytosine, and guanine; the OH group of phosphorus; and the 2′-OH group of ribose (in RNA synthesis). *Intermediary protecting groups* protect the 5′-OH group of the sugar and are cleaved after each condensation step.

The most important protecting groups are depicted in Figure 9. Adenine and cytosine are usually converted to the corresponding acid amides with excess benzoyl chloride, the amide of guanine is obtained with isobutyric anhydride. Base-protecting groups are introduced with a method developed by TI et al. (transient protection) [165]. After termination of the synthesis, the protecting groups are removed with 32 % aqueous ammonia at 50 – 60 °C in 4 – 5 h. In long, guanine-rich sequences, the O-6 position of guanine can also be protected [166–169], usually with a 4-nitrophenylethyl residue [170]. This is especially important for attaining good yields in the triester process because otherwise sulfonation with the condensing agent easily occurs. More labile amidine or acyl base-protecting groups are of interest for the synthesis of phosphate-modified oligonucleotides because they are more sensitive to hydrolysis than their diester analogues and treatment with ammonia at 60 °C can cause considerable strand breakage [171–173]. Important examples of such groups are the 4-phenoxyacetyl residue

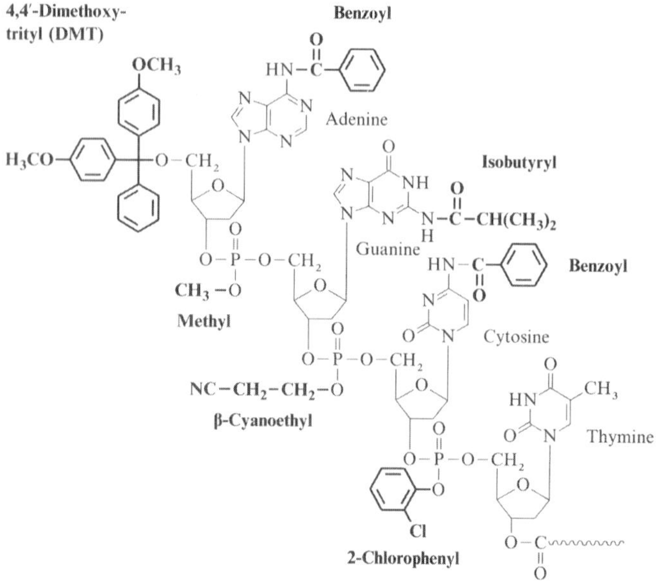

Figure 9. Protecting groups (bold type) used in the synthesis of oligonucleotides

(Pac) for adenine and guanine and the isobutyryl residue for cytosine.

The *tert*-butyldimethylsilyl residue is preferred as the protecting group for the sugar 2′-OH group in the solid-phase synthesis of oligoribonucleotides [174].

The 5′-OH group of deoxyribose or ribose is usually blocked with the 4,4′-dimethoxytrityl group [164], which can easily be removed after each coupling step by mild acid hydrolysis with 3 % di- or trichloroacetic acid in dichloromethane [175].

In the triester method the 2-chlorophenyl group is used to protect the phosphorus [176], while in the phosphoramidite method, the β-cyanoethyl group is preferred to the originally used methyl group [177, 178]. The β-cyanoethyl group can easily be removed at the end of the synthesis by β-elimination with aqueous ammonia [179]. No internucleotide phosphorus-protecting group is required in the H-phosphonate process.

7.3. Functionalization of the Support

Polystyrene [180], polyacrylamide [181], cellulose [182], and silica [183] have been used for solid-phase synthesis of oligodeoxyribonucleotides. The preferred support material is controlled pore glass (CPG) [184, 185] functionalized with a long-chain alkylamine spacer. The material is relatively rigid, does not swell, and is inert to all reagents used in the synthesis. In addition, it withstands the high mechanical stress encountered in the large number of reactions performed in automatic synthesizers. The narrow pore-size distribution gives more uniform diffusion parameters and, consequently, more reproducible flow rates and availability of functional support sites.

The support is functionalized with an alkylamine by silanization with 3-aminopropyltriethoxysilane [183, 186]. Heating and subsequent treatment with methanol result in cross-linkage of the silyl groups and hydrolysis of outer, less strongly fixed groups [187]. The nucleoside at the 3′-OH end of the DNA to be synthesized is linked to the support via a succinyl spacer [183, 188, 189]. Amino groups of the support that have not reacted are capped with acetic anhydride [188]. To determine the loading of the support with nucleoside material, an aliquot is treated with strong acid to remove the DMT residue. The resulting orange solution containing the dimethoxytrityl cation [190] can be determined by measuring the absorbance at 495 nm. The average loading is usually 15 – 25 µmol per gram of CPG.

7.4. Methods of Synthesis

The three main methods of oligonucleotide synthesis are the phosphate triester, the phosphoramidate (phosphite), and the H-phosphonate methods (see Fig. 8). The phosphoramidite method is the most widely used (Fig. 10, see next page). The reaction cycle involves

1) Cleavage of the 5′-OH DMT group
2) Condensation of the next monomer building block (chain elongation)
3) Masking of 5′-OH groups that have not reacted completely (capping)
4) In the phosphoramidite and H-phosphonate processes, an additional oxidation step is required after each condensation step or the end of the synthesis respectively to obtain the desired internucleotide phosphodiester.

Chain elongation in all processes occurs in the $3' \rightarrow 5'$ direction.

Phosphate Triester Method. This method was first used to synthesize oligonucleotides from suitable building blocks in solution [191–194]. It can also be employed for the solid-phase synthesis of nucleotides up to a chain length of about 20 bases. The monomer building block is a protected nucleoside phophodiester derivative (Fig. 8) [193] which reacts with the free 5′-end of a support-bound nucleoside to give a neutral phosphotriester. Various aromatic sulfonic acid derivatives [192, 195, 196], primarily mesitylene sulfonylnitrotriazole (MSNT), serve as condensing agents. The reaction can be accelerated by using catalysts such as N-methylimidazole [175]. This method is not widely used any longer due to disadvantages such as long reaction times, a large number of side reactions, and the frequently observed nicks caused by incomplete cleavage of the internucleotide protecting group.

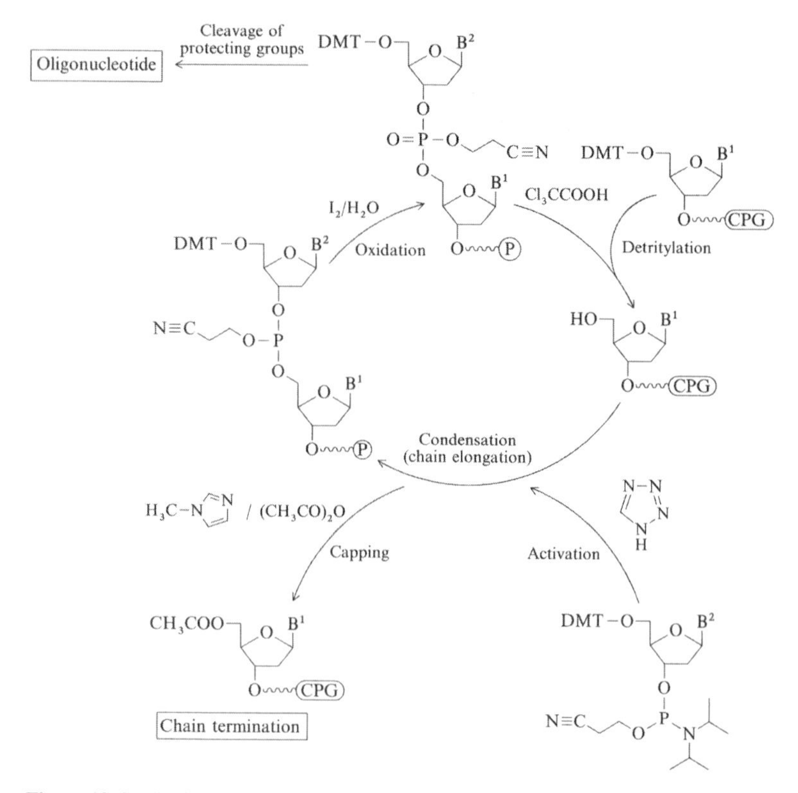

Figure 10. Synthesis cycle according to the phosphoramidite method
B = base; DMTr = 4,4′-dimethoxytrityl residue

Phosphoramidite (Phosphite) Method. This concept is similar to the triester method, but chain synthesis involves trivalent phosphite triester intermediates. The preferred monomer building blocks are the so-called nucleoside phosphoramidites introduced by CARUTHERS et al. [177, 178] that can be activated with tetrazole. After each condensation step, the phosphite triester is converted to the phosphate triester by oxidation with iodine solution. The exceptionally high reactivity of the activated phosphoramidites results in a condensation time of ca. 1 min and $\geq 99\ \%$ yield per chain elongation step.

***H*-Phosphonate Method.** This synthesis method was developed almost simultaneously by GARREGG et al. [197] and by FROEHLER and MATTEUCCI [198]. Condensation is achieved by activation of nucleoside 3′-*H*-phosphonate monomers with pivaloyl or adamantoyl chloride. Since the *H*-phosphonate bond is not cleaved during the synthesis cycle, only one oxidation step is required at the end of the synthesis. This considerably reduces the duration of the cycle. Another advantage over the phosphoramidite method is the higher stability of the monomers. Moreover, excess *H*-phosphonates can be regenerated after the synthesis, thus reducing costs; this makes this method particularly interesting for large-scale synthesis. Since the dinucleoside *H*-phosphonate bond can be easily attacked by nucleophilic reagents [199], oligonucleotide thiophosphates or phosphoamidate analogues can easily be prepared by oxidation with sulfur or amines [199, 200].

Synthesis of Oligoribonucleotides. The solid-phase synthesis of oligoribonucleotides can be conducted by the phosphoramidite [174] or the *H*-phosphonate method [201]. An additional protecting group is required for the 2′-OH group of the ribose [195]. The *tert*-butyldimethylsilyl group has gained acceptance for solid-phase synthesis with ribonucleoside phosphoramidites. It is stable under the acidic conditions required to remove the DMT group and can be cleaved with tetrabutylammonium fluoride at the end of the synthesis. However, steric hindrance due to the *tert*-butyldimethylsilyl group results in yields of ca. 95 % per chain elongation step. Hence, this method is suitable only for the synthesis of oligomers up to a chain length of about 40 nucleotides.

7.5. Cleavage of Protecting Groups and Purification of Oligonucleotides

Cleavage of Protecting Groups. After synthesis, the oligomer is still completely protected and attached to the support. Cleavage of protecting groups should always begin with the internucleotide protecting group on the phosphorus to prevent strand damage [196, 202].

If the triester method is used, the 2-chlorophenyl group is eliminated by treatment with an "aldoximate" [203]; the base-protecting groups and the support are subsequently removed with 32 % aqueous ammonia at 50 °C. Finally, treatment with 80 % acetic acid cleaves the DMT group, provided it is not required for purification by reversed phase high performance liquid chromatography (HPLC).

If methyl phosphoramidites are used for synthesis, the first deprotection step is treatment with thiophenol to cleave the methyl group [7]. If, however, β-cyanoethyl phosphoramidites are used, then ammonia treatment at 50 °C is sufficient to remove all protecting groups including the support.

Purification of Oligonucleotides. Since condensation does not give 100 % yields per coupling step, short-chain homologues are present in addition to the target sequence. For example, for an oligonucleotide with 70 bases and a 98.5 % yield per condensation step, the target sequence accounts for ≤ 35 % of the total yield. A purification step is therefore required for longer oligomers [204]. High performance liquid chromatography (HPLC) or polyacrylamide gel electrophoresis is commonly used for product isolation. Purification by *reversed phase HPLC* is ideal for the routine separation of oligonucleotides up to ca. 40 – 50 nucleotides [204, 205]. Here, the target sequence that is still tritylated is eluted last. Separation efficiency can be increased with special high-affinity protecting groups allowing purification of chains with 150 bases [206]. If the trityl group is removed before purification, the polyanionic character dominates. It is then more advantageous to separate the desired product by ion-exchange chromatography. *Polyacrylamide gel electrophoresis* under denaturing conditions is the most efficient method for oligonucleotide separation [207, 208]; the homologues are separated according to size. The target sequence, which is usually the longest sequence, has the lowest mobility. The individual fragments become visible on exposure to UV light (λ 254 nm) and are cut out of the gel matrix. Extraction with buffer or electroelution and subsequent desalting (dialysis or gel filtration) yield the desired oligonucleotide.

7.6. Synthesis of Modified Oligonucleotides

Oligonucleotides bind specifically to a defined target sequence under suitable hybridization conditions. Relatively short oligonucleotides (<20 bases) are used for new drug design strategies involving targeting interference of genetic expression at the level of transcription or translation [209].

For in vivo chemotherapeutic applications based on sequence-specific hybridization (antisense inhibition), nuclease-resistant oligonucleotides are required (the most important are shown in Fig. 11, see next page). They can be obtained by modifying the phosphate backbone of the oligonucleotides, by using α-anomers (**10**) or 2′-O-alkylribosides (**9**). In spite of the changes these oligonucleotides still hybridize with the target sequence and they can also pass through cell membranes more easily due to their lipophilic nature. Applications of these analogues as antisense oligonucleotides are summarized in [209–211].

Phosphothioate analogues (**1**), in which a nonbridging atom of the phosphate group is replaced by a sulfur atom either at a single specific position or at all positions within the chain, can be synthesized by following the phosphoramidite or *H*-phosphonate route [212–214]. The sulfurization reaction is usually carried out with a solution of sulfur in pyridine and carbon disulfide [215]. Other sulfurization reagents that can also be used in the synthesis are described in [216,217].

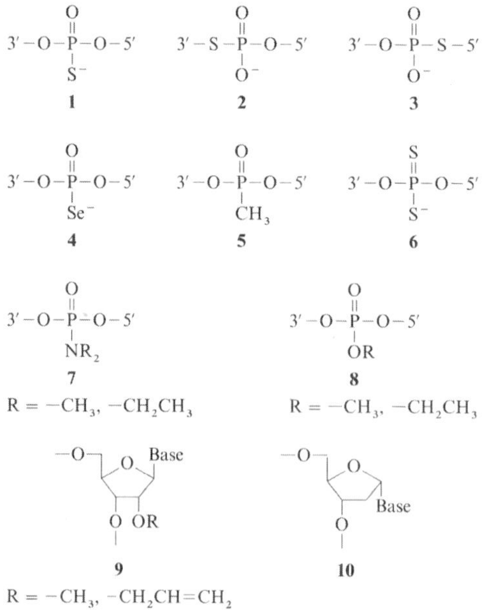

$3'{-}O{-}P(=O)(S^-){-}O{-}5'$ **1** $3'{-}S{-}P(=O)(O^-){-}O{-}5'$ **2** $3'{-}O{-}P(=O)(O^-){-}S{-}5'$ **3**

$3'{-}O{-}P(=O)(Se^-){-}O{-}5'$ **4** $3'{-}O{-}P(=O)(CH_3){-}O{-}5'$ **5** $3'{-}O{-}P(=S)(S^-){-}O{-}5'$ **6**

$3'{-}O{-}P(=O)(NR_2){-}O{-}5'$ **7**, $R = -CH_3, -CH_2CH_3$ $3'{-}O{-}P(=O)(OR){-}O{-}5'$ **8**, $R = -CH_3, -CH_2CH_3$

9, $R = -CH_3, -CH_2CH{=}CH_2$ **10**

Figure 11. Modified oligonucleotides

Like native DNA, phosphorodithioate analogues (**6**) are stereochemically uniform products and can be prepared by the phosphoramidite method with the corresponding thiophosphoramidite monomers [218].

Methylphosphonates (**5**), phosphoramidates (**7**), and phosphotriesters (**8**) are nonionic and have a more hydrophobic character than naturally occurring nucleic acids. The synthesis of oligonucleoside methylphosphonates is described in [219–222], and is best carried out with nucleoside 3′-methylphosphonous acid imidazolide, which in turn is prepared from methylphosphonous acid bisimidazolide and the fully protected nucleoside. Alternatively, a *H*-phosphonate is reacted with *tert*-butyldimethylsilyl chloride to give silyl phosphite and subsequently oxidized with alkyl or aryl halides [223]. All of the early investigations on the biological activity of these oligonucleotide analogues (which were named "matagen", an acronym for mashing tape for gene expression) were conducted by Ts' o and Miller [222].

Phosphoramidates (**7**) can be prepared by reacting *H*-phosphonate oligonucleotides with a solution of the appropriate amine in carbon tetrachloride [213].

The 2′-O-methylribonucleotides (**9**) are well established modified nucleosides that are used as monomer building blocks for oligonucleotide synthesis [224]. The synthesis of the corresponding nucleoside phosphoramidites is described in [225]. Oligonucleotides with α-2′-deoxynucleosides (α-DNA) form β-DNA with natural DNA. This form has parallel strands and is also nuclease resistant [226,227].

To increase the binding of an oligonucleotide to the target, an intercalating agent (e.g., 2-methoxy-6-chloro-9-aminoacridine [228,229] or phenazine [230]) can be covalently attached to the oligonucleotide. These agents insert "internally" between the base pairs of the double helix, increasing duplex stability.

A second method involves binding of photoreactive groups (e.g., psoralen [231–233], azidophenacyl or azidoproflavin derivatives, proflavins, and porphyrins [234–238]) that are activated by UV or visible light and can cross-link with the target DNA or RNA. "Sequence-specific artificial endonucleases" [239] are oligonucleotides that bear a group which can induce specific cleavage of the target nucleic acid after binding to the target. Such groups are the above-mentioned photoreactive groups: subsequent treatment with piperidine allows cleavage of the target sequence at the cross-linking site. Direct light-induced cleavage at neutral pH is achieved with ellipticine and diacapyrene derivatives [238]. Metal chelates attached to oligonucleotides (e.g., iron – EDTA [240–242], copper – phenanthroline [243–246], or iron – porphyrin [247]) lead to relatively specific cleavage of the target sequence. A nonspecific DNase or RNase bound to a complementary oligonucleotide can also destroy the target nucleic acid [248,249].

Nonradioactive labeling of synthetic oligonucleotides has become an important technique in molecular biology and several chemical procedures have been developed. Oligonucleotides are covalently attached to "marker groups" such as fluorescent dyes, biotin, and other biologically active groups (see also Section 8.2). These reporter groups can be attached to an oligonucleotide at the 5′-OH terminus, the 3′-OH terminus, the phosphate backbone, or one of the bases. Attachment at the 5′-OH terminus via protected aminoalkyl or thioalkyl phosphoramidites is most common. These amino- or thiolinkers are added in the last synthesis step and produce a free amino or thiol terminal after cleavage of the protecting groups [250–252]. Subsequent reaction with the activated marker molecule (e.g., biotin-*N*-hydroxysuccinimide) gives the labeled oligonucleotide.

A large number of bifunctional linkers (e.g., maleimidohexanoyl-*N*-hydroxysuccinimide ester) have been described which allow coupling of enzymes [253], peptides, and proteins to linked oligonucleotides via amino groups of the amino acids or thiol groups of cysteine. Attachment at the 3′-OH terminus can be carried out by linking a ribose unit with T 4 RNA ligase, followed by periodate cleavage of the ribose ring, and subsequent reductive amination [254,255]. Labeling via a 3′-terminal thiol group is described in [256]. Also a modified support is commercially available which produces an aliphatic primary amino group at the 3′-terminus [257]. At the internucleotide bond reporter groups can be attached, e.g., via a phosphoramidate linkage [258–260].

The attachment of reporter groups to modified bases usually occurs at the C-5 position of pyrimidines or the C-8 position of purines [261]. During oligonucleotide synthesis, a fully protected deoxyuridine phosphoramidite [262] modified with a trifluoroacetylaminopropenyl residue is inserted at the desired position in the sequence (usually at C-5); after cleavage of the protecting groups, the group to be coupled is bound via the resulting amino function. Numerous other labeling positions have been described (e.g., the C-4 position [263–265] of 2′-deoxycytidine and the N-6 position of the adenine ring [266] or the exocyclic amino function of the guanine residue [267]).

8. Uses

The direct use of nucleic acids in *nucleic acid probe technology* is becoming increasingly important. Here, labeled nucleic acids are used as probes to detect other nucleic acids by means of hybridization [21,23]. Identification of the specific nucleotide sequence of a pathogen (virus, bacterium, parasite) can be used to diagnose an infection [268,269]. More complex analyses are often necessary for the detection of changes in nucleic acids (point mutations, deletions, insertions, translocations, expression level or number of copies of the genetic information) in the diagnosis of hereditary diseases [24], or in the investigation of tumors [25,26]. This chapter will be limited to describing the uses of nucleic acids in probe technology but other applications will first be briefly summarized.

The second important field of application for nucleic acids is *genetic engineering* which allows the production of proteins in bacteria, fungi, animal or plant cell cultures, or in animals or plants. The nucleic acid that codes for the desired protein is introduced into the organism and expressed.

Antisense technology

is a third field of application of nucleic acids which, however, still only exists in model systems. This technology aims to prevent the expression of nucleic acid sequences that have a negative effect on an organism. This can be achieved by the interception and binding of mRNA with synthetic antisense oligonucleotides (see Section 7.6) or with an antisense RNA from a recombinant expression system. The mRNA forms double strands with the antisense RNA by specific base pairing and cannot be translated [270,271]. Another method of stopping expression is the specific cleavage of the target nucleic acid by catalytic RNA (ribozymes) [270,271]. The target nucleic acid is recognized by specific base pairing with the ribozyme. Antisense technology may be used to treat infections caused by viruses, bacteria, or parasites by cleaving or preventing the expression of the genetic information of these pathogens. Another possibility is the treatment of diseases resulting from overexpression of an endogenous gene or its expression in an erroneous target cell.

Finally, nucleic acid technology offers great promise in *research*. As a result of the dramatic improvements in cloning and sequencing large sections of DNA, DNA sequences or genes are increasingly being used as a starting point for the analysis of a biochemical mechanism. Nucleic acid sequences are also used to study evolutionary relationships between different organisms by comparing sequences of conserved genes.

8.1. Hybridization Techniques for Nucleic Acid Detection

Nucleic acid hybridization is based on the fact that two complementary nucleic acid strands (DNA – DNA, RNA – RNA, DNA – RNA) associate to form a double strand according to the rules of base pairing. The nucleic acid to be detected must be present as a single strand, i.e., denaturation of double strands is required before hybridization. The nucleic acid probe must contain a sequence complementary to that of the nucleic acid under investigation and have a label to permitidentification of the double strands formed by hybridization.

Solid-Phase Hybridization. The labeled hybrid strands must also be distinguishable from unhybridized probes. The simplest way of doing this is to use methods in which the nucleic acid of interest is bound to a solid phase (usually a nitrocellulose or nylon membrane). The potential nonspecific binding sites for the probe on the solid phase are blocked prior to the hybridization reaction with unlabeled nucleic acids of different origins, proteins, and polymers. Consequently, the labeled probe only binds to the solid phase via specific base pairing to the target nucleic acid. The unbound probe is subsequently washed away.

In the *dot blot process* nucleic acid samples are spotted on a membrane [272]. After hybridization the detection of bound label reveals which spots contain the nucleic acid of interest. Quantification is possible by comparison with the signal from a standard [273].

In *Southern blotting* the DNA to be analyzed is separated according to size by gel electrophoresis and subsequently transferred to a membrane without changing the spatial distribution of the DNA fragments [21,274]. After hybridization the size of the fragment containing the sequences of interest can be analyzed. Analogous separation and hybridization of RNA is referred to as *Northern blotting*.

In *in situ hybridization* the nucleic acid of interest is fixed in such a manner that its natural cellular compartmentalization is maintained [21,23,275]. As in histological studies, entire cells are applied and fixed to microscope slides or tissue is cut into ultrathin layers and fixed. Hybridization takes place on the slide between the labeled probes and the nucleic acids in the sample material. Examination under the microscope reveals which and how many cells contain the nucleic acid of interest, as well as its subcellular location. In situ hybridization permits differentiation between individual cells, whereas in solid-phase blotting methods the nucleic acids from many cells are mixed together. In situ hybridization also permits the assignment of a nucleic acid sequence to a chromosome by analyzing cells that are in the metaphase of cell division [276,277]. The areas of the chromosomes containing the complementary sequences are specifically labeled by the nucleic acid probe.

Hybridization in Solution. Hybridization reactions in which both reaction partners are in solution are more efficient. The hybridized probe, as part of a double strand, can be separated from the unreacted single-stranded probe by *electrophoresis* or *chromatography*.

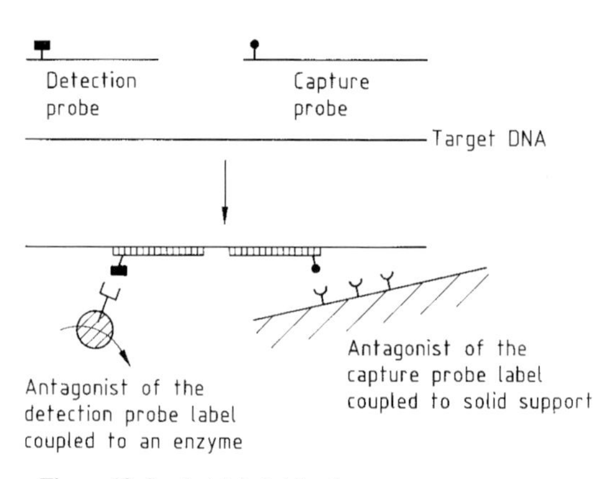

Figure 12. Sandwich hybridization
The oligonucleotide detection and capture probes are labeled with two different labels.

Sandwich hybridization offers an alternative way of separating hybrids formed in solution (Fig. 12) [278,279]. The nucleic acid of interest is hybridized with two different probes that recognize different sections of the nucleic acid. One of the probes (detection probe) is labeled to allow detection of the hybrid. The second probe acts as a capture probe to bind the hybrid consisting of the nucleic acid of interest and the detection probe to a solid phase. The detection probe only binds to the solid phase if the nucleic acid of interest is present in the solution and forms a bridge between the two probes. The capture probe can be covalently coupled to the solid phase [280], or can be labeled with a molecule that is distinct from the detection label and is bound to the solid phase via a fixed antagonist. Examples are biotin-labeled probes with a streptavidin- or avidin-loaded solid phase [281] or antigen-loaded probes with an appropriate antibody on the solid phase.

Strand displacement is another method for detecting hybrids formed in solution (Fig. 13) [282,283]. Here, the probe is a hybrid of a short, labeled oligonucleotide and a longer nucleic acid strand forming a partially single-stranded, partially double-stranded molecule. The longer strand is complementary to the nucleic acid of interest. If the nucleic acid of interest is present in the assay, double strands are formed with the single-stranded region of the probe. This double-stranded region of the unlabeled probe strand and the nucleic acid of interest is longer than the double-stranded region formed between the two probe strands. Therefore complete hybridization occurs between the longer probe strand and the nucleic acid of interest. As a result the short, labeled probe strand is released and can be detected.

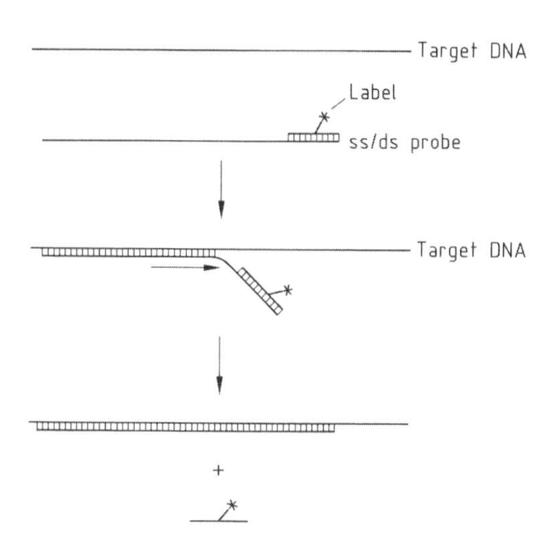

Figure 13. Strand displacement
ss = single-stranded; ds = double stranded

Finally, there are a series of modifications of a hybridization principle which does not require the separation of bound from unbound probe. Here, two labeled probes hybridize with the nucleic acid of interest at adjacent positions. Spatial proximity of the two labels is only guaranteed if the nucleic acid of interest is present and produces a signal that can be measured (e.g., enzymatic activity, light of a certain wavelength) [284].

8.2. Labeling and Detection Systems

Radioactive labeling is losing importance due to the development of sensitive, nonradioactive labels. Nucleic acids are usually radioactively labeled by enzymatic incorporation of nucleotides containing ^{3}H, ^{35}S, or ^{32}P isotopes with polymerases. Alternatively, the 5′-OH terminus of DNA can be labeled with ^{32}P by polynucleotide kinase [21]. Radioactive nucleic acids are detected by exposure to X-ray film or a photoemulsion (in situ hybridization) or by measurement in a scintillation counter.

In nonradioactive methods, the nucleic acid can be directly labeled with a fluorescent or luminescent dye [285], bound to an enzyme (e.g., β-galactosidase, horseradish peroxidase, alkaline phosphatase) or coupled with biotin or a hapten (antigenic determinant) [284]. The hapten (or biotin) is detected by the binding of an appropriate antibody (or streptavidin/avidin). The antibody and streptavidin/avidin are, in turn, directly coupled to a dye molecule or to an enzyme [21]. Additional chemical methods for labeling nucleic acids are described in [21,286–288] and in Section 7.6.

Photolabeling is a simple and fast labeling method [21,289]. Biotin or a hapten is coupled to a photoreactive group (photobiotin [289], photodinitrophenyl phosphate (DNP) [290], or photodigoxigenin [291]) which reacts with the amino groups of the nucleotides. On irradiation with light the photoreactive group is activated resulting in the formation of a covalent bond between biotin or the hapten (DNP, digoxigenin) and the nucleic acid.

Nucleic acids from biological sources (e.g., plasmids) are usually labeled with the aid of enzymes. Modified nucleotides are inserted with polymerases [21]; DNA polymerases are used to insert the deoxyribonucleotide triphosphate derivatives (e.g., biotin – dUTP, biotin-dATP, digoxigenin – dUTP, bromo – dUTP, and fluorescein – dUTP) in nick translation or by random primed labeling. The 3′-terminus of nucleic acids can be labeled with these nucleotides by terminal transferase. Modified ribonucleotides (e.g., biotin – UTP and digoxigenin – UTP) are inserted into RNA probes by in vitro transcription with RNA polymerases, particularly with phage-coded RNA polymerases [21,22].

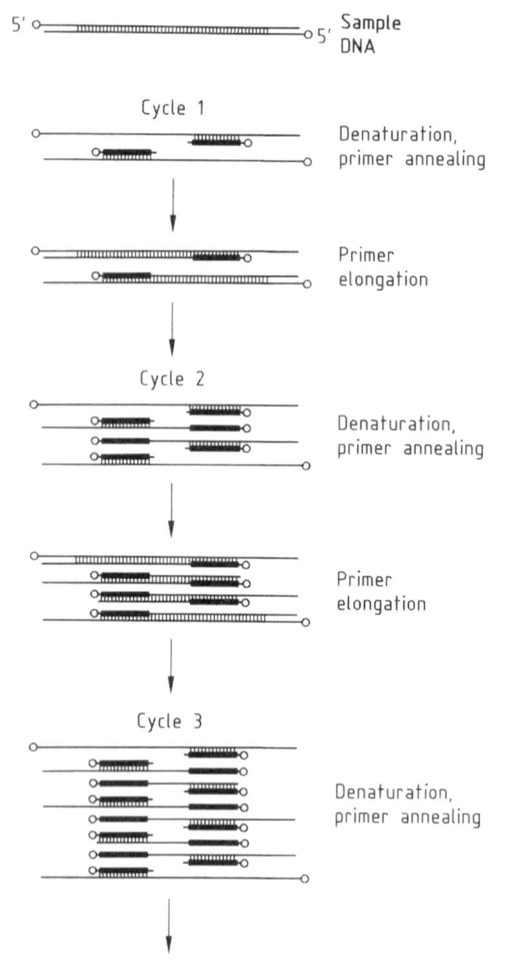

Figure 14. Polymerase chain reaction (PCR)

8.3. Amplification Systems

The sensitivity limit of nucleic acid probe technology can be increased by specifically amplifying (i.e., increasing the amount of) the nucleic acid of interest [21].

Polymerase Chain Reaction (PCR). The polymerase chain reaction is an established technique [15,27,292,293] which is used to amplify a particular nucleic acid from a complex mixture of sequences. The amplified DNA can be subsequently sequenced and/or cloned using genetic engineering techniques, or can be used to increase the sensitivity of pathogen detection. The sensitivity can be increased to such an extent that, for instance, one genome from the human immunodeficiency virus (HIV) integrated in the DNA of a single cell can be detected in the presence of many uninfected cells.

The PCR is based on the repetition of three partial reactions in 20 – 50 cycles (Fig. 14):

1) An oligonucleotide primer is annealed to each of the two denatured complementary strands of the DNA sequence to be amplified.
2) Primer extension catalyzed by a DNA polymerase converts the region to be amplified into two double strands.
3) The double strands are denatured by heating.

Theoretically the number of templates can be doubled in each cycle, so that 2^n templates would be created in n cycles; 25 cycles would result in an amplification rate of 3.4×10^7. However, the partial reactions are not perfectly efficient, and amplification rates of ca. $(1-3) \times 10^6$ are obtained in 25 cycles. Implementation of the PCR is facilitated by the use of thermostable DNA polymerases because otherwise new polymerase would have to be added at the start of each cycle. Typical PCR cycles consist of a few minutes of primer annealing at ca. 55 °C, a few minutes of elongation at ca. 70 °C, and 1 – 2 min of denaturation at 95 °C. These steps can be carried out automatically in programmable "thermocyclers".

RNA can also be amplified by PCR if a DNA template is created in the first step with the help of reverse transcriptase. The PCR product can be detected by electrophoresis with or without previous cleavage with restriction endonucleases, and with or without hybridization with a probe. The PCR product can be labeled directly by using labeled primers [294] or by incorporating labeled nucleotides (Section 8.2) [295,296]. Contamination is a serious problem in PCR because of the enormous sensitivity of the method. For example, contamination of a sample with a single molecule of the DNA of interest can lead to an incorrect positive result.

Other Amplification Systems. In *transcription-based amplification* oligonucleotide primers are annealed to the nucleic acid of interest which, in addition to complementary sequences to the target, also contain a promoter (binding site) for RNA polymerase [297]. After primer elongation RNA polymerases are used to produce RNA transcripts starting from the promoters. The transcripts can be converted to DNA with reverse transcriptase and used again as targets for the oligonucleotide primers.

The *ligase chain reaction* uses a ligase instead of a polymerase in temperature cycles analogous to the PCR [298, 299]. Two oligonucleotides are annealed adjacent to one another on the template strand and joined by a ligase. The ligated strand is then removed by denaturation and two oligonucleotides can anneal again in the next cycle.

Other amplification systems do not require cycles with different temperatures but proceed isothermally. In *self-sustained sequence replication* (3 SR) [300] and *nucleic acid sequence based amplification* (NASBA) [301], three enzymes act simultaneously at a constant temperature to amplify the target sequence. Reverse transcriptase synthesizes DNA on an RNA template by extending oligonucleotide primers that contain promoter sequences for RNA polymerases. The RNA strand of the resulting DNA–RNA hybrid is digested with RNase H allowing double-stranded DNA to be produced by reverse transcriptase with a second primer. RNA polymerase then transcribes the double-stranded DNA starting at the promoters, and the resulting RNA is again converted to DNA with reverse transcriptase.

Another isothermal amplification technique is based on the replication of RNA genomes with the help of *Qβ replicase*–a RNA-dependent RNA polymerase from phage Qβ [302, 58]. Natural substrates of Qβ replicase are characterized by RNA sequences forming a special secondary structure. For nucleic acid detection these sequences are modified by the insertion of short probe sequences. Unlike the amplification methods described above, the target sequence itself is not multiplied, instead the probe is amplified after it has located the target sequence (signal amplification). Nonspecifically bound and contaminating probe molecules cause background problems that can be alleviated by repeating the hybridization in cycles (reversible target capture) [303].

Another method of signal amplification involves the formation of probe networks in which the target-specific probe has target sequences for secondary probes. A "Christmas tree" of labeled probe sequences is produced for each original target molecule [304–306].

8.4. Applications of Probe Technology

Detection of Infectious Microorganisms. Detection of nucleic acids can be used to identify pathogenic viruses, bacteria, or parasites. In comparison with the detection of proteins or antigens, nucleic acid detection has the advantage that it allows identification of infectious organisms that frequently change their antigens (e.g., HIV) or that exist in a variety of serotypes. Furthermore, viral infections can be detected at the latent stage when no detectable synthesis of viral proteins occurs (e.g., HIV or herpes viruses). DNA probe technology can be more sensitive than immunological techniques due to the possibility of nucleic acid amplification. Although immunological tests are still easier, faster, and cheaper, competitive nucleic acid tests are to be expected in the future.

Detection of the nucleic acid of a pathogen can also serve as a direct measure of the infectiousness of a patient or test material [307]. Quantitative determination of the pathogenic nucleic acid can be used to follow the progress of treatment.

Detection of viruses is possible with a broad spectrum of nucleic acid probes, ranging from synthetic oligonucleotides to the complete cloned viral genome. Infections caused by some virus groups (e.g., papilloma viruses or enteroviruses [308]) can be diagnosed with a probe containing group-specific sequences identical in all members of the group and by using probes with nucleic acid sequences specific for each member.

For the detection of bacteria, the fact that each bacterium contains approximately 10 000 ribosomes and thus 10 000 copies of rRNAs can be used to increase the sensitivity level. The rRNA sequences are relatively well conserved and can used to establish relationships between different organisms in phylogenetic trees. Tests for individual bacterial species, for groups, or even for a wide range of bacteria can be carried out with appropriate oligonucleotide probes [309]. Another group of nucleic acid probes contains sequences of bacterial pathogenicity factors and allows differentiation between pathogenic and apathogenic types of a species (e.g., *Escherichia coli*) [268, 269]. Analogously, detection of resistance genes permits detection of antibiotic resistance and simultaneous identification of the pathogen [310]. Bacterial identification is also applied in the detection of pathogens in food (e.g., *Salmonella*).

Detection of Human DNA Sequences. Nucleic acid probe technology can be used for the *diagnosis of hereditary diseases* and *genetic counseling* [24]. If the genetic basis of the disease is known, the alteration in the nucleic acid can be detected by hybridization with the appropriate region. Such alterations include point mutations (sickle-cell anemia), deletions (Duchenne's muscular dystrophy), or variations in expression (forms of thalassemia). If the genetic cause of the disease is unknown, genetic markers that are inherited together with the defect gene are analyzed, e.g., recognition sites for restriction enzymes (restriction fragment length polymorphism, RFLP) [311].

The analysis of human sequences can also be applied in *tumor diagnosis* [25, 26]. Some human tumors are associated with characteristic chromosomal changes or alterations in genes (oncogenes) [312]. Suitable nucleic acid probes can be used for the highly sensitive detection of DNA sequences of different chromosomes connected by chromosomal translocations or for the detection of mutations in oncogenes that are associated with tumors. For instance, the Philadelphia chromosome (a translocation between chromosome 9 and 22) is characteristic for chronic myelogenous leukemia [313]. Point mutations in the *ras* oncogene are associated with several tumors, and amplification of the *N-myc* gene with neuroblastomas. In the case of recessive oncogenes (tumor suppressor genes) [314, 315], the loss of gene expression results in a tumor; for example, the loss of the retinoblastoma gene results in an eye tumor, retinoblastoma. Another area of tumor diagnosis is the detection of clonality of a leucocyte population caused by the transformation of a white blood cell [316]. Since all lymphocytes have their own characteristic rearrangement of immunoglobulin genes or T-cell receptor genes, the appearance of a defined pattern in many lymphocytes indicates proliferation of a single cell.

Methods that use the genetic diversity of humans to study individuals have wide applicability. Determination of the genes of the major histocompatibility locus (HLA) can be used both for diagnosis (transplantation medicine, prediction of risk of disease, autoimmune diseases) and for differentiation between individuals [317]. Probes which hybridize with repetitive sequences that occur in each individual in a slightly different, genetically determined arrangement can be used for finger printing [28]. Here nucleic acids cleaved by restriction enzymes are separated electrophoretically and hybridized with these probes in the Southern blot technique (Section 8.1). Several fragments of different sizes, each containing these repetitive sequences, produce a signal. Each individual inherits half the sequences from the father and half from the mother, and therefore has a unique fragment pattern [318–320]. This technique is used in forensic medicine to identify criminals by DNA isolated from cells (hair, blood etc.). Other fields of application are in the determination of paternity and family relationships.

9. Nucleosides and Nucleotides

9.1. Nucleosides

The term nucleosides was originally applied to the ribose derivatives of purines that can be isolated from an alkaline RNA hydrolysate. Later, this name was given to all purine and pyrimidine *N*-glycosides of D-ribose and 2-deoxy-D-ribose. Today, this term refers to all natural or synthetic compounds which consist of a heterocyclic nitrogen-containing base (aglycon) and a carbohydrate residue (glycon). The nitrogen atom (*N*-nucleoside) or carbon atom (*C*-

nucleoside) of the heterocycle is linked to the anomeric carbon atom of the sugar residue.

Ribonucleosides and deoxyribonucleosides are obtained from naturally occurring RNA and DNA by enzymatic or chemical hydrolysis. Most nucleosides are stable within a wide pH range. However, under strongly acidic or alkaline conditions at elevated temperature cleavage of the *N*-glycosidic bond may occur. The purine nucleosides are adenosine [*58-61-7*], deoxyadenosine [*958-09-8*], guanosine [*118-00-3*], and deoxyguanosine [*961-07-9*]. The pyrimidine nucleosides are cytidine [*65-46-3*], deoxycytidine [*951-77-9*], thymidine [*50-89-5*], and uridine [*58-96-8*].

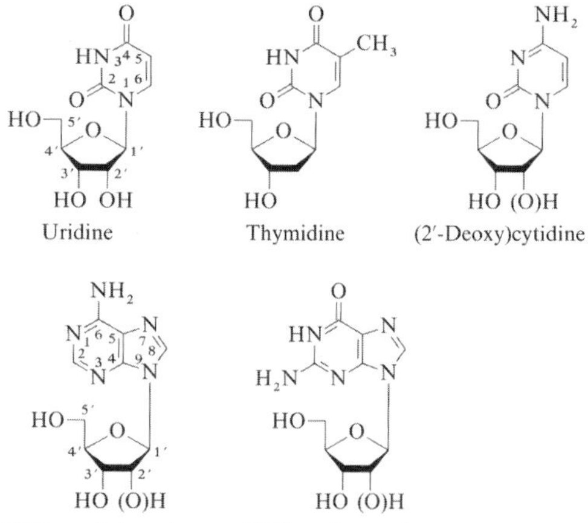

Uridine

Thymidine

(2′-Deoxy)cytidine

(2′-Deoxy)adenosine

(2′-Deoxy)guanosine

Figure 15. Common nucleosides

The furanoses D-ribose and 2-deoxy-D-ribose are linked to the bases by a β-glycosidic bond at C-1 (Fig. 15). A limited number of nucleosides in the α-D configuration (e.g., 5,6-dimethylbenzimidazole-α-D-ribofuranoside of vitamin B 12) can be isolated from natural substances.

Rare nucleosides and nucleoside antibiotics are found in biopolymers as fermentation products of microorganisms or as building blocks of nucleic acids [29]. They exhibit structural modifications of the nucleobase or the glycon residue.

An example of a rare building block of viral DNA is 2′-deoxy-5-hydroxymethylcytidine (**11**) and those of tRNA are the hypermodified nucleosides queuosine [*57072-36-3*] (**12**) and wyosine [*52662-10-9*] (**13**).

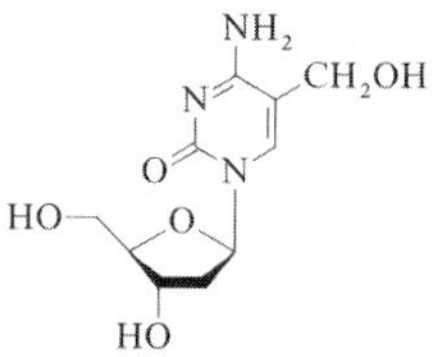

11

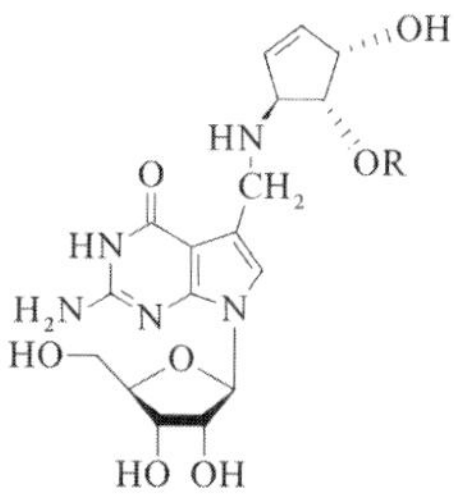

12

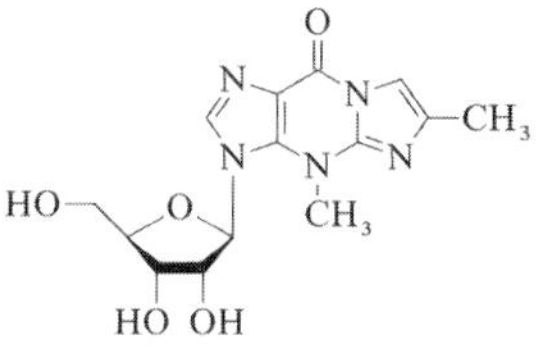

13

Examples of nucleosides formed by microorganisms are tubercidin [*69-33-0*] (**14**), formycin [*6742-12-7*] (**15**), cordycepin [*73-03-0*] (**16**), blasticidin S [*2079-00-7*] (**17**), vidarabin [*24356-66-9*] (**18**), oxetanocin [*103913-16-2*] (**19**), aristeromycin [*19186-33-5*] (**20**), neplanocin A [*72877-50-0*] (**21**), sinefungin [*58944-73-3*] (**22**), and tunicamycin A [*66081-37-6*] (**23**, $n=9$), tunicamycin B [*66081-36-5*] (**23**, $n=10$), tunicamycin C [*66081-37-6*] (**23**, $n=8$), and tunicamycin D [*66081-38-7*] (**23**, $n=11$) (Fig. 16).

Naturally occurring nucleosides show a wide variety of biological effects. Accordingly a large number of derivatives have been synthesized in order to find useful chemotherapeutics [30]. The 2′,3′-dideoxyribonucleosides are important synthetic nucleoside derivatives with modified sugar moieties. In the cell the 5′-triphosphate derivatives of these compounds inhibit the reverse transcriptase of retroviruses and thus are potentially important therapeutic drugs (e.g., 3′-azidothymidine [*30516-87-1*] (**24**) (AZT), the first approved drug against HIV infection.

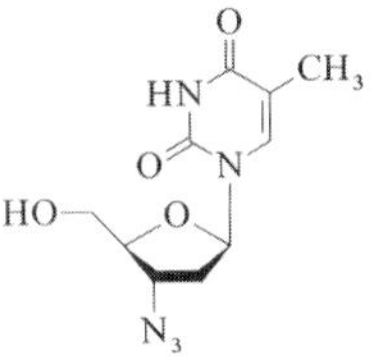

24

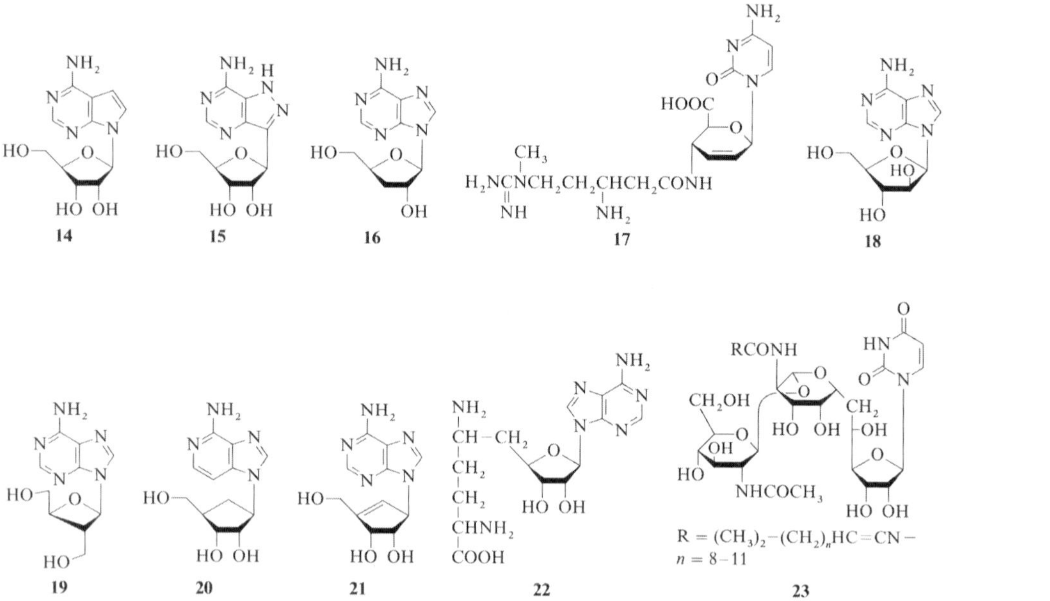

Figure 16. Nucleosides formed by microorganisms

Other derivatives which contain only a part of the ribofuranosyl residue (e.g., acyclonucleosides) also show antiviral activity. This class of substances includes acyclovir (**25**), ganciclovir (**26**) and phosphonylmethyl analogues [e.g., (*S*)-9-(3-hydroxy-2-phosphonylmethoxypropyl) adenine (**27**)]. Nucleosides with a halogen substituent in the heterocyclic moiety are important antiviral drugs (e.g., 5-iodo-2′-deoxyuridine, 5-bromovinyl-2′-deoxyuridine (**28**), or 2-chloro-2′-deoxyadenosine).

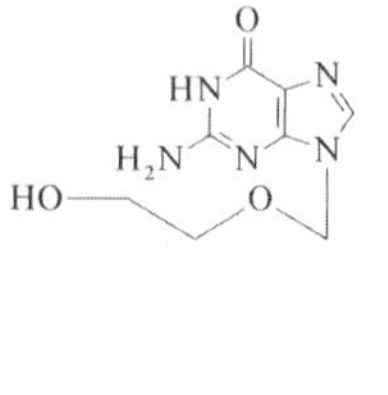

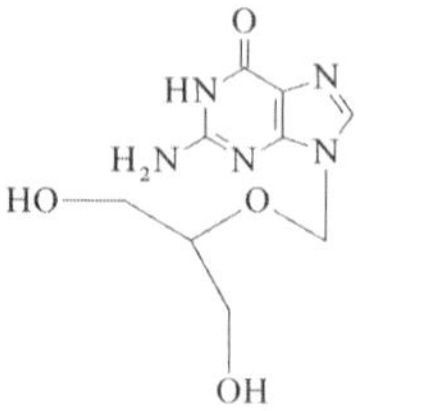

The nucleosides are obtained by modification of existing nucleosides or total synthesis. Total synthesis can be carried out in a linear or convergent manner. Due to the chiral centers on the glycon residue, preparation of optically pure nucleosides requires stereoselective methods of synthesis, especially for the glycosidation of the aglycons. The most important methods of total synthesis will be described briefly. Detailed surveys of the chemical synthesis of and on nucleoside derivatives are given in [31–35].

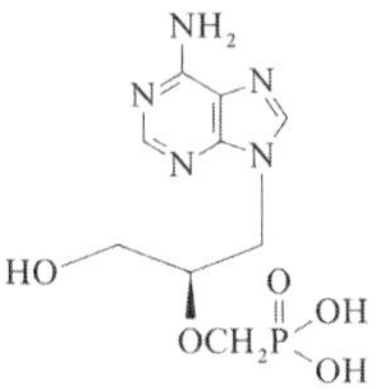

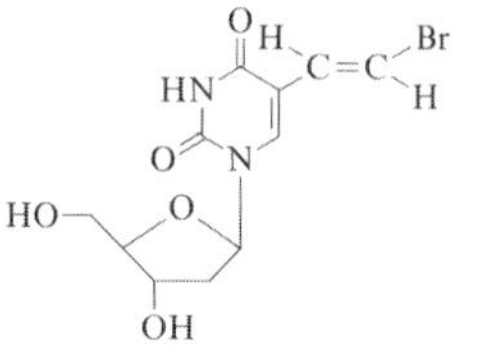

Synthesis of *N*-Nucleosides. The first syntheses date back to the work of FISCHER and HELFERICH who condensed acetyl-protected α-D-glucopyranosyl halides with the silver salts of heterocycles. The analogous chemical synthesis of the natural nucleosides adenosine and guanosine was described by TODD and coworkers [321]. This method was improved after the introduction of the chloromercury purine derivatives [322] and the use of mercury cyanide (which avoids the tedious preparation of the heavy metal salts of the nucleobases) [323]. The Hilbert–Johnson method was developed for the preparation of pyrimidine nucleosides [324]. This method starts with 2,4-dialkoxypyrimidines, which yield the corresponding nucleosides on reaction with O-acetyl-protected

1-halogenosugars (glycosyl halides) and subsequent ammonolysis or hydrolysis of the 4-alkoxy intermediates [325].

The introduction of the melt process according to SATO [326] represented another advance in nucleoside synthesis. Here, a polyacylated sugar is heated with the purine or pyrimidine base and a catalyst (e.g., 4-toluenesulfonic acid or zinc chloride) in vacuum. This method has the advantage that the free heterocyclic bases (instead of their heavy metal salts) and the 1-O-acetyl sugars (instead of the corresponding glycosyl halides) can be used.

Another synthesis, developed simultaneously by BIRKHOFER and coworkers [327] and NISHIMURA and coworkers [328], involves reaction of trialkylsilylated heterocycles with peracylated 1-halogenosugars in the presence of silver perchlorate as catalyst.

This method was improved by VORBRÜGGEN and is a silyl version of the Hilbert–Johnson method [329]. It is the method of choice for the synthesis of pyrimidine nucleosides. A similar process is based on transglycosylation, i.e., the transfer of the sugar residue of a nucleoside (donor) to a heterocyclic base (acceptor). Apart from chemical transglycosidation, enzymatic sugar transfer is also possible [330].

An elegant method for the synthesis of the nucleosides of purines or purine analogues, developed by SEELA and coworkers, is based on phase transfer catalysis [331]. It gives high yields of the desired β-anomers and the anomeric ratio can be controlled by the choice of catalyst. Finally, reference should be made to the synthesis of the heterocyclic moiety from the C-1 functionalized sugar molecule [332]. This is the method of choice for the synthesis of carbocyclic nucleosides [333].

Synthesis of *C*-Nucleosides. Two methods are available for synthesizing the C–C linkage between the glycosidic and the aglyconic parts of the molecule [334]:

1) Synthesis according to the Fischer–Helferich principle
2) Synthesis via a preformed function on C-1 of the sugar

Unlike *N*-glycosides, the glycosidic bond of *C*-nucleosides is stable to acids and these compounds are not attacked by nucleoside phosphorylases. Consequently *C*-nucleosides are more stable than *N*-nucleosides in biological systems.

9.2. Nucleotides

The term nucleotide generally refers to the phosphate esters of nucleosides. Only mononucleotides (i.e., nucleotides consisting of only one nucleotide unit) are considered here. Mononucleotides can be joined together to form an oligonucleotide (up to 200 nucleotide units) or a polynucleotide (> 200 nucleotide units).

Nucleotides are obtained by the chemical or enzymatic hydrolysis of nucleic acids and by the chemical or enzymatic phosphorylation of nucleosides [335–339]. They may be ribonucleotides (from RNA) or 2′-deoxyribonucleotides (from DNA). Depending on the position at which the sugar moiety is phosphorylated nucleotides can exist as 2′-, 3′-, or 5′-phosphoric acid monoesters. Cyclic 2′,3′- and 3′,5′-phosphoric acid diesters can be formed by intramolecular dehydration.

Cyclic adenosine monophosphate [*60-92-4*] (cAMP) is especially important as a mediator of the action of a large number of hormones (second messenger) [340].

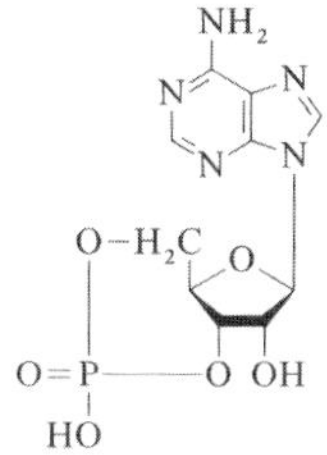

Adenosine
3′,5′-monophosphoric acid
(cyclic AMP)

Cyclic AMP is formed in the cell from ATP by the enzyme adenylate cyclase. It was first synthesized in the laboratory by COOK and coworkers [341]. Numerous attempts have been made to specifically alter the pharmacological effects of cAMP by chemical modification [342].

The nucleotides can contain more than one phosphate group. The 5′-triphosphates of the ribonucleosides and 2′-deoxyribonucleosides of the purines adenine (ATP or dATP) and guanine

(GTP or dGTP) and of the pyrimidines cytosine (CTP or dCTP) and thymine (dTTP) or uracil (UTP) are the most common natural nucleotides. They serve as building blocks for the polymerase-catalyzed synthesis of nucleic acids. Adenosine 5′-triphosphate [*56-65-5*] is also the main source of chemical energy in the living world.

Adenosine 5′-triphosphoric acid (ATP)

In contrast to many nucleosides, the corresponding phosphate esters are readily soluble in water as their alkali salts. Although the nucleoside monophosphates are stable compounds, the corresponding nucleoside di- and triphosphates undergo slow hydrolytic cleavage of the pyrophosphate bond, even at ambient temperature and neutral pH. Nucleoside mono- and polyphosphates are dephosphorylated by phosphatases (phosphomonoesterases) to yield the corresponding nucleosides.

Analogues containing modified phosphate groups (e.g., thiophosphate [343], imidophosphate [344], or methylene phosphonate functions [345]) are not or only very slowly attacked by phosphomonoesterases. The triphosphate derivatives of those analogues which are only slightly modified compared to the parent nucleotides are valuable tools for studying the reaction mechanisms of ATP-dependent enzyme systems.

The naturally occurring cyclic nucleoside 3′,5′-monophosphates are also very stable. They are converted to acyclic 5′-monophosphates by specific phosphodiesterases. Other important nucleotide building blocks occurring in nature are the adenosine phosphate residues in nicotin-amide adenine dinucleotide phosphate [*53-59-8*] (NADP), coenzyme A [*85-61-0*], and flavine adenine dinucleotide (FAD) [*146-14-5*].

β-Nicotinamide adenine dinucleotide phosphate (NADP)

Coenzyme A

Synthesis of Nucleotides. For a review of chemical nucleotide syntheses, see [346]. Nucleoside monophosphates are usually prepared by chemical phosphorylation of the nucleosides with phosphorus oxytrichloride. If the reaction is carried out with a trialkyl phosphate as solvent the 5′-monophosphates are obtained almost exclusively [347]. Protection of the 2′- and/or 3′-OH functions of the sugar is not required. Various methods are available for the chemical synthesis of nucleoside di- and triphosphates from nucleoside monophosphates.

In the *morpholidate method* the nucleoside monophosphate is reacted with morpholine and dicyclohexylcarbodiimide to give the corresponding phosphomorpholidate [348]. The latter is subsequently reacted with orthophosphate to give the nucleoside diphosphate or with pyrophosphate to yield the nucleoside triphosphate.

In the *imidazolidate method* the nucleoside monophosphates are reacted with 1,1′-carbonyldiimidazole to give the corresponding nucleoside monophosphate imidazolidates [349]. These compounds are in turn treated with inorganic ortho- or pyrophosphate to yield the nucleoside di- or triphosphates. In comparison with the morpholidate method, this process has the advantage of being a "one-pot" reaction.

In the *anhydride method* the nucleoside monophosphates are reacted with diphenyl chlorophosphate [350]. The resulting nucleoside diphenyl pyrophosphates can be subjected to nucleophilic exchange by treatment with anions such as ortho- or pyrophosphate.

These methods can be used analogously to prepare the nucleotides of the coenzyme type by replacing the orthophosphate with, for example, sugar phosphates. The possibility of reacting the nucleoside phosphodichloridate, obtained after treatment with phosphorus oxytrichloride, directly with pyrophosphate in a "one-pot" reaction has been described [351].

Enzymatic syntheses of nucleotides are described in [335–338].

9.3. Therapeutically Important Nucleoside and Nucleotide Derivatives

Antiviral Derivatives. The strategies involved in the development of antiviral nucleosides are reviewed in [352]. For the development and applications of antiviral active nucleotide analogues, see [353]. Of the large number of compounds that have been synthesized since the early 1970s, only a few derivatives have so far been approved. Among them are:

1) 5-iodo-2′-deoxyuridine [*54-42-2*] (IDU)
2) 5-trifluoromethyl-2′-deoxyuridine [*70-00-8*] (trifluorothymidine, TFT)
3) 9-β-D-arabinofuranosyladenine [*24356-66-9*] (Ara-A, vidarabin) (**18**)
4) 3′-azido-3′-deoxythymidine [*30516-87-1*] (AZT, zidovudine) (**24**)
5) 1-β-D-ribofuranosyl-1,2,4-triazole-3-carboxamide [*36791-04-5*] (ribavirin, virazol)
6) 9-[(2-hydroxyethoxy)methyl]guanine [*59277-89-3*] (acyclovir) (**25**)

The following nucleoside analogues are active against HIV infections. However, clinical trials of these compounds are still incomplete or they have not yet been approved partly because of serious side effects:

1) 2′,3′-dideoxycytidine [*7481-89-2*] (ddC), 2′,3′-dideoxyadenosine [*4097-22-7*] (ddA), and 2′,3′-dideoxyinosine [*69655-05-6*] (ddI)
2) 9-(1,3-dihydroxy-2-propoxymethyl)guanine [*82410-32-0*] (ganciclovir, DHPG) (**26**)
3) (*E*)-5-(2-bromovinyl)-2′-deoxyuridine [*73110-56-2*] (BVDU) (**28**)

For further details, see → Chemotherapeutics, Chap. 4.6.1. → Chemotherapeutics, Chap. 4.6.8.

Antitumor Derivatives. For details, see → Cancer Chemotherapy, Chap. 2.4.

10. References

General References

1. L. Stryer: *Biochemistry,* W. H. Freeman, New York 1988.
2. D. Freifelder: *Molecular Biology,* Jones and Bartlett Publishers, Boston 1987.
3. J. Darnell, H. Lodish, D. Baltimore: *Molecular Cell Biology,* Scientific American Books, W. H. Freeman, New York 1990.
4. B. Alberts: *Molecular Biology of the Cell,* Garland Publishing, New York-London 1983.
5. J. D. Watson et al.: *Molecular Biology of the Gene,* The Benjamin/Cummings Publishing Comp., Menlo Park, California 1987.
6. B. Lewin: *Gene,* VCH Verlagsgesellschaft, Weinheim 1988.
7. J. Sambrook, E. F. Fritsch, T. Maniatis: *Molecular Cloning. A Laboratory Manual,* 2nd ed., Cold Spring Harbor Laboratory Press 1989.
8. E.-L. Winnacker: *From Genes to Clones. Introduction to Gene Technology,* VCH Verlagsgesellschaft, Weinheim 1987.
9. A. Kornberg: *DNA Replication,* W. H. Freeman, San Francisco 1980.
10. E. C. Friedberg: *DNA Repair,* W. H. Freeman, New York 1985.
11. W. Saenger: *Principles of Nucleic Acid Structure,* Springer Verlag, New York 1984.
12. J. E. Dahlberg, J. N. Abelson (eds.): *Methods in Enzymology,* vol. 180, Part A, RNA processing, Academic Press, New York 1989.
13. P. M. Wassarman, R. D. Kornberg (eds.): *Methods in Enzymology,* vol. 170, Nucleosomes, Academic Press, New York 1989.
14. H. F. Noller, Jr., K. Moldave (eds.): *Methods in Enzymology,* vol. 164, Ribosomes, Academic Press, New York 1988.

15. M. A. Innis, D. H. Gelfand, J. J. Sninsky, T. J. White (eds.): *PCR Protocols – A Guide to Methods and Applications,* Academic Press, New York 1990.
16. B. S. Sproat, M. J. Gait in M. J. Gait (ed.): *Oligonucleotide Synthesis, A Practical Approach,* IRL Press, Oxford-Washington 1984.
17. H. G. Gassen, A. Lang (eds.): *Chemical and Enzymatic Synthesis of Gene Fragments: A Laboratory Manual,* Verlag Chemie, Weinheim 1982.
18. E. Sonveaux: "The Organic Chemistry Underlying DNA Synthesis," *Bioorg. Chem.* **14** (1986) 274 – 325.
19. J. W. Engels, E. Uhlmann in A. Fichter (ed.): *Gene Synthesis, Advances in Biochemical Engineering/Biotechnology,* vol. 37, Springer Verlag, Berlin – Heidelberg 1988.
20. G. Zon: "Oligonucleotide Analogues as Potential Chemotherapeutic Agents," *Pharm. Res.* 1988, no. 5, 539 – 549.
21. G. H. Keller, M. M. Manak: *DNA Probes,* Macmillan Publishers, New York 1989.
22. H.-J. Höltke, C. Kessler, *Nucleic Acid Res.* **18** (1990) 5843 – 5851.
23. B. D. Hames, S. J. Higgins (eds.): *Nucleic Acid Hybridization: A Practical Approach,* IRL Press, Oxford 1985.
24. K. E. Davies (ed.): *Human Genetic Diseases: A Practical Approach,* IRL Press, Oxford 1986.
25. J. Cossman (ed.): *The Molecular Genetics and the Diagnosis of Cancer,* Elsevier Science Publishing, New York 1990.
26. M. Furth, M. Greaves (eds.): *Molecular Diagnostics of Human Cancer, Cancer Cells 7,* Cold Spring Harbor Laboratory, New York 1989.
27. H. A. Erlich (ed.): *PCR Technology. Principles and Applications for DNA Amplification,* Stockton Press, New York-London-Tokyo-Melbourne-Hong Kong 1989.
28. L. T. Kirby: *DNA Fingerprinting, An Introduction,* Stockton Press, New York – London – Tokyo –Melborne – Hong Kong 1990.
29. R. J. Suhadolnik: *Nucleoside Antibiotics,* Wiley-Interscience, New York 1970.
30. R. J. Suhadolnik: *Nucleosides as Biological Probes,* Wiley-Interscience, New York 1979.
31. C. A. Decker, L. Goodman in W. Pigman, D. Horton (eds.): *The Carbohydrates,* vol. IIA, Academic Press, New York-London 1970.
32. W. Zorbach, R. S. Tipson: *Synthetic Procedures in Nucleic Acid Chemistry,* vol. I, Interscience Publishers, New York 1968.
33. L. Goodman in P. O. P. Ts'o (ed.): *Basic Principles in Nucleic Acid Chemistry,* **vol. I,** Academic Press, New York-London 1974, p. 93.
34. L. B. Townsend, R. S. Tipson: *Nucleic Acid Chemistry,* Wiley-Interscience, New York, Parts 1 and 2, 1978; Part 3, 1986.
35. Y. Mizuno; *The Organic Chemistry of Nucleic Acids,* Elsevier, Amsterdam 1986.

Specific References
36. G. F. Joyce, *Nature (London)* **338** (1989) 217.
37. A. I. Lamond, T. J. Gibson, *Trends Genet.* **6** (1990) 145.
38. J. D. Watson, F. H. C. Crick, *Nature (London)* **171** (1953) 737.
39. S. B. Zimmerman, *Annu. Rev. Biochem.* **51** (1982) 395.
40. R. Schleif, *Science (Washington, D. C.)* **241** (1988) 1182.
41. L. Joshua-Toret et al., *Nature (London)* **334** (1988) 82.
42. M. Miller et al., *Nature (London)* **334** (1988) 85.
43. A. Nordheim: *ISI Atlas of Science: Biochemistry* **1** (1988) 279.
44. H. R. Drew et al., *Annu. Rev. Cell Biol.* **4** (1988) 1.
45. J. Lebowitz, *TIBS* **15** (1990) 202.
46. J. Grinsted, P. M. Bennett, *Methods Microbiol.* **21** (1988) 129.
47. G. J. Pruss, K. Drlica, *Cell* **56** (1989) 521.
48. A. A. Travers, *Cell* **60** (1990) 177.
49. C. F. McAllister, E. C. Achberger, *J. Biol. Chem.* **264** (1989) 10 451.
50. A. A. Travers, *Annu. Rev. Biochem.* **58** (1989) 427.
51. D. M. Crothers et al., *J. Biol. Chem.* **265** (1990) 7093.
52. T. T. Eckdahl, J. N. Anderson, *Nucleic Acids Res.* **18** (1990) 1609.
53. A. Travers, A. Klug, *Nature (London)* **327** (1987) 280.
54. L. Bracco et al., *EMBO J.* **8** (1989) 4289.
55. M. S. Z. Horwitz, L. A. Loeb, *Science (Washington, D. C.)* **241** (1988) 703.
56. U. R. Müller, W. M. Fitch, *Nature (London)* **298** (1982) 582.
57. R. D. Wells, *J. Biol. Chem.* **263** (1988) 1095.
58. Y. Kohwi, T. Kohwi-Shigematsu, *Proc. Natl. Acad. Sci. USA* **85** (1988) 3781.
59. J. H. van de Sande et al., *Sience (Washington, D. C.)* **241** (1988) 551.

60. L. Perrouault et al., *Nature (London)* **344** (1990) 358.
61. C. Hélène, J.-J. Toulmé, *Biochim. Biophys. Acta* **1049** (1990) 99.
62. C. B. Harley et al., *Nature (London)* **345** (1990) 458.
63. G.-L. Yu et al., *Nature (London)* **344** (1990) 126.
64. V. A. Zakian et al., *Trends Genet.* **6** (1990) 12.
65. W. I. Sundquist, A. Klug, *Nature (London)* **342** (1990) 825.
66. J. D. Boeke, *Cell* **61** (1990) 193.
67. D. Sen, W. Gilbert, *Nature (London)* **344** (1990) 410.
68. F. Baudin, P. J. Romaniuk, *Nucleic Acids Res.* **17** (1989) 2043.
69. J. Normanly, J. Abelson, *Annu. Rev. Biochem.* **58** (1989) 1029.
70. A. Bhattacharyya et al., *Nature (London)* **343** (1990) 484.
71. R. M. W. Mans et al., *Nucleic Acids Res.* **18** (1990) 3479.
72. J. R. Wyatt et al., *BioEssays* **11** (1989) 100.
73. S. Arnott et al., *J. Mol. Biol.* **188** (1986) 631.
74. H. I. Elsner, E. B. Lindblad, *DNA* **8** (1989) 697.
75. H. Donis-Keller et al., *Nucleic Acids Res.* **4** (1977) 2527.
76. G. Krupp, H. J. Gross, *Nucleic Acids Res.* **6** (1979) 3481.
77. J. Parker, *Microbiol. Rev.* **53** (1989) 273.
78. M. Meselson, F. W. Stahl, *Proc. Natl. Acad. Sci. USA* **44** (1958) 671.
79. A. Kornberg, *Biochim. Biophys. Acta* **951** (1988) 235.
80. J. P. Vaughn et al., *Nucleic Acids Res.* **18** (1990) 1965.
81. D. Riesner, H. J. Gross, *Annu. Rev. Biochem.* **54** (1985) 531.
82. T. D. Fox, *Annu. Rev. Genet.* **21** (1987) 67.
83. H. M. Weintraub, *Sci. Am.* **262** (1990) 34.
84. J. Shine, L. Dalgarno, *Proc. Natl. Acad. Sci. USA* **71** (1974) 1342.
85. J. G. Belasco, C. F. Higgins, *Gene* **72** (1988) 15.
86. P. F. Johnson, S. L. McKnight, *Annu. Rev. Biochem.* **58** (1989) 799.
87. A. Baniahmad et al., *Cell* **61** (1990) 505.
88. D. R. Herendeen et al., *Science (Washington, D. C.)* **248** (1990) 573.
89. L. Gold, *Annu. Rev. Biochem.* **57** (1988) 199.
90. J. A. Steitz, *Sci. Am.* **258/6** (1988) 56.
91. M. R. Green, *Annu. Rev. Genet.* **20** (1986) 671.
92. S. M. Mount, *Nucleic Acids Res.* **10** (1982) 459.
93. A. M. Weiner, N. Maizels, *Cell* **61** (1990) 917.
94. V. Volloch et al., *Nature (London)* **343** (1990) 482.
95. D. Shippen-Lentz, E. H. Blackburn, *Science (Washington, D. C.)* **247** (1990) 546.
96. E. H. Blackburn, *J. Biol. Chem.* **265** (1990) 5919.
97. A. E. Dahlberg, *Cell* **57** (1989) 525.
98. L. Gold et al., *Annu. Rev. Microbiol.* **35** (1981) 365.
99. C. O. Gualerzi, C. L. Pon, *Biochemistry* **29** (1990) 5881.
100. W. Doerfler, *Annu. Rev. Biochem.* **52** (1983) 93.
101. E. U. Selker, *TIBS* **15** (1990) 103.
102. W. Zacharias et al., *J. Bacteriol.* **172** (1990) 3278.
103. N. D. F. Grindley, R. R. Reed, *Annu. Rev. Biochem.* **54** (1985) 863.
104. G. C. Walker, *Annu. Rev. Biochem.* **54** (1985) 425.
105. B. van Houten, *Microbiol. Rev.* **54** (1990) 18.
106. G. M. Myles, A. Sancar, *Chem. Res. Toxicol.* **2** (1989) 197.
107. L. A. Grivell, *Nature (London)* **344** (1990) 110.
108. D. L. Robertson, G. F. Joyce, *Nature (London)* **344** (1990) 467.
109. K. Kruger et al., *Cell* **31** (1982) 147.
110. C. Guerrier-Takada et al., *Cell* **35** (1983) 849.
111. O. C. Uhlenbeck, *Nature (London)* **328** (1987) 596.
112. T. R. Cech, *Science (Washington, D. C.)* **236** (1987) 1532.
113. J.-P. Perreault et al., *Nature (London)* **344** (1990) 565.
114. D. Herschlag, T. R. Cech, *Nature (London)* **344** (1990) 405.
115. J. J. Rossi, N. Sarver, *TIBTECH* **8** (1990) 179.
116. G. J. Raymond et al., *Anal. Biochem.* **173** (1988) 125.
117. J. Marmur, *J. Mol. Biol.* **3** (1961) 208.
118. U. Gubler, B. J. Hoffman, *Gene* **25** (1983) 263.
119. J. A. Thompson, R. D. Wells, *Nature (London)* **334** (1988) 87.
120. G. H. Hamm, G. N. Cameron, *Nucleic Acids Res.* **14** (1986) 5 – 9.
121. H. S. Bilofsky et al., *Nucleic Acids Res.* **14** (1986) 1 – 4.
122. E. Y. Chen, P. H. Seeburg, *DNA* **4** (1985) 165.
123. F. Sanger et al., *Proc. Natl. Acad. Sci. USA* **74** (1977) 5463.
124. A. M. Maxam, W. Gilbert, *Proc. Natl. Acad. Sci. USA* **74** (1977) 560.

125. M. M. Yang, D. C. Youvan, *Bio/Technology* **7** (1989) 576.
126. J. A. Brumbaugh et al., *Proc. Natl. Acad. Sci. USA* **85** (1988) 5610.
127. L. M. Smith, *Genet. Eng.* **10** (1988) 91.
128. W. Ansorge et al., *Nucleic Acids Res.* **15** (1987) 4593.
129. E. R. Mardis, B. A. Roe, *BioTechniques* **7** (1989) 840.
130. J. H. Jett et al., *J. Biomol. Struct. Dyn.* **7** (1989) 301.
131. J. A. Sorge, L. A. Blinderman, *Proc. Natl. Acad. Sci. USA* **86** (1989) 9208.
132. C. D. Carpenter, A. E. Simon, *BioTechniques* **8** (1990) 26.
133. E. P. Nikonowicz et al., *Biochemistry* **29** (1990) 4193.
134. F. J. M. van de Ven, C. W. Hilbers, *Eur. J. Biochem.* **178** (1988) 1.
135. C. de los Santos et al., *Biochemistry* **28** (1989) 7282.
136. S.-H. Chou et al., *Biochemistry* **28** (1989) 2435.
137. E. P. Geiduschek, G. P. Tocchini-Valentini, *Annu. Rev. Biochem.* **57** (1988) 873.
138. J. M. Sperrazza et al., *Gene* **31** (1984) 17.
139. M. I. Oakes, J. A. Lake, *J. Mol. Biol.* **211** (1990) 897.
140. M. I. Oakes et al., *J. Mol. Biol.* **211** (1990) 907.
141. D. D. Dunlap, C. Bustamante, *Nature (London)* **342** (1989) 204.
142. P. K. Hansma et al., *Science (Washington, D. C.)* **242** (1988) 209.
143. R. J. Driscoll et al., *Nature (London)* **346** (1990) 294.
144. C. Malvy, *Anal. Biochem.* **143** (1984) 158.
145. A. E. Dahlberg et al., *J. Mol. Biol.* **41** (1969) 139.
146. S. Peats, *Anal. Biochem.* **140** (1984) 178.
147. C. R. Merril, *Nature (London)* **343** (1990) 779.
148. N. C. Mills, J. Ilan, *Electrophoresis (Weinheim, Fed. Republ. Ger.)* **6** (1985) 531.
149. C. E. Willis, G. P. Holmquist, *Electrophoresis (Weinheim, Fed. Republ. Ger.)* **6** (1985) 259.
150. V. T. Kung et al., *Anal. Biochem.* **187** (1990) 220.
151. A. Glazer et al., *Proc. Natl. Acad. Sci. USA* **87** (1990) 3851.
152. J. Maddox, *Nature (London)* **345** (1990) 381.
153. N. C. Stellwagen, *Adv. Electrophor.* **1** (1987) 177.
154. C. R. Cantor et al., *Annu. Rev. Biophys. Chem.* **17** (1988) 287.
155. S. B. Smith et al., *Science (Washington, D. C.)* **243** (1989) 203.
156. S. M. Clark et al., *Sience (Washington, D. C.)* **241** (1988) 1203.
157. E. Lai et al., *BioTechniques* **7** (1989) 34.
158. M. V. Olson, *J. Chromatogr.* **470** (1989) 377.
159. A. M. Michelson, A. R. Todd, *J. Chem. Soc.* 1955, 2632 – 2638.
160. H. G. Khorana, G. M. Tener, J. G. Moffatt, E. H. Pol, *Chem. Ind. (London)* **34** (1956) 1523 – 1531.
161. H. G. Khorana, *Science (Washington, D. C.)* **203** (1979) 614.
162. R. B. Merrifield, *J. Am. Chem. Soc.* **85** (1962) 3821 –3827.
163. R. L. Letsinger, W. B. Lunsford, *J. Am. Chem. Soc.* **85** (1963) 3045 – 3046.
164. H. Schaller, G. Weimann, B. Lerch, H. G. Khorana, *J. Am. Chem. Soc.* **85** (1963) 3821 – 3827.
165. G. S. Ti, B. L. Gaffney, R. A. Jones, *J. Am. Chem. Soc.* **104** (1982) 1316 – 1319.
166. C. B. Reese, A. Ubasawa, *Tetrahedron Lett.* **21** (1980) 2265 – 2268.
167. A. Krazewski, J. Stawinski, M. Wiewiorowski, *Nucleic Acids Res.* **8** (1980) 2301 – 2305.
168. R. T. Pon, N. Usman, M. J. Damha, K. K. Ogilvie, *Nucleic Acids Res.* **14** (1985) 6453 – 6469.
169. K. K. Ogilvie, R. T. Pon, M. J. Damha, *Nucleic Acids Res.* **13** (1985) 6447 – 6465.
170. F. Himmelsbach et al., *Tetrahedron Lett.* **11** (1981) 59 – 72.
171. B. C. Froehler, M. D. Matteucci, *Nucleic Acids Res.* **22** (1983) 8031 – 8036.
172. J. C. Schulhof, D. Molko, R. Teoule, *Nucleic Acids Res.* **15** (1987) 397 – 416.
173. J. C. Schulhof, D. Molko, R. Teoule, *Nucleic Acids Res.* **16** (1988) 319 – 326.
174. N. Usman, K. K. Ogilvie, M.-Y. Jiang, R. J. Cedergren, *J. Am. Chem. Soc.* **109** (1987) 7845 – 7854.
175. B. S. Sproat, M. J. Gait in M. J. Gait (ed.): *Oligonucleotide Synthesis, A Practical Approach,* IRL Press, Oxford – Washington 1984, pp. 83 – 115.
176. C. B. Reese, R. C. Titmas, L. Yau, *Tetrahedron Lett.* **19** (1978) 2727 – 2730.
177. S. L. Beaucage, M. H. Caruthers, *Tetrahedron Lett.* **22** (1981) 1859 – 1862.
178. L. J. McBride, M. H. Caruthers, *Tetrahedron Lett.* **24** (1983) 245.
179. N. D. Sinha, J. Biernat, J. McMagnus, H. Köster, *Nucleic Acids Res.* **12** (1984) 4539 – 4557.

180. H. Ito, Y. Ike, S. Ikuta, K. Itakura, *Nucleic Acids Res.* **10** (1982) 1755.
181. K. Miyoshi, R. Arentzen, T. Huang, K. Itakura, *Nucleic Acids Res.* **8** (1980) 5507.
182. R. Crea, T. Horn, *Nucleic Acids Res.* **8** (1980) 2331.
183. M. D. Matteucci, M. H. Caruthers, *J. Am. Chem. Soc.* **103** (1981) 3171 – 3174.
184. V. A. Efimov, S. V. Reverdatto, O. B. Chakhmakhcheva, *Nucleic Acids Res.* **9** (1982) 6675 – 6694.
185. S. P. Adams et al., *J. Am. Chem. Soc.* **105** (1983) 661 – 663.
186. M. J. Nemer, K. K. Ogilvie, *Tetrahedron Lett.* **21** (1980) 4159 – 4162.
187. T. G. Wadell, D. E. Leyden, M. T. de Bello, *J. Am. Chem. Soc.* **103** (1981) 5303 – 5307.
188. M. H. Caruthers in H. G. Gassen, A. Lang (eds.): *Chemical and Enzymatic Synthesis of Gene Fragments: A Laboratory Manual,* Verlag Chemie, Weinheim 1982, pp. 71 – 79.
189. W. Bannwarth, P. Iaiza, *DNA* **5** (1986) 413 – 419.
190. F. Chow, T. Kempe, G. Palm, *Nucleic Acids Res.* **9** (1981) 2807.
191. J. C. Catlin, F. Cramer, *J. Org. Chem.* **38** (1973) 245 –250.
192. S. A. Narang, H. M. Hsiung, R. Brousseau, *Methods Enzymol.,* **68** (1979) 90 – 98.
193. S. S. Jones et al., *Tetrahedron* **36** (1980) 3075 – 3085.
194. J. H. van Boom et al., *J. Chem. Soc. Chem. Commun.* 1971, 869 – 871.
195. E. Ohtsuka, S. Iwai in S. Narang (ed.): *Synthesis and Application of DNA and RNA,* Academic Press, San Diego – New York – Berkley – Boston 1987.
196. J. H. van Boom, *J. Chem. Soc. Chem. Commun.* 1976, 167.
197. P. J. Garegg, T. Regberg, J. Stawinski, R. Strömberg, *Chem. Scr.* **25** (1985) 280 – 282.
198. B. C. Froehler, M. D. Matteucci *Tetrahedron Lett.* **27** (1986)469 – 472.
199. B. C. Froehler, *Tetrahedron Lett.* **27** (1986) 5575 –5578.
200. B. C. Froehler, P. Ng, M. D. Matteucci, *Nucleic Acids Res.* **16** (1988) 4831 – 4839.
201. J. Stawinski, R. Strömberg, M. Thelin, E. Westman, *Nucleic Acids Res.* **16** (1988) 9285 – 9298.
202. J. H. van Boom et al., *Nucleic Acids Res.* **4** (1977) 1047.
203. E. E. van Tamelen, S. V. Daub, *J. Am. Chem. Soc.* **99** (1977) 3526.
204. L. W. McLaughlin, J. U. Krusche in H. G. Gassen, A. Lang (eds.): *Chemical and Enzymatic Synthesis of Gene Fragments A Laboratory Manual,* : Verlag Chemie, Weinheim 1982.
205. L. W. McLaughlin, N. Piel in M. J. Gait (ed.): *Oligonucleotide Synthesis, A Practical Approach,* IRL Press, Oxford – Washington 1984, pp. 117 –132.
206. G. Schmidt, H. Seliger, *J. Chromatogr.* **397** (1987) 141 – 151.
207. R. Wu in M. J. Gait (ed.): *Oligonucleotide Synthesis, A Practical Approach,* IRL Press, Oxford – Washington 1984, pp. 135 – 151.
208. A. M. Maxam, W. Gilbert, *Methods Enzymol.* **65** (1980) 499 – 503.
209. G. Zon, *Pharmaceutical Research* **5** (1988) no. 9, 539 – 549.
210. C. Hélène, J.-J. Toulmé, *Biochim. Biophys. Acta* **1049** (1990) 99 – 125.
211. C. A. Stein, J. S. Cohen, *Cancer Research* **48** (1988) 2659 – 2668.
212. W. J. Stec, G. Zon, W. Egan, B. Stec, *J. Am. Chem. Soc.* **106** (1984) 6077 – 6079.
213. B. C. Froehler, *Tetrahedron Lett.* **27** (1986) 5565 – 5568.
214. A. Andrus, J. W. Efcavitch, L. J. McBride, B. Giusti, *Tetrahedron Lett.* **29** (1988) 861 – 864.
215. B. H. Dahl, K. Bjergarde, V. B. Sommer, O. Dahl, *Nucleosides & Nucleotides,* **8** (1989) 1023 – 1027.
216. S. L. Beaucage et al., *J. Am. Chem. Soc.* **112** (1990) 1254 – 1255.
217. P. C. J. Kamer et al., *Tetrahedron Lett.* **30** (1989) 6757 – 6760.
218. W. K.-D. Brill, J.-Y. Tang, Y.-X. Ma, M. H. Caruthers, *J. Am. Chem. Soc.* **111** (1989) 2321 – 2322.
219. P. S. Miller, K. N. Fang, N. S. Kondo, P. O. P. Ts'o, *J. Am. Chem. Soc.* **93** (1971) 6657 – 6665.
220. P. S. Miller et al., *Biochimie* **67** (1985) 769 – 776.
221. P. S. Miller et al., *Nucleic Acids Res.* **11** (1983) 6225 – 6242.
222. P. S. Miller et al., *Biochemistry* **25** (1986) 5092 – 5097.
223. E. de Vroom et al., *Recl. Trav. Chim. Pays-Bas* **106** (1987) 65 – 66.
224. H. Inoue et al., *Nucleic Acids Res.* **15** (1987) 6131 – 6148.
225. B. S. Sproat, B. Beijer, A. Iribarren, *Nucleic Acids Res.* **18** (1990) 41 – 49.
226. U. Sequin, *Experientia* **29** (1973) 1059 – 1062.

227. J. L. Imbach, B. Rayner, F. Morvan, *Nucleosides & Nucleotides* **8** (1989) 627–648.
228. C. Hélène et al., *Biochimie* **66** (1985) 777–783.
229. A. Zerial, N. T. Thuong, C. Hélène, *Nucleic Acids Res.* **15** (1987) 9909–9919.
230. V. V. Vlassov et al., *Nucleic Acids Res.* **14** (1986) 4065–4076.
231. H. B. Gamper, G. D. Cimino, J. E. Hearst, *J. Mol. Biol.* **197** (1987) 349–362.
232. Y. B. Shi, H. B. Gamper, J. E. Hearst, *J. Mol. Biol.* **263** (1988) 527–534.
233. P. S. Miller, P. O. P. Ts'o, *Anti-Cancer Drug Des.* **2** (1987) 117–128.
234. U. Asseline et al., *Proc. Natl. Acad. Sci USA* **81** (1984) 3297–3301.
235. Le Doan et al., *Nucleic Acids Res.* **15** (1987) 7749–7760.
236. D. Praseuth, L. Perrouault, T. Le Doan, *Proc. Natl. Acad. Sci USA* **85** (1988) 1349–1353.
237. D. Praseuth et al., *Biochemistry* **27** (1988) 3031–3038.
238. C. Hélène, T. Le Doan, N. T. Thuong in P. E. Nielsen (ed.): *Photochemical Probes in Biochemistry,* Kluwer Publ., Norwell MA, 1989, pp. 219–229.
239. C. Hélène, N. T. Thuong, T. Saison-Behmoaras, J. C. Francois, *Tibtech* **7** (1989) 310–315.
240. A. S. Boutorin et al., *FEBS Lett.* **172** (1984) 43–46.
241. B. C. F. Chu, L. E. Orgel, *Proc. Natl. Acad. Sci USA* **82** (1985) 963–967.
242. G. B. Dreyer, P. E. Dervan, *Proc. Natl. Acad. Sci. USA* **82** (1985) 968–972.
243. C. H. B. Chen, D. S. Sigman, *Proc. Natl. Acad. Sci. USA* **83** (1986) 7147–7151.
244. C. H. B. Chen, D. S. Sigman, *J. Am. Chem. Soc.* **110** (1988) 6570–6572.
245. J.-C. Francois et al., *Biochemistry* **27** (1988) 2272–2276.
246. J.-C. Francois et al., *J. Biol. Chem.* **264** (1989) 5891–5898.
247. T. Le Doan et al., *Nucleic Acids Res.* **15** (1987) 8643–8659.
248. D. Corey, P. Schultz, *Science (Washington, D. C.)* **238** (1987) 1401–1403.
249. R. N. Zuckermann, P. G. Schultz, *Proc. Natl. Acad. Sci. USA* **86** (1989) 1766–1770.
250. Applied Biosystems: DNA Synthesizer, User Bulletin, No. 49, Foster City 1988.
251. Beckman Instruments, Aminomodifiers and Easy Label Kits, User Manual, Palo Alto 1990.
252. B. C. F. Chu, G. M. Wahl, L. E. Orgel, *Nucleic Acids Res.* **11** (1985) 6513–6529.
253. E. Jablonski, E. W. Moomaw, R. H. Tullis, J. L. Ruth, *Nucleic Acids Res.* **14** (1986) 6115–6128.
254. M. Lemaitre, B. Bayard, B. Lebleu, *Proc. Natl. Acad. Sci. USA* **84** (1987) 648–652.
255. M. Lemaitre, C. Bisbal, B. Bayard, B. Lebleu, *Nucleosides & Nucleotides* **6** (1987) 311–315.
256. R. Zuckermann, D. Corey, P. Schultz, *Nucleic Acids Res.* **15** (1987) 5305–5321.
257. P. S. Nelson, R. A. Frye, E. Liu, *Nucleic Acids Res.* **17** (1989) 7187–7194.
258. S. Agrawal, J.-Y. Tang, *Tetrahedron Lett.* **31** (1990) 1543–1546.
259. R. L. Letsinger, M. E. Schott, US 4 547 569, 1985.
260. A. Jäger, M. J. Levy, S. M. Hecht, *Biochemistry* **27** (1988) 7237–7246.
261. R. P. Langer, A. A. Waldrop, D. C. Ward, *Proc. Natl. Acad. Sci. USA* **78** (1981) 6635–6637.
262. A. F. Cook, E. Vuocolo, C. L. Brakel, *Nucleic Acids Res.* **16** (1988) 4077–4095.
263. M. S. Urdea et al., *Gene* **61** (1987) 253–264.
264. I. C. Gillam, G. M. Tener, *Anal. Biochem.* **157** (1986) 199–207.
265. U. Pieles, B. S. Sproat, P. Neuner, F. Cramer, *Nucleic Acids Res.* **17** (1989) 8967–8978.
266. G. Gebeyehu et al., *Nucleic Acids Res.* **16** (1987) 4937–4534.
267. P. Tchen, R. P. P. Fuchs, E. Sage, M. Leng, *Proc. Natl. Acad. Sci. USA* **81** (1984) 3466–3470.
268. J. A. Washington, G. L. Woods in B. Swaminathan, G. Prakash (eds.): *Nucleic Acid and Monoclonal Antibody Probes,* Marcel Dekker, New York 1989, p. 319.
269. F. C. Tenover, *Clin. Microbiol. Rev.* **1** (1988) 82.
270. E. Uhlmann, A. Peyman, *Chemical Rev.* **90** (1990) 543–584.
271. C. Hélène, J.-J. Toulmé, *Biochim. Biophys. Acta* **1049** (1990) 99–125.
272. J. Brandsma, G. Miller, *Proc. Natl. Acad. Sci. USA* **77** (1980) 6851.
273. P. McIntyre, G. R. Stark, *Anal. Biochem.* **174** (1988) 209.
274. E. M. Southern, *J. Mol. Biol.* **98** (1975) 503.
275. T. R. Moench, *Mol. Cell. Probes* **1** (1987) 195.
276. J. E. Landegent et al., *Nature (London)* **317** (1985) 175.
277. P. Lichter et al., *Science (Washington, D. C.)* **247** (1990) 64.
278. M. Ranki et al., *Gene* **21** (1983) 77.

279. P. J. Nicholls, A. D. B. Malcolm, *J. Clin. Lab. Anal.* **3** (1989) 122.
280. M. Virtanen et al., *J. Clin. Microbiol.* **20** (1984) 1083.
281. A. C. Syvaenen et al., *Nucleic Acids Res.* **14** (1986) 5037.
282. M. S. Ellwood et al., *Clin. Chem.* **32** (1986) 1631.
283. C. P. H. Vary, *Nucleic Acids Res.* **15** (1987) 6883.
284. J. A. Matthews, L. J. Kricka, *Anal. Biochem.* **169** (1988) 1.
285. L. J. Arnold, Jr., et al., *Clin. Chem.* **35** (1989) 1588.
286. R. P. Viscidi et al., *J. Clin. Microbiol.* **23** (1986) 311.
287. A. H. N. Hopman et al., *Exp. Cell. Res.* **169** (1987) 357.
288. G. H. Keller et al., *Anal. Biochem.* **170** (1988) 441.
289. A. C. Forster et al., *Nucleic Acids Res.* **13** (1985) 745.
290. G. H. Keller et al., *Anal. Biochem.* **177** (1989) 392.
291. K. Mühlegger et al., *Biol. Chem. Hoppe-Seyler* **371** (1990) 953.
292. K. B. Mullis, F. A. Faloona, *Methods Enzymol.* **155** (1987) 335.
293. R. K. Saiki et al., *Science (Washington, D. C.)* **230** (1985) 1350.
294. H. Lee et al., *Science (Washington, D. C.)* **244** (1989) 471.
295. D. B. Schowalter, S. S. Sommer, *Analyt. Biochemistry* **177** (1988) 90 – 94.
296. T. Lion, O. A. Haas, *Analyt. Biochemistry* **188** (1990) 335 – 337.
297. D. Y. Kwoh et al., *Proc. Natl. Acad. Sci. USA* **86** (1989) 1173.
298. D. Y. Wu, R. B. Wallace, *Genomics* **4** (1989) 560.
299. K. J. Barringer et al., *Gene* **89** (1990) 117.
300. J. C. Guatelli et al., *Proc. Natl. Acad. Sci. USA* **87** (1990) 1874.
301. J. van Brunt, *Bio/Technology* **8** (1990) 291.
302. P. M. Lizardi et al., *Bio/Technology* **6** (1988) 1197.
303. H. Lomeli et al., *Clin. Chem. (London)* **35** (1989) 1826.
304. H. Wolf et al., *J. Virol. Methods* **13** (1986) 1.
305. M. S. Urdea et al., *Gene* **61** (1987) 253.
306. P. D. Fahrlander, *Bio/Technology* **6** (1988) 1165.
307. M. Berninger et al., *J. Med. Virol.* **9** (1982) 57.
308. H. A. Rotbart et al., *J. Clin. Microbiol.* **26** (1988) 2669.
309. J. J. Hogan in B. Swaminathan, G. Prakash (eds.): *Nucleic Acid and Monoclonal Antibody Probes,* Marcel Dekker, New York 1989.
310. S. Huovinen et al., *Antimicrob. Agents Chemother.* **32** (1988) 175.
311. D. N. Cooper, J. Schmidtke, *Hum. Genet.* **73** (1986) 1.
312. S. Nishimura, T. Sekiya, *Biochem. J.* **243** (1987) 313.
313. P. Benn et al., *Cancer Genet. Cytogenet.* **29** (1987) 1.
314. A. J. Levine, *BioEssays* **12** (1990) 60.
315. R. A. Weinberg, *Cancer Res.* **49** (1989) 3713.
316. M. A. Lovell, *Clin. Chem. (Winston-Salem, N.C.),* Suppl. 35 (1989) 1343.
317. H. A. Erlich et al., *Bio/Technology* **4** (1986) 975.
318. A. J. Jeffreys, *Biochem. Soc. Trans.* **15** (1987) 309.
319. H. Zischler et al., *Hum. Genet.* **82** (1989) 227.
320. A. H. Cawood, *Clin. Chem. (Winston-Salem, N.C.)* **35** (1989) 1832.
321. J. Davoll, B. Lythgoe, A. R. Todd, *J. Chem. Soc.* 1948, 1685.
322. J. Davoll, B. A. Lowy, *J. Am. Chem. Soc.* **73** (1951) 1650.
323. N. Yamaoka, K. Aso, H. Matsuda, *J. Org. Chem.* **30** (1965) 149.
324. G. E. Hilbert, T. B. Johnson, *J. Am. Chem. Soc.* **52** (1930) 4489.
325. J. Pliml, M. Prystas, *Adv. Heterocycl. Chem.* **8** (1967) 115.
326. T. Sato, T. Shimadate, Y. Ishido, *Chem. Abstr.* **56** (1962) 11 692 g.
327. L. Birkhofer, A. Ritter, H. P. Küelthau, *Angew. Chem.* **75** (1963) 209.
328. T. Nishimura, B. Shimizu, I. Iwai, *Chem. Pharm. Bull.* **11** (1963) 1470.
329. U. Niedballa, H. Vorbrüggen, *Angew. Chem., Int. Ed. Engl.* **9** (1970) 461.
330. A. Holy, I. Votruba, *Nucleic Acid Symp. Ser.* **18** (1987) 69.
331. F. Seela, H. D. Winkeler, *J. Org. Chem.* **47** (1982) 226.
332. D. H. Shannahoff, R. A. Sanchez, *J. Org. Chem.* **38** (1973) 593.
333. V. E. Marquez, M. I. Lim, *Med. Res. Rev.* **6** (1986) 1.
334. S. Hanessian, A. C. Parnet, *Adv. Carbohydr. Chem. Biochem.* **33** (1976) 111.
335. K. Ogata, *Adv. Appl. Microbiol.* **19** (1975) 209.
336. K. Ogata, S. Kinoshita, T. Tsunoda, K. Aida: *Microbial Production of Nucleic Acid-Related Substances,* Halsted Press, New York 1976.

337. S. L. Haynie, D. N. Whitesides, *Appl. Biochem. Biotech.* **23** (1990) 205.
338. D. J. Merkler, V. L. Schramm, *Anal. Biochem.* **167** (1987) 148.
339. L. A. Slotin, *Synthesis* 1977, 737.
340. G. A. Robinson, R. W. Butcher, E. W. Sutherland: *Cyclic AMP,* Academic Press, New York – London 1971.
341. W. H. Cook, D. Lipkin, R. Markham, *J. Am. Chem. Soc.* **79** (1957) 3607.
342. M. S. Amer in N. J. Harper, A. B. Simmonds (eds.): *Advances in Drug Research,* **vol. 12,** Academic Press, New York – London 1977, p. 1.
343. F. Eckstein, *Ann. Rev. Biochem.* **54** (1985) 367.
344. R. G. Yount, D. Babcock, D. Ojala, W. Ballantyne, *Biochemistry* **10** (1971) 2484.
345. T. C. Myers, K. Nakamura, J. W. Flesher, *J. Am. Chem. Soc.* **85** (1963) 3292.
346. G. R. Pettit: *Synthetic Nucleotides,* vol. I, Van Nostrand Reinhold, New York 1972.
347. M. Yoshikawa, T. Kato, T. Takenishi, *Tetrahedron Lett.* **50** (1967) 5065.
348. J. G. Moffatt, H. G. Khorana, *J. Am. Chem. Soc.* **83** (1961) 649.
349. D. E. Hoard, D. G. Ott, *J. Am. Chem. Soc.* **87** (1965) 1785.
350. A. M. Michelson, *Biochim. Biophys. Acta* **91** (1964) 1.
351. J. Ludwig, *Acta Biochim. Biophys. Acad. Sci. Hung.* **16** (1981) 131.
352. E. De Clercq, *Int. Congr. Ser. Excerpta Med.* **750** (1987) 631.
353. J. C. Martin: “Nucleotide Analogues as Antiviral Agents,” *ACS Symp. Ser.* **401** (1989).

Amino Acids

KARLHEINZ DRAUZ, Degussa-Hüls AG, Hanau-Wolfgang, Germany

BERND HOPPE, Degussa-Hüls AG, Frankfurt, Germany

AXEL KLEEMANN, Hanau, Germany

HANS-PETER KRIMMER, Degussa-Hüls AG, Hanau-Wolfgang, Germany

WOLFGANG LEUCHTENBERGER, Degussa-Hüls AG, Düsseldorf, Germany

CHRISTOPH WECKBECKER, Degussa-Hüls AG, Hanau-Wolfgang, Germany

1. Introduction and History

The proteins, although they occur in an almost infinite variety, are composed of a relatively small number of basic building blocks, all α-amino acids. In addition, the amino acids fulfill certain regulatory functions in the metabolism and are required for the biosynthesis of other functional structures. This review is limited, for the most part, to the protein-forming amino acids, because they are by far the most widely distributed in nature and are of considerable economic interest.

The ca. 20 different α-amino acids found in proteins are rather simple organic compounds, in which an amino group and a side chain (R) are attached alpha to the carboxyl function. The R group may be aliphatic, aromatic, or heterocyclic and may possess further functionality.

Ullmann's Industrial Toxicology

ISBN: 3-527-31247-1

$$\mathrm{R{-}CH(NH_2){-}COOH} \xrightarrow{-H_2O}$$

Amino acid

$$\mathrm{H_2N{-}CHR{-}CO{-}NH{-}CHR{-}COOH} \xrightarrow{-H_2O}$$

Dipeptide

$$\mathrm{H_2N{-}CHR{-}CO{-}(NH{-}CHR{-}CO)_{n-2}{-}NH{-}CHR{-}COOH}$$

Protein ($n > 50$)

At present over 200 naturally occurring α-amino acids are known [1–3], [11]. Table 1 shows the α-amino acids found in proteins, where they occur exclusively as the L-enantiomers. D-Amino acids have been found only in the cell walls of some bacteria, in peptide antibiotics, and in the cell pools of some plants [6], [12], [13].

History. The history of amino acid chemistry began in 1806, when two French investigators, VAUQUELIN and ROBIQUET, isolated asparagine from asparagus juice. It was not until 1925 that SCHRYVER and BURTON isolated threonine from oat protein, the last discovered of the ca. 20 protein-forming amino acids. STRECKER synthesized alanine in 1850 from acetaldehyde, ammonia, and hydrogen cyanide. ESCHER established the hypothesis of essential amino acids. EMIL FISCHER discovered that the amino acids were building blocks of the proteins. ABDERHALDEN synthesized threonine from acrylic acid derivatives and methanol. ROSE et al. recognized threonine as the last of the eight essential amino acids. D,L-Methionine was produced industrially in Germany in 1948, and in 1956 L-glutamic acid was produced by fermentation in Japan.

Origin of Amino Acids. The first amino acids were probably produced on the earth more than 3×10^9 years ago via "prebiotic synthesis" in the primordial atmosphere. The concept of prebiotic synthesis is based on laboratory experiments in which glycine, alanine, aspartic acid, glutamic acid, and other compounds were produced by the action of an electrical discharge on a simulated primordial atmosphere consisting of methane, hydrogen, water, and ammonia [14]. Since then, traces of amino acids have been detected in moon rocks, meteorites, and interstellar space.

2. Properties

2.1. Physical Properties and Structure

α-Amino acids are nonvolatile, white, crystalline compounds with no defined melting points. They are relatively stable on heating, generally decomposing at 250 – 300 °C. Both the low volatility and the thermal stability result from the low-energy dipolar structure (zwitterion, inner salt, betaine), which the amino acids assume in the solid state.

$$\mathrm{R{-}CH({}^{+}NH_3){-}COO^{-}}$$

Evidence for this structure is provided by infrared and Raman spectra in which the bands typical of $-NH_2$ and -COOH moieties are absent. Equilibrium in solution also lies almost exclusively on the side of the dipolar form; therefore, amino acids are insoluble in nonpolar solvents and usually not very soluble in polar ones. The only amino acids that exhibit any appreciable solubility in alcohol are proline and hydroxyproline. Solubility in water depends on the pH: the minimum is at the isoelectric point.

This solubility minimum at the isoelectric point is quite useful for purifying and recrystallizing amino acids. The analytical technique for separating amino acid mixtures by electrophoresis is based on the fact that a specific amino acid does not migrate in an electric field at its isoelectric point, pI, a physical constant for each amino acid.

The structures and physical properties of the most important α-amino acids are given in Section 2.3.1.

Stereochemistry. With the exception of glycine, the simplest amino acid (R = H), all natural α-amino acids are chiral compounds occurring in two enantiomeric (mirror-image) forms.

Table 1. Common amino acids of proteins

Trivial name	IUPAC abbreviation	One-Letter Code	Protein	Content, g/100 g	
				[4]	[6]
L-Alanine	Ala	A	silk fibroin	25	29.7
L-Arginine	Arg	B	salmin	87	
			edostin	17	
			wool	10	
			gelatin		8.3
			rat liver histone		15.9
L-Asparagine	Asn	N			
L-Aspartic acid	Asp	D	edestin	12	
			hemoglobin	9 – 10	
			barley globulin		10.3
L-Cysteine	Cys	C	wool keratin		11.9
			human hair keratin		14.4
			feather keratin		8.2
L-Cystine	Cys-Cys, $(Cys)_2$, Cyss	–			
L-Glutamic acid	Glu	E	gliadin	47	
			zein	31	
			wheat gliadin		39.2
			maize zein		22.9
L-Glutamine	Gln	Q			
Glycine	Gly	G	gelatin	26	
			silk fibroin	44	
L-Histidine	His	H	hemoglobin	7	
L-Hydroxyproline	Hyp	–	gelatin	15	
L-Isoleucine	Ile	I	edestin	21	
			hemoglobin	29	
			serum proteins	20	
			oat globulin		4.3
			beef serum albumin		2.6
L-Leucine	Leu	L	edestin	21	
			hemoglobin	29	
			serum proteins	20	
			maize zeins		19
L-Lysine	Lys	K	serum albumin	13	
			serum globulin	6	
			horse myoglobin		15.5
L-Methionine	Met	M	egg albumin	5	
			casein	3	4.1
			β-lactoglobulin		3.2
L-Phenylalanine	Phe	F	zein	8	
			egg albumin	5	7.7
			serum albumin		7.8
L-Proline	Pro	P	gelatin	17	16.3
			gliadin	13	
			salmin		6.9
			casein		10.6
L-Serine	Ser	S	silk fibroin	13	16.2
			trypsinogen		16.7
			pepsin		12.2
L-Threonine	Thr	T	casein	4	
			human hair keratin		8.5
			avidin		10.5
L-Tryptophan	Trp	W	fibrin	3	
			egg lysozyme		10.6
L-Tyrosine	Tyr	Y	fibrin	6	
			silk fibroin	13	
			papain		14.7
L-Valine	Val	V	casein	8	
			beef sinew		17.4
			beef aorta		17.6

$$\underset{\text{L-Amino acid}}{R\text{–}C(H)(\overset{+}{N}H_3)\text{–}COO^-} \qquad \underset{\text{D-Amino acid}}{{}^-OOC\text{–}C(H)(\overset{+}{N}H_3)\text{–}R}$$

L-Amino acid D-Amino acid

The prefixes L and D express the absolute configuration at the α-carbon atom by means of the formal stereochemical relationship to L- or D-glyceraldehyde, the reference substance introduced by EMIL FISCHER in 1891. In addition to the spacial representations shown above, the so-called Fischer projections are also universally recognized and used:

$$\begin{array}{c} COOH \\ H_2N\text{—}\!\!+\!\!\text{—}H \\ R \end{array} \qquad \begin{array}{c} COOH \\ H\text{—}\!\!+\!\!\text{—}NH_2 \\ R \end{array}$$

L-Amino acid D-Amino acid

Polarimetric determination of the specific rotation $[\alpha]_D^t$ can be used to differentiate between the two enantiomers and to check their optical purity. The molecular rotation $[M]_D^t$ is less common:

$$[M]_D^t = \frac{M_r}{100} \cdot [\alpha]_D^t$$

M_r molecular mass; t temperature; D 589.3 nm (wavelength of the sodium D line) Further methods for investigating the structure of amino acid enantiomers include the Cotton effect (change in molecular rotation as a function of the wavelength of plane-polarized light), as reflected in optical rotational dispersion (reversal of the direction of the molecular rotation at the wavelength of the absorption maximum), and circular dichroism (differing absorption for left- and right-handed circularly polarized light). L-Amino acids exhibit a positive carbonyl Cotton effect, D-amino acids a negative one.

Isoleucine, threonine, and hydroxyproline contain two chiral carbon atoms each; therefore, they appear in four stereoisomeric forms. Cystine, which likewise contains two chiral carbons, has only three stereoisomers: L-, D-, and *meso*-cystine, the meso form having a plane of symmetry. According to the Cahn – Ingold – Prelog rule (R, S rule) [17], all proteinogenic L-α-amino acids, with the exception of L-cysteine and L-cystine, are S; the D-α-amino acids are R. According to this system, L-threonine, for example, is termed (2S, 3R)-threonine. However, the R,S system has not attained wide acceptance for the simple L-α-amino acids.

Absorption Spectra. Aliphatic amino acids exhibit no absorption in the UV region above 220 nm, with the exception of cystine (240 nm). The aromatic amino acids, phenylalanine, tyrosine, and tryptophan, absorb between 250 and 300 nm [1, vol. 2]. The exact position of the maximum and the molar extinction coefficient ε are affected by the pH of the aqueous solution.

Two infrared bands are especially characteristic of amino acids: 1560 – 1600 cm^{-1} (–COO^-) and ca. 3070 cm^{-1} (–NH_3^+). These bands are also evidence of the bipolar nature of the amino acids.

2.2. Chemical Properties

Acidity and Basicity. The chemical properties of the α-amino acids are primarily the properties of the amino and carboxyl groups. Amino acids react with strong acids as proton acceptors (bases) and with strong bases as proton donors (acids). In acidic medium they are present predominantly as cations; in basic medium they are present predominantly as anions. Figure 1 shows the titration curves of glycine, glutamic acid, and lysine. The pK_1 and pK_2 values correspond to the inflection points of the titration curve, the pH value where the concentrations of the zwitterion form and the cationic or anionic form are equal. The pK_1, pK_2, and pI values can be calculated from the titration curves with the Henderson – Hasselbalch equation:

$$pK_1 = pH - \log \frac{[R\text{–}CH(\overset{+}{N}H_3)\text{–}COO^-]}{[R\text{–}CH(\overset{+}{N}H_3)\text{–}COOH]}$$

$$pK_2 = pH - \log \frac{[R\text{–}CH(NH_2)\text{–}COO^-]}{[R\text{–}CH(\overset{+}{N}H_3)\text{–}COO^-]}$$

$$pI = \frac{pK_1 + pK_2}{2}$$

Amino acids with additional basic or acidic groups (arginine, lysine, glutamic acid, cysteine) exhibit additional pK values. Amino acids act as buffers in the region of their pK values.

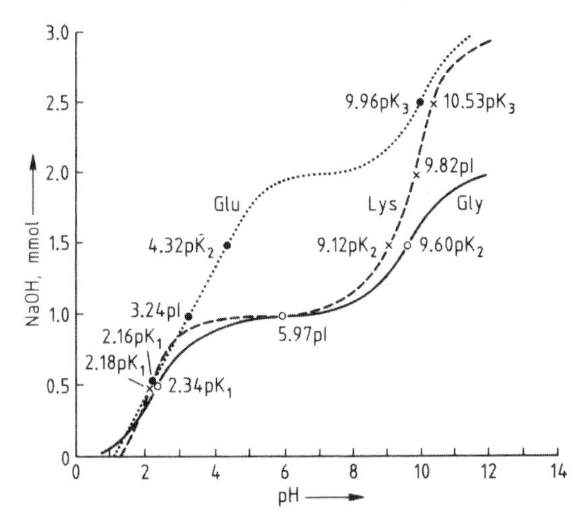

Figure 1. Titration curves of glycine, glutamic acid, and lysine [5]

The pK_1 values show the amino acids to be considerably stronger acids than acetic acid. However, because of intramolecular protonation of the amine moiety by the carboxyl group, aqueous solutions of amino acids are only weakly acidic. The pH values of aqueous monoaminomonocarboxylic acids lie between 5.5 and 6.0. Solutions of the acidic amino acids aspartic acid and glutamic acid have pH values of ca. 2. The weakly basic amino acid histidine has a pH of 7.5 in aqueous solution; the more strongly basic amino acids lysine and arginine have pH values of ca. 11 – 12. These differences in acidities and basicities are utilized in the separation of amino acid mixtures by ion-exchange chromatography and electrophoresis.

Reactions. Because of their bifunctional and sometimes trifunctional character, the α-amino acids are capable of taking part in a variety of chemical reactions. Comprehensive treatments of these may be found in monographs and reviews [1], [5–8], [18].

The introduction of groups protecting the amino, carboxyl, and side-chain functions is especially important for peptide synthesis [9]. Additionally, α-amino acids play a prominent role as intermediates in the synthesis of heterocycles [19]. The chiral α-amino acids and their derivatives are inexpensive and, for the most part, readily available synthons for numerous natural products and pharmaceuticals. In many cases optically active amino acids can also be used to induce chirality during the course of a synthetic process [20].

α-Amino acids form *chelate-like complexes* with heavy metal ions. The best known are the dark blue, easily crystallized copper chelates:

Bis(glycinato) copper(II) hydrate

This complex formation can be used to protect both the α-amino and the carboxyl group during synthesis of ε-N-acetyl derivatives or carbamates of lysine or other side-chain derivatives. In these copper or cobalt complexes the α-carbon atom is activated sufficiently to react with aldehydes. A well-known example is the alkaline condensation of the glycine – copper complex with acetaldehyde, resulting in threonine. Especially important is the reaction of α-amino acids with ninhydrin to form a blue-violet dye, the basis of a sensitive optical method for detecting amino acids (see Chap. 6).

Free amino acids react with *nitrous acid* to yield α-hydroxycarboxylic acids, with retention of configuration. Volumetric measurement of the nitrogen gas set free is the basis of Van Slyke's method for the quantitative analysis of amino acids. Reaction of amino acid esters with nitrous acid gives the acid-labile diazocarboxylic acid esters. Treatment of N-alkyl- or N-arylamino acids with nitrous acid yields the N-nitroso derivatives, which can be dehydrated to sydnones in the presence of acetic anhydride:

Oxidizing agents attack the amino group, converting the amino acid into an iminocarboxylic acid. These are unstable and either hydrolyze to α-oxocarboxylic acids or decompose, after decarboxylation, into ammonia and the aldehyde containing one carbon atom less. Diketones, triketones, *N*-bromosuccinimide, or silver oxide may serve as the oxidizing agent. The best known example is the reaction with ninhydrin. Oxidative deamination also can be carried out enzymatically with D- or L-amino acid oxidases. This also proceeds via the α-iminocarboxylic acids, which subsequently are hydrolyzed to α-oxo acids:

$$R{-}CH(NH_2){-}COOH \xrightarrow[-2\,H]{\text{amino acid oxidase}} R{-}C({=}NH){-}COOH \xrightarrow[-NH_3]{H_2O} R{-}C({=}O){-}COOH$$

This enantioselective enzymatic oxidative deamination is the basis of analytical methods for the determination of amino acid enantiomers.

The *N-acylation* of α-amino acids with acyl chlorides or anhydrides under Schotten-Baumann conditions produces *N*-acyl-α-amino acids, which have numerous uses. For example, the *N*-acetyl derivatives of D,L-amino acids (e.g., alanine, valine, methionine, phenylalanine, tryptophan) are intermediates in the production of L-amino acids by enzymatic resolution using aminoacylases (see Section 3.1). Amino acids acylated with naturally occurring fatty acid residues are used industrially as easily degradable surfactants.

If free amino acids are heated above 200 °C, especially in the presence of soda lime or metal ions, they readily *decarboxylate* to form amines. The enzymatic decarboxylation of amino acids gives biogenic amines. Some of these amines are physiologically active neurotransmitters (histamine, tyramine, dopamine, serotonin).

The *esters* usually are prepared by direct esterification, e.g., reaction of the α-amino acid with anhydrous alcohol in the presence of anhydrous hydrogen chloride. The initial product is the hydrochloride of the ester, which is set free by addition of base. On standing or warming, the free esters of α-amino acids eliminate alcohol to form 2,5-diketopiperazines:

$$2\; R{-}CH(NH_2){-}COOR \xrightarrow{-2\,ROH} \text{2,5-diketopiperazine (R-substituted)}$$

Esters of amino acids are useful for peptide synthesis because the carboxyl function is protected. The esters may be converted to amino alcohols by treatment with a strong reducing agent, such as lithium aluminum hydride.

Several cyclic derivatives of the α-amino acids are of industrial importance. The *hydantoins* or imidazolidine-2,4-diones, which have been discussed as raw materials for the synthesis of α-amino acids, are most conveniently prepared by treatment of aldehyde cyanohydrins with ammonium carbonate or urea. They also may be obtained by reacting amino acids with cyanates or isocyanates.

The initial product of this reaction is the ureidocarboxylic acid (hydantoic acid derivative):

$$R^1{-}CH(NH_2){-}COOH \xrightarrow{R^2{-}N{=}C{=}O} R^1{-}CH(COOH){-}NH{-}C({=}O){-}NH{-}R^2 \xrightarrow{H^+} \text{Hydantoin derivative}$$

Hydantoin derivative

The thiohydantoins are obtained by the reaction with isothiocyanates.

Dehydration of *N*-acylamino acids with acetic anhydride or carbodiimide yields *1,3-oxazolin-5-ones* (azlactones):

$$R^1{-}CH(COOH){-}NH{-}C({=}O){-}R^2 \xrightarrow{(CH_3CO)_2O} \text{4-}R^1\text{-2-}R^2\text{-1,3-oxazolin-5-one}$$

Racemization occurs readily during the reaction. The azlactones are intermediates in the synthesis of amino acids.

The *N-carboxylic acid anhydrides* (Leuchs anhydrides, 1,3-oxazolidine-2,5-diones) are often used in peptide synthesis, especially to prepare poly-α-amino acids. These reactive amino acid derivatives are obtained by treating the amino acid with phosgene in the presence of tertiary amines

or by elimination of benzyl chloride from *N*-benzyloxycarbonylamino acid chlorides.

2.3. Important Amino Acids

2.3.1. Proteinogenic Amino Acids

L-Alanine [*56-41-7*], 2-aminopropionic acid, $C_3H_7NO_2$, M_r 89.09, *mp* 314 °C (decomp.), $[\alpha]_D^{25}$ +14.47° (*c* = 10.03 in 6 M HCl), solubility 16.51 (25 °C) g/100 g H_2O, p*I* 6.01, dissociation constants: pK_1 2.34, pK_2 9.69.

L-Arginine [*74-79-3*], 2-amino-5-guanidinovaleric acid, 2-amino-5[(aminoiminomethyl)amino]pentanoic acid, $C_6H_{14}N_4O_2$, M_r 174.20, *mp* 244 °C (decomp.), $[\alpha]_D^{25}$ +27.58° (*c* = 2 in 6 M HCl), solubility 14.87 (20 °C) g/100 g H_2O, p*I* 10.76, dissociation constants: pK_1 2.01, pK_2 9.04 (α-NH_2), pK_3 12.48 (guanidyl). Hydrochloride: [*1119-34-2*], $C_6H_{15}CN_4O_2$, M_r 210.66, *mp* 220 °C (decomp.), $[\alpha]_D^{25}$ +11.7° (*c* = 5 in H_2O), solubility 75.1 (20 °C) g/100 g H_2O.

L-Asparagine [*70-47-3*], 2-aminosuccinamic acid, 2,4-diamino-4-oxobutanoic acid, $C_4H_8N_2O_3$, M_r 132.13, *mp* 236 °C (decomp.), $[\alpha]_D^{20}$ +32.6° (*c* = 1 in 0.1 M HCl), solubility 3.11 (28 °C) g/100 g H_2O, p*I* 5.41, dissociation constants: pK_1 2.02, pK_2 8.8.

L-Aspartic acid [*56-84-8*], aminosuccinic acid, 2-amino-1,3-butanedioic acid, $C_4H_7NO_4$, M_r 133.10, *mp* 270 °C (decomp., sealed tube), $[\alpha]_D^{25}$ +25.4° (*c* = 2 in 5 M HCl), solubility 0.5 (25 °C) g/100 g H_2O, p*I* 2.98, dissociation constants: pK_1 2.1, pK_2 3.86 (β-COOH), pK_3 9.82.

L-Cysteine [*52-90-4*], 2-amino-3-mercaptopropionic acid, 3-mercaptoalanine, $C_3H_7NO_2S$, M_r 121.16, *mp* 240 °C (decomp.), $[\alpha]_D^{25}$ +9.7° (*c* = 8 in 1 M HCl), solubility 28 (25 °C) g/100 mL solution, 16 (20 °C) g/100 g H_2O, p*I* 5.02, dissociation constants: pK_1 1.71, pK_2 8.27 (–SH), pK_3 10.78. Hydrochloride monohydrate: [*7048-04-6*], $C_3H_{10}ClNO_3S$, M_r 175.64, *mp* (anhydr.) 178 °C (decomp.), $[\alpha]_D^{25}$ +6.53° (calculated as cysteine) (*c* = 2 in 5 M HCl), solubility >100 (20 °C) g/100 g H_2O.

L-Cystine [*56-89-3*], 2, 2′-diamino-3, 3′-dithiobis(propionic acid), 3, 3′-dithiobis(2-aminopropanoic acid), $C_6H_{12}N_2O_4S_2$, M_r 240.30, *mp* 260 °C (decomp.), $[\alpha]_D^{25}$ −212° (*c* = 1 in 1 M HCl), solubility 0.011 (25 °C) g/100 g H_2O, p*I* 5.02, dissociation constants: pK_1 1.04, pK_2 2.05 (-COOH), pK_3 8.0 (–NH_2), pK_4 10.25 (–NH_2).

L-Glutamic acid [*56-86-0*], 2-aminoglutaric acid, 2-amino-1,4-pentanedioic acid, $C_5H_9NO_4$, M_r 147.13, *mp* 224 – 225 °C (decomp.), $[\alpha]_D^{25}$ +31.5° (*c* = 2 in 5 M HCl, 20 °C), solubility = 0.843 (25 °C) g/100 g H_2O, p*I* 3.08, dissociation constants: pK_1 2.1, pK_2 4.07, pK_3 9.47.

L-Glutamine [*56-85-9*], 2-aminoglutaramic acid, 2,5-diamino-5-oxopentanoic acid, $C_5H_{10}N_2O_3$, M_r 146.15, *mp* 185-6 °C (decomp.), $[\alpha]_D^{25}$ +31.8° ($c=2$ in 1 M HCl), solubility 3.6 (19 °C) g/100 g H_2O, p*I* 5.65, dissociation constants: pK_1 2.17, pK_2 9.13.

Glycine [*56-40-6*], aminoacetic acid, $C_2H_5NO_2$, M_r 75.07, *mp* 262, 292 °C (decomp.), solubility 24.99 (25 °C) g/100 g H_2O, p*I* 6.06, dissociation constants: pK_1 2.35, pK_2 9.13.

L-Histidine [*71-00-1*], α-amino-1*H*-imidazole-4-propionic acid, 1*H*-imidazole-4-alanine, $C_6H_9N_3O_2$, M_r 155.16, *mp* 277, 287 °C (decomp.), $[\alpha]_D^{25}$ +13.0° ($c=1$ in 6 M HCl), solubility 4.29 (25 °C) g/100 g H_2O, p*I* 7.64, dissociation constants: pK_1 1.77, pK_2 6.1 (imidazolyl), pK_3 9.18. Hydrochloride monohydrate: [*5934-29-2*], $C_6H_{12}ClN_3O_3$, M_r 209.63, *mp* 259 °C (decomp.), solubility 16.99 (20 °C) g/100 g H_2O.

L-Hydroxyproline [*51-35-4*], *trans*-4-hydroxy-2-pyrrolidinecarboxylic acid, $C_5H_9NO_3$, M_r 131.13, *mp* 274 °C (decomp.), $[\alpha]_D^{22}$ − 75.2° ($c=2$ in H_2O), solubility 36.11 (25 °C) g/100 g H_2O, p*I* 5.82, dissociation constants: pK_1 1.92, pK_2 9.73.

L-Isoleucine [*73-32-5*], 2-amino-3-methylvaleric acid, 2-amino-3-methylpentanoic acid, $C_6H_{13}NO_2$, M_r 131.18, *mp* 285-6 °C (decomp.), $[\alpha]_D^{25}$ +40.6° ($c=2$ in 6 M HCl), solubility 4.117 (25 °C) g/100 g H_2O, p*I* 6.02, dissociation constants: pK_1 2.36, pK_2 9.68.

L-Leucine [*61-90-5*], 2-amino-4-methylvaleric acid, 2-amino-4-methylpentanoic acid, 2-aminoisocaproic acid, $C_6H_{13}NO_2$, M_r 131.18, *mp* 293-5, 314-5 °C (decomp.), $[\alpha]_D^{25}$ − 10.9° ($c=0.52$ in H_2O), solubility 2.19 (25 °C) g/100 g H_2O, p*I* 5.98, dissociation constants: pK_1 2.36, pK_2 9.6.

L-Lysine [*56-87-1*], 2, 6-diaminohexanoic acid, 2, 6-diaminocaproic acid, $C_6H_{14}N_2O_2$, M_r 146.19, *mp* 225-5 °C (decomp.), $[\alpha]_D^{25}$ +25.9° ($c=2$ in 5 M HCl), solubility >100 (25 °C) g/100 g H_2O, p*I* 9.47, dissociation constants: pK_1 2.18, pK_2 8.95 (α-NH_2), pK_3 10.53. Hydrochloride: [*657-27-2*], $C_6H_{15}ClN_2O_2$, M_r 182.65, *mp* 253-6 °C (decomp.), solubility 72.5 (25 °C) g/100 mL solution.

L-Methionine [*63-68-3*], 2-amino-4-(methylthio)butyric acid, 2-amino-4-(methylthio)butanoic acid, $C_5H_{11}NO_2S$, M_r 149.21, *mp* 283 °C (decomp.), $[\alpha]_D^{28}$ +23.4° ($c=5$ in 3 M HCl), solubility 5.37 (20 °C) g/100 g H_2O, p*I* 5.74, dissociation constants: pK_1 2.28, pK_2 9.21.

L-Phenylalanine [*63-91-2*], 2-amino-3-phenylpropionic acid, α-aminobenzenepropanoic acid, $C_9H_{11}NO_2$, M_r 165.19, *mp* 283-4 °C (decomp.), $[\alpha]_D^{20}$ − 35.1° (c = 2 in H_2O), solubility 2.965 (25 °C) g/100 g H_2O, pI 5.48, dissociation constants: pK_1 1.83, pK_2 9.13.

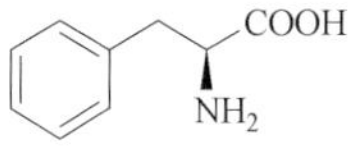

L-Proline [*147-85-3*], 2-pyrrolidinecarboxylic acid, 2-carboxypyrrolidine, $C_5H_9NO_2$, M_r 115.13, *mp* 220 – 222 °C (decomp.), $[\alpha]_D^{25}$ − 85.0° (c = 1 in H_2O), solubility 162.3 (25 °C) g/100 g H_2O, pI 6.3, dissociation constants: pK_1 2.0, pK_2 10.6.

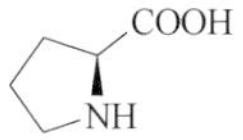

L-Serine [*56-45-1*], 2-amino-3-hydroxypropionic acid, 2-amino-3-hydroxypropanoic acid, $C_3H_7NO_3$, M_r 105.09, *mp* 228 °C (decomp.), $[\alpha]_D^{26}$ − 6.8° (c = 10 in H_2O), solubility 35.97 (20 °C) g/100 g H_2O, pI 5.68, dissociation constants: pK_1 2.21, pK_2 9.15.

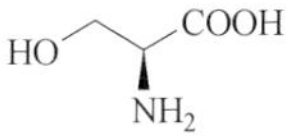

L-Threonine [*72-19-5*], 2-amino-3-hydroxybutyric acid, 2-amino-3-hydroxybutanoic acid, $C_4H_9NO_3$, M_r 119.12, *mp* 253 °C (decomp.), $[\alpha]_D^{26}$ − 28.6° (c = 2 in H_2O), solubility 9.03 (20 °C) g/100 g H_2O, pI 6.16, dissociation constants: pK_1 2.71, pK_2 9.62.

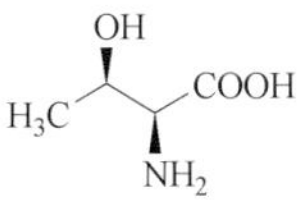

L-Tryptophan [*73-22-3*], 2-amino-3-(3′-indolyl)propionic acid, α-amino-1H-indole-3-propanoic acid, $C_{11}H_{12}N_2O_2$, M_r 204.23, *mp* 290 – 292 °C, 281 °C (decomp.), $[\alpha]_D^{25}$ − 32.15° (c = 1 in H_2O), solubility 1.14 (25 °C) g/100 g H_2O, pI 5.88, dissociation constants: pK_1 2.38, pK_2 9.39.

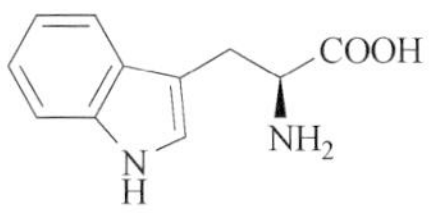

L-Tyrosine [*60-18-4*], 2-amino-3-(4-hydroxyphenyl)propionic acid, [3-(4-hydroxyphenyl)]alanine, 2-amino-3-(p-hydroxyphenyl)propionic acid, α-amino-4-hydroxybenzenepropanoic acid, $C_9H_{11}NO_3$, M_r 181.19, *mp* 342-4 (sealed tube), 297 – 298 °C (decomp.), $[\alpha]_D^{25}$ − 7.27° (c = 4 in 6 M HCl), solubility 0.045 (25 °C) g/100 g H_2O, pI 5.63, dissociation constants: pK_1 2.2, pK_2 9.11, pK_3 10.07 (–OH).

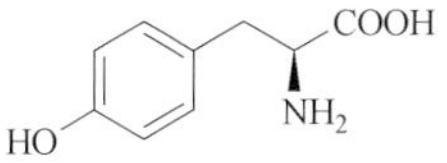

L-Valine [*72-18-4*], 2-amino-3-methylbutyric acid, 2-aminoisovaleric acid, $C_5H_{11}NO_2$, M_r 117.15, *mp* 315 °C (decomp.), $[\alpha]_D^{20}$ +26.7° (c = 3.4 in 6 M HCl), solubility 8.85 (25 °C) g/100 g H_2O, pI 5.96, dissociation constants: pK_1 2.32, pK_2 9.62.

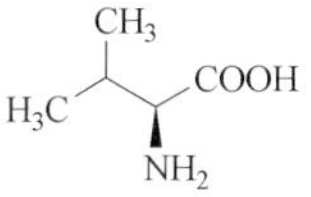

2.3.2. Other Important Amino Acids

β-Alanine [*107-95-9*], 3-aminopropionic acid, $C_3H_7NO_2$, M_r 89.09, dissociation constant: pK_1 3.6, occurrence: apple, constituent of pantothenic acid, carnosine, anserine.

D-Alanine [*338-69-2*], 2-aminopropionic acid, D-Ala, $C_3H_7NO_2$, M_r 89.09, application: LHRH-antagonists, e.g., Abarelix (antineoplastic) [21].

D,L-Alanine [*302-72-7*], 2-aminopropionic acid, $C_3H_7NO_2$, M_r 89.09, *mp* 295 °C (decomp.), solubility 16.72 (25 °C) g/100 g H_2O, p*I* 6.11, dissociation constants: pK_1 2.35, pK_2 9.87.

α-Aminoisobutyric acid [*62-57-7*], 2-amino-2-methylpropionic acid, Aib, $C_4H_9NO_2$, M_r 103.12, application: growth hormone secretagogues, e.g., MK-0677 [22].

H_3C, H_3C, COOH, NH_2

L-α-Aminobutyric acid [*1492-24-6*], 2-aminobutyric acid, L-Abu, $C_4H_9NO_2$, M_r 103.12, application: Levetiracetam (anticonvulsant) [23].

H_3C, COOH, NH_2

γ-Aminobutyric acid (GABA) [*56-12-2*], 4-aminobutyric acid, $C_4H_9NO_2$, M_r 103.12, occurrence: citrus fruits, sugar beet, brain.

H_2N, COOH

D,L-Aspartic acid [*617-45-8*], aminosuccinic acid, 2-amino-1,3-butanedioic acid, $C_4H_7NO_4$, M_r 133.10, *mp* 275 °C (decomp., sealed tube), solubility 0.775 (25 °C) g/100 g H_2O, p*I* 2.98, dissociation constants: pK_1 2.1, pK_2 3.86 (β-COOH), pK_3 9.82.

Betaine [*107-43-7*], carboxymethyl-trimethyl ammonium betaine, $C_5H_{11}NO_2$, M_r 117.15, occurrence: sugar beet.

H_3C, CH_3, H_3C, N^+, COO^-

L-Carnitine [*541-15-1*], (3-carboxy-2-hydroxy-propyl)-trimethyl ammonium betaine, $C_7H_{15}NO_3$, M_r 161.2, occurrence: Lys metabolite, muscle.

CH_3, OH, H_3C, N^+, H_3C, COO^-

L-Citrulline [*372-75-8*], 2-amino-5-ureidopentanoic acid, $C_6H_{13}N_3O_3$, M_r 175.19, *mp* 234 – 237 °C , 222 °C (decomp.), $[\alpha]_D^{25}$ +24.2° (c = 2 in 5 M HCl), solubility 10.3 (20 °C) g/100 g H_2O, p*I* 5.92, dissociation constants: pK_1 2.43, pK_2 9.41; occurrence: urea cycle, watermelon.

O, H_2N, N H, COOH, NH_2

D-Citrulline [*13594-51-9*], 2-amino-5-ureidopentanoic acid, D-Cit, $C_6H_{13}N_3O_3$, M_r 175.19, application: Cetrorelix (antineoplastic) [24].

Creatine [*57-00-1*], *N*-(aminoiminomethyl)-*N*-methyl-glycine, (*N*-methyl-guanetic acid, $C_4H_9N_3O_2$, M_r 131.14, occurrence: muscle of vertebrates.

CH_3, H_2N, N, COOH, NH

D-Cyclohexylalanine [*58717-02-5*], 2-amino-3-cyclohexylpropionic acid, D-Cha, $C_9H_{17}NO_2$, M_r 171.25, application: (thrombine inhibitors) [25].

COOH, NH_2

(3*S*,4a*S*,8a*S*)-Decahydroisoquinolinecarboxylic acid [*115238-58-9*], decahydroisoquinoline-3-carboxylic acid, $C_{10}H_{17}NO_2$, M_r 183.26, application: Nelfinavir (antiviral) [26].

COOH, NH

L-2,3-Diaminopropionic acid [*4033-39-0*], 2,3-diaminopropionic acid, L-Dap, $C_3H_8N_2O_2$, M_r 104.11, application: Imidapril (antihypertensive) [27].

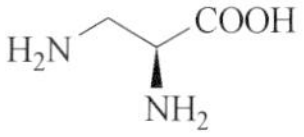

L-3,4-Dihydroxyphenylalanine (DOPA) [*59-92-7*], 2-amino-3-(3,4-dihydroxyphenyl)propionic acid, $C_9H_{11}NO_4$, M_r 197.17, occurrence: Tyr metabolite, bean.

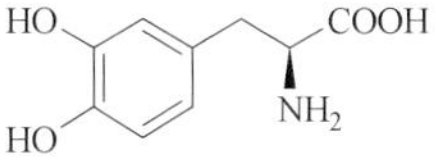

D-Glutamic acid [*6893-26-1*], 2-aminoglutaric acid, 2-amino-1,4-pentanedioic acid, D-Glu, $C_5H_9NO_4$, M_r 147.12, application: Spiroglumide (CCK-B-antagonist) [28].

L-Homocysteine [*6027-13-0*], 2-amino-4-mercaptobutyric acid, $C_4H_9NO_2S$, M_r 135.18, occurrence: Met metabolite, mushrooms, application: BMS-186716 [29].

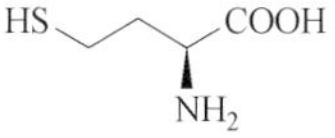

D-*p*-Hydroxyphenylglycine [*22818-40-2*], amino-(4-hydroxyphenyl)acetic acid, D-Phg(OH), $C_8H_9NO_3$, M_r 167.15, application: Amoxicillin (antibiotic) [30].

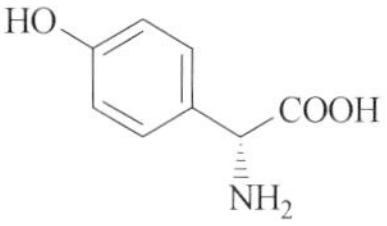

L-5-Hydroxytryptophan [*4350-09-8*], 2-amino-3-(5-hydroxy-1*H*-indol-3-yl)propionic acid, $C_{11}H_{12}N_2O_3$, M_r 220.22, occurrence: serotonin precursor.

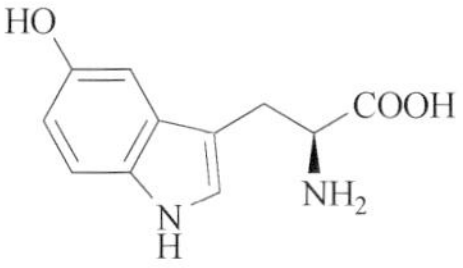

D,L-Isoleucine [*443-79-8*], 2-amino-3-methylvaleric acid, 2-amino-3-methylpentanoic acid, $C_6H_{13}NO_2$, M_r 131.18, *mp* 292 °C (decomp.), solubility 2.011 (25 °C) g/100 g H_2O, p*I* 6.04, dissociation constants: pK_1 2.32, pK_2 9.76.

D,L-Leucine [*328-39-2*], 2-amino-4-methylvaleric acid, 2-amino-4-methylpentanoic acid, 2-aminoisocaproic acid, $C_6H_{13}NO_2$, M_r 131.18, *mp* 293-5, 332 °C (decomp.), solubility 1.00 (25 °C) g/100 g H_2O, p*I* 6.04, dissociation constants: pK_1 2.33, pK_2 9.74.

L-*tert*-Leucine [*20859-02-3*], 2-amino-3,3-dimethylbutyric acid, L-Tle, $C_6H_{13}NO_2$, M_r 131.18, application: div. pharmaceuticals [31].

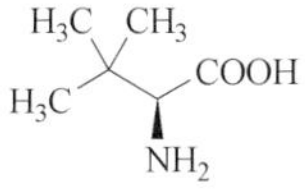

D,L-Lysine hydrochloride [*70-53-1*], $C_6H_{15}ClN_2O_2$, M_r 182.65, *mp* 264 °C (decomp.), solubility 35.98 (20 °C) g/100 g H_2O.

D,L-Methionine [*59-51-8*], 2-amino-4-(methylthio)butyric acid, 2-amino-4-(methylthio)butanoic acid, $C_5H_{11}NO_2S$, M_r 149.21, *mp* 281 °C (decomp.), solubility 3.35 (25 °C) g/100 g H_2O, p*I* 5.74, dissociation constants: pK_1 2.28, pK_2 9.21.

***S*-Methyl-L-cysteine** [*1187-84-4*],2-amino-3-methylthiopropionic acid L-Cys(Me), $C_4H_9NO_2S$, M_r 135.18, application: KNI-272 (antiviral) [32].

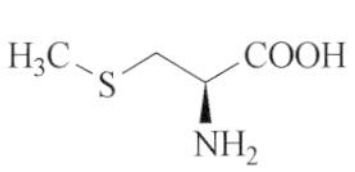

L-*S*-Methylmethionine (vitamin U) [*4727-40-6*], 3-amino-3-carboxy-propyl)-dimethyl sulfonium, $C_6H_{13}NO_2S$, M_r 163.24, occurrence: cabbage, asparagus.

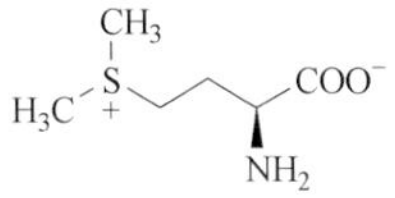

D-3-(2′-Naphthyl)-alanine [*76985-09-6*], 2-amino-3-naphthalen-2-ylpropionic acid, D-Nal, $C_{13}H_{13}NO_2$, M_r 215.25, application: LHRH-antagonists, e.g., Abarelix (antineoplastic) [33].

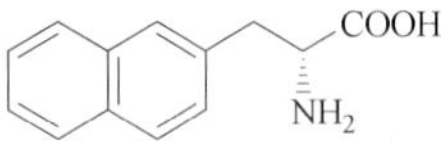

L-Ornithine [*70-26-8*], 2,5-diaminovaleric acid, $C_5H_{12}N_2O_2$, M_r 132.16, *mp* 140 °C (decomp.), $[\alpha]_D^{25}$ +16.5° (c=4.6 in H_2O), solubility, p*I* 9.7, dissociation constants: pK_1 1.94, pK_2 8.65 (a-NH), pK_3 10.76; occurrence: urea cycle, shark liver. Hydrochloride: [*3184-13-2*], $C_5H_{13}ClN_2O_2$, M_r 168.6, *mp* 215 °C (decomp.), $[\alpha]_D^{25}$ +28.3° (calculated as ornithine) (c=2 in 5 M HCl), solubility 54.36 (20 °C) g/100 g H_2O.

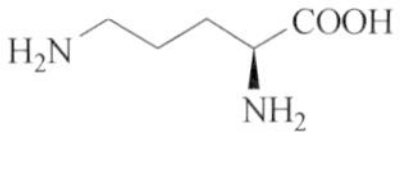

D-Penicillamine [*52-67-5*], 2-amino-3-mercapto-3-methylbutyric acid, $C_5H_{11}NO_2S$, M_r 149.21, occurrence: hydrolysis product of penicillin.

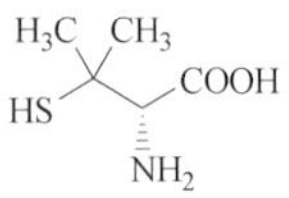

L-Penicillamine [*1113-41-3*], 2-amino-3-mercapto-3-methylbutyric acid, L-Pen, $C_5H_{11}NO_2S$, M_r 149.21, application: JE-2147 (antiviral) [34].

D-Phenylalanine [*673-06-3*], 2-amino-3-phenylpropionic acid, α-aminobenzenepropanoic acid, D-Phe, $C_9H_{11}NO_2$, M_r 165.19, application: AY-4166 (antidiabetic) [35].

D,L-Phenylalanine [*150-30-1*], 2-amino-3-phenylpropionic acid, α-aminobenzenepropanoic acid, $C_9H_{11}NO_2$, M_r 165.19, *mp* 284 – 288 °C, 320 °C (decomp.), solubility 1.29 (25 °C) g/100 g H_2O, p*I* 5.91, dissociation constants: pK_1 2.58, pK_2 9.24

D-*p*-Cl-Phenylalanine [*14091-08-8*], 2-amino-3-(4-chloro-phenyl)propionic acid, D-Phe(Cl), $C_9ClH_{10}NO_2$, M_r 199.63, application: LHRH antagonists, e.g., Abarelix (antineoplastic) [36].

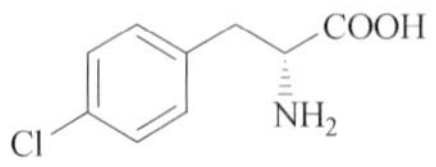

L-*p*-NO₂-Phenylalanine [*949-99-5*], 2-amino-3-(4-nitro-phenyl)propionic acid, L-Phe(NO_2), $C_9H_{10}N_2O_4$, M_r 210.17, application: Zolmitriptan (antimigraine) [37].

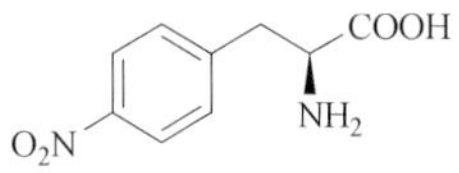

***S*-Phenyl-L-cysteine** [*34317-61-8*], 2-amino-3-phenylthiopropionic acid, L-Cys(Ph), $C_9H_{11}NO_2S$, M_r 197.61, application: Nelfinavir (antiviral) [38].

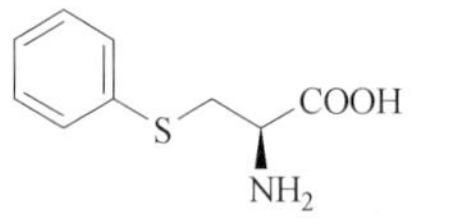

D-Phenylglycine [*875-74-1*], 2-aminophenyl-acetic acid, D-Phg, $C_8H_9NO_2$, M_r 151.16, application: Ampicillin (antibiotic) [39].

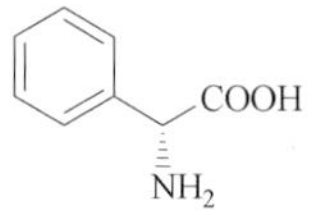

L-Pipecolic acid [*3105-95-1*], piperidine-2-carboxylic acid, L-Pec, $C_6H_{11}NO_2$, M_r 129.16, occurrence: legumes, metabolite of Lys, application: Ropivacaine (local anesthetic) [40].

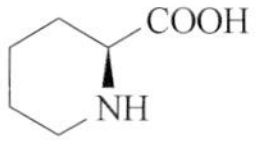

L-Piperazinecarboxylic acid [*147650-70-2*], piperazine-2-carboxylic acid, $C_5H_{10}N_2O_2$, M_r 130.15, application: Indinavir (antiviral) [41].

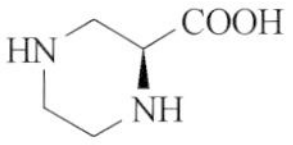

D-Piperidine-3-carboxylic acid [*25137-00-2*], $C_6H_{11}NO_2$, M_r 129.7, application: Tiagabine (anticonvulsant) [42].

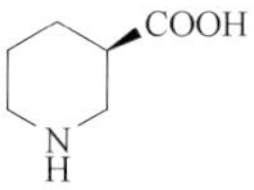

D,L-Proline [*609-36-9*], 2-pyrrolidine-carboxylic acid, 2-carboxypyrrolidine, $C_5H_9NO_2$, M_r 115.13, *mp* 205 °C (decomp.), p*I* 6.3, dissociation constants: pK_1 2.0, pK_2 10.6.

D-Proline [*344-25-2*], 2-pyrrolidine-carboxylic acid, 2-carboxypyrrolidine, D-Pro, $C_5H_9NO_2$, M_r 115.13, application: Eletriptan (antimigraine) [43].

D-3-(3′-Pyridyl)-alanine [*70702-47-5*], 2-amino-3-pyridin-3-ylpropionic acid, D-Pal, $C_8H_{10}N_2O_2$, M_r 166.18, application: LHRH antagonist, e.g. Abarelix (antineoplastic) [44].

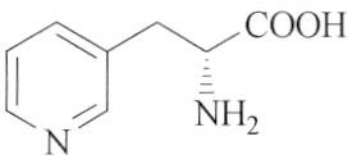

L-Saccharopine [*997-68-2*], 2-(5′-amino-5′-carboxy-pentylamino)pentanedioic acid, $C_{11}H_{20}N_2O_6$, M_r 276.27, occurrence: baker′s and brewer′s yeast.

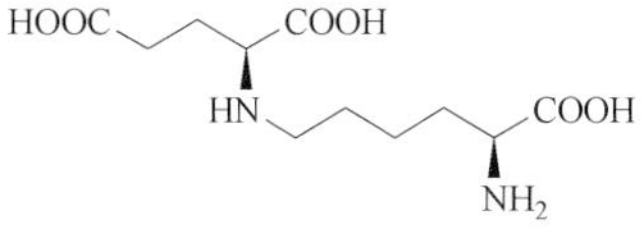

D,L-Serine [*302-84-1*], 2-amino-3-hydroxypropionic acid, 2-amino-3-hydroxypropanoic acid, $C_3H_7NO_3$, M_r 105.09, *mp* 246 °C (decomp., sealed tube), solubility 5.023 (25 °C) g/100 g H_2O, p*I* 5.68, dissociation constants: pK_1 2.21, pK_2 9.15.

D-Serine [*312-84-5*], 2-amino-3-hydroxy-propionic acid, 2-amino-3-hydroxypropanoic acid, D-Ser, $C_3H_7NO_3$, M_r 105.09, application: D-cycloserine (cognition enhancer) [45].

L-Thiazolidine-4-carboxylic acid [*34292-47-7*], thiazolidine-4-carboxylic acid, L-Tia, $C_4H_7NO_2S$, M_r 133.16, application: KNI-272 (antiviral) [32].

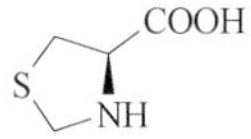

D,L-Threonine [*80-68-2*], 2-amino-3-hydroxybutyric acid, 2-amino-3-hydroxybutanoic acid, $C_4H_9NO_3$, M_r 119.12, *mp* 234-5 °C (decomp.), solubility 20.5 (25 °C) g/100 g H_2O, p*I* 6.16, dissociation constants: pK_1 2.71, pK_2 9.62.

L-Thyroxine [*51-48-9*], 2-amino-3-[4-(4-hydroxy-3,5-diiodophenoxy)-3,5,diiodo-phenyl)-propionic acid, $C_{15}H_{11}NO_4$, M_r 269.18, occurrence: thyroid gland.

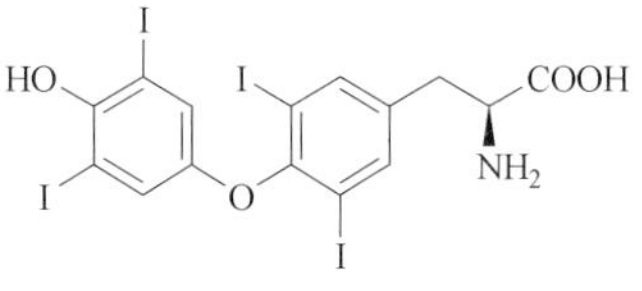

D,L-Tryptophan [*54-12-6*], 2-amino-3-(3′-indolyl)propionic acid, α-amino-1*H*-indole-3-propanoic acid, $C_{11}H_{12}N_2O_2$, M_r 204.23, *mp* 285 °C (decomp.), solubility 0.25 (30 °C) g/100 g H_2O, p*I* 5.88, dissociation constants: pK_1 2.38, pK_2 9.39.

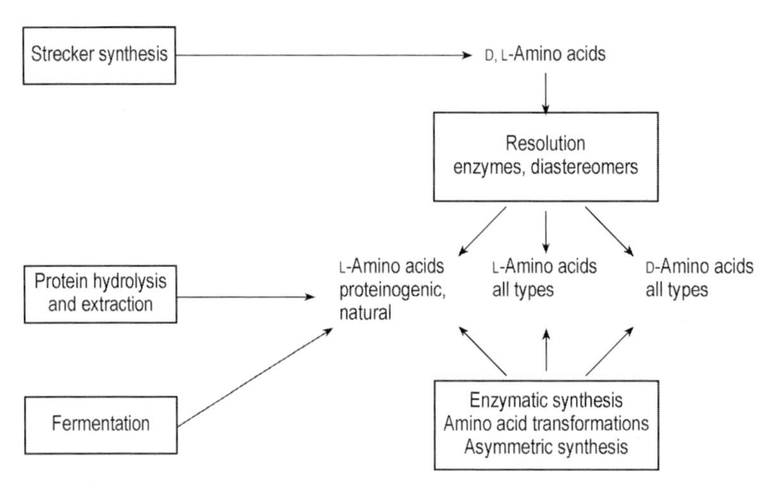

Figure 2. Routes for production of amino acids

D,L-Tyrosine [*556-03-6*], 2-amino-3-(4-hydroxyphenyl)propionic acid, [3-(4-hydroxyphenyl)]alanine, 2-amino-3-(*p*-hydroxyphenyl)propionic acid, α-amino-4-hydroxybenzenepropanoic acid, $C_9H_{11}NO_3$, M_r 181.19, *mp* 340 °C, 318 °C (decomp.), solubility 0.351 (25 °C) g/100 g H_2O, p*I* 5.63, dissociation constants: pK_1 2.2, pK_2 9.11, pK_3 10.07 (–OH).

D,L-Valine [*516-06-3*], 2-amino-3-methylbutyric acid, 2-aminoisovaleric acid, $C_5H_{11}NO_2$, M_r 117.15, *mp* 298 °C (decomp., sealed tube), solubility 7.09 (25 °C) g/100 g H_2O, p*I* 6.0, dissociation constants: pK_1 2.29, pK_2 9.72.

D-Valine [*640-68-6*], 2-amino-3-methylbutyric acid, D-Val, $C_5H_{10}NO_2$, M_r 116.14, application: Fluvalinate (insecticide) [46].

CH_3
COOH
H_3C
NH_2

3. Industrial Production of Amino Acids

Amino acids can be classified into natural and nonnatural ones. The natural amino acids are subdivided into proteinogenic, which means occurring in proteins. Twenty of the proteinogenic amino acids are genetically coded. For nutrition and feed additives the proteinogenic L-amino acids are of highest importance. The following sections will focus on production methods for the proteinogenic amino acids.

3.1. General Methods

Four basic processes (see Fig. 2) are suitable for the production of amino acids: chemical synthesis (including asymmetric synthesis), extraction, fermentation, and enzymatic routes. The *classical chemical synthesis* is applied to produce either the achiral amino acid glycine or racemic amino acids, for instance D,L-methionine. If L-amino acids are to be manufactured by this route, the chemical synthesis has to be followed by a resolution step. In some cases L-amino acids are directly produced from prochiral precursors by means of enzymes as chiral catalysts. For L-cysteine, amino acid (i.e., cystine) transformation is still very important. The *extraction process* has the advantage to offer access to nearly all proteinogenic L-amino acids by isolation from protein hydrolysates. Starting materials are protein-rich products, such as keratin, feathers, blood meal, or technical gelatin (see Fig. 3).

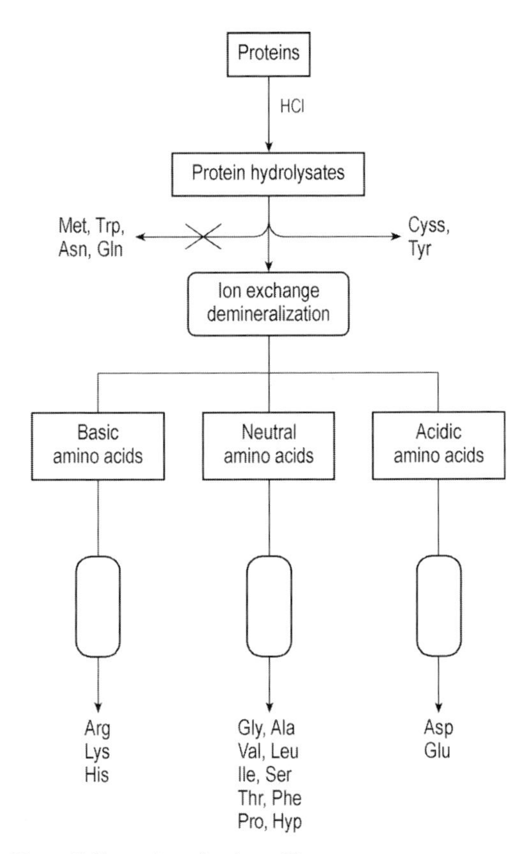

Figure 3. Extraction of amino acids

The use of overproducing microbial strains in *fermentation processes* with sucrose or glucose as carbon source is currently the most economic access to bulk amino acids such as monosodium L-glutamate, L-lysine hydrochloride, and L-threonine. Even the aromatic amino acids, L-phenylalanine and L-tryptophan – ten years ago only available by enzymatic processes – are now being produced by potent mutants of bacteria. Classical breeding and mutagenesis as well as recombinant DNA techniques are general methods of the strain development in order to influence the pathway in the biosynthesis of amino acids and to force the bacteria to overproduce the corresponding metabolites. Industrially used mutants are usually characterized through genetic markers introduced by methods of deregulation, e.g., auxotrophic mutants, in order to release key enzymes from strict regulation by metabolites (feedback inhibition and repression). A different way to get overproducing strains consists of screening on regulatory mutants, in that important enzymes are resistant to toxic compounds, i.e., analogs or derivatives of amino acids. In such cases high amino acid concentrations do not inhibit the enzymatic key in the pathway of biosynthesis and thus result in high production rates of the desired amino acid. The fourth method, *enzymatic catalysis*, uses whole cells or active cell components (enzymes) as such or, if possible, as immobilized biocatalysts in continuously operated reactors. The competitiveness of enzymatic processes depends on the availability and price of substrate, the activity and stability of the involved enzyme(s) and on the simplicity of product recovery.

In Table 2 the amino acids and their preferred production methods are listed, classified by the size of production volume. Table 3 specifies the amino acids produced by direct fermentation of carbohydrates.

Table 2. Production of amino acids, methods, and production volume

Type	Amino acid	Preferred production method
I	L-Alanine	enzymatic catalysis, fermentation
I	L-Asparagine	extraction
I	L-Glutamine	fermentation, extraction
I	L-Histidine	fermentation, extraction
I	L-Hydroxyproline	fermentation, extraction
I	L-Isoleucine	fermentation, extraction
I	L-Leucine	fermentation, extraction
I	L-Methionine	enzymatic resolution
I	L-Proline	fermentation, extraction
I	L-Serine	fermentation, extraction
I	L-Tyrosine	extraction
I	L-Valine	enzymatic catalysis, fermentation
II	L-Arginine	fermentation, extraction
II	L-Cysteine	reduction of L-cystine (extraction), enzymatic synthesis
II	L-Tryptophan	fermentation
III	Glycine	chemical synthesis
III	L-Aspartic acid	enzymatic catalysis
III	L-Phenylalanine	fermentation
III	L-Threonine	fermentation
IV	D,L-Methionine	chemical synthesis and enzymatic resolution
IV	L-Glutamic acid	fermentation
IV	L-Lysine	fermentation

Type I	100 – 1000	t/a
Type II	1000 – 8000	t/a
Type III	8000 – 100 000	t/a
Type IV	100 000 – 800 000	t/a

Table 3. Amino acids produced by direct fermentation from carbohydrates

Amino acid	Type of mutant	Amino acid yield, g/L	Reference
L-Alanine (95 % e.e.)	*Athrobacter oxydans*	75	HASHIMOTO and KATSUMATA (1994) [240]
D-Alanine (95 % e.e.)	*Brevibacterium lactofermentum*	46	YAHATA et al. (1993) [239]
L-Arginine	*Serratia marcescens* AU^r-1	100	CHIBATA et al. (1983) [242]
L-Glutamine	*Brevibacterium flavum* AJ 3409	57	YOSHIHARA et al. (1992) [272]
L-Histidine	*Serratia marcescens*	40	SUGIURA et al. (1987) [274]
L-Isoleucine	*Escherichia coli* H-8461	30	KINO et al. (1993) [286]
L-Leucine	*Brevibacterium flavum* AJ 3686	19.5	YOSHIHARA et al. (1992) [287]
L-Lysine	*Corynebacterium glutamicum*	> 120	OH et al. (1990) [303]
L-Methionine	*Pseudomonas putida* VKPM V-4167	3.5	University Odessa (1992) [355]
L-Proline	*Brevibacterium flavum* AP113	97.5	KOCARIAN et al. (1986) [325]
L-Threonine	*Escherichia coli* BKIIM B-3996	85	DEBABOV et al. (1990) [354]
L-Tryptophan	*Corynebacterium glutamicum* KY9218/pKW9901	50	IKEDA et al. (1994) [350]
L-Valine	*Brevibacterium lactofermentum* AJ 12341	39	KATSURADA et al. (1993) [353]

3.2. Production of Specific Amino Acids

3.2.1. L-Alanine

L-Alanine is industrially produced from L-aspartic acid by means of immobilized *Pseudomonas dacunhae* cells in a pressurized bioreactor [238]. In direct fermentation microorganisms usually accumulate D,L-alanine because of present alanine racemase. With a D-cycloserine resistant mutant selected from *Brevibacterium lactofermentum*, it is possible to obtain 46 g/L D-alanine with an enantiomeric excess (e.e.) of 95 % [239]. An alanine racemase-deficient mutant of *Arthrobacter oxydans* was reported, that produces 75 g/L L-alanine from glucose with a yield of 52 % and 95 % e.e. [240]. To a certain extent L-alanine is still isolated from protein hydrolysates.

3.2.2. L-Arginine

L-Arginine is produced by extraction from acidic hydrolysates of proteins such as gelatin or by fermentation. Suitable strains for fermentation are deregulated mutants derived from *Corynebacterium glutamicum* and *Bacillus subtilis* accumulating 25 – 35 g/L L-arginine from glucose [241]. Very potent strains of *Serratia marcescens* derived from mutants obtained by transduction, having feedback-insensitive and derepressive enzymes of arginine biosynthesis and 6-azauracil resistance, are able to produce 60 – 100 g/L L-arginine [242].

3.2.3. L-Aspartic Acid and Asparagine

L-Aspartic acid is industrially manufactured by an enzymatic process in which aspartase (L-aspartate ammonia lyase, EC 4.3.1.1) catalyzes the addition of ammonia to fumaric acid [243]. Advantages of the enzymatic production method are higher product concentration and productivity and the formation of fewer byproducts. Thus L-aspartic acid can be easily separated from the reaction mixture by crystallization.

A process involving continuous production of L-aspartic acid by means of carrier-fixed aspartase isolated from *Escherichia coli* was first commercialized in Japan [244]. Economically attractive processes for L-aspartic acid use resting or dried cells with high content of aspartase. A 4.5 % suspension of aspartase-containing *Brevibacterium flavum* cells could be recycled in seven repeated batches and concentrations up to 166 g/L L-aspartate could be achieved [245]. In 1973, an immobilized cell system based on *E. coli* cells entrapped in polyacrylamide gel lattice was introduced for large-scale production [246]. This example represents the first industrial application of immobilized microbial cells in a fixed-bed reactor. Further improvements, immobilization of the cells in κ-carrageenan, provided remarkably increased operational stability, resulting in biocatalyst half-lives of almost two years [247]. A column packed with the κ-carrageenan-immobilized system allows a theoretical productivity of $140\,g\,L^{-1}\,h^{-1}$ L-aspartate.

That L-aspartic acid gained importance as intermediate for the manufacture of the dipeptide sweetener aspartame (methyl ester of L-aspartyl-L-phenylalanine) prompted the industry to develop improved processes. For continuous production of L-aspartic acid *Escherichia coli* strains immobilized with polyurethane [248] or with polyethylenimine and glass fiber support [249] and κ-carrageenan-immobilized *Pseudomonas putida* [250] have been reported and protected, respectively. Less successful was the search for strains that are overproducing L-aspartate by fermentation on sugar basis. A pyruvate kinase-deficient mutant of *Brevibacterium flavum* accumulates up to 22.6 g/L L-aspartic acid in glucose-containing medium [251], not enough to be competitive with the enzymatic processes. On the other hand, genetic recombination techniques helped to improve aspartase-containing strains. An *asp*A gene bearing plasmid (pBR322::*asp*A-*par*) was able to elevate aspartase formation in *Escherichia coli* K 12 about 30-fold [252].

L-Asparagine can be isolated as a byproduct from the production of potato starch. A simple synthesis of L-asparagine starts from L-aspartic acid which is esterified to the β-methylester followed by treatment with ammonia [253].

3.2.4. L-Cystine and L-Cysteine

L-Cysteine is mainly produced via L-cystine from hydrolysates of hair or other keratins. L-Cystine is easily recovered because of its weak solubility in aqueous solutions. The electrolytic reduction of L-cystine leads to L-cysteine.

L-Cysteine can be prepared from β-chloro-D,L-alanine and sodium sulfide with cysteine desulfhydrase, an enzyme obtained from, e.g., *Citrobacterium freundii* [254]. An enzymatic process for L-cysteine has been successfully developed using microorganisms capable to hydrolyze 2-amino-Δ^2-thiazoline 4-carboxylic acid (ATC) which is readily available from methyl α-chloroacrylate and thiourea. A mutant of *Pseudomonas thiazolinophilum* converts D,L-ATC to L-cysteine in 95 % molar yield at product concentrations higher than 30 g/L [255].

3.2.5. L-Glutamic Acid

In 1957, a soil bacterium was discovered [256] which was able to excrete considerable amounts of L-glutamate. Fermentation processes using strains of this bacterium, later called *Corynebacterium glutamicum*, have been successfully commercialized not only for L-glutamic acid but also for the production of other economically important amino acids [257]. Numerous coryneform microorganisms have been isolated and found to be able to overproduce L-glutamic acid and other amino acids. Examples of these microorganisms are *Brevibacterium flavum*, *Brevibacterium lactofermentum*, and *Microbacterium ammoniaphilum*. Because of minor differences in the character of those bacteria [258] which are all gram-positive, non-spore-forming, nonmotile and all require biotin for growth, the name of genus *Corynebacterium* was suggested for these coryneform bacteria [259].

For industrial production of L-glutamic acid, molasses (sucrose), starch hydrolysates (glucose) and ammonium sulfate are generally used as carbon and nitrogen sources, respectively [260]. Key factors in controlling the fermentation are the presence of biotin in optimal concentration – to optimize cell growth and the excretion of L-glutamate – and sufficient supply of oxygen to reduce the accumulation of byproducts, such as lactic and succinic acid. In biotin-rich fermentation media the addition of penicillin or cephalosporin C favors the overproduction of L-glutamic acid due to effects on the cell membrane. The supplementation of fatty acids also results in an increased permeability of the cells thus enhancing glutamate excretion. In the past the mechanism of glutamate excretion was simply explained as a "leakage" or "overflow" phenomenon [261]. In the beginning of the 1990s it was reported that a specific carrier system exists [262], [263], which is responsible for active glutamate transport in *Corynebacterium glutamicum.* Generally the intracellular accumulation of glutamate does not reach levels sufficient for feedback control in glutamate overproducers due to rapid excretion of glutamate. However, the regulatory mechanisms of L-glutamic acid biosynthesis have been studied intensively to obtain mutants with increased productivity. Two enzymes have been shown to play key roles in the biosynthesis of L-glutamic acid [264].

1) Phosphoenolpyruvate carboxylase (PEPC) catalyzes carboxylation of phosphoenolpyruvate to yield oxaloacetate; it is inhibited by L-aspartic acid and repressed by both L-aspartic and L-glutamic acids.
2) α-Ketoglutarate dehydrogenase (KDH) converts α-ketoglutarate to succinyl-CoA. In L-glutamate overproducing strains KDH limits further oxidation of α-ketoglutarate to carbon dioxide and succinate, thus favoring formation of L-glutamic acid.

In L-glutamate overproducing strains the K_{m} value of KDH for α-ketoglutarate was nearly two magnitudes lower than that of L-glutamic acid dehydrogenase (GDH) which catalyzes the last step, the reductive amination of α-ketoglutarate to L-glutamate. Consequently, V_{max} of GDH was proven to be about 150 times higher than that of KDH. The *Corynebacterium glutamicum gdh* gene has been isolated and characterized [265]. A strain of *Microbacterium ammoniaphilum* cultured under biotin-deficient conditions produced 58 % of L-glutamic acid formed from glucose via phosphoenolpyruvate, citrate, and of α-ketoglutarate and the other 42 % via the tricarboxylic acid (TCA) or the glyoxylate cycle [266]. In large-scale production the formation of trehalose very often reduces the product yield. Trehalose consists of two α-1.1 bound glucose molecules and is excreted by the bacteria as a material with protects the cell against high osmotic pressure (osmoprotectant). A process was recently developed and successfully industrialized in which trehalose formation is controlled and decreased by culturing the overproducing mutant in media containing invert sugar from molasses [267], [268].

Today wild type isolates as well as mutants developed by classical breeding or even strains constructed by modern techniques, using cell fusion or recombinant DNA methods [269], are available for industrial L-glutamate production. In addition a new type of fermentation process was reported which uses a strain that overproduces L-glutamic acid and L-lysine simultaneously. The cultivation of an auxotrophic regulatory mutant of *Brevibacterium lactofermentum* in a medium, supplemented with polyoxyethylenesorbitan monopalmitate as surface-active agent, resulted in the accumulation of 162 g L-amino acids per liter (105 g L-lysine · HCl + 57 g L-glutamic acid). This corresponds to a production rate of 3.8 g (L-lysine · HCl + L-glutamic acid) per liter per hour [270].

The success of an economic production basically depends on the experience in fermentation technology, in up- and downstream processing and on the skill of employees in research, development, and production. A multistep inoculation procedure followed by the fed-batch mode (→ Biotechnology, Chap. 4.4.) in the main fermentation up to 500 m^3 scale is still the preferred technology for the production of L-glutamic acid.

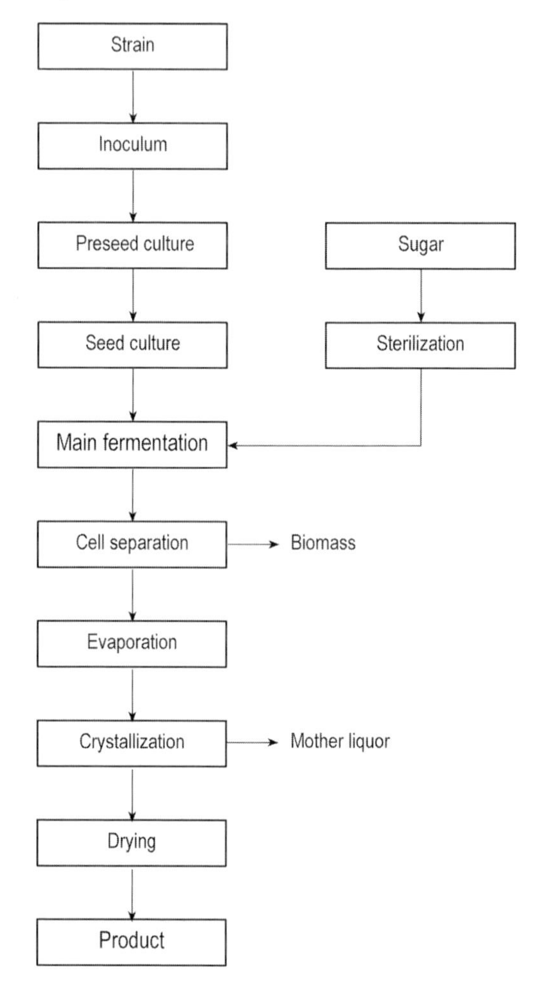

Figure 4. Flow diagram of fermentation and downstream processing of L-glutamic acid

Critical operations are the steam-forced sterilization of the fermenter and batchwise or continuous sterilization of the culture medium to prevent contamination by foreign microbes. During fermentation the temperature, pH, dissolved oxygen, and the sugar consumption have to be controlled as important process parameters. When the fermentation is completed after 40 – 60 h, the production strain may have accumulated more than 150 g/L L-glutamic acid. After deactivation of the broth, the recovery of the product is started by biomass separation using centrifuges or ultrafiltration units, concentration of the centrifugate or filtrate, followed by crystallization, filtration and drying. A scheme of production steps is given in Figure 4.

3.2.6. L-Glutamine

Microbial L-glutamine producers were selected from wild type glutamate-producing *coryneform* bacteria. A sulfaguanidine resistant mutant of *Brevibacterium flavum* accumulates 41 g/L L-glutamine in 48 h from 10 % glucose [271]. A yield of 44 % was achieved by the mutant *Brevibacterium flavum* AJ3409 [272].

3.2.7. L-Histidine

Efficient L-histidine fermentation can be performed with strains of *Corynebacterium glutamicum* and *Serratia marcescens*. A mutant of *Corynebacterium glutamicum* having resistance to 8-azaguanine, 1,2,4-triazole-3-alanine, 6-mercaptoguanine, 6-methylpurine, 5-methyltryptophan and 2-thiouracil, produces 15 g/L of L-histidine. Only after introduction of these resistance markers by mutagenesis is the mutant capable of releasing this amount of L-histidine into the nutrient solution. A strain of *Serratia marcescens* having both characters, feedback-insensitive and derepressed histidine enzymes combined with transductional techniques and 6-methylpurine resistance, accumulates 23 g/L L-histidine [273]. By amplification of the genes *his*G, *his*D, *his*B, *his*C in *Serratia marcescens* the final concentration of L-histidine could be elevated from 28 g/L to 40 g/L [274].

L-Histidine is also produced by isolation from blood-meal hydrolysates [275].

3.2.8. L-Hydroxyproline

L-Hydroxyproline (*trans*-4-hydroxy-L-proline, 2*S*,4*R*-hydroxyproline) is mainly obtained from hydrolysates of gelatin or other collagens. Collagen contains up to 15 % of L-4-hydroxyproline besides smaller amounts of the 3-isomer (2*S*,3*S*-hydroxyproline). An unusual feature of this amino acid is that it is not incorporated into collagen during biosynthesis at the ribosome, but is formed from L-proline by a posttranslational modification by an enzymatic hydroxylation reaction.

Another route for obtaining L-hydroxyproline is by a fermentative hydroxylation of L-proline with genetically modified microorganisms [276].

3.2.9. L-Isoleucine

An advantageous fermentation method for production of L-isoleucine is the use of chemically synthesized substrates that only require few steps to be converted to L-isoleucine. Among these the natural precursors 2-ketobutyrate [277] or D,L-2-hydroxybutyrate [278] have been used for the production with *Corynebacterium glutamicum.* A leucine requiring mutant of *Corynebacterium glutamicum*, with increased D-lactate utilization and consuming D,L-2-hydroxybutyrate, accumulates 13.4 g/L L-isoleucine. However, exploitation of this process is hampered by formation of byproducts [279]. Sugar based L-isoleucine processes have been developed with strains of *Corynebacterium glutamicum* [280], *Serratia marcescens* [281], and *Escherichia coli* [282–284]. The mutant *Escherichia coli* H-8285, being resistant to thiaisoleucine, arginine hydroxamate, and D,L-ethionine accumulates 26 g/L L-isoleucine in 45 h in a fed-batch process [285]. Introduction of resistance to 6-dimethylaminopurine in strain H-8285 resulted in a mutant *Escherichia coli* H-8461 that increased L-isoleucine accumulation to 30.2 g/L [286]. The biosynthesis of L-isoleucine has been investigated in detail on the level of involved genes [287], thus recombinant strains are being constructed with high productivity and selectivity. An appropriate balance of homoserine dehydrogenase and threonine dehydratase activities in the construct *Corynebacterium glutamicum* DR17/pECM3::*ilv*A(V323A) with feedback-resistant aspartate kinase create a specific productivity of 0.052 g L-isoleucine per gram dry biomass per hour [288]. The recombinant strain *Escherichia coli* AJ13100 produces L-isoleucine from glucose with high selectivity in 30 % yield [289].

3.2.10. L-Leucine

One method used for L-leucine production is the isolation from protein hydrolysates. As alternative method, a precursor fermentation was suggested: In a fed-batch culture of *Corynebacterium glutamicum* ATCC 13032 32 g/L 2-ketoisocaproate can be converted to 24 g/L L-leucine by the transaminase B reaction [290]. With the method of direct fermentation L-leucine can be produced by either α-aminobutyric acid-resistant mutants of *Serratia marcescens* or by 2-thiazole-alanine-resistant *coryneform* strains [291]. *Brevibacterium flavum* AJ3686 accumulates 19.5 g/L L-leucine (15 % yield) [292].

3.2.11. L-Lysine

The most potent microorganisms to overproduce L-lysine are mutants derived from *Corynebacterium glutamicum,* a gram-positive bacterium first introduced as an L-glutamate producing microbe; the wild strains themselves are not able to excrete L-lysine. Mainly auxotrophic and regulatory mutants of this bacterium have been developed by classical breeding methods and mutagenesis. The following techniques have been applied to channel the metabolic pathway in the biosynthesis of amino acids:

Screening for auxotrophic mutants, in order to release key enzymes, for instance aspartate kinase, from strict regulation by metabolites (feedback inhibition).

Screening for regulatory mutants, in which aspartate kinase is resistant to toxic analogues of L-lysine, such as *S*-(2-aminoethyl)-L-cysteine (AEC) or *O*-(2-aminoethyl)-L-serine. In such strains high lysine concentrations do not inhibit the enzymatic key step, the formation of 4-aspartylphosphate from L-aspartase catalyzed by aspartate kinase.

Screening for mutants having amino acid auxotrophy combined with deregulation.

Screening for regulatory mutants having additional enzyme defects (reduced pyruvate kinase) and low levels of other enzymes (citrate synthase).

As a modern technique, cell fusion with the method of protoplast fusion has been successively applied for breeding of industrial microorganisms [293], [294]. This technique allows the combination of positive characteristics of different strains such as high selectivity and high productivity. In fermentation with media of inhibitory osmotic stress (i.e., high substrate concentration results in high osmotic pressure on the cell that inhibits the performance of the mutant) the sugar consumption rate and L-lysine production rate of some mutants can be stimulated by the addition of glycine [295]. Another attractive approach is the development of thermophilic strains. A mutant of *Corynebacterium thermoaminogenes* was patented which is capable of growing at a temperature $>40\,°C$ and accumulating L-lysine in the culture medium [296].

Meanwhile, regulation of lysine excretion in overproducing strains is known in detail [297], [298]. Thus more attention should be paid to process development in the fields of molecular biology, biochemistry, and physiology and to finding new approaches for developing improved overproducing mutants [299]. The most specific and well-directed methods for strain development are offered by recombinant DNA techniques.

In principle all genes encoding the relevant enzymes in L-lysine biosynthesis have been isolated, characterized, and amplified in coryneform bacteria to enhance L-lysine formation [300]. As an example, amplification of *dap*A gene that codes for dihydrodipicolinate synthase in *Corynebacterium glutamicum* resulted in 35 % higher overproduction of L-lysine compared to the parent strain [301]. Another option for strain improvement is the transformation of *dap*A gene together with a *lys*C gene, coding for aspartate kinase with decreased feed back inhibition in *Corynebacterium glutamicum* [302].

In fed-batch culture and under appropriate conditions the favorable mutants for lysine production are able to reach final concentration of about 120 g/L L-lysine, calculated as hydrochloride [303]. Fermentation processes are performed in big tanks up to 500 m^3 size. An optimized feeding strategy practiced with a computer aided process control system may enable high conversion yield and productivity in large scale fed-batch cultivation [304]. Apart from the specific fermentation know how, inoculation, sterilization, and feeding strategy the recovery process and the quality of the product both can be decisive factors to guarantee competitiveness. The conventional route of lysine downstream processing is characterized by:

- Removal of the bacterial cells from fermentation broth by separation or ultrafiltration
- Absorbing and then collecting lysine in an ion exchange step
- Crystallizing or spray drying of lysine as L-lysine hydrochloride

An alternative process consists of biomass separation, concentration of the fermentation solution, and filtration of precipitated salts. The liquid product contains up to 50 % L-lysine base, that is stable enough to be marketed [305]. Recently, a new concept for lysine production was introduced. Here the lysine containing fermentation broth is immediately evaporated, spray-dried, and granulated to yield a feed-grade product, which contains lysine sulfate. Its lysine content correspond to that of a material which contains to at least 60 % of L-lysine hydrochloride [306], [307]. The process avoids any waste products usually present in the conventional L-lysine hydrochloride manufacture.

3.2.12. D,L-Methionine and L-Methionine

L-Methionine and its antipode D-methionine are of equal nutritive value, thus the racemate can directly be used as feed additive.

The most economic way for production of D,L-methionine is the chemical process based on acrolein, methyl mercaptan, hydrogen cyanide, and ammonium carbonate (see Fig. 5). β-Methylthiopropionaldehyde, formed by addition of methyl mercaptan to acrolein, is the intermediate that reacts with hydrogen cyanide to give α-hydroxy-γ-methylthiobutyronitrile. Treatment with ammonium carbonate leads to 5-(β-methylthioethyl)hydantoin that is saponified by potassium carbonate giving D,L-methionine in up to 95 % yield, calculated on acrolein [308].

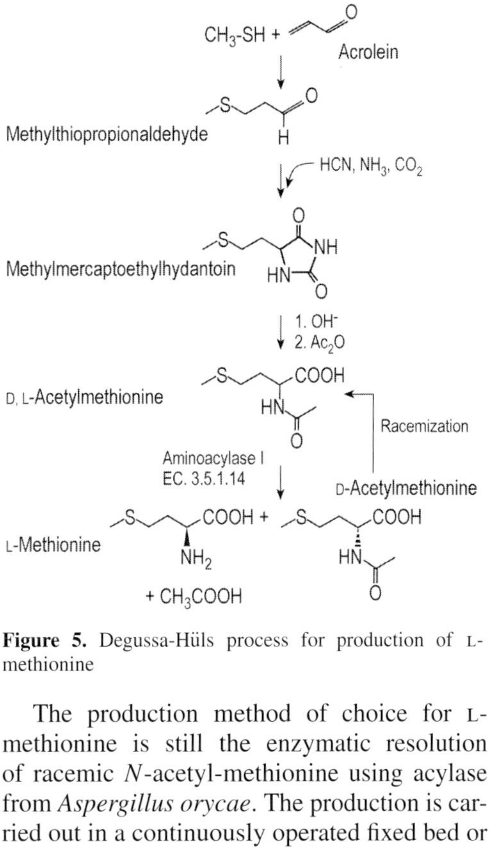

Figure 5. Degussa-Hüls process for production of L-methionine

The production method of choice for L-methionine is still the enzymatic resolution of racemic *N*-acetyl-methionine using acylase from *Aspergillus orycae*. The production is carried out in a continuously operated fixed bed or enzyme membrane reactor [309].

Alternatively, L-methionine may be produced by microbial conversion of the corresponding 5-substituted hydantoin. With growing cells of *Pseudomonas* sp. strain NS671, D,L-5-(2-methylthioethyl)hydantoin was converted to L-methionine; a final concentration of 34 g/L and a molar yield of 93 % have been obtained [310].

Biosynthesis of L-methionine and its regulation in bacteria is well known. Although some promising concepts, for example utilization of sulfate, sulfite, or thiosulfate as sulfur sources for microbes have been suggested [311], it was not possible so far to develop strains that are able

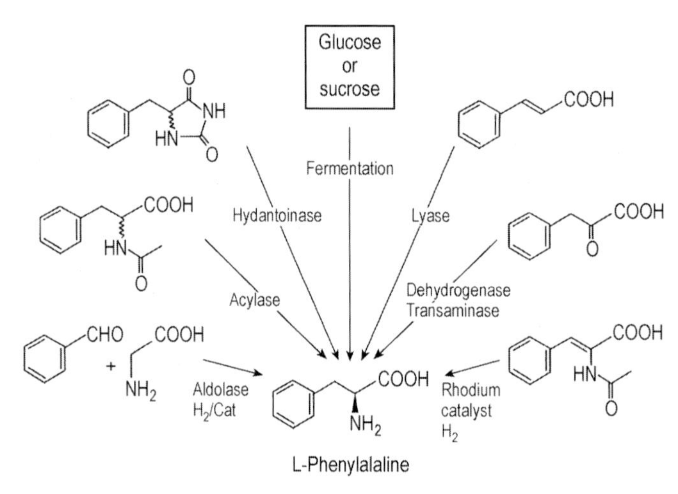

Figure 6. Asymmetric syntheses for production of L-phenylalanine

to excrete remarkable amounts of L-methionine into the culture medium.

3.2.13. L-Phenylalanine

Several methods for producing L-phenylalanine are known. Some of them have industrial significance (see Fig. 6).

In previous large-scale production processes for L-phenylalanine two enzymatic methods were applied:

1) Resolution of *N*-acetyl-D,L-phenylalanine by carrier-fixed microbial acylase: This process provided pharmaceutical-grade L-phenylalanine, but suffered from the disadvantage that the D-enantiomer had to be racemized and recycled.
2) Stereoselective and enantioselective addition of ammonia to *trans*-cinnamic acid, catalyzed by L-phenylalanine ammonia lyase (PAL, EC 4.3.1.5): PAL-containing *Rhodotorula rubra* was used in an industrial process [312] to supply L-phenylalanine for the first production campaign of the sweetener aspartame. When continuously operated in an immobilized whole cell reactor, the bioconversion reached concentration up to 50 g/L L-phenylalanine at a conversion of about 83 % [313]. Other processes started from phenylpyruvate with L-aspartic acid as amine donor using immobilized cells of *Escherichia coli* [314] or from α-acetamidocinnamic acid and immobilized cells of a *Corynebacterium equi* strain [315]. In both cases L-phenylalanine concentrations up to 30 g/L and more (molar yields as high as at least 98 %) were reached.

However, fermentation processes based on glucose-consuming L-phenylalanine overproducing mutants of *E. coli* and coryneform strains turned out to be more economical. The biosynthetic pathway for aromatic amino acids in bacteria is strictly regulated [316]. L-Phenylalanine is formed in ten enzymatic steps starting from erythrose-4-phosphate and phosphoenolpyruvate. The biosynthesis is governed by the first key enzyme 3-deoxy-D-arabinoheptulosonate-7-phosphate synthase (DAHPS) which is inhibited by both L-phenylalanine and L-tyrosine (by acting on the enzyme itself) and repressed by L-tyrosine (by acting on the according gene of the DNA). The other important enzyme is prephenate dehydratase (PDT) also inhibited by L-phenylalanine, but stimulated by L-tyrosine. To overcome these regulatory mechanisms either auxotrophs of *Corynebacterium glutamicum* have been constructed or L-phenylalanine analogues, e.g., 4-aminophenylalanine and 4-fluorophenylalanine, have been applied. The latter variant leads to resistant mutants of *Brevibacterium flavum* or *lactofermentum* [317]. These auxotrophic and regulatory mutants are able to produce more than 20 g/L of L-phenylalanine in a medium containing 13 % glucose. Similar results can be obtained by tyrosine auxotrophic regulatory mutants of *E. coli* [318]. With recombinant DNA techniques it was possible to improve overproducing strains of *coryneform* bacteria as well as of *E. coli.* Amplification of a deregulated DAHPS gene was achieved in a phenylalanine producer of *Brevibacterium lactofermentum* [319]. For optimal production of L-phenylalanine in fed-batch cultivation the critical specific glucose uptake rate has to be controlled. The specific feed rate during fermentation has to be adjusted below a critical limit, since otherwise the *E. coli* producer will be forced to excrete acetate [320]. A suitable profile of the specific glucose feed rate prevents acetate formation and leads to improved L-phenylalanine production with a final concentration up to 46 g/L and a corresponding yield of 18 %. L-Phenylalanine is recovered from the fermentation broth either by two-step crystallization or by an ion-exchange resin process. The preferred cell separation technique is ultrafiltration; and the filtrates may be treated with activated carbon for further purification. Instead of ion-exchange resins nonpolar, highly porous synthetic adsorbents are recommended to remove impurities [321], [322]. An alternative process in which a cell separator is integrated in the fermentation part, thus allowing cell recycling, was suggested for L-phenylalanine production and may lead to prospective developments [323].

3.2.14. L-Proline

L-Proline is still produced by isolation from protein hydrolysates. Direct fermentation using analogue-resistant mutants of *coryneform* bacteria or *Serratia marcescens* is proposed to be an economic alternative production method [324]. An isoleucine auxotrophic mutant of *Brevibacterium flavum* having resistance to sulfaguanidine and D,L-3,4-dehydroproline (DP) is able to accumulate 40 g/L L-proline. *Brevibacterium flavum* AP113 is claimed to produce 97.5 g/L L-proline; this mutant is characterized by isoleucine auxotrophy, resistance to DP, and osmotic pressure and incapable to degrade L-proline [325]. A proline oxidase-less strain of *Serratia marcescens*, having resistance to DP, thiazoline-4-carboxylate and azetidine-2-carboxylate, overproduces 58.5 g/L L-proline into the culture medium [326]. By amplification of the genes *pro*A and *pro*B in this type of regulatory mutant, a construct was obtained which yields 75 g/L L-proline [327].

3.2.15. L-Serine

L-Serine is obtained by extraction of protein hydrolysates or by microbial/ enzymatic conversion of glycine using immobilized resting cells or crude cell extracts. *Hyphomicrobium* strains possess the serine pathway and are able to produce L-serine from methanol and glycine. Methanol is oxidized by methanol dehydrogenase to formaldehyde which in turn is converted in an aldol-like reaction with glycine to L-serine. The reaction is catalyzed by serine hydroxymethyltransferase (SHMT) [328]. *Hyphomicrobium* sp. NCIB10099 was found to produce 45 g/L L-serine from 100 g/L glycine and 88 g/L methanol in three days [329]. In an enzyme bioreactor with a feedback control system a crude extract from *Klebsiella aerogenes* containing SHMT has been used to synthesize L-serine from glycine and formaldehyde in the presence of tetrahydrofolic acid and pyridoxal phosphate. In this bioreactor a serine concentration of 450 g/L with an 88 % molar conversion of glycine at a volumetric productivity of 8.9 g L^{-1} h^{-1} could be achieved under optimized conditions [330]. With whole cells of *Escherichia coli* MT-10350 L-serine is produced by treatment of an oxygenated aqueous glycine solution (485 g/L) with aqueous formaldehyde for 35 h at 50 °C in a molar yield of 89 % based on glycine [331].

3.2.16. L-Threonine

End of the 1980s, L-threonine was mainly used for medical purposes, in amino acid infusion solutions and nutrients. It was manufactured by extraction of protein hydrolysates or by fermentation using mutants of *coryneform* bacteria in amounts of several hundred tons per year worldwide. The production strains were developed by classical breeding. They were auxotrophic and resistant to threonine analogues such as α-amino-β-hydroxyvalerate (AHV), and reached product concentrations up to 20 g/L. These strains possessed deregulated

L-threonine pathways with feedback inhibition-insensitive aspartate kinase and homoserine dehydrogenase [332], [333]. In the 1990s strain developments, using both conventional methods and recombinant DNA techniques, have been very successful. Potent classically selected mutants suggested for industrial production are the species *Brevibacterium flavum*, *Providentia rettgeri*, *Serratia marcescens*, and *Escherichia coli.* However, in the competition between the favorable candidates, strains of *Escherichia coli* proved to be superior to other bacteria. Although the pathway of L-threonine biosynthesis in *Escherichia coli* is much more regulated than that in *Corynebacterium glutamicum*, new *Escherichia coli* strains with excellent yields and productivity in threonine formation could be constructed by genetic engineering. L-Threonine is successfully marketed as feed additive with a worldwide demand of more than 10 000 t/a. Production strains are based on *Escherichia coli* K-12 constructs harboring plasmids containing the *thr* operon that consists of the genes *thr*A, *thr*B, and *thr*C [334]. Further improvements resulted in strains capable to accumulate more than 80 g/L in about 30 h with a conversion yield of more than 40 % [335]. The strain stability could be further improved, for example by integrating the threonine operon into the chromosome [336]. The recovery of feed-grade L-threonine is rather simple. After fermentation is completed, cell mass is removed by centrifugation or ultrafiltration, the filtrate is concentrated, depigmentated and L-threonine isolated by crystallization [337].

3.2.17. L-Tryptophan

L-Tryptophan is one of the limiting essential amino acids required in the diet of pig and poultry. A mature and growing market for L-tryptophan as feed additive is still in development although many processes are available, that are even proven on a production scale. However, high production costs so far prevent a tolerable price level that would favor the introduction of L-tryptophan as bulk product. The most attractive production processes for tryptophan are based on microorganisms used as enzyme sources or as overproducers:

Enzymatic production from various precursors
Fermentative production from precursors
Direct fermentative production from carbohydrates by auxotrophic and analogue resistant regulatory mutants

L-tryptophan is synthesized from indole, pyruvate, and ammonia by the enzyme tryptophanase [338] or from indole and L-serine/D,L-serine by tryptophan synthase [339], [340]. Although production in enzyme bioreactors is quite efficient and concentrations of L-tryptophan up to 200 g/L could be achieved by condensation of indole and L-serine [341], these process variants were not economic due to the high costs of the starting materials. The microbial conversion of biosynthetic intermediates such as indole or anthranilic acid to L-tryptophan has also been considered as alternative for production. Whereas indole consuming mutants of *Corynebacterium glutamicum* produced about 10 g/L L-tryptophan [342], strains of *Bacillus subtilis* and *Bacillus amyloliquefaciens* reached final concentrations > 40 g/L L-tryptophan with anthranilic acid as carbon source [342], [344]. The process with anthranilic acid as precursor has been commercialized in Japan. However, the manufacturer using genetically modified strains derived from *Bacillus amyloliquefaciens* IAM 1521 was forced to stop L-tryptophan production. L-Tryptophan produced by this process was stigmatized because of side products found in the product causing a new severe disease termed eosinophilia-myalgia syndrome (EMS) [345]. One of the problematic impurities, "Peak E", was identified as 1,1′-ethylidene-*bis*-(L-tryptophan), a product formed by condensation of one molecule acetaldehyde with two molecules of tryptophan [346].

In other processes, i. e., direct fermentation using overproducing mutants and carbohydrates as carbon sources, formation of such impurities does not occur. In the 1990s striking progress has been made in the development of auxotrophic and deregulated mutants of *Brevibacterium flavum*, *Corynebacterium glutamicum*, and *Bacillus subtilis*. The biosynthesis of L-tryptophan and its regulation have been reviewed in detail for the different species [347–349]. The precise knowledge about the structure of the *trp* operon in *Escherichia coli* comprising the *trp* promoter and the genes *trp*E, *trp*D, *trp*C, *trp*B, and *trp*A which are coding for the enzymes anthranilate synthase (AS), phosphoribosyl anthranilate transferase (PRT), indole glycerol phosphate synthase (IGP) and tryptophan synthase (TS), respectively, was the benefit for further strain improvements.

Thus recombinant DNA techniques have been used to increase the capability of overproduction especially in strains of *Corynebacterium glutamicum* and *Escherichia coli*. One concept was realized successfully by amplification of *trp* operon genes together with *ser*A which codes for phosphoglycerate dehydrogenase. This key enzyme in L-serine biosynthesis should provide enough L-serine in the last step of L-tryptophan formation. Production strains are able to accumulate 30 – 50 g/L L-tryptophan with yields higher than 20 % based on carbohydrate. In general, crystals of tryptophan appear in the broth, when the concentrations are beyond the level of about 30 g/L [350]. Since L-tryptophan is sensitive to oxygen and heat [341], the recovery from the fermentation broth without considerable product losses is still a challenge and part of the sophisticated details of each specific process.

3.2.18. L-Tyrosine

L-Tyrosine is produced exclusively from protein hydrolysates. Its low solubility in water enables a quite simple isolation of the amino acid.

3.2.19. L-Valine

L-Valine is produced industrially in pharmaceutical quality by enzymatic resolution of *N*-acetyl-D,L-valine. Using direct fermentation the branched-chained amino L-valine can be produced by either α-aminobutyric acid-resistant mutants of *Serratia marcescens* or by 2-thiazolealanine-resistant *coryneform* strains [352]. *Brevibacterium lactofermentum* AJ12341 produces 39 g/L L-valine (28 % yield) [353].

4. Biochemical and Physiological Significance

The biosynthesis of amino acids begins with atmospheric nitrogen, which is reduced to ammonia by bacteria and plants. Ammonia is used by plants, by bacteria, and, to a limited extent, by ruminants as a raw material for amino acids. Amino acids, in turn, serve as starting materials for the synthesis of proteins and a variety of other nitrogen-containing compounds, such as the purine and pyrimidine bases in nucleic acids. Bacterial degradation leads, once again, to ammonia and nitrogen [150].

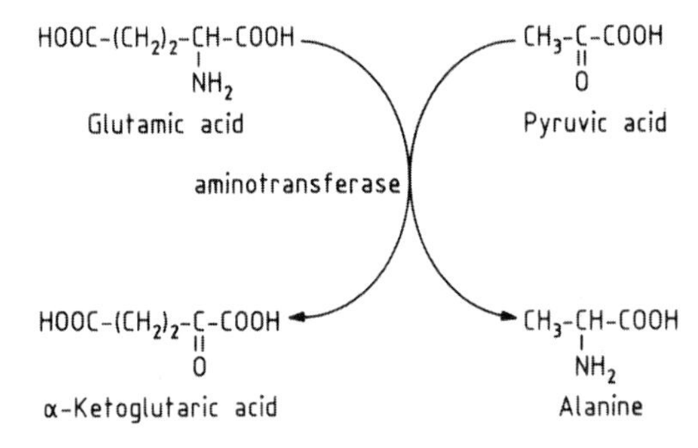

Figure 7. The transamination of amino acids

α-Ketoglutaric acid plays the central role in the assimilation of ammonia. Its transamination product, glutamic acid, in turn, can provide its amino group for the synthesis of other amino acids, e.g., alanine (Fig. 7).

Humans, animals, and some bacteria are incapable of synthesizing all the necessary amino acids in their own intermediary metabolism; i.e., they are heterotrophs and are therefore dependent on the biosynthetic capability of plants. Proteins that are consumed as foods by humans and animals are hydrolyzed to amino acids by the digestive enzymes. The amino acids are resorbed in the upper part of the intestine and enter the liver by way of the portal vein. The liver is the central organ for metabolism and homoeostasis of the plasma amino-acid level. The body's various requirements are met from the pool of free amino acids (Fig. 8), ca. 50 g in adult humans.

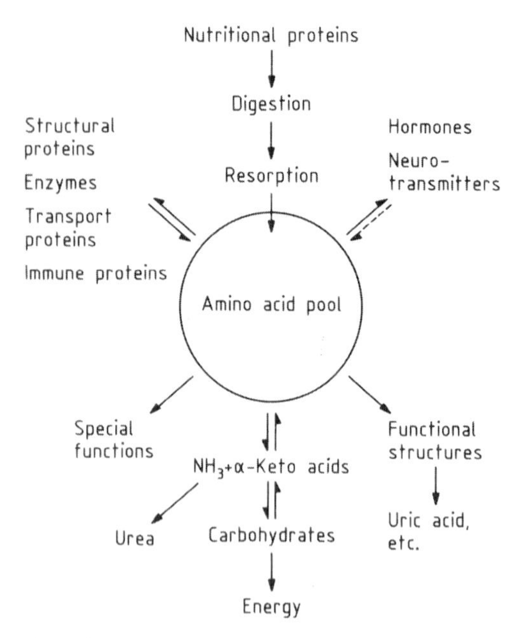

Figure 8. The amino acid pool and functions of amino acids in the intermediary metabolism

The lion's share of the amino acids (≈ 300 g/d for adults) is required for synthesis of proteins [151]: structural proteins, enzymes, transport proteins, and immune proteins. Additionally, amino acids are required for the synthesis of oligopeptides and polypeptides that fulfill regulatory functions in the body, i.e., hormones. Some amino acids or their metabolites are directly active as hormones or facilitate the transmission of nerve impulses (neurotransmitters), e.g., serotonin. Furthermore, there are amino acids that serve special functions, such as methionine, which is a methyl group donor. Finally, a series of amino acids serves as precursors for the biosynthesis of other structures. For example, glycine is used in the construction of the porphyrin skeleton. Amino acids are metabolized to produce energy in the case of a protein-deficient or a protein-excess diet.

The free amino acids in the amino acid pool undergo numerous transformations (Fig. 9), involving transamination and oxidative deamination, by which other amino acids can be synthesized. The α-keto acids are intermediates and in addition allow amino acids entrance into the carbohydrate (through pyruvate) and fatty acid (through acetylcoenzyme A) metabolisms. A distinction is therefore drawn between glucogenic and ketogenic amino acids.

D-Amino acids occur in the cell pool of plants and gram-positive bacteria and as building blocks in peptide antibiotics and bacterial cell walls [6], [12]. They do not occur in human or animal metabolism; proteins are made exclusively of L-amino acids. The traces of D-amino acids detected in metabolically inert protein (teeth, eye lenses) are believed to originate from racemization. Orally ingested D-amino acids are resorbed from the intestinal lumen more slowly than the L-form. The D-enantiomers cannot be utilized or can be utilized only to a slight extent as essential amino acids [152]. The one important exception is D-methionine. Animals and adult humans convert D-methionine into L-methionine by transamination. The α-keto acid of methionine is an intermediate. Otherwise, D-amino acids are degraded with the help of D-amino acid oxidases [153] to be used as an energy source [154].

The major end products of amino acid metabolism are urea, uric acid, ammonium salts, creatinine, and allantoin. The loss of nitrogen via these metabolites stabilizes at about 22 g protein per day after a few days on a protein-free diet.

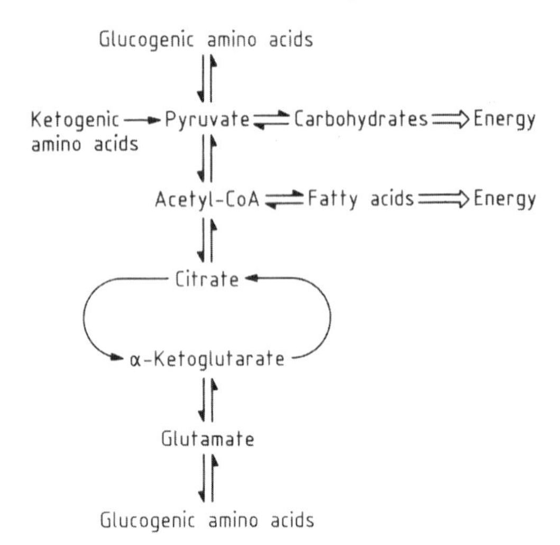

Figure 9. Intermediary metabolism of amino acids (simplified)

Inborn disorders in amino acid metabolism [155] can lead to marked alterations in the excretion profile. These disorders usually take the form of an enzyme or transport deficiency [150], [156]. The most common example is

Table 4. Essential (+) and semiessential (±) amino acids

	Baby	Adult	Rat	Chicken	Hen	Cat	Salmon
L-Arginine	±	±	±	+	±	+	+
L-Cysteine	+ (?)						
Glycine				+			
L-Histidine	+	±	+	+	±	±	+
L-Isoleucine	+	+	+	+	+	+	+
L-Leucine	+	+	+	+	+	+	+
L-Lysine	+	+	+	+	+	+	+
L-Methionine	+	+	+	+	+	+	+
L-Phenylalanine	+	+	+	+	+	+	+
L-Threonine	+	+	+	+	+	+	+
L-Tryptophan	+	+	+	+	+	+	+
L-Tyrosine	+ (?)						
L-Valine	+	+	+	+	+	+	+

phenylketonuria, a disruption of the normal metabolic pathway from phenylalanine to tyrosine caused by severe limitation in the activity of the phenylalanine hydroxylase [157].

Humans and animals are not capable of producing all the required L-amino acids in their intermediary metabolism. Therefore, they are dependent on an external source of these *essential* amino acids (Table 4). In situations of increased requirements (rapid growth, stress, trauma), histidine and arginine also become essential for humans. Cysteine and tyrosine may be essential for infants during their first few weeks, because their intermediary metabolism does not yet function well enough to produce these from methionine and phenylalanine in sufficient quantities.

5. Uses

The uses of amino acids have been treated in review articles [19], [158–160].

5.1. Human Nutrition

In addition to their nutritive value, amino acids are important flavor precursors and taste enhancers. In foods for humans, the flavor uses of amino acids represent the dominant factor in total market value. In animal nutrition, amino acids are used almost exclusively for their nutritive value.

Addition of small amounts of amino acids to improve the nutritive value of proteins is known as supplementation. Both supplementation and the combination of proteins with complementary amino acids are used to increase the biologic value of proteins. Usually the supply of at least one of the essential amino acids lies below the requirement. This, the limiting amino acid, determines what percentage of the protein (or, more precisely, its amino acids) can be used to meet the body's amino acid requirements. In most cases, methionine is the first limiting amino acid. Sometimes it is lysine; now and then it is both together.

The contents of essential amino acids found in several animal and vegetable foodstuffs are compiled in Table 5. Considerable variations may be present in the amino acid contents of a given foodstuff.

The published requirements for the individual essential amino acids differ. The values (Table 6) usually contain safety factors and therefore are higher than the minimum requirement. Requirement values were first determined by Rose [164]; those published by Hegsted [165] are considered the most reliable at present. The amino acid requirement pattern suggested by the FAO/WHO [163] is considered optimal for the greatest part of the population.

The "average safe level of daily protein intake for men and women," based on these amino acid requirement figures, is given as 0.55 g/kg body weight. The acute daily protein requirement, however, varies between 0.5 and 2.5 g/kg with age and constitution [166]. The Deutsche Gesellschaft für Ernährung (DGE) recommends a daily consumption of 0.9 g/kg body weight [167]. The Committee on Dietary Allowances, Food and Nutrition Board of the National Academy of Sciences (NAS), USA,

Table 5. Average amino acid content of some foodstuffs (mg/100 g) [a] [161]

Food	Ile	Leu	Lys	Cyss	Met	Phe	Tyr	Thr	Trp	Val	Arg	His	Protein, %	Moisture content, %
Maize, grain	350	1190	254	147	182	464	363	342	67 [M]	461	398	258	9.5	12.0
Rice, husked	300	648	299	84	183	406	275	307	98 [M]	433	650	197	7.5	13.0
Wheat, whole grain	426	871	374	332	196	589	391	382	142 [M]	577	602	299	12.2	12.0
Wheat, flour, 70 – 80 % extr. rate	435	840	248	304	174	581	277	321	128 [M]	493	422	248	10.9	12.0
Potato (*Solanum tuberosum*)	76	121	96	12	26	80	55	75	33 [M]	93	100	30	2.0	78.0
Bean (*Phaseolus vulgaris*)	927	1685	1593	188	234	1154	559	878	223 [M]	1016	1257	627	22.1	11.0
Soybean, milk	171	278	195	57	50	175	133	128	48 [M]	165	253	84	3.2	92.0
Soy protein, isolate [162]	4147	7119	5777	1008	1092	4644	3458	3211	1080	4210	6767	2378	75.7	4.7
Lettuce, leaves (*Lactuca sativa*)	50	83	50	24	–	67	35	54	10 [M]	71	59	21	1.3	94.8
Tomato (*Solanum lycopersicum*)	20	30	32	7	7	20	14	25	9 [M]	24	24	17	1.1	93.8
Apple (*Malus silvestris*)	13	23	22	5	3	10	6	14	3	15	10	7	0.4	84.0
Orange (*Citrus sinensis*)	23	22	43	10	12	30	17	12	6	31	52	12	0.8	87.4
Beef, veal, edible flesh	852	1435	1573	226	478	778	637	812	198 [M]	886	1118	603	17.7	61.0
Fish, fresh, all types	900	1445	1713	220	539	737	689	861	211 [M]	1150	1066	665	18.8	74.1
Milk, cows, untreated	162	328	268	28	86	185	163	153	48 [M]	199	113	92	3.5	87.3
Milk, human	48	104	81	16	19	41	39	53	20 [M]	54	46	30	1.2	87.6
Cheese, all types	956	1864	1559	76	530	950	973	725	217 [M]	1393	651	556	18.0	51.0
Egg, whole	778	1091	863	301	416	709	515	634	184 [M]	847	754	301	12.4	74.0

[a] Chemical determination. [M] Microbiological determination.

Table 6. Essential amino acid requirements of humans

Amino acid	Suggested patterns of requirements [163], g/100 g protein			Adult requirement, mg kg^{-1} d^{-1}			
	Infant child	School-age	Adult 1973	FAO/WHO 1980 (144)	NAS/NCR	Rose [a]	Hegsted [b]
His	1.4	–	–	–	–	–	–
Ile	3.5	3.7	1.8	10	12	10	10
Leu	8.0	5.6	2.5	14	16	11	13
Lys	5.2	7.5	2.2	12	12	9	10
Met + Cys [c]	2.9	3.4	2.4	13	10	14	13
Phe + Tyr [d]	6.3	3.4	2.5	14	16	14	13
Thr	4.4	4.4	1.3	7	8	6	7
Trp	0.85	0.46	0.65	3.5	3	3	3
Val	4.7	4.1	1.8	10	14	14	11
Total	37.3	32.6	15.2	83.5	91	81	80

[a] For men.
[b] For women.
[c] Cys can partly cover the total *S*-amino acid requirement.
[d] Tyr can partly cover the total aromatic amino acid requirement.

cites 0.8 g/kg as a desirable level of daily protein consumption [168].

5.1.1. Supplementation

In general, animal protein contains the essential amino acids in larger quantities and in a more favorable ratio than vegetable protein, which is often deficient in essential amino acids. Lysine is the first limiting amino acid in wheat, rye, barley, oats, maize, and millet, whereas methionine is the first limiting amino acid in meat, milk, soybeans, and other beans. The second limiting amino acids are usually threonine (wheat, rice) and tryptophan (maize, rice, casein). The limiting amino acids for several foodstuffs are listed in Table 7.

Table 7. Limiting amino acids in foodstuffs

Proteins	First limiting amino acid	Second limiting amino acid(s)
Peanut	Thr	Lys and Met
Fish	Met	Lys
Casein	Met	Trp
Torula yeast	Met	
Sesame	Lys	
Skim milk	Met	
Beans	Met	
Sunflower seed	Lys	Thr
Soy protein	Met	Lys
Wheat	Lys	Thr
Rice	Lys	Thr and Trp
Rye	Lys	Thr and Trp
Gelatin	Trp	
Maize	Lys	Trp and Thr

Improving the biologic value of vegetable protein in human nutrition is practiced for economic and dietary reasons. Combining different protein types is not always practical. Often a complementary protein is unavailable, too expensive, or not of acceptable taste. In these cases supplementation with amino acids is the simplest method of increasing the biologic value of proteins. There is a monumental volume of literature on the subject of amino acid supplementation [169–176].

The biologic value is an important criterion for the evaluation of proteins or amino acid mixtures. It can be determined experimentally [177], [178]. In principle, all methods measure the ability of the nutritional protein to replace body protein. Table 8 lists the biologic values of nutritional proteins as determined by the minimum requirement [166]. Whole egg protein is the reference in this scale.

Another scale of evaluation is the protein efficiency ratio (PER). This is the daily weight gain of young animals, usually rats, under standard feeding conditions (see Table 8). In an improved method, the animals are fed diets of various protein levels, and the protein efficiency is then determined using regressional analysis.

The net protein retention (NPR) method and the net protein utilization (NPU) method are more accurate than the PER method because they consider the protein (NPR) or total nitrogen (NPU) requirement for maintenance. The "chemical score," in which the availability of

Table 8. Protein quality of food and minimum requirements (human)

Food	Biologic value $g\ kg^{-1}\ d^{-1}$ [166]	Minimum requirement *, ratio (rat) [166]	Protein efficiency [161]
Whole egg	100	35	3.92
Beef	92	39	2.30
Cow's milk	88	40	3.09
Potato	86 [179]	41 [179] **	3.0 [180]
Fish			3.55
Casein			2.50
Soybean	84	42	2.32
Rice	81	44	2.18
Rye flour	76	46	
Maize	72	49	1.18
Beans	72	49	1.48
Wheat flour	56	63	0.6

* Of the protein part.
** Calculated.

the amino acids is not considered, is suitable for a gross estimation of the biologic protein quality. In this method the amino acid content is determined analytically and compared with the amino acid pattern of a reference protein, e.g., the FAO/WHO provisional scoring pattern (Table 9). This method provides an immediate picture of the size of amino acid gaps and the sequence of limiting amino acids.

$$\text{Score} = \frac{\text{Content of amino acid in test protein}}{\text{Content of amino acid in reference protein}} \times 100$$

Table 9. FAO/WHO provisional scoring pattern 1973 [163]

Amino acid	g/100 g protein
L-Isoleucine	4.0
L-Leucine	7.0
L-Lysine	5.5
L-Methionine + L-cystine *	3.5
L-Phenylalanine + L-tyrosine **	6.0
L-Threonine	4.0
L-Tryptophan	1.0
L-Valine	5.0
Total	36.0

* Cys can partly cover the total *S*-amino acid requirement.
** Tyr can partly cover the total aromatic amino acid requirement.

As can be seen in Table 10, addition of ca. 0.1 – 0.5 % of the limiting amino acid to such basic foodstuffs as wheat, rice, maize, and soybeans raises the protein efficiency in rat growth tests impressively.

Clinical studies [181–183] and field trials have solidified the evidence of the benefits to human nutrition of amino acid supplementation [184–187]. However, the general supplementation of basic foodstuffs, such as bread and rice, is not yet practiced extensively. In dietary nutrition, however, supplementation already plays an important role (Table 11).

Some infants exhibit lactose or cow's milk protein incompatibility. The formulas marketed for this condition often are based on isolated soybean protein and are supplemented with L-methionine to increase the biologic value. The advantageous effects of L-methionine supplementation on the physical development of infants has been demonstrated in a series of clinical studies [188], [189]. Additionally, the food for pregnant and nursing women, seniors, overweight persons, and athletes can also be supplemented. Extruded soy protein, which is already used in large quantities as a meat extender and vegetarian meat substitute, can be supplemented with *N*-acetyl-L-methionine [190].

5.1.2. Flavorings, Taste Enhancers, and Sweeteners

Free amino acids occur in almost all protein-based foods. In some foods their concentration is several percent. Foodstuffs having a relatively high concentration of free amino acids include fruit juices [191], cheese [192], [193], beer [194], and seafood [195]. Approximately 85 % of the free amino acids in orange juice is proline, arginine, asparagine, γ-aminobutyric acid, aspartic acid, serine, and alanine [196].

Table 10. Increase of the protein efficiency ratio (PER) by supplementation with amino acids [161], [172]

Food (protein content)	L-Lys·HCl, %	L-Thr, %	D,L-Trp, %	D,L-Met, %	PER *
Wheat flour (10 %)					0.65
	0.2				1.56
	0.4				1.63
	0.4	0.15			2.67
Rice (7.8 %)					1.50
	0.2	0.1			2.61
Maize (8.75 %)					1.41
	0.4		0.07		2.33
Soybean milk (10 %)					2.12
				0.3	3.01
Extruded soy protein (10%)					1.99
				0.23	2.62

*Reference protein: casein, PER=2.50.

Table 11. Amino acids in dietetic products

Protein/protein hydrolysate	Supplemented amino acid	Use	Indication
Cow's milk, casein, whey protein	L-Cyss and/or L-Lys · HCL	infant nutrition	adapted nutrition
Soy protein	L-Met	infant nutrition	lactose incompatibility and milk protein allergy
Casein/yeast	L-Lys · HCl	meal supplement	protein malnutrition, in place of conventional nutrition

Amino acids are relatively tasteless. Nonetheless, they contribute to the flavor of foods. They have characteristic synergistic flavor-enhancing and flavor-modifying properties, and they are precursors of natural aromas [197, 199]. Amino acids and protein hydrolysates are therefore useful additives in the food industry. The sodium salt of L-glutamic acid (MSG) exhibits a particularly pronounced flavor-enhancing effect [200] and has been recognized as a flavoring factor for seaweed, sake, miso, and soy sauce since 1908. The substance is used in concentrations of 0.1 – 0.4 % as an additive for spices, soups, sauces, meat, and fish, usually in combination with nucleotides [199].

L-Cysteine especially enhances the aroma of onion [201] and is therefore used to rearomatize dried onions. Glycine, which has a refreshing, sweetish flavor, occurs abundantly in mussels and prawns. It is considered to be an important flavor component of these products. When used as an additive for vinegar, pickles, and mayonnaise, it attenuates the sour taste and lends a note of sweetness to their aroma. D,L-Alanine is used for the same purpose in the Far East [202]. Glycine is used to mask the aftertaste of the sweetener saccharin [203], [204].

Table 12. Tastes of L- and D-amino acids* [207]

	L-Amino acid	D-Amino acid
Alanine	sweet (12 – 18)	sweet (12 – 18)
Arginine	bitter	neutral
Asparagine	neutral	sweet (3 – 6)
Aspartic acid	acidic/neutral	acidic/neutral
Cysteine	sulfurous	sulfurous
Glutamine	neutral	sweet (8 – 12)
Glutamic acid	acidic/"glutamate-like"	acidic/neutral
Glycine	sweet (25 – 35)	sweet (25 – 35)
Histidine	bitter	sweet (2 – 4)
Isoleucine	bitter	sweet (8 – 12)
Leucine	bitter	sweet (2 – 5)
Lysine	sweet/bitter	sweet
Methionine	sulfurous	sweet/sulfurous (4 – 7)
Phenylalanine	bitter	sweet (1 – 3)
Proline	sweet/bitter (25 – 40)	neutral
Serine	sweet (25 – 35)	sweet (30 – 40)
Threonine	sweet (35 – 45)	sweet (40 – 50)
Tryptophan	bitter	sweet (0.2 – 0.4)
Tyrosine	bitter	sweet (1 – 3)
Valine	bitter	sweet (10 – 14)

* Threshold values for sweet taste in parenthesis, µmol/mL; the threshold value for sucrose is 10 – 12 µmol/mL.

The L- and D-amino acids usually exhibit pronounced flavor differences. Many L-enantiomers taste weakly bitter, whereas their optical antipodes, the D-amino acids, taste sweet [205–207] (Table 12). For the most part, dipeptides and oligopeptides have bitter flavors. One of the few exceptions is the methyl ester of the dipeptide L-aspartyl-L-phenylalanine (Aspartame) [208,210], which is 150 – 200 times sweeter than sucrose.

The free amino acids are used widely in foodstuff technology as precursors for aromas and brown food colors [211]. The flavors are formed during foodstuff production, e.g., during the ripening of cheese [212], [213], the fermentation of alcoholic beverages [214], [215], or the leavening of dough [216], [217], or foodstuff cooking, e.g., frying, roasting, boiling, by the Maillard reaction between amino acids and reducing sugars (nonenzymatic browning) [197], [218], [219]. The Strecker degradation of amino acids plays a central role in this process. A broad spectrum of aroma-intensive, volatile compounds forms [220], [221]. The most important classes are aliphatic carbonyl compounds and heterocycles, such as furans, pyrones, pyrroles, pyrrolidines, pyridines, imidazoles, pyrazines, chinoxalines, thiophenes, thiolanes, trithianes, thiazoles, and oxazoles.

Table 13. Amino acids for Maillard flavors *

Meat, poultry [222]	Cys, Cyss, Gly, Glu, Ala, Met, His, Ser, Asp, Pro
Bread, cracker, biscuit [216], [223]	Pro, Lys, Arg, Val, His, Leu, Glu, Phe, Asp, Gly, Gln
Chocolate, cocoa [224]	Leu, Phe, Val, Glu, Ala
Honey [225–227]	Phe
Cream, butter [228]	Pro, Lys, Ala, Gly, His
Nut, peanut [229]	Leu, Val, Ile, Pro, Glu, Gln, His, Phe, Asp, Asn
Potato [230]	Met
Tobacco [226], [231]	Asn, Arg, GABA, Gln, Ala, Gly, Orn, Glu, Asp, Leu, Val, Thr, Pro, Tyr, Phe

* Key amino acids underlined.

It is often possible to assign certain aromas to specific amino acids [221]. For example, the sulfur-containing amino acid cysteine is primarily responsible for the formation of meat flavor. Proline seems to be important for the aroma of bread crust. Phenylalanine, as well as the branched-chain amino acids leucine and valine, is important for the characteristic flavor of chocolate. Valine and leucine are also involved in the aroma of roasted nuts. Methionine plays a key role in the aroma of french fries. The flavors of such products as precooked foods, snack articles, and spices may be improved by addition of the proper Maillard aromas. One variation is adding the precursors of the Maillard aromas, i.e., amino acid plus sugar, to the foodstuff and allowing the fragrance to form in situ. Some aroma profiles that can be prepared from amino acids are compiled in Table 13.

5.1.3. Other Uses in Foodstuff Technology

Amino acids are used in the foodstuff industry for purposes other than supplementation and flavoring. L-Cysteine, for example, is used by the baked goods and pasta industry as a flour additive [232], [233]. As a reducing agent, it relaxes wheat gluten proteins (by cleavage of the disulfide linkages), homogenizes the dough, accelerates dough development, and improves the structure of the baked product, while allowing shorter kneading times.

Because they are capable of forming complexes with metals, amino acids act as antioxidants for fats and fat-containing foodstuffs [234]. This effect is strengthened by primary antioxidants, such as α-tocopherol. Melanoidines, which are formed during the Maillard reaction, are stronger antioxidants than the amino acids themselves [235]. Maillard products are also reported to be preservatives [236]. Glycine apparently exhibits a special preservative effect [237].

5.2. Animal Nutrition

The use of amino acids for the nutrition of monogastric animals is based on the same foundation as the supplementation of human foodstuffs and the clinical experience with humans. In practice, the enrichment of animal feeds and formulated feeds with amino acids, especially methionine and lysine, represents far greater quantities than does human nutrition. By supplementing feeds or formulated feeds with the first limiting amino acid an obvious cost reduction can be achieved, while maintaining the quality of the ration.

Of the ca. 20 amino acids found in feed protein, about one half are essential for monogastric animals (see Table 4). Most natural feeds are relatively poor in methionine and lysine (Table 14). The requirement of our livestock, however, is comparatively high [357].

When formulating a feed mix for a given animal type, the manufacturer has two choices for

Table 14. Amino acid composition of feedstuffs, wt % [22]

Feedstuff	Dry matter	Crude protein	Met	Met+Cys	Lys	Thr	Trp	Arg	Leu	Ile	Val	His
Alfalfa	88	17.7	0.27	0.50	0.86	0.76	0.25	0.81	0.72	1.27	0.91	0.36
Barley	88	10.6	0.18	0.42	0.38	0.36	0.12	0.53	0.37	0.73	0.52	0.24
Beans, field	88	25.4	0.20	0.52	1.63	0.90	0.22	2.29	1.03	1.89	1.15	0.67
Blood meal	91	88.8	1.03	2.17	7.96	3.85	1.42	3.94	1.17	11.39	7.69	5.65
Corn	88	8.5	0.18	0.37	0.25	0.31	0.06	0.40	0.29	1.05	0.41	0.26
Corn gluten feed	88	19.0	0.32	0.71	0.58	0.68	0.11	0.85	0.59	1.69	0.89	0.59
Corn gluten meal	88	60.6	1.43	2.52	1.02	2.08	0.31	1.93	2.48	10.19	2.79	1.28
Feather meal	91	81.1	0.61	4.74	2.08	3.82	0.54	5.62	3.86	6.79	5.88	0.93
Fish meal	91	62.9	1.77	2.34	4.81	2.64	0.66	3.66	2.57	4.54	3.03	1.78
Meat and bone meal	91	49.1	0.68	1.18	2.51	1.59	0.28	3.45	1.34	2.98	2.04	0.91
Meat meal	91	48.8	0.68	1.24	2.44	1.63	0.30	3.42	1.40	2.99	2.13	0.93
Oat	88	12.6	0.22	0.57	0.53	0.44	0.14	0.87	0.48	0.92	0.66	0.31
Rapeseed meal,	88	34.8	0.70	1.59	1.95	1.53	0.45	2.15	1.37	2.47	1.77	0.97
Rice	88	7.3	0.20	0.37	0.26	0.26	0.09	0.60	0.29	0.59	0.40	0.18
Sesam meal	88	41.1	1.15	1.97	1.01	1.44	0.54	4.86	1.47	2.74	1.85	0.98
Sorghum	88	9.3	0.17	0.34	0.22	0.31	0.10	0.38	0.37	1.21	0.46	0.23
Soybean meal, 44 % CP*	88	44.0	0.64	1.31	2.75	1.76	0.57	3.28	2.01	3.44	2.09	1.21
Soybean meal, 48 % CP	88	47.6	0.69	1.41	2.98	1.89	0.61	3.52	2.16	3.71	2.23	1.31
Sunflower meal	88	33.5	0.77	1.36	1.19	1.25	0.40	2.75	1.37	2.15	1.66	0.88
Tapioca	88	3.3	0.04	0.09	0.12	0.11	0.04	0.18	0.11	0.19	0.14	0.08
Triticale	88	11.6	0.21	0.49	0.42	0.39	0.12	0.61	0.42	0.79	0.55	0.29
Wheat	88	13.3	0.21	0.50	0.38	0.38	0.15	0.64	0.44	0.87	0.56	0.32
Wheat bran	88	15.7	0.25	0.58	0.64	0.52	0.22	1.07	0.49	0.98	0.72	0.44
Wheat gluten feed	88	14.4	0.22	0.52	0.46	0.46	0.19	0.86	0.45	0.89	0.67	0.39
Wheat gluten meal	88	74.3	1.17	2.79	1.24	1.89	0.68	2.59	2.65	5.20	2.88	1.54

* Crude Protein

meeting the requirement of a particular amino acid. He may use either an excess of feed protein that contains large amounts of this amino acid or a minimum of natural protein and supplement it with synthetic amino acid. Because methionine, lysine, and threonine are commercially available and inexpensive, they are often used in formulated feed. L-tryptophan, which in many cases is the next limiting amino acids after methionine, lysine and threonine, is still too expensive to be regularly used for supplementation.

Amino Acids Content of Feedstuffs. Effective supplementation requires an exact knowledge of the natural amino acid content of both the individual feedstuffs and the formulated feed mix: the desired rates of supplementation must always be capable for being measured analytically. Ion-exchange and high-pressure liquid chromatography are reliable and proven methods for this.

The amino acid contents of individual feedstuffs are published internationally in a large series of tabulations (see, e.g., Table 14). However, such data must be current and reliable.

Amino Acid Requirements of Livestock. Determining amino acid requirement of animals requires difficult, time-consuming experiments. The values derived from these experiments are not constants valid for all times [263] but vary depending on environmental (heat stress, disease), genetic (sex, breed), and dietary factors (protein level, energy level, feed intake). There are essentially three methods for determining these requirements: carcass or milk analysis, synthetic rations, and semisynthetic rations.

For the first method, the amino acid content of a carcass is taken as a first-order approximation to the amino acid requirement of an animal. The same method can be applied for young suckling mammals by analyzing the amino acid content of the milk.

In the second method, the test animal is fed a synthetic mixture of all amino acids, along with an otherwise balanced ration (control group). The requirement for a single amino acid is determined by reducing its content in the diet to zero and from there on supplementing it stepwise up to an amount where the animal performs as well as the control group.

The third and most commonly used method is that a basal diet consisting of typically used feeds is formulated to be deficient in one amino acid but adequate in all other nutrients. To this basal diet graded levels of the amino acid in question are added until the performance of the animal approaches a maximum. The response in growth to stepwisely increasing amino acid supplementation follows the law of diminishing returns and can be described best by an exponential function.

Amino acid recommendations and requirements vary between species and for each species with age. There are several physiological reasons for that:

The various proteins deposited in the animal's body differ considerably in their amino acid composition. In poultry, for example, the proportion of lysine is much higher in muscle protein than in feather protein. In contrast, methionine and cystine are required in a higher percentage for the formation of feather protein and for maintaining metabolic functions than for muscle protein.

Not only the amino acid composition of the different proteins varies but also the quantity of each protein deposited in the animal changes with age. The daily accretion of muscle protein increases up to a certain age depending on the species and declines thereafter. The daily protein requirement for metabolic functions rises continually as the animal grows. Another reason for amino acid recommendations changing with age is the fact that young animals have a high potential for growth but at the same time a relatively small feed intake capacity which requires a high nutrient density in the diet.

The amino acid recommendations for domestic monogastric animals are listed in Table 15.

Economics of Amino Acid Supplementation. The purpose of modern, formulated feed mixes is to meet all nutritional requirements of the animal at a minimum cost, and amino acids and proteins are usually among the most expensive components of a feed mix. The performance of a feed mix is measured primarily by feed utilization and weight gain. Both feed utilization and weight gain, as well as other factors such as laying performance or feather or hair growth, are directly dependent on a sufficient supply of amino acids. Methionine and lysine play the major

Table 15. Amino acid recommendations, wt %, for poultry [268] and pigs [21]

Species	Metabolizable energy, MJ/kg	Crude protein	Met	Met+ Cys	Lys	Thr	Trp
Broiler							
Starter	13.2	21	0.56	0.96	1.24	0.77	0.21
Grower	13.4	20	0.52	0.92	1.12	0.70	0.19
Finisher	13.6	18	0.43	0.82	0.98	0.65	0.17
Laying hen (105 g feed intake per day)	12.1	16	0.40	0.74	0.84	0.55	0.15
Turkey (weeks of age)							
0 – 4	11.7	28	0.65	1.15	1.80	1.06	0.31
5 – 8	12.1	25	0.61	1.06	1.60	0.95	0.27
9 – 12	12.6	22	0.56	0.95	1.40	0.84	0.24
13 – 16	13.0	19	0.48	0.81	1.15	0.69	0.20
> 16	13.4	16	0.44	0.72	1.00	0.61	0.17
Pig (kg live weight)							
10 – 19	13.2	18	0.40	0.72	1.20	0.79	0.24
20 – 30	13.2	17	0.36	0.65	1.08	0.71	0.22
31 – 55	13.0	16	0.31	0.57	0.95	0.63	0.19
56 – 100	13.0	14	0.26	0.47	0.78	0.51	0.16
Sow							
Lactating	13.2	16.5	0.35	0.62	0.90	0.60	0.18
Pregnant	12.0	12.5	0.20	0.36	0.60	0.39	0.12

roles. Methionine in the form of D,L-methionine and lysine in the form of L-lysine · HCl are used in place of or as a supplement to natural methionine and lysine sources [358].

Linear programming is the method of choice for simultaneously optimizing a ration and minimizing cost [363]. This method allows simultaneous consideration of all demands that are made on the ration. A prerequisite is exact data on the nutrient content of all available feedstuffs and additives as well as their prices and availabilities. Additionally, the restrictions, i.e., the dietary requirements that the ration has to fulfill, must be known. Table 16 shows two examples of feed mixes formulated by the linear programming method commonly used.

Protein and Amino Acid Digestibilities of Feed Ingredients. The nutritive value of protein for monogastric animals is determined not only by the amino acid composition of the diet but also by digestibility of the individual amino acids, in particular the amino acids likely to be limiting. Over the last decades, considerable research has been carried out that demonstrates large differences in amino acid digestibilities between feeds.

Amino acid digestibilities can be determined according to the ileal or fecal analysis method. The fecal analysis method measures the amount of each amino acid consumed and excreted in feces. The amino acid digestibilities determined according to the ileal analysis method are calculated based on the intake and amount of each amino acid passing at the end of the distal ileum. The ileal analysis method should be considered as an improvement on the fecal analysis method, since the protein or amino acid absorbed in the large intestine make little or no contribution to the protein status of the animal. The ileal analysis method is also very sensitive to differences in amino acid digestibilities, as these result from processing conditions or from inherent differences between samples of the same feedstuff. Digestibilities measured over the entire digestive tract may be altered by resident bacteria. The bacteria may break down some of the amino acids, convert them to other amino acids, or even produce new amino acids. Measuring amino acids excreted in the feces will not reflect unabsorbed amino acid, but rather unabsorbed amino acids after possible alteration by the bacteria.

Amino acid digestibility values from the literature, determined with the ileal analysis method (Table 17) show large differences in amino acid digestibilities between feeds and among different samples of the same feed.

The differences in amino acid digestibility of feeds can be attributed to various factors. These include, e.g., fiber levels, heat damage during processing and the presence of antinutritional

Table 16. Two examples of formulated feeds

Broiler feed composition, wt %		Pig fattening feed composition, wt %	
Yellow corn	28.00	Feed grain (barley, wheat, corn)	35.00
Wheat	20.00	Soybean meal (44 % crude protein)	19.00
Soybean meal (48 % crude protein)	30.00	Tapioca meal	20.00
Tapioca meal	10.00	Corn gluten feed	15.00
Fat	7.00	Meat and bone meal (45 % crude protein)	3.00
Meat and bone meal (45 % crude protein)	3.00	Fat	3.00
Mineral premix	1.25	Beet molasses	2.00
Vitamin-trace element premix	0.50	Mineral premix	2.43
D,L-Methionine	0.25	Vitamin-trace element premix	0.50
	100.00	D,L-Methionine	0.07
			100.00

Table 17. Range of ileal digestibilities (%) of lysine, methionine, and threonine in different feeds for pigs modified from [22]

Amino acids	Lysine	Methionine	Threonine
Cereal grains			
Barley	65 – 79	72 – 88	64 – 76
Wheat	62 – 81	79 – 92	51 – 78
Corn	71 – 82	88 – 92	74 – 79
Protein supplements			
Soybean meal (44 % CP*)	85 – 89	77 – 90	73 – 81
Canola meal	74 – 76	81 – 87	66 – 67
Meat and bone meal	58 – 67	72 – 79	53 – 62
Cottonseed meal	53 – 70	65 – 82	55 – 69

* Crude Protein

factors that interfere with nutrient digestion and utilization.

Studies have clearly shown improvements in diet formulation practices, when diets are formulated on the basis of ileal digestible rather than on supply of limiting amino acids. This holds especially true when a good quality protein supplement is replaced by protein supplement(s) of lower quality combined with supplementary amino acids. In consequence, amino acid requirements should be expressed as ileal digestable, rather than as total amino acid requirements.

Amino Acids as a Measure to Reduce the N-Output from Livestock Production. Animal production accounts for a significant portion of nitrogen containing compounds released into the environment. In areas with intensive livestock production this might result in environmental problems, especially if N-requirements for crop fertilization and N-output from livestock production get out of balance. In this case nitrogen containing compounds are released into the surface and ground water where they accumulate.

The total amount of nitrogen produced is dependent on both, the average number of animals per unit available on and the efficiency of conversion of feed protein into body protein. This efficiency is often impaired due to diets, which are not balanced for the specific amino acid requirements of the animal fed. The portion of protein that is fed without supplying an adequate mix of amino acids is thus excreted without being utilized by the animals. Meanwhile the animal's requirements for amino acids are well known which allows to decrease the total protein content of diets as long as the diet is supplemented with amino acids in accordance with the animal's specific requirements. As a result, feed protein utilization is maximized and also water intake as a means to excrete excess nitrogen via the kidney is reduced.

In consequence the N-output from livestock production can be reduced down to 65 % when supplemental amino acids are used together with a reduction of the protein content of the diet.

5.3. Pharmaceuticals (→ Pharmaceuticals, General Survey and Development)

The pharmaceutical industry requires amino acids at a rate between 2000 and 3000 t/a worldwide. More than half of this is used for infusion solutions. During the last few years the potential of amino acids and their derivatives as active ingredients for pharmaceuticals has been recognized clearly, and considerable growth can be predicted.

5.3.1. Nutritive Agents

Infusion Solutions. Parenteral nutrition with L-amino acid infusion solutions is a well-established component of clinical nutrition therapy. A standard infusion solution contains the eight classical essential amino acids, the semiessential amino acids L-arginine and L-histidine, and several nonessential amino acids, generally glycine, L-alanine, L-proline, L-serine, and L-glutamic acid.

Also available are special infusion solutions tailored to the requirements of particular groups, such as newborn infants, seniors, or patients with an extreme negative nitrogen balance. Solutions rich in the branched-chained amino acids leucine, isoleucine, and valine and poor in methionine and aromatic amino acids are available for liver-disease patients. Solutions containing only essential amino acids are available for kidney patients. Enzymatic protein hydrolysates, which were used as infusion solutions until a few years ago, have disappeared almost completely from the market. They were not available in the optimal composition, and there were often compatibility problems. Only pure, crystalline L-amino acids are used in modern infusion solutions. The solutions (up to 10 %), which also contain electrolytes in addition to amino acids, are sterile and pyrogen-free.

The simultaneous administration of carbohydrates is necessary for optimal utilization of the amino acids. Glucose is normally a separate infusion. Some commercially available amino acid infusion solutions contain an energy source in the form of sugar alcohols (sorbitol, xylitol), which do not enter into a Maillard reaction with the amino acids.

Normally, parenteral nutrition is only practiced over a limited time. In principle, however, total parenteral nutrition over many years is possible. In such a case, all essential nutrients (unsaturated fatty acids, vitamins, and trace elements) must be provided.

Elemental Diets. Enteral nutrition is also a means of providing the essential nutrients [364]. Elemental diets, which were developed originally for the astronauts [365], contain chemically defined nutritive components. In addition to free amino acids the mixtures generally contain carbohydrates, fats, minerals, and vitamins in a combination adapted to the requirements. In many cases, elemental diets are used as an alternative and supplement to parenteral nutrition. They have high nutritional value and are totally resorbable. They are largely independent of the digestive function of the pancreas and reduce the intestinal bacteria flora. Amino acid elemental diets generally are used in cases of anatomic, functional, or enzymatic defects [366].

Formula diets based on peptides are gaining ground as an alternative to elemental diets based on L-amino acids. According to [367], short-chained peptides are resorbed rapidly via a peptide transport system in the gut, therefore in a process that is independent of amino acid transport. Compositions of nitrogen-free amino acid analogues (keto acids and hydroxy acids) have come into use for the special case of kidney insufficiency (chronic renal failure).

Elemental diets or formula diets are administered orally or via a nasogastric tube directly into the gastrointestinal tract.

5.3.2. Therapeutic Agents

Many therapeutic agents are derivatives of natural or nonnatural amino acids. Examples are benserazide, captopril, and dextrothyroxine. Only therapeutically useful amino acids and simple derivatives are treated here.

Amino Acids and Salts. The amino acids and their simple salts that are important therapeutic agents are compiled in Table 18. The proprietary names listed represent only a selection.

Table 18. Amino acids and their salts as drugs

Compound		Formula	M_r	Medical use	Trade names
L-Alanine	[*56-41-7*]	$C_3H_7NO_2$	89.09	parenteral nutrition, stimulant of glucagon secretion	
L-Arginine	[*74-79-3*]	$C_6H_{14}N_4O_2$	174.20	parenteral nutrition, stimulation of pituitary release of growth hormone and prolactin, stimulation of pancreatic release of glucagon and insulin	
L-aspartate	[*7675-83-4*]	$C_{10}H_{21}N_5O_6$	307.31	treatment of hyperammonemia	Argihepar, Laevil, Potentiator, Sargenor, Sorbenor
L-glutamate	[*4320-30-3*]	$C_{11}H_{23}N_5O_6$	321.34	treatment of hyperammonemia	Modumate
hydrochloride	[*1119-34-2*]	$C_6H_{15}ClN_4O_2$	210.66	treatment of hyperammonemia, electrolyte concentrate for i.v. solutions	Argivene, R-Gene, Polilevo
L-pyroglutamate		$C_{11}H_{21}N_5O_5$	303.27		
2-oxoglutarate	[*16856-18-1*]	$C_{11}H_{20}N_4O_7$	320.3	treatment of hepatic disorders	Leberam, Anetil, Argiceto, Eucol
L-Asparagine	[*70-47-3*]	$C_4H_8N_2O_3$	132.13	parenteral nutrition	
monohydrate	[*5794-13-8*]	$C_4H_{10}N_2O_4$	150.13		
D,L-Aspartic acid	[*617-45-8*]	$C_4H_7NO_4$	133.10		
magnesium, tetrahydrate	[*52101-01-6*]	$C_8H_{20}N_2O_{12}Mg$	360.54	cardiac agent, management of fatigue, mineral supplement	Magnesium Verla,
potassium, semihydrate	[*923-09-1*] (anhydrous)	$C_4H_6NO_4K \cdot 1/2\,H_2O$	180.20	cardiac agent, management of fatigue, mineral supplement	Trommcardin, Trophicard-Köhler
sodium, monohydrate		$C_4H_8NO_5Na$	173.10		
L-Aspartic Acid	[*56-84-8*]	$C_4H_7NO_4$	133.10	parenteral nutrition	
ferrous, tetrahydrate		$C_8H_2ON_2O_{12}Fe$	392.1	treatment of iron deficiency	Sideryl, Spartocine
magnesium, dihydrate	[*2068-80-6*]	$C_8H_{16}N_2O_{10}Mg$	324.52	management of fatigue and heart conditions	
potassium, semihydrate	[*1115-63-5*] (anhydrous)	$C_4H_6NO_4K \cdot 1/2\,H_2O$	180.20	management of fatigue and heart conditions	Magnesiocard,
sodium, monohydrate	[*3792-50-5*] (anhydrous)	$C_4H_8NO_5Na$	173.10	parenteral nutrition	Corroverlan
Betaine	[*107-43-7*]	$C_5H_{11}NO_2$	117.15		
citrate	[*17671-50-0*]	$C_{11}H_{19}NO_9$	309.27	lipotropic	Flacar, Panstabil
hydrochloride	[*590-46-5*]	$C_5H_{12}ClNO_2$	153.61	lipotropic, gastric acidifier	Aciventral, Acidol, Pesimuriat
monohydrate	[*17146-86-0*]	$C_5H_{13}NO_3$	135.15	lipotropic	Hepaderichol
L-Citrulline	[*372-75-8*]	$C_6H_{13}N_3O_3$	175.19	treatment of hepatic disorders	Polilevo
L-Cysteine	[*52-90-4*]	$C_3H_7NO_2S$	121.16	treatment of damaged skin, topically in ophthalmology, detoxicant	Reducdyn, Hepa-Loges, Irradian, Cicatrex, Felacomp

Table 18. (Continued)

Compound		Formula	M_r	Medical use	Trade names
hydrochloride monohydrate	*[7048-04-6]*	$C_3H_{10}ClNO_3S$	175.64		
hydrochloride	*[52-89-1]*	$C_3H_8ClNO_2S$	157.62		Cheihepar, Choldestal
L-Cystine	*[56-89-3]*	$C_6H_{12}N_2O_4S_2$	240.30	parenteral nutrition, lipotropic agent, treatment of hair and nail damage	Cystin "Brunner", Gerontamin, Pantovipar, Priorin
L-Glutamic acid	*[56-86-0]*	$C_5H_9NO_4$	147.13	parenteral nutrition, dietary supplement, treatment of hyperammonemia	Aciglut, Glutamin-Verla
calcium, dihydrate	*[5996-22-5]*	$C_{10}H_{20}N_2O_{10}Ca$	368.2		Vivacalcium
hydrochloride	*[138-15-8]*	$C_5H_{10}ClNO_4$	183.54	gastric acidifier	Bioprotein-Holzinger, Pansan, Pepsalara
monosodium, monohydrate	*[142-47-2]*	$C_5H_{10}NO_5Na$	187.13	flavoring and seasoning of food	
magnesium, tetrahydrate	(anhydrous) *[19238-50-7]*	$C_{10}H_{24}N_2O_{12}Mg$	388.62	tonics, mineral supplements	Magnesium Verla, Glutergen
potassium, monohydrate	*[19473-49-5]* (anhydrous)	$C_5H_{10}NO_5K$	203.24	tonics, mineral supplements	
magnesium, hydrobromide		$C_{10}H_{17}BrN_2O_8Mg$	397.5	tranquilizer	Psicosoma, Psychoverlan
L-Glutamine	*[56-85-9]*	$C_5H_{10}N_2O_3$	146.15	parenteral nutrition, treatment of mental disorders and alcoholism	
Glycine	*[56-40-6]*	$C_2H_5NO_2$	75.07	parenteral nutrition, antacid in conjunction with calcium carbonate	Cicatrex, Felacomp, Gastripan-K, Mirogastrin
aluminum, hydrate	*[13682-92-3]*	$C_2H_6NO_4Al\ (+x\ H_2O)$	135.1	antacid	Acidrine, Al-Glycin
L-Histidine	*[71-00-1]*	$C_6H_9N_3O_2$	155.16	parenteral nutrition, essential amino acid for infants	
acetate		$C_8H_{13}N_3O_4$	215.21		
hydrochloride monohydrate	*[5934-29-2]*	$C_6H_{12}ClN_3O_3$	209.63		Anti-rheuma, Rollkur-Ankermann, Laristine, Plexamine
L-Isoleucine	*[73-32-5]*	$C_6H_{13}NO_2$	131.18	parenteral nutrition, treatment of hepatic coma, dietary supplement	
L-Leucine	*[61-90-5]*	$C_6H_{13}NO_2$	131.18	parenteral nutrition, treatment of hepatic coma, dietary supplement	numerous combinations

Table 18. (Continued)

Compound		Formula	M_r	Medical use	Trade names
D,L-Lysine	*[70-54-2]*	$C_6H_{14}N_2O_2$	146.19	formation of salts with acidic drugs (to enhance solubility)	
monohydrochloride	*[70-53-1]*	$C_6H_{15}ClN_2O_2$	182.65	geriatric	Jestrosemin
acetylsalicylate	*[34220-70-7]*	$C_{15}H_{22}N_2O_6$	326.34	soluble form of acetylsalicylic acid (aspirin) for injection	Aspisol, Delgesic
L-Lysine	*[56-87-1]*	$C_6H_{14}N_2O_2$	146.19	parenteral nutrition, dietary supplement, prophylaxis of herpes simplex infection (?)	numerous combinations
acetate	*[57282-49-2]*	$C_8H_{18}N_2O_4$	206.24		
L-aspartate	*[27348-32-9]*	$C_{10}H_{11}N_3O_6$	279.30	dietary supplement	
L-glutamate	*[5408-52-6]*	$C_{11}H_{23}N_3O_6$	293.32	dietary supplement	
L-malate	*[71555-10-7]*	$C_{10}H_{20}N_2O_7$	280.28	dietary supplement	
monohydrate	*[39665-12-8]*	$C_6H_{16}N_2O_3$	164.21	dietary supplement	
monohydrochloride	*[657-27-2]*	$C_6H_{15}ClN_2O_2$	182.65	treatment of hypochloremia, alkaloses	Aktivanad, Athensa, Omnival, Vivioptal
D,L-Methionine	*[59-51-8]*	$C_5H_{11}NO_2S$	149.21	lipotropic and choleretic agent	numerous combinations
L-Methionine	*[63-68-3]*	$C_5H_{11}NO_2S$	149.21	parenteral nutrition, dietary supplement, lipotropic agent, treatment of paracetamol poisoning	
L-Ornithine					
acetate	*[60259-81-6]*	$C_7H_{16}N_2O_4$	192.22	parenteral nutrition	
L-aspartate	*[3230-94-2]*	$C_9H_{19}N_3O_6$	265.27	treatment of hepatic disorders	Hepa-Merz
monohydrochloride	*[3184-13-2]*	$C_5H_{13}ClN_2O_2$	168.62	parenteral nutrition	Ornitaine, Polilevo
2-oxoglutarate	*[5191-97-9]*	$C_{10}H_{18}N_2O_7$	278.14	treatment of hepatic disorders (hyperammonemia)	Ornicetil
D-Phenylalanine	*[673-06-3]*	$C_9H_{11}NO_2$	165.19	antidepressant	
L-Phenylalanine	*[63-91-2]*	$C_9H_{11}NO_2$	165.19	parenteral nutrition	
L-Proline	*[147-85-3]*	$C_5H_9NO_2$	115.13	parenteral nutrition, dietary supplement, starting material for captopril and enalapril	
L-Pyroglutamic acid	*[98-79-3]*	$C_5H_7NO_3$	129.07	formation of salts with basic drugs	
D,L-Serine	*[302-84-1]*	$C_3H_7NO_3$	105.09	starting material for benserazide	Aktiferrin
L-Serine	*[56-45-1]*	$C_3H_7NO_3$	105.09	parenteral nutrition, dietary supplement	Sulfolitruw
L-Threonine	*[72-19-5]*	$C_4H_9NO_3$	119.12	parenteral nutrition, dietary supplement	
L-Tryptophan	*[73-22-3]*	$C_{11}H_{12}N_2O_2$	204.23	parenteral nutrition, antidepressant, sleep inducer, dietary supplement	Optimax, Kalma, Pacitron
L-Tyrosine	*[60-18-4]*	$C_9H_{11}NO_3$	181.19	parenteral nutrition, dietary supplement	
L-Valine	*[72-18-4]*	$C_5H_{11}NO_2$	117.15	parenteral nutrition, dietary supplement, treatment of hepatic coma	

***N*-Acetylcysteine** [*616-91-1*], $C_5H_9NO_3S$, M_r 163.2, *mp* 109 – 110 °C, $[\alpha]_D^{20}$ +5 ° (c = 3, H_2O), is a mucolytic and secretolytic agent.

It is prepared by reaction of cysteine hydrochloride monohydrate with acetic anhydride in the presence of sodium acetate [368], [369].
Trade names:

FRG:
- ACC (Hexal)
- Acemuc (betapharm)
- Acetabs (Krewel Meuselbach)
- Acetyst (Pharbita)
- Azubrondün (Azupharma)
- Bisolvon (Boehringer Ingelheim)
- Brumoc (Klinge)
- durabronchal (durachemie)
- Fluimucil (Zambon)
- Muciteran (Farmasan)
- Mucocedyl (BM Medica)
- Muco-Perasthman (Polypharma)
- Muco Sanigen (Thiemann)
- Mucret (pharma-stern)
- Myxofat (Fatol)
- NAC (ct-Arzneimittel, ABZ-Pharma, Aliud Pharma, ratiopharm, Stada)
- Pulmicret (pharma-stern)
- Sigamucil (Kytta-Siegfried)
- Siran (Temmler)
- stas-akut (Stada)
- Tamuc (TAD)
- Tussiverlan (Verla)
- Vitemur (Orion Pharma)

F:
- Broncoclar (Oberlin)
- Codotussyl (Whitehall)
- Euronac (Europhta)
- Exomuc (Bouchara)
- Fluimucil (Zambon)
- Fluimucil Antibiotic 750 (Zambon)
- Genac (Génévrier)
- Mucolator (Abbott)
- Mucomyst (Bristol-Myers Squibb)
- Mucothiol (SCAT)
- Rhinofluimucil (Débat)-comb.
- Solmucol (Génévrier)
- Tixair (Byk)

GB:
- Ilube (Alcon)-comb.
- Parvolex (Evans)

I:
- Brunac (Bruschettini)
- Fluimucil (Zambon)
- Mucisol (Deca)
- Rinofluimucil (Zambon)-comp.

J:
- Acetein (Senju)
- Mucofolin Sol. (Eisai)

USA:
- Mucosit (Dey)

Carbocisteine (carbocysteine) [*638-23-3*], *S*-carboxymethyl-L-cysteine, $C_5H_9NO_4S$, M_r 179.2, *mp* 204 – 207 °C (decomp.), $[\alpha]_D^{20}$ −34.0 to −36.0 ° (c = 10, H_2O), is used to treat disorders of the respiratory tract associated with excessive mucus.

Synthesis involves *S*-alkylation of L-cysteine with chloroacetic acid in the presence of sodium hydroxide [370], [371].
Trade names:

FRG:
- Mucopront (Mack, Illert.)
- Sedotussin (Rodleben; UCB; Vedim)
- Transbronchin (ASTA Medica AWD)

F:
- Bronchathiol (Martin-Johnson & Johnson-MSD)
- Bronchocyst (SmithKline Beecham)
- Bronchokod (Biogalénique)
- Broncloclar (Oberlin)
- Broncorinol (Roche Nicholas)
- Bronkirex (Irex)
- Cadotussyl (Whitehall)
- Drill Expectorant (Pierre Fabre)
- Fluditec (Innotech International)
- Fluvic (Pierre Fabre)
- Médibronc (Elerté)
- Muciclar (Parke Davies)
- Mucotrophir (Sanofi Winthrop)
- Pectasan (RPR Cooper)
- Rhinathiol (Joullié)

GB:
- Mucodyne (Rhône-Poulenc Rorer)

I:
- Carbocit (CT)
- Fluifort (Dompé)
- Lisil (KBR)
- Lisomucil (Synthelabo)
- Mucocis (Crosara)
- Mucojet (Polifarma)
- Mucolase (Lampugnani)
- Mucosol (Tosi-Novara)
- Mucotreis (Ecobi)
- Polimucil (Poli)-comb.
- Reomucil (Astra-Simes)
- Solfomucil (Locatelli)
- Solucis (Magis)
- Superthiol (Francia Farm.)

J:
- Mucodyne (Kyorin)

Levodopa [*59-92-7*], (*S*)–3-(3,4-dihydroxyphenyl)alanine, $C_9H_{11}NO_4$, M_r 197, *mp* 285.5 °C (decomp.), $[\alpha]_D^{20}$ −12.15 ° ($c=4$ in 1 M HCl).

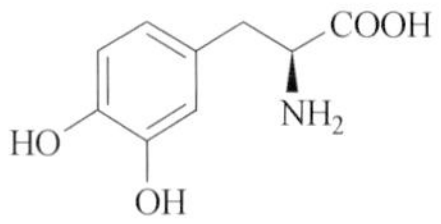

Levodopa is widely used for treatment of Parkinson's disease, most often in combination with peripheral decarboxylase inhibitors such as benserazide and carbidopa. For manufacture and trade names → Parkinsonism Treatment.

Trade names:

FRG:
- Dopaflex (medphano)
- Madopar (Roche)-comp. with benserazide
- Nacom (Du Pont Pharma)-comp. with carbidopa numerous generics

F:
- Modopar (Roche)-comp. with benserazide
- Sinemet (Du Pont Pharma)-comp. with carbidopa

GB:
- Madopar (Roche)-comp. with benserazide
- Sinemet (Merck Sharp & Dohme)-comp. with carbidopa

I:
- Larodopa (Roche)
- Madopar (Roche)-comp. with benserazide
- Sinemet (Du Pont Pharma)-comp. with carbidopa

J:
- Doparl (Kyowa)
- Dopasol (Daiichi)
- Dopaston (Sankyo)
- Larodopa (Roche)
- Neodopaston (Sankyo)-comp. with carbidopa

USA:
- Larodopa (Roche)
- Sinemet (Du Pont Pharma)-comp. with carbidopa generic

Mecysteine hydrochloride [*18598-63-5*], cysteine methylester hydrochloride, methyl L-2-amino-3-mercaptopropionatehydrochloride, $C_4H_{10}ClNO_2S$, M_r 171.66, *mp* 140 – 141 °C, is used in the treatment of disorders of the respiratory tract associated with excessive mucus. It is prepared by esterification of L-cysteine hydrochloride monohydrate with methanol in the presence of hydrogen chloride [372].

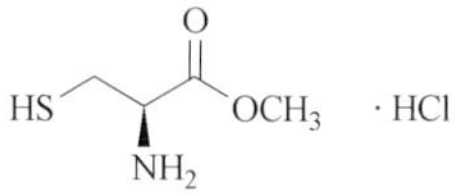

Trade names: Acthiol (Joullié, France), Actiol (Lirca, Italy), Visclair (Sinclair, UK), Aslos-C (Nissin, J), Epectan (Seiko, J), Moltanine (Tohok.-Tokyo Tanabe, J), Radcol (Nippon Universal, J), Sekinin (Tokyo Hosei, J).

Methiosulfonium chloride (iodide) [*1115-84-0*], L-methylmethionine sulfonium chloride, vitamin U, $C_6H_{14}ClNO_2S$, M_r 199.7; *mp* 134 °C (decomp.). Iodide [*3493-11-6*], $C_6H_{14}INO_2S$, M_r 291.1.

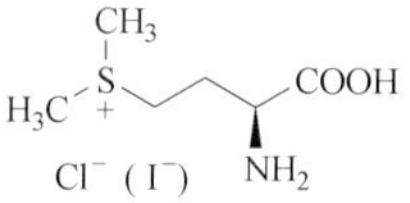

Methiosulfonium chloride is used for its protective effect on the liver and gastrointestinal mucosa, whereas the iodide finds use for rheumatic disorders. The compounds are made by heating L-methionine with methyl chloride or methyl iodide [373].

Trade names: Chloride: Ardesyl (Beytout, France), withdrawn from the market; Cabagin

(Kowa, Japan). Iodide: Lobarthrose (Opodex, France), withdrawn from the market.

Oxitriptan [*4350-09-8*], (*S*)-5-hydroxytryptophan, $C_{11}H_{12}N_2O_3$, M_r 220, *mp* 273 °C (decomp.), $[\alpha]_D^{22}$ −32.5 ° ($c=1$ in H_2O), $[\alpha]_D^{22}$ + 16.0 ° ($c=1$ in 4 M HCl).

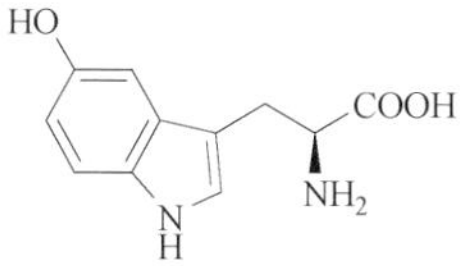

This intermediate in mammalian biosynthesis of serotonin is used as an antidepressant. It is produced either by total synthesis (analogous to L-tryptophan via the 5-benzyloxy derivative) [374–376] or by fermentation with *Chromobacterium violaceum* [377].
Trade names: Levothym (Promonta Lundbede, FRG), Lévotonine (Panpharma, F), Oxyfan (Coli, I), Tript-OH (Sigma-Tai, I).

D-Penicillamine [*52-67-5*], D-3-mercaptovaline, D-*β,β*-dimethylcysteine, $C_5H_{11}NO_2S$, M_r 149.21, *mp* 198.5 °C, $[\alpha]_D^{25}$ − 63 ° ($c=0.1$, pyridine). Hydrochloride [*2219-30-9*], $C_5H_{12}ClNO_2S$, M_r 185.7, *mp* 177.5 °C (decomp.) $[a]_D^{25}$ − 63 ° ($c=1$ in 1 M NaOH).

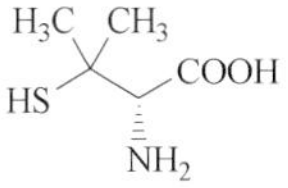

D-Penicillamine is a chelating agent that aids the elimination of toxic metal ions, e.g., copper in Wilson's disease. It is used, as an alternative to gold preparations, in the treatment of severe rheumatoid arthritis. It is useful in treating cystinuria because it reacts with cystine to form cysteine-penicillamine disulfide, which is much more soluble than cystine.

D-Penicillamine is produced by hydrolysis of benzylpenicillin via its Hg(II) complex [378] or by total synthesis. In the synthesis, isobutyraldehyde, sulfur, and ammonia are condensed to 5,5-dimethyl-2-isopropyl-Δ^3-thiazoline, which, on reaction with hydrogen cyanide, gives 5,5-dimethyl-2-isopropyl-thiazolidine-4-carbonitrile. Hydrolysis with boiling hydrochloric acid yields D,L-penicillamine hydrochloride. Cyclization with acetone and formylation leads to D,L-3-formyl-2,2,5,5-tetramethylthiazolidine-4-carboxylic acid, which can be resolved with (−)-phenylpropanolamine via the diastereomeric salts. Hydrolysis with hydrochloric acid leads to D-penicillamine hydrochloride [379].
Trade names: Cuprimine (Merck Sharp & Dohme, USA), Depen (Carter-Wallace, USA), Distamine (Dista, UK), Metalcaptase (Heyl/Knoll, FRG), Trolovol (ASTA Medica AWD, FRG), Trisorcin (Merdele, FRG), Trolovol (Bayer, F), Pendramine (ASTA Medica, UK), Pemine (Lilly, I).

5.4. Cosmetics

Amino acids are a component of the "natural moisturizing factor" (NMF) that protects the skin surface from dryness, brittleness, and a deleterious environment [380]. The epidermis of the skin contains about 15 % water, which, in the presence of amino acids, forms a stable water-in-oil emulsion with the skin lipids in the form of a thin layer. The amino acids simultaneously stabilize the pH of the skin (the acidic layer). Because of these properties, amino acids [381], [382], protein hydrolysates [383], and proteins [381] are widely utilized in skin and hair cosmetics, e.g., in mild skin creams, skin cleansing lotions, and hair shampoos.

The sodium salts of the reaction products of fatty acids with amino acids, such as glutamic acid [384–387], or short-chain peptides (protein hydrolysates) [388], [389] are surfactants. They are effective, skin-compatible cleaners and emulsifiers, which are used in shampoos, shower gels, baby baths, medicinal skin cleansers, cold-wave preparations, etc. [385], [390], [391]. The sulfur-containing amino acids exhibit a special normalizing effect on skin metabolism, e.g., in cases of excess skin lipid production (serborrhea), dandruff, or acne. Substances utilized for this purpose include derivatives of cysteine (e.g., *S*-carboxymethylcysteine), homocysteine (2-amino-4-mercaptobutyric acid), and methionine [392]. Amino acids are also used in hair lotions, where they are reported to have a nutritive effect [393]. Cysteine, which acts as a reducing agent, is gaining importance, especially in Japan, as a substitute for thioglycolic acid in permanent wave preparations that are less damaging to hair [394]. Use of aluminum, tin, and zirconium complexes of amino acids, especially glycine, as deodorants [395] and antiperspirants [395], [396] has been reported.

5.5. Agrochemicals

An increasing number of pesticides [397] are derivatives of natural or nonnatural amino acids. Important amino acids like Glyphosate or Glufosinate are major products in the agrochemical market. The synthesis of the active ingredient may start with an amino acid, but very often the amino acid moiety is formed during the chemical synthesis.

5.5.1. Herbicides

All pesticide segments (i.e., herbicides, fungicides, insecticides) contain compounds with amino acid substructures, but the major application of pesticidal amino acids is in the herbicide area [398].

Glyphosate [399] or Roundup [*1071-83-6*], $C_3H_8NO_5P$, M_r 169.1, *mp* 200 °C, is the isopropylamin salt of a phosphorus containing glycine derivative with a hybrid structure [400].

$(HO)_2OP$ N H COOH

Its mode of action is unique; it inhibits different enzyme systems that leads to the blocking of different amino acid biosyntheses. The substance is used as a nonselective contact herbicide to control deep rooted weeds. The sales [401] in 1997 grew to about 2180 million US dollar and reaching this figure it is the worlds top selling pesticide. The growth rates were double digits over the last five years (17 %) and for 1997 an annual production of 100 000 tons was estimated. Additional driving impact on the sales of glyphosate has been achieved by the introduction of glyphosate resistant crops (1996: soy beans, 1997: canola and cotton, 1998: maize) and the consequent distribution of genetically engineered seeds in combination with this herbicide.

The classical synthesis starts with iminodiacetic acid, phosphorous trichloride and formaldehyde, but also sequences using glycine and dimethyl phosphate have been applied. There are no confirmed reports of resistance. The expiration of the original patent and the advent of generic producers guarantee an increase of production of the substance becoming a bulk chemical.

Sulfosate [402] or Touchdown [*81591-81-3*], $C_6H_{16}NO_5PS$, M_r 245.2, is the trimesium salt of Glyphosate also applied as a non selective herbicide that was introduced 1989. As in the case of Glyphosate, sales are growing well.

Glufosinate [403] or Basta or Phosphinothricine [*53369-07-6*], $C_5H_{12}NO_4P$, M_r 181.1, *mp* 215 °C (for the ammonium salt). This phosphorous analogue of glutamic acid was indroduced in 1984 as a non selective herbicide in the speciality market, but meanwhile also resistant crops have been developed, which are distributed in combination with the herbicide (Liberty Link). In 1997 the sales were at 170 million US dollar.

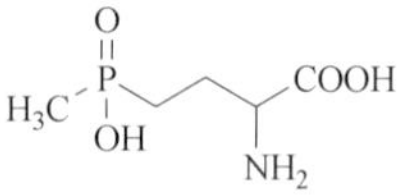

O P H_3C OH COOH NH_2

The contact herbicide inhibits the plant specific enzyme glutamine synthetase. The chemical synthesis requires acrolein, hydrocyanic acid and methyl phosphinous ester, resulting in a racemic product [404]. As the L-enantiomer is 40 times more active than the D-compound, many asymmetric synthetic routes are under investigation.

Bialofos [405] or Bilanafos [*35597-43-4*], $C_{11}H_{22}N_3O_6P$, M_r 323.3, *mp* 160 °C (decomp.), is a natural occuring tripeptide of the sequence L-phosphinothricine-L-alanine-L-alanine isolated from *Streptomyces viridochromogenes, Streptomyces hygroscopicus* and others. The compound was launched as a non selective contact herbicide in 1984. It has the same mode of action as Glufosinate, but has much lower sales.

Herbicides Based on 2-Methylvaline. In 1985, the first member of a new and very important class of selective herbicides based on 2-methylvaline was introduced. These imidazolinone herbicides are aceto lactate synthase (ALS) inhibitors [406] meaning that the synthesis of branched amino acids is blocked [407].

The synthesis of the racemic active incredients starts from methyl isopropylketone which is converted to 2-methylvaline and finally cyclized to an imidazolinone with an *ortho*-carboxy substituted aromatic or heteroaromatic ring system. The sales of the imidozolinone herbicides grew with 6.8 % per annum from 1992 to 1997. The sales for Imazethapyr, the star product against mono- and dicotyl weeds in soya were about 540 million US dollar in 1997.

Imazaquin [*81335-37-7*], 2-(4-isopropyl-4-methyl-5-oxo-4,5-dihydro-1*H*-imidazol-2-yl)-3-quinoline carboxylic acid, Scepter, $C_{17}H_{17}N_3O_3$, M_r 311.3, *mp* 219 – 224 °C, was launched in 1984.

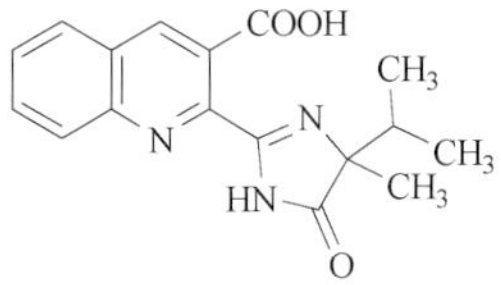

Imazapyr [*81334-34-1*], 2-(4-isopropyl-4-methyl-5-oxo-4,5-dihydro-1*H*-imidazol-2-yl)-nicotinic acid, Arsenal, $C_{13}H_{15}N_3O_3$, M_r 261.3, *mp* 169 – 173 °C, was launched in 1985.

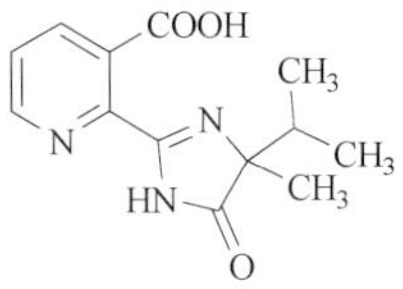

Imazamethabenz [*81405-85-8*], 2-(4-isopropyl-4-methyl-5-oxo-4,5-dihydro-1*H*-imidazol-2-yl)-5-methylbenzoic acid methyl ester, Assert, $C_{15}H_{18}N_2O_3$, M_r 274.3, *mp* 113 – 153 °C (of methyl ester), was launched in 1986.

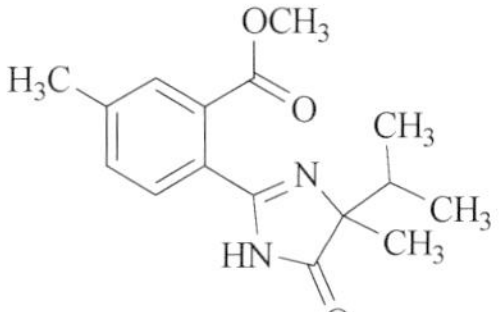

Imazethapyr [*81335-77-5*], 5-ethyl-2-(4-isopropyl-4-methyl-5-oxo-4,5-dihydro-1*H*-imidazol-2-yl)nicotinic acid, Pursuit, $C_{15}H_{19}N_3O_3$, M_r 289.3, *mp* 169 – 173 °C, was launched in 1987.

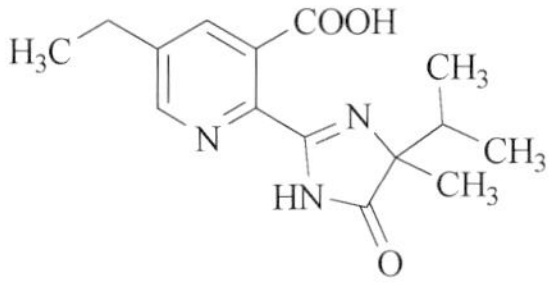

Imazapic [*104098-48-8*], 5-methyl-2-(4-isopropyl-4-methyl-5-oxo-4,5-dihydro-1*H*-imidazol-2-yl)-nicotinic acid, Cadre, $C_{14}H_{17}N_3O_3$, M_r 275.3, *mp* 204 – 206 °C, was launched in 1997.

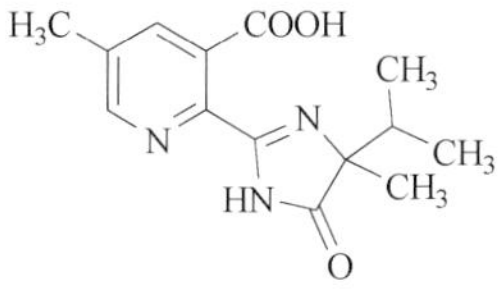

Imazamox [*114311-32-9*], 2-(4-isopropyl-4-methyl-5-oxo-4,5-dihydro-1*H*-imidazol-2-yl)-5-methoxymethylnicotinic acid, Raptor, $C_{15}H_{19}N_3O_4$, M_r 305.3, *mp*, was launched in 1997.

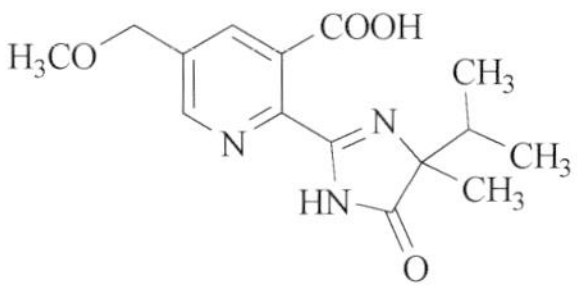

Herbicides derived from Acylphenyl Amino Acids. Acylphenyl amino acids are a group of herbicides with decreasing importance. The development of enantiomerically pure compounds are new concepts for improving the substance performance.

Benzoylprop [408] or Suffix [*22212-56-2*], $C_{16}H_{13}C_{12}NO_3$ (for ethyl ester), has an alanine subunit, which is formed by reacting 3,4-dichloroaniline with racemic ethyl 2-chloropropionate and benzoyl chloride. This herbicide is mainly used in wheat. The racemic active ingredient and the analogue with an 3-chloro-2-fluoroaniline aromat are more and more replaced by the D-alanine derivative (racemic switch).

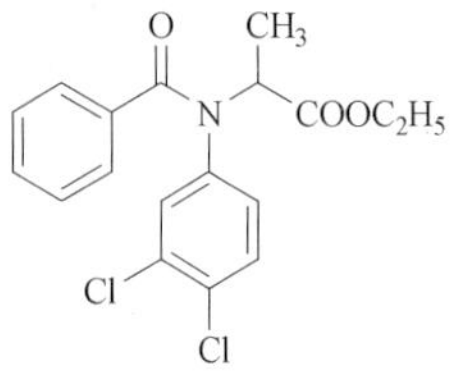

Flamprop-M [409] or Mataven L or Suffix BW [*63782-90-1*], $C_{19}H_{19}ClFNO_3$, M_r 363.8, *mp* 72.5 – 74.5 °C (data for isopropyl ester), is synthesized from 3-chloro-4-fluoroaniline, benzoyl chloride and methyl *S*-2-chloro propionate. This compound is used as selective grass herbicide in wheat and barley.

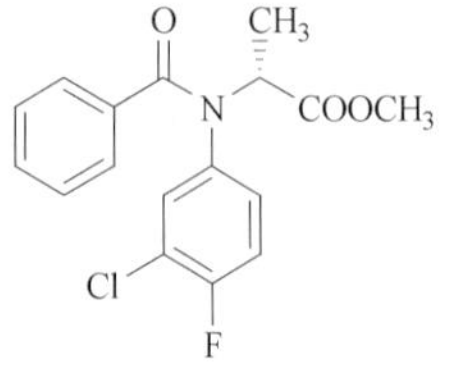

Amino Acids in Protox Herbicides. Inhibitors of the protoporphyrinogene oxidase [410], also known as Protox, are important herbicides blocking the biosynthesis of chlorophyll in plants. Phenyl substituted heterocycles have a strong impact on that enzyme [411]. These herbicides can contain amino acids in two ways. On the one side, amino acids like proline, 4-oxaproline [412] or pipecolinic acid [413] could be part of the heterocycle forming a hydantoin, on the other side amino acids like alanine can form the substituent at 5-position of the aromatic system. The preemergence activity and selectivity is increased using the R-enantiomer instead of the racemic alanine.

5.5.2. Fungicides

The fungicide market does not consist of many amino acid based active compounds, but patents filed from most of the major pesticide producers indicate that a new class of valine derivatives have a good chance to become market products in the future.

A lot of acylvaline anilides are claimed to be active against a broad spectrum of fungi, but especially the control of *Plasmopara* and *Phytophthora* [414] is pointed out. The literature indicates [415] that the *S*-compound is the more active enantiomer. This group of compounds may be divided into valine derivatives and non valine derivatives. The valine is usually acylated to form an urethane or a maleimide. The carboxylic group is condensated with a substituted aniline, an alkyl benzylamine or an alkyl homobenzylamine [416], [417].

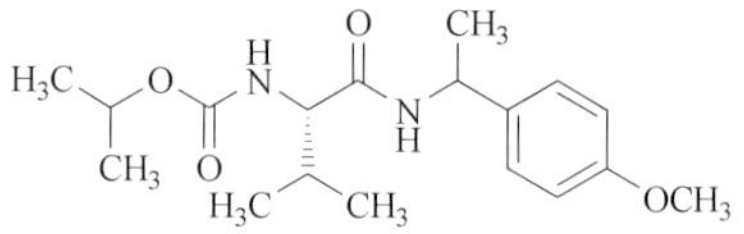

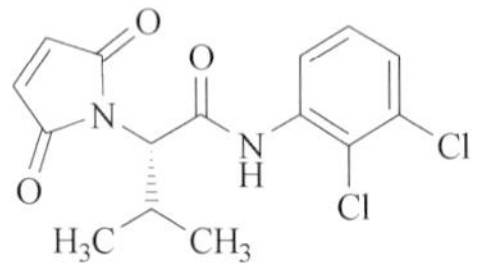

There are similar compounds described containing other amino acids. Examples are cyclopentylglycine, phenylalanine or alanine.

Anilide Fungicides. A number of anilide fungicides with a market volume of approx. 610 mio US dollar in 1997 have been established since several years. Three amino acid derivatives are sold as fungicides with systemic activity.

Metalaxyl [418] or Ridomil [*57837-19-1*], $C_{15}H_{21}NO_4$, M_r 279.3, *mp* 71.8 – 72.3 °C (data for isopropyl ester). This most important fungicide of its class synthesized by condensation of 2,6-dimethyl-aniline with methoxyacetyl chloride and subsequent coupling with racemic methyl 2-chloropropionate.

This fungicide is mainly used for the seed treatment business, in wine and in vegetables. The compound is sold as a stand-alone product or in a variety of mixtures with other fungicides. It was introduced in the market in 1978.

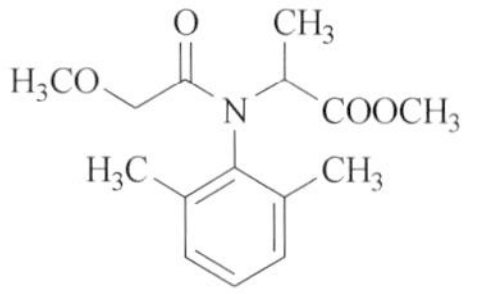

Furalaxyl or Fongarid [*57646-30-7*], $C_{17}H_{19}NO_4$, M_r 301.3, *mp* 70 and 84 °C (dimorphic) (data for isopropyl ester). Furalaxyl is a condensation product of furan-2-carboxylic acid and 2,6-dimethylaniline which is subsequently coupled with methyl 2-chloropropionate. The compound was introduced in 1984.

Benalaxyl [419] or Galben [*71626-11-4*], $C_{20}H_{23}NO_3$, M_r 325.4, *mp* 78 – 80 °C. In the commercial synthesis again 2,6-dimethylaniline is coupled with phenylacetyl chloride and subsequently methyl 2-bromopropionate.

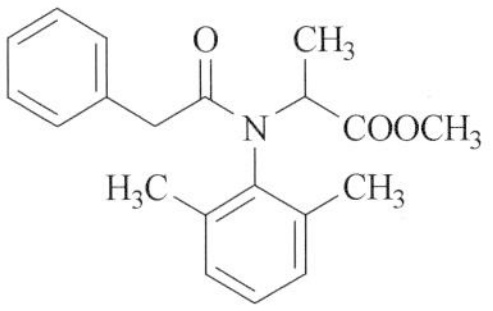

Hydantoins. Some amino acids being marketed as fungicides are sold as their hydantoins. One established product is iprodione.

Iprodione or Rovral [*36734-19-7*], $C_{13}H_{13}Cl_2N_3O_3$, M_r 330.2, *mp* 134 °C. The synthesis uses 3,5-dichloroaniline which is coupled with glycine and phosgene. The urea substructure is formed by reaction with isopropyl amine and phosgene. A lot of fungicide mixtures with Iprodione are on the market.

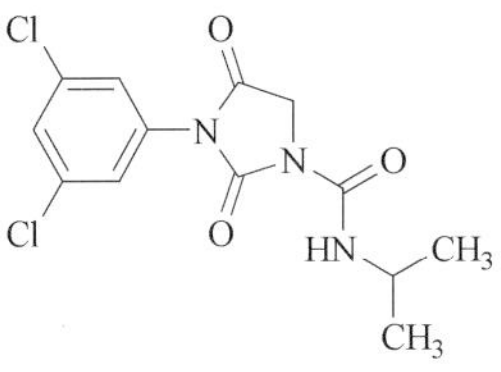

5.5.3. Insecticides

Only a few amino acid based insecticides are distributed in the market.

tau-Fluvalinate [420] or Klartan or Mavrik [*102851-06-9*], $C_{26}H_{22}ClF_3N_2O_3$, M_r 502.9, *bp* 164 °C (0.07 mm Hg) is the most important compound. This substance, preferably used in cotton fields, but also in corn, rape tomatoes and vegetables, is a pyrethroid with contact and stomach action. The market of that class had a volume of 1530 million US dollar in 1997. The chemical synthesis starts with 3,4-dichlorotrifluoromethylbenzene and D-valine. The carboxylic acid is esterified with the cyanohydrine of 3-phenoxy benzaldehyde. The compound is marketed with the defined stereochemistry at the valine moiety, but the cyanhydrine subunit is a *R/S* mixture.

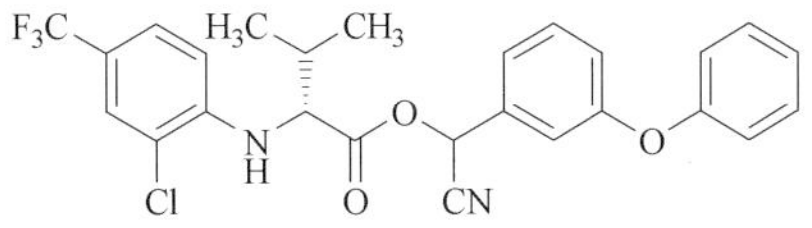

Imiprophin, $C_{17}H_{22}N_2O_4$, M_r 318.4, is a development candidate and also acting as a pyrethroid consisting of glycine hydantoin. The long synthetic pathway is leading to the vinyl substituted cyclopropane carboxylate, that forms an ester with 3-hydroxymethyl hydantoin. The resulting substance in *N*-alkylated with 3-bromopropine.

5.5.4. Plant Growth Regulators

Two compounds with amino acid structure are pointed out because of their direct chemical relationship to Glyphosate.

Glyphosine [421] or Polaris [*2439-99-8*], $C_4H_{11}NO_8P_2$ was introduced in 1973 and is preferably used in sugar beets. The synthesis starting with glycine uses two equivalents of fomaldehyde and of phosphorous trichloride under oxidative conditions.

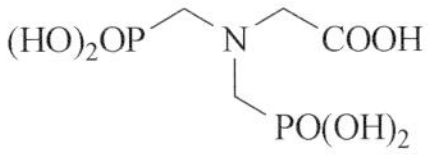

A second derivative of Glyphosate is a cyclic phosphaoxazole being formed from Glyphosate and formaldehyde. This compound is claimed to control the growth of some grasses without discoloration or phytotoxicity.

5.6. Industrial Uses

Amino acids and derivatives, as polyfunctional compounds, have potential for special industrial uses. The primary reason is their physicochemical properties, such as high thermal stability, low volatility, amphoterism, buffering capacity, and complex-forming capability. Increasingly important, however, are such aspects as environmental acceptability and low toxicity, properties that are characteristic for most amino acids and derivatives. Potential uses include acylamino acid monomers for epoxy resins [426]; amino acid dispersing agents for pigments in coloring polyester fibers [427]; *N*-acylamino acid dispersants for polyurethanes in water [428]; amino acid setting retarders for cement [429]; zinc salts of *N*-acyl-derivatives of basic amino acids and *N*-acylamino acids for the thermal stabilization of PVC [430]; polyglutamic acid esters and polyaspartic acid esters coatings for natural and synthetic leather [431]; amino acid hardening agents for methacrylate resins [432]; *N*-acylamino acid, amino acid ester, and amino acid amide gel-forming agents for oils [433]; basic amino acid vulcanization accelerators for natural rubber [434]; amino acid [435] and *N*-acylamino acid [436] corrosion inhibitors for metals; amino acids to stabilize the latent image of photographic emulsions [437]; and amino acid brighteners in galvanic baths [438], [439].

6. Chemical Analysis

Amino acids do not have defined melting points but decompose over a broad range between 250 and 300 °C. Therefore, they must be transformed into derivatives before melting points are useful for identification. Phenylisothiocyanate is used to yield the phenylthiohydantoin amino acid (PTH amino acid) [4], or 2,4-dinitrofluorobenzene (Sanger's reagent) is used to yield the dinitrophenyl amino acid [4]. Spectroscopic methods for the identification of amino acids include infrared [440], Raman, ^{1}H-NMR, and ^{13}C-NMR spectroscopy [441] of free amino acids or PTH derivatives and mass spectrometry of PTH derivatives [442]. Ultraviolet spectroscopy is important only for aromatic amino acids.

Various *staining methods* may be used for the qualitative identification of α-amino acids [443]. Some of these dye-forming reactions are suitable for quantitative analysis within the validity range of the Lambert–Beer law. By far the most important is the reaction with ninhydrin, which yields a red-violet to blue-violet dye (λ_{max} = 570 nm).

R–CH(NH$_2$)–COOH →

Ninhydrin

blue-violet dye

The imino acids proline and hydroxyproline form a structurally different dye, with an absorption maximum at 440 nm.

Fluorescent reagents also have been used successfully for quantitative analysis. The amino acid is converted into a strongly fluorescent derivative, which increases the sensitivity by orders of magnitude. Typical fluorescent reagents are *o*-phthalaldehyde – 2-mercaptoethanol [444], 1-dimethylaminonaphthalene-5-sulfonyl chloride (dansyl chloride) [445], and 4-phenylspiro[furan-2(3*H*)-1′-phthalan]-3,3′-dione (fluorescamine) [446].

The *separation* of amino acid mixtures is possible by electrophoresis or chromatography. The latter is especially useful, techniques including paper, thin layer, ion-exchange, high-pressure liquid, and gas chromatography. Paper chromatography is generally carried out in two dimensions. A number of eluents are available for this purpose. Quite often the amino acid is converted to its dinitrophenyl derivative.

More common than paper chromatography is *thin layer chromatography* (TLC) of either the free amino acids [447] or their PTH derivatives [448]. Silica gel, aluminum oxide, cellulose powder, or polyacrylamide may be used as carrier, but silica is preferred. An indicator reagent, most often ninhydrin, is used for detection [449]. For detection of very small quantities the amino acids separated on an TLC plate can be converted to the fluorescamine derivatives [450]. The detection limit is 10 pmol. The time required for the analysis of dinitrophenyl amino acids can be reduced by using high-pressure thin layer chromatography [451].

Ion-exchange chromatography [452], [453] on organic resins (Dowex, Amberlite, etc.) has proven to be the most exact and reliable method for the separation and quantitative analysis of amino acids. Before the automatic analysis technique was introduced [454], complete analysis of an amino acid mixture required 24 h. Today 2 h is the rule. Sodium citrate or lithium citrate buffer solutions are the eluents. The eluate is reacted with ninhydrin [452], [455] or *o*-phthalaldehyde [456]. With ninhydrin, 1 – 20 nmol amino acid can be measured with an accuracy of $\pm 1-5\,\%$. The detection limit with *o*-phthalaldehyde is in the picomole range.

Ion-exchange chromatography is the method of choice for analyzing amino acids in feeds, foodstuffs, and biologic fluids. In general, analysis is preceded by a hydrolysis, which degrades the proteins and peptides to their component amino acids. Figure 10 shows a sample aminogram of broiler feed.

Utilization of *high-pressure liquid chromatography* (HPLC) allows a further reduction in analysis time. However, not all amino acids can be separated cleanly. Furthermore, the amino acids generally must be converted to derivatives, e.g., to dansyl amino acids [457], PTH amino acids [458], or dinitrophenyl derivatives [459], before analysis. Reversed phases are the preferred stationary phases. Ninhydrin, *o*-phthalaldehyde – mercaptoethanol [460], or fluorescamine [461] are the usual reagents for detection. An HPLC method with direct UV detection has also been described [462] for analysis of infusion solutions.

Gas chromatographic methods [463] are useful for the analysis of complex amino acid mixtures. However, the amino acids must be converted into volatile derivatives, e.g., PTH amino acids [464], methyl esters of *N*-trifluoroacetylamino acids [465], *n*-butyl esters of *N*-trifluoroacetylamino acids [466], *N*-trimethylsilylamino acids [467], or *N,O*-bis-trimethylsilylamino acids [468].

Electrophoresis [469], which employs the differing rates of migration in an electric field, is relatively unimportant. Capillary isotachophoresis is a new high-resolution electrophoresis technique for amino acids.

The chromatographic separation of amino acid enantiomers is the subject of intensive investigation. Separation is currently possible by gas chromatography [470] and high-pressure liquid chromatography [471], using optically active phases or chiral solvents [472].

The *microbiological analysis* of amino acids is based on the fact that several L-amino acids are essential for certain bacteria strains. The growth of the bacteria cultures under standard conditions can be quantitatively evaluated (acidimetry or turbidimetry) and related to the amino acid concentration. Lactic acid bacteria [473] (Lactobacteriaceae) can be used to analyze 19 L-amino acids. Typical test microbes include *Leuconostoc mesenteroides* (ATCC 8042), *Lactobacillus arabinosus 17–5* (ATCC 8014), and *Streptococcus faecalis R* (ATCC 9790, 8043). The conventional microbiological methods are quite complicated but can be simplified by automation [474].

A series of L- or D-amino acids can be analyzed by *enzymatic methods.* L-Amino acids, or the enantiomeric purity of D-amino acids, can be determined with bacteria decarboxylases by measurement of CO_2 formed. D-Amino acids, or the enantiomeric purity of L-amino acids, can be determined with kidney D-amino acid oxidases by measurement of the O_2 consumption. Enzymes that react only with a single amino acid allow determination of that amino acid, e.g., arginine using liver arginase. An improved enzymatic method consists of the use of enzyme electrodes that contain the enzymes [475], or microorganisms having a special enzyme in a fixed form. Enzyme electrodes, however, are relatively unstable [476].

The *quantitative determination of crystallized amino acids* is carried out by acidimetric titration in nonaqueous medium [477], [478]. Glacial acetic acid is a suitable solvent. Formic acid may be added to improve solubility. The titrant is perchloric acid. Formol titration by the method of Sörensen can be used for aqueous solutions but is less accurate.

Standards of purity for individual L- and D,L-amino acids that are used in drugs or as food

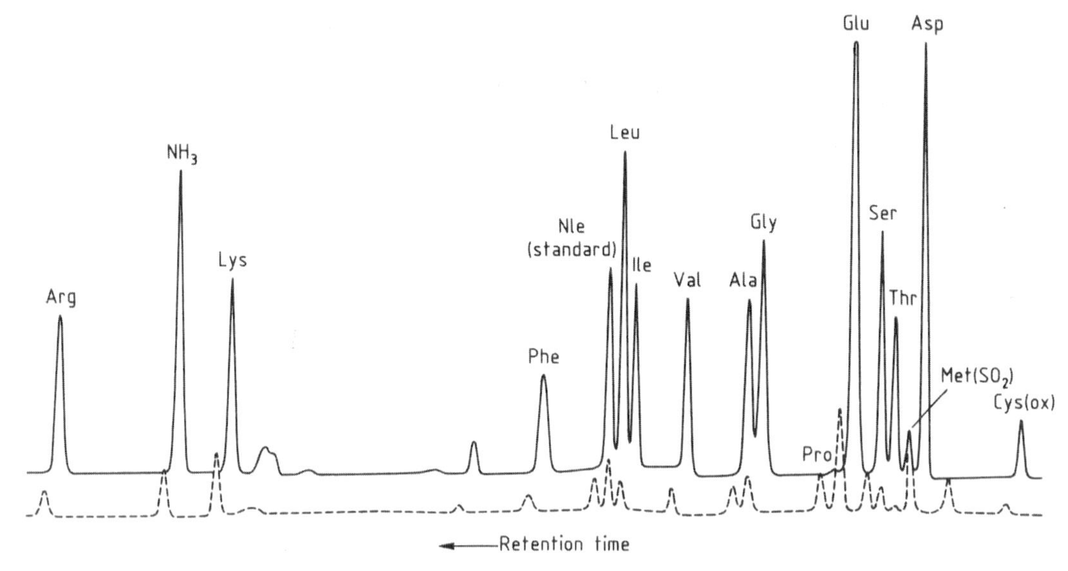

Figure 10. Amino acid chromatogram of a broiler feed
Internal standard: norleucine. Solid curve: UV detection at λ = 570 nm. Dotted curve: UV detection at λ = 440 nm for Pro (and Hyp).

additives are published in pharmacopeias [477], [479] and food codices [480].

Table 19. Amino acid production, 1999

Amino acid	Quantity, t/a [481] ([53])
L-Alanine	1200
L-Arginine *	1500
L-Asparagine	150
L-Aspartic acid	12 000
L-Cysteine **	3500
L-Glutamic acid *	700 000
L-Glutamine	1000
Glycine	15 000
L-Histidine	300
L-Hydroxyproline	100
L-Isoleucine	550
L-Leucine	800
L-Lysine *	430 000
D,L-Methionine	450 000
L-Methionine	400
L-Phenylalanine	11 000
L-Proline	800
L-Serine	300
L-Threonine	30 000
L-Tryptophan	1200
L-Tyrosine	150
L-Valine	1100
D-Phenylglycine + D-*p*-hydroxyphenylglycine	6000
Others	3000
Total	1 670 050

* Free amino acid and salts.
** Free amino acid and derivatives.

Table 20. Percentage of individual amino acids as a part of the total market, 1999

Amino acid	Quantity, %
L-Glutamic acid (Na)	58
D,L-Methionine	21
L-Lysine (HCl)	16
Other amino acids	5

Table 21. Market volume and value by field of use, 1999

Use	Volume, %
Human nutrition	62.2
Animal nutrition	37.2
Specialty *	0.6

* Pharmaceuticals, cosmetics, agrochemicals, and industrial uses.

7. Economic Significance

The 1999 world market for amino acids is estimated at more than 1.6×10^6 t/a (Table 19). The "big three" amino acids, sodium L-glutamate, D,L-methionine, and L-lysine · HCl, account for approximately 95 % of the volume (Table 20). The other amino acids play only a small role. The dominant amino acid, sodium glutamate, is used almost exclusively as a taste enhancer. D,L-Methionine and L-lysine · HCl are used almost exclusively to improve the nutritive value of animal feeds. The other amino acids have diversified applications. With the exception of glycine, they are more expensive than the big three amino acids. Table 21 shows the market volume and broken down by use.

The main manufacturers of amino acids are located in the Far East (e.g., Ajinomoto, Kyowa Hakko, Tanabe) and in Europe (Degussa, Alimentation Equilibrée de Commentry).

8. Toxicology

Excess amino acids are rapidly disposed of by increased metabolic degradation. Should the amino acid dose be suddenly increased, e.g., by extremely high protein consumption, within about two days the liver adaptively increases the levels of amino acid-catabolizing enzymes – transaminases, enzymes of the urea cycle, cystathionase, tryptophan pyrrolase, etc. The excess amino acids are to a large extent used to provide energy. The nitrogen is eliminated as urea. A smaller portion is used in protein synthesis, mainly liver protein and plasma albumin.

When too little protein or no protein is consumed or when the component amino acids are imbalanced, alteration of the ribosome profile occurs in the liver, and ribonucleic acids are catabolized. The manifestation of chronic protein deficiency is known as marasmus (slight deficiency) and kwashiorkor (extreme deficiency). Protein deficiency is usually coupled with calorie deficiency (protein – calorie malnutrition). This manifests itself on the biochemical level as a negative nitrogen balance, indicating a reduction in protein inventory. Initially, the labile enzyme and plasma proteins are consumed, the greatest losses occurring first in the liver, then in the musculature. Brain, heart, and kidneys suffer only minimal protein loss. Symptoms of protein synthesis disorders include disturbances in wound healing and bone growth, lowered resistance to infection and stress, loss of fertility and appetite, and anorexia. The urinary excretion of 3-methylhistidine is a common indicator of the catabolism of muscle protein.

The total absence of an essential amino acid in the diet is more serious than protein deficiency. In this case the proteins or amino acids in the diet are totally worthless because protein synthesis can occur only by degradation of body protein. A general interruption of the protein synthesis results. This manifests itself rapidly as a drop in enzyme activity and an impoverishment of the plasma proteins. Noticeable symptoms are loss of appetite and weight, alteration of cornea and lens, anatomic organ alterations, and an increased rate of mortality. In addition there appear specific deficiency symptoms characteristic of the missing amino acid or acids.

The metabolic disturbances brought about by gross divergences from the optimal amino acid pattern have three different causes [483], [484]: imbalance, antagonism, and toxicity.

Amino acid imbalance manifests as an appearance of deficiency symptoms for the first limiting amino acid when the other amino acids are consumed in great excess. The symptoms of imbalance are eliminated by administration of the first limiting amino acid in sufficient quantities. The main symptom is a severe reduction of food or feed assimilation and depression of growth. The depression of growth in rats has been investigated intensively by adding individual L- and D,L-amino acids to basal diets of various protein levels [485], [486]. *Amino acid antagonism*is caused by competition for common transport systems. An example is antagonism of the branched-chain amino acids isoleucine, leucine, and valine. The symptoms are reversible. In a study with young rats, addition of 5 % L-leucine to a low-protein diet (9 % casein) reduced the plasma levels of isoleucine and valine, depressed growth, and reduced feed consumption [487]. These effects were eliminated after a latent period of three days by small doses of L-isoleucine (0.16 %) and L-valine (0.15 %).

Amino acid toxicity occurs when very large quantities of one or more amino acids are consumed and is characterized by total failure of the adaptive mechanisms. The toxic level has been studied by adding increasing quantities of individual amino acids to a protein basal diet [484]. The toxicity of individual amino acids depends on the total protein consumption. Imbalance, antagonism, and toxicity are less pronounced when overall protein consumption is sufficient but become more severe at lower levels of protein consumption. The consumption of toxic amounts of amino acids increases their concentration in the plasma and brain. Because of the blood – brain barrier, however, the increase in the brain is not as great [488]. Failure of the adaption mechanisms during consumption of large excess of an amino acid can lead to accumulation of the amino acids or certain metabolites in the organism, leading directly or indirectly (e.g., by influencing hormone secretion) to anatomic or functional damage.

The toxicity of amino acids has been reviewed [484], [487]. The acute toxicities of most L-amino acids and some derivatives have been determined [489–491]. The toxicology of D-amino acids is discussed in review articles [484] and other publications [490], [491]. There is no evidence to date that the D-enantiomers of the α-amino acids found in proteins exhibit specific toxic effects. Their LD_{50} values are generally higher than those of the L-amino acids.

9. References

General References

1. J. P. Greenstein, M. Winitz: *Chemistry of the Amino Acids,* **vol. 1, 2, and 3,** J. Wiley & Sons, New York – London 1961.
2. A. Meister: *Biochemistry of the Amino Acids,* 2nd ed., **vol. 1,** Academic Press, New York 1965.
3. I. Wagner, H. Musso: “Neue natürliche Aminosäuren” *Angew. Chem.* **95** (1983), 827 – 920; *Angew. Chem. Int. Ed. Eng.* **22** (1983) 816 (review of the literature since 1956).
4. Th. Wieland et al.: “Methoden zur Herstellung und Umwandlung von Aminosäuren und Derivaten,” *Houben-Weyl* , 11/2.
5. H. D. Jakubke, H. Jeschkeit: *Aminosäuren, Peptide, Proteine,* Verlag Chemie, Weinheim 1982.
6. B. Weinstein (ed.): *Chemistry and Biochemistry of Amino Acids, Peptides and Proteins,* **vol. 1 – 5,** Marcel Dekker, New York 1971 – 1978.
7. D. Barton, W. D. Ollis: *Comprehensive Organic Chemistry,* **vol. 2, Chap. 9.6, p. 815; vol. 5** Pergamon Press, New York 1979.
8. G. Krügers: “Aminocarbonsäuren” in: *Methodicum Chimicum,* **vol. 6,** p. 611.
9. E. Wünsch: “Synthese von Peptiden,” *Houben-Weyl,* **15/1 and 15/2.**
10. I. C. Johnson: *Amino Acids Technology,* Noyes Data Corp., Parke Ridge, N. J., 1978.

Specific References

11. IUPAC Commission on Nomenclature of Organic Chemistry (CNOC) and IUPAC-IUB Commission on Biochemical Nomenclature (CBN), *Biochemistry* **14** (1975) 449.
12. S. K. Bhattacharyya, A. B. Banerjee, *Folia Microbiol. (Prague)* **19** (1974) 43.
13. T. Robinson, *Life Sci.* **19** (1976) 1097.
14. R. E. Dickerson, *Spektrum der Wissenschaften* 1979, no. 9, 98.
15. *Handbook of Chemistry and Physics,* 62nd ed., CRC Press, Boca Raton, Florida, 1981 – 1982.
16. D. M. Greenberg: *Amino Acids and Proteins,* Charles C. Thomas Publ., Springfield, Illinois, 1951.
17. R. S. Cahn, C. Ingold, V. Prelog, *Angew. Chem.* **78** (1966) 413; *Angew. Chem. Int. Ed. Engl.* **5** (1966) 385, 511.
18. K. Lübke, E. Schröder, G. Kloss: *Chemie und Biochemie der Aminosäuren, Peptide und Proteine,* **vol. I and II,** Thieme Verlag, Stuttgart 1975.
19. A. Kleemann, *Chem* . Ztg. 106 (1982) 151 – 167.
20. K. H. Drauz, A. Kleemann, J. Martens, *Angew. Chem.* **94** (1982) 590; *Angew. Chem. Int. Ed. Engl.* **21** (1982) 584.
21. *Drugs of the Future* **23** (1998) 1057.
22. R. P. Nargund, A. A. Patchett, M. A. Bach, M. G. Murphy, R. G. Smith, *J. Med. Chem.* **41** (1998) 3103.
23. *Drugs of the Future* **19** (1994) 111.
24. A. Müller, E. Busker, J. Engel, B. Kutscher, M. Bernd, A. V. Schally, *Int. J. Peptide Prot. Res.* **43** (1994) 264.
25. Europ. Pat. Appl. 0 530 167 A 1, 1993 (B. Atrash, D. M. Jones, M. Szelke).
26. *Drugs of the Future* **22** (1997) 371.
27. *Drugs of the Future* **17** (1992) 551.
28. *Drugs of the Future* **23** (1998) 751.
29. J. A. Robl et al., *J. Med. Chem.* **40** (1997) 1570.
30. S. Budavari (ed.): *The Merck Index,* 12th ed., Merck & Co., Inc., Whitehouse Station, NJ 1996, p. 97.
31. A. S. Bommarius, M. Schwarm, K. Stingl, M. Kottenhahn, K. Huthmacher, K. Drauz, *Tetrahedron: Asymmetry* **6** (1995) 2851.
32. *Drugs of the Future* **21** (1996) 1022.
33. *Drugs of the Future* **23** (1998) 1057.
34. T. Mimoto et al., *J. Med. Chem.* **42** (1999) 1789.

35. *Drugs of the Future* **18** (1993) 503.
36. *Drugs of the Future* **23** (1998) 1057.
37. *Drugs of the Future* **22** (1997) 260.
38. *Drugs of the Future* **22** (1997) 371.
39. S. Budavari (ed.): *The Merck Index,* 12th ed., Merck & Co., Inc., Whitehouse Station, NJ 1996, p. 99.
40. *Drugs of the Future* **14** (1989) 767.
41. *Drugs of the Future* **21** (1996) 600.
42. *Drugs of the Future* **18** (1993) 1129.
43. *Drugs of the Future* **22** (1997) 221.
44. *Drugs of the Future* **23** (1998) 1057.
45. *Drugs of the Future* **19** (1994) 988.
46. R. Meister (ed.): *Farm Chemical Handbook 1990,* Meister Publishing Co., Willoughby, OH 1990, p. C136.
47. M. S. Sadovnikova, V. M. Belikov, *Russ. Chem. Rev. (Engl. Transl.)* **47** (1978) 199.
48. Y. Izumi, J. Chibata, T. Itoh, *Angew. Chem.* **90** (1978) 187; *Angew. Chem. Int. Ed. Engl.* **17** (1978) 176.
49. S. Yamada, C. Hongo, R. Yoshioka, J. Chibata, *J. Org. Chem.* **48** (1983) 843.
50. Degussa, US 4356324, 1982 (W. Bergstein, A. Kleemann, J. Martens).
51. A. Collet, M.-J. Birenne, J. Jacques, *Chem. Rev.* **80** (1980) 215.
52. Nippon Kayaku, US 4224239, 1980 (Y. Tashiro, T. Nagashima, S. Aoki, R. Nishizawa).
53. *Finechemical (Jpn.)* 1982, no. 3, 4 – 26.
54. J. Chibata: *Immobilized Enzymes,* Kodansha – Halsted Press, Tokyo – New York 1978.
55. C. Wandrey: *Advances in Biochemical Engineering,* **vol. 12,** Springer Verlag, New York 1979.
56. Degussa, US 4304858, 1981 (C. Wandrey et al.).
57. Snamprogetti, DE 2621076, 1976 (F. Cecere et al.). Ajinomoto, US 4211840, 1980 (S. Nakamori et al.). AEC (Alimentation Equilibrée de Commentry), BE 883322, 1980. BASF, EP 46186, 1981 (R. Lungershausen et al.).
58. M. Guivarach et al., *Bull. Soc. Chim. Fr.* 1980, no. 1/2, II-91 – II-95.
59. M. Sugie, H. Suzuki, *Agric. Biol. Chem.* **44** (1980) 1089.
60. Dow Chemical, US 2700054, 1955 (H. C. White).
61. Ajinomoto, GB 908735, 1962 (J. Kato et al.).
62. Tanabe Seiyaku, US 3898128, 1975 (I. Chibata et al.).
63. K. Yamamoto et al., *Biotechnol. Bioeng.* **22** (1980) 2045.
64. Tanabe Seiyaku, US 3871959, 1975 (I. Chibata et al.).
65. Kyowa Hakko Kogyo, US 4086137, 1978 (K. Nakayama, K. Araki, H. Yoshida). Tanabe Seiyaku, US 3902967, 1975 (I. Chibata, M. Kisumi, J. Kato). Ajinomoto, GB 2084566, 1982 (K. Akashi et al.).
66. Prod. Organ. de Santerre Orsan, US 3933586, 1976 (Nguyen-Cong Puc). Ajinomoto, US 4000040, 1976 (T. Tsuchida, K. Kubota, Y. Hirose).
67. Tanabe Seiyaku, DE 2252815, 1972 (I. Chibata, T. Tosa, T. Sato)
68. Ajinomoto, US 4006057, 1977 (K. Sano et al.).
69. Showa Denko, DE 3021566, 1979 (K. Nakayasu, O. Furuya, C. Inouè). Showa Denko, JP-Kokai 80164669, 1980.
70. Mitsui Toatsu Chemicals, EP 79966, 1983.
71. J. Martens, H. Offermanns, P. Scherberich, *Angew. Chem.* **93** (1981) 680; *Angew. Chem. Int. Ed. Engl.* **20** (1981) 668.
72. Ajinomoto, DE 1196630, 1970 (J. Kato et al.). Ajinomoto, DE 1122538, 1962 (H. Kageyama, M. Sato, T. Inouè).
73. Shell Oil, US 3131211, 1964 (G. A. Kurhajec, D. S. La France).
74. T. Akashi, *Nippon Kagaku Zasshi* **81** (1960) 421.
75. Ajinomoto, US 3971701, 1976 (K. Takinami et al.). Ajinomoto, US 4347317, 1982 (M. Yoshimura et al.).
76. Kyowa Hakko Kogyo, US 3939042, 1976 (K. Nakayama et al.). Commercial Solvents Corp., US 3929575, 1975 (G. M. Miescher).
77. Chattan Drug & Chemical, US 3510515, 1970 (C. S. Colburn).
78. Degussa, DE 1134683, 1963 (H. Wagner et al.). Showa Denko, DE 2343599, 1982 (N. Mihara, O. Furuya, K. Wada).
79. Tanabe Seiyaku, US 3902966, 1975 (I. Chibata et al.).
80. Ajinomoto, US 3875001, 1975 (K. Kubota et al.).
81. Ajinomoto, EP 82637, 1982 (T. Tsuchida et al.).
82. Degussa, DE 2048790, 1976 (H. Wagner, F. Schäfer). Tanabe Seiyaku, JP 82156448, 1982.
83. Tanabe Seiyaku, JP-Kokai 7922513, 1979. W. R. Grace, US 4329427, 1982 (M. H. Updike, G. J. Calton).
84. Ajinomoto, EP 71023, 1982 (T. Tsuchida, K. Miwa, S. Nakamori).
85. T. Tosa, T. Mori, N. Fuse, I. Chibata, *Enzymologia* **32** (1967) 153.

86. Ethyl Corp., US 4259441, 1981 (D. P. Bauer).
87. Ajinomoto, US 3865690, 1975 (S. Okumura et al.).
88. Stamicarbon, DE 2005515, 1970 (I. A. Thoma, J. F. Klein, L. M. Geurts). Stamicarbon, DE 2010696, 1970 (I. A. Thoma, W. Reichrath).
89. Stamicarbon, DE 1949585, 1969 (W. K. van der Linden, G. M. Surerkropp).
90. Toray Ind., US 3770585, 1973, US 3796632, 1974 (T. Fukumura).
91. O. Tosaka, K. Takinami, Y. Hirose, *Agric. Biol. Chem.* **42** (1978) 745.
92. Kyowa Hakko Kogyo, DE 2531999, 1979 (K. Inuzuka, S. Hamada, Y. Hofu).
93. Ajinomoto, FR 2482622, 1981 (K. Miwa et al.) Ajinomoto, US 4346170, 1982 (K. Sano, T. Tsuchida).
94. Degussa, DE 1906405, 1969 (H. Wagner et al.).
95. Procter & Gamble, US 3963573, 1976 (C. E. Stauffer).
96. Kanegafuchi Kagaku Kogyo, US 4148688, 1979 (H. Yamada et al.).
97. R. H. Mazur, J. M. Schlatter, A. H. Goldkamp, *J. Am. Chem. Soc.* **91** (1969) 2684.
98. Y. Isowa et al., *Tetrahedron Lett.* 1979, 2611.
99. Ajinomoto, US 3867436, 1975 (M. Nakamura et al.).
100. Dynamit Nobel, DE 2931224, 1979 (H. aus der Fünten, K. Schrage).
101. Ajinomoto, DE 2364039, 1973 (T. Yukawa et al.).
102. Degussa, GB 2078218, 1981 (T. Lüssling, P. Scherberich).
103. Anic, EP 77099, 1982 (M. Fiorini, M. Riocci, M. Giongo).
104. S. Yamada et al., *Appl. Environ. Microbiol.* **42** (1981) 773.
105. Tanabe Seiyaku, JP 7020556, 1966.
106. Ajinomoto, US 3909353, 1975.
107. Ajinomoto, GB 2053906, 1980 (T. Tsuchida, K. Sano). Genex Corp., EP 77196, 1982 (R. A. Synenki).
108. Degussa, EP 52201, 1981 (K. Drauz, A. Kleemann, M. Samson).
109. V. E. Price et al., *Arch. Biochem.* **26** (1950) 92.
110. S. Yamada, C. Hongo, I. Chibata, *Agric. Biol. Chem.* **41** (1977) 2413.
111. Ajinomoto, US 4224409, 1980 (S. Nakamori, H. Morioka, F. Yoshinaga).
112. Tanabe Seiyaku, EP 76516, 1982 (I. Chibata et al.).
113. Schering, DE 3127361, 1981 (G. Siewert).
114. L. Bassignani et al., *Chem. Ber.* **112** (1979) 148.
115. Mitsui Toatsu Chemical, EP 30474, 1980 (R. Mita et al.).
116. Ges. für Biotechnol. Forschung, US 4060455, 1977 (F. Wagner, H. Sahm, W. H. Keune).
117. Ajinomoto, US 3880741, 1975 (K. Kageyama et al.).
118. Kyowa Hakko Kogyo, US 4183786, 1980 (K. Nakayama, K. Araki, Y. Tanaka).
119. M. Sato, K. Okawa, S. Akabori, *Bull. Chem. Soc. Jpn.* **30** (1957) 937.
120. Y. Ozaki et al., *Synthesis* 1979, 216.
121. Denki Kagaku Kogyo, DE 3247703, 1982 (M. Kato, T. Miyoshi, I. Kibayashi).
122. Tanabe Seiyaku, GB 2072185, 1981 (I. Chibata et al.).
123. V. G. Debabov et al., US 4321325, 1980. Ajinomoto, US 4347318, 1982 (K. Miwa et al.). Ajinomoto, EP 66129, 1982 (T. Tsuchida, K. Miwa, S. Nakamori).
124. O. T. Warner, O. A. Moe, *J. Am. Chem. Soc.* **70** (1948) 2765.
125. Degussa, DE 2647255, 1978 (H. Offermanns, H. Weigel).
126. I. Maeda, R. Yoshida, *Bull. Chem. Soc. Syn.* **41** (1968) 2975.
127. Degussa, EP 52199, 1981 (A. Kleemann, M. Samson).
128. Ajinomoto, GB 982727, 1965.
129. Ajinomoto, JP 7903021, 1979; JP 7903059, 1979; JP 7903010, 1979.
130. Degussa, EP 52200, 1981 (A. Kleemann, M. Samson).
131. Ajinomoto, DE 2819148, 1978 (S. Nakamori, K. Yokozeki, K. Mitsugi).
132. F. Masumi et al., *Chem. Pharm. Bull.* **30** (1982) 3831.
133. Tanabe Pharmaceutical, JP 78001836, 1978.
134. DE 2841642, 1978 (F. Wagner, J. Klein).
135. Mitsui Toatsu Chemicals, US 4335209, 1982 (Y. Asai, M. Shimada, K. Soda).
136. Asahi Kasei Kagyo, US 4349627, 1982; US 4360594, 1982 (A. Mimura et al.).
137. Ajinomoto, US 3929573, 1975 (S. Konosuke, K. Mitsugi).
138. AB Bofors, US 3963572, 1976 (S. V. Gatenbeck, P. O. Hedman).
139. Showa Denko, US 4363875, 1982 (T. Akashiba, A. Nakayama, A. Murata).
140. E. O. Denenu, A. L. Demain, *Appl. Environ. Microbiol.* **42** (1981) 497.
141. Mitsubishi Petrochemical, US 4271267, 1981 (H. Yukawa et al.).
142. Ajinomoto, EP 81107, 1982 (O. Kurahashi, M. Kamada, H. Enei).

143. Ajinomoto, US 4371614, 1983 (D. M. Anderson, K. M. Hermann, R. L. Somerville). Ajinomoto, EP 80378, 1982 (T. Tsuchida et al.).
144. Ajinomoto, DE 2411209, 1982 (T. Tsuchida, F. Yoshinaga).
145. Prod. Organ. de Santerre Orsan, GB 1578057, 1980.
146. W. D. Jefferson, *J. Am. Chem. Soc.* **43** (1978) 3980.
147. Zoecon Corp., US 4260633, 1981 (R. J. Anderson, K. G. Adams, C. A. Henrick).
148. Snamprogetti, CH 620943, 1980 (D. Dinelli, F. Morisi, D. Zaccardelli).
149. Kanegafuchi Kagaku Kogyo, DE 2757980, 1977 (H. Yamada, S. Takahashi, K. Yoneda). Kanegafuchi Kagaku Kogyo, JP-Kokai 8158493, 1981.
150. D. A. Bender: *Amino Acid Metabolism,* J. Wiley & Sons, London – New York – Sydney – Toronto 1975, p. 2.
151. H. N. Munro, *Drug Intell. Clin. Pharm.* **6** (1972) 216.
152. M. L. Sunde, *Poult. Sci.* **51** (1972) 44.
153. R. F. Barker, D. A. Hopkinson, *Ann. Hum. Genet.* **41** (1977) 27.
154. L. D. Stegink: *Clinical Nutrition Update: Amino Acids,* Am. Med. Assoc. Publ., Chicago 1977, pp. 198 – 205.
155. K. Schreier, *Ärztl. Fortbildung* **19** (1971) 107.
156. U. Porath, K. Schreier, *Med. Ernähr.* **11** (1970) 229.
157. H. Bickel, S. Kaiser-Grubel, *Dtsch. Med. Wochenschr.* **96** (1971) 1415.
158. Y. Izumi, I. Chibata, T. Itoh, *Angew. Chem.* **90** (1978) 187; *Angew. Chem. Int. Ed. Eng.* **17** (1978) 176.
159. M. S. Sadovnikova, V. M. Belikov, *Russ. Chem. Rev. (Engl. Transl.)* **47** (1978) 199.
160. K. Drauz, A. Kleemann, J. Martens, *Angew. Chem.* **94** (1982) 590; *Angew. Chem. Int. Ed. Engl.* **21** (1982) 584.
161. FAO: *Nutritional Studies No. 24, Amino Acid Content of Foods and Biological Data on Proteins,* Interprint (Malta), Rome 1970.
162. Degussa, unpublished
163. Report of a Joint FAO/WHO Ad Hoc Expert Committee: *Energy and Protein Requirements,* Rome 1973.
164. W. C. Rose, *Nutr. Abstr. Rev.* **27** (1957) 631.
165. D. M. Hegsted, *Fed. Proc. Fed. Am. Soc. Exp. Biol.* **22** (1963) 1424.
166. E. Kofrányi, *Ernähr. Umsch.* **23** (1976) 205.
167. Dtsch. Ges. für Ernährung: *Empfehlungen für die Nährstoffzufuhr,* 3rd ed., Umschau-Verlag, Frankfurt 1975.
168. *Recommended Dietary Allowances,* 9th ed., Nat. Acad. of Sci., Washington, D.C., 1980.
169. G. K. Parman, *J. Agric. Food Chem.* **16** (1968) 169.
170. H. H. Ottenheym, P. J. Jenneskens, *J. Agric. Food Chem.* **18** (1970) 1010.
171. N. S. Scrimshaw, A. M. Altschul: *Amino Acid Fortification of Protein Foods,* MIT Press, Cambridge, MA 1971.
172. E. E. Howe, G. R. Jansen, E. W. Gilfilian, *Am. J. Clin. Nutr.* **16** (1965) 315.
173. A. M. Altschul, *Nature (London)* **248** (1974) 643.
174. J. Kato, N. Muramatsu, *J. Am. Oil Chem. Soc.* **48** (1971) 415.
175. J. Mauron, *Z. Ernährungswiss. Suppl.* **23** (1979) 10.
176. D. M. Hegsted, *Am. J. Clin. Nutr.* **21** (1968) 688.
177. L. D. Satterlee, H. F. Marshall, J. M. Tennyson, *J. Am. Oil Chem. Soc.* **56** (1979) 103.
178. H. W. Staub, *Food Technol. (Chicago)* **32** (1978) 57.
179. F. Jekat, *Fette Seifen Anstrichm.* **79** (1977) 273.
180. J. C. Somogyi in: *Die Bedeutung der Eiweiße in unserer Ernährung,* Issue 48 a, Schriftenreihe der Schweizer Vereinigung für Ernährung, 1982, p. 3.
181. G. G. Graham et al., *Am. J. Clin. Nutr.* **22** (1969) 1459.
182. G. R. Jansen, C. F. Hutchison, M. E. Zanetti, *Food Technol. (Chicago)* **20** (1966) 323.
183. R. Bressani et al., *J. Nutr.* **79** (1963) 333.
184. J. L. Iwan, *Cereal Sci. Today* **13** (1968) 202.
185. D. Rosenfield, F. J. Stare, *Mod. Gov.* **11** (1970) 47.
186. S. N. Gershoff et al., *Am. J. Clin. Nutr.* **30** (1977) 1185.
187. H. N. Parthasarathy et al., *Can. J. Biochem.* **42** (1964) 385.
188. S. J. Fomon et al., *Am. J. Clin. Nutr.* **32** (1979) 2460.
189. A. L. Jung, S. L. Carr, *Clin. Pediatr. (Philadelphia)* **16** (1977) 982.
190. Procter & Gamble, US 3878305, 1975 (R. A. Damico, R. W. Boggs).
191. S. Wallrauch, *Flüss. Obst* **44** (1977) 386.
192. H. D. Pruss, I. P. G. Wirotama, K. H. Ney, *Fette Seifen Anstrichm.* **77** (1975) 153.
193. C. Ambrosino et al., *Minerva Pediatr.* **18** (1966) 759.

194. C. A. Masschelein, J. Van de Meerssche, *Tech. Q. Master Brew. Assoc. Am.* **13** (1976) 240.
195. T. Take, H. Otsuka, *Chem. Abstr.* **70** (1969) 46270 m.
196. J. Koch, *Flüss. Obst* **46** (1979) 212.
197. G. Baumann, K. Gierschner, *Dtsch. Lebensm. Rundsch.* **70** (1974) 273.
198. A. Askar, H. J. Bielig, *Alimenta* **15** (1976) 3.
199. W. Hashida, *Food Trade Rev.* **44** (1974) 21.
200. *Food Technol. (Chicago)* **34** (1980) 49.
201. S. Schwimmer, D. G. Guadagni, *J. Food Sci.* **32** (1967) 405.
202. Riken Kagaku, JP 7249707, 1972 (H. Watanabe et al.); *Chem. Abstr.* **79** (1973) 114219 q.
203. Pillsbury Comp., US 3510310, 1970.
204. C. Colburn, *Am. Soft Drink J.* **126** (1971) 16.
205. R. S. Shallenberger, T. E. Acree, C. Y. Lee, *Nature (London)* **221** (1969) 556
206. J. Solms, L. Vuataz, R. H. Egli, *Experientia* **21** (1965) 693.
207. H. Wieser, H. Jugel, H.-D. Belitz, *Z. Lebensm. Unters. Forsch.* **164** (1977) 277.
208. R. H. Mazur, J. M. Schlatter, A. H. Goldkamp, *J. Am. Chem. Soc.* **91** (1969) 2684
209. G. A. Crosby, *CRC Crit. Rev. Food Sci. Nutr.* **7** (1976) 297.
210. L. A. Pavlova et al., *Russ. Chem. Rev. (Engl. Transl.)* **50** (1981) 316.
211. W. J. Herz, R. S. Shallenberger, *Food Res.* **25** (1960) 491.
212. Lever Brothers, US 3922365, 1975 (K. H. Ney et al.).
213. H. Tanaka, Y. Obata, *Agric. Biol. Chem.* **33** (1969) 147.
214. M. Giaccio, L. Surricchio, *Quad. Merceol.* **16** (1977) 151.
215. H. Valaize, G. Dupont, *Ind. Agric. Aliment.* **68** (1951) 245.
216. E. L. Wick, M. DeFigueiredo, D. H. Wallace, *Cereal Chem.* **41** (1964) 300.
217. A. A. M. El-Dash, Dissertation, Kansas State University 1969 (Univ. Microfilms Inc., Ann Arbor, Michigan, No. 69-21123).
218. W. Baltes, *Ernähr. Umsch.* **20** (1973) 35.
219. M. Angrick, D. Rewicki, *Chem. Unserer Zeit* **14** (1980) 149.
220. H. E. Nurstein, *Food Chem.* **6** (1980) 263.
221. T. A. Rohan, *Food Technol. (Chicago)* **24** (1970) 29.
222. Maggi, DE 2246032, 1973 (R. J. Gasser). Z. Mielniczuk et al., *Acta Aliment. Pol.* **2** (1976) 213. Chas. Pfizer & Co., US 3365306, 1968 (M. A. Perret). Y.-P. C. Hsieh et al., *J. Sci. Food Agric.* **31** (1980) 943. R. A. Wilson, *J. Agric. Food Chem.* **23** (1975) 1032. P. van de Rovaart, J.-J. Wuhrmann, US 3930044, 1975. R. Schroetter, G. Woelm, *Nahrung* **24** (1980) 175.
223. H. Kisaki, *Chem. Abstr.* **69** (1968) 9824 d. Ajinomoto, FR 2005896 (1969). G. Rubenthaler, Y. Pomeranz, K. F. Kinney, *Cereal Chem.* **40** (1963) 658. I. R. Hunter, M. K. Mayo, US 3425840, 1969. Y. H. Liau, C. C. Lee, *Cereal Chem.* **47** (1970) 404. Y.-Y. Linko, J. A. Johnson, B. S. Miller, *Cereal Chem.* **39** (1962) 468. G. L. Bertram, *Cereal Chem.* **30** (1953) 126. Research Corp., US 3268555, 1966 (L. Wiseblatt). Hoffmann-La Roche, US 3547659, 1970 (W. Cort, L. Neck). M. Rothe, *Nahrung* **24** (1980) 185. A. A. El-Dash, A. A. Johnson, *Cereal Chem.* **47** (1970) 247. L. Wiseblatt, H. F. Zoumut, *Cereal Chem.* **40** (1963) 162. M. Rothe, *Ernährungsforschung* **5** (1960) 131.
224. G. Ziegleder, D. Sandmeier, *Dtsch. Lebensm. Rundsch.* **78** (1982) 315. W. Mohr, E. Landschreiber, Th. Severin, *Fette Seifen Anstrichm.* **78** (1976) 88. S. Turos, US 4346121, 1982.
225. E. Cremer, M. Riedmann, *Monatsh. Chem.* **96** (1965) 364.
226. Yuki Gosei Kogyo, US 3478015, 1969 (I. Onishi, A. Nishi, T. Kakizawa).
227. Fuji Oil Co., GB 1357511, 1974; GB 1488282, 1977.
228. Naarden Int., NL 7712745, 1979.
229. J. A. Newell, M. E. Mason, R. S. Matlock, *J. Agric. Food Chem.* **15** (1967) 767.
230. S.-C. Lee, B. R. Reddy, S. S. Chang, *J. Food Sci.* **38** (1974) 788. Research Corp., US 3814818, 1974 (S. S. Chang, B. R. Reddy). T. Y. Fan, M. H. Yueh, *J. Food Sci.* **45** (1980) 748. P. T. Arroyo, D. A. Lillard, *J. Food Sci.* **35** (1970) 769.
231. Yuki Gosei Kogyo, DE 1593733, 1972 (I. Onishi et al.). Japan Monopoly, Tanabe Seiyaku, US 3722516, 1973 (K. Suwa et al.). Philip Morris, US 4306577, 1981 (D. L. S. Wu, J. W. Swain). Reynolds Tobacco, US 3996941, 1976 (C. W. Miller, J. P. Dickerson, C. E. Rix).
232. R. G. Henika, N. E. Rodgers, *Cereal Chem.* **42** (1965) 397.
233. J. M Bruemmer, W. Seibel, H. Stephan, *Getreide Mehl Brot* **34** (1980) 173. Patent Technology, US 3803326, 1974. J. Geittner, *Gordian* **79** (1979) 202.
234. R. Marcuse, *Fette Seifen Anstrichm.* **63** (1961) 940.

235. H. Iwainsky, C. Franzke, *Dtsch. Lebensm. Rundsch.* **52** (1956) 129. H. Lingnert, C. E. Eriksson, *J. Food Process. Preserv.* **4** (1980) 161.
236. N. Watanabe et al., JP 7314042, 1973; *Chem. Abstr.* **79** (1973) 145052 j.
237. A. G. Castellani, *Appl. Microbiol.* **1** (1953) 195. Nisshin Flour Milling, JP 7319945, 1973(G. Ogawa, K. Taguchi); *Chem. Abstr.* **81** (1974) 76689 z. Nippon Kayaku, JP-Kokai 81109580, 1981; *Chem. Abstr.* **95** (1981) 202313 b.
238. M. Furui, K. Yamashita, *J. Ferment, Technol.* **61** (1983) 587 – 591.
239. S. Yahata, H. Tsutsui, K. Yamada, T. Konehara: "Fermentative production of D-alanine:" in: Proc. Ann. Meet. Agric. Chem. Soc. Jpn., (1993), 92.
240. S. Hashimoto, R. Katsumata, L-Alanine production by alanine racemase-deficient mutant of Athrobacter oxydans in: *Proc. Ann. Meet. (Agric. Chem. Soc. Jpn.)* (1994) 341.
241. H. Yoshida: "Arginine, Citrulline, and Ornithine," in K. Aida et al. (eds.): *Biotechnology of Amino Acid Production,* Kodansha, Tokyo/Elsevier, Amsterdam 1986, pp. 131 – 143.
242. Tanabe Seiyaku, JP-Kokai 58-9692, 1983 (I. Chibata, M. Kisumi, T. Tagaki).
243. I. Chibata, T. Tosa, T. Sato: "Aspartic Acid," in H. W. Blanch, S. Drew, D. I. C. Wang (eds.): *Comprehensive Biotechnology,* **vol. 3**, Pergamon Press, Oxford 1985.
244. Y. Yokote et al., *J. Solid-Phase Biochem.* **3** (1978) 247 – 261.
245. M. Terasawa, H. Yukawa, Y. Takayama, *Process Biochem.* **20** (1985) 124 – 128.
246. T. Sato et al., *Biotechnol. Bioeng.* **17** (1975) 1779 – 1804.
247. I. Chibata, T. Tetsuya, T. Sato, *Appl. Biochem. Biotechnol.* **13** (1986) 231 – 240.
248. M. C. Fusee, W. E. Swann, G. J. Calton, *Appl. Environ. Microbiol.* **42** (1981) 672 – 676.
249. Genex, EP-A 197 784, 1986 (W. E. Swann, A. C. Nolf).
250. J. Michelet, A. Deschamps, J. M. Lebeault, *3rd Eur. Congr. Biotechnol.,* **vol. 2,** Verlag Chemie, Weinheim 1984, pp. 133 – 138.
251. M. Mori, I. Shiio, *Agric. Biol. Chem.* **48** (1984) 1189 – 1197.
252. N. Nishimura, S. Komatsubara, M. Kisumi, *Appl. Environ. Microbiol.* **53** (1987) 2800 – 2803.
253. Maggi, DE 2 449 711, 1979 (P. Hirsbrunne, R. Bertholet).
254. Mitsubishi Chemical Ind., US 3974031, 1976 (H. Yamada, K. Gumagai, H. Ohkishi).
255. K. Sano, K. Mitsugi, *Agric. Biol. Chem.* **42** (1978) 2315 – 2321.
256. S. Kinoshita, S. Ukada, M. Shimono, *J. Gen. Appl. Microbiol.* **3** (1957) 139 – 205.
257. S. Kinoshita, K. Tanaka: "Glutamic acid," in K. Yamada, et al. (eds.): *The Microbial Production of Amino Acids,* John Wiley & Sons, New York 1972, pp. 263 – 324.
258. B. J. Eikmans, M. Kircher, D. J. Reinscheid, *FEMS Microbiol. Lett.* **82** (1991) 203 – 208.
259. W. Liebl, M. Ehrmann, W. Ludwig, K. H. Schleifer, *Int. J. Syst. Bacteriol.* **41** (1991) 225 – 235.
260. Y. Hirose, H. Enei, H. Shibai: "L-Glutamic acid fermentation," in H. W. Blanch, S. Drew, D. I. C. Wang (eds.): *Comprehensive Biotechnology,* **vol. 3,** Pergamon Press Ltd., Oxford 1985.
261. M. Kikuchi, Y. Nakao: "Glutamic acid," in K. Aida, I. Chibata, K. Nakayama, K. Takinami, H. Yamada (eds.): *Biotechnology of Amino Acid Production,* Kodansha, Tokyo/Elsevier, Amsterdam 1986, pp. 101 – 116.
262. R. Krämer, *BioEngineering* **9** (1993) 51 – 61.
263. R. Krämer, *FEMS Microbiol. Rev.* **13** (1994) 75 – 93.
264. I. Shiio, K. Ujigawa, *Agric. Biol. Chem.* **42** (1980) 1897 – 1904.
265. E. R. Börmann, B. J. Eikmanns, H. Sahm, *Molecular Microbiol.* **6** (1992) 317 – 326.
266. T. E. Walker et al., *J. Biol. Chem.* **257** (1982) 1189 – 1195.
267. H. Yoishii et al., *Nippon Nogei Kagaku Kaishi* **67** (1993) 949 – 954.
268. H. Yoishii, M. Yoshimura, S. Nakamori, S, Inoue, *Nippon Nogei Kagaku Kaishi* **67** (1993) 955 – 960.
269. Ajinomoto, JP 56 148 295, 1981 (T. Tsuchida, K. Miwa, S. Nakamori, H. Mimose).
270. M. Shirasuchi et al., *Biosci. Biotechnol. Biochem.* **59** (1995) 83 – 86.
271. T. Tsuchida et al., *Agric. Biol. Chem.* **51** (1987) 2089 – 2094.
272. Ajinomoto, US 5 164 307, 1992 (Y. Yoshihara, Y. Kawahara, Y. Yamada).
273. K. Araki: "Histidine," in K. Aida, I. Chibata, K. Nakayama, K. Takinami, H. Yamada (eds.): *Biotechnology of Amino Acid Production,* Kodansha, Tokyo/Elsevier, Amsterdam 1986, pp. 247 – 256.
274. M. Sugiura, S. Suzuki, M. Kisumi, *Agric. Biol. Chem.* **51** (1987) 371 – 377.

275. M. B. Vickery, *J. Biol. Chem.* **143** (1942) 77.
276. Kyowa Hakko, EP-A 759 472, 1996 (A. Ozaki, H. Mori, T. Shibasaki).
277. I. Eggeling, C. Cordes, L. Eggeling, H. Sahm, *Appl. Microbiol. Biotechnol.* **25** (1987) 346 – 351.
278. E. Scheer, L. Eggeling, H. Sahm, *Appl. Microbiol. Biotechnol.* **28** (1988) 474 – 477.
279. C. Wilhelm et al., *Appl. Microbiol. Biotechnol.* **31** (1989) 458 – 462.
280. Ajinomoto, Patent 5 164 307, 1992 (Y. Yoshihara, Y. Kawahara, Y. Yamada).
281. S. Komatsubara, M. Kisumi, I. Chibata, *J. Gen. Microbiol.* **119** (1980) 51 – 61.
282. Ajinomoto, EP-A 519 113, 1992 (V. A. Livshits et al.).
283. Kyowa Hakko, EP-A 595 163, 1994 (T. Nakano, T. Azuma, Y. Kuratsu).
284. Forsch. Z. Jülich, DE 44 00 926, 1995 (B. Möckel, L. Eggeling, H. Sahm).
285. Kyowa Hakko, US 5 362 637, 1994 (K. Kino, Y. Kuratsu).
286. Kyowa Hakko, EP-A 557 996, 1993 (K. Kino, K. Okamoto, Y. Takeda, Y. Kuratsu).
287. C. Keilhauer, L. Eggeling, H. Sahm, *J. Bacteriol.* **175** (1993) 5595 – 5603.
288. S. Morbach, H. Sahm, L. Eggeling, *Appl. Environ. Microbiol.* **61** (1995) 4315 – 4320.
289. Ajinomoto, EP-A 685 555, 1995 (K. Hashiguchi, H. Kishino, N. Tsujimoto, H. Matsui).
290. U. Groeger, H. Sahm, *Appl. Microbiol. Biotechnol.* **25** (1987) 352 – 236.
291. S. Komatsubara, M. Kisumi: "Histidine," in K. Aida, I. Chibata, K. Nakayama, K. Takinami, H. Yamada (eds.): *Biotechnology of Amio Acid Production,* Kodansha, Tokyo/Elsevier, Amsterdam 1986, pp. 233 – 246.
292. Ajinomoto, US 5 164 307, 1992 (Y. Yoshihara, Y. Kawahara, Y. Yamada).
293. M. Karasawa, O. Tosaka, S. Ikeda, H. Yoshi, *Agric. Biol. Chem.* **50** (1986) 339 – 346.
294. Kyowa Hakko, US 4 623 623, 1986 (T. Nakanashi et al.).
295. Kawahara et al., *Appl. Microbiol. Biotechnol.* **34** (1990) 87 – 90.
296. Ajinomoto, US 5 250 423, 1993 (Y. Murakami, H. Miwa, S. Nakamori).
297. A. Erdmann, B. Weil, R. Krämer, *Appl. Microbiol. Biotechnol.* **42** (1994) 604 – 610.
298. A. Erdmann, B. Weil, R. Krämer, *Biotechnol. Lett.* **17** (1995) 927 – 932.
299. R. Krämer, *J. Biotechnol.* **45** (1996) 1 – 21.
300. M. S. N. Jetten, A. J. Sinskey, *Crit. Rev. in Biol.* **15** (1995) 73 – 103.
301. Kyowa Hakko, EP-A 197 335, 1986 (R. Katsumata, T. Mitzukami, T. Oka).
302. Forsch. Z. Jülich, EP-A 435 132, 1991 (J. Cremer, L. Eggeling, H. Sahm).
303. Cheil Sugar, FR 2 645 172, 1990 (J. W. Oh, S. J. Kim, Y. J. Cho, N. H. Park, L. H. Lee).
304. Y.-C. Liu, W.-T. Wu, J.-H. Tsao, *Bioprocess. Eng.* **9** (1993) 135 – 139.
305. Eurolysine, EP-A 534 865, 1993 (P. Lucq, C. Domont).
306. Rhone-Polenc, EP 122 163, 1984 (N. Rouy).
307. Degussa, EP 533 039, 1993 (W. Binder et al.).
308. Degussa, DE 2 421 167, 1974 (T. Lüßling, K. Müller, G. Schreyer, F. Theissen).
309. W. Leuchtenberger, U. Plöcker: "Amino acids and hydroxycarboxylic acids," in W. Gerhartz (ed.): *Enzymes in Industry, Production and Applications,* VCH, Weinheim 1990.
310. T. Ishikawa et al., *Biosci. Biotech. Biochem.* **57** (1993) 982 – 986.
311. Genemcor, WO 93/17112, 1993 (J. C. Lievense).
312. Genex, US 4 598 047, 1986 (J. C. McGuire).
313. C. T. Evans, C. Coma, W. Petreson, M. Misawa, *Biotechnol. Bioeng.* **30** (1987) 1067 – 1072.
314. G. J. Calton et al., *Biotechnology* **4** (1986) 317 – 320.
315. C. T. Evans et al., *Biotechnology* **5** (1987) 818 – 923.
316. A. J. Pittard, *Amer. Soc. Microbiol. Conf. Proc.* (1987) 368 – 394.
317. I. Shiio: "Tryptophan, phenylalanine, and tyrosine," in K. Aida, I. Chibata, K. Nakayama, K. Takinami, H. Yamada (eds.): *Biotechnology of Amino Acid Production,* Kodansha, Tokyo/Elsevier, Amsterdam 1986, pp. 188 – 206.
318. S. O. Hwang et al., *Appl. Microbiol. Biotechnol.* **22** (1985) 108 – 113.
319. H. Ito et al., *Agric. Biol. Chem.* **54** (1990) 707 – 713.
320. B. K. Konstantinov, N. Nishio, T. Yoshida: "Glucose feeding strategy accounting for the decreasing oxidative capacity of recombinant Escherichia coli in fed-batch cultivation for phenylalanine production," *J. Ferment. Bioeng.* **70** (1990) 253.
321. Ajinomoto, US 4 584 400, 1986 (M. Ootani, S. Sano, I. Kusumotu).
322. A. M. Vasconcellos, A. L. Neto, D. M. Grassiano, C. P. De Oliveira, *Biotechnol. Bioeng.* **33** (1989) 1324 – 1329.
323. Ajinomoto, US 5 362 635, 1994 (T. Hirose et al.).

324. F. Yoshinaga: "Proline," in K. Aida, I. Chibata, K. Nakayama, K. Takinami, H. Yamada (eds.): *Biotechnology of Amino Acid Production,* Kodansha, Tokyo/Elsevier, Amsterdam 1986, pp. 117 – 120.
325. Sci. Res. Technol Inst. Amino Acids, USSR, DE-OS 3 612 077, 1986 (Š. M. Kocarian et al.).
326. M. Sugiura, T. Takagi, M. Kisumi, Abst. 31st Symp. Amino Acid and Nucleic Acid (Japan), 1982, p. 10.
327. Y. Imai, T. Takagi, M. Sugiura, M. Kisumi, Abst. Ann. Meet Agr. Chem. Soc. (Japan), 1984, p. 100.
328. Y. Izumi et al., *Appl. Microbiol. Biotechnol.* **39** (1993) 427 – 432.
329. T. Yoshida, T. Mitsunaga, Y. Izumi, *J. Fermentation Bioeng.* **75** (1993) 405 – 408.
330. H.-Y. Hsiao, T. Wie, *Biotechnol. Bioeng.* **28** (1986) 1510 – 1518.
331. Mitsui Toatsu, US 5 382 517, 1995 (D. Ura, T. Hashimukai, T. Matsumoto, N. Fukuhara).
332. K. Shimura: "Threonine," in H. W. Blanch, S. Drew, D. I. C. Wang (eds.): *Comprehensive Biotechnology,* **vol. 3,** Pergamon Press, Oxford 1985.
333. S. Nakamori: "Threonine and Homoserine," in K. Aida, I. Chibata, K. Nakayama, K. Takinami, H. Yamada (eds.): *Progress in Industrial Microbiology,* Kodansha, Tokyo/Elsevier, Amsterdam 1986, pp. 173 – 182.
334. Genetics Ind., US 4 278 765, 1981 (V. G. Debabov et al.).
335. Ajinomoto, EP-A 593 792, 1994 (V. G. Debabov et al.).
336. Eurolysine, FR 2 627 508, 1989 (F. Richaud et al.).
337. Ajinomoto, FR 2 588 016, 1987 (M. Ootani, T. Kitahara, K. Akashi).
338. A. Yokota, T. Takao, *Agric. Biol. Chem.* **48** (1984) 2663 – 2668.
339. Mitsui Toatsu, EP-A 341 674, 1989 (K. Ishiwata et al.).
340. Mitsui Toatsu, EP-A 438 591, 1991 (S. Ogawa, S. Iguichi, S. Morita, H. Kuwamoto).
341. B. K. Hamilton et al., *Trends Biotechnol.* **3** (1985) 64 – 68.
342. J. Plachy, S. Ulbert, *Acta Biotechnol.* **10** (1990) 517 – 522.
343. Showa Denko (1990) JP-Kokai 2 190 182.
344. Showa Denko, JP 62 186 786, 1987 (E. Takinishi, H. Takamatsu, K. Sakimoto, Y. Yajima).
345. A. N. Mayeno, G. J. Gleich, *TIBTECH* **12** (1994) 346 – 352.
346. K. Sakimoto, Y. Torigoe, *Curr. Prospects Med. Drug Saf.* (1994) 295 – 311.
347. K. Nakayama: "Tryptophan," in M. Moo-Young (ed.): *Comprehensive Biotechnology,* **vol. 3,** Pergamon Press, Oxford 1985, 621 – 631.
348. I. Shiio: "Tryptophan, Phenylalanine, and Tyrosine," in K. Aida, I. Chibata, K. Nakayama, K. Takinami, H. Yamada (eds.): *Biotechnology of Amino Acid Production,* Kodansha, Tokyo/Elsevier, Amsterdam 1986, pp. 188 – 206.
349. T. K. Maiti, S. P. Chatterje, *Hindustan Antibiotics Bulletin* **33** (1991) 26 – 61.
350. M. Ikeda, K. Nakanishi, K. Kino, R. Katsumata, *Biosci. Biotech. Biochem.* **58** (1994) 674 – 678.
351. J. L. Cuq, M. Gilot, *Sci. Aliments* **5** (1985) 687 – 697.
352. S. Komatsubara, M. Kisumi: "Isoleucine, Valine, and Leucine," in K. Aida, I. Chibata, K. Nakayama, K. Takinami, H. Yamada (eds): *Biotechnology of Amino Acid Production,* Kodansha, Tokyo/Elsevier, Amsterdam 1986, pp. 131 – 143.
353. Ajinomoto, EP 287 123, 1993 (N. Katsurada, H. Uchibori, T. Tsuchida).
354. All Union Sci.-Res. Inst. Genetics and Microorg. Gen. + Selec Univ. Odessa, WO 90/04636, 1990 (V. G. Debabov et al.).
355. University Odessa, SU 1 730 152, 1992 (I. I. Brown et al.).
356. H.-L. Bertram, H. Schmidtborn, *Feed International* 1984 (May) 37.
357. M. Kirchgeßner: *Tierernährung,* 5th ed., DLG-Verlag, Frankfurt 1982, p. 88.
358. M. L. Scott, M. C. Nesheim, R. J. Young: *Nutrition of the Chicken,* 3rd ed., M. L. Scott, Ithaka, New York, 1982.
359. *D,L-Methionine, The Amino Acid for Animal Nutrition,* Degussa, Frankfurt 1980.
360. *Nutrient Requirements of Poultry,* Nat. Res. Council, no. 1, 7th ed., Washington, D.C., 1977.
361. AEC (Alimentation Equilibrée de Commentry): *Alimentation Animale,* Doc. 4, Commentry, France, 1978.
362. *Nutrient Requirements of Swine,* Nat. Res. Council, 1979. Agricultural Research Council, Commonwealth Agricultural, Burlaux, London 1981. P. L. M. Berende, H.-L. Bertram, *Z. Tierphysiol. Tierernähr. Futtermittelk.* **49** (1983) 30. H.-L. Bertram, P. L. M. Berende, *Kraftfutter* (1983) 46.

363. W. Prinz, A. Becker, *Arch. Geflügelk.* **29** (1965) no. 2, 135.
364. R. Chernoff, *J. Am. Diet. Assoc.* **79** (1981) 426. M. R. Polk, *Am. Pharm.* **NS 22** (1982) 25.
365. M. Winitz et al., *Nature (London)* **205** (1965) 741.
366. R. J. Russell, *Gut* **16** (1975) 68. W. F. Caspary, *Dtsch. Ärztebl.* 1978 (2. Febr.) 243. H. Kasper, *Aktuel. Ernährung* **1** (1978) 22.
367. D. M. Matthews, S. A. Adibi, *Gastroenterology* **71** (1976) 151. M. T. Lis, R. F. Crampton, D. M. Matthews, *Br. J. Nutr.* **27** (1972) 159.
368. Mead Johnson, US 3091569, 1963; US 3184505, 1965.
369. H. A. Smith, G. Gorin, *J. Org. Chem.* **26** (1961) 820.
370. Rech. et Propagande Scientif., FR 1288907, 1962.
371. Degussa, US 4129593, 1978.
372. M. Bergmann, G. Michalis, *Ber. Dtsch. Chem. Ges.* **63** (1930) 987.
373. Degussa, DE 1239697, 1963.
374. May & Baker, GB 845 034, 1957.
375. B. Witkop et al., *J. Am. Chem. Soc.* **76** (1954) 5579.
376. A. J. Morris et al., *J. Org. Chem.* **22** (1957) 306.
377. E. A. Bell et al., *Nature (London)* **210** (1966) 529.
378. Distillers, GB 854339, 1957. Squibb, US 3281461, 1966. Heyl & Co., DE 2114329, 1971; DE 2413185, 1974.
379. Degussa, DE 1795299, 1968; DE 1795297, 1968; DE 2032952, 1970; DE 2123232, 1971; DE 2156601, 1971; DE 2335990, 1973; DE 2138122, 1971; DE 2258411, 1972; DE 2304055, 1973.
380. K. Schrader, *Am. Cosmet. Perfum.* **87** (1972) 49. S. Tatsumi, *Am. Cosmet. Perfum.* **87** (1972) 61. O. K. Jacobi, *Am. Cosmet. Perfum.* **87** (1972) 35. G. Hopf, J. König, G. Padberg, *Kosmetologie* **4** (1971) 132. A. Szakall, *Arch. Klin. Exp. Dermatol.* **201** (1955) 331. K. Laden, R. Spitzer, *J. Soc. Cosmet. Chem.* **18** (1967) 351. H. W. Spier, G. Pascher, *Klin. Wochenschr.* **31** (1953) 997.
381. Y. Kumano et al., *J. Soc. Cosmet. Chem.* **28** (1977) 285.
382. L'Oreál, DE 2807607, 1978 (J.-C. Ser et al.). Unilever, DE 2337342, 1973 (G. F. Johnston et al.). Orlane Paris, DE 2524297, 1975 (A. Meybeck, H. Noel). Shiseido, US 4035513, 1977 (Y. Kumano).
383. E. S. Cooperman, *Am. Cosmet. Perfum.* **87** (1972) 65. Colgate Palmolive, GB 1573529, 1980.
384. M. Takehara et al., *J. Am. Oil Chem. Soc.* **50** (1973) 227; **51** (1974) 419.
385. Ajinomoto, Kawaken Fine Chemicals, US 4273684, 1981 (T. Nagashima et al.)
386. Ajinomoto, GB 1483500, 1977.
387. Kawaken Fine Chemicals. JP-Kokai 75117806, 1975 (K. Nakazawa); *Chem. Abstr.* **84** (1976) 8858 r.
388. G. Schuster, H. Modde, E. Scheld, *Seifen Öle Fette Wachse* **38** (1965) 477. G. Schuster, H. Modde, *Parfüm + Kosmet.* **45** (1964) 337.
389. Estee Lauder, US 4005210, 1977 (J. Gubernick).
390. American Cyanamid, US 3988438, 1976. Ajinomoto, DE 2010303, 1970 (R. Yoshida et al.).
391. R. Yoshida, M. Takehara, *Chem. Abstr.* **84** (1976) 6843 h.
392. Dominion Pharmacal., US 4176197, 1979 (B. N. Olson). L'Oréal, US 4002634, 1977 (G. Kalopissis, C. Bouillon). L'Oréal, GB 1397623, 1975 (G. Kalopissis, G. Manoussos). L'Oréal, DE 1492071, 1965 (G. Kalopissis).
393. Mare Corp., US 4201235, 1980 (V. G. Ciavatta). Unilever, EP 8171, 1981 (G. P. Mathur et al.).
394. Kyowa Hakko Kogyo, US 4139610, 1979 (Y. Miyazaki et al.) Hans Schwarzkopf, DE 958501, 1957 (J. Saphir, E. Kramer). K. Yoneda et al., DE 2951923, 1979.
395. Schuylkill Chem., GB 1516890, 1978.
396. Procter & Gamble, EP 47650, 1982. Unilever, GB 1597498, 1981 (K. Gosling, M. R. Hyde).
397. C. Tomlin (ed.): *The Pesticide Manual,* 10th ed., Crop Protection Publications, Farnham 1994.
398. B. Hock, C. Fedtke, R. R. Schmidt: *Herbizide,* 1st ed., Georg Thieme Verlag Stuttgart 1995.
399. *Proc. North. Cent. Weed Control Conf.* **26** (1971) 64.
400. P. Knuuttila, H. Knuuttila, *Acta Chem. Scand.* **33** (1979) 623.
401. *Agrochemical Service,* Wood Mackenzie Consultans Limited, Edinburgh 1996.
402. Zeneca, US 4 315 765.
403. *Z. Pflanzenkr. Pflanzenschutz* (1981) Sonderheft IX, 431.
404. Hoechst, DE-OS 2 717 440, 1976.
405. K. Tachibana et al., *Abstr. 5th Int. Congr. Pestic. Chem.,* IVa Abstract 19.
406. D. W. Ladner, *Pestic. Sci.* **29** (1990) 341 – 356.
407. M. J. Muhitch, D. L. Shaner, M. A. Stidham, *Plant Physiol.* **83** (1987) 451 – 456.

408. T. Chapman et al., *Symp. New. Herbic. 3rd ed.* (1969) 40.
409. R. M. Scott et al., *Proc. Br. Crop. Prot. Conf. – Weeds* **2** (1976) 723.
410. S. O. Duke et al., *Pestic. Sci.* **40** (1994) 265 – 277.
411. N. Mito, R. Sato, M. Miyakado, H. Oshio, S. Tanaka, *Pestic. Biochem. Physiol.* **40** (1991) 128 – 135.
412. Degussa, WO 94/03458, 1994 (M. Schäfer, H. Baier, K. Drauz, H.-P. Krimmer, S. Landmann).
413. Sumitomo, JP 62 158 280.
414. BASF, DE 409 432, 1994.
415. BASF, DE 407 023, 1994.
416. American Cyanamid Co., EP 101 098, 1993.
417. Bayer, DE 030 062, 1990.
418. P. A. Urech, *Proc. Br. Crop. Prot. Conf. – Pests Dis.* **2** (1977) 623.
419. Garavaglia et al.: *Atti Simp. Chim. Antiparassitari, 3rd, Piacenza* 1981.
420. C. A. Henrick et al., *Pestic. Sci.* **11** (1980) 224.
421. C. A. Porter, L. E. Ahlrichs, *Hawaii Sugar Tech. Rep. 1971* **30** (1972) 71.
422. K. S. K. Prasad, K. G. H. Setty, H. C. Govindu, *Indian J. Plant Prot.* **5** (1977) 153.
423. Roussel-Uclaf, FR 2405650, 1979 (A. Boudet).
424. T. Kato, A. Iio, JP-Kokai 7629229, 1976; *Chem. Abstr.* **85** (1976) 88519 q.
425. A. Bouniols, J. Margara, *C. R. Hebd. Séances Acad. Sci. Ser. D* **273** (1971) 1193.
426. Ciba Geigy, EP 18948, 1980.
427. Emori Shoji, JP 78097022, 1978.
428. Dainichi Seika Kogy, JP 78079990, 1978.
429. L. Mueller, *Zem. Kalk Gips* **27** (1974) 69.
430. Ajinomoto, DE 2533136, 1975 (R. Yoshida et al.).
431. Kyowa Fermentation, GB 1400741, 1975. Honny Chemicals, GB 1402758, 1975 (Y. Nakagoshi). Kyowa Hakko Kogyo, DE 2229488, 1972 (Y. Fujimoto et al.). Toyo Cloth, US 3676206, 1972 (K. Nishitani et al.).
432. Sanyo Trading Co., DE 2716758, 1977 (M. Onizawa, S. Ohmiya).
433. Ajinomoto, JP 81035179, 1981. Ajinomoto, JP-Kokai 7522801, 1975 (T. Saito et al.); *Chem. Abstr.* **84** (1976) 33494 b.
434. Sanyo Trading Co., DE 2602988, 1976; US 4069213, 1978; DE 2604053, 1976 (M. Onizawa); DE 2658693, 1976 (M. Onizawa, S. Ohmiya).
435. R. M. Saleh, A. M. Shams El Din, *Corros. Sci.* **12** (1972) 688.
436. Ajinomoto, JP-Kokai 7426145, 1974 (Y. Kita et al.); *Chem. Abstr.* **81** (1974) 53195 w.
437. Ilford, GB 1378354, 1974 (A. D. Ezekiel). Ilford, DE 2316632, 1973 (R. Jefferson).
438. Fr. Blasberg & Co., DE 2050870, 1973 (W. Immel, W. Adams). Yokozawa Chemical Ind., JP 7410572, 1974 (K. Aoya et al.); *Chem. Abstr.* **81** (1974) 57631 h. Sony Corp., DE 2325109, 1973 (S. Fueki et al.).
439. A. Steponavicius, R. Visomirskis, *Electrodepos. Surface Treat. Lausanne* **1** (1972) 37.
440. F. S. Parker, D. M. Kirschenbaum, *Spectrochim. Acta* **16** (1960) 910. M. Tsuboi, T. Takenishi, A. Nakamura, *Spectrochim. Acta* **19** (1963) 271. J. F. Pearson, M. A. Slifkin, *Spectrochim. Acta Part A* **28** (1972) 2403.
441. C. S. Tsai et al., *Can. J. Biochem.* **53** (1975) no. 9, 1005.
442. H. Hagenmeyer et al., *Z. Naturforsch. B: Anorg. Chem. Org. Chem.* **25 B** (1970), 681.
443. E. Scoffone, A. Fontana, *Mol. Biol. Biochem. Biophys.* **8** (1970) 185.
444. M. Roth, *Anal. Chem.* **43** (1971) 880. J. R. Benson, P. E. Hare, *Proc. Natl. Acad. Sci. USA* **72** (1975) 619.
445. E. Bayer et al., *Anal. Chem.* **48** (1976) 1106.
446. S. Udenfried et al., *Science (Washington, D.C.)* **178** (1972) 871.
447. C. Haworth, R. W. A. Oliver, *J. Chromatogr.* **64** (1972) 305. E. Stahl: *Dünnschichtchromatographie,* Springer Verlag, Berlin1967, p. 701. A. R. Fahmy et al., *Helv. Chim. Acta* **44** (1959) 245.
448. M. Kubota et al., *Anal. Biochem.* **64** (1975) no. 2, 494. K. D. Kulbe, *Anal. Biochem.* **44** (1970) 548. P. A. Laursen, *Biochem. Biophys. Res. Commun.* **37** (1969) 663.
449. A. Wolf, *Prax. Naturwiss. Chem.* **23** (1974) no. 3, 74.
450. H. Nakamura, J. J. Pisano, *J. Chromatogr.* **121** (1976) 33.
451. K. Macek, Z. Deyl, M. Smrž, *J. Chromatogr.* **193** (1980) 421.
452. P. B. Hamilton, *Anal. Chem.* **35** (1963) 2055.
453. S. Blackburn: *Amino Acid Determination,* Marcel Dekker, New York1978.
454. D. H. Spackman, W. H. Stein, S. Moore, *Anal. Chem.* **30** (1958) 1190.
455. S. Moore, W. H. Stein, *J. Biol. Chem.* **211** (1954) 907.
456. H.-M. Lee et al., *Anal. Biochem.* **96** (1979) 298.
457. E. Bayer et al., *Anal. Chem.* **48** (1976) 1106. J. M. Wilkinson, *J. Chromatogr. Sci.* **16** (1978) 547. K.-T. Hsu, B. L. Currie, *J. Chromatogr.* **166** (1978) 555. T. Jamabe, N. Takei,

H. Nakamura, *J. Chromatogr.* **104** (1975) 359. A. Khayat, P. K. Redenz, L. A. Gorman, *Food Technol. (Chicago)* **36** (1982) 46.
458. A. P. Graffeo, A. Haag, B. L. Karger, *Anal. Lett.* **6** (1973) no. 6, 505. J. K. De Vries, R. Frank, C. Birr, *FEBS Lett.* **55** (1975) no. 1, 65. P. Frankhauser et al., *Helv. Chim. Acta* **57** (1974) 271. C. C. Zimmermann, E. Appella, J. J. Pisano, *Anal. Biochem.* **77** (1977) 569. G. Frank, W. Strubert, *Chromatographia* **6** (1973) no. 12, 522.
459. H. Beyer, U. Schenk, *J. Chromatogr.* **89** (1969) 483.
460. T. A. Kan, W. F. Shipe, *J. Food Sci.* **47** (1981) 338. H. Umagat, P. Kucera, L. F. Wen, *J. Chromatogr.* **239** (1982) 463.
461. W. Voelter, K. Zech, *J. Chromatogr.* **112** (1975) 643.
462. R. Schuster, *Anal. Chem.* **52** (1980) 617.
463. A. Darbre, *Biochem. Soc. Trans.* **2** (1974) 70. B. M. Nair, *J. Agric. Food Chem.* **25** (1977) 614.
464. J. J. Pisano, T. J. Bronzert, H. B. Brewer, *Anal. Biochem.* **45** (1972) 43.
465. A. Darbre, A. Islam, *Biochem. J.* **106** (1968) 923.
466. C. W. Gehrke, R. W. Zumwalt, K. Kuo, *J. Agric. Food Chem.* **19** (1971) 605.
467. K. Ruhlmann, W. Giesecke, *Angew. Chem.* **73** (1961) 113.
468. D. L. Stalling, C. W. Gehrke, R. W. Zumwalt, *Biochem. Biophys. Res. Commun.* **31** (1968) 4.
469. W. Grassmann, K. Hannig, *Houben-Weyl,* **1/1,** 708.
470. I. Abe, S. Musha, *J. Chromatogr.* **200** (1980) 195. G. J. Nicholson, H. Frank, E. Bayer, *HRC CC J. High Resolut. Chromatogr. Chromatogr. Commun.* **2** (1979) 411. W. A. König, G. J. Nicholson, *Anal. Chem.* **47** (1975) 951.
471. W. Lindner, *Chimia* **35** (1981) 294. V. A. Davankov et al., *Chromatographia* **13** (1980) no. 11, 677.
472. P. E. Hare, E. Gil-Av, *Science (Washington, D.C.)* **204** (1979) 1226.
473. L. M. Henderson, E. E. Snell, *J. Biol. Chem.* **172** (1947) 15. I. Grote, *Mühle Mischfuttertech.* **116** (1979) 465.
474. H. Itoh, T. Morimoto, I. Chibata, *Anal. Biochem.* **60** (1974) 573.
475. Ch. Calvot et al., *FEBS Lett.* **59** (1975) no. 2, 258. S. J. Updike, G. P. Hicks, *Nature (London)* **214** (1967) 986.
476. Ajinomoto, FR 2421380, 1979.
477. *Europäische Pharmacopoe,* **vol. 1 – 3,** Deutscher Apotheker-Verlag, Stuttgart, Govi-Verlag, Frankfurt 1974, vol. 1, p. 104.
478. W. Seaman, E. Allen, *Anal. Chem.* **23** (1951) no. 4, 592. H. P. Deppeler, G. Witthans, *Fresenius Z. Anal. Chem.* **305** (1981) 273.
479. *United States Pharmacopeia* USP XX, Convention, Rockville, Md., 1979.
480. *Food Chemicals Codex,* 3rd ed., National Academy Press, Washington, D.C., 1981.
481. T. Akashi: "Amino Acid Production and Use to Improve Nutrition of Foods and Feeds," *Chemrawn II Conference* Manila 1982.
482. *Chem. Eng. News* **61** (1983) Jan. 3, 18.
483. H. N. Munro, *Adv. Exp. Med. Biol.* **105** (1978) 119.
484. A. E. Harper, N. J. Benevenga, R. M. Wohlhueter, *Physiol. Rev.* **50** (1970) 428.
485. R. G. Daniel, H. A. Waisman, *Growth* **32** (1968) 255.
486. H. E. Sauberlich, *J. Nutr.* **75** (1961) 61.
487. K. Lang: *Biochemie der Ernährung,* 4th ed., Steinkopff, Darmstadt 1979.
488. Y. Peng et al., *J. Nutr.* **103** (1973) 608.
489. Degussa, unpublished, 1973 and 1983. R. J. Breglia, C. O. Ward, C. I. Jarowski, *J. Pharm. Sci.* **62** (1973) 49. P. Gullino et al., *Arch. Biochem. Biophys.* **58** (1955) 253. O. Strubelt, C.-P. Siegers, A. Schütt, *Arch. Toxicol.* **33** (1974) 55. W. Braun, *Strahlentherapie* **108** (1959) 262. H. Gutbrod et al., *Acta Hepatol.* **5** (1957) no. 1/2, 1. I. Petersone et al., *Eksp. Klin. Farmakoter.* **3** (1972) 5. D. G. Gallo, A. L. Sheffner, DE 2018599, 1971. Transbronchin, Homburg Pharma, Frankfurt 1971. L. Bonanomi, A. Gazzaniga, *Therapiewoche* **30** (1980) 1926. W. F. Riker, H. Gold, *J. Am. Pharm. Assoc.* **31** (1942) 306.
490. G. Maffii, G. Schott, M. G. Serralunga, *Res. Prog. Org. Biol. Med. Chem.* **2** (1970) 262.
491. Y. Kawaguchi et al., *Iyakuhin Kenkyu* **11** (1980) 635. Kaken Chemical, JP 52083940, 1976 (S. Suzuki). P. Gullino et al., *Arch. Biochem. Biophys.* **64** (1956) 319. Degussa, unpublished, 1981 and 1982. E.-J. Kirnberger et al., *Arzneim. Forsch.* **8** (1958) 72.

Cancer Chemotherapy

Bernhard Kutscher, ASTA Medica AG, Frankfurt am Main, Federal Republic of Germany

Gregory A. Curt, National Cancer Institute, Bethesda, Maryland 20205, United States

Carmen J. Allegra, National Cancer Institute, Bethesda, Maryland 20205, United States

Robert L. Fine, National Cancer Institute, Bethesda, Maryland 20205, United States

Hamza Mujagic, National Cancer Institute, Bethesda, Maryland 20205, United States

Grace Chao Yeh, National Cancer Institute, Bethesda, Maryland 20205, United States

Bruce A. Chabner, National Cancer Institute, Bethesda, Maryland 20205, United States

Ullmann's Industrial Toxicology

ISBN: 3-527-31247-1

1. Introduction

Malignant tumors represent one of the most common human diseases worldwide. Based on an estimation made in the United States, cancer will become the leading cause of death in the year 2000 [1].

Unfortunately, the subset of human cancer types that are amenable to curative treatment still is rather small. Although there is a tremendous progress in understanding the molecular events that lead to malignancy and many agents are known that effectively kill cancer cells, progress in development of clinically innovative drugs that can cure humans is slow [2], [3].

The heterogeneity of malignant tumors with respect to their genetics, biology, and biochemistry as well as primary or treatment-induced resistance to therapy hamper curative treatment [4], [5].

Searching for antineoplastic agents with improved selectivity to malignant cells remains the central task for drug discovery and development [5], [6].

According to a survey published in 1997 more then 315 drugs are under development in the United States for the treatment of cancer. This figure includes 42 drugs for treatment of lung cancer, 58 for breast cancer [8], [9], 60 for treatment of skin tumors (60), 36 for prostate cancer, and 35 for colon cancer.

In 1997 more than 1500 Americans are expected to die of cancer each day and more than a million new cases will be diagnosed with overall medical costs of $\$ 35 \times 10^9$. The total disease costs are estimated to sum up to more than $\$ 100 \times 10^9$ per year in treatment expenses and lost wages.

Increase in the incidence of cancer, mostly as a result of an aging population, is the driving growth in the marketplace. However, new medicines like hormone-analogs have lead to decrease in severity of the side effects of cancer therapy and have spurred wider use. According to a review [11] of ca. 90 approved anticancer drugs, more than 60 % are of natural origin or modeled on natural products parents. Cancer chemotherapeutics can be grouped according to their pharmacological and mechanistic profiles into

Antimetabolites
Alkylating agents
DNA-intercalating (agents) antibiotics
Mitose inhibitors
Signaltransduction-inhibitors

Two primary events in cell proliferation are DNA replication and cell division [12], [13]. The cell cycle has been divided into four sequential phases (G_0, G_1, S, M) and the cytostatics vary in the way they interfere with the cell cycle. Phasespecific agents/drugs, e.g., those interacting in G_1, S- or M-phase, are the mitose-inhibitors vincristine or vinblastine or the antimetabolites methotrexate or cytarabine, attacking in the S-phase. Alkylating drugs such as cyclophosphamide, cisplatin, or carboplatin inhibit and damage the cell in all phases and are thus phase unspecific. In general cells in the resting phase (G_0) are insensitive.

Approximately 70 % of patients diagnosed as having cancer have metastatic disease, i.e., disease that has spread beyond the primary site at the time of diagnosis [15]. However, the steady progress made in the treatment of cancer with drugs has contributed to the curing of an increasing proportion of patients with metastatic disease. The greatest change and improvements in cancer treatment have occurred because of the discovery and clinical development of drugs and the demonstration that metastatic cancer can be cured by these agents. Drugs, such as cytoxan, adriamycin, vincristine, *cis*-platinum, and bleomycin, all developed since 1960, are now regularly used by physicians to treat patients who would have been considered incurable a short time ago.

This article details the pharmacology and clinical use of the major classes of anticancer agents, including (1) antimetabolites, (2) alkylating agents, (3) anthracyclines and analogs, (4) other antitumor antibiotics, (5) antitubulin agents, (6) platinum complexes, (7) antihormones, and (8) signaltransduction-inhibitors. In each case, special emphasis is given to the progress made in developing clinically useful drugs and analogs that retain antitumor activity while decreasing host toxicity.

2. Antimetabolites

2.1. Methotrexate

Folic acid analogs comprise a class of antineoplastic agents of which methotrexate has gained the most widespread clinical use. These agents were the first to produce impressive remissions in acute leukemia [16] and cures in choriocarcinoma in women [17].

Reduced folates (tetrahydrofolates) are the biologically active form of folates required as cosubstrates in one-carbon transfer reactions. Included in these reactions are several important enzymatic steps in the de novo synthesis of purines and pyrimidines.

Methotrexate (NSC-740) [*59-05-2*], *N*-(4-{(2,4-diamino-6-pteridinyl)methyl]methylamino}benzoyl)-L-glutamic acid, $C_{20}H_{22}N_8O_5$, M_r 454.46, is a 2,4-diamino, N^{10}-methyl analog of folic acid that is capable of inhibiting certain folate-requiring reactions.

H2N N N CH3 N N N H N COOH NH2 O COOH

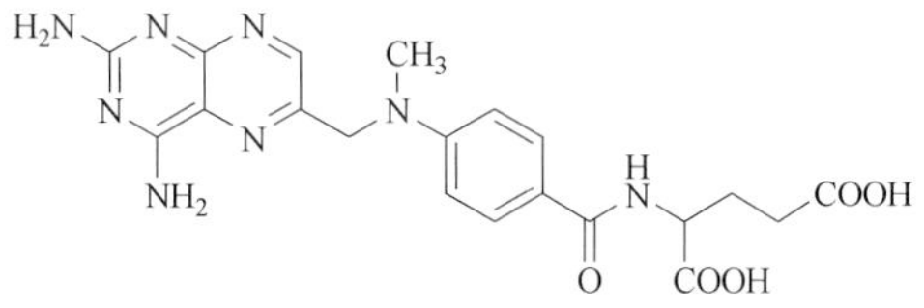

Methotrexate

Most importantly, methotrexate can inhibit dihydrofolate reductase (DHFR, $K_i = 10^{-11}$ M), a key enzyme for the maintenance of biologically active intracellular reduced folate pools. The folate-requiring reactions utilize reduced folates, and all reactions except that catalyzed by thymidylate synthase maintain folates in a reduced state during carbon transfer. Thymidylate synthase, which catalyzes the methylation of deoxyuridylate to thymidylate (required for DNA synthesis), requires the transfer of a carbon group from the folate cofactor N^{5-10}-methylene tetrahydrofolate with resultant oxidation of the folate to dihydrofolic acid. Oxidized folates must be reduced to the tetrahydro form by DHFR to be useful for intracellular metabolism. Inhibition of DHFR following methotrexate exposure ultimately leads to depletion of intracellular reduced folates. Cessation of first thymidylate and then purine nucleotide synthesis occurs as an indirect effect of methotrexate on reduced-folate levels. Methotrexate metabolites (polyglutamates) may also have direct inhibitory effects on folate-requiring enzymes, e.g., thymidylate synthase [18], and these effects may be important in inducing cytotoxicity.

2.1.1. Mechanism of Action and Mechanisms of Resistance

Transport. At concentrations less than 10 μM, methotrexate and reduced folates enter cells via an energy-dependent, temperature-sensitive carrier mechanism [19]. The affinity for this carrier has been variously reported to fall between 1 and 10 μM for tumor cell lines [19–21] and to be 87 μM for normal intestinal epithelial cells [22]. These differences in efficiency of transport may account for some of the selectivity of methotrexate for neoplastic cells. In addition to this carrier-mediated transport system, there exists a second, low-affinity transport mechanism that is poorly understood but appears to play a role in the transport of drug when concentrations exceed 20 μM [21], [23]. Methotrexate and reduced folates do not compete for uptake by this process, and it may represent a means for drug entry in cells resistant to low doses of drug by virtue of defective transport by the high-affinity mechanism. In some models, the sensitivity of a cell to methotrexate can be directly correlated with efficiency of drug transport, i.e., sensitive cells have a greater capacity for drug transport and longer intracellular retention of drug when compared to methotrexate-resistant cells [24]. Because decreased membrane transport may play a role in clinical drug resistance, a number of analogs with high lipid solubility have been developed that can circumvent a transport deficit. Methotrexate esters, diaminopyrimidines, and triazenates have been synthesized and used with success against experimental cell lines with defective methotrexate transport.

Intracellular Metabolism. Once inside the cell, naturally occurring folates may be metabolized to *polyglutamates;* that is, additional glutamyl moieties are added to the terminal glutamate present on the parent compound. This process allows for selective intracellular retention of the polyglutamated forms and an increased affinity for certain folate-requiring enzymes, such as thymidylate synthase, and for enzymes required for the de novo production of purine nucleotides [25]. Like the naturally occurring folates, intracellular methotrexate is also polyglutamated with the addition of from one to four additional glutamyl residues [26]. This process has been demonstrated in a variety of tissues, including human breast cancer cell lines [26], normal human liver [27], and murine leukemia cells [28]. The polyglutamates of methotrexate are selectively retained by the cells and appear to have an enhanced inhibitory potential for certain enzymes [18]. The inhibitory capacity of methotrexate polyglutamates for dihydrofolate reductase appears to be somewhat greater than that of the parent drug. The selective retention of methotrexate polyglutamates may be critical for the delayed cytotoxicity exhibited by cells capable of polyglutamate synthesis. In vitro experiments using MCF-7 human breast cancer cells have demonstrated that the intracellular retention and duration of binding to dihydrofolate reductase are directly related to the length of the polyglutamate tail [29].

Interaction with Dihydrofolate Reductase. The binding of methotrexate to dihydrofolate reductase has been extensively investigated by X-ray crystallographic and amino acid sequencing studies [30–34]. Methotrexate binds in a stoichiometric fashion to a hydrophobic pocket in the target enzyme DHFR [35]. The binding affinity depends on multiple factors, including pH, salt concentration, and NADPH concentration, and has been reported to be ca. 10 pM [36]. In the cell, methotrexate is a reversible inhibitor capable of being displaced by high concentrations of substrate. Thus, free intracellular drug in excess of the cellular dihydrofolate reductase binding capacity is required to maintain complete inhibition of the enzyme and thereby produce and maintain a state of reduced-folate depletion. If an excess of intracellular drug is not maintained, the intracellular reduced-folate pool recovers through enzymatic reduction of oxidized folates and cellular metabolism resumes.

Cellular resistance to methotrexate has been most commonly associated with an increase in dihydrofolate reductase activity. In general, the amplified enzyme is identical to the native protein in its affinity for methotrexate; however, altered methotrexate affinity has been reported to correlate with sensitivity to methotrexate in a series of murine leukemias [36]. Increased reductase activity and resistance to methotrexate has also been demonstrated in a number of cell lines made resistant in vitro by stepwise increases in drug concentration [37], [38], and in human tumor samples from clinically resistant tumors [39]. The increased enzyme levels can be correlated to gene amplification that may take the form of small new pieces of chromosomal material, called double minutes, or of large chromosomes, referred to as homogeneously staining regions (HSRs). The former variety of amplification imparts relatively unstable resistance, which requires the ongoing selective pressure of drug presence to be maintained [40], whereas the HSRs represent a more durable form of amplification and thus resistance. Several investigators [41–43] have successfully transvected amplified reductase genes into normal hematopoietic cells, allowing greater marrow resistance to methotrexate, an important dose-limiting toxicity of the drug.

Determinants of Cytotoxicity. Methotrexate is an S-phase-specific agent whose cytotoxic effects are determined by drug concentration and duration of cell exposure. These effects may be altered by the cellular milieu. The toxic effects of methotrexate can be completely reversed by exogenous administration of the end products (purines and thymidine) whose de novo synthesis is inhibited by methotrexate treatment. Also, the synthesis of these products may resume if an exogenous source of reduced folates is provided. These data provide the rationale for the treatment of patients with high-dose methotrexate and subsequent administration of a reduced folate in the form of leucovorin calcium (N^5-formyl tetrahydrofolic acid) as "rescue." The reversal of methotrexate cytotoxicity by reduced folates is a competitive process. The reasons for the competitive nature of this relationship are unclear but may be the result of a shared membrane transport system. In addition, methotrexate may have direct inhibitory effects on enzymes other than dihydrofolate reductase that require competitive levels of the folate cosubstrate to overcome the inhibition.

2.1.2. Analogs

Many new analogs to methotrexate have been developed in an effort to circumvent the cellular resistance that occurs with prolonged methotrexate exposure. As mentioned, drugs with increased lipid solubility have been successful in treating transport-resistant cells in vitro [44]. The lipophilic derivative metoprine and variations of the 10-deazaaminopterin series are the most interesting additions. Of the 10-deaza series, an ethyl sub-stitution at the 10-position imparts a marked increase in cytotoxicity when compared to methotrexate [45]. The analog possesses an improved membrane transport ability while retaining a high affinity for dihydrofolate reductase. A new antineoplastic agent is piritrexim isothionate [*79483-69-5*], 6-[(2,5-dimethoxyphenyl)methyl]-5-methylpyrido[2,3-d]pyrimidine-2,4-diamine mono-2-hydroxyethanesulfonate, $C_{17}H_{19}N_5O_2$ [46], that inhibits dihydrofolate reductase (DHFR). Further lipophilic DHFR-inhibitors are trimetrexate [*82952-645*] [47], and edatrexate [48] which are clinically studied and have shown activity, e.g., in non-small cell lung cancer. Thymidylate synthase (TS) is the rate-limiting enzyme in the anabolism of thymidine resulting in the incorporation into DNA. Raltitrexed (company codes: ZN-1694, D-1694, ICI-D1694) [*112887-68-0*] [49], is currently under clinical investigation with response rates in colon and breast cancers of up to 30 % [50], [51]. Myelosuppression seems to be the predominant dose limiting toxicity.

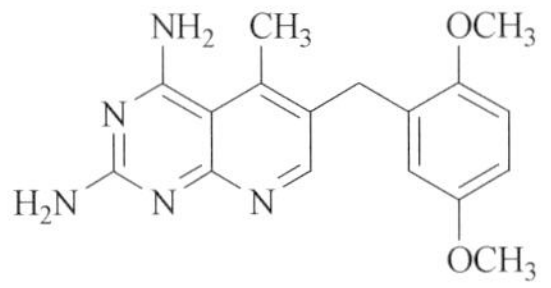

Piritrexim

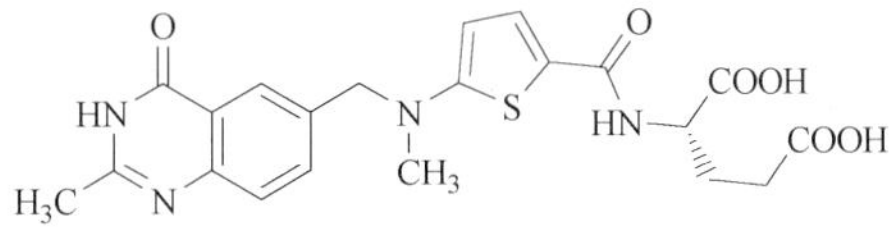

Raltitrexed

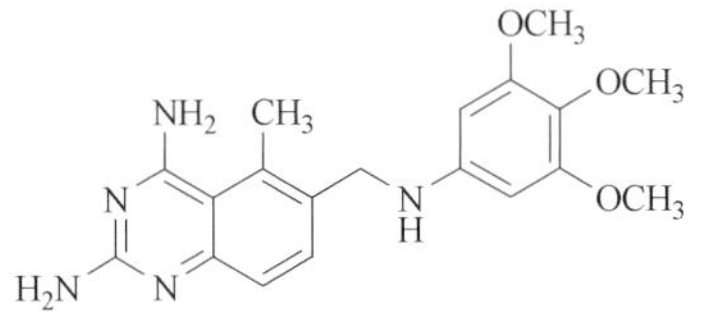

Trimetrexate

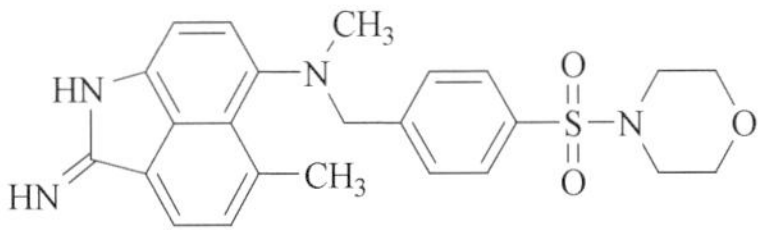

AG 331

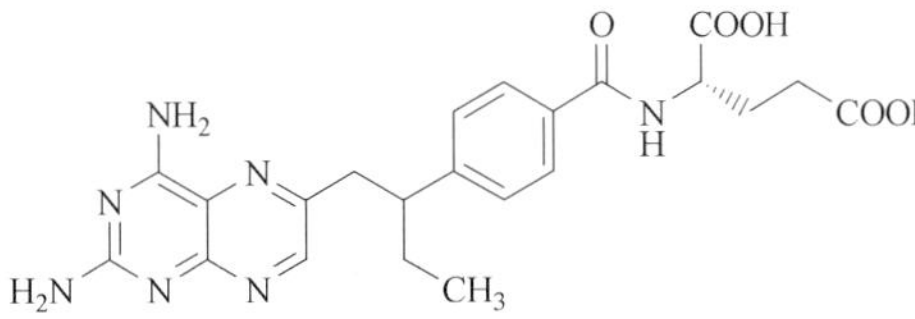

Edatrexate

Crystallographic data and computer-assisted drug design led to the development of thymidylate synthase (TS) inhibitors of the type of AG 331 [52].

Finally, CB 3717, a potent inhibitor of thymidylate synthase, is toxic for cell lines with altered dihydrofolate reductase; and homofolate, a de novo purine inhibitor requiring dihydrofolate reductase for activation, is effective in reductase-amplified lines [44].

2.2. Fluoropyrimidines

5-Fluorouracil (NSC-19893) [*51-21-8*], 5-fluoro-2,4-(1*H*, 3*H*)-pyrimidinedione, 5-FU, $C_4H_3FN_2O_2$, M_r 130.08, is a fluorinated pyrimidine whose structural formula resembles thymine; the hydrogen in the 5-position of the naturally occurring pyrimidine being replaced by fluorine. The *synthesis* of 5-FU in 1957 [15] represents the first successful effort in the rational design of anticancer drugs [53], and was predicated on the earlier observation that malignant cells selectively utilized uracil (and possibly toxic uracil analogs) in vivo [16], [54].

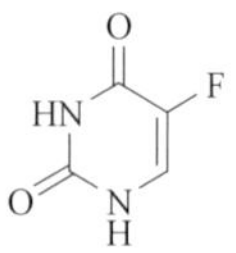

5-Fluorouracil

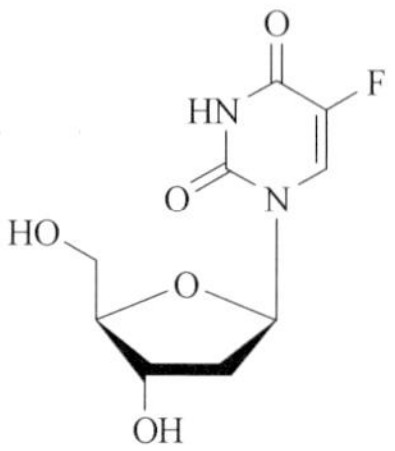

5-FUDR

Since the original synthesis of 5-FU and its nucleoside 5-FUDR (NSC-27640) [*50-91-9*], 2′-deoxy-5-fluorouridine, floxuridine, $C_9H_{11}FN_2O_5$, M_r 246.21, much has been learned about the mechanism of action of the fluoropyrimidines. These drugs are useful in the treatment of a wide range of human malignancies.

2.2.1. Mechanism of Action

Both 5-FU and 5-FUDR are prodrugs that require intracellular metabolism to their respective nucleotides for cytotoxicity. The pathways for fluoropyrimidine activation are shown in Figure 1. Each drug is enzymatically activated by different routes to FdUMP, FUMP, or FUTP, and each of these fluorinated nucleosides has different mechanisms of cytotoxicity.

Thymidine phosphorylase converts 5-FU to the deoxyribonucleotide 5-FUDR, which is then phosphorylated by thymidine kinase to yield 5-FdUMP. In the presence of methylene tetrahydrofolate, 5-FdUMP forms a stable ternary complex with thymidylate synthetase (TS), inhibiting this critical enzyme to cause "thymineless death." As expected, cytotoxicity is prevented in the presence of exogenous thymidine in these cells with intact salvage pathways. Inhibition of TS has long been considered the principal mechanism of 5-FU cytotoxicity, but it has also been demonstrated that 5-FU can be converted to 5-FUMP, either by orotic acid phosphoribosyl-transferase (OPRTase) in the presence of phosphoribosyl pyrophosphate (PRPP), or by stepwise conversion to the ribonucleotide 5-FUR (by uridine phosphorylase) followed by formation of 5-FUMP by uridine kinase. This intermediate can be converted to 5-FdUMP by ribonucleotide reductase to inhibit TS. Alternatively, 5-FUMP can be phosphorylated to 5-FUTP, which may be fraudulently incorporated into RNA to induce cytotoxicity. In a number of tumor models, loss of clonogenic capacity is directly correlated with the extent of incorporation of 5-FUTP into RNA [55], [56]. This RNA-specific toxicity is not reversed by thymidine. However, the precise mechanism of RNA-induced cell kill is speculative. The most consistent structural effect of 5-FU exposure is impaired processing of ribosomal RNA [57]. This concept is not supported by current evidence, however, as neither the synthesis nor the translation of messenger RNA (mRNA) appears affected by documented 5-FUTP incorporation into mRNA. Using human colon carcinoma cells propagated in vitro, neither quantitative nor qualitative differences in the translational products (polypeptides) of 5-FUTP-containing mRNA could be demonstrated [58]. However, small nuclear RNA species responsible for exon recognition during RNA splicing do contain significant quantities of uridylic acid [59]. Specific substitution of 5-FUTP into this RNA fraction may be critical for 5-FU toxicity.

As shown in Figure 1, 5-FdUMP can be further phosphorylated to 5-FdUTP, which can be incorporated into tumor cell DNA [60], [61]. This mechanism of drug action has been particularly difficult to appreciate, since the fraudulent base is quickly excised from DNA by the enzymes uracil-DNA glycosylase and dUTP nucleotidohydrolase. Thus, when tumor cells are incubated at low (0.1 μM) concentrations

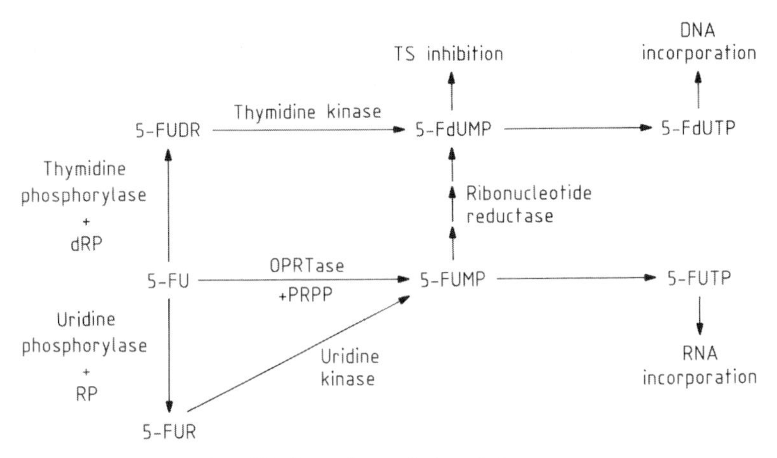

Figure 1. Pathways for fluoropyrimidine activation

of 5-FUDR, single-strand DNA shifts to lower molecular mass species, suggesting excision of the fluorinated base [62]. However, the actual presence of 5-FdUTP in cellular DNA cannot be detected until tumor cells are exposed to higher drug concentrations. Apparently, the importance of 5-FU incorporation into DNA varies from tumor to tumor, with human promyelocytic leukemia cells incorporating nearly 100-fold more 5-FdUTP into DNA than mouse leukemia cells [63]. The importance of this pathway both to tumor cell cytotoxicity and therapeutic index remains to be elucidated.

2.2.2. Mechanisms of Resistance

Tumor cells selected for in vitro resistance may demonstrate a deletion of critical drug-activating enzymes, including uridine kinase [64], orotic acid phosphoribosyltransferase [65], [66], and uridine phosphorylase [67]. Methylene tetrahydrofolate is required for 5-FdUMP inhibition of TS, and decreased availability of intracellular folates has been involved as a mechanism of 5-FU resistance [68].

In addition, alterations in the target enzyme TS can result in drug resistance. Resistant cells have been described with altered thymidylate synthetase having decreased affinity for FdUMP [69–71]. In addition, increased specific activity of TS has been reported in drug-resistant fibroblasts [72] and tumor cells [73], [74]. Whether elevated levels of target protein represent the end result of specific gene amplification (as has been documented for methotrexate resistance) remains to be determined.

2.2.3. Other Fluoropyrimidines

Attempts to develop fluoropyrimidines with improved therapeutic indexes have resulted in the synthesis and clinical trial of so-called *masked fluoropyrimidines.* The masked fluoropyrimidine 5′-deoxy-5-fluorouridine is a nontoxic prodrug that is converted to 5-FU by pyrimidine nucleoside phosphorylase. Because this enzyme may be present to a greater degree in some tumor cells than in normal human bone marrow, an improved therapeutic index can be demonstrated in vitro using breast, sarcoma, leukemia, and colon carcinoma cells [75]. Since 5′-deoxy-5-fluorouridine requires conversion to 5-FU and is in itself nontoxic, tumor cells that lack phosphorylase activity are resistant to the masked compound, while remaining cross-sensitive to 5-FU [76].

Ftorafur is a second masked fluoropyrimidine that is less myelosuppressive than 5-FU. However, the drug has a higher incidence of gastrointestinal and nervous system toxicity, which is probably due to organ-specific localization of activating enzymes [77]. In early clinical studies, ftorafur has shown antitumor activity in patients with 5-FU refractory colorectal and breast cancer [78].

2.3. 5-Azacytidine

5-Azacytidine (NSC-102816) [*320-67-2*], 4-amino-1-β-D-ribofuranosyl-1,3,5-triazine-2(1*H*)-one, 5-azacitidine, $C_8H_{12}N_4O_5$, M_r 244.21, is a pyrimidine analog that was first isolated as a fermentation product from *Streptoverticillium* cultures [79] and was chemically synthesized in Czechoslovakia in 1964 [80]. Structurally, 5-azacytidine differs from cytidine by the substitution of nitrogen in the 5 position of the pyrimidine nucleus.

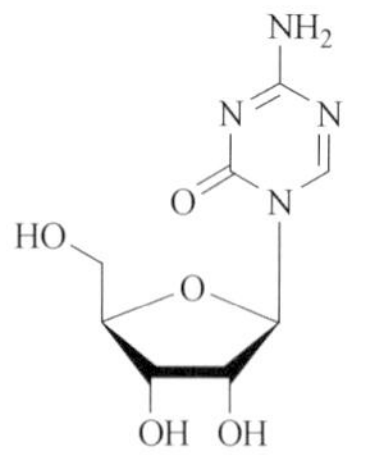

5-Azacytidine

2.3.1. Mechanism of Action

5-Azacytidine shares the facilitated transport system for cytidine for entry into cells [81] and must be phosphorylated to exert cytotoxic effects. Conversion to the monophosphate is catalyzed by the enzyme uridine – cytidine kinase [82], and this is likely the rate-limiting step for drug activation [83]. 5-Azacytidine monophosphate inhibits the enzyme orotidylate decarboxylase and interferes with de novo pyrimidine biosynthesis [84]. Subsequent metabolism of the monophosphate to azacytidine di- and triphosphate, catalyzed by cytidine monophosphate kinase and nucleoside diphosphate kinase, occurs rapidly and does not appear to be a rate-limiting step in drug activation [81].

The diphosphate of azacytidine is a substrate for ribonucleotide reductase and dAzaCTP (deoxyribonucleotide triphosphate of azacytidine) for DNA polymerase, allowing direct incorporation of drug into DNA [85]. This pathway may be critical for cytotoxicity because active DNA synthesis correlates with drug sensitivity in vitro [86]. In addition, incorporation of 5-azacytidine into DNA may affect gene expression. Mammalian DNA contains ca. 5 % of incorporated cytosine methylated in the 5 position [87]; methylation appears to inhibit gene expression. Thus the globin gene is hypomethylated in bone marrow as compared to other tissues [88–91]. The DNA containing even low levels of 5-azacytidine is a potent inhibitor of the enzyme responsible for cytosine methylation, i.e., DNA cytosine methyltransferase [92], [93]. The enzyme inhibition is disproportionately great compared with the small amount of incorporated fraudulent base, and appears to result from formation of a stable complex between 5-azacytidine residues and the methyltransferase, similar to the complex formed with thymidylate synthase and FdUMP [94]. Thus, in vitro treatment with 5-azacytidine can induce DNA hypomethylation and differentiation of murine cells [95], [96]. More recently these observations have been extended to clinical medicine. Azacytidine treatment in a patient with severe β-thalassemia could stimulate gamma-globin synthesis by inducing hypomethylation and expression of the gamma-globin gene [97].

After conversion to a triphosphate, azacytidine also competes with CTP for incorporation into RNA [98] and inhibits maturation of ribosomal and transfer RNA [99]. This causes disassembly of polyribosomes [100] and interferes with protein synthesis [101].

2.3.2. Mechanism of Resistance

Cellular resistance to 5-azacytidine may be the result of either decreased drug activation or possibly increased drug degradation. Drug metabolism by cytidine kinase appears to be rate-limiting in drug activation; deletion of this enzyme has been reported in 5-azacytidineresistant cells in vitro [81], [102]. Cytidine deaminase degrades 5-azacytidine to 5-azauridine; however, the role of this enzyme in drug resistance remains uncertain. For example, drug toxicity may be dependent on deamination; bacteria incapable of forming 5-azauridine from 5-azacytidine are resistant to the drug [103].

2.3.3. New Analogs

In clinical trials 5-azacytidine has demonstrated consistent antileukemic activity, inducing complete remissions in a significant number of heavily pretreated patients with acute myelogenous leukemia [104]. Acute dose-limiting toxicities associated with bolus drug administration (severe gastrointestinal symptoms, fever, life-threatening hypotension) may be ameliorated by administering the drug via constant intravenous infusion [105]. However, 5-azacytidine is chemically unstable, undergoing ring opening at the 5,6-imino double bond to form *N*-formylamidinoribofuranosylguanylurea, which further decomposes to ribofuranosylurea [106]. The halflife of this decomposition is 4 h, complicating the task of strict dosage control of prolonged infusions.

To circumvent the problem of aqueous instability, the hydrolytically susceptible 5,6-imino bond of 5-azacytidine was reduced to produce *dihydro-5-azacytidine* (NSC-264880) [*62402-31-7*], 4-amino-5,6-dihydro-1-β-D-ribofuranosyl-1,3,5-triazine-2(1*H*)one · monohydrochloride, $C_8H_{14}N_4O_5 \cdot HCl$, M_r 282.7 [107]. This compound has completed phase I trials in the 1980s with preliminary evidence of activity in lymphoma; in addition it exhibits the unusual dose-limiting toxicity of chest pain at the maximally tolerated dose [108].

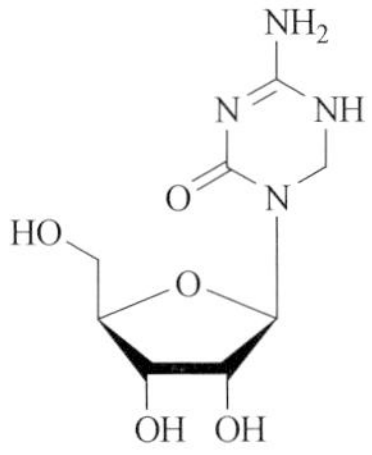

NH2 N NH O N HO O OH OH

Dihydro-5-azacytidine

2.4. Cytosine Arabinoside (Ara-C)

Ara-C (NSC-63878) [*147-94-4*], 4-amino-1-β-D-arabinofuranosyl-2(1*H*)-pyrimidone, cytarabine, cytosine arabinoside, $C_9H_{13}N_3O_5$, M_r 243.22, an antimetabolite that is a structural analog of cytidine, differs from the physiologic nucleoside by the epimeric configuration of the β-*trans*-hydroxyl group at the 2′ position of the sugar. The drug is transported into cells by a carrier-mediated process with shared affinity for deoxycytidine [109]. Once ara-C has entered the cell, cytotoxicity is dependent on formation of the triphosphate ara-CTP, which is responsible for inhibition of DNA synthesis.

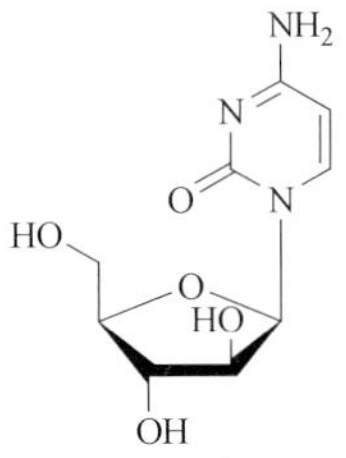

NH2 N O N HO O HO OH

Cytosine arabinoside

The precise *mechanism* by which ara-CTP inhibits DNA synthesis remains uncertain. Ara-CTP does inhibit both DNA polymerase α [110] and β [111]. The former is essential for DNA synthesis, the latter for DNA repair. Ara-CTP can be directly incorporated into DNA as well, and this pathway correlates strongly with cytotoxicity. The extent of drug incorporation into DNA is proportional to cell kill in both acute myelocytic and promyelocytic leukemia cells [112]; incorporation and cytotoxicity can be modulated by thymidine [113]. Also ara-C-substituted DNA is unstable under conditions of alkaline elution, suggesting drug-induced strand fragility and breakage [114]. Ara-CTP incorporation also directly blocks strand elongation [115] and causes premature strand termination [116]. This results in accumulation of DNA peaks of small sizes, suggesting that preexisting DNA may be nicked following exposure to ara-C [117]. A further effect of ara-C exposure is inhibition of DNA repair, as determined by alkaline elution studies in L 1210 cells [118]. As might be expected, cells pretreated with ara-C are more sensitive to ionizing radiation, suggesting potentiation of radiation damage by inhibition of DNA repair [119], [120–123].

2.4.1. Mechanisms of Resistance

A number of mechanisms of resistance to ara-C have been described. Ara-C itself is a prodrug that is metabolized first to ara-CMP by deoxycytidine kinase, then to ara-CDP by pyrimidine nucleoside monophosphate kinase, and finally to ara-CTP by nucleoside diphosphate kinase. Resistant cell lines lack the initial, rate-limiting activating enzyme [115], [124].

Since the 1980s, HPLC has been used to separate and semipurify deoxycytidine kinase, pyrimidine nucleoside monophosphate kinase, and pyrimidine nucleoside diphosphate kinase from cell extracts [125]. In murine leukemia cells selected for ara-C resistance, deoxycytidine kinase was found in lower specific activity, and similar results have been reported for other systems [126].

In addition, expansion of the intracellular pool of the physiologic substrate deoxycytosine triphosphate can inhibit ara-C activation by feedback inhibition of the initial activating enzyme [127]. Moreover, increased dCTP pools may also compete directly with ara-CTP for DNA polymerase. Experiments have shown a relationship between duration of leukemia remission and the ability of tumor cells obtained from patients to form and retain ara-CTP in vitro [128], but the mechanisms underlying this relationship remain to be established.

Although some workers have reported that increased drug catabolism by cytidine deaminase may underlie both de novo and acquired drug resistance in patients with leukemia [129], a definite clinical correlation between response and levels of this enzyme has not been established in several studies [130–132].

2.4.2. New Analogs

Several ara-C analogs have been rationally designed with the goal of overcoming specific mechanisms of resistance. Compounds with lipophilic side chains are relatively resistant to inactivation by cytidine deaminase. Another analog with lipophilic modifications in the side chain is enocitabine (NSC-239336, BHAC) [*055726-47-1*] [133], [134]. N^4-Behenoyl-ara-C (BHAC) has undergone clinical trial in patients with acute leukemia [135]. Despite its lipophilicity, the drug does not enter the cerebrospinal fluid and concentrates in bone marrow and red blood cells. N^4-Palmitoyl-ara-C can be administered orally and appears to be more active than the parent compound in preclinical models [136]. Although developed as an antiviral agent, 2′-fluoro-5-iodo-1-β-D-arabinofuranosylcytosine (FIAC) has significant antitumor activity [137]. Interestingly, this compound may be relatively cytotoxic for cells with high levels of cytidine deaminase because the catabolism product, FIA-uracil, is more toxic than the parent compound [138].

2.5. Deoxycytidine and Analogs

The pyrimidine antimetabolite gemcitabine (LY-188011, dFdC) [*095058-81-4*], 2’-deoxy-2’,2’-difluorocytidine, $C_9H_{11}F_2N_3O_4$ is an analog of deoxycytidine and a result of a program initiated at Lilly Research to synthesize fluorinated D-ribose and fluorinated nucleosides [139–142]. The difluorinated analog of cytarabine, gemcitabine was identified as novel antimetabolite with a broad spectrum of antitumor activity.

Trade names: Gemicitabine monohydrochloride [*122111-03-9*] is on the market as Gemzar in the indication of palliative treatment of locally advanced or metastatic non-small cell lung cancer.

Mechanism of Action. Gemcitabine shows good activity against human leukemic cell lines, a number of murine solid tumors, and human tumor xenografts [143–146]. Gemcitabine was significantly more cytotoxic than cytarabine in Chinese hamster ovary cells. The major cellular metabolite is the 5’-triphosphate of gemcitabine. The cytotoxicity was competitively reversed by deoxycytidine, suggesting that the biological activity required phosphorylation by deoxycytidine kinase [145].

Tumor-bearing mice were treated with either gemcitabine or cytarabine (20 mg/kg). DNA synthesis reached 1 % of control levels upon administration of gemcitabine. The greater accumulation of gemcitabine-5’-triphosphate compared with cytarabine-5’-triphosphate may cause greater cytotoxicity and therapeutic activity [146].

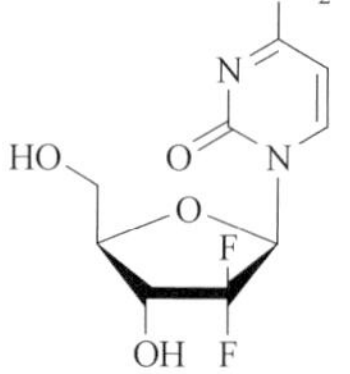

Gemcitabine

Further gemcitabine may enhance its own cytotoxic effects by self-potentiation mechanisms that act on, e. g., deoxycytidine monophosphate deaminase, deoxycytidine kinase or on DNA-synthesis [147], [148].

2.6. 2-Halopurines and Analogs

Cladribine (2-CdA, RWJ-26251) *004291-63-8*, is a purine deoxyribonucleoside with remarkable antileukemic activity. It represents a significant advance over existing therapy because it is given as a single 7-day continuous treatment, thus minimizing the side effects observed with multiple treatments. Remission rates of up to 89 % lasting for up to 25 months were observed in clinical trials in patients with hairy cell leukemia [149–152]. The antimetabolite used for first line treatment of hairy cell leukemia [153].

Trade name: Leustatin.

Fludarabine phosphate (NSC-312887, 2-F-ara-AMP) [*075607-67-9*] [154], 2-fluoro-9-(5-*O*-phosphono-β-D-arabino-furanosyl)-9*H*-purin-6-amine, $C_{10}H_{13}FN_5O_7P$, M_r 365.21, another cytotoxic purine antimetabolite, acts via inhibition of DNA synthesis. The product is used for treatment of patients with chronic lymphocytic leukemia.

Trade names: Fludara (Berlex), Benefluor (Schering AG).

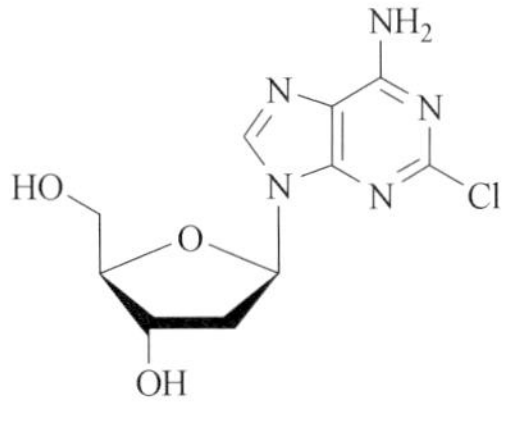

Cladribine

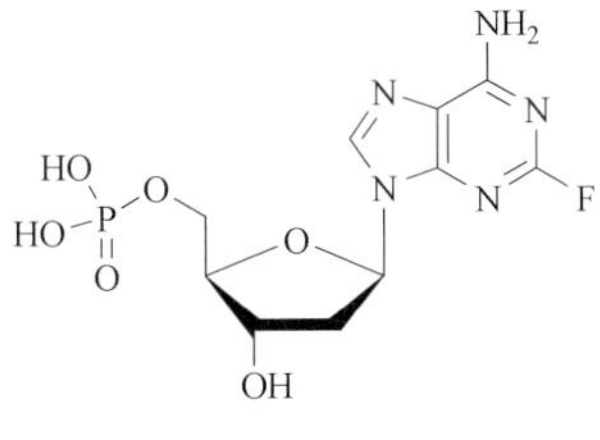

Fludarabine

Mechanism of Action. Fludarabine and its soluble derivatives interfere with phosphorylation, e.g., in L 1210 cells. Fludarabine behaves more like an analog of deoxycytidine than adenine or deoxyadenine as indicated by reports [155] demonstrating that the presence of fluorine in the 2-position of the adenine ring alters its function as a substrate for deaminase and nucleoside kinases. This results in differences in biological activity and metabolism. Halogenation does not simply block deamination, but also influences the enzyme that carries out the phosphorylation, as a result cytotoxicity is increased [156]. Fludarabine phosphate may selectively inhibit the incorporation of thymidine and uridine into the DNA molecule by inhibiting both ribonucleotide reductase [157] and DNA polymerase [158]. The maximum tolerated dose (MTD) in heavily pretreated patients with advanced malignancy/solid tumors on the daily regimen was about 15 mg/m^2. Granulocytopenia and thrombocytopenia were dose-limiting [159–161].

Pentostatin. The cytotoxic and immunosuppressant pentostatin (NSC-218321, CI-825, PD-81565, YK-176, 2-deoxycoformycin, 2'-dCF) [*063677-95-2*], (R)-3-(2-deoxy-β-D-erythro-pentofuranosyl)-3,6,7,8-tetrahydroimidazo[4,5-d][1,3]diazepin-8-ol, $C_{11}H_{16}H_4O_4$, M_r 268.13, *mp* 220 – 225 °C, can be isolated from the fermentation broth of *Streptomyces antibioticus* NRRL 3238 [162], [163] or *Aspergillus nidulanus* Y 176-2 or *Emericella* [164]. The adenosine nucleoside analog pentostatin, is the most potent inhibitor of adenosine deaminase, which is an important and ubiquitous cellular enzyme. The inhibition of this enzyme leads to accumulation of dATP which inhibits ribonucleotide reductase and thus DNA synthesis. Pentostatin was launched for treatment of hairy-cell leukemia refractory to α-interferon.

Trade names: Nipent (Parke-Davis, Lederle), Coforin (Katetsuken).

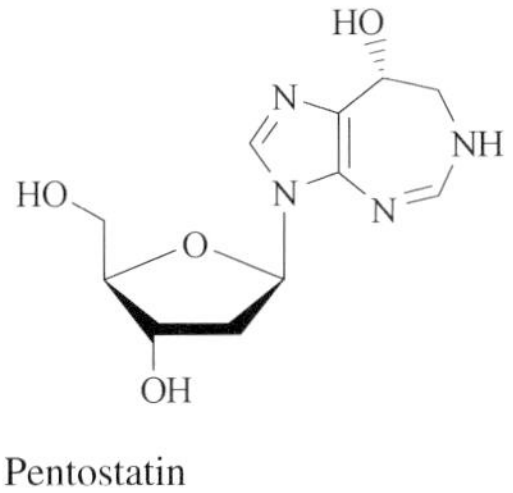

Pentostatin

Mechanism of Action. The highest activity of adenosine deaminase is found in lymphoid tissue as well as in the malignant cells of acute lymphoblastic leukemia [165]. Since pentostatin is the most potent inhibitor of adenosine deaminase it was expected to possess antitumor properties against certain malignancies, especially acute leukemias, chronic myelogenous leukemia, and lymphomas. Surprisingly, when tested against murine tumor models and also against various tumor cell cultures no antitumor effect was found [166]. However, in a phase I clinical trial the compound was found to produce a drop in lymphoblast count and showed antitumor activity in acute leukemia and lymphoma. The toxicity observed consisted mostly of CNS side effects including nausea, hepatic and renal dysfunction. In combination with vidarabin (Ara-A) (→ Chemotherapeutics, Chap. 4.6.1.) superiority to monotherapy was demonstrated in various studies [167], [168]. Durable complete remission was observed after pentostatin treatment in patients with hairycell leukemia resistant to α-interferon [169–174].

2.7. 6-Mercaptopurine and 6-Thioguanine

The purine antimetabolites, 6-mercaptopurine (NSC-755) [*50-44-2*], 1,7-dihydro-6*H*-purine-6-thione, 6-MP, $C_5H_4N_4S$, M_r 152.19; and 6-thioguanine (NSC-63878) [*154-42-7*], 2-amino-1,7-dihydro-6*H*-purine-6-thione, 6-TG, $C_5H_5N_5S$, M_r 167.21, have been used for several decades in the treatment of leukemia and certain other neoplastic diseases. The 6-thiopurine analog, 6-MP, was first synthesized in 1952 [175]; subsequently the same workers synthesized 6-TG in 1955 [176]. The clinical evaluation of 6-MP in the treatment of acute leukemia and chronic myelocytic leukemia was carefully studied in 1953 [177]. Shortly thereafter, the 2-amino analog of 6-MP, 6-TG, was submitted for clinical evaluation as an antitumor agent [178].

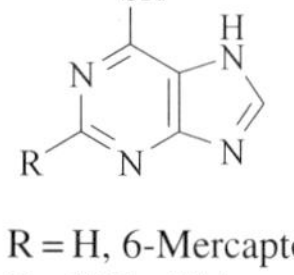

R = H, 6-Mercaptopurine
R = NH_2, Thioguanine

6-Mercaptopurine is used for maintenance therapy of acute lymphocytic and acute myelogenous leukemia, and 6-TG is used primarily for remission induction in acute myelogenous leukemia. Although 6-MP and 6-TG have an important role in chemotherapy of leukemia patients, the purine analogs produce low-response rates in patients with solid tumors, lymphomas, and chronic lymphocytic leukemia.

2.7.1. Mechanism of Action

6-Mercaptopurine has been used as an antineoplastic and immunosuppressive agent for decades, but the precise mechanism by which it exerts its cytotoxic effects has not yet been established. Both 6-MP and 6-TG must be converted to their thiol nucleotide form, which is the active cytotoxic moiety. The conversion is catalyzed by hypoxanthine guanine phosphoribosyltransferase (HGPRT) [179–181] and the reaction is dependent on the phosphoribosylpyrophosphate (PRPP) level in the cells [182–186]. 6-Thiouric acid (6-TU) is the major catabolic product of 6-MP. The rapid conversion of 6-MP to 6-TU by xanthine oxidase [187], [188] in leukemic cells may be a possible mechanism of 6-MP resistance [182], [189], [190]. The antineoplastic effect of 6-TG is similar to that of 6-MP. In its nucleotide form, 6-TG inhibits de novo purine biosynthesis and purine interconversions [191–195].

The cytotoxicity of 6-MP and 6-TG has been linked to (1) the interference with de novo purine biosynthesis and purine interconversions, (2) the inhibition of in vitro RNA synthesis, and (3) the incorporation into DNA during S phase, resulting in a deformation of the DNA.

The interference with de novo purine biosynthesis by 6-MP is regulated by 6-MP nucleotides. These nucleotides inhibit the enzyme 5-phosphoribosylpyrophosphate amidotransferase that catalyzes the initial reaction in the purine biosynthetic pathway [196]. The 6-MP nucleotides also inhibit conversion of inosine monophosphate (IMP) to adenine monophosphate (AMP) and to xanthine monophosphate (XMP), and limit the availability of XMP to form guanine monophosphate (GMP), thereby interfering with the supply of purine precursors for nucleic acid synthesis [181]. In the 1980s two new findings pertaining to the cytotoxicity mechanism in tumor cells have been reported. Studies of human lymphoma revealed that 6-MP was a potent inhibitor of cellular RNA synthesis and that 6-thioITP inhibited both the RNA polymerase I and RNA polymerase II activities of these cells [197]. These data suggested that direct inhibition of the enzymes mediating transcription by 6-thio-IMP may be one of the mechanisms for the cytotoxic action of 6-MP in human tumor cells. Using 6-TG as a cytotoxic agent resulted in severe chromosome damage in wild-type CHO cells [198]. Gross unilateral chromatid damage resulted, and the unilateral nature of this damage was probably due to malfunction of 6-TG-containing DNA as a replication template.

2.7.2. Mechanism of Resistance

Several mechanisms of resistance to these agents have been described in experimental tumors and relate to the pathways of antimetabolite activation and degradation. A decreased HGPRT activity in tumor cells diminishes antimetabolite activation, and this resistance pathway has been reported by several workers [199–202]. However, the HGPRT-regulated mechanism as a basis for drug resistance in human leukemic cells is relatively uncommon [203]. The resistance of 6-MP is related to an increase in alkaline phosphatase in sarcoma cells [204]. Increased alkaline phosphatase, a membrane-bound enzyme that converts the active mononucleotide to 6-thioinosine and inorganic phosphate, has also been reported in human leukemia patients resistant to drug treatment [205], [206]. Another enzyme in the degradation pathway of 6-MP that needs to be considered for drug efficacy and bioavailability is xanthine oxidase, which is responsible for converting 6-MP to 8-OH-6-MP and subsequently to thiouric acid, which is excreted through the urine.

To understand further the bioavailability and pharmacokinetics of thiopurines, the effect of allopurinol on 6-MP catabolism was studied. Allopurinol, an analog of hypoxanthine, enhances the therapeutic efficacy of 6-MP by inhibiting xanthine oxidase. The urinary excretion of 6-MP metabolites is markedly reduced in patients treated with allopurinol. Furthermore, allopurinol increased the plasma level of 6-MP in rabbits [207]. The data suggest that inhibition of 6-MP catabolism by allopurinol may contribute to a greater availability of 6-MP to tissues. Studies of the effect of allopurinol on the kinetics of oral and intravenous 6-MP in Rhesus monkeys and in humans demonstrated that allopurinol pretreatment resulted in a nearly 400 % increase in peak plasma concentration of 6-MP in monkeys and a 500 % increase in humans, but only when 6-MP was administered orally [208]. Allopurinol pretreatment had no effect on the kinetics of intravenously administered 6-MP. This difference is due to the action of allopurinol on liver or intestinal xanthine oxidase and inhibition of first-pass metabolism of oral 6-MP. This finding may explain the low and variable plasma levels of mercaptopurine in patients with acute lymphoblastic leukemia treated with oral 6-MP [209]. Although these studies emphasize the catabolic pathway for purines, all the purine and pyrimidine metabolic enzymes may be important to the bioavailability and activation of antimetabolites.

To clarify the mechanisms of resistance to thiopurines and potential drug interactions in tumor cells, studies have focused on the thiopurine-resistant cell lines deficient in HGPRT (L 1210) and the regulation of PRPP formation in thiopurine-resistant cell lines.

The thiopurines are inactive in the base form and must be converted to their respective nucleotides. This activation step requires PRPP as the cofactor and HGPRT as the enzyme to catalyze the conversion to nucleotide. A major biochemical effect of methotrexate is the suppression of purine biosynthesis and expansion of the PRPP pool. Studies of the cytotoxic and biochemical interaction of methotrexate and 6-TG in L 1210 mouse leukemia cells demonstrated that methotrexate can markedly enhance 6-TG activity [210]. Preexposure of cells to methotrexate resulted in a large increase in cytotoxic potency of 6-TG, whereas simultaneous exposure caused an antagonism of 6-TG cytotoxic activity. Although PRPP pools were not measured quantitatively, the effect of methotrexate preexposure was to increase PRPP pools, enhance the activation of 6-TG to 6-TG monophosphate, and thereby increase its incorporation into RNA.

2.7.3. New Analogs

Alkyl disulfide derivatives have been used as masked compounds of 6-mercaptopurines. The antitumor effect of these derivatives has been measured by their ability to decrease the degradation of 6-MP monophosphate. Of the seven 6-alkyl disulfide derivatives of 6-MP and 6-TG in L 1210 leukemia cells tested, decyl derivatives of both 6-MP and 6-TG were the most effective with therapeutic ratios as high as 50 and 48, while those of parent compounds were 6.2 and 5.0, respectively [211]. Partial circumvention of thiopurine resistance may have resulted from cellular uptake of intact acylated bis(6-MP-9-β-D-ribofuranoside-5′)pyrophosphate derivatives in 6-MP-resistant human cell lines deficient in HGPRT (L 1210) [212].

The 6-MP-resistant sublines of P 388 and L 1210 leukemia are also sensitive to two new *purine antagonists:* 5-carbamoyl-1*H*-imidazol-4-yl piperonylate and 4-carbamoylimidazolium-5-olate [213]. These two new purine analogs kill 6-MP-resistant cells by suppressive de novo purine synthesis. The activation of those new purine analogs is mediated by adenine phosphoribosyltransferase.

3. Alkylating Agents

Historically, alkylating agents were important in the early development of cancer chemotherapies. Victims of sulfur mustard gas exposure in World War I were found to have severe lymphoid aplasia as well as pulmonary irritation [214]. This led to clinical trials of the related, but less toxic, nitrogen mustard derivative, which produced tumor regressions in lymphoma patients [215].

Clinical use of nitrogen mustard today is mostly limited to the treatment of lymphomas, especially Hodgkin's disease, where it is used in a multidrug regimen called MOPP (nitrogen mustard, vincristine, procarbazine, and prednisone). Three widely used derivatives of nitrogen mustard used in patients with malignancies today include melphalan, cyclophosphamide, and chlorambucil (Sections 3.2, 3.3, 3.4).

The mechanism by which alkylating agents act can be classified as either S_N1 or S_N2. In the *S_N1 reaction,* a highly reactive intermediate forms initially and quickly reacts with a nucleophile to produce an alkylated product. This reaction follows first-order kinetics because the rate-limiting step is formation of the intermediate. The *S_N2 reaction* is a second-order reaction and thus is dependent on the concentration of both the alkylating agent and its target nucleophile [216]. In general, alkylating agents that react by an S_N1 mechanism, such as nitrogen mustard, are less selective in their reactions than S_N2 agents, but this rule is not always true. Selectivity also depends upon membrane permeability, charge, and reactivity of the drug.

3.1. Nitrogen Mustard

3.1.1. Mechanism of Action

Nitrogen mustard (NSC-762) [*51-75-2*], mechloroethamine, HN2, $CH_3N(CH_2CH_2Cl)_2$, $C_5H_{11}Cl_2N$, M_r 156.07, is activated through loss of one of its chlorines. The α carbon then reacts with the nucleophilic nitrogen to form the positively charged, highly reactive, cyclic aziridinium compound, which is attacked by nucleophiles to give the initial alkylated product. The second chlorine can also leave, initiating a second alkylation, which produces a cross-linked alkylation between two nucleophiles.

Because HN2 bonds covalently to many biologic molecules, such as DNA, RNA, and proteins, the alkylated sites responsible for its cytotoxicity are difficult to determine. However, studies have shown that cytotoxicity is likely to result from inhibition of DNA synthesis by damaging the DNA template [217–219]. The DNA molecule is rich in potential sites for alkylation, including the phosphate groups in the sugar phosphate backbone structure, and the oxygen and nitrogen sites in the purine and pyrimidine bases. However, the tendency for alkylation to occur in the N-7 position of guanine is enhanced. This may be mediated by the increased nucleophilic characteristics of the N-7 deoxyguanosine due to base stacking and charge transfer [220]. Other preferred sites of alkylation in decreasing order are the N-1 of adenosine, the N-3 of cytidine, and the N-3 of thymidine [221].

Bifunctional alkylating agents, such as HN2, produce intra- and interstrand cross-linking between DNA in the double-helix structure preferentially at the N-7 guanosine site. Thus, these bifunctional compounds are more effective antitumor agents than their monofunctional analogs; however, increasing the number of alkylating sites on the agent beyond two does not appear to increase antitumor activity [222]. This evidence suggests that DNA cross-linking is critical for alkylator activity. Further evidence for the importance of cross – linking to cytotoxicity comes from alkaline elution studies that can detect low levels of cross-linking in cells exposed to minimal doses of bifunctional alkylating agents [223].

In contrast, monofunctional agents, such as procarbazine and dacarbazine, do not produce DNA interstrand cross-links and appear to exert toxicity by producing single-strand DNA breaks. The increased carcinogenicity seen with some monofunctional alkylating agents may be due to incorrect base pair substitution by DNA repair enzymes that could result in malignant transformation.

3.1.2. Mechanisms of Drug Resistance

Several mechanisms have been elucidated for nitrogen mustard and other bifunctional alkylating agents. First, resistant cells with defective drug transport have been described. Nitrogen mustard enters cells by an active transport system that is physiologically utilized for choline transport. Lymphoma cells resistant to nitrogen mustard demonstrate decreased drug uptake by this specific active transport site, which also decreases its uptake of choline [224]. Second, cytosolic increase in nonprotein sulfhydryl levels [225] and higher nonprotein-bound thiol compounds that could inactivate the drugs before they reach the nucleus have been found in nitrogen mustard-resistant cells [226]. Third, enhanced repair of DNA cross-linking by repair enzymes has been demonstrated in vitro [227] and in vivo [228].

3.2. Melphalan

Melphalan (NSC-8806) [*148-82-3*], 4-[bis(2-chloroethyl)amino]-L-phenylalanine, L-phenylalanine mustard, $C_{13}H_{18}Cl_2N_2O_2$, M_r 305.20, was rationally designed and synthesized as a phenylalanine derivative of nitrogen mustard with the aim of obtaining increased specificity against melanoma tumor cells that utilize phenylalanine or tyrosine to produce melanin.

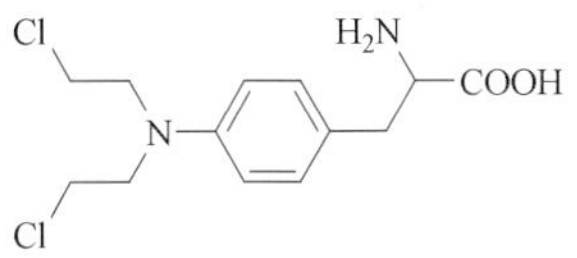

Melphalan

Although this compound does not exhibit specific antimelanoma activity, it has a broad-spectrum cytotoxicity in multiple myeloma, breast cancer, and lymphomas.

3.2.1. Mechanism of Action

Melphalan is a bifunctional alkylating agent but differs from HN2 mainly by the presence of an aromatic ring. This ring reduces the nucleophilicity of the nitrogen atom by withdrawing electrons, making the drug less reactive. Thus, it can be taken orally. It retains its alkylating activity but is more selective than nitrogen mustard because it is less likely to form the unstable and highly reactive aziridine intermediate indiscriminately. Like nitrogen mustard, it forms DNA cross-links that are critical for cytotoxic effect. Its cellular uptake is also mediated by an active, energy-dependent transport mechanism shared with leucine and glutamine uptake [218]. Thus, high concentrations of leucine or glutamine can reduce the cytotoxicity of melphalan in a marrow colony-forming unit assay [229]. Another transport mechanism, although less active, can be used by melphalan, and this mechanism is shared by the neutral amino acids alanine, cysteine, and serine [218].

3.2.2. Mechanism of Resistance

Decreased transport of melphalan into drug-resistant leukemia cells has been demonstrated and correlated to the melphalan-resistant phenotype. Specifically, a mutation in the higher velocity transport system has been suggested that results in a decreased affinity of the carrier protein for melphalan and leucine [230]. Evidence for the other two mechanisms of resistance seen with nitrogen mustard, i.e., increased intracellular thiol compounds and increased DNA cross-linking repair also, exists.

3.3. Cyclophosphamide

Cyclophosphamide (NSC-26271) [*50-18-0*], *N,N*-bis(2-chloroethyl)tetrahydro-2*H*-1,3,2-oxazaphosphorin-2-amine 2-oxide, $C_7H_{15}Cl_2N_2O_2P$, M_r 261.10.

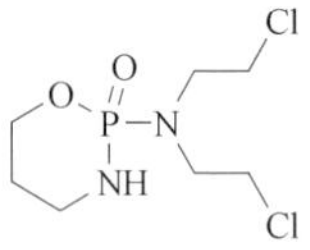

Cyclophosphamide

This widely used bifunctional, cyclic alkylating agent has important clinical use in lymphomas, leukemias, sarcomas, carcinomas of breast and ovary, as well as childhood malignancies. The compound was rationally designed based on data that tumor cells possess high concentrations of enzymes capable of cleaving the P–N bond. This reaction would activate drug by release of the potent antitumor agent phosphoramide mustard. Cyclophosphamide requires hepatic activation by oxidase enzymes. First, it is metabolized by liver microsomes to hydroxycyclophosphamide, which is spontaneously tautomerized to aldophosphamide. Aldophosphamide reaches peripheral tissues and tumors where it is hydrolyzed to yield the active antitumor agent phosphoramide mustard and acrolein. Acrolein has very weak antitumor activity and, when concentrated within the bladder by excretion, can cause hemorrhagic cystitis [216].

3.3.1. Mechanism of Action

Phosphoramide mustard can undergo similar bifunctional alkylation as nitrogen mustard. Also, because of its need for metabolic conversion for biologic activity, cyclophosphamide can be given orally or intravenously.

3.3.2. Mechanism of Resistance

Similar mechanisms of resistance occur for cyclophosphamide as for nitrogen mustard. Also, defective metabolic conversion by the hepatic microsomal system could serve to decrease the bioavailability of aldophosphamide to tumor tissue, but it is not known whether this is significant.

3.4. Chlorambucil

Chlorambucil (NSC-3088) [*305-03-3*], 4-[bis(2-chloroethyl)amino]benzenebutanoic acid, $C_{14}H_{19}Cl_2NO_2$, M_r 304.23. This drug is a close congener of melphalan and exhibits similar stability because of the electron-withdrawing properties of the aromatic ring. It is given orally and has proven efficacy in treating chronic lymphocytic leukemias, multiple myeloma, and lymphomas.

Chlorambucil is thought to have activation properties and mechanisms of resistance similar to melphalan.

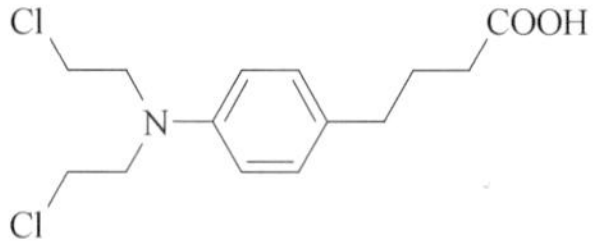

Chlorambucil

3.5. Thio-TEPA

Triethylenethiophosphoramide (NSC-6396) [*52-24-4*], 1,1′,1″-phosphinothioylidynetrisaziridine, thio-TEPA, $C_6H_{12}N_3PS$, M_r 189.23. This agent is representative of alkylating agents that have two or more aziridine rings. It has clinical activity against the same tumors as nitrogen mustard and has been used clinically in carcinomas of the breast and ovary. Thio-TEPA is also indicated for intrathecal therapy of meningeal carcinomas. The reactivity of the aziridine groups is increased by protonation; thus thio-TEPA is most active at low pH and has been used to cause sclerosis of malignant pleural effusions that often have an acidic pH.

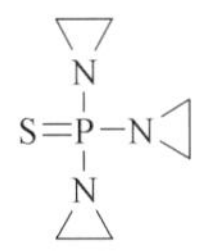

Thio-TEPA

The mechanisms of action and resistance are similar to those of nitrogen mustard.

3.6. Ifosfamide

Ifosfamide, or isofosphamide (NSC-109724) [*3778-73-2*], *N*,3-bis(2-chloroethyl)-tetrahydro-2*H*-1,3,2-oxazaphosphorin-2-amine 2-oxide, $C_7H_{15}Cl_2N_2O_2P$, M_r 261.07, is an analog of cyclophosphamide [231]. It has approximately one-third the alkylating activity of cyclophosphamide and requires hepatic microsomal conversion to its active form. Also, the rates of conversion by metabolism are similar in both drugs, but less biologically active alkylating moieties are formed in ifosfamide [232]. It has shown promising results in refractory pediatric bone and soft tissue sarcomas, refractory testicular tumors, and Wilms' tumor in children [233], [235].

Trade names: Ifex, Holoxan (ASTA Medica)

Ifosfamide

3.7. Estramustine

Estramustine

From the numerous analogs few found a way into the clinic or to the market. However, estramustine [*489-15-0*] was successfully launched for treatment of prostate cancer [234].

Trade name: Estracyt (Pharmacia/Upjohn)

3.8. Nitrosoureas

The clinically useful nitrosoureas include carmustine (NSC-409962) [*154-93-8*], *N,N′*-bis(2-chloroethyl)-*N*-nitrosourea, BCNU, $C_5H_9Cl_2N_3O_2$, M_r 214.04; lomustine (NSC-79037) [*13010-47-4*], *N*-(2-chloroethyl)-*N′*-cyclohexyl-*N*-nitrosourea, CCNU, $C_9H_{16}ClN_3O_2$, M_r 233.69; methylcyclohexylchloroethylnitrosourea (NSC-94941) [*52662-76-7*], 1-(2-chloroethyl)-3-(4-methylcyclohexyl)-1-nitrosourea, methyl-CCNU, $C_{10}H_{18}ClN_3O_2$, M_r 248; streptozotocin (NSC-85998) [*18883-66-4*], 2-deoxy-2{[(methylnitrosoamino)carbonyl]amino}-D-glucopyranose, $C_8H_{15}N_3O_7$, M_r 265.22; and chlorozotocin (NSC-178248) [*54749-90-5*], 2-({[(2-chloroethyl)nitrosoamino]carbonyl}amino)-2-deoxy-D-glucose, DCNU, $C_9H_{16}ClN_3O_7$, M_r 313.69. Streptozotocin is a naturally occurring nitrosourea derived from *Streptomyces acromogenes.*

This group of agents was developed by careful structure – function studies based on the antitumor activity of methyl-CCNU [235–237]. The chloroethyl derivatives were found to possess increased activity and a capacity to cross the blood – brain barrier because of their lipophilic nature. Each of these agents is capable of undergoing alkylating reactions with biologic molecules in a manner similar to the classic mustards through the formation of highly reactive chloroethyl carbocations. With BCNU, each molecule of drug may undergo two such reactions to produce nucleic acid strand breaks, and DNA – DNA and DNA – protein cross-links

[238]. Monofunctional CCNU and methyl-CCNU cross-link by initial carbocation formation and alkylation followed by loss of the chloride substituent and formation of a second reactive carbocation. These alkylation reactions appear to be the major mode of cytotoxicity for these agents [239], [240]. With the exception of chlorozotocin, the nitrosoureas may also form an isocyanate compound that may play a role in the toxic side effects of these agents [241], [242], but has little importance in the antitumor effect. In support of this view is the fact that chlorozotocin retains its potent cytotoxic capacity while producing little or no isocyanate compound and reportedly has less marrow toxicity. Experimental evidence raises additional questions concerning the role of isocyanates, which may enhance the antitumor activity of these compounds. Methylnitrosourea, which cannot alkylate DNA to produce cross-linking, produces alterations of nuclear protein in a manner similar to BCNU, and isocyanates are considered a possible explanation for these effects from both agents [243]. Also, sensitivity of a Walker tumor line made resistant to bifunctional alkylating agents can be restored by simultaneous treatment with an isocyanate-producing agent [244].

3.8.1. Mechanism of Action

Because of their lipophilicity, nitrosoureas enter cells by passive diffusion as opposed to an active transport mechanism common to the classic alkylating agents [245]. Once inside the cell, alkylating agent exposure results in pancellular covalent binding of drug to proteins, nucleic acids, and to a variety of smaller intracellular molecules. Which of these reactions is critical for cytotoxicity remains uncertain, but the majority of evidence points to interaction directly with DNA as the focal point of cytotoxicity [246–248]. The 7 position of guanine is particularly susceptible to alkylation, and accounts for the majority of the total alkylation of DNA [249], [250]. Since the chloroethyl nitrosoureas are each capable of two independent alkylations [238], DNA can be cross-linked by either interstrand or intrastrand processes [251]. Multiple studies using a variety of alkylating agents have confirmed that DNA cross-linking, which leads to inactivation of the DNA template, may well be the key mechanism of cytotoxicity [252–254]. Monofunctional DNA alkylations also occur following nitrosourea exposure and these must also be considered cytotoxic as monofunctional alkylating agents, incapable of cross-linking, retain cytotoxic activity. Monofunctional alkylations may produce single-strand DNA breaks by endonuclease cleavage at apurinic sites produced by the alkylation and repair process [255].

3.8.2. Mechanisms of Resistance

The mechanisms of cellular resistance to nitrosoureas remain unclear, although defective drug transport, as has been demonstrated for the classic alkylating agents, would be unlikely given the lack of need for an active membrane transport system. In human glioblastoma cells, increased activity of a specific excision enzyme, guanine-O^6-alkyltransferase, is correlated with in vitro resistance to nitrosoureas [256], and repair mechanisms would seem a likely mechanism since mammalian cells are capable generally of such repair [257].

3.8.3. Analogs

New agents have been synthesized that contain a sugar moiety similar to streptozotocin. Analogs that contain mannose, glucose, ribose, maltose, and galactose have all been produced, with the maltosyl derivative being exceptionally active in a variety of tumors tested [258], [259]. Sugar alcohols, such as mannitol, linked to nitrosoureas retain their cytotoxicity but appear to protect against the marrow toxicity induced by the parent nitrosourea compounds. A number of derivatives with di- and tripeptides containing alkyl nitrosoureas have also been introduced but have met with only moderate preclinical success [260].

There is continuous effort to obtain new nitrosoureas with higher efficiency and/or lower toxicity and several new compounds are undergoing clinical trials. In 1987 ranimustine (NSC-270561, MCNU) [*058994-96-0*], was marketed as Cymerin (Tokyo Tanabe) [261].

The phosphonoalanine derivative fotemustine (S-10036) [*092118-27-9*],

(±)-diethyl[1-[3-(2-chloroethyl)-3-nitroso-ureido]ethyl]phosphonate, $C_9H_{19}ClN_3O_5P$, M_r 315.69, *mp* 85 °C was approved as Muphoran (Servier) for treatment of disseminated malignant melanoma [262], [263]. Side effects of the related compound BCNU (carmustine) have been linked to the inhibitory effect on a major enzyme of the glutathione pathway, the cytosolic glutathione reductase. Fotemustine has no inhibitory effect on cytosolic glutathione reductase, indicating that fotemustine has a lower toxicity than carmustine [264–266]. In addition it has been shown to have lower mutagenicity in the Ames and micronucleus tests compared to BCNU [267]. In the clinical studies the major toxic effects of fotemustine were thrombocytopenia and leukopenia, which were delayed and reversible, nausea and vomiting being mild [268].

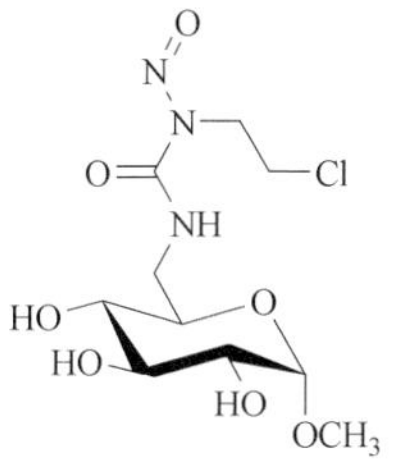

Ranimustine

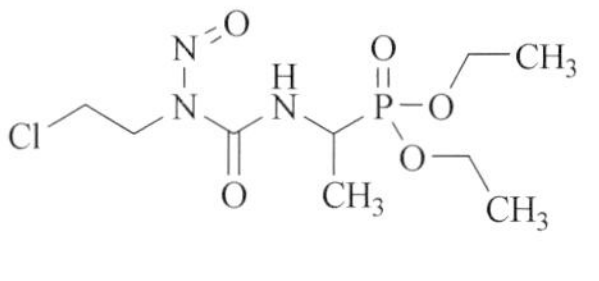

Fotemustine

3.9. Procarbazine

Procarbazine (NSC-77213) [*671-16-9*], *N*-(1-methylethyl)-4-[(2-methylhydrazino)methyl]-benzamide, $C_{12}H_{19}N_3O$, M_r 221.30, is one of a number of substituted hydrazine compounds originally synthesized as monoamine oxidase inhibitors in the early 1960s [269], and found to possess antineoplastic activity, particularly in the treatment of Hodgkin's disease [270]. Procarbazine is nontoxic as the parent compound but undergoes rapid chemical and metabolic degradation to intermediate products that, through a variety of mechanisms, are capable of cytotoxicity [271].

A crucial step in the activation of procarbazine appears to be the production of the azo analog *N*-isopropyl-α-(2-methyldiazeno)-*p*-toluamide, which is catalyzed by hepatic microsomal cytochrome P450 [272], [273]. Further metabolism by hepatic microsomes leads to the production of methyl- and benzyl-azoxy metabolites [272], [273]. Cytotoxic alkylating compounds may be formed from these metabolites via hydroxylation reactions, and these may play the major role in cytotoxicity. This sequence of metabolism and formation of the alkylating intermediates is consistent with the time course of appearance of the active species isolated in the serum and then excreted [274], [275]. In addition to the formation of these alkylating intermediates, methyl- and benzyl-azoxy metabolites may result in free radicals through the formation of diazenes [276], which in the presence of oxygen form free radicals and nitrogen. However, free-radical formation is not likely a major cause of cytotoxicity, because drug activity is preserved when cells are exposed to procarbazine under conditions that do not support the formation of free radicals [277]. The degree of toxicity induced by alkylation and by free-radical formation is unknown.

3.9.1. Mechanism of Action

The exact mechanism by which procarbazine produces cytotoxicity is unknown; however, the effects of its action have been well studied at the cellular level. Chromosome breaks and translocations have been demonstrated in vivo in Ehrlich ascites and L 1210 leukemia cells [278]. Inhibition of nucleic acid and protein synthesis and of a variety of enzymes has also been documented following procarbazine treatment. Inhibition of transfer and nuclear RNA synthesis occurs 2 h after procarbazine exposure and lasts for up to 24 h [279]. Thymidine incorporation into DNA is inhibited concomitantly with protein synthesis, reaching maximum inhibition in 12 – 16 h [271], [280]. Although many potentially cytotoxic events have been associated with procarbazine administration, it is not clear which of these effects causes cell death.

3.9.2. Mechanisms of Resistance

Resistance mechanisms for procarbazine are poorly understood, and no detailed studies have been reported that illustrate the typical cellular resistance encountered with alkylating agents. Since the drug enters cells by simple diffusion, resistance is unlikely to involve altered transport mechanisms [281]. New pieces of chromosomal material were found in Ehrlich ascites cells made resistant to procarbazine [271], and these may represent gene amplification, perhaps encoding for a target enzyme or detoxifying enzyme.

3.10. Dacarbazine

Dacarbazine (NSC-45388) [*4342-03-4*], 5-(3,3-dimethyl-1-triazenyl)-1*H*-imidazole-4-carboxamide, DTIC, $C_6H_{10}N_6O$, M_r 182.18, was synthesized in the late 1950s as an analog of 5-aminoimidazole-4-carboxamide, an intermediate in de novo purine synthesis. The drug is a product of rational synthesis, designed as a false intermediate capable of inhibiting de novo purine synthesis. Despite this theoretical basis for antitumor action, DTIC does not function as a purine analog; instead, it is extensively metabolized to a methylating agent [282]. Similar to procarbazine, DTIC must undergo activation by a microsomal oxidase to form a compound that can spontaneously produce a methyl diazonium ion intermediate that is probably the active metabolite. However, more recent evidence suggests a further metabolism to *N*-hydroxymethyl diazonium ion may be responsible for the selective antitumor effect of the drug [283]. Evidence for methylation of nucleic acids has been demonstrated in tissue culture [284]. An additional metabolic pathway involves the light sensitivity of the drug. Exposure to ultraviolet energy converts the parent compound to metabolites with moderate cytotoxicity in vitro [285], but this probably does not represent an important pathway for cytotoxicity in vivo.

Mechanism of Action. The mechanism of action of DTIC has not been systematically investigated, but it appears that the drug may act during any phase of the cell cycle [286] and can produce inhibition of RNA, DNA, and protein synthesis.

3.11. Hexamethylmelamine

Hexamethylmelamine (NSC-13875) [*645-05-6*], *N,N,N′,N′,N″,N″*-hexamethyl-1,3,5-triazine-2,4,6-triamine, altretamine, $C_9H_{18}N_6$, M_r 210.27, was synthesized in 1951 [287]. It has consistent antitumor activity in a variety of solid tumors, including ovarian, lung, and breast cancer.

The compound is almost insoluble in water, and thus, must be administered orally. Neither the mechanism of action nor the products of metabolic breakdown have been firmly established. Following administration of hexamethylmelamine, a spectrum of *N*-demethylation species has been isolated in the urine. The triazine ring appears unaffected by metabolism, as is evident by almost complete recovery of the intact ring in the urine using ring-labeled compound [288]. Existing evidence suggests possible formation of an alkylating species through *N*-demethylation [289] or the formation of *N*-methylol intermediates by hydroxylation of the parent compound [290] to account for the cytotoxic effects of the drug. *N*-Methylol derivatives have cytotoxic effects in vitro, but whether or not these derivatives are formed in vivo is unclear.

Analogs. Pentamethylmelamine (NSC-118742) [*35832-09-8*], *N,N,N′,N′,N″*-pentamethyl-1,3,5-triazine-2,4,6-triamine, $C_8H_{17}N_6$, M_r 196, is the most commonly used analog of hexamethylmelamine and differs by the absence of a single methyl group. Its major advantage is aqueous solubility, allowing an intravenous formulation. Its metabolism, toxicity, and antitumor activity parallel those of hexamethylmelamine [291], [292]. Other metabolites of hexamethylmelamine with varying numbers of methyl groups also possess antitumor activity that, in general, is directly proportional to the number of methyl groups on the triazene ring [292], [293].

3.12. Mitomycin-C

The mitomycins are a family of antibiotics isolated from *Streptomyces caespitosus*.

Mitomycin-C (NSC-26980, MIT-C) [*50-07-7*] [1aR-(1aα,8β,8aα,8bα)]-6-amino-8-[(aminocarbonyl)oxymethyl]-1,1a,2,8,8a,8b-hexahydro-8a-methoxy-5-methylazirino-[2',3':3,4]pyrrolo[1,2-a]indole-4,7-dione, $C_{15}H_{18}N_4O_5$, has DNA-alkylating properties [294]. Mitomycin is applied for treatment of stomach, breast, and gynecological cancers [295], [296]. Dose-limiting are leucopenia and thrombopenia.

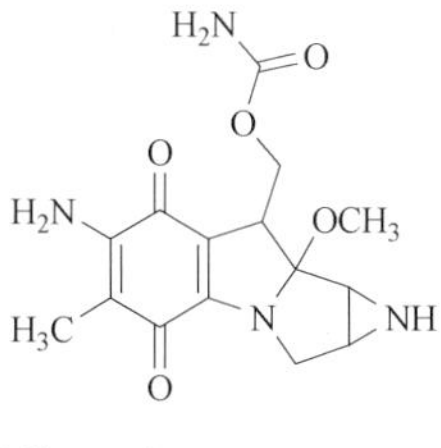

Mitomycin

4. Anthracyclines

Doxorubicin (NSC-123127) [*23214-92-8*], 10-[(3-amino-2,3,6-trideoxy-α-L-lyxo-hexopyranosyl)oxy]-7,8,9,10-tetrahydro-6,8,11-trihydroxy-8-(hydroxyacetyl)-1-methoxy-5,12-naphthacenedione, adriamycin, $C_{27}H_{29}NO_{11}$, M_r 543.54; and daunorubicin (NSC-82151 for HCl salt) [*20830-81-3*], 8-acetyl-10-[(3-amino-2,3,6-trideoxy-α-L-lyxo-hexopyranosyl)oxy]-7,8,9,10-tetrahydro-6,8,11-trihydroxy-1-methoxy-5,12-naphthacenedione, daunomycin, $C_{27}H_{29}NO_{10}$, M_r 527.51. Doxorubicin is used in the treatment of breast cancer, sarcoma, lymphoma, and small-cell lung cancer. Daunorubicin is used more commonly in acute myelocytic and lymphocytic leukemias.

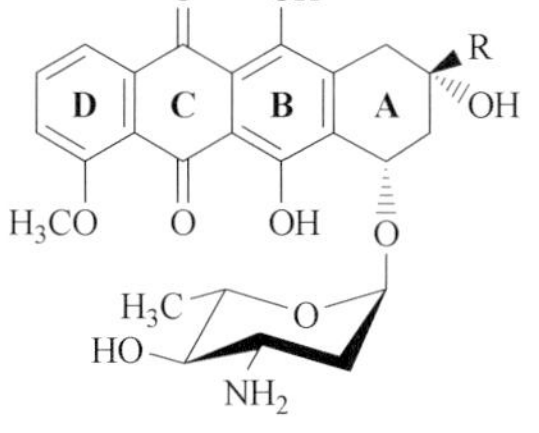

R = $COCH_2OH$, Doxorubicin
R = $COCH_3$, Daunorubicin

Meanwhile epirubicin (4'-epiadriamycin, pidorubicin, IMI-28) [*56420-45-2*], (8S-*cis*)-10-[(3-amino-2,3,6-trideoxy-β-L-arabino-hexopyranosyl)oxy]-7,8,9,10-tetrahydro-6,8,11-trihydroxy-8-(hydroxyacetyl)-1-methoxy-5,12-naphthacenedione, is the most commonly used antineoplastic antibiotic for breast cancer treatment. The total turnover exceeded \$ 200 $\times$ 10^6 in 1996. The compound can be synthesized by classical chemical synthesis [297–299].

Trade names: Farmorubicin and Pharmorubicin (Farmitalia).

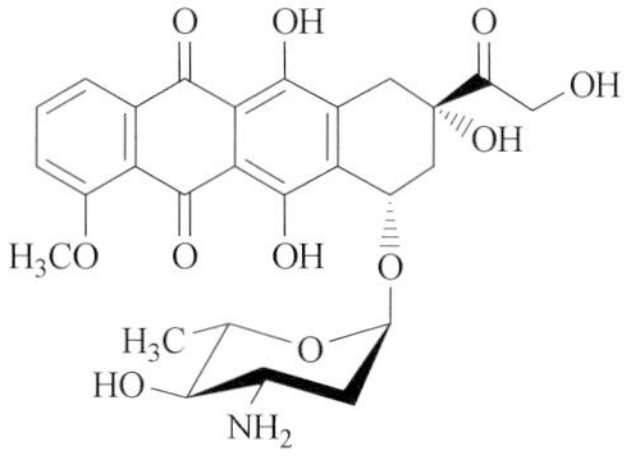

Epirubicin

In addition idarubicin (NSC-256439, IMI-30, DMDR) [*058957-92-99*] and pirarubicin (THP-ADM) [*072496-41-4*], (8S-*cis*)-10-[[3-amino-2,3,6-trideoxy-4-*O*-(tetrahydro-2H-pyran-2-yl]-α-L-lyxo-hexopyranosyl]oxy]-7,8,9,10-tetrahydro-6,8,11-trihydroxy-8-(hydroxyacetyl)-1-methoxy-5,12-naphthacenedione, $C_{32}H_{37}NO_{12}$, M_r 627,64 are available as idamycin (Adria) and pinorubicin (Nippon Kayaku), respectively.

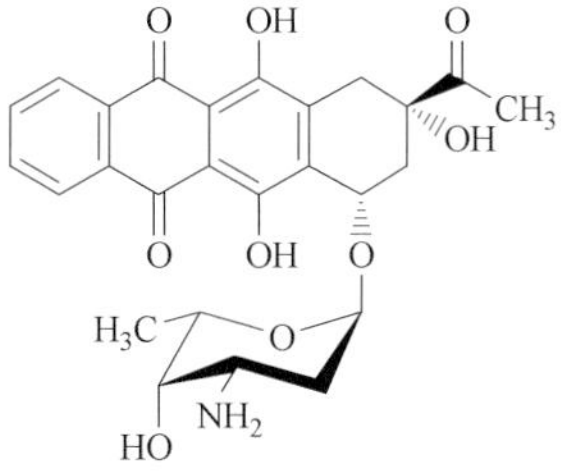

Idarubicin

Pirarubicin is the 4'-*O*-tetrahydropyranyl analog of adriamycin and can be synthesized from adriamycin [300], [301]. The acute cardiac toxicity was significantly less than that of adriamycin and general toxicity lower than that of other analogs [302,303,303]. The main indications for pirarubicin are cancer of the bladder, head and neck, and cervix.

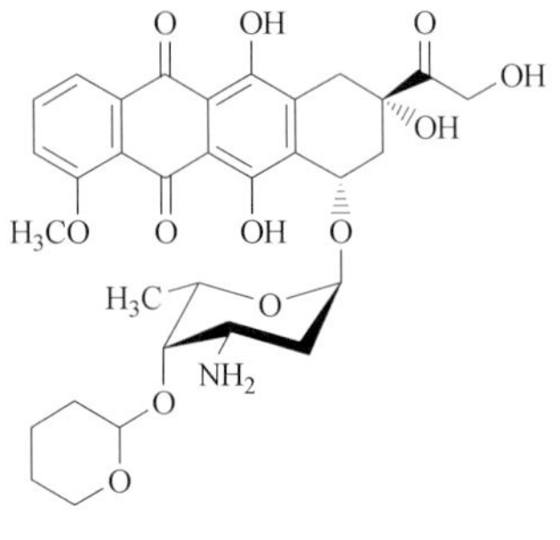

Pirarubicin

The anthracyclines are derived from *Streptomyces* species and are structurally tetracyclic chromophore antibiotics [305]. They are classified by their chromophore, otherwise known as aglycone, structure. The sugar most commonly attached to the aglycones is daunosamine, but other sugars may be involved, and these are mentioned in Section 4.3. The basic tetracyclic aglycone structure of the anthracyclines shares many characteristics with the hydroxyanthraquinones, which are ubiquitous in nature.

4.1. Mechanism of Action

DNA–RNA Binding. The exact mechanism of cytotoxicity by anthracyclines is unknown, but they do have multiple and distinct toxic effects that may kill a tumor cell in one or more ways [306].

Initially these drugs were found to bind DNA by intercalation between base pairs perpendicular to the long axis of the double helix, with the major binding occurring between the B and C rings of the drugs with the bases above and below [307]. The daunosamine sugar is thought to bind ionically with the sugar-phosphate backbone of DNA. The binding association constant is between 10^5 and 10^6 M^{-1}. Intercalation of the DNA causes a partial unwinding of the helix and thus disrupts DNA polymerases and transcription. However, these experiments were done with DNA in vitro; DNA in vivo is organized into chromatin, which is DNA wrapped around a series of histone core particles. Also, it has been established that the anthracyclines can affect every DNA function, including initiation, chain elongation, DNA synthesis, DNA repair, and RNA synthesis [308], [309]. Experimentally these compounds can cause sister chromatid exchanges, single- and double-strand breaks, and alkylation of DNA [310], [311]. Interestingly, some results strongly suggest that inhibition of DNA synthesis is not essential for cell kill. New anthracycline analogs, such as aclacinomycin A, which selectively inhibits preribosomal RNA synthesis and not DNA synthesis, retain cytotoxicity. This suggests other non-DNA-mediated mechanisms of cytotoxicity, such as preribosomal RNA synthesis [309].

Free-Radical Generation. Free-radical formation (highly reactive compounds with an unpaired electron) occurs during the metabolism of anthracyclines. When the microsomal enzyme P 450 reductase or xanthine oxidase interacts with and reduces the ketone oxygen in ring B to O^-, a semiquinone radical intermediate is formed. This interacts with oxygen to produce the superoxide radical with regeneration of the original anthracycline structure [312]. The superoxide radical can serve as a substrate for superoxide dismutase to form hydrogen peroxide, which can interact with the superoxide molecules to form hydroxyl radicals [313]. The superoxide and hydroxyl radicals can interact with and damage cells, especially the hydroxyl radical, which is one of the most reactive substances known. Hydroxyl radicals can also react with purine or pyrimidine bases, amines, and thiols. This free-radical formation is responsible for cardiac damage seen with chronic use of the anthracyclines in the treatment of human malignancies. Evidence exists in many animal models, as well as in humans, that superoxide and hydroxyl radical formation occurs in cardiac tissue, leading to lipid peroxidation of mitochondria and sarcosomes [314], [315]. Since mitochondria account for more than 40 % of cardiac muscle mass, as well as being the major source for ATP needed for contraction, and are coupled to calcium release during the action potential, one can easily visualize how these agents mediate cardiac toxicity. However, it now seems apparent that cardiac tissue lacks catalase and that doxorubicin destroys glutathione peroxidase [315], [316]. Agents that can scavenge free-radicals are under active investigation. Interestingly, there is no evidence to date to link anthracycline free-radical formation to its antitumor activity so that analogs incapable of free-radical formation may demonstrate improved therapeutic index.

Membrane Interactions. Anthracycline binding to cell membranes appears to be an important mechanism for cytotoxicity. Changes in membrane glycoproteins, transmembrane flux of ions, and membrane morphology have been demonstrated in a variety of cells. Doxorubicin binds most tightly to cardiolipin, a phospholipid found in high concentration in mitochondrial and tumor cell membranes but in low concentration in normal cell membranes [317]. Also, membrane redox potential changes occur with drug binding, and may promote free-radical generation. Perhaps the most interesting and important finding to date on this subject is that doxorubicin, covalently linked to beads to prevent cell entry of the drug, retains cytotoxic effects [318]. This suggests that doxorubicin does not need to enter cells or interact with DNA to mediate cell kill.

Another potential mechanism of action may be the dissociation of the cytochrome oxidase electron transport chain for ATP generation [319]. Cytochrome c oxidase requires cardiolipin for activity, and doxorubicin can remove cardiolipin from the enzyme complex, thus inactivating the enzyme.

Metal Ion Chelation. The anthracyclines are capable of chelating ions, including copper, calcium, magnesium, zinc, and iron. Of the metal ions, the tightest complex seems to be with iron(III). Doxorubicin – iron chelates can act as a redox catalyst for electron transfer from glutathione to oxygen, which leads to formation of cytotoxic oxygen radicals. This reaction can also utilize hydrogen peroxide and superoxide, leading to hydroxyl radical formation. Thus, evidence exists that the anthracycline – iron complex can mediate free-radical formation capable of lipid peroxidation and cell damage [320]. Whether this is a mechanism of tumor cell kill is unknown, but the phenomenon may be important to cardiac toxicity.

4.2. Mechanism of Resistance

Probably the most common and important mechanism by which tumor cells become resistant to the anthracyclines is decreased net intracellular accumulation. Many studies have shown that doxorubicin-resistant tumor cells are capable of effluxing the drug more efficiently than their parent sensitive cells [355]. In fact, this mechanism of drug resistance may be responsible for resistance to a variety of structurally unrelated compounds with different modes of antitumor action, such as the vinca alkaloids and actinomycin D. This has been termed pleiotropic drug resistance, and several laboratories have shown the reversal of this resistant phenotype by coincubation with various calcium channel blockers with doxorubicin [356]. These compounds increase the net intracellular drug accumulation in resistant cells, but the precise mechanism of this action is uncertain.

4.3. Analogs

As mentioned in Section 4.1, anthracycline tumor toxicity is probably related to intercalation with DNA, preribosomal RNA inhibition, and/or membrane binding effects, while cardiac toxicity may be due to free-radical formation by drug and drug – metal complex. Thus, analogs have been developed with less potential for free-radical formation. One such drug in early clinical development is aclacinomycin A (NSC-208734), which has an aklavinone aglycone structure and is derived as a fermentation product of *Streptomyces* [321]. This drug lacks the 11-hydroxyl, and therefore has about 10 % the potential to generate free-radicals in the P 450 reductase system. Although it does not bind DNA as well as doxorubicin, it retains significant antitumor activity.

The National Cancer Institute (USA) has screened hundreds of anthracycline analogs for antitumor activity. From these studies it can be concluded that (1) the amino sugar is not required for activity, (2) disaccharide analogs are generally more active than the parent saccharide in inhibiting RNA synthesis, and (3) in the aglycone, substituents in the 7 and 9 positions are important for activity. Studies have also shown that if the 4′-hydroxyl group is removed, cardiac toxicity is lessened significantly [322–324]. Thus, it seems that the amino sugar and/or the 4′-hydroxyl group are major determinants of cardiac toxicity in the anthracyclines.

5. Intercalating Anthracenes and Analogs

5.1. Mitoxantrone

Mitoxantrone hydrochloride (NSC-301739, DAD, CL-232315) [*070476-82-3*], 1,4-dihydroxy-5,8-bis[(2-(2-hydroxyethyl)amino)ethyl]amino-9,10-anthracenedione, is a new type of antineoplastic agent. Mitoxantrone is active against breast cancer, acute leukemia, lymphoma, cervix carcinoma, and liver cell cancer. Unlike the anthracyclines that have a red color, the anthracenediones are deep blue. Mitoxantrone is structurally similar to adriamycin but without the aminosugar at C9 and can be synthesized starting from 1,8-dihydroxyanthrachinone [325–329].

Trade names: Novanthrone (Lederle / American Cyanamide); Onkotrone (ASTA Medica AWD).

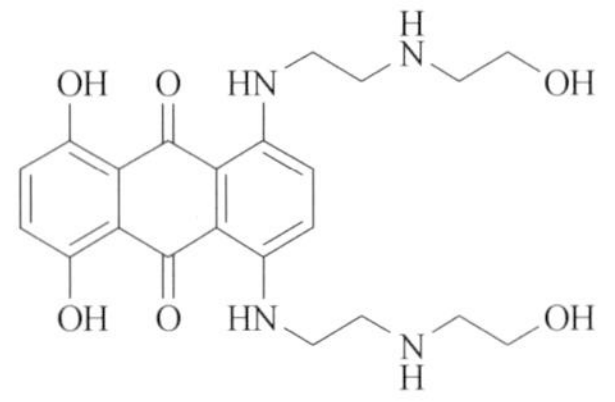

Mitoxantrone

Mechanism of Action. The quinone structure of mitoxantrone was recognized as being similar to that of adriamycin, having a lower cardiotoxic potential. However, its discovery was both a result of serendipity and of rational drug development [325]. The exact mechanism of action by which mitoxantrone exerts its cytotoxic effects has not been fully defined. The cytotoxicity is most likely associated with the action on chromosomal elements, resulting in DNA damage and leading to inhibition of nucleic acid synthesis and the eventual death of the cell. When compared to doxorubicin on an equimolar basis mitoxantrone proved to be six to seven times more potent in inhibiting the incorporation of uridine and thymidine into DNA [330]. Mitoxantrone is a cell phase nonspecific agent and has wide spectrum of activity against several experimental animal tumors. Cross-resistance to adriamycin was not always seen [331–333]. Adriamycin-like cardiac toxicity was not found in comparative studies using rats, dogs, rabbits, and monkeys maybe partially due to inhibition of free radical formation and due to the lack of the amino sugar moiety [334], [335]. Although mitoxantrone is not entirely free of cardiac toxicity in humans, minimal nausea and vomiting was observed [336–341].

5.2. Analogs

Bisantrene hydrochloride (NSC-337766, ADD, CL-216942) [*071439-68-4*], 9,10-anthracenedicarboxaldehyde bis[(4,5-dihydro-1H-imidazol-2-yl] dihydrochloride, $C_{22}H_{22}N_8$'2 HCl, M_r 471.39, is an intercalating anthracenebishydrazine cytostatic [342–344]. The product was launched for treatment of acute non-lymphocytic leukemia [345].

Trade names: Zantrene and Cyabin (Lederle).

As solubility is a problem, bisantrene prodrugs with enhanced water solubility (e.g., 199344, see below) were developed [346–348].

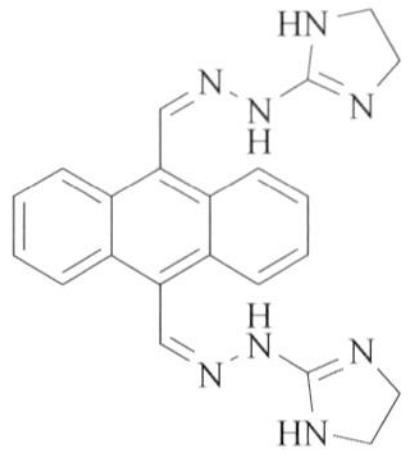

Bisantrene

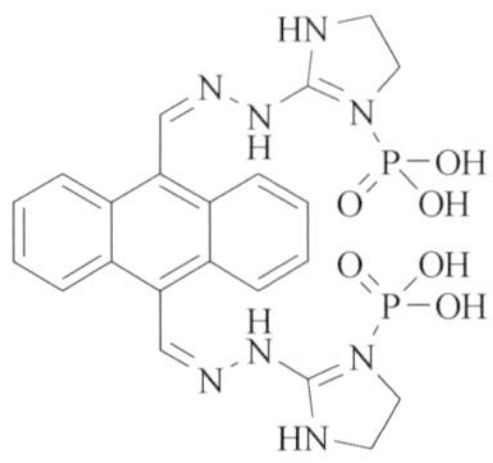

199344

Further intercalating agents such as amsacrine (NSC-156303/NSC-249992, m-AMSA, SN-11841) [*051264-14-3*] [349], [350], or nitracin [*4533-39-5*] [351] have the acridine structure in common.

Trade names: Amsacrine is marketed as Amsidine, Amecrin, and Ansidyl (Parke-Davis) [352–354].

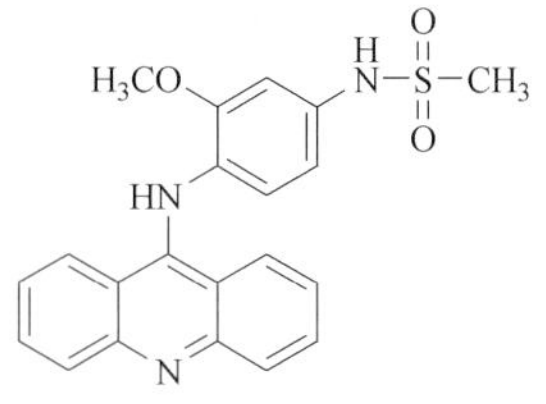

Amsacrine

6. Antitumor Antibiotics Other than Anthracyclines

The antitumor antibiotics are a group of microbial products capable of inhibiting tumor growth. This class of antitumor agents has been extensively studied as to mechanism of action and has a rather broad spectrum of activity. In contrast to antibacterial antibiotics, the therapeutic index of these drugs tends to be narrow and toxicity to normal host tissues is considerable.

6.1. Actinomycin D

The actinomycins are a family of antibiotics derived from *Actinomyces* broths during the 1940s [357]. Structurally, compounds in this class share a common phenoxone ring and two cyclic pentapeptides. The natural products differ among themselves in the amino acid composition of the peptide chains, but only actinomycin D is used in the clinical treatment of cancer, where it demonstrates reproducible activity against gestational choriocarcinoma and Wilms' tumor.

Actinomycin D (NSC-3053) [*50-76-0*], dactinomycin, $C_{62}H_{86}N_{12}O_{16}$, M_r 1255.47.

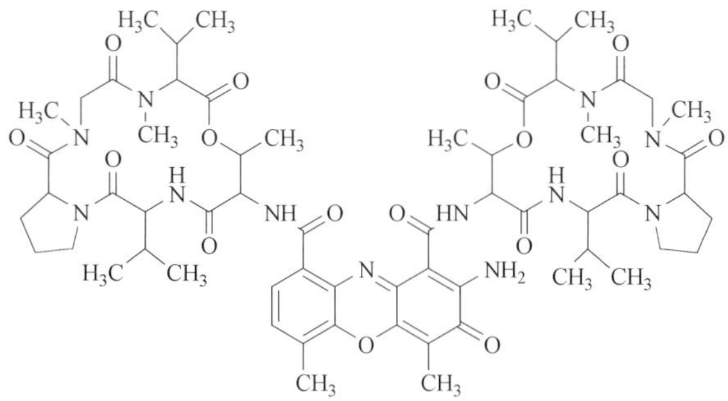

Actinomycin D

Mechanism of Action. Actinomycin interacts with DNA through "pseudo-intercalation" at deoxyguanylyl-3′,5′-deoxycytidine sequences [358], [359], resulting in inhibition of DNA-directed RNA synthesis and inhibition of protein synthesis [360]. Resistance to actinomycin D is linked to cross-resistance against other structurally unrelated amphiphilic drugs with dissimilar mechanisms of action as part of the phenomenon of pleiotropic drug resistance (see Section 4.2). Certainly it had been previously demonstrated that the accumulation of drug is greater in sensitive cells [361], [362], suggesting that alterations in transport at the membrane level might be important to resistance.

6.2. Bleomycin

Bleomycins (NSC-125066) are a group of antitumor antibiotics initially isolated from broths of *Streptomyces verticillus* [363].

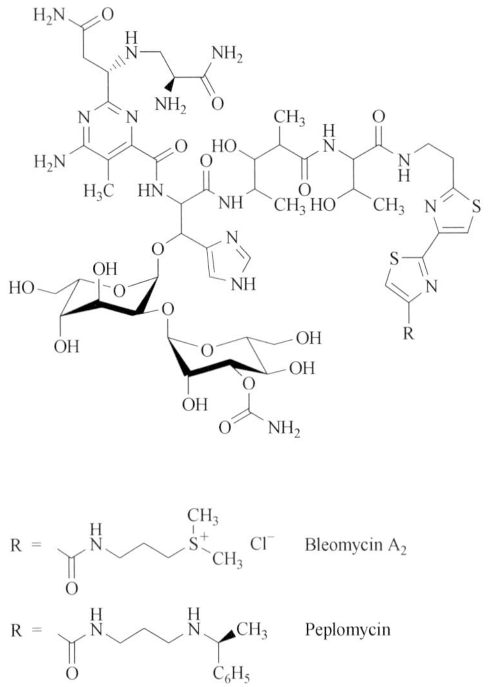

Bleomycin A_2

Peplomycin

The fermentation product consists of approximately 12 different components clinically marketed as bleomycin. Each is a peptide with a low relative molecular mass (ca. 1500), all containing bleomycinic acid but differing in terminal alkylamine groups. The clinical product is approximately 60 – 70 % bleomycin A_2 (NSC-146842), N^1-[3-(dimethylsulfonio)propyl]bleomycinamide, $C_{55}H_{84}N_{17}O_{21}S_3$, M_r 1416; and 20 – 30 % bleomycin B_2. Other analogs comprise about 5 % of the total. The drug is highly active against germ cell neoplasm of the testis.

6.2.1. Analogs

Peplomycin [*68247-85-8*], N^1-{3-[(1-phenylethyl)amino]propyl}bleomycinamide, $C_{61}H_{88}N_{18}O_{21}S_2$, M_r 1473.62, a biosynthetic bleomycin analog, demonstrates significantly reduced pulmonary toxicity in rodents [364]. The toxicology studies are supported by clinical trials in Japan (initiated in 1981) and France (1983), and suggest that peplomycin has a greater therapeutic index than the parent compound and may replace bleomycin in the clinic [365]. In addition, total synthesis of bleomycin described in the 1980s will likely lead to other analogs and a further understanding of the drug's mechanism of action [367]. A new drug delivery system which comprises peplomycin absorbed on to small activated carbon particles was introduced as slow release formulation suggesting to decrease the systemic toxicity [366].

6.2.2. Mechanism of Action

Bleomycin enters cells slowly. Labeling studies demonstrate that the drug is first detected on the cell membrane and reaches the nucleus only after several hours [368]. Bleomycin kills cells by producing single- and double-strand DNA breaks. Bleomycin first binds guanine bases in DNA through the amino terminal peptide of the drug [369]. Free-radical formation occurs through oxidation of a bleomycin – Fe(II) complex to Fe(III), which catalyzes the reduction of molecular oxygen to superoxides and free hydroxyl radicals [370]. In vitro resistance to the drug appears to be mediated either by defective accumulation [371] or by increased intracellular drug degradation by a specific bleomycin hydrolase [372–374].

The clinical toxicity of bleomycin is unusual. The drug has little hematopoietic toxicity; its major dose-limiting toxicities are to the lungs and skin. Acute and chronic pneumonitis followed by progressive pulmonary fibrosis appears to be dose dependent, with risk increasing significantly in patients receiving a cumulative dose of more than 450 mg. This toxic endpoint may be due to the terminal amines of the parent compound. This hypothesis is supported by preclinical studies [375] as well as by early clinical studies of peplomycin. The maximum-tolerated cumulative dose of this terminal amino acid-substituted bleomycin analog has yet to be determined, although during phase I analysis, pulmonary toxicity was not observed until weekly doses exceeded 10 mg/m^2 [376].

6.3. DNA Interactive Natural Products

Analogs related to the natural product CC-1065 like adozelesin, bizelesin, and carzelesin bind in the minor grove of the DNA and form a covalent adduct with adenine [377]. Adozelesin

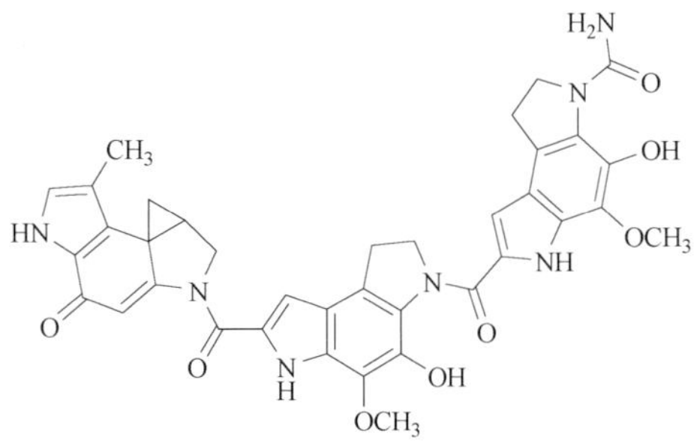

CC-1065

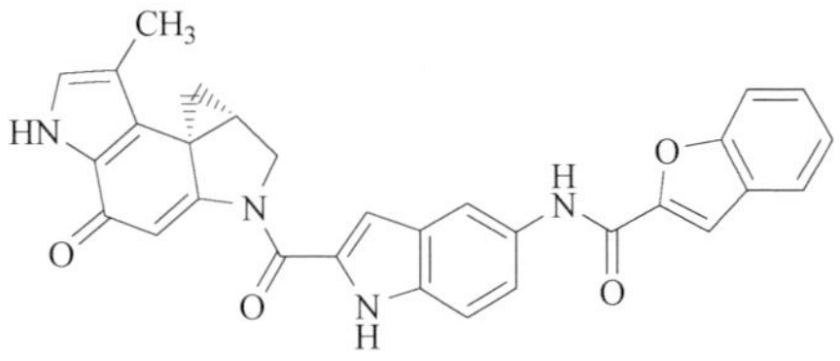

Adozelesin

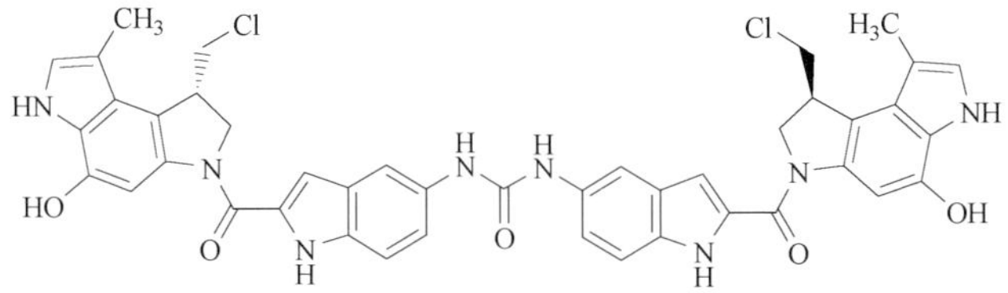

Bizelesin

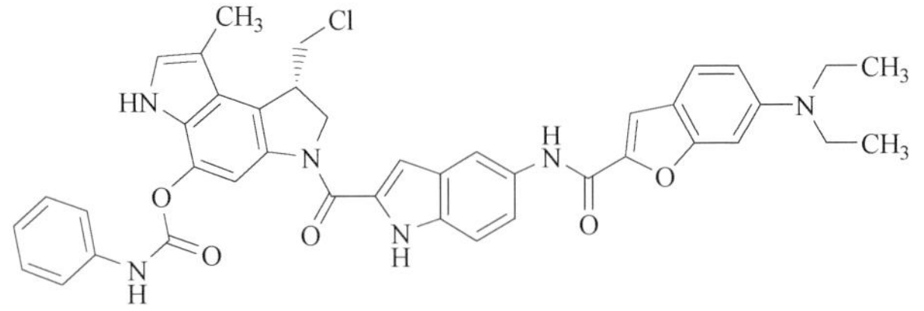

Carzelesin

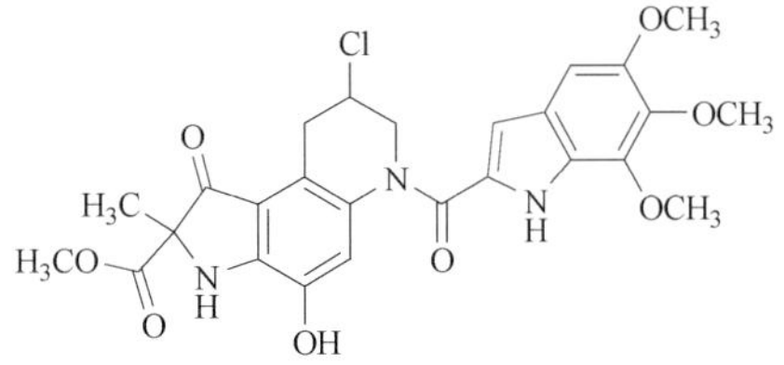

Duocarmycin C1

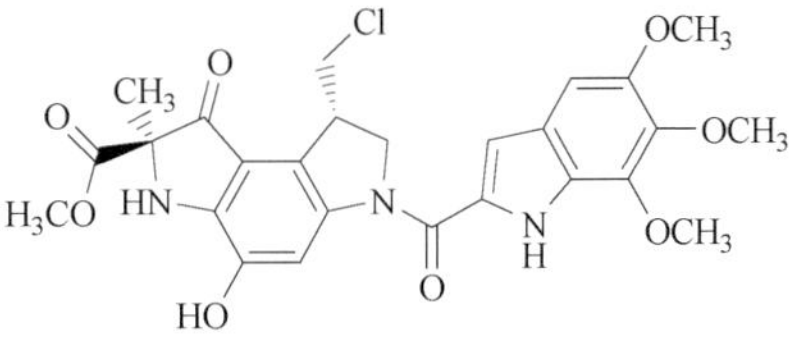

Pyrindamycin A

[*110314-48-2*] [378] is the clinically farthest advanced agent, while the prodrug carzelesin [*119813-10-4*] [379] and the dimer bizelesin [*129655-21-6*] [380] demonstrated impressive preclinical activity.

Further modified cyclopropylpyrroloindoles (CPI) forming covalent adducts with DNA are duocarmycin A [*118292-35-6*] [381], [382], pyrindamycin [*118292-36-7*] and FCE 24517 [383], [384].

A series of new DNA cleaving molecules based on the reactive enediyne moiety are the anticancer antibiotics calicheamicin [*113440-58-7*] [385], esperamicin [*114797-28-3*] [386], and dynemicin.

Simplified enediyne-type compounds damage DNA upon activation by chemical or biological means and are extremely potent cytotoxic agents in vitro [387], [388].

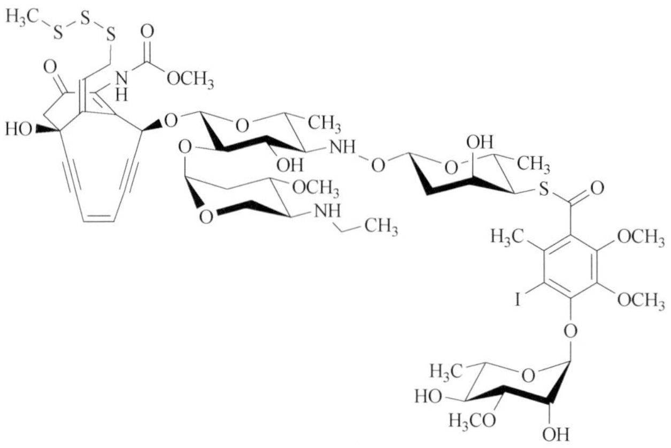

Esparamicin A1

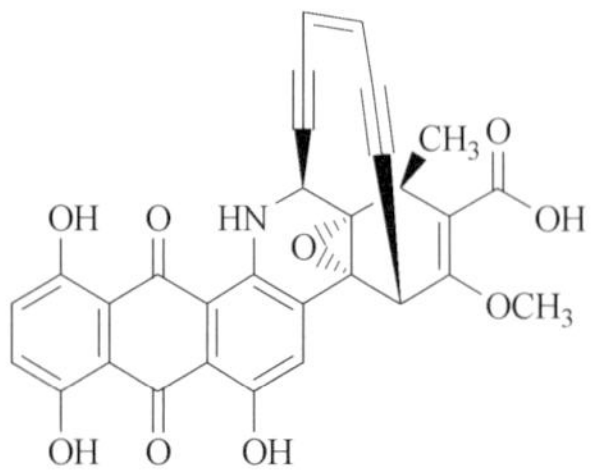

Dynemicin A

7. Antitubulin Agents

Tubulin-containing structures are important for diverse cellular functions, including chromosome segregation during cell division, intracellular transport, development and maintenance of cell shape, cell motility, and possibly distribution of molecules on cell membranes. The drugs that interact with tubulin are heterogeneous in structure. A common characteristic of these agents is binding to tubulin, causing its precipitation and sequestration to interrupt many important biologic functions that depend on the microtubular class of subcellular organelles. The tubulin – drug aggregates can be visualized by the indirect immunofluorescence technique as brightly stained cytoplasmic paracrystals. Of the tubulin binders, those that are important in cancer medicine include vinca alkaloids, podophyllotoxins, and taxoids.

7.1. Vinca Alkaloids

Vinca alkaloids are dimeric indole derivatives isolated from the periwinkle plant, *Catharantus roseus*. Of the whole family of more than 70 naturally occurring alkaloids, only a few have cytotoxic activity.

7.1.1. Vincristine and Vinblastine

The molecular structures of the dimeric *Catharantus roseus* alkaloids vincristine (NSC-67574) [*57-22-7*], 22-oxovincaleukoblastine, $C_{46}H_{56}N_4O_{10}$, M_r 824.94, Oncovin (sulfate); and vinblastine (NSC-49842) [*865-21-4*], vincaleukoblastine, $C_{46}H_{58}N_4O_9$, M_r 811.00, Velban (sulfate hydrate), are very similar. Both are formed of multiringed units, vindoline and catharantine, linked by a carbon – carbon bridge. They differ only in the nature of the substituent on the vindoline nitrogen atom.

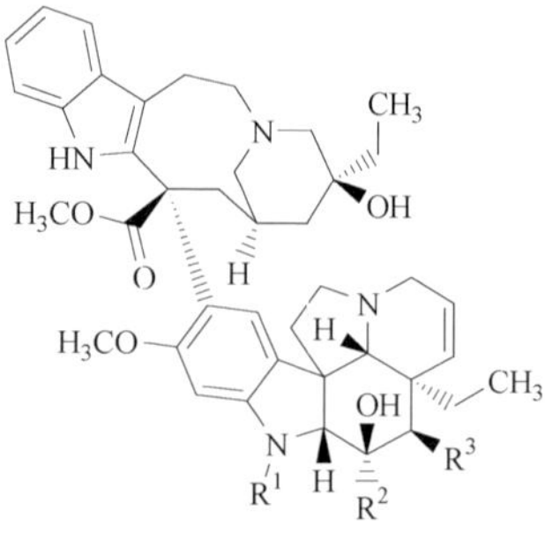

$R^1 = CHO$, $R^2 = CO(OCH_3)$, $R^3 = CO(CH_3)$, Vinristine
$R^1 = CH_3$, $R^2 = CO(OCH_3)$, $R^3 = CO(CH_3)$, Vinblastine
$R^1 = CH_3$, $R^2 = CO(NH_2)$, $R^3 = OH$, Vindesine

Cellular Pharmacology. As yet it is not clear how vinca alkaloids cross cell membranes. Some data suggest an energy-dependent transport system [389], while other data suggest simple diffusion across membranes [390]. However, passive diffusion is important only at drug concentrations exceeding 100 μM. Transport of vincristine is completely inhibited by vinblastine, suggesting a common carrier.

Vincristine and vinblastine exert their biologic effect through binding to tubulin. This occurs in interphase (late S and G_2), producing a transient G_2 block [391]. They do not affect DNA synthesis directly. The metabolic consequences of tubulin binding include polyploidy, nuclear fragmentation, and inhibition of cytokinesis, which occurs after prolonged drug exposure. Present studies suggest that sensitivity to these drugs increases progressively as cells approach mitosis and that cells at the end of the cycle at the time of exposure are likely to exhibit the greatest degree of mitotic disorganization [391].

Vincristine and vinblastine share a common binding site on each tubulin monomer, with the binding affinity (K_d) of about 1.6×10^{-6} M. The drug concentrations necessary to produce 50 % cell kill in vitro are ca. 4×10^{-8} M [392]. Malignant lymphocytes appear more susceptible than normal lymphocytes, presumably because of high tubulin content expressed on the surfaces of leukemic cells [393]. Other metabolic effects include inhibition of DNA and RNA and protein synthesis. However, these effects are exerted only at very high drug concentrations (1000 times greater than those achieved in vivo) and probably are secondary phenomena.

Drug Interactions. Both vincristine and vinblastine potentiate the effect of methotrexate through their blockade of methotrexate exit from cells [394], [395]. Some amino acids (glutamine, aspartic acid, ornithine, citrulline, and arginine) completely reverse the cytotoxic effect of vinblastine in tissue culture [396].

Mechanisms of Resistance. Vinca resistance may arise through mutations in tubulin, leading to decreased drug binding. Vinca-resistant cells may also share cross-resistance with antitumor antibiotics through a separate mechanism of resistance. The precise mechanism of this pleiotropic drug resistance remains to be clarified. However, tubulin is not responsible for that phenomenon because there is very little difference in affinity binding of colchicine to tubulin isolated from drug-sensitive and drug-resistant cells [397]. The appearance of a novel glycoprotein on the membrane of resistant cells and the accelerated drug efflux – leading to impaired drug accumulation – seem, at present, to be the possible mechanisms responsible for resistance [398].

7.1.2. Vindesine

Vindesine (NSC-245467) [*53643-48-4*], 3-(aminocarbonyl)-O^4-deacetyl-3-de(methoxy-carbonyl)vincaleukoblastine, deacetylvinblastineamide, $C_{43}H_{55}N_5O_7$, M_r 753.95 (structure, Section 7.1.1), is a vinblastine metabolite. It possesses antitumor activity that is similar to vincristine's rather than that of its parent compound, vinblastine. Vindesine demonstrates better activity than vinblastine in some tumor models (Gardner lymphosarcoma, Ridgway osteogenic sarcoma, and mammary carcinoma). In the past several years vindesine has attracted considerable attention among clinical investigators and has been intensively investigated. Response rates vary, with the highest responses observed in the highest lymphatic malignancies [399]. The drug is less neurotoxic than vincristine and appears to be active in vincristine-resistant tumors [400], [401].

7.1.3. Vinorelbine

Vinorelbine (KW-2307, NVB) [*071486-22-1*] 3',4'-didehydro-4'-deoxy-C'-norvincaleukoblastine, $C_{45}H_{54}N_4O_8$, M_r 778.94, is a semisynthetic vinca alkaloid. Starting from anhydrovinblastine the compound is obtained as 5'-noranhydrovinblastine in three steps [402], [403]. The product was launched for the treatment of non-small cell lung cancer.

Trade names: Navelbine (Pierre-Fabre/Glaxo), Ennades (Farmitalia).

Mechanism of Action. Vinorelbine was selected for drug development due to its high affinity for tubulin and its ability to prevent tubulin polymerization [403–405]. The compound induced total depolymerization of microtubules in P 388 murine leukemia cells, possibly via stimulation of microtubular protein synthesis [403]. Vinorelbine and vincristine were equally active against L 1210 leukemia in mice, while vinblastine had no significant effect. Vinorelbine exerted significant antitumor activity in the vincristine-resistant cell line P 377/VCR and low cross-resistance to other vinca alkaloids was observed [406], [407].

In clinical studies leukopenia was dose-limiting, no thrombocytopenia was observed [408]. Efficacy in non-small cell lung cancer was demonstrated [409–411].

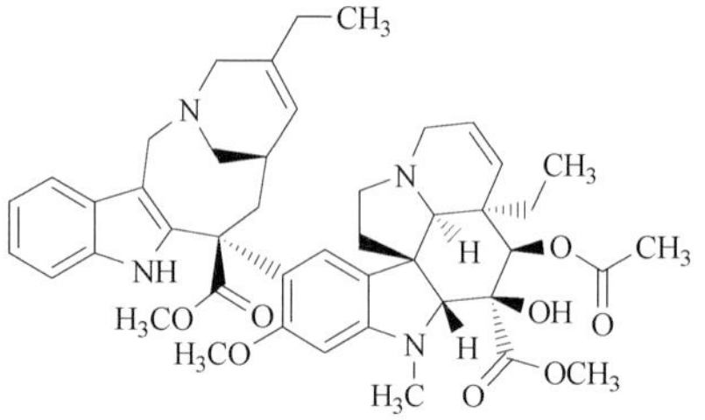

Vinorelbine

7.2. Podophyllotoxin and Its Derivatives

Podophyllin is a complex mixture of crystalline compounds derived from the mayapple plant. The active agent derived from this plant product, podophyllotoxin [*518-28-5*], shares a common binding to tubulin with colchicine, and the morphological effects of podophyllotoxin and colchicine exposure are indistinguishable [412]. The semisynthetic glycoside derivatives of podophyllotoxin, VP-16-213 (NSC-141540) [*33419-42-0*], etoposide, and VM-26 (NSC-122819) [*29767-20-2*], teniposide, have reproducible clinical activity against testicular cancer, small-cell lung cancer, and lymphomas [413].

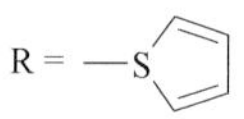

R = —S(thienyl) Teniposide (VM-26)

$R = CH_3$ Etoposide (VB-16-213)

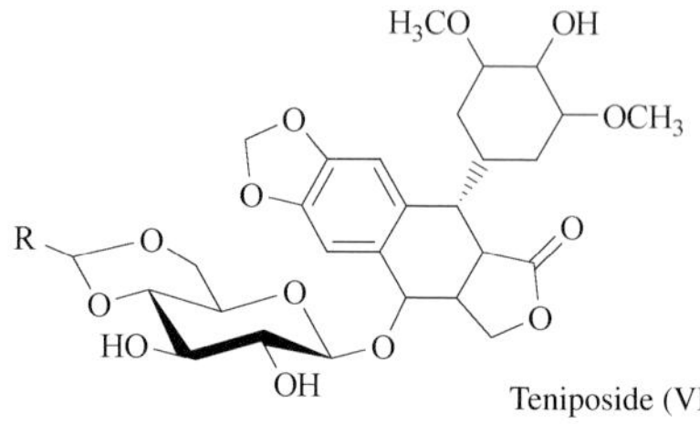

Despite structural similarities between the semisynthetic derivatives and parent natural product, neither VP-16 nor VM-26 binds tubulin. Evidence from a number of laboratories suggests that VP-16 exerts cytotoxic effects by causing DNA breakage [414], [415]. Selective double-strand DNA breaks, which are particularly lethal, appear to be caused by these agents [416].

The precise mechanism for VP-16 and VM-26 DNA damage is unknown, but the fact that the presence of a 4′-hydroxyl group is critical for cytotoxicity suggests formation of free-radical intermediates. Indeed, free-radical scavengers appear to be able to protect from podophyllin derivative cytotoxic effects in vitro [417]. Formation of DNA breaks may also be due to specific inhibition of the DNA repair enzyme topoisomerase II by both VP-16 and VM-26 [418]. It is likely that further characterization of topoisomerase II and the effects of its inhibition will result in the development of new leads in antitumor drug development.

7.3. Camptothecin and Analogs

Camptothecin (CPT) [*7689-03-4*], an alkaloid isolated from the Chinese plant *Camptotheca accuminata* in 1966 [419] was proved to have antineoplastic activity in various tumor systems [420]. The clinical use of CPT was, however, limited because of its high toxicity and low solubility in water. New derivatives have been synthesized carrying substitutents in 7,9,10- or 11-position of the ring A.

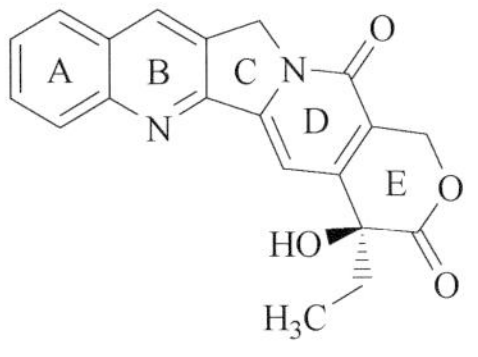

Campothecin

Irinotecan (NSC-616348, CPT-11, DQ-2805) [*097682-44-5*] [421], (+)-7-ethyl-10-[4-(1-piperidyl)-1-piperidyl]carbonyloxycamptothecin, $C_{33}H_{38}N_4O_6$, M_r 586.69 and topotecan (NSC-609699, SKF-S-104864-A, E-89/001) [*119413-54-4*] [422], (S)-10-dimethylaminomethyl-9-hydroxycamptothecin, $C_{23}H_{23}N_3O_5$, M_r 457.91 are camptothecin derivatives with topoisomerase I-inhibitory activity. As antineoplastic alkaloids, they can be synthesized semisynthetically starting from CPT [423]. In addition both agents have intercalating properties and are water soluble. Irinotecan hydrochloride was launched for the treatment of small cell and non-small cell lung cancer and cancers of the uterine, cervix, and ovaries.

Trade names: Campto (Rhône-Poulenc Rorer/Yakult Honsha) Topotecin (Daiichi Seiyaku).

Unlike irinotecan, topotecan is not a prodrug and does not require metabolic activation.

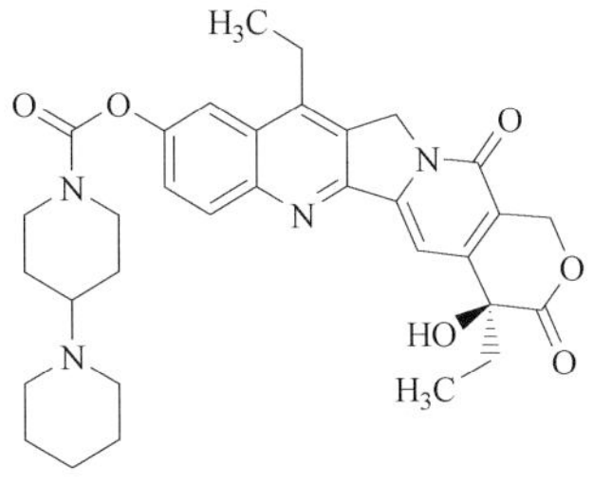

Irinotecan

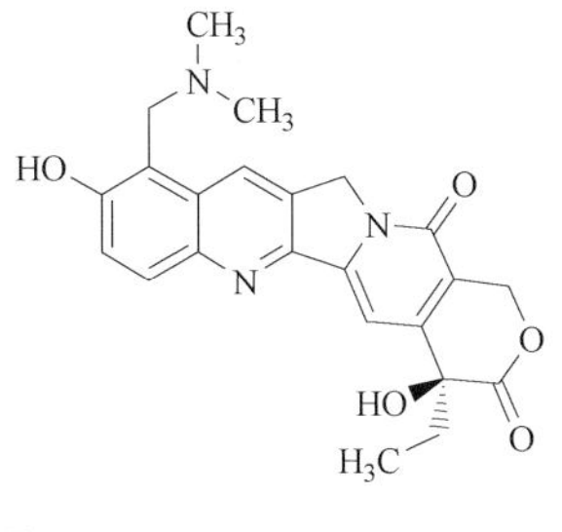

Topotecan

Mechanism of Action. Topoisomerase (topo I) is an ubiquitous nuclear enzyme involved in the regulation of essential cellular functions by relieving torsional DNA strain. Relaxation of supercoiled DNA is achieved through a series of topo I-mediated reactions. The main physiological role of topo-I is in sensing and subsequently releasing the positive supercoiling generated ahead of the moving transcription apparatus. A number of malignancies including acute leukemia, blasts, colon, esophageal, and ovarian cancers contain increased topo-I levels as compared to normal tissues [424], [425]. Topotecan binding to topo-I-DNA adducts results in markedly decreased rates of nick resealing and in delayed enzyme release, leading to increased numbers of strand breaks [426].

In animal tumor models including xenografts of human cancer lines topotecan and irinotecan exhibit a wide spectrum of antineoplastic activity [427–429]. Cell lines overexpressing P-glycoprotein display low levels of resistance to topotecan through decreased drug retention, more pronounced resistance is found in cell lines containing low levels of topo I [430], [431].

Dose-limiting toxicity in patients were stomatitis and esophagitis as well as neutropenia [432].

7.4. Taxoids

Paclitaxel (NSC-125973, BMS-181339) [*33069-62-4*], $C_{47}H_{51}NO_{14}$, M_r 853.9, is a natural antineoplastic taxane derivative originally isolated from the plant *Taxus brevifolia* [433] and subsequently synthesized due to limited supply from the bark of endangered yew trees. Starting from 10-deacetyl baccatin III (isolated from the renewable twigs and needles of *Taxus baccata*) the compound can be prepared in three steps utilizing protected *N*-benzoyl-(2R,3S)-3-phenylisoserine as key building block [434–436]. Various synthetic approaches are described and patented, including three total synthesis of paclitaxel [437–440].

Paclitaxel has a novel mechanism of action. Unlike vincristine or vinblastine, which bind tubulin to inhibit tubulin polymerization, paclitaxel stabilizes microtubular structures [441]. Indicated for the treatment of primary ovarian cancer in combination with cisplatin and for metastatic ovarian cancer where standard therapy has failed, paclitaxel has been marketed since 1993 as Taxol (Bristol-Myers Squibb) and was the best-selling anticancer agent in 1996. Taxol does not share a common binding site with other antitubulin agents and may instead bind to and stabilize polymerized tubulin [442].

A semisynthetic analog of paclitaxel, docetaxel (NSC-628503, RP-56976) [*114977-28-5*], (2R,3S)-*N*-carboxy-3-phenylisoserine, *N-tert*-butyl ester, 13-ester with 5β, 20-epoxy-1,2α-4,7β,10β,13α-hexahydroxytax-11-en-9-one 4-acetate 2-benzoate, $C_{43}H_{53}NO_{14}$, M_r 807.9, is a derivative with an *N-tert*-butyloxycarbonyl-(2R,3S)phenylisoserine C-13 side chain. Docetaxel is more potent than paclitaxel in vitro [443], [444]. Impressive clinical results have been reported for the treatment of ovarian, breast, and bronchial cancers with docetaxel [445]. The compound was launched in 1996 for the treatment of locally advanced breast cancer or relapse during anthracycline therapy of NSCLC and breast cancer.

Trade name: Taxotere (Rhône-Poulenc Rorer).

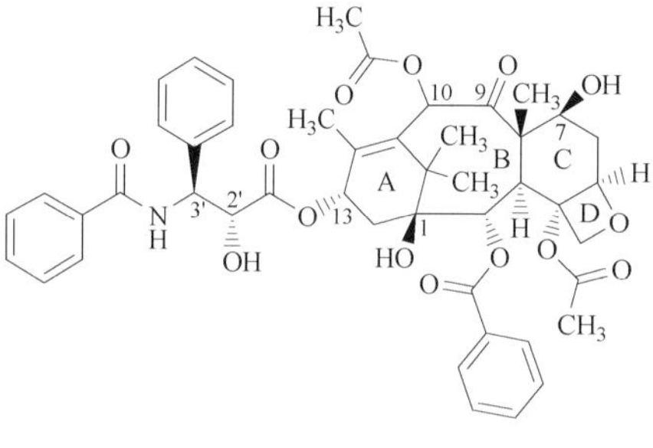

Paclitaxel

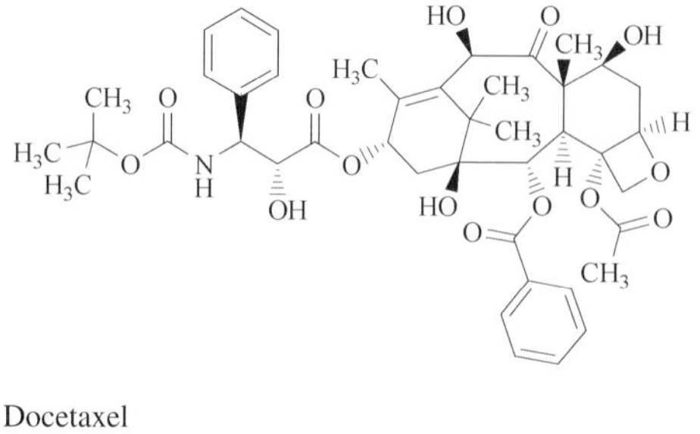

Docetaxel

Mechanism of Action. More than two decades after its isolation and the elucidation of its complex structure and cytotoxic activity [433] interest in paclitaxel raised again when S. HORWITZ et al. [446] reported on the novel mechanism of action. As spindle poison paclitaxel promotes the polymerization of tubulin to microtubules and stabilizes them against depolymerization, whereas vinca alkaloids induce microtubule disassembly (see Fig. 2). Thus, with paclitaxel the dynamic equilibrium of assembly and disassembly of microtubules is shifted in favor of the polymer, preventing cell division [447]. Paclitaxel binds preferentially to the β-tubulin subunit. This binding is reversible and the site is different from the binding sites of vinca alkaloids, colchicine and podophyllotoxin [448–450]. If paclitaxel is present, tubulin polymerizes without exogenous GTP and these stabilized, rigid microtubules cannot be disassembled. As a result, the dynamic organization of the cell is interrupted which leads to irreversible damage in rapidly dividing cells. Further evidence has been reported that the antiproliferative activity of paclitaxel is caused by additional effects [451–453].

A major impediment in the development of taxol as a drug was its poor water solubility [454]. With the solubility enhancer Cremophor EL hypersensitivity reactions, including hypotension, urticaria, and dyspnea, occurred in patients during rapid infusion. To cope these allergic reactions, 24-hour infusions and pretreatment with dexamethasone, diphenhydramine, or cimetidine was recommended [455].

The dose-limiting *toxicity* of paclitaxel is neutropenia. Several other toxic effects such as diarrhea, nausea, and emesis are less common. Docetaxel shares many toxic effects with paclitaxel such as dose-limiting neutropenia, alopenia, myalgias, and mucositis. In addition fluid retention and cutanous toxicities are observed [445], [460], [461]. Because of the low water solubility of paclitaxel and docetaxel, the synthesis of more soluble taxoid prodrugs and smaller analogs has become an interesting area of research (see Section 7.5).

Structure – Activity Relationship. A large number of taxoid analogs has been synthesized with emphasis to enhance biological activity and to improve the water solubility [436]. C-13

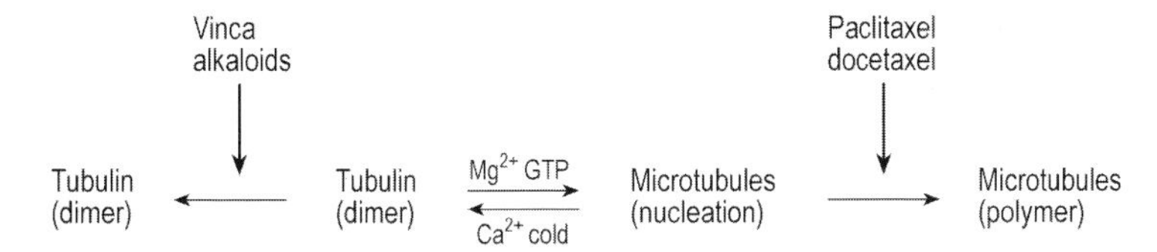

Figure 2. Mechanism of action of paclitaxel and vinca alkaloids

side chain depleted analogs such as baccatin III and its derivatives [456] as well as *N*-benzoyl-(2R,3S)-phenylisoserine are inactive. Simplified side chains at C-13 (like acetic, crotonic, or phenylacetic acid) possess reduced activity.

3'-Cyclohexyl-3'-dephenylpaclitaxel has a similar cytotoxicity to paclitaxel. Further modifications in the aromatic 3'-phenyl group gave compounds that were equipotent with paclitaxel. The compound with a 3'-(*p*-methoxyphenyl) has a slightly increased activity. Compounds with different substituents in the *N*-benzoyl part were similar to paclitaxel if these substituents were aromatic, aliphatic substituents reduced cytotoxicity.

As the 2'-hydroxyl group is essential for maximal biological activity [457] esterification leads to a total loss of activity in the microtubule assay, whereas cytotoxicity remains unchanged. Thus various amino acid esters were produced as prodrugs.

Notable loss of cytotoxicity was observed with A-ring modified analogs and oxidation at C-10 or C-7. 7-Acetylpaclitaxel and the C-7 epimer were similar in their ability to inhibit cell proliferation. All derivatives without the intact oxetane moiety are inactive [458]. Further, all C-4 modified analogs were devoid of activity underscoring the vital importance of these functional groups [459].

7.5. Epothilone A and B

The high cytotoxicity and good stabilization of microtubule raised interest in the natural products epothilone A [*152044-53-6*] and B [*152044-54-7*] originally isolated from myxobacteria *Sorangium cellulosum* [462]. Since BOLLAG et al. [463] reported on the mechanism of action, which resembles that of paclitaxel numerous reports were published on total synthesis [464–468] and biology of epothilones. Their unique capability to inhibit taxol resistant tumor cell lines [470] and their good solubility in water are the biggest advantages as compared to paclitaxel. Their in vivo activity is similar to that of paclitaxel [469].

R = H, Epothilone A
R = CH_3, Epothilone B

8. Heavy-Metal Complexes

8.1. *cis*-Platinum

cis-Platinum (NSC-119875) [*15663-27-1*], *cis*-diamminedichloroplatinum, cisplatin, $PtCl_2(NH_3)_2$, M_r 300.05, was the first heavy-metal compound to be introduced into clinical cancer chemotherapy. The discovery of this unique agent was predicated on the serendipitous observation that bacterial growth was inhibited when culture medium was subjected to an alternating current using platinum electrodes [471]. Moreover, the spent medium itself developed bacteriocidal characteristics, even in the absence of electrical current. Detailed analysis confirmed that, of the several platinum species produced by electrolysis, it was the cis isomer of $PtCl_2(NH_3)_2$ that had antibacterial activity. In 1969, ROSENBERG and co-workers reported that *cis*-platinum also had potent antitumor activity in murine tumors [472], and phase I clinical studies with the drug were initiated two years later [473], [474]. *Cis*-platinum, in combination with other agents, has led to highly active and often curative regimens in patients with testicular, ovarian, and head and neck cancer [475], [476].

8.1.1. Mechanism of Action

As with the alkylating agents, one of the major factors mediating the cytotoxicity of *cis*-platinum is probably the formation of cross-links between opposing strands of DNA (interstrand cross-links), linkage within a single strand of DNA (intrastrand cross-links), or the formation of linkages dependent on the hydrolysis of *cis*-platinum in solution. While the covalent stability of $Pt-NH_3$ bonds is high, both chlorides are good leaving groups and can be displaced by water or hydroxyl ions to form positively charged, aquated platinum species that avidly react with nucleophilic sites on macromolecules, especially the N^7 position of guanine and the N^3 position of cytosine [477]. Formation of the active intermediate is inhibited in the presence of Cl^-; in plasma the Cl^- concentration is sufficient to inhibit aquation of the drug, which has an in vitro plasma half-life of several hours [478]. However, this process occurs rapidly in the intracellular milieu, where Cl^- concentrations are low.

The cis configuration is central to cytotoxicity; the trans isomer is devoid of antitumor effects. This fact suggests that of the reactions caused by the aquated platinum species, formation of intrastrand cross-links is most important (intrastrand cross-links cannot be formed by the inactive trans compound).

As a result of nucleophilic attack on macromolecules, *cis*-platinum causes changes in the structural conformation of DNA [479] as well as intra- and interstrand cross-links. These changes inhibit RNA transcription from the DNA template [480] and, probably more important, directly inhibit DNA synthesis itself [481]. In addition, *cis*-platinum can react with tumor cell membrane to cause presentation of new antigenic determinants [482]. Whether this mechanism is related to tumor cell recognition and immune response remains to be determined.

As a heavy-metal-based compound, *cis*-platinum has unique clinical toxicities, including renal tubular damage, severe nausea and vomiting, high-tone hearing loss, and peripheral neuropathy. Less common are myelosuppression, hemolytic anemia, hypomagnesemic tetany, allergic reactions, and hepatotoxicity. Nephrotoxicity was dose limiting in early clinical trials and appeared to be due to tubular reabsorption of active platinum species causing proximal and distal tubule necrosis [483–485]. Tubular damage has been reported to cause defective reabsorption of magnesium, resulting in hypomagnesemia (which may result in tetany) [486], [487], as well as inappropriate renal loss of calcium, potassium, and phosphorus [488]. The acute renal toxicities of *cis*-platinum appear to be secondary to activation of the renin – angiotensin system, resulting in reduced renal blood flow and glomerular filtration [484], [489]. Thus, this toxicity can be mitigated by high-volume diuresis [490]. In 1984 it has been demonstrated that even high doses of *cis*-platinum are well tolerated when administered with high-volume chloresis, which not only dilutes urinary platinum levels but also prevents leaving of the chloride groups to form the toxic aquated molecule [491]. When administered by this regimen, the limiting side effects of *cis*-platinum become neurotoxicity and myelosuppression, while renal function is remarkably spared.

8.1.2. Mechanisms of Resistance

Although bacterial resistance to *cis*-platinum appears to be due to increased efficiency of DNA repair [492], the mechanisms of tumor cell resistance are less clear. Alkaline elution studies have demonstrated a direct relation between DNA cross-linking and tumor cell resistance [493–495], and a *cis*-platinum-resistant murine leukemia line has been described in which cross-links are formed at a reduced rate [496]. However, it remains uncertain whether this resistance is secondary to accumulation, impaired activation, or altered DNA repair processes. As with the alkylating agents in general, high levels of metallothionein have been reported in *cis*-platinum-resistant cells [497]. This sulfhydryl-rich protein is known to protect from metal toxicity by specifically binding to platinum, cadmium, and other heavy metals [498].

Because the cellular mechanisms of resistance to *cis*-platinum are so poorly characterized, new platinum analogs in clinical trial have been selected in an attempt to reduce host toxicity with the aim of improving therapeutic index [499].

8.2. Carboplatin

Carboplatin (NSC-241240, CBDCA) [*839805-03-3*], *cis*-diammine[1,1-cyclobutanedicarboxylato-(2)-*O,O'*]platinum(II), $C_6H_{12}N_2O_4Pt$, M_r 371.3, is a second generation cisplatin analog without significant nephrotoxicity or neurotoxicity and with less emetic potential than the parent compound [500–503]. Clinical trials show activity against several tumor types. Carboplatin is especially effective in treatment of ovarian and small-cell lung cancer [504], [505]. The minimal emetogenic dose of *cis*-platinum in dogs is 9 mg/m^2, whereas for carboplatin the dose is 624 mg/m^2. In addition, this compound retains antitumor activity in *cis*-platinum-restistant murine leukemia [506], demonstrating consistent antitumor activity in patients with ovarian cancer in the absence of either ototoxicity, nephrotoxicity, or neuropathy [507], [508]. Carboplatin entered the market in 1989 and is the leading platinum complex cancer drug with sales of \$ 373 × 10^6 worldwide in 1996. Carboplatin is indicated for the treatment of ovarian cancer and sales have benefited from the drug's use in combination with Taxol [509–512].

Trade names: Paraplatin (Bristol-Myers Squibb), Carboplat (BMS).

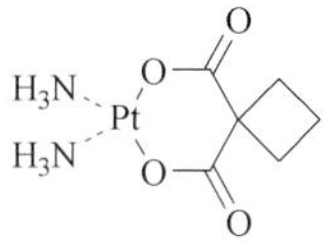

Carboplatin

8.3. Analogs

The synthesis and development of "second-generation" platinum compounds such as carboplatin has modified the problem of nephrotoxicity. Iproplatin (NSC-256927, CHIP) [*34348-60-2*], dichlorodihydroxy-bis-(2-propanamine)platinum $C_6H_{20}Cl_2N_2Pt$, M_r 418 [514], nedaplatin (NSC-375101D, 254-S), [*095734-82-0*], *cis*-diamine(glycolato-O^1,O^2)platinum, $C_2H_8N_2O_3Pt$, M_r 303.19 [515], and oxaliplatin (OHP) [*61825-94-3*], [Sp-4-2-(1R-*trans*)](1,2-cyclohexanediamine-*N,N'*)[ethanedioxato(2$^-$)-*O,O'*]-platinum, $C_8H_{14}N_2O_4Pt$ [516] were clinically investigated in depth. Nedaplatin entered the market in 1997 for the treatment of head and neck, small-cell lung, non-small cell lung, esophageal, bladder, testicular, ovarian, and uterine cervical cancers [517–520].

Trade name: Agupla (Shionogi).

Further studies of additional "third-generation" compounds are centering on the elimination of toxicity, enhanced therapeutic activity, non-cross-resistance and selective drug delivery [521]. Ormaplatin (NSC-363812, U-77233) [*62816-98-2*] [±(*trans*)]-tetrachloro(1,2-cyclohexanediamine-*N,N'*]platinum, $C_6H_{14}Cl_4N_2Pt$ and Lobaplatin (D-19466) [*135558-11-1*] are representatives of a series of platinum complexes in clinical development [522–524].

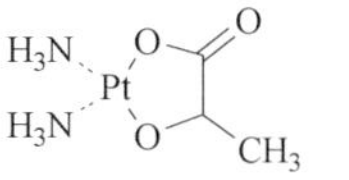

Nedaplatin

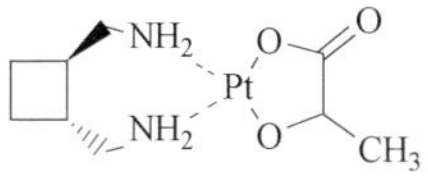

Lobaplatin

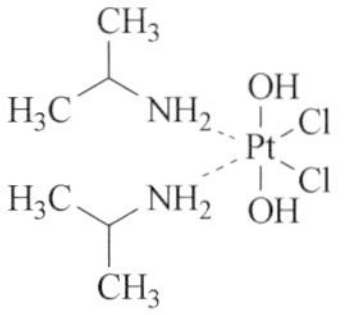

Iproplatin

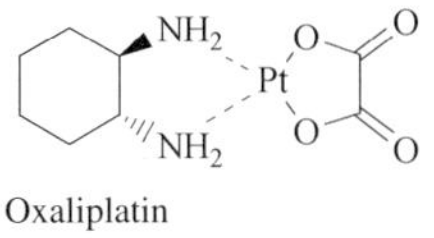

Oxaliplatin

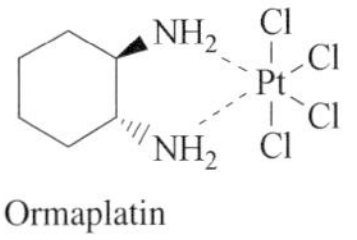

Ormaplatin

Orally active cisplatin-analogs with higher therapeutic ratio and specifically high anticancer activity are the challenge in the search for new "fourth-generation" derivatives. Platinum(IV)-complexes like JM 216, drug targeting approaches, e.g., the use of ligands with hormone receptor-binding affinity [525] or intercalating structures [526], and special formulation techniques are the major focus.

9. Hormonally Active Anticancer Drugs/Antihormones

Hormones and in particular, the sex hormones were the first growth factors discovered to be involuntary helpers of cancer. Female breast cancer and male prostate cancer are the best known examples of tumors acknowledged to be hormone-dependent. Shutting down the main production site of the sex hormones estrogen and testosterone either by removing the ovaries or by castration is a well-known and often effective therapy; however, these procedures can be problematic due to the concomitant psychological stress. Modern hormone therapy for advanced breast cancer and prostate cancer attempts to spare the patient such irreversible operative procedures for as long as possible by using hormone antagonists. Examples are the antiestrogens or LHRH antagonists, which hinder deployment of the hormone itself and thus its growth-promoting activity.

9.1. Antiestrogens

9.1.1. Antagonists

Estrogens can induce hormone-dependent human breast carcinoma and stimulate tumor growth. Reduced estrogen production is correlated with a lower risk of breast cancer and in particular with tumor regression. Hormonally active drugs are considered to be the treatment of first choice for advanced breast cancer, unless metastatic complications require immediate aggressive chemotherapy.

9.1.2. Tamoxifen, Toremifene

The antiestrogen tamoxifen (ICI 46474) [*10540-29-1*], (Z)-2-[*p*-(1,2-diphenyl-1-butenyl)phenoxy]-*N,N*-dimethylethylamine, $C_{26}H_{29}NO$, has become the standard first-line agent in postmenopausal patients [527–529]. The product is indicated for the treatment of breast cancer with worldwide sales of \$ 561 × 10^6 in 1996.

Trade name: Nolvadex (Zeneca).

However, primary and secondary resistance to tamoxifen treatment requires the introduction of new analogs with improved therapeutic activity for hormon-dependent neoplasia. Toremifene (NK-622, Fc-1157a) [*089778-26-7*] [530], [531], 2-[*p*-[(Z)-4-chloro-1,2-diphenyl-1-butenyl]phenoxy]-*N,N*-dimethylethylamine, $C_{26}H_{28}ClNO$, M_r 405.97, raloxifene [532], and droloxifene [533] were investigated clinically.

Toremifene is a novel triphenylethylene derivative structurally related to tamoxifen [527–529]. Almost all compounds related to tamoxifen contain an alkylaminoethoxy side chain, which seems to be essential for their binding to the estrogen-receptor (ER) and antiestrogenic activity [536].

Trade names: Toremifene has been launched as Fareston (Farmos, ASTA Medica) and Estrimex (Adria) for treatment of postmenopausal breast cancer.

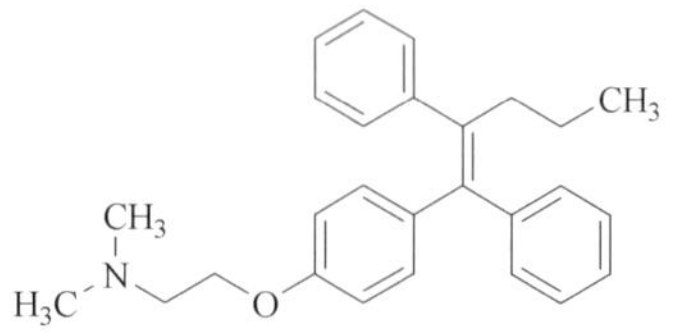

Tamoxifen

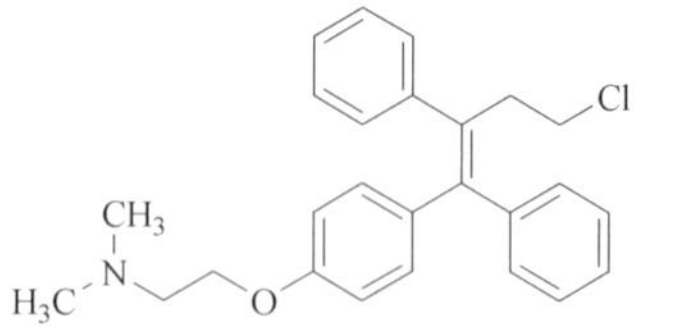

Toremifene

Mechanism of Action. Tamoxifen and toremifene bind to estrogen receptors in the cytosol, are translocated to the nucleus, and block estrogen-induced cell proliferation. However, specific differences in the drug profiles exist, indicating improved therapeutic properties of toremifene compared with tamoxifen. These differences include lower intrinsic estrogenic activity, longer nuclear retention, no retinal damage or neoplastic liver changes. Toremifene is more active against dimethylbenz[a]anthracene (DMBA)-induced rat mammary cancer and unlike tamoxifen, it inhibited the growth of an ER-negative transplantable mouse uterus sarcoma, although the antitumor effect of this compound was preferentially directed against estrogen-dependent tumors of the mammary gland and endometrium [534], [535]. In clinical trials no concrete side effects or pathological clinical chemistry values were observed in most patients [537], [538]. Some patients complained of light hot waves, sweating, nausea, and transient vertigo.

Cytotoxicity in vitro, dose related activity in ER-positive (i.e., cancer cells containing estrogen receptors) and ER-negative tumor models suggest that additional mechanism like growth-factor production may be triggered by toremifene [539].

9.1.3. Analogs

Raloxifene (LY-139481) [*82640-04-8*] is a benzo(b)thien-3yl-antiestrogen formerly under development for treatment of breast cancer. It mimics the effects of estrogen on the skeleton and is therefore effective in the prevention of postmenopausal osteoporosis [540]. Faslodex (ICI-182780, ZD-182780) [*129453-61-8*] is a pure, steroidal estrogen antagonist with oral anticancer activity [541], [542]. The in vivo antitumor activity of faslodex in xenografts of MCF-7 and Br 10 human breast cancers in mice was equivalent to that of tamoxifen. Miproxifene phosphate (Tat-59) [*115767-74-3*] [543], panomifene (Gyki-13504) [*77599-17-8*], idoxifene [544] (SB-223030, CB-7432) [*116057-75-1*] and droloxifene [*83647-29-4*] are more potent than tamoxifen against estrogen-dependent tumors in mice.

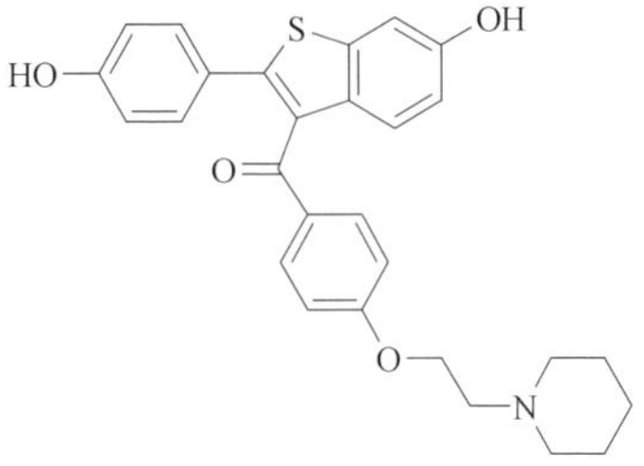

Raloxifene

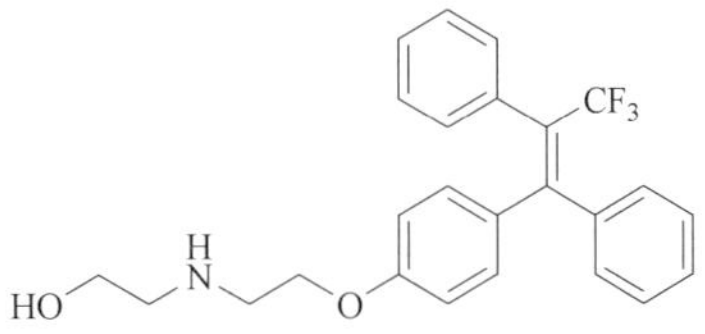

Panomifene

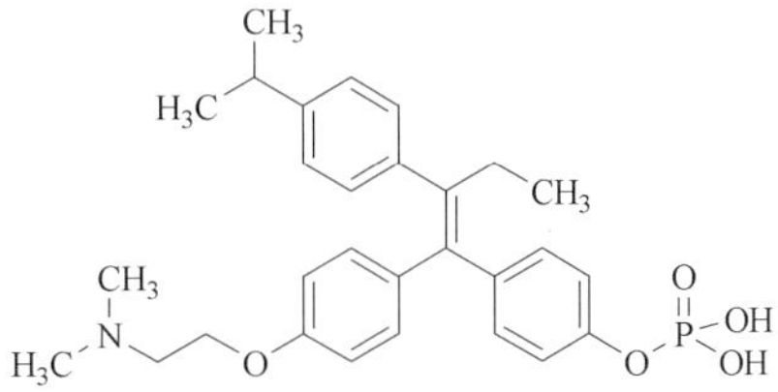

Miproxifene phosphate

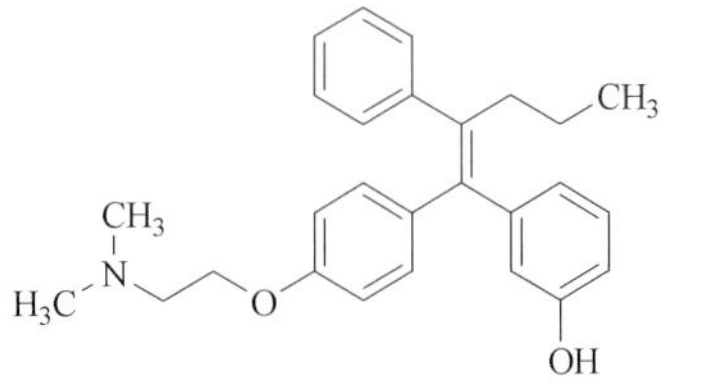

Droloxifene

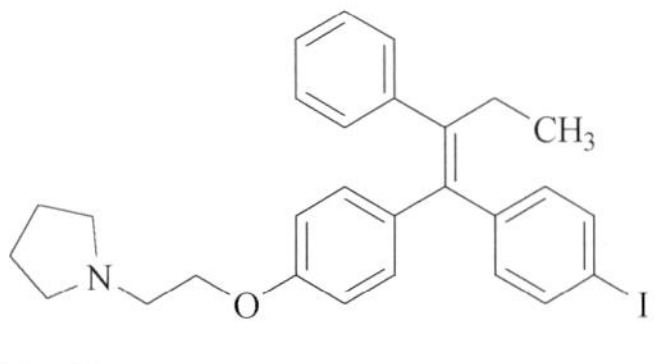

Idoxifene

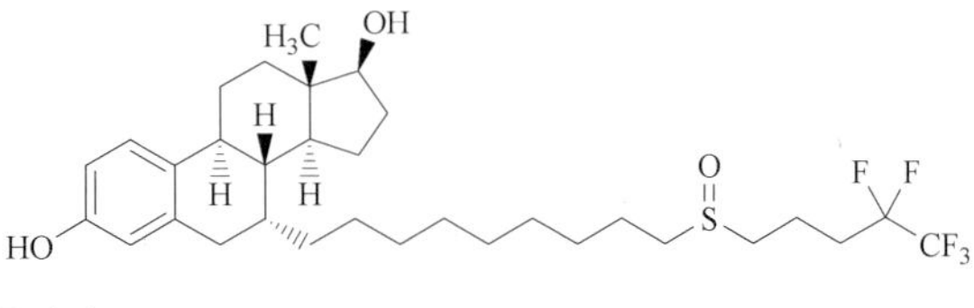

Faslodex

9.2. Aromatase Inhibitors

All endocrine therapies inhibit endogenous estrogen production or the interaction between estrogens and cellular estrogen receptors. Both ablative and additive hormone therapies are equal in efficiency.

Aromatase inhibitors block cellular estrogen synthesis and thus induce, particularly during postmenopause, a marked decrease in estrogen production. The inhibition of estrogen synthesis is caused by a suppression of the enzyme aromatase which converts androstenedione to estrone [545].

Aminoglutethimide [*125-84-8*], (±)-2-(4-aminophenyl)-2-ethylglutarimide, an unspecific aromatase inhibitor which also suppresses adrenal desmolase and 11-β-hydrolase, was the first aromatase inhibitor in the clinic and on the market.

Trade name: Orimeten (Ciba-Geigy).

Aminoglutethimide induces a decrease in cortisol, followed by an increase in ACTH [546]. Therefore new powerful specific aromatase inhibitors without influence on adrenal steroid synthesis were developed.

Formestane (4-OHA; CGP-32349) [*000566-48-3*], 4-hydroxyandrost-4-ene-3,17-dione, $C_{19}H_{26}O_3$, M_r 302.41, is an androstane derivative with highly specific aromatase inhibition [547]. The substance is used for the treatment of advanced breast cancer in postmenopausal women [548–553].

Trade name: Lentaron (Ciba-Geigy).

Atamestane, exemestane, and NKS-01 are orally active steroids.

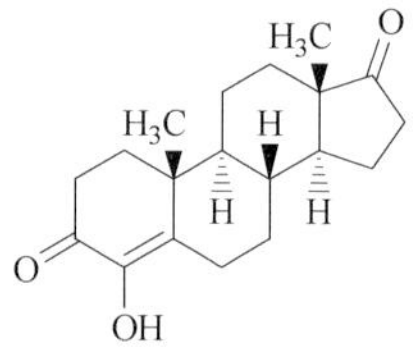

Formestane

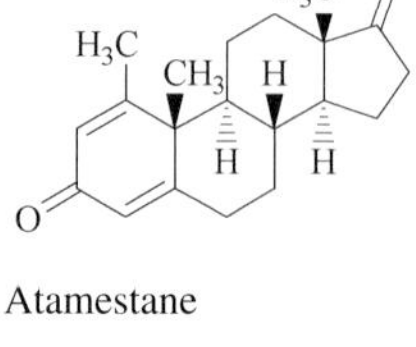

Atamestane

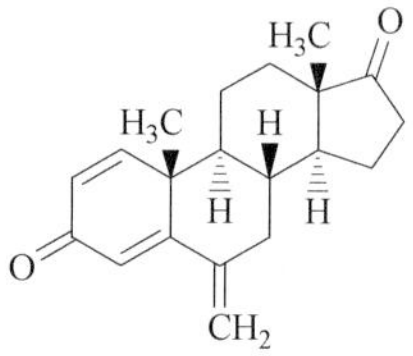

Exemestane

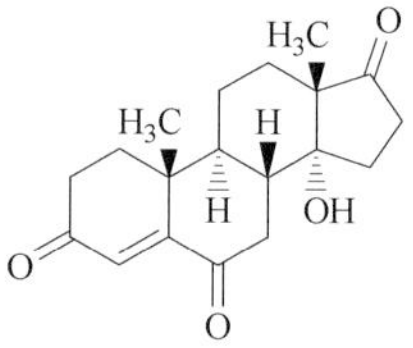

NSK-01

In addition to steroidal also nonsteroidal, in particular, imidazole/triazole derivatives were investigated as aromatase inhibitors. Anastrazole (ICI-D1033, ZD-1033) [*120511-73-1*] [554], 2,2'-[5-(1H-1,2,4-triazol-1-ylmethyl)-1,3-phenylene]-bis(2-methylpropionitrile), $C_{17}H_{19}N_5$, M_r 293.37, fadrozole (CGS-16949A) [*102676-96-0*] [555], (±)-4-(5,6,7,8-tetrahydroimidazo[1,5-a]pyridin-5-yl)benzonitrile, $C_{14}H_{13}N_3$, M_r 259.74, letrozole (CGS-20267) [*112809-51-5*] [556], 4,4'-(1H-1,2,4-triazol-1-ylmethylene)bis[benzonitrile], $C_{17}H_{11}N_5$, and vorozole (R83842), [*129731-10-8*] [557], (+)-6-[(4-chlorophenyl)-1H-1,2,4-triazol-1-ylmethyl]-1-methyl-1H-benzotriazole, $C_{16}H_{13}ClN_6$, are highly selective nonsteroidal aromatase inhibitors without intrinsic androgenic or estrogenic properties.

Trade names: Anastrazole and Fadrazole are marketed as Arimedex (Zeneca) and Arensin (Ciba-Geigy), respectively, for treatment of postmenopausal breast cancer.

Another aminoglutethimide analog in development is rogletimide [558] (pyridoglutethimide) [*121840-95-7*], (±)-3-ethyl-3(4-pyridinyl)-2,6-piperidinedione, $C_{12}H_{14}N_2O_2$.

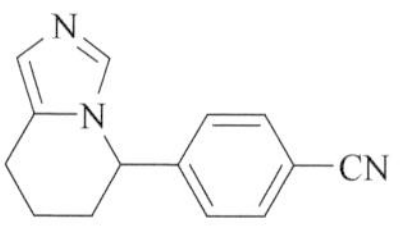

Fadrazole

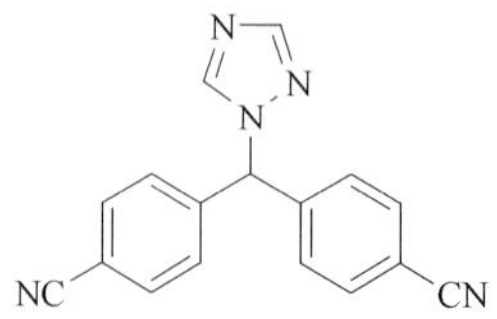

Letrozole

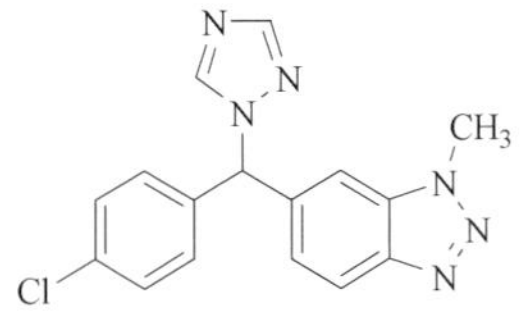

Vorozole

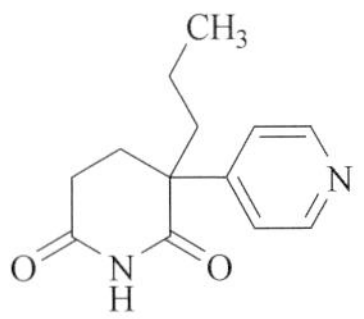

Rogletimide

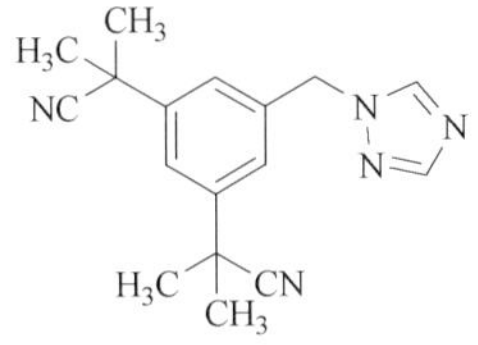

Anastrazole

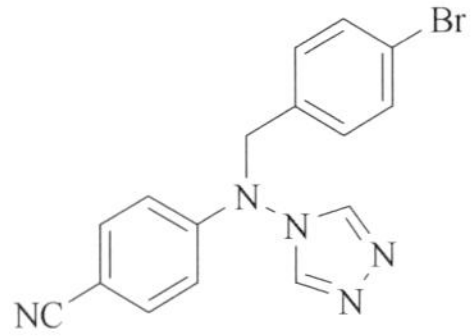

YM-511

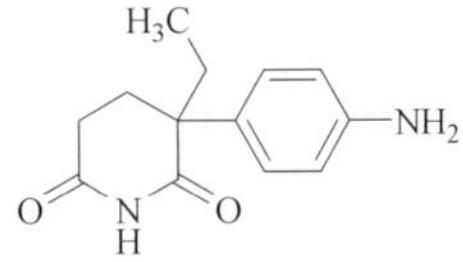

Aminoglutethimide

Mechanism of Action. Fadrazole [559] and anastrazole [560–567] for example, were found to be potent and specific aromatase inhibitors with neither androgenic nor estrogenic activity. Fadrazole was 180 times more potent than aminoglutethimide as an aromatase inhibitor. Anastrazole inhibited human placental aromatase with an IC_{50} = 15 nM (IC_{50} = inhibitory concentration). In animal studies fadrazole was found to lower serum estrogen, raise luteinizing hormone (LH) level, and reduce uterine weight as a result of aromatase inhibition. In addition, administration of 2 mg per kilogram body weight caused regression of DMBA-induced mammary tumors in female rats. In humans half-life of, e.g., anastrazole was more than 30 h. No serious side effects were reported and no significant effects on cortisol or aldosterone secretion was observed.

9.3. Antiandrogens

Prostate cancer is primarily a disease of the elderly and the second most common cancer in men in the United States. Ever since HUGGINS and HODGES demonstrated the partial androgen dependence of most prostatic tumors more than 50 years ago, androgen deprivation has become the commonly used initial treatment for prostate cancer [568], [569].

The major circulating androgen in man is testosterone, 90 % of which is produced in the testis. In addition, a small amount of androgen is produced by the adrenal gland under the control of ACTH. (see Fig. 3).

Four ways for androgen deprivation exist:

1) Removal of organs by surgery
2) Interference with control mechanism
3) Inhibition of biosynthesis
4) Competitive inhibition of androgens at the receptor site

The group 2 approach with, e.g., LHRH agonists and antagonists will be discussed in Section 9.4. Examples of group 3 include compounds (e.g., ketoconazole), which inhibit the synthesis of adrenal androgens [570] as well as inhibitors of the enzymes 5α-reductase [571] and aromatase [572]. "True" antihormones are compounds of group 4, based on the definition that an antiandrogen is a substance which binds to the target tissue androgen receptor and prevents the stimulatory effects of androgens. Although the exact mechanism of action of antiandrogens is not totally understood, an important feature is the competitive inhibition of the binding to the cytosol receptor. The first antiandrogen to be used clinically was the synthetic steroidal antiandrogen cyproterone acetate [*427-51-0*] [573]. However, cyproterone acetate also exhibits other steroidal activities, it is, e.g., a potent progestin, exhibits weak antigonadotrophic activity and has glucocorticoid like properties. Beyond the success in treating prostate cancer the steroidal properties are largely responsible for the fluid retention and thrombosis seen in patients [574]. Therefore, nonsteroidal antiandrogens are expected to have advantages by avoiding steroid-related side effects [575].

9.3.1. Flutamide

Flutamide (Sch-13521, NK-601, FTA) [*013311-84-7*], 2-methyl-*N*-[4-nitro-3(trifluormethyl)phenyl]-propanamide, $C_{11}H_{11}F_3N_2O_3$ was discovered in the early 1970s. Unlike the steroids it can easily be synthesized [576], [577] and is devoid hormonal activities. Total sales of the product were \$ 271×10^6 in 1996.

Trade names: Eulexin and Drogenil (Schering-Plough).

Mechanisms of Action. In a comparative study in castrated rats flutamide was shown to be equipotent to cyproterone acetate as an antiandrogen [578]. Several groups had observed that flutamide was a more potent antiandrogen in vivo than in vitro and suggested the involvement of an active metabolite [579], [580]. The major metabolite was identified as 2-hydroxyflutamide analog (Sch-16423) [*52806-53-8*] and high levels of 2-hydroxyflutamide in the plasma led to the conclusion that this was the active form of flutamide [581]. Although the plasma levels of testosterone increased on flutamide treatment, the levels of testosterone and dihydrotestosterone in androgen target tissue were reduced [582]. Flutamide exerts its antiandrogenic action by blocking the binding of androgens to the cytosolic androgen receptor and / or inhibiting the nuclear binding of androgens in the target tissue [581].

In patients improvements were seen in pain relief, prostatic enlargement and induration, reduction of metastases, and increase in body weight and phosphatase. Most common side effects of flutamide therapy are gynecomastia and breast tenderness [584]. More recent trials use a combination of flutamide with LHRH agonists (see Section 9.4). This is the concept of maximal androgen withdrawal in which the LHRH agonist wipes out androgens of testicular origin and the antiandrogen blocks the action of androgens of adrenal origin at the androgen receptor [585–587]. The therapeutic benefits seem to be greatest in patients with minimal disease at the start of treatment.

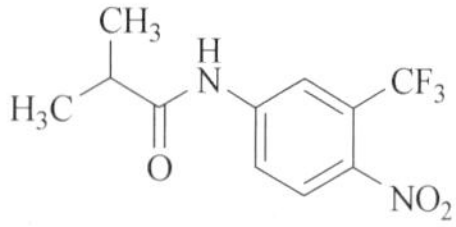

Flutamide

9.3.2. Nilutamide

Nilutamide has been discovered bei Roussel-Uclaf.

Trade name: Anandron (Roussel-Uclaf).

Mechanism of Action. Nilutamide competitively inhibits binding of androgens to the cytosolic androgen receptor. Administration over 7 days to immature, castrated male rats, nilutamide inhibited the increase of prostate weights induced by testosterone in a dose-dependent manner [588]. In rat pituitary cells nilutamide reverses the inhibition of LHRH-induced LH release elicited by dihydrotestosterone. It is probably because of these effects that nilutamide is recommended for use only in surgically or medically castrated males [589]. Nilutamide has demonstrated an antiandrogenic action in several animal tumor models [590].

Single dose kinetics of nilutamide in volunteers (100 mg) indicate a half-life of 43 ± 3 h, compared to 5.2 h for flutamide (200 mg) [591].

Clinical trials with nilutamide have concentrated on combination therapy with surgical or medical castration. The side effects observed in patients include hot flushes, nausea, vomiting, and visual problems [592].

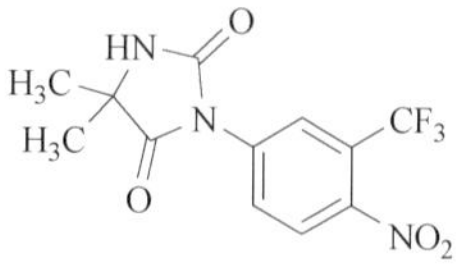

Nilutamide

9.3.3. Bicalutamide

Bicalutamide (ICI-176334) [*090357-06-5*], (±)-4-[3-(4-fluorophenylsulfonyl)-2-hydroxy-2-methylpropionamido]-2-(trifluoromethyl)benzonitrile, $C_{18}H_{14}F_4N_2O_4S$, M_r 430.37 was discovered at ICI / Zeneca and selected from more than 1000 compounds as having the desired properties of a pure, nonsteroidal, peripherally selective antiandrogen [593], [594]. The product was launched for the treatment of advanced prostate cancer in combination with LHRH analogs or surgical castration.

Trade name: Casodex (Zeneca).

Mechanism of Action. Bicalutamide inhibits the binding of the synthetic androgen [^{3}H]-R-1881 to both rat prostate and pituitary cytosol androgen receptors. The substance binds some 50 times less effectively than dihydrotestosterone and about 100 times less effectively than R-1881 to the prostate androgen receptor, Its affinity for the prostate receptor is about four-fold higher, that for the pituitary receptor ten times higher than that of hydroxyflutamide [595]. In vivo studies revealed that bicalutamide is about five times as potent as flutamide after oral application. Bicalutamide did not cause a significant elevation in LH or testosterone at any of the doses tested, whereas flutamide elicited increases. Half-life of bicalutamide in prostate cancer patients who received 10, 30 or 50 mg/d bicalutamide was around 6 d. The compound was well tolerated in all doses [596–598].

Nilutamide and bicalutamide offer advantages over flutamide because of their long half-lifes and sustained serum levels on once-daily dosing. The latter is essential to prevent androgen stimulation.

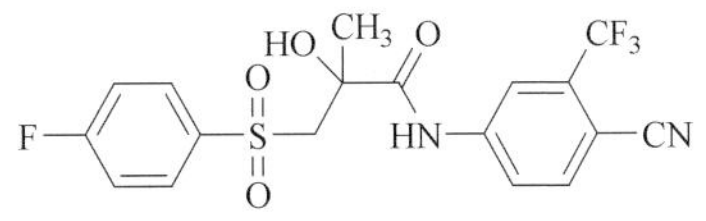

Bicalutamide

9.4. LHRH Analogs

The releasing hormone gonadorelin (GnRH, synonymous with LHRH, luteinizing hormone-releasing hormone or gonadoliberin), together with its specific receptor, plays a central role in neuroendocrinology [599]. The decapeptide LHRH is formed in the cell bodies of hypothalamic neurons and is secreted in pulses into the blood stream [600–603]. Ultimately it stimulates secretion of the sex-specific hormones in the testes and ovaries. Specific receptors for LHRH and synthetic analogs are also present in the pituitary gland and other tissues (for example, tumor cells) [604] and organs.

Three concepts for therapeutic application have emerged. The first is the restoration of normal physiology by administration of LHRH by infusion pump to promote fertility in men and women who are infertile due to defective endogenous LHRH secretion. Second, long-lasting LHRH agonists (so-called superagonists) are used in a depot form, which bring about desensitization of the pituitary receptors and thus interrupt the signal cascade. This results in a biochemical "castration", which opens up new therapeutic possibilities for hormone-dependent diseases such as prostate cancer, breast cancer, and endometriosis. Although superagonists are generally well tolerated, they have the disadvantage that hormone secretion (estrogen, testosterone) is initially stimulated before the depletion of receptors or "down" regulation can take place, and thus the illness temporarily worsens [605]. This has led to the development of the third concept: use of LHRH antagonists [606] (see Fig. 3). In the late 1980s about 5000 LHRH analogs had been synthesized worldwide and tested in vitro or in vivo [607–609]. Whilst LHRH agonists have been on the market for about ten years, the LHRH antagonists that have been developed farthest are still in clinical testing [610–614].

9.4.1. LHRH Agonists

The first years after the discovery of the gonadorelins were marked by the search for more active agonists, since the therapeutic potential of gonadorelins as, for example, antitumor agents or in gynecology, was apparent [606], [615]. Such superagonists bring about a very effective reversible inhibition of the release of steroidal sex hormones. The exchange of glycine at postion 6 ($glycine_6$) of the native LHRH for other, always D-configured, amino acids, is common to all modern superagonists; some have a *C*-terminal ethylamide (buserelin, leuprorelin) or azaglycinamide (goserelin) residue instead of glycinamide. Eight to ten amino acids of the LHRH sequence are thus conserved in all clinically relevant superagonists; by exchange at a maximum of two positions, the biological activity or hormone suppression in tumor patients, can be increased by a factor of up to 100 on subcutaneous application [606], [615]. Table 1 summarizes the most important derivatives.

Buserelin (Profact, Suprecur) [*57982-77-1*], leuprorelin (Lupron, Carcinil, Enatone) [*53714-56-0*], triptorelin (Decapeptyl) [*57773-63-4*], and goserelin (Zoladex) [*65807-02-5*] (trade names in Germany in parantheses) are the products on the market with the

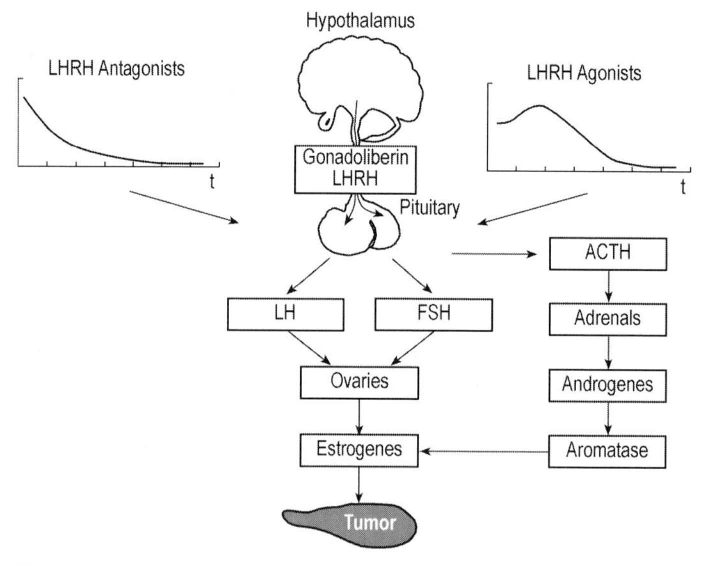

Figure 3. Antitumor activity of LHRH agonists and antagonists in femaleFSH = follicle stimulating hormone; ACTH = adrenocorticotropic hormone

Table 1. Structure of LHRH agonists on the market (given are only those amino acid residues that are different in LHRH).

Name (Co.)	Structure									
	1	2	3	4	5	6	7	8	9	10
LHRH	Glp[a]	His	Trp	Ser	Tyr	Gly	Leu	Arg	Pro	Gly-NH_2
Buserelin (Hoechst)						D-Ser(*t*Bu)				Gly-NHEt
Nafarelin						D-(2)Nal				
Leuprorelin (Abbott, Takeda)						D-Leu				Gly-NHEt
Goserelin (Zeneca)						D-Ser(*t*Bu)				Azagly-NH_2
Histrelin (Ortho)						D-His(Bzl)				Gly-NHEt
Triptorelin (Ferring)						D-Trp				

[a] Glp = pyroglutamic acid.

highest turnover. The Lupron line of products is the leading anticancer hormone drug with worldwide sales in 1996 of $\$ 810 \times 10^6$ followed by Zoladex with sales of $\$ 563 \times 10^6$.

Annual production is less than 100 kg for buserelin and significantly over 100 kg for the market leader, leuprorelin. At this order of magnitude, substances are only produced by classical organic preparative synthesis (fragment condensation in solution). Hoechst, for example, synthesizes the nonapeptide buserelin from the units pyroglutamic acid-histidine (Glp-His), tryptophan-serine-tyrosine (Trp-Ser-Tyr), and D-Serine-tert-butyl ether-leucine-arginine-proline-NHEt (D-Ser(*t*Bu)-Leu-Arg-Pro-NHEt); the tri- and tetrapeptide units are coupled to form the corresponding *C*-terminal heptapeptide, and then the *N*-terminal dipeptide is condensed with this to form the complete sequence. Control of the physicochemical process parameters, such as concentrations, precipitation, separations, reaction temperature profiles, and purification techniques, is important for successful scale-up of peptide synthesis to a technical scale [617].

9.4.1.1. Leuprorelin Acetate

Because leuprorelin inhibits the synthesis of androgen and estrogen, the drug blocks the growth of hormone-dependent tumors by shutting down testosterone production. Leuprorelin is indicated for the treatment of advanced prostate cancer, as an alternative to castration for the treatment of endometriosis, and for the presurgical management of patients with anemia caused by benign fibroid tumors.

Trade name: Lupron (Abbott, Takeda).

9.4.1.2. Goserelin

As certain prostate tumors grow in response to testosterone, goserelin prevents the production of testosterone in testes and is therefore indicated for the treatment of prostate cancer, advanced breast cancer, and endometriosis.

Trade name: Zoladex (Zeneca).

9.4.2. LHRH Antagonists

Common to the intrinsic activity of all superagonists is the initial temporary stimulation of gonadotropin release. Soon after the use of highly active agonists became an established therapy, a search began for corresponding antagonists, which do not bring about an initial hormone release, to avoid this therapeutically counterproductive effect. A final big hurdle for clinical use of highly active antagonists was the inherent anaphylactic potential of these peptides. Starting from the sequence of native LHRH, the individual positions of the peptide chain were examined in rapid succession for their contribution to biological activity. Particular attention was paid to side effects. The most effective early improvements in antagonistic activity were achieved by using D-phenylalanine (D-Phe) instead of histidine at position 2, by D-amino acids at position 6 instead of Gly, and the exchange of *C*-terminal glycine for D-alanine (D-Ala^{10}). Further stepwise optimization led to the sequence scheme now usual for all modern antagonists of D-Nal^{1}-D-Cpa^{2}-D-Pal^{3} (Nal = 2-naphthylalanine, Cpa = Phe(4-Cl), Pal = 3-pyridylalanine) as hydrophobic cluster, a D-configured aromatic or aliphatic, yet hydrophilic, aminocarboxylic acid at position 6, and the *C*-terminal hydrophilic sequence Xxx^{8}-Pro^{9}-D-Ala^{10} where Xxx is either arginine or isopropyllysine. Excellent documentation of the stepwise optimization can be found in [618], [619].

Antagonists of the second generation caused formation of temporary edemas of the face and extremities in animal experiments, due in part to massive histamine release by mast cell degranulation. Cyanosis and respiratory impairment were also observed [620]. The cause of these intolerable side effects is thought to be the combination of D-arginine at position 6 with the three aromatic amino acids at the N-terminus of the sequence [621]. For the desired biological antagonist potency, a D-configured basic amino acid is necessary at position 6. A. V. SCHALLY et al. achieved the breakthrough to highly active antagonists free of side effects with the derivatives SB-75 (cetrorelix) [*120287-85-6*] and SB-88 [*120287-93-6*], by incorporating hydrophilic, nonbasic amino acids with side-chain carbamoyl functions at position 6 (D-$Citrulline^{6}$, D-homocitrueline) [621], [622].

Today, e.g., cetrorelix is *manufactured* exclusively by classical fragment condensation on a kilogram scale. Two synthesis strategies proceed via either the *N*-terminal tripeptide (D-Nal^{1}-D-Cpa^{2}-D-Pal^{3}) or the *C*-terminal tripeptide (Arg^{8}-Pro^{9}-D-Ala^{10}) and the complementary heptapeptide (Ser^{4}-D-Ala^{10}) or (Nal^{1}-Leu^{7}) to the protected decapeptide with *tert*-butyl side chain protection. The deprotection with hydrochloric acid is followed by final purification by preparative HPLC. The *C*- and *N*-terminal functionalization of the acetylated decapeptide amide, necessary for biological activity and to avoid rapid enzymatic degradation, is introduced at the level of the terminal tripeptide. Functionalization is achieved by acetylation of the free α-amino group of $naphthylalanine^{1}$ with acetylhydroxysuccinimide, or amminolysis of the resulting $alanine^{10}$ methyl ester in alcoholic ammonia [623].

FOLKERS et al. were able to improve active antagonists successively by consistent and systematic modifications both at the relevant sequence positions and the side chain substituents; the most successful were complex substitutions at the positions 5, 6 and, in part, 8 [624–628]. An important contribution was made by RIVIER et al., who synthesized the decapeptide "azaline" with novel modifications at positions 5 and 6, where aminotriazole-substituted *p*-aminophenylalanine or lysine are positioned [629–631]. Azaline B [*134457-28-6*] is probably one of the most active antagonists presently available worldwide.

DEGHENGHI published a highly active decapeptide sequence with minimal histamine release and good water solubility. This structure, known as antarelix [*151277-78-5*], differs from SB-75 (cetrorelix) in that it has homocitrulline^6 instead of citrulline^6 and isopropylysine^8 instead of arginine^8 [633]. By using Lys(*i*Pr)^8, residual potential for histamine release can be further reduced. Organon is developing the antagonist ganirelix (RS26306) [*124904-93-4*] under license from Syntex; this is a decapeptide with novel alkyl-modified D- and L-homoarginine units at positions 6 and 8 [634]. With ramorelix (HOE 013) [*127932-90-5*], Hoechst (HMR) has a peptide antagonist with a sugar–amino acid unit (*O*-α-L-rhamnosyl-D-serine^6), that has improved water solubility [632]. Schering has also synthesized peptide antagonists, using nonproteinogenic amino acids, such as ε-dialkylated lysine or benzodiazepine aminocarboxylic acids [635], [636]. Abbott's A-76154 [*136989-30-5*] is an octapeptide antagonist with LHRH receptor affinity of a similar order of magnitude to that of active decapeptides such as A-75998 [*135215-95-1*] or "Nal-Glu" [*103733-02-4*] [644].

9.4.2.1. Receptor Assays

At present, selected peptides are being tested in vivo by measuring the testosterone concentration in male rat serum. Here, a single subcutaneous application of a potent LHRH antagonist leads to persistent testosterone suppression. The activity in animals generally correlates well with the binding affinity determined in vitro on human receptors. In cases where no in vivo activity was observed despite high binding affinity, the peptide probably had pharmacokinetic characteristics that played a decisive role. In the future, transgenic animals will also be very important for in vivo testing [645–647].

9.4.2.2. Peptidomimetics

For some years, increased efforts have also been made to find substances with affinity to the LHRH receptor that do not have the characteristic substance-specific properties and also disadvantages of peptides (short half-life, lack of bioavailability), yet have a high binding affinity. Ideally, such substances should be able to be administered orally, be sufficiently stable in the organism, and possess favorable pharmacological parameters comparable to peptide antagonists. Clinical investigations of the influence of the antimycotic ketoconazole on prostate cancer and testosterone suppression indicate that it may have a LHRH-mimetic effect [648], [649]. Further investigations by Abbott showed weak antagonistic activity both in vitro and in vivo for ketoconazole and its modified analogs; however, the biological activity of such derivatives is most probably not mediated mainly by LHRH antagonism [650], [651]. Complex, highly substituted nitrogen heterocycles seem to have LHRH antagonistic potential [652]. In patent literature benzodiazepines, benzodiazepinones, heterocyclic benzo-substituted alkylamines, and thienopyridine carboxylic acid derivatives are described as LHRH receptor antagonists with receptor-binding inhibition at submicromolar concentrations, suitable amongst other uses as antitumor agents for hormone-dependent tumors [647], [653]. A lead structure from the latter substance class is in extensive pharmacological trials [654–656].

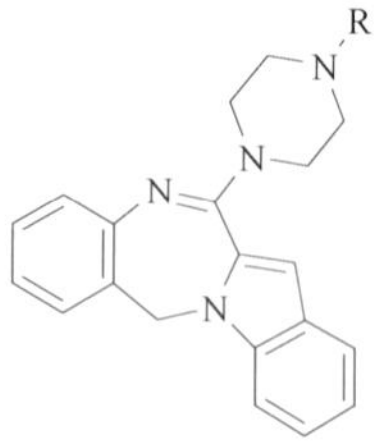

R N N N N

McNeil, US 4 678 784

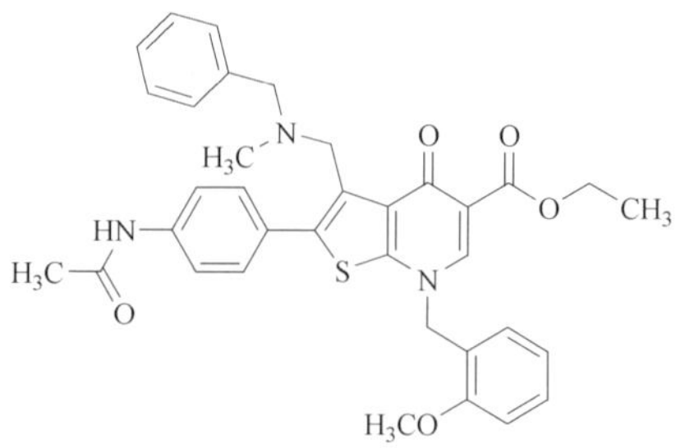

N H_3C O O O CH_3 HN H_3C O S N H_3CO

Takeda, WO 95/28405

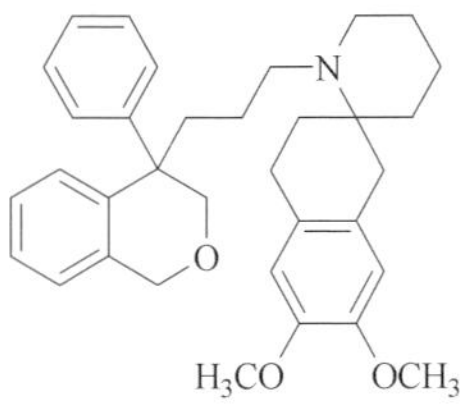

Takeda, EP 679 642 A1

10. Signal Transduction Inhibitors

Anticancer drug discovery has been directed away from agents that affect cells by producing DNA damage towards modulators of signal transduction pathways that have become unregulated or aberrant in malignant transformation [657].

10.1. Enzyme Inhibitors

Binding of growth factors [658], such as epidermal growth factor (EGF) [659] to a membrane bound receptor tyrosine kinase results in dimerization and autophosphorylation of tyrosine residues on the protein surface. As result the GTP-bound form of *ras* (→ Toxicology, Chap. 3.7.3.1.) [660] undergoes a conformational change on its surface, enabling it to bind to several effector molecules leading to the activation of transcription factors involved in DNA synthesis [661]. An overexpression of the EGF-receptor is observed in various types of human cancers.

Mechanism-based screens have identified the natural products bryostatin [662] [*83314-01-6*] as partial agonist of protein kinase C and fumagillin [663] [*23110-15-8*] as interfering with tumor-induced neovascularization thus inhibiting angiogenesis. FR-111142 as analog of fumagillin is reported to be less toxic and more active than the parent compound [664]. A potent inhibitor of *ras* protein farnesylation [665] is pepticinnamin E [666], and peptides related to the CAAX-tetrapeptide (where C is cysteine, A is valine, isoleucine, or leucine, and X is methionine or serine). In order to prepare completely nonpeptide and potentially more stable inhibitors a series of potent peptidomimetic inhibitors were designed [667].

Specific natural and synthetic kinase inhibitors are in clinical development as anticancer agents [668–670].

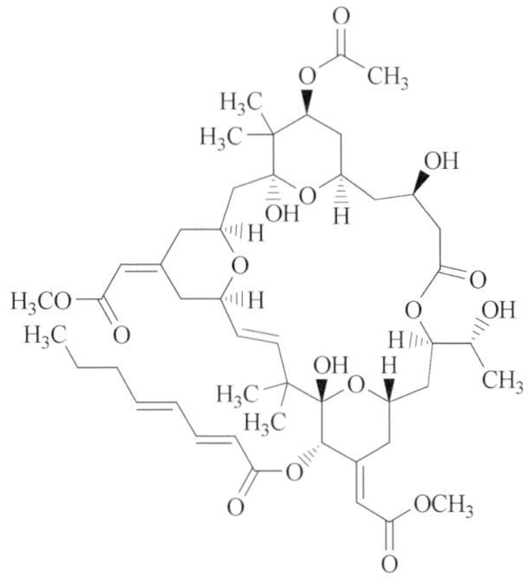

Bryostatin 1

10.2. Phospholipid – Based Antineoplastics

Ether phospholipids and lysophospholipids are naturally occurring derivatives of phospholipids of the cell membrane with many interesting properties. For example, the ether phospholipid 1-*O*-alkyl-2-acetyl-sn-glycero-3-phosphocholine (platelet activating factor, PAF) [671] causes platelet aggregation and dilation of blood vessels and lysophosphatidic acid (LPA), the simplest natural phospholipid, is a potent mitogen. The observation that alkyllysophospholipids exerted experimental antitumor activity in vitro and in vivo [672], [673] led to systematic modification of the structures and subsequent clinical investigation of, e.g., ilmofosine [*83519-04-4*] [675]. In the course of modification it was found that an analog without glycerol backbone was equally effective.

Miltefosine (D-18506, hexadecylphosphocholine), [*058066-85-6*] [676], 2-[[(hexadecyloxy)-hydroxyphosphinyl]oxy]-*N,N,N*-trimethylethanaminium hydroxide (inner salt); $C_{21}H_{46}NO_4P$, M_r 407.57 is a new phospholipid-based antineoplastic agent which exerts substantial antitumor activity in appropriate models. Because of its special pharmacology – high activity against mammary carcinoma and low toxicity to normal tissue – it is an ideal candidate for topical treatment of cutanous breast cancer metastases. The substance was launched in 1993 as a topical formulation for palliative treatment of refractory skin metastases of breast cancer [676].

Trade name: Miltex (ASTA Medica).

A variety of derivatives were synthesized and characterized, some of which appeared to be considerably less toxic to the gastrointestinal tract [677]. One of them, perifosine (D-21266) [*157716-52-4*] is currently in clinical trials.

Mechanism of Action. Miltefosine shows highly selective antineoplastic activity in DMBA- and *N*-methyl-*N*-nitrosourea (MNU)-mammary carcinoma of the rat. The activity is not due to an antiestrogenic effect in these estrogen-dependent tumor models [678]. The treatment was tolerated and no overt toxic symptoms were observed. The mechanism of action of miltefosine is most probably different from that of other chemotherapeutic agents [679]. Incorporation of miltefosine and metabolites into biological membranes was demonstrated, affecting the interactions of receptor proteins or membrane associated enzymes involved in growth control and cellular signalling [680]. The reduction in tumor mass was accompanied by morphological changes compatible with the induction of differentiation [675], [681]. Furthermore, the decrease of tumors was mainly due to cell loss by apoptosis.

$H_3C(CH_2)_{15}$ S O P O O^- O $\overset{+}{N}(CH_3)_3$ OCH_3

Ilmofosine

$H_3C(CH_2)_{15}$ O P O O^- O $\overset{+}{N}(CH_3)_3$

Miltefosine

In clinical trials mild dryness and flaking of the skin were noted, but no systemic toxicity was reported after topical administration. Several phase I and II studies in patients that were treated with oral capsules were terminated early due to gastrointestinal intolerance [682].

11. Economic Aspects

The world cancer drug market is expected to reach $\$ 14 \times 10^9$ by 2000 [10], [14]. Total sales of the top 100 anticancer drugs generated a turnover of $\$ 4.66 \times 10^9$ in 1996. Of the worlds eight top-selling anticancer drugs, four – the prostate cancer-drugs flutamide (Eulexin), bicalutamide (Casodex), leuprorelin (Lupron) and goserelin (Zoladex) – are merely palliative, yet have combined annual sales of $\$ 1.7 \times 10^9$, whereas sales of the breast cancer drugs tamoxifen (Nolvadex) and paclitaxel (Taxol) are approaching $\$ 500 \times 10^6$ and $\$ 850 \times 10^6$, respectively.

Taxol, BMS's drug for treating ovarian and breast cancers, was the biggest selling anticancer agent in 1996, with sales of $\$ 813 \times 10^6$.

12. References

1. I. B. Weinstein, *Cancer Res. (Suppl.)* **51** (1991) 5080.
2. R. Doll, *Eur. J. Cancer* **26** (1990) 5007.
3. C. LaVecchina, F. Levi, F. Lucchini, S. Garattini, *Anti-Cancer-Drugs* **2** (1991) 215.
4. D. Kessel, *In Vivo* **8** (1994) 829.
5. T. Tsurno, A. Toninda, *Anti-Cancer-Drugs* **6** (1995) 213.
6. G. Eisenbrand, S. Lauch-Birkel, W. C. Tang, *Synthesis* (1996) 1246.
7. G. B. Elion, *Cancer Res.* **45** (1985) 2943.
8. *Scrip* 1997, July 15, p. 20.
9. *Chem. Marketing Report,* July 14, 1997, p. 19.
10. *Med. Ad-News,* May 1996, p. 56.
11. G. M. Cragg, D. J. Newman, K. M. Snader, *J. Nat. Prod.* **60** (1997) 52–60.
12. K. Nasmyth, *Science* **274** (1996) 1643–1645.
13. G. Dratta, H. Pagano, *Annu. Rep. Med. Chem.* **31** (1996) 241–248.
14. *Pharma Business,* July/August 1997, p. 62.
15. S. A. Rosenberg: "Principles of Surgical Oncology," in V. T. DeVita, S. Hellman, S. A. Rosenberg (eds.): *Cancer: Principles and Practice of Oncology,* J. B. Lippincott Co., Philadelphia 1982, pp. 93–102.
16. S. Farber, L. K. Diamond, R. D. Mercer, R. F. Sylvester, V. A. Wolff, *N. Engl. J. Med.* **238** (1948) 787–793.

17. R. Hertz, G. T. Ross, M. B. Lipsett, *Am. J. Obstet. Gynecol.* **86** (1963) 808 – 814.
18. R. L. Kisliuk, Y. Gaumont, C. M. Baugh, J. M. Galivan, G. F. Maley, F. Maley: "Inhibition of Thymidylate Synthetase by Poly-gamma-glutamyl Derivatives of Folate and Methotrexate," in R. L. Kisliuk, G. M. Brown (eds.): *Chemistry and Biology of the Pteridines,* Elsevier/North Holland, New York 1979, pp. 431 – 435.
19. I. D. Goldman, N. S. Lichtenstein, V. T. Oliverio, *J. Biol. Chem.* **243** (1968) 5007 – 5017.
20. F. M. Sirotnak, R. C. Donsbach, *Cancer Res.* **36** (1976) 1151 – 1158.
21. R. D. Warren, A. P. Nichols, R. A. Bender, *Cancer Res.* **38** (1978) 668 – 671.
22. P. L. Chello, F. M. Sirotnak, D. M. Dorck et al., *Cancer Res.* **37** (1977) 4297 – 4303.
23. B. T. Hill, B. D. Bailey, J. C. White et al., *Cancer Res.* **39** (1979) 2440 – 2446.
24. F. M. Sirotnak, R. C. Donsbach, *Cancer Res.* **35** (1975) 1737 – 1744.
25. J. J. McGuire, J. R. Bertino, *Mol. Cell Biochem.* **38** (1981) 19 – 48.
26. R. L. Schilsky, B. D. Bailey, B. A. Chabner, *Proc. Natl. Acad. Sci. USA* **77** (1980) 2919.
27. S. A. Jacobs, C. J. Derr, D. G. Johns, *Biochem. Pharmacol.* **26** (1977) 2310 – 2313.
28. V. M. Whitehead, *Cancer Res.* **37** (1977) 408 – 412.
29. J. Jolivet, B. A. Chabner, *J. Clin. Invest.* **72** (1983) 773 – 778.
30. S. V. Gupta, N. J. Greenfield, M. Poe et al., *Biochemistry* **16** (1977) 3073 – 3079.
31. H. Nakamura, J. Littlefield, *J. Biol. Chem.* **247** (1972) 179 – 187.
32. D. A. Matthews, R. A. Alden, J. T. Bolin et al.:"X-ray Structural Studies of Dihydrofolate Reductase," in R. L. Kisliuk, G. M. Brown (eds.): *Chemistry and Biology of the Pteridines,* Elsevier/North Holland, New York 1979, pp. 465 – 470.
33. P. A. Charlton, D. W. Young, B. Birdsall et al., *Chem. Commun.* **20** (1979) 922 – 924.
34. D. A. Matthews, R. A. Alden, J. T. Bolin et al., *Science* **197** (1977) 452 – 455.
35. W. L. Werkheiser, *Cancer Res.* **23** (1963) 1277 – 1285.
36. B. A. Kamen, W. Whyte-Bauer, J. R. Bertino, *Biochem. Pharmacol.* **32** (1983) 1837 – 1841.
37. R. J. Kaufman, J. R. Bertino, R. T. Schimke, *J. Biol. Chem.* **253** (1978) 5852 – 5860.
38. E. W. Alt, R. E. Kellems, J. R. Bertino et al., *J. Biol. Chem.* **253** (1978) 1357.
39. G. A. Curt, J. Jolivet, B. D. Bailey, D. N. Carney, B. A. Chabner, *Biochem. Pharmacol.* **33** (1984) 1682 – 1685.
40. G. A. Curt, D. N. Carney, K. H. Cowan, J. Jolivet et al., *N. Engl. J. Med.* **308** (1982) 199.
41. M. J. Cline, H. Stang, K. Mercola et al., *Nature (London)* **284** (1980) 922 – 925.
42. M. Bar-Eli, H. D. Stang, K. E. Mercola, M. J. Cline, *Somatic Cell Genet.* **9** (1983) 55 – 67.
43. F. Carr, W. D. Medina, S. Dube, J. R. Bertino, *Blood* **62** (1983) 180 – 185.
44. H. Diddens, D. Niethammer, R. C. Jackson, *Cancer Res.* **43** (1983) 5286 – 5292.
45. F. M. Sirotnak, J. I. Degraw, F. A. Schmid, L. J. Goutas et al., *Cancer Chemother. Pharmacol.* **12** (1984) 26 – 30.
46. R. P. Hertzberg, R. K. Johnson, *Annu. Rep. Med. Chem.* **28** (1993) 167 – 176.
47. G. F. Fleming, R. L. Schilsky, *Semin. Oncol.* **19** (1992) 707.
48. K. Y. Shum et al., *J. Clin. Oncol.* **6** (1988) 446.
49. A. L. Lackman et al., *Cancer Res.* **51** (1991) 5579.
50. D. Cunningham, *Cancer Res.* **53** (1991) 810.
51. A. L. Jackman, D. C. Farrugia, W. Gibson, *Eur. J. Cancer* **13A** (1995) 1277 – 1282.
52. M. D. Varney et al., *J. Med. Chem.* **35** (1992) 663.
53. C. Heidelberger, N. K. Chandhavi, P. Dannenberg et al., *Nature (London)* **179** (1957) 663 – 666.
54. R. J. Rutman, A. Cantarow, K. E. Paschkis, *Cancer Res.* **14** (1954) 119 – 126.
55. R. L. Glazer, L. S. Lloyd, *Molec. Pharmacol.* **21** (1982) 468 – 473.
56. D. W. Kufe, P. P. Major, *J. Biol. Chem.* **256** (1981) 9802 – 9806.
57. K. V. Hadjiolova, Z. G. Naydenova, A. A. Hadjiolova, *Biochem. Pharmacol.* **30** (1981) 1861 – 1866.
58. R. I. Glazer, K. D. Hartman, *Molec. Pharmacol.* **23** (1983) 540 – 546.
59. Y. Oshima, M. Itoh, N. Okada, T. Miyata, *Proc. Natl. Acad. Sci. USA* **78** (1981) 4471 – 4474.
60. D. W. Kufe, P. P. Major, E. M. Egan, E. Leh, *J. Biol. Chem.* **256** (1981) 8885 – 8889.
61. P. V. Danenberg, C. Heidelberger, M. A. Mulkins, A. R. Peterson, *Biochem. Biophys. Res. Commun.* **102** (1981) 654 – 659.
62. Y. C. Cheng, K. Nakayama, *Molec. Pharmacol.* **23** (1983) 171 – 174.
63. M. Tanaka, K. Kimura, S. Yoshida, *Cancer Res.* **43** (1983) 5145 – 5150.
64. P. Reichard, O. Skold, G. Klein et al., *Cancer Res.* **22** (1962) 235 – 243.

65. P. Reyes, T. C. Hall, *Biochem. Pharmacol.* **18** (1969) 1587 – 1590.
66. D. K. Kasbeker, D. M. Greenberg, *Cancer Res.* **23** (1963) 818 – 825.
67. P. Reichard, O. Skold, G. Klein, *Nature (London)* **183** (1959) 939 – 941.
68. J. A. Houghton, S. J. Masada, J. O. Phillips et al., *Cancer Res.* **42** (1982) 144 – 149.
69. C. Heidelberger, G. Kaldos, K. L. Mukherjee et al., *Cancer Res.* **20** (1960) 903 – 909.
70. M. M. Lastieboff, B. Kedzioska, W. Rode, *Biochem. Pharmacol.* **32** (1983) 2259 – 2267.
71. A. R. Bapat, C. Zarow, P. V. Dannenberg, *J. Biol. Chem.* **258** (1983) 4130 – 4136.
72. C. C. Rossana, L. G. Rao, L. F. Johnson, *Mol. Cell. Biol.* **2** (1982) 1118 – 1125.
73. F. Baskin, S. C. Carlin, P. Kraus et al., *Molec. Pharmacol.* **11** (1975) 105 – 117.
74. D. G. Priest, B. E. Ledford, M. T. Doig, *Biochem. Pharmacol.* **29** (1980) 1549 – 1553.
75. R. D. Armstrong, E. Cadman, *Cancer Res.* **43** (1983) 2525 – 2528.
76. J. L. Au, Y. M. Rustum, J. Minonda, B. I. Sriva-stava, *Biochem. Pharmacol.* **32** (1983) 541 – 545.
77. Y. M. El Sayed, W. Sadee, *Cancer Res.* **43** (1983) 4039 – 4044.
78. E. J. Ansfield, G. J. Kallas, J. P. Simpson, *J. Clin. Oncol.* **1** (1983) 107 – 110.
79. F. Sorm, A. Piskala, A. Cihak et al., *Experientia* **20** (1964) 202 – 203.
80. A. Piskala, F. Sorm, *Collect. Czech. Chem. Commun.* **29** (1964) 2060 – 2075.
81. P. G. W. Plagemann, M. Behrens, D. Abraham, *Cancer Res.* **38** (1978) 2458 – 2466.
82. J. C. Drake, R. G. Stoller, B. A. Chabner, *Biochem. Pharmacol.* **26** (1977) 64 – 66.
83. T. Lee, M. Karon, R. L. Momparler, *Cancer Res.* **34** (1974) 2481 – 2488.
84. A. Cihak, *Collect. Czech. Chem. Commun.* **39** (1974) 3782 – 3792.
85. J. Doskel, V. Paces, F. Sorm, *Biochim. Biophys. Acta* **145** (1967) 771 – 779.
86. L. H. Li, E. J. Olin, T. J. Fraser et al., *Cancer Res.* **30** (1970) 2770 – 2775.
87. B. F. Vanyosbin, S. G. Tkacheug, A. W. Belozinsky, *Nature (London)* **225** (1976) 948 – 949.
88. J. L. Mandell, P. Chambon, *Nucleic Acids Res.* **7** (1979) 2081 – 2103.
89. J. D. McGhee, G. D. Gineler, *Nature (London)* **280** (1979) 419 – 420.
90. C. Scheir, T. Maniatis, *Proc. Natl. Acad. Sci. USA* **77** (1980) 6634 – 6638.
91. L. H. Van der Ploeg, R. A. Flavill, *Cell* **19** (1980) 947 – 958.
92. O. Niwa, T. Sugarhar, *Proc. Natl. Acad. Sci. USA* **78** (1981) 6290 – 6294.
93. S. Friedman, *Molec. Pharmacol.* **19** (1980) 314 – 320.
94. D. V. Santi, C. E. Garrett, P. J. Barr, *Cell* **25** (1983) 9 – 10.
95. P. G. Constantinides, P. A. Jones, W. Geness, *Nature (London)* **267** (1977) 364 – 366.
96. P. A. Jones, S. M. Taylor, *Cell* **20** (1980) 85 – 93.
97. T. J. Ley, J. DeSimone, N. P. Anagnov et al., *N. Engl. J. Med.* **307** (1982) 1469 – 1475.
98. J. Vesely, A. Chiak, *Pharmacol. Therap.* **2** (1978) 813 – 840.
99. T. Lee, M. R. Kovan, *Biochem. Pharmacol.* **25** (1976) 1737 – 1742.
100. A. Cihak, H. Vessela, F. Sorm, *Biochem. Biophys. Acta* **166** (1968) 277 – 279.
101. A. Cihak, J. Vesely, *Biochem. Pharmacol.* **21** (1972) 3257 – 3265.
102. J. Vesely, A. Cihak, F. Sorm, *Int. J. Cancer* **2** (1967) 639 – 646.
103. J. Doskacil, F. Sorm, *FEBS Lett.* **2** (1974) 30 – 32.
104. D. D. Von Hoff, M. Slavik, F. M. Muggia, *Ann. Intern. Med.* **85** (1976) 237 – 245.
105. Z. H. Israili, W. R. Vogler, E. S. Mingidi et al., *Cancer Res.* **36** (1971) 1453 – 2461.
106. J. A. Beisler, *J. Mol. Chem.* **21** (1978) 204 – 208.
107. J. A. Beisler, M. M. Abbasi, J. S. Driscoll, *Cancer Treat. Rep.* **60** (1976) 1671 – 1674.
108. G. A. Curt, J. A. Kelley, R. L. Fine et al., *Proc. Am. Assoc. Clin. Oncol.* **3** (1984) 37.
109. P. G. W. Plagemann, R. Marz, R. M. Wolhvete, *Cancer Res.* **38** (1978) 978 – 989.
110. A. Iwagaki, T. Nakamura, G. Wakisaka, *Cancer Res.* **29** (1969) 2169 – 2176.
111. M. Y. Chu, G. A. Fisher, *Biochem. Pharmacol.* **11** (1962) 423 – 430.
112. P. P. Major, E. M. Egan, G. P. Beardsley, M. D. Minden et al., *Proc. Natl. Acad. Sci. USA* **78** (1983) 3235 – 3239.
113. P. P. Major, L. Sargent, E. M. Egan, D. W. Kufe, *Biochem. Pharmacol.* **30** (1981) 2221 – 2224.
114. P. P. Major, E. M. Egan, D. Herrick, D. W. Kufe, *Biochem. Pharmacol.* **31** (1982) 861 – 866.
115. M. Y. Chu, G. A. Fisher, *Biochem. Pharmacol.* **14** (1965) 333 – 341.
116. M. R. Atkinson, M. Deutscher, A. Kornberg, A. Russel et al., *Biochemistry* **8** (1969) 4897 – 4904.

117. P. A. Diskwel, F. Wanka, *Biochim. Biophys. Acta* **520** (1978) 461 – 471.
118. R. J. Frann, C. M. Egan, D. W. Kufe, *Leuk. Res.* **7** (1983) 243 – 249.
119. J. F. Ward, E. I. Jones, W. F. Blakely, *Cancer Res.* **44** (1984) 59 – 63.
120. S. R. Bähring-Kuhlmey, *Drugs of Today* **13** (1977) 475.
121. A. A. Claesen et al., *Tetrahedron Lett.* (1966) 3499.
122. P. Major et al., *Proc. Natl. Acad. Sci. USA* **78** (1981) 3235.
123. R. Preston, *Treat. Carcinogen Mutagen* **1** (1980) 147.
124. P. Drahovsky, W. Kreis, *Biochem. Pharmacol.* **19** (1970) 940 – 944.
125. W. Kreis, J. Graham, L. A. Damin, *Biochem. Pharmacol.* **31** (1982) 3831 – 3837.
126. J. Balzarini, E. De Cleireg, *Mol. Pharmacol.* **23** (1983) 175 – 181.
127. A. W. Harris, E. C. Reynolds, L. R. Finch, *Cancer Res.* **39** (1979) 538 – 541.
128. Y. Rustum, H. Priesler, *Cancer Res.* **39** (1979) 42 – 49.
129. C. D. Stuart, P. J. Burke, *Nat. New Biol.* **233** (1971) 109 – 110.
130. P. Chang, P. H. Wiernik, S. D. Reich et al.: "Prediction of Response to Cytosine Arabinoside and Daunorubicin in Acute Nonlymphocytic Leukemia," in F. Mandelli (ed.): *Therapy of Acute Leukemias*, Lombardo-Editore, Rome 1977, pp. 148 – 159.
131. J. F. Smyth, A. B. Robins, C. L. Leese, *Eur. J. Cancer* **12** (1976) 567 – 573.
132. M. N. H. Tattersall, U. K. Ganeshaguru, A. V. Hoffbrand, *Br. J. Haematol.* **27** (1974) 39 – 46.
133. M. Aoshima et al., *Cancer Res.* **37** (1977) 1481.
134. F. Cabaninas, *Drugs of the Future* **5** (1980) 12, 603.
135. T. Ueda, T. Nakamura, S. Ando et al., *Cancer Res.* **43** (1983) 3412 – 3416.
136. K. Hoie, T. Tsuro, K. Naganuma, S. Tsukagoshi et al., *Cancer Res.* **44** (1984) 172 – 177.
137. C. W. Young, R. Schneider, B. Leyland-Jones et al., *Cancer Res.* **43** (1983) 5006 – 5009.
138. Y. C. Cheng, R. S. Tan, J. L. Ruth, G. Dutschman, *Biochem. Pharmacol.* **32** (1983) 726 – 729.
139. L. W. Hertel, J. S. Kroin, J. W. Misner, J. M. Tustin, *J. Org. Chem.* **53** (1988) 11, 2406 – 2409.
140. Eli Lilly, EP 122707, 1984 (L. W. Hertel).
141. Eli Lilly, EP 184365, 1986 (G. B. Grindey, L. W. Hertel).
142. Eli Lilly, EP 306190, 1989T. S. Chou, P. Heath). L. E. Patterson).
143. S. Chubb et al., *Proc. Am. Assoc. Cancer Res.* **28** (1987) Abst. 1282.
144. D. Y. Bouffard, L. F. Momparler, R. L. Momparler, *Eur. J. Pharmacol.* **183** (1990) 2, Abst. 032.
145. V. Heinemann, L. W. Hertel, G. B. Grindley, W. Phukett, *Cancer Res.* **48** (1988) 14, 4024 – 4031.
146. M. N. Serradel, J. Castañer, *Drugs of the Future* **15** (1990) 8, 794 – 797.
147. V. Heinemann et al., *Proc. Am. Assoc. Cancer Res.* **30** (1989) Abst 2204.
148. H. H. Hansen, *Ann. Oncol.* **120** (1994) 519 – 528.
149. E. Beutler, *Lancet* **340** (1992) 8825, 952.
150. H. M. Bryson, E. M. Sorkin, *Drugs* **46** (1993) 5, 872.
151. E. H. Estey et al., *Blood* **79** (1992) 4, 882.
152. S. P. Smith, *Hosp. Formul.* **28** (1993) 7, 621.
153. C. P. Robinson, *Drugs Today* **29** (1993) 6, 379.
154. US Dept. Of Health US 4357324, 1982 (J. A. Montgomery, A. T. Shortnacy).
155. R. W. Brockman, Y. C. Cheng, F. M. Schabel, J. M. Montgomery, *Cancer Res.* **40** (1980) 3610 – 3615.
156. J. A. Montgomery, *Cancer Res.* **42** (1982) 3911 – 3917.
157. W. C. Tseng et al., *Mol. Pharmacol.* **21** (1982) 474 – 477.
158. S. J. Hopkins, *Drugs of the Future* **10** (1985) 1, 20.
159. *Annu. Drug Data Rep.* (1992) 633 – 634.
160. J. S. Whelan et al., *Brit. J. Cancer* **64** (1991) 1, 120.
161. H. S. Hochster et al., *J. Clin. Oncol.* **10** (1992) 1, 28.
162. H. W. Dion et al., *Ann. NY Acad. Sci.* **284** (1977) 21 – 29.
163. Parke Davis, US 3923785, 1975 (A. Ryder et al.).
164. P. W. Woo et al., *J. Heterocycl. Chem.* **11** (1974) 4, 641 – 643.
165. J. F. Smith et al., *Cancer Chemother. Pharmacol.* **1** (1978) 49 – 51.
166. J. F. Smith et al., *Proc. Am. Assoc. Cancer Res.* **20** (1979).
167. G. A. Le Page et al., *Cancer Res.* **36** (1976) 1481 – 1485.
168. M. N. Seradel, J. Castañer, *Drugs of the Future* **6** (1981) 419 – 420.
169. M. Blick et al., *Am. J. Hematol.* **33** (1990) 3, 205 – 209.

170. S. Bernhard et al., *Med. Pediatr. Oncol.* **19** (1991) 4, 276.
171. A. D. Ho et al., *J. Natl. Canver Inst.* **82** (1990) 17, 1416 – 1420.
172. E. H. Kraut et al., *J. Clin. Oncol.* **7** (1989) 2, 168 – 172.
173. R. S. Witte et al., *Invest. New Drug* **10** (1992) 1,49.
174. B. J. Kane, J. G. Kuhn, M. K. Roush, *Ann. Pharmacother.* **26** (1992) 939 – 947.
175. G. B. Elion, E. Burgi, G. H. Hitchings, *J. Am. Chem. Soc.* **74** (1952) 411.
176. G. B. Elion, G. H. Hitchings, *J. Am. Chem. Soc.* **77** (1955) 1676.
177. J. H. Burchenal, M. L. Murphy, R. R. Ellison et al., *Blood* **8** (1953) 965.
178. G. H. Hitchings, G. B. Elion, *Ann. N.Y. Acad. Sci.* **60** (1954) 195.
179. R. W. Brockman, C. S. Debavadi, P. Stutts, D. J. Hutchison, *J. Biol. Chem.* **236** (1961) 1471 – 1479.
180. J. D. Davidson, *Cancer Res.* **20** (1960) 225 – 232.
181. G. B. Elion, *Fed. Proc.* **26** (1967) 898 – 904.
182. J. F. Henderson, M. K. Y. Khoo, *J. Biol. Chem.* **240** (1965) 2349 – 2357.
183. F. F. Snyder, J. F. Henderson, *Can. J. Biochem.* **51** (1973) 943 – 948.
184. P. C. L. Wong, J. F. Henderson, *Biochem. J.* **129** (1972) 1085 – 1094.
185. T. Fields, L. Brox, *Can. J. Biochem.* **52** (1974) 441 – 446.
186. C. D. Green, D. W. Martin Jr., *Proc. Natl. Acad. Sci. USA* **70** (1973) 3698 – 3702.
187. G. B. Elion, S. Callahan, W. Rundles, G. H. Hitchings, *Cancer Res.* **23** (1963) 1207 – 1217.
188. G. H. Hitchings, *Cancer Res.* **23** (1963) 1218 – 1225.
189. T. Nakamura, *Blood & Vessel (Jpn.)* **2** (1971) 237 – 248.
190. T. Higuchi, T. Nakamura, G. Wakisaka, *Blood &Vessel (Jpn.)* **3** (1972) 313 – 318.
191. P. W. Allan, L. L. Bennett Jr., *Biochem. Pharmacol.* **20** (1971) 847 – 852.
192. A. Hampton, *J. Biol. Chem.* **238** (1963) 3068 – 3074.
193. R. J. McCollister, W. R. Gilbert, D. M. Ashton, J. B. Wyngaarden, *J. Biol. Chem.* **239** (1964) 1560 – 1563.
194. R. P. Miech, R. E. Parks Jr., J. H. Anderson Jr., A. C. Sartorelli, *Biochem. Pharmacol.* **16** (1967) 2222 – 2227.
195. A. R. P. Paterson, D. M. Tidd: "6-Thiopurines," in A. C. Sartorelli, D. G. Johns (eds.): *Handbook of Experimental Pharmacology,* vol. 38, Springer-Verlag, Berlin 1975, Part 2, pp. 384 – 403.
196. D. L. Hill, L. L. Bennett Jr., *Biochemistry* **8** (1969) 122.
197. R. T. Kavahata, L. F. Chuang, C. A. Holmberg, B. I. Osburn et al., *Cancer Res.* **43** (1983) 3655 – 3659.
198. J. Maybaum, H. G. Mandel, *Cancer Res.* **43** (1983) 3852 – 3856.
199. J. S. Lazo, K. M. Huang, A. C. Sartorelli, *Cancer Res.* **37** (1977) 4250.
200. C. K. Carrico, A. C. Sartorelli, *Cancer Res.* **37** (1977) 1868.
201. K. Kim, W. J. Blechman, V. G. H. Riddle, A. B. Pardee, *Cancer Res.* **41** (1981) 4529.
202. J. D. Strobel-Stevens, S. M. El Dareer, M. W. Trader, D. L. Hill, *Biochem. Pharmacol.* **31** (1982) 3133.
203. M. Rosman, H. E. Williams, *Cancer Res.* **33** (1973) 1202.
204. M. K. Wolpert, S. P. Damle, J. E. Brown et al., *Cancer Res.* **31** (1971) 1620.
205. M. Rosman, M. L. Lee, W. A. Creasey et al., *Cancer Res.* **34** (1974) 1952.
206. E. M. Scholar, P. Calabresi, *Biochem. Pharmacol.* **28** (1979) 445.
207. L. Tterlikkis, J. L. Day, D. A. Brown, E. C. Schroeder, *Cancer Res.* **43** (1983) 1675 – 1679.
208. S. Zimm, J. M. Collins, D. O'Neill, B. A. Chabner et al., *Clin. Pharmacol. Therap.* **34** (1983) 810 – 817.
209. S. Zimm, J. M. Collins, R. Riccardi, D. O'Neill et al., *N. Engl. J. Med.* **308** (1983) 1005 – 1009.
210. R. D. Armstrong, R. Vera, P. Snyder, E. Cadman, *Biochem. Biophys. Res. Commun.* **109** (1982) 595.
211. M. Inomata, F. Fukuoka, A. Hoshi, K. Kuretani et al., *J. Pharmacobiodyn.* **4** (1981) 928.
212. D. M. Tidd, I. Gibson, P. D. G. Dean, *Cancer Res.* **42** (1982) 3769.
213. M. Inaba, M. Fukui, N. Yoshida, S. Tsukagoshi et al., *Cancer Res.* **42** (1982) 1103.
214. C. P. J. Adair, H. J. Bragg, *Ann. Surg.* **93** (1931) 190.
215. L. P. Jacobson, C. L. Spurr, E. S. O. Barron et al., *J. Am. Med. Assoc.* **132** (1946) 126.
216. M. Colvin: "The Alkylating Agents," in B. A. Chabner (ed.): *Pharmacologic Principles of Cancer Treatment,* W. B. Saunders, Philadelphia 1982, pp. 276 – 308.

217. N. O. Goldstein, R. J. Rutman, *Cancer Res.* **24** (1964) 1363.
218. R. W. Ruddon, J. M. Johnson, *Mol. Pharmacol.* **4** (1968) 258.
219. G. P. Wheeler, J. A. Alexander, *Cancer Res.* **29** (1969) 98.
220. P. Brookes, P. D. Lawley, *Biochem. J.* **80** (1961) 486.
221. C. C. Price, G. M. Gaucher, P. Koneru et al., *Biochim. Biophys. Acta* **166** (1968) 327.
222. A. Loveless, W. C. J. Ross, *Nature (London)* **166** (1950) 1113.
223. K. W. Kohn, L. C. Erickson, R. A. G. Ewig et al., *Biochemistry* **15** (1976) 4629.
224. R. J. Rutman, E. H. C. Chun, F. A. Lewis, *Biochem. Biophys. Res. Commun.* **32** (1968) 650.
225. I. Hirono, *Nature (London)* **186** (1960) 1059.
226. G. Calcutt, T. A. Connors, *Biochem. Pharmacol.* **12** (1963) 839.
227. R. A. G. Ewig, K. W. Kohn, *Cancer Res.* **37** (1977) 2114.
228. L. C. Ericson, G. Laurent, N. A. Sharkey et al., *Nature (London)* **288** (1980) 727 – 729.
229. D. T. Vistica, J. N. Toal, M. Rabinowitz, *Biochem. Pharmacol.* **27** (1978) 2865.
230. W. R. Redwood, M. Colvin, *Cancer Res.* **40** (1980) 1144.
231. W. P. Brade, J. Engel, *Drugs Today* **20** (1984) 491 – 496.
232. P. J. Creaven, L. M. Allen, D. A. Alford et al., *Clin. Pharmacol. Therap.* **16** (1974) 77.
233. T. A. Conners: "Alkylating Agents, Nitrosourea, and Alkyltriazines," in H. M. Pinedo, B. A. Chabner (eds.): *Cancer Chemotherapy Annual 5,* Elsevier, Amsterdam 1983, pp. 30 – 65.
234. A. Mittelman et al., *J. Urol. (Baltimore)* **115** (1976) 409.
235. W. A. Skinner, H. F. Gram, M. O. Green et al., *J. Med. Pharmaceut. Chem.* **2** (1960) 299.
236. K. A. Hyde, E. Acton, W. A. Skinner et al., *J. Med. Pharmaceut. Chem.* **5** (1962) 1.
237. F. M. Schabel, T. P. Johnston, G. S. McGaleb et al., *Cancer Res.* **23** (1963) 226.
238. K. W. Kohn, *Cancer Res.* **37** (1977) 1450.
239. R. J. Weinkam, D. F. Deen, *Cancer Res.* **42** (1982) 1008.
240. J. Mendel, R. Thust, H. Schwartz, *Arch. Geschwulstforsch.* **52** (1982) 371.
241. B. J. Bowdon, J. Grimsley, H. H. Lloyd, *Cancer Res.* **34** (1974) 194.
242. L. C. Panasci, D. Green, R. Nagourney et al., *Cancer Res.* **37** (1977) 2615.
243. J. M. Dornish, I. Smith-Kielland, *FEBS Lett.* **139** (1981) 41.
244. K. D. Tew, A. L. Wang, *Molec. Pharmacol.* **21** (1982) 729.
245. A. Begleiter, H. Y. P. Lam, G. J. Goldenberg, *Cancer Res.* **37** (1977) 1022.
246. N. G. Goldstein, R. J. Rutman, *Cancer Res.* **24** (1964) 1363.
247. R. W. Ruddon, J. M. Johnson, *Molec. Pharmacol.* **4** (1968) 258.
248. J. J. Roberts, T. P. Brent, A. R. Crathorn, *Eur. J. Cancer* **7** (1971) 515.
249. W. P. Tong, D. B. Ludlam, *Biochim. Biophys. Acta* **608** (1980) 174.
250. C. C. Price, G. M. Gaucher, P. Koneru et al., *Biochim. Biophys. Acta* **166** (1968) 327.
251. P. Brookes, P. D. Lawley, *Biochem. J.* **80** (1961) 486.
252. C. B. Thomas, R. Osieka, K. W. Kohn, *Cancer Res.* **38** (1978) 2448.
253. C. C. Erickson, M. O. Bradley, S. M. Ducore et al., *Proc. Natl. Acad. Sci. USA* **77** (1980) 467.
254. R. A. G. Ewig, K. W. Kohn, *Cancer Res.* **38** (1978) 3197.
255. W. G. Verly, Y. Paquette, *Cancer J. Biochem.* **50** (1972) 217.
256. L. C. Ericson, G. Laurent, N. A. Sharkey et al., *Nature (London)* **288** (1980) 727 – 729.
257. A. R. Crathorne, J. J. Roberts, *Nature (London)* **211** (1966) 150.
258. K. Tsujihara, M. Ozeki, T. Morikawa, M. Kawamori et al., *J. Med. Chem.* **25** (1982) 441.
259. Y. Akaike, Y. Arai, H. Takuchi, H. Satoh, *Gann* **73** (1982) 480.
260. W. J. Zeller, M. Berger, G. Eisenbrand, W. Tang et al., *Arzneim.-Forsch.* **32** (1982) 484.
261. J. R. Prous, *Drug News Perspect* **1** (1988) 1, 35.
262. ADIR, FR 2536075, 1984 (C. Cudennec, G. Lavielle).
263. M. N. Serradel, J. Castañer, *Drugs of the Future* **14** (1989) 11, 1042 – 1046.
264. S. Filippeschi et al., *Anticancer Res.* **14** (1988) 11, 1351 – 1354.
265. D. Khayat et al., *Cancer Res.* **47** (1987) 6782 – 6785.
266. J. A. Boutin et al., *Eur. Cancer Oncol.* (1989).
267. P. Deloffre, C. A. Cudennec, G. Lavielle, J. P. Bizzari, *15th Int. Cong. Chemother.,* Istanbul 1987, Abst. 755.
268. J. Jacquillat et al., *Proc. Am. Assoc. Cancer Res.* **30** (1989) Abst. 1088.
269. P. Zeller, H. Gutmann, B. Hegedus et al., *Experientia* **19** (1963) 129.
270. V. T. DeVita, A. Serpick, P. Carbone, *Ann. Intern. Med.* **73** (1970) 542.

271. J. Gutterman, A. Huang, P. Hochstein, *Proc. Soc. Exp. Biol. Med.* **130** (1979) 797.
272. R. J. Weinkam, D. A. Shiba, *Life Sci.* **22** (1978) 937.
273. D. L. Dunn, R. A. Lubet, R. A. Prough, *Cancer Res.* **39** (1979) 4555.
274. D. E. Schwartz, W. Bollag, P. Obrecht, *Arzneim.- Forsch.* **17** (1967) 1389.
275. D. Reed, F. Dost, *Proc. Am. Assoc. Cancer Res.* **7** (1965) 57.
276. M. Baggiolini, B. Dewald, H. Aebi, *Biochem. Pharmacol.* **18** (1969) 2187.
277. C. Pueyo, *Mutat. Res.* **67** (1979) 189.
278. E. Therman, *Cancer Res.* **32** (1972) 1111.
279. W. Kreis, *Proc. Am. Assoc. Cancer Res.* **7** (1966) 39.
280. A. Sartorelli, S. Tsunamura, *Proc. Am. Assoc. Cancer Res.* **6** (1965) 55.
281. H. Lam, A. Begleiter, W. Stein et al., *Biochem. Pharmacol.* **27** (1978) 1883.
282. J. L. Skibba, D. D. Beal, G. Ramirez et al., *Cancer Res.* **30** (1970) 147.
283. P. Farina, A. Gescher, J. A. Hickman, J. K. Horton et al., *Biochem. Pharmacol.* **31** (1982) 1887.
284. T. L. Loo, G. E. Housholder, A. H. Gerulath et al., *Cancer Treat. Rep.* **60** (1976) 149.
285. D. D. Beal, J. L. Skibba, K. K. Whitnable et al., *Cancer Res.* **36** (1976) 2827.
286. A. H. Gerulath, T. L. Loo, *Biochem. Pharmacol.* **21** (1972) 2335.
287. D. W. Kaiser, I. T. Thurston, J. R. Dudley et al., *J. Am. Chem. Soc.* **73** (1951) 2984.
288. J. F. Worzalla, B. D. Kaima, B. M. Johnson et al., *Cancer Res.* **34** (1974) 2669.
289. J. F. Worzalla, B. D. Kaima, B. M. Johnson et al., *Cancer Res.* **33** (1973) 2810.
290. B. C. V. Mitchley, S. A. Clarke, T. A. Connors et al., *Cancer Treat. Rep.* **61** (1977) 3.
291. M. Ames, G. Powis, J. S. Kovach et al., *Cancer Res.* **39** (1979) 5016.
292. L. M. Lake, E. E. Grunden, B. M. Johnson, *Cancer Res.* **35** (1975) 2858.
293. A. J. Cumber, W. C. J. Ross, *Chem. Biol. Interact.* **17** (1977) 349.
294. V. Iver, W. Szybalski, *Science (Washington)* **145** (1964) 55 – 58.
295. L. Littlefield et al., *Mut. Res.* **81** (1981) 377.
296. J. Mac Donald et al., *Ann. Intern. Med.* **93** (1980) 533.
297. *Drugs Today* **20** (1984) 489; **21** (1985) 420.
298. *Drugs of the Future* **8** (1983) 402; **9** (1984) 371; **10** (1985) 420.
299. F. Arcamone et al., *J. Med. Chem.* **18** (1975) 703.
300. Microbiochem. Res. Found., US 4303785, (H. Naganawa, T. Takenchi, H. Umezawa).
301. H. Umezawa et al., *J. Antibiot.* **32** (1979) 1082 – 1085.
302. M. N. Serradel, J. Castañer, *Drugs of the Future* **8** (1983) 7, 610 – 611.
303. T. Nishimura et al., *J. Antibiot.* **33** (1980) 737 – 743.
304. D. Dantchev et al., *J. Antibiot.* **32** (1979) 1085 – 1086.
305. H. Brockmann, *Fortschr. Chem. Org. Naturst.* **50** (1973) 121.
306. C. E. Myers: "Anthracyclines," in B. A. Chabner (ed.): *Pharmacologic Principles of Cancer Treatment,* W. B. Saunders, Philadelphia 1982, pp. 416–434.
307. R. Phillips, A. DiMarco, F. Zunino, *Eur. J. Biochem.* **85** (1978) 487.
308. R. B. Painter, *Cancer Res.* **38** (1978) 4445.
309. G. Daskal, C. Woodard, S. T. Crooke et al., *Cancer Res.* **38** (1978) 467.
310. W. E. Ross, L. A. Zwelling, K. W. Kohn, *Int. J. Radiat. Oncol., Biol. Phys.* **5** (1979) 1221.
311. B. Sinha, R. H. Sik, *Biochem. Pharmacol.* **29** (1980) 1867.
312. K. Handa, S. Sato, *Gann* **66** (1975) 43.
313. N. R. Bachur, S. L. Gordon, M. V. Gee, *Cancer Res.* **38** (1977) 1745.
314. R. Ogura, H. Toyama, T. Shimada et al., *J. Appl. Biochem.* **1** (1979) 325.
315. J. H. Doroshow, G. Y. Locker, C. E. Myers, *J. Clin. Invest.* **65** (1980) 128.
316. N. W. Revis, N. Marusic, *J. Mol. Cell. Cardiol.* **10** (1978) 945.
317. T. R. Tritton, S. A. Murphree, A. C. Sartorelli, *Biochem. Biophys. Res. Commun.* **84** (1978) 802.
318. T. R. Tritton, G. Yeh, *Science* **217** (1982) 248.
319. E. Goormaghtigh, R. Brasseur, J. M. Ruysschaert, *Biochem. Biophys. Res. Commun.* **104** (1982) 314.
320. C. E. Myers, C. Simone, L. Gianni et al., *Proc. Am. Assoc. Cancer Res.* **22** (1981) 112.
321. T. Oki, *Jpn. J. Antibiot.* **30** (1977) 570.
322. R. K. Y. Zee-Cheng, C. C. Cheng, *J. Med. Chem.* **21** (1978) 291.
323. L. H. Patterson, B. M. Gandecka, J. R. Brown, *Biochem. Biophys. Res. Commun.* **110** (1983) 399.
324. E. D. Kharasch, R. F. Novak, *Biochem. Biophys. Res. Commun.* **108** (1982) 1346.
325. S. S. Legha, *Drugs Today* **20** (1984) 12, 629 – 638.

326. *Drugs of the Future* **5** (1980) 234; **6** (1981) 316; **7** (1982) 357; **8** (1983) 64; **9** (1984) 380; **10** (1985) 429.
327. American Cyanamid, USP 4197249, 1980 (F. E. Durr, K. C. Murdock).
328. R. K.-Y. Zee-Cheng, C. C. Cheng, *J. Med.-Chem.* **22** (1979) 1024.
329. K. C. Murdock et al., *J. Med. Chem.* **22** (1979) 9, 1024 – 1030.
330. F. I. Durr, R. E. Wallace, R. V. Citarella, *Cancer Treat Rev.* **10** (1983) 3 – 11.
331. R. E. Wallace, K. C. Murdock, R. B. Angier, F. E. Durr, *Cancer Res.* **39** (1979) 1570 – 1574.
332. R. K. Johnson et al., *Cancer Treat. Rep.* **63** (1979) 425 – 439.
333. D. D. Von Hoff, C. A. Coltuan, B. Forseth, *Cancer Res.* **41** (1981) 1853 – 1855.
334. B. M. Henderson et al., *Cancer Treat. Rep.* **66** (1982) 1139 – 1143.
335. B. M. Sparano, G. Gordon, C. Hall, M. J. Iatropoulos, J. F. Noble, *Cancer Treat. Rep.* **66** (1982) 1145 – 1158.
336. K. C. Anderson et al., *Cancer Treat. Rep.* **67** (1983) 435 – 438.
337. F. C. Schell et al., *Cancer Treat. Rep.* **66** (1982) 1641 – 1643.
338. C. B. Pratt et al., *Cancer Treat. Rep.* **67** (1983) 85 – 88.
339. D. V. Unverferth et al., *Cancer Treat. Rep.* **67** (1983) 343 – 350.
340. M. S. Aapro, D. S. Alberts, *Invest. New Drugs* **2** (1984) 329 – 330.
341. S. A. Taylor, B. T. Tranum, D. D. Von Hoff, J. J. Constanzi, *Invest. New Drugs* **3** (1985) 67 – 69.
342. J. A. Elliott et al., *Anti-Cancer Drug Res.* **3** (1989) 4, 271 – 282.
343. R. F. Marschke et al., *Med. Pediat. Oncol.* **16** (1988) 4, 269 – 270.
344. K. Hillier, *Drugs of the Future* **6** (1981) 12, 762; **7** (1982) 896.
345. K. C. Murdock et al., *J. Med. Chem.* **25** (1982) 505.
346. American Cyanamid, EP 338372, 1989 (V. J. Lee, K. C. Murdock)
347. American Cyanamid, US 4900838, 1990 (K. C. Murdock)
348. K. C. Murdock et al., *J. Med. Chem.* **36** (1993) 15, 2098.
349. US Dept. Of Health, USP 25157, 1981 (H. Dubicki, J. L. Parsons, F. W. Starks).
350. B. F. Cain et al., *J. Med. Chem.* **18** (1975) 1110.
351. *Annu. Drug Data Report* (1981) 177.
352. *Drugs of the Future* **5** (1980) 277; **8** (1983) 535.
353. L. Steinherz et al., *Cancer Treat. Rep.* **66** (1982) 483.
354. J. Kolwas, *Drugs Today* **15** (1979) 200.
355. G. A. Curt, N. J. Clendeninn, B. A. Chabner, *Cancer Treat. Rep.* **68** (1984) 87 – 99.
356. T. Tsuruo, H. Lida, M. Nojiri, *Cancer Res.* **43** (1983) 2905.
357. S. Waksman, H. B. Woodruff, *Proc. Soc. Exp. Biol. Med.* **45** (1940) 609 – 611.
358. H. M. Sobell, S. C. Jam, T. D. Sakore et al., *Nat. New Biol.* **231** (1971) 200.
359. F. Takusagawa et al.: *12th Congress of the IUCR Associated Meeting on Molecular Structure and Biologic Activity,* Buffalo, New York, August 26 – 28, 1981, p. 55.
360. E. Reich, R. M. Franklin, A. J. Shatkin et al., *Proc. Natl. Acad. Sci. USA* **48** (1982) 1238.
361. H. S. Schwartz, E. Godoyien, R. Y. Ambaye, *Cancer Res.* **287** (1968) 192.
362. M. N. Goldstein, K. Hamm, E. Amrod, *Science* **151** (1966) 1555 – 1556.
363. H. Umezawa, K. Meada, T. Takeuchi et al., *J. Antibiot. Ser. A.* **19** (1966) 200 – 206.
364. B. T. Sikic, Z. H. Siddik, T. E. Gram, *Cancer Treat. Rep.* **64** (1980) 659 – 667.
365. T. Takita, Y. Muraoka, H. Umezawa: "Bleomycin and Peplomycin," in H. M. Pinedo, B. A. Chabner (eds.): *Cancer Chemotherapy Annual 6,* Elsevier, Amsterdam 1984, pp. 85 – 90.
366. A. Hagiwara et al., *Anti-Cancer Drug Res.* **2** (1988) 319 – 324.
367. S. Saito, Y. Umezawa, T. Yoshioka et al., *J. Antibiot. (Tokyo)* **36** (1983) 92 – 95.
368. J. Fugimito, H. Higashi, G. Kosaki, *Cancer Res.* **36** (1976) 2248 – 2251.
369. M. Chien, A. P. Grollman, S. B. Horwitz, *Biochemistry* **16** (1977) 3641 – 3645.
370. W. J. Caspay, C. Niziak, D. A. Lanzo et al., *Mol. Pharmacol.* **16** (1979) 256 – 260.
371. S. Biables, J. R. Warr, *Genet. Res.* **34** (1979) 269 – 279.
372. M. Mayaki, T. Ono, S. Hori et al., *Cancer Res.* **35** (1975) 2015 – 2018.
373. S. Akiyama, M. Kuwano, *J. Cell Physiol.* **107** (1981) 147 – 153.
374. S. Akiyama, K. Ikezaki, H. Kunamochi et al., *Biochem. Biophys. Res. Commun.* **101** (1981) 55 – 60.
375. W. E. G. Mueller, M. Geisort, R. K. Zahn et al., *Eur. J. Cancer Clin. Oncol.* **19** (1983) 665 – 669.
376. P. G. Sorensen, M. Dorth, H. H. Hansen, *Eur. J. Cancer Clin. Oncol.* **19** (1983) 319 – 323.

377. R. K. Johnson, R. P. Hertzberg, *Annu. Rep. Med. Chem.* **25** (1990) 129.
378. P. A. Aristoff et al., *Invest. New Drugs* **7** (1989) 364.
379. J. P. Mc Govren et al., *Invest. New Drugs* **7** (1989) 448.
380. M. A. Mitchell, P. D. Johnson, M. G. Williams, P. A. Aristoff, *J. Am. Chem. Soc.* **11** (1989) 6428.
381. K. Goni et al., *Jpn. J. Cancer Res.* **83** (1992) 113.
382. D. L. Boger, M. S. S. Palanki, *J. Am. Chem. Soc.* **114** (1992) 9318.
383. M. Fontana et al., *Anti-Cancer Drug Res.* **7** (1992) 131.
384. D. Volpe et al., *Invest. New Drugs* **10** (1992) 255.
385. N. Zein, M. Poncin, R. Nilakantan, G. A. Ellestad, *Science* **244** (1989) 697.
386. B. H. Long et al., *Proc. Natl. Acad. Sci. USA* **86** (1989) 2.
387. K. C. Nicolaou, *Science* **256** (1992) 1172.
388. K. C. Nicolaou, W. M. Dai, *J. Am. Chem. Soc.* **114** (1992) 8908.
389. W. A. Bleyer, S. A. Frisby, V. T. Oliverio, *Biochem. Pharmacol.* **24** (1979) 633 – 639.
390. R. A. Bender, W. D. Kornreich, *Proc. Am. Assoc. Cancer Res.* **22** (1981) 227.
391. H. Mujagic, S. S. Chen, R. Geist et al., *Cancer Res.* **43** (1983) 3591 – 3597.
392. D. V. Jackson, R. A. Bender, *Cancer Res.* **39** (1979) 4341 – 4349.
393. R. Schrek, *Am. J. Clin. Path.* **62** (1974) 1 – 7.
394. R. A. Bender, W. A. Bleyer, S. A. Frisby et al., *Cancer Res.* **35** (1975) 1305 – 1308.
395. R. F. Zager, S. A. Frisby, V. T. Oliverio, *Cancer Res.* **33** (1973) 1670 – 1676.
396. I. S. Johnson, H. F. Wright, G. H. Svoboda et al., *Cancer Res.* **20** (1980) 1016 – 1022.
397. G. A. Curt, B. D. Bailey, H. Mujagic et al., *Proc. Am. Assoc. Cancer Res.* **25** (1984) 337.
398. W. T. Beck, M. C. Cirtain, J. L. Lefko, *Cancer Treat. Rep.* **67** (1983) 875 – 879.
399. R. B. Sklaroff, D. Straus, C. Young, *Cancer Treat. Rep.* **63** (1979) 793 – 794.
400. G. Mathe, J. L. Misset, F. deVassal et al., *Cancer Treat. Rep.* **62** (1978) 805.
401. M. Bayssas, J. Gouveia, F. deVassal et al., *Cancer Res.* **74** (1980) 91 – 99.
402. R. Z. Andriamialisoa, N. Langlois, Y. Langlois, P. Potier, *Tetrahedron* **36** (1980) 3053 – 3060.
403. M. N. Serradel, J. Castañer, *Drugs of the Future* **11** (1986) 7, 575 – 577.
404. M. R. Paintrand, I. Pignot, *J. Electron. Microsc.* **32** (1983) 2, 115 – 124.
405. P. Mangeney et al., *J. Org. Chem.* **44** (1979) 3765 – 3768.
406. P. Maral, C. Bourut, E. Chenn, G. Mathe, *Cancer Chemother. Pharmacol.* **5** (1981) 3, 197 – 199.
407. R. Maral, C. Bourut, E. Chenn, G. Mathe, *Cancer Lett.* **22** (1984) 49 – 54.
408. G. Mathe, P. Reizenstein, *Cancer Lett.* **27** (1985) 285 – 293.
409. M. Besenval et al., *Semin. Oncol.* **16** (1989) 2, 37.
410. A. Depierre et al., *Semin. Oncol.* **16** (1989) 2, 26.
411. J. B. Sorensen, *Drugs* **44** (1992) 60.
412. P. B. Schiff, A. S. Kende, S. B. Horwitz, *Biochem. Biophys. Res. Commun.* **85** (1978) 737 – 740.
413. R. A. Bender, B. A. Chabner: "Tubulin Binding Agents: Epipodophyllotoxin," in B. A. Chabner (ed.): *Pharmacologic Principles of Cancer Treatment,* W. B. Saunders, Philadelphia 1982, pp. 263 – 266.
414. J. D. Loike, S. B. Horwitz, *Biochemistry* **15** (1976) 5435 – 5438.
415. D. K. Kalwinsky, A. T. Look, J. Ducore, A. Fridland, *Cancer Res.* **43** (1983) 1592 – 1596.
416. A. J. Wozniak, W. E. Ross, *Cancer Res.* **43** (1983) 130 – 135.
417. A. J. Wozniak, B. S. Glisson, K. R. Hande, W. E. Ross, *Cancer Res.* **44** (1984) 626 – 629.
418. B. H. Long, A. Minocha, *Proc. Am. Assoc. Cancer Res.* **24** (1983) 321.
419. M. E. Wall, M. C. Wani, C. E. Cook, K. H. Palmer, *J. Am. Chem. Soc.* **88** (1966) 3888.
420. J.-C. Cai, C. R. Hutchinson, *Chem. Heterocyl. Compd.* **25** (1983) 753.
421. Yakult Honsha, EP 137145, 1985 (T. Miyasaka et al.).
422. Smith Kline Beecham Corp., EP 321122, (J. C. Boehm, S. M. Hecht, K. G. Holden, R. K. Johnson, W. D. Kingsbury).
423. W. D. Kingsbury et al., *J. Med. Chem.* **34** (1991) 98 – 107.
424. N. Osheroff, *Pharmacol. Ther.* **41** (1989) 223 – 41.
425. L. Liu, J. Wang, *Proc. Natl. Acad. Sci. USA* **84** (1987) 7024 – 7027.
426. A. J. Ryan, S. Squires, H. L. Strutt, R. T. Johnson, *Nucl. Acid Res.* **19** (1991) 12, 3295 – 3300.
427. M. N. Serradel, J. Castañer, R. M. Castañer, *Drugs of the Future* **12** (1987) 3, 207.
428. C. H. Spiridonis, *Drugs of the Future* **20** (1995) 5, 483 – 489.

429. W. J. Slichenmyer, E. K. Rowinsky, R. C. Donehower, S. H. Kaufmann, *J. Natl. Cancer Inst.* **85** (1993) 4, 271 – 291.
430. C. B. Hendricks et al., *Cancer Res.* **52** (1992) 8, 2268 – 2278.
431. W. K. Eng et al., *Mol. Pharmacol.* **38** (1990) 4, 471 – 480.
432. D. Abigerges et al., *J. Natl. Cancer Inst.* **86** (1994) 446.
433. M. C. Wani et al., *J. Am. Chem. Soc.* **93** (1971) 2325 – 2327.
434. J.-N. Denis et al., *J. Am. Chem. Soc.* **110** (1988) 5917 – 5919.
435. J.-N. Denis, A. Correa, A. E. Greene, *J. Org. Chem.* **55** (1990) 1957 – 1959.
436. M. Hepperle, G. I. Georg, *Drugs of the Future* **19** (1994) 573 – 584.
437. K. C. Nicolaou, W. M. Dai, R. K. Guy, *Angew. Chem.* **106** (1994) 38 – 69.
438. K. C. Nicolaou et al., *Nature* **367** (1994) 630 – 634.
439. R. A. Holton et al., *J. Am. Chem. Soc.* **116** (1994) 1599 – 1600.
440. J. J. Masters et al., *Angew. Chem.* **107** (1995) 1883.
441. P. B. Schiff, S. B. Horwitz, *Proc. Natl. Acad. Sci. USA* **77** (1980) 1561 – 1564.
442. P. B. Schiff, S. B. Horwitz, *J. Cell Biol.* **91** (1981) 479 – 483.
443. M. A. Bissery, D. Guenard, F. Gueritte-Voegelein, F. Lavelle, *Cancer Res.* **51** (1991) 4845 – 4852.
444. I. Ringel, S. B. Horwitz, *J. Natl. Cancer Inst.* **83** (1991) 288 – 291.
445. R. Pazdur et al., *Cancer Treat. Rev.* **19** (1993) 351 – 386.
446. P. B. Schiff, J. Faut, S. B. Horwitz, *Pharmacol. Ther.* **25** (1984) 83 – 124.
447. S. Rao, J. J. Manfredi, S. B. Horwitz, I. Ringel, *J. Natl. Cancer Inst.* **84** (1992) 785 – 788.
448. S. Rao, S. B. Horwitz, I. Ringel, *J. Natl. Cancer Inst.* **84** (1992) 785 – 788.
449. J. Parness, S. B. Horwitz, *J. Cell. Biol.* **91** (1981) 479 – 487.
450. P. B. Schiff, S. B. Horwitz, *Biochemistry* **20** (1981) 3247 – 3252.
451. A. H. Ding, F. Porten, E. Sanchez, C. F. Nathan, *Science* **248** (1990) 370 – 372.
452. C. Bogdan, A. Ding, *J. Leukocyte Biol.* **52** (1992) 119 – 121.
453. M. E. Stearns, M. Wang, *Cancer Res.* **52** (1992) 3776 – 3781.
454. D. M. Vyas et al., *Biorg. Med. Chem. Lett.* **3** (1993) 1357 – 1360.
455. E. K. Rowinsky et al., *Semin. Oncol.* **20** (1993) 3, 1 – 15.
456. H. Lataste et al., *Proc. Natl. Acad. Sci. USA* **81** (1984) 4090 – 4094.
457. J. Kant et al., *Biorg. Med. Chem. Lett.* **3** (1993) 2471 – 2474.
458. G. Samaranayake, N. F. Magri, C. Jitransgri, D. G. Kingston *J. Org. Chem.* **56** (1991) 5114 – 5119.
459. A. Datta, L. Jayasinghe, G. I. Georg, *J. Med. Chem.* (1994) 4258 – 4260.
460. J. L. Fabre, D. Lolli-Tonelli, L. H. Spiridonidis, *Drugs of the Future* **20** (1995) 5, 464.
461. *Annual Data Report* (1995) 765.
462. G. Höfle et al., *Angew. Chem.* **108** (1996) 1671; *Int. Ed. Engl.* **35** (1996) 1567.
463. D. M. Bollag et al., *Cancer Res.* **55** (1995) 2325 – 2333.
464. A. Bolag et al., *Angew. Chem.* **108** (1996) 2976; *Int. Ed. Engl.* **35** (1996) 2801.
465. D. Meng et al., *J. Amer. Chem. Soc.* **119** (1997) 2733.
466. K. C. Nicolaou et al., *Angew. Chem.* **108** (1996) 2534.
467. K. C. Nicolaou et al., *Nature* **387** (1997) 268 – 272.
468. D. Schinzer et al., *Angew. Chem.* **109** (1997) 543 – 544.
469. D.-S. Su et al., *Angew. Chem.* **109** (1997) 2178.
470. R. J. Kowalski et al., *J. Biol. Chem.* **272** (1997) 2534.
471. B. Rosenberg, L. Van Camp, T. Krigas, *Nature (London)* **205** (1965) 698 – 700.
472. B. Rosenberg, L. Van Camp, J. E. Troska et al., *Nature (London)* **222** (1969) 385 – 386.
473. A. H. Rossof, R. E. Slayton, C. P. Perlia, *Cancer* **30** (1972) 1451 – 1455.
474. D. J. Higby, H. J. Wallace Jr., J. F. Holland, *Cancer Chemother. Rep.* **57** (1973) 459 – 463.
475. L. H. Einhorn, S. D. Williams, *N. Engl. J. Med.* **300** (1979) 289 – 292.
476. A. W. Prestayko, J. C. D'Aousst, B. F. Issel et al., *Cancer Treat. Rev.* **6** (1979) 17 – 24.
477. W. M. Scovell, T. O'Connor, *J. Am. Chem. Soc.* **99** (1977) 120 – 126.
478. A. F. LeRoy, R. J. Lutz, R. L. Dedrick et al., *Cancer Treat. Rep.* **63** (1979) 59 – 64.
479. G. L. Cohen, W. R. Bauer, J. K. Berton et al., *Science* **203** (1979) 1014 – 1016.
480. R. C. Srivastava, J. Froelich, G. L. Eichhorn, *Biochimie* **60** (1979) 879 – 881.
481. J. J. Roberts, A. J. Thomson, *Prog. Nucleic Acid Res. Mol. Biol.* **22** (1979) 71 – 133.

482. B. Rosenberg, *Naturwissenschaften* **60** (1973) 399 – 408.
483. T. F. Slater, M. Ahmed, S. A. Ibrahim, *J. Clin. Hematol. Oncol.* **7** (1977) 534 – 539.
484. N. E. Madias, J. T. Harrington, *Am. J. Med.* **65** (1978) 307 – 311.
485. D. C. Dobyan, J. Levi, C. Jacobs et al., *J. Pharmacol. Exp. Therap.* **213** (1980) 551 – 556.
486. R. L. Schilsky, T. Anderson, *Ann. Intern. Med.* **90** (1979) 929 – 931.
487. F. A. Hayes, A. A. Green, N. Jenzen et al., *Cancer Treat. Rep.* **63** (1979) 547 – 549.
488. S. Davis, W. Kessler, B. M. Haddad et al., *J. Med.* **11** (1980) 133 – 137.
489. M. Dentino, F. L. Luft, M. N. Yum et al., *Cancer* **41** (1978) 1274 – 1279.
490. K. K. Chang, D. J. Higby, E. S. Henderson et al., *Cancer Treat. Rep.* **61** (1977) 367 – 371.
491. R. F. Ozols, B. Corden, J. Jacobs et al., *Ann. Intern. Med.* **100** (1984) 19 – 24.
492. J. Drobnik, M. Urbankova, A. Krekulova, *Mut. Res.* **17** (1973) 13 – 20.
493. L. A. Zwelling, S. Michaels, H. Schwartz, *Cancer Res.* **41** (1981) 640 – 649.
494. J. Ducore, L. Zwelling, K. Kohn, *Proc. Am. Assoc. Cancer Res.* **21** (1980) 267.
495. L. C. Erikson, L. A. Zwelling, J. M. Ducore, *Cancer Res.* **4** (1981) 2791 – 2794.
496. K. Micetich, S. Michaels, G. Jude et al., *Proc. Am. Assoc. Cancer Res.* **22** (1981) 252.
497. A. Bakka, L. Endresen, A. B. S. Johnson et al., *Toxicol. Appl. Pharmacol.* **61** (1981) 215 – 226.
498. L. R. Beach, R. D. Palmiter, *Proc. Natl. Acad. Sci. USA* **78** (1981) 2110 – 2114.
499. C. R. Wilkenson, P. J. Cox, M. Jones et al., *Biochimie* **60** (1978) 851 – 857.
500. *Drugs of the Future* **8** (1983) 489; **9** (1984) 463; **10** (1985) 497; **11** (1986) 499.
501. *Drugs Today* **22** (1986) 255.
502. Research Corp., DE 2 329 485, 1973 (M. J. Cleare, J. D. Hoeschele, B. Rosenberg).
503. R. C. Harrison et al., *Inorg. Chim. Acta* **46** (1980) 215.
504. A. H. Calvert et al., *Cancer Chemother. Pharmacol.* **9** (1982) 140.
505. B. D. Evans et al., *Proc. Ass. Cancer Res.* **24** (1983) 154.
506. CBDCA, Clinical Brochure, Investigational Drug Branch, Division of Cancer Treatment, National Cancer Institute, Bethesda, Md., 1980.
507. B. D. Evans, K. S. Raju, A. H. Calvert, S. U. Harland et al., *Cancer Treat. Rep.* **67** (1983) 997 – 1001.
508. P. J. Creaven, S. Madajewicz, L. Pendyala et al., *Cancer Treat. Rep.* **67** (1983) 795 – 798.
509. C. Sternberg et al., *Cancer Treat. Rep.* **69** (1985) 1305 – 1307.
510. R. Canetta et al., *Cancer Treat. Rep.* **63** (1985)2107 – 2109.
511. I. N. Olver et al., *Cancer Treat. Rep.* **70** (1986) 421 – 422.
512. A. P. Kyriazis et al., *Cancer res.* **45** (1985) 2012 – 2015.
513. K. R. Harrap, M. Jones, C. R. Wilkenson et al.: "Antitumor Toxic and Biochemical Properties of *cis*-Platinum and Eight Other Platinum Analogs," in A. W. Prestayko, S. T. Crooke, S. K. Carter (eds.): *cis-Platinum: Current Status and New Developments (Pap Symposium),* Academic Press, New York 1980, pp. 193 – 212.
514. P. Chang, *Drugs of the Future* **8** (1983) 364; **10** (1985) 7, 561.
515. *Drugs of the Future* **12** (1987) 11, 1029 – 1031.
516. *Annu. Drug Data Rep.* **8** (1986) 6, 590.
517. Shionogi & Co., EP 216362 1987 (H. Kagawa, K. Shima, T. Tsukada).
518. K. Hirabayashi, E. Okada, *Cancer* **71** (1993) 9, 2769.
519. M. Koenuma et al., *Clin. Rep.* **29** (1995) 12, 259.
520. N. Uchida et al., *Clin. Rep.* **29** (1995) 12, 269.
521. I. H. Krakoff, 5th *Intl. Symp. Platinum Cancer Chemother.*, Abano Therme, 1987, Abst. L7.
522. B. K. Bhuyan et al., *Cancer Commun.* **3** (1991) 2, 53.
523. ASTA Medica AG, EP 324154 1989 (J. Engel et al.).
524. R. Voegeli, E. Günther, P. Aulenbacher, J. Engel, P. Hilgard, *Drugs of the Future* **17** (1992) 883 – 886.
525. A. M. Otto et al., *Pharm. Pharmacol. Lett.* **1** (1992) 103 – 106.
526. W. A. Denny et al., *J. Med. Chem.* **35** (1992) 2983 – 2987.
527. *Drugs Today* **10** (1974) 74.
528. ICI, GB 1013907 1962.
529. G. R. Bedford, D. N. Richardson, *Nature* **212** (1966) 733.
530. M. N. Serradel, J. Castañer, *Drugs of the Future* **11** (1986) 5, 398 – 400.
531. Farmos Group, EP 95875 (R. J. Tiovola et al.).
532. J. T. Pento et al., *Drugs of the Future* **9** (1984) 7, 5.
533. R. Löser, P.-St. Jamak, K. Seibel, *Drugs of the Future* **9** (1984) 3, 186.

534. S. Kallio et al., *Cancer Chemother. Pharmacol.* **17** (1986) 103 – 108.
535. L. Kangas et al., *Cancer Chemother. Pharmacol.* **17** (1986) 109 – 113.
536. V. C. Jordan, B. Gosden, *Mol. Cell. Endocrinol.* **27** (1982) 27, 921.
537. S. P. Robinson et al., *Eur. J. Cancer Clin. Oncol.* **24** (1988) 12, 1817.
538. R. Valavaara et al., *Eur. J. Cancer Clin. Oncol.* **24** (1988) 4, 785.
539. S. R. Ebbs et al., *Lancet II* (1987) 621.
540. R. F. Kauffmann et al., *J. Pharmacol. Exp. Ther.* **280** (1997) 146 – 153.
541. M. A. Ferreira, M. M. Caramona, L. M. Celeste, *Proc. Br. Pharmacol. Soc.* (1995) Abs. 267 P
542. M. Dukes, *Pharm. J.* **257** (1996) 176.
543. A. Hoshi, *Drugs of the Future* **16** (1991) 3, 217.
544. *Annu. Drug Data Report* **15** (1993) 4, 378.
545. A. Brodie, *J. Steroid Biochem. Mol. Biol.* **40** (1991) 1 – 3, 255.
546. S. A. Wells et al., *Cancer Res.* **Suppl. 42** (1982) 3454.
547. R. D. Burnett, D. N. Kirk, *J. Chem. Soc. Perkin Trans.* **I** (1973) 1830.
548. R. C. Coombes, *Eur. J. Cancer* **28A** (1992) 12, 1941.
549. D. Cunningham et al., *Brit. J. Cancer* **55** (1987) 331.
550. J. H. Davis et al., *Brit. J. Cancer* **66** (1992) 1, 139.
551. M. Dowsett et al., *Cancer Res.* **49** (1989) 1306.
552. M. Dowsett et al., *Eur. J. Cancer* **28** (1992) 2 – 3, 415.
553. R. C. Stein et al., *Cancer Chemother. Pharmacol.* **26** (1990) 75.
554. ICI, US 4935437, 1990 (P. N. Edwards, M. S. Lange).
555. Ciba Geigy AG, US 4617307, 1986 (L. J. Browne).
556. J. Prous, *Drugs of the Future* **19** (1994) 4, 335 – 337.
557. J. Castañer, *Annu. Drug Data Rep.* **13** (1991) 8, 716.
558. A. B. Foster et al., *J. Med. Chem.* **28** (1985) 200.
559. J. T. Pento, *Drugs of the Future* **14** (1989) 9, 843 – 845.
560. J. Prous et al., *Drugs of the Future* **20** (1995) 1, 30 – 32.
561. R. E. Steele et al., *Steroids* **50** (1987) 147.
562. M. Dukes et al., *Proc. Am. Assoc. Cancer Res.* **33** (1992) Abst. 1677.
563. P. V. Plourde et al., *Breast Cancer Res. Treat.* **30** (1994) 1, 103 – 111.
564. K. Schieweck, A. S. Bhatnagar, A. Matter, *Cancer Res.* **48** (1988) 834.
565. K. Schieweck et al., *Proc. Am. Assoc. Cancer Res.* **29** (1988) Abst. 968.
566. M. Dowsett et al., *Breast Cancer Res. Treat.* **27** (1993) 1 – 2, Abst. 87.
567. P. V. Plourde et al., *Proc. Am. Soc. Clin. Oncol.* **12** (1993) Abst. 165.
568. C. Huggins, C. V. Hodges, *Cancer Res.* **1** (1941) 293.
569. J. A. Smith, *J. Urol.* **137** (1987) 1 – 10.
570. J. Trachtenberg, *J. Urol.* **132** (1984) 61.
571. G. H. Rasmusson et al., *J. Med. Chem.* **27** (1984) 1690 – 1701.
572. M. R. Robinson, B. S. Thomas, *Brit. Med. J.* **4** (1971) 391 – 394.
573. F. Neumann, *J. Steroid Biochem.* **19** (1983) 391 – 402.
574. B. J. Furr, *Clin. Oncol.* **2** (1988) 581 – 590.
575. H. Tucker, *Drugs of the Future* **15** (1990) 3, 255 – 265.
576. *Drugs Today* **20** (1984) 296.
577. *Drugs of the Future* **1** (1976) 108; **8** (1983) 270.
578. R. O. Neri, E. A. Peets, *J. Steroid Biochem.* **6** (1975) 815 – 819.
579. W. I. P. Mainwaring, F. R. Mangan, P. A. Feherty, M. Freifeld, *Med. Cell. Endocrinol.* **1** (1974) 113 – 128.
580. S. Liao, D. K. Howell, T. M. Chang, *Endocrinology* **94** (1974) 1205 – 1209.
581. B. Katchen, S. Buxbaum, *J. Clin. Endocrinol. Metab.* **41** (1975) 373 – 379.
582. J. Geller et al., *Prostate* **2** (1981) 309 – 314.
583. E. A. Peets, M. F. Henson, R. O. Neri, *Endocrinology* **94** (1974) 532 – 540.
584. R. Neri, N. Kassem, *Prog. Cancer Res. Ther.* **31** (1984) 507 – 518.
585. F. Labrie et al., *Oncol.* **2** (1988) 597 – 619.
586. F. Labrie, A. Dupont, M. Ciguere, *Brit. J. Urol.* **61** (1988) 341 – 346.
587. A. Belanger, A. Dupont, F. Labrie, *J. Clin. Endocrinol. Metab.* **59** (1984) 422 – 426.
588. J.-P. Raynaud, C. Bonne et al., *Prostate* **5** (1984) 299 – 311.
589. T. Ojasoo, *Drugs of the Future* **12** (1987) 763 – 770.
590. L. Proulx, F. Labrie, *Prostate* **5** (1984) Abst. 429.
591. F. Labrie, A. Dupont, A. Belanger, *Prostate* **4** (1983) 579.
592. C. Harnois, A. Dupont, F. Labrie, *Brit. J. Opthalmol.* **70** (1986) 471 – 473.

593. H. Tucker, G. J. Chesterson, *J. Med. Chem.* **31** (1988) 885 – 887.
594. H. Tucker, J. W. Crook, G. J. Chesterson, *J. Med. Chem.* **31** (1988) 954 – 959.
595. B. J. A. Furr et al., *J. Endocrinol.* **113** (1987) R7 – 9.
596. S. N. Freeman, W. I. P. Mainwaring, B. J. A. Furr, *J. Endocrinol.* (1986) Suppl. III, 155.
597. C. J. Tyrell, *Prostate* (Suppl. 4) **20** (1992) 97.
598. G. Wilding et al., *Proc. Am. Assoc. Clin. Oncol.* **10** (1991) Abst. 593.
599. A. V. Schally, S. M. McCann, *Fertil. Steril.* **64** (1995) 452 – 453.
600. A. V. Schally, A. Arimura, A. J. Kastin, *Science* **173** (1971) 1036 – 1038.
601. H. Matsuo, Y. Baba, M. Nair, A. Arimura, A. V. Schally, *Biochem. Biophys. Res. Commun.* **43** (1971) 1334 – 1339.
602. K. Amoss et al., *Biochem. Biophys. Res. Commun.* **44** (1971) 205 – 210.
603. Y. Baba, H. Matsuo, A. V. Schally, *Biochem. Biophys. Res. Commun.* **44** (1971) 459 – 463.
604. G. Emons, A. V. Schally, *Hum. Reprod.* **9** (1994) 1364 – 1370.
605. P. M. Conn, W. Crowley, *New Engl. J. Med.* **324** (1991) 93 – 103.
606. G. F. Weinbauer, E. Nieschlag in K. Höffgen (ed.): *Peptides in Oncology* Springer, Heidelberg, 1992, pp. 113 – 136.
607. A. S. Dutta, *Drugs of the Future* **13** (1988) 43 – 57.
608. M. Filicori, C. Flamingi, *Drugs* **35** (1988) 63 – 82.
609. J. J. Nestor, B. H. Vickery, *Annu. Rep. Med. Chem.* **23** (1988) 211 – 220.
610. M. T. Goulet, *Annu. Rep. Med. Chem.* **30** (1995) 169 – 178.
611. T. Reissmann et al., *Hum. Reprod.* **20** (1995) 1974 – 1981.
612. M. J. Karten in W. F. Crowley, P. M. Conn (eds.): *Modes of Action of GnRH and GnRH Analogs,* Springer, Heidelberg 1992, pp. 277 – 297.
613. G. Flouret et al., *Pept. Sci.* **1** (1995) 89 – 105.
614. P. M. Conn, W. F. Crowley, *Annu. Rev. Med.* **45** (1994) 391.
615. A. V. Schally in J. F. Holand et al. (eds): *Cancer Medicine* 3rd ed., Lee & Febiger, Philadelphia, PA 1993, pp. 827 – 840.
616. A. V. Schally in P. Belfort, J. Pinotti, T. K. Eskes (eds.): *Advances in Gynecology and Obstetrics Vol. 6,* Parthenon, Cornforth 1989, pp. 3 – 22; b) Scrip 22, 1995, 2066.
617. A. Friedrich, G. Jaeger, K. Radscheit, R. Uhmann, *Pept. Proc. Eur. Pept. Symp. 22*nd 1992/1993, 47 – 49.
618. B. Kutscher et al., *Angew. Chem. Int. Ed. Engl.* **36** (1997) 2148 – 2161.
619. R. L. Barbieri, *Trends Endocrinol. Metab.* **3** (1992) 30 – 34.
620. R. Schmidt, K. Sundaram, R. B. Thau, C. W. Badrin, *Contraception* **29** (1984) 283 – 289.
621. S. Bajusz et al., *Int. J. Pept. Protein Res.* **32** (1988) 425 – 435.
622. J. Pinski et al., *Int. J. Pept. Protein Res.* **45** (1995) 410 – 417.
623. A. Kleemann et al., *Proc. Akabori Conf. Ger. Jpn. Symp. Pept. Chem. 4*th 1991, 96 – 101; b) F. R. Kunz, T. Müller, K. Drauz, *Proc. Akabori. Conf. Ger. Jpn. Symp. Pept. Chem.* 5th 1994, 15 – 16.
624. A. Ljugquist et al., *Proc. Natl. Acad. Sci. USA* **85** (1988) 8236 – 8240.
625. J. Leal et al., *Drugs of the Future* **16** (1991) 529 – 537.
626. A. Janecka, T. Janecki, C. Bowers, K. Folkers, *J. Med. Chem.* **37** (1994) 2238 – 2241.
627. A. Janecka et al., *Med. Chem. Res.* **1** (1991) 306 – 311.
628. A. Janecka, T. Janecki, C. Bowers, K. Folkers, *Int. J. Pept. Protein Res.* **44** (1994) 19 – 23.
629. P. Theobald et al., *J. Med. Chem.* **34** (1991) 2395 – 2402.
630. J. Rivier et al., *J. Med. Chem.* **35** (1992) 4270 – 4278.
631. J. E. Rivier et al., *J. Med. Chem.* **38** (1995) 2649 – 2662.
632. K. Stoeckemann, J. Sandow, *J. Cancer Res. Clin. Oncol.* **119** (1993) 457 – 462.
633. R. Deghenghi, F. Boutignon, P. Wüthrich V. Lenaerts, *Biomed. Pharmacother.* **47** (1993) 107 – 110.
634. J. Nester, et al., *J. Med. Chem.* **35** (1992) 3942 – 3948.
635. J. Mulzer in E. Ottow, U. Schöllkopf, B. G. Schulz (eds.): *Stereoselective Synthesis* , Springer, Heidelberg, 1994, pp. 37 – 61.
636. J. Mulzer et al., *Angew. Chem.* **106** (1994) 1813 – 1815; *Angew. Chem. Int. Ed. Engl.* **33** (1994) 1737 – 1739.
637. F. Haviv et al., *J. Med. Chem.* **32** (1989) 2340 – 2344.
638. F. Haviv et al., *J. Med. Chem.* **36** (1993) 928 – 933.
639. Abbott Laboratories, PCT/US 95/02410, 1995 (F. Haviv).
640. Abbott Laboratories, WO 95/04540, 1995 (F. Haviv).
641. Abbott Laboratories, WO 94/14841, 1994 (F. Haviv).

642. Abbott Laboratories, WO 94/13313, 1994 (J. Greer).
643. Tap Pharmaceuticals, US-A 5300492, 1994 (F. Haviv).
644. F. Haviv et al., *J. Med. Chem.* **37** (1994) 701 – 707.
645. T. Beckers, K. Marheineke, H. Reiländer, P. Hilgard, *Eur. J. Biochem.* **231** (1995) 535 – 543.
646. R. P. Millar, C. A. Flanagan, R. C. Milton, J. A. King, *J. Biol. Chem.* **264** (1989) 21007 – 21013.
647. G. A. McPherson, *J. Pharmacol. Methods* **14** (1985) 213 – 228.
648. J. Trachtenberg, A. Pont, *Lancet* **2** (1984) 433 – 435.
649. S. Bhasin et al., *Endocrinology* **118** (1986) 1229 – 1232.
650. B. De et al., *J. Med. Chem.* **32** (1989) 2036 – 2038.
651. Abbott Laboratories, US-A 4992421, 1991 (B. De).
652. McNeillab Inc., US-A 4678784, 1987 (C. Y. Ho).
653. Takeda Chemical Industries, WO 95/28405, 1995 (S. Furuya).
654. Takeda Chemical Industries, WO 96/34012A1, 1996 (C. Kitada).
655. Merck, WO 97/21435, 1997 (M. Goulet).
656. Takeda Chemical Industries, WO 97/14697, 1997 (S. Furuya).
657. C. Unger, *Drugs of the Future* **22** (1997) 12, 1337 – 1345.
658. G. Powis, *Pharmacol. Ther.* **62** (1994) 57 – 95.
659. W. J. Fantl, D. E. Johnson, L. T. Williams, *Annu. rev. Biochem.* **62** (1993) 453 – 481.
660. F. McCormick, *Nature* **363** (1993) 15 – 16.
661. C. A. Lang-Carter et al., *Science* **260** (1993) 315 – 319.
662. D. Rea et al., *Proc. 7th NCI-EORTC Symp. On New Drugs in Cancer Ther.* Amsterdam (1992), p. 62.
663. D. Ingber et al., *Nature* **348** (1990) 555.
664. T. Ozsuka et al., *J. Antibiotics* **45** (1992) 348.
665. Y. Reiss et al., *Cell* **62** (1990) 81 – 88.
666. S. Omura, D. van der Pyl, *Cell* **46** (1993) 222.
667. S. M. Sebti, A. D. Hamilton, *Drug Discovery Technol.* **3** (1998) 26 – 33.
668. A. Levitski, A. Gazit, *Science* **267** (1995) 1782 – 1788.
669. A. Levitski, *Eur. J. Biochem.* **226** (1994) 1 – 13.
670. R. T. Abraham, M. Aquarone, A. Anderson, *Biol. Cell.* **83** (1995) 105.
671. D. J. Hanahan, *Ann. Rev. Biochem.* **55** (1986) 483 – 509.
672. W. E. Berdel, *Onkologie* **13** (1990) 245 – 250.
673. W. E. Berdel et al., *Anticancer Res.* **1** (1981) 345 – 352.
674. G. Rodriguez et al., *Proc. Am. Assoc. Cancer Res.* **33** (1992) 262.
675. C. Unger, H. Eibl, *Lipids* **26** (1991) 1412.
676. P. Hilgard, J. Engel, *Drugs Today* **Suppl. B.,** (1994) 30.
677. P. Hilgard et al., *Cancer Chemother. Pharmacol.* **32** (1993) 90 – 95.
678. P. Hilgard, J. Stekar, C. Unger, *Proc. Annu. Meet. Am. Assoc. Cancer Res.* **31** (1990) A2457.
679. J. Engel et al., *Drugs of the Future* **13** (1988) 10, 948 – 951.
680. C. Geilen et al., *Eur. J. Cancer* **27** (1991) 12, 1650 – 1653.
681. R. Hass et al., *Cancer Res.* **52** (1992) 1445 – 1450.
682. R. Becher et al., *Onkologie* **16** (1993) 1, 11.

Immunotherapy and Vaccines

STANLEY J. CRYZ, JR., Swiss Serum and Vaccine Institute Berne, Switzerland (Chaps. 1 and2)

MARTA GRANSTROM, Departments of Clinical Microbiology and of Vaccine Production, National Bacteriological Laboratory, Karolinska Hospital, Stockholm, Sweden (Chap. 3)

BRUNO GOTTSTEIN, Institut für Parasitologie, Universität Zürich, Zürich, Switzerland (Section 4.1)

LUC PERRIN, Division d'Hematologie, Hôpital Cantonal Universitaire, Genève, Switzerland (Section 4.2)

ALAN CROSS, Department of Bacterial Diseases, Walter Reed Army Institute of Research, Washington D.C. 20307-5100, United States (Chap. 5)

JAMES LARRICK, Genelabs Incorporated, Redwood City, California, United States (Chap. 6)

Ullmann's Industrial Toxicology

ISBN: 3-527-31247-1

Abbreviations used in this article:

AIDS acquired immune deficiency syndrome
BCG acillus Calmette-Guerin
CMV Cytomegalovirus
CPS capsular polysaccharide
CS circumsporozoite
Da dalton
DNA deoxyribonucleic acid
DPT diphtheria–pertussis–tetanus
DPT-Pol. diphtheria–pertussis–tetanus–polio
DT diphtheria–tetanus
FHA filamentous hemagglutinin
HBIG human anti-HBV immune globulin
HBsAg hepatitis B surface antigen
HBV hepatitis B virus
HIV human immunodeficiency virus
HRIG human rabies immune globulin
humab human monoclonal antibodies
Ig immunoglobulin
IPV inactivated polio vaccine
ISG immune serum globulin
ITP idiopathic thrombocytopenic purpura
IU international units
IVIG intravenous immunoglobulin
Lf limit of flocculation
LPS lipopolysaccharide
MMR measles–mumps–rubella
NANP asparagine–alanine–asparagine–proline
NVDP asparagine–valine–aspartic acid–proline
OPV oral polio vaccine
PFU plaque forming units
PT pertussis toxin
RESA ring-infected erythrocyte surface antigen
RNA ribonucleic acid
TCID tissue culture infectious dose

1. Introduction [1–9]

1.1. Historical Aspects

Immunization is the most efficient, cost-effective means of preventing infectious diseases. The concept of preventing disease by vaccination is old: in China and India, the practice of "variolation," whereby small quantities of material from disease pustules were used to immunize people against smallpox, was practiced before 1000 B.C. The first "rational" approach to vaccination was taken by JENNER in 1798, who used naturally attenuated cowpox to immunize against smallpox. About 100 years later, PASTEUR introduced vaccines against anthrax and rabies based upon attenuated virulent organisms. The discovery by VON BEHRING in 1890, that serum antibodies could neutralize diphtheria toxin opened the door for a new avenue of vaccine development and passive therapy, whereby preformed antibodies were transferred to at-risk patients. By the beginning of the 20th century, serum obtained from immunized animals was used to treat a variety of diseases including diphtheria and tetanus. The use of human serum followed shortly in 1907.

Since the turn of the century, a wide range of vaccines and antisera have been introduced to manage infectious and noninfectious diseases (Table 1). The development of such agents has required expertise from a variety of disciplines including microbiology, biochemistry, immunology, and molecular biology.

Table 1. Vaccines and immunoglobulins

Type of vaccine or immunoglobulin	Disease
Bacterial vaccines	cholera
	diphtheria
	Hemophilus influenzae b
	meningococcal meningitis
	pertussis
	Streptococcus pneumoniae
	tetanus
	tuberculosis
	typhoid fever
Viral vaccines	hepatitis B
	influenza
	Japanese encephalitis
	measles
	mumps
	polio
	rabies
	Rift Valley fever
	smallpox
	tick-borne encephalitis
	varicella
	yellow fever
Immunoglobulins against bacterial diseases	diphtheria
	Hemophilus influenzae, meningococcal meningitis, *Streptococcus pneumoniae* (polyvalent preparation)
	pertussis
	tetanus
Immunoglobulins against viral diseases	cytomegalovirus
	hepatitis A
	hepatitis B
	human immunodeficiency virus (HIV) (normal intravenous immunoglobulin preparation administered to HIV-positive infants)
	measles
	mumps
	rabies
	rubella
	vaccinia
	varicella
Immunoglobulins against noninfectious diseases	hypogammaglobulinemia
	rhesus factor
	idiotypic thrombocytopenia purpura

1.2. Principles and Definitions

1.2.1. Antigens

An antigen is a molecule that can elicit an immune response (either humoral, i.e., antibody-mediated, or cellular, i.e., cell-mediated) or an immune reaction, such as an allergic reaction. An antigen that evokes an immune response is commonly referred to as an *immunogen*. Only foreign or "non-self" molecules are immunogenic. Usually the larger and more complex a molecule is, the more immunogenic it will be. For example, the gram-negative bacterial cell envelope shown in Figure 1 contains many different somatic (cell-associated) antigens, such as lipopolysaccharide (LPS), outer-membrane proteins, and phospholipids. Numerous factors determine the immunogenicity of a purified molecule. Size is of critical importance: proteins with a molecular mass of $\leq 20\,000$ are poorly immunogenic. Similarly, simple polysaccharides composed of a limited number of repeating monosaccharides are not immunogenic unless their molecular mass exceeds 500 000. Small molecules can be rendered immunogenic by covalently coupling them to larger molecules, forming conjugates.

A single antigen may contain many *epitopes,* which are specific areas of the molecule with a three-dimensional configuration that induces an immune response. Complex molecules, such as large proteins composed of many different amino acids, contain more epitopes than a comparatively simple polysaccharide composed of two or three monosaccharide repeats. The immune response to a given antigen can vary greatly among species and individuals within a species due to immune regulation genes.

1.2.2. Antibodies

Antibodies are proteins found primarily in the serum which are produced by B cells in response to contact with a foreign antigen. There are several different classes of antibodies with characteristic functions (see Chap. 5, Table 3). A schematic of an *immunoglobulin G* (IgG) molecule is shown in Figure 2. Immunoglobulins are composed of light and heavy chains held together by disulfide bonds. Each chain has a variable and a constant region. The tertiary structure of the variable region accounts for the specificity of antibody binding. An antibody produced from a given clone of B cells (Section 1.2.3) recognizes and binds to a given epitope or closely related epitopes. Antibodies which recognize more than one epitope are termed *cross-reactive.* The strength with which an antibody binds to an antigen is termed *affinity* and is determined by

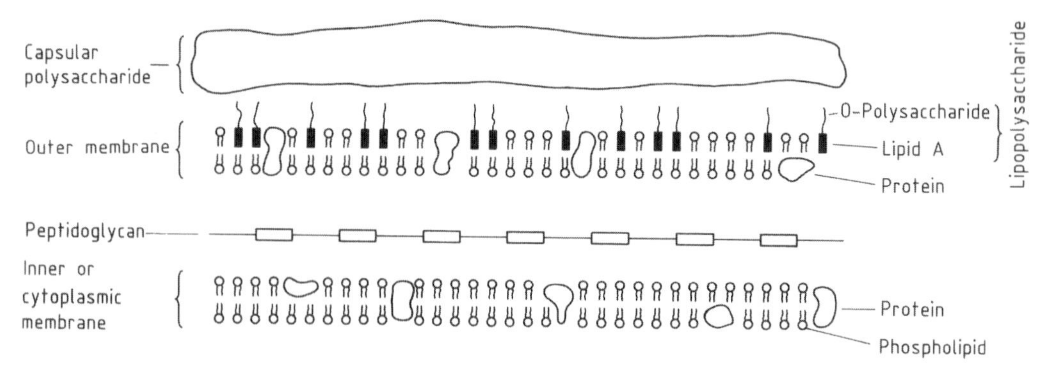

Figure 1. Schematic representation of the gram-negative bacterial cell envelope

the "fit" between the immunoglobulin binding site and the epitope.

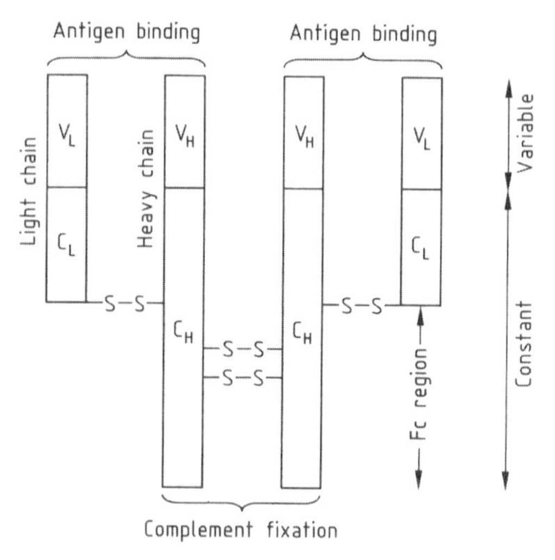

Figure 2. Schematic representation of an immunoglobulin G (IgG) molecule

Antibodies produced by a single clone of B cells are termed *monoclonal antibodies* and recognize only a single epitope (see Chap. 6). *Polyclonal antibodies* are produced by several B cell clones which recognize the same antigen but bind to different epitopes.

Immunoglobulins are divided into five classes termed IgG, IgM, IgA, IgD, and IgE based upon physical and structural differences; they are all glycoproteins. The vast majority of immunoglobulins circulate in the plasma fraction, but certain cells of the immune system can express immunoglobulin on their surface.

Immunoglobulin G (IgG) has a molecular mass of 150 000 and has two antigen binding sites; it represents approximately 80 % of all immunoglobulins in normal serum (8 – 16 mg/mL). Four IgG subclasses (IgG 1 – IgG 4) are found which account for ca. 70, 19, 8, and 3 % of total IgG, respectively; they differ with respect to their antigenic properties in the constant region of the heavy chains. The functional attributes of these subclasses are detailed elsewhere (see Section 5.1). Immunoglobulin G readily crosses the placenta and provides protection against a variety of infectious diseases in the neonate. For example, immunization of the mother against tetanus shortly before delivery ensures a protective level of antibody for the infant. Immunoglobulin G also diffuses into the extravascular tissue more readily than other immunoglobulin classes. The IgG antibody is thought to be responsible for neutralizing the majority of bacterial toxins (e.g., tetanus and diphtheria toxin) formed during an infection. Upon binding with invading bacteria, IgG activates the complement system (a group of interacting serum proteins) which attracts phagocytic cells. The binding of complement components to the Fc region of IgG permits the uptake and killing of the bacteria by phagocytes.

Immunoglobulin M (IgM) is a 900 000 dalton pentamer whose five IgG-like units are held together by intramolecular disulfide bonds and stabilized by a polypeptide "anchor" termed the "J-chain." IgM comprises 5 – 10 % of normal circulating immunoglobulins (1 – 2 mg/mL). Due to its large size, IgM is confined to the intravascular space. It binds complement but, unlike IgG, does not bind directly to phagocytic cells. Immunoglobulin M is usually the first antibody class formed in response to foreign antigen exposure. Due to its multivalency, IgM is extremely efficient at agglutinating bacteria. This phenomenon is important in the control of bacteremia and the neutralization of LPS released by gram-negative bacteria.

Immunoglobulin A (IgA) occurs as 160 000 dalton monomers or 320 000 dalton dimers. Monomers are found primarily in the intravascular space and comprise approximately 10 – 15 % (1.4 – 4 mg/mL) of total serum immunoglobulins. There are two subclasses, IgA 1 and IgA 2. Dimeric IgA, termed *secretory IgA*, is formed by noncovalent interaction with the "secretory piece," a glycoprotein of about 60 000 daltons. Secretory IgA is the predominant immunoglobulin found in mucous secretions, such as tears, colostrum, pulmonary, intestinal, and urinogenital fluids; IgA can bind complement and react with phagocytic cells.

Although parenteral immunization stimulates a vigorous serum IgA response, secretory IgA is considered to be of greater importance due to its "first line" defensive role in body secretions. Secretory IgA plays a critical role in providing protection against respiratory, intestinal, and urinogenital tract bacterial pathogens.

Immunoglobulin D (IgD) is a monomer of ca. 180 000 daltons; concentrations vary widely in humans ranging from 0 to ca. 0.4 mg/mL. Compared to other immunoglobulin classes, IgD is very susceptible to proteolysis and possesses a short half-life (about 3 days). Since IgD does not fix complement or bind to phagocytic cells, it is not considered to be a "protective" immunoglobulin as are IgA, IgG, and IgM. However, IgD is abundant on the surface of B lymphocytes and may play a central role in the activation process which leads to the development of antibody-secreting plasma cells.

Immunoglobulin E (IgE) is a monomer of 200 000 daltons; its average serum concentration (ca. 250 ng/mL) is the lowest of any immunoglobulin. Although IgE does not fix complement or react directly with phagocytic cells, it has a very high affinity for most cells and basophils. Immunoglobulin E appears to mediate both a protective and a detrimental immune response. At the mucoid surface, pathogens binding to IgE stimulate an acute inflammatory response by triggering the release of potent mediators from mast cells and basophils. A protective role for IgE is indicated in several chronic parasitic infections, most notably schistosomiasis (see Chap. 4) where high levels (> 100 µg/mL) of serum IgE have been noted. Immunoglobulin E induces an inflammatory response and "recruits" effector cells to the area. In contrast, IgE-mediated degranulation of effector cells leads to the release of vasoreactive molecules. Individuals suffering from certain allergies can also have elevated serum IgE levels.

1.2.3. Immune Response

The chain of events leading to an immune response is extremely complicated and not yet completely understood. A simple model is shown in Figure 3. The human is capable of forming an immune response to thousands of foreign antigens. The humoral (antibody-mediated) response depends on the *B cells* which are lymphocytes derived from bone marrow stem cells. Upon maturation, B cells form antibody-secreting *plasma cells.*

Each clone of B cells has an immunoglobulin molecule with a recognition site for a specific antigen on its surface. Binding of the appropriate antigen to the B cell causes proliferation of the clone whereby the progeny cells secrete antibody whose specificity is identical to that of the cell-surface immunoglobulin. A foreign antigen can be taken up by macrophages, processed, and presented upon the cell surface, or may remain in a soluble state. All antigens can be termed T-dependent or T-independent depending on whether they require or do not require the interaction of T cells for antibody synthesis (T cells are lymphocytes derived from the thymus). In the case of a *T-independent antigen* (usually polymers, such as bacterial capsular polysaccharides), the antigen can cross-link the B cell surface antibody molecules of the B cell; this initiates proliferation and antibody synthesis. A *T-dependent antigen* requires, in addition to binding to B-cell surface immunoglobulin, the release of T cell factors which act upon the B cell to initiate proliferation.

A human exposed to an antigen in this way is considered *primed*. This is a critical factor in immunization. After initial exposure to a T-dependent antigen by immunization, the induced serum antibody concentration is low and returns to near-basal levels comparatively quickly

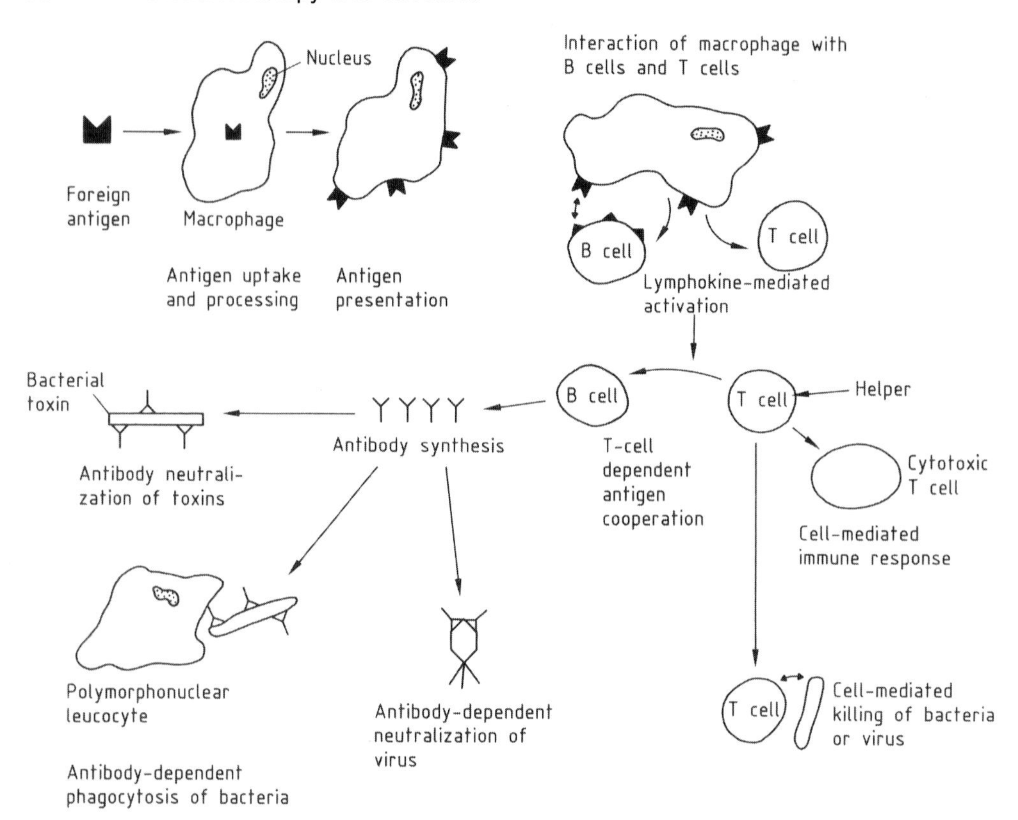

Figure 3. Immunological response cascade to a foreign antigen or vaccine, leading to antibody response or cell-mediated response

(the average half-life of IgG is ca. 22 days, whereas IgM is ca. 5 days). Upon revaccination (boosting) or natural exposure to the pathogen or toxin, a vigorous antibody response occurs in which high antibody levels are synthesized over a longer period of time. This is termed an *anamnestic response.* Therefore, high levels of serum antibodies do not have to be present for the individual to be protected by prior vaccination. A T-independent antigen does not evoke an anamnestic response.

The induction of a *cell-mediated immune (CMI) response* is rather more complicated and less well defined. The CMI response plays a critical role in immunity to pathogens able to live and proliferate within host cells. The first step is "activation" of T cells. Like B cells, T cells also recognize specific antigens. Recognition of an antigen on the surface of a macrophage by T cells results in the release of *interleukin 1* (IL-1), a lymphokine which activates them. One subpopulation of T cells synthesizes interleukin 2 (IL-2), also known as T-cell growth factor, which causes a second subpopulation of activated T cells expressing the IL-2 receptor to proliferate and become cytotoxic T cells. Cytotoxic T cells recognize the specific antigen they are activated against when expressed on the surface of infected cells; they thus kill these infected cells by lysis. A third subpopulation of T cells evolve to be "primed" memory cells which undergo rapid proliferation upon reexposure to the same antigen.

Immunization with a given vaccine may induce a CMI, a humoral antibody response, or both depending on the infecting pathogen. Both types of immunity are usually desirable and may act synergistically. For example, vaccine-induced immunity to many viral diseases, such as rabies, measles, mumps, and rubella, is confirmed by measuring serum antibody levels. However, infected cells can only be destroyed by cytotoxic T cells. Circulating antibody probably prevents the spread of the virus, while cytotoxic T cells eliminate already infected cells. Tuberculosis vaccines stimulate a good antibody response but are less proctective due to the suppression of the critical CMI response.

1.2.4. Active Immunization

Active immunization entails the administration of one or more antigens to a host in the form of a vaccine in an attempt to elicit a protective immune response. In this way an individual can be rendered immune to a variety of diseases. Most vaccines are administered to infants or young children. The trend is to combine several *monovalent vaccines* (vaccines containing a single antigen or vaccine strain) to form *multivalent vaccines* capable of simultaneously inducing immunity to several diseases. Infants are routinely immunized against diphtheria, tetanus, pertussis, and, in some countries, polio by administering the vaccines in a single injection. Similarly, young children are immunized simultaneously with a multivalent measles, mumps, and rubella vaccine. Administration of multivalent vaccines reduces the number of visits to healthcare centers, a critical point in developing countries.

Historically, *parenteral vaccines* have usually been used (i.e., injected with a syringe and needle). This approach is extremely successful for systemic diseases such as diphtheria or measles. However, in localized diseases (e.g., intestinal infections such as cholera), parenteral vaccines are of limited use since the immune system needs to be stimulated at the site of infection (e.g., local immune system). Therefore, effort is now being directed at controlling intestinal infectious diseases with orally administered vaccines. *Oral vaccines* can consist of

1) killed intact bacteria,
2) toxoids (i.e., toxins that are still immunogenic but are rendered biologically inactive by treatment with a chemical, heat, or mutation)
3) subunit vaccines, in which only the nontoxic portion of a molecule is used, or
4) live-attenuated vaccines, in which a viral or bacterial strain is rendered nonpathogenic (e.g., by passaging the virus in cell culture or deletion of bacterial genes), but is still able to multiply to a limited degree, thereby eliciting a protective immune response in the absence of disease symptoms.

One major problem associated with highly purified antigens or subunit vaccines is reduced immunogenicity. Unfortunately, as protective antigens are purified from either bacteria, viruses, or parasites in order to free them from nontoxic substances such as LPS, they lose their ability to evoke an immune response. Their immunogenicity can, however, be improved by using an *adjuvant.* Adjuvants function either to present the antigen to the immune system for a prolonged period of time, or to nonspecifically stimulate the immune system by releasing immune modulators such as lymphokines. To date, only aluminum gels have been licensed as adjuvants for human use. Tetanus or diphtheria toxoid is absorbed onto the gel which allows the toxoid to persist longer at the injection site. Biologically active peptides that can nonspecifically stimulate the immune system via lymphokine release or downregulate the suppression of the immune system, are being evaluated.

1.2.5. Passive Immunization

Passive immunization entails the transfer of preformed immunoglobulins to a host. Passive immunization is usually employed after known or suspected exposure to a given pathogen (for example, after being bitten by a rabid animal). It is used in cases were disease progress is rapid if the pathogen or toxin is not neutralized. The time needed for the host to mount a protective immune response following active immunization (7 – 14 days) would be too long. In some instances, globulin may be passively administered as a prophylactic measure, as in the case of travellers entering an area where hepatitis A is endemic.

Immunoglobulin preparations are assayed to confirm that they contain a high titer of neutralizing antibodies against the disease in question. Such *hyperimmune globulins* must contain at least five-fold higher levels than normal globulin and are obtained by screening plasma units (human) for a given antibody. Alternatively, volunteers may be vaccinated to enrich their plasma for a given antibody.

Immunoglobulin is administered either intramuscularly or intravenously. Only a limited volume of globulin can be given intramuscularly. Intravenous administration allows for comparatively large quantities (400 mg of immunoglobulin per kilogram body weight) of immunoglobulin to be administered quickly.

Several problems are associated with the preparation of immunoglobulin for passive therapy. Firstly, the identification of plasma donors who are "hyperimmune" to a given antigen requires expensive and tedious screening programs. Secondly, the administration of immunoglobulins entails the possibility of transmitting viral diseases, such as hepatitis B and non-A, non-B hepatitis. Additional screening procedures are therefore required.

An alternative to antibody obtained from donors is the use of human monoclonal antibodies (humabs, see Chap. 6). Humabs can be synthesized by hybridoma cell lines that are produced by fusing a human B cell secreting a desired antibody to a nonsecreting cell line. Hybridomas can be grown in fermentors containing up to 1000 L of serum-free medium. Antibody yields can reach 100 mg per liter of medium. The use of humabs can circumvent the need to obtain plasma from many donors and the risk of viral disease transmission, all at a lower cost. Humabs have been produced against a variety of infectious agents (e.g., cytomegalovirus, diphtheria and tetanus toxins) and will be tested in the near future.

1.2.6. Genetic Engineering

The application of recombinant DNA technology to the field of vaccine development has led to remarkable progress since the late 1970s. Critical protective antigenic determinants can be defined, cloned, produced on a large scale by using an appropriate expression system (vector and host), and purified. The first vaccine for human use to be produced by genetic engineering is that against hepatitis B; the surface antigen was cloned, then expressed in yeast, and purified to homogeneity. These vaccines have virtually replaced the first-generation, plasma-derived antigen vaccines due to increased safety and lower costs.

Numerous licensed, live-attenuated vaccines (measles, polio, mumps, rubella, varicella, and typhoid fever) have been obtained by rather empirical techniques, such as passage on tissue culture or chemical mutagenesis. Surprisingly, the precise genetic alterations responsible for avirulence are unknown. Several live-attenuated bacterial vaccine strains have been developed by using genetic engineering. The most advanced, in regards to clinical testing, are for cholera and typhoid fever. These strains have been attenuated by inactivating a gene or genes essential for virulence. A related approach has been to clone a given protective antigen and express it in a suitable, nonvirulent "carrier" strain. For example, surface glycoprotein from HIV-1 and rabies virus have been introduced into vaccinia virus (smallpox vaccine). In addition, antigens from several enteric pathogens such as *Shigella sonnei, S. dysenteriae,* and *Vibrio cholerae* have been introduced into the licensed live-oral typhoid vaccine strain, Ty21a. Such recombinant strains may prove useful as "bivalent" vaccines, conferring protection against two enteric pathogens.

2. Bacterial Vaccines

The chemotherapy of bacterial disease is treated elsewhere. (→ Chemotherapeutics, Chap. 2.).

2.1. Diphtheria Vaccine

Etiological Agent and Pathogenesis. The causative agent of diphtheria is *Corynebacterium diphtheriae,* first isolated by LOEFFLER in 1884. The disease is spread by inhalation of infected droplets and is characterized by the formation of a pseudo-membrane in the nasopharynx caused by localized bacterial replication. Disease symptoms are caused exclusively by the production of a lethal toxin which spreads from the site of infection via the bloodstream. Death is due to the inhibition of protein synthesis by the toxin in vital organs [10].

History of Immunization. Studies by ROUX and YERSIN in 1888 demonstrated that diphtheria is a "toxicosis." Two years later, BEHRING and KITASATO established that the disease could be prevented in animals by immunization with a crude diphtheria toxoid. Attempts to extend these findings to humans used a "toxoid" prepared by combining diphtheria toxin and antitoxin. The first large-scale immunization program with such a product was performed on New York school children in 1922. Soon after, crude toxoids prepared by formaldehyde treatment of *C. diphtheriae* filtrates replaced toxin-antitoxin toxoids. Routine mass vaccination against diphtheria was initiated in Western Europe and North America shortly after World War II.

Production and Properties of Diphtheria Vaccine. Diphtheria toxoid is prepared by detoxification of diphtheria toxin with formaldehyde. Derivatives of the hypertoxinogenic Park Williams 8 strain are used by most manufacturers. Large quantities of toxin are produced in fermentors (ca. 200 – 5000 L) and yields approach 500 mg/L. Although manufacturing processes vary, most employ the following steps. Formaldehyde [*50-00-0*] (0.4 – 0.6 %) is added to cell-free culture supernatants which are then stored at 35 – 37 °C for 3 – 5 weeks. Detoxification is accomplished first by reaction of formaldehyde with the ε-amino groups of lysine and then by formation of irreversible methylene bridges between aromatic amino acids. Upon confirmation of detoxification by animal testing, the toxoid is purified by diafiltration, ammonium sulfate or ethanol fractionation, or anion-exchange chromatography.

Immunization Recommendations. Diphtheria toxoid is rarely used as a monovalent vaccine, but is usually combined with tetanus toxoid (DT, see Section 2.2), pertussis vaccine (DPT, see Section 2.3), or inactivated polio vaccine (DPT-Pol, see Section 3.5) absorbed onto an aluminum salt adjuvant. Toxoid content is usually expressed in limits of flocculation, Lf (1 Lf $\approx$ 2 µg of toxoid). For primary immunization of infants, ca. 25 Lf of toxoid is administered intramuscularly starting at 6 – 12 weeks of age. Three doses are usually administered at 4 – 8 week intervals with a fourth dose 4 – 12 months later. To maintain life-long immunity, booster doses, together with tetanus toxoid, are recommended every 7 – 10 years.

Adverse Reactions. The absolute reactogenicity of diphtheria toxoid is difficult to determine since it is usually administered with tetanus toxoid or pertussis vaccines. Overall, the vaccine is considered to be safe; it primarily evokes mild, transient, local reactions.

Vaccine Efficacy. No large-scale field trials have been performed to determine vaccine efficacy. This is because it would be unethical to withhold the vaccine from control subjects in view of the dramatic decrease in disease following immunization in the 1920s. Furthermore, overwhelming data indicate that proper vaccination can provide absolute protection. For example, mass vaccination with diphtheria toxoid in Romania in 1958 showed that within seven years morbidity and mortality due to diphtheria declined by more then 99 %. In Western Europe and North America, where vaccination is universal, diphtheria has been virtually eradicated [11]. Rare cases occur almost exclusively in individuals lacking a proper history of vaccination.

Future Prospects. Given the combination of vaccine safety, efficacy, and cost, there has been little impetus to develop and test a "second generation" diphtheria toxoid. Any such vaccine would have to completely eliminate the possibility of toxic reversion and be more economical to produce. In another approach, synthetic peptides that express key epitopes of diphtheria toxin are conjugated to carrier proteins; these conjugates can elicit neutralizing antibodies [12]. The drawback with this approach is the cost of purifying the carrier protein.

The most promising approach to toxoid development is the use of nontoxic mutant proteins [13], obtained by recombinant DNA technology. Such proteins can be constructed by deletion of specific regions responsible for toxicity [14]. In addition, these "toxoids" can be synthesized in yields equal to those obtained by production strains of *C. diphtheriae.* The efficacy of such proteins is being evaluated.

2.2. Tetanus Vaccine

Etiological Agent and Pathogenesis. The causative agent of tetanus is the spore-forming bacillus, *Clostridium tetani*. The disease, first described by HIPPOCRATES, is spread via contamination of wounds or abrasions with soil containing *C. tetani* spores. Bacteria multiplying at the wound site release a potent neurotoxin which enters the bloodstream and acts upon the central nervous system evoking the "spastic paralysis" characteristic of tetanus [15]. The majority of tetanus cases occur in neonates due to the nonsterile severing of the umbilical cord.

History of Immunization. The identification of tetanus as a toxicosis by FABER in 1890 led to the development of crude tetanus toxoids as early as 1893. Initial attempts to vaccinate against tetanus were carried out during World War I. By 1926, parenteral immunization of humans with a safe, effective formalin toxoid was achieved. Until the mid-1960s, only such crude toxoids were available.

Production and Properties of Tetanus Vaccine. The presently used tetanus toxoid is a partially purified preparation. The hypertoxinogenic Harvard strain of *C. tetani* is most widely used for production of tetanus toxin. The bacteria are grown in fermentors until the cells autolyse, thereby releasing the toxin into the medium. Intact cells are removed by filtration. Formaldehyde is added to the filtrate to a final concentration of 0.4 – 0.6 %, the pH adjusted to 7.4 – 7.6, and the mixture held at 35 – 37 °C for about four weeks. After confirmation of detoxification by animal testing, tetanus toxoid is purified by diafiltration and ammonium sulfate fractionation.

Immunization Recommendations.
Tetanus toxoid is most frequently adsorbed to an aluminum salt adjuvant and administered together with diphtheria toxoid (DT, Section 2.1), pertussis vaccine (DPT, Section 2.3), or inactivated polio vaccine (DPT-Pol, Section 3.5). For primary immunization of infants, each dose of vaccine contains 20 – 30 Lf's of tetanus toxoid as DPT or DPT-Pol. Immunization starts at 6 – 12 weeks of age and consists of three doses given at 4 – 8 week intervals with a fourth dose 4 – 12 months later. To maintain life-long immunity (≥ 0.01 IU/mL of serum), booster doses, often combined with diphtheria toxoid, are recommended every 7 – 10 years.

Tetanus vaccine is often administered as a routine prophylactic measure following puncture wound trauma. This practice can lead to hyperimmunization and attendant reactions. Alternatively, an individual may be "tolerized" by repeated vaccination and, therefore, unable to mount a protective immune response. Vaccination of pregnant women is recommended with the final dose given about three weeks before expected delivery. The immune response of humans to tetanus vaccine varies widely [16]. The immunization schedule used appears to influence the magnitude of the immune response [17].

Adverse Reactions. Serious reactions following vaccination with tetanus toxoid are infrequent: swelling, pain, and/or redness at the injection site are the most common. However, immediate and delayed hypersensitivity, an Arthus-like reaction, and in rare instances (< 1 case in 2×10^6), neurological sequelae have been noted [15]. The majority of severe reactions are associated with high levels of pre-existing antitetanus antibody. Therefore, attention should be paid to the interval between booster doses to avoid these occurrences.

Vaccine Efficacy. Data supporting the efficacy of tetanus toxoid vaccine comes primarily from retrospective analysis of attack rates in immune versus nonimmune populations. The first such study was the evaluation of soldiers during World War II, where tetanus occurred primarily in nonimmune individuals. Immunization of pregnant women completely prevented neonatal tetanus [18]. Recently, cases of tetanus observed in the United States occurred exclusively in nonimmune individuals. Numerous studies have also shown that abbreviated one- or two-dose immunization regimens are able to induce long-lasting immunity [15]. This is of particular importance in developing countries where a multidose immunization regimen may not be feasible due to an inadequate health care system or economic constraints.

Future Prospects. As with diphtheria toxoid, tetanus vaccine is an extremely economic, effective, safe vaccine. Efforts directed to a "new generation" of tetanus vaccine are aimed at providing a toxoid more amenable to mass vaccination. The main thrust has been to produce a safer, more immunogenic toxoid capable of inducing long-lasting immunity after one or two administrations without reactogenic adjuvants. The substitution of glutaraldehyde for formaldehyde yields a safe, immunogenic vaccine not requiring adjuvants [19]. Production of toxoids by recombinant DNA techniques similar to those described for diphtheria toxin is also feasible [14].

2.3. Pertussis Vaccine

Etiological Agent and Pathogenesis. *Bordetella pertussis*, the cause of pertussis or "whooping cough," is a gram-negative rod that was first isolated by Bordet and Gengou in 1906. Humans are the only known host for *B. pertussis*. The disease is spread by infectious droplets with the vast majority of cases occurring in young children. *B. pertussis* shows a marked trophism for the ciliated epithelial cells of the upper respiratory tract. Initial symptoms are similar to those of the common cold, but worsen within 10 – 20 days. The paroxysmal stage, lasting an average of 15 – 20 days, is characterized by severe bouts of coughing. The convalescent stage can last for many months when secondary infections can present a major problem. *B. pertussis* remains localized within the initial disease stages which implies that the severe symptoms are toxin-induced [20]. This is supported by the fact that *B. pertussis* can synthesize a large number of toxic extracellular factors including pertussis toxin (also referred to as lymphocytosis promoting factor), adenylate cyclase, dermonecrotic toxin, and tracheal cytotoxin [21].

History of Immunization. Attempts at immunization against pertussis were initiated in the 1920s using killed whole-cell vaccines [22]. These studies demonstrated that vaccination could not only prevent a substantial proportion of disease, but that symptoms in vaccinated individuals who became infected were milder than in unvaccinated controls. Routine large-scale immunization against pertussis began shortly after World War II using standardized whole-cell vaccines of known potency. Even though the safety of these vaccines has been under attack, they are still used in most areas of the world except Japan, where acellular vaccines have been used since 1980 (see below).

Production and Properties of Pertussis Vaccine. At present, there is no standardized method, culture medium, or bacterial strain for producing pertussis vaccine. A given manufacturer's production procedure is designed to yield a vaccine which will meet the minimal requirements of the appropriate regulatory agency. *B. pertussis* is usually grown in fermentors on a synthetic or semisynthetic medium. The cells are harvested by centrifugation and resuspended to a given opacity using a reference standard. Inactivation is accomplished by heating and/or adding formaldehyde or thimerosal. In view of the diversity of production techniques employed, toxic components which escape inactivation can vary considerably from manufacturer to manufacturer.

Immunization Recommendations. Pertussis vaccine is used almost exclusively in combination with diphtheria and tetanus vaccines (see Sections 2.1 and 2.2). The vaccine is adsorbed onto an aluminum salt adjuvant. Immunization usually commences at 6 – 12 weeks of age and consists of three doses of vaccine given intramuscularly at 4 – 8 week intervals with a fourth dose given 4 – 12 months later. Vaccination with the acellular vaccine used in Japan consists of 2 – 3 doses of vaccine given at 4 – 8 week intervals starting at two years of age.

Adverse Reactions. The majority of children receiving pertussis vaccine experience an adverse reaction. Approximately 40 – 50 % of children have a local reaction and 30 – 40 % a systemic reaction (anorexia, vomiting, fretfulness, fever, or persistent crying) [23]. Of greater concern is the infrequent occurrence of convulsions and hypotonia. As a result of concern regarding vaccine safety, a large-scale study was initiated in England to determine the incidence of serious reactions in children aged 2 – 36 months [24]. The risk of vaccine-induced neurological illness was estimated to be 1/110 000 vaccinations and that for encephalopathy, 1/310 000 vaccinations. It is extremely difficult to establish a causal relationship in individual cases in such a study; incidence rates were evaluated on a temporal basis. Furthermore, neurological symptoms believed to be vaccine-related constitute only a small proportion of similar cases seen in this age group. Review of these data by health authorities has led to the conclusion that the benefit of vaccination outweighs the attendant risk in view of the fact that clinical pertussis can often result in neurological sequela.

Vaccine Efficacy. The first indication of vaccine efficacy came from a trial conducted in 1929 during an epidemic on the Faroe Islands; vaccination afforded 73 % protection against clinical disease [22]. Extensive information has been obtained in Japan, England, and Sweden where disease incidence correlates inversely with the overall immune status of the general population.

Routine immunization against pertussis in Japan was initiated in 1947 – 1949 with an acceptance rate of roughly 90 %. The number of pertussis cases declined from 152 072 in 1947 to less than 400 by 1971. In 1975, controversy concerning vaccine safety briefly halted vaccine usage. When it was resumed, the acceptance rate declined to ca. 25 – 30 %. Pertussis returned to epidemic proportions by 1979 with more than 13 000 cases reported nationally. At the urging of federal health authorities, vaccine acceptance rates increased to 65 – 70 % by 1982 with a concomitant decline in cases.

Similarly, concern about vaccine safety in England resulted in a dramatic decline in vaccine acceptance from ca. 80 % in 1973 to ca. 30 % in 1978. As in Japan, the disease became epidemic with more than 100 000 cases reported between 1977 and 1980. In Sweden, routine vaccination against pertussis was initiated in the early 1950s. One decade later, > 90 % of children were considered to be immune. Due to changes in the manufacturing technique, the pertussis vaccine used in the mid-1970s was not potent. Shortly after, a marked increase of pertussis cases was seen. When an effective vaccine was reintroduced, disease incidence was dramatically reduced.

Case-contact studies have shown the whole-cell vaccine to be 63 – 95 % effective at preventing overt disease. Efficacy of acellular vaccines is discussed below.

Future Prospects. Efforts to develop a "second generation" pertussis vaccine have centered on using purified detoxified antigen preparations containing a minimum amount of lipopolysaccharide. Several such acellular vaccines have been developed and clinically evaluated. They are composed of detoxified pertussis toxin (PT) alone or in combination with filamentous hemagglutinin (FHA), two key protective antigens [25]. The first generation acellular vaccines were produced in Japan using supernatant from static cultures as a source of antigen. Cell-free supernatants were subjected to ammonium sulfate precipitation followed by sucrose density gradient ultracentrifugation to simultaneously enrich PT and FHA and eliminate lipopolysaccharide; PT was inactivated by formaldehyde treatment. Preliminary testing in Japan showed the vaccine to be far better tolerated than whole-cell vaccine; case-contact studies revealed that it was ca. 90 % effective.

Second generation acellular vaccines of a greater purity have been produced on a large scale using fermentor-grown cultures. A monovalent formalin PT toxoid and a bivalent PT toxoid – FHA vaccine have been evaluated in a placebo-controlled trial in 5 – 11 month old children in Sweden [26]. Children received 2 doses of vaccine 8 – 12 weeks apart. The small number of children vaccinated (ca. 3000) does not allow evaluation of vaccine safety regarding the rare neurological reactions. However, the vaccine evoked far fewer local reactions in comparison to the whole-cell vaccine. After 15 months of observation, the PT toxoid vaccine was 54 % effective whereas the PT toxoid-FHA vaccine was 69 % effective against all forms of clinical pertussis. Both vaccines were equally effective (ca. 80 %) against severe disease.

2.4. Typhoid Fever Vaccine

Etiological Agent and Pathogenesis. *Salmonella typhi,* the causative agent of typhoid fever, is a gram-negative bacillus first isolated by GAFFKY in 1884. Infection is due to ingestion of the organisms in contaminated food or water. The bacteria penetrate the epithelium of the small intestine and are ingested by reticuloendothelial cells. Unlike most bacteria, the typhoid bacillus can survive and multiply within phagocytic cells and then spreads to the spleen, liver, lymph nodes, and gallbladder. Symptoms appear 1 – 2 weeks after exposure when the bacteria enter the bloodstream. Despite appropriate treatment, 1 – 2 % of those infected will become chronic asymptomatic carriers serving as an infectious reservoir. Typhoid primarily occurs in developing countries or in areas of poor sanitation. In endemic areas, the disease is primarily contracted by school-aged children, but travellers of all ages entering an endemic area are at risk.

History of Immunization. Attempts at immunizing against typhoid fever began in 1896. The first mass vaccination was conducted by the British Army during World War I using a killed whole-cell vaccine. The attack rate among vaccinated soldiers was far less than for nonimmunized soldiers. Consequently, the use of whole-cell vaccines became common practice in the armed forces.

In the 1960s and 1970s, attempts were made to use killed oral immunization. Killed bacteria were administered to volunteers at 1×10^{11} per dose; twelve doses resulted in a modest, but significant protection rate of 30 % [27]. Field trials in India showed that such killed preparations were ineffective at preventing disease.

An alternative approach to killed oral vaccines against typhoid was to use attenuated live-oral vaccine strains. The first candidate was a streptomycin-dependent mutant developed in 1967 [28]. When orally administered to volunteers in doses of up to 10^{11} bacteria, this strain did not evoke adverse reactions. Good levels of protection (66 – 78 %) were achieved with freshly harvested cultures. However, lyophilized preparations afforded little or no protection [28], which diminished interest in this strain.

The second live-oral vaccine candidate was a gal ε mutant termed Ty21a which lacked glucose 1,4-epimerase activity and therefore could not synthesize complete lipopolysaccharide [29]. Freshly harvested cultures of this strain were safe and effective in volunteer challenge studies [30]. A lyophilized Ty21a preparation afforded > 90 % protection for three years in a field trial in Egypt [31].

Production and Properties of Typhoid Vaccines. Killed whole-cell vaccines are usually produced from the Ty2 strain of *S. typhi.* Fermentor-grown cultures are inactivated either by heating in combination with phenol or formaldehyde or by acetone drying. Although the latter method yields a more efficacious vaccine (see below), only the vaccine produced by the former method is readily available. Each 0.5 mL dose of vaccine contains $(1-3)\times10^9$ killed organisms. Vaccine potency is estimated by a mouse-protection test.

The attenuated live-oral Ty21a vaccine strain is produced from fermentor-grown cultures. The harvested cells are suspended in a cryoprotective medium consisting of sugar and amino acids and lyophilized. The lyophilizate is placed in gelatin capsules which are then coated with an acid-resistant layer. Each capsule contains $(1-5)\times10^9$ viable bacteria. Vaccine potency is based upon the number of viable organisms per capsule [32].

Immunization Recommendations. People living in or travelling to an endemic area are vaccinated with two doses of whole-cell vaccine administered subcutaneously or intramuscularly 14 days apart. A single booster dose is advised for individuals travelling to an endemic area if 2 – 3 years have elapsed since primary immunization.

Three doses of attenuated live-oral vaccine are ingested, each dose on an alternate day. An additional complete immunization course is advised when entering an endemic area if more than 1 – 2 years have elapsed since primary immunization. The vaccine is for use in children six years of age or older and in all ages of non-immunocompromised adults.

Adverse Reactions. Reactions following immunization with the parenteral whole-cell vaccine are frequent and consist of local pain, swelling, and redness, often accompanied by fever, chills, and malaise. Transient debilitating reactions occur in ca. 10 % of vaccine recipients. Due to the high rate of adverse reactions, parenteral typhoid vaccine is not widely used.

Reactions following ingestion of the live oral Ty21a vaccine are rare (< 1 per 100 000 doses). Reactions are usually mild, transient, and resolve of their own accord. Gastrointestinal disturbances and rash are most frequently reported.

Vaccine Efficacy. To accurately assess the efficacy of parenteral typhoid vaccine, several field trials were conducted in the 1960s [33]. In most instances, two doses of either heat–phenol or acetone-dried vaccine were administered. Protection rates ranged from 51 – 77 % and 79 –93 %, respectively, for up to seven years. A further trial was conducted in Tonga in 1966 – 1973 where school-aged children received one or two doses of acetone-dried vaccine. Surprisingly, one dose gave no protection whereas two doses afforded only 40 % protection.

Efficacy of the live oral Ty21a vaccine administered in enteric-coated capsules has been evaluated in field trials in school-aged children in Chile [34]. Three doses of vaccine provided about 70 % protection over a three-year period.

Future Prospects. Two additional typhoid vaccines are undergoing clinical evaluation. The first is a double auxotrophic mutant of *S. typhi* requiring 4-aminobenzoate and adenine [35]. Although ingestion of up to 10^{10} organisms only evoked a poor humoral immune response to *S. typhi* cellular antigens, most vaccines manifested a specific cell-mediated immune response to *S. typhi* lipopolysaccharide.

The second candidate vaccine is a capsular polysaccharide termed V_i. Purified V_i antigen is safe upon parenteral immunization. A single dose of vaccine provided ca. 70 % protection for ca. one year in Nepal [36].

2.5. *Streptococcus pneumoniae* Vaccine

Etiological Agent and Pathogenesis. *Streptococcus pneumoniae,* often referred to as "pneumococcus," is a gram-positive diplococcus first isolated by STERNBERG and PASTEUR in 1881. The pneumococcus usually causes disease subsequent to a viral infection or in a compromised individual, i.e., is an opportunistic pathogen. Pneumonia is the most common disease syndrome caused by *S. pneumoniae*; bacteremia and meningitis are often sequelae to pneumococcal pneumonia. *Streptococcus pneumoniae* is also a leading cause of otitis media in young children. The pneumococcus often colonizes the nasopharynxes of healthy individuals which then serve as infectious foci. The organisms can then spread to another person by infected droplets or translocate to the inner ear or lungs following a viral infection.

History of Immunization. Early attempts at immunization against pneumococcal pneumonia employed heat-killed, whole-cell vaccines (reviewed in [37]). These trials and related studies led to the conclusion that immunity is mediated by serospecific antibody produced against the polysaccharide capsule surrounding the pneumococcus. A large-scale trial performed in the mid-1930s with a bivalent capsular polysaccharide (CPS) vaccine showed a modest decrease in disease. Conclusive proof of vaccine efficacy came from trials conducted with polyvalent vaccines; protection was serospecific. Further studies showed that vaccines comprising up to six distinct capsular antigens could be effectively used [38]. The first pneumococcal polysaccharide vaccines were hexavalent and were licensed in the early 1950s. However, owing to the belief that antibiotics could effectively control pneumococcal disease, vaccine acceptance was low, leading to a halt in production.

Production and Properties of *Streptococcus pneumoniae* Vaccine. This vaccine is composed of 23 types of purified CPS that have been selected on the basis of seroepidemiological surveillance studies of bacteremic isolates [39].

Pneumococcal CPSs are purified by a variety of techniques, depending upon serotype, which include precipitation with organic solvents, treatment with detergent, digestion with DNAse, RNAse, and protease, and ultracentrifugation. The vaccine contains only traces of protein and nucleic acids ($\leq$ 2 wt %) and is analyzed for its CPS constituents, serological purity, and pyrogenicity. Vaccine potency is based upon the determination of the molecular mass of the CPS, because a minimum antigen size is required for immunogenicity in humans. One human dose consists of 25 µg of each antigen in 0.5 mL administered intramuscularly or subcutaneously.

Immunization Recommendations. Pneumococcal vaccine is recommended for immunocompetent individuals with underlying clinical conditions that increase their risk of acquiring pneumococcal bacteremia or pneumonia. These conditions include alcoholism, asplenia, sickle cell anemia, reduced pulmonary function, congestive heart failure, diabetes, reduced renal function, and cirrhosis [40]. Routine vaccination of the elderly is also warranted due to the increased incidence of disease in this group. Patients with underlying conditions (leukemia, Hodgkin's disease, myeloma, and treatment with immunosuppressive drugs) that render them susceptible to pneumococcal disease may be considered candidates for vaccination even though a proportion may not respond to immunization [40]. Revaccination is not advised due to severe reactions.

Adverse Reactions. Pneumococcal vaccine is very safe [40]. Mild local reactions occur in ca. 30 – 40 % of vaccine recipients, fever and severe local reactions in ca. 1 %. Anaphylactic reactions are extremely rare with a frequency of $5/1\times10^6$ doses administered. Severe local and systemic reactions are more likely to occur upon revaccination.

Vaccine Efficacy. Considerable controversy exists concerning the degree of protection provided by pneumococcal vaccine. The vaccine was highly effective in reducing the incidence of pneumococcal pneumonia among South African gold miners [41]. However, this population differs substantially in age and in general health from those patients for whom the vaccine is now recommended. Several studies have attempted to determine vaccine efficacy among debilitated and/or elderly patients. The distribution of *S. pneumoniae* serotypes among vaccinated and nonvaccinated individuals with invasive pneumococcal disease has been compared [42]; efficacy was estimated to be 61 – 65 %. In contrast, a vaccine trial on high-risk patients, aged > 55, did not demonstrate any significant benefit [43], probably due to the inability of these patients to mount or maintain a protective antibody response. Similar findings were reported in a study on patients suffering from chronic pulmonary disease [44]. In spite of these divergent findings, health authorities still strongly recommend immunization based upon risk–benefit analysis.

Future Prospects. The current pneumococcal vaccine is poorly immunogenic in children under two years of age and in certain debilitated patient populations [40]. Attempts have been made to improve immunogenicity by covalently coupling pneumococcal CPS to carrier proteins [45]. Such conjugates evoked a more vigorous response in mice and rhesus monkeys when compared to native CPS [46]. However, a type 6A-tetanus toxoid conjugate administered to young adult volunteers was only slightly more immunogenic than native 6A-CPS, and a booster dose did not significantly increase anti CPS antibody levels [47]. However, these volunteers had been previously immunized with tetanus toxoid and possessed low, but detectable, levels of antibody to 6A-CPS.

2.6. Shigella Vaccines

Etiological Agent and Pathogenesis. Shigellosis or bacillary dysentery is caused by *Shigellae,* most notably *S. sonnei, S. flexneri* 2 a and 3, and *S. dysenteriae* 1. Disease results from ingestion of as few as 10 bacteria, making it the most infective bacterial enteric pathogen. In developing countries, where the attack rate can exceed $100\,000/1\times10^6$ population, shigellosis is primarily a disease of young children aged six months to six years.

Once ingested, the organisms penetrate the epithelium of the colon and rapidly multiply. This leads to localized inflammation and ulceration, possibly due to synthesis of a potent cytotoxin [48]. Symptoms include fever, bloody diarrhea, cramps, tenesmus, and shock. The mortality rate for untreated disease can exceed 10 %.

History of Immunization. Immunization of humans with killed whole-cell vaccines administered parenterally does not provide significant protection against shigellosis [49]. Subsequent attempts at immunization have therefore centered on use of attenuated live-oral vaccine strains. Vaccines based on streptomycin-dependent derivatives of *S. flexneri* and *S. sonnei* are well tolerated and provide significant immunity for 6 – 12 months [50], [51]. However, these vaccines were not pursued due to the necessity of administering four doses for primary immunization, yearly boosting to maintain efficacy, and genetic instability.

Attempts have been made to produce live-oral shigella vaccines by conjugal transfer of genes coding for *S. flexneri* surface antigens into *E. coli*. Multiple doses evoked no significant adverse reactions when fed to volunteers, but failed to protect against subsequent challenge [52].

The identification of critical *Shigellae*-protective antigens has allowed their expression in the attenuated live-oral typhoid vaccine strain, *S. typhi* Ty21a [53]. One strain, *S. typhi* Ty21a-5076-1C, carries a plasmid coding for the form I (O-polysaccharide) antigen of *S. sonnei*. This strain is safe and has afforded significant protection in volunteer studies [54].

Future Prospects. At present, no vaccines are available for use against shigellosis. As noted above, *S. typhi* 5076-1C that expresses *S. sonnei* form I antigen has shown promise in volunteer studies. However, efficacy varies from lot to lot and probably depends on the degree of flagellation of the bacteria. The construction of a *S. typhi* Ty21a strain expressing *S. flexneri* 2a type and group antigens has been described [55], but this strain has not yet been clinically evaluated.

Genes coding for the production of the O-antigen of *S. dysenteriae* 1 have been cloned and inserted in a plasmid [56]. The *S. dysenteriae* O-antigen can be expressed in *E. coli* K12 and *S. typhi* Ty21a by introduction of the recombinant plasmid. Volunteer studies to determine their safety are scheduled.

2.7. Cholera Vaccine

Etiological Agent and Pathogenesis. The causative agent of cholera is *Vibrio cholerae*, a gram-negative motile bacillus which is transmitted via the fecal–oral route. Upon passage through the stomach acid barrier, *V. cholerae* colonizes the ileum where it rapidly multiplies. The release of a potent enterotoxin (cholera toxin) causes massive watery diarrhea [57]. In most cases, oral or intravenous fluid replacement is sufficient treatment.

History of Immunization. Parenteral whole-cell cholera vaccines were employed by FENAN as early as 1885. Starting in the 1960s, several field trials were performed using inactivated whole-cell vaccines or cholera antigens (see "Vaccine Efficacy").

Numerous attempts have been made to orally vaccinate against cholera. Multiple oral doses of heat-killed *V. cholerae* evoked both a local and humoral antibody response [58]. Ten daily oral doses of 1.6×10^{10} killed vibrios provided protection against a homologous challenge [59]. However, parenteral vaccine gave slightly better protection. Although promising, the massive quantities of organisms used and the need for multiple doses made the oral approach expensive and cumbersome.

Several environmental isolates of *V. cholerae* with reduced virulence have been evaluated as live-oral vaccines. These strains failed to provide significant protection due to their inability to colonize the small intestine and evoke a protective immune response [60]. A hypotoxinogenic mutant of *V. cholerae* isolated by chemical mutagenesis afforded significant protection against diarrhea. However, due to its propensity to revert, it was not studied further [61]. Chemical mutagenesis has provided a strain of *V. cholerae* 3083 (Texas Star-SR), that produces only the B subunit of cholera toxin responsible for binding to cells and lacks the enzymatically active A subunit. Although vaccination with this strain afforded good levels of protection against cholera, about 25 % of vaccinees experienced mild to moderate diarrhea [62].

Production and Properties of Cholera Vaccine. The licensed cholera vaccine is composed of inactivated whole cells of *V. cholerae* and is administered parenterally. Fermentor-grown cultures are inactivated by a combination of heat and phenol or formaldehyde. Usually, two cultures of Ogawa-classical and Inaba-El Tor serotypes-biotypes are produced and then mixed. Cell concentration is adjusted to $(5-8)\times10^9$ cells/dose (0.5 mL). Vaccine potency is estimated by an intraperitoneal mouse challenge test. Two doses of vaccine given at two-week intervals are recommended. A booster dose is recommended whenever entering an endemic area.

Adverse Reactions. Approximately 20–30 % of vaccinees have a mild to moderate local reaction consisting of pain, redness, and/or swelling. Systemic reactions such as fever, headache, or malaise occur in ca. 5 % of vaccinees.

Vaccine Efficacy. Various parenteral vaccines, including inactivated whole-cell, purified lipopolysaccharide, cell-free Inaba antigen, and a whole-cell vaccine in adjuvant have been evaluated in field trials [58]. The results can be summarized as follows:

1) two doses of vaccine are superior to one;
2) protection afforded by vaccination is greater in adults than children;
3) vaccination is only moderately effective in nonendemic areas; and
4) protection lasts for only 3 – 6 months.

Therefore, the currently available cholera vaccine is not an effective tool to control endemic cholera, it is best employed by travellers entering an endemic area for a short time.

Future Prospects. Due to advances in understanding the pathogenesis and molecular genetics of *V. cholerae*, one or more safe and effective cholera vaccines will probably be introduced in the near future.

Several attenuated live-oral vaccine strains have been developed by deleting the cholera toxin gene or its enzymatically active A subunit [63], [64]. A derivative of *V. cholerae* 16961, in which the cholera toxin gene was deleted afforded 89 % protection against clinical disease although 50 % of the volunteers had mild to moderate diarrhea. A second vaccine strain was produced by deletion of the A subunit of cholera toxin from strain 395. Again, good protection was afforded by a single dose, but mild diarrhea was observed in 60 % of the subjects. The discovery, that strains of *V. cholerae* can produce a Shiga-like cytotoxin suggested that at least part of the diarrhea could be due to this toxin [65]. Therefore, KAPER and coworkers (personal communication) deleted the A subunit of cholera toxin from strain 569B which is naturally cytotoxin negative. This strain, called CVD-103, is far less reactinogenic than previously evaluated strains, eliciting mild diarrhea in $\sim$ 10 % of subjects. Vaccination with CVD-103 afforded excellent protection (87 %) against a homologous challenge. Furthermore, significant protection (67 – 78 %) was seen even when the challenge was of a different serotype or biotype.

Orally administered inactivated vaccines consisting of 1×10^{11} killed vibrios and 1 mg of the B subunit of cholera toxin or 1×10^{11} killed vibrios alone per dose have been evaluated [66]. After six months of surveillance, three doses of the combined vaccine afforded 85 % protection, whereas the cells alone were 58 % effective. After one year, protection had declined to 62 % for the combined vaccine and 53 % for the cells alone.

2.8. Vaccines Against Nosocomial Pathogenes

The routine use of broad-spectrum antibiotics, more frequent invasive surgery, increasing use of immunosuppressive agents in the management of cancer, and the ability to prolong the lives of critically ill patients have led to a dramatic increase of hospital-acquired (nosocomial) bacterial infections [67]. In the United States alone, nosocomial infections are a contributing factor in ca. 70 000 deaths per year; their treatment costs $> 1 \times 10^9$ \$ [68], [69]. The mortality rate for nosocomial bacteremia and pneumonia has remained unacceptably high (≥ 25 %) despite the introduction of numerous antibacterial agents [68]. Therefore, much effort has been devoted to developing immunological agents for controlling these infections [70].

Etiological agents and pathogenesis. *Pseudomonas aeruginosa*, *Escherichia coli*, *Klebsiella* spp., and *Staphylococcus aureus* account for the majority of serious nosocomial infections, such as bacteremia and pneumonia [71]. Infection with these agents usually arises due to the translocation of bacteria from the skin, intestine, or nasopharyngeal cavity following disruption of the host's defense systems by trauma, invasive surgical procedures, treatment with antibiotics, or immunosuppressive agents. Foci of infection are usually formed from which the bacteria enter the bloodstream. Death is most likely caused by shock.

Future Prospects. No vaccines are licensed against the four pathogens listed above. Attempts to control bacterial nosocomial infections by immunological means are complicated because of the diverse patient populations who are at high risk and the substantial number of bacterial pathogens of varying serotypes involved. Furthermore, vaccination of at-risk patients may not be feasible because of the time needed to mount a protective immune response and because a substantial portion of these patients are unable to mount a significant antibody response. A more effective approach is to passively immunize patients by using a hyperimmune globulin for intravenous use (see Chap. 5): vaccines developed against nosocomial pathogens could then be used to vaccinate healthy donors whose plasma would be processed into such a globulin [70].

Two different approaches are being taken to develop appropriate hyperimmune products. The first is based upon developing vaccines against the relevant serospecific bacterial antigens. The second is to evoke an antibody response to a common epitope expressed by the lipopolysaccharide of gram-negative bacteria.

Serospecific Vaccine against *Pseudomonas aeruginosa*. Human immunity to *P. aeruginosa* depends on the presence of humoral antibody directed against serospecific lipopolysaccharide (LPS) determinants and toxin A [72]. Several lipopolysaccharide-containing vaccines have been developed and clinically evaluated [73]. These vaccines have not gained wide acceptance due to their toxicity, poor immunogenicity, and poorly characterized antigenic content. Two serotypes of *P. aeruginosa* high molecular mass polysaccharides have been purified and found to be safe and immunogenic in humans [74]. In an attempt to produce both an antitoxin A and an antiLPS antibody response, toxin A – O-polysaccharide conjugate vaccines were synthesized. Such conjugates are nontoxic, nonpyrogenic, safe when administered to humans, and evoke an immune response to both vaccine moieties [75]. A polyvalent vaccine based upon O-polysaccharide serotypes is undergoing clinical evaluation; this vaccine would "cover" >90 % of *P. aeruginosa* bacteremic isolates.

Serospecific Vaccine against *Klebsiella spp.* Antibody to *Klebsiella* capsular polysaccharide (CPS) is highly protective against experimental infections [76]. Several experimental vaccines composed of purified *Klebsiella* CPS were found to be safe and immunogenic in human volunteers [70]. Based upon the seroepidemiology of *Klebsiella* blood isolates, a 24-valent CPS vaccine has been developed which is safe and immunogenic in humans. It also elicits production of antibody to 11 "cross-reactive" CPS serotypes not included in the vaccine and therefore, offers potential "coverage" against ca. 75 % of all *Klebsiella* bacteremic isolates.

Serospecific Vaccine against *Escherichia coli*. Antibodies to both O (LPS) and K (capsular) antigens can provide protection against experimental *E. coli* infections [77], [78]. However, ca. 40 % of *E. coli* bacteremic isolates do not have a polysaccharide capsule [79]. In addition, the two most frequently encountered K serotypes among blood isolates, K 1 and K 5, are nonimmunogenic in humans. Approximately 90 % of *E. coli* bacteremic isolates can be typed according to their O-antigen and most of them can be grouped within 11 serotypes, making a polyvalent formulation feasible [79]. The toxicity of native LPS precludes its use as a vaccine. However, nontoxic, serologically reactive O-polysaccharide can be isolated from *E. coli* LPS and O-polysaccharide – protein conjugates are being constructed in a manner similar to that described for *P. aeruginosa* (see above).

Serospecific *Staphylococcus aureus* Vaccine. *Staphylococcus aureus* produces numerous somatic antigens and extracellular toxins which have been implicated as virulence factors. These include teichoic acid, lipoteichoic acid, exopolysaccharide, capsular polysaccharide, α-toxin, and β-toxin. Patients with deep-seated *S. aureus* disease usually mount an antibody response to teichoic acid, α-toxin, and β-toxin [80]. However, the relative protective capacities of antibodies against these antigens are unknown.

A serotyping scheme has been described for *S. aureus* based upon cell-surface capsular polysaccharides [81]. Approximately 80 – 90 % of *S. aureus* clinical isolates belong to one of these serotypes [82]. Serospecific CPS is produced during experimental *S. aureus* infections and can be isolated from blood [83]. The protective capacity of antibody to CPS is being evaluated.

Crossreactive Anticore Glycolipid Vaccine. Antibody produced against the "core" glycolipid of *E. coli* strain J 5 crossreacts with the LPS of virtually all gram-negative bacteria [84]. Polyclonal or monoclonal antiJ 5 antibody (see also Section 6.1) can afford significant protection against several gram-negative pathogens [85]. Protective antibody is directed against the lipid A moiety of LPS and neutralizes the toxic activities of LPS. When used prophylactically or therapeutically, human plasma enriched with antiJ 5 antibody provides significant protection against death due to "gram-negative shock" [86], [87].

2.9. Meningococcal Meningitis Vaccine

Etiological Agent and Pathogenesis. The causative agent of meningococcal disease is *Neisseria meningitidis*, a gram-negative diplococcus. Humans are the only known reservoir for *N. meningitidis*. The meningococcus colonizes the nasopharynx; infrequently, it enters the bloodstream and infects the meninges. Meningococcal disease is most often seen in children less than 18 months of age and in settings such as military recruit camps where individuals from different geographical areas come into constant close contact [88]. The disease is endemic worldwide with an annual incidence of about 3/100 000 population. Numerous epidemics occurred in the 1970s and 1980s with attack rates exceeding 500/100 000. Eleven meningococcal capsular serotypes are known, with the majority of disease (ca. 95 %) caused by groups A, B, C, W 135, and Y [88], [89].

History of Immunization. The first attempts at immunizing against meningococcal disease used vaccines made of killed bacteria and were unsuccessful [90]. Attention then centered on using purified capsular antigens. In 1945 KABAT demonstrated that up to 1 mg of purified antigen could be administered to humans with no untoward reactions. However, these studies were not continued because the prophylactic treatment of meningococcal meningitis with antibiotics appeared to be highly effective. However, the appearance of resistant strains in the early 1960s revived interest in vaccine development. In the late 1960s, highly purified group A and group C capsular polysaccharides with a high molecular mass were shown to be safe and immunogenic in humans [91]. Human immune sera contained elevated levels of antibody capable of lysing meningococci in the presence of complement.

Production and Properties of Meningococcal Vaccines. Available vaccines contain serogroup A, C, W 135, and Y capsular antigens. The most widely used formulations are the tetravalent A, C, W 135, and Y, and the bivalent A and C vaccines. Capsular antigens are purified from fermentor-grown cultures by coprecipitation with detergents followed by ethanol fractionation, extraction with organic solvents, and ultracentrifugation. Preservation of the high molecular mass of the capsular antigens is of critical importance since this governs the immune response. Each vaccine dose contains 50 μg of capsular antigen per serotype in lyophilized form. Potency of the vaccine is based upon various physicochemical properties (size and purity). No animal test is required.

Immunization Recommendations. Meningococcal vaccine can be administered to any immunocompetent individual over 18 months of age. A single dose is given either subcutaneously or intramuscularly. The group A antigen is immunogenic in children $\geq$6 months of age, but two doses must be administered within a 2 – 3 month interval. Due to the low incidence of the disease in the general population, vaccine acceptance has been poor, it is usually used to immunize military recruits or travellers.

Adverse Reactions. The meningococcal vaccines are very safe–about 250×10^6 doses of vaccine have been administered worldwide with no vaccine-associated fatalities reported. Mild, transient local reactions occur in about 25 % of vaccinees.

Vaccine Efficacy. Conclusive proof that group C meningococcal vaccine prevents disease was presented in 1970 [92]. Subsequently, the group A vaccine was found to be effective in controlling endemic and epidemic disease in adults and children [93], [94]. Group C vaccine elicited a protective antibody response only in children $\geq$18 months of age, whereas the two doses of group A vaccine were immunogenic in children $\geq$6 months of age [95]. Efficacy for the W 135 and Y capsular antigens has not been clinically demonstrated, but is assumed because they engender appropriate levels of relevant functional antibodies in humans.

Future Prospects. Current meningococcal vaccines have two major limitations. First, they do not provide protection against group B organisms which may account for up to 50 % of endemic cases. Secondly, the group C vaccine is nonimmunogenic in children $\leq$18 months of age. The major difficulty in producing a group B meningococcal vaccine is that the group B capsular antigen, a homopolymer of α (2$\rightarrow$8)-linked scialic acid, is poorly immunogenic in humans, probably because similar structures are found in human gangliosides and fetal proteins. Even when coupled to carrier proteins [96] or mixed with group B outer membrane proteins [97], the group B antigen is a poor immunogen. An alternative approach is to use serotype-specific outer membrane proteins, 18 such serotypes have been identified. Outer membrane protein vaccines containing group B or C capsular antigens were found to be safe and immunogenic in humans [98], [99]. A large-scale field trial to determine the efficacy of these vaccines has been conducted.

In an attempt to construct vaccines capable of engendering a protective immune response in young children, capsular antigens have been coupled to carrier proteins to form conjugate vaccines [96]. Group A and C antigens coupled to tetanus toxoid are more immunogenic in animals than native capsular antigens. Studies are in progress to determine the safety and immunogenicity of group A and C conjugates in humans.

2.10. Tuberculosis Vaccine

Etiological Agents and Pathogenesis. Human pulmonary tuberculosis is caused by the bacilli, *Mycobacterium tuberculosis* and *M. bovis*, the latter causing disease primarily in children. *M. tuberculosis* is spread by infected droplets, whereas *M. bovis* is disseminated by ingestion of contaminated milk. The vast majority of clinical disease occurs in middle-aged individuals in developed countries and in children under 10 years of age in underdeveloped countries. The incidence of tuberculosis ranges from ca. 10/100 000 population in Europe and North America to 500/100 000 in parts of Africa. Although 95 – 98 % of people exposed to *M. tuberculosis* become infected, i.e., become tuberculin-positive, they do not manifest signs of clinical disease and are immune [100]. Only a small percentage of infections progress to a disease state. Once the inhaled bacilli reach the lower respiratory tract, they are ingested by alveolar macrophages. Rapid intracellular growth with inflammatory cell recruitment leads to granuloma formation. In the majority of cases, this is as far as the disease progresses. Active disease presents clinical symptoms ranging from self-limiting pulmonary involvement to acute disseminated disease which rapidly leads to death.

History of Immunization. Efforts to protect humans against tuberculosis have almost exclusively employed *Bacillus Calmette – Guérin* (BCG) vaccine (see below). Oral BCG vaccine was first evaluated in the 1920s in newborns. Subsequent studies showed BCG vaccine to be safe when given orally or parenterally. In the early 1930s, large-scale trials with BCG vaccine were started. Subsequent trials involving American Indians and Eskimos showed that intradermally administered BCG conferred protection against disease in infants. Additional proof of vaccine efficacy was obtained in a small-scale trial with Canadian Indians [101]. Even in the face of a high disease transmittance rate and a mortality rate of 800/100 000, the vaccine was more than 80 % effective over a 9-year observation period.

Production and Properties of Tuberculosis Vaccine. The original BCG vaccine strain was developed at the Pasteur Institute over 50 years ago by in vitro cultivation of *M. bovis* for 10 years. Numerous substrains are now used to produce commercial products. The bacteria are grown on Sauton's liquid media either in bottles with a high surface area or in submerged culture. They are then collected, suspended in a stabilizer solution, homogenized, and lyophilized. The bulk material is standardized by opacity, dry mass, and number of viable bacteria per unit mass. The vaccine is usually filled into glass ampules which are sealed under vacuum.

Controls of BCG vaccine are stringent. First, the culture is examined for purity, viability, and identity. Innocuity is tested by injection of vaccine into guinea pigs. The number of live organisms are determined by viability counting. General safety tests to confirm absence of unexpected toxicity are also performed. Resistance of the bacteria to elevated temperature is also documented. Vaccine potency is based upon the total viable bacteria per dose. However, this is not totally satisfactory because (1) different BCG strains may vary in their ability to evoke a protective cell-mediated immune response, and (2) the ratio of viable to nonviable bacteria can influence the magnitude of both the humoral and cell-mediated immune response to vaccination [102].

Immunization Recommendations. A single intradermal dose containing $(4-8)\times10^5$ viable organisms in 0.1 mL is normally administered. The preferred use of BCG vaccine depends on the disease incidence in the area in question. In developed areas where disease is infrequent, vaccination should be limited to individuals at high risk to exposure, i.e., family members of tuberculosis patients, health care workers expected to come into contact with tuberculosis patients or infected clinical specimens, or travellers entering an area of high endemicity for a prolonged period of time. In developing areas with a high rate of disease incidence, routine immunization of infants is a cost-effective method of controlling tuberculosis. In addition, all tuberculin-negative children upon entry into school and tuberculin-negative adults should be vaccinated.

Adverse Reactions. Correctly administered BCG vaccine is considered to be safe; one vaccination in 100 000 will result in a noticeable reaction [103]. A significant proportion of these reactions are ulcers caused by inadvertent subcutaneous administration of the vaccine. Of greater concern are the reports of mycobacterial disease following immunization [104]. Great care must be taken to insure that a potential vaccinee is not immunocompromised as the vaccine organisms can then rapidly multiply and lead to death [105].

Vaccine Efficacy. In numerous field trials the efficacy of BCG vaccine ranged from 0–80 % [100]. Although three trials failed to document significant vaccine-induced protection, considerable support still exists for large-scale vaccination in developing countries [106].

Future Prospects. At present, no "second-generation" tuberculosis vaccines are being clinically evaluated. Although the tuberculin skin test offers a practical method to measure "immunity" to disease, the molecular mechanism responsible for inducing a tuberculin-positive state is not known.

Several *M. tuberculosis* cell surface antigens have been identified which can induce a cell-mediated response in experimental animals [107]. The cloning of such antigens and their high-level expression in bacteria or yeast would permit economical large-scale purification. Alternatively, such cloned genes could be inserted into a suitable vector such as vaccinia virus [108]. Since most tuberculosis vaccine are used in underdeveloped nations, any new vaccine must not only be safe and effective, but also economical to produce and control.

2.11. *Escherichia coli* Vaccines

Etiological Agent and Pathogenesis. *Escherichia coli*, a gram-negative motile rod, is a major cause of diarrheal disease, urinary tract infections, neonatal meningitis, and nosocomial infections (see Section 2.8). *Escherichia coli* produces a variety of virulence factors including a polysaccharide capsule antigen (K-antigen), lipopolysaccharide antigen (O-antigen), fimbriae, enterotoxins, and cytotoxins. The majority of *E. coli* strains causing a given disease fall within a limited number of K or O serotypes.

Escherichia coli is the major cause of diarrhea worldwide. Three "categories" are recognized, i.e., enterotoxigenic, enteroinvasive, and enteropathogenic *E. coli* strains.

1) *Enterotoxigenic E. coli* invariably causes watery diarrhea which may be accompanied

by cramps, fever, or vomiting. The disease is acquired by ingesting contaminated food and affects infants, young children, and travellers to underdeveloped countries. The strains produce either a heat-stable enterotoxin and/or a heat-labile enterotoxin; they also possess fimbriae which mediate attachment to the ileal epithelium. Three distinct fimbriae (CFA/I, CFA/II, and E 8775) have been identified on enterotoxigenic *E. coli* strains isolated from human disease.
2) *Enteroinvasive E. coli* causes a dysentery-like disease characterized by fever, diarrhea, cramps, and stools containing blood and mucus. The disease is acquired by ingestion of contaminated food. The bacteria readily penetrate the ileal and colonic epithelia where they rapidly multiply resulting in tissue destruction. They can then infect adjacent cells or penetrate the underlying lamina propria.
3) *Enteropathogenic E. coli* affects primarily infants and young children after ingestion of contaminated food causing watery diarrhea often accompanied by fever and/or vomiting. These *E. coli* strains rarely produce a heat-labile or heat-stable enterotoxin and are unable to invade eukaryotic cells. They colonize the ileal epithelium with destruction of the surrounding brush border in the absence of invasion. Enteropathogenic *E. coli* strains can synthesize a cytotoxin similar to that produced by *Shigella dysenteriae* 1. This so-called Shiga-like toxin is thought to elicit the clinical symptoms.

Approximately 50 % of neonatal meningitis is due to *E. coli* transmitted from mother to infant. The bacteria apparently penetrate the intestinal epithelium of the neonate, enter the bloodstream, and are then translocated to the meninges. The majority of strains (>80 %) causing meningitis express the K 1 capsular antigen. However, the O-antigen also appears to play a role in virulence. *Escherichia coli* meningitis has a high attendant fatality rate.

Escherichia coli is the causative agent of ca. 90 % of nonobstructive urinary tract infections which are much more common among women than men. The clinical syndromes can include bacteriuria, cystitis, and pyelonephritis. Renal scarring associated with pyelonephritis occurs most often in children ≤5 years of age. The infecting organisms are thought to be acquired from the flora colonizing the vagina and periurethral areas. P and type I fimbriae are believed to play an important role by mediating tissue colonization [109].

History of Immunization. Initial attempts to vaccinate against *E. coli* diarrhea used crude cellular extracts derived from O types 111, 55, and 86 [110]. In one study, hospitalized infants up to one year of age received multiple doses of oral vaccine. Overall efficacy was 41 %. Efforts to develop vaccines against *E. coli* diarrhea center around the use of purified antigens or live-oral attenuated strains to elicit an antitoxic or antiadhesion response. Calves or neonatal piglets reared by mothers immunized with purified K 88, K 99, or 987 P fimbriae, procholeragnoid (heat-aggregated cholera toxin with reduced toxicity) or heat-labile enterotoxin are protected against *E. coli* bacillosis [111–113].

Graded doses (45 – 1800 μg) of purified type I fimbriae have been parenterally administered to human volunteers [114]. The vaccine was well tolerated and elicited both a humoral IgG and an intestinal secretory IgA antibody response. A moderate level of serospecific protein was obtained in volunteer challenge studies.

Oral administration of 1 mg of purified CFA/I evoked a significant secretory IgA response in about 50 % of vaccinated humans [115]. Responders were protected against challenge with an St^{+}, LT^{+}, CFA/I^{+} *E. coli* strain.

Future Prospects. No vaccines are available to prevent diarrhea urinary tract infections, or meningitis caused by *E. coli*. At present, both attenuated live-oral and inactivated vaccines are being evaluated as to their ability to prevent *E. coli* diarrhea. The production of a killed, inactivated, oral, whole-cell vaccine from a CFA/I^{+} strain has been described [115]. The majority of vaccinees responded with a good secretory IgA antibody response to CFA/I. The diarrhea attack rate for volunteers challenged with a toxinogenic CFA/I^{+} *E. coli* strain was 89 % for those in the placebo group and 20 % for those in the vaccine group.

A spontaneous mutant of *E. coli* has been isolated which has lost the genes for production of the heat-labile and heat-stable enterotoxins but still expresses CFA/II. This strain has been proposed as a likely carrier for cloned genes expressing relevant protective antigens such as colonization factors or "enterotoxoids" [116]. Several purified antigens show promise as oral vaccine candidates including conjugates of the two enterotoxins [117], the B subunit of the heat-labile toxin [118], and toxoid termed "procoligenoid" produced by heating the heat-labile enterotoxin [119].

Since the majority of *E. coli* strains associated with neonatal meningitis possess the K 1 capsule antigen, this would appear to be an excellent vaccine candidate. However, the K 1 antigen (which is identical to the group B meningococcal capsule) is not immunogenic in humans (see Section 2.9).

The most promising vaccines to prevent serious *E. coli* urinary tract infections (pyelonephritis) contain type I fimbrial antigens. Two fimbrial amino acid sequences have been shown to prevent colonization by a homologous *E. coli* strain in a murine pyelonephritis model [120]. Antibody to one sequence also afforded protection against an *E. coli* strain expressing a distinct fimbrial serotype.

2.12. *Neisseria gonorrhoeae* Vaccine

Etiological Agent and Pathogenesis. *Neisseria gonorrhoeae* was identified as the causative agent of gonorrhea by Bumm in 1885. It is a gram-negative diplococcus with humans serving as its only known natural host. Gonorrhea is a sexually transmitted disease of epidemic proportions worldwide. More than 500 000 cases are reported annually in the United States. The infection is usually restricted to the entry site, i.e., the genitalia, pharynx, and/or rectum. More than 90 % of infected males display symptoms characterized by pain upon urination and an urethral exudate [121]. However, up to 50 % of infected women remain asymptomatic and can serve as carriers. In approximately 10 % of women with genital gonorrhea, the bacteria invade the bloodstream and result in pelvic inflammatory disease. Disseminated gonococcal infections are far less frequent in male patients.

When cultured on solid medium, four colonial variants (T 1 – T 4) can be discerned [122]; however, only the piliated T 1 and T 2 variants can cause disease [123]. The first step in the infectious process is pili-mediated attachment of the gonococcus to mucus-secreting epithelial cells. Shortly thereafter, the bacteria are internalized and enter the submucosa [121]. Treatment of gonorrhea has been complicated by the occurrence of multiple antibiotic-resistant clinical isolates.

History of Immunization. The first attempt to prevent gonorrhea by immunization used a killed whole-cell vaccine administered parenterally [124], but did not provide protection in Eskimos. Subsequent human vaccine trials have used purified intact pili: antipili antibody blocks attachment of the gonococcus to epithelial cells, and, an antipili secretory IgA antibody response is mounted during natural infection [125]. Vaccination of human volunteers with purified *N. gonorrhoeae* pili elicited a humoral and a local vaginal antibody response. These antibodies blocked the attachment of *N. gonorrhoeae* to epithelial cells in vitro [125]. Two 2 mg doses of a monovalent pili vaccine resulted in significant protection against challenge with the homologous strain [126]. However, no protection was observed when the challenge strain possessed a pilus serotype modestly cross-reactive with the vaccine [121]. A subsequent field trial with this monovalent vaccine failed to show any protection [125].

Future Prospects. No vaccine has been licensed for the prevention of gonorrhea. The above human volunteer studies showed that protection against gonorrhea mediated by antipili antibody is serospecific [125]. Two different approaches are being taken to circumvent the problem of antigenic variation among pili:

1) Use of a polyvalent vaccine containing those types of pili which predominate in a given area [126].
2) Use of conserved antigenic domains from the antigenic pili molecule. A cyanogen bromide cleavage fragment can induce production of an antibody that can recognize various pili types [127]. Peptides corresponding to this conserved region block attachment of the gonococcus to epithelial cells in vitro [128]. If such peptides can induce a broadly protective immune response in

humans, synthetic peptides containing the proper sequence could be coupled with carrier proteins to produce a conjugate vaccine.

A possible alternative to pili as vaccine antigens are the outer membrane proteins of *N. gonorrhoeae*, the most promising being the PI protein. The PI protein has a sufficiently restricted antigenic variation to make construction of a polyvalent vaccine feasible. Furthermore, antiPI antibody is bactericidal and may block entry of the gonococcus into epithelial cells [129]. A parenterally administered, outer membrane preparation enriched with PI protein elicits an antibody response in humans [130].

2.13. *Hemophilus influenzae* Type b Vaccines

Etiological Agent and Pathogenesis. *Hemophilus influenzae* is a gram-negative bacillus first isolated by Pfeiffer in 1892. The vast majority (>90 %) of human disease is caused by type b capsular serotype organisms in children between the ages of three months to two years [131]. *Hemophilus influenzae* type b can cause several serious disease syndromes predominated by meningitis which is endemic worldwide and is one of the three leading causes of bacterial meningitis in developed countries. Infection is by inhalation of contaminated aerosols. After localized multiplication, the organisms enter the bloodstream and translocate to the meninges. Even though effective antibiotic treatment against *H. influenzae* type b has been available for many years, meningitis due to this organism still has a high attendant morbidity and mortality rate.

History of Immunization. Intensive efforts to understand human immunity to *H. influenzae* type b disease have largely replaced clinical trials with crude, poorly characterized vaccines. The capsular polysaccharide is a critical virulence factor because it confers resistance to lysis by serum [132]. Antibody-dependent, complement-mediated bacteriolysis is the basis for immunity against *H. influenzae* type b [132]. Antibodies directed against the type b capsular polysaccharide are bactericidal in the presence of complement and are believed to be the predominant protective antibody [131].

Production and Properties of *H. influenzae* Type b Vaccines. Two vaccines are licensed for use against *H. influenzae* type b: a purified capsular polysaccharide and a conjugate vaccine composed of capsular polysaccharide covalently coupled to diphtheria toxoid. The capsular antigen is purified by coprecipitation of the capsule with detergent, digestion with nucleases and protease, extraction in phenol, and ultracentrifugation. In the conjugate vaccine, the capsule is covalently linked to diphtheria toxoid by a bifunctional spacer molecule. The potency of *H. influenzae* type b vaccines is not determined in animals but by analysis of physicochemical characteristics which confirm that the capsular antigen is in an immunogenic form.

Immunization Recommendations. The capsular polysaccharide vaccine was first licensed for use in the United States in 1985. A single 25 μg dose, administerd parenterally is recommended for children 2 – 6 years of age; no booster dose is needed.

For the conjugate vaccine licensed in 1987, a single dose is recommended for children 18 months to 6 years of age; no booster dose is needed.

Vaccine Efficacy. The purified capsular polysaccharide vaccine was evaluated in a trial on Finnish children [133]. Vaccine efficacy was approximately 90 % for children who were vaccinated at 2 – 5 years of age or older. Vaccination provided no protection in children 3 – 18 months of age due to poor immunogenicity [132]. However, retrospective studies performed in the United States have found vaccine efficacy to range from 55 – 88 % [133]. In contrast, a capsular polysaccharide – diphtheria toxoid conjugate vaccine evoked protective levels of antibody in a majority of 18-month old children [135].

Future Prospects. A *H. influenzae* type b vaccine is clearly needed which is immunogenic in children <18 months of age, where approximately 75 % of serious disease occurs. Several candidate vaccines including the diphtheria toxoid conjugate mentioned above and a *Neisseria meningitides* outer membrane protein – capsular polysaccharide conjugate are being evaluated in clinical trials [135], [136]. A large-scale Finnish trial is being performed to evaluate the diphtheria toxoid conjugate vaccine. Vaccine efficacy was 83 % in children vaccinated at 3, 4, 6, and 14 months of age [137]. In contrast, this vaccine afforded insignificant protection when used to immunize Alaskan Eskimo children at 2, 3, and 4 months of age.

3. Viral Vaccines

3.1. Measles Vaccine

Etiological Agent and Pathogenesis. Measles (rubeola) is caused by a paramyxovirus first isolated by ENDERS and PEEBLES in 1954. This highly contagious disease is transmitted in an aerosol of virus-infected droplets from the respiratory tract. During the prevaccination era, children in industrialized countries generally had a self-limiting disease characterized by conjunctivitis, bronchitis, fever, and a rash; complications included pneumonia (mostly bacterial) and encephalitis. A late complication of measles in early childhood is the development of a progressive, fatal neurological disease, termed subacute sclerosing panencephalitis. Death due to complications following measles infection remains one of the leading causes of child mortality in developing countries.

History of Immunization. Two approaches to vaccine development were adopted, one was based on killed (inactivated) strains and the other on live, attenuated strains [138]. The formaldehyde-inactivated, alum-precipitated vaccine was licensed in the United States in 1963 and withdrawn in 1967 because its protective effect was insufficient. In addition, the vaccinees exposed to the wild virus developed atypical measles with fever and rash but also pneumonitis.

The attenuated live vaccine with the Edmonston B strain was also licensed in 1963 in the United States. A large-scale field trial in the United Kingdom showed high seroconversion rates, a long antibody response, and a 84 – 94 % decrease in attack rate among the 36 000 immunized children. Further attenuation of the vaccine strain in chick embryo cells yielded the Schwartz and the Moraten strains. Both strains are included in current measles (and combined measles – mumps – rubella) vaccines. Attenuation of the original Edmonston strain in human diploid cells yielded the Edmonston-Zagreb strain, also in current use [139]. With the introduction of general immunization against measles, the disease has almost disappeared in many countries (see below).

Production and Properties of Measles Vaccines. The attenuated measles vaccine strain is usually grown in primary culture using chick embryo cells or human diploid cells. The WHO test requirements for the seed virus and cell substrates are described in[140, pp. 52 – 73],[141, p. 179]. Tests for monitoring measles vaccine grown in chick embryo cells include tests for nonadsorbing viruses and avian leukosis viruses. For vaccines grown in human diploid cells, chromosomal monitoring requirements have been formulated. Potency is checked by titration of the virus in tissue culture, (minimum requirement, 1000 $TCID_{50}$ per human dose). To test for stability, a sample of the final freeze-dried vaccine is incubated at 37 °C for seven days: the sample must then contain at least 1000 $TCID_{50}$ in each human dose. The vaccine is lyophilized; reconstituted vaccine should be used immediately or stored at 0 – 10 °C for not more than 8 h.

Immunization Recommendations. The vaccine is given at the age of 15 – 24 months [142] or even younger in developing countries [139]. In the United States and many European countries, a combined measles – mumps – rubella (MMR) is usually given (see Section 3.4). As for all live vaccines, immunization of immuncompromised individuals is not generally recommended. Immunization of pregnant women should be avoided although no increased risks have been documented for measles vaccine.

Adverse Reactions. Current attenuated measles strains give a few mild reactions, mostly consisting of fever and rash 7 – 10 days after immunization. The neurologic reaction of the Guillian-Barré syndrome has been reported but is extremely rare.

Vaccine Efficacy. The rapid decrease of measles morbidity in the United States after introduction of general immunization is a good example of vaccine efficacy [142], [143]. Over a period of 20 years the rate of notified cases fell by more than 99 %. Subacute sclerosing panencephalitis also decreased. The overall protective efficacy of the current strains is 90 % but is lower in children of 12 months of age due to the presence of maternal antibody. Life-long immunity after one dose of vaccine has not been proven. Increased disease incidence in the United States has raised the question of the possible need for a two-dose schedule [142].

Future Prospects. Given the efficacy and acceptability of the currently used attenuated measles strains, there has been little incentive for vaccine development. However, current vaccines are not very stable at high temperature, which is a concern in many developing countries. Furthermore, a large proportion of morbidity and mortality in children in developing countries occurs before the recommended age for immunization. Research is centered on solving these two problems. More immunogenic strains and/or higher doses of attenuated vaccines are being investigated [139]. Another approach would be to renew work on the development of a killed (inactivated) vaccine. Inactivated vaccines have so far failed, this is probably the result of a lack of neutralizing antibodies to the viral fusion surface protein that is destroyed in the inactivation process [144]. Subunit vaccines based on the hemagglutinin and the fusion proteins of measles virus have successfully been tested in animals and may provide a vaccine especially suited for developing countries [145].

3.2. Mumps Vaccine

Etiological Agent and Pathogenesis. Mumps is caused by a member of the paramyxovirus group and is also known as (epidemic) parotitis due to its predominant symptom–infection of the salivary (parotid) gland. Transmission occurs via infected aerosol droplets. The disease is generally characterized by moderate fever and swelling of the salivary glands. Subclinical infection occurs in ca. one third of infected individuals. Most cases occur in children 5 – 10 years of age. The most frequent complication is meningoencephalitis, giving clinical symptoms in >10 % of patients. Orchitis, a less common but feared complication in postpubertal males, results in impairment of fertility in 10 – 20 % of cases but absolute sterility is rare.

History of Immunization. Isolation of the causative agent for mumps by Habel in 1945 from chick embryo was followed by attenuation studies for vaccine development. Development of an inactivated (killed) vaccine, was also initiated; a killed vaccine was licensed in the US in 1950 – 1978 but had low long-term protective efficacy [146]. An inactivated mumps vaccine was also produced in Finland for immunizing military recruits; it decreased disease incidence by 94 % [147]. Immunization with an attenuated mumps vaccine (Jeryl Lynn strain cultured in chick embryonic cells), licensed in 1967 in the United States resulted in a 97 % decline of disease incidence in the general population by 1981 [146], [148], [149]. Chick embryonic cells are used to propagate another highly attenuated strain, Urabe Am 9 developed in Japan [150]. The Rubini strain, developed in Switzerland, is used in a vaccine in which the virus is grown in human diploid cells [151]. Vaccines based on the Leningrad-3 strain have been used in the former Soviet Union and other countries since 1974.

Production and Characterization of Mumps Vaccines. The WHO requirements for the above-mentioned live mumps vaccine have been formulated [152, pp. 139 – 164]. Controls, thermostability tests, and requirements are as for measles vaccines, i.e., the minimum potency should be retained after inoculation for one week at 37 °C (see Section 3.1). No minimum requirement of potency has been established by the WHO but a commonly used minimum dosage is 5000 $TCID_{50}$ per human dose. The vaccine is lyophilized, reconstituted vaccine should be used without delay.

Immunization Recommendations. Mumps vaccine is recommended to be given to all susceptible individuals over the age of 12 months [153]. It is usually given in combination with measles and rubella vaccines in industrialized countries (see Section 3.4).

Vaccine Efficacy. Clinical efficacy is 75–90 % for the live vaccine containing the Jeryl Lynn strain [149]. Similar results have been shown for the other strains or inferred from serologic comparisons. The long-term protective efficacy of the inactivated mumps vaccine used in Finland has not been established but antibody responses indicate that it may give a shorter-term immunity than a live vaccine due to a lack of antibody response to the fusion protein [154] as in the case of the inactivated measles vaccine (see Section 3.1).

Adverse Reactions. Mumps vaccine is one of the least reactogenic attenuated vaccines. Side effects are rare and mild. No cases of atypical mumps were found after immunization of Finnish recruits with killed mumps vaccine [154].

Future Prospects. The present vaccine is highly satisfactory and little effort has been devoted to further development. Subunit vaccines based on the hemagglutinin neuraminidase and the fusion surface proteins of mumps virus are protective in animal models [155].

3.3. Rubella Vaccine

Etiological Agent and Pathogenesis. Rubella, also known as German measles, is caused by a member of the togavirus group. The virus was isolated in 1962 by two independent American groups. In 1941 GREGG reported that this mild disease could cause severe congenital cataracts in children born to mothers who were infected during the first trimester of pregnancy. Transmission of the virus occurs by the respiratory route. The clinical picture is a discrete rash and low-grade fever, 25–50 % of cases remain subclinical. Arthritis is a common complication; encephalitis is less common than in measles or varicellae (1/6000 cases). The congenital rubella syndrome is usually characterized by hearing impairment, ocular lesions, cardiac malformation, microcephaly, and mental retardation. The severity of defects is related to fetal age at the time of maternal infection, with the most severe damage seen after infection during the first month of pregnancy.

History of Immunization. Live attenuated rubella vaccines were licensed in several countries in 1969–1970. One of the first vaccines contained the Cendehill strain, grown in rabbit kidney cells [156]. Other vaccines were based on the HPV-77 strain, grown in either duck embryo cultures or in dog kidney cultures. The latter vaccine produced arthritis symptoms; 50–60 % of adult female vaccinees as compared to 10–20 % for the other two vaccines. A vaccine based on the RA 27/3 strain, isolated and propagated in human diploid cells, was licensed in the mid-1970s in Europe and in 1979 in the USA [157]. This vaccine gave a better immune response and had less side effects than the other vaccines; it has replaced other rubella strains in the current vaccines.

Production and Properties of Rubella Vaccines. The attenuated RA 27/3 strain is grown in human diploid cells. The WHO requirements for rubella vaccine include control of normal karology of the human diploid cells[158, pp. 54–84], [159, pp. 313–316]. The minimal potency requirement, determined by titration in tissue culture, is 1000 $TCID_{50}$ per human dose. The vaccine is lyophilized; reconstituted vaccine should be used immediately or stored at 2–8 °C for not more than 8 h.

Immunization Recommendations. Vaccination strategies range from protection of the individual as in the U.K. to indirect protection as in the US [160], [161]. Protection is achieved by general immunization of adolescent girls, usually combined with post partum vaccination of seronegative women. Indirect protection of adult women by immunization of young children is often combined with vaccination of seronegative women.

A combination of these strategies is used in some European countries, e.g., Sweden [162]. This program includes routine screening of all pregnant women, post partum vaccination of seronegatives, and two-dose immunization of children with combined measles–mumps–rubella vaccine. Although the risks for the fetus seem small, the vaccine should not be given to pregnant women and contraceptive measures are recommended for three months after immunization.

Adverse Reactions. The RA 27/3 strain has few and mild side effects. The joint manifestations, mainly arthralgia, are more common in adults. Intrauterine infections during pregnancy have been documented for all three strains (Cendehill, HPV-77, and RA 27/3) [163]. No abnormalities related to congenital rubella infections have been documented for any of the strains.

Vaccine Efficacy. The efficacy of the current strain is ca. 90 %. The incidence of congenital rubella virus infection decreased in countries with a routine immunization program against the disease. The duration of immunity has not yet been determined. The risks to the fetus upon a maternal reinfection are also unknown and may be very small.

Future Prospects. The current vaccine is considered safe and efficient and no efforts are devoted to further development.

3.4. Combined Measles – Mumps – Rubella Vaccine

History of Immunization. The first combined measles – mumps – rubella (MMR) vaccine was licensed in the US in 1971. It contains the Moraten measles strain (Section 3.1), the Jeryl Lynn mumps strain (Section 3.2), both grown in chick embryo cultures, and the RA 27/3 rubella strain (Section 3.3), grown in human diploid cells. Three other combined vaccines were licensed later. The triple vaccine has replaced the monovalent vaccines used in general immunization programs for children in the United States and in many European countries.

Production and Properties of Combined Vaccines. The minimum potencies for the attenuated strains of measles, mumps, and rubella virus, are the same as in the monovalent vaccines, i.e., 1000 $TCID_{50}$ for measles, >5000 $TCID_{50}$ for mumps, and >1000 $TCID_{50}$ for rubella. The vaccine is lyophilized and should be stored at 2 – 8 °C.

Immunization Recommendations. In many countries, one dose of vaccine is administered to children at the age of 15 – 24 months. Two doses are given in some European countries [162]. In Sweden, a first dose is administered at 18 months and a second at 12 years of age, replacing the monovalent programs for measles and rubella immunization respectively. In the United States, the combination of MMR with oral polio vaccine and diphteria – tetanus – pertussis (DTP) at 15 months has been recommended [164], [165].

Adverse Reactions. The side effects are the same as for the monovalent vaccines, the measles component being the most reactogenic. The rate of adverse reactions is <0.5 – 4 % [166].

Vaccine Efficacy and Future Prospects. The efficacy of the combined preparation is the same as for the monovalent vaccines with an overall efficacy of 90 – 95 % [143]. The life-long protective efficacy of a single injection has not been proven. Further combinations with an attenuated varicella component were investigated [167].

3.5. Polio Vaccine

Etiological Agent and Pathogenesis. Polio (infantile paralysis) is caused by a picornavirus of the genus enteroviruses. A poliovirus strain was first isolated in cell culture by Enders, Weller, and Robbins in 1949. In 1951, polio virus isolates were officially grouped into three serotypes, type 1 (Brunhilde), type 2 (Lansing), and type 3 (Leon). Transmission is mainly via the oral – fecal route and the virus reaches the central nervous system by way of the blood stream. The vast majority of infections are subclinical. The nonparalytic disease has a mild or minor form with fever and general malaise and a more severe or major form with additional symptoms of meningitis/meningoencephalitis. Paralytic polio, with its most severe bulbar form, is estimated to represent 5 – 10 % of clinical cases. The mortality rate is 5 – 10 % of clinical cases, i.e., 1 – 2 % of all infections. Severe paralysis is seen in 10 – 20 % of clinical cases, and mild or moderate paralysis in 30 %.

With improved sanitation and standards of living, a shift towards infection at a higher age was observed in industrialized countries in the prevaccination era. During the first half of this century, peak incidence was noted in the age group 5 – 14 years. A large proportion of cases occurred in young adults. In the developing countries, polio has maintained its character of infantile paralysis with the majority of children being infected during the first few years of life. Cases of paralytic polio occur only in the youngest age groups.

History of Immunization. The first large-scale field trial of an inactivated (killed) polio vaccine was launched in the US in the early 1950s, only a few years after successful propagation of the virus in tissue culture. Inactivated polio vaccine (IPV) was licensed for general use in the United States in 1955 [168], [169]. In 1955, cases of atypical paralytic polio were reported, most of them were associated with two lots of vaccine from CUTTER [170]. Although live poliovirus was recovered from vaccine supplied by other manufacturers, this failure of inactivation is known as the Cutter incident. The total toll was 269 cases, of which 192 were paralytic (with ten deaths). Clinical trials with a live attenuated oral polio vaccine (OPV) were started in 1958 and the vaccine was licensed in the United States in 1962 [171], [172]. Immunization with OPV alone has been used from the late 1950s – early 1960s in Europe. One of the largest immunization campaigns was launched in the former Soviet Union in 1960 when OPV was given to ca. 77×10^6 people. A few European countries have used only IPV [173–175]. In Sweden, clinical trials with IPV were started in 1955 and general immunization was introduced in 1957 [173]. All three strategies were effective in reducing the incidence of paralytic polio. With IPV alone, an 80 % reduction rate was achieved in the US with less than 50 % of the population immunized [168], [169], [171]. The decrease of polio continued at the same rate after introduction of OPV. Introduction of the massive IPV campaign in Sweden resulted in a 93 % decrease of paralytic disease after six years of immunization and elimination of the disease by 1964 [173].

Production and Properties of Polio Vaccines. Oral polio vaccines are manufactured with the attenuated Sabin strains. Inactivated polio vaccines are mostly produced with the type 1 Mahoney strain, type 2 MEF-1 strain, and type 3 Saukett strain. Primary and secondary monkey (Cynomolgus) kidney cultures or human diploid cells are most commonly used for culture. The WHO requirements for both OPV and IPV produced in primary monkey kidney cells include tests

1) for the absence of cytopathogenic viruses with the exception of some foamy viruses[141, pp. 40 – 84] , [159, pp. 157 – 173], [176],[182, pp. 108 – 110]
2) the absence of simian B virus in rabbits, and
3) the absence of SV 40 in sensitive cells.

(usually primary green monkey *Cercopitecus* kidney cultures). Minimal potency requirements for OPV were formulated by the WHO in 1987[152, pp. 165 – 166]. A single human dose of trivalent oral vaccine should contain approximately $10^{5.5} – 10^{6.5}$ infectious units of type 1, $10^{4.5} – 10^{5.5}$ of type 2, and $10^{5.0} – 10^{6.0}$ of type 3. For IPV, recommendations of 40 : 8 : 32 D-antigen units per human dose for types 1, 2, and 3 respectively were issued in 1981 [141]. The potency tests for OPV (and IPV prior to inactivation) are performed by titration in tissue culture. In vivo potency tests are also required for IPV but neither the animal species nor the number of injections is specified. Inactivation of polio vaccine by formaldehyde is controlled by titration of polio-sensitive cells (usually *Cynomolgus* or *Cercopitecus*) in tissue culture. Oral polio vaccines are supplemented with a stabilizer, usually magnesium chloride or sorbitol. Both OPV and IPV contain antibiotics, usually neomycin; OPV is best stored at −20 °C but can be stored at 2 – 4 °C for a variable length of time; IPV is stored at 2 – 8 °C.

Immunization Recommendations. Recommendations vary but at least three doses of OPV are given. In the US, five doses are recommended, in Sweden four, and in Finland six. The WHO recommends four doses for infants in developing countries.

Adverse Reactions. Serious adverse reactions of paralytic polio have only been reported for OPV with a predominance for type 3 [169], [177] (incidence = $1/1 \times 10^6$ vaccine recipients, including both nonimmune and immune individuals). No cases of IPV-induced paralytic polio have been reported after 30 years of use (with the exception of the Cutter incident). IPV is considered to be the least reactogenic of all the vaccines used for childhood immunization.

Vaccine Efficacy. Both OPV and IPV have proven highly efficient in eliminating polio from industrialized countries with general immunization programs. The protective efficacy is close to 100 % after at least three doses of vaccine with adequate immunogenicity. Serologic data reported for OPV from developing countries have been less than encouraging. Both vaccines have eliminated the circulation of wild type strains in industrialized countries with high immunization rates. Outbreaks among groups refusing immunization have occurred [168]. In 1984, nine cases of paralytic polio caused by type 3 occurred in Finland [177]. The low immunogenicity of the IPV used in Finland has been known since the late 1960s [178]. The new IPV developed in the Netherlands with its high antigen content is given in European countries in at least three doses [179]. Two doses or even one dose of the new IPV preparations have been claimed to be highly protective. A two dose-schedule has been used in Africa and protective efficiency was ca. 89 % [180]. Inactivated polio vaccine is more heat-stable and can be combined with diphteria – tetanus – pertussis vaccines.

Future Prospects. Further development of OPV is aimed at reducing production costs by use of microcarrier cultures (cell cultures on beads etc.) in fermentors [179]. Use of continuous cell lines is being investigated. An IPV produced in a continuous cell line is currently licensed in France [181]. The WHO requirements for production of vaccines in continuous cell lines are formulated in [182, pp. 93 – 107].

3.6. Hepatitis B Vaccine

Etiological Agent and Pathogenesis. Hepatitis B is caused by the hepatitis B virus (HBV), a member of the hepadna virus group. The presence of a new antigen called the Australia (Au) antigen or hepatitis B surface antigen (HBsAg) in the blood of patients with leukemia, Down's syndrome, and hepatitis was shown by BLUMBERG in 1964 – 67. Other antigens (HBeAg and HBcAg) were described later and correlated with infectivity; they are found in the 42-nm Dane particle, i.e., the infectious virion.

The HBV is a pathogen only for humans, but certain monkeys, in particular chimpanzees, are also susceptible to infection. Transmission in humans occurs mainly by inoculation with infected blood. The virus is excreted in body fluids, causing infection through saliva and sexual contact. A chronic carrier state is established in 5 – 10 % of adult cases; chronic active hepatitis develop in 25 – 30 % of the carriers, often leading to cirrhosis of the liver: 90 % of infants infected at birth become carriers. Hepatitis B occurs throughout the world; the incidence varies from <0.5 % in Western Europe and the US to 5 – 15 % in Southeast Asia and Southern Africa [183], [184].

Hepatitis B virus plays an important role in hepatocellular carcinoma although the exact mechanism has not yet been determined [183].

History of Immunization. Heat-inactivated HBsAg-positive serum was used by KRUGMAN in 1971 to immunize children. A 70 % protection rate was found upon challenge with active virus. The high protective efficacy of hepatitis B vaccine derived from purified, inactivated HBsAg from positive plasma was demonstrated in 1980 [185]. The vaccine, developed in the US, was licensed for general use in 1981. Studies in 1980 – 1983 also showed that the HBV vaccine in combination with human antiHBV immune globulin (HBIG) prevented development of the carrier state in infants born to carrier mothers. Hepatitis B vaccine was the first vaccine to be routinely produced by DNA recombinant technology in yeast (→ Genetic Engineering, Chap. 5.2.). Serologic studies indicate that the protective efficacy of the recombinant vaccines is the same as that of the plasma-derived vaccine. Genetically engineered vaccines were licensed in the US in 1986 and subsequently in several European countries.

Production and Properties of Hepatitis B Vaccines. The WHO requirements for plasma-derived HBV specify guidelines for selection of donors of HBsAg-positive plasma[186, pp. 70 – 101]. The plasma pool must be subjected to extensive tests in animals, fertile eggs and cell cultures for extraneous viruses and *Mycobacterium tuberculosis*. The purified HBsAg (>95 % pure) is inactivated by treatment with pepsin followed by urea (8 mol/L), and formaldehyde or by heat with or without formaldehyde. After controls for purity and antigen content, an alum adjuvant is added; the vaccine must be stored at 5 ± 3 °C. No minimal requirement of antigen content has been formulated but a commonly used vaccine contains 20 µg HBsAg per human dose.

The WHO requirements for HBV made by recombinant DNA technology state that such vaccines may contain the S gene products or the S/pre-S combination [152, pp. 106 – 138]; licensed recombinant vaccines contain only the S gene product. A full description of the host cell (*Saccharomyces cerevisiae*) and the expression vector is required.

The HBsAg is commonly purified by precipitation, ultrafiltration, and chromatography. Tests for HBsAg and residual cell or plasmid DNA are required before addition of adjuvant. No minimal antigen requirements have been formulated but a potency assay in mice has been outlined. Commonly used recombinant yeast vaccines contain 10 or 20 µg HBsAg per human dose.

Immunization Recommendations. Most Western European countries and the US have issued recommendations for vaccination of risk groups [187–189]. In general, pre-exposure prophylaxis is recommended for medical and other staff in frequent contact with high-risk groups or with blood from such groups. Immunization of patient groups frequently receiving blood or blood products, as well as patients in certain institutions is also recommended.

For pre-exposure prophylaxis, three 20 µg doses of plasma-derived vaccine is given at day 0, 30 and 6 months intramuscularly in the deltoid region. Children <11 years of age should receive 5 µg at each injection. Postexposure prophylaxis for infants to HBsAg, HBeAg-positive mothers is 0.5 mL intramuscular injection of human antiHBV immune globulin (HBIG, see also Section 5.3). Vaccine (10 µg) should be given at birth and then at one and six months of age. For postexposure prophylaxis of adults, 1 mL HBIG should be given immediately together with three doses of vaccine administrated as for preexposure prophylaxis.

Adverse Reactions. Side effects are mild and mainly local at the site of injection.

Vaccine Efficacy. The overall vaccine efficacy is 90 % with no response upon immunization in 5 – 10 % of the vaccines. In individuals with seroconversion to HBsAg, protective efficacy is almost 100 %. Studies on the duration of protection beyond 5 – 10 years are not yet available. The protective efficacy of HBIG and vaccine administered to newborns born to carrier mothers has been estimated to be more than 90 % [183].

Future Prospects. Current HBV vaccines are safe and have a high efficacy but also high production costs. Research is aimed at the development of cheaper, more immunogenic vaccines. The inclusion of the pre-S (P 31) region of the viral genome in addition to the present S (P 25) region is being investigated in yeast-derived recombinant vaccines [190], [191]. Other antigens such as HBcAg are also being studied. Polypeptide vaccines, based on the major determinants of HBsAg, are undergoing clinical trials. Synthetic peptides may provide the ultimate solution to the problem of immunogenic, readily available HBV vaccines.

3.7. Rabies Vaccine

Etiological Agent and Pathogenesis. Rabies is a lethal disease caused by a neutropic rhabdovirus affecting humans and warm-blooded animals. The most common route of transmission to humans is by infected saliva through the bite from a rabid animal. The virus ascends along peripheral nerves to the central nervous system. The incubation period varies from ten days to several months. Symptoms include a nonspecific prodromal stage followed by an acute neurological phase ending in coma and death. The disease is endemic in animals in most parts of the world with some exceptions such as the United Kingdom and parts of Scandinavia [192]. In the urban form, stray dogs and cats act as vectors, whereas the sylvatic form involves wild animals. Urban rabies is well-controlled in most countries, but progressive spread of the sylvatic form by foxes is a major problem in Europe. In the US, other animal species are involved including bats. In Europe and Africa, bats are the vector for an antigenic variant of the virus [193].

History of Immunization. In 1885 PASTEUR administered the first ever vaccination to a young boy who had been bitten by a rabid dog [194]. Serial injections containing a virus strain that had been propagated and attenuated in the spinal cord of rabbits ("fixed" virus) were given in an increasingly virulent form. The treatment was rapidly adopted as standard postexposure prophylaxis. The attenuated vaccine was later replaced by an inactivated (killed) vaccine also produced in neural tissue [194]. However, in some patients the myelin content of the vaccine caused sensitization resulting in neurological disease. An inactivated vaccine produced in duck embryos decreased the risk of neural tissue sensitization. The most widely used rabies vaccine in Europe and in the United States was developed in the 1960s and is produced in human diploid cells. A highly purified, concentrated duck embryo vaccine is also available. In Europe, other vaccines are produced in primary animal cells and continuous cell lines; non human primate diploid cells are used in the US.

Production and Properties of Rabies Vaccines. Attenuated strains of rabies virus are grown in tissue culture or embryonated duck eggs (see above). The virus suspension is usually concentrated and in some cases purified. Only inactivated vaccines are allowed for human use. Inactivation is mostly achieved by treatment with β-propiolactone, but phenol, formaldehyde, or ultraviolet irradiation are also used. The WHO has established requirements for controls of the cells used for virus propagation [152, pp. 167 – 194],[159, pp. 54 – 95]. The potency of the vaccine is determined in a mouse challenge assay (minimal requirement 2.5 IU per human dose). The vaccine is then usually lyophilized although adjuvanted preparations are also in use.

Immunization Recommendations. The recommendations vary by vaccine and country. As pre-exposure prophylaxis, three 1 mL doses are usually recommended on days 0, 7, and 28 or on days 0, 28, and 90 (or 365) as intramuscular injections in the deltoid or suprascapular region [192], [195]. As postexposure prophylaxis, recommended measures include local wound care, administration of vaccine, and administration of human rabies immune globulin (HRIG, 20 IU per kilogram of body weight; see also Section 5.3). Purified equine rabies immune globulin (40 IU/kg) is used in many developing countries with few side effects [196]. Postexposure vaccine prophylaxis is given in five 1 mL doses on days 0, 3, 7, 14, and 28 by intramuscular injection in previously unimmunized individuals. In individuals who have received pre-exposure prophylaxis, two doses are given on days 0 and 3. Children under four years of age receive 0.5 mL injections.

Vaccine Efficacy. The protective efficacy of the human diploid cell vaccine given (with HRIG) as postexposure prophylaxis was first documented in field trials conducted in Iran in 1974 – 75; the survival rate was 100 %. Other studies have confirmed the protective effect of pre- and postexposure immunization with this vaccine. The general opinion that postexposure vaccine prophylaxis in combination with immune globulin conveys 100 % protection has been challenged by two cases of vaccine failure [197]. In other vaccines, a high protective efficacy has been proven in field trials or inferred from comparative serologic studies.

Adverse Reactions. The most common side effects are local redness and induration at the injection site and fever. Desquiting episodes of a serum sickness-like reaction upon repeated immunization have been reported [198] and are possibly due to a sensitizing complex formation between β-propiolactone and human serum albumin [199] or to the high content of bovine serum residues in the vaccine [200].

Future Prospects. Work is centered on the development of large-scale vaccine production techniques to decrease the high costs and thereby increase availability in poorer countries. One such approach is the use of microcarrier culture systems in fermentors [179]. A vaccine produced in this manner in a continuous cell line has been evaluated. A subunit vaccine based on the surface glycoprotein of the virus is a further possibility. The peptide segment of the glycoprotein that induces the production of neutralizing antibodies has been identified and synthesized. A synthetic peptide vaccine may therefore be the most attractive future alternative [201].

3.8. Influenza Vaccine

Etiological Agent and Pathogenesis. Influenza is caused by two antigenically distinct members of the orthomyxovirus group–influenza viruses A and B. In 1934 ANDREWES managed to transfer the influenza A virus from human material to ferrets and later to mice. In 1940, FRANCIS and MAGILL independently isolated influenza B by transmission to ferrets. The virus was subsequently propagated in embryonated hen's eggs and tissue culture. The disease is transmitted by droplet infection from the respiratory tract; it is characterized by high fever, muscle pain, and a dry cough. Pneumonia is the main cause of mortality in elderly people and persons with chronic underlying diseases. Influenza occurs worldwide with regular epidemics during the winter months. The recurrent epidemics are caused by small changes (antigenic drift) in the main pathogenic determinants of the virus, i.e., hemagglutinin and neuraminidase. Large pandemics occur at 10 – 20 year intervals and are due to substantial changes (antigenic shift) in the pathogenic determinants. Antigenic drift is seen in both influenza A and B whereas antigenic shift has only been noted in influenza A [202].

History of Immunization. The first influenza vaccines produced in hens' eggs were tested in humans in the early 1940s. These inactivated *whole virus vaccines* had a low content of viral antigens and a high content of contaminating egg protein. Consequently, the protective effect was low and the rate of adverse reactions high. Whole virus vaccines purified by ultracentrifugation showed clearly improved immunogenicity and a decreased rate of side reactions. Zonal centrifugation further improved the vaccine for adults but still caused adverse reactions in children. Disruption of influenza vaccine with detergent was used to develop *split vaccines* in the mid 1960s. *Subunit vaccine* introduced in the mid 1970s contain concentrated, purified hemagglutinin and neuraminidase [202].

The development of live, attenuated virus strains was pursued in the former Soviet Union by serial passages in hens' eggs [202]. However, doubt was cast on their stability. More stable attenuated vaccines were obtained by isolation of temperature-sensitive and cold-adopted mutants [203]. These strains were successfully tested in humans in the late 1970s but are not in general use. Attenuated vaccines, based on avian – human recombinant strains, are being evaluated in clinical trials.

Production and Properties of Influenza Vaccines. The most commonly used influenza vaccines are of the whole or split virus type. The strains used are specified in annual WHO recommendations. The virus is grown in embryonated hens' eggs (usually pathogen-free) and the allantoic fluid is harvested. According to the WHO requirements, the inactivation method used (usually addition of formaldehyde or β-propiolactone) should inactivate avian leukosis viruses and mycoplasma[204, pp. 148 – 170]. The virus is concentrated and purified by high-speed centrifugation either before or after inactivation. Effective inactivation is controlled by inoculation of embryonated hens' eggs. Hemagglutinin content is usually checked by single radial immunodiffusion against a WHO standard. Whole virus and split virus vaccines usually contain 10 – 15 µg hemagglutinin per human dose. The inactivated vaccines are usually supplemented with a preservative and stored at 2 – 8 °C.

The WHO requirements for attenuated (live) vaccines stipulate that pathogen-free eggs must be used; the absence of other pathogens must be controlled in tissue cultures and animals[204, pp. 171 – 194]. Vaccine potency is determined by titration in embryonated eggs. No minimal requirements of infective dose have been formulated.

Immunization Recommendations. Most countries have recommendations for yearly immunization of high-risk groups. Some countries and the WHO recommend vaccination of all individuals over 65 years. In the United States, immunization of children with chronic pulmonary or cardiac disorders or with other chronic diseases residing in institutional care is recommended; 0.25 and 0.5 mL of split virus vaccine are given in the age groups 6 – 35 months and 3 – 12 years, respectively [205]. For individuals older than 12 years, 0.5 mL of vaccine is recommended.

Adverse Reactions. The incidence of systemic (febrile) and local reactions is <10 % and <20 % for the whole virus vaccines and split vaccines, respectively. An increased rate of the neurological reaction of Guillain – Barre's syndrome was reported in the United States after large-scale immunization against swine influenza in 1976.

Protective Efficacy. The protective efficacy of influenza vaccines is controversial. Discrepancies are probably due to different vaccines, number of injections given, the age groups under study, and occurrence of antigenic drift during the study period. The newer purified whole virus and split virus vaccines are considered to be 70 – 90 % effective in healthy adults. The duration of protection is largely dependent on the degree of antigenic drift. In the case of antigenic shift, little or no protection can be expected. The immune response and protective efficacy in children receiving chemotherapy and in debilitated elderly persons are lower [206]. The protective efficacy in preventing death has been estimated at 74 % [207].

Future Prospects. Research on influenza vaccines is directed toward the synthesis of the hemagglutinin and the neuraminidase antigens and to improvement of their immunogenicity [201]. Another line of development is the attenuated live vaccine approach [203].

3.9. Varicella Vaccine

Etiological Agent and Pathogenesis. The varicella-zoster virus, a member of the herpes virus group, causes two distinct clinical manifestations [208]–varicella and herpes zoster. *Varicella,* also known as chickenpox, is the primary infection. *Herpes zoster* is caused by the reactivation of the latent varicella virus in ganglion tissue. Varicella is usually a mild infection characterized by a vesicular rash. Transmission occurs by droplet infection from the respiratory tract and by direct contact with vesicle fluid. Complications are encephalitis (1/1000 – 1/3000 cases) and pneumonia. The disease occurs worldwide, with 90 – 95 % of cases occurring before the age of 15 years. Suspected cases of congenital varicella have been described mainly after varicella during the first trimester of pregnancy. Neonatal varicella with a 30 % mortality rate occurs with maternal varicella within one week prior to term without prophylaxis. Varicella in the compromised host is a severe disease with a 10 – 20 % mortality rate. Herpes zoster, a localized, often painful vesicular rash is most common in the elderly; it occurs in 10 % of the population.

History of Immunization. The varicella-zoster virus was first isolated and propagated in tissue culture by WELLER and STODDARD in 1952. Attenuated, live varicella vaccine was developed in Japan in 1970 using the Oka strain [209]. After serial passages, the attenuated strain was adapted to human diploid cell cultures. The immunogenicity and protective efficacy of the vaccine in healthy children has been demonstrated [209], [210]. Efficacy of the vaccine in children with malignant disease, in particular leukemia, has also been shown [211]. The Oka strain is used for vaccine production in Japan, Europe, and the United States.

Production and Properties of Varicella Vaccines. The attenuated, live Oka strain is propagated in human diploid cell cultures. The WHO requirements include control for the absence of adventitious agents and the usual conditions for culture in human diploid cells [186, pp. 102 – 133]. No minimal potency requirements have yet been formulated. Varying doses have been used in clinical trials. The most common formulation is ≥2000 plaque-forming units per human dose. The vaccine is lyophilized; reconstituted vaccine should be used without delay.

Immunization Recommendations. Varicella vaccine is not yet recommended for general immunization in Europe or the United States. It is given to children with leukemia during remission or when chemotherapy is withheld for one week prior to and after vaccination [211]. A second dose is often given to children who remain seronegative after the first injection.

Adverse Reactions. In healthy children and adults, local swelling and pain occurred in 1 – 5 % of healthy children and 20 % of adults. Systemic reactions with fever and a rash are reported in 5 – 10 % of both healthy children and adults.

Fever and rash occurred in 40 % of vaccine recipients with chemotherapy suspended for two weeks.

Vaccine Efficacy. Protective efficacy in healthy children is 95 – 100 % with persistence of immunity over a 5 – 10 year period. In adults, protective efficacy is 60 – 80 % and possibly of shorter duration. In children with malignancies, protective efficacy is 60 – 92 %. Spread from vaccine-induced vesicular rash has been documented in household contacts.

Future Prospects. The currently investigated, attenuated Oka strain vaccine is effective but long-term immunity and zoster incidence remain to be established. The problems of latency and of vaccine-induced rash could be overcome by the development of a subunit vaccine produced either by a recombinant DNA technique or by using synthetic peptides.

3.10. Yellow Fever Vaccine

Etiological Agent and Pathogenesis. Yellow fever is a hemorrhagic fever caused by a member of the toga virus group [212], [213]. The virus was transmitted to rhesus monkeys by Mathis and coworkers in 1927. The virus strain was then propagated by serial passages of intracerebral inoculations in white mice. Yellow fever is transmitted by mosquitoes of the genus *Haemagogus* in South America and of the genus *Aedes* in Africa. The animal reservoir is mainly monkeys. Subclinical infections are common. The incubation period in clinical cases is 3 – 6 days; symptoms range from transient fever and headache to high fever with meningoencephalitis, followed by jaundice and hemorrhagic manifestations. In the malignant form all these symptoms are present and death occurs within one week. Mortality rates are 40 – 50 % in the severe forms of the disease.

History of Immunization. Two types of attenuated vaccine were developed in the early 1930s cultured in neural tissue (Dakar vaccine) and in chick embryo [214]. Initially, both types were given together with human immune serum. In the late 1930s, the vaccine cultured in neutral tissue was used for mass immunization in Senegal and the chick embryo vaccine was tested in Brazil. The vaccines were administered by scarification using normal human serum as stabilizer. The chick embryo cultured vaccine (first the 17 E vaccine and later the 17 D vaccine) has been most widely used. Numerous problems and accidents were associated with both vaccines. The human serum used as stabilizer caused hepatitis affecting many vaccinees in the armed forces during World War II. Systemic reactions were common. Severe postvaccination encephalitis with a high mortality rate occurred mainly with the neural tissue cultured vaccine. This reaction (mostly in children $<$ one year of age) was also reported with some lots of the 17 D vaccine with increased neurotropism; furthermore, loss of protective efficacy was noted in tropical climates due to low thermostability. The 17 D vaccine is the result of developments aimed at careful definition of the properties of the seed virus and at improved thermostability [215].

Production and Properties of Yellow Fever Vaccines. Only certain institutes are approved by the WHO for production of yellow fever vaccine[159, pp. 34 – 53]. The seed lot virus, usually a substrain of 17 D-204, has to be shown to be free from neurotropism by testing in monkeys[159, pp. 34 – 53],[182, pp. 113 – 141],[216] . Most producers use seed lots that are free of leukosis virus for production but their use is not mandatory. Virus-infected embryos are harvested, homogenized, and the supernatant is used as vaccine. Several tests for adventitious agents are performed. Virus titrations are performed in a mouse assay with a minimal potency requirement of 1000 LD_{50} per human dose. The vaccines are lyophilized in the presence of stabilizer. Most current vaccines retain the minimal requirement for two weeks at 22 °C. A vaccine stable for two weeks at 37 °C is requested by WHO for use in tropical areas.

Immunization Recommendations. Vaccination of visitors to endemic areas in equatorial Africa and northern parts of South America is recommended. Immunization is mandatory in several countries for visitors from endemic areas. One 0.5 mL injection is given to both adults and children for both primary and booster immunization. Booster injections are recommended every ten years. The vaccine is not recommended for children under one year of age or for pregnant women.

Adverse Reactions. Yellow fever vaccines are safe and induce only minor local reactions. Transient headache can occur.

Protective Efficacy. Current vaccines are estimated to be 90 – 95 % protective. Immunization has decreased or eliminated the disease in many endemic areas.

3.11. Tick-Borne Encephalitis Vaccine

Etiological Agent and Pathogenesis. The tick-borne encephalitis virus, a member of the toga virus group, was first isolated in the former Soviet Union in 1937 [212]. Two antigenically distinct forms of the virus cause the disease in Europe and in the eastern former Soviet Union. The main vector for the European form is *Ixodes ricinus* and for the Eastern form *I. perulcatus*. Many wild and domestic animals can be infected; the main animal reservoirs are small mammals such as field mice. Transmission is usually by tick bite but infection can occur by drinking untreated cow milk. The incubation period is 7 – 14 days before onset of fever and malaise. After a 1 – 2 week recovery period, a second stage with fever and neurological symptoms can follow: meningitis (40 % of cases), meningoencephalitis (40 %), and severe meningoencephalomyelitis (20 %). Mortality rates are 1 – 2 % in the European form and 20 – 25 % in the Eastern form; neurological sequelae are seen in 15 – 40 % of cases. The disease is subclinical or abortive with only the first stage in 75 % of infections. Tick-borne encephalitis is endemic in Central Europe, in the Balkan countries, and in Finland and Sweden.

History of Immunization. The first inactivated (killed) vaccine was developed and used in humans in the former Soviet Union in 1939 followed by an attenuated, live vaccine in the 1960s [217]. An inactivated vaccine, developed in Europe with a virus strain isolated from a tick in Austria, was subjected to clinical trials in 1973 [218]. The vaccine used in Western Europe is a purified, concentrated version of this.

Production and Properties of Tick-Borne Encephalitis Vaccine. The inactivated vaccine is produced by propagation of the virus in hens' eggs, followed by purification by continuous flow zonal ultracentrifugation, and inactivation with formaldehyde. No WHO requirements have been formulated. The vaccine contains not less than 25 protective doses per human dose assayed in a mouse protection test. Human albumin is used as stabilizer and aluminum hydroxide as adjuvant.

Immunization Recommendations. In Austria and Bavaria, immunization of children older than one year is recommended. Other endemic countries recommend immunization of forest workers and other high-risk groups. Primary immunization consists of two 0.5 mL intramuscular injections at 1 – 3 month interval, followed by a third 0.5 mL dose 9 – 12 months after the second. Booster injections are recommended every three years.

Adverse Reactions. Local and systemic reactions are rare and mild. Low-grade fever is occasionally seen, mainly in children after the first injection.

Protective Efficacy. The vaccine is at least 95 % protective against all European virus strains in children and young adults. Seroconversion rates of about 90 % are reported for persons over 65 years of age.

3.12. Japanese Encephalitis Vaccine

Etiological Agent and Pathogenesis. The Japanese encephalitis virus, belonging to the toga virus group, was first isolated in 1935 in Japan [212]. The disease is transmitted by the mosquito *Culex tritaeniorhynchus*; the main animal reservoirs are pigs and birds. The incubation period is 5–15 days. The disease is subclinical in at least 95 % of infections. In clinical cases symptoms vary from mild febrile disease with headache to severe encephalitis. Paralytic forms more commonly affect the upper extremities. In endemic areas the disease affects mainly younger children but also elderly people with mortality rates of 50 %. Neurological sequelae have been reported in 30 – 40 % of survivors of the severe clinical forms.

History of Immunization. A formaldehyde-killed vaccine, consisting of a 5 % suspension of infected mouse brain tissue, was introduced for human use in Japan in 1954 [219]. The vaccine (Nakayama strain) has been purified by protamine sulfate precipitation since the late 1950s and by absorption with charcoal or kaolin since the early 1960s. A protective efficacy of 80 – 90 % for the vaccine was shown. The highly purified, mouse brain cultured vaccine produced in Japan is also used for immunizing travellers to endemic areas. In China, a vaccine produced in primary hamster kidney cells has been extensively used since the late 1950s.

Production and Properties of Japanese Encephalitis Vaccines. Several Japanese manufacturers produce the purified vaccine from culture of mouse neural tissue by similar methods. No WHO requirements have been formulated. A vaccine used for immunizing travellers to endemic areas is prepared by infecting mouse brain with the Nakayama strain. The brain homogenate is purified by protamine sulfate treatment and then inactivated with formaldehyde. Further purification involves ultracentrifugation on a sucrose density gradient. The vaccine is lyophilized; the reconstituted vaccine must be used immediately.

Immunization Recommendations. Primary immunization consists of subcutaneous injection of two 1 mL doses at a 1 – 2 week interval. A third 1 mL dose is recommended one month later as is a regular booster injection every 1 – 3 years. Extensive immunization in endemic areas has been considered by the WHO. Children less than three years of age should receive 0.5 mL doses. Immunization is generally recommended for health care workers and other people with extended stay in endemic areas.

Adverse Reactions. Only a few mild, local and systemic reactions have been reported.

Protective Efficacy. Immunization has drastically reduced disease incidence in Japan. The protective efficacy is estimated to be 90 – 95 % from serologic studies.

3.13. Smallpox Vaccine

Smallpox (variola) was caused by a member of the pox virus group, which also includes the vaccinia virus used for immunization. The disease was one of the most devastating infections in human history. Eradication of smallpox is the success story of immunization. In 1967 the WHO launched a massive eradication program–smallpox was still reported from 42 countries. In May 1980, WHO officially declared the world free from smallpox. No proven cases of variola have occurred in the past decade.

Two types of vaccine were manufactured, calf lymph vaccine and egg vaccine [220]. Both liquid and lyophilized forms were used. Requirements for manufacturing, control, and potency were formulated by the WHO. Immunization by multiple puncture with a bifurcated needle was most commonly used. Severe adverse reactions included postvaccination encephalitis and disseminated vaccinia.

General immunization against smallpox was withdrawn in most European countries in the mid 1970s. Requirement for smallpox vaccination was abandoned for international travel in 1982.

Recommendations for civilian immunization in the United States include only laboratory workers handling variola virus or other closely related orthopox viruses. Military personnel in the United States and the former Soviet Union are routinely vaccinated against smallpox [221].

3.14. Rift Valley Fever Vaccine

Rift Valley fever is an arthropod-borne disease known only in Africa [212], [213]. It is caused by a member of the Bunya virus group. The major vectors are mosquitoes (*Culex theileri* and *Aedes cabbalus*). The main natural hosts are sheep, cattle, and goats. The incubation period is 2 – 6 days. Rift Valley fever is a febrile disease with headache and abdominal pain lasting less than one week. Hemorrhagic fever with liver necrosis and encephalitis are the severe manifestations causing mortality and sequelae. Outbreaks occurred in Africa during the 1970s, with the largest outbreak in Egypt in 1977 – 1978.

An inactivated (killed) vaccine (NDBR-103) was produced by the US army in 1967 [222]. The Entebbe strain of the virus was grown in primary monkey kidney cells, inactivated with formaldehyde, and lyophilized. The adverse reactions are few and mild. One case of Guillain – Barré occurred in a Swedish military vaccinee [223]. Seroconversion rates after subcutaneous injection of three 1 mL doses given at 1 – 2 week intervals were over 95 %. A newer vaccine (GSD-200) is based on a cloned version of the original seed virus (Entebbe strain) and grown in diploid rhesus monkey cells. The WHO has formulated requirements for inactivated Rift Valley fever vaccines produced in primary monkey kidney cells and in human or non-human primate diploid cells[141, pp. 104 – 143]. No minimal potency requirements have been formulated.

4. Vaccines against Parasites

4.1. Vaccines against Helminths

Helminths represent one of the major causes of infectious diseases affecting humans and domestic animals. This results not only in a deleterious effect to the health of the hosts, but also in great economic losses. Although major advances have been made in the chemotherapy and epidemiology of diseases caused by helminths immunotherapy has produced only minor breakthroughs in the field of veterinary parasitology.

As a result of a long evolutionary development and a close parasite – host relationship, helminths have evolved strategies for circumventing complete elimination by the host immune response. Nevertheless, the immune response usually exerts deleterious effect upon their growth and proliferation. Thus, the use of a vaccine resulting in partial or complete protection might be one of the most cost-effective means of controlling helminth diseases.

Only a few reliable vaccines are available on a commercial scale for immunoprophylaxis of helminthoses in livestock. A vaccine against lungworms in cattle is a commercial success. Another vaccine against hookworms in dogs, although immunologically efficient, has failed commercially.

Important in the development of vaccines is the identification, isolation, and testing of putative protective antigens. Successful vaccination against helminth infections requires the priming of those responses which may subsequently be triggered during natural infection. Since antigen presentation plays a central role in the acquired immune response, the development of accessory cells and the activation of T and B lymphocytes has to be taken into account [224]. Developments have focussed on vaccines produced by recombinant DNA techniques. In vitro cultivation methods have been used for producing helminth antigens for immunoprophylaxis and combined with recombinant DNA technology [225].

The helminthoses described in this chapter were chosen on the basis of the parasites' importance to human health or because of their interesting biology.

4.1.1. Vaccines against Schistosoma

Etiologic Agents, Pathogenesis, and Epidemiology. Human schistosomiasis occurs primarily in tropical countries where it is one of the most threatening diseases. The infection affects about $(200-300)\times 10^6$ people throughout the world; more than 600×10^6 people live in *Schistosoma*-endemic areas. The importance of schistosomiasis has also increased in industrialized countries with intensive tourism and influx of high numbers of refugees from endemic areas [226].

The main causative agents of schistosomiasis are helminths of the genus *Schistosoma* which use aquatic or amphibious snails as intermediate hosts:

1) *Schistosoma haematobium* uses aquatic snails of the genus *Bulinus* as intermediate hosts and occurs mainly in Africa and some middle-eastern countries.
2) *S. mansoni* uses aquatic snails of the genus *Biomphalaria* as intermediate hosts; it occurs in Africa, parts of Arabia, northern and eastern parts of South America, and some Carribean islands.

3) *S. japonicum* uses amphibious snails of the genus *Oncomelania* as intermediate hosts; it occurs in Japan, the Philippines, and parts of China, Thailand, and Indonesia.

The life cycle of all three *Schistosoma* species is similar. The fully developed *miracidium* hatches from the egg in water and infects the intermediate host (i.e., the snail) where it multiplies asexually to produce numerous *cercariae*. The cercariae are released into the water and infect humans by penetration through the skin. They develop into immature worms (schistosomula) which migrate to the lungs and liver. The adult, sexually mature worms mate and migrate to their final destination which varies according to the species: *S. haematobium* migrates to the veins of the vesical plexus, *S. mansoni* and *S. japonicum* to the mesenteric veins. The adult paired worms produce eggs (300 – 3000 per pair per day) which pass through the vessels and tissues into the lumen of the gut and bladder. The eggs escape from the host in the feces and urine and the cycle is repeated.

Eggs from *S. haematobium* are mainly found in the bladder and urogenital tract causing hematuria and fibrosis of the bladder. In severe cases malignancies may develop. Eggs from *S. mansoni* and *S. japonicum* are trapped in the liver and bowels causing hepatomegaly. Subacute disease is probably due to the passage of worms through lungs leading to cough, pulmonary infiltrates, and fever [227]. Mainly in *S. japonicum*, acute schistosomiasis occurs 5 – 7 weeks after heavy primary infection. The illness if often associated with diarrhea, fever, hepatosplenomegaly, resulting in liver fibrosis and ascites.

Immunity and Vaccine Design. An age-dependent resistance to reinfection after chemotherapy was demonstrated with *S. haematobium* and *S. mansoni* [228], [229]. This is associated with an increased lymphocyte proliferation in response to egg, cercarial, and adult worm antigen [230], [231]. Lack of reinfection is also related to the eosinophil count [228], [232]; human eosinophils mediate antibody-dependent damage to the schistosomula of *S. mansoni* [233]. The main immune mechanisms involved in schistosomiasis are summarized in Table 2 and reviewed in [234].

Table 2. Main cells and antibodies active against *Schistosoma* in experimental animal models or in vitro

Developmental stage of parasite	Cells and antibodies involved	Mechanisms
Egg	effector T lymphocytes	initiation of granuloma formation (lymphokines)
Schistosomula	neutrophils, IgG	tegumental damage and killing of young larvae under certain conditions
Schistosomula	eosinophils, IgG	killing of young larvae in the presence of complement
Schistosomula	eosinophils, IgE	killing of young larvae
Schistosomula	macrophages, IgE	IgE-dependent cytotoxicity

The immune response may result in resistance to reinfection but not to simultaneous expulsion of an established parasite population from primary exposure [235]. In this way, the parasites evade the immune response. An alternative escape mechanism is that the worms may be covered by bound host IgG [236]. The immunoglobulin is partially cleaved by a parasite protease to produce peptides which may inhibit macrophage activation and thus depress macrophage-mediated, IgE-dependent destruction (cytotoxicity) of the schistosomula.

In the development of schistosomiasis vaccines, the findings regarding age-dependent host resistance will require application and efficacy at a very early age, before the child is exposed to natural infections [237]. The search for antigens that mediate protective immunity has concentrated on the exposed outer surface of the young schistosomula [238]. About 90 % of exposed epitopes consist of carbohydrates that crossreact with *Schistosoma* egg antigen. Antibodies to surface polypeptide antigen do not generally crossreact with egg antigen but are present on the surface membrane of adult worms. Monoclonal antibodies to some of these molecules seem to confer partial resistance when passively administered to animals and could thus be potentially protective antigens [239–241]. A range of candidate vaccine antigens of *S. mansoni* have been identified, several have been cloned and expressed in *Escherichia coli*, yeast, or vaccinia virus [238]. This important achievement will facilitate the production of large amounts of polypeptides for vaccination trials.

One of the most promising candidates is a schistosomula surface polypeptide with a molecular mass of 28 000. The gene coding for this polypeptide has been cloned [242] and expressed in *E. coli*. Immunization with this recombinant antigen induced a high level of serum cytotoxicity towards schistosomulas in the rat, hamster, and monkey. Significant protection against a natural challenge infection with live cercariae was obtained in rats and hamsters [242]. These results can be viewed with optimism for the development of a schistosomiasis vaccine.

4.1.2. Vaccines against Nematodes

4.1.2.1. Gastrointestinal Nematodes

The most prevalent and pathogenic gastrointestinal nematode parasites of humans belong to the genera *Ascaris, Strongyloides, Trichinella, Trichuris, Ancylostoma,* and *Necator.* Although many species parasitize deeper tissues of the body, the majority are intestinal. The intestine has been maintained as a site for adult stage development, whereas larval stages often invade other host tissues. The intestine consists of a series of distinct parasite habitats (gut sections, lumen, mucosa, etc.) each having its own characteristics. Large worms such as *Ascaris* must live within the lumen, smaller species such as hookworms are associated with the mucosa. For a long time research on immunity against intestinal worms was given low priority due to the lack of knowledge about local immune responses in the gut. Studies have shown that intestinal worms are indeed subject to protective immune responses, although these responses differ somewhat from classical immunity in the body because the worms live in the gut lumen. Special features are (1) macromolecular antigen uptake across the intact mucosal epithelium or by specialized epithelial cells overlying Peyer's patches and (2) complexation of the antigen by dimeric IgA secreted from the mucosa or intestinal IgG Fab fragments. Cells from the underlying lamina propria participate in cytotoxicity and hypersensitivity reactions and thus affect mucosal structure and function. Furthermore, a large variety of nonlymphoid effector cells occur within the intestine, including natural killer cells, macrophages, neutrophils, eosinophils and basophils; their numbers increase during parasite infections.

No vaccine against gastrointestinal nematodes is presently available for human application. However, promising, successful trials in veterinary parasitology have initiated interesting work in human gastrointestinal nematode infections and will probably result in vaccine supply in the near future.

Hookworm Disease. Various species of the family Ancylostomatidae are responsible for hookworm disease in humans, the most important being *Ancylostoma duodenale* and *Necator americanus*. Over a fifth of the world's population is afflicted by this disease, mainly in tropical and subtropical regions. Adult hookworms have a length of about 0.7 – 1.8 cm and a hook-like anterior end with a distinct mouth area. Females release eggs in the small intestine which are excreted in the feces of the host. In humid surroundings first-stage larvae hatch from the eggs and develop into infective larvae, which penetrate through the skin into new hosts. After migration through lymph or blood vessels, the larvae finally develop into mature adult worms in the small intestine, completing the parasite's life cycle. Hookworms suck blood from microlesions in the host's small intestine, thereby causing chronic gastrointestinal blood loss, anemia, and hypoalbuminemia. With a large worm burden death may occur. Patients usually suffer from extreme weakness, pallor, secondary respiratory tract infections, skin irritations, heart palpitations, and gastrointestinal distress.

Immunity and Vaccine Design. Nematode parasites present special problems for the host's protective immune response, because they possess a tough, protective, external cuticle. The cuticle is both antigenic and immunogenic, but it is doubtful whether responses directed against its surface play a major role in immunity against intestinal species. Protective responses are more likely initiated by antigens released through the orifices of the parasite [243]. One such antigen with a potential vaccine function is a secreted proteolytic enzyme [244]. Hookworms attached to the mucosa secrete the enzyme from glands in their mouths. The enzyme degrades the host proteins and inhibits blood coagulation which permits the hookworms to feed for an indefinite period of time. Immunizing the host against this enzyme would result in inhibition of any enzyme secreted by worms after a subsequent challenge infection. The worms would then be unable to feed. The gene coding for the enzyme in question has been cloned. Future investigations will have to demonstrate the applicability of this kind of vaccine.

A vaccine has been developed for the control of hookworm infections in dogs [245]. It was based upon irradiation-attenuated infective larvae and was administered parenterally. Although the vaccine was highly effective in preventing hookworm disease, it was withdrawn because it did not completely prevent infection, and effective anthelminthic chemotherapy was available. This example suggests that protective immunity may also occur in human hookworm infections.

4.1.2.2. Tissue-Invading Nematodes (Filariidae)

Many nematode species which live as adults in the intestine, (e.g., *Ascaris*, hookworms, and *Trichinella*) undergo development in parenteral tissues. Other species are wholly confined to these tissues and have no contact with the intestine (tissue-invading nematodes). This closed habitation site requires special conditions in order to obtain biological contact with the outside world, especially for reproduction. In the major group of tissue-invading nematodes, the Filariidae, the worms overcome this problem by using bloodfeeding arthropods as intermediate hosts. The female worms release microfilariae larvae which circulate in the blood or accumulate in the skin of the host. The arthropod feeds on the blood and takes up the microfilariae which develop into infective larvae. At a following blood meal, infective larvae are reinoculated into new human hosts.

Filariasis. The human disease filariasis comprises an extremely heterogeneous group of diseases. The main filiarial parasite species in humans are *Wuchereria bancrofti* and *Brugia malayi*, which are both transmitted by mosquitoes. The disease is widespread in tropical and subtropical regions affecting ca. 90×10^6 persons. Its early symptoms are fever, lymphangitis, and lymphadenitis. A following chronic stage is frequently characterized by more serious clinical manifestation including elephantiasis, hydrocele, and pulmonary eosinophilia. Adult worms are found in the lymphatic system, microfilariae may be found in blood.

Filiarial infections caused by *Onchocerca volvulus* occur in Africa and South and Central America, affects millions of patients, many of whom become blind. Adult worms are generally located under the skin, forming typical nodules; less often they penetrate deeply into the tissues. Pathology in onchocerciasis is due entirely to the microfilarial stage. Microfilariae are present in the skin and may penetrate into the eye, thus leading to severe eye lesions and blindness.

Loa loa is prevalent in the forest areas of West Africa. Adult worms penetrate into the tissues, provoking transient edema. Severe clinical symptoms are rare, loiasis being generally regarded as a benign infection [246]. *Dipetalonema perstans, D. streptocerca,* and *Mansonella ozzardi* infect humans (Africa, South America) but usually asymptomatically and rarely cause significant diseases [247].

Many of the changes associated with filarial infection are immunopathological in origin and hypersensitivity reactions are important in their development.

Immunity and Vaccine Design. Although filarial parasites provoke a strong immune response in the human host, the chronicity of these infections implies the absence of a protective response or the evasion of such responses by the worms. There is also no direct evidence that filarial infections in nature confer resistance to reinfection with the same parasite species.

No vaccines against filarial parasites are available. The development of new strategies for immunological control depends on a thorough understanding of immunological host – parasite relationship. Many studies on protective immunity in animals have concentrated upon responses directed against larval stages. A vaccine against microfilariae would inhibit transmission of the disease, a vaccine against infective larvae would provide protective immunity against primary infection of hosts. Inoculation with infective larvae attenuated by irradiation [248], [249], only confers partial protection to a subsequent challenge infection. Filarial vaccines based on irradiated larvae cannot be used in humans without first determining whether these attenuated larvae induce pathological changes [250].

The target antigens of antimicrofilarial immune responses are probably located on the surface of the microfilariae; they are currently being characterized [251]. Antibodies to surface antigens mediate adherence of host cells to the microfilariae; this can result in worm killing [243]. Using *Dipetalonema vitae* as a model, IgE was found to be the primary immunoglobulin involved in cell adherence to the worm cuticle. The first cell type to adhere is the eosinophil. Subsequent, adherence of macrophages is followed by release of lysosomal enzymes which degrade the cuticle. In humans, a comparatively long time is needed to establish such immunity. This may be due to pronounced immunosuppression or because the microfilariae cover their surface with host components and thus render their antigenic surface epitopes less accessible to the host's immune system. The search for filarial antigens that can safely and successfully be used in humans is still continuing; it is still not known how restricted series of antigens can be used to protect natural hosts against first or persistent infections with a complex, adaptable, genetically diverse parasite. Recombinant DNA technology and hybridoma technology may provide potential candidates for successful filarial vaccines.

4.1.3. Vaccines against Cestodes

Etiologic Agents, Pathogenesis, and Epidemiology. This section deals with the hydatidosis and cysticercosis disease complexes, which are caused by the larval stages of tapeworms belonging to the family Taeniidae. The most important causative agents of *hydatidosis* (echinococcosis) are *Echinococcus granulosus* and *E. multilocularis,* whose life cycles involve a definitive and an intermediate mammalian host. The definitive hosts are carnivores (mainly dogs for *E. granulosus* and foxes for *E. multilocularis*) in whose intestines the adult stage worms occur. Intermediate hosts are herbivorous and omnivorous species in which the larvae (metacestodes) develop. Humans and other intermediate hosts become infected by ingesting eggs passed in the feces of definitive hosts. The diseases caused by the metacestodes are referred to as cystic echinococcosis for *E. granulosus* and alveolar echinococcosis for *E. multilocularis. Echinococcus granulosus* is prevalent throughout the world and is a public health and economic problem in many areas. *Echinococcus multilocularis* only exists in the northern hemisphere and is relatively frequently seen in the former Soviet Union (Siberia), central Europe, northern China, Japan, and Alaska. The fully developed metacestode of *E. granulosus* is a typically unilocular, fluid-filled cyst, which is located in the liver, lungs, and other organs. *Echinococcus multilocularis* metacestode conforms a vesiculated parasitic mass in the liver of the host, it proliferates by continuous exogenous budding with possible metastasis formation in other organs.

The causative agent of *cysticercosis* is the tapeworm *Taenia solium*. Humans are the obligatory definitive hosts, pigs act as intermediate hosts. Metacestode stage infection can also occur in humans and may result in infection of the central nervous system by parasite larvae (neurocysticercosis). Morbidity includes intracranial hypertension, basal arachnoiditis, focal neurological deficits, and dementia. Hyperendemic areas are found in Latin America, Africa, and Asia; areas of lower endemicity occur in southern and eastern Europe.

Immunity and Vaccine Design. Host protective immunity is a striking feature of repeated infection with cestodes in mammalian intermediate hosts [225]. It plays a major role in regulating natural transmission of these parasites, and substantial research efforts have been undertaken towards development of vaccines against cestodes of veterinary importance [252]. For human cestode infections, it is debatable whether there is sufficient importance to warrant the research required to develop a vaccine. In areas of high egg contamination (e.g., the Turkana district (Kenya) for *E. granulosus*, St Lawrence Island for *E. multilocularis*, or Mexico for *T. solium*), vaccination should be given to very young persons, as patients usually become infected at very young age. An ideal human vaccine requires complete, long-lasting protection. Experiments in veterinary parasitology, however, demonstrated exactly opposite results. Vaccination of animals against *Taenia* species resulted only in marked reduction of cyst numbers which persisted only for a maximum of one year. In addition, strong adjuvants had to be employed which are not tolerated by humans. Other criteria may be important; a review of protective immune mechanisms is given in [253].

Little attention has been paid to the vaccination of definitive hosts (dogs and foxes for *Echinococcus*, and humans for *T. solium*). Experiments [254], [255] demonstrated an immune response after infection as well as highly significant suppression of egg production by *E. granulosus* after immunization of dogs with secretory antigens derived from adult tapeworms. This approach seems to be the most likely control measure for reducing infection risk in humans.

4.2. Malaria Vaccine

Malaria remains a major health problem in many tropical and subtropical countries and affects hundreds of million of people each year. A major effort was made to control malaria from 1950–1970 with insecticides (→ Insect Control) and antimalarial drugs (→ Chemotherapeutics, Chap. 3.3.). The initial remarkable results have been difficult to maintain. The advent of drug-resistant parasite strains and of insecticide-resistant mosquito vectors are major obstacles in the effort to control malaria. Since the early 1970s new approaches have been explored such as vector control through biological agents and control of malaria infection through vaccines.

The development of malaria vaccine has received considerable impetus: first, because immunization of animals with whole parasites can induce a degree of protection equal or superior to that induced following natural infection [256–258]; second, because in vitro culture systems have been developed for the blood and hepatic stages of the malarial parasite *Plasmodium falciparum* [259], [260]; and third, because monoclonal antibody and recombinant DNA techniques have been used to identify and produce the parasite polypeptides possibly involved in the development of protective immunity.

4.2.1. Strategy for Malaria Vaccine Development

More than 100 species of malarial plasmodia are known, but only four infect humans: *Plasmodium falciparum*, which is responsible for the majority of human deaths, *P. vivax, P. malariae*, and *P. ovale*.

The life cycle of the plasmodia is complex (Fig. 4). The female anopheline mosquitoes inoculate *sporozoites* into the blood of the vertebrate host. Within minutes the sporozoites invade the liver parenchymal cells (hepatocytes) where they divide asexually and develop into *merozoites* which rupture the hepatocytes and reenter the blood. In the subsequent *erythrocytic cycle*, the merozoites invade the red blood cells and mature into *schizonts* within 48–72 h depending on the species. The mature schizonts release merozoites which invade new erythrocytes. The erythrocytic cycle is responsible for the clinical manifestations of malaria. Some merozoites differentiate into sexual stages called *gametocytes* which are ingested by the mosquito. Fertilization of the gametes occurs solely in the midgut of mosquito. The resulting zygotes develop into *ookinetes* and then into *oocysts*. Sporozoites are released from mature oocysts and migrate to the mosquito salivary glands. The cycle is then repeated.

This complex life cycle involves continuous morphologic, enzymatic, and antigenic changes which are linked to the parasite's environmental adaptation to the host. The invasive stages of the parasite (sporozoites, merozoites, gametes) have unique, stage-specific surface determinants. Furthermore, immunologic crossreactivity exists between plasmodia species and between the developmental stages of a given species. However, immunization experiments in animal models have demonstrated that the antigenic determinants involved in protective responses are species- and stage-specific. Three types of malaria vaccine can therefore be devised, based on:

1) sporozoites,
2) asexual stages (merozoites, schizonts), and
3) sexual stages (gametes).

In addition, stagespecific parasitic antigens are expressed on liver cells containing merozoites [261] and may also be candidates for vac-

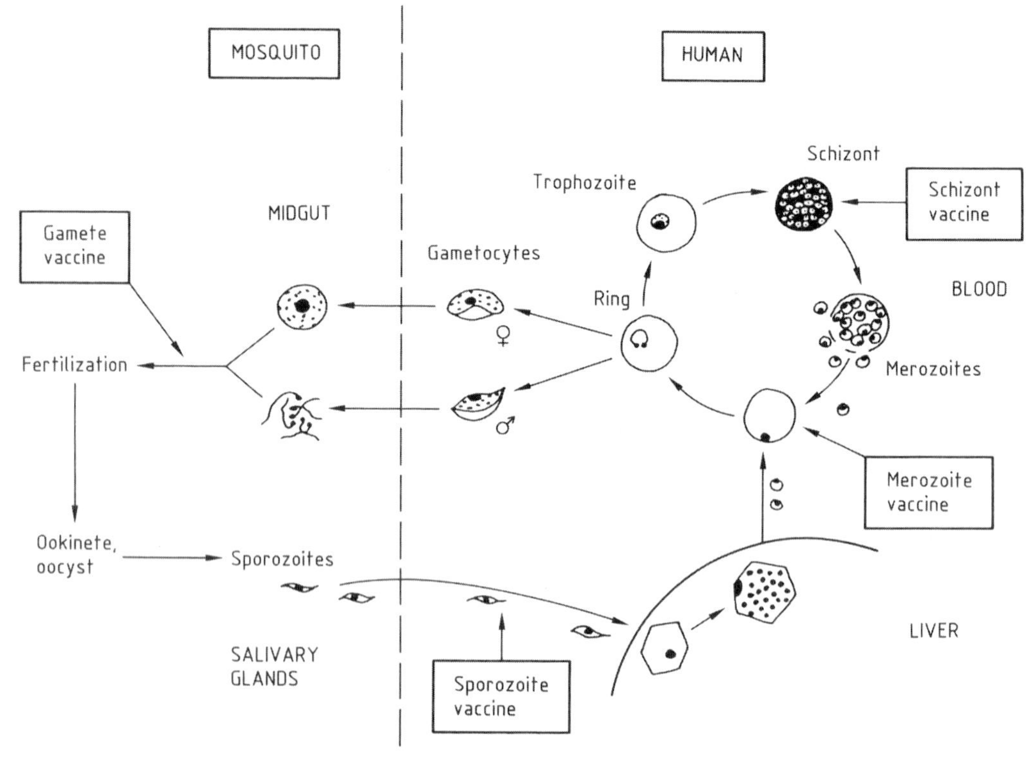

Figure 4. *Plasmodium falciparum* life cycle showing targets for vaccine development

cine development. A favored approach is the development of a multivalent vaccine containing components of several malaria stages. The selection of defined parasite components versus whole parasites (sporozoites, merozoites, gametes) for vaccine development is indicated by the following reasons:

1) Large-scale production and purification of whole parasites is not feasible.
2) Parasites cannot be obtained free of host components (e.g., mosquito salivary glands, erythrocyte membranes) which may induce adverse autoimmune reactions.
3) Most of the parasite components are irrelevant to the induction of protective responses and their inclusion may impair truly protective responses or induce immunopathologic lesions in the host.

The strategy for the development of malaria vaccine is as follows:

1) Identification and selection of malaria antigens that are the target of protective immune responses.
2) Functional, biochemical, and immunological characterization of these antigens, including identification of B and T cell epitopes.
3) Cloning of the genes coding for protective antigens, determination of DNA and amino acid sequences–what is the level of antigenic diversity?
4) Production of candidate protective antigens or epitopes by genetic engineering or chemical synthesis.
5) Evaluation of candidate protective antigens in terms of production of antigens, safety tests, adjuvants, carriers, etc.
6) Immunization trials in monkeys and human volunteers.

4.2.2. Sporozoite Vaccines

Following invasion of the hepatocytes, the sporozoites develop into thousands of merozoites, each of which may invade an erythrocyte. Obviously a vaccine which neutralizes sporozoites before their entry into liver cells or within liver cells would optimally prevent malaria infection. Vaccination with attenuated, irradiated sporozoites in rodents, monkeys, and humans induced complete protection against malaria [262], [263]. The induced immunity is species- and stage-specific but not strain-specific. It is at least partially mediated by antibodies as is shown by the protection against sporozoite challenge afforded by the passive transfer of antisporozoite monoclonal antibody [268] and by neutralization of sporozoite infectivity following incubation with the serum of protected animals. Deposition of specific antibodies on the sporozoite surface results in the formation of a tail-like precipitate; this is called the circumsporozoite (CS) reaction. Indirect evidence suggests that antisporozoite antibodies may play a role in human malaria [264], [265].

The control of sporozoite-induced infection also involves T cell-dependent mechanisms. In animal models, effector T cells can mediate antisporozoite immunity via antibody-independent mechanisms; for example, immunization with irradiated sporozoites can protect B cell-deficient mice against subsequent challenge infection with viable sporozoites [266]. Activation of malaria-specific T cells by malaria antigens induces the secretion of lymphokines which may act directly on the malaria parasite or indirectly by activating host effector systems [267].

The CS protein has been identified in several plasmodia species. Passive transfer of monoclonal antibodies directed against the CS protein protects mice from sporozoite challenge infection [268]. Indirect evidence suggests that CS protein is involved in the binding and penetration of sporozoites into liver cells. Similarly, fragments of monoclonal antibodies against CS prevent the attachment of sporozoites to hepatocytes in vitro [269].

The gene coding for the CS protein has been cloned in several plasmodial species [270], [271]. The protein contains a central block of tandemly repeated amino acids which vary in number and sequence among the different malaria species [272], [273]. For example, in *P. falciparum* the central area consists of 37 copies of the tetrapeptide asparagine – alanine – asparagine – proline (NANP) and four copies of the tetrapeptide asparagine – valine – aspartic acid – proline (NVDP). The regions flanking the repeats are more highly conserved between species than the repeats. Within a species, limited variations also occur outside the repeats [274].

The repeats cover the surface of mature sporozoites; for *P. falciparum* ca. 10^8 molecules of NANP are expressed on the membrane of mature sporozoites. The NANP repeats are the target of protective monoclonal antibodies passively transferred in vivo. The antibody response against sporozoites in humans is also mainly directed against the NANP repeats [275], [276].

In view of these findings, and because the NANP repeats are present on all the isolates of *P. falciparum* tested, two malaria vaccines based on NANP repeats have been prepared and tested on human volunteers. The first, produced by DNA recombinant technology, consisted of 32 repeats of NANP and NVDP fused to a 32 amino acid tail. The second was composed of three NANP repeats (NANP 3) conjugated to tetanus toxoid. Both formulations use aluminum hydroxide as adjuvant [277], [278]. The vaccines were safe and did not induce adverse reactions. Volunteers with high antisporozoite antibody titers were challenged with sporozoites and some were protected or presented a delay in appearance of parasitemia. Protection was shown in individuals with the highest antibody titers.

An antisporozoite vaccine must induce high antibody titers for a prolonged period and ideally a boosting effect should occur following exposure to sporozoites. Higher antibody responses can be obtained by changing the formulation and concentration of the immunogens and by using other adjuvants or live-attenuated vectors (vaccinia virus, salmonella) that carry and express the gene coding for the CS protein. Optimal antibody formation is dependent on the collaboration of T helper cells and B cells. In the two sporozoite vaccines tested, T cell help was provided by foreign protein (tetanus toxoid) but was unable to boost the antiNANP antibody response following exposure to sporozoites. More efficient sporozoite vaccines should contain T cell epitopes derived from sporozoites and ideally from the CS protein; in this context, proper help is provided by sporozoite-specific, primed T cells and can lead to optimal antiNANP antibody production by B cells. In a mouse model the response to some of the T cell epitopes of the CS protein is restricted by antigens of the major histocompatibility complex (MHC class II) [279–282]. In human populations, only three immunodominant epitopes are located in polymorphic regions of the CS protein outside the repetitive area. Since T cells have exquisitely specific reactivity, it follows that the polymorphism of T cell determinants may be responsible for a lack of proper help following exposure to sporozoites with T cell areas on CS that are different from those present on sporozoites responsible for previous infections in the same individual.

Therefore, it seems that more efficient vaccines should be based on either native malaria polypeptide(s) or cocktails of synthetic polypeptides containing multiple B and T cell epitopes. These epitopes should be selected in relation to constant and variant parasite components and in relation to epitope binding and recognition by components of the major histocompatibility complex of the human host.

4.2.3. Asexual Blood Stage Vaccine

The multiplication of asexual blood stages (merozoites and schizonts) is responsible for the morbidity and mortality associated with malaria. The level of parasitemia usually correlates with the severity of malaria infection. Immunity to malaria is mostly acquired but natural immunity also plays a role. Several single-gene disorders affecting erythrocytes, (e.g., sickle cell anemia, the thalassemias, and glucose phosphate deficiency) reduce the severity of malaria infection. Another genetic characteristic, the lack of the Duffy blood group antigens, is associated with complete resistance to *P. vivax* infection [283]; the Duffy blood group antigen or a closely associated antigen may be the receptor for *P. vivax* merozoites at the surface of erythrocytes.

The development of acquired resistance to malaria depends on the frequency and duration of the exposure to the parasite [284]. In endemic areas, babies born to immune mothers are resistant to malaria during the first three months of life due to the presence of maternal antibodies transferred during gestation. Later they suffer from severe, recurrent attacks; most deaths due to malaria occur in young children. From adolescence to adulthood there is a decrease in the severity and frequency of malaria attacks but sterile immunity is probably never achieved. In this context two types of vaccine based on asexual blood stages can be envisaged; (1) a vaccine which is more efficient than nature and leads to sterile immunity (i.e., infection no longer detectable) or (2) a vaccine capable of attenuating the parasite load by transforming the immune system of a non immune individual into that of an adult living in an endemic area.

The immune response to blood stages is complex and is directed against several antigens. Both antibody-mediated responses and cell-mediated, antibody-independent responses control asexual blood-stage infection. In humans, passive transfer of immunoglobulins purified from the sera of immune adults abort malaria infection in nonimmune infected children [285]. The antibodies may react with the surface of the merozoites and provoke their lysis upon addition of complement, or enhance their phagocytosis by mononuclear cells, or simply inhibit the binding of merozoites to erythrocytes. Other targets for antibodies are antigens on the surface of erythrocytes containing schizonts; binding of antibodies to schizonts may also lead to their destruction by phagocytosis [286] or induce the endothelial release of schizonts which may be later destroyed in the spleen [287]. Immunity to asexual blood stages also operates through a variety of antibody-independent mechanisms: T cell-dependent release of lymphokines, induction of oxidizing radicals leading to intracellular death of the malaria parasites, and activation of mononuclear cells in the spleen.

Immunization with merozoites and/or schizonts in a variety of plasmodia – host systems resulted in partial to almost complete protection [288]. Subsequent investigations were aimed at the characterization of components capable of inducing immunity. Several hundreds of asexual blood stage components can raise an immune response but only very few of the evoked responses are helpful to the host. Characteristics of some of the candidate antigens for asexual bloodstage vaccines are discussed in Sections 4.2.3.1, 4.2.3.2, 4.2.3.3, 4.2.3.4.

4.2.3.1. Merozoite Surface Antigens

A protein with a molecular mass of 190–200 kDa has been identified at the surface of *P. falciparum* schizonts and merozoites [289], [290]. During maturation of schizonts this polypeptide is processed into several components, one of them (M_r83 000) being the main surface component of the merozoites [290]. An important feature of the 190–200 kDa protein is its genetic polymorphism [291–294]. The gene coding for the protein can be divided into blocks ranging in homology among different *P. falciparum* isolates from 10–87 % at the amino acid level. A relatively short region of variable tripeptide repeats is found close to the N terminus. The blocks encoding for the N and C terminal sequences are highly conserved.

Immunization with the 190–200 kDa protein derived from *P. falciparum* in monkeys [295–297] and with an analogous protein from *P. yoelii* [298] can induce at least partial protection. Immunization with synthetic polypeptides corresponding to defined parts of the molecule (for example, the N terminus and amino acids 277–287) also confer partial protection in monkeys [299], [300].

An antigen with a molecular mass of 51 kDa is also expressed at the surface of *P. falciparum* merozoites and is the target of inhibitory monoclonal antibodies. It contains variant and constant epitopes for various *P. falciparum* isolates.

There is considerable antigenic diversity among *P. vivax* isolates as regards the components exposed at the surface of merozoites.

4.2.3.2. Rhoptry Antigens

Rhoptries are apical organelles of the merozoites which release their contents onto the erythrocyte membrane during invasion. A monoclonal antibody directed against a rhoptry protein of a rodent malaria, *P. yoelii*, reduced the virulence of the infection and a monoclonal antibody directed against 82 and 41 kDa components of *P. falciparum* inhibited the growth of *P. falciparum* in vitro [301–303]. The 82 kDa component is processed into 82 and 65 kDa components [304].

The 82 kDa polypeptide is membrane-bound through a glycosyl–phosphatide–inositol anchor; hydrolysis of its anchor activates the proteolytic activity of the 76 kDa polypeptide and may play a role in the invasion of erythrocytes by merozoites [305]. The 41 kDa polypeptide displays aldolase activity [306]. Interestingly, both the 76 kDa and the 41 kDa components can induce at least partial protection against *P. falciparum* infection in monkeys [307], [308]. The gene coding for the 41 kDa protein has been cloned and presents two interesting characteristics in terms of vaccine development; absence of variable amino acid repeats and almost complete conservation of the amino acid sequences among isolates from *P. falciparum* [306].

Another rhoptry antigen of *P. falciparum* with a molecular mass of 225 kDa has been identified in the peduncle of the rhoptries. It is synthesized as a 240 kDa polypeptide which is processed into a 225 kDa protein during schizogony and is quantitatively recovered in the culture supernatant following merozoite invasion [309].

A third set of rhoptry-associated proteins is the 105–130–140 kDa complex composed of three coprecipitating but unrelated proteins [310]. The 225 kDa proteins and the 105–130–140 kDa complex have not been evaluated in immunization trials.

4.2.3.3. Antigens Associated with the Membrane of Infected Erythrocytes

The ring-infected erythrocyte surface antigen (RESA) is a *P. falciparum* antigen with a molecular mass of 155 kDa. It is synthesized in trophozoites, accumulates in the merozoite, and following invasion becomes associated with the membrane of erythrocytes containing ring forms of the parasite [311], [312] but is not directly accessible on the external erythrocyte surface. AntiRESA antibodies inhibit the multiplication of asexual blood stages in vitro and may interfere with the invasion process [313]. The gene coding for RESA has been cloned and sequenced [314]. It contains two blocks of repetitive amino acid sequences which are the immunodominant regions of the molecules in terms of antibody response. Antibodies directed against the RESA repeats crossreact with at least six other asexual blood stage components. Aotus monkeys have been immunized with fusion proteins corresponding to various areas of the RESA and with synthetic polypeptides corresponding to the repetitive sequences [315]. Partial protection was observed in some groups of animals; work is in progress to optimize the efficacy of immunization based on RESA-derived molecules.

Erythrocytes containing mature asexual blood stages of *P. falciparum* attach to endothelial cells lining the venules of deep tissues. This cytoadherence of mature parasites prevents their passage through the spleen and thus their exposure to localized destructive mechanisms. Electron-dense protuberances (knobs) on the plasma membranes of infected erythrocytes are implicated in cytoadherence [316]. The genes coding for two knob components (knob-associated histidine-rich protein M_r85 – 105 kDa) and mature parasite-infected erythrocyte surface antigen (M_r240 – 300 kDa) have been cloned [317], [318]. The two proteins differ antigenically among isolates and contain repeated amino acid sequences. Cytoadherence can be inhibited by antisera in a strain-specific manner [319]. However, an antigenically invariant epitope has also been identified on the surface of infected erythrocyte isolates and may be an important antigen for vaccine development [320].

4.2.3.4. Other Proteins and Synthetic Peptides

A number of other antigens are also candidates for vaccine development. *P. falciparum* requires exogenous iron in the form of ferrotransferrin. A malaria transferrin receptor at the surface of infected erythrocytes transports bound ferrotransferrin to the parasite and may be used as a target for the vaccine [321]. Glycophorins exposed at the erythrocyte surface may act as ligands for *P. falciparum* merozoites and *P. falciparum* proteins have been identified which either bind to glycophorins or to human erythrocytes [322], [323]. A prominent antigen of *P. falciparum* with an apparent molecular mass of 126 – 140 kDa is associated with merozoite release. The gene coding for this protein has been cloned and contains at least two stretches of amino acid repeats, one being composed of polyserine repeats [324]. Monkeys immunized with this protein are protected from a lethal challenge infection [295].

Several immunization trials have been conducted in monkeys using synthetic peptides derived from asexual blood stages of *P. falciparum* coupled to carrier proteins [298], [299], [315]. A partial protective response was observed with peptides corresponding to various areas of the 190 – 200 kDa protein, to RESA, and to fragments of parasite components identified by their molecular mass of 55 and 35 kDa [299]. Synthetic hybrid polymer – proteins containing several peptides corresponding to epitopes of 195 – 200 kDa, RESA, 55 kDa, 35 kDa, and CS protein have been used for immunization of human volunteers [324]. The vaccine was well tolerated and no adverse effects were observed. All the immunized and control volunteers had patent parasitemia but the majority of the immunized volunteers were able to control their parasitemia in the absence of drug therapy. The immune response was low in terms of specific antibody production, and cell mediated responses as measured by proliferation assays was undetectable.

4.2.4. Sexual Stages–Transmission Blocking Immunity

The transmission of malaria from the vertebrate host to the mosquito vector is effected by sexual parasite stages–the gametocytes–which develop from merozoites. Within the vertebrate host the gametocytes are surrounded by the erythrocyte membrane; following ingestion by the mosquito vector, the gametes become extracellular. The female gametes are fertilized by the male gametes in the midgut of the mosquito to produce zygotes which develop into ookinetes. The ookinetes penetrate the midgut wall where they remain to form oocysts in which the sporozoites develop.

In the vertebrate host, the sexual stages do not produce illness, and their intracellular localization prevents direct attack by host effector mechanisms. Within the midgut of the mosquito the extracellular gametes are exposed to antigamete and/or antizygote antibodies from the vertebrate host that are ingested with the mosquito's blood meal. The antibodies partially or completely prevent the development of sexual stages and subsequent production of sporozoites. Transmission of the parasite is therefore blocked. This phenomenon is termed *transmission blocking immunity*.

The development of vaccines based on sexual blood stages could have an important impact on the epidemiology of malaria in endemic areas by reducing the level of malaria transmission. Ideally, transmission blocking vaccines have to be used in combination with vaccines based on sporozoite and/or asexual blood stages (see Sections 4.2.2 and 4.2.3).

Transmission blocking immunity has been induced by immunization with extracellular gametes in several animals [325–328]. The induced antigamete response is long lasting and in some cases is boosted by malaria infection, probably due to the gametocyte antigens in the circulation of the vertebrate host [328], [329]. There is also evidence that in *P. vivax* malaria in humans the antigamete response is boosted during natural infection [330]. Addition of sera of previously infected individuals to gametes can prevent fertilization and development of oocysts in mosquitos.

Specific targets for antigamete immunity have been identified using monoclonal antibodies in species including the human parasites *P. falciparum* and *P. vivax* [331], [332]. The antibodies act against the gametes by preventing fertilization and against the zygotes and ookinetes by preventing further development. In *P. falciparum* the target antigens for inhibition of fertilization are polypeptides with a molecular mass of 230 kDa and 45 – 48 kDa [331], [332]. Some of the epitopes on the 45 – 48 kDa antigen have been defined and are the targets of inhibitory monoclonal antibodies [333]. New antigens are expressed at the surface of the zygote and ookinete and one of them (M_r= 25 kDa) is a probable target of inhibitory monoclonal antibodies [331].

5. Immunotherapy

Since the late nineteenth century considerable progress has been made in our concepts of passive immunotherapy and in the development of preparations safe for human use; however, the proper role for such therapy in clinical medicine still needs to be defined.

By 1900 immune serum from various animal species had been used to treat pneumonia, tetanus, diphtheria, and rabies. Human serum was first used in 1907 by CENCI for the modification of measles and later for mumps and pertussis [334]. Placental extracts prepared by ammonium sulfate precipitation [335], [336] were also employed and may be considered as the first immunoglobulins prepared for human therapy [336]. Placental material is still used as a source of immunoglobulin.

The serious hypersensitivity reactions associated with animal serum proteins and the risk of viral hepatitis with convalescent human serum limited the use of serum therapy to life-threatening infections. In the 1920s attempts were made to separate the immune substances from animal serum by alcohol or acetone treatment. One such preparation, Huntoon's antibody solution, was administered intravenously to over 400 patients without the occurrence of anaphylaxis or serum sickness; however, pyrexia, cyanosis, and dyspnea did occur and were implicated in the deaths of three patients [337]. The introduction of antibiotics in the 1930s decreased the demand for serum therapy [338]. During this period, however, the experimental basis for combination therapy with antimicrobials and hyperimmune animal serum was established [339], [340].

A wide variety of biological products is now available for immunotherapy, the most important being purified gamma globulin for intramuscular injection (standard immune serum globulin, ISG) and globulin for intravenous injection (standard intravenous immunoglobulin, IVIG). Hyperimmune globulins with a high antibody titer against specific pathogens are also used. Additional preparations include antitoxins, plasma, and other blood products.

5.1. Gamma Globulin Preparations

5.1.1. Standard Immune Serum Globulin

Historical Aspects. In 1936 ARNE TISELIUS separated serum proteins into four major fractions by electrophoresis; subsequently he and KABAT found that immunoglobulin occurred predominantly in the gamma electrophoretic fraction [341]. COHN and colleagues devised a procedure for recovering immunoglobulins (gamma globulins) from serum on a large scale.

The serum proteins were fractionated by precipitation with ethanol under carefully controlled conditions of pH, temperature, protein concentration, and ionic strength [336], [342]. This process enriched and stabilized the gamma globulin from plasma at a relatively uniform antibody content while also denaturing most viruses [334]. Such gamma globulin prepared from large pools (>500 donors) of donor plasma were used in the treatment of infectious diseases during World War II [343].

Shortly after World War II gamma globulin was shown to contain antibody titers adequate for the prevention or attenuation of measles [344], infectious hepatitis [345], and polio [346]. Since the report of hypogammaglobulinemia in 1952, and the demonstration that gamma globulin administered on a monthly basis decreased the incidence of infection [347], antibody replacement of this deficiency has been routine.

Properties of Gamma Globulins. Gamma globulins occur at a serum concentration of 600 – 1200 mg/100 mL in adults, they represent approximately 11 – 14 % of total serum proteins [341] and 80 % of serum antibody [348]. Immunoglobulin G has a half-life in the normal circulation of approximately 25 days (35 – 40 days in patients with agammaglobulinemia) and is synthesized by adults at a daily rate of 35 mg/kg body weight [348]. Its rate of synthesis is regulated by serum IgG levels. Immunoglobulin M and IgA have lower serum concentrations and shorter half-lifes than IgG. The molecular mass of IgG has been estimated to be 145 000 [341] with 2.5 wt % being carbohydrate that is associated with the heavy chain (see also Fig. 5).

The IgG isotype can be divided into four subclasses (Table 3). The most abundant is IgG 1 (60 – 70 % of total IgG) which binds to the Fc receptors of neutrophils and mononuclear cells and to the first component of complement [349]. The IgG 2 subclass (20 – 30 % of total IgG) activates the classical complement pathway poorly but can activate complement by the alternate pathway. It is more resistant to proteolysis than the other subtypes [348]. The antibody response to pneumococcal and hemophilus polysaccharides may be related to the preimmune levels of this subclass [350]. The subclasses IgG 3 and IgG 4 bind to the Fc receptors of phagocytes (mononuclear cells and neutrophils) and basophils, respectively [349]. The IgG 3 has a short half-life (nine days); IgG 4 is unable to bind complement.

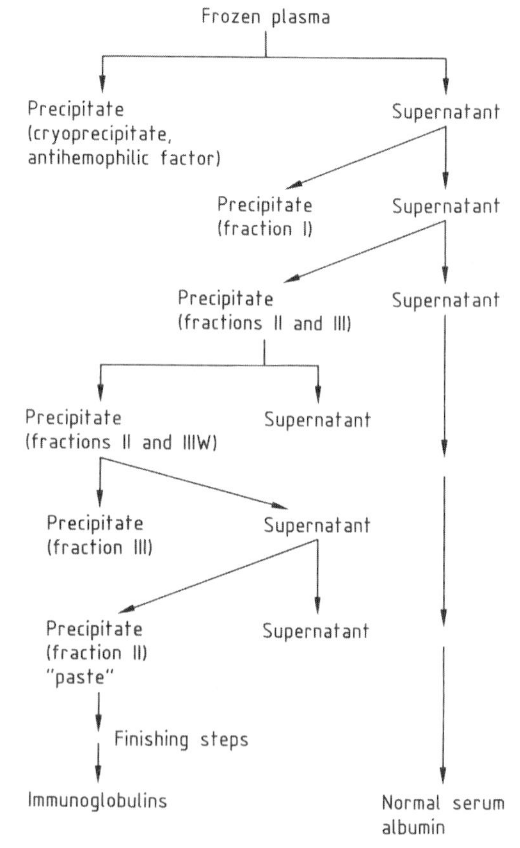

Figure 5. Preparation of immunoglobulin by Cohn – Oncley fractionation ("6/9 method")
Cohn fractionation of plasma by cold – alcohol procedures yields a fraction II precipitate or "paste". This is followed by applying the Oncley procedure to Cohn fraction II material to give immunoglobulin. The procedure is based on solubility differences that depend on ethanol concentration, ionic strength, pH, temperature, and protein concentration. Precipitates are usually collected by centrifugation.

Preparation of Immune Serum Globulin. Standard immune serum globulin (ISG, gamma globulin) is prepared from the plasma of pools of >1000 donors (see Fig. 5); → Blood, Chap. 2.2.5.). This minimizes differences between individual antibody levels to specific antigens and ensures a broad range of antibody specificities.

Table 3. Immunoglobulin G subclasses *

Property	IgG 1	IgG 2	IgG 3	IgG 4
Percentage of total IgG (adult)	60 – 70	20 – 30	5 – 10	<5
Molecular mass (daltons)	146 000	146 000	170 000	146 000
No. of interchain disulfide bonds	2	4	11	2
No. of amino acids in hinge region	15	12	62	12
Half-life (days)	21 – 23	20 – 23	7 – 8	21 – 23
Fc receptor binding				
mononuclear cells **	++	+	++	±
neutrophils	++	±	++	+

* Adapted from [239].
** ++ strong binding; + detectable binding; ± binding detected in some, but not all studies.

Standard immune serum globulin is usually prepared as a 16.5 % injectable solution (165 mg per mL) and represents a ca. 20 – 25-fold concentration of plasma IgG [336], [343]. This provides an effective dose of antibody in a relatively small volume [343]. Because of the viscosity of the preparation, it can only be given intramuscularly or subcutaneously [343], usually through a large gauge needle (16 – 18 ga). It has been estimated that 1 g of ISG contains 4×10^{18} antibody molecules with $>10^7$ specificities [334]. The product, although highly stable at 4 °C, can still undergo proteolysis, presumably due to plasmin contamination [336]. The ISG may also contain blood group substances, IgA, and IgG dimers. The presence of IgA can result in anaphylactic reactions in patients who lack IgA [336].

5.1.2. Immunoglobulin for Intravenous Use

Intravenously administered gamma globulin is needed to allow greater patient comfort, increase acceptability, and provide larger quantities of antibody. This is particularly important for patients who have a small muscle mass (children) or insufficient skin surface (burns), who are at risk from uncontrollable bleeds (Wiskott-Aldrich or other bleeding diatheses), or who need either large repeated doses (immunodeficient) or rapid onset of peak levels (intoxicated patients). Furthermore, intravenous administration avoids the local degradation of ISG antibody that may occur at the injection site [351].

In 1948 Cohn administered 25 – 50 mL of ISG intravenously and saw no adverse effects [336], [352]. Janeway and Cohn repeated the experiment with a new preparation and induced severe reactions after only 2 mL. This reaction was attributed to contamination of the preparation by staphylococcal enterotoxin [352]. The intravenous administration of ISG was repeated by Janeway in 1970 [336], and induced severe cardiovascular (tachycardia, arrhythmias, hypotension, severe chest pain), tachypneic, and pyrogenic (fever, chills, malaise) reactions [343]. These adverse effects were attributed to the presence of immunoglobulin aggregates that activated complement [353], [354].

To avoid the above-mentioned severe reactions, methods were introduced to rid the preparations of the high molecular mass immunoglobulin aggregates. Although they could be removed by centrifugation, reaggregation usually occurred and this method was not practical on a large scale [355]. Consequently, attention was turned to chemical modification.

Various methods of chemical modification have been tried. Degradation with the enzymes pepsin and plasmin resulted in the formation of antibody fragments with immunologic activity [336], [356]. In addition, since the proteolytic enzyme was not removed, it may have continued to be active in the absence of lyophilization; porcine pepsin also induced antibody formation [354]. Reaggregation was also prevented by acidification with hydrochloric acid [357] and reduction – oxidation with dithiothreitol, iodoacetamide, sulfite, or tetrathionite. Alkylation and acylation of immunoglobulin has also been accomplished with β-propiolactone.

The IVIG preparations that have been reduced and alkylated or otherwise modified may have impaired complement binding and altered subclass distribution [358], impaired opsonic activity in vitro [359], shortened serum half-life, and decreased protective efficacy [360]. Consequently, a new generation of native or intact IVIG preparations has been produced without modification by methods which include adjustment to pH 4, use of poly(ethylene glycol), ethanol precipitation, ultra- or diafiltration, and ion-exchange chromatography [353]. Four unmodified products are licensed in the United States: one prepared by adjustment to pH 4 in the presence of traces of pepsin, one prepared by ion-exchange chromatography and ultrafiltration, and a third by diafiltration, ultrafiltration, and adjustment of pH to 4 – 4.5. A fourth product (Venoglobulin-1 from Alpha Therapeutic Corporation) licensed in the United States but produced in Japan is prepared by poly(ethylene glycol) fractionation and ion-exchange adsorption.

Commercially prepared IVIG is usually stabilized with a mono- or disaccharide, such as 10 % maltose [338] or glucose. This considerably decreases the incidence of side effects that accompany the infusion of IVIG [336], [357], [361], perhaps by minimizing precipitation or aggregation. Maltose can cause a mild diuretic effect [362].

5.1.3. Hyperimmune Globulins and Antitoxins

Hyperimmune globulins are high-titered preparations of ISG against a specific antibody. They are prepared from the plasma of patients who have recently recovered from the disease or who have a high titer as a result of a previous vaccination or natural infection. Antitoxins for use in intoxication with botulinus or diphtheria toxin are prepared from snake venom or in horses [363].

5.1.4. Production Requirements

The WHO requirements stipulate that current immunoglobulin products

1) are prepared from pools of >1000 donors;
2) are free of kinins, plasmin, and prekallikrein activity;
3) have a low IgA content;
4) are as free as possible from immunoglobulin aggregates;
5) contain at least 90 % intact IgG without fragments;
6) are as unmodified as possible so that opsonic, complement, and other biologic activities are maintained;
7) contain all IgG subclasses;
8) have high levels of antibody to at least two bacterial species or toxins and two viruses (to be ascertained by neutralization tests), and
9) contain at least 0.1 International Units (IU) of antibody to hepatitis B and have a 1 : 1000 titer to hepatitis A [364].

In addition, manufacturers should specify the diluent and any chemical modifications used. Finally a product can only be classified as a hyperimmune immunoglobulin if the antibody level is five times that of standard ISG preparations.

There is no consensus on which laboratory tests should be used to predict the safety of IgG (contact activation, anticomplementary activity) [336]. Safety requirements should, however, consider the IgG half-life and virus transmission (particularly of human immunodeficiency virus HIV, but also of hepatitis B and non-A, non-B hepatitis viruses) [353].

5.2. Prophylaxis with Immune Serum Globulin

Passive immunization with ISG has been recommended for the short-term prevention of disease when vaccines for active immunization are unavailable or when active immunization was not given before disease exposure; it should be given before expected contact or early in the disease incubation. In these situations active immunization is always preferable because the antibodies subsequently formed by the vaccine provide longterm protection. Immune serum globulin is also used for antibody replacement in patients who lack adequate serum levels of immunoglobulin (i.e., hypogammaglobulinemia) [335].

The efficacy of ISG for the short-term prophylaxis of specific infections was established shortly after its widespread availability in the 1940s. These include:

Hepatitis. Stokes and Neefe showed that ISG could prevent or modify the course of hepatitis A [335], [336], [345], but Krugman found that ISG modified, but did not prevent this disease [365]. The ISG dose for treating hepatitis A is 0.02 mL/kg body weight [336]. A hyperimmune antibody preparation is also available commercially. In the case of hepatitis B, ISG is not recommended. Instead, hyperimmune globulin is given (see Section 5.3). Although ISG can decrease the incidence of post-transfusion hepatitis [366], its use for non-A, non-B hepatitis is considered optional [364].

Measles. The ability of convalescent serum to modify the course of measles was demonstrated in 1907 [335]. The ISG has also been shown to prevent or attenuate the disease [343]. The introduction of active immunization with a measles vaccine in 1963 (see Section 3.1) significantly decreased the incidence of this disease in the United States. Standard ISG is now recommended for infants under one year of age or for immunodeficient patients within six days of acute exposure to a case of measles [364].

Polio. If given early, ISG can modify the paralytic complications of polio [335], [346]. An unexposed individual should receive 0.15 mL/kg body weight.

Prevention of Infection in Patients with Hypogammaglobulinemia. An intramuscular ISG dose of 100 mg/kg given every 3 – 4 weeks is recommended for patients with hypogammaglobulinemia. This maintains a serum level of circulating IgG above 200 mg/mL [336] and confers protection against a wide variety of infections [342]. This minimum recommended dose is based on a study by the British Medical Research Council conducted between 1956 – 66 on 176 patients [367]. Although serum IgG levels at this dose rarely rise to the normal range [368], total replacement of IgG does not appear necessary for preventing infection [369]. Although a monthly dose of 200 mg/kg was found superior to 100 mg/kg, most patients did not tolerate more than the 100 mg/kg [361]. Two other studies also showed that higher doses of ISG decreased the incidence of acute infections [334]. Individualization of doses and their frequency has also been suggested [368].

Rubella. Standard ISG is unreliable in the modification of rubella [335], [337], [365]. High titers of antibody are needed [365]. Nevertheless, the use of ISG is recommended for women exposed to the disease during early pregnancy [364].

Clinical Studies. On the basis of data available in 1980, a committee of the WHO observed that it was "inappropriate" to use standard immunoglobulin for the prevention of infection in premature infants, during the physiologic hypogammaglobulinemia of infancy, or for malnutrition. Its use is contraindicated in patients with selective IgA deficiency [364]. Previous attempts in the 1960s showed that ISG did not prevent infection in a variety of clinical situations [370], [371] including multiple myeloma [372]. Although ISG was unable to prevent upper or lower respiratory tract infections in children or institutionalized, elderly adults, a significant decrease was seen in the incidence of mumps in children and fevers of unknown origin in adults [373]. Its use in the prevention of infection in burned patients has yielded conflicting data [335]. In a study on Peruvian children with burns over at least 10 % of their body surface area, there was a 41 % incidence of septicemia and 15 % mortality in control patients compared to a 21 % incidence of septicemia and 6 % mortality among those who received either plasma or ISG [374]. However, in another study in the United States administration of ISG did not beneficially affect either the rate of septicemia or mortality in burned patients [375]. More recently, patients admitted to a burn unit in India were randomized into groups that received active immunization against *Pseudomonas aeruginosa*, passive immunization with a cold ethanol precipitate of plasma from normal immunized volunteers, both immunologic treatments, or neither. In children, but not adults, there was a significant reduction in mortality following a daily dose of just over 20 mg protein for three days. There was a decrease in the incidence of not only *P. aeruginosa*, but also other gram-negative bacilli among those treated immunologically [376].

The ability of ISG to prevent infections has been most extensively studied in high-risk, premature infants. This population has a hypogammaglobulinemia that exposes them to a high incidence of bacterial infection [377]. The more premature the infant the lower the IgG, since most of the transplacental transfer of IgG occurs during the last six weeks of gestation [370]. Children do not attain adult levels of immunoglobulin until two years of age. In two studies using 0.5 – 3 mL/kg, no prophylactic effect was observed [371], [378], although there was a suggestion that the high dose regimen may have had a beneficial effect.

5.3. Prophylaxis with Hyperimmune Globulins

Several hyperimmune globulins are commercially available or undergoing development. Hyperimmune IVIG preparations are discussed in Section 5.5.

Diphtheria [342]. Diphtheria antitoxin is obtained from the blood of horses immunized against diphtheria toxin. Individuals should be tested for sensitivity to horse serum before being given the product. For prophylaxis, 1000 – 5000 IU of antitoxin is administered to Schick-positive individuals exposed to diphtheria. Higher doses (20 000 – 80 000 IU) are given for treatment.

Hepatitis. A hyperimmune globulin is available for hepatitis A. In the case of hepatitis B, a hyperimmune globulin is administered after mucosal or percutaneous exposure (including sexual contact) to an antigen-positive individual [336]. This is also recommended for newborns of antigen-positive mothers [365].

Mumps. Immune serum globulin has no efficacy in the modification of this disease, but hyperimmune globulin does [343].

Pertussis. Hyperimmune globulin modifies the disease. For example, 2.5 mL of hyperimmune antipertussis gamma globulin with a followup dose at 5 – 7 days led to a 75 % reduction in disease among exposed, nonimmune individuals [343]. However, use of hyperimmune globulin has been superseded by antibiotics [379].

Rabies. Hyperimmune globulin is recommended in addition to active immunization following exposure to a possible or proven case of rabies [364] (see Section 3.7). Combined active and passive immunization against rabies have been shown to be superior to active immunization alone [336].

Rho(D) immune globulin. This hyperimmune globulin is recommended for rhesus-negative mothers who deliver rhesus-positive infants [364].

Tetanus. The efficacy of tetanus immune globulin in either the prophylaxis or treatment of tetanus has not been clearly shown in clinical study. Current recommendations are for 250 IU injected intramuscularly for individuals whose tetanus immunization is not known and whose wound is of a sufficiently serious nature. For treatment of clinical tetanus, doses of 3000 – 6000 units are recommended.

Vaccinia. Hyperimmune globulin is used for prophylaxis against smallpox and for treatment of the dermal complications of vaccination [343].

Varicella. The efficacy of standard ISG in the prophylaxis of this disease is not well-established [335], [365]. Use of the hyperimmune product, however, is indicated in individuals who have never had chicken pox, who are exposed to acute cases and belong to a high risk group (newborns or immunocompromized patients), and women who are pregnant [364].

Hyperimmune Globulins under Development. Globulins with high titers to cytomegalovirus, and *Pseudomonas aeruginosa* have been tested. A product (bacterial polysaccharide immune globulin) has been tested that is prepared from the sera of donors immunized with licensed vaccines against *Hemophilus influenzae* (type b), pneumococci, and meningococci.

Certain subpopulations of children, such as native American Indians and Eskimos, are at particularly high risk of acquiring serious infection with encapsulated bacteria. Standard ISG preparations have failed to show a beneficial prophylactic effect for many types of infection. Immune serum globulin made hyperimmune to polysaccharide antigens by immunizing volunteers with licensed vaccines against pneumococci, *H. influenzae*, and meningococci has been used to immunize Apache infants. There was a significant reduction in the incidence of systemic disease caused by *H. influenzae* and pneumococci during the first six followup months as well as a significant decrease in the incidence of bacteremia [380].

5.4. Therapy with Immune Serum Globulin

Data suggest ISG is active in bacterial infections in animal models [379], [381] and may be synergistic with antibiotics [339], [340], [382]. However in 1968 SCHLESS and HARRELL [379] and others [336] observed that there was little evidence to support its therapeutic efficacy in established infection in humans; they suggested the need for a controlled clinical trial of ISG in the treatment of systemic infection in patients that were not deficient in antibody [379]. Indeed, since the most functionally-active antibody to gram-negative bacteria is IgM and not the IgG found in ISG, little benefit was to be expected [342]. The ISG is active against a wide variety of human pathogens in animal models of infection [379]. When used with antimicrobial agents, ISG has possible benefits in experimental infection with a wide variety of organisms [382], [383] and in clinical infections in humans [339], [340], [384], [385]. Large daily doses of ISG administered intravenously for 10 days to patients with leukemia and fever was well tolerated [386]. However, no benefit was found for patients who received the ISG in addition to antibiotics (compared to antibiotics alone). The administration of ISG to shorten the course of infection in children under two years of age was ineffective [387].

5.5. Prophylaxis and Therapy with Intravenous Immunoglobulin (IVIG)

In studies on the prophylactic or therapeutic efficacy of ISG, the possibility that higher doses of standard immunoglobulin might improve efficacy was a recurring theme. With the availability of IVIG, larger volumes of immunoglobulin could be administered directly into the bloodstream (see Section 5.1.2). In addition, with the ability to screen large numbers of samples for antibody levels or to immunize volunteers with an increasing number of vaccines, an increasing number of publications have examined the efficacy of passive immunotherapy, particularly with hyperimmune IVIG preparations, in the treatment and prophylaxis of infectious diseases.

Intravenous immunoglobulin G is effective in the prevention of infection in patients with hypogammaglobulinemia [358]. It has also been shown to be effective for treating chronic infection in such patients who developed sinopulmonary infection despite ISG maintenance therapy [351].

5.5.1. Viral Infection

Since cytomegalovirus (CMV) is a frequent cause of infection in patients undergoing organ transplantation and effective antiviral therapy is lacking, there has been considerable interest in the use of IVIG that is hyperimmune in CMV antibody for both the prophylaxis and treatment of CMV infections.

Prophylaxis. Hyperimmune CMV-IVIG (total dose 550 mg/kg) was shown to decrease the attack rate of symptomatic CMV infection among CMV-negative patients who received kidneys from CMV antibody-positive donors from 60 – 21 % [388]. This treatment also decreased the incidence of fungal and parasitic infection. Hyperimmune CMV-IVIG (3×200 mg/kg doses) also prevented mortality and interstitial pneumonia from CMV for 120 days in leukemic patients who underwent bone marrow transplantation [389]; however in another study it did not prevent acquisition of infection [390]. In another trial, nonimmune IVIG with a high antiCMV titer did not decrease the incidence of CMV seroconversion in bone marrow transplantation patients [391]; however, the incidence of symptoms and interstitial pneumonia decreased. In contrast, CMV-IVIG given to CMV-negative patients undergoing bone marrow transplantation had no effect on either the prevention of new disease or the amelioration of established disease [392]. Therapy of bacterial infections following infection with human immunodeficiency virus are discussed in Section 5.5.2.

Therapy. Immunoglobulin has been used sporadically in the treatment of viral disease. High-titered CMV-IVIG from screened donors did not show efficacy in the treatment of bone marrow transplant patients with documented CMV infection [393]. IVIG has been shown to alter the course of echovirus encephalitis infection in three patients with hypogammaglobulinemia. Although no benefit occurred from giving the IVIG intravenously in two of these patients, intraventricular administration resulted in clinical cures [394], [395]. The use of IVIG was unable to alter the lethal course of polymyositis secondary to echovirus in another patient with hypogammaglobulinemia [396]. IVIG has also been used experimentally in the successful treatment of herpes infection in mice [397].

5.5.2. Bacterial Infection

Numerous studies have demonstrated the efficacy of hyperimmune IVIG in the prophylaxis and, if used early after infection, the treatment of bacterial infection. In 1943 ALEXANDER showed that the combination of a sulfa drug and animal hyperimmune sera was more effective in reducing mortality than either agent alone [339], [340]. A number of studies have demonstrated the efficacy of IVIG in both the prevention and treatment of experimental infection with *H. influenzae* in neonatal rat models [398]; with *E. coli* [399] and *Klebsiella* [400] in mouse models; and with *P. aeruginosa* in neutropenic and burned rodent models [401], [402]. Intravenous immunoglobulin hyperimmune to group B streptococcal surface antigens can prevent and treat serious bacteremia in experimental infection in monkeys [403].

Prophylaxis. Earlier data with ISG in both experimental bacterial infection in animals and in clinical infection in humans indicated that in some situations (e.g., patients with hypogammaglobulinemia), exogenous standard gamma globulin could prevent the acquisition of serious bacterial infection. Similar studies with IVIG have established the efficacy of these preparations [358].

Since earlier investigators believed that larger doses of ISG might be effective in the prevention of bacterial infection in high-risk neonates, it is not surprising that similar studies have now been reported with IVIG. In one nursery with a high rate of infection, a single dose of IVIG (120 mg/kg within 2 h of birth) decreased acquisition of infection and mortality in preterm, lowbirthweight neonates [404]. A second dose at eight days, conferred no further advantage. Prophylaxis with IVIG (0.5 mg/kg/week for four weeks) was significantly better in preventing infection and death in neonates, but only in the subpopulation that weighed less than 1500 g and had a gestational age of less than 34 weeks [370]. In another study, antibiotics were given either alone or with IVIG to women 27 – 36 weeks pregnant who had chorioamnionitis. Only high doses (24 g/d for five days) of IVIG given after the 32nd week of pregnancy prevented infection in the delivered babies [405]. The investigators concluded that little transplacental transfer of IgG occurred before the 32nd week of gestation.

Data on the use of standard IVIG for the prevention of bacterial infections in patients without hypogammaglobulinemia is limited. High levels of specific antibody are needed; these may not be found in standard, nonimmune IVIG, despite the possibility of delivering larger amounts of IVIG than was the case with ISG. Hyperimmune products have been shown to prevent specific infections in experimental models (see Sections 5.4, 5.5.2, and 5.6). The administration of nonimmune IVIG (1000 mg/kg) before bone marrow transplantation and weekly for 17 weeks thereafter had no effect on the acquisition of either bacterial or fungal infection [406]. Clinical trials are being performed to test the efficacy of IVIG in the prophylaxis of bacterial infection in adults with chronic lymphocytic leukemia, a condition which may be complicated by hypogammaglobulinemia [407].

Therapy. Patients with human immunodeficiency virus (HIV) infection (i.e., acquired immune deficiency syndrome, AIDS) have a dysfunction in their humoral immune system which includes both a decreased antibody response to bacterial antigens and an altered distribution of IgG subclasses [408]. Unlike adults, who tend to acquire opportunistic infections (infections in which immunodeficient individuals are infected by organisms that are normally withstood by immunocompetent individuals), young children and particularly infants infected with

HIV often resemble patients with primary humoral immunodeficiency and tend to suffer from bacterial infections. Consequently, prophylaxis with monthly doses of IVIG has been used in the management of these patients. In one pilot study, a decreased incidence in episodes of fever and bacteremia was noted. This was accompanied by clinical improvement, prolongation of life, and improvements in other immunologic parameters [408]. In a 37 month old child with AIDS, monthly doses of IVIG produced increases in IgG 2 antibody and antibody to 12 pneumococcal serotypes, and prevented subsequent episodes of bacteremia [409]. On the basis of these preliminary data, IVIG prophylaxis has been advocated in the treatment of childhood AIDS [408]. Periodic administration of IVIG decreased lactic dehydrogenase activity (a proposed indicator of pulmonary interstitial inflammation) in adults and children with HIV infection [410].

Attempts to demonstrate a significant effect in the therapy of bacterial infections with IVIG have been much less successful than with its use in prophylaxis. This may be attributable to the shorter half-life of IVIG in the blood during infection [368], [381]. The addition of IVIG to standard regimens of antibiotics decreased the mortality from documented bacteremia, particularly among preterm infants. The number of subjects studied was too small for statistical analysis, however [411].

In adults, little data is available to support the use of IVIG in the treatment of bacterial infections. This may be due to the need for high levels of antibody specific for the invading organism (rather than simply high levels of nonspecific antibody) as well as the need for prompt initiation of therapy. In most experimental studies with hyperimmune IVIG, little benefit can be shown if the exogenous immunoglobulin is given more than 8 h after infection (see [381], [383]). In clinical medicine, however, identification of the time of onset of infection is often difficult.

5.5.3. Noninfectious Diseases

5.5.3.1. Therapeutic Effect of IVIG

In 1981 it was reported that a patient who received IVIG for hypogammaglobulinemia also recovered from bleeding secondary to a coincidental platelet deficiency (*thrombocytopenia*) following the infusion [412]. Following this fortuitous observation, experiments showed that a high dose of IVIG could indeed reverse thrombocytopenia in the absence of hypogammaglobulinemia; furthermore, a regimen of 400 mg/kg/d for five days was effective treatment for idiopathic thrombocytopenic purpura (ITP) in children [413]. The use of IVIG and oral steroids was also compared. Among those who responded rapidly to treatment (62 %), IVIG was as efficacious as steroids; however the slower responders responded better to the IVIG treatment. Administration of IVIG was accompanied by a doubling of serum IgG; IgM levels rose under both treatment regimens. The IVIG may possibly reduce clearance of antibody-coated platelets via an Fc-mediated mechanism (see Section 5.5.3.2) [414]. The dosage regimen used in the above study (400 mg/kg/d for five days) has been used in many other studies on the noninfectious uses of IVIG. However, in another study 800 – 1000 mg/kg was given as a single infusion to children with ITP with similar results and no reported untoward effects [415]. Experience with IVIG for ITP has not been uniformly successful however [416].

IVIG has been used in the treatment of *immunologically-mediated blood disorders* in both pediatric and adult populations. These include autoimmune hemolytic anemia [417], [418], autoimmune neutropenia [419], posttransfusion purpura [420], ITP in adults [421], [422], chronic ITP in adults and children [423], [424], thrombocytopenia secondary to alloimmunization in leukemic patients receiving platelet transfusions who became refractory to subsequent platelet transfusions [425], thrombocytopenia secondary to transplacental passage of antiplatelet antibodies [426], ITP of pregnancy [427], antibody-mediated red cell aplasia [428], during pregnancy in women with severe rhesus immunization [429], and in conjunction with cyclophosphamide used to treat antibody to factor VIII in hemophilia [430].

In addition to these hematologic disorders, two large trials conducted in patients with *Kawasaki's disease* have compared the use of aspirin alone to that of aspirin and IVIG (400 mg/kg/d for 4 and 3 days) [431], [432].

The IVIG reduced the fever and the incidence of coronary artery disease [431], [432]. It suppressed the T and B cell activation characteristic of patients with this disease and decreased the levels of spontaneous immunoglobulin synthesis in vitro [433].

Placentally derived immunoglobulin has been administered to patients with severe *rheumatoid arthritis* because the symptoms of some patients improved during pregnancy [434]. None of five patients given IVIG derived from control plasma improved whereas 3/5 improved under the placentally derived globulin. An antihuman leukocyte antigen – DR surface antigen (anti HLA-DR) antibody in the placental preparation was presumed to be a possible mechanism for this improvement.

High dose immunoglobulin treatments have also been used in patients with *Felty's syndrome* [435], *myasthenia gravis* [436], *epilepsy* [437], and *multiple sclerosis* [438]. In the latter study one-third of patients worsened with IVIG therapy.

5.5.3.2. Mechanism of Action

Many immunologic mechanisms have been invoked to explain the beneficial effects of IVIG in noninfectious illnesses. IVIG is a potent immune modulator; the role of the Fc portion of the immunoglobulin molecule in this respect is still a subject of active investigation.

Antibody-coated material is removed by phagocytes in a process known as *Fc-receptor-mediated clearance.* The Fc portion of the antibody coating the bacterium cell, protein etc. binds to the Fc receptor of the phagocyte. The material is then taken up by the phagocyte and digested. Initially, the Fc receptors in the reticuloendothelial system were thought to become saturated by the high dose of immunoglobulins in IVIG resulting in a decreased clearance of IgG-coated particles due to competitive inhibition [417], [439]. This is consistent with studies in which radiolabelled, autologous erythrocytes were cleared more slowly after IVIG infusion than before [414], [423]; this treatment altered Fc receptor affinity, not receptor number [440]. Other suggested mechanisms include decreased synthesis of autoantibodies [441], the clearance of persistent, occult viral infections [396], [441], the protection of platelets or megakaryocytes against antiplatelet antibodies [441], antiidiotypic suppression of antibody synthesis [436], [442], blocking of the Fc receptor by antibodies, and the production of antilymphocyte antibody [443]. However, patients with ITP have responded to IVIG without demonstrable alteration in Fc-receptor mediated clearance [441]. In addition, preincubation of erythrocytes in IVIG failed to inhibit the phagocytosis of sensitized erythrocytes by cultured macrophages [444]. Thus, IVIG may lead to improvement in such patients by multiple mechanisms.

IVIG is a potent modulator of antibody production both in vitro and in vivo. It inhibits lectin-driven B cell differentiation in vitro [445], [446] and immunoglobulin production by peripheral blood mononuclear cells stimulated with pokeweed mitogen [446]. For these effects the Fc portion of immunoglobulin must be present: the Fc portion alone was 100 fold more effective than the intact IgG. The observation that IgM antibody rises after IVIG infusion has raised the possibility that it may stimulate some immunoglobulin-producing cell populations [423]. Interestingly, the monthly administration of ISG to premature infants during the first year of life resulted in a significantly lower level of immunoglobulin compared to the control group [371], [447].

In adults with ITP, the infusion of IVIG (400 mg/kg) led to a decrease in T4 (helper) lymphocytes and elevation in T8 (suppressor) lymphocytes (and decrease in T4/T8 ratio) [421]. Similarly, the administration of IVIG to patients with hypogammaglobulinemia led to increased suppressor T cell activity, decreased total T cells, a significant decrease in the T4/T8 ratio, and a decrease in lectin-induced immunoglobulin syntheses in vitro [448].

The ability of IVIG to decrease antibody production may be desirable in patients producing autoantibody, but dangerous in those with infection [445]. In patients with acute otitis, repeated monthly [379] infusions of IVIG resulted in higher IgG levels if the initial antipneumococcal antibodies were low. However, in the presence of high initial levels of antibody, specific antipneumococcal antibody levels decreased [449]. The suggestion has also been made that high levels of nonspecific antibody could lead to a reduction in the survival of specific antibody [450],

[451]. Evidence suggests that IVIG may even exacerbate infections. A patient with neutropenia died due to acceleration of the yeast infection shortly after infusion of IVIG; the IVIG may have blocked the normal Fc-mediated clearance mechanism [452].

Immunoglobulin G can bind both native C 3 and C 3 b during complement activation by soluble immune complexes and by bacteria [453], [454]. Finally since immunoglobulin has unique antigenic determinants, antibodies to the immunoglobulin may develop [455], [456]. These antibodies may play a role in the anti-idiotype network.

In summary, exogenous immunoglobulin has a diverse, potent effect on a wide variety of immune regulatory mechanisms. These interactions may often work to the patient's benefit, but may also result in previously unsuspected adverse effects.

5.6. Prophylaxis and Therapy with Plasma and Other Blood Products

Although the regular, large-scale, clinical use of plasma and other blood products is considered impractical (both for production and safety reasons), passive immunotherapy with these agents may provide some efficacy in the prophylaxis and treatment of infections in humans.

Plasma has been used to treat infections, particularly those caused by *Pseudomonas* [374], [457]. Plasma was superior to ISG in the prevention of infection among patients who suffered >30 % burns [457]. Similar observations were made during experimental infection [458]. One possible explanation for the higher efficacy of plasma is that it contains IgG, IgA, and IgM (ISG contains only IgG). This may improve the distribution of antibody at different anatomic sites. Since the isotypes have functional differences, the antibacterial activity is also increased (natural antibody against gram-negative bacilli is thought to be predominantly of the IgM isotype). However, plasma infusions carry the risk of transmitting hepatitis (there is no manufacturing process for plasma that inactivates the virus as is the case with ISG).

Further efforts in the use of passive immunotherapy have also focused on septic shock. The rapid infusion of large volumes of lyophilized human plasma from blood with >40 µg/mL of antibody to a mixture of lipopolysaccaride serotypes resulted in nearly 7-fold decrease in mortality in South African women treated in an obstetrical–gynecological ward for septic shock [459]. Equine plasma similarly screened for antilipopolysaccaride antibodies has also been used routinely in veterinary practice in South Africa [460].

Opsonins. The defective neonatal antibody (opsonic) response that mediates the uptake and killing of bacteria by phagocytes can be partially corrected with gamma globulin [461]. The ability passively administered opsonins to prevent infection was first suggested in one study in which fresh whole blood was administered to infants to prevent group B streptococcal sepsis. All nine infants transfused with blood having antibody to group B streptococci lived versus 3/6 transfused with blood having undetectable antibody titers. Protection was correlated with opsonic antibody titers in the infants' blood and with having >40 % of their blood volume replaced [462].

Postimmune Serum. Donors were immunized with a J 5 mutant of *E. coli* to elicit antibodies to widely shared core epitopes in the lipopolysaccharides of gram-negative bacteria. The passive administration of postimmune serum from these individuals to patients in bacterial endotoxic shock decreased the incidence of death as well as the need for pressor therapy in patients with profound shock. However, since this protection did not correlate with the presence of antibody to the core epitopes, it is not clear whether the protection was conferred by the antibody [463]. In a follow-up study, the prophylactic administration of postimmune plasma appeared to significantly prevent shock and death from gram-negative infections in patients admitted to a surgical intensive care unit. Such treatment had no effect on infection rate, however [464]. In both studies the protective moiety in the plasma and serum was presumed to be antibody to the endotoxin. In any event, for this to be a practical therapy, the protective portion of the postimmune product must be mass produced in the form of a safe, standardized preparation (e.g., made into an IVIG from a large pool of donors immunized with the J 5 vaccine or into a monoclonal antibody preparation). Attempts to show a protective effect from an IVIG prepared from immunized donors have been unsuccessful to date [465]. Nevertheless, monoclonal antibody preparations directed toward a J 5 epitope were evaluated in human volunteers (see also Section 6.1).

5.7. Adverse Effects of Gamma Globulin Preparations

Immune serum globulin G is one of the safest biological products available with an overall incidence of reactions of 3 – 12 % [334]. Rare systemic (anaphylactic) reactions may occur during or within minutes of administration of ISG (ca. 1/500 – 1/1000 injections). Late-occurring systemic reactions (within hours or days) include arthralgias, pyrexia, and diarrhea and are not uncommon in immunodeficient patients [364]. Local reactions occur with intramuscular administration of ISG and are related primarily to the large volume administered. Local reactions to ISG can be prevented by using small volumes or by pretreatment with analgesics. Systemic reactions may be modified by the use of aspirin, hydrocortisone, or antihistamines.

Systemic reactions with IVIG are more common than with ISG and often depend on the rate of infusion and the type of preparation. Patients receiving IVIG for hypogammaglobulinemia have a reported incidence of 2.5 % reactions which are usually mild [358]; the incidence of these reactions decreases after the first two months of therapy. The cause of reactions to IVIG are not clear but may depend on aggregate formation, IgA contamination, and activated enzymes that initiate the release of inflammatory mediators [364].

The experimental work discussed above and clinical experience give reason for caution in the use of immunoglobulin preparations. Clinical experience primarily involves concern about viral transmission, anaphylactic reactions among patients with IgA deficiency, and isolated case reports of unusual occurrences. The indications for the use of gamma globulin should be well founded [363].

Viral Transmission. Immune serum globulin G has had an outstanding safety record with regard to lack of transmission of virus to recipients [466]. For example, ISG is free of serum hepatitis, even if the virus occurs in the original plasma. Ten children who received plasma for prophylaxis against measles in England became jaundiced and three died. In contrast, when ISG prepared from the same plasma was given to 56 children, only one child became jaundiced [343]. In the United States, 4 out of 15 volunteers who were inoculated subcutaneously with hepatitis-positive plasma developed hepatitis. Five further volunteers received large doses of ISG and none developed hepatitis [343]. The ability of the Cohn – Oncley fractionation (see Fig. 5) to inactivate virus may in part be due to the high concentration of antivirus antibody [353]. Non-A, non-B hepatitis has been reported in pastes produced in this procedure: in one, the lyophilization step was omitted, and in two others an ion-exchange step was added to lower IgA but it lowered IgG 4 levels as well [353].

In contrast, non-A, non-B (NANB) hepatitis has been reported following IVIG infusion [467–469]. In one such episode 16 out of 77 patients who received IVIG for immunodeficiency acquired hepatitis with death occurring in 5 cases. Interestingly, no hepatitis was observed following administration of ISG prepared from the same serum pool [467]. Since neither of the IVIG products was subjected to a recognized virucidal finishing step, such as treatment with acid, β-propiolactone, or ultraviolet radiation, the virus may still have been viable. Thus ethanol fractionation by itself might not fully inactivate putative virus [470], though it is virucidal for enveloped viruses [471]. Supplementary virucidal procedure could make IVIG safe from NANB hepatitis transmission [471]. Unlike plasma, IVIG has not been known to transmit hepatitis B [472].

There have been no confirmed cases of transmission of HIV virus in IVIG but antibody to HIV has been isolated from two patients with primary hypogammaglobulinemia [471]. When HIV was added to plasma and then processed into IVIG, $>10^4$ PFU/mL (PFU denotes plaque-forming units) of HIV were inactivated during alcohol fractionation and poly-(ethylene glycol) fractionation to IVIG [473]. Antibody to HIV was detected by an enzyme-linked immunoabsorbent assay (ELISA) in all 10 lots of a reduced, alkylated IVIG and in 4/8 lots of a pH 4/pepsin IVIG; 8/10 and 3/8 lots were also positive by Western blot analysis. The HIV antibody-positive lots, which were negative for HIV by culture and reverse transcriptase activity, were infused into patients. Comparison of pre- and post-infusion sera demonstrated that antibody to HIV was detectable for up to one month before converting to a negative antibody status [472].

One woman who had received hyperimmune Rho (D) immunoglobulin subsequently was found to be infected with HIV but belonged to a group with an increased risk of acquiring infection with HIV [474].

Anaphylaxis. Anaphylaxis has been documented following the administration of IVIG to some patients with hypogammaglobulinemia. Indeed, patients with antibody deficiency have an increased incidence of adverse reactions to IVIG [342]. Low levels of IgA are associated with an increased risk of such anaphylactic reactions [475]; IVIG products differ markedly in their IgA content [476]. In two patients with common variable immunodeficiency, IgE antibody to IgA in the infused IVIG was felt to account for these reactions [475]. In addition, IgE has been detected in commercial preparations of IVIG [477].

Other clinical problems have been described which involve the use of IVIG. IVIG prophylaxis precipitated cryoglobulinemic nephropathy in a patient with hypogammaglobulinemia secondary to B cell neoplasm [478].

The passive transfer of antibodies against specific blood types in IVIG may also be associated with hemolytic anemia or cause problems in the crossmatching of blood prior to surgery [479]. Immunologically normal patients may receive antibody to immunoglobulin that is of a different genotype.

5.8. Future Prospects

Large amounts of immunoglobulin can now be safely delivered directly into the bloodstream by using IVIG. This need was recognized by Cohn in the 1940s. Since immunoglobulin is now known to be a potent modulator of the immune system, IVIG has been used as a form of therapy for many noninfectious diseases. However, the impact of such treatment on the ability to respond to infectious diseases often complicates these conditions and still has to be assessed. Studies to date do not clearly show that the indications for the use of IVIG in infectious diseases should be expanded beyond those already shown for ISG. In some conditions involving functional hypogammaglobulinemia (e.g., HIV infection in young children, chronic lymphocytic leukemia), prophylaxis with standard IVIG may ultimately demonstrate efficacy. In certain subpopulations that are at risk of infection with specific organisms over a finite period of time, immunoglobulin preparations enriched in antibody to those organisms (i.e., hyperimmune preparations) may prove useful. The most promising use appears to be in the prophylaxis of specific infections; this usually involves hyperimmune serum and, as shown by studies in neonates, the substantial replacement of blood volume.

The use of IVIG to treat established infections with pharmacological (not replacement) doses of immunoglobulin, (with the possible exception of antitoxin treatment of septic shock), still has no proven role. This may be related to the need for prompt intervention and a large amount of a specific antibody, requiring rapid identification of the microorganisms involved. Preparations that are known to be immunotherapeutically effective are less efficacious when given 8 h or more after the onset of infection. Substantial evidence suggests that passive immunotherapy may provide an enhanced therapeutic benefit; however, this still remains to be proved in a prospective, controlled therapeutic trial.

Finally, more recent studies that show some benefit from passive immunotherapy have not been performed with IVIG preparations, e.g., [376], [463], [464], [459]. Thus despite the substantial progress made in the development of IVIG preparations for human use, their firm recommendation for widespread use in infectious disease requires further well-designed studies.

6. Immunotherapeutic Uses of Monoclonal Antibodies

When a vertebrate is vaccinated with a foreign antigen, it produces a mixture of antibodies that can bind to the antigen. Each antibody binds to a different epitope on the immunogen and is made by an individual clone of B lymphocytes. Industrial quantities of a single (i.e., monoclonal) antibody recognizing a specific epitope can be made by fusing a B lymphocyte producing the antibody of interest with an immortal tumor cell line to give hybridomas. For a detailed description of monoclonal antibodies, see → Monoclonal Antibodies.

Infectious diseases continue to cause much morbidity and mortality despite the discovery of a wide variety of novel antimicrobial agents and improved methods for immunizing at-risk populations. Therapy with human or animal serum is an effective adjunct to conventional drug therapy and, prior to the development of antimicrobials, was the only effective therapy for many infectious diseases (see Chap. 5). The efficacy of serum treatment for various infectious diseases suggests that this approach could be a useful adjunct to drug therapy if safety, efficacy, and production problems could be overcome. Human monoclonal antibodies (humabs) may solve many of the problems associated with serum therapy. Therapeutic humabs have been developed against many target organisms and are being tested in clinical trials.

6.1. Introduction

Despite the fact that the first reports describing the generation of antigen-specific murine monoclonal antibodies were published in the mid 1970s, routine production of humabs has been more difficult. Methods to generate and produce these molecules have improved and serious efforts are underway to develop them as therapeutic products for which passively administered human antisera against specific target antigens are in use in a number of clinical settings:

1) Red cell antigens: Rh (hemolytic disease of the newborn)
2) White cell antigens: antilymphocyte/thymocyte globulin
3) Viral antigens: hepatitis A and B, rabies, CMV, HSV, and varicella zoster
4) Bacterial antigens: tetanus, endotoxins, and pneumococcus
5) Elimination of circulating drugs (overdoses)
6) Antisnake venom
7) Fertility control (e.g., anti-β-human chorionic gonadotropin)

Humabs that recognize many of these antigens have been produced, and will probably augment or replace pooled antisera in the near future. Concerns, which include the possible contamination of pooled globulins by infectious agents (e.g., human immunodeficiency virus HIV, various hepatitis viruses, and cytomegalovirus) and the diminished availability of serum donors, as well as the relative ease of reproducible humab manufacture, will accelerate this trend.

Humabs will be preferred for the diagnosis and treatment of a variety of diseases because they minimize the problems encountered when a foreign animal monoclonal antibody is administered (e.g., anaphylaxis, clinical manifestations of immune complex formation, and reduction of efficacy by anti-antibodies). In well over half of the patients treated to date with murine monoclonal antibodies, the human antimouse immunoglobulin response has limited their usefulness [480]. Only a fraction of the antimouse immune response is directed to the variable region (idiotype) of the rodent immunoglobulins. This suggests that humabs will be more effective than murine monoclonal antibodies. Furthermore, humabs are more likely to have species-specific carbohydrates which may be important in a number of effector functions, such as Fc receptor-mediated, antibody-dependent, cellular cytotoxicity, complement activation, and phagocytosis [481]. Serum half-life and effector functions of immunoglobulin subclasses are very important for designing the optimal anti-infectious agent therapeutic. There are a number of reasons why passively administered humabs will be preferred for therapeutic use. A good summary and technical background are to be found in [482], [483].

6.2. Bacterial Targets

Various bacterial targets are recognized by humabs:

Tetanus toxoid [484–488]
- Diphtheria toxoid [489]

Gram-negative endotoxins [491–494]
- *Pseudomonas aeruginosa* lipopolysaccharide [496] and exotoxin A [497]
- *Hemophilus influenzae* [498], [499]
- *Mycobacterium leprae* [500]
- *Neisseria meningitides* [501]
- *Pneumococcus* [502], [503]

Tetanus-neutralizing humabs have been frequently generated because of the ease of obtaining immune B cells from vaccinated persons [484–488].

Because immunization against tetanus in the United States is universal and most individuals have relatively high titers, it is unlikely that any of these monoclonals will be scaled up and used in this country. The same is true of antidiphtheria monoclonal antibodies [489]. In other countries, vaccination against tetanus and diphtheria is not widespread and administration of immune human or animal sera is still practiced. Concern about HIV-contaminated serum has stimulated efforts to generate humabs against tetanus for use in India, South America, and southern Europe.

Gram-negative bacterial infections account for 1 – 2 % of hospital admissions and up to 100 000 deaths each year in the USA. Despite therapies including antibiotics and various support measures, mortality rates remain as high as 50 – 70 %. Much interest has been generated by the idea that lipopolysaccharide, the lethal component of gram-negative bacteria, might have antigenic determinants that are shared by many bacterial species. The outer carbohydrate domain of lipopolysaccharide varies highly from one species to the next whereas the inner carbohydrate core and lipid A regions appear to be much more conserved. In a clinical trial antisera were obtained from firemen vaccinated with a rough mutant *E. coli* strain called J 5 that lacks the outer carbohydrate. This antiserum gave significant protection to patients with gram-negative sepsis [490]. Several laboratories have reported the generation of humabs that are highly cross-reactive with bacterial lipopolysaccharide [491–494]. They are all of the IgM class and protect mice against infection when given at high doses. Two of these and a murine monoclonal of the same specificity are in clinical trials for the treatment of gram-negative sepsis. At present it is too early to determine whether they are successful, however several points should be made regarding this approach:

Firstly, it is unclear whether antibody was the active factor in the antisera in the original study [490]. Secondly, no documentation on the binding of the antisera to the bacteria causing a specific case of sepsis was made, i.e., no correlation could be made between response and nonresponse. Thirdly, immune serum is known to contain many other factors that might have benefited the treated patients. Fourthly, although cross-reactive antisera and monoclonal antibodies appear to bind to a wide number of bacteria and blotted lipopolysaccharide, questions still exist regarding the accessibility of the core and lipid A region for antibody binding. Fifthly, several trials with slightly different design have given inconsistent results [495]. Thus for a number of reasons the so-called core concept is questionable.

Type-specific horse antibodies recognizing typhoid and pneumococcus immunotypes were used from the turn of the century up to the beginning of the antibiotic era in the 1940s. The generation of humabs recognizing type-specific determinants on gram-negative bacteria is a reasonable alternative to the above-mentioned highly cross-reactive monoclonals. Although the diversity of possible immunotypes of gram-negative bacteria is large, a particular subset is probably responsible for most of the invasive bacteremias. Several groups are trying to generate a number of type-specific humabs and administer a cocktail of antibodies. This approach has already been tried for *Pseudomonas aeruginosa* [496]. In model systems, murine monoclonal antibodies recognizing a single immunotype were much more effective than those recognizing a core determinant that was shared between Pseudomonas species. If the cocktail approach is to prove effective, it may be necessary to add humabs that neutralize various virulence factors. In the case of *Pseudomonas* infections, humabs have also been generated to exotoxin A [497].

Humabs have been reported that recognize *Hemophilus influenza* type B capsular polysaccharide [498]. The successful introduction of an *H. influenza* type B capsular polysaccharide vaccine [499] (see Section 2.13) has made the clinical use of these humabs less attractive than originally anticipated. However, a safe humab for this life-threatening disease may still be developed for clinical use in conjunction with the vaccines.

Anti-mycobacterium leprae humabs have been reported [500]. In view of the central role of cell-mediated immunity in resistance and recovery from this disease, these reagents are unlikely to find clinical application. Nevertheless, these antibodies and those made by patients suffering from other infectious diseases can be used to probe the human humoral immune response, to clone antigens recognized by the humabs, and to investigate antibody-mediated autoimmunity initiated by microorganisms.

6.3. Viral and Chlamydial Targets

Antiviral humabs are especially attractive therapeutic targets. Immunoprophylaxis of cytomegalovirus (CMV) infections in immunosuppressed patients (e.g., transplantation) is of major clinical interest. Several groups have generated CMV-neutralizing humabs [504], [505] and clinical trials were scheduled to begin in 1989.

Humabs that neutralize hepatitis B virus [506–508] will replace the antisera currently used after acute exposure to this agent. These humabs may also be useful for "active – passive" immunization of at-risk infants. A very large population of infants, particularly in less developed countries, become chronically infected in the perinatal period. Evidence suggests [509] that administration of immune sera will prevent lifelong infection and could have an impact on the prevalence of the most common human cancer in the world, hepatitis-B-positive hepatoma [509]. The widespread use of anti-hepatitis A antisera suggests that the humabs generated to this target may also find clinical application.

A minor yet important target for humabs is the treatment of varicella zoster infections [510]. These humabs may have to be developed as an orphan drug given the small numbers of patients. (The development of drugs for rare diseases is supported by the US government; these pharmaceuticals are called orphan drugs.)

The acquired immune deficiency syndrome (AIDS) is caused by the human immunodeficiency viruses (HIV) and is currently the major infectious viral disease problem. Many groups are generating humabs that recognize one or more HIV strains. Many of these humabs recognize gene product gp 160 that is found in the viral envelope [511], [512]. At present, there is no evidence that a humoral immune response can prevent or change the pattern of HIV infection in people or animals. The viruses undergo rapid mutation (five times the rate of influenza). Furthermore, they can spread by cell to cell contact and reside permanently out of reach of humoral immunity inside the mononuclear phagocytes scattered throughout the body. Trials are underway to test the efficacy of a high-titer anti-HIV antiserum in the prevention of neonatal AIDS. The dismal failure of all attempted vaccines and passive serum to delay the spread of the virus in subhuman primate models is very discouraging. Antiviral humabs attached to toxins or engineered to bring cytotoxic cells into contact with infected cells might improve the potency of the antibody approach.

Several humabs have been generated against chlamydia [513] and other viruses:

Epstein Barr Virus [514]
Herpes simplex [515–517]
Human T cell leukemia virus I [518]
Measles (SSPE) virus [519], [520]
Rabies [521]
Rubella [522]
X 31 influenza virus [523]

However, clinical interest in the therapeutic purposes and use of these antibodies is limited.

6.4. Parasite Targets

The major therapeutic target of interest among parasitic diseases is malaria (see Section 4.2). Much less is known about the humoral immune response of humans suffering from other parasitic diseases (Section 4.1). Passively administered humabs may have therapeutic potential in acute *Falciparum malaria*. Immunoglobulin

M and IgG humabs have been generated [524], [525]. Monoclonal antibodies are also useful research tools in the development of malaria vaccines (Chap. 4).

7. References

1. I. Roitt: *Essential Immunology,* 5th ed., Blackwell Scientific Publications, Oxford, United Kingdom 1984.
2. B. D. Davis, R. Dubecco, H. N. Eisen, H. S. Ginsberg: *Microbiology: Including Immunology and Molecular Genetics,* 3rd ed., Harper and Row, Hagerstown, MD, 1980.
3. R. H. Schwartz, A. Yano, W. E. Paul: "Interaction between antigen-presenting cells and primed T-lymphocytes," *Immunol. Rev.* **40** (1978) 153.
4. W. Stroler, L. A. Hanson, K. W. Sell (eds.): *Recent Advances in Mucosal Immunity,* Raven Press, New York 1982.
5. T. Yamamura, T. Tada: *Progress in Immunology V,* Academic Press, Tokyo 1984.
6. J. B. Robbins, J. C. Hill, J. C. Sadoff (eds.): *Bacterial Vaccines, Seminars in Infectious Disease,* vol. IV, Thieme-Stratton, New York 1982.
7. U. E. Nydegger (ed.): *Immunotherapy, A Guide to Immunoglobulin Prophylaxis and Therapy,* Academic Press, London 1981.
8. R. E. Black, M. M. Levine, M. L. Clements, G. Losonsky et al.: "Prevention of shigellosis by a Salmonella typhi – Shigella sonnei bivalent vaccine," *J. Infect. Dis.* **155** (1987) 1260 – 1265.
9. M. M. Levine, J. B. Kaper, D. Herrington, J. Ketley et al.: "Safety, immunogenicity, and efficacy of recombinant live oral cholera vaccines, CVD 103 and CVD 103-HgR," *Lancet* **2** (1988) 467 – 470.
10. R. J. Collier: "Diphtheria toxin: mode of action and structure," *Bacteriol. Rev.* **39** (1975) 54 – 85.
11. A. M. Pappenheimer, Jr. in R. Germanier (ed.): "Diphtheria," *Bacterial Vaccines,* Academic Press, Orlando, FL 1984, pp. 1 – 36.
12. F. Audibert et al.: "Active antitoxic immunization by a diphtheria toxin synthetic oligopeptide," *Nature (London)* **289** (1981) 593 – 594.
13. T. Uchida, D. M. Gill, A. M. Pappenheimer, Jr.: "Mutation in the structural gene for diphtheria toxin carried by temperate phage *β*," *Nature (London)* **283** (1971) 8 – 11.
14. L. Greenfield, H. L. Dovey, F. C. Lawyer, D. H. Gelfand: "High level expression of diphtheria toxin peptides in Escherichia coli," *Bio/Technology* **4** (1986) 1006 – 1011.
15. B. Bizzinin in R. Germanier (ed.): "Tetanus," *Bacterial Vaccines,* Academic Press, Orlando, FL 1984, pp. 37 – 68.
16. M. C. Hardegree et al.: "Titration of tetanus toxoids in international units; relationship to antitoxin responses of rhesus monkeys," *6th Proc. Int. Conf. Tetanus 1981,* 1982, pp. 409 – 423.
17. M. Huet: "La standardisation des vaccins tetaniques," *6th Proc. Int. Conf. Tetanus 1981,* 1982 pp. 425 – 433.
18. K. W. Newell, A. Dueñas-Lehmann, D. R. Leblanc, N. Garces-Osorio: "The use of toxoid for the prevention of tetanus neonatorum; final report of a double-blind controlled field trial," *Bull. W.H.O.* **35** (1966) 863 – 871.
19. N. Guerin, C. Fillastre: "Vaccin D.T.C. Polan," *6th Proc. Int. Conf. Tetanus 1981,* 1982, pp. 477 – 479.
20. M. Pittman: "Pertussis toxin; the cause of the harmful effects and prolonged immunity of whooping cough," *Rev. Infect. Dis.* **1** (1979) 401 – 412.
21. C. R. Manclark, J. L. Cowell in R. Germanier (ed.): "Pertussis," *Bacterial Vaccines,* Academic Press, Orlando, FL 1984, pp. 69 – 106.
22. T. Madsen: "Vaccination against whooping cough," *JAMA J. Am. Med. Assoc.* **101** (1933) 187 – 188.
23. C. L. Cody et al.: "Nature and rates of adverse reactions associated with DTP and DT immunizations in infants and children," *Pediatrics* **68** (1981) 650 – 660.
24. R. Aldersdale et al.: *Whooping Cough. Reports from the Committee on Safety of Medicines and the Joint Committee on Vaccination and Immunization,* **vol. 4,** H. M. Stationary Office, London, England 1981, pp. 79 – 169.
25. M. Oda, J. L. Cowell, D. G. Burstyn, R. Manclark: "Protective activities of the filamentous hemagglutinin and the lymphocytosis-promoting factor of Bordetella pertussis in mice," *J. Infect. Dis.* **150** (1984) 823 – 833.
26. Ad Hoc Group for the study of Pertussis Vaccines: "Placebo controlled trial of two acellular pertussis vaccines in Sweden–protective efficacy and adresse events." *Lancet* **1** (1988) 955 – 960.

27. C. S. Chuttani et al.: "Controlled field trial of a high-dose oral killed typhoid vaccine in India," *Bull. W.H.O.* **55** (1977) 643 – 644.
28. M. Reitman: "Infectivity and antigenicity of streptomycin-dependent Salmonella typhosa," *J. Infect. Dis.* **117** (1967) 101 – 107.
29. R. Germanier, E. Fürer: "Isolation and characterization of gal ε mutant Ty 21 a of Salmonella typhi; a candidate strain for a live, oral typhoid vaccine," *J. Infect. Dis.* **131** (1975) 553 – 558.
30. R. H. Gilman, R. B. Hornick, W. E. Woodward, H. L. Dupont et al.: "Evaluation of a UDP-glucose-4-epimeraseless mutant of Salmonella typhi as a live oral vaccine," *J. Infect. Dis.* **136** (1977) 717 – 723.
31. M. H. Wahdan, C. Sérié, Y. Cerisier, S. Sallam et al.: "A controlled field trial of live Salmonella typhi Ty 21 a oral vaccine against typhoid; three year results," *J. Infect. Dis.* **145** (1982) 292 – 295.
32. World Health Organization Committee on Biological Standardization: Requirement for typhoid vaccine (live-attenuated, Ty21 a, oral). World Health Organization Technical Report, series 700; Geneva, Switzerland: 34th report, 1984, pp. 48 – 68.
33. R. Germanier in R. Germanier (ed.): "Typhoid fever," *Bacterial Vaccines,* Academic Press, Orlando, FL 1984, pp. 137 – 165.
34. M. M. Levine, C. Ferreccio, R. E. Black, R. Germanier, Chilean Typhoid Committee: "Large-scale field trial of Ty21 a live oral typhoid vaccine in enteric-coated capsule formulation," *Lancet* **ii** (1987) 1049 – 1052.
35. M. M. Levine, D. Herrington, J. R. Murphy, J. G. Morris et al.: Safety, infectivity, immunogenicity, and in vivo stability of two auxotrophic mutant strains of *Samonella typhi,* 541 Ty and 543 Ty, as live oral vaccine in humans, *J. Clin. Invest.* **79** (1987) 888 – 902.
36. I. L. Acharya, C. V. Lowe, R. Thapa, V. L. Gurubacharya et al.: "Prevention of typhoid fever in Nepal with the Vi capsular polysaccharide of *Salmonella typhi," N. Engl. J. Med.* **317** (1987) 1101 –1104.
37. R. Austrian: "Of gold and pneumococci; a history of pneumococcal vaccines in South Africa," *Trans. Am. Clin. Climatol. Assoc.* **89** (1977) 141 – 161.
38. C. M. MacLeod, M. R. Krauss: "Stepwise intratype transformation of pneumococcus from R to S by way of a variant intermediate in capsular polysaccharide production," *J. Exp. Med.* **86** (1947) 439 – 453.
39. J. B. Robbins, R. Austrian, C.-J. Lee, S. C. Rastogi et al.: "Considerations for formulating the second-generation pneumococcal capsular polysaccharide vaccine with emphasis on the cross-reactive types," *J. Infect. Dis.* **148** (1983) 1136 – 1159.
40. Health and Public Policy Committee, American College of Physicians: "Pneumococcal vaccine," *Ann. Intern. Med.* **104** (1986) 118 – 120.
41. P. Smit, D. Oberholzer, S. Hayden-Smith, H. J. Koornhof et al.: "Protective efficacy of pneumonococcal polysaccharide vaccines," *JAMA J. Am. Med. Assoc.* **238** (1977) 2613 – 2616.
42. G. Bolan, C. V. Broome, R. R. Facklam, B. D. Plikaytis et al.: "Pneumococcal vaccine efficacy in selected populations in the United States," *Ann. Intern. Med.* **104** (1986) 1 – 6.
43. M. S. Simberkoff, A. P. Cross, M. Al-Ibrahim, A. L. Baltch et al.: "Efficacy of pneumococcal vaccine in high-risk patients," *N. Engl. J. Med.* **315** (1986) 1318 – 1327.
44. J. A. Leech, A. Gervais, F. L. Ruben: "Efficacy of pneumococcal vaccine in severe chronic obstructive pulmonary disease," *Can. Med. Assoc. J.* **136** (1987) 361 – 365.
45. C. Chu, R. Schneerson, J. B. Robbins, S. C. Rastogi: "Further studies on the immunogenicity of Haemophilus influenzae type b and pneumococcal type 6 A polysaccharide-protein conjugates," *Infect. Immun.* **40** (1983) 245 – 256.
46. R. Schneerson, J. R. Robbins, C. Chu, A. Sutton et al.: "Serum antibody response of juvenile and infant rhesus monkeys injected with Haemophilus influenzae type b and pneumococcus type 6 A capsular polysaccharide-protein conjugates," *Infect. Immun.* **45** (1984) 582 – 591.
47. R. Schneerson, J. B. Robbins, J. C. Parke, Jr., C. Bell et al.: "Quantitative and qualitative analyses of serum antibodies elicited in adults by Haemophilus influenzae type b and pneumococcus type 6 A capsular polysaccharide-tetanus toxoid conjugates," *Infect. Immun.* **52** (1986) 519 – 528.
48. A. D. O'Brien, M. R. Thompson, P. Gemski, B. P. Doctor et al.: "Biological properties of Shigella flexneri 2 A toxin and its serological relationship to Shigella dysenteriae 1 toxin," *Infect. Immun.* **15** (1977) 796 – 798.
49. A. R. Higgins, T. M. Floyd, M. A. Kader: "Studies in shigellosis. III. A controlled

evaluation of a monovalent Shigella vaccine in a highly endemic environment," *Am. J. Trop. Med. Hyg.* **4** (1955) 281 – 288.
50. D. M. Mel, E. J. Gangarosa, M. L. Radovanović, B. L. Arsić et al.: "Studies on vaccination against bacillary dysentery. 6. Protection of children by oral immunization with streptomycin-dependent Shigella strains," *Bull. W.H.O.* **45** (1971) 457 – 464.
51. M. M. Levine, P. A. Rice, E. J. Gangarosa, G. K. Morris et al.: "An outbreak of Sonnei shigellosis in a population receiving oral attenuated shigella vaccines," *Am. J. Epidemiol.* **99** (1973) 30 – 36.
52. S. B. Formal, M. M. Levine in R. Germanier (ed.): "Shigellosis," *Bacterial Vaccines,* Academic Press, Orlando, FL 1984, pp. 167 – 186.
53. S. B. Formal, L. S. Baron, D. J. Kopecko, O. Washington et al.: "Construction of a potential bivalent vaccine strain: introduction of Shigella sonnei Form I antigen genes into the gal ε Salmonella typhi Ty 21 a typhoid vaccine strain," *Infect. Immun.* **34** (1981) 746 – 750.
54. R. E. Black, M. M. Levine, M. L. Clements, G. Losonsky et al.: "Prevention of shigellosis by a Salmonella typhi – Shigella sonnei bivalent vaccine," *J. Infect. Dis.* **155** (1987) 1260 – 1265.
55. L. S. Baron, D. J. Kopecko, S. B. Formal, R. Seid et al.: "Introduction of Shigella flexneri 2 a type and group antigen genes into oral typhoid vaccine strain Salmonella typhi Ty 21 a," *Infect. Immun.* **55** (1987) 2797 – 2801.
56. S. Sturm, K. Timmis: "Cloning of the rfb gene region of Shigella dysenteriae 1 and construction of an rfb-rfp gene cassette for the development of lipopolysaccharide-based live anti-dysentery vaccines," *Microb. Pathog.* **1** (1986) 289 – 297.
57. R. A. Finkelstein: "Immunology of cholera," *Immunol.* **69** (1975) 137 – 195.
58. R. A. Finkelstein: "Cholera," *CRC Crit. Rev. Microbiol.* **2** (1973) 553 – 623.
59. R. A. Cash, S. I. Music, J. P. Libonati, J. P. Craig et al.: "Response of man to infection with Vibrio cholerae. II. Protection from illness afforded by previous disease and vaccine," *J. Infect. Dis.* **130** (1974) 325 – 333.
60. R. A. Cash, S. I. Music, J. P. Libonati, A. R. Schwartz et al.: "Live oral cholera vaccine; evaluation of the clinical effectiveness of two strains in humans," *Infect. Immun.* **10** (1974) 762 – 764.
61. W. E. Woodward, R. H. Gilman, R. B. Hornick, J. P. Libonati et al.: "Efficacy of a live oral cholera vaccine in human volunteers," *Dev. Biol. Stand.* **33** (1976) 108 – 112.
62. M. M. Levine, R. E. Black, M. L. Clements, C. Lanata et al.: "Evaluation in humans of attenuated *Vibrio cholerae* El Tor Ogawa strain Texas Star-SR as a live oral vaccine, *Infect. Immun.* **43** (1984) 515 – 522.
63. J. B. Kaper, M. M. Levine: "Cloned cholera enterotoxin genes in study and prevention of cholera," *Lancet* **ii** (1981) 1162 – 1163.
64. J. J. Mekalanos, D. J. Suartz, G. D. N. Pearson, N. Harford et al.: "Cholera toxin genes: nucleotide sequence, deletion analysis, and vaccine development," *Nature* (London) 306 (1983) 551 – 557.
65. A. D. O'Brien, M. E. Chen, R. K. Holmes et al.: Environmental and human isolates of *Vibrio cholerae* and *Vibrio parahaemolyticus* produce a *Shigella dysenteriae* 1 (Shiga)-like cytotoxin," *Lancet* **i** (1984) 77 – 78.
66. J. D. Clements, J. R. Harris, M. R. Khan, B. A. Kay et al.: "Field trial of oral cholera vaccine in Bangladesh," *Lancet* **ii** (1986) 124 – 127.
67. J. E. McGowan, M. W. Barnes, M. Finland: "Bacteremia at Boston City Hospital; occurrence and mortality during 12 selected years (1935 – 1972), with special reference to hospital-acquired cases," *J. Infect. Dis.* **132** (1975) 316 – 335.
68. C. S. Bryan, K. L. Reynolds, E. R. Brenner: "Analysis of 1186 episodes of gram-negative bacteremia in non-university hospitals; the effects of antimicrobial therapy," *Rev. Infect. Dis.* **5** (1983) 629 – 638.
69. R. W. Haley: *Preliminary cost-benefit analysis of hospital infection control programs (The SENIC Project),* Proc. Internat. Workshop, Baiersbromm, Germany (1977) pp. 93 – 95.
70. S. J. Cryz, Jr.: "Prospects for the prevention and control of gram-negative nosocomial infections," *Vaccine* **5** (1987) 261 – 265.
71. W. R. Jarvis, J. W. White, V. P. Munn, J. L. Mosser et al.: "Nosocomial infection surveillance," 1983. Centers for Disease Control Surveillance Summaries **33**, no. 299, pp. 955 – 2155 (1985).
72. M. S. Pollack, L. S. Young: "Protective activity of antibodies to exotoxin A and lipopolysaccharide at the onset of Pseudomonas aeruginosa septicemia in man," *J. Clin. Invest.* **63** (1979) 276 – 286.

73. S. J. Cryz, Jr. in R. Germanier (ed.): "Pseudomonas aeruginosa infections," *Bacterial Vaccines,* Academic Press, Orlando, FL 1984, pp. 317–351.
74. G. B. Pier, S. E. Bennett: "Structural analysis and immunogenicity of Pseudomonas aeruginosa immunotype 2 high molecular weight polysaccharide," *J. Clin. Invest.* **77** (1986) 491–495.
75. S. J. Cryz, Jr., E. Fürer, A. S. Cross, A. Wegmann et al.: "Safety and immunogenicity of a Pseudomonas aeruginosa O-polysaccharide-toxin A conjugate vaccine in humans," *J. Clin. Invest.* **80** (1987) 51–56.
76. S. J. Cryz, Jr., A. S. Cross, E. Fürer, N. Chariatte, et al.: "Activity of intravenous immune globulin against *Klebsiella,*" *J. Lab. Clin. Med.* **108** (1986) 182–189.
77. W. D. Welch, W. J. Martin, P. Stevens, L. S. Young: "Relative opsonic and protective activities of antibodies against Kl, O and lipid A antigens of *Escherichia coli,*" *Scand. J. Infect. Dis.* **11** (1979) 291–298.
78. B. Kaijser, S. Ahlstedt: "Protective capacity of antibodies to Escherichia coli O and K antigens," *Infect. Immun.* **17** (1977) 286–289.
79. A. S. Cross, P. Gemski, J. C. Sadoff, F. Ørskov et al.: "The importance of the Kl capsule in invasive infections caused by *Escherichia coli,*" *J. Infect. Dis.* **149** (1984) 184–193.
80. M. Granström, I. Julander, R. Möllby: "Serological diagnosis of deep Staphylococcus aureus infections by enzyme-linked immunosorbent assay (ELISA) for staphylococcal hemolysin and teichoic acid," *Scand. J. Infect. Dis. Suppl.* **41** (1983) 132–139.
81. W. W. Karakawa, J. M. Fournier, W. F. Vann, R. Arbeit et al.: "Method for the serological typing of the capsular polysaccharides of *Staphylococcus aureus,*" *J. Clin. Microbiol.* **22** (1985) 445–447.
82. H. K. Hochkeppel, D. G. Braun, W. Visher, A. Imm et al.: "Serotyping and electron microscopy studies of Staphylococcus aureus clinical isolates with monoclonal antibodies to capsular polysaccharide types 5 and 8," *J. Clin. Microbiol.* **25** (1987) 526–530.
83. R. D. Arbeit, J. M. Nelles: "Capsular polysaccharide antigenemia in rats with experimental endocarditis due to *Staphylococcus aureus,*" *J. Infect. Dis.* **155** (1987) 242–246.
84. L. M. Mutharia, G. Crockford, W. C. Bogard, Jr., R. E. W. Hancock: "Monoclonal antibodies specific for Escherichia coli J5 lipolysaccharide: cross-reaction with other gram-negative bacterial species," *Infect. Immun.* **45** (1984) 631–636.
85. E. J. Ziegler, J. A. McCutchan, H. Douglas, A. I. Braude: "Treatment of E. coli and Klebsiella bacteremia in agranulocytopenic animals with antiserum to a UDP-GAL epimerase-deficient mutant," *J. Immunol.* **111** (1973) 433–438.
86. E. J. Ziegler, J. A. McCutchan, J. Fierer, M. P. Glauser et al.: "Treatment of gramnegative bacteremia and shock with human antiserum to a mutant of *Escherichia coli,*" *N. Engl. J. Med.* **307** (1982) 1225–1230.
87. J. D. Baumgartner, M. P. Glauser, J. A. McCutchan, E. J. Ziegler et al.: "Prevention of gramnegative shock and death in surgical patients by antibody to endotoxin core glycolipid," *Lancet* **ii** (1985) 59–63.
88. J. D. Band, M. E. Chamberland, T. Platt, R. E. Weaver et al.: "Trends in meningococcal disease in the United States, 1975–1980," *J. Infect. Dis.* **148** (1983) 754–758.
89. E. C. Gotschlich in R. Germanier (ed.): "Meningococcal meningitis," *Bacterial Vaccines,* Academic Press, Orlando, FL 1984, pp. 237–255.
90. L. Lapeyssonnie: "La meningite cerebro-spinale en Afrique," *Bull. W.H.O.* **28** (1963) 1–114.
91. E. C. Gotschlich, I. Goldschneider, M. S. Artenstein: "Human immunity to the meningococcus. IV. Immunogenicity of group A and group C meningococcal polysaccharides in human volunteers; *J. Exp. Med.* **129** (1969) 1367–1384.
92. M. S. Artenstein, R. Gold, J. G. Zimmerly, F. A. Wyle et al.: "Prevention of meningococcal disease by group C polysaccharide vaccine," *N. Engl. J. Med.* **282** (1970) 417–420.
93. P. H. Mäkelä et al.: "Effect of group-A meningococcal vaccine in army recruits in Finland," Lancet **ii** (1975) 883–886.
94. M. H. Wahdan, S. A. Sallam, M. N. Hassan, A. A. Gawad et al.: "A second controlled field trial of a serogroup A meningococcal vaccine in Alexandria," *Bull. W.H.O.* **55** (1977) 645–651.
95. H. Peltola, P. H. Mäkelä, H. M. Käyhty, H. Jousimies et al.: "Clinical efficacy of meningococcus group A capsular polysaccharide vaccine in children three

months to five years of age," *N. Engl. J. Med.* **297** (1977) 686 – 691.
96. H. J. Jennings, C. Lugowski: "Immunochemistry of groups A, B and C meningococcal polysaccharide-tetanus toxoid conjugates," *J. Immunol.* **127** (1981) 1011 – 1018.
97. C. E. Frasch: "Prospects for the prevention of meningococcal disease: special reference to group B," *Vaccine* **5** (1987) 3 – 4.
98. W. D. Zollinger, R. E. Mandrell, J. M. Griffis in J. B. Robbins, J. C. Hill, J. C. Sadoff (eds.): "Enhancement of immunological activity by monovalent complexing of meningococcal group B polysaccharide and outer membrane protein," *Seminars in Infectious Disease,* vol. 4, Thieme-Stratton, New York 1982, pp. 254 – 262.
99. C. E. Frasch, G. Coetzee, D. M. Zahradrik, L. Y. Wang in G. K. Schoolnik (ed.): "New developments in meningococcal vaccines," *Pathogenic Neisseria,* American Society for Microbiology, Washington, D.C. 1985, pp. 633 – 640.
100. F. M. Collins in R. Germanier (ed.): "Tuberculosis," *Bacterial Vaccines,* Academic Press, Orlando, FL 1984, pp. 373 – 418.
101. R. G. Ferguson, A. B. Simes: "BCG vaccination of Indian infants in Saskatchewan," *Tubercle* **30** (1949) 5 – 11.
102. S. V. Boyden: "The effect of previous injections of tuberculoprotein on the development of tuberculin sensitivity following B.C.G. vaccination in guinea pigs," *Br. J. Exp. Pathol.* **38** (1957) 611 – 617.
103. A. Lotte, O. Wasz-Höckert, F. Lert, N. Dumitrescu et al.: "Balance of risks and balance of cost regarding tuberculosis and antituberculosis vaccination," *Dev. Biol. Stand.* **43** (1979) 111 – 119.
104. R. Torriani, A. Zimmermann, A. Morell: "Die BCG-Sepsis als letale Komplikation der BCG-Impfung," *Schweiz. Med. Wochenschr.* **109** (1979) 708 –713.
105. W. Hennessen, H. Freudenstein, H. Engelhardt: "Observations on a BCG vaccine causing adverse reactions in newborns," *J. Biol. Stand.* **5** (1977) 139 – 146.
106. D. B. Travers: "BCG vaccination," *Lancet* **i** (1981) 1001 – 1002.
107. M. C. Lu, M. H. Lien, R. E. Becker, H. C. Heine et al.: "Genes for immunodominant protein antigens are highly homologous in Mycobacterium tuberculosis, Mycobacterium africanum, and the vaccine strain Mycobacterium bovis BCG," *Infect. Immun.* **55** (1987) 2378 – 2382.
108. B. Moss, C. Flexner: "Vaccinia virus expression vectors," *Ann. Rev. Immunol.* **5** (1987) 305 – 324.
109. H. Leffler, C. Svanborg-Eden: "Glycolipid receptors for uropathogenic E. coli attaching to human urinary tract epithelial cells and agglutinating human erythrocytes," *FEMS Microbiol. Lett.* **8** (1980) 127 – 134.
110. K. Rauss, I. Kétyi, E. Matusovits, L. Szendrei et al.: "Specific oral prevention of infantile enteritis. III. Experiments with corpuscular vaccine," *Acta. Microbiol. Acad. Sci. Hung.* **19** (1972) 19 – 28.
111. E. Fürer et al.: "Protection against colibacillosis in neonatal piglets by immunization of dams with procholeragenoid," *Infect. Immun.* **35** (1982) 887 – 894.
112. L. Doberescu, C. Huygelen: "Protection of piglets against neonatal *E. coli* enteritis by immunization of the sow with a vaccine containing heat-labile enterotoxin (LT), *Zentralbl. Veterinärmed. Reihe A* **23** (1976) 79 – 88.
113. R. L. Morgan, R. E. Isaacson, H. W. Micon, C. C. Brinton et al.: "Immunization of suckling pigs against enterotoxigenic Escherichia coli-induced diarrheal disease by vaccinating dams with purified 987 or K 99 pili; protection correlates with pilus homology of vaccine and challenge," *Infect. Immun.* **22** (1978) 771 – 777.
114. M. M. Levine, R. E. Black, C. C. Brinton, Jr., M. L. Clements et al.: "Reactogenicity, immunogenicity and efficacy studies of *Escherichia coli* type 1 somatic pili parenteral vaccine in man, *Scand. J. Infect. Dis. Suppl.* **33** (1982) 83 – 95.
115. D. G. Evans, D. J. Evans, Jr., A. R. Opekun, D. Y. Graham in A. Tagliabue, R. Rappuoli, S. E. Piazzi (eds.): "Oral whole cell vaccine protective against enterotoxigenic Escherichia coli diarrhea," *Bacterial Vaccines and Local Immunity,* Edita Da Sclavo Sp.A., Siena, Italy 1986, pp. 155 – 156.
116. M. M. Levine in R. Germanier (ed.): *Escherichia coli* infections," Bacterial Vaccines, Academic Press, Orlando, FL 1984 pp. 187 – 235.
117. F. A. Klipstein, R. F. Engert, R. A. Houghton: "Protection in rabbits immunized with a vaccine of Escherichia coli heat-stable toxin cross-linked to the heat-labile toxin B subunit," *Infect. Immun.* **40** (1983) 888 – 893.

118. J. Holmgren: "Actions of cholera toxin and the prevention and treatment of cholera," *Nature (London)* **292** (1981) 413 – 417.
119. R. A. Finkelstein, C. V. Sciortino, L. C. Rieke, M. F. Burks et al.: "Preparation of procholigenoids from Escherichia coli heat-labile enterotoxins," *Infect. Immun.* **45** (1984) 518 – 521.
120. M. A. Schmidt, P. O'Hanley, G. K. Schoolinik in A. Tagliabue, R. Rappuoli, S. E. Piazzi (eds.): Synthetic peptides protect against infection in a model of experimental pyelonephritis in mice," *Bacterial Vaccines and Local Immunity,* Edita Da Sclavo Sp.A., Siena, Italy 1986, pp. 389 – 398.
121. E. C. Gotschlich in R. Germanier (ed.): ''Gonorrhea," *Bacterial Vaccines,* Academic Press, Orlando, FL 1984, pp. 353 – 371.
122. D. S. Kellogg, Jr., I. R. Cohen, L. C. Norins, A. L. Schroeter et al.: "Neisseria gonorrhoeae. II. Colonial variation and pathogenicity during 35 months in vitro," *J. Bacteriol.* **96** (1968) 596 – 605.
123. J. Swanson, S. J. Kraus, E. C. Gotschlich: "Studies on gonococcus infection. I. Pili and zones of adhesion; their relation to gonococcal growth patterns," *J. Exp. Med.* **134** (1971) 886 – 906.
124. L. Greenberg, B. B. Diena, F. A. Ashton, R. Wallace et al.: "Gonococcal vaccine studies in Inuvik," *Can. J. Public Health* **65** (1974) 29 – 33.
125. E. C. Tramont, J. W. Boslego: "Pilus vaccines," *Vaccine* **3** (1985) 3 – 10.
126. C. C. Brinton, S. W. Wood, A. Brown, A. M. Labik et al. in : J. B. Robbins, J. Hill, J. C. Sadoff (eds.): "The development of a Neisseria pilus vaccine for gonorrhea and meningococcal meningitis," *Seminars in Infectious Disease,* vol. 4, Bacterial Vaccines, Thieme-Stratton, Inc., New York 1982, pp. 140 – 159.
127. G. K. Schoolinik, J. Y. Tai, E. C. Gotschlich: "A pilus peptide vaccine for prevention of gonorrhea," *Prog. Allergy* **33** (1983) 314 – 331.
128. J. B. Rothbard, R. Fernandez, L. Wang, N. N. H. Teng et al.: "Antibodies to peptides corresponding to conserved sequence of gonococcal pilins block bacterial adhesion," *Proc. Nat. Acad. Sci. U.S.A* (2) (1985) 915 – 919.
129. M. S. Blake in G. G. Jackson, H. Thomas (ed.): "Functions of the outer membrane proteins of *Neisseria gonorrhoeae," The Pathogenesis of Bacterial Infections*, Springer-Verlag, Berlin 1985, pp. 51 – 66.
130. T. M. Buchanan, M. S. Siegel, K. C. S. Chen, W. A. Pearce in J. B. Robbins, J. Hill, J. C. Sadoff (ed.): "Development of a vaccine to prevent gonorrhea," *Seminars in Infectious Disease,* vol. 4, Bacterial Vaccines, Thieme-Stratton, Inc., New York 1982, pp. 160 – 164.
131. J. B. Robbins, R. Schneerson, M. Pittman in R. Germanier (ed.): *Haemophilus influenzae type b infections," Bacterial Vaccines,* Academic Press, Orlando, FL 1984, pp. 289 – 316.
132. A. Sutton, R. Schneerson, S. Kendall-Morris, J. B. Robbins: "Differential complement resistance mediates virulence of Haemophilus influenzae type b," *Infect. Immun.* **35** (1982) 95 – 104.
133. H. Peltola, H. Käyhty, A. Sivonen, P. H. Mäkelä: "Haemophilus influenzae type b capsular polysaccharide vaccine in children; a double-blind field study of 100 000 vaccines 3 months to 5 years of age in Finland," *Pediatrics* **60** (1977) 730 – 737.
134. E. A. Mortimer, "Efficacy of Haemophilus b Polysaccharide Vaccine: an Enigma," *J. Am. Med. Assoc.* **260** (1988) 1454 – 1455.
135. H. Käyhty, J. Eskola, H. Peltola, M. G. Stout et al.: "Immunogenicity in infants of a vaccine composed of Haemophilus influenzae type b capsular polysaccharide mixed with DPT or conjugated to diphtheria toxoid," *J. Infect. Dis.* **155** (1987) 100 – 106.
136. A. A. Lenoir, P. D. Granoff, D. M. Granoff: "Immunogenicity of Haemophilus influenzae type b polysaccharide-Neisseria meningitidis outer membrane protein conjugate vaccine in 2- to 6-month-old infants," *Pediatrics* **80** (1987) 283 – 287.
137. J. Eskola, H. Peltola, A. K. Takala, H. Käyhty et al.: "Efficacy of Haemophilus influenzae type b polysaccharide-diphtheria toxoid conjugate vaccine in infancy," *N. Engl. J. Med.* **317** (1987) 717 –722.
138. C. D. Mitchell, H. H. Balfour, Jr.: "Measles control: So near and yet so far," *Prog. Med. Virol.* **31** (1985) 1 – 42.
139. L. E. Markowitz, R. H. Bernier: "Immunization of young infants with Edmonston-Zagreb measles vaccine," *Pediatr. Infect. Dis. J.* **6** (1987) 809 – 812.
140. WHO Expert Committee on Biological Standardization: *WHO Tech. Rep. Ser.* **329** (1966) 52 – 73.

141. WHO Expert Committee on Biological Standardization: *WHO Tech. Rep. Ser.* **673** (1982).
142. Centers for Disease Control: "Measles – United States, 1986," *Morb. Mort. Wkly. Rep.* **36** (1987) 301 – 305.
143. K. J. Bart, W. A. Orenstein, A. R. Hinman: "The virtual elimination of rubella and mumps from the United States and the use of combined measles, mumps and rubella vaccines (MMR) to eliminate measles," *Dev. Biol. Stand.* **65** (1986) 45 – 52.
144. E. Norrby, G. Enders-Ruckle, V. ter Meulen: "Differences in the appearance of antibodies to structural components of measles virus after immunization with inactivated and live virus," *J. Infect. Dis.* **132** (1975) 262 – 269.
145. T. M. Varsanyi, B. Morein, A. Löve, E. Norrby: "Protection against lethal measles virus infection in mice by immune-stimulating complexes containing the hemagglutinin or fusion protein," *J. Virol.* **61**(1987) 3896 – 3901.
146. M. R. Hilleman: "Mumps vaccination," in: *Modern Trends in Medical Virology,* vol. 2, Butterworths, London (1970) pp. 241 – 261.
147. K. Penttinen, K. Cantell, P. Somer, A. Poikolainen: "Mumps vaccination in the Finnish defense forces," *Am. J. Epidemiol.* **88** (1968) 234 – 244.
148. M. R. Hilleman, R. E. Weibel, E. B. Buynak, J. Stokes, Jr. et al.: "Live, attenuated mumps-virus vaccine; 4. Protective efficacy as measured in a field evaluation," *N. Engl. J. Med.* **276** (1967) 252 – 258.
149. Centers for Disease Control: "Efficacy of mumps vaccine – Ohio," *Morb. Mort. Wkly. Rep.* **32** (1983) 391–398.
150. F. E. Andre, J. Peetermans: "Effect of simultaneous administration of live measles vaccine on the "take rate" of live mumps vaccine," *Dev. Biol. Stand.* **65** (1986) 101 – 107.
151. R. Glück, J. M. Hoskins, A. Wegmann, M. Just et al.: "Rubini, a new live attenuated mumps vaccine virus strain for human diploid cells," *Dev. Biol. Stand.* **65** (1986) 29 – 35.
152. WHO Expert Committee on biological standardization: *WHO Tech. Rep. Ser.* **760** (1987).
153. Recommendation of the Immunization Practices Advisory Committee (ACIP): "Mumps vaccine," *Morb. Mort. Wkly. Rep.* **31** (1982) 617 – 625.
154. K. Penttinen, E.-P. Helle, E. Norrby: "Differences in antibody response induced by formaldehyde inactivated and live mumps vaccines," *Dev. Biol. Stand.* **43** (1979) 265 – 268.
155. A. Löve, R. Rydbeck, G. Utter, C. örvell et al.: ''Monoclonal antibodies against fusion protein are protective in necrotizing mumps meningoencephalitis," *J. Virol.* **58** (1986) 220 – 222.
156. F. T. Perkins: "Licensed vaccines," *Rev. Infect. Dis.* **7** (1985) Suppl. 1, 73 – 76.
157. S. A. Plotkin, F. Buser: "History of RA 27/3 rubella vaccine," *Rev. Infect. Dis.* **7** (1985) Suppl. 1, 77 – 78.
158. WHO Expert Committee on Biological Standardization: *WHO Tech. Rep. Ser.* **610** (1977) 54 – 84.
159. WHO Expert Committee on Biological Standardization: *WHO Tech. Rep. Ser.* **658** (1981).
160. J. A. Dudgeon: "Selective immunization; Protection of the individual," *Rev. Infect. Dis.* **7** (1985) Suppl. 1, 185 – 190.
161. Recommendation of the Immunization Practices Advisory Committee (ACIP): "Rubella prevention," *Morb. Mort. Wkly. Rep.* **33** (1984) 301 – 318.
162. B. Christenson, M. Böttiger, L. Heller: "Mass vaccination program aimed at eradicating measles, mumps, and rubella in Sweden; first experience," *Br. Med. J.* **287** (1983) 389 – 391.
163. Centers for Disease Control: "Rubella vaccination during pregnancy – United States, 1971 – 1986," *Morb. Mort. Wkly. Rep.* **36** (1987) 457 – 461.
164. Recommendation of the Immunization Practices Advisory Committee (ACIP): "New recommended schedule for active immunization of normal infants and children," *Morb. Mort. Wkly. Rep* **35** (1986) 577 – 579.
165. A. Deforest, S. S. Long, H. W. Lischner, J. A. C. Girone et al.: "Simultaneous administration of measles-mumps-rubella vaccine with booster doses of diphtheria-tetanus-pertussis and poliovirus vaccines," *Pediatrics* **81** (1988) 237 – 246.
166. H. Peltola, O. P. Heinonen: "Frequency of true adverse reactions to measles-mumps-rubella vaccine," *Lancet* **i** (1986) 939 – 942.
167. A. M. Arbeter, L. Baker, S. E. Starr, S. A. Plotkin: "The combination measles, mumps, rubella and varicella vaccine in healthy children," *Dev. Biol. Stand.* **65** (1986) 89 – 93.

168. L. B. Schonberger, J. Kaplan, R. Kim-Farley, M. Moore et al.: "Control of paralytic poliomyelitis in the United States," *Rev. Infect. Dis.* **6** (1984) Suppl. 2, 424 – 426.
169. J. Salk: "Commentary: Poliomyelitis vaccination – choosing a wise policy," *Pediatr. Infect. Dis. J.* **6** (1987) 889 – 893.
170. G. S. Wilson: *The hazards of immunization,* University of London, The Athlone Press 1967.
171. A. B. Sabin: "Oral poliovirus vaccine: History of its development and use and current challenge to eliminate poliomyelitis from the world," *J. Infect. Dis.* **151** (1985) 420 – 436.
172. A. B. Sabin: "Commentary: Is there a need for a change in poliomyelitis immunization policy?" *Pediatr. Infect. Dis. J.* **6** (1987) 887 – 889.
173. M. Böttiger: "Experiences of vaccination with inactivated poliovirus vaccine in Sweden," *Dev. Biol. Stand.* **47** (1981) 227 – 232.
174. K. Lapinleimu, M. Stenvik: "Experiences with polio vaccination and herd immunity in Finland," *Dev. Biol. Stand.* **47** (1981) 241 – 246.
175. H. Bijkerk: "Poliomyelitis in the Netherlands," *Dev. Biol. Stand.* **47** (1981) 233 – 240.
176. WHO Expert Committee on Biological Standardization: *WHO Tech. Rep. Ser.* **687** (1983) 107 – 174.
177. A. M. Mcbean, J. F. Modlin: "Rationale for the sequential use of inactivated poliovirus vaccine and live attenuated poliovirus vaccine for routine poliomyelitis immunization in the United States," *Pediatr. Infect. Dis. J.* **6** (1987) 881 – 887.
178. K. Lapinleimu: "Sero-immune pattern of poliomyelitis type I and III in a population vaccinated with inactivated polio vaccine," in: *Proceedings of the 11th Symposium of the European Association Against Poliomyelitis and Allied Diseases,* Rome 1966, Brussels; European Association Against Poliomyelitis, 1967, pp. 119 – 25.
179. A. L. van Wezel, G. van Steenis, Ch. A. Hannik, H. Cohen: "New approach to the production of concentrated and purified inactivated polio and rabies tissue culture vaccines," *Dev. Biol. Stand.* **41** (1978) 159 – 168.
180. S. E. Robertson, H. P. Traverso, J. A. Drucker, E. Z. Rovira et al.: "Clinical efficacy of a new, enhanced-potency, inactivated poliovirus vaccine," *Lancet* **i** (1988) 897 – 899.
181. B. J. Montagnon, B. Fanget, A. J. Nicolas: "The large-scale cultivation of vero cells in micro-carrier culture for virus vaccine production. Preliminary results for killed poliovirus vaccine," *Dev. Biol. Stand.* **47** (1981) 55 – 64.
182. WHO Expert Committee on Biological Standardization: *WHO Tech. Rep. Ser.* **745** (1987).
183. Prevention of liver cancer: *WHO Tech. Rep. Ser.* **691** (1983).
184. A. A. McLean: "Development of vaccines against Hepatitis A and Hepatitis B," *Rev. Infect. Dis.* **8** (1986) 591 – 598.
185. W. Szmuness, C. E. Stevens, E. J. Harley, E. A. Zang et al.: "Hepatitis B vaccine; Demonstration of efficacy in a controlled clinical trial in a high-risk population in the United States," *N. Engl. J. Med.* **303** (1980) 833 – 841.
186. WHO Expert Committee on Biological Standardization: *WHO Tech. Rep. Ser.* **725** (1985) 102 – 133.
187. Recommendation of the Immunization Practices Advisory Committee (ACIP): "Recommendations for protection against viral hepatitis," *Morb. Mort. Wkly. Rep.* **34** (1985) 313 – 335.
188. Recommendation of the Immunization Practices Advisory Committee (ACIP): "Update on Hepatitis B prevention," *Morb. Mort. Wkly. Rep.* **36** (1987) 353 – 366.
189. G. Eder, J. L. McDonel, F. Dorner: "Hepatitis B vaccine," in: *Progress in Liver Diseases,* vol. 8, Grune and Stratton, New York 1986, pp. 367 – 394.
190. A. Zuckerman: "Tomorrow's hepatitis B vaccines," *Vaccine* **5** (1987) 165 – 167.
191. D. N. Standring, W. J. Rutter: "The molecular analysis of hepatitis B virus," in: *Progress in Liver Diseases,* vol. 8, Grune and Stratton, New York 1986, pp. 311 – 333.
192. WHO Expert Committee on Rabies: *WHO Tech. Rep. Ser.* **709** (1984).
193. Centers for Disease Control: "Bat-rabies – Europe," *Morb. Mort. Wkly. Rep.* **35** (1986) 430 – 432.
194. P. Sureau: "Rabies vaccine production in animal cell cultures," *Adv. Biochem. Eng. Biotechnol.* **34** (1987) 111 – 128.
195. B. D. Perry: "Rabies," *Vet. Clin. North. Am. Small Anim. Pract.* **17** (1987) 73 – 89.
196. H. Wilde, P. Chomchey, S. Prakongsri, P. Punyaratabandhu: "Safety of equine rabies immune globulin," *Lancet* **ii** (1987) 1275.

197. Centers for Disease Control: "Human rabies despite treatment with rabies immune globulin and human diploid cell rabies vaccine – Thailand," *Morb. Mort. Wkly. Rep.* **36** (1987) 759 – 765.
198. Centers for Disease Control: "Systemic allergic reactions following immunization with human diploid cell rabies vaccine," *Morb. Mort. Wkly. Rep.* **33** (1984) 185 – 187.
199. M. C. Swanson, E. Rosanoff, M. Gurwith, M. Deitch et al.: "IgE and IgG antibodies to β-propiolactone and human serum albumin associated with urticarial reactions to rabies vaccine," *J. Infect. Dis.* **155** (1987) 909 – 913.
200. M. Granström, M. Eriksson, G. Edevåg: "A sandwich ELISA for bovine serum in viral vaccines," *J. Biol. Stand.* **15** (1987) 193 – 197.
201. B. Morein, K. Simons: "Subunit vaccines against enveloped viruses; virosomes, micelles and other protein complexes," *Vaccine* **3** (1985) 83 – 93.
202. C. H. Stuart-Harris, G. C. Schild: "Influenza," *The Viruses and Disease,* E. Arnoldt Ltd, London 1976.
203. P. F. Wright, D. T. Karzon: "Live Attenuated Influenza Vaccines," *Prog. Med. Virol.* **34** (1987) 70 – 88.
204. WHO Expert Committee on Biological Standardization: *WHO Techn. Rep. Ser.* **638** (1979).
205. Recommendations of the Immunization Practices Advisory Committee (ACIP): "Prevention and Control of Influenza," *Morb. Mort. Wkly. Rep.* **36** (1987) 373 – 387.
206. P. A. Gross, A. L. Gould, A. E. Brown: "Effect of cancer chemotherapy on the immune response to influenza virus vaccine: Review of published studies," *Rev. Infect. Dis.* **7** (1985) 613 – 618.
207. M. A. Strassburg, S. Greenland, F. J. Sorvillo, L. E. Lieb et al.: "Influenza in the elderly: report of an outbreak and a review of vaccine effectiveness reports," *Vaccine* **4** (1986) 38 – 44.
208. J. M. Ostrove, G. Inchauspe: "The biology of varicella-zoster virus" in: S. E. Straus "Varicella-zoster virus infections: biology, natural history, treatment, and prevention," *Ann. Intern. Med.* **108** (1988) 221 –237.
209. M. Takahashi: "Clinical overview of varicella vaccine; Development and early studies," *Pediatrics* **78** (1986) Suppl., 736 – 741.
210. R. E. Weibel, B. J. Neff, B. J. Kuter, H. A. Guess et al.: "Live attenuated varicella virus vaccine. Efficacy trial in healthy children," *N. Engl. J. Med.* **310** (1984) 1409 – 1415.
211. A. A. Gershon, S. P. Steinberg, L. Gelb, and The National Institute of Allergy and Infectious Diseases Varicella Vaccine Collaborative Study Group: "Live attenuated varicella vaccine use in immunocompromised children and adults," *Pediatrics* **78** (1986) Suppl., 757 – 762.
212. Arthropod-borne and rodent-borne viral diseases: *WHO Tech. Rep. Ser.* **719** (1985).
213. Viral haemorrhagic fevers: *WHO Tech. Rep. Ser.* **721** (1985).
214. K. C. Smithburn, C. Durieux, R. Koerber, H. A. Penna et al.: *Yellow Fever Vaccination,* World Health Organization, Geneva 1956.
215. J. C. Roche, A. Jouan, B. Brisou, R. Rodhain et al.: "Comparative clinical study of a new 17 D thermostable yellow fever vaccine," *Vaccine* **4** (1986) Suppl., 163 – 165.
216. WHO Expert Committee on Biological Standardization: *WHO Tech. Rep. Ser.* **594** (1976) 23 – 49.
217. A. A. Smorodintsev, A. V. Dubov, V. I. Ilyenko, V. G. Platonov: "A new approach to development of live vaccine against tick-borne encephalitis," *J. Hyg.* **67** (1969) 13 – 20.
218. C. Kuntz, F. X. Heinz, H. Hofmann: "Immunogenicity and reactogenicity of a highly purified vaccine against tick-borne encephalitis," *J. Med. Virol.* **6** (1980) 103 – 109.
219. A. Oya in H. Fukumi (ed.): "Japanese encephalitis vaccine," *The vaccination. Theory and practice,* International Medical Foundation of Japan (1975) pp. 69 – 82.
220. WHO Expert Committee on Biological Standardization: *WHO Tech. Rep. Ser.* **323** (1966) 56 – 71.
221. Recommendation of the Immunization Practices Advisory Committee (ACIP): *Smallpox Vaccine Morb. Mort. Wkly. Rep.* **34** (1985) 341 – 342.
222. R. Randall, L. N. Binn, V. R. Harrison: "Immunization against Rift Valley fever virus," *J. Immunol.* **93** (1964) 293 – 299.
223. B. Niklasson: "Rift Valley fever virus vaccine trial; Study of side effects in humans," *Scand. J. Infect. Dis.* **14** (1982) 105 – 109.
224. P. M. Kaye, *Parasitol. Today* **3** (1987) 293 – 299.
225. M. D. Rickard, M. J. Howell in E. E. R. Taylor, J. R. Baker (ed.): *In Vitro Methods for Parasite Cultivation,* Academic Press, London 1987, pp. 407 – 451.

226. G. Piekarski: *Medizinische Parasitologie,* Springer Verlag, Berlin 1987.
227. V. Houba in V. Houba (ed.): *Immunological Investigation of Tropical Parasitic Diseases,* Churchill Livingstone, Edinburgh 1980, pp. 130 – 147.
228. P. Hagan, P. J. Moore, A. B. Adjukiewicz, B. M. Greenwood et al.: *Parasite Immunol.* **7** (1985) 617 – 624.
229. A. E. Butterworth, M. Capron, J. S. Cordingley, P. R. Dalton, *Trans. R. Soc. Trop. Med. Hyg.* **79** (1985) 393 – 408.
230. G. Gazzinelli, J. R. Lambertucci, N. Katz, R. S. Rocha, *J. Immunol.* **135** (1985) 2121 – 2127.
231. C. W. Todd, R. W. Goodgame, D. Colley, *J. Immunol.* **122** (1979) 1440 – 1446.
232. R. F. Sturrock, R. Kimani, B. I. Cottrell, A. E. Butterworth, *Trans. R. Soc. Trop. Med. Hyg.* **77** (1983) 363 – 371.
233. A. E. Butterworth, R. F. Sturrock, V. Houba, A. A. F. Mahmoud, Nature *(London)* **256** (1975) 727 – 729.
234. A. E. Butterworth, *Adv. Parasitol.* **23** (1984) 143 – 235.
235. S. R. Smithers, R. J. Terry, *Adv. Parasitol.* **14** (1976) 399 – 422.
236. A. Capron, *Fortschr. Zool.* **27** (1982) 259 – 264.
237. A. E. Butterworth, P. Hagan, *Parasit. Today* **3** (1987) 11 – 16.
238. A. J. G. Simson, D. Cioli, *Parasit. Today* **3** (1987) 26 – 28.
239. C. Dissous, I. M. Crzych, A. Capron, *J. Immunol.* **129** (1982) 2232 – 2234.
240. Q. D. Bickle, B. I. Andrews, M. G. Taylor, *Parasit Immunol.* **8** (1986) 95 – 107.
241. C. Kelly, A. I. G. Simpson, E. Fox, S. M. Phillips et al.: *Parasit Immunol.* **8** (1986) 193 – 198.
242. J. M. Balloul, P. Sondermeyer, D. Dreyer, M. Capron, *Nature (London)* **326** (1987) 149 – 153.
243. D. Wakelin: *Immunity to Parasites,* Edward Arnold Ltd., London 1984.
244. S. Friedmann, *Gen Eng. News* (1985) 24 – 25.
245. V. K. Vinayak, N. K. Gupta, A. K. Chopra, G. L. Sharma et al., *Parasitology* **82** (1981) 375 – 382.
246. P. Ambroise-Thomas in V. Houba (ed.): *Immunological Investigation of Tropical Parasitic Diseases,* Churchill Livingstone, Edinburgh 1980, pp. 84 – 103.
247. B. M. Ogilvie, M. J. Worms in S. Cohen, E. H. Sadun (eds.): *Immunology of Parasitic Infections,* Blackwell Scientific Publ., Oxford 1976, 397 – 402.
248. M. M. Wong, H. I. Fredericks, C. P. Ramachandran, *Bull. W.H.O.* **40** (1969) 493 – 501.
249. C. P. Ramachandran, *Southeast Asian J. Trop. Med. Public Health* **1** (1970) 78 – 92.
250. A. Haque, A. Capron in T. W. Pearson (ed.): *Parasite Antigens,* Marcel Dekker Inc., New York 1986, pp. 317 – 402.
251. R. M. Maizels, F. Partono, Sri Oemijati, D. A. Denham et al., *Parasitology* **87** (1983) 249 – 263.
252. M. D. Rickard in C. Arme, P. W. Pappas (eds.): *Biology of the Eucestoda,* Academic Press, London 1983, pp. 539 – 579.
253. M. D. Rickard, F. F. Williams, *Adv. Parasitol.* **21** (1982) 229 – 296.
254. R. P. Herd, R. I. Chappel, D. Biddell, *Int. J. Parasitol.* **5** (1975) 395 – 399.
255. D. J. Jenkins, M. D. Rickard, *Aust. Vet. J.* **63** (1986) 40 – 42.
256. R. S. Nussenzweig, J. Vanderberg, H. Most, C. Orton, *Nature (London)* **216** (1967) 160 – 162.
257. W. A. Siddiqui, *Science* **197** (1977) 388 – 389.
258. G. H. Mitchell, W. H. G. Richards, G. A. Butcher, S. Cohen, *Lancet (i)* (1977) 1335 – 1358.
259. W. Trager, J. B. Jensen, *Science* **193** (1976) 673 – 675.
260. D. Mazier, R. L. Beaudoin, S. Mellouk, P. Druilhe et al., *Science* **227** (1985) 440 – 442.
261. C. Guerin-Marchand, P. Druilhe, B. Galey, A. Londono et al., *Nature (London)* **329** (1987) 164 – 167.
262. A. H. Cochrane in J. P. Kreier (ed.): *Malaria, Immunology and Immunization* vol. 3, Chap. 4, Academic-Press, New-York, 1980.
263. D. F. Clyde, *Am. J. Trop. Med. Hyg.* **24** (1975) 397 – 401.
264. E. H Nardin, R. S. Nussenzweig, I. A. McGregor, J. H. Bryan, *Science* **206** (1979) 597 – 599.
265. G. Del Giudice, H. D. Engers, C. Tougne, S. S. Biro et al., *Am. J. Trop. Med. Hyg.* **36** (1987) 203 – 212.
266. D. H. Chen, R. E. Tigelaar, F. I. Weinbaum, *J. Immunol.* **118** (1977) 1322 – 1327.
267. A. Ferreira, L. Schofield, V. Enea, H. Schellekins et al., *Science* **232** (1986) 881 – 884.
268. P. Potocnjak, N. Yoshida, R. S. Nussenzweig, V. Nussenzweig, *J. Exp. Med.* **151** (1980) 1504 – 1513.
269. D. Mazier, S. Mellouk, R. L. Beaudoin, B. Texier et al., *Science* **231** (1986) 156 – 159.

270. J. B. Dame, J. L. Williams, T. F. McCutchan, J. L. Weber et al., *Science* **225** (1984) 593 – 599.
271. V. Enea, J. Ellis, F. Zavala, D. E. Arnot et al., *Science* **225** (1984) 628 – 630.
272. D. J. Kemp, R. L. Coppel, R. F. Anders, *Ann. Rev. Microbiol.* **41** (1987) 181 – 208.
273. V. F. De la Cruz, A. A. Lal, T. F. McCutchan, *J. Biol. Chem.* **262** (1987) 11935 – 11941.
274. M. J. Lockyer, R. T. Schwarz, *Mol. Biochem. Parasitol.* **22** (1987) 101 – 107.
275. G. Del Giudice, A. S. Verdini, M. Pinori, J. P. Verhave et al., *J. Clin. Microbiol.* **25** (1987) 91 – 96.
276. S. L. Hoffman, R. Wistar, Jr., W. R. Ballou, M. R. Hollingdale et al., *New-Engl. J. Med.* **315** (1986) 601 – 603.
277. D. A. Herrington, D. F. Clyde, G. Losonsky, M. Cortesia et al., *Nature (London)* **328** (1987) 257 – 259.
278. W. R. Ballou, S. L. Hoffman, J. A. Sherwood, M. R. Hollingdale et al., *Lancet* **I** (1987) 1277 – 1279.
279. G. Del Giudice, J. A. Cooper, J. Merino, A. S. Verdini et al., *J. Immunol.* **137** (1986) 2952 – 2955.
280. M. F. Good, J. A. Berzofsky, W. L. Maloy, Y. Hayashi et al., *J. Exp. Med.* **164** (1986) 655 –660.
281. A. R. Togna, G. Del Giudice, A. S. Verdini, F. Bonetti et al., *J. Immunol.* **137** (1986) 2956 – 2960.
282. M. F. Good, D. Pombo, I. A. Quakyi, E. M. Riley et al., *Proc. Natl. Acad. Sci. USA* **85** (1988) 1199 – 1203.
283. L. H. Miller, S. J. Mason, D. F. Clyde, M. H. McGinis, *New-Eng. J. Med.* **295** (1976) 302 – 304.
284. M. J. Miller, *Trans. Roy. Soc. Trop. Med. Hyg.* **52** (1958) 152 – 158.
285. S. Cohen, I. A. McGregor, S. C. Carrington, *Nature (London)* **192** (1961) 733 – 735.
286. A. Celada, A. Cruchau, L. H. Perrin, *Clin. Exp. Immunol.* **47** (1982) 635 – 641.
287. P. H. David, *Proc. Natl. Acad. Sci. USA* **80** (1983) 5075 – 5081.
288. L. H. Miller, R. J. Howard, R. Carter, M. F. Good et al., *Science* **234** (1986) 1349 – 1356.
289. L. H. Perrin, E. Ramirez, L. Er-Hsiang, P.-H. Lambert; *Clin. Exp. Immunol.* **41** (1980) 91 – 106.
290. A. A. Holder, R. R. Freeman, *J. Exp. Med.* **156** (1982) 1528 – 1538.
291. J. S. McBride, D. Walliker, G. Morgan, *Science* **217** (1982) 254 – 256.
292. A. A. Holder, M. J. Lockyer, K. G. Odink, J. S. Sandhu et al., *Nature (London)* **317** (1985) 270 – 273.
293. K. G. Oding, M. J. Lockyer, S. C. Nicholls, Y. Hillman et al., *FEBS Lett.* **173** (1984) 108 – 112.
294. M. Mackay, M. Goman, N. Bone, J. E. Hyde et al., *EMBO J.* **4** (1985) 3823 – 3829.
295. L. H. Perrin, B. Merkli, M. Loche, C. Chizzolini et al., *J. Exp. Med.* **160** (1984) 441 – 447.
296. R. Hall, J. E. Hyde, M. Goman, D. L. Simmons et al., *Nature (London)* **311** (1984) 379 – 382.
297. W. A. Siddiqui, L. Q. Tam, K. J. Kramer, G. S. N Hui et al., *Proc. Natl. Acad. Sci. USA* **84** (1987) 3014 – 3018.
298. A. A. Holder, R. R. Freeman, *Nature (London)* **294** (1981) 361 – 363.
299. M. E. Patarroyo, P. Romero, M. L. Torres, P. Clavijo et al., *Nature (London)* **328** (1985) 629 – 631.
300. A. Cheung, J. Leban, A. R. Shaw, B. Merkli et al., *Proc. Natl. Acad. Sci. USA* **83** (1986) 8323 – 8332.
301. L. H. Perrin, R. Dayal, *Immunol. Rev.* **61** (1982) 245 – 268.
302. R. R. Freeman, A. J. Trejosiewicz, G. A. Cross, *Nature (London)* **284** (1980) 366 – 369.
303. L. H. Perrin, E. Ramirez, P.-H. Lambert, P. A. Miescher, *Nature (London)* **289** (1981) 301 – 303.
304. C. Braun-Breton, M. Jendoubi, E. Brunet, L. Perrin et al., *Mol. Bioch. Parasitol.* **20** (1986) 33 – 38.
305. C. Braun-Breton, T. L. Rosenberg, L. Pereira Da Silva, *Nature (London)* **332** (1988) 457 – 459.
306. U. Certa, P. Ghersa, H. Dobeli, H. Matile et al., *Science* **240** (1988) 1036 – 1038.
307. P. Dubois, J. P. Dedet, T. Fandeur, C. Roussilhon et al., *Proc. Natl. Acad. Sci. USA* **81** (1984) 229 – 335.
308. L. H. Perrin, B. Merkli, M. S. Gabra, J. Stocker et al., *J. Clin. Invest.* **75** (1985) 1718 – 1725.
309. J. F. Dubremetz, P. Delplace, B. Fortier, G. Tronchin et al., *Mol. Biochem. Parasitol.* **27** (1988) 135 – 142.
310. R. L. Coppel, J. G. Culvenor, A. E. Bianco, P. E. Crewther et al., *Mol. Biochem. Parasitol.* **25** (1987) 73 – 81.
311. H. Perlamnn, K. Berzins, M. Wahlgren, J. Carlsson et al., *J. Exp. Med.* **159** (1984) 1686 – 1693.

312. G. V. Brown, I. G. Culvenor, P. E. Crewther, A. E. Bianco et al., *J. Exp. Med.* **162** (1985) 774–779.
313. B. Wahlin, M. Wahlgren, H. Perlamnn, H. Berzins et al., *Proc. Natl. Acad. Sci. USA* **81** (1984) 7912–7916.
314. J. M. Favarolo, R. L. Coppel, L. M. Corcoran, S. J. Foote et al., *Nucl. Acid. Res.* **14** (1986) 8265–8277.
315. W. E. Collins, R. F. Anders, M. Pappaioanou, G. H. Campbell et al., *Nature (London)* **323** (1986) 259–262.
316. S. A. Luse, L. H. Miller, *Am. J. Trop. Med. Hyg.* **20** (1971) 655–670.
317. T. Triglia, H. D. Stahl, P. E. Crewther, D. Scanlon et al., *EMBO J.* **6** (1987) 1413–1419.
318. M. Koenen, A. Scherf, O. Mercereau, G. Langsley et al., *Nature (London)* **311** (1984) 382–385.
319. I. J. Udeinya, L. H. Miller, I. A. McGregor, J. B. Jensen, *Nature (London)* **231** (1983) 429–431.
320. K. Marsh, R. J. Howard, *Science* **231** (1986) 150–153.
321. S. Pollack, J. Fleming, *Brit. J. Haematol.* **58** (1984) 289–294.
322. M. E. Perkins, *J. Exp. Med.* **160** (1984) 788–797.
323. D. Camus, T. J. Hadley, *Science* **230** (1985) 553–557.
324. P. Delplace, B. Fortier, G. Tronchin, J. F. Dubremetz et al., *Mol. Biochem. Parasitol.* **23** (1987) 193–202.
325. M. E. Patarroyo, R. Amador, P. Lavijo, A. Moreno et al., *Nature (London)* **332** (1988) 158–161.
326. R. W. Gwadz, *Science* **193** (1976) 1150–1153.
327. K. N. Mendis, G. A. T. Targett, *Nature (London)* **277** (1979) 289–391.
328. R. W. Gwadz, L. C. Koontz, *Infect. Immun.* **44** (1984) 137–145
329. N. Kumar, R. Carter, *Mol. Biochem. Parasitol.* **13** (1984) 333–341.
330. K. Mendis, P. Udayama, R. Carter, P. H. David, *J. Cell Biochem. Suppl.* **10 A** (1986) 149–156.
331. A. N. Vermuelen, T. Ponnudurai, P. J. A. Beckers, J. P. Verhave et al., *J. Exp. Med.* **162** (1985) 1460–1476.
332. J. Rener, P. M. Graves, R. Carter, J. L. Williams et al., *J. Exp. Med.* **158** (1983) 976–980.
333. P. M. Graves, R. Carter, T. R. Burkot, J. Rener et al., *Infect. Immun.* **48** (1985) 611–619.
334. E. R. Stiehm, E. Ashida, K. S. Kim, D. J. Winston et al., *Ann. Intern. Med.* **107** (1987) 367–82.
335. E. R. Stiehm, *Pediatrics* **63** (1979) 301–19.
336. J. A. Finlayson, in C. S. F. Easmon, J. Jeljaszewics (eds.): *Medical Microbiology,* **vol. 1,** Academic Press, London 1982, pp. 129–82.
337. R. Heffron: *Pneumonia, With Special Reference to Pneumococcus Lobar Pneumonia,* The Commonwealth Fund, New York 1939.
338. R. H. Rousell, M. S. Collins, M. B. Dobkin, R. E. Louie et al., *Am. J. Med. Suppl.* **76** (1984) 40–145.
339. H. E. Alexander, *Am. J. Dis. Child.* **66** (1943) 172–187.
340. H. E. Alexander, *Am. J. Dis. Child.* **66** (1943) 160–171.
341. E. Merler, F. S. Rosen, *N. Engl. J. Med.* **275** (1966) 480–86; 536–542.
342. C. A. Janeway, F. S. Rosen, *N. Engl. J. med.* **275** (1966) 826–831.
343. P. A. M. Gross, D. Gitlin, C. A. Janeway, *N. Engl. J. Med.* **260** (1959) 170–178.
344. J. Stokes, E. P. Maris, S. S. Gellis, *J. Clin. Invest.* **23** (1944) 531–540.
345. J. Stokes, J. R. Neefe, *JAMA J. Am. Med. Assoc.* **127** (1945) 531–540.
346. W. M. Hammon, L. L. Coriell, J. Stokes et al., *JAMA J. Am. Med. Assoc.* **150** (1950) 739–49; 750–756.
347. C. A. Janeway, L. Apt, D. Gitlin, *Trans. Assoc. Am. Physician* **66** (1953) 200–202.
348. P. A. M. Gross, D. Gitlin, C. A. Janeway, *N. Engl. J. Med.* **260** (1959) 121–125.
349. R. G. Hamilton, *Clin. Chem. (Winston Salem N.C.)* **33** (1987) 1707–1725.
350. G. R. Siber, P. H. Schur, A. C. Weitzman, G. Schiffman, *N. Engl. J. Med.* **303** (1980) 178–182.
351. C. M. Roifman, H. M. Lederman, S. Lavi, L. D. Stein et al., *Am. J. Med.* **79** (1985) 171–174.
352. S. Barandun, H. Isliker, *Vox Sang.* **51** (1986) 157–160.
353. A. Hassig, *Vox Sang.* **51** (1986) 10–17.
354. D. D. Schroeder, M. L. Dumas, *Am. J. Med. Suppl.* **76** (1984) 33–39.
355. C. A. Janeway, E. Merler, F. S. Rosen, S. Salmon et al., *N. Engl. J. Med.* **278** (1968) 919–923.
356. J. Passwell, F. S. Rosen, E. Merler in B. Alving, J. Finlayson (eds.): *Immunoglobulins; characteristics and uses of intravenous preparations,* U.S.Dept. H.H.S., Washington 1979, pp. 139–142.

357. H. D. Ochs, S. H. Fischer, R. J. Wedgwood, *J. Clin. Immunol Suppl.* **2** (1982) 22 S – 29 S.
358. C. Cunningham-Rundles, F. P. Siegal, E. M. Smithwick, A. Lion-Boule et al., *Ann. Intern. Med.* **101** (1984) 435 – 439.
359. M. S. Collins, J. H. Dorsey, *J. Infect. Dis.* **151** (1985) 1171 – 1173.
360. J. R. Schreiber, V. A. Barrus, G. R. Siber, *Infect. Immun.* **47** (1985) 142 – 148.
361. H. D. Ochs et al., *Lancet* **2** (1980) 1158 – 1159.
362. *Medical Letter* **24** (1982) 81 – 82.
363. R. H. Johnson, R. J. Ellis, *Ann. Intern. Med.* **81** (1974) 61 – 67.
364. C. Cunningham-Rundles, L. A. Hanson, W. H. Hitzig, W. Knapp et al., *Bull. W.H.O.* **60** (1982) 43 – 47.
365. S. Krugman, *N. Engl. J. Med.* **269** (1963) 195 – 201.
366. E. B. Grosman, S. G. Stewart, J. Stokes, *JAMA J. Am. Med. Assoc.* **129** (1945) 991 – 994.
367. Medical Research Council Workingparty, *Lancet* **1** (1969) 163 – 168.
368. C. L. S. Leen, P L. Yap, D. B. L. McClelland, *Vox Sang.* **51** (1986) 278 – 286.
369. *Lancet* **1** (1983) 105 – 106.
370. G. Chirico, G. Rondini, A. Plebani, A. Chiara et al., *J. Pediatr.* **110** (1987) 437 – 442.
371. J. Amer, E. Ott, F. A. Ibbott, D. O'Brien et al., *Pediatrics* **32** (1963) 4 – 9.
372. A. A. Hertler, S. C. Ross, *J. Clin. Lab. Immunol.* **21** (1986) 177 – 181.
373. S. Baron, E. V. Barnet, R. S. Goldsmith, S. Silbergeld et al., *Am. J. Hyg.* **79** (1964) 186 – 195.
374. N. A. Kefalides, J. A. Arana, A. Bazan, M. Bocanegra et al., *N. Engl. J. Med.* **267** (1962) 317 – 324.
375. H. H. Stone, C. D. Graber, J. D. Martin, L. Kolb, *Surgery St. Louis* **58** (1965) 810 – 814.
376. R. J. Jones, E. A. Roe, J. L. Gupta, *Lancet* **2** (1980) 1263 – 1265.
377. R. L. Wasserman, *Pediatric Infectious Diseases* **5** (1986) 620 – 621.
378. J. A. Steen, *Arch. Pediatr.* **77** (1960) 291 – 304.
379. A. P. Schless, G. S. Harell, *Amer. J. Med.* **44** (1968) 325 – 329.
380. M. Santosham, R. Reid, D. M. Ambrusino, M. C. Wolff et al., *Engl. J. Med.* **317** (1987) 923 – 9.
381. S. M. Rosenthal, R. C. Millican, J. Rust, *Proc. Soc. Exp. Biol. Med.* **94** (1957) 214 – 217.
382. M. W. Fisher, *Antibiot. Chemother. (Washington D.C.)* **7** (1957) 315 – 321.
383. M. W. Fisher, M. C. Manning, *Antibiot. Chemother. (Washington D.C.)* **8** (1958) 29 – 31.
384. B. A. Waisbren, *Antibiot. Chemother. (Washington D.C.)* **7** (1957) 322 – 333.
385. B. A. Waisbren, D. Lepley, *Arch. Intern. Med.* **109** (1962) 712 – 716.
386. G. P. Bodey, B. A. Nies, N. R. Mohberg, E. J. Freireich, *JAMA J. Am. Med. Assoc.* **190** (1964) 1099 – 1102.
387. K. C. Finkel, J. C. Haworth, *Pediatrics* **25** (1960) 798 – 806.
388. D. R. Snydman, B. G. Werner, B. Heinzer-Lacey, V. P. Berardi et al., *N. Engl. J. Med.* **317** (1987) 1049 – 1054.
389. R. M. Condie, R. J. O'Reilly, *Am. J. Med.* **76** (Suppl. 30 March 1984) 134 – 141.
390. A. Hagenbeek, H. G. J. Brummelhuis, A. Donkers, A. M. Dumas et al., *J. Infect. Dis.* **155** (1987) 897 – 902.
391. D. J. Winston, W. G. Ho, C.-H. Lin, M. D. Budinger et al., *Am. J. Med.* **76** (Suppl. 30 March 1984) 128 – 133.
392. R. A. Bowden, M. Sayers, N. Flournoy, B. Newton et al., *N. Engl. J. Med.* **314** (1986) 1006 – 1010.
393. E. C. Reed et al., *J. Infect. Dis.* **156** (1987) 641 –644.
394. K. Erlendsson, T. Swartz, J. M. Dwyer, *N. Engl. J. Med.* **312** (1985) 351 – 353.
395. P. J. Mease, H. D. Ochs, R. J. Wedgood, *N. Engl. J. Med.* **304** (1981) 1278 – 1281.
396. J. M. Crennan, R. E. Van Scoy, C. H. McKenna, T. F. Smith, *Am. J. Med.* **81** (1986) 35 – 42.
397. K. S. Erlich, J. Mills, *Rev. Infect. Dis.* **8** (1986) S 439 – 445.
398. D. Ambrosino, J. R. Schreiber, R. S. Daum, G. R. Siber, *Infect. Immun.* **39** (1983) 709 – 714.
399. T. E. Harper, R. D. Christensen, G. Rothstein, *Pediatr. Res.* **22** (1987) 455 – 460.
400. S. J. Cryz, A. S. Cross, E. Furer, N. Chariatte et al., *J. Lab. Clin. Med.* **108** (1986) 182 – 189.
401. I. A. Holder, J. G. Naglich, *Am. J. Med.* **76** (1984) 161 – 167.
402. J. E. Pennington, G. J. Small, *J. Infect. Dis.* **155** (1987) 973 – 978.
403. V. G. Hemming, W. T. London, G. W. Fischer, B. L. Curfman et al., *J. Infect. Dis.* **156** (1987) 655 – 658.
404. K. N. Haque, M. H. Zaidi, S. K. Haque, H. Bahakim et al., *Pediatric Infectious Diseases* **5** (1986) 622 – 625.
405. D. Sidiropoulos, U. Herrmann, A. Morell, G. von Muralt et al., *J. Pediatr. St. Louis* **109** (1986) 505 – 508.

406. W. G. Ho, D. J. Winston, K. Bartoni, R. E. Champlin et al., Abstr. 456, 1983 Interscience Confer. Infect. Dis. Chemother., Las Vegas, Nevada.
407. C. Bunch, H. Chapel, K. R. Rai, R. P. Gale, *Blood* **70** (1987) 224 a (Abstr); *N. Engl. J. Med.* **319** (1988) 902–907.
408. H. D. Ochs, *Pediatr. Infect. Dis.* **6** (1987) 509–511.
409. C. C. Wood, J. G. McNamara, D. F. Schwarz, W. W. Merrill et al., *Pediatr. Infect. Dis.* **6** (1987) 564–566.
410. B. A. Silverman, A. Rubinstein, *Am. J. Med.* **78** (1985) 728–736.
411. D. Sidiropoulos, U. Boehme, G. von Muralt, A. Morell et al., *Pediatr. Infect. Dis.* **5** (1986) 193–194.
412. S. Barandun, P. Imbach, A. Morrell, H. P. Wagner in U. E. Nydegger (ed.): *Immunohemotherapy. A guide to immunoglobulin prophylaxis and therapy*, Academic Press, New York 1981, pp. 275–282.
413. P. Imbach, H. P. Wagner, W. Berchtold, G. Gaedicke et al., *Lancet* **2** (1985) 464–468.
414. J. Fehr, V. Hoffmann, U. Kappeler, *N. Engl. J. Med.* **306** (1982) 1254–1258.
415. S. Rosthoj, G. K. Steffensen, T. K. Guld, *Acta Paediatr. Scand.* **76** (1987) 631–635.
416. R. Bohm, C. Hofstaetter, R. C. Briel, *Blut* **48** (1984) 469–470.
417. F. E. Leickly, R. H. Buckley, *Am. J. Med.* **82** (1987) 159–162.
418. G. W. Richmond, I. Ray, A. Korenblitt, *J. Pediatr. (St. Louis)* **110** (1987) 917–919.
419. M. W. Hilgartner, J. Bussel, *Am. J. Med.* **83** (1987) 35–39.
420. T. Becker, S. Panzer, D. Maas, V. Kiefel et al., *Br. J. Haematol.* **61** (1985) 149–155.
421. F. Dammacco, G. Iodice, N. Campobasso, *Br. J. Haematol.* **62** (1986) 125–135.
422. A. C. Newland, J. G. Treleaven, R. M. Minchinton, A. H. Waters, *Lancet* **1** (1983) 84–87.
423. J. B. Bussel, R. P. Kimberly, R. D. Inman, I. Schulmann et al., *Blood* **62** (1983) 480–486.
424. J. R. Duran-Suarez, A. Martin, M. C. Botella, S. de la Torre et al., *Haematologica* **68** (1983) 564–566.
425. C. A. Schiffer et al., *Blood* **64** (1984) 937–940.
426. G. Chirico, M. Duse, A. G. Ugazio, G. Rondini, *J. Pediatr. (St. Louis)* **103** (1983) 654–655.
427. E. C. Besa, M. W. McNab, A. J. Solan, M. J. Lapes et al., *Am. J. Hematol.* **18** (1985) 373–379.
428. W. A. McGuire, H. H. Yang, E. Bruno, J. Brandt et al., *N. Engl. J. Med.* **317** (1987) 1004–1008.
429. C. De la Camara, R. Arrieta, A. Gonzalez, E. Iglesias et al., *N. Engl. J. Med.* **318** (1988) 519–520.
430. I. M. Nilsson, E. Berntorp, O. Zettervall, *N. Engl. J. Med.* **318** (1988) 947–950.
431. J. W. Newburger, M. Takahashe, J. C. Burns, A. S. Beiser et al., *N. Engl. J. Med.* **315** (1986) 341–347.
432. M. Nagashima, M. Matsushima, H. Matsuoka, A. Ogawa et al., *J. Pediatr. (St. Louis)* **110** (1987) 710–712.
433. D. Y. M. Leung, J. C. Burns, J. W. Newburger, R. S. Geha, *J. Clin. Invest.* **79** (1987) 468–472.
434. B. Combe, B. Cosso, J. Clot, M. Bonneau et al., *Am. J. Med.* **78** (1985) 920–928.
435. F. C. Breedveld, A. Brand, W. G. van Aken, *J. Rheumatol.* **12** (1985) 700–702.
436. E. L. Arsura, A. S. Bick, N. G. Brunner, T. Namba et al., *Arch. Intern. Med.* **146** (1986) 1365–1368.
437. K. Kawada, P. I. Terasaki, *Exp. Hematol (N.Y.)* **15** (1987) 133–136.
438. E. Schuller, A. Govaerts, *Eur. Neurol.* **22** (1983) 205–212.
439. A. Salama, C. Mueller-Eckhardt, V. Kiefel, *Lancet* **2** (1983) 193–195.
440. R. P. Kimberly, J. E. Salmon, J. B. Bussel, M. K. Crow et al., *J. Immunol.* **132** (1984) 745–750.
441. U. Budde et al., *Scand. J. Haematol.* **37** (1986) 125–129.
442. A. Etzioni, S. Pollack, A. Benderly, *N. Engl. J. Med.* **318** (1988) 994.
443. G. P. Sandilands, H. I. Atrah, G. Templeton, J. E. Cocker et al., *J. Clin. Lab. Immunol.* **23** (1987) 109–115.
444. T. W. Jungi, S. Barandun, *Vox Sang.* **49** (1985) 9–19.
445. W. Stohl, *Clin. Exp. Immunol.* **62** (1985) 200–207.
446. F. Hashimoto, Y. Sakiyama, S. Matsumoto, *Clin. Exp. Immunol.* **65** (1986) 409–415.
447. H. H. Hodes, *Pediatrics.* **32** (1963) 1–3.
448. W. B. White, C. R. Desbonnet, M. Ballow, *Am. J. Med.* **83** (1987) 431–434.
449. K. Prellner, P. Christensen, O. Kalm, K. Offenbartl, *Acta Pathol. Microbiol. Scand. Sect.* **94 C** (1986) 207–211.

450. K. K. Christensen, P. Christensen, *Pediatric Infectious Diseases* **5** (1986) S 189 – 192.
451. J. G. Kelton, C. J. Carter, C. Rodger, G. Bebenek et al., *Blood* **63** (1984) 1434 – 1438.
452. A. S. Cross, B. M. Alving, J. C. Sadoff, P. Baldwin et al., *Lancet* **1** (1984) 912.
453. M. Berger, P. Rosencranz, C. Y. Brown, *Clin. Immunol. Immunopathol.* **34** (1985) 227 – 236.
454. J. Kulics, E. Rajnavolgyi, G. Fust, J. Gergely, *Mol. Immunol.* **20** (1983) 805 – 810.
455. E. F. Ellis, C. S. Henney, *J. Allergy* **43** (1969) 45 – 54.
456. R. C. Williams, O. J. Mellbye, G. Kronvall, *Infect. Immun.* **6** (1972) 316 – 323.
457. D. S. Feingold, F. Oski, *Arch. Intern. Med.* **116** (1965) 226 – 228.
458. R. C. Millican, J. D. Rust, *J. Infect. Dis.* **107** (1960) 389 – 394.
459. E. Lachman, S. B. Pitsoe, S. L. Gaffin, *Lancet* **1** (1984) 981 – 983.
460. M. A. Thomson, *J. S. Afr. Vet. Assoc.* **54** (1983) 279 – 281.
461. M. L. Forman, R. Stiehm, *N. Engl. J. Med.* **281** (1969) 926 – 931.
462. A. S. Shigeoka, R. T. Hall, H. R. Hill, *Lancet* **1** (1978) 636 – 638.
463. E. J. Ziegler, J. A. McCutchan, J. Fierer, M. P. Glauser et al., *N. Engl. J. Med.* **307** (1982) 1225 – 1230.
464. J.-D. Baumgartner, M. P. Glauser, J. A. McCutchan, E. J. Ziegler et al., *Lancet* **2** (1985) 59 –63.
465. T. Calandra, J. Schellekens, J. Verhoef, M. P. Glauser, *Program and Abstracts for the 4th International Symposium on Infections in the Immunocompromised Host,* Ronneby Brunn, Sweden, 1986. Abstract No. 128.
466. A. H. Levy, *J. Chronic Dis.* **15** (1962) 589 – 598.
467. J. Bjorkander, C. Cunningham-Rundles, P. Lundin, R. Olsson et al., *Am. J. Med.* **84** (1988) 107 – 111.
468. R. S. Lane, *Lancet* **2** (1983) 974 – 975.
469. H. D. Ochs, S. H. Fischer, F. S. Virant, M. L. Lee et al., *Lancet* **1** (1985) 404 – 405.
470. C. L. S. Leen, P. L. Yap, G. Neill, D. B. L. McClelland et al., *Vox Sang.* **50** (1986) 26 – 32.
471. B. Cuthbertson, R. J. Perry, P. R. Foster, K. G. Reid, *J. Infect.* **15** (1987) 125 – 133.
472. C. C. Wood, A. E. Williams, J. G. McNamara, J. A. Annunziata et al., *Ann. Intern. Med.* **105** (1986) 536 – 538.
473. Y. Hamamoto, S. Harada, N. Yamamoto, Y. Uemura et al., *Vox Sang.* **53** (1987) 65 – 69.
474. *Morb. Mort. Wkly. Rep.* **36** (1987) 728 – 729.
475. A. W. Burks, H. A. Sampson, R. H. Buckley, *N. Engl. J. Med.* **314** (1986) 560 – 564.
476. R. Apfelzweig, D. Piskiewicz, J. A. Hooper, *J. Clin. Immunol.* **7** (1987) 46 – 50.
477. R. Paganelli, I. Quinti, G. P. D'Offizi, C. Papetti et al., *Vox Sang.* **51** (1986) 87 – 91.
478. J. C. Barton, G. A. Herrera, J. H. Galla, L. F. Bertoli et al., *Am. J. Med.* **82** (1987) 624 – 629.
479. A. G. Brox, D. Cournoyer, M. Sternbach, G. Spurll, *Am. J. Med.* **82** (1987) 633 – 635.
480. R. Schroff, K. Foon, S. Beatty, R. Oldham et al.: "Human anti-murine immunoglobulin responses in patients receiving monoclonal antibody therapy," *Can. Res.* **45** (1985) 879 – 885.
481. M. Nose, H. Wigzell: "Biological significance of carbohydrate chains on monoclonal antibodies," *Proc. Natl. Acad. Sci. U.S.A.* **80** (1983) 6632 – 6636.
482. E. Englemann, S. Foung, J. Larrick, A Raubitschek (eds.): *Human Hybridomas and Monoclonal Antibodies,* Plenum Press, New York 1985.
483. A. Strelkelkaus (ed.): *Human Hybridomas; Diagnostic and Therapeutic Applications,* Marcel Dekker, New York 1986.
484. D. Kozbor, J. Roder, T. Chang, Z. Stplewski et al.: "Human antitetanus toxoid monoclonal antibody secreted by EBV-transformed human B cells fused with murine myeloma," *Hybridoma* **1** (1982) 323 – 328.
485. J. Larrick, K. Truitt, A. Raubitschek, G. Senyk et al., "Characterization of human hybridomas secreting antibody in tetanus toxoid," *Proc. Natl. Acad. Sci. U.S.A.* **80** (1983) 6376 – 6380.
486. N. Chiorazzi, R. Wasserman, H. Kunkel: "Use of Epstein-Barr virus-transformed B cell lines for the generation of immunoglobulin-producing human B cell hybridomas," *J. Exp. Med.* **156** (1982) 930 – 935.
487. F. Gigliotti, R. Insel: "Protective human hybridoma antibody to tetanus toxin," *J. Clin. Invest.* **70** (1982) 1306 – 1309.
488. L. Olsson, T. Mazauric, J. Vincent-Falquet, J. Armand: "A human monoclonal antibody specific for tetanus toxoid," *Dev. Biol. Stand.* **57**(1984) 87 – 91.
489. S. Tsuchiya, S. Yokoyama, O. Yoshie, Y. Ono: "Production fo diphtheria antitoxin antibody in Epstein-Barr virus induced lymphoblastoid cell lines." *J. Immunol.* **124** (1980) 1970 – 1976.

490. E. Ziegler, J. McCutchan, J. Fierer et al.: "Treatment of Gram-negative bacteremia and shock with human anti-serum to a mutant Escherichia coli," *N. Engl. J. Med.* **307** (1982) 1225–1230.
491. W. Bogard, E. Hornberger, P. Kung in E. Englemann, S. Foung, J. Larrick, A. Raubitscheck (eds.): "Production and characterization of human monoclonal antibodies against Gram-negative bacteria," *Human Hybridomas and Monoclonal Antibodies,* Plenum Press, New York 1985, pp. 95–112.
492. N. Teng, H. Kaplan, J. Hebert et al.: "Protection against Gram-negative bacteremia and endotoxemia with human monoclonal IgM antibodies," *Proc. Natl. Acad. Sci. U.S.A.* **82** (1985) 179–184.
493. J. Larrick, M. Jahnsen, G. Senyk, S. Weiss et al. in H. Friedman (ed.): Production of human monoclonal antibodies recognizing cross-reactive determinants on lipopolysaccharides," *The Immunology and Immunopharmacology of Endotoxins,* 1986, pp. 75–81.
494. M. Pollack, A. Raubitschek, J. Larrick: "Cross-reactive human monoclonal antibodies that recognize conserved epitopes in the core-lipid A region of endotoxin," *J. Clin. Invest.* **79** (1987) 1421–1430.
495. J. Baumgartner, M. Glauser, J. McCutchan, E. Zeigler et al.: "Prevention of gramnegative shock and death in surgical patients by antibody to endotoxin core glycolipid." *Lancet* **2** (1985) 59–63.
496. J. Larrick, S. Hart, D. Lippman, M. Glembourtt et al. in A. Strelkelkaus (ed.): "Generation and characterization of human monoclonal anti-Pseudomonas aeruginosa antibodies," *Human Hybridomas; Diagnostic and Therapeutic Applications,* Marcel Dekker, New York 1986, pp. 65–80.
497. J. Larrick, B. Dyer, G. Senyk et al. in E. Engleman, S. Foung, J. Larrick, A. Raubitschek (eds.): "In vitro expression of human B cells for the production of human monoclonal antibodies," *Human Hybridomas and Monoclonal Antibodies,* Plenum Press, New York 1985, pp. 149–165.
498. K. Hunter, Jr., G. Fischer, V. Hemming, S. Wilson et al.: "Antibacterial activity of a human monoclonal antibody to Haemophilus influenzae type B capsular polysaccharide," *Lancet* **2** (1982) 789–799.
499. H. Peltola, H. Kayhty, M. Virtanen, P. Makela: "Prevention of Hemophilus influenza type b bacteremic infections with the capsular polysaccharide vaccine," *N. Engl. J. Med.* **310** (1984) 1561–1565.
500. T. Atlaw, D. Kozbor, J. Roder: "Human monoclonal antibodies against Mycobacterium Leprae," *Infect. Immun.* **49** (1985) 104–110.
501. B. Brodeur, L. Lagace, Y. Larose, M. Martin et al. in B. Schook (ed.): "Mouse-human myeloma partner for the production of heterohybridomas," *Monoclonal Antibodies* , Marcel Dekker, 1986, p. 51.
502. M. Steinitz, S. Tamir, A. Goldfarb: "Human anti-pneumococci antibody produced by an Epstein-Barr virus (EBV)-immortalized cell line," *J. Immunol.* **132** (1984) 877–882.
503. J. Schwaber, M. Posner, S. Schlossman, H. Lazarus: "Human-human hybrids secreting pneumococcal antibodies," *Hum. Immunol.* **9** (1984) 137–142.
504. D. Emanuel, J. Gold, J. Colacino, C. Lopez et al.: "A human monoclonal antibody to cytomegalovirus (CMV)," *J. Immunol.* **133** (1984) 2202–2205.
505. C. Amadei, S. Michelson, J. Frot, M. Fruchart et al.: "Human anticytomegalovirus (CMV) immunoglobulins secreted by EBV-transformed B-lymphocytes cell lines," *Dev. Biol. Stand.* **57** (1984) 283–286.
506. K. Burnett, J. Leung, J. Marinis in E. Engleman, S. Foung, J. Larrick, A. Raubitschek (eds.): "Human monoclonal antibodies to defined antigens," *Human Hybridomas and Monoclonal Antibodies,* Plenum Press, New York 1985, pp. 113–133.
507. Y. Ichimori, K. Sasano, H. Itoh, S. Hitosumachi et al.: "Establishment of hybridomas secreting human monoclonal antibodies against tetanus toxin and hepatitis B virus surface antigen," *Biochem. Biophys. Res. Commun.* **129** (1985) 26–38.
508. E. Stricker, R. Tiebout, P. Lelie, W. Zeijlemaker: "A human monoclonal IgG antihepatitis B surface antibody; production, properties and applications," *Scand. J. Immunol.* **22** (1985) 337–345.
509. R. Beasley et al.: "Hepatitis B immune globulin (HBIG) efficacy in the interruption of perinatal transmission of hepatitis B virus carrier state," *Lancet* **2** (1981) 388–393.
510. S. Foung et al.: "Human monoclonal antibodies neutralizing Varicella Zoster virus," *J. Infect. Dis.* **52** (1985) 280–285.

511. L. Evans, J. Homsy, W. Morrow, I. Gaston et al.: "Human monoclonal antibody directed against gag gene products of the human immunodeficiency virus," *J. Immunol.* **140** (1988) 941 – 943.
512. B. Banapour et al.: "Characterization and epitope mapping of a human monoclonal antibody reactive with the envelope glycoprotein of human immunodeficiency virus," *J. Immunol.* **138** (1987) 4027 – 33.
513. A. Rosen, K. Persson, G. Klein: "Human monoclonal antibodies to a genus-specific chlamydial antigen, produced by EBV-transformed B cells," *J. Immunol.* **130** (1983) 2899 – 2902.
514. S. Koizumi, S. Fujiwara, H. Kikuta et al.: "Production of human monoclonal antibodies against Epstein-Barr virus specific antigens by the virus immortalized lymphoblastoid cell lines," *Virology* **150** (1986) 161 – 170.
515. J. Seigneurin et al.: "Herpes simplex virus glycoprotein D; human monoclonal antibody produced by bone marrow cell line," *Science* **221** (1983) 173 – 175.
516. Y. Masuho et al.: "Generation of hybridomas producing human monoclonal antibodies against herpes simplex virus after in vitro stimulation," *Biochem. Biophys. Res. Commun.* **135** (1986) 495 – 505.
517. L. Evans, C. Maragos, J. May: "Human lymphoblastoid cell lines established from peripheral blood lymphocytes secreting immunoglobulins directed against herpes simplex virus," *Immunol. Lett.* **8** (1984) 39 – 50.
518. S. Matsushita, M. Robert-Gurhoff, J. Trepel, J. Cossman et al.: "Human monoclonal antibodies directed against an envelope glycoprotein of human T-cell leukemia virus type 1," *Proc. Natl. Acad. Sci. U.S.A.* **83** (1986) 2672 – 2677.
519. C. Croce, A. Linnenbach, W. Hall, Z. Steplewski et al.: "Production of human hybridomas secreting antibodies to measles virus," *Nature (London)* **288** (1980) 488 – 489.
520. R. Ritts, Jr., A. Ruiz-Arguelles, K. Weyl et al.: "Establishment and characterization of a human non-secreting plasmacytoid cell line and its hybridization with human B cells," *Int. J. Cancer* **31** (1983) 133 – 151.
521. J. Hilfenhaus, E. Kanzy, R. Kohler, W. Willems: "Generation of human anti-rubella monoclonal antibodies from human hybridomas constructed with antigen-specific Epstein-Barr virus transformed cell lines," *Behring Inst. Mitt.* **80** (1986) 31 – 40.
522. F. van Meel, P. Steenbakkers, J. Oomen: "Human and chimpanzee monoclonal antibodies," *J. Immunol. Methods* **80** (1985) 267 – 280.
523. D. Crawford et al.: "Production of human monoclonal antibody to X 31 influenze virus nucleoprotein," *J. Gen. Virol.* **64** (1983) 697 – 700.
524. R. Schmidt-Ullrich, J. Brown, R. Whittle, P. Lin: "Human-human hybridomas secreting monoclonal antibodies to the M.W. 195 000 Plasmodium falciparum blood stage antigen," *J. Exp. Med.* **163** (1986) 179 – 189.
525. R. Udomsangpetch et al.: "Human monoclonal antibodies to Pf-155, a major antigen of malaria parasite Plasmodium falciparum," *Science* **231** (1986) 55 – 59.

Pharmaceuticals, General Survey and Development

JOHN L. MCGUIRE, Johnson & Johnson, New Brunswick, New Jersey 08933, United States (Chaps. 1 and 2)
HORST HASSKARL, Ludwigshafen, Federal Republic of Germany (Chap. 3)
ROLF KRETZSCHMAR, Knoll AG, Ludwigshafen, Federal Republic of Germany (Section 4.1)
KLAUS-JÜRGEN HAHN, Knoll AG, Ludwigshafen, Federal Republic of Germany (Section 4.2)
MANUEL ZAHN, Knoll AG, Ludwigshafen, Federal Republic of Germany (Section 4.3)

1. Introduction

The pharmaceutical industry is largely a research-based industry. Not only has it been historically successful in providing valuable new drugs for treatment of diseases, but its research and development laboratories have made significant contributions to our basic understanding of those diseases. In significant part because of the success of this industry, people today look forward to longer and healthier lives than ever before. Excellent reviews of this industry have been published [1,2].

Historically, today's pharmaceutical industry emerged from a number of different sciences and approaches to finding therapeutics. Examples are: (1) the early discovery that extracts of plants such as salicylates from willow tree bark could provide products with medicinal value; (2) the discovery that proteins could be used as drugs, i.e., BANTING and BEST's demonstration that insulin could reverse the ravages of diabetes in 1938 and commercial expression of

Ullmann's Industrial Toxicology

ISBN: 3-527-31247-1

recombinant proteins such as tissue plasminogen activator (TPA) by Genentech; (3) Lister's small pox vaccine and the development of the first monoclonal antibody therapeutic, OKT3, by Johnson & Johnson for prevention of tissue transplant rejection; (4) research in gene therapy; (5) use of genomics to find new targets for pharmaceutical intervention; and (6) the scientific approach which spawned most of today's pharmaceutical armamentarium – modification of organic molecules to provide therapeutically active agents through medicinal chemistry.

Pharmaceutical agents based on organic molecules which have been synthesized and modified to provide medicinal products comprise the largest segment of available drugs today. This is true irrespective of whether the overall pharmaceutical market based on currency volume, the numbers of prescriptions, or the numbers of products sold over-the-counter is considered. Biotechnology is part of the pharmaceutical industry today, but drugs based on organic chemistry remain the largest part of pharmaceutical R & D and comprise the largest percentage of new drugs launched annually.

Relatively few of the medicines used today were available before World War II [2]. Morphine, digitalis, and quinine existed. Insulin and small pox vaccine were used. A few synthetic organic compounds, such as aspirin and barbiturates, were available, and the first sulfonamide antibacterial agents were known. Some of these were considered miracle drugs, but the fact is that there were few drugs available.

World War II provided an important stimulus for development of new drugs, primarily because the combatants needed drugs to treat wounded servicemen, and other drugs such antimalarials and drugs for treatment of venereal diseases to keep troops healthy so they could fight. Much of the early pre-war effort in medicinal chemistry had taken place in the German chemical industry. The war meant that the search for new drugs became more international. This blossomed into a new industry after the war, the research-based pharmaceutical industry which has grown into one of the most productive industries in the world. The primary emphasis was production of new heterocycle-based pharmaceuticals, but vaccine production became a strength of some pharmaceutical companies, and fermentation-based antibiotic product research and insulin production developed in others. The biotechnology industry innovated commercial production of recombinant proteins and gene based research, and these technologies too were rapidly incorporated into the pharmaceutical industry.

In 1998, global sales for pharmaceutical products were estimated to be approximately $\$ 300 \times 10^9$ [3]. This includes both drugs sold via prescription and drugs sold over-the-counter (OTC), much like other consumer products.

The pharmaceutical industry has features that distinguish it from other industries. First, this industry is research-intensive, having close interaction with universities and other cutting edge research facilities. R & D is expensive, and times needed for the discovery and development of new drugs typically long. Secondly, although there are exceptions, most companies in this industry tend to be either global or regional in nature, not national companies. Because of the enormous cost of pharmaceutical R & D, companies strive to develop and sell their products worldwide in order to recoup their R & D costs. Thirdly, this industry is heavily regulated, both in terms of requiring regulatory approval to sell products and, in many countries, requiring price approvals. Price regulation is important not only because it affects profits and the ability to reinvest in new drug research within the pharmaceutical industry, but also because it relates to the cost of healthcare generally. Finally, physicians tend to be the customers rather than patients who actually use the pharmaceutical industry's products. As a result, a complex interaction exists with the medical profession, governments and insurance groups which actually pay much of the cost of this industry's products throughout the world.

Many studies show that pharmaceuticals are one of the most efficient forms of medicine available today. Because of this success and the aging of the population, however, the percentage share of healthcare costs in GNP has grown. Many governments provide healthcare to their citizens, so healthcare costs can impact taxes. Even though drugs represented only 8 % of total 1997 healthcare costs, as governments try to find ways to contain healthcare costs, the costs of pharmaceuticals have frequently become a political issue around the world.

Like many other industries, the pharmaceutical industry is undergoing change. In addition to constantly assimilating new technologies into its research and adjusting to changing market and regulatory environments, a number of pharmaceutical company mergers are taking place. As the cost of R & D in this industry continues to grow and companies try to continue to fill their "pipelines" with new products, ways are sought to help with the costs. Coupled with this is the fact that, in marketing terms, the pharmaceutical industry is highly fragmented. The largest companies have less than 5 % of the worldwide market share for pharmaceuticals. Perhaps as a result, mergers and acquisitions have become more frequent. Examples are the merger of the two British companies Glaxo and Wellcome; the life sciences operations of Hoechst, Marion Merril Dow, Rousell, and Rorer merged in a series of transactions to form Aventis. Sanofi merged with Synthelabo; Novartis was formed by a merger of the Swiss companies Ciba Geigy and Sandoz, and Astra and Zeneca merged to form Astra Zeneca.

2. Companies, Markets, and Products

Of the worldwide pharmaceutical market in 1998, of $ 300×10^9 [3] North American sales were $ 118×10^9 (38 % of worldwide sales), 95 % of that being US sales. The remainder was approximately equally divided among Europe and the rest of the world. It has been predicted that global sales will surpass $ 400×10^9 by 2002. Readers are referred to the many excellent references on the pharmaceutical market including Scrip reports, PharmaVitae by Datamonitor and IMS Health World Drug Monitor.

The global pharmaceutical market is divided into two segments. One segment are ethical products that are not promoted to the consumer. Most of them are available, at least in developed countries, only on prescription by a physician or other healthcare professional. The second segment is OTC products. These are generally well established treatments for minor conditions, ones where safety and the ability to self-diagnose are not concerns. The distribution between these two categories is quite clear in countries that have a well-developed healthcare system. In this discussion of the pharmaceutical market, both segments are combined.

On a geographical basis, the sales figures for the major markets of ethical products are shown in Table 1. These data show that the top six countries account for about 70 % of the worldwide market. This is not surprising, for they are the most populous of the industrially advanced nations.

Table 1. World pharmaceutical market by country in 1997 (sources: Scrip reports [4]and IMS Health World Drug Monitor [5])

Country	Sales, $ 10^9
United States	66.5
Japan	41.7
Germany	14.7
France	13.7
Italy	8.6
United Kingdom	7.7
Spain	4.9
Canada	4.1
Netherlands	1.9
Belgium	1.8

If a breakdown of worldwide sales is made among companies, it can be seen that the top ten companies account for just under a third of world pharmaceutical sales (Table 2). The remainder is supplied by a large number of smaller companies. Currently, the largest company in terms of worldwide sales covers about 5 % of world sales and the 20th just over 1 %. Obviously, many smaller companies still represent very significant businesses.

Table 2. Top ten Pharmaceutical companies (ethical sales) worldwide in 1998 (source: Datamonitor [6])

Rank	Company	Country	Pharmaceutical sales, $ 10^9
1	Merck, Sharp & Dohme	USA	15.3
2	Glaxo-Wellcome	United Kingdom	13.4
3	Pfizer	USA	12.2
4	Bristol Myers Squibb	USA	10.4
5	Novartis	Switzerland	9.5
6	Lilly	USA	8.6
7	Johnson & Johnson	USA	8.6
8	Roche	Switzerland	8.4
9	Smith Kline Beecham	USA	7.8
10	Hoechst Marion Roussel	Germany	7.6

The top ten pharmaceutical companies maintain research facilities in a number of countries in North America, Europe and Asia and have multiple commercial operations throughout the world.

The global ethical pharmaceutical market ($\approx \$ 280 \times 10^9$ in 1998) is divided into many therapeutic segments. The largest segment (Table 3) is cardiovascular drugs which accounted for approximately 23 % of the global market. Antiinfectives is the next largest segment, accounting for $\approx$ 14 % of the global market. Gastrointestinal drugs and those directed toward treatment of CNS disorders, immunological/inflammatory disease, and cancer follow.

Table 3. Top five therapeutic segments of the pharmaceutical market in 1998 (source: Datamonitor [6])

Therapeutic segments	1998 Sales, $\$ 10^9$	% of 1998 Sales
Cardiovascular	64	23
Antiinfectives	38	14
Gastrointestinals	35	13
CNS	31	11
Immunological/inflammatory	20	7
Oncology	14	5

Within each therapeutic area different drug classes exist. They vary in degree of market fragmentation. For example, within the cardiovascular market, the top five products within the antihypertensive market accounted for only 24 % of 1998 sales in this class, whereas in the lipid-lowering market, the top five products accounted for nearly 75 % of total global sales [7].

The degree of market domination by individual companies varies substantially between product classes. For example, within the antiinfective market, Glaxo Wellcome dominated the antiviral market in 1998, with three of its products accounting for nearly a third of the market, whereas the antibacterial market was not dominated by any single company.

The top 25 branded drugs worldwide by sales in 1998 are shown in Table 4. Several points should be made. First, these sales figures are worldwide sales. The leading products in individual countries vary considerably because of differences in the incidence of diseases, medical practice, and the particular strengths of international and local companies in that particular market. Secondly, if the top 25 drugs were ranked by prescription number rather than sales, a somewhat different picture would emerge. This is because drug prices may vary. Older products tend to be available at lower prices than newly launched ones because each product is generally priced to bear the research costs involved in their discovery and development, and R & D costs continue to rise. Additionally, the patents on older products have often expired, and generic products or copies of the original drugs, frequently compete with the originator in the marketplace on the basis of price.

Table 4. Top 30 branded products by sales in 1998 [8]

Products	Generic name
Losec/Prilosec	omeprazole
Zocor	simvastatin
Prozac	fluoxetine
Norvasc	amlodipine
Renitec/Vasotec	enalapril
Clarityne/Claritin	loratadine
Lipitor	atorvastatin
Zoloft	sertraline
Seroxat/Paxil	paraxetine
Premarin/Prempro/Premphase	conjugated estrogens
Lipostat/Pravachol	pravastatin
Augmetin	amoxicillin and clavulante
Zyprexa	olanzapine
Ciproxin/Cipro	ciprofloxacin
Epoxen/Procrit	epoetin
Prevacid	lansoprazole
Neoral/Sandimmun	cyclosporine
Zantac	ranitidine
Klacid/Biaxin	clarithromycin

Finally, most of the sales volumes discussed above relate to sales of ethical pharmaceutical products. There are also OTC generic pharmaceutical products that are sold without a prescription. These products are commonly used for relatively minor ailments and have not only been shown over time to be safe and simple to use, but it has been concluded by regulatory authorities that no medical guidelines beyond instructions on the package are needed by the patient to diagnose the ailment or to use the product correctly. The *OTC company* typically does not do innovative research to discover new active pharmacological agents, but rather simply manufactures and distributes well-established patent-free compositions of long established safety. New research on the part of generic manufacturers usually relates to formulations rather than trying to discover new drugs. OTC manufacturers primarily formulate the product into an acceptable dosage form, manufacture, and market it with a well recognized brand name through retailers. Advertising to the general consumer is important for the success of the OTC product.

In the case of *ethical pharmaceuticals*, however, the research-based pharmaceutical company is the primary innovator of totally new pharmacologically active compounds known as new chemical entities (NCEs). The research to discover a useful new drug is expensive and lengthy, and the probability of success on any single project is low. Once a compound has been found, lengthy development studies are required to prove to regulatory agencies that the compound is safe and effective for use (Chap. 4) and that it can be consistently manufactured. The time from initiation of a new research project to approval for sale can easily take 12 – 15 years and consume $\$ 600 \times 10^9$ [9]. This business requires high research investment over a prolonged period of time and will see a long delayed return on investment. The risks of failure are high, but the rewards for a successful product can also be high. These drugs are promoted primarily to medical professionals, citing claims approved by regulatory agencies, although direct to consumer advertising is sometimes being used more recently.

The *generic pharmaceutical industry* is another part of the ethical pharmaceutical industry. Products that are the subject of current valid patents can be made and sold only by the patent holders or by a licensee. When the patents have expired, after 20 years of patent life in most countries, other manufacturers are free to make and sell the product if approved by regulatory agencies. Companies can sell these off-patent products as generic drugs. The same active ingredient can be used in generic products, but not with the same trade name because that will typically be owned by the innovator. The generic business involves little research since it relies upon the research of others and concentrates only on commercially successful products whose patents have expired. These products are still sold on prescription, usually at prices lower than the innovator since generic companies do not have to recoup extensive R & D expenses.

Finally, there are *chemical supply companies* which supply active substances and excipients to the pharmaceutical industry. The industry also can be said to include drug distributors and retail outlets. These supply and distribution centers are beyond the scope of this review, however.

3. Legal Requirements

3.1. General

According to the German Drug Law (Arzneimittelgesetz, AMG), drugs (pharmaceuticals, pharmaceutical products, medicinal products) are a special kind of – normally – industrial finished products consisting of chemical elements and compounds as well as their naturally occurring mixtures and solutions, plants, parts of plants, plant constituents, bodies of animals, body constituents and metabolic products of humans or animals, and microorganisms including viruses, their constituents, and metabolic products.

Drugs are intended to treat or prevent diseases, to make diagnoses, or to restore, correct, or modify physiological functions in humans or animals.

In almost all countries throughout the world the development of drugs, their manufacture, and placing on the market depend on governmental approval. Normally a comprehensive national legal framework exists to avoid risks in connection with the development, manufacture, sale, and use of drugs. Drugs are regarded as being potentially dangerous. Drug law is therefore the comprehensive legal attempt to safeguard drug safety for the benefit of humans and animals.

The following chapter gives an overview of internationally and generally accepted legal requirements. It is based on the main laws of the United States, the European Community (EC), and Germany.

In the following, law means legal provisions adopted by the legislative organ (i.e., parliament), regulation means legal provisions made by an institution of the executive (i.e., the government) on the basis and within the limits of an authorization given by the law. Administrative regulations are governmental measures directly binding not the individual or a company but the competent authorities.

United States Law. The basic law in the United States is the Federal Food, Drug, and Cosmetic Act, as amended, the latest amendment being the Food and Drug Administration Modernization Act of 1997 (Public Law no. 105 – 115). Many regulations have been issued relating to drugs and based on this Act. The main regulations (all revised as of April 1, 1998) are:

1) Investigational New Drug Application (IND) [21 Code of Federal Regulations (CFR) Part 312]
2) Applications for FDA Approval to Market a New Drug or an Antibiotic Drug (NDA) (21 CFR Part 314)
3) Protection of Human Subjects (21 CFR Part 50)
4) Institutional Review Boards (21 CFR Part 56)
5) Good Laboratory Practice for Non-Clinical Laboratory Studies (21 CFR Part 58)
6) Current Good Manufacturing Practice for Finished Pharmaceuticals (GMP) (21 CFR Part 211)

European Community Law. The legal system of the European Community (EC) differentiates between regulations and directives, both issued by the Council of Ministers and occasionally by the Commission. Regulations have direct and binding effect in the member states, whereas directives have to be transferred separately into national law. The main regulations and directives are:

1) Council Regulation (EEC) 1768/92 of June 18, 1992 concerning the creation of a supplementary protection certificate for medicinal products.
2) Council Regulation (EEC) no. 2309/93 of July 22, 1993 laying down Community procedures for the authorization and supervision of medicinal products for human and veterinary use and establishing a European Agency for the Evaluation of Medicinal Products (EMEA).
3) Council Directive 65/65/EEC of January 26, 1965 on the approximation of provisions laid down by law, regulation or administrative acts relating to medicinal products, as amended.
4) Council Directive 75/319/EEC of May 20, 1975 on the approximation of provisions laid down by law, regulation or administrative acts relating to medicinal products, as amended.
5) Council Directive 75/318/EEC of May 20, 1975 on the approximation of the laws of member states relating to analytical, pharmacotoxicological and clinical standards and protocols in respect of the testing of medicinal products, as amended.

These and additional directives, guidelines, and so-called Notices to Applicants for marketing authorizations for medicinal products for human use in the member states of the European Community have been published by the European Commisson under the title "The Rules Governing Medicinal Products in the European Union" in 13 volumes.

German Law. The main law governing drugs in Germany is the Drug law (Arzneimittelgesetz), published in its latest version of December 11, 1998 (Federal Law Gazette part I, p. 3586), hereinafter called AMG.

International Harmonization. A very important attempt has been made to harmonize the legal requirements for registration of drugs world-wide. This attempt is called International Conference on Harmonization of Technical Requirements for Registration of Pharmaceuticals for Human Use (ICH). ICH has elaborated numerous recommendations for the regulatory bodies of the European Union, Japan and USA. ICH recommendations do not legally bind.

3.2. Definitions

Apart from the drug definition (Section 3.1) additional definitions are important. *Quality* is determined by identity, content, purity etc. (§ 4 Subparagraph 15 AMG); *safety* means that, in the light of currently prevailing scientific knowledge, the drug has no harmful effects exceeding medical acceptability (§ 5 Subparagraph 2 AMG). *Effectiveness* (*efficacy*) means a reasonable expectation that, in a significant proportion of the target population, the pharmacological effect of the drug, when used under adequate directions for use and warnings against unsafe use, will provide clinically significant relief of the type claimed [21 CFR 330.10 (a) (4)]. Effectiveness is not defined in German law. *Side effects* means the occurrence of undesired reactions (§ 4 Subparagraph 13 AMG).

3.3. Requirements Relating to the Development of Drugs

Before a drug can be put on the market various prerequisites have to be fulfilled notwithstanding the necessary final governmental approval. These prerequisites relate to the testing of the drug. The performance of drug testing is not within the free discretion of the pharmaceutical company or the investigator. Internationally accepted standards and methods have to be applied. Drug development is divided into analytical tests, pharmacological and toxicological tests (see also Section 4.1), and clinical trials (see also Section 4.2).

The main specifications for the performance of analytical, pharmacological – toxicological, and clinical trials for the EU and Germany are contained in the Council Directive of May 20, 1975 as amended (75/318/EEC) and in the German Drug Testing Guidelines (Arzneimittelprüfrichtlinien) of May 5, 1995, respectively.

Analytical Tests. Analytical tests comprise physicochemical, biological, and microbiological tests including qualitative and quantitative properties of the constituents, description of method of preparation, control of starting materials, characteristics affecting the bioavailability, tests on the final product, identification of active ingredients and excipient constituents, safety tests (e.g., abnormal toxicity test), and stability tests.

Pharmacological and Toxicological Tests. The function of these tests is to show the potential toxicity of the drug and its pharmacological properties, both qualitatively and quantitatively. *Pharmacological tests* comprise pharmacodynamic and pharmacokinetic studies. *Toxicological tests* consist of single-dose (acute) toxicity, repeated (subacute or chronic) toxicity, fetal toxicity, as well as the investigation of reproduction function, mutagenic potential, and carcinogenic potential. Some of these tests are performed in vitro, but many are carried out in vivo. Animals tests have to comply with additional legal requirements. The investigator also has to adhere to the principles of Good Laboratory Practice for Non-Clinical Laboratory Studies (GLP) which ensure reliable and comparable results under high quality with respect to safety studies. GLP pertains to personnel, testing facility management, study director, independent quality assurance unit, animal care facilities, standing operation procedures (SOP), protocols, records and reports, and disqualification of testing facilities.The necessity to comply with the GLP and its wording is laid down in the Council Directive of Dec. 18, 1986 (87/18/EEC), in § 19 a of the German Chemical Act

(Chemikaliengesetz) amended as of July 25, 1994 in conjunction with Annex 1 of this Act (Federal Law Gazette Part I p. 1703), and in 21 CFR Part 58. The second additional legal requirement concerning in vivo tests is compliance with animal protection provisions for ethical reasons. Details are contained in the Council Directive of 24 Nov., 1986 (86/609/EEC) and in the German Animal Protection Act (Tierschutzgesetz) published in the version of May 25, 1998 (Federal Law Gazette Part I p. 1105).

In Germany the performance of drug tests in animals is required by the Drug Development Guidelines (Arzneimittelprüfrichtlinien). To this extent no governmental permission is necessary for performing these tests, a notification to the competent authority is sufficient. If animal tests are not required by law, special governmental permission is necessary. Permission is only given if the animal experiment is unavoidable because of the nonexistence of alternative methods, is ethically justified, and serves the purpose of diagnosis or treatment of diseases, of identifying environmental risks, of performing safety tests with chemicals, or performing basic research.

Clinical Trials. The performance of clinical trials constitutes a substantial part of drug development. These trials are aimed at evaluating the safety and effectiveness of the drug in humans. The tests must normally be carried out as controlled clinical trials consisting of a test group and a control group of patients. The test group is treated with a test drug, whereas the control group is treated with a placebo or standard drug. Controlled clinical trials may be single blind or double blind and normally consist of three phases. Clinical phase I is performed in healthy volunteers, whereas phases II and III are performed in patients (see Section 4.2.1).

One of the most crucial ethical and legal problems of drug development concerns clinical trials because these trials involve experimentation in humans. Fundamental legal precautions have to be observed. The three main conditions are as follows:

1) Medical and ethical justification of the trial
2) Information of the human subject participating on the risk involved
3) His or her freely given consent (informed consent)

Thus, the goal of the special legal requirements is the protection of the human subject. Special parts of the U.S. Code of Federal Regulations have been devoted to this aim [21 CFR Part 50 (Protection of Human Subjects) and Part 312 (Investigational New Drug Application)]. Basic elements of informed consent are designed to guarantee a free and self-responsible decision of the trial subject. This involves a statement that the study involves research, an explanation of the purpose of the research, and the expected duration of the subject's participation. A description of benefits and any reasonably foreseeable risks or discomforts must be given to the subject. Benefits obviously do not exist for trials in voluntary healthy subjects. The subject must be informed of appropriate alternative procedures or courses of treatment, especially if these might be to his or her advantage. Existing confidentiality and exceptions thereto also have to be described; an exception, for example, is that the records may be inspected by the competent German authority or by the FDA. Furthermore information on possible financial compensation in case of injury or medical treatment must be provided. The fact that participation in the trial is voluntary and that the subject can refuse to participate without penalty must be included in the information.

Further elements pertaining to informed consent may include information on currently unforeseeable risks, on possible termination of the investigation, and the consequences of a subject's decision to withdraw from the trial. The patients should be notified of the number of subjects participating in the trial and also of significant new findings to let them decide whether or not to continue their participation. The subject must give written consent to participation in the trial. In addition the sponsor of an investigation has to submit an investigational new drug application (IND) to the FDA, if a clinical investigation is to be conducted with a new drug. Unless the FDA notifies the sponsor of any objections, IND goes into effect and the trial may be started thirty days after the FDA has received the IND.

German law (§§ 40, 41 AMG) contains additional requirements. The clinical trial must be supervised by a physician who can prove to have at least two years experience in the field of clinical drug trials. Appropriate pharmacological – toxicological tests have to be carried out before the first test is performed in humans. The documents relating to these pharmacological – toxicological tests have to be deposited with the competent federal health office. In addition, the trial protocol including the names of the investigators and the test sites have to be submitted to the authority. A written positive opinion of an ethical committee has to be added. Contrary to U.S. law and some national European laws, however, this does not mean that permission has to be given by the authority; no governmental permission is necessary for the performance of clinical trials in Germany. The participants in the clinical trial must be given compulsory insurance. The coverage must be in adequate proportion to the risks involved in the clinical trial; insurance has to cover an injury of at least € 500 000 (ca. $ 600 000).

In the United States the ethical committee is called the Institutional Review Board (IRB), (21 CFR Part 56). The respective regulations describe the general standards for the composition, operation, and responsibility of an IRB that reviews clinical investigations. The term IRB denotes any board, committee, or other group formally designated by an institution to review, to approve the initiation of, and to conduct periodic review of biomedical research involving subjects. The primary purpose of such a review is to assure the protection of the rights and welfare of the human subjects. Clinical investigations relating to new drugs can only be initiated after the investigation has been reviewed by, approved by, and remains subject to continuing review by an IRB. An IRB has to have at least five members with varying professional backgrounds necessary to review specific research activities. The persons must be knowledgeable in these areas. No IRB may consist entirely of men or women or entirely of members of one profession. One member has to work in a nonscientific area (e.g., lawyer, ethicist, member of the clergy). One member of the IRB is not to be affiliated with the institution whereas the other members may belong to the institution. No IRB member may have a conflicting interest in the research project (e.g., because the IRB member is also an investigator).

The IRB has to approve, require modification of, or disapprove all research activities covered by the investigation. The Informed Consent of the participant in the trial is subject to inspection by IRB. The criteria for IRB approval of research are contained in 21 CFR Part 56.111. The institution responsible for the trial or the IRB must prepare and maintain adequate documentation of IRB activities. The parent institution is presumed to be responsible for the operation of an IRB.

In Germany physicians are obliged to ask an ethics committee for its opinion (not for its permission !) before they start a clinical trial. The ethics committees are normally associated with the official chambers of physicians that exist in each state of the Federal Republic of Germany. The rights and obligations of these ethics committees are by no means as clearly defined as those described in the U.S. regulations. German ethics committees are not entitled to give permission for or to disapprove a clinical trial.

The World Medical Association has adopted and published recommendations guiding physicians in biomedical research on human subjects. These recommendations are known as the Declaration of Helsinki (amended in Hong Kong 1989). They are directed to the international community of physicians and are not legally binding. The Declaration of Helsinki consists of three parts:

1) Basic Principles
2) Medical Research Combined with Professional Care (Clinical research)
3) Non-Therapeutic Biomedical Research Involving Human Subjects (Non-Clinical Biomedical Research)

The United States has made the Declaration of Helsinki a part of the national law relating to foreign clinical studies (21 CFR Part 312.120). Insofar the Declaration of Helsinki is binding. In Germany the Declaration of Helsinki has been made binding for physicians, but only in their capacity as members of the chamber of physicians.

Finally the EC Commission has published " Good Clinical Practice for Trials on Medicinal Products in the European Community (GCP)" as of June 26, 1990. The GCP guidelines recommend the investigator to request the opinion of an ethics committee. The EC Commission, however, wants to convert this recommendation into a directive. A proposal for such a directive relating to the implementation of Good Clinical Practice in the conduct of clinical trials on medical products for human use has been published (Official Journal no. C 306/9 of October 8, 1997). As a result of the adoption of that directive, GCP would become legally binding in the member states.

3.4. Marketing Approval

The placing of drugs on the market needs governmental approval as defined in relevant national laws. Such an approval generally needs proof of quality, safety, and effectiveness of the drug. The competent authority has to give its approval on the basis of evidence provided during drug development (article 5 of 65/65/EEC; § 25 AMG; 21 CFR Part 314.105 and Part 314.125). See also Section 4.3. Within the EC, a distinction has to be made between two procedures: a centralized procedure which exclusively applies for medicinal products derived from biotechnology. The approval to be given by EMEA is valid for all member states of the EC. In addition, the procedure of mutual recognition exists meaning that normally a member state is bound by an approval of another member state in the sense that it recognizes that approval as basis for its own identical decision. The approval, however, remains national (The Rules governing Medicinal Products in the European Union, Volume IIA, Notice to Applicants – Procedures for marketing authorization, 1998).

Relevant details of the new drug have to be submitted to the competent authority. These include the name and address of the applicant and the manufacturer, the name of the drug, a qualitative and quantitative description of the drug constituents, the pharmaceutical form, the pharmacological effects, the fields of application (indications), contraindications, the side effects, interactions with other products, dosage, a brief summary of the drug manufacture, method and duration of administration, package size, method of preservation, shelf life, results of stability tests, and methods of quality control. The results of analytical, pharmacological, toxicological, and clinical tests have also to be submitted. Normally this authority has to make its decision within seven months of receiving the application. In practice, however, applications unfortunately require a much longer time to be processed. In Germany, for example, it sometimes takes more than two years.

If minor changes are made to the documentation already submitted to the health authority, the applicant simply has to notify the authority of these changes. More important changes (e.g., dosage, duration of application, limitation of contraindications, side effects, or interactions) have to be approved by the authority. The authority is deemed to have given its approval if it does not react within a period of three months.

If changes are made in the composition of the active ingredients or mode of drug application, the indications are extended, or a genetic engineering production process is implemented, the applicant has to reapply for market approval.

The competent authority is entitled to withdraw, revoke, or suspend the authorization given at any time if there are special reasons relating to the quality, especially to the safety or to the effectiveness of the drug.

Unless previously prolonged, the marketing authorization normally expires after a period of five years as from the date of its granting (§ 31 AMG).

3.5. Requirements for the Manufacture and Quality Control of Drugs

The manufacture of drugs may itself involve risks. The quality of the drug product has to be identical to that accepted by the authority responsible for marketing approval. Consequently the drug manufacturer needs special governmental authorization to produce drugs. The requirements to be met by the applicants are suitable and sufficient premises, technical equipment, and control facilities as well as, according to EC law, at least one appropriately qualified person responsible for production, i.e., the production director. The qualified person is usually a pharmacist, but chemists, biologists, physicians, and veterinarians are also admitted provided that they have acquired additional knowledge and experience specified in the relevant law. The qualified person has to be permanently present during drug manufacture. In addition German law requires an independent quality-control director who has to have the same qualifications as the production director, and a marketing director (Chap. IV of 75/319/EEC; §§ 13 et sequentes AMG).

An additional, very important requirement for drug manufacture is observance of the rules of Good Manufacturing Practice (GMP). These rules relate to quality assurance system, personnel, premises, equipment, documentation, production, quality control, contract manufacture and analysis, complaints, product recall, and self-inspection. These principles are laid down in the "EC-Guide to Good Manufacturing Practice for Medicinal Products" of 1998 (Rules Governing Medicinal Products in the European Community Volume IV) and in 21 CFR Parts 210 and 211. Drug manufacture is also subject to supervision by the competent authority. This authority varies from country to country, it may be a central governmental authority or the authority of a federal state.

The manufacture, testing, storing, packaging, and marketing of drugs are subject to supervision and inspection by the competent authority. The drug manufacturer has to give the inspectors of the authority access to the premises in which drugs are manufactured. This also holds for the manufacture of active ingredients. The development of drugs and their clinical testing are also supervised by the competent authority which has to check whether the legal requirements for clinical trials have been met. The law gives the competent authority various authorizations to ensure adherence to the drug law.

3.6. Requirements for the Placing of a Drug on the Market

The most important legal prerequisite for placing a drug on the market is the marketing approval. However, many other stringent provisions also have to be observed. Most drugs can only be sold to the patient by a pharmacy. This applies to drugs prescribed by a physician and to nonprescription (over-the-counter) drugs. The over-the-counter (OTC) drugs are generally recognized as safe and effective and not misbranded if certain conditions are met, especially with respect to any applicable monograph (21 CFR Part 330). Some information has to be declared on the drug package (e.g., name and address of the drug manufacturer, name of the drug, number of approval, batch number, type and amount of active ingredients, expiry date). The most important source of information for the patient, however, is the package insert (e.g., § 11 AMG) which contains instructions for use and a description of indications, contraindications, side effects, and interactions. Comprehensive scientific written information must also be available in written form for doctors and physicians (e.g., § 11a AMG). If the pharmaceutical company wants to promote the drug, compulsory information on the drug has to be given (e.g., Heilmittelwerbegesetz in Germany). This information is almost identical to that contained in the package insert.

All these and additional information instruments are designed to avoid risks in connection with the administration of drugs and to guarantee maximum safety. The competent authority is entitled and authorized to control and inspect the company to make sure that it complies with the law.

An additional legal obligation of the pharmaceutical company is to report all possible side effects of the drug, whether previously known or unknown, to the competent authority without undue delay. Every pharmaceutical company has to have a special drug safety monitor (e.g., Stufenplanbeauftragter, § 63 a AMG) whose function is to collect and to evaluate reports on side effects and to coordinate the necessary measures. The drug safety monitor is personally responsible for the fulfilment of the notification obligations toward the health authority and must be a physician, veterinary physician, or pharmacist with at least two years professional experience. The name of the person must be notified to the competent authority. Another specialist (Informationsbeauftragter, § 74 a AMG) to be appointed by the pharmaceutical company has to guarantee for the authenticity of the product information used with the data contained in the registration file.

3.7. Drug Liability

If, as a result of the administration of a drug, a person is killed by the drug or the body or the health of a person is considerably injured, then the pharmaceutical company that has placed the drug on the market is obliged to compensate for the harm caused by the injury. This liability is independent of fault and not imposed on the drug manufacturer, but on the company placing the product on the market; it is a strict liability (§ 84 AMG). Liability only exists if the drug, despite correct use, produces harmful effects that exceed the bounds considered justifiable in the light of scientific knowledge available in medical science and which have their origin in the development of the drug or the manufacturing processes, or if damage has occurred as a result of labeling or instructions for use that do not comply with the knowledge available in medical science. The responsibility for damage in Germany is limited to 100×10^6 €. In general a liability insurance is compulsory to cover the potential financial risk.

Finally many penal provisions and administrative fines complete the instrumentarium of the drug law; in many cases violation of the law leads to personal responsibility and liability (e.g., §§ 95 – 97 AMG).

3.8. International Non-Proprietary Names (INN)

International Non-Proprietary Names (INN) were introduced in 1950 for the purpose of creating an unambiguous single non-proprietary name for drug substances worldwide. (WHO Drug Information vol. 13, no. 1, 1999, pp. 13 – 14). The WHO is in charge of selecting a single name which is acceptable all over the world for each active substance which is to be marketed as a pharmaceutical product. Trademarks for medicines are not allowed to be similar to INNs so as not to confuse the patients or doctors.

In order to select an INN, a request is usually submitted to the World Health Organisation on an INN request form. In certain countries where national nomenclature commissions exist, this is done through the national nomenclature authority, e.g., in the United States it is the United States Adopted Names (USAN)

Council. Precise information on the substance's chemistry, pharmacological action and therapeutic use must be provided. The proposed nonproprietary names and the name and address of the manufacturer are also entered in the form.

The names proposed by the originator are examined and one name is selected in accordance with the "Procedure for the selection of recommended international non-proprietary names (INN) for pharmaceutical substances" and the "General principles for guidance in devising INNs". All the members of the WHO Expert Panel on International Pharmacopeia and Pharmaceutical Preparations designated to select non-proprietary names have to agree to the name selected. Then it is first published as a *proposed INN*. During a four-month period, any person can forward comments, or lodge a formal objection to a name, e.g., on grounds of similarity with a trade name. If no objection is raised, the name will be published a second time as a *recommended INN*.

Almost 7000 INNs have been published to date in the "WHO Drug Information" journal. The names are given in English, French and Spanish, as well as Latin. A cumulative list of all INNs is published periodically and includes INNs in Russian, as well as references to other common names (www.who.int/medicines/organization/qsm/qualityassurance/inn/orginn.shtml).

4. Drug Testing

4.1. Preclinical Testing

4.1.1. Introduction

Modern drug development can be divided into three major stages:

1) Finding the substance
2) Pharmacological – toxicological and pharmaceutical studies
3) Clinical studies

Rational methodology dominates from the very first phase of this process. The search for a new substance starts with the chemical synthesis of a compound or with its isolation from natural products or a fermentation broth of natural or genetically engineered microorganisms. In products produced by genetic engineering, the desired primary effect is usually known. This is, however, not the case with chemically synthesized products. Although a definite effect can often be predicted from structural data or from molecular biological concepts based on receptor models, this effect still has to be checked in pharmacological tests. These tests are normally carried out in a specific screening process that incorporates models for both the desired effect and possible side effects. The tests are performed on standardized models (animals, animal organs, cellular systems, microorganisms, or enzymes), and the most suitable substances are selected (screened) from a large number of analogous substances. In the 1950s, a 1 : 1000 chance of success for the discovery of a promising substance was assumed. Today, a chance of 1 : 7000 – 1 : 10 000 is expected for chemically synthesized substances because of the demand for increased selectivity of a desired effect.

The purpose of the preclinical evaluation of a new drug is to establish its complete spectrum of pharmacological and toxicological properties as well as its pharmaceutical quality. A detailed description of all experiments and results is required by the drug-approval authorities in all countries as evidence of drug efficacy and safety. Animal experiments play a central role in the elucidiation of the pharmacology and toxicology of the active agent or mixtures in preclinical drug development. They are designed to identify all the effects of an active drug candidate which may justify its subsequent administration to humans or which are to be regarded as undesirable or harmful. A therapeutic dose range can be determined from a quantitative comparison of desirable and undesirable effects and the benefit – risk ratio can then be estimated. Chemical analysis of body fluids and organs should give insight into the chemical behavior of the substance in the animal. This includes its absorption, distribution, metabolism, and excretion. Pharmacokinetic and metabolic studies of this type characterize the animal species used in the pharmacological and toxicological experiments with regard to its suitability for predicting the corresponding characteristics in humans.

In toxicological tests, administration of often excessively high doses should produce local or systemic damage of the animal organism so that tolerance in humans can be predicted and possible target organs for acute or chronic poisoning can be determined.

Guidelines and Laws. Investigation of the pharmacological and toxicological properties of a new drug in accordance with the latest scientific knowledge is stipulated in all national and supranational (e.g., EC) drug approval procedures. See also Section 3.3. Most of the pharmacological and toxicological data have to be existent when clinical trials commence. In many countries (e.g., United States, Canada, United Kingdom, and Japan), clinical trials are officially permitted only after these results have been submitted. These tests should qualitatively and quantitatively establish the toxicity limits of the substance, as well as any unwanted or adverse effects, and the desired activity potential. Furthermore, as initiated by the FDA (USA), almost every country now stipulates that the safety testing of drugs be carried out under Good Laboratory Practice (GLP) guidelines (e.g., Council Directive 87/18/EEC). GLP requires that the reported results are comprehensible and that the quality of the data is secured and reviewable. Detailed recording of the planning and performance of experiments, precisely defined instructions (standard operating procedures, SOPs) for the staff, validation of methods, and independent quality control (quality assurance) are therefore necessary. Moreover, laws in many countries regulate the use of animals and the number and quality of animal experiments (e.g., Protection of Animals, Council Directive 86/609/EEC). The necessity of animal experiments for establishing the pharmacological and toxicological profile of a drug is stipulated in national and supranational guidelines. For instance:

1) *EC:* Pharmacotoxicological Standards (Council Directive 75/318/EEC) and several Notes for Guidance concerning single tests
2) *United States*: Guidelines for the Format and Content of the Nonclinical/Pharmacology/Toxicology Section of the Application, US Department of Health and Human Services, Public Health Service, Food and Drug Administration, Washington, D.C./USA, 1987
3) *Japan*: General Toxicology Guidelines, Ministry of Health and Welfare, Tokyo, Japan, 1989

Several guidelines have been compiled by the WHO [10].

4.1.2. Animal Experiments

Animal experiments should allow prediction of the possible reaction of the human organism after single or repeated administration of a drug. The results of animal experiments are applicable to humans with some reservations. Neither the physiology nor the metabolism of commonly used experimental animals completely corresponds to that of humans. The information value can be improved by testing at least two different animal species (preferably one rodent and one nonrodent).

The physical reactions of humans do not generally differ from those of animals. Reactions vary between humans and animals to the same extent as between individual animal species. Thus, good or moderate agreement or even incompatibility is possible. It would, however, be incorrect to reject animal experiments on the basis of the rare cases where no relevance is observed, at least until other, more informative methods have been found.

Methodological difficulties in establishing drug activity or the emergence of unknown and, therefore, untested reactions (e.g., as in the thalidomide disaster of the 1960s) do not legitimate that subsequent drug-induced damage be blamed solely on the shortcomings of animal experiments. Numerous observations have shown that there is generally a good correlation between the reactions shown by animals (primarily vertebrates) and humans to a drug. Differences are frequently due to variations in the pharmacokinetics and metabolism of the active substance. These properties must therefore be carefully checked in all animal species. Animal experiments are certainly not a substitute for careful clinical trials and for the subsequent long-term observation of patients who have been treated with the new drug. However, they are still essential for the estimation of risk in the development period. Restriction to experiments involving important problems and based on validated methods should be an ethical obligation.

A special problem with animal experiments is that many human diseases do not occur spontaneously in animals. Many animal models have therefore been developed with the aim of "experimentally producing a comparable diseased condition" [11]. The applicability of results obtained from these systems is naturally less than that of results obtained in tests on animals with spontaneous pathological conditions. These limitations mainly apply to experiments for finding active substances that counteract specific diseases. For instance, it is easier to find an antihypertensive or antiarrhythmic agent using animal experiments than a psycho-pharmacologically active drug. Experiments for establishing drug safety are always carried out on healthy, purebred experimental animals.

The standardization of methods is another important factor for transferability of the results. Even minor changes in the test procedure (changing the breed of experimental animal, the time of testing, or temperature) can cause substantial deviations of results. All reports on pharmacological and toxicological testing must therefore be accompanied by a detailed description of the methodology used. Standard pharmacological methods are described in appropriate handbooks (e.g., [12, 13]). Comparative testing of a standard substance, if available, is of great importance. Ring experiments with standard substances involving several laboratories have proved useful for toxicological tests.

4.1.3. Pharmacological Testing

Studies of Special and General Pharmacodynamic Properties. Actual drug development begins when the signs of a therapeutically interesting effect found in the preliminary tests (in vitro screening tests or specific in vivo tests) can be reproduced and biometrically confirmed in comparison with standard substances. The aim of this phase of drug development is to analyze the special (i.e., therapeutically desired effect) and general pharmacodynamic (i.e., sum of all biological activities of a drug) profiles. The methods should be varied, for example by using different animal species. Tests include the determination of dose – response relationships, calculation of mean effective doses (ED_{50} = the dose that causes a defined effect in 50 % of the animals or changes a quantifiable reaction in the animals by 50 %), and the determination of time – response relationships. Different routes of administration should be used (generally oral and intravenous, but also subcutaneous, intraperitoneal, and intraduodenal). The duration and onset of the effect and the time of maximum response should be established.

Comparison of the mean effective doses for the special pharmacodynamic effect and the toxic doses determined in the toxicological experiments (e.g., LD_{50} = lethal dose in 50 % of an animal group) furnishes important information on the "therapeutic range", i.e., whether a desired response can be achieved in animal experiments with doses that do not produce side effects or lethality.

The responses obtained in the acute experiments must be checked again after repeated dosage to determine whether the response decreases (tolerance) or increases (accumulation).

The general pharmacodynamic characterization of an active agent should show whether it also affects functions of the organism that are not necessarily connected with the primary, therapeutic effect. The functioning of all important organs (e.g., heart and circulation, lungs, nervous system, kidneys, endocrine glands, and gastrointestinal system) must therefore be tested in separate pharmacological experiments and changes in function must be quantitatively assessed by biometric methods. Experiments on isolated organs, isolated cell systems (e.g., receptor binding assays), and cell-free enzyme systems are of primary importance in this respect. If substance-dependent changes in functions are observed, the relevance of the results should be checked in the intact animal. If essential physiological reactions are inhibited and can no longer be evaluated as a result of anesthesia (e.g., central nervous reactions, central cardiovascular regulation), experiments must be performed on animals without anesthesia. Dose – response relationships must be determined if pharmacodynamic effects occur. The dosage level must be based on the dose required to give the therapeutically desired response. Furthermore, substance-dependent effects must be checked after repeated administration.

Some experiments are designed to find antagonistic drugs if poisoning by an overdose of the developing drug is the result of an exaggerated pharmacodynamic response. This antagonist should be used as an antidote for humans in the case of overreaction or poisoning. Studies on interaction with other drugs that are used frequently to treat the same indications should also be performed.

Pharmacokinetic Studies. Investigations which evaluate the "biological fate" of the new drug in the animal organism are an important part of its pharmacological characterization. The species used for these pharmacokinetic studies are the same as those employed for the pharmacodynamic and toxicological experiments. The path of the active substance is followed from the site of administration through the organism to the excretory organs. The elucidation of the pharmacokinetics of a drug, known as ADME, is divided into the following parts:

1) *Absorption or Application* (*A*) *Phase*. Determination of the degree and rate of absorption of the substance into the blood stream from the site of administration. The substance may be applied in a dissolved or suspended form, but also in a pharmaceutically prepared form.
2) *Distribution* (*D*) *Phase*. Determination of the concentration of the substance in body fluids and tissues after its distribution in the organism, including binding to special components (e.g., plasma proteins).
3) *Metabolism* (*M*) *Phase*. Determination of the metabolism of the substance, i.e., its biological degradation in the organism by the action of enzymes, primarily those in the liver, intestine, plasma, or kidneys.
4) *Elimination* (*E*) *Phase*. Determination of the routes and amounts by which the unchanged or metabolized substance is excreted.

Chemicoanalytical, immunological, and radiological methods are used to determine the substance in body fluids and organs. Radioactive substances (i.e., ^{3}H- or ^{14}C-labeled) or deuterium-labeled substances serve to identify metabolites.

Knowledge of the pharmacokinetics of a substance obtained in the animal species that are used for the pharmacodynamic and toxicological tests is essential for the interpretation of drug safety data. It is the most important criterion for estimating the relevance of the data for humans. It permits the association of a defined drug concentration in the plasma or at the site of action with a pharmacodynamic effect in animals and humans. Special features of the pharmacokinetic behavior of an active substance can be recognized in animal experiments and allow specific analysis in humans. Examples of such behavior include rapid metabolism during initial passage of the substance through the liver or the intestinal wall after enteral absorption (first pass effect), enterohepatic circulation (reabsorption of substances excreted into the intestine with the bile), and storage of the substance in certain tissues.

Conclusions about the dosage interval needed in long-term pharmacological and toxicological studies can be drawn from the retention time of a substance in the organism if, as in humans, a continuous plasma level of the drug and prevention of its accumulation are required.

Pharmacokinetic studies on animals are helpful for preparation of protocols for experiments on humans when the drug is first administered.

4.1.4. Toxicological Testing

Investigation of the toxicological properties of a potential drug is of considerable importance for its further development. Information about possible toxic effects and dose limitation must be obtained before performing the first trials on humans. In addition to acute toxicity tests, experiments are performed with repeated drug administration (varying duration) in different animal species. The duration of these animal tests depend on the planned duration of tests and subsequent administration in humans. Special toxicological tests are used to exclude local intolerance to parenterally administered drugs, reproductive damage, mutagenic, and, in the case of long-term human application for more than six months, carcinogenic effects. Tests for allergization and phototoxicity are required for dermatological drugs. As a rule, experiments must be conducted in such a manner that statistical analyses are possible. Toxicological tests must comply with GLP regulations and internationally standardized guidelines.

Single-Dose Acute Toxicity. Almost all countries require the testing of systemic tolerance after a single oral and parenteral administration of high doses to two animal species (mouse and rat), with a postobservation period of 14 days. Systemic toxic symptoms should be described separately for male and female animals, data on the dose dependency of symptoms and lethality (LD_{50}) should be recorded. The value of acute toxicity experiments is disputed. Their primary importance is the fact that appropriate animals deliver information for the experimental design of the more important toxicity tests involving repeated drug administration. Also, information is obtained on the poisoning profile of an acute overdosage in humans. Calculation of the therapeutic range from ED_{50} and LD_{50} values is discussed in Section 4.1.3.

It is generally sufficient if the approximate acute toxicity is determined on a small number of experimental animals (i.e., three per sex and species for each dose tested) without statistical evaluation (cf. EC Note for Guidance: single dose toxicity). This allows estimation of the mean, minimum, and maximum lethal doses.

In the case of new drug combinations, acute toxicity tests should establish whether the combination of active agents potentiates their action or leads to new toxic effects. If this is the case, extended tests are required to clarify the toxic properties in long-term application (full toxicological development program as with a single new drug).

Toxicity on Repeated Administration. Repeated administration of a drug (usually at daily intervals) should show which functional and morphological changes are produced by the active agent. The test must include a control group of animals that receives a preparation without an active ingredient (i.e., solvent alone). Doses range from therapeutic (i.e., in the range of the pharmacodynamic effective dose) to toxic. The highest dose should produce harmful effects and the lowest dose should be free of toxic effects (i.e., the results in the test group should not differ from those of the control group). As a rule, two series of experiments of varying duration are set up. In order to determine the subchronic toxicity (14 days to 3 months, daily administration) and chronic toxicity (3 – 12 months, daily administration) each series is performed on two animal species. A rodent and a nonrodent species should be chosen (usually rat and dog or hamster and monkey). In many countries, the appropriate animal species, the number of groups of experimental animals exposed to varying dosages, the number of animals in each group and of each sex, and the duration of treatment are stipulated by guidelines. In applications for drug approval, the required duration of drug administration to animals depends primarily on the expected maximum duration of the clinical treatment of patients. A survey of the requirements in some countries is given in Table 5.

Table 5. Toxicity study requirements in the EC, Japan, and USA

Duration of	Duration of toxicity studies		
patient therapy (oral or parenteral administration)	EC	Japan	USA*
≤ 1 day	2 weeks	28 days	14 days
2 – 7 days	4 weeks	90 days	90 days
8 – 30 days	3 months	26 weeks	26 weeks
> 30 days	6 months	52 weeks	26** or 52 weeks

* Depending on the stage of drug development, considerably shorter durations are sufficient for the performance of clinical studies before drug approval (IND phase).

** Duration of therapy up to three months.

The following investigations are used to evaluate tolerance or toxicity: functional tests (behavior, course of growth and food intake, acoustic and optic reactions, measurement of blood pressure, and electrocardiogram), clinical chemistry of blood and urine, status, autopsy including organ observation and weights, and light microscopy supplemented by electron microscopic investigations of all organs. Studies on the reversibility of the changes complete the toxicological tests.

Reproduction Studies. Reproduction tests have been an essential requirement since the thalidomide disaster in 1962. The effect of the test substance on fertility and mating behavior, as well as effects that lead to abortion of the fetus, fetal abnormalities, or damage of further progeny are studied. These tests are divided into three segments:

1) *Fertility Testing*. The influence of the substance on male and female fertility is generally tested in rats.
2) *Embryo Toxicity Testing*. These tests are carried out on at least two species, usually rat (or mouse) and rabbit. The test substance is administered to pregnant animals during the

critical phase of organogenesis (day 6 – 15 of pregnancy in rats, day 6 – 18 in rabbits).

3) *Peri- and Postnatal Investigation.* The test substance is usually administered to pregnant rats in the last third of gestation and continued during the four-week lactation period. The influence on the size of the litter and breeding of progeny is recorded.

Mutagenicity tests (→ Mutagenic Agents) are designed to detect any changes in genetic properties. As a rule, a series of at least four tests is conducted for each drug. Various types of mutation are investigated in different systems. Point mutations (i.e., gene mutations) are investigated in vitro in bacterial test systems (e.g., Ames test) and eucaryotic (e.g., mammalian) cells, predominantly lymphocytes, that are cultured in a suitable nutrient medium. The point mutation tests are also used as precarcinogenicity tests. Chromosomal changes are studied in vitro in mammalian bone marrow cells and in vivo in a rodent species. A series of other models can be used to supplement the investigations in individual cases [14].

Carcinogenicity tests (→ Carcinogenic Agents) are designed to establish whether a drug has cancer-producing (carcinogenic) effects. These tests are required if the drug is to be taken either regularly for more than six months or with interruptions, but frequently enough to produce the same total burden. They are also necessary if the chemical structure of the substance or previous chronic toxicity and mutagenic tests suggest a possible carcinogenic potential. The investigations should be performed on two rodent species that are treated with varying daily doses of the test substance for 18 – 24 months (mouse) and 24 – 30 months (rat).

Special Toxicity Tests. Pharmaceutical agents that are to be injected or applied to the skin or mucous membranes must be tested for local irritant effects. These tests are carried out on animals (dog, rabbit) by single and repeated application of different amounts of the final preparations at these sites. Special investigations should be conducted with dermatological drugs after combined systemic and topical application to reveal possible sensitization (Magnussen test), allergization, or phototoxicity. Systemic toxicity testing is also required for repeatedly administered dermatological drugs that penetrate the skin in relevant amounts (generally >1 %). This test is most effectively carried out by subcutaneous application.

Safety Testing of Biotechnological Products. Human peptides, proteins, and monoclonal antibodies produced by biotechnological methods require preclinical evaluation of their safety for use in humans. This evaluation differs from that required for conventional, chemically synthesized substances. The development of antibodies to these products in test animals influences the test results (allergic reactions or neutralization) and limits the duration of administration. Exaggerated pharmacodynamic responses are usually responsible for the most serious toxicological problems caused by these proteins (e.g., recombinant insulin, erythropoietin, or tissue plasminogen activator). Sufficient information can therefore still be obtained under such restricted testing conditions provided the tests include as much pharmacological and biochemical data as possible (see Section 4.1.3). Toxic effects can, however, also occur after overdosage, i.e., when the concentrations of agents normally present in the body are greatly exceeded (e.g., with cytokines) during therapy or when mutants (homologues of human proteins) are tested.

If antibody development is detected in the animal species usually used for testing, studies on nonhuman primates are sometimes required to determine the most likely toxic profile of the products in humans. Another possibility would be to test a biotechnological product that has an identical amino acid sequence to the endogenous substance in the animal species used. This would, however, greatly increase development costs and would still only be a model.

4.2. Clinical Trials

The term clinical trials comprises the systematic, controlled and/or documented administration of potential drugs to a limited number of healthy or sick human subjects in order to provide proof of effectiveness (efficacy) and safety and to allow the subsequent, wide-scale, safe, application of the medicinal product in patient therapy. These studies must be performed in a group of sick subjects who are representative of the population for which the drug is intended. The animal studies described in Section 4.1 are an indispensable prerequisite for clinical studies: knowledge of the pharmacological and toxicological profile is needed to assess benefits and risks in future therapy, to estimate risks for initial administration in humans, and to design suitable clinical trials. However, the restricted applicability of data from animal experiments to humans must be taken into consideration (cf. Section 4.1.2). The effectiveness and tolerability of a substance in humans can only be determined in patients in clinical trials. The necessity for clinical studies is recognized throughout the world and is established in national laws. Before initiation of any clinical trial certain regulatory requirements have to be met, ranging from simple notification to the competent authority to official trial permission based on review of submitted data.

4.2.1. Phases of Clinical Drug Development

Clinical studies can be divided into four phases:

1) Human pharmacological testing (Phase I)
2) Tolerability and dose finding in patients (Phase II)
3) Proof of safety and effectiveness (Phase III)
4) Studies and follow-up checks after approval (Phase IV)

Although the phases are performed in consecutive order, wide overlap may occur. The definition of the individual phases is therefore not always standardized; Phases II and III are sometimes further subdivided.

Phase I. In human pharmacological testing, a new active substance (NAS) is administered to humans for the first time. The effect of the substance on the human body, namely pharmacodynamics (including effects and adverse drug reactions), is recorded with as much detail as possible. The effect of the body on the substance, namely pharmacokinetics (absorption, distribution, metabolism, and elimination: ADME), is also investigated; the four steps involved in the elucidation of the pharmacokinetic system are defined in Section 4.1.3. Sensitive analytical methods are used to determine the concentration of the substance in body fluids. For methodological and ethical reasons, Phase I investigations are usually carried out in a group of ca. 60 healthy volunteers. Studies in patients are only performed under special circumstances (e.g., in the case of cytostatic and immunologic agents).

The first application to subjects generally starts with the administration of a dose that is only a fraction of the dose effective in animals. This single dose is gradually increased until it exceeds the expected therapeutic range and/or adverse drug reactions occur. Statements made by the subject about his well-being are recorded. Physical and chemical data (e.g., vital functions, laboratory values, and the concentration of the substance in the blood and excretory products) are also measured. Important basic pharmacokinetic parameters (e.g., the distribution volume, half-life, and metabolism) are determined from the chemical data. The absolute bioavailability is calculated by comparing plasma levels after systemic (e.g., intravenous) and enteral (oral) administration. Since the metabolism of foreign substances is genetically determined and varies from person to person, the pharmacogenetics of the test persons must be defined.

The results of Phase I investigations provide important information for establishing the limiting dose and the dosing intervals for the first administration of the drug to patients in Phase II. Phase I investigations are completed by determining pharmacokinetics behavior on repeated administration, the bioavailability of the final dosage form, interactions with other drugs, and, finally, the influence of food, age, disease, and of impaired liver and kidney function on kinetics.

Phase II. In Phase II the potential drug is for the first time administered to patients suffering from the disease intended for treatment. The objectives are to obtain information on tolerability and to find the optimal dose and dosing scheme. Usually, several hundred patients are necessary. Since these investigations can only be carried out in patients, the accompanying conditions must be taken into account. As a result of the variety and variability of diseases and their symptoms, these studies demand high requirements and performance as regards planning, execution, diagnostic methodology, documentation, biometric evaluation, and clinical interpretation of the results. The choice of method depends on the indication and the test parameters; it may range from establishing influence of a drug on a clinically relevant, objectively measurable parameter (e.g., blood pressure) to the detection of effects that are only perceptible after long-term treatment and are based on subjective parameters such as patient's statements and physician's impressions (e.g., psychopharmacological drugs). The internationally standardized scales used for this purpose are validated in the patient's language (national language). Most of the relevant diagnostic procedures for a given disease are also suitable for demonstrating effects or efficacy of treatment; they are utilized selectively in clinical trials. Such efficacy parameters are used to determine the minimal and fully effective dosages. Suitable means of long-term recording methods provide information about the dosage intervals. The investigations should be performed on an adequate number of patients who are representative of the population to be treated later as regards the spectrum of the illness, age, sex, and race. The clinical situation comprises the common cold treated by the family doctor as well as the life-threatening arrhythmia of an acute myocardial infarction in the intensive care unit. Phase II investigations must comply with valid scientific standards which are defined in national and international guidelines for some groups of therapeutic agents. Furthermore, the value of the new therapy should be compared with established therapies, as regards recovery rates, incidence range of side effects, quality of life, secondary diseases, or rates of survival. Depending on the disease this information can be obtained from studies in as many as several hundred patients.

Phase III. The objectives of Phase III are to establish scientific proof of efficacy of the drug in question and to confirm the optimal dose in controlled randomized trials (see Section 4.2.2) in comparison to placebo and/or an established comparator drug. For these studies the considerations for clinical trials of Phase II also apply. In addition, special attention must be paid to drug safety and rarely occurring adverse drug reactions. Influences of accompanying therapy, interactions with other drugs and stimulants, effect on driving fitness and working ability should also be clarified. The investigations should extend over a period corresponding to the planned duration of application. In the case of drugs used for long-term treatment, the investigations should be performed for at least one year. The cross section of the diseased population should be as representative as possible and include widely differing forms of the disease. Since a large number of patients is required (several hundred to a thousand) the tests must be performed by a number of physicians according to a standard protocol. These multicenter studies are usually carried out on an international basis. A claim of prevention can usually be verified only after several thousand patients have been subjected to long-term therapy (several years) in multinational studies. Many of these intervention studies are therefore carried out after approval in Phase IV.

Phase IV. A new drug is granted marketing authorization by the relevant authorities on the basis of the results of pharmaceutical/quality data, preclinical testing, and clinical studies, particularly Phase I – III. Phase IV involves medicinal products that have already been approved. These studies are required to update information about the drug in keeping with scientific developments. Drug monitoring or pharmacovigilance (i.e., the collection and evaluation of adverse drug reactions) is often included in this phase. Phase IV begins with drug approval and ends with the discontinuation of sales.

4.2.2. Methods for Proving Effectiveness

As with all scientific experiments, clinical trials should provide generally valid evidence from a limited number of individual observations. The reproducibility of a result can be determined after repetition of the test and the probability of general validity can be calculated with statistical methods. Clinical tests must be therefore designed and evaluated on the basis of biostatistical criteria.

The basis for clinical trials for proof of efficacy is the therapeutic comparison in which the new drug is compared with an established drug (standard therapy) or placebo (dummy or sham therapy). Ideally, the two groups of subjects should differ only as regards the form of therapy they receive. In order to obtain a high degree of uniformity, factors affecting the groups (e.g., type, degree, and stage of illness; demographic characteristics) are defined by inclusion criteria. The structural uniformity can be further increased by stratification. Known interfering factors but also risk factors are eliminated by the definition of exclusion criteria. The patients must be randomly assigned to the treatment groups (randomization) to increase their comparability with respect to unknown factors and to lay the basis for the application of mathematical probability analysis.

In addition to structural uniformity, observational uniformity has to be established to eliminate subjective influences resulting from expectations or bias; this is achieved by performing blind studies. In a single-blind test only the patients are not informed about the treatment they receive. In a double-blind test the assignment of the patient to the test groups also remains unknown to the investigator. Coded test preparations with the same appearance and taste are randomly allocated to the test subjects. If a therapy is not to be compared with another treatment, a dummy or sham preparation (placebo) is used that looks exactly like the active preparation but does not contain an active agent. When comparing different galenic forms, blind testing is achieved by administering the two forms to both groups. Each form is administered as an active preparation and as a placebo (double dummy technique). For reasons of safety, provision for the decoding of the preparations must be made in case of an adverse event.

Noncontrolled, open studies do not generally permit scientifically valid conclusions to be made about the efficacy of a medication. A rare exception is if the drug has a striking effect on a previously incurable illness. These observation studies are, however, performed in other circumstances, e.g., if a drug is being administered for the first time, as a pilot experiment to check methodology, to establish dosages by dose titration, for long-term observation, and to document drug safety. Statistically valid results can only be obtained in comparative studies which are carried out in two main ways: as a comparison between different subjects in two groups (interindividual) or as a crossover experiment on a single subject (intraindividual).

Group (interindividual) comparison (parallel design) can be used to evaluate most medical problems, especially for dose confirmation and determination or comparison of drug effectiveness. However, this design is occasionally confronted with ethical limitations. If the treatment produces results that vary widely between different individuals or if small differences have to be verified, relatively large numbers of patients are required.

Variability and thus the number of observations can be significantly reduced if the study is conducted as an intraindividual comparison in the same patient. Each patient receives each of the test preparations in a random order for a sufficiently long period. This design can only be used in patients suffering from chronic illnesses with stable characteristics. Methodological problems sometimes restrict the use of this procedure, especially when the effect of one therapy continues into the subsequent treatment period (carry-over effect).

4.2.3. Adverse Drug Reactions

Very few therapeutic measures are completely free of risk. Adverse drug reactions occur unintentionally at the dosages normally used in diagnostics, therapy, or prophylaxis and are unpleasant or even harmful for the patient. Some adverse reactions can be expected from the pharmacological and toxicological properties of the drug and mainly occur after an overdose or as a result of hyperreactions in certain susceptible individuals (genetic origin). These reactions must be distinguished from immunological reactions which may result in a true drug allergy via sensitization or in a pseudoallergy via the release of mediator substances of the immune response. Since the value of a drug depends not only on its therapeutic benefit but also on its potential risks, proof of its tolerability must be substantiated with the same care as its efficacy. All adverse events are therefore recorded, irrespective of their causal relationship with drug administration because this relationship sometimes only becomes apparent after reoccurrence of these events. Each adverse event is recorded with respect to its symptomatology, onset, duration, and the degree of severity. Although the investigator evaluates the probability of a causal relationship, correct assignment can only be made later during retrospective overall evaluation. Standardized algorithms are available for determining the probability of a causal relationship. The entire range of adverse reactions of the test substance is compared with those of the control preparation, including the placebo. In addition to adverse drug reactions occurring during clinical development, physicians and pharmaceutical companies are obliged to report adverse drug reactions occurring during treatment with marketed drugs. The forms to be used, the time factor to be observed, and the authorities to be informed in the relevant notification system vary from country to country. The national centers report to the Collaborating Center of International Drug Monitoring of the WHO.

4.2.4. Good Clinical Practice

Quality standards and the assurance of their maintenance in drug production and in preclinical and clinical development are defined in guidelines such as Good Manufacturing Practice (GMP), Good Laboratory Practice (GLP), and Good Clinical Practice (GCP), see also Section 3.3. Compliance with these guidelines is subject to official control. The International Conference of Harmonization (ICH) has developed over the years binding guidelines on major aspects of drug development. The ICH guideline on good clinical practice (GCP) was adopted in 1997 by the regulatory authorities of the EU, Japan, and the United States. It represents the international ethical and scientific quality standard for clinical trials in humans. These guidelines have three main objectives:

1) The protection of the trial subjects
2) Definition of responsibilities
3) Data quality assurance

Written Standard Operating Procedures (SOP) ensure the adherence to these rules during the many activities involved in the planning, performance, monitoring, and evaluation of clinical studies.

Protection of Trial Subjects. The Declaration of Helsinki (last revision of Hong Kong 1989) of the World Medical Association gives guidance to physicians in biomedical research involving human subjects (see Section 3.3) and is the accepted basis for clinical trial ethics.

Biomedical research must improve diagnostic, therapeutic, and prophylactic procedures and the understanding of the etiology and pathogenesis of disease. However, the right of the research subject to safeguard his or her integrity always has priority. Among the many principles to be regarded are careful benefit – risk assessment, adequate information and voluntary participation (informed consent), and insurance for the case of injury. A specially appointed committee independent of the investigator in conformity with the national laws (Ethics Committee, Institutional Review Board) should give its opinion as to whether these principles, legal requirements etc. are adequately taken in consideration.

Responsibilities of the persons involved in clinical trials (e.g., sponsor, monitor, investigator) are defined and must be laid down in written Standard Operating Procedures and in a trial protocol. All these measures serve the purpose of protecting the trial subjects/patients, complying with relevant national laws, and validating the data obtained from the study.

Data Quality Assurance in Clinical Trials. Since a new drug is granted approval by the authorities on the basis of the submitted data, considerable attention is paid in GCP guidelines to the careful collection, documentation, and validation of the data. The basic condition is that each result of a clinical report must be verifiable and must be retraceable to individual observations in individual patients. For pivotal studies, the Federal Code of Regulations of the United States specifies FDA inspections of the sponsor and of the testing location. The GCP guidelines of the EU also require audits at the site of the investigator and sponsor which are carried out either by an external institute or by an independent data quality assurance unit of the sponsor. The Japanese regulatory authority performs site and sponsor audits. Data audits are carried out on random samples and according to biometric criteria. The entire data handling must take into account the national data protection laws. The period of time for which the data must be kept at the individual locations is also regulated. Data must be kept for up to five years after termination of the marketing of the drug.

In Germany and some other countries, compliance with regulations and legal provisions that apply to the execution of clinical studies are subject to official surveillance.

4.2.5. Clinical Application Dossier

Although the formal requirements to be met by an application dossier vary in different countries, they are being harmonized. These requirements are contained in the "Notice to Applicants" in the EU and in the "Guideline for the Format and Content of the Clinical and Statistical Sections of a New Drug Application" in the United States. Japan has its own regulations. The official evaluation of a dossier and possible time limits are still regulated very differently at a national level. In addition to their own expertise, the authorities generally also make use of external professional commissions, advisory boards, or external expert opinions. After a drug has been granted approval and put on the market, it is still subject to official surveillance with respect to production quality, information of professional circles and patients, notification of adverse reactions, and promotional information.

4.3. Quality Control of Pharmaceutical Products

The manufacturers of pharmaceuticals products strive to achieve quality of their products both by controlling the quality of the outcome and also by manufacturing according to the strict rules and high standards laid down in the GMP guidelines (Section 3.5).

Established quality standards are reached by agreement between the manufacturer and the health authorities prior to marketing pharmaceutical products. This agreement is similar to a legal contract in which the company producing and controlling the product is committed to meeting quality standards in future, i.e., prospectively, which are based on experience gained in the past while developing the product, i.e., retrospectively.

Quality standards are published by national regulatory authorities. Since the adoption of the first European Directive on pharmaceuticals (65/65/EEC) in 1965, the Member States of the European Union (EU) have been working towards harmonization of their requirements. These requirements have been published in "The Rules Governing Medicinal Products in the European Union" Volume III – Part 1 "Guidelines on the Quality, Safety, and Efficacy of Medicinal Products for Human Use" January 1996, European Commission III/5380/96, to be found at http://dg3.eudra.org.

4.3.1. Manufacturing of Drug Substances – Test of Starting Materials

All starting materials have to be tested according to specifications applying suitable analytical procedures before manufacturing or synthesizing a drug substance. A chain of documents is generated as chemical synthesis proceeds. This documentation includes, amongst others, a written description of the process and appropriate production records, records of the raw materials used, records of batch numbers, records of the critical processing steps accomplished, and intermediate test results with meaningful standards (FDA Guide to Inspections of Bulk Pharmaceutical Chemicals, September 1991, reformatted in May 1994 with editorial changes.)

4.3.2. Control of Drug Substances: Impurities

Impurities, i.e. process and drug-related organic and inorganic impurities, may arise during the manufacturing process and/or storage of the drug substance. They may be identified or unidentified, volatile or nonvolatile, and include starting materials, byproducts, intermediates, degradation products, reagents, and catalysts. Inorganic impurities are normally known and identified and include heavy metals, inorganic salts, and other materials, such as, filter aids and charcoal. The amount of impurities, identified or unidentified, must be limited to ensure the safety of patients. The procedure used to establish acceptable upper limits is described in detail in the "International Conference on Harmonisation of Technical Requirements for the Registration of Pharmaceuticals for Human Use" (ICH) *Harmonised Tripartite Guideline "Impurities in New Drug Sub-*

stances", published in 1995 (to be found at http://www.ifpma.org/ich5q.html).

Solvents are used in the manufacture of drug substances and excipients or in the preparation of medicinal products. The levels of residual solvents in medicinal products should be no higher than can be supported by the safety data. Solvents that are known to cause unacceptable toxicity, such as, benzene, should be avoided in the production of drug substances, excipients, or medicinal products unless their use can be strongly justified in a risk-benefit assessment. Solvents associated with less severe toxicity, e.g., chloroform, methanol, or toluene, should be limited in order to protect patients from potential adverse effects. Ideally, less toxic solvents, such as acetone, ethanol, or ethyl ether, should be used where viable.

A method for classifying residual solvents by risk assessment is provided in the *ICH Guideline "Note for Guidance on Impurities: Residual Solvents" (CPMP/ICH/283/95), published 17 July 1997* (to be found at http://www.ifpma.org).

4.3.3. Control of Drug Products: Impurities

As stated in an ICH guideline, "Impurities in drug products may be produced by degradation of the active ingredient or by reaction of the active ingredient with an excipient and/or with the immediate container/closure system." Consequently, the amount of degradation products allowed in a finished product should be such that the therapeutic efficacy of the product is not affected. The toxicity of degradation products must also be evaluated and assessed.

Levels of degradation products can be measured, for example, by comparison of an analytical response for a degradation product to that of an appropriate reference standard or to the response of the active ingredient itself. Methods for setting up limits for degradation products are described in detail in the *ICH Guideline "Note for Guidance on Impurities in New Medicinal Products" (CPMP/ICH/282/95)* of 6 November 1996 (to be found at http://www.ifpma.org).

4.3.4. Validation of Analytical Procedures

The objective of analytical procedure validation is to demonstrate that the method in question is suitable for its intended purpose. The following analytical procedures have to be validated in particular:

- Identification tests intended to ensure the identity of an analyte in a sample by comparing a property of the sample to that of a reference standard
- Quantitative tests for the content of impurities
- Limit tests for the control of impurities
- Quantitative tests of the active moiety in the drug substance or drug product
- Dissolution testing for drug products
- Particle size determination for drug substances

Relevant information on validation of analytical procedures is summarized in *ICH Guideline "Validation of Analytical Procedures: Definitions and Terminology"* of November 1994 (to be found at http://www.ifpma.org/ich5q.html).

Guidance and recommendations on how to consider the various validation characteristics for each analytical procedure are provided in the *ICH Guideline "Note for Guidance on Validation of Analytical Procedures: Methodology" (CPMP/ICH/281/95)* of 6 November 1996 (to be found at http://www.ifpma.org).

4.3.5. Process Validation and Qualification

Since it is important to ensure that the defined quality standards of the pharmaceutical product are met, the manufacturers conduct "Process Validation and Qualification" on the product prior to marketing thereby establishing and documenting that:

- The facilities, equipment and processes have been designed in compliance with the requirements of the current GMP guidelines.
- The facilities and equipment have been built and installed as specified. This constitutes the Installation Qualification (IQ).

- The facilities and equipment are operated as specified. This constitutes the Operational Qualification (OQ).
- The facilities and equipment operate as specified and repeatedly and reliably produce a finished product of the requisite quality. This constitutes the Process Validation (PV).

Any aspect of the premises, facilities, equipment or processes, which may affect the quality of the product, either directly or indirectly, is subject to qualification and validation (*Convention for the Mutual Recognition of Inspections: "Principles of Qualification and Validation in Pharmaceutical Manufacture," Document PH 1/96*, January 1996).

4.3.6. Batch Release

Some types of pharmaceutical products, e.g., vaccines, sera, toxins, antitoxins, antigens, blood products, etc., have to be tested and released on a batch-by-batch basis by an independent state laboratory specializing in this particular analytical area. These products contain substances whose purity and potency cannot be measured by chemical methods alone. The batch release system is used in many countries, and the independent laboratory responsible for testing in Germany is the Paul Ehrlich Institut. In France, it is the Institut Pasteur.

4.3.7. Stability Testing

Stability studies are a major expense when developing new products. This applies in particular to drug products that are to be marketed in several strengths and package types. Multiple strengths and package types combined with multiple batches, various storage conditions, test parameters, and test intervals require a great number of samples to be tested at considerable cost. Further to this, the different requirements of the regulatory agencies must be taken into account. As a result, an enormous amount of stability testing, much of it redundant, was performed by multinational pharmaceutical companies seeking approvals in several countries. Therefore, the compilation of a common set of stability requirements for marketing authorizations was considered to be a top priority for the pharmaceutical industry when the International Conference on Harmonisation of Technical Requirements for the Registration of Pharmaceuticals for Human Use (ICH) was formed in 1991. Regulators and pharmaceutical industry representatives from the EU, Japan, and the United States with observers from the Canadian and Swiss Health Authorities and the WHO chose stability testing as one of the first issues to be discussed and harmonized. An ICH Guideline on Stability Testing was subsequently developed and published in October 1993, after which it was adopted throughout the ICH region [15]. This guideline describes the stability testing requirements for a registration application within the "ICH territory": EU, Japan, and the USA *(ICH Harmonized Tripartite Guideline "Stability Testing of New Drug Substances and Products")* of 27 October 1993, as amended (to be found at http://www.ifpma.org).

As this guideline was limited to new active substances and products for human use sold in ICH countries in which climatic conditions are moderate, the WHO developed its own guideline to include established substances, and also hot or hot and humid climatic zones. The WHO guideline was published in 1996 (*"Guidelines for stability testing of pharmaceutical products containing well-established drug substances in conventional dosage forms" WHO Technical Report Series, No. 863*, 1996) [16]. Meanwhile, a variety of additional ICH stability guidelines have been developed, for example:

- *Photostability Testing* and a guideline concerning the *Stability of Biotechnology Products*, http://www.ifpma.org/ich5q.html
- *Reduced Stability Testing Plan – Bracketing and Matrixing (CPMP/QWP/157/96),* dated 22 October 1997
- *Note for Guidance on Stability Testing of Existing Active Substances and Related Finished Products* (CPMP/QWP/556/96)

In the USA, the Code of Federal Regulations (CFR) provides guidance on stability testing, e.g., the GMP Guideline for Pharmaceutical Products, 21 CFR 211, contains a subpart 166 describing stability testing requirements, and 21 CFR 314.50 (d) (1) (ii) (a) requires that the stability of a drug substance and product be studied after approval as per the stability protocols approved [17].

4.3.8. Documentation

Format of Registration Documentation. Regulatory authorities in various regions require applicants to submit marketing authorization applications according to their national guidelines and regulations. In *Europe*, the format for presenting quality related documents is outlined in detail in *The Rules governing Medicinal Products in the European Union, Volume 2B: Notice to Applicants,* January 1997, to be found at http://dg3.eudra.org/eudralex/index.htm. The format for applications in the *USA* is described in the *Code of Federal Regulations, Title 21, Chapter I, Subchapter D, Part 314, Subpart B* (http://www.fda.gov). In *Japan,* the Ministry of Health and Welfare regularly publishes updates of their Technical Requirements for New Drug Applications, updated versions are available from Yakuji Nippo, Ltd., Japan, editorial supervision by the Pharmaceuticals and Cosmetics Division, Pharmaceutical Affairs Bureau, Japanese Ministry of Health and Welfare.

The Common Technical Document – Quality. The ICH topic "Common Technical Document (CTD) Quality" proposes a common structure for the chemical/pharmaceutical section of dossiers for marketing authorization applications. The final ICH Guideline (expected to be released in November 2000) will be a major achievement and the result of a concerted effort on the part of the reviewers working in the health authorities and the quality control/regulatory affairs departments in the pharmaceutical companies. The aim of the Common Technical Document project is to harmonize both the structure (or Table of Contents) and data of regulatory dossiers submitted to the health authorities. It encompasses all documents except for administrative application forms and raw data. It includes Common Tabulated Summaries based on the structure of the documentation, as well as Common Written Summaries.

The following data have to be supplied to the regulatory authorities when submitting an application for marketing authorization:

Drug Master Files. Drug Master Files (DMFs) are documents that are submitted to the U.S. FDA providing confidential detailed information on the facilities, processes, or articles used in the manufacturing, processing, packaging, and storage of medicinal products intended for human use. DMFs allow parties other than the holder of the DMF to reference information not disclosed to them (Center for Drug Evaluation and Research, FDA *Guideline for Drug Master Files,* September 1989, to be found at http://www.fda.gov).

In the EU, the use of Drug Master Files is restricted to active drug substances only (*European Drug Master File Procedure for Active Substances, The Rules governing medicinal products in the EU, Vol. III – Part 1,*, January 1996 to be found at http://dg3.eudra.org).

4.3.9. Variations – Changes

Drug manufacturers are required to amend and update manufacturing and control methods within the life-time of a medicinal product. In Europe, this is detailed in Council Directive 65/65/EEC, Article 11. Since the level of scientific development is constantly subject to change, modifications in a product's composition, manufacturing methods and control methods are unavoidable, and even desirable as they improve quality. However, every change that is made entails revision and amendments to the documentation. Sometimes approval has to be obtained from the health authorities before the proposed changes can be implemented (Commission Regulation (EC) No. 541/95, dated 10 March 1995, as amended by Commission Regulation (EC) 1146/98, EC Commission Guideline on Dossier Requirements for Type I Variations (III/5783/93), *Note for Guidance on Stability Testing for a Type II Variation to a Marketing Authorisation, CPMP/QWP/576/96*, dated 22 April 1998).

In the USA, regulation 21 CFR 314.70 provides a guideline for changes subject to FDA approval prior to implementation and for changes that can be reported to the FDA in an annual report. In addition, the FDA has developed various guidelines for testing requirements in the case of post-approval changes under the general titles of Scale-Up and Post-Approval Changes (SUPAC) and Bulk Active Chemicals Post-Approval Changes (BACPAC).

Drug Substance

General information	
Nomenclature	Recommended INN[a], pharmacopoeial name, if relevant, chemical name(s), other name(s), company or laboratory code, regional name, national name, e.g., BAN[b], USAN[c], JAN[d], and CAS registry number.
Structure	NCE: structural formula, including relative and absolute stereochemistry, the molecular formula, and relative molecular mass.
	Biotech: schematic amino acid sequence indicating glycosylation sites where available and relative molecular mass.
General properties	Brief description of the physicochemical and other relevant properties of the drug substance.
Manufacture	
Manufacturer(s)	Name, address, and responsibility of each manufacturer and each site or facility involved.
Manufacturing process description and process controls	Information on manufacturing route including, for example, a flow diagram and a description of the synthetic process. Also, a description of process parameters such as temperature, pressure, pH, and time. Tests at the appropriate control points.
	Additionally for Biotechnology: description of the manufacturing process including the cell culture, harvest, isolation, purification, concentration, filling, storage and shipping conditions.
Control of materials	Starting materials, solvents, reagents, catalysts, and any other materials used in the manufacture indicating where in the process each material is used. Acceptance criteria and testing.
	Additionally for biotech products produced from cell banks: description of genetic construct for recombinant cell substrates, development and stability, and cell bank system, and materials used for cell banking, with information on their origin.
Control of critical steps and intermediates	Critical steps: tests and acceptance criteria, with justification including experimental data, performed at critical steps of the manufacturing process to assure that the process is controlled.
	Intermediates: specifications and analytical procedures, if any, on intermediates including validation of analytical procedures, where appropriate.
Characterization	
Elucidation of structure and/or biological characterization	NCE: Confirmation of structure based on synthetic route and spectral analyses, for example. Information on the potential for isomerism and the identification of stereochemistry.
	Biotech: details on primary, secondary and higher order structure and information on biological activity, purity, and immunochemical properties (where relevant).
Impurities	
Control of Drug Substance	
Specification	
Analytical Procedures	
Validation	
Batch Analyses	
Justification	
Reference standards or materials	Information on the reference standards or reference materials used for testing of the drug substance and drug product.
Container closure system	Description of the container closure system suitable for the storage and shipment of the drug substance, including components, composition, and specifications.
Stability	
Stability summary and conclusions	Summary of the types of studies conducted, protocols used, and the results obtained; conclusions with respect to storage conditions and retest date or expiry period.
Stability data	Results of the stability studies conducted in an appropriate format such as tabular, graphical, narrative; information on analytical procedures used and their validation.

[a]INN = International Non-Proprietary Names. [b]BAN = British Approved Names. [c]USAN = United States Adopted Names. JAN = Japanese Approved Names.

4.3.10. Change Control

To ensure that the systems concerned are continuously validated, the manufacturer must monitor all changes with respect to facilities, materials, equipment and processes used in the manufacture of drug substances and medicinal products. This commitment should be declared in the Validation Master Plan. As part of its Quality Management System, the manufacturer must establish a defined, formalized change control procedure (*Convention for the Mutual Recognition of Inspections: "Principles of Qualification and Validation in Pharmaceutical Manufacture," Document PH 1/96*, January 1996).

Drug product	
Composition	Description of pharmaceutical form, list of all components, and their amounts on a per unit basis, their function and reference to their specifications, type of container and closure used.
Pharmaceutical development report	Description and, if necessary, justification of differences between clinical formulation(s) and the formualtion proposed for marketing. Pharmaceutical development studies on key parameters which might have an influence on the performance of the drug product. NCE: examples include physicochemical characteristics of the drug substance which affect drug product dissolution, e.g., polymorphic form, particle size. Biotech: an example is the selection of a stabilizing excipient. Identification and discussion of critical steps in the manufacturing process and justification of non-standard sterilization process. Justification of overages, if any. Microbiological attributes, suitability of container closure system.
Manufacture	
Manufacturer(s)	Name, address, and responsibility for each manufacturer and each site and facility involved.
Batch formula	Names and amounts of all components to be used in the manufacturing process.
Manufacturing process description and process controls	Information on manufacturing process including, for example, a flow diagram, a description of the process, and type of equipment; a description of process parameters and their control.
Control of intermediates	Critical steps: tests and acceptance criteria, with justification, including experimental data, performed at critical steps. Intermediates: specifications and analytical procedures, if any, for intermediates including validation of analytical procedures, where appropriate.
Process validation or evaluation	Process evaluation or validation protocol, based on experimental data on pilot or production scale batches, or validation data for production scale batches, where appropriate.
Control of excipients	
Specifications	
Analytical procedures	
Validation	
Justification	
Control of drug product	
Specifications	
Analytical procedures	
Validation	
Batch analyses	
Justification	
Container closure system	Description of the container closure system suitable for the storage and shipment of the drug product, including components, composition, and specifications.
Stability	
Stability summary and conclusions	A summary discussing the types of studies conducted, protocols used and the results obtained; conclusions with respect to storage conditions and expiry period, and, if applicable, in-use storage conditions and expiry period.
Stability data	Results of the stability studies conducted in an appropriate format such as tabular, graphical, narrative; information on analytical procedures used and their validation.

4.3.11. Compliance

Manufacturers must make sure that the documentation approved by the health authorities complies with the methods applied in the manufacturing and quality control of the drug substances and products.

4.3.12. Changes Requiring New Applications – Line Extension

Certain changes to a marketing authorization are considered to fundamentally change the terms of the authorization in question and therefore cannot be considered as a variation. For example:

- Adding or removing of one or more active ingredient(s)
- Changing the amount of active ingredient(s)
- Replacing the active ingredient(s) with a different salt/ester complex/derivative (with the same therapeutic moiety)
- Replacing an isomer, or using a different mixture of isomers, replacing a mixture with a single isomer (e.g., replacing a racemate with a single enantiomer)

- Replacing a biological substance or a biotechnology product with one of a different molecular structure; modification of the vector used to produce the antigen/source material, including a master cell bank from a different source
- Using a new ligand or coupling mechanism for a radiopharmaceutical

An abridged application for a new marketing authorization must be submitted for these changes and, in certain cases, the existing marketing authorization may have to be withdrawn. Otherwise, the new product is regarded as a "line extension" to the existing formulations and dosage forms.

4.3.13. Inspections

Manufacturers of both drug substances and drug products are inspected by the national authorities. The inspectors must establish whether the control methods are applied and the facilities adequate, the personnel properly trained and manufacturing procedures followed.The inspectors also have to check whether cross-contamination can be excluded, i.e. whether the substance or product is subject to contamination by material stored or used in the same building. Other factors to be considered are:

- The degree of exposure of the material to adverse environmental conditions
- Relative ease and thoroughness of clean-up
- Sterile vs. non-sterile operations

(*FDA Guide to Inspections of Bulk Pharmaceutical Chemicals*, September 1991, reformatted May 1994, with editorial changes). The inspections may take place during the clinical development of a new substance, prior to granting marketing authorization (pre-approval inspection), prior to the first launch of a drug product, or during the marketing phase.

4.3.14. The Mutual Recognition Agreement (MRA)

To cut down on the need for GMP inspections by regulatory authorities in foreign countries, agreements have been negotiated between the EU and the USA, the EU and Canada, the EU and Australia, the EU and New Zealand, and Switzerland and Japan. The aim is to achieve mutual recognition of GMP inspections conducted by local inspectors, with the results documented in an "Inspection Report".

The main advantage of an MRA would be the assurance that imported pharmaceutical products meet GMP requirements without resource expenditures on the part of the authorities of the importing country. Another important benefit of an MRA would be a reduction of costs for the pharmaceutical manufacturer, mainly as a result of inspection reports being accepted by more than one regulatory authority. This would eliminate the need for duplication of inspections and imply fewer inspections of the manufacturing sites. An MRA could also reduce regulatory review times as it would increase the efficiency of pre-marketing approval inspection activities. The current status of the MRAs between the EU and other countries is reported on the web site of the European Commission's DGIII (http://dg3.eudra.org/).

Agreement Between the USA and the EU. On June 20, 1997, the United States and the EU concluded negotiations on an agreement titled "Agreement on Mutual Recognition between the United States of America and the European Community". The MRA was signed on 18 May 1998 and came into force on 1 December 1998. The transition period will end on 30 November 2001. The complete text of this MRA is available on the Internet at the FDA's web site http://www.fda.gov. It has also been published in the US Federal Register April 10, 1998 (Volume 63, Number 69) Page 17744 – 17771.

4.3.15. Pharmaceutical Inspection Convention (PIC)

Another initiative to cut down on foreign GMP inspections is the "Convention for the Mutual Recognition of Inspections in respect of the Manufacture of Pharmaceutical Products" (shortened to Pharmaceutical Inspection Convention or PIC). This convention was signed in October 1970 by all EFTA countries at the time (i.e., Austria, Denmark, Finland, Iceland, Liechtenstein, Norway, Portugal, Sweden,

Switzerland, and the United Kingdom). Since coming into force the convention has been extended to include Hungary, Ireland, Romania, Germany, Italy, Belgium, France, and Australia.

Similar to the MRAs mentioned above, the "PIC provides that an inspection of a pharmaceutical manufacturer, whose intention it is to export products in one of the Contracting States, is carried out by its national authority and shall be regarded and assessed by the health authority of the country of importation as if it had been carried out by its own inspectors."

A Committee of Officials supervises the operation of the PIC concerning the technical aspects of the manufacturing and control of pharmaceuticals. In the meantime, the convention's original scope has been extended to the training of laboratory staff, the mutual recognition of the quality standards of inspectorates, and also to cooperation concerning blood products, radiopharmaceuticals, medicinal gases, etc. Working groups meet regularly to develop guides, guidelines, and recommendations. Also seminars on technical subjects are organized, and the proceedings of these seminars and experts' meetings are published (http://www.efta.int/structure/EFTA/efta-sec.cfm).

4.3.16. Pharmaceutical Inspection Cooperation Scheme (PIC/S)

Some of the contracting states within the PIC are also Member States of the European Union with the result that certain rules and regulations were incompatible. Therefore, a new arrangement, which may eventually replace the PIC, called "Pharmaceutical Inspection Cooperation Scheme" (or PIC Scheme) has been set up. This new scheme is meant to be less formal and more flexible concerning cooperation between the inspectorates of the present PIC contracting states. Also the PIC Scheme allows for inspectorates of other countries to join. In addition to the recognition of foreign inspections, the PIC Scheme is in harmony with the main tasks of the PIC, namely, "the exchange of information, networking and confidence building between the national inspection authorities, the development of quality standards systems, the training of inspectors and other related experts and its work towards the global harmonisation of Good Manufacturing Practice (GMP)".

The contracting members of the new PIC Scheme are the following inspectorates: Australia, Canada (since 1999), the Czech Republic, Denmark, Finland, Hungary, Iceland, Ireland, Liechtenstein, the Netherlands, Norway, Romania, the Slovak Republic, Sweden, Switzerlan, and the United Kingdom. The remaining PIC inspectorates have confirmed their intention to join the PIC Scheme sooner or later. In addition, the following applications for membership are being considered: Estonia, Latvia, Singapore, South Africa, and Taiwan.

4.3.17. Pharmacopeias

National Pharmacopeias. In order to control drug quality, lists of drugs, their preparation and uses were developed and published in a "pharmacopeia". The earliest of these was probably the "New Compound Dispensatory" issued in 1498 by the "Florentine Guild of Physicians and Pharmacists". Various pharmacopeias were generated in the ensuing centuries: Barcelona (1535), Nürnberg (1546), Cologne (1565), Rome (1583), and London (1618). The first attempt to harmonize different standards and monographs was made by the Pharmacopeia Committees of London, Edinburgh and Dublin. As the result of their efforts, the first "British Pharmacopeia" documenting the agreement on harmonized international quality standards was published in 1864 [18].

International Pharmacopeia. The project of developing an "International Pharmacopeia" started in 1929 with an agreement signed by 19 countries. The next step was not made until 1948, when the first World Health Assembly established an expert committee to adapt and harmonize the existing monographs of major pharmacopeias, and also to develop new standards. As the committee also kept developing countries in mind, some of the drug substances included in the first edition of the International Pharmacopeia were for the treatment of tropical diseases. Analytical methods are adapted for the practice of quality control testing in countries that need inexpensive methods, such as on-the spot identification of counterfeit pharmaceuticals (WHO Drug Information Vol. 13, No. 1, 1999, pp. 13–14).

European Pharmacopeia. The need to unify the various national pharmacopeias in Europe arose as a result of the EU proposal to allow the free movement of medicinal products. Both public health and international trade require common manufacturing and quality control standards concerning drug substances for human and veterinary use. It also requires these standards be updated to stay in line with scientific progress. In 1964, the "European Pharmacopeia Convention" was signed by six Member States of the Council of Europe. It is open to European countries (members and, under certain conditions, non-members of the Council of Europe), as well as to an international European organization. Observer status is also possible.

Twenty-seven parties are members of the Convention with 25 member states belonging to the Council of Europe (including the 15 EU member states): Austria, Belgium, Croatia, Cyprus, Czech Republic (since June 1998), Denmark, Finland, the former Yugoslav Republic of Macedonia, France, Germany, Greece, Iceland, Ireland, Italy, Luxembourg, the Netherlands, Norway, Portugal, the Slovak Republic, Slovenia, Spain, Sweden, Switzerland, Turkey and the United Kingdom; plus one state which is not a member of the Council of Europe, Bosnia-Herzegovina; and the Commission of the European Union.

A further 17 parties participate as observers: Albania, Bulgaria, Estonia, Hungary, Latvia (since 1998), Lithuania, Poland, Romania and the Ukraine, as well as the non-European countries Australia, Canada, China, Malaysia, Morocco, Syria, Tunisia, and the WHO. Some of these observers have implemented all or parts of the European Pharmacopoeia in their national pharmaceutical regulations.

In marketing authorization application dossiers submitted to regulatory authorities in the EU Member States, the monographs of the European Pharmacopoeia are obligatory. This requirement is based on Directive 75/318/EEC of 20 May 1975 of the Council of the Communities. Cooperation between the Council of Europe and the EU was further advanced on 26 May 1994, when the "European Pharmacopeia Secretariat" took over additional objectives, e.g., setting up a European network of laboratories involved in the quality control of pharmaceutical products. Consequently, the "European Pharmacopeia Secretariat" changed its name to the "European Department for the Quality of Medicines" (EDQM).This European network, called the Official Medicines Control Laboratories (OMCLs), is open not only to EU countries but also to members and observers of the European Pharmacopoeia Commission. Its main objectives are to achieve the mutual recognition of tests carried out at the national level for countries that belong to the EU, and the sharing of expertise, standardization and international collaboration for the other countries.

The European Pharmacopoeia is one of the cofounders of the Pharmacopoeial Discussion Group (PDG) that was set up in 1990 with Japan and the United States. Monographs and general methods of analysis proposed by national associations of manufacturers of pharmaceutical products are selected for harmonization in the three pharmacopoeias. The European Pharmacopeia Commission has set the goals of rapidly adding to the number of existing European monographs (1500 so far), reducing the time taken to compile them, and, if necessary, generating monographs on recent substances still protected by patent. The European Pharmacopeia (EP or Ph. Eur.) features more than 250 general analytical methods on which the specifications prescribed in the monographs are based. The methods range from physical, chemical and biological to microbiological, phytochemical, and so on. The EP also includes technical methods used to test the functional parameters of pharmaceutical forms such as the dissolution test for tablets.

The texts are updated at regular intervals to take into account scientific progress and any changes in the commercial products. The principles of monograph compilation are adapted to accommodate regulatory requirements in the public health area (licensing, control and inspection authorities), industrial constraints and recent advances in technology and science (http://www.pheur.org/).

The monographs and methods published in the European Pharmacopoeia are normally transferred to the national pharmacopoeias, e.g., the *Deutsches Arzneibuch* (DAB), the British Pharmacopoeia (BP), the Pharmacopeé Francaise, the Farmacopeia Ufficiale della Republica Italiana.

4.3.18. The EP Certificate of Suitability

Global expansion of international trade has given rise to a wide range of qualities of medicinal substances within the territory covered by the Convention on the Elaboration of a European Pharmacopoeia. The great variety of supply sources has made it necessary to take additional measures to protect public health so that the European Pharmacopoeia can continue to play its role as a reliable reference in defining the quality of medicines. It is essential to have a procedure that allows the drug manufacturer to prove that the purity of the substance is suitably controlled by the EP monograph. The procedure described in Resolution AP-CSP (98) 2 of the Council of Europe meets this need with a certificate of suitability of EP monographs issued by the EDQM.

As specified in Resolution AP-CSP (98) 2, the manufacturer has to submit a full dossier on the manufacturing method for the substance and its associated impurities for validation. The dossier is assessed according to a procedure that guarantees its confidentiality and, if the information received is pertinent, a certificate of suitability is issued. This certificate can be included in the dossier for marketing authorization application for all medicinal products containing the substance in question (http://www.pheur.org/).

5. References

1. J. S. Bindra, D. Lednicer: *Chronicles of Drug Discovery,* John Wiley, New York 1983.
2. C. Thromber: "The Pharmaceutical Industry," in Alan Heaton (ed.): *The Chemical Industry,* Blackie Academic and Professional, London 1986.
3. *Scrip 1999 Yearbook,* **vol. 1,** PJB Publ. Ltd., Richmond, Surrey 1999.
4. http://www.pjbpubs.com/scrip/
5. http://www.imshealth.com/html/wdm/-index.htm
6. http://www.datamonitor.com/
7. *PharmaVitae* 1999, April, 6
8. *Pharma Business* 1999, May, 42.
9. *Scrip* **2416** (1999) March 3, 27.
10. "Principles for Pre-clinical Testing of Drug Safety" *WHO Tech. Rep., Ser.* **341,** 1966. "Principles for Testing of Drugs for Teratogenicity" *WHO Tech. Rep. Ser.* **364,** 1967. "Principles for Testing of Drugs for Carcinogenicity" *WHO Tech. Rep. Ser.* **426,** 1969. "Evaluation and Testing of Drugs for Mutagenicity" *WHO Tech. Rep. Ser.* **482,** 1971.
11. O. Eichler (ed.): "Erzeugung von Krankheitszuständen durch das Experiment," in *Handbook of Experimental Pharmacology,* vol. XVI, Parts 1 – 15, Springer Verlag, Berlin-Heidelberg-New York, 1962 – 1966.
12. R. A. Turner: *Screening Methods in Pharmacology,* **vol. 1, vol. 2,** Academic Press, NewYork-London, 1965, 1971.
13. L. Ther: *Grundlagen der experimentellen Arzneimittelforschung,* Wissenschaftliche Verlagsges. mbH, Stuttgart, 1965.
14. J. Ashby: "The Prospects for a Simplified and Internationally Harmonized Approach to the Defection of Possible Human Carcinogens and Mutagens," *Mutagenesis* **1** (1986) 3 – 16.
15. P. Jeffs: "The Importance of Stability Testing in the Registration of Pharmaceutical Products" in D. Mazzo (ed.): *International Stability Testing,* Interpharma Press, Inc., Buffalo Grove, IL 1999.
16. S. Kopp-Kubel, M. Zahn: "The WHO Stability Guideline" in D. Mazzo (ed.): *International Stability Testing,* Interpharma Press, Inc., Buffalo Grove, IL 1999.
17. D. Shah: ″Postapproval FDA Stability Requirements″ in D. Mazzo (ed.): *International Stability Testing,* Interpharma Press, Inc., Buffalo Grove, IL 1999.
18. A. Cartwright, B. Matthews (eds.): *Pharmaceutical Product Licensing – Requirements for Europe,* Ellis Horwood Ltd., Chichester 1991.

Chemotherapeutics

See also → *Cancer Chemotherapy;* → *Pharmaceuticals*

PAUL ACTOR, Smith Kline & French Laboratories , Philadelphia, Pennsylvania 19101, United States (Chaps. 1, 3)

ALFRED W. CHOW, Smith Kline & French Laboratories , Philadelphia, Pennsylvania 19101, United States (Chaps. 1, 3)

FRANK J. DUTKO, Sterling-Winthrop Research Institute, Rensselaer, New York 12144, United States (Chap. 4)

MARK A. MCKINLAY, Sterling-Winthrop Research Institute , Rensselaer, New York 12144, United States (Chap. 4)

Ullmann's Industrial Toxicology

ISBN: 3-527-31247-1

1. Introduction

This article discusses the synthetic agents that are effective against pathogenic bacteria, protozoa, and viruses. Although a large percentage of the agents employed to treat these infections, especially bacterial infections, are natural products, i.e., antibiotics, many synthetic compounds continue to provide useful alternatives, and some are the agents of choice for the treatment of specific clinical entities.

2. Chemotherapy of Bacterial Infections

2.1. Classification of Bacteria Causing Disease

A broad variety of organisms are primary human pathogens. No universal agreement exists for the classification of these bacteria, and it is beyond the scope of this chapter to discuss the arguments inherent in attempting to create a universally acceptable and useful system. The definitive work in bacterial classification is Bergy's *Manual of Determinative Bacteriology* Bacterial species still are arbitrarily defined by descriptive features. In the approximate sequence of importance, some of the major features employed include the following:

1) *morphological appearance of the cell,* including shape, size, flagellar pattern if motile, capsule occurrence, colonial morphology, and pigmentation
2) *staining properties,* including the gram capsule, spore, and acid fast
3) *metabolic patterns*
4) *macromolecular composition* and *structure*
5) *ecological habitat*
6) ability to be *pathogenic*

The major families of bacteria causing human infection are shown in Table 1. However, many bacteria that are not primary pathogens are capable of causing clinical infections in immunocompromised hosts. Furthermore, transfer of genetic material between bacterial species, e.g., via plasmids, allows saprophytic bacteria to acquire virulence factors and become pathogenic.

2.2. Emergence of New Bacterial Pathogens

The emergence and recognition of organisms as important human pathogens are strongly influenced by three major factors: (1) the transfer of virulence factors and antibiotic resistance between bacterial species by extrachromosomal elements, (2) opportunistic infection of immunocompromised hosts by saprophytic bacteria, and (3) improvement and widespread use in the microbiological clinical laboratory of new diagnostic procedures for isolating and identifying pathogens.

Opportunistic infections, particularly in hospitalized individuals, have become major infectious disease problems. This problem is likely to continue because of the increasing use of instrumentation, antibiotics, and drugs that either

Table 1. The major groups of pathogenic bacteria

Morphological type	Family	Genus	Gram stain	Oxygen utilization *
Cocci	Micrococcaceae	*Micrococcus*	+	A
		Staphylococcus	+	A/F – AN
	Streptococcaceae	*Streptococcus*	+	F – AN
	Peptococcaceae	*Peptococcus*	+	AN
		Peptostreptococcus	+	AN
	Neisseriaceae	*Neisseria*	–	A
	Veillonellaceae	*Veillonella*	–	AN
Rods	Enterobacteriaceae	*Escherichia*	–	F – AN
		Shigella	–	F – AN
		Salmonella	–	F – AN
		Citrobacter	–	F – AN
		Klebsiella	–	F – AN
		Enterobacter	–	F – AN
		Erwinia	–	F – AN
		Serratia	–	F – AN
		Hafnia	–	F – AN
		Edwardsiella	–	F – AN
		Proteus	–	F – AN
		Providencia	–	F – AN
		Morganella	–	F – AN
		Yersinia	–	F – AN
	Vibrionaceae	*Vibrio*	–	F – AN
		Aeromonas	–	F – AN
	Pasteurellaceae	*Pasteurella*	–	F – AN
		Haemophilus	–	F – AN
	Pseudomonadaceae	*Pseudomonas*	–	A
	Legionellaceae	*Legionella*	–	A
	Neisseriaceae	*Moraxella*	–	A
		Acinetobacter	–	A
		Brucella	–	A
		Bordetella	–	A
	Bacteroidaceae	*Bacteroides*	–	AN
		Fusobacterium	–	AN
		Leptotricha	–	AN
	Bacillaceae	*Bacillus*	+	AN
		Clostridium	+	AN
		Listeria	+	A
		Erysipelothrix	+	A
Actinomycetes andrelated organisms		*Corynebacterium*	+	A/F – AN
	Propionibacteriaceae	*Propionibacterium*	+	AN
		Eubacterium	+	AN
	Actinomycetaceae	*Actinomyces*	+	FA
	Mycobacteriaceae	*Mycobacterium*	+	A
	Nocardiaceae	*Nocardia*	+	A
Rickettsias and Chlamydias	Rickettsiaceae	*Rickettsia*	–	P
		Coxiella	–	P
	Bartonellaceae	*Bartonella*	–	P
	Anaplasmataceae	*Grahamella*	–	P
		Anaplasma	NR**	P
		Haemobartonella	–	P
		Eperythrozoan	–	P

bypass or reduce the level of natural resistance or the specific immune mechanisms of the host. Thus, microorganisms previously considered as innocuous commensals or contaminants are now able to invade and cause disease. Some of the more important bacterial pathogens that have created significant problems include *Staphylococcus epidermidis, Bacteroides fragilis* and other *Bacteroides* species, *Clostridium difficile,* the gram-negative rods including *Acinetobacter* species, *Serratia, Citrobacter* species, *Yersinia, Moraxella,* and the atypical mycobacteria.

Since the mid-1960s, the role of *S. epidermidis* has changed from a bothersome contami-

Table 1. (Continued)

Morphological type	Family	Genus	Gram stain	Oxygen utilization *
	Chlamydiaceae	*Chlamydia*	–	P
Mycoplasmas	Mycoplasmataceae	*Mycoplasma*	–	FA
		Ureaplasma	–	FA
	Acholeplasmataceae	*Acholoplasma*	NR**	FA
Spirochetes	Spirochaetaceae	*Treponema*	–	AN
		Borrelia	–	AN
		Leptospira		A

* A= Aerobic; AN = Anaerobic; F – AN = Facultative anaerobic; A/F – AN = Aerobic or facultative anaerobic growth; P = Parasitic usually require host cells for growth.
** NR = Gram stain not revealed.

nant to a major pathogen causing infections associated with foreign bodies (prosthetic valvular endocarditis and infections of CNS shunts and of joint and vascular prostheses). In addition, this organism can cause infections not associated with foreign bodies, such as endocarditis and urinary tract infections. *Bacteroides fragilis,* the preeminent anaerobic human pathogen, has evolved as a problem. Information as to the importance and widespread incidence of this organism only became available when techniques were established for its facile isolation and rapid diagnosis. Another important anaerobe is *Clostridium difficile,* a component of the normal gut flora, which overgrows to large populations in the presence of antibiotics. Toxigenic stains can cause pseudomembranous colitis, a severe necrotizing disease of the large intestine. With the waning of *M. tuberculosis* infections in the United States, other mycobacterioses, the so-called atypical mycobacteria, have become increasingly important.

2.3. Antimicrobial Resistance

Mechanisms of Resistance. The widespread clinical use of antimicrobics has resulted in the emergence of many strains of bacteria resistant to one or more of these agents. In most cases in which adequate studies have been done, the role of antimicrobial agents apparently is to exert selective pressure, resulting in the emergence of resistant organisms. In some instances, the organisms are *naturally resistant* to the antibiotic used. In other cases, the resistant bacteria may have acquired *R-factors* or *plasmids.* These extrachromosomal agents may provide the organism with the ability to synthesize enzymes that modify or inactivate the antimicrobial agent. The plasmids may also cause changes in the organism's ability to accumulate the antimicrobial agent, or they may stimulate the cell to produce or overproduce metabolic enzymes resistant to inhibition by the antimicrobial agent. Additionally, alterations in the permeability of the bacterial cell envelope could result in drug resistance.

Chromosomal resistance develops as a result of spontaneous mutation in a locus that controls susceptibility to a given antimicrobial agent. The presence of the drug serves as a selection mechanism for drug-resistant mutants. Chromosomal mutants are most commonly resistant by virtue of an alteration in a structural receptor for the drug. Cross resistance is frequently observed between chemically related drugs showing a similar mechanism of action, but may also exist for unrelated chemicals.

Clinical Implications of Resistance. Bacterial resistance problems have resulted in the continuing need for new antimicrobics or modification in the ways in which we treat patients with the available antimicrobials. Important examples with the synthetic antimicrobials include the sulfonamides, which were only active against a small percentage of gonococcal strains 6 years after they were first successfully employed for the treatment of gonorrhea. The sulfonamides also have lost their usefulness in the prevention and treatment of meningococcal infection because of drug resistance. Drug-resistant mutants have arisen in tuberculosis, and naturally resistant species of mycobacteria have become clinically important especially in immunocompromised hosts. The emergence of drug-resistant bacteria in the hospital setting has led to restricted use of certain valuable drugs in hospitals. Other strategies that have been employed to minimize the drug resistance problem include treatment regimes to maintain high drug levels in tissues and the simultaneous administration of two drugs, each of which delays emergence of resistance to the other drug (e.g., sulfonamides and trimethoprim).

2.4. Pathogenesis and Virulence Factors

All organ systems are subject to the pathogenic effects of bacteria, and the resulting infections range from the trivial to the fatal. The major organ systems involved are skin and soft tissues, urinary and reproductive tissues, the respiratory tract, and the central nervous, digestive, and cardiovascular systems. Infections accompanying other medical problems that result in a breach of the anatomical and immunological barriers to infection are particularly difficult to treat, e.g., *Pseudomonas* infection in burn patients or those with cystic fibrosis. Mixed infections involving more than one organism are common.

Although humans are continually exposed to many different environmental microorganisms, only a small percentage of these have the capacity to produce disease. The production of *virulence factors* by microorganisms is the important determinant in disease. Virulence factors are those components of the microbe that are essential for the establishment of infection and the development of disease in the host. Usually more than one factor is involved in the disease process. The adherence of microorganisms to host tissues is determined by highly specific host receptors for bacterial surface components and is a necessary prerequisite for infection. Some bacteria, e.g., *Vibrio cholerae* and *Escherichia coli*, colonize mucosal surfaces and cause damage by elaboration of a toxin. Other organisms, e.g., *Salmonella* and *Shigella,* can invade following attachment and either enter host cells or disseminate throughout the body. The invasion is facilitated by the production of enzymatic substances that circumvent anatomical barriers. To survive, these organisms may have specialized virulence factors that enable them to avoid or disarm host defenses. Survival and continued proliferation of the invading organism are often accompanied by the production of *toxins.* These toxins are proteinaceous substances capable of producing adverse biological effects.

2.5. Selection of an Appropriate Antimicrobial Agent

The selection of a specific agent for treating infection involves consideration of the infecting organism, the status of the host, and the specific attributes of the antimicrobic. The key factors to be considered are the drug's antimicrobial spectrum and potency, its physical characteristics (solubility and stability in body fluids), safety profile, pharmacokinetics, compatibility with other drugs, and cost. The choice of an appropriate agent involves identifying the infecting organism, obtaining information as to its susceptibility, and assessing various host factors.

Identification of the Microorganism. Rapid identification can often be made by *microscopic examination* of gram-stained specimens. A number of immunologic procedures can be employed, such as *latex agglutination tests* or the *enzyme-linked immunoabsorbent assay* (ELISA). Much progress has been made with assays employing *DNA probes* but these are not generally available to the clinical laboratory. *Cultural techniques* are the major means for identification of microbial pathogens. A number of automated and semiautomated tests and kits facilitate this identification and may combine the capability of antimicrobial susceptibility testing.

Determination of Antimicrobial Susceptibility. Several methods for determination of antimicrobial susceptibility are available. The *disk-diffusion method* is widely employed and gives semiquantitative information that is clinically useful. Methods employing *serial dilution* of the drug in culture broth or agar give quantitative data. This is usually expressed as the minimal inhibitory concentration (MIC). A subculture onto an antibiotic-free medium of broth cultures from the MIC tube showing no visible growth can allow for the determination of the minimal bactericidal concentration (MBC), which is usually defined as the concentration that causes decline in colony count of 99.9 % or more.

Host Factors. The clinician must consider a number of host factors in selecting the drug. A history of adverse reactions to antimicrobials, the patient's age, genetic or metabolic abnormalities, pregnancy, renal or hepatic function, and the site of infection all influence this choice.

Combination Therapy. At times combination therapy offers advantages over treatment with a single antimicrobial agent; however, most infections can be treated with one drug. Combination therapy may be valuable in prevention of the emergence of resistance or for polymicrobial infections. When the initial diagnosis is unclear, a combination of antimicrobial agents may be needed.

2.6. Chemoprophylaxis

Chemoprophylaxis is the administration of drugs prior to or shortly after exposure to an infectious agent to prevent infection. For example, isoniazid may be given to individuals who show recent change from negative to positive skin tests for tuberculosis. The risk of infection must be balanced against toxicity, cost, efficacy, and inconvenience of taking the proposed chemoprophylaxis.

Antibacterial chemoprophylaxis is an accepted clinical procedure for preventing group A *Streptococcus* and *Meningococcus,* and plague infections. Prophylactic drugs have also been used to prevent infective endocarditis, postcoital cystitis, and exacerbations in chronic bronchitis in high-risk individuals. Immunosuppressed patients are often given antimicrobials to prevent the complications resulting from the spread of endogenous bacteria. In some surgical procedures, prophylactic administration of antimicrobials is valuable in preventing infection.

2.7. Chemotherapy

2.7.1. Biochemical Targets for Antimicrobial Agents

A chemical affects cell growth if it hinders the cell's endogenous biosynthetic processes by restricting the availability of building materials or catalytic enzymes or interferes with the supply of usable energy. Considerable advances have been made in identifying several general biochemical targets (i.e., peptidoglycan synthesis, biosynthetic enzymes, and ribosomal function), although in many cases the final molecular site of action is uncertain. At the cellular and subcellular level, most known antimicrobial agents function in one of four major ways:

1) inhibition of bacterial cell wall synthesis
2) alteration of cell membrane permeability or inhibition of active transport across cell membranes
3) inhibition of nucleic acid synthesis
4) inhibition of protein synthesis

The specific mechanisms of action for the individual classes of antimicrobials are discussed in the following sections [2].

2.7.2. Quinolone Antibacterial Agents

Historical Background. Following the discovery of nalidixic acid, numerous antibacterial quinolones were synthesized [3]. These early quinolone carboxylic acids were absorbed fairly well, but suffered from poor pharmacokinetics and tissue penetration, which relegated them to use only in urinary tract infections caused by gram-negative bacteria. In addition to poor pharmacokinetics, these compounds also showed CNS side effects and rapid development of resistant microorganisms. Since the initial introduction of the early compounds, many second- and third-generation compounds have been discovered with increased potency and antibacterial spectrum against gram-negative bacteria, including *P. aeruginosa.* In addition, some of the newer compounds are active against gram-positive bacteria and anaerobes and appear to have fewer adverse effects, as well as improved pharmacokinetics and tissue penetration. Several have shown promise when administered parenterally and thus offer therapeutic potential outside the urinary tract. The early compounds have been reviewed comprehensively, especially with respect to synthetic methods, microbiology, and structure activity relationships [4]. The newer compounds have been reviewed [5–7].

2.7.2.1. Structure Function

Nalidixic acid, the prototype of the series, is a naphthyridine derivative; however, the newer, more potent congeners are not derived from the naphthyridine skeleton. The prototype compound of the quinoline class is oxolinic acid. Although all congeners of nalidixic acid can be named as derivatives of quinoline, such a system is cumbersome because the common skeleton is termed 4-oxo-1,4-dihydroquinoline. A simpler term, 4-quinolone (**1**), has been suggested as a generic name for the agents.

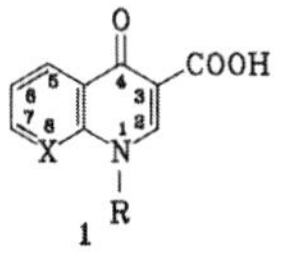

1

4-Quinolones

X = N, Naphthyridine

X = CH, Quinolone

The second- and third-generation 4-quinolones have been modified at C-6, C-7, and C-8 (**1**). The position 1 is vital to antimicrobial activity. The presence of a two-carbon fragment or a spatial equivalent is essential and almost all marketed compounds in this class have an ethyl group at position 1. Ciprofloxacin has a cyclopropyl group at this site, which fills the spatial requirements of the two-carbon chain. Very few modifications at C-2 have been studied, although cinoxacin, which substitutes a nitrogen for the carbon at this position, has improved absorption properties. Relative to its parent compound, oxolinic acid, cinoxacin is somewhat less active in vitro.

Positions C-3 and C-4 are the most structurally critical, and substitution of other groups at these positions results in loss of activity. Modification of C-5 also offers few advantages. Addition of a nitrogen within the ring may result in an improved use profile. Cinoxacin, mentioned previously, is an example of improved pharmacokinetics resulting from such a change at C-2; however, substitution of a ring nitrogen for C-5, C-6, C-7, and C-8 generally decreases or abolishes activity.

Addition of a fluorine atom to C-6 results in a dramatic increase in the antibacterial spectrum and potency of these compounds [8]. Consequently, all of the newer quinolones have C-6-substituted fluorine as a component of the molecule. These structures have been given the name fluoroquinolones [5].

An aromatic group at C-7 may tend to increase CNS side effects (e.g., rosoxacin), but the piperazine ring may not produce this enhanced effect. The piperazine ring at C-7 also seems to achieve the necessary balance of enhancing penetration capabilities and in vivo activity. The activity of the 4-quinolones against gram-positive and anaerobic bacteria is enhanced through substitutions at C-8 with short one-to-three-atom chains (e.g., ofloxacin). Increased activity against gram-positive bacteria is achieved by the addition of a fluorine atom at C-8 (e.g., CI – 934).

2.7.2.2. Mechanism of Action

The quinolone antibacterial agents act by inhibiting bacterial DNA-gyrase (topoisomerase II). This essential bacterial enzyme, discovered in 1976 in *Escherichia coli* [9], contains two A-subunits and two B-subunits. The *B-subunit,* which is the site of action of coumermycin – novobiocin type antibiotics, is responsible for supplying the energy necessary for the supercoiling of DNA. The *A-subunit,* the site of action for the quinolone type antibiotics, performs the physical act of supercoiling the bacterial DNA molecule in conjunction with the B-subunit [10].

2.7.2.3. Nalidixic Acid and First-Generation Quinolones

Nalidixic acid and the related first-generation 4-quinolones are orally administered agents that concentrate primarily in the urinary tract. Because effective plasma concentrations are not obtained with safe doses of these agents, they cannot be used for the treatment of systemic infections; thus, their use is for the most part limited to infections of the urinary tract.

Nalidixic acid (International Nonproprietary Name – INN, *United States Pharmacopeial – U.S.P.*) [*389-08-2*], 1-ethyl-1,4-dihydro-7-methyl-4-oxo-1,8-naphthyridine-3-carboxylic acid, $C_{12}H_{12}N_2O_3$, M_r 232.24, *mp* 229 – 230 °C, is a pale buff crystalline powder.

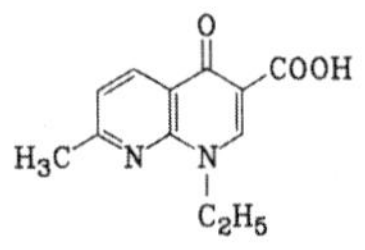

Nalidixic acid is more active against gram-negative bacteria than against gram-positive bacteria. It is active against most of the members of the Enterobacteriaceae, including ca. 99 % of the strains of *Escherichia coli,* 98 % of *Proteus mirabilis,* 75 – 97 % of other *Proteus* species, 92 % of *Klebsiella* and *Enterobacter* species, and 80 % of other coliform bacteria.

These organisms are susceptible to a urinary concentration of 16 μg/mL or less [3]. Some strains of *Salmonella* and *Shigella* are also sensitive. *Pseudomonas* species are resistant, as are most of the important species of gram-positive clinical pathogens, including *Staphylococcus* species, *Streptococcus pneumoniae,* and *Streptococcus faecalis.* Resistance can be induced by in vitro passage [11]; however, surveys of resistance in clinically isolated uropathogens have shown that the incidence of isolates resistant to nalidixic acid remained surprisingly low despite extensive clinical use [12].

Nalidixic acid administered orally is 96 % absorbed from the gastrointestinal tract [13]. It is rapidly metabolized in the liver. Plasma levels of 20 – 50 μg/mL may be obtained 2 h after a 1-g dose [14]; however, the drug does not accumulate in the tissues even after prolonged administration, and the kidney is the only organ in which tissue concentrations may exceed plasma levels [13]. Approximately 85 % of the drug excreted in the urine is the inactive conjugated form; the remainder is the hydroxynalidixic acid metabolite, which is 16 times more active than the parent compound [15]. Urinary concentrations range from 25 – 250 μg/mL following a single oral dose of 0.5 – 1 g and remain between 100 and 500 μg/mL with a 1-g dose administered every 6 h [14], [16].

Oral nalidixic acid is generally well tolerated; however, various *adverse reactions* have been reported. Gastrointestinal side effects include nausea, vomiting, diarrhea, and abdominal pain. Dermatological reactions and photosensitivity have also been reported. Additionally, a range of reversible central nervous system reactions are observed. The drug is contraindicated in infants and during early pregnancy; it should not be used in children.

Synthesis: Condensation of 2-amino-6-methylpyridine with diethyl ethoxymethylenemalonate and thermal cyclization gives 4-hydroxy-7-methyl-1,8-naphthyridine-3-carboxylate. Alkaline saponification of the ester group and alkylation of the nitrogen atom in one step yields nalidixic acid [17].

Trade names: Cybis (Sterling-Winthrop), NegGram, and Wintomylon (Sterling-Winthrop).

Oxolinic acid (INN) [*14698-29-4*], 5-ethyl-5,8-dihydro-8-oxo-1,3-dioxolo[4.5-*g*]quinoline-7-carboxylic acid, $C_{13}H_{11}NO_5$, M_r 261.23, *mp* 314 – 316 °C (decomp.), occurs as crystals.

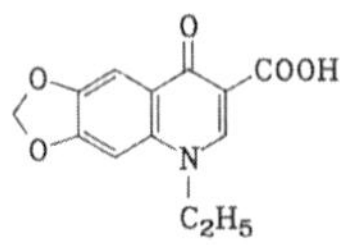

Oxolinic acid shares a similar antibacterial spectrum of activity with nalidixic acid, i.e., it is active against gram-negative rods with the exception of *Pseudomonas* species. Its in vitro potency, however, is significantly greater and it is more active in vivo [18]. Except for *S. aureus* strains that are inhibited at concentrations of 6.25 μg/mL, oxolinic acid does not inhibit gram-positive bacteria. Cross resistance with nalidixic acid is observed [19]. As with nalidixic acid, emergence of resistance during treatment of patients with bacteriuria has been reported.

Oxolinic acid achieves good urinary concentrations within 4 h following oral administration. An oral dosage of 2 g/d produces an average urinary concentration range of 16 – 64 μg/mL in 24-h urinary collections. Elimination of the drug is via urine and feces with low or borderline plasma levels against susceptible bacteria.

Adverse Effects: Central nervous system toxicity has been observed with this drug, and the potential is increased in elderly patients, in which CNS side effects are more common than with nalidixic acid.

Synthesis: Condensation of 3,4-methylenedioxyaniline with diethyl ethoxymethylenemalonate in boiling diphenyl ether and alkylation of the resultant 1,4-dihydroxy-6,7-methylenedioxy-4-oxoquinoline-3-carboxylate with ethyl iodide in the presence of caustic soda – DMF (*N,N*-dimethylformamide) gives the precursor ester. Saponification with dilute sodium hydroxide and neutralization yields oxolinic acid [20].

Trade name: Utibid (Parke-Davis).

Cinoxacin (INN, U.S.P.) [*28657-80-9*], 1-ethyl-1,4-dihydro-4-oxo[1,3]dioxolo[4,5-*g*]cinnoline-3-carboxylic acid, $C_{12}H_{10}N_2O_5$, M_r 262.22, *mp* 261 – 262 °C (decomp.), occurs as tan crystals.

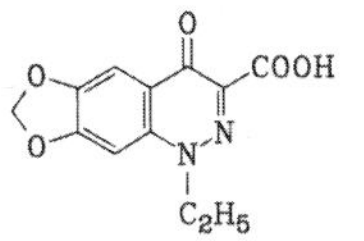

The antibacterial activity spectrum of cinoxacin resembles that of nalidixic acid, although it is more potent against selected species [21]. Generally its potency lies between that of nalidixic acid and oxolinic acid. Cinoxacin is most active against *E. coli*, of which > 90 % of the strains are inhibited at 16 µg/mL. Its activity is mainly against gram-negative rods; a broad spectrum of these organisms are susceptible to cinoxacin [21], [22]. Strains that are resistant to cinoxacin are cross-resistant with nalidixic acid and oxolinic acid [22], [23]. Similar to observations with the other quinolones, no evidence exists that plasmids play a role in resistance.

Excellent absorption is observed on oral administration. Peak plasma levels occur at 2 – 3 h after administration of a 250- or 500-mg dose. Serum binding is ca. 70 % and the serum half-life is ca. 1 h. Food may delay and reduce absorption. Peak urine concentrations range from 88 to 925 µg/mL within 4 – 6 h after dosing. Cinoxacin concentrates in renal tissue, where levels may exceed serum levels [24]. Approximately 60 % of the parent compound is excreted in the urine along with at least four microbiologically inactive metabolites [25].

Low frequencies of *adverse effects* (4.4 %) are reported on oral administration of cinoxacin with gastrointestinal reactions being the most common.Central nervous system side effects are < 1 %.

Synthesis: 2-Nitro-4,5-methylenedioxy-acetophenone is catalytically reduced to the corresponding amine, which is cyclized by treatment with $NaNO_2$ and HCl in water to give 6,7-methylenedioxycinnolin-4-ol. Bromination followed by treatment with cuprous cyanide in refluxing DMF gives 4-hydroxy-6,7-methylenedioxycinnolin-3-carbonitrile; hydrolysis with HCl in refluxing acetic acid gives cinoxacin [26].

Trade name: Cinubac (Eli Lilly).

Pipemidic Acid and Piromidic Acid. Both pipemidic and piromidic acid have been marketed outside of the United States as urinary tract antiseptics. These compounds previously have been reviewed in detail [27], [28]. Generally they offer no significant advantages over the other available 4-quinolones and are cross-resistant with these quinolones. They are of historical interest because they are the *forerunners of the second-generation fluoroquinolones.* The addition of a piperazine ring at C-7 of the 4-quinolone skeleton (**1**) in pipemidic acid seems to enhance in vivo performance through improved pharmacokinetics. Pipemidic acid has been reported to be orally effective against experimental *P. aeruginosa* infections in mice; however, its clinical utility against this organism is not established.

Pipemidic acid (INN, dénomination commune Francaise, French approved name – DCF, *Merck Index* – MI) [*51940-44-4*], 8-ethyl-5,8-dihydro-5-oxo-2-(1-piperazinyl)pyrido[2,3-*d*]-pyrimidine-6-carboxylic acid, $C_{14}H_{17}N_5O_3$, M_r 303.32, *mp* 253 – 255 °C, is a yellowish-white, odorless, bitter-tasting crystal.

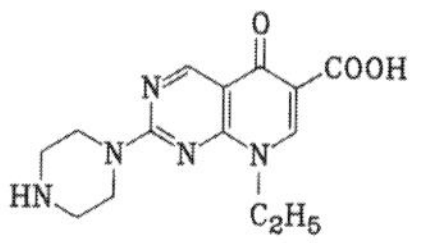

Synthesis: Ethyl 8-ethyl-5,8-dihydro-2-methylthio-5-oxopyrido[2,3-*d*]pyrimidine-6-carboxylate is heated with excess piperazine at 90 – 110 °C in DMSO (dimethyl sulfoxide) for 3 h [29].

Piromidic acid (INN, National Formulary Name – NFN, MI 9) [*19562-30-2*], 8-ethyl-5,8-dihydro-5-oxo-2-(1-pyrrolidinyl)pyrido[2,3-*d*]-pyrimidine-6-carboxylic acid, $C_{14}H_{16}N_4O_3$,

M_r 288.31, *mp* 314 – 316 °C, is crystalline in form.

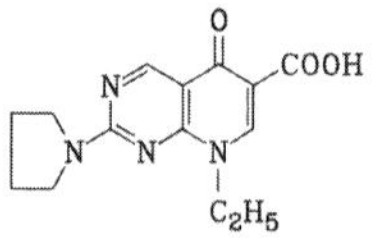

Synthesis: Ethyl 8-ethyl-5,8-dihydro-2-methylthio-5-oxopyrido[2,3-*d*]pyrimidine-6-carboxylate is heated in a pressure tube with pyrrolidine at 95 °C in ethanol for 6 h to yield piromidic acid [30].

Rosoxacin (INN) [*40034-42-2*], 1-ethyl-1,4-dihydro-4-oxo-7-(4-pyridyl)-3-quinolinecarboxylic acid, $C_{17}H_{14}N_2O_3$, M_r 294.31, *mp* 290 °C, forms yellow crystals, is stable in dry heat at 70 °C, and is sensitive to light.

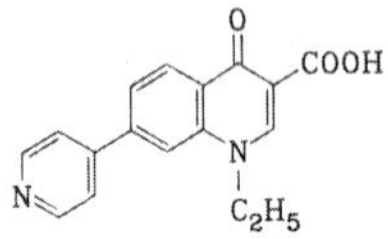

Rosoxacin is a potent first-generation quinolone with minimum inhibitory concentrations (MIC) ranging from 0.1 μg/mL for the most susceptible *E. coli* to 1.6 μg/mL for *Providencia* and *Klebsiella* species [31]. Some strains of *Pseudomonas* are susceptible. *Neisseria gonorrhoeae, Neisseria meningitidis,* and *Haemophilus influenzae* are particularly susceptible to the action of rosoxacin (MIC, 0.02 μg/mL); low plasma levels are achieved in humans after oral administration. A dose of 250 mg produces a peak plasma level of 6.4 μg/mL 2 h after administration.

This quinolone has been marketed as an antigonorrheal agent. Single oral doses are reported to produce high cure rates in patients with acute uncomplicated gonorrhea. *Adverse effects* include dizziness and itching.

Synthesis: A five-step synthesis starting with 3-nitrobenzaldehyde yields 4-(3-aminophenyl)-pyridine, which is condensed with diethyl ethoxymethylenemalonate and then cyclized thermally to yield ethyl 1,4-dihydro-4-oxo-7-(4-pyridyl)quinoline-3-carboxylate. Ethylation followed by saponification and neutralization yields rosoxacin [32].

Trade name: Roxadyl (Sterling-Winthrop).

Miloxacin (INN) [*37065-29-5*], 5,8-dihydro-5-methoxy-8-oxo-1,3-dioxolo[4,5-*g*]quinoline-7-carboxylic acid, $C_{12}H_9NO_6$, M_r 263.21, *mp* 264 °C (decomp.), occurs as colorless prisms from DMF.

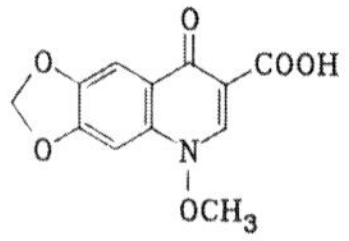

Miloxacin, marketed in Japan, has an antimicrobial spectrum and potency similar to that of oxolinic acid. Clinically, it does not seem to have any advantages over oxolinic acid; however, it appears to offer somewhat superior absorption and higher urinary excretion over oxolinic acid [33], [34]. A peak serum concentration of 7.7 μg/mL is achieved in humans 1 – 2 h following a 500-mg oral dose [35]. The drug is highly metabolized (87 %) to a biologically inactive glucuronide form, and only 5.4 % is found in the urine 8 h after dosing.

Synthesis: [36].

2.7.2.4. Second-Generation Fluoroquinolones

The second-generation fluoroquinolones are distinguished by their excellent and broad antibacterial activity, their relatively fewer adverse effects when compared with nalidixic acid, and their low propensity for inducing bacterial resistance. Some of these compounds can be administered parenterally and are effective in systemic infections outside of the urinary tract.

Their *activity* seems scarcely affected by inoculum size, type of medium, or presence of serum [37–40]. Some of the compounds lose activity in the presence of urine and when the pH of the medium is < 5.0 [38–41]. Generally these drugs are highly active against the enteric gram-negative bacilli and cocci. They are also active against other gram-negative bacteria, including *P. aeruginosa* (but less so against other *Pseudomonas* species), *Aeromonas hydrophila, H. influenzae,* and *Legionella pneumophilia.* They have excellent activity against pathogens of the gastrointestinal tract, including *E. coli, Salmonella* species, *Shigella* species, *Yersinia enterocolitica, Camplobacter jejuni,* and *Vibro* species. Their activity against gram-positive species is poorer than against gram-negative ones, but it is still within a potential therapeutic range especially for ofloxacin and ciprofloxacin. Activity against anaerobic bacteria is marginal. Cross resistance among the various quinolones is observed [41], [42]. The fluoroquinolones are active in vitro against multiple antibiotic-resistant bacteria. They are rapidly bactericidal. Other organisms reported to be susceptible to one or more of these compounds include *Chlamydia trachomatis, Ureaplasma urealyticum, Mycoplasma hominis, Mycoplasma pneumoniae,* and *Mycobacterium tuberculosis.* Fungi are not susceptible to the fluoroquinolones.

Although differences in potency between the fluoroquinolones exist for the various pathogens, their clinical usefulness will be predicted best in conjunction with key parameters, such as pharmacokinetic properties and toxicities of individual drugs [5].

Norfloxacin (INN, BAN – British Approved Name) [*70458-96-7*], 1-ethyl-6-fluoro-1,4-dihydro-4-oxo-7-(1-piperazinyl)-3-quinolinecarboxylic acid, $C_{16}H_{18}FN_3O_3$, M_r 319.34, *mp* 227–228 °C, is crystalline in form.

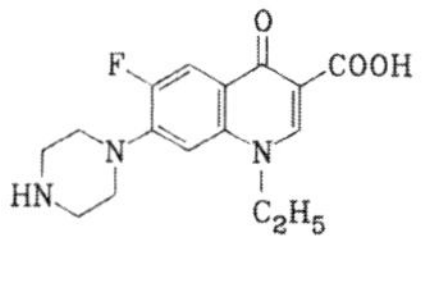

Norfloxacin has an in vitro spectrum of antibacterial activity that includes most gram-negative organisms with MIC_{90} values in the range of 1 µg/mL or less [43], [44]. Norfloxacin is about 100 times more active than nalidixic acid, and its spectrum of activity includes enterococci and staphylococci as well as *Pseudomonas* [43]. Norfloxacin is also active against *H. influenzae* at a concentration of 0.12 µg/mL and *N. gonorrhoeae* at 0.016 µg/mL. The MIC_{90} for *S. aureus* strains is 1.6 µg/mL, with the streptococci being even less sensitive [44]. Members of the *Bacteroides fragilis* group of anaerobes are relatively resistant to norfloxacin (MIC 8 – 128 µg/mL), as are other anaerobic bacteria. Norfloxacin is more potent against sexually transmitted diseases than available quinolone chemotherapeutics [45]. It is less active than other fluoroquinolones against *Ureaplasma* and shows only moderate activity against *Chlamydia trachomatis* [46].

Absorption of norfloxacin is poorer than that of enoxacin or ciprofloxacin [47–50]. Prostatic tissue levels in humans exceed serum levels, whereas blister fluid levels are 70 % of serum levels [48], [51]. Approximately 30 % of the administered dose is excreted in urine; 80 % of the excreted dose is parent compound [52]. Oral administration of 400 mg BID (twice daily) for 10 d was as effective as trimethoprim – sulfamethoxazole in treatment of patients with urinary tract infections or acute pyelonephritis [53]. *Adverse effects,* which included dizziness and nausea, were low. Norfloxacin administered in a single oral dose of 600 mg gave 100 % cure of gonococcal infections in men [45].

Synthesis: by condensation of 1-ethyl-6-fluoro-7-chloro-1,4-dihydro-4-oxo-3-quinolinecarboxylic acid with piperazine at 170 °C in water in a pressure tube [54].

Trade (or code) *names:* Noroxin, Baccidal, MK – 0366, and AM – 715 (Kyorin, Merck).

Ciprofloxacin (INN, BAN) [*85721-33-1*], 1-cyclopropyl-6-fluoro-1,4-dihydro-4-oxo-7-(1-piperazinyl)-3-quinolinecarboxylic acid, $C_{17}H_{18}FN_3O_5$, M_r 331.35, *mp* 255 – 257 °C, is crystalline in form.

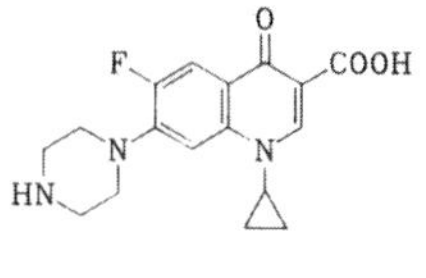

Ciprofloxacin has the most potent in vitro activity of the fluoroquinolones reported to date. It is active against most bacterial strains that cause urinary tract infections at concentrations well in excess of those observed in the urine [55–57]. The minimum inhibitory concentrations of ciprofloxacin that inhibit 90 % of clinical isolates (MIC_{90}), in µg/mL, are as follows.

Pathogen	MIC_{90}, µg/mL
Escherichia coli	0.06
Klebsiella species	0.25
Salmonella species	0.015
Shigella species	0.008
Citrobacter diversus	0.03
Citrobacter freundii	0.125
Enterobacter cloacae	0.03
Enterobacter aerogenes	0.06
Proteus mirabilis	0.06
Proteus vulgaris	0.06
Morganella morganii	0.016
Providencia stuartii	0.50
Pseudomonas aeruginosa	0.25
Pseudomonas maltophilia	4.0
Pseudomonas cepacia	8.0
Serratia marcescens	0.13
Acinetobacter calcoaceticus	0.5
Yersinia enterocoliticia	0.06
Seromonas hydrophilia	0.008
Pasteurella multocida	0.016
Haemophilus influenzae	0.015
Neisseria gonorrhoeae	0.004
Bacteroides fragilis	4.00
Staphylococcus aureus	0.50
Staphylococcus epidermidis	0.25
Streptococcus pyogenes	0.20
Streptococcus agalactiae	1.00
Streptococcus faecalis	2.00
Streptococcus pneumoniae	2.00
Streptococcus viridans	4.00
Listeria monocyogenes	1.00
Chlamydia trachomatis	1.00

As can be seen from these values, most organisms are highly susceptible to ciprofloxacin, although the gram-positive bacteria are much less susceptible than the gram-negative organisms.

Oral doses of ciprofloxacin are rapidly absorbed and peak plasma levels of 2 – 3 µg/mL are observed 1 – 1.5 h after a 500-mg dose. The serum half-life is 3.9 – 4.9 h [50], [58]. Approximately 20 % of the administered dose can be recovered in the urine as active drug during the first 4 h, with a total of 30 – 40 % recovered in 24 h. Ciprofloxacin can also be recovered from blister fluid, where it achieves 57 % of the serum concentration [50]. A parenteral formulation is available. An intravenous dose results in a serum half-life of 4 h, with ca. 76 % of the dose recovered in the urine [49]. Clinical trials have shown the drug to be efficacious for the treatment of urinary tract infections as well as systemic bacterial infections.

Synthesis: Condensation of 2,4-dichloro-5-fluorobenzoylchloride with diethyl malonate by means of magnesium ethoxide in ether gives diethyl 2,4-dichloro-5-fluorobenzoylmalonate, which is partially hydrolyzed and decarboxylated with *p*-toluenesulfonic acid – water, yielding ethyl 2,4-dichloro-5-fluorobenzoylacetate. Condensation of this with triethyl orthoformate in refluxing acetic anhydride affords ethyl 2-(2,4-dichloro-5-fluorobenzoyl)-3-ethoxyacrylate, which is treated with cyclopropylamine in ethanol to give ethyl 2-(2,4-dichloro-5-fluorobenzoyl)-3-cyclopropylaminoacrylate. Cyclization with NaH in refluxing dioxane then yields 7-chloro-1-cyclopropyl-6-fluoro-1,4-dihydro-4-oxoquinoline-3-carboxylic acid, which is finally condensed with piperazine in hot DMSO to yield ciprofloxacin [59].

Trade name: Bay 09867 (Bayer).

Enoxacin (INN) [*74011-58-8*], 1-ethyl-6-fluoro-1,4-dihydro-4-oxo-7-(1-piperazinyl)-1,8-naphthyridine-3-carboxylic acid, $C_{15}H_{17}FN_4O_3$, M_r 320.32, *mp* 220–224 °C, occurs as crystals; HCl salt *mp* 300 °C.

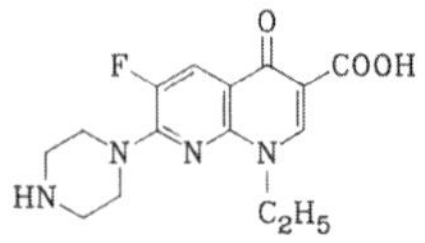

Enoxacin has a spectrum of activity similar to that of norfloxacin, but it appears to be somewhat less potent [41]. The antibiotic is well-absorbed in humans and animals [47], [49], [60]. In experimental animals, enoxacin produced greater tissue concentrations than did norfloxacin when administered orally. In humans, a 600-mg oral dose of enoxacin shows superior oral absorption to norfloxacin, a higher peak serum concentration c_{max} (3.7 µg/mL), a longer serum half-life (6.2 h), greater urinary recovery (67 % in 0 – 24 h), and greater blister fluid penetration [49]. This quinolone is effective in treating urinary tract and respiratory infections in humans, with a low incidence of side effects [49]. Clinical success has also been reported in the treatment of uncomplicated gonorrhea and systemic *P. aeruginosa* infections [61], [62]. A parenteral formulation is available and is currently being tested clinically.

Synthesis: Condensation of 2,3-difluoro-6-nitrophenol with chloroacetone yields 2-acetonyloxy-3,4-difluorobenzene, which is cyclized (H_2/Raney Ni) to give 7,8-difluoro-2,3-dihydro-3-methyl-4*H*-benzoxazine. Condensation with diethyl ethoxymethylenemalonate followed by cyclization (ethyl polyphosphate, 145 °C) yields 9,10-difluoro-3-methyl-7-oxo-2,3-dihydro-7*H*-pyrido[1,2,3-*d*]1,4-benzoxazine-6-carboxylate. Hydrolysis with HCl gives enoxacin [63].

Trade (or code) *names:* CI – 919, AT – 2266, Flumark, and PD – 107779 (Dainippon).

Pefloxacin (INN) [*70458-92-3*], 1-ethyl-6-fluoro-7-(4-methyl-1-piperazinyl)-4-oxo-1,4-di-hydroquinoline-3-carboxylic acid, $C_{17}H_{20}FN_3O_3$, M_r 333.36, *mp* 270 – 272 °C. The mesylate salt, *mp* 284 – 286 °C (decomp.), is a yellowish-white crystalline powder.

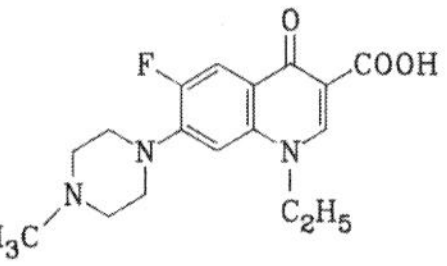

Pefloxacin is marketed in France. The in vitro spectrum and potency of this fluoroquinolone are similar to norfloxacin but pefloxacin appears to have improved activity against gram-positive bacteria [64]. In experimental immunocompromised animals with *Pseudomonas* infection, pefloxacin showed better activity than did ciprofloxacin. In humans, renal plasma levels were 3.8 μg/mL following a 400-mg oral dose, and the plasma half-life (9 – 10 h) exceeded that for most quinolones currently under development [65]. Approximately 10 % is recovered in the urine after an 800-mg oral dose. One of the active metabolites of this compound is norfloxacin. Pefloxacin has been reported to be efficacious in treating meningitis in humans [66].

An intravenous preparation is available, and significant bone levels are achieved after i.v. dosing. Pefloxacin has been reported active clinically in the treatment of staphylococcal osteomyelitis when administered along with rifampicin [67].

Synthesis: 3-Chloro-4-fluoroaniline is condensed with diethyl ethoxymethylenemalonate and then cyclized thermally to give ethyl 6-fluoro-7-chloro-4-hydroxyquinoline-3-carboxylate. Subsequent alkylation with ethyl iodide leads to the ester, which is then saponified and neutralized to give pefloxacin [68].

Trade (or code) *names:* 1589 – RB, Eu – 5306 (Rhône – Poulenc).

Ofloxacin (INN) [*83380-47-6*], (±)9-fluoro-3-methyl-10-(4-methyl-1-piperazinyl)-7-oxo-2,3-dihydro-7*H*-pyrido[1,2,3-*de*]-1,4-benzoxazine-6-carboxylic acid, $C_{18}H_{20}FN_3O_4$, M_r 361.16, *mp* 250 – 257 °C (decomp.), occurs as colorless needles.

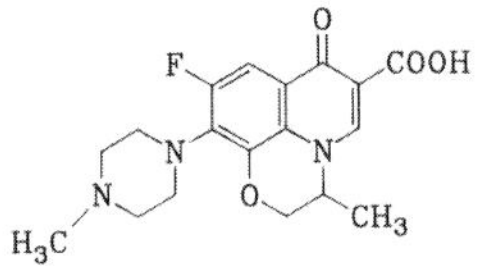

Ofloxacin has no spectrum or potency advantages over the other newer quinolones. Its in vitro microbiological activity is comparable to that of norfloxacin, but it has somewhat better activity against gram-positive cocci [37], [64]. The major attribute of this fluoroquinolone lies in its superior pharmacokinetics. A peak serum level of 10.7 μg/mL at 1.2 h after a 600-mg oral dose was achieved in human volunteers [69]. A long serum half-life (7.0 h) and excellent urinary recovery (73 %) make this compound one of the more interesting members of the group of fluoroquinolones under worldwide development. Ofloxacin has been reported to be efficacious in a variety of clinical indications after oral dosing of 100 – 600 mg/d [60].

Synthesis: prepared from 2,3,4-trifluoronitrobenzene by a seven-step synthesis [70].

Trade (or code) *names:* DL – 8280, Hoe – 280, Ru – 43 – 280, and Tarivid (Daiichi Seiyaku).

Amifloxacin (INN) [*86393-37-5*], 6-fluoro-1,4-dihydro-1-(methylamino)-7-(4-methyl-1-piperazinyl)-4-oxo-3-quinolinecarboxylic acid, $C_{16}H_{19}FN_4O_3$, M_r 334.35, *mp* 299 – 301 °C (decomp.), occurs as crystals.

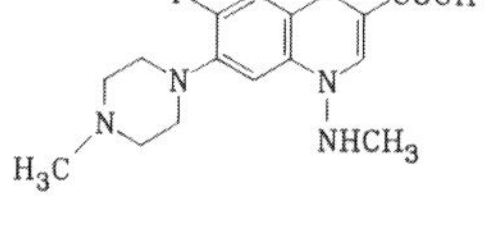

Amifloxacin has broad spectrum in vitro activity similar to that of the other quinolones, but with poorer potency than ciprofloxacin and little activity against anaerobic bacteria [71], [72]. The MIC_{90} for various Enterobacteriaceae, including *E. coli, Klebsiella, Proteus, Enterobacter, Citrobacter, Salmonella,* and *Shigella,* was comparable to that of norfloxacin and enoxacin, but was 2- to 8-fold less than that of ciprofloxacin [73]. Amifloxacin was more active than norfloxacin and enoxacin against *P. aeruginosa,* but less active than ciprofloxacin.

Amifloxacin is well-absorbed on oral administration and is excreted at high urinary levels in experimental animals [74]. A parenteral dosage form is available for humans, and early phase 1 studies have been initiated.

Synthesis: prepared from ethyl 6-fluoro-7-chloro-1,4-dihydro-4-oxo-3-quinolonecarboxylate through a five-step synthesis [75].

Trade (or code) *name:* WIN 49375 (Sterling-Winthrop).

Other Fluoroquinolones. A number of interesting fluoroquinolones are currently under study in various countries. They include flumequine (R – 802), A – 56620, AM – 833, and CI – 934. The most advanced of these is flumequine, which is available for human and animal use in Europe.

Compound A – 56620 (difloxacin), like the other fluoroquinolones, has potent antibacterial activity, but also has good activity against staphylococci, streptococci, and *B. fragilis* [76], [77]. Its potency is not appreciably affected by inoculum size. In dogs, the half-life is extended (8.2 h), and the compound undergoes enterohepatic circulation. The compound is in phase I development.

Fluoroquinolone AM – 833 has in vitro microbiological activity similar to that of norfloxacin, but improved activity against staphylococci. It is more active in animal infections when administered orally than is norfloxacin, probably because of superior pharmacokinetics [78]. Peak serum levels and serum half-life in dogs are clearly superior to the values obtained with norfloxacin [79].

Compound CI – 934 has broad gram-negative activity, but is less potent against these organisms than is norfloxacin or ciprofloxacin. It does have superior activity against gram-positive cocci when compared with the other quinolones [80–82].

Flumequine (INN, BAN) [*42835-25-6*], 9-fluoro-6,7-dihydroxy-5-methyl-1-oxo-1*H*,5*H*-benzo[*i,j*]quinolizine-2-carboxylic acid, $C_{14}H_{12}FNO_3$, M_r 261.15, *mp* 253 – 255 °C, is a white crystalline powder.

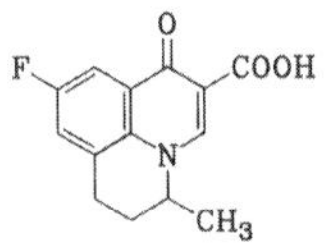

O F COOH N CH3

Synthesis: Condensation of 5-fluoro-2-methyltetrahydroquinoline with diethyl ethoxymethylenemalonate followed by thermal cyclization gives ethyl 6,7-dihydro-9-fluoro-5-methyl-1-oxo-1*H*,5*H*-benzo[*i,j*]quinolizine-2-carboxylate,which is saponified with sodium hydroxide to give flumequine [83].

A – 56620 [*98105-99-8*], 1-(*p*-fluorophenyl)-6-fluoro-1,4-dihydro-4-oxo-7-(1-piperazinyl)-quinoline-3-carboxylic acid hydrochloride, $C_{20}H_{17}F_2N_3O_3 \cdot HCl$, M_r 421.83, *mp* 275 °C.

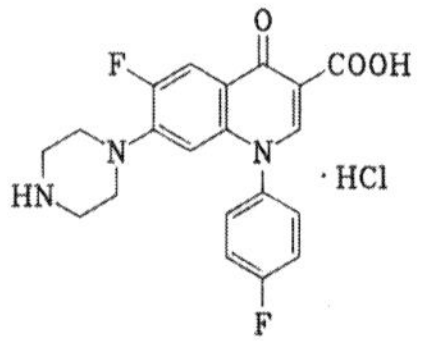

O F COOH N HN N · HCl F

Synthesis: Condensation of 7-chloro-1-(*p*-fluorophenyl)-6-fluoro-1,4-dihydro-4-oxoquinoline-3-carboxylic acid with *N*-carboethoxypiperazine in hot 1-methyl-2-pyrrolidinone yields the *N*-carboethoxy derivative of A – 56620. Hydrolysis of the derivative with sodium hydroxide in aqueous ethanol followed by treatment with dilute hydrochloric acid gives A –56620 [84].

AM – 833 [*79660-72-3*], 6,8-difluoro-1-(2-fluoroethyl)-1,4-dihydro-7-(4-methyl-1-piperazinyl)-4-oxo-3-quinolinecarboxylic acid, $C_{17}H_{18}F_3N_3O_3$, M_r 369.34; HCl salt [*79660-53-0*], *mp* 269 –271 °C.

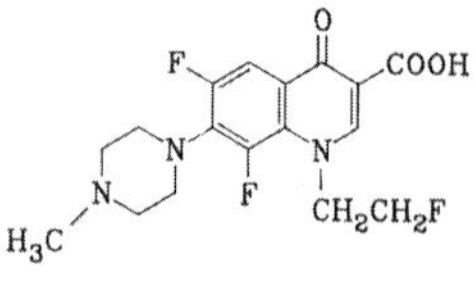

O F COOH N N N F CH2CH2F H3C

Synthesis: The reaction of 6,7,8-trifluoro-1,4-dihydro-4-oxoquinoline-3-carboxylic acid with 1-bromo-2-fluoroethane by means of NaI in DMF gives 6,7,8-trifluoro-1-(2-fluoroethyl)-1,4-dihydro-4-oxoquinoline-3-carboxylicacid, which is then condensed with *N*-methylpiperazine in refluxing pyridine to give AM – 833.

Manufacturer: Kyorin.

CI – 934 *[91188-00-0]*, 1-ethyl-7-{3-(ethylamino)methyl-1-pyrrolidinyl}-6,8-difluoro-1,4-dihydro-4-oxo-3-quinolinecarboxylic acid, $C_{19}H_{23}F_2N_3O_3$, M_r 379.41, *mp* 208 – 210 °C.

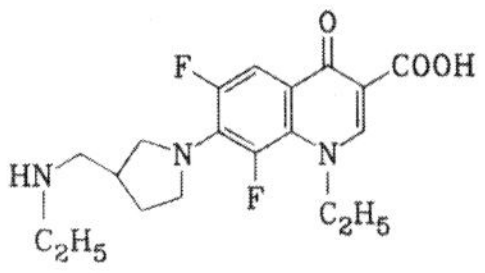

Synthesis: Refluxing 1-ethyl-1,4-dihydro-4-oxo-6,7,8-trifluoro-3-quinolinecarboxylic acid and *N*-ethyl-3-pyrrolidinemethanamine and 1,8-diazabicyclo[5.4.0]undec-7-ene in acetonitrile gives CI – 934 [85].

2.7.3. Sulfa Drugs

The sulfonamides continue to maintain a niche in human and animal medicine even 50 years after their discovery. More than 5000 sulfonamides have been synthesized with a broad diversity of antimicrobial activity and pharmacokinetic profiles. Approximately 13 compounds continue to be used in human medicine; however, their use is limited by the emergence of resistant organisms. The sulfonamide prontosil, one of a group of dyes synthesized in 1935, was the first clinically effective compound. Prontosil protected mice and rabbits against bacterial infections, but had no in vitro activity. Subsequently, prontosil was found to be, in effect, a prodrug, which on metabolic cleavage produced 4-aminobenzenesulfonamide (sulfonalamide) [86]. Because of their ease of synthesis, low cost, relative safety, and broad efficacy in humans, many research organizations embarked on broad synthesis programs designed to produce compounds with improved potency, spectrum of activity, and pharmacokinetic properties. Interest shifted from the sulfonamides in 1945 after the introduction of penicillin, but was revived again in the late 1950s.

2.7.3.1. Biological Activity and Medical Uses

Sulfonamides have broad antimicrobial activity that includes many species of bacteria and protozoa, such as the following:

Gram-negative bacteria
Escherichia coli
Proteus species
Klebsiella species
Salmonella species
Shigella
Vibrio cholerae
Neisseria species
Haemophilus species
Pseudomonas (some strains)
Calymmatobacterium granulomatis
Legionella pneumophila

Gram-positive bacteria	Others
Bacillus anthracis (some strains)	*Chlamydia trachomatis*
	Trachoma virus
Staphylococcus aureus	Lymphogranuloma venereum virus
Streptococcus pyogenes	
Clostridium welchii	*Plasmodium falciparum*
Clostridium tetani	*Plasmodium malariae*
	Toxoplasma
	Nocardia species

The common gram-positive and gram-negative bacteria are susceptible, as are the protozoan organisms that cause malaria and toxoplasmosis. The lymphogranuloma venereum and trachoma viruses are reported to be susceptible as well. A high percentage of bacterial organisms have developed resistance to sulfonamides; and thus, these drugs are indicated only in a few diseases, including urinary tract infections, chancroid, inclusion conjunctivitis, trachoma, and nocardia infections.

The sulfonamides have been classified on the basis of their serum half-lifes in humans and their topical or gastrointestinal uses. Those compounds with a half-life of < 10 h are termed short-acting, and those with a half-life between 10 and 24 h are medium-acting; long-acting sulfonamides have a half-life > 24 h.

2.7.3.2. Mechanism of Action and Antimicrobial Resistance

4-Aminobenzoic acid (**3**) is an essential metabolite for many microorganisms.

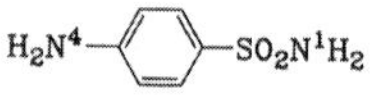

H_2N^4–C$_6$H$_4$–$SO_2N^1H_2$ **2** 4-Aminobenzenesulfonamide [*63-74-1*]

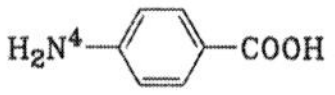

H_2N^4–C$_6$H$_4$–COOH **3** 4-Aminobenzoic acid [*150-13-0*]

This compound is used by bacteria as a precursor in the synthesis of folic acid [*59-30-3*], which, in turn, serves as an important intermediate in nucleic acid synthesis. *Sulfonamides,* as structural analogues of **3**, interfere with microbial folic acid synthesis by inhibiting the adenosine triphosphate dependent condensation of a pteridine with **3** to yield dihydropteroic acid, which is subsequently converted to folic acid [87–89]. As a result, nonfunctional analogues of folic acid are formed that do not allow bacterial cells to grow. Because mammalian cells cannot synthesize folic acid, the activity of sulfonamides is selective for bacteria that are capable of producing it. Tubercule bacilli are poorly inhibited by sulfonamides, but their growth is inhibited by 4-aminosalicylic acid [*65-49-6*], whereas most sulfonamide-sensitive bacteria are resistant to this acid. Thus, the receptor site for 4-aminobenzoic acid apparently differs in different types of organisms.

Trimethoprim [*738-70-5*] inhibits dihydrofolic acid reductase 10 000 times more efficiently in bacteria than in mammalian cells. This enzyme is important in the folic acid pathway leading to the synthesis of purines and ultimately of DNA. The combination of sulfonamides with trimethoprim acts synergistically to inhibit bacteria by sequential blockage (see Section 2.7.3.6).

Pyrimethamine [*58-14-0*] (Daraprim) also inhibits dihydrofolate reductase, but it is more active against the enzyme in mammalian cells, and, thus more toxic than trimethoprim. Toxoplasma and other protozoal infections can be successfully treated with pyrimethamine – sulfonamide combinations.

The usefulness of sulfa drugs has been limited by the development of *bacterial resistance.* At least three mechanisms are well documented: (1) alteration of cell wall permeability, (2) increased production of an essential metabolite, and (3) a mechanism involving dihydropteroic acid synthetase. Two enzymes have been reported for *E. coli* with less affinity for sulfonamides than enzymes from susceptible strains, but with no change in 4-aminobenzoic acid affinity. Resistant bacteria, e.g., staphylococci, synthesize excess 4-aminobenzoic acid, which can antagonize sulfa drugs. Plasmids may code for drug-resistant enzymes [90] or decreased bacterial cell permeability to sulfonamides [91]. Multiple resistance mechanisms in the same organism are possible [92]. Sulfonamide resistance is found in 20 – 40 % of community and nosocomial strains of bacteria, including *Staphylococcus,* Enterobacteriaceae, *Neisseria meningitidis,* and *Pseudomonas* species [93].

2.7.3.3. Structure – Function Relationships

All clinically useful sulfa drugs are derived from sulfonalamide (**2**), an analogue of 4-aminobenzoic acid (**3**). The synthesis of thousands of derivatives of **2** has allowed for excellent structure – function determinations, which have been reviewed in detail [94]. A free amino group at N-4 is essential for microbiological activity. Acyl substitution, such as phthalyl or succinyl, at N-4 is acceptable, provided the substituent is hydrolyzed in the body to the free amine (e.g., phthalylsulfathiazole). Replacement of the benzene ring by a heterocycle leads to loss in activity; thus, only the acid portion of the molecule is amenable to substitution. Substitution at the sulfonyl radical can result in marked changes in pharmacologic properties, such as absorption, solubility, pharmacokinetics, and gastrointestinal tolerance.

Monosubstitution at N-1, especially when the substituent is a heterocycle, has led to most of the clinically useful sulfa drugs. Derivatives containing five- or six-membered heterocyclic rings, including oxazole, thiazole, pyridine, pyrimidine, and many others, have been most successful.

2.7.3.4. Pharmacokinetics

Sulfa drugs are commonly administered orally; absorption is generally rapid. Systemic derivatives, such as sulfisoxazole and sulfadiazine, are available, but are rarely used in clinical medicine. An intravenous formulation of trimethoprim and sulfamethoxazole is used to treat *Pneumocystis carinii* pneumonitis, shigellosis, and urinary tract infections caused by susceptible organisms. In addition various topical preparations are available for treating ophthalmic and vaginal infections and burn patients.

Generally the available sulfa drugs are rapidly absorbed from the gastrointestinal tract. Their excretion rates vary widely, depending on their physical and chemical properties. Several derivatives of **2** substituted at N-4 are poorly absorbed and are employed for diseases of the gastrointestinal tract. Peak blood levels following oral absorption are usually observed at 2 – 4 h. A wide variation in half-lifes from 2.5 to 150 h for the commercially available agents has been reported, but most agents are excreted over a 24-h period.

Sulfonamides are usually well-distributed in the body and achieve tissue levels of ca. 80 % of those observed in the serum [95]. The degrees of protein binding and lipid solubility play a major role in determining serum half-life and distribution in the body. The primary route of metabolism is via the liver, where acetylation or glucuronidation occur. Excretion is mainly via the kidney, where most drugs are removed by glomerular filtration, although partial reabsorption and active tubular secretion also are involved. Sulfa drugs bind reversibly to serum proteins, primarily albumin, which tends to slow metabolism by the liver [96].

2.7.3.5. Toxicity and Drug Interactions

Significant hypersensitivity reactions have been associated with the sulfonamides; the long-acting derivatives, especially, have caused fatal hypersensitivity reactions. Other hypersensitivity reactions include erythema nodosum and erythema multiforme, including Stevens – Johnson syndrome. Photosensitization, itching, and rash usually require cessation of therapy. With earlier drugs, crystalluria and related renal damage occurred, primarily with the less water-soluble compounds, such as sulfadiazine, sulfamerizine, and sulfapyridine. Other adverse effects observed less commonly are nausea, vomiting, diarrhea, hepatic toxicity, and a syndrome that resembles serum sickness. Serious adverse reactions include hemolytic anemia, aplastic anemia, agranulocytosis, thrombocytopenia, and leukopenia.

Sulfonamides interact with other drugs and compete with drugs, such as warfarin and methotrexate, for albumin-binding sites resulting in increased toxicity of the displaced drugs. Concomitant administration of probenecid, which results in decreased renal tubular secretion, prolongs sulfa drug levels and may increase toxicity.

2.7.3.6. Combination Therapy

Both oral and intravenous formulations of sulfamethoxazole in combination with trimethoprim are available. This synergistic mixture is used orally to treat urinary tract and certain other infections. In the United States, trimethoprim is also available as a single agent for the treatment of urinary tract infections. Combinations of sulfa drugs and various antibiotics for both oral and topical use are also available.

Systemic formulations in combination with 2,6-diamino-3-phenylazopyridine · hydrochloride, an azo dye that is a mild local analgesic, have been employed to reduce pain of inflammation that results from urinary tract infections.

2.7.3.7. Rapidly Absorbed Short- and Medium-Acting Sulfa Drugs

Sulfamethoxazole (INN, U.S.P.) [*723-46-6*], N^1-(5-methyl-3-isoxazolyl)sulfanilamide, $C_{10}H_{11}N_3O_3S$, M_r 253.28, *mp* 170 °C, is an odorless and colorless crystalline powder.

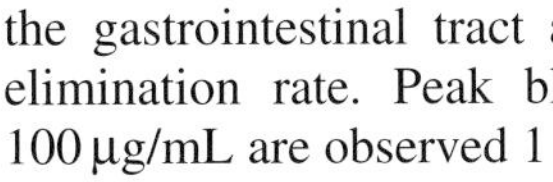

H_2N–C6H4–SO_2NH–(isoxazolyl, N, O, CH_3)

Sulfamethoxazole is rapidly absorbed from the gastrointestinal tract and has a moderate elimination rate. Peak blood levels of 80 – 100 μg/mL are observed 1 – 4 h following a 2-g oral dose. The plasma half-life is 9 – 11 h and the drug is 70 % protein-bound; 80 – 90 % of the unbound compound remains unacylated and biologically active in the blood. About 60 % of the drug is excreted in the urine, and half of this is N-4 acylated or conjugated. Urine levels are 3 times blood levels.

The major indication for sulfamethoxazole is treatment of urinary tract infections caused by susceptible strains of *E. coli, Klebsiella, Proteus* species, and staphylococci. Although the compound has broader microbiological activity, it generally is not used for infections outside of the urinary tract. Sulfamethoxazole is also indicated in the treatment of nocardiosis, inclusion conjunctivitis, trachoma, chancroid, and malaria caused by chloroquine-resistant strains, and is used in combination with pyrimethamine for toxoplasmosis.

The usual dose for mild infections is 2 g followed by 1 g every 12 h for 4 days. For severe infections, the drug is administered 3 times a day at twice this dose. Pediatric dose is 50 – 60 mg/kg followed by 12-h doses of 25 – 30 mg/kg.

Synthesis: prepared from ethyl 5-methylisoxazole-3-carbamate [97].

Trimethoprim (INN, U.S.P.) [*738-70-5*], 2,4-diamino-5-(3,4,5-trimethoxybenzyl)pyrimidine, $C_{14}H_{18}N_4O_3$, M_r 290.32, *mp* 199 – 203 °C, is a white-to-cream-colored bitter, crystalline powder.

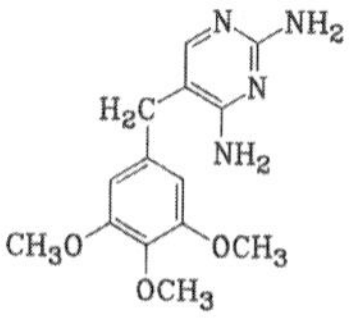

Synthesis: prepared from guanidine and β-ethoxy-3,4,5-trimethoxybenzylbenzalnitrile [98].

Trade names: Cofrim (Lemmon), Bactrim (Hoffman-La Roche), Septa (Burroughs Wellcome), Proloprim (Burroughs Wellcome), and Trimpex (Hoffman-La Roche, trimethoprim only).

Sulfamethoxazole – Trimethoprim. Sulfamethoxazole is usually administered along with trimethoprim in a fixed ratio of 5 : 1. This synergistic combination results in a sequential blockage of folic acid synthesis. Synergistic antibacterial activity has been demonstrated in both in vitro and in vivo systems [99], [100].

The pharmacokinetic profiles of sulfamethoxazole and trimethoprim fit well, both compounds showing peak blood levels 1 – 4 h after oral administration. With administration at 12-h intervals, a 20 : 1 ratio of unbound sulfa drug: trimethoprim is achieved in the plasma, an optimal ratio for antibacterial activity. Trimethoprim is 44 % protein bound and has a serum half-life of 8 – 10 h. Both drugs are metabolized by the liver and excreted by the kidneys.

Trimethoprim has a broad spectrum of antimicrobial activity and extends the spectrum of sulfamethoxazole to include many urinary and systemic pathogens untreatable by this sulfa drug alone. The major use of the drug combination is for urinary tract infections, in which case the drugs are administered orally for 10 – 14 days every 12 h. Additionally, the combination is indicated for chronic bronchitis in adults and otitis media in children for susceptible strains of *Streptococcus pneumoniae* and *Haemophilus influenzae.* Bacillary dysentery caused by *Shigella* species also responds to oral treatment.

An intravenous preparation is available for treatment of severe infections, as well as pneumonitis caused by the protozoan, *Pneumocystis carinii.* For treatment of *Pneumocystis* pneumonia, the recommended i.v. dose is 15 – 20 mg/kg, based on the trimethoprim component, given in 3 or 4 equally divided doses every 6 – 8 h for up to 14 days [101]. For severe urinary tract infections and shigellosis, the total daily dose is 8 – 10 mg/kg, based on the trimethoprim component, given in 2 – 4 equally divided doses every 6, 8, or 12 h for up to 14 days for urinary tract infections and 5 days for shigellosis.

Sulfisoxazole [*127-69-5*], sulfafurazole (U.S.P., INN), N^1-(3,4-dimethyl-5-isoxazolyl)-sulfanilamide, $C_{11}H_{13}N_3O_3S$, M_r 267.30, *mp* 192 °C, occurs as colorless and odorless prisms.

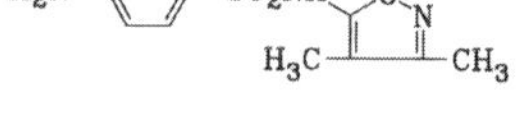

Sulfisoxazole has antimicrobial activity somewhat less than that of sulfadiazine; however, it is more easily tolerated because of its high solubility and rapid excretion. It is clinically useful in urinary tract infections and is especially potent against *E. coli* and *Prot. vulgaris.* Later derivatives of this compound, i.e., sulfamethoxazole, have superior solubility properties. At 2 h after oral administration of a 2-g dose, peak serum levels of ca. 150 µg/mL are achieved. The drug is 85 % protein-bound, and about 30 % of the unbound portion is in the acetylated form.

Excretion is primarily by the kidneys with ca. 95 % of a single oral dose eliminated within 24 h; 70 % of the excreted material is in the active form. The acetyl sulfisoxazole is metabolized in the gastrointestinal tract to release sulfisoxazole, which results in delayed absorption.

Although this sulfa drug is still used clinically, its effectiveness is limited mainly because of high protein binding. The usual adult dose is 2 – 4 g in a loading dose followed by 4 – 8 g/d divided in 4 – 6 doses. The compound is also formulated in vaginal preparations for the treatment of vaginitis caused by *Haemophilus vaginalis.* This sulfa drug is also available as the water soluble diolamine salt in 40 % aqueous solution for dilution with sterile water as a parenteral preparation.

Synthesis: p-Acetaminobenzenesulfonyl chloride is reacted with 3,4-dimethyl-5-aminoisoxazole followed by deacetylation [102].

Trade names: Gantrisin (Hoffmann-La Roche), SK-Soxazole (Smith Kline & French).

Sulfadiazine (INN, U.S.P.) [*68-35-9*], N^1-2-pyrimidinylsulfanilamide, $C_{10}H_{10}N_4O_2S$, M_r 250.27, *mp* 252 – 256 °C, is a white or pale yellow crystalline powder.

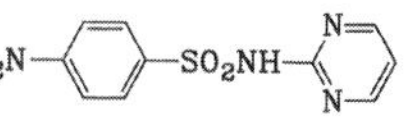

Sulfadiazine is classified as a rapidly absorbed agent with a moderate rate of excretion. It is the sulfa drug of choice for treating CNS infections caused by antibiotic-resistant organisms and for treating nocardia infections. Sulfadiazine is also used to treat urinary tract infections; however, crystalluria is a problem because of its poor water solubility. The usual precautions of urine alkalinization and forced fluid intake are recommended for patients taking sulfadiazine.

Sulfadiazine is absorbed rapidly from the gastrointestinal tract after oral administration. The peak blood level of free drug following a 2-g dose is 30 – 60 µg/mL with about 45 % protein-binding. The drug achieves excellent cerebrospinal fluid levels, which are 5 – 13 % of the blood levels. The plasma half-life is 17 h. About 20 % of an oral dose is excreted in the urine as the free form. The major metabolite is the N-4-acetylated compound.

This sulfa drug can be administered parenterally; however, because its alkaline solution is irritating, the i.v. route is preferred. The usual oral dose for adults is 2 – 4 g initially, followed by 2 – 4 g/d in 3 – 6 divided doses. When given intravenously for adults and children over 2 months of age, the drug is administered at a loading dose of 50 mg/kg or 1.25 g/m^2 followed by a total daily dose of 100 mg/kg or 2.25 g/m^2 administered 4 times a day.

A silver salt of sulfadiazine is available for topical treatment of burn patients. In this case the sulfonamide acts primarily as a vehicle for release of silver ions, which exert an antibacterial effect. Resistant organisms have been reported in burn units.

Synthesis: prepared by condensing 2-aminopyrimidine with acetylsulfaninyl chloride followed by hydrolysis with NaOH [103].

Trade name: Suladyne (Reid-Provident).

Sulfacytine (INN) [*17784-12-2*], N^1-(1-ethyl-1,2-dihydro-2-oxo-4-pyrimidinyl)sulfanilamide, $C_{12}H_{14}N_4O_3S$, M_r 294.33, *mp* 166.5 – 168 °C, is crystalline in form.

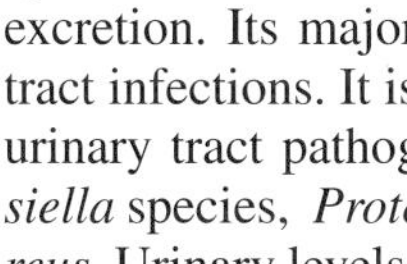

Sulfacytine is absorbed rapidly from the GI (gastrointestinal) tract with a high rate of urinary excretion. Its major use is for treating urinary tract infections. It is active against the common urinary tract pathogens, such as *E. coli, Klebsiella* species, *Proteus* species, as well as *S. aureus.* Urinary levels far in excess of the MIC are found following a daily dose of 1 g. The usual dose is 500 mg initially followed by 250 mg 4 times a day for 10 days. Peak blood levels occur within 2 – 3 h after administration, with ca. 90 % of the oral dose being recovered in the urine as the free form. High urinary concentrations are observed, but the drug is bound 86 % to serum proteins. The serum half-life is 4.5 h.

Synthesis: [104].

Trade Name: Renoquid (Parke-Davis).

Sulfamethizole (INN, U.S.P.) [*144-82-1*], N^1-(5-methyl-1,3,4-thiadiazol-2-yl)sulfanilamide, $C_9H_{10}N_4O_2S_2$, M_r 270.32, *mp* 208 °C, is a colorless crystal.

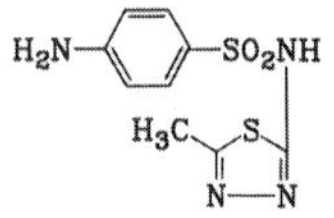

Sulfamethizole is well-absorbed and rapidly excreted. Its major indications are confined to therapy of urinary tract infections caused by susceptible organisms. It has the same therapeutic indications as sulfisoxazole; however, it is less bound to serum proteins than is sulfisoxazole. The usual adult dose is 2 – 4 g followed by 2 – 4 g/d administered in 6 divided doses.

Synthesis: prepared by the reaction of acetaldehyde thiosemicarbazone with *p*-acetylaminobenzenesulfonyl chloride in pyridine [105].

Trade names: Thiosnifil-A (Ayerst), Proklar (O'Neal, Jones & Feldman), component of Azotrex (Bristol), and component of Suladyne (Reid-Provident).

2.7.3.8. Long-Acting Sulfonamides

Sulfadoxine (INN, U.S.P.) [*2447-57-6*], N'-(5,6-dimethoxy-4-pyrimidinyl)sulfanilamide, $C_{12}H_{14}N_4O_4S$, M_r 310.33, *mp* 201 – 202 °C, forms colorless crystals.

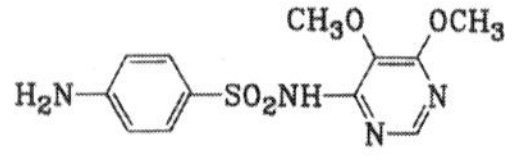

Sulfadoxine is the only long-acting agent available for clinical use in the United States. All of the long-acting sulfonamides are associated with serious hypersensitivity reactions, such as Stevens – Johnson syndrome. Sulfadoxine, originally known as sulfamethoxine, is combined with pyrimethamine (Fansidar). It has a half-life of 200 – 230 h and a peak serum level of 51 – 76 µg/mL at 2.5 – 6 h following an oral dose of 0.5 g. The major indication is for the treatment and prophylaxis of chloroquine-resistant *Plasmodium falciparum* malaria.

Synthesis: [106].

Trade name: Fanasil (Hoffmann-La Roche International).

2.7.3.9. Sulfonamides for Use in the Gastrointestinal Tract

Sulfasalazine (INN, U.S.P.) [*599-79-1*], 5-{[*p*-(2-pyridylsulfamoyl)phenyl]azo}salicylic acid, $C_{18}H_{14}N_4O_5S$, M_r 398.39, *mp* 240 – 245 °C, is a brownish-yellow powder.

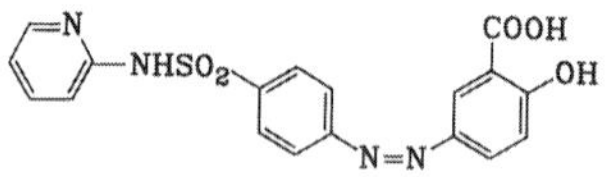

Sulfasalazine, used for the treatment of ulcerative colitis, is a prodrug of sulfapyridine. The compound is a combination of sulfapyridine and 5-aminosalicylic acid joined by an azo link, which is cleaved in the lower intestinal tract by bacterial enzymes. The release of 5-aminosalicylic acid is believed to produce a local antiinflammatory action in the colon; this is the mechanism by which the drug is effective in ulcerative colitis.

Sulfasalazine is absorbed only 10 – 15 % from the upper gastrointestinal tract and gives peak blood levels of 14 µg/mL 2 – 4 h after a 2-g oral dose. The drug is excreted in the bile and reabsorbed from the intestine as its metabolite, sulfapyridine. Peak blood levels of sulfapyridine have been reported at 13 µg/mL 6 – 24 h after dosing. The major urinary metabolite of sulfasalazine is sulfapyridine and its metabolites (60 %); however, small amounts of the unchanged drug (15 %) and 5-aminosalicylic acid and its metabolites (20 – 33 %) are also observed.

Synthesis: prepared by coupling diazotized 2-sulfanilamidopyridine with salicylic acid [107].

Sulfaguanidine, Sulfasuxidine, Sulfathalidine. Sulfaguanidine, sulfasuxidine, and sulfathalidine are all absorbed poorly from the GI tract following an oral dose. They have been employed as prophylactic agents prior to bowel surgery.

Sulfaguanidine (INN, *National Formulary* – NF **11,** MI) [*57-67-0*], N^1-(diaminomethylene)-sulfanilamide, $C_7H_{10}N_4O_2S$, M_r 214.24, *mp* 190 – 193 °C, occurs as colorless crystals.

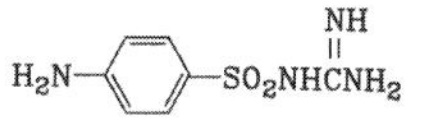

Synthesis: prepared by fusing N^4-acetylsulfanilamide and dicyanodiamide [108].

Sulfasuxidine, succinylsulfathiazole (U.S.P. **18,** MI) [*116-43-8*], 4′-(2-thiazolylsulfamoyl)-succinanilic acid monohydrate, $C_{13}H_{13}N_3O_5S_2 \cdot H_2O$, M_r 373.41, *mp* reported as 184 – 186 °C and as 192 – 195 °C, occurs as crystals.

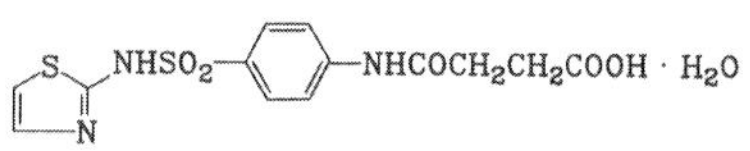

Synthesis: prepared by refluxing sulfathiazole with a slight excess of succinic anhydride in alcohol [109].

Sulfathalidine [*85-73-4*], phthalylsulfathiazole (INN, U.S.P.), 4′-(2-thiazolylsulfamoyl)-phthalanilic acid, $C_{17}H_{13}N_3O_5S_2$, M_r 403.43, *mp* 270 °C, is a white crystalline powder.

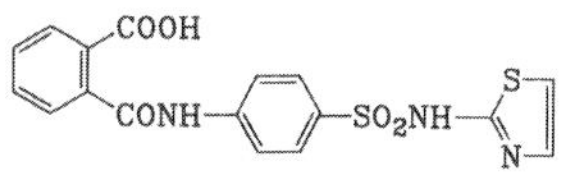

Synthesis: prepared by refluxing sulfathiazole with a slight excess of succinic anhydride in alcohol [110].

2.7.4. Agents for Treating Mycobacterial Infections

The mycobacteria are a group of rod-shaped bacteria that include many pathogenic species [111–113]. They do not stain readily, but once stained, they resist decoloration by acid or alcohol; therefore, they are called acid-fast bacilli. They cause chronic diseases producing granulomatous lesions. The major organisms causing human disease are *Mycobacterium tuberculosis, M. leprae,* and the so-called atypical mycobacteria.

Mycobacterium tuberculosis and *M. bovis,* the causative agents of tuberculosis, produce no known toxins. The organisms are usually taken in via the respiratory tract, where they establish pulmonary infections. Resistance and hypersensitivity of the host generally influence the development of the disease. Once established in the tissues, they reside in phagocytic cells, where their intracellular location makes chemotherapy difficult. The organisms spread in the host by direct extension, through the lymphatics and bloodstream, and via bronchi and the gastrointestinal tract. Every organ system can be involved. Chronic pulmonary infections are usually established; however, meningitis or urinary tract involvement can occur in the absence of other symptoms of tuberculosis. Dissemination via the bloodstream leads to miliary tuberculosis involving many organs and a high fatality rate.

Treatment of *M. tuberculosis* infections usually involves combinations of chemotherapeutic agents. The most widely used antituberculosis drugs are isoniazid, pyrazinamide, ethambutol, and the antibiotics rifampin and streptomycin (→ Antibiotics). Other drugs, such as ethionamide, 4-aminosalicylic acid, viomycin, and cycloserine, have severe adverse effects and are employed less frequently. Treatment is for long periods of time and clinical cure can be achieved in 6 – 18 months. Drug-resistant strains to all of these agents emerge rapidly; thus, greater success is achieved when the drugs are administered concomitantly.

There has been a increase in the incidence of atypical mycobacterial infection usually observed in patients whose immunity has been compromised by other factors. Many of these species of *Mycobacterium* cause severe disease closely resembling tuberculosis. These atypical mycobacteria, e.g., *M. avium – intracellularis* group, are generally much less sensitive to the available antituberculosis chemotherapeutics, especially streptomycin and isoniazid. Other organisms respond to the antitubercular chemotherapeutic agents, whereas many are sensitive to antibiotics employed for treating other bacterial infections (aminoglycosides and tetracyclines). A summary of the chemotherapeutic agents used to treat atypical mycobacterial infections is shown in Table 2 [114].

Mycobacterium leprae, the etiological agent of *leprosy*, is a disease involving the cooler tissues of the body, i.e., skin, superficial nerves, nose, pharynx, larynx, eyes, and testicles. In un-

Table 2. Antimicrobials used to treat atypical mycobacteria *

Mycobacterial species	Primary drug	Secondary drug
M. kansasii	isoniazid with rifampin, with or without ethambutol	ethambutol, streptomycin, pyrazinamide, PAS, cycloserine, ethionamide, kanamycin, and capreomycin
M. fortuitum complex	amikacin and doxycycline	cefoxitin, rifampin, erythromycin, and a sulfonamide
M. avium – intracellularis – scrofulaceum	isoniazid, rifampin, ethambutol, and streptomycin	clofazimine, capreomycin, ethionamide, cycloserine, ansamycin, imipenem, and amikacin
M. marinum (balnei)	minocycline	trimethoprim – sulfamethoxazole, rifampin, and cycloserine

* Adapted from [114].

treated cases, severe disfiguration is observed because of skin infiltration and nerve involvement. There are two distinct types of leprosy, lepromatous and tuberculoid. The more severe, lepromatous type is progressive and leads to lesions involving skin and nerves and to bacteremia. The tuberculoid type is benign and nonprogressive. Treatment usually is lengthy and requires administration of one of the sulfones and the semisynthetic antibiotic, rifampin. Drug resistance to the sulfones has emerged.

2.7.4.1. Antituberculosis Agents

Isoniazid (INN, U.S.P.) [*54-85-3*], isonicotinic acid hydrazide, $C_6H_7N_3O$, M_r 137.14, *mp* 171.4 °C, occurs as crystals (alc).

O=CNHNH$_2$

In 1952 isoniazid was demonstrated to be valuable in treating tuberculosis [115]. It kills *M. tuberculosis* organisms by inhibiting mycolic acid synthesis [116]. At higher concentrations, it may have another mechanism of action against atypical bacteria. The MIC of isoniazid for susceptible *M. tuberculosis* strains ranges from 0.025 to 0.05 µg/mL. Emergence of resistance is observed when the drug is administered alone [115].

Isoniazid is well-absorbed after oral or intramuscular administration and is subsequently distributed throughout the body. The drug is acetylated in the liver, with the rate of acetylation being separated into genetic populations of slow and normal acetylators [116].

Isoniazid is generally well-tolerated. An infrequent major toxicity is hepatitis [117]. Elevations in serum glutamic-oxaloacetic transaminase (SGOT), which are observed in 15 % of patients, disappear with continued therapy. Peripheral neuropathy occurs in 17 % of people receiving 6 mg kg^{-1} d^{-1} of isoniazid. Manifestations of central nervous system toxicity have been reported, as have hypersensitivity reactions. Isoniazid potentiates dilantin toxicity, especially in slow acetylators.

Isoniazid is indicated for all forms of tuberculosis. It is administered in combination with one or more other antitubercular agents, the usual dose being 5 – 10 mg kg^{-1} d^{-1}. Higher doses of the drug (15 mg/kg, orally) can be administered twice weekly along with other agents. This twice-weekly regime can be started after an initial period of daily drug administration and serves to encourage compliance and reduce cost [118].

Synthesis: prepared by condensing pyridine-4-carboxyethylate and hydrazine [119].

Trade names: Cotinazin (Pfizer), Dinacrin (Sterling-Winthrop), INH (Ciba-Geigy), and Nydrazid (Squibb).

Ethambutol hydrochloride (INN, U.S.P.) [*1070-11-7*], [*74-55-5*], (+)-2,2′-(ethylenediimino)-di-1-butanol dihydrochloride, $C_{10}H_{24}N_2O_2 \cdot 2\,HCl$, M_r 277.23, *mp* 194.5 °C.

$$CH_3CH_2-\underset{H}{\overset{CH_2OH}{C}}-NHCH_2CH_2NH-\underset{CH_2OH}{\overset{H}{C}}-CH_2CH_3 \cdot 2\ HCl$$

Ethambutol is a tuberculostatic agent active against most human strains of tubercle bacilli at 8 µg/mL, with ca. 75 % of the strains susceptible at 1 µg/mL. It is not cross-resistant with isoniazid-resistant strains. Resistance develops stepwise when the drug is administered alone. The major use of the drug is in combination with other antitubercular agents to limit the emergence of resistance. Ethambutol has replaced 4-aminosalicylic acid as the companion drug for isoniazid because of its greater potency and lower incidence of adverse effects.

The usual dose is 15 – 25 mg kg^{-1} d^{-1} initially, followed after 60 days by 15 mg kg^{-1} d^{-1} as a single dose. Ethambutol is 75 – 80 % absorbed after an oral dose with peak plasma levels of 5 µg/mL after a 25-mg/kg dose. It is well-distributed in the body and reached CNS levels of 10 – 50 % of serum levels in patients with meningeal inflammation. The drug is excreted mainly unchanged in the urine with ca. 15 % being converted to inactive metabolites following absorption. Neuropathic toxicities are the major *adverse effects* of this drug, especially retrobulbar neuritis, which is common at a dose of 50 mg kg^{-1} d^{-1}.

Synthesis: prepared by warming (+)-2-amino-1-butanol with ethylene bromide at 110 – 115 °C, liberation of the base with potassium hydroxide, and conversion to the dihydrochloride [120].

Trade name: Myambutol (Lederle).

Pyrazinamide (INN, U.S.P.) [*98-96-4*], pyrazinecarboxamide, $C_5H_5N_3O$, M_r 123.11, *mp* 189 – 191 °C, occurs as crystals.

$CONH_2$

Pyrazinamide is a bactericidal antitubercular agent that is used in combination with other antitubercular agents, especially in developing nations [121]. It is a nicotinamide analogue with an unknown mechanism of action. When used alone, resistance develops rapidly. Because of its toxicity, it is used for short-course therapy regimes. Treatment with pyrazinamide can result in hepatotoxicity, especially in the extended dose regimes employed in the early clinical trials, i.e., 40 mg kg^{-1} d^{-1}, but a recommended dose of 20 – 35 mg kg^{-1} d^{-1} appears to be safe [122].

In the United States, pyrazinamide is generally reserved for combination use in patients infected with drug-resistant strains. Mainly because of low cost and convenient once-weekly administration, its use is favored in developing nations, especially in areas of high incidence of drug resistance [123].

Pyrazinamide is well-absorbed when administered orally and reaches levels in body fluids in excess of the MIC for susceptible organisms. It is metabolized by the liver and excreted in the urine. The usual total daily dose is 1.5 – 2.0 g divided into 2 – 4 dosing intervals. Higher levels administered once or twice weekly appear to be clinically effective with no observable liver toxicity.

Synthesis: prepared by ammonolysis of methyl pyrazinoate (from quinoxaline) [124].

Manufacturer: Lederle.

Ethionamide (INN, U.S.P.) [*536-33-4*], 2-ethylthioisonicotinamide, $C_8H_{10}N_2S$, M_r 166.24, *mp* 164 – 166 °C (decomp.), occurs as yellow crystals.

C_2H_5

$S{=}C{-}NH_2$

This nicotinic acid analogue is considered a secondary agent for treating drug-resistant tuberculosis. It is tuberculostatic for most susceptible strains at 0.6 – 2.5 µg/mL. The usual dose is 1 g/d, starting at 250 mg in divided doses and increasing by 125 mg kg^{-1} d^{-1} until the desired dose is reached. A frequent *adverse effect* is gastrointestinal irritation accompanied by nausea and vomiting. Central nervous system side effects, including psychiatric disturbances and peripheral neuropathy, have been reported. In about 5 % of the patients, a reversible hepatotoxicity is observed. Ethionamide is absorbed well from the gastrointestinal tract, giving plasma concentrations of 20 µg/mL. It penetrates both normal and inflamed meninges, giving high CNS levels. Ethionamide metabolism is in the liver, and the metabolites are excreted in the urine.

Synthesis: prepared by addition of hydrogen sulfide to 2-ethylnicotinonitrile in the presence of triethanolamine [125].

Trade name: Trecator – SC (Ives).

***p*-Aminosalicylic acid** [*65-49-6*], PAS (U.S.P.), 4-aminosalicylic acid, $C_7H_7NO_3$, M_r 153.14, *mp* 150 – 151 °C (effervescence), is in the form of minute crystals (alcohol).

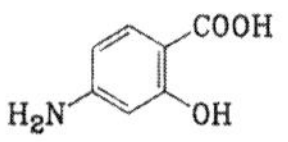

4-Aminosalicylic acid is an inhibitor of mycobacterial growth by impairing of folate synthesis. It has largely been replaced by ethambutol in the standard combination therapy for *M. tuberculosis.* The compound is poorly absorbed when administered orally; thus, a 4-g dose results in peak plasma concentrations of only 7 – 8 μg/mL. The metabolites of PAS are excreted in the urine. The major *adverse effect* is gastrointestinal intolerance, which frequently results in poor patient compliance. 4-Aminosalicylic acid has been reported to produce a lupus-like syndrome. Hypersensitivity reactions are frequent, occurring at a 5 – 10 % rate. The usual adult dose is 10 – 12 g/d in three or four divided doses and 200 – 300 mg kg^{-1} d^{-1} in divided doses for children.

Synthesis: prepared by carboxylation of 3-aminophenol with ammonium carbonate solution at 110 °C under pressure [126].

Trade name: PAS – Heyl (Heyl).

Amithiozone [*104-06-3*], thiacetazone (INN, MI), 4′-formylacetanilide thiosemicarbazone, $C_{10}H_{12}N_4OS$, M_r 236.29, *mp* 225 – 230 °C (decomp.), occurs as minute, pale yellow crystals that darken on exposure to light.

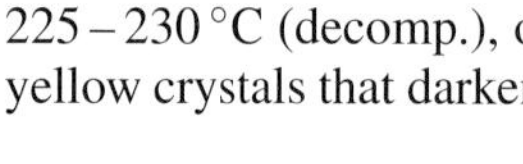

Amithiozone is a second-line drug for the treatment of tuberculosis. This thiosemicarbazone is active against *M. tuberculosis,* inhibiting most strains at 1 μg/mL [127]. The drug is administered orally at 150 mg/d or 450 mg twice weekly. Peak serum concentrations are 1 – 2 μg/mL. Amithiozone should be used in combination with other drugs because resistance develops readily when the drug is used alone. The drug is not available in the United States and has only limited use in Europe because of gastrointestinal irritation and bone-marrow suppression. It is, however, used as a first-line drug in East Africa, where apparent lower incidence of adverse effects in Africans and low cost favor its use.

Synthesis: prepared by treating 4-acetamidobenzaldehyde with thiosemicarbazide in alcohol [128].

2.7.4.2. Antileprosy Agents

Dapsone (U.S.P.) [*80-08-0*], 4,4′-sulfonyldianiline, $C_{12}H_{12}N_2O_2S$, M_r 248.30, *mp* 175 – 176 °C (also recorded 180.5 °C), occurs as crystals (alc).

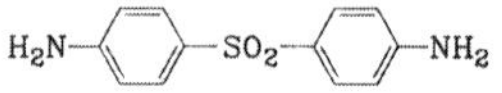

Dapsone is the basic therapeutic agent for the treatment of leprosy. It is used either alone or more frequently as a component of multidrug programs. Dapsone has been employed, primarily as a single chemotherapeutic agent, since the mid-1940s with satisfactory clinical efficacy. Both secondary and primary drug resistance have emerged and are of worldwide concern; thus, combination therapy with the antibiotic rifampin or with clofazimine is recommended. Many dapsone derivatives, all of which share many pharmacological properties, have been synthesized; however, dapsone remains the most clinically useful agent for the treatment of leprosy.

Sulfones share a mechanism of action similar to that of the sulfonamides and are bactericidal for *M. leprae* [129]. Dapsone is slowly and completely absorbed from the gastrointestinal tract with peak levels achieved in 1 – 3 h. The half-life varies over a broad range with a mean of 28 h. The drug is 50 % bound to plasma proteins [130]. Tissue distribution is broad and reabsorption of bile-secreted drug from the gastrointestinal tract tends to extend the residence time of the drug in the circulation. The drug is excreted in the urine.

Daily therapy with 50 mg of dapsone has been successful in adults. The daily dose can be increased to 100 mg if necessary, and twice-weekly doses of 100 – 400 mg have been clinically successful. Therapy is gradually increased to the effective level and should continue for at least 2 years, but may be necessary for the lifetime of the patient. The most common *adverse effect* is hemolysis, generally observed in patients dosed with levels greater than 100 mg/d. Anorexia, nausea, and vomiting may result from oral administration.

Synthesis: [131].

Trade name: Avlosulfon (Ayerst).

Sulfoxone, sodium (INN, U.S.P.) [*144-75-2*], [*144-76-3*] free acid, disodium sulfonylbis(*p*-phenyleneimino)dimethanesulfinate, $C_{14}H_{14}N_2Na_2O_6S_3$, M_r 448.43.

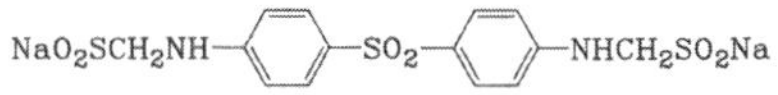

Sulfoxone has been substituted for dapsone in cases in which poor gastrointestinal tolerance restricts dapsone therapy. The maximum daily oral dose of sulfoxone is 660 mg. Sulfoxone is incompletely absorbed from the gastrointestinal tract and large amounts are excreted in the feces. Generally the distribution and excretion of sulfoxone in humans after oral absorption is similar to that observed for dapsone.

Synthesis: prepared by combining 4,4′-sulfonyldianiline with sodium formaldehyde sulfoxylate in acetic acid or alcohol [132].

Trade name: Diasone Sodium Enterab (Abbott).

Acedapsone (INN) [*77-46-3*], 4,4′-sulfonylbis(acetanilide), $C_{16}H_{16}N_2O_4S$, M_r 332.37, *mp* 289 – 292 °C, is a crystalline solid.

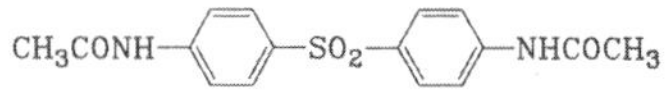

Acedapsone is a long-acting injectable repository derivative of dapsone. It is slowly absorbed when administered intramuscularly. Peak serum concentrations occur 3 – 5 weeks after administration. Acedapsone is metabolized to dapsone in the body, where it has a serum half-life exceeding 40 d [133]. Injections administered 5 times yearly have been employed with promising results [134].

Synthesis: [135].

Clofazimine (INN) [*2030-63-9*], 3-(*p*-chloroanilino)-10-(*p*-chlorophenyl)-2,10-dihydro-2-(isopropylimino)phenazine, $C_{27}H_{22}Cl_2N_4$, M_r 473.40, *mp* 210 – 212 °C, occurs as dark red crystals.

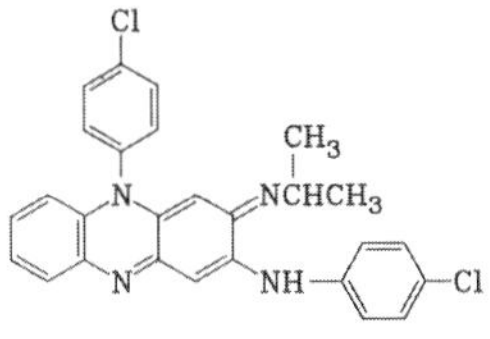

Clofazimine is a phenazine dye with weak bactericidal activity against *M. leprae* [136]. Its main use has been for treating sulfone-resistant infections and people who cannot tolerate sulfones. It also exerts an antiinflammatory effect and prevents the development of erythema nodosum leprosum. The drug is absorbed from the gastrointestinal tract and appears to accumulate in the tissues. Appreciable clinical effects are not observed until 50 days after initiation of therapy. The dose of clofazimine used is 100 – 300 mg spaced at 2-week intervals [137–139]. The major *adverse effects* are skin pigmentation and mild gastrointestinal intolerance; other effects are negligible.

Synthesis: by heating the hydrochloride of the corresponding 2-imino compound with isopropylamine at 80 °C [140].

Trade name: Lampren (Ciba-Geigy).

2.7.5. Miscellaneous Nitroheterocycles Used to Treat Bacterial Infection

Metronidazole (INN, U.S.P.) [*443-48-1*], 2-methyl-5-nitroimidazole-1-ethanol, $C_6H_9N_3O_3$, M_r 171.16, *mp* 158 – 160 °C, occurs as cream-colored crystals.

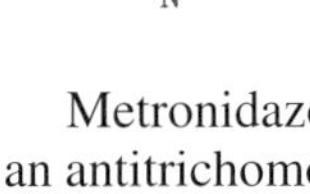

Metronidazole was originally introduced as an antitrichomonal agent and was later found to be useful in treating infections caused by several other protozoal organisms as well as by anaerobic bacteria.

Metronidazole inhibits most anaerobic bacteria at in vitro concentrations of 16 μg/mL or less; however, some organisms are less susceptible [141], [142].

This drug is useful clinically for treating the majority of anaerobic infections with the exception of actinomycosis. A major use of metronidazole is to treat infections caused by the *Bacteroides fragilis* group, the most common etiological agent of serious anaerobic infections. The emergence of resistance to clindamycin, an antibiotic useful in the treatment of anaerobic bacteria, makes metronidazole an even more important agent.

Metronidazole is absorbed rapidly after oral dosing and the serum levels obtained during the elimination phase are similar to those observed when an equivalent dose is administered intravenously. The drug is distributed widely in the body and diffuses well into all tissues including the central nervous system. Five major metabolites are found in the urine, with the hydroxy metabolite being the most important. The usual intravenous treatment regime for the susceptible anaerobic bacteria is a loading dose of 15 mg/kg followed by 7.5 mg/kg 4 times a day. The oral dose is 1 – 2 g/d in 2 – 4 doses at intervals of 6 or 12 h [143]. The maximum daily dosage is 4 g. Generally the drug is well-tolerated; however, major *adverse reactions* involving the central nervous system have been reported. In patients ingesting alcohol, metronidazole may cause reactions similar to those observed with disulfiram.

Synthesis: prepared by nitrating 2-methyl-5-nitroimidazole followed by alkylation with 2-chloroethyl alcohol [144].

Trade names: Flagyl (Searle), Metro I.V. (American McGaw), Satric (Savage), Metronid (Asher), Metryl (Lemmon), and SK-Metronidazol (Smith Kline & French).

Nitrofurantoin (INN, U.S.P.) [*67-20-9*], [*17140-81-7*], monohydrate, 1-[(5-nitrofurfurylidene)amino] hydantoin, $C_8H_6N_4O_5$, M_r 238.16, *mp* 270 – 272 °C (decomp.), occurs as orange-yellow needles (dilute acetic acid).

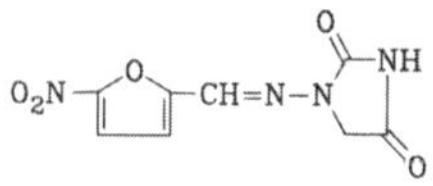

O_2N–(furan)–CH=N–N(hydantoin: O, NH, O)

Nitrofurantoin is used to prevent and treat urinary tract infections caused by susceptible bacteria. The drug is bacteriostatic for most susceptible organisms at 32 μg/mL or less. Nitrofurantoin is active against many strains of *E. coli;* however, most *Proteus* and *Pseudomonas* species and many of the *Enterobacter* and *Klebsiella* species are resistant [145]. Other bacteria commonly found outside the urinary tract are susceptible in vitro to nitrofurantoin, but their susceptibility is of little practical significance [146]. Microorganisms initially sensitive to nitrofurantoin generally do not become resistant during therapy.

Nitrofurantoin is absorbed rapidly and completely after oral administration with therapeutically active concentrations found only in the urinary tract. Approximately one-third of the drug is rapidly excreted in the urine by both glomerular filtration and tubular secretion, with significant reabsorption when the urine is acid [147]. The macrocrystalline form of the drug is absorbed and excreted more slowly. The average dose gives urinary concentrations of ca. 200 μg/mL. The usual adult dose for treating acute and recurrent uncomplicated urinary tract infections is 50 or 100 mg 4 times per day for 2 weeks. The most common *adverse effects* are nausea, vomiting, and diarrhea. Various hypersensitivity reactions have been reported. Other rare, but serious side effects involving various organ systems occur.

Synthesis: prepared by reaction of 5-nitrofurfural diacetate with 1-aminohydantoin in acid solution [148].

Trade names: Furadantin (Norwich Eaton), Macrodantin (Norwich Eaton), and the sodium salt is Dantrirem (Norwich Eaton).

3. Chemotherapy of Protozoan Infections

3.1. Classification of Pathogenic Protozoa

The human parasites that are members of the phylum Protozoa are a diverse group of eukaryotic single-cell organisms. They can be divided into four principal groups:

1) flagellates (Mastigophora)

2) sporozoans (Sporozoa),
3) ciliates (Ciliata)
4) amebas (Sarcodina)

These organisms continue to be among the leading causes of infectious diseases in underdeveloped countries and are now encountered more frequently throughout the world. The ease of travel and the increased use of immunosuppressive drugs have increased the incidence of protozoal infections in the temperate climates. Table 3 lists the major parasitic protozoal human diseases and their causative agents. Excellent reviews and textbooks, which detail the epidemiology, pathogenesis, and chemotherapy of protozoal infections, are available [149–163].

Table 3. Major protozoan organisms causing human disease

Disease	Causative organism
Flagellates	
African trypanosomiasis (sleeping sickness)	*Trypanosoma gambiense* *Trypanosoma rhodesiense*
American trypanosomiasis (Chagas' disease)	*Trypanosoma cruzi*
Visceral leishmaniasis	*Leishmania donovani*
Mucocutaneous leishmaniasis	*Leishmania braziliensis*
Cutaneous leishmaniasis	*Leishmania tropica* *Leishmania mexicana*
Trichomoniasis	*Trichomonas vaginalis*
Lambliasis	*Giardia lamblia*
Sporozoans	
Malaria	*Plasmodium falciparum* *Plasmodium vivax* *Plasmodium malariae* *Plasmodium ovale*
Babesiosis	*Babesia* species
Isosporiasis (human coccidiosis)	*Isospora belli*
Toxoplasmosis	*Toxoplasma gondii*
Cryptosporidium	*Cryptosporidium* species
Pneumocystosis	*Pneumocystis carinii*
Ciliates	
Balantidiasis	*Balantidium coli*
Amebas	
Amebiasis	*Entamoeba histolytica* *Entamoeba polechi* *Dientamoeba fragilis*

3.2. Flagellates

3.2.1. Hemoflagellates

The hemoflagellates of humans include the genera *Trypanosoma* and *Leishmania* (Table 3). There are two distinct types of human trypanosomiasis: (1) African (or sleeping sickness), caused by *T. rhodesiense* and *T. gambiense*, and transmitted by tsetse flies; and (2) American, caused by *T. cruzi,* which is the agent of Chagas' disease and is transmitted by cone-nosed bugs. The *Leishmania* include four species and several subspecies, all of which are transmitted by sand flies.

Trypanosomes appear in the blood as trypanomastigotes, an elongated form with a free flagellum that is an extension of a lateral undulating membrane. Other developmental forms among the hemoflagellates include the amastigote, a leishmanial rounded intracellular stage; the promastigote (formerly called a leptomonad), an elongated flagellated extracellular form without an undulating membrane; and an epimastigote (formerly called the crithidial stage), a flagellated form with a short undulating membrane.

In *Leishmania* only the amastigote and promastigote stages are found, the latter being restricted to the insect. In African trypanosomes only the trypanomastigote is found in humans, whereas the other flagellated stages appear in the tsetse fly. In *Trypanosoma cruzi* all three stages appear in humans, whereas only the trypanomastigote and epimastigote stages are found in the insect.

3.2.1.1. African Trypanosomiasis

3.2.1.1.1. Biology and Epidemiology

The two human parasites causing African trypanosomiasis, i.e., *Trypanosoma rhodesiense* and *T. gambiense,* are believed to have evolved from *T. brucei,* a pathogen of livestock and game animals, and are considered by some to be subspecies of this animal trypanosome. The three forms are indistinguishable morphologically but differ ecologically and epidemiologically.

These parasites are introduced through the bite of the tsetse fly (*Glossina*) and the disease is generally restricted to tsetse fly areas. *Trypanosoma gambiense,* transmitted primarily by *Glossina palpalis,* extends from west to central Africa and produces a relatively chronic infection with progressive central nervous system disease. *Glossina morsitans* transmits *T. rhodesiense* in more restricted geographical areas, primarily to the south and west of Lake Tanganyika (Africa). This disease is generally more acute, leading to death in a matter of weeks to months.

After introduction of the parasite by the tsetse fly bite, a primary lesion is formed; the parasite may spread to lymph nodes and the bloodstream and in terminal stages to the central nervous system, where it produces the syndrome of sleeping sickness. Trypanosomes appear in the blood and tissues as trypanomastigotes. When trypanosomes are ingested with a blood-meal, they undergo a developmental cycle in the tsetse fly, producing metacyclic forms in the salivary glands that are reintroduced with the next blood-meal.

3.2.1.1.2. Chemotherapy

The chemotherapy of African trypanosomiasis centers on three key drugs: (1) pentamidine for chemoprophylaxis, (2) suramin for treatment of the early stages of the disease, and (3) melarsoprol for treatment of the late stages when trypanosomes are present in the central nervous system [164]. A polyamine biosynthesis inhibitor, α-difluoromethylornithine, has been reported to be effective clinically in the treatment of African trypanosomiasis [165]. Detailed reviews have been published on the chemotherapy of the African trypanosomes [166–172].

Drug resistance is not a serious clinical problem despite the fact that producing resistance to the available agents in the laboratory is relatively easy [166–169]. Suramin and pentamidine do not penetrate the central nervous system, a serious defect that limits the clinician to the more toxic arsenic compounds for treating late-stage disease. In addition lack of oral activity, adverse effects, and lack of efficacy against all stages of disease and all strains and species of pathogens limit the available agents.

Suramin (see Section 4.6.9). Suramin is a sulfated naphthylamine that is the drug of choice for the treatment of early stage hemolymphatic African trypanosomiasis. The usual adult dose is a test dose of 100 – 200 mg followed by 5 intravenous doses of 1 g administered on days 1, 3, 7, 14, and 21. Initial plasma concentrations fall rapidly and the drug, tightly bound to serum proteins, remains in low concentration for up to 3 months. Suramin does not penetrate the central nervous system and, thus, does not cure infections of the central nervous system. A transient albuminuria is often observed during treatment, as are shock, febrile reactions, skin lesions, and other toxic effects.

Trade names: Antrypol, Bayer 205, Belganyl, Fourneau 309, Germanin, Moranyl, Naganol, and Naphuride.

Pentamidine (INN, BAN, DCF, NFN) [*100-33-4*], 4,4′-(pentamethylenedioxy)dibenzamidine, $C_{19}H_{24}N_4O_2$, M_r 340.42, *mp* 186 °C (decomp.); pentamidine isethionate (M & B 800) [*140-64-7*], $C_{23}H_{36}N_4O_{10}S_2$, M_r 592.28, *mp* ca. 180 °C, occurs as very bitter crystals.

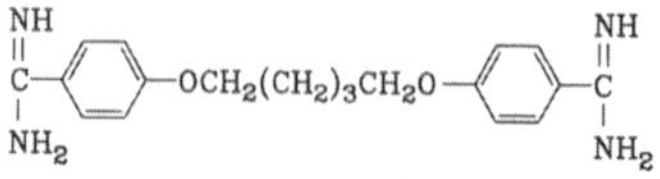

NH=C(NH$_2$)–C$_6$H$_4$–$OCH_2(CH_2)_3CH_2O$–C$_6$H$_4$–C(NH$_2$)=NH

Two salts of this aromatic diamidine are in use: pentamidine isethionate and pentamidine methanesulfonate (lomidine). The doses and adverse effects of the two salts are similar and are discussed together. Treatment usually involves an intramuscular dose of 4 mg kg^{-1} d^{-1} for 10 days. For chemoprophylaxis, the usual dose is 4 mg/kg given intramuscularly every 3 – 6 months. Generally, this drug is more effective against the Gambian disease than the Rhodesian disease. Abdominal cramping may occur during treatment. Pentamidine can cause hypoglycemia and renal damage.

Pentamidine has been shown to be effective in the treatment of pneumonia due to *Pneumocystiscarinii,* an opportunistic protozoal pathogen, and as a secondary drug for treating visceral leishmaniasis.

Synthesis: Saturating an anhydrous alcoholic solution of 4,4′-dicyanodiphenoxypentane with dry hydrogen chloride and allowing it to stand gives pentamidine [173].

Trade names: Lomidine, dimethanesulfonate salt (Specia, France); pentamidine isethionate B.p., diisethionate salt (May & Baker, United Kingdom).

Melarsoprol (INN, BAN, DCF, NFN, MI 9) [*494-79-1*], 2-[4-(4,6-diamino-1,3,5-triazin-2-

yl-amino)phenyl]-1,3,2-dithiarsolane-4-methanol, $C_{12}H_{15}AsN_6OS_2$, M_r 398.33.

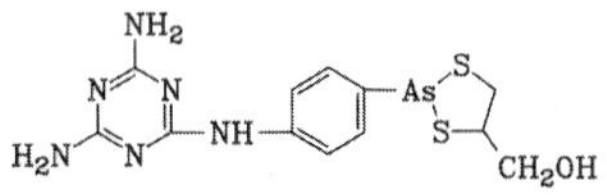

Melarsoprol and the other arsenic compounds discussed in this section are rarely used because of their extreme toxicity. The use of these arsenic compounds had been restricted to patients with overt central nervous system symptoms; however, it is now realized that invasion of the central nervous system occurs early in the course of infection and that late relapses after treatment with drugs other than the arsenic compounds are due to failure to kill organisms present in the central nervous system. Thus, a regime of intravenous suramin followed by one or more courses of intramuscular injections of melarsoprol has been proposed to prevent relapse [155]. The recommended treatment for central nervous system infections is a daily intravenous dose of 2 – 3.6 mg/kg for 3 days followed after a week with 3.6 mg/kg for 3 days and repeated after 10 – 21 days.

Adverse effects with melarsoprol and other arsenic compounds are common and can be quite severe. These include various neuropathies, including the optic nerve, skin rashes, and a syndrome that resembles acute encephalitis and is believed to be due to a reaction that destroys parasites in the brain, which occurs in up to 10 % of the patients. Treatment with arsenic compounds is also complicated by the emergence of drug-resistant strains.

Synthesis: [174].

Tryparsamide (INN; U.S.P. **17,** MI 9) [*554-72-3*], monosodium *N*-(carbamoylmethyl)-arsanilate, $C_8H_{10}AsN_2NaO_4 \cdot 1/2\,H_2O$, M_r 305.10, occurs as hemihydrate, platelets, is slowly affected by light and stable to air.

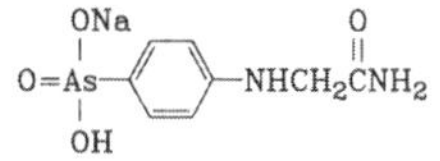

Tryparsamide is a pentavalent arsenic compound that has no in vitro activity, but is most probably metabolized to the trivalent arsenic form after injection. This drug is active only against *T. gambiense,* which limits its use. It crosses the blood – brain barrier, and, thus can be used to treat late-stage trypanosomiasis; however, adverse effects are such that melarsoprol is preferred. Tryparsamide has been employed as an alternate drug of choice in tandem with suramin for treating late-stage disease with central nervous system involvement. The regime employed for tryparsamide is intravenous administration of 30 mg/kg given every 5 days for a total of 12 injections. The regime may be repeated after 1 month. Suramin should be administered intravenously at 10 mg/kg at the same dose regime employed for tryparsamide.

Synthesis: reaction of arsanilic acid with chloracetamide in the presence of sodium hydroxide and sodium carbonate [175].

Trade names: Tryparsam, Tryparsamidium, Glyphenarsine, and Tryparsone.

Berenil [*908-54-3*], diminazene aceturate (INN, BAN, NFN), 4,4′-(1-triazene-1,3-diyl)bis(benzenecarboximidamide)bis(*N*-acetylglycinate), $C_{22}H_{29}N_9O_6$, M_r 515.54, *mp* 217 °C (decomp.), is a yellow solid; free base [*536-71-0*].

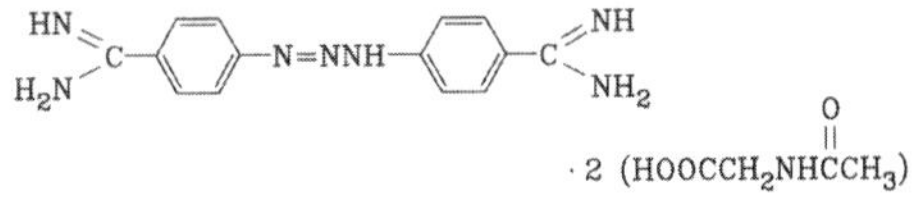

This aromatic diamidine, closely related to pentamidine, was developed originally as a cattle trypanocide. It has been employed effectively in an intravenous dose to treat both Gambian and Rhodesian forms of trypanosomiasis. Recently, the drug was also found to be active on oral administration, with the effective dose being 5 mg/kg taken at 2-day intervals for a total of 3 doses [164]. The drug appears to be relatively well tolerated; however, a persistent albuminuria has been observed in humans following therapy.

Synthesis: [176].

Mel W [*13355-00-5*], melarsonyl potassium (INN, BAN, DCF), melarsenoxide potassium dimercaptosuccinate, dipotassium 2{4-[(4,6-diamino-1,3,5-triazin-2-yl)amino]phenyl}-1,3,2-dithiarsolane-4,5-dicarboxylate, $C_{13}H_{11}AsK_2N_6O_4S_2$, M_r 532.51; free acid [*37526-80-0*].

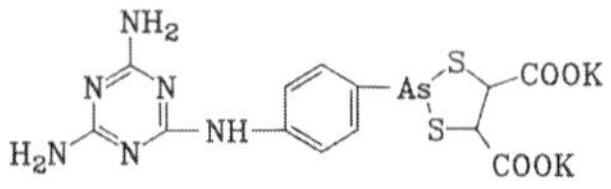

Mel W is an arsenic compound that was developed as a possible replacement for melarsoprol. Similar to other arsenic compound, it crosses the blood – brain barrier and thus can be employed to treat late-stage Gambian and Rhodesian trypanosomiasis. Although it is more water soluble than melarsoprol, Mel W has similar adverse effects to those of melarsoprol. It is almost never employed clinically because of lack of obvious efficacy and lack of safety advantages.

3.2.1.2. American Trypanosomiasis

3.2.1.2.1. Biology and Epidemiology

Trypanosoma cruzi, which causes Chagas' disease, is a parasite of humans and many small animals. This disease is found mainly in Central and South America, but several cases have been reported as far north as the southern United States. This organism is transmitted by triatomid bugs of the family Reduviidae, of which at least 36 species have been found to be naturally infected.

The trypanosomes are ingested with a bloodmeal and undergo development in the intestine of the bug. They eventually give rise to infective metacyclic trypanosomes, which resemble those in vertebrate blood. Transmission usually results from contamination of mucous membranes or skin with infected insect excreta.

After the organisms gain access to the host, they usually multiply in lymphoid tissue. From there, all cells of the body may be infected, but the reticuloendothelial system, the central nervous system, and cardiac muscle are most likely involved. The digestive tract also may be infected, especially the esophagus and colon, leading to development of megaesophagus or megacolon.

The amastigote (leishmania) form is found intracellularly where it multiplies, leading to the formation of pseudocysts, which rupture and release trypanosomes capable of invading other cells or circulating in the blood. Inflammatory reactions resulting from cell damage and infected cells produce many of the disease symptoms.

Acute Chagas' disease occurs mainly in children and only in a small percentage of those people who become infected in endemic areas. Chronic disease develops slowly, with cardiac involvement observed in 20 – 40 % of patients.

3.2.1.2.2. Chemotherapy

A great need exists for agents to treat Chagas' disease. Only two drugs have been reported to have any significant efficacy for treatment (nifurtimox and benznidazole) and none are available for chemoprophylaxis [169], [177–181].

Nifurtimox [*23256-30-6*], 4-[(5-nitrofurfurylidene)amino]-3-methyl-4-thiomorpholine-1,1-dioxide, $C_{10}H_{13}N_3O_5S$, M_r 287.29, *mp* 180 – 182 °C, occurs as orange red crystals.

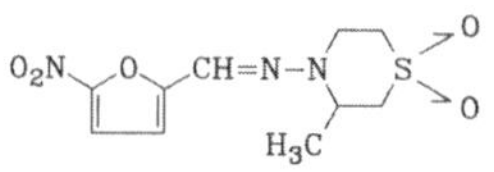

This nitrofuran has been available for 10 years for treating the acute stage of the disease. Its efficacy in chronic Chagas' disease is open to question. Apparently there are variations in response by various strains of the parasite. The usual dose is 8 – 10 mg kg^{-1} d^{-1} in 4 divided oral doses given for 120 days. It is considered by some to be the only available agent for the treatment of Chagas' disease. Nifurtimox is not well tolerated by patients over the 120-day recommended treatment period. *Adverse effects*, such as nausea, weight loss, and memory and sleep disorders, are common, and few patients complete the full treatment period.

Synthesis: prepared from 5-nitrofurfural and 5-amino-3-methyltetrahydro-1,4-thiazine-1,1-dioxide [182].

Trade names: Bayer 2502, Lampit.

Benznidazole [*22994-85-0*], *N*-benzyl-2-nitroimidazole-1-acetamide, $C_{12}H_{12}N_4O_3$, M_r 260.26, *mp* 188.5 – 190 °C, is crystalline in form.

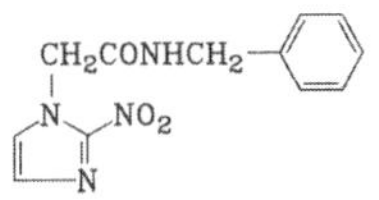

```
CH2CONHCH2—⟨benzene⟩
 |
 N  NO2
(imidazole ring)
 N
```

This orally administered nitroimidazole is claimed to be effective for both acute and chronic Chagas' disease. The recommended dose is 5 mg kg^{-1} d^{-1} for 60 days. No evidence has shown that this drug has any significant advantages over nifurtimox. In doses employed in the initial clinical trials in excess of 5 mg kg^{-1} d^{-1}, serious *adverse effects*, such as polyneuropathy, were observed.

Synthesis: [183].

Trade names: R07-1051, Radanil and Rochagan.

3.2.1.3. Leishmaniasis

3.2.1.3.1. Biology and Epidemiology

Leishmania infections are caused by four species of *Leishmania* (*L. donovani, L. tropica, L. braziliensis,* and *L. mexicana*). These are obligate intracellular parasites during their amastigote (nonflagellar) stage in the human host. One species, *L. donovani,* mainly invades the internal organs whereas the others invade the skin. The Central and South American forms of leishmaniasis, caused by *L. braziliensis* and *L. mexicana* involve the mucous membranes of the nose, mouth, and pharynx, in addition to the skin. *Leishmania* are transmitted by sand flies of the genus *Phlebotomus.* The sand fly acquires the parasite with a blood-meal or from infected skin. In the sand fly, the organism is transformed into the flagellated promastigote stage, which multiplies and is introduced into the skin of a host when the fly feeds.

A broad spectrum of pathology is observed, partly dependent on the site of invasion of the specific parasite and the host's inflammatory response to the organism. The visceral form of the disease, kala-azar, caused by *L. donovani,* produces a progressive disease mainly of the reticuloendothelial system (spleen, liver, and bone marrow). The disease usually results in death if untreated. The cutaneous forms vary from self-healing sores to diffuse progressive disfiguring lesions that can lead to a disseminated disease in hosts with defective cellular immunity.

3.2.1.3.2. Chemotherapy

Two groups of compounds are used to treat visceral and cutaneous forms of leishmaniasis: pentavalent organic antimony compounds and aromatic diamidines. The antifungal antibiotic, amphotericin B, has been shown to be effective in selected situations.

Pentavalent antimony compounds have been used to treat kala-azar since the 1920s. Their introduction was preceded, for a few years, by the use of trivalent antimony compounds, especially tartar emetic. With correct use, cure rates exceeding 90 % have been obtained with pentavalent antimony compounds [184]. Response to therapy for visceral leishmaniasis varies in different geographical locations, with more resistance to therapy being observed in the Sudan than in India.

Pentamidine, a drug used for treating African trypanosomiasis, has been employed for the therapy of visceral leishmaniasis after initial treatment with pentavalent antimony compounds has failed. Although this compound, which now is the only readily available diamidine for use in humans, is effective, the patients may have a high relapse rate. The recommended dose of pentamidine is 2 – 4 mg/kg intramuscular or slow intravenous injections administered weekly or, at most, 3 times weekly, until a clinical and parasitological cure is achieved, usually involving many months of therapy.

Sodium stibogluconate (INN, BAN, DCF) [*16037-91-5*], antimony(V) derivative of sodium gluconate, 2′,4′-*O*-(oxydistibylidyne)bis(D-gluconic acid)-2,4-Sb,Sb′-dioxide trisodium salt · nonahydrate, $C_{12}H_{17}Na_3O_{17}Sb_2 \cdot 9\,H_2O$, M_r 907.6, water-soluble amorphous powder.

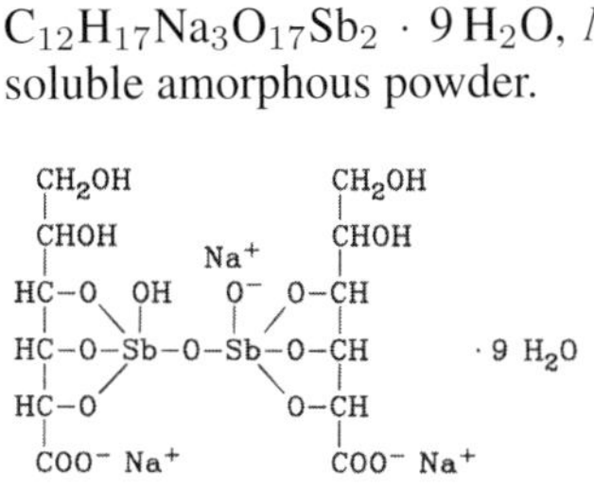

```
 CH2OH                   CH2OH
 |                       |
 CHOH           Na+      CHOH
 |                       |
HC-O   OH       O-   O-CH
 |   \ |        |  /    |
HC-O-Sb-O-Sb-O-CH              · 9 H2O
 |   /           \    |
HC-O              O-CH
 |                    |
 COO- Na+             COO- Na+
```

Sodium stibogluconate is the drug of first choice for the treatment of all forms of leishmaniasis. Some evidence suggests that the antiprotozoal activity of this compound may depend on its reduction to the trivalent antimony compound after treatment. The drug can be administered by either the intravenous or the intramuscular routes.

A large percentage of patients are cured with a single course of therapy consisting of 20 mg kg^{-1} d^{-1} up to a maximum of 800 mg/d. Sodium stibogluconate is tolerated relatively well; however, a disturbing aspect of therapy is the occurrence of sudden death, although it is difficult to determine whether the deaths are related to the clinical disease, the drug therapy, or the interaction of disease and drug.

Synthesis: heating gluconic acid with freshly made antimonic acid paste until it is completely dissolved and then neutralized with sodium hydroxide.

Trade name: Pentostam (Burroughs Wellcome, United Kingdom).

Glucantime [*133-51-7*], meglumine antimonate, 1-deoxy-1-(methylamino)-D-glucitol trioxoantimonate, $C_7H_{18}NO_8Sb$, M_r 365.91.

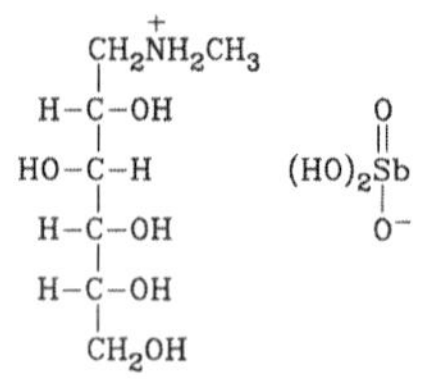

```
          +
    CH2NH2CH3
    |
  H-C-OH               O
    |                  ||
 HO-C-H            (HO)2Sb
    |                  |
  H-C-OH               O-
    |
  H-C-OH
    |
    CH2OH
```

Glucantime is a pentavalent antimony compound related to pentostam. The former drug has no obvious advantage over the latter one and is used interchangeably with pentostam as the drug of choice in some countries. It can be administered intramuscularly or intravenously. The usual dose is 10 or 20 mL of a 30 % solution on alternate days for a total of 200 – 250 mL. As has been observed with pentostam, viable organisms can be recovered from treated patients both during and even after therapy.

Trade names: Protosib, 2168-RP (Rhône-Poulenc).

3.2.2. Intestinal and Urogenital Flagellates

Although the human alimentary and urogenital tracts are colonized by seven species of flagellate protozoa, only two, *Trichomonas vaginalis* and *Giardia lamblia*, are generally considered to be pathogens. Reviews covering broad biological data for these two protozoans are available [185–191].

3.2.2.1. Trichomonas Vaginalis

3.2.2.1.1. Biology and Epidemiology

Of the three species of *Trichomonas* that infect humans, only *T. vaginalis* is pathogenic, causing trichomoniasis. *Trichomonas vaginalis* is a common pathogen in the female genitourinary tract, where it exists only as the trophozoite form. In the female, the infection is normally limited to the vulva, vagina, and cervix. In the male, where infection occurs less often, the prostate, seminal vesicles, and urethra may be involved. Signs and symptoms in the female include profuse vaginal discharge in addition to local tenderness, vulval pruritus, and burning. About 10 % of infected males have a thick, white urethral discharge.

Trichomonas infection is typically transmitted during sexual intercouse, although nonvenereal routes of transmission cannot be excluded. Infection rates vary greatly, but may be quite high in some populations, e.g., 38 – 56 % of symptomatic women attending a venereal disease clinic. Assessing the prevalence of trichomoniasis in men is difficult because most infections in men are asymptomatic. An estimated 3×10^6 American women contract trichomoniasis every year.

3.2.2.1.2. Chemotherapy

Successful treatment of vaginal infections requires the destruction of the trichomonads. Numerous topical preparations are available for treatment; however, they suffer from failure to completely eliminate the parasite in female infections and cannot be applied to infected males. Systemically active drugs are, therefore, the normal treatment of choice. Treatment is necessary both in symptomatic patients and simultaneously in asymptomatic sexual consorts to prevent reinfection or spread of infection.

The effective agents for the systemic treatment of trichomoniasis are all related to the 5-nitroimidazole, metronidazole (see Section 2.7.5). The drugs include tinidazole, nimorazole, ornidazole, secnidazole, and carnidazole. None of these compounds are approved for use in the treatment of trichomoniasis in the United States, and not all are available in every country. No clear clinical advantage has been shown for any of these 5-nitroimidazoles over the others. Cure rates for trichomoniasis are ca. 85 – 95 %, with symptoms usually relieved within a few days. The original recommended regimes for the various compounds for therapy called for oral doses of up to 1 g/d in divided doses for 4 – 10 days; e.g., the recommended dose for metronidazole is 250 mg given 3 times each day for 7 days. Patients now generally receive a shortened regime or a single large oral dose of 2 g. A single dose is as efficacious as multiple-dose treatment and may be superior with some imidazoles. Treatment failures with the 5-imidazoles may be associated with poor serum and local tissue levels or resistant isolates. Cross resistance with metronidazole and the other nitroimidazoles, has been reported. Generally, the 5-nitroimidazoles are free of acute side effects. The most common *adverse effects* are gastrointestinal, and these generally are transient and mild. This group of compounds is mutagenic in bacteria, and tumors have been observed in rodents fed high doses of metronidazole for long periods.

Metronidazole: see Section 2.7.5.

Tinidazole (INN) [*19387-91-8*], 1-[2-(ethylsulfonyl)ethyl]-2-methyl-5-nitroimidazole, $C_8H_{13}N_3O_4S$, M_r 247.27, *mp* 127 – 128 °C, occurs as colorless crystals (benzene).

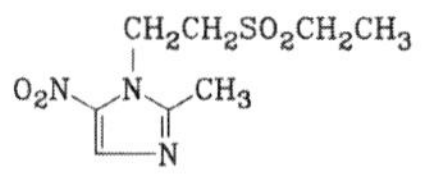

Synthesis: heating 2-methyl-5-nitroimidazole with 2-ethylsulfonylethyl-4-toluenesulfonate for 4 h [192].

Trade names: Fasigyn (Pfizer, USA), Simplotan (Pfizer, FRG).

Nimorazole (INN, BAN, NFN, MI) [*6506-37-2*], 4-[2-(5-nitroimidazolyl-1-yl)-ethyl]morpholine, $C_9H_{14}N_4O_3$, M_r 226.23, *mp* 110 – 111 °C, occurs as crystals (water).

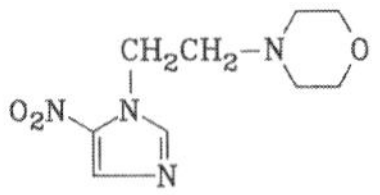

Synthesis: condensing the sodium salt of 5-nitroimidazole with 4-(2-chloroethyl)morpholine in boiling acetone [193].

Trade names: Naxogin (Carlo Erba, Italy).

Ornidazole (INN) [*16773-42-5*], α-(chloromethyl)-2-methyl-5-nitro-1*H*-imidazole-1-ethanol, $C_7H_{10}ClN_3O_3$, M_r 219.63, *mp* 77 – 78 °C, is crystalline in form.

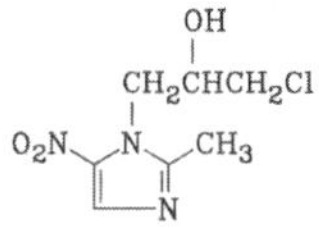

Synthesis: [194].

Secnidazole (INN, DCF, NFN, BAN) [*3366-95-8*], α-2-dimethyl-5-nitro-1*H*-imidazole-1-ethanol, $C_7H_{11}N_3O_3$, M_r 185.18, *mp* 76 °C, occurs as crystals.

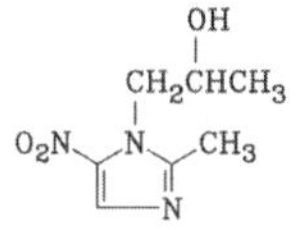

Synthesis: [195].

Code names: PM 185184, RP 14539.

Carnidazole (INN) [*42116-76-7*], *O*-methyl [2-(2-methyl-5-nitro-1*H*-imidazol-1-yl)ethyl]thio-carbamate, $C_8H_{12}N_4O_3S$, M_r 244.27.

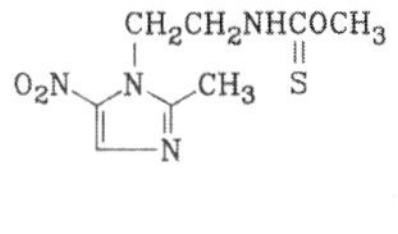

Synthesis: [196].

Code names: R28096 (as hydrochloride), R25831 (as the free base).

3.2.2.2. *Giardia Lamblia*

3.2.2.2.1. Biology and Epidemiology

Giardia lamblia is a flagellated protozoan residing in humans in the upper portion of the small intestine. The trophozoite form of the parasite is converted to the infective cyst stage on its passage to the colon. Ingestion of cysts and the subsequent formation of motile trophozoite forms in the upper small intestine complete the life cycle. This parasite, which was long considered a harmless commensal, is now regarded as the most common intestinal parasite in the United States and United Kingdom.

Endemic infections with *Giardia* are found in every country of the world, but the incidence varies from one country to another. Giardiasis occurs more commonly in children. Transmission is either by direct fecal – oral contamination or by indirect transfer of cysts in food or water. Diarrhea is the most frequent symptom. The disease may be acute or chronic. Anorexia, abdominal cramps, bloating, and weight loss are common symptoms. Most infected individuals do not present evidence of the disease.

3.2.2.2.2. Chemotherapy

Several reviews describe in detail the treatment regimes for *Giardia* infection [191], [197], [198]. Although there is not general agreement as to the optimal chemotherapeutic agent, all three of the following chemical compounds have been reported to give cure rates $\geqq 90$ % with oral administration: metronidazole (and other 5-nitroimidazoles), quinacrine, and furazolidone. In the United States, quinacrine is the most commonly used drug, whereas in many other countries the 5-nitroimidazoles are preferred. Furazolidone is not widely used except as a suspension for children. This agent is generally considered less efficacious than the nitroimidazoles.

Metronidazole has been used to treat giardiasis since 1961, but the original dose regime has been modified. The recommended oral dose for metronidazole is 250 mg 3 times a day for 5 days, although shortened dosing regimes similar to those employed in the treatment of trichomoniasis have been reported to be successful. Other 5-nitroimidazoles that are employed include tinidazole, nimorazole, and ornidazole. Similar to metronidazole, these compounds are generally used with good clinical success in the same dose regimes as have been employed for treating trichomoniasis.

Metronidazole: see Section 3.2.2.1.2.

Tinidazole: see Section 3.2.2.1.2.

Nimorazole: see Section 3.2.2.1.2.

Ornidazole: see Section 3.2.2.1.2.

Quinacrine hydrochloride (U.S.P.) [*6151-30-0*], 6-chloro-9-{[4-(diethylamino)-1-methylbutyl]-amino}-2-methoxyacridine dihydrochloride dihydrate, $C_{23}H_{30}ClN_3O \cdot 2\,HCl \cdot 2\,H_2O$, M_r 508.91, *mp* 248 – 250 °C (*mp* poorly discernable), is a bitter, bright yellow crystal; anhydrous quinacrine [*83-89-6*], atabrine, mepacrine.

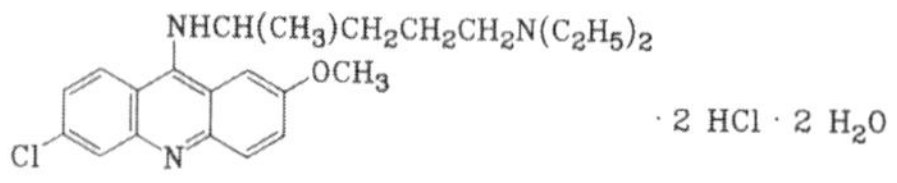

$NHCH(CH_3)CH_2CH_2CH_2N(C_2H_5)_2$; OCH_3 ; Cl ; N ; $\cdot\,2\,HCl \cdot 2\,H_2O$

Quinacrine is an acridine derivative previously employed to treat malaria and tapeworm infections. The usual adult dose is 100 mg 3 time a day given for 5 days. The pediatric dose is 2 mg/kg 3 times a day for 5 days with a maximum of 300 mg/d. Gastrointestinal *adverse effects* as well as malaise and headache, occur with administration of this compound. Quinacrine is excreted in the urine, imparting a deep yellow color. Approximately 4 – 5 % of the patients develop yellow skin staining at the dose used for giardiasis. Quinacrine can induce red blood cell hemolysis in G-6-PD deficient patients. Prolonged high-dose therapy with quinacrine has caused rare instances of retinopathy similar to that observed with chloroquine.

Synthesis: condensing 1-(diethylamino)-4-aminopentane with 3,9-dichloro-7-methoxyacridine [199].

Furazolidone (INN, BAN, U.S.P.) [*67-45-8*], furoxone, 3-[(5-nitrofurfurylidene)amino]-2-oxazolidinone, $C_8H_7N_3O_5$, M_r 225.16, *mp* 275 °C (decomp.), occurs as yellow crystals.

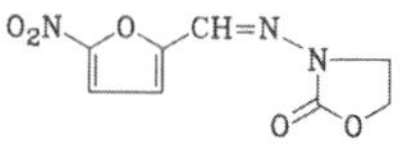

Furazolidone is a nitrofuran that is active against *G. lamblia* and various gram-negative bacteria, including *Salmonella, Shigella,* and *Vibrio cholerae.* The recommended adult oral dose is 100 mg 4 times a day for 7 – 10 days. The pediatric dose is 1.25 mg/kg 4 times a day for 7 – 10 days. Compared with other agents used for treating *Giardia* infections, furazolidone is somewhat less active. Furazolidone treatment can result in gastrointestinal and central nervous system *adverse effects.* Furazolidone turns the urine brown and can cause red blood cell hemolysis and mild, reversible anemia in individuals with G-6-PD deficiency. Furazolidone, similar to other nitrofurans, has carcinogenic potential. Adenocarcinoma of the lung in mice and of the mammary gland in female rats has been reported with long-term administration of furazolidone.

Synthesis: [200].

Trade names: Furox (Smith Kline Beckman), Furoxone (Norwich Eaton).

3.3. Sporozoans

Sporozoans have a complex life cycle that often involves two hosts (e.g., arthropod and human). The Coccidia, a subclass of essentially intestinal Sporozoa, and the hemosporidians, which include the malaria parasite, are animal and human parasites. *Toxoplasma,* a parasite of cats, is a common human parasite. *Babesia,* a tick-borne protozoan and a common animal parasite, is a rare human pathogen (these organisms are no longer considered sporozoans, but will be covered in this section). *Pneumocystis carinii* and *Cryptosporidium* have emerged as important human pathogens, especially in immunocompromised patients.

3.3.1. Plasmodia

3.3.1.1. Biology and Epidemiology

Malaria infections are distributed widely in countries of Africa, Asia, and Latin America, with an estimated worldwide prevalence of 100×10^6 cases associated with ca. 1×10^6 deaths each year. At least five species of *Plasmodium* can infect humans: *P. vivax, P. ovale, P. malaria, P. falciparum,* and *P. knowlesi* (only in Malaysia). In addition, at least two species of nonhuman primate plasmodia are transmissible to humans experimentally. Diagnosis of infection is made on the basis of morphological characteristics of specific species. The identification of the infecting organism is important in determination of the chemotherapeutic approach. Equally important in the therapy and prophylaxis of infection is an understanding of the life cycle and course of infection of the malaria parasites.

Transmission to humans is by the blood-sucking bite of various species of mosquitoes of the genus *Anopheles,* in which the sexual or sporogonic cycle of development occurs. The asexual cycle (schizogany) takes place in humans. The mosquito introduces the infective sporozoites, which quickly invade the liver parenchymal cells (the preerythrocytic cycle). After further development, numerous asexual progeny (the merozoites) enter the bloodstream and invade the erythrocytes. Multiplication in the red blood cell is characteristic for each species, resulting in synchronous destruction of host cells (the erythrocytic cycle). Successive production of merozoites occurs every 48 h (*P. vivax, P. ovale,* and *P. falciparum*) or at 72-h intervals (*P. malariae*). The *P. falciparum* infections are generally confined to the red blood cells after the first liver cycle; thus, untreated infections will terminate spontaneously, usually in 6 – 8 months, or end fatally. The other three species continue to multiply in liver cells, and persistence of the parasite in these cells long after the parasites have disappeared from the bloodstream is observed. During the erythrocytic cycle, certain merozoites become differentiated as male or female gametocytes. Thus, the sexual cycle begins in humans, but for continuation of the cycle, the gametocytes must be ingested by mosquitoes. Development time of the various stages in the mosquito's stomach wall ranges from 8 to 14 days, depending on the malaria species. The infective stage is the sporozoite, which migrates to the salivary gland of the mosquito and is injected with a blood-meal.

The *pathogenic mechanisms* resulting in clinical illness in humans can be divided into four processes: (1) fever, (2) anemia, (3) tissue hypoxia, and (4) immunopathologic events. Cyclic fever and its physiological consequences involve rupture of erythrocytes and release of schizonts. Anemia is a common complication of malaria and mainly results from the rupture of red blood cells during schizogony, but other factors, such as autoimmune mechanisms, may contribute to the anemia. Tissue hypoxia, resulting from alterations in microcirculation and anemia, may cause serious complications including renal failure, pulmonary edema, and cerebral dysfunction in *P. falciparum* infections. Immunologic response to malaria infection can result in clinical disease. Immune-complex glomerulonephritis and greatly enlarged spleens are two examples of this phenomenon.

3.3.1.2. Chemotherapy

Antimalarial drugs can be divided into two groups on the basis of their mechanism of action: (1) The aminoquinolines, such as chloroquine, apparently exert their effect by intercalation into parasite DNA. This is not the only mechanism of action of these compounds, because mefloquine does not intercalate DNA. (2) The other group of compounds inhibits the synthesis of folic acid from 4-aminobenzoic acid. This group includes chloroguanide, pyrimethamine, and their derivatives, as well as the sulfonamides and sulfones. Quinine, a natural product and a mainstay of therapy for more than 60 years, still remains an important chemotherapeutic agent, especially for treating malaria parasites that are resistant to the newer synthetic agents.

Antimalarial drugs have been grouped in at least six different categories according to use, including (1) causal prophylaxis, (2) suppressive treatment, (3) clinical cure or treatment of the acute attack, (4) radical cure, (5) suppressive cure, and (6) gametocidal therapy [201]. True *causal prophylactic agents* should be capable of killing sporozoites prior to their entry into red blood cells. Although no such agent is available, several agents, e.g., primaquine and chloroguanide, have activity against the preerythrocytic stages of *P. falciparum. Suppressive therapy* inhibits development of the erythrocytic stages, thus preventing clinical symptoms. Chloroquine, chloroguanide, and pyrimethamine are suppressive; however, insensitivity or drug resistance in malarial strains in certain localities has created problems. A *clinical cure* can be achieved by agents that interrupt the development of the intracellular schizont in the red blood cell. The 4-aminoquinoline derivatives, chloroquine and amodiaquin, are the major drugs in this category, although the slower-acting chloroguanide and pyrimethamine are also highly active schizonticides. *Radical cure* involves elimination of both erythrocytic and exoerythrocytic parasites. Vivax malaria can be treated with the 8-aminoquinoline derivatives, of which only primaquine is used currently. Radical cure of falciparum malaria is relatively easy to achieve by continuation of treatment. A *suppressive cure* involves complete elimination of the malaria parasites by treatment that exceeds the life span of the parasite. *Drugs that kill gametocytes* directly are not available, but chloroguanide and pyrimethamine prevent the development of gametocytes in mosquitoes.

Drug resistance to all of the synthetic agents used in the treatment and prophylaxis of malaria has been observed in many areas of the world. *Plasmodium falciparum* is resistant to amodiaquine and chloroquine and to the combination of sulfadoxine and pyrimethamine. This therapeutic problem extends to South America, 13 African nations, south of the Equator, and to the South Asian continent, from Pakistan to eastern India. When used alone, quinine has lost some of its activity against falciparum malaria in those areas where it has been used indiscriminately or with poor compliance. Treatment of the other malarial species with 4-aminoquinoline derivatives, such as amodiaquine and chloroquine, still remains effective.

Chloroquine (INN, U.S.P.) [*54-05-7*], 7-chloro-4-(4-diethylamino-1-methylbutylamino)-quinoline; $C_{18}H_{26}ClN_3$, M_r 319.88, *mp* 87 °C; diphosphate, *mp* 193 – 195 °C or 215 – 218 °C (two modifications); sulfate, *mp* ca. 207 °C.

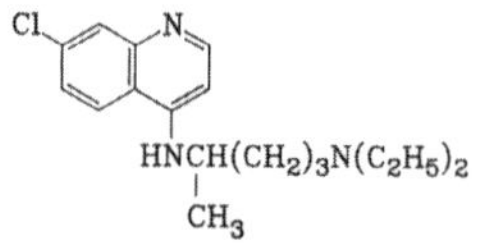

Cl N HNCH(CH$_2$)$_3$N(C$_2$H$_5$)$_2$ CH$_3$

Chloroquine, a 4-aminoquinoline, is highly effective against the asexual erythrocytic forms of *P. vivax* and *P. falciparum,* and gametocytes of *P. vivax.* In human vivax malarias, chloroquine has no prophylactic or radically curative value; however, it is effective in terminating acute attacks of vivax malaria. When administered continuously for long periods, it acts as a suppressive agent. It is highly effective in controlling acute attacks of falciparum malaria and generally cures the disease. In certain parts of the world, drug-resistant strains of *P. falciparum* limit the use of chloroquine.

Chloroquine is well absorbed when administered orally. It is highly bound to tissues, especially liver, spleen, kidney, and lung; about 55 % of the drug is bound to plasma proteins. The drug is excreted slowly in the urine, with ca. 70 % being recovered as parent compound. Because of high tissue binding, a loading dose is necessary to achieve adequate plasma concentration. After a single dose, the half-life of the drug in plasma is 3 days, with a longer half-life achieved on multiple doses. Chloroquine at the dose employed for prophylaxis of acute malarial attacks causes relatively few *adverse effects*, mainly pruritus and gastrointestinal discomfort.

The oral dose for suppression of sensitive malarial strains in adults is 300 mg of the free base weekly, starting 2 weeks prior to and while present in a malarious area and for 6 weeks after leaving the area. For oral treatment of acute malaria in adults, 600 mg of base is given initially, followed by 300 mg 6 h later and 300 mg at 24 and 48 h. Primaquine should be given after chloroquine doses when treating malaria due to *P. vivax* and *P. ovale* or for radical cure following exposure.

Chloroquine is of use in treating extraintestinal amebiasis, and is also reported to be of some value in the treatment of *Giardia* and *Babesia* infections.

Synthesis: condensation of 4,7-dichloroquinoline with 1-(diethylamino)-4-aminopentane [202].

Trade names: Aralen hydrochloride, Aralen phosphate (Sterling Winthrop), Resochin (diphosphate, Bayer), Nivaquine (sulfate, Specia, France).

Amodiaquine (INN, U.S.P.) [*86-42-0*], 4-[(7-chloro-4-quinolyl)amino]-α-(diethylamino)-*o*-cresol, $C_{20}H_{22}ClN_3O$, M_r 358.87, *mp* 208 °C, is crystalline in form; dihydrochloride dihydrate, *mp* 150 – 160 °C (decomp.), is a yellow, bitter crystal.

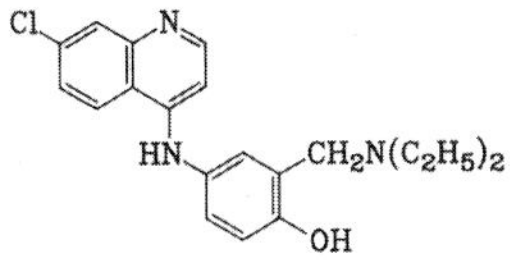

Amodiaquine, a congener of chloroquine, is claimed to have superior activity against some strains of *P. falciparum* that are partially resistant to chloroquine. It is used for treating overt malaria attacks and for suppression. The observed *adverse effects* and the frequency of these effects are similar to chloroquine, i.e., diarrhea, vomiting, and vertigo. Dosing also is similar to that employed for chloroquine. For suppressive therapy, the unit dose is 400 mg of the free base (520 mg of dihydrochloride). For treatment of acute attacks, 600 mg of the base is given initially with subsequent daily doses of 400 mg for 2 days.

Synthesis: Condensation of 4,7-dichloroquinoline and 4-acetamido-α-diethylamino-*o*-cresol gives amodiaquin dihydrochloride dihydrate [203].

Trade name: Camoquin hydrochloride (Parke-Davis).

Hydroxychloroquine sulfate (U.S.P.), hydroxyquinine (INN) [*747-36-4*], 2-({4-[(7-chloro-4-quinolyl)amino]pentyl}ethylamino)ethanolsulfate (1 : 1 salt), $C_{18}H_{26}ClN_3O \cdot H_2SO_4$, M_r 439.95, *mp* (usual form) ca. 240 °C, *mp* (other form) ca. 198 °C, is a white crystalline odorless powder with a bitter taste; hydroxychloroquine [*118-42-3*].

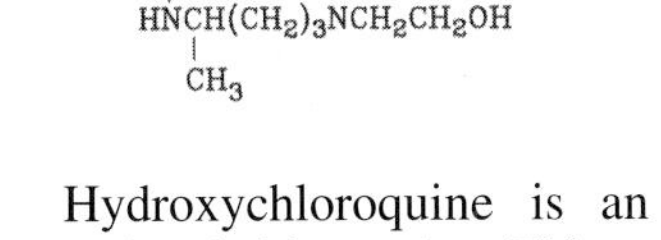

Hydroxychloroquine is an *N*-ethyl-β-hydroxylated chloroquine. This compound has an activity and safety profile that is similar to chloroquine. A dose of 400 mg of hydroxychloroquine sulfate is equivalent to 500 mg of chloroquine phosphate.

Synthesis: prepared by reacting a mixture of 4,7-dichloroquinoline, phenol, and N'-ethyl-N'-β-hydroxyethyl-1,4-pentadiamine at 125 – 130 °C [204].

Trade name: Plaquenil sulfate (Sterling-Winthrop, USA).

Primaquine (INN) [*90-34-6*]; primaquine phosphate (U.S.P.) [*63-45-6*], 8-[(4-amino-1-methylbutyl)amino]-6-methoxyquinoline phosphate (1 : 2 salt), $C_{15}H_{21}N_3O \cdot 2\,H_3PO_4$, M_r 455.34, *mp* 197 – 198 °C, occurs as yellow crystals (90 % ethanol).

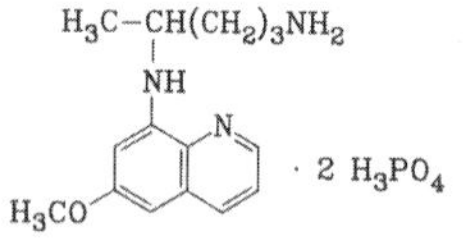

$H_3C-CH(CH_2)_3NH_2$ / NH / H_3CO · 2 H_3PO_4

Primaquine is an α-aminoquinoline used for prevention of malarial relapses and for radical cure of *P. vivax* and *P. ovale* malaria by acting on the exoerythrocytic stage of the parasite. It also destroys the gametocytes of these species. It has no significant activity against the asexual blood forms of *P. falciparum.*

Primaquine, similar to all of the other α-aminoquinolines, is absorbed rapidly when administered orally and is rapidly metabolized. Plasma levels peak at 6 h. Minor *adverse effects* include nausea, abdominal discomfort, and headache. Severe hemolytic reactions in people with a glucose-6-phosphate dehydrogenase deficiency of the type found in those of Mediterranean ancestry have been observed. Most blacks have this enzyme deficiency, but it is usually confined to older erythrocytes and thus the hemolytic reaction is less severe.

Primaquine is supplied as the phosphate salt with 26.3 mg being equivalent to 15 mg of free base. For prevention of attack after departure from areas where *P. vivax* and *P. ovale* are endemic, a daily oral dose of 15 mg of base along with the last 2 weeks of chloroquine prophylaxis is recommended. For prevention of relapses to vivax and ovale malaria, a daily dose of 15 mg/d for 14 days or a weekly dose of 45 mg of base for 8 weeks is employed.

Synthesis: 6-Methoxy-8-aminoquinoline is condensed with 1-phthalimido-4-bromopentane, and the phthalyl group is cleaved by heating with hydrazine in ethanol and hydrochloric acid [205].

Pyrimethamine (INN, U.S.P.) [*58-14-0*], 2,4-diamino-5-(4-chlorophenyl)-6-ethylpyrimidine, $C_{12}H_{13}ClN_4$, M_r 248.71, *mp* 233 – 234 °C, is crystalline in form.

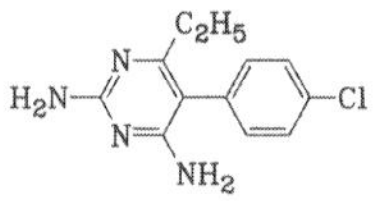

C_2H_5 / H_2N / N / N / Cl / NH_2

Pyrimethamine is a 2,4-diaminopyrimidine that has high affinity for the enzyme dihydrofolate reductase from the malaria parasite and thus interferes with folate synthesis. Trimethoprim, a related compound with good antibacterial activity, also has antimalarial activity. Pyrimethamine by itself has little value in treating a primary attack of malaria. However, combined with sulfadoxine (Fansidar), malaria has been prevented in subjects in areas where there is drug-resistant falciparum malaria. The combination oral product of 25 mg of pyrimethamine and 500 mg of sulfadoxine is given once weekly during exposure. It is also employed in combination with quinine sulfate for treatment of chloroquine-resistant falciparum malaria. The dose employed for the combination is 650 mg of quinine 3 times daily and 25 mg of pyrimethamine twice daily for 3 days. Pyrimethamine is well absorbed when administered orally. It is eliminated slowly with a plasma half-life of 4 days. The dose of 25 mg weekly produces few adverse effects. Excessive doses may result in a reversible megaloblastic anemia resembling folic acid deficiency.

Synthesis: prepared by condensing 3-isobutoxy-2-(4-chlorophenyl)pent-2-enonitrile with guanidine nitrate in the presence of sodium methylate [206].

Trade name: Daraprim (Burroughs Wellcome).

Chloroguanide [*500-92-5*], chloroguanide hydrochloride (U.S.P. **14**) [*637-32-1*], proguanil (INN), 1-(4-chlorophenyl)-5-isopropylbiguanide hydrochloride, $C_{11}H_{16}ClN_5 \cdot HCl$, M_r 290.20, *mp* 243 – 244 °C, is a white powder with a bitter taste.

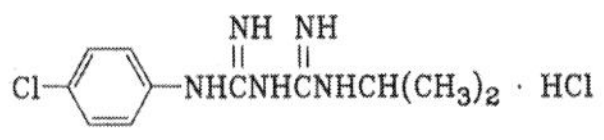

NH NH
Cl–C6H4–NHC(=NH)NHC(=NH)NHCH(CH3)2 · HCl

Chloroguanide is used to treat overt clinical vivax and falciparum malaria; however, response to treatment is slower than that observed with most other antimalarial agents. Chloroguanide is not recommended for treatment of an acute attack of falciparum malaria. Although it is of use in acute vivax malaria, it offers no advantages over other available agents. It is a causal prophylactic, suppressive, and radical cure agent in falciparum malaria. Chloroguanide is active against developing preerythrocytic stages of some malarias and is reported to sterilize gametocytes. Resistance to chloroguanide greatly compromises its usefulness.

Chloroguanide is converted metabolically to a triazine derivative that inhibits the enzyme dihydrofolate reductase. It is slowly absorbed from the gastrointestinal tract with peak serum concentrations achieved at 2 – 4 h after oral administration. About 50 % of the drug is excreted in the urine and 10 % directly into the feces. The prophylactic dose is 100 – 200 mg daily for nonimmune subjects. Less drug (300 mg weekly) is employed for partly immune individuals. For treatment of acute attack of vivax malaria, an initial dose of 300 – 600 mg is followed by a daily dose of 300 mg, usually for 5 – 10 days. Chloroguanide is a relatively safe drug at the doses employed for malaria treatment. Large daily doses of 1 g may cause vomiting, abdominal pain, and diarrhea.

Synthesis: prepared by heating 4-chlorophenyldicyandiamide with isopropylamine [207].

Trade name: Paludrine (ICI, United Kingdom).

Mefloquine (INN) [*53230-10-7*], (DL-*erythro*-α-2-piperidyl-2,8-bis(trifluoromethyl)-4-quinolinemethanol, $C_{17}H_{16}F_6N_2O$, M_r 378.32, *mp* 178 – 178.5 °C; mefloquine · HCl [*51773-92-3*], *mp* 259 – 260 °C.

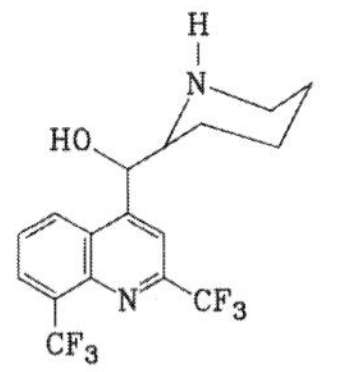

Mefloquine is a 4-quinolinemethanol that is currently well along into clinical trials, especially in Asia and South America, where multiple drug-resistant *P. falciparum* is a major problem [208, 209]. This compound has an unusually long half-life in humans (17 days) and may be efficacious for use as a chemoprophylactic when given weekly. It has proved highly effective in curing infections caused by *P. falciparum* when administered orally. Liver concentrations are high and prolonged, and the drug is excreted mainly in the feces. A quinate salt is available with a markedly increased water solubility and an improved pharmacokinetic profile.

Synthesis: [210].

Trade name: Mefloquine Quinate, $C_{17}H_{16}F_6N_{20}$ · $C_7H_{12}O_6$ (Smith Kline Beckman).

Halofantrine [*69756-53-2*], [±]-halofantrine [*66051-63-6*]; halofantrine hydrochloride (INN) [*36167-63-2*], 1,3-dichloro-α-[2-(dibutylamino)-ethyl]-6-(trifluoromethyl)-9-phenanthrenemethanol hydrochloride, $C_{26}H_{30}Cl_2F_3NO \cdot HCl$, M_r 536.89, *mp* 93–96 °C, *mp* 203–204 °C, occurs as crystals.

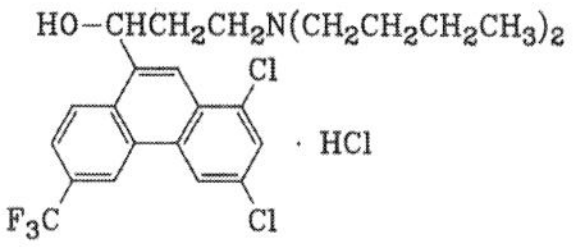

HO–CHCH2CH2N(CH2CH2CH2CH3)2 · HCl

Halofantrine is a 9-phenanthrene methanol currently under development for the treatment of drug-resistant falciparum malaria. It is curative with minimal adverse effects in nonimmune subjects infected with Vietnam or Cambodian strains of *P. falciparum* or Chesson strain of *P. vivax*. The oral dose employed in these studies was 250 mg every 6 h for 12 doses [211]. The compound is rapidly but incompletely absorbed when administered orally.

Synthesis: [212].

Trade name: Halofantrine-β-glycerophosphate, $C_{26}H_{30}Cl_2F_3NO \cdot C_3H_9O_6P$ (Smith Kline Beckman).

3.3.2. Babesia

Babesia are intraerythrocytic protozoan parasites that are transmitted by ticks. They are mainly parasites of domestic and wild animals, although on occasion they can be transmitted to humans. Infections tend to be self-limiting and are characterized by fever, hemolytic anemia, and hemoglobinuria. No effective therapy exists for these infections.

3.3.3. Isosporiasis

The sporozoan order Coccidia contains a number of parasites that invade the intestinal mucosa at one stage in their life cycle. *Isospora belli,* usually acquired by fecal contamination of food or water, is a member of this group. Although a rare human pathogen, *T. belli* can cause serious gastrointestinal disease. Treatment is usually with antifolates, such as pyrimethamine and sulfadiazine. The drug of choice is trimethoprim – sulfamethoxazole. The usual oral dose of this combination is 160 mg of trimethoprim and 800 mg of sulfamethoxazole four times daily for 10 days then twice a day for 3 weeks. *Isospora hominis,* now called *Sarcocystis hominis,* is a related parasite that can cause intestinal infection in humans. This organism, similar to toxoplasma, has an asexual stage in the muscles of many mammals. Human infection is acquired from eating improperly cooked beef or pork containing *Sarcocystis.* Treatment is similar to that employed for *I. belli.*

3.3.4. Toxoplasmosis

Toxoplasma gondii is a sporozoan of the order Coccidia. The parasite is an obligate intracellular organism that exists in three forms including the tachyzoites (formerly trophozoites), tissue cysts, and oocysts. *Toxoplasma gondii* is a ubiquitous parasite with a worldwide distribution and a capability of infecting a wide range of animals and birds. The final hosts are members of the cat family. Infection is by ingestion of either tissue cysts or oocysts, releasing viable organisms that invade the epithelial cells of the intestine, where they undergo an asexual cycle and a sexual cycle in the cat. Many oocysts are shed in the feces and sporulate outside the cat.

Ingestion of primarily lamb or pork containing tissue cysts or other food products contaminated with oocysts is the major means of transmission. Active infection during gestation can result in congenital infection. Toxoplasma infection can be acute or chronic, symptomatic or asymptomatic. Most infected adults show no symptoms. Chronic infections can be latent and later exacerbate in immunocompromised individuals, resulting in severe disease, such as encephalitis, myocarditis, and pneumonitis. Congenital infection may lead to subsequent disease, such as impaired vision, hearing loss, or neurologic disorders.

The most effective treatment for toxoplasmosis is a combination of pyrimethamine and sulfadiazine. Although the tachyzoites are susceptible to these agents, the tissue cyst is resistant to available agents. A loading regime of pyrimethamine at 100 mg kg^{-1} d^{-1} twice a day for 2 days followed by 25 mg every other day is the recommended treatment for adults with significant infection. A synergistic sulfa drug, such as sulfadiazine, sulfamethazine, or sulfamerazine, is usually given in combination with pyrimethamine. The antibiotics spiramycin and clindamycin or trimethoprim – sulfamethoxazole are less active than the combination of pyrimethamine and a sulfa drug.

3.3.5. Cryptosporidium

This coccidian protozoa is a significant cause of death in immunocompromised patients. The parasite, long known in domestic animals, is the same organism that causes disease in humans. Infection is acquired by ingestion of oocysts, which excyst and release sporozoites that invade and replicate in the intestinal microvilli. Male and female gametes are produced, initiating the sexual cycle. Oocysts are then formed and can be passed in the feces or can release sporozoites to cause autoinfection. In individuals with normal host defenses, the organism may not cause symptoms or may cause a transient diarrhea and gastrointestinal disease that is generally self-limiting. The disease is more serious in immunocompromised individuals and is a major contribution to death in patients with acquired immune deficiency syndrome (AIDS). No effective chemotherapy exists for cryptosporidiosis.

3.3.6. Pneumocystis Carinii

This protozoan causes infection in the immunocompromised host and is the most common pathogen of AIDS patients, accounting for 43 % of reported opportunistic diseases. The disease is usually limited to the lungs, where it causes a diffuse pneumonitis. The parasite has been found in the lungs of a wide variety of animals and is distributed globally. Subclinical infections in normal individuals are probably common. The untreated infection in the immunocompromised patient is 90 – 100 % fatal. The drug of choice for treating pneumocystis infection is the combination of trimethoprim and sulfamethoxazole. The recommended dose is 20 mg kg^{-1} d^{-1} of trimethoprim with 100 mg kg^{-1} d^{-1} of sulfamethoxazole given orally or intravenously in 4 doses for 14 days. This combination therapy results in a high incidence of rash, neutropenia, fever, and diarrhea in AIDS patients. Pentamidine isethionate is an alternative choice, but is associated with significant adverse effects. The recommended dose is 4 mg kg^{-1} d^{-1} administered intramuscularly for 12 – 13 days.

3.4. Ciliates

Infections in humans caused by ciliated protozoans are rare. Only one parasite, *Balantidium coli*, causes disease in humans. The ciliated form of this parasite penetrates the mucosa of the colon and multiplies in the submucosal tissues. This results in colitis, with mucus and blood in the feces. The parasite can form cysts when conditions are not ideal for penetration and it is the cyst that allows for transmission of this disease, usually from animals (swine) to humans. The disease is treated with the antibiotics tetracycline or paromomycin or with metronidazole (750 mg twice a day for 5 days).

3.5. Amebas

3.5.1. Biology and Epidemiology

The disease amebiasis is worldwide in distribution and is almost always associated with poor sanitary conditions. The etiological agent, *Entamoeba histolytica*, occurs in three stages: (1) the inactive cyst, (2) the intermediate precyst, and (3) the ameboid trophozoite, which is the only stage found in the tissues. The ameboid form is also found in liquid feces during amebic dysentery. Multiplication among trophozoites occurs by binary fission. After ingestion of infective cysts in food or water contaminated with feces, the cysts are activated in the stomach, and development takes place during passage to the large intestine.

A population of lumen-dwelling trophozoites, capable of invading the intestinal epithelium, emerge from the cyst. Most infections are without symptoms, although when tissue invasion occurs (in ca. 10 % of the cases), there is disease associated with the infection. Asymptomatic infected persons harbor lumen-dwelling amebas that produce cysts, which are passed in the feces. Diseased individuals usually pass trophozoites and cysts in the feces. Extraintestinal amebiasis is observed in ca. 4 % or more of clinical infections, usually taking the form of amebic hepatitis or liver abscess. Abscesses may also occur rarely in other areas, including the lungs, brain, and spleen.

Entamoeba histolytica must be distinguished from four other amebas that are also intestinal parasites of humans. These include the common ameba *Ent. coli; Dientamoeba fragilis,* the only intestinal ameba other than *Ent. histolytica* suspected of causing diarrhea, but not of invading the tissues; *Iodamoeba butschlii;* and *Endolimax nana.*

Primary amebic meningoencephalitis has been reported in less than 100 cases with amebas living free in soil and in water. Two genera of ameba, *Naegleria* and *Acanthamoeba,* have been associated with pathogenicity. Most cases of infection have developed in children who were swimming in contaminated outdoor pools. *Naegleria* meningoencephalitis infection is rapidly fatal and does not respond to the available amebicides. A few patients have been reported to respond to intravenous and intrathecally administered amphotericin B.

3.5.2. Chemotherapy

The approach to treating amebiasis is in part based on the location of the amebic organism and includes treatment of the asymptomatic individual who is passing cysts. The criterion for cure in intestinal amebiasis is the elimination of the organism. This is usually achieved with tetracycline and an 8-hydroxyquinoline derivative or diloxanide furoate. The antibiotic paromomycin also may be employed. For those infections involving only the bowel wall and resulting in acute amebic dysentery, a variety of agents are employed. Symptomatic amebiasis involving the intestine can be treated with a nitroimidazole, such as metronidazole or tinidazole, a course of therapy usually resulting in 90 % cure. Dehydroemetine, a semisynthetic derivative of the alkaloid emetine, can be used for the rapid relief of symptoms in severely ill patients. As an alternative to metronidazole, combination therapy of dehydroemetine with tetracycline or paromomycin has been employed. For extraintestinal infections, use of metronidazole is recommended, usually in combination with diiodohydroxyquin or diloxanide furoate, to prevent continued intraluminal infection. In seriously ill patients with complicated amebic infections, parenteral emetine in combination with iodoquinol may be employed.

8-Hydroxyquinolines. A number of 8-hydroxyquinolines are available for treating amebiasis. Iodoquinol, the most widely used drug, is described in this chapter; however, other derivatives have been applied in specific areas of the world. These include clioquinol (iodochlorhydroxyquin), broxyquinoline, chlorquinaldol, and chiniofon.

Iodoquinol [*83-73-8*], diiodohydroxyquinoline (INN, U.S.P.), 5,7-diiodo-8-quinolinol, $C_9H_5I_2NO$, M_r 396.95, *mp* 200 – 215 °C (extensive decomp.), occurs as crystals or yellowish brown powder.

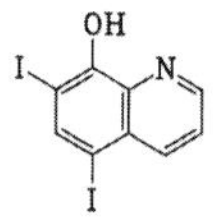

OH I N I

Iodoquinol is an 8-hydroxyquinoline that is directly amebicidal. It is active only on amebas in the intestinal tract and, like the other members of this class of compounds, is ineffective against amebic abscess and hepatitis. The drug is partly absorbed when it is administered orally, with ca. 25 % of the drug recovered in the urine as the glucuronide. Although the 8-hydroxyquinolines were originally thought to be of low toxicity, a number of *toxic reactions* are now known to result from their use, the most significant being a subacute myelooptic neuropathy, which is particularly observed with clioquinol. The usual dose of iodoquinol for treating asymptomatic carriers of *Ent. histolytica* is 650 mg 3 times a day for 20 days. This same dose may be employed in combination therapy with metronidazole in treating frank intestinal disease or in hepatic abscess, where it serves to limit reoccurrence of the intestinal forms.

Synthesis: prepared by the action of iodine monochloride on 8-hydroxyquinoline [213].

Trade name: Yodoxin (Glenwood).

Clioquinol [*130-26-7*], iodochlorhydroxyquin (INN, U.S.P., BAN), 5-chloro-8-hydroxy-7-iodoquinoline, C_9H_5ClINO, M_r 305.50, *mp* ca. 178 – 179 °C (decomp.), is a brownish-yellow bulky powder.

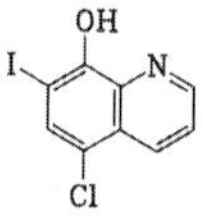

OH I N Cl

Synthesis: [214].

Trade names: Vioform (Ciba-Geigy), Rheaform, veterinary (Squibb).

Broxyquinoline (INN, DCF, MI) [*521-74-4*], 5,7-dibromo-8-hydroxyquinoline, $C_9H_5Br_2NO$, M_r 302.95, *mp* 196 °C, occurs as monoclinic needles.

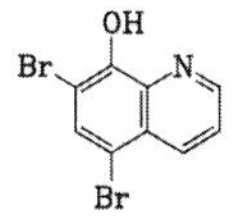

Synthesis: prepared by bromination of 8-quinolinol [215].

Chiniofon (INN, DCF, NFN, NF **11**) [*8002-90-2*], mixture of four parts (by mass) 8-hydroxy-7-iodo-5-quinolinesulfonic acid and one part sodium hydrogencarbonate.

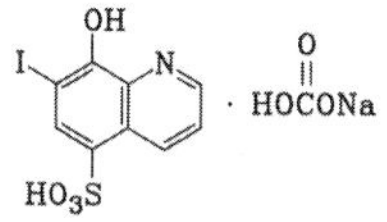

Synthesis: of 8-hydroxy-7-iodo-5-quinolinesulfonic acid [216].

Diloxanide furoate [*3736-81-0*], 4-[(dichloroacetyl)methylamino]phenyl-2-furancarboxylate, $C_{14}H_{11}Cl_2NO_4$, M_r 328.15, *mp* 112.5 – 114 °C.

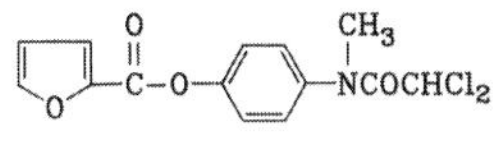

The furoate ester of diloxanide is one of the agents of choice in the treatment of persons who are asymptomatic passers of cysts or for invasive and extraintestinal amebiasis (administered with other appropriate drugs). The recommended oral dose is 500 mg 3 times daily given for 10 days; a second course may be necessary. When administered orally, the ester is hydrolyzed in the lumen or mucosa of the intestine, resulting in diloxanide and furoic acid. Most of the oral dose is excreted in the urine within 48 h, and peak drug concentrations appear in the plasma in 1 h. There are few *adverse effects* reported with this compound, the most common being flatulence or mild abdominal discomfort.

Diloxanide (INN, BAN, DCF, NFN, MI) [*579-38-4*], 2,2-dichloro-4′-hydroxy-*N*-methylacetanilide, $C_9H_9Cl_2NO_2$, M_r 234.08, *mp* 175 °C, is crystalline in form.

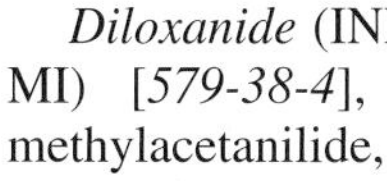

Synthesis: Diloxanide is prepared from 3,3-dianisyl-4-hexanone [217]. Diloxanide furoate is obtained by reacting diloxanide with 2-furoyl chloride in pyridine.

4. Chemotherapy of Viral Infections

4.1. Physical and Biological Characteristics

Obligate Intracellular Parasites. Viruses are among the smallest of all life forms. Unlike bacteria, viruses are obligate intracellular parasites that are metabolically inert in the extracellular state. Even though viruses depend to varying degrees on the host cell's metabolic machinery, the differences existing between the metabolic processes of the host cell and those specified by viruses have been exploited in the search for antiviral drugs that selectively inhibit virus replication.

Nucleic Acid. Unlike other microorganisms, viruses contain only one form (double-stranded or single-stranded) of only one type (RNA or DNA) of nucleic acid as their genetic material. The polarity of the genetic material determines whether transcriptase enzymes (enzymes not found in eukaryotic cells) must be present within virions. For example, a negative-stranded RNA genome is complementary to messenger RNA (mRNA) and must be transcribed by a virion transcriptase enzyme to produce viral mRNA necessary for initiation of the infection process. Other virion transcriptases, such as the reverse transcriptase present in retroviruses, catalyze the formation of DNA from viral RNA. The unique reverse transcriptase enzyme of the human T-cell lymphotrophic virus type III, lymphadenopathy associated virus, AIDS-related virus, or human immunodeficiency virus (HIV), which are various names for the virus implicated in AIDS (acquired immune deficiency syndrome), is the target of several antiviral drugs.

Viral Proteins. The functions of the structural virion proteins are to protect the genetic material from degradation and also to determine what types of host cells can be infected by a particular virus (host range). The latter function is determined by envelope proteins, which may be glycosylated, because these proteins are the first to contact the host cell. In viruses that lack an envelope, capsid proteins protect the genetic material and determine host range. The capsid is the protein coat that is complexed with the nucleic acid. Nonstructural proteins, such as the virion transcriptases, are also present within the virion.

Lipid Envelope. The lipids present in enveloped viruses resemble, in composition, those lipids of the host-cell membranes. Certain viral proteins may be embedded in the lipid bilayer of virions.

Laboratory Diagnosis of Viral Infections. Rapid detection of viral infections is critical for rational chemotherapeutic intervention [218]. Detection methods include serological tests, as well as direct examination of clinical specimens for virus. Serological tests measure the rise in serum antibody late in the disease course (convalescent) vs. serum antibody levels found early in the course of the disease (acute). The presence of antibodies specific for a particular virus may be difficult to assess if antibodies to related viruses are present or if the virus infection represents reactivation or reinfection of the host with the same virus. The immunoglobulin antibody of class IgM, which is produced early in infections, can indicate the presence of active virus replication. This IgM antibody can be detected by using IgM capture techniques in which serum interacts with anti-IgM on an immobilized solid surface. Other serological techniques use enzyme- (peroxidase), radioactive-, or fluorescent-labeled antibodies.

Direct observation using an electron microscope can rapidly detect some virus infections. This technique works well for examining fecal contents in acute gastroenteritis caused by rotaviruses, caliciviruses, astroviruses, and the Norwalk agent, in which sufficiently large numbers of morphologically distinct virions are present. In these cases, the viruses do not grow well in cell culture.

The classic method, and still the standard, is to isolate and identify viruses by growing them from clinical specimens in a cell culture. Several different cell cultures must be used because no one culture can support the replication of all viruses. Preliminary virus identification can be obtained by examination of infected cells for the cytopathic effect (CPE) that is characteristic for

Table 4. Virus families

Virus family	Distinguishing characteristics *	Human pathogens *
Poxviridae	DS DNA, enveloped	smallpox, vaccinia
Herpesviridae	DS DNA, enveloped	HSV, VZV, CMV, EBV
Adenoviridae	DS DNA, nonenveloped	human adenoviruses
Papovaviridae	DS DNA, nonenveloped	papilloma, BK and JC
Hepadnaviridae **	DS DNA, nonenveloped	hepatitis B virus
Parvoviridae	SS DNA, nonenveloped	parvoviruses, Norwalk (?)
Reoviridae	DS RNA, nonenveloped	reoviruses, rotaviruses
Togaviridae	SS RNA, enveloped, no DNA step, (+) sense	EEE, WEE, VEE, rubella, yellow fever, dengue
Coronaviridae	SS RNA, enveloped, no DNA step, (+) sense	coronaviruses
Paramyxoviridae	SS RNA, enveloped, no DNA step, (−) sense, nonsegmented	parainfluenza, measles, mumps, RSV
Rhabdoviridae	SS RNA, enveloped, no DNA step, (−) sense, nonsegmented	VSV, rabies
Filoviridae **	SS RNA, enveloped, no DNA step, (−) sense, nonsegmented	Marburg, Ebola
Orthomyxoviridae	SS RNA, enveloped, no DNA step, (−) sense, segmented	influenza A and B
Bunyaviridae	SS RNA, enveloped, no DNA step, segmented, ambisense	Bunyamwera Rift Valley fever
Arenaviridae	SS RNA, enveloped, no DNA step, segmented, ambisense	Lassa, Machupo, Junin, LCMV
Retroviridae	SS RNA, enveloped, DNA step	human T-lymphotrophic viruses
Picornaviridae	SS RNA, nonenveloped	poliovirus, Coxsackievirus, echovirus, hepatitis A, rhinovirus
Caliciviridae	SS RNA, nonenveloped	Norwalk (?), caliciviruses

* DS = double-stranded; SS = single-stranded; HSV = herpes simplex virus; VZV = varicella – zoster virus; CMV = cytomegalovirus; EBV = Epstein – Barr virus; EEE = eastern equine encephalitis virus; WEE = western equine encephalitis virus; VEE = Venezuelan equine encephalitis virus; RSV = respiratory syncytial virus; VSV = vesicular stomatitis virus; LCMV = lymphocytic choriomeningitis virus; (+) sense = genomes containing nucleotide sequences translated by ribosomes; (−) sense = genomes composed of nucleotide sequences complementary to (+) sense; ambisense = genome composed of (−) sense covalently attached to (+) sense.
** Not yet approved by ICTV.

a virus group. In the absence of a characteristic CPE, serological tests are used for final identification of viruses.

Many viruses have been cloned by using recombinant DNA techniques. Probes produced from these cloned virus DNAs can be used to detect certain viruses in clinical samples. Eventually, nonradioactive probes that use biotin or fluorescent tags may be widely used in clinical laboratories.

4.2. Classification of Viruses

The International Committee on Taxonomy of Viruses (ICTV) published a report in 1982 [219] that is summarized in Table 4. A total of 16 of the 55 virus families recognized by the ICTV contain viruses that are human pathogens. The distinguishing characteristics are the type of nucleic acid, the presence or absence of an envelope, the replication strategy of the genome, whether it is a positive-, negative-, or ambisense genome, and genome segmentation.

4.3. Assessment of Antiviral Activity in Cell Culture

In vitro cell culture systems for determining the antiviral effect of compounds have been useful, even though in vitro activity does not always predict in vivo or clinical efficacy. Plaque reduction [220], [221], yield reduction, cell growth, and macromolecular synthesis assays are among of the methods used to quantify the effects of antiviral candidates on virus replication and host cell metabolism. In the plaque reduction assay, the concentration of compound that reduces the number of plaques to a defined end point (usually 50 % or 90 %) is determined. In the yield reduction assay, the concentration of compound that reduces the yield of infectious virus by 50 % or 90 % is determined. In the cell growth or macromolecular synthesis assay, the effect of a compound on the increase in cell number or the incorporation of radioactive precursors into DNA, RNA, and protein, respectively, is determined. A compound is a viral-specific inhibitor if the antiviral concentration in the plaque reduction or virus yield assay does not affect cell growth or macromolecular synthesis.

4.4. Animal Models of Virus Infection

In addition to the in vitro cell culture systems described in Section 4.3, animal models of human virus infections are used to assess the systemic efficacy of candidate antiviral compounds. A compound that is a potent inhibitor of virus replication in a cell culture is commonly completely ineffective in preventing disease when administered to a virus-infected animal. These results are usually explained by poor absorption or rapid metabolism of the compound, resulting in a concentration of the compound that is too low to inhibit virus replication in the critical tissues.

Results in animal models of human viral infections often differ significantly from the natural course of the human infection, even when the identical virus is used. For example, mice infected intracerebrally with a mouse-adapted human poliovirus develop flaccid limb paralysis and succumb to the infection within 24 h of the onset of paralysis [222]. The severity of the poliovirus infection in mice is greater than that usually observed in humans, presumably because the route of inoculation in mice does not allow the animal to develop a protective immune response prior to the onset of paralysis. In humans, poliovirus replicates first in the Peyer's patches in the gut, which stimulates an immune response prior to the spread of virus to the central nervous system.

Mouse models of herpes simplex virus (HSV) infections are used to assess the efficacy of antiherpetic compounds; however, the disease in mice is more severe than the acute infection in humans. Mice generally succumb to encephalitis following intraperitoneal or intravaginal inoculation of virus. The severity of the disease in the animal model used in evaluating an antiviral compound should be considered when assessing potency and projecting dose levels in humans.

4.5. Rationale for Chemotherapy of Viral Infections

The rationale for the discovery of specific chemotherapeutic agents effective against viruses is based on the knowledge that certain virus functions are distinct from cellular functions (for reviews of antiviral chemotherapeutics, see [223]). A number of unique virus-coded enzymes are known to be critical for virus replication. These enzymes (e.g., deoxypyrimidine kinase, also known as thymidine kinase, and ribonucleotide reductase from HSV) have different substrate specificities compared to cellular enzymes and, therefore, can be used as targets for antiviral chemotherapy. The antiherpetic drug, acyclovir (see Section 4.6.1), is an excellent example of the exploitation of the differences between the viral and cellular enzyme. Acyclovir is activated by deoxypyrimidine kinase to eventually form acyclo-GTP, which inhibits the viral DNA polymerase. This process does not occur to a significant extent in uninfected cells. A second example is the reverse transcriptase enzyme (from the retroviruses implicated in AIDS), which presumably does not have a normal cell counterpart. This enzyme has been the target of several agents that specifically inhibit the replication of the AIDS viruses.

Virus replication can also be inhibited by agents, such as rhodanine or arildone, that specifically bind with high affinity to the picornaviral capsid proteins and stabilize virion capsid conformation. This stabilization prevents cell-induced uncoating of the virion nucleic acid, thereby halting the virus infection at the uncoating step.

The host's immune response to a viral infection works in concert with antiviral drugs to clear the infection. The virus infection of a host can be viewed as a race between replication of the virus and the mounting immune response to the virus. If the extent of virus replication and resultant tissue destruction exceeds a threshold value, disease occurs. If the extent of virus replication is limited by the immune response or by chemotherapeutic intervention, then the disease course will be milder or asymptomatic. Thus, the role of antiviral chemotherapy in the treatment of viral disease is to inhibit virus replication sufficiently to enable the host's immune system to overcome the virus infection.

4.6. Chemotherapeutic Agents

4.6.1. Nucleoside Analogues

Acyclovir (INN) [*59277-89-3*], 9-[(2-hydroxyethoxy)methyl]guanine, $C_8H_{11}N_5O_3 \cdot 1/2\,H_2O$, M_r 234.2, *mp* > 200 °C.

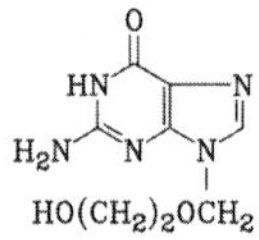

The antiherpetic activity of acyclovir, the acyclic analogue of guanosine, was first reported along with its synthesis in 1978 [224]. The rationale for this synthesis was the previous observation that the acyclic analogue of adenosine was a substrate for adenosine deaminase [225]. Acyclovir has potent activity against several herpesviruses (herpes simplex virus types 1 and 2, Epstein – Barr virus, and varicella – zoster virus), but is not as potent against human cytomegalovirus.

Acyclovir is clinically effective in the treatment of herpesvirus infections in humans. Acyclovir was first reported to be clinically efficacious in the topical treatment of herpetic keratitis [226]. Acyclovir was subsequently shown to be as effective as vidarabine [227] or as trifluorothymidine [228] against herpetic keratitis. Topical acyclovir cream was found to be effective against herpes labialis [229] and primary genital herpes [230], but was not effective against recurrent genital herpes [231]. Unlike the topical cream, orally administered acyclovir did reduce the duration of viral shedding, as well as the time to healing in recurrent genital herpes. Oral acyclovir, however, had no significant effect on symptoms of recurrent genital herpes [232–234]. Treatment protocols using daily administration of prophylactic oral acyclovir for 4-month periods have been described for patients with frequent recurrences [235], [236]. During treatment with acyclovir, the patients had significantly fewer recurrences than those receiving placebo. However, after treatment with acyclovir was stopped, the patients' recurrence rates returned to pretreatment frequencies. While on treatment with acyclovir, many patients reported the symptoms of a prodrome that did not progress to a complete lesion [237].

Adverse Effects. Several problems exist with the chronic use of acyclovir in patients. The major question is one of long-term safety. First, because acyclovir is a nucleoside analogue, there is concern it may be mutagenic. Thus far, however, acyclovir has not been shown to cause cancer in laboratory animals. Furthermore, it is not known whether long-term prophylactic use of acyclovir is more likely to result in development of resistant viruses. Other minor concerns with acyclovir include reversible toxicity to the bone marrow, reversible kidney damage due to acyclovir crystals, and diarrhea. Because only 15 – 20 % of acyclovir is absorbed after oral administration, a prodrug of acyclovir activated by xanthine oxidase has recently been developed [238].

Synthesis: Reaction of 2,6-dichloro-9-(2-benzoyloxyethoxymethyl) purine with methanolic ammonia, followed by treatment with nitrous acid and then with methanolic ammonia yields acyclovir [224].

Trade name: Zovirax (Burroughs Wellcome).

Idoxuidine [*54-42-2*], IDU (INN, U.S.P.), 2′-deoxy-5-iodouridine, $C_9H_{11}IN_2O_5$, M_r 354.12, *mp* 160 °C (decomp.).

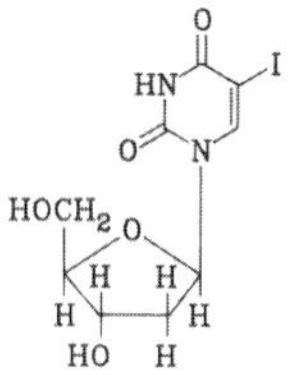

Idoxuidine was the first clinically effective nucleoside analogue. It was synthesized in 1959 [239] as part of a program aimed at developing synthetic nucleosides as DNA synthesis inhibitors to treat cancer. In effect, antiviral chemotherapy became a reality in 1962 when IDU was reported to cure herpes simplex keratitis in humans [240]. Although IDU was effective topically against herpes simplex keratitis, it was not active when administered systemically.

Adverse Effects. Because of the insolubility and toxicity of IDU, other drugs, such as trifluorothymidine, are used for the topical treatment of ocular herpetic infections [241].

Synthesis: obtained by refluxing uracil-deoxyriboside, iodine, chloroform, and HNO_3 [242].

Trade names: Stoxil (Smith Kline).

Trifluridine (INN) [*70-00-8*], 5′-trifluoromethyl-2′-deoxyuridine, trifluorothymidine, TFT, $C_{10}H_{11}F_3N_2O_5$, M_r 296.2, *mp* 186 – 189 °C.

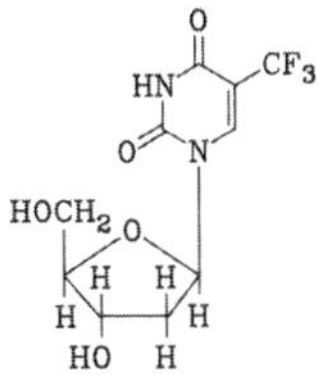

Trifluridine was first synthesized in 1964 [243]. Its use is limited to the topical treatment of herpetic keratitis. It was more effective than IDU in patients with dendritic or ameboid herpetic keratitis [244].

The *antiviral activity* and the *mechanisms of action* of TFT have been reviewed [245], [246]. It is converted to the active triphosphate form by cellular and viral thymidine kinases. The triphosphate form of TFT is then preferentially incorporated into viral DNA, which results in inhibition of the transcription of late viral mRNAs.

Adverse Effects. Because TFT can also be incorporated into cellular DNA, TFT is toxic to uninfected cells; thus, TFT cannot be used systemically.

Synthesis: prepared by treatment of 5-trifluoromethyluracil with a bacterial enzyme [243].

Trade name: Viroptic (Burroughs Wellcome).

Vidarabine [*24356-66-9*], ara-A (INN, U.S.P.), 9-β-D-arabinofuranosyladenine, $C_{10}H_{13}N_5O_4 \cdot 4\,H_2O$, M_r 339.36, *mp* 257 °C.

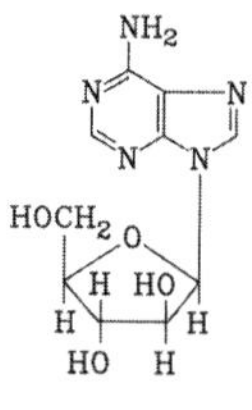

Ara-A was initially synthesized in the 1960s as a potential anticancer agent [247], [248]. Ara-A was reported to inhibit herpesvirus, vaccinia virus, cytomegalovirus, and varicella – zoster viruses [249], [250]. Topical ara-A was as effective against herpetic keratitis as IDU [251]. Intravenous ara-A was also effective against herpes zoster in immunosuppressed patients [252]. Ara-A was sufficiently nontoxic to uninfected cells to permit systemic administration to patients with herpes encephalitis. Ara-A reduced mortality from 70 to 28 % in a study of 28 patients with herpes encephalitis [253]. Recently, acyclovir was shown to be superior to ara-A for herpes encephalitis [254]. Topical ara-A and ara-A monophosphate have no effect in patients with genital or labial herpes [255], [256].

Mechanism of Action. Clearly, ara-A has multiple sites of inhibition of HSV. Ara-A is phosphorylated to give ara-A triphosphate (ara-ATP). Ara-ATP is a selective inhibitor of ribonucleoside diphosphate reductase and HSV DNA polymerase, as well as of the addition of poly-A to viral mRNA. Ara-A also slows viral DNA elongation, inhibits terminal deoxynucleotidyl transferase, and inhibits *S*-adenosylmethionine-dependent methylation (capping) of viral mRNA [257].

Synthesis: Ara-A is prepared by treating 9-(3′,5′-*O*-isopropylidene-β-D-xylofuranosyl)adenine with methanesulfonyl chloride. The crystalline 9-(3′,5′-*O*-isopropylidene-2′-*O*-methanesulfonyl-β-D-xylofuranosyl)adenine is exposed to 90 % aqueous acetic acid at 100 °C for 5 h. The epoxide is formed with the use of methanolic sodium hydroxide. The epoxide is converted to the arabinoside by reaction with sodium benzoate or sodium acetate in 95 % aqueous *N,N*-dimethylformamide [248].

Trade name: Vira-A (Parke-Davis).

Bromovinyldeoxuridine [*73110-56-2*], BVDU, (*E*)-5-(2-bromoethenyl)-2′-deoxyuridine, $C_{11}H_{13}BrN_2O_5$, M_r 331.6.

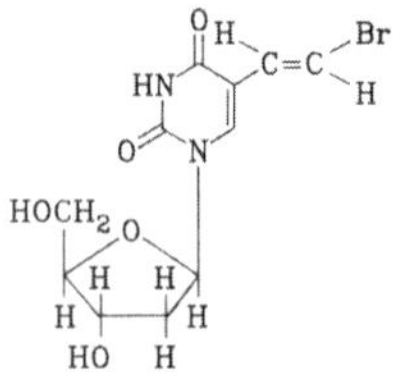

The antiherpetic activity of BVDU was first reported in 1979 [258]. This pyrimidine derivative is about 100 times more active against HSV – 1 in cell culture than against HSV – 2 and is also active against varicella – zoster virus.

The clinical efficacy of BVDU has been demonstrated in several uncontrolled (open) trials. First, because BVDU was well-absorbed when given orally, the progression of varicella – zoster virus in immunocompromised patients was retarded by BVDU [259]. Second, topical administration of BVDU was effective against herpetic keratitis [260]. Placebo-controlled, double-blinded clinical trials have been initiated.

The *mechanism of action* of BVDU is similar to that of other selective nucleoside analogues having antiherpetic activity. It is converted to the triphosphate by the viral and cellular thymidine kinases. The 5′-triphosphate is then incorporated into viral DNA in virally infected cells.

Adverse Effects. There are several problems with the use of BVDU. First, BVDU is rapidly degraded by thymidine phosphorylase. Second, herpesviruses can readily develop resistance to BVDU by lowering the virus content of thymidine kinase activity. Finally, the development of BVDU by G. D. Searle has been halted because of increased incidence of tumors in animals dosed for long periods [261].

Synthesis: Condensation of the trimethylsilyl derivative of (*E*)-5-(2-bromovinyl)uracil with 2′-deoxy-3,5-di-*O*-*p*-toluoyl-α-D-erythro-pentofuranosyl chloride gives a mixture of the α- and β-anomers of the protected nucleoside. The *p*-toluoyl protecting groups are removed by treatment with sodium methoxide in methanol. The β-anomer, but not the α-anomer, is biologically active [258].

FIAC [*69124-05-6*], 1-(2′-fluoro-2′-deoxy-β′-D-arabinofuranosyl)-5-iodocytosine hydrochloride, $C_9H_{11}FIN_3O_4 \cdot HCl$, M_r 407.58, *mp* 177 – 181 °C.

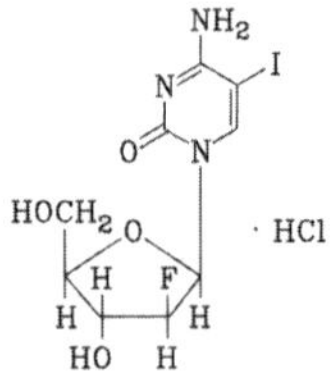

The synthesis and antiherpetic activity of a series of 2′-fluoro-5-substituted arabinofuranosylcytosines and uracils have been recently reported [262], [263]. Several compounds in this series are potent inhibitors of HSV – 1, HSV – 2, varicella – zoster virus, cytomegalovirus, and Epstein – Barr virus. For example, 1-(2′-fluoro-2′-deoxy-β-D-arabinofuranosyl)-5-iodocytosine (FIAC) was active systemically after i.v. administration to immunocompromised patients infected with varicella – zoster virus [264].

The mode of action of these 2′-fluoro-5-substituted nucleoside analogues is similar to that of acyclovir. The analogues are substrates for the herpesvirus-coded thymidine kinase (deoxypyrimidine kinase) [265]. Viral DNA polymerase is more sensitive to inhibition by the 5′-triphosphate of FIAC than is the host cell DNA polymerase [266].

A related compound, 1-(2′-fluoro-2′-deoxy-β′-D-arabinofuranosyl)-5-iodouracil (FIAU), has recently been reported to have some toxic effects in humans. FIAU caused cardiac fibrosis at a dose of 150 mg/kg in a long-term study [267].

Synthesis: FIAC is prepared by condensation of 3-*O*-acetyl-5-*O*-benzoyl-2′-deoxy-2′-fluoro-D-arabinofuranosyl bromide with trimethylsilylated cytosines to yield the blocked nucleosides. After deprotection by saponification to the 2′-fluoroarabinofuranosylcytosine nucleoside, the 5-iodo analogue (FIAC) is obtained by iodination [262].

Ribavirin [*36791-04-5*], Virazole (INN, British Approved Name – BAN), 1-β-D-ribofuranosyl-1*H*-1,2,4-triazole-3-carboxamide, $C_8H_{12}N_4O_5$, M_r 244.2, *mp* 166–168 °C.

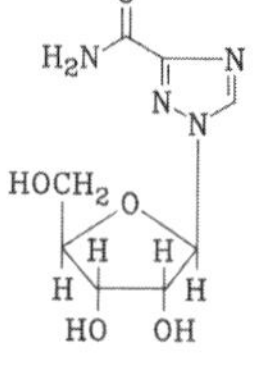

Ribavirin was initially described in 1972 as the result of a program to synthesize broad spectrum antiviral agents [268]. Ribavirin is active in cell culture against most DNA and RNA viruses [269].

Ribavirin probably has several *mechanisms of action.* Ribavirin is phosphorylated by cellular enzymes, and both the monophosphate (RMP) and the triphosphate (RTP) have antiviral activity. The monophosphate inhibits synthesis of guanosine 5′-monophosphate (GMP) [270]. Thus, the supply of guanosine triphosphate is depleted and nucleic acid synthesis is inhibited. In addition RTP inhibits the virus-specific mRNA capping enzymes of vaccinia virus [271]. The ability of RTP to inhibit the capping of viral-specific mRNA could explain how ribavirin could be active against both DNA and RNA viruses. An interesting finding is that ribavirin was active in cell culture against the virus which causes AIDS [272].

Clinical trials with ribavirin have been performed with respiratory syncytial virus (RSV), Lassa virus, and influenza viruses. Aerosolized ribavirin was effective in reducing fever and symptoms in adults experimentally infected with RSV [273]. Similar findings were observed in infants treated with aerosolized ribavirin after natural RSV infection [274]. Intravenous ribavirin reduced mortality in comatose patients infected with Lassa virus [275]. Conflicting results have been found with influenza virus and oral ribavirin. Several studies demonstrated decreased severity of illness and decreased viral shedding with oral ribavirin [276], [277]. Other studies demonstrated no therapeutic effect of oral ribavirin against influenza virus [278], [279]. However, aerosolized ribavirin has been shown to reduce fever and systemic illness in patients infected with influenza virus [280], [281].

Synthesis: Acid-catalyzed fusion of methyl 1,2,4-triazole-3-carboxylate and 1,2,3,5-tetra-*O*-acetyl-β-D-ribofuranose or 1-*O*-acetyl-2,3,5-tri-*O*-benzoyl-β-D-ribofuranose yields blocked methyl ester nucleosides. Ribavirin is obtained after treatment of the blocked nucleoside with methanolic NH_3 [268].

Trade name: Virazole (ICN).

Cyclaradine [*69979-46-0*], carbocyclic arabinofuranosyladenine, [±]-9-[2α,3β-dihydroxy-4α-(hydroxymethyl)cyclopent-1α-yl]adenine, $C_{11}H_{15}N_5O_3 \cdot H_2O$, M_r 283.33, *mp* 253 – 255 °C.

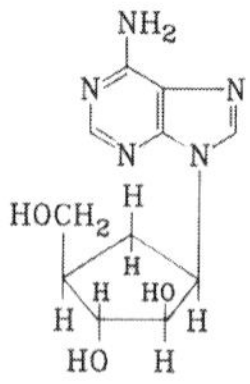

NH2
HOCH2
HO H

Cyclaradine is a carbocyclic derivative of ara-A. Cyclaradine was synthesized in 1977 in an effort to obtain a derivative of ara-A that is resistant to adenosine deaminase [282]. The deamination product of ara-A is 9-β-D-arabinofuranosylhypoxanthine, which is considerably less active than ara-A against herpesviruses. Cyclaradine is resistant to adenosine deaminase.

Cyclaradine is active in cell culture against herpes simplex virus type 1 [282]. Clinical studies in humans have not yet been conducted with cyclaradine. However, cyclaradine is effective against HSV – 1 induced encephalitis in mice [283] and HSV – 2-induced genital infections in guinea pigs [284].

Synthesis: Reaction of 5-amino-4-*N*-[2α,3β-dihydroxy-4α-(hydroxymethyl)cyclopent-1α-yl]-amino-6-chloropyrimidine with diethoxymethyl acetate gives the 6-chloropurine. Treatment of the 6-chloropurine with liquid ammonia gives cyclaradine [282].

DHPG [*82410-32-0*], 2-amino-1,9-dihydro{[2-hydroxy-1-(hydroxymethyl)ethoxy]-methyl}-6*H*-purin-6-one, BIOLF – 62 [285], BW B759U, 2′-nor-2′-deoxyguanosine (2′-NDG) [286], $C_9H_{13}N_5O_4$, M_r 255.3, *mp* > 300 °C.

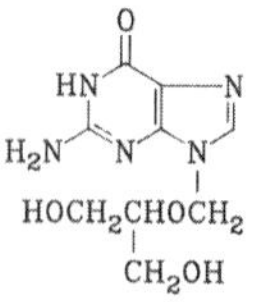

HOCH2CHOCH2
CH2OH

DHPG is an analogue of acyclovir; its potency is equal to or greater than that of acyclovir against some herpesviruses. In comparison with acyclovir, DHPG is at least 10-fold more active against CMV and Epstein – Barr virus in cell culture and about 50-fold more active than acyclovir in animal models of herpesvirus infection [287].

Several preliminary clinical studies with DHPG have been reported. It was effective against CMV pneumonia in transplant recipients [288] and against CMV retinitis [289]. DHPG has also been used for CMV infections in AIDS patients [290].

Synthesis: 2-*O*-(Acetoxymethyl)-1,3-di-*O*-benzylglycerol is condensed with N^2,9-diacetylguanine in the presence of a catalytic amount of *p*-toluenesulfonic acid in sulfolane. The desired isomer is crystallized from toluene, debenzylated over 20 % palladium hydroxide, and deacetylated with concentrated NH_4OH/methanol [291].

Buciclovir [*86304-28-1*], (*R*)-9-(3,4-dihydroxybutyl)guanine, (*R*)-DHBG, (*R*)-2-amino-9-(3,4-dihydroxybutyl)-1,9-dihydro-6*H*-purin-6-one, $C_9H_{13}N_5O_3$, M_r 239.23, *mp* 260 – 261 °C (decomp.).

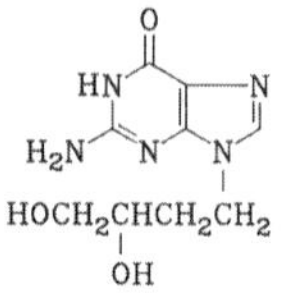

HOCH2CHCH2CH2
OH

(*RS*)-9-(3,4-Dihydroxybutyl)guanine, which is a close analogue of acyclovir and DHPG, has recently been shown to inhibit HSV – 1 and HSV – 2 both in cell culture and in animal models of virus infection [292]. The (R)-enantiomer of DHBG is about 5-fold more active against HSV – 2 in plaque reduction tests than the (S)-enantiomer. The mode of action of DHBG is similar to that of acyclovir [293], [294]. DHPG is phosphorylated by the HSV-induced thymidine kinase; presumably it is phosphorylated by cellular enzymes to give the 5′-triphosphate derivative of DHPG, which selectively inhibits viral DNA synthesis.

No clinical data on DHBG are available at the present time.

Synthesis: Condensation of 2-amino-6-chloropurine with 4-bromo-2-hydroxybutyrate and subsequent hydrolysis yields 4-(9-guanyl)-2-hydroxybutyric acid, which is esterified and reduced to buciclovir [295].

4.6.2. Phosphonoacetate and Phosphonoformate

Phosphonoacetic acid (PAA) [*4408-78-0*], $C_2H_5O_5P$, M_r 136.01, was discovered by routine screening methods in 1973 to inhibit HSV [296]. The sodium salt is called phosphonoacetate. PAA has been shown to inhibit the replication in cell culture of HSV, cytomegalovirus, Epstein–Barr virus, varicella – zoster virus, vaccinia virus, avian herpesvirus, and equine abortion virus. PAA inhibits the replication of herpesvirus by interacting directly with the virus-induced DNA polymerase at the pyrophosphate binding site [297]. Host cell polymerases are less sensitive to PAA [298]. Preclinical studies in rats showed that intravenous PAA was deposited in bone [299]. This result prevented clinical studies from being performed with PAA. Structure – activity studies have been reported for 100 analogues of PAA, but all analogues were less active than PAA [300].

Phosphonoformate [*63585-09-1*], foscarnet sodium, phosphonoformate sodium, PFA (INN), phosphonoformic acid (trisodium salt), foscarnet, CNa_3O_5P, M_r 191.95, *mp* >250 °C.

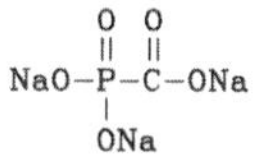

```
    O  O
    ‖  ‖
NaO–P–C–ONa
    |
    ONa
```

Phosphonoformate is closely related to PAA and also directly inhibits the virus-induced DNA polymerase [301]. Phosphonoformate is deposited in bone, but to a lesser extent than PAA [302]. Clinical trials have indicated that topical treatment of patients with recurrent genital herpesvirus infection was effective in shortening the time to healing, but this effect was only observed in male patients [303]. Intravenous phosphonoformate was effective against CMV in allograft recipients [304].

Phosphonoformate has also shown activity in cell culture against HTLV – III (the virus implicated in AIDS) [305].

Synthesis: Phosphonoformate is prepared by saponification of the triethyl ester of phosphonoformate with NaOH. Trisodium phosphonoformate hexahydrate is recrystallized from H_2O [301].

Trade name: Foscarnet (Astra).

4.6.3. Amantadine and Rimantadine

Amantadine is presently the only chemotherapeutic agent accepted for clinical use against influenza virus. Amantadine was discovered in 1964 and was found to be active against influenza A viruses [306]. Administered orally, subcutaneously, or intraperitoneally, amantadine reduced the mortality in mice infected intranasally with influenza A virus. The precise *mechanism of action* of amantadine is not defined although most studies indicate that amantadine inhibits uncoating (the release of infectious nucleic acid into the cell's cytoplasm) [307]. The first clinical trials in volunteers lacking antibody showed that prophylactic amantadine reduced the incidence of influenza [308]. These results were confirmed in a natural outbreak of influenza virus [309]. The use of amantadine did not become widespread because until September, 1976; the FDA would not approve its use against any strain of influenza except for the original A2/Asian strain. Thus, amantadine could not be used against the major Hong Kong influenza virus epidemic in the late 1960s.

In comparison studies with amantadine, rimantadine is equally effective prophylactically with fewer *adverse effects* (nervousness, difficulty in concentrating, and insomnia) [310]. Rimantadine has similar therapeutic efficacy as amantadine with respect to faster resolution of symptoms in patients with influenza [311]. A randomized, placebo-controlled, double-blind trial demonstrated the therapeutic efficacy of oral rimantadine given once daily to patients with influenza A virus infection [312].

Amantadine (INN) [*665-66-7*], 1-adamantanamine hydrochloride, $C_{10}H_{17}N \cdot HCl$, M_r 187.74, *mp* 206 – 208 °C.

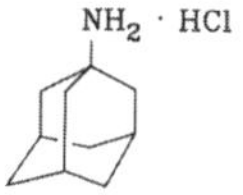

$NH_2 \cdot HCl$

Synthesis: 1-Bromoadamantane is reacted with acetonitrile in the presence of concentrated sulfuric acid. The resulting 1-acetaminoadamantane is hydrolyzed with NaOH in boiling diethylene glycol.

Trade name: Symmetrel (Du Pont).

Rimantadine (INN) [*13392-28-4*], [*1501-84-4*], α-methyl-tricyclo[3.3.1.1^{3,7}]decane-1-methan-amine, $C_{12}H_{21}N$, M_r 179.31, *mp* 373 – 375 °C, an analogue of amantadine, was also described in 1964 [306]; HCl salt [*1501-84-4*].

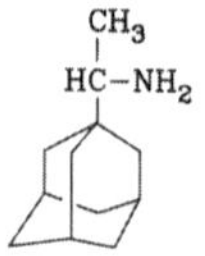

4.6.4. Enviroxime

Enviroxime (INN) [*72301-79-2*], (*E*)-2-amino-6-benzoyl-1-(isopropylsulfonyl)benzimidazoleoxime, 6-{[(hydroxyimino)phenyl]methyl}-1-[(1-methylethyl)sulfonyl]-1*H*-benzimidazol-2-amine, $C_{17}H_{18}N_4O_3S$, anti-isomer, M_r 358.45, *mp* 198 – 199 °C.

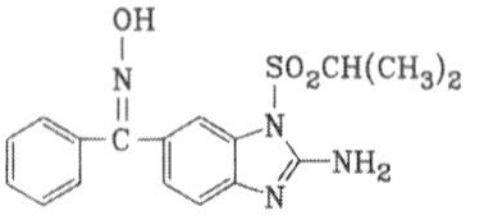

Enviroxime is an extremely potent inhibitor of human rhinoviruses in cell culture. A total of 15 different rhinovirus serotypes were inhibited in plaque reduction assays by enviroxime concentrations ranging from 0.04 µg/mL to < 0.01 µg/mL [313], [314]. However, the results of five clinical trials showed that enviroxime had only a modest, clinically significant, beneficial effect on rhinovirus infections. Prophylactic intranasal enviroxime administered prior to virus challenge in volunteers resulted in a statistically significant reduction in rhinorrhea, with no significant difference detected in infection rate or in the quantity of secreted virus. In addition, oral administration of enviroxime caused nausea and vomiting [315–317]. Therapeutically, intranasal administration of enviroxime lessened symptoms, but only on the fifth (last) day of treatment [318]. An open-field trial of intranasally applied enviroxime against natural rhinovirus infections showed no statistically significant differences compared to placebo [319]. Eli Lilly has terminated all studies on enviroxime because of the marginal efficacy observed in the clinic.

Synthesis: Reaction of cyanogen bromide and 3,4-diaminobenzophenone gives 2-amino-1-(isopropylsulfonyl)-6-benzimidazolyl phenylketone. The benzimidazole is sulfonated by using isopropylsulfonyl chloride and sodium hydride in dimethoxyethane. The anti- and syn-isomers of enviroximes are separated by high-performance liquid chromatography or by fractional crystallization [313].

4.6.5. 4′,6-Dichloroflavan

4′,6-Dichloroflavan [*73110-56-2*], BW-683C, 6-chloro-2-(4-chlorophenyl)-3,4-dihydro-4′,6-di-chloroflavan, $C_{15}H_{12}Cl_2O$, M_r 279.18.

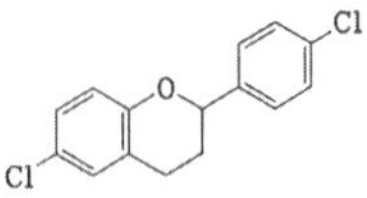

Dichloroflavan, like enviroxime, is a potent inhibitor of the replication of human rhinoviruses in cell culture [320]. Dichloroflavan appears to bind to virion capsid proteins, thereby inhibiting uncoating of the viral RNA [321], [322]. Only one clinical trial with oral dichloroflavan has been reported. In this double-blind, placebo-controlled trial, orally administered dichloroflavan was given to volunteers before and after challenge with rhinovirus 9. Dichloroflavan had no effect on the incidence or course of rhinovirus infection in this study [323].

Chemical synthetic details are not available.

4.6.6. Chalcone Ro 09 – 0410

Chalcone Ro 09 – 0410 [*76554-66-0*], 1-(4-ethoxy-2-hydroxy-6-methoxyphenyl)-3-(4-methoxyphenyl)-2-propen-1-one, 4′-ethoxy-2′-hydroxy-4,6′-dimethoxychalcone, Ro 09 – 0410, $C_{19}H_{20}O_5$, M_r 328.37.

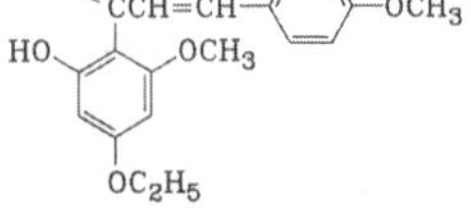

Chalcone Ro 09 – 0410 is a potent inhibitor of human rhinovirus in cell culture. Only 0.002 µg of Ro 09 – 0410 per mL is sufficient to inhibit some serotypes of human rhinovirus. Other serotypes of human rhinovirus require as much as 2 µg of the chalcone per mL for inhibition to occur [324]. The mode of action of Ro 09 – 0410 is similar to dichloroflavan in that it prevents uncoating by irreversible binding to viral capsid proteins. The chalcone could, however, be removed from virus particles by extraction with a nonpolar solvent [325].

A phosphate ester prodrug of the chalcone has been tested in volunteers because the chalcone itself was not well-absorbed orally and was irritating intranasally. The prodrug did not have any effect on the course of the rhinovirus infection, probably because even though the prodrug was well-absorbed, the chalcone did not reach the nasal mucosa [326].

Chemical synthetic details are not available.

4.6.7. Arildone and Disoxaril

Arildone [327] and disoxaril [328] have been shown to be active in cell culture, as well as in experimental animals infected with human picornaviruses [329–331]. *Mechanism of action* studies indicated that arildone blocked poliovirus uncoating, but did not inhibit adsorption or penetration [332]. Further work has shown that both arildone and disoxaril interact directly with picornavirion capsid proteins to prevent uncoating of virions.

Arildone (INN) [*56219-57-9*], 4-[6-(2-chloro-4-methoxy)phenoxy]hexyl-3,5-heptanedione, $C_{20}H_{29}ClO_4$, M_r 368.94, *bp* 180 °C.

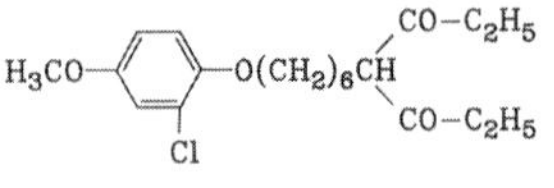

H_3CO–$C_6H_3(Cl)$–$O(CH_2)_6CH(CO{-}C_2H_5)_2$

Synthesis: Treatment of 1-(2-chloro-4-methoxyphenoxy)-6-iodohexane and lithio-3,5-heptanedione in dimethylformamide for 48 h at 60 °C gives arildone [327].

Disoxaril (INN), Win 51711 [*87495-31-6*], 5-{7-[4-(4,5-dihydro-2-oxazolyl)phenoxy]heptyl}-3-methylisoxazole, $C_{20}H_{26}N_2O_3$, M_r 342.42, *mp* 86 – 89 °C.

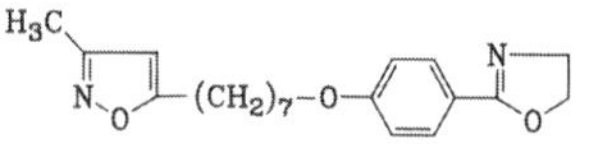

H_3C–isoxazole–$(CH_2)_7$–O–C_6H_4–oxazoline

4.6.8. 3′-Azidothymidine

3′-Azidothymidine [*30516-87-1*], AZT, 3′-azido-3′-deoxythymidine, $C_{10}H_{13}N_5O_4$, M_r 267.3, *mp* 119 – 121 °C.

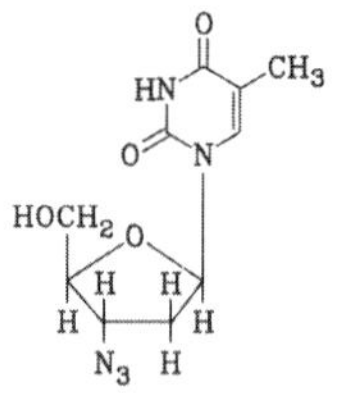

(Structure: thymine (HN, O, CH_3, N) attached to deoxyribose ring with $HOCH_2$, H, H, H, H, N_3, H)

3′-Azidothymidine was initially synthesized in 1979 [333]. This compound is active against the viruses responsible for AIDS (HTLV – III); it blocks the cytopathic effects of HTLV – III in cell culture [334]. 3′-Azidothymidine entered phase II clinical trials against AIDS in January, 1986 [335]; it has shown some positive results in preventing mortality because of the AIDS virus in patients treated with AZT for a few months.

Synthesis: prepared by detritylation of the 3′-azidoderivative of 1-(2′-deoxy-5′-*O*-trityl-β-D-lyxosyl)thymine [333], [336].

4.6.9. Suramin

Suramin (INN, U.S.P. **12**, DCF) [*145-63-1*], 8,8′{carbonylbis[imino-3,1-phenylenecarbonyl-imino(4-methyl-3,1-phenylene)-carbonylimino]}-bis(1,3,5-naphthalene)trisulfonic acid; suramin sodium (hexasodium salt) [*129-46-4*], $C_{51}H_{34}N_6Na_6O_{23}S_6$, M_r 1429.15.

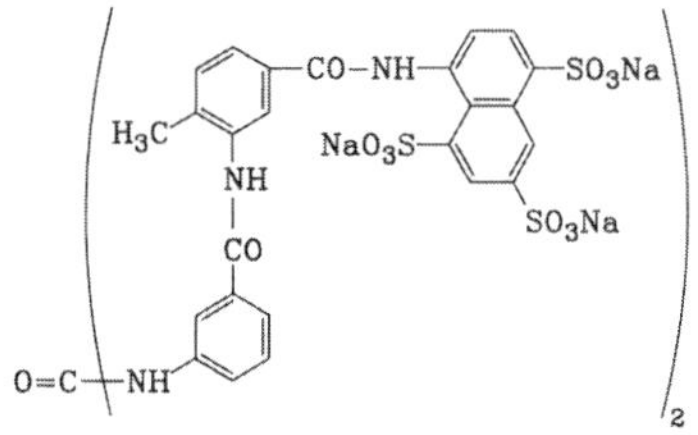

(Structure: $O{=}C{-}[NH{-}C_6H_4{-}CO{-}NH{-}C_6H_3(CH_3){-}CO{-}NH{-}C_{10}H_4(SO_3Na)_3]_2$; labels: H_3C, NH, CO, CO–NH, SO_3Na, NaO_3S, SO_3Na, O=C, NH, 2)

Suramin sodium is the hexasodium salt of 8,8′-{carbonylbis[imino-3,1-phenylenecarbonyl-imino(4-methyl-3,1-phenylene)carbonylimino]}-bis(1,3,5-naphthalene)trisulfonic acid. It is the drug of choice for the therapy of early African trypanosomiasis

[337], and has been the subject of renewed interest as an inhibitor of the reverse transcriptase of the virus that causes AIDS [338], [339]. Preliminary findings in a few AIDS patients showed that suramin sodium was beneficial after short-term intravenous treatment [340].

Synthesis: Suramin sodium is prepared by condensing 1-naphthylamine-4,6,8-trisulfonic acid with 3-nitro-4-methylbenzoyl chloride, reducing the product, condensing with 3-nitrobenzyl chloride, reducing again, and then treating with carbonyl chloride and neutralizing with sodium hydroxide [341].

4.6.10. HPA 23

HPA 23 [*89899-81-0*], ammonium 21-tungsto-9-antimonate, 5′-tungsto-2-antimonate, $(NH_4)_{17}Na/NaW_{21}Sb_9O_{86} \cdot 14\,H_2O$, M_r 6937.39.

HPA 23 is a tungsto-antimonate compound that is a competitive inhibitor of the reverse transcriptase of murine oncornavirus [342] and of the AIDS viruses. In a preliminary uncontrolled clinical trial, HPA 23 was able to inhibit AIDS virus replication in patients. The virus, however, reappeared when therapy was discontinued [343].

Synthesis: Addition of a hot hydrochloric acid solution of antimony trioxide to an aqueous solution of sodium tungstate and NH_3 gives HPA 23 [344].

5. References

1. R. E. Buchanan, N. E. Gibbons, (eds.): *Bergey's Manual of Determinative Bacteriology,* 8th ed., The Williams & Wilkins Co., Baltimore 1974.
2. E. F. Gale, E. Cundliffe, P. E. Reynolds, M. H. Richmond, M. J. Waring in *The Molecular Basis of Antibiotic Actions,* 2nd ed., J. Wiley & Sons, London 1972.
3. G. Y. Lesher, E. J. Froelich, M. D. Gruett, J. H. Bailey et al., *J. Med. Pharm. Chem.* **5** (1962) 1063.
4. R. Albrecht, *Prog. Drug Res.* **21** (1977) 9.
5. J. S. Wolfson, D. C. Hooper, *Antimicrob. Agents Chemother.* **28** (1985) 581 – 585.
6. M. P. Wentland, J. B. Cornett in D. M. Bailey (ed.): *Annual Reports in Medicinal Chemistry,* **vol. 20,** Academic Press, New York 1985, pp. 145 – 153.
7. D. C. Hooper, J. S. Wolfson, *Antimicrob. Agents Chemother.* **29** (1986) in press.
8. H. Koga, A. Itoh, S. Murayama, S. Suzae, T. Irikura, *J. Med. Chem.* **23** (1980) 1358 – 1364.
9. M. Gellert, K. Mizuuchi, M. H. O'Dea, H. A. Nash, *Proc. Natl. Acad. Sci.* **73** (1976) 3872 – 3878.
10. A. Sugino, C. L. Peebles, K. N. Kruezer, N. R. Cozzar, *Proc. Natl. Acad. Sci.* **47** (1977) 4767 – 4771.
11. A. M. Barlow, *Br. Med. J.* **2** (1963) 1308.
12. T. A. McAllister, A. Percival, J. G. Alexander, J. G. Boyce et al., *Postgrad. Med. J.* **47** (1971, Sept.) Suppl., 7.
13. E. W. McChesney, E. J. Froelich, G. Y. Lesher et al., *Toxicol. Appl. Pharmacol.* **6** (1964) 292.
14. T. A. Stamey, N. J. Nemoy, M. Higgins, *Invest. Urol.* **6** (1969) 582.
15. G. A. Portmann, E. W. McChesney, H. Stander et al., *J. Pharm. Sci.* **55** (1966) 72.
16. M. Buchbinder, J. C. Webb, L. Anderson, W. R. McCabe, *Antimicrob. Agents Chemother.* 1962, 308.
17. Sterling Drug, US 3 149 104, 1964 (G. Y. Lesher, M. D. Gruett).
18. F. J. Turner, S. M. Ringel, J. F. Martin et al., *Antimicrob. Agents Chemother.* 1967, 475.
19. S. M. Ringel, F. J. Turner, F. L. Lindo et al., *Antimicrob. Agents Chemother.* 1967, 480.
20. D. Kaminsky, R. I. Meltzer, *J. Med. Chem.* **11**(1968) 160.
21. W. E. Wick, D. S. Preston, W. A. White et al., *Antimicrob. Agents Chemother.* **4** (1973) 415.
22. R. G. Gordon, L. I. Stevens, C. E. Edmiston et al., *Antimicrob. Agents Chemother.* **10**(1976) 918.
23. W. A. Goss, W. H. Dietz, T. M. Book, *J. Bacteriol.* **10** (1965) 918.
24. R. A. P. Burt, T. Morgan, J. P. Payne et al., *Br. J. Urol.* **49** (1977) 147.
25. H. R. Black, K. S. Israel, R. L. Wolen et al., *Antimicrob. Agents Chemother.* **15** (1979) 165.
26. Eli Lilly, DE 2 065 719, 1975 (W. A. White).
27. J. R. Prous (ed.), *Ann. Drug Data Rep.* **2** (1979/1980)177 – 180.
28. J. R. Prous (ed.), *Ann. Drug Data Rep.* **2** (1979/1980)181 – 184.
29. J. Matsumoto, S. Minami, *J. Med. Chem.* **18** (1975) 74.
30. Dainippon, JP 25 912, 1967 (Minami et al.).

31. J. R. O'Connor, R. A. Dobson, P. E. Came, R. B. Wagner in J. D. Nelson, C. Grassi (eds.): "Current Chem. and Infect. Dis.," *Proc. 11th ICC and 19th ICAAC,* vol. 1, The American Society for Microbiology, Washington, D.C., 1980, pp. 440 – 442.
32. Sterling Drug, US 3 753 993, 1973 (G. Y. Lesher, P. M. Carabateas).
33. A. Izawa, Y. Kisaki, K. Irie, Y. Eda et al., *Antimicrob. Agents Chemother.* **18** (1980) 37 – 40.
34. A. Izawa, A. Yoshitake, T. Komatsu, *Antimicrob. Agents Chemother.* **18** (1980) 41 – 44.
35. A. Yoshitake, K. Kawahara, F. Shono, I. Umeda et al., *Antimicrob. Agents Chemother.* **18** (1980) 45 – 49.
36. Sumitomo, US 3 799 930, 1974 (T. Nakagome et al.).
37. K. Sato, Y. Matsuura, M. Inoue, T. Une et al., *Antimicrob. Agents Chemother.* **22** (1982) 548 – 553.
38. N. X. Chin, H. C. Neu, *Antimicrob. Agents Chemother.* **25** (1984) 319 – 326.
39. D. S. Reeves, M. J. Bywater, H. A. Holt, L. O. White, *J. Antimicrob. Chemother.* **13** (1984) 333 – 346.
40. J. B. Cornett, R. B. Wagner, R. A. Dobson, M. P. Wentland et al., *Antimicrob. Agents Chemother.* **27** (1985) 4 – 10.
41. N. X. Chin, H. C. Neu, *Antimicrob. Agents Chemother.* **24** (1983) 754 – 763.
42. A. L. Barry, R. N. Jones, *Antimicrob. Agents Chemother.* **25** (1984) 775 – 777.
43. A. Ito, K. Hirai, M. Inoue et al., *Antimicrob. Agents Chemother.* **17** (1980) 103.
44. H. C. Neu, P. Labthavikul, *Antimicrob. Agents Chemother.* **22** (1982) 23.
45. S. R. Crider, S. D. Colby, L. K. Miller, W. O. Harrison et al., *N. Eng. J. Med.* **311** (1984) 137.
46. P. Cantet, H. Ranaudin, C. Quentin, C. Babear, *Pathol. Biol. (Paris)* **31** (1983)501.
47. S. Nakamura, N. Kurobe, S. Kashimoto, T. Ohue et al., *Antimicrob. Agents Chemother.* **24** (1983) 54.
48. Z. N. Adhami, R. Wise, D. Wetson, B. Crump, *Antimicrob. Chemother.* **13** (1984) 87.
49. R. Wise, R. Lockley, J. Dent, M. Webberly, *Antimicrob. Agents Chemother.* **26** (1984) 17.
50. B. Crump, R. Wise, J. Dent, *Antimicrob. Agents Chemother.* **24** (1983) 784.
51. M. Bologna, L. Vaggii, C. M. Forchetti, E. Martine, *Lancet* **2** (1983) 280.
52. B. N. Swanson, V. K. Boppana, P. H. Vlasses, H. M. Rotmensch et al., *Antimicrob. Agents Chemother.* **23** (1983) 284.
53. G. Panichi, A. Pantosti, G. P. Testore, *J. Antimicrob. Chemother.* **11** (1983) 589.
54. Kyorin, BE 870 917, 1979.
55. R. Wise, J. M. Andrew, L. J. Edwards, *Antimicrob. Agents Chemother.* **23** (1983)559.
56. H. L. Muytjens, J. van der Ros-van de Repe, G. van Veldhuizen, *Antimicrob. Agents Chemother.* **24** (1983) 302.
57. R. J. Fass, *Antimicrob. Agents Chemother.* **24** (1983) 568.
58. W. Brumfitt, I. Franklin, D. Grady, J. M. T. Hamilton-Miller et al., *Antimicrob. Agents Chemother.* **26** (1984) 757.
59. Bayer, EP-A 78 362, 1983 (K. Grohe, H. J. Zeiler, K. G. Metzger).
60. S. Nakamura, K. Nakata, H. Katae, A. Minami et al., *Antimicrob. Agents Chemother.* **23** (1983) 742.
61. A. Notowicz, E. Stolz, B. van Klingeren, *J. Antimicrob. Chemother.* **14** (1984) Suppl. C, 91.
62. J. M. Hubrechts, R. Vanhoof, J. Servais, R. Toen et al., *Lancet* **1** (1984) 860.
63. Dainippon, EP-A 9 425, 1980 (J. I. Matsumoto, Y. Takase, Y. Nishimura).
64. D. L. van Caekenberghe, S. R. Pattyn, *Antimicrob. Agents Chemother.* **25** (1984) 518.
65. G. Montay, Y. Gouffon, R. Roquet, *Antimicrob. Agents Chemother.* **25** (1984) 463.
66. M. Wolff, B. Regnier, C. Daldoss, M. Nkam et al., *Antimicrob. Agents Chemother.* **26** (1984) 289.
67. N. Desplaces, L. Gutmann, J. F. Acar, *24th ICAAC,* Washington, D.C., 1984, article no. 279.
68. F. B. McGillion, *Drugs of the Future* **7** (1982) 946.
69. M. R. Lockley, R. Wise, J. Dent, *Antimicrob. Agents Chemother.* **14** (1984) 647.
70. Daiichi Seiyaku, EP-A 47 005, 1982 (I. Hayakawa, T. D. S. Hiramitsu, Y. Tanaka).
71. N. V. Jacobus, F. P. Tally, M. Barza, *Antimicrob. Agents Chemother.* **26** (1984) 104 – 107.
72. I. Garcia, G. P. Bodey, V. Fainstein, D. Hsitto et al., *Antimicrob. Agents Chemother.* **26** (1984) 421 – 423.
73. H. C. Neu, P. Labthavikul, *24th ICAAC,* Washington, D.C., 1984, article no. 402.
74. R. B. Wagner, R. A. Dobson, M. P. Wentland, D.M. Bailey et al., *23rd ICAAC,* Las Vegas, Nev., 1983, article no. 378.

75. Sterling Drug, EP-A 90 424, 1983 (M. P. Wentland, D. M. Bailey).
76. P. B. Fernandes, D. Chu, R. Bowen, N. Shipkowitz et al., *24th ICAAC,* Washington, D.C., 1984, article no. 78.
77. P. B. Fernandes, N. Shipkowitz, D. Chu, L. Doen et al., *24th ICAAC,* Washington, D.C., 1984, article no. 79.
78. N. X. Chin, D. C. Brittain, H. C. Neu, *24th ICAAC,* Washington, D.C., 1984, article no. 399.
79. K. Takagi, M. Hosaka, H. Kusajima, Y. Oomori et al., *23rd ICAAC,* Las Vegas, Nev., 1983, article no. 659.
80. M. A. Cohen, P. A. Bien, T. J. Griffin, C. L. Heifetz, *24th ICAAC,* Washington, D.C., 1984, article no. 81.
81. J. C. Sesnie, M. A. Shapiro, C. L. Heifetz, T. F. Mich, *24th ICAAC,* Washington, D.C., 1984, article no. 82.
82. J. M. Domagala, J. B. Nichols, C. L. Heifetz, T. F. Mich, *24th ICAAC,* Washington, D.C., 1984, article no. 80.
83. Riker Lab, US 3 896 131, 1975 (J. F. Gerster).
84. D. T. W. Chu, G. R. Granneman, P. B. Fernandes, *Drugs of the Future* **10** (1985) 546.
85. Warner Lambert, AU 8 318 698, 1984 (T. P. Culbertson, J. M. Domatgala, T. F. Michh, J. B. Nickols).
86. J. Trefouël, J. Trefouël, F. Nitti, D. Bovet, *C. R. Seances Soc. Biol. Ses Fil.* **120**(1935) 756.
87. D. D. Woods, *Br. J. Exp. Pathol.* **21** (1940)74.
88. P. Fildes, *Lancet* **1** (1940) 955.
89. A. K. Miller, P. Bruno, R. M. Berglund, *J. Bacteriol.* **54** (1947) 9.
90. O. Sköld, *Antimicrob. Agents Chemother.* **9** (1976)49.
91. S. A. Kabins, M. V. Panse, S. Cohen, *J. Infect. Dis.* **123** (1971) 158.
92. R. L. Then, *Rev. Infect. Dis.* **4** (1982) 261.
93. J. M. J. Hamilton-Miller, *J. Antimicrob. Chemother.* **5 Suppl. B** (1979) 61.
94. J. K. Seydel, *J. Pharm. Sci.* **57** (1968) 1455.
95. N. Anand: "Sulfonamides and Sulfones," in M. E. Wolff (ed.): *Burger's Medicinal Chemistry,* 4th ed., J. Wiley & Sons, New York 1979, Part II, Chap. 13.
96. T. Fujita, C. Hanch, *J. Med. Chem.* **10** (1967) 991.
97. H. Kano, US 2 888 455, 1959.
98. B. Roth et al., *J. Med. Chem.* **23** (1980) 535.
99. S. R. M. Bushby, *J. Infect. Dis.* **128 Suppl.** (1973) S443.
100. E. Grunberg, *J. Infect. Dis.* **128 Suppl.** (1973) S478.
101. D. J. Winston, W. K. Lau, R. P. Gale, L. S. Young, *Ann. Intern. Med.* **92** (1980) 762.
102. J. Wüst, *Antimicrob. Agents Chemother.* **11** (1977) 631.
103. R. O. Roblin, *J. Am. Chem. Soc.* **62** (1940) 2002.
104. L. Doub, *J. Med. Chem.* **13** (1970) 242.
105. H. Iundbeck & Co., US 2 447 702, 1948 (O. Hübner).
106. H. Bretschneider, US 3 132 139, 1962.
107. Aktiebolaget Pharmacia, US 2 396 145, 1946 (A. Askëlof).
108. ICI, GB 551 513, 1943 (E. Haworth).
109. M. L. Moore, *J. Am. Chem. Soc.* **64** (1942) 1572.
110. Sharpe & Dohme, US 2 324 013 – 15, 1943 (M. L. Moore).
111. R. M. Des Prez, R. A. Goodwin, Jr. in G. L. Mandell, R. G. Douglas, Jr., J. E. Bennett (eds.): *Principles and Practice of Infectious Diseases,* 2nd ed., J. Wiley & Sons, New York 1985, p. 1383.
112. W. E. Bullock, [111], p. 1406.
113. W. E. Sanders, Jr., [111], p. 1413.
114. *Med. Lett.* **28** (1986, Mar. 28) 710.
115. Tuberculosis Chemotherapy Trials Committee: Interim Report to the Medical Research Council, "The Treatment of Pulmonary Tuberculosis with Isoniazid," *Br. Med. J.* **2** (1952) 735.
116. W. B. Pratt in *Chemotherapy of Infection,* Oxford University Press, New York 1977, p. 231.
117. D. E. Kopanoff, D. E. Spider, G. J. Caras, *Am. Rev. Respir. Dis.* **117** (1978) 991.
118. American Thoracic Society, *Am. Rev. Respir. Dis.* **110** (1974) 374.
119. Distillers, US 2 280 994, 1958.
120. R. G. Wilkinson et al., *J. Am. Chem. Soc.* **83** (1961) 2212.
121. W. W. Stead, A. K. Dutt, *Am. Rev. Respir. Dis.* **125 Suppl.** (1982) no. 3, 94.
122. D. J. Girling, *Tubercule* **59** (1978) 13.
123. A. K. Dutt, W. W. Stead, *J. Infect. Dis.* **146** (1982) 698.
124. Merck, US 2 149 279, 1939 (O. Dalmer).
125. D. Lieberman, *Bull. Soc. Chim. Fr.* (1958) 687.
126. J. T. Sheehan, *J. Am. Chem. Soc.* **70** (1948) 1665.
127. G. L. Mandell, M. A. Sande in A. G. Gilman, L. S. Goodman, A. Gilman (eds.): *Drugs Used in the Chemotherapy of Tuberculosis and Leprosy, The Pharmacological Basis of Therapeutics,* 6th ed., Macmillan Publ. Co., New York 1980, p. 1200.

128. G. Domagk et al., *Naturwissenschaften* **33** (1946) 315.
129. C. C. Shepard, L. Levy, P. Fasal, *Am. J. Trop. Med. Hyg.* **18** (1969) 258.
130. R. W. Riley, L. Levy, *Proc. Soc. Exp. Biol. Med.* **142** (1973) 1168.
131. I.G. Farbenind, FR 829 926, 1938.
132. H. Bauer, *J. Am. Chem. Soc.* **61** (1939) 617.
133. J. H. Peters, J. F. Murray, G. R. Gordon et al., *Am. J. Trop. Med. Hyg.* **26** (1977) 127.
134. D. A. Russell, C. C. Shepard, D. H. McRae et al., *Am. J. Trop. Med. Hyg.* **24** (1975) 485.
135. E. Eslarger et al., *J. Med. Chem.* **12** (1969) 357
136. L. Levy, *Am. J. Trop. Med. Hyg.* **23** (1974) 1097.
137. J. Convit, S. G. Browne, J. Languillon et al., *WHO Bull.* **42** (1970) 667.
138. C. C. Shepard, L. L. Walker, R. M. Van Lindingham et al., *Proc. Soc. Exp. Biol. Med.* **137** (1971) 725.
139. C. C. Shepard, L. L. Walker, R. M. Van Lindingham et al., *Proc. Soc. Exp. Biol. Med.* **137** (1971) 728.
140. V. C. Barry, *J. Chem. Soc.* (1958) 859.
141. V. L. Sutter in S. M. Finegold, W. L. George, R. D. Rolfe (eds.): *First United States Metronidazole Conference, Proc. Symp.,* Tarpon Springs, Fla., Biomedical Information Corp., New York 1982, p. 61.
142. Hoffmann La Roche, US 2 430 094, 1947 (H. M. Wuest).
143. S. M. Finegold, [111], p. 220.
144. R. M. Jacob, Rhône-Poulenc, US 2 944 061, 1960.
145. M. Turck, A. R. Ronald, R. G. Petersdorf, *Antimicrob. Agents Chemother.* 1966, 446.
146. A. Kucers, N. M. Bennett in A. Kucers, N. M. Bennett (eds.): *The Use of Antibiotics,* J. B. Lippincott Co., Philadelphia 1979, p. 749.
147. H. K. Reckendorf, R. G. Castringius, H. K. Spingler, *Antimicrob. Agents Chemother.* 1962, 531.
148. Eaton Labs., US 2 610 181, 1952 (K. J. Hayes).
149. R. A. Marcial-Rojas (ed.): *Pathology of Protozoal and Helminth Diseases, With Clinical Correlation,* Williams & Wilkins, New York 1971.
150. H. Van de Bossche, *Nature London* **273** (1978) 626.
151. T. C. Jones, [111], pp. 1505, 1513, 1560.
152. D. J. Wyler, [111], p. 1514.
153. D. R. Hill, [111], p. 1552.
154. J. J. Marr, [111], p. 286.
155. [111], p. 1537.
156. J. I. Ravdin, T. C. Jones, [111], p. 1506.
157. W. T. Hughes, [111], p. 1549.
158. L. V. Kirchoff, F. Neva, [111], p. 1531.
159. M. R. Rein, [111], p. 1556.
160. T. K. Ruebush, II, [111], p. 1559.
161. H. W. Brown, *Basic Clinical Parasitology,* 4th ed., Appleton-Century-Crofts, New York 1976.
162. R. D. Pearson, A. de Queiroz Sousa, [111], p. 1522.
163. R. E. McCabe, J. E. Remington, [111], p. 1540.
164. W. E. Gutteridge, *Br. Med. Bull.* **41** (1985) 162.
165. A. Sjoerdsma, J. A. Golden, P. J. Schechter et al., *Clin. Res.* **32** (1984) 559 A.
166. J. Williamson, *Exp. Parasitol.* **12** (1962) 274.
167. J. Williamson in H. W. Mulligan (ed.): *The African Trypanosomiases,* Allen & Unwin, London 1970, p. 125.
168. J. Williamson, *Trop. Dis. Bull.* **73** (1976) 531.
169. F. Hawking in R. J. Schnitzer, F. Hawking (eds.): *Experimental Chemotherapy,* **vol. 1,** Academic Press, New York 1963, p. 129.
170. B. B. Waddy, [167], p. 711.
171. F. I. C. Apted, *Pharmacol. Ther.* **11** (1980) 391.
172. D. A. Evans, *Antibiot. Chemother.* **30** (1981)272.
173. May & Baker, US 2 410 796, 1946 (G. Neuberry).
174. E. Friedheim, *Chem. Abstr.* **47** (1953) 144.
175. W. A. Jacobs, *J. Am. Chem. Soc.* (1919) 1590.
176. Hoechst, US 2 838 485, 1958 (R. Brodersten).
177. Z. Brenner, *Adv. Pharmacol. Chemother.* **13** (1975)1.
178. Z. Brenner, *Pharmacol. Ther.* **7** (1979) 71.
179. W. E. Gutteridge, *Trop. Dis. Bull.* **73** (1976) 699.
180. W. E. Gutteridge in J. R. Baker (ed.): "Perspectives in Trypanosomiasis Research," *Proceedings of the Twenty-first Trypanosomiasis Seminar,* Research Studies Press, Cinchester 1982, p. 47.
181. R. Ribeiro-dos-Santos, A. Rassi, F. Koberle, *Antibiot. Chemother.* **30** (1980) 115.
182. Bayer, US 3 262 930, 1964 (H. Herlinger).
183. Hoffmann-La Roche, *Chem. Abstr.* **71** (1969) 3383.
184. P. H. Rees, P. A. Kager, T. Ogada, J. K. M. Eeftinck Schattenkerk, *Trop. Geogr. Med.* **37** (1985) 37.
185. L. L. Branborg, R. Owen, R. Fogel, H. Goldberg et al., *Gastroenterology* **78** (1980) 1602.
186. B. M. Honigberg in L. Kreiser, *Parasitic Protozoa,* **vol. 2,** Academic Press, New York 1978, p. 275.

187. O. Jirovec, M. Petru, *Adv. Parasitol.* **6** (1968) 117.
188. R. Knight, *Clin. Gastroenterol.* **7** (1978) 47.
189. E. A. Meyer, S. Radulescu, *Adv. Parasitol.* **17** (1979) 1.
190. J. W. Smith, M. S. Wolfe, *Ann. Rev. Med.* **31** (1980) 373.
191. J. G. Meingassner, P. G. Heyworth, *Antibiot. Chemother.* **30** (1981) 163.
192. K. Butler, *Chem. Abstr.* **71** (1964) 3384 e.
193. Carlo Erba, US 3 399 193, 1965 (P. N. Giraldi).
194. Hoffmann La Roche, US 3 435 049, 1966 (M. Hoffer).
195. Rhône-Poulenc, *Chem. Abstr.* **63** (1965) 11 571 d.
196. J. Heeres, *Eur. J. Med. Chem.-Chim. Ther. (Paris)* **11** (1976) 237.
197. S. J. Lerman, R. A. Walker, *Clin. Pediatr. (Phila.)* **21** (1982) 409.
198. M. S. Wolfe, *J. Am. Med. Assoc.* **233** (1975)1362.
199. B. R. Brown, *J. Chem. Soc.* 1948, 99.
200. G. D. Drake, *Chem. Abstr.* **51** (1957) 2051 e.
201. I. M. Rollo in A. F. Goodman, L. S. Goodman, A. Gilman (eds.): *The Pharmacological Basis of Therapeutics,* 6th ed.,Macmillan Publ. Co., New York 1980, p. 1061.
202. A. R. Surrey, *J. Am. Chem. Soc.* **68** (1946) 113.
203. J. H. Burckhalter, *J. Am. Chem. Soc.* **70** (1948) 1363.
204. A. R. Surrey, *J. Am. Chem. Soc.* **72** (1950) 1814.
205. R. C. Elderfield, *J. Am. Chem. Soc.* **68** (1946) 1525.
206. P. B. Russel, *J. Am. Chem. Soc.* **73** (1951) 3763.
207. J. Souza, *WHO Bull.* **61** (1983) 815.
208. F. H. S. Curd, *J. Chem. Soc.* (1948) 1630.
209. T. Harinasuta, D. Bunnag, W. H. Wernsdorfer, *WHO Bull.* **61** (1983) 299.
210. F. I. Carroll, *J. Med. Chem.* **17** (1974) 210.
211. J. Rinehart, *Am. J. Trop. Med. Hyg.* **25** (1976) 769.
212. K. Killer, *Drugs of the Future* **5** (1980) 547.
213. V. Papesch, *J. Am. Chem. Soc.* **58** (1936) 1314.
214. A. Das, *J. Org. Chem.* **22** (1957) 1111.
215. A. Luis, *Chem. Abstr.* **47** (1953) 10 533 d.
216. I.G. Farbenind, DE 545 915, 1930 (K. Schranz).
217. Boots Pure Drug, US 2 912 438, 1959 (P. Oxley).
218. T. H. Flewett, *Br. Med. Bull.* **41** (1985) 315 – 321.
219. R. E. F. Matthews, *Intervirology* **17** (1982) 1 – 199.
220. B. Rada, D. Blaskovic, F. Sorm, J. Skoda, *Experientia* **16** (1960) 487.
221. E. C. Herrmann, *Proc. Soc. Exp. Biol. Med.* **107** (1961) 142 – 145.
222. B. Jubelt, O. Narajan, R. T. Johnson, *J. Neuropathol. Exp. Neurol.* **39** (1980) 149 – 159.
223. B. Clement, *Pharm. Unserer Zeit* **15** (1986) 72 – 84.
224. H. J. Schaeffer, L. Beauchamp, P. deMiranda, G. B. Elion et al., *Nature (London)* **272** (1978) 583 – 585.
225. H. J. Schaeffer, S. Gurwara, R. Vince, S. Bittner, *J. Med. Chem.* **14** (1971) 367 – 369.
226. B. R. Jones, D. J. Coster, P. N. Fison, G. M. Thompson et al., *Lancet* **1** (1979) 243 – 244.
227. P. R. Laibson, D. Pavan-Langston, W. R. Yeakley, J. Lass, *Am. J. Med.* **73 A** (1982) 281 – 285.
228. C. La Lau, J. A. Oosterhuis, J. Versteeg, G. Van Rij et al., *Am. J. Med.* **73 A** (1982) 305 – 306.
229. J. M. Yeo, A. P. Fiddian, *J. Antimicrob. Chemother.* **12** (1983) Suppl. B, 95 – 103.
230. L. Corey, A. J. Nahmias, M. E. Guinan, J. K. Benedetti et al., *New. Engl. J. Med.* **306** (1982) 1313 – 1319.
231. L. Corey, J. K. Benedetti, C. W. Critchlow, M. R. Remington et al., *Am. J. Med.* **73** (1982) 326 – 334.
232. A. E. Nilsen, T. Aasen, A. M. Halsos, B. R. Kinge et al., *Lancet* **2** (1982) 571 – 573.
233. O. P. Salo, A. Lassus, T. Hovi, A. P. Fiddian, *Eur. J. Sex. Trans. Dis.* **1**(1983) 95 – 98.
234. R. C. Reichman, G. J. Badger, G. J. Mertz, L. Corey et al., *J. Am. Med. Assoc.* **251** (1984) 2103 – 2107.
235. J. M. Douglas, C. Critchlow, J. Benedetti, G. J. Mertz et al., *New. Engl. J. Med.* **310** (1984) 1551 – 1556.
236. S. E. Straus, H. E. Takiff, M. Seidlin, S. Bachrach et al., *New Engl. J. Med.* **310** (1984) 1545 – 1550.
237. D. Brigden, *Br. Med. Bull.* **41** (1985) 357 – 360.
238. T. A. Krenitsky, W. W. Hall, P. deMiranda, L. M. Beauchamp et al., *Proc. Natl. Acad. Sci. U.S.A.* **81** (1984) 3209 – 3213.
239. G. R. Revankar, J. H. Huffman, L. B. Allen, R. W. Sidwell et al., *J. Med. Chem.* **18** (1975) 721 – 726.
240. H. E. Kaufman, *Proc. Soc. Exp. Biol. Med.* **109** (1962) 251 – 252.
241. H. E. Kaufman, E. D. Varnell, Y. M. CentifantoFitzgerald, J. G. Sanitato, *Antiviral Res.* **4** (1984) 333 – 338.

242. W. H. Prusoff, *Biochim. Biophys. Acta* **32** (1959) 295 – 296.
243. P. C. Heidelberger, D. G. Parsons, D. C. Remy, *J. Med. Chem.* **7** (1964) 1 – 5.
244. C. Wellings, P. N. Awdry, F. H. Bors, B. R. Jones et al., *Am. J. Ophthalmol.* **73** (1972) 932 – 942.
245. P. C. Heidelberger, D. King, *Pharmacol. Ther.* **6** (1979) 427 – 442.
246. P. C. Heidelberger, *Ann. N. Y. Acad. Sci.* **255** (1975) 317 – 325.
247. W. W. Lee, A. Benitez, L. Goodman, B. R. Baker, *J. Am. Chem. Soc.* **82** (1960) 2648 – 2649.
248. E. J. Reist, A. Benitez, L. Goodman, B. R. Baker et al., *J. Org. Chem.* **27**(1962) 3274 – 3279.
249. J. DeRudder, M. Privat de Garlihe, *Antimicrob. Agents Chemother.* **5** (1965) 578 – 584.
250. F. M. Schabel, *Chemotherapy* **13** (1968) 321 – 338.
251. D. Pavan-Langston, C. H. Dohiman, *Am. J. Ophthalmol.* **74** (1972) 81 – 88.
252. R. J. Whitley, L. T. Ch'ien, R. Dolin, G. J. Galasso et al., *New. Engl. J. Med.* **294** (1976) 1193 – 1199.
253. R. J. Whitley, S.-J. Soong, R. Dolin, G. J. Galasso et al., *New. Engl. J. Med.* **297** (1977) 289 – 294.
254. B. Sköldenberg, M. Forsgren, K. Alestig, T. Bergström et al., *Lancet* **2** (1984) 707 – 711.
255. H. G. Adams, E. A. Benson, E. R. Alexander, L. A. Vontver et al., *J. Infect. Dis.* **133** (1976) Suppl. A, 151 – 159.
256. S. L. Spruance, C. S. Crumpacker, H. Haines, C. Bader et al., *New. Engl. J. Med.* **300** (1979) 1180 – 1184.
257. W. H. Prusoff, M. Zucker, W. R. Mancini, M. J. Otto et al., *Antiviral Res.* (1985) Suppl. 1, 1 – 10.
258. E. DeClercq, J. Descamps, P. DeSomer, P. J. Barr et al., *Proc. Natl. Acad. Sci.* **76** (1979) 2947 – 2951.
259. E. DeClercq, H. Degreef, J. Wildiers, G. DeJonge et al., *Br. Med. J.* **281** (1980)1178.
260. P. C. Maudgal, E. De Clercq, L. Missotten, *Antiviral Res.* **4** (1984) 281 – 291.
261. R. K. Robins, *Chem. Eng. News* **64** (1986) 28 – 40.
262. K. A. Watanabe, U. Reichman, K. Hirota, C. Lopez et al., *J. Med. Chem.* **22** (1979) 21 – 24.
263. K. A. Watanabe, T.-L. Su, R. S. Klein, C. K. Chu, *J. Med. Chem.* **26** (1983) 152 – 156.
264. C. W. Young, R. Schneider, B. Leyland-Jones, D. Armstrong et al., *Cancer Res.* **43** (1983) 5006 –5009.
265. Y.-C. Cheng, G. Dutschman, J. J. Fox, K. A. Watanabe et al., *Antimicrob. Agents Chemother.* **20** (1981) 420 – 423.
266. H. S. Allaudeen, J. Descamps, R. K. Sehgal, J. J. Fox, *J. Biol. Chem.* **257** (1982) 11 879 – 11 882.
267. C. McLaren, M. S. Chen, R. H. Barbhaiya, R. A. Buroker et al. in R. Kono (ed.): *Herpesviruses and Virus Chemotherapy,* Elsevier Science Publ., New York 1985, pp. 57 – 61.
268. J. T. Witkowski, R. K. Robins, R. W. Sidwell, L. N. Simon, *J. Med. Chem.* **15** (1972) 1150 – 1154.
269. R. W. Sidwell, R. K. Robins, I. W. Hillyard, *Pharmacol. Ther.* **6** (1979) 123 – 146.
270. D. G. Streeter, J. T. Witkowski, G. P. Khare, R. W. Sidwell et al., *Proc. Natl. Acad. Sci.* **70** (1973) 1174 – 1178.
271. B. B. Goswami, E. Borek, O. K. Sharma, J. Fujitaki et al., *Biochem. Biophys. Res. Commun.* **89** (1979) 830 – 836.
272. J. B. McCormick, J. P. Getchell, S. W. Mitchell, D. R. Hicks, *Lancet* **2** (1984) 1367 – 1369.
273. C. B. Hall, J. T. McBride, E. E. Walsh, D. M. Bell et al., *New Engl. J. Med.* **308** (1983) 1443 – 1447.
274. C. B. Hall, J. T. McBride, C. L. Gala, S. W. Hildreth et al., *J. Am. Med. Assoc.* **254** (1985) 3047 – 3051.
275. J. B. McCormick, I. J. King, P. A. Webb, C. L. Scribner et al., *New Engl. J. Med.* **314** (1986) 20 – 26.
276. C. R. Magnussen, J. R. Douglas, R. F. Betts, F. K. Roth et al., *Antimicrob. Agents Chemother.* **12** (1977) 498 – 502.
277. F. Salido-Rengell, H. Nasser-Quinones, B. Briseno Garcia, *Ann. N.Y. Acad. Sci.* **284** (1977) 272 –277.
278. C. B. Smith, R. P. Charette, J. P. Fox, M. K. Cooney et al. in R. A. Smith, W. Kirkpatrick (eds.): *Ribavirin: A Broad Spectrum Antiviral Agent,* Academic Press, New York 1980, pp. 147 – 164.
279. Y. Togo, E. A. McCracken, *J. Infect. Dis.* **133** (1976) Suppl. A, 109 – 113.
280. V. Knight, S. Wilson, J. M. Quarles, S. E. Greggs et al., *Lancet* **2** (1981) 945 – 949.
281. H. W. McClung, V. Knight, B. E. Gilbert, S. Z. Wilson et al., *J. Am. Med. Assoc.* **249** (1983) 2671 – 2674.
282. R. Vince, S. Daluge, *J. Med. Chem.* **20** (1977) 612 – 613.

283. W. M. Shannon, L. Westbrook, G. Arnett, S. Daluge et al., *Antimicrob. Agents Chemother.* **24** (1983) 538 – 543.
284. R. Vince, S. Daluge, H. Lee, W. M. Shannon et al., *Science* **221** (1983) 1405 – 1406.
285. K. K. Ogilvie, U. O. Cheriyan, B. K. Radatus, K. O. Smith et al., *Can. J. Chem.* **60** (1982) 3005 – 3010.
286. W. T. Ashton, J. D. Karkas, A. K. Field, R. L. Tolman, *Biochem. Biophys. Res. Commun.* **108** (1982) 1716 – 1721.
287. A. K. Field, M. E. Davies, C. DeWitt, H. C. Perry et al., *Proc. Natl. Acad. Sci.* **80** (1983) 4139 – 4143.
288. D. H. Shepp, P. S. Dandliker, P. deMiranda, T. C. Burnette et al., *Ann. Intern. Med.* **103** (1985) 368 – 373.
289. D. Felsenstein, D. J. D'Amico, M. S. Hirsch, D. A. Neumeyer et al., *Ann. Intern. Med.* **103** (1985) 377 – 380.
290. M. C. Bach, S. P. Bagwell, N. P. Knapp, K. M. Davis et al., *Ann. Intern. Med.* **103** (1985) 381 – 382.
291. J. C. Martin, C. A. Dvorak, D. F. Smee, T. R. Matthews et al., *J. Med. Chem.* **26** (1983) 759 – 761.
292. A. Larsson, B. Öberg, S. Alenius, C.-E. Hagberg et al., *Antimicrob. Agents Chemother.* **23** (1983) 664 – 670.
293. A. Larsson, P.-Z. Tao, *Antimicrob. Agents Chemother.* **25** (1984) 524 – 526.
294. A. Larsson, S. Alenius, N.-G. Johansson, B. Öberg, *Antiviral Res.* **3** (1983) 77 – 86.
295. Astra, EP-A 55 239, 1982 (C. E. Hagberg, K. N. G. Johansson, Z. M. I. Kovacs, G. B. Stening).
296. N. L. Shipkowitz, R. R. Bower, R. N. Appell, C. W. Nordeen et al., *Appl. Microbiol.* **26** (1973) 264 – 267.
297. J. C.-H. Mao, E. E. Robishaw, *Biochemistry* **14** (1975) 5475 – 5479.
298. C. L. K. Sabourin, J. M. Reno, J. A. Boezi, *Arch. Biochem. Biophys.* **187** (1978) 96 – 101.
299. B. A. Bopp, C. B. Estep, D. J. Anderson, *Fed. Proc.* **36** (1977) 939.
300. J. C.-H. Mao, E. R. Otis, A. M. von Esch, T. R. Herrin et al., *Antimicrob. Agents Chemother.* **27** (1985) 197 – 202.
301. J. M. Reno, L. F. Lee, J. A. Boezi, *Antimicrob. Agents Chemother.* **13** (1978) 188 – 192.
302. E. Helgstrand, H. Flodh, J.-O. Lernestedt, J. Lundstrom et al. in L. H. Collier, J. Oxford (eds.): *Developments in Antiviral Therapy,* Academic Press, London 1980, pp. 63 – 83.
303. J. Wallin, J.-O. Lernestedt, S. Ogenstad, E. Lycke, *Scand. J. Infect. Dis.* **17** (1985) 165 – 172.
304. G. Klintmalm, B. Lönngvist, B. Öberg, G. Gahrton et al., *Scand. J. Infect. Dis.* **17** (1985) 157 – 163.
305. E. G. Sandstrom, J. C. Kaplan, R. E. Byington, M. S. Hirsch, *Lancet* **1** (1985) 1480 – 1482.
306. W. L. Davies, R. R. Grunert, R. F. Haff, J. W. McGahen et al., *Science* **144** (1964) 862 – 863.
307. N. Kato, H. J. Eggers, *Virology* **37** (1969) 632 – 641.
308. G. G. Jackson, R. L. Muldoon, L. W. Akers, *Antimicrob. Agents Chemother.* **3** (1963) 703 – 707.
309. A. W. Galbraith, J. S. Oxford, G. C. Schild, G. I. Watson, *Lancet* **2** (1969) 1026 – 1028.
310. R. Dolin, R. C. Reichman, H. P. Madore, R. Maynard et al., *New Engl. J. Med.* **307** (1982) 580 – 584.
311. L. P. VanVoris, R. F. Betts, F. G. Hayden, W. A. Christmas et al., *J. Am. Med. Assoc.* **245** (1981) 1128 – 1131.
312. F. G. Hayden, A. S. Monto, *Antimicrob. Agents Chemother.* **29** (1986) 339 – 341.
313. J. H. Wikel, C. J. Paget, D. C. DeLong, J. D. Nelson et al., *J. Med. Chem.* **23** (1980) 368 – 372.
314. D. C. DeLong, S. E. Reed, *J. Infect. Dis.* **141** (1980) 87 – 91.
315. R. J. Phillpotts, D. C. DeLong, J. Wallace, R. W. Jones et al., *Lancet* **1** (1981) 1342 – 1344.
316. F. G. Hayden, J. M. Gwaltney, *Antimicrob. Agents Chemother.* **21** (1982) 892 – 897.
317. R. A. Levandowski, C. T. Pachucki, M. Rubenis, G. G. Jackson, *Antimicrob. Agents Chemother.* **22** (1982) 1004 – 1007.
318. R. J. Phillpotts, J. Wallace, D. A. J. Tyrrell, V. B. Tagart, *Antimicrob. Agents Chemother.* **23** (1983) 671 – 675.
319. F. D. Miller, A. S. Monto, D. C. DeLong, A. Exelby et al., *Antimicrob. Agents Chemother.* **27** (1985) 102 – 106.
320. D. J. Bauer, J. W. T. Selway, J. F. Batchelor, M. Tisdale et al., *Nature (London)* **292** (1981) 369 – 370.
321. M. Tisdale, J. W. T. Selway, *J. Gen. Virol.* **64** (1983) 795 – 803.
322. M. Tisdale, J. W. T. Selway, *J. Antimicrob. Chem.* **14** (1984) Suppl. A, 97 – 105.
323. R. J. Phillpotts, J. Wallace, D. A. J. Tyrrell, D. S. Freestone et al., *Arch. Virol.* **75** (1983) 115 – 121.

324. H. Ishitsuka, Y. T. Ninomiya, C. Ohsawa, M. Fujiu et al., *Antimicrob. Agents Chemother.* **22** (1982) 617 – 621.
325. Y. Ninomiya, C. Ohsawa, M. Aoyama, I. Umeda et al., *Virology* **134** (1984) 269 – 276.
326. R. J. Phillpotts, P. G. Higgins, J. S. Willman, D. A. J. Tyrrell et al., *J. Antimicrob. Chemother.* **14** (1984) 403 – 409.
327. G. D. Diana, U. J. Salvador, E. G. Zalay, R. E. Johnson et al., *J. Med. Chem.* **20** (1977) 750 – 756.
328. G. D. Diana, M. A. McKinlay, M. J. Otto, V. Akullian et al., *J. Med. Chem.* **28** (1985) 1906 – 1910.
329. M. A. McKinlay, J. V. Miralles, C. J. Brisson, F. Pancic, *Antimicrob. Agents Chemother.* **22** (1982) 1022 – 1025.
330. M. J. Otto, M. P. Fox, M. J. Fancher, M. F. Kuhrt et al., *Antimicrob. Agents Chemother.* **27** (1985) 883 – 886.
331. M. A. McKinlay, B. A. Steinberg, *Antimicrob. Agents Chemother.* **29** (1986) 30 – 32.
332. J. J. McSharry, L. A. Caliguiri, H. J. Eggers, *Virology* **97** (1979) 307 – 315.
333. T.-S. Lin, W. H. Prusoff, *J. Med. Chem.* **21** (1978) 109 – 112.
334. C. Norman, *Science* **230** (1985) 1355 – 1358.
335. *FDC Reports,* Mar. 3, 1986.
336. J. P. Horwitz, J. Chua, M. Noel, *J. Org. Chem.* **29** (1964) 2076 – 2078.
337. R. S. Goldsmith in B. G. Katzung (ed.): *Basic Clinical Pharmacology,* 2nd ed., Lange Medical Publications, Los Altos, Calif., 1984, pp. 648 – 675.
338. E. DeClercq, *Cancer Lett.* **8** (1979) 9 – 22.
339. H. Mitsuya, M. Popovic, R. Yarchoan, S. Matsushita et al., *Science* **226** (1984) 172 – 174.
340. D. Rouvroy, J. Bogaerts, J.-B. Habyarimana, D. Nzaramba et al., *Lancet* **1** (1985) 878 – 879.
341. M. Windholz, S. Budavari, R. F. Blumetti, E. S. Otterbein (eds.): *The Merck Index,* 10th ed., Merck & Co., Rahway, N.J., 1983, p. 1294.
342. J. C. Chermann, F. C. Sinoussi, C. Jasmin, *Biochem. Biophys. Res. Commun.* **65** (1975) 1229 – 1236.
343. W. Rozenbaum, D. Dormont, B. Spire, E. Vilmer et al., *Lancet* **1** (1985) 450 – 451.
344. M. Michelon, G. Herve, *C. R. Acad. Sci. Paris* **274** (1972) 209 – 212.

Antimycotics

See also → Pharmaceuticals; → Chemotherapeutics; → Disinfectants

AXEL SCHMIDT, Bayer AG, Wuppertal, Federal Republic of Germany

FRANK-ULRICH GESCHKE, Bayer AG, Wuppertal, Federal Republic of Germany

1. Introduction

The significant advances in antimycotic chemotherapy in the last decades are mainly due to the introduction of new, potent, and broadly active antimycotics to the market. The substances can be classified according to their mode of application into topical and systemic antimycotics. Topical antimycotics are applied to the skin, mucous membranes, and intravaginally in order to treat superficial mycoses locally. Peroral therapy and prophylaxis of intraluminal colonializations or infections of the gastrointestinal tract, e.g., with drops, syrups, or lozenges containing antimycotics that are not absorbed or are absorbed to an extremely small extent from the gastrointestinal tract (no or extremely low bioavailability) must also be regarded as topical therapy. Systemic antimycotics are applied parenterally and/or orally and show moderate to good bioavailability.

At present, numerous drugs for the topical treatment of superficial mycoses such as cutaneous and mucocutaneous infections are available, but there are only a limited number of drugs for the treatment of systemic mycoses. Additionally, resistance problems are arising, especially in the antimycotic chemotherapy of systemic mycoses, so that the medical need for new substances for this indication is great. Several "historical" compounds that show a relatively unselective antifungal activity, such as dyes, organic acids, mercury salts, and quaternary ammonium compounds, are sometimes still used in the treatment of superficial mycoses. Their therapeutic efficacy, tolerance, and patient acceptance are rated as low nowadays, and these substances are of decreasing relevance as much more selective antimycotics are available.

Ullmann's Industrial Toxicology

ISBN: 3-527-31247-1

Modern antimycotics are expected to be effective in vitro, in animal models, and in infected humans and animals against one or more of the following classes of pathogenic fungi: dermatophytes (e.g., *Trichophyton* and *Microsporum* species, *Epidermophyton floccosum*), yeasts (e.g., *Candida* and *Torulopsis* species, *Cryptococcus neoformans*), moulds (e.g., *Aspergillus fumigatus* and other *Aspergillus* species, *Fusarium* species, zygomycetes such as *Mucor, Rhizopas,* and *Absidia* species), and dimorphic (biphasic) fungi, which can express a yeast phase, which is mostly the pathogenic form in the infected host, and a mould phase, which is mostly the infective, saprophytic form in nature, e.g., in soils [1].

In addition to the very common superficial mycoses of the skin and mucous membranes, yeasts, moulds, and dimorphic fungi can also induce serious, life-threatening generalized opportunistic infections, especially in immunocompromised hosts. As the number of patients at high risk of systemic mycoses (e.g., patients after transplantations, under aggressive antineoplastic chemotherapy or in intensive care, AIDS patients) is increasing steadily, these infections are much more common nowadays [1]. Dermatophytes solely cause infections of the skin, hair, and nails, and are unable to induce systemic, generalized mycoses, even in immunocompromised hosts.

The in vitro antimycotic activity of the substances is assessed by the determination of MIC values (minimal inhibitory concentrations). This is the lowest concentration of a substance that inhibits the growth of a fungal isolate in vitro. The MICs of antimycotics can vary considerably – up to as much as several decades – depending on culture conditions such as medium composition, incubation temperature, inoculum size and preparation method. Therefore, a high degree of standardization must be maintained to obtain comparable results. Furthermore, there is often only a poor or no in vitro/in vivo correlation of MIC values with the clinical outcome.

Some antimycotics, including azole agents, also show a moderate to mostly low in vitro antibacterial activity which does not seem to be relevant for the clinical situation during therapy. Therefore, these minor antibacterial properties of the substances are not discussed in this chapter. Many formulations, especially of topically applicable antimycotics, contain other compounds such as urea, antibiotics, and steroids. These combinations are not extensively discussed here.

2. Azole Antimycotics

Azoles with antimycotic properties can be devided into imidazole and triazole derivatives. Both groups have the same mechanism of antifungal action. Imidazole derivatives with antimycotic activity were first synthesized at Bayer and Janssen and described in 1968/1969. Serval important commercial products have been developed for human and veterinary medicine and plant protection [2].

Low doses of azole antimycotics produce morphological changes in sensitive fungi. These are reflected in changes in the cell volume and plasma membrane [3] and are probably due to the interaction of the azoles with ergosterol biosynthesis [4–7]. Biochemically, azoles specifically inhibit fungal ergosterol biosynthesis by interference with fungal specific isoenzymes of the cytochrome P450-dependent 14α-demethylation of lanosterol or 24-methylenedihydrolanosterol, which is a key step in the biosynthesis of ergosterol. The accumulation of membrane-disturbing 14α-methylsterols seems to be the origin of the antifungal properties of azole compounds, as known from in vitro studies [8]. Ergosterol is an essential constituent of the fungal plasma membrane, helping to maintain the integrity of the cell membrane which has a barrier function. Ergosterol deficiency leads to loss of essential cytoplasmatic constituents through the plasma membrane, and results in cessation of growth and even death of the fungus. In all studies, an accumulation of sterols with a methyl group on C14, which are unsuitable as membrane components, was observed [9], [10]. At higher doses, azoles also interfere with synthesis of fatty acids and triglyceride [4]. The weak antibacterial activity of these substances seems to be based on another mechanism. However, MIC values of azole antimycotics strongly depend on the test method and are mostly of limited use in predicting clinical efficacy.

2.1. Azole Antimycotics for Systemic Treatment

2.1.1. Fluconazole

Fluconazole [*86386-73-4*], α-(2,4-difluorophenyl)-α-(1*H*-1,2,4-triazol-1-ylmethyl)-1*H*-1,2,4-triazole-1-ethanol, $C_{13}H_{12}F_2N_6O$, M_r 306.27, *mp* 138 – 140 °C, crystals from ethyl acetate/hexane.

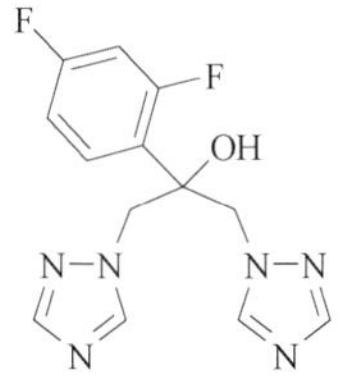

Synonym(s). UK 49858.

Solubility. Soluble in water.

Stability. The pure solid form is stable under normal storage conditions.

Description. The triazole antimycotic fluconazole was first synthesized at Pfizer Pharmaceuticals [11], [12].

Formulations. Capsules à 0.05, 0.1, 0.15, and 0.2 g. Suspensions (0.5 and 1 %) for oral application. Infusion bottles à 0.1, 0.2, and 0.4 g.

Trade Names. Biozolene (Bioindustria), Diflucan (Mason, Pfizer, Roerig), Dimycon (Alkaloid), Elazor (Sigma-Tau), Flavizol (Gador), Fungata (Mack), Mutum (Raffo), Triflucan (Pfizer).

Antimycotic Properties. Fluconazole is a broad spectrum triazole antimycotic with good in vitro and in vivo activity against pathogenic *Candida* species (apart from *Candida krusei*, which is a primary resistant *Candida* species), *Cryptococcus neoformans*, and dermatophytes [13]. *Aspergillus* species mostly show primary resistance towards fluconazole. Secondary resistance development can be observed after prolonged or repeated therapy of *Candida*-associated mycoses, especially in AIDS patients. In vitro susceptibilities of yeasts against fluconazole are reported in Table 1 [14].

Table 1. MIC values for fluoconazole

Fungus	No. isolates tested	MIC range, µg/mL
Candida albicans	443	0.125 – > 80
Candida kefyr	22	32 – 64
Candida lusitaniae	28	0.25 – 32
Candida parapsilosis	68	0.5 – 4
Candida tropicalis	88	0.5 – > 64
Trichosporon beigelii	15	1 – 64

Pharmacokinetics. The bioavailability after oral application ranges around 90 % [15]. The plasma concentrations and AUC (Area Under the Curve) are dose-proportional in the dose range of 50 – 400 mg/d [15]. After a single oral dose of 100 mg, a plasma concentration of 1.7 µg/L is achieved (6.7 µg/L after 400 mg) [16]. A steady state is mostly achieved after a therapy duration of 6 – 10 d. The serum protein binding of fluconazole is 12 % [17], and the plasma half-life 25 h. In *Cryptococcus neoformans* meningitis in AIDS patients, the liquor concentration was 80 % of the plasma concentration [18]. In vaginal secretions, the concentration of fluconazole was similar to the plasma concentration [19]; 60 – 75 % of the substance is eliminated by the kidneys in unmetabolized form. The elimination half-life is between 22 and 37 h [17].

Indications. Vaginal candidiasis [20] and systemic and mucocutaneous candidiasis [21–24]. In *Cryptococcus neoformans* meningitis the substance is mostly used in combination with amphotericin B and flucytosine [25]. Prophylaxis and preemptive therapy in AIDS patients, patients with leukemia, and patients with bone marrow transplants [26], [27].

Side Effects. Fluconazole is well tolerated.

CNS. Vertigo: 3.7 %, headache: 1.9 %.

GIT. Stomachache: 1.7 %, vomiting: 1.7 %, diarrhoea: 1.5 %.

Liver. Rise in transaminases (> 8 % of normal level): 1 %.

Skin. Exanthema: 1.8 %.

A Stevens – Johnson syndrome was observed after fluconazole therapy in an AIDS patient.

2.1.2. Itraconazole

Itraconazole [*84625-61-6*], 4-[4-[4-[4-[[2,4-Dichlorophenyl)-2-(1*H*-1,2,4-triazol-1-ylmethyl)-1,3-dioxolan-4-yl]-methoxy]phenyl]-1-piperazinyl]phenyl]-2,4-dihydro-2-(1-methylpropyl)-3*H*-1,2,4-triazol-3-one, $C_{35}H_{38}Cl_2N_8O_4$, M_r 705.64, *mp* 166.2 °C, crystals from toluene.

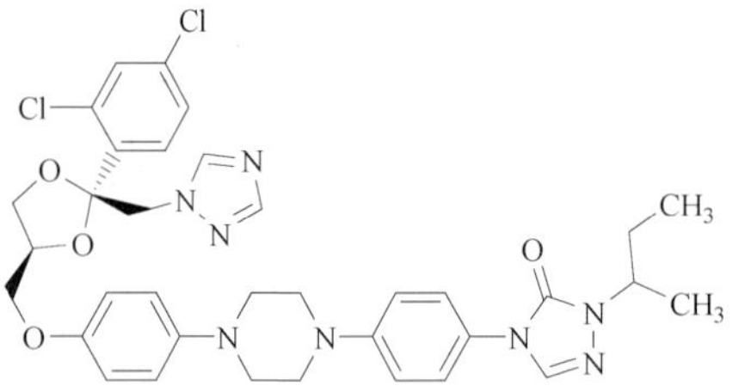

Synonym(s). R 51211.

Solubility. Itraconazole is a lipophilic substance which is practically insoluble in water and dilute acidic solutions. Partition coefficient in *n*-octanol/aqueous buffer of pH 8.1: 5.66.

Stability. The pure solid compund is stable under normal storage conditions.

Synthesis. Itraconazole is synthesized as follows [28], [29] (see top of next page):

Description. The triazole antimycotic itraconazole was first synthesized at Janssen Pharmaceuticals [30], [31].

Formulations. Capsules à 0.1 g.

Trade Names. Itrizole (Cilag), Sempera (Janssen), Siros (Cilag), Sporanox (Edward Keller, Janssen), Triasporin (Lifepharma).

Antimycotic Properties. Itraconazole is a broad-spectrum triazole antimycotic with good in vitro and in vivo activity against pathogenic dermatophytes, moulds (including *Aspergillus* species), *Candida* species, and dimorphic fungi such as *Blastomyces dermatitidis, Coccidioides immitis, Histoplasma capsulatum, Cryptococcus neoformans,* and *Sporothrix schenckii* [32–40]. MIC values of itraconazole are given in Table 2 [14].

Table 2. MIC values for itraconazole

Fungus	No. isolates tested	MIC range, µg/mL
Aspergillus fumigatus	96	0.001 – 10
Candida albicans	1605	0.063 – 128
Candida (Torulopsis) glabrata	4	2 – 128
Candida krusei	15	0.125 – 0.5
Candida lusitaniae	15	≤ 0.03 – 0.25
Candida parapsilosis	46	0.063 – > 128
Candida tropicalis	50	0.13 – > 128
Cryptococcus neoformans	118	0.001 – 0.5
Epidermophyton floccosum	5	0.063
Fusarium species	44	8 – > 32
Histoplasma capsulatum	10	0.063
Malassezia furfur	35	0.001 – 1
Microsporum species	10	0.063 – 0.25
Sporothrix schenckii	33	0.001 – 4
Trichophyton species	15	0.063 – 64
Trichosporon beigelii	15	≤ 0.03 – 0.25

Pharmacokinetics. The substance has a bioavailability after oral application of 40 % if taken on an empty stomach to 55 % if taken simultaneously with food [41]. The absorption of itraconazole after oral application is slow; peak plasma concentrations are reached 3 – 6 h after intake [41], [42]. Plasma concentrations are low, and a peak plasma concentration of 0.1 µg/L was observed after a single oral dose of 100 mg [42]. The serum protein binding of itraconazole is high; more than 99 % of the substance is bound to plasma proteins, mostly to albumin. The substance shows good tissue penetration, especially into the skin and nails [41]. Itraconazole is extensively metabolized in the liver, and more than 30 metabolites have been identified so far [43]. The elimination of the substance mostly occurs by the biliary-fecal route and renally in a metabolized form [44]. The elimination half-life ranges between 15 and 42 h, depending on the applied dose [41].

Indications. Therapy of dermatophytoses and onychomycoses [45–47] with a dosing regimen of 100 to 200 mg per day. Vaginal candidosis with 200 mg/d for 3 days [48]. Therapy of systemic infections such as candidosis and aspergillosis including disseminated infections, blastomycosis, coccidioidomycosis, histoplasmosis, and cryptococcosis with a dosing regimen of 100 to 400 (or up to 600) mg/d [49–51].

cis-[2-(2,4-Dichlorophenyl)-2-(1*H*-1,2,4-triazol-1-ylmethyl)-1,3-dioxolan-4-yl]-methylmethanesulfonate

1-Acetyl-4-(4-hydroxyphenyl)piperazine

NaH (50%) / DMSO, 80 °C, 5h

1. NaOH / BuOH, 12 h reflux
2. $Cl-C_6H_4-NO_2$, K_2CO_3 / DMSO

1. H_2, 5% Pt / C
2. $ClCO_2C_6H_5$ / pyridine / $CHCl_3$
3. H_2NNH_2, H_2O / dioxane

n-BuOH / $NaOCOCH_3$

KOH / DMSO

Itraconazole

Especially in aspergillosis, an early onset of therapy is important for the clinical outcome.

Side Effects. Itraconazole is generally well tolerated.

CNS. Vertigo and headache: 10 %.

GIT. Stomachache, vomiting, and diarrhoea were observed.

Liver. Rise in transaminases and other reversible liver affections: 7 %.

Metabolism. Hypokaliemia and rise in serum triglycerides [52], [53].

2.1.3. Ketoconazole

Ketoconazole [*65277-42-1*], *cis*-1-acetyl-4-[4-[[2-(2,4-dichlorophenyl)-2-(1*H*-imidazol-1-ylmethyl)-1,3-dioxolan-4-yl]methoxy]phenyl]-piperazine, $C_{26}H_{28}Cl_2N_4O_4$, M_r 531.4, *mp* 146 °C, colorless crystals.

Synonym(s). R 41400.

Solubility. Ketoconazole is soluble in methanol, ethanol, and chloroform; slightly soluble in dilute acids; and barely soluble in water.

Stability. The pure solid compund is stable under normal storage conditions.

Synthesis. The synthesis of ketoconazole begins with the ketalization of 2,4-dichloroacetophenone (**1**) with glycerol. The intermediate (**2**) is not isolated but treated with bromine to form the bromide (**3**). The benzoxylation of (**3**) gives (**4**) in the form of a *cis – trans* mixture. The *cis* form can be separated by crystallization. The reaction of (**4**) with imidazole in dimethylacetamide gives (**5**), which is then hydrolyzed to (**6**), which in turn is converted with methanesulfonyl chloride into the sulfonate (**7**). The reaction of (**7**) with the sodium salt of *N*-acetyl-*N*-*p*-hydroxyphenylpiperazine in DMSO gives ketoconazole.

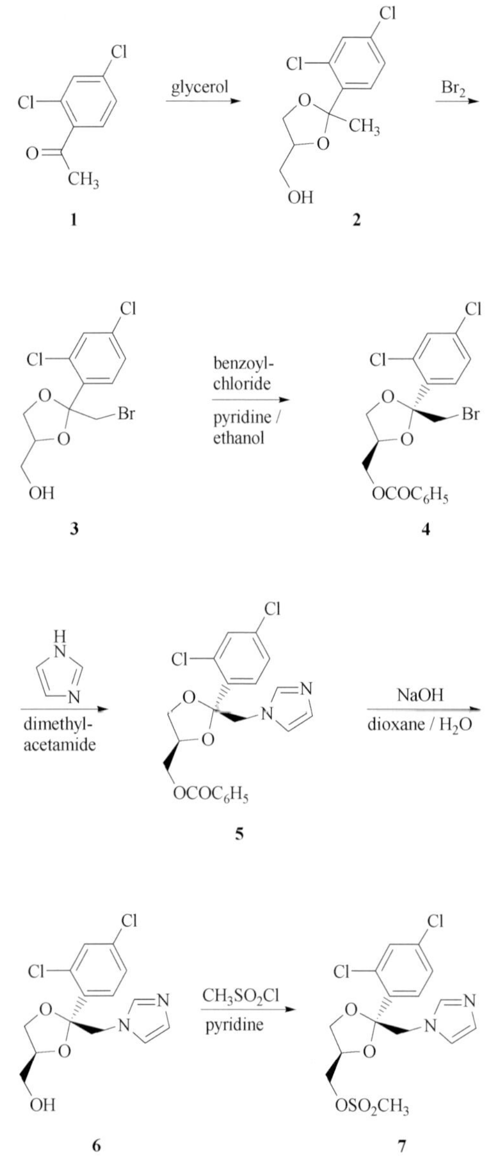

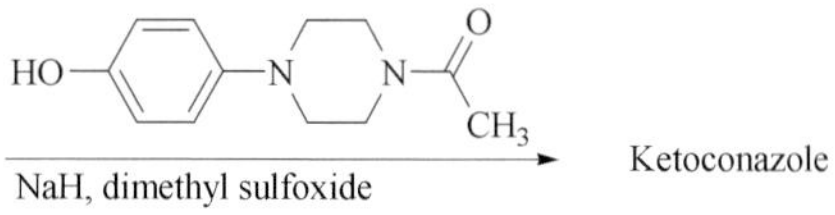

Description. The imidazole antimycotic ketoconazole was first synthesized at Janssen Pharmaceuticals [54–58].

Formulations. Tablets à 0.2 g; 2 % solutions, ointments, and cream preparations.

Trade Names. Candoral (Aché), Cetonax (Cilag), Fitonal (Disprovent), Fungarest (Janssen), Fungicil (Labinca), Fungo-Hubber (Hubber), Fungoral (Ilsan, Janssen, Leo), Ketazol (Exa), Ketoderm (Janssen), Ketoisdin (Isdin), Ketonan (IMA), Ketoral (Bilim), Micoral (Cassara), Micotek (Kressfor), Micoticum (Vita), Nizoral (Abic, Edward Keller, Janssen), Nizshampoo (Janssen), Oromycosal (Alkaloid), Oronazol (Krka), Panfungol (Esteve), Rofenid (Rhône-Poulenc), Terzolin (Janssen).

Antimycotic Properties. Ketoconazole shows a good efficacy against dermatophytes, dimorphic fungi (*Blastomyces dermatitidis, Coccidioides immitis, Histoplasma capsulatum*), and pathogenic *Candida* species [59], [60]. The activity against hyphomycetes and *Cryptococcus neoformans* is limited. MIC values of ketoconazole are summarized in Table 3. The in vitro activity of ketoconazole, like that of the other azoles, is strongly dependent on the test method used.

Table 3. MIC values for ketoconazole

Fungus	MIC range, µg/mL
Dermatophytes	0.04 – 10
Candida species	0.04 – 40
Cryptococcus neoformans	0.08 – 5
Aspergillus species	1.25 – 40
Dimorphic fungi	0.04 – 5
Chromomycetes	0.04 – 20

Pharmacokinetics. After oral administration, ketaconazole shows a variable degree of absorption of up to 70 %. The degree of absorption depends on the pH of the gastric fluid [61], [62]. Antacids and H_2-antagonists reduce the absorption of the drug. Peak serum concentrations are reached 2 – 4 h after administration and are between 2.0 and 3,5 μg/mL for a dose of 200 mg (3 mg/kg), and between 3.5 and 7.0 μg/mL for a 400 mg dose. Serum concentrations of 10 – 20 μg/mL can be reached after administration of a single dose of 1200 mg [63]. The substance shows a binding potency to plasma proteins and erythrocytes of 99 % [61]. After vaginal application, 1 % of the substance is absorbed; after cutaneous application, no absorption could be measured [64]. Ketoconazole penetrates well into the stratum corneum of the skin and vaginal secretions [61]. The concentration of ketoconazole in urine is low, with less than 1 μg/mL after a daily dose of 400 mg. The drug diffuses only slowly into the cerebrospinal fluid, where it reaches about 5 % of the serum concentration. The elimination half-life ranges from 4 to 9 h. The substance is mainly eliminated by the biliary – fecal route in a metabolized, inactive form.

Indications. Systemic therapy of dermatomycoses, blastomycosis, coccidioidomycosis, and histoplasmosis. Because of the side effects of ketoconazole, the newer azoles such as fluconazole and itraconazole should be considered instead of ketoconazole for systemic treatment. Ketoconazole is especially indicated in topical therapy of dermatomycoses, pityriasis versicolor, seborrheoic dermatitis, and dandruff.

Side Effects. After oral administraition of ketaconazole, side effects are seen at up to 10 % (400 mg/d) to > 50 % (800 mg/d) [65].

CNS. Vertigo, dizziness, and headache.

GIT. Somachache, loss of appetite, vomiting, and diarrhoea.

Liver. Rise in transaminases and other liver affections which are mostly reversible, although fulminant hepatitis was observed in 1 in 10 000 cases.

Vessels. Angioedema. Further, skin reactions, insomnia, and lethargy can be observed. Gynaecomastia, oligospermia, loss of libido, and alopecia, which may occur, are due to an inhibition of testosterone and cortisol biosynthesis [66–69]. Daily doses of higher than 400 mg reduce the testosterone serum concentration.

2.1.4. Miconazole

Miconazole [*22916-47-8*], 1-[2-(2,4-dichlorophenyl)-2-[(2,4-dichlorophenyl)methoxy]ethyl]-1*H*-imidazole, $C_{18}H_{14}Cl_4N_2O$, M_r 415.92.

Miconazole nitrate, [*22832-87-7*], $C_{18}H_{15}Cl_4N_3O_4$, M_r 479.15, *mp* 184 – 185 °C, colorless crystals.

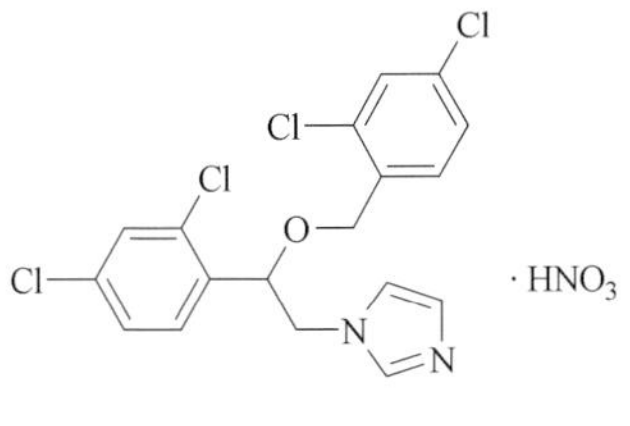

Synonym(s). R 18134; R 14889 (mononitrate).

Solubility. Miconazole is barely soluble in water (0.3 mg/mL) and slightly soluble to soluble in the many organic solvents and dilute inorganic acids.

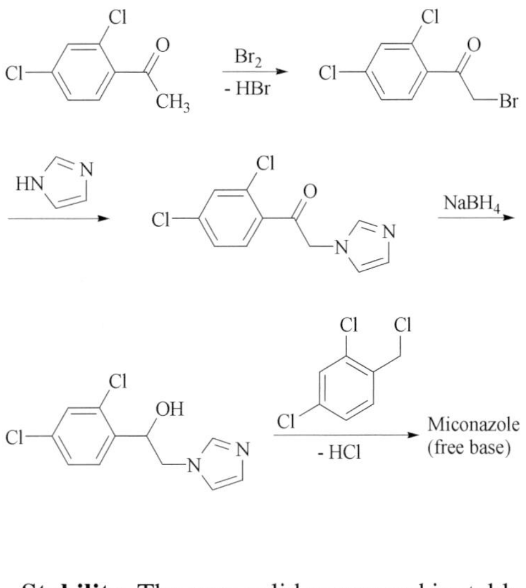

Stability. The pure solid compound is stable under normal storage conditions.

Synthesis. The synthesis [70], [71] begins with 2,4-dichloroacetophenone, which is first brominated to form the phenacyl bromide and then treated with imidazole to form the phenacylimidazole. Reduction with sodium borohydride and reaction of the resulting alcohol with 2,4-dichlorobenzyl chloride yields miconazole.

Description. The imidazole antimycotic miconazole was first synthesized at Janssen Pharmaceuticals [72], [73].

Formulations. Ampulles à 0.2 g; lozenges à 0.25 g; 2 % solutions, lotions, powders, gels, ointments, and cream preparations; vaginal ovula.

Trade Names (Miconazole). Andergin (Pierell), Brentan (Janssen), Daktar (Janssen), Daktarin (Abic, Edward Keller, Janssen), Dumicoat (Dumex), Femeron (Janssen), Micofim (Elofar), Micotar (Dermapharm), Micotef (LPB), Monistat (Cilag, Ethnor, Janssen), Monistat I.V. (Janssen), Vodol (Searle).

Trade Names (Miconazole Nitrate). Aflorix (Gerardo Ramon), Albistat (Cilag), Aloid (Janssen), Andergin (Pierrel), Anotit (Janssen), Brenazol (DuraScan), Brentan (Janssen), Britane (Janssen), Canofite (Janssen), Conoderm (C-Vet), Conofite (Janssen, Pitman-Moore), Daktar (Janssen), Daktarin (Abic, Janssen), Decomyc (Merck), Deralbine (Andromaco), Dermacure (Janssen), Dermonistat (Ortho), Epi-Monistat (Cilag, Janssen), Florid (Janssen, Mochida), Fungiderm (Janssen), Funginazol (Morgens), Fungisdin (Esteve, Janssen), Fungucit (Iltas), Fungur (Salutas), Gyno-Daktar (Janssen), Gyno-Daktarin (Edward Keller, Janssen, Johnson & Johnson), Gyno-Monistat (Cilag, Janssen), Ipec (Byk-Gulden), Micatin (McNeil, Ortho), Micoderm (Italsuisse), Micogel (Cipla), Micogyn (Elofar), Miconal (Ecobi), Micotef (LPB), Monistat (Cilag, Janssen, Ortho), Monostat 7 (Ortho), Mykotral (Chephasaar), Prilagin (Gambar), Surolan (Janssen), Vodol (Searle).

Antimycotic Properties. Miconazole is a broad-spectrum imidazole antimycotic that generally shows activity against pathogenic dermatophytes (*Epidermophyton floccosum; Trichophyton* and *Microsporum* species), *Candida* species, moulds including *Aspergillus* species, and pathogenic dimorphic fungi [74–78]. MIC values are summarized in Table 4.

Table 4. MIC values for miconazole

Fungus	No. isolates tested	MIC range µg/mL
Aspergillus fumigatus	74	0.5 – 64
Blastomyces dermatitidis	57	0.001 – 2
Candida albicans	1815	0.016 – 100
Candida (Torulopsis) glabrata	224	0.016 – 64
Candida kefyr	33	0.016 – 4
Candida krusei	40	< 0.063 – 6.25
Candida lusitaniae	7	0.18 – 6.3
Candida parapsilosis	54	0.016 – 32
Candida tropicalis	172	0.016 – 33
Coccidioides immitis	99	0.063 – 6
Cryptococcus neoformans	69	0.063 – 25
Epidermophyton floccosum	19	0.06
Histoplasma capsulatum	48	0.05 – 2
Malassezia furfur	44	0.2 – 50
Microsporum canis	18	0.1 – 4
Pseudoallescheria boydii	42	< 0.016 – 4
Sporothrix schenckii	65	0.5 – 16
Trichophyton mentagrophytes	28	0.1 – 4
Trichophyton rubrum	63	0.001 – 32

Phamacokinetics. Miconazole is incompletely absorbed after oral application with an average bioavailability of 27 % [79]. The active drug is dissolved in Cremophor EL for intravenous administration. After intravenous administration of a single dose of 200 mg, a peak plasma concentration of 1.6 mg/L was observed. Infusion of 500 mg of miconazole briefly produces serum concentrations of 2 – 9 µg/mL, which, however, rapidly fall to under 0.2 µg/mL. Miconazole is barely absorbed after cutaneous and/or mucocutaneous application. The substance shows a plasma protein binding of 90 % and poor penetration into the liquor cerebrospinalis and other body fluids. Miconazole is extensively metabolized in the liver. The substance is eliminated predominantly by the biliary – fecal route as inactive metabolites. Elimination occurs in phases with a terminal elimination half-life of 24 h [79].

Indications. Especially topical therapy of skin infections caused by dermatophytes and yeasts, and mucocutaneous infections including *Candida* vulvovaginitis. Systemic, intravenous therapy especially of infections caused by *Pseudoallescheria boydii* (dosage up to 1.2 g 3 × per day) [80]. For treatment of other generalized mycoses, mostly the newer, more effective, and less toxic azole antimycotics for systemic use, such as fluconazole and itraconazole, are indicated.

Side Effects. Multiple side effects have been observed after intravenous infusion of miconazole, which seem to be partly due to the formulation of the solvent, which contains the detergent Cremophor EL; mostly anaphylactic reactions, pruritus, dizziness, tachyarrhythmia, hyperaggregation of erythrocytes, anemia, thromocytosis, phlebitis, nausea, vomiting, hyperlipidemia, hyponatriemia, arthralgia, and flush [81].

2.2. Azole Antimycotics for Topical Treatment

2.2.1. Bifonazole

Bifonazole [*60628-96-8*], 1-([1,1'-biphenyl]-4-ylphenylmethyl)-1*H*-imidazole, $C_{22}H_{18}N_2$, M_r 310.4, *mp* 147 – 148 °C, colorless, odorless, tasteless crystals.

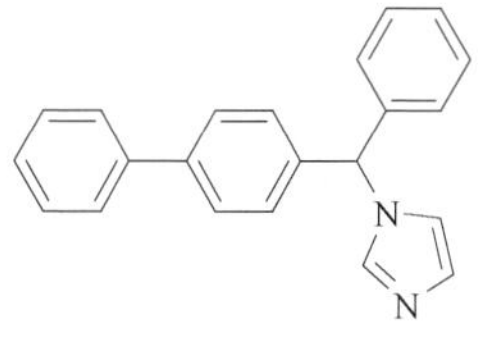

Synonym(s). Trifonazole, Bay h 4502.

Solubility. Bifonazole is soluble in glacial acetic acid (833 mg/mL), chloroform (344 mg/mL), benzene (38.2 mg/mL), methanol (36.5 mg/mL), ethanol (24.6 mg/mL), ethyl acetate (17.1 mg/mL), and acetonitrile (10.4 mg/mL). It is practically insoluble in water, but in Walpole's buffer, solubility is 3.75 mg/mL at pH 2.

Stability. Pure bifonazole is stable under normal storage conditions. In solution, the substance is stable under neutral or slightly alkaline conditions. It slowly decomposes in strongly alkaline or strongly acidic media. Bifonazole solutions are sensitive to light.

Synthesis. Bifonazole is synthesized as follows:

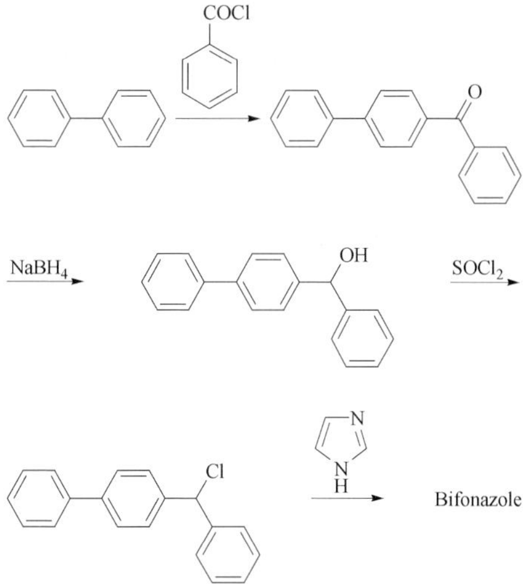

COCl

O

$NaBH_4$

OH

$SOCl_2$

Cl

N

N H

Bifonazole

Description. The imidazole antimycotic bifonazole was first synthesized at Bayer Pharmaceuticals in 1974 [82–86].

Formulations. 1 % solutions, powders, gels, ointments, and cream preparations.

Trade Names. Amycor (Lipha), Azolmen (Menarini), Bifonazol (Bayropharm), Micofun (Incobra), Mycospor (Bayer, Kai Cheong), Mycosporin (Bayer).

Antimycotic Properties. Bifonazole is a broad-spectrum imidazole antimycotic with excellent in vitro activity against pathogenic dermatophytes, yeasts, moulds, and dimorphic fungi. MIC values of bifonazole are summarized in Table 5.

Table 5. MIC values for clotrimazole and bifonazole

Fungus	MIC range, µg/mL	
	clotrimazole	bifonazole
Dermatophytes	<0.01 – 2	0.01 – 1
Candida species	0.04 – 10	0.08 – 10
Cryptococcus neoformans	0.04 – 10	0.04 – 2.5
Pityrosporum species	2 – 4	0.5 – 1
Aspergillus species	0.1 – 1.25	0.5 – 1
Chromomycetes	0.04 – 2.5	<0.04 – 0.31
Dimorphic fungi	0.04 – 0.31	<0.04 – 0.08

Pharmacokinetics. Bifonazole is absorbed adequately after oral administration to humans and animals. The substance quickly induces ribosomal liver enzymes and is primarily metabolized to metabolites without antimycotic activity. Therefore, bifonazole can only be used as a topical antimycotic [87], [88]. The substance is minimally absorbed after topical application of 1 % formulations. Concentrations in blood serum remain below 5 ng/mL, even after long term treatment of a 200 cm^2 skin area. Concentrations in the stratum papillare of the skin reach 2 – 3 μg/g [88].

Indications. Topical treatment of mycoses of the skin induced or sustained by fungi such as dermatophytes, yeasts, and also chromomycetes.

Side Effects. Topical applications, including intravaginal application, are well tolerated. Skin irritations and allergic reactions are rarely observed and seem to be mainly due to the galenic formulation (e.g., fatty preparations on acute, inflammative lesions). Allergic reactions are rarely observed [89].

2.2.2. Butoconazole

Butoconazole [*64872-76-0*], (±)-1-[4-(4-chlorophenyl)-2-[(2,6-dichlorphenyl)thio]butyl]-1*H*-imidazole, $C_{19}H_{17}Cl_3N_2S$, M_r 411.78, *mp* 68 – 70.5 °C, crystals from cyclohexane.

Butoconazole nitrate [*64872-77-1*], $C_{19}H_{18}Cl_3N_2O_3S$, M_r 475.01, *mp* 162 – 163 °C, colorless blades from acetone/ethyl acetate.

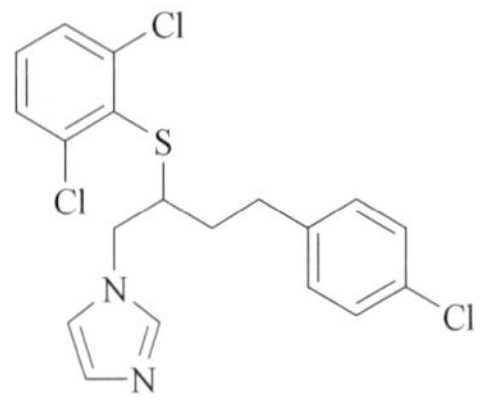

Solubility. For further information, see [90].

Stability. The pure solid compound is stable under normal storage conditions.

Synthesis [91]:

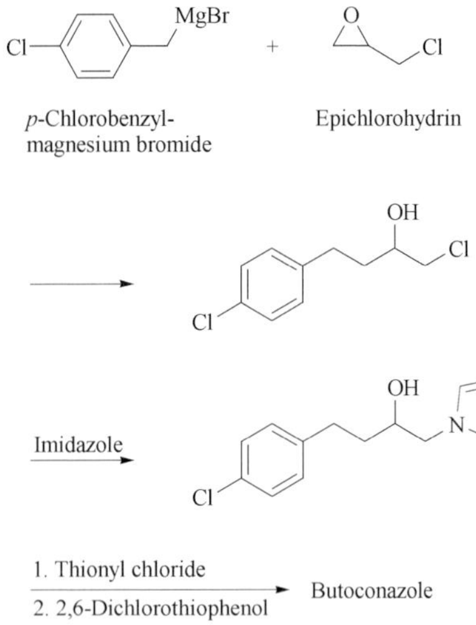

p-Chlorobenzylmagnesium bromide + Epichlorohydrin → (OH, Cl intermediate) —Imidazole→ (imidazolyl alcohol) —1. Thionyl chloride, 2. 2,6-Dichlorothiophenol→ Butoconazole

Description. The imidazole antimycotic butoconazole was first synthesized at Syntex Pharmaceuticals [92].

Formulations. 2 % cream preparations.

Trade Names (Butoconazole Nitrate): Femcosyn (Protochemie), Femstat (Syntex), Gynomyk (Cassenne).

Antimycotic Properties. Broad-spectrum imidazole antimycotic with in vitro and in vivo activitiy against pathogenic yeasts, dermatophytes, moulds, and dimorphic fungi [93], [94].

Pharmacokinetics. After intravaginal application of 100 mg of butoconazole nitrate, 5 % of the dose was found in the urine and feces [95].

Indications. Butoconazole is generally indicated for therapy of vulvovaginal mycoses caused by *Candida* species and *Candida (Torulopsis) glabrata* [96], [97].

Side Effects. Local irritations such as itching and burning sensations and allergic reactions may occur in rare cases and are mainly due to the galenic formulation [98].

2.2.3. Clotrimazole

Clotrimazole [*23593-75-1*], 1-[(2-chlorophenyl)diphenylmethyl]-1*H*-imidazole, $C_{22}H_{17}ClN_2$, M_r 344.8, *mp* 143 – 144 °C, colorless, tasteless crystals.

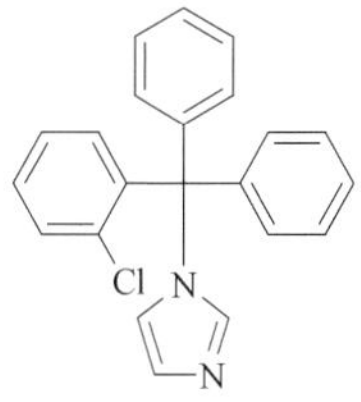

Synonym(s). Chlortritylimidazole, Bay 5097, FB b 5097, PCPIM.

Solubility. Clotrimazole is readily soluble in methanol, chloroform, and DMF (all > 100 mg/mL); slightly soluble in acetone (50 mg/mL), ethyl acetate (45 mg/mL), ethanol (95 mg/mL), and PEG 400 (60 mg/mL); and practically insoluble in water (< 0.01 mg/mL).

Stability. Pure clotrimazole is stable under normal storage conditions. The stability of the solution is pH dependent. Above pH 7 the solution is stable. In acid solution, clotrimazole undergoes gradual hydrolysis to form 2-chlorophenylbisphenylmethanol and imidazole. The hydrolysis half-life at pH 1 is about 170 h in ethanol/water (1/1) at room temperature.

Synthesis. The synthesis of clotrimazole begins with 2-chlorobenzyl chloride, which is converted to benzotrichloride. Friedel – Crafts reaction with benzene followed by reaction with imidazole forms clotrimazole [99].

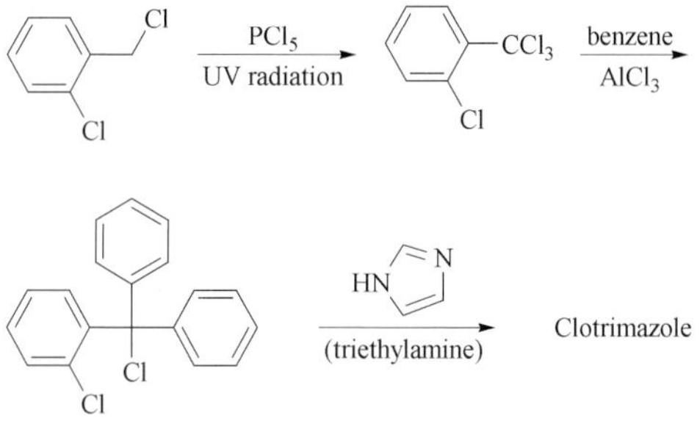

PCl5 / UV radiation; benzene / AlCl3; (triethylamine); Clotrimazole

Description. The imidazole antimycotic clotrimazole was first synthesized at Bayer Pharmaceuticals in 1967 [100]. Ist antimycotic activity was first reported in 1969 [101], [102–104].

Formulations. Solutions, powders, gels, ointments, and cream preparations; lozenges and vaginal tablets.

Trade Names. Agisten (Agis), Akneclor (Spirig), Antifungol (Hexal), Antimicotico (Savoma), Antimyk Neu (Pfleger), Apocanda (Apogepha), Azutrimazol (Azuchemie), Canesten (Bayer, Bayropharm, Kai Cheong), Canifug (Wolf), Clazol (Ilsan), Clocim (Cimex), Clot-basan (Sagitta), Clotrifug (Wolff), Clotrimaderm (Taro), Clotrimix (Hosbon), clotri OPT (Optimed), Clotrizol (Jossa), Clozole (Jean-Marie), Contrafungin (PharmaGalen), Cutistad (Stada), Dignotrimazol (Dignos), Dolexalan (Wölfer), Durafungol (Durachemie), Empecid (Bayer, Yoshitomi), Eurosan (Mepha), Femcare (Schering-Plough), Fungiframan (Oftalmiso), Fungi-med (Permamed), Fungizid-ratiopharm (ratiopharm), Fungosten (Mulda), Fungotox (Mepha), Gilt (Lyssia), Gino-Canesten (Bayer), Gino-Clotrimix (Hosbon), Gino-Lotramina (Schering Corp.), Gromazol (Grossman), Gyne-Lotremin (Mason, Schering Corp., Schering-Plough), Jenamazol (Jenapharm), KadeFungin (Kade), Localicid (Dermapharm), Lotramina (Schering Corp.), Lotremin (Mason, Schering Corp.), Lotrimin (Schering Corp., Schering-Plough), Micomisan (Hosbon), Micoter (Cusi), Mono-Baycuten (Bayropharm), Mycelex (Bayer, Dome), Mycelex-7 (Bayer), Mycelex-G (Bayer, Dome), Myclo (Boehringer Ingelheim), Mycofug (Hermal), Myco-Hermal (Hermal, Jebsen), Mycoril (Remedica), Mycosporin (Bayer), Mycotrim (Lagap), Myko Crdes (Ichthyol), Mykofungin (Wyeth), Mykohaug C (Salutas), Ovis Neu (Aldenylchemie), Peckle (Hisamitsu), Pedisafe (Sagitta), Plimycol (Pliva), SD-Hermal (Hermal), Stiemazol (Stiefel), Tibatin (DAK), Tricosten (Farmion), Trimysten (Bellon), Uromykol (Hoyer).

Antimycotic Properties. Broad-spectrum imidazole antimycotic with good activity against yeasts, dermatophytes, moulds, and dimorphic fungi. MIC values of clotrimazole are summarized in Table 5.

Pharmacokinetics. Clotrimazole is absorbed adequately after oral administration to humans and animals. The substance quickly induces ribosomal liver enzymes and is primarily metabolized to metabolites without antimycotic activity. Therefore, clotrimazole can only be used as a topical antimycotic [105], [106]. The substance is minimally absorbed after the topical application of 1 % formulations. Concentrations in blood serum remain below 5 ng/mL, even after long-term treatment of a 200 cm^2 skin area. Concentrations in the stratum papillare of the skin reach 2 – 3 μg/g [106].

Indications. Topical treatment of mycoses of the skin induced or sustained by fungi such as dermatophytes, yeasts, and chromomycetes. In addition, clotrimazole is indicated for therapy of vulvovaginal mycoses caused by *Candida* species and *Candida (Torulopsis) glabrata.*

Side Effects. Topical applications, including intravaginal application, are well tolerated. Skin irritations and allergic reactions are rarely observed and seem to be mainly due to the galenic formulation (e.g. fatty preparations on acute, inflammative lesions). Allergic reactions are rarely observed [107].

2.2.4. Croconazole

Croconazole [*77175-51-0*], 1-[1-[2-(3-chlorobenzyloxy)phenyl]vinyl]imidazole, $C_{18}H_{15}ClN_2O$, M_r 310.78, *mp* 72 – 73 °C, white crystals.

Croconazole hydrochloride [*77174-66-4*], $C_{18}H_{16}Cl_2N_2O$, M_r 347.24, *mp* 148.5 – 150 °C, white crystals.

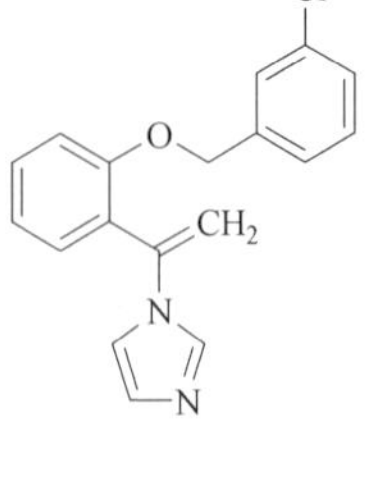

Synonym(s). Croconazole (free base); S 710674 (hydrochloride).

Solubility. Soluble in ethyl acetate.

Stability. The pure solid compound is stable under normal storage conditions.

Description. The imidazole antimycotic croconazole was first synthesized at Shionogi Pharmaceuticals [108], [109].

Formulations. 1 % creams and gel preparations.

Trade Names (Croconazole Hydrochloride). Pilzcin (Merz, Merz + Schoeller, Shionogi).

Antimycotic Properties. Broad-spectrum imidazole antimycotic with activity against almost all pathogenic fungi.

Pharmacokinetics. Croconazole is not absorbed after cutaneous application.

Indications. Topical treatment of mycoses of the skin induced or sustained by fungi such as dermatophytes and yeasts.

Side Effects. Local irritations and allergic reactions after topical treatment may occur in rare cases and are mainly due to the galenic formulation.

2.2.5. Econazole

Econazole [*27220-47-9*], 1-[2-[(4-chlorphenyl)-methoxy]-2-(2,4-dichlorophenyl)ethyl]-1*H*-imidazole, $C_{18}H_{15}Cl_3N_2O$, M_r 381.69, *mp* 86.8 °C, colorless crystals.

Econazole nitrate [*68797-31-9*], $C_{18}H_{16}Cl_3N_3O_4$, M_r 444.7, *mp* 164 – 165 °C, colorless crystals.

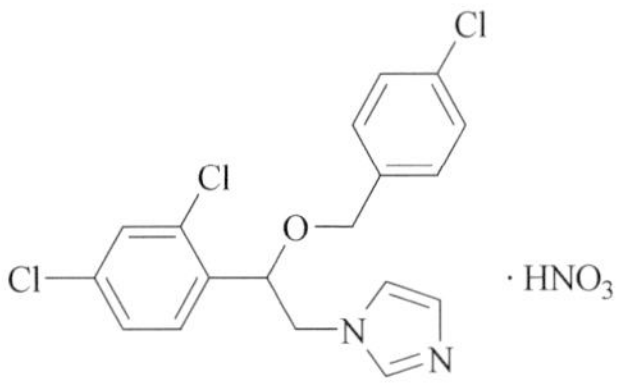

Synonym(s). R 14827 (mononitrate).

Solubility. Econazole is soluble in methanol, acetic acid, and PEG 400; less soluble in ethanol, acetone, chloroform, and butanol; and barely soluble in water, ether, cyclohexane, and hexane [110].

Stability. The solid substance is stable under usual storage conditions but should be stored in the dark.

Synthesis. The synthesis is almost the same as that of miconazole (see Section 2.1.4) [70], [71], [111], differing only in the last stage, in which the 1-[2,4-dichloro[β-hydroxy)phenethyl]imidazole intermediate is treated with 4-chlorobenzyl chloride.

Description. The imidazole antimycotic econazole was first synthesized at Janssen Pharmaceuticals [111] in 1969 [112], [113].

Formulations. 1 % solutions, powders, ointments, and cream preparations; vaginal ovula (150 mg).

Trade Names (Econazole). Pevalip (Janssen), Pevaryl (Cilag), Pevaryl Lipogel (Cilag, Janssen), Pevaryl lotion (Cilag), Pevaryl P.V. (Cilag).

Trade Names (Econazole Nitrate). Amicel (Salus), Chemionazolo (Brocchieri), Dermazol (CT), Dermazole (Lision Hong), Ecalin (JAKA-80), Ecodergin (Leben's), Eco Mi (Geymonat), Ecorex (Tosi), Ecostatin (Squibb, Westwood-Squibb), Ecotam (Alacan), Epi-Pevaryl (Cilag), Etramon (Johnson & Johnson), Fitonax (Cilag), Gyno-Pevaryl (Cilag, Edward Keller), Ifenec Zilliken), Micocert (Dexter), Micoespec (Centrum), Micofungal (Labopharma), Micogin (Crosara), Micogyn (Crosara), Micos (AGIPS), Micosten (Bergamon), Micostyl (Stiefel), Mitekol (Lek), Mykopevaryl (Cilag), Pargin (Gibipharma), Pevaryl (Cilag, Edward Keller, Smith Kline Beecham), Pevaryl P.V. (Edward Keller), Polycain (Taiho), Skilar (Bonomelli), Spectazole (Ortho).

Antimycotic Properties. Broad-spectrum imidazole antimycotic with activity against almost all species of pathogenic fungi [114], [115]. MIC values of econazole are summarized in Table 6 [14].

Table 6. MIC values for econazole

Fungus	No. isolates tested	MIC range µg/mL
Aspergillus fumigatus	24	0.15 – 25
Candida albicans	283	0.016 – 25
Candida (Torulopsis) glabrata	38	0.016 – 25
Candida kefyr	33	0.016 – 6.25
Candida krusei	34	0.125 – 12.5
Candida parapsilosis	32	0.016 – 25
Candida tropicalis	52	0.016 – 25
Malassezia furfur	19	0.0125 – 25

Pharmacokinetics. After vaginal application, 1 – 27 % of the applied dose was found in the urine [116], [117].

Indications. Topical treatment of mycoses of the skin induced or sustained by fungi such as dermatophytes and yeasts [118]. In addition, econazole is indicated for therapy of vulvovaginal mycoses caused by *Candida* species and *Candida (Torulopsis) glabrata* [119], [120].

Side Effects. Local irritations such as itching and burning sensations and allergic reactions may occur in rare cases after topical therapy and are mainly due to the galenic formulation [121].

2.2.6. Fenticonazole

Fenticonazole [*72479-26-6*], 1-[2-(2,4-dichlorophenyl)-2-[[4-(phenylthio)phenyl]-methoxy]ethyl]-1*H*-imidazole, $C_{24}H_{20}Cl_2N_2OS$, M_r 455.4.

Fenticonazole nitrate [*73151-29-8*], $C_{24}H_{21}Cl_2N_3O_4S$, M_r 518.63, *mp* 136 °C, odorless white crystalline powder.

Synonym(s). Rec 15/1476 (mononitrate).

Solubility. Solubility at 20 °C: water < 0.1 mg/mL, diethyl ether < 0.1 mg/mL, ethanol 30 mg/mL, methanol 100 mg/mL, chloroform 300 mg/mL, DMF 600 mg/mL [122].

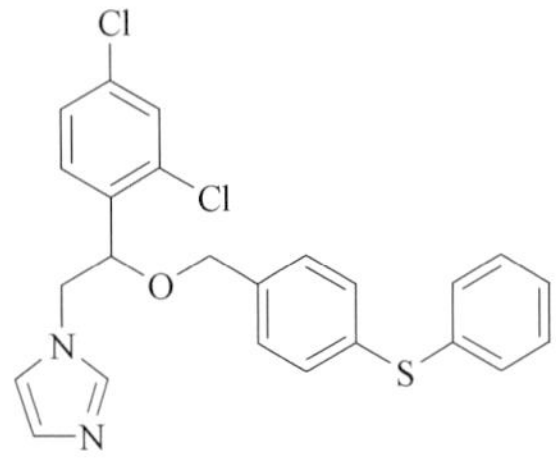

Stability. The pure solid compund is stable under normal storage conditions.

Description. The imidazole antimycotic fenticonazole was first synthesized at Recordati Pharmaceuticals [123], [124].

Formulations. 2 % solutions, ointments, and cream formulations; vaginal ovula (600 mg).

Trade Names (Fenticonazole Nitrate). Falvin (Farmades), Fenizolan (Organon), Fentiderm (Zyma), Fentigyn (Novartis), Lomexin (Grünenthal, Recordati), Mycodermil (Vifor Fribourg).

Antimycotic Properties. Broad-spectrum imidazole antimycotic with activity against almost all species of pathogenic fungi.

Pharmacokinetics. After topical application, the substance is not absorbed.

Indications. Topical treatment of mycoses of the skin induced or sustained by fungi such as dermatophytes and yeasts [125], [126]. In addition, fenticonazole is indicated for therapy of vulvovaginal mycoses caused by *Candida* species and *Candida (Torulopsis) glabrata* [127].

Side Effects. Local irritations such as itching and burning sensations and allergic reactions may occur in rare cases and are mainly due to the galenic formulation.

2.2.7. Isoconazole

Isoconazole [*27523-40-6*], 1-[2-(2,4-dichlorophenyl)-2-[(2,6-dichlorophenyl)methoxy]ethyl]-1*H*-imidazole, $C_{18}H_{14}Cl_4N_2O$, M_r 416.1, white crystals.

Isoconazole nitrate [*24168-96-5*], $C_{18}H_{15}Cl_4N_3O_4$, M_r 479.1, *mp* 182 – 183 °C, colorless crystals.

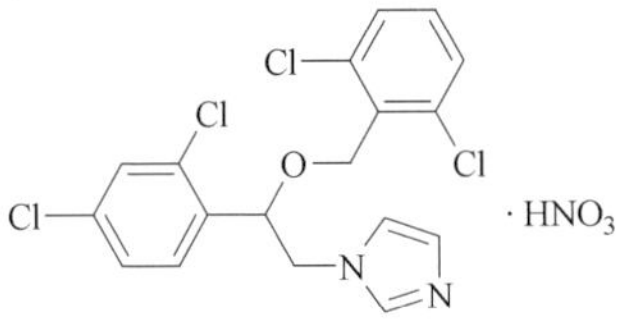

Synonym(s). R 15454 (mononitrate).

Solubility. Isoconazole is soluble in methanol, acetic acid, and PEG 400; less soluble in ethanol, acetone, chloroform, and butanol; and barely soluble in water, diethyl ether, cyclohexane, and hexane.

Stability. The solid substance is stable under normal storage conditions.

Synthesis. The synthesis of isoconazole [128] differs from that of miconazole (see Section 2.1.4) only in the last stage, in which 1-[2,4-dichloro(β-hydroxy)phenethyl]imidazole is treated with 2,6-dichlorobenzyl chloride in hot benzene – DMF in the presence of sodium hydride.

Description. The imidazole antimycotic isoconazole was first synthesized at Janssen Pharmaceuticals [128–131].

Formulations. 1 % solutions, ointments, and cream formulations.

Trade Names (Isoconazole Nitrate). Fazol (Bellon), Gyno-Travogen (Jebsen, Schering), Icaden (Schering), Isogyn Ginecologico (Crosara), Mupaten (Schering-Plough), Travogen (Jebsen, Schering), Travogyn (Schering).

Antimicrobial Properties. Broad-spectrum azole antimycotic with activity against almost all species of pathogenic fungi [132], [133].

Pharmacokinetics. After topical application, isoconazole remains practically unabsorbed. Application in an ethanol/propyleneglycol vehicle increases penetration into the stratum corneum of the skin significantly [134].

Indications. Topical treatment of mycoses of the skin induced or sustained by fungi such as dermatophytes and yeasts. In addition, isoconazole is indicated for therapy of vulvovaginal mycoses causes by *Candida* species and *Candida (Torulopsis) glabrata* [135].

Side Effects. Local irritations such as itching and burning sensations and allergic reactions may occur in rare cases and are mainly due to the galenic formulation.

2.2.8. Omoconazole

Omoconazole *[74512-12-2]*, (Z)-1-[2-[2-(4-chlorophenoxy)ethoxy]-2-(2,4-dichlorophenyl)-1-methylethenyl]-1*H*-imidazole, $C_{20}H_{17}C_{13}N_2O_2$, M_r 423.73, *mp* 89 – 90 °C, crystals from ethyl acetate/hexane (1/4).

Omoconazole nitrate *[83621-06-1]*, $C_{20}H_{18}Cl_3N_3O_5$, M_r 485.96, *mp* 118 – 120 °C (Büchi) / 122.5 (Mettler), crystals from ethyl acetate/ethanol.

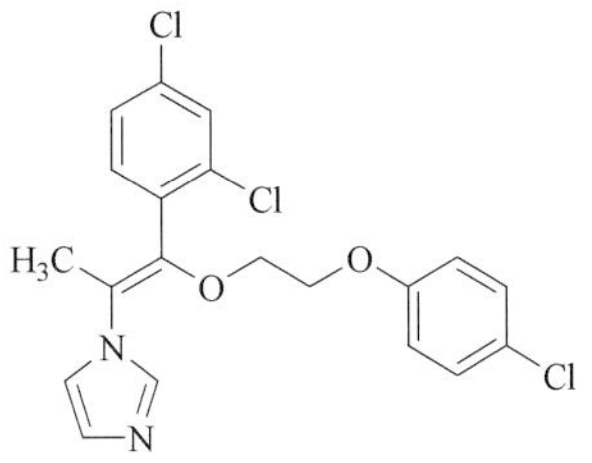

Synonym(s). CM 8282 (mononitrate).

Stability. The pure solid substance is stable under normal storage conditions.

Description. The imidazole antimycotic omoconazole was first synthesized at Siegfried Pharmaceuticals [136], [137]. Stereospecific synthesis [138].

Formulations. 1 % cream formulations.

Trade Names (Omoconazole Nitrate). Azameno (Kwizda, Wyeth), Fongamil (Biorga), Fongarex (Sanofi), Fongorex (Siegfried), Melur (Siegfried).

Antimycotic properties. Broad-spectrum imidazole antimycotic with activity against almost all pathogenic fungi [139].

Pharmacokinetics. After topical application, omoconazole is not absorbed.

Indications. Topical treatment of mycoses of the skin induced or sustained by fungi such as dermatophytes and yeasts.

Side Effects. Local irritations such as itching and burning sensations and allergic reactions may occur in rare cases and are mainly due to the galenic formulation.

2.2.9. Oxiconazole

Oxiconazole *[64211-45-6]*, (Z)-1-(2,4-dichlorophenyl)-2-(1*H*-imidazol-1-yl)ethanone *O*-[(2,4-dichlorophenyl)methyl]oxime, $C_{18}H_{13}Cl_4N_3O$, M_r 429.92.

Oxiconazole nitrate *[64211-46-7]*, $C_{18}H_{14}Cl_4N_4O_4$, M_r 492.15, *mp* 137 – 138 °C, crystals from ethanol.

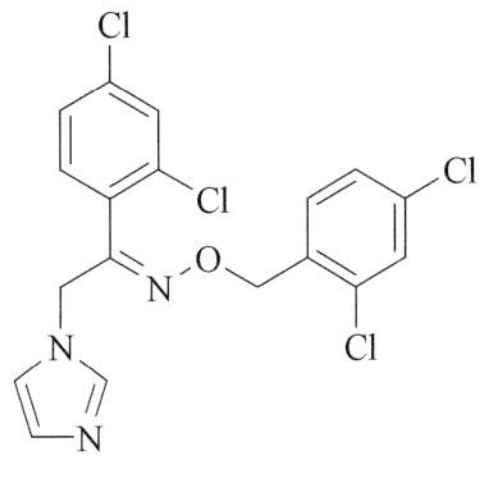

Synonym(s). Ro 13-8996/000 (free base); Ro 13-8996/001, Sgd 301-76 (mononitrate).

Stability. The pure solid substance is stable under normal storage conditions.

Synthesis. Oxiconazole is synthesized as follows [140], [141]:

Description. The imidazole antimycotic oxiconazole was synthesized at Roche and Siegfried Pharmaceuticals [142], [143].

Formulations. 1 % solutions, powders, ointments, and cream formulations. Vaginal tablets (600 mg).

Trade Names (Oxiconazole Nitrate). Gyno-Myfungar (Klinge), Myfungar (Klinge, Siegfried, Wyeth), Oceral (Roche), Oxistat (Glaxo).

Antimycotic Properties. Broad-specktrum imidazole antimycotic with activity against almost all species of pathogenic fungi [144].

Pharmacokinetics. After local application, oxiconazole penetrates well into the stratum corneum of the skin and the hair follicle [145].

Indications. Topical treatment of mycoses of the skin induced or sustained by fungi such as dermatophytes and yeasts. In addition, oxiconazole is indicated for therapy of vulvovaginal mycoses caused by *Candida* species and *Candida (Torulopsis) glabrata* [146].

Side Effects. Local irritations such as itching and burning sensations, and allergic reactions may occur in rare cases and are mainly due to the galenic formulation.

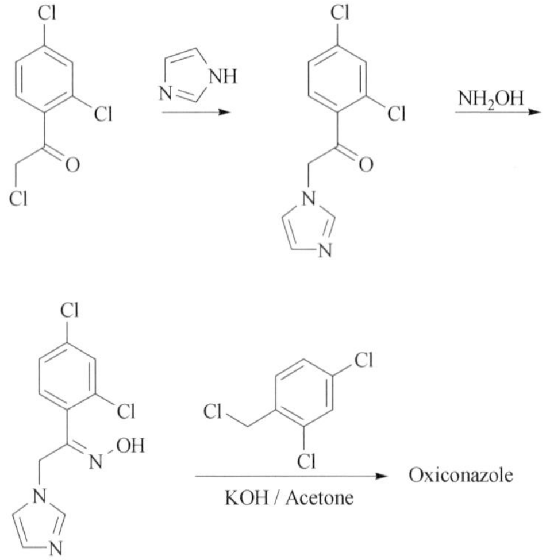

NH_2OH

KOH / Acetone

Oxiconazole

2.2.10. Sertaconazole

Sertaconazole [*99592-32-2*], (±)-1-[2,4-dichloro-*β*-[(7-chlorobenzo[*b*]thien-3-yl)methoxy]phenethyl]imidazole, $C_{20}H_{15}Cl_3N_2OS$, M_r 437.78.

Sertaconazole nitrate [*99592-39-9*], $C_{20}H_{15}Cl_3N_3O_4S$, M_r 501.01, *mp* 158 – 160 °C, odorless, white crystalline powder.

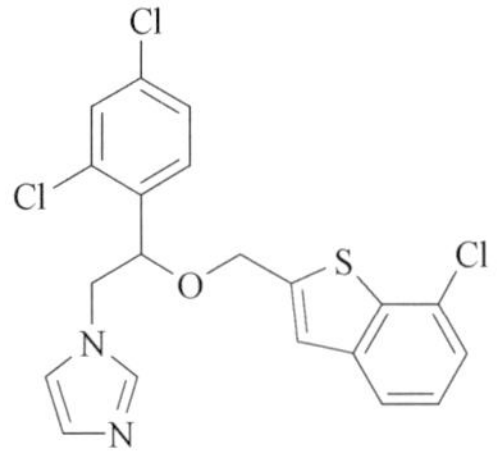

Synonym(s). FI-7045 (free base); FI-7056 (mononitrate).

Solubility. Fairly soluble in ethanol (1.7 %), chloroform (1.5 %); slightly soluble in acetone (0.95 %); very slightly soluble in *n*-octanol (0.069 %). Practically insoluble in water (< 0.01 %).

Stability. The pure solid substance is stable under normal storage conditions.

Synthesis. The Synthesis is described in [147].

Description. See [148], [149].

Formulations. 1 % cream formulations, gels, and ointments. New formulations for the treatment of vaginal candidiasis and oral formulations for the prophylaxis and treatment of mucocutaneous buccopharyngeal candidiasis are being developed.

Trade Names (Sertaconazole Nitrate). Dermofix (Ferrer), Dermoseptic (Smith Kline Beecham), Zalain (Robert).

Antimycotic Properties. Sertaconazole is a rather new broad-spectrum imidazole antimycotic with activity against almost all species of pathogenic fungi. It also has excellent activity against pathogenic yeasts [150–153].

Parmacokinetics. The substance is not absorbed after topical administration.

Indications. Topical treatment of mycoses of the skin induced or sustained by fungi such as yeasts and dermatophytes [154]. New formulations for the treatment of vaginal mycoses are in development.

Side Effects. Local irritations and allergic reactions may occur in rare cases and are mainly due to the galenic formulation.

2.2.11. Sulconazole

Sulconazole [*61318-90-9*], 1-[2-[[(4-chlorophenyl)methyl]thio]-2-(2,4-dichlorophenyl)ethyl]-1*H*-imidazole $C_{18}H_{15}Cl_3N_2S$, M_r 460.8.

Sulconazole nitrate [*61318-91-0*], $C_{18}H_{15}Cl_3N_3O_3S$, M_r 524.03, *mp* 130.5–132.0 °C, white or off-white crystalline powder.

Synonym(s). RS-44872, RS-44872-00-10-3 (both mononitrate).

Solubility. 1 part in 3,333 (water), 1 part in 100 (ethanol), 1 part in 130 (acetone), 1 part in 333 (chloroform), 1 part in 2000 (dioxan), 1 part in 71 (methanol), 1 part in 286 (chloromethane), 1 part in 10 (pyridine), 1 part in 2000 (toluene).

Stability. The pure solid substance is stable under normal storage conditions, but shoud be protected from light.

Synthesis. [155]:

4-Chlorothiobenzoic acid

THF →

Thiobenzoate

$NaBH_4$ →

HNO_3 → Sulconazole nitrate

Description. See [156–158].

Formulations. 1 % solutions, creams, and ointments.

Trade Names (Sulconazole Mononitrate). Exelderm (Schwarz), MYK (Cassenne), MYK-1 (Will), Sulcosyn (Syntex), Suldicyn (Syntex).

Antimycotic Properties. Sulconazole is a broad-spectrum imidazole antimycotic with activity against almost all pathogenic fungi [159–161].

Pharmacokinetics. The substance is not absorbed after topical administration.

Indications. Topical treatment of mycoses of the skin induced or sustained by fungi such as dermatophytes and yeasts.

Side Effects. Local irritations and allergic reactions may occur in rare cases and are mainly due to the galenic formulation.

2.2.12. Terconazole

Terconazole [*67915-31-5*], *cis*-1-[4-[[2-(2,4-dichlorophenyl)-2-(1*H*-1,2,4-triazol-1-ylmethyl)-1,3-dioxolan-4-yl]methoxy]phenyl]-4-(1-methylethyl)piperazine, $C_{26}H_{31}Cl_2N_5O_3$,

M_r 532.48, *mp* 126.3 °C, crystals from isopropyl ether.

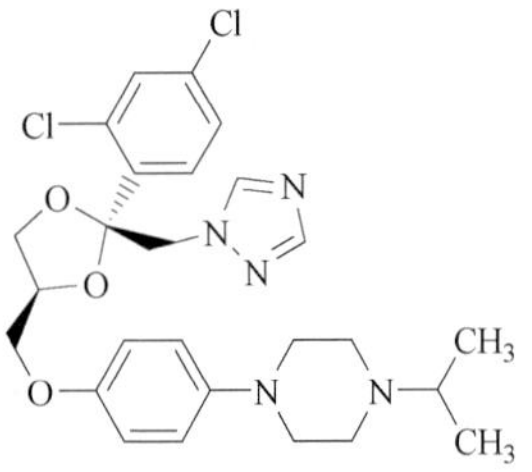

Synonym(s). Triaconazole, R 42470.

Stability. The pure solid substance is stable under normal storage conditions.

Synthesis. [162], [163]:

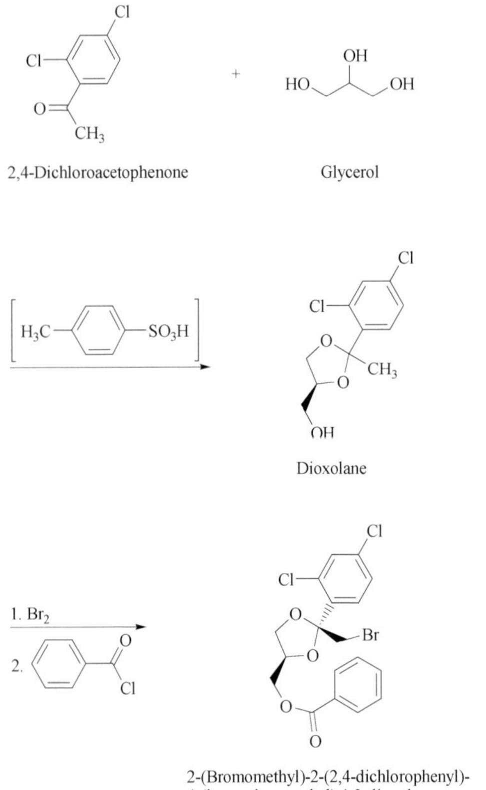

2,4-Dichloroacetophenone

Glycerol

Dioxolane

2-(Bromomethyl)-2-(2,4-dichlorophenyl)-4-(benzoyloxymethyl)-1,3-dioxolane

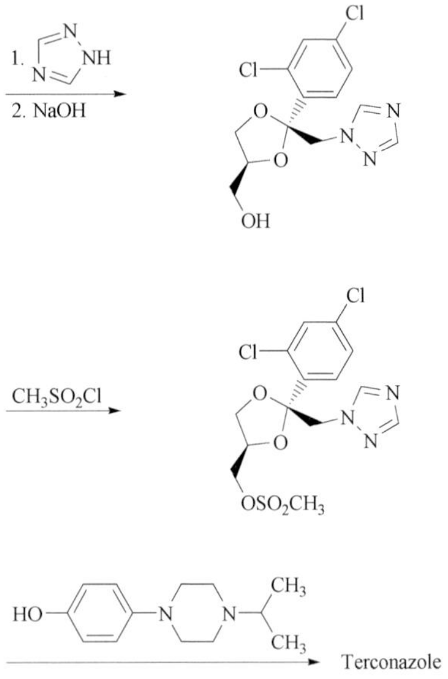

1. Br$_2$ 2. (benzoyl chloride)

1. (1,2,4-triazole) 2. NaOH

CH_3SO_2Cl

Terconazole

Description. The triazole antimycotic terconazole was first synthesized at Janssen Pharmaceuticals [164–166].

Formulations. 1 % solutions, powders, ointments, and cream formulations; vaginal pessar (80 mg).

Trade Names. Fungistat (Janssen), Gyno-Terazol (Cilag), Panlomyc (Janssen), Terazol (Cilag, Organon, Ortho), Terconal (Fisons), Tercospor (Cilag).

Antimycotic Properties. Terconazole is a broad-spectrum triazole antimycotic with activity against most species of pathogenic fungi [163].

Pharmacokinetics. After intravaginal application, 5 – 16 % of the dose is aborbed. The substance is extensively metabolized and eliminated via the urine and feces [163].

Indications. Topical treatment of mycoses of the skin induced or sustained by fungi such as dermatophytes and yeasts. In addition, terconazole is indicated for theraphy of vulvovaginal mycoses caused by *Candida* species and *Candida (Torulopsis) glabrata* [167], [168].

Side Effects. *CNS.* After intravaginal application of terconazole, a flu syndrome (fever, headache, hypotension) was observed [169].

Local irritations and allergic reactions may occur in rare cases and are mainly due to the galenic formulation.

2.2.13. Tioconazole

Tioconazole [*65899-73-2*], 1-[2-[(2-chloro-3-thienyl)methyloxy]-2-(2,4-dichlorophenyl)ethyl]-1*H*-imidazole, $C_{16}H_{13}Cl_3N_2OS$, M_r 387.70, crystals.

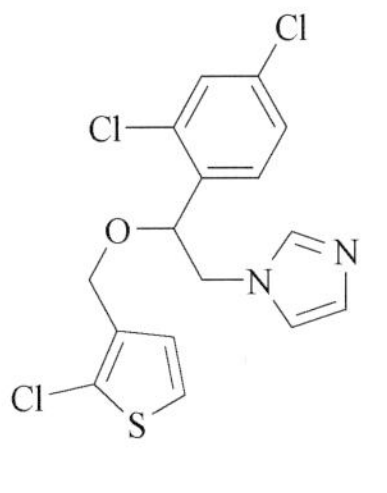

Synonym(s). TIO, IK 20 349.

Stability. The pure solid substance is stable under normal storage conditions.

Synthesis. [170]:

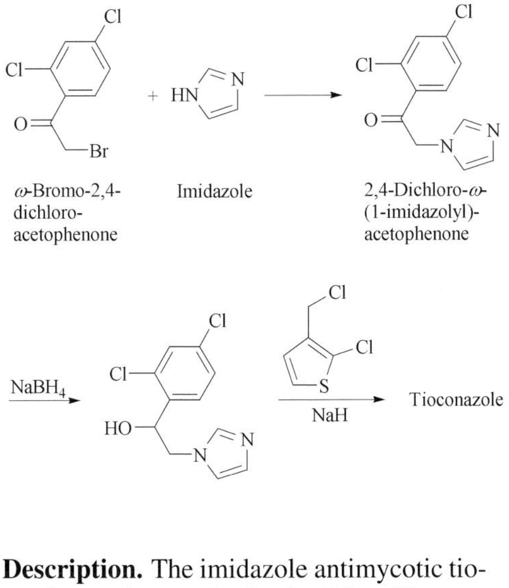

ω-Bromo-2,4-dichloroacetophenone + Imidazole → 2,4-Dichloro-ω-(1-imidazolyl)-acetophenone

$NaBH_4$ → ; + 2-chloro-3-(chloromethyl)thiophene, NaH → Tioconazole

Description. The imidazole antimycotic tioconazole was first synthesized at Pfizer Pharmaceuticals [171], [172].

Formulations. 1 % solutions, lotions, ointments, and cream preparations for therapy of dermatomycoses; 28 % solutions for onychomycoses; 2 – 6 % cream preparations for vaginal candidiasis.

Trade Names. Dermo-Trosyd (Pfizer), Fungibacid (Asche), Gino-Tralen (Pfizer), Gyno-Trosyd (Mason, Pfizer), Tralen (Pfizer), Trosid (Pfizer), Trosyd (Mason, Pfizer, Roerig), Trosyl (Roerig, Novex, Pfizer), Vagistat (Bristol-Myers-Sqibb), Zoniden (Irbi).

Antimycotic Properties. Tioconazole is a broad-spectrum imidazole antimycotic with activity against almost all species of pathogenic fungi [173–175].

Pharmacokinetics. Tioconazole is normally not absorbed after topical application to the skin [173]. On vaginal application of an ovulum containing 300 mg tioconazole, a serum concentration of 21 µg/mL was observed after 8 h [174].

Indications. Topical treatment of mycoses of the skin induced or sustained by fungi such as dermatophytes and yeasts [176]. In addition, tioconazole is indicated for therapy of vulvovaginal mycoses caused by *Candida* species and *Candida (Torulopsis) glabrata.*

Side Effects. Local irritations such as itching, burning sensations, erythema, and dermatitis, and allergic reactions may occur in rare cases and are mainly due to the galenic formulation [177], [178].

3. Polyene Antimycotics

The mechanism of action of polyene antimycotics is based on the formation of a complex with ergosterol in the fungal plasma membrane. This complex causes changes in the permeability of the plasma membrane, loss of ions and low molcular mass cytoplasmatic components, and inhibition of glycolysis [179]. The effect of polyenes on fungi is antagonized by Ca^{2+}, Mg^{2+}, and sterols in vitro. Further, an antagonism between polyene and azole antimycotics has been observed in several cases, both in vitro and in vivo. This antagonism can be explained in terms of both substance classes interfering with ergosterol biosynthesis.

The polyenes are primarily fungistatic at concentrations around the MIC level. In vitro fungicidal effects are achieved with 2, 4 or up to 10 times the MIC. The three relevant polyene antimycotics amphotericin B, natamycin (pimaricin), and nystatin show broad-spectrum in vitro antimycotic activity. Amphotericin B is the most potent of these three substances. However, in vivo activity is confined to yeasts, dimorphic (biphasic) fungi, and moulds, including *Aspergillus* species. Generally, the polyenes are less active to mostly ineffective and therefore generally not indicated in the treatment of dermatophytoses.

Polyenes are not absorbed sufficiently, if at all, on oral administration. Parenteral administration causes serious local inflammations. Nystatin and pimaricin are therefore unsuitable for intravenous administration.

3.1. Amphotericin B

Amphotericin B [*1397-89-3*], natural product, 38-membered cyclic lactone, isolated, e.g., from culture filtrates of *Streptomyces nodosus*. There are seven double bonds in the molecule. The keto group in position 1 is cyclized with the hydoxl group in position 35 to form a hemiacetal. The carbon atom at position 33 is atached to the dideoxy sugar mycosamine by a β-glucoside bond. $C_{47}H_{73}NO_{17}$, M_r 924.1, *mp*: does not have an exact melting point and decomposes above 170 °C, amorphous yellow powder.

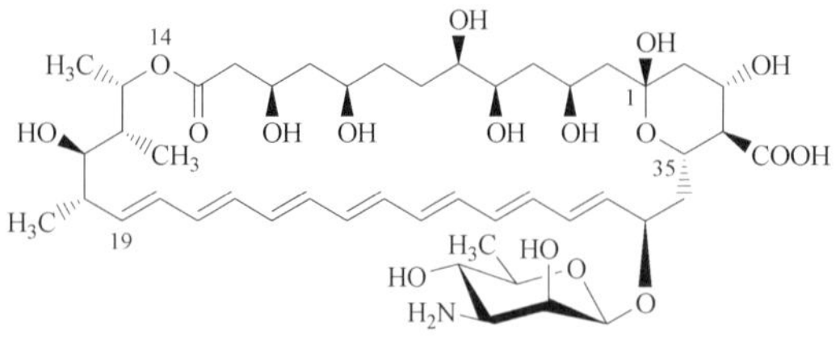

Synonym(s). RP 17774.

Solubility. Insoluble in water at pH 6 – 7. Solubility in water at pH 2 or pH 11: 0.1 mg/mL. Soluble in aqueous ethanol; DMF: 2 – 4 mg/mL; DMF + HCl: 60 – 80 mg/mL; DMSO: 30 – 40 mg/mL; insoluble in ether, ethyl acetate, or benzene. Soluble in aqueous sodium desoxycholate.

Stability. Amphotericin B is unstable to heat and prolonged exposure to light. It should be stored in the dark and kept refrigerated. In solution it decomposes at pH < 4 and > 10 [180]. Solutions show an optimum stability in a citrate – phosphate buffer at pH 5 – 7 [181–183].

Synthesis. Mostly fermentative from *Streptomyces* species [184]. First isolated from a culture of *Streptomyces nodosus* M4575, an isolate obtained from soil of the Orinoco river region of Venezuela.

Description. The first gereal characterization appeared in 1961 [185]. After some partial structures were determined, the complete structure was published in 1970 [186], and the total synthesis in 1987 [187], [188].

Formulations. For systemic treatment: Ampules à 0.05 g. Liposomal formulations with reduced toxicity [189–191] are available for systemic treatment. For local therapy, amphotericin B formulations such as cream preparations and ointments, lozenge formulations (à 0.1 and 0.01 g), and suspensions (100 mg/mL) are used. Also topical cream formulations additionally containing antibiotics (e.g., neomycine) and/or steroids (e.g., triamcinolone) are on the market.

Trade Names. AmBisome (Vestar), Ampho-Moronal (Bristol-Myers Squibb, Heyden, Squibb), Funganiline (Med. y Prod. Quim.), Fungilin (Bristol-Myers Squibb, Novo Nordisk, Squibb), Fungizone (Bristol-Myers Squibb, Squibb).

Antimycotic Properties. Amphotericin B is a broad-spectrum polyene antimycotic with activity against pathogenic yeasts, moulds, dimorphic fungi, and in vitro also against dermatophytes [192], [193]. MIC values of amphotericin B are summarized in Table 7.

Table 7. MIC values for amphotericin B, nystatin, and pimaricin

Fungus	MIC range, µg/mL		
	amphotericin B	nystatin	pimaricin
Candida species	0.1 – 0.5	1 – 5	1.5 – 10
Cryptococcus neoformans	0.1 – 0.5	0.5 – 4	6 – 12.5
Aspergillus species	0.1 – 0.5	1 – 10	3 – 12
Coccidioides immitis	0.1 – 0.5	0.5 – 1.5	2.5 – 25
Histoplasma capsulatum	0.1 – 1	0.5 – 1	3
Blastomyces species	0.1 – 0.5	0.5 – 1	1.5
Dermatophytes	2 – 4	2 – 10	1.2 – 100

Pharmacokinetics. Amphotericin B is scarcely absorbed after oral application. Therefore, amphtericin B has to be applied by the intravenous route for the treatment of systemic fungal infections [194]. Intravenous administration of 5 – 75 mg/d leads to serum concentrations of 0.14 – 2.39 mg/L 4 h after infusion [194]. The extravascular tissue distribution of amphotericin B is minimal because of cell membrane and lipoprotein binding [195]. The highest amphotericin B tissue concentrations are achieved in the liver and spleen, and also in the lungs and kidneys [196], [197]. Because of its poor penetration of the blood – brain barrier, meningitis and infections of the central nervous system can only be treated intralumbally and/or intrathecally [194]. The metabolization of amphotericin B is still unclear, and no metabolites have been charcterized so far [196]. Only 3 % of the administered dose is found in the urine [198], and 1 – 15 % of the daily dose is eliminated by the biliary – fecal route [197]. Much of the substance is trapped in the organs. Elimination occurs in two phases. The initial half-life is between 1 and 2 d, with a terminal half-life of 15 d [194].

Indications. Amphotericin B is effective in therapy of *Candida* infections, also including generalized and disseminated forms, caused by *Candida albicans* and other pathogenic *Candida* species [199]. Further, amphotericin B is indicated in the therapy of mycotic diseases such as histoplasmosis, sporotrichosis, cryptococcosis, blastomycosis, aspergillosis, zygomycosis, and coccidioidomycosis [200]. Although the substance shows an in vitro activity against dermatophytes, amphotericin B is not indicated in the therapy of dermatophytoses, mostly due to toxicity and stability problems and high cost. In the treatment of generalized mycoses, often a synergistic effect between amphotericin B and fluorocytosine [201], [202], and an antagonistic effect between amphotericin B and azole antimycotics is observed (the latter with the exception of CNS cryptococcosis due to *Cryptococcus neoformans*). Indications and efficacy of amphotericin B are summarized in Table 8. Amphotericin B can be given as a continuous drip infusion, even though serious, and in some cases, irreparable side effects, predominantly to the kidney tubulus, have been observed. Therefore, systemic treatment should be initiated only after exact diagnosis, and the patient should be kept under strict supervision. The optimum dosage for systemic therapy is 1 mg/kg every 24 to 48 h. Despite serious side effects, amphotericin B still remains a drug of choice for treating systemic mycoses.

Table 8. Indications and efficacy of amphotericin B at optimal dosage

Indication	Pathogen	Clinical efficacy
Candidosis	*Candida* species, mostly *C. albicans*	effective in > 50 % of cases
Aspergillosis	*Aspergillus* species except *A. nidulans*	uncertain
Zygomycosis	*Mucor* and *Absidia* species	uncertain
Blastomycoses	*Blastomyces dermatidis*, *Bl. brasiliensis*	effective in ≈ 50 % of cases
Coccidioidomycosis	*Coccidioides immitis*	effective
Histoplasmosis	*Histoplasma capsulatum*	usually effective

Side Effects. The following side effects have been observed under systemic therapy:

CNS. Fever, vomiting, and chills are observed in about 50 % and seem to be mediated by prostaglandines.

Heart. Rarely, arrythmia is observed under infusion [203].

Circulation. Collapse.

Vessels. Thrombophlebitis at the injection site is requently observed, liposomal formulations show a lower phlebotoxicity [204].

Blood/Spleen. Anemia and thrombocytopenia [204] are rarely observed.

Liver. In some cases, hepatotoxicity was observed.

Kidney. Amphotericin B shows a high degree of nephrotoxicity. Cumulative doses of 4 – 5 g generally cause reversible damage to the kidneys [205]. Higher doses mostly cause an irreversible kidney damage with persisting hypopotassemia, hematuria, proteinuria, and azotemia [203–206]. Therefore, serum potassium, magnesium, and creatinine levels have to be frequently monitored.

In the case of intralumbal and/or intrathecal administration, severe side effects of the central nervous system are common. Liposomal formulation of amphotericin B can reduce toxicity problems [207].

Topical application of amphotericin B is highly effective and well tolerated in the treatment of cutaneous and mucocutaneous candidoses. This is also true for inhalation and instillation. Local irritations and allergic reactions may occur in rare cases and are partly due to the galenic formulation.

3.2. Natamycin (Pimaricin)

Natamycin [*7681-93-8*], natural product isolated from culture filtrates of *Streptomyces natalensis* (original isolate from soil of Pietermaritzburg, South Africa) and *Streptomyces chattanoogensis* [208], [209], contains mycosamine as sugar component, like amphotericin B and nystatin. $C_{33}H_{47}NO_{13}$, M_r 665.7, *mp*: does not have an exact melting point and decomposes above 200 °C, colorless substance.

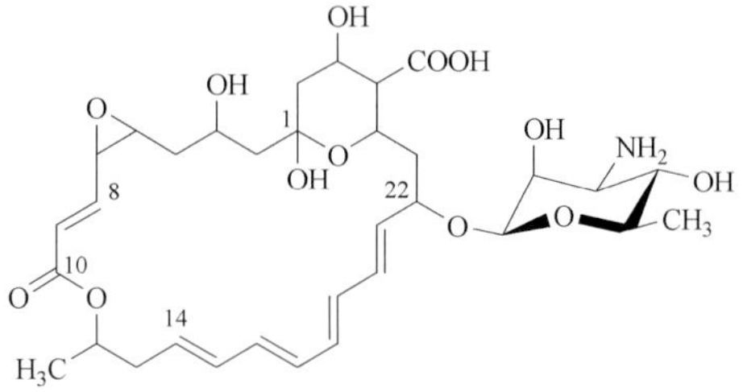

OH COOH OH O 1 O OH OH NH$_2$ OH 8 22 O O CH$_3$ O 10 O 14 H$_3$C

Synonym(s). Pimaricin, antibiotic A 5283, CL 12625, tennecetin.

Solubility. Natamycin is soluble in propylene glycol (20 mg/mL), DMF (50 mg/mL), and *N*-methylpyrrolidone (120 mg/mL); slightly soluble in methanol (2 mg/mL); and practically insoluble in water (0.05 mg/mL). Natamycin is amphoteric, dissolving in dilute acids and bases. However, these solutions are unstable.

Stability. Solutions or suspensions of natamycin are stable for several weeks at pH 5 – 7. The solutions can even be sterilized by heating at 110 °C for 20 min. However, they must be protected from light, which causes rapid decomposition [208], [210].

Synthesis. Fermentative from *Streptomyces* species.

Description. First isolated in 1957 [208], [211], [212].

Formulations. For therapy of human infections: Ointments, powders, creams, lotions, lozenges, dragees, suspensions, and vaginal tablets of different concentration. Also in combination with antibiotics and/or steroids (e.g., neomycine, hydrocortisone).

Trade Names. Deronga (Basotherm), Mycophyt (Gist-Brocades), Myprozine (Lederle), Natacyn (Alcon), Natafucin (Brocades), Pima-Biciron (Basotherm), Pimafucin (Basotherm, Beytout, Byk Procienx, Gist-Brocades), Pimagyn (Doetsch-Grether), Synogil (Basotherm).

Antimycotic Properties. Natamycin is a broad-spectrum polyene antimycotic, MIC values are summarized in Table 7.

Phamacokinetics. After topical application, no systemic absoption was detectable.

Indications. Cutaneous and mucocutaneous candidosis; in some cases, the substance was also successful for therapy of human dermatophytoses [213–215]. In veterinary medicin, natamycin is used for the therapy of dermatophytoses in cows (generally caused by *Trichophyton verrucosum*) and horses (generally caused by *Trichophyton equinum*) [216], [217]. As natamycin also has antibacterial properties, it is also used as a food preservative.

Side Effects. When applied topically, natamycin is highly effective and well tolerated in the treatment of cutaneous and mucocutaneous candidoses. Local irritations and allergic reactions may occur in rare cases and are mostly due to the galenic formulation. Especially in veterinary medicine, phototoxic effects were observed.

3.3. Nystatin A1

Nystatin (A1) [*1400-61-9*], three biologically active components – Nystatin A1, A2, and A3 – have been described. Nystatin A1 is a natural product isolated from culture filtrates of *Streptomyces noursei* [218], *Streptomyces aureus*, and other *Streptomyces* species. Like amphtericin B it is a 38-membered macrocyclic lactone, from which it differs solely in the position of the hydroxyl groups and in the absence of a double bond at C23 [219]. $C_{47}H_{75}NO_{17}$, M_r 926.1, *mp*: does not have a melting point, begins to decompose above 160 °C and decomposes above 250 °C without melting, yellow powder.

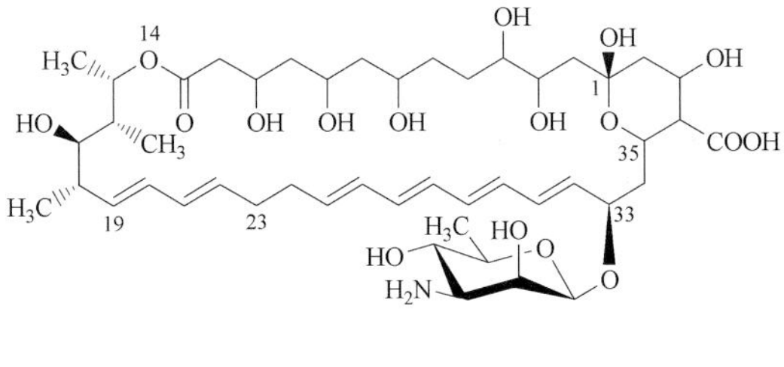

Synonym(s). Polyfungin A1.

Solubility. Nystatin A1 is soluble in pyridine, DMSO, DMF, ethylene glycol (8.7 μg/mL), methanol (11 μg/mL), ethanol (1.2 μg/mL), butanol, dioxane, and water (4.0 μg/mL). The substance is hygroscopic [220].

Stability. The substance is amphoteric, but aqueous and alkaline solutions are unstable. Nystatin shows optimum stability in phosphate – citrate buffers at pH 5 – 7. If kept refrigerated, the pure substance can be stored for several months without loss of activity. For stability, see [221].

Synthesis. Fermentative from *Streptomyces* species.

Description. See [222], [223].

Formulations. Tablets, dragées, lozenges, suspensions, drops, ointments, powders, gels, creams, ovulas, and vaginal tablets of different concentrations. There are also formulations available in combination with antibiotics and/or steroids such as tetracycline, neomycin, gramicidin, and cortisone. Dosages are often expressed in IE (100 000 IE = 22.73 mg nystatin A1).

Trade Names. Adiclair (Ardeypharm), Biofanal (Pfleger), Candex (Dome), Candio-Hermal (Hermal, Merck), Canstat (Lederle), Diastatin (Pfizer), Fungicidin (Spofa), Fungireduct (Azupharma), Herniocid (Mayrhofer), Korostatin (Holland-Rantos), Lederlind (Lederle), Lystin (Mekim), Mikostatin (Squibb), Moronal (Heyden, Squibb), Multilind (Bristol-Myers Squibb, Fair), Mycostatin (Bristol-Myers Squibb, Heyden, Sanofi Winthrop, Westwood-Squibb), Mykinac (NMC), Myco-Posterine N (Kade), Mykundex (Jossa), Nadostine (Pan-Well), Nilstat (Lederle), Nyaderm (Taro), Nysert (Norwich Eaton), Nystacid (Farmos Group), Nystaderm (Dermapharm), Nysta-Dome (Dome), Nystan (Squibb), Nystat-Rx (Pharma-Tek), Nystavescent (Squibb), Nystex (Savage), Oranyst (Taro), O-V Statin (Squibb), Restatin (Remedica), Rivostatin (Rivopharm) Stereomycin (Medica).

Antimycotic Properties. Nystatin A1 is a broad-spectrum polyene antimycotic with in vitro activity against pathogenic yeasts, moulds, dermatophytes, and dimophic fungi. MIC values of nystatin are summarized in Table 12.

Pharmacokinetics. After topical application, no systemic absorption was detectable.

Indications. Nystatin A1 is especially indicated in the therapy of cutaneous and mucocutaneous infections caused by pathogenic yeasts.

Side Effects. When applied topically, nystatin A1 is highly effective and well tolerated in the treatment of cutaneous and mucocutaneous candidoses. Local irritations and allergic reactions may occur in rare cases and are mostly due to the galenic formulation. If topically applicated in the gastrointestinal tract, nausea, vomiting, and diarrhoea are rarely observed.

4. Flucytosine

Flucytosine [*2022-85-7*], 4-amino-5-fluoro-2(1*H*)-pyrimidinone, $C_4H_4FN_3O$, M_r 129.03, *mp* 297 °C (decomp.), colorless crystals.

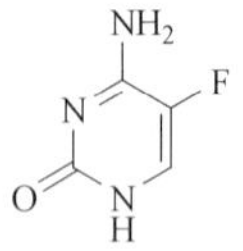

Synonym(s). Fluorocytosine, 5-Flurocytosine, 5-FC, Ro 2-9915.

Solubility. Flucytosine is readily soluble in water. It is basic and forms salts with acids.

Stability. Pure flucytosine as well as aqueous solutions, both acidic and basic, are stable. The substance should be stored in the dark.

Synthesis. The synthesis begins with *S*-ethylisothioureahydrobromide and ethyl-2-fluoro-3-methoxy-acrylate [224], [225]:

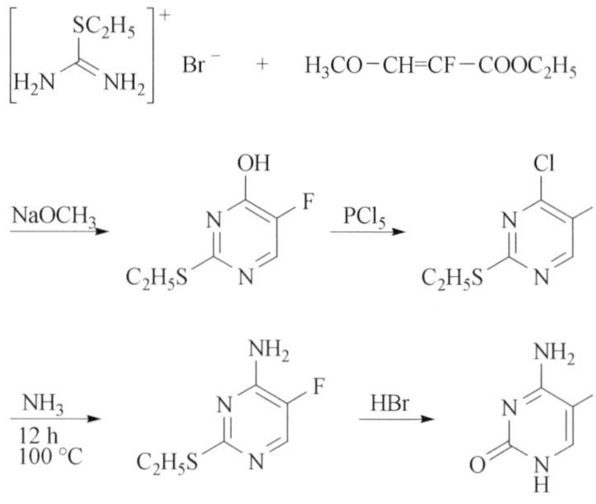

Description. The substance was initially synthesized at Hoffmann-La Roche Pharmaceuticals as an antimetabolite [224], [226–228].

Formulations. Infusion bottles à 0.1, 0.2, and 0.4 g. Capsules à 0.05, 0.1, 0.15, and 0.2 g. 0.5 and 1 % suspensions for oral application, 10 % suspension creams.

Trade Names. Alcoban (Roche), Alcobon (Roche), Ancobon (Roche), Ancotil (Edward Keller, Hoffmann-La Roche, Roche), Cocol (Horita, Tobishi).

Mechanism of Action. The mechanism of action of flucytosine is based on the incorporation of 5-fluorouracil, which is relatively selectively formed in fungal cells by deamination of flucytosine by the fungal cytosine-desaminase, into the messenger RNA and ribosomal RNA of the fungal cell. Fungal RNA containing 5-fluorouracil-riboside increasingly inhibits fungal protein biosynthesis [231–234].

Antimycotic Properties. Flucytosine has a narrow in vitro antimycotic activity spectrum that includes yeasts, *Cryptococcus neoformans*, as well as some *Aspergillus* strains and some chromomycetes. The activity spectrum and MIC values of flucytosine are summarized in Table 9 [231], [232]. Because the antimycotic action of flucytosine is antagonized by purines, pyrimidines, peptides, and some amino acids in vitro, only media free from these antagonists, e.g., yeast – nitrogen base medium containing dextrose, can be used for in vitro testing. Of the fungus strains affected by flucytosine, more than 10 % can be expected to show primary resistance. The development of secondary resistance by sensitive fungi during treatment also occurs, especially in cases of underdosage or prolonged treatment [232], [234]. In vitro fungicidal effects can be obtained with > 10 times the MIC and a contact time of > 150 h.

Pharmacokinetics. Flucytosine can be administered orally, parenterally, and topically. After oral administration, the drug is absorbed rapidly and almost completely in the intestine (80 – 90 %) [235]. Average serum concentrations of 30 – 60 µg/mL are normally reached with a single oral dose of 35 mg/kg. In patients with normal kidney function, optimal continuous serum levels between 25 and 100 mg/L are achieved with a dose of 50 mg/kg every 6 h [236]. Flucytosine penetrates readily into the tissues, body fluids, and cerebrospinal fluid (CSF) [237]. The CSF level reaches approximately 70 – 80 % of blood levels [236]. The plasma protein binding of flucytosine is low (ca. 4 %) [238]. More than 90 % of the dose is eliminated via the kidneys in unchanged form. The elimination half-life is 3 – 6 h. In patients with severely impaired kidney function, the elimination half-life can rise to 4 d. In patients with restricted renal function, the daily dose of flucytosine therefore must be adjusted according to creatinine elimination. The compound can be readily eliminated by hemodialysis [231–233]. A small amount of flucytosine is metabolized to 5-fluorouracil, e.g., by the bacterial flora of the gut, which seems to be a reason for the hematotoxicity of flucytos e [239], [240].

Table 9. MIC values for flucytosine

Fungus	MIC range, µg/mL
Candida albicans	0.1 – 10 (> 100)
Candida tropicalis	0.1 – 100
Candida parapsilosis	0.1 – 0.5
Candida pseudotropicalis	0.1
Candida krusei	0.5 – 1
Candida (Torulopsis) glabrata	0.1 – 1
Cryptococcus neoformans	0.5 – 4 (16)
Aspergillus fumigatus	0.5 – 10
Aspergillus niger	0.5 – 1
Aspergillus flavus	1
Aspergillus nidulans	100
Phialophora species	1 – 10
Cladosporium species	1 – 10

Indications. Flucytosine has proved very effective in the treatment of *Candida* fungemia and systemic candidoses (but not mucocutaneous candidiasis) in combination with amphotericin B. Also, *Cryptococcus* infections, including *Cryptococcus neoformans* meningoencephalitis in immunocompromized patients (e.g., AIDS patients), can be treated with flucytosine in combination with amphotericin B (and in the case of *Cryptococcus neoformans* associated infections, additionally fluconazole). Further, flucytosine is indicated in the therapy of chromomycoses [241]. There can also be some activity of flucytosine in disseminated aspergilloses. Because of the rapid development of resistance, the substance should only be given in combination with other antimycotic agents, mostly amphotericin B [242].

Side Effects. Flucytosine is normally well tolerated provided dosage instructions are observed and precautionary measures are taken [232]. The following side effects have been observed during treatment with flucytosine: *Blood/spleen.* Neutropenia, leukopenia, thrombopenia [243].

GIT. Nausea, vomiting, and diarrhoea were observed in rare cases [243].

Liver. Liver malfunctions with reversible hepatomegalia and increase in serum transaminases; an extensive liver cell necrosis was reported after flucytosine therapy in two patients [243].

Skin. Allergic skin reactions were observed under therapy. Changes in blood counts can have serious consequences. Blood counts and liver function should therefore be checked regularly during therapy. Hematotoxic side effects were especially observed if serum levels exceeded 100 mg/L. Flucytosine can prolong engraftment in bone marrow transplantation patients.

5. Griseofulvin

Griseofulvin [*126-07-8*], 7-chloro-4,6-trimethoxy-6′-methylspiro[benzofuran-2(3*H*),1′-[2]cyclohexene]-3,4′-dione, $C_{17}H_{17}ClO_6$, M_r 352.8, *mp* 222 °C, colorless crystals.

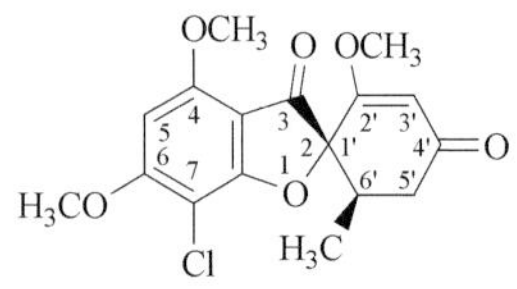

Synonym(s). Amudane, Curling Factor.

Solubility. Griseofulvin is soluble in methanol (2.5 g/mL), ethanol (1.5 g/mL), ether, and glacial acetic acid; quite soluble in DMF (150 mg/mL); slightly soluble in chloroform (25 mg/mL), ethyl acetate, acetone (33 mg/mL), and toluene; and insoluble in petroleum ether. The solubility in water is 40 µg/mL.

Stability. Griseofulvin is stable under normal storage conditions.

Synthesis. Griseofulvin is mostly produced by fermentation with strains of *Penicillium patulum*. After the first total synthesis in 1960 [244], other processes were published, but none of them has acquired industrial significance up to now.

Description. Griseofulvin was isolated from culture broths of *Penicillium griseofulvum* in 1938 and later also found in culture filtrates of *Penicillium nigricans* (= *P. janczewskii*), *P. patulum*, and other *Penicillium* species. The structure was determined in 1952. Griseofulvin was first used in plant protection and was introduced to medicine as an antimycotic in 1958 [245].

Formulations. Tablets à 125 and 500 mg. 5 % cream formulations.

Trade Names. B-GF (Biokema), Delmofulvina (Coli), Fulcin (ICI, Mason, Zeneca), Fulcin S (ICI), Fulvicin (Schering), Fulvicina (Interpharma), Fulvicin P/G (Schering-Plough), Fungivin (Nycomed), Gefulvine (Dif-Dogu), Greosin (Glaxo), Gricin (Arzneimittelwerke Dresden), Grifulin (Teva), Grifulvin V (McNeil, Ortho), Grisactin (Ayerst, Wyeth), Griséfuline (Sanofi Winthrop), Griseo (Ayerst), Griseoderm (Trinity), Griseomed (Waldheim), Griseostatin (Mason, Schering Corp.), Griseo von ct (ct-Arzneimittel), Grisfulvin (Protea), Grisol (Gebro), Grisovin (Glaxo), Grisovina Fp (Glaxo, Teofarma), Grisovin-FP (Glaxo), Grivate (Fujisawa), Grysio (Ayerst), Lamoryl (Leo, Lövens), Likuden (Hoechst), Norofulvin (Norbrook), Polygris (Schering Corp.), Sulvina (Dibios), Ultragris (Sidmak).

Mechanism of Action. In addition to the inhibition of protein synthesis resulting from the interference of griseofulvin with guanine during RNA synthesis, there may also be a specific inhibition of chitin synthesis. Chitin is an essential cell-wall constituent in dermatophytes. Inhibition of its synthesis would explain the characteristic morphological changes in the hyphae, called curling, that occur in dermatophytes in the presence of griseofulvine [231], [232], [246].

Antimycotic Properties. Griseofulvin has a narrow spectrum of antimycotic activity, including only dermatophytes. *Trichophyton* and *Microsporum* species and *Epidermophyton floccosum* are inhibited by concentrations of 0.1 – 0.5 µg/mL. The in vitro effect of griseofulvin depends primarily on the test system used and is confined to proliferating hyphae. Griseofulvin is fungistatic. Purines and pyrimidines are antagonistic in vitro [232], [247]. Acquired in vitro resistance towards griseofulvin develops gradually following a number of fungal passages [248], [249]. However, increases in resistance during therapy are rarely observed, even during long-term treatment. Although dermatophyte variants showing primary resistance have been described, they are very rare [250]. MIC values for griseofulvin are summarized in Table 10 [14].

Table 10. MIC values for griseofulvine

Fungus	No. isolates tested	MIC range µg/mL
Epidermophyton floccosum	6	3 – > 5
Microsporum canis	19	0.1 – > 18
Microsporum gypseum	6	0.5 – 3.13
Trichophyton mentagrophytes	57	0.1 – > 30
Trichophyton rubrum	558	0.1 – > 30
Trichophyton schoenleinii	170	0.1 – 25
Trichophyton tonsurans	25	0.1 – > 18
Trichophyton violaceum	345	0.1 – 21

Pharmacokinetics. The degree of absorption of orally administered griseofulvin varies from individual to individual from 25 to 75 % of the administered dose. The absorption is strongly dependent on the degree of micronization of the drug, a high fat content in the food also increases absorption. After the administration of a single dose of 0.25 – 0.5 g, serum concentrations reach 0.6 to 1.5 µg/mL. The serum half-life is about 20 h [231], [232], [251]. Griseofulvin accumulates during keratinization in human epidermis cells, hair, and nails. In the course of long-term griseofulvin treatment, concentrations up to 25 µg/g can be detected in the stratum corneum of the skin [232]. Griseofulvin concentrations of 0.2 to 0.3 µg/mL are found in the sweat [252]. Further, the substance shows high concentrations in liver, body fat, and muscles. Griseofulvin is metabolized in the liver and is mostly eliminated by the kidneys in a metabolized, inactive form. Up to 75 % of the quantity absorbed is excreted in the urine as inactive 6-demethylgriseofulvin. The concentration of unchanged griseofulvin in the urine is 1 – 2 µg/mL, corresponding to approximately 1 % of the dose [232], [253]. Elimination occurs biphasically with a terminal elimination half-life of 10 – 40 h [254].

Indications. The indications for griseofulvin are dermatophytoses and onychomycoses which do not respond to topical therapy. The optimal daily dose is 5 – 10 mg/kg body weight. The duration of treatment ranges from 3 to 5 weeks for dermatophytoses of the trunk and extremities, and from 5 to 10 weeks for infections of palms, soles, and scalp, but is more than 6 months and even up to 18 months for onychomycoses [231], [232]. Additional topical therapy increases the cure rate and reduces duration of treatment. Due to better tolerability and efficacy, newer antimycotic compounds such as azole antimycotics and allylamines have mostly replaced griseofulvin in the therapy of dermatomycoses and onychomycoses. Griseofulvin is inactive in mycoses caused by fungi other than dermatophytes.

Side Effects. *CNS.* Headache, depression, dizziness, sleeplessness.

Nerves. Paresthesia and neuropathia [255].

GIT. Dryness of the mouth, changes in taste, and other unspecific gastrointestinal disorders.

Skin. Minor allergic skin reactions, Lyell syndrome [256], erythema multiforme, exfoliative dermatitis, angioedema, photosensitization.

Occasionally, reversible hematopathies, proteinuria, and, especially in case of previous hepatopathy, increases in the liver enzyme values in the blood may be observed. A carcinogenic and embryotoxic potential of griseofulvin was observed in laboratory animals such as rats.

6. Ciclopirox

Ciclopirox [*29342-05-0*], 6-cyclohexyl-1-hydroxy-4-methyl-2(1*H*)-pyridinone, $C_{12}H_{17}NO_2$, M_r 207.27, *mp* 144 °C, colorless crystals.

Ciclopirox olamine [*41621-49-2*], 6-cyclohexyl-1-hydroxy-4-methyl-2(1*H*)-pyridinone olamine, $C_{14}H_{24}N_2O_3$, M_r 268.4, *mp* 143 °C, colorless crystals.

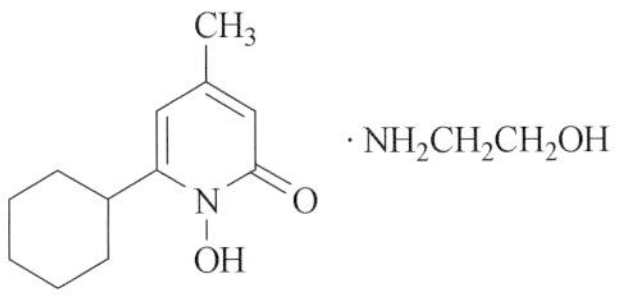

Synonym(s). HOE 296b (free base); Ciclopirox ethanolamine, Hoe 296 (olamine salt).

Stability. The substance is stable under normal storage conditions.

Synthesis. Base **9** is produced by heating 4-methyl-6-cyclohexyl-2-pyrone (**8**) [257] for 8 h at 80 °C with hdroxylamine hydrochloride and 2-aminopyridine [258] or from 5-cyclohexyl-3-methyl-5-oxopentenoate (**10**) [258] by reaction with hydroxylamine hydrochloride via the oxime **11**, followed by cyclization. The final step is to form the salt with ethanolamine.

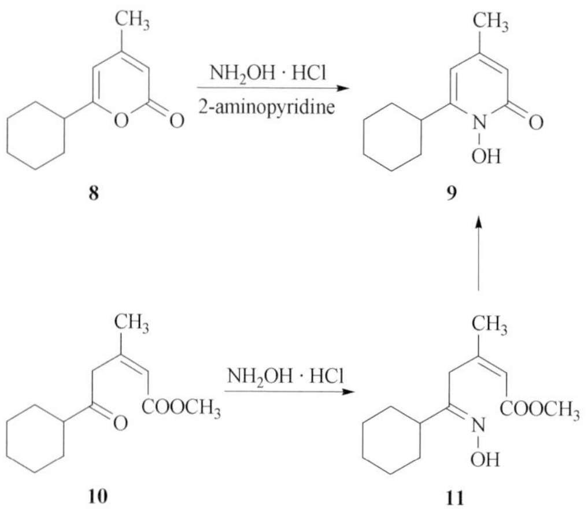

Description. The substance was first synthesized by Hoechst Pharmaceuticals in 1973 [258–262].

Formulations. 1 % solutions, ointments, and cream preparations.

Trade Names (Ciclopirox). Nagel Batrafen (Casella-Riedel).

Trade Names (Ciclopiroxolamine). Batrafen (Casella-Riedel, Hoechst, Knoll), Brumixol (Bruschettini), Ciclochem (Novag), Dafnegin (Poli), Fungiderm (Heilmittelwerke Wien), Fungowas (Wassermann), Loporox (Hoechst-Roussel), Miclast (Ellem), Micomicen (Vita), Micoxolamin (Delalande), (Sinbio), Obytin (Hoechst, Jugoremedija).

Mechanism of Action. Ciclopirox is a substituted pyridione derivative. Experiments show that the mechanism of action involves a disruption of transport mechanisms into the cell plasma [263], [264], especially the uptake of leucine [265]. Further, the uptake of other amino acids such as phenylalanine and lysine, as well as the uptake of potassium and phosphate ions is impaired. This all probably involves a change in membrane permeability, as reported for some other antimycotics, although ciclopirox does not show lytic activity [263], [264]. Consecutively, the substance significantly reduces fungal growth rates [266].

Antimycotic Properties. Ciclopirox has a broad in vitro antifungal activity spectrum, including dermatophytes, yeasts, and moulds. The most important fungi within the activity spectrum of ciclopirox and the corresponding MIC values are listed in Table 11 [263], [264], [267]. Ciclopirox is primarily fungistatic, but if the contact time is long enough and the concentration exceeds 20 µg/mL, it exerts a fungicidal effect on yeasts and dermatophytes [264]. In vitro, the effect is limited to proliferating fungi and is greatly dependent on the nutrient medium.

Table 11. MIC values for ciclopirox olamine

Species	MIC range, µg/mL
Dermatophytes	0.5 – 5
Candida species	0.5 – 5
Moulds	0.5 – 30

Pharmacokinetics. Ciclopirox penetrates rapidly into and through the skin and nail keratin. After dermal application of a 1 % cream formulation, about 1.3 % of the dose is absorbed. Serum levels range around 10 µg/L after application of a dose of 37 mg of ciclopirox [268]. After vaginal administration, 15 – 20 % of the dose is absorbed [269]. Ciclopirox penetrates well into the skin and nail matrix. Absorbed ciclopirox is mainly eliminated by the kidneys; 1.1 – 1.7 % of the topical dose is eliminated in the urine within four days. Only a fraction thereof is unaltered ciclopirox. Approximately 96 % of the absorbed drug is bound to human serum proteins [270].

Indications. Ciclopirox is indicated for the topical treatment of dermatophytoses and vaginal infections caused by *Candida* species [271].

Side Effects. Topically applied, ciclopirox is well tolerated. The incidence of local irritations such as itching and burning sensations range between 1 and 4 % and are often due to the galenic formulation.

7. Thiocarbamates

Thiocarbamates are inhibitors of fungal squalene epoxidase (see also Chap. 8). By this mechanism, tolciclate and tolnaftate inhibit biosynthesis of fungal ergosterol [272].

7.1. Tolciclate

Tolciclate [*50838-36-3*], methyl (3-methlphenyl)carbamothioic acid *O*-(1,2,3,4-tetrahydro-1,4-methanonaphtalen-6-yl) ester, $C_{20}H_{21}NOS$, M_r 323.46, *mp* 120 – 121 °C, white crystals from ethanol – diethylether.

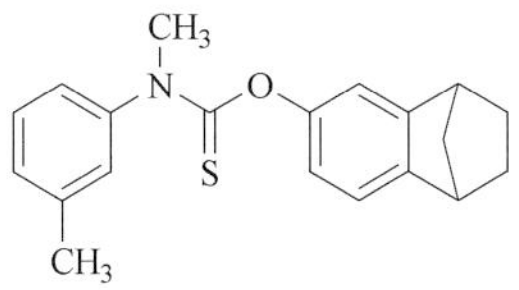

Synonym(s). KC 9147.

Solubility. The substance shows high liposolubility. Soluble in *n*-hexane (15 mg/mL) and *n*-octanol (24 mg/mL).

Stability. The substance is stable under normal storage conditions, but should be protected from light.

Synthesis. [273]:

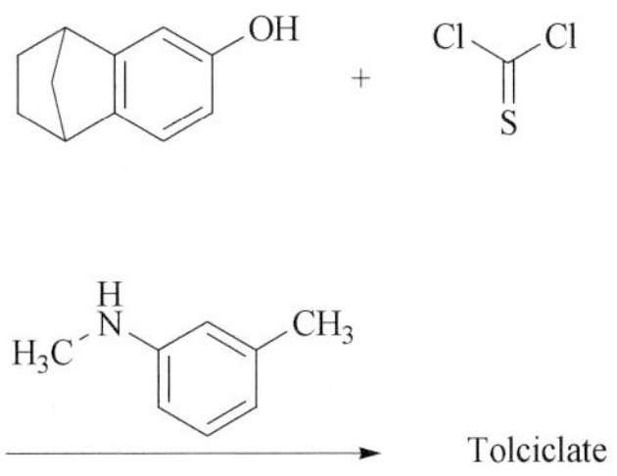

Tolciclate

Description. Tolciclate was first synthesized at Carlo Erba Pharmaceuticals [274], [275].

Formulations. 1 % solutions, ointments, and cream formulations.

Trade Names. Fungifos (Basotherm), Kilmicen (Carlo Erba), Tolmicen (Carlo Erba, Wing Yee), Tolmicol (Carlo Erba).

Antimycotic Properties. In vitro, tolciclate has a narrow activity spectrum with good efficacy only towards dermatophytes. Isolates of susceptible species showing primary resistance or developing secondary resistance are rarely observed.

Indications. The indications for tolciclate are defined by its narrow spectrum and limited to the treatment of dermatophytoses. Because of inadequate penetration into the nail keratin, it is unlikely to be effective for the treatment of onychomycoses. On average, the efficacy of tolciclate in topical therapy of dermatomycoses is lower than that of most other antimycotics for topical treatment.

Side Effects. Tolciclate is well tolerated on the skin.

7.2. Tolnaftate

Tolnaftate [*2398-96-1*], (3-methylphenyl)carbamothioic acid *O*-2-naphthalenylmethylester, $C_{19}H_{17}NOS$, M_r 307.4, *mp* 110.5 – 111.5 °C, colorless crystals.

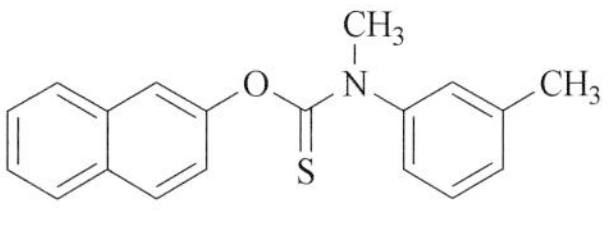

Synonym(s). Naphthiomate-T.

Solubility. The substance shows a good solubility in ethanol, ether, and glacial acetic acid; is quite soluble in DMF, chloroform, dichloromethane, and benzene; and practically insoluble in water.

Stability. Tolnaftate is stable under normal storage conditions, but schould be protected from light.

Synthesis. Tolnaftate is synthesized from β-naphthol, thiophosgene, and *N*-methl-*m*-tolulidine by the following pathway [276]. Another route, which does not use thiophosgene, has been described [277].

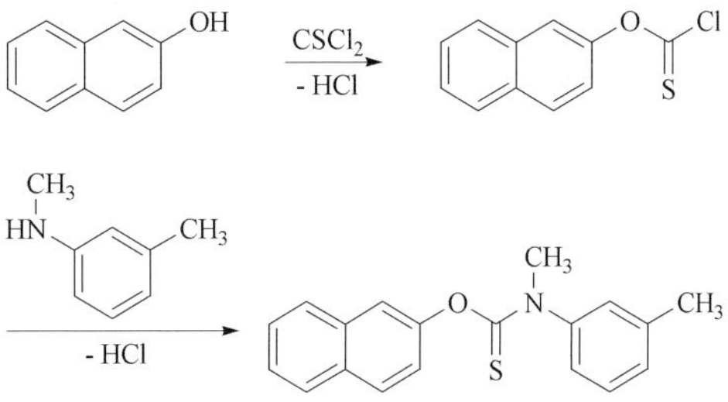

$CSCl_2$ – HCl

– HCl

Description. Tolnaftate was first synthesized by Japan Soda and first described as a topical antimycotic in 1962 [278–281].

Formulations. 1 % solutions, ointments, and cream preparations.

Trade Names. Aftate (JDH, Plough), Alber-T (Hing Yip), Chinofungin (Chinoin), Chlorisept (Chinosol), Ezon-T (Yamanouchi), Focusan (Lundbeck), Footwork (Lederle), Hi-Alarzin (Hing Yip, Yamanouchi), Mikoderm (Akdeniz), NP-27 (Thompson), Pedesal (Krka), Pediderm (Nelson), Pedimycose (Scholl), Pitrex (Taro), Separin (Sumitomo), Sorgoa (Scheurich), Sporiderm (Cétrane), Sporiline (Schering-Plough), Tinacidin (Schering Corp.), Tinactin (Schering-Plough), Tinatox (Brenner), Tinavet (Schering), Tineafax (Wellcome), Ting (Fisons), Tonoftal (Schering), Zeasorb-AF (Stiefel).

Antimycotic Properties. In vitro, tolnaftate has a narrow activity spectrum with fungistatic effect on towards dermatophytes and a few moulds such as some isolates of selected *Aspergillus* species. The MIC values range between 0.1 and 3 μg/mL for susceptible fungi. Isolates of susceptible species showing primary resistance or developing secondary resistance are rarely observed. Tolnaftate is almost ineffective against yeasts and moulds.

Pharmacokinetics. Tolnaftate combines good penetration into the stratus corneum of the skin with poor penetration into nail keratin.

Indications. The indications for tolnaftate are defined by its narrow spectrum and limited to the treatment of dermatophytoses. Tolnaftate is effective when topically administered several times daily over a period of several weeks [231], [232]. Because of its inadequate penetration into nail keratin, tolnaftate is unlikely to be effective for the treatment of onychomycoses. On the average, the efficacy of tolnaftate in topical therapy of dermatomycoses is lower than that of most other antimycotics for topical treatment [282].

Side Effects. Tolnaftate is well tolerated on the skin. Erythemas and allergic reactions are very rare and might also be due to the galenic formulation. No other side effects are known [232], [283].

8. Allylamines

The allylamines are inhibitors of the enzyme squalene epoxidase. This results in an accumulation of squalene, which exhibits lethal effects towards the fungal organism at high concentrations. Thereby, the inhibition of squalene epoxidase blocks ergosterol biosynthesis in the fungal cell [284]. The essential portion of the allylamine molecule is the conjugated enzyme group with a *trans* configuration in the side chain [285]. The enzyme squalene epoxidase is not linked to the cytochrome P450 complex.

8.1. Naftifine

Naftifine [*65472-88-0*], (*E*)-*N*-Methyl-*N*-(3-phenyl-2-propenyl)-1-naphthalenemethenamine, $C_{21}H_{21}N$, M_r 287.40, *bp* 162 – 167 °C, colorless, viscous oil.

Naftifine hydrochloride [*65473-14-5*], $C_{21}H_{22}ClN$, M_r 323.9, *mp* 177 °C, crystals from propanol.

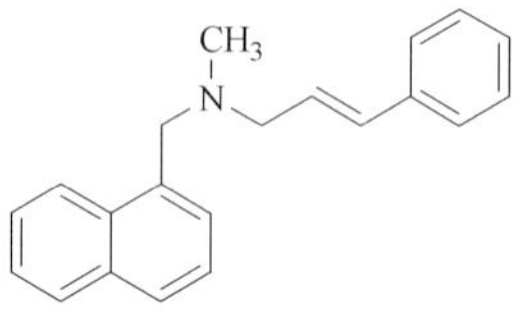

Synthesis. [286], [287]:

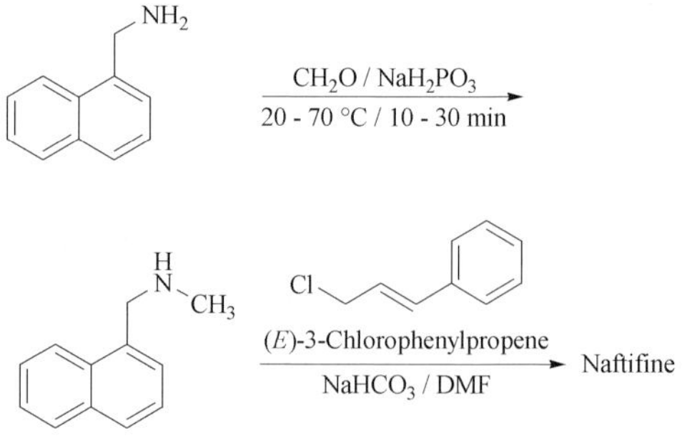

Description. See [288–290].

Formulations. 1 % creams, ointments, gels, and solutions.

Trade Names (Naftifine Hydrochloride). Exoderil (Allergan, Novartis, Rentschler, Schering), Fotimin (KRKA), Nafteryl (Schering), Naftin (Allergan).

Antimycotic Properties. Naftifine is fungicidal against dermatophytes such as *Epidermophyton floccosum, Trichophyton* and *Microsporum* species. Against pathogenic yeasts such as *Candida* species and moulds, it shows only an intermediate fungistatic activity in vitro [291], [292].

Pharmacokinetics. Naftifine penetrates well into the stratum corneum of the skin; 2 – 4 % of the topically administered dose were absorbed after administration of a 1 % gel preperation. After occlusion, an absorption of 6 % of the administered dose was observed [293].

Indications. Naftifine is primarily indicated in the therapy of superficial dermatomycoses [294–296]. It also shows some clinical efficacy in superficial infections caused by yeasts and moulds.

Side Effects. Naftifine is well tolerated although rare cases of local skin irritations and contact dermatitis have been found, which might also be due the galenic formulation [293].

8.2. Terbinafine

Terbinafine [*91161-71-6*], (*E*)-*N*-(6,6-dimethyl-2-hepten-4-ynyl)-*N*-methyl-1-naphthalenemethanamine, $C_{21}H_{25}N$, M_r 291.44, colorless crystals.

Terbinafine hydrochloride [*78628-80-4*], $C_{21}H_{26}ClN$, M_r 327.90, *mp* 195 – 198 °C, colorless crystals from 2-propanol diethyl ether, change in crystal structure begins at 150 °C.

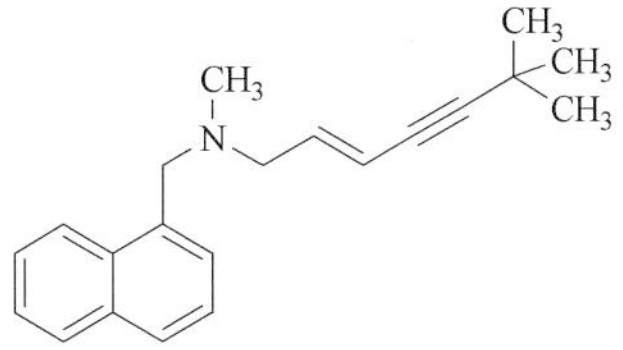

CH_3 N CH_3 CH_3 CH_3

Synonym(s). SF 83627, SF 83-627 (both hydrochloride).

Solubility. Terbinafine is a highly lipophilic substance.

Stability. The substance is stable under normal storage conditions.

Synthesis. [297] (see top of next page):

Description. The substance was first synthesized at Sandoz Pharmaceuticals [298].

Formulations. 1 % cream and ointment preparations, and tablets à 250 mg.

Trade Names (Terbinafine). Lamisil (Edward Keller).

Trade Names (Terbinafine Hydrochloride). Lamisil (Sanabo, Sandoz-Wander).

Antimcotic Properties. Terbinafine shows a good and mostly fungicidal in vitro antimycotic activity against dermatophytes such as *Epidermophyton floccosum, Trichophyton,* and *Microsporum* species. The substance also shows a high but often fungistatic in vitro activity against isolates of various *Aspergillus* species. The activity against pathogenic yeasts is quite variable. Many isolates of *Cryptococcus neoformans* and *Candidia parapsilosis* show an in vitro sensitivity towards terbinafine, although isolates of *Candida albicans, Candida (Torulopsis) glabrata,* and *Candida tropicalis* are mostly resistant or have rather high MIC levels [299].

Pharmakocinetics. Terbinafine is well absorbed from the gastrointestinal tract with a bioavailability of 70 – 80 %. Peak plasma concentrations of 0.97 µg/mL were observed following a single oral dose of 250 mg within 2 h

Terbafin hydrochloride (Synthesis 1):

CuCl

CuCl

$(CH_2O)_n$

Diisobutylaluminum hydride

Terbafin hydrochloride (Synthesis 2):

1. C_4H_9OLi

2. H_2O

HBr

Na_2CO_3 / DMF

(*E* / *Z* ~ 3 / 1)

(*E*)-Isomer

(*Z*)-Isomer

HCl / EtOH

Terbinafin hydrochloride

of administration. Steady state concentrations are reached within two weeks of therapy [300]. The substance is highly bound to plasma proteins. After oral administration, terbinafine distributes well into tissues, including skin, the nail plate, sweat, sebum, and hair. Terbinafine is metabolized in the liver to inactive metabolites which are excreted in the urine. Elimination occurs biphasically. Plasma elimination half-lifes of 11 – 17 h have been reported, as has a slower elimination half-life of 90 – 100 h [301]. After topical administration, the lipophilic substance shows a good penetration into the skin and nail matrix and is hardly absorbed (< 5 % of the administered dose) [302].

Indications. Oral and topical treatment of dermatophytoses and onychomycoses [303]. The substance also shows activity in the topical treatment of pityriasis versicolor and cutanous candidiasis. 250 mg of terbinafine are given once daily by mouth for 2 to 4 weeks for therapy of tinea cruris and tinea corporis. Especially for tinea pedis, treatment may have to be continued for up to 6 weeks [292].

Side Effects. The following side effects were observed after oral administration:

CNS. Headache.

GIT. Nausea, diarrhoea, anorexia, mild abdominal pain (5 %) [301].

Arthralgia, myalgia, and skin reactions such as rush and urticaria rarely occur. Severe skin reactions including erythema multiforme (1:100,000), Stevens–Johnson syndrome and Lyell syndrome have occurred occasionally. Loss or disturbance of taste, which is mostly reversible [304], [305], and liver disfunction have been observed [306]. After topical administration, the substance is well tolerated, although side effects such as skin irritations (e.g., itching and burning sensations) and allergic reaction may occur, which may also often be due to the galenic formulation.

9. Amorolfine

Amorolfine *[78613-35-1]*, (±)-*cis*-2,6-dimethyl-4-[2-methyl-3(*p-tert*-pentylphenyl)propyl]morpholine, $C_{21}H_{35}NO$, M_r 317.60, *bp* 134 °C.

Amorolfine hydrochloride *[78613-38-4]*, $C_{21}H_{36}ClNO$, M_r 354.06.

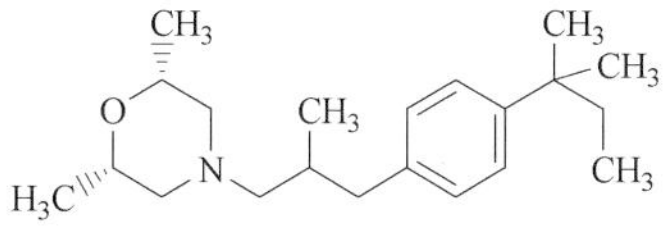

Synonym(s). Ro 14-4767/000 (free base); Ro 14-4767/002 (hydrochloride).

Description. The substance was first synthesized at Roche Pharmaceuticals [307], [308].

Formulations. 0.25 % cream preparations, 5 % nail lacquers, 50 mg vaginal tablets.

Trade Names (Amorolfine Hydrochloride). Loceryl (Roche, Sauter), Locéryl (Roche).

Mechanism of Action. Amorolfine is a morpholine derivative and selectively inhibits Δ^{14}-reduction and Δ^{8}-Δ^{7}-isomerase, which are required for Δ^{14}-reductase and Δ^{8}-Δ^{7}-isomerization, following C14 demethylation of lanosterol or 24-methlenedihydrolanosterol in the ergosterol biosnthesis pathway of the fungal cell [309].

Antimycotic Properties. Amorolfine is a broad-spectrum antimycotic [310] with in vitro activity against dermatophytes, pathogenic *Candida* species, and a variety of pathogenic dimophic fungi such as *Blastomyces dermatitidis, Histoplasma capsulatum*, and *Sporothrix schenckii*. The substance also shows some activity against various isolates of *Aspergillus* species.

Pharmacokinetics. Amorolfine is not absorbed after topical administration on normal skin. After application of a single vaginal tablet containing 50 mg of amorolfine, plasma concentrations between 27 and 83 ng/mL were observed [311].

Indications. For the treatment of skin infections, a 0.25 % cream preparation is applied once daily until clinical cure is achieved and then for a further 3–5 d [312]. For the treatment of nail mycoses caused by dermatophytes, yeasts, and moulds, a lacquer preparation containing 5 % amorolfine is painted onto the affected nail once or twice a week until the nail has regenerated. Treatment of onychomycoses normally needs to be continued for up to 1 year. Amorolfine is also active in the treatment of vaginal candidiasis.

Side Effects. Skin irritations, such as erythema, pruritus, or a burning sensation, rarely occur.

10. Unspecific Topical Antimycotics

Other unspecific antimycotics for topical use are listed in Table 12. These somewhat "historical" substances are of minor importance in clinical use nowadays, as highly effective and safe specific substances have become available.

11. Recent Developments and Outlook

The antimycotic market is clearly dominated by azoles, especially for the treatment of systemic mycoses. The latter market is led by fluconazole and itraconazole. Improved azoles are in clinical development (mostly phase I and II); pradimicins, aureobasidins, echinocandins, and pneumocandins for the treatment of systemic mycoses. The new azoles are reported to have

Table 12. Other unspecified topical antimycotics

Generic name	Structural formula	Activity spectrum
Quaternary ammonium compounds		
Benzalkonium chloride [*8001-54-5*]	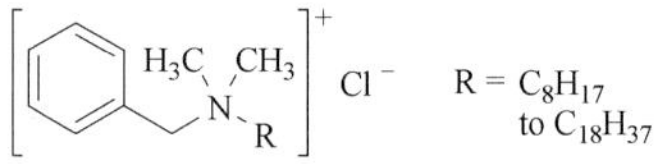 $[C_6H_5CH_2N(CH_3)_2R]^+ Cl^-$ R = C_8H_{17} to $C_{18}H_{37}$	Dermatophytes, yeasts, moulds
Domiphen bromide [*13900-14-6*]	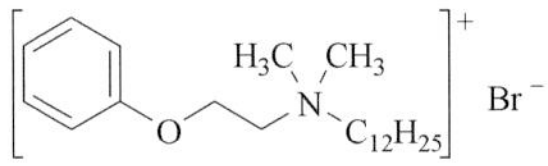 $[C_6H_5OCH_2CH_2N(CH_3)_2C_{12}H_{25}]^+ Br^-$	Dermatophytes, yeasts, moulds
Dibromosalane [*87-12-7*]	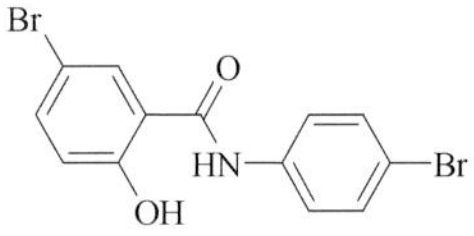	Dermatophytes, yeasts, gram-positive bacteria
Chloroquinaldol [*72-80-0*]	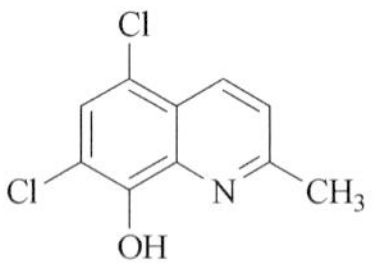	Dermatophytes, yeasts
Dichlorophene [*97-23-4*]	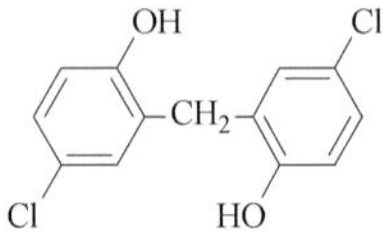	Dermatophytes, yeasts
Hexachlorophene [*70-30-4*]	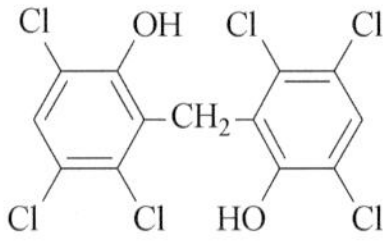	Dermatophytes, yeasts, gram-positive bacteria
Dibenzthione [*350-12-9*]	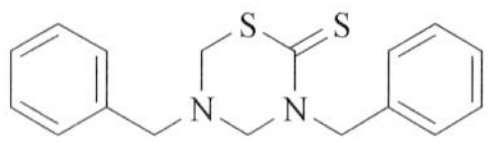	Dermatophytes, yeasts
Chlormidazole [*3689-76-7*]	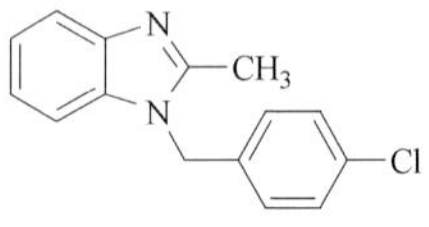	Dermatophytes, yeasts
Mercury compounds		
Merbromin (Mercurochrome) [*129-16-8*]	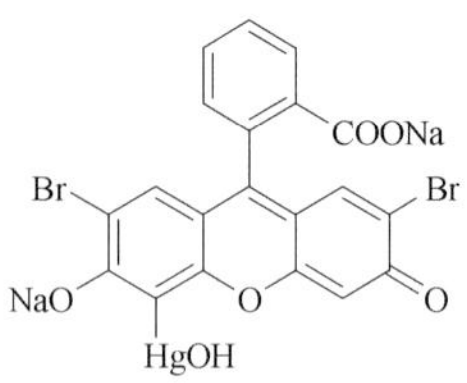	Dermatophytes, yeasts, bacteria
Thimerosal [*54-64-8*]	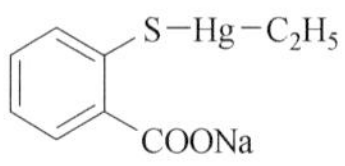	Dermatophytes, yeasts, bacteria
Triphenylmethane dyes		
Fuchsin (basic violet) [*632-99-5*]	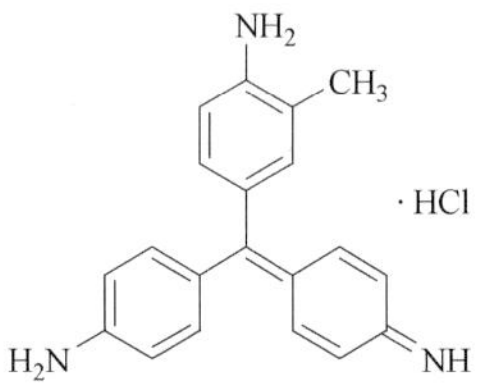 · HCl	Dermatophytes, yeasts
Undecylenic acid [*112-38-9*]	$CH_2 = CH(CH_2)_8COOH$	Dermatophytes, yeasts

activity against fluconazole-resistant strains and species as well as a broader spectrum of antifungal activity. Except the azoles, all other reported compounds habe a relatively low bioavailability and/or tolerability on the average which may limit their use to therapy of life threatening infections within the hospital segment. Production costs of all these compounds seem to be rather high as compared, e.g., to fluconazole.

Significant increase in research activities in antifungals for systemic application has been observed within the last few years in many pharmaceutical companies and institutes worldwide due to the high medical need and the fast growing market. Therefore, the chances for the development of new antifungal agents with new chemical structures and new mechanisms of action are improving.

12. References

1. K. J. Kwon-Chung et al.: *Medical Mycology,* Lea & Febiger, Philadelphia 1992.
2. K. H. Büchel et al., in J. S. Bindra et al. (eds.): *Chronicles of Drug Discovery,* **vol. 2,** J. Wiley, New York 1984.
3. H. Yamaguchi et al., *Sabouraudia* **17** (1979) 311.
4. H. Van den Bossche et al., *Biol. Interact.* **21** (1978) 59.
5. H. Van den Bossche et al., in W. Siegenthaler et al. (eds.): *Current chemotherapy,* Am. Soc. Microbiol., Washington, D.C. 1978 p. 228.
6. J. Haller, *Abstracts XII – Intl. Congr. Microbiol.,* Munich Sept. 1978, no. 38/3.
7. H. Van den Bossche et al., *Arch. Int. Physiol. Biochim.* **87** (1979) 849.
8. H. Van den Bossche in M. R. McGinnis (ed.): *Current Topics in Medical Mycology,* **vol. 1** Springer, New York 1985, p. 313.
9. W. R. Ness et al., *J. Biol. Chem.* **253** (1978) 6218.
10. K. E. Bloch, *CRC Crit. Rev. Biochem.* **7** (1979) 1.
11. Pfizer, GB 2 099 818, 1982.
12. Pfizer, US 4 404 216, 1983.
13. T. E. Rogers et al., *Antimicrob. Agents Chemother.* **30** (1986) 418.
14. M. R. McGinnis et al. in V. Lorian (ed.): *Antibiotics in Laboratory Medicine,* Williams & Wilkins, Baltimore 1996.
15. K. W. Brammer et al., *Rev. Infect. Dis.* **12** (1990) Suppl. 1, S 318.
16. J. E. Thorpe et al., *Antimicrob. Agents Chemother.* **34** (1990) 2032.
17. M. J. Humphrey et al., *Antimicrob. Agents Chemother.* **28** (1985) 648.
18. E. T. Houang et al., *Antimicrob. Agents Chemother.* **34** (1990) 909.
19. R. M. Tucker et al., *Antimicrob. Agents Chemother.* **32** (1988) 369.
20. K. W. Brammer, *Br. J. Obstet, Gyn.* **96** (1989) 226.
21. B. Dupont et al., *J. Med. Vet. Mycol.* **26** (1988) 67.
22. J. A. Como et al., *N. Engl. J. Med.* **33** (1994) 263.
23. J. E. Mangino et al., *Lancet* **345** (1995) 6.
24. J. H. Rex et al., *Antimicrob. Agents Chemother.* **39** (1995) 1.
25. J. E. Bennet et al., *N. Engl. J. Med.* **301** (1979) 126.
26. W. G. Powderly et al., *N. Engl. J. Med.* **326** (1992) 793.
27. M. Rozenberg-Arska et al., *J. Antimicrob. Chemother.* **27** (1991) 369.
28. J. Heeres et al., *J. Med. Chem.* **27** (1984) 894.
29. F. von Bruchhausen et al. (eds.): *Hagers Handbuch der Pharmazeutischen Praxis,* **vol. 8,** Springer, Berlin – Heidelberg – New York 1993, p. 634.
30. Janssen, EP 6 711, 1980.
31. Janssen, US 4 267 179, 1981.
32. D. A. Borelli, *Rev. Infect. Dis.* **9** (1987) 57.
33. M. Bogers et al., *Rev. Infect. Dis.* **9** (1987) 33.
34. G. Cauwenbergh et al., *Drug Dev. Res.* **8** (1986) 317.
35. J. D. Cleary et al., *Ann. Pharmacother.* **26** (1992) 502.
36. D. W. Denning et al., *Arch. Intern. Med.* **149** (1989) 2301.
37. T. S. Jennings et al., *Ann. Pharmacother.* **27** (1993) 1206.
38. R. Negroni et al., *Rev. Infect. Dis.* **9** (1987) 47.
39. J. W. Van't Wout et al., *J. Infect.* **22** (1991) 45.
40. J. W. Van't Wout et al., *J. Infect.* **20** (1990) 147.
41. J. Heykants et al., *Mycoses* **32** (1989) Suppl. 1, 67.
42. T. C. Hardin et al., *Antimicrob. Chemother.* **32** (1988) 1310.
43. J. Heykants et al. in R. A. Fromtling (ed.): *Recent Trends in the Discovery, Development and Evaluation of Antifungal Agents,* J. R. Prous, Barcelona 1987, p. 223.
44. S. M. Grant et al., *Drugs* **37** (1989) 310.
45. J. Faegermann, *Mycoses* **31** (1988) 377.

46. E. Van Hecke et al., *Mycoses* **31** (1988) 641.
47. T. Piepponen et al., *J. Antimicrob. Chemother.* **29** (1992) 195.
48. G. E. Stein et al., *Antimicrob. Agents Chemother.* **37** (1993) 89.
49. W. E. Dismukes et al., *Am. J. Med.* **93** (1992) 489.
50. J. S. Hostetler et al., *Chemotherapy* **38** (1992) Suppl. 1, 12.
51. J. R. Graybill et al., *Am. J. Med.* **89** (1990) 282.
52. A. P. Lavrijsen et al., *Lancet* **340** (1992) 251.
53. R. M. Tucker et al., *J. Antimicrob. Chemother.* **26** (1990) 561.
54. Janssen Pharmaceutica, DE-OS 2 804 094, 1978 (J. V. Heeres et al.).
55. J. V. Heeres et al., *J. Med. Chem.* **22** (1979) 1003.
56. Janssen, DE 2 804 096, 1978.
57. Janssen, US 4 144 346, 1979.
58. Janssen, US 4 223 036, 1980.
59. N. R. Blatchford et al., *Lancet* **320** (1982) 770.
60. A. Tavihan et al., *Gastroenterology* **90** (1986) 443.
61. T. K. Daneshmend et al., *Clin. Pharmacokin.* **14** (1988) 13.
62. G. Lake-Bakaar et al., *Ann. Int. Med.* **109** (1988) 471.
63. T. K. Daneshmend et al., *Antimicrob. Agents Chemother.* **25** (1984) 1.
64. M. D. Ene et al., *Br. J. Clin. Pharmacol.* **17** (1984) 173.
65. A. M. Sugar et al., *Antimicrob. Agents Chemother.* **31** (1987) 1874.
66. G. Cauwenbergh, *Mycoses* **32** (1989) Suppl. 2, 59.
67. R. De Felice et al., *Antimicrob. Agents Chemother.* **19** (1981) 1073.
68. A. Pont et al., *Arch. Int. Med.* **142** (1982) 2137.
69. J. H. Lewis et al., *Gastroenterology* **86** (1984) 503.
70. Janssen Pharmaceutica, US 3 717 655, 1969 (E. F. Godefroi et al.).
71. E. F. Godefroi et al., *J. Med. Chem.* **12** (1969) 784.
72. Janssen, DE 1 940 388, 1970.
73. Janssen, US 3 717 655, 1973.
74. L. Brugmans et al., *Arch. Dermatol.* **102** (1970) 428.
75. R. Godts et al., *Arzneim. Forsch. / Drug Res.* **21** (1971) 256.
76. R. C. Heel et al., *Drugs* **19** (1980) 7.
77. J. R. Graybill et al., *Drugs* **25** (1983) 41.
78. A. Kucers et al., in A. Kucers et al., (eds.): *The Use of Antibiotics,* Lippincott, Philadelphia 1987, p. 1492.
79. T. K. Daneshmend et al., *Clin. Pharmacokin.* **8** (1983) 17.
80. D. A. Stevens, *Drugs* **26** (1983) 347.
81. V. Feinstein et al., *Ann. Intern. Med.* **93** (1980) 432.
82. Bayer, DE-OS 2 461 406, 1976 (E. Regel et al.).
83. Bayer, DE 2 461 406, 1976.
84. Bayer, FR 2 295 747, 1976.
85. Bayer, DE 2 643 563, 1978.
86. Bayer, US 4 118 487, 1978.
87. W. Ritter, *Proc. VII Int. Congr. ISHAM Palmerston North* (1982) 105.
88. H. Weber, *Proc. VII Int. Congr. ISHAM Palmerston North* (1982) 105.
89. S. Stettendorf, *Arzneim. Forsch. / Drug Res.* **33** (1983) 750.
90. T. Anik et al., *J. Pharm. Sci.* **70** (1981) 8.
91. F. von Bruchhausen et al. (eds.): *Hagers Handbuch der Pharmazeutischen Praxis,* **vol. 7,** Springer, Berlin – Heidelberg – New York 1993, p. 581.
92. Syntex, US 4 078 071, 1978.
93. K. A. M. Walker et al., *J. Med. Chem.* **21** (1978) 840.
94. F. C. Odds et al., *J. Antimicrob. Chemother.* **14** (1984) 105.
95. W. Droegemueller et al., *Obstst. Gynecol.* **64** (1984) 530.
96. J. B. Jacobson et al., *Acta. Obstet. Gynecol. Scand.* **64** (1985) 241.
97. C. S. Brandbeer et al., *Genitourin. Med.* **61** (1985) 270.
98. W. A. Van Dyk, *J. Reproduct. Med.* **31** (1986) Suppl. 7, 662.
99. K. H. Büchel et al., *Arzneim. Forsch. / Drug Res.* **22** (1972) 1260.
100. Bayer, DE-OS 1 617 481, 1976 (K. H. Büchel et al.).
101. M. Plempel et al., *Dtsch. Med. Wochenschr.* **94** (1969) 1356.
102. Bayer, DE 1 940 626, DE 1 940 627, 1969.
103. Bayer, ZA 6 805 392, ZA 6 900 039, 1969.
104. Bayer, US 3 657 445, US 3 705 172, 1972.
105. W. Ritter, *Proc. VII Int. Congr. ISHAM Palmerston-North* (1982) 105.
106. H. Weber, *Proc. VII Int. Congr. ISHAM Palmerston-North* (1982) 105.
107. S. Stettendorf et al., *Arzneim. Forsch. / Drug Res.* **33** (1983) 750.
108. Shionogi, BE 883 665, 1980.
109. Shionogi, US 4 328 348, 1982.

110. F. Toshiaka et al., *Iyakuhin Kenkyu* **9** (1978) 448.
111. E. F. Godefroi et al., *J. Med. Chem.* **12** (1969) 784.
112. Janssen, DE 1 940 388, 1970.
113. Janssen, US 3 717 655, 1973.
114. V. Tullio et al., *Mycoses* **33** (1990) 257.
115. R. C. Heel et al., *Drugs* **16** (1978) 177.
116. W. Rindt et al., *Arzneim. Forsch. / Drug Res.* **29** (1979) 697.
117. A. Vukovich et al., *Clin. Pharmacol. Ther.* **21** (1977) 121.
118. L. E. Millikan et al., *J. Am. Acad. Dermatol.* **18** (1988) 52.
119. G. Gabriel et al., *Br. J. Vener. Dis.* **59** (1983) 56.
120. J. S. Bingham et al., *Br. J. Vener. Dis.* **57** (1981) 204.
121. J. E. Benett in L. S. Goodman et al. (eds.): *The Pharmacological Basis of Therapeutics*, Pergamon Press, New York 1990 p. 1165.
122. A. Tajana et al., *Arzneim. Forsch. / Drug Res.* **31** (1981) 2127.
123. Recordati, DE 2 917 244, 1979.
124. Recordati, US 4 221 803, 1980.
125. A. Finzi et al., *Mykosen* **29** (1986) 41.
126. E. M. Kokoschka et al., *Mykosen* **29** (1986) 45.
127. A. Gastaldi, *Curr. Ther. Res.* **38** (1985) 489.
128. E. F. Godefroi et al., *J. Med. Chem.* **12** (1969) 784.
129. Janssen, DE 1 940 388, 1970.
130. Janssen, US 3 717 655, 1973.
131. Janssen, US 3 839 574, 1974.
132. H. J. Kessler, *Arzneim. Forsch. / Drug Res.* **29** (1979) 1344.
133. H. J. Kessler et al., *Arzneim. Forsch. / Drug Res.* **29** (1979) 1352.
134. U. Täubler, *Arzneim. Forsch. / Drug Res.* **37** (1987) 461.
135. H. Wendt et al., *Arzneim. Forsch. / Drug Res.* **29** (1979) 846.
136. Siegfried, DE 2 839 388, 1980.
137. Siegfried, US 4 210 657, 1980.
138. Siegfried, US 4 554 356, 1985.
139. M. Mosse et al., *Pathol. Biol.* **34** (1986) 684.
140. G. Mixich et al., *Arzneim. Forsch. / Drug Res.* **29** (1979) 1510.
141. F. von Bruchhausen et al. (eds.): *Hagers Handbuch der Pharmazeutischen Praxis,* **vol. 8,** Springer, Berlin – Heidelberg – New York 1993, p. 1262.
142. Siegfried, DE 2 657 578, 1977.
143. Siegfried, US 4 124 767, 1978.
144. A. Polak, *Arzneim. Forsch. / Drug Res.* **32** (1982) 17.
145. G. Stüttgen et al., *Mykosen* **28** (1985) 138.
146. W. H. Beggs, *IRCS Med. Sci.* **11** (1983) 677.
147. M. Raga et al., *J. Med. Chem. Chim. Ther.* **21** (1986) 329.
148. Ferrer, EP 151 477, 1985.
149. Ferrer, US 5 135 943, 1992.
150. E. Martin-Mezuelos et al., *Chemotherapy* **42** (1996) 112.
151. E. Drouhet et al., *Arzneim. Forsch. / Drug Res.* **42** (1992) 705.
152. C. Palacin et al., *Arzneim. Forsch. / Drug Res.* **42** (1992) 699.
153. C. Palacin et al., *Arzneim. Forsch. / Drug Res.* **42** (1992) 714.
154. O. Azcona et al., *Curr. Ther. Res.* **49** (1991) 1046.
155. F. von Bruchhausen et al. (eds.): *Hagers Handbuch der Pharmazeutischen Praxis,* **vol. 9,** Springer, Berlin – Heidelberg – New York 1993, p. 691.
156. Syntex, DE 2 541 833, 1976.
157. Syntex, FR 2 285 126, 1976.
158. Syntex, US 4 038 409, US 4 039 677, 1977.
159. P. Benfield et al., *Drugs* **35** (1988) 143.
160. A. Yoshida et al., *Chemotherapy* **32** (1984) 477.
161. F. Iwata et al., *Shinkin to Shinkinsho* **25** (1984) 147.
162. J. Heeres et al., *J. Med. Chem.* **26** (1983) 611.
163. F. von Bruchhausen et al. (eds.): *Hagers Handbuch der Pharmazeutischen Praxis,* **vol. 9,** Springer, Berlin – Heidelberg – New York 1993, p. 808.
164. Janssen, DE 2 804 096, 1978.
165. Janssen, US 4 144 346, 1979.
166. Janssen, US 4 223 036, 1980.
167. R. A. Fromtling, *Clin. Microbiol. Rev.* **1** (1988) 187.
168. A. Kjaeldgaard, *Pharmacotherapeutica* **4** (1986) 525.
169. U. M. Moebius, *Lancet* **332** (1988) 966.
170. F. von Bruchhausen et al. (eds.): *Hagers Handbuch der Pharmazeutischen Praxis,* **vol. 9,** Springer, Berlin – Heidelberg – New York 1993, p. 944.
171. Pfizer, BE 841 309, 1976.
172. Pfizer, US 4 062 966, 1977.
173. S. P. Clissold et al., *Drugs* **31** (1986) 29.
174. S. Jevons, *Antimicrob. Agents Chemother.* **15** (1979) 597.
175. F. C. Odds, *J. Antimicrob. Chemother.* **6** (1980) 749.
176. Y. M. Clayton et al., *Clin. Exp. Dermatol.* **7** (1982) 543.

177. D. Brunelli et al., *Contact Dermatitis* **27** (1992) 120.
178. R. Izu et al., *Contact Dermatitis* **26** (1992) 130.
179. R. W. Holz in F. E. Hahn (ed.): *Antibiotics,* **vol. 5,** Springer, Berlin – Heidelberg – New York 1979, p. 313.
180. J. Vandeputte et al., *Antibiot. Anm.* (1955/56), 587.
181. J. M. T. Hamilton-Miller, *J. Pharm. Pharmacol.* **25** (1973) 401.
182. J. F. Gallelli, *Drug Intell.* **1** (1967) 102.
183. E. Borowski et al.: *Vth Intl. Congr. Biochem. Moscow 1961,* Pergamon Press, London 1961, p. 3.
184. F. von Bruchhausen et al. (eds.): *Hagers Handbuch der Pharmazeutischen Praxis,* **vol. 7,** Springer, Berlin – Heidelberg – New York 1993, p. 237.
185. C. P. Schaffner et al., *Antibiot. Chemother.* **11** (1961) 724.
186. E. Borowski et al., *Tetrahedron Lett.* (1970) 3909.
187. K. C. Nicolaou et al., *Chem. Commun.* (1986) 413.
188. O. Mathieson, US 2 908 611, 1959.
189. R. Janknegt, *Clin. Pharmacokinet.* **23** (1992) 279.
190. H. Lackner et al., *Pediatrics* **89** (1992) 1259.
191. A. Zoubek et al., *Pediat. Hematol. Oncol.* **9** (1992) 187.
192. E. D. Ralph et al., *Antimicrob. Agents Chemother.* **35** (1991) 188.
193. C. E. Hughes et al., *Antimicrob. Agents Chemother.* **25** (1984) 560.
194. T. K. Daneshmend et al., *Clin. Pharmacokin.* **8** (1983) 17.
195. J. Brajtburg et al., *J. Infect. Dis.* **149** (1984) 986.
196. K. J. Christiansen et al., *J. Infect. Dis.* **152** (1985) 1037.
197. N. Collette et al., *Antimicrob. Agents Chemother.* **33** (1989) 362.
198. A. J. Atkinson et al., *Antimicrob. Agents Chemother.* **13** (1978) 271.
199. A. Kucers et al. in A. Kucers et al. (eds.): *The Use of Antibiotics,* J. B. Lippincott, Philadelphia 1987, p. 1441.
200. D. J. Drutz, *Drugs* **26** (1983) 337.
201. G. Medoff et al., *Proc. Natl. Acad. Sci. USA* **69** (1972) 196.
202. J. E. Bennett et al., *N. Engl. J. Med.* **301** (1979) 126.
203. H. A. Gallis et al., *Rev. Infect. Dis.* **12** (1990) 308.
204. F. Gigliotti et al., *J. Infect. Dis.* **156** (1987) 784.
205. M. S. Maddux et al., *Drug Intell. Clin. Pharm.* **14** (1980) 177.
206. W. T. Butler et al., *Ann. Intern. Med.* **61** (1964) 175.
207. G. Lopez-Berenstein et al., *J. Infect. Dis.* **151** (1985) 704.
208. A. P. Struyk et al., *Antibiot. Ann.* (1957/58) 878.
209. W. E. Meyer, *Chem. Commun.* (1968) 470.
210. T. Strittmatter, *Chem. Abstr.* **94** (1981) 214469.
211. Koninklijke Nederlandsche Gist en Spiritus-Fabriek, GB 844 289, 1960.
212. American Cyanamid, GB 846 933, 1960.
213. E. Skytte Christensen et al., *Acta Obstst. Gynecol. Scand.* **61** (1982) 325.
214. D. Kerridge, *Ad. Microb. Physiol.* **27** (1986) 1.
215. T. Wegemann, *Internist* **30** (1989) 46.
216. M. I. Maurad, *Assiut. Vet. Med.* **11** (1984) 236.
217. D. J. Sutton et al., *Vet. Rec.* **118** (1986) 27.
218. E. L. Hazen et al., *Proc. Soc. Exp. Biol. Med.* **76** (1951) 93.
219. C. N. Chong et al., *Tetrahedron Lett.* (1970) 5145.
220. A. C. Moffat: *Ckarke's Isolation and Identification of Drugs,* Pharmaceutical Press, London 1986.
221. J. R. Carlson et al., *Antibiot. Chemother.* **9** (1959) 139.
222. O. Mathieson, US 2 832 719, 1958.
223. American Canamid, US 3 517 100, 1970.
224. R. Duschinsky et al., *J. Am. Chem. Soc.* **79** (1957) 4559.
225. Hoffmann-La Roche, US 2 945 038, 1960 (R. Duschinsky et al.).
226. Hoffmann-La Roche, US 2 802 005, 1957.
227. Hoffmann-La Roche, US 2 945 038, 1960.
228. Hoffmann-La Roche, US 3 040 026,1962.
229. Hoffmann-La Roche, BE 628 615, 1963.
230. Hoffmann-La Roche, US 3 368 938, 1968.
231. D. C. E. Speller (ed.): *Antifungal Chemotherapy,* J. Wiley & Sons, New York 1980.
232. H. Otten et al.: *Antibiotika-Fibel,* Thieme, Stuttgart 1975, p. 666.
233. A. Polak et al., *Eur. J. Biochem.* **32** (1973) 276.
234. J. Schönbeck et al., *Sabouraudia* **11** (1973) 10.
235. R. E. Cutler et al., *Clin. Pharmacol. Ther.* **24** (1978) 333.
236. A. Polak, *Postgrad. Med. J.* **55** (1979) 667.

237. F. von Bruchhausen et al. (eds.): *Hagers Handbuch der Pharmazeutischen Praxis,* **vol. 8,** Springer, Berlin – Heidelberg – New York 1993, p. 226.
238. T. K. Daneshmend et al., *Clin. Pharmacokin.* **8** (1983) 17.
239. R. B. Diasio et al., *Antimicrob. Agents Chemother.* **14** (1978) 903.
240. B. E. Harris et al., *Antimicrob. Agents Chemother.* **29** (1986) 44.
241. K. F. Wagner in G. L. Mandell et al. (eds.): *Principles and Practice of Infectious Diseases,* Churchill Livingstone, New York 1990, p. 1975
242. D. Amstrong in J. F. Ryley (ed.): *Chemotherapy of Fungal Diseases,* Springer, Berlin – Heidelberg – New York 1990, p. 439.
243. A. Kucers et al. in A. Kucers et al. (eds.): *The Use of Antibiotics,* J. B. Lippincott, Philadelphia 1987, p. 1534.
244. A. Brossi et al., *Helv. Chim. Acta.* **43** (1960) 1444.
245. Glaxo, US 3 069 328, US 3 069 329, 1962.
246. P. W. Brian, *Trans. Br. Mycol. Soc.* **43** (1960) 1.
247. W. Dittmar, *Mykosen* **9** (1966) 104.
248. R. Brehme et al., *Pharmazie* **34** (1979) 372.
249. T. Nogichi et al., *Antimicrob. Agents Chemother.* **1962** (1961 – 70) 259.
250. W. Adam et al., *Ärztl. Forsch.* **14** (1960) 144.
251. M. Rowland, *J. Pharm. Sci.* **57** (1968) 984.
252. M. Kraml, *Antibiot. Chemother.* **12** (1962) 239.
253. V. Shah, *J. Pharm. Sci.* **61** (1972) 634.
254. C. Lin et al., *Drug Metabol. Rev.* **4** (1975) 75.
255. B. R. Lecky, *Lancet* **336** (1990) 230.
256. G. Mion et al., *Lancet* **334** (1989) 1331.
257. G. Lohhaus et al., *Chem. Ber.* **100** (1967) 658.
258. Hoechst, DE-AS 2 214 608, 1973 (G. Lohhaus et al.).
259. Hoechst, ZA 6 906 039, 1970.
260. Hoechst, DE 2 214 608, 1973.
261. Hoechst, US 3 883 545, 1975.
262. Hoechst, DE 2 795 831, 1978.
263. W. Dittmar, *Proc. III. Intern. Congr. Trop. Der.,* Sao Paulo 1974.
264. K. Iwata et al., *Arzneim. Forsch. / Drug Res.* **31** (1981) 1323.
265. K. Sakurai et al., *Chemotherapy* **24** (1978) 68.
266. K. Iwata et al., *Arzneim. Forsch. / Drug Res.* **31** (1981) 1323.
267. W. Dittmar et al., *Arzneim. Forsch. / Drug Res.* **31** (1981) 1317.
268. H. M. Kellner et al., *Arzneim. Forsch. / Drug Res.* **31** (1981) 1337.
269. S. G. Jue et al., *Drugs* **29** (1985) 330.
270. H. G. Alpermann et al., *Arzneim. Forsch. / Drug Res.* **31** (1981) 1328.
271. H. G. Peil, *Arzneim. Forsch. / Drug Res.* **31** (1981) 1366.
272. N. S. Ryder et al., *Antimicrob. Agents Chemother.* **29** (1986) 858.
273. I. De Carneri et al., *Arzneim. Forsch. / Drug Res.* **26** (1976) 769.
274. Carlo Erba, DE 2 313 845, 1973.
275. Carlo Erba, US 3 855 263, 1974.
276. Nippon Soda K.K. Tokio, DE-AS 1 468 388, 1962 (M. K. Taoka et al.).
277. R. Brehme et al., *Pharmazie* **34** (1979) 372.
278. T. Nogichi et al., *Antimicrob. Agents Chemother.* (1962) 259.
279. Japan Soda, FR 1 337 797, 1963.
280. Japan Soda, BE 627 322, 1962.
281. Japan Soda, US 3 334 126, 1967.
282. S. C. Harvey in L. S. Goodman et al. (eds.): *The Pharmacological Basis of Therapeutics,* Macmillan Publ., New York 1985 p. 959.
283. A. H. Gould, *South. Med. J.* **59** (1966) 176.
284. G. Petranyi et al., *Science* **224** (1984) 1239.
285. A. Schütz et al., *J. Med. Chem.* **27** (1984) 1539.
286. F. von Bruchhausen et al. (eds.): *Hagers Handbuch der Pharmazeutischen Praxis,* **vol. 8,** Springer, Berlin – Heidelberg – New York 1993, p. 1068.
287. H. Loibner et al., *Tetrahedron Lett.* **25** (1984) 2535.
288. Sandoz, BE 853 976, 1977.
289. Sandoz, US 4 282 251, 1981.
290. Lab. Frumtost-Prem S.A., ES 504 432, 1982.
291. A. Georgopoulos et al., *Antimicrob. Agents Chemother.* **19** (1981) 386.
292. G. Petranyi et al., *Antimicrob. Agents Chemother.* **19** (1981) 390.
293. J. P. Monk et al., *Drugs* **42** (1991) 659.
294. S. Nolting et al., *Mykosen* **28** (1985) 69.
295. L. E. Millikan et al., *J. Am. Acad. Dermatol.* **18** (1988) 52.
296. I. Effendy et al., *Therapiewoche* **36** (1986) 848.
297. F. von Bruchhausen et al. (eds.): *Hagers Handbuch der Pharmazeutischen Praxis,* **vol. 9,** Springer, Berlin – Heidelberg – New York 1993, p. 802.
298. Sandoz, EP 24 587, 1981.
299. Y. M. Clayton et al., *Br. J. Dermatol.* **130** (1994) Suppl. 43, 7.
300. J. C. Jensen, *Clin. Exp. Dermatol.* **14** (1989) 110.
301. J. A. Balfour et al., *Drugs* **43** (1992) 259.

302. P. J. Dykes et al., *J. Dermatol. Treat.* **1** (1990) Suppl. 2, 19.
303. T. C. Jones, *Br. J. Dermatol.* **132** (1995) 683.
304. L. Juhlin, *Lancet* **339** (1992) 1483.
305. J. P. Ottervanger et al., *Lancet* **340** (1992) 728.
306. G. Lowe et al., *B.M.J.* **306** (1993) 248.
307. Hoffmann-La Roche, DE 2 752 096, 1978.
308. Hoffmann-La Roche, US 4 202 894, 1980.
309. A. Polak in J. F. Ryley (ed.): *Chemotherapy of Fungal Diseases,* Springer, Berlin – Heidelberg – New York 1990, p. 153.
310. A. Polak in J. F. Ryley (ed.): *Chemotherapy of Fungal Diseases,* Springer, Berlin – Heidelberg – New York 1990, p. 505.
311. E. Rhode et al. in R. A. Fromtling (ed.): *Recent Trends in the Discovery, Development and Evaluation of Antifungal Agents,* J. R. Prous, Barcelona 1987, p. 575.
312. M. Haria et al., *Drugs* **49** (1995) 103.

Neuropharmacology

Victor P. Whittaker Arbeitsgruppe Neurochemie, Max-Planck-Institut für biophysikalische Chemie, Göttingen, Federal Republic of Germany Johannes Gutenberg Universität , Mainz, Federal Republic of Germany

1. Introduction

Most drugs or toxins act on the nervous system in a highly specific way [1–7]. In many cases the particular subsystem or component (e.g., enzyme, ion channel, or transmitter receptor) with which the drug interacts is known precisely. In other cases, the target may be unidentified, but research into the mode of action of the drug, toxin, or toxic substance is best planned on the assumption that the action is specific, and a clue will often be provided by the symptoms that the agent produces in animals or humans. The nonmetabolized gaseous anesthetics are one of the few classes of pharmacological agents to act in a more generalized way on nerve cells but here too their mode of action is to selectively induce biophysical changes in nerve cell membranes which block their excitability, rather than to exert a general effect on the whole cell.

The purpose of this article is to summarize what is known about the organization and function of the nervous system, its chief subsystems and components down to the molecular level in order to provide a background for the following articles on specific topics in neuropharmacology:

Analeptics
Analgesics and Antipyretics
Anesthetics, General
Anti-Obecity Drugs
Antiarrhythmic Drugs
Antiasthmatic Agents
Antiemetics
Antiepileptics
Antihypertensives
Antihypotensives
Calcium Antagonists
Cardiac Glycosides and Synthetic Cardiotonic Drugs
Coronary Therapeutics
Cough Remedies
Hypnotics
Local Anesthetics
Parasympatholytics and Parasympathomimetics
Parkinsonism Treatment
Psychopharmacological Agents
β-Receptor Blocking Agents
Sedatives
Skeletal Muscle Relaxants

Ullmann's Industrial Toxicology

ISBN: 3-527-31247-1

Spasmolytics
Sympatholytics
Sympathomimetics

2. Types of Neural Cells

The cells making up nervous tissue can be divided broadly into two types: *nerve cells* (*neurons*) whose task is the transmission and processing of information, and *glia*, to which an insulating and supporting role is assigned (Fig. 1). Within these two categories are many subtypes whose classification is based both on structure and function.

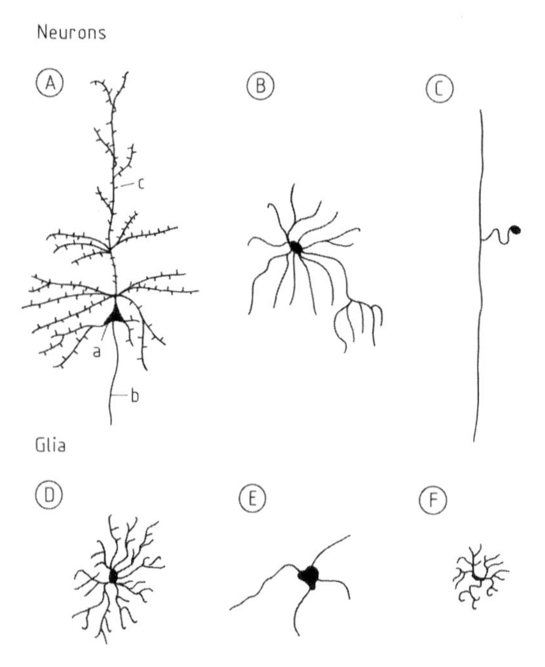

Figure 1. Types of neurons (A – C) and glia (D – F)
Neurons (Golgi staining): A) Pyramidal neuron with cell body (a), axon (b), and dendrites (c); B) Basket cell; C) Bipolar cell
Glia: D) Astrocyte; E) Oligodendrocyte; F) Microglia

2.1. Neurons

A primary distinction can be made between sensory, motor, and interneurons. In the peripheral nervous system (Section 3.1), *sensory neurons* are linked to a variety of sensors (sensory receptors) by which different sensations (touch, pain, heat, cold, pressure, tension, sound, light, smell, taste) and certain internal parameters (such as blood oxygen and carbon dioxide tension) are converted into trains of nerve impulses that convey information about the internal and external environment to the central nervous system (CNS), often through ascending sensory or afferent pathways involving several neurons. By contrast, *motor neurons*, usually under the control of descending or efferent pathways, convey signals to and stimulate effector cells (muscle or gland cells) causing muscles to contract or gland cells to secrete physiologically active substances (hormones, enzymes).

The constant flow of information conveyed by afferent impulses needs to be processed before it can result in effective action. This is achieved by *interneurons* that are often arranged in complex repetitive patterns which resemble the logic units of computer circuits. However, although the nervous system may be thought of as a biological computer regulating and optimizing the responses of the organism to changes in its environment, any close comparison with man-made computers reveals a fundamental difference: the nervous system is not digitalized. It performs numerical computations slowly and with difficulty but is excellent at pattern recognition, association, learning and memory, though again, not of the computational type.

The morphology of nerve cells is elegantly displayed at the light-microscopic level by silver staining, a technique originally devised by C. Golgi (1843 – 1926). In this method only a few cells are stained in a given section of tissue, but the stain spreads until the whole neuron is displayed with all its ramifications (Fig. 1). Cells can then be classified according to their morphology: *pyramidal cells* (Fig. 1 A) are a type of efferent cell with prominent triangular cell bodies; *basket* (Fig. 1 B) and *stellate* (star-shaped) cells are interneurons with bushy processes resembling the wickerwork of a basket or the rays of a star, respectively.

All neurons possess a cell body, which contains a prominent *nucleus* surrounded by *cytoplasm* (perikaryon); in addition almost all neurons send out several or many processes known as *dendrites* (from their resemblance to the branches of trees) and a single process known as an axon. The dendrites and cell body collectively form a receptive field which receives contacts from the axons of neighboring cells. The axon conveys impulses arising from the convergence of this receptive field to the target cell–another nerve cell or effector (muscle or gland) cell. In sensory cells a dendrite and axon may fuse to form a bipolar cell; impulses generated by receptors attached to the dendrite are picked up by the axon, bypassing the peripherally placed cell body (Fig. 1 C). Towards its terminal the axon may branch several times before terminating on another cell; it may also form swellings or *varicosities* along its length which are in close contact with the target cells. Within the neuron, information is conducted by nerve impulses–self-propagating waves of ion flux that can be detected by minute accompanying fluctuations in membrane potential (Section 4.2). The transmission of information between neurons is brought about by the release of chemical transmitter substances (*neurotransmitters*) from stores in the axon terminal or varicosity (Chap. 5). Neurotransmitters are released by the arrival of a nerve impulse and are of two types–low molecular mass compounds such as acetylcholine and noradrenaline (Section 5.2.1) or neuropeptides (Section 5.2.2); they interact with *receptors* on the cell membrane of the target cell and induce nerve impulses in it. Many drugs owe their effect to their ability to selectively bind to these receptors (Section 5.3.3). The complex of nerve axon termination together with the region of the cell membrane of the target cell immediately apposed to it is known as a *synapse*, a word coined by C.S. SHERRINGTON (1857 – 1952) from two Greek words meaning a "clasping together".

Because so many drugs and toxins act by interfering with synaptic transmission, the synapse has sometimes been called "the Achilles heel" of the nervous system.

2.2. Glia

Glial cells can also be classified on the basis of morphology and function. Three main types are recognized in the CNS: astrocytes, oligodendrocytes, and microcytes. In the PNS, Schwann cells are the peripheral counterparts of oligodendrocytes.

Glial cells were originally regarded as the neural counterpart of connective tissue and as the glue (Greek glía) holding the tissue together. Embryological studies showed that astrocytes and oligodendrocytes, like neurons, are of ectodermal, not mesodermal origin; microglia, however, are derived from mesoderm, like connective tissue.

Astrocytes are star-shaped cells with many radiating processes (Fig. 1 D). The processes may form end-feet applied to brain capillaries and so contribute to the structural and functional barrier between the brain and the blood known as the *blood – brain barrier* or they may surround nerve cells in a manner suggesting a supportive or service function. Much evidence indicates that astrocytes regulate the environment of nerve cells by controlling the extracellular concentrations of hydrogen and potassium ions and chemical transmitters, and by releasing trophic factors essential for the survival of neurons.

Oligodendrocytes. The primary function of these cells (Fig. 1 E) in the CNS and of their counterparts, the *Schwann cells*, in the PNS is to provide an insulating wrapping around axons. This wrapping known as *myelin* is applied in segments (Fig. 2); the short, poorly insulated spaces between segments are known as nodes of Ranvier after their discoverer, L.A. RANVIER (1835 – 1922). The segmental character of this insulation confers an important advantage on the neuron; propagated nerve impulses are forced to jump from one node to the next, a phenomenon known as *saltatory conduction*, and this greatly speeds up nerve conduction.

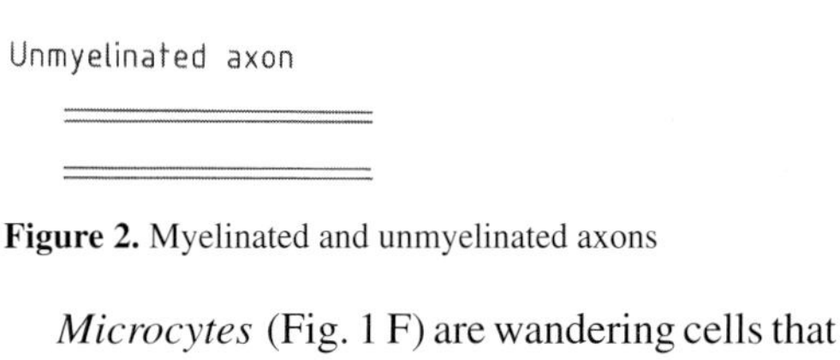

Myelinated axon

Node of Ranvier

Myelin sheath

Unmyelinated axon

Figure 2. Myelinated and unmyelinated axons

Microcytes (Fig. 1 F) are wandering cells that invade injured or inflamed regions of the nervous system from the blood stream and remove dead cells and tissue debris by phagocytosis. They are not present in healthy nervous tissue in significant numbers.

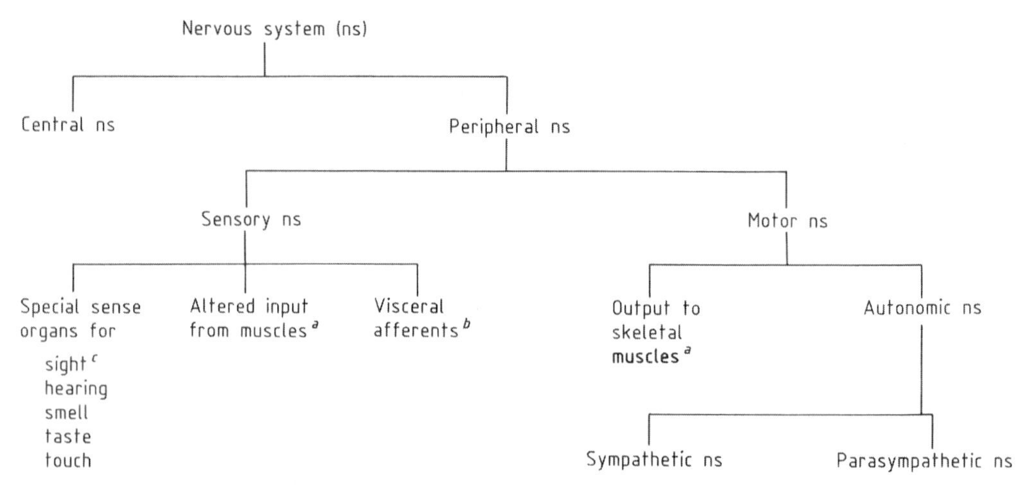

Figure 3. Subdivisions of the vertrebrate nervous system
[a] These portions of the peripheral nervous system are collectively referred to as the somatic nervous system. [b] Often classified as the afferent portion of the autonomic nervous system. [c] The retina is often regarded as a peripheral extension of the CNS.

3. Evolution and Organization of the Vertebrate Nervous System

3.1. Evolution and Main Subdivisions

The simplest vertebrate nervous system is seen in fish, amphibia, and reptiles. Birds and mammals represent separate developments from primitive reptilian forms, mainly characterized by a relative expansion of the parts of the brain concerned with sensory processing and motor agility. Nevertheless, the basic plan of all vertebrate nervous systems is similar (Fig. 3) being divided into the *central nervous system* (CNS), comprising the brain and spinal cord, and the *peripheral nervous system* (PNS), comprising the *somatic nerves* (those innervating the voluntary skeletal musculature), the *autonomic nerves* (those innervating glands and the smooth involuntary musculature of the viscera and blood vessels), and the nerves from the organs of special senses. An alternative, more functional classification of the PNS distinguishes between sensory (afferent) and motor (efferent) nerves. The *sensory nerves* mediate information from the special senses, the somatic (skeletal) musculature, and the viscera. The *motor nerves* induce the contraction of skeletal muscle and control the hollow organs of the body and the glands, the latter being termed *autonomic nerves*. These nerves are further subdivided into *sympathetic* and *parasympathetic*. The *sympathetic nervous system* prepares the body for emergencies: the blood supply to the muscles, heart, and brain is increased by dilatation of their blood vessels, while that to the skin and viscera is reduced by vasoconstriction. The heart beats faster, the urinary and rectal outlets are blocked by contraction of their sphincters, intestinal movement and alimentary secretion are stopped, and blood sugar is raised. By contrast the *parasympathetic system* restores bodily function when the crisis is over. Wastes are eliminated by the relaxation of the appropriate sphincters, blood pressure and heart-rate fall, and alimentary function is resumed. Many drugs block (antagonists) or stimulate (agonists) the action of the transmitters involved.

The anatomy of the sensory and somatic motor systems is relatively simple (Fig. 4 A). Sensory nerves derived from bipolar neurons (Section 2.1) enter the spinal cord through the dorsal (posterior) roots; their cell bodies are located in the dorsal root ganglia. The motor nerves leave the spinal cord through the ventral (anterior) roots; their cell bodies are located in the ventral columns of the spinal cord, protuberances of neuron-rich grey matter jutting out from the central mass of grey matter in the spinal cord.

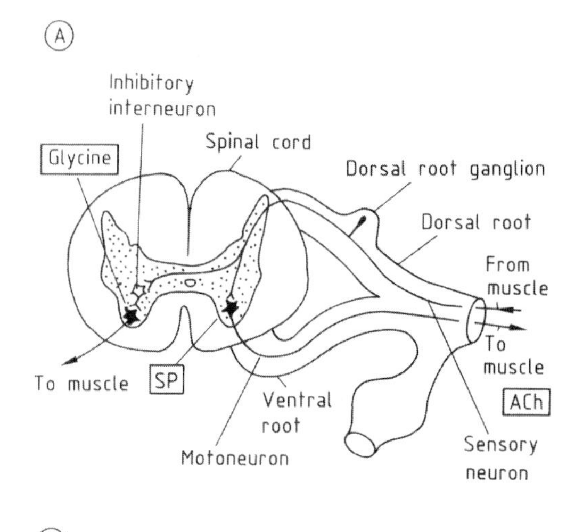

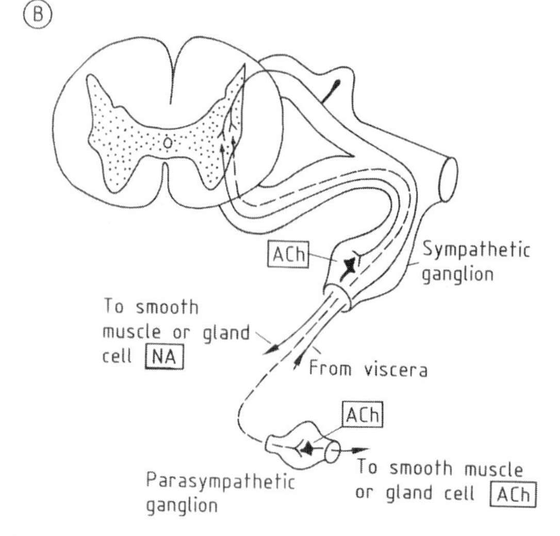

Figure 4. The relationship of the peripheral nervous system to the spinal cord
A) Somatic system: sensory nerves from muscle receptors form a two-neuron arc with a motor neuron (motoneuron) on the same side and inhibit a motor neuron on the opposite side via an inhibitory interneuron, thus forming a three-neuron arc.
B) Autonomic system: continuous lines indicate visceral afferents, preganglionic sympathetic, and postganglionic; dashed lines indicate preganglionic parasympathetic.
The transmitters utilized (Section 5.2) are shown in boxes; ACh = acetylcholine, NA = noradrenaline, SP = substance P. In reality both systems run together in each root and bilaterally; they are shown separately and unilaterally for clarity.

The autonomic outflow is more complex (Fig. 4 B). Motor nerves issuing from the ventral roots enter ganglia arranged in chains on either side of the backbone. Some (the *preganglionic sympathetic* nerves) make contact with nerve cell bodies in these ganglia. The axons of these ganglion cells (the *postganglionic sympathetic* nerves) provide the sympathetic innervation to the heart, viscera, and blood vessels. Other preganglionic nerves (the *preganglionic parasympathetic* nerves) travel through the ganglia, usually accompanying the postganglionic sympathetic innervation to the heart, viscera, and blood vessels, and eventually making contact with ganglion cells nearer or in the target organs themselves. These cells provide the *postganglionic parasympathetic* innervation. How these connections were worked out is explained in Section 5.1. Brain weight is not a good guide to the degree of evolution of the nervous system since large animals have large brains but may show a behavior no more complex than that of their smaller congeners. The larger brains of large members of a particular family or order contain more glial cells, not necessarily more neurons. To accommodate the increased surface area of larger brains, the outer layers (the *cortex*) are thrown into folds; this is particularly obvious in the human brain, where, compared to other animals, there is in addition a net increase (though not as large as is often supposed) in cortical mass relative to whole brain mass.

Another criterion of development is the relative volume of the hind-brain (cerebellum), a part of the brain concerned with the performance of skilled acts. Originally evolved to control swimming movements in fish, it shows marked development in birds and mammals, especially primates.

3.2. Patterns of Neuronal Integration

3.2.1. The Reflex Arc

The unit of integrated action in the nervous system is the *reflex*. The study of reflexes was begun by SHERRINGTON [8] in decapitated (spinal) cats, maintained by artificial respiration and warmth. Such preparations show integrated responses to stimuli of a quasi-mechanical kind– the foot is sharply withdrawn in response to a prick on the foot pad while the contralateral limb extends to support the body (nociceptive reflex), a limb offers resistance to being pulled (stretch reflex) or jerks when a tendon at the knee or ankle is suddenly stretched by tapping (knee or ankle jerk); stepping movements can be elicited. Muscular movements involve the relaxation of antagonists (muscles opposing a limb movement) when agonists (muscle causing a movement) contract and, very often, compensating movements on the opposite side.

SHERRINGTON grasped that the simplest integrative unit consisted of a *two-neuron reflex arc*, comprising an afferent sensory neuron synapsing with a motor neuron (Fig. 4 A). Thus, in the stretch reflex, stretching a muscle generates impulses from muscle stretch receptors which are conducted to synapses on motor neurons in the spinal cord and cause a reflex contraction of the muscle from which the impulses originate.

To explain the relaxation of antagonists, SHERRINGTON introduced the concept of *central inhibition*, whereby some synapses activated by incoming impulses inhibit firing of a motor neuron. Normally all motor nerves to muscles generate impulses at a low rate, thus generating the tension (muscle tone) needed to maintain posture; inhibition of a motor neuron to an antagonist muscle thus causes it to relax. Reciprocal excitation and inhibition were invoked to explain the compensatory movements of limbs on the opposite side of the body in the nociceptive and stepping reflexes: sensory impulses would activate agonists and inhibit antagonists on one side of the cord, but activate antagonists and inhibit agonists on the other side.

Careful analysis of the speed of reflex activation and inhibition of motor neurons showed that inhibitory reflex action involves a three-neuron arc with an inhibitory interneuron inserted between the sensory input and motor output (Fig. 4 A). This interneuron is activated by the afferent sensory signal and inhibits the motor neuron. The rate of firing of stretch receptors in muscle can be adjusted by the amount of contraction occurring in the fine muscle fibers (muscle spindles) to which they are attached. These can be adjusted independently of surrounding muscle fibers by the so-called γ-motor neurons (Fig. 5).

Such systems are now familiar to engineers as servomechanisms embodying negative feedback loops, but were first discovered by physiologists.

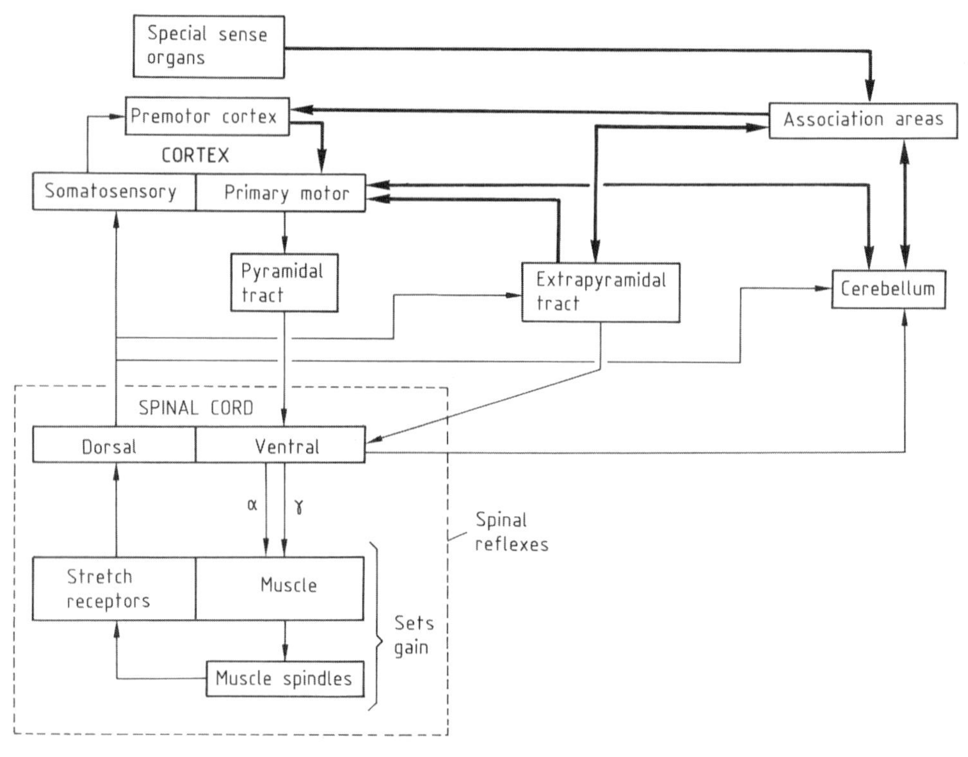

Figure 5. Simplified scheme showing main systems involved in posture and movement

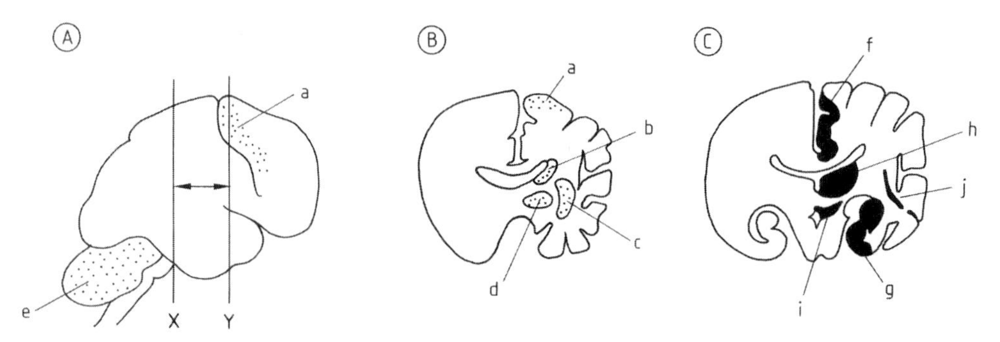

Figure 6. Sketches of human brain anatomy to illustrate the location of pyramidal, extrapyramidal, and limbic systems described in text
A) Side view of the human brain; B, C) Transverse sections along the lines in A marked X and Y, respectively
A, B) Shaded areas show structures involved in voluntary movement: a) Motor cortex containing pyramidal cells; (b – d) Basal ganglia (extrapyramidal tract); b) Caudate nucleus; c) Putamen; d) Globus pallidus; e) Cerebellum
C) Black regions are structures involved in the limbic system: f) Olfactory cortex; g) Hippocampus; h) Thalamus; i) Hypothalamus; j) Part of basal ganglia (pallidum)
For clarity, structures are shown on one side only but are actually on both sides. Arrows in A show view-points of sections.

3.2.2. Voluntary Movement

In the conscious animal muscular movement is under voluntary control, and occurs smoothly and effortlessly in spite of the vast number of adjustments that have to be made. Figures 5 and 6 summarize the main pathways involved. Playing onto the repertoire of spinal reflexes in this systems analysis are three pathways:

1) the direct pathway of the *pyramidal tract* (Fig. 6 B, a) (so called because it originates in large neurons in the part of the cortex controlling movement whose cell bodies are shaped like a pyramid; Fig. 1 A)
2) the *extrapyramidal tract* operating through a set of well defined nuclei (conglomerations of neurons) below the cortex (Fig. 6 B, b – d), and
3) the hindbrain or *cerebellum* (Fig. 6 A, e), which performs the "staff work" for a skilled movement.

Broadly, the extrapyramidal tract is involved in the control of posture and the cerebellum in the programming of skilled action. Volition seems to act through the premotor cortex. The unit of integrated activity is the spinal reflex (Fig. 4 A). Thick lines in Figure 5 indicate the circuits involved in the preparation for voluntary movement, thin lines the evolving muscular response. Diseases (Parkinsonism, Huntington's chorea) which affect the extrapyramidal system cause postural defects, difficulty in initiating movements, tremor at rest (Parkinsonism), or uncontrolled movements at rest (chorea). Damage to the cerebellum causes intention tremor and clumsy execution of movement.

Variations in the repertoire of skilled acts between animals (e.g., a cat and a chimpanzee) are largely caused (1) by the differing extents to which individual muscles or muscle groups are represented in the cortex, and (2) the degree of control over individual spinal motor neurons exerted by the cortical tracts. Strychnine is a specific blocker for inhibitory synapses (which are described in detail in Section 5.2). The painful cramps caused by its injection are due to the failure of central inhibition as antagonists pull against agonists.

The extrapyramidal tract seems particularly sensitive to degenerative diseases and drugs, some of which mimick the symptoms of such diseases while others alleviate them. Such phenomena require an understanding of the synaptic transmitters involved and will be discussed in Section 5.2.1. The transmitter dopamine plays a key role in this system.

3.2.3. The Limbic System

The limbic system is the term often used to denote those (often phylogenetically old) parts of the brain concerned with feeling and emotion

(Fig. 6 C). It is much less clearly defined and well understood than the motor system described in Sections 3.2.1 and 3.2.2. Two well-defined cortical areas, the olfactory cortex (a phylogenetically old part of the cortex concerned with smell) and the hippocampus (also concerned with memory), are involved, as well as various subcortical nuclei located in the thalamus, hypothalamus (also concerned with nervous control of many bodily functions), and the basal ganglia (also concerned with movement). Probably all the brain structures concerned with visceral control, together with their links to the parts of the brain concerned with cognition, learning, memory, and voluntary movement can be regarded as part of the limbic system. Integrated reactions indicative of rage, fear, aggression, or pleasure can be elicited in animals by local stimulation of certain subcortical limbic structures, particularly the hypothalamus, thus providing experimental justification for the concept of the limbic system. Derangements in the limbic system are believed to be involved in psychoses such as schizophrenia and in various forms of depression (→ Psychopharmacological Agents).

3.2.4. The Ascending Reticular Activation System

This system is phylogenetically old and is located in the brain stem with projections to almost all the subcortical nuclei and the brain cortex. It is believed to "activate" higher centers involved in attention, awareness, and the conscious state but is also thought to be active in the rapid eye movement phase of sleep (REM sleep). Many of the neurons forming this system use acetylcholine as their transmitter and they activate largely by suppressing inhibition (Section 5.2.1). Other important central cholinergic pathways are those arising in the basal nucleus of Meynert (nucleus basalis magnocellularis) projecting to the cortex, and the cholinergic input to the hippocampus which arises in the septum [4]. In several degenerative diseases of the nervous system (e.g., Alzheimer's disease, Parkinsonian dementia) the cholinergic neurons of the nucleus basalis are among the first to degenerate.

4. Techniques for Investigating the Nervous System

The nervous system may be investigated morphologically, physiologically, biochemically, and pharmacologically. Each approach has its advantages and limitations and only a combined approach can give the detailed knowledge needed for complete understanding.

4.1. Neuroanatomy

Neuroanatomy provides a basically static view of the nervous system; only with difficulty can series of static images be assembled into sequences from which the underlying dynamic events can be inferred.

Dissection. Gross dissection of the nervous systems of vertebrates reveals many similarities and homologies of structure with an increasing complexity of organization as the evolutionary tree is ascended, culminating in humans. The macroscopic components of the brain with their areas of grey (neuron-rich) and white (myelin-rich) matter are given fanciful names based on their appearance (e.g., hippocampus, caudate nucleus) or more logical ones based on location (e.g., temporal lobe) or known function (e.g., motor cortex).

Light Microscopy. Histology at the light microscopic level reveals individual neuronal and glial cells and gives some clues as to connections. Histochemical methods, based on staining for specific enzymes or antigens, reveal differences within the broad classes of neurons or glia which may have important functional significance. Here, a powerful new technique, *in situ* hybridization with a DNA preparation (cDNA) complementary to a messenger RNA (mRNA) coding for a specific functional protein, is providing valuable information unobtainable in other ways. Microinjection of a strongly fluorescent dye (e.g., procyon yellow) into a neuron often vividly reveals connexions, thus supplementing older methods based on sectioning fiber tracts and tracing their subsequent degeneration.

Electron Microscopy. In the CNS, presynaptic nerve terminals often appear under the electron microscope as club-like expansions (ca.

0.5 μm in diameter) of the fine (0.2 μm diameter) terminal axonal branches (Fig. 7 A). There is a distinct electron-lucent gap between the pre- and postsynaptic membranes, the purpose of which may be to shunt the potentials associated with the arrival of a nerve impulse and to prevent them invading the postsynaptic membrane, thus altering its excitability. This allows synaptic transmission to be purely chemical in character. In some rare synapses where there is evidence of electrotonic transmission, this "synaptic gap" does not appear and pre- and postsynaptic membranes are closely apposed.

The presynaptic terminal cytoplasm is packed with vesicular organelles of equal size (ca. 0.05 μm in diameter). These are referred to as *synaptic vesicles*. Their function is to store transmitter and to release it when the nerve terminal is invaded by a nerve impulse, by temporarily fusing with the terminal plasma membrane and opening out to discharge their contents, a process known as *exocytosis*. This release process is Ca^{2+}-dependent. Synaptic vesicles that have released their contents are recovered by the reverse process, known as *endocytosis*, pulled back into the terminal and recycled. The process is similar to and is possibly an evolutionary adaptation of the general process of cell secretion, with the important difference that whereas in normal secretion vesicles are used only once before being returned to the Golgi region of the cell body for reprocessing, synaptic vesicles are recovered, refilled, and reused.

The synthesis, storage, and release of transmitter demands considerable amounts of metabolic energy so it is not surprising that presynaptic nerve terminals contain local concentrations of the organelles responsible for cellu-

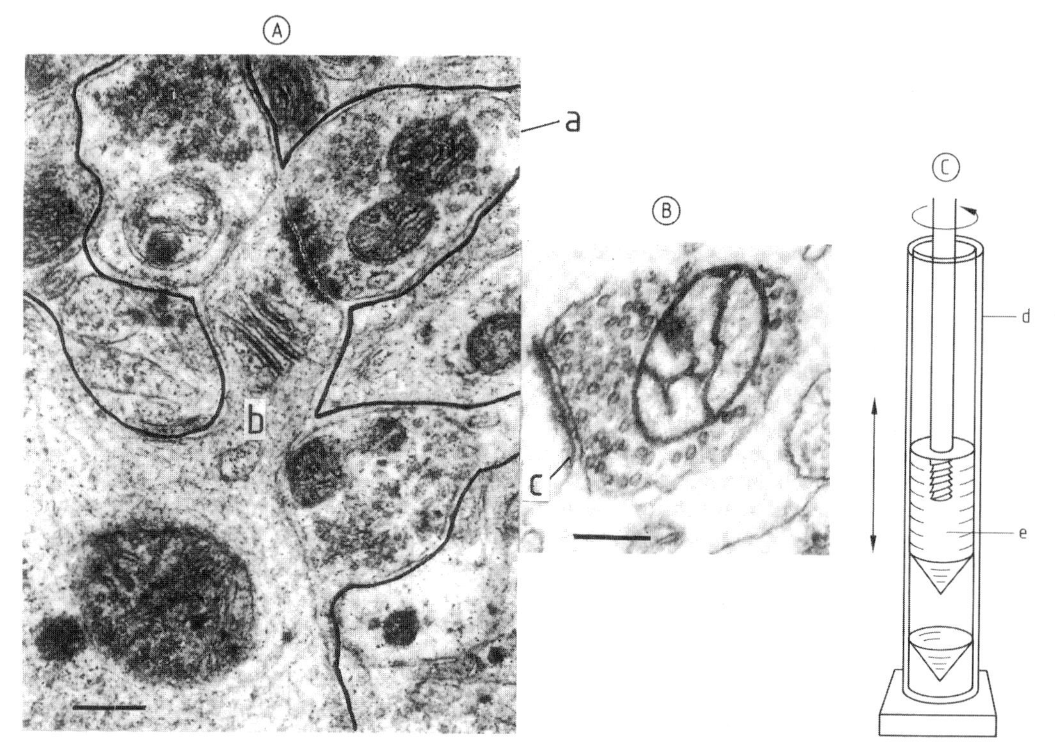

Figure 7. An axodendritic cortical synapse in situ (A) and a synaptosome derived from it (B) with the aid of an homogenizer (C)

A) Cortical synapse, three synapses are outlined: a) Presynaptic terminal bulb filled with synaptic vesicles and containing mitochondria and presynaptic dense projections; b) Dendritic spine with dense postsynaptic membrane and spine apparatus; Bar in lower left-hand corner represents 0.2 μm

B) Synaptosome: c) Postsynaptic membrane still adhering to the synaptosome; Bar in lower left-hand corner represents 0.2 μm

C) Homogenizer: d) Precision-bore glass tube (mortar); e) Rotating (900 rpm) plastic pestle, tissue suspended in iso-osmotic sucrose is forced between the pestle and mortar.

lar respiration and energy production, the *mitochondria*.

The postsynaptic region of the synapse sometimes shows specializations and provides a basis for classification. Thus, in the CNS a broad distinction can be made between *axoaxonic*, *axosomatic*, and *axodendritic* synapses according as to whether the presynaptic axonal termination is made with another nerve terminal, or the cell body or dendrites of the target cell, respectively. Axodendritic synapses are often made, not directly onto the dendrite, but onto a small projection from it called a spine (Fig. 7 A). The numerous spines on dendrites of large neurons of the CNS give them a "furry" appearance in silver staining (Fig. 1 A). There is some evidence that spines increase in size with continued use of the synapse concerned. This is one of the few morphological indications of the elusive synaptic remodeling that must accompany learning and memory.

4.2. Neurophysiology

Physiological techniques have the advantage that they can be applied to the intact animal or to surviving tissue so that the results are often easier to interpret than those obtained by other methods. Against this, physiological methods often leave us with a model of a vital process in which basic mechanisms are concealed within "black boxes".

The simplest preparations for the physiological investigation of the nervous system are single neurons or single synapses. Conduction down axons or transmission across synapses is detected by recording the transient electrical signals that accompany nerve impulses with fine microelectrodes coupled to amplifiers and oscilloscopes (Fig. 8 A). Use is often made of invertebrate preparations (e.g., the giant axon and giant synapse of the squid) that can be penetrated by the microelectrode. Such intracellular recordings, in which the potential across the cell membrane is measured, reveal that in neurons, as in other cells, a resting potential exists across the membrane whose magnitude can be inferred from the difference in K^+ concentrations (strictly, activities) between the inside and outside of the cell. Living cells generally maintain K^+ concentrations in their cytoplasm higher, and Na^+ concentrations lower than those of the extracellular medium. If the membrane is permeable solely to potassium ions, a potential E_m arises across the membrane of a magnitude close to that predicted by the Nernst equation

$$E_m = \frac{RT}{F} \ln \frac{[K_0^+]}{[K_i^+]} = 58 \log_{10} \frac{[K_0^+]}{[K_i^+]} \quad \text{at } 20\,^\circ\text{C}$$

where R is the gas constant, T the absolute temperature, F the Faraday constant, and $[K_0^+]$ and $[K_i^+]$ the outside and inside K^+ concentrations respectively; under normal resting conditions E_m is close to -75 mV. When a nerve impulse is generated, usually by stimulating the nerve cell electrically, an intracellular electrode registers a transient reversal of the membrane potential due to a short-lasting rise in permeability to Na^+. The Na^+ potential temporarily superimposes itself on the resting potential. Recovery ensues as the Na^+ channel closes and is assisted by a transient increase in K^+ permeability.

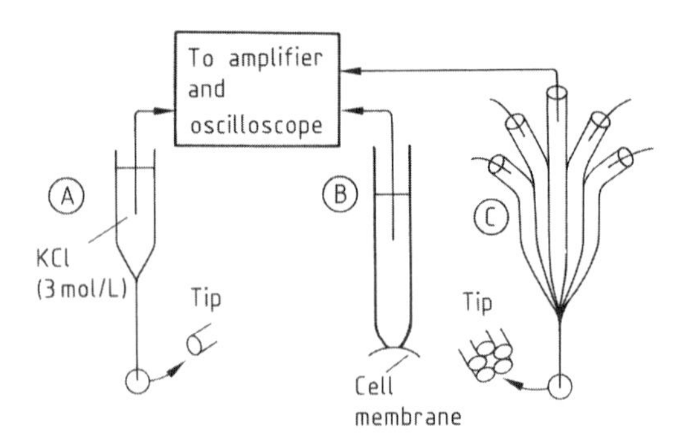

Figure 8. Types of glass electrodes
A) Simple microelectrode for intracellular recording; B) Patch-clamp electrode; C) Multibarrelled electrode (not to scale)

Permeability to ions is brought about by selective, voltage-sensitive ion channels embedded in the membrane. These are complex membrane-spanning proteins whose structure is now known as a result of their isolation and sequencing by the complementary DNA (cDNA) technique. However the fine conformational changes that accompany the opening and closing of the channels are not yet fully understood. Local anesthetics (e.g., cocaine) act by blocking influx of Na^+.

The opening and closing of single ion channels in the membrane can be detected by means of "patch-clamp" electrodes (Fig. 8 B). These are polished glass electrodes with a small hole in the tip onto which a patch of cell membrane is drawn by suction. A good seal is essential for the detection of the very small electrical signals caused by the spontaneous opening and closing of single ion channels and requires a "clean" membrane. The method works well with certain isolated cell preparations and a large range of ion channels, some of them "ligand-gated" (i.e., opened by specific substances such as transmitters interacting chemically with the channel protein), have been detected and studied.

At the synapse, the signal is transformed from an electrical to a chemical one by the inflow of Ca^{2+} ions through voltage-sensitive calcium channels that are activated by the arrival of a nerve impulse. This causes the exocytosis of numerous vesicles and the release of transmitter, which rapidly diffuses across the 25 to 100-nm synaptic gap and triggers a response in the postsynaptic target cell. Responses are of several kinds but are easiest to detect when they involve the opening of transmitter-controlled ion channels in the postsynaptic membrane and the generation of an excitatory postsynaptic potential (EPSP). This in turn, if large enough, will generate a propagated impulse in the postsynaptic cell. In muscle cells, this propagated impulse is accompanied by muscle contraction.

Analysis of the magnitude of EPSPs reveals them to be made up of subunits or quanta, whose release is synchronized by the arrival of the nerve impulse [9]. The morphological counterparts of these quanta are the synaptic vesicles, the exocytosis of a single vesicle generating a single quantum.

An example of a more complex electrophysiological investigation is the use of the multibarrelled electrode (Fig. 8 C) which may be inserted into the CNS with its tip in proximity to a nerve cell. One "barrel" of the electrode records the response of the cell to chemical substances or drugs released from other barrels by iontophoresis. A complete pharmacological experiment can be performed in this way on a single central neuron. This technique has been important in revealing the responses of central neurons to drugs and putative transmitters.

Intracellular recordings from neurons in specialized areas of the CNS, often using indwelling electrodes in animals free to move, has led to the discovery that neurons are involved in a hierarchy of organization and has given important insights into the functioning of the CNS. This is a complex subject which can only be touched upon in this article. To give one striking example: certain neurons in the striatal vocal center of birds respond when the bird hears the call of its own species but not when it hears those of other species or sounds in general. Such cells are at the apex of a complex process of neural analysis.

Physiological methods have also revealed the existence of conditioned reflexes. The pioneer in this field was I. P. Pavlov (1849 – 1936) who noted that dogs who regularly heard a bell each time they were fed quickly salivated at the sound of the bell alone, in the absence of the sight or smell of food. This approach with its many variants and model systems, ranging from invertebrates to primates, has shed light on such important higher neural functions as learning and memory.

Perhaps the chief failure of neurophysiology is to provide a convincing theory of how the "logic units"–repetitive groups of neurons whose connections and individual properties are known in some detail– combine to achieve the pattern recognition, control of skilled movements, or other quasi-computer-controlled activities of the nervous system. This may well have to await a better understanding of the properties of networks of interacting components.

4.3. Neurobiochemistry (Neurochemistry)

The biochemical investigation of the nervous system has given rise to the subdiscipline of neurobiochemistry (neurochemistry) which includes the investigation of the nervous system by cell- and molecular-biological, and molecular genetical methods. Such methods may reveal little about the way the nervous system functions as a whole, but they can help in prying open the physiologists' "black boxes" to reveal what is going on inside. They are also of great importance in understanding long-term phenomena such as development and aging, and the interaction between the nervous system and the immune and hormonal control systems.

Such investigations often begin by the application of *tissue fractionation techniques*. These involve dispersion of the tissue and fractionation by centrifugal density gradient separation or particle exclusion chromatography. The technique was originally developed for use with tissues with a relatively homogeneous and easily disrupted cellular structure like liver, but in the early 1960s the author of this article and his colleagues pioneered their application to the much more complex mammalian brain [10]. The application of mild liquid shear in a "homogenizer" (Fig. 7 C) breaks open nerve and glial cells, fragments axons and dendrites, and "pinches off" the club-like presynaptic nerve terminals to form sealed structures called *synaptosomes* (Fig. 7 B). A combination of moving-boundary (differential) centrifugation, in which particulate structures are sedimented in progressively more intense centrifugal fields, and density gradient centrifugation, in which such structures are centrifuged until they reach a region of the gradient whose density matches their own, allows synaptosomes to be separated from other organelles and cell fragments such as nuclei, mitochondria, myelin fragments, and vesiculated membrane fragments (microsomes) derived from external and internal cell membranes.

Synaptosomes are sealed structures and conserve to a remarkable degree the properties of the synapse. Morphologically, they retain the characteristic structure of the presynaptic nerve terminal, with its small mitochondria and synaptic vesicles (Fig. 7 B). Physiologically, they are capable of respiring and generating ATP when incubated in a balanced saline medium containing an energy-yielding substrate such as glucose. Such metabolizing synaptosomes take up glucose, transmitters or their precursors by carrier-mediated mechanisms and provide a convenient in vitro system on which to test the action of drugs on such processes. Anorexiants are an example of a class of drugs which may act by controlling monoamine uptake. Synaptosomes also release transmitter in a Ca^{2+}-dependent manner when depolarized with high concentrations of K^+ or Na^+ plus veratridine, a drug that depolarizes by opening up Na^+ channels. Fluorescent probes that report membrane potentials confirm that metabolizing synaptosomes possess a resting membrane potential of the same magnitude as that of intact neurons and analysis of their internal K^+ and Na^+ concentrations shows that they are able to maintain ionic gradients of the requisite magnitude.

Synaptosomes disrupted by osmotic shock release their contents: density gradient separation of these yields fractions of presynaptic terminal cytosol, synaptic vesicles, presynaptic plasma membranes and intraterminal mitochondria for further biochemical analysis. Transmitters are specifically associated with the synaptic vesicles, in the case of acetylcholine, in the amount required by the vesicle theory [11].

Milder disruptive procedures, involving teasing tissue through stainless steel or nylon sieves of decreasing mesh size, yield intact cell bodies. Astrocytes, oligodendrocytes and neurons can be isolated in this way, usually after a short further fractionation of the initial cell dispersion. If cells are prepared from embryonic or neonatal brain, they are often viable in tissue culture.

One class of neuronal cytoplasmic constituents of great importance is comprised by the mRNAs (Section 4.1) that code for the numerous proteins needed for neuronal structure and function. In general, neurons are rich in mRNA, because their cell bodies have to maintain an enormously expanded peripheral cytoplasm enclosed within the dendrites and axon (Section 5.4). A mRNA coding for a functional protein (e.g., an ion channel or receptor) may be tested for activity by transfecting a frog oocyte (egg cell) with it. The transfected oocyte is not only able to synthesize the appropriate protein, it can effect any necessary posttranslational modification and the required protein is then expresssed on the oocyte's cell membrane. The protein can be sequenced by making cDNA (Section 4.1) to the neuron's mRNA, inserting this into the DNA of the bacterium *Escherichia coli* and detecting bacterial clones bearing the appropriate cDNA sequence either by hybridization with a synthetic oligonucleotide whose sequence is based on a partial sequence of the protein in question or by immunochemical means. In the latter procedure, the cDNA is inserted into the host cell along with an "expression vector" which directs the synthesis of protein according to the code supplied by the cDNA insert.

Although cellular or subcellular preparations are preferred for neurochemical studies, a variety of more integrated preparations are available. *Tissue slices*, thin enough (0.3 – 0.5 mm) to allow for adequate oxygenation and rapid penetration and diffusion of metabolites, are viable for several hours in a suitable oxygenated medium and permit metabolic processes, electrophysiological recordings, and drug responses to be studied. In the form of a *sterile explant*, tissues and cells may survive for days or weeks; neurites may grow out from the neurons present and form synapses with neighboring cells, thus providing a test system for detecting endogenous *neuronotrophic factors* released by glial or target cells that are essential for neuronal survival and differentiation.

An even higher level of organization is made experimentally accessible by *perfusion of the brain*. "Push – pull cannulae" allow perfusion of a restricted brain region by injecting fluid through the annular space between concentric cannulae and withdrawing it through the inner cannula, or perfusion of portions of the hollow spaces (ventricles) within the brain. Such cannulae have proved useful in detecting transmitter release or investigating the metabolism of transmitter precursors associated with specific brain subsystems. In a later variant (microdiffusion) the cannula ends in a semipermeable membrane.

The highest organizational level of all is reached by the introduction of metabolites or markers into the blood just before it enters the brain. Neural activity in response to stimuli causes local increases in blood circulation or metabolism which can be followed by the local accumulation of label. In an earlier form of the technique, radioactive deoxyglucose was used; the brains were fixed, sectioned, and subjected to autoradiography to determine the local accumulation of the label. In *positron-emission tomography* (PET), which can be applied to humans, the label is a short-lived positron – emitting radionuclide (^{15}O, ^{13}N, ^{11}C, or ^{18}F) generated in a nearby cyclotron. An array of radiation detectors picks up the γ-rays generated by the annihilation of the positrons by electrons in the environment, and computer tomography is used to convert the response of the array to an image of the positron distribution in the tissue in real time. Cerebral blood flow may be detected with $[^{15}O]CO_2$ or $[^{15}O]H_2O$ and the distribution of a variety of metabolites and drugs may also be studied. In spite of the huge capital and running costs of the equipment, this method is now widely used in neuropharmacological research and in clinical investigation. The dynamic character of the results obtained with living patients gives the method a unique power.

Another noninvasive technique, less well developed but of great potential, is *magnetic resonance imaging* (MRI). In its simplest form nuclear magnetic resonance (NMR) signals from hydrogen atoms are transformed by computer tomography to provide an image of the distribution of the resonating atom in any required plane. Most of the signals come from the hydrogen of water, but the image of the tissue so obtained is superior in contrast and resolving power to that obtained by X-ray computed tomography (CT); brain cancers, plaques, vascular damage and other brain abnormalities can be detected. In principle, MRI could also be used to detect the compartmentation of metabolites and the kinetics of their reactions, changes in intracellular pH and the distribution of drugs in small tissue volumes using ^{31}P, ^{13}C, or fluorine labeling.

In spite of the advantages of these new noninvasive techniques, the picture of the nervous system furnished by neurochemistry remains stubbornly molecular. It has much to tell us about development, aging and the function and modus operandi of the units which make up the system but does not reveal how it works as a whole.

4.4. Neuropharmacology

The ideas and concepts of pharmacology, especially that of specific tissue receptors for drugs and toxins, have greatly influenced the understanding of the nervous system. Transmitters and other endogenous neuroactive substances may be described as *autopharmacological agents*. Until chemical or immunochemical methods are devised they must be estimated by *bioassays*, in which the response of the assay system – often the contraction of a muscle registered by a recorder coupled to a strain gauge – is matched by a "standard". The standard may be the endogenous substance itself (if known), a stable preparation of the endogenous substance whose activity is expressed in arbitrary units, or another, known substance with pharmacologically similar or identical properties. Further characterization may involve parallel assays in which comparisons are made between unknown and standard in two or more bioassay systems with different ligand specificities. Bioassay–although sensitive, rapid, and accurate in experienced hands and requiring inexpensive equipment–has largely been superseded by other methods such as fluorimetry, gas chromatography, high performance liquid chromatography (HPLC), or radioimmunoassay.

The existence of a drug with a powerful action on the nervous system prompts the search for a receptor. This is greatly helped by the use of radiolabeled drugs and a theoretical analysis of the binding data which leads to the calculation of affinity constants and other parameters. At this stage a crude membrane preparation may be all that is required as a source of receptor. The uncritical application of the technique has, however, led to the pejorative description "grind and bind" research.

If well-founded evidence for a specific receptor is obtained, its presence suggests the possible existence of endogenous ligands. Thus, when morphine-specific receptors were found in the brain, a search was instituted for endogenous morphine-like substances (opioids). Some 14 such substances are now known and include short-chain peptides (the *enkephalins*) and the longer chain *endorphins*, all of which contain the common sequences tyr-gly-gly-phe-leu or tyr-gly-gly-phe-met. Centrally acting analgesics and antitussives are among those pharmaceuticals interacting with the brain opioid system.

Specificity studies of receptors for endogenous ligands, using synthetic analogues, often enable apparently identical receptors to be resolved into a number of subtypes (Chap. 5).

Drugs and toxins often have profound effects on the nervous system in extremely low doses. Thus as little as 10 μg of the hallucinogen lysergic acid diethylamide (LSD) causes disturbance of consciousness in humans, while 1 ng doses of botulinum toxin evoke toxic symptoms. Such extreme sensitivity to drugs illustrates the importance of chemical processes in the working of the nervous system.

5. Synaptic Transmission

5.1. Historical Background

The preceding chapters have shown that the synapse is a specialized region of contact and chemical secretion between the presynaptic neuron and its target cell (neuron, muscle, or gland cell), that transmitters are stored in the synaptic vesicles of the nerve terminals or varicosities, that release is brought about by vesicle exocytosis, that vesicles can be recovered and reutilized, that release is quantized, and that the response of the postsynaptic membrane to transmitter release is frequently the opening of a ligand-gated ion channel in the postsynaptic membrane induced by the transient binding of the transmitter to the channel at a specialized receptor site. Estimates of the number of different transmitters vary, but may well exceed 20.

The observation by C. BERNARD (1813–1878) that the South American arrow poison, curare, blocked neuromuscular transmission without affecting the excitability of either the nerve or the muscle, showed that excitation changes its character at the junction between nerve and muscle and becomes particularly sensitive to chemical blockade. It did not however immediately lead to the concept of chemical transmission at synapses: at the turn of this century, resemblances were seen on the one hand between the physiological effects of adrenaline and stimulation of sympathetic nerves and on the other between the effects of muscarine and stimulation of the parasympathetic system. This led to the suggestion that adrenaline, or something similar, mediated sympathetic stimulation and muscarine, or an endogenous analogue, mediated parasympathetic stimulation. Another alkaloid, nicotine, was found first to stimulate and then to block synaptic transmission through autonomic ganglia and this property enabled the anatomy of the autonomic system (Section 3.1 and Fig. 4) to be worked out. U. S. VON EULER (1905–1983) discovered that the transmitter at most sympathetic nerve terminals is noradrenaline rather than adrenaline and the work of O. LOEWI (1873–1961) and H.H. DALE (1875–1968) showed that the transmitter at autonomic ganglia, most parasympathetic nerve endings, and at neuromuscular junctions is acetylcholine. DALE coined the convenient–if not quite accurate–terms adrenergic and cholinergic to describe the two types of synapse and this terminology has been extended subsequently to cover other transmitters.

The fact that muscarine and nicotine do not have identical actions on cholinergic synapses, indicates that the acetylcholine receptors in the target cells have markedly different *chemical specificities* as well as differing physiological roles. This was the first example of an important development in pharmacology: the classification of receptors having a common endogenous ligand by means of agonists or antagonists of known but different structures. Thus acetylcholine receptors may be classified as muscarinic (these are found at postganglionic parasympathetic nerve terminals and in the CNS) and nicotinic (found in autonomic ganglia, at the neuromuscular junction, and also in the CNS). That there is more than one subtype of nicotinic receptor is shown by the fact that the muscular relaxant decamethonium is an excellent blocker of neuromuscular transmission, whereas hexamethonium, not very effective at this synapse, is much more so in autonomic ganglia. There are at least two subtypes of muscarinic receptor (designated M_1 and M_2) and at least four types of adrenergic receptors (α_1, α_2, β_1, and β_2). Multiple types of glutamate (probably three) and γ-aminobutyrate (at least two, $GABA_A$ and $GABA_B$) are also known.

The isolation and sequencing of receptors from different parts of the nervous system has fully confirmed the existence of subtypes. In situ hybridization shows that these subtypes are not mingled together indiscriminately but may be specific for particular classes of neuron. This opens up the possibility of very specific drug targeting if variants can be synthesized which specifically recognize the subtypes thus far known only from their molecular composition.

5.2. Survey of Transmitters

5.2.1. Low Molecular Mass Transmitters

The first transmitters to be identified were the low molecular mass compounds noradrenaline and acetylcholine.

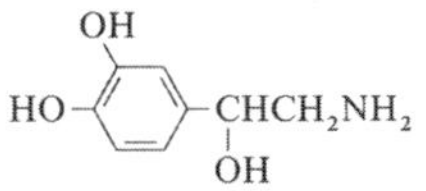

Noradrenaline
[51-41-2]

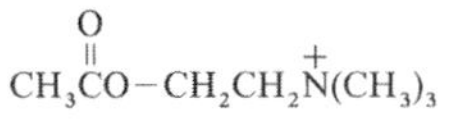

$CH_3CO-CH_2CH_2\overset{+}{N}(CH_3)_3$

Acetylcholine
[51-84-3]

Others followed: dopamine, 5-hydroxytryptamine (5-HT, also known as serotonin), glutamate, γ-aminobutyrate (GABA), glycine, and, less certainly, histamine, aspartate, and taurine. Such transmitters are often nowadays referred to as "classical" transmitters, in contradistinction to the neuropeptides (Section 5.2.2).

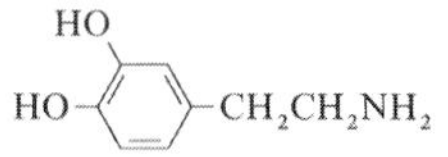

Dopamine
[51-61-6]

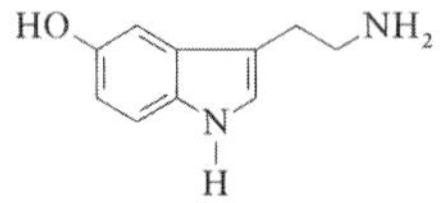

5-Hydroxytryptamine
[50-67-9]

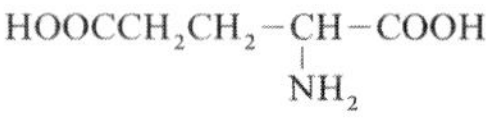

$HOOCCH_2CH_2-CH(NH_2)-COOH$

Glutamate
[56-86-0]

$CH_2(NH_2)CH_2CH_2COOH$

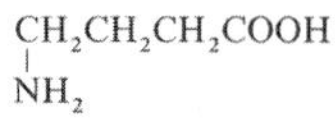

γ-Aminobutyrate (GABA)
[56-12-2]

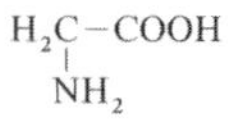

$H_2C(NH_2)-COOH$

Glycine
[56-40-6]

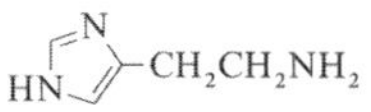

Histamine
[51-45-6]

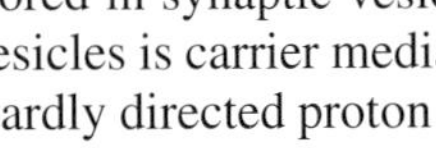

$HOOCCH_2-CH(NH_2)-COOH$

Aspartate
[56-84-8]

$NH_2CH_2CH_2SO_3H$

Taurine
[107-35-7]

The classical transmitters are usually synthesized in the nerve terminal cytoplasm and are stored in synaptic vesicles. Uptake by synaptic vesicles is carrier mediated and driven by an inwardly directed proton gradient across the vesicle membrane. Release is via Ca^{2+}-activated exocytosis of synaptic vesicles; the vesicles are recovered and refilled with the locally synthesized or recovered transmitter, a process known as vesicle recycling. Transmitter release is thus not immediately dependent on axonal transport (Section 5.4). The released transmitter is salvaged by a high affinity, carrier-mediated uptake system in the presynaptic terminal, either as such or in the form of its hydrolysis products. Thus acetylcholine is hydrolyzed to choline and acetate by the enzyme acetylcholinesterase, present in high concentration in the synaptic cleft; both hydrolysis products are taken up and resynthesized to acetylcholine.

The hydrolysis of acetylcholine is also an inactivating mechanism, since choline has only one ten-thousandth of the activity of acetylcholine on acetylcholine receptors and acetate is not a ligand for such receptors. Inactivation mechanisms involving oxidation, methylation, and other reactions also exist for other transmitters, but these usually come into play only when the concentration of transmitter in the synaptic cleft is very high; the usual form of inactivation is the presynaptic uptake of the transmitter. This applies to the catecholamines, 5-HT, and the amino acids, but not histamine. In the case of the amino acids, glial cells may also play a part in the uptake process.

Acetylcholine and noradrenaline are transmitters at numerous peripheral and central synapses. Glutamate is an excitatory transmitter widely distributed in the CNS of mammals; in many invertebrates (arthropods, cephalopods) it replaces acetylcholine as the excitatory transmitter at the neuromuscular junction. Dopamine and 5-HT are excitatory transmitters in the mammalian central nervous system; GABA and glycine are inhibitory transmitters in the brain and spinal cord, respectively. In arthropods inhibition is brought about not centrally but by specific inhibitory endings in the muscles where GABA is a transmitter; indeed, GABA was first discovered by using the inhibitory nerve to the stretch receptor of the crayfish as an assay system.

Inhibition is a widespread and important phenomenon in the CNS. Without it, the smooth relaxation of an antagonist muscle when an agonist contracts would be impossible, and painful cramps would accompany any movement. Thoughts and behavior would become wild and disordered. Inhibitory synaptic transmission involves the release of transmitter, the transmitter opening chloride channels instead of channels for sodium and potassium; chloride ions move into and transiently *hyperpolarize* the postsynaptic neuron. Hyperpolarization raises or stabilizes the neuron's resting potential, thereby partially or completely neutralizing excitatory postsynaptic potentials and preventing them from achieving the membrane depolarization which would trigger off a propagated impulse. Epilepsy has been defined as "a disorder of brain function characterized by the episodic recurrence of paroxysmal neurological or behavioral manifestations caused by abnormal synchronous and excessive discharges of large groups of neurons" [1]. The cause is obscure but may be related to an abnormal accumulation of the excitatory transmitter glutamate, a reduced availability of the inhibitory transmitter GABA, or an excessive accumulation of extracellular K^+ which would cause massive depolarization of neurons and extensive discharges. Two other classes of drugs exert a general effect on neuronal excitability in the CNS, the analeptics, which stimulate and the sedatives which depress it. A fourth class, the neuroleptics, which include some sedatives, seem to act primarily on D_2 dopamine receptors (Section 5.3.3).

5.2.2. Neuropeptides

The discovery of neuroactive peptides in the nervous system grew out of the discovery of peptide and protein hormones. The first hormone to be discovered was secretin which is released from the intestine and evokes pancreatic secretion. Many others followed; among the most important are insulin and glucagon both of which are released from the pancreas and regulate blood sugar, but in opposite directions. The intestine continued to yield hormones, among them substance P, vasoactive intestinal polypeptide (VIP) and cholecystokinin (CK).

Another potent source of physiologically active peptides is the pituitary, or hypophysis, a gland at the base of the brain. Here are found hormones controlling diuresis and water balance (vasopressin), milk ejection (oxytocin), growth (growth hormone), reproductive function (the two gonadotropins, namely luteinizing hormone and follicle-stimulating hormone), lactation (prolactin), fat mobilization (β-lipotropin), thyroid activity (thyrotropin), and adrenal cortical activity (adrenocorticotropic hormone). These in turn are controlled by peptides located in the region at the base of the brain known as the hypothalamus. Such hormones either stimulate (releasing hormones) or inhibit (releaseinhibiting hormones) the release of pituitary hormones. Releasing hormones for the gonadotropins, thyrotropin, adrenocorticotropic hormone, and growth hormone, and a release-inhibiting hormone for growth hormone (somatostatin) have been characterized. The hypothalamus also links up with the autonomic nervous system; thus the hypothalamic – hypophyseal axis is extremely important for central nervous control of bodily function and metabolism.

Peptides are found in bewildering variety at all levels of the phylogenetic tree down to *Hydra*, a simple animal related to the sea anemone. Some peptides have definite functions, thus one peptide in *Hydra* is essential for the differentiation of the head region and has been detected in the rat hypothalamus; another, in *Aplysia* (a sea-slug) regulates egg laying; others may act as neurotoxins. Frog skin has yielded at least a dozen peptides, many resembling those found in the mammalian hypothalamus and elsewhere.

Immunological techniques have been extremely important in studying the neuropeptides. Specific antisera or monoclonal antibodies (mABs) raised to them have enabled them to be estimated in tissues and subcellular fractions by means of *radioimmunoassay*. Such sera or antibodies have also been extensively used to localize the peptides they recognize in tissues, including nervous tissue, by applying the techniques of immunocytochemistry. Usually, the specific antiserum or mAB is applied after permeabilizing tissue sections with a detergent. After washing out unbound antibody, a second antibody, labeled with a fluorescent group and directed against the immunoglobulin fraction of the animal providing the antiserum or mAB is applied. The section is washed again and the location of the original antigen is revealed by the pattern of fluorescence observed under the microscope.

Such studies have given two remarkable results: (1) peptides including many already identified as hormones are widely but not uniformly distributed in the nervous system; and (2) the neurons staining positively for peptides often utilize low molecular mass, classical transmitters. Certain peptides tend to be associated with certain classical transmitters such as GABA, acetylcholine, and noradrenaline. Thus it has been estimated that 50 % of all central cholinergic terminals also contain VIP and that 75 % of all VIP-containing terminals are also cholinergic [12]. Although peptides (neuropeptides) are present throughout the neuron, they tend to be highly concentrated in the terminals and can often only be identified in cell bodies by using colchicine to block axonal transport (Section 5.4), the process whereby substances or organelles synthesized in the neuronal cell body are conveyed to the terminal.

It thus seems reasonable to conclude that neurons primarily utilize neuropeptides as transmitters. When a classical transmitter coexists with the neuropeptide, the latter may be regarded as a cotransmitter. Neuropeptides are indeed released from nerve terminals on stimulation in a Ca^{2+}-dependent manner and can induce or modify electrical responses in postsynaptic membranes. A good example is substance P which is released on stimulation of afferent neurons entering the spinal cord of the type mediating pain. Neuropeptides differ considerably from classical transmitters in the dynamics of their synthesis, storage, and release [13, 14]. Thus they are not stored in synaptic vesicles as the classical transmitters are but in larger vesicles, which sometimes have electron-dense cores (i.e., have black-looking cores in electron micrographs). They cannot be synthesized in the nerve terminals or salvaged after release so there is no recycling vesicular pool. This means that release is completely dependent on axonal transport.

These differences in intracellular dynamics suggest that the pattern of release of neuropeptide cotransmitters reflects the past intensity and duration of stimulation of the neurons containing them. This could result in a subtle regulation of the rate of synthesis and pattern of release of the classical cotransmitter, especially if the neuropeptide combines with receptors to release "second messengers" (Section 5.3.2) able to regulate intracellular transmitter metabolism.

Neuropeptides are rarely if ever synthesized as such; they are the products of the posttranslational modification of much larger molecules, known as propeptides (prohormones) or prepropeptides (preprohormones). These contain the sequences of neuroactive peptides as segments separated by lysine – lysine or lysine – arginine, points at which the propeptide is easily cleaved by proteolytic enzymes. Thus the hypophyseal hormones adrenocorticotropic hormone, γ-melanotropin and β-lipotropin are synthesized as parts of a single much larger prohormone; adrenocorticotropic hormone itself contains the sequence of α-melanotropin while β-

lipotropin breaks down to yield the opioid peptide β-endorphin and γ-lipotropin; the latter in turn yields β-melanotropin.

The discovery of the neuropeptides has opened up a baffling yet exciting new era in neurobiological research. In many cases the function of these peptides in the brain is not known; some may act as transmitters, but others may have a modulating role or link the nervous system with the autonomic system, the immune system or tissue repair after injury. The implications for neuropharmacology are also immense, if stable analogues with antagonistic or agonist properties can be successfully prepared.

5.2.3. Evolutionary Significance of the Two Types of Transmitter

The low molecular mass transmitters are not found in the simplest living organisms that possess nervous systems; thus acetylcholine is not found in coelenterates and first appears in the echinoderms and platyhelminths. Their appearance may coincide with an evolutionary demand for fast, sustained movement. It is probable that neuropeptides were the first transmitters and that chemical transmission is an adaptation of humoral secretion. Neuropeptide secretion is dependent on de novo synthesis and axonal transport, these processes are wasteful and relatively inefficient especially as the size of organisms increased and axons grew longer, separating the sites of synthesis and release by ever greater distances.

Chemical transmission, especially when speeded up by recycling, remained however an attractive mechanism. The store of transmitter in the presynaptic terminal guarantees unidirectional synaptic transmission–an essential prerequisite for the orderly processing of information. If an antidromic impulse (i.e., one travelling in the reverse direction) should arise in an axon, it would be unable to effect reversed transmission due to lack of a store of transmitter in the neuronal cell body or dendrites.

The existence of more than one transmitter and specific receptors for them may be a device to suppress "cross-talk" between interdigitated but functionally distinct systems. The retention of neuropeptides as transmitters or cotransmitters alongside the classical transmitters is harder to explain. Some may represent the vestigial survival of phylogenetically old transmitters which evolutionary pressures were not strong enough to eliminate. Others clearly retain a transmitter function in circuits where short synaptic delays and continuous action may not be so important, as in pain afferents. Yet others have a regulating role or have important hormonal functions linking the nervous system to other body systems.

All chemical synapses– even those using fastacting, easily mobilized transmitters–impose a short delay of not less than 0.5 – 1 msec. It is significant that where fast action has evolutionary survival value (e.g., the escape reactions of decapods), chemical transmission has been abandoned for the even faster electrical (electrotonic) transmission in the corresponding neuronal circuits. Electrical transmission is also less temperature dependent than chemical transmission, which may explain why electrotonic synapses are more frequently encountered in cold-blooded animals. Such synapses show apposition of pre- and postsynaptic membranes and are devoid of synaptic vesicles.

5.3. Transmitter Action

5.3.1. Fast Ionotropic Action

The term ionotropic action denotes the action of transmitters which operate by opening channels in the postsynaptic membrane. Such action is swift and terminated by spontaneous closing of the channel and the inactivation of the transmitter by hydrolysis or reuptake.

Excitatory Ionotropic Action . Both acetylcholine (when acting on nicotinic receptors) and glutamate are examples of fast-acting excitatory transmitters. The nicotinic acetylcholine receptor (nAChR) was first isolated in a pure form from an exceptionally rich source: the electric organ of the electric ray, *Torpedo*. Such organs consist of a honeycomb array of columns of flattened cells, called electrocytes, which are embryologically derived from muscle with loss of the contractile machinery; each electrocyte is 80 % covered on its under surface with cholinergic nerve terminals. Electric organ contains weight for weight, 500 – 1000 times more synaptic material and nAChR than muscle and isolation of the latter becomes a feasible proposition.

Isolation of the receptor was made possible by the discovery that one of the main proteins of cobra venom is a powerful neuromuscular blocking agent and has a high affinity for the receptor. The receptor was solubilized in detergent and adsorbed onto columns of immobilized venom; after washing the column, the receptor could be eluted in concentrated form. It consists of four peptides of molecular masses 40 (α), 50 (β), 60 (γ), and 65 (δ) kDa; the complete receptor is a complex of all four in which the α-peptide is repeated twice. In negative staining the complex appears as a pentagon with a central hole presumed to be the ion channel.

All four peptides have been sequenced [15]: each contains four hydrophobic transmembrane segments and the complete molecule is an assembly of 20 such segments forming an ionic channel through the membrane. When two molecules of acetylcholine bind to the two α-subunits an allosteric change takes place and the channel opens. The channel opening time varies with the ligand. Each package (quantum) of transmitter opens 5000 – 6000 such channels and generates a transient change of about 0.2 mV in the resting membrane potential of the postsynaptic membrane. The *Torpedo* electrocyte is not electrically excitable and cannot conduct an impulse but in an excitable nerve or muscle cell, the synchronized release of several hundred such quanta will generate a postsynaptic potential change sufficient to initiate a conducted response. Reversible neuromuscular blocking agents (muscular relaxants) usually contain two quaternary nitrogen atoms separated by a bridge of carbon atoms instead of one as in acetylcholine. Such compounds block by straddling the two α-subunits of the receptor. One such muscular relaxant, succinylcholine, is readily hydrolyzed and provides a transient, easily controlled block very useful in abdominal surgery. By contrast the snake toxins work by blocking the ion channel directly.

The characterization of the nAChR has been followed by that of the glutamate receptor, which operates on similar lines.

Inhibitory Ionotropic Action. The nature of spinal inhibition was worked out by J.C. ECCLES (b. 1903) using microelectrodes inserted into motor neurons and the reflex activation of inhibitory interneurons (Section 3.2.1) [2]. His observation that minute amounts of a chemical substance, strychnine, could block inhibition convinced him that an inhibitory chemical transmitter, subsequently identified as glycine, must be involved. To identify the ligand-gated ionic channel involved, different ions were injected into motor neurons and the effect on the inhibitory (hyperpolarizing) postsynaptic potential (IPSP) generated by the reflexly activated interneuron was observed. The IPSP could be abolished and then converted to a depolarizing potential by increasing the internal Cl^- concentration of the cell, showing that the ionic channel opened by the transmitter is a Cl^- channel. Hydrated univalent anions smaller or only a little larger than Cl^- also reduce the IPSP and are thus also able to pass through the Cl^- channel; other ions (e.g. organic anions, di- or trivalent inorganic anions) are ineffective.

The high affinity of strychnine for the glycine receptor has enabled the latter to be isolated and sequenced using affinity chromatography on immobilized strychnine.

In the brain and the dorsal horn of the spinal cord, GABA (not glycine) is the inhibitory transmitter. It acts in a similar way to glycine but its receptor has a different pharmacological specificity: it is blocked by picrotoxin and bicuculline (ineffective at glycine receptors) but not strychnine. Presynaptic inhibition (i.e., inhibition exerted on excitatory presynaptic nerve terminals by inhibitory synapses present on them) is also mediated by this transmitter.

5.3.2. Slow Metabolotropic Action

Acetylcholine at muscarinic receptor sites, the monoamine transmitters (noradrenaline, dopamine, 5-HT, histamine), and probably some neuropeptides do not operate directly through ligand-gated ion channels but via *second messengers*. These messengers are intracellular components capable of regulating a variety of intracellular functions (including that of ion channels) which in turn are mobilized by binding the transmitter (or an agonist analogue) to its receptor. An intricate series of reactions has been implicated (Fig. 9). The basic features are:

1) the identification of a family of G (or N) proteins which link the receptor to adenylate (or guanidylate) cyclase,

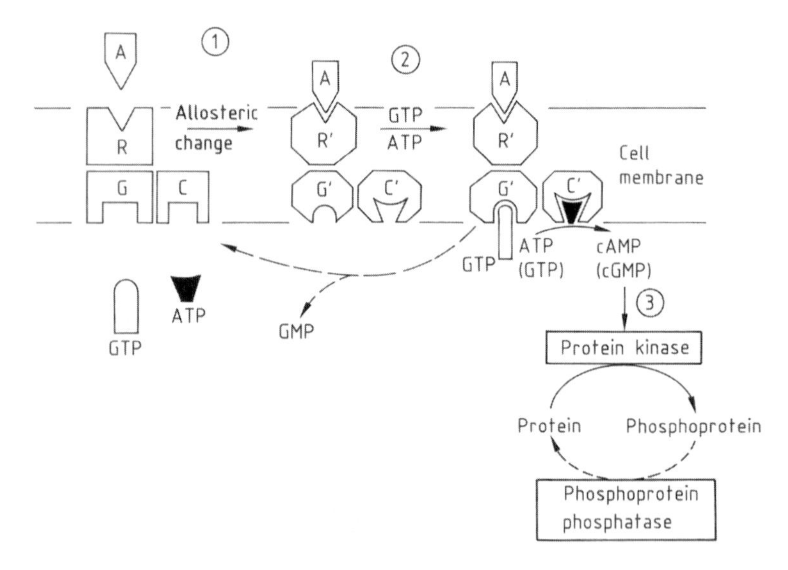

Figure 9. Scheme showing sequence of events regulated by metabolotropic receptor (R)
Allosteric changes are shown schematically by different geometric figures.
1) Agonist A combines with complex of R, G-protein (G) and adenylate (guanidylate) cyclase (C).
2) The ligand-induced conformational change in R communicates itself to G which binds GTP and through G to C which converts ATP (GTP) to cyclic AMP (cyclic GMP).
3) Cyclic AMP (cyclic GMP) activates a protein kinase which regulates functional proteins by phosphorylating them.

2) the regulation of protein kinases by the cyclic adenosine 3′,5′-monophosphate (cAMP) or cyclic guanosine 3′,5′-monophosphate (cGMP)
3) the regulation of functional proteins by phosphorylation catalyzed by such kinases.

Most receptor-regulated cyclases are adenylate cyclases but those regulated by H_1 (histamine receptor) and mAChR (acetylcholine receptor) are guanidylate cyclases. In some cases the functional proteins so regulated are ion channels. Thus an important effect of the activation of α_2 receptors by noradrenaline is a long-lasting inhibition achieved by an increase in K^+ conductance following channel phosphorylation. In other cases, proteins controlling gene expression may be so regulated. A phospholipase C capable of generating triphosphoinositide and diacylglycerol from a membrane phospholipid is yet another protein to be thus regulated. Both of these substances can release Ca^{2+} from intracellular membranes. The Ca^{2+} ions can react with another protein calmodulin to activate another functionally important class of protein kinases. Calcium antagonists are used therapeutically in coronary heart disease, arrhythmia, and hypertension. However, in many cases the link between protein phosphokinase activity induced by the second messenger– cAMP, cGMP–or Ca^{2+} and the physiological response has yet to be worked out.

The involvement of so many allosteric and enzymic steps may explain the relatively slow onset and long duration of such responses. This justifies the classification of these transmitter actions as metabolotropic.

5.3.3. Receptors of Importance in Drug Action

Drugs may act on receptors as *agonists* (mimetics) that produce the same action as the transmitter, or as *antagonists* (lytics, blockers, blocking agents). Several examples have already been given, notably the neuromuscular blocking agents which act by blocking nicotinic acetylcholine receptors in muscle subtypes (Section 5.3.1). The receptors for glutamate, GABA, noradrenaline, and dopamine are also important targets for drug action and pharmacological analysis has revealed many subtypes.

Many clinically important drugs are used to regulate the autonomic nervous system by acting as either agonists or antagonists at the acetylcholine or noradrenaline receptors in smooth and cardiac muscle or glands. β-Blockers do however gain entrance to the CNS and produce depression apparently by blocking adrenergic pathways in the limbic system. By contrast, iproniazid and other monamine oxidase inhibitors, raise the intracellular concentration of noradrenaline and are antidepressive. This strongly suggests that depression involves a disturbance in CNS (possibly limbic) function mediated by adrenergic neurons, which could result either from a failure to synthesize enough noradrenaline or from a desensitization of noradrenergic receptors.

Dopaminergic transmission though not widespread is also of critical importance in the CNS. Dopaminergic cell bodies are present in the substantia nigra, a nucleus of dark hue between the hypothalamus and thalamus (Fig. 5 C) and project to the caudate nucleus (Fig. 5 B). In Parkinson's disease these neurons lose their ability to produce enough dopamine. Administration of a precursor, L-dopa, or of dopamine agonists is therapeutic. Conversely, agents which displace dopamine, e.g., the neurotoxin *N*-methylphenyltetrahydropyridine, induce Parkinsonian symptoms.

Dopamine pathways appear to be affected in schizophrenia. All three main classes of neuroleptic (antipsychotic) drugs, phenothiazines, butyrophenones, and thioxanthenes, have side effects indicating the involvement of extrapyramidal neurons: Parkinsonian rigidity and tremor or dyskinesia (convulsive movements). D_2 receptor blockade correlates well with neuroleptic activity and one of the symptoms of schizophrenia, catatonia, may be related to Parkinson rigidity.

A third type of receptor of great importance for drug action is the $GABA_A$ receptor which is the main target for the anxiolytic and anticonvulsant benzodiazepines. These compounds bind to a receptor which allosterically modifies the GABA receptor to increase the efficacy of its activator GABA, an inhibitory transmitter. There is an excellent correlation between binding affinity and potency in animal and clinical tests.

5.4. Axonal Transport

The presynaptic nerve terminal cannot synthesize structural and functional proteins and lipoprotein membranes; it is entirely dependent on the synthetic activity of the cell body (perikaryon). The electromotor cell body has been estimated to maintain over 30 times its volume of presynaptic nerve terminal; similar figures must apply to other neurons. Most neuronal perikarya show corresponding signs of intense synthetic activity: they are rich in polysomes (collections of ribosomes synthesizing proteins from mRNA) and membranes, the latter often budding off synaptic vesicle-like structures.

The process whereby soluble proteins and membrane-bound organelles are conveyed from the cell body to the terminals is referred to as axonal transport. It involves more than one mechanism and rate of transport. Paradoxically, synaptic vesicles are conveyed at the fastest rate (400 mm/d), probably with the participation of microtubules (a kind of scaffolding) using a mechanism akin to muscular contraction. Larger particles (e.g., mitochondria) travel more slowly, apparently as a result of increased mechanical resistance, but soluble cytoplasmic proteins travel at the slowest rate (ca. 1 mm/d). Here the mechanism of transport is uncertain but may involve the streaming of cytoplasm within spaces between bundles of microtubules. Colchicine stops transport by causing the disassembly of microtubules.

There is also a need for retrograde transport– the conveyance of spent or worn-out organelles to the cell body for salvage–but the mechanisms involved are still unclear.

6. References

General References

1. G. Adelman (ed.): *Encyclopedia of Neuroscience,* Birkhäuser, Boston – Basel – Stuttgart 1987.
2. P. L. McGeer, J. C. Eccles, E. G. McGeer: *Molecular Biology of the Mammalian Brain,* Plenum Press, New York – London 1978.
3. F. Hucho: *Neurochemistry, Fundamentals and Concepts,* VCH Verlagsgesellschaft, Weinheim 1986.
4. R. Nieuwenhuys: *Chemoarchitecture of the Brain,* Springer Verlag, Berlin 1985.

5. J. R. Cooper, E. F. Bloom, R. H. Roth: *The Biochemical Basis of Neuropharmacology,* Oxford University Press, New York 1986.
6. S. Iversen, L. L. Iversen: *Behavioral Pharmacology,* Oxford University Press, Oxford 1981.
7. E. R. Kandel, J. H. Schwartz: *Principles of Neural Science,* 2nd ed., Elsevier, New York – Amsterdam – Oxford 1985.

Specific References

8. C. S. Sherrington: *Integrative Action of the Nervous System,* University Press, Cambridge 1948.
9. B. Katz: *Nerve, Muscle, and Synapse,* McGraw Hill, New York 1966.
10. V. P. Whittaker in A. Lajtha (ed.): *Handbook of Neurochemistry,* 2nd ed., **vol. 7,** Plenum Press, New York – London 1984, p. 1.
11. V. P. Whittaker in [10]p. 41.
12. D. V. Agoston, E. Borroni, P. J. Richardson, *J. Neurochem.* **50** (1988) 1659-1662.
13. D. V. Agoston, J. M. Conlon, V. P. Whittaker, *Exp. Brain Res.* **72** (1988) 535-542.
14. D. V. Agoston, V. P. Whittaker, *J. Neurochem.* **52** (1989) 1474-1480.
15. M. Noda et al., *Nature (London)* **302** (1983) 528-532.

Author Index

Author Index

Actor, Paul, Smith Kline & French Laboratories, Philadelphia, Pennsylvania 19101, United States (Chaps. 1, 3) *Chemotherapeutics*, **1013**
Allegra, Carmen J., National Cancer Institute, Bethesda, Maryland 20205, United States *Cancer Chemotherapy*, **839**
Aschenbrenner, Hans, Bayer AG, Leverkusen, Federal Republic of Germany (Section 6.4.2) *Transport, Handling, and Storage*, **487**

Bartels, Klaus, Berufsgenossenschaft der chemischen Industrie, Heidelberg, Federal Republic of Germany (Section 4.1) *Plant and Process Safety*, **361**
Bender, Herbert, BASF Aktiengesellschaft, Ludwigshafen, Federal Republic of Germany (Section 2.1) *Plant and Process Safety*, **361**
Berner, Sibylle, Boehringer Mannheim GmbH, Tutzing, Federal Republic of Germany (Chap. 7) *Nucleic Acids*, **739**
Berson, Emile, Ministère des Transports, Paris, France (Chap. 2) *Transport, Handling, and Storage*, **487**
Berson, Nadia, Paris, France (Chap. 2) *Transport, Handling, and Storage*, **487**
Berthold, Werner, BASF Aktiengesellschaft, Ludwigshafen, Federal Republic of Germany(Section 2.2.3) *Plant and Process Safety*, **361**
Binetti, Roberto, Istituto Superiore di Sanità, Rome, Italy (Chap. 3) *Chemical Products: Safety Regulations*, **553**
Bino, Goffredo Del, European Commission, Brussels, Belgium (Chap. 1) *Chemical Products: Safety Regulations*, **553**
Bosatra, Dan, Dow Europe S. A., Horgen, Switzerland (Section 3.1– 3.9) *Occupational Health and Safety*, **299**
Broecker, Bernard, formerly Hoechst Aktiengesellschaft, Frankfurt/Main, Federal Republic of Germany (Chaps. 2 and 4) *Chemical Products: Safety Regulations*, **553**
Burtscher, Helmut, Boehringer Mannheim GmbH, Penzberg, Federal Republic of Germany (Chaps. 1–6) *Nucleic Acids*, **739**

Chabner, Bruce A., National Cancer Institute, Bethesda, Maryland 20205, United States *Cancer Chemotherapy*, **839**
Chow, Alfred W., Smith Kline & French Laboratories, Philadelphia, Pennsylvania 19101, United States (Chaps. 1, 3) *Chemotherapeutics*, **1013**
Conrad, Dietrich, Berlin, Federal Republic of Germany (Section 2.2.2) *Plant and Process Safety*, **361**
Cross, Alan, Department of Bacterial Diseases, Walter Reed Army Institute of Research, Washington D.C. 20307-5100, United States (Chap. 5) *Immunotherapy and Vaccines*, **899**
Cryz, Stanley J. Jr., Swiss Serum and Vaccine Institute Berne, Switzerland (Chaps. 1 and 2) *Immunotherapy and Vaccines*, **899**
Curt, Gregory A., National Cancer Institute, Bethesda, Maryland 20205, United States *Cancer Chemotherapy*, **839**

Dekant, Wolfgang, Institute of Toxicology, University of Wuerzburg, Germany *Toxicology*, **3**
Devos, Jean-Marie, CEFIC, Brussels, Belgium (Chaps. 1, 3) *Legal Aspects*, **585**
Döhrn, Klaus Jochem, Germanischer Lloyd, Hamburg, Federal Republic of Germany (Section 6.2.1.3) *Transport, Handling, and Storage*, **487**
Doktor, Kurt-Joseph, Bayer AG, Leverkusen, Federal Republic of Germany (Sections 6.2.1.1, 6.2.1.2, 6.3, 6.4) *Transport, Handling, and Storage*, **487**
Dosch, Rainer, Bayer AG, Leverkusen, Federal Republic of Germany (Section 4.7) *Occupational Health and Safety*, **299**
Drauz, Karlheinz, Degussa-Hüls AG, Hanau-Wolfgang, Germany *Amino Acids*, **777**
Drees, Stefan, Bayer AG, Leverkusen, Federal Republic of Germany (Section 3.5.1) *Plant and Process Safety*, **361**
Droste, Bernhard, Bundesanstalt für Materialforschung und -prüfung, Berlin, Federal Republic of Germany (Chap. 4) *Transport, Handling, and Storage*, **487**
Dutko, Frank J., Sterling-Winthrop Research Institute, Rensselaer, New York 12144, United States (Chap. 4) *Chemotherapeutics*, **1013**

Eberz, Albert, Bayer AG, Leverkusen, Federal Republic of Germany(Section 2.3) *Plant and Process Safety*, **361**

Fine, Robert L., National Cancer Institute, Bethesda, Maryland 20205, United States *Cancer Chemotherapy*, **839**
Förster, Hans, Physikalisch-Technische Bundesanstalt, Braunschweig, Federal Republic of Germany (Section 3.5.3 in part) *Plant and Process Safety*, **361**
Fränzle, Otto, Christian-Albrechts-Universität, Kiel, Federal Republic of Germany [(Sections1.1, 1.2, 2.1 (in part), 2.2, 2.3 (in part), 2.4, 3.1, 3.2 (in part)] *Ecology and Ecotoxicology*, **151**

Geiger, Adrian, Sandoz, Basel, Switzerland (Section 4.2.1) *Plant and Process Safety*, **361**
Geschke, Frank-Ulrich, Bayer AG, Wuppertal, Federal Republic of Germany *Antimycotics*, **1075**
Glor, Martin, Ciba-Geigy AG, Basel, Switzerland (Section 2.2.3) *Plant and Process Safety*, **361**
Gottstein, Bruno, Institut für Parasitologie, Universität Zürich, Zürich, Switzerland (Section 4.1) *Immunotherapy and Vaccines*, **899**
Granstrom, Marta, Departments of Clinical Microbiology and of Vaccine Production, National Bacteriological Laboratory, Karolinska Hospital, Stockholm, Sweden (Chap. 3) *Immunotherapy and Vaccines*, **899**
Grenner, Dieter, Bayer AG, Dormagen, Federal Republic of Germany (Sections 4.4, 4.5) *Plant and Process Safety*, **361**
Gruber, Joachim, Bayer AG, Leverkusen, Federal Republic of Germany (Section 4.4) *Occupational Health and Safety*, **299**

Hagen, Hans, Bayer AG, Leverkusen, Federal Republic of Germany (Sections 3.4.1, Section 5.1 in part) *Plant and Process Safety*, **361**
Hahn, Klaus-Jürgen, Knoll AG, Ludwigshafen, Federal Republic of Germany (Section 4.2) *Pharmaceuticals, General Survey and Development*, **981**
Harbordt, Jürgen, Bayer AG, Leverkusen, Federal Republic of Germany (Section 5.1 in part) *Plant and Process Safety*, **361**
Harke, Hans-P., Schülke & Mayr, Hamburg/Norderstedt, Federal Republic of Germany *Disinfectants*, **721**
Hasskarl, Horst, Ludwigshafen, Federal Republic of Germany (Chap. 3) *Pharmaceuticals, General Survey and Development*, **981**
Hawighorst, Josef, Bayer AG, Leverkusen, Federal Republic of Germany (Sections 6.2.1.1, 6.2.1.2) *Transport, Handling, and Storage*, **487**
Hesse, Günter, Bayer AG, Brunsbüttel, Federal Republic of Germany (Section 3.5.2) *Plant and Process Safety*, **361**

Ullmann's Industrial Toxicology

ISBN: 3-527-31247-1

Hoppe, Bernd, Degussa-Hüls AG, Frankfurt, Germany *Amino Acids*, **777**
Hübner, Heinz W., Bundesanstalt für Materialforschung und -prüfung, Berlin, Federal Republic of Germany (Section 6.1) *Transport, Handling, and Storage*, **487**

Itoda, Muneto, JCIA, Tokyo, Japan (Chap. 7) *Legal Aspects*, **585**

Jørgensen, Sven Erik, Danmarks Farmaceutiske Højskole, Copenhagen, Denmark (Section 3.3) *Ecology and Ecotoxicology*, **151**

Kiesselbach, Niko, Bayer AG, Leverkusen, Federal Republic of Germany (Section 4.3) *Occupational Health and Safety*, **299**
King, Norman, Department of the Environment, London, United Kingdom (Chaps. 5 and 6) *Chemical Products: Safety Regulations*, **553**
Kleemann, Axel, Hanau, Germany *Amino Acids*, **777**
Klemchuk, Peter P., CIBA-GEIGY Corporation, Ardsley, New York 10502, United States *Antioxidants*, **699**
Klusacek, Hans, Bayer AG, Dormagen, Federal Republic of Germany (Section 6.3) *Transport, Handling, and Storage*, **487**
Kolk, Jan J., Arnhem, The Netherlands (Sections 2.2, 2.3 in part, 4.1, 4.2, 4.5) *Occupational Health and Safety*, **299**
Krämer, Ludwig, Head of Legal Services, DG XI, European Commission, Brussels, Belgium (Chap. 2, Chap. 5 in part) *Legal Aspects*, **585**
Kretzschmar, Rolf, Knoll AG, Ludwigshafen, Federal Republic of Germany (Section 4.1) *Pharmaceuticals, General Survey and Development*, **981**
Krimmer, Hans-Peter, Degussa-Hüls AG, Hanau-Wolfgang, Germany *Amino Acids*, **777**
Kromarek, Pascale, Elf Aquitaine, Direction Environment, Paris-La-Defense, France (Chap. 5) *Legal Aspects*, **585**
Kutscher, Bernhard, ASTA Medica AG, Frankfurt am Main, Federal Republic of Germany *Cancer Chemotherapy*, **839**

Lamb, Donald W., Miles Inc., Pittsburgh, United States (Sections 2.4, 3.10, 4.8) *Occupational Health and Safety*, **299**
Larrick, James, Genelabs Incorporated, Redwood City, California, United States (Chap. 6) *Immunotherapy and Vaccines*, **899**
Lersner, Heinrich Freiherr von, Umweltbundesamt, Berlin, Federal Republic of Germany (Chap. 8) *Transport, Handling, and Storage*, **487**
Leuchtenberger, Wolfgang, Degussa-Hüls AG, Düsseldorf, Germany *Amino Acids*, **777**
Lipinski, Gert-Wolfhard von Rymon, formerly Hoechst Aktiengesellschaft, Frankfurt, Federal Republic of Germany *Food Additives*, **677**
Lück, Erich, formerly Hoechst Aktiengesellschaft, Frankfurt, Federal Republic of Germany *Food Additives*, **677**
Lühr, Hans-Peter, Technische Universität, Institut für wassergefährdende Stoffe, Berlin, Federal Republic of Germany (Chap. 7) *Transport, Handling, and Storage*, **487**

McGuire, John L., Johnson & Johnson, New Brunswick, New Jersey 08933, United States (Chaps. 1 and 2) *Pharmaceuticals, General Survey and Development*, **981**
McKinlay, Mark A., Sterling-Winthrop Research Institute, Rensselaer, New York 12144, United States (Chap. 4) *Chemotherapeutics*, **1013**
Meyer, Gunter, IG-Chemie-Papier-Keramik, Hannover, Federal Republic of Germany (Section 2.1) *Occupational Health and Safety*, **299**
Miksche, Leopold W., Bayer AG, Leverkusen, Federal Republic of Germany (Chap. 1, Section 4.6) *Occupational Health and Safety*, **299**
Mühlegger, Klaus, Boehringer Mannheim GmbH, Tutzing, Federal Republic of Germany (Chap. 9) *Nucleic Acids*, **739**
Müller, Edmund, BASF Aktiengesellschaft, Ludwigshafen, Federal Republic of Germany (Section 4.3) *Plant and Process Safety*, **361**
Müller, Michael, Bayer AG, Leverkusen, Federal Republic of Germany (Section 2.2.1, Section 3.5.3 in part) *Plant and Process Safety*, **361**
Mix, Karl-Heinz, Bayer AG, Leverkusen, Federal Republic of Germany (Section 2.2.4) *Plant and Process Safety*, **361**
Mujagic, Hamza, National Cancer Institute, Bethesda, Maryland 20205, United States *Cancer Chemotherapy*, **839**

Noha, Klaus, Hoechst Aktiengesellschaft, Frankfurt, Federal Republic of Germany (Section 4.2.2) *Plant and Process Safety*, **361**
Noordam, Philip, The Hague, The Netherlands (Sections 2.2, 2.3 in part) *Occupational Health and Safety*, **299**

Perrin, Luc, Division d'Hematologie, Hôpital Cantonal Universitaire, Genève, Switzerland (Section 4.2) *Immunotherapy and Vaccines*, **899**
Pfau, Wolfgang, Department of Toxicology and Environmental Medicine of the Fraunhofer Society and Department of Toxicology, Hamburg University Medical School, Hamburg, Germany *Carcinogenic Agents*, **651**
Pfau, Wolfgang, Umweltmedizin Hamburg e.V. and Institute of Toxicology, Hamburg University Medical School, Hamburg, Germany *Mutagenic Agents*, **667**
Pilz, Volker, Bayer AG, Leverkusen, Federal Republic of Germany (Chap. 1, Sections 3.1–3.3) *Plant and Process Safety*, **361**

Rennoch, Detlef, Bundesanstalt für Materialforschung und -prüfung, Berlin, Federal Republic of Germany (Chap. 3) *Transport, Handling, and Storage*, **487**
Rolle, Annette, Bundesanstalt für Materialforschung und -prüfung, Berlin, Federal Republic of Germany (Section 6.2.2) *Transport, Handling, and Storage*, **487**

Schacke, Helmut, Bayer AG, Leverkusen, Federal Republic of Germany (Section 3.4.2) *Plant and Process Safety*, **361**
Schmidt, Axel, Bayer AG, Wuppertal, Federal Republic of Germany *Antimycotics*, **1075**
Schrörs, Bernd, Bayer AG, Leverkusen, Federal Republic of Germany (Section 3.4.3) *Plant and Process Safety*, **361**
Schulz, Nikolaus, Bayer AG, Dormagen, Federal Republic of Germany (Sections 4.4, 4.5) *Plant and Process Safety*, **361**
Seibl, Rudolf, Boehringer Mannheim GmbH, Penzberg, Federal Republic of Germany (Chap. 8) *Nucleic Acids*, **739**
Smith, Turner T. Jr., Hunton & Williams, Washington, D.C., United States (Chap. 6) *Legal Aspects*, **585**
Steinbach, Jörg, Schering AG, Berlin, Federal Republic of Germany(Section 2.3) *Plant and Process Safety*, **361**
Stoessel, Francis, Ciba-Geigy AG, Basel, Switzerland(Section 2.3) *Plant and Process Safety*, **361**
Straškraba, Milan, Akademie věd České Republiky, Biomatematická laboratoř, Entomologický institut, České Budějovice, Czech Republic (Sections 2.1.4, 2.3.5, 3.2.3) *Ecology and Ecotoxicology*, **151**
Stroetmann, Clemens, Bad Honnef, Federal Republic of Germany (Chap. 1) *Transport, Handling, and Storage*, **487**

Vamvakas, Spiridon, Institute of Toxicology, University of Wuerzburg, Germany *Toxicology*, **3**
Viard, Richard, Bayer AG, Leverkusen, Federal Republic of Germany (Section 3.4.2) *Plant and Process Safety*, **361**
Visser, Wieger Johannes, Nederlandse Spoorwegen, Utrecht, The Netherlands (Sections 5.1–5.2.1) *Transport, Handling, and Storage*, **487**

Walper, Matthias, Bayer AG, Brunsbüttel, Federal Republic of Germany (Sections 5.2, 5.3) *Plant and Process Safety*, **361**
Walther, Claus-Diether, Bayer AG, Leverkusen, Federal Republic of Germany (Section 2.2.2) *Plant and Process Safety*, **361**
Wandrey, Peter-Andreas, Bundesanstalt für Materialforschung und –prüfung, Berlin, Federal Republic of Germany (Section 2.2.4) *Plant and Process Safety*, **361**
Weckbecker, Christoph, Degussa-Hüls AG, Hanau-Wolfgang, Germany *Amino Acids*, **777**
Weidlich, Stephan, Hoechst Aktiengesellschaft, Frankfurt, Federal Republic of Germany (Section 3.4.3) *Plant and Process Safety*, **361**
Whittaker, Victor P., Arbeitsgruppe Neurochemie, Max-Planck-Institut für biophysikalische Chemie, Göttingen, Federal Republic of Germany Johannes Gutenberg Universität, Mainz, Federal Republic of Germany *Neuropharmacology*, **1115**
Widmer, Ulrich, Sandoz, Basel, Switzerland (Section 4.2.1) *Plant and Process Safety*, **361**
Williamson, Gabrielle H., LeBoeuf, Lamb, Greene & MacRae, Brussels, Belgium (Chap. 4) *Legal Aspects*, **585**
Wybenga, Frits, International Standards Coordinator, Office of Hazardous Materials Safety, Washington D.C. 20590, United States (Section 5.2.2) *Transport, Handling, and Storage*, **487**

Yasogawa, Yoshio, Nippon Kaiji Kentei Kyokai, Tokyo, Japan (Section 5.2.3) *Transport, Handling, and Storage*, **487**
Yeh, Grace Chao, National Cancer Institute, Bethesda, Maryland 20205, United States *Cancer Chemotherapy*, **839**

Zahn, Manuel, Knoll AG, Ludwigshafen, Federal Republic of Germany (Section 4.3) *Pharmaceuticals, General Survey and Development*, **981**
Zimmermann, Jürgen, Bayer AG, Dormagen, Federal Republic of Germany (Sections 4.4, 4.5) *Plant and Process Safety*, **361**

Subject Index

Subject Index

Ullmann's Industrial Toxicology

ISBN: 3-527-31247-1

O